Proceedings

1999 IEEE INTERNATIONAL CONFERENCE ON

May 10-15, 1999
Marriott Hotel, Renaissance Center
Detroit, Michigan

Sponsored by

IEEE Robotics and Automation Society

Volume 4

Pages 2495-3286

ICRA99 PROCEEDINGS

Additional copies may be ordered from:

IEEE Service Center
445 Hoes Lane
P.O. Box 1331
Piscataway, NJ 08855-1331 U.S.A.

IEEE Catalog Number 99CH36288
ISBN 0-7803-5180-0 (Softbound)
ISBN 0-7803-5181-9 (Casebound)
ISBN 0-7803-5182-7 (Microfiche)
Library of Congress Catalog Number 90-640158
ISSN 1050-4729
IEEE Catalog Number (CD-ROM): 99CH36288C
ISBN 0-7803-5183-5 (CD-ROM)

Copyright and Reprint Permission:

Abstracting is permitted with credit to the source. Libraries are permitted to photocopy beyond the limit of U.S. copyright law for private use of patrons those articles in this volume that carry a code at the bottom of the first page, provided the per-copy fee indicated in the code is paid through Copyright Clearance Center, 27 Congress Street, Salem, MA 01970. Instructors are permitted to photocopy isolated articles for non-commercial classroom use without fee. For other copying, reprint, or republication permission, write to IEEE Copyrights Manager, IEEE Service Center, 445 Hoes Lane, P.O. Box 1331, Piscataway, NJ 08855-1331. All rights reserved. Copyright © 1999 by the Institute of Electrical and Electronics Engineers, Inc.

The papers in this book comprise the proceedings of the meeting mentioned on the cover and title page. They reflect the author's opinions and, in the interests of timely dissemination, are published as presented and without change. Their inclusion in this publication does not necessarily constitute endorsement by the editors or the Institute of Electrical and Electronics Engineers, Inc.

Printed in the United State of America by OMNIPRESS.......The Proceedings Printer.

The Institute of Electrical and Electronics Engineers, Inc.

Welcome Note from the Conference Chairman

I would like to extend a warm welcome to all participants in this year's International Conference on Robotics and Automation, ICRA99. This conference carries the distinction of being held in Detroit, the heart of the automotive industry in America. It ascertains the strong bond between R&D in academia and research laboratories, and the practice of robotics in industry.

More than any time in the history of this technology, this is an era for effective exchange of information and ideas between the two entities. ICRA99 provides a unique chance for industry engineers to meet with and exploit the minds of some of the most prominent researchers in the world. Reciprocally, this is an unequaled opportunity for researchers to pick the brains of industry practitioners seeking ideas for problems to solve through research, and avenues for industry support.

Only at ICRA conferences is there such a gathering of prominent robotics researchers in one place. This is particularly obvious this year; look up the names, affiliations, and topics of the presentations and you will recognize the wealth of information that will be presented at the conference. The panel sessions explore the frontiers of robotics research while the Banquet speaker Dr. Gustaf Olling of DaimlerChrysler symbolizes the renaissance of the American automotive industry through robotic and information technologies.

Detroit is experiencing a spectacular revival. After decades of relative dormancy, Greater Detroit is now bustling with activity, social, cultural, professional and economic. You are invited to participate in the many cultural and industrial tours available through the conference to see the new Detroit. Across the Detroit River, in Windsor, there is another country and a different culture.

This is a conference organized and managed by volunteers who deserve special thanks. Much hard work and sleepless nights have gone into making this conference a reality. Please let them know that we all appreciate their efforts.

Welcome and please enjoy this great conference.

Hadi Akeel
General Chairman

Foreword

The 1999 IEEE International Conference on Robotics and Automation (ICRA99) has two unique features. First, the conference is held in the final year of the 20th century which has witnessed the dramatic growth of information technology in recent years. The conference therefore emphasizes the robotics and automation technologies for the next century with integration of information. This leads to the theme of the conference as **Mega-Information Integration for Robotics and Automation in the 21st Century.** Secondly, the conference is held in Detroit, Michigan, the international center for automobile manufacturing which has a constant need of new robotics and automation technologies and enables the conference to attract many industry practitioners, while traditionally most ICRA participants are academicians. In addition, ICRA99 overlaps the annual Robotics Industries Association (RIA) Robot Show and Conference which is held in Cobo Hall, a walking distance from the conference hotel. This overlap further symbolizes the cooperation between academia and industry.

Eight-hundred thirty-five papers covering a wide scope of robotics and automation were submitted to the conference. The Program Committee had a very difficult time to select 521 papers for inclusion in the conference proceedings. While selecting the papers, the Program Committee focused on quality, and broad coverage of the areas as stressed by the conference theme. We regret that many good papers were not selected by the Program Committee because of the limited space. The conference program also includes 5 tutorials and 9 workshops, 3 video sessions, 2 plenary sessions, and a banquet speech. These supplementary programs further enhance the theme of the conference.

I am grateful for the assistance of Peggy Gerds of the Department of Electrical Engineering at The Ohio State University in handling the submitted papers and responses with the authors, and in sending thousands of email messages and letters. I also like to extend my sincere thanks to the General Chair Hadi Akeel, and other members of the Organizing Committee: Michael Bridges, Nikolaos Papanikolopoulos, T.J. Tarn, Rajiv Dubey, Daniel Koditschek, Elsayed Orady, Gary Rutledge, Guy Potok, Jianming Tao, Jason Tsai, Bruno Siciliano and Shin'ichi Yuta as well as the Program Committee members, Local Arrangements Committee members, and Video Proceedings Committee members. Special thanks should go to C.S. George Lee for assisting the publication of the Advance and Final Programs.

Finally we should thank all the authors and participants of the conference. Without them, there will not be a successful ICRA99.

Yuan F. Zheng
Program Chair

ORGANIZING COMMITTEE

General Chair
Hadi A. Akeel
FANUC Robotics, North America, USA

Program Chair
Yuan F. Zheng
Ohio State University, USA

Program Vice Chairs
Bruno Siciliano
Universita' degli Studi di Napoli Federico II, Italy

Shin'ichi Yuta
University of Tsukuba, Japan

Local Arrangements Chair
Michael Bridges
University of Michigan, Ann Arbor, USA

Publicity Chair
Nikolaos Papanikolopolous
University of Minnesota, USA

Publications Chair
T. J. Tarn
Washington University, USA

Workshops and Tutorials Chair
Daniel Koditschek
University of Michigan, Ann Arbor, USA

Video Proceedings Chair
Rajiv Dubey
University of South Florida, USA

Registration Chair
Elsayed Orady
University of Michigan, Dearborn, USA

Exhibits Chair
Gary Rutledge
FANUC Robotics, North America, USA

Fund Raising Chair
Guy Potok
GP Associates, USA

Finance Chair
Jianming Tao
FANUC Robotics, North America, USA

At Large
Jason Tsai
FANUC Robotics, North America, USA

Program Committee

Arvin Agah, Univ. of Kansas, USA
Omar Al-Jarrah, Jordan Univ. of Science and Technology, Jordan
Fumihito Arai, Nagoya Univ., Japan
Beno Benhabib, Univ. of Toronto, Canada
Matthew Berkemeier, Boston Univ., USA
Antonio Bicchi, Univ. di Pisa, Italy
John Billingsley, Univ. of Southern Queensland, Australia
Martin Buehler, McGill Univ., Canada
Hegao Cai, Harbin Inst. of Technology, China
Alicia Casals, Univ. Politencnica Catalunya, Spain
Raja Chatila, LAAS-CNRS, France
Yilong Chen, General Motors, USA
Fan-Tien Cheng, National Cheng Kung Univ. R.O.C.
Pasquale Chiacchio, Univ. di Napoli Federico II, Italy
James Clark, McGill Univ., Canada
Paolo Dario, Scuola Superiore Sant'Anna, Italy
Pierre Dauchez, LIRMM, France
Alessandro DeLuca, Univ. di Roma La Sapienza, Italy
Rüdiger Dillmann, Univ. of Karlsruhe, Germany
Gregory Dudek, McGill Univ., Canada
Olav Egeland, Norwegian Univ. of Science and Technology, Norway
Li-Chen Fu, National Taiwan Univ. R.O.C.
Toshio Fukuda, Nagoya Univ. Japan
Alessandro Giua, Univ. di Cagliari, Italy
Ken Goldberg, Univ. of California, Berkeley, USA
Andrew Goldenberg, Univ. of Toronto, Canada
David Greenstein, General Motors, USA
Kamal Gupta, Simon Fraser Univ., Canada
Gregory Hager, Yale Univ., USA
William R. Hamel, Univ. of Tennessee, USA
Hideki Hashimoto, Univ. of Tokyo, Japan
Hooshang Hemami, Ohio State Univ., USA
Shigeo Hirose, Tokyo Inst. of Technology, Japan

Gerd Hirzinger, Inst. of Robotics and System Dynamics, Germany
Steve Hsia, Univ. of California Davis, USA
Han-Pang Huang, National Taiwan Univ., R.O.C.
Seth Hutchinson, Univ. of Illinois, USA
Makoto Kaneko, Hiroshima Univ., Japan
Lydia Kavraki, Rice Univ., USA
Kazuhiko Kawamura, Vanderbilt Univ., USA
Wisama Khalil, Ecole Contrale de Nantes, France
Farshad Khorrami, Polytechnic Univ., USA
Pradeep Khosla, Carnegie Mellon Univ., USA
Richard Klafter, Temple Univ., USA
Antti Koivo, Purdue Univ., USA
Kazuhiro Kosuge, Tohoku Univ., Japan
Krzysztof Kozlowski, Pozan Univ. of Technology, Poland
Reid Kress, Oak Ridge National Lab., USA
Kostas Kyriakopoulos, National Technical Univ. of Athens, Greece
Dave Lane, Heriot-Watt Univ., Scotland
Jean-Paul Laumond, LAAS-CNRS, France
Steven M. LaValle, Iowa State Univ., USA
Beom-Hee Lee, Seoul National Univ., Korea.
C.S.George Lee, Purdue Univ., USA
Kok-Meng Lee, Georgia Inst. of Technology, USA
Sukhan Lee, Samsung Advanced Inst. of Technology, Korea
Jadran Lenarcic, The Jozef Stefan Inst., Slovenia
Zexiang Li, Hong Kong Univ. of Science and Technology, Hong Kong, China
John Luh, Clemson Univ., USA
Peter Luh, Univ. of Connecticut, USA
Vladimir Lumelsky, Univ. of Wisconsin, USA
Ren Luo, National Chung Cheng Univ., R.O.C.
Anthony Maciejewski, Purdue Univ., USA
Claudio Melchiorri, Univ. of Bologna, Italy
Max Meng, Univ. of Alberta, Canada
Giuseppe Menga, Politecnico di Torino, Italy
Howard Moraff, National Science Foundation, USA
Charles Nguyen, Catholic Univ. of America, USA
David Orin, Ohio State Univ., USA
Angel del Pobil, Univ. Jaume-I, Spain

Elon Rimon, Technion, Israel
Imre Rudas, Banki Donat Polytechnic, Hungary
Art Sanderson, National Science Foundation, USA
Joris De Schutter, Katholieke Univ. Leuven, Belgium
Kevin Smith, Brigham Young Univ., USA
Tarek Sobh, Univ. of Bridgeport, USA
Sergey Sokolov, Russian Academy of Sciences, Russia
Il Hong Suh, Hanyang Univ. Korea
Gaurav Sukhatme, Univ. of Southern California, USA
Satoshi Tadokoro, Kobe Univ., Japan
Kazuo Tanie, Mechanical Engineering Lab., Japan
Russell Taylor, Johns Hopkins Univ., USA
Spyros Tzafestas, National Technical Univ. of Athens, Greece
Kimnon Valavanis, Univ. SW Louisiana, USA
Qidi Wu, Tongji Univ., China
Ning Xi, Michigan State Univ., USA
Jing Xiao, National Science Foundation, USA
Yangsheng Xu, Chinese Univ. of Hong Kong, China
Keon Young Yi, Kwangwoon Univ., Korea
Junku Yuh, Univ. of Hawaii, USA
Xiaoping Yun, Naval Postgraduate School, USA

Video Proceedings Committee

Rajiv V. Dubey (Chair)
I-Ming Chen
T. Kesavada
Oussama Khatib
Heikki Koivo
Peter B. Luh
Mark Spong
Shigeki Sugano

Local Arrangements Committee

Chair: Michael Bridges (Chair)
Robert N.K. Loh
Edward Y.L. Gu
Feng Lin
Ning Xi

1999 IEEE International Conference on Robotics and Automation
Technical Sessions Schedule

May 12, 1999

	1	2	3	4	5	6	7	8	9	10	11	12	13
8:15-8:30	OPENING CEREMONY												
WAI 8:30-10:10	PLENARY SESSION: Intelligent Transportation												
WAII 10:35-12:15	Mobile Robot Maneuvering	Navigation in Unknown Environment	Biped Robot I	Underwater Vehicles	Robot Planni. & Program. for Assembly	Discrete Even. Control of Manu. Syst.	Motion Planning I	Robot Control I	Actuator	Teleop. I Force & Pos. Control	Contact and Grasping Control	Visual Servo Control I	Tactile Sensing
WPI 1:30-3:10	Mobile Robots and Applications	Sensor Based Navigation	Biped Robot II	Underwater Robotics: Sensing, etc.	Flexible Manipulator I	Task Scheduling	Motion Planning II	Robot Control II	Actuators and Joint Actuation	Teleop. II Sensor Based Teleopera.	Contact Geometry	Visual Servo Control II	Sonar Based Sensing
WPII 3:35-5:15	Mobile Robot Environment Interaction	Mobile Robot Motion Planning I	Biology-Inspired Methods	Service and Underwater Robots	Flexible Manipulator II	Manufactur. Planning and Scheduling	Constraint & Nonholono. Systems	Robot Control III	Fault Tolerant Robot	Parallel Manipulators	Dexterous Manipulation	Computer Vision in Manufactu.	Contact Sensing

May 13, 1999

	1	2	3	4	5	6	7	8	9	10	11	12	13
TAI 8:30-10:10	PLENARY SESSION: Robot Art												
TAII 10:35-12:15	Mobile Robot Field Applications	Mobile Robot Motion Planning II	Humanoid and Walk. Robots	New Robotic Technol. & Applications	Flexible Robots	Manufactu. Process Control	Control Architectures	Fuzzy Control I	Study of Robot Kinematics	Multiple Manipulator	Grasping Analysis	New Geometry Method in Comp. Vision.	Robotic Sensing & Its Applications
TPI 1:30-3:10	Mobile Robot Localization I	Mobile Robot Motion Planning III	Legged Locomotion I	Space Robots	Calibrat. and Tolerances	Production Planning and Control	Force Control	Fuzzy Control II	Kinematics	Cooperative Robots	Fixture Desi. and Manipu. Planning	Calibration-Free Visual Servo	Sensing and Sensor Design
TPII 3:35-5:15	Mobile Robot Localization II	Mobile Robot Motion Planning IV	Legged Locomotion II	Medical Robotics	Calibrat. and Friction Modeling	Process Planning and Manufactu.	Impedance Control	Robot Control IV	Distance and Contact Calculations	Teleop. III Experiments and Control	Grasping Computation	Real-Time Computer Vision	Sensor Fusion I
6:30	Conference Banquet - Banquet Speech - Dr. Gustav Olling, Daimler Chrysler Corp. "CAD/CAM/CAE Applications"												

Note: There are three video sessions on May 13, 1999: TAII-VS 10:35-12:15: Video Session I; TPI-VS 1:30-3:10: Video Session II; TPII-VS 3:35-5:15: Video Session III.

May 14, 1999

	1	2	3	4	5	6	7	8	9	10	11	12	13
FAI 8:30-10:10	Mobile Robot Localization III	Mobile Robot Systems I	Legged Locomotion III	Medical Robotics II	Learning and Identification	Manufacturing Automation	Compliance Control	Robotic Stiffness Control	Micro-Manipulator	Tele-Manipulation	Parts Manipulation	Vision-Based Mobile Robotics	Sensor Fusion II
FAII 10:35-12:15	Mobile Inspect. and Reconnaiss. Systems	Mobile Robot Systems II	Quadruped Locomotion	Medical Robotics III	Computational Intelligence	Virtual Factory	Genetic Algorithm	Intelligent and Fuzzy Control	Robot Mechanism	Sensor-Based Human/Mac Interaction	Part Feeding and Orienting	Vision Based Navigation I	Sensor Selection and Placement
FPI 1:30-3:10	Mobile Robot Motion Control I	Mobile Robot Sensor-Based Control	Small Scale Mobile Robots	Mobile Robot Mobility & Locomotion	Micro/Nano Manipulation	Assembly Planning	Learning Control	Manipulator Control	Hyper Redundant Robots	Human Robot Interaction I	Reasoning and Handling of Objects	Vision-Based Navigation II	Haptic Display
FPII 3:35-5:15	Mobile Robot Motion Control II	Mobile Robot Sensing	Articulated Locomotion	Multi-Finger Hands	Industr. Appl. of Robot and Auto. Tech.	Comp.-Aided Assembly Planning	Robot Programming	Neural Network Applications	Redundant Robots	Human Robot Interaction II	Sensor-Based Grasping	Visual Tracking	Haptic Interface

Table of Contents

Welcome iii
Foreword iv
Organizing Committee v
Program at a Glance vii
Author Index li

Volume 1 (pages 1-830)

WAII-1
MOBILE ROBOT MANEUVERING

Control of a Car-Like Mobile Robot for Parking Problem 1
    Sungon Lee, Youngil Youm, and Wan Kyun Chung

A Generalization of Path Following for Mobile Robots 7
    F. Díaz del Río, G. Jiménez, J. L. Sevillano, S. Vicente, and A. Civit Balcells

On a Manipulative Difficulty of a Mobile Robot with Multiple Trailers for Pushing and Towing? 13
    Wen Li, Takashi Tsubouchi, and Shin'ichi Yuta

Experiments of Backward Tracking Control for Trailer System 19
    Doh-Hyun Kim and Jun-Ho Oh

WAII-2
NAVIGATION IN UNKNOWN ENVIRONMENT

A Neural Network Approach to Real-Time Collision-Free Navigation of 3-D.O.F. Robots in 2D 23
    Xianyi Yang and Max Meng

Navigation in Partially Unknown, Narrow, Cluttered Spaces 29
    Matthias Strobel

Coastal Navigation: Mobile Robot Navigation with Uncertainty in Dynamic Environments 35
    Nicholas Roy, Wolfram Burgard, Dieter Fox, and Sebastian Thrun

Optimal Motion Planning for a Wheeled Mobile Robot 41
    Weiguo Wu, Ping Jiang, and Huitang Chen

WAII-3
BIPED ROBOTS I

Foot Rotation Indicator (FRI) Point: A New Gait Planning Tool to Evaluate
Postural Stability of Biped Robots 47
    Ambarish Goswami.

Point to Point Motion of Skeletal Systems with Multiple Transmission Delays 53
    Abhay Kataria, Hitay Özbay, and Hooshang Hemami

Energy Analysis During Biped Walking 59
    Filipe M. Silva and J. A. Tenreiro Machado

A High Stability, Smooth Walking Pattern for a Biped Robot 65
    Qiang Huang, Shuuji Kajita, Noriho Koyachi, Kenji Kaneko, Kazuhito Yokoi,
    Hirohiko Arai, Kiyoshi Komoriya, and Kazuo Tanie

WAII-4
UNDERWATER VEHICLES

Hovering and Altitude Control for Open-Frame UUVs 72
    M. Caccia, G. Bruzzone, and G. Veruggio

Real-Time Path Planning and Obstacle Avoidance for an Autonomous Underwater Vehicle 78
    Gianluca Antonelli, Stefano Chiaverini, Roberto Finotello, and Emanuele Morgavi

Fault Tolerant Decomposition of Thruster Forces of an Autonomous Underwater Vehicle 84
    Tarun Kanti Podder and Nilanjan Sarkar

Design and Simulation of Amoebot-A Metamorphic Underwater Vehicle 90
    I-Ming Chen, Hsi-Shang Li, and Arnaud Cathala

WAII-5
ROBOT PLANNING AND PROGRAMMING FOR ASSEMBLY

Distributed Control of Flexible Transfer System (FTS) Using Learning Automata 96
    Toshio Fukuda, Kosuke Sekiyama, Yoshiaki Hasebe, Yasuhisa Hasegawa,
    Susumu Shibata, Hironobu Yamamoto, and Yuji Inada

An Integration of Robot Programming and Sequence Planning 102
    Xiaobu Yuan and Yuqing Gu

An Integrated Intelligent Approach and System for Rapid Robotic
Assembly Prototyping, Planning and Control 108
    X. F. Zha

A Geometric Algorithm for Hybrid Localization/Inspection/Machinability Problem 114
    Y. X. Chu, J. B. Gou, and Z. X. Li

WAII-6
INVITED SESSION: DISCRETE EVENT CONTROL OF MANUFACTURING SYSTEMS

Control of Flexible-Manufacturing Workcells Using Extended Moore Automata 120
    A. Ramirez, C. Sriskandarajah, and B. Benhabib

A Computer Implementable Algorithm for the Synthesis of an Optimal
Controller for Acyclic Discrete Event Processes 126
    Satya Ranjan Mohanty, Vigyan Chandra, and Ratnesh Kumar

Discrete Event Control with Active Events 131
    Michael Heymann, Feng Lin, and George Meyer

Performance Analysis of Machining Systems with Modular Logic Controllers 137
Euisu Park, Dawn M. Tilbury, and Pramod P. Khargonekar

WAII-7
MOTION PLANNING I

Proportional Navigation Guidance in Robot Trajectory Planning for Intercepting Moving Objects 145
M. Mehrandezh, M. N. Sela, R. G. Fenton, and B. Benhabib

Path Planning for Elastic Plates Under Manipulation Constraints 151
Florent Lamiraux and Lydia E. Kavraki

Control of Dynamics and Sensor Based Motion Planning for a Differential Drive Robot 157
S. Bobyr and V. Lumelsky

Range-Sensor Based Navigation in Three Dimensions 163
Ishay Kamon, Ehud Rivlin, and Elon Rimon

WAII-8
ROBOT CONTROL I

Position Control of a Robot Manipulator Using Continuous Gain Scheduling 170
M. A. Jarrah and Omar M. Al-Jarrah

A Muscular-Like Compliance Control for Active Vehicle Suspension 3275
Shih-Lang Chang, Chi-haur Wu, and D. T. Lee

A Taxonomy for Robot Control 176
D. M. Miljanovic and E. A. Croft

The Extremal Properties of Spatial Stiffness Matrices 182
Shuguang Huang and Joseph M. Schimmels

WAII-9
ACTUATOR

High Stiffness Control of Direct-Drive Motor System by a Homogeneous ER Fluid 188
Naoyuki Takesue, Guoguang Zhang, Junji Furusho, and Masamichi Sakaguchi

Development of X-Screw: A Load-Sensitive Actuator Incorporating a Variable Transmission 193
Shigeo Hirose, Craig Tibbetts, and Tetsuo Hagiwara

Robust Force Controller Design for an Electro-Hydraulic Actuator Based on Nonlinear Model 200
N. Niksefat and N. Sepehri

Distributed Control Scheme for Motor Networks with Communication-Constrained Channels 207
Kyunghwan Kim and Nicola J. Ferrier

WAII-10
TELEOPERATION I: FORCE AND POSITION CONTROL

The Distributed Controller Architecture for a Masterarm and Its Application
to Teleoperation with Force Feedback 213
    Sooyong Lee, Jangwook Lee, Dae-Sung Choi, Munsang Kim, and Chong-Won Lee

Action Synchronization and Control of Internet Based Telerobotic Systems 219
    Ning Xi and T. J. Tarn

Free and Constrained Motion Teleoperation via Naturally-Transitioning Rate-to-Force Control 225
    Robert L. Williams II, Jason M. Henry, Mark A. Murphy, and Daniel W. Repperger

Teleoperation with Adaptive Motion/Force Control 231
    Wen-Hong Zhu and S. E. Salcudean

WAII-11
CONTACT AND GRASPING CONTROL

Automatic Generation of High-Level Contact State Space 238
    Xuerong Ji and Jing Xiao

Planning of Regrasp Operations 245
    Sascha A. Stoeter, Stephan Voss, Nikolaos P. Papanikolopoulos, and Heiko Mosemann

Experimental Investigation into Contact Transition Control with Joint Acceleration Feedback Damping 251
    J. D. Han, Y. C. Wang, and W. L. Xu

Analytical Conditions for the Rotational Stability of an Object in Multi-Finger Grasping 257
    M. M. Svinin, K. Ueda, and M. Kaneko

WAII-12
VISUAL SERVO CONTROL I

Visual Servoing with Linearized Observer 263
    Koichi Hashimoto and Toshiro Noritsugu

A Novel Visual Servoing Approach Involving Disturbance Observer 269
    Joon-Soo Lee, Il Hong Suh, Bum-Jae You, and Sang-Rok Oh

Performance of a Partitioned Visual Feedback Controller 275
    Paul Y. Oh and Peter K. Allen

Neural Network-Based Vision Guided Robotics 281
    Kevin Stanley, Q. M. Jonathan Wu, Ali Jerbi, and William A. Gruver

WAII-13
TACTILE SENSING

An Integrated Tactile/Shear Feedback Array for Stimulation of Finger Mechanoreceptor 287
    Darwin G. Caldwell, N. Tsagarakis, and C. Giesler

Tactile Sensing with Minimal Wiring Complexity 293
    Martin Nilsson

A Micromachined Piezoelectric Teeth-like Laparoscopic Tactile Sensor:
Theory, Fabrication and Experiments 299
    J. Dargahi, S. Payandeh, and M. Parameswaran

The Role of Contact Area Spread Rate in Haptic Discrimination of Softness 305
    Guilio Ambrosi, Antonio Bicchi, Danilo De Rossi, and Enzo Pasquale Scilingo

WPI-1
MOBILE ROBOTS AND APPLICATIONS

A Sensor Guided Autonomous Parking System for Nonholonomic Mobile Robots 311
    K. Jiang and L. D. Seneviratne

Kinodynamic Motion Planning for All-Terrain Wheeled Vehicles 317
    Moëz Cherif

The Formula One Tire Changing Robot (F-T.C.R.) 323
    Raul Mihali and Tarek Sobh

The Path Planning of Mobile Manipulator with Genetic-Fuzzy Controller in Flexible Manufacturing Cell 329
    Xiaowei Ma, Yili Fu, Yufei Yuan, Wang Wei, Yulin Ma, and Hegao Cai

WPI-2
SENSOR-BASED NAVIGATION

On-Line Sub-Optimal Obstacle Avoidance 335
    Zvi Shiller

High-Speed Navigation Using the Global Dynamic Window Approach 341
    Oliver Brock and Oussama Khatib

An Autonomous Sensor-Based Path-Planner for Planetary Microrovers 347
    S. L. Laubach and J. W. Burdick

On the Near-Optimality of Sensor-Based Navigation in a 2-D Unknown Environment with Simple Shape 355
    Iliroshi Noborio and Kenji Urakawa

WPI-3
BIPED ROBOTS II

Physical Interaction Between Human and a Bipedal Humanoid Robot:
Realization of Human-Follow Walking 361
    Samuel Agus Setiawan, Sang Ho Hyon, Jin'ichi Yamaguchi, and Atsuo Takanishi

Development of a Bipedal Humanoid Robot: Control Method of
Whole Body Cooperative Dynamic Biped Walking 368
    Jin'ichi Yamaguchi, Eiji Soga, Sadatoshi Inoue, and Atsuo Takanishi

Biped Robot Locomotion in Scenes with Unknown Obstacles 375
    M. Yagi and V. Lumelsky

Blind Walking of a Planar Bipedal Robot on Sloped Terrain 381
    Chee-Meng Chew, Jerry Pratt, and Gill Pratt

WPI-4
INVITED SESSION: UNDERWATER ROBOTICS: SENSING, NAVIGATION, AND CONTROL

Motion Coordination of Underwater Vehicle-Manipulator Systems Subject to Drag Optimization 387
    Nilanjan Sarkar and Tarun Kanti Podder

Experimental Study on Adaptive Control of Underwater Robots 393
    J. Yuh, Jing Nie, and C. S. G. Lee

Advances in Doppler-Based Navigation of Underwater Robotic Vehicles 399
    Louis Whitcomb, Dana Yoerger, and Hanumant Singh

Passive Arm Based Dynamic Positioning System for Remotely Operated Underwater Vehicles 407
    Liu Hsu, Ramon R. Costa, Fernando C. Lizarralde, and José Paulo Vilela Soares da Cunha

WPI-5
FLEXIBLE MANIPULATORS I

A Robust Controller Design Method for a Flexible Manipulator with
a Time Varying Payload and Parameter Uncertainties 413
    Jee-Hwan Ryu, Dong-Soo Kwon, and Youngjin Park

An Infinite-Dimensional Analysis of a PD-Controlled Single Flexible Link in Collision 419
    Francis Ching and David Wang

Scaling Laws for the Dynamics and Control of Flexible-Link Manipulators 427
    Milind Ghanekar, David W. L. Wang, and Glenn R. Heppler

Modeling and Control of a New Three-Degree-of-Freedom Flexible Arm with Simplified Dynamics 435
    J. A. Somolinos, V. Feliu, L. Sánchez, and J. A. Cerrada

WPI-6
TASK SCHEDULING

Task Scheduling on Spacecraft by Hydrid Genetic Algorithms 441
    Il-Jun Jeong, George Papavassilopoulos, and David S. Bayard

Elevator Group Control with Accurate Estimation of Hall Call Waiting Times 447
    Young Cheol Cho, Zavarin Gagov, and Wook Hyun Kwon

Dynamic Vehicle Routing Using Hybrid Genetic Algorithms 453
    Wan-rong Jih and Jane Yung-jen Hsu

A New Generation of Evaluation Tool for On-Line Design and
Scheduling in an Advanced Manufacturing System 459
    Ming-Hung Lin and Li-Chen Fu

WPI-7
MOTION PLANNING II

CONTROLAB MUFA: A Multi-Level Fusion Architecture for Intelligent Navigation of a Telerobot 465
    E. P. L. Aude, G. H. M. B. Carneiro, H. Serdeira, J. T. C. Silveira, M. F. Martins, and E. P. Lopes

Randomized Kinodynamic Planning 473
    Steven M. LaValle and James J. Kuffner, Jr.

Weightlifting Motion Planning for a Puma 762 Robot 480
    Chia-Yu E. Wang, Wojciech K. Timoszyk, and James E. Bobrow

Computing Fault Tolerant Motions for a Robot Manipulator 486
    Scott K. Ralph and Dinesh K. Pai

WPI-8
ROBOT CONTROL II

PID Control of Robotic Manipulator with Uncertain Jacobian Matrix 494
    C. C. Cheah, S. Kawamura, S. Arimoto, and K. Lee

Stabilization of the Pendulum on a Rotor Arm by the Method of Controlled Lagrangians 500
    Anthony M. Bloch, Naomi Ehrich Leonard, and Jerrold E. Marsden

Real-Time Motion Control in the Neighborhood of Singularities: A Comparative
Study Between the SC and the DLS Method 506
    Dragomir N. Nenchev, Yuichi Tsumaki, and Masaru Uchiyama

Task-Space Tracking Control Without Velocity Measurements 512
    Fabrizio Caccavale, Ciro Natale, and Luigi Villani

WPI-9
ACTUATORS AND JOINT ACTUATION

Control Performance of an Air Motor: Can Air Motors Replace Electric Motors? 518
    S. R. Pandian, F. Takemura, Y. Hayakawa, and S. Kawamura

Development of a Pneumatic Muscle Actuator Driven Manipulator
Rig for Nuclear Waste Retrieval Operations 525
    Darwin G. Caldwell, N. Tsagarakis, G. A. Medrano-Cerda, J. Schofield, and S. Brown

Joint Actuation Switching in Closed-Chain Mechanisms for High Task Adaptability 531
    Sungbok Kim

Development of Dual Mode X-Screw: A Novel Load-Sensitive Linear
Actuator with a Wide Transmission Range 537
    Tetsuo Hagiwara and Shigeo Hirose

WPI-10
TELEOPERATION II: SENSOR-BASED TELEOPERATION

Vision-Based End-Effector Alignment Assistance for Teleoperation 543
    S. E. Everett, Y. Isoda, R. V. Dubey, and C. Dumont

Image Based Predictive Display for Tele-Manipulation 550
    Martin Jägersand

Task-Based Data Exchange for Remote Operation System Through a Communication Network 557
    Takafumi Matsumaru, Syun'ichi Kawabata, Tetsuo Kotoku, Nobuto Matsuhira,
    Kiyoshi Komoriya, Kazuo Tanie, and Kunikatu Takase

Collision Control in Teleoperation by Virtual Force Reflection: An Application to the ROBTET System 565
    M. Hernando, E. Gambao, E. Pinto, and A. Barrientos

WPI-11
CONTACT GEOMETRY

A Simple Characterization of the Infinitesimal Motions Separating General Polyhedra in Contact 571
    Ernesto Staffetti, Lluís Ros, and Federico Thomas

Contact Analysis of Spatial Fixed-Axes Pairs Using Configuration Spaces 578
    Iddo Drori, Leo Joskowicz, and Elisha Sacks

Identification of Contact Conditions from Contaminated Data of Contact Moment 585
    Tetsuya Mouri, Takayoshi Yamada, Yasuyuki Funahashi, and Nobuharu Mimura

Interactive Manipulation of Articulated Objects with Geometry Awareness 592
    Min-Hyung Choi and James F. Cremer

WPI-12
VISUAL SERVO CONTROL II

A Two Loops Direct Visual Control of Direct-Drive Planar Robots with Moving Target 599
    Rafael Kelly, Fernando Reyes, Javier Moreno, and Seth Hutchinson

Exponentially Stable Positioning of a Rigid Robot Using Stereo Vision 605
    Wen-Chung Chang and A. S. Morse

Planar Image Based Visual Servoing as a Navigation Problem 611
    Noah J. Cowan and Daniel E. Koditschek

Visual Servoing with Dynamics: Control of an Unmanned Blimp 618
    Hong Zhang and James P. Ostrowski

WPI-13
SONAR-BASED SENSING

A Diffuse Reflection Model for Time of Flight Sonar 624
    Paul Gilkerson and Penelope Probert

A Fast and Accurate Sonar-Ring Sensor for a Mobile Robot 630
    Teruko Yata, Akihisa Ohya, and Shin'ichi Yuta

Localization and Classification of Target Surfaces Using Two Pairs of Ultrasonic Sensors 637
    Youngjoon Han and Hernsoo Hahn

The Arc-Transversal Median Algorithm: An Approach to Increasing Ultrasonic Sensor Accuracy 644
    Keiji Nagatani, Howie Choset, and Nicole Lazar

WPII-1
MOBILE ROBOT-ENVIRONMENT INTERACTION

Marker-Augmented Robot-Environment Interaction 652
    Amol D. Mali

Spontaneous, Short-Term Interaction with Mobile Robots 658
    J. Schulte, C. Rosenberg, and S. Thrun

Interactive Task Training of a Mobile Robot through Human Gesture Recognition 664
    Paul E. Rybski and Richard M. Voyles

A Person Following Behaviour for a Mobile Robot 670
    H. Sidenbladh, D. Kragic, and H. I. Christensen

WPII-2
MOBILE ROBOT MOTION PLANNING I

Efficient Topological Exploration 676
    Ioannis M. Rekleitis, Vida Dujmovic, and Gregory Dudek

EquiDistance Diagram: A New Roadmap Method for Path Planning 682
    S. S. Keerthi, C. J. Ong, E. Huang, and E. G. Gilbert

Probabilistic Roadmap Methods are Embarrassingly Parallel 688
    Nancy M. Amato and Lucia K. Dale

Perception-Based Motion Planning for Indoor Exploration 695
    Peter Leven, Seth Hutchinson, Darius Burschka, and Georg Färber

WPII-3
BIOLOGY-INSPIRED METHODS

ARMAR: An Anthropomorphic Arm for Humanoid Service Robot 702
    K. Berns, T. Asfour, and R. Dillmann

Gesture-Based Programming: A Preliminary Demonstration 708
    Richard M. Voyles and Pradeep K. Khosla

Dynamics Computation of Structure-Varying Kinematic Chains for Motion Synthesis of Humanoid 714
    Katsu Yamane and Yoshihiko Nakamura

Fault Tolerance via Component Redundancy for a Modularized Sensitive Skin 722
    D. Um and V. Lumelsky

WPII-4
SERVICE AND UNDERWATER ROBOTS

Cobots for the Automobile Assembly Line 728
    Prasad Akella, Michael Peshkin, Ed Colgate, Wit Wannasuphoprasit, Nidamaluri Nagesh,
    Jim Wells, Steve Holland, Tom Pearson, and Brian Peacock

Feedforward Modulation of Dynamic Behaviour in Anthropomorphic Robots via Force Redundancies 734
    B.-J. Yi and J. H. Lee

ProVAR Assistive Robot System Architecture 741
    H. F. M. Van der Loos, J. J. Wagner, N. Smaby, K. Chang, O. Madrigal, L. J. Leifer, and O. Khatib

Vision Based Assisted Operations in Underwater Environments Using Ropes 747
    Josep Amat, Alícia Casals, and Josep Fernández

WPII-5
FLEXIBLE MANIPULATORS II

Modeling of Kineto-Elastodynamics of Robots with Flexible Links 753
    Xiang-Rong Xu, Won-Jee Chung, and Young-Hyu Choi

Simulation of Flexible Manipulators with Elastic Non-Linearities 759
    F. Boyer and N. Glandais

Deformation Modeling of Viscoelastic Objects for Their Shape Control 767
    Shinichi Tokumoto, Yoshiaki Fujita, and Shinichi Hirai

Propulsion and Control of Deformable Bodies in an Ideal Fluid 773
    Richard Mason and Joel Burdick

WPII-6
MANUFACTURING PLANNING AND SCHEDULING

An Operational Approach for the Control of Manufacturing Processes 781
    Angela Di Febbraro, Riccardo Minciardi, and Simona Sacone

An Effective Method to Reduce Inventory in Job Shops 787
    Peter B. Luh, Xiaohui Zhou, and Robert N. Tomastik

A Translation Method for Ladder Diagram with Application to a Manufacturing Process 793
    Hyung Seok Kim, Naehyuck Chang, and Wook Hyun Kwon

Methodology of Generating Recovery Procedures in a Robotic Cell 799
    Hsien-Jung Wu

WPII-7
CONSTRAINT AND NONHOLONOMIC SYSTEM

Robot Calibration with Planar Constraints 805
    Hanqi Zhuang, Shui H. Motaghedi, and Zvi S. Roth

Curve Fitting Approach to Motion Planning of Nonholonomic Chained Systems 811
    W. L. Xu, B. L. Ma, and S. K. Tso

Nonholonomic Deformation of a Potential Field for Motion Planning 817
    S. Sekhavat and M. Chyba

Steering Nonholonomic Systems via Nilpotent Approximations: The General Two-Trailer System 823
    Marilena Vendittelli, Jean-Paul Laumond, and Guiseppe Oriolo

Volume 2 (pages 831-1664)

WPII-8
ROBOT CONTROL III

Formally Non-Exact Analytical Modeling of Mechanical Systems and
Environmental Interactions in an Adaptive Control 831
    József K. Tar, Imre J. Rudas, Okyay M. Kaynak, and János F. Bitó

New Results in NPD Control: Tracking, Integral Control, Friction
Compensation and Experimental Results 837
    Brian Armstrong, David Neevel, and Todd Kusik

Decentralized Learning Control for Robot Manipulators 843
    Weiqing Huang, Lilong Cai, and Xiaoqi Tang

Study on Piezoelectric Actuators in Control of a Single-Link Flexible Manipulator 849
    Dong Sun and James K. Mills

WPII-9
FAULT-TOLERANT ROBOTS

Failure Recovery in Redundant Serial Manipulators Using Nonlinear Programming 855
    Christopher Cocca, Daniel Cox, and Delbert Tesar

Fault Detection and Robust Fault Recovery Control for Robot Manipulators with Actuator Failures 861
    Jin-Ho Shin and Ju-Jang Lee

The Design of Control Strategies Tolerant to Undetected Failures in Kinematically Redundant Manipulators 867
    M. Goel, A. A. Maciejewski, and V. Balakrishnan

Failure Tolerant Teleoperation of a Kinematically Redundant Manipulator: An Experimental Study 874
    M. Goel, A. A. Maciejewski, V. Balakrishnan, and R. W. Proctor

WPII-10
PARALLEL MANIPULATORS

Dynamic Modeling of a Class of Spatial Statically-Balanced Parallel Platform Mechanisms 881
    Imme Ebert-Uphoff and Clément M. Gosselin

Singularity Analysis of 6-DOF Manipulators with the Analytical Representation of the Determinant 889
    Doik Kim, Wan Kyun Chung, and Youngil Youm

On Design of a Redundant Wire-Driven Parallel Robot WARP Manipulator 895
    Kiyoshi Maeda, Satoshi Tadokoro, Toshi Takamori, Motofumi Hattori,
    Manfred Hiller, and Richard Verhoeven

Hybrid Micro-Gravity Simulator Consisting of a High-Speed Parallel Robot 901
    Toshifumi Akima, Susumu Tarao, and Masaru Uchiyama

WPII-11
DEXTEROUS MANIPULATION

The Robonaut Hand: A Dexterous Robot Hand for Space 907
    C. S. Lovchik and M. A. Diftler

Goldfinger: A Non-Anthropomorphic, Dextrous Robot Hand 913
    Ann M. Ramos, Ian A. Gravagne, and Ian D. Walker

Grasping with Elastic Finger Tips 920
    K. K. Choi, S. L. Jiang, and Z. Li

A Skeletal Framework Artificial Hand Actuated by Pneumatic Artificial Muscles 926
    Y. K. Lee and I. Shimoyama

WPII-12
COMPUTER VISION IN MANUFACTURING

Invariant Features and the Registration of Rigid Bodies 932
    Gregory C. Sharp, Sang W. Lee, and David K. Wehe

Determining Surface Orientation from Fixated Eye Position and Angular Visual Extent 938
    Nicola J. Ferrier

3-D Cueing: A Data Filter for Object Recognition 944
    Owen Carmichael and Martial Hebert

The Design of a 3-D Surface Geometry Acquisition System for Highly Irregular
Shaped Objects: With Application to CZ Semiconductor Manufacture 951
    Vivek A. Sujan and Steven Dubowsky

WPII-13
CONTACT SENSING

Telemetric Robot Skin 957
    Mitsuhiro Hakozaki, Hideki Oasa, and Hiroyuki Shinoda

Photo-Plethysmograph Nail Sensors for Measuring Finger Forces Without
Haptic Obstruction: Modeling and Experimentation 962
    Stephen Mascaro, Kuo-Wei Chang, and H. Harry Asada

Body Parts Positions and Posture Estimation System Based on Pressure Distribution Image 968
    Tatsuya Harada, Taketoshi Mori, Yoshifumi Nishida, Tomohisa Yoshimi, and Tomomasa Sato

Active Haptic Perception and Control of Its Fixation 976
    Satoru Emura and Susumu Tachi

TAII-1
MOBILE ROBOT FIELD APPLICATIONS

Automated Planning and Scheduling for Planetary Rover Distributed Operations 984
    Paul G. Backes, Gregg Rabideau, Kam S. Tso, and Steve Chien

Kinematic Modeling of a High Mobility Mars Rover 992
    M. Tarokh, G. McDermott, S. Hayati, and J. Hung

Evaluation of Internal Navigation Sensor Suites for Underground Mining Vehicle Navigation 999
    Raj Madhavan, Eric Nettleton, Eduardo Nebot, Gamini Dissanayake, Jock Cunningham,
    Hugh Durrant-Whyte, Peter Corke, and Jonathan Roberts

A Fuzzy Genetic Based Embedded-Agent Approach to Learning & Control
in Agricultural Autonomous Vehicles 1005
    Hani Hagras, Victor Callaghan, Martin Colley, and Malcom Carr-West

TAII-2
MOBILE ROBOT MOTION PLANNING II

Shortest Path Planning for a Tethered Robot or an Anchored Cable 1011
    Patrick G. Xavier

The Gaussian Sampling Strategy for Probabilistic Roadmap Planners 1018
    Valérie Boor, Mark H. Overmars, and A. Frank van der Stappen

MAPRM: A Probabilistic Roadmap Planner with Sampling on the Medial Axis of the Free Space 1024
    Steven A. Wilmarth, Nancy M. Amato, and Peter F. Stiller

Motion Planning in an Unknown Polygonal Environment with Bounded Performance Guarantee 1032
    Amitava Datta and Subbiah Soundaralakshmi

TAII-3
HUMANOID AND WALKING ROBOTS

Action Acquisition Framework for Humanoid Robots Based on Kinematics and Dynamics Adaptation 1038
    Fumio Kanehiro, Masayuki Inaba, and Hirochika Inoue

Making Feasible Walking Motion of Humanoid Robots from Human Motion Capture Data 1044
    Anirvan Dasgupta and Yoshihiko Nakamura

Stable Adaptive Control of a Bipedal Walking Robot with CMAC Neural Networks 1050
    Jianjuen Hu, Jerry Pratt, and Gill Pratt

Development of "MEL HORSE" 1057
    Hiroki Takeuchi

TAII-4
NEW ROBOTIC TECHNOLOGY AND APPLICATIONS

Keeping the Analog Genie in the Bottle: A Case for Digital Robots 1063
    Ian D. Walker, Joseph R. Cavallaro, and Martin L. Leuschen

A Multi-Disciplinary Approach in Evaluating and Facilitating the Use of the Manus Robot 1071
    M. Mokhtari, N. Didi, and A. Roby-Brami

Task Verification Facility for the Canadian Special Purpose Dextrous Manipulator 1077
    J.-C. Piedboeuf, J. de Carufel, F. Aghili, and E. Dupuis

Laser Garment Cut Robot System: Design and Realization 1084
    Yuntao Wang, Qingshan Bao, Shuguo Wang, and Lin Wu

TAII-5
FLEXIBLE ROBOTS

Experimental Results on Tracking Control of a Flexible-Link Manipulator:
A New Output Re-Definition Approach 1090
    H. A. Talebi, K. Khorasani, and R. V. Patel

Modeling and Vibration Control with Neural Network for Flexible Multi-Link Structures 1096
    Masahiro Isogai, Fumihito Arai, and Toshio Fukuda

Tip-Trajectory Tracking Control of Single-Link Flexible Robots via Output Redefinition 1102
    H. Yang, H. Krishnan, and M. H. Ang, Jr.

Synthesis of Bounded-Input Nonlinear Predictive Controller for Multi-Link Flexible Robots 1108
    H. Yang, H. Krishnan, and M. H. Ang, Jr.

TAII-6
MANUFACTURING PROCESS CONTROL

A Multi-Contract Net Protocol for Dynamic Scheduling in Flexible Manufacturing Systems (FMS) 1114
    D. Ouelhadj, C. Hanachi, A. Moualek, and A. Farhi

A Genetic Algorithm for Flexible Job-Shop Scheduling 1120
    Haoxun Chen, Jürgen Ihlow, and Carsten Lehmann

Optimal Rate Allocation in Unreliable, Assembly/Disassembly Production Networks with Blocking 1126
    Vassilis S. Kouikoglou

Design and Implementation of a Process Control Language for an Automated Pager Manufacturing System 1132
    Keon Young Yi and Se Joong Jeon

TAII-7
CONTROL ARCHITECTURES

Using Control Networks for Distributed Robotic Systems 1138
    J. A. Janét, W. J. Wiseman, R. D. Michelli, A. L. Walker, and S. M. Scoggins

A Language for Declarative Robotic Programming 1144
    John Peterson, Gregory D. Hager, and Paul Hudak

Open Architecture Controller Software for Integration of Machine Tool Monitoring 1152
    Shige Wang, C. V. Ravishankar, and Kang G. Shin

A Real-Time Software Architecture for Robotics and Automation 1158
    Jonathan M. Roberts, Peter I. Corke, Robin J. Kirkham, Frederic Pennerath, and Graeme J. Winstanley

TAII-8
FUZZY CONTROL I

Appearance-Based Visual Learning in a Neuro-Fuzzy Model for Fine-Positioning of Manipulators 1164
    Jianwei Zhang, Ralf Schmidt, and Alois Knoll

Robot Skill Transfer Based on B-Spline Fuzzy Controllers for Force-Control Tasks 1170
    Markus Ferch, Jianwei Zhang, and Alois Knoll

Target Tracking by Grey Prediction Theory and Look-Ahead Fuzzy Logic Control 1176
    Ren C. Luo and Tse Min Chen

Fuzzy Selection of Fuzzy-Neuro Robot Force Controllers in an Unknown Environment 1182
    Kazuo Kiguchi and Toshio Fukuda

TAII-9
STUDY OF ROBOT KINEMATICS

Efficient Simulation of a Multilayer Viscoelastic Beam Using an Equivalent Homogeneous Beam 1188
    J.-C. Piedboeuf, L.-L. Pagé, I. Tremblay, and M.-J. Potvin

Kinematically Dual Manipulators 1194
    Herman Bruyninckx

Unified Analysis on Mobility and Manipulability of Mobile Manipulators 1200
    Yoshio Yamamoto and Xiaoping Yun

Development & Application of Wall-Climbing Robots 1207
    Yan Wang, Shuliang Liu, Dianguo Xu, Yanzheng Zhao, Hao Shao, and Xueshan Gao

TAII-10
MULTIPLE MANIPULATORS

Experimental Verification of Distributed Event-Based Control of Multiple Unifunctional Manipulators 1213
    Khalid Munawar and Masaru Uchiyama

Agent-Based Planning and Control of a Multi-Manipulator Assembly System 1219
    Juan-Carlos Fraile, Christiaan J. J. Paredis, and Pradeep K. Khosla

A Multiple Robot System for Cooperative Object Transportation
with Various Requirements on Task Performing 1226
    Zhi Dong Wang, Majid Nili Ahmadabadi, Eiji Nakano, and Takayuki Takahashi

M+: A Scheme for Multi-Robot Cooperation Through Negotiated Task Allocation and Achievement 1234
    S. C. Botelho and R. Alami

TAII-11
GRASPING ANALYSIS

Examples of 3D Grasp Quality Computations 1240
    Andrew T. Miller and Peter K. Allen

Interpretation of Grasp and Manipulation Based on Grasping Surfaces 1247
    Fuminori Saito and Kazuyuki Nagata

A Criteria-Based Approach to Grasp Synthesis 1255
    R. D. Hester, M. Cetin, C. Kapoor, and D. Tesar

Grasp Analysis as Linear Matrix Inequality Problems 1261
    Li Han, Jeffrey C. Trinkle, and Zexiang Li

TAII-12
NEW GEOMETRIC METHOD IN COMPUTER VISION

Self-Calibration of Eye-Hand System Based on Geometric Method with Perception Net 1269
    Sukhan Lee and Sookwang Ro

Geometric Understanding of Rigid Body Transformations 1275
    Y. Liu and M. A. Rodrigues

The Effect of Measurement Noise on Intrinsic Camera Calibration Parameters 1281
    Sunil Kumar Kopparapu and P. Corke

Calibration of the Omnidirectional Vision Sensor: SYCLOP 1287
    Cyril Cauchois, Eric Brassart, Cyril Drocourt, and Pascal Vasseur

TAII-13
ROBOTIC SENSING AND ITS APPLICATIONS

Integrated Photo and Acceleration Sensing Module for Robot Planning and Control 1293
    Fathi M. Salam and Ning Xi

Vision Based Navigation System for Autonomous Mobile Robot with Global Matching 1299
    Yasunori Abe, Masaru Shikano, Toshio Fukuda, Fumihito Arai, and Yoshio Tanaka

Tactile Mapping of Palpable Abnormalities for Breast Cancer Diagnosis 1305
    Y. Wang, C. Nguyen, R. Srikanchana, Z. Geng, and M. T. Freedman

Mathematical Model for Acquiring a Cone Axis from X-Ray Images 1310
    Y. Zhu, R. Phillips, W. Viant, A. Mohsen, M. Bielby, and J. G. Griffiths

TPI-1
MOBILE ROBOT LOCALIZATION I

Simultaneous Localisation and Map Building Using Millimetre Wave Radar to Extract Natural Features 1316
    S. Clark and G. Dissanayake

Monte Carlo Localization for Mobile Robots 1322
    Frank Dellaert, Dieter Fox, Wolfram Burgard, and Sebastian Thrun

Mobile Robot Localization Based on an Omnidirectional Stereoscopic Vision Perception System 1329
    Cyril Drocourt, Laurent Delahoche, Claude Pegard, and Arnaud Clerentin

A New Estimator for Mixed Stochastic and Set Theoretic Uncertainty
Models Applied to Mobile Robot Localization 1335
    Uwe D. Hanebeck and Joachim Horn

TPI-2
MOBILE ROBOT MOTION PLANNING III

Safe Actions and Observations Planning for Mobile Robots 1341
    Alain Lambert and Nadine Le Fort-Piat

Real Time Motion Planning for Autonomous Mobile Robot Using Framework of Anytime Algorithm 1347
    Kae Fujisawa, Soichiro Hayakawa, Takeshi Aoki, Tatsuya Suzuki, and Shigeru Okuma

Planning Tracking Motions for an Intelligent Virtual Camera 1353
    Tsai-Yen Li and Tzong-Hann Yu

Efficient Generation of Object Hierarchies from 3D Scenes 1359
    Miguel Angel García, Angel Domingo Sappa, and Luis Basañez

TPI-3
LEGGED LOCOMOTION I

Hybrid Control for Biped Robots Using Impedance Control and Computed-Torque Control 1365
    Jong H. Park and Ho A. Chung

A New Free Gait Generation for Quadrupeds Based on Primary/Secondary Gait 1371
    Shaoping Bai, K. H. Low, Gerald Seet, and Teresa Zielinska

Design of a Spider-Like Robot for Motion with Quasistatic Force Constraints 1377
    Shraga Shoval, Elon Rimon, and Amir Shapira

Prescribed Synergy Method Based Hybrid Intelligent Gait Synthesis for Biped Robot 1384
    Changjiu Zhou, Kanniah Jagannathan, and Than Myint

TPI-4
SPACE ROBOTS

Space Robot Experiments on NASDA's ETS-VII Satellite 1390
    Mitsushige Oda

Path Planning and Control for AERCam, a Free-Flying Inspection Robot in Space 1396
    Howie Choset, Ross Knepper, Joleen Flasher, Sean Walker, Andrew Alford,
    Dean Jackson, David Kortenkamp, Jaime Fernandez, and Robert Burridge

New Mobility System for Small Planetary Body Exploration 1404
    Tetsuo Yoshimitsu, Ichiro Nakatani, and Takashi Kubota

Automated Planning for a Deep Space Communications Station 1410
    T. Estlin, F. Fisher, D. Mutz, and S. Chien

TPI-5
CALIBRATION AND TOLERANCES

The Use of Kinematic Model to Analyze Positional Tolerances in Assemblies 1418
    L. Pino, F. Bennis, and C. Fortin

A Generic Algorithm for CAD-Directed CMM Dimensional Inspection Planning 1424
    Yueh-Jaw Lin and Rahul Mahabaleshwarkar

Development of a Non-Linear psi-Angle Model for Large Misalignment Errors and Its
Application in INS Alignment and Calibration 1430
    Xiaoying Kong, Eduardo Mario Nebot, and Hugh Durrant-Whyte

A Geometric Approach to Establishment of Datum Reference Frames 1436
    J. B. Gou, Y. X. Chu, and Z. X. Li

TPI-6
PRODUCTION PLANNING AND CONTROL

Production Planning and Control in Flexibly Automated Manufacturing Systems:
Current Status and Future Requirements 1442
    Spyros A. Reveliotis

Dynamic Lead Time Modeling for JIT Production Planning 1450
    Michael C. Caramanis and Osman M. Anli

Message-Oriented Decomposition for Modeling Supervisory Control in Manufacturing System 1456
    X. Hua Du and Chen Zhou

Optimal Control of Production Systems with Unreliable Machines and Finite Buffers 1462
    F. Balduzzi, G. Menga, and A. Giua

TPI-7
FORCE CONTROL

Six-Axis Force Control for Walking Robot with Serial/Parallel Hybrid Mechanism 1469
    Yusuke Ota, Kan Yoneda, Shigeo Hirose, and Fumitoshi Ito

A Force Control for a Robot Finger Under Kinematic Uncertainties 1475
    Zoe Doulgeri and Suguru Arimoto

Hard Contact Surface Tracking for Industrial Manipulators with (SR) Position Based Force Control 1481
    R. Maass, V. Zahn, M. Dapper, and R. Eckmiller

A General Contact Model for Dynamically-Decoupled Force/Motion Control 3281
    Roy Featherstone, Stef Sonck Thiebaut, and Oussama Khatib

TPI-8
FUZZY CONTROL II

Visual Servoing of a Robotic Manipulator Based on Fuzzy Logic Control 1487
    R. De Guiseppe, F. Taurisano, C. Distante, and A. Anglani

A Robust Fuzzy Controller for Multiple Tentacle Cooperative Robots 1495
    Mircea Ivanescu and Nicu Bizdoaca

Modeling Product Quality in a Machining Center Using Fuzzy Petri Nets with Neural Networks 1502
    M. Hanna

Fuzzy Admission Control and Scheduling of Production Systems 1508
    Runtong Zhang and Yannis A. Phillis

TPI-9
KINEMATICS

Inverse Kinematic Solution Based on Decomposed Manipulability 1514
    Jihong Lee and K. T. Won

Redundant Actuation for Improving Kinematic Manipulability 1520
    John O'Brien and John T. Wen

Completeness Results for a Point-to-Point Inverse Kinematics Algorithm 1526
    Juan Manuel Ahuactzin and Kamal Gupta

Dualities Between Serial and Parallel "321" Manipulators 1532
    Herman Bruyninckx

TPI-10
COOPERATIVE ROBOTS

Multiple Cooperating Mobile Manipulators 1538
    Thomas Sugar and Vijay Kumar

A Decentralized Approach to the Conflict-Free Motion Planning for Multiple Mobile Robots 1544
    Chun Li, Zhiqiang Zheng, and Wensen Chang

Group Behavior Control for MARS (Micro Autonomous Robotic System) 1550
    Toshio Fukuda, Hiroo Mizoguchi, and Kosuke Sekiyama

Control of Changes in Formation for a Team of Mobile Robots 1556
    Jaydev P. Desai, Vijay Kumar, and James P. Ostrowski

TPI-11
FIXTURE DESIGN AND MANIPULATION PLANNING

A Task-Dependent Approach to Minimum-Deflection Fixtures 1562
    Qiao Lin and Joel W. Burdick

Folding Cartons with Fixtures: A Motion Planning Approach 1570
    Liang Lu and Srinivas Akella

Automated Fixture Layout Design for 3D Workpieces 1577
    Michael Y. Wang

Dexterity Through Rolling: Manipulation of Unknown Objects 1583
    Antonio Bicchi, Alessia Marigo, and Domenico Prattichizzo

TPI-12
CALIBRATION-FREE VISUAL SERVO

Model-Based 3D Object Tracking Using Projective Invariance 1589
    Sung-Woo Lee, Bum-Jae You, and Gregory D. Hager

A Dynamic Quasi-Newton Method for Uncalibrated Visual Servoing 1595
    Jenelle Armstrong Piepmeier, Gary V. McMurray, and Harvey Lipkin

Matching of Affinely Invariant Regions for Visual Servoing 1601
    T. Tuytelaars, L. Van Gool, L. D'haene, and R. Koch

Task Specification and Monitoring for Uncalibrated Hand/Eye Coordination 1607
    Z. Dodds, G. D. Hager, A. S. Morse, and J. P. Hespanha

TPI-13
SENSING AND SENSOR DESIGN

Fluorescent Dye Based Optical Position Sensing for Planar Linear Motors 1614
    Gregory A. Fries, Alfred A. Rizzi, and Ralph L. Hollis

What Kind of Haptic Perception Can We Get with a One-Wire Interface? 1620
    C. Melchiorri, G. Vassura, and P. Arcara

An Implementation of a Gyro Actuator for the Attitude Control of an Unstructured Object 1626
    Keon Young Yi and Young Gu Chung

Analysis of Parallel Beam Gyroscope 1632
    Hiroshi Sato, Toshio Fukuda, Fumihito Arai, Hitoshi Iwata, and Kouichi Itoigawa

TPII-1
MOBILE ROBOT LOCALIZATION II

Fusion of Fixation and Odometry for Vehicle Navigation 1638
    Amit Adam, Ehud Rivlin, and Héctor Rotstein

Using Infrared Sensors and the Phong Illumination Model to Measure Distances 1644
    P. M. Novotny and N. J. Ferrier

Real Time Position Estimation for Mobile Robots by Means of Sonar Sensors 1650
    C. Urdiales, A. Bandera, R. Ron, and F. Sandoval

Circumventing Dynamic Modeling: Evaluation of the Error-State Kalman Filter
Applied to Mobile Robot Localization 1656
    Stergios I. Roumeliotis, Gaurav S. Sukhatme, and George A. Bekey

Volume 3 (pages 1665-2494)

TPII-2
MOBILE ROBOT MOTION PLANNING IV

Real-Time Control of a Mobile Robot by Using Visual Stimuli 1665
    Nicola Ancona and Antonella Branca

A Probabilistic Roadmap Approach for Systems with Closed Kinematic Chains 1671
    Steven M. LaValle, Jeffery H. Yakey, and Lydia E. Kavraki

Visibility-Based Pursuit-Evasion: The Case of Curved Environments 1677
    Steven M. LaValle and John Hinrichsen

A Parallel Algorithm to Construct Voronoi Diagram and Its VLSI Architecture 1683
    N. Sudha, S. Nandi, and K. Sridharan

TPII-3
LEGGED LOCOMOTION II

The ARL Monopod II Running Robot: Control and Energetics 1689
    M. Ahmadi and M. Buehler

Control of Hopping Height in Legged Robots Using a Neural-Mechanical Approach 1695
    Matthew D. Berkemeier and Kamal V. Desai

Passive Walking with Leg Compliance for Energy Efficient Multilegged Vehicles 1702
    Eric Y. Raby and David E. Orin

Increasing the Locomotive Stability Margin of Multilegged Vehicles 1708
    Fan-Tien Cheng, Hao-Lun Lee, and David E. Orin

TPII-4
MEDICAL ROBOTICS I

Virtual Endoscope System with Force Sensation 1715
    Koji Ikuta, Masaki Takeichi, and Takao Namiki

Haptic Control of the Master Hand Controller for a Microsurgical Telerobot System 1722
    Dong-Soo Kwon, Ki Young Woo, and Hyung Suck Cho

Force Display Method to Improve Safety in Teleoperation System for Intravascular Neurosurgery 1728
    Mitsutaka Tanimoto, Fumihito Arai, Toshio Fukuda, and Makoto Negoro

On Inverse Kinematics and Trajectory Planning for Tele-Laparoscopic Manipulation 1734
    Ali Faraz and Shahram Payandeh

TPII-5
CALIBRATION AND FRICTION MODELING

Determination of Viscous and Coulomb Friction by Using Velocity Responses to Torque Ramp Inputs 1740
    Rafael Kelly and Jesús Llamas

A Pressure-Based, Velocity Independent, Friction Model for Asymmetric Hydraulic Cylinders 1746
    Adrian Bonchis, Peter I. Corke, and David C. Rye

Geometric Algorithms for Closed Chain Kinematic Calibration 1752
    C. C. Iurascu and F. C. Park

Calibration of Parallel Robots with Two Inclinometers 1758
    S. Besnard and W. Khalil

TPII-6
PROCESS PLANNING AND MANUFACTURING

Modeling Process Planning Problems in an Optimization Perspective 1764
    Y. F. Zhang, G. H. Ma, and A. Y. C. Nee

Mobility Analysis for Feasibility Studies in CAD Models of Industrial Environments 1770
    C. Van Geem, T. Siméon, J-P. Laumond, J-L. Bouchet, and J-F. Rit

Development of a Collaborative and Event-Driven Supply Chain Information
System Using Mobile Object Technology 1776
    Eric Huang, Fan-Tien Cheng, and Haw Ching Yang

Remote Rapid Manufacturing with "Action Media" as an Advanced User Interface 1782
    Mamoru Mitsuishi, Katsuya Tanaka, Yasuyoshi Yokokohji, and Takaaki Nagao

TPII-7
IMPEDANCE CONTROL

Spatial Impedance Control of Redundant Manipulators 1788
    Ciro Natale, Bruno Siciliano, and Luigi Villani

Adaptive Impedance Modification of a Master-Slave Manipulator 1794
    A. Rubio, A. Avello, and J. Florez

Adaptive Force Tracking Impedance Control of Robot for Cutting Nonhomogeneous Workpiece 1800
    Seul Jung and T. C. Hsia

Stable Neuro-Adaptive Control for Robot with Unknown Dynamics 1806
    Fuchun Sun, Zengqi Sun, Kab-Il Kim, Yunyue Zhu, and Lu Wenjuan

TPII-8
ROBOTIC CONTROL IV

Closed-Loop Control of a Base XY Stage with Rotational Degree-of-Freedom
for a High-Speed Ultra-Accurate Manufacturing System 1812
    Hemant Melkote and Farshad Khorrami

Adaptive Control of Time-Varying Mechanical Systems: Modeling, Controller Design and Experiments 1818
    Prabhakar R. Pagilla, Kiu Ling Pau, and Biao Yu

A General Framework for Cobot Control 1824
    R. Brent Gillespie, J. Edward Colgate, and Michael Peshkin

Global Convergence of the Adaptive PD Controller with Computed Feedforward for Robot Manipulators 1831
    Víctor Santibañez and Rafael Kelly

TPII-9
DISTANCE AND CONTACT CALCULATIONS

Computing Form-Closure Configurations 1837
    A. Frank van der Stappen, Chantal Wentink, and Mark H. Overmars

Bound Coherence for Minimum Distance Computations 1843
    David Johnson and Elaine Cohen

Green's Function Contact Maps for Accurate Real Time Collisions 1849
    C. Ullrich and D. K. Pai

Benefits of Applicability Constraints in Decomposition-Free Interference
Detection Between Nonconvex Polyhedral Models 1856
    P. Jiménez and C. Torras

TPII-10
TELEOPERATION III: EXPERIMENTS AND CONTROL

On the Use of Local Force Feedback for Transparent Teleoperation 1863
    K. Hashtrudi-Zaad and S. E. Salcudean

Experiments with Transparent Teleoperation Under Position and Rate Control 1870
    Wen-Hong Zhu, S. E. Salcudean, and Ming Zhu

Automatic Property Identification via Parameterized Constraints 1876
    Thomas Debus, Pierre Dupont, and Robert Howe

Teleoperation via Bilateral Behavior Media: Control, Accumulation, and Assistance 1882
    Stephen Palm, Taketoshi Mori, and Tomomasa Sato

TPII-11
GRASPING COMPUTATION

A Fast and Robust Grasp Planner for Arbitrary 3D Objects 1890
    Ch. Borst, M. Fischer, and G. Hirzinger

Computing Parallel-Jaw Grips 1897
    Gordon Smith, Eric Lee, Ken Goldberg, Karl Böhringer, and John Craig

Constructing 3D Frictional Form-Closure Grasps of Polyhedral Objects 1904
    Yun-Hui Liu, Dan Ding, and Shuguo Wang

Segmenting Manipulative Hand Movements by Dividing Phase Plane Trajectories 1910
    M. Zacksenhouse and Thomas Moestl

TPII-12
REAL-TIME COMPUTER VISION

Self Windowing for High Speed Vision 1916
    Idaku Ishii and Masatoshi Ishikawa

Panoramic Video with Predictive Windows for Telepresence Applications 1922
    Jonathan Baldwin, Anup Basu, and Hong Zhang

Frame-Rate Stereopsis Using Non-Parametric Transforms and Programmable Logic 1928
    Peter I. Corke, Paul A. Dunn, and Jasmine E. Banks

Fast 3D Boundary Computation from Occluding Contour Motion 1934
    A. Bendiksen and G. D. Hager

TPII-13
SENSOR FUSION I

Contact Localization by Multiple Active Antenna 1942
    Naohiro Ueno and Makoto Kaneko

Stereo Based Registration of Multi-Sensor Imagery for Enhanced Visualization of Remote Environments 1948
    Mark D. Elstrom and Philip W. Smith

Sensor Fusion and Calibration for Motion Captures Using Accelerometers 1954
    Jihong Lee and Insoo Ha

Achieving Efficient Data Fusion Through Integration of Sensory Perception Control and Sensor Fusion 1960
    Tomasz Celinski and Brenan McCarragher

FAI-1
MOBILE ROBOT LOCALIZATION III

Estimation of Vehicle Pose and Road Curvature Based on Perception-Net 1966
    Sukhan Lee, Jae-Won Lee, Dongmok Shin, Woong Kwon, Dong-Yoon Kim,
    Kyoungsig Roh, and Kwang S. Boo

Learning Visual Landmarks for Pose Estimation 1972
    Robert Sim and Gregory Dudek

Smoother Based 3-D Attitude Estimation for Mobile Robot Localization 1979
    Stergios I. Roumeliotis, Gaurav S. Sukhatme, and George A. Bekey

Subpixel Localization and Uncertainty Estimation Using Occupancy Grids 1987
    Clark F. Olson

FAI-2
MOBILE ROBOT SYSTEMS I

Two Compact Robots for Remote Inspection of Hazardous Areas in Nuclear Power Plants 1993
    J. Savall, A. Avello, and L. Briones

MINERVA: A Second-Generation Museum Tour-Guide Robot 1999
    Sebastian Thrun, Maren Bennewitz, Wolfram Burgard, Armin B. Cremers, Frank Dellaert, Dieter Fox,
    Dirk Hähnel, Charles Rosenberg, Nicholas Roy, Jamieson Schulte, and Dirk Schulz

Traversability Index: A New Concept for Planetary Rovers 2006
    Homayoun Seraji

Adaptive Intelligent Asitance Control of Electrical Wheelchairs by Grey Fuzzy Decision-Making Algorithm 2014
    Ren C. Luo, Tse Min Chen, Zu Hung Hsiao, and Chi-Yang Hu

FAI-3
LEGGED LOCOMOTION III

Analysis and Design Approach to Inchworm Robotic Insects 2020
    Nicolae Lobontiu, Michael Goldfarb, and Ephrahim Garcia

Gait Generation for Inchworm-Like Robot Locomotion Using Finite State Model 2026
    I-Ming Chen, Song Huat Yeo, and Yan Gao

Study on Roller-Walker: System Integration and Basic Experiments 2032
    Gen Endo and Shigeo Hirose

An Investigation into Non-Smooth Locomotion 2038
    Milos Zefran, Francesco Bullo, and Jim Radford

FAI-4
MEDICAL ROBOTICS II

Design and Control of an Active Mattress for Moving Bedridden Patients 2044
    William H. Finger and H. Harry Asada

A Semi-Active, Flexible, Beaded Support Surface for Tangential Transport and Tissue
Therapy of Bedridden Patients 2051
    Joseph Spano and Haruhiko H. Asada

A Dexterous Manipulator for Minimally Invasive Surgery 2057
    Mark Minor and Ranjan Mukherjee

General Evaluation Method of Safety for Human-Care Robots 2065
    Koji Ikuta and Makoto Nokata

FAI-5
LEARNING AND IDENTIFICATION

A Novel Computationally Intelligent Architecture for Identification and Control of Nonlinear Systems 2073
    M. Önder Efe, Okyay Kaynak, and Imre J. Rudas

A Heuristic Q-Learning Architecture for Fully Exploring a World and Deriving
an Optimal Policy by Model-Based Planning 2078
    Gang Zhao, Shoji Tatsumi, and Ruoying Sun

Learning Accurate Path Control of Industrial Robots with Joint Elasticity 2084
    Friedrich Lange and Gerhard Hirzinger

Multiple Hypothesis Testing Method for Decision Making 2090
    Xiao-gang Wang and Helen C. Shen

FAI-6
MANUFACTURING AUTOMATION

Force-Responsive Robotic Assembly of Transmission Components 2096
    Wyatt S. Newman, Michael S. Branicky, H. Andy Podgurski, Siddharth Chhatpar,
    Ling Huang, Jayendran Swaminathan, and Hao Zhang

Clustering of Qualitative Contact States for a Transmission Assembly 2103
    Marjorie Skubic, Benjamin Forrester, and Brent Nowak

A Hierarchical Method to Improve the Productivity of a Multi-Head Surface Mounting Machine 2110
    S. H. Lee, B. H. Lee, and T. H. Park

Manufactuability Analysis in 5-Axis Sculptured Surface Machining 2116
    Wenyu Yang, Han Ding, and Youlun Xiong

FAI-7
COMPLIANCE CONTROL

Position Control of a Compliant Mechanism Based Micromanipulator 2122
    Kevin Fite and Michael Goldfarb

Robust Robot Compliant Motion Control Using Intelligent Adaptive Impedance Approach 2128
    Dragoljub Surdilovic and Zarko Cojbasic

Intelligent Compliance Control for Robot Manipulators Using Adaptive Stiffness Characteristics 2134
    Byoung-Ho Kim, Nak Young Chong, Sang-Rok Oh, Il Hong Suh, and Young-jo Cho

An Independent Joint-Based Compliance Control Method for a
Five-Bar Finger Mechanism via Redundant Actuators 2140
    B. R. So, B.-J. Yi, S.-R. Oh, and I. H. Suh

FAI-8
ROBOTIC STIFFNESS CONTROL

Synthesis of Cartesian Stiffness for Robotic Applications 2147
    Namik Ciblak and Harvey Lipkin

Minimal Realization of a Spatial Stiffness Matrix with Simple Springs Connected in Parallel 2153
    Rodney G. Roberts

Ideal Motion Control Using Pre-Shape Sliding Mode Controller 2159
    Ying Ma, Geok Soon Hong, and Aun Neow Poo

A Limitation of Position Based Impedance Control in Static Force Regulation: Theory and Experiments 2165
    B. Heinrichs and N. Sepehri

FAI-9
MICROMANIPULATOR

Development of a New Type of Capsule Micropump 2171
    Shuxiang Guo, Seiji Hata, Koichi Sugumoto, Toshio Fukuda, and Keisuke Oguro

Multi-DOF Device for Soft Micromanipulation Consisting of Soft Gel Actuator Elements 2177
    Satoshi Tadokoro, Shinji Yamagami, Masahiro Ozawa, Tetsuya Kimura, Toshi Takamori, and Keisuke Oguro

Design and Analysis of a 3-DOF Micromanipulator 2183
    James Nielsen, Tamio Tanikawa, and Tatsuo Arai

Micro Object Handling Under SEM by Vision-Based Automatic Control 2189
    Takeshi Kasaya, Hideki Miyazaki, Shigeki Saito, and Tomomasa Sato

FAI-10
TELEMANIPULATION

Model-Based Variable Position Mapping for Telerobotic Assistance in a Cylindrical Environment 2197
    S. E. Everett and R. V. Dubey

Tele-Control of Rapid Prototyping Machine via Internet for Automated Tele-Manufacturing 2203
    Ren C. Luo, Wei Zen Lee, Jyh Hwa Chou, and Hou Tin Leong

Distributed Generic Control for Multiple Types of Telerobot 2209
    Alan R. Graves and Chris Czarnecki

Operating in Configuration Space Significantly Improves Human Performance in Teleoperation 2215
    I. Ivanisevic and V. Lumelsky

FAI-11
PARTS MANIPULATION

A Miniature Mobile Parts Feeder: Operating Principles and Simulation Results 2221
    Arthur E. Quaid

On Manipulating Polygonal Objects with Three 2-DOF Robots in the Plane 2227
    Attawith Sudsang, Jean Ponce, Mark Hyman, and David J. Kriegman

Discrete Actuator Array Vectorfield Design for Distributed Manipulation 2235
    Jonathan E. Luntz, William Messner, and Howie Choset

Open-Loop Orientability of Objects on Actuator Arrays 2242
    Jonathan E. Luntz, William Messner, and Howie Choset

FAI-12
VISION-BASED MOBILE ROBOTICS

Motion Control Using Visual Servoing and Potential Fields for a Rover-Mounted Manipulator 2249
    Ricardo Swain Oropeza and Michel Devy

Probabilistic Localization by Appearance Models and Active Vision 2255
    B. J. A. Kröse and R. Bunschoten

A Real-Time Occupancy Map from Multiple Video Streams 2261
    Adam Hoover and Bent David Olsen

A Trinocular Stereo System for Highway Obstacle Detection 2267
    Todd Williamson and Charles Thorpe

FAI-13
SENSOR FUSION II

A New Algorithm for the Alignment of Inertial Measurement Units Without
External Observation for Land Vehicle Applications 2274
    Gamini Dissanayake, Eduardo Nebot, Salah Sukkarieh, and Hugh Durrant-Whyte

Semi-Automatic Acquisition of Symbolically-Annotated 3D-Models of Office Environments 2280
    Michael Beetz, Markus Giesenschlag, Roman Englert, Eberhard Gülch, and Armin B. Cremers

Rapid Physics-Based Rough-Terrain Rover Planning with Sensor and Control Uncertainty 2286
    Karl Iagnemma, Frank Genot, and Steven Dubowsky

Online Self-Calibration for Mobile Robots 2292
    Nicholas Roy and Sebastian Thrun

FAII-1
MOBILE INSPECTION AND RECONNAISSANCE SYSTEMS

Cyclops: Miniature Robotic Reconnaissance System 2298
    Brian Chemel, E. Mutschler, and H. Schempf

ROMA: A Climbing Robot for Inspection Operations 2303
    M. Abderrahim, C. Balaguer, A. Giménez, J. M. Pastor, and V. M. Padrón

Design of In-Pipe Inspection Vehicles for φ25, φ50, φ150 Pipes 2309
    Shigeo Hirose, Hidetaka Ohno, Takeo Mitsui, and Kiichi Suyama

Pandora: Autonomous Urban Robotic Reconnaissance System 2315
    Hagen Schempf, E. Mutschler, C. Piepgras, J. Warwick, B. Chemel, S. Boehmke,
    W. Crowley, R. Fuchs, and J. Guyot

FAII-2
MOBILE ROBOT SYSTEMS II

A Mobile Manipulator 2322
    Matthew T. Mason, Dinesh Pai, Daniela Rus, Lee R. Taylor, and Michael A. Erdmann

A Control System Development Environment for AURORA's Semi-Autonomous Robotic Airship 2328
    Ely Carneiro de Paiva, Samuel S. Bueno, Sergio B. Varella Gomes,
    Marcel Bergerman, and Josué J. G. Ramos

Fundamental Performance of 6-Wheeled Off-Road Vehicle "HELIOS-V" 2336
    Yasuyuki Uchida, Kazuya Furuichi, and Shigeo Hirose

Experimental Study of a Cable-Driven Suspended Platform 2342
    M. A. Rahimi, H. Hemami, and Yuan F. Zheng

FAII-3
QUADRUPED LOCOMOTION

Stable Open Loop Walking in Quadruped Robots with Stick Legs 2348
    M. Buehler, A. Cocosco, K. Yamazaki, and R. Battaglia

Adaptive Periodic Movement Control for the Four Legged Walking Machine BISAM 2354
    W. Ilg, J. Albiez, H. Jedele, K. Berns, and R. Dillmann

Towards Efficient Implementation of Quadruped Gaits with Duty Factor of 0.75 2360
    Vincent Hugel and Pierre Blazevic

A Quadruped Robot Which Can Take Various Postures 2366
    Toru Omata and Eiichiro Tanaka

FAII-4
MEDICAL ROBOTICS III

Development of an Electric Heating Type SMA Active Forceps for Laparoscopic Surgery 2372
    Minoru Hashimoto, Tsuyoshi Tabata, and Takahiro Yuki

Two-Lead-Wire Drive for Multi-Micro Actuators 2378
    Koji Ikuta and Makoto Nokata

Shape Memory Alloy Actuated Robot Prostheses: Initial Experiments 2385
    Charles Pfeiffer, Kathryn Delaurentis, and Constantinos Mavroidis

Development of a Myoelectric Discrimination System for a Multi-Degree Prosthetic Hand 2392
    Han-Pang Huang and Chun-Yen Chen

FAII-5
COMPUTATIONAL INTELLIGENCE

Synthesis, Learning and Abstraction of Skills Through Parameterized
Smooth Map from Sensors to Behaviors 2398
    Y. Nakamura, T. Yamazaki, and N. Mizushima

Research on Holographic Virtual Manufacturing Basis 2406
    Zhiyan Wang, Chengxiang Guan, and Zhichao Zhang

Multi-Layer Template Correlation Neural Network for Recognition of Lane Mark
Based on Pipelined Image Processing Structure 2410
    Xiangjing An, Wensen Chang, and Xiangdong Chen

An On-Line Self-Organizing Neuro-Fuzzy Control for Autonomous Underwater Vehicles 2416
    Jeen-Shing Wang, C. S. George Lee, and Junku Yuh

FAII-6
VIRTUAL FACTORY

Virtual Factory: A Novel Testbed for an Advanced Flexible Manufacturing System 2422
    Ming-Hung Lin, Li-Chen Fu, and Teng-Jei Shih

Interactive Design of a Virtual Factory Using Cellular Manufacturing System 2428
    M. Ernzer and T. Kesavadas

Development of a Virtual Factory Emulator Based on Three-Tier Architecture 2434
    Chien-Fa Yeh and Han-Pang Huang

A Customizable Software Infrastructure for Virtual Factories Development 2440
    D. Brugali, D. Dragomirescu, S. Galarraga, and G. Menga

FAII-7
GENETIC ALGORITHMS

Geometrical Synthesis of Star-Like Parallel Manipulators Topology with a Genetic Algorithm 2446
    Alain Tremblay and Luc Baron

Nonlinear MBPC for Mobile Robot Navigation Using Genetic Algorithms 2452
    D. R. Ramírez, D. Limón, J. Gómez-Ortega, and E. F. Camacho

Robot Hand Manipulation by Evolutionary Programming 2458
    Toshio Fukuda, Kenichiro Mase, and Yasuhisa Hasegawa

Motion Planning Based on Hierarchical Knowledge Using Genetic Programming 2464
    Kentarou Kurashige, Toshio Fukuda, and Haruo Hoshino

FAII-8
INTELLIGENT AND FUZZY CONTROL

Intuitionistic, 2-Way Adaptive Fuzzy Control 2470
    Evren Gurkan, A. M. Erkmen, and I. Erkmen

A New Adaptive Fuzzy Hybrid Force/Position Control for Intelligent Robot Deburring 2476
    Feng-Yih Hsu and Li-Chen Fu

Steering Fuzzy Logic Controller for an Autonomous Vehicle 2482
    Neil Eugene Hodge and Mohamed B. Trabia

Share Control in Intelligent Arm/Hand Teleoperated System 2489
    Song You, Tianmiao Wang, Jun Wei, Fenglei Yang, and Qixian Zhang

Volume 4 (pages 2495-3286)

FAII-9
ROBOT MECHANISM

Dynamic Analysis of the Cable Array Robotic Crane 2495
    Wei-Jung Shiang, David Cannon, and Jason Gorman

Design and Kinematic Analysis of Modular Reconfigurable Parallel Robots 2501
    Guilin Yang, I-Ming Chen, Wee Kiat Lim, and Song Huat Yeo

A Robust Control Scheme for Dual-Arm Redundant Manipulators: Experimental Results 2507
    Haipeng Xie, Iain J. Bryson, Farshid Shadpey, and R. V. Patel

Self-Reconfiguration Planning with Compressible Unit Modules 2513
    Daniela Rus and Marsette Vona

FAII-10
SENSOR-BASED HUMAN/MACHINE INTERACTION

What Can Be Learned from Human Reach-to-Grasp Movements for the
Design of Robotic Hand-Eye Systems? 2521
    Alexa Hauck, Michael Sorg, Thomas Schenk, and Georg Färber

Development of a Visual Space-Mouse 2527
    Tobias Peter Kurpjuhn, Alexa Hauck, Kevin Nickels, and Seth Hutchinson

Virtual Switch Human-Machine Interface Using Fingernail Touch Sensors 2533
    Stephen Mascaro and H. Harry Asada

Arm-Manipulator Coordination for Load Sharing Using Predictive Control 2539
    Kamran Iqbal and Yuan F. Zheng

FAII-11
PART FEEDING AND ORIENTING

Part Orienting with a Force/Torque Sensor 2545
    Shawn Rusaw, Kamal Gupta, and Shahram Payandeh

Toppling Manipulation 2551
    Kevin M. Lynch

Trap Design for Vibratory Bowl Feeders 2558
    Robert-Paul Berretty, Ken Goldberg, Lawrence Cheung, Mark H. Overmars,
    Gordon Smith, and A. Frank van der Stappen

Testing and Analysis of a Flexible Feeding System 2564
    Greg C. Causey, Roger D. Quinn, and Michael S. Branicky

Planning and Control of Indirect Simultaneous Positioning Operation of Deformable Objects 2572
    Takahiro Wada, Sinichi Hirai, and Sadao Kawamura

FAII-12
VISION-BASED NAVIGATION I

Image-Based Robot Navigation Under the Perspective Model 2578
    Ronen Basri, Ehud Rivlin, and Ilan Shimshoni

Autonomous Dirigible Navigation Using Visual Tracking and Pose Estimation 2584
    Mário Fernando Montenegro Campos and Lúcio de Souza Coelho

Robotic Wheelchair with Three Control Modes 2590
    Yoshinori Kuno, Satoru Nakanishi, Teruhisa Murashima, Nobutaka Shimada, and Yoshiaki Shirai

Experiments of Decision Making Strategies for a Lane Departure Warning System 2596
    Woong Kwon, Jae-Won Lee, Dongmok Shin, Kyoungsig Roh, Dong-Yoon Kim, and Sukhan Lee

FAII-13
SENSOR SELECTION AND PLACEMENT

A Robust Approach for Sensor Placement in Automated Vision Dimensional Inspection 2602
    Xiangrong Gu, Michael M. Marefat, and Frank W. Ciarallo

Improving Sensory Perception Through Predictive Correction of Monitoring Errors 2608
    Tomasz Celinski and Brenan McCarragher

Sensor Selection by Reliability Based on Possibility Measure 2614
    F. Kobayashi, F. Arai, and T. Fukuda

The Sensor Selection Task of The Gaussians Mixture Bayes' with Regularised EM
(GMB-REM) Technique in Robot Position Estimation 2620
    Takamasa Koshizen

FPI-1
MOBILE ROBOT MOTION CONTROL I

Point Stabilization of Mobile Robots via State Space Exact Feedback Linearization 2626
    KyuCheol Park, Hakyoung Chung, and Jang Gyu Lee

Trajectory Tracking Control of a Four-Wheel Differentially Driven Mobile Robot 2632
    Luca Caracciolo, Alessandro De Luca, and Stefano Iannitti

Tracking Control of Mobile Robots Using Saturation Feedback Controller 2639
    Ti-Chung Lee, Kai-Tai Song, Ching-Hung Lee, and Ching-Cheng Teng

Tracking Control of Wheeled Mobile Robots with Unknown Dynamics 2645
    Wenjie Dong and Wei Huo

FPI-2
MOBILE ROBOT SENSOR-BASED CONTROL

Unifying Exploration, Localization, Navigation, and Planning Through a Common Representation 2651
    Alan C. Schultz, William Adams, Brian Yamauchi, and Mike Jones

Cooperative Robot Localization with Vision-Based Mapping 2659
    Cullen Jennings, Don Murray, and James J. Little

Motion Control of Multiple Autonomous Mobile Robots Handling a Large Object in Coordination 2666
    Kazuhiro Kosuge, Yasuhisa Hirata, Hajime Asama, Hayato Kaetsu, and Kuniaki Kawabata

Analysis of Deformable Object Handling 2674
    Herbert G. Tanner and Kostas J. Kyriakopoulos

FPI-3
SMALL SCALE MOBILE ROBOTS

Development of BEST Nano-robot Soccer Team 2680
    Sung Ho Kim, Jong Suk Choi, and Byung Kook Kim

The Miniature Omni-Directional Mobile Robot OmniKity-I (OK-I) 2686
    Myung-Jin Jung, Hyun-Sik Shim, Heung-Soo Kim, and Jong-Hwan Kim

Microgravity Experiment of Hopping Rover 2692
    Tetsuo Yoshimitsu, Takashi Kubota, Ichiro Nakatani, Tadashi Adachi, and Hiroaki Saito

Ant Trails -- An Example for Robots to Follow? 2698
    R. Andrew Russell

FPI-4
MOBILE ROBOT MOBILITY AND LOCOMOTION

Dynamics and Distributed Control of Tetrobot Modular Robots 2704
    Woo Ho Lee and Arthur C. Sanderson

Force Distribution Equations for General Tree-Structured Robotic Mechanisms with a Mobile Base 2711
    Min-Hsiung Hung, David E. Orin, and Kenneth J. Waldron

Brachiation on a Ladder with Irregular Intervals 2717
    Jun Nakanishi, Toshio Fukuda, and Daniel E. Koditschek

Robust Sensing for a 3,500 Tonne Field Robot 2723
    Jonathan M. Roberts, Frederic Pennerath, Peter I. Corke, and Graeme J. Winstanley

FPI-5
MICRO/NANO MANIPULATION

Two-Dimensional Fine Particle Positioning Using a Piezoresistive
Cantilever as a Micro/Nano-Manipulator 2729
    Metin Sitti and Hideki Hashimoto

Pick and Place Operation of a Micro Object with High Reliability and Precision
Based on Micro Physics Under SEM 2736
    Shigeki Saito, Hideki Miyazaki, and Tomomasa Sato

Micro Tri-Axial Force Sensor for 3D Bio-Micromanipulation 2744
    Fumihito Arai, Tomohiko Sugiyama, Toshio Fukuda, Hitoshi Iwata, and Kouichi Itoigawa

Modelling of a Piezohydraulic Actuator for Control of a Parallel Micromanipulator 2750
    Quan Zhou, Pasi Kallio, and Heikki N. Koivo

FPI-6
ASSEMBLY PLANNING

Identification of Assembly Process States Using Polyhedral Convex Cones 2756
    H. Mosemann, A. Raue, and F. Wahl

Disassembly Sequencing and Assembly Sequence Verification Using Force Flow Networks 2762
    Sukhan Lee and Hadi Moradi

Adaptive Accommodation Control for Complex Assembly: Theory and Experiment 2768
    Sungchul Kang, Munsang Kim, Chong W. Lee, and Kyo-Il Lee

Compliant-Motion Planning and Execution for Robotic Assembly 2774
    Jan Rosell, Luis Basañez, and Raúl Suárez

FPI-7
LEARNING CONTROL

Learning Friction Estimation for Sensorless Force/Position Control in Industrial Manipulators 2780
    V. Zahn, R. Maass, M. Dapper, and R. Eckmiller

Learning of Robot Tasks via Impedance Matching 2786
    Suguru Arimoto Tomohide Naniwa, and P. T. A. Nguyen

Releasing Manipulation with Learning Control 2793
    Chi Zhu, Yasumichi Aiyama, and Tamio Arai

Learning Method for Hierarchical Behavior Controller 2799
    Yasuhisa Hasegawa and Toshio Fukuda

FPI-8
MANIPULATOR CONTROL

Feedback Control of a Planar Manipulator with an Unactuated Elastically Mounted End Effector 2805
    Mahmut Reyhanoglu, Sangbum Cho, and N. Harris McClamroch

Posture Control of Casting Manipulation 2811
    Hitoshi Arisumi and Kiyoshi Komoriya

Achieving Fine Absolute Positioning Accuracy in Large Powerful Manipulators 2819
    Marco A. Meggiolaro, Peter C. L. Jaffe, and Steven Dubowsky

Performance of Linear Decentralized H-Infinity Optimal Control for Industrial Robotic Manipulators 2825
    Jonghoon Park, Wan Kyun Chung, and Youngil Youm

FPI-9
HYPER-REDUNDANT ROBOTS

Development of a Lightweight Torque Limiting M-Drive Actuator for
Hyper-Redundant Manipulator Float Arm 2831
    Shigeo Hirose and Richard Chu

The Shape Jacobian of a Manipulator with Hyper Degrees of Freedom 2837
    Hiromi Mochiyama and Hisato Kobayashi

A Geometric Approach to Anguilliform Locomotion: Modelling with an Underwater Eel Robot 2843
    Ken McIsaac and James P. Ostrowski

Continuum Robots: A State of the Art 2849
      G. Robinson and J. B. C. Davies

FPI-10
HUMAN-ROBOT INTERACTION I

New Sensors for New Applications: Force Sensors for Human/Robot Interaction 2855
      Andy Lorenz, Michael A. Peshkin, and J. Edward Colgate

Construction of a Human/Robot Coexistence System Based on a Model of
Human Will – Intention and Desire 2861
      Yoji Yamada, Yoji Umetani, Haruyoshi Daitoh, and Takayuki Sakai

Emergence of Emotional Behavior Through Physical Interaction Between Human and Robot 2868
      Takanori Shibata, Toshihiro Tashima, and Kazuo Tanie

A Human-Robot Interface Using an Extended Digital Desk 2874
      Maho Terashima and Shigeyuki Sakane

FPI-11
REASONING AND HANDLING OF OBJECTS

Automatic Orienting of 3D Shapes by Using a New Data Structure for Object Modeling 2881
      A. Adán, C. Cerrada, and V. Feliu

Toward Development of a Generalized Contact Algorithm for Polyhedral Objects 2887
      Christopher A. Tenaglia, David E. Orin, Robert A. LaFarge, and Chris Lewis

Estimation of Mass and Center of Mass of Graspless and Shape-Unknown Object 2893
      Yong Yu, Kenro Fukuda, and Showzow Tsujio

4-Axis Electromagnetic Microgripper 2899
      Arianna Menciassi, Blake Hannaford, Maria Chiara Carrozza, and Paolo Dario

FPI-12
VISION-BASED NAVIGATION II

Integration of Two Complementary Depth Mapping Processes: Motion Parallax and Focus Principle .... *
      P. Bonzom and B. Jouvencel

Goal Directed Reactive Robot Navigation with Relocation Using Laser and Vision 2905
      J. R. Asensio, J. M. M. Montiel, and L. Montano

Vision-Based Self-Localization of a Mobile Robot Using a Virtual Environment 2911
      M. Schmitt, M. Rous, A. Matsikis, and K.-F. Kraiss

Continuous Mobile Robot Localization: Vision vs. Laser 2917
      J. A. Pérez, J. A. Castellanos, J. M. M. Montiel, J. Neira, and J. D. Tardós

FPI-13
HAPTIC DISPLAY

Assistance System for Crane Operation with Haptic Display: Operational
Assistance to Suppress Round Payload Swing 2924
    Mitsunori Yoneda, Fumihito Arai, Toshio Fukuda, Keisuke Miyata, and Toru Naito

Haptic Exploration of Fine Surface Features 2930
    Allison M. Okamura and Mark R. Cutkosky

Passive Implementations for a Class of Static Nonlinear Environments in Haptic Display 2937
    Brian E. Miller, J. Edward Colgate, and Randy Freeman

A Three-Dimensional Touch/Force Display System for Haptic Interface 2943
    Tsuneo Yoshikawa and Akihiro Nagura

FPII-1
MOBILE ROBOT MOTION CONTROL II

Numerically Efficient Trajectory Tracking Control of Polynomic Nonlinear Systems 2952
    Raj Madhavan

Quasi-Time-Optimal Motion Planning of Mobile Platforms in the Presence of Obstacles 2958
    Motoji Yamamoto, Makoto Iwamura, and Akira Mohri

Analysis and Design of Non-Time Based Motion Controller for Mobile Robots 2964
    Wei Kang, Ning Xi, and Jindong Tan

Sensor Fusion Based on Fuzzy Kalman Filtering for Autonomous Robot Vehicle 2970
    J. Z. Sasiadek and Q. Wang

FPII-2
MOBILE ROBOT SENSING

Environmental Support Method for Mobile Robots Using Visual Marks with Memory Storage 2976
    Jun Ota, Masakazu Yamamoto, Kazuo Ikeda, Yasumichi Aiyama, and Tamio Arai

A Real Time Detection Algorithm for Direction Error in Omnidirectional Image Sensors for Mobile Robots ... 2982
    Dong Sung Kim, Young Shin Kim, and Wook Hyun Kwon

Modeling and Classification of Rough Surfaces Using CTFM Sonar Imaging 2988
    Z. Politis and P. J. Probert

Laser Based Pose Tracking 2994
    Patric Jensfelt and Henrik I. Christensen

FPII-3
ARTICULATED LOCOMOTION

Limbless Locomotion: Learning to Crawl 3001
    Kevin Dowling

Analysis of Snake Movement Forms for Realization of Snake-Like Robots 3007
    Shugen Ma

GMD-SNAKE2: A Snake-Like Robot Driven by Wheels and a Method for Motion Control 3014
    Bernhard Klaassen and Karl L. Paap

Why to Use an Articulated Vehicle in Underground Mining Operations? 3020
    Claudio Altafini

FPII-4
MULTI-FINGER HANDS

Simulating Dextrous Manipulation of a Multi-fingered Robot Hand Based on a Unified Dynamic Model 3026
    Joseph C. Chan and Yun-Hui Liu

Coordinated Motion Generation for Multifingered Manipulation Using Tactile Feedback 3032
    S. L. Jiang, K. K. Choi, and Z. X. Li

A Multi-Fingered End Effector for Unstructured Environments 3038
    R. M. Crowder, V. N. Dubey, P. H. Chappell, and D. R. Whatley

An Off-Line Iterative and On-Line Analytical Force Distribution Approach for Soft Multi-Fingered Hands 3044
    Bing-Ran Zuo, Wen-Han Qian, and Günther Seliger

FPII-5
INDUSTRIAL INVITED SESSION: INDUSTRIAL APPLICATION OF ROBOT AND AUTOMATION TECHNOLOGIES

Assessment of Feedback Variable for Through-the-Arc Seam Tracking in Robotic Gas Metal Arc Welding 3050
    Jianming Tao and Peter Levick

"Through-Arc" Process Monitoring Techniques for Control of Automated Gas Metal Arc Welding 3053
    Darren Barborak, Troy Paskell, Chris Conrady, and Bruce Madigan

Dynamic Modeling of GMAW Process 3059
    Zafer Bingul and George E. Cook

Development of Impulsive Object Sorting Device with Air Floating 3065
    Shinichi Hirai, Masaaki Niwa, and Sadao Kawamura

FPII-6
COMPUTER-AIDED ASSEMBLY PLANNING

Integrated Computer Tools for Top-Down Assembly Design and Analysis 3071
    R. Mantripragada, J. D. Adams, S. H. Rhee, and D. E. Whitney

The Development of a Rapid Prototyping Machine System for Manufacturing Automation 3079
    Ren C. Luo and Wei Zen Lee

Qualification of Standard Industrial Robots for Micro-Assembly 3085
    Michael Höhn and Christian Robl

Understanding of Mechanical Assembly Instruction Manual by Integrating
Vision and Language Processing and Simulation 3091
    Kazuaki Tanaka, Norihiro Abe, and Hirokazu Taki

FPII-7
ROBOT PROGRAMMING

An Integrated Interface Tool for the Architecture for Agile Assembly 3097
    Jay Gowdy and Zack Butler

Programming in the Architecture for Agile Assembly 3103
    Jay Gowdy and Alfred A. Rizzi

An Approach to Automated Programming of Industrial Robots 3109
    Donald R. Myers

An Object-Oriented Realtime Framework for Distributed Control Systems 3115
    R. D. Schraft and A. Traub

FPII-8
NEURAL NETWORK APPLICATIONS

Neural Adaptive Control of Two-Link Manipulator with Sliding Mode Compensation 3122
    Wen Yu, Alexander S. Poznyak, and Edgar N. Sanchez

How to Compensate Stick-Slip Friction in Neural Velocity Force
Control (NVFC) for Industrial Manipulators 3128
    M. Dapper, V. Zahn, R. Maass, and R. Eckmiller

Transfer of Human Control Strategy Based on Similarity Measure 3134
    Jingyan Song, Yangsheng Xu, Michael C. Nechyba, and Yeung Yam

Modeling of Human Strategy in Controlling Light Source 3140
    Jiong Zhang and Yangsheng Xu

FPII-9
REDUNDANT ROBOTS

Using Redundancy to Reduce Accelerations Near Kinematic Singularities 3146
    K. O'Neil and Y. C. Chen

An Improved Trajectory Planner for Redundant Manipulators in Constrained Workspace 3153
    Tzu-Chen Liang and Jing-Sin Liu

On the Quantification of Robot Redundancy 3159
    Jadran Lenarcic

Learning of Inverse Kinematics Behavior of Redundant Robot 3165
    Goran S. Dordevic, Milan Rasic, Dragan Kostic, and Dragoljub Surdilovic

FPII-10
HUMAN-ROBOT INTERACTION II

Interactions and Motions in Human-Robot Coordination 3171
    J. Y. S. Luh and Shuyi Hu

Emotional Communication Between Humans and the Autonomous Robot Which Has the Emotion Model 3177
    Tetsuya Ogata and Shigeki Sugano

Development of Human Symbiotic Robot: WENDY 3183
    Toshio Morita, Hiroyasu Iwata, and Shigeki Sugano

Reflexive Behavior of Personal Robots Using Primitive Motions 3189
    Li Xu and Yuan F. Zheng

FPII-11
SENSOR-BASED GRASPING

High Speed Grasping Using Visual and Force Feedback 3195
    Akio Namiki, Yoshihiro Nakabo, Idaku Ishii, and Masatoshi Ishikawa

Vision-Aided Object Manipulation by a Multifingered Hand with Soft Fingertips 3201
    Yasuyoshi Yokokohji, Moriyuki Sakamoto, and Tsuneo Yoshikawa

Human Visual Servoing for Reaching and Grasping: The Role of 3-D Geometric Features 3209
    Y. Hu, R. Eagleson, and M. A. Goodale

Sensor Based Control for the Execution of Regrasping Primitives on a Multifingered Robot Hand 3217
    Mohammad Asim Farooqi, Takashi Tanaka, Yukio Ikezawa, Toru Omata, and Kazuyuki Nagata

FPII-12
VISUAL TRACKING

VISP: A Software Environment for Eye-in-Hand Visual Servoing 3224
    Éric Marchand

Measurement Error Estimation for Feature Tracking 3230
    Kevin Nickels and Seth Hutchinson

Visual Servoing of a 6-DOF Manipulator for Unknown 3D Profile Following 3236
    Jacques A. Gangloff, Michel de Mathelin, and Gabriel Abba

Accurate Image Overlay on Head-Mounted Displays Using Vision and Accelerometers 3243
    Yasuyoshi Yokokohji, Yoshihiko Sugawara, and Tsuneo Yoshikawa

FPII-13
HAPTIC INTERFACE

Design of a 3R Cobot Using Continuously Variable Transmissions 3249
    Carl A. Moore, Michael A. Peshkin, and J. Edward Colgate

Development of an Anthropomorphic Head-Eye Robot WE-3RII with an
Autonomous Facial Expression Mechanism 3255
    Atsuo Takanishi, Hideaki Takanobu, Isao Kato, and Tomohiko Umetsu

Interaction with a Realtime Dynamic Environment Simulation Using a
Magnetic Levitation Haptic Interface Device 3261
    Peter J. Berkelman, Ralph L. Hollis, and David Baraff

Guaranteed Convergence Rates for Five Degree of Freedom In-Parallel Haptic Interface Kinematics 3267
    Christopher Lee, Dale A. Lawrence, and Lucy Y. Pao

* Paper not available at time of printing.

Author Index

Abba, Gabriel 3236
Abderrahim, M. 2303
Abe, Norihiro 3091
Abe, Yasunori 1299
Adachi, Tadashi 2692
Adam, Amit 1638
Adams, J. D. 3071
Adams, William 2651
Adán, A. 2881
Aghili, F. 1077
Ahmadabadi, Majid Nili 1226
Ahmadi, M. 1689
Ahuactzin, Juan Manuel 1526
Aiyama, Yasumichi 2793, 2976
Akella, Prasad 728
Akella, Srinivas 1570
Akima, Toshifumi 901
Al-Jarrah, Omar M. 170
Alami, R. 1234
Albiez, J. 2354
Alford, Andrew 1396
Allen, Peter K. 275, 1240
Altafini, Claudio 3020
Amat, Josep 747
Amato, Nancy M. 688, 1024
Ambrosi, Guilio 305
An, Xiangjing 2410
Ancona, Nicola 1665
Ang, M. H. Jr. 1102, 1108
Anglani, A. 1487
Anli, Osman M. 1450
Antonelli, Gianluca 78
Aoki, Takeshi 1347
Arai, F. 2614
Arai, Fumihito 1096, 1299, 1632,
                      1728, 2744, 2924
Arai, Hirohiko 65
Arai, Tamio 2793, 2976
Arai, Tatsuo 2183
Arcara, P. 1620
Arimoto, Suguru 494, 1475, 2786
Arisumi, Hitoshi 2811
Armstrong, Brian 837
Asada, H. Harry 962, 2044, 2051, 2533
Asama, Hajime 2666
Asensio, J. R. 2905
Asfour, T. 702

Aude, E. P. L. 465
Avello, A. 1794, 1993
Backes, Paul G. 984
Bai, Shaoping 1371
Balaguer, C. 2303
Balakrishnan, V. 874, 867
Balcells, A. Civit 7
Balduzzi, F. 1462
Baldwin, Jonathan 1922
Bandera, A. 1650
Banks, Jasmine E. 1928
Bao, Qingshan 1084
Baraff, David 3261
Barborak, Darren 3053
Baron, Luc 2446
Barrientos, A. 565
Basañez, Luis 1359, 2774
Basri, Ronen 2578
Basu, Anup 1922
Battaglia, R. 2348
Bayard, David 441
Beetz, Michael 2280
Bekey, George A. 1656, 1979
Bendiksen, A. 1934
Benhabib, B. 120, 145
Bennewitz, Maren 1999
Bennis, F. 1418
Bergerman, Marcel 2328
Berkelman, Peter J. 3261
Berkemeier, Matthew D. 1695
Berns, K. 702, 2354
Berretty, Robert-Paul 2558
Besnard, S. 1758
Bicchi, Antonio 305, 1583
Bielby, M. 1310
Bingul, Zafer 3059
Bitó, János F. 831
Bizdoaca, Nicu 1495
Blazevic, Pierre 2360
Bloch, Anthony M. 500
Bobrow, James E. 480
Bobyr, S. 157
Boehmke, S. 2315
Böhringer, Karl 1897
Bonchis, Adrian 1746
Bonzom, P. .... *
Boo, Kwang S. 1966

Boor, Valérie	1018	Chan, Joseph C.	3026
Borst, Ch.	1890	Chandra, Vigyan	126
Botelho, S. C.	1234	Chang, K.	741
Bouchet, J-L.	1770	Chang, Kuo-Wei	962
Boyer, F.	759	Chang, Naehyuck	793
Branca, Antonella	1665	Chang, Shih-Lang	3275
Branicky, Michael S.	2096, 2564	Chang, Wen-Chung	605
Brassart, Eric	1287	Chang, Wensen	1544, 2410
Briones, L.	1993	Chappell, P. H.	3038
Brock, Oliver	341	Cheah, C. C.	494
Brown, S.	525	Chemel, Brian	2298, 2315
Brugali, D.	2440	Chen, Chun-Yen	2392
Bruyninckx, Herman	1194, 1532	Chen, Haoxun	1120
Bruzzone, G.	72	Chen, Huitang	41
Bryson, Iain J.	2507	Chen, I-Ming	90, 2026, 2501
Buehler, M.	1689, 2348	Chen, Tse Min	1176, 2014
Bueno, Samuel S.	2328	Chen, Xiangdong	2410
Bullo, Francesco	2038	Chen, Y. C.	3146
Bunschoten, R.	2255	Cheng, Fan-Tien	1708, 1776
Burdick, Joel W.	347, 773, 1562	Cherif, Moëz	317
Burgard, Wolfram	35, 1322, 1999	Cheung, Lawrence	2558
Burridge, Robe	1396	Chew, Chee-Meng	381
Burschka, Darius	695	Chhatpar, Siddharth	2096
Butler, Zack	3097	Chiaverini, Stefano	78
Bshringer, Karl	1897	Chien, Steve	984, 1410
Caccavale, Fabrizio	512	Ching, Francis	419
Caccia, M.	72	Cho, Hyung Suck	1722
Cai, Hegao	329	Cho, Sangbum	2805
Cai, Lilong	843	Cho, Young Cheol	447
Caldwell, Darwin G.	287, 525	Cho, Young-jo	2134
Callaghan, Victor	1005	Choi, Dae-Sung	213
Camacho, E. F.	2452	Choi, Jong Suk	2680
Campos, Mário Fernando Montenegro	2584	Choi, K. K.	920, 3032
Cannon, David	2495	Choi, Min-Hyung	592
Caracciolo, Luca	2632	Choi, Young-Hyu	753
Caramanis, Michael C.	1450	Chong, Nak Young	2134
Carmichael, Owen	944	Choset, Howie	644, 1396, 2235, 2242
Carneiro, G. H. M. B.	465	Chou, Jyh Hwa	2203
Carr-West, Malcom	1005	Christensen, Henrik I.	670, 2994
Carrozza, Maria Chiara	2899	Chu, Richard	2831
Casals, Alícia	747	Chu, Y. X.	114, 1436
Castellanos, J. A.	2917	Chung, Hakyoung	2626
Cathala, Arnaud	90	Chung, Ho A.	1365
Cauchois, Cyril	1287	Chung, Wan Kyun	1, 889, 2825
Causey, Greg C.	2564	Chung, Won-Jee	753
Cavallaro, Joseph R.	1063	Chung, Young Gu	1626
Celinski, Tomasz	1960, 2608	Chyba, M.	817
Cerrada, C.	2881	Ciarallo, Frank W.	2602
Cerrada, J. A.	435	Ciblak, Namik	2147
Cetin, M.	1255	Clark, S.	1316

Clerentin, Arnaud	1329	Di Febbraro, Angela	781
Cocca, Christopher	855	Díaz del Río, F.	7
Cocosco, A.	2348	Didi, N.	1071
Coelho, Lúcio de Souza	2584	Diftler, M. A.	907
Cohen, Elaine	1843	Dillmann, R.	702, 2354
Cojbasic, Zarko	2128	Ding, Dan	1904
Colgate, Ed	728	Ding, Han	2116
Colgate, J. Edward	1824, 2855, 2937, 3249	Dissanayake, Gamini	999, 1316, 2274
		Distante, C.	1487
Colley, Martin	1005	Dodds, Z.	1607
Conrardy, Chris	3053	Dong, Wenjie	2645
Cook, George	3059	Dordevic, Goran S.	3165
Corke, Peter I.	999, 1158, 1281, 1746, 1928, 2723	Doulgeri, Zoe	1475
		Dowling, Kevin	3001
Costa, Ramon R.	407	Dragomirescu, D.	2440
Cowan, Noah J.	611	Drocourt, Cyril	1287, 1329
Cox, Daniel	855	Drori, Iddo	578
Craig, John	1897	Du, X. Hua	1456
Cremer, James F.	592	Dubey, R. V.	543, 2197
Cremers, Armin B.	1999, 2280	Dubey, V. N.	3038
Croft, E. A.	176	Dubowsky, Steven	951, 2286, 2819
Crowder, R. M.	3038	Dudek, Gregory	676, 1972
Crowley, W.	2315	Dujmovic, Vida	676
Cunningham, Jock	999	Dumont, C.	543
Cutkosky, Mark R.	2930	Dunn, Paul A.	1928
Czarnecki, Chris	2209	Dupont, Pierre	1876
D'haene, L.	1601	Dupuis, E.	1077
da Cunha, José Paulo V. Soares	407	Durrant-Whyte, Hugh	999, 1430, 2274
Daitoh, Haruyoshi	2861	Eagleson, R.	3209
Dale, Lucia K.	688	Ebert-Uphoff, Imme	881
Dapper, M.	1481, 2780, 3128	Eckmiller, R.	1481, 2780, 3128
Dargahi, J.	299	Efe, M. Önder	2073
Dario, Paolo	2899	Elstrom, Mark D.	1948
Dasgupta, Anirvan	1044	Emura, Satoru	976
Datta, Amitava	1032	Endo, Gen	2032
Davies, J. B. C.	2849	Englert, Roman	2280
de Carufel, J.	1077	Erdmann, Michael A.	2322
De Guiseppe, R.	1487	Erkmen, A. M.	2470
De Luca, Alessandro	2632	Erkmen, I.	2470
de Mathelin, Michel	3236	Ernzer, M.	2428
de Paiva, Ely Carneiro	2328	Estlin, T.	1410
De Rossi, Danilo	305	Everett, S. E.	543, 2197
De Schutter, Joris	1194	Faraz, Ali	1734
Debus, Thomas	1876	Farber, Georg	695, 2521
Delahoche, Laurent	1329	Farhi, A.	1114
Delaurentis, Kathryn	2385	Farooqi, Mohammad Asim	3217
Dellaert, Frank	1322, 1999	Featherstone, Roy	3281
Desai, Jaydev P.	1556	Feliu, V.	435, 2881
Desai, Kamal V.	1695	Fenton, R. G.	145
Devy, Michel	2249	Ferch, Markus	1170

Fernandez, Jaime	1396	Goel, M.	874, 867
Fernández, Josep	747	Goldberg, Ken	1897, 2558
Ferrier, N. J.	1644	Goldfarb, Michael	2020, 2122
Ferrier, Nicola J.	207, 938	Gomes, Sergio B. Varella	2328
Finger, William H.	2044	Gómez-Ortega, J.	2452
Finotello, Roberto	78	Goodale, M. A.	3209
Fischer, M.	1890	Gorman, Jason	2495
Fisher, F.	1410	Gosselin, Clément M.	881
Fite, Kevin	2122	Goswami, Ambarish	47
Flasher, Joleen	1396	Gou, J. B.	114, 1436
Florez, J.	1794	Gowdy, Jay	3097, 3103
Forrester, Benjamin	2103	Gravagne, Ian A.	913
Fortin, C.	1418	Graves, Alan R.	2209
Fox, Dieter	35, 1322, 1999	Griffiths, J. G.	1310
Fraile, Juan-Carlos	1219	Gruver, William A.	281
Freedman, M. T.	1305	Gu, Xiangrong	2602
Freeman, Randy	2937	Gu, Yuqing	102
Fries, Gregory A.	1614	Guan, Chengxiang	2406
Fu, Li-Chen	459, 2422, 2476	Gülch, Eberhard	2280
Fu, Yili	329	Guo, Shuxiang	2171
Fuchs, R.	2315	Gupta, Kamal	1526, 2545
Fujisawa, Kae	1347	Gurkan, Evren	2470
Fujita, Yoshiaki	767	Guyot, J.	2315
Fukuda, Kenro	2893	Ha, Insoo	1954
Fukuda, T.	2614	Hager, G. D.	1607, 1934
Fukuda, Toshio	96, 1096, 1182, 1299, 1550, 1632, 1728, 2171, 2458, 2464, 2717, 2744, 2799, 2924	Hager, Gregory D.	1144, 1589
		Hagiwara, Tetsuo	193, 537
		Hagras, Hani	1005
Funahashi, Yasuyuki	585	Hahn, Hernsoo	637
Furuichi, Kazuya	2336	Hähnel, Dirk	1999
Furusho, Junji	188	Hakozaki, Mitsuhiro	957
Gagov, Zavarin	447	Han, J. D.	251
Galarraga, S.	2440	Han, Li	1261
Gambao, E.	565	Han, Youngjoon	637
Gangloff, Jacques A.	3236	Hanachi, C.	1114
Gao, Xueshan	1207	Hanebeck, Uwe D.	1335
Gao, Yan	2026	Hanna, M.	1502
Garcia, Ephrahim	2020	Hannaford, Blake	2899
García, Miguel Angel	1359	Harada, Tatsuya	968
Geng, Z.	1305	Hasebe, Yoshiaki	96
Genot, Frank	2286	Hasegawa, Yasuhisa	96, 2458, 2799
Ghanekar, Milind	427	Hashimoto, Hideki	2729
Giesenschlag, Markus	2280	Hashimoto, Koichi	263
Giesler, C.	287	Hashimoto, Minoru	2372
Gilbert, E. G.	682	Hashtrudi-Zaad, K.	1863
Gilkerson, Paul	624	Hata, Seiji	2171
Gillespie, R. Brent	1824	Hauck, Alexa	2521, 2527
Giménez, A.	2303	Hayakawa, Soichiro	1347
Giua, A.	1462	Hayakawa, Y.	518
Glandais, N.	759	Hayati, S.	992

Hebert, Martial	944	Iannitti, Stefano	2632
Heinrichs, B.	2165	Ihlow, Jürgen	1120
Hemami, Hooshang	53, 2342	Ikeda, Kazuo	2976
Henry, Jason M.	225	Ikezawa, Yukio	3217
Heppler, Glenn R.	427	Ikuta, Koji	1715, 2065, 2378
Hernando, M.	565	Ilg, W.	2354
Hespanha, J. P.	1607	Inaba, Masayuki	1038
Hester, R. D.	1255	Inada, Yuji	96
Heymann, Michael	131	Inoue, Hirochika	1038
Hiller, Manfred	895	Inoue, Sadatoshi	368
Hinrichsen, John	1677	Iqbal, Kamran	2539
Hirai, Shinichi	767, 2572, 3065	Ishii, Idaku	1916, 3195
Hirata, Yasuhisa	2666	Ishikawa, Masatoshi	1916, 3195
Hirose, Shigeo	193, 537, 1469, 2032, 2309, 2336, 2831	Isoda, Y.	543
		Isogai, Masahiro	1096
Hirzinger, Gerhard	1890, 2084	Ito, Fumitoshi	1469
Hodge, Neil Eugene	2482	Itoigawa, Kouichi	1632, 2744
Höhn, Michael	3085	Iurascu, C. C.	1752
Holland, Steve	728	Ivanescu, Mircea	1495
Hollis, Ralph L.	1614, 3261	Ivanisevic, I.	2215
Hong, Geok Soon	2159	Iwamura, Makoto	2958
Hoover, Adam	2261	Iwata, Hiroyasu	3183
Horn, Joachim	1335	Iwata, Hitoshi	1632, 2744
Hoshino, Haruo	2464	Jackson, Dean	1396
Howe, Robert	1876	Jaffe, Peter C. L.	2819
Hsia, T. C.	1800	Jagannathan, Kanniah	1384
Hsiao, Zu Hung	2014	Jägersand, Martin	550
Hsu, Feng-Yih	2476	Janét, J. A.	1138
Hsu, Jane Yung-jen	453	Jarrah, M. A.	170
Hsu, Liu	407	Jedele, H.	2354
Hu, Chi-Yang	2014	Jennings, Cullen	2659
Hu, Jianjuen	1050	Jensfelt, Patric	2994
Hu, Shuyi	3171	Jeon, Se Joong	1132
Hu, Y.	3209	Jeong, Il-Jun	441
Huang, E.	682	Jerbi, Ali	281
Huang, Eric	1776	Ji, Xuerong	238
Huang, Han-Pang	2392, 2434	Jiang, K.	311
Huang, Ling	2096	Jiang, Ping	41
Huang, Qiang	65	Jiang, S. L.	920, 3032
Huang, Shuguang	182	Jih, Wan-rong	453
Huang, Weiqing	843	Jiménez, G.	7
Hudak, Paul	1144	Jiménez, P.	1856
Hugel, Vincent	2360	Johnson, David	1843
Hung, J.	992	Jones, Mike	2651
Hung, Min-Hsiung	2711	Joskowicz, Leo	578
Huo, Wei	2645	Jouvencel, B.	*
Hutchinson, Seth	599, 695, 2527, 3230	Jung, Myung-Jin	2686
Hyman, Mark	2227	Jung, Seul	1800
Hyon, Sang Ho	361	Kaetsu, Hayato	2666
Iagnemma, Karl	2286	Kajita, Shuuji	65

Kallio, Pasi	2750	Koivo, Heikki N.	2750
Kamon, Ishay	163	Komoriya, Kiyoshi	65, 557, 2811
Kanehiro, Fumio	1038	Kong, Xiaoying	1430
Kaneko, Kenji	65	Kopparapu, Sunil Kumar	1281
Kaneko, Makoto	257, 1942	Kortenkamp, David	1396
Kang, Sungchul	2768	Koshizen, Takamasa	2620
Kang, Wei	2964	Kostic, Dragan	3165
Kapoor, C.	1255	Kosuge, Kazuhiro	2666
Kasaya, Takeshi	2189	Kotoku, Tetsuo	557
Kataria, Abhay	53	Kouikoglou, Vassilis S.	1126
Kato, Isao	3255	Koyachi, Noriho	65
Kavraki, Lydia E.	151, 1671	Kragic, D.	670
Kawabata, Kuniaki	2666	Kraiss, K.-F.	2911
Kawabata, Syun'ichi	557	Kriegman, David J.	2227
Kawamura, Sadao	494, 518, 2572, 3065	Krishnan, H.	1102, 1108
Kaynak, Okyay M.	831, 2073	Kröse, B. J. A.	2255
Keerthi, S. S.	682	Kubota, Takashi	1404, 2692
Kelly, Rafael	599, 1740, 1831	Kuffner, James J. Jr.	473
Kesavadas, T.	2428	Kumar, Ratnesh	126
Khalil, W.	1758	Kumar, Vijay	1538, 1556
Khargonekar, Pramod P.	137	Kuno, Yoshinori	2590
Khatib, Oussama	341, 741, 3281	Kurashige, Kentarou	2464
Khorasani, K.	1090	Kurpjuhn, Tobias Peter	2527
Khorrami, Farshad	1812	Kusik, Todd	837
Khosla, Pradeep K.	708, 1219	Kwon, Dong-Soo	413, 1722
Kiguchi, Kazuo	1182	Kwon, Wook Hyun	447, 793, 2982
Kim, Byoung-Ho	2134	Kwon, Woong	1966, 2596
Kim, Byung Kook	2680	Kyriakopoulos, Kostas J.	2674
Kim, Doh-Hyun	19	LaFarge, Robert A.	2887
Kim, Doik	889	Lambert, Alain	1341
Kim, Dong Sung	2982	Lamiraux, Florent	151
Kim, Dong-Yoon	1966, 2596	Lange, Friedrich	2084
Kim, Heung-Soo	2686	Laubach, S. L.	347
Kim, Hyung Seok	793	Laumond, Jean-Paul	823, 1770
Kim, Jong-Hwan	2686	LaValle, Steven M.	473, 1671, 1677
Kim, Kab-Il	1806	Lawrence, Dale A.	3267
Kim, Kyunghwan	207	Lazar, Nicole	644
Kim, Munsang	213, 2768	Le Fort-Piat, Nadine	1341
Kim, Sung Ho	2680	Lee, B. H.	2110
Kim, Sungbok	531	Lee, C. S. George	393, 2416
Kim, Young Shin	2982	Lee, Ching-Hung	2639
Kimura, Tetsuya	2177	Lee, Chong-Won	213, 2768
Kirkham, Robin J.	1158	Lee, Christopher	3267
Klaassen, Bernhard	3014	Lee, D. T.	3275
Knepper, Ross	1396	Lee, Eric	1897
Knoll, Alois	1164, 1170	Lee, Hao-Lun	1708
Kobayashi, F.	2614	Lee, J. H.	734
Kobayashi, Hisato	2837	Lee, Jae-Won	1966, 2596
Koch, R.	1601	Lee, Jang Gyu	2626
Koditschek, Daniel E.	611, 2717	Lee, Jangwook	213

Lee, Jihong	1514, 1954	Low, K. H.	1371
Lee, Joon-Soo	269	Lu, Liang	1570
Lee, Ju-Jang	861	Luh, J. Y. S.	3171
Lee, K.	494	Luh, Peter B.	787
Lee, Kyo-Il	2768	Lumelsky, V.	157, 375, 722, 2215
Lee, S. H.	2110	Luntz, Jonathan E.	2235, 2242
Lee, Sang W.	932	Luo, Ren C.	1176, 2014, 2203, 3079
Lee, Sooyong	213	Lynch, Kevin M.	2551
Lee, Sukhan	1269, 1966, 2596, 2762	Ma, B. L.	811
Lee, Sung-Woo	1589	Ma, G. H.	1764
Lee, Sungon	1	Ma, Shugen	3007
Lee, Ti-Chung	2639	Ma, Xiaowei	329
Lee, Wei Zen	2203, 3079	Ma, Ying	2159
Lee, Woo Ho	2704	Ma, Yulin	329
Lee, Y. K.	926	Maass, R.	1481, 2780, 3128
Lehmann, Carsten	1120	Maciejewski, A. A.	874, 867
Leifer, L. J.	741	Madhavan, Raj	999, 2952
Lenarcic, Jadran	3159	Madigan, Bruce	3053
Leonard, Naomi Ehrich	500	Madrigal, O.	741
Leong, Hou Tin	2203	Maeda, Kiyoshi	895
Leuschen, Martin L.	1063	Mahabaleshwarkar, Rahul	1424
Leven, Peter	695	Mali, Amol D.	652
Levick, Peter	3050	Mantripragada, R.	3071
Lewis, Chris	2887	Marchand, Éric	3224
Li, Chun	1544	Marefat, Michael M.	2602
Li, Hsi-Shang	90	Marigo, Alessia	1583
Li, Tsai-Yen	1353	Marsden, Jerrold E.	500
Li, Wen	13	Martins, M. F.	465
Li, Z.	920	Mascaro, Stephen	962, 2533
Li, Z. X.	114, 1436, 3032	Mase, Kenichiro	2458
Li, Zexiang	1261	Mason, Matthew T.	2322
Liang, Tzu-Chen	3153	Mason, Richard	773
Lim, Wee Kiat	2501	Matsikis, A.	2911
Limón, D.	2452	Matsuhira, Nobuto	557
Lin, Feng	131	Matsumaru, Takafumi	557
Lin, Ming-Hung	459, 2422	Mavroidis, Constantinos	2385
Lin, Qiao	1562	McCarragher, Brenan	1960, 2608
Lin, Yueh-Jaw	1424	McClamroch, N. Harris	2805
Lipkin, Harvey	1595, 2147	McDermott, G.	992
Little, James J.	2659	McIsaac, Ken	2843
Liu, Jing-Sin	3153	McMurray, Gary V.	1595
Liu, Shuliang	1207	Medrano-Cerda, G. A.	525
Liu, Y.	1275	Meggiolaro, Marco A.	2819
Liu, Yun-Hui	1904, 3026	Mehrandezh, M.	145
Lizarralde, Fernando C.	407	Melchiorri, C.	1620
Llamas, Jesús	1740	Melkote, Hemant	1812
Lobontiu, Nicolae	2020	Menciassi, Arianna	2899
Lopes, E. P.	465	Meng, Max	23
Lorenz, Andy	2855	Menga, G.	1462, 2440
Lovchik, C. S.	907	Messner, William	2235, 2242

Meyer, George	131	Nakabo, Yoshihiro	3195
Michelli, R. D.	1138	Nakamura, Yoshihiko	714, 1044, 2398
Mihali, Raul	323	Nakanishi, Jun	2717
Miljanovic, D. M.	176	Nakanishi, Satoru	2590
Miller, Andrew T.	1240	Nakano, Eiji	1226
Miller, Brian E.	2937	Nakatani, Ichiro	1404, 2692
Mills, James K.	849	Namiki, Akio	3195
Mimura, Nobuharu	585	Namiki, Takao	1715
Minciardi, Riccardo	781	Nandi, S.	1683
Minor, Mark	2057	Naniwa, Tomohide	2786
Mitsui, Takeo	2309	Natale, Ciro	512, 1788
Mitsuishi, Mamoru	1782	Nebot, Eduardo	999, 1430, 2274
Miyata, Keisuke	2924	Nechyba, Michael C.	3134
Miyazaki, Hideki	2189, 2736	Nee, A. Y. C.	1764
Mizoguchi, Hiroo	1550	Neevel, David	837
Mizushima, N.	2398	Negoro, Makoto	1728
Mochiyama, Hiromi	2837	Neira, J.	2917
Moestl, Thomas	1910	Nenchev, Dragomir N.	506
Mohanty, Satya Ranjan	126	Nettleton, Eric	999
Mohri, Akira	2958	Newman, Wyatt S.	2096
Mohsen, A.	1310	Nguyen, C.	1305
Mokhtari, M.	1071	Nguyen, P. T. A.	2786
Montano, L.	2905	Nickels, Kevin	2527, 3230
Montiel, J. M. M.	2905, 2917	Nie, Jing	393
Moore, Carl A.	3249	Nielsen, James	2183
Moradi, Hadi	2762	Niksefat, N.	200
Moreno, Javier	599	Nilsson, Martin	293
Morgavi, Emanuele	78	Nishida, Yoshifumi	968
Mori, Taketoshi	968, 1882	Niwa, Masaaki	3065
Morita, Toshio	3183	Noborio, Hiroshi	355
Morse, A. S.	605, 1607	Nokata, Makoto	2065, 2378
Mosemann, Heiko	245, 2756	Noritsugu, Toshiro	263
Motaghedi, Shui H.	805	Novotny, P. M.	1644
Moualek, A.	1114	Nowak, Brent	2103
Mouri, Tetsuya	585	O'Brien, John	1520
Mukherjee, Ranjan	2057	O'Neil, K.	3146
Munawar, Khalid	1213	Oasa, Hideki	957
Murashima, Teruhisa	2590	Oda, Mitsushige	1390
Murphy, Mark A.	225	Ogata, Tetsuya	3177
Murray, Don	2659	Oguro, Keisuke	2171, 2177
Mutschler, E.	2298, 2315	Oh, Jun-Ho	19
Mutz, D.	1410	Oh, Paul Y.	275
Myers, Donald R.	3109	Oh, Sang-Rok	269, 2134, 2140
Myint, Than	1384	Ohno, Hidetaka	2309
Nagao, Takaaki	1782	Ohya, Akihisa	630
Nagata, Kazuyuki	1247, 3217	Okamura, Allison M.	2930
Nagatani, Keiji	644	Okuma, Shigeru	1347
Nagesh, Nidamaluri	728	Olsen, Bent David	2261
Nagura, Akihiro	2943	Olson, Clark F.	1987
Naito, Toru	2924	Omata, Toru	2366, 3217

Ong, C. J. 682
Orin, David E. 1702, 1708, 2711, 2887
Oriolo, Guiseppe 823
Ostrowski, James P. 618, 1556, 2843
Ota, Jun 2976
Ota, Yusuke 1469
Ouelhadj, D. 1114
Overmars, Mark H. 1018, 1837, 2558
Ozawa, Masahiro 2177
Özbay, Hitay 53
Paap, Karl L. 3014
Padrón, V. M. 2303
Pagé, L.-L. 1188
Pagilla, Prabhakar R. 1818
Pai, Dinesh K. 486, 1849, 2322
Palm, Stephen 1882
Pandian, S. R. 518
Pao, Lucy Y. 3267
Papanikolopoulos, Nikolaos P. 245
Papavassilopoulos, George 441
Parameswaran, M. 299
Paredis, Christiaan J. J. 1219
Park, Euisu 137
Park, F. C. 1752
Park, Jong H. 1365
Park, Jonghoon 2825
Park, KyuCheol 2626
Park, T. H. 2110
Park, Youngjin 413
Paskell, Troy 3053
Pastor, J. M. 2303
Patel, R. V. 1090, 2507
Pau, Kiu Ling 1818
Payandeh, Shahram 299, 1734, 2545
Peacock, Brian 728
Pearson, Tom 728
Pegard, Claude 1329
Pennerath, Frederic 1158, 2723
Pérez, J. A. 2917
Peshkin, Michael A. 728, 1824,
                                  2855, 3249
Peterson, John 1144
Pfeiffer, Charles 2385
Phillips, R. 1310
Phillis, Yannis A. 1508
Piedboeuf, J.-C. 1077, 1188
Piepgras, C. 2315
Piepmeier, Jenelle Armstrong 1595
Pino, L. 1418
Pinto, E. 565

Podder, Tarun Kanti 84, 387
Podgurski, H. Andy 2096
Politis, Z. 2988
Ponce, Jean 2227
Poo, Aun Neow 2159
Potvin, M.-J. 1188
Poznyak, Alexander S. 3122
Pratt, Gill 381, 1050
Pratt, Jerry 381, 1050
Prattichizzo, Domenico 1583
Probert, P. J. 2988
Probert, Penelope 624
Proctor, R. W. 874
Qian, Wen-Han 3044
Quaid, Arthur E. 2221
Quinn, Roger D. 2564
Rabideau, Gregg 984
Raby, Eric Y. 1702
Radford, Jim 2038
Rahimi, M. A. 2342
Ralph, Scott K. 486
Ramirez, A. 120
Ramírez, D. R. 2452
Ramos, Ann M. 913
Ramos, Josue J. G. 2328
Rasic, Milan 3165
Raue, A. 2756
Ravishankar, C. V. 1152
Rekleitis, Ioannis M. 676
Repperger, Daniel W. 225
Reveliotis, Spyros A. 1442
Reyes, Fernando 599
Reyhanoglu, Mahmut 2805
Rhee, S. H. 3071
Rimon, Elon 163, 1377
Rit, J-F. 1770
Rivlin, Ehud 163, 1638, 2578
Rizzi, Alfred A. 1614, 3103
Ro, Sookwang 1269
Roberts, Jonathan 999, 1158, 2723
Roberts, Rodney G. 2153
Robinson, G. 2849
Robl, Christian 3085
Roby-Brami, A. 1071
Rodrigues, M. A. 1275
Roh, Kyoungsig 1966, 2596
Ron, R. 1650
Ros, Lluís 571
Rosell, Jan 2774
Rosenberg, C. 658

Rosenberg, Charles	1999	Sekiyama, Kosuke	96, 1550
Roth, Zvi S.	805	Sela, M. N.	145
Rotstein, Héctor	1638	Seliger, Günther	3044
Roumeliotis, Stergios I.	1656, 1979	Seneviratne, L. D.	311
Rous, M.	2911	Sepehri, N.	200, 2165
Roy, Nicholas	35, 2292	Seraji, Homayoun	2006
Rubio, A.	1794	Serdeira, H.	465
Rudas, Imre J.	831, 2073	Setiawan, Samuel Agus	361
Rus, Daniela	2322, 2513	Sevillano, J. L.	7
Rusaw, Shawn	2545	Shadpey, Farshid	2507
Russell, R. Andrew	2698	Shao, Hao	1207
Rybski, Paul E.	664	Shapira, Amir	1377
Rye, David C.	1746	Sharp, Gregory C.	932
Ryu, Jee-Hwan	413	Shen, Helen C.	2090
Sacks, Elisha	578	Shiang, Wei-Jung	2495
Sacone, Simona	781	Shibata, Susumu	96
Saito, Fuminori	1247	Shibata, Takanori	2868
Saito, Hiroaki	2692	Shih, Teng-Jei	2422
Saito, Shigeki	2189, 2736	Shikano, Masaru	1299
Sakaguchi, Masamichi	188	Shiller, Zvi	335
Sakai, Takayuki	2861	Shim, Hyun-Sik	2686
Sakamoto, Moriyuki	3201	Shimada, Nobutaka	2590
Sakane, Shigeyuki	2874	Shimoyama, I.	926
Salam, Fathi M.	1293	Shimshoni, Ilan	2578
Salcudean, S. E.	231, 1863, 1870	Shin, Dongmok	1966, 2596
Sanchez, Edgar N.	3122	Shin, Jin-Ho	861
Sánchez, L.	435	Shin, Kang G.	1152
Sanderson, Arthur C.	2704	Shinoda, Hiroyuki	957
Sandoval, F.	1650	Shirai, Yoshiaki	2590
Santibañez, Víctor	1831	Shoval, Shraga	1377
Sappa, Angel Domingo	1359	Siciliano, Bruno	1788
Sarkar, Nilanjan	84, 387	Sidenbladh, H.	670
Sasiadek, J. Z.	2970	Silva, Filipe M.	59
Sato, Hiroshi	1632	Silveira, J. T. C.	465
Sato, Tomomasa	968, 1882, 2189, 2736	Sim, Robert	1972
Savall, J.	1993	Siméon, T.	1770
Schempf, Hagen	2298, 2315	Singh, Hanumant	399
Schenk, Thomas	2521	Sitti, Metin	2729
Schimmels, Joseph M.	182	Skubic, Marjorie	2103
Schmidt, Ralf	1164	Smaby, N.	741
Schmitt, M.	2911	Smith, Gordon	1897, 2558
Schofield, J.	525	Smith, Philip W.	1948
Schraft, R. D.	3115	So, B. R.	2140
Schulte, J.	658	Sobh, Tarek	323
Schultz, Alan C.	2651	Soga, Eiji	368
Schulz, Dirk	1999	Somolinos, J. A.	435
Scilingo, Enzo Pasquale	305	Song, Jingyan	3134
Scoggins, S. M.	1138	Song, Kai-Tai	2639
Seet, Gerald	1371	Sorg, Michael	2521
Sekhavat, S.	817	Soundaralakshmi, Subbiah	1032

Spano, Joseph	2051	Tang, Xiaoqi	843
Sridharan, K.	1683	Tanie, Kazuo	65, 557, 2868
Srikanchana, R.	1305	Tanikawa, Tamio	2183
Sriskandarajah, C.	120	Tanimoto, Mitsutaka	1728
Staffetti, Ernesto	571	Tanner, Herbert G.	2674
Stanley, Kevin	281	Tao, Jianming	3050
Stiller, Peter F.	1024	Tar, József K.	831
Stoeter, Sascha A.	245	Tarao, Susumu	901
Strobel, Matthias	29	Tardós, J. D.	2917
Suárez, Raúl	2774	Tarn, T. J.	219
Sudha, N.	1683	Tarokh, M.	992
Sudsang, Attawith	2227	Tashima, Toshihiro	2868
Sugano, Shigeki	3177, 3183	Tatsumi, Shoji	2078
Sugar, Thomas	1538	Taurisano, F.	1487
Sugawara, Yoshihiko	3243	Taylor, Lee R.	2322
Sugiyama, Tomohiko	2744	Tenaglia, Christopher A.	2887
Sugumoto, Koichi	2171	Teng, Ching-Cheng	2639
Suh, Il Hong	269, 2134, 2140	Tenreiro Machado, J. A.	59
Sujan, Vivek A.	951	Terashima, Maho	2874
Sukhatme, Gaurav S.	1656, 1979	Tesar, Delbert	855, 1255
Sukkarieh, Salah	2274	Thiebaut, Stef Sonck	3281
Sun, Dong	849	Thomas, Federico	571
Sun, Fuchun	1806	Thorpe, Charles	2267
Sun, Ruoying	2078	Thrun, Sebastian	35, 658, 1322, 1999, 2292
Sun, Zengqi	1806		
Surdilovic, Dragoljub	2128, 3165	Tibbetts, Craig	193
Suzuki, Tatsuya	1347	Tilbury, Dawn M.	137
Svinin, M. M.	257	Timoszyk, Wojciech K.	480
Swain Oropeza, Ricardo	2249	Tokumoto, Shinichi	767
Swaminathan, Jayendran	2096	Tomastik, Robert N.	787
Tabata, Tsuyoshi	2372	Torras, C.	1856
Tachi, Susumu	976	Trabia, Mohamed B.	2482
Tadokoro, Satoshi	895, 2177	Traub, A.	3115
Takahashi, Takayuki	1226	Tremblay, Alain	2446
Takamori, Toshi	895, 2177	Tremblay, I.	1188
Takanishi, Atsuo	361, 368, 3255	Trinkle, Jeffrey C.	1261
Takanobu, Hideaki	3255	Tsagarakis, N.	287, 525
Takase, Kunikatu	557	Tso, Kam S.	984
Takeichi, Masaki	1715	Tso, S. K.	811
Takemura, F.	518	Tsubouchi, Takashi	13
Takesue, Naoyuki	188	Tsujio, Showzow	2893
Takeuchi, Hiroki	1057	Tsumaki, Yuichi	506
Taki, Hirokazu	3091	Tuytelaars, T.	1601
Talebi, H. A.	1090	Uchida, Tasuyuki	2336
Tan, Jindong	2964	Uchiyama, Masaru	506, 901, 1213
Tanaka, Eiichiro	2366	Ueda, K.	257
Tanaka, Katsuya	1782	Ueno, Naohiro	1942
Tanaka, Kazuaki	3091	Ullrich, C.	1849
Tanaka, Takashi	3217	Um, D.	722
Tanaka, Yoshio	1299	Umetani, Yoji	2861

Umetsu, Tomohiko	3255	Whatley, D. R.	3038
Urakawa, Kenji	355	Whitcomb, Louis	399
Urdiales, C.	1650	Whitney, D. E.	3071
Van der Loos, H. F. M.	741	Williams, Robert L. II	225
Van der Stappen, A. Frank	1018, 1837, 2558	Williamson, Todd	2267
Van Geem, C.	1770	Wilmarth, Steven A.	1024
Van Gool, L.	1601	Winstanley, Graeme J.	1158, 2723
Vasseur, Pascal	1287	Wiseman, W. J.	1138
Vassura, G.	1620	Won, K. T.	1514
Vendittelli, Marilena	823	Woo, Ki Young	1722
Verhoeven, Richard	895	Wu, Chi-haur	3275
Veruggio, G.	72	Wu, Hsien-Jung	799
Viant, W.	1310	Wu, Lin	1084
Vicente, S.	7	Wu, Q. M. Jonathan	281
Villani, Luigi	512, 1788	Wu, Weiguo	41
Vona, Marsette	2513	Xavier, Patrick G.	1011
Voss, Stephan	245	Xi, Ning	219, 1293, 2964
Voyles, Richard M.	664, 708	Xiao, Jing	238
Wada, Takahiro	2572	Xie, Haipeng	2507
Wagner, J. J.	741	Xiong, Youlun	2116
Wahl, F.	2756	Xu, Dianguo	1207
Waldron, Kenneth J.	2711	Xu, Li	3189
Walker, A. L.	1138	Xu, W. L.	251, 811
Walker, Ian D.	913, 1063	Xu, Xiang-Rong	753
Walker, Sean	1396	Xu, Yangsheng	3134, 3140
Wang, Chia-Yu E.	480	Yagi, M.	375
Wang, David	419	Yakey, Jeffery H.	1671
Wang, David W. L.	427	Yam, Yeung	3134
Wang, Jeen-Shing	2416	Yamada, Takayoshi	585
Wang, Michael Y.	1577	Yamada, Yoji	2861
Wang, Q.	2970	Yamagami, Shinji	2177
Wang, Shige	1152	Yamaguchi, Jin'ichi	361, 368
Wang, Shuguo	1084, 1904	Yamamoto, Hironobu	96
Wang, Tianmiao	2489	Yamamoto, Masakazu	2976
Wang, Xiao-gang	2090	Yamamoto, Motoji	2958
Wang, Y.	1305	Yamamoto, Yoshio	1200
Wang, Y. C.	251	Yamane, Katsu	714
Wang, Yan	1207	Yamauchi, Brian	2651
Wang, Yuntao	1084	Yamazaki, K.	2348
Wang, ZhiDong	1226	Yamazaki, T.	2398
Wang, Zhiyan	2406	Yand, Haw Ching	1776
Wannasuphoprasit, Wit	728	Yang, Fenglei	2489
Warwick, J.	2315	Yang, Guilin	2501
Wehe, David K.	932	Yang, H.	1102, 1108
Wei, Jun	2489	Yang, Wenyu	2116
Wei, Wang	329	Yang, Xianyi	23
Wells, Jim	728	Yata, Teruko	630
Wen, John T.	1520	Yeh, Chien-Fa	2434
Wenjuan, Lu	1806	Yeo, Song Huat	2026, 2501
Wentink, Chantal	1837	Yi, B.-J.	734, 2140

Yi, Keon Young	1132, 1626
Yoerger, Dana	399
Yokoi, Kazuhito	65
Yokokohji, Yasuyoshi	1782, 3201, 3243
Yoneda, Kan	1469
Yoneda, Mitsunori	2924
Yoshikawa, Tsuneo	2943, 3201, 3243
Yoshimi, Tomohisa	968
Yoshimitsu, Tetsuo	1404, 2692
You, Bum-Jae	269, 1589
You, Song	2489
Youm, Youngil	1, 889, 2825
Yu, Biao	1818
Yu, Tzong-Hann	1353
Yu, Wen	3122
Yu, Yong	2893
Yuan, Xiaobu	102
Yuan, Yufei	329
Yuh, J.	393
Yuh, Junku	2416
Yuki, Takahiro	2372
Yun, Xiaoping	1200
Yuta, Shin'ichi	13, 630
Zacksenhouse, M.	1910
Zahn, V.	1481, 2780, 3128
Zefran, Milos	2038
Zelinsky, Alex	2620
Zha, X. F.	108
Zhang, Guoguang	188
Zhang, Hao	2096
Zhang, Hong	618, 1922
Zhang, Jianwei	1164, 1170
Zhang, Jiong	3140
Zhang, Qixian	2489
Zhang, Runtong	1508
Zhang, Y. F.	1764
Zhang, Zhichao	2406
Zhao, Gang	2078
Zhao, Yanzheng	1207
Zheng, Yuan F.	2342, 2539, 3189
Zheng, Zhiqiang	1544
Zhou, Changjiu	1384
Zhou, Chen	1456
Zhou, Quan	2750
Zhou, Xiaohui	787
Zhu, Chi	2793
Zhu, Ming	1870
Zhu, Wen-Hong	231, 1870
Zhu, Y.	1310
Zhu, Yunyue	1806
Zhuang, Hanqi	805
Zielinska, Teresa	1371
Zuo, Bing-Ran	3044

Dynamic Analysis of the Cable Array Robotic Crane

Wei-Jung Shiang
David Cannon
Department of Industrial and Manufacturing Engineering
Jason Gorman
Department of Mechanical Engineering
Pennsylvania State University
University Park, Pennsylvania

Abstract

Offshore loading and unloading of cargo vessels and on board cargo relocation during conditions of Sea State 3 or greater have been found to be difficult with existing crane technology due to oscillation of the payload. A new type of crane which uses four actuated cables to control the motion of the payload is presented. The closed chain configuration will intuitively provide more stability with respect to the motion of the sea compared to existing cranes. The kinematics and dynamics are derived using cable coordinates. Since there are four cables and three degrees of freedom, the system is redundant. This problem is solved by applying a geometric constraint to the equations of motion such that the reduced number of equations equals the degrees of freedom. The force distribution method is applied using linear programming to solve for the required cable tensions. Simulation results showing cable tensions and cable lengths during a typical crane operation are presented.

1 Introduction

Offshore loading and unloading of cargo vessels and on board cargo relocation during conditions of Sea State 3 or greater have been found to be difficult with existing crane technology. The sea state may directly cause large motions of the cargo ship or indirectly introduce large motions of the hoisted container with the effect of excitation of parametric instability.[4] Traditional cranes are stable only in the vertical direction while they load are free to rotate in all directions and sway in the horizontal plane under the slightest side disturbance.[5]

Many investigators have introduced robotics concepts such as parallel manipulators to design stiffer crane systems [1], [5], and [10]. They argued that the most efficient structural form for carrying large steady loads is a set of tension members and compression members configured to hold the load with the lowest possible loads in each member.[10] The advantages of parallel manipulators are high precision positioning capability, fast motion, large lifting capacity, and light weight.[9]

The four-cable array robot shown in Fig. 1 is proposed to manipulate a container load in an efficient way on the vessel deck by controlling the length of all four cables simultaneously. This closed chain system has three constrained degrees of freedom while the remaining three degrees of freedom are controlled by an end effector.

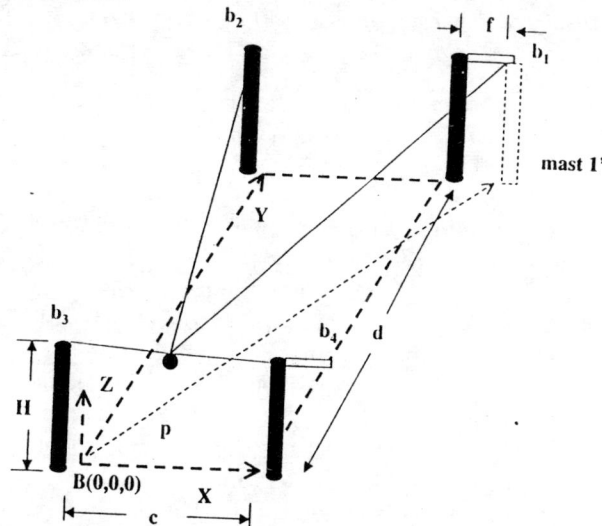

Figure 1: Schematics of the cable array robot

Four cables are used rather than three so that the workspace on deck is maximized. However, this advantage increases the complexity of the problem by causing the system to be kinematically redundant.

Therefore, the cable tensions can not be uniquely determined by the geometric configuration. Similar situations can be found in force/position control cases and multiple-chain robotic systems [2], [3], and [8].

The equations of motion are derived using Lagrange's equations in cable space which is analogous to the the standard joint space. A systematic procedure based on [8] is applied to the dynamic equations to eliminate the redundancy and thereby reduce the equations of motion to a form which is well suited to force distribution methods.

In the cable array robot, the dynamic equations for the rigid cables under-specify the system. The compression forces and limitations of the cable tensions are not included in the dynamic equations. In order to find proper tensions, optimal force distribution,[3], is used with adequate constraints for the cable tensions and a suitable objective function. The equations are then formulated in the standard linear programming (LP) form.

2 Model Definition and Kinematics

The cable array robot is shown in Fig. 1. The boom allows offloading cargo to a dock or other vessel. There is a mast/winch pair at each corner of the workspace which actuate four cables that are linked to a loading module used to lift and carry containers.

The coordinates of the joint b_i on the mast i with respect to (w.r.t.) the frame B can be written as

$$b_1 = [c+f, d, H]$$
$$b_2 = [0, d, H]$$
$$b_3 = [0, 0, H]$$
$$b_4 = [c+f, 0, H]$$

It is assumed that the cables act as rigid bodies both axially and transversely and they are considered to be massless. The length of the cable can, however, be changed by rotating the winch to reel the cable in or let it out. The container is assumed to be a point mass and the cable array robot base is considered to be stationary.

The inverse kinematics of the cable array robot for any number of links can be formulated fairly easily since the system is closed chain. The required cable lengths for a given position of the container, suppose that the coordinate of the container w.r.t. the frame B is $[x, y, z]$ and the coordinates of the joint on the mast i w.r.t. the frame B is $[x_i, y_i, z_i]$, the cable length l_i will be

$$l_i = ((x-x_i)^2 + (y-y_i)^2 + (z-z_i)^2)^{\frac{1}{2}} \quad (1)$$

$i = 1, 2, 3, 4, \cdots$

The forward kinematic relationship between the Cartesian system and the cable lengths can be directly formulated, e.g. by assuming the container is a point mass, but the relationship becomes progressively more complex as the number of cables increases. The forward kinematic equations are expressed as functions of the four cable lengths. One geometric compatibility equation is also required to guarantee the connection of all four cables. The forward kinematic equations are

$$x = \frac{(l_1^2 - l_2^2 + x_2^2 - x_1^2)}{2(x_2 - x_1)} \quad (2)$$

$$y = \frac{(l_2^2 - l_3^2 + y_3^2 - y_2^2)}{2(y_3 - y_2)} \quad (3)$$

$$z = -(l_4^2 - (x-x_4)^2 - (y-y_4)^2)^{\frac{1}{2}} + z_4 \quad (4)$$

The geometric compatibility constraint equation is:

$$l_4^2 - l_1^2 - l_3^2 + l_2^2 = 0 \quad (5)$$

From the forward kinematic equations, the dynamic model of the cable array robot can be written in cable link space and cable tensions are considered as generalized forces in the model. This cable link space approach can handle the dynamic problem in terms of cable link lengths, linear link velocities, and linear link accelerations.

3 Dynamic Equations

The general dynamic equations of motion can be obtained from the Lagrangian formulation. Lagrange's equation can be written in the form of potential energy, $\mathbf{P}$, kinetic energy, $\mathbf{K}$, and generalized forces or torques in the form

$$\frac{d}{dt}\left(\frac{\partial K(\mathbf{q},\dot{\mathbf{q}})}{\partial \dot{\mathbf{q}}}\right) - \frac{\partial K(\mathbf{q},\dot{\mathbf{q}})}{\partial \mathbf{q}} + \frac{\partial P(\mathbf{q})}{\partial \mathbf{q}} = \tau \quad (6)$$

According to [7], the dynamic equations for a system can be rewritten by expanding and regrouping Equation (6) in matrix form as follows:

$$\mathbf{M}(\mathbf{q})\ddot{\mathbf{q}} + \mathbf{V}(\mathbf{q},\dot{\mathbf{q}})\dot{\mathbf{q}} + \mathbf{G}(\mathbf{q}) = \tau \quad (7)$$

where the mass matrix $\mathbf{M}(\mathbf{q})$ is directly obtained from the kinetic energy $\mathbf{K}$, $\mathbf{G}(\mathbf{q})$, derived from the potential energy $\mathbf{P}$ and $\mathbf{V}(\mathbf{q},\dot{\mathbf{q}})$ represents Coriolis and centrifugal forces. According to [6], the following equations define the matrices $\mathbf{M}(\mathbf{q})$ and $\mathbf{G}(\mathbf{q})$.

$$\mathbf{M}(\mathbf{q}) = m\mathbf{J}(\mathbf{q})^{\mathbf{T}}\mathbf{J}(\mathbf{q})$$
$$\mathbf{G}(\mathbf{q}) = \frac{\partial P(\mathbf{q})}{\partial \mathbf{q}} \quad (8)$$

where $\mathbf{J(q)}$ is the Jacobian matrix transforming velocity from cable space to Cartesian space.

The Coriolis and centrifugal matrix $\mathbf{V(q,\dot{q})}$ is obtained by the Christoffel symbols. The kth row, jth column element of the matrix $\mathbf{V(q,\dot{q})}$ is defined as

$$\begin{aligned} v_{kj} &= \sum_{i=1}^{n} \mathbf{Ch}_{ijk}\dot{\mathbf{q}}_i \\ &= \sum_{i=1}^{n} \frac{1}{2}(\frac{\partial m_{kj}}{\partial q_i} + \frac{\partial m_{ki}}{\partial q_j} - \frac{\partial m_{ij}}{\partial q_k})\dot{\mathbf{q}}_i \end{aligned} \quad (9)$$

where $\mathbf{Ch}$ is the Christoffel symbol, and m_{ij} is the ith row, jth column element of $\mathbf{M(q)}$. The symmetric positive definite mass matrix and coriolis and centrifugal matrix in link space (l_1, l_2, l_3, l_4) take the form:

$$\mathbf{M(q)} = \frac{1}{h}\mathbf{M'(q)}, \ \mathbf{V(q,\dot{q})} = \frac{1}{h^2}\mathbf{V'(q,\dot{q})}$$

where

$$\begin{aligned} h =\ & (2cf + d^2 + c^2 + f^2)l_2^4 + (-2d^2l_1^2 \\ & + (-4cf - 2f^2 - 2c^2)l_3^2 - 8d^2cf - 4d^2f^2 \\ & -4d^2c^2)l_2^2 + d^2l_1^4 + (2d^2f^2 + 2d^2c^2 \\ & +4d^2cf)l_1^2 + (2cf + c^2 + f^2)l_3^4 + (2d^2f^2 \\ & +d^2c^2 + 4d^2cf)l_3^2 + (-8d^2cf - 4d^2f^2 - 4d^2c^2) \\ & l_4^2 + 4d^2c^3f + d^4f^2 + 4d^2cf^3 + d^2f^4 \\ & +d^4c^2 + 6d^2c^2f^2 + 2d^4cf + d^2c^4 \end{aligned}$$

the sample entries of $\mathbf{M'(q)}$ and $\mathbf{V'(q,\dot{q})}$ are

$$\begin{aligned} m_{11} =\ & ml_1^2(-4l_4^2d^2 + l_2^4 + l_3^4 + d^4 - 2l_2^2d^2 \\ & -2l_2^2l_3^2 + 2l_3^2d^2) \\ v_{11} =\ & ml_1\dot{l}_1((f^2 + 2fc + c^2 + d^2)l_2^8 + (f^2 + 2fc \\ & +c^2)l_3^8 + ((-4f^2 - 8fc - 4c^2 - 2d^2)l_3^2 + \cdots \end{aligned}$$

4 Reduction of Variables

The number of generalized coordinates must be reduced by one due to the redundancy of the cables. The concept of applying constrained motion in force/position control can be applied to the cable array robot by making three cables work as actuators to perform tasks, and the forth cable is conceptually treated as the desired constraint force for the tasks. This concept preserves the advantage of cable-space dynamic equations in this application.

A decoupling transformation, based on a similar approach proposed by [8] and [2], is applied to reduce the dynamic equations to three variables with a kinematic constraint equation (5). The nonlinear constraint equation is from equation (5) and is redefined as

$$\phi_1(\mathbf{q}) = 0 \quad (10)$$

with the virtual constraint force represented by

$$\mathbf{f} = \mathbf{J_1}^T \lambda \quad (11)$$

where λ is a vector of generalized multipliers associated with the constraints and in this case a scalar. The Jacobian matrix $\mathbf{J_1}$ is defined as $\partial\phi_1(\mathbf{q})/\partial\mathbf{q}$. The generalized coordinates are divided into two groups. One group of three for position constraints and a group of one force constraint. There are four different combinations in which to divide these four link variables. Therefore, there are four different sets of transformed dynamic equations which are generated from four different combinations of link variables.

Note that, l_4 is arbitrarily assigned as the group $\mathbf{q_1}$, and (l_1, l_2, l_3) is assigned as the group vector $\mathbf{q_2}$ without loss of generality. A function $\mathbf{\Omega}$ exists, which is $(l_1^2 + l_3^2 - l_2^2)^{.5}$ in this example, such that a nonlinear transformation can be defined as

$$\mathbf{w} = \mathbf{W(q)} = \begin{bmatrix} \mathbf{q_1} - \mathbf{\Omega(q_2)} \\ \mathbf{q_2} \end{bmatrix} \quad (12)$$

and an inverse nonlinear transformation can be defined as

$$\mathbf{q} = \mathbf{Q(w)} = \begin{bmatrix} \mathbf{w_1} + \mathbf{\Omega(q_2)} \\ \mathbf{w_2} \end{bmatrix} \quad (13)$$

The Jacobian matrix of Equation(13) is defined as

$$\mathbf{T(w)} = \frac{\partial \mathbf{Q(w)}}{\partial \mathbf{w}} = \begin{bmatrix} \mathbf{I_1} & \frac{\partial \mathbf{\Omega(w)}}{\partial \mathbf{w_2}} \\ 0 & \mathbf{I_3} \end{bmatrix} \quad (14)$$

Transforming Equation (7) to the new generalized coordinates $\mathbf{w}$ with premultiplying $\mathbf{T(w)}$ and adding virtual constraint forces $\mathbf{f}$ inside the model, the transformed dynamic equations are derived as follows

$$\begin{aligned} & T(w)^T(M(Q(w))T(w)\ddot{w} + M(Q(w))\dot{T}(w)\dot{w} \\ & +T(w)^T(V(Q(w), T(w)\dot{w})T(w)\dot{w} + G(Q(w))) \\ & = T(w)^T(\tau + f) \quad (15) \\ & \bar{M}(w)\ddot{w} + \bar{V}(w,\dot{w})\dot{w} + \bar{G}(w) \\ & = T(w)^T(\tau + f) \quad (16) \end{aligned}$$

where

$$\begin{aligned} \bar{M}(w) &= T(w)^T M(Q(w)) T(w) \\ \bar{V}(w,\dot{w}) &= T(w)^T(M(Q(w))\dot{T}(w) \\ &\quad + V(Q(w), T(w)\dot{w})T(w)) \\ \bar{G}(w) &= T(w)^T G(Q(w)) \end{aligned}$$

The transformed dynamic equations can be partitioned into two parts by the partitioning matrix $\mathbf{E_1}$ and $\mathbf{E_2}$, where $[\mathbf{E_1^T}|\mathbf{E_2^T}]$ is an identity matrix of rank 4, E_1 is a 1×4 matrix and E_2 is a 3×4 matrix. Equation (16) can be written in a reduced form as

$$E_1 \bar{M}(w_2) E_2^T \ddot{w}_2 + E_1 \bar{V}(w_2, \dot{w}_2) \dot{w}_2 + E_1 \bar{G}(w_2)$$
$$= E_1 T(w_2)^T (\tau + f) \quad (17)$$
$$E_2 \bar{M}(w_2) E_2^T \ddot{w}_2 + E_2 \bar{V}(w_2, \dot{w}_2) \dot{w}_2 + E_2 \bar{G}(w_2)$$
$$= E_2 T(w_2)^T \tau \quad (18)$$

In the transformed coordinates the constraint equation is

$$\mathbf{w_1} = 0 \quad (19)$$

and the virtual constraint force satisfies

$$\mathbf{f} = \mathbf{J_1^T}(\mathbf{w_2}) \lambda \quad (20)$$

Equation (17) can be viewed as an algebraic equation for the virtual constraint force, and Equation (18) characterizes the motion of the container on the constraint space. The upper left 3×3 sub-matrix of $\bar{M}(\mathbf{w})$ is the equivalent of the mass matrix for a three cable configuration. This is similarly true for $\bar{G}(\mathbf{w})$. The relation between the transformed equations and the three cable configuration show the major advantage in using this form of nonlinear transformation.

Using Equation (11), the right hand side of Equation (16) is

$$\begin{bmatrix} 1 & 0 & 0 & \frac{l_1}{(l_1^2 + l_3^2 - l_2^2)^{\frac{1}{2}}} \\ 0 & 1 & 0 & \frac{-l_2}{(l_1^2 + l_3^2 - l_2^2)^{\frac{1}{2}}} \\ 0 & 0 & 1 & \frac{l_3}{(l_1^2 + l_3^2 - l_2^2)^{\frac{1}{2}}} \\ 0 & 0 & 0 & 1 \end{bmatrix} \tau + \begin{bmatrix} 0 \\ 0 \\ 0 \\ 1 \end{bmatrix} \lambda \quad (21)$$

The new variable or the generalized multiplier λ is introduced into this transformed dynamic equations, and this situation shows that multiple tension solutions exist for a specific four-cable array robot motion.

5 Inverse Dynamic Analysis

The inverse dynamic analysis of robot manipulators computes generalized forces which are required to produce specified kinematics, such as position, velocity, and acceleration.[7] In many control applications, the desired position, velocity, and acceleration of the moving robot are known or predetermined. Therefore, the required forces to actuate links along a predetermined trajectory can be obtained from the dynamic equations by plugging in kinematic data. The same procedure could be applied to the cable array robot but equation (16) is under-specified. Therefore the resultant forces might include compression forces which are not realistic since cables cannot carry compressive loads. Such forces could appear if sufficient constraints are not applied in Equation (16) to eliminate compression forces. The redundancy of the cables also introduces multiple tension solutions. This could result in dangerously high tension solutions unless the redundancy is properly addressed.

To avoid these problems, an optimal force distribution method is applied.[3] The left hand side, or the resultant forces R of Equation (16), are fixed for a specified motion, but the right hand side can be adjusted by changing the magnitude of the generalized multiplier λ. In order to find proper tensions, more adequate constraints for the cables and a suitable objective function need to be defined. Then the problem can be formulated in the standard linear programming (LP) form as follows

Maximize : objective function
Subject to :

$$T(w)^T (\tau + \mathbf{J_1^T}(\mathbf{w_2}) \lambda) = R$$
$$A\tau \leq v_A$$
$$\lambda \text{ unconstrained}$$

where A is the constraint matrix and v_A is the right hand side vector.

In order to avoid the cables becoming slack and prevent overloading, the physical limits of the tensions can be defined as

$$\tau_{\min} \leq \tau \leq \tau_{\max}$$

which for linear programming purposes can be rewritten as

$$\tau \leq \tau_{\max}$$
$$-\tau \leq -\tau_{\min}$$

where $\tau_{\max}$ is the vector of the maximum cable tensions and $\tau_{\min}$ is the vector of the minimum cable tensions. This limit constraint will eliminate compression forces from the feasible region of this LP problem. For the rigid body assumption, the minimum cable tension vector ($\tau_{\min}$) can be a constant vector, but for a more realistic model, the minimum cable tensions can be characterized as functions of cable configurations. The next constraint for the LP problem is based on the fact that in the absence of disturbance forces, the longest cable will provide the least tension when supporting the container. For different cable configurations, different constraint equations may be included in the LP model. For example, if the cable 3 is the longest cable in a specific configuration, the constraint equations will be defined as

$$\begin{bmatrix} 1 \\ 1 \\ 1 \end{bmatrix} \tau_3 \leq \begin{bmatrix} \tau_1 \\ \tau_2 \\ \tau_4 \end{bmatrix} \quad (22)$$

The first objective function candidate minimizes the sum of all the tensions. The physical meaning of this is to find the minimum total effort to achieve the specified task. The objective function for this case is as follows

$$\text{Minimize}: T_1 + T_2 + T_3 + T_4$$

However, a more relevant objective function for the cable array robot maximizes tension on the two longest cables. This will ensure that the cables remain as taut as possible to insure greatest protection from becoming slack. The objective function of this criterion can be expressed as maximizing total tension of the two longest cables and it is as follows

$$\text{Maximize}: T_i + T_j \quad (23)$$

were T_i and T_j are tensions in the two longest cables for a specific configuration.

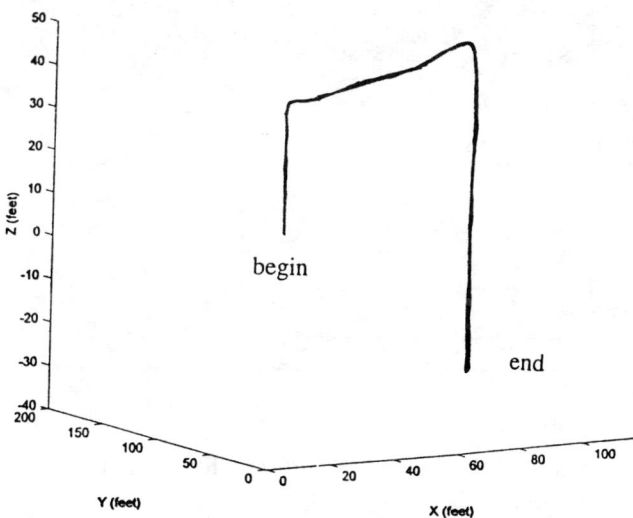

Figure 2: The specified trajectory in Cartesian space

A desired trajectory shown in Figure 2 is selected for the cable array robot to test the two objective functions discussed. The motion consists of vertically lifting up from the left lower corner of the deck, moving sideways towards the other side of the deck, and then landing on the other vessel. When the trajectory was applied using both objective functions, it was found that the solutions were identical. The basic reason is constraint equations (22) firmly restrict the feasible region of the LP problem and those two objective functions will generate the same tension solution. Fig. 3 shows the cable tensions and lengths during the prescribed motion. For this particular smooth trajectory, there is no temporal discontinuity problem in the solution sequence. It is important that the results are smooth and continuous since this reduces the chance of vibration in the cables.

6 Conclusion

Although the cable array robot is redundant, it is possible to find an optimal set of cable tensions for a specified trajectory using a nonlinear transformation and the optimal force distribution method. The objective function and solution constraints may be adjusted to alter performance. Multi-objective programming can be applied to this method which would allow more complex and effective objective functions.

The assumption of rigid body cables is not realistic but it allowed the application of force distribution without the complication of a higher order model due to the modes of vibration. Future research will involve the introduction of flexible cables and motion of the cable array robot base. The dynamic model of distributed weight system needs to describe the motion of elastic and inclined travelling cables. Also, the complexity of robotic crane will increase if the sea wave motion is included in the dynamic model. With more complete model, the analysis of cable catenary, sagging and loading effects, and the excitation of instability can be further addressed. In addition to the trajectory control demonstrated in the rigid cable case, the force distribution method must also be able to control the vibrations due to the flexible cables.

References

[1] Albus, J, R. Bostelman, and N. Dagalakis, "The NIST Spider, a Robot Crane," *Journal of Research of the NIST*, Vol. 97, pp. 373-385, 1992.

[2] Carignan, C. R., and D. L. Akin, "Optimal Force Distribution for Payload Positioning Using a Planar Dual-Arm Robot," *Journal of Dynamic Sys-*

tems, Measurement, and Control, Vol. 111, pp. 205-210, 1989.

[3] Cheng, F-T, and D.E. Orin, "Optimal Force Distribution in Multiple-Chain Robotic Systems," *IEEE Transactions on Systems, Man, and Cybernetics*, Vol. 21, pp. 13-24, 1991.

[4] Chin, C. and Nayfeh, A. H., "Nonlinear Dynamics of Crane Operation at Sea," *37th AIAA Structure, Structural Dynamics, and Material Conference*, Part 3, pp. 1536-1546, 1996.

[5] Dagalakis, N., J. Albus, B. Wang, J. Unger, and J.D. Lee, "Stiffness Study of a Parallel Link Robot Crane for Ship-Building Applications," *ASME Transactions of Offshore Mechanics Arctic Engineering*, Vol. 111, No. 3, pp. 183-193, 1989.

[6] Lew, J.Y. and W.J. Book, "Dynamics of Two Serially Connected Manipulators," *ASME Advanced Control Issues for Robot Manipulators*, DSC-Vol. 39, pp. 17-22, 1992.

[7] Lewis, F., C.T. Fitzgerald, and D.M. Dowson, *Control of Robot Manipulators*, Macmillan, 1993.

[8] McClamroch, N. H., and DanWei Wang, "Feedback Stabilization and Tracking of Constrained Robots," *IEEE Transactions on Automatic Control* Vol. 33, pp. 419-426, 1988.

[9] Pang, H. and M. Shahinpoor "Inverse Dynamics of a Parallel Manipulator," *Journal of Robotic Systems*, Vol. 11 pp. 693-702, 1994.

[10] Yang L.F, M Mikulas, Jr., J.C. Chiou, "Stability and 3-D Spatial Dynamics Analysis of a Three Cable Crane," AIAA-92-2456-CP, pp 2096-2076, 1992.

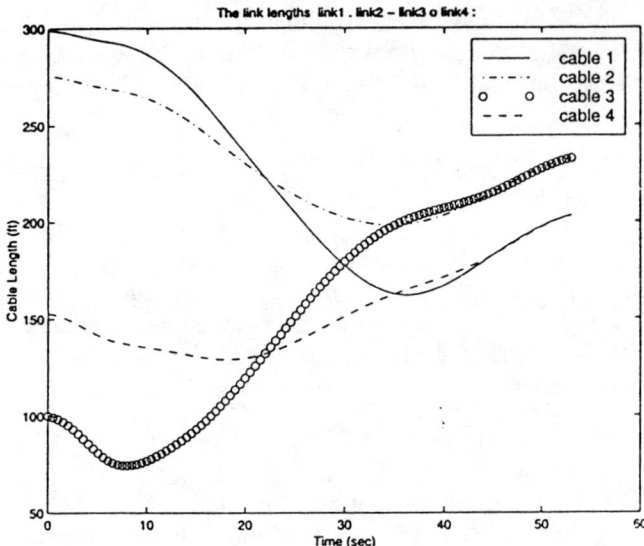

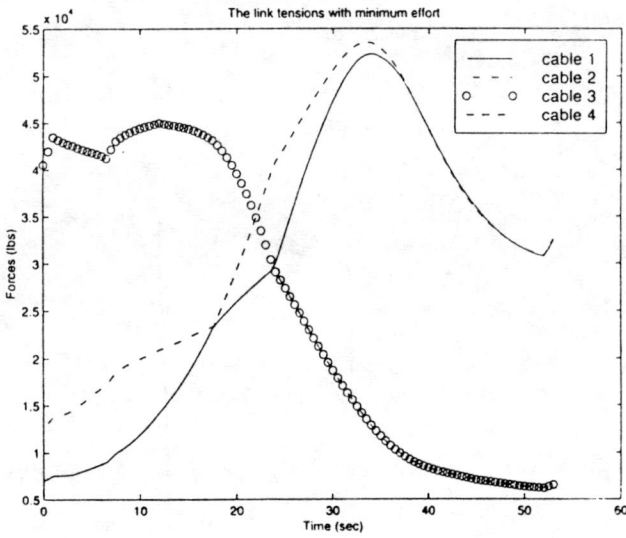

Figure 3: Cable tensions and lengths for a specified trajectory

Design and Kinematic Analysis of Modular Reconfigurable Parallel Robots

Guilin Yang[†] I-Ming Chen[‡] Wee Kiat Lim[‡] Song Huat Yeo[‡]

[†]Automation Technology Division
Gintic Institute of Manufacturing Technology, Singapore 638075

[‡]School of Mechanical & Production Engineering
Nanyang Technological University, Singapore 639798

ABSTRACT— A modular parallel robotic system consists of a collection of individual standard units that can be assembled into various robot configurations for a diversity of task requirements. This paper is focused on the design and kinematic analysis of modular reconfigurable parallel robots. A set of fundamental modules is considered. A local frame representation of the Product-Of-Exponentials (POE) formula, i.e., the local POE formula, is employed for the kinematic analysis of modular parallel robots. Two forward displacement analysis algorithms and a workspace visualization scheme are presented for a class of 3-leg modular parallel robots. Computation examples are also given to demonstrate the effectiveness of the proposed algorithms. The kinematic formulation shows that the local POE formula is a systematic and well-structured method for the kinematic analysis of parallel robots.

1 Introduction

A parallel robot is a closed-loop mechanism in which the mobile platform is connected to the base by at least two serial kinematic chains (legs). Because a parallel robot usually has a limited workspace, trajectory planning and application development become difficult. In this paper, the modularity concept is introduced in the design of parallel robot configurations. The modularly designed parallel robot system consists of a collection of individual standard actuators, passive joints, rigid links (connectors), mobile platforms, and end-effectors. A modular parallel robot configuration thus can be rapidly constructed and its workspace can be adjusted by changing the leg positions, the joint types, and the link lengths for a diversity of task requirements.

Because of the reconfigurability, a modular parallel robot system has unlimited configurations. To find general and systematic algorithms for the kinematic analysis of various possible parallel robots are of primary importance. In this paper, we mainly focus on the forward displacement analysis and the workspace visualization of a class of 3-leg parallel robots. For the forward displacement analysis, no general closed-form solution algorithm has been discovered for parallel robots except for some special robot configurations [5], [6], [7]. Most of the closed-form solution algorithms follow an algebraic analytical approach, and result in solving high order polynomial equations, which requires extensive computation effort. The sensor-based algorithm and the iterative numerical algorithm hence become more feasible and practical. Previous works on such algorithms usually follow a pure geometrical (vector computation) modeling approach [10], [3]. Here, based on the local POE formula, a sensor-based solution algorithm and an iterative numerical algorithm are re-formulated. We show that the local POE formula is a systematic and well-structured method for the kinematic analysis of parallel robots.

For a parallel robot with fixed configuration, it is possible to develop a specific analytical algorithm to determine the workspace boundary [4], [8]. For a modular reconfigurable parallel robot, a general analytical algorithm is more effective for the workspace analysis. Since such an algorithm is difficult to formulate, a general and simple visualization scheme is employed to generate the workspace of the mobile platform.

2 Design Considerations

2.1 Robot modules

A set of commercial grade and custom designed actuator modules, passive joint modules, rigid link modules (connectors), and mobile platforms are considered as the basic parallel robot modules. Off-the-shelf intelligent mechatronic drives, PowerCube, from Amech GmbH, Germany, are selected as actuator modules for rapid deployment. Both revolute (Fig. 1(a)) and prismatic (Fig. 1(b)) actuator modules are considered. Such a compact actuator module is a self-contained drive unit with a built-in motor, a controller, an amplifier, and the communication interface. It has a cubic or double-cube design with multiple connecting sockets so that two actuator modules can be connected in many different orientations. Three types of passive joint modules (without actuators) are custom designed and fabricated: the rotary joint (Fig. 2(a)), the pivot joint (Fig. 2(b)), and the spherical joint (Fig. 2(c)). An angular displacement sensor is built into each of the passive rotary and pivot joint modules. The rigid link modules (connectors) are used to connect joint modules through bolts. A set of links with various geometrical shapes and dimensions has been designed for different robot configurations (Fig. 3(a)). A hexagonal mobile platform is designed (Fig. 3(b)). This hexagonal platform has one connecting socket on each of its six edges so that it can be used with parallel robots with 2, 3, or 6 legs. For a variety of applications, each type of modules has several different sizes.

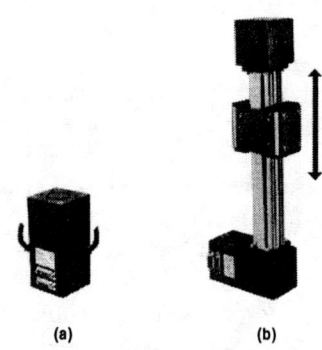

Fig. 1: Actuator modules

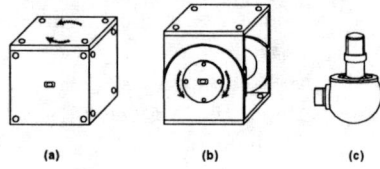

Fig. 2: Passive joint modules

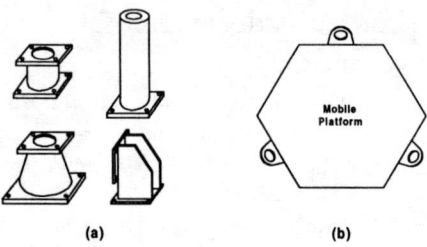

Fig. 3: Links (connectors) and the mobile platform

2.2 Possible robot configurations

Based on the module designs, many possible parallel robot configurations can be constructed. Here, however, we mainly focus on the 6-DOF, nonredundant, parallel robot configurations. An enumeration scheme for such parallel robot topological structures is presented in [12]. In this work, a class of 3-leg, nonredundant, parallel robots is identified as having simple kinematics and desirable characteristics. Such a parallel robot consists of three legs. And each leg has two actuator joints, one passive revolute joint, and one passive spherical joint. The actuator joint modules in each leg are always placed nearby the base because they are heavier than the passive joint modules. Base on this fact, we can generate all of the possible robot configurations in this class. Fig. 4 shows two such possible robot configurations.

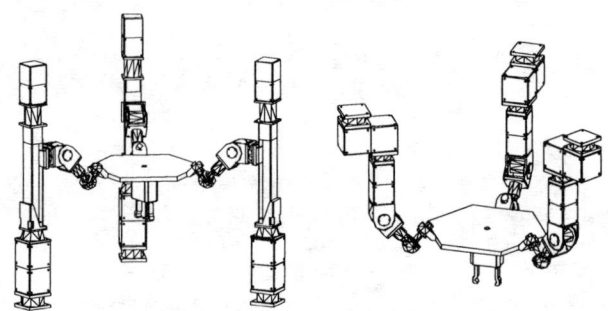

Fig. 4: Two modular 3-leg parallel robot configurations

3 Local POE Formula

In [1], Brocket shows that forward kinematic equation of an open chain robot containing either revolute or prismatic joints can be uniformly expressed as a product of matrix exponentials. Because of its compact representation and its connection with Lie groups and Lie algebras, the POE formula has proven to be a useful modeling tool in robot kinematics [11]. According to the coordinate frames used for expressing the joint axes, the POE formula can be written in different forms such as the base frame, the local frame, and the tool frame representation of the POE formula [2], [9], [11]. For our purpose, only the local frame representation of the POE formula is introduced in this article.

3.1 Dyad kinematics

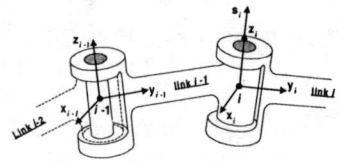

Fig. 5: Two consecutive links: a dyad

Let link $i-1$ and link i be two adjacent links connected by joint i, as shown in Fig. 5. Link i and joint i are termed as *link assembly i*. If we denote the body coordinate frame on link assembly i by frame i, then the relative pose (position and orientation) of frame i with respect to frame $i-1$, under a joint displacement, q_i, can be described by a 4×4 homogeneous matrix, an element of $SE(3)$, such that

$$T_{i-1,i}(q_i) = T_{i-1,i}(0)e^{\widehat{s_i}q_i}, \qquad (1)$$

where $\hat{s}_i \in se(3)$ is the twist of joint i expressed in frame i, and $T_{i-1,i}(0) \in SE(3)$ is the initial pose of frame i relative to frame $i-1$.

3.2 Local POE formula for open chains

Based on the dyad kinematics, the forward kinematic transformation for an open kinematic chain can be easily derived. Consider an open kinematic chain with $n+1$ links, sequentially $0, 1, \ldots, n$ (from the base 0 to the end link n). The forward kinematic transformation thus can be given by:

$$T_{0,n}(q_1, q_2, \ldots, q_n) = T_{0,1}(q_1) T_{1,2}(q_2) \ldots T_{(n-1),n}(q_n)$$
$$= \prod_{i=1}^{n} (T_{(i-1),i}(0) e^{\hat{s}_i q_i}) \quad (2)$$

4 Forward Displacement Analysis

We consider a class of modular 3-leg (6-DOF) parallel robots as shown in Fig. 6. Each leg contains 4 joint modules, i.e., two actuator modules, one passive revolute (rotary or pivot) joint module, and one passive spherical joint module which is at end of the leg. We assume that joint ij ($\hat{s}_{ij}$) are actuating joints ($i = 1, 2, 3; j = 1, 2$), while joint $i3$ ($\hat{s}_{i3}$) are passive revolute joints ($i = 1, 2, 3$). Define frame A as the local frame attached to the mobile platform and frame B as the base frame. The coordinate of point A_i ($i = 1, 2, 3$) relative to frame A and frame B_{i3} are given by $p'_i = (x'_{ai}, y'_{ai}, z'_{ai})^T$ and $p''_i = (x''_{ai}, y''_{ai}, z''_{ai})^T$ respectively. The forward displacement analysis becomes to determine the pose of frame A with respect to the base frame B when the joint displacements of the six actuating joints, q_{ij} ($i = 1, 2, 3; j = 1, 2$), are known.

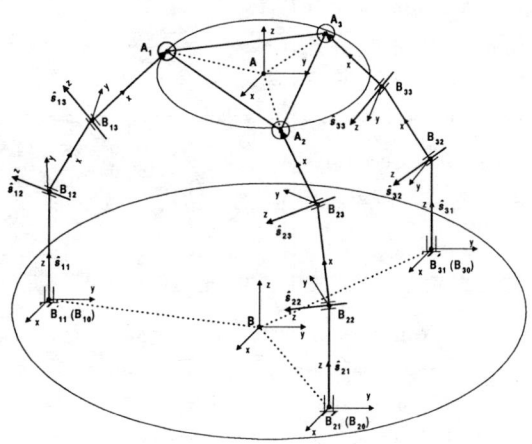

Fig. 6: A 3-leg parallel robot

4.1 Sensor-based solution approach

The sensor-based method is a simple and practical approach for the forward displacement analysis of parallel robots. The basic idea is to install a sensor in each of the passive joint modules to measure its corresponding joint displacement. In this case, the position of point $A_i (i = 1, 2, 3)$ (with respect to the base frame B) can be directly determined as a function of the actuating joint displacements and the sensed passive joint displacement in leg i. For each leg, the computation is identical to the forward kinematics of serial chains. Based on the local POE formula (Eq. (2)), p_i - the positional vector of point A_i, can be given by

$$\begin{bmatrix} p_i \\ 1 \end{bmatrix} = T_{B,B_{i0}} T_{B_{i0},B_{i1}}(0) e^{\hat{s}_{i1} q_{i1}} T_{B_{i1},B_{i2}}(0) e^{\hat{s}_{i2} q_{i2}}$$
$$T_{B_{i2},B_{i3}}(0) e^{\hat{s}_{i3} q_{i3}} \begin{bmatrix} p''_i \\ 1 \end{bmatrix}, \quad (3)$$

where $T_{B,B_{i0}}$ is the fixed kinematic transformation from frame B (the base platform frame) to frame B_{i0} (the base frame of leg i). Note that the homogeneous coordinate representation is employed in Eq. (3). Denote the forward kinematic transformation from the base frame B to the mobile platform frame A as $T_{B,A}$. Then

$$T_{B,A} \begin{bmatrix} p'_i \\ 1 \end{bmatrix} = \begin{bmatrix} p_i \\ 1 \end{bmatrix} \quad (i = 1, 2, 3). \quad (4)$$

We also have

$$T_{B,A} \begin{bmatrix} p'_{12} \times p'_{23} \\ 0 \end{bmatrix} = \begin{bmatrix} p_{12} \times p_{23} \\ 0 \end{bmatrix}, \quad (5)$$

where, $p'_{12} = p'_2 - p'_1$, $p'_{23} = p'_3 - p'_2$, $p_{12} = p_2 - p_1$, and $p_{23} = p_3 - p_2$.

Combining Eq. (4) and Eq. (5), we obtain

$$T_{B,A} \begin{bmatrix} p'_1 & p'_2 & p'_3 & p'_{12} \times p'_{23} \\ 1 & 1 & 1 & 0 \end{bmatrix} = \begin{bmatrix} p_1 & p_2 & p_3 & p_{12} \times p_{23} \\ 1 & 1 & 1 & 0 \end{bmatrix}. \quad (6)$$

Therefore, the pose of the mobile platform, $T_{B,A}$, can be given by

$$T_{B,A} = \begin{bmatrix} p_1 & p_2 & p_3 & p_{12} \times p_{23} \\ 1 & 1 & 1 & 0 \end{bmatrix} \begin{bmatrix} p'_1 & p'_2 & p'_3 & p'_{12} \times p'_{23} \\ 1 & 1 & 1 & 0 \end{bmatrix}^{-1}. \quad (7)$$

Note that the matrix $\begin{bmatrix} p'_1 & p'_2 & p'_3 & p'_{12} \times p'_{23} \\ 1 & 1 & 1 & 0 \end{bmatrix}$ is a constant matrix for a specific robot configuration, and it is always invertible if point A_1, A_2, and A_3 neither coincide with each other nor fall on the same line.

4.2 Numerical solution approach

The limitation of the sensor-based algorithm is that it can only be implemented on the actual parallel robot in which each of the passive revolute joints has a displacement sensor. Hence, in situations where the passive joint displacements are unable to be sensed, e.g., off-line computations and simulations, the iterative numerical solution method could be more practical because it does not require the passive joint displacements as the input variables.

For simplicity, Eq. (3) can also be written as

$$P_i = T_i e^{\hat{s}_{i3} q_{i3}} P_i'' \quad (i = 1, 2, 3), \quad (8)$$

where $P_i = \begin{bmatrix} p_i \\ 1 \end{bmatrix}$, $P_i'' = \begin{bmatrix} p_i'' \\ 1 \end{bmatrix}$, and
$T_i = T_{B,B_{i0}} T_{B_{i0},B_{i1}}(0) e^{\hat{s}_{i1} q_{i1}} T_{B_{i1},B_{i2}}(0) e^{\hat{s}_{i2} q_{i2}} T_{B_{i2},B_{i3}}(0)$.
Since

$$\|\overrightarrow{A_1 A_2}\|^2 = (P_2 - P_1)^T (P_2 - P_1), \quad (9)$$

The differential form of Eq. (9) can be written as

$$d\|\overrightarrow{A_1 A_2}\| = \frac{(P_2 - P_1)^T}{\|\overrightarrow{A_1 A_2}\|}(dP_2 - dP_1). \quad (10)$$

According to Eq. (8), we find

$$dP_i = T_i e^{\hat{s}_{i3} q_{i3}} \hat{s}_{i3} P_i'' dq_{i3} \quad (i = 1, 2, 3). \quad (11)$$

The actual distance from point A_1 to point A_2 is a constant, denoted by a_{12}. Then,

$$d\|\overrightarrow{A_1 A_2}\| = a_{12} - \|\overrightarrow{A_1 A_2}\|$$
$$= \frac{(P_2 - P_1)^T}{\|\overrightarrow{A_1 A_2}\|}(dP_2 - dP_1). \quad (12)$$

similarly, we have

$$a_{23} - \|\overrightarrow{A_2 A_3}\| = \frac{(P_3 - P_2)^T}{\|\overrightarrow{A_2 A_3}\|}(dP_3 - dP_2); \quad (13)$$

$$a_{31} - \|\overrightarrow{A_3 A_1}\| = \frac{(P_1 - P_3)^T}{\|\overrightarrow{A_3 A_1}\|}(dP_1 - dP_3). \quad (14)$$

Combining Eq. (12), (13), and (14), we get

$$dq = J^{-1} da, \quad (15)$$

where $dq = column[dq_{13}, dq_{23}, dq_{33}]$,
$da = column[a_{12} - \|\overrightarrow{A_1 A_2}\|, a_{23} - \|\overrightarrow{A_2 A_3}\|, a_{31} - \|\overrightarrow{A_3 A_1}\|]$,

$$J = \begin{bmatrix} -\frac{(P_2-P_1)^T}{\|\overrightarrow{A_1 A_2}\|}\dot{P}_1 & \frac{(P_2-P_1)^T}{\|\overrightarrow{A_1 A_2}\|}\dot{P}_2 & 0 \\ 0 & -\frac{(P_3-P_2)^T}{\|\overrightarrow{A_2 A_3}\|}\dot{P}_2 & \frac{(P_3-P_2)^T}{\|\overrightarrow{A_2 A_3}\|}\dot{P}_3 \\ \frac{(P_1-P_3)^T}{\|\overrightarrow{A_3 A_1}\|}\dot{P}_1 & 0 & -\frac{(P_1-P_3)^T}{\|\overrightarrow{A_3 A_1}\|}\dot{P}_3 \end{bmatrix},$$

in which $\dot{P}_i = T_i e^{\hat{s}_{i3} q_{i3}} \hat{s}_{i3} P_i''$.

Eq. (15) can be written as an iterative form, i.e.,

$$q^{(k+1)} = q^{(k)} + (J^{-1} da)^{(k)}, \quad (16)$$

where k represents the number of iterations. Based on the standard iterative form of Eq. (16), the Newton Raphson method is employed to derive the numerical solution of $q(= [q_{13}, q_{23}, q_{33}]^T)$. Since the Jacobian matrix, J, is a 3×3 matrix, the inverse of the Jacobian matrix J can be found in a symbolic form. From computational point of view, however, the proposed algorithm is equivalent to the method given by Cleary and Brooks[3]. After the joint displacement q is derived, the pose of the mobile platform can be easily determined by using the sensor-based algorithm.

4.3 Computation examples

In the following computation examples, we wish to employ the proposed forward displacement analysis algorithms for the kinematic analysis of a modular 3-leg parallel robot, i.e., to determine the pose of the mobile platform when the joint displacements of the actuating joints are known. As shown in Fig. 6, the given geometric parameters are as follows.

$$T_{B,B_{10}} = \begin{bmatrix} 0.5 & -0.866 & 0 & -250 \\ 0.866 & 0.5 & 0 & -433 \\ 0 & 0 & 1 & 80 \\ 0 & 0 & 0 & 1 \end{bmatrix}, T_{B,B_{20}} = \begin{bmatrix} -1 & 0 & 0 & 500 \\ 0 & -1 & 0 & 0 \\ 0 & 0 & 1 & 80 \\ 0 & 0 & 0 & 1 \end{bmatrix},$$

$$T_{B,B_{30}} = \begin{bmatrix} 0.5 & 0.866 & 0 & -250 \\ -0.866 & 0.5 & 0 & 433 \\ 0 & 0 & 1 & 80 \\ 0 & 0 & 0 & 1 \end{bmatrix};$$

$$T_{B_{10},B_{11}}(0) = T_{B_{20},B_{21}}(0) = T_{B_{30},B_{31}}(0) = \begin{bmatrix} 0 & 0 & 1 & 90 \\ 0 & -1 & 0 & 0 \\ 1 & 0 & 0 & 0 \\ 0 & 0 & 0 & 1 \end{bmatrix};$$

$$T_{B_{11},B_{12}}(0) = T_{B_{21},B_{22}}(0) = T_{B_{31},B_{32}}(0) = \begin{bmatrix} 0 & 0 & 1 & 195 \\ 0 & 1 & 0 & 0 \\ -1 & 0 & 0 & 0 \\ 0 & 0 & 0 & 1 \end{bmatrix};$$

$$T_{B_{12},B_{13}}(0) = T_{B_{22},B_{23}}(0) = T_{B_{32},B_{33}}(0) = \begin{bmatrix} 1 & 0 & 0 & 0 \\ 0 & 0 & -1 & 0 \\ 0 & 1 & 0 & 105 \\ 0 & 0 & 0 & 1 \end{bmatrix};$$

$s_{ij} = (0, 0, 0, 0, 0, 1)^T$ (i,j=1,2,3);
$p_1' = (-150, -259.81, 0)^T$, $p_2' = (300, 0, 0)^T$, and
$p_3' = (-150, 259.81, 0)^T$; $p_1'' = p_2'' = p_3'' = (0, 500, 0)^T$.

TABLE 1: Computation Results

$q_a = (0, 0, 0, 0, 0, 0)^T$ $q_p = (0.2218, 0.2218, 0.2218)^T$ $N = 3$	$\begin{bmatrix} 1 & 0 & 0 & 0 \\ 0 & 1 & 0 & 0 \\ 0 & 0 & 1 & 867.75 \\ 0 & 0 & 0 & 1 \end{bmatrix}$
$q_a = (0, 0.1, 0.2, 0, 0.1, 0.2)^T$ $q_p = (0.1386, 0.4121, 0.1735)^T$ $N = 3$	$\begin{bmatrix} 0.9737 & -0.2195 & 0.0616 & -81.42 \\ 0.2170 & 0.9752 & 0.0436 & -9.32 \\ -0.0696 & -0.0291 & 0.9972 & 857.21 \\ 0 & 0 & 0 & 1 \end{bmatrix}$
$q_a = (0, 0.2, 0.4, 0, 0.2, 0.4)^T$ $q_p = (0.0605, 0.6486, 0.1882)^T$ $N = 4$	$\begin{bmatrix} 0.9084 & -0.4038 & 0.1089 & -158.52 \\ 0.3840 & 0.9085 & 0.1651 & -35.25 \\ -0.1656 & -0.1081 & 0.9802 & 826.15 \\ 0 & 0 & 0 & 1 \end{bmatrix}$
$q_a = (0, 0.3, 0.6, 0, 0.3, 0.6)^T$ $q_p = (0.0011, 0.9069, 0.2650)^T$ $N = 4$	$\begin{bmatrix} 0.8367 & -0.5313 & 0.1328 & -217.24 \\ 0.4673 & 0.8191 & 0.3327 & -71.67 \\ -0.2855 & -0.2163 & 0.9336 & 780.98 \\ 0 & 0 & 0 & 1 \end{bmatrix}$
$q_a = (0, 0.4, 0.8, 0, 0.4, 0.8)^T$ $q_p = (-0.0280, 1.1732, 0.4047)^T$ $N = 4$	$\begin{bmatrix} 0.7843 & -0.5987 & 0.1623 & -249.90 \\ 0.4532 & 0.7317 & 0.5091 & -109.10 \\ -0.4236 & -0.3258 & 0.8452 & 731.63 \\ 0 & 0 & 0 & 1 \end{bmatrix}$

The entire computation scheme consists of two parts. The first part is to solve for the displacements of passive joints $q_p(= (q_{13}, q_{23}, q_{33})^T)$ with the known actuating joint displacements $q_a(= (q_{11}, q_{21}, q_{31}, q_{12}, q_{22}, q_{32})^T)$ by using the iterative numerical algorithm. The second part, based on the derived solution of q_p, is to determine the pose of the mobile platform $T_{B,A}$ by using the sensor-based algorithm. The computation results are listed in Table 1. In the right column: q_a– the input actuating joint displacements; q_p the derived passive joint displacements; and N–the number of iterations. The left column represents the derived poses of the mobile platform–$T_{B,A}$. The initial guess solutions for the passive joints, q_p, are identically taken as $(0,0,0)^T$. The desired solutions in all five cases can be obtained in 3 to 4 iterations. Nevertheless, the iterative numerical algorithm cannot guarantee that a solution will be derived within a fixed number of iterations. The computation efficiency depends on the initial guess solutions. Note that in the computation examples, the joint angles are expressed in radians (for revolute joints), and the position vectors are in millimeters.

5 Workspace Visualization

5.1 Geometrical description of workspace

The complete workspace of a 6-DOF parallel robot is embedded in a 6-D manifold which cannot be represented in a 3-D space. Hence, the workspace determined here is the reachable workspace, i.e., the region of the 3-D Cartesian space that can be reached by the mobile platform with a given orientation of the platform [4].

Since the pose of the mobile platform frame A, $T_{B,A}$, has the form of $T_{B,A} = \begin{bmatrix} R_{B,A} & p_{B,A} \\ 0 & 1 \end{bmatrix}$, Eqn.(4) can also be written as:

$$p_{B,A} = p_i - R_{B,A} p'_i \quad (i=1,2,3). \quad (17)$$

Eqn.(17) is an alternative expression of leg i's inverse kinematic equation, where each leg is considered as an independent serial chain. For a given mobile platform orientation $-R_{B,A}$, the platform position, $p_{B,A}$, can be derived by translating point A_i through a fixed vector $(-R_{B,A} p'_i)$. In other words, the reachable workspace of the mobile platform determined by leg i is a fixed translation of point A_i's workspace. However, the motion of a 3-leg parallel robot is constrained by its three legs such that the position of the mobile platform must satisfy the three inverse kinematic equations, i.e., Eqn.(17), simultaneously. The reachable workspace of the mobile platform is, therefore, the intersection of the individual ones determined by each of the three legs.

5.2 Workspace visualization scheme

The workspace visualization scheme is also focused on the class of 3-leg modular parallel robots mentioned in Section 2.2. For each of the three legs, substituting Eqn.(3) into Eqn.(17), the individual workspace can be theoretically generated through infinitely varying the three joint displacements within their joint limitations. For simplicity, we directly employ the 3-D parametric surface plotting function provided by the *Maple V* software to generate each of the individual workspaces. However, this 3-D plotting function only takes two parametric variables, i.e., two joint variables, at a time and the resulting plot is a 3-D surface. We, therefore, evenly discretize the working range of one joint variable into many intermediate values; then for each of them, with the other two joint variable to generate a 3-D surface. Plotting all of these surfaces together, a 3-D workspace can be visualized. Finally, plotting the three individual workspaces in the same graph, the intersection of the three workspaces can be obtained, which is the actual reachable workspace of the mobile platform. In addition, the 2-D cross-section view provided by *Maple V* can be used to investigate the internal structure of the workspace.

5.3 Visualization example

In this example, we employ the workspace visualization scheme developed above to graphically represent the workspace of the parallel robot given in Section 4.3. The orientation of the mobile platform is specified as a 3×3 identity matrix. The workspace visualization results are shown in Fig.7 and Fig.8.

Fig.7(b) show the 3-D workspace (the intersection part) of the platform generated by the three legs. This 3-D workspace can be graphically studied by plotting it in different Maple functions, such as PATCH, CONTOUR, and WIREFRAME, and observing it from different view points and angles. In order to understand it better, the individual workspace generated by Leg 1 is shown in Fig.7(a). To further study the internal structure of the workspace, several sectional views of both Fig.7(a) and (b) in the planes parallel to the xy plane are shown in Fig.8. The left and right column represent the cross-sections of Fig.7(a) and Fig.7(b) respectively. Two graphs in the same row have the same z coordinates. With the help of the graphs on the left, we can easily identify the working regions (intersection areas) in the composite workspaces based on the plotting density of the graphs on the right. It can be seen from these figures that the workspace generated by each leg is a part of a solid torus. There is a void in the lower part of the intersection of these three toruses indicating that the platform has a poor motion capability in the lower z coordinates. This example shows that the workspace can be clearly visualized although the derived workspace is not a true 3-D solid model. Most importantly of all, this visualization scheme is simple and general for various 3-leg modular parallel robot configurations.

6 Summary

In this paper, we employ the modularity design concept in the design of parallel robots. The robot modules considered can be categorized into the actuator modules, passive joint modules, link (connector) modules, and mobile platforms. With such an inventory of modules, many possible parallel robot configurations can be rapidly constructed for a diversity of tasks. Based on the

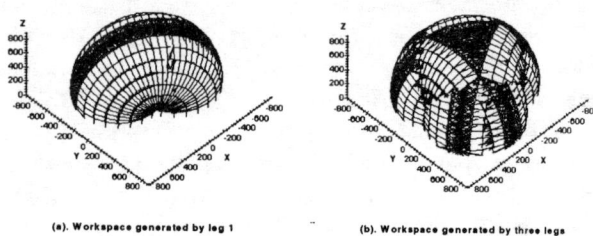

Fig. 7: 3-D workspace of the platform

local POE formula, two simple and systematic forward displacement analysis algorithms, namely the sensor-based algorithm and the iterative numerical algorithm, are derived for a class of 3-leg modular parallel robots. With the POE formula, these two algorithms can uniformly deal with various 3-leg parallel robots regardless of the joint types and the degrees of freedom. For the workspace analysis, a simple and general workspace visualization scheme is also presented. As demonstrated by the examples, the kinematic analysis algorithms and the workspace visualization scheme are suitable for the design and analysis of modular reconfigurable parallel robots.

Acknowledgment

The authors thank Prof. Frank Chongwoo Park of Seoul National University, Korea, for his valuable advice on modular reconfigurable parallel robots. This work is supported by GINTIC Institute of Manufacturing Technology, Singapore, under Upstream project U-97-A006.

References

[1] R. Brockett. Robotic manipulators and the product of exponential formula. In *International Symposium in Math. Theory of Network and Systems*, pages 120–129, Beer Sheba, Israel, 1983.

[2] I.M. Chen and G. Yang. Configuration independent kinematics for modular robots. In *Proceedings of IEEE Conference on Robotics and Automation*, pages 1440–1445, 1996.

[3] K. Cleary and T. Brooks. Kinematic analysis of a novel 6-DOF parallel manipulator. In *Proceedings of IEEE Conference on Robotics and Automation*, pages 708–713, 1993.

[4] C.M. Gosselin. Determination of the workspace of 6-DOF parallel manipulators. *ASME Journal of Mechanical Design*, 112:331–336, Sept. 1990.

[5] C. Innocenti and V. Parenti-Castelli. Direct position analysis of the Stewart platform mechanism. *Journal of Mechanism and Machine Theory*, 25(6):611–621, 1990.

[6] H.Y. Lee and B. Roth. A closed-form solution of the forward displacement analysis of a class of in-parallel mechanisms. In *Proceedings of IEEE Conference on Robotics and Automation*, pages 720–724, 1993.

[7] W. Lin, J. Duffy, and M. Griffis. Forward displacement analysis the 4-4 Stewart platforms. In *Proceedings of ASME 21nd Biennial Mechanism Conference*, De-25, pages 263–269, 1990.

[8] J.P. Merlet. Trajactory verification in the workspace for parallel manipulators. *International Journal of Robotic Research*, 13(4):326–333, 1990.

[9] R. Murray, Z. Li, and S.S. Sastry. *A Mathematical Introduction to Robotic Manipulation*. CRC Press, 1994.

[10] L. Notash and R.P. Podhorodeski. Complete forward displacement solutions for a class of three-branch parallel manipulators. *Journal of Robotic System*, 11(6):471–485, 1994.

[11] F.C. Park. Computational aspect of manipulators via product of exponential formula for robot kinematics. *IEEE Transactions on Automatic Control*, 39(9):643–647, 1994.

[12] R.P. Podhorodeski and K.H. Pittens. A class of parallel manipulators based on kinematically simple branches. *ASME Journal of Mechanical Design*, 116:908–914, 1994.

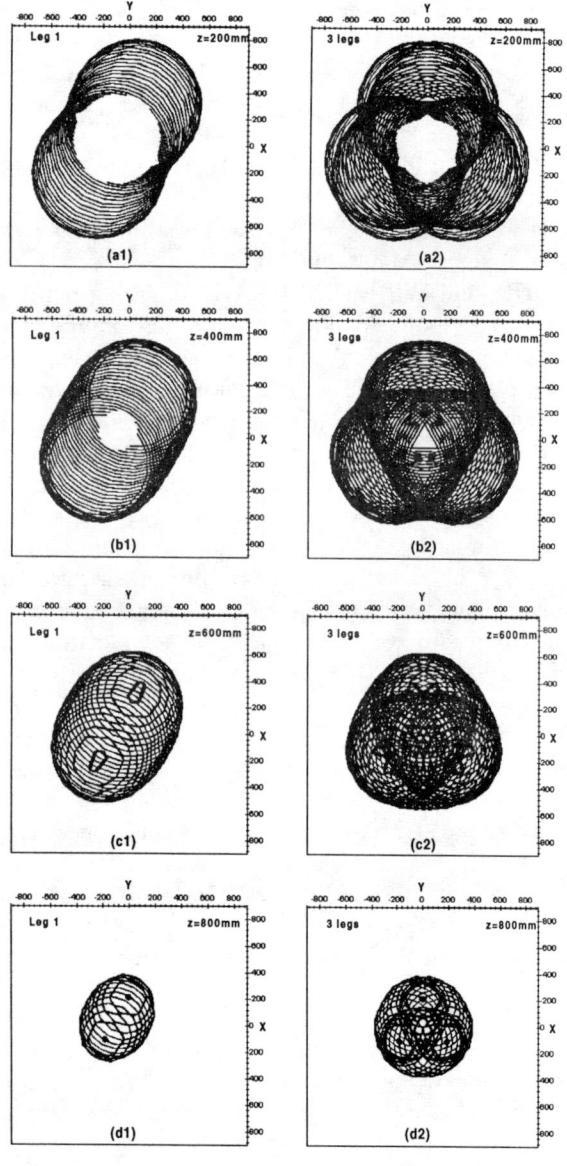

Fig. 8: Sections of the workspace parallel to the xy plane

A Robust Control Scheme for Dual-Arm Redundant Manipulators: Experimental Results

Haipeng Xie[*], Iain J. Bryson[*], Farshid Shadpey[+] and R.V. Patel[*]

[*]Department of Electrical & Computer Engineering, University of Western Ontario
London, Ontario, Canada N6A 5B9

[+]Bombardier Services, Defence, Mirabel, Quebec, Canada J7N 1H3

1. Abstract

The problem of tracking a Cartesian space trajectory for an object held by two redundant robots while controlling the contact force and the object's internal force is investigated in this paper. A Two-Level Impedance Control (TLIC) algorithm incorporating an Error Reference Controller (ERC) is developed. This algorithm is robust to system and environmental kinematic and dynamic uncertainties. Joint redundancy is used to fulfill additional tasks such as singularity robustness. The control algorithm is applicable to dual independent as well as closed-chain robot control. The algorithm has been extensively tested using a computer simulation of the full dynamic model of a dual-arm experimental redundant system in our Robotics and Control Systems Laboratory. The algorithm has also been implemented on the dual-arm experimental system. Experimental results that illustrate various features of the dual-arm control algorithm are presented in this paper.

2. Introduction

Multi-robot systems are useful in transporting massive and/or bulky objects. Redundant manipulators have been successfully used in tasks involving singularity and obstacle avoidance. When a task involves contact with an environment, controlling the contact force as well as the interacting (squeezing) force and torque applied by the robots on the object are major issues. Impedance control algorithms have been successfully developed for single robot systems by Hogan [1], Liu and Goldenberg [2], Shadpey [3] and for dual-robot system by Schneider and Cannon [4] and Yoshikawa and Zheng [5] in handling contacts with an unknown environment. An intensive analysis of common object internal loading was done by Walker et al. [6]. Internal force-based impedance control was proposed by Bonitz and Hsia [7]. This method enforces an impedance relationship between internal force error and end-effector tracking deviation. Ramadorai et al. [8] addressed the multi-redundant robot control problem by means of non-linear feedback. However, this approach is based on perfect knowledge of the environment and the system's kinematics and dynamics. In this paper, a two-level impedance control approach is proposed which results in significant robustness for the closed-loop system to uncertainties in the robot dynamics and kinematics. An object-level impedance is controlled between the object and the environment, while a second level impedance is achieved between the internal force error and the end-effector tracking deviation. The approach not only gives high robustness to system kinematic and dynamic uncertainties, but also gives an approach that is applicable for dual independent as well as closed-chain operations. As a result, this control approach allows smooth control switching from independent robot control to closed-chain dual-arm control. To the best of acknowledge, these issues have not been well addressed to date in the robotics literature.

3. Dynamic Model

The following notation is used in this paper:

$\{W\}$: World frame which is attached at some point in the robots' work space; $\{R_i\}$: Robot i base frame which is attached to the base of the ith robot; $\{T_i\}$: Robot i tool frame which is attached to the end-effector of the ith robot; $\{O\}$: Object frame which is attached to the center of mass of the object; ${}^wR_{R_i}, {}^wP_{R_i}$: Orientation and position of robot i in the world frame; ${}^oR_{T_i}, {}^oP_{T_i}$: Grasp orientation and position of robot i in the object frame; I_o, m_o: Moment of inertia and mass of the object (payload) in the object frame. A closed kinematic chain is formed (Figure 1) when two robots hold a common object.

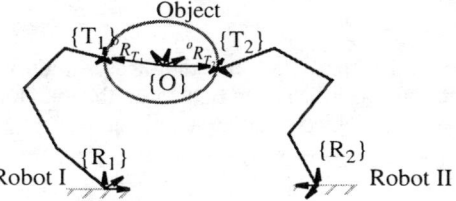

Figure 1: Dual-arm system

3.1 Closed-chain constraints

The position and the orientation constraints are defined as

$${}^wP_{R_1} + {}^wR_{R_1}{}^{R_1}P_{T_1} - {}^wR_o{}^oP_{T_1} = {}^wP_{R_2} + {}^wR_{R_2}{}^{R_2}P_{T_2} - {}^wR_o{}^oP_{T_2} = {}^wP_o \quad (1)$$

$${}^wR_{R_1}{}^{R_1}R_{T_1}{}^{T_1}R_o = {}^wR_{R_2}{}^{R_2}R_{T_2}{}^{T_2}R_o = {}^wR_o \quad (2)$$

and the velocity constraints as

$$G_1^{-T}\begin{bmatrix}{}^w v_1\\{}^w \omega_1\end{bmatrix} = G_2^{-T}\begin{bmatrix}{}^w v_2\\{}^w \omega_2\end{bmatrix} = \begin{bmatrix}{}^w v_o\\{}^w \omega_o\end{bmatrix} \quad (3)$$

where G_i is the grasp matrix of robot i:

$$G_i = \begin{bmatrix}I_3 & O_3\\{}^w\hat{P}_{oT_i} & I_3\end{bmatrix} \text{ with } \hat{P} = \begin{bmatrix}0 & -P_z & P_y\\P_z & 0 & -P_x\\-P_y & P_x & 0\end{bmatrix} \text{ and } {}^w P_{oT_i} = {}^w R_o{}^o P_{T_i} \quad (4)$$

I_3, O_3 are the (3×3) identity and zero matrices respectively. The joint velocity constraint is

$$\begin{bmatrix}G_1^{-T}{}^w\tilde{R}_{R_1}J_{R_1} & -G_2^{-T}{}^w\tilde{R}_{R_2}J_{R_2}\end{bmatrix}\begin{bmatrix}\dot{q}_1\\\dot{q}_2\end{bmatrix} = 0 \text{ with } {}^w\tilde{R}_{R_i} = \begin{bmatrix}{}^w R_{R_i} & O_3\\O_3 & {}^w R_{R_i}\end{bmatrix} \quad (5)$$

3.2 Forward dynamics of the dual-arm system

3.2.1 Object Dynamics

The dynamics of the object in frame $\{O\}$ are defined by

$$^w M_o {}^w\ddot{x}_o + {}^w h_o = G^w F - {}^w F_e \quad (6)$$

where ${}^w M_o = \begin{bmatrix}m_o O_3\\O_3 & {}^w I_o\end{bmatrix}$, and ${}^w h_o = \begin{bmatrix}-m_o g\\{}^w\omega_o \times ({}^w I_o {}^w\omega_o)\end{bmatrix}$, with ${}^w I_o = {}^w R_o I_o {}^w R_o^T$. The term ${}^w F_e$ denotes the forces and moments exerted by the object on the environment. The grasp matrix $G = [G_1\ G_2]$ and ${}^w F = {}^w\tilde{R}_R {}^R F$ where ${}^w\tilde{R}_R = \begin{bmatrix}{}^w\tilde{R}_{R_1} & O_6\\O_6 & {}^w\tilde{R}_{R_2}\end{bmatrix}$ and ${}^R F = \begin{bmatrix}{}^{R_1}F\\{}^{R_2}F\end{bmatrix}$. The term ${}^{R_i}F = \begin{bmatrix}{}^{R_i}R_{T_i} & O_3\\O_3 & {}^{R_i}R_{T_i}\end{bmatrix}\begin{bmatrix}{}^{T_i}f_i\\{}^{T_i}n_i\end{bmatrix}$

denotes the force and moments exerted by robot i on the object measured at the origin of the tool frame and expressed in the robot's base frame.

3.2.2 Robot Dynamics

The dynamic equation of the motion of a manipulator is given by

$$M\ddot{q} + h = \tau - J_R^T {}^R F = \tau - J_w^T {}^w F \quad (7)$$

where the inertia matrix, the joint acceleration vector, the centrifugal, Coriolis and gravity term, the joint torque vector and the Jacobian matrices are given by:

$$M = \begin{bmatrix}M_1 & O_7\\O_7 & M_2\end{bmatrix},\ \ddot{q} = \begin{bmatrix}\ddot{q}_1\\\ddot{q}_2\end{bmatrix},\ h = \begin{bmatrix}h_1\\h_2\end{bmatrix},\ \tau = \begin{bmatrix}\tau_1\\\tau_2\end{bmatrix},\ J_R = \begin{bmatrix}J_{R_1} & O_{6\times 7}\\O_{6\times 7} & J_{R_2}\end{bmatrix} \text{ and }$$

$J_w = {}^w\tilde{R}_R J_R$ respectively.

3.2.3 Dynamic model of the complete system

Differentiating (3) with respect to time yields ${}^w\ddot{x} = G^T {}^w\ddot{x}_o + \dot{G}^T {}^w\dot{x}_o$. This can be written as

$$^w\tilde{R}_R(J_R\ddot{q} + \dot{J}_R\dot{q}) = G^T {}^w\ddot{x}_o + \dot{G}^T {}^w\dot{x}_o \quad (8)$$

From (6), (7) and (8), we obtain the complete dynamic equation of the dual-arm system:

$$\begin{bmatrix}M & O & J_w^T\\O & {}^w M_o & -G\\J_w & -G^T & O\end{bmatrix}\begin{bmatrix}\ddot{q}\\\ddot{x}_o\\{}^w F\end{bmatrix} = \begin{bmatrix}\tau - h\\-{}^w F_e - {}^w h_o\\\dot{G}^T {}^w\dot{x}_o - {}^w\tilde{R}_R \dot{J}_R \dot{q}\end{bmatrix} \quad (9)$$

4. Two-level impedance controller (TLIC)

As shown in Figure 2, the controller consists of two levels of impedance control. The first level is implemented at the object level while the second is between the end-effector tracking error and the internal force error.

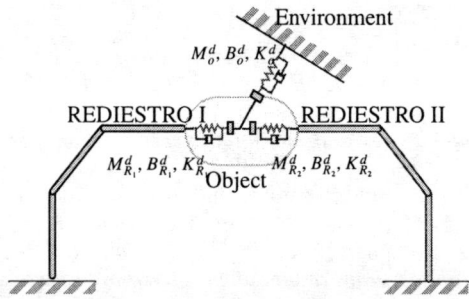

Figure 2: Two-level impedance controller

4.1 Object-Level Augmented Hybrid Impedance Controller (OAHIC)

An augmented hybrid impedance control scheme is adopted here to achieve robust control when in contact with an unknown environment. A desired impedance is specified between the object and the environment via the parameters M_o^d, B_o^d, K_o^d:

$$^w\ddot{x}_o^t = (M_o^d)^{-1}\left\{{}^w F_e - (I-S){}^w F_e^d - B_o^d({}^w\dot{x}_o^t - S^w\dot{x}_o^d) - K_o^d S\begin{bmatrix}{}^w x_{o_r}^t - {}^w x_{o_r}^d\\-e_o\end{bmatrix}\right\} + S^w\ddot{x}_o^d$$

$$(10)$$

where ${}^w\ddot{x}_o^t$ is the object target acceleration trajectory; a "1" in the selection matrix S implies a position controlled direction and a "0" implies a force controlled direction; ${}^w x_o^d, {}^w\dot{x}_o^d, {}^w\ddot{x}_o^d$ denote the desired object trajectory; M_o^d, B_o^d, K_o^d denote the desired object impedance parameters; and e_o denotes the orientation error which is defined as

$$e_o = \Lambda(R_{o_{err}}) = \begin{bmatrix}R_{o_{err}}(3,2) - R_{o_{err}}(2,3)\\R_{o_{err}}(1,3) - R_{o_{err}}(3,1)\\R_{o_{err}}(2,1) - R_{o_{err}}(1,2)\end{bmatrix} \text{ where } R_{o_{err}} = {}^w R_o^d({}^w R_o^t)^T \quad (11)$$

The terms ${}^w F_e$ and ${}^w F_e^d$ are the estimated and the desired external forces exerted on the environment by the object, where

$$^w F_e = G^w F - ({}^w M_o {}^w\ddot{x}_o^t + {}^w h_o) \quad (12)$$

From ${}^w\ddot{x}_o^t$, we can determine ${}^w x_o^t, {}^w\dot{x}_o^t$. Using constraints (1), (2) and (3), we obtain the desired robot end-effector trajectory ${}^w x_R^d, {}^w \dot{x}_R^d, {}^w \ddot{x}_R^d$ from the object target trajectories ${}^w x_o^t, {}^w \dot{x}_o^t, {}^w \ddot{x}_o^t$:

$$ {}^w x_{R_i}^d = {}^w x_{O_T}^t - {}^w R_{O_i}{}^o P_{T_i} \qquad (13) $$

with $i = 1$ or 2 whichever has less kinematic error and

$$ {}^w R_o = {}^w R_{R_i}{}^{R_i} R_{T_i}{}^{T_i} R_o \qquad (14) $$

The desired rotation matrix is:

$$ {}^w R_{R_i}^d = {}^w R_o^t {}^o R_{t_i} \qquad (15) $$

The desired velocity and acceleration are given by

$$ {}^w \dot{x}_R^d = G^{Tw} \dot{x}_o^t \quad \text{and} \quad {}^w \ddot{x}_R^d = G^{Tw} \ddot{x}_o^t + \dot{G}^{Tw} \dot{x}_o^t \qquad (16) $$

4.2 Internal force control

Here, we use a weighted pseudo-inverse definition of the grasp matrix G:

$$ G^\dagger = P G^T (G P G^T)^{-1} \qquad (17) $$

where $P = \begin{bmatrix} \Delta & 0 \\ 0 & \Delta \end{bmatrix}$ with $\Delta = \begin{bmatrix} O_3 & I_3 \\ I_3 & O_3 \end{bmatrix}$. Therefore

$$ G^\dagger = \frac{1}{2} \begin{bmatrix} G_1^{-1} \\ G_2^{-1} \end{bmatrix} = \begin{bmatrix} I_3 & O_3 \\ -{}^w \hat{P}_{oT_1} & I_3 \\ I_3 & O_3 \\ -{}^w \hat{P}_{oT_2} & I_3 \end{bmatrix} \qquad (18) $$

The commanded internal force must be chosen to lie in the range of the internal force projection operator $(I - G^\dagger G)$. If the desired squeezing force at the center of mass is specified as $F_{int_o}^d$, then

$$ F_{int}^d = \begin{bmatrix} G_1^{-1} \\ -G_2^{-1} \end{bmatrix} F_{int_o}^d \qquad (19) $$

Let us define the internal force error $e_{int} = F_{int}^d - F_{int}$, where

$$ F_{int} = (I - G^\dagger G)^w F \qquad (20) $$

An impedance control scheme is used to control the internal force:

$$ {}^w \ddot{x}_{R_i}^t = (M_{R_i}^d)^{-1} \left\{ e_{int} + B_{R_i}^d ({}^w \dot{x}_{R_i}^d - {}^w \dot{x}_{R_i}^t) + K_{R_i}^d \begin{bmatrix} {}^w x_{R_{ip}}^t - {}^w x_{R_{ip}}^d \\ -e_{R_i} \end{bmatrix} \right\} + {}^w \ddot{x}_{R_i}^d \qquad (21) $$

where ${}^w \ddot{x}_{R_i}^t$ is the ith robots' target acceleration trajectory, ${}^w x_{R_i}^d, {}^w \dot{x}_{R_i}^d, {}^w \ddot{x}_{R_i}^d$ the ith robot's desired trajectory, $M_{R_i}^d, B_{R_i}^d, K_{R_i}^d$ the desired impedance parameters for internal force control of the ith robot, and $e_{R_i} = \Lambda(R_{R_{ierr}})$ where $R_{R_{ierr}} = {}^w R_{R_i}^d ({}^w R_{R_i}^t)^T$

4.3 Error correction

A PD loop is implemented on each robot to obtain the error reference trajectory ${}^w \ddot{x}_{R_i}^r$.

$$ {}^w \ddot{x}_{R_i}^r = {}^w \ddot{x}_{R_i}^t + K_{v_{R_i}} ({}^w \dot{x}_{R_i}^t - {}^w \dot{x}_{R_i}) + K_{p_{R_i}} \left(\begin{bmatrix} {}^w x_{R_{ip}}^t - {}^w x_{R_{ip}} \\ e_{R_i} \end{bmatrix} \right) \qquad (22) $$

where $e_{R_i} = \Lambda(R_{R_{ierr}})$ with $R_{R_{ierr}} = {}^w R_{R_i}^t ({}^w R_{R_i})^T$.

4.4 Redundancy Resolution (RR)

Joint redundancy is used to avoid singularities and can be used to achieve joint limit avoidance or obstacle avoidance with very little modification. The singularity avoidance scheme requires the minimization of the cost function:

$$ L_i = \ddot{E}_i^T W \ddot{E}_i + (\ddot{q}_i^d + \lambda \dot{q}_i)^T W_v (\ddot{q}_i^d + \lambda \dot{q}_i) \qquad (23) $$

where $\ddot{E}_i = {}^w \tilde{R}_{R_i}^{-1} {}^w \ddot{x}_{R_i}^r - J_{R_i} \ddot{q}_i^d - \dot{J}_{R_i} \dot{q}_i$. The solution is given by

$$ \ddot{q}_i^d = (J_{R_i}^T W J_{R_i} + W_v)^{-1} (J_{R_i}^T W^w \tilde{R}_{R_i}^{-1} {}^w \ddot{x}_{R_i}^r - J_{R_i}^T W J_{R_i} \dot{q}_{R_i} - W_v \lambda \dot{q}_i) \qquad (24) $$

However, joint redundancy can also be used to achieve robot-to-robot collision avoidance, robot self-collision avoidance and robot-to-environment collision avoidance by adding additional tasks into (23).

4.5 Complete controller

The required joint torque is calculated as

$$ \tau = \tau_a + \tau_e \qquad (25) $$

where $\tau_a = M \ddot{q}^d + h$ and $\tau_e = J_w^{Tw} F$.

5. Independent Dual-Arm Robot Control

By modifying the internal force impedance control law (21) as shown below, the independent robot Augmented Hybrid Impedance Controller is obtained:

$$ {}^w \ddot{x}_{R_i}^t = (M_{R_i}^d)^{-1} \left\{ (I - S_i)^w F_{ei}^d - {}^w F_{ei} + B_{R_i}^d ({}^w \dot{x}_{R_i}^d - {}^w \dot{x}_{R_i}^t) + K_{R_i}^d S_i \begin{bmatrix} {}^w x_{R_{ip}}^t - {}^w x_{R_i}^d \\ -e_{R_i} \end{bmatrix} \right\} + S_i {}^w \ddot{x}_{R_i}^d \qquad (26) $$

where ${}^w F_{ei}^d$ is the desired contact force/torque exerted on the environment by robot i, ${}^w F_{ei}$ is the measured contact force, ${}^w x_{R_i}^d$ is the desired Cartesian space trajectory and s_i is the hybrid selection matrix. Because of the consistency of the controller, after the two robots working independently under impedance control grasp an object, the controller can switch smoothly to closed-chain control.

6. Experiments

To demonstrate the performance and robustness of the proposed controller, extensive simulations have been done based on a full dynamic model of a dual-arm redundant manipulator system, consisting of two 7-DOF manipulators (called REDIESTRO (Red) 1 and 2) [9]. Here, a dual-arm cooperative task is designed to demonstrate the key features of this control scheme. This includes independent to closed-chain control switching, dual-arm cooperation for maneuvering the common object to track desired position and orientation trajectories while regulating the internal force/torque under closed-chain control and object-level external force control to insert the common object back into its base.

6.1 Control scheme switching

In the experimental task, the two robotic arms start from the rest configuration, move to the common object (Figure 3) and "grasp" it by inserting the square-shape end-effector into each side of the object under independent force control (Figure 4). Then, the controller switches the control scheme from independent control (26) to closed-chain dual-arm control (25). The second level of the impedance control loop in the closed-chain control scheme is consistent with the independent impedance control scheme. This feature results in continuous and smooth switching between independent and closed-chain control. Figures 10 and 11 plot the end-effector force measurements obtained from Red1 and Red2 respectively.

6.2 Robustness to kinematic and dynamic uncertainties

The kinematic uncertainties come mostly from joint reading offsets. Without a sophisticated calibration procedure, the robots have end-effector kinematic errors in the level of several centimeters. When the robots complete the "grasping" by the peg insertion operation, the distance between the two robot end-effectors is constrained by the structure of the common object and is known to be $[-0.15, 0, 0]m$. But the distance obtained according to the kinematics and the joint readings is $[-0.149, 0.390, -0.042]m$, this is plotted in Figure 9 which shows a kinematic error of about $4cm$. The second level impedance control is essential in balancing the priority of the internal force control and the object level tracking via the control parameter K_{R_l} in (21). When the kinematic error is large, a smaller value of K_{R_l} is required to ensure better internal force tracking by sacrificing object-level tracking. This results in an easy way of applying the same controller to robots with different levels of calibration. The joint friction is not modeled in the controller and is around 30 % of the total dynamics during normal operation. The error correction is essential to ensure that the required end-effector motion is achieved when large unmodeled dynamics exist. Higher error correction control gains result in better performance but the gain magnitudes are limited by the servo capacity of the robots.

6.3 Object position and orientation tracking

During the dual-arm cooperating maneuver, the common object is required to track a desired position and orientation trajectory. This trajectory includes three parts: After lifting up the object, both robots move it from $[0, -0.4, 0.38]m$ with orientation $[0, 0, 0]$ in fixed angle representation to the new position $[0.2, -0.2, 0.6]m$ (Figures 5 and 6) with orientation $[17.2^0, 17.2^0, 17.2^0]$; then, the object is moved to position $[-0.2, -0.2, 0.6]m$ with orientation $[0, 0, 0]$ (Figures 7); finally the object is brought back to the position $[0, -0.4, 0.38]m$ (Figure 8). Figures 14 and 15 show the object position and orientation tracking calculated from Red1 kinematics. To avoid damaging the object during the closed-chain operation, the robots are commanded to apply $15N$ along the squeezing direction and 0 in all other directions including the twisting torque. Figures 12 and 13 show the object's internal force and torque obtained using the force sensors mounted near the end-effectors of the two robots. There is a relatively larger error in tracking the internal torque. By increasing the value of K_{R_l}, the performance can be improved, but the Red1-Red2 system cannot deliver the required joint torque without saturation.

6.4 Object external force control

The common object is inserted back into the base by external force control. A $-40N$ pressing force toward the table and 0 in all other directions is specified. Figure 16 shows the achieved object external force via the commanded force profile, and Figure 17 shows the achieved object external torque.

7. Conclusions

The dual-arm controller has been shown to be significantly robust to environmental and system kinematic and dynamic uncertainties while achieving object position and orientation tracking, external force control and internal force regulation. A two-level impedance control scheme has been used. The scheme gives a consistent control approach for independent as well as closed-chain dual-arm operations. Experimental results for a dual-arm systems (consisting of two 7-DOF manipulators) in our Robotics and Control Systems Laboratory have been presented to illustrate the performance of the controller. The problem of choosing appropriate values of the control parameters has also been addressed.

Acknowledgment

This work was support in part by the Natural Sciences and Engineering Research Council (NSERC) of Canada under Grant CRD 172884 (Principal Investigator: R.V. Patel) and by the Canadian Space Agency's STEAR-5 program under Contract 9F006-5-0219/01-SW with the Department of Public Works and Government Services Canada.

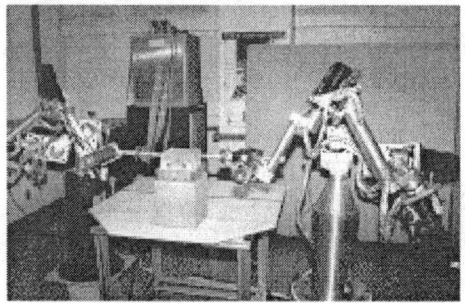

Figure 3: Open-chain dual-arm maneuver

Figure 4: Dual-arm object grasping

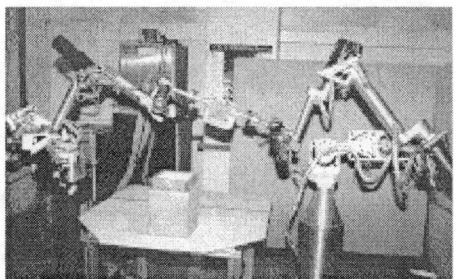

Figure 5: Maneuvering the object

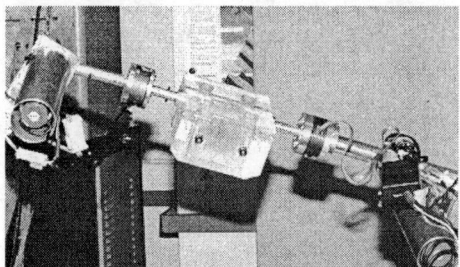

Figure 6: Zoom-in of object maneuver

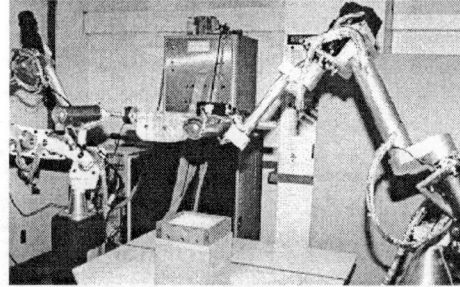

Figure 7: Maneuvering the object to the left

Figure 8: Final landing back into the base

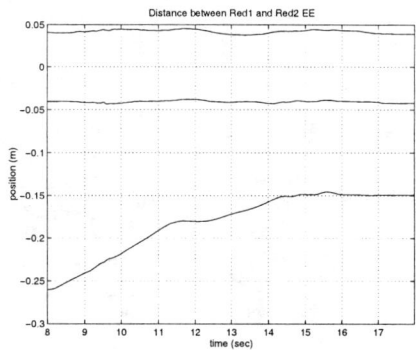

Figure 9: Kinematic uncertainty: End-effector position difference between Red1 and Red2

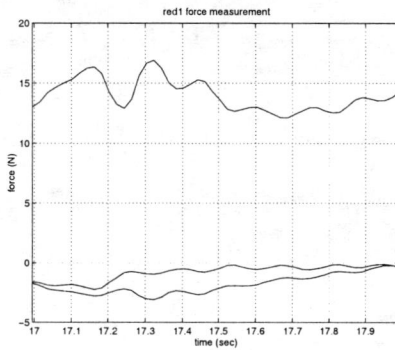

Figure 10: Switching - Red1 EE force measurement

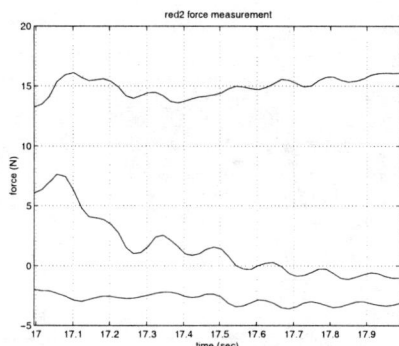

Figure 11: Switching - Red2 EE force measurement

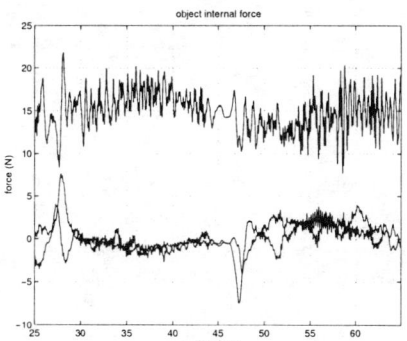

Figure 12: Object internal force

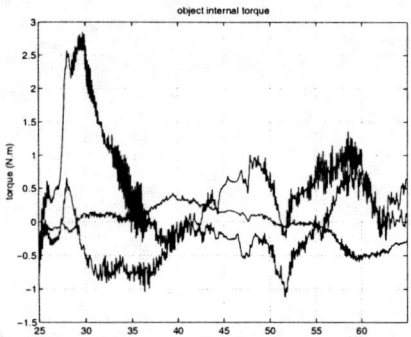

Figure 13: Object internal torque

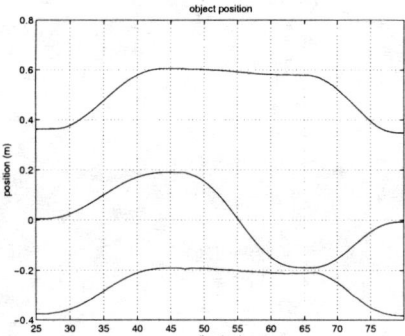

Figure 14: Object position tracking

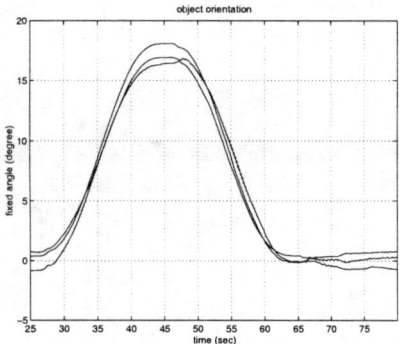

Figure 15: Object orientation tracking

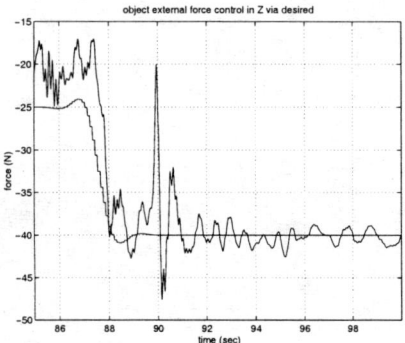

Figure 16: Object external force via commanded profile

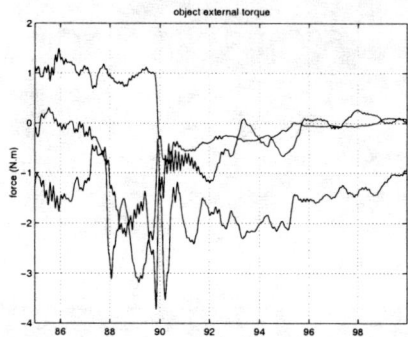

Figure 17: Object external torque

Reference

[1] N. Hogan, "Impedance Control: An approach to Manipulation", Transactions of the ASME J. Dyn. Sys., Meas., Contr., Vol.107, 1985, pp.1-24.

[2] G.J. Liu & A.A. Goldenberg, "Robust Hybrid Impedance Control of Robot Manipulators", Proceedings of the 1992 IEEE International Conference on Robotics and Automation, May 1991, pp.287-292.

[3] Farshid Shadpey, "Force Control and collision avoidance strategies for kinematically redundant manipulators", Ph.D thesis, Concordia University, Montreal, Canada, 1997.

[4] S.A. Schneider and R. Cannon, "Object Impedance Control for Cooperative Manipulation: Theory and Experimental Results", IEEE Transactions on Robotics and Automation, Vol.8 No.3, February 1993, pp. 383-394.

[5] T. Yoshikawa and X-Z. Zheng, "Coordinated Dynamic Hybrid Position/Force Control for Multiple Robot Manipulators Handling One Constrained Object", The International Journal of Robotics Research, Vol.12, No.3, June 1993, pp.219-230.

[6] I.D. Walker, R.A. Freeman and S.I. Marcus, "Analysis of Motion and Internal Loading of Objects Grasped by Multiple Coorperating Manipulators", International Journal of Robotics Research, Vol.10, 1991, pp.396-409.

[7] R.G. Bonitz and T.C. Hsia, "Internal Force-Based Impedance Control for Cooperating Manipulators", IEEE Transactions on Robotics and Automation, Vol.12, No.1, February 1996, pp. 78-89.

[8] A.K. Ramadorai, T.J. Tarn and A.K. Bejczy, "Task Definition, Decoupling and Redundancy Resolution by Nonlinear Feedback in Multi-Robot Object Handling", Proceedings of the 1992 IEEE International Conference on Robotics and Automation, May 1992, pp.467-474.

[9] H.P. Xie, F. Shadpey and R.V. Patel, "A Two-level Impedance Control Scheme for Dual-Arm Redundant Manipulators", 1998 World Automation Congress, Anchorage, Alaska, ISORA 116.1 - 116.6, 1998.

Self-reconfiguration Planning with Compressible Unit Modules

Daniela Rus   Marsette Vona

Department of Computer Science
Dartmouth College
Hanover, NH 03755

{rus,mav}@cs.dartmouth.edu

Abstract

We discuss a robotic system composed of Crystalline *modules. Crystaline modules can aggregate together to form distributed robot systems. Crystalline modules can move relative to each other by expanding and contracting. This actuation mechanism permits automated shape metamorphosis. We describe the crystalline module concept and show the basic motions that enable a crystalline robot system to self-reconfigure. We present an algorithm for general self-reconfiguration and describe simulation experiments.*

1 Introduction

We wish to develop versatile and extensible massively-parallel distributed robotic systems. We believe that versatility can be obtained by developing self-reconfigurable systems. Self-reconfiguring robots have the ability to adapt to the operating environment and the required functionality by changing shape. They consist of a set of identical robotic modules that can autonomously and dynamically change their aggregate geometric structure to suit different locomotion, manipulation, and sensing tasks. For example, a self-reconfiguring robot system could self-organize as a snake shape to pass through a narrow tunnel and re-organize as a multi-legged walker upon exit to traverse rough terrain.

Self-reconfiguration poses new challenges to designing and controlling distributed robot systems. Because the connectivity topology of these systems changes dynamically, new models of synchronization, communication, control, and planning are needed. Solutions to these problems will impact more broadly computation in distributed systems. For example, new models of distributed computing are emerging due to the proliferation of wireless mobile computers. As with self-reconfiguring robot systems, topology and reachability changes dynamically in wireless networks of mobile computers, requiring new solutions to communication and routing[1].

In our previous work [13, 14, 18] we discuss a small and simple robotic module we call a *Robotic Molecule*. This module has already been prototyped. Our experiments demostrate that it is capable of self-reconfiguration in three dimensions. We have also demonstrated how systems composed of robotic molecules can use self-reconfiguration to increase their locomotive versatility [15, 16]. The robotic molecule uses 4 rotational degrees of freedom to accomplish motion relative to a structure that consists of identical modules.

In this paper we propose a different approach to self-reconfiguring robot systems. The new approach uses an actuation mechanism inspired by that that of muscles and amoebas. The idea is to create a mechanical module that can expand and contract by a constant factor. In addition, this module should be capable of making and breaking connections with neighboring modules. By expanding and contracting the neighbors in a connected structure, an individual module can be moved relative to the structure. This basic operation leads to new algorithms for global self-reconfiguration. For example, Figure 2 shows three snapshots from a simulation of a four-unit robot.

The module we propose is called the *Crystalline* module. It has cubic shape with connectors to other modules in the middle of each face. This module is activated by three binary actuators that permit the cube to shrink and expand by a factor of two. This actuation scheme allows an individual module to relocate to arbitrary positions on the surface of a convex structure of modules in constant time (unlike previous systems that required linear time in the number of modules on

---

[1]The solutions employed by cellular phone systems rely on global broadcast, which is inadequate for our applications.

the surface [31, 24, 15, 16]). In this paper we focus on algorithms for shape metamorphosis planning for the class of robot systems consisting of crystalline modules, called *crystals*. We present an algorithm called *melt-grow* for self-reconfiguration where the initial and the goal structures have the same volume. The melt-grow planner achieves shape morphing by using an intermediate structure, which is a projection of the robot modules into a pool on the ground. This approach leads to reconfiguration algorithms that maintain stability in real-world environments where gravity is present. The planner runs in $O(n^2)$ time, where n is the number of modules in the crystal. We also describe a simulator we have built to study crystalline robot systems and give examples from our experience using this simulator.

2 Related work

Related work in self-reconfiguring robots includes robots in which modules are reconfigurable using external intervention [5]. In [6] a cellular robotic system is proposed to coordinate a set of specialized modules. Several specialized modules and ways of composing them were proposed. [30] studies multiple modes of locomotion that are achieved by manually composing a few basic elements in different ways. This work also presents extensive examples of locomotion and reconfiguration in simulation. [19, 31, 29, 20] consider a system of modules that can achieve planar motion by walking over one another. The reconfiguration motion is actuated by varying the polarity of electromagnets that are embedded in each module. More recently [21] this group developed a twelve DOF module capable of three-dimensional motion. [24] describes metamorphic robots that can aggregate as two-dimensional structures with varying geometry. The modules are deformable hexagons. This work also examines theoretical bounds for planning the self-reconfiguring motion of such modules. In [18] we have shown a constant-time reduction between robotic molecule structures our group has designed to support self-reconfiguration [15, 16] and metamorphic robots [24].

The robot proposed in this paper is different than the previously proposed modules in its actuation capabilities, which lead to new types of self-reconfiguration planning algorithms. The high-level idea of a shrinkable module that can be a cell in a reconfigurable system has been presented as the patent [28].

3 The Crystalline Module
3.1 Concept

The idea behind the crystalline module is to create a mechanism that has some of the motive properties of muscles, that can be closely packed in 3D space, and that can attach itself to similar units. We chose a design based on cubes with connectors to other modules in the middle of each face. Our idea is to build a cube that can contract by a factor of two and expand to the original size (see Figures 2 and 8). We wish to effect compression along all three principal directions (e.g., x, y, z) individually or in parallel. We call the module an *atom*, and each connector a *bond*. Figure 1 shows a design for the mechanics of a two-dimensional (square) implementation of the atom. Our idea is to use complimentary rack and pinion mechanisms to implement the shrinking and expansion. In three dimensions, the rack and pinion mechanisms could be replaced with lead screws. We are currently constructing this design, which will be the subject of a different paper.

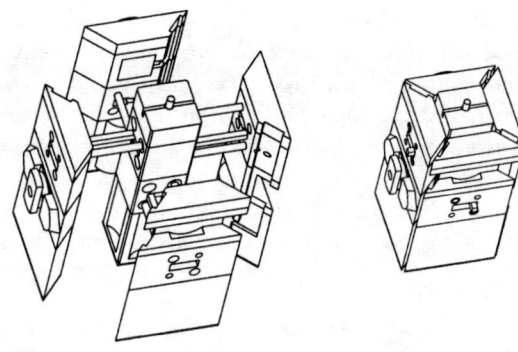

Figure 1: The mechanics of a 2D atom actuated by complimentary rack-and-pinion mechanisms. The atom is 4 inches tall (not including electronics, which are not shown). When expanded (left), the atom occupies a 4 inch square; when contracted (right) the atom occupies a 2 inch square.

3.2 Motion relative to a structure

A crystalline module can connect with identical modules to create crystalline robot systems. Only lattices whose faces are normal to the $x, y,$ and z axes can be created using crystalline robots. By manipulating the size of the atom, it is possible to approximate any finite solid shape to an arbitrary precision using crystalline modules[2].

Crystalline robot systems are dynamic structures: (1) they can move using sequences of reconfigurations to implement locomotion gaits; and (2) they can undergo shape metamorphosis. The dynamic nature of

---

[2]The aliasing error for any shape on a raster display can be arbitrarily reduced by increasing the resolution of the display.

Figure 2: Three snapshots from a simulation using crystalline robots. The left image shows the initial state. The middle image shows the robot after shrinking two modules. The right image shows the robot after relaxing the shrunk modules. Notice that the entire structure moved forward one unit, in an inchworm-like fashion.

these systems is supported by the ability of individual modules to move globally relative to the structure.

The basic operations in a crystalline robot system are:

- (expand <atom, dimension>) - expand a compressed atom in the desired dimension (x, y, or z)
- (contract <atom, dimension>) - compress an expanded atom in the desired dimension
- (bond <atom, dimension>) - activate one of the atom's connectors to bond with a neighboring atom in the structure
- (free <atom, dimension>) - deactivate one of the atom's connectors to break a bond with a neighboring atom in the structure

Figure 2 illustrates the use of these primitives for generating a linear propagation algorithm called the *inchworm propagation* algorithm for crystals. The robot consists of four connected crystalline modules. The modules rest on a substrate of other crystalline modules[3]. We assume that each module can compress by a factor of 2. In the first phase of the algorithm, the rightmost module attaches to the substrate and the middle modules compress. This operation causes the leftmost module to advance by one unit (where the unit is denoted by the size of the module). In the second phase of the algorithm the leftmost module makes a connection to the substrate, the rightmost module disconnects and the middle two modules expand. The net effect of this algorithm is a global translation of one unit for the crystal. It is possible to describe similar algorithms for effecting global translations and 90 degree concave and convex transitions about crystalline structures.

[3] Note that the substrate is not necessary for all locomotion gaits.

3.3 Relocating a Module on a Crystal

An interesting property of these robots is illustrated in Figure 3. The problem is to find a way to move the crystalline module on the surface of the cubic crystal to any other location. Instead of propagating the module along the surface of the large cube, which would require time linear in the side length of the cube, it is possible to reach the goal using a constant number of primitive operations. The number of operations remains constant no matter where the start and goal locations are oriented relative to each other. First, the module is pulled inside the cube by compresing two internal modules, selected depending on the goal location. The two compressed modules are at the intersection of the supporting line for the starting location and the supporting line for the final location. Second, the compressed modules are expanded in the direction of the goal location so as to pop out a crystalline module in the desired place. If the two supporting lines do not intersect, two transitions will be required instead of one. Note that this popped module is not identical to the original module on the face of the structure. Using this algorithm, a module can be relocated in constant time on any convex substrate. This algorithm assumes that the actuators are strong enough to pull or push any number of modules during these two operations. If the actuators are not strong enough an inchworm-like propagation can be used instead, but this will no longer be a constant time operation. Alternatively, parallelism can be employed to retain constant-time performance.

When the crystalline robot structure is non-convex, a similar algorithm effects the module relocation operation in $O(k)$-time, where k is the number of concave angles in the structure. The idea is to iterate the algorithm for convex substrates at each concave angle between the starting location and the goal location.

Figure 4 shows the details of the algorithm for relocating a module on the surface of a crystal with two

Figure 3: A crystalline module can be pushed into the large cube and popped out at any location on the surface of the cube in constant time. The three images are snapshots from a simulation. The left image shows the initial configuration (with the extra cube located on the side face) and the right image shows the final configuration (where the extra cube is on the top face). The middle image shows the base cube where two internal modules are compressed (not visible in the figure).

concave corners. The algorithm iterates compression and expansion steps.

We define a *scrunch* to be two adjacent, connected atoms that are compressed in the dimension normal to their connected faces. We define an *axis* to be a connected string of at least two atoms along one dimension. Two axes intersect if they have one atom in common. It follows that:

Theorem 1 *If an axis contains a scrunch, that scrunch can be moved to any position on the axis by the inchworm propagation algorithm. If one of two intersecting axes contains a scrunch, that scrunch can be transfered to the other axis, provided there exists sufficient surrounding structure to maintain connectedness throughout the operation.*

Proof: We have described the intuition behind these results. The technical proof is omitted for space considerations. □

To obtain a general algorithm for relocating crystalline modules on the surface of three-dimensional crystals the following two steps are sufficient:

1. Find a path with segments oriented along the $x, y,$ and z directions from the starting location to the target location.

2. Iterate the algorithm in Figure 4 to travel around each turn in the path.

The complexity of the algorithm is $O(t)$, where t is the number of turns in the path. Note that $t = O(k)$, with k the number of concave angles in the structure.

4 A Planner for Shape Metamorphosis

In this section we describe a planner for self-reconfiguration in crystalline robot systems. More precisely, given a pair of crystals (S, G), each composed of n atoms, find a feasible reconfiguration plan P that transforms S into G. A reconfiguration plan P is a partially ordered sequence of primitive operations. A reconfiguration plan is feasible iff at no time during the execution of the plan does the crystal become disconnected.

The key observation for planning is that our crystalline systems consist of identical modules. Since all the modules are identical and interchangeable, it is not necessary to compute goal locations for each element. Thus, self-reconfiguration is different from the related warehouse problem (where modules are assigned unique ids and have to be placed at desired locations), which is intractable.

We have developed a centralized planning algorithm called the *melt-grow* planner that is complete over a useful subset *Grain(4)* of crystals and runs in $O(n^2)$ time, where n is the number of atoms in the crystal:

1. *Melt* S into an intermediate crystal I[4].

2. *Grow* G out of I.

The set *Grain(n)* contains the crystals that can be tiled by cubes of $n \times n \times n$ atoms (or squares of $n \times n$ atoms in 2D) called *grains*, so that the set of planes (or edges in 2D) that coincide with all sides of all grains intersect only at grain edges and corners. Figure 5 gives an example of a 2D crystal in Grain(4) and a 2D crystal not in Grain(4). The subset Grain(4) is useful because we can argue that by manipulating the scale of the atom, it is possible to approximate any finite solid shape to an arbitrary precision with a crystal in Grain(4).

---

[4]I is the projection of a 3D structure onto the ground plane or the projection of a 2D structure onto a line.

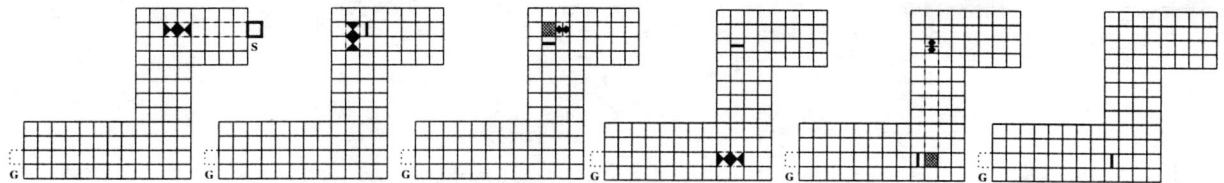

Figure 4: This figure illustrates the algorithm for relocating a crystalline module on a concave substrate of crystalline modules. The left figure shows the initial configuration. The relocating atom is in the upper right corner of the structure. The goal location is in the bottom left corner. Large dark diamonds mark two atoms about to be be compressed. Small dark diamonds mark two compressed atoms about to be expanded. Dark lines mark compressed pairs. The second figure shows the structure after the compression of the first pair of candidate atoms, and two atoms preparing for the next compression. The third figure shows two pairs of compressed atoms and a hole. The fourth figure shows the first compressed pair expanded into the hole and a candidate pair of atoms for the next compression. The fifth figure shows the state of the structure after this compression, with the resulting hole. The right-most figure shows the structure after an expansion into the hole. At this point, the remaining compressed pair can be expanded into the goal location.

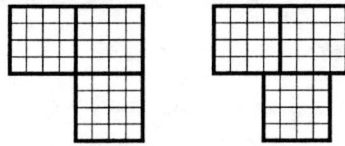

Figure 5: (Left): a 2D crystal in Grain(4). (Right): a 2D crystal not in Grain(4).

We use the intermetiate crystal I both to maintain stability during reconfiguration and in order to avoid backtracking[5]. We project the 3D structures S and G to and from a planar I. Physical stability can be guaranteed throughout the reconfiguration in environments where gravity is present by transforming S into I from the top down and I into G from the bottom up.

At a high level, the Melt algorithm works by finding a *mobile grain* g in S, transporting g to a place in I, and repeating until all grains are in I. Similarly, the Grow algorithm works by selecting mobile grains from I and transporting them to locations in G until all grains are in G. A grain is *mobile* iff it can be removed without disconnecting the crystal.

Figure 6 shows the details of the melt-grow algorithm. The first step in the algorithm is to compute the location of the intermediate structure I. This is done by the function Locate-Stem. The function

```
Melt-Grow(Crystal S, Crystal G)
    Grain stem <- Locate-Stem(S)
    Crystal I <- Melt(S, stem)
    Grow(I, stem, G)

Melt(Crystal S, Grain stem)
    Crystal I <- stem
    S <- S-stem
    Crystal C <- Design-Pool(Volume(S), stem)
    While !empty(S)
        Grain mover <- Find-Mobile(S)
        Grain parent <- Find-Parent(I, C)
        Transport(mover, parent, union(S,I))
    return I

Grow(Crystal I, Grain stem, Crystal G)
    Crystal C <- stem
    I <- I-stem
    While !empty(I)
        Grain mover <- Find-Mobile(I)
        Grain parent <- Find-Parent(C, G)
        Transport(mover, parent, union(C,I))
```

Figure 6: The implementation of the melt-grow algorithm.

---

[5]Another planner might generate intermediate states in which no out-of-place grain is mobile, which would require backtracking.

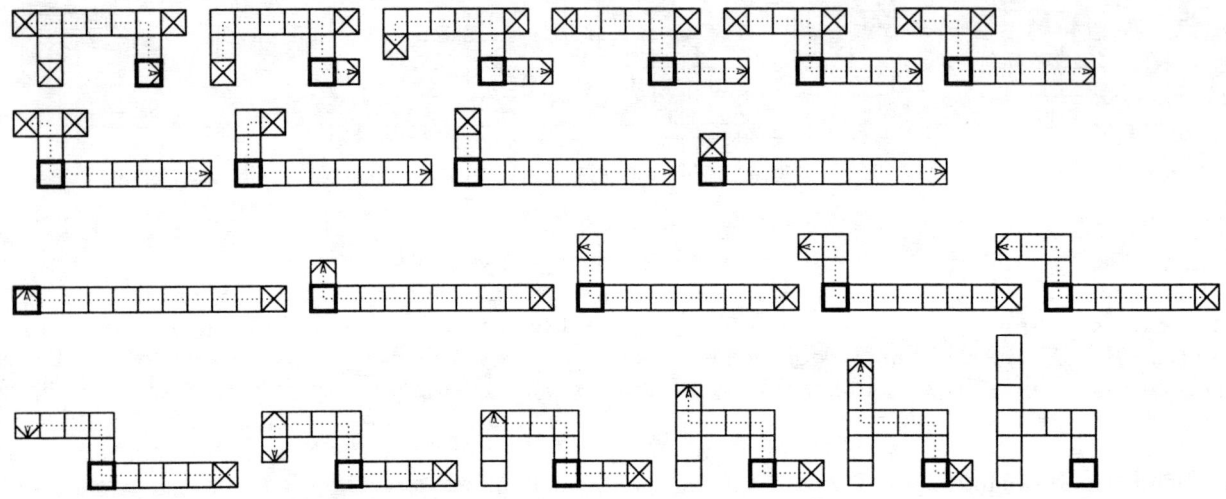

Figure 7: Reconfiguring a *Table* to a *Chair* using the melt-grow algorithm. Each square represents one grain (16 atoms). Grains marked ⊠ are mobile, the stem grain is marked ▫, and candidate parent grains are marked ⊠. This figure represents schematically the output of a simulation where the table and chair are composed of 176 crystals each.

`Design-Pool` generates an intermediate crystal with suitable volume. Many types of intermediate crystal are possible; for simplicity, we use a planar spiral of grains (or a one-dimensional line of grains for 2D systems). Such an intermediate crystal design makes locating mobile grains in I trivial.

The second step of the algorithm is to melt S into I; the third step is to grow I into G. These steps require locating a mobile grain, locating a good destination for this grain, and finding a path for the grain to the destination. Mobile grains in crystal C can be located by searching for vertices which are not articulation points in $GCG(C)$, the Grain Connectivity Graph of crystal C. $GCG(C)$ is an undirected graph whose vertices represent grains in C and whose edges represent active connections between neighboring grains. While Melting, any mobile grain still in S is a suitable mover; while Growing, any mobile grain still in I is a suitable mover. Parent grains can be located by searching for grains that are adjacent to yet-to-be-filled vacancies. While Melting, any such grain in I is a suitable parent; while Growing, any such grain in G is a suitable parent.

After locating a mobile grain and a parent, the mobile grain is transported to a space adjacent to the parent by (1) Finding a route through the crystal (i.e. by Depth-First-Search from parent in $GCG(C)$); (2) Decomposing the route into a sequence of grain motion primitives, and from there into a sequence of atom motion primitives; and (3) Executing the atom motion primitives. The following are the grain motion primitives:

- (`scrunch <grain, dimension, sense>`) - create a planar (linear in 2D) compression in a mobile grain at one of its six faces ($+x$, $-x$, $+y$, $-y$, $+z$, $-z$)

- (`relax <grain>`) - expand a compression at one face of a grain into a grain-in-progress

- (`transfer <grain>`) - transfer a compression at one face of a grain to the adjacent grain

- (`propagate <grain>`) - move a compression at one face of a grain to the opposing face of the grain

- (`convert <grain, dimension, sense>`) - move a compression at one face of a grain to one of the faces in an orthogonal dimension

Any mobile grain in a Grain(4) crystal can always be `scrunched`. A scrunch can always be transported to any parent grain by a sequence of transfer, propagate, and convert operations. Finally, any candidate parent containing a scrunch can always `relax` into an adjacent grain-in-progress. The algorithmic simplicity afforded by these guarantees is the reason for introducing the Grain(4) abstraction.

Figure 7 shows an example where the melt-grow algorithm was used to automatically transform a 2D *table* into a 2D *chair*. Since the melt-grow algorithm operates at the grain level, not the atom level, each square in Figure 7 represents one grain (16 atoms).

Theorem 2 *The melt-grow algorithm is complete for the subset Grain(4) of crystals. The running time of the melt-grow algorithm is $O(n^2)$, where n is the number of atoms in the crystal.*

Proof: Omitted for space considerations. □

5 Simulation

We have developed a simulator for 3D crystalline robots called *xtalsim*. We implemented the melt-grow algorithm in xtalsim, and we have performed experiments in simulation which verify that crystalline robots can self-reconfigure. The entire *table* to *chair* metamorphosis in Figure 7 was generated automatically by the melt-grow planner. Each object consists of 176 atoms. Figure 8 shows several snapshots from a smaller 3D simulation in which a dog-shaped object is made to metamorphose into a couch-shaped object via a hand-written plan.

Simulations are specified to xtalsim in a high-level language, either as initial and goal Grain(4) crystals, as sequences of grain motion primitives, or as sequences of atom motion primitives. The language includes constructs to allow parallel and serial combinations of actions and supports reusable, modular algorithms.

Xtalsim facilitates high-level experimentation with crystalline robots since it (1) efficiently extrapolates the global effects of individual atom movements, (2) verifies that all specified actions are feasible, and (3) displays the resulting atom actions as a 3D animation.

6 Discussion and Future Work

We presented the crystalline robotic module which is capable of self-reconfiguration by using a muscle-like actuation mechanism. The module is capable of making and breaking connections with identical modules, and of compressing by a constant factor. This actuation mechanism supports very fast algorithms for relocating one module on the surface of a crystal, which leads to an efficient $O(n^2)$ planner for shape metamorphosis. We are currently prototyping this module.

Acknowledgements

We are grateful to Keith Kotay for extensive discussions about the crystalline module. We are also grateful to Michael Shin for his help in implementing the simulator.

This paper describes research done in the Dartmouth Robotics Laboratory. Support for this work was provided through the NSF CAREER award IRI-9624286 and the NSF award IRI-9714332.

References

[1] Y. Cao, A. Fukunaga, A. Kahng, and F. Meng. Cooperative mobile robots: Antecedents and directions. Technical report, UCLA Department of Computer Science, 1995.

[2] I. Chen and J. Burdick. Enumerating the Non-Isomorphic Assembly Configurations of a Modular Robotic System. To appear in the *International Journal of Robotics Research*.

[3] P. Chew and K. Kedem. Getting around a lower bound for the minimum Hausdorff distance. In *Third Scandinavian Workshop on Algorithm Theory*, eds. O Nurmi and E. Ukkonen, Lecture Notes in Computer Science 621, pp 318–325, Springer Verlag 1992.

[4] G. Chirikjian and J. Burdick. Kinematics of a hyper-redundant robot locomotion with applications to grasping. In *Proceedings of the IEEE International Conference on Robotics and Automation*, 1991.

[5] R. Cohen, M. Lipton, M. Dai, and B. Benhabib. Conceptual design of a modular robot. In *Journal of Mechanical Design*, pp. 117-125, March 1992.

[6] T. Fukuda and Y. Kawauchi. Cellular robotic system (CEBOT) as one of the realization of self-organizing intelligent universal manipulator. In *Proceedings of the IEEE Conference on Robotics and Automation*, pp. 662-667, 1990.

[7] G. Hamlin and A. Sanderson. Tetrabot modular robotics: prototype and experiments. In *Proceedings of the IEEE/RSJ International Symposium of Robotics Research*, pp 390-395, Osaka, Japan, 1996.

[8] Kazuo Hosokawa, Isao Shimoyama, and Hirofumi Miura. Dynamics of self-assembling systems — analogy with chemical kinetics. *Artificial Life*, 1(4), 1995.

[9] D. Huttenlocher, G. Klanderman, and W. Rucklidge. Comparing images using the Hausdorff distance. *IEEE Transactions on Pattern Matching and Machine Intelligence*, 1993.

[10] S. Kelly and R. Murray, Geometric phases and robotic locomotion. CDS Technical Report 94-014, California Institute of Technology, 1994.

[11] K. Kotay and D. Rus. Navigating 3d steel web structures with an inchworm robot. In *Proceedings of the 1996 International Conference on Intelligent Robots and Systems*, Osaka, 1996.

[12] K. Kotay and D. Rus. Task-reconfigurable robots: navigators and manipulators. In *The 1997 International Conference on Intelligent Robots and Systems*, 1997.

Figure 8: Five snapshots from a simulation using crystalline robots. The initial configuration (on the left) is a dog-shaped object. The final configuration (on the right) is a couch-shaped object. The middle images show intermediate steps in the transformation from dog to couch. The planning for this transformation was done manually. Note that some atoms are left in a compressed state so that the volume of the final shape is less than the volume of the initial shape.

[13] K. Kotay, D. Rus, M. Vona, and C. McGray. The self-reconfigurable robotic molecule. In *Proceedings of the 1998 International Conference on Robotics and Automation*, 1998.

[14] K. Kotay, D. Rus, M. Vona, and C. McGray. The self-reconfiguring robotic molecule: design and control algorithms. In *the 1998 Workshop on Algorithmic Foundations of Robotics*, 1998.

[15] K. Kotay and D. Rus. Motion Synthesis for the Self-reconfiguring Robotic Molecule. In *Proceedings of the 1998 International Conference on Intelligent Robots and Systems*, 1998.

[16] K. Kotay and D. Rus. Locomotion Versatility through Self-reconfiguration In *Robotics and Autonomous Systems*, 1998 (to appear).

[17] J. C. Latombe. *Robot Motion Planning*. Kluwer Academic Publishers 1991.

[18] C. McGray and D. Rus. Motion Self-reconfiguring Molecules as 3D Metamorphic Squares In *Proceedings of the 1998 International Conference on Intelligent Robots and Systems*, 1998.

[19] S. Murata, H. Kurokawa, and Shigeru Kokaji. Self-assembling machine. In *Proceedings of the 1994 IEEE International Conference on Robotics and Automation*, San Diego, 1994.

[20] S. Murata, H. Kurokawa, K. Tomita, and Shigeru Kokaji. Self-assembling method for mechanical structure. In *Artif. Life Robotics*, 1:111–115, 1997.

[21] S. Murata, H. Kurokawa, E. Yoshida, K. Tomita, and S. Kokaji. A 3-D Self-Reconfigurable Structure. In *Proceedings of the 1998 IEEE International Conference on Robotics and Automation*, Leuven, 1998.

[22] R. Murray. Trajectory generation for underactuated systems with applications to robotic locomotion. In *Workshop on Algorithmic Foundations of Robotics*, eds. P. Agrawal, L. Kavraki, and M. Mason, A. K. Peters, 1998.

[23] B. Neville and A. Sanderson. Tetrabot family tree: modular synthesis of kinematic structures for parallel robotics. In *Proceedings of the IEEE/RSJ International Symposium of Robotics Research*, pp 382-390, Osaka, Japan, 1996.

[24] A. Pamecha, C-J. Chiang, D. Stein, and G. Chirikjian. Design and implementation of metamorphic robots. In *Proceedings of the 1996 ASME Design Engineering Technical Conference and Computers in Engineering Conference*, Irvine, CA 1996.

[25] C. Paredis and P. Khosla. Kinematic Design of Serial Link Manipulators from Task Specifications. In *International Journal of Robotic Research*, Vol. 12, No. 3, pp 274–287, 1993.

[26] C. Paredis and P. Khosla. Design of Modular Fault Tolerant Manipulators. In *The First Workshop on the Algorithmic Foundations of Robotics*, eds. K. Goldberg, D. Halperin, J.-C. Latombe, and R. Wilson, pp 371-383, 19 95.

[27] D. Rus. Self-Reconfiguring Robots. *IEEE Intelligent Systems*, 13(4), 2-5, July/August 1998

[28] K. Tanie and H. Maekawa. Self-reconfigurable cellular robotic system. US Patent 5361186, 1993.

[29] K. Tomita, S. Murata, E. Yoshida, H. Kurokawa, and S. Kokaji. Reconfiguration method for a distributed mechanical system. In *Distributed Autonomous Robotic Systems 2*, pp 17–25, Springer Verlag 1996.

[30] M. Yim. A reconfigurable modular robot with multiple modes of locomotion. In *Proceedings of the 1993 JSME Conference on Advanced Mechatronics*, Tokyo, Japan 1993.

[31] E. Yoshida, S. Murata, K. Tomita, H. Kurokawa, and S. Kokaji. Distributed Formation Control of a Modular Mechanical System. In *Proceedings of the 1997 International Conference on Intelligent Robots and Systems*, 1997.

What Can Be Learned from Human Reach-To-Grasp Movements for the Design of Robotic Hand-Eye Systems?*

Alexa Hauck, Michael Sorg, Georg Färber
Lab. Process Control & Real-Time Systems
Technische Universität München
Munich, Germany
Email: {a.hauck,m.sorg}@ei.tum.de

Thomas Schenk
Department of Neurology
Ludwig-Maximilians-Universität München
Munich, Germany
Email: tschenk@nefo.med.uni-muenchen.de

Abstract

In the field of robot motion control, visual servoing has been proposed as the suitable strategy to cope with imprecise models and calibration errors. Remaining problems such as the necessity of a high rate of visual feedback are deemed to be solvable by the development of real-time vision modules. However, human grasping, which still outshines its robotic counterparts especially with respect to robustness and flexibility, definitely requires only sparse, asynchronous visual feedback. We therefore examined current neuroscientific models for the control of human reach-to-grasp movements with the emphasis lying on the visual control strategy used. From this, we developed a control model that unifies the two robotic strategies look-then-move *and* visual servoing, *thereby compensating the problems that each strategy shows when used alone.*

1 Introduction

In the field of visually controlled robot manipulators, two strategies have been proposed: *look-then-move systems* try to determine the object's pose from the visual input as accurately as possible and then move the robot appropriately. This approach lends itself very well to integration in manufacturing; the visual input replaces the a priori knowledge about the exact pose of the parts to handle. As the trajectory of the manipulator is computed in advance, this approach scales well in the case of more complex environments where additional constraints for the robot motion exist, such as obstacles. Unfortunately, its accuracy heavily depends on the accuracy of the sensor/robot calibration and of the sensor itself.

To overcome these problems, *visual servoing* has been proposed as an alternative approach (see [12] for a tutorial). Here, visual information about the current position of the manipulator is used in a feedback control loop to guide the robot. The main advantage of visual feedback control is that the accuracy is relatively insensitive to calibration errors. However, to close the robot control loop with visual information, image processing has to be fast. This is difficult to realize without the use of specialized vision hardware at the moment. Furthermore, the trajectory of the manipulator cannot be computed in advance easily. This can lead to a situation in which the system loses sight of the gripper and thus loses its feedback.

For both approaches, there exists a large number of successfully realized hand-eye systems which cope very well with a specific problem (for a extensive survey see [3], for a collection of articles on state-of-the-art results [11, 19]). Yet, in comparison with the human example they all show a considerable lack of robustness and flexibility. The main difference is that humans can grasp successfully using only little visual information, for example looking at the target once with only one eye. More visual informtion, for example a view of the hand or stereo vision, results in a more precise and efficent grasp [6, 7]. Looking at the hand at the beginning or during motion allows to react to large disturbances and to recalibrate the internal models [13, 5].

One has to conclude that there might be something to learn for roboticists by taking a closer look at the human example. In Sec. 2, we therefore analyze and discuss current neuroscientific models that try to explain human reach-to-grasp movements. We then derive a motion control module for the implementation on a robot.

---

*The work presented in this paper was supported by the *Deutsche Forschungsgemeinschaft* as part of the Special Research Program "Sensorimotor – Analysis of biological systems, modelling, and medical-technical application" (SFB 462).

2 Models of human reaching

The main problems roboticists encounter when trying to learn from neuroscience are the different "hardware" addressed, and the different "languages" spoken by researchers. The former rules out a direct copying of results, the second impedes their transfer. In order to mitigate these problems we limited ourselves to examining current *analytical* models of human reaching, as they facilitate a later implementation on a robot. The models are analyzed with the aim of answering the following questions:

- What is the underlying control structure and how is (visual) sensor information integrated?

- In which space are the movements planned and which constraints are applied?

In 1899 already, Woodworth proposed that a reaching movement consists of two components: an *"initial impulse propelling the hand towards the target"* and a *"current control to home in on the final position via successive approximations"* [20]. The former was found to be dependent on visual information only at the beginning, to generate a trajectory; the latter depends on primarily visual feedback during motion.

Woodworth's results stem from experiments that analyzed the accuracy of reaching movements with varying availability of visual information. The notion that the required accuracy of a movement affects the corresponding trajectory was supported by Milner [16] who measured the trajectories of human subjects inserting a pin into a hole. For small holes and therefore high precision requirements, the velocity profile showed small oscillations at the end, corresponding to a sequence of submovements.

Concerning the form of human multijoint arm movements, it has been consistently reported [15] that the invariant features are that

- the path of the hand is a roughly straight line in Cartesian coordinates and that

- the profile of tangential (Cartesian) hand velocity is bell-shaped.

Thus, one has to conclude that human reaching movements are planned in spatial coordinates[1], not in joint space.

Concerning the control structure, a number of different models has been developed.

---

[1] The variable measured is the position of the hand; orientation, i.e. the shape of the hand, is thought to be controlled by a process running in parallel [14].

2.1 Feedforward models

As known from the literature on motion planning, the problem of trajectory generation is underconstrained. A classical approach in optimal control theory is to specify an *objective function* that is to be optimized. For example, the trapezoid joint velocity profile provided by many joint controllers minimizes peak velocity and thereby total impulse [17].

An optimization criterion that has been shown to match experimental data of human reaching very well is *minimum jerk* [4], "jerk" being the first derivative of (Cartesian) acceleration. Eq. 1 gives the objective function C for a 2D-movement, Eq. 2 the corresponding hand path

$$C = \frac{1}{2} \int_{t=0}^{t=T} \{(\frac{d^3x}{dt^3})^2 + (\frac{d^3y}{dt^3})^2\} dt \quad (1)$$

$$\mathbf{x}_1(t) = \mathbf{x}_0 + (\mathbf{x}_{T1} - \mathbf{x}_0) \cdot (10\tau_1^3 - 15\tau_1^4 + 6\tau_1^5) \quad (2)$$

with $\mathbf{x}_0$ and $\mathbf{x}_{T1}$ being the start and target position respectively, T being the movement duration, and $\tau_1 = t/T$. The resulting paths are straight Cartesian lines, the tangential velocity profiles are bell-shaped.

By the so-called *superposition scheme*, this model can also cope with the case of a double-step target, i.e. the case that the target "jumps" during the movement from $\mathbf{x}_{T1}$ to $\mathbf{x}_{T2}$: At the time $t = t_2$, a second trajectory (Eq. 3) is superimposed on the first:

$$\mathbf{x}_2(t \geq t_2) = (\mathbf{x}_{T2} - \mathbf{x}_{T1}) \cdot (10\tau_2^3 - 15\tau_2^4 + 6\tau_2^5) \quad (3)$$

with $\tau_2 = (t - t_2)/(T - t_2)$.

This pure feedforward control structure has already been successfully implemented on a robot [9].

In [2], the minimum jerk superposition model is extended to a full-scale motion planner by adding a method to compute the amplitude and duration of the submovements, and an adaptive control structure to learn and compensate modelling errors. This model, though, has not been implemented on a robot yet.

2.2 Feedback models

A feedback version of the minimum jerk trajectory generator is described in [10]. The control variable used is the derivative of the acceleration, $\dot{\mathbf{a}}$. For any time t, the current value of $\dot{\mathbf{a}}$ can be computed as follows:

$$\dot{\mathbf{a}}(t) = -9\frac{\mathbf{a}(t)}{T} - 36\frac{\mathbf{v}(t)}{T^2} + 60\frac{(\mathbf{x}_T - \mathbf{x}(t))}{T^3} \quad (4)$$

with $\mathbf{v}(t) = \dot{\mathbf{x}}(t), \mathbf{a}(t) = \ddot{\mathbf{x}}(t)$ and T being the desired duration of the movement.

All the models described up to now constrained the trajectory by specifying a criterion to optimize. In [17] it was shown, though, that criteria such as minimum jerk, minimum energy, and minimum force all lead to similar (bell-shaped) velocity profiles. One could therefore argue that human arm trajectories are not optimizing one concrete criterion, but that by exhibiting a bell-shaped velocity profile an almost optimal performance regarding more than one criterion results. Following this line of reasoning, in [8] a feedback scheme was proposed that, similarly to visual servo control, computes the current velocity from the remaining distance to the target as described by the differential equation given in Eq. 5:

$$\dot{\mathbf{x}}(t) = \frac{1}{T_1} \cdot \dot{g}(t) \cdot (\mathbf{x}_{T1} - \mathbf{x}(t)) \qquad (5)$$

with $g(t) = \frac{1}{\tau^2} \cdot t^3$, τ being a variable discretizing time. $g(t)$ models a nonlinear time perception and assures a bell-shaped velocity profile and a smooth, two-phasic acceleration profile[2]. Eq. 6 shows the solution of Eq. 5:.

$$\mathbf{x}_1(t) = \mathbf{x}_0 + (\mathbf{x}_{T1} - \mathbf{x}_0) \cdot (1 - e^{-\frac{1}{T_1} g(t)}) \qquad (6)$$

In the case of a double-step target, a second differential equation given in Eq. 7 is superimposed

$$\dot{\mathbf{x}}_2(t \geq t_2) = \frac{1}{T_2} \cdot \dot{g}(t - t_2) \cdot (\mathbf{D}_2 - \mathbf{x}_2(t)) \qquad (7)$$

with $\mathbf{D}_2 = (\mathbf{x}_{T2} - \mathbf{x}_{T1})$. This leads to the solution

$$\begin{aligned}\mathbf{x}(t \geq t_2) &= \mathbf{x}_0 + (\mathbf{x}_{T1} - \mathbf{x}_0) \cdot (1 - e^{-\frac{1}{T_1} g(t)}) + \\ & (\mathbf{x}_{T2} - \mathbf{x}_{T1}) \cdot (1 - e^{-\frac{1}{T_2} g(t - t_2)}) \end{aligned} \qquad (8)$$

Unfortunately, Eq. 5 and Eq. 7 cannot be merged into a single feedback equation due to the nonlinearity in time.

2.3 Discussion

Due to the (visual) feedforward structure, the models described in Sec. 2.1 cannot cope with perturbations in the arm's state or with an inaccurate model of the spatial relation between eye and hand. The corresponding feedback version does not show these problems, but raises two new ones: First, robot control on the level of joint acceleration increments is more difficult to realize than on the level of joint positions;

---
[2]Note, that T does not directly correspond to movement duration here; it scales the velocity.

secondly, the current position, velocity, and acceleration would have to be measured at a high rate which is problematic in the case of vision.

The second feedback model shows a remarkable similarity to visual servo control; the nonlinear time function corresponds to a gain that increases with time. However, no feedback version is presented for the case of a double-step target, nor for a perturbation of the hand position.

3 Proposed control model

The former sections showed that for the planning and the control of visually-guided motion of a robot manipulator a control structure combining visual feedforward and feedback parts would be advantageous. None of current models of human reaching fulfills this requirement. As the model described in Eqs. 5-8 exhibits such a hybrid structure at least for the non-disturbed case, we extended it by generalization.

3.1 *Generalization of Goodman's model*

In the case of n target jumps, Eq. 8 can be written as

$$\begin{aligned}\mathbf{x}(t \geq t_n) &= \mathbf{x}_0 + (\mathbf{x}_{T1} - \mathbf{x}_0) \cdot (1 - e^{-\frac{1}{T_1} g(t)}) + \\ & \sum_{i=2}^{n} (\mathbf{x}_{Ti} - \mathbf{x}_{Ti-1}) \cdot (1 - e^{-\frac{1}{T_i} g(t - t_i)}) \\ &= \mathbf{x}_0 + \sum_{i=1}^{n} \mathbf{D}_i \cdot (1 - e^{-\frac{1}{T_i} g(t - t_i)}) \\ &= \mathbf{x}_{Tn} - \sum_{i=1}^{n} \mathbf{D}_i \cdot e^{-\frac{1}{T_i} g(t - t_i)} \end{aligned} \qquad (9)$$

with $\mathbf{D}_i = \mathbf{x}_{Ti} - \mathbf{x}_{Ti-1}$, $\mathbf{x}_{T0} = \mathbf{x}_0$, $t_1 = 0$.
Differentiating Eq. 9 yields

$$\dot{\mathbf{x}}(t \geq t_n) = \sum_{i=1}^{n} \frac{1}{T_i} \cdot \dot{g}(t - t_i) \cdot \mathbf{D}_i \cdot e^{-\frac{1}{T_i} g(t - t_i)} \qquad (10)$$

Rearranging Eq. 9 as

$$\mathbf{D}_n e^{-\frac{1}{T_n} g(t - t_n)} = \mathbf{x}_{Tn} - \mathbf{x}(t) - \sum_{i=1}^{n-1} \mathbf{D}_i \cdot e^{-\frac{1}{T_i} g(t - t_i)} \qquad (11)$$

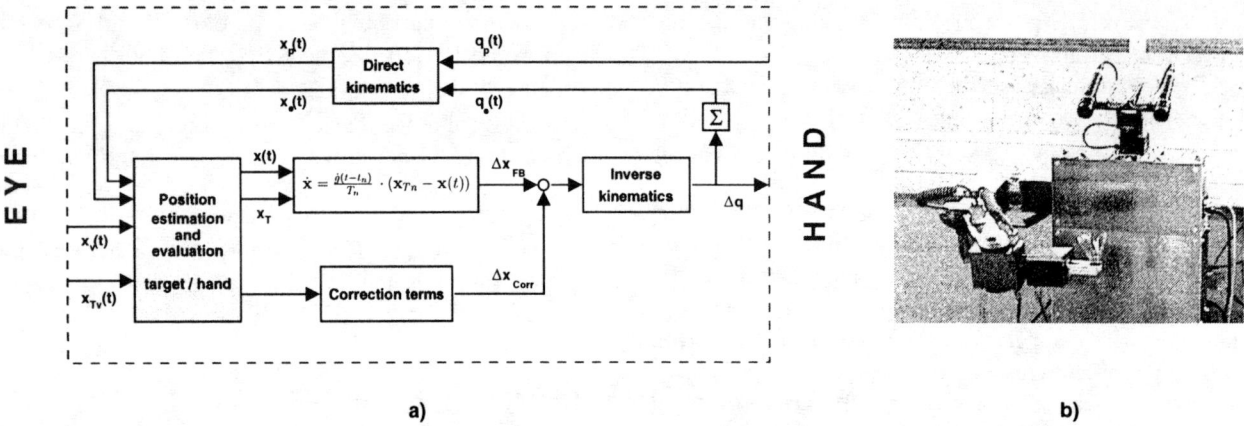

Figure 1: a) Block diagram of the motion control   b) Hand-eye system MINERVA

and substituting Eq. 11 into Eq. 10 yields

$$\dot{\mathbf{x}}(t \geq t_n) = \frac{1}{T_n} \cdot \dot{g}(t - t_n) \cdot (\mathbf{x}_{Tn} - \mathbf{x}(t)) +$$
$$\sum_{i=1}^{n-1} \mathbf{D}_i \cdot e_i(t) \cdot \left(\frac{\dot{g}(t - t_i)}{T_i} - \frac{\dot{g}(t - t_n)}{T_n} \right)$$
$$= \dot{\mathbf{x}}_{FB}(t) + \sum_{i=1}^{n-1} \dot{\mathbf{x}}_{Corr,i}(t) \qquad (12)$$

with $e_i(t) = e^{-\frac{1}{T_i} g(t - t_i)}$. Thus, the control is split into a feedback unit $\dot{\mathbf{x}}_{FB}$ and a set of feedforward terms $\dot{\mathbf{x}}_{CJ}$ which correct the target jumps. Note, that the time-dependent gain of $\dot{\mathbf{x}}_{FB}$, $\frac{1}{T_n} \dot{g}(t - t_n)$, is "reset" on each jump; therefore, stability can be ensured independent of the number of target jumps.

As the original differential equation Eq. 5 exhibits a symmetry regarding target and hand position, visually or proprioceptively measured differences between expected hand position $\mathbf{x}_e$ and actual one $\mathbf{x}_{act}$ can be integrated as a virtual target jump $\mathbf{D}_{n+1} = \mathbf{x}_e - \mathbf{x}_{act}$. Thus, the feedback unit $\dot{\mathbf{x}}_{FB}$ can also be employed in a feedforward fashion by estimating the current position of the hand from the history of Cartesian or joint angle increments computed, and superimposing corrective movements only when the current position can be measured and differs too much from the estimated one.

Fig. 1a illustrates this mixed feedforward/feedback control structure in a block diagram.

3.2 Simulation

To validate the control module, we simulated robot movements using *Matlab*. Fig. 2 shows the path projections and the Cartesian velocity profiles for a target jump using a model of our robot (Fig. 1b). Path and velocity profiles show the smooth adaptation known from human reaching movements. This property would enable a smooth tracking of moving objects without enforcing hard real-time constraints on the visual estimation of the target's position.

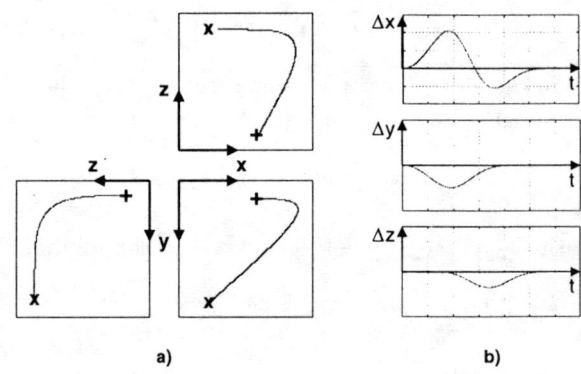

Figure 2: Double-step target: a) projections of the path,   b) velocity profiles

Next, we simulated the case of a disturbed internal kinematic model of a 2D manipulator: "In reality", link 2 is 25% longer than assumed in the model (see Fig. 3). In human reaching experiments, this would correspond to the case that the subject points with a tool in his hand.

Fig. 3a and Fig. 3b show the assumed and the real path of the manipulator. Without feedback about the position of the hand, the manipulator does not reach the target position. In Fig. 3c, feedback about the hand position was available once at $t = 0.5s$ (duration of the undisturbed movement $\approx 2s$); even with only one corrective movement, end-point precision is

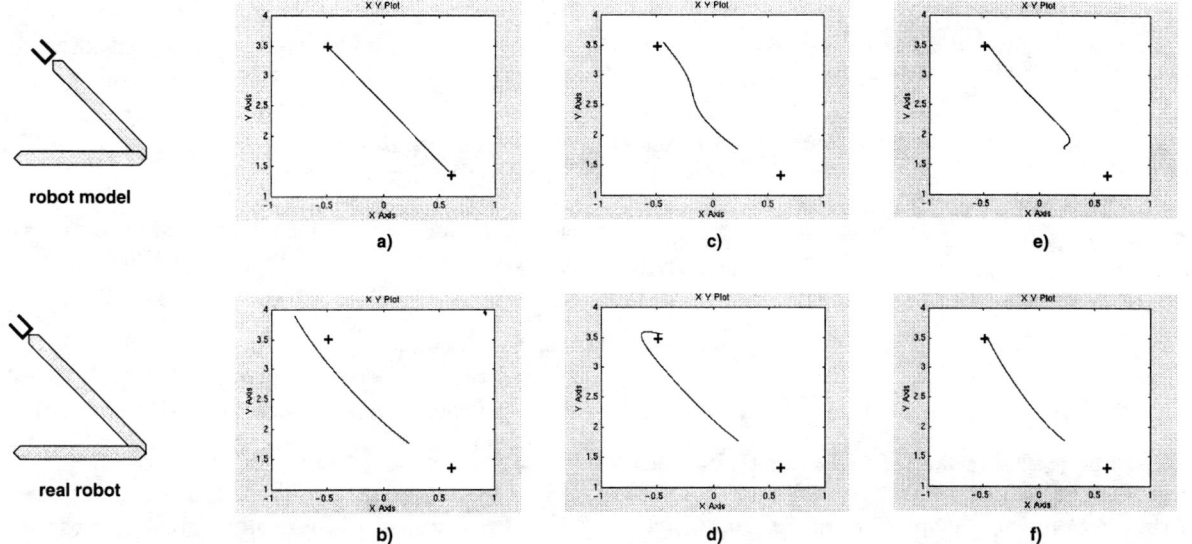

Figure 3: Internal model error: a) assumed path, b) real path, c) real path with corrective movement at $t = 0.5s$, d) real path with corrective movement at $t = 1.0s$, e) real path with corrective movement at $t = 0.1s$, f) real path with slower corrective movement at $t = 0.1s$

considerably improved in spite of the large model error. Fig. 3d plots the path for the case of a later onset ($t = 1.0s$), Fig. 3e for the case of an earlier onset of the corrective movement ($t = 0.1s$). To illustrate the effect of the choice of parameter T_2, Fig. 3f depicts the same situation as Fig. 3e, but with $T_{2e} = 0.8 \cdot T_1$.

This strategy also works in the case of errors in the internal or external camera parameters. Purely translational errors can be compensated completely by one submovement following a look at the hand. For rotational errors, there is a remaining error due to the difference in the observed positions (target/hand); to keep this error low, the hand should be observed when it is close to the target.

4 Conclusion

The previous sections showed that robotics can actually learn by taking a closer look at the human example: The model for the visual control of reach-to-grasp movements presented in Sec. 3 was developed by examining current analytical models for human reaching movements and extending a suitable one to meet the needs of a robotic hand-eye system. The resulting control model combines feedforward and feedback structures, thereby combining the advantages of "look-then-move" and "look-and-move" strategies.

Due to its hybrid structure, the model allows a very flexible control of robotic reaching movements:

If no visual or proprioceptive information about the position of the hand is available, a pure feedforward control results, with end-point precision depending on the accuracy of the internal visuomotor and kinematic models. Yet, whenever information about the position of the hand is available, it can be integrated asynchronously and smoothly by superimposing a corrective movement. Thereby, errors in the visuomotor or kinematic models can be compensated. If visual feedback about the position of the hand is available at a high rate, it can be integrated continuously, thus leading to a classical visual servoing structure, which additionally can deal with target jumps.

The control module is currently being implemented on our experimental hand-eye system MINERVA (Manipulating Experimental Robot with Visually guided Actions), which consists of a 6 dof manipulator arm and a stereo camera system on a pan-tilt head, mounted in an anthropomorphic fashion (see Fig. 1b). To test the control strategy with real visual information, image interpretation modules for the recognition of the objects to grasp [1] and for the determination of grasping points [18] are being integrated.

The flexibility of the motion control regarding the use of visual information naturally calls for a planning module that decides when to look at the hand, depending on the situation and the requirements for the grasping task. This is especially important in the case of a mobile robot that uses vision for other tasks as well.

In the future, the control model is to be further validated by analyzing the kinematics of human reaching with restricted visibility of the hand and comparing them to the predictions of the model. This will close the loop by transferring knowledge from robotics back to neuroscience.

Eventually, we want to "learn how to learn", that is not only react to detected disturbances by superimposing submovements, but also use the measured differences to adapt the internal models.

References

[1] T. Bandlow, A. Hauck, T. Einsele, and G. Färber. Recognising Objects by their Silhouette. In *IMACS Conf. on Comp. Eng. in Systems Appl. (CESA'98)*, pages 744–749, Apr. 1998.

[2] E. Burdet. *Algorithms of Human Motor Control and their Implementation in Robotics*. PhD thesis, ETH Zürich, 1996.

[3] P. I. Corke. Visual Control of Robot Manipulators – A Review. In K. Hashimoto, editor, *Visual Servoing*, pages 1–31. World Scientific Publishing Company, 1993.

[4] T. Flash and E. Henis. Arm trajectory modification during reaching towards visual targets. *J. Cogn. Neurosci.*, 3:220–230, 1991.

[5] C. Ghez, J. Gordon, M. F. Ghilardi, and R. Sainburg. Contributions of Vision and Proprioception to Accuracy in Limb Movements. In M. S. Gazzaniga, editor, *The Cognitive Neuroscience*, pages 549–564. MIT Press, 1995.

[6] M. A. Goodale, D. Pelisson, and C. Prablanc. Large adjustments in visually guided reaching do not depend on vision of the hand or perception of target displacement. *Nature*, 320:748–750, 1986.

[7] M. A. Goodale and P. Servos. Visual Control of Prehension. In H. N. Zelaznik, editor, *Advances in Motor Learning and Control*, chapter 5. Human Kinetics, 1996.

[8] S. R. Goodman and G. G. Gottlieb. Analysis of kinematic invariances of multijoint reaching movement. *Biol. Cybern.*, 73:311–322, 1995.

[9] E. Henis. *Strategies Underlying Arm Trajectory Modification During Reaching Toward Visual Targets*. PhD thesis, Weizmann Institute of Science, 1991.

[10] B. Hoff and M. A. Arbib. A Model of the Effects of Speed, Accuracy, and Perturbation on Visually Guided Reaching. In R. Caminiti, editor, *Control of Arm Movement in Space: Neurophysiological and Computational Approaches*, pages 285–306. Springer-Verlag, 1991.

[11] R. Horaud and F. Chaumette, editors. *Workshop on New Trends in Image-Based Visual Servoing*. In association with IROS'97, Sept. 1997.

[12] S. Hutchinson, G. D. Hager, and P. I. Corke. A Tutorial on Visual Servo Control. *IEEE Trans. on Robotics and Automation*, 12(5):651–670, Oct. 1996.

[13] L. S. Jakobson and M. A. Goodale. Trajectories of reaches to prismatically displaced targets: evidence for "automatic" visuomotor recalibration. *Experimental Brain Research*, 78:575–587, 1989.

[14] M. Jeannerod, Y. Paulignan, C. Mackenzie, and R. M. Marteniuk. Parallel Visuomotor Processing in Human Prehension Movements. In R. Caminiti, editor, *Control of Arm Movement in Space: Neurophysiological and Computational Approaches*, pages 27–44. Springer-Verlag, 1991.

[15] M. Kawato. Trajectory Formation in Arm Movements: Minimization Principles and Procedures. In H. N. Zelaznik, editor, *Advances in Motor Learning and Control*, chapter 9. Human Kinetics, 1996.

[16] T. E. Milner. A model for the generation of submovements requiring endpoint precision. *J. Neuroscience*, 49(2):487–496, 1992.

[17] W. L. Nelson. Physical Principles for Economies of Skilled Movements. *Biol. Cybern.*, 46:135–147, 1983.

[18] J. Rüttinger. Visual Determination of Grasping Points for Hand-Eye Coordination. Master's thesis, TU München, Nov. 1998.

[19] M. Vincze and G. D. Hager, editors. *Workshop on Robust Vision for Vision-Based Control of Motion*. In association with ICRA'98, May 1998.

[20] R. Woodworth. The accuracy of voluntary movement. *Psychological Review*, 3:1–114, 1899.

Development of a Visual Space-Mouse

Tobias Peter Kurpjuhn
kurpjuhn@lpr.e-technik.tu-muenchen.de
Technische Universität München
Munich, Germany

Alexa Hauck
hauck@lpr.e-technik.tu-muenchen.de
Technische Universität München
Munich, Germany

Kevin Nickels
knickels@trinity.edu
Trinity University
San Antonio, TX, USA

Seth Hutchinson
seth@uiuc.edu
University of Illinois at Urbana-Champaign
Urbana, IL, USA

Abstract

The pervasiveness of computers in everyday life coupled with recent rapid advances in computer technology have created both the need and the means for sophisticated Human Computer Interaction (HCI) technology. Despite all the progress in computer technology and robotic manipulation, the interfaces for controlling manipulators have changed very little in the last decade.

Therefore Human-Computer interfaces for controlling robotic manipulators are of great interest. A flexible and useful robotic manipulator is one capable of movement in three translational degrees of freedom, and three rotational degrees of freedom. In addition to research labs, six degree of freedom robots can be found in construction areas or other environments unfavorable for human beings.

This paper proposes an intuitive and convenient visually guided interface for controlling a robot with six degrees of freedom. Two orthogonal cameras are used to track the position and the orientation of the hand of the user. This allows the user to control the robotic arm in a natural way.

1 Introduction

In many areas of our daily life we are faced with rather complex tasks that have to be done in circumstances unfavorable for human beings. For example heavy weights may have to be lifted, or the environment is hazardous to humans. Therefore the assistance of a machine is needed. On the other hand, some of these tasks also need the presence of a human, because the complexity of the task is beyond the capability of an autonomous robotic system or the decisions that have to be made demand complicated background knowledge.

This leads to the demand for a comfortable 3D control and manipulation interface. A very special kind of controlling device is the space-mouse that has been developed recently by [1] and [2]. The space-mouse is a controlling device similar to the standard computer mouse, but instead of moving it around on a table, which causes plane related reactions on a computer program, it consists of a chassis that holds a movable ball. This ball is attached to the chassis in such a way that it can be moved with six degrees of freedom: three translations and three rotations. This makes it possible for the user to control six degrees of freedom with one hand. Because of this, even complex robotic devices can be controlled in a very intuitive way.

One step further to a more intuitive and therefore more effective controlling device would be a system that can be instructed by watching and imitating the human user, using the hand of the user as the major controlling element. This would be a very comfortable interface that allows the user to move a robot system in a natural way. This is called the *visual space mouse*.

The purpose of this project was to develop a system that is able to control a robotic system by observing the human and directly converting intuitive gestures into movements of the manipulator. The hand serves as the primary controller to affect the motion and position of a robot gripper. For the observation of the user, two cameras are used. A precise calibration is not required for our method. In fact, the only calibration that is required is the approximate knowledge of the directions "up", "down", "left" and "right". If a translation or rotation of the camera moves the controlling hand of the human out of the view of the camera, the system will fail. The gripper of a PUMA 560 robot arm with six degrees of freedom is used as a manipulator. One camera is placed on the ceiling providing a vertical view of the controlling hand and one camera is placed on a tripod on the floor to provide a horizontal view (see Figure 1). Together, the cameras create a 3D work-space in which the user is allowed to move.

Figure 1: Structure of the Visual Space-Mouse.

The structure of the system yields some very powerful advantages. The first group of advantages is determined by the structure itself: the system provides a quantitative and cheap control unit without any aids or moving parts. That means there is no physical wear in the controlling system. This eliminates one potential source of failure, thereby making the system more robust.

The second group of advantages is determined by the basic concept of the system which provides upgrading possibilities. One possibility would be the use of sensor data combined with an intelligent robot control system. This would lead to a robotic manipulator with teleassistance and all its advantages [3]. Another possibility would be the implementation of hand gestures as a communication language with the system to produce a high level control-interface [4]. Additionally the work-space created by the two cameras is determined by the angle of view of the two cameras. Therefore it can be individually adjusted to the needs of the controlling task. A movement guided initialization sequence tells the system which hand serves as the primary input device. It is possible to keep track of several objects to perform more complex tasks. The scaling factor that translates the hand movement of the user into manipulator motion is fixed, but freely adjustable.

The remainder of the paper is organized as follows. In Section 2 we describe the major components of the system. We begin in Section 2.1 by giving a brief overview of the image processing unit. In Section 2.2 we describe the robot control portion of the system.

Section 3 discusses two different approaches realizing the implementation of the visual space-mouse.

2 System Overview

The system of the visual space mouse can be divided into two main parts: image processing and robot control. The role of image processing is to perform operations on a video signal, received by the video cameras. The aim is to extract desired information out of the video signal. The role of robot control is to transform electronic commands into movements of the manipulator.

2.1 Image Processing

Our image processing unit consists of two greyscale CCD (charge coupled device) cameras connected to an image processing device, the Datacube MaxVideo 20. This Datacube performs operations on the video output. The operations of the Datacube are controlled by a special image processing language: VEIL [5], which is run on a host computer. In this way the data collected by the camera can be processed in a convenient way. This makes it possible to extract the desired information from the video output. In our case we identify, track and estimate the position of the hand of the user.

A special feature of VEIL is the use of blobs. A *blob* in VEIL is defined as a connected white region within a darker environment. The use of blobs makes it possible to detect and track special regions in the image.

The image processing unit is supposed to detect and track the hand. To achieve this task, within the environmental constraints imposed on the project, an image processing task was set up as following. The video output of the camera is convolved with a blurring filter and then thresholded. After thresholding the image blob-detection is applied.

Since the purpose of this project is to generate a prototype validating the benefits of the visual space-mouse, extra constraints were placed on the image processing. In particular, a black background in combination with a dark clothing is used. By doing so, the output image of the threshold operation gives a black and white image of the camera view where objects such as the hands or the face are displayed as pure white regions. This image is then imported to the blob-routine, which will mark every white region as a blob and choose one of the blobs to be the control blob.

To gain tracking of the desired hand, a motion guided initialization sequence was added. In the initialization sequence, the user waves the controlling hand in the workspace. The image processing unit records the image differences for several successive images and creates a map of these differences. After applying a blurring filter to suppress pixel-noise, resulting in spikes in this map, the image processing unit chooses the control blob. The blob in the current image that is closest to the region that changes most is chosen to be the control blob.

The values of this blob are stored in a global data structure to make it accessible to the robot control unit. Blobs other than the control blob are ignored in the controlling process. To ensure tracking of the hand after initialization without any sudden changes in the control blob, the bounding box is only allowed to change up to S_{cb} pixels each cycle in each direction. In the current implementation, this threshold value is set to 5 pixels. This causes the blob to get stuck to the hand and not to jump to other objects that are near the hand. One advantage to this is some measure of robustness to occlusion of the hand. If an object (either dark or bright) passes between the camera and the hand, the bounding box for that object will not match the bounding box for the controlling blob. Thus, the controlling blob will remain in the position it was before the occluding object appeared.

The orientation of the major axis of the object is computed by using the centered second moments of the object as follows (see [6]):

$$\varphi = \frac{1}{2} \arctan \frac{\widehat{m}_{xy}}{\widehat{m}_{xx} - \widehat{m}_{yy}} \quad (1)$$
$$\widehat{m}_{ab} = m_{ab} - m_a \cdot m_b, \quad a,b \in \{x,y\}$$

where $\widehat{m}_{xx}$, $\widehat{m}_{yy}$, and $\widehat{m}_{xy}$ are the centered second moments about the horizontal, vertical, and 45° axes, respectively. The second equation is used to compute these centered second moments, with m_{ab} and m_a representing the non-centered second and first moments about the appropriate axes, respectively. By using these equations, angles between $+\pi/4$ and $-\pi/4$ can be measured. When the real object oversteps an angle of $\pm\pi/4$ the result of Equation (1) will change its sign. The routine to measure the orientation of the hand takes care of this effect by causing the angle returned to be clipped to $\pm\pi/4$ when the orientation of the blob oversteps this border.

As there are two cameras each providing a different view, those computations have to be done for each image source. The image processing task is set up in such a way that it processes first the horizontal, and after that the vertical view. Both views are treated in one cycle.

Because the blob search is run on a host computer, the image has to be transmitted from the Datacube to a workstation over a bus network. The bus network is the bottleneck of the whole image processing unit, as illustrated in Figure 2. By shrinking the image to $\frac{1}{16}$ of the original size, much transfer time can be saved with an acceptable loss of accuracy.

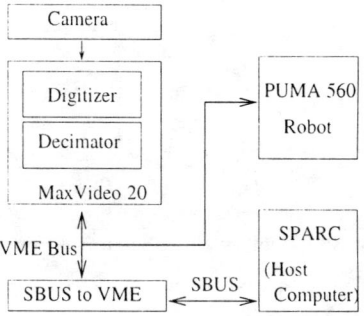

Figure 2: Bus structure and data-flow.

2.2 Robot Control

The second unit of the visual space mouse system is the robot control unit. The main elements of the robot control unit are the controller task, written in RCCL (Robot Control C Library) as a task level robot control language [7] [8], run on the host computer, and the manipulator itself.

Describing a manipulator task requires specifying positions to be reached in space (the *where*) as well as specifying aspects of the trajectory (the *how*). RCCL describes target positions using either Cartesian position equations or sets of joint angles. [9].

Cartesian position equations consist of several transform matrices that are multiplied. Each transform matrix describes a rotation and translation of the coordinate system. Together they form two systems of coordinate transformations: one on the right side and one on the left side of a position equation. Equation (2) describes the relationship of the two coordinate transformations.

$$\mathbf{T}_{start} \cdot \mathbf{T}_{variable} = \mathbf{T}_{base} \cdot \mathbf{T6} \cdot \mathbf{T}_{tool}, \quad (2)$$

where $\mathbf{T}_{base}$, $\mathbf{T6}$ and $\mathbf{T}_{tool}$ represent the homogeneous coordinate transformations from the world frame to the robot's base frame, from the robot's base frame to a frame attached to link 6 of the robot, and from, and from a frame attached to link 6 of the robot to the tool frame. The transform $\mathbf{T}_{start}$ represents a homogeneous coordinate transform from the world coordinate frame to the initial position and orientation of the tool, and $\mathbf{T}_{variable}$ is a variable homogeneous coordinate transformation matrix that is continuously updated, thereby causing the manipulator to move to a goal position and orientation.

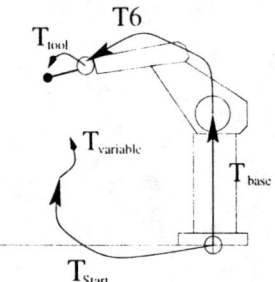

Figure 3: Effect of the position equation.

As both sides of the equations are said to be equal, both coordinate frame transformations have the same effect. That means that both sides of the equation start shifting from the same point and reach the same destination point. Equation (2) is solved for the matrix $\mathbf{T6}$, describing the desired position of the manipulator arm:

$$\mathbf{T6} = \mathbf{T}_{base}^{-1} \cdot \mathbf{T}_{start} \cdot \mathbf{T}_{variable} \cdot \mathbf{T}_{tool}^{-1} \qquad (3)$$

To reach a point in space with the manipulator, you have to create the position equation, solve for $\mathbf{T6}$ in Cartesian space and transform the solution into joint space to achieve the desired values of the joint angles of the manipulator. With these joint angles the manipulator is able to reach the destination point. The trajectory generator in RCCL will then plan a path to the desired joint angles and update it as necessary.

The only inputs for the control unit are the two blob data-structures, described in Section 2.1. These data-structures represent the spatial position and orientation of the object being tracked. The controlling unit looks at the center of the blob rectangles in the image planes, which each contain a pixel-coordinate-system. The center of the camera views are said to be the origin of the coordinate systems. These pixel coordinates are translated into global coordinates for the manipulator. This is done by directly mapping the movements of the blobs into movements of the manipulator: blob motion in the image causes the manipulator to move in the corresponding direction.

The orientations of the hand can also be observed (see Section 2.1) and are transformed into manipulator movements. Two orientations, the rotation of the hand about the optical axis of each camera, can be observed directly from the images. The third orientation, the rotation of the hand about the horizontal axis, has to be computed from the image data. The value of the orientation of the (constant sized) hand could be computed, for example, from knowledge of the size, position, and orientation of the projections of the hand in the two images.

Every time these new values are passed to the control unit a new transform matrix is created with respect to the movement of the hand. This matrix is included in the coordinate transformation equation used to control the robot. The equation is solved for $\mathbf{T6}$, the joint values of the manipulator are computed and the results are passed to the trajectory generator.

3 The Visual Space-Mouse

In some cases it is not possible to use two cameras to watch the controlling hand of the user. This can be caused by limitations on free space or on accessible hardware.

In Section 3.1 we discuss the space-mouse proposed previously, but we also suggest an approach to solve the dilemma of limited resources in Section 3.2.

3.1 Two Camera Space-Mouse

In our laboratory, only one image processing hardware device was available. Both camera views had to be processed by switching between two video channels. Combined with the transfer time via bus system (see Figure 2) this was a very time consuming procedure. So the biggest problem with the two-camera version was the speed of the image processing unit. The whole network slowed down the performance to 3 fps (frames per second). This forced the user to slow down hand motion in an unnatural way.

The use of a second image processing device would increase the image processing performance to the level seen in the one-camera version described below, allowing the user to move the hand at a natural speed. Nevertheless it could be shown that the tracking of the hand and controlling of the manipulator worked quite nicely in all six dimensions.

3.2 Space-Mouse with one Camera

In some cases limitations have to be applied to the structure of the visual space-mouse, as described above. The solution of this obstacle leads to a one-camera-version of the visual space-mouse.

By removing the overhead camera, any information about the depth of the controlling hand is lost. Any rotation with the rotation axis parallel to the image plane will just change the height and the width of the object. The sign of the rotation can not be determined easily. There are three dimensions of an object in a plane that are easily and robustly detectable: height, width and rotation in the image plane. The controlling task of a manipulator with six degrees of freedom is therefore very difficult with just 3 values. To handle this problem but keep the user interface intuitive and simple, a state machine was implemented in the controller.

The state machine consists of three different levels: two control levels and one transition level. The control levels are used to move the manipulator. The transition level connects the two control levels and affects the gripper of the robot arm.

When the palm of the hand is facing toward or away from the camera, the state machine of the controlling unit is in one of two control levels. In each control level the manipulator can be moved in a plane, by moving the hand in the up-down direction or forward-backward direction. The control levels differ in the orientation of the planes the manipulator can be moved in. The plane of control level 2 is orthogonal to the plane of control level 1 (see Figure 4). The planes intersect at the manipulator.

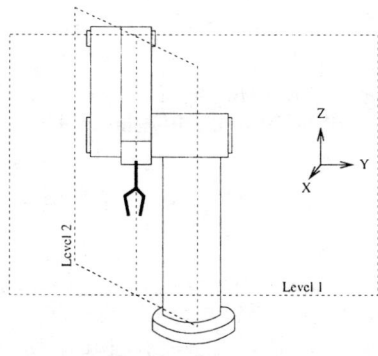

Figure 4: Motion planes of the two control levels.

To change the control levels the hand is turned so that the palm is facing down. In this mode the hand can be moved within the workspace without effecting the position of the manipulator. This mode is called the transition level (see Figure 5).

The transition level gets its name from its position between the two control levels, which are the actual steering levels. The task of the transition level is to connect both control levels and to perform additional actions on the workspace managed by the control levels. Those action are actuating the gripper and rotating the whole manipulator along its vertical axis.

By the use of the two planes, described previously, only a cubic space in front of the arm can be accessed. With the rotation along the z-axis this cube can be rotated and so the whole area around the manipulator is accessible. The rotation is initiated by rotating the hand in the plane of the image. This causes the robot to turn in steps of 10 degrees.

The gripper movement controls the opening and closing of the gripper. This movement is initiated by rotating the hand in the horizontal plane as shown in Figure 5. Placing the gesture for the gripper in the transition level has the advantage that any movement of the hand has no effect on the position of the manipulator, which will keep the gripper fixed during actuation.

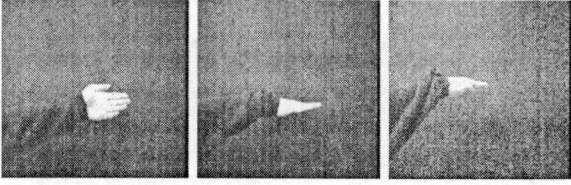

Figure 5: Control level gesture (left), transition level gesture (middle) and gripping gesture (right).

To determine when the state machine is supposed to change state, two threshold levels are computed and stored during initialization of the program. The first threshold level is set to $\frac{3}{4}$ of the height of the original blob($= height\_3\_4$), the second one is set to $\frac{3}{4}$ of the width ($= width\_3\_4$) of the original blob. The state machine goes into the transition level when the height of the actual blobs falls below $height\_3\_4$. It goes into the opposite control level when the actual height exceeds $height\_3\_4$. In the transition level, the gripper is actuated when the width of the hand is reduced below $width\_3\_4$. The structure of the state machine is illustrated in Figure 6.

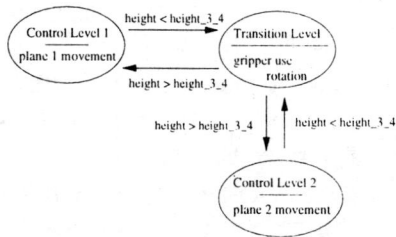

Figure 6: Structure of the state machine.

The origin of a control level plane is reset to the current position of the manipulator when the state machine enters that control level. This has the advantage that positions that are out of reach within the first attempt can be reached in the second attempt just by going into the transition mode, moving the hand to a more convenient position and then returning to the same control level.

The state machine always starts in control level 1. To visualize the state of the state machine of the control unit, the rectangle around the hand on the monitor is shown in a state-dependent color.

3.3 Discussion

An experiment was performed to validate the functions of the one-camera space-mouse. The task was to assemble a house out of three randomly placed wooden pieces. Several persons have been chosen to perform this experiment with minimal training, and each was able to successfully finish the task. The experiment showed that the state machine described above was usable. The biggest problem was that the gesture for the gripping movement was found to be unnatural. Most of the candidates not only turned the hand in the horizontal image plane, reducing the width of the hand below *width_3_4* as shown in Figure 5, but also turned their wrist. By doing so, they overstepped *height_3_4* and inadvertently transitioned into a control level.

One solution for this problem would be to introduce a third control level, as both control levels have been exhausted in terms of robustly detectable intuitive gestures. But this would require a significant change in the state machine, and the control would become less intuitive. Thus, it has not been implemented. Another possibility would have been to change the gripping gesture. This was not possible because of the limited possibilities of gestures that were bound to the robustly detectable dimensions: position, size, and orientation of the controlling blob.

3.4 Future Work

Both versions of the space mouse have several areas for improvement. Some of them are as follows:

1. Implementation of a motion model of the hand
2. Segmentation of the hand blob, for higher resolution control of the robot
3. Including sensor data for achieving teleassistance [3] (e.g. collision avoidance)
4. Implementation of a high-level gesture language
5. Adding a routine for filtering out the background
6. Implementation of a state machine in the two-camera version of the space-mouse

4 Conclusions

The objective of developing a high-level visually guided interface has been realized. As the experiment described in Section 3.3 showed, simple remote tasks can be performed with minimal training or teaching. This is a good demonstration of intuitive and convenient way in which a 3D interface can be operated. Additionally, several possible extensions to this implementation of the visual space-mouse have been proposed that would make it an even more powerful interface for control and manipulation.

References

[1] G. Hirzinger, B. Gombert, J. Dietrich, and V. Senft, "Die Space Mouse - Eine neue 3D-Mensch-Maschine-Schnittstelle als Spin-Off-Produkt der Raumfahrt," in *Jahrbuch für Optik und Feinmechanik* (W. D. Prenzel, ed.), Berlin: Schiele und Schön, 1996.

[2] G. Hirzinger, B. Gombert, and M. Herrmann, "Space mouse, the natural man-machine interface for 3D-CAD comes from space," in *ECUA'96 Conference for CATIA and CADAM users in Europe*, (Göteborg, Schweden), 1996.

[3] P. K. Pook, *Teleassistance: Using Deictic Gesture to Control Robot Action*. PhD thesis, University of Rochester, 1995.

[4] V. I. Pavlovic, R. Sharma, and T. S. Huang, "Visual interpretation of hand gestures for human-computer interaction: A review," *IEEE Transactions on Pattern Analysis and Machine Intelligence*, vol. 19, pp. 384–401, July 1997.

[5] T. J. Olson, R. J. Lockwood, and J. R. Taylor, "Programming a pipelined image processor," *Journal of Computer Vision, Graphics and Image Processing*, Sept. 1996.

[6] B. K. P. Horn, *Robot Vision*. Massachusetts Institute of Technology, 1986.

[7] J. Lloyd and V. Hayward, *Multi-RCCL User's Guide*. Montréal, Québec, Canada, Apr. 1992.

[8] P. Leven, "A multithreaded implementation of a robot control C library," Master's thesis, University of Illinois at Urbana-Champaign, 1997.

[9] M. Spong and M. Vidyasagar, *Robot Dynamics and Control*. John Wiley & Sons, 1989.

Virtual Switch Human-Machine Interface Using Fingernail Touch Sensors

Stephen Mascaro and H. Harry Asada

d'Arbeloff Laboratory for Information Systems and Technology
Department of Mechanical Engineering
Massachusetts Institute of Technology
Cambridge, MA 02139, e-mail: smascaro@mit.edu, asada@mit.edu

Abstract

A novel human-machine interface using wearable sensors is presented. Fingernail sensors that detect touch forces at the fingertip are used as a means to acquire human intentions of pressing buttons and switches. Combined with a magnetic tracker detecting the position of the human hand, the nail sensor system can identify which switch the human wishes to push and when the switch has been pushed. This allows traditional physical switches, embedded in a wall, control panel, etc., to be replaced by "virtual switches" that contain no electrical circuits or mechanical parts, but are merely pictures showing the location of the switch. In the virtual switch system, the location and functionality of switches are determined by software and can thereby be changed flexibly depending on the progress of task performance, environmental conditions, and context. The virtual switch method is combined with a "hyper manual" that stores task-procedure and operational information on a computer. Monitoring the human task performance allows the digital hyper manual to better guide the human, detect errors, and provide a safer environment.

1. Introduction

Glove-based input has been used increasingly in the last decade for teleoperation and other forms of human-machine interaction. Sturman and Zeltzer provide a comprehensive review of glove-based input [1], including applications to teleoperation and robotic control. Postural gesture recognition has been applied to teleoperation of robots and to teaching of robots by demonstration and guiding [2][3][4]. Recently, Voyles and Khosla developed a system for teaching-by-guiding by inferring human intentions from tactile gestures, which were measured by force sensors on the robot [5]. However, such a system is not very flexible, as it requires modification of the hardware on the robot.

By instrumenting the human rather than the machine, we can achieve a much greater flexibility in measurement. However, few electronic gloves collect touch-force data from the human fingers as the human interacts with the environment. The ones that do make use of pressure sensing pads consisting of conductive rubber, capacitive sensors, optical detectors, or other devices which are placed between the fingers and the environment surface [6][7][8][9]. These sensor pads inevitably deteriorate the human haptic sense, since the fingers cannot directly touch the environment surface. Recently, a new type of touch force sensor has been developed, which works by measuring the color change of the fingernail caused by touch forces [10]. This new sensor does not obstruct the natural haptic sense of the human, but rather is attached to the human fingernail. The sensor uses miniaturized optics and electronics and thus can be disguised as a decorative plastic nail. By combining new touch sensor technology, existing data glove capabilities, a vision for understanding human intentions through instrumenting the human, and finally a graphical user interface for feedback to the human, a new level of human-machine interaction can be achieved.

First, this paper describes the principle of this virtual switch system and discusses its features and utility. Practical embodiment of the concept by using fingernail sensors and magnetic tracker is presented. The method is applied to a human-robot interface for task programming and manual control. Effectiveness of the virtual switch method is quantitatively evaluated through experiments using human subjects. Lastly, the virtual switch method is combined with a "hyper manual" storing task-procedure and operational information on a computer. Monitoring the human task performance allows the digital hyper manual to better guide the human, detect errors, and provide a safer environment. Application to the operation of a hybrid bed/wheelchair system for bedridden patients demonstrates the features of the combined virtual switch and hyper manual system.

2. Concept of Virtual Switches

2.1 Physical Switch vs. Virtual Switch

In the home as well as work environments, humans constantly supervise, control, and communicate with devices, computers and machines using a multitude of switches. Switches are rudimentary means for the human to communicate his/her intentions to machines. The wearable finger touch sensor would replace the traditional switches and enhance the human-machine interface. Figure 1 depicts the functionality of a traditional switch and shows how the wearable finger touch sensor provides the same functionality and replaces the physical switch. As shown in Figure 1(a), the traditional switch works by means of:

1. The human movement of his/her finger to physically push some button of the switch
2. The detection of the human intention by electrical contact in the switch, and
3. Transfer of the detected signal to a specific part of the machine to change its state

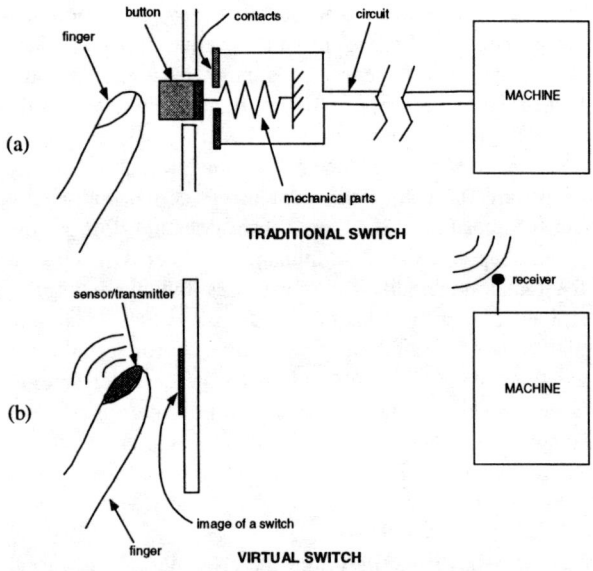

Figure 1: Traditional switch vs. virtual switch

In the traditional switch, the detection of human intention is performed by a device that is physically attached and wired to a specific part of the machine. Traditional switches have the following limitations:

1. They generally cannot be reconfigured without reinstallation
2. They can be installed only on a solid wall, or on a certain class of surfaces appropriate for securing the body of a switch
3. They take up space that could be used for some other purpose
4. They can be activated accidentally if bumped with something other than a finger
5. They can be damaged in hazardous environments – chemical, mechanical, etc.

With the new wearable fingernail touch sensor, the ability now exists to measure human intention through touch without affecting a change on the environment.

Human intention can be detected by a device worn by the human rather than a device attached to the machine. To this end, we propose the concept of a "virtual switch." As shown in Figure 1(b), the virtual switch is not a mechanism, but is an image on a surface that represents a switch. The intention of activating the switch is detected by the fingernail touch sensor, coupled with 3-D finger position measurements from an electronic glove. The signal is transmitted wirelessly from the human to the machine. When a touch is detected on a finger whose position measurement corresponds to a certain virtual switch, that switch is activated. This will eliminate many of the problems listed above. To summarize, the virtual switch panel offers the following advantages over traditional switches:

1. Virtual switches can be rearranged and reconfigured without reinstallation
2. Virtual switches can be placed on diverse surfaces including soft surfaces, furniture, the body of the machine, and even the human body
3. Virtual switches can move around in space and change in functionality as a task progresses
4. Virtual switches can share the workspace and do not monopolize a work surface
5. Virtual switches can have different functions for different fingers
6. Virtual switches cannot be activated accidentally if bumped with something other than finger

The concept of the virtual switch panel opens up numerous possibilities for human-machine communication, and can be anything from a simple virtual on-off button to an entire virtual computer keyboard.

2.2 Human-Machine Interface

Figure 2 shows a sketch of one embodiment of the virtual switch panel. In this scenario, the human is working alongside a robot to accomplish a task. The human is wearing some form of electronic glove with open fingertips, which tracks the position of his fingers in 3-D space, and his fingertips are instrumented with the fingernail sensors to measure finger touch force. Virtual switches are painted on the surfaces around his workspace as well as the surface of his workspace. Some switches may even be painted on the robot or human himself. Whenever the fingernail sensor detects a sudden touch force, it relays the signal to the computer or robot controller along with the position of the finger that committed the touch. If the computer recognizes that the position corresponds to a certain virtual switch, then that switch is declared "activated." The function associated with the switch is performed, and the computer provides feedback to the human audibly or otherwise to confirm the activation of the switch. In this way the human can activate the robot, the computer, or other devices in his work area without affecting any change on the environment. Furthermore, the functions of each of the switches can be reprogrammed by the human at any time without having to do any work mechanically. The virtual switches can even take on different functions automatically at different stages of a task, or have different functions depending on which finger activates them. Like a computer mouse with two buttons, different actions can be recognized by using multiple finger touch sensors. Finally, the human can work on top of the virtual switches and use his desk for other tasks because the switches do not get in the way.

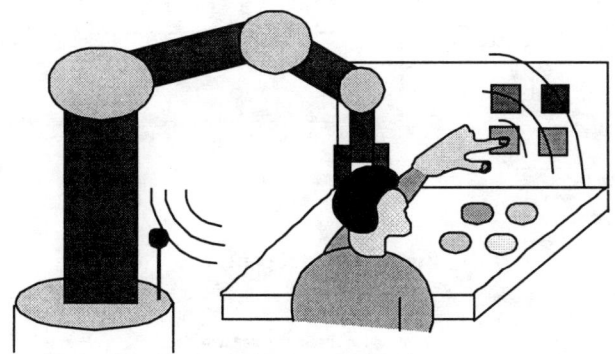

Figure 2: Virtual Switch Panel

Figure 3 shows the next level of this embodiment, which is the "totally virtual switch panel." This embodiment has all the features of the original virtual switch panel, only in this case the switches are not even painted or drawn on the surfaces of the workspace. Instead, the switches are either projected onto the workspace, or the human wears a head-mounted, head-up-display, which superimposes computer images of the switches on his view of the workspace. By tracking head motion, the images can be made to appear stationary on a particular surface or move around in a desired fashion. Looking in different places can cause different switch panels to be displayed. Switches can be rearranged and reconfigured completely by software.

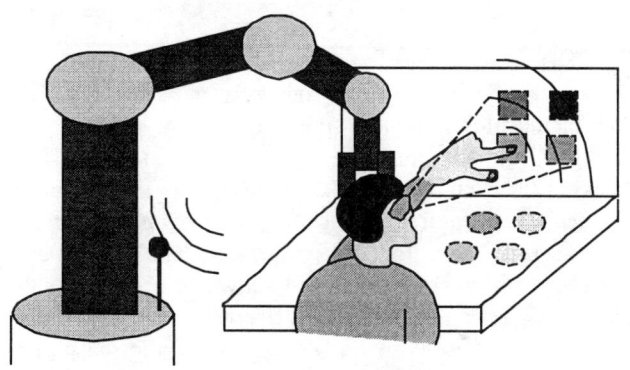

Figure 3: Totally virtual switch panel

3. Manual Robot Control and Programming Using a Virtual Switch Panel

3.1 System Construction

Since the virtual switch increases flexibility in terms of positioning and functionality, it seems reasonable to expect that increased performance can be achieved by making the virtual switch panel more intuitive than a traditional switch panel. A good case study for this hypothesis is the teaching of an industrial robot such as the PanaRobo KS-V20 manufactured by Panasonic. Figure 4 shows a picture of the robot and its teaching pendant. The robot has five rotational joints in series and can be controlled in joint angle mode or linear Cartesian mode. As seen in Figure 4, the top three pairs of buttons on the right are used for linear Cartesian moves, and the top five pairs are used for joint angle moves.

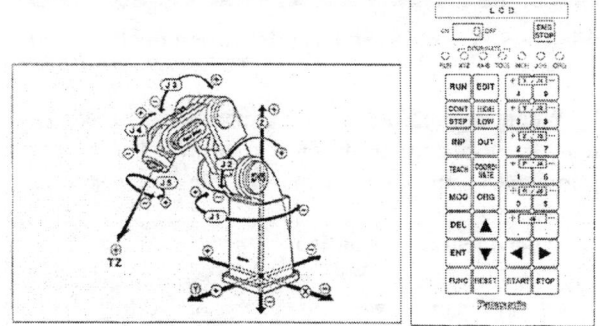

Figure 4: PanaRobo KS-V20 and teaching pendant

A major limitation of this standard method of control is the lack of intuition between the buttons on the pendant and the resultant actuation of the robot. Depending on which way the person is facing the robot, a right button for example could cause the robot to move to the left. A button for a particular joint has no intuitive connection to the particular joint of the robot. Intuition can be greatly enhanced by using a virtual switch panel. In particular, by using virtual switches, we now have the option of mounting switches on the workspace and the robot itself. Because the switches are virtual, they are not at risk of being damaged within the workspace environment, and they do not require any physical modifications to the robot hardware or the workspace.

Figure 5 shows an obvious configuration for the virtual switch panel for the KS-V20 robot. Virtual switches are located on each joint in order to perform joint moves, intuitively connecting each switch with its resulting actuation. The human is equipped with two touch force sensors on the index and middle fingers. The index finger causes counter-clockwise rotations while the middle finger causes clockwise rotations. Virtual switches for linear Cartesian control are located on the surface of the workspace in the shape of a set of coordinate axes, corresponding to the coordinate frame of the robot. The operator can use two switches for each axis, or one switch for each axis with separate fingers for positive and negative directions.

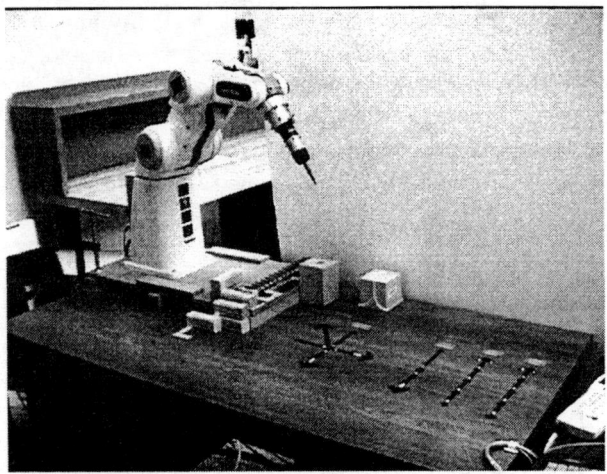

Figure 5: Virtual switch configuration for the robot

We now present specific experiments to compare human performance using the virtual switch panel with performance using the traditional teaching pendant.

3.2 Experimental Evaluation

In this paper, we will focus on testing human performance using linear Cartesian control. An experiment was set up in which a human operator is required to move the tip of the robot tool along a path in 3-dimensional Cartesian space. Figure 6 shows a drawing of the experimental setup. The operator is asked to maintain a constant gap of approximately 5-7 mm between the tip of the tool and the path surface. The total path length is 1.42 m. The path is comprised of 27 distinct moves: 11 in the x-direction, 9 in the y-direction, and 7 in the z-direction. The operator must therefore make at least 27 decisions in order to choose the correct switches, either on the teaching pendant or on the virtual switch panel. The robot is programmed to run at a linear speed of 25 mm/sec for both cases. The time it takes the operator to complete the course will be used as a measure of human performance for each method of control.

Figure 7 shows the results of this experiment for four human operators. As one might expect, there is certain variation between the operators. However, the learning curve trends are quite consistent. Compared to using the teaching pendant, the learning curve is noticeably faster when using the virtual switches. The operators converge more quickly on their peak performance when using the virtual switches. This suggests that the virtual switches are a better match for the learning pattern of a human. Also, for some of the operators, the steady-state performance is better when using the virtual switches as compared to the teaching pendant. This suggests non-intuitiveness cannot always be overcome by repetitive experience. The intuitiveness of virtual switches can allow some people to operate at a consistently higher performance level in tasks of this nature.

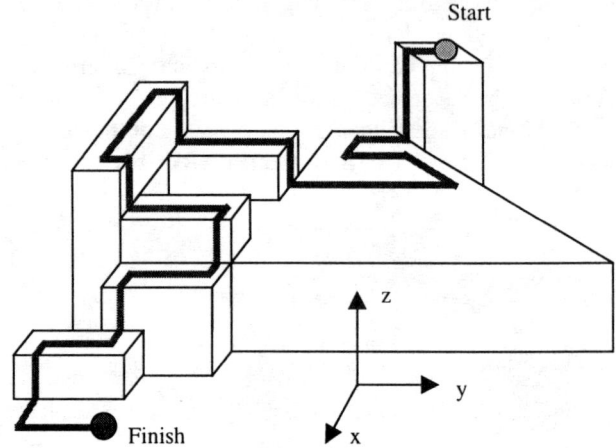

Figure 6: Tool course

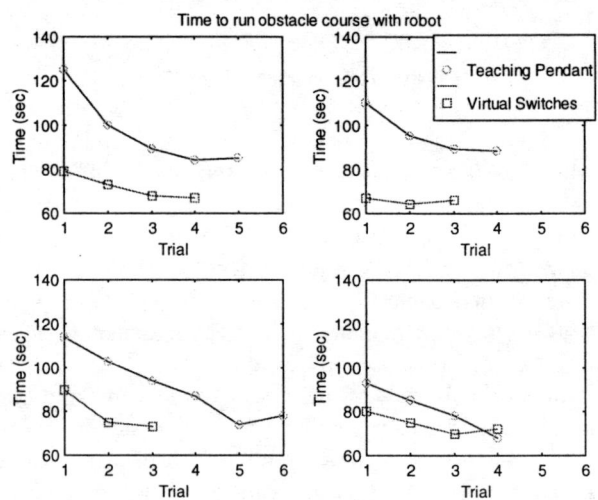

Figure 7: Performance data for teaching pendant and virtual switches

Based on the speed of the robot, the minimum time to cover the distance of the course is 57 seconds. By subtracting this minimum time required for moving, we can get the approximate time taken up by mental decision making for each operator. Dividing by the total number of decisions required, we get the approximate average time per decision for the operators. The results are averaged for five operators and shown in Figure 8 with standard deviation. The difference here is quite pronounced. The virtual switch seems to offer an initial advantage of taking less than half the time per decision compared to the teaching pendant. At most, about half the number of trials is required for the learning curve to level off. To make any general comparisons about the final steady-state performance, this experiment needs to be conducted with a larger number of trials on a larger pool of human operators.

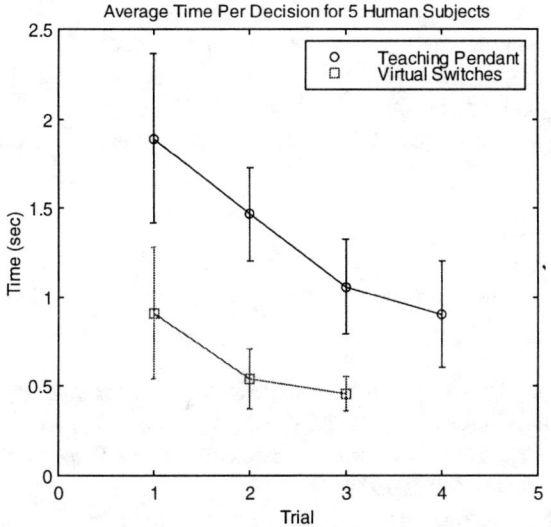

Figure 8: Composite average time per decision for teaching pendant and virtual switches

4. Digital Manuals with the Virtual Switch Human-Machine Interface

4.1 Integration of Manual Information with Human Behavior Understanding

The flexibility provided by the virtual switch system would allow us to assist the human in performing complex tasks that require interactive control and coordination between the human and a machine. In this section we explore the possibility of using virtual switches for performing a complex procedure described in a manual. Different portions of virtual switches are to be activated and presented to the human following the procedure described in the manual.

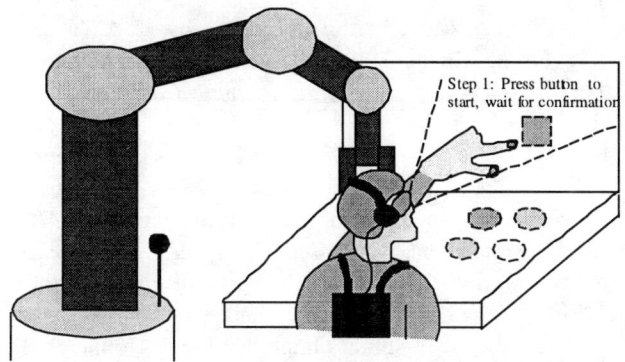

Figure 9: Wearable Hyper Manual

In general, a manual contains step-by-step instructions for operating a machine in order to perform a certain task. Typically a human follows a procedure described in a manual, which includes a sequence of operations such as pressing buttons and knobs. In the past decade, many manuals have been computerized, i.e. digital manuals, for better service and easier use. The functionality and features of digital manuals can be enhanced by combining them with wearable sensors such as the finger touch sensors and hand position sensors that monitor human behavior. Namely, combining virtual switches with a digital manual would create a powerful aid for guiding a user through a complex operational procedure. Figure 9 illustrates such a combined system, called a "wearable hyper manual."

The wearable hyper manual consists of a wearable computer that stores manual information, wearable and/or stationary sensors monitoring human behavior, and a display and/or headset for providing instructions to the human. Unlike traditional digital manuals, where the user must retrieve items of information needed for each step of operation, the wearable digital manual system monitors the human behavior, identifies the stage of procedure in which the human is currently involved, and provides the right items of instruction needed for that stage. Furthermore, inputs from the human are acquired from the virtual switch panel described above. The virtual switch panel presented to the human would be varied depending on the stage of the procedure and the relevance to the context of the task. The execution of this entire process entails a task programming and process control engine. Such an engine would represent the task, code the procedure, recognize each task stage, observe human behavior, retrieve manual information, present instructions, display control panels, and acquire human inputs to coordinate a target machine with the human inputs.

4.2 Application to a Hybrid Bed/Wheelchair System

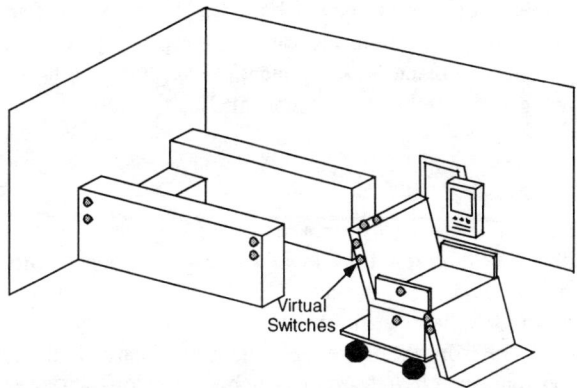

Figure 10: Distributed Virtual Switch System

As an example, consider the application of a virtual switch system to a hybrid bed/wheelchair system for bedridden patients, as shown in Figure 10. This hybrid bed/wheelchair, called RHOMBUS (Reconfigurable Holonomic Omnidirectional Mobile Bed with Unified

Seating), was developed to eliminate transfer between a bed and wheelchair [11]. This requires a certain procedure for converting the bed to a chair and vice versa. To alleviate difficulties and ensure safety, virtual switches are placed at various surfaces on the chair and bed. The system is operated by a caregiver wearing the finger touch sensors and hand position sensor. Numerous virtual switches can be embedded in the bed/chair surface for acquiring the caregiver's intention. For example, when the caregiver wishes to raise the back leaf of the bed/chair system, he/she touches the back side of the back leaf and tends to push it upward. The virtual switch embedded in the back leaf recognizes the human motion by detecting the hand location and the pressure increase at his/her fingertips.

The detected signal is then transmitted to the powered bed/chair to activate the actuator that raises the back leaf. Let us suppose that, after raising the back leaf, the caregiver wants to push the bed (now reconfigured to a wheelchair) forward. The virtual switch embedded in the wheelchair handle detects the caregiver's intention when he/she touches the wheelchair handle and tends to push it forward. The position sensor recognizes that the caregiver places his/her hand on the handle, and the finger touch sensors detect that the forward button printed on the handle is pressed by his/her fingers.

The assignment of virtual switches to individual portions of the bed/chair system can be changed depending on the context, situation, and stage of the task. For example, virtual switches can be altered between the bed-mode operations and the chair-mode operations by simply changing the "map" relating sensor signals to the control actions. Although the same point of the machine is pressed, different actions can be generated. Pressing the back leaf, for example, is recognized as the intention to change the back leaf angle only when the system is in the bed mode. Pressing the same back leaf in the wheelchair mode creates no action, thus avoiding erratic operations. This context-dependent assignment can be extended to the concept of "wearable digital manuals."

5. Conclusions

In this paper we proposed the concept of a "virtual switch," whereby a traditional switch is replaced with merely an image of a switch on a surface. Measurements of a human's hand position, along with measurements of touch force from the fingernail sensor are used to activate the virtual switch. Virtual switches offer an increased flexibility in placement and functionality over traditional switches. The virtual switch concept was tested against a traditional teaching pendant for teleoperated teaching of a robot. Experiments show that the flexibility in designing virtual switch panels can result in increased intuition for the human operator. Initial decision-making time as well as learning time can be reduced by over 50%.

We also presented the concept of a wearable hyper manual, which can be combined with the virtual switch method to prompt a human operator with instructions while interacting with the machine. This entails a task programming and process control engine that would represent the task, code the procedure, recognize each task stage, observe human behavior, retrieve manual information, present instructions, display control panels, and acquire human inputs to coordinate actuation of the machine with the human inputs.

References

[1] D.J. Sturman and D. Zelzer, "A Survey of Glove-base Input", *IEEE Computer Graphics & Applications*, January, pp. 30-39, 1994.

[2] M. Brooks, "The DataGlove as a Man-Machine Interface for Robotics," *2nd IARP Workshop on Medical and Healthcare Robotics*, Newcastle upon Tyne, UK, Sept., pp. 213-225, 1989.

[3] S.B. Kang, and K. Ikeuchi, "Grasp Recognition and Manipulative Motion Characterization from Human Hand Motion Sequences", *IEEE Int. Conf. on Robotics and Automation* **2**, pp.1759-1764, 1994.

[4] C.P. Tung, and A.C. Kak, "Automatic Learning of Assembly Tasks using a DataGlove System", *IEEE/RSJ Int. Conf. on Intelligent Robots and Systems* **1**, pp. 1-8, 1995

[5] R.M. Voyles, and P.K. Khosla, "Tactile Gestures for Human/Robot Interaction," *IEEE/RSJ Int. Conf. on Intelligent Robots and Systems* **3**, pp. 7-13, 1995.

[6] D.J. Beebe, D.D. Denton, R.G. Radwin, and J.G. Webster, "A Silicon-Based Tactile Sensor for Finger-Mounted Applications", *IEEE Transactions on Biomedical Engineering* **45:2**, pp. 151-159, 1988.

[7] J.G. Webster, Ed., *Tactile Sensors for Robotics and Medicine*, New York: Wiley, 1988.

[8] T.R. Jenson, R.G. Radwin, and J.G. Webster, "A Conductive Polymer Sensor for Measuring External Finger Forces", *Journal of Biomechanics* **24:9**, pp. 851-858, 1991.

[9] J.S. Son, A. Monteverde, and R.D. Howe, "A Tactile Sensor for Localizing Transient Events in Manipulation", *Proceedings of the IEEE International Conference on Robotics and Automation*, pp. 471-476, 1994.

[10] S. Mascaro, K.W. Chang, and H.H. Asada, "Finger Touch Sensors using Instrumented Nails and their Application to Human-Robot Interactive Control", *ASME IMECE, Symposium on Haptic Interfaces for Virtual Environment and Teleoperator Systems*, November, 1998.

[11] Mascaro, S., Spano, J., and Asada, H. "A Reconfigurable Holonomic Omnidirectional Mobile Bed with Unified Seating (RHOMBUS) for Bedridden Patients," *IEEE Int. Conf. on Robotics and Automation*, Albuquerque, New Mexico, April 1997.

Arm-Manipulator Coordination For Load Sharing Using Predictive Control

Kamran Iqbal
Programs in Physical Therapy
Northwestern University
Email: k-iqbal@nwu.edu

and

Yuan F. Zheng
Department of Electrical Engineering
The Ohio State University, Columbus, Ohio
E-mail: zheng@ee.eng.ohio-state.edu

Abstract

The coordination problem of a human arm and a robot manipulator is explored using compliant motion and predictive control. The problem arises when a human arm and a robot manipulator coordinate for the execution of a task in unscheduled task environment. In such scenarios the arm, by virtue of its intelligence, is assumed to lead the task while the manipulator is required to comply with the motion of the arm and support the object load. Such a scheme is superior to the better known multiple manipulator coordination problem, which normally assumes known trajectories and a structured task environment. By coordinating manipulator with the arm of its operator the uncertainty due to the environment can be reduced while load sharing can help relieve the arm of the physical strain. This paper addresses the problem in the framework of model-based predictive control. The transfer function from the manipulator position command to the wrist sensor force output is defined. The desired set point for the manipulator force is set to equal the gravitational force. A predictive control scheme is then used to design a two-degree of freedom controller for the problem. The simulation results indicate that the manipulator effectively takes over the object load and the arm force stays close to zero. Moreover, the manipulator is seen to be highly responsive to the arm movement and relatively small arm force can effectively initiate the manipulation task.

Introduction

In recent years, coordination of multiple robot manipulators has received much attention by the researchers [1,2]. The coordination is aimed to tackle those manipulation tasks that would be difficult for a single manipulator to execute. Many industrial and service applications fall in this category, e.g., sharing heavy loads, bending metal sheets, etc. Such coordination results in an increase the efficiency of task execution. Although the problem has received considerable attention in recent years, the scope of its application is essentially limited, due to complexity of coordinating mechanisms, to simple tasks in well designed environments. The present robot technology does not address the execution of complex coordination tasks in unstructured environments.

The arm-manipulator coordination problem has been recently introduced in the field of robotics. The problem visualizes a task situation where two coordinating manipulators are needed, but lack the necessary intelligence to perform the task. An alternative approach is to replace one of the manipulators with the human arm (Fig. 1). The intelligence of the arm helps perform complex functions, e.g., task planning, obstacle avoidance, etc., while the manipulator performs the load sharing function. Due to its ability to tackle tasks requiring intelligence, the arm-manipulator coordination is superior to manipulator-manipulator coordination, which can perform only low intelligence tasks.

Arm-manipulator coordination for load sharing has been studied earlier by [3,4]. In the previous work two different control strategies, i.e., compliant motion control and reflexive motion control have been proposed and tested in the laboratory. In [3] it was shown that the manipulator was required to possess negative stiffness to ensure compliance in load sharing tasks involving arm-manipulator scenario. In that study it was also discovered that in situations where the human arm exerted large forces to change the motion of the object, the manipulator acted as load to the arm. In [4] reflexive motion control was proposed to solve the loading problem. In reflexive control scheme a supervisory loop was added to the compliant controller, that acted like a force control. Reflexive control improved the speed of the manipulator in response to the motion of the arm. In a way the controller anticipated the movement of the arm and applied the required corrections in advance. Reflexive control thus assisted the manipulator in comprehending the intentions of the human arm.

Encouraged by the success of the earlier studies, we now propose model based predictive control in the arm-manipulator coordination problem. In such a scheme the control action is aimed to reduce the error between the projected and the desired motion of the object. As the human arm moves the object, the wrist sensor on the manipulator picks up the forces exerted by the environment. The controller then predicts the manipulator force over a finite horizon and commands the manipulator to cancel the environmental forces. Since the motion of the inertial object does not require excessive arm forces, the control strategy would ensure that the arm force stays at a low value, and the load is effectively taken over by the manipulator.

The objective of this paper, similar to the previous work, is to develop a coordination mechanism for the manipulator to work with its operator's natural arm in relatively unknown and unstructured task environment. This paper is organized as follows: Model-based predictive control is discussed in the following section. The dynamics of arm-manipulator coordination problem are formulated next, followed by results, discussion, and a conclusion.

Model-Based Predictive Control

Predictive control [5,6,7] is a model based optimal control strategy, in which a quadratic cost function comprising error among the desired and estimated set-point trajectories and controller inputs is minimized at future points in time utilizing the available information at the present time. The controller output can be structured in a desired manner and suitable performance criteria can be included into the cost function. The resulting control law is very similar to other model based design methods, such as pole placement or LQ control.

The use of the model based predictive control is attractive in the field of robotics due to its ability to conveniently handle desired set point trajectories. Its disadvantage is that the stability of the feedback system can not be guaranteed up front and usually a number of attempts are made before arriving at a successful design. The tradeoffs associated with a number of tuning parameters involved in the design further diffuse the choice of the controller.

As with other model-based control schemes, the heart of the model based predictive control (MBPC) is the process model itself. A good design includes mechanisms for obtaining the 'best' model and for making the controller robust against inevitable plant/model mismatch. The model has to be rich enough to capture the essential dynamics of the plant. It has been recognized [5], that it is important to include a disturbance model in a MBPC design. The internal model principle can then effectively lead to disturbance annihilating controllers [6]. The assumed model can thus increase the robustness of the system.

The predictive control law can be most conveniently applied to discrete time system models (although a continuous time version is also available). In order to compute the predictive control law the following process model is normally considered [6]:

$$y(k) = q^{-d-1} \frac{B(q^{-1})}{A(q^{-1})} u(k) + \xi(k) \quad (1)$$

where q^{-1} represents the backward shift operator that is normally used to get causal solutions, d denotes the process delay, and $\xi(k)$ denotes a disturbance input of the following type:

$$\xi(k) = \frac{C(q^{-1})}{D(q^{-1})} e(k) \quad (2)$$

where $e(k)$ represents a sequence of zero mean white noise. A popular choice for the disturbance model is given as $D(q)=\Delta(q^{-1})=1-q^{-1}$, which is effective against step disturbances and Brownian motion type noise [6]. The use of $\Delta(q^{-1})$ in the determination of noise term leads to integral/incremental contol laws that are clearly desirable. When written in this form the model is also known as the Controlled Autoregressive and Moving Average (CARIMA) model. The i-step ahead predictor for the process in (1) is given as [7]:

$$\hat{y}(k+i) = G_i u(k+i-1) + \frac{H_i}{A} u(k-1) + \frac{F_i}{C}[y(k) - \hat{y}(k)] \quad (3)$$

where the polynomials F, G and H are solved from the following set of Diophantine recursion equations:

$$PC = DE_i + q^{-i} F_i$$
$$BDE_i = CG_i + q^{-i} H_i \quad (4)$$

An important step in designing predictive controllers involves the choice of the criterion function. The following unified criterion function has been proposed by [7]:

$$J = \sum_{H_m}^{H_p} [P\hat{y}(k+i) - P(1)w(k+i)]^2 + \rho \sum_{1}^{H_c} [\frac{Q_n}{Q_d} u(k+i-1)]^2 \quad (5)$$

where H_m, H_p and H_c represent the minimum, prediction and the control horizons; the polynomial P defines the desired set point behavior; and, polynomials Q_n and Q_d, together with the scalar ρ, define the controller output weighting. The above function represents a generalization of cost functions used in a number of predictive controllers proposed in the literature, e.g., generalized predictive control (GPC) [5,6] and minimum variance control [8].

The above cost function can be augmented by end point state weighting or terminal constraints [6]. In order to define a control structure the cost function is minimized subject to a set of input constraints. For example, the following constraints ensure that the control signals for the given set point trajectory become steady:

$$\varphi Pu(k+i-1)=0, \quad 1 \leq H_c < i \leq H_p \qquad (6)$$

where polynomial φ governs the steady state behavior of the set point and the disturbances. For a constant disturbance P can be chosen as $P=\Delta$, where $\Delta=1-q^{-1}$.

Successful application of the MBPC involves selection of a number of tuning parameters: the minimum, prediction and control horizons H_m, H_p and H_c; the weighting polynomials φ, P, Q_n, and Q_d; the disturbance model (given by polynomials C and D); and the controller input weighting ρ. The rules for selection of these parameters are given in the literature (e.g., [5,7]) and will not be repeated here.

It is shown by [7] that the control law derived by minimization of the above unified cost function can be represented in the form of a polynomial solution, representing a two degree-of-freedom controller, similar to that of the pole-placement design, given as:

$$Ru(k) = -Sy(k) + Tw(k+H_p) \qquad (7)$$

where $w(k+H_p)$ represents the future set point sequence. The polynomial approach is shown in block diagram form in Fig. 2. Using polynomial approach the process and the controller output, $y(k)$ and $u(k)$, are given as:

$$y(k) = \frac{q^{-1}BCw(k+H_p) + AR\xi(k)}{AR + q^{-1}BS}$$
$$u(k) = \frac{ACw(k+H_p) + AS\xi(k)}{AR + q^{-1}BS} \qquad (8)$$

where in order to ensure stability the characteristic polynomial, $AR+q^{-1}BS$, should have all its roots inside the unit circle. The servo behavior of the resulting closed-loop system is governed by the zeros of polynomials C, and can be tuned independently from the regulator behavior (governed by the zeros of polynomial R). For detailed computations and a code for the solution algorithm the reader is referred to [7].

Formulation of the Arm-Manipulator Coordination Problem

In order to formulate the arm-manipulator coordination problem, we consider an object jointly manipulated by a human arm and a robot manipulator. The forces acting on the object include the arm force (F_a), the manipulator force (F_m), and the force of gravity (F_g). There are no restrictions placed on the motion of the object, and no prior knowledge of the trajectory is assumed.

Following other researchers (e.g., [9]) we consider the manipulator to be a positioning device, which decouples the dynamics of the robot and provides position tracking in Cartesian coordinates. A second order linear system is used to model the end-effector position control system along each coordinate. In such formulation $X_m = G X_c$, where X_m represents the position and the orientation of the end-effector, X_c represents the commanded position and orientation, and G is a 6x6 diagonal transfer matrix of second order transfer functions.

The manipulator is assumed to be equipped with a wrist sensor that can sense environmental forces. Let given by Z_f denotes the sensor impedance, then the manipulator force output F_m is given as:

$$F_m = Z_f(X_c - X_o) \qquad (9)$$

where X_o represents the position and orientation of the object handled by the end-effector. In the context of this paper the term `force' has a generalized meaning that also include torques.

Similarly, following other researchers [10], we represent the human arm with a black box model neglecting the behavior of the musculo-skeletal system. The input to the arm model is the desired position and orientation, and the output is the arm force, F_a, (including moment) that is applied to the object under manipulation. The arm similarly possesses impedance, Z_a, which may arise from the elastic and viscous properties of the biological muscles. The force generated by the arm is then given as:

$$F_a = Z_a(X_d - X_o) \qquad (10)$$

where Z_a is a 6x6 transfer matrix of the arm impedances, and X_d represents the desired arm trajectory in terms of its position and orientation. The desired trajectory is assumed to be planned in the central nervous system, and no *apriori* knowledge of it is assumed. The arm stiffness is generally unknown, but it is much smaller than the wrist sensor stiffness, that has a high value that is known.

It is assumed that the speed of manipulation task is small, such that the Coriolis and other nonlinear effects can be neglected. It is further assumed that the only forces acting on the object consist of the arm force, F_a, the manipulator force, F_m, and the force of gravity, F_g. In the absence of any friction, the dynamics of the object can be written as:

$$M_o \ddot{X}_o = F_a + F_m + F_g \qquad (11)$$

where M_o represents a 6x6 matrix of inertial properties of the object, F_m (6x1) and F_a (6x1) represents the force and moment exerted by the manipulator and the human arm, and F_g represents the gravitational force. A simplified block diagram of the arm-manipulator coordination problem is shown in Fig. 3.

In the following analysis we initially neglect the force of gravity to compute the transfer function for the coordination problem. The same can be considered as constant disturbance acting on the object, and is later included in the computation for the predictive controller. Also, for simplicity, we only consider a one-dimensional view of the problem, replacing complex impedances Z_a and Z_f by their stiffness components K_a and K_f.

Ignoring gravity and the robot dynamics, the manipulator force output is given as:

$$F_m = \frac{-k_a k_g}{M_o s^2 + k_a + k_g} X_d + \frac{(M_o s^2 + k_a) k_g}{M_o s^2 + k_a + k_g} G X_c \quad (12)$$

where X_d represents the desired movement initiated by the human arm, X_c represents the manipulator command, and $G(s)$ represents the transfer function of the positioning system. At slow speeds of manipulation the first component in the above expression can be approximated as $-k_a X_d$. This component is unknown and can not be directly measured or controlled, however, its magnitude stays small. The second component, which represents the transfer function from the manipulator position command to the wrist sensor force output, represents the dominant component, and is used for the controller computations. This component can be suitably modified using a feedback controller to reflect desired dynamic response properties. Thus, for design purposes the plant transfer function is given as:

$$F_m = \frac{(M_o s^2 + k_a) k_g}{M_o s^2 + k_a + k_g} G X_c \quad (13)$$

In order to compute the predictive control law for the arm-manipulator coordination problem, a discrete-time equivalent of equation (13) was first obtained. A relatively small sampling time (1-2 msec) was deemed necessary due to the high natural frequency of the response modes introduced by the high impedance (10^5 oz-in^{-1}) of the force sensor. It can be verified that the denominator polynomial of the discrete-time transfer function is given by:

$$[z^2 + 2\cos\{(k_a + k_g)T/m\}z + 1] \cdot [z^2 - (e^{-aT} + e^{-bT})z + e^{-(a+b)T}] \quad (14)$$

where $G = ab/[(s+a)(s+b)]$. To simplify the computations a low fixed value of arm stiffness (100 oz-in^{-1}) was assumed. Then, the bandwidth of the continuous-time model is given as $\omega_b = 2\sqrt{(k_a + k_g)/m}$, and the minimum sampling rate is given by $\omega_s = 2\sqrt{(k_a + k_g)/m}$. A sampling rate of 500Hz was accordingly selected. A 2nd order manipulator transfer function with time constants of 0.04 sec and 0.25 sec was used [3]. The object mass was assumed to be 1Kg.

As stated earlier, the gravitational force, F_g, was assumed to be as a constant disturbance during manipulation task. The polynomials in the disturbance model (Eq. 2) were accordingly selected as: $C = 1 - \mu q^{-1}$, and $D = \Delta$; where $\mu = 0.9$, and $\Delta = 1 - q^{-1}$. The polynomials in the criterion function were selected as: $\varphi = \Delta$, $P = 1$, $Q_n = \Delta$, and $Q_d = 1$. The parameter ρ was selected as $\rho = 10^{-5}$. Finally, the following values for the horizon parameters were selected: $H_m = 1$, $H_p = 10$, and $H_c = 2$ ($H_c = 1$ could not be selected due to unstable poles of the process, see [7]). The computation results for the predictive controller are given in the appendix.

Results

The arm-manipulator coordination problem with the predictive control law derived above was simulated on the computer in Matlab/Simulink environment. The desired movement trajectory was prescribed to be a step in the vertical direction. A time of two seconds was allowed for simulation. The simulation results are presented as follows: The commanded and the achieved object position are plotted in Figure 4. The arm force for the manipulation task is plotted in Figure 5, and the manipulator force is plotted in Figure 6.

From the simulation results it is seen that the manipulator is highly responsive to the movement initiated by the human arm. A small force applied by the arm results in a large matching manipulator response. Except for the difference in magnitude, the trajectories for F_a and F_m are almost identical. The manipulator thus emulates the arm effort with a high gain that is determined by the ratio of the manipulator stiffness to the arm stiffness.

The performance of the predictive controller would, however, depend on the unknown arm stiffness during any manipulation task. For simplicity, a low constant value of the arm stiffness was assumed in the computer simulations. In practice it would be highly desirable to observe/estimate the arm stiffness in real-time and accordingly update the controller parameters. A method to estimate the arm stiffness on-line using Recursive Least Squares algorithm was suggested in [11].

Conclusion

A model predictive control scheme for the arm-manipulator coordination problem is proposed. The simulation results indicate that such a strategy is effective in minimizing the arm effort to move an inertial object. The resulting object movement is smooth and requires minimum arm force to initiate the movement.

Acknowledgement

The first author acknowledges the support of the Ohio State University, Department of Electrical Engineering for the completion of this work.

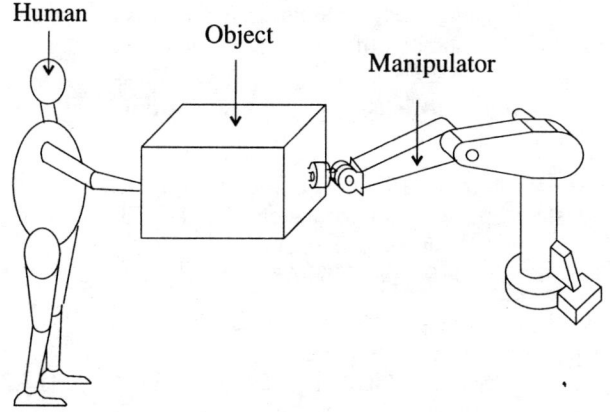

Fig. 1. The conceptual visualization of the arm-manipulator coordination problem: a human arm and a robot manipulator jointly handle an inertial object. (Reproduced from Ref. [3] with permission)

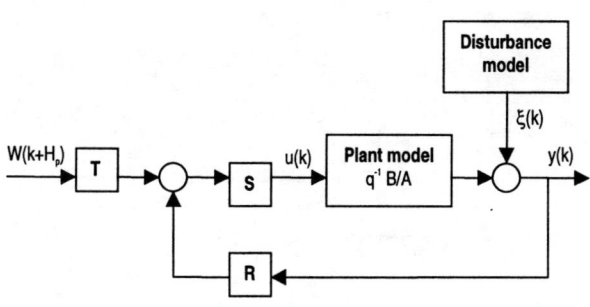

Fig. 2: Bock diagram representation of the two degree-of-freedom model-based predictive controller. T, R, and S represent polynomials in backward shift operator.

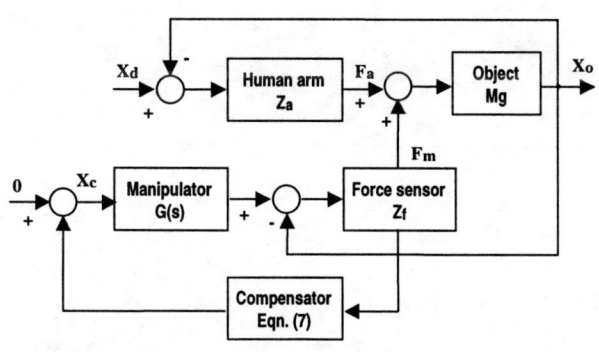

Fig. 3: Block diagram representation of the arm-manipulator coordination problem.

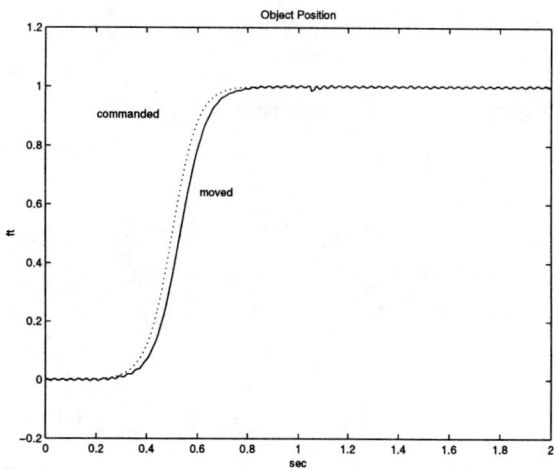

Fig. 4: The commanded and the achieved object trajectory during the manipulation task.

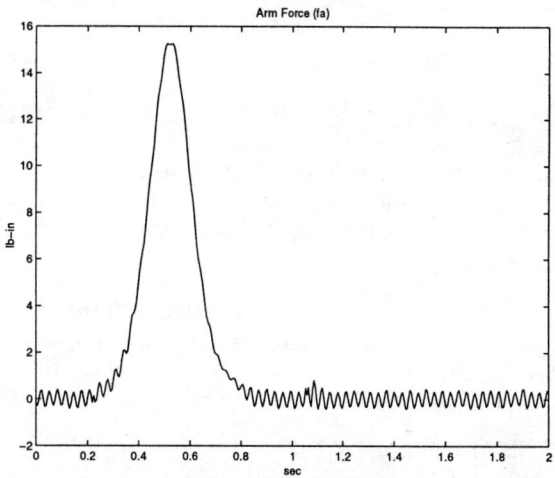

Fig. 5: The arm force applied to initiate the manipulation task.

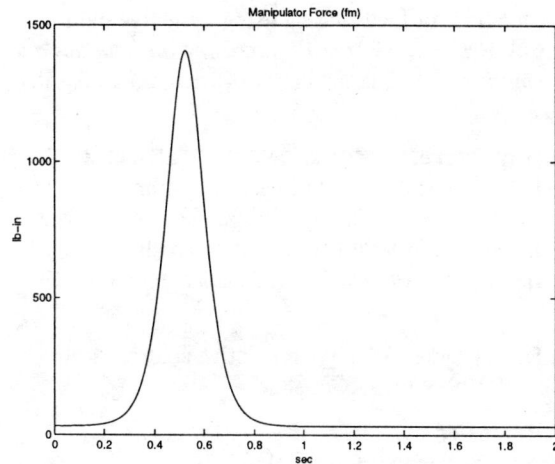

Fig. 6: The manipulator response measured by the wrist sensor during the manipulation task.

Appendix

The continuous-time plant model is given as:

n = 10000 0 100000

d = 1 25 10014 250250 40040

The discre-time plant model is given as:

b = 0 0.0050 -0.0050 -0.0049 0.0049

a = 1.0000 -3.9653 5.9062 -3.9162 0.9753

The polynomials used in the predictive control law for the plant (Fig. 1) are given as:

r = 1.0000 0.4191 -2.0456 -0.7800 1.4064

s = 123.8425 -302.0060 248.5781 -69.3824

t = 10.3213 -9.2892

References

[1] J.Y.S. Luh and Y.F. Zheng, Constrained relation between two coordinated industrial robots for motion control. *Int. J. Robotic Research,* Vol. 6, No, 3, Fall 1987, pp.60-70.

[2] Y.F. Zheng and J.Y.S. Luh, Optimal load distribution for two industrial robots handling a single object. *ASME J. Dynamic System, Meas. Control,* Vol. 111, June 1989, pp.232-237.

[3] O.M. Al-Jarrah and Y.F. Zheng, Arm-manipulator coordination for load sharing using compliant control. *Proceed. 1996 IEEE Int. Conf. on Robotics and Automation,* Minneapolis, Minnesota, 22-28 April, 1996, pp. 1000-1005.

[4] O.M. Al-Jarrah and Y.F. Zheng, Arm-manipulator coordination for load sharing using reflexive motion control. *Proceed. 1997 IEEE Int. Conf. on Robotics and Automation,* Albuquerque, New Mexico, 20-25 April, 1997.

[5] D.W.Clarke, C. Mohtaki and P.S. Tuffs, Generalized predictive control, Pt. I: The basic algorithm. *Automatica*, 23, No. 2, pp. 137-148, and Generalized Predictive Control. Part II, Extensions and interpretations. *Automatica*, 23, No. 2, 1987, pp. 149-160.

[6] D.W. Clarke, Adaptive predictive control. *A. Rev. Control,* Vol. 20, 1996, pp. 83-94

[7] R. Stoeterboek, *Predictive Control A Unified Approach,* Prentice Hall Int. (UK), 1992.

[8] K.J. Astrom and Wittenmark, *Computer Controlled Systems* (3rd Ed.), Prentice Hall Inc. 1997.

[9] Y. Xu and R.P.Paul, On position compensation and force control stability of a robot with compliant wrist. *Proceed. 1988 IEEE Int. Conf. on Robotics and Automation,* Philadelphia, Pennsylvania, April 14-19, 1988, pp. 1173-1178.

[10] H. Kazerooni, "Human-robot interaction via the transfer of power and information signals," IEEE Trans. On Systems, Man and Cybernetics, Vol. 20, No.2, March/April, 1990, pp. 450-463.

[11] K. Iqbal and Y.F. Zheng, "Predictive control application in arm-manipulator coordination," Proc. IEEE Int. Symp. on Intelligent Control, Istanbul, Turkey, July 16-18, 1997.

Part Orienting with a Force/Torque Sensor

Shawn Rusaw   Kamal Gupta   Shahram Payandeh

Experimental Robotics Laboratory
School of Engineering Science
Simon Fraser University
Burnaby, British Columbia, Canada

Abstract

This paper examines sensor-based orientation of polygonal shaped objects using a 6-axis force/torque sensor which is used to supply rotation and force information during the orientation process. Sensor-based algorithms utilizing rotation sense and force information are presented which offer improvements over the best sensorless orienting techniques, and are shown to be better than previous sensor-based techniques for parts with the same diameter value for every stable edge. Plans generated by our algorithm were tested and verified using a unique conveyor/robotic car testbed.

1 Introduction

A long term goal of the part orienting research community is the design of a *flexible part feeder* that can orient a large variety of parts without redesign for each new part. The flexible feeder will take a computer/CAD model of the part and automatically reconfigure based on a knowledge of the part mechanics and the part's interaction with its environment. Until recently, part orienting research has focused on orienting techniques that do not require sensors (see Section 2). The main justifications for this being (i) current techniques like bowl feeders are sensorless and (ii) sensorless apparatus do not have the added complications introduced by sensors like signal processing and extra time to process data. Since sensor-based orienting systems are more complicated to implement than sensorless techniques, we need to examine the potential benefits that outweigh this complexity.

In this paper we examine the use of a 6-axis force/torque sensor, a sensor commonly used in industry, in the part orienting process. This sensor has the advantage of being able to work in obscured or smoky environments, unlike a typical vision system. The force/torque sensor is used to detect rotation sense of the part and measure the forces exerted on the part during the orientation process. The sensor is mounted to a flat pusher (see Figure 2) and the part is oriented through a sequence of push-align operations, where the angle of rotation for each push-align operation is selected based on previous rotation sense and force readings.

The best sensorless orienting algorithms generate orienting plans no longer than $2n - 1$ steps [6], where n is the number of stable edges of the part. We shown that using only rotation sense, which is a simple binary measurement, reduces the worst case plan length to n. When force is used in conjunction with rotation sense, the worst case plan length is further reduced to $m + 1$, where m is the largest set of edges resulting in indistinguishable force measurements.

2 Related Work

Sensorless orientation of objects has been examined by Mason [11] who used pushing to reduce the uncertainty in an object's orientation. Peshkin and Sanderson [12] developed a planner for orienting parts with fences suspended above a conveyor belt. Peshkin and Goldberg [4] improved on this idea by designing curved fences which further reduce the uncertainty in part orientation to a specific single orientation. Rusaw, Gupta and Payandeh [15] examined the effect of allowing subsequent fences to have different coefficients of friction. Recently, a complete algorithm for planning $O(n^2)$ length curved fence assemblies has been presented by Berretty, Goldberg, Overmars and van der Stappen [3] for the frictionless case. Brost [5] showed how to plan parallel jaw grasping motions to orient a part in the presence of uncertainty. Goldberg [7] developed the complete backchaining algorithm to plan a series of parallel jaw motions to orient a part to symmetry. Akella,Huang, Lynch and Mason [2] have developed a system that orients an object on a conveyor using a single fence with 1 DOF. Akella [1] developed a planner that can take a part from any initial configuration to any final goal position and orientation using a sensorless sequence of pushes.

Sensor-based orienting was examined by Lynch, Maekawa and Tanie [10] who utilized the limit surface model [8] of quasi-static motion to develop a closed loop system that localized the position of a part. Jia and Erdman [9] developed a system to determine the orientation of a part by pushing with a round, tactile sensor equipped fingertip. Salvarinov and Payandeh [16] utilize a single joint, strain gauge equipped fence above a conveyor and detect the contact signature of a part interacting with the fence. Rusaw, Gupta and Payandeh [14] used the limit surface model to detect the final edge in contact with a pusher at the end of a linear normal push.

This paper is most closely related to the research of Akella [1]. Akella showed that utilizing sensor input during the orientation process reduces the number of manipulation steps required over the sensorless case. In Akella's work, the sensor readings corresponded to measurements of the part diameter at the stable orientations. So, the sensor reading is dependent only on the final orientation. In our work, the final state may be reached via clockwise or counterclockwise rotation, each leading to different sensor readings. For this reason, we make a distinction between Akella's resting ranges, actions ranges and representative

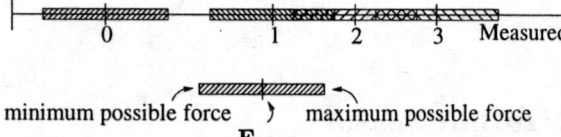

Figure 1: Distinguishability of force measurements. $\{1, 2, 3\}$ form an indistinguishable set.

actions and our *sensor resting ranges, sensor action ranges* and *sensor representative actions*.

Akella introduced the notion of *state distinguishability* to specify which stable edges cannot be distinguished from the diameter measurement. Rotation sense is not able to distinguish anything since all edges can result in both possible binary values (clockwise/counterclockwise). Although rotation sense cannot give any direct information regarding the final orientation after a particular push-align operation, it is still able to reduce the number of orienting steps over the sensorless case. Incorporating force information requires the concept of state distinguishability. Since different edges require different forces to bring them into alignment with the fence, final stable edges may be distinguished if the forces differ enough.

3 Assumptions

The orienting techniques outlined in the remainder of this paper require the following assumptions:

1. Parts are polygons. Since the pusher is flat, non-convex polygons are considered by using the convex hull.
2. Part vertices and center of mass are known.
3. Motion is quasi-static (conveyor moves slowly).
4. Coefficient of friction μ is constant at support surface.
5. Friction is described by Coulomb model.
6. There is zero friction between part and pusher.
7. All pushes are perpendicular to the conveyor motion.
8. The objects and fence are perfectly rigid.

4 Description of Sensor

The sensor used in this paper is an ATI 15/50 6-axis force/torque sensor. The force/torque sensor is used to determine the rotation sense and force during a push-align operation. Rotation sense is determined by observing the motion of the contact point as the part rotates from vertex contact to edge contact (see [14] for details). Our force/torque sensor is able to consistently determine rotation sense for edges that are greater than 2cm in length. Shorter edges are also detectable, but failures are more common due to sensor noise.

4.1 Distinguishability of Force and Rotation Sense Data

As the part rotates into a final stable state, both a rotation sense and a force measurement will be taken. Since different force measurements are possible for clockwise and counterclockwise rotation, distinguishability is dependent on the sense measurement. Figure 1 shows several ranges of force measurements corresponding to several final stable states. The states with force ranges that overlap are considered an indistinguishable set. The center of each range $\mathbf{F}_{ideal}$ is calculated based on the limit surface model (see [14] for details).

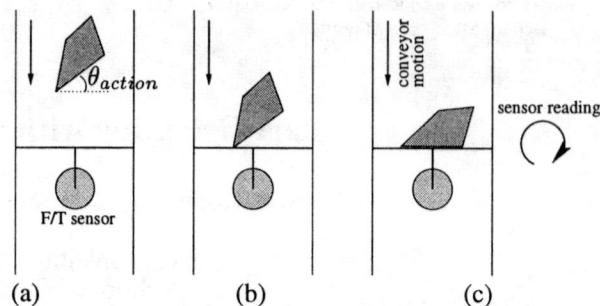

Figure 2: The push-align operation. (a) part is placed upstream after being rotated through some angle θ_{action}, (b) the part moves toward the fence and (c) the part rotates to a stable edge which is followed by a sensor reading

Final stable states are considered distinguishable if the range of possible force readings for the states do not overlap. The range of possible forces represents all possible force measurements given (i) sensor noise, (ii) uncertainty in physical parameter measurement and (iii) deviations from ideal uniform pressure distribution. Bounds for (i) can be determined from sensor noise. Bounds for (ii) are easily determined from measurement error when determining the constants—i.e. inclined plane angle measurement error when determining the coefficient of friction μ. Bounds for (iii) cannot be explicitly determined since the pressure distribution is indeterminate, and we are not aware of any studies that examine the *probability* of having any particular pressure distribution. Bounds for (iii) must be determined empirically based on experimentation. If the bounds are too small, leading to algorithm failures, simply increase the range until the algorithm works consistently.

5 Analyzing The Push-Align Operation

This section outlines the modifications made to Akella's algorithm [1] in order to facilitate searching with force and/or rotation sense data. Sections 5.1 through 5.3 discuss the modifications when using only rotation sense and Section 5.4 discusses incorporating force data with rotation sense

5.1 Sensor Resting Ranges

Definition 5.1 *A* **sensor resting range** *of a stable orientation is the set of initial orientations which have the same sensor reading after coming to rest in that stable orientation.* ∎

A sensor resting ranges is the range of initial angles θ that result in a given final state and sensor reading. The *sensor resting range diagram* (see Figure 3) is used to represent this information. $\{s_1, s_2, s_3\}$ are the final stable *state* (orientation) of the part. The stable orientation of a sensor resting range is indicated by an ×. The sensor resting range is delimited by a vertical line at each end. For rotation sense, the stable orientation coincides with a limit of the range. The range has a ↻ or ↺ to signify the sensor reading for that range. For a sensor measurement corresponding to the diameter, Akella's resting ranges are identical to the sensor resting ranges. There are twice as many sensor resting ranges when using rotation sense as opposed to diameter.

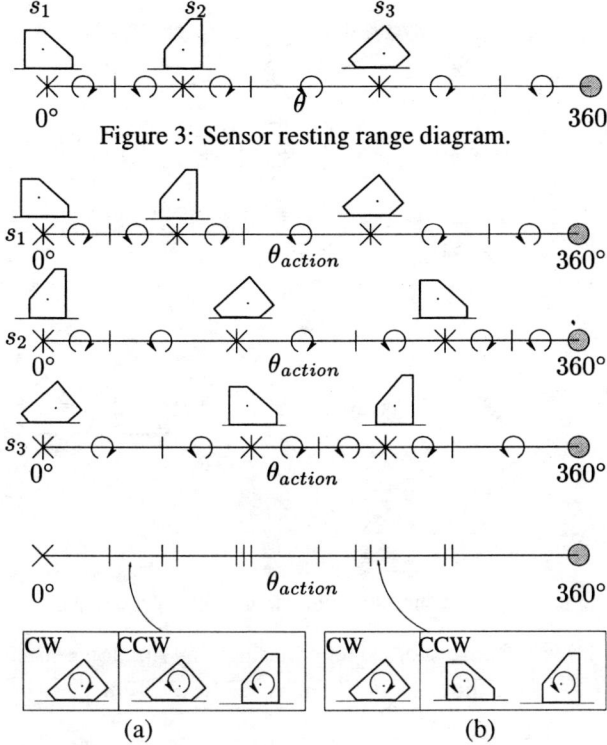

Figure 3: Sensor resting range diagram.

Figure 4: Overlap ranges for all sensor action ranges (top of figure)

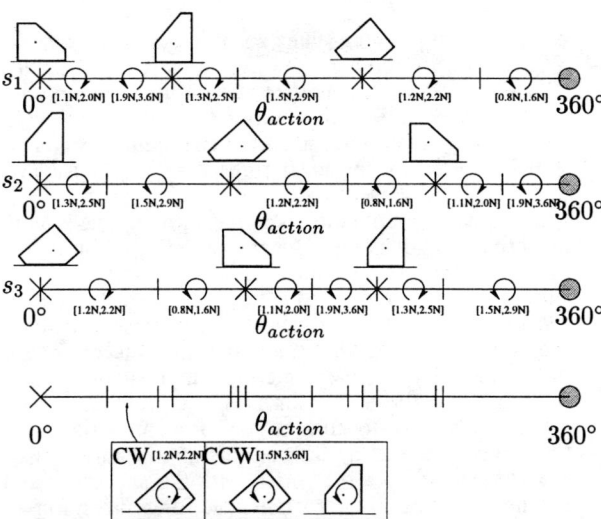

Figure 5: Sensor representative actions using both force and rotation sense.

5.2 Sensor Action Ranges

Definition 5.2 *A sensor action range is a contiguous interval of rotations that forms an equivalence class of actions and sensor readings. That is, every action that belongs to a sensor action range results in the same final state and sensor measurement for a given initial state.* ∎

The sensor action ranges represent the result of a push-align operation, and are constructed from the sensor resting ranges in the same manner action ranges are constructed from resting ranges in [1]. Note however that the final stable orientation may be a limit of the sensor action range. Figure 4 shows an example of the sensor action ranges. $\{s_1, s_2, s_3\}$ are the initial states (each line corresponds to an initial state), θ_{action} is the angle of rotation and the polygons above the line show the final orientation.

5.3 Sensor Representative Actions

We now examine the effect of an action on a group of initial stable states. Merging the endpoints of k sensor action ranges results in a set of *overlap ranges* [1]. For a given set of initial orientations, any action chosen from these overlapping ranges will result in the same set of final orientations and sensor data. To stay with the tone of previous sections we will call the center of each overlapping range a *sensor representative action*. Figure 4 (bottom) shows the set of overlap ranges with sensor information for all initial stable states.

5.4 Incorporating Force Data

In addition to rotation sense, the sensor action ranges may further modified to contain force information. Since the force measurements are dependent on the rotation sense, the sensor action ranges have the same limits as when using only rotation sense. Each sensor action range now has a corresponding range of possible force sensor readings to go along with the rotation sense data. Figure 5 shows the sensor representative actions.

6 The Sensor-Based Planner

We have modified Akella's AND/OR tree-based algorithm, making it suitable for use with rotation sense and force data. The algorithm must be run offline once for each part, generating a plan which can be followed based on the sensor data collected during the orientation process.

Each level of the search tree corresponds to a push-align operation. The root of the tree is the set of all possible orientations of the part. A node in the tree corresponds to the set of possible states after the sequence of push-align operations and sensor readings up to that node. All links are AND links (a solution exists only if all child nodes are solved) corresponding to the set of child nodes generated from the same sensor representative action θ_{action}. Each node consists of a set of nodes with different and/or distinguishable sensor readings.

The following descriptions make reference to sensor reading being 'same/indistinguishable' or 'different/distinguishable'. These terms must be used in the context of the particular sensor data being used. When using only rotation sense, issues of distinguishability are not used, so use 'same/different'. When using force and rotation sense together, use 'indistinguishable/distinguishable'. The following set of points (1-4) describes how to generate the branches from a given base node.

1. Generate a list of sensor representative actions for the constituent states of the base node. Steps 2-4 will be done for every representative action θ_{action} in this list to generate child nodes with same/indistinguishable sets of final states.
2. Apply the sensor representative action θ_{action} to the constituent states of the base node to obtain the final possible state and sensor readings.

2547

3. Sort the final possible states according to sensor value. When using rotation sense alone, this results in a list of states with the first half corresponding to clockwise and the latter half corresponding to counterclockwise. When using rotation sense and force together, each half will be sorted by the ideal force data $\mathbf{F}_{ideal}$ for that half's rotation sense.
4. We now group the states into nodes with same/indistinguishable sensor readings and remove duplicate states from each node. The first node is initialized to have the first state on the list. Traversing the list, if a state is identical to its successor, discard the successor. If the state and the successor are not identical but have same/indistinguishable sensor readings add the successor to the current node and set the successor to current state for comparison. If the successor has a different/distinguishable sensor reading, initialize a new node with the successor and set the successor to the current state for comparison.

The search process utilizes an OPEN and CLOSED list. OPEN corresponds to nodes which have been generated but not fully explored (all child nodes created) and CLOSED is the list of nodes which have been fully explored. Initialize the OPEN list to contain the root node and the CLOSED list to empty. The first push-align operation and sensor reading leads to the sets of final stable states with different/distinguishable sensor readings. If a node has a single state it is labeled solved. The search is breadth-first and proceeds by popping a node from the head of OPEN list and generating the set of child nodes for the constituent states (steps 1-4). Each node that is generated is placed at the tail of the OPEN list if it is not already in OPEN or CLOSED. When a node is solved, pass this information to it's parent who can then check if itself is now solved. A node is solved if it has a single state or if all children are labeled solved. Fully explored nodes are placed on the CLOSED list. The search succeeds when the root is labeled solved and fails if OPEN becomes empty and root is not solved. An example plan when using only rotation sense is given in Figure 6 and an example plan when using rotation sense and force is given in Figure 7.

The algorithm is complete and has a time complexity of $O(n^4 2^n)$ [1]. For a part with n stable states, there are C_k^n subsets with k states. Worst case, the algorithm will have to examine every representative action for every set of k states. When using force and/or rotation sense, the algorithm takes about twice as long as when using diameter—there are twice as many branches to examine.

7 Naive Shortest Plan Length

The *naive shortest plan length* is the worst case shortest plan for a part. We refer to the plan length as naive shortest length for a part since shortest length plans for many parts are typically shorter than the naive shortest length. The straightforward proofs are detailed in [13].

7.1 Using Only Rotation Sense

Table 1 gives the naive shortest plan length for parts with n stable edges and p periods in their radius function. For the asymmetric case, after the first push-align operation, the number of unknown states is equal to n. Each push-align operation reduces the number of unknown states by at least one, leading to plan no longer than n.

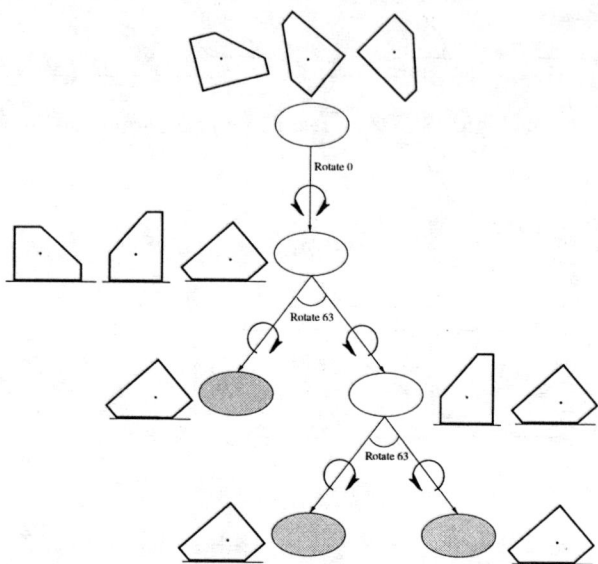

Figure 6: Plan generated when using rotation sense data.

Table 1: Naive shortest length using only rotation sense.

Radius function	naive shortest length
asymmetric	n
symmetric	n/p

7.2 Collecting Valid Data

On the first push, if the part contacts the fence on a stable edge, no contact location step will be observed, and no valid sense or force data will be available. When planning using only rotation sense, collecting valid data in the first push is not important, since rotation sense cannot determine anything in the first step anyway—the first node after the root node is always the set of all stable states. However, if no valid force data is taken, an additional push-align operation, corresponding to a small rotation, may be required to ensure collection of valid force data.

7.3 Using Rotation Sense and Force

Given a part with n stable edges, let m be the largest indistinguishable set of final stable states. For example, for a part with $n = 5$, if clockwise rotation leads to indistinguishable sets of size 2 and 3 (based on force), and counterclockwise rotation leads to indistinguishable sets of size 4 and 1, then $m = 4$. After the first push-align operation, the worst case number of unknown states is m. Table 2 gives the naive shortest length for orienting parts with force and sense data. Note the extra step required.

8 Experimental Trials

8.1 Experimental Apparatus

An experimental testbed was built to test the algorithm. The conveyor was constructed from a pair of hand cranked mylar rollers from an overhead projector (see Figure 8(a)). An ATI 15/50 force/torque sensor with a T-shaped fence was used to collect rotation sense and force data (see Figure 8(b)).

A cable driven car guided by tracks is used to perform the push-align operations (see Figure 8(c)). Polygons are

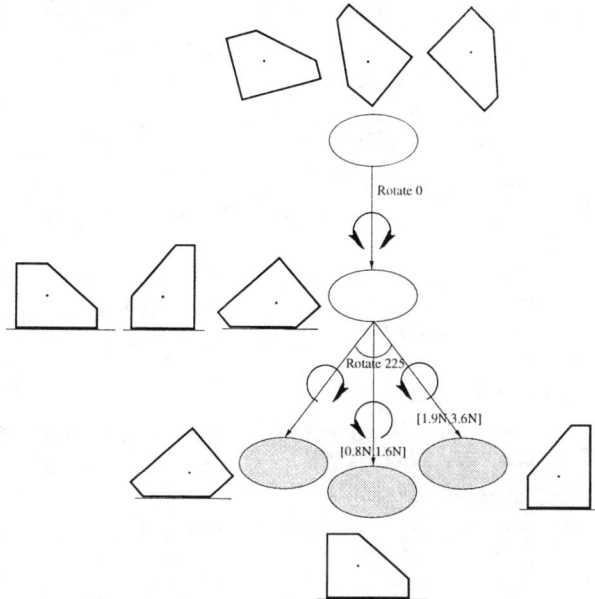

Figure 7: Plan when using force and rotation sense data.

Table 2: Naive shortest length using both force and sense.

Radius function	naive shortest length
asymmetric	$m + 1$
symmetric	$m/p + 1$

picked up and rotated using a rotating suction cup on the bottom of the car (see Figure 8(d)). The suction cup is attached to a vacuum pump, and turned on/off using a relay driven valve. The suction cup is shaped like an accordion, so when the air is removed, it actually picks the part several centimeters off the conveyor. To keep the part level and adjust the lift height, a set of four limit screws are placed around the suction cup (see Figure 8(f)).

The parts were made of aluminum, and had masses on the order of 200 grams. Our force/torque sensor has a resolution of 0.04 N/bit, so these parts only register about 20 bits of force, which is unusable since the contact location noise is extremely high. In order to provide useful force readings, a 2 kg mass was added to the top of the parts. In order to pick this mass up with the suction cup, a flat surface was attached to the top of the mass. Figure 8(e) shows a part with extra mass attached and Figure 8(f) shows the suction cup picking up the part and extra mass—notice how the limit screws adjust the height and help level the part. In Akella's work, a suction cup picked up the parts directly, which we could have done if we had used a sensor with finer resolution.

8.2 Experiments using Only Rotation Sense

Plans generated by the algorithm were tested using the parts shown in Figure 9. For each part, 20 trials were run. The number of successful trials is given in the table in Figure 9. Out of 80 trials, only two failures were observed. One failure occurred when the conveyor became stuck and its controller compensated by jerking it. The subsequent vibrations of the force/torque sensor lead to an invalid step detection. The second failure occurred when the suction cup failed to pick up the part. After the height of the cup

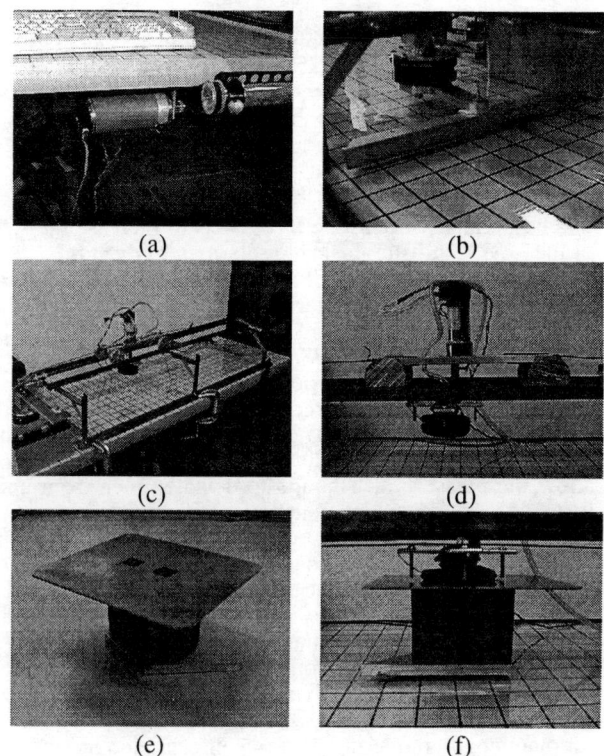

Figure 8: Photographs of experimental apparatus.

Polygon	Number Successful
(a)	18/20
(b)	20/20
(c)	20/20
(d)	20/20

Figure 9: Experiments using only rotation sense.

was adjusted, this type of failure was not observed again.

8.3 Experiments using Rotation Sense and Force

For each push-align operation the rotation sense is determined from the direction of the contact location step and the force is that just *before* the step occurs. Forces during the step must not be used as they don't fit our model of vertex contact.

To determine the coefficient of friction between the conveyor and the parts, the force/torque sensor was used to measure the normal force between the fence and the part during a 50 second push. The resultant coefficient of friction was in the range [0.30,0.44]. When determining the ranges of possible force measurements, to establish the distinguishability of final stable states, we need to include the effect of measurement uncertainty, sensor noise and deviations from a uniform pressure distribution. The range of friction values adds ±20% to the range, sensor noise adds about ±2-5% (0.1N/5N) and deviations from limit surface model is placed at another ±20%. In terms of planning,

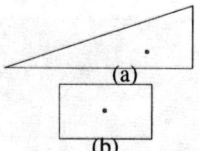

Polygon	Number Successful
(a)	19/20
(b)	19/20

Figure 10: Experiments using rotation sense and force.

this large range of possible forces acts to increase the size of the largest indistinguishable set, often to $m = n$.

The two parts shown in Figure 10 were used for the experimental trials. When a force range of ±40% is used, the shortest length plans are no better than when using only rotation sense (assuming all pushes collect valid data). For this reason, the allowable deviation from a uniform pressure distribution was lowered until the shortest plan length was improved. For part (a) this range was ±30% and for part (b) this range was ±20%. Since these ranges do not allow the pressure distribution to deviate much from uniform, one expects to encounter a few failures.

For each part, 20 trials were run using both rotation sense and force information. The results of the tests is shown in the lower table of Figure 10. The single failures for each part were caused by a force measurement that was slightly larger than any of the possible ranges at that stage of the plan predicted. The forces could have easily been accommodated with a slightly larger force range. However, increasing the force ranges would have increased the shortest plan length, rendering the force information redundant.

9 Conclusions and Future Work

This paper presented a sensor based algorithm that generates orienting plans with less steps that the best sensorless techniques. When using rotation sense as the sensor data, shortest plan length is no longer than n, the number of stable edges. Using force data in conjunction with the sense data reduces the worst case shortest length plan to $m + 1$ steps, where m is the size of the largest indistinguishable set of final stable states, based on force data.

Compared to Akella's [1] sensor based orienting work, our shortest length plans are better when all final states of the part have the same diameter value—the diameter sensor provides no information. Figure 11 gives optimal plans for orienting a number of parts with symmetric shapes and offset centers of mass using diameter (sensorless) and rotation sense.

9.1 Future Work

The force/torque sensor can provide several other pieces of useful data that may act to further reduce the number of manipulation steps. In the case of parts with unstable edges, the number of steps in the contact location profile may be useful. As well, the 'height' of the steps in the contact location offer insight into the length of the corresponding edge.

Better planning algorithms using rotation sense and force should be developed. The additional algorithms presented by Akella [1] are a starting point, but algorithms with improvements directly related to the data type may also be a possibility.

Some formalization of the information provided by different types of sensors should be examined. For example, a stable edge can have only a single range of possible diameter values but can have multiple rotation sense values

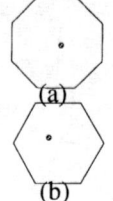

part	shortest plan length	
	diameter	sense
(a)	8	4
(b)	6	4

Figure 11: Case where sense outperforms diameter.

and/or force ranges. As well, diameter and force are able to distinguish between stable edges but rotation sense provides no such information yet is still better than the sensorless case.

Acknowledgments

This work was supported by the NSERC grants of Drs. Gupta and Payandeh.

References

[1] S. Akella. *Robotic Manipulation for Parts Transfer and Orientation: Mechanics Planning and Shape Uncertainty*. PhD thesis, Carnegie Mellon University, 1996.

[2] S. Akella, W. H. Huang, K. M. Lynch, and M. T. Mason. Sensorless parts orienting with a one-joint manipulator. In *Proc. IEEE Int. Conf. on Robotics and Automation*, pages 2383–2390, 1997.

[3] R. P. Berretty, K. Goldberg, M. H. Overmars, and A. F. van der Stappen. Algorithms for fence design. In *Proc. 4th Int. Workshop on Algorithmic Foundations of Robotics (pre-prints)*, 1998.

[4] M. Brokowski, M. A. Peshkin, and K. Goldberg. Curved fences for part alignment. *ASME J. of Mechanical Design*, 117(1):27–34, 1995.

[5] R. C. Brost. Automatic grasp planning in the presence of uncertainty. *Int. J. of Robotics Research*, 7(1):3–17, 1988.

[6] Y. B. Chen and D. J. Ierardi. Oblivious plans for orienting and distinguishing polygonal parts. In *Proc. 4th Canadian Conf. on Computational Geometry*, 1992.

[7] K. Y. Goldberg. Orienting parts without sensors. *Algorithmica*, 10(2/3/4):201–225, 1993.

[8] S. Goyal, A. Ruina, and J. Papadopoulos. Planar sliding with dry friction. part 1. limit surface and moment funtion. *Wear*, 143:307–330, 1991.

[9] Y. B. Jia and M. Erdmann. Pose from pushing. In *Proc. IEEE Int. Conf. on Robotics and Automation*, pages 165–171, 1996.

[10] K. M. Lynch, H. Maekawa, and K. Tanie. Manipulation and active sensing by pushing using tactile feedback. In *Proc. IEEE/RSJ Int. Conf. on Intelligent Robots and Systems*, pages 416–421, 1992.

[11] M. Mason. Mechanics and planning of manipulator pushing operations. *Int. J. of Robotics Research*, 5(3):53–71, 1986.

[12] M. A. Peshkin and A. C. Sanderson. Planning robotic manipulation strategies for workpieces that slide. *IEEE J. of Robotics and Automation*, 4(5):524–531, 1988.

[13] S. Rusaw. Orienting polygons: Experiments in automated part orienting. M.A.Sc. Thesis, Simon Fraser Univ., 1998.

[14] S. Rusaw, K. Gupta, and S. Payandeh. Determining polygon orientation using model based force interpretation. In *Proc. IEEE Int. Conf. on Robotics and Automation*, 1998.

[15] S. Rusaw, K. Gupta, and S. Payandeh. Orienting polygons with fences over a conveyor belt: Empirical observations. In *To Appear: Proc. IEEE/RSJ Int. Conf. on Intelligent Robots and Systems*, 1998.

[16] A. Salvarinov and S. Payandeh. An application of a 1 dof fence for active part detection, manipulation and alignment on a conveyor belt. In *Proc. IEEE Int. Conf. on Robotics and Automation*, pages 428–434, 1997.

Toppling Manipulation

Kevin M. Lynch
Laboratory for Intelligent Mechanical Systems
Department of Mechanical Engineering
Northwestern University
Evanston, IL 60208-3111

Abstract

This paper describes a robotic manipulation primitive called toppling—knocking a part over. We derive the mechanical conditions for toppling, express these as constraints on robot contact locations and motions, and describe an application of toppling to minimalist parts feeding of 3D objects on a conveyor with a 2 joint robot.

1 Introduction

In the spirit of minimalist robotics, we are studying a manipulation primitive called *toppling*. Toppling occurs when a robot knocks a part over to a new face. This work has two motivations: (1) it adds to the repertoire of manipulation primitives (which includes grasping, pushing, throwing, tapping, and tumbling; see Lynch and Mason [9] for an overview) available to a robot in manipulation planning; and (2) it can be accomplished using very simple, low-degree-of-freedom robot motions. One natural application for toppling is reorienting parts on conveyors. Since the conveyor provides linear motion for the part, the toppling "robot" can in fact be a fixed overhang (for example, the overhangs above a bowl feeder track). Thus robot motion is reduced to fixed automation. When combined with previous work on orienting planar parts on conveyors (Akella *et al.* [3]; Peshkin and Sanderson [14]; Brokowski *et al.* [6]; Wiegley *et al.* [20]), toppling permits full 3D parts feeding on a conveyor.

In this paper we study the mechanics of toppling and derive an algorithm for finding the set of contact points on a part from which it can be toppled to a new face. The results of this algorithm provide the basis for automatic motion planning to find a sequence of topples to a desired goal face, or for finding a sequence of fixed overhangs. Based on our results, we have constructed the 2JOC, a two joint robot which can position and orient 3D parts on a moving conveyor (Figure 1).

This work is inspired by K. Goldberg's [8] suggestion of the utility of toppling in minimalist manipulation and parts feeding. Our mechanics analysis is related to Erdmann's [7] work on two-palm nonprehensile manipulation. Erdmann constructed a planner to find motions for two independent three degree-of-freedom planar palms manipulating an object without grasping it. The planner uses a quasistatic model of frictional mechanics and finds motions that utilize a variety of slipping and rolling motions between the object and the palms. In our work we consider a much simpler set

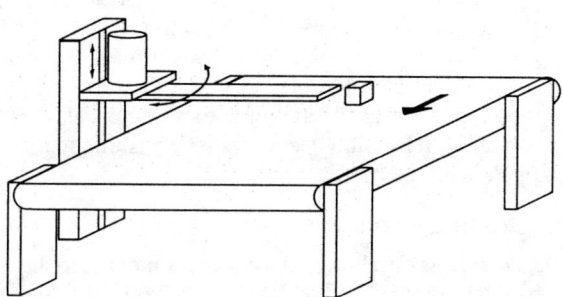

Figure 1: The 2JOC feeds parts on a conveyor by pushing and toppling.

of robot motions and focus on the conditions for inducing toppling of an object on a moving conveyor.

Other related work includes pivoting and tumbling parts using a robot hand (Brock [5]; Sawasaki *et al.* [16]; Aiyama *et al.* [2]; Trinkle [18, 19]). By grasping a part with a pivoting gripper that allows the part to reorient under gravity, Rao *et al.* [15] showed that a 4 DOF SCARA robot could induce out-of-plane rotations. Zhang and Gupta [21] recently presented an approach to orienting parts on a conveyor by allowing them to fall over steps. Toppling can also be effected by controlled acceleration of the conveyor. A similar idea for toppling parts was presented by Singer and Seering [17].

Our 2JOC robot controls the position and orientation of parts on a conveyor by pushing and toppling. The motivations are closely related to those of Bicchi and Sorrentino [4] who demonstrated control of the position and orientation of an object by rolling it between two grasping palms. Because spatial rolling constraints are nonholonomic, the three degree-of-freedom robot is able to control the object's position and orientation. With the 2JOC, the conveyor-plane position and orientation of a polyhedral part is controlled by pushing the part over the conveyor, while Marigo *et al.* [10] have studied the set of reachable configurations for polyhedra on a planar surface when the parts can only roll about edges.

In Section 2 we derive the contact conditions for toppling and the toppling transition directed graph, which indicates the new resting face once the part has toppled. In Section 3 we describe some applications of toppling to parts feeding, including the 2JOC. We conclude in Section 4.

2 Toppling

Toppling consists of two phases: rolling (Section 2.1) and settling (Section 2.2). During rolling the robot pushes the part up onto a *toppling edge*, which is perpendicular to the motion of the conveyor, until the center of mass of the part is directly above the edge. During settling the part falls under gravity, lands on a new face, and perhaps continues to roll onto another face before coming to rest.

We define two planes: the *toppling plane* and the *conveyor plane*. The toppling plane is a plane orthogonal to the toppling edge. In our analysis of toppling we project the part onto this plane, and the toppling edge projects to a toppling, or pivot, vertex. We assume the projection of the part to the toppling plane is polygonal. The conveyor plane is the plane of the conveyor, and it is orthogonal to the toppling plane. All toppling analysis occurs in the toppling plane; the conveyor plane is relevant when we include pushing motions in this plane with the 2JOC.

2.1 Rolling Conditions

The part rests on a horizontal conveyor moving to the right on the page (Figure 2). We define a frame fixed to the conveyor with origin at the pivot vertex of the part (which moves with the conveyor) with the x-axis aligned with the direction of motion of the conveyor and the y-axis vertical. The center of mass of the part in this frame is at a distance r from the origin at an angle η. The friction coefficients μ_c and μ_f correspond to friction between the part and the conveyor and between the part and the fence, respectively. The corresponding friction cone half-angles are $\alpha_c = \tan^{-1}\mu_c$ and $\alpha_f = \tan^{-1}\mu_f$.

The fence contacts a part edge with one endpoint at (x,y) in the conveyor frame. The angle of the edge is ψ from (x,y) and the inward-pointing contact normal for the edge is at $\psi + \pi/2$. The friction cone between the part and the fence is bounded by the angles of the left friction cone edge $\beta_l = \psi + \pi/2 + \alpha_f$ and the right friction cone edge $\beta_r = \psi + \pi/2 - \alpha_f$.

The question is, what fence contact points along this edge will result in the part initially rolling over the pivot vertex?

The key construction is shown in Figure 2. Draw a vertical line through the center of mass, extend the right edge of the friction cone at the pivot until it intersects this line, and extend the left edge of the friction cone backward until it intersects this line. This defines a triangle with vertices P_1 at $(r\cos\eta, (r\cos\eta)/\mu_c)$, P_2 at $(0,0)$ (the pivot), and P_3 at $(r\cos\eta, (-r\cos\eta)/\mu_c)$ in the conveyor frame. If the fence is rigid, and the contact force the fence applies to the part makes positive moment about every point in this triangle (i.e., the contact force passes around the triangle in a counterclockwise fashion), then the only quasistatic solution is that the part rolls about the pivot vertex.[1] To guarantee rolling, every force in the fence friction cone must make positive moment about every point in the $P_1P_2P_3$ triangle. In Figure 2, the fence friction cone shown barely satisfies this condition.

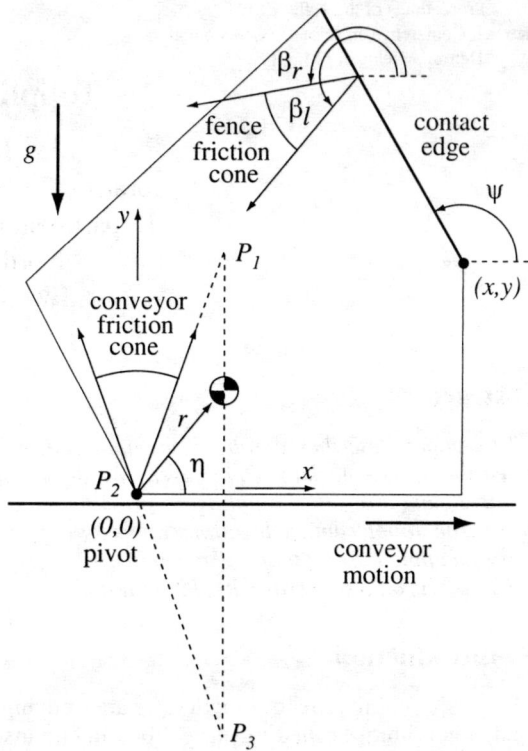

Figure 2: Notation for rolling.

The contact friction cone in Figure 2 marginally satisfies the rolling condition, and it is apparent that any higher contact point will also satisfy the condition. We can state this formally. Parameterize the contact edge by w, so that points on the contact edge are given by $(x,y) + w(\cos\psi, \sin\psi)$, $w \in [0, w_{max}]$. Then if a contact force (any force with a positive component in the direction of the inward-pointing contact normal) through $(x,y) + w_0(\cos\psi, \sin\psi)$ makes positive moment about a point P, it is easy to show that any other parallel force through $(x,y) + w(\cos\psi, \sin\psi)$, $w > w_0$, also makes positive moment about P. Therefore, if contact at w_0 causes rolling, then contact at $w > w_0$ also causes rolling. Thus the construction of Figure 2 confirms these intuitive properties of rolling:

- "higher" fence contacts tend to produce rolling, while lower contacts result in slipping on the conveyor;
- a larger conveyor friction coefficient μ_c results in a smaller $P_1P_2P_3$ triangle, increasing the set of contact points that produce rolling;
- a center of mass further to the left results in a smaller $P_1P_2P_3$ triangle, increasing the set of contact points that produce rolling.

The construction also shows that the height of the center of mass plays no role in the quasistatic, dry friction rolling conditions.

The analysis of Figure 2 only addresses the instantaneous initial condition for rolling. As the part rolls, it may become wedged or begin slipping on the conveyor. To analyze

---

[1] We state this without proof. See (Mason [11]) for related examples.

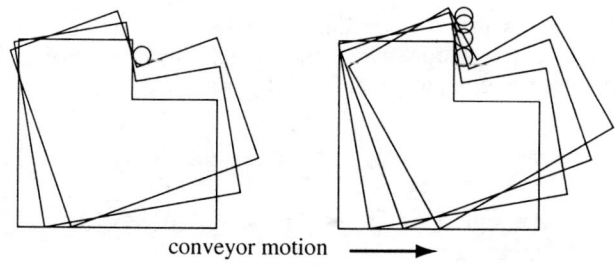

Figure 3: The part on the left becomes wedged or begins slipping with a fence of fixed height. The fence on the right lifts up to maintain a constant contact point on the part, allowing it to topple over.

the gross motion of the part after it begins rolling, we consider two models of the motion of the fence. In the first model, the fence "complies" to the shape of the part in a position-controlled manner, so that the contact point on the part remains constant during rolling. In the second model, the fence remains motionless. The first model is easier to analyze, and it increases the set of parts that can be toppled. The second model requires no motion by the fence.

2.1.1 Position-Controlled Fence

To simplify analysis and to be able to topple parts that might otherwise slip or become wedged, we raise or lower the fence so *the contact point on the part remains constant* as the part rolls. The velocity of the conveyor is known, and given the geometry and initial position of the part, the fence can simply "comply" to the shape of the part in a position-controlled manner. The fence maintains contact until the angle η to the center of mass becomes greater than $\pi/2$. The part then falls to a new stable edge. See Figure 3.[2]

The goal is to find contact points that maintain the rolling condition as the center of mass rolls from its initial angle η_i to its final angle $\eta_f = \pi/2$, at which point the part topples over. As the part rolls, the fence friction cone moves in the conveyor frame. (The $P_1 P_2 P_3$ triangle also shrinks.) As a result, a contact that causes rolling at η_i may not cause rolling as η increases—the part may begin to slide or wedge. So, for a given pivot vertex and contact edge, we must find w_0 such that all contact points $w > w_0$ result in rolling at all angles $\eta \in [\eta_i, \eta_f]$.

Consider the top triangle vertex P_1, at $(r\cos\eta, (r\cos\eta)/\mu_c)$. At $\eta = \eta_i$ the contact edge is at an angle ψ from the vertex at (x, y). Consider the right edge of the fence contact friction cone, at an angle β_r. We would like to find the function $w_{1r}(\eta)$ which gives the edge contact point where the right edge of the friction cone passes exactly through P_1. We define w_{1r}^* to be the maximum value of $w_{1r}(\eta)$ for $\eta \in [\eta_i, \eta_f]$. Then the positive moment rolling condition for the triangle vertex P_1 and the right edge of the friction cone is satisfied for all contact points $w > w_{1r}^*$. (Similarly we can define w_{1l}^* for vertex P_1 and the left edge

of the friction cone; for vertex P_2 we have w_{2r}^*, w_{2l}^*, and for vertex P_3 we have w_{3r}^*, w_{3l}^*.)

The line of the contact edge during rolling can be expressed as a function of the angle η and the contact parameter w:

$$(x\cos\Delta\eta - y\sin\Delta\eta + w\cos(\psi + \Delta\eta),$$
$$x\sin\Delta\eta + y\cos\Delta\eta + w\sin(\psi + \Delta\eta)), \quad (1)$$

where $\Delta\eta = \eta - \eta_i$ is the amount of rolling the part has undergone. The contact force is at an angle $\beta_r + \Delta\eta$, and the line of action of the force through the triangle vertex P_1 can be parameterized by v:

$$(r\cos\eta + v\cos(\beta_r + \Delta\eta), (r\cos\eta)/\mu_c + v\sin(\beta_r + \Delta\eta)). \quad (2)$$

We solve for the contact point that provides a force through P_1 by equating (1) and (2) and solving for w:

$$w_{1r}(\eta) = (2\mu_c y \cos\beta_r - r\cos\lambda - r\cos v - 2\mu_c x \sin\beta_r$$
$$+ \mu_c r \sin\lambda + \mu_c r \sin v)/2\mu_c \sin(\beta_r - \psi), \quad (3)$$

where $\lambda = \beta_r - \eta_i$ and $v = \beta_r + 2\eta - \eta_i$. A necessary condition for $w_{1r}(\eta)$ to reach its maximum is $dw_{1r}(\eta)/d\eta = 0$:

$$\frac{dw_{1r}(\eta)}{d\eta} = \frac{2\mu_c r \cos v + 2r\sin v}{2\mu_c \sin(\beta_r - \psi)} = 0. \quad (4)$$

Solving (4) for η, we get $\eta_{1r} = (\tan^{-1}(-\mu_c) - \beta_r + \eta_i)/2$. Then w_{1r}^* must occur at either η_i, η_f, or η_{1r}. To find w_{1r}^*, we need only evaluate (3) at a discrete set of angles.

Similarly for the bottom triangle vertex P_3, we get

$$w_{3r}(\eta) = (2\mu_c y \cos\beta_r + r\cos\lambda + r\cos v - 2\mu_c x \sin\beta_r$$
$$+ \mu_c r \sin\lambda + \mu_c r \sin v)/2\mu_c \sin(\beta_r - \psi), \quad (5)$$

$$\frac{dw_{3r}(\eta)}{d\eta} = \frac{2\mu_c r \cos v - 2r\sin v}{2\mu_c \sin(\beta_r - \psi)} = 0, \quad (6)$$

and $\eta_{3r} = (\tan^{-1}(\mu_c) - \beta_r + \eta_i)/2$. For the triangle vertex P_2 (the pivot point):

$$w_{2r} = \frac{y\cos\beta_r - x\sin\beta_r}{\sin(\beta_r - \psi)}, \quad (7)$$

and w_{2r} (hence w_{2r}^*) is independent of η.

Substituting β_l for β_r in the equations above, we find $w_{1l}^*, w_{2l}^*, w_{3l}^*$. Define w_{max}^* to be the maximum of the six values $w_{\{1,2,3\}\{r,l\}}^*$. If w_{max}^* is greater than w_{max} defining the end of the edge segment, then there are no toppling contacts on this edge. Otherwise, the range of toppling contacts is given by the range $(\max(0, w_{max}^*), w_{max})$.

We have implemented an algorithm in Lisp which, for each stable resting configuration of the polygon, finds the toppling contacts on each edge. An example is shown in Figure 5. As expected, increasing conveyor friction increases the range of toppling contacts.

This analysis ensures that any force inside the fence friction cone causes rolling, and is therefore conservative. Instead, we could simply verify that, at all times, there exists

---

[2] If the next clockwise edge contacts the conveyor before η reaches $\pi/2$, that edge is unstable. In this case the next vertex clockwise on the part's convex hull becomes the pivot vertex. For simplicity, we will ignore this case.

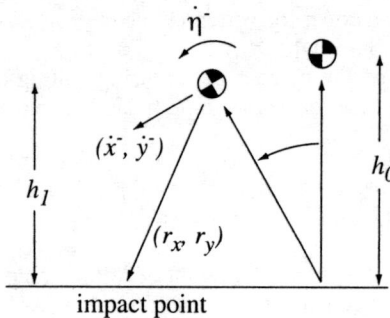

Figure 4: Notation for settling analysis.

a contact force inside the fence friction cone that produces rolling. The fence could move such that it barely slips on the part as it moves; the contact point is not quite maintained. This allows the fence to choose a force at the edge of its friction cone. Therefore, the rolling analysis does not have to be satisfied for all possible forces in the friction cone, but just for the chosen edge of the friction cone.

2.1.2 Fixed Fence

If the fence remains a fixed height as the part moves on the conveyor, we must find a range of fence heights y that guarantees toppling. To do that we again equate (1) and (2) for a particular contact edge, but now solve for $y(\eta)$ instead of $w(\eta)$. There is no closed-form solution to $dy(\eta)/d\eta = 0$, however, so the minimum y value that ensures toppling must be determined numerically.

The following conditions must also be satisfied: the fence cannot lose contact with the part as it rolls, and rolling must continue if the fence switches contact to a new edge during rolling (the part cannot wedge in a concavity or begin to slip).

2.2 Settling

When the center of mass passes the vertical with respect to the pivot point, the part begins to free fall. The part may simply come to rest on the next edge, or it may continue past this edge and come to rest on a subsequent edge.

The part falls like a pendulum, as shown in Figure 4, until impacting at the next vertex clockwise of the pivot vertex on the part's convex hull. The pre-impact linear velocity of the part's center of mass is $(\dot{x}^-, \dot{y}^-)$, and the angular velocity is $\dot{\eta}^-$, given by

$$\dot{\eta}^- = \sqrt{\frac{2mg(h_0 - h_1)}{I_p}},$$

where m is the mass of the part, g is the gravitational constant, and h_0 and h_1 are the height of the center of mass above the conveyor at the beginning and end of the free fall, respectively. I_p is the inertia of the part about the pivot, where $I_p = m(\rho^2 + h_0^2)$, and ρ is the radius of gyration of inertia of the part measured about its center of mass. The vector from the center of mass to the impact vertex is (r_x, r_y).

To determine the settling edge, we assume a perfectly plastic impact between the impact vertex and the conveyor.

This places three constraints on the post-impact velocity. The first two are kinematic, and the third indicates that the impulse passes through the impact point:

$$\dot{x}^+ = r_y \dot{\eta}^+$$
$$\dot{y}^+ = -r_x \dot{\eta}^+$$
$$r_x(\dot{y}^+ - \dot{y}^-) - r_y(\dot{x}^+ - \dot{x}^-) = \rho^2(\dot{\eta}^+ - \dot{\eta}^-).$$

Solving yields

$$\dot{\eta}^+ = \frac{\rho^2 \dot{\eta}^- + r_y \dot{x}^- - r_x \dot{y}^-}{\rho^2 + r_x^2 + r_y^2}.$$

The pendulum now begins a new free fall stage about the new impact vertex with this initial angular velocity. The part has settled when the post-impact velocity causes immediate re-impact with the previous vertex on successive impacts.

2.3 Toppling Transition Directed Graph

With the rolling conditions and the settling analysis, we can construct the *toppling transition directed graph* for a planar part. Each node of the graph corresponds to a stable resting edge for the part. From each node there is a single arc that leads to the node the part reaches after toppling. This arc is tagged with the contact points on the part (in the case of a position-controlled fence) or the fence heights (in the case of fixed fences) that result in toppling. If no fence contacts can result in toppling from this node, the arc is eliminated. Increasing conveyor friction μ_c can result in the addition of arcs to the graph. Figure 5 shows an example for a position-controlled fence.

Finding a fence plan amounts to searching this graph for a sequence of actions leading from the start node to the goal node.

3 Applications

3.1 Sensorless Parts Feeding

A sequence of stationary fences over a conveyor can be used to reduce uncertainty in the orientation of a part in the toppling plane. This is similar to the work of Zhang and Gupta [21], who showed that uncertainty in the orientation of a part can be reduced by having the part fall over a series of steps on the conveyor. These sensorless strategies for reducing uncertainty in the toppling plane can be combined with sensorless approaches to conveyor-plane parts feeding (Akella *et al.* [3]; Peshkin and Sanderson [14]; Brokowski *et al.* [6]; Wiegley *et al.* [20]) to construct sensorless 3D parts feeding devices on conveyors.

An example sequence of fixed fences is shown in Figure 6. The part is a 3-4-5 triangle, with edges labeled as shown in Figure 6 and vertices at $(-2, -0.5)$, $(1, -0.5)$, and $(1, 3.5)$ with respect to an origin at the center of mass. The friction coefficients are $\mu_f = 0$ and $\mu_c = \tan 30° = 0.577$. A fixed-height peg at a height y above the conveyor, $3.467 < y < 3.881$, causes a triangle at rest on edge 1 to topple to rest on edge 3, while a triangle at rest on edges 2 or 3 passes under the peg. (If $y > 4$, the triangle passes under the peg; if

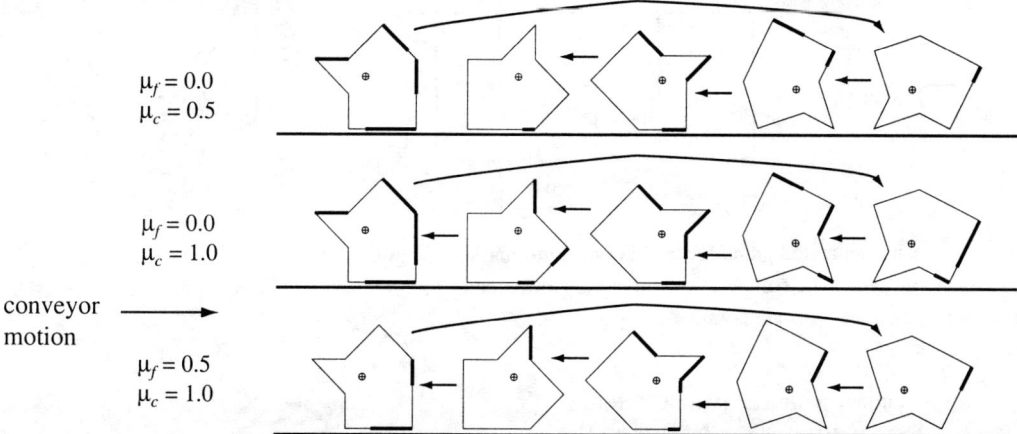

Figure 5: Toppling contacts for a position-controlled fence and two different conveyor friction coefficients μ_c. The toppling contacts are indicated by heavy lines. As we increase μ_c, the ranges of toppling contacts increase. The arrows indicate a toppling transition directed graph. In the case $\mu_c = 0.5$, there is a single resting configuration from which the part cannot be toppled. The fence would have to contact the bottom edge of the part to cause toppling.

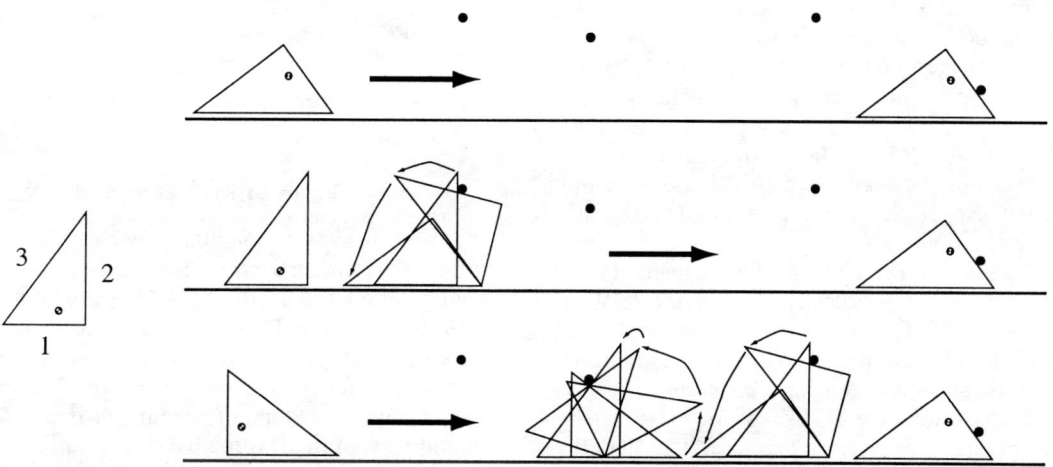

Figure 6: A sequence of fixed-height pegs that eliminates uncertainty in the part's toppling-plane orientation. At each peg, the part either passes underneath it or topples to a new edge.

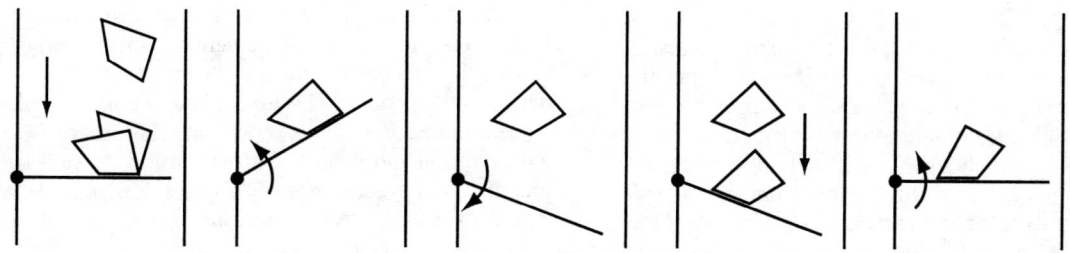

Figure 7: The 1JOC positions and orients parts in the conveyor plane by pushing.

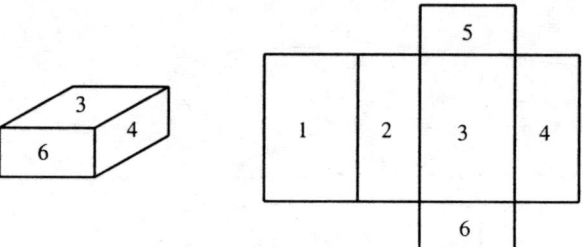 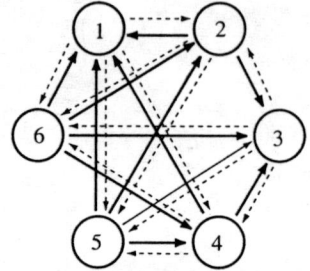

Figure 8: The toppling transition directed graph for a 3x2x1 uniform mass rectangular prism where $\mu_f = 0$ and $0.333 < \mu_c \leq 1.0$. Each node corresponds to a support face. The solid arrows correspond to transitions with feasible toppling contacts; the dotted transitions are unachievable. For $\mu_c > 1.5$, all transitions are possible.

$y > 3.881$, the peg loses contact with the triangle before it has finished rolling.) A peg at $2.361 < y < 2.698$ causes the triangle to topple from edge 2 to edge 1. No peg location will cause toppling from edge 3 to edge 2. Combining these constraints, we find that three fixed-height pegs are necessary and sufficient to bring this triangle to a unique resting edge at the end of the sequence.

In general, fixed-height fences cannot remove all toppling-plane uncertainty. These fences are analogous to the overhangs used to topple parts in bowl feeders.

3.2 2JOC

The 1JOC (1 Joint Over Conveyor) [3] performs parts feeding in the conveyor plane by using a single revolute robot joint to push parts on the conveyor (Figure 7). By combining part motion on the conveyor with a series of pushing motions, we have shown that it is possible for the 1JOC to take any polygonal part from any configuration on the conveyor (upstream of the robot) to a single desired configuration in the conveyor plane. We have also demonstrated a sensorless variant of the 1JOC.

We have augmented the 1JOC with a prismatic joint that allows the fence to move vertically, as in Figure 1. We call this system the 2JOC. The goal is to combine toppling with the ability of the 1JOC to perform conveyor-plane feeding, resulting in full 3D parts feeding on a conveyor.

For example, consider the 3x2x1 uniform mass rectangular block of Figure 8. Set $\mu_f = 0$ and $\mu_c = 0.5$. With these values, the toppling transition directed graph is shown. The part can be toppled from face A to face B when the edge AB (adjacent to faces A and B) is perpendicular to the motion of the conveyor and furthest upstream on the conveyor. Then the spatial toppling problem reduces to the planar problem above, with edge AB acting as the pivot vertex. During toppling, the fence uses its vertical prismatic motion to maintain a constant contact point on the part, but does not rotate in the conveyor plane. By sequencing 1JOC pushing and toppling, we can control the 3D configuration of the part.

If $\mu_f = 0, \mu_c > 1.5$ for the block of Figure 8, then any topple is possible; any face is reachable from any other face by toppling. In fact, for any rectangular prism with $\mu_f = 0$, there exists a μ_c^{crit} such that this property holds for all $\mu_c > \mu_c^{crit}$. This property, coupled with the 1JOC feeding property,

Figure 9: The experimental 2JOC.

indicates that any rectangular prism can be taken from any initial 3D configuration on the conveyor to a desired goal configuration, provided μ_c is sufficiently high.

Although the analysis in this paper is sufficient to analyze the 2JOC manipulating rectangular prisms, a number of issues remain for more general 3D parts. Determining which face a part will settle on is one major issue. Another is determining the full 3D force-balance conditions for rolling; in some cases, the part may undergo conveyor-plane rotation during the rolling phase of toppling. Another mechanics issue is studying the effect of out-of-plane forces in pushing (Mason and Salisbury [12]; Mayeda and Wakatsuki [13]), which are not considered in the 1JOC. We must also address automatic planning, and the existence of feasible plans as a function of the part geometry, center of mass, and μ_c and μ_f.

We have recently finished construction of a 2JOC under computer control (Figure 9), and we have implemented some simple 2JOC plans on rectangular prisms (see [1] for a description and video). Current work is toward automatic planning and execution given a general 3D part description, a goal state, and initial conditions from sensory data.

3.3 Feeding Efficiency

The Adept Flex Feeder uses a system of conveyor belts to present parts in randomized orientations to an overhead vision system. If a part in the visual field is in a graspable configuration, i.e., it is resting on the desired face, then a SCARA robot grasps the part and places it in a pallet. The conveyor remains motionless until all graspable parts in the visual field have been processed. It then advances, bringing new parts into the visual field.

When the robot moves to grasp a part, it could "bump" other parts on the way, causing them to topple onto the desired face (Goldberg [8]). The part could then be processed in the next step. This makes dual use of the robot's motion (processing a part and preparing to process a part) which could improve the overall throughput of the system. The idea of knocking a part over to put it into a graspable configuration is similar to letting gravity reorient a part in a pivot grasp (Brock [5]; Rao et al. [15]).

4 Conclusions

Toppling is a mechanically simple manipulation primitive that can increase the dexterity of a simple minimalist robot. In this paper we have used a quasistatic analysis to reduce the toppling conditions to geometric contact conditions for a fixed-height fence or a one degree-of-freedom position-controlled robot operating above a conveyor. We have shown that toppling can be used in conjunction with pushing to allow a two joint robot to manipulate 3D parts on a conveyor. Remaining work includes deriving toppling transition directed graphs for 3D parts using impact simulation for the settling phase, and automatic motion planning and implementation on the 2JOC for 3D parts which are not rectangular prisms.

Acknowledgments

Thanks to Ken Goldberg for suggesting the toppling problem, Tao (Mike) Zhang for a careful reading of the paper, and Tom Scharfeld for building the 2JOC.

References

[1] http://lims.mech.nwu.edu/~lynch/research/2JOC/

[2] Y. Aiyama, M. Inaba, and H. Inoue. Pivoting: A new method of graspless manipulation of object by robot fingers. In *IEEE/RSJ International Conference on Intelligent Robots and Systems*, pages 136–143, Yokohama, Japan, 1993.

[3] S. Akella, W. Huang, K. M. Lynch, and M. T. Mason. Parts feeding on a conveyor with a one joint robot. *Algorithmica*, to appear.

[4] A. Bicchi and R. Sorrentino. Dexterous manipulation through rolling. In *IEEE International Conference on Robotics and Automation*, pages 452–457, 1995.

[5] D. L. Brock. Enhancing the dexterity of a robot hand using controlled slip. In *IEEE International Conference on Robotics and Automation*, pages 249–251, 1988.

[6] M. Brokowski, M. Peshkin, and K. Goldberg. Curved fences for part alignment. In *IEEE International Conference on Robotics and Automation*, pages 3:467–473, Atlanta, GA, 1993.

[7] M. A. Erdmann. An exploration of nonprehensile two-palm manipulation. *International Journal of Robotics Research*, 17(5):485–503, May 1998.

[8] K. Y. Goldberg. Personal communication, 1998.

[9] K. M. Lynch and M. T. Mason. Dynamic nonprehensile manipulation: Controllability, planning, and experiments, 1998. *International Journal of Robotics Research*, to appear.

[10] A. Marigo, Y. Chitour, and A. Bicchi. Manipulation of polyhedral parts by rolling. In *IEEE International Conference on Robotics and Automation*, pages 2992–2997, 1997.

[11] M. T. Mason. Two graphical methods for planar contact problems. In *IEEE/RSJ International Conference on Intelligent Robots and Systems*, pages 443–448, Osaka, Japan, Nov. 1991.

[12] M. T. Mason and J. K. Salisbury, Jr. *Robot Hands and the Mechanics of Manipulation*. The MIT Press, 1985.

[13] H. Mayeda and Y. Wakatsuki. Strategies for pushing a 3D block along a wall. In *IEEE/RSJ International Conference on Intelligent Robots and Systems*, pages 461–466, Osaka, Japan, 1991.

[14] M. A. Peshkin and A. C. Sanderson. Planning robotic manipulation strategies for workpieces that slide. *IEEE Journal of Robotics and Automation*, 4(5):524–531, Oct. 1988.

[15] A. Rao, D. Kriegman, and K. Y. Goldberg. Complete algorithms for feeding polyhedral parts using pivot grasps. *IEEE Transactions on Robotics and Automation*, 12(6), Apr. 1996.

[16] N. Sawasaki, M. Inaba, and H. Inoue. Tumbling objects using a multi-fingered robot. In *Proceedings of the 20th International Symposium on Industrial Robots and Robot Exhibition*, pages 609–616, Tokyo, Japan, 1989.

[17] N. C. Singer and W. P. Seering. Utilizing dynamic stability to orient parts. *Journal of Applied Mechanics*, 54:961–966, Dec. 1987.

[18] J. C. Trinkle and R. P. Paul. Planning for dexterous manipulation with sliding contacts. *International Journal of Robotics Research*, 9(3):24–48, June 1990.

[19] J. C. Trinkle, R. C. Ram, A. O. Farahat, and P. F. Stiller. Dexterous manipulation planning and execution of an enveloped slippery workpiece. In *IEEE International Conference on Robotics and Automation*, pages 2: 442–448, 1993.

[20] J. Wiegley, K. Goldberg, M. Peshkin, and M. Brokowski. A complete algorithm for designing passive fences to orient parts. In *IEEE International Conference on Robotics and Automation*, pages 1133–1139, 1996.

[21] R. Zhang and K. Gupta. Automatic orienting of polyhedra through step devices. In *IEEE International Conference on Robotics and Automation*, pages 550–556, 1998.

Trap Design for Vibratory Bowl Feeders *

Robert-Paul Berretty [‡†]   Ken Goldberg [§¶]   Lawrence Cheung [§¶]   Mark H. Overmars [†]
Gordon Smith [§¶]   A. Frank van der Stappen [†]

Abstract

The vibratory bowl feeder is the oldest and still most common approach to the automated feeding (orienting) of industrial parts. In this paper we consider a class of vibratory bowl filters that can be described by removing polygonal sections from the track; we refer to this class of filters as *traps*. For an n-sided convex polygonal part and m-sided convex polygonal trap, we give an $O((n+m)\log(n+m))$ algorithm to decide if the part will be rejected by the trap, and an $O((nm(n+m))^{1+\epsilon})$ algorithm which deals with non-convex parts and traps. We then consider the problem of designing traps for a given part, and consider two rectilinear subclasses, *balconies* and *gaps*. We give linear and $O(n^2)$ algorithms for designing feeders and have tested the results with physical experiments using a commercial inline vibratory feeder. Our algorithms can be tested using online java applets: http://ford.ieor.berkeley.edu/trap-design.

1 Introduction

Part feeders, which singulate and orient parts prior to packing and insertion, are critical components of an automated assembly line. Although there is a substantial body of research in analytic feeder design, it has not yet produced a science base for practitioners, who still rely on instinct and rules-of-thumb [16]. Thus feeder design remains one of the biggest obstacles to automated manufacturing.

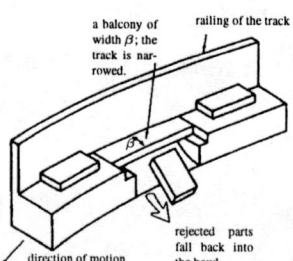

Figure 1: Vibratory bowl feeder track [7].

The oldest and still most common approach to automated feeding is the *vibratory bowl feeder* which consists of a bowl filled with parts surrounded by a helical metal track [7]. The bowl and track undergo an asymmetric helical vibration that causes parts to move up the track, where they encounter a sequence of mechanical devices such as wiper blades, grooves, gaps, and balconies. Most of these devices are filters that serve to reject (force back to the bottom of the bowl) parts in all orientations except for the desired one. Thus a stream of oriented parts emerges at the top after successfully running the gauntlet (See Figure 1).

In this paper we consider a class of vibratory bowl filters that can be described by removing polygonal sections from the track; we refer to this class of filters as *traps*. We first give algorithms to decide if a polygonal part will be rejected by the trap. We then consider the problem of designing traps for a given part, and consider two rectilinear subclasses, *gaps* and *balconies*. Proofs of some theorems can be found in a technical report [5].

2 Related Work

Space does not permit an adequate review of research in part feeding. An excellent introduction to mechanical parts feeders can be found in Boothroyd's book [7], which describes vibratory bowl feeders in detail as well as non-vibratory feeders such as the magnetic and revolving hook feeders. Sony introduced a novel approach using random

---

*Research is supported by NATO Collaborative Research Grant CRG 951224.

[†] Department of Computer Science, Utrecht University, PO Box 80089, 3508 TB Utrecht, The Netherlands.

[‡] Berretty's research is supported by the Dutch Organization for Scientific Research (N.W.O.).

[§] Industrial Engineering and Operations Research, University of California at Berkeley, Berkeley, CA 94720, USA.

[¶] Goldberg and his students are also supported by the National Science Foundation under Presidential Faculty Fellow Award IRI-9553197 and CDA-9726389.

motion of parts over part-specific pallets [22, 20]. A variety of sensor-based alternatives to mechanical bowl feeders have been proposed. For example, Carlisle *et al.* [9] describe a system that combines machine vision with a high-speed robot arm and [18] use an optical silhouette sensor with air nozzle to reject parts on a feeder track.

Specific to vibratory bowls, researchers have used simulation [14, 4, 19], heuristics [16], and genetic algorithms [10] to design traps. Perhaps closest in spirit to our work is M. Caine's PhD thesis which develops geometric analysis tools to help designers by rendering the C-space for a given combination of part, trap, and obstacle [8].

Consider a part feeding system that accepts as input a set of part orientations Θ. Based on a definition by Akella *et al* [1], we might say that a system has the *feeding property* if there exists some orientation $\theta \in \Theta$ such that the system outputs parts only in orientation θ. This paper reports on algorithms that design traps with the feeding property. We are not aware of any previous algorithms for the systematic design of vibratory bowl traps.

3 Geometric Modeling

Throughout this paper, we focus on planar motion of the part as it slides across the track while maintaining contact with the vertical railing. In the figures, the railing is coincident with the x axis and the part moves in the positive x direction. We denote the part by P, its center of mass by c, and the trap by R. We focus on cases where P is a polygonal part with n vertices. The boundary of the trap belongs to the track, and therefore supports the part. The interior of the trap cannot support the part. We denote the interior of a shape by $\text{int}(\cdot)$. The part of the track underneath P, supporting it is $S = P - \text{int}(R)$.

The track is slightly tilted toward the railing so that the part remains in contact with the railing as it moves along the railing. The radius function for the part identifies stable orientations of the part against the railing [13].

Definition 3.1 *The radius of a part at an angle θ is the distance from the center of mass to the line tangent to the part, and orthogonally intersecting the ray from the center of mass in the direction of θ.*

Each stable orientation of P corresponds to a local minimum in the radius function. The stable orientations of the part can easily be computed in linear time from the description of the part [17]. There are only $O(n)$ stable orientations of the part against the railing.

We investigate the conditions that cause a part to drop through a trap. Let us for the sake of simplicity first fix position of the part with respect to the trap. Even when the part is partially supported, i.e. $S \neq \emptyset$, it might still drop.

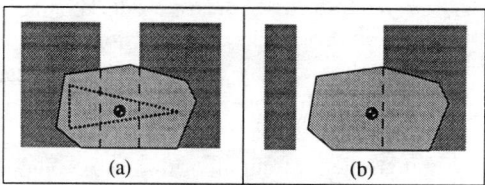

Figure 2: (a) A safe pose. The triangle is evidence of safeness. (b) An unsafe pose of the same part above a different trap.

Definition 3.2 *Let P be a part with center of mass c. Let R be a trap in the track of the bowl. The part P is* safe *if and only if there exists a triangle $\Delta = \{t_1, t_2, t_3\}$ with $c \in \Delta$, and $t_1, t_2, t_3 \in S$. Otherwise, P is* rejected.

In other words, there are three points of the part around the center of mass, which are supported by the track. Note, that the center of mass itself need not be supported.

We now state a lemma that gives a relation between the convex hull of the intersection of the part and the trap and the safeness of the part. Let $\mathcal{CH}(\cdot)$ denote the convex hull of a shape.

Lemma 3.3 *P is safe if and only if $c \in \mathcal{CH}(S)$.*

By a standard map overlay algorithm, $S = P - \text{int}(R)$ can be computed in $O((n + m + k) \log(n + m))$ time. Here, k is the complexity of the intersection of the part and the trap. The convex hull can be computed in $O(k)$ time. The number of points, k, can be reduced to $O(n+m)$ by only taking into account the leftmost and rightmost intersection of the part and the trap, per edge of the trap. These intersections are computed using an $O((n + m) \log(n + m))$ sweepline algorithm. Hence, S is computed in $O((n+m) \log(n+m))$ time. (For these standard computational geometry algorithms we refer to the book of De Berg *et al.* [12]).

4 Analyzing a Trap

The analysis problem is to decide whether or not a part in a given orientation will be rejected (fall through) as it moves in the $+x$ direction along the railing and over the trap.

To analyze the safeness of the part, we consider the set of intersections between part and gap edges. This set changes as the part moves across the trap. We first determine which edges of the part are intersected by the edges of the gap as the part moves over the track. Second, we determine in which order the edges intersect each other. Therefore, we first sort the vertices of the part and the trap by their y-coordinate. Now, we have the y-coordinates of every edge, either from the trap, or the part.

We merge the edges of the part with the edges of the trap, and determine each x position of the part where a part edge intersects with a trap edge. These "events" characterize the combinatorial changes in the intersection of the trap and the part during the motion. This preprocessing step takes $O(n + m)$ time in the convex case and $O(nm \log(nm))$ time otherwise.

A basic ingredient of our algorithm is to compute when the center of mass crosses a line defined by two intersecting pairs of trap and part edges. The position of an intersection point of a trap and a part edge is linearly dependent on the position of the part. This implies that the equation which describes the colinearity of the center of mass and the two intersection points is quadratic, and has at most two solutions.

The running time of the analysis algorithm is dependent on the shape of the part and the shape of the trap. If both the part and the trap are convex, then the problem is considerably easier to solve.

4.1 Convex Part, Convex Traps

For a convex part with a convex trap, the condition which describes the safeness of the part can be reformulated in terms of the vertices defined by intersections of the part and the trap edges.

Lemma 4.1 *Given a convex part P with center of mass c, and a convex trap R. P is safe if and only if c is in the convex hull of the vertices of $S \cap R$ or $c \notin R$.*

We restrict ourselves to the part of the motion when $c \in R$, which is a necessary condition for rejection of the part. We maintain $\mathcal{CH}(S \cap R)$, and check whether the center of mass is always inside $\mathcal{CH}(S \cap R)$.

As mentioned, during the motion of the part, intersections between the part and the trap edges may appear and disappear. Also, intersection points move. Fortunately, in our case, there are only four types of events at which the combinatorial structure of the convex hull changes. Details are in the technical report [5]. The four events occur when (1) a vertex of P moves across an edge of R, introducing or deleting a vertex of $\mathcal{CH}(S \cap R)$; (2) an edge of P moves across an vertex of R, introducing or deleting a vertex of $\mathcal{CH}(S \cap R)$; (3) a vertex of P moves across an edge of R, changing the defining edges of a vertex; (4) an edge of P moves across an vertex of R, changing the defining edges of a vertex.

Every event only requires a constant complexity update of the convex hull. By appropriately storing the convex hull, we can locate the place where the update is necessary in logarithmic time. Hence, the events can be handled in logarithmic time. After preprocessing, we know at which positions of the part, edges of the trap coincide with vertices of the part and vice versa. Therefore, maintaining the convex hull takes $O((n + m) \log(n + m))$ time.

During part motion, the center of mass always has the same distance to the edge of the track. Therefore, at any moment during the motion, there are only two edges of $\mathcal{CH}(S \cap R)$ that the center of mass can possibly cross. The intersecting edges of the trap and the part defining these edges might change, though. Every time the description of a relevant edge changes, a new event is generated for the position at which the center of mass (C.O.M.) will cross the new edge. This is accomplished without increasing the asymptotic running time. From the motion of the center of mass, and the motion of the relevant edges we derive poses of the part at which the center of mass leaves the convex hull. We add these poses as extra events. We handle such events as follows. We first check if the event is still valid, by checking if the edge associated with the event still is a relevant edge. If so, we report dropping of the part, otherwise, we discard the event. This gives no extra overhead for the algorithm. The following theorem summarizes the result.

Theorem 4.2 *Testing one orientation of a convex part against a convex trap can be accomplished in time $O((n + m) \log(n + m))$.*

4.2 Non-convex Part, Non-Convex Traps

In this section we briefly discuss how to test a pose of a not-necessarily-convex part against a not necessarily convex trap in the track. An approach with a promising time bound is the extension of the method of the previous section. Unfortunately, Lemma 4.1 is no longer valid. We must take into account supports of the part outside the trap. We have to maintain the convex hull of a set of nm moving and appearing and disappearing points. It turns out, that we can use an adaptation of the algorithm of Basch *et al.* [3, 2], leading to an algorithm with $O((nm(n + m))^{1+\epsilon})$ running time for any constant $\epsilon > 0$. We do not discuss this algorithm in this version of the paper, the interested reader is referred to the full version [5].

Theorem 4.3 *Testing one orientation of a polygonal part against a polygonal trap can be accomplished in $O((nm(n + m))^{1+\epsilon})$ time.*

5 Designing Traps

In this section we discuss how to design one trap such that the track satisfies the feeder property, i.e. only one orientation of the part survives the trap.

In the first subsection, we will discuss "gaps" as illustrated in Figure 3. In the second subsection we will discuss "balconies". The edges of gaps and balconies are orthogonal to the railing.

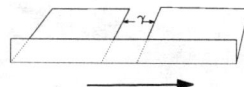

Figure 3: A gap in the track. The edges of the gap are orthogonal to the railing

5.1 The Gap

We assume, for a start, that P is convex. To get some flavor of the problem, we first show what happens if we move the part along a gap of arbitrary width. We start with the part to the left of the gap; the part is safe. During the motion, there might be a rejected position, if the gap is wide enough. At the end, when the center of mass is to the right of the right edge of the gap, the part is safe again. The position of the part at which the safeness of the part changes we call a *critical pose* (See Figure 5). If the gap is small enough, then the part remains safe throughout the whole motion, and there are no critical poses.

We focus on the *critical gap-width*, γ: the part passes safely over this gap but is rejected for gap-widths $\gamma + \epsilon$, for any $\epsilon > 0$. We compute the critical gap-width for a part in a given orientation along the railing.

The part is safe if and only if there is a supported triangle around the center of mass. This implies that if the part is rejected, then the supported area of the part is contained in a half-plane that does not contain the center of mass. We distinguish two different types of rejected poses of the part: (1) the part is only supported to the left (or the right) of the center of mass; (2) the supports are contained in a half-plane below (or above) the center of mass.

In the first type of rejected poses, the part can only be supported by one side of the gap, either the left or the right side. The second type of rejected poses correspond to poses in which the part is supported by both sides of the gap. See the last two frames of Figure 4.

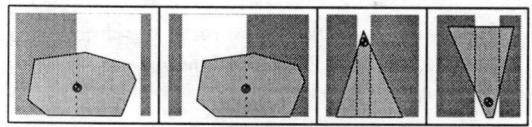

Figure 4: The types of rejected poses.

The critical gap-widths related to the first type of poses are relatively easy to compute by considering the radius function of the part at $\sigma - \frac{\pi}{2}$ and $\sigma + \frac{\pi}{2}$. Clearly, if one of these radii is less than the gap-width, then the part will fall either forward or backward. Thus, the critical gap-width γ is at most $\min\{\text{radius}(\sigma - \frac{\pi}{2}), \text{radius}(\sigma + \frac{\pi}{2})\}$.

The critical gap-width for the second type of critical poses is a bit harder to compute. Let us investigate how the supports of the part can be contained in a half-plane below the center of mass (the case for the supports above the center of mass is similar). The line defining the half-plane plays a crucial role in the analysis. Let us picture a part that is supported by two sides of the gap. The supported area of the part now consists of two convex regions, one to the left of the gap, and one to the right of the gap. The center of mass is contained in the gap. A half-plane extending downward and containing the supported area will always contain the entire lower half of the convex hull of the part. Therefore, the center of mass has to lie in the upper hull of the part to obtain an rejected pose.

A half-plane corresponding to a critical pose of the part is tangent to both the supported regions, as well as to the center of mass. In other words, the intersection points of the upper hull of the part and the gap, and the center of mass are colinear. Figure 5 shows an example of a critical pose of a convex part.

Computing the critical gap-width is accomplished by rotating a line around the center of mass, hereby sweeping over all critical poses of the part which in general have different gap-widths. The gap-widths corresponding to a critical pose is the horizontal distance between the intersection points of the line and the edges of the (upper hull of the) part. During the sweep, a linear number of pairs of edges are intersected by the line. For each such pair of edges of the upper hull of the part we compute the smallest gap-width such that there is a critical pose during the motion. The maximum of these gap-widths is the critical gap-width corresponding to the second type of rejected poses of the gap. This computation takes linear time.

Theorem 5.1 *For any orientation of the part, the critical gap-width can be computed in $O(n)$ time.*

Corollary 5.2 *Let P be a convex, polygonal part with n edges. In $O(n^2)$ time we can design a feeder with a gap, if such feeder exists, or report failure otherwise.*

If we drop the assumption that P is convex, the analysis is a bit more complex. But, it turns out that the critical gap-width can be computed in $O(n \log n)$ time, leading to an $O(n^2 \log n)$ algorithm to compute the feeder gap-width.

5.2 The Balcony

In this section, we treat a trap called a *balcony* which narrows the supporting surface of the track. Like a gap, a balcony is rectilinear. We can define this trap by giving the width, β, of the track. The trap is at least a long as the length of the part. See Figure 1 for an example.

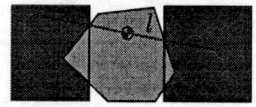

Figure 5: A critical pose.

For a balcony, this is also a sufficient condidtion.

Theorem 5.3 *Let P be a part with center of mass c. Let R be a balcony trap having width β. If at some moment during the motion, $c \in \text{int}(R)$, then the part will be rejected.*

Recall that the distance from c to the railing is exactly the radius of the part in the orientation it is traveling in. Hence, the theorem tells us that orientations with radius greater than the balcony-width drop off the track, and orientations with radius smaller than the balcony-width remain stable. The critical balcony-width for a given orientation is exactly its radius. Therefore, using a balcony, the orientation with the smallest radius can be selected by the bowl feeder. Clearly, this minimum can be computed in linear time.

Theorem 5.4 *Let P be a polygonal part with n vertices. In $O(n)$ time we can design a feeder using a balcony, if such feeder exists, or report failure otherwise.*

Note that the railing of the track always touches the part at the convex hull. Therefore, the given analysis holds for both convex and non-convex parts. A balcony can select orientations with a unique smallest radius. The only parts we cannot feed using a balcony are parts for which the minimal radius is not unique.

5.3 Parameterized Traps

In the previous two sections, we discussed simple traps such as gaps and balconies that can be described with only one parameter. More complicated traps can be specified with more parameters.

One way to look at the problem of designing a trap which is specified by k parameters, is to consider it as an arrangement of algebraic surfaces which divides a higher dimensional space into cells for which the part is safe, and cells for which the part is rejected. The space is spanned by the parameters of the trap, and the position of the part above the trap. The algebraic surfaces are derived from the higher dimensional boundaries of the convex hull of the part and the trap in different configurations. This is very much in the flavor of general robot motion planning using a cell decomposition approach. We refer the reader to Latombe's book [15] for an overview of robot motion planning, and to the paper of Schwartz and Sharir [21] for a solution to the general motion planning problem. Computing and processing the cells can be done by Collins' cylindrical algebraic decomposition [11], and is therefore doubly exponential in k. Details and possible improvements are in [5].

6 Experimental Results

To facilitate study, we implemented a java applet that allows users on the internet to experiment with our trap anal-

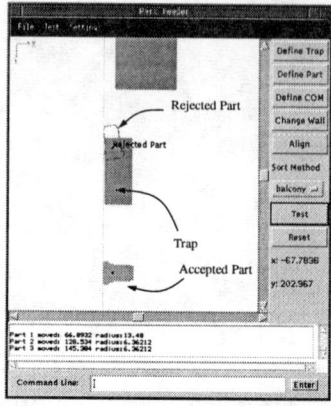

Figure 6: Trap Design Applet.

ysis and design algorithms. With a mouse, users can create a polygonal part. The applet automatically designs traps to feed the given part and demonstrates the results with animation: http://ford.ieor.berkeley.edu/trap-design.

One function in the applet is an analytic test to decide if a polygonal part, in its current position and orientation, will fall through a given trap. For a polygonal part sliding across a trap, the applet predicts when, if ever, the part reaches a point where it will fall in.

Instead of using the algorithm of Section 4.2, we currently use a simple algorithm which does a repeated search for support triangles. Even though the theoretical running time of this algorithm is rather poor ($O((nm)^5)$), it runs fast in practice.

The applet is also able to design balconies using the algorithm of Section 5.2. It computes the radius function of the user's part. If there is a unique minimum value, a balcony with this width will ensure that all other poses with a larger radius will fall prey to the trap.

The applet can also design a type of trap we did not discuss in the paper. This trap consists of a "bridge" or pair of inner and outer balconies. Two parameters define this trap: the width of the bridge and width of the balcony at the side of the railing.

After designing the traps, the applet simulates the behavior of the parts as they are randomly rotated and fall against the railing, and then move along the track. We are currently working to incorporate gap design into the applet.

Also, we built a physical inline vibratory feeder and succesfully tested various traps. See Figure 7 for a picture of our feeder.

7 Conclusion

In this paper, we report algorithms for the planar analysis and design of traps for polygonal parts moving across a

Figure 7: Part on track and railing mounted on Model 5300A.1 (T-18) adjustable inline vibratory feeder from Automation Devices, Inc. Approximate length 18 inches. The traps were designed by the algorithm and cut with a milling machine. The feeders successfully feeds a stream of these parts.

feeder track. We are not aware of any previous algorithms for the systematic design of vibratory bowl traps.

Treating parts and motion in all of their three-dimensional glory will require additional work, for example consider the common technique used for feeding screws, where a slot is used such that the screw shaft drops in, but the screw head prevents the screw from falling back into the bowl. We are working to extend our analysis to 3D.

We are also investigating systematic algorithms for the design of other commonly used feeder devices such as blades, wipers, and ramps. Armed with these tools, we can then address the planning problem, which is to automatically design an optimal sequence of such devices to deliver a desired part orientation.

Our goal is to develop complete algorithms for synthesizing feeder tracks: given part geometry, a complete algorithm will either find a track if one exists or report that no track exists for this part. Complete algorithms have been reported for feeding with parallel-jaw grippers [13] and for fences on a conveyor belt [6].

References

[1] S. Akella, W. Huang, K. Lynch, and M. Mason. Sensorless parts feeding with a one joint robot. *Algorithms for Robotic Motion and Manipulation*, 1996.

[2] J. Basch. Private communication. 1998.

[3] J. Basch, Leonidas J. Guibas, and J. Hershberger. Data structures for mobile data. In *Proc. 8th ACM-SIAM Sympos. Discrete Algorithms*, pages 747–756, 1997.

[4] D. Berkowitz and J. Canny. Designing parts feeders using dynamic simulation. In *IEEE International Conference on Robotics and Automation (ICRA)*, Minneapolis, MN, 1996.

[5] R.-P. Berretty, K. Goldberg, L. Cheung, M.H. Overmars, G. Smith, and A.F. van der Stappen. Trap design for vibratory bowl feeders. Technical report, Department of Computer Science, Utrecht University, 1999. To appear.

[6] R.-P. Berretty, K. Goldberg, M. Overmars, and A.F. van der Stappen. Computing fence designs for orienting parts. *Computational Geometry, Theory and Applications*, 10(4):249–262, 1998.

[7] G. Boothroyd, C. Poli, and L. Murch. *Automatic Assembly*. Marcel Dekker, Inc., New York, 1982.

[8] Mike Caine. The design of shape interactions using motion constraints. In *IEEE International Conference on Robotics and Automation*, 1994.

[9] Brian Carlisle, Ken Goldberg, Anil Rao, and Jeff Wiegley. A pivoting gripper for feeding industrial parts. In *International Conference on Robotics and Automation*. IEEE, May 1994.

[10] A. Christiansen, A. Edwards, and C. Coello. Automated design of parts feeders using a genetic algorithm. In *IEEE International Conference on Robotics and Automation (ICRA)*, Minneapolis, MN, 1996.

[11] G. E. Collins. Quantifier elimination for real closed fields by cylindrical algebraic decomposition. In *Proc. 2nd GI Conference on Automata Theory and Formal Languages*, volume 33 of *Lecture Notes Comput. Sci.*, pages 134–183. Springer-Verlag, Berlin, West Germany, 1975.

[12] Mark de Berg, Marc van Kreveld, Mark Overmars, and Otfried Schwarzkopf. *Computational Geometry: Algorithms and Applications*. Springer-Verlag, Berlin, 1997.

[13] Ken Goldberg. Orienting polygonal parts without sensors. *Algorithmica*, 10(2):201–225, August 1993.

[14] M. Jakiela and J. Krishnasamy. Computer simulation of vibratory parts feeding and assembly. In *2nd International Conference on Discrete Element Methods*, March 1993.

[15] J.-C. Latombe. *Robot Motion Planning*. Kluwer Academic Publishers, Boston, 1991.

[16] L. Lim, B. Ngoi, S. Lee, S. Lye, and P. Tan. A computer-aided framework for the selection and sequencing of orientating devices for the vibratory bowl feeder. *International Journal of Production Research*, 32(11), 1994.

[17] M. Mason. *Manipulator grasping and pushing operations*. PhD thesis, MIT, 1982. published in *Robot Hands and the Mechanics of Manipulation*, MIT Press, Cambridge, 1985.

[18] G. Maul and N. Jaksic. Sensor-based solution to contiguous and overlapping parts in vibratory bowl feeders. *Journal of Manufacturing Systems*, 13(3), 1994.

[19] G. Maul and M. Thomas. A systems model and simulation of the vibratory bowl feeder. *Journal of Manufacturing Systems*, 16(5), 1997.

[20] Paul Moncevicz, Mark Jakiela, and Karl Ulrich. Orientation and insertion of randomly presented parts using vibratory agitation. In *ASME 3rd Conference on Flexible Assembly Systems*, September 1991.

[21] J. T. Schwartz and Micha Sharir. On the "piano movers" problem II: general techniques for computing topological properties of real algebraic manifolds. *Adv. Appl. Math.*, 4:298–351, 1983.

[22] Sony. Advanced parts orienting system (apos). Technical Report 801-8902-11, Sony Corporation, 1989.

Testing and Analysis of a Flexible Feeding System

Greg C. Causey[†], Roger D. Quinn[†], Michael S. Branicky[☆]

Case Western Reserve University (CWRU), Cleveland Ohio, 44106

Abstract

Flexible parts feeding techniques have recently begun to gain industry acceptance. However, one barrier to effective flexible feeding solutions is a dearth of knowledge of the underlying dynamics involved in flexible part feeders. This paper presents the results of testing the CWRU flexible parts feeder. Data was collected for extended periods while feeding a variety of parts. The data was examined to determine throughput and statistical properties. In addition, tests were performed to examine other aspects of the system. A new metric for specifying the throughput of vision-based flexible feeders is presented, interesting system phenomena is examined, and a statistical analysis of the data is performed.

1 Introduction

Automated manufacturing is undergoing a paradigm shift. Previously it was equated with dedicated systems producing large volumes of a single product at high speed. In the past few years it has changed to one which demands a flexible, reconfigurable system capable of producing a wide variety of products in small lot sizes as well as being capable of accepting the rapid introduction of new products. A major component of such a system is a flexible parts feeder.

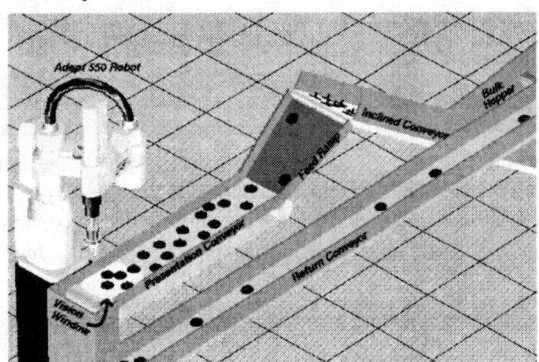

Figure 1: Schematic of the CWRU Flexible Feeder

While a small number of reconfigurable parts feeders have been brought to market [1,2,3,4,5,6,7,8,9] or patented [10,11,12,13,14], there has been little research in the underlying principles governing the dynamics of such systems. Several papers have discussed methods of static and dynamic simulation [15,16,17]. All these efforts estimate throughput by simulating the tumbling of parts rather than gaining an understanding of why the system performs as it does. Goldberg and Gunmundsson [18] attempted to determine the best setting for relative conveyor speeds in an Adept FlexFeeder by examining a 2-D model of the system coupled with statistical part distribution. Goldberg et al. [19] outline how the throughput of a feeder may be estimated using the probability of stable part poses, conveyor speed, and robot cycle time. Again, their research did not examine the underlying dynamics governing the operation of the feeder.

This paper presents the results of testing the CWRU flexible parts feeder (shown schematically in Figure 1), whose design has been published previously [20,21,22,23]. First, a new metric for specifying the throughput of a flexible feeder is presented. Second, the throughput of the system for a variety of parts is presented. Next, interesting system behavior observed in the test data is discussed. Finally, a statistical analysis of the data is performed. For all following sections, see [23] for more details.

2 System Testing

During automated testing, the system was programmed to run continuously to allow data to be collected for extended, unattended periods. Parts which were retrieved from the feeder were simply dropped onto the return feeder and recirculated. Data collected included the time duration of all portions of the feeding cycle (feeder advance time, vision processing time, robot motion time), the total run time, the number of parts fed, and (when multiple parts were being fed at once) the type of part retrieved. The entire operation proceeded serially during testing. First the vision system would take a picture and identify a valid part. Next the robot would retrieve the part, then the vision system would look for another part. If a new part was not found, the feeder would advance and the vision system would again search for a part.

System throughput tests of six hour duration were performed on a variety of different parts. During this testing, no physical parameters of the feeder were altered. Control software was changed depending on the parts being fed. *Angle dependence* tests were performed to examine the effect of the angle of the inclined conveyor on the overall system throughput. A *"slow robot"* test was conducted to determine the effect of running one component of the feeder at a different rate. Lastly, an *endurance* test was performed to determine the reliability of the system over time and to determine the robustness of the system to mechanical jams.

Parts used for testing consisted of a mixture of ⅜" and ⁵⁄₁₆" hex nuts, plastic snap rings (approximately

[†]Department of Mechanical and Aerospace Engineering
[☆]Department of Electrical Engineering and Computer Science

1¾" in diameter and ⅛" in cross-section), plastic sockets (approximately ¾" on a side by 1" long), and clear plastic disks (approximately 2⅛" in diameter and ⅛" thick) with black rims (see Figure 2).

After the data was collected, it was post-processed before further examination. The data was converted from a time per part form to parts per minute (PPM). This was accomplished by moving a five-minute-wide window over the data and determining the average PPM inside that window. The act of converting the data from time per part to parts per time had a smoothing effect. Five minutes was chosen as a compromise window size so that short term system behavior would not be missed due to undue smoothing.

3 Results

3.1 A Metric for Feeder Evaluation

Reporting results on the throughput of a flexible parts feeder requires a redefinition of the standard parameters used previously to qualify feeders. Unlike a bowl feeder, a typical flexible feeder is composed of several major components (a mechanism to present quasi-singulated parts, a vision system to determine the location of graspable parts, and a mechanism for removing those parts from the system). Simply stating a number as the throughput of the feeder doesn't indicate the relative speed of each system component or pinpoint possible bottlenecks.

To report on the throughput of the feeder the following four different parameters will be used.

1. Overall throughput
2. Throughput of the parts presentation system
3. Throughput of the vision system
4. Throughput of the parts removal system

The overall throughput is the standard feeder parameter. This is a measure of how many parts are removed from the system in a certain amount of time. The throughput of the parts presentation system is a measure of the physical capability of the system to present singulated parts to the workcell. The throughput of the vision system is an indication of the speed of the vision system in locating candidate parts. The throughput of the part removal system is an indication of how fast the parts can be removed from the vision window.

3.2 Throughput Results

3.2.1 Snap Rings, Disks, Sockets

In the first three tests, snap rings, disks, and sockets were fed (individually) respectively. The average and standard deviation of each parts' throughput is shown in Tables 1-3. All values are in parts per minute.

	Overall	Conveyor	Vision System	Robot
Average	12.65	51.51	31.94	35.60
Std Dev.	0.49	4.80	1.13	0.53

Table 1: Throughput of Snap Rings

	Overall	Conveyor	Vision System	Robot
Average	10.52	31.71	27.40	37.57
Std Dev.	0.80	4.00	2.52	0.30

Table 2: Throughput of Plastic Disks

	Overall	Conveyor	Vision System	Robot
Average	18.56	111.07	69.40	33.15
Std Dev.	0.69	17.43	3.80	0.33

Table 3: Throughput of the Plastic Sockets

3.2.2 Nuts: 5/16" and 3/8"

Two tests were performed using a combination of 5/16" and 3/8" hex nuts. Tables 4 and 5 show the results of the first and second test. Table 6 shows the results from the second test, broken out by individual part throughput (while still feeding both parts).

In Table 6, the overall and robot motion throughputs are shown. It was impossible to determine the throughput for the vision and conveyor sub-systems for individual parts. For example, while a single advance may have brought multiple parts into the vision window, only the first part retrieved was assigned the time for the conveyor advance.

	Overall	Conveyor	Vision System	Robot
Average	19.50	189.34	59.73	34.92
Std Dev.	0.95	48.46	5.67	0.41

Table 4: Throughput of Hex Nuts - Test 1

	Overall	Conveyor	Vision System	Robot
Average	21.37	237.58	77.78	33.89
Std Dev.	0.43	34.62	4.56	0.31

Table 5: Throughput of Hex Nuts - Test 2

Figure 3 shows the average throughput for the feeder and its sub-systems during the first test. Figure 4 shows the overall system throughput and the individual part throughput during the second test. The individual throughputs are shown for a shortened time period and were generated using a window of 30 minutes length to more clearly show the relationship between the feeding rates of the two parts.

An interesting feature can be seen in the overall throughput graph (Figure 3). A drastic drop in the throughput of the system can be seen at approximately 40 minutes. This was due to a lack of sufficient parts in the

Figure 2: Parts used for testing

system. Over time, the parts were distributed throughout the system more evenly, therefore the overall throughput of the system stabilized. This also had the effect of causing the throughput of the conveyor sub-system to vary, as seen in the upper right plot.

	Overall	Robot
3/8" Nuts		
Average	9.89	33.95
Std Dev.	0.46	0.21
5/16" Nuts		
Average	11.48	33.84
Std Dev.	0.48	0.16

Table 6: Individual Throughput when Feeding Mixed Hex Nuts

The second test occurred approximately 60 hours of operation after the first, which had the effect of further reducing the variation in system throughput. The right hand graph of Figure 4 shows the interaction of the two different-sized parts (top line: 5/16" nuts, bottom line: 3/8" nuts). As one of the part's throughput increases, the corresponding throughput of the other part decreases. Since the size of the presentation window is constant, it follows that if more of one part is present then less of the other part will be present.

Since there were an equal number of parts placed in the bin, it was expected that the average feed rate for each part would be the same. However, as seen in the right-hand plot of Figure 4, this was not the case. A possible explanation for this behavior is the relative stability of each part (nuts resting on their sides are undesirable). So, a third test was conducted using the same parameters with the addition of a pause programmed into the controller to halt the system after each part was retrieved which allowed the number of parts on the horizontal conveyor and the number of parts retrieved to be manually counted.

Mixed Nuts	Overall	3/8"	5/16"
Percentage Parts on Edge	15.6 %	21.0 %	10.4 %
Percentage Parts Fed	73.4 %	69.5 %	77.1 %

Table 7: Statistical Data from Manual Nut Test

From the data, it is clear that the 3/8" nuts are more prone to standing on edge (21% vs. 10.4%) and that the percentage of parts removed by the robot is higher for the 5/16" nuts (77.1% vs. 69.5%). These results agree with those seen in Test 2. Finally, Table 8 shows a comparison of the results of Test 2 and the manual test.

% Parts Presented	Manual Test	Test 2
3/8" Nuts	48.3 %	45.7 %	46.3 %
5/16" Nuts	51.7 %	54.3 %	53.7 %

Table 8: Percentage of Each Type of Nut

From this table, it is clear that approximately the same number of parts are being spilled onto the horizontal conveyor from the inclined conveyor (48.3% vs. 51.7%). The percentages of parts fed during both the manual test and test 2 correlate well. Therefore, the differences in the throughput of each size nut can be attributed to their relative stability.

3.2.3 Mixed Nuts and Plastic Sockets

Finally, two tests were performed using a combination of hex nuts and plastic sockets. These tests were interesting because the parts themselves were of different material and geometry; the nuts were steel and relatively flat while the sockets were plastic and more box-like.

Table 9 shows overall system throughput and Table 10 shows individual part throughput. Figure 5 shows the

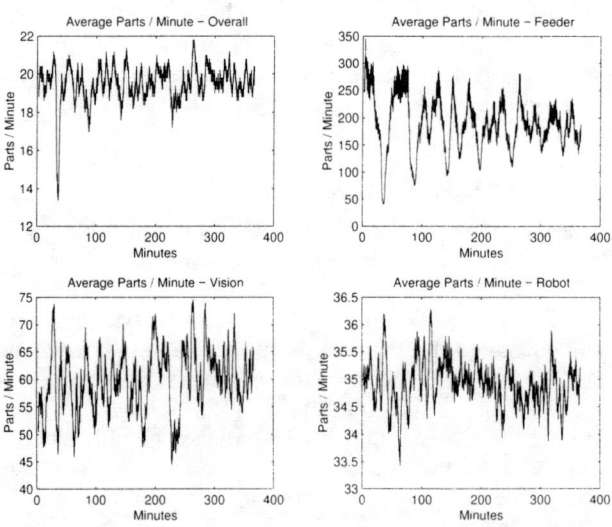

Figure 3: Feeder Throughput for the 5/16" and 3/8" Hex Nuts

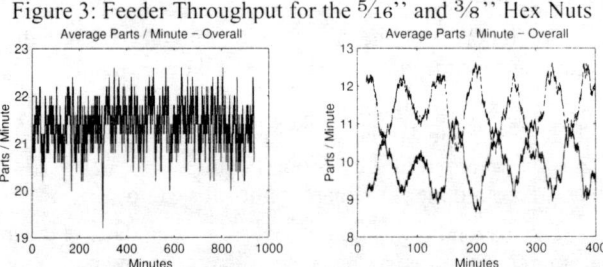

Figure 4: Feeder Throughput for the 5/16" and 3/8" Hex Nuts

overall system and individual part throughputs for the two tests. As with the previous test, a 30-minute window was used to generate the individual throughput plots (top line: 5/16" nuts, middle line: 3/8" nuts, bottom line: sockets).

The lower plots show that further mixing of the parts occurred between the tests (performed about a week apart).

	Overall	Conveyor	Vision System	Robot
TEST 1				
Average	20.82	254.85	68.90	34.03
Std Dev.	0.52	33.61	4.70	0.35
TEST 2				
Average	20.79	291.74	66.98	33.87
Std Dev.	0.55	35.68	5.14	0.34

Table 9: Throughput of the System Feeding All Parts

	Overall	Robot	Overall	Robot
	TEST 1		TEST 2	
3/8" Nuts				
Average	7.64	34.05	7.42	33.84
Std Dev.	0.39	0.24	0.50	0.21
5/16" Nuts				
Average	8.50	34.13	8.22	33.78
Std Dev.	0.99	0.18	0.67	0.25
Plastic Sockets				
Average	4.66	33.74	5.15	34.00
Std Dev.	0.90	0.28	0.74	0.26

Table 10: Individual Part Throughput When All Parts Fed

To reflect a typical assembly requirement, two tests were conducted: one in which the parts were fed in a particular order (in this case, 3/8" nut, 5/16" nut, socket, 3/8" nut, etc.) and another in which only one part was fed with all the parts in the bin. Tables 11 and 12 show the respective results.

	Overall	Conveyor	Vision System	Robot
Average	15.47	119.40	37.81	34.25
Std Dev.	1.38	26.31	5.44	0.35

Table 11: System Throughput Feeding Parts in a Specific Order

	Overall	Conveyor	Vision System	Robot
3/8" Nuts				
Average	13.62	89.93	31.03	33.68
Std Dev.	0.87	13.74	3.06	0.34
5/16" Nuts				
Average	14.07	97.24	32.33	33.78
Std Dev.	0.69	12.37	2.39	0.37
Plastic Sockets				
Average	10.61	58.99	21.61	32.20
Std Dev.	1.31	13.16	3.75	0.38

Table 12: System Throughput Feeding Individual Parts

3.3 Angle Dependence Test

One of the many parameters that contribute to the overall throughput of the system is the angle of the inclined conveyor with respect to the horizontal.

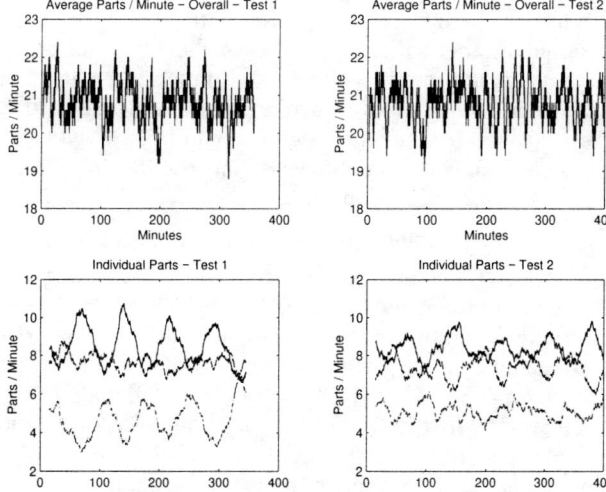

Figure 5: Overall and Indiviual Throughput with Multiple Parts in the Feeder

Intuitively, it would be expected that altering the angle of the inclined conveyor would cause system throughput to range from 0 (the conveyor too steep for any parts to advance) to some nominal value (the conveyor set to an angle of 0°) with the maximum at some angle between these two extremes. If the conveyor is at too shallow of an angle, too many parts will be spilled onto the presentation conveyor, thereby reducing part singulation. A mixture of nuts and sockets were fed and data was collected for 2.5 hours per angle.

The angles tested were 32°, 34°, 36°, and 38°. Testing of other angles was not straightforward due to mechanical limitations of the construction of the current feeder. Table 13 shows system throughput for the tests.

	Overall	Conveyor	Vision System	Robot
Angle = 32°				
Average	19.20	243.54	54.64	33.99
Std Dev.	0.80	29.20	5.35	0.32
Angle = 34°				
Average	19.66	242.16	57.58	34.29
Std Dev.	0.67	25.23	5.09	0.30
Angle = 36°				
Average	18.98	224.60	52.95	34.28
Std Dev.	0.62	25.18	3.84	0.31
Angle = 38°				
Average	19.14	185.50	57.55	34.25
Std Dev.	0.74	23.97	5.62	0.38

Table 13: Throughput for Angles of the Inclined Conveyor

It is apparent from the data that these small changes in the angle of the inclined conveyor has little impact on overall system throughput, but it does affect the throughput of the conveyor sub-system.

3.4 Slow Robot Test

In all previous tests the throughput of the robot was consistent. The purpose of this test was to determine if the throughput of the robot is unrelated to the rest of the system (this is also discussed in Section 3.7.3). This is important because it allows the throughput of the feeding system to be determined for a wide variety of robots.

The test was identical to the one described in Section 3.2.3. The only change was in the speed of the robot (from 130% to 10% of its nominal speed).

	Overall	Conveyor	Vision System	Robot
Average	5.10	245.48	67.26	5.46
Std Dev.	0.15	120.07	7.01	0.14

Table 14: Throughput of the System with the Robot Slowed

Examining the results, it can be seen that while the overall and robot throughputs are greatly reduced in comparison to the previous test (Table 9), the throughput of the conveyor and vision system are nearly identical.

3.5 Endurance Test

An endurance test was performed to determine the ability of the system to feed parts for a long period of time without jamming or human intervention. Data was

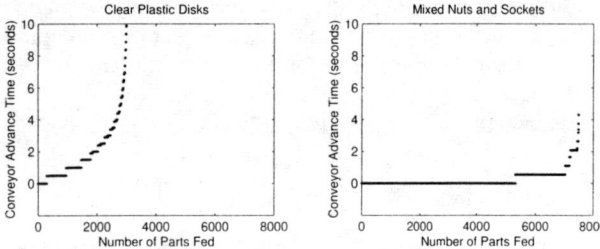

Figure 6: Time Spent Moving the Conveyors

analyzed in discrete 30-minute windows to reduce computation time.

A mixture of nuts and plastic sockets were fed. Total run time for the test was about 3 days 9 hours. During that time the system fed over 150,000 parts without intervention. Table 15 shows the throughput of the system and its components.

	Overall	Conveyor	Vision System	Robot
Average	31.38	424.60	107.80	49.30
Std Dev.	0.20	22.10	2.07	0.16

Table 15: Throughput of the System During Extended Testing

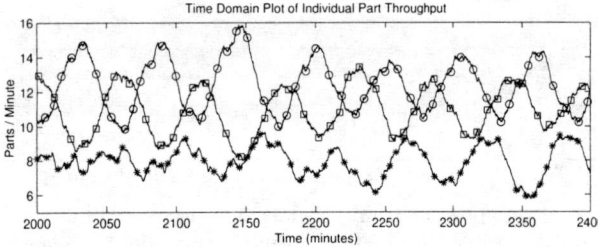

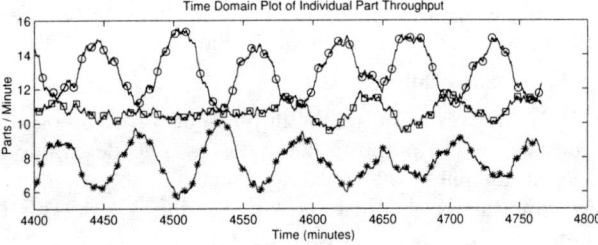

Figure 7: Individual Part Throughput - Extended Test

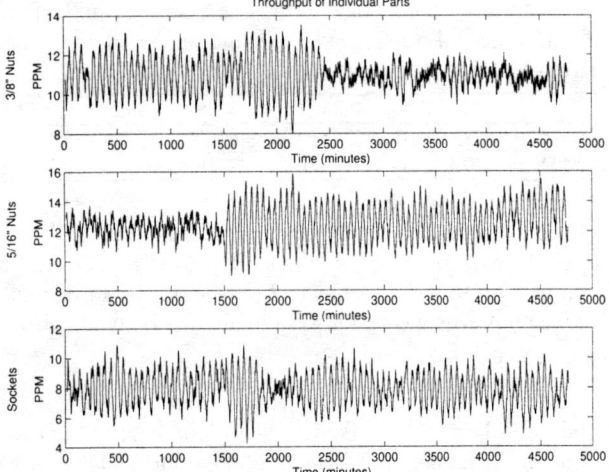

Figure 8: Individual Part Throughput - Extended Test

When comparing the results with those reported in Table 9, a major improvement (about a 50% increase) in the throughput of the system can be seen, which was obtained by altering system parameters and control code.

3.6 Other Observations

3.6.1 Effects of Part Size on Throughput

System throughput versus part size is examined in Table 16.

Part Type	Major Dimension	Throughput
Nuts and Sockets	0.850''	20.7 ppm
Mixed Nuts	0.640''	19.4 ppm
Plastic Sockets	0.850''	18.5 ppm
Plastic Snap Rings	1.700''	12.6 ppm
Clear Plastic Disks	2.100''	10.5 ppm

Table 16: Major Part Dimension and System Throughput

Clearly, higher throughputs can be expected for smaller parts. A better indicator of the effect of part size on the throughput of the system may be to examine the number of conveyor increments made during operation. Figure 6 shows this for plastic disks and mixed nuts and sockets. A zero indicates that no conveyor move was necessary to acquire the next part (i.e. it was already in the vision window). When examining the graph for the mixed nuts and retainers, it can be seen that no conveyor move was necessary when retrieving over 5000 parts and only a single move was necessary when retrieving approximately 2000 additional parts.

3.6.2 System Throughput Oscillations

While examining the extended test data, interesting oscillations in individual part throughput were noticed. Figure 7 shows the individual part throughput for two 6½ hour windows during the test (○-5/16'' nuts, □-3/8'' nuts, *-sockets). The top plot begins by showing a strong anti-phase relationship between the two sizes of nuts. By the end of the window, however, the phase between the two sizes of nuts is beginning to align while the sockets have begun oscillating out of phase with the 5/16'' nuts. Examining the bottom plot shows the 5/16'' nuts and sockets are anti-phase while the 3/8'' nuts are feeding at a consistent rate. However, by the end of the window, the 3/8'' nuts are beginning to oscillate out of phase with the 5/16'' nuts and the amplitude of the oscillations of the sockets seems to be diminishing.

3.6.3 Jumps in the Variation of System Throughput

Another phenomenon noticed while examining the test data from Sections 3.2.2 and 3.2.3 was sudden jumps in the variation of system throughput. Figure 8 shows the throughput of each part versus time for the extended test (Section 3.2.3). There is a jump in the magnitude of the oscillations of the 3/8'' nuts at approximately 2200 minutes and a jump in the oscillations of the 5/16'' nuts at approximately 1500 minute, however, no jump was seen in socket throughput.

Next, the data from Section 3.2.2, second test was further examined as shown in Figure 9. Again, a reduction in the oscillations of the throughput of each part is seen (at about 500 minutes). In this case, however, both the parts display the jump at the same time.

The phenomena of jumping between two seemingly stable operating regions is a characteristic of a nonlinear system. It is important to determine the underlying dynamics which are causing this to happen so that the variation in the throughput of the system may be reduced.

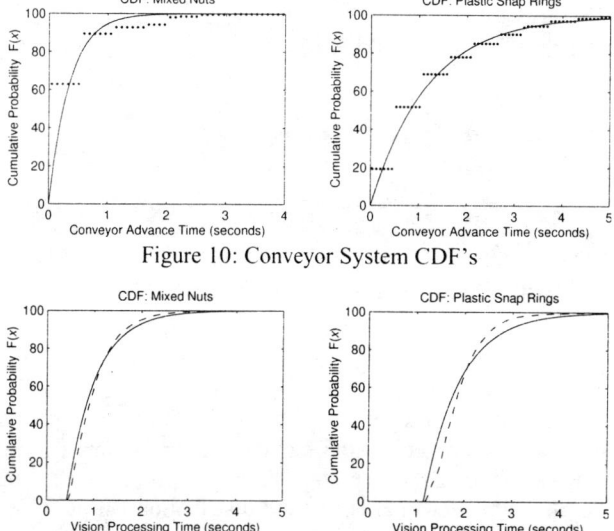
Figure 10: Conveyor System CDF's

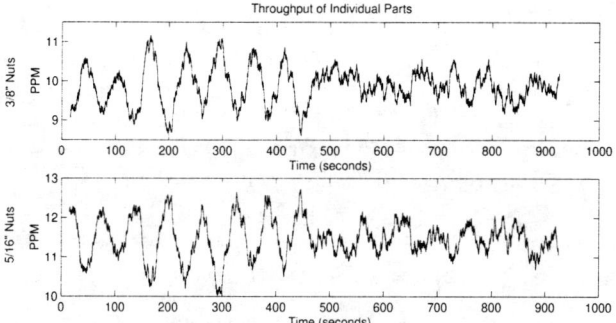
Figure 9: Individual Part Throughput - Second Mixed Nut Test

3.7 Statistical Properties of the Feeder

Statistical distributions associated with the test data are first examined followed by a discussion of determining system properties when only sub-system properties are known. Lastly, the correlation between the sub-systems is investigated. In all statistical distribution figures, a solid line represents the ideal distribution while a dashed line or histogram represents experimental data.

3.7.1 Statistical Distribution

The Poisson distribution is a common model of discrete arrival processes. Hence, it (shifted from 0 by the minimum part retrieval time) appears a logical choice for modeling the *Overall System Throughput*. In this case, the times between retrievals (also called interarrival times) are exponentially distributed. Therefore, a test of the fit of an exponential distribution to these interarrival times must be examined. Figure 12 shows the cumulative distribution function (CDF) for several test parts.

Examining the graphs, it is clear that the exponential distribution does fit the data, which can therefore be modeled by a time-shifted Poisson process. This agrees with the distribution used in [18].

The *Conveyor Sub-system* is unique in that its data is very discrete. It takes a repeatable amount of time to advance the conveyor, therefore the possible move intervals can only be combinations of those times. Again, an exponential distribution was fit to the data. In this case, there was no minimum part retrieval time since it was possible for the next part which was removed from the system to require no conveyor advance. Figure 10 shows an exponential distribution fit to the mixed nuts and plastic snap rings. As with the overall system case, the data fit an exponential well. Therefore the throughput of the conveyor can also be modeled as a Poisson process.

The *Vision Sub-system* is unique in that there is a minimum process time (for an empty window), to which is added the time required to find a part. The vision processing time depends on many factors including the complexity of the part and associated vision algorithms, the number of parts in the window, the location of the part in the window, and the speed of the vision system. An exponential distribution was fit to the data to see if the process could be modeled as Poisson. Figure 11 shows the mixed nuts and plastic snap rings fit with this distribution. As can be seen, the exponential again fits the data well. It can therefore be concluded that the vision system may also be modeled as a time-shifted Poisson process.

It was expected that the times for the *Part Manipulator* (robot) to retrieve parts would be normally distributed. Figure 13 shows the mixed nuts and sockets

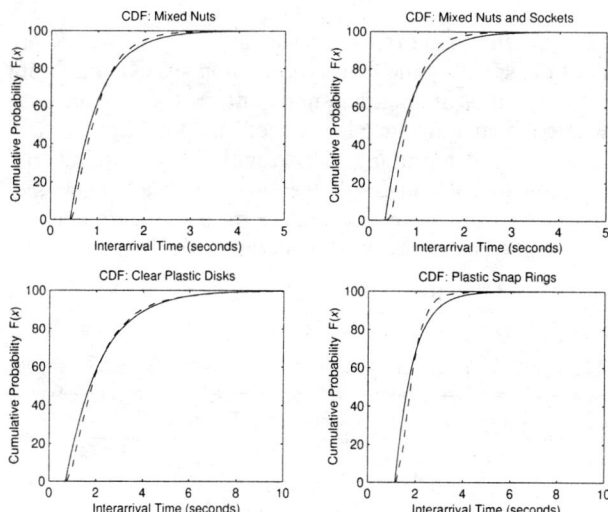
Figure 12: CDF for Overall throughput vs. Exponential Dist.

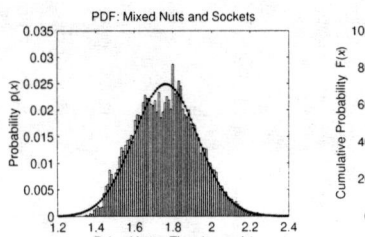
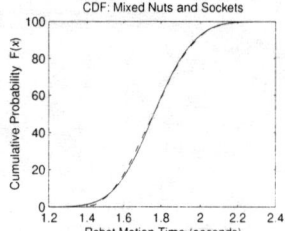

Figure 13: Normal Distribution and Mixed Nuts and Sockets test with a normal distribution, which fits the data quite well.

Since the ultimate goal is to be able to specify the throughput of the feeder in parts per minute it is important to examine what happens (statistically) to the data after it has been averaged by the moving window. This can be answered by the central limit theorem [24], the implication of which is that the distribution (in parts per unit time) of all the sub-systems can be approximated as normal. Figure 15 shows that both overall system and conveyor data (after averaging) are well modeled by a normal distribution. Even conveyor system throughput, which when plotted as first order arrival time is very discrete (Figure 10), is well modeled.

3.7.2 Computing New System Parameters

As both computer vision systems and robots increase in speed and decrease in price, it is expected that those sub-systems of a flexible parts feeder will be upgraded. It is then desired to know apriori what the effect of the upgrade will have on the overall throughput of the system. Four properties will be examined (for derivations, see [23]).

Mean Part Interarrival Time: Figure 14 is a time line of the sequences which must occur for parts to be retrieved. From this, it can be seen that the total time to acquire a part is simply the sum of the times required by each subsystem.

$$F_i = C_i + V_i + R_i \qquad (1)$$

F_i is the time require to retrieve part i. C_i, V_i, and R_i are the time spent by the conveyors, vision system, and robot in presenting, finding, and retrieving part i. For the case where a feeder advance is not needed, a time of 0 is used. Using (1), determining the overall mean requires the addition of each mean of the sub-system average time. The subscript "ppm" denotes parts per minute and the subscript "iat" denotes 1st order interarrival time.

$$\mu_{F_{iat}} = \mu_{C_{iat}} + \mu_{V_{iat}} + \mu_{R_{iat}} \qquad (2)$$

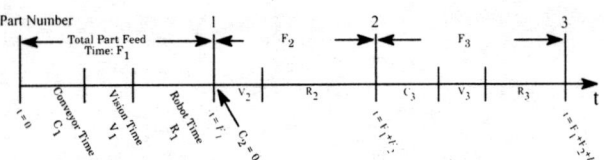

Figure 14: Time Line of Serial Part Retrieval

Average System Throughput: First, the mean of the overall system in terms of the 1st order interarrival times (in PPM) can be expressed as

$$\mu_{F_{ppm}} = n \bigg/ \sum_{i=0}^{n} F_i \qquad (3)$$

After writing the average throughput of each sub-system and substituting those and (1) into (3), the equation for the average system throughput in terms of the average sub-system throughputs is

$$\mu_{F_{ppm}} = 1 \bigg/ (\mu_{C_{ppm}}^{-1} + \mu_{V_{ppm}}^{-1} + \mu_{R_{ppm}}^{-1}) \qquad (4)$$

Variance of the 1st Order Interarrival Time: The population variance of the system is determined using the standard formula [24]. Substituting (1) and (2) into the standard formula and rearranging yields the system variance in terms of the sub-system variances:

$$\sigma_{F_{iat}}^2 = \sigma_{C_{iat}}^2 + \sigma_{V_{iat}}^2 + \sigma_{R_{iat}}^2 \\ - 2\big[Cov(C,V) + Cov(C,R) + Cov(V,R)\big] \qquad (5)$$

Variance of the Average Throughput: In this case, one can write an equation for the system variance, but it cannot be partitioned into an equation depending only on the sub-system variances and covariances (as in (5)).

3.7.3 Correlation Between Sub-systems of the Feeder

There are three sub-systems in the feeder (conveyors, vision system, robot) and therefore there are three possible sub-system interactions. A correlation coefficient is determined for each, shown in Table 17.

It is clear that there is a fairly strong relationship between the conveyor and the vision system while there is very little correlation between the robot and either the conveyor or vision system.

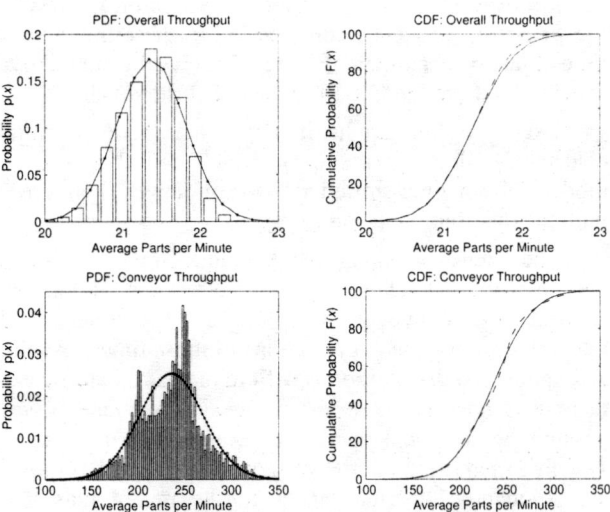

Figure 15: PDF and CDF for the Overall and Conveyor Sub-system Throughput

	Snap Rings	Disks	Sockets	Mixed Nuts	Mixed Nuts and Sockets
Vision vs. Feeder	0.8210	0.8841	0.8696	0.5696	0.4494
Vision vs. Robot	-0.0205	-0.0460	-0.1257	-0.1452	-0.1329
Feeder vs. Robot	0.0139	-0.0555	-0.1297	-0.1435	-0.1102

Table 17: Correlation Coefficients from Feeder Tests

By examining the serial mode of operation of the feeder, shown in pseudo code:

```
While (Part Found == False) {Advance the Feeder}
Return the Part Location
```

an explanation of the dependence between the feeder and the vision system is possible. The advancement of the feeder is signaled by the vision system, but to determine if an advance is necessary, the vision system must take a picture. This leads to two pictures for each conveyor advance. Hence, in general, when the vision time is large, the conveyor advance time will also be large.

4   Conclusions

Preliminary testing has been performed on the CWRU flexible feeder using several different parts. A new metric has been constructed to better describe the throughput of a vision-based flexible parts feeder. Testing has shown throughputs between 10 to 30 parts per minute for a variety of parts. The system was shown to be nonlinear and it displayed jump phenomena. A statistical analysis was also performed. An exponential distribution fit the overall system, conveyor sub-system, and vision processing sub-system while a normal distribution fit the robot sub-system. When the data was converted into average parts per minute, all the data was fit by a normal distribution. It was shown that the average throughput and 1st order interarrival times could be determined from knowing the properties of the sub-systems. However, the variances of the average throughput and 1st order interarrival times could not be determined from knowing the properties of the sub-systems. Finally, it was determined that the conveyor and vision sub-systems are inter-related while the robot is mostly independent of other sub-systems.

Acknowledgments

This work was funded through the Center for Automation and Intelligent Systems Research (CAISR) by the Cleveland Advanced Manufacturing Program (CAMP).

References

[1] Comtech. MRW Robo-Pot. Company Product Literature. Farmington Hills, MI 48333

[2] Davis, William F. Centrifugal Feeders The Flexible Approach. Custom Systems Integration Co. Presented at the 1994 RIA Flexible Parts Feeding for Automated Handling and Assembly Workshop, 1994

[3] Mahoney and Marshell, Tape and Reel Technology for Automatic Packaging and Feeding of Connectors. GPAX International, Inc. Company Product Literature. Columbus, Ohio.

[4] SonyFA, Advanced Parts Orientating System, T-APOS. Company product Literature. 560 Route 303 Rangeburg, New York 10962.

[5] Daniel Boehlke. Smart Design for Flexible Feeding. Machine Design, December 12, 1994.

[6] Adept Robotics, Adept FlexFeeder 250, Company Product Literature, 150 Rose Orchard Way, San Jose, CA, Sept. 1995.

[7] Steve Gordon. Flexible Feeding of Small Parts. Intelligent Automation Systems. Proceedings of the 1996 RIA Flexible Parts Feeding for Automated Handling and Assembly Workshop. Minneapolis, Minnesota, October 1996

[8] Terrence Lynch, Smart Parts Feeder Speeds Agile Assembly. Design News, 8/14/95

[9] E.M. Ross. Flexible Feeding Devices for Robotic Assembly. Robotic Production Methods Inc (RPM). Proceedings of the 1994 RIA Flexible Parts Feeding for Automated Handling and Assembly Workshop. Cincinnati, Ohio, October 1994

[10] US Patent #4333558. Nonaka, Takeyoshi and Otsuka, Kazuo. Photoelectric Control System for Parts Orientation. June 8, 1982

[11] US Patent #4608646. Goodrich, Jerry L. and Devlin, William L. Programmable Parts Feeder. August 26, 1986

[12] US Patent #5687831. Carlisle, Brian R. Flexible Parts Feeder. November 18, 1997.

[13] US Patent #5314055. Gordon, Steven J. Programmable Reconfigurable Parts Feeder. May 24, 1994

[14] US Patent #4909376. Herndon, Donnie, et al. Robotically Controlled Component Feed Mechanism Visually Monitoring Part Orientation. March 20, 1990

[15] Mirtich, B. et al., "Estimating Pose Statistics for Robotic Parts Feeders", Proceeding of the 1996 IEEE International Conference on Robotic and Automation.

[16] Craig, J. "Simulation-Based Robot Cell Design in AdeptRapid", Proceeding of the 1997 IEEE International Conference on Robotic and Automation.

[17] Berkowitz, D. & Canny, J. "Designing Parts Feeders Using Dynamic Simulation", Proceeding of the 1996 IEEE International Conference on Robotic and Automation.

[18] Gudmundsson, D. & Goldberg, K. "Tuning Robotic Part Feeder Parameters to Maximize Throughput", Proceeding of the 1997 IEEE International Conference on Robotic and Automation.

[19] Goldberg, K. et al., "Estimating Throughput for a Flexible Parts Feeder". International Symposium on Experimental Robotics. June 1995.

[20] R.D. Quinn et al. Design of an Agile Manufacturing Workcell for Light Mechanical Applications. IEEE International Conference on Robotics and Automation Proceedings, 858-863, 1996.

[21] F.L. Merat et al. Advances in Agile Manufacturing. CAISR Technical Report TR96-104. IEEE Conference on Robotics and Automation Proceedings, 1997.

[22] Causey, G. & Quinn, R. "Design of a Flexible Parts Feeding System". IEEE International Conference on Robotics and Automation. Proceedings, 1997.

[23] Greg C. Causey. Elements of Agility in Manufacturing. Ph.D. Dissertation, Case Western Reserve University, 1999.

[24] Jay L. Devore. Probability and Statistics for Engineers and the Sciences. Brooks/Cole, Monterey CA, 1987.

Planning and Control of Indirect Simultaneous Positioning Operation for Deformable Objects

Takahiro Wada, Shinichi Hirai, and Sadao Kawamura
Department of Robotics
Ritsumeikan University
Kusatsu, Shiga, 525-8577, Japan
e-mail: wachan@robot.club.or.jp , {hirai, kawamura }@se.ritsumei.ac.jp

Abstract

A novel control method for positioning operations of deformable objects will be presented. In many manipulative operations of deformable objects, it is required to guide multiple points on an object simultaneously. Moreover, the points often cannot be manipulated directly. A model of the manipulated deformable object is essential to perform these operations. It is, however, difficult to build a precise model of a deformable object. Thus, we need a new control scheme that allows us to perform the operations successfully despite of discrepancy between a manipulated deformable object and its model.

In this paper, we will introduce a coarse model of a deformable object and will develop a new control law for its positioning operation. First, we will propose a mathematical model of deformable objects for their positioning operations. Second, indirect simultaneous positioning operations of deformable objects is formulated. Then, we will propose a novel iterative control method to realize a given positioning operation. The validity of the proposed method and the effect of model errors to the operation will be examined through experiments using textile fabrics. Experimental results show that coarse model of deformable objects is effective to their positioning operations. Finally, we will discuss on the locations of robotic fingers, which apply forces to a manipulated object.

1 Introduction

Many researches on object manipulations have been studied for a past decade while few result has been applied to actual manipulative operations. Model-based approaches have been applied to the manipulation of objects. Object models are requisite for complex manipulative operations. One problem in this approach is that it is difficult to build an exact model of a manipulated object. For example, it is not easy to build friction model between fingers and an object. It is difficult to build a model of deformable object. Due to this discrepancy between a manipulated object and its model, model-based approaches fail to be applied to actual manipulative operations.

Wakamatsu et al. [1] have analyzed grasping of deformable objects and have introduced bounded force closure. Their approach is static and control of manipulative operations is out of consideration. Howard et al. [2] have proposed a method to model elastic objects as the connections of springs and dampers, and have developed a method to estimate the coefficients of the springs and dampers by recursive learning method for grasping. This study has focused on model building and control law for manipulative operations has not been investigated. Sun et al. [3] have studied on the positioning operation of deformable objects using two manipulators. They have focused on the control of the object position and deformation control is not discussed.

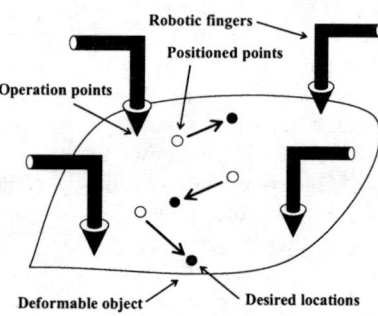

Figure 1: Indirect positioning operation of deformable object

Object models are requisite for object manipulation while precise models cannot be built. To solve this dilemma, we will propose a new control method using coarse object models. In our approach, we will first build a coarse model of a manipulated object and will derive a control law robust to discrepancy between the object and its coarse model. Especially, we will apply this approach to the indirect simultaneous positioning operation of a deformable object [4]. In this operation, multiple points on an object should be guided to their desired locations simultaneously, as illustrated in Fig.1. The positioned points cannot be operated directly. In our previous work [4], we have defined indirect simultaneous positioning operations of deformable objects and have proposed a control method for these operations based on their coarse models. We have assumed that locations of robotic fingers are determined in advance. Namely, planning of robotic finger locations have been out of consideration. In this paper, we will propose planning methods of robotic finger locations considering forces exerted on the object by the fingers. In this paper, we will first propose a coarse model of deformable objects. Secondly, positioning operations is formulated using the coarse model. Then, simple control law is derived based on the formulation. Experimental results illustrate the validity of the proposed method and the effects of model errors to the convergence. Further, we will discuss on planning problem of robotic fingers locations based on the coarse model for the positioning operations.

2 Coarse modeling of deformable objects for positioning operations

In this section, coarse model of deformable objects is proposed for their positioning operations.

For simplicity, we deal with two dimensional deformable objects such as textile fabrics, in this paper. Force-displacement characteristics of deformable objects can be nonlinear. Additionally, many objects show hysteresis property. However, we model the objects as mesh points connected by linear springs each other as shown in Fig.2. Let l_c, l_w, l_s be natural length of horizontal, vertical, and diagonal springs, respectively. Coefficients k_c, k_w, k_s are spring constants of horizontal, vertical, and diagonal springs. Each mesh point is connected by vertical, horizontal, and diagonal springs as shown in Fig.2. In the model, we assume that the object deforms in a two-dimensional plane. Position vector of the (i,j)-th mesh point $\boldsymbol{p}_{i,j} \triangleq [x_{i,j}, y_{i,j}]^T$ $(i = 0, \cdots, M \; ; \; j = 0, \cdots, N)$ is utilized in order to describe translations, rotations, and deformations of the object. The resultant force exerted on mesh point $\boldsymbol{p}_{i,j}$ can be described as follows (Fig.2):

$$F_{i,j} = \sum_{k=1}^{8} F_{i,j}^k = -\frac{\partial U}{\partial \boldsymbol{p}_{i,j}} \qquad (1)$$

where

$$\begin{aligned}
\boldsymbol{F}_{i,j}^1 &= k_c(|\boldsymbol{p}_{i,j+1} - \boldsymbol{p}_{i,j}| - l_c) e_{i,j+1}^{i,j} \\
\boldsymbol{F}_{i,j}^2 &= k_s(|\boldsymbol{p}_{i+1,j+1} - \boldsymbol{p}_{i,j}| - l_s) e_{i+1,j+1}^{i,j} \\
\boldsymbol{F}_{i,j}^3 &= k_w(|\boldsymbol{p}_{i+1,j} - \boldsymbol{p}_{i,j}| - l_w) e_{i+1,j}^{i,j} \\
\boldsymbol{F}_{i,j}^4 &= k_s(|\boldsymbol{p}_{i+1,j-1} - \boldsymbol{p}_{i,j}| - l_s) e_{i+1,j-1}^{i,j} \\
\boldsymbol{F}_{i,j}^5 &= k_c(|\boldsymbol{p}_{i,j-1} - \boldsymbol{p}_{i,j}| - l_c) e_{i,j-1}^{i,j} \qquad (2)\\
\boldsymbol{F}_{i,j}^6 &= k_s(|\boldsymbol{p}_{i-1,j-1} - \boldsymbol{p}_{i,j}| - l_s) e_{i-1,j-1}^{i,j} \\
\boldsymbol{F}_{i,j}^7 &= k_w(|\boldsymbol{p}_{i-1,j} - \boldsymbol{p}_{i,j}| - l_w) e_{i-1,j}^{i,j} \\
\boldsymbol{F}_{i,j}^8 &= k_s(|\boldsymbol{p}_{i-1,j+1} - \boldsymbol{p}_{i,j}| - l_s) e_{i-1,j+1}^{i,j}
\end{aligned}$$

and

$$e_{k,l}^{i,j} = \frac{\boldsymbol{p}_{k,l} - \boldsymbol{p}_{i,j}}{|\boldsymbol{p}_{k,l} - \boldsymbol{p}_{i,j}|}. \qquad (3)$$

At an end of the objects, there exist no springs along some directions, then the force of the corresponding spring is replaced by zero. U denotes whole potential energy of the object. In other words, function U can be calculated by sum of all energies of springs [4]. Here, we assume that the shape of the object is dominated by eq.(1). Then, we can calculate the deformation of the object by solving eq.(1) under given constraints. Note that the following discussions are valid even if the object has an arbitrary three-dimensional shape by modeling the object similarly.

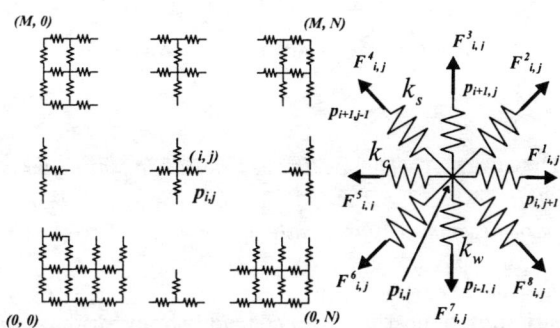

Figure 2: Spring model of two dimensional deformable object

3 Formulation of indirect simultaneous positioning operations

Here, we classify parameters $\boldsymbol{p}_{i,j} = [x_{i,j}, y_{i,j}]^T$ into the following three categories(see Fig.3) in order to formulate indirect simultaneous positioning operations.

Operation points: are defined as the points that can be manipulated directly by robotic fingers. ($\triangle$ in Fig.3)

Positioned points: are defined as the points that should be positioned indirectly by manipulating operation points appropriately, while we cannot manipulate directly. ($\bigcirc$ in Fig.3)

Non-target points: are defined as the all points except the above two points. (others in Fig.3)

Let the number of operation points and that of positioned points be m and p, respectively. Then the number of non-target points is $n = (M+1) \times (N+1) - m - p$ since the object has $(M+1) \times (N+1)$ meshe points. Note that the operation points and the positioned points are mesh points on the object. Then, $\boldsymbol{r}_m$ is defined as a vector that consists of coordinate values of the operation points, which is referred to as an operation point parameter. Positioned point parameter $\boldsymbol{r}_p$ and non-target point parameters $\boldsymbol{r}_n$ are also defined in the similar way.

Let us consider the following problem:

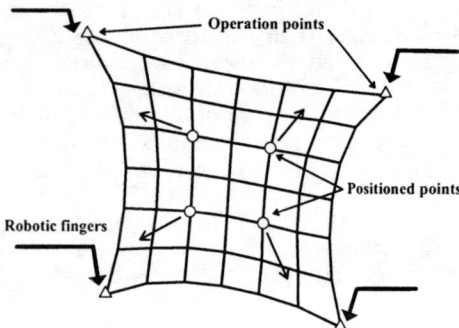

Figure 3: Classification of mesh point

Problem *The positioned points r_p are guided to their desired location r_p^d by controlling operation point parameter r_m appropriately.*

In this paper, we will assume that the goal to a positioning operation is to coincide the positioned point parameter r_p with its desired location r_p^d, say, $r_p = r_p^d$. We also assume that each robotic finger contacts with an object at one point.

Let $U(r_m, r_n, r_p)$ be a potential energy of an object. The equations of equilibrium of springs can be written as follows:

$$\frac{\partial U(r_m, r_n, r_p)}{\partial r_m} - \lambda = \mathbf{0}, \quad (4)$$

$$\begin{bmatrix} \frac{\partial U(r_m, r_n, r_p)}{\partial r_p} \\ \frac{\partial U(r_m, r_n, r_p)}{\partial r_n} \end{bmatrix} = \mathbf{0} \quad (5)$$

where a vector λ denotes a set of forces exerted on the object at the operation points r_m by robotic fingers.

It should be noted that the external forces λ can appear only in eq.(4), not in eq.(5). This implies that no external forces are exerted on positioned points and at non-target points. These equations represent characteristics of indirect simultaneous positioning operations of deformable objects.

Here, the given desired locations of positioned points should be realized. Namely, a condition $r_p = r_p^d$ should be realized by controlling the operation points r_m appropriately. However, in general, it is difficult to identify physical parameters in the model exactly. Thus, a model inversion approach is not effective. Therefore, it is important to develop a control method that is robust to the error between an object and its model.

4 Control method for indirect positioning operations of deformable objects

4.1 Infinitesimal relation among positioned points and operation points

In this section, we will propose a novel control method for indirect positioning operations of deformable objects. Applying this method, the desired location of positioned points can be achieved by modifying the position of operation points. Errors between positioned point parameter and its desired value decrease with an iterative manner based on the infinitesimal relation among positioned points and operation points. A vision sensor is used to measure the position of positioned points.

Let us derive infinitesimal relation among positioned points and operation points. Now, consider a neighborhood around an equilibrium point $r_0 = [r_{m0}^T, r_{p0}^T, r_{n0}^T]^T$. We can obtain the following equation by linearizing eq.(5) around the equilibrium point.

$$A\delta r_m + B\delta r_n + C\delta r_p = 0 \quad (6)$$

where

$$A \triangleq \begin{bmatrix} \frac{\partial^2 U}{\partial r_m \, \partial r_p} \\ \frac{\partial^2 U}{\partial r_m \, \partial r_n} \end{bmatrix}\bigg|_{r_0} \in R^{(2p+2n)\times 2m}$$

$$B \triangleq \begin{bmatrix} \frac{\partial^2 U}{\partial r_n \, \partial r_p} \\ \frac{\partial^2 U}{\partial r_n \, \partial r_n} \end{bmatrix}\bigg|_{r_0} \in R^{(2p+2n)\times 2n} \quad (7)$$

$$C \triangleq \begin{bmatrix} \frac{\partial^2 U}{\partial r_p \, \partial r_p} \\ \frac{\partial^2 U}{\partial r_p \, \partial r_n} \end{bmatrix}\bigg|_{r_0} \in R^{(2p+2n)\times 2p}.$$

Vector δr_m is defined as a deviation of the operation points from their equilibrium points. Vectors δr_n and δr_p are defined in the similar way.

We obtain the following equation from eq.(6).

$$F \begin{bmatrix} \delta r_m \\ \delta r_n \end{bmatrix} = -C\delta r_p \quad (8)$$

where $F = [A \; B]$.

The following theorems can be proved [4].

Result 1 *There exist displacements of the operation points δr_m corresponding to any displacements of positioned points δr_p, if and only if, $\mathrm{rank} F = 2p + 2n$ is satisfied.*

Result 2 *The number of the operation points must be greater than or equal to that of the positioned points in order to realize any arbitrary displacement δr_p.*

In this paper, we deal with the case of $m = p$. If there exists F^{-1}, eq.(8) can be rewritten as follows:

$$\delta r_m = -S_U F^{-1} C \delta r_p, \quad (9)$$
$$\delta r_n = -S_D F^{-1} C \delta r_p, \quad (10)$$

where $S_U = [I \mid 0]$, $S_D = [0 \mid I]$.

4.2 Control method

In this section, we will propose a novel control method to achieve a given indirect positioning operation by controlling the operation points based on the infinitesimal relation among the positioned points and operation points, which are given by eq.(8).

First, the position of operation points in the k-th trial is updated based on eq.(9) as follows:

$$r_m^k = r_m^{k-1} - k_p S_U F_{k-1}^{-1} C_{k-1}(r_p^d - r_p^{k-1}) \quad (11)$$

where F_k and C_k are functions of r_m^k, r_n^k, and r_p^k. Superscript and subscript k on variables mean their values obtained in the k-th trial. A scalar k_p describes a gain. Positions r_p^k can be measured by a vision sensor, and positions of the operation points can be computed from the motion of end effector of robot fingers as long as robot fingers grasp the object firmly. On the other hand, it is cannot be effective to measure the non-target points r_n^k due to large number of the points. Thus, we will estimate r_n^k as follows:

$$r_n^k = r_n^{k-1} - k_p S_D F_{k-1}^{-1} C_{k-1}(r_p^{k-1} - r_p^{k-2}) \quad (12)$$

Operation points r_m^k is desired positions of robotic fingers. Position control of individual fingers is achieved by a simple control law.

After robot fingers converged to r_m^k, new positions of positioned points r_p^k are measured again by a camera, as the k-th trial. Then, the same procedure is iterated.

In this paper, we deal with the case that the matrix F_k is non-singular for any k, for simplicity. We can show that the positioned points r_p^d can be converged to the desired ones by control law (11) even if the model includes some errors [4]. If $\|I - Q_k \tilde{Q}_k\| < 1$ is satisfied, $r_p^k \to r_p^d$ as $k \to \infty$ can be guaranteed, where $Q_k = -S_U F_k C_k$, $\tilde{Q}_k = -k_p S_U \tilde{F}_k \tilde{C}_k$. $\tilde{F}_k$ and $\tilde{C}_k$ denote matrices F_k and C_k including errors. Decreasing k_p yields the reduction of each element of estimated matrix $\tilde{Q}_{k-1}$. This implies that we can maintain the convergence to the desired positions by decreasing k_p despite large errors of matrix $\tilde{Q}_{k-1}$. The speed of convergence is, however, reduced when gain k_p decreases.

5 Experiments

In this section, we will show experimental results in order to illustrate the validity of the proposed control method and to investigate the effect of model errors to the convergence. In the experiments, knitted fabrics of the acrylic 85[%] and wool 15[%] (100[mm]×100[mm]) are utilized. The fabric is descritized into 3×3 meshes. The numbers of the operation and positioned points are three, respectively. Their locations on the object at initial condition are shown in Fig.4. Marks are plotted on the positioned points of the fabric. Then, their positions are measured by an image sensor. The measured positions are utilized for calculating the desired positions of the operation points using eq.(11) at each trial. Then, robot fingers grasping the operation points are controlled so that their positions converge to the desired ones. Three 2DOF robots with stepping motors are utilized as robotic fingers. The configurations of the operation and positioned points are as follows:

$$r_m \triangleq [x_{0,3}, y_{0,3}, \; x_{1,0}, y_{1,0}, \; x_{3,2}, y_{3,2}]^T,$$
$$r_p \triangleq [x_{1,1}, y_{1,1}, \; x_{1,2}, y_{1,2}, \; x_{2,2}, y_{2,2}]^T.$$

The desired positioned points used in the experiments are

$$r_p^d = [30, 40, \; 65, 50, \; 53.57, 90]^T.$$

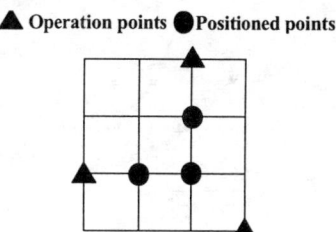

Figure 4: Configuration of points in experiments

We have identified spring constants $(k_c, k_w, k_s) = (4.17, 13.2, 3.32)$ [gf/mm] coarsely for control law in the experiments, through tensile tests. Note that the ratio of the spring constants is important in our control method. Then, we define $\alpha = k_c/k_s$ and $\beta = k_w/k_s$. From coarsely identified spring constants, $\alpha = 1.256$ and $\beta = 3.976$ are obtained. In experiments, various values of α and β including errors were utilized in the control method, in order that the effects of the deviations of α and β from the identified ones are investigated. Moreover, values 0.1 and 0.5 of gains k_p are used in eq. (11) to show that the effects of the gains. Fig.5 illustrates the experimental results of $k_p = 0.1$ and α is fixed. Fig.6 shows the results of $k_p = 0.1$ and β is fixed. Similarly, Fig.7 and 8 show the results with $k_p = 0.5$, and α and β is fixed, respectively. In all of these figures, we can find that the positioned points can converge to the desired positions if the coefficients (α, β) is near their identified values, while they are gradually oscillatory or diverge if (α, β) is far from the identified values. Fig.5 and 6 show that the positioned points converge to the desired ones despite of 100 times or 0.01 times of deviations of parameters α and β. Fig.7 and 8 show that the positioned points diverge for 100 times and 0.01 times deviations of the parameters. On the other hand, the speed of convergence is high with $k_p = 0.5$.

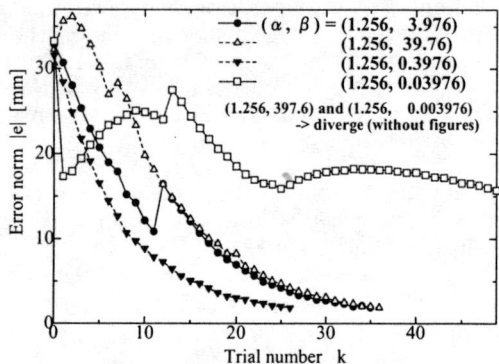

Figure 5: Experimental results ($k_p = 0.1$, $\alpha = 1.256$)

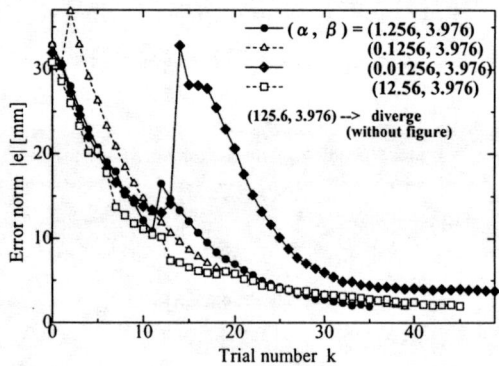

Figure 6: Experimental results ($k_p = 0.1$, $\beta = 3.976$)

As an example, Fig.9 shows behaviors of operation and positioned points with $k_p = 0.5$, $(\alpha, \beta) = (1.256, 3.976)$. The accuracy of convergence to the desired ones can be reached to a resolution level of the visual sensor (about 1[mm]).

According to the experimental results, we can conclude that very coarsely estimated parameters can be utilized in the proposed control method in a practical operations of the fabrics. Gain k_p should be chosen carefully.

6 Planning locations of robot fingers

So far, we discuss on positioning operations assuming that the locations of fingers are determined in advance. However, if the locations are not appropriate, determinant of matrix F become 0, then the operations cannot be achieved. In addition, excessive deformation of the object may occur.

In this section, we propose two methods for the planning of finger locations. In one approach, forces exerted on the object by robotic fingers are evaluated for an arbitrary displacements of the positioned points. The locations are determined so that minimal forces are realized. In another approach, forces required for a given displacements are evaluated. Now,

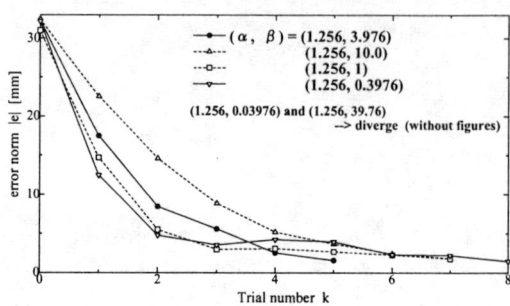

Figure 7: Experimental results ($k_p = 0.5$, $\alpha = 1.256$)

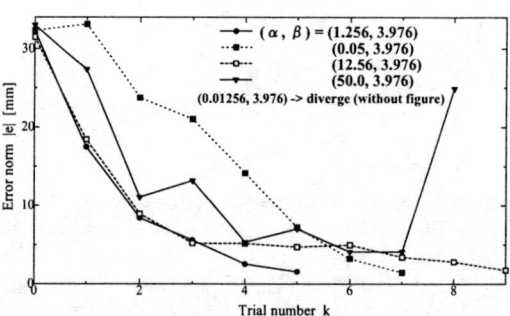

Figure 8: Experimental results ($k_p = 0.5$, $\beta = 3.976$)

we assume that meshes are generated and the locations of positioned points are given in advance.

Here, we derive an infinitesimal relationship between the positioned points and forces exerted on the object. Linearlinzing eq.(4) at their natural conditions, eq.(13) is obtained.

$$\frac{\partial U}{\partial \boldsymbol{r}_m}\bigg|_{\boldsymbol{r}_0} + \overline{F}\begin{bmatrix} \delta \boldsymbol{r}_m \\ \delta \boldsymbol{r}_n \end{bmatrix} + \overline{C}\delta \boldsymbol{r}_p = \lambda \qquad (13)$$

where matrices $\overline{F}$ and $\overline{C}$ are defined as follows:

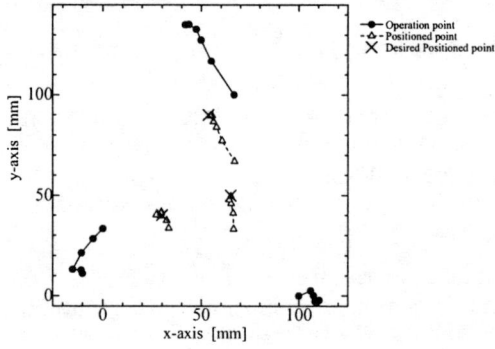

Figure 9: Behavior of positioned and operation points in experiments

$$\overline{F} \triangleq [\overline{A}\ \overline{B}],\ \overline{C} \triangleq \left[\frac{\partial^2 U}{\partial r_p \partial r_m}\right]\bigg|_{r_0} \quad (14)$$

and

$$\overline{A} \triangleq \left[\frac{\partial^2 U}{\partial r_m \partial r_m}\right]\bigg|_{r_0}, \overline{B} \triangleq \left[\frac{\partial^2 U}{\partial r_n \partial r_m}\right]\bigg|_{r_0} \quad (15)$$

Substituting eq.(8) to eq.(13), vectors r_m and r_n are eliminated, then the following equations are derived.

$$\frac{\partial U}{\partial r_m}\bigg|_{r_0} + G\delta r_p = \lambda \quad (16)$$

$$G \triangleq -\overline{F}F^{-1}C + \overline{C} \quad (17)$$

Suppose that the planning is performed before beginning of the operations, and locations of the operation points are not changed during the operations. Then, the object is in its natural shape at the planning. Thus,

$$\frac{\partial U}{\partial r_m}\bigg|_{r_0} = \lambda|_{r_0} = 0 \quad (18)$$

is satisfied. As the result, eq.(16) can be rewritten as follows:

$$\lambda = G\delta r_p \quad (19)$$

(a) For an arbitrary infinitesimal displacement
In this case, a cost function is defined as follows:

$$J_1 = max\frac{\|G\delta r_p\|}{\|\delta r_p\|} = \|G\|. \quad (20)$$

(b) For a given displacement
The locations of the fingers are determined so that the maximal force is minimized for an error with respect to their desired positions $e_p = r_p^d - r_p$. In this case, a cost functions is defined as follows:

$$J_2 = \left\|G\frac{e_p}{\|e_p\|}\right\| \quad (21)$$

In eqs.(20)and(21), in the case that $\det F = 0$, J_1 and J_2 are ∞.

In both cases, the locations are determined by minimizing cost functions.

[Example]
Suppose that $k_c = k_w = k_s = 1$, $r_p = [p_{1,1}^T, p_{1,2}^T, p_{2,2}^T]^T$, and $m = p = 3$, that is, both of the numbers of operation and positioned points are three. By calculating function J_1, we obtain the locations of operation points with $minJ_1$ as

$$r_m = [p_{0,3}^T, p_{1,0}^T, p_{3,2}^T]^T \quad (22)$$

as shown in Figure 10

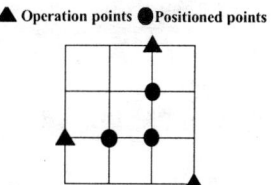

Figure 10: Configuration of operation points minimizing function J_1

7 Conclusions

In this paper, we have proposed a new control method for positioning operations of deformable objects based on their coarse model. Especially, we have focused on indirect simultaneous positioning operation of a deformable object, where multiple points on the object must be guided to the desired locations simultaneously despite the points cannot be guided directly. First, we have proposed a mathematical model of deformable objects for their positioning operations. Second, indirect simultaneous positioning operations of deformable objects have been formulated. Then, we have found that a coarse model of the object can be used in our method. Further, we discuss on planning of locations of robotic fingers based on a relationship between forces exerted on the object by fingers and infinitesimal displacement of the positioned points. Future problems are (1) an appropriate method to generate meshes, (2) extension to 3D deformable objects, and (3) learning control to improve the performance of positioning operations.

References

[1] Wakamtatsu, H., Hirai, S., Iwata, K., "Static Analysis of Deformable Object Grasping Based on Bounded Force Closure", Proc. IEEE Int. Conf. on Robotics and Automation, pp.3324-3329, 1996

[2] Howard, A.M. and Bekey, G.A., "Recursive Learning for Deformable Object Manipulation", Proc. of Int. Conf. on Advanced Robotics, pp. 939–944, 1997.

[3] Sun, D., Liu, Y., Mills, J.K., "Cooperative Control of a Two-Manipulator System Handling a General Flexible Object", Proc. of IROS'97, pp. 5–10, 1997

[4] Wada, T., Hirai, S., Kawamura, S., "Indirect Simultaneous Positioning Operations of Extensionally Deformable Objects", Proc. of IROS'98, 1998

Image-Based Robot Navigation Under the Perspective Model

Ronen Basri*
Dept. of Applied Math
The Weizmann Inst. of Science
Rehovot 76100 Israel

Ehud Rivlin
Dept. of Computer Science
The Technion
Haifa 32000 Israel

Ilan Shimshoni[†]
Dept. of Ind. Eng. and Mgmt.
The Technion
Haifa 32000 Israel

Abstract

In a recent paper we have presented a method for image-based navigation by which a robot can navigate to desired positions and orientations in 3-D space specified by single images taken from these positions. In this paper we further investigate the method and develop robust algorithms for navigation assuming the perspective projection model. In particular, we develop a tracking algorithm that exploits our knowledge of the motion performed by the robot at every step. This algorithm allows us to maintain correspondences between frames and eliminate false correspondences. We combine this tracking algorithm with an iterative optimization procedure to accurately recover the displacement of the robot from the target. Our method for navigation is attractive since it does not require a 3-D model of the environment. We demonstrate the robustness of our method by applying it to a six degree of freedom robot arm.

1 Introduction

In a recent paper [1] we proposed an approach to the problem of guiding a robot to desired positions and orientations in space. In this method the target pose is specified by an image taken from that pose (the *target image*). The task given to the robot is to move to a position where an image taken by a camera mounted on the robot will be identical to the target image. During the execution of this task the robot is allowed to take pictures of the environment, compare them with the target image, and use the result of this comparison to determine its subsequent steps. This method is attractive since it requires no advance measurement of the environment. Only the target image is given to the robot as an input. In this paper we further investigate the

---

*This research was supported in part by the Israeli Ministry of Science, Grant No. 9766. Ronen Basri is an incumbent of Arye Dissentshik Career Development Chair at the Weizmann Institute.

[†]Ilan Shimshoni is supported in part by the Goldschmidt Foundation, and is a David and Ruth Moskowitz Academic Lecturer.

method and focus on constructing robust algorithms to perform navigation under perspective projection.

Our method can be compared to studies such as [10, 8] in which the path of the robot is predetermined, and a pre-storage of the entire path is needed. This is particularly problematic if the starting position of the robot may vary. "On-line" methods (e.g [3]) exist, but are commonly limited to the 2-D plane, or need the storage of a 3-D model of the environment (e.g [2, 4]). Also of relevance is work on image-based visual servoing (see reviews in [7, 6]).

Below we present a method for visual navigation under the perspective imaging model. In our method the robot is instructed to reach a desired pose specified by a single image taken from that pose. The method then proceeds by comparing the target image to images taken by the robot as it moves toward the target. At every steps pairs of corresponding feature points are found and used to estimate the remaining displacement of the robot to the target position. A robust tracking technique that exploits our knowledge of the motion performed by the robot between frames is developed, and an iterative, least-square estimation is used to accurately determine the target position and orientation. These algorithms enable us to maintain correspondences throughout the frames and eliminate false matches. This is important in order to obtain a fast and efficient convergence to the desired pose.

Our navigation method is attractive for several reasons. The method does not require a 3-D model of the environment. The path taken by the robot is not determined in advance. Its starting position is allowed to vary as long that the initial and target images contain sufficiently many correspondences to determine the displacement between the corresponding positions. Furthermore, the path determined at every step is almost independent of the previous steps taken by the robot. Because of this property the robot may be able, while moving toward the target, to perform auxiliary tasks or to avoid obstacles, without this impairing its ability to converge to the target position. Finally, by using visual feedback the robot can overcome motion calibration errors. Applications to our method exist in almost every domain of robot navigation and manipulation. In addition, the method offers a convenient and natural relay for human-robot interface.

The paper is divided as follows. Section 2 reviews the principles of our navigation method under the perspective projection model. Section 3 introduces our tracking method and the iterative solution for the pose problem. Experimental results are shown in Section 4.

2 Perspective visual navigation

In this section we review the principles of our navigation method under perspective projection. Additional details can be found in [1]. We wish to move a robot to an unknown target position and orientation S, which is specified by an image I taken from that position. Denote the current unknown position of the robot by S', our goal then is to lead the robot to S. Below we assume that the internal parameters of the camera are all known. The external parameters, that is, the relative position and orientation of the camera in these pictures, is unknown in advance.

To determine the motion of the robot we would like to recover the relative position and orientation of the robot S' relative to the target pose S. Given a target image I taken from S and given a second image I' taken from S', by finding sufficiently many correspondences in the two images we may estimate the motion parameters separating the two images using the algorithm described in [5, 12]. This algorithm requires at least eight correspondences in the two images.

The algorithm proceeds by first recovering the essential matrix E relating corresponding points in the two images. Once the essential matrix is recovered, it can be decomposed into a product of two matrices $E = RT$, the rotation matrix R and a matrix T which contains the translation components. The rotation matrix, which determines the orientation differences between the two images, can be fully recovered. The translation components, in contrast, can be recovered only up to an unknown scale factor. These recovered translation components determine the position of the epipole in the current image, which indicates the direction to the target position. In Section 2.1 below we show how to determine whether the target position is in front or behind the current position of the robot. However, the distance to the target position cannot be determined from two images only. Note that in the presence of noise this procedure does not guarantee that the recovered matrix R would in fact represent a rotation. In Section 3 we outline an iterative, non-linear minimization procedure to compute the rigid displacement of the robot from the target. In our experiments we use the recovered rotation and translation as a starting point for this procedure.

2.1 Resolving the ambiguity in the direction to the target

Using the current and target images we have completely recovered the rotation matrix relating the two images. Since a rotation of the camera is not affected by depth we may apply this rotation to the current image to obtain an image that is related to the target image by a pure translation. After applying this rotation the two image planes are parallel to each other and the epipoles in the two images fall exactly in the same position. Denote this position by $(v_x, v_y, f)^T$. We may now further rotate the two image planes so as to bring both epipoles to the position $(0, 0, f)^T$. Denote this rotation by R_0. Notice that there are many different rotations that can bring the epipoles to $(0, 0, f)^T$, all of which are related by a rotation about $(0, 0, f)^T$. For our purpose it will not matter which of these rotations is selected.

After applying R_0 to the two images we now have the two image planes parallel to each other and orthogonal to the translation vector. The translation between the two images, therefore, is entirely along the optical axis. Denote the rotated target image by I and the rotated current image by I'. Relative to the rotated target image denote an object point by $P = (X, Y, Z)$. Its coordinates in I are given by $x = fX/Z, y = fY/Z$ and its corresponding point $(x', y', f)^T \in I'$,

$$x' = \frac{fX}{Z+t}, \qquad y' = \frac{fY}{Z+t}. \qquad (1)$$

t represents the magnitude of translation along the optical axis (so $|t| = \|(t_x, t_y, t_z)\|$), and its sign is positive if the current position is in front of the target position, and negative if the current position is behind the target position. We can therefore resolve the ambiguity in the direction by recovering the sign of t. To do so we divide the coordinates of the points in the target image with their corresponding points in the current image, namely

$$\frac{x}{x'} = \frac{y}{y'} = \frac{Z+t}{Z} = 1 + \frac{t}{Z}. \qquad (2)$$

This implies that $t = Z(x/x' - 1)$. Unfortunately, the magnitude of Z is unknown. Thus, we cannot fully recover t from two images. However, its sign can be determined since

$$sign(t) = sign(Z)\,sign(\frac{x}{x'} - 1). \qquad (3)$$

Notice that since we have applied a rotation to the target image Z is no longer guaranteed to be positive. However, we can determine its sign since we know the rotation R_0, and so we can determine for every image point whether it moved to behind the camera as a result of this rotation. Finally, the sign of $x/x' - 1$ can be inferred directly from the data, thus the sign of t can be recovered. The sign of t is determined by the sign computed by The majority of pairs of points.

2.2 Recovering the distance to the target

Computing the distance to the target is important in order to perform smooth motion in which the robot gradually translates and rotates in the same rate toward the target position and orientation. Such a gradual motion is important particularly in order to roughly maintain the same part of the scene visible throughout the motion. Unfortunately, the distance to the target cannot be determined by comparing a single image acquired by the robot to the target image. Instead, we let the robot move one step and take a second image. We then use the changes in the position of feature points due to this motion to recover the distance.

Using the current and target images we have completely recovered the rotation matrix relating the two images. Since a rotation of the camera is not affected by depth we may apply this rotation to the current image to obtain an image that is related to the target image by a pure translation. Below we refer by I' and I'' to the current and previous images taken by the robot after rotation is compensated for so that the image planes in I, I', and I'' are all parallel.

Given a point $\mathbf{p} = (x, y, f)^T \in I$, suppose the direction from the current image I' to the target position is given by $\mathbf{t} = (t_x, t_y, t_z)^T$, and that between the previous image I'' and the current image the robot performed a step $\alpha \mathbf{t}$ in that direction. Denote by n the remaining number of steps of size $\alpha \mathbf{t}$ separating the current position from the target (so that $n = 1/\alpha$). The x coordinate of a point in the target, current, and previous images are

$$x = \frac{fX}{Z}, \quad x' = \frac{f(X + t_x)}{Z + t_z}, \quad x'' = \frac{f(X + (1+\alpha)t_x)}{Z + (1+\alpha)t_z} \quad (4)$$

respectively. Eliminating X and Z and dividing by t_z we obtain that

$$n = \frac{(x' - x)(x'' - v_x)}{(x'' - x')(x - v_x)}. \quad (5)$$

The same computation can be applied to the y coordinate of the point. In fact, we can obtain a better recovery of n if we replace the coordinates by the position

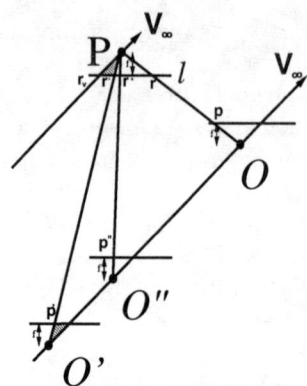

Figure 1: Geometric interpretation of the cross-ratio.

of the point along the epipolar line in the three images. (Thus, n is obtained as a cross-ratio along this line.)

The computation of the distance can be interpreted geometrically as follows. Consider Figure 1. Given three images whose centers of projection O, O' and O'' lie along a straight line leading to a point at infinity V_∞. The number of steps to the target is given by the cross-ratio of these four points. However, the positions of the centers of the cameras is not given to us explicitly. Instead, for some 3D point $\mathbf{P}$ we have its projection in the three images from which the cross ratio in (5) has been computed. These two cross ratios, however, are the same. To see this, consider the line l which lies in the plane O, O', P, is parallel to the image planes of the images, and whose distance from $\mathbf{P}$ is f (the focal length). Clearly, the cross-ratio obtained for the points on l ($\mathbf{r}, \mathbf{r'}, \mathbf{r''}, \mathbf{r}_v$) is the same as the cross-ratio we are seeking. As can be readily shown the shaded triangles in the figure are congruent. Consequently, the value obtained by the distances in the image from the points to the epipole yields the same cross-ratio as the value obtained for l, which is equal to cross-ratio we are seeking. It is obvious from the figure that the cross-ratio obtained is invariant to the choice of $\mathbf{P}$.

3 Motion recovery and tracking

When we perform visually guided navigation in the presence of extreme perspective distortions our method outlined in Section 2 may often poorly recover the relative position of the robot with respect to the target. Furthermore, maintaining correct correspondences throughout the robot's motion may be difficult, and this may prevent the robot from converging to the target position. To improve the performance of our navigation procedure we developed and implemented two algorithms. The first algorithm recovers exactly the rotation matrix R

and the epipole **v** given correct feature correspondences. The second algorithm tracks feature points from frame to frame exploiting our knowledge about the robot's motion between the frames. This algorithm is used in addition to robustly estimate distance of the robot to the target. Below we describe the two algorithms.

3.1 Motion parameters estimation

As is mentioned in the previous section, we use Hartley's algorithm [5] to recover the essential matrix E and separate it into the rotation matrix R and the epipole **v**. We then use R and **v** as initial values for a non-linear minimization procedure that finds the R and **v** which minimize a least-squares error function.

The error function that has to be chosen must minimize the squares of errors in values measured in the image (i.e., errors in image position). We first describe our procedure in the case that only translation separates the robot from the target. Later, we generalize the procedure to account for rotation as well.

Suppose that the robot is posed such that only translation separates it from the target. In this case every pair of corresponding points is collinear with the epipole **v**. However, because of noise the lines connecting all pairs of corresponding points may not intersect at a single point. One way, then, to determine the epipole **v** is to use a linear least squares algorithm which finds the point that is closest to all the lines going through the pairs of points (see Figure 2, the dashed line represents a line through two corresponding points, **p** and **p'**, and **ls** denotes the distance of the epipole **v** from this line). A better measure is as follows. Given a candidate epipole **v**, for each pair of points compute the line through **v** that is closest to the two points (the solid line in Figure 2). Now, measure the distance of the points from this line (d and d' in the figure), and add the square of each of these two distances to the error function.

Given an epipole **v** and a pair of points **p** and **p'** we find the optimal line as follows. Denote by θ the angle between the line and the x-axis, and denote $\mathbf{q} = \mathbf{p} - \mathbf{v}$ and $\mathbf{q'} = \mathbf{p'} - \mathbf{v}$. Then

$$d^2 = (\mathbf{q} \cdot (\sin\theta, -\cos\theta))^2, \quad d'^2 = (\mathbf{q'} \cdot (\sin\theta, -\cos\theta))^2.$$

To find the angle θ that minimizes $d^2 + d'^2$ we differentiate this expression by θ:

$$((\mathbf{q}_x^2 + \mathbf{q'}_x^2) - (\mathbf{q}_y^2 + \mathbf{q'}_y^2))2\sin\theta\cos\theta + 2(\mathbf{q}_x\mathbf{q}_y + \mathbf{q'}_x\mathbf{q'}_y)(\cos^2\theta - \sin^2\theta) = 0.$$

Thus,

$$\theta = \frac{1}{2}\tan^{-1}\frac{2(\mathbf{q}_x\mathbf{q}_y + \mathbf{q'}_x\mathbf{q'}_y)}{(\mathbf{q}_x^2 + \mathbf{q'}_x^2) - (\mathbf{q}_y^2 + \mathbf{q'}_y^2)}.$$

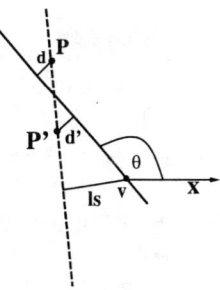

Figure 2: An illustration of the error function: given a pair of corresponding points **p** and **p'** and a candidate epipole **v** the distance between **v** and the line through the two points is **ls**. A better estimate is given by taking the distance from the two points and the solid line (the nearest line through **v** to the points). In this case the squared error value is given by $d^2 + d'^2$.

In the general case when the two images are related by both rotation and translation, we first rotate one of the images by R and then compute the function described above. This measure produces suboptimal results because the errors minimized in the rotated image are of rotated points and not of the original points measured in the image. This small penalty in accuracy is the price we pay for being able to compute the closest line going through **v** analytically.

After recovering the epipole **v** we can improve our estimate of the rotation matrix R. At this point we seek a matrix that satisfies the non-linear constraints $RR^T = I$ and $\det R = 1$. For this purpose we apply a Levenberg-Marquardt non-linear minimization procedure which estimates the rotation matrix R and translation vector **t** that bring the corresponding points to a best fit. To ensure that R satisfies the non-linear constraints we specify it using the Euler angles.

3.2 Tracking feature points

In order to recover the relative position of the robot with respect to the target it is essential to find correspondences between the images taken by the robot and the target image. In this paper we do not address the problem of finding correspondences between the initial image of the robot and the target image. In the experiments below (Section 4) we supplied these correspondences manually. However, once the robot begins moving it acquires new images, and we wish to maintain as many possible correspondences to accurately estimate the relative displacement of the robot from the target. To achieve this we track feature points between frames exploiting our knowledge of the motion performed by the robot in every step. Below we describe this tracking procedure in detail.

Let I_i and I_{i+1} denote the two images acquired by the robot in the i'th step, and let R_i and t_i denote the rotation and translation performed by the robot at this step. Given feature points in the previous image we apply the camera rotation R_i to the points to obtain the new positions of the feature points in the new image had only rotation been applied between the frames. The remaining translation component of the motion between the frames causes the points to move along epipolar lines. Denote by p'_j the position of a feature point after applying R_i and by p_j its yet unknown position in I_{i+1}. The epipolar line connecting p'_j and p_j also passes through v and so can be computed. Using the computed epipolar line we choose a candidate corresponding point and by estimating robustly the distance to the target we determine which of the corresponding points is correct.

To select the candidate correspondences we apply a corner detector to the new image I_{i+1}. Then, for every feature point p'_j in the rotated image we look for a corner point that lies close to p'_j and very close to the epipolar line through p'_j. The nearest point according to this criterion is selected to be a candidate corresponding feature to p'_j in the new image. Then, we apply Equation (5) to p_j and p'_j, which estimates the number of steps n to the goal. For all correct matches we expect to obtain approximately the same value for n, whereas for incorrect matches we expect to obtain random values. In order to find a robust estimate for n, we look for the interval where the values are densest. This is found using a shortest-window mode estimator. I.e., the values are sorted and the middle of the window of size k which is shortest is chosen as the estimate.

Once the value for n has been estimated, we have an estimate for the target position. In addition to that Equation (5) now can be inverted, and the actual position of the feature point in the new image can be **computed**:

$$x' = \frac{nx''(x - v_x) + x(x'' - v_x)}{n(x - v_x) + x'' - v_x}.$$

When the computed position is far from the position found in the previous step (a false match), a corner in the vicinity of the computed position is searched for, and if such a corner is not found the computed position is used. The result of this procedure is a set of corresponding feature points in the new image.

Once the corresponding features are obtained we re-estimate the rotation and translation to the target position. This is important due to errors in the estimates of the relative displacement of the robot in previous steps. Denote the rotation and epipole computed in the previous step by R and v, we apply the iterative procedure described in Section 3.1 using the initial values $R_i^{-1}R$ and v, which are expected to be very close to the correct rotation matrix and epipole. This process of tracking and improving the estimates of the displacement of the robot is repeated at every step of the robot until the target position is reached.

4 Experimental results

In the following experiment we used an Eshed-Robotec ER9 six degree of freedom robot arm. Throughout the sequence we extracted feature points using a variant of the SUSAN corner detector [11]. This algorithm extracted about 200 corners in each image. In the source and target images we manually selected 32 pairs of points. We then recovered the location of the epipole and the rotation that separates the source from the target using Hartley's algorithm (described in Section 2). To further improve the estimated parameters we used them as a starting point for a Levenberg-Marquardt non-linear minimization procedure (we used the MINPACK library [9]). This algorithm was described in Section 3.1.

After recovering the motion parameters we instructed the robot to perform a step toward the target pose. The magnitude of the rotation of the robot was set to a fraction of the angular difference between the source and the target. The magnitude of the translation was set arbitrarily since it could be recovered only to within a scale factor.

We then obtained a new image. To maintain correspondences we tracked the feature points between consecutive frames using the method described Section 3.2. Thus we were able not to lose the matched points, detect false correspondences, and compute the distance to the target. In addition to the geometric constraints we also used color SSD to choose between competing possible correspondences. In the eight steps of the experiment we only lost two correspondences. One because it fell outside the boundaries of the image, and the other due to a false match.

Figure 3 shows the pose estimates of the robot relative to the target and the errors in these estimates obtained in each of the steps . As can be seen the robot manages to proceed to the target almost along a straight line and to rotate to the desired orientation along a great circle. Figure 4 shows the images acquired by the robot along its path to the target (Fig. 4(a)-(g)) along with the target image (Fig. 4(h)).

5 Conclusions

In this paper we have presented a robust method for visual navigation under the perspective camera model.

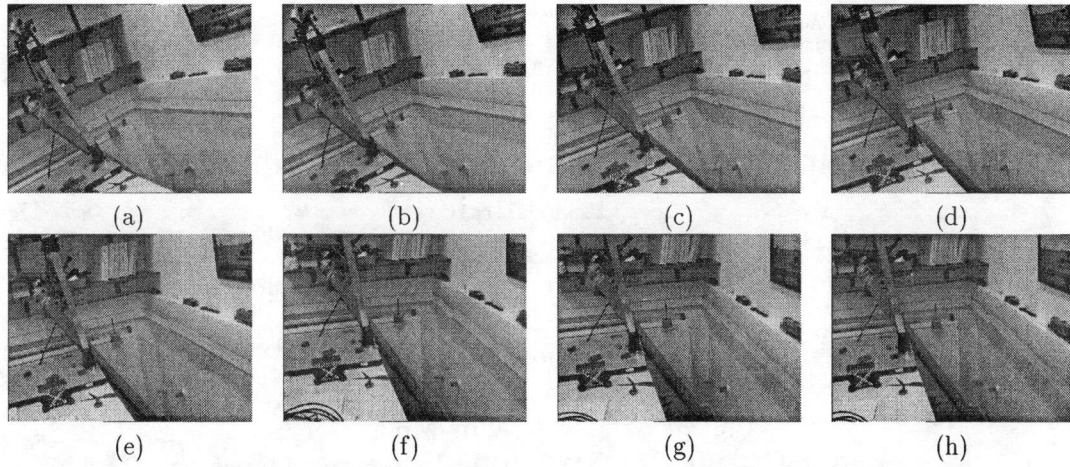

Figure 4: Real experiment: (a) The initial image; (b-f) Intermediate images; (g) Final image; (h) Target image;

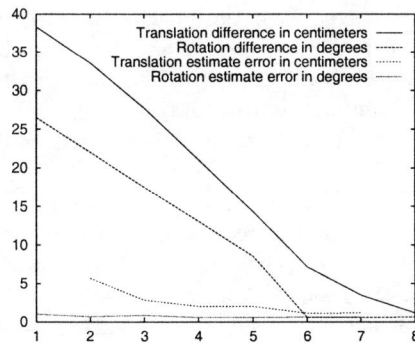

Figure 3: Experimental results: The top two curves show the difference in translation and orientation between the pose of the robot and the target pose at every step. The bottom two curves show the error in the estimate of the pose. Note that there is no estimate for the error in the translation in the first image because the translation is recovered at that point only to within a scale factor.

Using this method a robot can be sent to desired positions and orientations specified by images taken from these positions. The method requires the pre-storage of the target image only. It then proceeds by comparing the target image to images taken by the robot while it moves, one at a time. In this paper we have focused on constructing robust algorithms for performing the visual navigation by tracking the feature points from frame to frame exploiting our knowledge of the motion of the robot between the frames. This enabled us to eliminate false matches and to obtain accurate estimates of the relative displacement of the robot from the target position and orientation. Future research will be devoted to finding the initial correspondences between feature points in the source and target images.

References

[1] R. Basri, E. Rivlin, and I. Shimshoni, *Visual homing: surfing on the epipoles* ICCV-98, Bombay: 863-869, 1998.

[2] R. Basri and E. Rivlin, *Localization and homing using combinations of model views.* AI, **78**: 327-354, 1995.

[3] G. Dudek and C. Zhang, *Vision-based robot localization without explicit object models.* IEEE Int. Conf. on Robotics and Automation: 76-82, 1996.

[4] C. Fennema, A. Hanson, E. Riseman, R. J. Beveridge, and R. Kumar. *Model-directed mobile robot navigation.* IEEE Trans. on Systems, Man and Cybernetics, **20**: 1352-1369, 1990.

[5] R.I. Hartley. *In defense of the eight-point algorithm.* PAMI, **19**(6): 580-593, 1997.

[6] K. Hashimoto (Editor). *Visual Servoing* World Scientific, Singapore, 1993.

[7] S. Hutchinson, G.D. Hager, and P.I. Corke. *A tutorial on visual servo control.* IEEE Transaction on Robotics and Automation, **12**(5): 651-670, 1996.

[8] Y. Matsumoto, I. Masayuki and H. Inoue, *Visual navigation using view-sequenced route representation.* IEEE Int. Conf. on Robotics and Automation: 83-88, 1996.

[9] J.J. Moré, B.S. Garbow, and K.E. Hillstrom, *User guide for MINPACK-1.* ANL-80-74, Argonne National Laboratories, 1980.

[10] R. C. Nelson. *Visual homing using an associative memory.* DARPA Image Understanding Workshop: 245-262, 1989.

[11] S.M. Smith and J.M. Brady. *SUSAN - a new approach to low level image processing.* IJCV, **23**(1): 45-78, 1997.

[12] J. Weng, T.S. Huang, and N. Ahuja. *Motion and structure from two perspective views: Algorithms, error analysis, and error estimation.* PAMI, **11**(5): 451-476, 1989.

Autonomous Dirigible Navigation Using Visual Tracking and Pose Estimation

Mário Fernando Montenegro Campos   Lúcio de Souza Coelho

Laboratório de Robótica, Visão Computacional e Percepção Ativa
Departamento de Ciência da Computação
Universidade Federal de Minas Gerais
Av. Antônio Carlos 6627 – 31270-010 - Belo Horizonte - MG - Brazil
{mario,omni}@dcc.ufmg.br

Abstract

This paper describes a methodology for visual navigation of dirigibles based on tracking and pose estimation of visual beacons – artificial landmarks with known geometrical and visual properties. Fast pose estimation for the dirigible is necessary since the 3D position and orientation data may be used for attitude control. Robust tracking of the beacons is also an important feature. The methodology is validated through different experiments. Results on precision of position and velocity parameters, robustness in tracking the beacon in unstructured environments and time performance of the system are discussed. Finally, future improvements and applications to external airships are discussed.

1 Introduction

Autonomous navigation of unmanned aerial vehicles (UAVs) cannot be always done by using inertial, GPS and other kinds of dead-reckoning sensors. In the case where an aerial robot is used to directly coordinate a team of unmanned ground vehicles (UGVs) in a dynamical environment, such as in surface mining [1], a vision system would be necessary. In the same case, the use of structure from motion techniques by the UAV could accomplish the desired task more efficiently and with lower cost than by using the conventional approach of range sensing and map fusion by the ground vehicles. Also, visual navigation alone can be a cheaper substitutive to high precision dead-reckoning navigation, such as DGPS, when it comes to tasks as docking, clearance verification and target tracking [2].

In robot navigation, the goal is to safely reach a destination, without collisions or other harmful problems. The three subtasks involved in navigation are mapping, planning and driving [3]. The subtasks of mapping and driving are directly dependent on pose estimation.

There has been important developments in the area of visual navigation in recent years. Among these efforts are those of map based, artificial landmark-based and landmark-based. However, there has been fewer examples when it comes to aerial vehicles and 3D motion with little or no restrictions. In general, current approaches consist of a basically 2D-level visual navigation, such as the "visual odometer" of the CMU autonomous helicopter [4], leaving three-dimensional issues for dead-reckoning and inertial systems.

The authors intend to develop a completely visual 3D navigation system for UAVs, based on natural landmarks. This paper presents a first step towards that goal: a visual six degrees of freedom pose estimation system designed for use onboard of unmanned aerial vehicles (UAVs). In particular, it is intended for being used on board of an indoor autonomous blimp under development by the authors. The system is based on the tracking of artificial landmarks, or *visual beacons*. The visual system underwent preliminary tests for pose estimation of a terrestrial mobile robot. Results showing the efficiency of the system under pursued criteria are shown and future directions for improvement of the system and its evaluation experiments are pointed out.

1.1 Related Work

Visual pose estimation for terrestrial (or surface) robots is usually simplified due to constraints of 2D motion, as seen in [5]. Nevertheless, the advent of UAVs raises the subject of visual pose estimation for airborne mobile robots, such as autonomous LTAs [6]. A possible way to accomplish that is the use of image projections of artificial landmarks of a visual beacon as input for the Perspective N-Point Problem (PNPP) [7]. However, recognizing a visual beacon in unstructured environment, as well as deciding between possible solutions of the PNPP, are problematic tasks. They can in some extent be tackled by using tracking both at visual [8] and geometric levels.

1.2 Problem Characterization

The autonomous blimp discussed here is a small indoor dirigible equipped with motors for propulsion and a microcamera for visual sensing. The gondola of the blimp sends video signal from the microcamera to a commander computer and receives control signals from the computer to the propelling motors. The navigation system utilizes an off-board visual beacon – a calibration object with known visual and geometric properties – to calculate the position and orientation of the blimp in 3D space; that 6 DOF estimation is done without any previous assumption for the orientation, such as constant zero pitch with respect to the ground. Figure 2 depicts the indoor dirigible facing the beacon used in this work. A close-up of the gondola can be seen in Figure 3.

A mission starts with take-off from the platform where the mast that holds the mooring device of the blimp is mounted. That platform and mast are known objects to which visual beacons are attached. The blimp is docked to the mast in such a way that the visual beacon and the platform surface are on the field of view of the microcamera and can be easily identified by

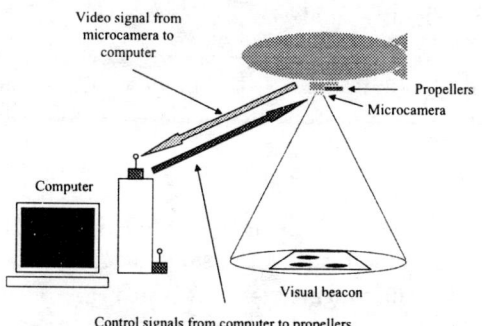

Figure 1: overall vision of the system associated to the autonomous blimp.

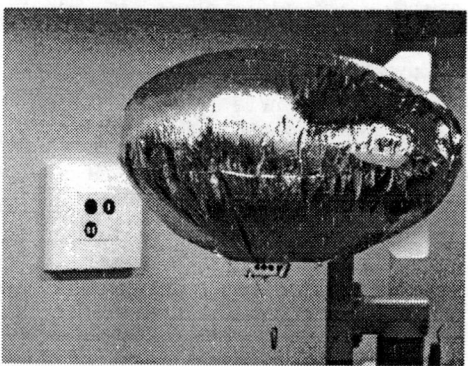

Figure 2: view of the blimp floating next to the visual beacon.

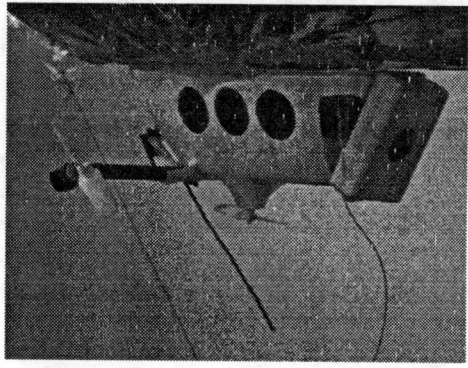

Figure 3: close up of the blimp gondola, with its motors and camera.

2 Methodology

A brief summary of the basic methodologies of image processing and computational geometry used by the system, described in more detail in [9], is presented here. A visual beacon with triangular topology, assigned by three black circular markers, is identified in the image by simple threshold segmentation and contour following. The rectangular boundaries of the image projections of each circular marker are then used as input to a hybrid method – analytical but with numerical refinement – for solving the PNPP, thus calculating pose.

The correct operation of those methodologies precludes that the frames captured by the camera always show the beacon lying on an white surface occupying all the image, allowing easy threshold segmentation. Also, the camera has to be always close enough to the beacon to produce appreciable perspective distortions in the image, allowing a closed solution for the PNPP.

Clearly the assumptions above are very strict and difficult to attain in real world. Therefore, in order to surpass those limitations, a simple *tracking* strategy was adopted in addition to those described in [9]. The following sections describe how that tracking methodology operates at both visual and geometrical levels.

2.1 Visual Tracking

The tracking methodology is based on the assumption that the camera's position and orientation does not change significantly between two consecutive frame acquisitions. It is also assumed that the *rate* of change of pose is constant. Thus, given two locations containing the beacon in two consecutive images, it is possible to predict the new location in the third frame.

The image of the beacon location is defined by a window – a rectangular region of the image containing the beacon. Formally, a window is expressed as a quadruple (x, y, w, h), where (x, y) is the lower left corner pixel of the window and w and h are the width and the height of the window, respectively. Let $W = (x_W, y_W, w_W, h_W)$ be the window describing the beacon location in the image. The image rectangular boundaries of the markers of the beacon can also be seen as windows. Let R be the set of such rectangular boundaries. W contains the image of the beacon according to the definitions below:

$x_W = x_L \mid ((x_L, y_L, w_L, h_L) \in R) \wedge (x_L \leq x \forall (x, y, w, h) \in R)$ (x_L is the x coordinate of the leftmost rectangular boundary in R).

$y_W = y_B \mid ((x_B, y_B, w_B, h_B) \in R) \wedge (y_B \leq y \forall (x, y, w, h) \in R)$ (y_B is the y coordinate of the bottommost rectangular boundary in R).

$w_W = x_R + w_R - x_L \mid ((x_R, y_R, w_R, h_R) \in R) \wedge (x_R + w_R \geq x + w \forall (x, y, w, h) \in R)$ ($x_R + w_R - 1$ is the column where the rightmost rectangular boundary side lies).

$h_W = y_T + h_T - y_B \mid ((x_T, y_T, w_T, h_T) \in R) \wedge (y_T + w_T \geq x + w \forall (x, y, w, h) \in R)$ ($y_T + w_T - 1$ is the line where the uppermost rectangular boundary side lies).

Let $W_1 = (x_1, y_1, w_1, h_1)$ and $W_2 = (x_2, y_2, w_2, h_2)$ be the locations of the image of the beacon in two successive camera frames F_1 and F_2, respectively. A predicted third window W_3' is given by

$$W_3' = (2x_2 - x_1, 2y_2 - y_1, 2w_2 - w_1, 2h_2 - h_1)$$

It can be seen from the equation above that the coordinates of W_3' are computed by simply adding the two-dimensional image displacement vector $[\, x_2 - x_1 \;\; y_2 - y_1 \,]$ to (x_2, y_2). Since the width and the height of these tracking windows can change between frames, they are also estimated using the same assumption of smooth variation. This can be done by computing a dimensional vector $[\, w_2 - w_1 \;\; h_2 - h_1 \,]$ and adding it to (w_2, h_2), yielding the dimensions of the predicted window.

The search for the beacon markers, using threshold segmentation and contour following, is then performed on the pixels defined by the predicted window. Nevertheless, the chain code algorithm for contour following is allowed to go beyond the window boundaries, thus defining the whole contour of the markers even if they are not perfectly contained in the predicted window. Once the contours are defined, the actual beacon image location W_3 is calculated and stored, as well as W_2, for the calculation of the next predicted window.

2.2 Geometrical Tracking

The above methodology presents a problem when the camera is sufficiently far from the beacon. In this case, the projected area of the visual markers on the visual plane is about the same, and the corresponding calculated relative distance differences between them tend to vanish. In such cases, the system is prone to provide inadequate initial conditions for the numerical module, which then converges into an incorrect solution for the PNPP problem.

This limitation can be overcome by incorporating the geometrical results in a frame-to-frame technique somewhat similar to that used for visual tracking. Assuming that the system starts tracking the target in a position sufficiently close to the beacon, the transform describing the beacon for the first frame can be straightforwardly calculated using the geometric procedures from [9]. From the second frame on, the estimation of the initial tentative solution is modified. Instead of using the ratios of the areas of projections of markers, the initial solution for a frame $n > 1$ is generated from the final vectors $\vec{V}_{0,n-1} = \epsilon_{0,n-1}\vec{e}_{0,n-1}$, $\vec{V}_{1,n-1} = \epsilon_{1,n-1}\vec{e}_{1,n-1}$ and $\vec{V}_{2,n-1} = \epsilon_{2,n-1}\vec{e}_{2,n-1}$ calculated for the frame $n - 1$. The new initial values of $\epsilon_{0,n}$, $\epsilon_{1,n}$ and $\epsilon_{2,n}$ are then calculated as follows:

$$\epsilon_{i,n} = \left|\left[\; x_{i,n}\frac{z_{i,n-1}}{f} \;\; y_{i,n}\frac{z_{i,n-1}}{f} \;\; z_{i,n-1} \;\right]\right|, 0 \leq i \leq 2,$$

where f is the focal distance of the camera. That is, $\epsilon_{i,n}$ is such that $\vec{V}_{i,n}$ has the same magnitude in Z of $\vec{V}_{i,n-1}$, but the magnitudes in X and Y are scaled in a way that the projections of the point $P_{i,n} = F + \vec{V}_{i,n}$ are the points p_i that were determined during the image processing methodology. In other words, the vertices of the visual beacon are assumed to have the same Z coordinates obtained in the previous frame, but with X and Y coordinates that guarantee the projections of the vertices actually found in the image plane.

The $\epsilon_{i,n}$ are very close to their real values, since displacements and rotations of the beacon between frames are assumed to be very small. Hence, the $\epsilon_{i,n}$ calculated as shown above are used as the initial condition for the numerical module, ensuring that a correct final solution based on the previously obtained data is computed. The use of previous solutions as just initial conditions for searching a new point in the solution space of the problem ensures that there is no error accumulation along the process.

3 Experimental Results

The experiments presented in this section were planned in order to evaluate the performance of the system under the following three requirements:

Spatial Accuracy: the system should provide pose estimation, as well as estimation of other related measurements such as trajectories and velocities, accurately enough to enable robot navigation.

Temporal Efficiency: pose estimation has to happen at a rate compatible with the expected velocities of the mobile robot used.

Robustness: the system should be able to track a beacon, hence it should be able determine locating the beacon even in frames containing other objects from the environment.

The methodology was evaluated in two different situations called Type I and Type II. Those were investigated to compare the performance of the system under the three criteria above.

In Type I experiments a Sony XC77-CE camera, with a $25mm$ lens was used. This camera was mounted on the top of a Nomad 200 mobile robot. In each Type I experiment, the robot was programmed to move one meter in a straight line trajectory along the Z axis of the beacon coordinate system. (Both the beacon and the robot wheels were manually aligned to rectilinear features on the floor of the lab, ensuring the movement described.) The translations were performed at a constant speed of $v = 101.6mm/s$ (or 4 inches per second). The starting point of all translations was at a distance of $78cm$ from the beacon. Figure 4 shows a picture of this experimental assembly.

Figure 5 summarizes the results for ten Type I experiments. The distance computed by $d(t) = d_i + vt$, where d_i is the initial Z coordinate and t is the elapsed time since the beginning of the translation, was used

Figure 4: experimental assembly for the experiments.

as a "ground-truth" comparative measure. Then, a percent error was calculated by $Error = 100\frac{z(t)-d(t)}{d(t)}$, where $z(t)$ is the value estimated by the system for the Z coordinate at time t. The graph in Figure 5 shows that error for 527 data points collected along the experiments. For purposes of visualization, that data set was divided horizontally in intervals of $100mm$. The vertical bars show the standard deviation for the subset corresponding to each interval.

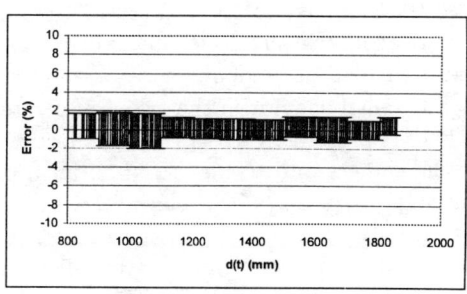

Figure 5: error of Z estimation *versus* "ground-truth" distance along Z for ten Type I experiments.

Errors are within the range between -8 and $+8\%$, however, the standard deviation is quite smaller than the magnitude of that range. That implies that system accuracy in general is about 2% of the distances in the Type I experiments. That seems to be reasonable

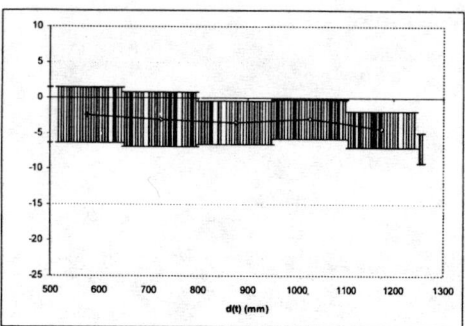

Figure 6: error of Z estimation *versus* "ground-truth" distance along Z for ten Type II experiments.

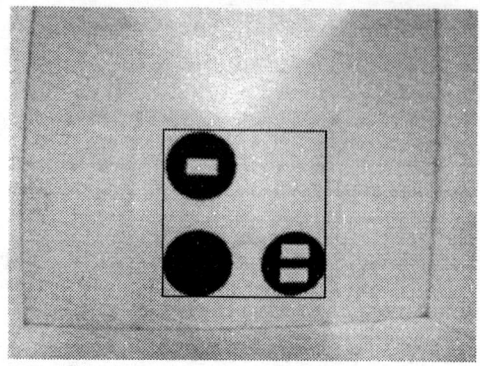

Figure 7: image captured by the microcamera at the start position of a Type II translation.

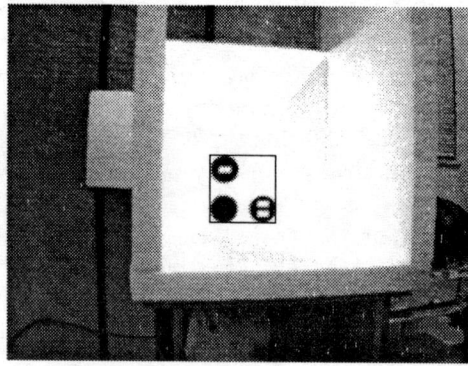

Figure 8: images captured by the microcamera at the finish position of a Type II translation.

for all the applications envisioned for the system but docking. Also, contrary to intuitive expectations, the error does not seem to have any direct relation with the distance to the beacon in the range encompassed by the experiments.

A Type II experiment is in some ways similar to a Type I. However, the camera used is the on board microcamera of the blimp. It has a focus of $6.8mm$ and the total translation is of just $75cm$, starting with an initial distance between the beacon and the camera of $50cm$. Type II experiments showed a different behavior.

The results of ten Type II experiments are shown in the graph of Figure 6. The same conventions of Type I experiments, shown in Figure 5, are used here, but length intervals are $150mm$ long. Moreover, the light square markers inside each interval represent the average error for that interval. It can be seen that in this case the error is directly related to the distance from the beacon. Nevertheless the error seems to be systematic instead of random.

The different behavior in the two study cases presented here is probably related to the different geometry of the camera and to the use of an ideal pinhole camera model by the system. The relatively large focal distance greatly reduces radial lens distortion in the case of the Sony camera. In the microcamera, the focus is small and the radial distortion is large. This is clearly due to the inadequacy of a simple pinhole camera model, which upsets the accuracy of the system. Until a distance of about a meter from the beacon, the accuracy for both cameras is about the same, but beyond that their performance start to be quite different.

The geometric model would also explain part of the systematic error in Type II experiments. Other part could be caused by some lack of synchronism between the system and the robot. Inaccurate measurement of the initial distance is very likely to occur due to odometric problems.

Type I and Type II experiments also allowed the measurement of the system performance in terms of time. The computer used to run the system has a Pentium II processor with a clock of $233MHz$ and $64Mbytes$ of RAM memory, running Windows98 with only the typical interactive load. The time to process each frame was measured and the average frame rate for a data set of 54 processing times was of $5Hz$. Although this frequency is far below a state-of-art frame rate of $60Hz$, it appears to be sufficient for the visual navigation of a blimp. Indeed, the speed of approximately $10cm/s$ used by the mobile robot in experiments Type I and II is roughly the maximum expected speed for the indoor autonomous blimp under construction.

The evaluation of the robustness of the system was also performed during experiments of types I and II.

Unfortunately, in this case the evaluation can be only qualitative. Figure 7 and 8 shows the images captured by the microcamera at the start and finish positions, respectively, of a Type II translation. The fovea window as it would be calculated by the system is shown in both images as a black rectangle. It can be seen that the beacon appears in an unstructured environment in the image after, and certainly that was also the situation along a considerable part of the translation. But the correct functioning of the tracking technique allowed the pose estimation even in those unstructured frames.

4 Conclusions and Future Directions

The experimental results suggest that the tracking system designed for the autonomous blimp is able to perform acceptable pose estimation inside a radius of one meter around a visual beacon with dimensions in the order of one decimeter. That workspace radius could be enlarged in direct proportion with the enlargement of the beacon dimensions. However, results with a camera with large focus, as well as related work [10], indicates that the workspace could also be expanded by improving the camera model and calibration. The sources of systematic errors identified need further study. Exhaustive experiments specially dedicated to evaluation of sensitivity, as described in [11], plus more accurate ground-truth comparisons should help identify and predict those error sources. The system will soon be ported to the AS800 dirigible of the Automation Institute of the Fundação Centro Tecnológico para Informática. This dirigible currently has only INS and GPS navigation systems. Experiments will be performed in the integration and use of the different sensor modalities in order to evaluate better configurations for different flight phases.

Acknowledgments

The authors wish to thank Vijay Kumar and Hong Zhang from the GRASP Lab, University of Pennsylvania, and Alberto Elfes, from the AI/CTI for invaluable discussions and suggestions. This research was partially supported by CNPq 522618/96-0NV and 170645/97-5NV-BSP and FAPEMIG TEC-609/96.

References

[1] Cork, P.I. et al, "Applications of Robotics and Robot Vision to Mining Automation", in *Proceedings of Workshop on Robotics and Robot Vision*, Gold Coast, Australia, 1996.

[2] Rock, S. M. et al, "Combined CDGPS and Vision-Based Control of a Small Autonomous Helicopter", in *Proceedings of the 1998 American Control Conference*, Philadelphia, USA, 1998.

[3] McKerrow, J. P., *Introduction to Robotics*, Addison-Wesley, Sidney, 1998.

[4] Miller, R., Mettler, B. and Amidi, O., "Carnegie Mellon University's 1997 International Aerial Robotics Competition Entry", in *Proceedings of the AUVSI'97 - Association for Unmanned Vehicle Systems: International*, 1997.

[5] D. Jung, J. Heinzmann and A. Zelinsky, "Range and Pose Estimation for Visual Servoing of a Mobile Robot", in *Proceedings of the 1998 IEEE International Conference on Robotics and Automation*, Leuven, Belgium, May 1998, pp. 1226-31.

[6] A. Elfes et al, "Project AURORA: Development of an Autonomous Unmanned Remote Monitoring Airship", in *Journal of the Brazilian Computer Society - Special Issue on Robotics*, Number 3, Vol. 4, April 1998, pp. 70-8.

[7] D. DeMenthon and L. S. Davis, "Exact and Approximate Solutions of the Perspective-Three-Point Problem", in *IEEE Transactions on Pattern Analysis and Machine Intelligence*, Vol. 14, Number 11, November 1992, pp. 1100-5.

[8] H. Wang et al, "Real-time object tracking from corners", in *Robotica*, 1998, Vol. 16, pp. 109-16.

[9] L. S. Coelho and M. F. M. Campos, "Computer Vision-Based Navigation for Autonomous Blimps", in *Proceedings of the SIBGRAPI'98 - Brazilian Symposium on Computer Graphics and Imagem Processing*, Rio de Janeiro, Brazil, 1998.

[10] R. Y. Tsai, "An Efficient and Accurate Camera Calibration Technique for 3D Machine Vision", in *Proceedings of IEEE Conference on Computer Vision and Pattern Recognition*, Miami Beach, FL, 1986, pp. 364-374.

[11] C. B. Madsen, "A Comparative Study of the Robustness of Two Pose Estimation Techniques", in *Machine Vision and Applications - Special Issue on Performance Characterization*, Vol. 9, February 1997, pp. 291-303.

Robotic Wheelchair with Three Control Modes

Yoshinori Kuno, Satoru Nakanishi, Teruhisa Murashima,
Nobutaka Shimada and Yoshiaki Shirai

Department of Computer-Controlled Mechanical Systems, Osaka University
2-1, Yamadaoka, Suita, Osaka 565-0871, Japan
kuno@mech.eng.osaka-u.ac.jp

Abstract

With the increase of senior citizens, there is a growing demand for human-friendly wheelchairs as mobility aids. This paper proposes the concept of a robotic wheelchair to meet this need. It has three motion control modes. One is an autonomous control mode. The wheelchair has autonomous navigation capabilities, such as avoiding obstacles and moving straight toward a target object or along a corridor using ultrasonic and vision sensors. Another is a manual control mode with minimum intentional operations. The wheelchair observes the user's face by vision, moving in the direction indicated by the use's face movements. The third is a remote control mode. When the user is not riding, it can find him/her using face recognition. The user can make it come or go by hand gestures. We have developed an experimental robotic wheelchair based on the above concept. Experimental results prove our approach promising.

1 Introduction

As the number of senior citizens has been increasing year by year, demand for human-friendly wheelchairs as mobility aids has been growing. Recently, robotic wheelchairs have been proposed to meet this need [1][2][3]. These wheelchairs help humans with the aid of ultrasonic, vision, and other sensors to avoid obstacles, to go to pre-designated places, and to pass through narrow or crowded areas.

In this paper, we propose a robotic wheelchair that has three motion control modes.

One is an autonomous control mode, which has been investigated in conventional robotic wheelchair researches.

Another is a manual control mode with minimum intentional operations. Conventional researches have pointed out the necessity of user-friendly interfaces for robotic wheelchairs. However, users need to continue intentional operations with these systems although the operations may be easier than usual power wheelchairs. We propose a manual control method that reduces intentional operations by observing human behaviors. If the wheelchair system could guess the user's intentions only from the behavior observations, the user would not need intentionally to give commands to the system. However, this is difficult to be achieved. Thus, we consider to use human behaviors which are done intentionally but so naturally that the user may not feel any burden. We use face direction. The wheelchair moves in the direction of the user's face. For example, it turns left if the user turns the face to the left. The user should know this fact and needs to move the face intentionally. However, it is a natural behavior to look in the direction where he/she intends to go. It is also a natural behavior turning back the face to the frontal position as the turn is getting completed. Although this behavior can be done almost unconsciously, it can control the wheelchair appropriately. When we use a steering lever or wheel to control the vehicle motion, we need to operate it intentionally all the time. Using face direction can remove a considerable part of such intentional operations, realizing a user-friendly interface.

The third is a remote control mode, which has not been considered in conventional systems. This is the control mode when the user is not riding. People who need a wheelchair may find it difficult to walk to their wheelchair. Thus, our wheelchair has the capability to find the user by face recognition. Then, It can recognize his/her hand gestures. The user can make it come or go by hand gestures.

This paper describes the design concept of our robotic wheelchair and shows our experimental system with operational experiments.

2 System Design

Figs. 1 and 2 show the configuration of our robotic wheelchair system and its overview, respectively. As a computing resource, the system has a PC (Pentium II 266 MHz) with a real-time image processing board consisting of 256 processors developed by NEC [4]. The system has sixteen ultrasonic sensors and two video cameras. One camera is set up to see the outside of the system; and the other to see the inside of the system, that is, the camera is set to look at the user's face. Using these sensor data, we realize the following three modes to control the wheelchair motion.

Autonomous control

The system uses the ultrasonic sensor data to detect and avoid obstacles. This obstacle-avoidance behavior overrides other behaviors except those done manually with the joystick to ensure the safety.

With conventional power wheelchairs, we need to hold the joystick to control the motion all the time even when we just want to go straight. In addition to the obstacle-avoidance behavior, the current system has two autonomous behaviors to reduce the burden of users in such cases.

One is a target-tracking behavior. It is initiated by the face direction observation result in the manual control mode. If the face is looking straight for a while, the system considers that the user wants to go straight, starting this behavior. If it can find a certain feature that can be tracked around the image center, it controls the wheelchair motion to keep the feature in the image center. When it cannot find such a feature, it detects several features, calculating geometrical relations among them to control the wheelchair motion.

The other is a wall-following behavior. When the wheelchair moves along a corridor, it keeps its distance from the wall using the ultrasonic sensor data. We realize this behavior using the method developed for behavior-based mobile robots [5].

Manual control with minimum intentional operations

We need to express our intentions such as turning right or left to the system. We use a joystick and other devices to do this with conventional power wheelchairs. However, using such devices requires us continuous intentional operations. Our wheelchair system has the camera to observe the user's behaviors. It uses the results of observations to reduce the burden of operations.

Remote control

This is the control mode when the user is not riding. People who need a wheelchair may find it difficult to walk to their wheelchair. Thus, our wheelchair has the capability to find the user by face recognition. Then, It can recognize his/her hand gestures. The user can make it come or go by hand gestures.

Although the first control mode is important, it has been investigated in conventional robotic wheelchair researches. Thus, we concentrate on the second and third modes in this paper.

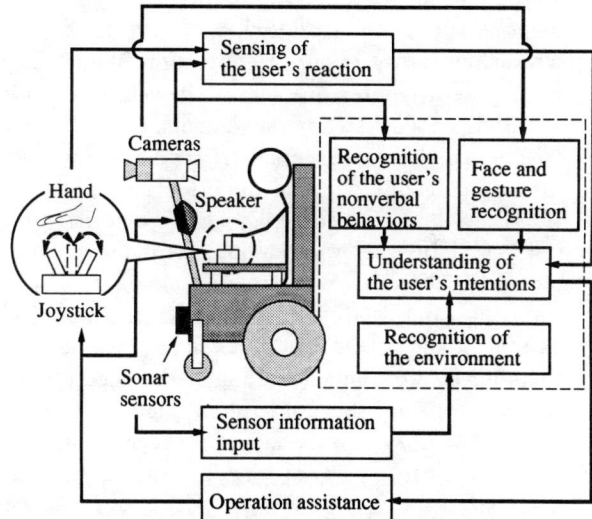

Figure 1: System configuration.

Figure 2: System overview.

3 Manual Control Using Minimum Intentional Operations

3.1 Use of Face Direction

We use face direction for motion control. As mentioned in Introduction, this will make a user-friendly interface.

The problem, however, is that we move our head in various other occasions than in controlling the wheelchair's direction. The robotic wheelchair system needs to separate wheelchair-control behaviors from others. This is the intention understanding problem in the current system. We assume that we move our head slowly and steadily when we are told that the wheelchair moves in our face direction. Thus, the current system ignores quick head movements, only responding to slow steady movements.

There are, however, cases that we look in a certain direction steadily without any intention of controlling the wheelchair. For example, when we notice a poster on a corridor wall, we may look at it by turning our face in its direction while moving straight. The environmental information around the wheelchair can be helpful in such cases. In the example case, even if we turn our face to the left to look at the poster, the system can consider this not to show a left-turn intention if it knows that a wall exists on the left and that it cannot turn left. We have made a simple experiment on this.

3.2 Face Direction Computation

We do not need precise face direction computation because the system uses the face direction data only to determine the turning direction where the user wants to go. Thus, we use the following simple method.

Fig.3 shows each step of the process. Fig.3 (a) is an original image. First, we extract skin color regions in the image (Fig.3 (b)), choosing the largest region as the face region (Fig.3 (c)). Then, we extract dark regions inside the face region, which contain face features such as the eyes, eyebrows, and mouth (Fig.3 (d)). We compare the centroid of the face region and that of the all face features combined. In Figs. 3 (c) and (d), the vertical line passing the centroid of the face region is drawn in a thick line; and in Fig. 3 (d), that passing the centroid of the face features is drawn in a thin line. If the latter lies to the right/left of the former (to the left/right in the image), the face can be considered as turning right/left. The output of the process is this horizontal distance between the two centroids, which roughly indicates the face direction.

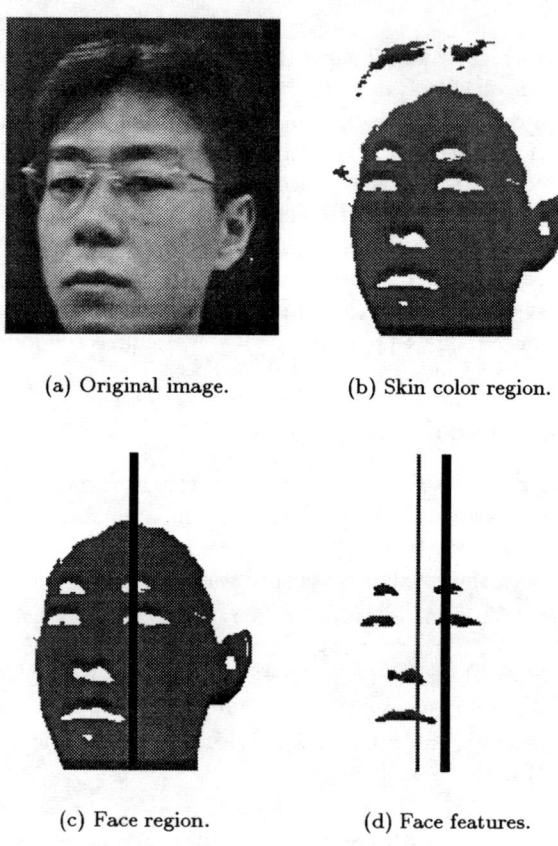

(a) Original image.   (b) Skin color region.

(c) Face region.   (d) Face features.

Figure 3: Face direction computation.

3.3 Experiments

The system can compute face direction 30 frames per second. We apply a filter to the direction data to separate wheelchair-control behaviors from others. The filter that we use is a simple smoothing filter by averaging the values over a certain number of frames. If this number is large, the system will not be affected by quick unintentional head movements. However, the user may feel uneasy at the slow response of the system. We made actual running experiments by changing the number of frames n used in the filter.

First, we examined the degree of uneasiness about the slow response. We used six subjects. They were male students at our university and not regular wheelchair users. We told them that the wheelchair would move in their face direction. They drove the wheelchair without any problem. This confirms that face direction can be used for the wheelchair opera-

Table 1: Experimental evaluation for response time by six subjects (A-F)

Frames(n)	A	B	C	D	E	F	Total (shown by the number of *'s)
1	1	2	3	3	2	3	14 **************
3	2	2	3	3	3	2	15 ***************
5	3	3	2	3	3	3	17 *****************
6	1	3	2	2	2	2	12 ************
7	1	3	2	2	2	2	12 ************
8	1	3	3	2	1	2	12 ************
9	1	2	2	1	2	1	9 *********
10	1	2	2	1	2	2	10 **********
12	1	1	2	1	2	2	9 *********
14	1	1	1	1	1	2	7 *******
16	1	1	1	1	1	2	7 *******
18	1	1	1	1	1	1	6 ******
20	1	1	1	1	1	1	6 ******

tion. They were asked to give their subjective evaluation score for each smoothing filter condition from the viewpoint of the system's response to their head movements: 1 for not good, 2 for moderate, and 3 for good. Table 1 shows the result. When n is small, the wheelchair responds sensitively to any head movements. Thus, the scores are a little small. When n is large, the wheelchair does not respond soon even when the user turns the head intentionally. The scores in these cases are considerably small. The highest score is obtained for $n = 5$.

Second, we examined whether the system would be affected by quick unintentional head movements. We consider three levels of movements: quick movements with duration less than 0.5 second (level 1), moderate speed movements with duration from 0.5 to 1 second (level 2), and slow movements with duration from 1 to 1.5 seconds (level 3). At the level 3, we turn our head and can read characters. Simple filtering cannot discriminate such movements from wheelchair-control movements. Thus, the purpose here is to make the system not to be affected by up to the level 2 movements. We asked a subject to move his head five times for each level while the wheelchair was moving straight. Then, we examined whether or not the wheelchair motions were affected. Table 2 shows the result. It suggests that n should be equal to or larger than 15 for the system not to be affected by movements at the levels 1 and 2.

The subjects commented as follows after the experiments. When they turned the wheelchair to the left/right, they did not mind the slow response. However, when the turn was getting completed and they turned back their head to the frontal position, they felt uneasy if the response was slow. In the former case, they did turning-head behaviors intentionally. Thus, they did not mind the slow response because their movements themselves were slow and steady. However, in the latter case, their behaviors were almost unconscious and quick. Thus, the slow response caused their uneasiness.

Based on this observation, we have modified the system to use 5 frames for smoothing for center-oriented face directions and 15 for left/right directions. The six subjects have given the same high scores as in the case of n=5 to this modified version wheelchair. Also, experimental runs have shown that it has the same degree of stability against unintentional quick head movements as in the case of n=15.

Although the filtering process solves the problem of quick unintentional movements, there are times that we look in a direction for a while even though we do not intend to turn in the direction. An example case is that we notice a poster on a wall and examine it. In such cases, the environmental information around the wheelchair can be useful. In the example case, even if we turn our face to the left to look at the poster, the system can consider this not to show a left-turn intention if it knows that a wall exists on the left and that it cannot turn left.

We made a simple experiment on this idea. We gave map data to the system. The system was able to tell the existence of walls and other obstacles using the dead reckoning data and the map data. The experiment was successful. In this experimental run, the user turned his face to the right twice: to look at a poster on the wall and to show his intention of turning right. The wheelchair responded only to the second movement of the user's head.

We made several other running experiments. These

Table 2: Experimental evaluation for unintentional head movements

Frames(n)	Level 1					Level 2					Level 3				
5	×	×	×	×	×	×	×	×	×	×	×	×	×	×	×
10	○	○	○	×	×	×	×	×	×	×	×	×	×	×	×
15	○	○	○	○	×	○	○	○	○	×	○	×	×	×	×
20	○	○	○	○	○	○	○	○	○	○	○	○	○	○	×
30	○	○	○	○	○	○	○	○	○	○	○	○	○	○	○

○: The wheelchair motion was not affected. ×: The wheelchair motion was affected.

experimental results confirm that we can control the wheelchair by face direction. Owing to smoothing, the system does not respond to quick head movements. Thus, if we look for a while in the direction where we want to go, we can move the wheelchair to meet our intention. However, this slow response means that we cannot make quick precise control of the wheelchair. We may not be able to avoid an obstacle if it suddenly comes close. Thus, we program the system so that the autonomous control to avoid obstacles using ultrasonic sensors can override the manual control.

The sensor information around the wheelchair can also be helpful to choose intentional head movements. Instead of map data, we are working on the use of ultrasonic and vision sensors.

4 Remote Control Using Face and Gesture Recognition

In various occasions, people using wheelchairs have to get off. They need to move their wheelchairs where they do not bother other people. Then, when they leave there, they want to make their wheelchairs come to them. It is convenient if we can do these operations by hand gestures, since hand gestures can be used in noisy conditions. However, computer recognition of hand gestures is difficult in complex scenes. In our typical cases, many people are moving with their hands moving. Thus, it is difficult to distinguish a coming-here or any other command gestures by the user from other movements in the scene.

We propose to solve this problem by combining face recognition and gesture recognition. Our system first extracts face regions, detecting the user's face. Then, it tracks the user's face and hands, recognizing hand gestures. Since the user's face is located, a simple method can recognize hand gestures. It cannot be distracted by other movements in the scene.

The system extracts skin color regions by color seg-

(a) Original image. (b) Detection result.

Figure 4: Face recognition to detect the user (Person A).

mentation. It also extracts moving regions by subtraction between consecutive frames. Regions around those extracted in both processes are considered as face candidates. The system zooms in each face candidate one by one, checking whether or not it is really a face by examining the existence of face features. Then, the selected face region data are fed to the face recognition process. We use the face recognition method proposed by Moghaddam and Pentland [6]. Images of the user from various view angles are compressed in an eigenspace in advance. The observed image is projected onto the eigenspace and the distance-in-feature-space (DIFS) is computed. The method obtains the distance-from-feature-space (DFFS) by the sum of squared distance between the input image and the reconstructed image from the eigenspace data. Using both data, the system identifies whether or not the current face is the user's face. Fig. 4 shows an example of face recognition.

After the user's face is detected, it is tracked based on the simple SAD (sum of absolute difference) method. Moving skin color regions around and under the face are recognized as the hands. They are also tracked. Fig. 5 shows a tracking result. The relative hand positions with respect to the face are computed.

The spotting recognition method [7] based on continuous dynamic programming carries out both segmentation and recognition simultaneously using the position data.

Gesture recognition in complex environments cannot be perfect. Thus, the system improves the capability through interaction with the user. If the matching score for a particular registered gesture exceeds a predetermined threshold, the wheelchair moves according to the command indicated by this gesture. Otherwise, the gesture with the highest matching score, although it is smaller than the threshold, is chosen. Then, the wheelchair moves a little according to this gesture command. If the user continues the same gesture after seeing this small motion of the wheelchair, it considers that the recognition result is correct, carrying out the order. If the user changes his/her gesture, it begins to recognize the new gesture to iterate the above process. In the former case, the gesture pattern is registered so that the system can learn it as a new variation of the gesture command.

Experiments in our laboratory environments where several people are walking have confirmed that we can move the wheelchair by the same hand gestures as we use between humans.

Figure 5: Tracking the face and the hands.

5 Conclusion

We have proposed a robotic wheelchair that has three motion control modes: an autonomous control mode, a manual control mode with minimum intentional operations, and a remote control mode. Experimental results prove our approach promising. Recognition results of the environments by the video camera and the ultrasonic sensors can be useful to improve the capability of the second mode, that is, they will help the system to understand the user's intentions. We are working on this as well as quantitative evaluation of the system.

Acknowledgments

This work has been supported in part by the Ministry of Education, Science, Sports and Culture under the Grant-in-Aid for Scientific Research (09555080), the Kurata Foundation under the Kurata Research Grant, and the Kayamori Foundation of Informational Science Advancement.

References

[1] D. P. Miller and M. G. Slack, "Design and testing of a low-cost robotic wheelchair prototype," *Autonomous Robotics*, vol. 2, pp. 77-88, 1995.

[2] R. C. Simpson and S. P. Levine, "Adaptive shared control of a smart wheelchair operated by voice control," *Proc. 1997 IEEE/RSJ International Conference on Intelligent Robots and Systems*, vol. 2, pp. 622-626, 1997.

[3] H. A. Yanco and J. Gips, "Preliminary investigation of a semi-autonomous robotic wheelchair directed through electrodes," *Proc. Rehabilitation Engineering Society of North America 1997 Annual Conference*, pp. 414-416, 1997.

[4] S. Okazaki, Y. Fujita, and N. Yamashita, "A compact real-time vision system using integrated memory array processor architecture," *IEEE Trans. on Circuits and Systems for Video Technology*, vol.5, no.5, pp. 446-452, 1995.

[5] I. Kweon, Y. Kuno, M. Watanabe, and K. Onoguchi, "Behavior-based mobile robot using active sensor fusion," *1992 IEEE International Conference on Robotics and Automation*, pp.1675-1682, 1992.

[6] B. Moghaddam and A. Pentland, "Maximum likelihood detection of faces and hands," *International Workshop on Automatic Face- and Gesture-Recognition*, pp.122-128, 1995.

[7] T. Nishimura, T. Mukai, S. Nozaki, and R. Oka, "Spotting recognition of gestures performed by people from a singe time-varying image using low-resolution features," *Trans. IEICE D-II*, vol.J80-D-II, no.6, pp.1563-1570, 1997 (in Japanese).

Experiments on Decision Making Strategies for a Lane Departure Warning System

Woong Kwon, Jae-Won Lee, Dongmok Shin, Kyoungsig Roh, Dong-Yoon Kim, and Sukhan Lee

System and Control Sector, Samsung Advanced Institute of Technology(SAIT),
Yongin-si, Kyungki-do, 449-712, KOREA.
E-mail:wkwon@sait.samsung.co.kr

Abstract

This paper describes two decision making strategies of the lane departure warning system for a vehicle and provides experimental results to assess the performance of each strategy. The overview of the lane departure warning system mounted on the vehicle is presented. Lateral offset(LO) based strategy and time to lane crossing(TLC) based strategy are proposed to be used in the warning module which detects unintended lane departure so as to warn the driver. The performance criteria of the warning strategies are defined as: 1) false alarm rate, 2) alarm triggering time(ATT). The proposed strategies are incorporated in our lane departure warning system to be tested in real expressway experiments. The experimental results for the performance optimization of both strategies are discussed and compared with each other.

1 Introduction

The importance of the lane departure warning system[1]-[3] and the automatic lane keeping system[4]-[7] has been increasing for safety of a car driver. While automatic lane keeping system keeps the vehicle near a lane without the driver's control, lane departure warning system only warns the driver of danger. They prevent the driver from drifting off the lane by sleeping or being inattentive in maneuvering the vehicle in a monotone environment such as quiet expressway. Since vehicle departure accidents lead to fatal results, many researchers have been interested in the intelligent vehicle systems such as the lane departure warning system and the automatic lane keeping system.

But, currently it is known that the lane departure warning system is more commercially useful than the automatic lane keeping system since it is cost effective, can be easily implemented and, in the most important

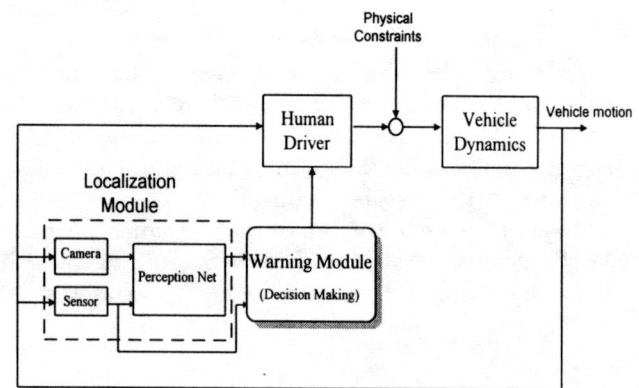

Fig. 1 Lane departure warning system: localization module and warning module

point of view, is relatively safe in case of malfunction or fault of the system[1]. Hence, although many automatic lane keeping systems have been proposed and exhibited their superior performance, they are not yet commercialized but demonstrated only in laboratory level.

The lane departure warning system is composed of two modules, the localization module and the warning(decision making) module as shown in Fig. 1. The *localization module* observes the lane in the road and solves current lateral position of the vehicle and the lane geometry. The *warning module* decides on giving an alarm using the information from sensory data and the localization module. This paper deals with the lane departure warning system which is mounted on our test automobile in Fig. 2, a 1998 Samsung Motors SM520V model including a CCD camera on the rear view mirror and a Pentium PC in the trunk. It has been shown that the localization module of our system can determine the lateral position and the lane curvature successfully[8]. In this paper, two strategies of the warning(decision making) module are proposed

Fig. 2 The test automobile where lane departure warning system is mounted

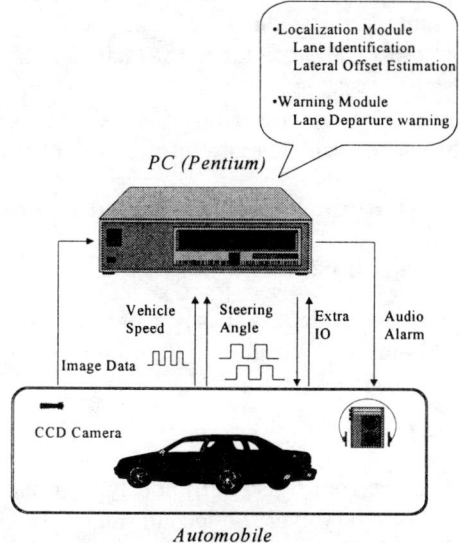

Fig. 3 Overview of the system mounted on the vehicle

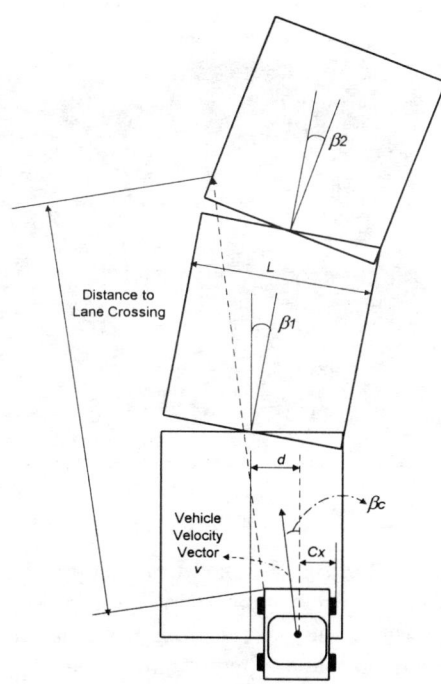

Fig. 4 Lane identification using the Perception Net based method

and compared with each other. Performance criteria of a warning strategy are defined and used to assess the proposed strategies. Experimental results from our test vehicle are provided and analyzed.

This paper is organized as follows. In Section 2, the overview of our lane departure warning system is outlined. In Section 3, two warning strategies are proposed. Performance criteria are defined and investigated in Section 4, and experimental results and conclusions are discussed in the following sections.

2 System Overview

The lane departure warning system mounted on the test vehicle is illustrated in Fig. 3. A CCD camera is equipped on the rear view mirror to sense the roadway up to about 60m in front of the vehicle at 15Hz. A Pentium PC receives data from the CCD camera and the other sensors to estimate lane geometry and gives an alarm when lane departure is detected. For convenience, we used a pulse output from the active damper suspension(ADS) unit, instead of speedometer, to compute the speed of the vehicle. The steering angle is acquired from the steering encoder pulse output in the vehicle. The output of the lane departure warning system is provided by a voice alarm using the audio system of the vehicle. Extra IO systems are also equipped to attain other purposes such as the indicator of the direction light of the car.

In the PC all the sensor data and the image data are aggregated to be used in the localization module and the warning module. The localization module utilizes the Perception Net based method developed by Lee et. al[8]. In this method the road is modeled by three connected plates moving with the vehicle, as shown in Fig. 4. In order to remove the effect of disturbance which may exist near the center of the road, each plate is partitioned into a pair of plates with planar gap. Fig. 5 shows the monitor screen of the system where three pairs of connected plates, identified lane, sensory data and the alarm sign are displayed. The Perception Net based method is based on sensor fusion of the image data and the other sensor data with uncertainty propagation and guarantees computational efficiency and robustness[8]. By this method the lane geometry is identified and the lateral offset of the vehicle is estimated under all weather conditions.

The warning module utilizes the lane geometry and

Fig. 5 Monitor screen of the system

the lateral offset obtained from the localization module to make decision on giving a low-level warning and an alarm. The next section describes in detail the warning module applied to our system. The notations used in Fig. 4 are applied throughout this paper and have meanings as follows.

- d : lateral offset of the vehicle center from the lane center line
- L : lane width which is assumed to be constant
- C_x : distance between center and side of the vehicle
- β_c : angle between the velocity vector of the vehicle and the first plate
- β_1 : angle between the first plate and the second plate
- β_2 : angle between the second plate and the third plate
- v : speed of the vehicle

3 Warning Module

To give a low-level warning and a voice alarm the warning module requires some warning criteria based on the lane information and vehicle pose relative to the lane. Literature surveys show that there are two warning criteria: lateral offset(LO)[2][3] and time to lane crossing(TLC)[1][2].

LO, described as d in Fig. 4, means the relative distance between the lane edge and the vehicle. The warning strategy based on LO gives an alarm when the vehicle approaches to or remains in the vicinity of the lane edge. However, it has its own limits such that LO does not consider the vehicle speed and the information of the lane geometry in front of the vehicle.

TLC is the time remaining before the vehicle will cross over the lane edge if it is assumed that the steering angle is fixed so that the moving direction of the vehicle is maintained. It is computed as the distance to lane crossing in Fig. 4 divided by the speed of the vehicle.

Although LO and TLC are reliable warning criteria, some additional decision logics and very fine tunings are required to cope with various situations in real highway navigation. Also, the performance criteria of warning strategies should be provided to improve their performance and to tune some decision parameters.

In this paper, we propose two warning strategies based on LO and TLC respectively, compensated with additional decision logics so as to improve the performance of the system. The performance criteria will be explained in Section 4 and the numerical values of the parameters in the strategies are tuned and shown in Section 5. The fundamental logics of the two strategies are described as follows. For convenience, we distinguish low-level warning from alarm to monitor the state of the vehicle more intuitively in the experiment.

A. Lateral offset(LO) based strategy

Give a low-level warning:

- if the vehicle is in the neighborhood of the lane boundary within the first safety margin L_{s1}, i.e. if $L/2 - L_{s1} < d + C_x$.

Give an alarm:

- if the vehicle is continuously approaching to lane boundary over the second safety margin L_{s2}, for a predetermined interval, i.e. if $L/2 - L_{s2} < d + C_x$ and $\Delta d > 0$ for N_l consecutive samples

No alarm or cancel the existing alarm :

- if none of the conditions for an alarm and a low-level warning are satisfied.

- if N_a sampling times elapse after the alarm was triggered.

- if the vehicle still remains in the region near the lane boundary where an alarm was previously given so as to prevent duplicate alarms.

- if the vehicle speed is lower than a threshold, i.e. if $v < v_{th}$. We set v_{th} as 40km/h.

- if there exists an extreme value in lane curvature (regarding it as a lane identification error), i.e. if $\beta_1 \geq \bar{\beta}_1$.

- if the moving direction of the vehicle, β_c, is extremely large (also regarding it as a lane identification error), i.e. if $\beta_c \geq \bar{\beta}_c$.

- if the propagated uncertainty in the Perception Net exceeds a predetermined threshold (regarding it as an estimation error), i.e. if $\sigma_d^2 > \bar{\sigma}_d^2$ where σ_d^2 denotes variance of d.

- if points that can be presumed as lane boundary are insufficiently detected.

- if a correct direction light is on indicating pertinent lane change instead of lane departure. We do not use this logic in the experiment.

B. Time to lane crossing(TLC) based strategy

Give a low-level warning:

- if TLC is continuously lower than a first threshold T_{th1} during a predetermined interval, i.e. if $T \leq T_{th1}$ for N_{t1} consecutive samples where T means TLC.

Give an alarm:

- if TLC gets lower than a second threshold T_{th2} within a predetermined interval after a low-level warning was triggered, i.e. as soon as $T \leq T_{th2}$ during N_{t2} samples after N_{t1} samples which led to a low-level warning.

No alarm or cancel the existing alarm:

- if the same conditions of the LO based strategy for no alarm or canceling the existing alarm are satisfied.

4 Performance of a Warning Strategy

We choose two main criteria to evaluate the performance of a warning strategy as follows.

Criterion 1 *The number of triggering a false alarm and missing a true alarm must be minimized.*

The false alarm is defined as the alarm which is triggered in normal situation when the vehicle remains in the lane and also be expected not to departure from the lane. Since a driver may be annoyed at such frequent false alarms, the number of false alarms must be minimized for a commercially wide use of the system. But, emphasis on minimization of false alarm rate may lead to missing an alarm due to more strict decision process. The number of missed alarms should also be minimized. We also regard a missed alarm as a kind of false alarms for convenience.

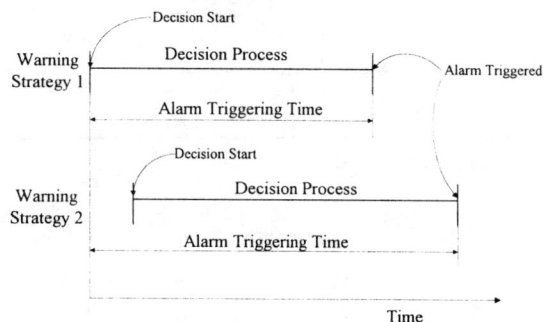

Fig. 5 Definition of alarm triggering time(ATT)

Criterion 2 *An alarm must be triggered as soon as possible guaranteeing proper human response time.*

In order to observe the Criterion 1 and reduce the false alarm rate, it is required that the judgment of triggering alarm is carried out more precisely. However, more precise judgment tends to require more decision process time such as more sampling times. Since human driver must have sufficient time to react the alarm and steer the vehicle, the alarm should be triggered as soon as possible when the lane departure nearly happens. Thus the warning strategy should have an additional criterion, the Criterion 2, as well as the Criterion 1. In order to define a metric for the Criterion 2, we define the *alarm triggering time* (ATT) of a warning strategy as the time between the earliest start of decision process of all the warning strategies and final decision of the corresponding warning strategy on triggering an alarm, as shown in Fig. 6.

5 Experimentation

The lane departure warning system mounted on our test automobile in Fig. 2 was used for the experiments. The objectives of the experiments are outlined as: 1) the tuning parameters of both strategies such as L_{s1}, L_{s2}, $\bar{\beta}_1$, $\bar{\beta}_c$, T_{th1}, T_{th2}, N_l, N_{t1}, N_{t2}, and $\bar{\sigma}_d^2$ are determined to optimize the performance, 2) after the determination of the parameters the resultant false alarm rate and the alarm triggering time (ATT) of both strategies are measured, and 3) the two strategies are compared with each other taking the performance criteria into consideration in such a way as to assess the ability of lane departure detection.

The experiments are repeatedly executed in real expressway between two points with a distance 15km under speed limit 120km/h. The driver of the car intentionally change the lane to imitate lane departure situation. The width of the lane is assumed to be 3.3m throughout the test road to be used in the localization module. In this lane change experiment, the

Table 1: Tuned parameters

LO-based		TLC-based	
$\bar{\beta}_1$ (rad)	0.09	$\bar{\beta}_1$ (rad)	0.10
$\bar{\beta}_c$ (rad)	0.07	$\bar{\beta}_c$ (rad)	0.08
$\bar{\sigma}_d^2$ (m^2)	1/5000	$\bar{\sigma}_d^2$ (m^2)	1/300
L_{s1} (m)	0.16	T_{th1} (sec)	0.50
L_{s2} (m)	0.08	T_{th2} (sec)	0.40
N_l	5	N_{t1}	8
		N_{t2}	5

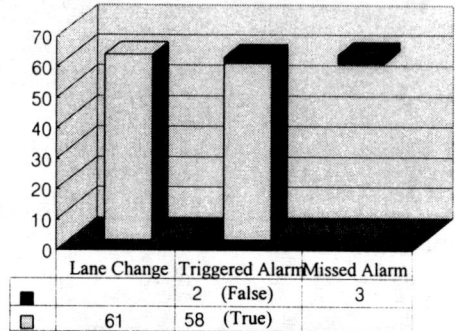

Fig. 6 Alarm state in LO based method

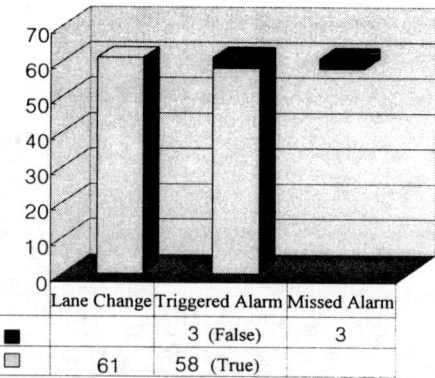

Fig. 7 Alarm state in TLC based method

system mounted on the vehicle aggregates and stores sensory data such as speed data, steering angle and image of the lane in front of the vehicle. Off-line simulation using the stored data was performed to optimize the performance of the warning strategy by tuning the parameters. Since there exists a trade-off between the number of false alarms and the delayed time till an alarm was triggered(ATT), i.e. minimizing false alarms tends to induce long ATT and vice versa, the off-line tuning was iteratively performed to produce minimum false alarm rates with reasonable ATT. The tuned parameters of both strategies from the off-line simulation are illustrated in Table I.

The second phase of the experiments are carried out by on-line decision making of the warning algorithm during real expressway navigation using the optimally tuned parameters in Table I obtained from the off-line simulation. In Fig. 6 and 7, the resultant number of imitated lane departure accidents, correctly triggered alarms, false alarms and missed alarms of both strategies are illustrated. As shown in these figures, 58 lane changes of total 61 lane changes were successfully detected and only three lane changes were missed in both strategies, while two(three) false alarms took place in LO(TLC) based method. The false alarm rates were greatly reduced when compared with the result of the warning algorithms using intuitively selected decision parameters. In the repetitive experiments we could observe that there are some factors that give rise to false alarms. One of the most important factors is the fault of lane parameter estimation algorithm which is followed by the computational error of LO and/or TLC. The faults result from image processing error due to bad pavement condition, bad lane painting in the road, lane-like objects that can cause optical illusion, etc. Since they are unavoidable, some additional decision logics and low pass filtering in line extraction process are needed to compensate these faults. Another factor is the unpredicted change of lane width in driving real road such as a ramp in the highway interchange and a wider road in an abrupt curve. False alarm is also produced by inherent drawbacks of the warning strategy. If a driver had a habit of driving nearly stick to lane boundary, there would be frequent false alarms. To minimize false alarm rate arising from various reasons as mentioned above, many additional decision logics and tuning processes are needed as shown in the previous section. In particular, this paper's Perception-Net based method has a merit in that the uncertainty level, variance, of the estimated parameter can be used in the decision logics. This considerably reduced the number of false alarms. In the experiment, mean alarm triggering time(ATT), the other performance criteria, was about 9.65(8.40) sampling times, i.e. 0.643(0.560)sec at 15Hz, in LO(TLC) based method, which is believed to be sufficiently fast.

We cannot clearly find out the difference in false alarm rates between both strategies. However, it is clear that the ATT of TLC based strategy is faster than that of LO based strategy. This is because TLC

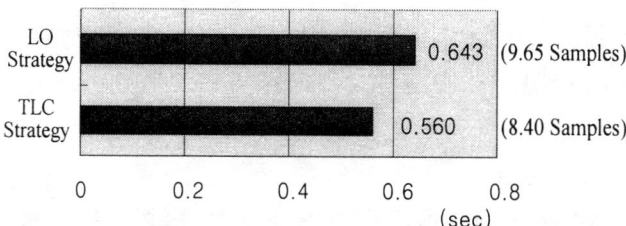

Fig. 8 Mean alarm triggering time(ATT)

strategy utilizes the lane geometry data in front of the vehicle as well as vehicle pose data.

Fig. 9 illustrates one example of time history of LO and TLC, and the alarm states in the same lane departure situation using both strategies. 'Left side of the vehicle' labeled in Fig. 9(a) means the distance between the left side of the vehicle and the center line of the lane. In Fig. 9(b), TLC is bounded above for convenience by limiting the lane crossing point inside the connected plates of Fig. 4.

6 Conclusions

It was proposed that the LO based method and the TLC based method may be used as an warning strategy of a lane departure warning system with additional decision logics and parameter tunings. The performance criteria of an warning strategy were proposed as: 1) the false alarm rate and 2) the delayed time to judge an alarm, defined as alarm triggering time(ATT). This paper's main contribution is that the proposed warning strategies were tuned and compared with each other by real expressway experiments taking the performance criteria into cosideration. The results will help the lane departure warning system to be practically used and commercialized.

References

[1] D.J. LeBlanc et al., "CAPC: A Road-Departure Prevention System," *IEEE Control Systems Magazine*, vol. 16, no. 6, pp.61-71, 1996.

[2] M. Chen, T. Jochem, and D. Pomerleau, "AURORA: A Vision-Based Roadway Departure Warning Systems," *Proc. of IEEE IROS95*, August 1995, pp.243-248.

[3] S. Bajikar et al., "Evaluation of In-Vehicle GPS-Based Lane Position Sensing for Preventing Road Departure," *IEEE Conf. on Intel. Transport. Systems*, Nov 1997, pp.397-402.

[4] T. Zimmermann et al., "VECTOR-A Vision Enhanced/Controlled Truck for Operational Research," *Proc. Int. Truck & Bus Exposition*, Nov 1994, SAE942284.

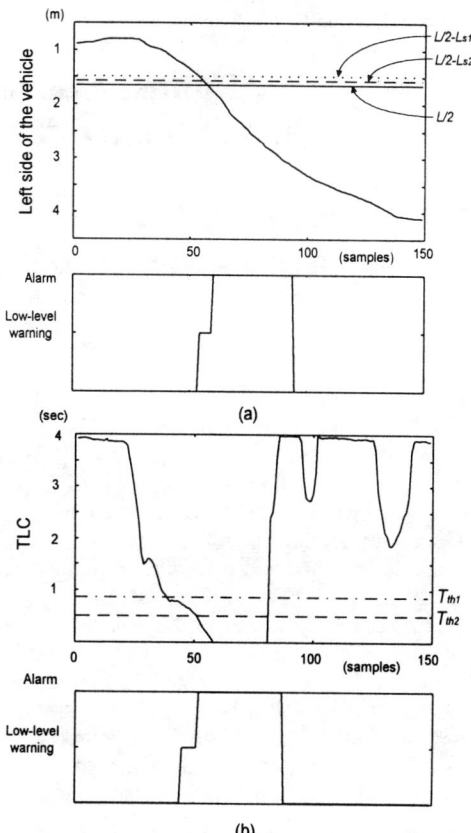

Fig. 9 Time history of both strategies
(a) LO based (b) TLC based

[5] T. Jochem, D. Pomerleau, B. Kumar, and J. Armstrong, "PANS: A Portable Navigation Platform", *IEEE Symposium on Intelligent Vehicles*, September 25-26, 1995.

[6] B. Brumitt, M. Hebert, and the CMU UGV Group, "Experiments in Autonomous Driving With Concurrent Goals And Multiple Vehicles", *Proc. of IEEE Int. Conf. on Robot. and Automat.*, May 1998, pp.1895-1902.

[7] J. Kosěcka, R. Blasi, C.J. Taylor, and J. Malik, "A Comparative Study of Vision-Based Lateral Control Strategies for Autonomous Highway Driving", *Proc. of IEEE Int. Conf. on Robot. and Automat.*, May 1998, pp.1903-1908.

[8] S. Lee, K.S. Boo, D. Shin and D.H. Lee, "Automatic Lane Following with a Single Camera," *Proc. of IEEE Int. Conf. on Robot. and Automat.*, May 1998, pp.1689-1694.

A Robust Approach for Sensor Placement in Automated Vision Dimensional Inspection

Xiangrong Gu*, Michael M. Marefat †, and Frank W. Ciarallo ‡

*Teradyne, Inc., 880 Fox Lane, San Jose, CA 95131-1685
†Department of Electrical and Computer Engineering, University of Arizona, Tucson, AZ 85721
‡Department of Systems and Industrial Engineering, University of Arizona, Tucson, AZ 85721

Abstract

This paper introduces a novel sensor placement approach for dimensional inspection with active visual inspection systems. For a given set of entities, an effective sensor plan includes parameter values that allow some entities to be observed simultaneously. Two kinds of approaches have generally been used for sensor planning in visual inspection systems: the generate-and-test method and the synthesis method. In order to have high performance and satisfactory accuracy, uncertainty must be considered in the sensing model of the visual system. Furthermore, in order to avoid sharp deterioration of the performance of the visual inspection system for very small disturbances of the parameters, an appropriate definition of robustness is proposed. In an experiment a new sensor planning approach based on robustness resolves some problems with earlier approaches.

1 Introduction

Given a set of geometric part entities, such as line segments, to be inspected, the function of visual sensor planning is to determine the position and orientation of the camera (often called a *sensor setting* or *camera pose*) that can observe the set simultaneously with a specified inspection quality [5]. The function of inspection planning is to determine a set of sensor settings of minimal size that can effectively observe all the needed entities and that satisfy all sensor constraints such as focus, resolution, field of view, etc.

Uncertainty is intrinsic in machine vision, both in the sensing process and the images derived. Effective models that describe the uncertainty can help to dramatically improve the performance of these systems. In this work, two main errors in vision inspection are considered: displacement error and quantization error. These errors are due to the erroneous placement of the sensor and digitization of the image. The models for the displacement error and the quantization error that will be used are introduced in Section 3 [8].

In a visual inspection system, the sensor (camera) is constrained by the requirements of visibility, resolution, field-of-view, and focus. To obtain an acceptable image of an object, all of these sensor constraints [5, 6, 7] must be satisfied.

In the remainder of this paper, Section 2 provides a summary of previous work on two main approaches to sensor planning in automated visual inspection systems: generate-and-test and synthesis. Section 3 formulates the problem of determining sensor settings in automated dimensional inspection. Section 4 casts the problem of uncertainty in inspection as one of robustness. Section 5 describes a new optimization procedure for sensor planning that searches for a balance between accuracy and robustness. Finally, in section 6, an experiment demonstrates the effectiveness of this new synthesis approach.

2 Summary of existing works
2.1 Generate-and-test methods

The most widely applied method in previous vision sensor planning systems uses a geodesic dome or tessellated sphere surrounding an object to determine candidate camera positions. The camera is oriented toward an object reference point, and a generate-and-test method based on the tessellated sphere is employed to determine the final camera placement. Quite a few sensor planning systems have been developed using the generate-and-test method such as the HEAVEN system by Sakane, et al [10, 11], the VIO (Vision Illumination Object) system by Niepold [12], et al, the ICE (Illumination Control Expert) system by Yi, et al [13], and the GASP (General Automatic Sensor Planning) by Trucco et al [7]. The generate-and-test method draws on well-known techniques for discretizing and efficiently searching spaces. This method is useful for finding an initial solution for other methods, such as the synthesis method. Some potential problems with this method include:

1. The computational cost of finely tessellating and searching high dimensional parameter spaces.
2. Sensor constraints including focus, resolution and field-of-view constraints are ignored.
3. The solution space is limited to one viewpoint and view direction at the center of each facet.
4. Usually, uncertainties are not considered.

2.2 Synthesis methods

The synthesis methods use analytical relationships between sensor parameters and the goals to be achieved to directly determine the parameters that satisfy the task constraints. The synthesis methods attempt to determine the "best" generalized viewpoint in a high dimension space. The measures of the "best" generalized viewpoint can be minimum error, maximum robustness or simply satisfying sensor and visibility constraints. This approach requires an understanding of the causal relationships between the parameters to be planned and the goals to be achieved. In its most successful applications, the synthesis method provides insight into the structure of the sensor planning problem.

However, the dimension of the search space (which consists of camera location, orientation, optical and illumination parameters) is typically too large to determine a solution efficiently. Therefore, considering the parameters in subsets has the advantage of reducing the dimensionality of the space and providing bounds for solutions in the entire admissible space.

Cowan et al [1] represents the resolution, field-of-view and focus constraints as a three dimensional region where all constraints can be realized. But this approach requires an iterative computational procedure and it only considers three degrees of freedom.

MVP (machine vision planner) developed by Tarabanis et al [6] utilizes more parameters. In MVP, robustness is defined as a weighted sum of the sensing constraints and the goal of MVP is to maximize robustness, as follows:

Max Robustness = $\alpha_1 g_1 + \alpha_{2a} g_{2a} + \alpha_{2b} g_{2b} + \alpha_3 g_3 + \alpha_4 g_4$

where g_1, g_{2a}, g_{2b}, g_3 and g_4 are the resolution, field-of-view, depth-of-field and visibility constraints and $\alpha_1, \alpha_{2a}, \alpha_{2b}, \alpha_3$ and α_4 are the corresponding weights. g_1, g_{2a}, g_{2b}, g_3 and g_4 are each mathematically an expression whose value determines whether the corresponding constraint is satisfied. g_1, g_{2a}, g_{2b}, g_3 and g_4 are derived based on the geometry of the object and capabilities of the sensor. For a given sensor setting, when each of g_1, g_{2a}, g_{2b}, g_3 and g_4, have values that are greater than or equal to zero, the generalized viewpoint will satisfy each of the feature constraints of visibility, resolution, focus, and field of view. The greater g_1, g_{2a}, g_{2b}, g_3 and g_4 are, the better the constraints are satisfied.

Crosby, et al. [9] use nonlinear programming with nonlinear constraints for sensor planning. The formulation tries to minimize, $E[\varepsilon^2]$, the mean squared error. The error ε is defined as the sum of the displacement and quantization errors for the geometric entities (i.e. line segments) to be inspected and all sensor constraints must be satisfied. The mean squared error depends on the camera position t_x, t_y and t_z and camera orientation ϕ, θ and τ, as follows,

Minimize $E[\varepsilon^2(t_x, t_y, t_z, \phi, \theta, \tau)]$

Subject to: Sensor Constraints (e.g. resolution, field-of-view, depth-of-field and visibility).

3 Problem Formulation

In sensor planning one sensor setting is chosen so that a given set of entities can be observed simultaneously with a specified level of quality, including the satisfaction of all constraints. The quality of a sensor setting is determined by its ability to measure all of the required entities to within desired tolerances with a high level of certainty.

The error model presented below is similar to the model in [8, 9]. Given a set of K geometric entities (line segments) to be dimensioned, consider two kinds of errors, displacement error ε_d and quantization error ε_q:

1. Displacement error ε_d.

The camera setting includes six degrees of freedom for a typical end-effector, i.e. the camera position $t = (t_x, t_y, t_z)$ and orientation (ϕ, θ, τ) in the world coordinate system. Define the position of the origin of the world coordinate system in the camera coordinate system as $p = (p_u, p_v, p_w)$ and the coordinates of an object point (x, y, z) projected on the image plane as (u, v).

The transformation matrix relating the world coordinate system to the camera coordinate system is

$$\begin{bmatrix} cu \\ cv \\ c \end{bmatrix} = PQ \begin{bmatrix} x \\ y \\ z \\ 1 \end{bmatrix} = P \begin{bmatrix} r_{11} & r_{12} & r_{13} & p_u \\ r_{21} & r_{22} & r_{23} & p_v \\ r_{31} & r_{32} & r_{33} & p_w \\ 0 & 0 & 0 & 1 \end{bmatrix} \begin{bmatrix} x \\ y \\ z \\ 1 \end{bmatrix} \quad (1)$$

where $[r_{ij}]$ sub matrix is a 3×3 rotation matrix relating the camera coordinate system to the world coordinate system, i.e., $r_{11} = \cos\phi\cos\theta$, $r_{12} = \sin\phi\cos\theta$, $r_{13} = -\sin\theta$, etc., and the perspective matrix P as:

$$P = \begin{bmatrix} 1 & 0 & 0 & 0 \\ 0 & 1 & 0 & 0 \\ 0 & 0 & -1/f & 1 \end{bmatrix} \quad (2)$$

where f is the lens focal length of the camera. Hence,

$$u = fC_1/(f - C_3), v = fC_2/(f - C_3) \quad (3)$$

where
$$\begin{aligned} C_1 &= r_{11}x + r_{12}y + r_{13}z + p_u \\ C_2 &= r_{21}x + r_{22}y + r_{23}z + p_v \\ C_3 &= r_{31}x + r_{32}y + r_{33}z + p_w \end{aligned} \quad (4)$$

Usually there are displacement errors in camera placement, represented as errors in the six degrees of freedom by $dx, dy, dz, d\phi, d\theta$ and $d\tau$. The displacement errors result in errors, du and dv, in image plane coordinates of a projected object point. Usually $dx, dy, dz, d\phi, d\theta$ and $d\tau$ are very small, so du and dv can be computed as follows:

$$\begin{aligned} du &= (\partial \tfrac{fC_1}{f-C_3}/\partial x)dx + (\partial \tfrac{fC_1}{f-C_3}/\partial y)dy + (\partial \tfrac{fC_1}{f-C_3}/\partial z)dz + \\ &\quad (\partial \tfrac{fC_1}{f-C_3}/\partial \phi)d\phi + (\partial \tfrac{fC_1}{f-C_3}/\partial \theta)d\theta + (\partial \tfrac{fC_1}{f-C_3}/\partial \tau)d\tau \\ dv &= (\partial \tfrac{fC_2}{f-C_3}/\partial x)dx + (\partial \tfrac{fC_2}{f-C_3}/\partial y)dy + (\partial \tfrac{fC_2}{f-C_3}/\partial z)dz + \\ &\quad (\partial \tfrac{fC_2}{f-C_3}/\partial \phi)d\phi + (\partial \tfrac{fC_2}{f-C_3}/\partial \theta)d\theta + (\partial \tfrac{fC_2}{f-C_3}/\partial \tau)d\tau \end{aligned} \quad (5)$$

where all of the partial derivatives are functions of (x, y, z) and (ϕ, θ, τ).

It is typical to assume $dx, dy, dz, d\phi, d\theta$ and $d\tau$ are random variables with Gaussian distributions. $dx, dy, dz, d\phi, d\theta$ and $d\tau$ all have zero mean and their variances are $\delta_x^2, \delta_y^2, \delta_z^2, \delta_\phi^2, \delta_\theta^2$ and δ_τ^2 respectively.

The errors in the six degrees of freedom, $dx, dy, dz, d\phi, d\theta$ and $d\tau$ are usually assumed to be independent. Therefore du and dv are also random variables with Gaussian distributions and means of zero and variances as follows:

$$\begin{aligned} \delta_u^2 &= (\partial \tfrac{fC_1}{f-C_3}/\partial x)^2 \delta_x^2 + (\partial \tfrac{fC_1}{f-C_3}/\partial y)^2 \delta_y^2 + (\partial \tfrac{fC_1}{f-C_3}/\partial z)^2 \delta_z^2 + \\ &\quad (\partial \tfrac{fC_1}{f-C_3}/\partial \phi)^2 \delta_\phi^2 + (\partial \tfrac{fC_1}{f-C_3}/\partial \theta)^2 \delta_\theta^2 + (\partial \tfrac{fC_1}{f-C_3}/\partial \tau)^2 \delta_\tau^2 \\ \delta_v^2 &= (\partial \tfrac{fC_2}{f-C_3}/\partial x)^2 \delta_x^2 + (\partial \tfrac{fC_2}{f-C_3}/\partial y)^2 \delta_y^2 + (\partial \tfrac{fC_2}{f-C_3}/\partial z)^2 \delta_z^2 + \\ &\quad (\partial \tfrac{fC_2}{f-C_3}/\partial \phi)^2 \delta_\phi^2 + (\partial \tfrac{fC_2}{f-C_3}/\partial \theta)^2 \delta_\theta^2 + (\partial \tfrac{fC_2}{f-C_3}/\partial \tau)^2 \delta_\tau^2 \end{aligned} \quad (6)$$

For a line entity, there are two endpoints. Define the image plane coordinates of these two endpoints as $(u1, v1)$ and $(u2, v2)$, respectively. Then the errors in the measured dimension of a line in each direction (horizontal and vertical) are:

$$\varepsilon_{dx} = du1 - du2$$
$$\varepsilon_{dy} = dv1 - dv2 \qquad (7)$$

These are also Gaussian random variables with means of zero and variances as follows:

$$\text{Variance of } \varepsilon_{dx} = 2\delta_u^2$$
$$\text{Variance of } \varepsilon_{dy} = 2\delta_v^2 \qquad (8)$$

A graphical depiction of the displacement error is shown in Fig. 1, using the geometric approximation in [8] which assumes that nominal and displaced entities are approximately parallel. The displacement error ε_d is then:

$$\varepsilon_d = \varepsilon_{dx} \cos\gamma + \varepsilon_{dy} \sin\gamma \qquad (9)$$

where $\gamma = \tan^{-1}((v1-v2)/(u1-u2))$ and $(u1, v1)$, $(u2, v2)$ are the coordinates of the end-points of the line entity to be dimensioned.

Since the displacement error, ε_d, is a linear combination of random variables with Gaussian distributions, ε_d is also a random variable with Gaussian distribution, mean zero and variance as follows:

$$\text{Variance of } \varepsilon_d = \delta_d^2 = 2(\delta_u^2 \cos^2\gamma + \delta_v^2 \sin^2\gamma) \qquad (10)$$

2. Quantization error ε_q.

Quantization error for a line entity is introduced because of the digitization of the image. The pixel dimensions are r_x and r_y in the x and y directions, respectively. The probability density function of the dimensional error due to quantization error of a line entity in two dimensions has been derived by Yang, et al [8]. Line entity endpoints are assumed to have a uniform distribution within a pixel in both the horizontal and vertical directions. Assume that nominal and displaced entities are approximately parallel. Define L and L_q to be the actual and quantized lengths of a line entity on the image plane, respectively. Then the dimensional error due to quantization for a entity, $\varepsilon_q = L_q - L$, has the following properties:

$$\text{Mean of } \varepsilon_q = 0$$
$$\text{Variance of } \varepsilon_q = \delta_q^2 = (r_x^2 \cos^2\gamma + r_y^2 \sin^2\gamma)/6 \qquad (11)$$

When $r_x = r_y = r$, the variance $\delta_q^2 = r^2/6$.

3. Integrating errors

The integrated error ε is the sum of the displacement error, ε_d and quantization error, ε_q. The displacement error, ε_d and quantization error, ε_q are assumed to be independent. As a result, ε, has the following two properties:

$$\text{Mean of } \varepsilon = 0$$
$$\text{Variance of } \varepsilon = \delta^2 = \delta_d^2 + \delta_q^2 \qquad (12)$$

In addition to representing these errors for a given camera pose, all the sensor constraints have to be checked so that the resolution, field-of-view, depth-of-field and visibility restrictions of the sensor are satisfied. This ensures that the result is practically useful. In order to ensure that dimensional tolerances (which are defined in the design of an object) are satisfied, the probability that the dimensional inspection error for an entity with length L is within a specified deviation ΔL can be computed and then compared with an acceptable threshold. If the proba-

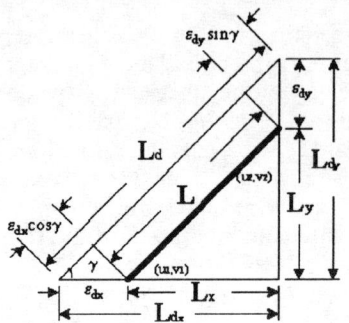

Figure 1: Graphical depiction of the approximation representing the dimensional error for a line entity L.

bility is greater than or equal to the threshold, then the camera pose is acceptable:

$$\int_{-\Delta L}^{+\Delta L} f_\varepsilon(\varepsilon) d\varepsilon \geq Threshold \qquad (13)$$

4 Robust Sensor Placement

In [9], the approach to automated vision inspection planning is based on nonlinear programming with nonlinear constraints with an objective function based on accuracy.

Given an entity to be inspected, define the maximum permissible inspection variance δ^{*2}, as the maximum variance such that the probability that the error in dimensioning for an entity with length L is within a specified deviation ΔL is not less than an acceptable threshold. When the actual inspection variance, δ^2 of an entity equals the maximum permissible inspection variance, δ^{*2}, the probability that the dimensioning error for that entity with length L is within a specified deviation ΔL equals the acceptable threshold, i. e.,

$$\text{If } \delta^2 = \delta^{*2}, \text{ then } \int_{-\Delta L}^{+\Delta L} f_\varepsilon(\varepsilon) d\varepsilon = Threshold \qquad (14)$$

where $f_\varepsilon(\varepsilon)$ is the probability density function of the dimensional inspection error for a single line entity.

Given a set of entities to inspect, E_i (i=1,2...n), the acceptable tolerances ΔL_i and thresholds T_i, and a camera pose to observe all the entities, the maximum permissible inspection variance δ_i^{*2} and the actual inspection variance δ_i^2 can be computed. The robustness, R_i, for entity i is defined as the ratio between the maximum permissible entity inspection variance, δ_i^{*2} and the actual entity inspection variance, δ_i^2:

$$R_i = \delta_i^{*2} / \delta_i^2 \qquad (15)$$

This new definition of robustness considers the accuracy of an inspection in a sensor planning system. While the definition of robustness in [6] is derived from a weighted sum of constraints, here the variance is used to define robustness.

If R_i is not less than 1, the inspection certainty is not less than the threshold. Therefore, the entity E_i can be observed effectively using this camera pose. The greater R_i, the more believable and accurate the inspection result for entity E_i using the camera pose.

Given a camera position (t_x, t_y, t_z) and orientation (ϕ, θ, τ) in a world coordinate system and a line entity E_i whose dimension is to be determined, numerical integration is used to compute the maximum permissible inspection variance δ_i^{*2} on the basis of the acceptable tolerance ΔL_i and corresponding threshold T_i. Using (10), (11) and (12), the mean and variance of displacement error ε_{di}, quantization error ε_{qi}, and total error ε for this entity can be computed.

Two kinds of inspections can be performed, single entity inspection and multiple entities inspection, as follows:

- Single Entity Inspection.

Given a line entity E, the acceptable tolerance ΔL and corresponding threshold T, search for a final camera pose so that the robustness, is maximized, i. e., the reciprocal of the robustness is minimized, as follows:

$$\text{Minimize } 1/R = \delta^2/\delta^{*2} \qquad (16)$$

Subject to: Sensor constraints.

It can be shown that for a single entity inspection, this method is identical to the minimum mean squared error approach of [11,12, 13].

- Multiple Entities Inspection

Given a set of entities to be inspected, E_i (i=1, 2,..., n), the acceptable tolerances ΔL_i and corresponding thresholds T_i, search for a final camera pose so that the minimum robustness of all the entities, is maximized, i.e., the maximum reciprocal of the robustness is minimized, as follows:

$$\text{Minimize } \{ \text{Maximum } \{1/R_1, 1/R_2, ..., 1/R_n\} \} \qquad (17)$$

Subject to: Sensor constraints.

If the sum of the robustness is maximized, the inspection certainty of any single entity is not guaranteed. For example, the maximum total robustness of all entities may occur with a camera pose where all entities except for one have a very small error. The remaining entity may have an unacceptably large error.

Given a camera pose, it is computationally expensive to compute the maximum permissible inspection variance because of the numerical integration procedure. When the set of entities to be inspected is large, that process is especially time-consuming. In order to devise a simplified method, Theorems 4.1 and 4.2, and their corollaries, are useful:

Theorem 4.1 Suppose L_i and L_j are the lengths of entities E_i and E_j in the image coordinate system. Let A_i and A_j be the accuracy and let T_i and T_j be the thresholds for E_i and E_j. If (1) The quantization errors of entities E_i and E_j are negligible and (2) $T_i = T_j$, then

$$\delta_i^{*2}/(A_i L_i)^2 = \delta_j^{*2}/(A_j L_j)^2 \qquad (18)$$

where δ_i^{*2} and δ_j^{*2} are the maximum permissible inspection variances.

Proof: Suppose ΔL_i is the acceptable tolerances for E_i. Then,

$$A_i = \Delta L_i / L_i$$

(and similarly for E_j). If the displacement errors, ε_{di} and ε_{dj}, are much greater than the quantization errors, ε_{qi} and ε_{qj}, then the total errors ε_i and ε_j can be approximated as follows:

$$\varepsilon_i \approx \varepsilon_{di} \quad \text{and} \quad \varepsilon_j \approx \varepsilon_{dj}$$

In this case, because the displacement errors are random variables with Gaussian distribution, the total errors can also be regarded as random variables with the same Gaussian distribution as the displacement errors.

Thus, for entity E_i, we have

$$T_i = \frac{1}{\sqrt{2\pi}\delta_i^*} \int_{-\Delta L_i}^{\Delta L_i} e^{-\frac{\varepsilon_i^2}{2\delta_i^{*2}}} d\varepsilon_i$$

Because of the symmetry of the Gaussian distribution,

$$T_i = \frac{1}{\sqrt{2\pi}\delta_i^*} \int_{-\Delta L_i}^{\Delta L_i} e^{-\frac{\varepsilon_i^2}{2\delta_i^{*2}}} d\varepsilon_i = \frac{2}{\sqrt{2\pi}\delta_i^*} \int_{-\infty}^{\Delta L_i} e^{-\frac{\varepsilon_i^2}{2\delta_i^{*2}}} d\varepsilon_i - 1$$

$$= \frac{2}{\sqrt{2\pi}} \int_{-\infty}^{\Delta L_i/\delta_i^*} e^{-\frac{\varepsilon_i^2}{2}} d\varepsilon_i - 1$$

Similarly for entity E_j. Because $T_i = T_j$, i.e.,

$$\frac{2}{\sqrt{2\pi}} \int_{-\infty}^{\Delta L_i/\delta_i^*} e^{-\frac{\varepsilon_i^2}{2}} d\varepsilon_i - 1 = \frac{2}{\sqrt{2\pi}} \int_{-\infty}^{\Delta L_j/\delta_j^*} e^{-\frac{\varepsilon_j^2}{2}} d\varepsilon_j - 1$$

we have

$$\Delta L_i / \delta_i^* = \Delta L_j / \delta_j^*$$

and

$$\delta_i^{*2}/(A_i L_i)^2 = \delta_j^{*2}/(A_j L_j)^2 \qquad \textbf{Q E D.}$$

Corollary 1 Suppose L_i and L_j are the lengths of entities E_i and E_j in the image coordinate system. Let A_i and A_j be the accuracies, and T_i and T_j be the thresholds of E_i and E_j. If (1) The quantization errors of entities E_i and E_j are negligible, (2) $T_i = T_j$, and (3) $A_i = A_j$ then

$$\delta_i^{*2}/\delta_j^{*2} = L_i^2/L_j^2 \qquad (19)$$

where δ_i^{*2} and δ_j^{*2} are the maximum permissible inspection variances for entities E_i and E_j.

Theorem 4.2 Given a set of entities E_i (i=1, 2, ..., n) to be inspected, suppose that for entity E_i: L_i is the length in the image coordinate system, A_i is the accuracy, and T_i is the threshold. If
(1) For all entities in the set, the displacement error is much greater than the quantization error,
(2) The threshold T_i (i=1, 2, n) of all entities are identical, then

$$\begin{bmatrix} 1/R_1 \\ 1/R_2 \\ \vdots \\ 1/R_i \\ \vdots \\ 1/R_n \end{bmatrix} = \begin{bmatrix} \delta_1^2/\delta_1^{*2} \\ \delta_2^2/\delta_2^{*2} \\ \vdots \\ \delta_i^2/\delta_i^{*2} \\ \vdots \\ \delta_n^2/\delta_n^{*2} \end{bmatrix} = k \begin{bmatrix} \delta_1^2/(A_1 L_1)^2 \\ \delta_2^2/(A_2 L_2)^2 \\ \vdots \\ \delta_i^2/(A_i L_i)^2 \\ \vdots \\ \delta_n^2/(A_n L_n)^2 \end{bmatrix} \qquad (20)$$

where δ_i^{*2} and δ_i^2 are the maximum permissible inspection variance and the actual inspection variance of entity E_i, k is a constant, and R_i is the robustness in inspection of entity E_i.

Proof: Please see [15].

Based on Theorem 4.2, $R_i' = (A_i L_i)^2/\delta_i^2$ is a scaled version of the robustness of entity E_i, R_i, which is equal to δ_i^{*2}/δ_i^2.

Corollary 2 Given a set of entities E_i (i=1, 2,..., n) to be inspected, Suppose that for entity E_i: L_i is the length in the image coordinate system, A_i is the accuracy and T_i is the

threshold. If
(1) For all entities in the set, the displacement error is much greater than the quantization error,
(2) The thresholds T_i ($i=1, 2,..., n$) of all entities are identical,
then there exists a single camera pose that optimizes both objectives 1 and 2, below:
1. Minimize (Maximum ($1/R_1'$, $1/R_2'$, ..., $1/R_i'$, ..., $1/R_n'$))
2. Minimize (Maximum ($1/R_1$, $1/R_2$, ..., $1/R_i$, ..., $1/R_n$))

where R_i is the robustness and R_i' is the scaled version of the robustness of entity E_i, i. e.,

$$R_i = \delta_i^{*2} / \delta_i^2$$
$$R_i' = (A_i L_i)^2 / \delta_i^2 \quad (21)$$

where δ_i^{*2} is the maximum permissible inspection variance and δ_i^2 is the actual inspection variance of entity E_i.

5 Optimization

The following alternative optimization approach avoids the computationally expensive numerical integration used to compute the maximum permissible inspection variance. The second step of the procedure is much simpler. Assume the quantization error is negligible, relative to displacement error. In order to dimensionally inspect a desired set of entities E_i ($i=1,2,..,n$):

Step 1: Nonlinear programming.

$$\text{Minimize} \left\{ \text{Maximum} \left\{ 1/R_1', 1/R_2', ..., 1/R_n' \right\} \right\}$$

Subject to: Sensor constraints.

Where for the entity E_i, R_i' is defined in (21), L_i is the length in the image coordinate system and A_i is the desired accuracy.

Step 2: Check dimensional tolerances.

Using (13), compute the probability that the error for the entity with the smallest robustness is within a specified deviation ΔL. Compare this value with an acceptable threshold. The camera pose is acceptable for all the entities to be inspected simultaneously if the probability for the entity with the least robustness is greater than or equal to the threshold.

In step 1 of this procedure, the burden of the expensive computation is reduced. Using the scaled version of the robustness, the same camera setting can be found without the numerical integration to get the maximum permissible variance. In step 2, generally only the probability of the entity with the smallest robustness needs to be computed to judge the camera setting when the displacement error is much greater than the quantization error. When the quantization error is significant, then the probabilities of all entities must to be checked.

Given k desired geometric entities to be dimensioned and m half spaces in the visibility constraints, the set of k resolution constraints for the k entities can be represented by g_{1j} for $j = 1$ to k. The set of half spaces for all m visibility constraints can be represented by g_{4i} for $i = 1$ to m [9]. Suppose that for entity E_i, δ_i^2 is the variance of errors, L_i is the measured length and A_i is the accuracy. The focus constraint ensures that the positions of entities are within an interval between the far and near limits of the depth of field. The largest angle between an endpoint of an entity and the viewing direction of the camera can not be greater than the field-of-view angle of the camera. Therefore, the problem can be formulated as follows, (for detailed descriptions of g_1, g_2, g_3, and g_4, please see [9]).

Minimize $\left\{ F(t_x, t_y, t_z, \phi, \theta, \tau) = \text{Maximum} \left\{ 1/R_1', 1/R_2', ..., 1/R_n' \right\} \right\}$

Subject to: $g_{1j} \geq 0$ (resolution) for $j = 1$ to k
$g_{2a} \geq 0$ (focus a)
$g_{2b} \geq 0$ (focus b) (22)
$g_3 \geq 0$ (field-of-view)
$g_{4i} \geq 0$ (visibility) for $i = 1$ to m

where $1/R_j' = \delta_j^2 / (A_j L_j)^2$ for $j=1$ to k and (t_x, t_y, t_z) is the position of the camera in the world coordinate system and (ϕ, θ, τ) is the orientation of the camera.

6 Experimental Results

A general non-linear programming package called GRG2 was used for the examples solved in this paper. The generalized reduced gradient (GRG) with quasi-Newton method was chosen at each iteration because it is a fast method if there are not too many variables in the problem.

6.1 Experiments

Consider the object in Figure 2, where each dimension is in millimeters. This object is also considered in [9]. Assume the data below in for the pixel dimensions, focal length, and the moments of the distributions for the displacement errors in the six degrees of freedom:

$r_x = 0.01$ mm (horizontal pixel dim.)
$r_y = 0.013$ mm (vertical pixel dim.)
$f = 25$ mm (focal length)
$\eta_{dx} = \eta_{dy} = \eta_{dz} = 0$ mm (mean of dx, dy and dz)
$\sigma^2_{dx} = \sigma^2_{dy} = \sigma^2_{dz} = 0.3$ mm^2 (variance of dx, dy, dz)
$\eta_{d\phi} = \eta_{d\theta} = \eta_{d\tau} = 0$ rad (mean of $d\phi$, $d\theta$ and $d\tau$)
$\sigma^2_{d\phi} = \sigma^2_{d\theta} = \sigma^2_{d\tau} = 0.003$ rad^2 (variance of $d\phi$, $d\theta$, $d\tau$)

Here, the variances of the errors in the six degrees of freedom (t_x, t_y, t_z, ϕ, θ, τ) are ten times greater than in [9]. Also, for the sake of example, a particular camera pose is chosen to be (t_x, t_y, t_z, ϕ, θ, τ) = (0, 200, 15, 0.7854, 0, 1.571), where the position coordinates are in millimeters and the angles are in radians. Suppose we would like to observe the entities E4 and E5 simultaneously.

Table 1: Experimental results for simultaneously inspecting E4 and E5 in Experiment 1.1

Initial setting: (0, 200, 15, 0.7854, 0, 1.571)
Final setting: (0.0436, 200.04, 14.95, 0.0891, 0.5958, 1.57)
Var. of displacement error for E4: 8.95×10^{-5} mm^2
Var. of displacement error for E5: 3.67×10^{-4} mm^2
Var. of quantization error for E4: 2.454453×10^{-5} mm^2
Var. of quantization error for E5: 2.454488×10^{-5} mm^2
Total variance for entity E4: 1.14×10^{-4} mm^2
Total variance for entity E5: 3.92×10^{-4} mm^2
Length of E4 in world coordinate system: 10mm
Length of E5 in world coordinate system: 30mm
Optimization time: 58 seconds.
Error-free length of E4 in image: 1.592mm
Error-free length of E5 in image: 4.682mm

Experiment 1.1: Attempt to simultaneously inspect entities E4 and E5 for the object in Figure 2. The initial and final data from the optimization in (22) are provided in Table 1. Variances are computed using (10) and (11).

Experiment 1.2: Check of dimensional tolerances.
Assume we want to determine the probabilities that the lengths of entity E4 and E5 for the object in Figure 2 are within tolerances after the optimization.
E4: Accuracy = 97.5%, Threshold = 99%
E5: Accuracy = 97.5%, Threshold = 99%

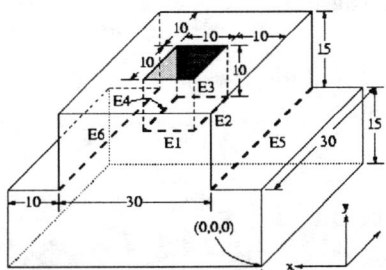

Figure 2: Object to be inspected, including six line entities E1 through E6. This object has 1 pocket and 2 steps. All coordinates are in millimeters.

After numerical Integration, the probabilities are determined to be 99.94% and 99.99% for entities E4 and E5, respectively. Both of them are greater than the desired threshold of 99%. Therefore, using the final camera setting, entities E4 and E5 can be inspected simultaneously within the specified dimensional tolerances with the required level of certainty.

In [9], E4 and E5 can not be effectively inspected simultaneously because the inspection probabilities are determined to be 96.73% and 99.99% for E4 and E5, respectively and the desired thresholds are 99%. Although the variances of the errors in the six degrees of freedom (t_x, t_y, t_z, ϕ, θ, τ) in this experiment are ten times greater than in [12], the probabilities for entities E4 and E5 are both greater than the required thresholds, 99%.

7. Conclusion

In a dimensional inspection system, it is desirable for the system to have both certainty and robustness. Certainty assures the high accuracy of the system and robustness ensures adaptability. But usually these two properties are in conflict with each other. A compromise between these determines the performance of the system.

In this work, a new definition of robustness for dimensional vision inspection has been proposed. A novel synthesis approach for automated dimensional vision inspection using this definition is introduced. To simplify the expensive computation, an equivalent but much simpler optimization procedure is also presented. An experiment has been carried out using the new synthesis approach. The result of this experiment points out that the new method is very promising for simultaneously inspecting several entities. Together with the sensor constraint graph (SCG) approach in [9], this new synthesis approach using robustness can be an effective method for sensor planning and inspection plan generation.

References

[1] Cowan, C. K. and Kovesi, P. D., "Automatic Sensor Placement from Vision Task Requirements," *IEEE Trans. on Pattern Analysis and Machine Intelligence*, Vol.10, No.3, pp. 407-416, May 1988.

[2] Griffin, P. M. and Villalobos, J. R., "Process Capability of Automated Visual Inspection Systems," *IEEE Trans. on Systems, Man, and Cybernetics*, Vol. 22, No. 3, pp. 441-448, May/June 1992.

[3] Smith, R. C. and Cheeseman, P., "On the Representation and Estimation of Spatial Uncertainty," *The Int. Journal of Robotics Research*, Vol. 5, No. 4, pp. 56-68, winter 1986.

[4] Shen, R. C. and Duffie, N. A., "An Uncertainty Analysis Method for Coordinate Referencing in Manufacturing Systems," *Trans. of the ASME Journal of Engineering for Industry*, Vol. 117, No.1, pp. 42-8, February 1995.

[5] Tarabanis, K., Tsai, R. Y., and Allen, P. K., "A Survey of Sensor Planning in Computer Vision," *IEEE Trans. on Robotics and Automation*, Vol.11, No.1, pp. 86-104, February 1995.

[6] Tarabanis, K., Tsai, R. Y., and Allen, P. K., "The MVP Sensor Planning System for Robotic Vision Tasks," *IEEE Trans. on Robotics and Automation*, Vol. 11, No. 1, pp. 72-85, February 1995.

[7] Trucco, E., Umasuthan, M., Wallace A. M. and Roberto, V., "Model-Based Planning of Optimal Sensor Placements for Inspection," *IEEE Trans. on Robotics and Automation*, Vol. 13, No. 2, pp. 182-194, April 1997.

[8] Yang, C. C., Marefat, M. M., and Ciarallo, F. W., "Error Analysis and Planning Accuracy for Dimensional Measurement in Active Vision Inspection," *IEEE Trans. on Robotics and Automation*, Vol. 14, No. 3, pp. 476-487, June 1998.

[9] Crosby, K. L., Yang, C. C., Ciarallo, F. W., Marefat, M. M., "Camera Settings for Dimensional Inspection Using Displacement and Quantization Errors," *IEEE Int. Conf. on Robotics and Automation*, Albuquerque, New Mexico, April 21-25, 1997.

[10] Sakane, S., Ishii, M., and Kakikura, M., "Occlusion Avoidance of Visual Sensors Based on a Hand Eye Action Simulator System: HEAVEN," *Advanced Robotics*, Vol. 2, No. 2, pp. 149-165, 1987.

[11] Sakane, S., Sato, T., "Automatic Planning of Light Source and Camera Placement for an Active Photometric Stereo System," *Proc. of IEEE Int. Conf. on Robotics and Automation*, pp. 1080-1087, Sacramento, California, April 1991.

[12] Niepold, R., Sakane, S., Sato, T., and Shirai, Y., "Vision Sensor Set-up Planning for a Hand-eye System using Environmental Model," *Proc. Society Instrum. Control Engineering*, Japan, pp. 1037-1040, 1988.

[13] Yi, S., Haralick, R. M., and Shapiro, L. G., "Automatic Sensor and Light Source Positioning for Machine Vision," *Proc. 10th Int. Conf. on Pattern Recognition*, pp.55-59, 1990.

[14] Ho, C., "Precision of Digital Vision Systems," *IEEE Trans. on Pattern Analysis and Machine Intelligence*, Vol. PAMI-5, No. 6, pp. 593-601, November 1983.

[15] Gu, X., Marefat, M., and Ciarallo, F. W., "Robust Sensor Placement in Active Vision Dimensional Inspection", submitted to *IEEE Trans. on Robotics and Automation*, 1999.

Improving Sensory Perception through Predictive Correction of Monitoring Errors

Tomasz Celinski and Brenan McCarragher
Department of Engineering
Faculty of Engineering and Information Technology
The Australian National University
Canberra, Australia
Fax: int + 61 6 249 0506
E-mail: {tomasz,brenan}@faceng.anu.edu.au

Abstract

We present a novel approach to managing the quality and cost of perception in a multi-sensor robotic system. The approach involves prediction of monitoring errors of low-performance process monitors using a weighted least squares algorithm, and detection of instances when high-performance process monitoring is necessary. Two significant characteristics of the approach are (1) dynamic, real-time management of process monitors, and (2) the ability to deliver high quality information while keeping the cost of perception low.

1 Introduction

The perception system is one of the essential components of an autonomous robot. Such a system typically consists of a number of process monitors[1] some of which may be able to provide the same type of information. Although some process monitors may output the same type of information, the quality (precision) of the information can vary significantly between the monitors. Since the quality of information can strongly affect the overall robot performance it is important to be able to control it, as shown in Figure 1. Ideally the robot's task controller should be able to specify an information quality level and the robot's perception controller should attain that level at the lowest possible cost (shortest possible time). For example, a mobile robot traversing a large empty room may only require position information precise to

[1] a process monitor is a combination of sensors and a data processing algorithm

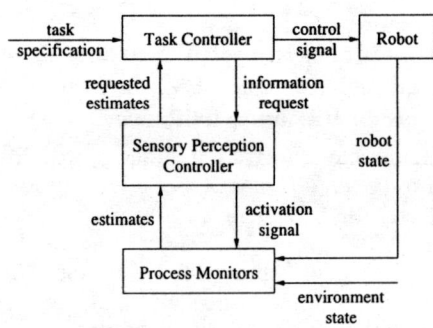

Figure 1: A robotic system with perception control capabilities. The Task Controller responsibilities range from high-level task planning and supervision to low-level control, such as PID or adaptive control. The Perception Controller is responsible for efficient acquisition of high-quality information for the Task Controller.

within 1 meter, but the same robot traversing a narrow door may require positioning precision of 0.1 meter, but in both cases perception should be performed as efficiently as possible. This leads to the following question which is addressed in this paper: How can process monitors of varying levels of performance be controlled to ensure sufficiently high quality of perceived information at minimum cost?

Past approaches to managing multiple process monitors have mainly been concerned with sensor fusion where information from multiple, concurrently operating sensors is combined in order to improve estimates of robot state or the state of the environment. Examples include mobile robot localization [10, 2] and recognition and localization of assembly parts [4, 11]. The traditional sensor fusion approaches usually assume that all process monitors are activated whenever information is required. Such systems tend to produce high (but fixed) information quality levels, but at a high cost (also fixed). A different approach, based on stochastic dynamic programming, was proposed in [6, 5]. Process monitors are dynamically selected during task execution in a way which optimizes a criterion reflecting monitoring confidence and cost. The approach was specialized to monitoring of discrete events in a discrete event control framework. Another interesting approach is the instrumented sensor approach [8, 7]. It is based on a sensori-computational model which defines the sensing system in terms of the functionality, accuracy, robustness and efficiency of its components. The approach has been shown to be effective in mobile robot applications.

The approach to be presented in this paper is significantly different. It centers around an error correction algorithm which attempts to predict and correct monitoring errors of cheap, low-performance process monitors using a weighted least-squares scheme. The accuracy of the predictions is assessed based on a prediction interval. If the predictions are not sufficiently accurate then high-performance process monitors are employed. The approach allows a desired information quality level to be specified and it delivers the information at a low cost by actively managing process monitors in real time.

2 Problem Formulation

The scenario under consideration is illustrated in Figure 2. The figure shows three components: two process monitors, m_{lo} and m_{hi}, and an Error Correction Algorithm (ECA). The process monitors are assumed to have significantly different performance characteristics. Process monitor m_{lo} has low-precision but is cheap to use (the cost of using a process monitor is measured by the time it takes to complete it's operation). Process monitor m_{hi} is a high-precision process monitor but it is expensive to use. The third component is the Error Correction Algorithm (ECA) which has four inputs (estimates $\hat{z}_{hi}$ and $\hat{z}_{lo}$, an error bound δ as well as a probability p), and one output $\hat{z}$. The algorithm is responsible for ensuring that the output ($\hat{z}$) is of higher quality than ($\hat{z}_{lo}$) but obtained at a lower cost than m_{hi}. Ideally one would like the error correction algorithm to:

- minimize the cost of producing estimates $\hat{z}(l)$ (l is a discrete time) subject to the constraint that

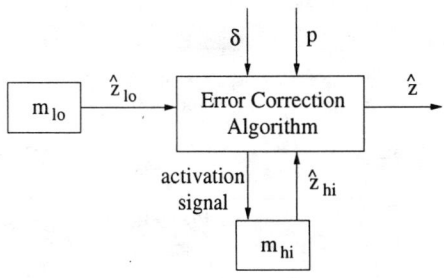

Figure 2: A schematic view of the system. The Error Correction Algorithm is given the desired bound δ and probability p as design parameters (see equation (1). It receives estimates from m_{lo} at every time step but activates m_{hi} only when necessary. An estimate, $\hat{z}$, is produced at every time step.

the probability of $\hat{z}(l)$ being within δ of the true value $z(l)$ is p:

$$P(|\hat{z}(l) - z(l)| < \delta) \geq p \quad (1)$$

3 Error Correction Algorithm

The problem formulated in the previous section may be approached in the way illustrated in Figure 3. The ECA aims to correct the output z_{lo} of monitor m_{lo} based on predictions of the monitoring error, e_{lo}, associated with z_{lo}. If the quality of the error predictions deteriorates to a level which is likely to exceed the specified bound (δ) then the high-precision monitor m_{hi} is employed to determine the error e_{lo} based on the difference between z_{lo} and z_{hi}. The main components of the approach may therefore be identified as follows:

- error prediction
- prediction assessment
- error correction

The components will now be discussed in detail.

3.1 Error Prediction

The error prediction part of the approach is concerned with estimation of monitoring error $\hat{e}_{lo}(l)$ at time l based on past values $e_{lo}(k_i)$[2], where the discrete time k_i is one of n times when the true monitoring error e_{lo} was measured. The error prediction

---
[2]prediction times are denoted by l; observation times when m_{lo} and m_{hi} are both used are denoted by k_i

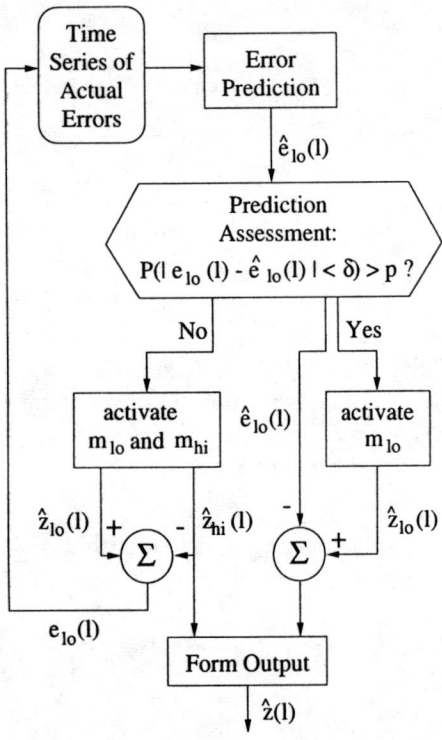

Figure 3: The Error Correction Algorithm

may therefore be viewed as a problem of time series analysis and prediction constrained by the following conditions:

1. *Irregular sampling intervals*: The data consists of monitoring errors $e_{lo}(k_i)$ which are obtained only at those times k_i when monitors m_{lo} and m_{hi} are both used; the details of when m_{hi} is to be used and how $e_{lo}(k_i)$ are obtained will be discussed in Section 3.2.

2. *Small sample size*: The number of available data points is likely to be low. The amount of available data will depend on how many times m_{lo} and m_{hi} are both employed; it will also depend on how many times m_{lo} is employed during task execution.

3. *Non-stationarity*: The data may be non-stationary because the stochastic process generating the monitoring errors of m_{lo} is likely to be non-stationary.

4. *No a priori models*: The nature of the error may be affected by unpredictable factors such as environmental conditions, sensor damage, etc. The error dynamics may therefore be very difficult, or even impossible, to model a priori.

In addition to the above constraints imposed by the nature of the problem we also impose the following constraint:

5 *Low computation time*: The time required for the data analysis and prediction should be insignificant compared to the time of monitoring operations of m_{lo}.

The prediction approach which we propose to use is regression based on weighted least squares [9, 3]. The approach is not affected by constraint 1 and it's simplicity makes it compatible with 5. It is able to cope with 2 because, in our case, it involves estimation of only two parameters. The approach copes with constraint 3 through exponential weighting of data (exponential forgetting) and deals with 4 by using a linear local approximation to the actual error dynamics. Most other approaches to time series analysis, such as [1], are not able to cope with the above constraints.

The basic linear regression model adapted for our purposes is

$$e_{lo} = \beta_0 + \beta_1 k + \epsilon \qquad (2)$$

where the monitoring error e_{lo} is the dependent variable, the time k is the independent variable, β_0 and β_1 are parameters to be estimated, and ϵ is a random noise (not to be confused with the *monitoring error* e_{lo}). Let $e_{lo}(k_i)$ be the monitoring error observed at time k_i, and let ϵ_i be given by

$$\epsilon_i = e_{lo}(k_i) - (\beta_0 + \beta_1 k_i)$$

Standard approach to regression aims to find β_0 and β_1 which minimize the sum of square errors. Instead we are interested in minimizing the weighted sum of square errors given by

$$S^2 = \underline{\epsilon}^T \mathbf{W} \underline{\epsilon} \qquad (3)$$

where $\underline{\epsilon}^T = [\epsilon_1 \ \epsilon_2 \ ... \ \epsilon_n]$ and $\mathbf{W}$ is a diagonal matrix of weights. The weighted least squares estimate $\hat{\underline{\beta}}$ of the parameters $\underline{\beta} = [\beta_0 \ \beta_1]^T$ is given by

$$\hat{\underline{\beta}} = (\underline{k}^T \mathbf{W} \underline{k})^{-1} \underline{k}^T \mathbf{W} \underline{e}_{lo} \qquad (4)$$

The parameter estimates may be used for prediction of monitoring errors at time l via

$$\hat{e}_{lo}(l) = \hat{\beta}_0 + \hat{\beta}_1 l \qquad (5)$$

The estimation of parameters is carried out only at those times when new values of e_{lo} become available (ie. when m_{lo} and m_{hi} are both employed). After n

values of e_{lo} are recorded the weighting matrix $\mathbf{W}$ is diagonal and of size $n \times n$. The purpose of this matrix is to restrict the influence of old data on the parameter estimates through exponential forgetting. Recent data are assumed to be more informative than old data and for that reason old data are exponentially discarded through a discounting factor ϕ which satisfies $0 \leq \phi \leq 1$. The matrix $\mathbf{W}$ therefore consists of diagonal entries

$$w_{ii} = \phi^{n-i} \quad (6)$$

3.2 Assessment of Prediction Accuracy

The accuracy of predicted values $\hat{e}_{lo}$ may be assessed using a prediction interval for e_{lo} which, instead of a single prediction, gives a range of predicted values which has a certain probability of containing the true e_{lo}. A prediction interval for weighted regression is given by

$$\hat{e}_{lo}(l) \pm t_{\alpha/2,n-2} S \sqrt{1 + \underline{l}(\underline{k}^T \mathbf{W} \underline{k})^{-1} \underline{l}^T} \quad (7)$$

In this expression $t_{\alpha/2,n-2}$ is the $\alpha/2 \times 100$ percentile of the Student's t distribution with $n-2$ degrees of freedom, and $\underline{l} = [1 \ l]^T$.

The prediction interval may be related to the specifications expressed in equation (1). By setting $\alpha = 1 - p$ in (7), letting

$$\Psi(l) = t_{\alpha/2,n-2} S \sqrt{1 + \underline{l}(\underline{k}^T \mathbf{W} \underline{k})^{-1} \underline{l}^T}$$

and using the definition of prediction interval one obtains

$$P(\hat{e}_{lo}(l) - \Psi(l) < e_{lo}(l) < \hat{e}_{lo}(l) + \Psi(l)) = p$$

where $e_{lo}(l)$ is the actual monitoring error at time l. It follows that

$$P(|\hat{e}_{lo}(l) - e_{lo}(l)| < \Psi(l)) = p \quad (8)$$

Observing that, when m_{hi} is not used, in (1)

$$\hat{z}(l) = \hat{z}_{lo}(l) - \hat{e}_{lo}(l)$$

and noting also that

$$\hat{z}_{lo}(l) = z(l) + e_{lo}(l)$$

leads to

$$z(l) - \hat{z}(l) = \hat{e}_{lo}(l) - e_{lo}(l)$$

This, combined with (8) gives

$$P(|z(l) - \hat{z}_{lo}(l)| < \Psi(l)) = p \quad (9)$$

It follows immediately that condition (1) is satisfied if $\Psi(l) \leq \delta$ or, equivalently, if

$$t_{(1-p)/2,n-2} S \sqrt{1 + \underline{l}(\underline{k}^T \mathbf{W} \underline{k})^{-1} \underline{l}^T} \leq \delta \quad (10)$$

Inequality (10) is the basis for assessment of accuracy of the predicted monitoring errors $\hat{e}_{lo}(l)$. If the inequality is satisfied then the predicted value is judged to be satisfactory and only monitor m_{lo} is consulted (the flow chart in Figure 3 follows the "Yes" branch). If the inequality is not satisfied than the predicted value is unsatisfactory and monitors m_{lo} and m_{hi} are both consulted (the flow chart in Figure 3 follows the "No" branch).

4 Experiments

The experiments involved two process monitors capable of producing estimates of the position of the part, P, shown in Figure 4 (a). The object is a part of an axle assembly from a children's bicycle. Monitor m_{lo} assumes that the only object in the image is the part of interest. The position of the part is then determined using binary vision techniques. Monitor m_{hi} also relies on binary vision techniques, but it doesn't make the assumption that P is the only object visible in the picture. Instead it makes the somewhat less restrictive assumption that P is the biggest visible object. The improved performance of m_{hi} over m_{lo} comes at a cost, as shown in Table 1.

Monitor	Cost (CPU time)
m_{lo}	0.37
m_{hi}	2.92

Table 1: The computational cost of using m_{lo} and m_{hi} (measured in seconds)

The evaluation of the error correction algorithm was based on two sequences of images. The first sequence ($S1$) contained 15 images of the part P as well as a second object (a small spanner) positioned at various points in the field of view.. The second sequence ($S2$) contained 15 images of part P without the spanner. Examples of images from the two sequences are shown in Figure 4 (b) and (c).

The parameter settings used in the evaluation were: $\delta = 3$, $p = 0.95$ (see equations (1) and (10)). The

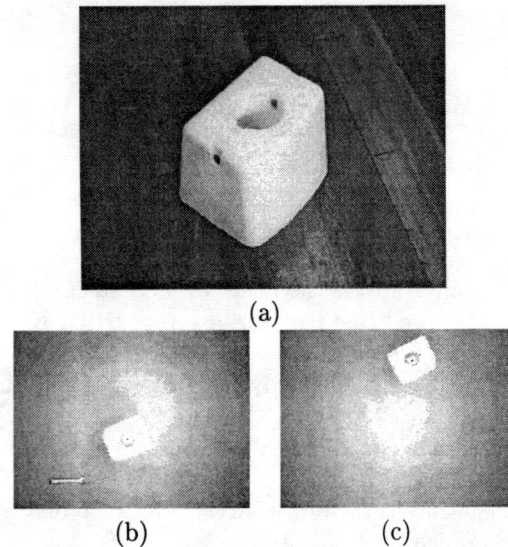

Figure 4: (a) The part used in the experiments, (b) image from $S1$, (c) image from $S2$

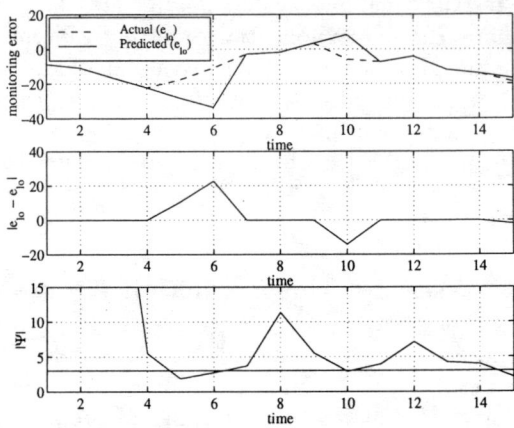

Figure 5: Results of processing sequence $S1$

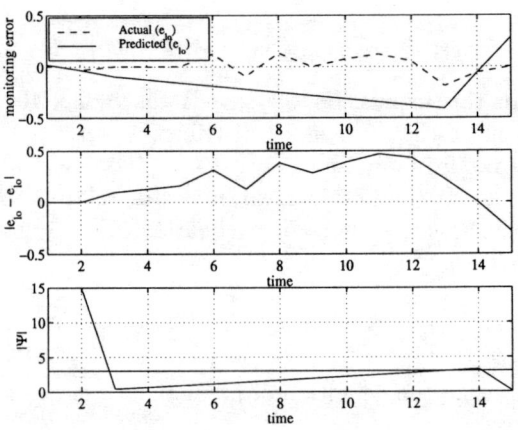

Figure 6: Results of processing sequence $S2$

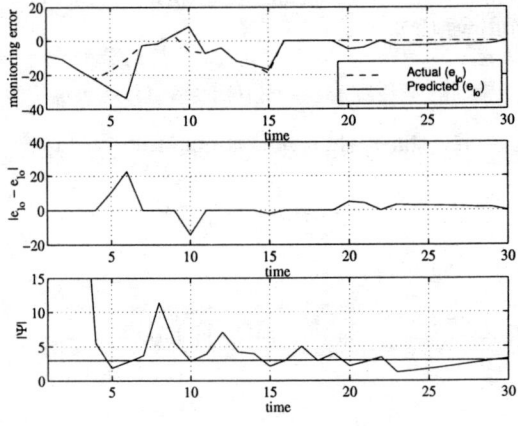

Figure 7: Results of processing the sequence $\{S1, S2\}$

discounting factor ϕ, of equation (6), was set to 0.3. The error correction algorithm was applied to the sequences $S1$, $S2$ and the combination of these sequences $\{S1, S2\}$. The results of processing the three image sequences are shown in Figures 5, 6, 7 and 8. Figures 5, 6 and 7 show the actual and predicted monitoring errors of the low-performance monitor m_{lo}, the magnitude of the difference between the actual and predicted errors, and the behavior of the Ψ function in inequality (10) (the inequality is satisfied whenever Ψ is above the horizontal line corresponding to $\delta = 3$). Figure 8 shows the average cost and error magnitude obtained using the ECA, using m_{hi} only, and using m_{lo} only.

The following observations can be made regarding the results. The exclusive use of m_{lo} resulted in the lowest possible costs but high errors. The use of m_{hi} provided the lowest errors but at the highest cost. This is not surprising since the two methods represent opposite extremes of the possible performance range. The ECA delivered errors within the specified limit ($\delta = 3$) for sequences $S1$, $S2$, and $\{S1, S2\}$. The error bound was achieved at a much lower cost than by using m_{hi} exclusively. Exclusive use of m_{lo} exceeded the bound on two of the three sequences.

The ECA invoked both monitors 73% of the time for $S1$, 20% of the time for $S2$, and 53% of the time for $\{S1, S2\}$. The relatively high monitoring activity for $S1$ is due to a combination of small sample size and large variation of the error series. The low activity for $S2$ is due to the small variation of the series relative to the bound ($\delta = 3$). The activity in $\{S1, S2\}$ is a combination of the the behaviors for $S1$ and $S2$.

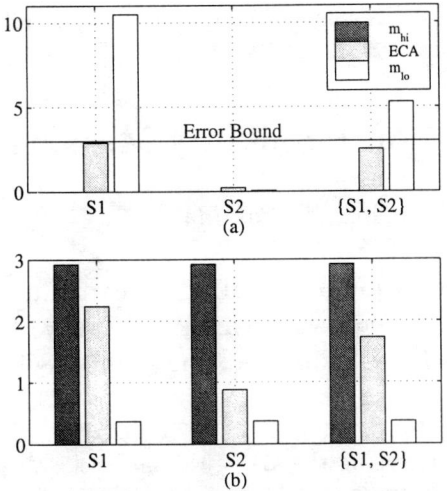

Figure 8: Comparison of performance: (a) Error is the average error magnitude per image (b) Cost is the average CPU time (seconds) per image

5 Discussion

The experiments have highlighted the following performance aspects:

- The exclusive use of m_{hi} resulted in the lowest errors but highest cost. The exclusive use of m_{lo} had the opposite outcome, with error bounds exceeded in two out of three experiments.

- The ECA delivered low errors (within specification) at much lower costs than m_{hi}. The ECA was therefore more efficient than exclusive use of m_{hi} in terms of cost, and superior to m_{lo} in terms of error levels.

Although the experiments compared the ECA to exclusive use of each process monitor, they provide some indication about the ECA's performance relative to traditional data fusion methods. A typical data fusion approach would attempt to combine the output of all process monitors at each time step. This means that, in the above experiments, both m_{hi} and m_{lo} would be used at each step. Such a data fusion scheme would produce very low errors but at a very high cost (the cost of using m_{hi} plus the cost of using m_{lo}). Traditional data fusion would therefore lack the efficiency advantages of the ECA. Because of its fixed structure, it would also lack the ECA's ability to cope with changing performance specifications.

The overall aim of this work was to develop a sensory perception methodology for autonomous robotic systems, which typically have dynamically changing information requirements, and whose performance depends on the efficiency of perception. The experiments showed that the ECA has the ability to attain low perceptual errors at low costs, and is able to dynamically adapt its behavior to changing performance specifications (changing δ and p). The ECA is therefore well suited to autonomous robotic applications.

References

[1] G.E.P. Box and G.M. Jenkins. *Time Series Analysis*. Holden-Day, 1976.

[2] M. Bozorg, E.M. Nebot, and H.F. Durrant-Whyte. Optimal data fusion in decentralised systems with flexible observation rates: Estimation for a land vehicle. In *Proceedings of the International Conference on Field and Service Robotics*, pages 306–312, 1997.

[3] R.J. Carroll and D. Ruppert. *Transformation and Weighting in Regression*. Chapman and Hall, 1988.

[4] B.K. Ghosh, D. Xiao, N. Xi, and T.J. Tarn. Multi-sensor based robotic manipulation in an uncalibrated manufacturing workcell. Preprint submitted to Elsevier Science.

[5] G.E. Hovland and B.J. McCarragher. Controlling sensory perception for indoor navigation. Proceedings of the 1998 IEEE Int. Conf. Robotics and Automat.

[6] G.E. Hovland and B.J. McCarragher. Dynamic sensor selection for robotic systems. In *Proc. 1997 IEEE Int. Conf. Robotics and Automat.*, pages 272–277, 1997.

[7] T.C. Henderson M. Dekhil. Instrumented logical sensor systems - practice. In *Proceedings of the 1998 IEEE International Conference on Robotics and Automat.*, 1998.

[8] T.C. Henderson M. Dekhil. Instrumented sensor system architecture. *The International Journal of Robotics Research*, April 1998.

[9] T.P. Ryan. *Modern Regression Methods*. John Wiley & Sons, 1997.

[10] J. van Dam. *Environment Modelling for Mobile Robots: Neural Learning for Sensor Fusion*. PhD thesis, University of Amsterdam, 1998.

[11] B.J. You, H.J. Kim, S.R. Oh, and C.W. Lee. Fusion of gray-level images and laser stripe images for three dimensional posture determination of polyhedral objects. In *Proceedings of the World Automation Congress WAC 98*, 1998.

Sensor Selection by Reliability Based on Possibility Measure

F. Kobayashi, F. Arai

Dept. of Micro System Eng.
Nagoya Univ.
Furo-cho, Chikusa-ku, Nagoya 464–8603

T. Fukuda

Center for Cooperative Research in
Advanced Science & Technology, Nagoya Univ.
Furo-cho, Chikusa-ku, Nagoya 464–8601

Abstract

Robotic or manufacturing systems become more and more complex for adapting to various environmental conditions. In these systems, sensor fusion methods for estimating states of a system from multiple sensor information have received much attention. Also, it is necessary to select the sensor information for adapting to various situation flexibly. In these days, various methods of fusing multiple information have been proposed so far, but these methods cannot select the sensor information. In this paper, we propose a sensor selected fusion system using a recurrent neural network. The sensor selection method is based on the production system, considering with the reliability calculated by the possibility measure. The effectiveness of the proposed method is shown through a simulation of a mobile robot.

1 Introduction

Human beings recognize surroundings or oneself and decide a action by using some sense organs, such as eyes, ears, tactile organs and so on. Although these sensations have ambiguity or contradiction, human beings can recognize and decide them correctly by selecting the useful sense organs or fusing them.

Recently, in robotics and manufacturing fields, a system has many various kinds of sensors like human for measuring the states of the system. As the states become complex and numerous more and more, the measurement methods for its state become also complex. Also, The real time measurements of all states is impossible by equipped sensors. So, sensor fusion method for inferring states which cannot be measured in realtime from multiple sensor information have received much attention. Many sensor fusion methods have been proposed so far. The previous methods may be divided into three types: The first is the research work concerning the statistical analysis[1, 2, 3]; The second is the research work concerning the artificial intelligence[4, 5]; The third is the research work concerning the the neural networks[6, 7]. We also have proposed a sensor fusion system using recurrent fuzzy inference which can fuse some sensor information with different sampling time and accuracy[8]. However, in these methods, the system have only some sensors which is selected by an operator in advance and recognize surroundings by all sensor information. On the other hand, there are research works which the system select the sensor information considering with the sensor error[9]. In these works, the system has some sensors for measuring a specific state. Then, the system selects a sensor with high accuracy among equipped sensors or integrates some sensor information. However, these methods are nothing but switching the sensor information or weighted average of some sensor information. So, it is difficult for the system to recognize in various environmental conditions because the system cannot acquire the needed sensor information and the sensor information have an error and ambiguity. Therefore, the system needs a sensor selection method according to an environmental condition for recognizing it accurately.

In this paper, we propose a sensor fusion system which can select the sensor by the reliability of sensor information. The reliability represents the degree how reliable the information is and is defined by the possibility measure which is a kind of the fuzzy measure[10]. The sensor selection method is based on the production system. In this method, rules of the production system are supplied by an operator in advance. Also, we use a recurrent neural network for fusing the selected sensor information because the recurrent neural network can predict the required information from time series data. For showing the effectiveness of our sensor fusion system, we apply a simulation which the system estimates the position of a robot with some sensors.

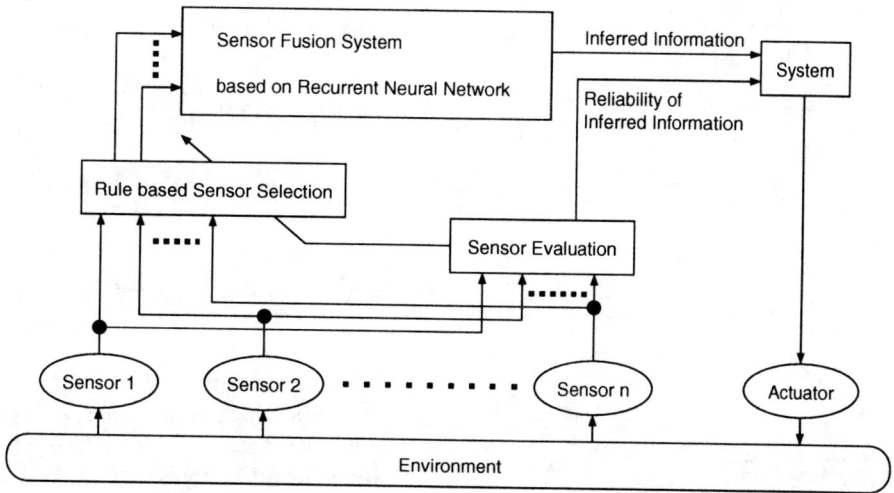

Figure 1: Sensor Fusion System with Sensor Selection

2 Sensor Selected Fusion System

2.1 Sensor Fusion System

Figure 1 shows the concept of a sensor selected fusion system. This system consists of three module; fusion module, evaluation module and selection module.

In this system, sensor information are used as the inputs of evaluation and selection module. Firstly, in evaluation module, the system calculates the reliability of each sensor information considering with sensor characteristics (sensing range and accuracy). Secondly, in selection module, the system selects the sensor information which are used in fusion module according to the reliability. Finally, in fusion module, the system estimates states from some selected sensor information by a recurrent neural network. At the same time, in evaluation module, the system calculates the reliability of estimated states considering with the structure of the recurrent neural network and the reliability of each sensor information. In the following section, we describe the detailed explanation of each module.

2.2 Reliability of Sensor Information

A sensor information belongs to a subset into which the measuring range is divided by the required accuracy. However, when the information and boundary value of the subset are same perfectly or nearly, the information has the ambiguity which it belongs to the subset or not because it has the error. In this paper, we use the possibility measure which is a kind of the fuzzy measures as the reliability how reliable the information is.

The sensor information, the estimated value, and the accuracy of them are normalized. We assume that the required accuracy of the sensor information is nearly equal to the accuracy of the estimated value which is determined by an operator in advance. The measuring range S_i of the sensor information s_i is divided into some subsets as shown in Fig. 2 by the output's accuracy dy. As the sensor information s_i belongs to one subset among some subsets, we assume that s_i belongs to the subset S_{ik} as expressed by Eq. (1).

$$S_{ik} = \left\{ s_i \mid (k-1)dy \leq s_i < kdy \right\} \quad (1)$$
$$(k = 1, 2, \cdots, K)$$

However, the sensor information s_i belongs to a subset E determined by the peculiar accuracy ds_i as expressed by Eq. (2).

$$E = \left\{ s_i \mid s_i - \frac{ds_i}{2} \leq s_i \leq s_i + \frac{ds_i}{2} \right\} \quad (2)$$

Then, the system transforms the subset S_{ik} into the section Ω as expressed by Eq. (3).

$$\Omega = \left\{ \omega \mid 0 \leq \omega < 1 \right\} \quad (3)$$

As the same way, the system transforms the subset E into a subset E' expressed by Eq. (4).

$$E' = \left\{ \omega \mid \omega_i - \frac{ds_i/2}{dy} \leq \omega \leq \omega_i + \frac{ds_i/2}{dy} \right\} \quad (4)$$

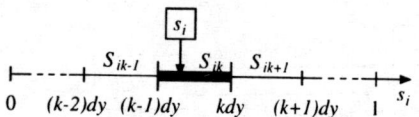

Figure 2: Division by output's accuracy

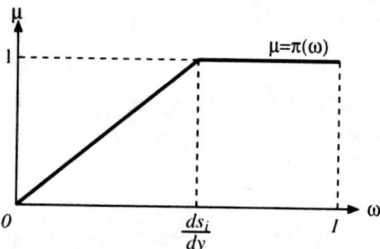

Figure 3: Possibility distribution function $\pi(\omega)$

where, ω_i is expressed by Eq. (5).

$$\omega_i = \begin{cases} \dfrac{s_i - (k-1)dy}{dy} & (s_i \le (k-\frac{1}{2})dy) \\ \dfrac{kdy - s_i}{dy} & (s_i > (k-\frac{1}{2})dy) \end{cases} \quad (5)$$

provided that the lower limit of the set E' is 0 and the upper limit of that is 1. Here, if the boundary value of the set E' is equal to the boundary value of the set Ω, the sensor information s_i has the ambiguity. So, we define the possibility distribution function as a function expressed by Eq. (6) as shown in Fig. 3.

$$\pi(\omega) = \begin{cases} \dfrac{dy}{ds_i} & \left(0 \le \omega \le \dfrac{ds_i}{dy}\right) \\ 1 & \left(\dfrac{ds_i}{dy} < \omega \le 1\right) \end{cases} \quad (6)$$

Consequently, the possibility measure λ_{s_i} of the sensor information s_i is calculated by Eq. (7).

$$\lambda_{s_i} = \sup\left\{\pi(\omega) \mid \omega \in E'\right\} \quad (7)$$

2.3 Sensor Select according to Reliability

Human beings recognize surroundings by using many sense organs, such as eyes, ears and so on. Although some of sensations are vague or inconsistent, they can recognize environments precisely by selecting the suitable sensations or fusing the sensations. Here, when they select the sense organs, they use an index and a rule including the knowledge acquired in the past.

In this paper, we use the reliability described in 2.2 as the index. Then, the system select the sensor information by the production system which have the rule determined by the operator in advance. The rule for the production system is expressed by Eq. (8).

$$\begin{aligned} \text{IF} \quad & \lambda_{s_1} \text{ is } \{ \text{ Low, Middle, High } \}, \\ & \lambda_{s_2} \text{ is } \{ \text{ Low, Middle, High } \}, \\ & \qquad\qquad \vdots \\ & \lambda_{s_I} \text{ is } \{ \text{ Low, Middle, High } \}, \\ \text{THEN} \quad & \text{using sensors are } \{ 1, 2, \cdots, s \} \end{aligned} \quad (8)$$

where, the terms "Low", "Middle" and "High" are the sets which represent the degree of the reliability. Also, we produce the rule for the sensor selection by the concept as follows:

- If all reliabilities belong to the set of the same kind, the system uses all sensor information.

- If there is the reliability which belongs to the "High" set, the system selects the sensor information whose reliability belongs to the "High" set.

- If all reliability don't belongs to the "High" set and there is the reliability which belongs to the "Middle" set, the system selects the sensor information whose reliability belongs to the "Middle" set.

Here, the boundary value of each set is determined by the operator in advance as follows:

$$\begin{aligned} \text{Low} &: \{0 \le r_{s_i} < 0.35\} \\ \text{Middle} &: \{0.35 \le r_{s_i} < 0.65\} \\ \text{High} &: \{0.65 \le r_{s_i} < 1\} \end{aligned}$$

2.4 Sensor Fusion by Neural Network

In human beings, sensations from sense organs are passed to the specific neuron in the brain. Then, in the brain, sensations are transformed into the required information. In this paper, we use a recurrent neural network for fusing the selected sensor information. The neural network is helpful in fusing the sensor information because it is defined as the model of processing information in the human brain. Especially, the recurrent neural network is very useful for the sensor fusion system which estimates the states of the system from time series data. Also, as the neural network can change the structure of oneself by learning, the system can construct the sensor fusion system efficiently and flexibly.

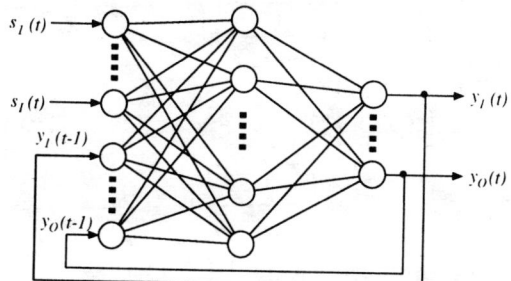

Figure 4: Recurrent Neural Network

In this paper, we use the Jordan type of the recurrent neural networks which have three layers (input layer, hidden layer and output layer) and feedback from output layer to input layer as shown in Fig. 4. Then, we use the back propagation method as the learning method of the recurrent neural network.

2.5 Reliability of Estimated Result

The reliability of sensor information is calculated by the possibility measure which is a kind of the fuzzy measure as described in 2.2. In this section, we describe the method of calculating the reliability of the estimated result considering with the reliability of sensor information and the structure of the recurrent neural network.

In the neural network, signals propagate from the input layer to the hidden layer and output layer. In this paper, we assume that the reliabilities of sensor information are propagate each layer in the same way.

The signal of the neuron j in the layer h is calculated by the summation of signals in the layer $h-1$ which are weighted down. As the same way, the reliability of the neuron j in the layer h is calculated by the weighted average of the reliability of neurons in the layer $h-1$ as expressed by Eq. (9).

$$R_{(h,j)} = \frac{\sum_i R_{(h-1,i)} \cdot |w_{ij}|}{\sum_i |w_{ij}|} \quad (9)$$

where $R_{h,j}$ represents the reliability of the neuron j of the layer h, w_{ij} represents the weight between the neuron j of the layer h and the neuron i of the layer $h-1$.

Thus, in each neuron, the reliability is propagated like the signal. Then, in the output layer, the reliability of the estimated result is calculated.

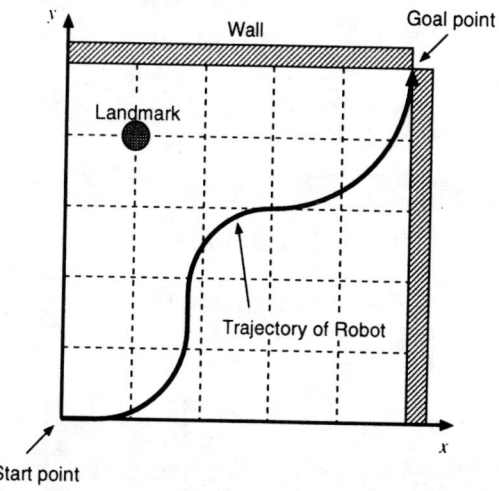

Figure 5: Simulation Environment

3 Simulation

We apply the proposed sensor selected fusion system to a numerical simulation of a mobile robot self-localization. The purpose of this simulation is to estimate with accuracy and high reliability the position of the robot which follows the trajectory described in Fig. 5.

In this simulation, the sensor information which used in sensor selected fusion system are the rotation angle of right and left tire, the size of the landmark on image, the rotation angle of a camera and the distance between the robot and the wall. The sensor parametrics are described in Table 1. Here, for showing the effectiveness of the proposed system, we compare the estimated position by the sensor selected fusion system to that by the sensor nonselected fusion system.

Figure 6 shows the sensor information and the reliability of each sensor information. The recurrent neu-

Table 1: Sensor Parametrics

sensor	range		accuracy
	min	max	
angle of right tire	0.00°	90.00°	0.18°
angle of left tire	0.00°	90.00°	0.18°
distance	0.000m	6.000m	0.006m
size of landmark	0.000	1.000	0.004
angle of camera	-180.00°	180.00°	0.36°

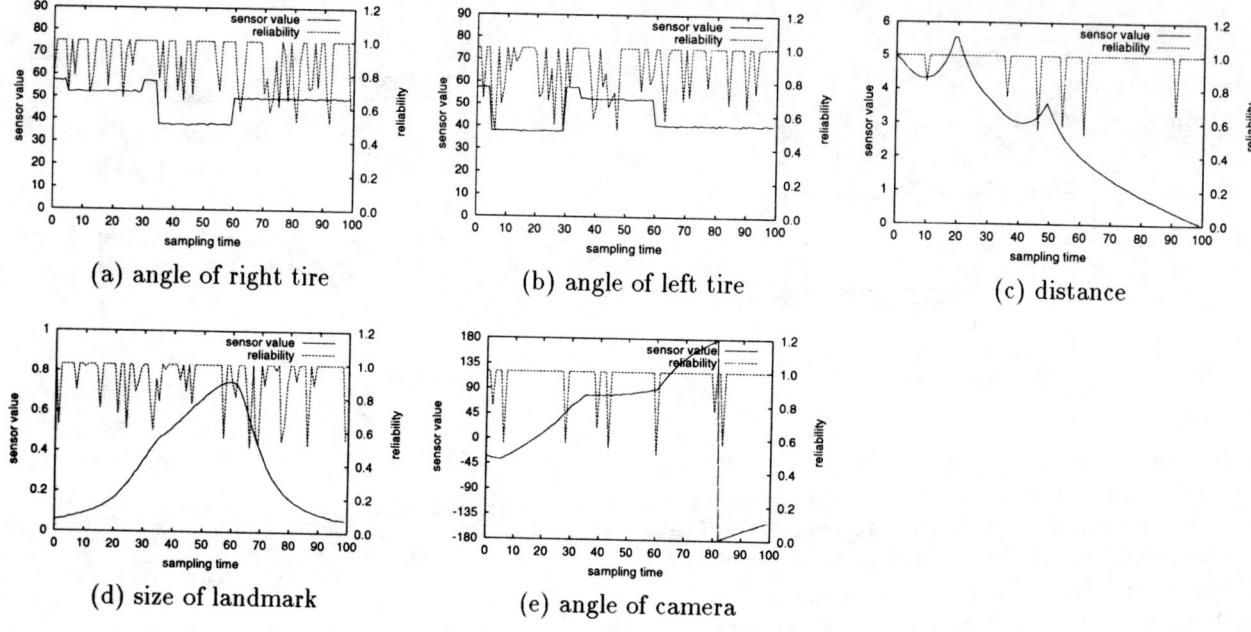

Figure 6: Sensor Output / Reliability

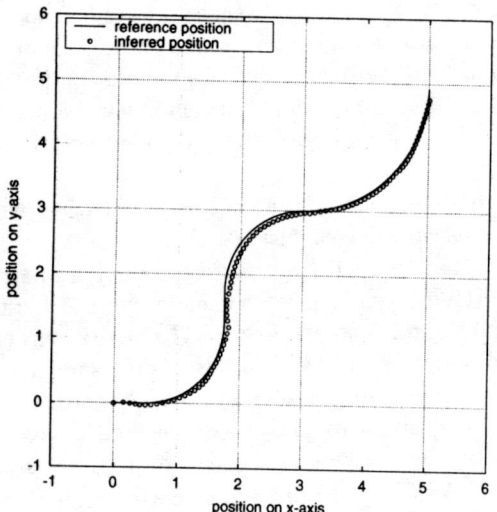

Figure 7: Result in Sensor Nonselected

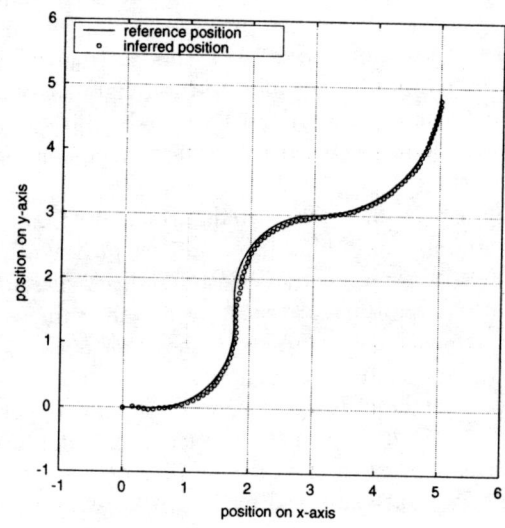

Figure 8: Result in Sensor Selected

ral network is learned until the the position error is less than 10^{-4}. The result of estimating the position by the fusion system without sensor selection shows in Fig. 7. Then, the result of it by the sensor selected fusion system shows in Fig. 8. Here, in these figures, the line represents the trajectory of the robot and the point represents the estimated position of the robot. Figure 9 shows which the sensor information is selected or not at each sampling time. Table 2 shows the mean square error and the reliability of the robot's position.

As shown in these figure and table, the proposed sensor selected fusion system can estimate the position of the robot with accuracy and high reliability. In the nonselected fusion system, the system has to use the sensor information with low reliability. On the other

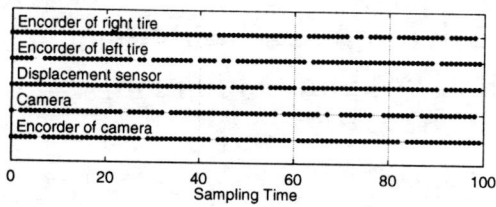

Figure 9: Selected Sensor

Table 2: Square Error

	average error	reliability
selected	1.082×10^{-4}	0.973
nonselected	6.043×10^{-4}	0.945

hand, in the proposed system, the system can eliminate it in selection module. Therefore, the system can enhance the accuracy and reliability of the position even if all sensor information have low reliability.

4 Summary

Human beings select the useful sense organs according to a environmental condition. In this paper, we propose the sensor fusion system which can select the sensor information considering with the reliability of sensor information. The reliability of sensor information represents the degree how reliable the information is and is based on the possibility measure which is a kind of the fuzzy measure. The system selects the sensor information by the production system which the rules are supplied by an operator in advance. Then, the system fuse the selected sensor information by the recurrent neural network. For showing the effectiveness of proposed method, we compare the sensor selected fusion system with the nonselected fusion system in a simulation of a mobile robot self-localization problem. By this simulation, although the learning speed of the selected method is slower than that of the nonselected method, the selected method can estimate the position with accuracy and high reliability.

In this paper, we use the rule which an operator arranges in advance as the sensor selection rule in the production system. Our next challenge is that the system automatically arranges the rule for selecting the sensor information.

Acknowledgments

This research was supported in part by a grant from "IMS-SIMON: Sensor Fused Intelligent Monitoring System for Machining".

References

[1] S. Shekhar, O. Khatib, M. Shimojo, "Object Localization with Multiple Sensors", *Int. J. of Robotics Research*, Vol. 7, No. 6, pp. 34–44, 1988.

[2] L. Sukhan, R. Sookwang, "Uncertainty Self-Management with Perception Net Based Geometric Data Fusion", *Proc. of the IEEE International Conference on Robotics and Automation*, pp. 2075–2081, 1997.

[3] S. V. R. Nageswara, "Nadaraya-Watson Estimator for Sensor Fusion Problems", *Proc. of the IEEE International Conference on Robotics and Automation*, pp. 2069–2074, 1997.

[4] P. K. Allen, "Integrating Vision and Touch for Object Recognition Tasks", *Int. J. of Robotics Research*, Vol. 7, No. 6, pp. 15–33, 1988.

[5] A. R. Hanson, "Sensor and information fusion from knowledge-based constraints", *Proc. of SPIE*, pp. 186–196, 1988.

[6] D. Zipser, R. A. Anderson, "A back-propagation programmed network that simulates response properties of a subset posterior partial neurons", *Nature*, Vol. 331, pp. 679–684, 1988.

[7] Y. Sakaguchi, "Topographic organization of nerve field with teacher signal", *Neural Networks*, Vol. 3, pp. 411–421, 1990.

[8] F. Kobayashi, F. Arai, T. Fukuda, M. Onoda, N. Marui, "Online Estimation of Surface Roughness by Recurrent Fuzzy Inference in Grinding Process", *Manufacturing Science and Engineering - 1997, MED-Vol. 6-1*, pp. 49–54, 1997.

[9] K. Shimojima, T. Fukuda, F. Arai, H. Matsuura, "Fuzzy Inference Integrated 3-D Measuring System with LED Displacement Sensor and Vision System", *Intelligent and Fuzzy Systems*, Vol. 1, No. 1, pp. 63–72, 1993.

[10] L. A. Zadeh, "Fuzzy sets as a basis for a theory of possibility", *Fuzzy Sets and Systems*, Vol. 1, pp. 3–28, 1978.

The Sensor Selection Task of The Gaussians Mixture Bayes' with Regularised EM (GMB-REM) Technique in Robot Position Estimation

Takamasa Koshizen

Department of Systems Engineering
Research School of Information Sciences and Engineering
The Australian National University
Canberra, 0200, Australia
takamasa@syseng.anu.edu.au

Abstract

Modelling and reducing uncertainty are two essential problems of mobile robot localisation. Our previous work has been to develop a robot localisation system, namely the Gaussian Mixture Bayes with Regularised Expectation Maximisation (GMB-REM), using a single sensor. It allows a robot position to be modelled as a probability distribution, and uses the Bayes' theorem to reduce the uncertainty of a robot's location. In this paper, a new system, which is enhanced from the GMB-REM system in order to perform a sensor selection task, is introduced. Empirical results show the proposed new system outperforms the GMB-REM system with sonar alone. That is, the new system can deal with multiple sensors and further minimise the average localisation error of the robot by performing the sensor selection task.

1. Introduction

In this paper, we focus on our new real mobile robot localisation technique in terms of a sensor selection issue for a robot's position estimation problem. Position estimation is one of the crucial issues for robot navigation and goal-reaching as a robot must be able to avoid several static/dynamic obstacles with different forms by utilising its perception to estimate its position. The position estimation problem is generally called a mobile robot localisation problem, i.e. a robot is located at an unknown position for which it has a map. It "looks" around from its position, and based on these observations it must infer the place (or set of places) on a map where it can be located [1].

Essentially, the problem of robot localisation can be summarised as *how "uncertainty" can be modelled and then reduced*. To cope with this uncertainty, we previously employed the Gaussian Mixture Bayes' with Regularised Expectation Maximisation (GMB-REM) technique with a sonar sensor. The GMB-REM utilised the probability distribution across a robot's possible location, and employed the Bayes' theorem to reduce uncertainty.

Note that there are key differences between the other works and our GMB-REM localisation system. The first is the method introduced by Kuc and Barshan [2], it lead to a unimodal density functions, in contrast to the GMB-REM that is multimodal. The second is the map generation technique, represented by Moravec et al. [3], whose work has been to mainly construct a map of the (indoor) environment using probabilistic occupancy grids. Our approach provides a way of making 'combined' use of map and sensor readings to compute the position of the robot within the map. In other words, the received sensory information can be used *immediately* to update the estimate of robot location. Another is that we allow a mobile robot to learn explicitly the pattern of high dimensional sensor signatures rather than to learn implicitly the features or landmarks of the environment, as has been done by Thrun [4]. Furthermore, in our localisation system, the Expectation Maximisation (EM) Algorithm [5] is employed to learn the relationship between sensory information and the robot's position. In practice, the EM is regularised to prevent both *singular* covariance matrices and *overfitting* problems since the Maximising Likelihood (ML) can lead to overfitting. The problem of overfitting is typically more severe in density estimation due to singularities in the log-likelihood function in high-dimensional spaces.

A mixture of Gaussians is well suited to approximate continuous probability densities of sensor readings given a robot's true location. A severe problem, however, comes up particularly when only sonar sensor is used in the system. That is, the sonar model tends to be somehow noised as the robot approaches particular locations in the cluttered room [6]. Therefore, we hereby introduce a sensor selection scheme into the previous GMB-REM with sonar alone.

This paper is organised as follow. In the next section, we present the enhanced GMB-REM localisation with its sensor selection task. In Section 3, we show that compared to the previous GMB-REM with sonar sensor alone, the new approach yields better results, subject to minimising (average) robot's localisation error. Section 4 outlines the conclusions and some remarks.

2. The New Gaussian Mixture Bayes with Regularised EM (GMB-REM) Localisation System with Its Sensor Selection Task.

2.1. EM Algorithm and Density Estimation

Generally, the *Expectation Maximization* (EM) algorithm is an iterative algorithm for the computation of maximum likelihood parameter estimation. Furthermore, the EM algorithm is widely used for parameter estimation of mixture models, in particular for *a mixture of Gaussians* model.

Initially, it is assumed that the probability density function of an n-dimensional random vector X is a mixture of Q multivariate Gaussians,

$$p(\chi|\theta) = \sum_{j=1}^{Q} \frac{a_j}{(2p)^{n/2}\sqrt{|\Sigma_j|}} \exp\left[-\frac{1}{2}(\chi-\mu_j)^T \Sigma_j^{-1}(\chi-\mu_j)\right]$$

where, $\theta = \{\alpha_j, \mu_j, \Sigma_j \mid j=1,...,Q\}$ is the set of parameters of the model, α_j are the mixing proportions, μ_j and Σ_j denote the means and the covariance matrices respectively. We denote $\chi = \{x^1,...,x^n\}$ as a sample data set (training set) of values of X, and assume that elements of χ are independent and identically distributed. The ultimate goal of the EM algorithm is to obtain parameter values $\hat{\theta}$ which maximise the likelihood of χ given the data, i.e.

$$\hat{\theta} = \arg\max_{\theta} p(\chi|\theta)$$
$$= \arg\max_{\theta} \prod_{i=1}^{N} p(x^i|\theta)$$

The EM algorithm can be summarised as follows:

1. Initialise the means, μ_j, to randomly picked data points from χ and the covariance matrices, Σ_j, to unit matrices. Set $\alpha_j = 1/Q$ for all j. Set the iteration counter $t=0$.

2. Expectation (**E**)-step: Compute the posterior probabilities $h_{ij}(t) \equiv E[z_{ij} \mid x^i, \theta]$ of membership of x^i to the jth Gaussians, for all i and j,

$$h_{ij}(t) = \frac{\frac{\alpha_j(t)}{(2\pi)^{n/2}\sqrt{|\Sigma_j(t)|}} \exp\left[-\frac{1}{2}(x^i-\mu_j(t))^T \Sigma_j(t)^{-1}(x^i-\mu_j(t))\right]}{\sum_{k=1}^{Q} \frac{\alpha_k(t)}{(2\pi)^{n/2}\sqrt{|\Sigma_k(t)|}} \exp\left[-\frac{1}{2}(x^i-\mu_k(t))^T \Sigma_k(t)^{-1}(x^i-\mu_k(t))\right]}$$

3. Maximisation (**M**)-step: Re-estimate the mixing proportions $\alpha_j(t+1)$, means $\mu_j(t+1)$, and covariances $\Sigma_j(t+1)$ of the Gaussians using the data set weighted by $h_{ij}(t)$. Increment the iteration counter $t=t+1$ and go to the E-step by the following manner:

$$\alpha_j(t+1) = \frac{1}{N}\sum_{i=1}^{N} h_{ij}(t)$$

$$\mu_j(t+1) = \frac{\sum_{i=1}^{N} h_{ij}(t) x^i}{\sum_{i=1}^{N} h_{ij}(t)}$$

$$\Sigma_j(t+1) = \frac{\sum_{i=1}^{N} h_{ij}(t)(x^i - \mu_j(t+1))(x^i - \mu_j(t+1))^T}{\sum_{i=1}^{N} h_{ij}(t)}$$

4. Iterate the above two steps until the change in the means μ_j is below some specified threshold.

2.2. The Gaussian Mixture Bayes with Regularised EM (GMB-REM) Localisation System

An overview of our robot localisation system with sensory data is given in Fig. 1. It mainly consists of 4 phases:

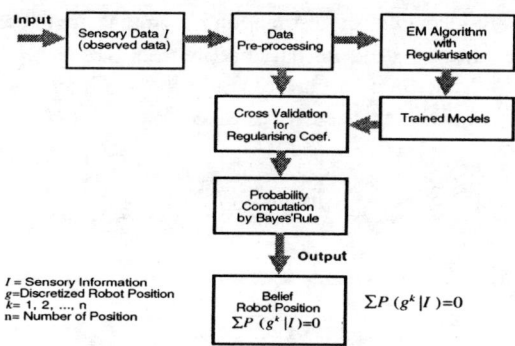

Figure 1. The GMB-REM Localisation System

1. *Data Collection* The robot is moved on a planar workspace $W \subset \Re^2$ which is a darker area as shown in Fig.2. A sensor signature is then taken for some position. After correction, the sonar data are normalised to *zero mean* and *unit variance*. Further, these are divided into several data sets, namely training, validation and test sets respectively. It must be noted that test data is obtained separately from training and validation data sets.

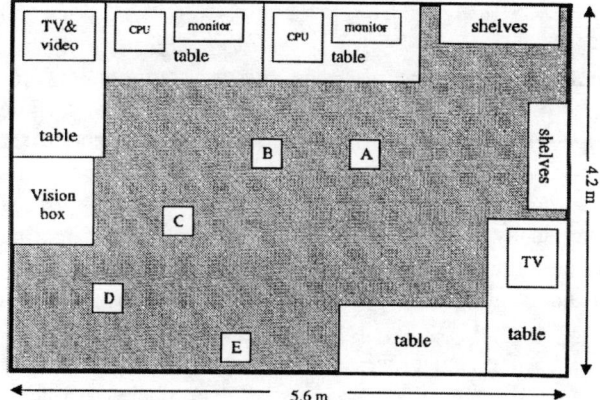

Figure 2. Top Down View of the Cluttered Room

2. *Training*: A set of given training data is constructed into a model to predict the probabilistic conditional distribution $P(\mathbf{r}|\mathbf{x})$ by using the *Regularised* EM algorithm.

Moreover, the EM algorithm is widely applied for parametric *density estimation* for a mixture of Gaussian models as described in section 2.1. In fact, the Gaussian mixture densities are able to approximate any continuous probability function with an appropriately chosen mixture [7]. In our localisation system, the training data are separated into each different robot's position $\mathbf{x}$ estimating the parameters of the conditional probability.

Fig. 3 shows the prediction of the conditional probability distribution, which is approximated by constructing a model, given an arbitrary robot position $\mathbf{x} \in \Re^2$. The model must produce a density estimation f such that $f(\mathbf{r}; \theta) = P(\mathbf{r}|\mathbf{x})$. If we choose f as normal distribution like a Gaussian function, and that is correct, then the parameter vector $\theta = (\mu, \sigma)$ is sufficient to fully describe $P(\mathbf{r}|\mathbf{x})$.

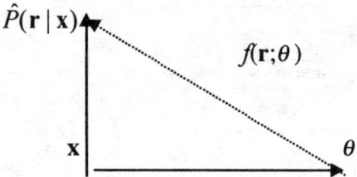

Figure 3. From Position x to Probability Density Function

μ and σ are called the mean and the standard deviation respectively. In our approach, a true distribution of sonar readings is approximated by *a mixture of Gaussian models*, which model probability densities by weighted sums of normal distributions. The mixture model is also parameterised by $\theta_\mathbf{x} = (\alpha_{x j}, \mu_{x j}, \Sigma_{x j})$. Once the parameters $\theta_\mathbf{x}$ are found, the conditional density function f of a certain position $\mathbf{x}$ can be written as:

$$P(\mathbf{r}|\mathbf{x}) = \sum_{j=1}^{Q_\mathbf{x}} \frac{\alpha_{j\mathbf{x}}}{(2\pi)^{16/2} \sqrt{|\Sigma_{j\mathbf{x}}|}} \exp\left[-\frac{1}{2}(\mathbf{r} - \mu_{j\mathbf{x}})^T \Sigma_{j\mathbf{x}}^{-1}(\mathbf{r} - \mu_{j\mathbf{x}})\right]$$

where we have used $Q_\mathbf{x}$ as mixture components for modelling the function of a robot's position $\mathbf{x}$. The probabilistic generative model parameter $\theta_\mathbf{x}$ can be adjusted to maximise the log-likelihood for the given set of training data via the EM. Certainly, the parameter $\theta_\mathbf{x}$ could also be obtained by Neural Networks in terms of approximating any continuous functions [8]. Furthermore, in our system, we also regularise Gaussian mixture models by adding a penalty term Σ_p to the log-likelihood cost function. Actually, the penalty we use is realised by a small value of $\varepsilon = \max\{|\Sigma_{j\mathbf{x}}|, \varepsilon_{j\mathbf{x}}\}, (0.0 < \varepsilon_{j\mathbf{x}}, \varepsilon < 1.0)$ for each covariance matrix in the M-step of every EM iteration. We also add ε to the *posteriors* to prevent the Gaussians from being too far from all the data points. Thus, we name the EM as a *Regularised EM* algorithm. Use of the regularisation technique has become widespread in many areas of science, including statistics [9]. The most important role of the regularisation technique is to prevent both singular covariance matrices and overfitting [10]. Particularly, in high-dimensional spaces, these two problems frequently cause the computed density estimates to possess only relatively limited generalisation capabilities with respect to predicting the densities of new data.

3. *Cross Validation* (*CV*): A suitable value for regularising coefficient ε can be found by seeking the value which gives the best performance on a validation set.

4. *Calculating the Posterior Probability* (or *belief*) *of the Robot's position*: Initially, a robot position $\mathbf{x}$ is *dicretised* into a grid cell g^k, and the prior distribution of each position is *uniform*. In our system, it is set to 0.5. For a given test set, the posterior probability of robot position $P(g^k|\mathbf{r})$ is calculated by using the GMB approach. The GMB assigns a vector from R^{16} (16 dimensional sonar data space) to one of the grid cells. The input vector $\mathbf{r}$ is called sonar vector or observation vector and consists of sets of measurements that distinguish among grid cells. In our framework the likelihood of $\mathbf{r}$ at each position is multiplied by the robot's prior probability of being in the cell. The products are then renormalised to obtain a *posterior* probability distribution, which incorporates the information in the most current sonar readings. The probability distribution $P(g^k|\mathbf{r})$ is alternatively called the *belief* of the robot's position. The belief actually can provide a variance of its position, which also corresponds to an existence probability of a robot in a grid cell.

2.3. The Sensor Selection Task and the GMB Localisation System

In this section, we will describe the GMB-REM system in terms of the sensor selection task. An overview of the GMB-REM localisation system and its sensor selection task is represented in Figure 4.

Mainly, it consists of two stages; sensor perception and sensor selection tasks respectively. The perception stage is exactly the same as the one previously described in Section 2.2. In addition, the system is capable of inputting multiple sensor information such as sonar, infrared and camera-photo sensors. The system also allows us to choose an option whether we perform the sensor selection task or not. Fig.4 represents the sensor selection task for a robot's position estimation, subject to minimising a robot's localisation error.

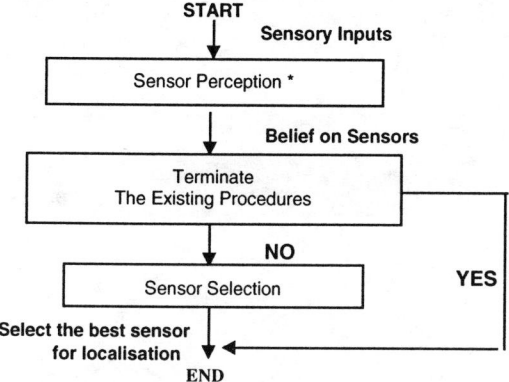

Figure 4. The Extended GMB-REM System

The basic idea is that sensor selection allows us to choose the sensor which works best at a certain position, considering the minimum (metric) error of the localisation. In other words, by comparing the error for different positions, the robot can decide which one yields more accurate localisation.

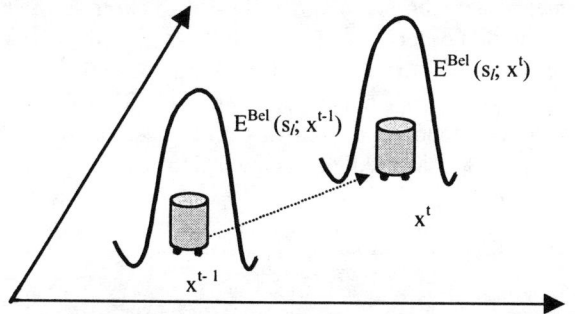

Figure 5. Localisation Error for Each Robot's Position

In addition, Fig.5 shows that error associated with different types of sensors for each position of the robot. Where $x^t \in x$ denotes a robot's position at a certain time, and s_l is a type of sensor (l: the number of different types of sensors).

Mathematical definitions of localisation metric error on beliefs E^{Bel} and sensor selection are as follows,

Let X be a robot's position space and S be a sensor space.

<u>Definition1</u> (Localisation error E^{Bel}): The metric error is calculated or measured at an arbitrary robot's position, i.e.,

$$E^{Bel} = \{ s \in S | E^{Bel}(s;x') \forall x' \in X \}$$

<u>Definition2</u> (Sensor Selection): There is a sensor that minimises the metric error of robot position, i.e.,

$$\exists_{s' \in S}; s' = \text{argmin } E^{Bel}$$
$$= \text{argmin} \{ s \in S | E^{Bel}(s;x') \forall x' \in X \}$$

In practice, there are possibly various ways to determine the form of localisation error. However, in our case, the localisation error was simply a metric distance, which can be measured by comparing the predicted position of the robot based on a sensor signature, and the true position at which the reading was taken.

3. Empirical Results

In our research, A Nomad200 mobile robot produced by Nomadic Technologies was used. It has a cylindrical shape with an approximate radius of 0.23 m and a kinematic model equivalent to a unicycle. It also has a base (lower part) and a turret (upper part) orientation. The base is formed by 3 wheels and a tactile sensor. The turret is a uniform 16 sided polygon, which entirely corresponds to a heading direction of the robot. On each side, there are sonar (alternatively called ultrasonic), infrared and bumper (touch) sensors. In our experiment, for the sensor selection task with the GMB-REM localisation system, sonar and infrared sensors were used to implement the sensor selection scheme, in order to cope with the disadvantage of the long-distance sensor (sonar), infrared sensor which has the nature of a short-distance. The sonar sensor was outlined in the previous section; here the infrared sensor is described briefly. The infrared sensor measures the difference between emitted light and reflected light. It is also very sensitive to the ambient light. It is normally assumed that for short distances and for the areas shorter than about 0.5 m, the range information of the infrared is acceptable; however the use of sonar sensor here is improper.

In the experiment, the room was represented into grids, which contained a matrix of cells representing a room of 4.2m by 5.6m. In addition, the size of each grid cell was just 0.1m by 0.1m. Fig.2 above shows the top-down view of our robot's room, used in our research. In the cluttered room, there were chair legs, feet, desks and computers, a trash bin, as well as a vision box, which was located fairly close to the corner of the room. In addition, to evaluate the performance of our sensor selection scheme, we chose the location where robot moved from a centre of the room to a corner with nearby some obstacles. Our previous work suggested that the sonar model was somehow noised around corner [6]. This *noised* area can be used to show how our sensor selection scheme reduces the uncertainty.

Initially, sonar and infrared data were collected for each position of the robot. Each set of sensor data constituted a vector of 16 sensor readings, because these were obtained at a point of 16 sensor readings of the Nomad200 robot. In addition, each set was comprised of 400 training and 200 validation data points for each position of the robot. For both sensors, 70 samples of the test data were taken separately from the training and validation data sets. In particular, test data obtained by successive movements of the robot approximately 0.02 to 0.04 m away from the point where the training data were collected. The number of mixture Gaussians we applied here was from 1 to 20. Moreover, the range of the regularising coefficients was within 0.3 to 1.0, and the numerical singularity for covariance vectors actually occurred between the range of 0.0 and 0.2.

Fig.6 represents an example of the relationship between the log-likelihood and the regularising coefficients. The range of the regularising coefficient was within 0.3 to 1.0. In this case, the maximised log-likelihood was highest when the coefficient was 0.3 for both sensors.

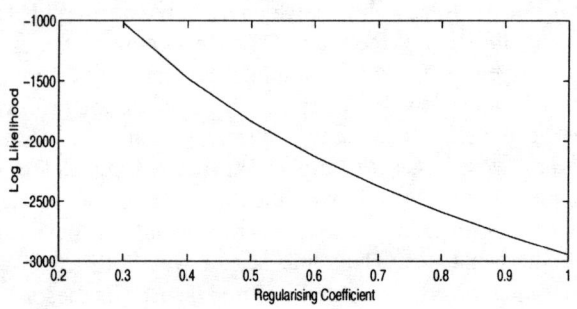

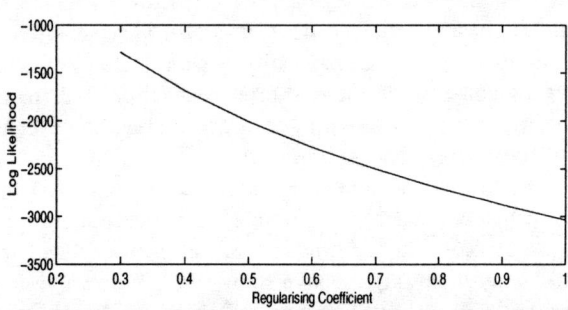

Figure 6. Log Likelihood and Regularising Coefficient

Fig.7 describes the time-consuming on (internal) robot's belief (0.0<belief<1.0) perceived by sonar sensor when the robot was located at point B in the Fig. 2. In our experiment, a single sensor reading was usually not enough to determine the position of the robot *uniquely*. Reports from the sensor readings were combined until the belief converged to be equal or greater than 0.99. This figure also indicates that the variance of the belief apparently decreased after the second sonar signal was combined by the Bayes' rule. What we must note here is the importance of the size of the variance, as smaller variance indicates the reduced uncertainty of the robot's position. Furthermore, the size of variance actually depends on the size of each grid cell. As a result, the measurements of the robot's position were very accurate when the size of each grid cell was small enough.

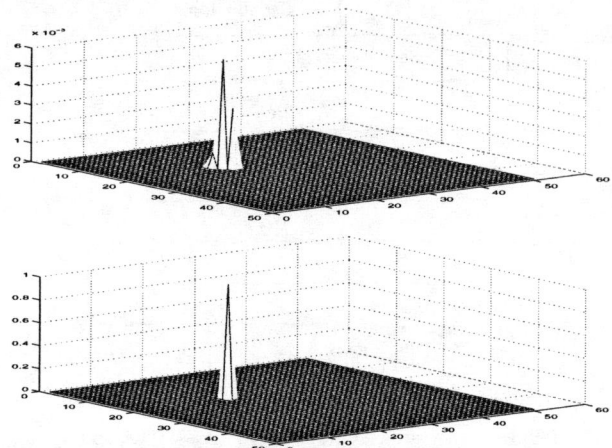

Figure 7. Time Consuming on Belief (Sonar Sensor)

Fig. 8 shows the localisation error resulting from sonar and infrared sensor models. The solid line represents the sonar's error, while the dash represents the infrared's error. In this experiment, the robot was moved by joy-sticking through the path), in this order: A → B → C → D → E in the Fig 2. From path A to C, the infrared sensor was not able to be used because the distance between the robot and an obstacle such as the vision box is more than 1 m. Hence, the infrared sensor was actually applied into the path C to E.

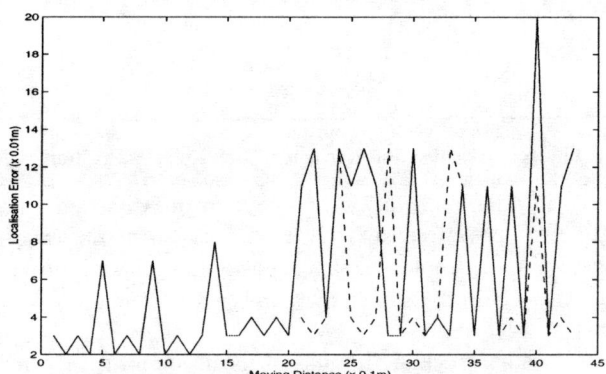

Figure 8. Localisation error of sonar and infrared models

**Picture 1. (left)-Robot is located at near corner of the room.
(right)-Robot is located at the centre of the room.**

Fig.9 represents the normalised average localisation error, which was calculated by the following equation;

$$\frac{1}{\tau * K} \sum_{k=1}^{K} E^{Bel}(k)$$

where, k denotes k-th test sample, and τ denotes the total number of test samples. The solid line denotes the error of sonar alone, while the dash denotes that of sensor selection using both sonar and infrared. This figure shows a considerable decrease in the error in the noised area with the use of new system, whereas the reverse is true of the previous GMB-REM with sonar alone.

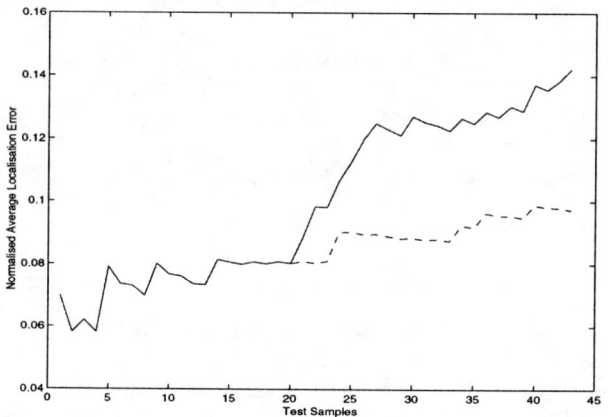

Figure 9. Normalised Average Localisation Error of Sonar alone and Sensor Selection

4. Conclusion

In this paper, we proposed a new robot localisation system, which is an enhancement of our previous system, namely the Gaussian Mixture Bayes with Regularised Expectation Maximisation (GMB-REM) with sonar alone. The enhanced localisation system for a sensor selection task is capable of dealing with multiple sensor inputs and constructing their probabilistic models by using a density estimation technique. The principal idea of the sensor selection task is to enable us to choose the sensor, which works best at a particular position, considering the minimum (metric) localisation error. In short, by comparing the error for different positions of a robot, the robot can determine which sensor model comes out more accurate for localisation.

Our experiment was implemented by locating the robot close to a corner with nearby obstacles because obstacle avoidance in this sort of area is very important for a robot's navigation. In meet points (corners and dead ends), a large dead-reckoning error often occurs, for instance [11]. Our results indicate the sensor selection task can further minimise the robot's average localisation error where the robot is located close to a corner with nearby some obstacles.

In summary, our approach can be regarded as a mean of reducing uncertainty of sensor information. Further research can be extended by applying the proposed method to more complex environments.

References

1. I.J. Cox, "Blanche-An Experiment in guidance and navigation of an Autonomous Robot Vehicle", IEEE Trans. Robotics and Automation, Vol.7, pp.193-204, 1991.
2. B. Barshan and R. Kuc, "Differentiating Sonar Reflections from Corners and Planes by Employing an Intelligent Sensor", IEEE Transactions on Pattern Analysis and Machine Intelligence, pp.560-569, June 1990.
3. H. Moravec, "Sensor Fusion in Certainty Grids for Mobile Robots", AI Magazine, pp. 61-74, 1988.
4. S. Thrun, "Bayesian Landmark Learning for Mobile Robot Localisation", to appear in Machine Learning, 1998.
5. A.P. Dempster, N.M. Laird, D.B. Rubin, "Maximum Likelihood from Inclomplete Data via the EM Algorithm", Journal of Royal 38, 1987. Statistical Society, Series B, Vol. 39, pp.1-38, 19878.
6. T. Koshizen, "The Gaussian Mixture Bayes with Regularised EM Algorith (GMB-REM) for Real Mobile Robot Position Estimation Technique", In Proc International Conference of Automation, Robotics, Control and Vision (ICARCV), Singapore, 1998.
7. L. Xu and M.I. Jordan, "On Convergence Properties of the EM algorithm for Gaussian Mixtures", Neural Computation, 8, 129-151, 1996.
8. K. Hornik, M Stinchcombe, H. White, "Universal Approximation of an Unknown Function and its Derivatives using Multi layer Feedforward Networks", Neural Networks, Vol. 3, pp. 551-560, 1990.
9. V. Vapnik, "The Nature of Statistical Learning Theory", Springer, NewYork, 1995.
10. T. Koshizen and Y. Rosseel,, "A New EM Algorithm with a Class of Thikonov Regularisers", To be submitted to International Joint Conference on Neural Networks, Washington, DC, USA, 1999.
11. K. Nagatani, H. Choset, S. Thrun, "Towards Exact Localization without Explicit Localization", Proc. of IEEE International Conference on Robotics and Automation, 1998.

Point Stabilization of Mobile Robots Via State Space Exact Feedback Linearization

KyuCheol Park[1], Hakyoung Chung[2], and Jang Gyu Lee[1]

[1] Automatic Control Research Center and School of Electrical Engineering
Seoul National University, Seoul, 151-742, Korea

[2] Department of Control and Instrumentation Engineering
Seoul National University of Technology, Seoul, 139-743, Korea

Abstract

In this paper, the point stabilization of mobile robots via state space exact feedback linearization is presented. The state space exact feedback linearization has not been possible for the point stabilization of mobile robots due to the restricted mobility caused by nonholonomic constraints. Under our proposed coordinates, however, the point stabilization problem can be exactly transformed into the problem of controlling a linear time invariant system. Thus, using well-established linear control theory, the point stabilization of robots can be easily formulated.

1. Introduction

The controllers of mobile robots aim to stabilize a robot at a desired point — point stabilization of the robot — or to control the robot to follow a desired trajectory. Since the mobile robot has restricted mobility, wheel driven mobile robots have nonholonomic constraints that arise from constraining the wheels of the mobile robot from rotating without slipping [1]. Furthermore, the linearized system of mobile robot with nonholonomic constraints has a controllability deficiency (Brockett theorem [2]). Therefore, linear control approaches are not applicable for mobile robot control. Path tracking and path following problems have been solved via standard nonlinear control approaches such as input/output feedback linearization [3, 4]. Since the point stabilization problem has not been solved using the standard nonlinear control approach, a variety of approaches have been attempted [5, 6, 7].

The point stabilization problems are concerned with obtaining feedback control laws that guarantee an equilibrium point of the entire closed loop system to be asymptotically stable. For a linear time invariant system, if all unstable eigenvalues are controllable, then the system can be asymptotically stabilized by a linear time invariant state feedback. However, nonholonomic systems such as mobile robots can not be asymptotically stabilized by a linear time invariant static state feedback since the linearization of the nonholonomic system about any equilibrium point is not asymptotically stabilizable, nor can the robot be stabilized with smooth state feedback. In spite of this limitation, the complete controllability of the nonholonomic system guarantees that there exist feedback control laws that asymptotically stabilize the system about an equilibrium point. Several nonstandard nonlinear control approaches for the nonholonomic system have been studied [1].

Those controllers can be categorized into discrete time invariant controllers, time varying controllers and hybrid controllers. All of the controllers are based on Lyapunov control theory. Thus, one can guarantee that the robot moves to the equilibrium point and consequently the convergence rate to the equilibrium point has been the performance measure of the designed controller. Therefore, to increase the convergence rate, several variant norms are adopted instead of conventional norms.

In this paper, the point stabilization of mobile robots is investigated using standard nonlinear control approaches such as the state space exact feedback linearization. The objective of the controller is to stabilize a mobile robot at a given target point. Three possible cases are investigated according to the condition of the robot at the target point. First, the robot moves to a target point without heading angle constraints. Secondly, the robot is stabilized at the target point satisfying a given heading angle constraint. Finally, we have considered the situation in which the target point is not fixed, *i.e.*, the target point itself maneuvers. In this case, the situation is transformed into a target and pursuer problem. The objective is to find the hitting conditions. The heading angle constraint at the hitting time is also considered. For the three cases, conditions for the state space exact feedback linearization are derived under a unified framework.

This paper is organized as follows. In section 2, the state space exact feedback linearization for mobile robots is developed. In section 3, the state space exact feedback linearization for the pursuer and target problem is developed. Finally, in section 4, conclusions of the feedback linearization for the mobile robot are addressed.

2. State space exact feedback linearization for mobile robots

The kinematic equation of a mobile robot in the two dimensional Cartesian space is

$$\dot{x} = V\cos\phi$$
$$\dot{y} = V\sin\phi \qquad (1)$$
$$\dot{\phi} = \omega.$$

However, there is a nonholonomic constraint that should be satisfied. The equation (2) shows a nonholonomic constraint of a robot: that is the wheel of the robot is rotating without slipping.

$$\dot{x}\sin\phi - \dot{y}\cos\phi = 0. \qquad (2)$$

The control problem of a robot is difficult since the control inputs have to be generated satisfying the constraint (2) while the robot moves.

Consider a mobile robot position and orientation with respect to a target point in polar frame as shown in Fig. 1. In Fig. 1, γ is the distance to go. α is the line of sight angle. ϕ is the heading angle. The control variables of a mobile robot, V and ω are the linear velocity and the angular velocity, respectively. In this coordinate, the target point is assumed to be $(0,0,0°)$. The control objective is how to move the robot to the target point. The mobile robot motion is expressed by

$$\dot{\gamma} = -V\cos\theta$$
$$\dot{\theta} = -\omega + V\frac{\sin\theta}{\gamma} \qquad (3)$$
$$\dot{\alpha} = V\frac{\sin\theta}{\gamma}.$$

The nonholonomic constraints do not appear explicitly in (3), and the controllability of (3) is as follows.

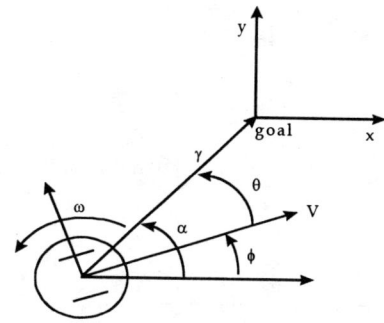

Fig. 1 Mobile robot coordinate in the polar frame

Lemma 1
The system expressed in (3) is completely controllable.
Proof
The driftless system (3) is completely controllable if the rank of the iterated Lie bracket generated by input vector space g_i is equal to the system dimension. The rank of the iterated Lie bracket is

$$rank\{g_1, g_2, [g_1, g_2]\} = rank\begin{bmatrix} -\cos\theta & 0 & \sin\theta \\ \frac{\sin\theta}{\gamma} & -1 & \frac{\cos\theta}{\gamma} \\ \frac{\sin\theta}{\gamma} & 0 & \frac{\cos\theta}{\gamma} \end{bmatrix} = 3.$$

Therefore, the system (3) is completely controllable.
Q. E. D.

The necessary and sufficient condition for state space exact feedback linearization for a driftless system is that $rank\{g_1(0), g_2(0), \cdots, g_m(0)\} = n$ [3]. The equation (3) does not satisfy the condition of the feedback linearization. To avoid this problem, an auxiliary variable $a = \dot{V}$ is introduced — that is one of the control inputs is modified to the linear acceleration from the linear velocity. With the augmentation of the auxiliary variable, the system (3) is no longer a driftless system. Thus, the feedback linearization methods are applicable to the nonlinear kinematic motion of a robot.

In the following sections 2.1 and 2.2, it is shown that the kinematic equations of a mobile robot can be state space exact feedback linearizable. The kinematic system of the mobile robot expressed in polar coordinate are completely transformed into a linear time invariant system by introducing an auxiliary variable related with the linear velocity.

2.1 Free heading angle at the equilibrium state

Let the state variables be $x_1 = \gamma, x_2 = \theta, x_3 = \alpha, x_4 = V$, then kinematic motion of (3) is expressed as (4) in the form of $\dot{x} = f(x) + g(x)u$ where $\varpi = -\omega$ and $a = \dot{V}$.

$$\dot{x} = \begin{bmatrix} -x_4 \cos x_2 \\ x_4 \dfrac{\sin x_2}{x_1} \\ x_4 \dfrac{\sin x_2}{x_1} \\ 0 \end{bmatrix} + \begin{bmatrix} 0 & 0 \\ 0 & 1 \\ 0 & 0 \\ 1 & 0 \end{bmatrix} \begin{bmatrix} a \\ \varpi \end{bmatrix} \quad (4)$$

Proposition 1

Let the output functions be $\lambda_1 = x_1 \cos x_3, \lambda_2 = x_1 \sin x_3$, then the conditions of state space exact feedback linearization are satisfied where $x_1 \neq 0, x_4 \neq 0$. Thus the state space exact feedback linearization problem of a mobile robot can be solved.

Proof

It can be easily shown that each distribution $G_i = \text{span}\{ad_f^i g_1, ad_f^i g_2 \, 0 \leq i \leq 4, ad_f^0 g = g\}$ associated with f and g_i has a constant dimension and is involutive. Therefore, there exist a neighborhood U of x^0 and m real valued functions $\lambda_1(x), \lambda_2(x), \cdots, \lambda_m(x)$ defined on U. The system has relative degree $\{r_1, r_2, \cdots, r_m\}$ at x^0, with $r_1 + r_2 + \cdots + r_m = n$. As the output functions are $\lambda_1 = x_1 \cos x_3, \lambda_2 = x_1 \sin x_3$, then the input transformation matrices are defined as follows.

$$A(x) = \begin{bmatrix} -\cos(x_2 - x_3) & x_4 \sin(x_2 - x_3) \\ \sin(x_2 - x_3) & x_4 \cos(x_2 - x_3) \end{bmatrix}, b(x) = 0$$

$\det(A(x)) = -x_4 \neq 0$, where $x_4 \neq 0$.

The local diffeomorphism and the state transformation are defined as follows.

$\xi_1 = \phi_1^1 = \lambda_1 = x_1 \cos x_3, \xi_2 = \phi_2^1 = L_f \lambda_1 = -x_4 \cos(x_2 - x_3)$
$\xi_3 = \phi_1^2 = \lambda_2 = x_1 \sin x_3, \xi_4 = \phi_2^2 = L_f \lambda_2 = x_4 \sin(x_2 - x_3)$

Then, the nonlinear system (4) is transformed to a linear system shown in (5).

$$\dot{\xi} = \begin{bmatrix} 0 & 1 & | & 0 \\ 0 & 0 & | & \\ \text{---} & \text{---} & | & \text{---} \\ & & | & 0 & 1 \\ 0 & & | & 0 & 0 \end{bmatrix} \begin{bmatrix} \xi_1 \\ \xi_2 \\ \xi_3 \\ \xi_4 \end{bmatrix} + \begin{bmatrix} 0 & 0 \\ 1 & 0 \\ 0 & 0 \\ 0 & 1 \end{bmatrix} \begin{bmatrix} v_1 \\ v_2 \end{bmatrix}$$

$$y = \begin{bmatrix} 1 & 0 & 0 & 0 \\ 0 & 0 & 1 & 0 \end{bmatrix} \begin{bmatrix} \xi_1 \\ \xi_2 \\ \xi_3 \\ \xi_4 \end{bmatrix} \quad (5)$$

Q. E. D.

The input transformation matrix $A(x)$ is nonsingular, where $x_4 \neq 0$ which means that the speed of the mobile robot is not zero. As the distance to go is assumed to be nonzero, *i.e.*, $\gamma \neq 0$, the nonzero speed of the robot is acceptable. The nonlinear system is decomposed into X and Y axis motion via the state space exact feedback linearization. Moreover, each axis motion is perfectly decoupled.

The control inputs v_1, v_2 are expressed with the states $x_1 \sim x_4$ and u_1, u_2 as shown in (6).

$$\begin{aligned} v_1 &= -\cos(x_2 - x_3)u_1 + x_4 \sin(x_2 - x_3)u_2 \\ &= -a\cos\phi + V\sin\phi\omega = -a_x + V_y\omega \\ v_2 &= \sin(x_2 - x_3)u_1 + x_4 \cos(x_2 - x_3)u_2 \\ &= -a\sin\phi - V\cos\phi\omega = -a_y - V_x\omega \end{aligned} \quad (6)$$

In equation (6), the inputs v_1, v_2 are expressed with X and Y axis linear velocity components, linear acceleration components, and angular velocity. The first channel of the linear system (5) is related with the X axis velocity since the states ξ_1, ξ_2 are $\xi_1 = \gamma \cos\alpha, \xi_2 = -V\cos\phi$. The second channel is related with the Y axis velocity since the states ξ_3, ξ_4 are $\xi_3 = \gamma \sin\alpha, \xi_4 = -V\sin\phi$. Therefore, the inputs v_1, v_2 can be regarded as computed accelerations with respect to X axis and Y axis, respectively.

2.2 Fixed heading angle at the equilibrium state

In The Proposition 1, the heading angle at an equilibrium state is free since the heading angle is not considered explicitly in the feedback linearization and the kinematic equation of the robot. To consider the heading angle of the robot, the kinematic equations of the robot are modified to equation (7). with the form $\dot{x} = f(x) + g(x)u$, where $a = \dot{V}$.

$$\dot{x} = \begin{bmatrix} -x_4 \cos x_2 \\ x_4 \dfrac{\sin x_2}{x_1} \\ 0 \\ 0 \end{bmatrix} + \begin{bmatrix} 0 & 0 \\ 0 & -1 \\ 0 & 1 \\ 1 & 0 \end{bmatrix} \begin{bmatrix} a \\ \omega \end{bmatrix} \quad (7)$$

In (7), the state variables are defined as $x_1 = \gamma$, $x_2 = \theta$, $x_3 = \phi, x_4 = V$.

Proposition 2

Let the output functions be $\lambda_1 = x_1, \lambda_2 = x_3 + x_2$, then the conditions of the state space exact feedback linearization are satisfied. Therefore, the state space exact feedback linearization problem of the mobile robot can be solved. Furthermore, the heading angle at the equilibrium state is fixed to zero.

Proof

It can be easily shown that each distribution G_i associated with f and g has a constant dimension and is involutive. As the output functions are $\lambda_1 = x_1, \lambda_2 = x_3 + x_2$, the input transformation matrices are defined as follows.

$$A(x) = \begin{bmatrix} -\cos x_2 & -x_4 \sin x_2 \\ \dfrac{\sin x_2}{x_1} & -x_4 \dfrac{\cos x_2}{x_1} \end{bmatrix}, \quad b(x) = \begin{bmatrix} \dfrac{x_4^2 \sin^2 x_2}{x_1} \\ 2\dfrac{x_4^2 \sin x_2 \cos x_2}{x_1^2} \end{bmatrix}.$$

The local diffeomorphism and the transformed states are defined as follows.

$$\xi_1 = \phi_1^1 = \lambda_1 = x_1, \xi_2 = \phi_2^1 = L_f \lambda_1 = -x_4 \cos x_2$$

$$\xi_3 = \phi_1^2 = \lambda_2 = x_2 + x_3, \xi_4 = \phi_2^2 = L_f \lambda_2 = x_4 \dfrac{\sin x_2}{x_1}.$$

Then, the nonlinear system (7) is transformed into a linear system (8). The linear system has a linear controllable and observable form.

$$\dot{\xi} = \begin{bmatrix} 0 & 1 & & \\ 0 & 0 & \multicolumn{2}{c}{\mathbf{0}} \\ \hline & & 0 & 1 \\ \multicolumn{2}{c}{\mathbf{0}} & 0 & 0 \end{bmatrix} \begin{bmatrix} \xi_1 \\ \xi_2 \\ \xi_3 \\ \xi_4 \end{bmatrix} + \begin{bmatrix} 0 & 0 \\ 1 & 0 \\ 0 & 0 \\ 0 & 1 \end{bmatrix} \begin{bmatrix} v_1 \\ v_2 \end{bmatrix}$$

$$y = \begin{bmatrix} 1 & 0 & 0 & 0 \\ 0 & 0 & 1 & 0 \end{bmatrix} \begin{bmatrix} \xi_1 \\ \xi_2 \\ \xi_3 \\ \xi_4 \end{bmatrix} \quad (8)$$

Q. E. D.

As $\det(A(x))$ is $\dfrac{x_4}{x_1}$, the input transformation matrix $A(x)$ is nonsingular where $x_1 \neq 0, x_4 \neq 0$ which means that the speed of the mobile robot is not zero when the distance to go is nonzero, i. e., $\gamma \neq 0$. This condition is also acceptable due to the same reason mentioned in section 2.1. The linear system (8) has four states and the equilibrium is achieved when $\xi = 0$. Thus, $\xi_1 = \gamma = 0$,

$$\xi_2 = -V \cos \theta = 0 \quad \xi_3 = \alpha = 0, \quad \xi_4 = \dfrac{V \sin \theta}{\gamma} = 0. \text{ That is}$$

$\gamma = 0$, $V = 0$, $\alpha = 0$, and $\theta = 0$. As a consequence, the equilibrium point in the XY coordinate is (0,0), and the heading angle of the mobile robot is fixed to zero at the equilibrium point since $\alpha = 0$ and $\theta = 0$. In this case, it is difficult to investigate physical meanings of the control inputs v_1 and v_2.

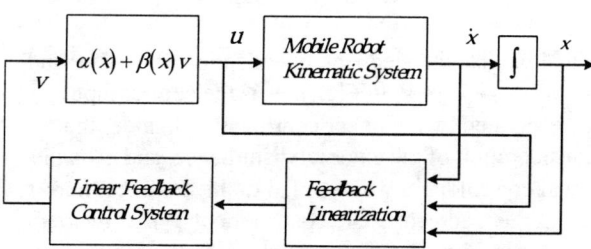

Fig. 2 The schematic diagram of the mobile robot control

However, the control energy of the robot can be expressed by inputs and state variables as shown in (9).

$$a^2 + (V\omega)^2 = (v_1 - \xi_1 \xi_4^2)^2 + (\xi_1 v_2 - \xi_4 \xi_2 - \xi_1 \xi_4^2)^2$$

$$= \left(v_1 - \dfrac{V^2 \sin^2 \theta}{\gamma}\right)^2 + \left(\gamma v_2 + \dfrac{V^2 \sin \theta \cos \theta}{\gamma} - \dfrac{V^2 \sin^2 \theta}{\gamma}\right)^2 \quad (9)$$

Since γ is always greater than zero, equation (10) is satisfied.

$$a^2 + (V\omega)^2 \leq v_1^2 + \left(\gamma v_2 + \dfrac{V^2 \sin \theta \cos \theta}{\gamma}\right)^2 = v_1^2 + (\gamma v_2 + \dot{\gamma}\dot{\alpha})^2 \quad (10)$$

Thus, the control energy of the robot is upper bounded by the control inputs of the linear system (8).

In Fig. 2, v is the control input generated by the system, and u is the actual control input which is fed to the nonlinear kinematic system of the mobile robot. The control input v is converted to the actual control input u using the Proposition 1 and 2. According to the design of the linear feedback control system, the mobile robot can be controlled satisfying several performance indices. Furthermore, it is easily shown that the stability of the mobile robot system can also be guaranteed since the developed system is linear time invariant.

3. Feedback linearization for pursuer and target problem

In the Proposition 1 and 2, the target or goal point is fixed. In this section, we have considered the situation when the target is not fixed. This condition usually arises in missile guidance or chasing problems. The pursuer and target engagement situation is depicted in Fig. 3. In Fig. 3, γ indicates the distance from the pursuer to the target, α is the line of sight angle, and ϕ is the heading angle of the pursuer. The variables V_m and ω are the linear velocity and the angular velocity of the pursuer. Let the states be $x_1 = \gamma, x_2 = \theta_m, x_3 = \phi, x_4 = V_m$,

$x_5 = V_t \cos \theta_t$, $x_6 = -\dfrac{V_t \sin \theta_t}{\gamma}$, then the system is expressed in $\dot{x} = f(x) + g_1(x)u_1 + g_2(x)u_2$ with control inputs ω and a where $\dot{\phi} = \omega$ and $\dot{V}_m = a$.

$$\dot{x} = \begin{bmatrix} x_5 - x_4 \cos x_2 \\ x_4 \dfrac{\sin x_2}{x_1} + x_6 \\ 0 \\ 0 \\ 0 \\ 0 \end{bmatrix} + \begin{bmatrix} 0 \\ 0 \\ 0 \\ 1 \\ 0 \\ 0 \end{bmatrix} u_1 + \begin{bmatrix} 0 \\ -1 \\ 1 \\ 0 \\ 0 \\ 0 \end{bmatrix} u_2 + \begin{bmatrix} 0 \\ 0 \\ 0 \\ 0 \\ 1 \\ 0 \end{bmatrix} w_1 + \begin{bmatrix} 0 \\ 0 \\ 0 \\ 0 \\ 0 \\ 1 \end{bmatrix} w_2 \quad (11)$$

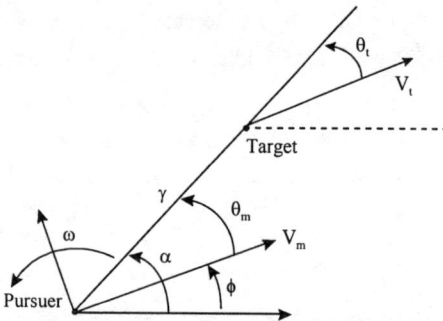

Fig. 3 The coordinate in the polar frame with moving target.

In equation (11), it is assumed that $\dot{x}_5 = w_1$ and $\dot{x}_6 = w_2$. The states x_5 and x_6 are associated with the target states. If the pursuer is a missile, then the target has a tendency to escape from the pursuer. As a consequence, the target will move to increase γ and θ_t. The target motion may or may not be known to the pursuer. Thus, w_1 and w_2 can be considered as disturbances to the given system. With this configuration, the feedback linearization method is applied using the same concept in the Proposition 1 and 2.

Proposition 3

Let the output functions be $\lambda_1 = x_1, \lambda_2 = x_3 + x_2$, then the conditions of state space exact feedback linearization are satisfied. Thus the state space exact feedback linearization problem of the pursuer and target problem can be solved. Also, the heading angle of the pursuer at the equilibrium state is fixed to zero.

Proof

It can be easily shown that each distribution G_i associated with f and g_i has a constant dimension and is involutive. As the output functions are $\lambda_1 = x_1, \lambda_2 = x_3 + x_2$, the input transformation matrices are defined as follows.

$$A(x) = \begin{bmatrix} -\cos x_2 & -x_4 \sin x_2 \\ \dfrac{\sin x_2}{x_1} & -x_4 \dfrac{\cos x_2}{x_1} \end{bmatrix},$$

$$b(x) = \begin{bmatrix} \dfrac{x_4^2 \sin^2 x_2}{x_1} + x_4 x_6 \sin x_2 \\ 2\dfrac{x_4^2 \sin x_2 \cos x_2}{x_1^2} - \dfrac{x_4 x_5 \sin x_2}{x_1^2} + \dfrac{x_4 x_6 \cos x_2}{x_1} \end{bmatrix}$$

The local diffeomorphism and the transformed states are defined as follows.

$$\xi_1 = \phi_1^1 = \lambda_1 = x_1$$
$$\xi_2 = \phi_2^1 = L_f \lambda_1 = x_5 - x_4 \cos x_2$$
$$\xi_3 = \phi_1^2 = \lambda_2 = x_3 + x_2$$
$$\xi_4 = \phi_2^2 = L_f \lambda_2 = \dfrac{x_4 \sin x_2}{x_1} + x_6$$
$$\dot{x}_5 = w_1, \dot{x}_6 = w_2$$

Then, the nonlinear system is transformed into a linear system (12). The linear system has a controllable and observable form with the disturbances w_1 and w_2.

$$\dot{\xi} = \begin{bmatrix} 0 & 1 & & \\ 0 & 0 & & \mathbf{0} \\ \hline & & 0 & 1 \\ \mathbf{0} & & 0 & 0 \end{bmatrix} \begin{bmatrix} \xi_1 \\ \xi_2 \\ \xi_3 \\ \xi_4 \end{bmatrix} + \begin{bmatrix} 0 & 0 \\ 1 & 0 \\ 0 & 0 \\ 0 & 1 \end{bmatrix} \begin{bmatrix} v_1 \\ v_2 \end{bmatrix} + \begin{bmatrix} 0 & 0 \\ 1 & 0 \\ 0 & 0 \\ 0 & 1 \end{bmatrix} \begin{bmatrix} w_1 \\ w_2 \end{bmatrix}$$

$$y = \begin{bmatrix} 1 & 0 & 0 & 0 \\ 0 & 0 & 1 & 0 \end{bmatrix} \begin{bmatrix} \xi_1 \\ \xi_2 \\ \xi_3 \\ \xi_4 \end{bmatrix}$$

(14)

Q. E. D.

The disturbances are related with the target motion. The disturbance w_1 is related with the range between the pursuer and the target. The disturbance w_2 is related with the line of sight angle between the pursuer and the target. Hence, the pursuer controller design problem can be formulated with the linear control theory under the effect of the disturbances. The actual control inputs to the pursuer is given by (13). The inputs of the transformed linear system v_1 and v_2 are given by (14). If the states x_5 and x_6 are equal to the zero, i.e., the target is fixed, then (13) and (14) satisfy the condition expressed in (9).

$$a = u_1 = -\cos x_2 v_1 + x_1 \sin x_2 v_2 - \dfrac{x_4^2 \sin^2 x_2 \cos x_2}{x_1} + \dfrac{x_4 x_5 \sin^2 x_2}{x_1}$$

$$\omega = u_2 = -\dfrac{\sin x_2}{x_4} v_1 - \dfrac{x_1 \cos x_2}{x_4} v_2 + \dfrac{x_4 \sin x_2 (1 + \cos^2 x_2)}{x_1} - \dfrac{x_5 \sin x_2 \cos x_2}{x_1} + x_6$$

(13)

$$v_1 = -u_1 \cos x_2 - u_2 x_4 \sin x_2 + \dfrac{x_4^2 \sin^2 x_2}{x_1} + x_4 x_6 \sin x_2$$

$$v_2 = u_1 \dfrac{\sin x_2}{x_1} - u_2 \dfrac{x_4 \cos x_2}{x_1} - \dfrac{x_4 x_5 \sin x_2}{x_1^2} + \dfrac{x_4 x_6 \cos x_2}{x_1} + \dfrac{2 x_4^2 \sin x_2 \cos x_2}{x_1^2}$$

(14)

Fig. 4 depicts the feedback linearization of pursuer and target geometry. In Fig. 4, v is the control input generated by the linear control system under the consideration of the external disturbance, and u is the actual control input which is fed to the nonlinear purser and target geometry. The control input v is converted to the actual control input u using the Proposition 3. This diagram is similar to the mobile robot control with fixed target point except that the disturbance w_1 and

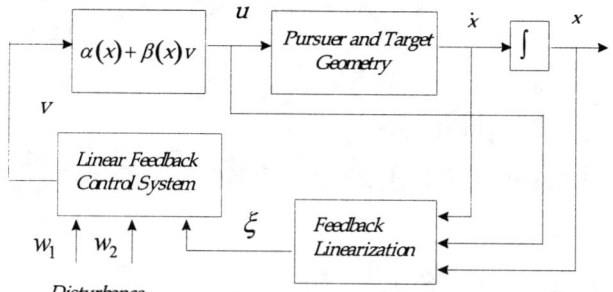

Fig. 4 A schematic of the pursuer and the target geometry with feedback linearization

w_2 are newly introduced. As a consequence, the disturbance can be rejected according to the design of the linear feedback control system. Furthermore, as the equilibrium state is the same to (8), the pursuer can hit the target with a fixed heading angle.

4. Conclusions

In this paper, the point stabilization of mobile robots is formulated by utilizing the state space exact feedback linearization. A kinematic system of a mobile robot under polar coordinates is perfectly transformed into a linear time invariant system. Moreover, nonholonomic constraints do not appear in the polar coordinates explicitly. With the introduction of an auxiliary variable, the kinematic system of the mobile robot is no longer a driftless system. Hence, standard nonlinear control theory, the state space exact feedback linearization method, is applied to the control problem of the robot, which has not been achieved in the past. It is shown that the developed kinematic system of the mobile robot satisfies the necessary and sufficient conditions for the state space exact feedback linearization.

According to the conditions at a target point, three possible cases are considered. First of all, the case in which the mobile robot moves to a target point without heading angle constraints is investigated. Secondly, the control of the robot to a target point with heading angle constraint is formulated. Under the suggested framework in this paper, both cases have unified characteristics, *i.e.*, the derived linear systems have the same form as the aggregate of a double integrator system. Furthermore, the derived linear system is a decoupled one. As a consequence, the nonlinear control problem of the mobile robot is converted into the control problem of a simple linear time invariant decoupled MIMO system. Additionally, one can design a controller which minimizes certain performance indices with the help of the relationships between the control energy of the nonlinear kinematics of the robot and the inputs of the transformed linear system. Finally, we have considered the situation when the target is not fixed. In this case, the pursuer and target problem is defined as a guidance problem. As shown in section 3, the guidance problem is converted into a linear system under the effect of disturbances, which are caused by the maneuvering of the target. Therefore, in this case, the control objective is to reject the disturbances and to stabilize the given linear system.

One can design a point stabilization controller in the context of linear control theory by using the developed linear system model.

References

[1] I. Kolmanovsky and N. H. McClamroch, "Developments in Nonholonomic Control Problems," *IEEE Control Systems*, December, 1995.

[2] R. W. Brockett, "Asymptotic stability and feedback Stabilization," *Differential Geometric Control Theory*, Birkhauser, Boston, pp. 181-191, 1983.

[3] B. d'Andrea-Novel, G. Campion and G. Bastin, "Control of Nonholonomic Wheeled Mobile Robots by State Feedback Linearization," *The International Journal of Robotics Research*, vol. 14, no. 6, pp. 543-559, 1995.

[4] A. Isidori, *Nonlinear Control Systems*, 2nd edition, Springer, 1989.

[5] C. Canudas de Wit and O. J. Sordalen. "Exponential Stabilization of mobile robot with nonholonomic constraints," *IEEE transaction on Automatic Control*, vol. 13, no. 11, pp. 1791-1797, 1992.

[6] C. Samson, "Control of chained system: Application to path following and time-varying point-stabilization of mobile robots," *IEEE transaction on Automatic Control*, vol. 40, no. 1, pp. 64-77, 1995.

[7] M. Aicardi, "Closed Loop Steering of Unicycle-like Vehicles via Lyapunov Techniques," *IEEE robotics & Automation Magazine*, March, 1995.

Trajectory Tracking Control of a Four-Wheel Differentially Driven Mobile Robot

Luca Caracciolo   Alessandro De Luca   Stefano Iannitti

Dipartimento di Informatica e Sistemistica
Università degli Studi di Roma "La Sapienza"
Via Eudossiana 18, 00184 Roma, Italy
{deluca}@labrob.ing.uniroma1.it

Abstract

We consider the trajectory tracking control problem for a 4-wheel differentially driven mobile robot moving on an outdoor terrain. A dynamic model is presented accounting for the effects of wheel skidding. A model-based nonlinear controller is designed, following the dynamic feedback linearization paradigm. An operational nonholonomic constraint is added at this stage, so as to obtain a predictable behavior for the instantaneous center of rotation thus preventing excessive skidding. The controller is then robustified, using conventional linear techniques, against uncertainty in the soil parameters at the ground-wheel contact. Simulation results show the good performance in tracking spline-type trajectories on a virtual terrain with varying characteristics.

1 Introduction

Robotic autonomous navigation tasks in outdoor environments can be effectively performed by skid-steering vehicles [1]. Typically, these are 4-wheel differentially driven (4wdd) vehicles in which rotational motion is achieved by a differential thrust on wheel pairs at opposite sides. One commercial example is the ATRV-2 mobile robot by RWI (see Fig. 1). The absence of a steering system makes 4wdd vehicles mechanically robust and able to move on rough terrains with ease and good maneuverability. In applications where the robot needs to move also on paved grounds, skid-steering vehicles are preferred to tracked vehicles because they do not corrupt the contact surface. Although the dynamic behavior of 4wdd skid-steering vehicles is similar to that of vehicles with tracks, the literature on autonomous navigation has focused mostly on the latter class [2, 3].

Figure 1: ATRV-2 mobile robot

When considering the problem of accurate trajectory tracking, 4wdd vehicles are quite difficult to control. In fact, in order to follow a curved path, the wheels need to skid laterally and cannot be tangent to the desired path. Moreover, the instantaneous center of rotation (ICR) of 4wdd vehicles may move out of the robot wheelbase, causing loss of motion stability. This is different from car-like vehicles, whose ICR is always theoretically fixed along the rear wheels axis [4]. From the modeling point of view, the equilibrium equation of the forces orthogonal to the wheels should be taken into account and this prescribes the use of a dynamic model for control design purposes, instead of a simpler kinematic one.

In this paper, we present a robust trajectory tracking control system for 4wdd vehicles. After deriving a dynamic model of a 4wdd vehicle, we design a model-based tracking controller, by borrowing an approach used for motion planning and control of nonholonomic wheeled mobile robots [4, 5]. The basic idea is to specify the longitudinal coordinate of the ICR, in order to force it to remain inside the robot wheelbase. Adding this operational kinematic constraint, we proceed with full linearization of the system via dynamic state feedback. The obtained closed-loop system is linear and input-output decoupled, thus stabilization to a desired trajectory is easily achieved by means of linear control techniques. However, the overall nonlinear control law depends on soil parameters [3]. We study the effects of uncertain soil parameters on the dynamics of the closed-loop system with respect to a specific class of trajectories and propose then a robust control scheme that reject these disturbances. Simulation results on tracking of robot trajectories over a virtual terrain characterized by varying parameters are finally presented.

2 Dynamic Modeling

We develop a vehicle dynamic model useful for control design, neglecting some side effects introduced, e.g., by suspensions and tire deformation (see, e.g., [6]). In particular, we make the following assumptions:

1. Rigid vehicle moving on a horizontal plane.
2. Vehicle speed below 10 km/h (about 6 mph).
3. Longitudinal wheel slippage neglected.
4. Tire lateral force function of its vertical load.

2.1 Skid-steering motion analysis

Define a fixed reference frame $F(X, Y)$ and a moving frame $f(x, y)$ attached to the vehicle body, with origin at the vehicle center of mass G (see Fig. 2). The center of mass is located at distances a and b (usually, $a < b$) from the front and rear wheels axes, respectively, and is symmetric with respect to the vehicle sides (at distance t).

Let $\dot{x}, \dot{y}, \dot{\theta}$ be, respectively, the longitudinal, lateral, and angular velocity of the vehicle in frame f. In the fixed frame F, the absolute velocities are

$$\begin{bmatrix} \dot{X} \\ \dot{Y} \end{bmatrix} = \begin{bmatrix} \dot{x}\cos\theta - \dot{y}\sin\theta \\ \dot{x}\sin\theta + \dot{y}\cos\theta \end{bmatrix} = R(\theta) \begin{bmatrix} \dot{x} \\ \dot{y} \end{bmatrix}.$$

Differentiation with respect to time gives

$$\begin{bmatrix} \ddot{X} \\ \ddot{Y} \end{bmatrix} = R(\theta) \begin{bmatrix} \ddot{x} - \dot{y}\dot{\theta} \\ \ddot{y} + \dot{x}\dot{\theta} \end{bmatrix} = R(\theta) \begin{bmatrix} a_x \\ a_y \end{bmatrix},$$

where a_x and a_y are the absolute accelerations expressed in the moving frame f. At each instant the vehicle motion is a pure rotation around a point C, the instantaneous center of rotation, in which the linear velocity components in f vanish. Its coordinates are

$$\begin{bmatrix} x_c \\ y_c \end{bmatrix} = \begin{bmatrix} -\dot{y}/\dot{\theta} \\ \dot{x}/\dot{\theta} \end{bmatrix}.$$

The angular velocity $\dot{\theta}$ and the lateral velocity $\dot{y}$ both vanish during straight line motion, and the ICR goes to infinity along the y-axis. On a curved path, the ICR shifts (forwards) by an amount $|x_c|$. When $\dot{y} = 0$, there is no lateral skidding. If x_c goes out of the robot wheelbase, the vehicle skids dramatically with loss of motion stability.

Finally, note that the longitudinal velocity $\dot{x}_i$ and the lateral (skidding) velocity $\dot{y}_i$ of each wheel ($i = 1, \ldots, 4$) are given by

$$\begin{aligned} \dot{x}_1 = \dot{x}_4 &= \dot{x} - t\dot{\theta} \quad \text{(left)} \\ \dot{x}_2 = \dot{x}_3 &= \dot{x} + t\dot{\theta} \quad \text{(right)} \\ \dot{y}_1 = \dot{y}_2 &= \dot{y} + a\dot{\theta} \quad \text{(front)} \\ \dot{y}_3 = \dot{y}_4 &= \dot{y} - b\dot{\theta} \quad \text{(rear)}. \end{aligned} \quad (1)$$

2.2 Equations of motion

The free-body diagram of forces and velocities is shown in Fig. 2, with the vehicle having instantaneous positive velocity components $\dot{x}$ and $\dot{\theta}$ and negative velocity $\dot{y}$. Wheels develop tractive forces F_{xi} and are subject to longitudinal resistance forces R_{xi}, for $i = 1, \ldots, 4$. We assume that wheel actuation is equal on each side so as to reduce longitudinal slip. Thus, it will always be $F_{x4} = F_{x1}$ and $F_{x3} = F_{x2}$. Lateral forces F_{yi} act on the wheels as a consequence of lateral skidding. Also, a resistive moment M_r around the center of mass is induced in general by the F_{yi} and R_{xi} forces.

For a vehicle of mass m and inertia I about its center of mass, the equations of motion can be written in frame f as:

$$\begin{aligned} ma_x &= 2F_{x1} + 2F_{x2} - R_x \\ ma_y &= -F_y \\ I\ddot{\theta} &= 2t(F_{x1} - F_{x2}) - M_r. \end{aligned} \quad (2)$$

To express the longitudinal resistive force R_x, the lateral resistive force F_y, and the resistive moment M_r, we should consider how the vehicle gravitational load mg is shared among the wheels and introduce a Coulomb friction model for the wheel-ground contact. We have

$$\begin{aligned} F_{z1} = F_{z2} &= \frac{b}{a+b} \cdot \frac{mg}{2} \\ F_{z3} = F_{z4} &= \frac{a}{a+b} \cdot \frac{mg}{2}. \end{aligned}$$

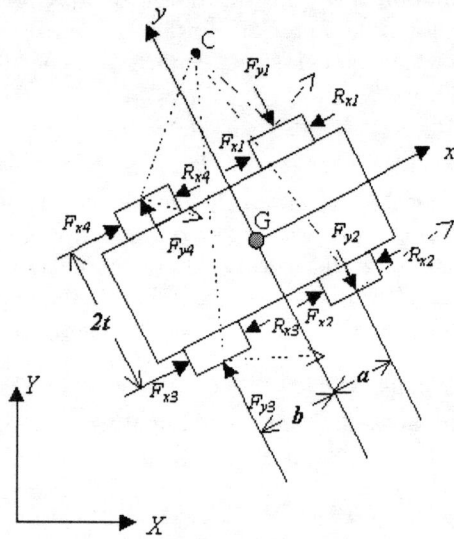

Figure 2: Free-body diagram

At low speed, the lateral load transfer due to centrifugal forces on curved paths can be neglected. In case of hard ground, we can assume that the contact patch between wheel and ground is rectangular and that the tire vertical load produces an uniform pressure distribution. In this condition, $R_{xi} = f_r F_{zi} \text{sgn}(\dot{x}_i)$, where f_r is the coefficient of rolling resistance, assumed independent from velocity [7]. The total longitudinal resistive force is then

$$R_x = \sum_{i=1}^{4} R_{xi} = f_r \frac{mg}{2} \left(\text{sgn}(\dot{x}_1) + \text{sgn}(\dot{x}_2) \right). \quad (3)$$

Introducing a lateral friction coefficient μ, the lateral force acting on each wheel will be $F_{yi} = \mu F_{zi} \text{sgn}(\dot{y}_i)$. The total lateral force is thus

$$F_y = \sum_{i=1}^{4} F_{yi} = \mu \frac{mg}{a+b} \left(b\,\text{sgn}(\dot{y}_1) + a\,\text{sgn}(\dot{y}_3) \right), \quad (4)$$

while the resistive moment is

$$\begin{aligned}
M_r &= a(F_{y1}+F_{y2}) - b(F_{y3}+F_{y4}) \\
&\quad + t\left[(R_{x2}+R_{x3}) - (R_{x1}+R_{x4})\right] \\
&= \mu \frac{abmg}{a+b} \left(\text{sgn}(\dot{y}_1) - \text{sgn}(\dot{y}_3)\right) \\
&\quad + f_r \frac{tmg}{2} \left(\text{sgn}(\dot{x}_2) - \text{sgn}(\dot{x}_1)\right).
\end{aligned} \quad (5)$$

The dynamic model can be rewritten in frame F, introducing the generalized coordinates $q = (X, Y, \theta)$ and matrix notation

$$M\ddot{q} + c(q,\dot{q}) = E(q)\tau, \quad (6)$$

with

$$M = \begin{bmatrix} m & 0 & 0 \\ 0 & m & 0 \\ 0 & 0 & I \end{bmatrix}, \quad c(q,\dot{q}) = \begin{bmatrix} R_x \cos\theta - F_y \sin\theta \\ R_x \sin\theta + F_y \cos\theta \\ M_r \end{bmatrix}$$

and

$$E(q) = \begin{bmatrix} \cos\theta/r & \cos\theta/r \\ \sin\theta/r & \sin\theta/r \\ t/r & -t/r \end{bmatrix}, \quad \tau_i = 2r F_{xi}\ (i=1,2),$$

being r the wheel radius, τ_1 and τ_2 the torques produced by the left and right side motors at the load side, respectively. An ideal transmission factor is also assumed.

3 Trajectory Control

3.1 Operative nonholonomic constraint

We start by observing that x_c (the x-axis projection of the instantaneous center of rotation) cannot be larger than a. If this happens, the vehicle would skid along the y-axis thus losing control. In order to have the vehicle move properly, one should have then

$$\left| -\frac{\dot{y}}{\dot{\theta}} \right| < a.$$

Therefore, we can introduce the following operative constraint

$$\dot{y} + d_0 \dot{\theta} = 0, \quad 0 < d_0 < a,$$

or, in terms of generalized coordinates,

$$\begin{bmatrix} -\sin\theta & \cos\theta & d_0 \end{bmatrix} \begin{bmatrix} \dot{X} \\ \dot{Y} \\ \dot{\theta} \end{bmatrix} = A(q)\dot{q} = 0. \quad (7)$$

This relation represents a nonholonomic constraint that can be attached to the dynamic model (6) for control design purposes. When this constraint is enforced, the robot dynamics becomes

$$M\ddot{q} + c(q,\dot{q}) = E(q)\tau + A^T(q)\lambda, \quad (8)$$

where λ is the vector of Lagrange multipliers corresponding to eq. (7).

Admissible generalized velocities $\dot{q}$ can be expressed as

$$\dot{q} = S(q)\eta, \quad \eta \in \mathbb{R}^2, \quad (9)$$

where η is a pseudo-velocity and $S(q)$ is a 3×2 full rank matrix, whose columns are in the null space of $A(q)$, e.g.,

$$S(q) = \begin{bmatrix} \cos\theta & -\sin\theta \\ \sin\theta & \cos\theta \\ 0 & -\frac{1}{d_0} \end{bmatrix}.$$

We can differentiate (9) and eliminate λ from eq. (8) so as to obtain the reduced dynamic model (dropping dependencies)

$$\begin{aligned} \dot{q} &= S\eta \\ \dot{\eta} &= (S^T M S)^{-1} S^T \left(E\tau - M\dot{S}\eta - c \right). \end{aligned} \quad (10)$$

3.2 Partially linearizing static feedback

Following [8], if we apply the nonlinear static state-feedback law

$$\tau = (S^T E)^{-1} \left(S^T M S u + S^T M \dot{S} \eta + S^T c \right), \quad (11)$$

where $u = (u_1, u_2)$ is the vector of new control variables, system (10) becomes a purely (second-order) kinematic model

$$\begin{aligned} \dot{q} &= S\eta \\ \dot{\eta} &= u. \end{aligned}$$

In our case, the control law (11) has the explicit form

$$\begin{bmatrix} \tau_1 \\ \tau_2 \end{bmatrix} = \begin{bmatrix} \frac{r}{2}\left(mu_1 + \frac{m}{d_0}\eta_2^2 + R_x\right) \\ -\frac{rd_0}{2t}\left(\left(m + \frac{I}{d_0^2}\right)u_2 - \frac{m}{d_0}\eta_1\eta_2 + F_y - \frac{M_r}{d_0}\right) \\ \frac{r}{2}\left(mu_1 + \frac{m}{d_0}\eta_2^2 + R_x\right) \\ +\frac{rd_0}{2t}\left(\left(m + \frac{I}{d_0^2}\right)u_2 - \frac{m}{d_0}\eta_1\eta_2 + F_y - \frac{M_r}{d_0}\right) \end{bmatrix} \quad (12)$$

and gives

$$\begin{aligned} \dot{X} &= \cos\theta\, \eta_1 - \sin\theta\, \eta_2 \\ \dot{Y} &= \sin\theta\, \eta_1 + \cos\theta\, \eta_2 \\ \dot{\theta} &= -\frac{1}{d_0}\eta_2 \\ \dot{\eta}_1 &= u_1 \\ \dot{\eta}_2 &= u_2. \end{aligned} \quad (13)$$

3.3 Fully linearizing dynamic feedback

We show next that, by choosing a particular output, eqs. (13) can be fully linearized and input-output decoupled by means of a dynamic state feedback.

For, we choose as linearizing outputs the position of a point D placed on the x-axis at a distance d_0 from the vehicle frame origin

$$z = \begin{bmatrix} X + d_0 \cos\theta \\ Y + d_0 \sin\theta \end{bmatrix}, \quad (14)$$

and add one integrator on the input u_1 (dynamic extension)

$$\begin{aligned} u_1 &= \xi \\ \dot{\xi} &= v_1 \\ u_2 &= v_2, \end{aligned} \quad (15)$$

where ξ is the controller state and v_1 and v_2 are the new control inputs.

By applying the standard input-output decoupling algorithm (see, e.g., [9]), we differentiate eq. (14) until the input v explicitly appears. We obtain

$$\begin{aligned} \ddot{z} &= \begin{bmatrix} \cos\theta & \frac{1}{d_0}\eta_1 \sin\theta \\ \sin\theta & -\frac{1}{d_0}\eta_1 \cos\theta \end{bmatrix} v \\ &+ \begin{bmatrix} \frac{2}{d_0}\xi\eta_2 \sin\theta - \frac{1}{d_0^2}\eta_1\eta_2^2 \cos\theta \\ -\frac{2}{d_0}\xi\eta_2 \cos\theta - \frac{1}{d_0^2}\eta_1\eta_2^2 \sin\theta \end{bmatrix} \\ &= \alpha(q,\eta)v + \beta(q,\eta). \end{aligned}$$

Since

$$\det[\alpha(q,\eta)] = -\frac{1}{d_0}\eta_1,$$

we have that the decoupling matrix α is nonsingular iff the vehicle longitudinal velocity η_1 is different from zero. Whenever defined, the control law

$$v = \alpha^{-1}(q,\eta)\left[r - \beta(q,\eta)\right], \quad (16)$$

where r is the trajectory jerk reference, yields

$$\dddot{z} = r, \quad (17)$$

i.e., two independent input-output chains of three integrators.

Combining eqs. (15) and (16) gives the following input-output decoupling and fully linearizing dynamic controller

$$\begin{aligned} \dot{\xi} &= \cos\theta\, r_1 + \sin\theta\, r_2 + \frac{1}{d_0^2}\eta_1\eta_2^2 \\ u_1 &= \xi \\ u_2 &= \frac{d_0}{\eta_1}(\sin\theta\, r_1 - \cos\theta\, r_2) - \frac{2}{\eta_1}\xi\eta_2. \end{aligned} \quad (18)$$

We note that the limitation $\eta_1 \neq 0$ does not avoid to achieve good tracking performance by means of controller (18), as long as the trajectory is persistent.

3.4 Linear stabilization for tracking

It is easy to complete the control design for eq. (17) using a exponentially stabilizing state feedback for each integrator chain with input r_i. For $i = 1, 2$, we choose

$$r_i = \dddot{z}_{di} + k_{ai}(\ddot{z}_{di} - \ddot{z}_i) + k_{vi}(\dot{z}_{di} - \dot{z}_i) + k_{pi}(z_{di} - z_i), \quad (19)$$

where the gains are such that $\lambda^3 + k_{ai}\lambda^2 + k_{vi}\lambda + k_{pi}$ ($i = 1, 2$) are Hurwitz polynomials, $z_d(t)$ is the desired smooth reference trajectory, and z, $\dot{z}$ and $\ddot{z}$ can be evaluated in terms of q, η and ξ.

The state-feedback control law (19) can be seen as an output-feedback linear controller having two (realizable) minimum-phase zeros, characterized by the

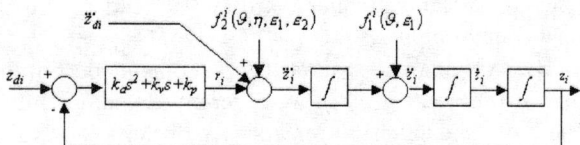

Figure 3: Linear control scheme for tracking

gain ratios k_v/k_a and k_p/k_a, and a feedforward action depending on $\ddot{z}_d$ (see Fig. 3). The resulting control scheme has the open-loop transfer function

$$F(s) = C(s) \cdot P(s) = (k_a s^2 + k_v s + k_p) \cdot \frac{1}{s^3}. \quad (20)$$

4 Robust Control Design

In case of perfect knowledge of the ground-wheel contact parameters, the control law (12) perfectly compensates for the terrain factors and leads to the kinematic system (13) for which trajectory tracking is easily achieved by means of the dynamic controller (18) and the stabilizer (19).

When R_x, F_y, and M_r are unknown (because of μ and f_r) or just incorrectly estimated, the vehicle closed-loop dynamics will still be nonlinear and coupled. The influence of an error on the terrain factors estimate is a torque disturbance that should be rejected by a proper robust control design.

4.1 Disturbance analysis

Let $\hat{R}_x$, $\hat{F}_y$, $\hat{M}_r$ be the estimated values of the terrain factors. Then, the implementable control law is still given by (12) with these estimates in place of the real values. The resulting system, in place of (13), is

$$\begin{aligned}
\dot{X} &= \cos\theta\,\eta_1 - \sin\theta\,\eta_2 \\
\dot{Y} &= \sin\theta\,\eta_1 + \cos\theta\,\eta_2 \\
\dot{\theta} &= -\frac{1}{d_0}\eta_2 \quad (21)\\
\dot{\eta}_1 &= u_1 + \epsilon_1 \\
\dot{\eta}_2 &= u_2 + \epsilon_2,
\end{aligned}$$

where

$$\begin{aligned}
\epsilon_1 &= \frac{1}{m}\left(\hat{R}_x - R_x\right) \\
\epsilon_2 &= \frac{d_0^2}{md_0^2 + I}\left(\left(\hat{F}_y - \frac{\hat{M}_r}{d_0}\right) - \left(F_y - \frac{M_r}{d_0}\right)\right).
\end{aligned}$$

If we suppose that the vehicle moves on a (unknown) uniform terrain, the above disturbances will be constant.

When we apply the dynamic controller (18), ϵ_1 and ϵ_2 will affect the resulting nominal integrator chains (17) in a special way. Beside the output (14), we have

$$\dot{z} = \begin{bmatrix} \cos\theta\,\eta_1 \\ \sin\theta\,\eta_1 \end{bmatrix}$$

$$\ddot{z} = \begin{bmatrix} \xi\cos\theta + \frac{1}{d_0}\eta_1\eta_2\sin\theta \\ \xi\sin\theta - \frac{1}{d_0}\eta_1\eta_2\cos\theta \end{bmatrix} + \begin{bmatrix} \cos\theta \\ \sin\theta \end{bmatrix}\epsilon_1,$$

and

$$\dddot{z} = r + \begin{bmatrix} \frac{2}{d_0}\eta_2\sin\theta & \frac{1}{d_0}\eta_1\sin\theta \\ -\frac{2}{d_0}\eta_2\cos\theta & -\frac{1}{d_0}\eta_1\cos\theta \end{bmatrix} \begin{bmatrix} \epsilon_1 \\ \epsilon_2 \end{bmatrix}.$$

The way each integrator chain is affected by the disturbances is depicted in Fig. 3, with $f_1^i(\theta,\eta_1)$ and $f_2^i(\theta,\eta,\epsilon_1,\epsilon_2)$ following from the above expressions for $\ddot{z}_i$ and $\dddot{z}_i$ ($i = 1, 2$).

From standard linear system analysis and Fig. 3, it follows that, even for constant disturbances f_1^i and f_2^i, the steady-state error of the controlled output will be different from zero.

4.2 A modified linear controller for disturbance rejection

Consider a desired trajectory made of a straight line path with a kth order (canonical) polynomial timing law, i.e.,

$$z_{d1}(t) = \sum_{h=0}^{k} c_h \frac{t^h}{h!}, \quad z_{d2}(t) = c_0 + \rho z_{d1}(t),$$

where ρ is a proportionality factor.

Suppose that the closed-loop control system is of type k, i.e., is asymptotically stable and has zero steady-state error for a $(k-1)$th order canonical input (see, e.g., [10]). As a result, when $t \to \infty$, the vehicle orientation θ will converge to a constant value $\bar{\theta}$, its steady-state longitudinal velocity η_1 will be a $(k-1)$th polynomial function of time, while η_2 will go to zero. Therefore, disturbances f_1^i will become constant whereas disturbances f_2^i will be at most $(k-1)$th polynomial functions of time. In order to reject those disturbances at steady state, the linear stabilizer $C(s)$ in eq. (20) (equivalently, eq. (19)) should be modified so as to include k cascaded integrators.

If we focus on trajectories built up with 3rd order splines, we can design a controller that accomplish the goal of robust trajectory tracking by using three integrators. Taking advantage of a state feedback from vehicle position and velocity (which is equivalent to the presence of one realizable zero), and selecting two pairs of complex zeros, a real zero, and two real poles the resulting closed-loop system can be stabilized. Summarizing, the transfer function of the robust stabilizing controller will have, for each input-output channel, the structure

$$C'(s) = (k_v s + k_p) \frac{(s+\alpha)\prod_{i=1}^{2}(s^2 + 2\omega_{ni}\zeta_i s + \omega_{ni}^2)}{s^3(s+p_1)(s+p_2)}. \quad (22)$$

Note that the above analysis holds only in the case of a straight line path, while for other kind of paths the disturbance equations are quite difficult to analyze. Nevertheless, the performance of the linear stabilizer (22) is satisfactory also for more general trajectories as illustrated in the following section.

5 Simulation Results

Numerical simulations of the tracking controller made of eqs. (12), (18), and (22) were performed with SIMULINK^{TM}, using the mechanical data characterizing the ATRV-2 robot. The vehicle dimensions are $a = 0.37$ m, $b = 0.55$ m, $2t = 0.63$ m, while the wheel radius is $r = 0.2$ m. The vehicle mass is $m = 116$ kg and its inertia is $I = 20$ kgm^2. The maximum achievable velocity is 2 m/s, while torque saturation for each motor occurs at 125 Nm, so that the maximum allowed torques τ_1 and τ_2 for each side of the vehicle are 250 Nm.

The desired trajectory is $z_{d1} = t$, $z_{d2} = -0.002t^3 + 0.1t^2 + 0.1t$ for $t \in [0, 50]$ sec, assigned to a point with $d_0 = 0.18$ m in eq. (14). The vehicle starts from the origin of frame F with a heading $\theta(0) = 10°$, i.e., it is initially out of the desired trajectory. Initial longitudinal and lateral (skidding) velocities are $\dot{x}(0) = \dot{y}(0) = 0.5$ m/s, with angular velocity $\dot{\theta}(0) = -2.2$ rad/s.

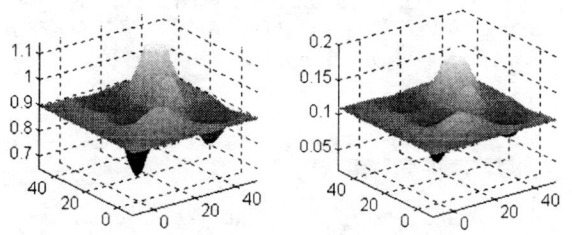

Figure 4: Fields of soil parameters μ and f_r

In order to validate the robustness of the tracking controller, we have simulated a virtual square (planar) terrain of side 50 m, with varying values of the friction coefficient μ and of the rolling resistance coefficient f_r, as shown in Fig. 4. The nonlinear controller uses instead constant estimates of soil parameters $\hat{\mu} = 0.895$ and $\hat{f}_r = 0.1$. The real values of μ and f_r encountered by the vehicle during its motion are plotted in Fig. 5.

The parameters of the robust linear stabilizer (22) are $p_1 = 9$, $p_2 = 5$, $\alpha = 1.5$, $\omega_{n1} = 1.5$, $\zeta_1 = 0.8$,

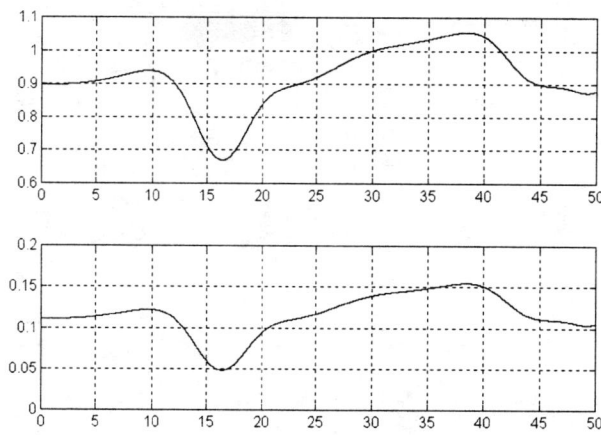

Figure 5: Actual values of μ and f_r during motion

$\omega_{n2} = 1$, $\zeta_2 = 1$, $k_v = 65$, and $k_p = 1.5k_v$, while $\xi(0) = 0$ in eq. (18).

Position and velocity errors are shown in Fig. 6, while the acceleration error and control torque behavior are presented in Fig. 7. Figure 8 shows a stroboscopic view of the vehicle tracking the desired trajectory on the virtual terrain. Gray levels indicate different values of soil parameters.

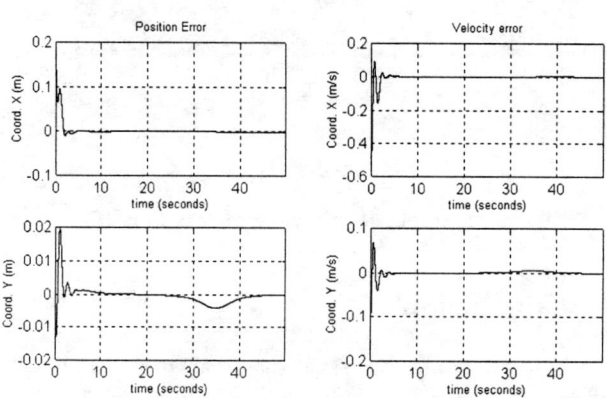

Figure 6: Position and velocity errors

The controller is able to recover from the initial error and stabilizes the vehicle to the desired trajectory, even if soil parameters are variable during motion. While position and velocity errors are rapidly compensated and remain very small, changes in soil parameters are more visible in the acceleration error, as expected. Note finally that the torques behavior is rather smooth, with initial peak values well within their feasible limits.

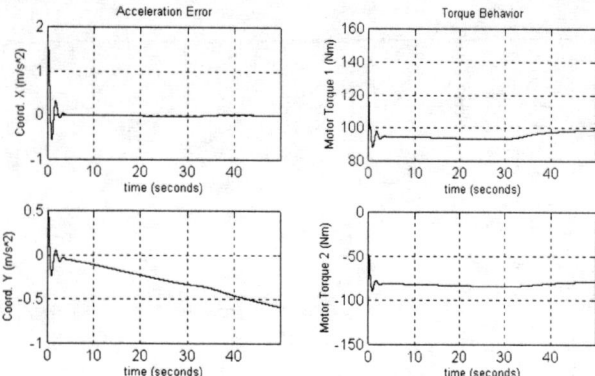

Figure 7: Acceleration error and torque behavior

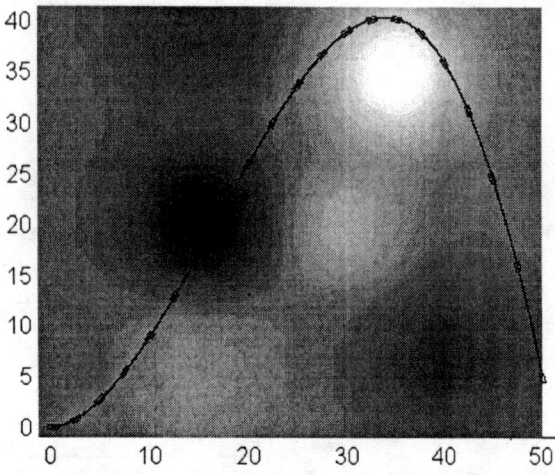

Figure 8: Stroboscopic view of robot motion

6 Conclusions

A robust controller for trajectory tracking of 4-wheel differentially driven vehicles has been presented. The control design minimizes unwanted lateral skidding, by imposing the longitudinal position of the instantaneous center of vehicle rotation. The controller is able to handle uncertain soil parameters.

The reported work constitutes one module of a more complex planning and control architecture that is going to be implemented for an ATRV-2 mobile robot. Among the extensions that should be included, we mention the control of 3D motion on uneven terrains and the use of global positioning sensors for measuring the actual robot state. For the latter problem, we have already simulated the presence of noise and/or transmission delays on state measurements, as would happen when using a GPS or a gyroscope for this purpose. Tracking performance remains satisfactory up to 60 ms delay and 10 cm peak position noise.

Acknowledgements

This work is supported by *Italian Space Agency*, under contract *ASI ARS-98-135*.

References

[1] Y. Fuke and E. Krotkov, "Dead reckoning for a lunar rover on uneven terrain," *1996 IEEE Int. Conf. on Robotics and Automation*, Minneapolis, MN, pp. 411–416, 1996.

[2] Z. Shiller, W. Serate, and M. Hua, "Trajectory planning of tracked vehicles," *1993 IEEE Int. Conf. on Robotics and Automation*, Atlanta, GA, pp. 796–801, 1993.

[3] A.T. Le, D.C. Rye, and H.F. Durrant-Whyte, "Estimation of track-soil interactions for autonomous tracked vehicles," *1997 IEEE Int. Conf. on Robotics and Automation*, Albuquerque, NM, pp. 1388–1393, 1997.

[4] A. De Luca, G. Oriolo, and C. Samson, "Feedback control of a nonholonomic car-like robot," in J.-P. Laumond (Ed.) *Robot Motion Planning and Control*, Lecture Notes in Control and Information Sciences, vol. 229, pp. 171–253, Springer-Verlag, London, 1998.

[5] B. d'Andrea-Novel, G. Campion, and G. Bastin, "Control of nonholonomic wheeled mobile robots by state feedback linearization," *Int. J. of Robotics Research*, vol. 14, no. 6, pp. 543–559, 1995.

[6] A.B. Will and S.H. Zak, "Modelling and control of an automated vehicle," *Vehicle System Dynamics*, vol. 27, pp. 131-155, 1997.

[7] J.Y. Wong, *Theory of Ground Vehicles*, John Wiley, New York, 1978.

[8] G. Campion, G. Bastin, and B. d'Andrea-Novel, "Structural properties and classification of kinematic and dynamic models of wheeled mobile robots," *IEEE Trans. on Robotics and Automation*, vol. 12, no. 1, pp. 47–62, 1996.

[9] A. Isidori, *Nonlinear Control Systems*, 3rd Edition, Springer-Verlag, London, 1995.

[10] J.J. D'Azzo and C.H. Houpis, *Linear Control System Analysis and Design*, 3rd Edition, McGraw-Hill, New York, 1988.

Tracking Control of Mobile Robots Using Saturation Feedback Controller

Ti-Chung Lee
Department of Electrical Engineering
Ming Hsin Institute of Technology
Hsinchu 304, TAIWAN

Kai-Tai Song   Ching-Hung Lee   Ching-Cheng Teng
Department of Electrical and Control Engineering
National Chiao Tung University
Hsinchu, TAIWAN

Abstract: *A general tracking control problem with saturation constraint for nonholonomic mobile robots is proposed and solved using the backstepping technique. A global result is given in which some artificial assumptions about the linear and the angular velocities of mobile robots from recent literature are dropped. The proposed controller can simultaneously solve both the tracking problem and the regulation problem of mobile robots. With the proposed control laws, mobile robots can now globally follow any path such as a straight line, a circle and the path approaching to the origin using a single controller. Computer simulations are presented, which confirm the effectiveness of the tracking control laws. Moreover, practical experimental results concerning the tracking control are reported with saturation constraint for mobile robots.*

Keywords: Mobile robots, tracking, time-varying systems, nonlinear systems, backstepping.

1. Introduction

The control problem of nonholonomic mechanical systems (e. g., mobile robots) has attracted considerable attention among the control community recently. Attempts to control such systems are, however, deceptively simple. The challenge of this problem is reflected in the fact that a mobile robot in the plane possesses three degrees of freedom of motion, which have to be controlled by only two control inputs and under the nonholonomic constraint. Several researchers have shown, based on Brockett's theorem [5], that such a system is open-loop controllable, but not stabilizable by pure smooth time-invariant feedback, see [2,17]. That is, there does not exist a smooth, or continuous, feedback. Recently, methods for solving this problem can be put into two categories. The first one is based on the motion planning by using the vision guided, environment prediction, artificial neural network, or other sensory information, [7,8,13,23,24].

The second category is using the nonlinear system theorems for solving this problem, e.g., the hybrid control, the backstepping technique, the sliding model control, the fuzzy logic control, and the adaptive control, etc., [1,3,4,9,11,20]. In this paper, our approach categorizes the second one. In this category, two main research directions have been adopted for mobile robots. The first direction, started from Bloch et al. in [3], uses the discontinuous feedback whereas the second one uses the time-varying continuous feedback, which was first investigated by Samson in [21]. Subsequent to these investigations, Pomet [20] then proposed the "smooth" feedback control laws. But they were to solve the regulation problem found to yield a slow asymptotic convergence. In order to obtain faster convergence (e.g., exponential convergence), an alternative approach was proposed by M'Closkey and Murray in [15] initially and taken up in several studies [15,16] subsequently.

The tracking problem for mobile robots has also attracted as much attention among researches [6,10,17,19,21]. Using Barbalet lemma or the backstepping method, some controllers have been proposed such that the mobile robots could globally follow special paths such as circles and straight lines. Despite the apparent advancement of the above methods, there remains however, several main restrictions on their applications: (a) In the tracking problem, only some special cases (straight lines or circles) are solved. (b) In other cases, for example, the tracking problem with the linear and the angular velocities approaching to zero remains unsolved.

In applications, it is preferred to solve the tracking problem and the regulation problem simultaneously using a single controller, otherwise, there will be a need to switch between two controllers of different types. In this paper, both the tracking problem and the regulation problem of mobile robots will be solved simultaneously without any further assumptions except for some regular assumptions. Moreover, for the saturation constraint in control inputs (the linear and angular velocities) are also taken into consideration in the tracking problem. For practical applications, the bounds on the velocities of wheels must be attended to avoid the high-gain control signal.

The organization of this paper is as follows. Some notations and the problem formulation are presented in Section 2. Main results including the control law and the stability analysis are given in Section 3. In Section 4, some results from computer simulation and practical experiment for mobile robots are given to verify the proposed tracking controller. Section 5 presents the discussion for experimental environment, results, extension of the proposed approach, and future research. The conclusion remark is given in Section 6.

2. Preliminaries
2.1 Notation
1. $\Re$ denotes the set of all real numbers and $\Re^+$ denotes the set of all nonnegative real numbers.
2. A function $f: \Re_+ \to \Re$ is uniformly continuous if for any $\varepsilon > 0$, there exists a $\delta(\varepsilon) > 0$ such that if $|x_1 - x_2| < \delta$, with $x_1, x_2 \in \Re^+$, then $|f(x_1) - f(x_2)| < \varepsilon$.
3. Define the saturation function $sat_\delta(x)$ with $\delta > 0$ as

$$sat_\delta(x) = \begin{cases} x & |x| \leq \delta \\ sgn(x)\delta & |x| > \delta. \end{cases} \quad (1)$$

2.2 Problem formulation
A. A simplified mobile robot model and its properties

The car-like robots considered in this paper, are a simplified model of the constraint on the movement of a rear car, see Fig. 1. The nonholonomic constraint for this simplified model is

$$\dot{y}\cos\theta - \dot{x}\sin\theta = 0, \quad (2)$$

that specifies the tangent direction along any feasible path for the robot and a bound on the curvature of the path. From the driver's viewpoint, there are two degrees of freedom for a general car: the accelerator and the steering wheel. Herein, we assume that the reference point lies in the midpoint of the rear wheels. We denote by v the linear velocity and by w the angular velocity of the mobile robot. Moreover, (x, y) are the Cartesian coordinates of the center of mass of the vehicle, and θ is the angle between the heading direction and the x-axis. Let us consider a simplified mobile robot model [13,19]

$$\begin{aligned} \dot{x} &= v\cos\theta \\ \dot{y} &= v\sin\theta \\ \dot{\theta} &= w. \end{aligned} \quad (3)$$

Notice that, in the plane, the mobile robot (3) possesses three degrees of freedom of motion, which have be controller by only two control input and under nonholonomic constraint. When we consider the state vector (x, y, θ), it is rather straightforward to show that system (3) is controllable in both position and orientation, see [20]. Unfortunately, we must confront a system where linearization results in a loss of controllability, although the nonlinear system is controllable. In particular, there are many researches, based on Brockett's theorem [5], showed that such a system is open-loop controllable, but not stabilizable by pure smooth time-invariant feedback, see [2,17].

B. Tracking control problem with saturation constraint

Suppose the reference trajectory is described by

$$\begin{aligned} \dot{x}_r &= v_r \cos\theta_r \\ \dot{y}_r &= v_r \sin\theta_r \\ \dot{\theta}_r &= w_r \end{aligned} \quad (4)$$

where the desired linear velocity satisfying $0 \leq v_r (v_r = \sqrt{\dot{x}_r^2 + \dot{y}_r^2}) < v_{max}$ and the desired angular velocity satisfying $\sup_{t \geq 0}|w_r(t)| < w_{max}$ are bounded and uniformly continuous, with v_{max} and w_{max} representing saturation bounds from practical restriction for control inputs $v(t)$ and $w(t)$, respectively. Note that in the perfect tracking case, i.e., $x \equiv x_r, y \equiv y_r, \theta \equiv \theta_r$, the equation (4) always holds. In this paper, our main purpose is to solve the following tracking problem for (3).

Tracking Problem with Saturation Constraint (TPSC):
To find saturation control laws for v and w with $|v| < v_{max}, |w| < w_{max}$ such that the mobile robot with the dynamic model (3) follows a reference trajectory $(x_r(t), y_r(t), \theta_r(t))$. That is, $\lim_{t\to\infty}|x(t) - x_r(t)| = 0$, $\lim_{t\to\infty}|y(t) - y_r(t)| = 0$, and $\lim_{t\to\infty}|\theta(t) - \theta_r(t)| = 0$.

Using the body frame, we then define the error coordinates by (see [12])

$$\begin{bmatrix} x_e \\ y_e \\ \theta_e \end{bmatrix} = \begin{bmatrix} \cos\theta & \sin\theta & 0 \\ -\sin\theta & \cos\theta & 0 \\ 0 & 0 & 1 \end{bmatrix} \begin{bmatrix} x_r - x \\ y_r - y \\ \theta_r - \theta \end{bmatrix}.$$

Therefore, the tracking error model is obtained as

$$\begin{aligned} \dot{x}_e &= wy_e - v + v_r \cos\theta_e \\ \dot{y}_e &= -wx_e + v_r \sin\theta_e \\ \dot{\theta}_e &= w_r - w. \end{aligned} \quad (5)$$

For convenience, we choose new coordinates and inputs as

$$\begin{bmatrix} x_0 \\ x_1 \\ x_2 \end{bmatrix} = \begin{bmatrix} \theta_e \\ y_e \\ -x_e \end{bmatrix}, \quad \begin{bmatrix} u_0 \\ u_1 \end{bmatrix} = \begin{bmatrix} w_r - w \\ v - v_r \cos x_0 \end{bmatrix}.$$

System (5) can be rewritten as (see [16])

$$\dot{x}_0 = u_0 \quad (6a)$$
$$\dot{x}_1 = (w_r - u_0)x_2 + v_r \sin x_0 \quad (6b)$$
$$\dot{x}_2 = -(w_r - u_0)x_1 + u_1. \quad (6c)$$

With the new-coordinates (x_0, x_1, x_2), the tracking problem is now transformed to a stability problem. Throughout this paper, these coordinates (x_0, x_1, x_2) will be used to solve the tracking problem. More explicitly, by the invariance of coordinate transformations, if the (x_0, x_1, x_2) converge to zero then the TPSC is solved. That is, $\lim_{t\to\infty}|x(t) - x_r(t)| = 0$, $\lim_{t\to\infty}|y(t) - y_r(t)| = 0$, and $\lim_{t\to\infty}|\theta(t) - \theta_r(t)| = 0$.

3. Globally tracking control laws

In this section, a globally tracking control law with saturation constraint is presented using the backstepping method and the idea from the standard stability theory. The stability of the closed-loop system is guaranteed without any further assumptions relating v_r and w_r.

First, let us define a positive-definite function
$$V_1(x_1, x_2) = x_1^2 + x_2^2. \qquad (7)$$
Note that, it is the square of the distance between the mobile robot and the desired trajectory. Taking the derivative of V_1 along the trajectory of (3) yields
$$\dot{V}_1 = 2(x_2 u_1 + x_1 v_r \sin x_0). \qquad (8)$$
Choosing the saturation control law
$$u_1 = -sat_a(k_0 x_2) \qquad (9)$$
where $k_0 > 0$ and $0 < a < v_{max} - \sup_{t \geq 0} v_r$. Note that
$$|v| \leq |u_1| + |v_r \cos x_0| < v_{max} \qquad (10)$$
satisfying the saturation constraint of the line velocity.

Using the saturation control (9), we have
$$\dot{V}_1 = -2 sat_a(k_0 x_2) x_2 + 2 x_1 v_r \sin x_0. \qquad (11)$$
Next, introduce a new variable
$$\bar{x}_0 = x_0 + \frac{\varepsilon h(t) x_1}{1 + V_1^{\frac{1}{2}}} \qquad (12)$$
with $h(t) = 1 + \gamma \cos(t - t_0), 0 < \gamma < 1$, and $0 < \varepsilon < \frac{1}{1+\gamma}$, where t_0 will be specified in the following theorem. Note that if $(\bar{x}_0, x_1, x_2)$ converges to zero, then the stability of (x_0, x_1, x_2) will be guaranteed. With (12), system (6a) is transformed into
$$\dot{\bar{x}}_0 = \alpha(x_1, x_2, t) u_0 + \beta(x_0, x_1, x_2, t) \qquad (13)$$
where
$$\alpha(x_1, x_2, t) = 1 - \frac{\varepsilon h x_2}{1 + V_1^{\frac{1}{2}}},$$
$$\beta(x_0, x_1, x_2, t) = \varepsilon[\frac{\dot{h} x_1 + h w_r x_2 + h v_r \sin x_0}{1 + V_1^{\frac{1}{2}}}$$
$$- \frac{h x_1}{(1 + V_1^{\frac{1}{2}})^2 V_1^{\frac{1}{2}}} (x_1 v_r \sin x_0 - sat_a(k_0 x_2) x_2)].$$
It is easy to check that $0 < 1 - \varepsilon(1 + \gamma) \leq \alpha(x_1, x_2, t) < 2$.

Choose the control law with saturation constraint u_0 as
$$u_0 = \frac{-\beta(x_0, x_1, x_2, t)}{\alpha(x_1, x_2, t)} - sat_b(k_1 \bar{x}_0) \qquad (14)$$
with $k_1 > 0$ and $b > 0$. Then,
$$\dot{\bar{x}}_0 = -\alpha(x_1, x_2, t) sat_b(k_1 \bar{x}_0). \qquad (15)$$
Note that $\lim_{\varepsilon \to 0} \frac{-\beta(\cdot)}{\alpha(\cdot)} = 0$. So, it is possible to choose a small $\varepsilon > 0$ and $b > 0$ such that $\sup_{t \geq 0} |u_0| < w_{max} - \sup_{t \geq 0} |w_r(t)|$. Then $|w(t)| \leq |u_0| + |w_r| < w_{max}$ satisfying the saturation constraint of the angular velocity. Define $V_2(\bar{x}_0) = \bar{x}_0^2/2$. We then obtain

$$\dot{V}_2 = -\alpha(x_1, x_2, t) sat_b(k_1 \bar{x}_0) \bar{x}_0 = \begin{cases} -\alpha(x_1, x_2, t) k_1 \bar{x}_0^2 & |k_1 \bar{x}_0| \leq b \\ -\alpha(x_1, x_2, t) b \bar{x}_0 & |k_1 \bar{x}_0| > b \end{cases} \qquad (16)$$
$$\leq 0 \quad \forall \bar{x}_0 \in \Re.$$

This means that $\lim_{t \to \infty} \bar{x}_0(t) = 0$ (by the Lyapunov stability theorem, see [21]).

Now, we are in a position to present main results.

Theorem 1 *Consider a simplified model (3) of mobile robots. Then the tracking problem with saturation constraint can be solved using control laws (9) and (14) with $h(t) = 1 + \gamma \cos(t - t_0)$, where t_0 will be defined in the proof of theorem.*

Proof: Due to limited space, the proof is omitted here. A similar proof can be found in the result of [14].

Remark: Although the LaSalle invariance principle can not be applied directly to non-autonomous systems. Its ideal can be applied here using some modification results, for details see [14].

4. Experimental results

The experimental mobile robot developed in our laboratory has two driving wheels and one casters for balance. The physical relation of experimental mobile robot (two wheels mobile robot) and the mobile robot is briefly described here. Let v_L and v_R denote the velocities of the left wheel and the right wheel, respectively. In our case, the center of mass is equal to the center point between the left wheel and the right wheel. Thus, its kinematic equation can be described as the equation (3) with $v = \frac{v_L + v_R}{2}$ and $w = \frac{v_R - v_L}{E}$ where E is the distance between the two driving wheels. So, the proposed controller can be applied to this case with $v_R = v + \frac{wE}{2}$ and $v_L = v - \frac{wE}{2}$. Therefore,
$$\sup_{t \geq 0} \max(v_L, v_R) \leq \sup_{t \geq 0} v + \frac{E}{2} \sup_{t \geq 0} w = v_{max} + \frac{E}{2} w_{max}. \qquad (17)$$
In our case, $E = 54$ cm and the velocities of the left wheel and the right wheel satisfying the constraint $\max(v_L, v_R) \leq 60$ cm/s. Thus, the values of v_{max} and w_{max} can be chosen according to the constraint
$$v_{max} + 27 w_{max} \leq 60 \qquad (18)$$
due to the practical consideration. For v_L and v_R, the above inequality is a conservative result. In practical application, the above condition can be chosen by simulation results. In general, the constraint for v_{max} and w_{max} are larger than the condition (18).

The proposed algorithms have been verified by using computer simulation and practical experiments on an experimental mobile robot. The robot's hardware structure has two independent drive wheels and a free caster for balance. The motion of the robot is controlled by changing the velocities of the left and right wheels. Two dedicated motion control chips HCTL-1100 from HP were employed for servo control of the two drive wheels. The integral velocity control mode was used in the experiments. The control computer only needs to send commands to the chips, which take care of the motor servo control. The position estimation of the robot is conducted using an odometer, which samples the left and right wheel velocities to calculate the current posture of the robot. The formula of this estimator is presented in (19) ~ (23).

$$v = \frac{v_L + v_R}{2} \quad (19)$$

$$\omega = \frac{v_R - v_L}{E} \quad (20)$$

$$x_{new} = x_{old} + \Delta T \cdot v \cdot \cos(\theta_{old}) \quad (21)$$

$$y_{new} = y_{old} + \Delta T \cdot v \cdot \sin(\theta_{old}) \quad (22)$$

$$\theta_{new} = \theta_{old} + \Delta T \cdot \omega \quad (23)$$

where E is the distance between two drive wheels and T is the sampling period. The desired velocity calculated in the tracking algorithm is transferred into the left and right wheel velocities using

$$v_r = v_{out} + \frac{L \cdot w_{out}}{2} \quad (24)$$

$$v_l = v_{out} - \frac{L \cdot w_{out}}{2} \quad (25)$$

Two trajectories have been used to show the performance of the control law. Computer simulations were first conducted for these trajectories and the parameters (k_0, k_1, γ, ε) are determined by examining the desirable tracking performance. In the following the simulation and experimental results of tracking as well as docking performance are presented respectively for circular and parallel parking trajectories. The model of these trajectory generators are illustrated below:

■ Circular trajectory:

$$x_r = x_c + R \cdot \sin(c \cdot t)$$
$$y_r = y_c - R \cdot \cos(c \cdot t)$$
$$\theta_r = c \cdot t$$
$$v_r = c \cdot R$$
$$w_r = c$$

where (x_c, y_c) is the center coordinate, R is the radius。

■ Parallel parking :

$$x_r(t) = \begin{cases} 2a\cos(c(t+\frac{\pi}{4c})) & 0 \leq t \leq \frac{\pi}{2c} \\ -a\sqrt{2} & \frac{\pi}{2c} \leq t \end{cases},$$

$$y_r(t) = \begin{cases} b\sin(2c(t+\frac{\pi}{4c})) & 0 \leq t \leq \frac{\pi}{2c} \\ -b & \frac{\pi}{2c} \leq t \end{cases},$$

$$\theta_r(t) = \begin{cases} \pi - \tan^{-1}\left(\frac{b\cos(2c(t+\pi/4c))}{a\sin(c(t+\pi/4c))}\right) & 0 \leq t \leq \frac{\pi}{2c} \\ \pi & \frac{\pi}{2c} \leq t \end{cases},$$

$$v_r(t) = \begin{cases} 2c\sqrt{a^2\sin^2(c(t+\frac{\pi}{4c})) + b^2\cos^2(2c(t+\frac{\pi}{4c}))} & 0 \leq t \leq \frac{\pi}{2c} \\ 0 & \frac{\pi}{2c} \leq t \end{cases}$$

$$w_r(t) = \begin{cases} \frac{abc\cos(c(t+\frac{\pi}{4c}))\cos(2c(t+\frac{\pi}{4c})) + 2\sin(2c(t+\frac{\pi}{4c}))}{(a\sin(c(t+\frac{\pi}{4c})))^2 + (b\cos(2c(t+\frac{\pi}{4c})))^2} & 0 \leq t \leq \frac{\pi}{2c} \\ 0 & \frac{\pi}{2c} \leq t \end{cases}$$

An "8" shappped trajectory was employed, where $2a$ and b represent the long and short axis respectively。

Because of the experimental mobile robot hardware structure, the outputs of the controller were limited as (k_0, k_1, γ, ε, v_{max}, w_{max})=(1,1,0.5,0.5,30 cm/s,1 cm/s). For comparison, the simulation and experimental results are depicted together in the figures.

By theorem 1, the tracking errors will converge to zero. These simulations illustrated the effectiveness of the tracking control laws and note that in all cases $h(t)$ can be chosen as $h(t) = 1 + \gamma \cos t$.

Simulation and experimental results with saturation constraint are shown in Figs. 3-4. The circular trajectory and the parallel parking problem are presented in Figs. 3 and 4, respectively. Experimental results corresponding to the same conditions are also shown in these plots for comparison. The dashed-line presents the computer simulation and the solid-line presents the experimental results. These results demonstrate the ability of the proposed controllers in producing nice convergence behavior.

Moreover, in problems associated with the parking case which v_r and w_r are approaching to zero, has remained unsolved in present literature using the nonlinear theory approach. This research effort has proven, that the problem can be solved, using the proposed controllers (9) and (14). Moreover, with our approach, the assumptions in [10,21] are unnecessary. This means that the tracking control problem without any constraint on velocities is solved using a single controller.

5. Experimental discussion

In the practical environment, we must consider many real-restrictions for the mobile robot, e.g., the limitation of velocities, time-delay of actuator, etc. With rolling and no slipping assumption for wheels, this paper only proposes saturation control laws for practical consideration. Unfortunately, in practical, there are many problems must be considered. For instance, the mobile robot must have high mobility and high payload capacity on rugged terrain and soft soil, which makes them most suitable for military and nuclear applications, such as clearing mine fields and excavation contaminated soil in decommissioned nuclear sites. Besides, the friction of wheels, variation of system parameters, or external disturbance will affect a fair of control. In the future, we would like to investigate the robust control and adaptive control of the mobile robot. This means that the computed torque controller must be used for tracking control. On the other word, how to guide a mobile robot to avoid the static (or dynamic) obstacle and reach the desired target is another interesting problem.

6. Conclusions

A novel control law has been proposed to solve the tracking problem of mobile robots. The tracking control and the regulation problems were solved simultaneously. Simulation results show that mobile robots globally follow any special paths such as a straight line, a circle, and a path approaching to zero. Moreover, the technique presented in this paper can be generalized to the control problem of other systems, e.g., the control of knife edge using steering and pushing inputs, four wheels mobile robot, and the general chained-form system. This is the first step towards achieving a more complex control systems, e. g., the chained-form systems. The next interesting challenge will be to simplify the proposed control law (e.g., using a linear controller) and extend the result to general chained-form systems.

Acknowledgements: This paper was supported by the National Science Council, Republic of China, under Grant NSC-88-2212-E-159-003 and NSC-88-2213-E-009-121.

References

[1] S. Bentalba, A. E. Hajjaji, and A. Rachid, "Fuzzy control of a mobile robot: a new approach," in *Proc. IEEE Conf. Control Applications*, Hartford, CT, Oct. pp. 69-72, 1997.

[2] A. M. Bloch, M. Reyhanoglu, and N. H. McClamroch, "Control and stabilizability of nonholonomic dynamic systems," in *IEEE Trans. On Automatic Control*, Vol. 37, No. 11, pp. 1746-1757, 1992.

[3] A. M. Bloch and S. Drakunov, "Stabilization of a nonholonomic system via sliding modes," in *Proc. 33rd IEEE Conf. On Decision and Control*, Lae Buena Vista, FL, pp. 2961-2963, 1994.

[4] A. M. Bloch and S. Drakunov, "Tracking in nonholonomic dynamic system via sliding modes," in *Proc. 34rd IEEE Conf. On Decision and Control*, New Orleans, LA, pp. 2103-2106, 1995.

[5] R. W. Brockett, "Asymptotic stability and feedback stabilization," in: R. W. Brockett, R. S. Millman, and H. H. Sussmann Eds., *Differential Geometric Control Theory*, pp. 181-191, 1983.

[6] C. de Wit Canudas, B. Siciliano, and G. Bastin, eds *Theory of Robot Control*, Springer-Verlag, London, 1996.

[7] C. C. Chang and K. T. Song, "Environment prediction for a mobile robot in a dynamic environment," in *IEEE Trans. on Robotics and Automation*, Vol. 13, No. 6, pp.862-872, 1997.

[8] P. Fiorini and Z. Shiller, "Motion planning in dynamic environments using velocity obstacles," in the *Int. J. of Robotics Research*, Vol. 17, No. 7, pp. 760-772, 1998.

[9] J. Guldner and V. I. Utkin, "Stabilization of nonholonomic mobile robots using Lyapunov functions for navigation and slide model control," in *Proc. 33rd IEEE Conf. On Decision and Control*, Lae Buena Vista, FL, pp. 2967-2972, 1994.

[10] Z. P. Jiang and H. Nijmeijer, "Tracking control of mobile robots: a case study in backstepping," in *Automatica* Vol. 33, pp. 1393-1399, 1997.

[11] Z. P. Jiang and J. B. Pomet, "Combining backstepping and time-varying techniques for a new set of adaptive controllers," in *Proc. IEEE Conf. On Decision and Control*, New Orlean, LA, Dec. pp. 2207-2212, 1994.

[12] Y. J. Kanayama, Y. Kimura, F. Miyazaki, and T. Noguchi, "A stable tracking control scheme for an autonomous mobile robot," in *Proc. IEEE Int. Conf. on Robotics and Automation*, pp. 384-389, 1990.

[13] J. P. Laumond, P. E. Jacobs, M. Taix, and R. M. Murray, "A motion planner for nonholonomic mobile robots," in *IEEE Trans. on Robotics and Automation*, Vol. 10, No. 5, pp. 577-593, 1994.

[14] T. C. Lee, C. H. Lee, and C. C. Teng, "Tracking control of mobile robots using backstepping technique," 5th International Conference on Control, Automation, Robotics and Vision, pp. 1715-1719, Dec. 8-11, 1998, Mandarin Singapore, Singapore

[15] R. T. M'Closkey, and R. M. Murray, "Exponential stabilization of driftless control systems using homogeneous feedback," in *IEEE Trans. Automatic Control* 42 No.5, pp. 614-628, 1997.

[16] R. T. M'Closkey and R. M. Murray, "Exponential stabilization of nonlinear driftless control systems via time-varying homogeneous feedback," in *IEEE Conf. On Decision and Control*, Lake Buena Vista, pp. 1317-1322, 1994.

[17] R. M. Murray, G. Walsh, and S. S. Sastry, "Stabilization and tracking for nonholonomic control systems using time-varying state feedback," in *IFAC Nonlinear Control Systems Design*, Bordeaux, ed. M. Fliess, pp. 109-114, 1992.

[18] R. M. Murray and S. S. Sastry, "Nonholonomic motion planning: steering using sinusoids," in *IEEE Trans. on Automatic Control*, Vol. 38, No. 5, pp. 700-716, 1993.

[19] W. Oelen and J. van Amerongen, "Robust tracking control of two-degree-freedom mobile robots," in *Control Engng Practice* 2, pp. 333-340, 1994.

[20] J. B. Pomet, "Explicit design of time-varying stabilizing control laws for a class of controllable systems without drift," in *Syst, and Contr. Letters* 18, pp. 467-473, 1992.

[21] C. Samson and K. Ait-Abderrahim, "Feedback control of a nonholonomic wheeled cart in cartesian space," in *Proc. IEEE Int. Conf. on Robotics and Automation*, Sacramento, CA, pp. 1136-1141, 1991.

[22] J. J. Slotine and W. Li, *Applied Nonlinear Control*, Prentice-Hall Inc, 1991.

[23] K. T. Song, C. Y. Chen, and C. C. Chang, "On-line motion planning of an autonomous mobile robot based on sensory information," in *Proc. IROS'91*, pp. 423-428, 1991.

[24] J. Taylor and D. J. Kriegman, " Vision-based motion planning and exploration algorithms for mobile robots," in *IEEE Trans. on Robotics and Automation*, Vol. 14, No. 3, pp. 417-426, 1998.

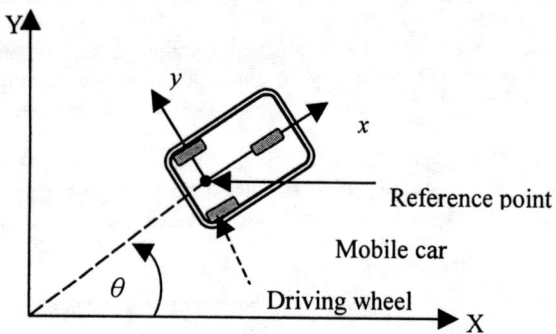

Figure 1: A nonholonomic mechanical system: simplified mobile car

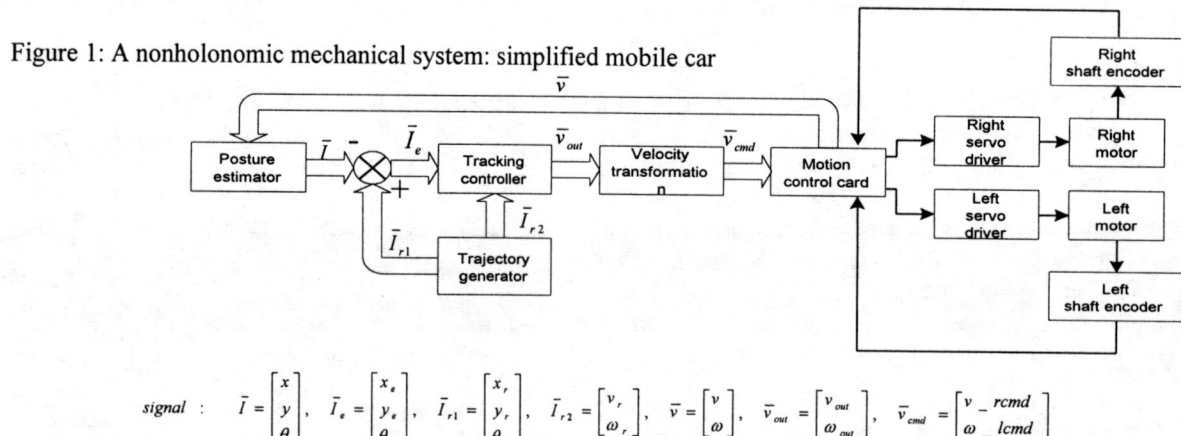

$$signal: \quad \bar{I} = \begin{bmatrix} x \\ y \\ \theta \end{bmatrix}, \quad \bar{I}_e = \begin{bmatrix} x_e \\ y_e \\ \theta_e \end{bmatrix}, \quad \bar{I}_{r1} = \begin{bmatrix} x_r \\ y_r \\ \theta_r \end{bmatrix}, \quad \bar{I}_{r2} = \begin{bmatrix} v_r \\ \omega_r \end{bmatrix}, \quad \bar{v} = \begin{bmatrix} v \\ \omega \end{bmatrix}, \quad \bar{v}_{out} = \begin{bmatrix} v_{out} \\ \omega_{out} \end{bmatrix}, \quad \bar{v}_{cmd} = \begin{bmatrix} v\_rcmd \\ \omega\_lcmd \end{bmatrix}$$

Figure 2: Block diagram of the tracking control system

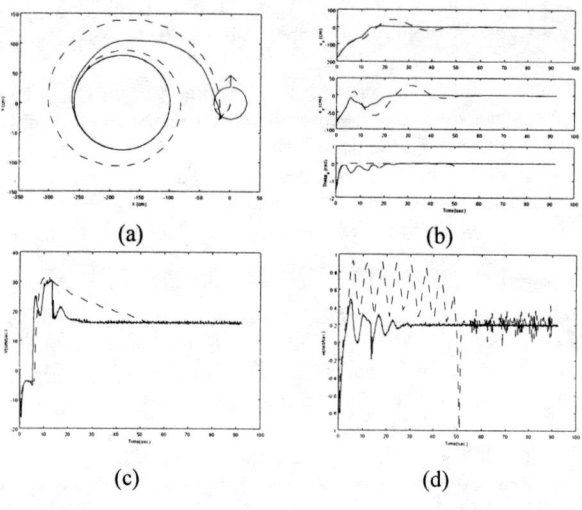

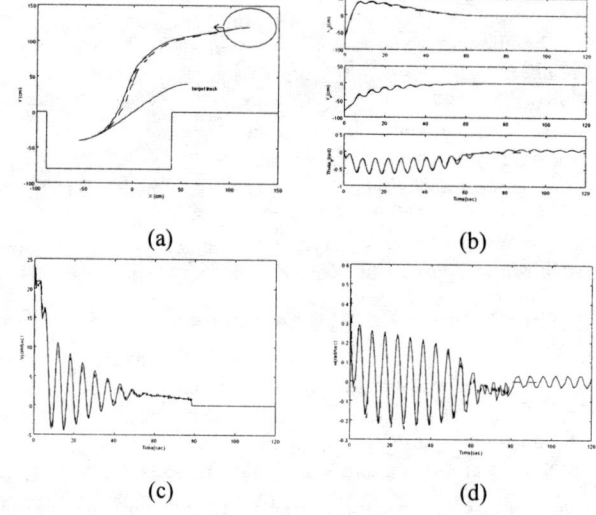

Figure3: Experimental and simulation results of tracking a circular trajectory: (a) position variations, (b) tracking errors, (c) linear velocity, (d) angular velocity (solid-line: experiment; dotted-line simulation).

Figure4: Experimental and simulation results of parallel parking: (a) position variations, (b) tracking errors, (c) linear velocity, (d) angular velocity (solid-line: experiment; dotted-line :simulation).

Tracking Control of Wheeled Mobile Robots with Unknown Dynamics

Wenjie Dong   Wei Huo

The seventh Research Division,
Beijing Univ. of Aero. and Astro., 100083 Beijing, P.R. China

Abstract

This paper considers the tracking problem of nonholonomic wheeled mobile robots with unknown dynamics. A new adaptive robust global dynamic controller is presented based on the canonical form of the wheeled mobile robots. This novel controller has low dimension and no singular points. Simulations show the effectiveness of the control scheme.

1 Introduction

In recent years, there has been a growing interest in design of feedback control laws for wheeled mobile robots (WMR) subjected to nonholonomic constraints. Due to Brockett's theorem, it is well-known that the nonholonomic WMR cannot be asymptotically stabilized to a rest configuration by differentiable pure-state feedback laws [1]. However, several approaches have been proposed for stabilizing such systems. For details, see the survey paper [14].

Another control problem of WMR is trajectory tracking problem. For the tracking problem of the kinematic model (i.e. control inputs are generalized velocities), several authors have presented some results. In [3][16][13][11], the tracking problem of individual WMR is solved by designing a controller which makes the robot follow a virtual reference robot. However, these results do not fit all nonholonomic WMR. Since the kinematic model of WMR can be put into the chained system locally or globally [15], some authors have discussed the tracking problem of the chained system. A solution based on a linear approximation of the kinematic nonholonomic system around sufficiently exciting reference trajectories was described and analyzed in [18]. Using results on "differentially flat" nonlinear systems [9], an n-dimensional single-generator, one-chain system can be dynamically linearized by adding $(n-2)$ integrators. Therefore, the tracking problem of WMR can be solved by an $(n-2)$-dimensional dynamic controller [7]. Unfortunately, the control law has singular points, and the dimension of the controller is high if n is large. With change of the time scale, the singular points may be avoided [10], but the dimension of the controller can not be reduced. With aid of backstepping technique, a static controller was also presented for the tracking problem of the chained system in [12]. Since there is a unknown large enough number λ in the controller, it may be difficult to implement the control in practice.

In many cases, tracking control of the dynamic model of WMR with uncertainty are important for practical applications. Few results on the subject are presented now. [17][5][4] studied the tracking problem of dynamic nonholonomic systems with uncertainty. However, in these papers configuration tracking for the given trajectory still has not been solved, since only partial state (i.e. generalized velocity) tracking was achieved.

Our purpose in this paper is to investigate how to construct a controller which can make all states of the closed-loop dynamic model of WMR with unknown dynamics globally track a given trajectory. To this end, generic structures of WMR model equations are reviewed, and the canonical form is introduced based on them. With aid of the canonical form and the well-known backstepping technique, a new adaptive robust tracking controller is presented. Our result is novel in the following aspects: 1) The tracking problem of the nonholonomic WMR with unknown dynamics is solved, which has not been addressed in the existing literature to our knowledge. 2) Our global controller can make all states of the closed loop system asymptotically track the given desired trajectory, while in paper [17][5][4] only partial states (i.e. generalized velocities) can track the desired trajectory. 3) Our controller is simple in structure and easy to be implemented, since no information and calculation on the system dynamics are involved.

2 Model Equations and Problem Statement

Consider the nonholonomic *WMR* discussed in [2][7], let $\xi = [x, y, \theta]^T$, where x, y are the coordinates of a reference point P on the frame in a fixed orthonormal inertial basis $\{O, \vec{I}_1, \vec{I}_2\}$, and θ is the orientation of an arbitrary basis $\{\vec{x}_1, \vec{x}_2\}$ attached to the frame with respected to the inertial basis $\{\vec{I}_1, \vec{I}_2\}$, β is the angle of the orientation wheel. It is shown in [2] that the mobility of any *WMR* can be characterized by two integers δ_m and δ_s. The interesting nonholonomic *WMR* identified by (δ_m, δ_s) are type (2,0), (2,1), (1,1), (1,2) robots. Generally, the dynamic model of the nonholonomic *WMR* can be expressed in the following form [2]:

$$J(X)\dot{X} = 0 \quad (1)$$
$$M(X)\ddot{X} + C(X,\dot{X})\dot{X} = A(X)\tau + J^T(X)\lambda \quad (2)$$

where the generalized coordinate $X = \xi$ if $\delta_s = 0$ or $X = (\xi, \beta)$ if $\delta_s \neq 0$, $M(X)$ is a bounded positive definite symmetric inertia matrix, $C(X,\dot{X})\dot{X}$ presents the vectors of centripetal and Coriolis torques, $A(X)$ is input transformation matrix, $J(X)$ is a full rank matrix, λ is Lagrange multiplier, τ is $(\delta_m + \delta_s)$-dimensional control input and the superscript T denotes the transpose. The constraint (1) is assumed to be completely nonholonomic, and (2) satisfies the following two properties [6]:

Property 1: $\dot{M} - 2C$ is skew-symmetric for a suitable definition of C.

Property 2: There exist positive constants $c_i > 0 (1 \leq i \leq 3)$ such that $\forall X$ and $\forall \dot{X}$, $\|M(X)\| \leq c_1$, $\|C(X,\dot{X})\| \leq c_2 + c_3\|\dot{X}\|$.

Given a desired differentiable trajectory $X^*(t)$ which satisfies the nonholonomic constraint

$$J(X^*)\dot{X}^* = 0 \quad (3)$$

the dynamic tracking problem is defined as follows: For the system (1)-(2) with $M(X)$ and $C(X,\dot{X})$ unknown and a given desired trajectory $X^*(t)$ satisfying (3), find a feedback law τ such that $\lim_{t \to \infty}(X - X^*) = 0$ and $\lim_{t \to \infty}(\dot{X} - \dot{X}^*) = 0$. To make this problem resolvable, the following assumptions about the desired trajectory X^* are made:

Assumption 1: $x^*(t)$ and $y^*(t)$ are bounded, X^* does not contain singular points listed in Table 2.

Assumption 2: For type (2,0), (1,1) and (1,2) robots, $\dot{\theta}^*(t)$ and $\ddot{\theta}^*(t)$ are bounded. For type (2,1) robot, $\dot{\theta}^*(t) + \dot{\beta}^*(t)$ and $\ddot{\theta}^*(t) + \ddot{\beta}^*(t)$ are bounded.

Assumption 3: There exists a time diverging sequence $\{t_i\}_{i \in N}(N = 1, 2, \cdots)$, and $|t_i - t_{i-1}| \leq T_0 <$ ∞, such that for type (2,0), (1,1) and (1,2) robots $\lim_{i \to \infty} \inf |\dot{\theta}^*(t_i)| = \epsilon > 0$, for type (2,1) robot $\lim_{i \to \infty} \inf |\dot{\theta}^*(t_i) + \dot{\beta}^*(t_i)| = \epsilon > 0$,

Table 1. The kinematic models of nonholonomic WMR

Type	Model equation: $\dot{X} = B(X)u$
(2,0)	$\begin{bmatrix} \dot{x} \\ \dot{y} \\ \dot{\theta} \end{bmatrix} = \begin{bmatrix} -\sin\theta & 0 \\ \cos\theta & 0 \\ 0 & 1 \end{bmatrix} \begin{bmatrix} \eta_1 \\ \eta_2 \end{bmatrix}$
(2,1)	$\begin{bmatrix} \dot{x} \\ \dot{y} \\ \dot{\theta} \\ \dot{\beta} \end{bmatrix} = \begin{bmatrix} -\sin(\theta+\beta) & 0 & 0 \\ \cos(\theta+\beta) & 0 & 0 \\ 0 & 1 & 0 \\ 0 & 0 & 1 \end{bmatrix} \begin{bmatrix} \eta_1 \\ \eta_2 \\ \zeta_1 \end{bmatrix}$
(1,1)	$\begin{bmatrix} \dot{x} \\ \dot{y} \\ \dot{\theta} \\ \dot{\beta} \end{bmatrix} = \begin{bmatrix} -L\sin\theta\sin\beta & 0 \\ L\cos\theta\sin\beta & 0 \\ \cos\beta & 0 \\ 0 & 1 \end{bmatrix} \begin{bmatrix} \eta_1 \\ \zeta_1 \end{bmatrix}$
(1,2)	$\begin{bmatrix} \dot{x} \\ \dot{y} \\ \dot{\theta} \\ \dot{\beta}_1 \\ \dot{\beta}_2 \end{bmatrix} = \begin{bmatrix} -L[\sin\beta_1\sin(\theta+\beta_2) + \sin\beta_2\sin(\theta+\beta_1)] & 0 & 0 \\ L[\sin\beta_1\cos(\theta+\beta_2) + \sin\beta_2\cos(\theta+\beta_1)] & 0 & 0 \\ \sin(\beta_2-\beta_1) & 0 & 0 \\ 0 & 1 & 0 \\ 0 & 0 & 1 \end{bmatrix} \begin{bmatrix} \eta_1 \\ \zeta_1 \\ \zeta_2 \end{bmatrix}$

Following [1], it is shown in [2] that, after eliminating Lagrange multiplier, (1)-(2) can be written as

$$\dot{X} = B(X)u \quad (4)$$
$$M_1(X)\dot{u} + C_1(X,\dot{X})u = A_1(X)\tau \quad (5)$$

where (4) is listed in Table 1 for each type of WMR, $M_1(X) = B^T(X)M(X)B(X)$, $C_1(X,\dot{X}) = B^T(X)M(X)\dot{B}(X) + B^T(X)C(X,\dot{X})B(X)$, $A_1(X) = B^T(X)A(X)$. Similarly, (3) can be written as

$$\dot{X}^*(t) = B(X^*(t))u^*(t) \quad (6)$$

where $u^*(t)$ is a known virtual input, (6) is called the virtual reference system. Since (4)-(5) describes the motion of the system (1)-(2), the dynamic tracking problem can be discussed based on (4)-(6).

3 Controller Design

To solve the dynamic tracking problem, the system (4)-(5) are converted into the extended chained form:

$$\begin{cases} \dot{q}_1 = v_1, \quad \dot{q}_{i,j} = v_1 q_{i+1,j} \quad (2 \leq i \leq n_j - 1) \\ \dot{q}_{n_j,j} = v_{1+j} \quad (1 \leq j \leq m) \end{cases} \quad (7)$$
$$M_2(q)\dot{v} + C_2(q,\dot{q})v = A_2(q)\tau \quad (8)$$

By the diffeomorphic state transformation:

$$\begin{cases} q = [q_1, q_{2,1}, \ldots, q_{n_1,1}, \ldots, q_{2,m}, \ldots, q_{n_m,m}]^T \\ \quad = T_1(X) \\ v = [v_1, \ldots, v_{m+1}]^T = T_2^{-1}(X)u \end{cases} \quad (9)$$

where $M_2(q) = g^T(X)M(X)g(X)|_{X=T_1^{-1}(q)}$, $C_2(q,\dot{q}) = [g^T(X)M(X)\frac{d}{dt}g(X) + g^T(X)C(X,\dot{X})\cdot$

Table 2. Canonical Forms of Kinematic Models of restricted mobility robots and Corresponding Transformations

Type	State transformation $q = T_1(X)$	Input transformation $v = T_2^{-1}(X)u$	Canonical form	Singular points
(2,0)	$q_1 = \theta$ $q_{2,1} = x\cos\theta$ $\quad + y\sin\theta$ $q_{3,1} = -x\sin\theta$ $\quad + y\cos\theta$	$v_1 = u_2$ $v_2 = u_1 - (x\cos\theta$ $\quad + y\sin\theta)u_2$	$\dot{q}_1 = v_1$ $\dot{q}_{2,1} = q_{3,1}v_1$ $\dot{q}_{3,1} = v_2$	
(2,1)	$q_1 = \theta + \beta$ $q_{2,1} = -x\cos(\theta+\beta)$ $\quad -y\sin(\theta+\beta)$ $q_{3,1} = x\sin(\theta+\beta)$ $\quad -y\cos(\theta+\beta)$ $q_{2,2} = \beta$	$v_1 = u_2 + u_3$ $v_2 = -u_1 + (u_2$ $\quad +u_3)(x\cos(\theta+\beta)$ $\quad +y\sin(\theta+\beta))$ $v_3 = u_3$	$\dot{q}_1 = v_1$ $\dot{q}_{2,1} = q_{3,1}v_1$ $\dot{q}_{3,1} = v_2$ $\dot{q}_{2,2} = v_3$	$\theta + \beta$ $= 0 (\text{mod}\pi)$
(1,1)	$q_1 = \theta$ $q_{2,1} = x\cos\theta$ $\quad + y\sin\theta$ $q_{3,1} = -x\sin\theta$ $\quad + y\cos\theta$ $q_{4,1} = L\tan\beta$ $\quad -q_{2,1}$	$v_1 = u_1\cos\beta$ $v_2 = u_1\cos\beta(x\sin\theta$ $\quad -y\cos\theta) + \frac{Lu_2}{\cos^2\beta}$	$\dot{q}_1 = v_1$ $\dot{q}_{2,1} = q_{3,1}v_1$ $\dot{q}_{3,1} = q_{4,1}v_1$ $\dot{q}_{4,1} = v_2$	$\beta = \frac{\pi}{2}$ $(\text{mod}\pi)$
(1,2)	$q_1 = \theta$ $q_{2,1} = x\cos\theta$ $\quad + y\sin\theta$ $q_{3,1} = -x\sin\theta$ $\quad + y\cos\theta$ $\quad -2L\frac{\sin\beta_1\sin\beta_2}{\sin(\beta_2-\beta_1)}$ $q_{2,2} = x\sin\theta$ $\quad -y\cos\theta$ $q_{3,2} = x\cos\theta$ $\quad +y\sin\theta$ $\quad -L\frac{\sin(\beta_1+\beta_2)}{\sin(\beta_2-\beta_1)}$	$v_1 = u_1\sin(\beta_2-\beta_1)$ $v_2 = -q_{3,2}u_1\sin(\beta_2$ $\quad -\beta_1) - \frac{2Lu_2\sin^2\beta_2}{\sin^2(\beta_2-\beta_1)}$ $\quad + \frac{2Lu_3\sin^2\beta_1}{\sin^2(\beta_2-\beta_1)}$ $v_3 = q_{3,1}u_1\sin(\beta_2$ $\quad -\beta_1) - \frac{Lu_2\sin(2\beta_2)}{\sin^2(\beta_2-\beta_1)}$ $\quad + \frac{Lu_3\sin(2\beta_1)}{\sin^2(\beta_2-\beta_1)}$	$\dot{q}_1 = v_1$ $\dot{q}_{2,1} = q_{3,1}v_1$ $\dot{q}_{3,1} = v_2$ $\dot{q}_{2,2} = q_{3,2}v_1$ $\dot{q}_{3,2} = v_3$	$\beta_1 = \beta_2$ $(\text{mod}\pi)$ $\beta_1 = 0$ $(\text{mod}\pi)$ $\beta_2 = 0$ $(\text{mod}\pi)$

$g(X)|_{X=T_1^{-1}(q)}$, $A_2(X) = g^T(X)A(X)|_{X=T_1^{-1}(q)}$, and $g(X) := B(X)T_2(X)$, $T_1(X)$ and $T_2(X)$ are given in [15] and listed in Table 2 for each type of WMR.

Remark 1: Canonical form of (2,1) robot is a special case of (7). From Table 2, there are singular points in some transformations. If a robot rests on a singular point at the initial time, a disturbance should be exerted on it such that it leaves the point before applying the transformation.

By the same transformation as (9), i.e.

$$\begin{cases} q^* = [q_1^*, q_{2,1}^*, \ldots, q_{n_1,1}^*, \ldots, q_{2,m}^*, \ldots, \\ \qquad q_{n_m,m}^*]^T = T_1(X^*) \\ v^* = [v_1^*, \ldots, v_{m+1}^*]^T = T_2^{-1}(X^*)u^* \end{cases} \quad (10)$$

the given reference system (6) can be put into

$$\begin{cases} \dot{q}_1^* = v_1^*, \quad \dot{q}_{i,j}^* = v_1^* q_{i+1,j}^* \quad (2 \le i \le n_j - 1) \\ \dot{q}_{n_j,j}^* = v_{1+j}^* \quad (1 \le j \le m) \end{cases} \quad (11)$$

where v^* is a known vector. Additionally, Assumption 1~3 about $X^*(t)$ can be rephrased as follows:

Assumption 4: Q_1^* is bounded, Q_1^* denotes the remainder vector of q^* after element q_1^* is removed.

Assumption 5: v_1^* and $\dot{v}_1^*$ are bounded.

Assumption 6: There exists a time diverging sequence $\{t_i\}_{i\in N}$, and $|t_i - t_{i-1}| \le T_0 < \infty$, such that $\lim_{i\to\infty}\inf|v_1^*(t_i)| = \epsilon > 0$.

With the transformations (9) and (10), it is easy to verify that the dynamic tracking problem is equivalent to find a feedback law τ such that $\lim_{t\to\infty}(q(t) - q^*(t)) = 0$ and $\lim_{t\to\infty}(\dot{q}(t) - \dot{q}^*(t)) = 0$. Let $e = \Psi(q - q^*) := [e_1, e_{2,1}, \ldots, e_{n_1,1}, \ldots, e_{2,m}, \ldots, e_{n_m,m}]^T$, where $\Psi = \text{diag}[\Psi^1, \Psi_1^2, \ldots, \Psi_1^m]$, $\Psi_1^l (2 \le l \le m)$ is the resulting matrix eliminating the first row and the first column of the matrix $\Psi^l = \{\psi_{i,j}^l\} \in R^{n_l \times n_l}$, $\psi_{i,j}^l$ is defined as follows.

$$\psi_{i,i}^l = 1(1 \le i \le n_l), \quad \psi_{i,j}^l = 0(i < j; 1 \le i,j \le n_l)$$
$$\psi_{i,1}^l = 0(2 \le i \le n_l), \quad \psi_{i,j}^l = 0(i \ne j(\text{mod}2))$$
$$\psi_{i,j}^l = k_{i-3,l}\psi_{i-2,j} + \psi_{i-1,j-1}(4 \le i \le n_l; 2 \le j \le n_l)$$

and constants $k_{i,l} > 0(1 \le i \le n_l - 3)$, the following lemma can be proved.

Lemma: Consider the system (7) and a given desired trajectory q^* in (11), under Assumption 4~6, the control law

$$\dot{p} = -\mu_2 p - \mu_1 e_1 - \sum_{l=1}^{m}\sum_{j=2}^{n_l-1}\sum_{i=2}^{j}\frac{e_{j,l}\psi_{j,i}^l q_{i+1,l}}{k_{0,l}k_{1,l}\cdots k_{j-2,l}}$$
$$- \sum_{l=1}^{m}\sum_{j=2}^{n_l-1}\frac{e_{n_l,l}\psi_{n_l,j}^l q_{j+1,l}}{k_{1,l}k_{2,l}\cdots k_{n_l-2,l}} \quad (12)$$

$$v = \begin{bmatrix} v_1^* + p \\ v_2^* - \mu_{3,1}e_{n_1,1} - k_{n_1-2,1}v_1^* e_{n_1-1,1} \\ \quad -v_1^*\sum_{i=2}^{n_1-1}\psi_{n_1,i}^1(q_{i+1,1} - q_{i+1,1}^*) \\ \vdots \\ v_{m+1}^* - \mu_{3,m}e_{n_m,m} - k_{n_m-2,m}v_1^* e_{n_m-1,m} \\ \quad -v_1^*\sum_{i=2}^{n_m-1}\psi_{n_m,i}^m(q_{i+1,m} - q_{i+1,m}^*) \end{bmatrix}$$
$$=: \sigma \quad (13)$$

makes $q(t)$ and $\dot{q}(t)$ asymptotically converge to $q^*(t)$ and $\dot{q}^*(t)$ respectively, where $\mu_1 > 0$, $\mu_2 > 0$, $\mu_{3,l} > 0$, $k_{0,l} = 1$, $k_{n_l-2,l} > 0$ $(1 \le l \le m)$.

Proof: The closed-loop system of (7), (12) and (13) can be written as

$$\dot{e}_1 = p, \quad \dot{e}_{2,l} = v_1^* e_{3,l} + pq_{3,l}$$
$$\dot{e}_{j+3,l} = v_1^*(-k_{j+1,l}e_{j+2,l} + e_{j+4,l}) + p\sum_{i=2}^{j+3}\psi_{j+3,i}^l q_{i+1,l}$$
$$(0 \le j \le n_l - 4; 1 \le l \le m)$$
$$\dot{e}_{n_l,l} = -\mu_{3,l}e_{n_l,l} - k_{n_l-2,l}v_1^* e_{n_l-1,l} + p\sum_{i=2}^{n_l-1}\psi_{n_l,i}^l q_{i+1,l}$$

$$\dot{p} = -\mu_2 p - \mu_1 e_1 - \sum_{l=1}^{m}\sum_{j=2}^{n_l-1}\left[\sum_{i=2}^{j}\frac{e_{j,l}\psi_{j,i}^l q_{i+1,l}}{k_{0,l}k_{1,l}\cdots k_{j-2,l}}\right.$$
$$\left.+\frac{e_{n_l,l}\psi_{n_l,j}^l q_{j+1,l}}{k_{1,l}k_{2,l}\cdots k_{n_l-2,l}}\right]$$

Let $V = 0.5[p^2 + \mu_1 e_1^2 + \sum_{l=1}^{m}\sum_{j=2}^{n_l} e_{j,l}^2/(k_{0,l}k_{1,l}\cdots k_{j-2,l})]$, differentiating V along the closed loop system yields

$$\dot{V} = -\mu_2 p^2 - \sum_{l=1}^{m}\frac{\mu_{3,l}e_{n_l,l}^2}{k_{1,l}k_{2,l}\cdots k_{n_l-2,l}} \leq 0 \quad (14)$$

thus V is non-increasing and has limit $V_{lim} \geq 0$. Noting the expression of V, so p and e are bounded. By Assumption 4~5, $q_{i,l}(3 \leq i \leq n_l; 1 \leq l \leq m)$, $\dot{e}$ and $\dot{p}$ are bounded. Since $\frac{d}{dt}\dot{V} = -2\mu_2 p\dot{p} - \sum_{l=1}^{m}(2\mu_{3,l}e_{n_l,l}\dot{e}_{n_l,l})/(k_{1,l}k_{2,l}\cdots k_{n_l-2,l})$ is bounded, $\dot{V}$ is uniformly continuous. By Barbalat's lemma, $\dot{V} \to 0$, hence p and $e_{n_l,l}(1 \leq l \leq m)$ tend to zero, respectively.

Since v_1^* is bounded, $v_1^{*2}e_{n_l,l}(1 \leq l \leq m)$ tend to zero. Differentiating $v_1^{*2}e_{n_l,l}$ yields $\frac{d}{dt}(v_1^{*2}e_{n_l,l}) = -k_{n_l-2,l}v_1^{*3}e_{n_l-1,l} + [2v_1^*e_{n_l,l}\dot{v}_1^* - \mu_{3,l}v_1^{*2}e_{n_l,l} + v_1^{*2}p\sum_{i=2}^{n_l-1}\psi_{n_l,i}^l q_{i+1,l}]$, where the first term is uniformly continuous, since its derivative $\frac{d}{dt}(-k_{n_l-2,l}v_1^{*3}e_{n_l-1,l}) = -3k_{n_l-2,l}\dot{v}_1^* v_1^{*2}e_{n_l-1,l} - k_{n_l-2,l}v_1^{*3}\dot{e}_{n_l-1,l}$ is bounded. The other terms tend to zero (since v_1^*, $\dot{v}_1^*$, and $q_{i,l}(3 \leq i \leq n_l; 1 \leq l \leq m)$ are bounded, $v_1^*e_{n_l,l}$ and p tend to zero). By Barbalat's lemma, $\frac{d}{dt}(v_1^{*2}e_{n_l,l}) \to 0$, thus $v_1^{*3}e_{n_l-1,l} \to 0$. Furthermore, $v_1^{*2}e_{n_l-1,l}$ and $v_1^*e_{n_l-1,l}$ converge to zero.

Differentiating $v_1^{*2}e_{j,l}(j = n_l-1,\ldots,2)$ and repeating the above procedure, it can be proved $v_1^{*2}e_{j,l}$ and $v_1^*e_{j,l}$ $(j = n_l-1,\ldots,2)$ converge to zero, respectively.

Again v_1^* is bounded and p tends to zero, so $v_1^{*2}p$ converges to zero. Differentiating $v_1^{*2}p$, yields

$$\frac{d}{dt}(v_1^{*2}p) = -\mu_1 v_1^{*2}e_1 + 2v_1^*\dot{v}_1^* p - \mu_2 v_1^{*2}p$$
$$-\sum_{l=1}^{m}\sum_{j=2}^{n_l-1}\left[\sum_{i=2}^{j}\frac{v_1^{*2}e_{j,l}\psi_{j,i}^l q_{i+1,l}}{k_{1,l}k_{2,l}\cdots k_{j-2,l}} + \frac{v_1^{*2}e_{n_l,l}\psi_{n_l,j}^l q_{j+1,l}}{k_{1,l}k_{2,l}\cdots k_{n_l-2,l}}\right]$$

where the first term is uniformly continuous (since its time derivative is bounded), the other terms tend to zero (since v_1^* and $q_{j,l}^*(3 \leq j \leq n_l; \leq l \leq m)$ are bounded, p and $v_1^*e_{j,l}(2 \leq j \leq n_l; 1 \leq l \leq m)$ tend to zero). By Barbalat's lemma, $\frac{d}{dt}(v_1^{*2}p)$ tends to zero, so $v_1^{*2}e_1$ tends to zero. Furthermore $v_1^*e_1$ tends to zero.

Considering $v_1^*e_1$, $v_1^*e_{j,l}(2 \leq j \leq n_l, 1 \leq l \leq m)$ and p tend to zero, so $v_1^{*2}V$ tends to zero. Since V has limit $V_{lim} \geq 0$ and v_1^* does not tend to zero (by Assumption 6), V_{lim} is necessarily equal to zero. Therefore e_1, $e_{j,l}(2 \leq j \leq n_l; 1 \leq l \leq m)$ and p tend to zero. Since Ψ is a nonsingular constant matrix, q and $\dot{q}$ asymptotically converge to q^* and $\dot{q}^*$, respectively. $\diamondsuit$

Remark 2: By the proof, in the Lemma Assumption 6 can be relaxed as: $v_1^* \not\to 0$ as $t \to \infty$. By the inverse state transformation, Assumption 3 can be replaced by: For type $(2,0)$, $(1,1)$ and $(1,2)$ robots, $\dot{\theta}^*(t) \not\to 0$ as $t \to \infty$. For type $(2,1)$ robot, $\dot{\theta}^*(t) + \dot{\beta}^*(t) \not\to 0$ as $t \to \infty$.

With aid of the Lemma and the well-known backstepping technique, the following theorem can be proved.

Theorem: Consider the system (7)-(8) with unknown dynamics and the virtual reference system (11), under Assumption 4~6, the control law (12),

$$\tau = A_2^\#\left[-K_p(v-\sigma) - \frac{\hat{a}\gamma_2^2\chi^2(\sigma,\dot{\sigma})(v-\sigma)}{\gamma_2\chi(\sigma,\dot{\sigma})\|v-\sigma\| + \gamma(t)} - \Lambda\right] \quad (15)$$

and the adaptive law

$$\dot{\hat{a}} = \frac{\gamma_1\gamma_2^2\chi^2(\sigma,\dot{\sigma})\|v-\sigma\|^2}{\gamma_2\chi(\sigma,\dot{\sigma})\|v-\sigma\| + \gamma(t)} \quad (16)$$

make $q(t)$ and $\dot{q}(t)$ asymptotically converge to $q^*(t)$ and $\dot{q}^*(t)$ respectively, and $\hat{a}$ is bounded, where $\#$ is any left inverse, K_p is a positive matrix, constants $\gamma_2 \geq 1$ and $\gamma_1 > 0$, $\gamma(t) > 0$ and such that

$$\int_0^\infty \gamma(t)dt = d_1 < \infty \quad (17)$$

σ is defined in Lemma and $\Lambda = [\Lambda_1,\ldots,\Lambda_{m+1}]^T$

$$\Lambda = \begin{bmatrix} \mu_1 e_1 + \sum_{l=1}^{m}\sum_{j=2}^{n_l-1}\left[\sum_{i=2}^{j}\frac{e_{j,l}\psi_{j,i}^l q_{i+1,l}}{k_{0,l}k_{1,l}\cdots k_{j-2,l}}\right. \\ \left.+\frac{e_{n_l,l}\psi_{n_l,j}^l q_{j+1,l}}{k_{1,l}k_{2,l}\cdots k_{n_l-2,l}}\right] \\ \frac{e_{n_1,1}}{k_{1,1}k_{2,1}\cdots k_{n_1-2,1}} \\ \vdots \\ \frac{e_{n_m,m}}{k_{1,m}k_{2,m}\cdots k_{n_m-2,m}} \end{bmatrix}$$

$$\chi(\sigma,\dot{\sigma}) := \|g(T_1^{-1}(q))\| \cdot \|[\tfrac{d}{dt}(g(T_1^{-1}(q))\sigma)\| + (1 + \|\tfrac{d}{dt}T_1^{-1}(q)\|) \cdot \|g(T_1^{-1}(q))\sigma\|$$

where $\mu_1 > 0$, $\mu_2 > 0$, $\mu_{3,j} > 0$, $k_{0,l} = 1$, $k_{n_l-2,l} > 0$ $(1 \leq l \leq m)$.

Proof: Let $w = [w_1, w_2]^T = v - \sigma$, $a = \max\{c_1, c_2, c_3\}/\gamma_2$, $\tilde{a} = \hat{a} - a$, the closed-loop system of (7), (8), (12), (15) and (16) can be written as

$$
\begin{cases}
\dot{e}_1 = p + w_1, \quad \dot{e}_{2,l} = v_1^* e_{3,l} + (p + w_1)q_{3,l} \\
\dot{e}_{j+3,l} = v_1^*(-k_{j+1,l}e_{j+2,l} + e_{j+4,l}) + (p \\
\quad + w_1)\sum_{i=2}^{j+3}\psi_{j+3,i}^l q_{i+1,l}, (0 \leq j \leq n_l - 4; 1 \leq l \leq m) \\
\dot{e}_{n_l,l} = -\mu_{3,l}e_{n_l,l} - k_{n_l-2,l}v_1^* e_{n_l-1,l} + w_2 + (p \\
\quad + w_1)\sum_{i=2}^{n_l-1}\psi_{n_l,i}^l q_{i+1,l} \\
M_2(q)\dot{w} = -C_2(q,\dot{q})w - K_p w - \Lambda - \Phi(\sigma,\dot{\sigma}) \\
\quad - \dfrac{\hat{a}\gamma_2^2 \chi^2(\sigma,\dot{\sigma})w}{\gamma_2\chi(\sigma,\dot{\sigma})\|w\| + \gamma} \\
\dot{\hat{a}} = \dfrac{\gamma_1\gamma_2^2 \chi^2(\sigma,\dot{\sigma})\|w\|^2}{\gamma_2\chi(\sigma,\dot{\sigma})\|w\| + \gamma}, \quad \dot{p} = -\mu_2 p - \Lambda_1
\end{cases}
$$
(18)

where $\Phi(\sigma,\dot{\sigma}) = M_2(q)\dot{\sigma} + C_2(q,\dot{q})\sigma$. Let

$$V = \frac{1}{2}[p^2 + \mu_1 e_1^2 + \sum_{l=1}^{m}\sum_{j=2}^{n_l}\frac{e_{j,l}^2}{k_{0,l}k_{1,l}\cdots k_{j-2,l}} + w^T M_2 w + \tilde{a}^2/\gamma_1]$$

Differentiating V along (18) yields $\dot{V} = -\mu_2 p^2 - \sum_{l=1}^{m}(\mu_{3,l}e_{n_l,l}^2)/(k_{1,l}k_{2,l}\cdots k_{n_l-2,l}) - w^T K_p w + R$, where

$$
\begin{aligned}
R &= -w^T \Phi - \frac{\hat{a}\gamma_2^2 \chi^2\|w\|^2}{\gamma_2\chi\|w\| + \gamma} + \frac{\tilde{a}\gamma_2^2 \chi^2\|w\|^2}{\gamma_2\chi\|w\| + \gamma} \\
&\leq a\gamma_2\chi\|w\| - \frac{a\gamma_2^2 \chi^2\|w\|^2}{\gamma_2\chi\|w\| + \gamma} = \frac{a\gamma_2\chi\gamma\|\tilde{v}\|}{\gamma_2\chi\|\tilde{v}\| + \gamma} \leq a\gamma
\end{aligned}
$$

therefore

$$\dot{V} \leq -\mu_2 p^2 - \sum_{l=1}^{m}\frac{\mu_{3,l}e_{n_l,l}^2}{k_{1,l}k_{2,l}\cdots k_{n_l-2,l}} - w^T K_p w + a\gamma$$

Since γ satisfies (17), integrating both sides of the above inequation gives

$$V(t) - V(0) = \int_0^t \left[-\mu_2 p^2(s) - w^T(s)K_p w(s) + a\gamma(s) - \sum_{l=1}^{m}\frac{\mu_{3,l}e_{n_l,l}^2(s)}{k_{1,l}k_{2,l}\cdots k_{n_l-2,l}}\right]ds \leq ad_1 \quad (19)$$

thus V is bounded, which implies that $p \in L_\infty$, $e \in L_\infty$, $w \in L_\infty$, and $\tilde{a} \in L_\infty$. By Assumption 4-5, $q_{i,l}(3 \leq i \leq n_l; 1 \leq l \leq m)$, $\dot{e}$, and $\dot{p}$ are bounded. So M_2, C_2, and Λ are bounded, furthermore $\dot{w}$ is bounded. From (19),

$$\int_0^t \left[-\mu_2 p^2(s) - \sum_{l=1}^{m}\frac{\mu_{3,l}e_{n_l,l}^2(s)}{k_{1,l}k_{2,l}\cdots k_{n_l-2,l}} - w^T(s)K_p w(s)\right]ds \leq V(0) - V(t) + ad_1$$

so $p \in L_2$, $e_{n_l,l} \in L_2(1 \leq l \leq m)$, and $w \in L_2$. Therefore, $p \to 0$, $e_{n_l,l} \to 0(1 \leq l \leq m)$, and $w \to 0$ as $t \to \infty$, respectively.

Mimicking Proof of the Lemma, by differentiating $v_1^{*2}e_{i,l}(i = n_l, n_l - 1, \ldots, 2; 1 \leq l \leq m)$ and $v_1^{*2}p$, we can prove $v_1^* e_{i,l}(i = n_l - 1, n_l - 2, \ldots, 2; 1 \leq l \leq m)$ and $v_1^* e_1$ tend to zero step by step. With respect to (18), $\dot{e}_1$ and $\dot{e}_{i,l}(2 \leq i \leq n_l; 1 \leq l \leq m)$ tend to zero, therefore $\dot{q}$ asymptotically converges to $\dot{q}^*$. Noting Assumption 6, the sequence $\{e_{i,l}(t_j)\}_{j\in N}$ and $\{e_1(t_j)\}_{j\in N}$ tend to zero respectively. Using Taylor expansion, $e_{i,l}(t) = e_{i,l}(t_s) + (t - t_s)\dot{e}_{i,l}(t')$, where $|t - t_s| \leq T_0$, t' is some time between t and t'. When $t \to \infty$, then $t_s \to \infty$, $e_{i,l}(t) \to 0$, and $\dot{e}_{i,l}(t') \to 0$. Therefore, $e_{i,l}(t) \to 0$ as $t \to \infty$. Similarly, we can prove e_1 tends to zero. Thus q asymptotically converges to q^*. Additionally, $\hat{a}$ being bounded is guaranteed by boundedness of $\tilde{a}$. ◇

Remark 3: In the control law, $\gamma(t)$ may be $1/(1+t)^{d_2}(d_2 \geq 2)$, $e^{-d_2 t}(d_3 > 0)$, or anything else which satisfies (17). By inverse transformation, the controller for the original system can be easily obtained, so it is omitted here.

4 Simulation

Consider the tracking problem of type $(2,0)$ robot moving on a horizontal plane whose dynamic model equation is described in [8], given a desired trajectory $X^* = [X_1^*, X_2^*, X_3^*]^T$, where $X_1^* = \cos t$, $X_2^* = \sin t$, X_3^* is determined by the nonholonomic constraint $\dot{X}_1^* \cos X_3^* + \dot{X}_2^* \sin X_3^* = 0$. Following Section 2, the control law can be easily derived step by step, due to space limit, it is omitted here.

In order to simulate, suppose in (2) $M(X) = \text{diag}[5,5]$ and $C(X,\dot{X}) = 0$. In the simulation, let $[x(0), y(0), \theta(0)] = [1.2, -0.3, 0.2]$, $[\dot{x}(0), \dot{y}(0), \dot{\theta}(0)] = [-0.002, 0.01, 0.1]$, $p(0) = 0.5$, $\hat{a}(0) = 0$. Select $\mu_1 = 12.25$, $\mu_2 = 7$, $\mu_{3,1} = 6$, $k_1 = 5$, $K_p = \text{diag}(50, 50)$ and $\gamma = 1/(1+t)^2$, $\gamma_1 = 1$, $\gamma_2 = 1$ in the feedback law and the adaptive law. Simulation results are depicted in Figure 1-2 respectively.

Acknowledgments

This paper was supported by the National Science Foundation of China.

References

[1] A. M. Bloch, R. Reyhanoglu, and N. H. McClamroch, "Control and stabilization of nonholonomic dynamic systems," *IEEE Trans. on Automat. Contr.*, Vol.37, pp.1746-1757, 1992.

[2] G. Campion, G.Bastin, and B. d'Andrea-Novel, "Structure properties and classification of kinematic and dynamic models of wheeled mobile robots," *IEEE Trans. Robotics and Automation*, Vol.12, pp.47-62, 1996.

[3] C. Canudas de Wit, H. Khennouf, C. Sammson, and O. J. Sordalen, "Nonlinear control design for mobile robots," In Y. F. Zheng (Ed.), *Recent Trends in Mobile Robots*, World Scientific, 1993.

[4] Y. C. Chang and B. S. Chen, "Adaptive tracking control design of nonholonomic mechanical systems," *Proc. of the IEEE Conf. on Decision and Control*, pp.4379-4744, 1996.

[5] B. S. Chen, T. S. Lee, and W. S. Chang, "A robust H^∞ model reference tracking design for non-holonomic mechanical control systems," *Int. J. Control*, Vol.63, pp.283-306, 1996.

[6] J. J. Craig, *Adaptive Control of Mechanical Manipulators*, New York, Addison-wesley, 1988.

[7] B. d'Andrea-Novel, G. Campion, and G. Bastin, "Control of nonholonomic wheeled mobile robots by state feedback linearization," *Int. J. Robotics Research*, Vol.14, pp.543-559, 1995.

[8] Wenjie Dong and Wei Huo, "Adaptive Stabilization of Dynamic Nonholonomic Chained Systems with Uncertainty", *Proc. of the IEEE Conf. on Decision and Control*, pp.2362-2367, 1997.

[9] M. Fliess, J. Levine, P. Martin, and P. Rouchon, "Flatness and defect of nonlinear systems: introductory theory and examples," *Int. J. Control*, Vol.61, pp.1327-1361, 1995.

[10] M. Fliess, J. Levine, P. Martin, and P. Rouchon, "Design of trajectory stabilizing feedback for driftless flat systems," *Proc. European Control Conf.*, pp.1882-1887, 1995.

[11] Z.-P. Jiang and H. Nijimeijer, "Tracking control of mobile robots: a case study in backstepping," *Automatica*, Vol.33, pp.1393-1399, 1997.

[12] Z.-P. Jiang, and N. Nijmeijer, "Backstepping-based tracking control of nonholonomic chained systems" *Proc. European Control Conf.*, 1997.

[13] Y. Kanayama, Y. Kimura, F. Miyazaki, and T. Noguchi, "A stable tracking control method for an autonomous mobile robot," *IEEE Proc. of the IEEE Conf. on Robotics and Auotomation*, pp.384-389, 1990.

[14] I. Kolmanovsky and N. H. McClamroch, Developments in nonholonomic control problem. *IEEE Contr. Syst.*, Dec., pp.20-36, 1995.

[15] W. Leroquais and B. d'Andrea-Novel, "Transformation of the kinematic models of restricted mobility wheeled mobile robots with a single platform into chained forms," *Proc. of the IEEE Conf. Decision and Control*, pp.3811-3816, 1995.

[16] A. Micaelli and G. Samson, "Trajectory tracking for two-steering-wheels mobile robots," *Proc. Symp. Robot Control'94,* Capri, 1994.

[17] C. Y. Su, and Y. Stepanenko, "Robust motionforce control of mechanical systems with classical nonholonomic constraints," *IEEE Trans. Automat. Contr.*, Vol.39, pp.609-614, 1994.

[18] G. Walsh, D. Tilbury, S. S. Sastry, R. M. Murray, and J. P. Laumond, "Stabilization of trajectories for systems with nonholonomic constraints," *IEEE Trans. Automat. Contr.*, Vol.39, pp.216-222, 1994.

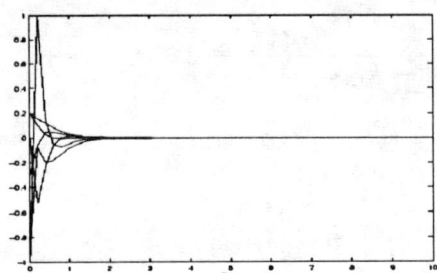

Figure 1. Response of $X - X^*$ and $\dot{X} - \dot{X}^*$.

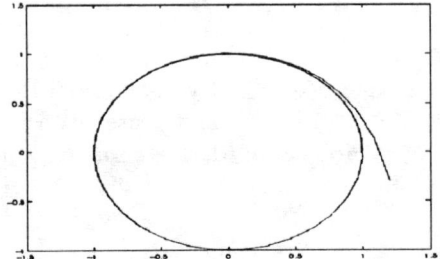

Figure 2. Geometric trajectory of x^*-y^* and x-y

Unifying Exploration, Localization, Navigation and Planning Through a Common Representation*

Alan C. Schultz, William Adams, Brian Yamauchi, and Mike Jones

Navy Center for Applied Research in Artificial Intelligence
Naval Research Laboratory, Washington, DC 20375-5337, U.S.A.
schultz@aic.nrl.navy.mil

Abstract

The major themes of our research include the creation of mobile robot systems that are robust and adaptive in rapidly changing environments and the view of integration as a basic research issue. Where reasonable, we try to use the same representations to allow different components to work more readily together and to allow better and more natural integration of and communication between these components. In this paper, we describe our most recent work in integrating mobile robot exploration, localization, navigation, and planning through the use of a common representation, evidence grids.

1 Introduction

A central theme of our research is the view of integration as a basic research issue, studying the combination of different, complementary capabilities. One principle that allows integration is the use of unifying representations. Where reasonable, we try to use the same representations to allow different components to work more readily together and to allow better and more natural integration of and communication between these components. In the work reported here, the unifying representation is the evidence grid, a probabilistic metric map. In this paper, we describe how using evidence grids as a unifying representation not only allows for better integration across techniques, but also allows reuse of data in learning and adaptation.

We have developed and integrated techniques for autonomous exploration, map building, and continuous self-localization. Further, we have integrated these techniques with methods for navigation and planning. In addition, this integrated system includes methods

*This work was sponsored by the Office of Naval Research.

for adapting maps to allow for robust navigation in dynamic environments. As a result, a robot can enter an unknown environment, map it while remaining confident of its position, and robustly plan and navigate within the environment in real time.

In the next section, we describe the common representation we use for integrating the various techniques. In Section 3, we review our previous results in localization and exploration, along with our integration of these techniques and mechanisms to make them adaptive to changes in the environment. In Sections 4 and 5 we introduce the new components for reactive navigation and planning, and show how they integrate into the system using our representation. In the remaining sections, we describe experiments to verify that the resulting system works robustly and repeatably, and present the results of these experiments.

2 Unifying Representation

We use evidence grids [7] as our spatial representation. An evidence grid is a probabilistic representation which uses Cartesian grid cells to store evidence that the corresponding region in space is occupied. Evidence grids have the advantage of being able to fuse information from different types of sensors. To update an evidence grid with new sensor readings, the sensor readings are interpreted with respect to a sensor model that maps the sensor datum at a given pose to its effect on each cell within the evidence grid [1] The interpretation is then used to update the evidence in the grid cells using a probabilistic update rule. Evidence grids have been created that use different updating methods, most notably, Bayesian [7], and Dempster-Shafer [4]. In the results reported here, Bayesian updating is used.

---
[1]These sensor models may be learned or may be explicitly modeled. Our results use a simple, untuned, explicit model.

In this study, we use sonar sensors in combination with a planar structured light rangefinder. In order to reduce the effect of specular reflections, we have developed a technique we call laser-limited sonar. If the laser returns a range reading less than the sonar reading, we update the evidence grid as if the sonar had returned the range indicated by the laser, in addition to marking the cells actually returned by the laser as occupied.

We create two types of representations with the evidence grids: short-term perception maps, and long-term metric maps. The short-term maps store very recent sensor data that does not contain significant odometry error, and these maps can be used for obstacle avoidance and for localization. The long-term maps are used to represent the environment over time, and can be used for navigation and path-planning.

3 Previous Results in Exploration and Localization

3.1 Learning Where You Are

Evidence grids provide a uniform representation for fusing temporally and spatially distinct sensor readings. However, the use of evidence grids requires that the robot be localized within its environment. Due to odometric drift and non-systematic errors such as slippage and uneven floors, odometry errors typically accumulate over time making localization estimates degrade. This can introduce significant errors into evidence grids as they are built. We have addressed this problem by developing a method for *continuous localization*, in which the robot corrects its position estimates incrementally and on the fly [9].

Continuous localization builds short-term perception maps of the robot's local environment. These maps typically contain very small amounts of error, and are used to locate the robot within a global, long-term map via a registration process. (In the next section we will describe how these long-term maps are created.) The results from this process are used to correct the robot's odometry.

Fig. 1 shows the process of continuous localization. The robot builds a continuous series of short-term perception maps of its immediate environment, each of which is of brief duration and contains only a small amount of dead reckoning error. After several time intervals, the oldest (most "mature") short-term map is used to position the robot within the long-term map by registering the two maps.

The registration process consists of sampling the

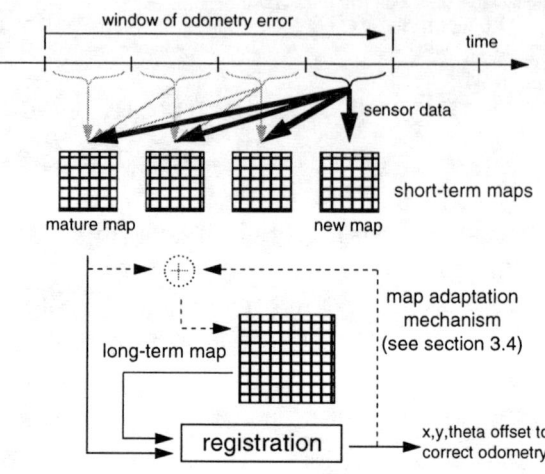

Figure 1: Continuous Localization

possible poses within a small area around the robot's current estimated pose. For each tested pose, the mature short-term map is rotated and translated by the difference in pose (the offset) and a match score is calculated based on agreement between the cell values of the short-term map and the long-term map, summed across all cells. The match scores for all tested poses are then used to determine the offset that is likely to have the highest match score. This offset is applied to the robot's odometry, placing it at the pose which causes its local perceptions to best match the long-term map. After the registration takes place the most mature map is discarded, and a new short-term perception map is created. See [2] and [9] for more details, experimental results, and comparisons with other techniques.

3.2 Learning New Environments

In order for mobile robots to operate in unknown environments, they need the ability to explore and build maps that can be used for navigation. We have developed an exploration strategy based on the concept of frontiers, regions on the boundary between open space and unexplored space. When a robot moves to a frontier, it can see into unexplored space and add the new information to its map. As a result, the mapped territory expands, pushing back the boundary between the known and the unknown. By moving to successive frontiers, the robot can constantly increase its knowledge of the world. We call this strategy *frontier-based exploration*[11].

A process analogous to edge detection and region extraction in computer vision is used to find the boundaries between open space and unknown space in the

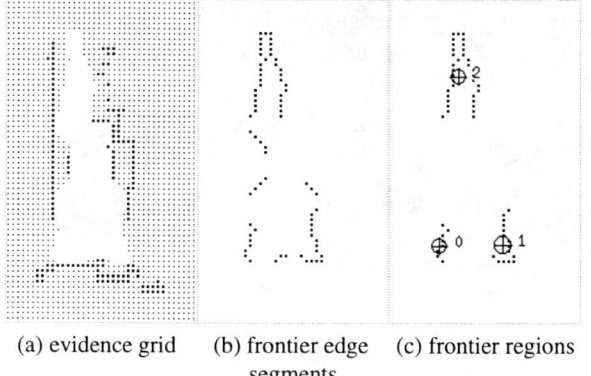

(a) evidence grid (b) frontier edge segments (c) frontier regions

Figure 2: Frontier detection

3.3 Integrated Exploration and Localization

Frontier-based exploration provides a way to explore and map an unknown environment, given that a robot knows its own location at all times. Continuous localization provides a way for a robot to maintain an accurate estimate of its own position, as long as the environment is mapped in advance. The question of how to combine exploration with localization raises a "chicken-and-egg" problem: the robot needs to know its position in order to build a map, and the robot needs a map in order to determine its position. By integrating continuous localization and frontier-based exploration, we can solve this problem, allowing the robot to explore and build a map while maintaining an accurate estimate of its position [13].

This works because the exploration strategy will only take the robot as far as the edge of its "known world," such that about half of its sensors can still see the old, known environment, which can be used to localize, while its other sensors are building up the map in the unknown environment. Frontier-based exploration and continuous localization run in parallel. Whenever the robot arrives at a new frontier, it adds to the map of the environment and passes this map to continuous localization. Continuous localization uses this map of the known world as its long-term map. As the robot navigates to the next frontier, continuous localization constructs short-term maps based on the robot's recent perceptions, and compares them to the long-term map to correct the robot's position estimate. When the robot arrives at the new frontier, its position estimate will be accurate, and new sensor information will be integrated at the correct location within the map.

While other systems have been developed for mobile robot exploration, they have been limited to constrained environments, e.g. where all walls are either parallel or perpendicular to each other [5], [10] or where the entire environment can be explored using wall-following [6]. Our system differs in being able to explore unstructured environments where walls and obstacles may be in any orientation.

3.4 Dynamic Environments: Adaptive Long-Term Maps

We are also interested in explicitly modeling changes that occur in the world after the robot has finished exploration. This is useful for learning and representing changes in the environment. We have extended the continuous localization algorithm to allow the long-term map to be updated with recent sensor data from

evidence grid. Any open cell adjacent to an unknown cell is labeled a frontier edge cell. Adjacent edge cells are grouped into frontier regions. Any frontier region above a certain minimum size (roughly the size of the robot) is considered a frontier. Fig.2a shows an evidence grid built by a real robot in a hallway adjacent to two open doors. Fig.2b shows the frontier edge segments detected in the grid. Fig.2c shows the regions that are larger than the minimum frontier size. The centroid of each region is marked by crosshairs. Frontier 0 and frontier 1 correspond to open doorways, while frontier 2 is the unexplored hallway.

Once frontiers have been detected within a particular evidence grid, the robot attempts to navigate to the nearest accessible, unvisited frontier. When the robot reaches its destination (or if the navigation routine determines that the robot cannot get to the frontier), it performs a sensor sweep using laser-limited sonar, and adds the new information to the evidence grid. The robot then detects frontiers in the updated grid, and navigates to the nearest remaining accessible, unvisited frontier.

We have demonstrated that frontier-based exploration can successfully map real-world office environments [11], and that this technique scales well for use in multi-robot environments [12]. In relatively small environments, such as a single office, frontier-based exploration was capable of mapping accurately using dead reckoning for position estimation. However, for larger environments, dead reckoning errors would generate large errors in the generated maps. In the next section, we show how we integrated continuous localization and frontier-based exploration.

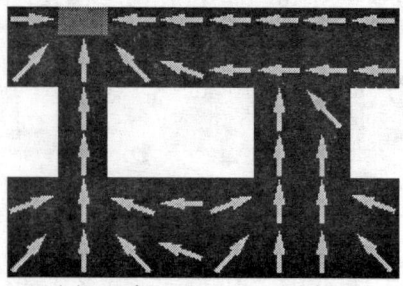  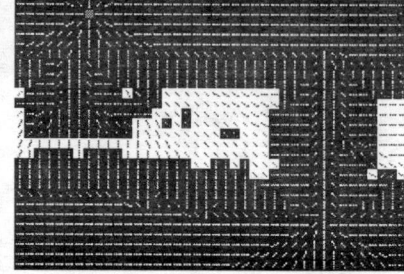

(a) native representation (b) paths using long term map (c) paths after adaptation

Figure 3: Paths generated by Trulla

the short-term perception maps, making the long-term map adaptive to the environment [2]. After the most mature short-term map is used to correct the robot's dead reckoning, the odometry correction from the continuous localization process is also applied to the short-term perception map, and then its cells are combined with the corresponding cells of the long-term map using Bayesian updating (dashed lines in Fig. 1). The cells are weighted by a learning rate that controls the effect the short-term map has on the long-term map.

Previous results demonstrated that this allows the robot to recognize and respond to changes in the environment without introducing excessive errors in odometry [2]. In the next section we will describe how this capability can be used in conjunction with planning to allow adaptation to changing environments.

4 Planning: Trulla

While previous versions of our system used a simple path planner, we have extended our system to use Trulla, a propagation-based path planner [3]. Trulla uses a navigability grid to describe which areas in the environment are navigable (considering floor properties, obstacles, etc). In order to integrate Trulla into our system, we note that Trulla's notion of a navigability grid is similar to our long-term metric map.

Trulla works as follows: beginning from the cell containing the goal, the neighboring cells are explored outward, and each is assigned its own subgoal. Each newly tested cell is assigned the closest subgoal of its already-tested neighbors, if that subgoal is visible from the new cell. If none of the neighbors' subgoals are visible, then the new cell lies around the corner of an obstacle, and the neighbor with the closest subgoal is itself assigned as the subgoal of the new cell. In this manner, the shortest paths to the goal are propagated out to all cells. Since each cell can only point to a closer subgoal, the paths that Trulla produces do not suffer from local minima. Once the subgoals are determined, each cell is assigned the direction to its subgoal, resulting in a field of vectors that point in the direction of the shortest path to the goal. See [3] for more details on Trulla.

We have replaced Trulla's navigability grid with our long-term map – cell occupancy probabilities are mapped to navigability values. As our long-term map adapts to changes in the environment, as described in Section 3.4, Trulla can update its paths to reflect the robot's current knowledge about the world. Trulla is capable of replanning quickly, and we have reached speeds in excess of one hertz.

Fig. 3a shows an example of a native Trulla navigability grid and the vectors to get from any grid cell to the goal, located in the upper, left-hand corner. Fig. 3b shows the the same area as represented by the long-term map. Fig. 3c shows the vectors produced for the same goal after a change has occured to the environment and the long-term map has been updated by continuous localization.

Although the long-term map can adapt to somewhat rapid and persistent changes in the environment, very fast changes, such as a person walking through the room, will not appear in the long-term map. Paths generated by Trulla will avoid persistent obstacles but are not sufficient to prevent collisions with transient obstacles. In related work, Trulla has previously been combined with reactive navigation to avoid collisions with unmodeled obstacles [8]. In the work reported here, Trulla is combined with Vector Field Histogram navigation to avoid transient obstacles and to perform reactive navigation.

5 Reactive Navigation: VFH

Vector Field Histogram (VFH) is a reactive navigation method which uses recent, local sensor perception to drive a robot towards a specified goal [1]. It was chosen over other methods because of its performance and similar representation of the environment, making integration easier. VFH uses the Histogrammic In-Motion Mapping (HIMM) method to construct an occupancy grid from sensor readings filtered through a simple sensor model. The area of the HIMM grid immediately surrounding the robot is divided into arcs, and for each arc an object density is computed as the weighted sum of the occupancy values of the grid cells contained by the arc. Given a goal, VFH searches for the contiguous set of arcs with sufficiently low object density which best matches the direction to the goal. Because the method models the robot as a point object, the free path cannot be blindly followed – the robot's body would collide with the edges and corners of obstacles. To compensate for this assumption, the HIMM grid is also used to compute a potential field. The resulting repulsion vector is added to the vector from the chosen set of arcs to provide a force away from nearby obstacles while generally heading in the chosen direction. The robot is steered in the direction of this summed heading vector.

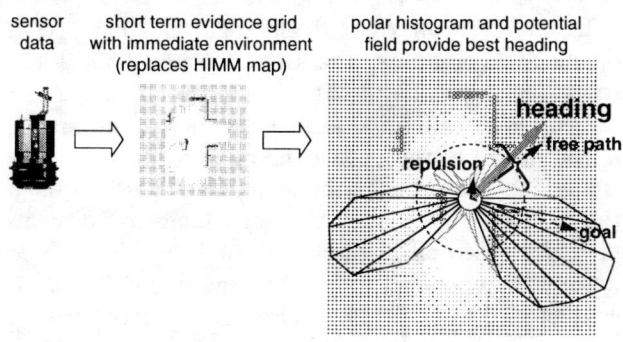

Figure 4: Integration of Vector Field Histogram

In our integration, illustrated in Fig. 4, we replace the HIMM occupancy grid with the short-term perception map produced by continuous localization. The short-term perception map allows VFH to consider all sensors, and yields a more consistent and less noisy picture of the robot's immediate environment.

6 Integrated Architecture

Fig. 5 illustrates the complete architecture. When heading into an unknown environment, the robot autonomously maps the environment, producing the initial long-term map [2]. Continuous localization runs in parallel, regularly correcting the odometry of the robot. While continuous localization maintains the robot's odometry, it regularly produces the short-term perception maps and updates the long-term map, both of which are sent to a separate Map Server process. The Map Server allows the sensor-fused perceptions of the immediate environment to be shared among the various processes, reducing the sensor bottleneck and replicated sensor data gathering and fusion code.

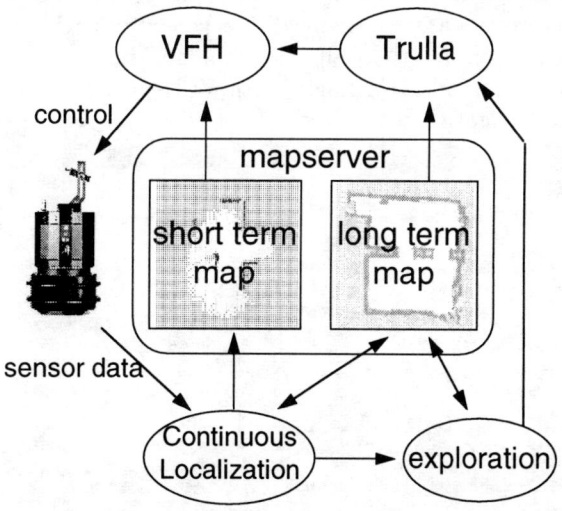

Figure 5: Architecture of integrated system

The user (or possibly some other high-level process) specifies a navigation goal to Trulla, which consults the Map Server for the current long-term map and computes the vector field describing the best path from each cell to the goal. Trulla sends the vector field to VFH, which uses the robot's current position to index the vector field and get the direction to the goal. VFH then retrieves the short-term map from the Map Server, computes the object density and potential field, and steers the robot. VFH repeats this seqeunce until the goal is reached.

While VFH is steering the robot, continuous localization continues to correct odometry and produce short-term and adapted long-term maps. When a new

---
[2] We are currently extending the system to recognize previously explored environments in which case the map is simply retrieved.

long-term map is available Trulla replans and sends the new vector field to VFH. When new vector fields or a new short-term map is available, VFH uses them to reactive navigate along the current path to the goal.

7 Experiment Design

To demonstrate the capability of our integrated system to plan and navigate reliably in environments with unexpected changes, we conducted four experiments with a Nomad200 mobile robot. All four experiments used an environment characterized by a wall between two rooms with one or two passages in which the robot could move between rooms. The robot was required to navigate from one room to the other starting with a long-term map learned through exploration. One of the passages was then changed (blocked or unblocked), requiring continuous localization to adapt the map and Trulla to replan accordingly, with VFH providing reactive navigation.

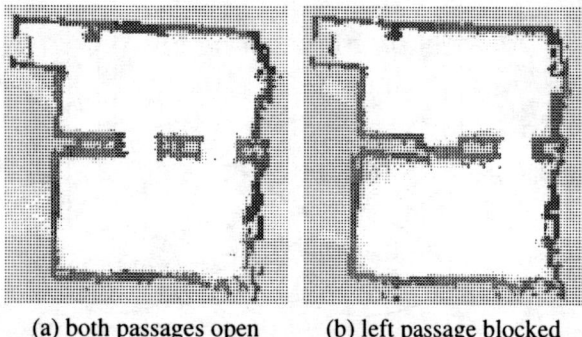

(a) both passages open   (b) left passage blocked

Figure 6: Initial room maps

8 Results

In the first experiment, the system was given the long-term map shown in Fig. 6a, with both passages open. However, the left passage was physically blocked as shown in Fig. 6b. This was the "unexpected blockage" configuration. In the second experiment, the robot was given the long-term map from Fig. 6b, which showed the left passage blocked, but the environment was actually configured as shown in Fig. 6a, with both passages open. This was the "unexpected opening" configuration.

Each experiment was repeated using learning rates of 0.1 and 0.5. Ten runs were performed for each experiment, with varying start and goal locations chosen near the left side of the environment to ensure the robot would have an opportunity to sense the changes.

We expected that the higher learning rate would yield faster adaptation and replanning and more fuzziness or blurring around the edges of the map, while the lower learning rate would take longer to adapt but cause less blurring of the map edges. The match of the left passage area of the adapted map with the a priori map for the actual configuration (how well it learned the change) was expected to be roughly the same with either learning rate.

For the unexpected blockage experiment, the robot, as expected, planned a path through the left opening, which its map indicated was open. Approaching the blockage, VFH detected and tried to navigate around the blockage. Continuous localization accumulated evidence of the blockage and updated the long term map. When the long-term map sufficiently represented the blockage, Trulla replanned its next path through the right passage, which VFH then followed to the goal. The run ended when the robot reached the goal. For the unexpected opening experiment, the robot planned a path through the right passage according to its map, unaware of the shortcut. As the robot passed by the closer opening on its way to the planned passage, sensor readings showing that the left passage was in fact open were obtained as chance permitted, and the long-term map updated. After one or more traversals past the opening, the long-term map indicated the left passage was open and Trulla planned a path through it as the shorter route.

In both the unexpected blockage and unexpected opening experiments, the runs continued until the robot actually traversed the unexpected opening. In the two unexpected blockage experiments, the change is considered learned when the planned paths change enough to cause the robot to follow a path through the right passage, even if the left passage is not completely blocked off in the long-term map. In the two unexpected opening experiments, the change is considered learned when Trulla can first plan a path through the opening in the current direction of travel which has a significant effect on the overall vector field, even if the robot's current position at that time causes it to instead follow a path through the right passage.

All runs were completed without collisions. During one run of the unexpected opening experiment with learning rate 0.1, the robot's odometry was corrupted (due to a communication error) and the robot was unable to complete the run. All results for that experiment are based on the nine successful runs.

Fig. 7 shows the effectiveness of learning in terms

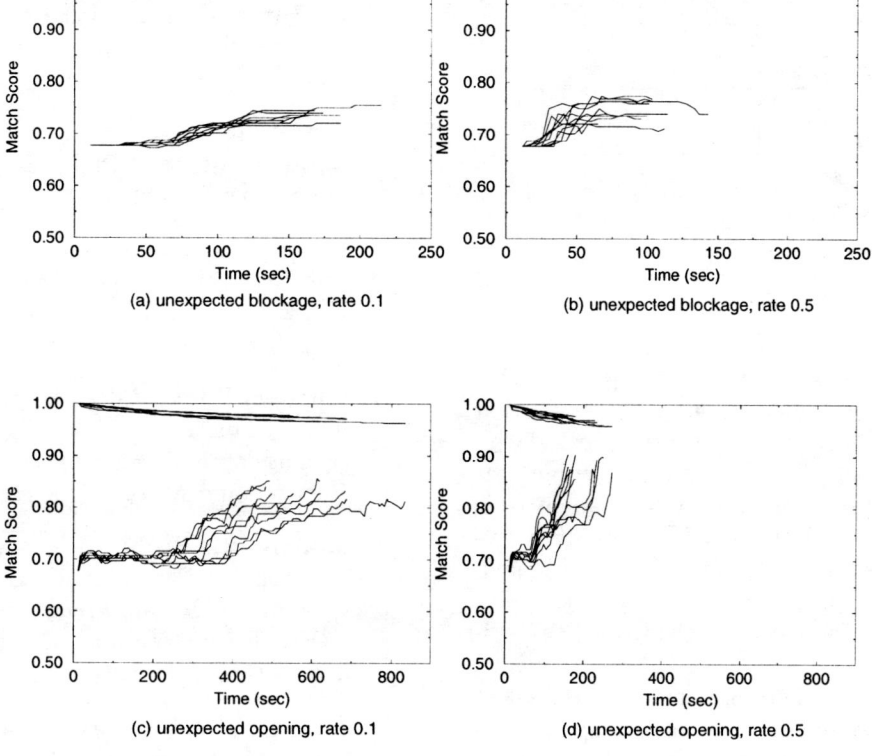

Figure 7: Effect of learning on long-term maps

of the match between the learned maps and the actual environment configuration as represented by the initial maps. Values shown are the percentage of cells in agreement – occupied, empty, or unkown. The average time to learn the change in the environment (as defined above) and the average error in the robot's pose (periodically measured during each run) are shown in Table 1.

The lower set of lines in each graph illustrates the percentage of matching cells in the local area around the left passage between the adapted map and the initial long-term map which included the change. Initially there is a low match because the robot started with a map that did not match the environment, but the match improves over time as the long-term map adapts to the true state of the environment. Although the match score would ideally rise to 100 percent, it does not because of blurring and incomplete learning. The blockage is incompletely learned because the robot can only see the front until it replans through the alternate opening and passes to the rear of the blockage. The upper set of lines in each graph shows the match between the remainder of the adapted map and the initial long-term map. Before learning has had any effect the match is perfect, but over time the edges blur from the inaccuracies in pose.

As shown in Table 1, for a given learning rate, learning the blocked passage case was faster than learning in the unexpected opening case because the robot could gather a lot of sensor data while VFH was trying to navigate the blocked passage prior to the replanning. Learning that the passage was open took longer because it was dependent on getting occasional readings of the area while the robot followed its path through the other passage. As expected, the learning rate had a significant effect on the ability to quickly adapt to changes. A higher learning rate results in a faster ability to learn the changes in the environment. In addition, there are no significant differences in the pose error as corrected by continuous localization.

By examining the differences across the 10 runs for each of the four experiments, we could examine the ability of the system to perform reliable and repeatably. As can be seen within each graph in Fig. 7, the difference among the runs was very small. The shape of the curves is almost identical, with the main difference being in the length of time required to notice the difference.

		Learning Rate	
		0.1	0.5
Unexpected Blockage	avg time:	123 sec	46 sec
	avg pose error:	10.3 in	10.3 in
Unexpected Opening	avg time:	493 sec	120 sec
	avg pose error:	7.8 in	6.4 in

Table 1: Effects of learning rate: summary

9 Conclusion

We have created a system where a robot can enter a previously unknown indoor environment, map that environment while maintaining accurate position information, and robustly plan and navigate within that environment. The system is designed to be adaptive to rapid changes in the environment. Using a unified representation for localization, exploration, reactive navigation and planning components enhanced the ability to integrate these components, allowing for more efficient data reuse.

Experimental results were presented for the effect of the learning rate on adaptation to changing environments, and also to show that the system performs reliably and repeatable. Work continues on a method for storing, identifying and using previously learned environments, using a topological representation for the overall world in which the robot works. In addition, we are enhancing the algorithms to eliminate an assumption that the robot is on level ground.

References

[1] Borenstein, J. and Koren, Y. (1991). "The Vector Field Histogram - Fast Obstacle Avoidance for Mobile Robots," IEEE Transactions on Robotics and Automation, IEEE: New York, 7(3):278-288.

[2] Graves, K., Adams, W., and Schultz, A., (1997). "Continuous Localization in Changing Environments," proc. of 1997 IEEE Int. Symp. on Computational Intelligence in Robotics and Automation, IEEE, Monterey, CA, July, 28-33.

[3] Hughes, K. Tokuta, A., and Ranganathan, N., (1992). "Trulla: An Algorithm for Path Planning Among Weighted Regions by Localized Propagations," In Proc. of the 1992 IEEE/RSJ International Conference on Intelligent Robots and Systems, Raleigh, NC, 469-475.

[4] Hughes, K. and Murphy, R. (1992). "Ultrasonic Robot Localization using Dempster-Shafer Theory," In 1992 SPIE Stochastic Methods in Signal Processing and Computer Vision, invited session on applications for vision and robotics.

[5] Lee, W. (1996). "Spatial Semantic Hierarchy for a Physical Robot," Ph.D. Thesis, Dept. of Computer Sciences, The University of Texas at Austin.

[6] Mataric, M. (1992). "Integration of Representation Into Goal-Driven Behavior-Based Robots," IEEE Transactions on Robotics and Automation, 8(3), IEEE: New York, 304-312.

[7] Moravec, H. and Elfes, A., (1985). "High Resolution Maps From Wide Angle Sonar," In Proceedings of the IEEE International Conference on Robotics and Automation, 116-121.

[8] Murphy, R., Hughes, K., and Noll, E., (1996). "An Explicit Path Planner to Facilitate Reactive Control and Terrain Preferences," In Proc. of the 1996 IEEE International Conference on Robotics and Automation, 2067-2072.

[9] Schultz, A. and Adams, W. (1998). "Continuous localization using evidence grids," In Proc. of the 1998 IEEE International Conference on Robotics and Automation, Leuven, Belgium, 2833-2839.

[10] Thrun, S. and Bücken, A., (1996). "Integrating Grid-Based and Topological Maps for Mobile Robot Navigation," In Proc. of the Thirteenth National Conference on Artificial Intelligence (AAAI-96), Portland, OR, 944-950.

[11] Yamauchi, B., (1997). "A Frontier-Based Approach for Autonomous Exploration," In Proc. of the 1997 IEEE International Symposium on Computational Intelligence in Robotics and Automation, 146-151.

[12] Yamauchi, B., (1998). "Frontier-Based Exploration Using Multiple Robots," In Proc. of the Second International Conference on Autonomous Agents (Agents'98), Minneapolis, MN.

[13] Yamauchi, B., Schultz, A., and Adams, W., (1998). "Mobile Robot Exploration and Map-Building with Continuous Localization," In Proc. of the 1998 IEEE International Conference on Robotics and Automation, Leuven, Belgium, 3715-3720.

Cooperative Robot Localization with Vision-based Mapping *

Cullen Jennings   Don Murray   James J. Little

Computer Science Dept.
University of British Columbia
Vancouver, BC, Canada V6T 1Z4
{jennings,donm,little}@cs.ubc.ca

Abstract

Two stereo vision-based mobile robots navigate and autonomously explore their environment safely while building occupancy grid maps of the environment. A novel landmark recognition system allows one robot to automatically find suitable landmarks in the environment. The second robot uses these landmarks to localize itself relative to the first robot's reference frame, even when the current state of the map is incomplete. The robots have a common local reference frame so that they can collaborate on tasks, without having a prior map of the environment.

Stereo vision processing and map updates are done at 5Hz and the robots move at 200 cm/s. Using occupancy grids the robots can robustly explore unstructured and dynamic environments. The map is used for path planning and landmark detection. Landmark detection uses the map's corner features and least-squares optimization to find the transformation between the robots' coordinate frames.

The results provide very accurate relative localization without requiring highly accurate sensors. Accuracy of better than 2cm was achieved in experiments.

1 Introduction

Most robots that successfully navigate in unconstrained environments use sonar transducers or laser range sensors as their primary spatial sensors (for example, [1, 2]). While computer vision is often used with mobile robots, it is usually used for feature tracking or landmark sensing and seldom for occupancy grid mapping or obstacle detection.

In this paper, we present a working implementation of a multi-robot navigation system that uses stereo vision-based robots. The robots use correlation-based stereo vision to map their environment with occupancy grids, which are continuously updated to reflect the robot's changing environment. The robots use the maps to navigate and autonomously explore unknown, dynamic indoor environments. We have also developed a method for detecting corners in the map and using them as landmarks. These landmarks are used to localize the robots with respect to one another. The robots can then safely share maps and cooperate within the same local environment.

In earlier work our previous robot, *Spinoza*[3], demonstrated stereo-based mapping and navigation. In [4], we presented algorithms for path planning and exploration using the generated map. In [5], we described in detail how to generate occupancy grid maps from stereo vision and how to compensate for various systematic errors specific to stereo vision in order to have a robust mapping system.

This paper focuses on the localization problem of robust navigation. Localization is the problem of identifying the robot's current position and orientation with respect to some common coordinate system. A robot in an unknown position must refer to its environment to determine its location. Even after localization, drift in odometry causes the estimated position to incrementally deviate from the true position. We present a landmark-based method for localizing within a local coordinate system.

Localization is a difficult problem that has inspired several different approaches. One is to place distinctive, identifiable landmarks at known locations in the robot's environment. This engineering solution is prohibitively expensive and generally difficult to justify outside hospitals or factories[6].

Another approach is to search the local environment for visually distinctive locations and record as

---

* This research was supported by grants from the Natural Sciences and Engineering Research Council of Canada and the Networks of Centres of Excellence Institute for Robotics and Intelligent Systems.

landmarks these locations and a method to identify them (usually their visual appearance)[7]. Some systems concentrate on known types of visual events—finding vertical lines associated with doors, for example, or using range sensing, such as laser stripe systems, sonar, or active stereo vision to find locally salient geometric locations, such as corners, doors, or pillar/floor junctions. One can also track "corners", local 2D image features, over sequences of image frames[8], from which the 3D location of the corner points and the motion of the sensor can be determined[9].

Several working systems have been reported. Borthwick and Durrant-Whyte [10] base their system on detecting corner and edge features in 2D maps, using Extended Kalman Filtering to estimate both feature locations and robot position. Weckesser et al. [11] use *a priori* landmarks at known positions in the global coordinate frame and particular models for landmarks (such as door jambs). Their system uses active stereo cameras and can effectively solve for pose of the robot with respect to landmarks. Thrun and Bucken [12] base their system on sensing regular landmarks, such as overhead lights or doorways, rather than distinctive ones. It uses a Bayesian approach implemented in neural nets and learns which landmarks are most salient.

Our system uses landmarks within the 2D map, as do Borthwick and Durrant-Whyte. Since the occupancy grids are the result of the integration of several sensor readings, features within the map tend to be more robust and less subject to noise than features in a single sensor reading. We divide our map into a series of local coordinate frames, each with a set of landmarks for localization within the local frame. When two robots are within the same "node" of the map, they can localize with the same landmarks and operate in a common coordinate system.

Corner landmarks in the map are found with a least-squares model fitting approach that fits corner models to the edge data in the map. These landmarks are found for both the reference map and the current map. A least-squares approach then finds the transformation between the two maps based on matching the landmarks between the two.

Section 2 describes the robots and how they build maps with stereo vision using occupancy grids. Section 3 shows how we detect corner features in the occupancy grids. Section 4 describes how we localize the robots given the landmark information. In Section 5, we show the results of our localization approach. The final section concludes and describes future work.

2 Mapping

We used two RWI B-14 mobile robots to conduct our experiments. They used a Triclops trinocular stereo vision camera module.[1] The stereo vision module has three identical wide angle (90° field-of-view) cameras. The environment is like a normal office: it is not highly textured and it has many right-angled corners. The hallway that produced the localization data is shown in Figure 1.

Figure 1: Hallway near where map was localized

2.1 Occupancy grid mapping and stereo vision

Occupancy grid mapping, pioneered by Moravec and Elfes [13, 14], is simple and robust, flexible enough to accommodate many kinds of spatial sensors, and adaptable to dynamic environments. It divides the environment into a discrete grid and assigns to each grid location a value related to the probability that the location is occupied by an obstacle. Sensor readings are used to determine regions where an obstacle is anticipated. The grid locations that fall within these regions have their values increased, while locations in the sensing path between the robot and the obstacle have their probabilities of occupancy decreased. Grid locations near obstacles thus tend to have a higher probability of being occupied than do other regions.

[1]See *www.ptgrey.com*.

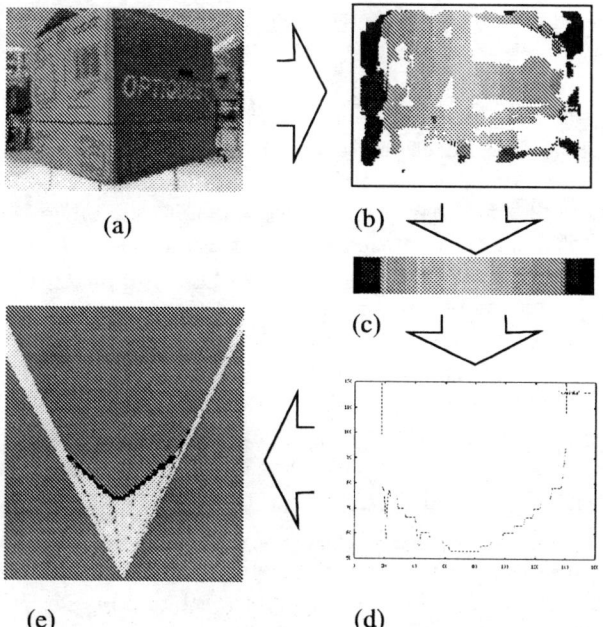

ing the closest obstacle in each column. Figure 2(d) shows these column values converted into distance, and (e) shows these distance values converted into an occupancy grid representation, with black indicating the uncertainty region around the object and white indicating regions that are clear. The two "spikes" in Figure 2(d) were caused by mismatches in the stereo algorithm; their causes and removal are discussed in [5].

The process illustrated in Figure 2 generates the input into our stereo vision occupancy grid. The mapping system then integrates these values over time, to expand the map and keep it current in the changing world. Although the map in Figure 3 looks like directions to buried treasure, it actually depicts our lab and nearby hallways.

Figure 2: From stereo images to top-down views: (a) grayscale image (b) disparity image (white indicates invalid, otherwise brighter indicates closer to the cameras) (c) maximum disparity in each column of the disparity image (d) depth versus column (depth in centimeters) (e) resultant estimate of clear, unknown and occupied regions (white is clear, black is occupied and gray is unknown)

Although occupancy grids may be implemented in any number of dimensions, most mobile robotics applications (including ours) use 2D grids. The stereo data provides 3D information that is lost in the construction of a 2D occupancy grid map. This reduction in dimension is justified because indoor mobile robots inhabit a fundamentally 2D world. The robot possesses 3 DOF (X, Y, heading) within a 2D plane corresponding to the floor. By projecting all sensed obstacles to the floor, we can uniquely identify free and obstructed regions in the robot's space.

Figure 2 shows the construction of the 2D occupancy grid sensor reading from a single 3D stereo image. Figure 2(a) shows the reference camera grayscale image (160x120 pixels). The resulting disparity image is shown in Figure 2(b). White regions indicate areas of the image which were invalidated and thus not used. Darker areas indicate lower disparities, and are farther away from the camera. Figure 2(c) shows a column-by-column projection of the disparity image, taking the maximum valid disparity in each column. The result is a single row of maximum disparities, represent-

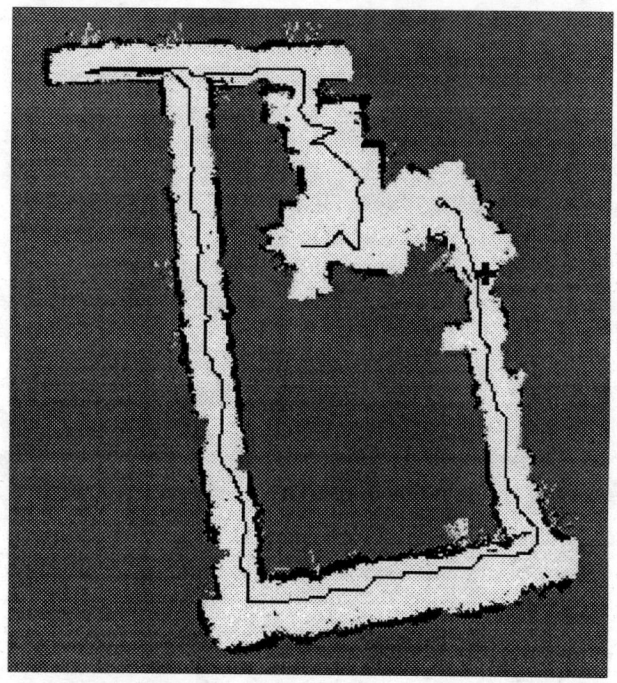

Figure 3: Map generated over single run

3 Landmarks in Occupancy Grids

We do localization using visual information stored in an occupancy grid, so we need to detect features in the map. Corners are appealing because they are distinctive, each constrain two degrees of freedom, occur frequently in indoor environments, and stand out well in occupancy grids despite quantization. There are many ways to detect features (correlation, Hough transform, Karhunen-Loeve Transform), but we chose

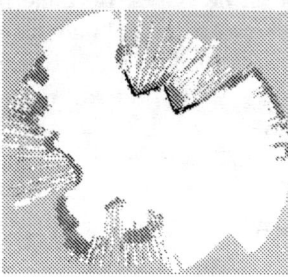

Figure 4: Edges found in the occupancy grid (left), and the occupancy with landmarks found (right).

a least-squares model matching approach because it is capable of subpixel resolution and is fast and trackable, meaning that if we have an estimate of the solution, we can use it to find the real solution.

Once the landmarks are found, we match the landmarks in the two maps being considered to find the local coordinate transformation between the two maps.

3.1 Landmark Detection

First we find boundaries between empty and solid spaces in the occupancy grid—we call these edges. Next, we find possible locations for corner models and use these as initial estimates. Third, we do a least-squares fit of a corner model to the map edge data to find all the corner landmarks.

3.1.1 Finding edges in map

Finding edges in occupancy maps is complicated by the fact that there is often a region of uncertainty between a region that is certainly occupied and one that is not. To overcome this we grow regions of certainty in the map into regions of uncertainty. Where we are highly confident that a certain region is empty or occupied, we mark these accordingly. We preserve the regions that have never been seen. Next we find occupied regions that are adjacent to empty regions and mark them as "edge" pixels. Models are fit directly to edge grid locations. Figure 4 shows the edges found by this process.

3.2 Model initialization

Our problem is finding initial estimates of possible locations that might fit a corner model. We start with the edge image. We dilate this several times and then compute an orientation map for each grid location—the ith bit is set if a line in the dilated edge map can extend for a fixed number of grid locations from this location in a direction of $i \times 45°$. Then we determine the corner map by looking within the orientation map for grid locations that have the ith and the $(i+2)^{th}$ bits set. The search for a model fit is performed in the dilated edge map in order to save processing time. The dilated map has been dilated enough that a line at $22.5°$ dilates to an area that overlays a line at $0°$. This guarantees that the orientation map will have a bit set if the model fits the edge data at any rotation.

3.3 Model fitting

The corner model is just two line segments of a fixed length ($500mm$ in our case) that intersect at a $90°$ angle. The problem is to fit a model of a corner that is parameterized by X, Y, Θ to the edge grid locations, given an initial estimate of position and orientation.

The model of the corner is discretized into n points, labeled $m_1, m_2, ..m_n$, which are evenly distributed along the model. Let $x = [X, Y, \theta]^T$. Now, the n points can be determined as a function of x, so $m_i = f_i(x)$. Let x_0 be the initial estimate of x. The error of the model at each point m_i is the distance from m_i to the nearest edge location. A Euclidean distance metric is used. These distances form the vector

$$e = \begin{bmatrix} \text{dist}(\mathtt{m_1, nearestedge}), \\ \text{dist}(\mathtt{m_2, nearestedge}), \\ \vdots, \\ \text{dist}(\mathtt{m_n, nearestedge}), \end{bmatrix}$$

The vector e has the distance from each point to the nearest edge. The Jacobian, J, of e with respect to x is computed by evaluating the partial derivative of

$$\frac{\partial \text{dist}(\mathtt{f_i(x), nearestedge})}{\partial x_j}$$

around the current estimate of x. Most of the time in localization is spent finding the nearest edge location to a given point while computing the Jacobian. Following the technique of Newton, the change in x from this estimate x_i to the next estimate x_{i+1} is given by:

$$J(x_{i+1} - x_i)$$

This is an over-constrained system, since the model is discretized into n points, which make it much larger than the 3 DOF of the model. The least-squares solution to this is given by

$$\min \|J(x_{i+1} - x_i) - e\|^2$$

This is the same as solving the normal equations

$$J^T J(x_{i+1} - x_i) = J^T e$$

This model fitting works much better when this system of equations is stabilized by adding the constraint

$$x_{i+1} - x_i = 0$$

This constraint needs to be appropriately scaled in the least-squares solution, as shown by Lowe[15].

We measure position in grid cells and angle in degrees, and we scale all constraints equally by 0.1. This scaling indicates that we are willing to change the solution by 10 grid locations to avoid a 1 grid location error. Let the matrix W be $\frac{1}{10}I$. The system to which we wish to find the least-squares solution becomes:

$$\begin{bmatrix} J \\ W \end{bmatrix}(x_{i+1} - x_i) = \begin{bmatrix} e \\ 0 \end{bmatrix}$$

and the normal equations become

$$(J^T J + W^T W)(x_{i+1} - x_i) = J^T e$$

These are solved with Gaussian elimination with pivoting on the optimal row. The system is run for ten iterations or until the change from the previous iteration is very small. Convergence is fairly good, normally occurring after three iterations.

4 Localization

4.1 Corner Matching

The goal at this stage is to take two maps, the reference and current maps, and match the corners between them. First we eliminate all corner models with an RMS error above 1.5 pixels. (This is one of the few "magic numbers" in the system.) A corner will suppress all corners with a higher RMS error that are within $500mm$ of the suppressing corner. This non-maximal suppression reduces the chances of the corner being mismatched.

Finally each corner is matched to the nearest corner in the other map. If no matching corner is found within $1000mm$ of a particular corner, the corner is left unmatched and not used for localization. Checking that the orientations of the corners matches does not seem to change the system's robustness much. If there had been more mismatches, it would likely have helped.

4.2 Localization

The goal here is to find the transformation that maps the corners in one map to the other. This stage uses the same least-squares model matching technique that we used to match the corner models to the edges. The transform is parameterized by a $2D$ translation and rotation about the robot. The three parameters are found by setting up equations for each pair of landmarks that match between the maps. The rotation error between the maps is usually quite small (less than 15°), so the linearization of the transform in the least-squares matching process introduces minimal error and the method quickly converges. The stabilization technique mentioned in Section 3.3 is important for nice convergence at this stage.

5 Experiments

The first robot toured around our building creating the map shown in Figure 3. The intersection of two hallways (Figure 1) was selected as a localization node. The local occupancy map for this area is shown in Figure 5. The map edges are shown in Figure 6 with the detected landmarks drawn as black corners and overlaid on the edges. The location of the robot

Figure 5: Reference map from first robot

when it acquired this map was marked so that it could be used as ground truth for comparing the location of the second robot. The second robot was run out to a nearby location and created the map shown in Figure 7. The associated edges and landmarks are shown in Figure 8.

The odometry estimates have drifted, as shown in Figure 9, which overlays the edges from the two robots. By comparing the locations of the robots as marked on the floor, we know that the length of the transformation between the reference frames should be

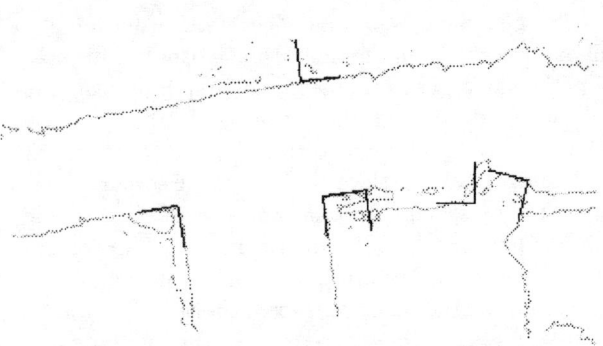

Figure 6: Edges and landmarks in reference map

Figure 7: Current map from second robot

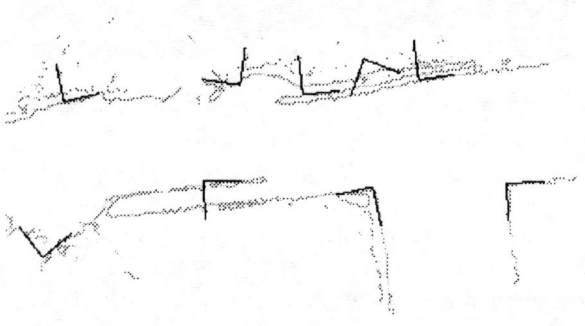

Figure 8: Edges and landmarks in current map

$80\pm20mm$. The localization routine produced a transform of length $92mm$. The transformation found was applied to the edges in the second map, and they were overlayed on the reference map from the first robot to get the image shown in Figure 10.

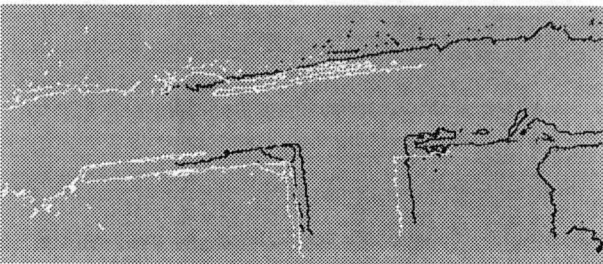

Figure 9: Edges of two maps overlayed before localization

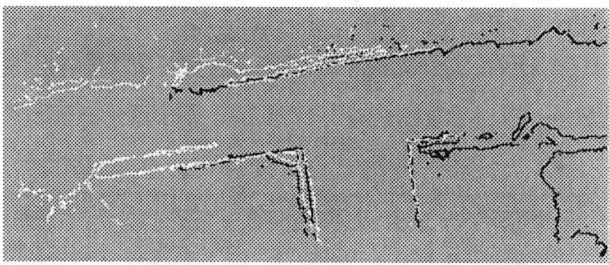

Figure 10: Edges of two maps overlayed after correction for localization

The system is quite robust and correctly localizes the robot in a wide variety of situations. It starts to fail where the odometry estimate is so far out that the corners are mismatched. By using fairly large corners ($500mm$ per side) and non-maximal suppression we ensure that we do not detect multiple corners very close together. Most of our corners are more than $1000mm$ apart. This ensures that for up to $500mm$ of odometry error, we are highly likely to match to the correct corner.

Since the system finds naturally occurring corners in the environment, it can localize fairly often and have much smaller odometry errors between localizations. The technique can be used to localize a robot to a location that it has previously visited or that has been visited by another robot. It can also be used to localize two robots that are in the same space and need to share coordinate systems to collaborate on a task.

The system is fast enough for real time robotics. The landmark detection for the image shown in this paper took 14 seconds running on a 266MHz Pentium II. Computing the localization took only a few milliseconds. The speed of this localization would allow for many possible transformation models to be tested and the best one chosen if needed. We did not find a need for this.

6 Conclusions and Future work

This paper demonstrates a working system in which robots can localize themselves relative to other robots or to themselves at other times. The system is very accurate and works with incomplete maps of environments that are being explored. It was shown that localization using features in stereo vision occupancy grids is a feasible solution to the mobile robot localization problem. Accuracies better than a few centimeters are attainable in real time.

Our corner localization works well because we base the localization on data integrated over several views, a method more robust than a single reading would be. We can also extend the landmark detection method to use 'multi-level' 2D maps. These are collections of 2D slices of the 3D environment. Near ceilings, room corners are rarely obstructed and are readily apparent. These can be used as commonly occurring landmarks.

Further, a more comprehensive cooperative mobile robot environment could be built relying on the groundwork presented here. A group of landmarks can form a local coordinate frame, and locally consistent frames can be linked so as to create nodes useful for developing topological maps. Exploration of the environment can also be a cooperative task, with one robot altering its behavior in response to its partner's activities. These and other related ideas would be suitable topics for future work.

Acknowledgments

Special thanks to Rod Barman, Stewart Kingdon, David Lowe, and the rest of the Robot Partners Project.

References

[1] G. Dudek, M. Jenkin, E. Milios, and D. Wilkes, "On building and navigating with a local metric environmental map," in *ICAR-97*, 1997.

[2] W. Burgard, A. B. Cremers, D. Fox, D. Hähnel, G. Lakemeyer, D. Schulz, W. Steiner, and S. Thrun, "The interactive museum tour-guide robot," in *AAAI-98*, pp. 11–19, 1998.

[3] V. Tucakov, M. Sahota, D. Murray, A. Mackworth, J. Little, S. Kingdon, C. Jennings, and R. Barman, "Spinoza: A stereoscopic visually guided mobile robot," in *Proceedings of the Thirteenth Annual Hawaii International Conference of System Sciences*, pp. 188–197, Jan. 1997.

[4] D. Murray and C. Jennings, "Stereo vision based mapping for a mobile robot," in *Proc. IEEE Conf. on Robotics and Automation, 1997*, May 1997.

[5] D. Murray and J. Little, "Interpreting stereo vision for a mobile robot," in *IEEE Workshop for Perception for Mobile Agents*, pp. 19–27, June 1998.

[6] S. B. Nickerson *et al.*, "Ark: Autonomous navigation of a mobile robot in a known environment," in *IAS-3*, pp. 288–293, 1993.

[7] S. Oore, G. Hinton, and G. Dudek, "A mobile robot that learns its place," *Neural Computation* **9**, pp. 683–699, Apr. 1997.

[8] J. Shi and C. Tomasi, "Good features to track," in *Proc. IEEE Conf. Computer Vision and Pattern Recognition, 1994*, pp. 593–600, 1994.

[9] C. Tomasi and T. Kanade, "Factoring image sequences into shape and motion," in *Proc. IEEE Workshop on Visual Motion, 1991*, pp. 21–28, 1991.

[10] S. Borthwick and Durrant-Whyte, "Simultaneous localisation and map building for autonomous guided vehicles," in *IROS-94*, pp. 761–768, 1994.

[11] P. Weckesser, R. Dillmann, M. Elbs, and S. Hampel, "Multiple sensor processing for high-precision navigation and environmental modeling with a mobile robot," in *IROS-95*, 1995.

[12] S. Thrun and A. Bucken, "Learning maps for indoor mobile robot navigation," tech. rep., CMU-CS-96-121, Apr. 1996.

[13] H. Moravec and A. Elfes, "High-resolution maps from wide-angle sonar," in *Proc. IEEE Int'l Conf. on Robotics and Automation*, (St. Louis, Missouri), Mar. 1985.

[14] A. Elfes, "Using occupancy grids for mobile robot perception and navigation," *Computer* **22**(6), pp. 46–57, 1989.

[15] D. G. Lowe, "Fitting parameterized three-dimensional models to images," *IEEE Transactions on Pattern Analysis and Machine Intelligence* **13**, pp. 441–450, May 1991.

Motion Control of Multiple Autonomous Mobile Robots Handling a Large Object in Coordination

Kazuhiro Kosuge*, Yasuhisa Hirata*,
Hajime Asama**, Hayato Kaetsu** and Kuniaki Kawabata**

*Department of Machine Intelligence and Systems Engineering,
Tohoku University
Aoba-yama01, Sendai 980-8579, JAPAN
**Biochemical Systems Laboratory
The Institute of Physical and Chemical Research, RIKEN
Hirosawa 2-1, Wako, Saitama 351-0198, JAPAN

Abstract

In this paper, we discuss a problem relating to the force/moment transformation for the handling of a large object by multiple mobile robots in coordination. We propose a control algorithm using geometrical constraints among the grasping points and the representative point of the object, which reduce the effect of noise amplified by force/moment transformation. We extend this algorithm to the decentralized control algorithm of multiple robots handling an object in coordination. The proposed control algorithm is experimentally applied to the mobile robots. Experimental results illustrate the validity of the proposed control algorithm.

1 Introduction

The coordination of multiple robots has some advantages similar to the case of a task executed by humans in coordination. Multiple robots in coordination can execute tasks which could not be done by a single robot. Many control algorithms of robots have been proposed for the handling of a single object by multiple robots in coordinations [1]-[4], etc.

Most of the control algorithms proposed so far have been designed under the assumption that the force/moment applied to a representative point of the object is available. The force/moment applied to the object is usually calculated from the force/moment detected by a force sensor attached to each robot. When the representative point is located far from the force sensor, the calculation amplifies the sensor noise included in the force/moment information from the force sensor and lower the equivalent sensor resolution at the representative point especially when we consider the problem of handling a large object.

In this paper, we propose a control algorithm for handling of a large object in coordination. In the following part of this paper, we briefly review the compliance-based control algorithm of multiple robots, which we proposed in [5]. And We discuss the problem relating to the force/moment transformation. We propose a control algorithm using geometrical constraints among the grasping points and the representative point of the object which reduces the effect of sensor noise. We extend this algorithm to a decentralized control algorithm of multiple mobile robots

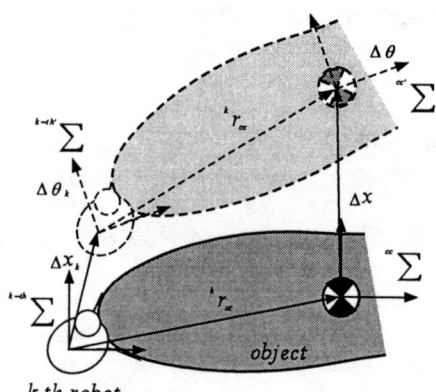

Figure 1: Coordinate system

handling a single object in coordination. The control algorithm is experimentally implemented in the autonomous omnidirectional mobile robots, ZEN. The experimental results illustrate the effectiveness of the proposed control algorithm.

2 Coordinate systems

Consider a problem to handle a rigid object by n mobile robots in coordination. We define coordinate systems as shown in Figure 1; the object coordinate system $^{cc}\Sigma$, and the k-th mobile robot coordinate system $^{k-th}\Sigma$. The object coordinate system is attached to a representative point of the object and moves together with the object. The k-th mobile robot coordinate system is attached to the grasping point of the k-th mobile robot and its coordinate axes are defined parallel to those of the object coordinate system. Under the assumption that each robot holds the object firmly and no relative motion between the object and the robot occurs, the k-th mobile robot coordinate system also moves together with the object coordinate system without changing their geometrical relation.

Assume that external force/moment is applied to the object. $\Delta x \in R^2$ and $\Delta \theta \in R$ is position deviation and orientation deviation of the object coordinate system with respect to the original object coordinate system as shown in Figure 1. $\Delta x_k \in R^2$ and $\Delta \theta_k \in R$ is position deviation and orientation deviation of the

k-th mobile robot coordinate system, with respect to $^{cc}\Sigma$, corresponding to the motion of the object coordinate system Δx and $\Delta\theta$. $^{k-th'}\Sigma$ and $^{cc'}\Sigma$ express the mobile robot coordinate system and the object coordinate system respectively, after the deviational motion, and correspond to $^{k-th}\Sigma$, $^{cc}\Sigma$, respectively before the deviation.

3 Compliance-based coordinated motion control algorithm

First, let us review the compliance-based coordinated motion control algorithm we proposed in [5]. In [5], we assumed that each robot is driven by velocity-controlled actuators and its impedance around the representative point attached to the object is controlled so as to have the following dynamics;

$$\begin{bmatrix} _k f^{ext} \\ _k n^{ext} \end{bmatrix} = \begin{bmatrix} _k D & 0 \\ 0 & _k D_\theta \end{bmatrix} \begin{bmatrix} \Delta \dot{x} \\ \Delta \dot{\theta} \end{bmatrix} + \begin{bmatrix} _k K & 0 \\ 0 & _k K_\theta \end{bmatrix} \begin{bmatrix} \Delta x \\ \Delta \theta \end{bmatrix} \quad (1)$$

where, $\begin{bmatrix} _k D & 0 \\ 0 & _k D_\theta \end{bmatrix}$, and $\begin{bmatrix} _k K & 0 \\ 0 & _k K_\theta \end{bmatrix}$ are a 3×3 damping matrix, and a 3×3 stiffness matrix, respectively. In eq.(1) we assumed that the translational motion and the rotational motion are decoupled each other. $_k f^{ext} \in R^2$, $_k n^{ext} \in R$ represent the external force and moment born by the k-th mobile robot respectively at the representative point with respect to the nominal object coordinate system.

Next, we derive the equivalent impedance of each mobile robot at the grasping point. To simplify the discussion, we assume that the force sensor of each mobile robot is located at its grasping point and directly detects force/moment applied to the robot at its grasping point. Considering the homogeneous coordinate transformation from $^{cc'}\Sigma$ to $^{k-th}\Sigma$, we have the following relation;

$$\begin{bmatrix} E & ^k r_{cc} \\ 0 & 1 \end{bmatrix} \begin{bmatrix} Rot(\Delta\theta) & \Delta x \\ 0 & 1 \end{bmatrix}$$
$$= \begin{bmatrix} Rot(\Delta\theta_k) & \Delta x_k \\ 0 & 1 \end{bmatrix} \begin{bmatrix} E & ^k r_{cc} \\ 0 & 1 \end{bmatrix} \quad (2)$$

where $Rot(\Delta\theta)$ and $Rot(\Delta\theta_k)$ are 2×2 rotation matrices corresponding to $\Delta\theta$ and $\Delta\theta_k$ respectively. E is the 2×2 identity matrix. $^k r_{cc} \in R^2$ is the position vector from the origin of $^{k-th}\Sigma$ to the origin of $^{cc}\Sigma$ with respect to $^{cc}\Sigma$. The left-hand side of eq.(2) represents the homogeneous coordinate transformation from $^{cc'}\Sigma$ to $^{k-th}\Sigma$ via $^{cc}\Sigma$, and the right-hand side of eq.(2) represents the homogeneous coordinate transformation from $^{cc'}\Sigma$ to $^{k-th}\Sigma$ via $^{k-th'}\Sigma$. Under the assumption that each robot grasps the object firmly and no relative motion between the grasping point and the object occurs, we have

$$\Delta\theta = \Delta\theta_k \quad (3)$$

From eq.(2), eq.(3), and Figure 1, we have

$$\begin{aligned} \Delta x_k &= \Delta x - (Rot(\Delta\theta) - E)^k r_{cc} \\ &\cong \Delta x - \Delta\theta \, ^k R^T \end{aligned} \quad (4)$$

where, we assumed that the rotational deviation $\Delta\theta$ is small. $^k R, ^k R'$ are matrices defined by the element of the position vector $^k r_{cc} = (r_x, r_y)^T$ as follows;

$$^k R = \begin{bmatrix} -r_y & r_x \end{bmatrix}, \quad ^k R' = \begin{bmatrix} r_y & -r_x \end{bmatrix}^T \quad (5)$$

We also have the following relations for force and moment;

$$_k f^{ext} = f_k^{ext}, \quad _k n^{ext} = -^k r_{cc} \times f_k^{ext} + n_k^{ext} \quad (6)$$

where f_k^{ext} and n_k^{ext} are equivalent force and moment vectors at the origin of $^{k-th}\Sigma$, or the grasping point, corresponding to $_k f^{ext}$, $_k n^{ext}$.

Substituting eq.(3) eq.(4), and eq.(6) into eq.(1), we have the impedance dynamics of the k-th mobile robot around its grasping point as follows;

$$\begin{bmatrix} f_k^{ext} \\ n_k^{ext} \end{bmatrix} = \begin{bmatrix} _k D & -_k D \, ^k R' \\ ^k R_k D & _k D_\theta - ^k R_k D \, ^k R' \end{bmatrix} \begin{bmatrix} \Delta\dot{x}_k \\ \Delta\dot{\theta}_k \end{bmatrix} + \begin{bmatrix} _k K & -_k K \, ^k R' \\ ^k R_k K & _k K_\theta - ^k R_k K \, ^k R' \end{bmatrix} \begin{bmatrix} \Delta x_k \\ \Delta\theta_k \end{bmatrix} \quad (7)$$

Eq.(7) represents the impedance dynamics of each mobile robot at its grasping point when the compliance-based coordinated motion control algorithm is implemented. Translational motion and rotational motion are strongly coupled in eq.(7) through off-diagonal block matrices $-_k D \, ^k R'$, $^k R_k D$, $-_k K \, ^k R'$ and $^k R_k K$, which involve terms concerned with a vector cross product relating to the distance $^k r_{cc}$ between the coordinate systems $^{cc}\Sigma$ and $^{k-th}\Sigma$. When a large object is handled, these terms lead to the amplification of the sensor noise and the reduction of the force/moment resolution for the implementation of the control algorithm. A numerical example will be given in the section 5. Note that this problem is inevitable for any control algorithm which requires force/moment information at a representative point attached to the object.

To solve the above problem, suppose that each mobile robot has the following decoupled impedance at its grasping point.

$$\begin{bmatrix} f_k^{ext} \\ n_k^{ext} \end{bmatrix} = \begin{bmatrix} D_k & 0 \\ 0 & D_{\theta k} \end{bmatrix} \begin{bmatrix} \Delta\dot{x}_k \\ \Delta\dot{\theta}_k \end{bmatrix} + \begin{bmatrix} K_k & 0 \\ 0 & K_{\theta k} \end{bmatrix} \begin{bmatrix} \Delta x_k \\ \Delta\theta_k \end{bmatrix} \quad (8)$$

Substituting eq.(3), eq.(4) and eq.(6) into eq.(8), we can derive the apparent impedance of the object around its representative point as follows;

$$\begin{bmatrix} \sum D_k & \sum D_k \, ^k R' \\ \sum -^k R D_k & \sum (D_{\theta k} - ^k R D_k \, ^k R') \end{bmatrix} \begin{bmatrix} \Delta\dot{x} \\ \Delta\dot{\theta} \end{bmatrix}$$
$$+ \begin{bmatrix} \sum K_k & \sum K_k \, ^k R' \\ \sum -^k R K_k & \sum (K_{\theta k} - ^i R K_k \, ^k R') \end{bmatrix} \begin{bmatrix} \Delta x \\ \Delta\theta \end{bmatrix}$$
$$= \begin{bmatrix} f^{ext} \\ n^{ext} \end{bmatrix} \quad (9)$$

where, f^{ext} and n^{ext} are external force and moment applied to the object and born by all of the mobile robots;

$$f^{ext} = \sum_{k=1}^{n} {}_k f^{ext} = \sum_{k=1}^{n} f_k^{ext} \quad (10)$$

$$n^{ext} = \sum_{k=1}^{n} {}_k n^{ext} = \sum_{k=1}^{n} (-{}^k r_{cc} \times f_k^{ext} + n_k^{ext}) \quad (11)$$

As shown by eq.(9), the behavior of the object is very complex. We could not use the decoupled impedance expressed by eq.(8) as it is to specify the apparent impedance of the handled object, although the decoupled impedance, eq.(8), is not affected by the distance between the grasping point and the representative point. In the following, we propose a control algorithm for multiple mobile robots handling a large object in coordination. To lessen the effect of the force/moment transformation problems, we will use geometrical constraints as well as the decoupled impedance.

4 Algorithm with geometrical constraints

In this section, we propose a control algorithm for the handling of an object by multiple mobile robots in coordination. The algorithm utilizes geometrical constraints as well as the decoupled impedance at grasping points so that the apparent impedance of the handling object is specified.

Consider to specify a decoupled impedance of the handling object at the representative point of the object using decoupled impedance expressed by eq.(8). As mentioned in the previous section, in general, we could not specify the apparent impedance of the object as long as each mobile robot has a decoupled impedance at its grasping point. In the following part of this section, we will show that the apparent impedance will be specified for a special case satisfying certain geometrical constraints. Consider the case where D_k and K_k are expressed as follows;

$$D_k = d_k E, \quad K_k = k_k E \quad (12)$$

where $d_k \in R$, $k_k \in R$ are positive real numbers and E is the 2×2 identity matrix. Assume that the following conditions on ${}^k r_{cc}$, d_k and k_k are satisfied;

$$\sum_{k=1}^{n} d_k {}^k r_{cc} = 0, \quad \sum_{k=1}^{n} k_k {}^k r_{cc} = 0 \quad (13)$$

Then, the resultant impedance of the object expressed by eq.(9), is rewritten as follows;

$$\begin{bmatrix} \sum D_k & 0 \\ 0 & \sum(D_{\theta k} - {}^k R D_k {}^k R') \end{bmatrix} \begin{bmatrix} \Delta \dot{x} \\ \Delta \dot{\theta} \end{bmatrix}$$
$$+ \begin{bmatrix} \sum K_k & 0 \\ 0 & \sum(K_{\theta k} - {}^k R K_k {}^k R) \end{bmatrix} \begin{bmatrix} \Delta x \\ \Delta \theta \end{bmatrix}$$
$$= \begin{bmatrix} f^{ext} \\ n^{ext} \end{bmatrix} \quad (14)$$

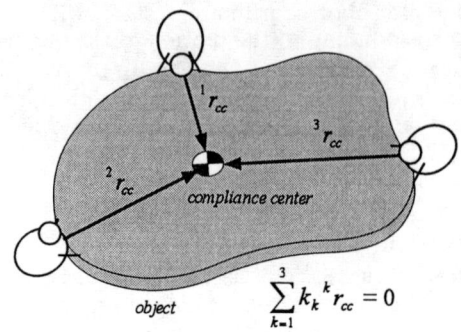

Figure 2: Handling using three mobile robots

To specify apparent impedance of the rotation, we modify eq.(8) as follows;

$$\begin{bmatrix} f_k^{ext} \\ n_k^{ext} \end{bmatrix} = \begin{bmatrix} D_k & 0 \\ 0 & D_{\theta k} + {}^k R D_k {}^k R' \end{bmatrix} \begin{bmatrix} \Delta \dot{x}_k \\ \Delta \dot{\theta}_k \end{bmatrix}$$
$$+ \begin{bmatrix} K_k & 0 \\ 0 & K_{\theta k} + {}^k R K_k {}^k R' \end{bmatrix} \begin{bmatrix} \Delta x_k \\ \Delta \theta_k \end{bmatrix} \quad (15)$$

where $D_{\theta k} + {}^k R D_k {}^k R'$, $K_{\theta k} + {}^k R K_k {}^k R'$ are a positive real numbers to control the robot in stable manner. The resultant apparent impedance of the object is expressed by

$$\begin{bmatrix} \sum D_k & 0 \\ 0 & \sum D_{\theta k} \end{bmatrix} \begin{bmatrix} \Delta \dot{x} \\ \Delta \dot{\theta} \end{bmatrix}$$
$$+ \begin{bmatrix} \sum K_k & 0 \\ 0 & \sum K_{\theta k} \end{bmatrix} \begin{bmatrix} \Delta x \\ \Delta \theta \end{bmatrix} = \begin{bmatrix} f^{ext} \\ n^{ext} \end{bmatrix} (16)$$

Since eq.(15) does not have off-diagonal blocks, the translational motion caused by the external force f_k^{ext} and the rotational motion caused by the external moment n_k^{ext} is decoupled completely. The elimination of direct coupling largely reduces the vibration caused by the force/moment transformation. The geometrical constraints expressed by eq.(13) says that the representative point should be located at the centroid of the system which consists of n mobile robots. The centroid could be specified by selecting the impedance parameters as shown by eq.(13). To satisfy eq.(13), the following relation should be satisfied.

$$\alpha d_k = \beta k_k \quad (k = 1, 2 \cdots, n) \quad (17)$$

where α and β are constant real numbers. Figure 2 illustrates the relation among the grasping points and the representative point.

5 Effect of noise included in force/moment

In this section, we calculate the effect of sensor noise on the mobile robot velocity to illustrate the validity of the proposed control algorithm. We evaluate the velocity $[\Delta \dot{x}_k \ \Delta \dot{\theta}_k]^T$ of the grasping point of the k-th mobile robot for a given sensor noise expressed by

$$\begin{bmatrix} f_k^{ext} \\ n_k^{ext} \end{bmatrix} \leq 1 \quad (18)$$

Let us express the relation between the sensor noise and the grasping point velocity by

$$F = Dv \quad (19)$$

where $F = [f_k^{ext}\ n_k^{ext}]^T$, $v = [\Delta \dot{x}_k\ \Delta \dot{\theta}_k]^T$ and D is a damping matrix. From eq.(18) and eq.(19), we have

$$|F|^2 = F^T F = v^T D^T D v \leq 1 \quad (20)$$

Let λ_i and e_i ($i = 1, 2, 3$) be the eigenvalue and the corresponding eigne vector of the matrix $D^T D$. As is well known, eq.(20) represents an ellipsoid whose principal axes are expressed by the eigne vectors e_i with the magnitude of $1/\sqrt{\lambda_i}$ ($i = 1, 2, 3$).

Using this ellipsoid, we will show how much the effect of sensor noise will be decreased for the experimental system as shown in Figure 6(a). In case of the proposed control algorithm, the relation between the sensor noise F and the velocity v is expressed by eq.(15). In case of the conventional algorithm, the relation is expressed by eq.(7). The ellipsoids for both cases are shown in Figure 3. This figure shows that the effect of sensor noise included in the force/moment from a force sensor is decreased by the proposed algorithm.

6 Decentralized motion control algorithm in coordination

6.1 Design of controller for each robot

In this section, we extend the proposed algorithm explained in the previous section to a decentralized control algorithm of mulitple mobile robots handling a single object in coordination. We assume that each robot is controlled by its own controller in a decentralized way. The desired trajectory of the object is given to the leader, and the follower estimates the desired motion of the object commanded to the leader to transport the object in coordination with the leader.

We assume that each robot has the dynamics as shown in eq.(15), where $k = l, i$. The subscripts l and i indicate that the leader robot and the i-th follower robot respectively. Let $x_{ld}, x_{ie} \in R^2$, $\theta_{ld}, \theta_{ie} \in R$ be the desired trajectories of the leader and the i-th follower respectively. $x_{dl}, x_{ei} \in R^2$, $\theta_{dl}, \theta_{ei} \in R$ indicate the trajectories of the representative point attached to the object which are calculated by transfoming the

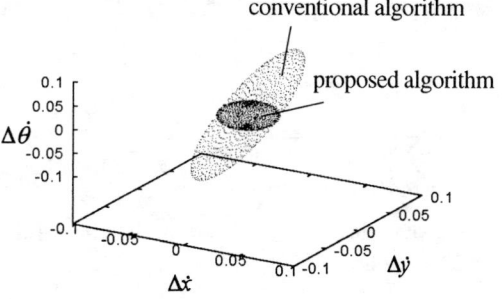

Figure 3: Effect of noise for mobile robot

coordination of the desined trajectories of the leader and the i-th follower respectively. $x \in R^2, \theta \in R$ is the real trajectory of the object. Under the assumption that each robot holds the object firmly and no relative motion between the object and each robot occurs, the relation between the deviation of each robot and the deviation of representative point is expressed as follows;

$$\Delta x_l = \Delta x_{ol} - \Delta \theta_{ol}\ {}^l R^T, \quad \Delta \theta_l = \Delta \theta_{ol} \quad (21)$$
$$\Delta x_i = \Delta x_{oi} - \Delta \theta_{oi}\ {}^i R^T, \quad \Delta \theta_i = \Delta \theta_{oi} \quad (22)$$

where $\Delta x_{ol}, \Delta \theta_{ol}, \Delta x_{oi}, \Delta \theta_{oi}$ are deviation relating to the position and the orientation of representative point of the object corresponding to the motion of each robot, and are expressed as follows;

$$\Delta x_{ol} = x - x_{dl} \quad (23)$$
$$\Delta \theta_{ol} = \theta - \theta_{dl} \quad (24)$$
$$\Delta x_{oi} = x - x_{ei} \quad (25)$$
$$\Delta \theta_{oi} = \theta - \theta_{ei} \quad (26)$$

${}^l R^T, {}^i R^T$ are defined by eq.(5) and indicate matrices expressed by the element of the position vectors with respect to the leader and the i-th follower respectively.

For the simplicity of discussions, we rewrite the parameters of the leader and the follower simply as follows;

$$\begin{bmatrix} D_k & 0 \\ 0 & D_{\theta k} + {}^k R D_k {}^k R' \end{bmatrix} = D_k' \quad (27)$$

$$\begin{bmatrix} K_k & 0 \\ 0 & K_{\theta k} + {}^k R K_k {}^k R \end{bmatrix} = K_k' \quad (28)$$

$$\begin{bmatrix} \Delta x_k \\ \Delta \theta_k \end{bmatrix} = \Delta X_k \quad (29)$$

$$\begin{bmatrix} f_k^{ext} \\ n_k^{ext} \end{bmatrix} = F_k^{ext} \quad (30)$$

where $k = l, i$, $(x_{ld}, \theta_{ld})^T = X_{ld}$, $(x_{ie}, \theta_{ie})^T = X_{ie}$, $(x_{dl}, \theta_{dl})^T = X_{dl}$, $(x_{ei}, \theta_{ei})^T = X_{ei}$, and $(x, \theta)^T = X$ Then, the resultant impedance of the leader robot and i-th follower robot expressed by eq.(15) is rewritten as follows.

$$D_l' \Delta \dot{X}_l + K_l' \Delta X_l = F_l^{ext} \quad (31)$$
$$D_i' \Delta \dot{X}_i + K_i' \Delta X_i = F_i^{ext} \quad (32)$$

6.2 Handling of a single object by two robots in coordination

For the simplicity of explanation, we consider the case of two autonomous omnidirectional mobile robots, that is $i = 1$ in eq.(32). Under the assumption that the external force applied to the object is negligible, consider the geometric constraints expressed by eq.(13), and the leader and the followers are controlled using the same parameters, that is,

$$D_l' = D_i' = D \quad (33)$$
$$K_l' = K_i' = K \quad (34)$$

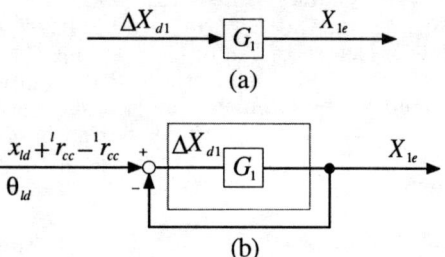

Figure 4: Estimator

We have the following relations with respect to the forces/moment applied to each robot;

$$F_l^{ext} + F_1^{ext} = 0 \quad (35)$$

From eq.(31), eq.(32), eq.(33), eq.(34) and eq.(35), we obtain the following relation;.

$$D(\Delta\dot{X}_l + \Delta\dot{X}_1) + K(\Delta X_l + \Delta X_1) = 0 \quad (36)$$

As time tends to infinity, we obtain the following relationship from eq.(36) and the positive definiteness of the damping matrix and the stiffness matrix, even if the initial values of $\Delta X_l + \Delta X_1$ is not zero.

$$\Delta X_l + \Delta X_1 = 0 \quad (37)$$

Subtracting eq.(21) from eq.(22) and eliminating X, we have following relation;

$$\Delta X_1 - \Delta X_l = X_{dl} - X_{e1} \quad (38)$$

Let ΔX_{d1} be the difference between the trajectories of the object X_{dl} with respect to the desired trajectories of the leader X_{ld} and the trajectory of the object X_{e1} with respect to estimated trajectories of the follower X_{1e}. From eq.(37) and eq.(38) ΔX_{d1} is expressed as

$$\Delta X_{d1} = X_{dl} - X_{e1} = 2\Delta X_1 \quad (39)$$

It should be noted that the follower can calculate ΔX_{d1} using observable variable ΔX_1.

Let us consider how X_{ld} is estimated using ΔX_{d1}. Let G_1 be the transfer function, which estimates X_{ld}, as X_{1e}, based on ΔX_{d1} as shown in Figure 4(a). From eq.(39), Figure 4(a) can be rewritten as a feedback system as shown in Figure 4(b). To eliminate the steady-state position and velocity estimation errors, the transfer function G_1 is designed as follows;

$$G_1 = \frac{a_1 s + b_1}{s^2} \quad (40)$$

6.3 Handling by $n+1$ mobile robots in coordination

6.3.1 Forces and trajectory deviations

We extend the result in the previous section to a general case. First, we consider the relationship among trajectory deviations. We assumed that the external force applied to the object is negligible and consider the geometric constraints expressed by eq.(13). We have the following relations with respect to the forces applied to each robot;

$$F_l^{ext} + \sum_{j=1}^{n} F_j^{ext} = 0 \quad (41)$$

As time tends to infinity, we obtain the following relationship from eq.(31), eq.(32), eq.(41) and the positive definite of the damping matrix and the stiffness matrix, even if the initial values of $K'_l \Delta X_l + \sum_{j=1}^{n} K'_j \Delta X_j$ is not zero.

$$K'_l \Delta X_l + \sum_{j=1}^{n} K'_j \Delta X_j = 0 \quad (42)$$

6.3.2 Dynamics of Virtual Leader

It is impossible for the i-th follower to estimate the desired trajectory of the leader because the trajectory deviation of the i-th follower, which was used for the estimation of the desired trajectory of the leader in previous case, is affected by motions of all of the robots. Therefore, for the i-th follower, the robots is classified into two groups as shown in Figure 5(a); one is the i-th follower itself and the other is the rest of the robots including the leader. In this paper, we referred to the rest of the robots as the i-th virtual leader. The i-th virtual leader consists of the leader, and j-th followers ($j = 1, \ldots, i-1, i+1, \ldots, n$). For the i-th follower, the i-th virtual leader behaves as if it is a real leader as shown in Figure 5(b). Using the concept of the virtual leader, the i-th follower estimate the desired trajectory of the i-th virtual leader based on the estimation algorithm in the previous section.

We can derive the dynamics of the i-th virtual leader as follows;

$$D'_l \Delta \dot{X}_l + \sum_{j=1(j\neq i)}^{n} D'_j \Delta \dot{X}_j$$
$$+ K'_l \Delta X_l + \sum_{j=1(j\neq i)}^{n} K'_j \Delta X_j$$
$$= F_l^{ext} + \sum_{j=1(j\neq i)}^{n} F_j^{ext} \quad (43)$$

where

$$\sum_{j=1(j\neq i)}^{n} c_j = \sum_{j=1}^{i-1} c_j + \sum_{j=i+1}^{n} c_j \quad (44)$$

The trajectory deviation of the i-th virtual leader $\Delta X_{li}, \Delta \dot{X}_{li}$ are expressed as follows;

$$\Delta \dot{X}_{li} = \frac{1}{D'_i}\{D'_l \Delta \dot{X}_l + \sum_{j=1(j\neq i)}^{n} (D'_j \Delta \dot{X}_j)\} \quad (45)$$

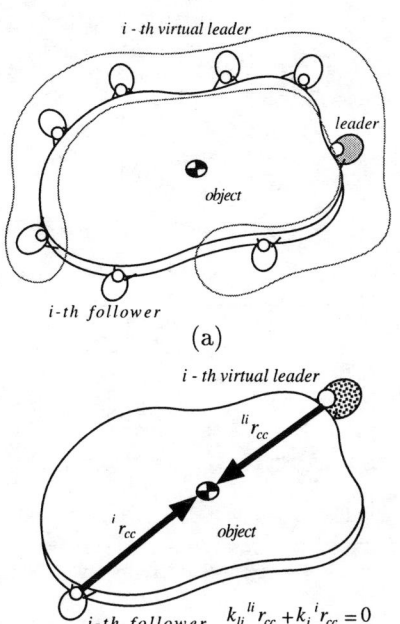

Figure 5: Virtual Leader

$$\Delta X_{li} = \frac{1}{K'_i}\{(K'_l \Delta X_l + \sum_{j=1(j\neq i)}^{n}(K'_j \Delta X_j)\} \quad (46)$$

As time tends to infinity, from eq.(42), we obtain the following relationship, regardless of the desired trajectory given to the leader X_{ld}.

$$\Delta X_{li} + \Delta X_i = 0 \quad (47)$$

The force and moment applied to the i-th virtual leader F_{li}^{ext} is expressed as follows;

$$F_{li}^{ext} = F_l^{ext} + \sum_{j=1(j\neq i)}^{n} F_j \quad (48)$$

From eq.(41), we also have

$$F_{li}^{ext} + F_i^{ext} = 0 \quad (49)$$

From eq.(45), eq.(46) and eq.(48), (43) is rewritten as

$$D'_i \dot{\Delta X}_{li} + K'_i \Delta X_{li} = F_{li}^{ext} \quad (50)$$

It should be noted that eq.(50) expresses the behavior of the i-th virtual leader. Let X_{di} be the desired trajectory of the representative point attached to the object which are calculated by transforming the coordination of the desired trajectories of the i-th virtual leader. Then, $\Delta X_{li} = (\Delta x_{li}, \Delta \theta_{li})^T$ are expressed as

$$\Delta x_{li} = \Delta x_{oli} - \Delta \theta_{oli}{}^{li}R^T \quad \Delta \theta_{li} = \Delta \theta_{oli} \quad (51)$$

where Δx_{oli}, $\Delta \theta_{oli}$ are deviation relating to the position and the orientation of representative point of the object corresponding to the motion of the i-th virtual leader, and are expressed as follows;

$$\Delta x_{oli} = x - x_{di} \quad \Delta \theta_{oli} = \theta - \theta_{di} \quad (52)$$

6.3.3 Estimation

Using the concept of the virtual leader, the i-th follower can estimate the desired trajectory of the i-th virtual leader based on the estimation algorithm which we discussed in the previous section because eq.(50) and eq.(51) have the form as (31) and (21).

We define ΔX_{di} as the estimation error of the i-th virtual leader estimated by the follower. ΔX_{di} is expressed as

$$\Delta X_{di} = X_{di} - X_{ei} = 2\Delta X_i \quad (53)$$

It should be noted that ΔX_{di} is calculated by each follower based on the observable state of each follower ΔX_i. Using this ΔX_{di} and the transfer function matrix G_i which we design in previous section, the i-th follower can estimate the desired trajectory of the i-th virtual leader. We design the transfer function matrix G_i similar to the case of previous section. The transfer function matrix G_i is expressed as follows;

$$G_i = \frac{a_i s + b_i}{s^2} I_3 \quad (54)$$

The stability of the resultant system is shown similar to [5].

7 Experiments

First, we did two types of experiments to compare the conventional algorithm proposed in [5] and the proposed algorithm explained in the previous section to handle a large object in coordination. We did these experiments using two autonomous omnidirectional mobile robots as shown in Figure 6(a). Second, we did the experiment using three autonomous omnidirectional mobile robots with the proposed algorithm explained in the previous section as shown in Figure 6(b). The control algorithm was implemented in the autonomous omnidirectional mobile robots, ZEN, developed by RIKEN[7]. Each mobile robot has three degrees of freedom of motion and equipped with the Body Force Sensor[8]. The control algorithm is implemented using VxWorks. The sampling rate is 1024Hz.

7.1 Experimental results based on conventional algorithm

In this experiment with the conventional algorithm, the leader was given a desired trajectory along y-axis which was calculated by a fifth order function. The orientations of all of the robots were kept constant during the transportation of the object. The results are shown in Figure 7. You can see that the moment detected by each robot include vibratory part. The vibratory part caused estimation error and the follower could not follow the leader accurately.

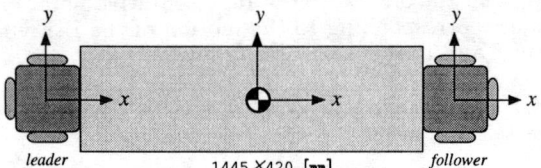

(a) Experiment by two mobile robots

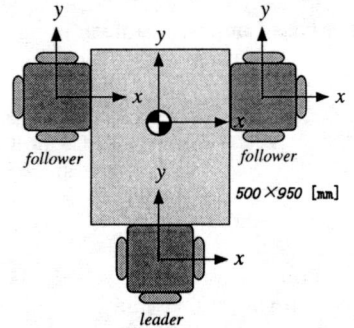

(b) Experiment by three mobile robots

Figure 6: Experimental system

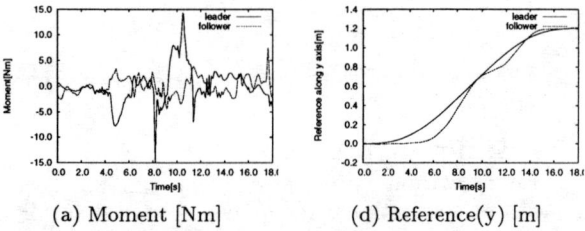

(a) Moment [Nm] (d) Reference(y) [m]

Figure 7: Experiment using conventional algorithm

7.2 Experimental results using proposed control algorithm

This experiment was done using the same desired trajectory of the object and the same impedance parameters as the previous experiment. The results are shown in Figure 8. You can see that the moment of each robot include little vibration. The less estimation error was observable than the previous experiment and the transportation of the large object was successfully achieved.

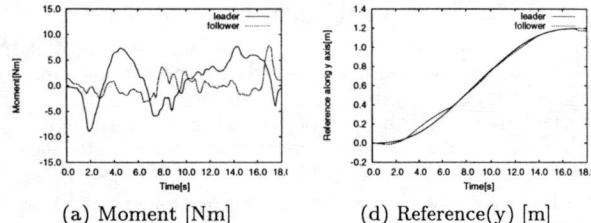

(a) Moment [Nm] (d) Reference(y) [m]

Figure 8: Experiment using proposed algorithm

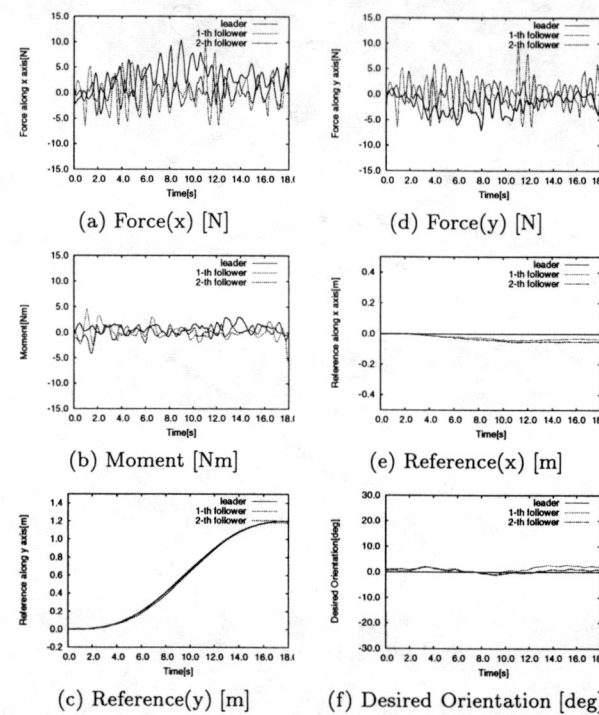

(a) Force(x) [N] (d) Force(y) [N]

(b) Moment [Nm] (e) Reference(x) [m]

(c) Reference(y) [m] (f) Desired Orientation [deg]

Figure 9: Experimental results by three mobile robots

7.3 Experimental results using 3 mobile robot

In the experiment using three autonomous omnidirectional mobile robots, the leader was given the desired trajectory along y-axis which was calculated by a fifth order function and each follower estimated the desired trajectory of its own virtual leader using the algorithm proposed in the previous section. The results are shown in Figure 9. The three robots transported the large object in coordination successfully.

Figure 11 shows an example of the experiments. In this figure, the desired trajectory of the object was commanded to the leader as shown in Figure 10

8 Conclusions

In this paper, we pointed out a problem relating to the force/moment transformation for the coordinated motion control problem of multiple mo-

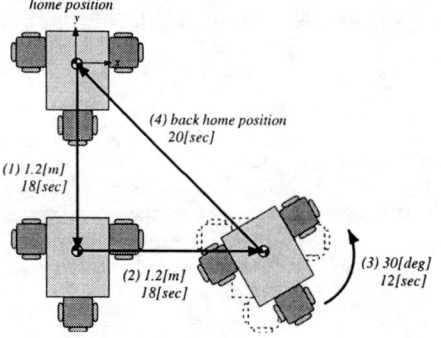

Figure 10: Desired trajectory in Figure11

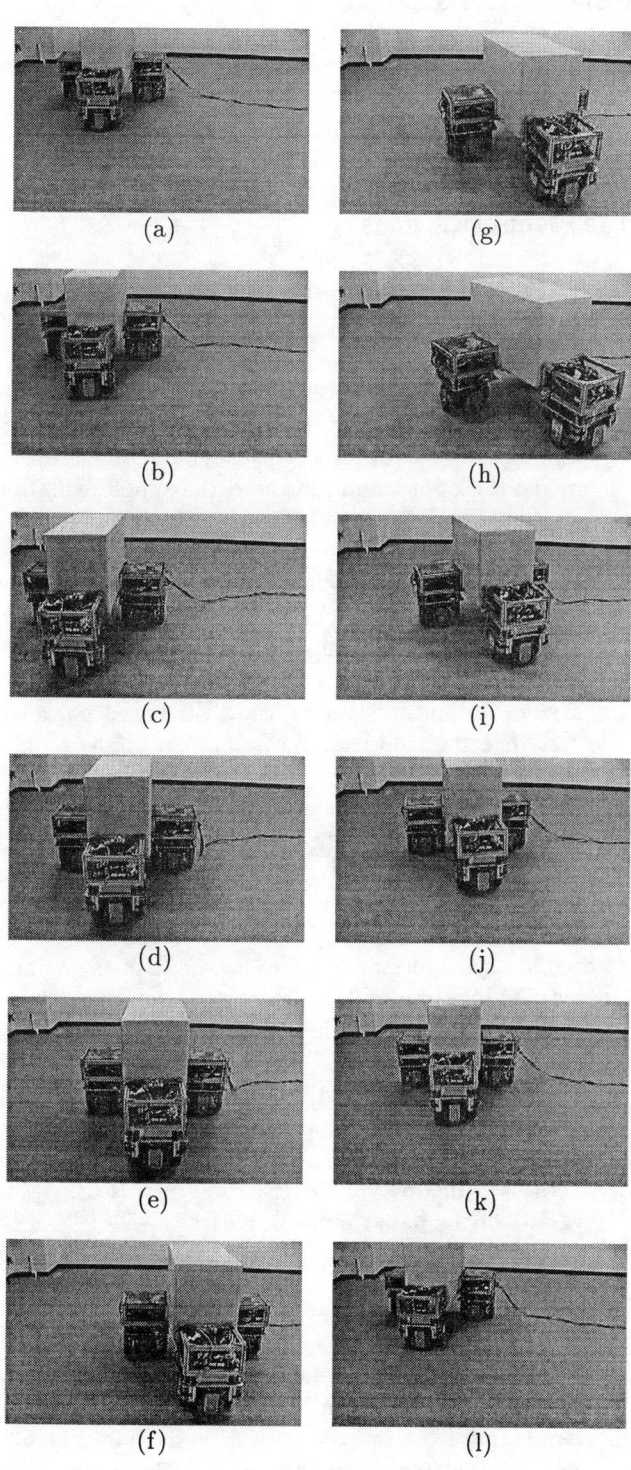

Figure 11: Example of experiment

bile robots handling a large object in coordination. The control algorithm with geometrical constraints expressed by eq.(15) was proposed to overcome this problem. This algorithm is efficient to lessen the effect of noises caused by force/moment transformation, since most of the impedance parameters do not include the vector cross product terms with respect to the distance between the grasping point and the representative point. With the algorithm, we could specify the apparent impedance of the handling object completely.

We extend this algorithm to the decentralized control algorithm of multiple robots handling a single object in coordination. The proposed decentralized control algorithm is experimentally applied to the omnidirectional mobile robots, ZEN. Experimental results illustrate the validity of the proposed control algorithm.

References

[1] E. Nakano, S. Ozaki, T. Ishida, I. Kato, "Co-operational Control of the Anthropomorphous Manipulator "MELARM", *Proc. of 4th International Symposium on Industrial Robots*, pp.251-260, Tokyo 1974.

[2] Y.F. Zheng, J.Y.S. Luh, "Optimal Load Distribution for Two Industrial Robots Handling a Single Object", *Proc. of IEEE International Conference on Robotics and Automation*, pp.344-349, 1988.

[3] Y. Nakamura, K. Nagai, T. Yoshikawa, "Mechanics of Cooperative Manipulation by Multiple Robotic Mechanism", *Proc. of 1987 International Conference on Robotics and Automation*, pp.991-998, 1987.

[4] J. Ota, Y. Buei, T. Arai, H. Osumi, K. Suyama, "Transferring Control by Cooperation of Two Mobile Robots", *Journal of the Robots Society of Japan*, Vol. 14, pp.263–270, 1996.(In Japanese).

[5] K. Kosuge, T. Oosumi, "Decentralized Control of Multiple Robots Handling an Object", *Proc. of 1996 IEEE Int. Conf. on Intelligent Robots and Systems*, pp.318–323, 1996.

[6] K. Kosuge, S. Hashimoto, K. Takeo "Coordinated Motion Control of Multiple Robots Manipulating a Large Object",*Proc. of 1997 IEEE/RSJ Int. Conf. on Intelligent Robots and Systems*, pp.208-213, 1997.

[7] H. Asama, M. Sato, H. kaetsu, K. Ozaki, A. Matsumoto, I.Endo, "Development of an Omni-Directional Mobile Robot with 3 DoF Decoupling Drive Mechanism" *Journal of the Robots Society of Japan*, Vol. 14, pp.249–254, 1996.(In Japanese)

[8] K. Kosuge, T. Oosumi, Y. Hirata, H. Asama, H. Kaetsu, K. Kawabata "Handling of a Single Object by Multiple Autonomous Mobile Robots in Coordination with Body Force Sensor", *Proc. of 1998 IEEE Int. Conf. on Intelligent Robots and Systems*, to appear.

Analysis of Deformable Object Handling

Herbert G. Tanner, Kostas J. Kyriakopoulos

Control Systems Laboratory,
Division of Machine Design and Control,
Dept. of Mechanical Engineering,
National Technical University of Athens, NTUA
e-mail: {htanner,kkyria}@central.ntua.gr

Abstract

A manipulated deformable object is viewed as an underactuated mechanical system. In this context controllability issues are discussed and results on the nature of the constraints and the controllability properties of an important class of deformable objects being modeled with finite elements are stated. For this class of deformable objects the results permit to circumvent the usual procedure of calculating Lie brackets to establish a base for the associated Lie algebra, and answers the question of determining the kind of constraints imposed on the system in a straightforward algebraic way. Inequality constraints associated to material strength limitations are also included.

1 Introduction

Among all kinds of material that a robot is called to manipulate, very few are actually rigid. Our world is formed mainly of deformable materials, the flexibility of which varies significantly. It becomes clear that different kinds of material can not allow a uniform treatment. Therefore, there should be a way to distinguish between manipulated objects that have different properties and follow different handling strategies, bearing in mind the individual characteristics.

Previous approaches to deformable object handling focus mainly on formulating general continuous dynamic equations for the object in a way to enhance computation or allow for certain control strategies. Sun et al. [9] followed Terzopoulos' [11] hybrid approach to deformable objects. Kosuge et al. [2] used finite elements but ignored the dynamics of the object and consentrated only on the static conditions. Wu et al. [12] approximated the distributed parameter system with a lumped parameter model. Yukawa et al. [13] investigated a vibrating flexible object and modeled it using model reduction theory.

The authors have previously modeled a deformable object being manipulated by multiple mobile manipulators [10]. They used elastodynamic equations to model the object and indicated the simplest finite element grid structure able to describe the object being handled by multiple manipulators. In this paper we generalize this approach, by introducing a framework that both includes a rich variety of mechanical systems and allows more detailed description. We consider the deformable object as an underactuated mechanical system [8] and 'discretize' the distributed parameter system using finite elements [6]. As underactuated systems, a great variety of mechanical systems can be included in this framework e.g. chains, structures with passive joints, systems with rolling contact, etc. This broadens considerably the prespective of the approach to object handling.

The rest of the paper is organized as follows: In section 2 a deformable object is described as an underactuated system and the state equations are derived. In section 3 the dynamic constraints imposed on the system are classified. The controllability properties are investigated in section 4. Constraints related to material strength limitations are included in section 5. In section 6, some examples are presented which verify the theoretical results. Finally, in section 7 the results of the paper are summarized.

2 The Underactuated System

Strictly speaking, a deformable object has infinite degrees of freedom. An attempt to simplify the problem is to 'discretize' the structure, reducing the number of its degrees of freedom to a finite countable set. A popular way is finite elements.

The extend of discretization can depend on the individual characteristics of the material. Almost rigid materials do not require dense discretization grids; flexible ones do. Adjusting the grid, completely rigid to flexible materials can be described.

We consider the discetized deformable object as an underactuated mechanical system. Underactuated systems have less inputs than degrees of freedom [8]. The system at hand is underactuated since only a few degrees of freedom are directy controlled, namely the ones coinciding with the grasp points. The rest are staticaly and dynamically coupled to the actuated and can be regarded as passive.

The class of underactuated mechanical systems is very broad and includes systems with passive joints, flexible link robots, mobile robots, flexible link robots, space robots and a variety of other systems out of many robotics fields. This broad class motivates the investigation of new manipulation tasks, including systems of rigid bodies or combinations of rigid and deformable objects, modeled as underactuated systems.

Following the Lagrangian formulation of the dynamics of mechanical systems with n generalized coordinates $\mathbf{q} = (q_1, \ldots, q_n)^T \in \mathbf{Q}$, the equations of motion can be derived

$$\frac{d}{dt}\left(\frac{\partial \mathcal{L}}{\partial \dot{q}_i}\right) - \frac{\partial \mathcal{L}}{\partial q_i} = \mathbf{F}_i \qquad i = 1, \ldots, n$$

It is well known that the above can take the form:

$$\mathbf{M}(\mathbf{q})\ddot{\mathbf{q}} + \mathbf{C}(\mathbf{q},\dot{\mathbf{q}})\dot{\mathbf{q}} + \mathbf{K}(\mathbf{q}) = \mathbf{B}(\mathbf{q})\mathbf{u} \qquad (1)$$

where $\mathbf{M}$ is the symmetric and positive definite inertia matrix, $\mathbf{C}$ contains the Coriolis and centrifugal terms, $\mathbf{K}$ is formed by the terms associated with gravity and elastic forces, and u is input. In the case of underactuated systems the space of generalized coordinates can be partisioned into an actuated and passive part: $\mathbf{q}^T = (\mathbf{q}_1^T, \mathbf{q}_2^T)$, and (1) can be written [8]

$$\mathbf{m}_{11}\ddot{\mathbf{q}}_1 + \mathbf{m}_{12}\ddot{\mathbf{q}}_2 + \mathbf{c}_1(\mathbf{q},\dot{\mathbf{q}}) + \mathbf{k}_1(\mathbf{q}) = 0 \qquad (2a)$$
$$\mathbf{m}_{21}\ddot{\mathbf{q}}_1 + \mathbf{m}_{22}\ddot{\mathbf{q}}_2 + \mathbf{c}_2(\mathbf{q},\dot{\mathbf{q}}) + \mathbf{k}_2(\mathbf{q}) = \mathbf{b}(\mathbf{q})\mathbf{u} \qquad (2b)$$

where $\mathbf{b}(\mathbf{q}) \in \Re^{m \times m}$ is assumed nonsingular.

Viewing (2a) as a set of $n-m$ dynamic constraints, a natural question to ask what kind are they. We will name them *intrinsic constraints* to distinguish them from any external imposed constraints related to object material strength or obstacle avoidance. Intrinsic constraints can either be holonomic, first order nonholonomic or second order nonholonomic. The kind of constraints imposed determines the controllability properties of the system. Therefore, it is quite important to classify these constraints before proceeding to investigating controllability properties.

2.1 Collocated Linearization and State Space Description

Certain forms of system description enhance analysis. A valuable tool for analysis is feedback linearization, which unveils the system structure. An important property of system (2a) - (2b) is that it can be partially feedback linearized with respect to the actuated degrees of freedom [8]. Indeed, by examination of (2a) it can be seen that $\mathbf{m}_{11}$ is square and nonsingular, since the original inertia matrix is positive definite. Therefore (2a) can be solved for $\ddot{\mathbf{q}}_1$

$$\ddot{\mathbf{q}}_1 = -\mathbf{m}_{11}^{-1}[\mathbf{m}_{12}\ddot{\mathbf{q}}_2 + \mathbf{c}_1(\mathbf{q},\dot{\mathbf{q}}) + \mathbf{k}_1(\mathbf{q})]$$

and then substitute in (2b)

$$\bar{\mathbf{m}}(\mathbf{q})\ddot{\mathbf{q}}_2 + \bar{\mathbf{c}}(\mathbf{q},\dot{\mathbf{q}}) + \bar{\mathbf{k}}(\mathbf{q}) = \mathbf{b}\mathbf{u}$$

where

$$\bar{\mathbf{m}}(\mathbf{q}) = \mathbf{m}_{22}(\mathbf{q}) - \mathbf{m}_{21}(\mathbf{q})\,\mathbf{m}_{11}^{-1}(\mathbf{q})\,\mathbf{m}_{12}(\mathbf{q})$$
$$\bar{\mathbf{c}}(\mathbf{q},\dot{\mathbf{q}}) = \mathbf{c}_2(\mathbf{q},\dot{\mathbf{q}}) - \mathbf{m}_{21}(\mathbf{q})\,\mathbf{m}_{11}^{-1}(\mathbf{q})\,\mathbf{c}_1(\mathbf{q},\dot{\mathbf{q}})$$
$$\bar{\mathbf{k}}(\mathbf{q}) = \mathbf{k}_2(\mathbf{q}) - \mathbf{m}_{21}(\mathbf{q})\,\mathbf{m}_{11}^{-1}(\mathbf{q})\,\mathbf{k}_1(\mathbf{q})$$

Now, using the linearizing feedback

$$\mathbf{u} = \mathbf{b}^{-1}[\bar{\mathbf{m}}(\mathbf{q})\mathbf{v} + \bar{\mathbf{c}}(\mathbf{q},\dot{\mathbf{q}}) + \bar{\mathbf{k}}(\mathbf{q})] \qquad (3)$$

the system (2a) - (2b) takes the form

$$\ddot{\mathbf{q}}_2 = \mathbf{v} \qquad (4a)$$
$$\ddot{\mathbf{q}}_1 = \mathbf{J}(\mathbf{q})\ddot{\mathbf{q}}_2 + \mathbf{R}(\mathbf{q},\dot{\mathbf{q}}) \qquad (4b)$$

where

$$\mathbf{J}(\mathbf{q}) = -\mathbf{m}_{11}^{-1}(\mathbf{q})\mathbf{m}_{12}(\mathbf{q})$$
$$\mathbf{R}(\mathbf{q},\dot{\mathbf{q}}) = -\mathbf{m}_{11}^{-1}(\mathbf{q})\mathbf{c}_1(\mathbf{q},\dot{\mathbf{q}}) - \mathbf{m}_{11}^{-1}(\mathbf{q})\mathbf{k}_1(\mathbf{q})$$

Then setting

$$\mathbf{x}_1 = \mathbf{q}_2 \in \Re^m, \quad \mathbf{x}_2 = \mathbf{q}_1 \in \Re^{n-m}, \quad \mathbf{x}_3 = \dot{\mathbf{q}}_2, \quad \mathbf{x}_4 = \dot{\mathbf{q}}_1,$$

equations (4a)-(4b) come into state space form [7]

$$\dot{\mathbf{x}}_1 = \mathbf{x}_3 \qquad (5a)$$
$$\dot{\mathbf{x}}_2 = \mathbf{x}_4 \qquad (5b)$$
$$\dot{\mathbf{x}}_3 = \mathbf{v} \qquad (5c)$$
$$\dot{\mathbf{x}}_4 = \mathbf{J}(\mathbf{x}_1, \mathbf{x}_2)\mathbf{v} + \mathbf{R}(\mathbf{x}_1, \mathbf{x}_2, \mathbf{x}_3, \mathbf{x}_4) \qquad (5d)$$

3 Constraint Classification

When the manipulated object is considered completely rigid, intrinsic constraints permit integration yielding algebraic expressions of the form

$$\mathbf{q}_1 = g(\mathbf{q}_2)$$

In this case they are **holonomic** and they can be used to eliminate a number of generalized coordinates and reduce the dimension of the state space by $2(n-m)$:

$$\dot{\mathbf{x}}_1 = \mathbf{x}_3 \qquad \dot{\mathbf{x}}_3 = \mathbf{v}$$

The second case is when the constraints are **first-order nonholonomic**. The dimension of the configuration space remains the same, but the dimension of its tangent bundle is reduced by $n-m$. The constraints can be expressed in the form

$$\mathbf{A}(\mathbf{q}) \begin{bmatrix} \dot{\mathbf{x}} \\ \mathbf{x} \end{bmatrix} = 0 \quad \mathbf{A} \in \Re^{(n-m) \times 2n}. \quad (6)$$

A base $\mathbf{S}(\mathbf{q}) \in \Re^{(n+m) \times 2n}$ for the annihilator of $\mathbf{A}$ is formed and (1) is 'projected' into the space generated by the columns of $\mathbf{S}^T$. This results in $n+m$ state equations. The existence of $\mathbf{S}$ implies a relation [1]:

$$\begin{bmatrix} \dot{\mathbf{x}} \\ \mathbf{x} \end{bmatrix} = \mathbf{S}^T \eta \qquad \eta \in \Re^{m+n} \quad (7)$$

which can be used to obtain

$$\dot{\eta} = \mathbf{D}(\mathbf{q}) + \mathbf{G}(\mathbf{q}, \eta) \mathbf{u} \quad (8)$$

where $\mathbf{D}$ and $\mathbf{G}$ are derived by differentiation and substitution of (7) into (1) and 'projection' of the resulting equations onto the distribution spanned by $\mathbf{S}^T$.

The last case is when these constraints are completely nonintegrable, forming a set of $n-m$ **second-order nonholonomic constraints**. A formal definition of second order nonholonomic constraints is given in [7] by defining the distribution $\mathbf{\Delta} = span\{\tau_0, \tau_j\}$ where

$$\tau_0 = \sum_{j=1}^{m} \dot{q}_{2,j} \frac{\partial}{\partial q_{2,j}} + \sum_{k=1}^{n-m} (\dot{q}_{1,k} \frac{\partial}{\partial q_{1,k}} + R_k \frac{\partial}{\partial \dot{q}_{1,k}}) + \frac{\partial}{\partial t}$$

$$\tau_j = \frac{\partial}{\partial \dot{q}_{2,j}} + \sum_{i=1}^{n-m} J_{ij} \frac{\partial}{\partial \dot{q}_{1,i}}, \quad j = 1, \ldots, m.$$

Definition. *(from [7]): Consider the distribution $\mathbf{\Delta}$ and $\tilde{\mathcal{C}}$ its accessibility algebra. Let $\tilde{\mathbf{C}}$ the accessibility distribution generated by $\tilde{\mathcal{C}}$. The system (4a)-(4b) is said to be completely second-order nonholonomic if $dim\tilde{\mathbf{C}}(\mathbf{x}, t) = 2n+1, \forall (\mathbf{x}, t) \in \mathbf{M} \times \Re$.*

Here, the dimension of the state space is not reduced. We show that finite element models of deformable objects exhibit second-order nonholonomic constraints. The presence of second-order nonholonomic constraints is typical in the dynamic description of underactuated robots [5].

4 Controllability Issues

When k intrincic constraints are holonomic, then the system motion is restricted to an $n-k$ dimensional manifold. The system has $n-k$ degrees of freedom; the remaining k can be eliminated. Control must be restricted to this manifold of admissible motions.

If the constraints turn out to be completely first order nonholonomic then the system configuration space is not confined. Proving that these constraints are completely nonholonomic is equivalent to evaluating the dimension of the accessibility distribution of the system, $\mathbf{C}$, which should be $2n$. Then the system is accessible. Due to the presence of a drift term $\mathbf{D}$ in (8), accessibility do not imply controllability. For nonholonomic systems with drift there is no available general necessary and sufficient result for establishing complete controllability [4]. One has to resort to other forms of controllability, such as strong accessibility and STLC (small-time local controllability). For the latter there exist only sufficient conditions, but once it has been established one can utilize the manifold of equilibrium points of the drift vector to reach an arbitrary small neighborhood of the desired configuration. Systems which are STLC are not asymptotically stabilizable via time-invariant feedback; piecewise analytic, however, may be used.

If the system (4a)-(4b) is proved to be second-order nonholonomic, then it is has been proved that it is strongly accessible [7]. Moreover, there is a chance for smooth stabilization, provided that a sufficient condition for non-existence of a smooth stabilizing control law is not satisfied.

Theorem. *([7]): Assume that $R_i(\mathbf{q}, 0) = 0, \forall \mathbf{q} \in \mathbf{Q}$, for some $i \in I_{n-m} = \{1, \ldots, n-m\}$. Let $n - m \geq 1$ and let $(\mathbf{q}^e, 0)$ denote an equilibrium solution. Then the second-order nonholonomic system, defined by (4a)-(4b), is not asymptotically stabilizable to $(\mathbf{q}^e, 0)$ using time-invariant continuous (static or dynamic) state feedback law.*

If $\forall i, \exists \mathbf{q} \in \mathbf{Q} | R_i(\mathbf{q}, 0) \neq 0$, then the system could perhaps be stabilizable by continuous control law.

In the remaining of this section we will show that a class of deformable objects being modeled with finite elements share some interesting properties if they satisfy a certain condition. The discussion concentrates on finite element models for which (1) takes the form:

$$\mathbf{M\ddot{q} + C\dot{q} + Kq = Bu} \quad (9)$$

i.e. the characteristic matrices are independent of the generalized coordinates and speeds. The above model is standard and can be found in many finite elements textbooks [6]. The state equations derived from the above model have the form

$$\ddot{\mathbf{q}}_2 = \mathbf{v} \quad (10a)$$
$$\ddot{\mathbf{q}}_1 = \mathbf{J}\ddot{\mathbf{q}}_2 + \mathbf{R}(\mathbf{q}, \dot{\mathbf{q}}) \quad (10b)$$

We proceed with the following Lemma:

Lemma 1. *For equations (10a - 10b) it holds:*

1. $[\tau_k, ad_{\tau_0}^r \tau_j] = 0 \quad \forall r \geq 0, \quad k, j \in \{1, \ldots, m\}$

2. $ad_{\tau_0}^{1+r} \tau_j = \begin{bmatrix} A_r \begin{bmatrix} \mathbf{J}_j & e_j \end{bmatrix}^T \\ 0_{m \times 1} \\ B_r \begin{bmatrix} \mathbf{J}_j & e_j \end{bmatrix}^T \\ 0_{(m+1) \times 1} \end{bmatrix}$ *for $r > 0$ and*
$A_r, B_r \in \Re^{(n-m) \times 1}$, *where $\mathbf{J}_j$ is the jth column of $\mathbf{J}$ and e_j is the jth base vector of $\Re^m$.*

3. *The vectors that form the vector field $ad_{\tau_0}^{1+i} \tau_j$ can be expressed as*

$$B_r = (-1)^{r-1} \left[\frac{\partial \mathbf{R}}{\partial \mathbf{q}_1} A_{r-1} + \frac{\partial \mathbf{R}}{\partial \dot{\mathbf{q}}_1} B_{r-1} \right]$$
$$A_r = (-1)^{r-1} B_{r-1} \quad with$$
$$B_1 = \frac{\partial \mathbf{R}}{\partial \mathbf{q}} + \frac{\partial \mathbf{R}}{\partial \dot{\mathbf{q}}_1} \cdot \frac{\partial \mathbf{R}}{\partial \dot{\mathbf{q}}} \qquad A_1 = \frac{\partial \mathbf{R}}{\partial \dot{\mathbf{q}}}$$

Proof. (1) can be proved by noticing that if we set $\bar{q} = \begin{bmatrix} \mathbf{q}^T & \dot{\mathbf{q}}^T & t \end{bmatrix}^T$ then $\frac{\partial \tau_j}{\partial \bar{q}} = 0$. (2) can be proved by straightforward calculation, taking into account that $\mathbf{J}$ is independent of $\bar{q}$ and $\mathbf{R}$ does not contain quadratic terms in neither $\mathbf{q}$ or $\dot{\mathbf{q}}$. For (3), note the special structure of $\frac{\partial \tau_0}{\partial \bar{q}}$ and that $\frac{\partial ad_{\tau_0}^i \tau_j}{\partial \bar{q}} = 0$. □

Our main result follows:

Proposition 1. *If for system (9) all matrices in the sequence*

$$\begin{bmatrix} A_1 \\ 0 \\ B_1 \\ 0 \end{bmatrix} \begin{bmatrix} \mathbf{J} \\ \mathbf{I}_{m \times m} \end{bmatrix} \cdots \begin{bmatrix} A_{\frac{2(n-m)}{m}} \\ 0 \\ B_{\frac{2(n-m)}{m}} \\ 0 \end{bmatrix} \begin{bmatrix} \mathbf{J} \\ \mathbf{I}_{m \times m} \end{bmatrix}$$

have full rank, (9) is second-order nonholonomic.

Proof. The proof follows immediately from the above Lemma and from Definition, by noting that this sequence is directly accossiated with the filtration of the accessibility distribution $\tilde{C}$. The filtration is regular with relative growth vector $s = (m+1, m, \underbrace{m, \ldots, m}_{r})$.

Therefore the dimension of the distribution is finally $2m + 1 + \frac{2(n-m)}{m}m = 2n$ □

Being second-order nonholonomic, the finite element model (9) is also **strongly accessible** [7]. Moreover, a careful investigation shows that it **does not satisfy the sufficient condition for STLC** presented in [3], since $\mathbf{J}$ is constant and $\mathbf{R}$ does not have a term which is quadratic in velocities. This of course does not mean that the system is not STLC. Finally, at the equilibrium the system does not satisfy the condition $R(\mathbf{q}^e, 0) = 0 \ \forall \mathbf{q}$ which is sufficient for nonexistence of a time-invariant continuous state feedback law. Therefore, such a control law might exist.

5 Material Constraints

In practice, rarely do we need to control all degrees of freedom [10]. Steering only the directly controlled degrees of freedom is usually sufficient, provided that the rest are confined within specific limits. These limits naturally arise from material strength limitations and are written in the form:

$$\boldsymbol{\sigma} \leq \bar{\boldsymbol{\sigma}} \quad (11)$$

where $\boldsymbol{\sigma}$ is the stress tensor of the structure and $\bar{\boldsymbol{\sigma}}$ is the maximum admissible stress for the particular material and object. These constraints can be included in (9) through Kuhn-Tucker multipliers:

$$\mathbf{M\ddot{q} + C\dot{q} + Kq = Bu} + (\frac{\partial \mathbf{r}}{\partial \mathbf{q}})^T \boldsymbol{\mu}$$
$$\boldsymbol{\mu}^T \mathbf{r}(\mathbf{q}) = 0, \quad \boldsymbol{\mu} \geq 0$$

Note that in this case, the multipliers cannot be eliminated because they correspond to inequality conditions which do not reduce the dimension of the state space. Then the Kuhn-Tucker term can be included into the potential terms, yielding equations (2a)-(2b)

$$\mathbf{m}_{11} \ddot{\mathbf{q}}_1 + \mathbf{m}_{12} \ddot{\mathbf{q}}_2 + \mathbf{c}_1 \dot{\mathbf{q}} + \mathbf{k}_1(\mathbf{q}, \boldsymbol{\mu}) = 0$$
$$\mathbf{m}_{21} \ddot{\mathbf{q}}_1 + \mathbf{m}_{22} \ddot{\mathbf{q}}_2 + \mathbf{c}_2 \dot{\mathbf{q}} + \mathbf{k}_2(\mathbf{q}, \boldsymbol{\mu}) = \mathbf{bu}$$
$$\boldsymbol{\mu}^T \mathbf{r}(\mathbf{q}) = 0, \quad \boldsymbol{\mu} \geq 0$$

When the stress conditions are satisfied, then $\boldsymbol{\mu}$ vanish and the equations describe the motion of an unconstrained system.

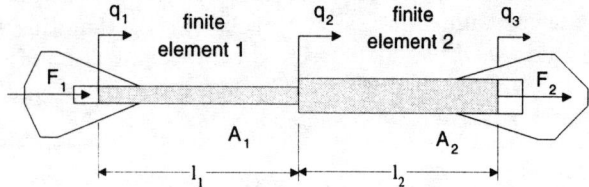

Figure 1: A deformable object under axial load

6 Examples

6.1 Rod under axial load

Consider a beam under axial load. The beam is divided into two finite elements (Figure 1). The system has three degrees of freedom, two of which are directly controlled. The element characteristic matrices are:

$$\mathbf{M} = \frac{\rho A \ell}{6} \begin{bmatrix} 2 & 1 \\ 1 & 2 \end{bmatrix} \quad \mathbf{C} = \frac{\mu A \ell}{6} \begin{bmatrix} 2 & 1 \\ 1 & 2 \end{bmatrix}$$
$$\mathbf{K} = \frac{AE}{\ell} \begin{bmatrix} 1 & -1 \\ -1 & 1 \end{bmatrix}$$

Assembling the element equations to form the complete equations and rearranging the terms to single out the actuated part from the unactuated, yields the dynamic equations

$$\begin{bmatrix} \frac{\rho(A_1\ell_1 + A_2\ell_2)}{3} & \frac{\rho A_1 \ell_1}{6} & \frac{\rho A_2 \ell_2}{6} \\ \frac{\rho A_1 \ell_1}{6} & \frac{\rho A_1 \ell_1}{3} & 0 \\ \frac{\rho A_2 \ell_2}{6} & 0 & \frac{\rho A_2 \ell_2}{3} \end{bmatrix} \begin{bmatrix} \ddot{q}_2 \\ \ddot{q}_1 \\ \ddot{q}_3 \end{bmatrix}$$
$$+ \begin{bmatrix} \frac{\mu(A_1\ell_1 + A_2\ell_2)}{3} & \frac{\mu A_1 \ell_1}{6} & \frac{\mu A_2 \ell_2}{6} \\ \frac{\mu A_1 \ell_1}{6} & \frac{\mu A_1 \ell_1}{3} & 0 \\ \frac{\mu A_2 \ell_2}{6} & 0 & \frac{\mu A_2 \ell_2}{3} \end{bmatrix} \begin{bmatrix} \dot{q}_2 \\ \dot{q}_1 \\ \dot{q}_3 \end{bmatrix}$$
$$+ \begin{bmatrix} \frac{A_1 E}{\ell_1} + \frac{A_2 E}{\ell_2} & -\frac{A_1 E}{\ell_1} & -\frac{A_2 E}{\ell_2} \\ -\frac{A_1 E}{\ell_1} & \frac{A_1 E}{\ell_1} & 0 \\ -\frac{A_2 E}{\ell_2} & 0 & \frac{A_2 E}{\ell_2} \end{bmatrix} \begin{bmatrix} q_2 \\ q_1 \\ q_3 \end{bmatrix} = \begin{bmatrix} 0 \\ F_1 \\ F_2 \end{bmatrix}$$

Where A_1, A_2 and ℓ_1, ℓ_2 are the elements cross sections and lengths, respectively. Applying the linearizing feedback (3) yields the familiar form (10a-10b)

$$\ddot{q}_1 = v_1$$
$$\ddot{q}_3 = v_2$$
$$\ddot{q}_2 = \mathbf{J} \begin{bmatrix} v_1 & v_2 \end{bmatrix}^T + R$$

where

$$\mathbf{J} = -\frac{1}{2(A_1\ell_1 + A_2\ell_2)} \begin{bmatrix} A_1\ell_1 & A_2\ell_2 \end{bmatrix}$$
$$R = -\frac{3}{\rho(A_1\ell_1 + A_2\ell_2)} \left[\frac{\mu}{3}(A_1\ell_1 + A_2\ell_2)\dot{q}_2 + \frac{\mu A_1 \ell_1}{6}\dot{q}_1 \right.$$
$$\left. + \frac{\mu A_2 \ell_2}{6}\dot{q}_3 + E\left(\frac{A_1}{\ell_1} + \frac{A_2}{\ell_2}\right)q_2 - \frac{A_1 E}{\ell_1}q_1 - \frac{A_2 E}{\ell_2}q_3 \right]$$

Using Proposition 1 we can conclude that the system is second-order nonholonomic. Indeed, in this case $r = 1$ and the matrix

$$L = \begin{bmatrix} \frac{\partial R}{\partial \dot{q}} \\ \frac{\partial R}{\partial q} + \frac{\partial R}{\partial \dot{q}_2} \cdot \frac{\partial R}{\partial \dot{q}} \end{bmatrix} \begin{bmatrix} J_1 & J_2 \\ 1 & 0 \\ 0 & 1 \end{bmatrix}$$

has a determinant:

$$det[L] = -\frac{9(A_1\ell_1 + 2A_2\ell_2)E\mu A_2}{4\rho^2 \ell_2 (A_1\ell_1 + A_2\ell_2)^2} \neq 0$$

Therefore it is nonsingular and the system is second-order nonholonomic. This can be verified by taking the vector fields:

$$\tau_0 = \begin{bmatrix} \dot{q}_1 & \dot{q}_2 & \dot{q}_3 & 0 & R & 0 & 1 \end{bmatrix}^T$$
$$\tau_1 = \begin{bmatrix} 0 & 0 & 0 & 1 & J_1 & 0 & 0 \end{bmatrix}^T$$
$$\tau_2 = \begin{bmatrix} 0 & 0 & 0 & 0 & J_2 & 1 & 0 \end{bmatrix}^T$$

and calculating a base for the accessibility algebra:

$$[\tau_0, \tau_1] = \begin{bmatrix} -1 & \frac{A_1\ell_1}{2(A_1\ell_1 + A_2\ell_2)} & 0 & 0 & 0 & 0 \end{bmatrix}^T$$
$$[\tau_0, \tau_2] = \begin{bmatrix} 0 & \frac{A_1\ell_1}{2(A_1\ell_1 + A_2\ell_2)} & -1 & 0 & 0 & 0 \end{bmatrix}^T$$
$$ad_{\tau_0}^2 \tau_1 = \begin{bmatrix} 0 & 0 & 0 & 0 & G & 0 & 0 \end{bmatrix}^T$$
$$ad_{\tau_0}^2 \tau_1 = \begin{bmatrix} 0 & -G & 0 & 0 & 0 & 0 & 0 \end{bmatrix}^T$$

where $G = \frac{3A_1 E(3A_1\ell_1\ell_2 + 2A_2\ell_2^2 + A_2\ell_1^2)}{2\rho(A_1\ell_1 + A_2\ell_2)^2 \ell_1 \ell_2} \neq 0$. Clearly the dimension of the accessibility algebra is $2n + 1$ and therefore the system is second-order nonholonomic by definition.

6.2 Beam under bending load

Consider a beam resisting bending moments applied on its plane (Figure 2). Here the finite element model is more complex. There are six degrees of freedom four of which, the linear and angular displacements at the grasp points, are actuated. Displacements q_3 and q_4 are not directly controlled. This system is also second-order nonholonomic. The matrix

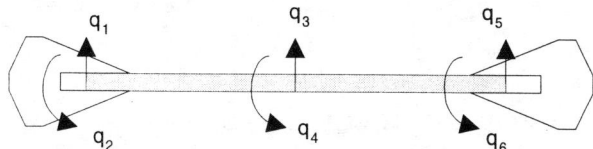

Figure 2: A deformable beam resisting bending

sequence has $\frac{2(n-m)}{m} = 1$ elements. Therefore we only need to calculate the matrix:

$$L = \begin{bmatrix} \frac{\partial \mathbf{R}}{\partial \mathbf{q}} + \frac{\frac{\partial \mathbf{R}}{\partial \dot{\mathbf{q}}}}{\frac{\partial \mathbf{R}}{\partial \dot{\mathbf{q}}_k}} \cdot \frac{\partial \mathbf{R}}{\partial \dot{\mathbf{q}}} \end{bmatrix} \begin{bmatrix} \mathbf{J} \\ \mathbf{I} \end{bmatrix} \quad k = 3, 4.$$

Its determinant turns out to be:

$$det[L] = \frac{4254529125}{562432} \frac{\mu^2 A^2 E^2 I^2}{m^4 \ell^4} \neq 0$$

indicating that the system is second-order nonholonomic. A direct calculation of a base for the accessibility algebra verifies the result.

7 Conclusion

A deformable object being manipulated can be discretized using finite elements and regarded as an underactuated mechanical system. In this framework many different kind of mechanical systems and manipulating tasks can be studied. We presented a sufficient condition for second-order nonholonomy, for an important class of deformable objects modeled with finite elements. This result permits to circumvent the usual procedure of calculating Lie brackets to establish a base for the associated Lie algebra, and answers the question of determining the kind of constraints imposed on the system in a straightforward algebraic way. Moreover, constraints associated with material strenght limitations have also been included.

References

[1] B. d'Andrea-Novel G. Campion and G. Bastin. Modelling and state feedback control of nonholonomic mechanical systems. In *Proceedings of the 1991 IEEE Conference on Decision and Control*, December 1991.

[2] K. Kosuge, M. Sakai, K. Kanitani, H. Yoshida, and T. Fukuda. Manipulation of a flexible object by dual manipulators. In *IEEE Int. Conf. on Rob. and Autom.*, pages 318–322, 1995.

[3] A. De Luca, R. Mattone, and G. Oriolo. Dynamic mobility of redundant robots using end-effector commands. In *1996 Int. Conf. on Robotics and Automation*, pages 1760–1767, Minneapolis, MN, April 1996.

[4] H. Nijmeiher and A.J. van der Schaft. *Nonlinear Dynamical Control Systems*. Springer-Verlag, 1990.

[5] Giuseppe Oriolo and Yoshihiko Nakamura. Control of mechanical systems with second-order nonholonomic constraints: Underactuated manipulators. In *Proceedings of the 1991 IEEE Conference on Decision and Control*, December 1991.

[6] S.S. Rao. *The Finite Element Method in Engineering*. Pergamon, 1989.

[7] M. Reyhanoglu, A. van der Schaft, N.H. McClamroch, and I. Kolmanovsky. Nonlinear control of a class of underactuated systems. In *35th IEEE Conf. on Decision and Control*, pages 1682–1687, Kobe, Japan, December 1996.

[8] M. W. Spong. Underactuated mechanical systems. In B. Siciliano and K.P. Valavanis, editors, *Control Problems in Robotics and Automation, Lecture Notes in Control and Information Sciences 230*, pages 135–150. Springer, 1998.

[9] D. Sun, X. Shi, and Y. Liu. Modeling and cooperation of two-arm robotic system manipulating a deformable object. In *1996 IEEE Int. Conf. on Rob. and Autom.*, pages 2346–2351, Minneapolis, Minnesota, April 1996.

[10] H.G. Tanner and K.J. Kyriakopoulos. Modeling of multiple mobile manipulators handling a common deformable object. *Journal of Robotic Systems*. to appear.

[11] D. Terzopoulos and K. Fleischer. Deformable models. *The Visual Computer*, 4:306–331, 1988.

[12] C. Wu and C. Jou. Design of a controlled spatial curve trajectory for robot manipulators. In *Proceedings of the 27th Conference on Decision and Control*, pages 161–166, December 1988.

[13] T. Yukawa, M. Uchiyama, and H. Inooka. Cooperative control of a vibrating flexible object by a rigid dual-arm robot. In *IEEE Int. Conf. on Rob. and Autom.*, pages 1820–1826, 1995.

Development of BEST Nano-Robot Soccer Team

Sung Ho Kim, Jong Suk Choi, and Byung Kook Kim*
Department of Electrical Engineering
Korea Advanced Institute of Science and Technology
Taejon 305-701 Korea

Abstract

In this paper, we describe the development of our BEST nano-robot soccer team composed of 5 robots, a vision system, and communication modules. Each nano-robot is designed with an all-in-one type microcontroller and two DC motors fixed in parallel due to its small size $4(cm) \times 4(cm) \times 4.5(cm)$. A vision system capable of 30 recognitions/sec provides each robot's position feedback via RF communication. A three level hierarchical software control scheme is adopted: role allocation, target generation, and visual servoing. Dynamic role allocation is done in the highest layer(the role allocation layer) regarding the context and role importance. In the target generation layer, target positions of robots are generated based on 6 roles, 2-2-1 formation, and libero system. In the visual servoing layer, path planning is achieved using a newly proposed line-circle algorithm with linear and circular paths as our basic strategy. Normally, 2-2-1 formation is adopted. When the both attackers are blocked, one of defenders is assigned to continue the attack as libero(temporary 3-1-1 formation).

1 Introduction

NAROSOT(NAno-RObot SOccer Tournament) is a robot soccer category with the smallest soccer robots. The word "nano" means that the robots are the smallest in the robot soccer categories of FIRA(Federation of International Robot-soccer Association). Each robot must be smaller than $4(cm) \times 4(cm) \times 5.5(cm)$ hexahedron[3]. There are no rooms for internal sensors such as encoders in these robots, and hence vision is the only feedback available. This makes it hard to control robots precisely and fast. Since 5 robots in each team play a game, coordination and cooperation among robots become more important and difficult than in MIROSOT(MIcro RObot SOccer Tournament, 3 robots in each team with 8 cm cube).

We make small but self-contained nano-robot and use only vision data to control two DC motors of each robot. We also make robot control scheme with expansible structure, where new functions can be easily added. A three level hierarchical scheme for visual servoing, target generation, and role allocation constitutes its control structure, where each layer is divided according to its function.

In Section 2, we present hardware design of our system. A hierarchical controller for soccer robot is proposed in Section 3, where visual servo layer, target generation layer, and role allocation layer are presented. In Section 4, simulation results of the proposed controller and experimental results are shown. Section 5 summarizes this paper and shows further works.

2 Hardware Description

Our system is composed of 5 nano-robots, an image processing board, a host PC and a camcoder. RF modules are used to communicate between host PC and robots. Each subsystem is independent and self-contained and can be tested independently for debugging its functions. Each subsystem also has its own restrictions which can limit the performance of overall system. Hence, it is considered in hardware design that a specific subsystem should not deteriorate the performance of overall system.

Nano-robots of the BEST is $4 \times 4 \times 4.5$ cm^3 in size(1 cm less height than regulation: currently the smallest NAROSOT robot) and has a Phillips 87C51 all-in-one type micro-controller, two DC motors without encoder and a RF module for link to host PC. For motor-driver chip, TC4428 is used. Two DC micromotors with 11.8:1 gear heads from MINIMOTOR are kept charge of locomotion of robots. The picture of robot are shown in Fig. 1.

We use Cognachrome vision board from Newton Labs. This board is capable of field-rate(60Hz) pro-

---
*Professor to whom all correspondence should be addressed

cessing, which can give centroids, aspect ratios and angles for color patches of robots. The results transfer to our host PC by serial port at 20 or 30 Hz rate. To implement our controller, we use PC with Pentium

Figure 1: BEST Nano-robot

166 MHz and Windows 95. We use Rx/Tx FM RF modules from Radiometrix (BiM-418-F, BiM-433-F), which can operate stably at 19600 bps and transmits 14 bytes(140 bits) during one sampling period in the experiments.

3 Control Structure

Our robot controller has a three level hierarchical scheme so that the controller may be easy to debug, and new functions can be easily added. The proposed hierarchical controller consists of a visual servoing, a target generation, and a role allocation layer, which are divided according to their function. The overall block diagram of our controller is shown in Fig. 2.

3.1 Visual Servo Layer

We design visual servo layer using visual servoing concept[1] so that our robots can dribble a ball, pass a ball, and play some kind of expertise like human soccer players.

In our robot, two motors are placed in parallel but there exists some offset between two axes to satisfy the size limit. We set a simplified dynamic model of first order into our plant of robot intuitively. Then a simple velocity control can be achieved by taking the inverse model.

$$v^i(t_0 + T_s) = v^i(t_0)e^{-\frac{T_s}{\tau^i}} + u^i K^i(1 - e^{-\frac{T_s}{\tau^i}}), \quad i = 1, 2 \quad (1)$$

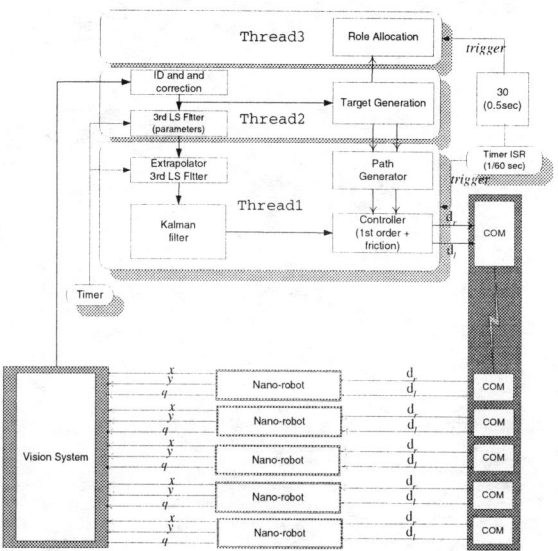

Figure 2: Overall Block diagram of Controller

Given the reference of velocity v^i_{ref}, control input u^i can be derived from Eq. (1).

$$u^i = \frac{v^i_{ref} - v^i(t_0)e^{-\frac{T_s}{\tau^i}}}{K^i(1 - e^{-\frac{T_s}{\tau^i}})} \quad (2)$$

In robot soccer game, it is important to control robots to satisfy the desired angle as well as desired position for kicking, passing, or dribbling. Let's define a posture composed of position and angle as $P = (p, \theta) = (x, y, \theta)$, start posture as $P_s = (p_s, \theta_s)$, and desired target posture as $P_t = (p_t, \theta_t)$. Given P_s and P_t, we want to draw a shortest path composed of start circle, target circle and a line tangential to both circles. Each radius of the circles(R_c) should be selected properly considering the characteristic of robot. If there are any obstacles on the tangential line, the robot can perform obstacle-avoidance by modifying path with thought of the nearest obstacle on the line as new target. Fig. 3 shows the planned path and flow-chart of the planning algorithm. P_{t2} is an added posture for stable tracking to P_t with larger velocity and P_{tm} is the modified target posture in the presence of obstacle.

After planning the path from start posture to target posture, path-following is performed by line-control or circle-control. If we select proper circle and control-direction(Counter-ClockWise or ClockWise), circle-control can be performed on-line as follows. First define direction-flow d composed of two components of

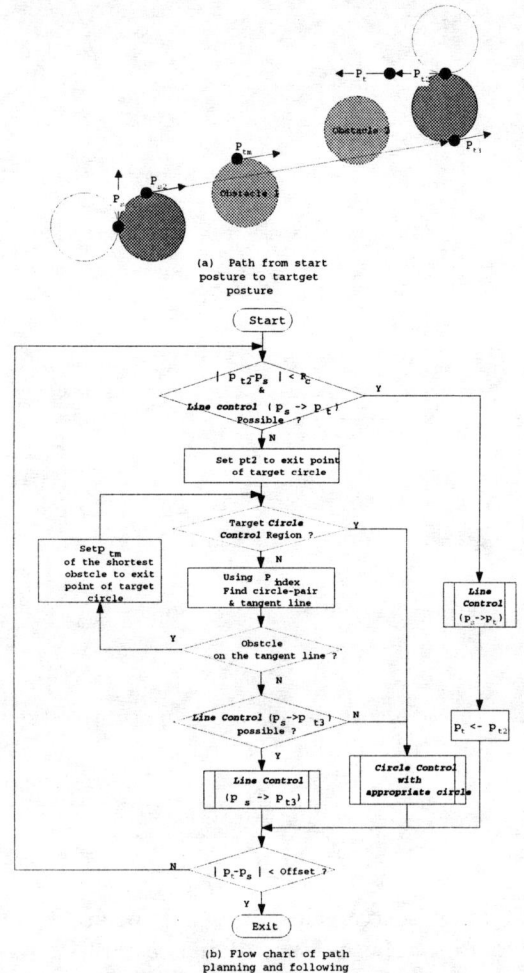

Figure 3: Path Planning using Line and Circles

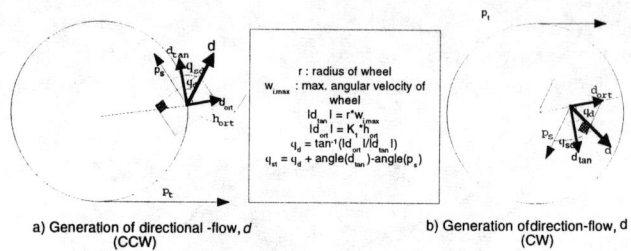

Figure 4: Path Following using Circle-Control

where
$$\theta_{sd} = \begin{cases} \angle \vec{d} - \angle \vec{p}_s = (\angle \vec{d}_{tan} - \theta_d) - \angle \vec{p}_s, & CCW \\ \angle \vec{d} - \angle \vec{p}_s = (\angle \vec{d}_{tan} + \theta_d) - \angle \vec{p}_s, & CW \end{cases}$$
$$\theta_d = \tan^{-1}(\frac{|d_{ort}|}{|d_{tan}|})$$

3.2 Target Generation Layer

Six roles - goalie, left-defender, right-defender, kicker, assistant and libero - are available for 5 robots in the target generation layer. A target position of each role is decided according to the positions of a ball, our robots, and opponent robots.

3.2.1 Formation

A 2-2-1 formation or a 3-1-1 formation can be adopted selectively: 2-2-1 formation means 2 attackers, 2 defenders, and 1 goalie; 3-1-1 formation means 3 mid fielders, defenders, and 1 goalie. It is important to

vectors: tangential one d_{tan} and orthogonal one d_{ort} as Fig. 4. After finding of d, the next procedure is generation of curvature σ which can be considered as control variable.

$$w^{ref} = \sigma v^{ref} \qquad (3)$$

where
$$\sigma = \begin{cases} \frac{1}{R_c} + K_2\theta_{sd}, & CCW \\ -\frac{1}{R_c} + K_2\theta_{sd}, & CW \end{cases}$$
$$\theta_{sd} = \begin{cases} \angle \vec{d} - \angle \vec{p}_s = (\angle \vec{d}_{tan} - \theta_d) - \angle \vec{p}_s, & CCW \\ \angle \vec{d} - \angle \vec{p}_s = (\angle \vec{d}_{tan} + \theta_d) - \angle \vec{p}_s, & CW \end{cases}$$
$$\theta_d = \tan^{-1}(\frac{|d_{ort}|}{|d_{tan}|})$$

In Eq.(3), constant value of K_2 is used for proportional gain. The line-control is similar to the circle-control except generation of curvature.

$$\sigma = K_2 \theta_{sd} \qquad (4)$$

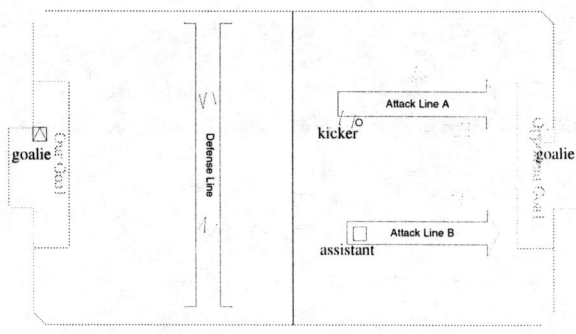

Figure 5: 2-2-1 Formation

maintain efficient formation for preparing attacks and defenses. In our system, 2-2-1 formation is implemented in considering batteries, defense enforcement and stable games. The main advantage of this formation is that uniform distribution of robots, which causes to use all robots evenly and to avoid flocking of robot. These can help games to be smooth. In this paper, we describe details about only this formation.

This formation sets up one lateral defense line and two vertical attack lines[Fig. 5]. In this formation, goalie moves laterally in the front of our goal and defend near the goal. Two defenders move on defense line, defend the opponent attacks, intercept the ball, and then change to attackers. The kicker among two attackers drives the ball, shoots and passes according to tactics. The other attacker, assistant, cooperates with the kicker, occupies the good position to attack, and is ready to attack when the kicker passes the ball. These can improve success probability of our attacks.

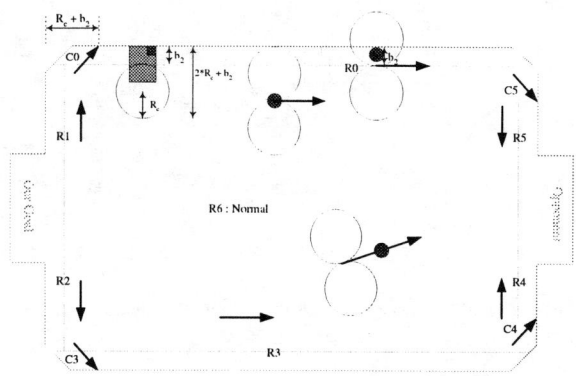

Figure 7: Target Generation of Kicker

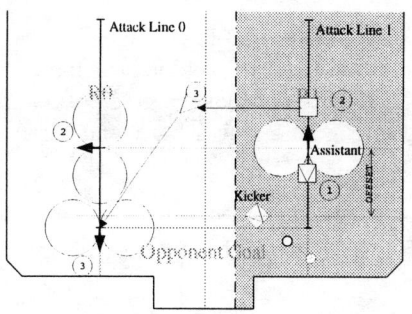

Figure 8: Target Generation of Assistant

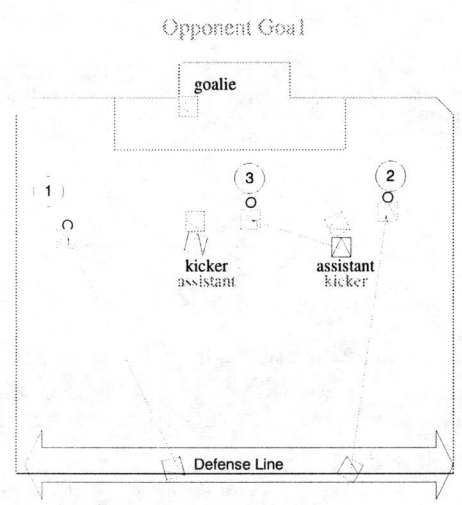

Figure 6: Libero System

The formations can be compensated with libero system[Fig. 6]. In case that one or two attackers are blocked in our attack situations, one of the defenders or an assistant is assigned as libero to help the attack. This makes attacks to be continued against interruptions.

3.2.2 Roles

A. Kicker

Target of kicker is determined to keep the ball toward the opponent goal. There may be the case that ball is near wall or corner. In this case, it is impossible to generate target to keep the ball toward opponent goal, so the target is set to parallel to wall or corner toward the opponent goal[Fig. 7]. Intermediate targets are needed when the ball is near a wall or corner and back of the kicker(direction of our goal), which are used for moving around the ball to avoid back pass, which causes a dangerous situation.

B. Assistant

Target of assistant moves on one of two attack lines which is farther to the kicker than the other. When the kicker becomes near to assistant, the assistant change attack line to avoid colliding with and blocking the kicker[Fig. 8]. In the figure, two sets of targets are drawn. One is the case where assistant is near to kicker vertically. When the ball becomes near, the assistant changes to kicker and continues to attack. At the same time, the kicker changes to assistant and assists the new kicker.

C. Defenders

The defenders defend opponent attacks on a defense line. If the ball is inside of the defense line(toward our goal) and our attacker catches the ball inside and

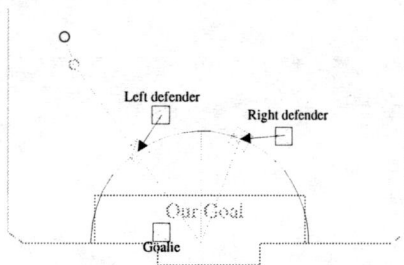

Figure 9: Target Generation of Defenders

begins to attack, the defenders open a path on the defense line. When a defender intercepts the ball, the defender changes to kicker and begins to attack[Fig. 9].

D. Goalie
Goalie moves on the goal line in lateral direction and predicts the ball position to block the ball. In the real game, goalie is often blocked by goal posts. To prevent this situation, goalie moves back and forth near the goal posts.

E. Libero
Target of libero is generated by the same algorithm as the kicker.

3.3 Role Allocation Layer

Dynamic role allocation is done in the highest layer, that is in the role allocation layer, as per the context and role importance. In this layer, connections between robots and roles are changed dynamically according to the context. In the present state, the most appropriate robot to a role can be selected and connected to the role.

In this scheme, when some of our robots are failed, the other robots play the games normally by filling the spaces of the failed robots automatically. Hence, this can make overall control algorithm more robust and reliable. It must be considered that toggling of roles is avoided in the allocation algorithm. Priority rules of the dynamic allocation are as follows.

1. Goalie: the nearest robot to the goal line.
2. Kicker: the nearest robot to the ball except above.
3. Assistant: the robot in the forefront except above.
4. Left defender: the left robot except above.
5. Right defender: the robot except above.

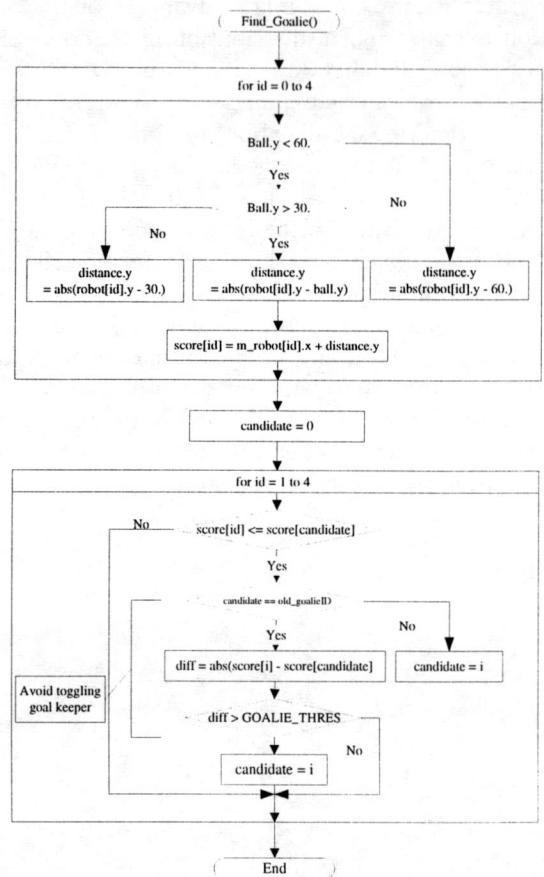

Figure 10: Allocation Algorithm for Goalie

Figs. 10 and 11 show the flow chart of the detail allocation algorithm for goalie and a result of the dynamic role allocation algorithm respectively. Detailed allocation algorithms for the other roles are established using similar scheme.

4 Simulation and Experimental Results

In this section, we show some results of simulations and experiments. First, we show the result of line-circle algorithm for path-planning and path-following in presence of two obstacles[Fig. 12]. We can show that our algorithm does also perform obstacle-avoidance by modifying path with adding new target posture P_{tm}.

Next, Fig. 13 shows a simulator for target generation and role allocation layers, which shows the situation that one of the defenders continues to attack as libero when the 2 attackers are blocked. Last, Fig. 14

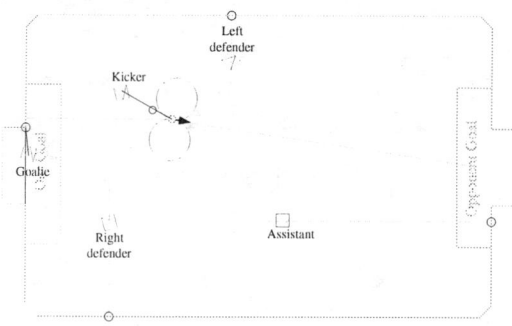

Figure 11: Result of Dynamic Role Allocation

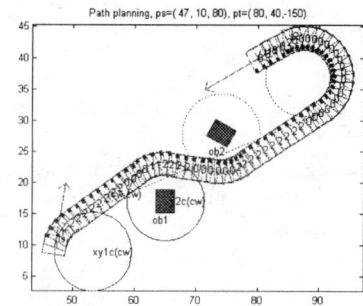

Figure 12: Simulation of Line-Circle Algorithm with Obstacles

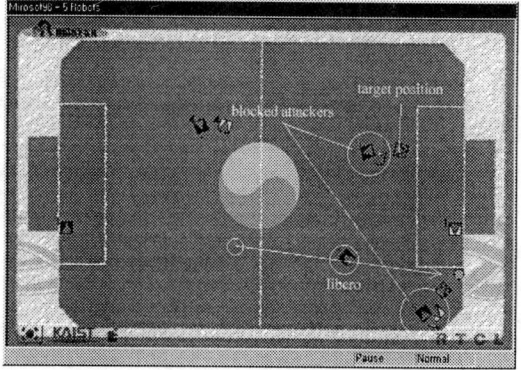

Figure 13: Simulator for Target Generation and Role Allocation

Figure 14: Final Game of NAROSOT in FIRA'98

shows a photo of the final game between the yellow colored BEST and the blue colored Y2K2 teams in FIRA'98 in Paris. In this game, our team won the game with score 6-2, and got the first prize. Although the tournament did not give a direct performance evaluation, it showed fitness and excellence of our three layers concept and robot system.

5 Conclusion

In this paper, we presented the NAROSOT system of our BEST team. We made the smallest NAROSOT size robots($4 \times 4 \times 4.5 cm$) and total system(vision and communication). We proposed the three-layered hierarchical control scheme which is composed of visual servo layer, target generation layer and role allocation layer. In the visual servo layer, we use simplified model and line-circle path algorithm. In the target generation layer, we generates target positions of robots based on 6 roles, 2-2-1 formation and libero system. Dynamic role allocation and libero strategies are used in the role allocation layer.

We think the NAROSOT system is a good benchmark for multi-robotic system research. More versatile formation and strategy will be implemented as further works.

References

[1] S. Hutchinson, G. D. Hager and P. I. Corke, "A Tutorial on Visual Servo Control", *IEEE transactions on Robotics and Automation*, Vol. 12, No. 5, October 1996, pp. 651-670

[2] A. Bicchi, G. Casalino and C. Santilli, "Planning Shortest Bounded-Curvature Paths for a Class of Nonholonomic Vehicles among Obstacles", *J. of Intel. and Robotic Systems*, vol. 16, no. 4, 1996, pp. 387-405

[3] FIRA home page, http://www.fira.net/fira/

The Miniature Omni-directional Mobile Robot OmniKity-I (OK-I)

Myung-Jin Jung, Hyun-Sik Shim, Heung-Soo Kim and Jong-Hwan Kim

Dept. of Electrical Engineering, KAIST, Taejon-shi, 305-701,

Republic of Korea

{mjjung, hsshim, hskim, and johkim}@vivaldi.kaist.ac.kr

http://vivaldi.kaist.ac.kr/~iclab

Abstract

The structure of OmniKity-I (OK-I), the miniature omni-directional mobile robot, and its kinematics and dynamics models are presented. OK-I has three DOFs in the plane. A conventional two wheeled mobile robot (2-WMR) serves as the base of OK-I for translational motion. The robot body is on top of the base. The third motor controls angular motion of the body as the base turns according to the translational motion. The circular surface connector (CSC) is devised to transfer signals from the base to the body without constraining angular motion of the body. Experimental results show proper omni-directional features of OK-I.

1 Introduction

A wheeled mobile robot (WMR) has three coordinates in the plane. Position x, y and robot heading angle θ in an absolute frame. An omni-directional WMR can be defined as a WMR that can move along any path that originates from the current posture (x, y, θ) in the configuration space.

For the past decade, many omni-directional robot structures have been suggested. Most of them used specially designed wheels, e.g., universal wheels, Swedish wheels, ball wheels or roller wheels [1]-[5]. These wheels are designed in such a manner that each wheel has two degrees of freedom in motion but only one of them is active-driving. Three or more of these wheels are combined to guarantee the omni-directional motion of the body. Some structures have the advantage that special wheel locations allow slippage detection [1], [5].

In recent years new structures have been introduced. Compared with one degree of active driving of the wheels, two degrees of active-driving concept can be found in [6]-[7]. It is straightforward that a pair of wheels which are active-driving in two dimension can be combined to make the omni-directional motion. In [7], the velocity factor perpendicular to the robot heading angle is generated by the angular motion of the offset axis. It is known that an offset steered wheel is omni-directional [8]. It is very robust to slippage between the actuators and the wheels and between the wheels and the ground. One point is that a pair of two dimensionally active-driving wheels cost four actuators. Still the redundancy may be compensated by little slippage.

The principle of generating the perpendicular motion factor can be applied in the very same manner to 2-WMR's by modeling a two wheeled mobile robot as an offset steered uni-cycle, not as a conventional uni-cycle (Figure 1). While the uni-cycle model is active in two dimension in the sense of having active v and ω, the offset steered model is in the sense of having active v_x and v_y. A big difference between the offset steered uni-cycle and a 2-WMR modeled as an offset steered uni-cycle is that, to be active in two dimension, the former needs to be coupled with another, but the latter does not; a offset steered uni-cycle cannot translate the body with active v_x and v_y, but a 2-WMR is sufficient to translate the body. From this, one 2-WMR modeled as an offset steered uni-cycle is an independent active wheel in two dimension in a manner that its v_x and v_y are active. In this model the angular velocity of the 2-WMR is dependent on the translational velocities.

In the omni-directional robot OmniKity-I (OK-I), the 2-WMR serves as the base that controls translational motions of the body. To make the robot heading angle independent of the translational motions, the robot body is put on the base with a revolute joint. The base contains two geared DC motors with encoders and a caster. Since the body revolves most of time on the base to keep its desired angular velocity as the base translates, the circular surface connector (CSC) is devised to flow signals from the base to the body where most of the electronic circuitry is contained without limiting the angular motion of the body.

OK-I is a miniature robot less than 8.5 cm in each side. The name OmniKity comes from Kity, the 1 $inch^3$ micro robot equipped with a CPU, sensors, geared motors, and batteries on-board and specially designed for running in a maze [9]. OK-I is equipped with actuators and control circuitry on-board. Future generations of OK will be with some proximity sensors and on-board cameras for MiroSot, Micro-Robot World Cup Soccer Tournament (www.fira.net), and will be a little bit bigger than the prototype (the first micro omni-directional robot with the ball-wheel mechanism was shown at MiroSot'97 [10]).

In Section 2 the kinematics of OK-I is presented. Control scheme is descrived in Section 3 and experimental results are reported in Section 4. Concluding remarks follow in Section 5.

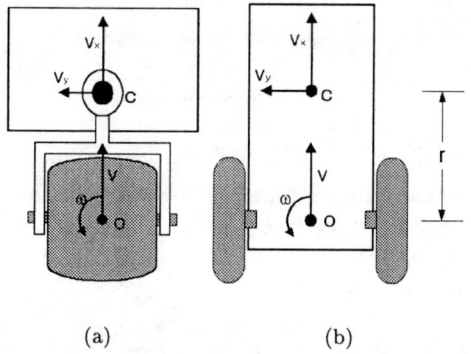

Figure 1: (a) Offset steered uni-cycle model (b) 2-WMR modeled as offset steered uni-cycle

2 Kinematic Equations

A conventional 2-WMR is modeled as a uni-cycle as

$$\begin{pmatrix} v_x \\ v_y \\ \omega \end{pmatrix} = \begin{pmatrix} \cos\theta & 0 \\ \sin\theta & 0 \\ 0 & 1 \end{pmatrix} \begin{pmatrix} v \\ \omega \end{pmatrix} \quad (1)$$

where v and ω are the control inputs of the 2-WMR. In (1) point o is assumed to be the center of the 2-WMR. On the other hand, if point c is assumed to be the center, the kinematic equation becomes

$$\begin{pmatrix} v_x \\ v_y \\ \omega \end{pmatrix} = \begin{pmatrix} \cos\theta & -r\sin\theta \\ \sin\theta & r\cos\theta \\ 0 & 1 \end{pmatrix} \begin{pmatrix} v \\ \omega \end{pmatrix}. \quad (2)$$

From (2), we have

$$\begin{pmatrix} v_x \\ v_y \end{pmatrix} = \begin{pmatrix} \cos\theta & -r\sin\theta \\ \sin\theta & r\cos\theta \end{pmatrix} \begin{pmatrix} v \\ \omega \end{pmatrix}. \quad (3)$$

Since the 2 × 2 matrix on the right-hand side is invertible, the position(x, y) of point c can follow any path as long as r is not zero. It has been noted from the nonholonomic problem point of view that such a point c, r apart from the wheel axis, can be exponentially point stabilizable provided that r is not zero [11]-[12]. This is drawn very naturally from the system equation (3).

A rather interesting form of this equation is

$$\begin{pmatrix} v_x \\ v_y \end{pmatrix} = \begin{pmatrix} \cos\theta & -\sin\theta \\ \sin\theta & \cos\theta \end{pmatrix} \begin{pmatrix} v \\ r\omega \end{pmatrix} \quad (4)$$

from which it is seen that the global translational velocity factors v_x and v_y are just the rotational transformation of v, the velocity factor along the robot heading and $r\omega$, the velocity factor perpendicular to the robot heading angle as shown in Figure 2. Since v

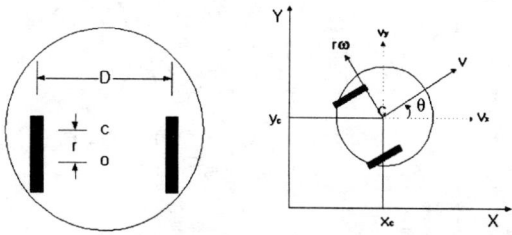

Figure 2: Active-driving wheel in two dimension

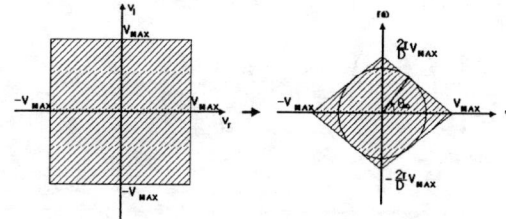

Figure 3: Feasible region of translational velocity

and ω are represented by

$$\begin{pmatrix} v \\ \omega \end{pmatrix} = \begin{pmatrix} \frac{1}{2} & \frac{1}{2} \\ -\frac{1}{D} & \frac{1}{D} \end{pmatrix} \begin{pmatrix} v_l \\ v_r \end{pmatrix} \quad (5)$$

where v_l and v_r are the left and right wheel velocities, respectively, the translational velocity is limited to the shaded region as in Figure 3.

Thus the minimum instantaneous translational velocity v_m is obtained at the angle $\pm\theta_m$ or $\pi \pm \theta_m$ to the robot heading

$$v_m = 2V_{MAX}\frac{r}{D} \cdot \sin\theta_m \qquad (6)$$

where the angle θ_m is a 2-WMR's geometric parameter defined by

$$\theta_m = \tan^{-1}\frac{V_{MAX}}{\frac{2r}{D} \cdot V_{MAX}} = \tan^{-1}\frac{D}{2r}. \qquad (7)$$

As in (4), the angular velocity of the base ω becomes a subordinate variable to the translational velocities v_x and v_y. Therefore by putting the body on point c and by introducing an additional control input ω_3 for the body, the angular velocity ω_B of the robot body becomes independently controllable:

$$\omega_B = \omega + \omega_3. \qquad (8)$$

Combining (4) and (8) together, the forward kinematic equations are obtained as

$$\begin{pmatrix} v_x \\ v_y \\ w_B \end{pmatrix} = \begin{pmatrix} \cos\theta & -r\sin\theta & 0 \\ \sin\theta & r\cos\theta & 0 \\ 0 & 1 & 1 \end{pmatrix} \begin{pmatrix} v \\ \omega \\ \omega_3 \end{pmatrix}. \qquad (9)$$

The determinant of the above 3×3 matrix in (9) is r. Since r is not zero, the inverse kinematics is obtained as

$$\begin{pmatrix} v \\ \omega \\ \omega_3 \end{pmatrix} = \begin{pmatrix} \cos\theta & \sin\theta & 0 \\ -\frac{1}{r}\sin\theta & \frac{1}{r}\cos\theta & 0 \\ \frac{1}{r}\sin\theta & -\frac{1}{r}\cos\theta & 1 \end{pmatrix} \begin{pmatrix} v_x \\ v_y \\ w_B \end{pmatrix}. \qquad (10)$$

Replacing v and ω by v_l and v_r, we have

$$\begin{pmatrix} v_l \\ v_r \\ w_3 \end{pmatrix} = J^{-1} \begin{pmatrix} v_x \\ v_y \\ w_B \end{pmatrix} \qquad (11)$$

where

$$J^{-1} = \begin{pmatrix} \cos\theta + \frac{D}{2r}\sin\theta & \sin\theta - \frac{D}{2r}\cos\theta & 0 \\ \cos\theta - \frac{D}{2r}\sin\theta & \sin\theta + \frac{D}{2r}\cos\theta & 0 \\ \frac{1}{r}\sin\theta & -\frac{1}{r}\cos\theta & 1 \end{pmatrix}.$$

As the base revolves below the body during most of the operation, ordinary wires cannot be used for transferring the signals from the base to the body. For OK-I, the circular surface connector is devised to flow signals without restricting the revolute motion of the body. In the next section the structure of OK-I and the control scheme are described.

3 Control Scheme and Prototype

The control loop of OK-I is twofold. One is for the low level PID control of the wheels and body. There are three loops of this type as shown in Figure 4. In the figure subscript d means *desired* and the value with index (k-1) is of one sampling interval before. Three dedicated PID controllers are used which

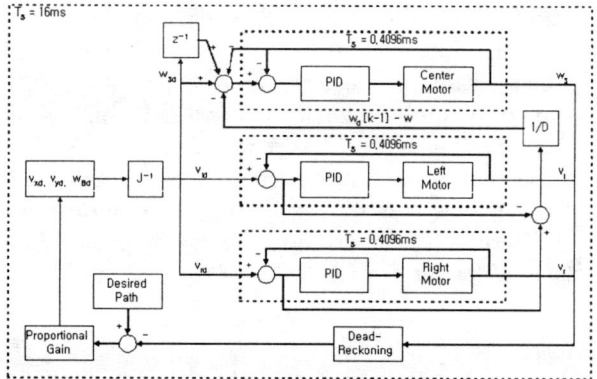

Figure 4: Control loop

have the 0.4096 ms sampling interval. The other is the outer loop for angular speed control of the body. The error, e_{w_3}, is fed back to calculating w_{3d} to keep the body's angular velocity at every 16 ms (the outer loop sampling interval):

$$e_{w_3} = (w_{3d}[k-1] - w_3) - (w_d[k-1] - w). \qquad (12)$$

The onboard CPU calculates the desired velocity of each motor as per the desired path at every sampling time of 16 ms. It is highly desired that the base has little odometry error to the ground as the body has to keep the angular velocity of w_B.

The prototype OK-I, sized 7.5 cm × 7.5 cm × 8.5 cm, has a 16 bit controller 80C196KC as a main CPU with 32KB RAM and ROM, three PID controllers on the body and three geared motors with encoders and a caster on the base. Owing to the three dedicated PID controllers, the main CPU is fully devoted to dead-reckoning and calculating desired velocities for the given path. The maximum wheel speed V_{MAX} is 40 cm/s and r and D are 1 cm and 5.9 cm, respectively. Figure 5 shows the prototype standing by Kity. The space in the left of the body is reserved for batteries, but for the moment power is provided from an external power supply.

The circular surface connector in Figure 6 is implemented with the printed circuit board on the base and 10 commuters on the body. It flows motor and encoder signals of the base wheels to the body where most circuitries are located. There are ten signals from the two base wheels: two power lines for the encoders, two pairs of the encoder signals (A_l, B_l, A_r, B_r) and two pairs of the motor signals (M_l+, M_l-, M_r+, M_r-). Since CSC can flow up to only several hundreds of mili-amperes, for large scale robots the signals sent though this connecter should be changed. For example, instead of being supplied power from the body, the base would be provided with a separate battery. In this case the revolute axis can serve as a ground connector between the body and the base circuits. And the direct motor signals (M+, M-) would be replaced by a PWM and a direction signal that are in much less current. By putting supporters around the circular surface connectors, the OK-I is expected to carry a high payload.

Figure 5: OK-I

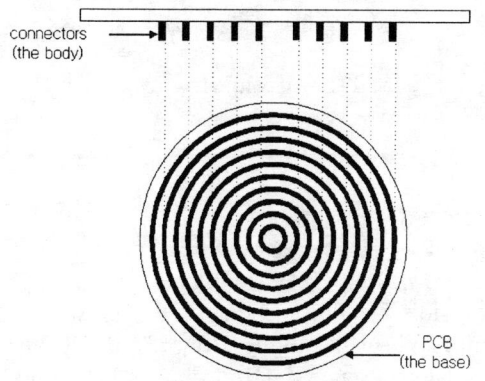

Figure 6: The circular surface connector (CSC)

4 Experimental Results

For testing omni-directional mobility and dead-reckoning precision, two pen drawings are performed by OK-I (Figure 7). One is for the translational motion (Figure 8) and the other is for the angular motion while translating (Figure 9). In both tests, no external locating sensors are used. Since the path following

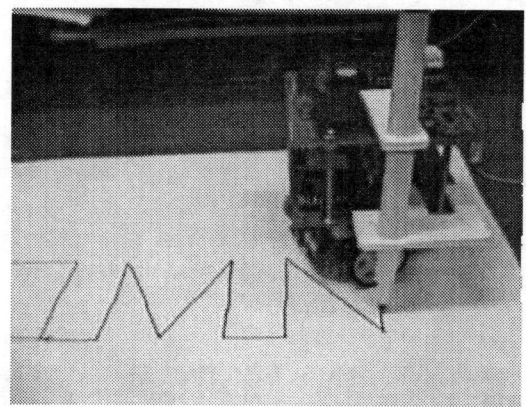

Figure 7: Pen drawing by OK-I

is purely based on the kinematic equation, the speed along the path is limited to 1 cm/s. Owing to the 2-WMR characteristics of the base, simple proportional feedback to the two wheels is enough for making the robot trajectory asymptotically close to the desired path. The height of each letter is 5 cm, the width of

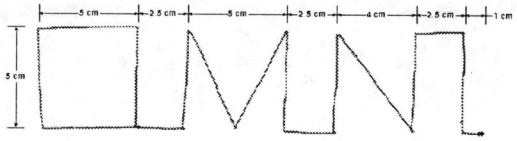

Figure 8: Translational motion

letters 'O' and 'M' is 5 cm, and that of 'N' is 4 cm. The last letter 'I' is completed with a 1 cm horizontal line. All the letters are placed 2.5 cm apart. So the path consisting of 16 line segments sums up approximately 73.6 cm traveling distance and 15 turnings. The initial base angle is 180 degrees (or facing to the left side of the page), and the maximum translational velocity along the path is 1 cm/s. Of the total 73.6 cm traveling with 15 turnings, the final distance error was less than 3 mm all the time. As one of the wheels changes the direction twice when drawing the angles in the figure, the back-lash caused most of the odometry error. Ticks on the middle of some straight lines are due to the back-lash, where the base turns less or stops while the body still keeps turning to the opposite direction. Error modeling techniques or attaching encoders on the wheel axis will be considered to reduce the back-lash errors in the future OK series.

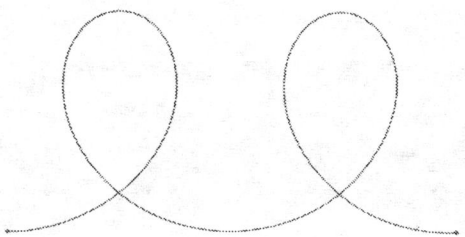

Figure 9: Angular motion with translation

The angular motion of OK-I is quite smooth once the base heads towards the direction of the traveling. To demonstrate smooth angular motion even in the case when the base is angled with the traveling direction, the initial angle of the base is 90 degrees to the direction of the traveling as in Figure 9. The translational velocity along the path keeps 1 cm/s and the angular velocity is 36 deg./s. During the 20 cm (or 20 s) traveling, smooth angular motion of the body was obtained even when the base was turning to the direction of traveling at the very initial stage. In this test, the distance between the left and right cross points measured 10 cm with less than 0.5 mm error.

5 Conclusions

The omni-directional mobile robot OmniKity-I (OK-I) was developed and its kinematic equation was derived. By introducing two dimensionally active-driving wheel concept to the conventional 2-WMR, it can track any desired path. The 2-WMR serves the base of the structure and by putting the robot body on the center of the base, the robot can follow any path in the configuration space (x, y, θ).

The circular surface connector (CSC) was devised so that ten signals (two power lines for encoders, two pairs of encoder signals (A, B), two pairs of motor signals (M+, M-)) can flow between the body and the base without restricting the revolute motion of the body.

OK-I, the prototype, showed proper omni-directional mobility by drawing various sharp angles and smooth circles. Similarities of the base to the conventional 2-WMR make the control of path following or point stabilization even simple. Still it is critical that the base should have little slippage so that the body can keep the desired angular motion with respect to the ground. The 2-WMR base takes the advantage of little slippage compared with other specially designed omni-directional wheels.

Figure 10: Right side of OK-I

Figure 11: Bottom view of OK-I

References

[1] P. F. Muir and C. P. Neuman, "Kinematic Modeling for Feedback Control of an Omnidirectional Wheeled Mobile Robot," IEEE International Conference on Robotics and Automation, pp.1772-1778, 1987.

[2] M. West, Haruhiko Asada, "Design of a Holonomic Omnidirectional Vehicle," IEEE Interna-

tional Conference on Robotics and Automation, pp.97-103, 1992.

[3] S. M. Killough, F. G. Pin, "Design of an Omnidirectional and Holonomic Wheeled Platform Prototype," IEEE International Conference on Robotics and Automation, pp.84-90, 1992.

[4] H. Asama, M. Sato, L. Bogoni, "Development of an Omni-Directional Mobile Robot with 3 DOF Decoupling Drive Mechanism," IEEE International Conference on Robotics and Automation, pp.1925-1930, 1995.

[5] M. West, Haruhiko Asada, "Design and Control of Ball Wheel Omnidirectional Vehicles," IEEE International Conference on Robotics and Automation, pp.1931-1938, 1995.

[6] J. Borenstein, "The OmniMate: A Mobile Robot With Omnidirectional Motion, Internal Odometry Error Correction, and Recovery After Actuator Failure," Proceedings of the World Automation Congress, Vol.3, pp.119-124, 1996.

[7] M. Wada, S. Mori, "Holonomic and Omnidirectional Vehicle with Conventional Tires," IEEE International Conference on Robotics and Automation, pp.3671-3676, 1996.

[8] P. F. Muir and C. P. Neuman, "Kinematic Modeling of Wheeled Mobile Robots," Journal of Robotic Systems 4(2), pp. 281-340, 1987.

[9] J.-H. Kim, M.-J. Jung, H.-S. Shim, S.-W. Lee, "Autonomous Micro-Robot 'Kity' for Maze Contest," Proceedings of International Symposium on Artificial Life and Robotics, pp.261-264, 1996.

[10] R. Blank, J. M. Koller, M. Lauria, F. Conti, "Omnidirectional robot and fast positioning systems," Proceedings of Micro-Robot World Cup Soccer Tournament (MiroSot), pp.85-86, 1997.

[11] C. Samson, K. Ait-Abderrahim, "Feedback Control of a Nonholonomic Wheeled Cart in Cartesian Space," IEEE International Conference on Robotics and Automation, pp.1136-1141, 1991.

[12] X. Yun, Y. Yamamoto, "Stability Analysis of the Internal Dynamics of a Wheeled Mobile Robot," Journal of Robotic Systems 14(10), pp. 697-709, 1997.

Microgravity Experiment of Hopping Rover

Tetsuo YOSHIMITSU *, Takashi KUBOTA **, Ichiro NAKATANI **,
Tadashi ADACHI***, Hiroaki SAITO***

* Department of Electrical Engineering, The University of Tokyo
** The Institute of Space and Astronautical Science (ISAS)
*** Nissan Motor Co. Ltd.
3-1-1 Yoshinodai, Sagamihara, Kanagawa 229-8510, Japan

Abstract

Under the micro gravity environment such as on the surface of small planetary bodies (asteroids or comets), traditional wheeled rovers are not expected to move effectively due to the low friction and inevitable detachment from the surface. Therefore the authors have proposed a hopping rover suitable for exploration on the surface of small bodies. This paper describes the micro-gravity experiments of a proposed hopping rover which is driven by an internal DC motor torquer. The proposed rover could hop under a micro gravity environment. Experimental results are compared with the computational simulations.

1. Introduction

There have been several spacecraft launched for the exploration of small planetary bodies such as asteroids and comets. So far they only approached close to the target body. In upcoming asteroid or comet missions, it is increasingly important to make science-equipped rovers "hang around" the surface of such a small body whose surface gravity is very weak[1].

A wheeled nano-rover[2], one of the friction-based mobilities, is proposed for the ISAS's asteroid sample return mission to be launched in 2002. The novel mechanism proposed in [3] is a hopping rover equipped with an internal drive, which also moves based on a friction but overcomes the difficulties of mobility around the small bodies.

This paper reports the experimental results of the proposed hopping rover under the micro gravity environment obtained by the free falls. The prototype rover uses a DC motor for the torquer.

This paper is structured as follows. Section 2 briefly explains the hopping rover proposed in [3]. The model of DC motor driven rovers is also described. Section 3 details the prototype rover, experimental parameters, computational simulations and experimental results.

2. Hopping Rover

2.1. Hopping Mechanism

The authors have proposed a hopping rover for the exploration of small planetary bodies[3]. This rover includes a torquer inside. Turning the rover by rotating the torquer, the reaction force against the surface makes the rover hop with significant horizontal velocity(Figure 1). Once hopped into the air, the rover moves ballistically. This rover moves by an innovative mechanism in that the contact force between the surface and the body is increased with help of the intentional pushing force made by the torquer, which makes the limit of friction larger and provides the ability of faster horizontal speed.

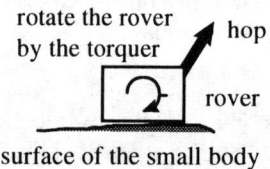

Figure 1: Hopping rover with a torquer[3]

2.2. Rover Model

The 2D simulation model is detailed in [3]. This paper extends the model considering an actual torquer. In [3], the simulations are made on the assumption of a constant torque imposed which can be attained not so easily. To make the rover simple, a DC motor is used for the torquer, whose output torque changes as the states of the rotor and stator of the motor. The stator of the motor is connected to the rover body. The output torque is dependent on the rover attitude.

Here, the rover is expressed by a rectangle, and the surface of the small body is flat. A DC motor is located at the center of the rover, around which a torque $T(t)$ is imposed. Figure 2 shows the coordinate system. The x-axis is along the flat surface, and the y-axis is vertical to the surface.

The rover state is expressed by the following four

parameters.

- x : horizontal coordinate of the center of gravity (point C)
- y : vertical coordinate of the center of gravity
- θ : attitude of the rover (DC motor stator)
- θ' : attitude of the DC motor rotor

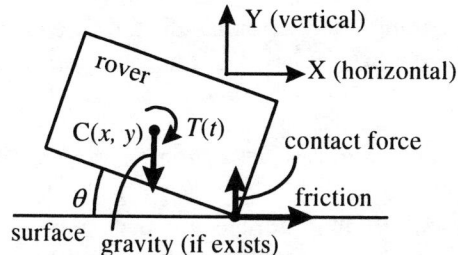

Figure 2: Coordinate system used in the 2D simulation

The rover state variables with respect to time are calculated numerically using Runge-Kutta method. Simulation starts when the DC motor starts. Note that the initial value of the state variables are $(x, y, \theta, \theta')_{t=0} = (0,0,0,0)$. The differential equations which describe the motion of the rover are described in [3].

The output torque of a DC motor is given by:
$$T(t) = k \cdot I(t)$$
$$V = I(t) \cdot R + k \cdot \omega(t)$$

where k denotes the motor torque constant and $I(t)$ denotes the current. V and R are the applied voltage to the motor and the resistance of the motor respectively. $\omega(t)$ denotes the rotor speed relative to the stator which is given by :

$$\omega(t) = \frac{d\theta(t)}{dt} - \frac{d\theta'(t)}{dt}$$

3. Experimental Study

Experiments under the micro gravity environment obtained by the free falls are conducted. This section explains the prototype rover, experimental parameters, simulations under the same conditions as the experiments, and the experimental results.

3.1. Prototype Rover

The prototype rover specification is shown in Table 1(a). Table 1(b) shows the DC motor specification. The DC motor is driven by PWM of 8.1[V] pulses, the pattern of which can be programmed on the following two modes.

(a) step drive: the constant duty ratio pulses are imposed on the DC motor. The duty ratio d_m can be programmed to an arbitrary value.

(b) ramp drive: At $t=0$, the duty ratio of the pulse is 0%, increased proportionally to the passed time, consequently comes to 100%. Time t_m when the duty ratio arrives at 100% can be programmed to an arbitrary value.

A 30[g] aluminum fly-wheel is attached to the rotor of the DC motor for increasing the inertia of the rotor.

In order to keep the rover still before the start of the free fall, the rover is locked to the surface physically and magnetically. In practice these locks are needed because the capsule with the experimental instrumentation are being conveyed to the entrance of the deep hole waiting for the free falling, many large shocks sufficient for the rover to shift attack to the capsule. After the free fall starts, both locks are sequentially unlocked. First, the physical locks (two pins) are unlocked after t_l[sec] have passed since the start of the fall. 1[sec] after the physical unlock, the magnetic lock by an electromagnet is unlocked and the rover becomes free. The motor starts 1[sec] after the last unlock and the rover starts to move t_l+2[sec] after the start of the free fall. In each experiment, t_l can be adjusted and is usually set to 3 or 4[sec].

3.2. Experimental Parameters

The micro gravity of 10[sec] is obtained by the free fall of 490[m] with help of jets thrusting upward in order to compensate for the resistance of the air[4]. With this facility, three experiments are conducted under different conditions in (1) the friction between the rover and the surface and (2) the motor driver method. Table 1(c) shows the simulation conditions in each experiment.

Table 1: Parameters in experiments

(a) rover spec.

width × height	120[mm] × 96.5[mm]
total mass	0.55 [kg]
rover inertia	1.0×10^{-3}[kg m^2]
DC motor rotor inertia	2.3×10^{-5}[kg m^2]

(b) DC motor spec. (in 9[V] measuring voltage)

Stall torque	500[gf cm]
No-load speed	7650 [rpm]
Torque constant	11.2[mNm/A]

(c) different parameters in each experiment

experiment	friction	motor drive method
#1	∞	step($d_m = 100$[%])
#2	$\mu \approx 1$	
#3		ramp($t_m = 1.71$[sec])

In experiment #2 and #3, the surface is made up of a flat piece of wood, where the coefficient of static friction is about 1.0. In experiment #1, a board is inserted between the rover and the surface. Also a needle sticks out from the bottom of the rover. With these items, the rover never slides leftward, making the friction ∞.

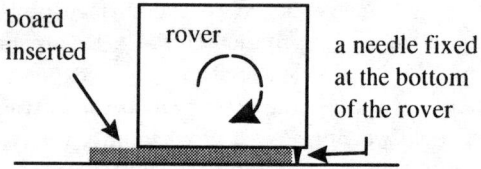

Figure 3: Inserted board and needle to make friction ∞

3.3. Computational Simulations

The experiments of step motor drives are simulated in Figure 4(a)(b) and (c). In Figure 4, the following parameters are used.

Hop time : the passed time since the DC motor started.

hop speed : v_x denotes the horizontal speed in the instant of hop.
V_y denotes the vertical speed in the instant of hop.
$$v = \sqrt{v_x^2 + v_y^2}$$

hop angle θ_h : $\tan \theta_h = v_x / v_y$

μ : coefficient of static friction between the rover and the surface

μ' : coefficient of dynamic friction between the rover and the surface which is given by $\mu' = \mu \times 0.8$

The first and second experiments correspond to $\mu=1$ and $\mu=\infty$ respectively in Figure 4. Figure 5(a)(b) and (c) show simulation results of ramp motor drives of $t_m = 1.71$[sec]. The third experiment corresponds to $\mu=1$ in Figure 5. Table 2 summarizes the expectations of the experiments. Note that all the simulations are done with gravity equal to zero.

3.4. Experimental Results

Figure 6 shows the video frames of the first experiment, the camera of which is set in front of the rover. These series of figures show that the rover hops with a significant horizontal velocity. The proposed hopping mechanism proves to work well not only in numerical simulation[3], but also in experiment.

1[sec] after the magnetic lock unlocked, the rover moved slightly in spite of no imposed torque (observed in Figure 6(b)). This is because the gravity in the free-fall capsule is not strictly at zero. The gravity profile during 10 [sec] of the first free fall is shown in Figure 7. At first and toward the end of the free fall, the obtained gravity is disturbed, and the experiment is not conducted in these periods. From 0.5[sec] to 9[sec], the gravity is fairly constant. The gravity from 1 to 8[sec] ranges from -0.34×10^{-2} to 0.38×10^{-2}[m/s^2]. Minimum and maximum accelerations of the three falls are summarized in Table 3. The peak-to-peak change during the falls are less than 1[mG]($= 10^{-2}$[m/s^2]). In this way, because the gravity during a free fall does not face in a single direction, zero-gravity is assumed in simulations.

The images in Figure 6 are distorted by optical aberrations in the camera lens. The distortions are corrected theoretically and coordinate transformations are made in order to calculate the rover states (x, y, θ) in each frame. Figure 8(a),(b) and (d) show the experimental transitions of rover's states (x, y, θ) with respect to time. Figure 8(c) shows the spacial movement of the center of gravity of the rover. Each figure in Figure 8 is overlaid the corresponding simulation in gray. In the same way, Figure 9 and Figure 10 show the second and third experiments respectively. The hop angle, speed and time are calculated by Figure 8, Figure 9 and Figure 10, which are summarized in Table 4. Table 2 of the simulations and Table 4 of the experimental results are consistent except for the hop angle and hop time of the first experiment. So the simulations are concluded to reflect the actual movement. In the future design of the hopping rover, we are able to make the most use of the simulations.

The hop angle and hop time of the first experiment have differences from those in the simulation. In the first experiment, compared with the corresponding simulation, the contact duration time between the rover and the surface in the experiment(0.53[sec]) is longer than in the simulation(0.26[sec]). This makes the actual rover's hop angle larger. In experiment #1, to prevent the rover slide leftward making the friction ∞, a board is inserted. There should be no force from outside after the ascending speed of the rover surpasses the descending speed of the contact point with the surface. But in the case of experiment #1, the needle of the rover contacts the side of the inserted board after the detachment from the surface. This contact force from the side of the board increases the rover's horizontal speed and makes the contact duration time longer and the hop angle larger.

4. Conclusion

The experiments of the proposed hopping rover are

conducted under micro gravity environment. Experimental results show the validity of the proposed hopping mechanism. The computational simulations and the experimental results are consistent in many cases.

The authors are now seeking a sophisticated DC motor PWM profile through the simulations. Also the 3D simulations can be made to control the hop direction of the rover with some motors and to design a actual rover targeted for an existing small asteroid.

References

[1] T.Kubota et al, "A Collaborative Micro-Rover Exploration Plan on the Asteroid Nereus in MUSES-C Mission", 48th International Astronautical Congress, IAF-Q.5.06, 1997.

[2] B.Wilcox et al, "Nanorover Technology and the MUSES-CN Mission", Proc. of i-SAIRAS '97, pp.445-450, 1997.

[3] T.Yoshimitsu et al, "Hopping Rover for MUSES-C Asteroid Exploration Mission", Proc. of ISAS 8th Workshop on Astrodynamics and Flight Mechanics, 1998.

[4] "JAMIC USER's GUIDE", Japan Microgravity Center

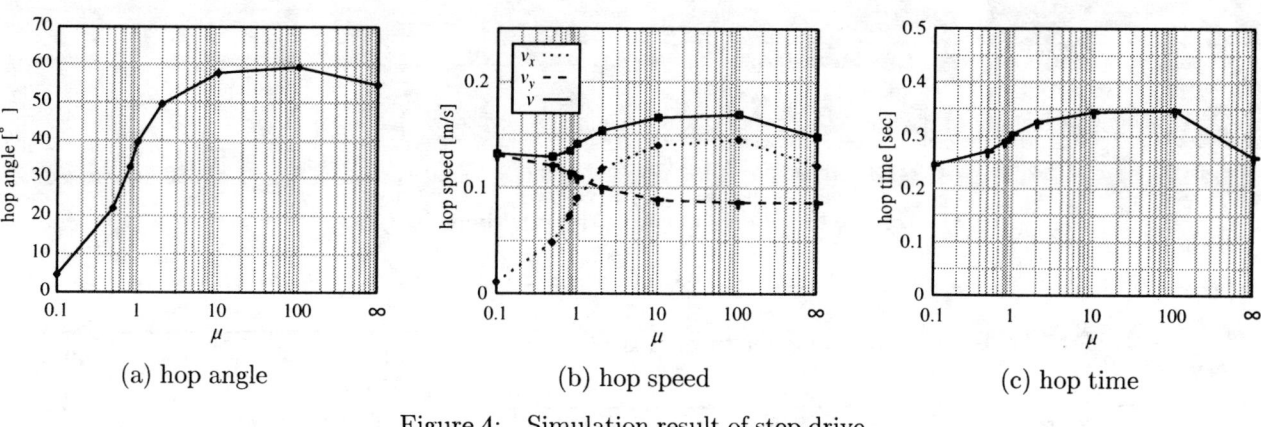

Figure 4: Simulation result of step drive

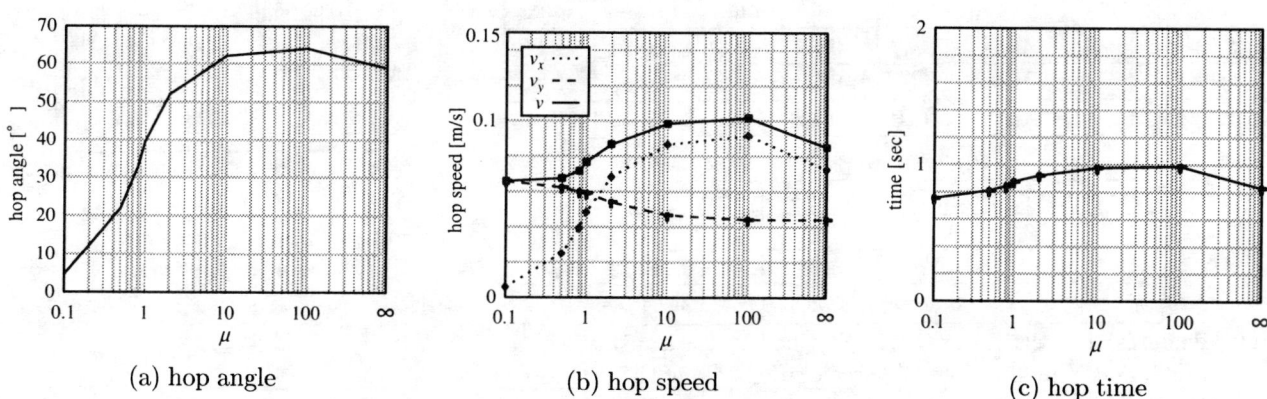

Figure 5: Simulation result of ramp drive

Table 2: Simulation result (hop angle, speed and time)

experiment	hop angle [°]	hop speed			hop time [sec]
		v_x [mm/s]	v_y [mm/s]	v [mm/s]	
#1	55	120	86	148	0.26
#2	40	90	109	141	0.30
#3	39	48	59	76	0.87

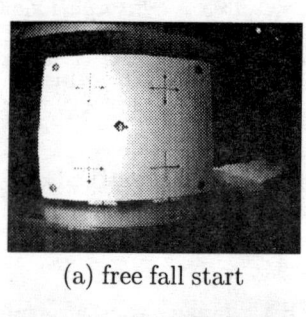 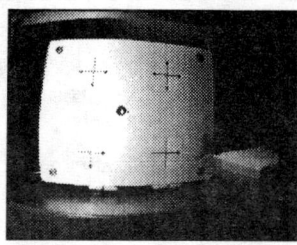 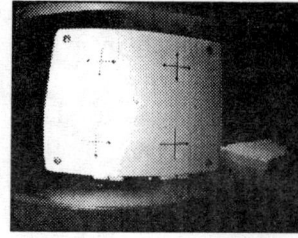 

(a) free fall start (b) 0.667 sec passed since magnetic lock unlocked (c) DC motor starts (d) 0.167 sec passed since DC motor started

 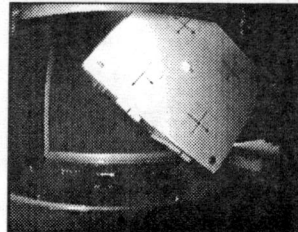 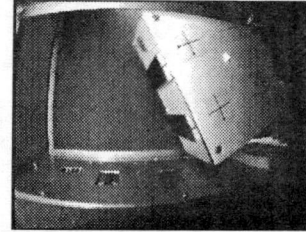 

(e) 0.333 sec passed (f) 0.500 sec passed (g) 0.667 sec passed (h) 0.833 sec passed

Figure 6: Video images of the experiment #1

Table 3: Acceleration during the three falls.

(a) minimum and maximal values

experiment	time [sec]	X[mG] min	X[mG] max	Y[mG] min	Y[mG] max	Z[mG] min	Z[mG] max
#1	1∼8	-0.17	0.09	-0.07	0.08	-0.34	0.38
#2	1∼7	-0.08	0.09	-0.07	0.07	-0.36	0.43
#3	1∼8	-0.18	0.20	-0.11	0.15	-0.49	0.43

(b) peak to peak

experiment	X[mG]	Y[mG]	Z[mG]
#1	0.26	0.15	0.72
#2	0.17	0.14	0.79
#3	0.38	0.26	0.92

Z[mG] demotes the vertical gravities obtained in the capsule in each table. X[mG] and Y[mG] denote the accelerations along the orthogonal horizontal directions.

Figure 7: Gravity acceleration profile of the first fall.

Table 4: Hop angle, speed and time of three experiments

experiment	hop angle [°]	hop speed v_x [mm/s]	hop speed v_y [mm/s]	hop speed v [mm/s]	hop time [sec]
#1	73	144	44	151	0.53
#2	43	83	91	123	0.43
#3	46	57	58	82	1.10

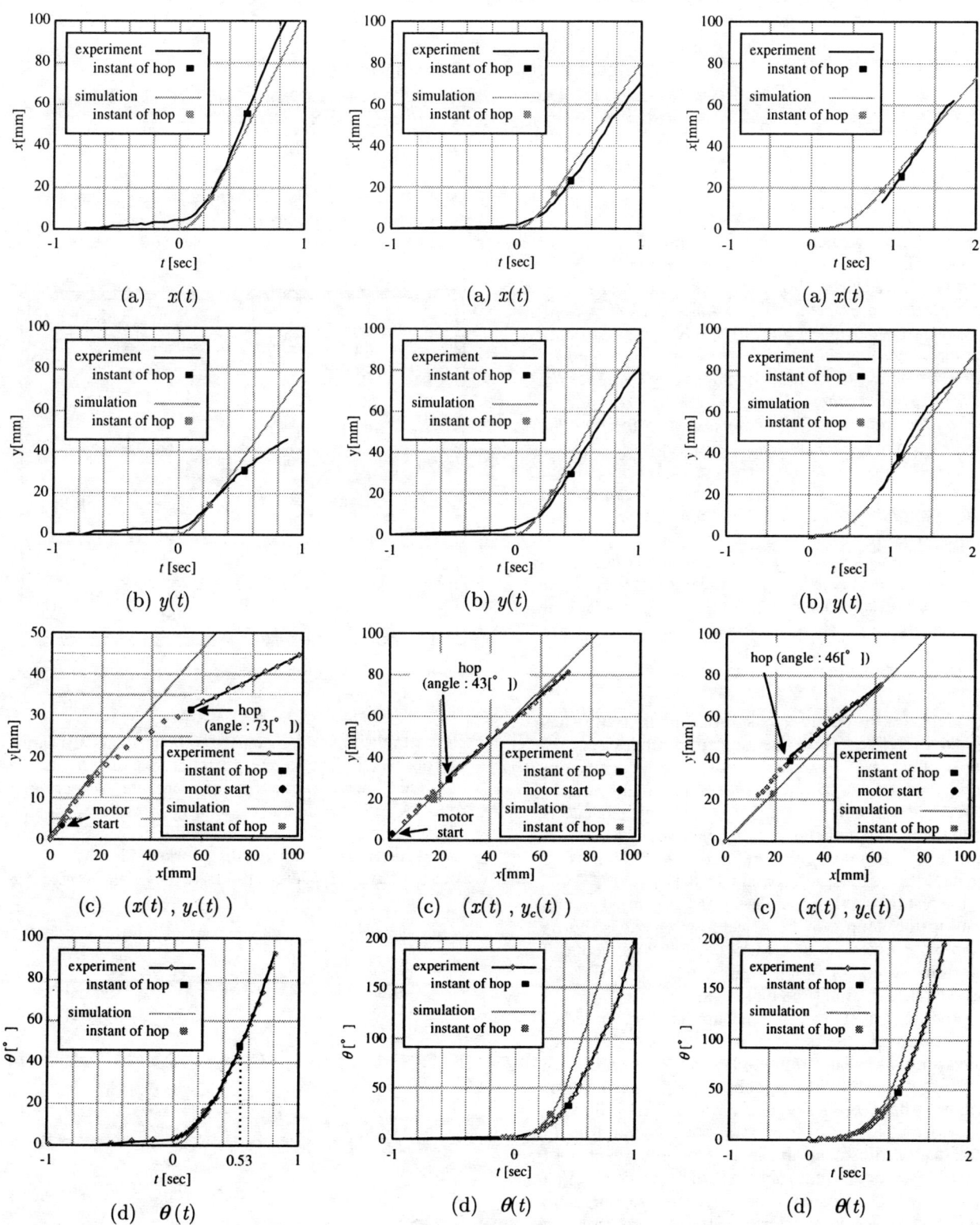

Figure 8. Transition of the rover states in experiment #1

Figure 9. Transition of the rover states in experiment #2

Figure 10. Transition of the rover states in experiment #3

Ant Trails – an Example For Robots to Follow?

R. Andrew Russell
Intelligent Robotics Research Centre
Department of Electrical and Computer Systems Engineering
Monash University, Clayton, VIC 3168
AUSTRALIA

Abstract

Sensitivity to chemical signals was the first sense to evolve and it is crucial for the survival of many biological creatures. However, there have been few attempts to transfer this sensory modality to robotic systems. In order to operate effectively in their variable and unstructured real world environment insects employ many strategies. Generating and detecting odours is the basis of several of these strategies. It seems certain that similar techniques can be implemented to improve the competence of robotic systems. This paper describes a project to investigate transferring the pheromone trail tracking capabilities of ants to robotic systems. A robotic ant has been built and equipped with a pair of odour sensing 'antennae'. This system has been programmed with a trail following algorithm observed in the ant Lasius fuliginosus. The design of this robot, the control algorithm and results of odour trail following experiments are presented in this paper.

1. INTRODUCTION

The detection of odours is a sensory modality that has received relatively little attention from roboticists. As humans our olfactory sense has a great influence on our moods and actions but almost at a subliminal level. In insects and other simpler creatures detecting chemicals is of immediate and crucial importance to many aspects of their lives. Chemical clues are used to find food, avoid danger, locate a mate and communicate [1].

Interest in robotic odour detection has centred around locating the source of both waterborne and airborne chemical plumes [2..6]. Many of the robot control algorithms used for plume source location have been based on information from biological examples.

Despite the many potential applications in robot navigation less attention has been paid to odour trail following [7, 8]. The ability to lay and follow trails can be a useful capability for a mobile robot. A number of scenarios have been suggested where trail following behaviour would be of use to a robotic system [9]. In the situation illustrated in Figure 1 a large quantity of materiel is to be moved from the start to the goal. 'A', a relatively sophisticated robot, has the sensors and intelligence to find the goal. Simple load carrying robots 'B' and 'C' follow an odour trail layed by 'A' to deliver the bulk of the materiel to the goal. Once an odour trail is established it can also be used as a guide for robots to return to their starting position. These uses of odour trails mimic the pheromone trails of ants [10].

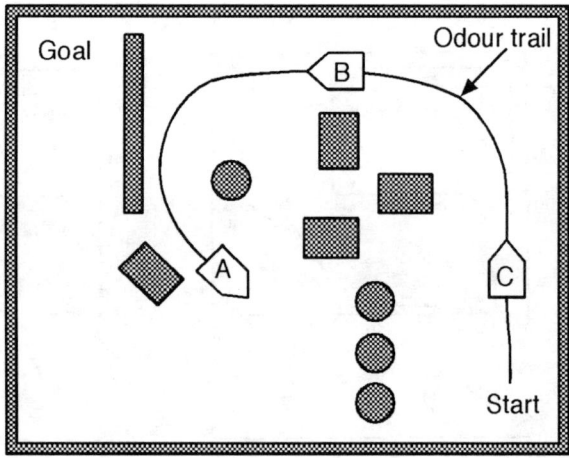

Figure 1: Laying odour trails to guide outgoing or returning robots.

Odour trails could also be used to help organise swarms of robots. In Figure 2 specialist worker robots are required to perform a task at the work site. A recruiting robot 'A' lays odour trails radiating from the work site. The chemical used to mark the trail represents a request for a specific kind of specialist worker robot 'B'. When robots of type 'B' detect the trail they follow it to the work site.

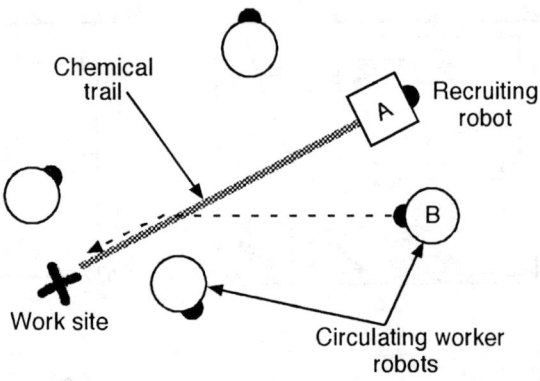

Figure 2: Recruiting specialist worker robots from a circulating robot swarm.

All of these uses of odour markings require robots to follow trails on the floor. In this application conventional control schemes do not function well, particularly in the face of sensor errors and imperfections in the trail. Millions of years of evolution have produced ant trail tracking algorithms which are remarkably simple and robust. This paper reports on a project to investigate robotic applications of one of the trail tracking algorithms observed in biological ants.

2. TRAIL FOLLOWING BY ANTS

Ant trails are markings of pheromone on the ground that provide a guide path for worker ants to follow. The initial trail is usually layed by a foraging ant returning to the nest after discovering a source of food. Worker ants navigate along this trail to transport the food back to the nest and at the same time they reinforce the trail with more pheromone. An interesting example of distributed optimisation takes place which leads to an initially winding and haphazard trail being converted into a straight, direct route [11]. This autocatalytic effect has been taken as a model for the distributed solution of some optimisation problems. Algorithms incorporating such ideas have been proposed for routing phone calls through a telecommunications network [12] and for the solution of difficult optimisation problems such as the travelling salesman problem [13].

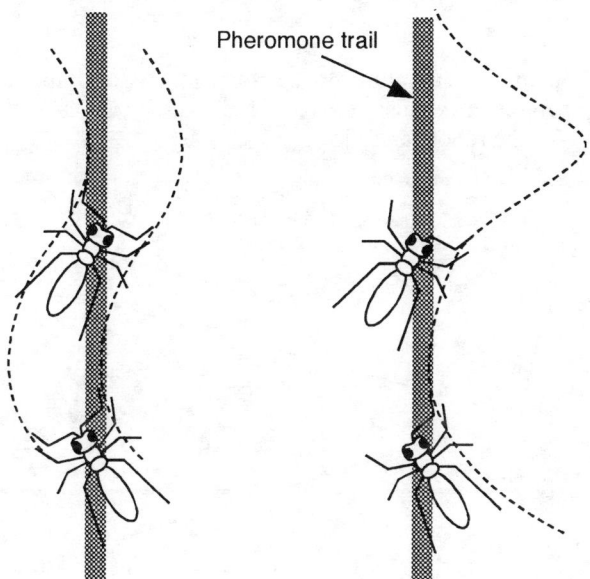

Figure 3: Trail following by the ant *Lasius fuliginosus* (adapted from [14]).

The technique that ants use to follow pheromone markings appears very simple. However, it does exhibit remarkable fault tolerance. Hangartner [14] describes the trail following behaviour of the ant *Lasius fuliginosus*. The normal mode of trail following for the ant is to curve to the left until the right antenna detects the trail and then to curve to the right until the left antenna contacts the trail (Figure 3). In this way the ant maintains the trail between its two antennae. Experiments have shown that if the ant has its left antenna removed then the remaining right antenna still triggers a swing to the right when it detects the pheromone trail. Even though a signal is not received from the left antenna the ant will turn back towards the trail after swinging rather wide to the right of the trail. In this way the ant can still follow the trail, albeit in a slightly less efficient manner. Adaptation of the trail following technique of *Lasius fuliginosus* to the control of a robotic ant is the subject of this paper.

3. A SENSOR FOR ODOUR MARKINGS

In order for a robot to follow a trail of robotic pheromone it must be able to detect it. Even in places where airflow is predominantly turbulent there is a layer of air close to the ground where airflow is laminar. Chemicals can only move through this region and away from the floor surface by diffusion [1]. Ants penetrate this laminar sublayer by tapping the ground with their antennae and sense the pheromone by contact or at very close proximity within the sublayer. Current robotic devices are built on a much larger scale compared to a biological ant. It is not possible to measure chemical concentration a fraction of a millimetre above the ground. In practice robotic odour sensors must have a ground clearance of at least 5 mm. Here they respond to plumes of chemical carried from some distance away and this reduces their ability to localise patches of odour on the ground.

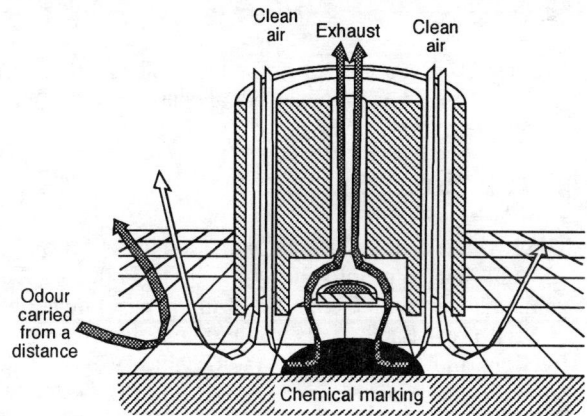

Figure 4: A cross-section view of the odour sensor.

For this project a sensor has been developed which uses controlled airflow to overcome the problem of localising odour markings. The sensor is based upon the quartz crystal microballance that uses a quartz crystal to weigh odour molecules [15]. A quartz crystal is coated with a substance that has an affinity for the chemical to be detected. In these experiments silicone OV-17 was used as a coating to detect trails of camphor. Odour molecules temporarily 'stick' to the coating material, increasing the mass of the crystal, and lowering its resonant frequency. This varying crystal frequency is the output of the sensor. Figure 4 shows a cross-section view of the odour sensor. A small circular printed circuit board fan blows air towards the floor through an annular opening around the edge of the sensor. Most of this air is forced outwards to produce an air curtain excluding external odour. Some of the air moves inward pressurising the area under the quartz crystal and causing an airflow which carries odour from the floor, over the crystal and then away to be vented to the atmosphere

As demonstrated in Figure 5 the odour sensor provides a much more localised response to odour markings on the ground compared to an unenclosed crystal.

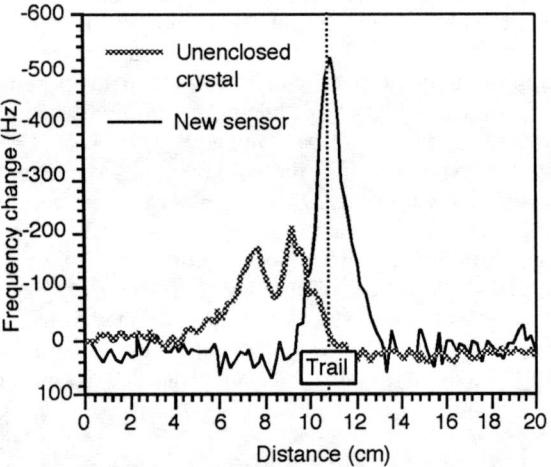

Figure 5: Response of an unenclosed crystal 5 mm from the ground and the new odour sensor passing over a camphor trail at 0.2 cm/sec.

4. THE ROBOT VEHICLE

To provide a reasonable approximation to the form of an ant a six-legged robot vehicle was built (Figure 6). This vehicle is based on Genghis, a walking robot developed at MIT [16]. Each of its six legs has two degrees of freedom and is actuated by radio control servos. Two odour sensors positioned 10 cm ahead of the robot and 13 cm apart simulate the odour sensing capabilities of the ant's antennae. Each leg of the robot is controlled by two actuators. These actuators are revolute and therefore, even for straight line walking, there is slip as the tip of each leg describes an arc. In order to turn the front-to-back range of movement of the legs down one side of the robot is restricted. To analyse the turning behaviour of the robot it is convenient to consider that each pace is divided into a straight line movement when legs on both sides of the robot are moving back together and a turn on the spot through angle τ where legs on one side of the robot are halted and the other set keep moving as illustrated in Figure 7.

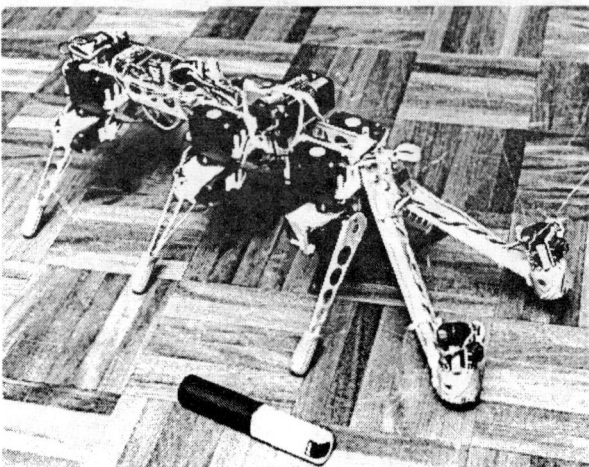

Figure 6: The robot ant with an odour pen.

This is a simplification of the actual walking action of the robot. However, it does allow us to write a simple equation for the radius of curvature r of the robot turn:

$$r = \frac{t}{2\cos(\tau)\tan\left(\frac{\tau}{2}\right)} \quad (1)$$

Radius of curvature of the robot turn is an important parameter because it gives a limit to the trail curvature that the robot can follow.

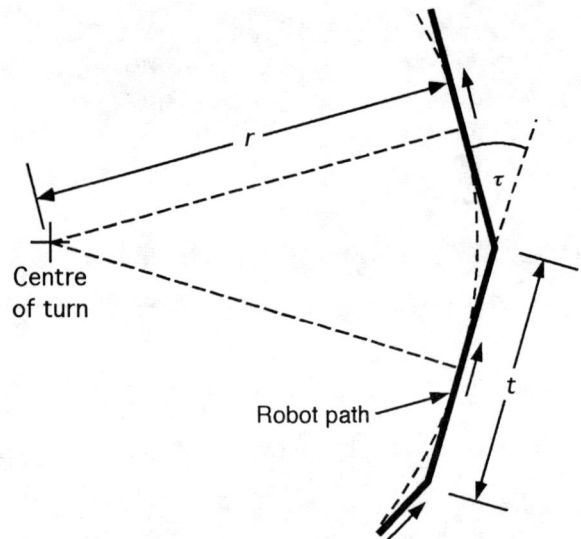

Figure 7: Robot turn made up of straight moves followed by rotation on-the-spot.

The structure of the robot's control program is based on Brooks' subsumption architecture [17]. A key feature of subsumption architecture is the direct connection of sensory inputs to control outputs with a minimum of data processing. This reflex level control provides very rapid response to changing sensory input. More sophisticated levels of control can be added to take over (subsume) control when appropriate. The basic walking and turning action of the robot is derived from material published about Genghis. To this has been added an additional layer of control based on data from the odour sensors.

5. THE CONTROL ALGORITHM

The ant control algorithm outlined in Section 2 was implemented as follows:

- Take 8 paces bearing to the right until completed or until trail detected by the left antenna.
- Take 8 paces bearing to the left until completed or until trail detected by the right antenna.
- repeat

(If the left sensor detects odour then the sequence of 8 paces to the left is started or restarted and stimulation of the right sensor starts or restarts the sequence of 8 paces to the right)

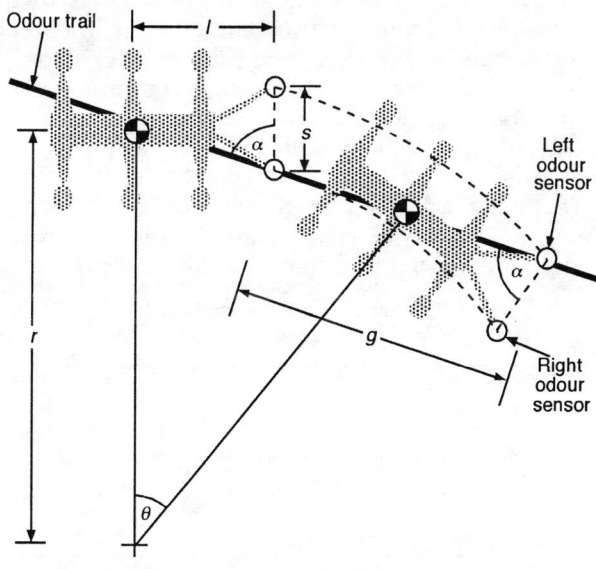

Figure 8: Trail following trajectory of the six-legged robot.

During trail following the path of the robot is determined by the location of the sensors and radius of turn of the robot. When the robot has settled to a stable trajectory while following a straight trail the turning centre of the robot will be positioned over the trail at the points when either sensor detects the trail (Figure 8).

Therefore:

$$\tan(\alpha) = \frac{l}{\left(\frac{s}{2}\right)} \quad (2)$$

where:

- l = distance between the sensors and the turning centre of the robot,
- s = the separation between the sensors, and
- α = the angle between the trail and a line through the sensors.

Now the angle θ that the robot turns through between one sensor detecting the trail and the other is:

$$\theta = \pi - 2\alpha \quad (3)$$

Therefore, the distance between successive points where the robot samples the trail g is:

$$g = \sqrt{2r^2(1 - \cos(\theta))} \quad (4)$$

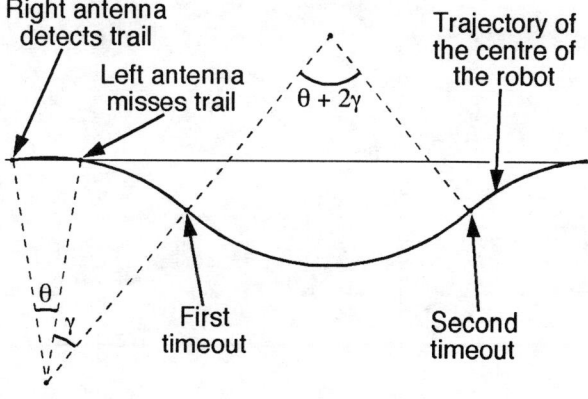

Figure 9: Reacquiring a lost trail.

Thus the distance g between points at which the robot samples the trail increases as the sensor separation s is increased. This distance also increases as the radius of the robot turn r is increased and reduces as the sensors are moved further away from the turning centre of the robot (*l*). All other things being equal the robot will advance along the trail faster the fewer times it samples the trail. The ability of the robot to follow sharp deviations in the trail is directly related to the radius of curvature of the robot turn

6. REACQUIRING A LOST TRAIL

Without any prior knowledge it is safest to assume that the trail will keep going in the same direction that it took before contact was lost. When the robot fails to detect the trail a timeout causes the robot to swing back towards the expected location of the trail. As illustrated in Figure 9 the robot turns through an angle γ following the point at which an antenna would normally detect the trail (a total turn of $\theta + \gamma$). In order for the robot to head in the expected direction of the trail the algorithm described in Section 5 has been modified so that subsequent turns are $\theta + 2\gamma$. In practice, to allow for a left turn in the trail or a failed left antenna sensor the turn resulting from a timeout should the slightly larger than $\theta + 2\gamma$.

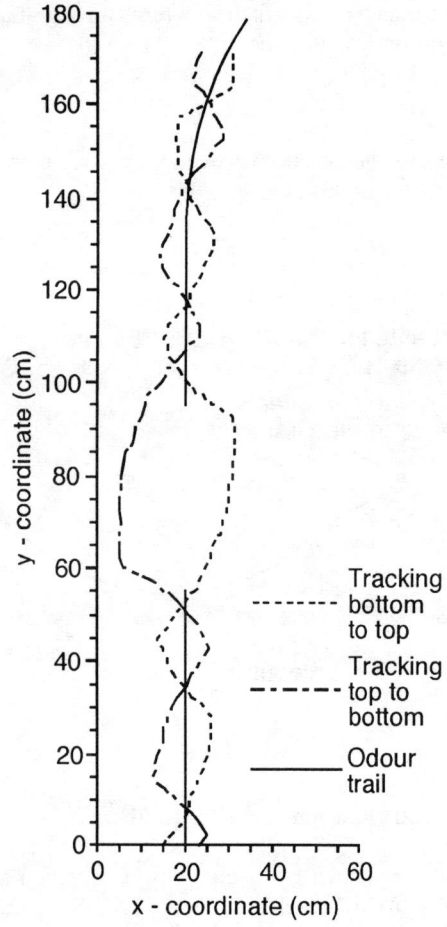

Figure 10: Track of the robot following an odour trail.

7. EXPERIMENTAL RESULTS

Ants lay pheromone trails by touching their gaster on the ground and transferring a patch of chemical. For this reason the trail will be inherently discontinuous. It is reasonable to assume that the ant's tracking algorithm can accommodate breaks in the trail. To assess the capabilities of the algorithm a camphor trail was layed on the floor consisting of a straight section, a curved section and a large gap. As shown in Figure 10 the robot ant tracked the trail in both directions and showed that the algorithm can robustly follow an odour trail even when contact with the trail is lost for a time. Turning of the robot utilises slipping of the robot's feet on the ground. Because of differing friction between the feet it was found that the robot turned right quicker than it turned left. This tendency will be corrected when new rubber tips are installed on all the feet. However, robust trail following behaviour occurred in spite of this imperfection in the robot.

8. ROBUSTNESS

Details of the operation of the ant trail tracking algorithm are shown by recordings of a simulation of the ant robot. In Figure 11 the trajectory of a point located between the sensors is plotted as the robot follows a straight trail, crosses a gap and then turns to the right with a reducing radius of curvature. The simulated robot negotiates the break in the trail and then follows the curved path until the path curves too sharply. At this point the robot losses contact and wanders away. Timing diagrams at the bottom of Figure 11 show how the signal from left and right sensors and the timeout pulse affect the turning direction of the robot.

The second simulation result shown in Figure 12 demonstrates the effectiveness of the control algorithm in accommodating sensor failure. With one sensor disabled the simulated robot is still able to negotiate the same trail as presented to the intact ant robot. The only difference is that the robot swings more widely and therefore progress along the trail is slowed.

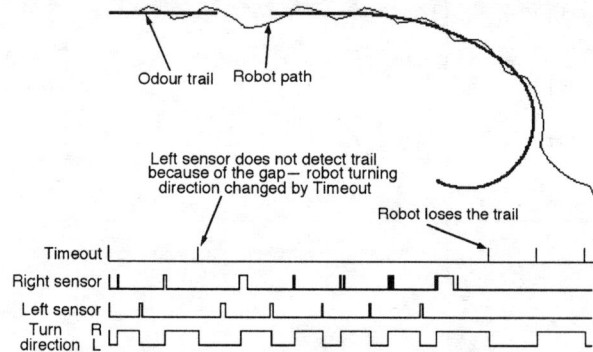

Figure 11: A simulation of the ant robot negotiating a trail with a missing section and a tight curve.

Figure 12 shows that an ant missing its left antenna is able to follow a trail curving sharply to the right. However, the ability to follow a left turn is much more impaired.

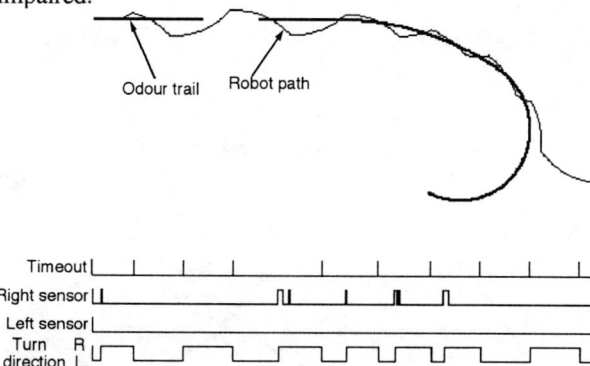

Figure 12: A simulation of the ant robot with a damaged left antenna negotiating a trail.

9. CONCLUSIONS

This paper has described the development of a robotic ant equipped with odour sensing 'antennae'. This device was used to investigate the robotic implementation of a trail following algorithm observed in the ant *Lasius fuliginosus*. The algorithm was found to provide robust trail tracking and will help to establish the viability of odour trail laying and tracking as aids to robot navigation.

10. ACKNOWLEDGMENTS

I would like to acknowledge the valuable assistance of Professor David Thiel who helped with the design of the olfactory sensor. This project was supported by grants from the Australian Research Council.

REFERENCES

[1] Dusenbery, D.B., *Sensory Ecology: How Organisms Acquire and Respond to Information*, W.H. Freeman and Company, N.Y., 1992

[2] Consi, T.R., Grasso, F., Mountain, D., and Atema, J., 'Explorations of turbulent odour plumes with an autonomous underwater robot', *The Biological Bulletin*, Vol. 189, October/November 1995, pp. 231-232.

[3] Ishida, H., Hayashi, K., Takakusaki, M., Nakomoto, T., Moriizumi, T., and Kanzaki, R., 'Odour-source localization system mimicking behaviour of silkworm moth', *Sensors and Actuators A*, Vol. 51, 1996, pp. 225-230.

[4] Kuwana, Y., Shimoyama, I., Sayama, Y. and Miura, H., 'Synthesis of pheromone-oriented emergent behaviour of a silkworm moth', *Proceedings of the IEEE/RSJ International Conference on Intelligent Robots and Systems*, 1996, pp. 1722-1729.

[5] Sandini, G., Lucarini, G. and Varoli, M., 'Gradient driven self-organising systems', *Proc. IEEE/RSJ International Conference on Intelligent Robots and Systems*, Yokohama, Japan, July 26-30, 1993, pp. 429-432.

[6] Rozas, R., Morales, J., and Vega, D., 'Artificial smell detection for robotic navigation', *Fifth International Conference on Advanced Robotics*, 1991, pp. 1730-1733.

[7] Russell, R.A., 'Laying and sensing odor markings as a strategy for assisting mobile robot navigation tasks', *IEEE Robotics & Automation Magazine*, September 1995, pp. 3-9.

[8] Stella, E., Musio, F., Vasanelli, L. and Distante, A., 'Goal-oriented mobile robot navigation using an odour sensor', *Proc. 1995 Intelligent Vehicles Symposium*, pp. 147-151.

[9] Deveza, R., Russell, R.A., Thiel D. and Mackay-Sim A., 'Odour sensing for robot guidance', *The International Journal of Robotics Research*, Vol. 13, No. 3, June 1994, pp. 232-239.

[10] Agosta, W.C., *Chemical Communication: the Language of Pheromones*, Scientific American Library, N.Y., 1992.

[11] Beckers. R., Deneubourg, J.L. and Goss, S., 'Trails and U-turns in the selection of a path by the ant *Lasius niger*', *Journal of Theoretical Biology*, Vol. 159, 1992, pp. 397-415.

[12] Ward, M., 'There's an ant in my phone...', *New Scientist*, 24th January 1998, pp. 32-35.

[13] Colorni, A., Dorigo, M., and Maniezzo, V., 'Distributed optimization by ant colonies', *Towards a Practice of Autonomous Systems: Proc. of the First European Conference on Artificial Life*, MIT Press, 1992, pp.134-142.

[14] Hangartner, W., 'Spezifität und Inaktivierung des Spurpheromons von *Lasius fuliginosus* Latr. und Orientierung der Arbeiterinnen im Duftfeld', *Zeitschrift für Physiologie*, Vol. 57, 1967, pp. 103-136.

[15] King, W.H., 'Piezoelectric sorption detector', *Anal. Chem.*, Vol. 36, No. 9, 1964, pp. 1735-1739.

[16] Brooks, R.A., 'A robot that walks: emergent behaviour from a carefully evolved network', *Neural Computation*, No. 1, 1989, pp. 253-262.

[17] Brooks, R.A., 'A robust layered control system for a mobile robot', *IEEE Journal of Robotics and Automation*, Vol. RA-2, No.1, March 1986, pp.14-23.

Dynamics and Distributed Control of Tetrobot Modular Robots

Woo Ho Lee

Department of Mechanical Engineering,
Aeronautical Engineering, and Mechanics
Rensselaer Polytechnic Institute
Troy, New York, 12180

Arthur C. Sanderson

Department of Electrical, Computer,
and Systems Engineering
Rensselaer Polytechnic Institute
Troy, New York, 12180

Abstract

Reconfigurable robotic systems can be easily adapted to different tasks or environments by reorganizing their mechanical configurations. In order to implement these new generations of robotic system, new approaches must be considered for design, analysis, and control. This paper proposes a distributed control scheme for a highly redundant parallel Tetrobot mechanism. The architecture of the proposed distributed control is composed of processors dedicated to each module and a network used to communicate the information between the modules. Each processor computes the kinematics, dynamics and control input for the dedicated module using the subsystem dynamic model and the information communicated only from adjacent modules. The simulation results of set-point and tracking control are provided to demonstrate the feasibility of the proposed scheme and compared to centralized control. The results show that the controlled node reaches the desired position without a steady state error even though the convergence rate is slower than for the centralized scheme. To improve the problem caused by a local optimization technique, an iteration scheme of local optimization was also applied to obtain a global solution to generate the paths of uncontrolled nodes.

1 Introduction

Robotics research in the fields of remote, unstructured, and hazardous environments such as space and undersea exploration, mining, and construction is growing rapidly as recent technology provides more intelligence and autonomy to the robots. A variety of tasks might be assigned for robots to complete a given mission in these fields. It will be very attractive if one robot can complete all the tasks by reconfiguring its structure according to the different tasks or environments.

Reconfigurable robotic systems based on the modular approach were proposed for this goal by several researchers in recent years. Fukuda and Nakagawa [1] and Pamecha and Chirikjian [14] proposed a dynamically self-reconfigurable robotic systems which can reorganize their

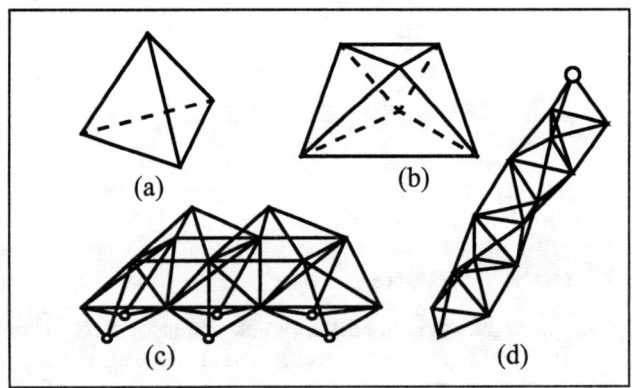

Figure 1: Examples of Tetrobot Designs. Each strut is an independent linear actuator, while each node is a multi-link (CMS [4]) joint. (a) Tetrahedral module (b) Octahedral module (c) Six-legged Walker (d) Tetrahedron-based Tetrobot Manipulator

shapes to meet the given tasks. Murata et al. [12] extended the concept to a 3-D self-reconfigurable robotic system. They designed and implemented the 3-D self-reconfigurable homogeneous unit and also discussed a distributed self-reconfiguration algorithm.

Hamlin and Sanderson [3, 4, 5] proposed the Tetrobot, a class of modular reconfigurable parallel robotic systems. The Tetrobot family of robots is a type of variable geometry truss that uses a novel joint mechanism design and a cooperating system of hardware and software components to build a wide variety of parallel or hybrid series-parallel robots. As shown in Fig. 1, Tetrobots can be constructed in any configuration of struts and nodes which are consistent with structural rigidity (statically determinate structures). Since only the lengths of the struts are actuated, the topology of each Tetrobot remains constant throughout the range of motion of all of its actuated links. Fig. 2 shows a Tetrobot prototype system configured as a six-legged walking robot [4].

This paper proposes a distributed dynamic scheme for

Figure 2: Photo of Six-legged Walker.

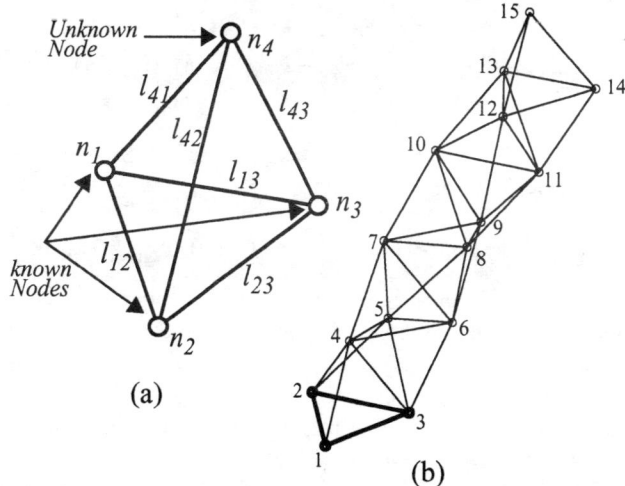

Figure 3: (a) A Unit Tetrahedral Module (b) Tetrahedron-Based Tetrobot

the Tetrobot reconfigurable robotic system. The distributed control architecture is based on a modular approach for the real-time control of a large scale system, and has been investigated for multiple redundant manipulators [10] and multiple robotic system [2]. It can offer significant benefits compared to a centralized control including application to all the configurations of a Tetrobot, parallel computing, modularity, robustness, and fault tolerance.

In this paper, the architecture of the proposed distributed control is composed of processors dedicated to each module and a network which is used to communicate the information between the modules. Each processor computes the kinematics, dynamics and control input for one dedicated module using the subsystem dynamic model and the information communicated only from adjacent modules.

2 Kinematics of Tetrobot

The tetrahedron-based Tetrobot is composed of tetrahedral modules with actuated links and multilink spherical (CMS [4]) nodes, and is constructed by appending additional modules to the triangular faces of adjacent modules. Fig. 3(a) shows a unit tetrahedral module that consists of four nodes connected together with six struts. The manipulator type Tetrobot composed of 12 tetrahedral modules is shown in Fig. 3(b). The first module is connected to a support by the three "fixed" nodes, and the "last" module in the chain acts as the end-effector. The kinematics of the Tetrobot relates the positions of the nodes in the truss and the strut lengths. For example, the forward kinematics may be expressed as,

$$\boldsymbol{n}_{y\{y \in N_C \cup N_U\}} = F_f(\boldsymbol{n}_{x\{x \in N_F\}}, l_{ij\{(i,j) \in S\}}) \quad (1)$$

where, N_C is the set of controlled nodes for which a particular path is desired, N_F is the set of fixed nodes, N_U, is the set of unconstrained nodes whose positions do not directly affect the current task, and S denotes the set of all struts. The kinematic problem for the entire Tetrobot is based on the propagation of modular kinematic solutions through the structure and is discussed in reference [4].

3 Dynamic Model of Tetrobot

In this section, the dynamic model of the tetrahedron-based Tetrobot is summarized based on the previous work [9]. Fig. 4 shows the free body diagram of a Tetrobot composed of 4 tetrahedral modules. Applying Newton's

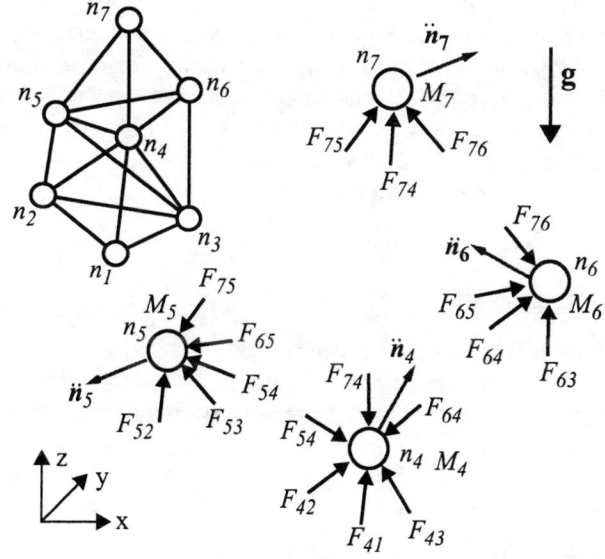

Figure 4: Free body diagram of 4 tetrahedral modules.

equation of motion under the assumption of a lumped mass model gives,

$$\boldsymbol{F}_{74} + \boldsymbol{F}_{75} + \boldsymbol{F}_{76} = M_7(\ddot{\boldsymbol{n}}_7 + \boldsymbol{g})$$

$$F_{63}+F_{64}+F_{65}-F_{76} = M_6(\ddot{n}_6+g) \quad (2)$$
$$F_{52}+F_{53}+F_{54}-F_{65}-F_{75} = M_5(\ddot{n}_5+g)$$
$$F_{41}+F_{42}+F_{43}-F_{54}-F_{64}-F_{74} = M_4(\ddot{n}_4+g)$$

The actuating forces of each node point can be expressed in general form as,

$$\boldsymbol{F}_{node} = [\hat{v}]_{node}^{-1}\{M_{node}\ddot{\boldsymbol{n}}_{node}+M_{node}\boldsymbol{g}+\sum \hat{v}_{ij}\boldsymbol{f}_{ij}\} \quad (3)$$

where $\sum \hat{v}_{ij}\boldsymbol{f}_{ij}$ are the interacting forces between the adjacent modules. The matrices denoted by $[\hat{v}]_{node}$ represent the configuration of each module and contains a unit vector along the strut. All of the terms in eq(3) can be obtained from the information of each module except the interacting forces. The actuating forces can be calculated by propagating from the end-effector to the base. This principle of propagation may be applied to any tetrahedron-based Tetrobot, where any systematic traversal of the modules may be used.

4 Distributed Control

4.1 The Architecture of Distributed Control

Most robot control schemes have been focused on the centralized control approach. However, the centralized control scheme may not be appropriate to a system with many degrees of freedom because of the computational burden and lack of flexibility. There is a need to distribute the computational burden among many processors for the real time control and flexibility of a large scale robotic system. The proposed distributed control architecture is shown in Fig. 5. The processors dedicated to each module calculate all the information for each module and communicate the information between adjacent modules. The task of each processor may be classified into the following functions: sensing, forward kinematics, controller, communication, and path planning. As an example, the task of the $i-th$ processor dedicated to the $i-th$ node (n_i) is described as follows.

- Sensing: $l_{i,b}, \dot{l}_{i,b}$

where, $b = \{b_1, b_2, b_3\}$ is the set of base nodes and (i,b) represents the struts of $i-th$ module. The lengths and velocities of struts located between the nodal point (n_i) and base nodal points of $i-th$ module are sensed and locally used to calculate the updated position and velocity of $i-th$ nodal point.

- Forward kinematics: $n_i, \dot{n}_i$

The position and velocity of nodal point of $i-th$ module is calculated and updated using the local sensing signals and the positions and velocities of base nodal points.

$$n_i = f(n_b, l_{i,b}) \quad (4)$$
$$\dot{n}_i = f(n_b, \dot{n}_b, l_{i,b}, \dot{l}_{i,b})$$

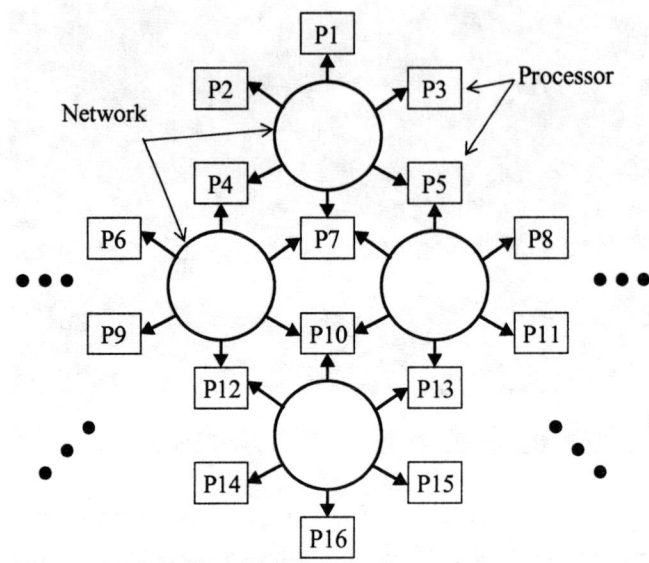

Figure 5: The architecture of Tetrobot distributed dynamic control.

where, n_b is three base nodes of $i-th$ module.

- Controller: Actuating force F_i

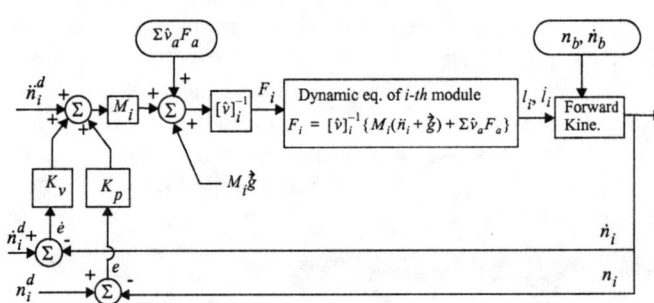

Figure 6: The block diagram of $i-th$ module.

The control input is composed of 2 parts. The first part (F_i^c) is used to compensate the nonlinear terms of the dynamic equation using the dynamic model of $i-th$ module. The second term (F_i^s) is the PD feedback controller which is driven by the error terms. The controller is based on the computed torque technique in task space [7].

$$F_i = F_i^c + F_i^s \quad (5)$$
$$F_i^c = [\hat{v}]_i^{-1}\{\sum \hat{v}_{i,a}F_a + M_i\vec{g}\}$$
$$F_i^s = M_i[\hat{v}]_i^{-1}\{\ddot{n}_i^d + K_v(\dot{n}_i^d - \dot{n}_i) + K_p(n_i^d - n_i)\}$$

where, K_p and K_v are 3×3 diagonal position and velocity control gain matrices. The block diagram of a controller is shown in Fig. 6. n_i^d is the desired nodal position of $i-th$ node in the task space. The information of adjacent modules is utilized to compensate the interacting forces from the adjacent modules and to solve the forward kinematics of the nodal point of the $i-th$ module.

- Communication: $n, \dot{n}, F$

Each processor computes the nodal positions and velocities, and actuating forces of each module using the information of its modules obtained from the network. The new values computed and updated at each processor are communicated through the network. Therefore, the information through the network is not the current value, but the previous one. For this reason, the errors caused by the computation time delay exist and accumulate as the information passes through the adjacent modules.

- Path planning of unconstrained nodal points

The nodal points of a Tetrobot can be divided into three categories: the controlled, fixed and unconstrained nodal points. The paths of controlled nodal points can be uniquely described to follow the given task by path planning or user input. On the other hand, the unconstrained nodal points can be arbitrarily located within some ranges. The path planning of unconstrained nodal points is used to resolve the redundancy which can be utilized for certain purposes such as obstacle avoidance [11], joint range limits [8], task priority based redundant control [13], singularity avoidance [16] and joint torque optimization [6, 9]. Most researchers utilized the inverse Jacobian method [15] which is quite general, but computationally inefficient for a large scale system. In this paper, the paths of unconstrained nodal points are resolved to avoid the singular configuration of a Tetrobot using the virtual force method [4]. The virtual force method is a potential field approach which can be applied to the distributed computation. The virtual force method uses the potential function which is composed of local values, nodal points of adjacent and base module in each module.

$$P_i = \sum_i \sum_j \frac{(|n_i^d - n_j| - l_{ave})^2}{l_{st}^2} \quad |j \in S_i \quad (6)$$

where, l_{ave} is the center position for the struts and l_{st} is the stroke length of the actuator. n_i^d is a desired nodal point and S_i is the set which contains the base (n_b) and adjacent (n_a) nodal points of $i-th$ node. The minimum of a potential function can be obtained by differentiating with respect to n_i^d.

$$\text{Min. } P_i = \frac{\partial P_i}{\partial n_i^d} = 0 \quad (7)$$

The solution n_i^d of eq(7) represents the desired unconstrained nodal points which try to avoid the singular configuration of the $i-th$ module based on the local information.

4.2 Error analysis

This section presents the error analysis caused by computation time delay in the distributed control system. Each processor utilizes the information of the adjacent and base nodes obtained at the previous time step, not the current time information. For this reason, the errors caused by computation time delay are accumulated and should not be ignored for the analysis and implementation. First, let us consider the value of the $i-th$ nodal point computed from the $i-th$ processor at the $k-th$ sampling time.

$$^k n_i = f(^{k-1}n_b, {}^k l_{i,b}) \quad (8)$$

In order to calculate the current position of the nodal point $^k n_i$, the current strut length values $^k l_{i,b}$, previous values of base nodal points $^{k-1}n_b$ are utilized. The real position can be expressed as,

$$^k n_i^* = f(^k n_b^*, {}^k l_{i,b}) \quad (9)$$

where, $(\cdot)^*$ represents the real value. The error between the real value $(^k n_i^*)$ and computed value $(^k n_i)$ can be obtained as,

$$\begin{aligned} ^k e_i &= {}^k n_i^* - {}^k n_i \\ &= f(^k n_b^*, {}^k l_{i,b}) - f(^{k-1}n_b, {}^k l_{i,b}) \\ &= f(^k n_b^*, {}^k l_{i,b}) - f(^{k-1}n_b^* - {}^{k-1}e_b, {}^k l_{i,b}) \end{aligned} \quad (10)$$

The error equation (10) shows that the error is composed of the time delay part $(^{k-1}n_b^*)$ and the previous time error $(^{k-1}e_b)$ of base nodes. The second term is caused by the lower nodes of base nodes and accumulates proportional to the number of nodes which were passed through.

5 Distributed Control Simulation

The simulation of the proposed distributed control has been performed and compared to the centralized control scheme. In order to illustrate the differences of the two schemes, the same strategy is applied to the centralized control. For the centralized control, the global virtual force method is applied to generate the desired paths of uncontrolled nodes using all the nodes in the truss, and the errors caused by computation time delay are assumed to zero. The fourth order Runge-Kutta method is used to integrate the dynamic equation of motion. The masses of nodes are 1kg and the initial strut lengths are 0.45m. It is assumed that all the simulation parameters are known.

5.1 Set-point Control

The set-point control simulation is performed to explore the proposed distributed controller for the Tetrobot composed of 6 tetrahedra modules (18 DOF). The Tetrobot is composed of three base nodes, n_1, n_2, n_3, one controlled node, n_9, and five unconstrained nodes, $n_4, \cdots, n_8$. The desired displacement of the controlled node is $\delta N_x = -0.5$m, $\delta N_y = 0.3$m, and $\delta N_z = 0.3$m. The position (K_p) and velocity gain (K_v) were chosen to have a critical damping. The controlled node gains $K_{p,c} = 50, K_{v,c} = 14.1$ and uncontrolled node gains $K_{p,u} = 400, K_{v,u} = 40$ are used and sampling time t_s is 0.005 sec. As an optimization method of minimizing a potential function of virtual force method, the Newton's gradient search (MATLAB) is used. Fig. 7 illustrates the entire motion of a Tetrobot during the simulation.

The controlled node reaches to the desired position without a steady state error and overshoot, and the response is close to the centralized control scheme as shown in Fig. 8. (Thick line: distributed control, thin line: centralized control). On the other hand, the responses of struts and actuating forces show the large differences between the two schemes shown in Fig. 9–10. (solid line = (i, b_1), center line = (i, b_2), hidden line = (i, b_3)). The responses of the distributed control scheme converge slowly compared to the centralized control and also show an overshoot. The differences of the two schemes originate from the computation time delay and local optimization scheme. To improve the problem caused by a local optimization technique, an iteration scheme is proposed. The solution of a local optimization technique is fed into the optimization problem again and iterates the local optimization several times for each sampling time. The desired paths of unconstrained nodes approach the global solution as the iteration rate increases.

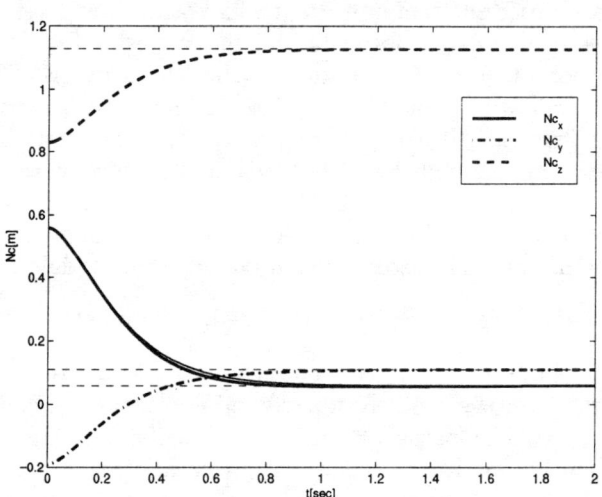

Figure 8: The position of the controlled node, n_9. Centralized and distributed controlled trajectories are almost identical.

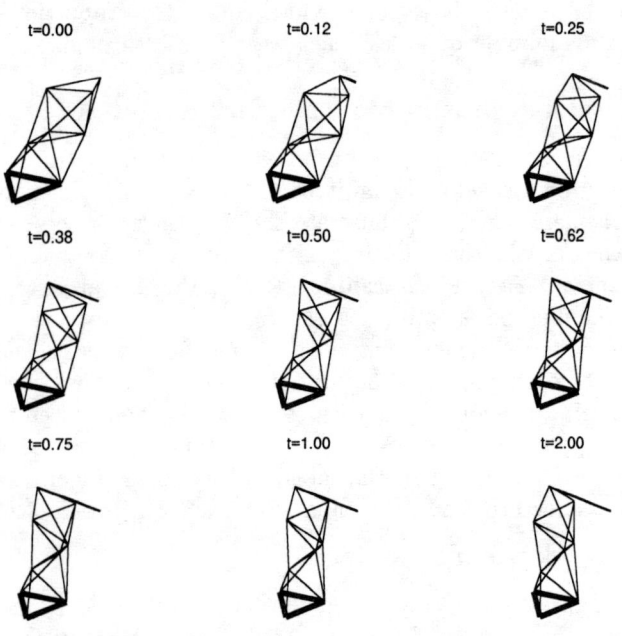

Figure 7: The dynamic motion of 6 module Tetrobot executing a displacement of the controlled node.

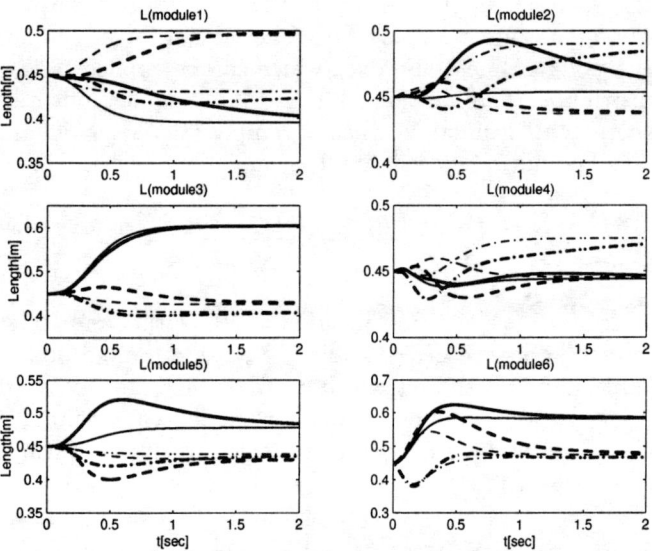

Figure 9: The strut lengths during execution shown in Fig. 7. Centralized and distributed control trajectories are quite different.

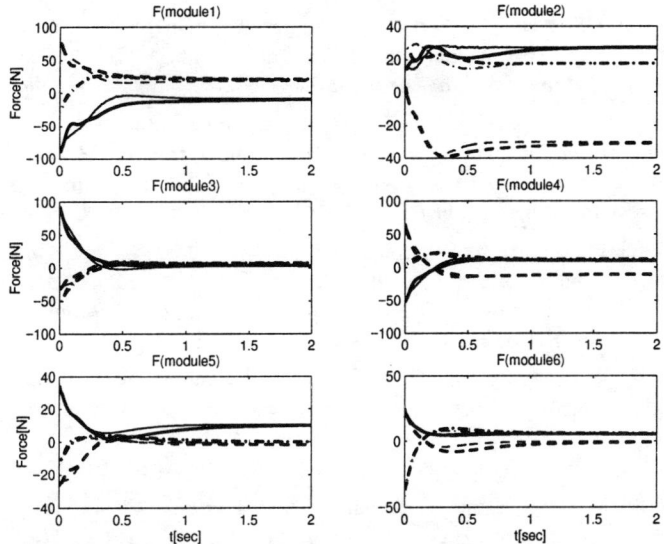

Figure 10: The actuating forces for motion in Fig. 7. Centralized and distributed control are different.

5.2 Tracking Control

The distributed control scheme is also applied to the tracking problem for the 7 module Tetrobot (21 DOF). The prescribed trajectory is a circular motion and the radius of a circle is 0.3m. The sampling time is 0.002 sec and control gains of all the nodes are $K_p = 200, K_v = 28.3$.

First, the simulation is performed without the iteration of local optimization. It is observed that the motion is not smooth for the entire time in this trial. The struts which are close to the base do not actuate much compared to those of the controlled node. The strokes of struts of the controlled node (module #7) are about 0.4m and the strokes of the base module (module #1) are about 0.05m. This big difference of strokes between modules is mainly due to the local optimization technique which is applied to calculate the desired trajectory of unconstrained nodes.

As a second case, the Tetrobot motion applying 5 iterations of local optimization in shown in Fig. 11. The results show that all the struts are actuating uniformly for the entire time. The tracking errors of the two schemes are compared in Fig. 12 (Thick line:5 iteration, thin line:No iteration). The results show that the tracking performance is not sensitive to the path planning of uncontrolled nodes when multiple iterations of the local optimization algorithm are used. The actuating forces are illustrated in Fig. 13.

6 Conclusions

This paper proposes a distributed dynamic control scheme for the Tetrobot reconfigurable parallel robot. A

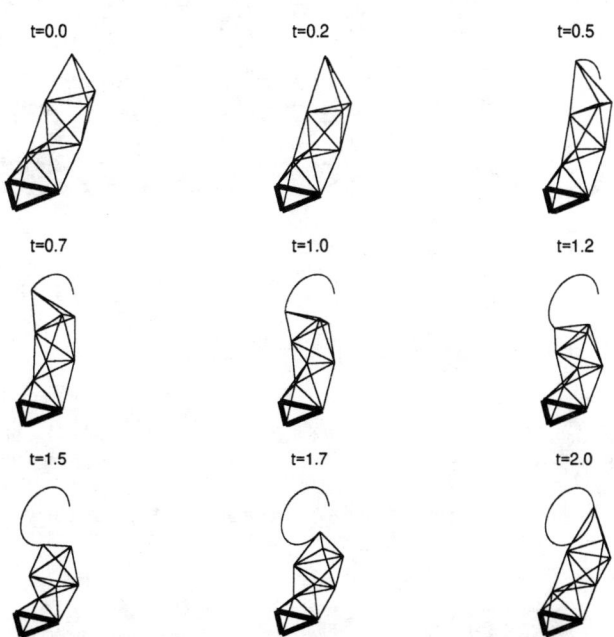

Figure 11: The motion of 7 module Tetrobot.

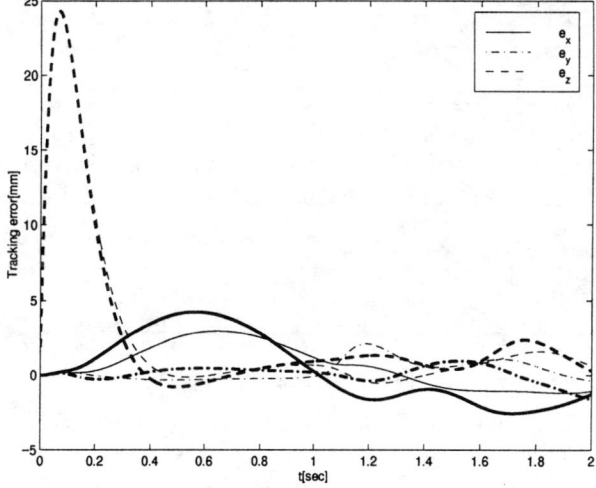

Figure 12: The tracking error of the controlled node for motion in Fig. 11.

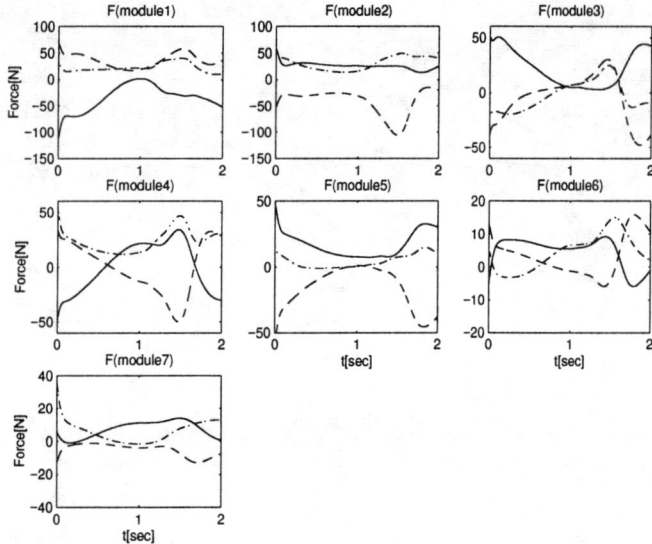

Figure 13: The actuating forces for motion in Fig. 11.

set-point control simulation was performed to investigate the proposed distributed controller and compared it to centralized control. It is shown that the controlled node reaches the desired position without a steady state error even though the convergence rate is slower than the centralized scheme. An iteration scheme of local optimization was also applied to obtain a global solution to generate the paths of uncontrolled nodes. While the examples shown in this paper demonstrate the approaches for a serial chain of tetrahedral modules, the control scheme applies to any Tetrobot such as the six-legged walker shown in Fig. 2.

Acknowledgments

We are pleased to acknowledge the use of computing facilities of the Electronics Agile Manufacturing Research Institute at Rensselaer. The EAMRI is partially funded by Grant number DMI-9320955 from the NSF and ARPA.

References

[1] T. Fukuda and S. Nakagawa, "Dynamically Reconfigurable Robotic System," *IEEE Int'l Conf. on Robotics and Automation*, pp. 1581-1586, May, 1988.

[2] T. Fukuda and T. Ueyama, *Cellular Robotics and Micro Robotic Systems*, World Scientific, New Jersey, 1994.

[3] G. J. Hamlin and A. C. Sanderson, "A Novel Concentric Multilink Spherical Joint with Parallel Robotics Applications," *IEEE Int'l Conf. on Robotics and Automation*, pp. 1267-1272, May, 1994.

[4] G. J. Hamlin and A. C. Sanderson, *Tetrobot: A Modular Approach to Reconfigurable Parallel Robotics*, Kluwer Academic Publishers, Newton, MA, 1997.

[5] G. J. Hamlin and A. C. Sanderson, "Tetrobot: A Modular Approach to Parallel Robotics," *IEEE Robotics and Automation Magazine*, March, 1997.

[6] J. M. Hollerbach and K. C. Suh, "Redundancy resolution of manipulators through torque optimization," *IEEE Journal of Robotics and Automation*, vol. RA-3, no. 4, pp. 308-316, August, 1987.

[7] O. Khatib, "A Unified Approach for Motion and Force Control of Robot Manipulators: The Operational Space Formulation," *IEEE Journal of Robotics and Automation*, vol. RA-3, no. 1, pp. 43-53, February, 1987.

[8] C. A. Klein and C. H. Huang, "Review of pseudoinverse control for use with kinematically redundant manipulators," *IEEE Trans. on Systems, Man, and Cybernetics*, vol. SMC-13, no. 3, pp. 245-250, March/April, 1983.

[9] W. H. Lee and A. C. Sanderson, "Dynamic Simulation of Tetrahedron-Based Tetrobot," *IEEE/RSJ Int'l Conf. on Intelligent Robots and Systems*, pp. 630-635, Oct., 1998.

[10] Y. Liu and S. Arimoto, "Decentralized Adaptive and Nonadaptive Position/Force Controllers for Redundant Manipulators in Cooperations," *Int'l Journal of Robotics Res.*, vol. 17, no. 3, pp. 232-247, 1998.

[11] A. A. Maciejewski and C. A. Klein, "Obstacle avoidance for kinematically redundant manipulators in dynamically varying environments," *Int'l Journal of Robotics Res.*, vol. 4, no. 3, pp. 109-117, Fall, 1985.

[12] S. Murata, H. Kurokawa, E. Yoshida, K. Tomita, and S. Kokaji, "A 3-D Self-Reconfigurable Structure," *IEEE Int'l Conf. on Robotics and Automation*, pp. 432-439, May, 1998.

[13] Y. Nakamura and H. Hanafusa, "Task-priority based redundancy control of robot manipulators," *Int'l Journal of Robotics Res.*, vol. 6, no. 2, pp. 3-15, 1987.

[14] A. Pamecha and G. Chirikjian, "A Useful Metric For Modular Robot Motion Planning," *IEEE Int'l Conf. on Robotics and Automation*, pp. 442-447, April, 1996.

[15] D. Whitney, "Resolved motion rate control of manipulators and human prostheses," *IEEE Trans. Man-Machine Syst.*, vol. MMS-10, no. 2, pp. 47-53, June, 1969.

[16] T. Yoshikawa, "Analysis and control of robot manipulators with redundancy," *Robotics Research: 1st Int'l Symposium*, eds. M. Brady and R. Paul, MIT Press, pp. 439-446, 1984.

Force Distribution Equations for General Tree-Structured Robotic Mechanisms with a Mobile Base

Min-Hsiung Hung and David E. Orin

Dept. of Electrical Engineering
The Ohio State University
Columbus, Ohio

Kenneth J. Waldron

Dept. of Mechanical Engineering
The Ohio State University
Columbus, Ohio

Abstract

An efficient formulation of the force distribution equations for actively-coordinated vehicles is presented. The applicable platforms include not only systems with star topologies, such as walking machines that have multiple legs with a single body, but also general tree-structured mechanisms, such as variably-configured wheeled vehicles having multiple modules. Based on this formulation, several standard optimization techniques, such as linear programming or quadratic programming, can be applied to obtain the solution. The efficiency of the formulation is demonstrated with results showing real-time execution on a Pentium PC.

1 Introduction

Actively-coordinated vehicle systems usually have a larger number of independently-controlled actuators than that of degrees of freedom of mobility. These vehicles have the ability to use the redundancy of actuation to improve the vehicle-terrain contact conditions. In turn, various aspects of system performance can be enhanced or optimized. Thus, such vehicles are very suitable for off-road terrain-adaptive applications, such as those in the following fields: mining, agriculture, forestry, military locomotion, or exploration of planetary surfaces. Examples of actively-coordinated vehicles include legged walking machines [1], usually with multiple legs and a single body, and variably-configured wheeled systems [2], usually with multiple modules.

Force distribution is one of the most important issues to utilize actively-coordinated vehicles. It deals with the allocation of desired contact forces among legs/wheels to equilibrate a given inertial force/moment (wrench) of a vehicle for a specified motion trajectory. This is an under-constrained problem. That is, there are infinite sets of contact forces to balance a given inertial wrench of a system. Therefore, an optimal set of contact forces can be found for a specific motion based on some optimization criteria of system performance, such as minimization of power consumption, minimization of slip, maximization of stability, or load balance.

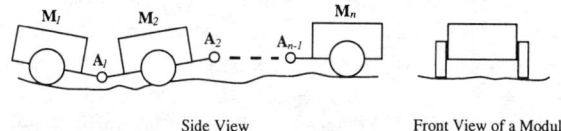

Figure 1: Actively-Articulated Wheeled Vehicle [2]

Past work has been directed to the problem of formulating the force distribution equations for actively-coordinated vehicles [3]. However, the previous formulations were only applicable to simple star-topology systems, such as multilegged vehicles with a single, nonarticulated body structure. It is necessary to have other approaches to formulate the force distribution equations for actively-coordinated systems with more complex kinematic configurations, such as the actively-articulated wheeled vehicles shown in Figure 1 [2]. This figure shows an actively-articulated wheeled vehicle with n modules. Each module M_i consists of a module body and two wheels that are mounted on a solid axle through revolute joints. The consecutive modules are connected by articulations with three degrees of freedom (DOF) that can be kinematically modeled as a spherical joint or three consecutive revolute joints. All actuators of the wheels and articulations are independently controlled.

Several prototypes of such a vehicle have been constructed [2, 4]. A preliminary formulation of the force distribution equations of a three-module vehicle was presented in [5, 6]. In the approach taken for balancing all forces on the master module of the vehicle, quantities, such as each link's spatial inertia, spatial acceleration, velocity, or velocity-dependent terms, are first expressed in the master module's coordinate system. The quantities are then used to compute the resultant inertial force. This method of calculation involves a considerable number of spatial screw transforms such that it is computationally very expensive.

In this paper, we will present a systematic and efficient approach to formulate the force distribution equations for actively-coordinated vehicle systems, including legged vehicles with articulated bodies and wheeled systems. In our formulation, both the joint torque constraints and the contact friction constraints will be efficiently incorporated. Also, we will consider each link's dynamic effect, which can be optionally included in the formulation according to the requirements of the application. In addition, in our formulation, the legs or wheels of the vehicle are allowed to possess multiple contact points with the environment.

With regard to systematization, we will use a general tree-structured mechanism to represent the vehicle system. Then, a set of link parameters and functions are defined and used to systematically specify the relations among the links in the tree. To achieve efficiency, we use recursive computation to derive the force balance and other constraint equations.

The remainder of the paper is organized as follows. Section 2 describes the relevant background and the notation used in this work. Also, a general tree-structured mechanism is described. In Section 3, force balance equations are derived. The joint torque constraints and the contact friction constants are incorporated in Sections 4. In Section 5, a concise matrix form of the force distribution equations is presented. Section 6 presents an example mechanism with its associated computational cost given. The paper ends with a summary and conclusions of the work.

2 Background and Notation

In this section, the relevant background used in deriving our formulation is briefly described. Also, a review of the spatial dynamic equations for a rigid body, along with the notation used in this work, will be presented.

Notation We use an italic bold variable, such as $\boldsymbol{a}$, to denote a Cartesian (three-dimensional) vector, and a block bold variable, such as $\mathbf{a}$, to refer to a spatial (six-dimensional) vector. If a vector for a link or contact is expressed in its own coordinate system, no direct indication of this is given. Otherwise, a leading superscript is used to indicate the coordinate system. For example, $\mathbf{f}_i$ denotes a spatial force vector acting on link i with components expressed in frame $\{i\}$, whereas $^j\mathbf{f}_i$ denotes the same vector, but expressed in frame $\{j\}$.

Also, for the transformation notation, $^i\boldsymbol{R}_j$ is a 3×3 matrix which is used to transform a Cartesian vector expressed in frame $\{j\}$ to one expressed in frame $\{i\}$. The 6×6 matrix $^i\mathbf{X}_j$ is used to transform spatial vectors between coordinate systems $\{i\}$ and $\{j\}$.

General Tree-Structured Mechanism An example of a general tree-structured mechanism is shown in Figure 2. The tree structure has a reference member, which is the root of the tree, and indexed as 0, and N links. The reference member is assumed to be mobile and refers to the main body of a mobile vehicle, while

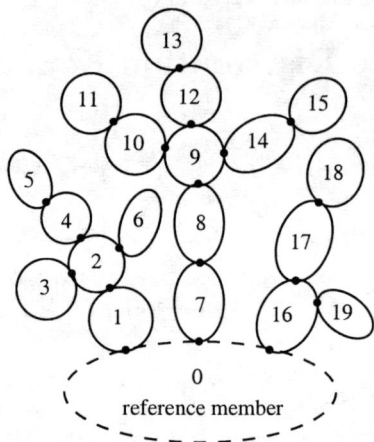

Figure 2: Example of a general tree-structured mechanism

the N links of the tree compose hybrid serial/parallel chains branching off of the reference member. Like the link indexing system used in a serial open chain, the links in the tree are also indexed from 1 to N. Also, to facilitate development of the dynamic equations, a link's index-number must always be greater than those of its predecessors. Indexing the links in either a depth-first or a breadth-first way will satisfy this requirement [7], and the former is adopted in Figure 2. In addition, in this paper, we will use single DOF revolute or prismatic joints to represent the joints between the links. This will simplify the analysis of the computational requirements, while it is still able to account for most robotic systems. Furthermore, most multiple-DOF joints can be modeled by concatenation of several single-DOF joints.

To describe the relationship between the links in a tree, two functions are defined. The first one is the parent function, $p(i)$, which denotes the index-number of the parent link of link i. The other one is the descendent function, $d(i)$, which is the set of index-numbers of all descendent links of link i. In a tree-structured mechanism, each link can have only one parent link; however, it may have many descendent links. For example, referring to Figure 2, we have the following relationships: $p(1) = p(7) = p(16) = 0$, $p(2) = 1$, $p(3) = p(4) = p(6) = 2$, $p(5) = 4$, $d(0) = \{1, \cdots, 19\}$, $d(1) = \{2, \cdots, 6\}$, $d(7) = \{8, \cdots, 15\}$, $d(16) = \{17, 18, 19\}$, and so on.

For efficient computation, the Modified Denavit-Hartenberg (MDH) parameters are used to relate link coordinate systems. Also, in each branch, in addition to MDH parameters, a pair of constant parameters, are involved to specify the transformation between link coordinate systems [7].

Dynamic Equations The derivation of the force distribution equations of a vehicle begins with the set of dynamic equations for the force balance on each link using spatial notation. The force balance equation for link i can be expressed as follows [7]:

$$\mathbf{f}_i = \mathbf{f}_i^* + \mathbf{f}_i^+, \tag{1}$$

where $\mathbf{f}_i$ is the spatial force exerted onto link i by its parent link, $\mathbf{f}_i^*$ is link i's resultant inertial force with the gravitational effect, and $\mathbf{f}_i^+$ is the sum of the spatial force exerted by link i onto all of its outboard links and transformed to the ith coordinate frame. Each of these 6×1 spatial force vectors is composed of three moment components at the top and three force components at the bottom.

The spatial force $\mathbf{f}_i^*$ can be computed as follows:

$$\mathbf{f}_i^* = \mathbf{I}_i \mathbf{a}_i' - \boldsymbol{\beta}_i, \tag{2}$$

where $\mathbf{I}_i$ is the 6×6 spatial inertia matrix of link i, $\mathbf{a}_i' = \mathbf{a}_i - [\mathbf{0}^T \ ^i\mathbf{a}_g^T]^T$, which is the spatial acceleration $\mathbf{a}_i$ of link i biased by gravitational acceleration, and $\boldsymbol{\beta}_i$ is a spatial bias force vector which is velocity-dependent. Here, the spatial acceleration is composed of the angular acceleration vector at the top and the translational acceleration vector at the bottom.

Since the outboard links' equations would be computed first, $\mathbf{f}_i^+$ is calculated by accumulation using the parent function as follows [7]:

$$\mathbf{f}_{p(i)}^+ := \mathbf{f}_{p(i)}^+ + {}^i\mathbf{X}_{p(i)}^T \mathbf{f}_i, \tag{3}$$

where ${}^i\mathbf{X}_{p(i)}$ is the spatial transform matrix transforming spatial vectors between the frames of link i and its parent link. The equation for each link i is computed in a backward recursive manner; that is, for $i = N, \cdots, 1$.

In this paper, we assume that all velocities and accelerations of each member of the vehicle are known and have been calculated through a kinematic analysis. Our focus will be on the force balance analysis.

3 Force Balance Equations

To derive the force balance equations for a tree-structured vehicle, we will balance all forces on the reference member. We assume that the mechanism has a mobile base numbered as 0, N links in a tree structure numbered from 1 to N, and m contact points numbered from 1 to m. Also, it is assumed that only the outermost links (leaves) possess contact points with the environment.

From Eq. (1), we know that the force exerted on a link by its parent is equal to the sum of the resultant inertial force of the link and the forces exerted by the link itself onto all of its outboard links. In a similar point of view, we can suppose that the reference member connects with the ground through a virtual 6-DOF joint [8], and the entire vehicle system is seen as a whole. Then, if the reference member experiences an external force $\mathbf{f}_0^e$ (which may come, for instance, from supporting a manipulator installed on the reference member with interaction with the environment), this external force will support the resultant inertial forces of each link in the vehicle and the forces exerted onto the environment, at the contact points, by the vehicle.

In order to efficiently form the force balance equations for the vehicle, we consider the resultant inertial forces and contact forces separately and use the superposition principle. With no contact forces, i.e., by initially setting $\mathbf{f}_k^+ = \mathbf{0}$, for $k = 0, \cdots, N$, we can compute the net inertial forces ($\mathbf{f}_i$) by using Eqs. (1) and (3) as follows:

$$\begin{aligned} &\text{for } i = N, \cdots, 0 \\ &\quad \mathbf{f}_i = \mathbf{f}_i^* + \mathbf{f}_i^+ \\ &\quad \text{if } i \neq 0 \\ &\quad\quad \mathbf{f}_{p(i)}^+ := \mathbf{f}_{p(i)}^+ + {}^i\mathbf{X}_{p(i)}^T \mathbf{f}_i \\ &\quad \text{end if} \\ &\text{end for } i \end{aligned} \tag{4}$$

Next, the contact forces, expressed in frame $\{0\}$, are computed. In this paper, multiple contact points are allowed on the leaf links. Also, the total number of contact points may vary with time, and the contact points are indexed from 1 to m. In order to effectively trace the contact points in the tree, the function, $c(i)$, is defined, which is a set of all the indices of contact points on link i. As shown in Figure 3, contact coordinate systems are assigned by placing their origins at the contact points with their $\hat{\mathbf{z}}_{c_j}$ unit vector along the normal directions to the contact surface. In this example, two leaf links, link 11 and link 13, are shown. Link 11 has two contact points, c_4 and c_5, with the environment, whereas link 13 has only one contact point, c_6. Therefore, the following holds: $c(11)=\{4,5\}$ and $c(13)=\{6\}$.

In this paper, we assume that only force components, $\mathbf{h}_j$, exist at each contact point, and any contact moments are negligible. With $c(i)$ as defined above and with the contact coordinate system assignment, we can then calculate the spatial contact forces ${}^0\mathbf{h}_j$. First, the spatial transform between frames $\{c_j\}$ and $\{0\}$ is calculated and expressed as follows [9]:

$${}^{c_j}\mathbf{X}_0 = {}^{c_j}\mathbf{X}_E \, {}^E\mathbf{X}_0 = \begin{bmatrix} \mathbf{R} & \mathbf{0} \\ \mathbf{C} & \mathbf{R} \end{bmatrix}, \tag{5}$$

where E represents the earth (inertial) coordinate sys-

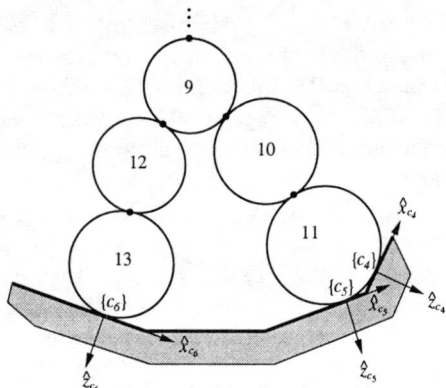

Figure 3: Example of contact point coordinate systems of the leaf links in a tree structure

tem. Then, $^0\mathbf{h}_j$ may be computed as follows:

$$^0\mathbf{h}_j = {}^{c_j}\mathbf{D}_0^T\, h_j, \qquad (6)$$

where

$$^{c_j}\mathbf{D}_0 = \begin{bmatrix} C & R \end{bmatrix}. \qquad (7)$$

Since only force components are transformed, only part of the spatial transformation is needed in the above computation.

Now, the force balance equations for the system in frame $\{0\}$ can be expressed as follows:

$$\mathbf{f}_0^e = \mathbf{f}_0 + \sum_{j=1}^m {}^0\mathbf{h}_j = \mathbf{f}_0 + \sum_{j=1}^m {}^{c_j}\mathbf{D}_0^T\, h_j. \qquad (8)$$

Since the contact forces in the force distribution problem are unknown and to be determined, these quantities will be set as the primary variables in the formulation [3]. With the known parts calculated as above, the force balance equation, Eq. (8), can be expressed in a compact matrix-vector form:

$$\mathbf{W}\mathbf{x} = \mathbf{F}_0, \qquad (9)$$

where

$$\mathbf{W} = \begin{bmatrix} {}^{c_1}\mathbf{D}_0^T & {}^{c_2}\mathbf{D}_0^T & \cdots & {}^{c_m}\mathbf{D}_0^T \end{bmatrix}, \qquad (10)$$

$$\mathbf{x} = \begin{bmatrix} h_1^T & h_2^T & \cdots & h_m^T \end{bmatrix}^T, \text{ and} \qquad (11)$$

$$\mathbf{F}_0 = \mathbf{f}_0^e - \mathbf{f}_0. \qquad (12)$$

4 Joint Torque and Contact Friction Constraints

The basic problem of force distribution in an actively-coordinated vehicle is to find the desired contact forces and joint torques given the desired motion. Thus, it is necessary to derive the relationship between the contact forces and the joint torques in a general tree-structured mechanism. Also, the desired joint torques are bound by the actuator limits. Therefore, it is important that these joint torques constraints be incorporated in the formulation of the force distribution problem. This section derives these sets of equations.

A. Relations Between Joint Torque and Contact Force: To derive the relations between joint torques and contact forces in a tree structure, we use the same concept as was used to derive the force balance equation, Eq. (8). That is, the spatial force exerted onto link i by link i's parent will balance all spatial inertial forces of link i and its descendents, and contact forces exerted onto the environment by leaf links among link i and link i's descendents. The torque of joint i is then computed as follows:

$$\tau_i = \boldsymbol{\phi}_i^T \left(\mathbf{f}_i + \sum_{\substack{\forall j \in c(k) \\ \left\{\begin{array}{c} \forall k \in \{i, d(i)\} \\ \text{and} \\ k \in \text{leaf links} \end{array}\right\}}} {}^i\mathbf{h}_j \right), \qquad (13)$$

where $\boldsymbol{\phi}_i$ is the spatial joint axis connecting link i with its parent [8].

The first term in the above equation is just picking up one element from the known vector $\mathbf{f}_i$:

$$\kappa_i = \boldsymbol{\phi}_i^T \mathbf{f}_i. \qquad (14)$$

The second term relating the contact forces to the joint torque in Eq. (13) is computed as follows:

$$\boldsymbol{\phi}_i^T\, {}^i\mathbf{h}_j = (\boldsymbol{\phi}_i^T\, {}^0\mathbf{X}_i^T)\, {}^{c_j}\mathbf{D}_0^T\, h_j, \qquad (15)$$

where h_j is first transformed to frame $\{0\}$ before being transformed to link i's coordinate system with $^0\mathbf{X}_i^T$.

Combining Eqs. (14) and (15), Eq. (13) can be rearranged as follows:

$$\tau_i = \kappa_i + \boldsymbol{\lambda}_i \mathbf{x}, \qquad (16)$$

where κ_i is a constant calculated using Eq. (14), and $\boldsymbol{\lambda}_i$ is a $1 \times 3m$ vector whose jth slot is filled by $(\boldsymbol{\phi}_i^T\, {}^0\mathbf{X}_i^T)\, {}^{c_j}\mathbf{D}_0^T$, a 1×3 vector, where j is for indices of all of the contact points on leaf links among link i and its descendent links. The remaining elements of $\boldsymbol{\lambda}_i$ are filled with zeros.

B. Joint Torque Limit Constraints: To incorporate the joint torque limit constraints in the formulation of the force distribution equations, Eq. (16) is computed for all of the links in the tree, $(i = 1, \cdots, N)$. This gives

$$\boldsymbol{\tau} = \boldsymbol{\kappa} + \boldsymbol{\lambda}\mathbf{x}, \qquad (17)$$

where τ is the $N \times 1$ joint torque vector, κ is an $N \times 1$ vector, with $\kappa = [\ \kappa_1\ \cdots\ \kappa_N\ ]^T$, and λ is an $N \times 3m$ matrix, with $\lambda = [\ \lambda_1^T\ \cdots\ \lambda_N^T\ ]^T$.

When the joint torque vector has an upper bound, τ_{max}, and lower bound, τ_{min}, then the joint torque limit constraints are given as follows:

$$\tau_{min} \leq (\tau = \kappa + \lambda \mathbf{x}) \leq \tau_{max} \tag{18}$$

The above equation can be rearranged:

$$\mathbf{A}_1 \mathbf{x} \geq \mathbf{B}_1, \tag{19}$$

where $\mathbf{A}_1 = \begin{bmatrix} \lambda \\ -\lambda \end{bmatrix}$ and $\mathbf{B}_1 = \begin{bmatrix} \tau_{min} - \kappa \\ -\tau_{max} + \kappa \end{bmatrix}$.

C. Contact Friction Constraints During the locomotion of an actively-coordinated vehicle, avoiding slippage and maintaining contact are very desirable. Combining the friction constraints at all of the contact points, the full set of contact-friction force constraints may be written as [3]:

$$\mathbf{A}_2 \mathbf{x} \geq \mathbf{B}_2. \tag{20}$$

5 Concise Matrix Form of the Force Distribution Equations

The force balance equations, Eq. (9), joint torque constraints, Eq. (19), and friction force constraints, Eq. (20), define an under-constrained problem which has an infinite number of solutions. An objective function may be defined as below to set criteria for improving system performance:

$$\Phi = f(\mathbf{x}), \tag{21}$$

where Φ is a scalar objective function which is a function of the contact forces, $\mathbf{x}$. This objective function may be set to be linear or quadratic. Now, we can combine Eqs. (9), (19), (20), and (21) to formulate the force distribution problem in a concise matrix-vector form:

$$\begin{aligned} \min\quad & \Phi = f(\mathbf{x}) \\ \text{subject to}\quad & \mathbf{W}\mathbf{x} = \mathbf{F}_0 \\ & \mathbf{A}_1 \mathbf{x} \geq \mathbf{B}_1 \\ & \mathbf{A}_2 \mathbf{x} \geq \mathbf{B}_2 \end{aligned} \tag{22}$$

This problem can be effectively solved by standard optimization techniques, such as linear programming [10] and quadratic programming [11].

6 Example Mechanism

An n-module actively-articulated wheeled vehicle (AAWV) is used to test the formulation. We model

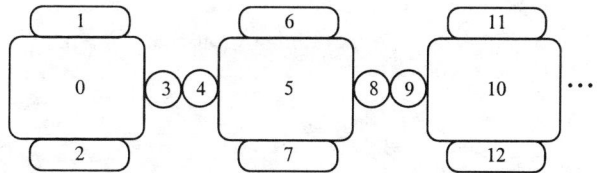

Figure 4: Numbering system in an AAWV by breadth first rule

Figure 5: The Wheeled Actively-Articulated Vehicle (WAAV) [12]

each articulation as concatenated single-DOF revolute joints. If we treat the vehicle as a general tree structure and number the tree by breadth first rule, the system is shown in Figure 4 from which we know the relationship between the number of modules, n, and the number of links, N (excluding the reference member): $N = 5(n-1) + 2$. Also, if each wheel is assumed to possess only one contact point with the environment, the relationship between the number of contact points and the number of modules is: $m = 2n$.

A three-module actively-articulated wheeled vehicle, called WAAV which was designed and built in the Department of Mechanical Engineering at The Ohio State University, is shown in Figure 5 [12].

Computational Requirements Efficient methods developed in [9] are employed to implement the formulation. The computational cost of the formulation developed for an n-module AAWV in terms of the number of multiplications ($\times$) and additions/subtractions ($+$) required is determined and listed in Table 1. The results show that the computational cost is a quadratic function of the number of modules, n. However, the coefficient in the quadratic term is relatively small in comparison with that in the linear term, so that the computational cost is nearly linear with the number of modules when n is small. For WAAV with $n = 3$, the total cost is

Table 1: Computational cost

Computation	×	+/−
Inertial forces ($\mathbf{f}_i$)	$305n - 128$	$309n - 138$
Spatial transf. ($^{c_j}\mathbf{D}_0$)	$162n$	$126n$
Force balance eq.	0	6
Spatial transf. ($^i\mathbf{X}_0$)	$330n - 279$	$198n - 198$
Joint torque constr.	$54n^2 - 18n$	$45n^2 - 5n - 6$
Total	$54n^2 + 779n - 407$	$45n^2 + 628n - 336$
(n=3)	2416	1953

[2416×, 1953+]. A more detailed description of the computation can be found in [13].

The formulation has been implemented in C++ and run on a PC with a Pentium-MMX 233 CPU. The execution times for a AAWV with 2 to 6 modules are 0.11 ms, 0.22 ms, 0.36 ms, 0.54 ms, and 0.74 ms, respectively. All of the execution times are less than 1 ms, which can be computed in real time.

7 Summary and Conclusions

In this paper, the force distribution problem for actively-coordinated vehicles is efficiently formulated. In addition to the force balance equations, the proposed formulation involves both the joint torque-limit and contact-friction force constraints. The final form of the force distribution equations is very suitable to be solved by standard optimization techniques, such as linear programming or quadratic programming. The applicable platforms for this formulation include not only systems with star topologies, such as walking machines that have multiple legs with a single body, but also general tree-structured mechanisms, such as variably-configured wheeled vehicles having multiple modules.

The developed formulation has been implemented in C++ and run on a PC with a Pentium-MMX 233 CPU. The execution times for a AAWV with a small number of modules are very fast; in particular, they are less than 1 ms. Also, the force distribution problem for a three-module actively-articulated wheeled vehicle, called WAAV, has been implemented and solved using a quadratic programming algorithm. The total computation of both the formulation part, given in this paper, and the solution part may be completed in less than 5 ms. This should be adequate for real-time operation and will hopefully be applied to a variety of actively-coordinated terrain-adaptive vehicles in the near future.

8 Acknowledgments

This work was supported by a fellowship from The Chung Cheng Institute of Technology, Taiwan, R.O.C.

9 References

[1] D. R. Pugh et al., "Technical Description of the Adaptive Suspension Vehicle," *Int. Journal of Robotics Research*, vol. 9, no. 2, pp. 24–42, 1990.

[2] S. V. Sreenivasan and P. Nanua, "Force Distribution Characteristics of Actively Reconfigurable Wheeled Vehicle Systems," in *1996 ASME Design Engineering Tech. Conf. and Computers in Engineering Conf.*, (Irvine, California), pp. 1–10, Aug. 1996.

[3] F. T. Cheng and D. E. Orin, "Efficient Formulation of the Force Distribution Equations for Simple Closed-Chain Robotic Mechanisms," *IEEE Trans. on Systems, Man, and Cyber.*, vol. 21, pp. 25–32, Jan./Feb. 1991.

[4] A. J. Spiessbach and W. R. Woodis, "Final Report for Mars Rover/Sample Return (MRSR) Rover Mobility and Surface Rendezvous Studies," tech. rep., Martin Marietta Space Systems, Denver, March 1989.

[5] V. R. Kumar and K. J. Waldron, "Actively Coordinated Vehicle Systems," *Journal of Mechanisms, Transmissions, and Automation in Design*, vol. 111, pp. 223–231, June 1989.

[6] S. C. Venkataraman, *Active Coordination of Wheeled Vehicle Systems for Improved Dynamic Performance*. PhD thesis, The Ohio-State University, Columbus, Ohio, 1997.

[7] S. McMillan, D. E. Orin, and R. B. McGhee, "A Computational Framework for Simulation of Underwater Robotic Vehicle Systems," *Journal of Autonomous Robots*, Kluwer Academic Publishers, Boston, pp. 253–268, 1996.

[8] S. McMillan, D. E. Orin, and R. B. McGhee, "Efficient Dynamic Simulation of an Underwater Vehicle with a Robotic Manipulator," *IEEE Trans. on Systems, Man, and Cyber.*, vol. 25, pp. 1194–1206, Aug. 1995.

[9] S. McMillan and D. E. Orin, "Efficient Computation of Articulated-Body Inertias Using Successive Axial Screws," *IEEE Trans. on Robotics and Automation*, vol. 11, pp. 606–611, Aug. 1995.

[10] F. T. Cheng and D. E. Orin, "Efficient Algorithm for Optimal Force Distribution–The Compact-Dual LP Method," *IEEE Trans. on Robotics and Automation*, vol. 6, pp. 178–187, April 1990.

[11] M. A. Nahon and J. Angeles, "Real-Time Force Optimization in Parallel Kinematic Chains under Inequality Constraints," *IEEE Trans. on Robotics and Automation*, vol. 8, pp. 439–450, Aug. 1992.

[12] C. J. Hubert, "Coordination of Variably Configured Vehicles," Master's thesis, The Ohio-State University, Columbus, Ohio, 1998.

[13] M. H. Hung, D. E. Orin, and K. J. Waldron, "Efficient Formulation of the Force Distribution Equations for General Tree-Structured Robotic Mechanisms with a Mobile Base," tech. rep., The Ohio State University, Columbus, Ohio, 1998.

Brachiation on a Ladder with Irregular Intervals

Jun Nakanishi[†], Toshio Fukuda[‡] and Daniel E. Koditschek[*]

[†] Dept. of Micro System Engineering, Nagoya University, Nagoya, Aichi 464-8603, Japan
[‡] Center for Cooperative Research in Advanced Science and Technology, Nagoya University, Nagoya, Aichi 464-8603, Japan
[*] Dept. of Electrical Engineering and Computer Science, The University of Michigan, Ann Arbor, MI 48109-2110, USA

Abstract

We have previously developed a brachiation controller that allows a two degree of freedom robot to swing from handhold to handhold on a horizontal ladder with evenly space rungs as well as swing up from a suspended posture using a "target dynamics" controller. In this paper, we extend this class of algorithms to handle the much more natural problem of locomotion over irregularly spaced handholds. Numerical simulations and laboratory experiments illustrate the effectiveness of this generalization.

1 Introduction

This paper presents a control strategy for brachiation on a ladder with irregular intervals. Our interest in this problem arises from the general concern about how dynamically dexterous robotic tasks can be achieved by combining physical insight into the designated task and the intrinsic dynamics of the robot in its environment. The study of brachiation has design implications for other tasks involving dynamical dexterity such as legged locomotion, [8, 12], dexterous manipulation [1, 2, 4] and underactuated systems [15].

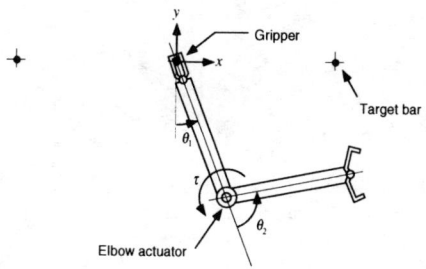

Figure 1: A two-link brachiating robot

For the last few years, we have been studying the control of the two degree of freedom brachiating robot depicted in Figure 1, which dynamically moves from handhold to handhold like a long armed ape. We initially proposed a new control algorithm based upon what we termed the "target dynamics" method. Motivated by the desire to have the robot's trajectories mimic the pendulous motion of an ape's brachiation, this method enabled us to force a one degree of freedom virtual composite of the physical 2 dof revolute-revolute kinematics to oscillate as if governed by the equations of motion of a harmonic oscillator [6]. Preliminary analysis, extensive simulation [6] and subsequent experimental studies [7] confirmed the proposed algorithm could achieve brachiation as well as swing up from a one hand to a two hand grip on a level ladder with uniform intervals.

The question remains whether this approach is likely to yield a flexible enough repertoire of behaviors to motivate its further analytical and experimental exploration. In this paper we take the modest step of increasing the behavioral repertoire to include the "irregular ladder problem" — brachiation on a ladder with irregularly spaced rungs placed at the same height. This addition seems to be essential, if only from the point of view of our initial biomechanics motivation, since very few unstructured environments confront an ape with equally spaced branches. The original robot brachiation studies by Saito et al. [3, 9, 10, 11] considered brachiation on bars with different distances and heights using heuristic learning and neural networks [10]. However, experimental implementation of their control algorithms were not carried out in the irregular ladder problem because of the enormous experimental burden and parametric iterations required of the physical robot[1]. Here, we employ a deadbeat style control strategy to solve the irregular ladder problem by extending the results in our previous studies. Numerical simulation and experimental results illustrate the effectiveness of our approach.

2 Experimental Setup[2]

2.1 Physical Apparatus

This section briefly describes our experimental system. We use the two-link brachiating robot originally developed by Saito [11] having updated the controller hardware (computer, input-output devices and motor driver circuits). Figure 2 depicts the experimental setup. The length of each arm is 0.5m and the total weight of the robot is about 4.8kg. The details of the description of the robot can be found in [7].

2.2 Model

The dynamical equations used to model the robot depicted in Figure 3 take the form of a standard two-link planar manipulator

$$\dot{T}q = \mathcal{L}(Tq, v_r) \qquad (1)$$

where

$$\mathcal{L}(Tq, v_r) = \begin{bmatrix} \dot{q} \\ M^{-1}\left(-V - k - B\dot{q} - C\,\text{sgn}(\dot{q}) + \begin{bmatrix} 0 \\ K v_r \end{bmatrix}\right) \end{bmatrix},$$

$q = [\,\theta_1,\,\theta_2\,]^T \in \mathcal{Q}$, $Tq = [\,q^T,\,\dot{q}^T\,]^T \in T\mathcal{Q}$, M is the inertia matrix, V is the Coriolis/centrifugal vector, and k is the gravity vector. C and B denote the coulomb and viscous friction coefficient matrices respectively. We assume

---

[1] They did implement the learning algorithm on the physical two-link robot in the uniform ladder problem [11].
[2] Portions of this section are excerpted from [7].

*This work was supported in part by NSF IRI-9510673

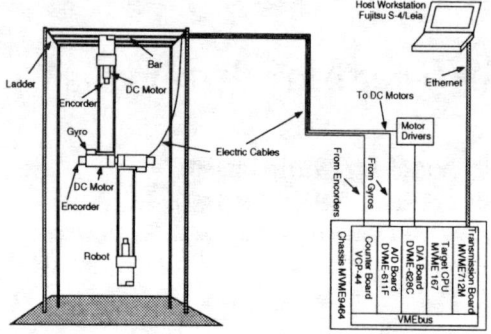

Figure 2: The experimental setup of the two-link brachiating robot.

that the elbow actuator produces torque proportional to a voltage command, v_r, sent to a driver as $\tau = Kv_r$, where K is a positive constant. It is generally known that DC

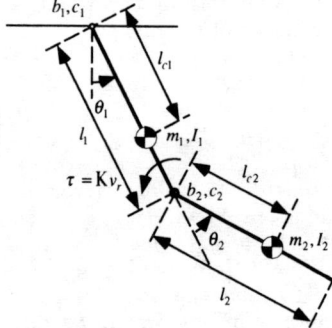

Figure 3: The mathematical model of the two-link brachiating robot used in this paper.

motors with harmonic gear mechanisms bear complicated nonlinear characteristics [16]. However, for simplicity, we model the dynamics of the elbow actuator using only viscous and coulomb friction and rotor inertia. As the results of parameter identification presented in [7] suggest, the model we offer here fits the dynamics of the physical system fairly well. The dynamical parameters of the robot are shown in Table 1.

Description		i=1	i=2
Mass	m_i(kg)	3.499	1.232
Moment of inertia	I_i(kgm^2)	0.090	0.033
Link length	l_i(m)	0.50	0.50
Location of CG	l_{ci}(m)	0.414	0.333
Viscous friction	b_i(Nm/s)	0.02	0.14
Coulomb friction	c_i(Nm)	0.02	0.45
Torque constant	K(Nm/V)	1.752	

Table 1: The dynamical parameters of the robot obtained by the procedure described in [7].

3 Irregular Ladder Problem[3]

In this section, we present our control strategy for brachiation on a ladder with irregular intervals. First, we review our "target dynamics" strategy and its application to the uniform ladder problem and the rope problem discussed in [6]. Next, we introduce a deadbeat style strategy which extends these ideas to the present problem setting.

3.1 Review of the Uniform Ladder Controller

A detailed development of the target dynamics controller can be found in [6]. The strategy is a particular instance of input/output linearization Specifically, brachiation is encoded as the output of a target dynamical system—a harmonic oscillator determined by a "virtual frequency", ω, which we will force the robot to mimic. At the end of [6], we define the controller for the lossless model where $B, C \to 0$

$$\begin{aligned}\tau &:= \left(D_q h \begin{bmatrix} n_{12} \\ n_{22} \end{bmatrix}\right)^{-1} \left[-\omega^2 \theta - (D_q h)\dot{q} + D_q h M^{-1}(V+k)\right] \\ &= \frac{1}{n_{12} + \frac{1}{2}n_{22}} \left[-\omega^2(\theta_1 + \frac{1}{2}\theta_2) + (n_{11} + \frac{1}{2}n_{21})(V_1 + k_1)\right] \\ &\quad + V_2 + k_2 \end{aligned} \quad (2)$$

where, $h(q) := \theta = \theta_1 + \frac{1}{2}\theta_2$, n_{ij} denotes each component of M^{-1}, and $v_r = \frac{1}{K}\tau$. In the subsequent simulations and experiments, we use the lossy model, and the friction terms are added in the controller to cancel them. Note that

$$D_q h \begin{bmatrix} n_{12} \\ n_{22} \end{bmatrix} = \frac{m_1 l_{c1}^2 + m_2(l_1^2 - l_{c2}^2) + I_1 - I_2}{2 \det(M)} \neq 0, \quad (3)$$

i.e., the invertibility condition of the first term in (2) is satisfied in the particular setting of concern.

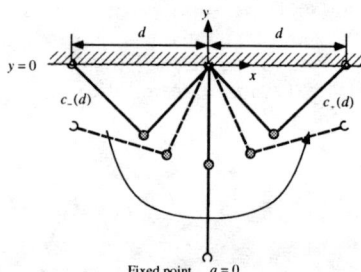

Figure 4: The ceiling is parametrized by the distance between the grippers d. A left branch $c_-(d)$ and right branch $c_+(d)$ are defined in this manner.

The ladder problem arises when an ape transfers from one branch to another and the control of arm position at the next capture represents the control task requirement. Here, we restrict our attention to brachiation on a set of evenly spaced bars at the same height. We showed in [6] how a symmetry property of an appropriately chosen target system — determined by ω in (2) — can solve this problem. The next question concerns the choice of ω in the target dynamics to achieve the desired motion. This is determined according to the principle of "neutral orbits" defined in [6, 13, 14]. Namely, following Raibert [8], a "neutral orbit" is one with a reverse time symmetry whereby orbits forward in time from the bottom are horizontal reflections of orbits backward in time from the same bottom state[4]. In the sequel, we will denote the integral curve of a vector field f by the notation f^t.

Define the "ceiling" to be those configurations where the hand of the robot reaches the height $y = 0$ as depicted

---
[3]Portions of sections 3.1 and 3.2 are excerpted from [6].

[4]The "bottom states" are those characterized by both joint angles at zero: i.e. the arm is hanging straight down.

in Figure 4,
$$\mathcal{C} = \{q \in Q | \cos\theta_1 + \cos(\theta_1 + \theta_2) = 0\}. \quad (4)$$

Note that $\mathcal{C}$ can be parameterized by two branches,
$$\mathcal{C} = \text{Im } c_- \cup \text{Im } c_+ \quad (5)$$

of the maps
$$c_\pm(d) = \begin{bmatrix} \pm \arcsin\left(\frac{d}{2l}\right) \\ \pm \left[\pi - 2\arcsin\left(\frac{d}{2l}\right)\right] \end{bmatrix}, \quad (6)$$

where $l = l_1 = l_2$. Suppose we have chosen a feedback law, $\tau(q, \dot{q})$, denote the closed loop dynamics of the robot as
$$\dot{Tq} = \mathcal{L}_\tau(q, \dot{q}) = \mathcal{L}(Tq, \tau(q, \dot{q})). \quad (7)$$

In the sequel, we will be particularly interested in initial conditions of (7) originating in the zero velocity sections of the ceiling that we denote $T\mathcal{C}_0$ in (13). We conclude in [6] that any feedback law, τ, which respects the reverse time symmetry solves the ladder problem, assuming we can find d such that $[c_-(d)^T, 0, 0]^T$ is in a neutral orbit. Note that finding such a ceiling point requires solving the equation
$$\Phi(d, t_N) = [\, I, \quad 0\,] \mathcal{L}_\tau^\nu \left(\begin{bmatrix} c_-(d) \\ 0 \\ 0 \end{bmatrix} \right) = \begin{bmatrix} 0 \\ 0 \end{bmatrix}, \quad (8)$$

for d and t_N simultaneously, where $\nu = \frac{t_N}{4}$ and I is a 2×2 identity matrix. Of course solving this equation is very difficult: it requires a "root finding" procedure that entails integrating the dynamics, $\mathcal{L}$.

The feedback law to achieve the desired target dynamics is given by (2). We show in [6] that the choice of the target dynamical system — a harmonic oscillator — has a very nice property relative to the difficult root finding problem (8). Namely, using this control algorithm, t_N is given by
$$t_N(\tau_\omega) = \frac{2\pi}{\omega} \quad (9)$$

because θ follows the target dynamics $\ddot{\theta} = -\omega^2 \theta$. In this light, then, we need merely solve (8) for d. More formally, we seek an implicit function $d^* = \lambda^{-1}(\omega)$ such that $\Phi\left(\lambda^{-1}(\omega), \frac{2\pi}{\omega}\right) = 0$. In practice, we are more likely to take an interest in tuning ω as a function of a desired distance between the bars, d^*. Thus, we are most interested in determining
$$\omega = \lambda(d^*). \quad (10)$$

In general, we can expect no closed form expression for λ or λ^{-1}, and we resort instead to a numerical procedure for determining an estimate, $\hat{\lambda}$. The details of the numerical procedure is discussed in [5]. In Figure 5 we plot a particular instance of $\hat{\lambda}$ for the case where the robot parameters are as specified in Table 1.

3.2 Rope Problem: Review

In this section, we consider the rope problem discussed in [6]: brachiation along a continuum of handholds such as afforded by a branch or a rope. First, the average horizontal velocity is characterized as a result of the application of the target dynamics controller, τ_ω, introduced above. Then, we consider the regulation of horizontal velocity using this controller. An associated numerical "swing map"

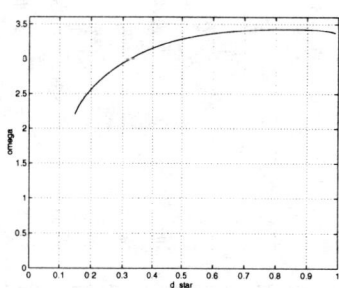

Figure 5: Numerical approximation $\omega = \hat{\lambda}(d^*)$. Target dynamics controller, τ_ω, is tuned according to this mapping, $\hat{\lambda}$, that is designed to locate neutral orbits originating in the ceiling.

suggests that we indeed can achieve good local regulation of the foward velocity through the target dynamics method.

Supposing that the robot starts in the ceiling with zero velocity, then it must end in the ceiling under the target dynamics controller since θ follows the target dynamics $\ddot{\theta} = -\omega^2 \theta$. However, if d and ω are not "matched" as $\omega = \lambda(d)$, then the trajectory ends in the ceiling, $Tq \in T\mathcal{C}_+$, with $\dot{\theta} = 0$ but $r \neq d$ and $\dot{r} \neq 0$. Here, (r, θ) denots the position of the gripper in polar coordinates arising from the change of coordinates from joint space. This leads to the definition of the swing map.

When a gripper moves a distance $2d^*$ in the course of the ladder trajectory, and if the trajectory is immediately repeated, then the body will also move a distance of d^* each swing, hence, its average horizontal velocity will be
$$\bar{h} = \frac{d^* \omega}{\pi} = \frac{d^* \lambda(d^*)}{\pi} := \tilde{V}(d^*) \quad (11)$$

according to the previous discussion.

Consider now the task of obtaining the desired forward velocity $\bar{h}^*$ of brachiation. If $\tilde{V}$ is invertible, then $d^* = \tilde{V}^{-1}(\bar{h}^*)$ and we can tune ω in the target dynamics as
$$\omega = \lambda \circ \tilde{V}^{-1}(\bar{h}^*) \quad (12)$$

to achieve a desired $\bar{h}^*$ where λ is again the mapping (10). Consider the ceiling condition with zero velocity
$$T\mathcal{C}_{0\pm} = \{[c_\pm(d)^T, 0, 0]^T \in T\mathcal{C} \mid d \in [0, 2l]\} \quad (13)$$

Define the maps, $C_\pm$, relating d and the initial state of the robot, and Π which "kills" any velocity in the ceiling as
$$C_\pm : [0, 2l] \to T\mathcal{C}_{0\pm} : d \mapsto [c_\pm(d)^T, 0, 0]^T \quad (14)$$
$$\Pi : T\mathcal{C}_\pm \mapsto T\mathcal{C}_{0\pm}. \quad (15)$$

We now define a "swing map" [6], σ_ω, as a transformation of $[0, 2l]$ into itself,
$$\sigma_\omega(d) := C_+^{-1} \circ \Pi \circ \mathcal{L}_{\tau_\omega}^{2\nu} \circ C_-(d) : [0, 2l] \to [0, 2l] \quad (16)$$

Note that if $\omega = \omega^* = \lambda(d^*)$, then
$$\sigma_\omega(d^*) = d^* \quad (17)$$

that is, d^* is a fixed point of the appropriately tuned swing map. Suppose we iterate by setting the next initial condition in the ceiling to be

$$Tq_0[k+1] = C_- \circ \sigma_\omega(d[k]). \quad (18)$$

This yields a discrete dynamical system governed by the iterates of σ_ω,

$$d[k+1] = \sigma_\omega(d[k]). \quad (19)$$

Numerical evidence suggests that the iterated dynamics converges, $\lim_{k \to \infty} \sigma_{\omega^*}^k(d) = d^*$ when d is in the neighborhood of d^* [6].

3.3 Deadbeat Control Strategy for Irregular Ladder Problem

This section presents a deadbeat style control strategy for the irregular ladder problem which extends the ideas discussed in the previous sections. Now, we consider brachiation on a ladder with irregularly spaced rungs placed at the same height as depicted in Figure 6. Using the target dynamics, a single parameter, ω, in the controller characterizes the full range of the swing motion of the robot. Now, we seek the tuning rule for ω which locates the desired orbit from $C_-(d[k])$ to $C_+(d[k+1])$.

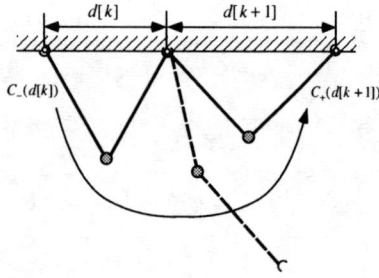

Figure 6: The irregular ladder problem. The robot moves from the left branch to the right branch with the intervals $d[k]$ and $d[k+1]$.

Define a new function

$$\tilde{\lambda} : [0, 2l] \times [0, 2l] \to \mathbb{R} \quad (20)$$

to solve the implicit function in ω by (19):

$$\tilde{\lambda}(d_1, d_2) := \text{solve}_{\omega \in \mathbb{R}} [d_2 - \sigma_\omega(d_1) = 0], \quad (21)$$

where d_1 and d_2 are the intervals between the bars of the left branch and the right branch respectively. This function is computed numerically and involves integrating the Lagrangian dynamics as in (18). In practice, we find that $\tilde{\lambda}$ is well defined only on a subset $\mathcal{D} \subseteq [0, 2l]$ whose extent depends upon the dynamical parameters of the robot as

$$\omega = \tilde{\lambda}(d[k], d[k+1]) : \mathcal{D} \times \mathcal{D} \to \mathbb{R}, \quad (22)$$

where $\mathcal{D} \subseteq [0, 2l]$. We plot in Figure 7 a particular instance of $\tilde{\lambda}$ for the case where the robot parameters are as specified in Table 1. The target dynamics controller is tuned according to this mapping to locate the orbit which achieves the desired gait of locomotion. Note that the mapping, $\omega = \lambda(d^*)$, in (10) is the intersection of the surface, $\tilde{\lambda}$, and the plane $d[k] - d[k+1] = 0$.

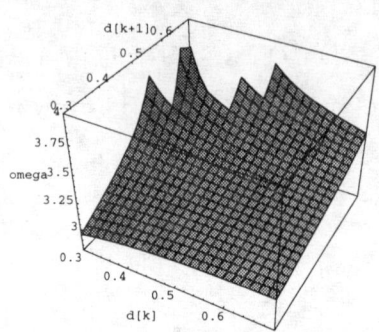

Figure 7: Numerical approximation of $\omega = \tilde{\lambda}(d[k], d[k+1])$. Target dynamics controller, τ_ω, is tuned according to this mapping, $\tilde{\lambda}$, that is designed to locate the desired orbit.

4 Simulation

Consider the following three cases of the intervals between the bars as specified in Table 2. The initial condition of the robot is $Tq_0 = [c_-(d[k])^T, 0, 0]$. From the numerical solution to the mapping (22) depicted in Figure 7, ω is tuned for each case as shown in Table 2.

Case	$d[k]$	$d[k+1]$	ω
1	0.4	0.6	3.66
2	0.5	0.6	3.47
3	0.6	0.5	3.255

Table 2: Intervals between the bars and ω considered in numerical simulation and experiments.

In this simulation, we use the lossy model with the dynamical parameter as specified in Table 1. Note that discontinuity of the voltage command observed in Figures 9 and 11 results from the coulomb friction terms added in the controller. These simulation results suggest the effectiveness of the proposed strategy.

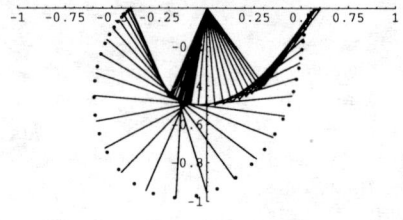

Figure 8: Movement of the robot (simulation), where $d[k] = 0.4, d[k+1] = 0.6$.

Case 1: $d[k] = 0.4, d[k+1] = 0.6$ Figure 8 depicts the movement of the robot, and Figure 9 shows the joint trajectories and the voltage command to the motor driver.

Case 2: $d[k] = 0.5, d[k+1] = 0.6$ Figure 10 depicts the movement of the robot, and Figure 11 shows the joint trajectories and the voltage command to the motor driver.

Case 3: $d[k] = 0.6, d[k+1] = 0.5$ Figure 12 depicts the movement of the robot, and Figure 13 shows the joint trajectories and the voltage command to the motor driver.

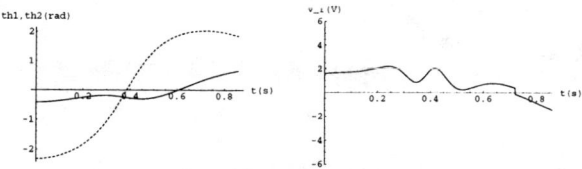

Figure 9: The simulation results, where $d[k] = 0.4, d[k+1] = 0.6$. Left: Joint trajectories (solid: θ_1, dashed: θ_2), Right: Voltage command to the motor driver.

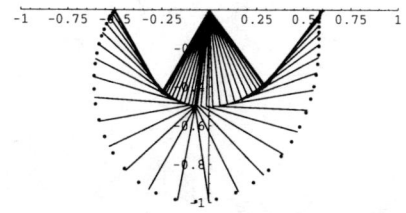

Figure 10: Movement of the robot (simulation), where $d[k] = 0.5, d[k+1] = 0.6$.

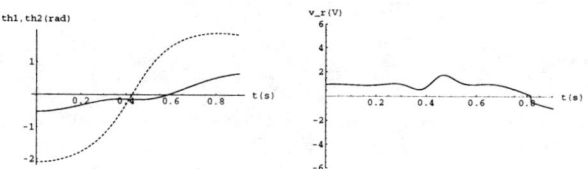

Figure 11: The simulation results, where $d[k] = 0.5, d[k+1] = 0.6$. Left: Joint trajectories, (solid: θ_1, dashed: θ_2), Right: Voltage command to the motor driver.

5 Experiments

This section presents the experimental implementaion of the proposed control strategy. We consider the same ladder intervals as specified in Table 2.

As we have experienced in our previous experimental work [7], we refine the dynamical parameters in the controller and the timing of bar release manually so that the robot successfully achieves the desired brachiation because of the parameter mismatch and a delay in the actuator mechanism the gripper. The command to close the gripper is sent and the voltage command to the motor driver is turned off simultaneously when the gripper approaches the target bar. Some experience is helpful in these refinements.

Case 1: $d[k] = 0.4, d[k+1] = 0.6$ The typical movement of the robot is depicted in Figure 14, while the joint trajectories and the voltage commands sent to the driver are shown in Figure 15. We choose to use the dynamical parameters, $m_1 = 3.39, m_2 = 1.30, c_2 = 0.65, b_2 = 0.9$, instead of the values shown in Table 1. The mean time of ten runs at which the robot reaches the ceiling is 0.949 seconds with ±0.04 second error, which is close to its analytical value, $t = \frac{\pi}{\omega} = 0.854$ seconds.

Case 2: $d[k] = 0.5, d[k+1] = 0.6$ The typical movement of the robot is depicted in Figure 16, while the joint trajectories and the voltage commands sent to the driver are shown in Figure 17. We choose to use the dynamical parameters, $m_1 = 3.39, m_2 = 1.30, c_2 = 0.73, d_2 = 0.6$, instead of the values shown in Table 1 and send the com-

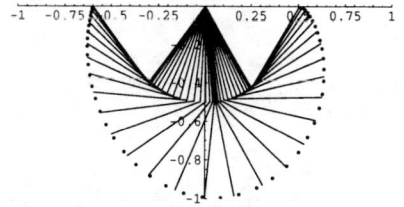

Figure 12: Movement of the robot (simulation), where $d[k] = 0.6, d[k+1] = 0.5$.

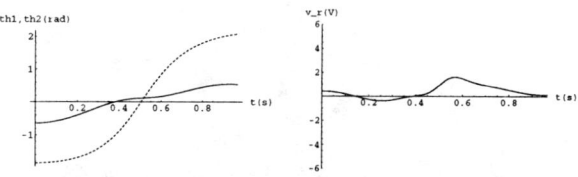

Figure 13: The simulation results, where $d[k] = 0.6, d[k+1] = 0.5$. Left: Joint trajectories, (solid: θ_1, dashed: θ_2), Right: Voltage command to the motor driver.

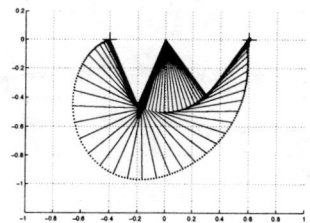

Figure 14: Movement of the robot (experiment), where $d[k] = 0.4, d[k+1] = 0.6$.

mand to open the gripper 0.01 seconds before the controller is turned on. The mean locomotion time of ten runs is 0.870 seconds with ±0.03 second error, which is close to its analytically calculated value, $t = \frac{\pi}{\omega} = 0.905$ seconds.

Case 3: $d[k] = 0.6, d[k+1] = 0.5$ The typical movement of the robot is depicted in Figure 18, while the joint trajectories and the voltage commands sent to the driver are shown in Figure 19. We choose to use the dynamical parameters, $m_1 = 3.39, m_2 = 1.30, c_2 = 0.73, b_2 = 0.33$, instead of the values shown in Table 1 and send the command to open the gripper 0.08 seconds before the controller is turned on. The mean locomotion time of ten runs is 0.841 seconds with ±0.08 second error, which is very close to its analytical value, $t = \frac{\pi}{\omega} = 0.965$ seconds.

As we have begun to investigate the discrepancy between the simulation and experiments seen above, numerical studies suggest that this seems to be due to the model mismatch of the friction and unmodelled torque saturation of the elbow actuator.

6 Conclusion

We present a deadbeat style control strategy which increases the behavioral repertoire of a brachiating robot to handle irregularly space handholds by an appropriate modification of our earlier "target dynamics controller." Numerical simulation and experimental results illustrate the effectiveness of this strategy. More analytical work will be required to completely understand the effect of this

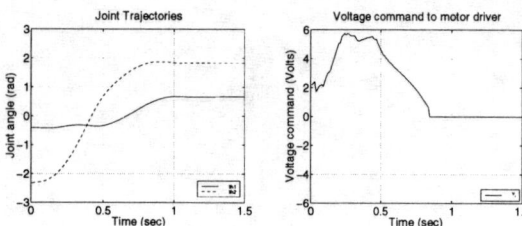

Figure 15: The experimental results, where $d[k] = 0.4, d[k+1] = 0.6$. Left: Joint trajectories (solid: θ_1, dashed: θ_2), Right: Voltage command to the motor driver.

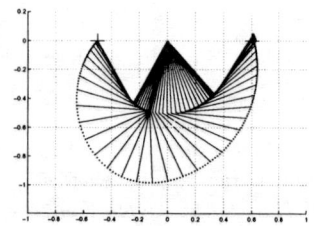

Figure 16: Movement of the robot (experiment), where $d[k] = 0.5, d[k+1] = 0.6$.

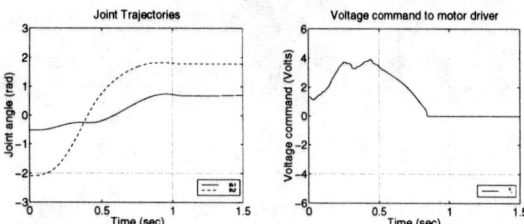

Figure 17: The experimental results, where $d[k] = 0.5, d[k+1] = 0.6$. Left: Joint trajectories (solid: θ_1, dashed: θ_2), Right: Voltage command to the motor driver.

style of controller on such underactuated mechanisms.

Motivated by the close analogy between brachiation and legged locomotion, future directions of work in this area suggest the desirability of "passive" or somewhat less model dependent approaches. In the more distant future, we are interested as well in "leaping" gaits analogous to an ape's fast brachiation that include a nonholonomic flight phase. We are hopeful that the ideas presented here may still have wider applications to other problems in the study of dynamical dexterous robotics.

References

[1] Andersson, R. L. *A Robot Ping-Pong Player: Experiment in Real-Time Intellgent Control*. MIT Press, 1988.

[2] Bühler, M., Koditschek, D. E., and Kindlmann, P. J. Planning and control of robotic juggling and catching tasks. *International Journal of Robotics Research*, 13(2):101–118, April 1994.

[3] Fukuda, T., Saito, F., and Arai, F. A study on the brachiation type of mobile robot (heuristic creation of driving input and control using CMAC. In *IEEE/RSJ International Workshop on Intelligent Robots and Systems*, pages 478–483, 1991.

[4] Lynch, K. M. *Nonprehensile Robotic Manipulation: Controllability and Planning*. PhD thesis, Carnegie Mellon University, The Robotics Institute, March 1996.

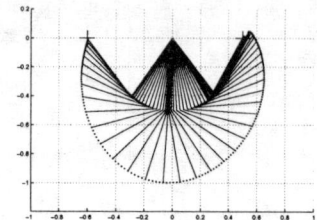

Figure 18: Movement of the robot (experiment), where $d[k] = 0.6, d[k+1] = 0.5$.

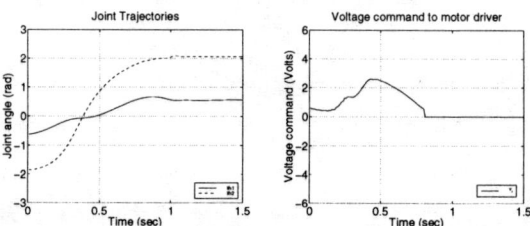

Figure 19: The experimental results, where $d[k] = 0.6, d[k+1] = 0.5$. Left: Joint trajectories (solid: θ_1, dashed: θ_2), Right: Voltage command to the motor driver.

[5] Nakanishi, J., Fukuda, T., and Koditschek, D. E. Preliminary analytical approach to a brachiation robot controller. Technical report: CGR 96-08 / CSE TR 305-96, The Univ. of Michigan, EECS Depertment, August 1996.

[6] Nakanishi, J., Fukuda, T., and Koditschek, D. E. Preliminary studies of a second generation brachiation robot controller. In *IEEE International Conference on Robotics and Automation*, pages 2050–2056, April 1997.

[7] Nakanishi, J., Fukuda, T., and Koditschek, D. E. Experimental implementation of a "target dynamics controller on a two-link brachiating robot". In *IEEE International Conference on Robotics and Automation*, pages 787–792, May 1998.

[8] Raibert, M. H. *Legged Robots that Balance*. MIT Press, 1986.

[9] Saito, F. *Motion Control of the Brachiation Type of Mobile Robot*. PhD thesis, Nagoya University, March 1995. (in Japanese).

[10] Saito, F., Fukuda, T., and Arai, F. Movement control of brachiation robot using CMAC between different distance and height. In *IMACS/SICE International Symposium on Robotics, Mechatronics and Manufacturing Systems*, pages 35–40, 1992.

[11] Saito, F., Fukuda, T., and Arai, F. Swing and locomotion control for a two-link brachiation robot. *IEEE Control Systems Magazine*, 14(1):5–12, February 1994.

[12] Schwind, W. J. *Spring Loaded Inverted Pendulum Running: A Plant Model*. PhD thesis, The University of Michigan, Dept. of EECS, September 1998.

[13] Schwind, W. J. and Koditschek, D. E. Control the forward velocity of the simplified planner hopping robot. In *IEEE International Conference on Robotics and Automation*, pages 691–696, 1995.

[14] Schwind, W. J. and Koditschek, D. E. Characterization of monoped equilibrium gaits. In *IEEE International Conference on Robotics and Automation*, pages 1986–1992, 1997.

[15] Spong, M. The swing up control problem for the acrobot. *IEEE Control Systems Magazine*, 15(1):49–55, February 1995.

[16] Tuttle, T. D. and Seering, W. P. A nonlicar model of a harmonic drive gear transmission. *IEEE Transaction on Robotics and Automation*, 12(3):368–374, June 1996.

Robust Sensing for a 3,500 tonne Field Robot

Jonathan M. Roberts[1,2], Frederic Pennerath[1], Peter I. Corke[1,2] & Graeme J. Winstanley[1,2]
[1]CSIRO Manufacturing Science & Technology
[2]CRC for Mining Technology and Equipment
PO Box 883, Kenmore, Qld 4069, Australia
http://www.cat.csiro.au/cmst/automation

Abstract

The mining industry is highly suitable for the application of robotics and automation technology since the work is arduous, dangerous and often repetitive. This paper discusses a robust sensing system developed to find and track the position of the hoist ropes of a dragline. Draglines are large 'walking cranes' used in open-pit coal mining to remove the material covering the coal seam. The rope sensing system developed uses two time-of-flight laser scanners. The finding algorithm uses a novel data association and tracking strategy based on pairing rope data.

1 Introduction

Draglines are large machines used for overburden stripping in open-cut coal mining(Figure 1). Details of their operation are given in [1]. They typically weigh up to 5000 tonnes, and can employ up to sixteen 373kW (500hp) motors on the slew drive. The bucket capacity is of the order of 100 tonnes. The fundamental goal of the project was to reduce the average cycle time of the machine, leading to immediate reductions in overburden removal costs. We aim to achieve this by reducing the time of the dominant part of the cycle — swinging the bucket through free space. Our work is based on the observation that control of the bucket moving through free space is a robotics problem.

This paper describes a hoist rope angle sensing system developed for a prototype swing cycle automation system for a full-scale production dragline (a 3,500 tonne Bucyrus-Erie 1370W). The system uses two time-of-flight laser scanners. Data from the scanners is applied to a rope finding algorithm that uses a novel data association and tracking strategy based on pairing rope data.

The remainder of this paper is structured as follows: Section 2 discusses the problem of robust load position

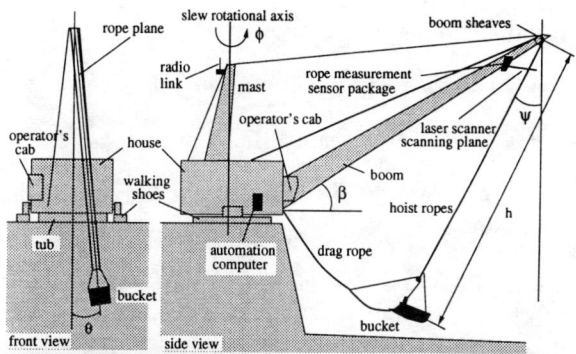

Figure 1: Top: a schematic of a dragline showing the main elements of the machine. Bottom: a dragline.

sensing. Section 3 then describes the laser based rope angle measurement system developed. Section 4 discusses the rope tracker and Section 5 talks about a Kalman filter designed to fuse data and generate velocity estimates. Section 6 shows some results, and Section 7 lists some conclusions.

2 Robust load position sensing

The motion of a dragline bucket with respect to the boom can be considered as a pendulum with the rope's internal friction and aerodynamic forces providing damping. Control of such a system requires a knowledge of instantaneous bucket position and velocity.

Existing encoders on all drives provide hoist and drag rope lengths as well as house angle, ϕ (Figure 1). Whilst this information is sufficient to locate the bucket on the rope plane, to determine the swing angle, θ, another sensor must be employed. This sensor's performance criteria are stringent: it must be capable of operating:

- 24 hours a day in all weather conditions
- at a rate of at least 3 Hz (control constraint)
- over a range of 5 to 100 m and invariant to bucket pose.

The extremely harsh nature of the bucket's interaction with the ground precludes instrumentation on the bucket itself.

At first glance, computer vision appears to be an ideal sensor for this application: the sensors are low cost, processing power is becoming cheaper, it is able to measure the state of the load without being in contact, and it mimics the human operator's sensing modality. Experiments with computer vision were reported in [2, 3]. Strong shadows, background texture, motion and lack of contrast combined to thwart all the machine vision approaches that we evaluated.

3 A laser based approach

The machine vision experiments did teach us that hoist ropes may be a easier to locate than the bucket itself. The approach selected for the automation system is instead based on a scanning infra-red laser range-finder that looks for the ropes. This retains the key advantage of non-contact position sensing, the devices are rugged and low in cost, and the difficult problem of robust scene segmentation is side stepped.

The bucket's position is inferred from the drag and hoist rope lengths, the hoist rope swing angle (θ) (Figure 1). Once calibrated, both the drag and hoist rope lengths can be measured using encoders located on the drag and hoist winch drums. The problem is: how to measure the swing angle of the hoist rope (θ)? Another useful measure is the hoist angle (ψ) of the hoist

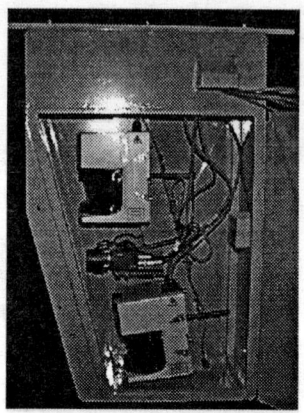

Figure 2: The sensor package housing showing the two scanners and the camera.

ropes (Figure 1). This angle is used for on-line bucket weight estimates and bucket pose recovery.

Figure 1 shows the location of the rope sensing system on the dragline. The aim of the rope finding system is to calculate the swing and hoist angles. Note that there are actually two hoist ropes separated by about one metre. The angles computed are therefore the angles for an imaginary *average* hoist rope located mid-way between the left and right hoist ropes (Section 4.4).

The system uses two Proximity Laser Scanners (PLS) manufactured by Sick Opto-electronics, Germany. The PLS returns range, R, and bearing, ω, data across a 180° field-of-view, giving a reading every 0.5° and a range accuracy of 20 mm and have been extensively tested under the range of conditions expected on a dragline (night, day, rain and dust). Two laser scanners are used for redundancy, and to address the problem of blinding due to direct sunlight. The two scanners have different look-down angles (Figure 2) and so can not be blinded by the sun simultaneously. With some knowledge of the geometry of the dragline boom tip, it is possible to calculate swing and hoist angles directly from the laser scanner data.

The raw data from the laser scanners passes to the rope finding element (Figure 3), which extracts the range and bearing of the two ropes with respect to the scanner in the scanning plane (Section 4). These positions are passed to a transformation element that transforms the positions of the ropes in the scanning plane to swing and hoist angles of the hoist rope. The swing and hoist angles for each rope are averaged, and the average is passed to a Kalman filter. Finally, the Kalman filter produces estimates for swing and hoist angles and the respective velocities (Section 5).

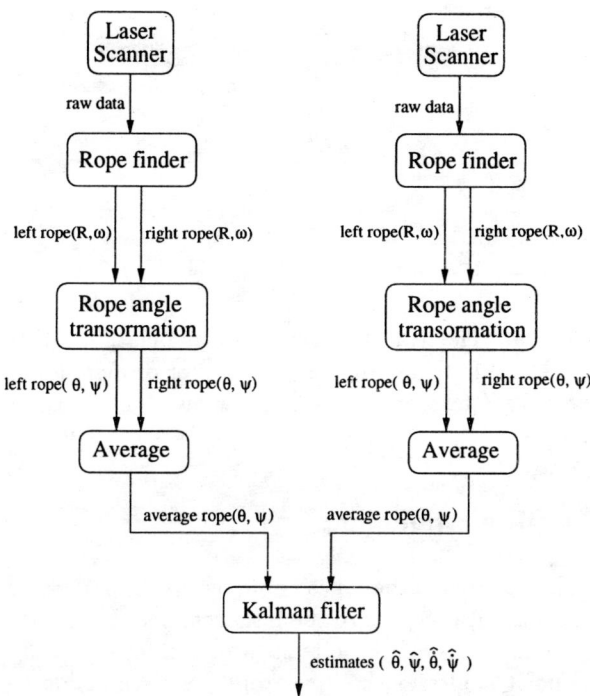

Figure 3: Schematic of the hoist rope angle measurement system.

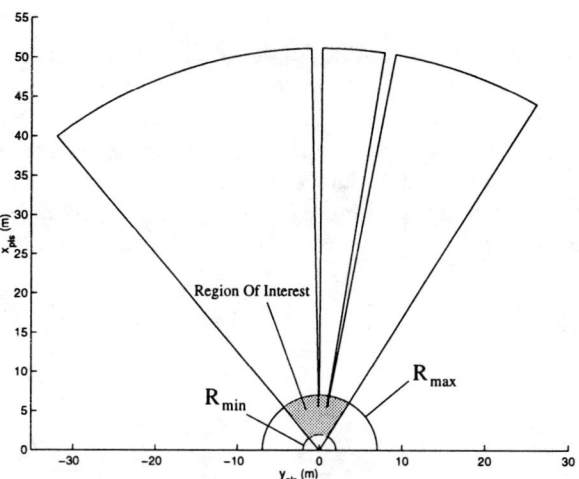

Figure 4: Range data from a laser scanner showing the region-of-interest produced by range gating. The two large spikes indicate the position of the two hoist ropes.

Rain rate	Probability (%)
'light'	0.02
'medium'	0.28

These probabilities are very low and therefore medium and light rain should not cause a significant problem. False targets may also be generated by insects and birds. Insects tend to swarm around the bright lights that are used to illuminate the pit at night.

4 Finding the ropes

Each laser scanner produces 361 measurements per scan. Each measurement represents the range to the first object in the corresponding 0.5° segment of the semi-circular scan. The positions of the two hoist ropes must be extracted from this data. False targets (targets that are not the ropes) complicate the task of tracking the ropes. The rope finding algorithm therefore must be able to locate the positions of the hoist ropes reliably in the presence of these false targets.

4.1 False targets

Rain drops may cause false targets. The laser scanners on the dragline are mounted below the boom structure and boom sheaves (Figure 1). The extent to which rain drops can be seen in this region, when the boom is moving through the air due to swinging, is unknown, and no data has yet been captured to indicate the scale of the problem. Some experiments were therefore carried out using an unmounted scanner at our laboratory in rain and in the open. The important measure, for the rope tracking system to work reliably, is the actual probability of seeing a rain drop in a given scan segment of the laser scanner, which for the two experiments conducted were:

4.2 The rope finding algorithm

4.2.1 Range gating

The first stage of the rope finding algorithm is to *range gate* the raw laser scanner data. This means that all targets found between a range R_{min}, and R_{max} are considered potential hoist rope targets. We can do this because we know that the ropes can only be in a certain region, which is constrained by the mechanical design. Figure 4 shows the region-of-interest created by these range gate limits combined with the restricted view from the sensor housing.

4.2.2 Rope discrimination (pairing data)

The most significant thing that we know about the ropes is that there are two ropes, and that they are constrained in the way they move relative to one another. This allows us to *pair up* data points into possible rope pairs (an n^2 search task). Experiments have shown that the hoist ropes are always within 0.5m of one another in the x-direction and are between 0.5 and 1.0m of one another in the y-direction. Figure 5 shows some data captured of the ropes when it was not raining, with some randomly added false target data added. The circles show the range gated data points. The squares show the mid-point between possible rope pairs. In Figure 5 three possible rope pairs can be seen.

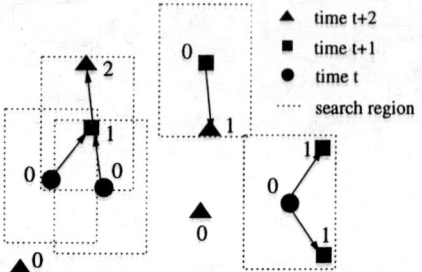

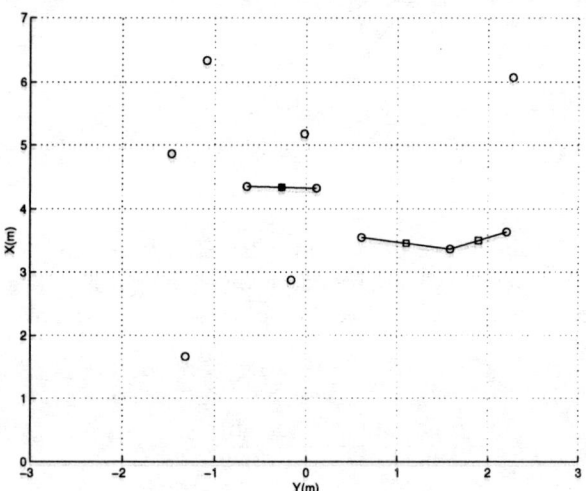

Figure 5: Data from the scanner clearly showing paired data points (squares) that represent the position of possible rope pairs.

4.2.3 Tracking and data association

The next problem encountered is that of data association and tracking, i.e., how do we match pairs between successive scans, and hence identify the hoist ropes and reject the false target pairs? As observed by the laser scanners, which operate at a low frequency (3 Hz), the ropes move with an almost random walk type motion (Figure 7) which makes directly tracking the hoist ropes using model based tracking techniques, such as Kalman filters, impossible.

The most practical (in this case) and the simplest form of tracking and data association is to look for a target at time t in the area immediately around where it was seen at time $t-1$. The size and shape of the search area is based on observed maximum motions between samples from experimental data.

The number of consecutive matches of a particular target pair can be counted and the pair with the highest count, assuming it is above some pre-defined threshold, should be the hoist ropes (Figure 6).

Figure 6: The tracking and data association strategy. The numbers indicate the number of matches of each point (the search area is only shown for successful matches).

4.3 Rope modelling

The position of the ropes found by the rope tracking system is with respect to the laser scanner. These positions must therefore be transformed into real swing and hoist angles of the hoist ropes. It is possible to derive an expression for both swing and hoist angles as a function of the sensed hoist rope position with respect to the laser scanner and the geometry of the sensing/boom structure. These transformations have been derived, but are not shown here.

4.4 Rope vibration and angle averaging

Observations of the hoist ropes using the boom mounted camera revealed that the ropes vibrate significantly while the dragline is both digging and swinging. It was also observed that the ropes vibrate 180° out of phase with one another (Figure 7). This allows us to take the average of the swing and hoist angles in an attempt to remove the high frequency component of the rope motion. We are only interested in the low frequency component of the motion from a control point-of-view.

5 Kalman filtering

The rope angle system consists of two laser scanners for robustness, and hence averaged swing and hoist angles are calculated for both scanners. A Kalman filter was used to fuse the measurements from the separate scanners, and to estimate the rate of change of swing and hoist angles, which are needed by the dragline control algorithm.

The angle data from each scanning system is not synchronous. The Kalman filter provides estimates

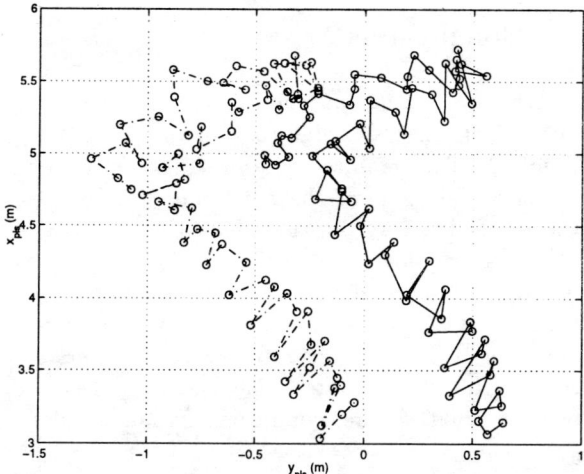

Figure 7: The motion of the two hoist ropes in the laser scanning plane.

at a fixed frequency (5 Hz) for the control system, whereas the measurements from the laser scanners arrive irregularly (on average 3/second).

By its nature, the Kalman filter gives the covariance of the estimates which in turn can be used for two things:

1. To reject a laser measurement if this measure is incompatible with the current estimate. In other words, it implements sensor failure detection.

2. To validate the estimates. The estimates are valid once their covariance fall under a given threshold depending on the desired accuracy.

5.1 Model of rope movement

The model of the bucket/rope system is extremely simple. The swing angle (θ) is assumed to follow the equation of an un-forced pendulum:

$$\ddot{\theta} + \omega^2 \theta = 0 \qquad (1)$$

where

$$\omega^2 = \frac{g}{h} \frac{cos(\beta)}{cos(\beta + \psi)} \qquad (2)$$

The hoist angle (ψ) is assumed to follow a constant velocity:

$$\ddot{\psi} = 0 \qquad (3)$$

5.2 Filter implementation

The measurements from the two laser scanners should be treated as a measurement from only one virtual laser scanner, since both laser scanners measure the same angles. The filter was implemented as a discrete Kalman filter with a variable time step. Notation used below is consistent with [4]. The state ($\hat{\mathbf{x}}$) vector is as follows:

$$\hat{\mathbf{x}} = \begin{bmatrix} \hat{\theta} \\ \hat{\dot{\theta}} \\ \hat{\psi} \\ \hat{\dot{\psi}} \end{bmatrix} \qquad (4)$$

Note that the state vector includes four components, actually two independent sub-vectors of two components. The first sub-vector is the swing angle and its derivative whereas the second sub-vector is the hoist angle and its derivative.

The conditional state propagation relation is:

$$\hat{\mathbf{x}}(t_i^-) = \boldsymbol{\Phi}(t_i, t_{i-1}) \hat{\mathbf{x}}(t_i^+) \qquad (5)$$

The state transition matrix ($\boldsymbol{\Phi}(t_i, t_{i-1})$) is given by:

$$\boldsymbol{\Phi}(t_i, t_{i-1}) = \begin{bmatrix} cos(\omega dt) & \frac{sin(\omega dt)}{\omega} & 0 & 0 \\ -\omega sin(\omega dt) & cos(\omega dt) & 0 & 0 \\ 0 & 0 & 1 & dt \\ 0 & 0 & 0 & 1 \end{bmatrix} \qquad (6)$$

where

$$dt = t_i - t_{i-1} \qquad (7)$$

Note that the first row in Equation 6 is obtained from the standard solution to the pendulum differential equation (Equation 1), and the second row from its derivative.

The observation model is given by:

$$\mathbf{z}(t_i) = \mathbf{H}(t_i)\mathbf{x}(t_i) + \mathbf{v}(t_i) \qquad (8)$$

where

$$\mathbf{H}(t_i) = \begin{bmatrix} 1 & 0 & 0 & 0 \\ 0 & 0 & 1 & 0 \end{bmatrix} \qquad (9)$$

Note that the observation vector has two components, the measured swing angle and the measured hoist angle. These angles come from either of the two laser scanners.

In addition to the two classical steps of every Kalman filter (integration of the model followed by a correction with the measure), the filter includes a third independent step, called prediction. This prediction step is simply an integration of the model from the time of the last received measure to the current time. This is needed because data loggers and/or control algorithms that ultimately use the estimated angle data require fixed time interval estimates. The filter is used in two different ways:

1. Every time a measure is received (from either scanner), the state vector is integrated from the time of the last measure to the current time. The correction is then made with the current measure and the state is replaced by its new value.

2. Every time a request from an external *client* is received (e.g., from a data logger or controller), the current state of the filter is integrated according to the model from the time of the last correction to the current time. The output state is not stored as the new state of the filter but is provided to the client. The covariance is also integrated to determine if the estimates are valid. If they are invalid, the client is told.

6 Results

The swing and hoist angle estimation system described here is currently being used in the prototype dragline swing automation system being trialled on a production dragline at a mine in Queensland, Australia. Figure 8 shows the results from the angle sensing system during a typical automatic cycle.

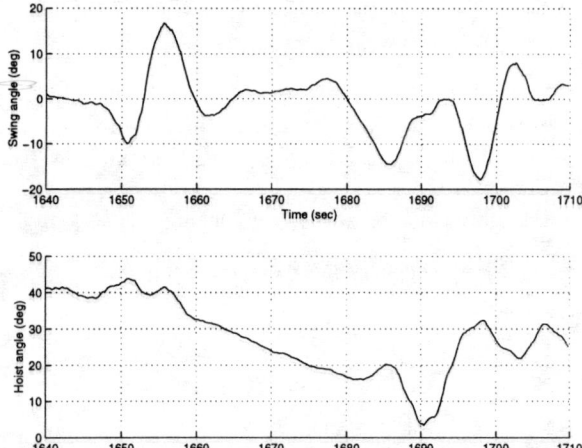

Figure 8: Swing angle (top), and hoist angle (bottom) during an automatic cycle.

The swing and hoist angle system has so far been used to demonstrate:

- effective bucket swing damping at dumping and digging points;
- online estimation of bucket weight, which is used to determine when the bucket has emptied;
- bucket pose recovery.

7 Conclusions

We have successfully automated the swing cycle of a 3,500 tonne dragline (one of the world's largest robots). This required a robust sensing system to measure the angle of the hoist rope with respect to the vertical. A system based on time-of-flight laser scanners was developed which implemented a tracking algorithm based on a novel data association and tracking strategy. A Kalman filter was used to fuse data from the two asynchronous laser scanners and to cope with sensor failure. The system is designed to operate in all weather conditions, 24 hours a day.

Acknowledgements

The authors gratefully acknowledge the help of their colleagues Stuart Wolfe and David Hainsworth. This work was funded by the Australian Coal Association Research Program (ACARP) as Project C5003, Rio Tinto, BHP Australia Coal Pty Ltd and the Cooperative Research Centre for Mining Technology and Equipment (CMTE). Tarong Coal's Meandu mine have made their BE1370W dragline available as a platform for development. Thanks also to Bucyrus (Australia) and Tritronics Pty Ltd who have provided invaluable in-kind support. The first phase of this project was funded by ACARP as Project C3007.

References

[1] G. Winstanley, P. Corke, and J. Roberts, "Dragline swing automation," in *IEEE Conf. Robotics and Automation*, 1997.

[2] D. W. Hainsworth, P. I. Corke, and G. J. Winstanley, "Location of a dragline bucket in space using machine vision techniques," in *Proc. Int. Conf. on Acoustics, Speech and Signal Processing (ICASSP-94)*, vol. 6, (Adelaide), pp. 161–164, Apr. 1994.

[3] G. J. Winstanley, "Dragline bucket identification," in *Proceedings of Image and Vision Computing*, (Christchurch, New Zealand), August 1995.

[4] P. S. Maybeck, *Stochastic Models, Estimation and Control*. Academic Press, New York, 1982.

Two-Dimensional Fine Particle Positioning Using a Piezoresistive Cantilever as a Micro/Nano-Manipulator

Metin Sitti and Hideki Hashimoto
Institute of Industrial Science, University of Tokyo
Roppongi, 7-22-1, Minato-ku, Tokyo, 106-8558, Japan

Abstract

In this paper, a fine particle positioning system using a pizoresistive cantilever, which is normally utilized in Atomic Force Microscopy, as the manipulator has been proposed. Modeling and control of the interaction forces among the manipulator, particle and surface have been realized for moving particles with sizes less than 3 μm on a Si substrate in 2-D. Optical Microscope (OM) is utilized as the vision sensor, and the cantilever behaves also as a force sensor which enables contact detection and surface alignment sensing. A 2-D OM real-time image feedback constitutes the main user interface, where the operator uses mouse cursor and keyboard for defining the tasks for the cantilever motion controller. Particle manipulation experiments are realized for 2.02 μm goal-coated latex particles, and it is shown that the system can be utilized in 2-D micro particle assembling.

1 Introduction

By the recent advances on micro-mechatronics technology, micro sensors and actuators, high precision positioners, micro robots inside the nuclear plant pipes, etc. have become possible. Especially, imaging devices such as Scanning Probe Microscopes (SPM) and Near-Field Optical Microscopes can provide imaging down to submicron and atomic scale at 2-D or 3-D. However, for constructing more complex micro/nano machines or devices by assembling micro/nano parts, necessary fabrication and manipulation technologies are still very immature at the scales less than 10 μm. Because, at these scales, micro/nano sticking forces become dominant to the inertial force, and a new robotics approach is indispensable. At this point, many researchers are trying to find new strategies for micro/nano assembly.

For the assembly of fine particles, different tools or control approaches have been proposed. For the particles sizes at the micro scale, Miyazaki et al. [1] proposed a two probe-based directly teleoperated assembly of spherical latex particles with the diameter of approximately 2 μm in 3-D, and tried to construct a 3-D pyramid where they had problems of assembling the last top particle due to adhesive forces. Tanikawa et al. [2] developed a directly teleoperated two-finger micro hand like a chopstick for moving glass spheres in 3-D with the size of 2 μm. Pappas et al. [3] proposed a robotics system where a micro/nano-tool is driven automatically for realizing simple tasks using visual servoing. They achieved positioning of 50 μm diameter diamond particles using a glass pipette with air pressure controlled picking and placing. For the nano scale particle manipulation, SPM probes such as Atomic Force Microscope (AFM) and Scanning Tunneling Microscope (STM) are utilized [4], [5], [6].

In this paper, a task-based semi-autonomous 2-D fine particle assembly system which utilizes piezoresistive AFM cantilever with its tip as the micro/nano manipulator and force sensor is proposed. As different from other works, the proposed system has the potential of also manipulating the nano objects by replacing the optical microscope imaging with the AFM non-contact imaging [7], the manipulator-particle contact is detected automatically using the real-time force feedback where the deformation of the tip or particle is prevented, even using one optical microscope, the depth (distance between the substrate/particle and cantilever) information is obtained through the cantilever substrate contact feedback, and manipulator-particle interaction forces during the manipulation are analyzed.

2 Problem Definition and Approach

Spherical polyvinyl gold-coated latex particles with sizes around 2.02μm (JEOL Datum Ltd.) are semi-fixed/absorbed to a Si substrate such that they are to be positioned by changing their xy positions [8], [9], [10]. The particles are also called as *absorbates*. The operations are to be realized in open air conditions with high relative humidity ($20-60\%$). Thus, the sticking forces such as capillary and van der Waals are strong. Here, it is assumed that the electrostatic

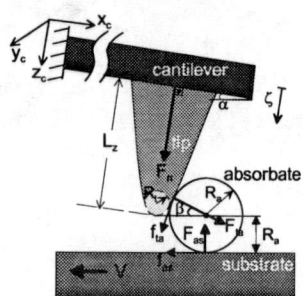

Figure 1: Positioning of the micro particles using the AFM tip as the manipulator.

forces are negligible with respect to other forces.

The cantilever tip which has a radius $R_t \approx 30nm$ is being utilized in contact pushing and pulling of the particles. Assuming the tip is pushing from a contact point passing through the absorbate center, the 3-D system can be simplified to 2-D as shown in Figure 1. The extension of the formulation to the 3-D case without assumption is direct. In the figure, F_{ta} and f_{ta} correspond to the tip-absorbate attractive/repulsive interaction force and friction force respectively. F_{as} and f_{as} are the interaction and friction forces for the absorbate and the substrate. The angle between the contact point of the tip and particle center is represented as β. The tip is aligned with an angle α. The tip is placed above the substrate with the parking height equals to the radius of the particle R_a. Thus, it is assumed not to touch to and interact with the substrate. By this way, the tip is not deformed, and vertical sticking force between the tip and particle is reduced.

Here, not the tip base but the substrate is moved with a constant speed V, and the manipulation strategy is as follows: *use the fixed cantilever as a stopper while moving the substrate under the particle with a uniform speed.* For fixing the particles, following conditions should be held:

- *at non-contact:* when the tip approaches or retracts from the particle, the particle should not stick to the tip and not move such that

$$F_{ta}sin\beta \leq F_{as},$$
$$f_{as} \geq F_{ta}cos\beta, \quad (1)$$

- *at contact:* when the tip contacts with the particle the resulting pushing or pulling force should be enough to fix the particle without breaking the tip, and flipping away the particle such that

$$f_{ta}cos\beta + F_{ta}sin\beta \geq F_{as},$$
$$f_{as} \leq F_{ta}cos\beta - f_{ta}sin\beta. \quad (2)$$

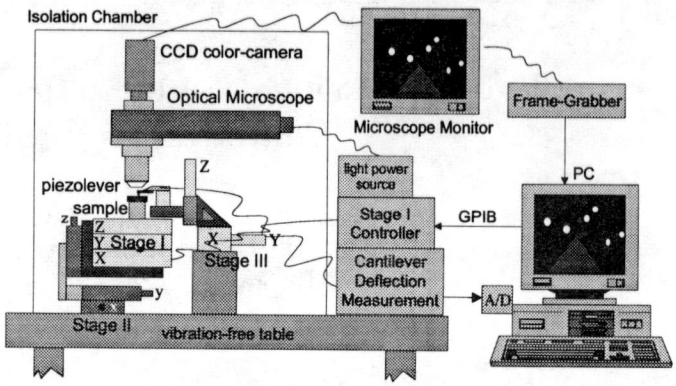

Figure 2: Overall system setup for the micro particle assembly.

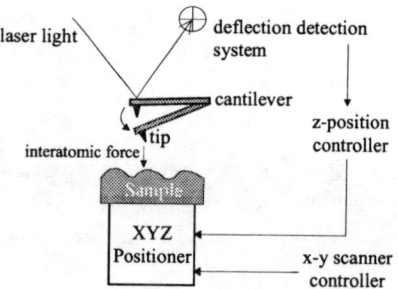

Figure 3: The basic structure of a conventional AFM.

The aim is to design the control parameters such as the contact point and angle, cantilever properties such as stiffness and tip radius and shape, particle-substrate friction, adhesion forces, humidity level and motion resolution and speed so that the above conditions can be held for a reliable manipulation.

3 System Setup

A task-based teleoperation control approach is selected where an operator determines the tasks to be realized, and a controller realizes these tasks automatically. The overall system is shown in Figure 2.

3.1 AFM as the Manipulator and Sensor

The basic structure of a conventional AFM system is shown in Figure 3. The very sharp cantilever tip atoms are interacted with the sample atoms by moving the sample/cantilever in the z-direction. The interatomic force $F_n(t)$ is attractive or repulsive, and the resulting typical deflection curve of the cantilever $\zeta(t)$ depending on the tip-sample distance $h(t)$ is shown in Figure 4. If the sample is moved slowly, i.e. the cantilever is assumed to be at equilibrium at each point:

$$F_n(t) = k_c\zeta(t), \quad (3)$$

where k_c is the previously known cantilever spring constant. Thus, the tip-sample force can be measured

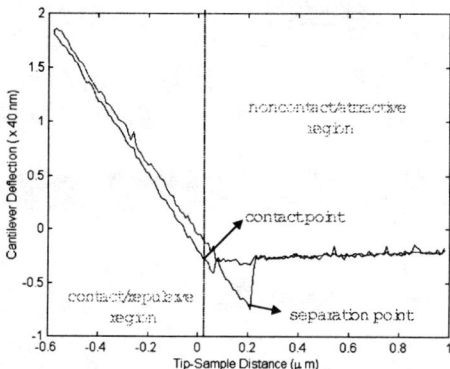

Figure 4: The typical cantilever deflection, i.e. interatomic force, and tip-sample distance relation for Si sample (experimental).

by measuring $\zeta(t)$. Instead of a laser interferometry-type deflection detection system, $\zeta(t)$ is measured by a Wheastone bridge-based deflection measurement electronics in our system since a piezoresistive cantilever (Park Scientific Instruments Co.) [11] is used. Thus, the output of the bridge is a voltage difference $V_{out}(t)$, and the nanometer value of the $\zeta(t)$ is computed from the below equation:

$$\zeta(t) = SG_2(G_1 V_{out}(t) + V_{off}), \quad (4)$$

where $G_1 = G_2 = 100$ are the amplification gains, V_{off} is the offset voltage which is needed when there is an offset depending on the different resistance values of the cantilevers, $V_0 = 2.5V$ is the bridge voltage, and S is the constant scaling ratio which is calibrated for each cantilever previously.

Using the Eq. (3) and $F_n(t)$ vs. $h(t)$ relation curve, if a reference ζ^* is set at the contact linear region, then the z-stage is moved until detecting this point, and the (x, y) positions are scanned and at each point the same reference is tracked. Thus, the surface 3-D topology image can be held. This method is called *Contact Imaging Mode*.

Besides of being the 2-D contact push-pull manipulation tool, sensory functions of the AFM are as follows:

- $F_n(t)$ force feedback is held in real-time,
- the alignment error of the substrate can be compensated by getting a contact 3-D topology planar image of the substrate along single x and y lines where there is no particle,
- the contact between the tip and the particle can be detected by measuring $\zeta(t)$.
- the depth information can be obtained through the cantilever substrate contact feedback.

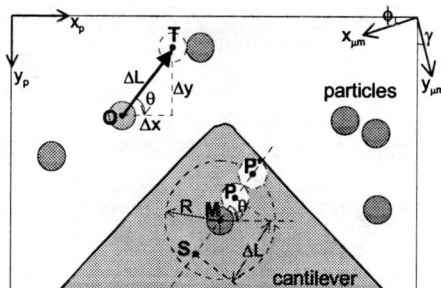

Figure 5: Automatic pushing or pulling operation, and image and positioner coordinates.

3.2 Vision Sensor: Optical Microscope

A reflecting light-type Optical Microscope (OM) (Olympus Co.) is used as the top-view vision sensor. It has approximately ×5000 monitor magnification with ×80 lens magnification, and $25mm$ working distance. A color camera (Sony Co.) on the OM is connected to a Matrox Co. Meteor framegrabber which enables real-time color image viewing of the micro world on the PC screen with the range of 640×480 pixels in the image frame, (x_p, y_p), and approximately $57 \times 45 \mu m^2$ in the world coordinates, $(x_{\mu m}, y_{\mu m})$. Hereafter, the world coordinates mean the coordinate frame for the AFM positioner x-y motion space. The image and world coordinate frames are given in Figure 5. In the case of linear mapping between both spaces, $x_{\mu m} = p_x x_p$ and $y_{\mu m} = p_y y_p$ where $p_x = p_y \approx 95 nm/pixel$ are constant x and y scaling constants. However, during the experiments, it is observed that there is also some rotation in the coordinates due to the orientations of the camera, stage and sample surface. Therefore, a calibration process is needed before the experiments in order to map (x_p, y_p) to $(x_{\mu m}, y_{\mu m})$ which is explained later.

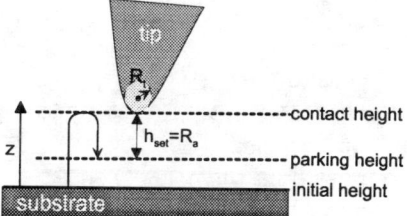

Figure 6: Parking height setting for the tip.

3.3 User Interface

The user interface constitutes of the real-time display of the top-view images from the camera mounted on the OM. The operator uses the mouse cursor and keyboard for defining the tasks for the AFM position controller. At present there are three main tasks:

- *Task 1 (T1)*: position the tip above the substrate,

- *Task 2 (T2)*: point-to-point relative motion,
- *Task 3 (T3)*: automatic pushing or pulling.

At the task T1, the tip is set to the height of R_a above from the substrate automatically. For this, the substrate is moved along the z-direction until touching to the tip, i.e. tip deflection is measured as ζ_{set}, then retracted back to the z-position given as $z = z_{final} - h_{set}$ where $h_{set} = R_a$ as shown in Figure 6.

In the task T2, using the mouse cursor, the operator clicks a point and its target point to be moved. Then the positon controller automatically moves with the amount of relative distance between the two points. This operation is needed for searching and initially positioning the particles on the screen.

The task T3 is used for semi-autonomous particle pushing or pulling. Freezing the continuous image, the operator selects the particle to be moved by clicking on the particle. Using image processing, the particle center **O** is automatically located. Next, the target position **T** is clicked by the operator. Then the pushing or pulling operation is realized automatically as shown in Figure 5 by the following strategy:

- *Step 1*: the tip radius R_t and center position **M**, particle center position **O** and radius R_a, and the target position **T** are computed or known,
- *Step 2*: compute the relative distance and orientation between **O** and **T** such that

$$\Delta x = x_T - x_O, \; \Delta y = y_T - y_O,$$
$$\theta = tan^{-1}(|\Delta y|/|\Delta x|),$$
$$s_x = \Delta x/|\Delta x|, \; s_y = \Delta y/|\Delta y|, \quad (5)$$

- *Step 3*: find the point **P'** which is the initial set point before tip-particle contact such that

$$x_{P'} = x_M + s_x R cos\theta,$$
$$y_{P'} = y_M + s_y R sin\theta,$$
$$R = R_t + 3R_a, \quad (6)$$

- *Step 4*: move the tip $3R_a$ upward from the substrate for avoiding any possible collision between the tip and particle,
- *Step 5*: move from the point **O** to **P'**,
- *Step 6*: move back the tip to its parking height using T1,
- *Step 7*: find the contact point **P** by moving from **P'** through **M** by measuring the cantilever deflection,
- *Step 8*: find the point **S** where

$$x_S = x_P - s_x \Delta L cos\theta,$$
$$y_S = y_P - s_y \Delta L sin\theta,$$
$$\Delta L = (\Delta x^2 + \Delta y^2)^{1/2}, \quad (7)$$

- *Step 9*: move from **P** to **S**,
- *Step 10*: move from **P** to **T**.

In above operations, since the tip cannot be seen from the top view (it is approximately 7 μm inside from the end of the cantilever), the tip center **M** is computed roughly by using the Si substrate as a mirror, and then precisely by several particle contact tests.

3.4 Position Control

For the manipulation of the particles and initial settings, three different XYZ stages are utilized as shown in Figure 2. Stage I, the fine positioning XYZ piezoelectric stage (Physick Instrumente Co.) with 10 nm resolution, integrated LVDT (Linear Variable Differential Transformer) sensors, 100 μm range in all axes, %0.1 linearity error, and closed-loop PI control is utilized during the automatic particle assembly control. The other two stages are used for initial alignment and particle search on the surface. The main motion of the stage is point to point motion. Depending on the desired speed V, stairwise point-to-point motion is utilized for controlled manipulation speed where for moving from (x_1, y_1) to (x_2, y_2) point:

$$x_i = x_1 + s_x \Delta cos\theta i,$$
$$y_i = y_1 + s_y \Delta sin\theta i \quad (8)$$

where $i = 1, \ldots, N$, $N = \Delta L/\Delta$, Δ is the predetermined motion step resolution, s_x, s_y, θ and ΔL are computed as the equations (5) and (7) for the case of $\Delta x = x_2 - x_1$ and $\Delta y = y_2 - y_1$. Thus, at each ith step, firstly x_i then y_i is moved.

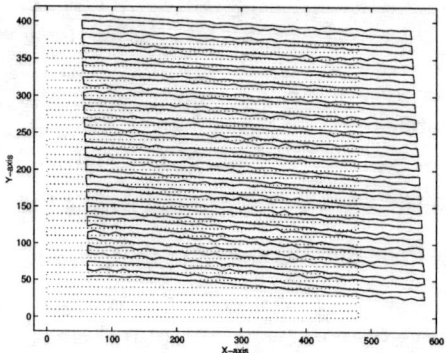

Figure 7: Controlled x-y scanning of the Stage I (dotted lines with 10 times magnification), and resulting detected particle center positions (solid lines).

3.5 Coordinate Frame Calibration

Image to world coordinates transformation constitutes of scaling and rotation transformations. Rotation transformation is due to the misalignment of the substrate surface and OM image plane, and pan and tilt errors due to the rough manual positioner. For this purpose, a selected particle is tracked automatically during a controlled x-y motion. Particle detection image processing part is the same with the selection of the particle procedure. As an example, a scanning motion with 1 μm steps and 0.1% hystheresis error results in the particle center positions as shown in Figure 7.

From the calibration tests, the rotation angles ϕ and γ for the x and y axes, and scaling factor $p = p_x = p_y$ are computed. Then, the pixel to μm position trasformation is as follows:

$$\mathbf{x_{\mu m}} = \mathbf{A} \mathbf{x_p},$$
$$\mathbf{A} = \begin{bmatrix} -p\cos\phi & p\sin\gamma & 640 p\cos\phi \\ p\sin\phi & p\cos\gamma & -640 p\sin\phi \\ 0 & 0 & 1 \end{bmatrix},$$
$$\mathbf{x_{\mu m}} = [x_{\mu m} \quad y_{\mu m} \quad 1]^T,$$
$$\mathbf{x_p} = [x_p \quad y_p \quad 1]^T. \quad (9)$$

4 Analysis of Interaction Forces

For getting the conditions of the equations (1) and (2), non-contact and contact forces are analyzed. In this analysis, it is assumed that the contacting point of the cantilever tip behaves as a sphere of radius R_t during the force interaction while the whole tip shape is taken as a cone (that is the real case in our cantilever), electrostatic forces are negligible, and objects can be deformable. Furthermore, as the notation, '+' forces mean attractive, and '−' ones do the repulsive forces. Those conditions can be analyzed in two parts:

4.1 At Non-Contact

In this case, the tip is approaching to the particle. Then $f_{ta} = 0$, and F_{ta}, F_{as} and f_{as} can be modeled using the capillary and van der Waals forces, and JKR contact deformation model [12] as follows:

$$F_{as} = 4\pi\gamma R_a,$$
$$F_{ta}(h(t)) = \frac{H\widetilde{R}}{6h^2} + \frac{2\pi\gamma\widetilde{R}\cos\delta}{1 + \frac{(h(t)-a_0)}{2r_1}},$$
$$f_{as} = \mu F_L + cA_{as},$$
$$A_{as}^3 = \frac{R_a}{K}\left(F_L + 6\pi R_a\gamma + \sqrt{12\pi R_a\gamma F_L + (6\pi R_a\gamma)^2}\right),$$
$$F_L = F_{as} - F_{ta}\sin\beta, \quad (10)$$

where γ is the water surface energy, $h(t)$ is the tip-absorbate distance, r_1 is the meniscus curvature radius, H is the Hamacker constant (for the case of a liquid layer on the sample, $H = \{(H_{tip} - H_{liquid})(H_{sample} - H_{liquid})\}^{1/2}$ [12]), $\widetilde{R} = R_a R_t/(R_a +$ $R_t)$, $a_0 = 30^{-1/6}\sigma$ is the contact point where σ is the interatomic distance, $K = 4/(3\pi)((1 - \nu_t^2)/E_t + (1 - \nu_p^2)/E_p)^{-1}$ is the equivalent Young modulus where E_t and E_p are the Young modulus and ν_t and ν_t are the Poisson's coefficients of the tip and particle respectively, δ is the liquid contact angle, and μ and c are the scaling constants for the load and adhesion part of the friction force respectively [13], [8]. In the case of micro scale particle size, $R_a/R_t >> 1$, and $\widetilde{R} \approx R_t$. Furthermore, F_{ta} becomes maximum at $h = a_0$ such that $F_{ta}(max) \approx 2\pi\gamma R_t$. Assuming the vertical forces are balanced, i.e. F_L is negligible, the conditions become as:

$$2R_a \geq R_t \sin\beta,$$
$$R_t \cos\beta \leq c\widetilde{K} R_a^{2/3}, \quad (11)$$

where $\widetilde{K} = (4\pi^2\gamma^2 K/3)^{-1/3}$. The first equality is correct for any β where $R_a >> R_t$. The second one depends on c, K, β and R_t and R_a. For the case of $R_a = 30R_t = 1\mu m$, $\gamma = 0.072 J/m^2$, $K = 10 GPa$, approximate result is $\cos\beta \leq 2c$. For this case, $c \geq \cos\beta/2$ for not moving the particle on the substrate.

4.2 At Contact

At contact pushing, the forces are:

$$F_{ta}(x(t)) = 2\pi\gamma R_t - 4\pi K'\sqrt{R_t}(a_0 + x(t)/\cos\beta)^{3/2},$$
$$F_{as}(x(t)) = 4\pi\gamma R_a - 4\pi K\sqrt{R_a}(a_0 + x(t)\sin\beta/\cos\beta)^{3/2},$$
$$f_{ta}(t) = \mu' F_{ta}(t) + c' A_{ta}(t),$$
$$f_{as}(t) = \mu F_{as}(t) + cA_{as}(t), \quad (12)$$

where $x(t)$ is the horizontal sample motion which depends on the motion speed V and resolution Δ, K' is the modulus for the tip-particle interaction, and $A_{ta}(t)$ and $A_{as}(t)$ are the contact areas. Putting these forces inside the conditions, the below equality is obtained:

$$(\mu'\sin\beta - \cos\beta - \mu\mu'\cos\beta)F_{ta}(t) \leq -cA_{as}(t) + c'A_{ta}(t)(\mu\cos\beta - \sin\beta). \quad (13)$$

Here, using any deformation models such as Hertz, JKR, DMT or Maugis-Dugdale [7] models $A_{as}(t)$ and $A_{(t_a t)}$ can be written in terms of adhesive forces and loads. For example, using the Hertz model, $A_{as}(t) = (R_a F_{as}(t)/K)^{1/3}$ and $A_{ta}(t) = (R_t F_{ta}(t)/K')^{1/3}$. Putting these contact areas to the Eq. (13), the relation is nonlinear and complex. However, assuming not smooth tip and particle surface geometry, the friction terms due to the contact area can be neglected with respect to the Amonton's load-based friction such that $c, c' \to 0$. Then, the following simplified condition is held:

$$\tan\beta \leq \mu + \frac{1}{\mu'}, \quad (14)$$

where $\beta \in [0, \pi/2]$. This condition implements that μ' should be small and relatively μ should be large.

During the contact pushing, the cantilever bending force is important for understanding the pushing force and cantilever dynamics. At the equilibrium points, the contact forces deflect the cantilever along the z_c axis as:

$$k_c \zeta(t) = f_{ta}(t)\cos\psi + F_{ta}(t)\sin\psi + \lambda(-F_{ta}(t)\cos\psi + f_{ta}(t)\sin\psi), \quad (15)$$

where $\lambda = 2L_x/L_z$ is the structural contstant of the cantilever with the cantilever length L_x and tip height L_z respectively, and $\psi = \beta - \alpha$ is the tip half angle ($\psi = 16.5^o$ in our conic tip). Assuming $f_{ta}(t) = \mu' F_{ta}(t)$, the following equality held from the Eq. (15):

$$F_{ta}(t) = \frac{k_c}{\mu'(\cos\psi + \lambda\sin\psi) + \sin\psi - \lambda\cos\psi}\zeta. \quad (16)$$

Thus, $\zeta(t)$ curves are directly proportional to the $F_{ta}(t)$ curves. Furthermore, there is also a torsional force τ_c on the cantilever due to the contact force along the y_c axis where it can not be measured by the present hardware.

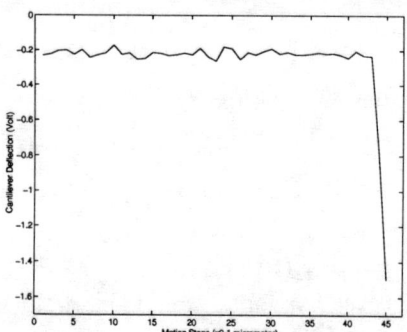

Figure 8: The cantilever deflection during automatic contact detection.

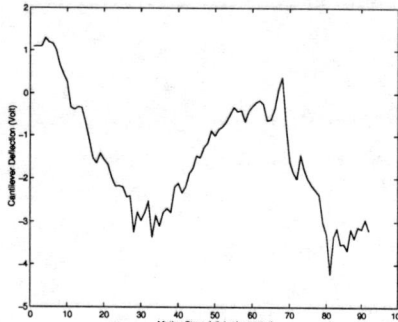

Figure 9: The cantilever deflection during pushing.

5 Experiments

As the first experiment, the contact point detection is tested. As the initial calibrations, the co-

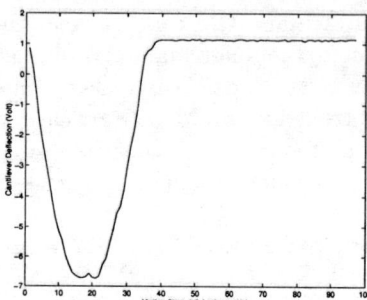

Figure 10: The tip and particle can lose the contact during pushing due to initial misalignment.

ordinate transformation parameters are computed as $p = 0.095$, $\phi = 9^o$, $\gamma = 1^o$, and the tip center is $(320, 304) pixel$. The cantilever parameters are $R_t = 30nm$, $S = 20nm/V$, $V_{off} = 3.62V$, and $k_c = 20N/m$, $L_x = 155\mu m$, and $L_z = 4\mu m$. The motion speed of the Stage I is around $V = 2\mu m/sec$ with $\Delta = 0.1\mu m$ steps. A particle is moved along a line that passes through the tip center, and $\zeta(Volt)$ is observed as in Figure 8. In the figure, $-0.2V$ is the no deflection line where around the 43th step of motion the particle contacts and bends until to $-1.5V$ which is the ζ_{set} for automatic contact detection.

As the next experiment, the automatic pushing of two particles to form a line shape is realized as shown in Figure 11. The automatic pushing results can be seen in the figure where the particles are positioned to the user-defined target points. During pushing, one of the bending data of the cantilever, which also correspond to $F_{ta}(t)$, can be seen from Figure 9. From the figure, F_{ta} has some maximum points, F_{ta}^*, which correspond to the the separation points of the particle from the substrate. This phenomenon is caused by the shearing of the contact points [14] during the separation such that at these maxima:

$$f_{as}^* = -k_c S\zeta^* = \kappa A_{as}^* = F_{ta}^*(\cos\beta - \mu'\sin\beta), \quad (17)$$

where κ is the shear stregth of the contact points, and ζ^* is the measured minima of the curve in Figure 9. In the figure, after breaking the bonds, since the cantilever contacts with the particle and breaks its bonds again, there are many peaks in the deflection curve such that $f_{as}^* = 1.2\mu N$ and $f_{as}^* = 1.6\mu N$ respectively. Also, the maximum contact areas can be calculated using the Eq. (17) such that $A_{as}^* = 0.0083\mu m^2$ and $0.011\mu m^2$ with the contact radii $51.5nm$ and $59.5nm$ respectively where $\kappa = 144 \times 10^6 N/\mu^2$ in the case of water layer between the particle and substrate.

In some cases, if the contact point of the particle and tip is not well centered, i.e. does not pass through the particle center, the particles can be pushed only once, and then the tip can lose its contact with the particle as shown in Figure 10. For avoiding this problem,

Δ is decreased for reducing positioning errors. Moreover, pushing can break the cantilever tip sometimes due to very large F_{ta}. This possibility is minimized by checking the ζ during pushing such that if it exceeds $-10V$, which is also the saturation voltage of the amplifier, then the pushing is stopped, and the design of the particle and substrate should be changed according to the force analysis by changing the environment parameters such as adding lubricants between the particle and substrate, realizing the manipulation in a liquid environment, etc.

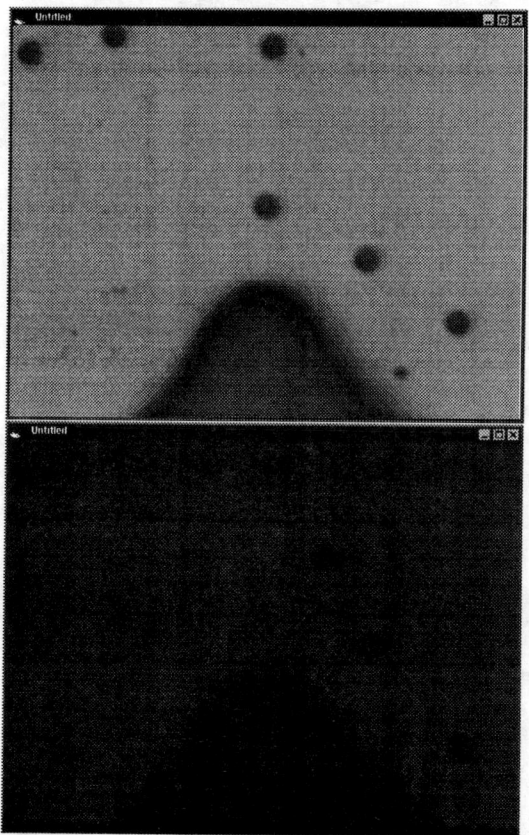

Figure 11: Pushing particle result: the initial positions (upper image), and the last configuration with the moved two particles (bottom).

6 Conclusion

In this paper, a micro particle manipulation system using a piezoresistive AFM cantilever as the manipulator and force sensor has been proposed. Modeling and control of the AFM cantilever tip and particle interaction has been realized for moving particles with sizes less than 3 μm on a Si substrate in 2-D. Particle manipulation experiments are realized, and it is shown that the system can be utilized in 2-D micro particle assembling. As the future work, the size of particles will be reduced down to 10s of nanometer, and the manipulation operations will be realized in a liquid environment where the capillary and electrostatic forces are reduced. Also, 3-D manipulation is aimed where special electromechanical MEMS or biochemical grippers can be used.

Acknowledgements

Authors would like to thank Park Scientific Instruments Co. for providing the piezoresistive cantilevers, and all other Hashimoto Lab. members for their various helps.

References

[1] H. Miyazaki and T. Sato, "Pick and place shape forming of three-dimensional micro structures from fine particles," in *Proc. of the IEEE Int. Conf. on Robotics and Automation*, pp. 2535–2540, 1996.

[2] T. Tanikawa, T. Arai, and T. Masuda, "Development of micro manipulation system with two-finger micro hand," in *Proc. of the IEEE/RSJ Int. Conf. on Intelligent Robots and Systems*, pp. 850–855, 1996.

[3] I. Pappas and A. Codourey, "Visual control of a microrobot operating under a microscope," in *Proc. of the IEEE/RSJ Int. Conf. on Intelligent Robots and Systems*, pp. 993–1000, 1996.

[4] D. Schafer, R. Reifenberger, A. Patil, and R. Andres, "Fabrication of two-dimensional arrays of nanometric-size clusters with the atomic force microscopy," *App. Physics Letters*, vol. 66, pp. 1012–1014, Feb. 1995.

[5] A. A. Requicha, C. Baur, and et al., "Nanorobotics assembly of two-dimensional structures," in *Proc. of the Workshop on Precision Manipulation at the IEEE Robotics and Autom. Conf.*, pp. 136–150, 1998.

[6] J. Stroscio and D. Eigler, "Atomic and molecular manipulation with the scanning tunneling microscope," *Science*, vol. 254, pp. 1319–1326, Nov. 1991.

[7] M. Sitti and H. Hashimoto, "Tele-nanorobotics using atomic force microscope," in *Proc. of the IEEE/RSJ Int. Conf. on Intelligent Robots and Systems*, pp. 1739–1746, Canada, Oct. 1998.

[8] M. Sitti and H. Hashimoto, "Macro to nano telemanipulation through nanoelectromechanical systems," in *Proc. of the IEEE Ind. Electronics Conf.*, pp. 98–103, Germany, Sept. 1998.

[9] M. Sitti, K. Hirahara, and H. Hashimoto, "2-d micro particle assembly using atomic force microscope," in *Proc. of the IEEE Int. Symp. on Micro Machine and Human Science*, pp. 143–148, Nagoya, Nov. 1998.

[10] M. Sitti and H. Hashimoto, "Two-dimensional fine particle positioning under optical microscope using a piezoresistive cantilever as a manipulator," *J. of Micromechatronics*, 1999 (submitted).

[11] F. J. Giessibl and B. M. Trafas, "Piezoresistive cantilevers utilized for scanning tunneling and scanning force microscope in ultrahigh vacuum," *Rev. Sci. Instrum.*, vol. 65, pp. 1923–1929, June 1994.

[12] J. Israelachvili, *Intermolecular and Surface Forces*. Academic Press, 1992.

[13] A. Berman and J. Israelachvili, "Control and minimization of friction via surface modification," *Micro/Nano Tribology and its Application*, Kluwer Academics Pub., pp. 317–329, 1997.

[14] C. M. Mate, "Force microscopy studies of the molecular origin of friction and lubrication," *IBM J. Res. Develop.*, vol. 39, pp. 617–627, Nov. 1995.

Pick and Place Operation of a Micro Object with High Reliability and Precision based on Micro Physics under SEM

Shigeki SAITO, Hideki MIYAZAKI, and Tomomasa SATO
Research Center for Advanced Science and Technology, the University of Tokyo
4-6-1 Komaba, Meguro-ku, Tokyo, 153-8904, JAPAN

Abstract

Recently, techniques of arranging micrometer-sized objects with high reliability and precision are required to construct micro devices such as quantum optics devices. Since micro objects tend to adhere to other objects mainly by electrostatic force, it is possible to pick them easily with a needle tip instead of grasping by a gripper. On the other hand, it is difficult to place them on a substrate precisely, and furthermore even picking them is not achieved with enough reliability. To solve this problem, in this study, basic strategy for reliable pick and precise place operation is indicated based on the actual measurement of adhesion under SEM first. Secondly dynamics for pick and place operation of a microsphere by a needle-shaped tool is analyzed. Thirdly a new method of reliable pick and precise place is proposed on the basis of the basic strategy and the dynamics. An experimental system for executing the proposed operation is constructed based on the numeric-control of the position measurement of the target object. On this system, a lot of numeric-controlled experiments with 2 μm polystyrene spheres reveal that under the optimal conditions, pick operation has never failed once out of over 20 trials and place operation has 140nm precision in position. Consequently, this paper concludes that the proposed method realizes pick operation with high reliability and place operation with high precision under SEM based on micro physics.

1 Introduction

Recently techniques of arranging micrometer-sized objects are required in the fields of basic research and applied research of electronic information equipment. In the field of physics, matrix array of dielectric microspheres as small as 1μm or less is expected to work as a quantum optics device. The key point for the array to work as expected is the precision of the structure of the array [1, 2].

To realize the precise array-structure, techniques for arranging objects by operation with high reliability and precision are indispensable. Therefore in this paper, a micro-manipulation system for numeric-controlled operation under scanning-electron-microscope(SEM) is constructed with the aim of realizing the automation of arraying micrometer-sized objects through SEM image-processing.

Typical procedure of arraying microspheres for a quantum optics device is described as follows [2]:

1. prepare microspheres on a substrate.
2. select a target microsphere
3. **pick** the target
4. transfer the target to the worktable
5. **place** the target to the desired position

The desired task is completed by repetition of step 2 to step 5. Among these steps, pick and place are critical operations compared to the others.

Micro objects generally behave differently from macro objects because of the scale effect. The adhesion of a micro object is strong compared to its gravity force [3]. The tendency is stronger especially in SEM environment because charging is remarkable due to electron beam injection [4].

Since micro objects tend to adhere to other objects, it is possible to pick them easily by contact with a needle tip instead of grasping by a gripper. On the other hand, it is difficult to place them precisely on a substrate and furthermore even picking them is not successful with enough reliability [2,5].

Because the dynamics for pick and place operation in micro world is not yet clear, formerly results of pick and place operation by a needle-shaped tool depend on the probability.

Therefore, for reliable and precise pick and place operation, Tanikawa has developed a-pair-of-needles manipulator with a 6-DOFs-parallel-link structure. He aims at reliable and precise operation through a skillful technique like the use of chopsticks [6]. Koyano has also developed a manipulator with two needle-shaped tools which rotates at a needle-tip-concentrated configuration. He has been able to pick a micro object with a large-tip-area tool and to place it with a small-tip-area tool after switching the tools [7]. Kasaya has constructed a system which enables visual-feedback manipulation with a needle-shaped tool under SEM. He has realized pick and place operation by trial-and-error or repetition method based on the observation through SEM-image-processing [5]. These solutions are successful in some cases, however, some problems may still occur if the size of an object is a few micrometer or less. For example, Tanikawa's skillful technique would be difficult to control for precise operation because quite often dextrous tip-motion is severely demanded. Since Koyano's needle-tip-concentrated configuration is realized with a mechanism, it would be unrealistic to operate a micrometer-sized object. Kasaya's system might not work as expected, because reliable and precise operation by the trial-and-error method requires the observation for a long time in high magnification under SEM although it should be avoided for such reasons as follows [8]:

- The adhesion of micro object increases with elapse time under SEM due to electron beam irradiation.
- The adhesion of micro object tends to be larger because the volume of electron-beam current becomes larger at observation in higher magnification.

In this study, a new method of reliable pick and precise place operation by numeric control is developed; Pick and place operation with high reliability and precision based on micro physics under SEM is proposed. In addition, the effectiveness of the proposed operation is proved through experiments in which objects are micrometer-sized spheres of dielectric material(polymer).

In the following, adhesion of a micro object under SEM, basic strategy for pick and place operation, analysis of its dynamics, and proposition of new operations based on the dynamics are described in section 2. Section 3 describes an experimental system which consists of a SEM-image-based measurement system and a numeric-controlled micro manipulator. Experimental results about reliability and precision for the proposed operation are discussed in section 4. In section 5, obtained results are summarized and necessary future studies are discussed.

2 Reliable Pick and Precise Place Operation

2.1 Adhesion of a micro object under SEM

As described in section 1, some dispositions of adhesion acting on a micro object under SEM are elucidated [8]. Out of these dispositions, possible large obstacles for reliable pick and precise place are summarized as follows:

Time Dependency: The relationship between time elapse and adhesion is investigated as shown in Figure 1. X-axis shows period of contact after stick-

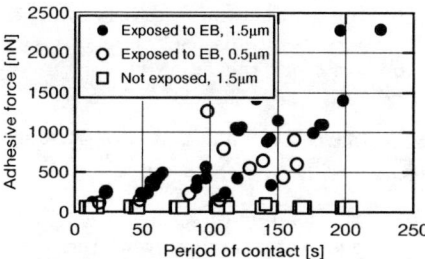

Figure 1: Time Dependency of Adhesion

ing the tool-tip surface on the top of an object and Y-axis shows the measured force. Two series of data with a black dot '●' and a white dot '○' show the measured adhesions when the object is exposed to electron beam(EB) in the case of tool-tip diameter of $1.5\mu m$ and $0.5\mu m$ respectively. A series of data with a box dot '□' show the measured adhesions when the object is not exposed to EB in the case of tool-tip diameter of $1.5\mu m$. From these data, it is proved that the adhesion of micro object increases with time elapse under SEM due to EB irradiation.

EB Condition Dependency: Figure 2 shows the relationship between SEM-image magnification at observation and adhesion. Each graph of Figure 2 is a histogram about the magnitude of measured force. X-axis shows the magnitude of the measured adhesion and Y-axis shows the frequency. From these data, it is confirmed that the adhesion becomes larger at higher-magnification observation. This means that the adhesion increases mainly depending upon the EB current because the volume of EB-current injection is proportional to the square of magnification.

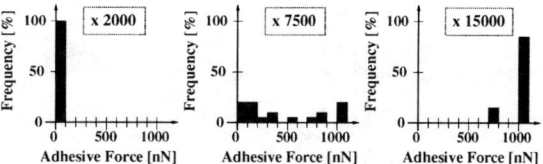

Figure 2: EB Condition Dependency of Adhesion

In addition to these knowledges about adhesion under SEM, it is also widely known from experience of electron microscopists that more contamination accumulates under SEM as the magnification at observation is higher or as the period of observation is longer.

2.2 Basic Strategy

The discussion mentioned above concludes that observation under SEM in high magnification for a long time should be avoided, although high magnification is indispensable for precise operation of a micro object. Thus if the size of an object is a few micrometer or less, former ways of micro-manipulation might never be effective because they often require trial-and-error or repetition method which is based on the observation for a long time. Koyano proposes the way of place operation which does not depend upon the trial-and-error or repetition. Nevertheless the proposed way would not be realistic for a micrometer-sized object since the operation relies on mechanical accuracy.

Consequently pick and place operation with reliability and precision should be simple one which needs neither a special mechanism nor observation for a long time. In other words, it is the most desirable that pick and place operation can be completed by simple trajectory-control of a tool-tip without visual-feedback. Therefore, in this study, dynamics of a micro object on which adhesion acts is analyzed. Based on the dynamics, simple pick and place operation with reliability and precision is proposed.

2.3 Dynamics Analysis

This section describes the dynamics analysis for pick and place operation, of which the target objects are micro spheres, in order to realize them by controlling the tool-tip trajectory.

To simplify the analysis, it is assumed that pick and place operation can be executed by loading force to a micro sphere gradually when the tool-tip is contacting a sphere on a substrate.

Definition of Coordinate

Figure 3 shows the definition of coordinate for dynamics analysis of pick and place operation. The value r is radius of the sphere. The value ϕ is contact angle. The contact interface between a tool-tip and a sphere and that between a substrate and a sphere are defined as S_1 and S_2, respectively. Every parameter with a certain suffix corresponds to the contact interface with the same suffix in the following. The values N_1, N_2 are normal reactions. The values μ_1, μ_2 are coefficients of maximum friction. The values f_1, f_2 are friction forces along with tangential lines at contact interfaces.

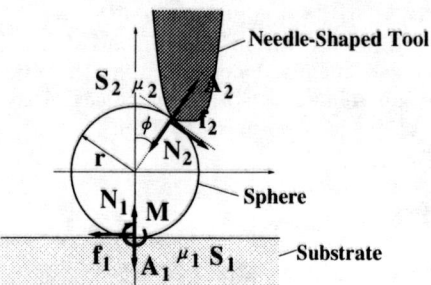

Figure 3: Definition of Coordinate

Besides the values A_1, A_2 are introduced to this system as adhesions and the value M is introduced as a rolling-resistance moment. These are specific parameters in the micro world.

Furthermore the parameter k regarding the loading-direction is defined by $k \equiv (A_2 - N_2)/f_2$. The value of $k = 0$ shows the direction along with the tangential line. The value of $k < 0$ shows the direction in which the tool-tip gets large reaction from the sphere by loading to it. And the value of $k < 0$ shows the direction in which the tool-tip gets only attractive force.

Then equilibrium equations are formulated utilizing these parameters as follows:

$$\begin{aligned}
0 &= N_1 - A_1 + A_2 \cos\phi - f_2 \sin\phi - N_2 \cos\phi \\
0 &= N_2 \sin\phi - A_2 \sin\phi - f_2 \cos\phi + f_1 \\
0 &= r(f_1 + f_2) + M
\end{aligned} \quad (1)$$

The third equation (1) describes the equilibrium about rotation. This equation can be established as long as external moment $|M_{ex}|$ does not exceed the maximum rolling-resistance M_{max} discussed in the following.

Maximum Rolling-Resistance Moment

Rolling-resistance moment and the existence of maximum rolling-resistance are pointed out here.

In micro world, adhesion necessarily causes deformation to some extent at the contact point and generates contact interface with finite area. When surface energy of a microsphere is considered, the ratio of the contact-interface size to the sphere size is proportional to the sphere size to the minus 1/3 th power [9]. Thus as the sphere becomes smaller, it becomes more difficult to neglect the size of the contact interface.

The stress distribution in the near-field of the contact-interface edge is similar to that in the near-field of the crack tip. As Maugis pointed out, separation process of micro objects can be considered equivalent to the fracture process of crack [10].

As it is generally known in fracture mechanics, the stress more than certain threshold is needed to cause the fracture of the crack [11]. This knowledge concludes that separation of micro sphere also needs stress, more than certain threshold, and that this threshold determine the maximum rolling-resistance moment.

Since it is naturally assumed that the maximum rolling-resistance moment M_{max} is proportional to the area of contact interface, M_{max} is written in the form $M_{max} = c \cdot \pi a_1^2$ where c is a function of the surface energy. a_1 is radius of contact interface S_1.

According to the equilibrium equations (1) and lin-

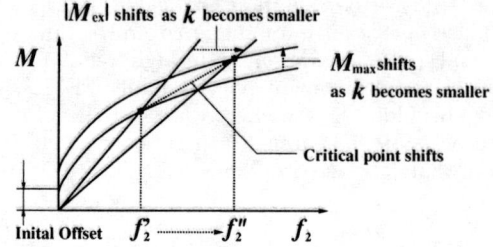

Figure 4: Relationship between M_{max} and $|M_{ex}|$

ear elastic theory. M_{max} is expressed in the form of

$$\begin{aligned}
M_{max} &= c \cdot \pi a_1^2 = c' \cdot N_1^{\frac{2}{3}} \\
&= c' \{A_1 + (\sin\phi - k\cos\phi)f_2\}^{\frac{2}{3}}
\end{aligned} \quad (2)$$

where c' is a function of surface energies, sphere radius, and elastic moduluses.

Similarly external moment M_{ex} is given by

$$M_{ex} = -r(1 + \cos\phi + k\sin\phi)f_2 \quad (3)$$

The qualitative relationship between M_{max} and $|M_{ex}|$ is shown in Figure 4.

Because loaded force f_L is expressed in the form $f_L = \sqrt{1+k^2}\, f_2$, some knowledge about difficulty of rolling is acquired from Figure 4 as follows:

- In case that the direction of loading is fixed (i.e. k is fixed):

  1. When the loaded force f_L is quite large (i.e. f_L exceeds the critical loaded force $\sqrt{1+k^2}\, f_2'$), the micro sphere will start rolling.

  2. When the loaded force f_L is small enough, the micro sphere never starts rolling because there is an initial offset value of M_{max} which derives from the adhesion on the substrate side A_1.

- In case that the value of loaded force is fixed (i.e. $\sqrt{1+k^2}\, f_2$ is fixed):

  1. For the loading-direction in which the tool-tip gets the reaction from the sphere (i.e. $k < 0$), the sphere has more difficulty to start rolling as the direction in which the tool-tip gets a larger reaction is taken (i.e. k becomes smaller).

  2. For the loading-direction in which the tool-tip gets only attractive force (i.e. $k > 0$), control of the sphere rolling is unstable.

Analysis through Equilibrium Equations

To simplify the analysis of the results after executing pick and place operation, it is assumed that a sphere does not start rolling (i.e. always $M_{max} > |M_{ex}|$).

After equilibrium equations are solved on this assumption, conditions for rolling-start are considered in the range of its solution for more precise analysis.

Conditions for slipping-start at contact interfaces S_1, S_2 are given by

$$f_1 > \mu_1 N_1, \quad f_2 > \mu_2 N_2 \quad (4)$$

respectively. The equations (4) are expressed from equation (1) as follows:

$$\{k(\mu_1\cos\phi+\sin\phi)+(\cos\phi-\mu_1\sin\phi)\}f_2 > \mu_1 A_1 \quad (5)$$

$$(1+k\mu_2)f_2 > \mu_2 A_2 \quad (6)$$

Equation (5) shows the condition for slipping-start at contact interface S_1. Equation (6) shows the condition at S_2.

From these equations (5) and (6), some critical values of the loading-direction parameter k are given by

$$k_1 = -\frac{1}{\mu_2} \quad (7)$$

$$k_2 = \frac{\mu_1\sin\phi-\cos\phi}{\mu_1\cos\phi+\sin\phi} \quad (8)$$

$$k_3 = \frac{(\mu_1\sin\phi-\cos\phi)\mu_2 A_2+\mu_1 A_1}{(\mu_1\cos\phi+\sin\phi)\mu_2 A_2-\mu_1\mu_2 A_1} \quad (9)$$

The meaning of these critical values is as follows:

k_1 : critical value determining
whether slip can be caused on S_2

k_2 : critical value determining
whether slip can be caused on S_1

k_3 : critical value determining on which
slip can be caused earlier, S_1 or S_2

When the ratio of adhesion on the substrate side A_1 to the adhesion on tool-tip side A_2 is defined as $\alpha \equiv A_1/A_2$, k_3 is transformed into the form

$$k_3 = -\frac{1}{\mu_2}+\frac{(\mu_1-\mu_2)\cos\phi+(1+\mu_1\mu_2)\sin\phi}{(\mu_1\cos\phi+\sin\phi)\mu_2-\mu_1\mu_2\alpha} \quad (10)$$

To give a typical example, the relationship between α and critical value of k in the case of $\mu_1=\mu_2=0.8$, $\phi=30°$ is shown in Figure 5.

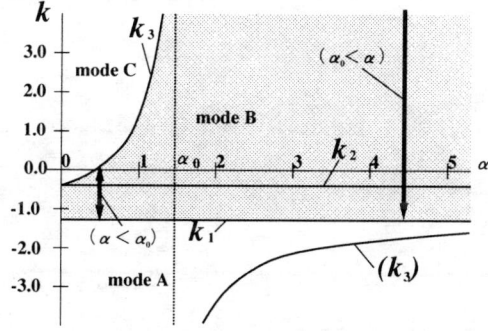

Figure 5: Relationship between α and critical k

In this case α_0 is defined by

$$\alpha_0 \equiv \frac{\mu_1\cos\phi+\sin\phi}{\mu_1} \simeq 1.5$$

As recognized from Figure 5, by the use of these critical k, a phenomenon caused by operation is classified into 2 modes for $\alpha_0 < \alpha$ and 3 modes for $\alpha < \alpha_0$ as follows:

- In case that adhesion on substrate side is larger than adhesion on tool-tip side (i.e. $\alpha_0 < \alpha$):

  - **mode A:** $k < k_1$
    Slip is caused on neither of contact interfaces. Therefore as loaded force becomes larger, deformation or drastic movement of sphere may be caused finally.
  - **mode B:** $k_1 < k$
    Slip is caused on contact interface of the tool-tip side S_2.

- In case that the adhesion on the tool-tip side is larger than adhesion on substrate side (i.e. $\alpha < \alpha_0$):

  - **mode A:** $k < k_1$
    Slip is caused on neither of contact interfaces. Therefore as the loaded force becomes larger, deformation or drastic movement of the sphere may be caused finally.
  - **mode B:** $k_1 < k < k_3$
    Slip is caused on contact interface of the tool-tip side S_2.
  - **mode C:** $k_3 < k$
    Slip is caused on contact interface of the substrate side S_1.

Figure 5 also shows a map of modes due to both the ratio of adhesions α and the loading-direction k. The sketch of predicted phenomenon depending on the several modes is shown in Figure 6.

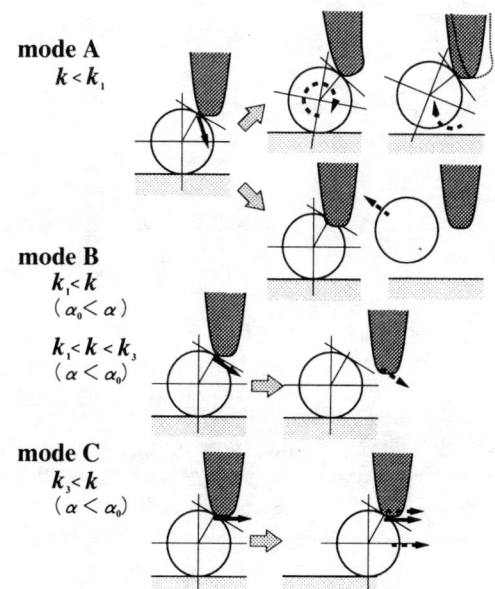

Figure 6: Predicted Phenomenon

2.4 Proposition of Reliable Pick and Precise Place Operation

The data presented in section 2.1 reveals that adhesion acting on micro object increases with time elapse after contact because of EB irradiation. It is naturally presumed that adhesion depends on the history of irradiation, and therefore the adhesion of a sphere on substrate side is larger than that on tool-tip side in pick operation. And also for place operation adhesion of a sphere on the tool-tip side is larger than that on the substrate side. In this section, a new method of pick and place based on the condition and analysis stated above is proposed.

Reliable Pick Operation

Adhesion between a sphere and a substrate is larger than that between a sphere and a tool-tip in pick operation, because the sphere on the substrate has been exposed to EB irradiation before pick operation. For this reason, the key point of realizing reliable pick operation is to cause the fracture of the contact interface on a substrate side without the fracture on the tool-tip side.

The fracture on the substrate side results in the reduction of adhesion of contact interface on the same side. Adhesion on the tool-tip side, thereby, becomes relatively larger than that on the substrate side. By lifting up the tool-tip soon after the fracture, pick operation is completed with high reliability.

From knowledge described in section 2.2, we know that to cause the fracture of interface on the substrate without that on the tool-tip side, mode A and furthermore rolling-start must be caused by taking loading-direction of optimal k and ϕ (i.e. $k < k_1$ and $M_{max} < |M_{ex}|$).

In addition to this, the best loading-direction is axial (i.e. normal in the configuration of this study) because the rigidity of a needle-shaped tool in axial direction is much higher than that in bending direction.

From the discussion above, offset-loading method is proposed as reliable pick operation. This method enables reliable pick due to the substrate-side fracture caused by loading to a sphere along with offset normal axis as shown in Figure 7.

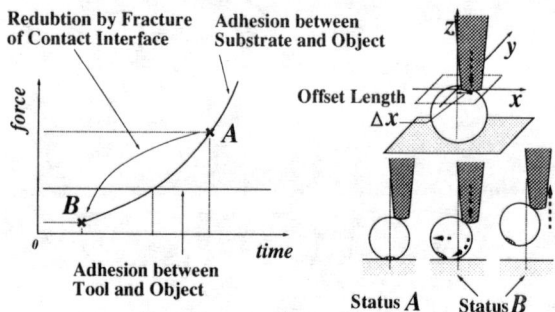

Figure 7: Offset-Loading Method

To simplify, it is assume that loading-direction is collinear to actual tool-tip trajectory. Equation (2) and (3) are then written for $k = -\cot\phi$ in the form

$$M_{max} = c'\{A_1 + f_L\}^{\frac{2}{3}} \quad (11)$$
$$M_{ex} = -r \cdot \sin\phi \cdot f_L \quad (12)$$

where offset length Δx corresponds to $r \cdot \sin\phi$.

In this method, external moment M_{ex} is applied by the loaded force in the direction of normal with offset length. Offset length is, thus, desirable to be larger as long as slip of tool-tip does not occur.

Precise Place Operation

Adhesion on the tool-tip side is larger than that on the substrate side in place operation because a sphere sticking to a tool-tip has been exposed to EB irradiation before place operation. For this reason, the key point to realize precise place operation is to cause the fracture of contact interface on a tool-tip side without the fracture on a substrate side.

Section 2.2 indicates that to cause the fracture of interface on a tool-tip without that on a substrate side, mode B must be caused without rolling-start of a sphere by taking loading-direction of optimal k (i.e. $k_1 < k < k_3$ and $M_{max} > |M_{ex}|$).

In general smaller k (i.e. the direction in which a tool-tip gets larger reaction from a sphere) is desired because rolling-start is less likely to occur.

Nevertheless as recognized from equation (6), when slipping starts on contact interface S_2, necessary loaded force $f_L(=\sqrt{1+k^2}f_2)$ becomes infinity for $k = k_1$, $1.36 \times A_2$ for k_2, and $0.75 \times A_2$ for k_3, where $\alpha = 0.5$ and the other parameters are the same as in section 2.2. Thus the value of k near to k_1 should be avoided because for $k \sim k_1$ rolling inevitably starts before slipping starts on the interface of tool-tip side. As a conclusion, optimal k is in the range of $k_1 < k < k_2$ except the near range of k_1. From the discussion above,

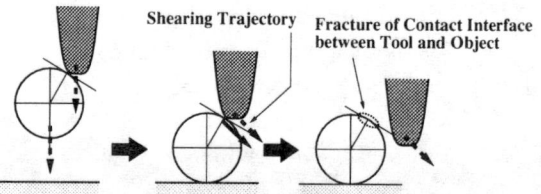

Figure 8: Shearing-Trajectory Method

shearing-trajectory method is proposed as precise place operation. This method enables precise place due to the tool-tip-side fracture caused by taking almost tangential trajectory at contact point and by letting the tool-tip get small reaction from the sphere as shown in Figure 8.

3 Experimental System

Experimental System for executing the proposed operation is shown in Figure 9. The system consists of a micro manipulator, a position measurement system, and a computer for system integration.

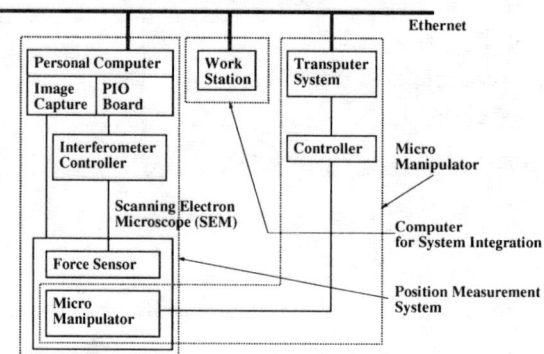

Figure 9: Configuration of Experimental System

In experiments, the tip size of a needle-shaped tool is about $0.7\mu m$ and the material is glass coated with Au. The diameter of a sphere is $2.0\mu m$ and the material is polystyrene. The material of a substrate is glass coated with Au.

3.1 Micro Manipulator

In this study, the micro manipulator named Nano Robot II is utilized [1]. This manipulator consists of a tool unit, a substrate unit, and a base unit. Micro objects are put on the substrate of the substrate unit and operated by a needle-shaped tool attached to the tool unit under the observation by SEM from upside.

Each unit has only translational DOFs. A tool unit has three DOFs with resolution of $15nm$ to operate a micro object. A substrate unit has three DOFs with coarse motion to adjust the tool-tip to the target object. A base unit has two DOFs with coarse motion to keep the tool-tip in the field of SEM view. Each DOF can be controlled numerically by the transputer system through the actuator controller. Figure 10 shows a micro manipulator and the configuration of a tool, a sphere, and a substrate.

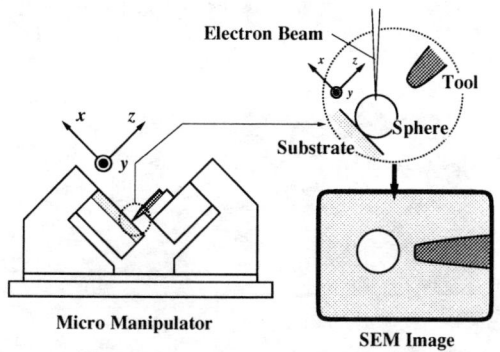

Figure 10: Micro Manipulator and Configuration of a Tool, a Sphere, and a Substrate

3.2 Position Measurement System

Position measurement system is composed of an SEM image-processing system and the microforce-sensing touch sensor. The SEM image-processing system measures the two-dimensional position of an object in the SEM monitoring screen coordinates. Microforce-sensing touch probe measures the position in depth direction.

SEM Image-Processing
SEM image-processing system consists of an SEM, an image-capture board, and a computer for image-processing. The SEM with a field emission electron gun (Hitachi, S-4200) is used as the microscope which observes the tasks from upside. In the experiments of this study, SEM image is acquired at video rate(30 image/sec) under the following conditions: acceleration voltage of $10kV$ and pressure of 10^{-6} Torr. The practical resolution of this SEM is about $10nm$ for video rate observation. The magnifications of 5,000 and 25,000 are used. The sizes of one pixel at these magnifications correspond to $6.8nm$ and $34nm$ respectively. From this SEM image, the two-dimensional position of the tool and the objects are measured. The SEM outputs image as monochrome NTSC video signal. The SEM image is captured as monochrome 8-bit digital images with 640×480 pixels by the PC (Dell, Dimension XPS with Pentium II, 266MHz) equipped with an image capture board (ArgoCraft, VideoCapture). Processes such as filtering and image recognition are executed by the PC.

SEM image is generally noisy especially when observed at the video rate of 30 images/sec. And edges are naturally emphasized in the SEM image. SEM images have also imbalance of brightness. Furthermore, there is a problem related to images that since the task is executed by the tool contacting the object, sometimes a part of the object is hidden by the tool. In SEM image-processing, features and problems mentioned above must be considered.

The image-processing techniques Gaussian filter, Sobel filter, and Hough transform are implemented in this system to measure the two-dimensional position of a tool and a object [5].

To realize precise measurement, the following steps are taken. First, acquired SEM images are averaged in the time domain by image-improvement function equipped with the SEM. Secondly, Gaussian filter spatially smoothes SEM image processed above. Thirdly, Sobel filter derives edge fragments from smoothed SEM image. Finally, Hough transform detects given shapes from an image. Given shapes are geometry models drawn by segments. Hough transform is considered as a good method for pattern-matching between geometry models and image made of edge fragments [12].

From the result of estimations about image-processing techniques stated above, it has turned out that the precision of SEM image-processing measurement is about $40nm$ in the case of the magnification of 5,000 and is about $8nm$ in the case of the magnification of 25,000.

Microforce-Sensing Touch Probe
The constructed microforce-sensing touch probe is composed of a needle-shaped tool, a cantilever with parallel structure of plate springs, and a laser interferometer. A tool is attached to a cantilever and also works as a touch probe. Force on a tool-tip is sensed by measuring the displacement of a cantilever through the use of a laser interferometer. The rigidity of the cantilever is about $23N/m$. The resolution of a laser interferometer is about $8nm$. Therefore the force sensitivity of touch probe is about $180nN$. By detecting contact on a tool-tip through the sense of microforce, touch probe measures the position in depth direction with resolution of less than $\pm 150nm$.

3.3 Computer for System Integration

A workstation(Sun, SPARK station IPX) is used as a computer for system integration. This workstation has data-management function for position data measured by the position measurement system and the operation data for generating actual tool trajectory. In addition, it also has management function for experimental software utilities. Both transputer and PC are connected to this workstation through Ethernet.

4 Experimental Results
4.1 Reliable Pick Operation

In order to execute the proposed pick operation (i.e. offset-loading method), the tool-tip approach to the standard position of the target object must be taken first. Procedure of the approach is realized as follows:

1. Position measurement in depth direction by detecting the contact between the tool-tip and the substrate

2. Lifting-up movement of the tool-tip in Z-axis by the diameter of the sphere

3. Position measurement of the tool-tip and the standard position of the sphere in two-dimensional coordinate on the SEM screen by SEM image-processing

4. Horizontal movement of the tool-tip in XY-plane to the initial position

In this experiment, the standard position is set $1.0\mu m$ above the top of a sphere.

In order to prove the effectiveness of the proposed pick operation (i.e. offset-loading method), the relationship between offset length Δx and possibility of success are examined by judging whether the result is successful or not when offset length varies from $0.0\mu m$ to $0.5\mu m$ by $0.1\mu m$. The number of trials is over 10 in every offset length. Figure 11 shows SEM image during execution of offset-loading method ($\Delta x = 0.2\mu m$).

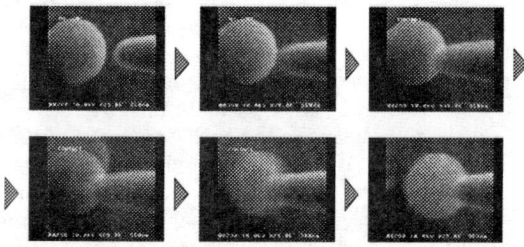

Figure 11: Execution of Offset-Loading Method

The obtained result is shown in Figure 12. This data shows that pick operation by offset-loading method has never failed once out of over 20 trials in the case of $\Delta x = 0.1\mu m$ and $\Delta x = 0.2\mu m$, and that there is optimal value of offset length in the middle of the range. When the operation is successful, the fracture of contact interface between a sphere and a substrate can be detected by monitoring the force output of the touch probe.

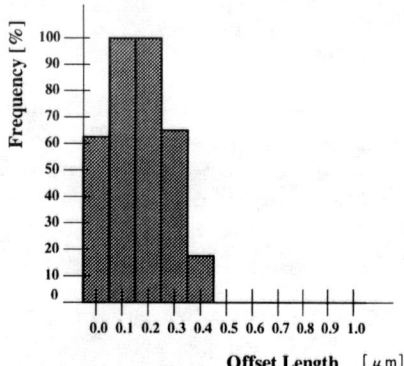

Figure 12: Relationship between offset length and success

An experiment of comparison with the proposed pick operation is executed by lifting up tool-tip soon after the contact detection. This method does not cause the fracture of the contact interface between a sphere and a substrate. This has never succeeded even once out of 10 trials. This result also shows the most important factor to realize reliable pick is that to fracture the contact interface between the sphere and the substrate.

4.2 Precise Place Operation

In order to execute the proposed place operation (i.e. shearing-trajectory method), the contact position of the tool-tip on the surface of the sphere must be measured first. The contact position is expressed by angles of ϕ and θ. ϕ is the same in section 2. θ is rotational angle in a plane parallel to XY-plane. The angle of tool-tip trajectory is defined by ϕ'. For example, $\phi' = 0°$ means horizontal trajectory and $\phi' = 90°$ means vertical trajectory. Besides the difference of angles is defined by $\Delta\phi = \phi' - \phi$.

In order to prove the effectiveness of the proposed place operation, the relationship between difference of angles $\Delta\phi$ and mode as shown in section 2 were examined by judging which mode was caused by the operation when difference of angles $\Delta\phi$ varied from $-10°$ to $+30°$ by $10°$. The number of trials was over 10 in every difference of angles. Figure 13 shows SEM images during execution of shearing-trajectory method.

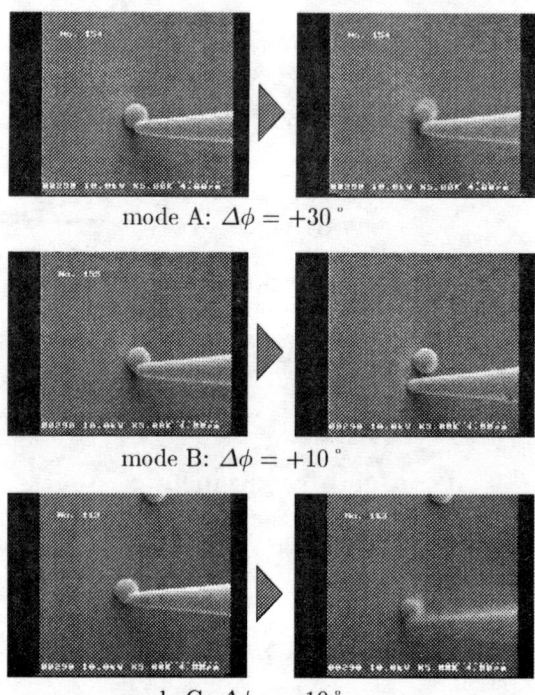

mode A: $\Delta\phi = +30°$

mode B: $\Delta\phi = +10°$

mode C: $\Delta\phi = -10°$

Figure 13: Execution of Shearing-Trajectory Method

The obtained result is shown in Figure 14. This data shows that place operation by shearing-trajectory method is the most successful in mode B in the case of $\Delta\phi = +10°$ and $\Delta\phi = +20°$, and that there is an optimal value of difference of angles in the middle of the range.

As for the mode B, precision in position was evaluated by measuring the positioning errors of the trials though the use of the SEM image-processing. As a result, the maximum positioning error of mode B is about $140nm$.

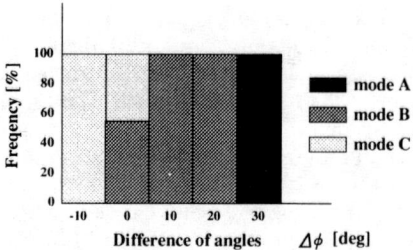

Figure 14: Relationship between difference of angles $\Delta\phi$ and mode

5 Conclusion

This paper proposes a new method of reliable pick and precise place operation for a micro object, based on micro physics under SEM.

Basic strategy is that pick and place operation must be completed simple trajectory-control of a tool-tip without visual-feedback, i.e. the proposed operation should be based on the unique disposition of a micro object. This strategy is derived from the knowledge that SEM observation in high magnification for a long time should be avoided for reliable and precise operation. This knowledge is based on the facts from the actual measurement (1)that adhesion on a micro object under SEM increases with time elapse after its contact because of EB irradiation and (2)that larger adhesion is generated at observation in higher magnification because EB current injection into an object becomes larger in proportion to the square of magnification at observation.

Dynamics for reliable and precise operation is analyzed. The dynamics system has a configuration of a needle-shaped tool, a sphere, and a substrate. The key point of this analysis is the introduction of rolling-resistance moment and adhesion which are unique factors in micro world. Regarding the rolling-resistance moment, the existence of maximum rolling-resistance moment is discussed. As the result of analysis, the phenomenon after executing operation has turned out to be classified into 2 or 3 modes. These classified modes depend upon the direction of loading by a tool and the ratio of adhesions on both sides (tool-tip side and substrate side).

Based on the discussion stated above, a new method of pick and place operation is proposed. As for reliable pick even in case that the substrate-side adhesion is larger than the other, offset-loading method is proposed. This method enables reliable pick due to the substrate-side fracture caused by loading to a sphere along with offset normal axis. Concerning precise place even in case that tool-side adhesion is larger than the other, shearing-trajectory method is proposed. This method enables precise place due to the tool-tip-side fracture caused by taking almost tangential trajectory at contact point and by letting tool-tip get slight reaction from the sphere.

Experimental system for the proposed operation is constructed and the proposed pick and place operation has turned out to be effective through experiments on the system. In these experiments, the object is a polystyrene sphere of $2.0 \mu m$ diameter, the substrate is a glass plate coated with Au, and the tool is a glass needle coated with Au. As the result of the experiments, offset-loading method has proved to enable pick operation with high reliability and shearing-trajectory method has turned out to enable place operation with $140 nm$ precision in position. This result shows the feasibility of highly precise arrangements to realize photonic crystals and suggests that manipulation under SEM would contribute to the realization of manufacturing micromachine and microelectronics device in the future.

Several problems, however, still have to be solved. The proposed operation techniques should be generalized to applied to many kinds of material, shape, and size. This theme would be the basis of micro automation. Besides maximum rolling-resistance moment must be actually measured and be quantitatively analyzed to establish more completed dynamics model.

Acknowledgements

The authors would like to thank Dr. Taketoshi Mori for insightful comments and valuable discussion, Mr. Yasushi Tomisawa for the measurement of adhesion data, Mr. Takeshi Kasaya for the development of the software library for SEM image-processing, and Mr. Hiroshi Morishita for the advice upon the experimental setup. This research was supported by the Grant-in-Aid for Scientific Research of the Ministry of Education, Science, Sports, and Culture, by the Grant-in-Aid for JSPS Research Fellows(No. 00097026), and by the Core Research for Evolutional Science and Technology of the Japan Science and Technology Corporation.

References

[1] T. Sato, T. Kameya, H. Miyazaki, and Y. Hatamura, "Hand-eye System in Nano Manipulation World", Proc. of 1995 IEEE International Conference on Robotics and Automation, Vol.1, pp.59-66, 1995.

[2] H. Miyazaki and T. Sato, "Mechanical Assembly of Three-Dimensional Microstructures from Fine Particles", Advanced Robotics, Vol.11, No.2, pp.169-185, 1997.

[3] R. S. Fearing, "Survey of Sticking Effects for Micro Parts Handling", The Proceedings of IEEE Micro Electro Mechanical Systems, pp. 212 - 217, January 29 - February 2, 1995, Amsterdam, the Netherlands.

[4] H. Miyazaki, Y. Tomisawa, K. Koyano, and T. Sato, "Adhesive forces acting on micro objects in manipulation under SEM", Proc. of SPIE Microrobotics and Microsystems Fabrication, Vol.3202, pp.197-208, 1997.

[5] T. Kasaya, H. Miyazaki, S. Saito, and T. Sato, "Micro Object Handling under SEM by Vision-based Automatic Control", Proc. SPIE Microrobotics and Micromanipulation, Vol. 3519, to be published, November, 1998.

[6] T. Tanikawa, T. Arai, T. Masuda, "Development of Micro Manipulation System with Two-Finger Micro Hand", Proc. IEEE/RSJ International Conference on Intelligent Robotics and Systems, pp. 850-855, Nov. 4-8, 1996.

[7] K. Koyano and T. Sato, "Micro Object Handling System with Concentrated Visual Fields and New Handling Skills", The Proc. of IEEE International Conference on Robotics and Automation, pp. 2541 - 2548, 1996.

[8] H. Miyazaki, Y. Tomisawa, S. Saito, and T. Sato, "Adhesive force acting on micro objects under scanning electron microscope", Proc. of Conferecne on Robotics and Mechatronics (CDROM), 1BIV1-7, June, 1998 (in Japanese).

[9] K. L. Johnson, K. Kendall and A. D. Roberts, "Surface Energy and the contact of elastic solids", Proc. R. Soc. Lond. A. 324, pp.301-313, 1971.

[10] D. Maugis, "Adhesion of Spheres: The JKR-DMT Transition Using a Dugdale Model", Journal of Colloid and Interface Science, Vol. 150, No. 1, pp.243-269, 1991.

[11] Kobayashi, "Fracture Mechanics", Kyouritu Press, 1997(in Japanese).

[12] D. H. Ballard, "Generalizing the Hough Transform to Detect Arbitary Shapes", Pattern Recognition, Vol.13, No.2, pp.111-122, 1981.

Micro Tri-axial Force Sensor for 3D Bio-Micromanipulation

Fumihito ARAI*, Tomohiko SUGIYAMA*, Toshio FUKUDA**,
Hitoshi IWATA***, and Kouichi ITOIGAWA***

* Department of Micro System Engineering, Nagoya University
Furo-cho, Chikusa-ku, Nagoya 464-8603, JAPAN, http://www.mein.nagoya-u.ac.jp
** Center for Cooperative Research in Advanced Sci. & Tech., Nagoya University
*** Tokai Rika Co., LTD., Oguchi, Niwa-gun, Aichi 480-0195, JAPAN

Abstract

It is important to manipulate a biological small object, such as a cell and embryo. Three-dimensional high speed micromanipulation is needed as a fundamental technology for biology and bio engineering application. In this paper, we focus on the contact type micromanipulation in the liquid. We designed and made a contact type micromanipulation system which has a 3DOF narrow range positioning system on a 3 DOF wide range positioning system. We developed a micro tri-axial force sensor which can be installed near the tip of the endeffector. Performance of this micro force sensor is presented.

1. Introduction

Recently, biology is progressed so much with the advancement of bio-technology. Bio-technology can be classified into gene engineering, cell engineering, and development engineering. In these research fields, the micro and nano-manipulation, mass production, repetitive processing, and high speed and high precision processing in the liquid are required. For the breakthrough in these fields, integration of the distributed research fields and system technologies is important. At this moment, conventional robotics has potentiality for the breakthrough in the bio-engineering. But, it is not enough. Main problem is caused by the size of the object. The size of the object is too small. So, the system must consists of fine mechanisms and have a fine motion control system. Thus, we have to take a new approach to improve the system performance.

Key technologies are manipulation technology, microsystem technology, visualization technology, human interface technology, automation technology, and so on. Especially, manipulation of an animal or plant cell, and embryo is important. Anatomic operations are frequently performed. Most of the cases, operators manage to manipulate the biological objects by the micromanipulators with the two-dimensional image from the optical microscope. However, operation of the micromanipulator in the three-dimensional micro/nano world is quite difficult. From these points of view, three-dimensional high speed micromanipulation is needed. In this paper, we focus on the contact type micromanipulation in the liquid. We designed and made a contact type micromanipulation system which has a 3DOF narrow range positioning system on a 3 DOF wide range positioning system. Moreover, we developed a micro tri-axial force sensor which can be installed near the tip of the endeffector.

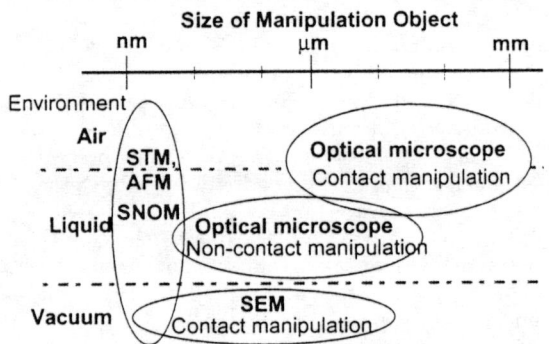

Fig. 1 Classification of micromanipulation based on environment

2. What is Necessary for Bio-Micromanipulation?

2.1 Target of contact type micromanipulation

Figure 1 shows the classification of micromanipulation based on the environment[1,10]. Micromanipulation is broadly classified into the following two types:
(a) contact type;
(b) non-contact type.
In the case of biological application, most of the objects are kept in the liquid, and observed by the optical microscope. If the object is smaller than 2 or 3 micrometer, it is difficult to manipulate by the contact

type manipulation under the microscope[1]. In such cases, non-contact type manipulation method, such as the optical tweezers[3,6] and dielectrophoresis[4-6], is suitable. Force produced by these non-contact type method is sub-nN order or less, and it is not strong enough for manipulation of much more bigger object. If the size of object is bigger than about 10 micron, contact type manipulation is suitable. In this paper, we focus on the contact type micromanipulation in the liquid. As an example of the contact tasks in the liquid, anatomical operations such as nuclear transplantation and electrophysiological inspection are considered.

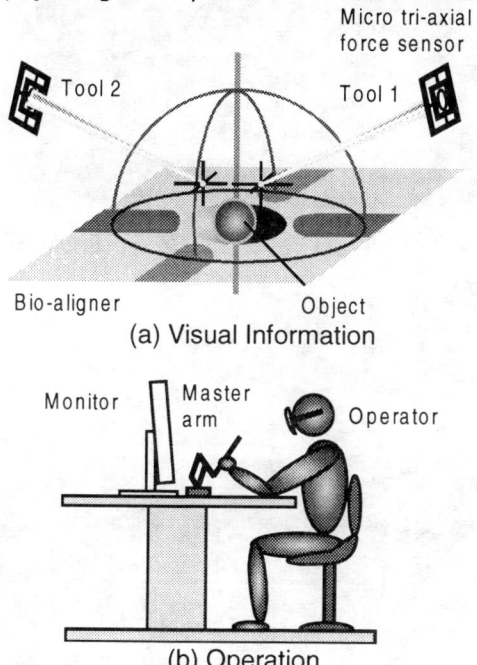

Fig. 2 Concept of the bio-micromanipulation system

2.2 Basic strategies to improve the operability

Most of the bio-micromanipulation tasks are performed in water solution. Thus, compared with the general manipulation tasks in the air, we have to consider not only the gravitational force but also the buoyancy, fluid force from flow, Brownian motion, interactive forces in the micro/nano world such as Van der Waals force and electrostatic force depending on the surface electrons. Moreover, size of each manipulation object is in micron order, it is quite difficult to recognize its configuration and to manipulate it freely. Automation of manipulation is quite difficult and skills are required because most of the objects are not rigid but fragile. For the contact type manipulation, it is important to control the internal force precisely. The operability of the contact type manipulation depends on the fixation of the object. From these point of view, we summarize the basic strategies to improve the operability of the bio-micromanipulation as follows (See details in ref. [2]).

(1) Utilization of the fixation device (Bio-aligner[2,7])
(2) Presentation of visual information
(3) Force measurement and presentation
(4) Realization of autonomous function

Based on these strategies, Fig. 2 shows the concept of the bio-micromanipulation system.

3. Three-Dimensional Micromanipulation System

3.1 Problem of the Position Accuracy

In case of manipulating a micro object, such as a cell(Ex. animal cell: 20 – 30 μm), manipulation area for one object is micron order. When we array the multiple objects in plane, the moving area of the micromanipulator is more than millimeter order. Moreover, if we change the endeffector and transport the objects, it will become more than centimeter order. Thus, micromanipulator must have moving mechanism more than centimeter order and position accuracy of nanometer order. The scale difference between the work distance and position accuracy is $10^5 - 10^7$. The micromanipulation system must achieve this requirements in the three-dimensional space.

Generally, most of the conventional micromanipulators have fine mobile resolution(from sub-micron to nano meter). However, absolute position accuracy for all mobile range is not guaranteed. As a method to extend the mobile range, elastic deformation or slide mechanism are considered. The elastic deformation must be less than about 10 % of the total length, so this method is not suitable for relatively long work distance. To realize long work distance of more than 20 or 30 μm, an inchworm mechanism or slide mechanism is used. In the case of inchworm mechanism, piezoelectric actuator(Ex. PZT) is frequently used. In this case, repeatability of one step is important to realize high positioning accuracy for all working area. In the case of slide mechanism, linear bearing with the steel balls is used to reduce the friction. However, we must increase the pushing load to avoid the backlash, and this will cause the deformation which will degrade the absolute accuracy. Moreover, thermal deformation is not negligible. From these reasons, it is difficult to improve the absolute position accuracy when we employ the slide mechanism. To improve the absolute position accuracy, we have to contrive the bearing mechanism, consider the heat transfer characteristics, and utilize the high precision position sensors.

3.2 Improvement of Position Accuracy

Absolute position accuracy in the three-dimensional space is quite important to perform micromanipulation tasks safely and smoothly. However, it is very difficult to guarantee the absolute position accuracy in sub-micron order for all working area, because of the alignment error, elastic and thermal deformation, and slide. These errors can be classified into the repeatable error and unrepeatable error. Basically, we should design the system so that these errors are as small as possible. In this point of view, mechanical design (rigidity, configuration, motion etc.), thermal design, material design, sensing system design, and environment condition are all important to reduce these errors. However, these errors are not perfectly evaded, and are essential to some extent. So, we should consider how to cope with these errors.

Repeatable errors can be estimated and should be compensated in advance. For example, when the endeffector of the micromanipulator is changed, the length and orientation of it may change. In such case, we have to calibrate the positioning system in advance to guarantee the absolute position accuracy. Calibration method against the miss alignment of the system components and tool exchange is important. It must be simple and accurate. If we can calibrate the system in the three-dimensional space easily, we can realize automation in high speed and safe manipulation.

For the unrepeatable errors, we should integrate the micromanipulator and sensing system. If the error changes rapidly, we need a real-time feedback compensation. If the error changes relatively slow compared with the speed of manipulation task execution, feedback compensation or online calibration is needed.

We can summarize the methods to improve the position accuracy of the micromanipulator as follows:
(1) calibration
   (kinematic parameters[1,10], vision system);
(2) visual feedback[8,9].

We have proposed the three-dimension calibration method with the micromanipulation system[1,10]. The accuracy of the calibration depends on the accuracy of the global sensor. We proposed to use the microscope as the image sensor which can measure the tip of the endeffector with high accuracy. The proposed method is simple and can calibrate the unknown kinematic parameters fast. The reference point(Ex: tip of the endeffector) is measured fast by simple image processing[10].

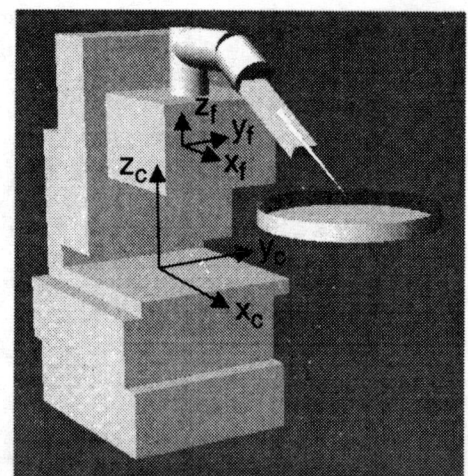

Fig. 3 3DOF Micromanipulator in simulator

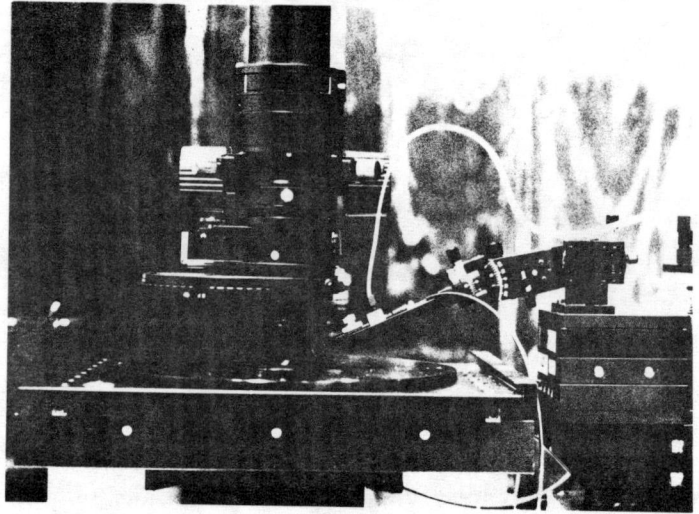

Fog. 4 Photo of 3DOF Micromanipulator

3.3 Configuration of micromanipulator

Main task of the micromanipulator is to move the endeffector tip to the goal position. The goal position is determined based on the state of the manipulation object. It is important for the system to know the global potion of the endeffector tip and object. To realize the simple calibration of the positioning system with high accuracy, configuration of the micromanipulator is important. So we designed the Cartesian type micromanipulator, which has two xyz translation stages. The system has high rigidity and kinematic calculation is simplified. The narrow range(micron order) positioning mechanism was set on the wide range(centimeter order) positioning mechanism. Both positioning mechanisms are controlled by the high precision position feedback system. The narrow range positioning mechanism has three degrees of freedom (xf, yf, zf), and the wide range positioning mechanism has three degrees of freedom (xc, yc, zc).

Figure 3 shows the simulator of this micromanipulator which is made by the computer graphics on the SGI. We made a prototype of this micromanipulator as shown in Fig. 4.

The narrow range positioning mechanism is actuated by the three PZT actuators. Position of each xf, yf, and zf axes is measured by the magnetic sensor. Mobile range of each xyz axes is 100 μm, and has absolute position accuracy of 40nm by the position feedback along each axis. The resonance frequency of the narrow range stage is about 300 Hz. The mobile range of this stage is narrow, but the position of this stage is controlled very fast.

The wide range positioning mechanism is actuated by the three stepping motors. Mobile range of each xc, yc, and zc axes is 25 mm. Step resolution of the stepping motor is 26nm(lead: 1mm, 0.0094 deg/pulse). Position of each axes is measured by the linear scale. The resolution of this linear scale is less than 26nm. This stage has wide range working area, but the response speed of the stage is not so fast compared with the narrow range stage.

4. Micro Tri-axial Force Sensor for Three-Dimensional Micromanipulation

4.1 Why we need a micro force sensor?

In the case of anatomical operation, there is force interaction between the object and endeffector. At present, operation of the micromanipulator is quite difficult, since the operator controls its tip position based on the image from the microscope, and force information is not available or is limited in only one direction[11].

A multi-axial force sensor is demanded to improve operability and to protect the tool tip and the fragile object. However, there is no practical micro force sensor which can measure multi-axial force information. We can find many multi-axial force sensors, however, most of the conventional force sensors are big and heavy[12,13] for micromanipulation. Thus, the resonant frequency of the force sensor tends to be decreased if we keep the accuracy of the force measurement. In such case, the mechanical vibration of the finger tip will degrade the operation or we must slow down the motion control of the micromanipulator. We have developed the variable stiffness control system, but the mechanism was complicated[2].

For the three dimensional high speed micromanipulation, it is desirable to install a miniaturized multi-axial force sensor near the tool tip as shown in Fig. 2. The sensor must be accurate, stiff, small, and light weighted. If the size of the sensor is miniaturized, the mass of the sensor reduces dramatically and the resonant frequency of the elastic body increases. From the scaling analysis, it can be explained as follows.

Sensor mass: $m \propto L^3$

Resonant frequency: $\omega \propto L^{-1}$

L : characteristic length

So, we miniaturized the multi-axial force sensor to reduce the mass and to increase the resonant frequency while keeping the accuracy of force measurement.

4.2 Design and performance

Figure 5 shows the force sensor chips which are made by the silicon micromachining. The chip size is designed small (4.5mm×5.0mm×0.525mm). Figure 6 shows the gauge allocation on the chip (Type A). The center mass of both types is suspended by four cantilevers. Each cantilevers has two piezoresistive strain gauges. Gauge resistance is 1050 Ω each. The force sensing principle is based on the piezoresistivity effect, and it has good linearity[14]. The piezoresistive force sensor is suitable for real time force feedback control.

The cantilever structure is made by the anisotropic etching of <100> silicon. Cantilever size is 200μm×500μm×50μm and its Young's modulus is 130 GPa. The spring coefficient of the sensor is estimated as 1600 N/m. If we install a glass tool (weight: 67 mg), the resonance frequency of the sensor tip is estimated as 730 Hz. It is bigger than the resonance frequency of the narrow range xyz stage(300 Hz), so we will be able to operate the micromanipulator fast.

We made four pairs of bridge circuits as shown in Fig. 7. Here, V1, V2, V3, and V4 are output voltage in each bridge circuits, respectively. We calculate Vx, Vy, and Vz as follows.

$$Vx = V1 - V3 \quad (1)$$
$$Vy = V4 - V2 \quad (2)$$
$$Vz = V1 + V2 + V3 + V4 \quad (3)$$

The output voltage vector V is written as follows by the

applied moments Mx and My, force Fz, and compliant matrix C.

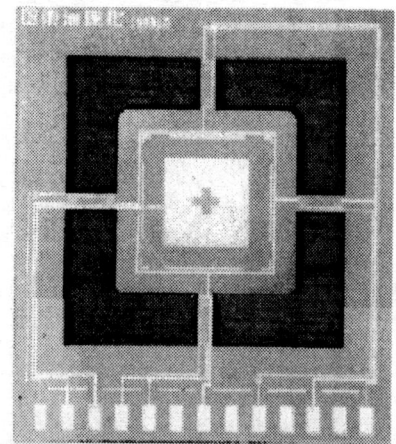

(a) Type A: Without a through hole

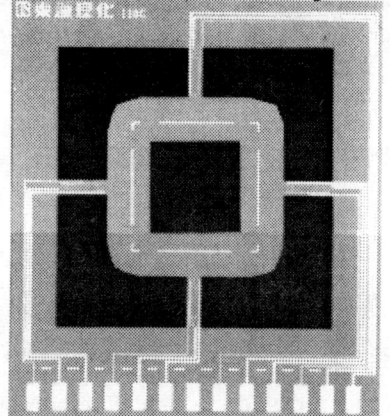

(b) Type B : With a through hole at the center mass
Fig. 5 Micro tri-axial force sensor chip

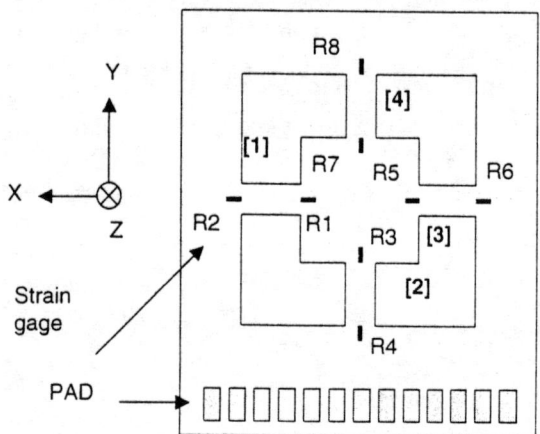

Fig. 6 Piezoresistive strain gauge allocation

$$V = CF \quad (4)$$

where

$$V = [V_x, V_y, V_z]^T, \quad F = [M_y, M_x, F_z]^T$$

$$C = \begin{bmatrix} C_{xx} & C_{xy} & C_{xz} \\ C_{yx} & C_{yy} & C_{yz} \\ C_{zx} & C_{zy} & C_{zz} \end{bmatrix}$$

Here, the compliant matrix C can be identified by a simple calibration experiment by measuring different kinds of loads. Then, the force on the tip is calculated by the following equation.

$$Ft = MC^{-1}V \quad (5)$$

where

$$Ft = [F_x, F_y, F_z]^T, \quad M = diag\left[\frac{1}{l}, \frac{1}{l}, 1\right]$$

l : length of tool

We calibrated the compliant matrix by the experiment as follows.

$$C^{-1} = \begin{bmatrix} 1.240 & -0.058 & 0.543 \\ 0.120 & 0.771 & 0.660 \\ 0.038 & 0.001 & 0.621 \end{bmatrix}, \quad l = 10\,mm$$

Figure 8 (a), (b), and (c) shows the sensor outputs when the force components Fx, Fy, and Fz are applied at the tool tip, respectively. Three axial forces are measured. However, coupling in the Fz force measurement is relatively big. We plan to improve the calibration method and sensor performance in future.

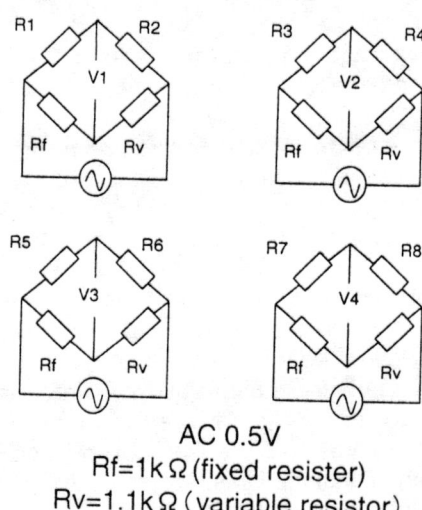

AC 0.5V
Rf=1kΩ (fixed resister)
Rv=1.1kΩ (variable resistor)

Fig. 7 Bridge circuits to measure the tri-axial forces

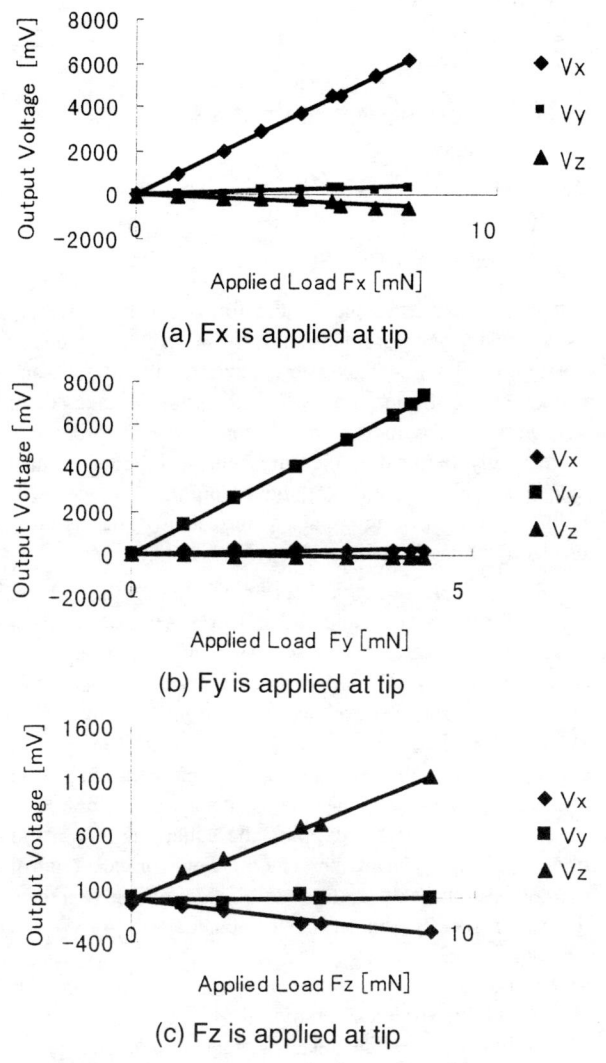

Fig. 8 Output voltage of force sensor

5. Conclusions

Here we focused on the contact type bio-micromanipulation in the liquid. We introduced basic strategies to improve the bio-micromanipulation. For three-dimensional high speed micromanipulation, we developed a Cartesian type micromanipulation system which has the narrow range positioning system on the wide range positioning system.

Moreover, we developed two types of micro tri-axial force sensor which can be installed near the endeffector. The basic performance of this micro force sensor is presented. It has fine resolution for bio-micromanipulation. The proposed sensor can be modified to realize a six-axis force sensor[12].

In future, we will integrate the basic components to realize safe and dexterous 3D micromanipulation.

Acknowledgement

This research work was supported in part by a grant from NEDO (New Energy and Industrial Technology Development Organization).

References

[1] F. Arai and T. Fukuda: " Micromanipulation – Three-Dimensional High Speed Micromanipulation –", Tutorial notes, Tutorial on Micro Mechatronics and Micro Robotics, IEEE ICRA'98, 1998
[2] F. Arai, et al.: "Bio-Micro-Manipulation (New Direction for Operation Improvement)", Proc. of IEEE/RSJ Int. Conf. Intelligent Robots and Systems (IROS), VOL. 3, 1997, pp.1300-1305.
[3] A. Ashkin: " Acceleration and Trapping of Particles by Radiation Pressure", Physical Review Letters, 24 (4), p.156-15
[4] H. A. Pohl: " Dielectrophoresis", Cambridge University Press, Cambridge, 1978
[5] T. Muller, et al.: " Trapping of Micrometre and Sub-micrometre Particles by High-frequency Electric Fields and Hydrodynamic Forces", J. Phys. D: Appl. Phys. 29, p. 340-349, 1996
[6] K. Morishima, F. Arai, T. Fukuda, and K. Yoshikawa: " Screening of Single Escherichia coli by Electric Field and Laser Tweezers", IEEE Proc. Int'l Symp. on Micro Machine and Human Science(MHS), 155-160, 1997
[7] F. Arai, T. kasugai, and T. Fukuda: " 3D Positioning and Orientation Control Method of Micro Object by Dielectrophoresis", Proc. MHS'98, 1998, p.149-154
[8] I. Pappas, A. Codourey: " Visual Control of a Microrobot Operating under a Microscope", Proc. IROS, VOL. 2, pp. 993-1000, 1996.
[9] B. Vikramaditya and B.J. Nelson: " Visually Guided Microassembly Using Optical Microscopes and Active Vision Techniques ", Proc. ICRA'97, Vol. 4, p.3172-3177.
[10] F. Arai, et al.: " 3D Micromanipulation System under Microscope ", IEEE Proc. MHS'98, 1998, p.127-134
[11] T. Fukuda, K. Tanie, and T. Mitsuoka, A Study on Control of a Micromanipulator", Proc. of Micro Robots and Teleoperators Workshop, Hyannis, MA, Nov. 1987
[12] M. Kaneko and T. Nishihara: "Basic Study of Six-Axis Force Sensor Design Based on Combination Theory", J. Robotics Society Japan, Vol. 11, No. 8, 1993, pp.1261-1271(in Japanese)
[13] K. Nagai, Y. Eto, A. Asai, and M. Yazaki: " Development of a Three-fingered Robotic Hand-Wrist ", Proc. IROS, VOL. 1, 1998, pp.476-481.
[14] F. Arai, et al.: "Integrated Microendeffector for Micromanipulation", IEEE/ASME Trans. Mechatronics, VOL. 3, No.1, March 1998, p.17-23.

Modelling of a Piezohydraulic Actuator for Control of a Parallel Micromanipulator

Quan Zhou*, Pasi Kallio* and Heikki N. Koivo**

*Tampere University of Technology, Automation and Control Institute, P.O. Box 692, 33101 Tampere, Finland;
Email: quan@ad.tut.fi, kallio@ad.tut.fi
**Helsinki University of Technology, Control Engineering Laboratory, P.O. Box 3000, 02015TKK, Finland;
Email: heikki.koivo@hut.fi

ABSTRACT

Piezoelectric actuators have such disadvantageous properties as hysteresis nonlinearity and drift. This paper discusses a hybrid model to describe those properties. The model consists of a dynamic linear part and a static nonlinear part. The dynamic part uses two poles to describe the dynamic response and the drift property. The static nonlinear part modifies the parameters of the linear dynamic model to adapt the model to hysteresis nonlinearity and gain nonlinearity. The model is generated for a piezohydraulic actuator based on a multi-stage step signal. Simulation and experimental results show that the model is able to predict the drift and the nonlinear properties of the actuator.

1 INTRODUCTION

Miniaturization has been one of the most important technological trends in recent decades. Microelectronics has paved the way. The successful fabrication and operation of microactuators and micromechanical devices provides an opportunity to produce microminiature machines and mechanical systems. Such systems are called microsystems in Europe, microelectromechanical systems (MEMS) in the United States and micromachines in Japan.

Research on microsystems began to gather momentum in 1980's and currently it is one of the most prominent research areas all over the world. The study of microrobots and micromanipulators is an essential part of microsystems. Structures vary from etched silicon implementations to miniature mechanisms. Nevertheless, the objective of all micromanipulators is to manipulate micrometer sized objects. The size of the micromanipulator is not as important in such applications as biotechnological operations, the assembly of microelectro-mechanical systems, and the testing of microelectronics circuits [1]. Small-scale manipulators are needed in catheter types of applications, such as medical diagnostics and therapy, for example.

Micromanipulator is a precision positioning device. It requires actuators having displacement resolution of nanometers, high stiffness and fast frequency response. These requirements can be fulfilled by using piezoelectric actuators. Therefore, piezo actuators have been widely used in many micromanipulators and other precision positioning devices. However, piezo actuators have such properties as hysteresis and drift, which make them nonlinear. As the micromanipulation applications become increasingly demanding, requirements for better control design are also raised. A good manipulator model will help the control design. Such a model, in turn, requires proper dynamic models for the piezo actuators.

The properties of piezoceramic actuators have been studied for many years and different approaches have been used. IEEE ultrasonics, Ferroelectrics and Frequency Control Society formulated linearized constitutive relations describing piezoelectric continua, published in 1966 and revised in 1987 [2]. Leigh and Zimmerman discussed an implicit algorithm for predicting the hysteresis behaviour of piezoceramic actuators, employing the trapezoidal rule for stepping the equations forward in time [3]. Jung and Kim presented a feed-forward control method with three different deterministic models to reduce the hysteresis effect in piezoceramic actuators [4]. Ge and Jouaneh modelled the hysteresis nonlinearity of a piezoceramic actuator using numerical Preisach models [5], [6]. Goldfarb and Celanovic developed a generalized Maxwell model of a piezoelectric stack actuator where the static hysteresis was identified as energy storage coupled to rate-independent dissipation and was represented by a generalized elasto-slip model [7]. The emphasis has been on hysteresis nonlinearity: models are often static, and drift modelling is normally ignored.

In this paper, we use a hybrid approach to model the piezohydraulic actuators of a micromanipulator developed at Tampere University of Technology and Helsinki University of Technology. The dynamic part of the model describes the dynamic behaviour of the actuator including the drift property. In the mean while, the hysteresis and gain nonlinearities are modelled statically using neural networks. In the next section, the actuator and the micromanipulator under study will be briefly discussed. Section 3 presents the structure of the hybrid nonlinear model of the actuators. The simulation and experimental results are presented and discussed in section 4. Section 5 concludes the paper.

2 THE MICROMANIPULATOR

A micromanipulator is a device that facilitates the remote handling of microscopic objects under computer-assisted human control. The operator obtains visual information about the end-effector and micro objects using a microscope and a CCD camera, as shown in Fig. 1. The micromanipulator can be controlled either by using a joystick or a PC keyboard. Automatic operations, such as automatic injections, can be activated using the keyboard.

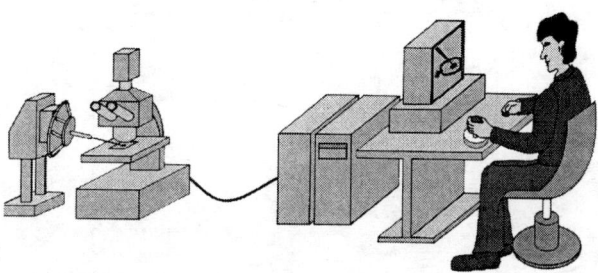

Figure 1: Concept of micromanipulation.

2.1 Actuation system

The actuation system of the micromanipulator consists of a piezoelectric actuator, a small tank and a metallic bellows, as illustrated in Fig. 2 [8]. The piezo actuator is placed in the tank filled with hydraulic oil. When a voltage is applied to the piezo actuator, it deforms. When the actuator buckles, oil flows from the tank to the bellows which elongates, and vice versa: when the actuator gets straightened, oil flows from the bellows to the tank. Since the effective area of the bellows is smaller than that of the actuator, the displacement is magnified. The results of displacement experiments have shown that the movement range of the actuator is about ± 250 μm.

The piezoelectric actuator (RAINBOW™) is structurally similar to unimorph type of elements. The RAINBOW™ actuator is a single element structure consisting of a PZT side and a metallic side [9]. When a voltage is applied opposite to the poling field, the wafer buckles and when the voltage is parallel to the poling field, the wafer straightens.

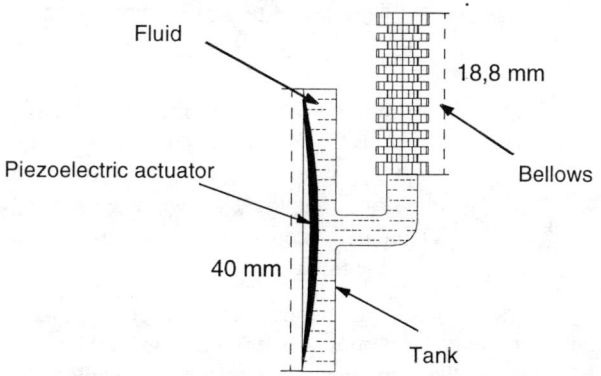

Figure 2: Overview of the actuation system.

The bellows is a spring type of passive component where the force required to deform the bellows is directly proportional to the displacement. The bellows can be considered as a linear system.

1. End-effector 2. Mobile platform 3. Bellows
4. Magnet mechanism 5. Piezo actuator

Figure 3: Overview and photograph of the micromanipulator.

2.2 Manipulator structure

The manipulator has a Stewart platform [10] type parallel structure consisting of three identical piezohydraulic actuation systems as described in [11]. The actuators are connected by a mobile platform resulting in a tripod-like parallel configuration, Fig. 3. The mobile platform is fixed to each bellows using a venting screw and a small nut. The end-effector (needle) is mounted on the platform.

This manipulator differs from the conventional Stewart platform type of structures, since the bending of the bellows facilitates a joint-free structure. Using the bending character of the bellows instead of spherical joints simplifies the structure and the manufacturing process, since especially the manufacture of miniaturised spherical joints is difficult. By changing the lengths of the bellows the orientation of the mobile platform, and thus the position of the end-effector can be controlled.

3 MODELLING OF THE PIEZOHYDRAULIC ACTUATION SYSTEM

In the paper, we will develop a dynamic model of the piezohydraulic actuator for its z-axis displacement. Since the bellows and the hydraulic system can be considered as a linear amplifier, the specific properties of the piezohydraulic actuator inherit mainly from the properties of the piezoelectric actuator – hysteresis and drift, as shown in Fig. 4 and Fig. 5, respectively. From Fig. 4, it can be noticed that the gain of the actuator is nonlinear. Due to the hysteresis and drift, a linear dynamic model is not sufficient to describe the system behaviours. Therefore, the model parameters have to be changed to

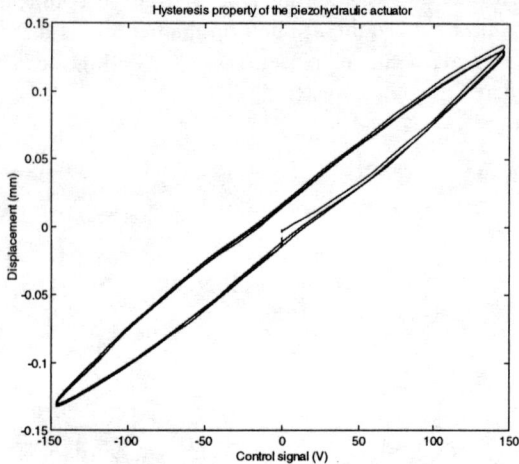

Figure 4: Hysteresis and gain nonlinearity properties of the piezohydraulic actuator.

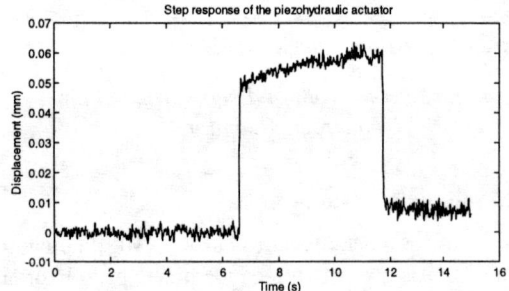

Figure 5: Step response of the piezohydraulic actuator, the drift being obvious.

cope with the nonlinear features. This paper proposes a model structure where parameters of a properly formulated linear dynamic model are scheduled by neural networks.

3.1 The dynamic part of the actuator model

The step response shown in the Fig. 5 can be described by a linear model containing two time constants, one describing the fast behaviour and the other describing the slow behaviour — the drift — of the system. As shown in Fig. 6, the model consists of two first order transfer functions having time constants T_1 and T_2, and respectively two gains k_1 and k_2. The gains determine the contribution of the two parts to the system output. The system can be described by a transfer function having two poles and one zero:

$$G(s) = \frac{(k_1 a + k_2 b)s + ab(k_1 + k_2)}{(s+a)(s+b)} \quad (1)$$

where $a = 1/T_1$ and $b = 1/T_2$.

A simulated output of the model is shown in Fig. 7. However, the linear transfer function with fixed parameters is only locally valid due to the nonlinearities of the system.

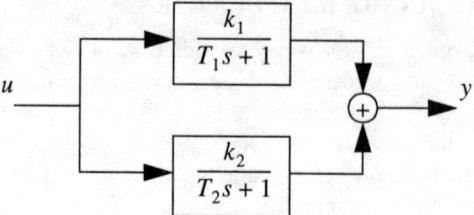

Figure 6: Structure of the dynamic model.

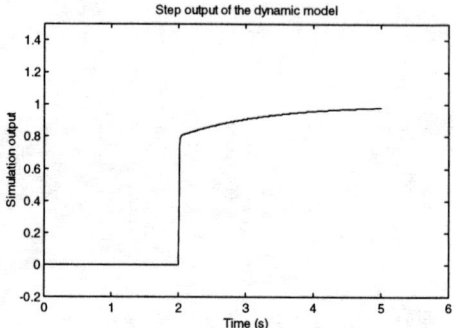

Figure 7: Simulated output of the dynamic model.

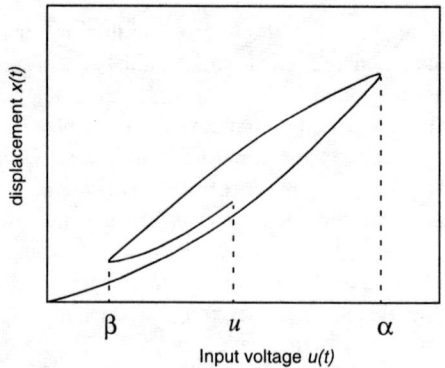

Figure 8: Hysteresis loops.

3.2 The Static nonlinearity part of the actuator model

The main responsibility of the static nonlinearity part of the hybrid model is to compensate for the hysteresis nonlinearity of the actuator, In the mean while, it also takes into account the gain nonlinearity. One of the classical models for hysteresis is the Preisach model [12]:

$$x(t) = \iint_{\alpha \geq \beta} \mu(\alpha, \beta) \gamma_{\alpha\beta}[u(t)] d\alpha d\beta \quad (2)$$

where $x(t)$ is the output response of a piezoceramic actuator; $\mu(\alpha, \beta)$ is a weighing function in Preisach model; α and β correspond to up and down switching values of the input, as shown in Fig. 8; and $\gamma_{\alpha\beta}[u(t)]$ is a binary hysteresis operator whose value is determined by the input operation.

The classical Preisach model can represent hysteresis nonlinearity which satisfies two properties: the wiping-out property and the congruency property. The wiping-out property means that the hysteresis curve does not depend on how it was approached [5]. The congruency property refers that the change in displacement should be the same for the same control input change. This property is not satisfied in e.g. piezoceramic actuators. In such a case, generalized Preisach model can be used [6].

If we regard the Preisach model as a nonlinear function, it could be generalized as:

$$x(t) = f(\alpha, \beta, u(t)) \tag{3}$$

where α and β correspond to up and down switching values of the input, $u(t)$ is the input signal of the actuator.

Thus, nonlinear mappings such as feed forward neural networks can be used to describe the hysteresis nonlinearity. Moreover, the dynamic model is only locally valid since it depends on the input voltage and history of the inputs. In Fig. 9, we can observe changes in the step responses when the actuator was driven by a multi-stage step signal. Therefore, the parameters of the dynamic model should be adapted to the hysteresis and gain nonlinearities.

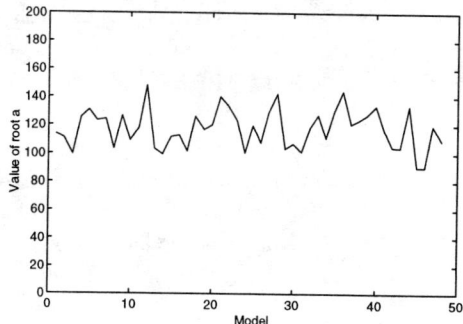

Figure 10: Variation of the parameter a in different models.

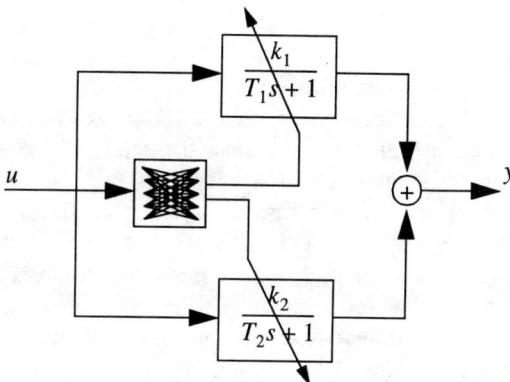

Figure 11: Principle of the hybrid model structure.

The parameter a is kept constant, since the response time of the piezohydraulic actuator remains unchanged despite the hysteresis nonlinearity. This has been proven in experiments and is demonstrated in Fig. 10 which shows the values of a for 50 different models. Those models are determined from 50 step responses covering the whole workspace of the actuator. The parameters b, k_1 and k_2 are functions of up and down switching values α, β, and input signal $u(t)$. The functions are nonlinear and implemented using neural networks. Naturally, other methods can also be used. The hybrid model structure is shown in Fig. 11.

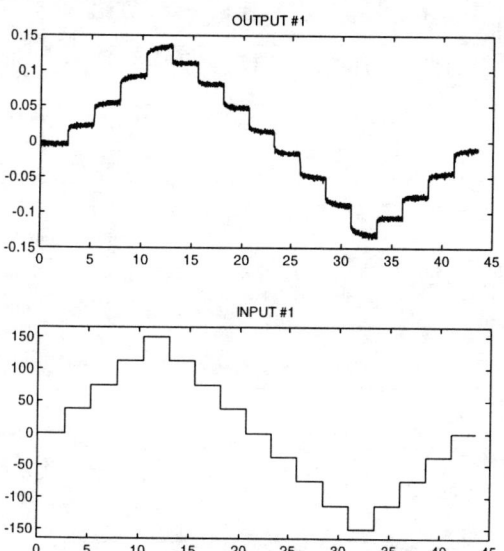

Figure 9: The system response of multi-stage step signal.

3.3 The hybrid model structure

The dynamic model (1) can be modified to take into account the system nonlinearities by replacing the parameters with functions:

$$G(s) = \frac{\lambda_1(\alpha, \beta, u(t))s + \lambda_2(\alpha, \beta, u(t))}{(s+a)(s+b(\alpha, \beta, u(t)))} \tag{4}$$

where
$$\lambda_1 = k_1(\alpha, \beta, u(t))a + k_2(\alpha, \beta, u(t))b(\alpha, \beta, u(t))$$
and
$$\lambda_2 = ab(\alpha, \beta, u(t))[k_1(\alpha, \beta, u(t)) + k_2(\alpha, \beta, u(t))]$$

4 SIMULATION AND EXPERIMENTAL RESULTS

In general, the system response can be divided into multiple windows to determine how parameters of the linear dynamic models change. Determination of the dynamic model for each window results in a group of models $G_i(s)$ and groups of model parameters $(a_i, b_i, k_{1i}, k_{2i})$.

The system response for the multi-stage step signal (Fig. 9) can be divided into multiple windows such that each window contains one step signal. 17 windows are obtained for one signal period. The actual test signal has 3 periods resulting in totally 49 windows. An ARX model with one zero and two poles is identified for each step response. From the parameters a_i, b_i, k_{1i}, k_{2i} an estimation for the fast pole a and nonlinear functions ($k_1(\ )$, $k_2(\ )$

and $b(\ )$) approximating the other parameters are determined. The a parameter is relatively constant for all ARX models and has a mean value of 117.4 and a standard deviation of 13.9, as shown in Fig. 10. Thus, the mean value can simply be chosen to represent a.

The gains k_{1i} are obtained by computing the amplitude average of each step response between time instants T_1 and $2T_1$. The gains k_{2i} are determined based on the DC gain of the ARX models and the gain k_{1i}:

$$k_{2i} = dcgain_i - k_{1i} \qquad (5)$$

After the model parameters for each window have been computed, they are used as the training data for neural networks. Three two-layer feedforward neural networks are used for computing the functions $k_1(\ )$, $k_2(\ )$ and $b(\ )$. Input variables for the networks are α, β and $u(t)$. The resulting functions $k_1(\ )$, $k_2(\ )$ and $b(\ )$ generated by the neural networks are shown as functions of input signal $u(t)$ in Fig. 12. Since the step size is rather large, the generated nonlinear mappings are not very smooth but as can be seen basically symmetric. The areas (around $u = -100\ \text{V}$ and $u = 100\ \text{V}$), where the parameters change rapidly, correspond to the turning points in the hysteresis curves. The initial parts of the mappings are different from the later parts, which also coincides the starting part of the hysteresis curve shown in Fig. 4.

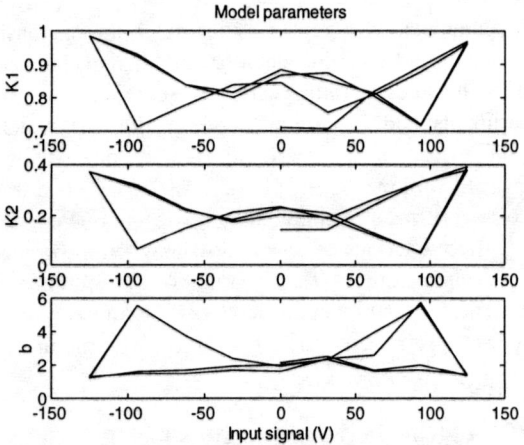

Figure 12: Nonlinear mapping for k_1, k_2 and b.

As shown in Fig. 13, the simulated output of the model matches the test signal at each step. Fig. 14 compares simulated and experimental results for two verification signals. For the large multi-stage step verification signal, the difference between the simulated output and the experimental result is distinguishable in the first up-down part of the signals. However, the error decreases as the time goes. For the ramp verification signal, the difference is almost indistinguishable. When the results are plotted as a hysteresis curve in Fig. 15, the small difference between the experimental results and simulated outputs can be observed. Overall, the model

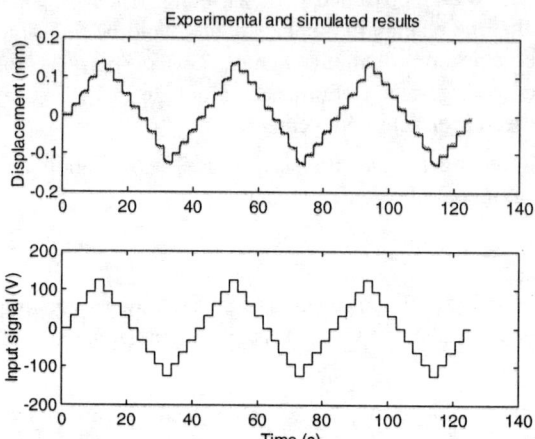

Figure 13: Model matching: the measurement signal (gray) and the model output (black).

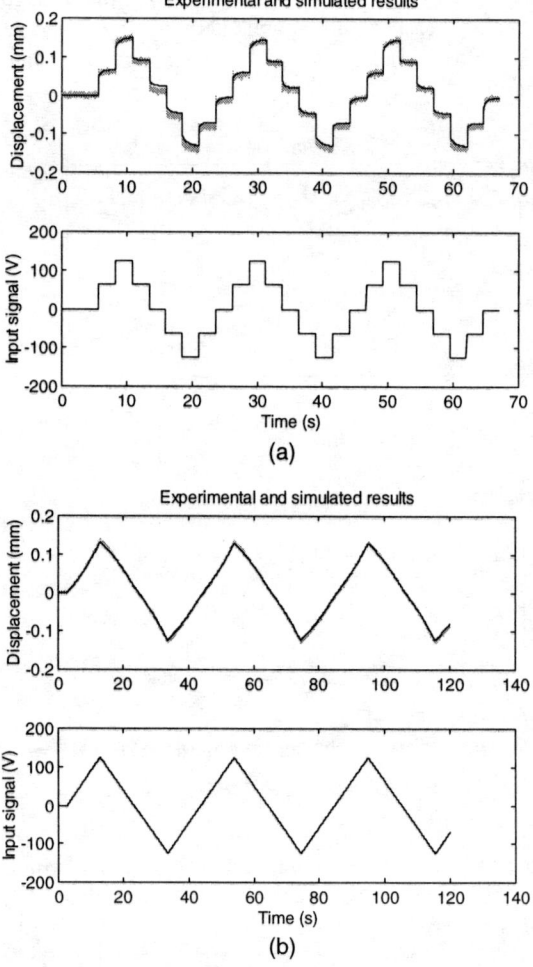

Figure 14: Verification of the hybrid model: (a) simulated output (black) and experimental measurement (gray) of a multi-stage step signal; (b) simulated output (black) and experimental measurement (gray) of a ramp signal.

built on a small number of discrete windows describes rather accurately the dynamics, the drift and the nonlinear properties of the actuator.

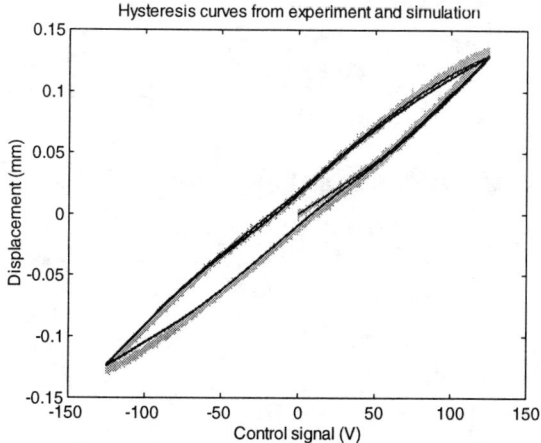

Figure 15: Verification of the model: the hysteresis curves from the simulation is close to the experimental ones.

The model can help to linearize the piezohydraulic actuator using a model-based control strategy, as shown in Fig. 16. The simulated result is shown in Fig. 17. Even though only a simple PI controller was used, the simulated result is quite linear. This encourages us to use model based controllers in a control scheme presented in [13].

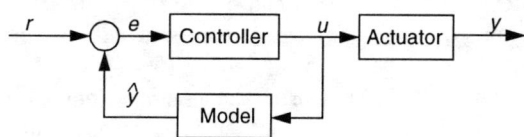

Figure 16: Model-based controller for linearizing the piezohydraulic actuator.

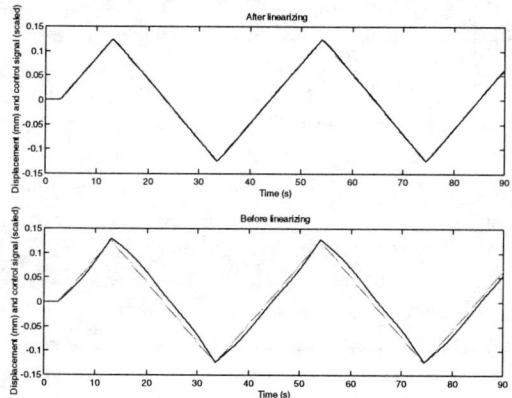

Figure 17: Simulation results with and without model-based controller. Gray lines are scaled control signal and the black lines are simulated actuator output.

5 CONCLUSIONS

A hybrid nonlinear dynamic model for a piezohydraulic actuator has been presented in this paper. The model structure can conceptually be separated in two parts: a linear dynamic part to model the dynamics and the drift property of the actuator and a static nonlinear part to adapt the dynamic model to the hysteresis and gain nonlinearities. The hybrid model was generated using a multi-stage step signal. The resulted model was verified for another multi-stage step signal and a ramp signal. The results have shown that the nonlinear dynamic model is capable of compensating the hysteresis and gain nonlinearities of the actuator as well as the drift property.

The hybrid model presented here has been tested on the major loop of the hysteresis. With its flexible nonlinear structure, the model can be extended to deal with minor loops and more complex verification signals, by training the neural networks with more data and using more complex patterns than the multi-stage step signal as perturbation signal.

ACKNOWLEDGMENT

The authors wish to thank the Technology Development Centre of Finland (TEKES) and the Academy of Finland for funding the project.

REFERENCES

[1] Kallio, P. and Koivo, H. N., "Microtelemanipulation: a Survey of the Application Areas", Proceedings of the International Conference on Recent Advances in Mechatronics, ICRAM'95, Istanbul, Turkey, August 1995, 365 - 372.

[2] Standards Committee of the IEEE Ultrasonics, Ferroelectrics, and Frequency Control Society, "An American National Stand: IEEE Standard on Piezoelectricity", The Institute of Electrical and Electronic engineers, ANSI/IEEE Std. 176-1987, New York, 1987.

[3] Leigh, T. and Zimmerman, D., "An Implicit Method for the Nonlinear Modelling and Simulation of Piezoceramic Actuators displaying Hysteresis", Smart Structures and Materials, AD-Vol. 24/AMD-Vol. 123, ASME, 1991, pp. 57-63.

[4] Jung, S. and Kin, S., "Improvement of Scanning Accuracy of PZT Piezoelectric Actuators by Feed-forward Model-reference Control", Precision Engineering 16, 1994, pp. 49-55.

[5] Ge, P. and Jouaneh, M., "Modeling Hysteresis in Piezoceramic Actuators", Precision engineering 17, 1995, pp. 211-221.

[6] Ge, P. and Jouaneh, M., "Generalized Preisach Model for Hysteresis nonlinearity of Piezoceramic Actuators", Precision Engineering 20, 1997, pp. 99-111.

[7] Goldfarb, M. and Celanovic, N., "Modeling Piezoelectric Stack Actuators for Control of Micromanipulation", IEEE Control Systems, VOL. 17, NO. 3, June 1997, pp. 69-79.

[8] Kallio, P., Lind, M., Kojola, H., Zhou, Q., Koivo, H., "An Actuation System for Parallel Link Micromanipulators", Proceedings of the Intelligent Robots and Systems, IROS'96, Osaka, Japan, November 1996, 856 - 862.

[9] Haertling, G., "Rainbow Ceramics - a New Type of Ultra-high-displacement Actuator". American Ceramic Society Bulletin, Vol. 73, No. 1, January 1994, 93 - 96.

[10] Stewart, D., "A Platform with 6 Degrees of Freedom". Proceedings of the Institution of Mechanical Engineers. Vol. 180, Part 1, No. 15, 1965, pp. 371 - 386.

[11] Kallio, P., Lind, M., Zhou, Q., Koivo, H. "A 3 DOF Piezohydraulic Parallel Micromanipulator". Proceedings of the IEEE ICRA'98, Leuven, Belgium, May 1998.

[12] Mayergoyz, I., "Mathematical Models of Hysteresis", New York: SPringer-Verlag, 1991.

[13] Kallio, P., Zhou, Q., Lind, M. and Koivo, H. N., "Position Control of A 3 DOF Piezohydraulic Parallel Micromanipulator". Proceeding of IEEE/RSJ International Conference on Intelligent Robots and Systems, IROS'98, Victoria Conference Center, Victoria B. C., Canada, October 1998, 770 - 775.

Identification of Assembly Process States Using Polyhedral Convex Cones

H. Mosemann, A. Raue, and F. Wahl
Institute for Robotics and Process Control
Technical University of Braunschweig
Hamburger Str. 267, 38114 Braunschweig (Germany)
email: H.Mosemann@tu-bs.de

Abstract

The execution of automatically generated assembly plans by robots is one of the key technologies of modern and flexible manufacturing. During the execution of assembly sequences the robot comes into contact with the environment. Since there are positional and geometrical uncertainties from object representation, robot motion, and sensor information, compliance is used typically to prevent excessive contact forces. The contact forces and the resulting torques provide information about the contact geometry which is used to guide the assembly operation. In this paper we present a new and fast algorithm to identify assembly process states considering static friction under uniform gravity. This identification of assembly process states enables a robot to select and modify its motion strategies adequately according to the state. We give a symbolic representation of contact states and analyze the static properties of assemblies at each contact state by using the theory of polyhedral convex cones.

1 Introduction

The increasing expectation of assembly processes to be automated leads to a strong desire for high level assembly planning systems. These systems accept CAD descriptions of assembly components and high level assembly specifications as input and generate as well as evaluate assembly sequences automatically. One basic idea in most of the work on assembly planning is the *assembly by disassembly strategy* in which assembly sequences are generated by starting with the complete mechanical product and decomposing it by disassembly operations. The concepts of assembly planning are described for example in [9], [14]. We implemented the **H**igh **L**evel **A**ssembly **P**lanning system ^{High}LAP [12], [13] which takes CAD descriptions of assembly components and high level assembly specifications using symbolic spatial relationships as input. The output of ^{High}LAP are sequence plans which specify the order in which parts are to be assembled to form the desired product and provide the trajectories to bring the parts together. Here, in each step of an assembly sequence only one active and one passive subassembly are joined. This constraint results in a two-handed assembly sequence plan.

When the robot performs the assembly sequences in real-world it comes into contact with the environment during the execution of the assembly sequences. Compliance is used typically to prevent excessive contact forces, due to positional and geometrical uncertainties from object representation, robot motion, and sensory information. On the other hand contact forces and the resulting torques provide information about the contact geometry and can be used to guide the assembly operation. In this paper we present a new algorithm to synthesize and recognize contact states between subassemblies considering static friction. The sequence plans generated by ^{High}LAP are the basis for robot assembly. But for the execution of the sequences in real-world the synthesis of the contact states between subassemblies is indispensable.

We are using the theory of polyhedral convex cones for the recognition of the contact states of the assembly sequence during execution. Therefore, force-torque sensor information acquired in the assembly process is interpreted using state classifiers which are formulated by polyhedral convex cones. The classifiers map the sensor data to a contact state. Subsequently a state transition to the goal of the assembly process is computed.

This paper is based on the following assumptions:

- The objects are rigid polyhedra with given CAD descriptions.

- The position and orientation of the robot end-effector can be obtained from the robot internal sensors.
- The force torque sensor is calibrated and measures only the contact forces and torques.
- We consider static friction only.

2 Previous Work

In this paper we present an approach to the automatic recognition and synthesis of contact states of subassemblies for force guided assembly. There are several bodies of work relevant to this goal. The first deals with fine-motion strategies for robots; [11] describes a formal approach to the synthesis of compliant-motion strategies from geometric descriptions of assembly operations and explicit estimates of errors in sensing and control. Erdmann [3] extended this approach by using back-projections. Laugier [10] gives a planning strategy by reasoning on an explicit representation of the contact space.

Hirai [7, 8] use the theory of polyhedral convex cones to recognize contact states based on a geometric model for manipulative operations. He proposes a method to estimate the contact states by using force information acquired in the mating process and develops an algorithm for generating the state classifiers based on geometric models of the objects. But Hirai ignores friction in his model. Bicchi et al. [1] discuss the problem of resolving the location of a contact, the force at the interface and the torque about the contact normals. In [2] a systematic and general approach to identify topological transitions in contact situations during compliant motion is presented; the identification is based on energy considerations. Vougioukas and Gottschlich [16] discuss compliance synthesis for force guided assembly.

3 Identification of Assembly Process States

To realize the identification of *assembly process states* it is necessary to be able to identify a single contact formation. We propose to do this by using the theory of polyhedral convex cones (PCCs) including the consideration of static friction.

3.1 Contact States

The contacts between the active and passive subassemblies are described by principal contacts:

Definition 3.1 (Principal Contact) *A principal contact (PC) is a single contact between a pair of topological elements vertex, edge, face [7]. Here, it is assumed that each topological element is in one contact only. A PC is described by*

$$PC = <i,j> \in T \times T, \; i,j \in T := \{face, edge, vertex\} \quad (1)$$

Definition 3.2 (Contact Formation)
A contact formation (CF) between two polyhedra is a set of PCs.

Definition 3.3 (Fundamental Contact)
Assuming that all objects are polyhedra, it is possible to express each PC by a set of the following pairs of topological elements [7]:

$$\text{vertex - face, edge - edge, face - vertex}$$

These pairs of topological elements are called Fundamental Contacts (FC).

Since each *PC* can be described by a set of *FCs* it is also possible to describe each *CF* by a set of *FCs*.

Definition 3.4 (Configuration) *A Configuration C is defined by a certain position and orientation of all workpieces.*

3.2 Polyhedral Convex Cones

We use the theory of PCCs [5] to map the measured forces and torques to the contact states. How to calculate all contact states of an assembly operation is described in section 4.1.

Definition 3.5 (Polyhedral Convex Cone) *Let $\vec{a}_1$ through $\vec{a}_n$ be n real vectors. Then the set given by*

$$\mathcal{P} = \{\vec{x} \mid \vec{a}_i^T \vec{x} \leq 0, \; \forall i \in [1,\ldots,n]\} \quad (2)$$

represents a semi-infinite region surrounded by hyperplanes. $\mathcal{P}$ is called a Polyhedral Convex Cone.

Definition 3.6 (Convex Sum) *Let $\mathcal{P}_X$ and $\mathcal{P}_Y$ be two PCCs. The Convex Sum of these PCCs is then given by:*

$$\mathcal{P}_X + \mathcal{P}_Y = \{\vec{x}+\vec{y} \mid \vec{x} \in \mathcal{P}_X, \vec{y} \in \mathcal{P}_Y\} \quad (3)$$

Definition 3.7 (Union of PCCs) *Let $\mathcal{P}_X$ and $\mathcal{P}_Y$ be two PCCs. The Union of these PCCs is then given by:*

$$\mathcal{P}_{X \cup Y} = \mathcal{P}_X \cup \mathcal{P}_Y \quad (4)$$

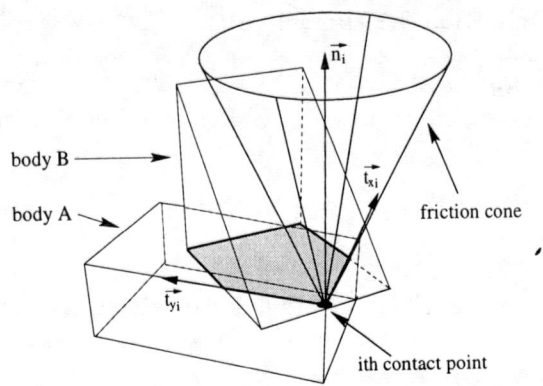

Figure 1: Contact frame of the ith contact point of two objects with its corresponding friction cone.

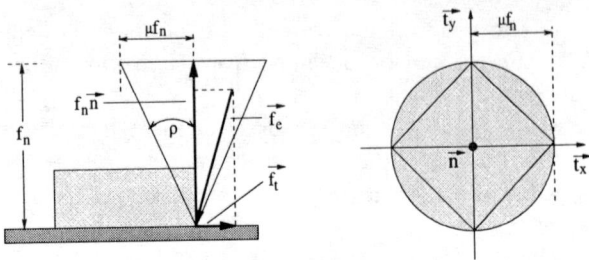

Figure 2: Projected approximation of the friction cone with $l = 4$

3.3 Modeling Friction

Implementation of a friction model which reflects the precise physical behavior is quite difficult if not impossible. We adopted a deterministic static friction model commonly known as *Coulomb friction* (see e. g. [4]). Our model has been successfully implemented, evaluated and described in [6] and [12]. The Coulomb model of friction gives an accepted empirical relationship between the normal force magnitude f_n and the tangential force magnitude f_t. The proportionality constant μ is known as the coefficient of static friction. Considering Coulomb's law of friction, the following inequality must be satisfied by the tangential force magnitudes $f_{t_{x_i}}, f_{t_{y_i}}$ at each contact point i of the assembly in order to prevent sliding[1]:

$$(f_{t_{x_i}}^2 + f_{t_{y_i}}^2)^{\frac{1}{2}} \leq \mu f_{n_i} \quad (5)$$

Since we are using polyhedral convex cones we approximate inequality 5 by an adaptively chosen number l ($l = 2^n, n \geq 2$) of linear inequalities. Such an approximation can also be found in [15]. Thus, considering friction, l linear constraints are induced by the approximated friction cone which is depicted in Figure 1 and Figure 2.

3.4 Describing Assembly Process States using PCCs considering Friction

A certain stable assembly process state given by contact formation CF_m is described by a PCC $\mathcal{P}_{\mathcal{F}m}$ which defines the corresponding admissible force set $\mathcal{F}_m$.

[1] For simplicity of notation we assume an identical coefficient of static friction at each contact.

Definition 3.8 (Admissible Force Set) *The Admissible Force Set $\mathcal{F}_m$ is the range of forces and torques which satisfy the static equilibrium condition for CF_m. $\mathcal{F}_m$ can be defined by a PCC $\mathcal{P}_{\mathcal{F}m}$ as described below.*

In order to obtain $\mathcal{P}_{\mathcal{F}m}$ the union of all PCCs $\mathcal{P}_{\mathcal{F}ml}$ has to be calculated:

$$\mathcal{P}_{\mathcal{F}m} = \bigcup_{l=1}^{L} \mathcal{P}_{\mathcal{F}ml} \quad (6)$$

where L is the number of configurations belonging to CF_m. The PCC $\mathcal{P}_{\mathcal{F}ml}$ describes the admissible force set $\mathcal{F}_{ml}$ of configuration C_l belonging to CF_m.

Let CF_m be described by N fundamental contacts FC_n. Assuming that the origin is fixed to the moving object, C_l is defined by the fundamental contacts FC_n and one normal vector of a face of a fixed object.

Thus, $\mathcal{P}_{\mathcal{F}ml}$ can be computed by the convex sum of all $\mathcal{P}_{\mathcal{F}nl}$:

$$\mathcal{P}_{\mathcal{F}ml} = \sum_{n=1}^{N} \mathcal{P}_{\mathcal{F}nl} \quad (7)$$

where $\mathcal{P}_{\mathcal{F}nl}$ is the PCC that describes the admissible force set $\mathcal{F}_{nl}$ corresponding to FC_n considering C_l. Assuming that there is no friction, $\mathcal{F}_{nl}$ is given by:

$$\mathcal{F}_{nl} = \{\, f_{nl} \vec{w}_{nl} \mid f_{nl} \geq 0 \,\} \quad (8)$$

where f_{nl} denotes the magnitude of force and torque. Let FC_n be a vertex - face contact; then $\vec{w}_{nl}$ called the wrench-vector is given by:

$$\vec{w}_{nl} = \begin{bmatrix} \vec{n}_{nl} \\ \vec{x}_{nl} \times \vec{n}_{nl} \end{bmatrix} \quad (9)$$

where $\vec{n}_{nl}$ is the outward normal vector of the face corresponding to C_l and $\vec{x}_{nl}$ denotes the coordinates of the vertex corresponding to C_l. The origin is located at the point where the forces and torques take effect.

Thus, $\mathcal{F}_{nl}$ can be defined by a PCC $\mathcal{P}_{\mathcal{F}nl}$ described by the single span-vector $\vec{w}_{nl}$.

In order to consider static friction the Coulomb model of friction mentioned in section 3.3 has to be applied. Therefore, the forces represented by $f_{nl}\vec{n}_{nl}$ and the resulting torques represented by $\vec{x}_{nl} \times \vec{n}_{nl}$ which may appear have to be modified. According to Figure 2, $\vec{n}_{nl}$ has to be replaced by a cone with angel ρ, where ρ can be obtained by:

$$\rho = \arctan(\mu) \qquad (10)$$

Let $\vec{s}_1 \neq \vec{0}$ be a vector perpendicular to $\vec{n}_{nl}$ and let $\vec{s}_2 \neq \vec{0}$ be a vector perpendicular to $\vec{n}_{nl}$ and $\vec{s}_1$; then the friction cone can be approximated by a linear combination of the following four vectors with positive coefficients :

$$\vec{n}_{nl1} = \cos(\rho) \cdot \vec{n}_{nl} + (1-\cos(\rho)) \cdot \frac{\vec{n}_{nl}\vec{s}_1}{|\vec{s}_1|^2} \cdot \vec{s}_1 +$$
$$\frac{\sin(\rho)}{|\vec{s}_1|} \cdot \vec{s}_1 \times \vec{n}_{nl} \qquad (11)$$

$$\vec{n}_{nl2} = \cos(-\rho) \cdot \vec{n}_{nl} + (1-\cos(-\rho)) \cdot \frac{\vec{n}_{nl}\vec{s}_1}{|\vec{s}_1|^2} \cdot \vec{s}_1 +$$
$$\frac{\sin(-\rho)}{|\vec{s}_1|} \cdot \vec{s}_1 \times \vec{n}_{nl} \qquad (12)$$

$$\vec{n}_{nl3} = \cos(\rho) \cdot \vec{n}_{nl} + (1-\cos(\rho)) \cdot \frac{\vec{n}_{nl}\vec{s}_2}{|\vec{s}_2|^2} \cdot \vec{s}_2 +$$
$$\frac{\sin(\rho)}{|\vec{s}_2|} \cdot \vec{s}_2 \times \vec{n}_{nl} \qquad (13)$$

$$\vec{n}_{nl4} = \cos(-\rho) \cdot \vec{n}_{nl} + (1-\cos(-\rho)) \cdot \frac{\vec{n}_{nl}\vec{s}_2}{|\vec{s}_2|^2} \cdot \vec{s}_2 +$$
$$\frac{\sin(-\rho)}{|\vec{s}_2|} \cdot \vec{s}_2 \times \vec{n}_{nl} \qquad (14)$$

Using these four vectors $\mathcal{F}_{nl}$ can be described as follows:

$$\mathcal{F}_{nl} = \left\{ \sum_{i=1}^{4} f_{nli}\,\vec{w}_{nli} \mid f_{nli} \geq 0 \right\} \qquad (15)$$

with

$$\vec{w}_{nli} = \begin{bmatrix} \vec{n}_{nli} \\ \vec{x}_{nl} \times \vec{n}_{nli} \end{bmatrix} \qquad (16)$$

Thus, the PCC $\mathcal{P}_{\mathcal{F}nl}$ is defined by the four span-vectors $\vec{w}_{nli}$, $i = 1, \dots, 4$.

3.5 Identifying Assembly Process States using PCCs

A measured force-torque vector belongs to contact formation CF_m if this vector is element of $\mathcal{P}_{\mathcal{F}m}$. This

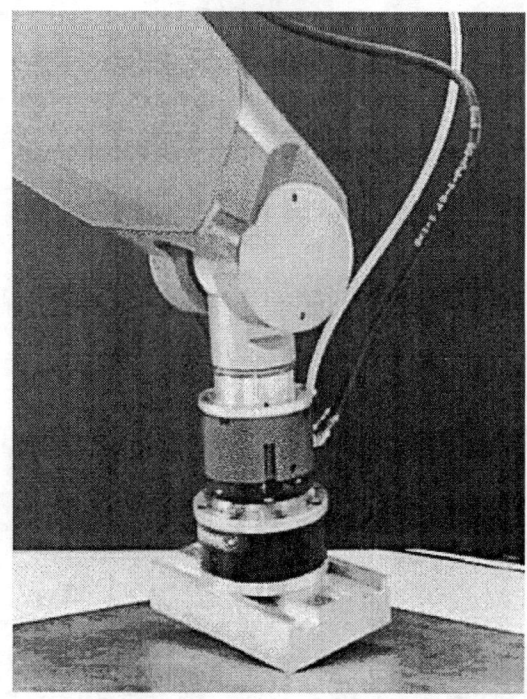

Figure 3: Rectangular parallelepiped fixed to a force-torque sensor mounted on a Stäubli RX60 robot.

can be checked by using the PCC-tools described in [7]. Since we consider friction there are some force-torque vectors, which can be mapped to different contact formations at a time. We map these force-torque vectors to those detected CFs which consist of the least number of FCs.

4 Results

In order to demonstrate the identification of assembly process states, we implemented the necessary tools to handle $PCCs$ on a workstation with an Intel Pentium Pro 200 MHz. A rectangular parallelepiped of aluminum was fixed to a force-torque sensor mounted on a Stäubli RX60 robot as shown in Figure 3. The operation in our experiment was to put this workpiece down on a plate using the PCC-tools to identify the assembly process states. Before doing this, we had to precompute the $PCCs$ for all possible contact states. We did this once considering friction and once setting the proportionality constant μ to zero, in order to compare the identification of process states with and without friction.

4.1 Computation of PCCs

In case of a rectangular parallelepiped there are nine possible contact states, namely four vertex-face-contacts, four edge-face-contacts, and one face-face-contact. For calculating the *PCCs*, the outward normal vector of the plate and the coordinates of each vertex must be given. Measured in millimeters the coordinates of the vertices in the force-torque sensor-frame are given by:

$$\begin{aligned} \vec{x}_1 &= [\ \ 39,\ -60,\ 35\ ] \\ \vec{x}_2 &= [\ \ 39,\ \ 60,\ 35\ ] \\ \vec{x}_3 &= [-39,\ \ 60,\ 35\ ] \\ \vec{x}_4 &= [-39,\ -60,\ 35\ ] \end{aligned} \quad (17)$$

The outward normal vector of the plate for a face-contact is given by:

$$\vec{n} = [\ 0,\ 0,\ -1\ ] \quad (18)$$

Since it is impossible to evaluate all configurations belonging to one CF, we sample over each CF using a constant δ. In order to illustrate this, we take e.g. an edge-face-contact. Such a contact is specified by the two vertex-face-contacts of the vertices bounding the edge and by the angle α between the inner normal vector of the lower face and the outer normal vector of the plate. α is specified by:

$$\epsilon \le \alpha \le (90° - \epsilon)\ |\ \epsilon > 0,\ \epsilon \to 0 \quad (19)$$

It easily can be seen that there is an infinite number of possible α. Thus, we use a constant δ to split the given range up into a finite number of configurations.

Next a *PCC* is calculated for each configuration as described in section 3.4. Using the methods of interpolating *PCCs* described in [8] we get nine *PCCs*, one for each CF.

4.2 Experiment

While the workpiece shown in Figure 3 is put down on the plate, the sensor data is gathered using the force-torque sensor type manufactured by Schunk. Getting one force-torque vector from the sensor takes 6 ms. In order to measure the contact forces and torques at the contact points only, the force-torque sensor has been calibrated prior to experimentation. We used the method explained in section 3.5 to map the gathered sensor data to the contact state which takes only 130 μs for each mapping. The further movement of the workpiece depends on the detected contact state. Figure 4 shows an example of measured force and torque. In addition, the corresponding contact state considering friction is given. The assembly process state changes as follows:

$$V1 \to E41 \to E12 \to FACE \quad (20)$$

Comparing the mapping of sensor data with and without considering friction we found, that without friction nearly no edge-face-contact and no face-face-contact has been detected correctly. On the other hand all configurations have been mapped to the correct contact state when friction has been considered considered.

Let us take a look at Figure 4 with time $t = 2000ms$ to illustrate this. The current CF is an edge-face-contact $E12$. Since this edge is colinear with the x-axis, there could not be any force along this axis unless friction is considered. From Figure 4 we see that a force of $-1\ N$ is measured along this axis. Thus there is no way to map this vector to the correct state without considering friction. Therefore, our approach is a major contribution for assembly process identification.

5 Conclusion

We presented and implemented an algorithm to classify and recognize contact states between assemblies considering friction. The application of this algorithm is force guided assembly. In order to identify assembly process states the theory of polyhedral convex cones is used. Force-torque sensor information acquired in the assembly process is interpreted using state classifiers which are formulated by polyhedral convex cones which were calculated considering friction. The classifiers map the sensor data to a contact state. The movement of the workpieces can be modified depending on the identified contact state.

References

[1] A. Bicchi, J. K. Salisbury, and D. L. Brock. Contact sensing from force measurment. *The International Journal of Robotics Research*, 12(3):249–262, June 1993.

[2] S. Dutre, H. Bruyninckx, and J. De Schutter. Contact identification and monitoring based on energy. In *IEEE International Conference on Robotics and Automation*, pages 1333–1338, April 1996.

[3] M. Erdmann. Using backprojections for fine motion planning with uncertainty. *The International Journal of Robotics Research*, 5(1):19–45, 1986.

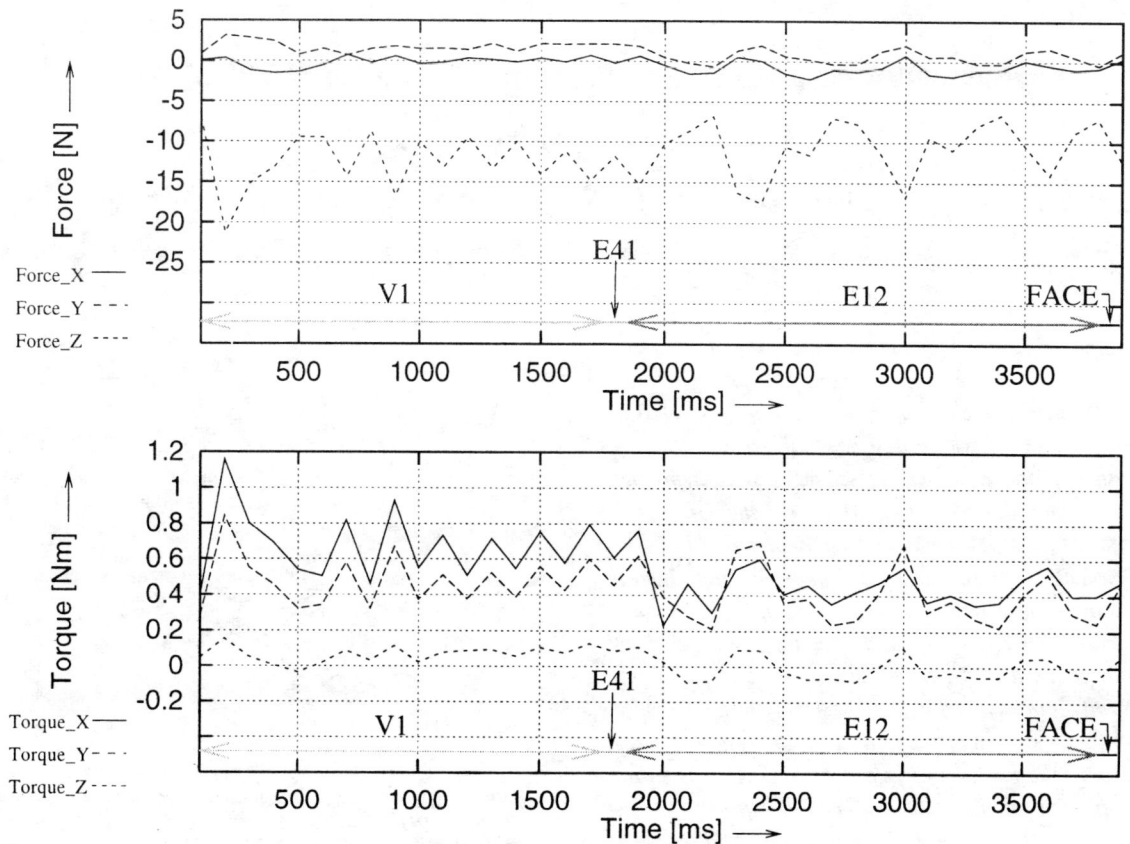

Figure 4: Measured forces and torques during the put down of the parallelepiped.

[4] M. Erdmann. On a Representation of Friction in Configuration Space. *International Journal of Robotics Research*, 13(3):240–270, 1994.

[5] A. J. Goldman and A. W. Tucker. Polyhedral convex cones. *Annals of Math. Studies*, 38:19–39, 1956.

[6] H. Mosemann, F. Röhrdanz and F. M. Wahl. Stability analysis of assemblies considering friction. *IEEE Transactions on Robotics and Automation*, 13(6):805–813, December 1997.

[7] S. Hirai. Identification of contact states based on a geometric model for manipulative operations. *Advanced Robotics*, 8(2):139–155, 1994.

[8] S. Hirai and H. Asada. Kinematcs and statics of manipulation using the theory of polyhedral convex cones. *The International Journal of Robotics Research*, 12(5):434–447, October 1993.

[9] S. G. Kaufman, R. H. Wilson, R. E. Jones, and T. L. Calton. The archimedes 2 mechanical assembly planning system. In *IEEE International Conference on Robotics and Automation*, pages 3361–3368, 1996.

[10] C. Laugier. Planning fine motion strategies by reasoning in the contact space. In *IEEE International Conference on Robotics and Automation*, pages 653–659, 1989.

[11] T. Lozano-Pérez, M. T. Mason, and R. H. Taylor. Automatic synthesis of fine-motion strategies for robots. *The International Journal of Robotics Research*, 3(1):3–24, Spring 1984.

[12] H. Mosemann, F. Röhrdanz, and F. M. Wahl. Assembly stability as constraint for assembly sequence planning. In *IEEE International Conference on Robotics and Automation*, pages 233–238, Leuven, Belgium, May 1998.

[13] F. Röhrdanz, H. Mosemann, and F. M. Wahl. Generating und Evaluating Stable Assembly Sequences. *Journal of Advanced Robotics*, 11(2):97–126, 1997.

[14] B. Romney, C. Godard, M. Goldwasser, and G. Ramkumar. An efficient system for geometric assembly sequence generation and evaluation. In *ASME International Computers in Engineering Conference*, pages 699–712, 1995.

[15] J. Trinkle, J. S. Pang, S. Sudarsky, and G. Lo. On dynamic multi-rigid-body contact problems with coulomb friction. Technical report, Department of Computer Science, Texas A&M University, 1995.

[16] S. G. Vougioukas and S. N. Gottschlich. Compliance synthesis for force guided assembly. In *IEEE International Conference on Robotics and Automation*, pages 976–981, 1995.

Disassembly Sequencing and Assembly Sequence Verification Using Force Flow Networks

Sukhan Lee[1,2]   Hadi Moradi[1]

Abstract

The design of an assembly line for a mechanical assembly has to satisfy all the geometrical and the physical constraints presented in the assembly such as blocking relation and contact forces. Given a desired final assembly, determining an appropriate assembly sequence which satisfies not only the blocking constraints but also the force constraints needs a proper model to directly provide the information needed to reason about assembly. We introduce a representation, the Directional Force Graph, which explicitly describes how parts interact in the assembly. This representation is used to determine disassemblable parts that are naturally broken apart from an assembly by a decomposition force. A decomposition force is an external force applied to a held part(s) against a fixtured part(s). The proposed approach provides a mechanism to explicitly incorporate the real-world constraints of assembly directions, fixturing, and tooling requirements into planning. We describe a complete planning algorithm and report the results.

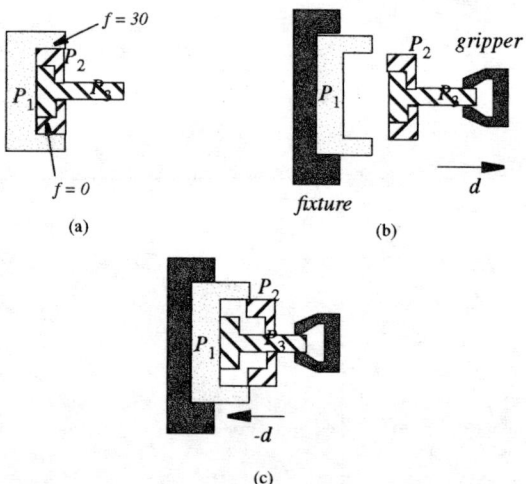

Figure 1: (a) An assembly with irreversible assembly sequence. (b) The assembly during disassembly operation, and (c) the assembly during assembly which cannot be the reverse of the disassembly sequence due to the contact forces.

1. Introduction

Assembly planning has emerged as an important engineering discipline due to its impact on the quality and efficiency in product design and manufacturing cycles. However the current manual practice of assembly planning in factories is error prone and labor intensive, requiring a number of iterations to complete improvement in designs, assembly methods and configurations, work instructions, and adjustments in part supply. Thus there is a need for CAD tools that can assist the designers of assembly plans in ordering the assembly operations and selecting the required tools and motions, as well as giving rapid feedback to product designers to facilitate Design for Assembly. CAD tools for assembly planning are expected to help in upgrading the quality of assembly plans and significantly reducing the heavy indirect cost incurred by manual assembly planning.

---

1. Department of Computer Science-EE Systems, Institute for Robotics and Intelligent Systems, University of Southern California, Los Angeles, CA, 90089.

2. Jet Propulsion Laboratory, California Institute of Technology, Pasadena, CA 91109

In the past decade, several researches have been conducted studies leading to some semi-automatic assembly planners such as Archimedas2 [5] or STAAT [1]. In particular, substantial progress has been made on the problem of efficiently determining all geometrically feasible directions of assembling a product [4]. However, there exist many factors other than the existence of mating paths under ideal geometric constraints that are crucial for real-world assembly planning. For instance, although the majority of the assembly planners assume that an assembly sequence is the reverse of a disassembly sequence, Fig. 1 shows an assembly sequence generated by a conventional geometrical assembly sequence planner which fails to realize the irreversibility of the disassembly sequence.

Lee and Yi [7] introduced the Force Flow Network to incorporate the physical contact forces, between parts in an assembly, into planning. However, they limit their study to assemblies with only orthogonal contacts. In this paper, we generalize the approach by introducing the Directional Force Graph (DFG) which explicitly describes how parts interact in the assembly. This representation is used to determine disassemblable parts that are naturally broken apart from an assembly by a decomposition force. A decomposition force is an external force applied to a gripped part(s) against a fixtured part(s). DFG is then converted to a Force Flow Network (FFN) by selecting a source (grasped part) node and a sink (fixtured part) node. The identification of the minimal cut-sets of an FFN network, based on a max-flow, min-cut algorithm of polynomial time complexity, show the disassemblable parts that are naturally broken apart from the assembly by the decomposition force. Therefore, the proposed force-based approach is applicable to assemblies with large number of parts. Furthermore, the approach provides a natural mechanism to incorporate real-world assembly constraints into planning. For instance, the current fixturing and tooling capabilities can be incorporated into the selection of grasped and fixtured parts.

2. Related Work

A fully automatic planner takes a CAD model of an assembly as input and computes the set of all plans that separate the parts. The geometric approach has been one of the leading trends in assembly sequencing. Homem De Mello and Sanderson [3] use a generate-and-test approach for generating all possible subassemblies and testing the feasibility of a given operation to remove each subassembly from the rest of the assembly. The exponential nature of the generate-and-test approach makes this method infeasible when the number of parts grows in an assembly. The first step to overcome this exponential nature is to identify all ranges of translation directions that allow two parts to separate without collision. Wilson and Latombe [11] introduced the notion of a Blocking Graph based on a critically-based decomposition of motion directions. In this way, the uncountable set of motion directions can be partitioned into a finite number of equivalence classes to yield a set of blocking graphs (the "Non-Directional Blocking Graph", NDBG) that captures all possible directions of motion separating two polyhedra. However, like all geometrical assembly planners, this approach does not incorporate more than pure geometrical factors and does not consider other realistic factors such as friction.

Srinivasan and Gadh [10] addressed the importance of optimal disassembly planning and proposed a wave propagation method to determine the locally optimal disassembly plan using the geometric CAD model of a product.

Now, the critical issue in assembly planning is how to bring it closer to reality by incorporating a variety of real-world assembly constraints. There exist many factors other than the existence of mating paths that are crucial for real-world assembly planning. Examples include the requirement of fixturing, tooling, and reorientation for the stability of assembly operations[2,4,6].

In this paper, we introduce Directional Force Graph to incorporate the force constraints into assembly planning. Our Directional Force Graph is similar to Directional Blocking Graph (DBG) introduced by Wilson-Latombe, but force information is added. Consequently, the assembly sequences verified by DFG are closer to reality because it considers the friction force between mating features. Moreover, DFG has an explicit mechanism to incorporate real-world assembly constraints such as limitation on fixturing, tooling, grasping capabilities, into planning. Our planning algorithm uses DFG to determine disassembly plans or verify assembly plans, which are determined by NDBG or other approaches. It is necessary to mention that not all disassembly sequences are the reverse of a feasible assembly sequence [7]. Thus we have to verify the reversibility of a disassembly plan to be able to use it as an assembly plan. We further show that DFG is not suitable to directly determine assembly plans because it may overlook some of possible assembly plans.

3. Assumptions

Incorporating force constraints into planning requires modeling of the forces involved in part separations (due to frictions, forced fits, etc.) as well as the force propagations through interconnected parts. However, the modeling of forces involved in disassembly may not be simple [12] because: (a) the separation forces due to frictions and forced fits depend on various mechanical properties, such as stiffness, rigidity, (b) the forces may vary during separations due to the change in contact areas as well as the change in the mechanism of friction, (c) the separation forces may vary according to the direction in which the force is applied. In this study, we consider the following assumptions to ease up the force calculation.

1. There is no uncertainty in part position, orientation, and dimension.

2. Assembly consists of polyhedral parts.

3. Parts are not perfectlyy rigid, but sufficiently rigid such that a small amount of surface deformation is allowed to generate a squeezing force while allowing uniform propagation of forces to the contact surfaces so that force summation becomes valid.

4. The force assigned to two mating parts is constant during infinitesimal translation.

5. The force assigned to two mating parts is resulted from the connection type (force fit, loose fit). The effect of gravity has not been considered in this report.

6. Only one part is grasped by the gripper and one part is held by the fixture.

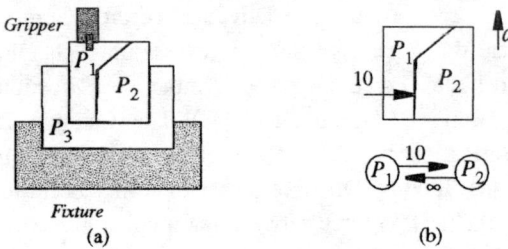
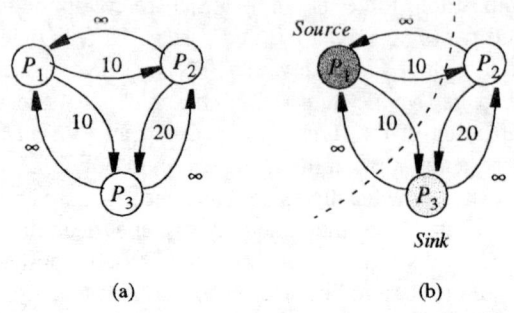

Figure 2: *An assembly of 3 parts. The decomposition force for translating P_1 against P_2 in direction d is shown with an arc from P_1 to P_2. Consider P_1 is grasped for infinitesimal translation and P_2 is fixtured, then the decomposition force is 10. The decomposition force for translating P_2 against P_1, P_2 is grasped and P_1 is fixtured, in direction d is infinite because P_1 blocks P_2.*

Figure 3: *The DFG and FFN for the assembly in Fig. 2. (a) The DFG for the assembly in direction d, (b) the FFN for the assembly when part P_3 is fixtured and part P_1 is grasped.*

We also consider linear and monotone [9,13] assembly planning in which one part is added (or removed) from the assembly at a time and it is placed in its final pose with respect to the other parts in the assembly.

The decomposition forces between each two parts in an assembly is calculated based on their mating forces and their contact surfaces using the above assumptions. For example, Fig. 2 shows an assembly of 3 parts. P_1 cannot move against P_2 in direction d unless the decomposition force overcomes the friction. On the other hand P_2 cannot move against P_1 in direction d because it is completely blocked by P_1 (infinite force needed).

4. Directional Force Graph

Consider an assembly A consists of N parts $\{P_1, ..., P_N\}$. In a given direction d, a directed graph, called Directional Force Graph (DFG) and represented as $G(d, A)$, is assigned with N nodes representing $P_1, P_2, ..., P_N$. A directed arc from P_i to P_j in an DFG with weight f_{ij} represents the maximum static force that needs to be overcome to infinitesimally translate part P_i is direction d while P_i is grasped and P_j is fixtured. Figure 3 (a) shows the DFG for the assembly in Fig. 2.

DFG can be tailored to a more realistic model of an assembly during assembly operation by considering a fixture and a gripper which are used to hold and manipulate parts respectively. The resulting model is called Force Flow Network (FFN) with a source as the grasped part and a sink as the fixtured part (Fig. 3 (b)). The resulting flow network can then be used to directly determine which decompositions are actually feasible. This is analogous to a human operator who fixtures the assembly and grabs a part to remove it by increasing the force gradually. At a certain time, the assembly naturally breaks into two subassemblies (or the assembly cannot be broken apart because he needs infinite force or he will break the assembly). This makes DFG more promising than a pure geometrical approach to better model the actual world.

5. Finding Locally Free Subassembly

The geometric reasoning of LFM, Local Freedom of Motion, assumes ideal part connections with no frictions and mating forces involved. However, we can define the force space of a decomposable subassembly as the dual of the motion space spanned by LFMs. That is, a static force can be defined in the space dual to the motion space spanned by LFMs. It is the duality between the motion and force space of a decomposable subassembly that allows to transform the problem defined in the motion space into a dual problem in the force space, and vice versa. To be more specific, let us consider the problem of identifying the LFM for a subassembly in a given disassembly direction. In other words we have to answer the following question:

How much force do we have to apply to remove a subassembly from the assembly and which subassembly will be removed?

On the other hand, DFG represents the amount of maximum force that can be exerted in a subassembly when a part is grasped and another part is fixtured. Consequently, the above LFM problem can be transformed into a dual problem of determining a minimal cut of the Force Flow Network constructed for the given disassembly direction based on corresponding DFG. We use max-flow, min-cut algorithm to determine the amount of flow and the cut set of FFN [7]. In the following we briefly explain a flow network and the theorem about max-flow, min-cut algorithm [8].

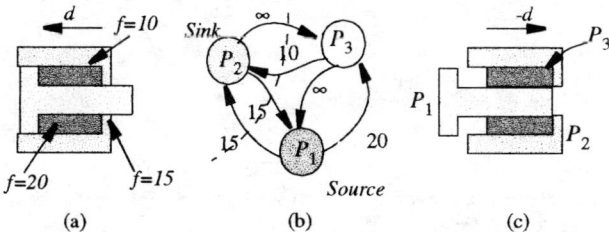
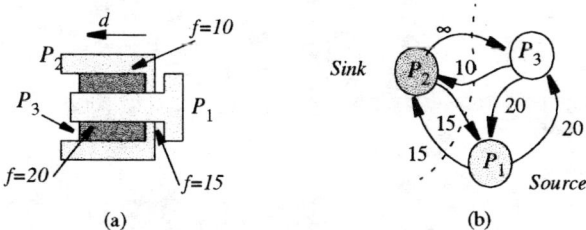

Figure 4: (a) An assembly in its final state, (b) the FFN for the assembly as P_1 is grasped and P_2 is fixtured. The FFN shows that P_1 cannot be disassembled linearly. (c) Although the linear disassembly of P_1 is impossible, but it is possible to assemble part P_1 into subassembly consisting of (P_2, P_3).

Figure 5: An assembly during assembly operation. (a) Part P_1 is inserted in direction d. (b) The FFN for the assembly which shows that the insertion of P_1 pushes P_3 and changes the state of the assembly. Dashed line show the cut point of the FFN.

A flow network $N = (s, t, V, E)$ is built on top of a directed graph, $G = (V, E)$, in which each edge $(u, v) \in E$ has a non-negative *flow* $f(u, v) \geq 0$. If $(u, v) \notin E$, then we assume that $f(u, v) = 0$. Source (s), and sink (t) are distinguished from other vertices. An *s-t* cut is a partition of nodes in N into two sets W and $\overline{W}$ such that $s \in W$ and $t \in \overline{W}$. The capacity of an *s-t* cut is:

$$C(W, \overline{W}) = \sum_{\substack{(i,j) \in E \\ \text{such that} \\ i \in W, j \in \overline{W}}} b(i,j)$$

in which $b(i, j)$ is the maximum flow that can be passed through arc $(i, j) \in E$. The max-flow, min-cut theorem says: *The value f of any s-t flow cannot be greater than the capacity $C(W, \overline{W})$ of any cut-set. Moreover, the value of the maximum flow equals the capacity of the minimum cut.* Note that it is equivalent to saying that the maximum force that can be exerted to an assembly is equal to the minimum cut in the assembly which makes the assembly break naturally into two subassemblies. Figure 3 shows an example of FFN in which part P_1 is gripped and part P_3 is fixtured. It can be seen that part P_1 can be removed with 20 unit of force. However, if part P_2 was gripped, then parts P_1 and P_2 together could be removed with 30 units of force.

6. Disassembly Sequence Planning

After constructing DFGs for desired directions, first, we need to select a part to be fixtured and a part to be grasped by available fixtures, tools, and grippers. For the selection of parts to be fixtured and grasped, the accessibility, fixturability, and graspability of parts, as well as the cost-effectiveness associated with fixturing and grasping, at the current stage of the assembly should be taken into consideration. Although the subject of selecting parts to be fixtured or grasped is interesting on its own, we use two simple criteria to select fixtured and grasped parts in our disassembly sequence planner which have been discussed in section 9.

Then for each direction and for each selection of grasped and fixtured parts, we use FFN to determine which parts can be removed from the assembly. We then recurs on the remaining subassembly. In other words, this is as building a tree with A, the whole assembly, at the root and a single part at each leaf. Each path from A to a leaf represents a disassembly sequence.

As mentioned before, assembly by disassembly using the above algorithm may overlook some of the possible plans. For instance, Fig. 4 shows an assembly with the valid assembly sequence of assembling P_3 to P_2 in direction -d and then adding P_1 to the subassembly in direction -d. In order to generate this assembly sequence by reversing a disassembly sequence, we have to be able to disassemble P_1 in direction d from the assembly first, then P_3 from P_2. However, as the FFN for the assembly shows, removing only P_1 in direction d is not possible unless an extra fixture is used to keep P_3 in its place. Consequently, the above assembly sequence will be overlooked using the reverse of FFN disassembly sequences.

7. Assembly Sequence Verification

As shown before, a disassembly sequence is not always guaranteed to be the reverse of an assembly sequence (Fig 1). We use FFN to determine if the infinitesimal translation for assembling a part to a subassembly changes the state of the assembly. Consider an assembly A consists of N parts $\{P_1, ..., P_N\}$ in which P_i is about to be inserted into the assembly in direction d. Build the FFN for the current assembly in which P_i is the source and another part (e.g. P_j) should be selected as the sink. Then use the max-flow, min-cut algorithm to determine its cut sets ($P_i \in W, P_j \in \overline{W}$). If $W = \{P_i\}$, W includes only P_i, then the assembly oper-

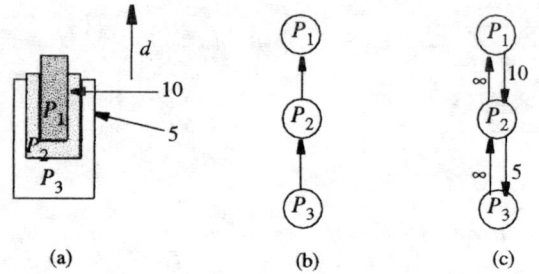

Figure 6: (a) an assembly with the contact forces, (b) the DBG for the given direction, (c) the DFG for the assembly in the given direction. DBG can be converted to its corresponding DFG and vice versa.

ation is valid. Otherwise, the assembly operation is not valid and the assembly is not stable (another part moves during insertion). For instance, Fig. 5 shows an assembly during assembly operation. The FFN shows that the assembly operation is not valid because part P_3 moves during insertion operation. Thus the assembly operation is not valid unless a fixture is used to part P_3.

8. DFG vs. DBG

As mentioned earlier, DFG is similar to DBG introduced by Wilson [11]. In fact, it is very easy to show that a DFG can be transformed into its corresponding DBG by erasing all arcs with finite force value and assigning "1" to all arcs with infinite force value (Fig. 6 (c) converted to Fig. 6 (b)). This shows that DBG is an abstract version of DFG. More precisely, DBG models an assembly into motion space while DFG models an assembly into force space.

Also we can use the DBG and the liaison graph for a given direction of an assembly to construct the corresponding DFG as follows:

1. For each pair of parts, P_i and P_j, with mating force f_{ij} (in liaison graph), assign the force flow capacity $b(i,j) = b(j,i) = f_{ij}$ in DFG,

2. For each arc in the DBG, replace the corresponding arc in the DFG with infinite force flow capacity.

9. Experimental Results

We implemented the max-flow, min-cut algorithm to solve force flow network in parallel with our 2D assembly planner [2] on an UltraSparc Creator. The planner determines valid Vertical Assembly (VA plans) plans for infinitesimal translations in which each subassembly is stable before and after an insertion operation on a flat surface. By definition, VA plans are linear, monotone, two-handed, and contact-coherent [9,13].

We consider a flat surface as the simplest possible fixture to hold the parts in a stable orientation. We considered only one fixtured part which was chosen as the most left part on the surface. Also, because of our linearity assumption, any part with an infinite outgoing arc cannot be a candidate for a grasped part. Thus we used this criterion to determine graspable parts.

The results for a model aircraft engine (Fig 7) is given in Table1. The set of best plans is considered as plans with minimum number of rotations. Figure 8 shows a disassembly plan generated using a pure geometrical disassembly planner (based on DBG) which was rejected by DFG approach. It is obvious that removing the screws on the cap of crank shaft will loosen the cap and the crank shaft. Thus removing the screws on the crank shaft will not be possible (using linear disassembly planning).

TABLE 1. The number of plans, the number of best plans and the running times for the model air craft engine.

Assembly planning	t (sec)	plans	best plans
Using DBG	<1.0	155,424	137,904
Using DFG	<1.0	11,520	4,032

10. Conclusion

The problem of assembly sequencing, limited to linear and monotone assembly sequences, using force-based reasoning is presented. Directional Force Graph has an explicit mechanism to incorporate real-world assembly constraints, such as the limitations in fixturing, tooling, and grasping capabilities, into planning. The primary results on generating assembly plans using this method is given which needs more investigation. The proposed algorithm can be easily extended to include multiple grasped and multiple fixtured parts.

Further study is needed for modeling the force constraints in assembly, especially, by taking into consideration the possible variation and uncertainty associated with mating/separation forces due to the flexibility of parts as well as the dynamics involved in part mating/separation. Also generation of non-monotonic and non-sequential plans using DFG need to be investigated.

Acknowledgments

The authors wishes to thank Ken Goldberg for initiating and supporting the VA planner at USC.

References

[1] B. Romney, C. Godard, M. Goldwasser, G. Ramkumar. "An Efficient System for Geometric Assembly Sequence Generation and Evaluation" ASME Proc Int. Conf. in Computers in Engineering. pp 699-712, 1995.

[2] Ken Goldberg and Hadi Moradi. "Compiling Assembly plans into Hard Automation". Proc. of ICRA 1996.

[3] .S. Homem de Mello and A. C. Sanderson, "A Correct and Complete Algorithm for the Generation of Mechanical Assembly Sequences," IEEE Transactions on Robotics and Automation, vol. 7, pp. 228-240, 1991.

[4] Luis Homem De Mello and S.Lee, *Computer Aided Mechanical Assembly Planning*. Kluwer Academic Publisher, 1991.

[5] S. G. Kaufman, R.H. Wilson, R.E. Jones, T.L. Calton, and A. L. Ames. "The Archimedes 2 Assembly Planning System". Proc of ICRA, vol 4. pp 3361-3368, 1996.

[6] H. Moseman et. al. "Assembly Stability as a constraint for Assembly Sequence". in Proc. of ICRA, Belgium 1998.

[7] Lee and C. Yi, "Force-Based reasoning in Assembly Planning", in Proc. of ASME 1994.

[8] C. H. Papadimitriou and K. Steiglitz "Combinatorial Optimization, algorithms and complexity", Prentice-hall NJ.

[9] B.K. Natarajan. "On Planning Assemblies". Proc. of the 4th ACM Symp. on Computational Geometry. pp-299-308, 1988.

[10] H. Srinivasan and R. Gadh. "Complexity Reduction in geometric Selective Disassembly Using the Wave Propagation Abstraction'. in Proc. of the IEEE Int. Conf. on Robotics and Automation, Belgium 1998.

[11] R. H. Wilson and Jean-Claude Latombe. "Geometric reasoning about mechanical assembly". AI journal, Dec. 1994.

[12] D. Whitney and O. Gilbert, "Representation of Geometric Variations Using Matrix Transforms for Statistical Tolerance Analysis in Assemblies," in Proc. of ICRA, Atlanta, Georgia, pp. 314-321, May 1993.

[13] . Jan D. Wolter. *On the automatic generation of plans for Mechanical Assembly*. PhD thesis, The Univ. of Michigan. 1988.

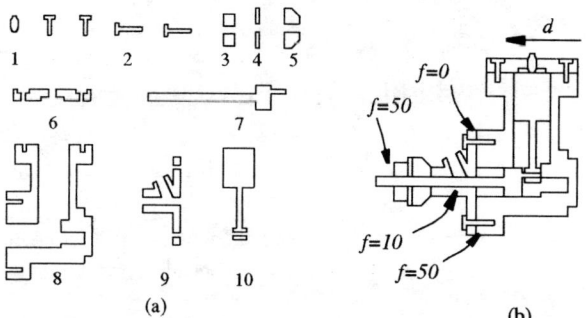

Figure 7: The model aircraft engine. (a) Parts that are included in the assembly, (b) the final desired assembly with contact force information in direction d. By using these force information, the assembly planning using DFG rejects the plan found by the pure geometrical approach (Fig 8).

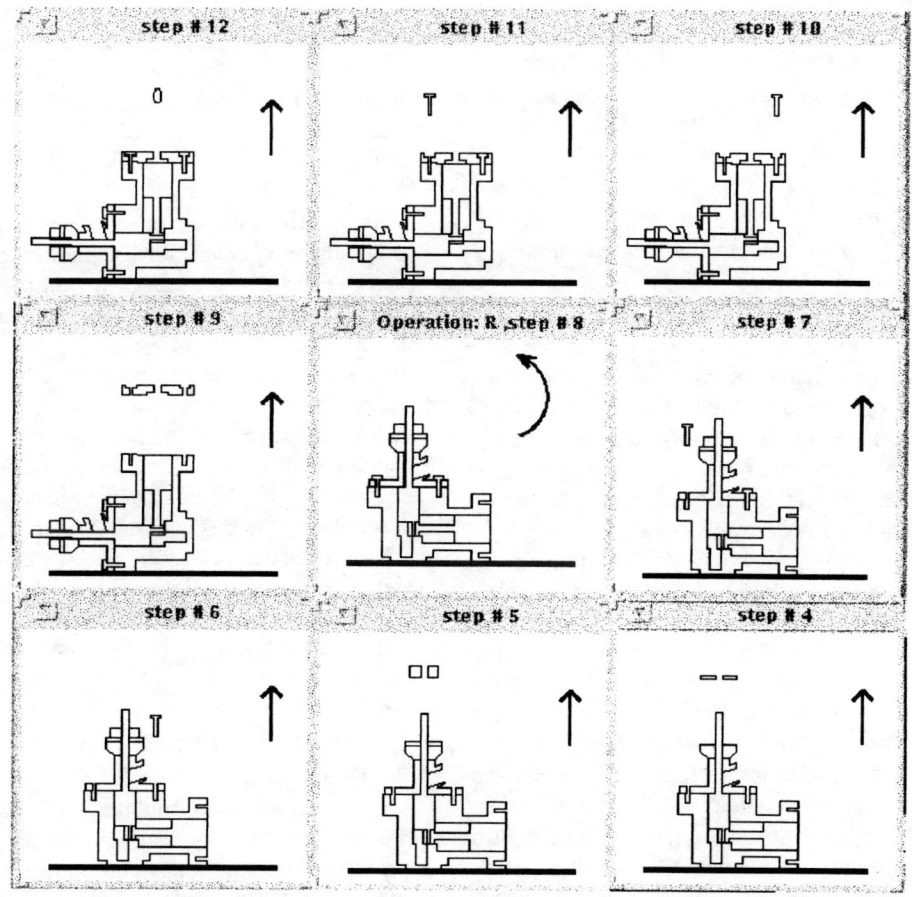

Figure 8: The disassembly plan generated using the pure geometrical reasoning (DBG). The removal of the screw on the crank shaft (step 5) cannot be accomplished because the screw on the cap of crank shaft are removed (steps 7 and 6) and there is no force to hold the cap and the crank shaft in their place. Consequently, this disassembly plan is rejected by force-based planning.

Adaptive Accommodation Control for Complex Assembly : Theory and Experiment

Sungchul Kang   Munsang Kim   Chong W. Lee   Kyo-Il Lee†

Korea Institute of Science and Technology
Seoul, Korea

†Seoul National University
Seoul, Korea

ABSTRACT

In this paper, an adaptive accommodation control for complex assembly is presented. The complex assembly (CA) is defined as a task whose assembled parts have complex geometry including concavity. The concept of the adaptive accommodation, which is inspired by the insertion operation of a blindfold human, is to adaptively change the accommodation property depending on the sensed contact wrench and the current target twist. Both the bounded wrench condition and the target approachability condition can be satisfied simultaneously by applying the adaptive accommodation control law. By using the convex optimization technique, an optimum target approaching twist with adaptive accommodation property can be determined at each instantaneous contact state as a global minimum solution. Incorporated with an admissible perturbation method, a new adaptive accommodation control law, which is independent of part geometry, is developed without motion planning nor contact analysis procedure. A VME-bus based real-time control system is built to experiment various CA tasks. T-insertion task as a planar CA and double-peg assembly task as a spatial assembly were successfully executed by implementing the adaptive accommodation control law.

1 Introduction

Recently, various robot applications, e.g., assembly (or fixturing) lines in manufacturing, construction, component repairing in space or in hazardous environment and robotic surgery, require an insertion of a complicated object into an positionally uncertain environment. In these applications, the shape of the part inserted is not simple or the environment geometry is not exactly known.

In this paper, a complex assembly (CA) is defined as a task which deals with complex shaped parts including concavity or whose environment is so complex that unexpected contacts occur frequently during insertion. Different from simple peg-in-hole tasks, CA has features that the dimension of the nominal insertion path can usually be more than 1 and the contact states occurring during the insertion for CA are various and complicated. Therefore the conventional compliant motion planning approaches based on various existing compliant control schemes have limitation to be applied to CA due to their computational complexity in generating a multi-dimensional motion plan. Several results related to CA have been presented by extending the existing active compliance approaches. Shimmels and Peshkin[1-3] proposed an admittance control law for force-guided assembly applicable to a typical fixture assembly task with or without friction. The method requires the contact analysis for every possible contact occurring in fixture assembly. Due to the limitation of linear compliance mapping[4], although all possible contact states can be generated from the geometry of the part and environment[5], the method is hard to be applied to a CA tasks which deals with complex parts with concavities and whose contact states are complicated and change severely. McCarragher[6,7] presented a discrete event controller by imitating human decision making mechanism for assembly task. Also, the control system includes a qualitative matching process[8] to monitor the current contact state during assembly. However, similar to the work of Shimmels and Peshkin's, since it requires the geometrical classification procedure for all possible contact states to build a petri net as a discrete event system model, the more complex the geometry of the parts becomes, the more complicated and time-consuming to build the petri net becomes.

As described above, most existing compliant control and planning approaches have limitation to be easily applicable to CA due to their dependency on the part geometry or on the dimension of the motion since they requires contact kinematic analysis procedures for every possible contact or planning procedures to produce a compliant path. In our recent previous work on CA[9], a target-approachable force-guided control has been proposed and verified by CA simulation. By refining and developing the control algorithm for CA, an adaptive accommodation control law is exploited in this work. Also an *accommodation index* which indicates the assemblability at each contact state is derived from the solution of the Kuhn-Tucker conditions for convex optimization. As a result of experiments of a variety of CA tasks including T-insertion and double-peg-assembly, its feasibility and applicability are verified.

2 Approach to complex assembly
2.1 Convex optimization for CA

As a first step to formulate the CA algorithm in this work, we assume that there is no nominal motion plan nor prescribed contact states determined from the geometry of the part and environment. This assumption corresponds to the situation where a blindfold human inserts a part into a hole without any geometric information of the part and the hole. In this situation contact states occur unexpectedly due to the geometric complexity or the misalignment between the part and the environment. When this situation is implemented to a robotic assembly system shown in Fig.1, the information usually available to the robot are 1) the end point position/orientation, 2) the sensed *resultant* contact wrench(forces/torques) and 3) target position/orientation with uncertainty.

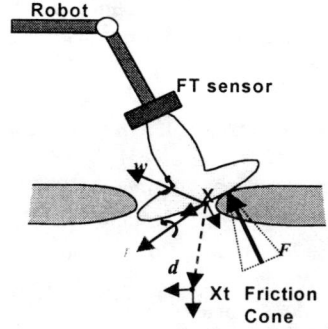

Fig.1 Target approaching twist in contact

Based on the previous work[9], the objective function $f(t)$ is defined as a half of the square of the normed deviation between the *target twist d* and the differential twist *t*. The magnitude of the target twist is given as a desired translational and angular velocities V_d and Ω_d. Thus the target twist d, computed from the deviation between a target pose $X_t = (O_t, P_t)$ and a sensed current pose $X = (O, P)$ is represented as

$$d = (\omega_{dx}, \omega_{dy}, \omega_{dz}, v_{dx}, v_{dy}, v_{dz})^T$$
$$= (\omega_d^T, v_d^T)^T \quad (1)$$

where $\omega_d = (O_t - O)/\|O_t - O\| \cdot \Omega_d$

and $v_d = (P_t - P)/\|P_t - P\| \cdot V_d$

Therefore, the objective function is described as

$$f(t) = \frac{1}{2}\|d - t\|^2 \quad (2)$$

where $t = (\omega_x, \omega_y, \omega_z, v_x, v_y, v_z)^T$

To balance the magnitude of the translational twist and the rotational twist in $f(t)$, the rotational twist vector is scaled by multiplying a *weighting radius of rotation*. The weighting radius of rotation is defined as a maximum radius of rotation about an axes of the inserted object.

The constraint function $g(t)$ is constructed by the reciprocal and the repelling constraints[10] representing the admissible motion space in contact. It is described by a contact wrench *w* and a differential twist *t* as an inequality form as

$$g(t) = -w^T t \leq 0 \quad (3)$$

where $w = (\tau_x, \tau_y, \tau_z, f_x, f_y, f_z)$.

Note that both wrench and twist are described in TCP (Tool Center Point) frame X which is described as X in Fig.2. The resultant wrench measured in a FT sensor can be transformed to that in the TCP frame by using the static forces and moments equilibrium condition. Note that the constraint function $g(t)$ is coordinates invariant while the objective function is not invariant with the change of the reference frame[11]. Thus the optimization problem for complex assembly does not hold the coordinates invariance property.

Both the quadratic objective function $f(t)$ and constraint function $g(t)$ are convex functions since the Hessian matrix of each function is positive-semi-definite[12]. Therefore, this is a convex optimization problem which guarantees a global minimum at an instantaneous quasi-static contact state. Finally, the CA problem can be formulated as a simple quadratic convex optimization to find a target approachable twist t^* as follows :

Find a twist t^* which minimizes $f(t) = \frac{1}{2}\|d - t\|^2$ **subject to** $g(t) = -w^T t \leq 0$.

By applying the Kuhn-Tucker conditions for the optimization problem with inequality constraints, the Lagrange function L is described as

$$L = f(t) + u \cdot g(t) \quad (4)$$

Then there exists a Lagrangian multiplier u^* such that the Lagrangian is stationary with respect to each twist component t_i, i.e. $v_x, v_y, v_z, \omega_x, \omega_y$ or ω_z and u as follows.

$$\frac{\partial L}{\partial t_i} \equiv \frac{\partial f}{\partial t_i} + u^* \frac{\partial g}{\partial t_i} = 0, \quad (5)$$

$$g(t^*) \leq 0, \quad (6)$$

$$u^* \cdot g(t^*) = 0 \text{ and} \quad (7)$$

$$u^* \geq 0. \quad (8)$$

2.2 Adaptive accommodation law

It is important to note that the condition (7) can be divided into two cases, which are $u^* = 0$ and $g(t^*) = 0$. When $u^* = 0$, the inequality constraint is inactive. Thus the optimum target approaching twist is simply same as d. In this case, the contact wrench does not constrain the part to be inserted to the target pose. This case corresponds to the situation illustrated in Fig.2. Fig.2(a) shows a physical situation during a planar CA task. The optimum twist lies in the repelling motion space. Fig.2(b) shows the optimum within the differential twist vector space in 3 dimensional planar CA problem. The optimum twist t^* is determined from (5)-(8) as follows.

$$t^* = d \quad (9)$$

In the other case that $g(t^*) = 0$, in which the inequality constraint is active at the contact state such as shown in Fig.3(a), the optimum twist is in the reciprocal motion space as shown in Fig.3(b). In this case, the optimum twist t^* can be solved from (5)–(8) as follows.

$$t^* = d + u^* \cdot w \quad (10)$$

where $u^* = -w^T d / \|w\|^2 \geq 0$.

Note that the Lagrangian multiplier u^* maps the sensed resultant wrench into the target approaching twist using the dynamically updated target twist d.

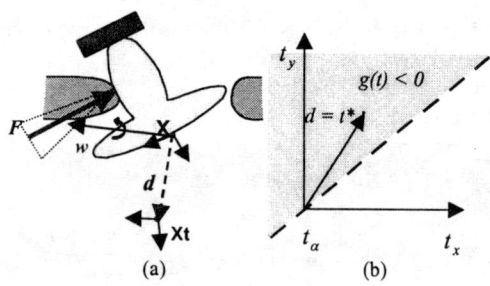

Fig.2 Optimum by repelling motion
(a) Corresponding contact state in planar CA task
(b) Optimum twist t^* in the repelling motion space

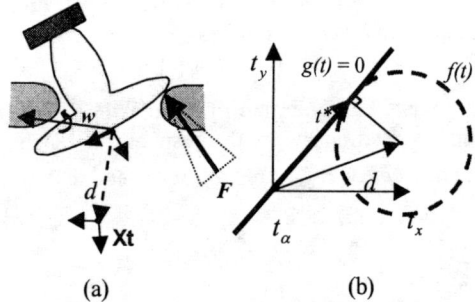

Fig.3 Optimum by reciprocal motion
(a) Corresponding contact state in CA task
(b) Optimum twist t^* in the reciprocal motion space

Thus the u^* has accommodation characteristics similar to the conventional damping control law. However, In contrast to the conventional accommodation matrices, which are usually linear mappings and determined in the least-square sense, the Lagrangian multiplier u^*, which is a scalar, determines an optimum target approaching twist dynamically using the sensed current resultant wrench and the current target twist d. Therefore, the optimum twist equation (10) has the characteristics of an adaptive accommodation, which dynamically changes the mapping depending on the current contact wrench and the target twist. Thus the new control law (9) and (10) is called *adaptive accommodation law* in this work.

The condition of u^* that should be greater than or equal to 0, is a necessary condition for determining whether a target approaching twist exists. If $u^* < 0$, there is no optimum solution and thus a target approaching twist does not exist. Therefore, the u^* is an important index to check if the target approachability condition is satisfied at a contact state. Thus it is called *"accommodation index."* The accommodation index plays an important role in determining whether the optimum twist is feasible or not in the CA operation.

2.3 Perturbation Twist in the Virtual Admissible Motion Space

In a multi-contact case shown in Fig.4, the optimum twist determined from the resultant wrench may not exist in the AMS(admissible motion space) but in the VAMS(virtual admissible motion space) as shown in Fig.5. Therefore, if it does not exist in the AMS, the optimum twist obtained from the Kuhn-Tucker conditions can not guarantee the target approachability. Rather, it can even make the part exert a large wrench to the environment making a dangerous contact situation. To prevent this problem, the bounded wrench condition should be added to the optimization. That is, when the magnitude of the sensed resultant wrench exceeds the wrench limit in case of an ill-conditioned contact state, a twist perturbation in a random direction within the VAMS is generated to find a direction making the sensed wrench bounded. As a result of repetitive perturbations in VAMS, the part can be moved to a different contact state(or a non-contact state) which may be possible to proceed insertion. Especially in the case of a jamming (or wedging) situation due to a multi-contact with large friction, the rotational perturbation in the same direction as that of the sensed torque vector is effective since it tends to change the contact state to a different one or to a non-contact state. This concept is similar to the conventional damping control law.

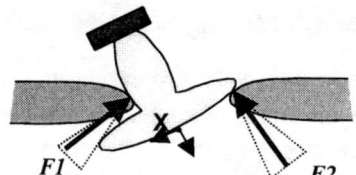

Fig.4 Multi-contact example in CA process

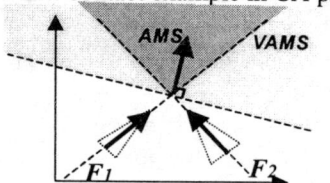

Fig.5 Difference between AMS and VAMS

The translational perturbation twist vector δ_V is modeled as

$$\delta_V = diag(C_V, C_V, C_V) \cdot f \quad (11)$$

In the equation (11), the $diag(C_V, C_V, C_V)$ is a 3 by 3 diagonal matrix whose diagonal components are all C_V's. Based on the damping control method, C_V is a constant accommodation parameter offering a translational compliance in contact. The magnitude of the C_V can be determined from the inverse of the tolerable translational stiffness K_V at the robot end-point and the sampling time T_s for the controller as follows,

$$C_V = \frac{1}{K_V T_S} \quad (12)$$

The tolerable stiffness K_V can be estimated from the desired robot end point stiffness and the measurable range of the FT sensor. On the other hand, the rotational perturbation twist vector δ_ω is also modeled by the sensed resultant torque vector $\tau = (\tau_x, \tau_y, \tau_z)$ as

$$\delta_\omega = diag(C_\omega, C_\omega, C_\omega) \cdot \tau \quad (13)$$

Similar to the equation (11), the $diag(C_\omega, C_\omega, C_\omega)$ is a diagonal matrix whose diagonal components are all C_ω's. The C_ω is also a constant accommodation parameter for rotational compliance. The magnitude of the C_ω can be designed in the same way as (12),

$$C_\omega = \frac{1}{K_\omega T_S} \quad (14)$$

Finally, six dimensional perturbation twist vector δ is determined in VAMS as follows,

$$\delta = (\delta_V^T, \delta_\omega^T)^T \quad (15)$$

Consequently, the perturbation twist δ is generated from the sensed resultant wrench to proceed insertion maintaining the tolerable magnitude of the resultant contact wrench, when the value of the accommodation index is less than 0 or the sensed resultant wrench is larger than the prescribed wrench bound during CA.

The proposed adaptive accommodation controller does not guarantee the target approachability at every contact state during insertion depending on the initial/target pose as well as the geometry of the assembled parts. This problem can be prevented with a help of a gross motion planner which generates a non-contact motion plan by using the geometry information of the part and environment. The issue of the alignment of the initial pose correctly is not a compliant motion planning problem but a gross motion planning problem.

3 Adaptive Accommodation Controller

The adaptive accommodation(AA) controller is designed by using the conventional accommodation control approach[2] which maps the contact wrench into the twist to produce an accommodation (or an admittance) effect. When contact is detected during insertion, target-approachable force-guided (TAF) module is activated and computes a target approaching twist in either optimum twist generator or perturbation twist generator. The block diagram of the TAF module is as shown in Fig.6.

The TAF module handles three control modes, i.e., the non-contact mode, the optimum twist mode and perturbation twist mode. In the non-contact twist mode, the controller performs like a free-space position controller to approach the commanded target pose. When contact is detected, the optimum twist mode or perturbation twist mode is activated. If the accommodation index is zero or positive and the bounded resultant wrench condition are satisfied, the optimum twist mode is activated in the TAF module. On the other hand, either the accommodation index or the bounded resultant wrench condition is not satisfied, the perturbation twist mode is activated to determine the perturbation twist vector.

Unlike the accommodation matrix[1] based on linear mapping, the mapping in our damping controller has nonlinear characteristics as represented in (9) and (10). However, the AA control system can be simply implemented to an existing high stiffness position controller by adding an external wrench feedback loop. The algorithm for solving an optimum twist which is analytically derived in (9) and (10), and the perturbation generation module can be simply implemented as an external feedback loop. The accommodation parameters C_V and C_ω used to generate perturbations should be designed carefully by taking into account the tolerable stiffness of the part and environment, and the bandwidth of the robot. The overall block diagram of the proposed AA control system for CA is shown in Fig.7.

4 CA Experiment

4.1 Robot control system for CA

A variety of CA tasks are experimented by using the VME based real-time robot control system shown in Fig.8. This control system, which runs in a real-time OS (VxWorks), is a general-purpose robot control system aiming at building a humanoid robot called CENTAUR in Korea Institute of Science and Technology.

4.2 T assembly/disassembly

As a typical example of planar assembly, which needs a rotational motion to insert a T-shaped part into the target pose, T-assembly/disassembly task is implemented and

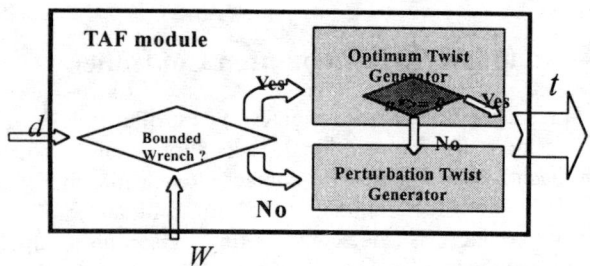

Fig.6 Target-approachable force-guided (TAF) module

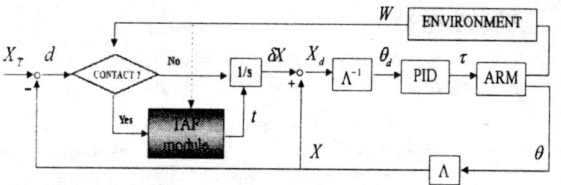

Fig.7 Block diagram of adaptive accommodation control

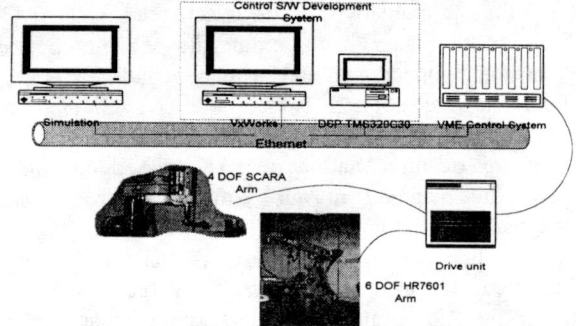

Fig.8 VME-based real time robot control system

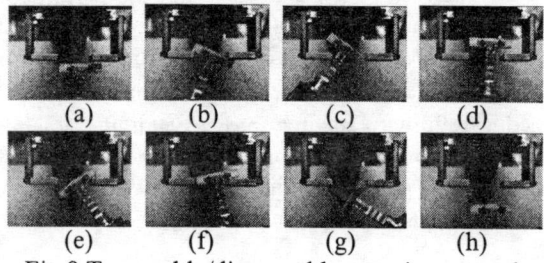

Fig.9 T-assembly/disassembly experiment result

experimented by the robot control system described above. The trajectory(twist) update rate is assigned by considering the computation time for the twist generation and the inverse kinematics of the robot. The servo rate in the PID controller is also set by considering the steady state response of the robot dynamics.

The result of the T-assembly/disassembly experiment is shown sequentially from Fig.9(a) to Fig.9(h). Fig.9(a) shows an initial position of T. The initial position is set to ensure an approaching direction to the target position where the insertion is completed. There is no limitation to the initial position unless it does make a sticking state which the target approachability is not satisfied in view of the accommodation index. From the contact mode plot shown in Fig.11, almost contact modes are optimum twist mode "1" thanks to the adaptive accommodation property. In the perturbation twist mode "2", which is usually corresponds to a multi-contact case, the perturbed motion helps to disengage the jamming condition by changing the current contact state to another state. Fig.10 show the resultant forces and torque measured by F/T sensor during insertion. From the tolerable stiffness specified, the measured forces and torques satisfy the bounded wrench property which offers a compliance in contact.

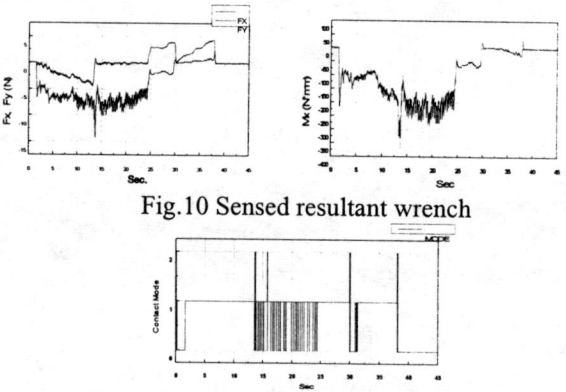

Fig.10 Sensed resultant wrench

Fig.11 Contact modes during insertion

4.2 Double-peg assembly

As a second example of the complex assembly to verify the adaptive accommodation controller, a double-peg assembly task has been experimented. This double-peg assembly experiment focuses on the robustness against the initial misalignment for the successful insertion. If the initial pose is given so that the two pegs are partially inserted into the holes, and thus the resultant wrench can easily find an admissible motion to the target position, the assembly operation is successful regardless the geometric complexity of the parts. As a result of generating the optimum twist and the perturbation twist to approach the target position, the assembly was finally succeeded in spite of the misalignments. Fig.12 shows the sequential result of the assembly operation and Fig.13 shows the bounded wrench result subject to the prescribed tolerable stiffness. Fig.14 shows the contact modes monitored during the insertion. Due to the small tolerance (0.1 mm), the free non-contact modes are scarcely seen once the mating stage overcoming the misalignment is completed. The double hole disk tries to be inserted against the misalignment by interchanging its contact mode among free, optimum and perturbation mode. This motion can be seen from 0 until 5 sec. Once the disk is inserted after 5 sec, the inserting motion interchanges between the optimum mode and the perturbation mode. It means that during the time the disk can be inserted by a simple compliant motion just like a single peg-in-hole task.

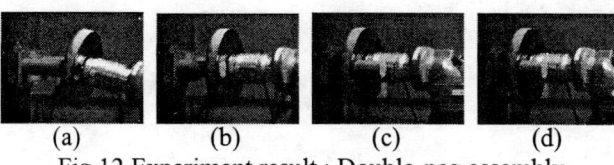

Fig.12 Experiment result : Double-peg assembly

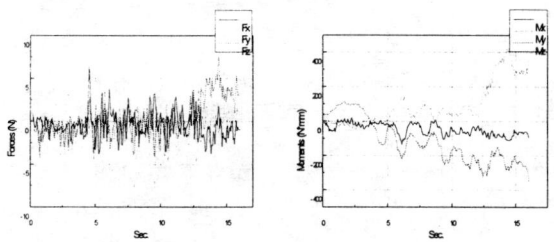

Fig.13 Sensed resultant wrench

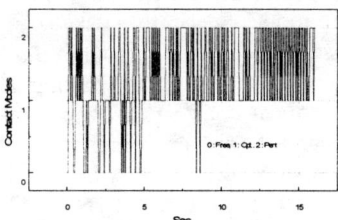

Fig.14 Contact modes during insertion

5 Conclusions

In this work, the complex assembly(CA) is newly defined as a result of the investigation of various assembly tasks encountered in industry and the other automation fields. By examining the operation of a blindfold human for CA, a simple assembly rule called adaptive accommodation law, which imitates human operation is exploited and mathematically formulated using twist and wrench. To attempt automation for the CA tasks using a robot with a force/torque sensor, a new geometry-independent assembly algorithm called adaptive accommodation control is developed by modeling the complex assembly operation as a convex optimization problem. The goal of the optimization is to achieve both the target approachability and the bounded wrench property at each contact state during insertion. Additionally, the admissible perturbation method which generates a twist in the virtual admissible motion space constructed from the sensed resultant wrench is proposed to make the contact wrench bounded and to disengage an unexpected jamming (or wedging) state. Since both the optimum twist generator and the perturbation twist generator does not require any geometry information of the parts or the environment, the control algorithm can be easily implemented to various complex assembly tasks.

This work is expected to be easily extended to a cooperative assembly task using two arms, which is more human-like and efficient in terms of the measure of time and energy, since the adaptive accommodation control algorithm is independent of the geometry of the assembled parts, the number of the contact states and the structure of the robot kinematics.

References

[1] Peshkin, M.A., "Programmed Compliance for Error Corrective Assembly", *IEEE Trans. Robotics and Automation,* Vol.6, No.4, 1990, pp.473-482

[2] Shimmels, J.M. and Peshkin, M.A., "Admittance Matrix Design for Force-Guided Assembly", *IEEE Trans. Robotics and Automation*, Vol.8, No.2, 1992, pp.213-227

[3] Shimmels, J.M. and Peshkin, M.A., "Force Assembly with Friction", *IEEE Trans. Robotics and Automation*, Vol.10, No.4, 1994, pp.465-479

[4] Asada, H., "Representation and Learning of Nonlinear Compliance Using Neural Nets", *IEEE Trans. Robotics and Automation*, Vol.9, No.6, 1993, pp.863-867

[5] Xiao, J. and Zhang, L., "Contact Constraint Analysis and Determination of Geometrically Valid Contact Formations from Possible Contact Primitives," *IEEE Trans. of Robotics and Automation*, Vol.13, No.3, 1997, pp.456-466

[6] McCarragher, B. J. and Asada, H., "The Discrete Event Control of Robotic Assembly Tasks," *ASME Journal of Dynamic systems, Meas. And Control*, Vol.117, 1995, pp.384-393

[7] McCarragher, B. J. and Asada, H., "The Discrete Event Modeling and Trajectory Planning of Robotic Assembly Tasks," *ASME Journal of Dynamic systems, Measurement and Control*, Vol.117, 1995, pp.394-400

[8] McCarragher, B. J. and Asada, H., "Qualitative Template Matching Using Dynamic Process Models for State Transition Recognition of Robotic Assembly", *ASME Journal of Dynamic Systems, Measurement and Control,* Vol.115, 1993, pp.261-269

[9] Kang, S., Kim, M., Lee, K-Y and Lee, C-W, "A Target Approachable Force-Guided Control for Complex Assembly", Proc. IEEE Int. Conf. Robotics and Automation, May-1998, pp.826-831,

[10] Ohwovoriole, M.S. and Roth, B., "An Extension of Screw Theory," *Journal of mechanical design*, vol.103, October, 1981, pp.725-735

[11] Kang, S., A Study on Target-Approachable Force-Guided Control with Adaptive Accommodation for Complex Assembly, *Ph.D. Thesis,* Dept. Mechanical design and Production engineering, Seoul National University, Korea, 1998

[12] Arora, J.S., *Introduction to Optimum Design*, McGraw-Hill, Singapore, 1989

COMPLIANT-MOTION PLANNING AND EXECUTION FOR ROBOTIC ASSEMBLY*

Jan Rosell   Luis Basañez   Raúl Suárez

Institut d'Organització i Control de Sistemes Industrials (UPC), Diagonal 647, 08028 Barcelona, SPAIN
Phone: +34 (93) 4016653, Fax: +34 (93) 4016605, e-mails: {rosell,basanez,suarez}@ioc.upc.es

Abstract: This paper presents a method for the planning and execution of compliant motions within the scope of a two-phase fine-motion planner for the performance of planar assembly tasks with robots. Algorithms are provided to find a nominal solution path in both free and contact configuration space which is feasible in spite of the uncertainties affecting the task. Compliant-motion commands based on the generalized damping control mode are synthesized to follow this path, allowing to maintain a constant bounded force.

1 Introduction and overview

Fine-motion planning usually gives rise to an active-compliance strategy that describes geometric trajectories as a function of the current actual situation during the task execution. Three main approaches have been presented following this research line: a) the LMT approach [5] describes the synthesis of compliant motions as the backchaining of preimages from the goal region to the initial region, the preimage for a given velocity command being the set of configurations that guarantee that the goal is reachable and recognizable, taking into account uncertainty in sensing and control; b) two-phase planners first generate a nominal plan assuming no uncertainty, and then consider uncertainty and replan the steps of the path pruning possible errors (e.g. [10]); c) contact-space approaches represent the task as a graph of contact states and synthesize a plan by searching in this graph, considering the uncertainty in the states definition and in the state transition operators (e.g. [9]).

This paper presents a method for the planning and execution of compliant motions (velocities are modified according to reaction forces), which is used in a two-phase fine-motion planner for planar assembly tasks (two degrees of freedom of translation and one of rotation). Section 2 introduces the basic motion planning phase which analyzes all the geometric constraints of a planar assembly task and uses them to obtain an exact cell partition of the free configurations in configuration space ($\mathcal{C}_{free}$). A graph representation of this partition allows the search for a nominal solution path using graph searching techniques. Section 3 introduces the motion synthesis phase, which evaluates the nominal solution path in free configuration space taking into account all the uncertainties affecting the task. When necessary, the arcs of the path are patched in contact configuration space ($\mathcal{C}_{contact}$), where the paths are found analogously to those in $\mathcal{C}_{free}$. The synthesis of motion commands is performed in Section 4. The generalized damping control mode is assumed and compliant-motion commands are synthesized with two components, one devoted to follow the solution path previously found, and the other devoted to maintain the contact taking into account the effect of friction. Task execution issues are tackled in Section 5, where task execution experiments are reported. Finally, Section 6 concludes the work.

2 Basic Motion planning

Let $\mathcal{A}$ and $\mathcal{B}$ be two polygons describing a manipulated object and an static object, respectively. Let $\{W\}$ and $\{T\}$ be the reference frames attached to the workspace and to the manipulated object $\mathcal{A}$, respectively. $\{T\}$ has the origin at the manipulated object reference point, and an orientation ϕ with respect to $\{W\}$. Each vertex of $\mathcal{A}$ is described with respect to $\{T\}$ by a vector $\vec{h}$, with module h and orientation γ. The vertices of $\mathcal{B}$ are described with respect to $\{W\}$ by their coordinates x and y.

Two types of basic contacts can take place: an edge of $\mathcal{A}$ against a vertex of $\mathcal{B}$ (Type-A) and a vertex of $\mathcal{A}$ against an edge of $\mathcal{B}$ (Type-B). The following

---
*This work was partially supported by the CICYT projects TAP96-0868 and TAP98-0471.

subsections analyze the geometric constraints of contact situations involving different combinations of basic contacts, considered only those imposed by the edges and vertices of the basic contacts involved.

2.1 One Basic Contact

Let us define:

ψ_W, ψ_T: the orientation of the normal to the contact edge with respect to $\{W\}$ and $\{T\}$, respectively.

d_W, d_T: the signed distances between the straight line that supports the contact edge and the origins of $\{W\}$ and $\{T\}$, respectively. If (x_e, y_e) is a point of the contact edge, then:

$$d_W = x_e \cos\psi_W + y_e \sin\psi_W \quad (1)$$
$$d_T = x_e \cos\psi_T + y_e \sin\psi_T \quad (2)$$

A $\mathcal{C}$-face is defined as the set of contact configurations involving only one basic contact, and a $\mathcal{C}'$-face its parametrized projection into the xy-plane [7, 8]. The $\mathcal{C}'$-face represents the contact positions for each possible contact orientation. For a given orientation ϕ the $\mathcal{C}$-face is a segment with the following features:

- Its supporting line is

$$x \cos\psi_W + y \sin\psi_W = D \quad (3)$$

where, for type-A basic contacts

$$D = x_v \cos\psi_W + y_v \sin\psi_W + d_T \quad (4)$$
$$\psi_W = \psi_T + \phi + \pi \quad (5)$$

(x_v, y_v) being the coordinates of the contact vertex; and for type-B basic contacts

$$D = h_v \cos(\psi_W + \pi - \gamma_v - \phi) + d_W \quad (6)$$

h_v and γ_v being the module and orientation of the vector defining the contact vertex, respectively, and ψ_W being independent of ϕ.

- Each extreme is on a circumference:

$$x = x_v + h_v \cos(\pi + \phi + \gamma_v)$$
$$y = y_v + h_v \sin(\pi + \phi + \gamma_v)$$

where (x_v, y_v) and $(h_v \cos\gamma_v, h_v \sin\gamma_v)$ are, respectively, the contact vertex and a vertex of the contact edge for a type-A basic contact, and vice versa for a type-B basic contact.

The range of nominal contact orientations is determined by the parallelism condition between the contact edge and the edges adjacent to the contact vertex, which is given by:

$$\psi_W^{sta} = \phi + \psi_T^{mob} + \pi \quad (7)$$

where ψ_T^{mob} and ψ_W^{sta} represent, respectively, the orientation of the contact edge and the orientation of an adjacent edge of the contact vertex for a type-A basic contact, and vice versa for a type-B basic contact.

2.2 Two Basic contacts

Let a $\mathcal{C}$-edge be the set of contact configurations for a contact situation involving two basic contacts i and j, and a $\mathcal{C}'$-edge its parametrized projection into the xy-plane [7, 8]. The $\mathcal{C}'$-edge is an arc of a curve obtained from the system of equations of the supporting lines of each basic contact given by equation (3). The solution for the $\mathcal{C}'$-edge is:

(a) If $\sin(\psi_{Wj} - \psi_{Wi}) \neq 0$:

$$x = \frac{D_i \sin(\psi_{Wj}) - D_j \sin(\psi_{Wi})}{\sin(\psi_{Wj} - \psi_{Wi})}$$
$$y = -\frac{D_i \cos(\psi_{Wj}) - D_j \cos(\psi_{Wi})}{\sin(\psi_{Wj} - \psi_{Wi})} \quad (8)$$

(b) Otherwise it is the straight line:

$$x \cos(\psi_{Wi}) + y \sin(\psi_{Wi}) = D_i \quad (9)$$

for the orientation that satisfies $|D_i| = |D_j|$.

In case (a), each extreme of the $\mathcal{C}'$-edge occurs for a value of the orientation such that one of the contacts satisfies any of the following two constraints:

- *orientation constraint*: the contact edge is parallel to an adjacent edge of the contact vertex.
- *finite length constraint*: the contact vertex coincides with a vertex of the contact edge.

The expression of these values, as well as the value of the orientation in case (b), can be found in [8].

2.3 Three Basic contacts

Let a $\mathcal{C}$-vertex be the contact configuration of a three basic contact situation, involving contacts i, j and k. Then, if $\sin(\psi_{Wj} - \psi_{Wk}) \neq 0$, $\sin(\psi_{Wk} - \psi_{Wi}) \neq 0$ and $\sin(\psi_{Wi} - \psi_{Wj}) \neq 0$ the orientation ϕ of the $\mathcal{C}$-vertex satisfies:

$$D_i \sin(\psi_{Wj} - \psi_{Wk}) + D_j \sin(\psi_{Wk} - \psi_{Wi}) +$$
$$D_k \sin(\psi_{Wi} - \psi_{Wj}) = 0 \quad (10)$$

otherwise, it is the unique orientation where one of the contact situation involving only two of the contacts occurs. This condition is obtained by solving the system of equations describing the geometric constraints of the $\mathcal{C}'$-edge of contacts i and j and that of the $\mathcal{C}'$-face of contact k. Equation (10) is a second order equation when all basic contacts are of the same type, or a fourth order equation, otherwise [8].

For more than three basic contacts, the corresponding $\mathcal{C}$-vertex coincides with the $\mathcal{C}$-vertex of any subset of three non-redundant basic contacts. These situation is tackled in the following subsection.

2.4 Considering all the Constraints

There can be other constraints than those imposed by the edges and vertices of the involved basic contacts, either due to concave objects or due to the existence of several static objects. The $\mathcal{C}$-space considering these additional constraints can be built by first generating, as described above, the $\mathcal{C}$-faces, the $\mathcal{C}$-edges and the $\mathcal{C}$-vertices, in this order, and by pruning then these sets considering all the constraints:

- **$\mathcal{C}$-vertex pruning**: Eliminate the $\mathcal{C}$-vertices that correspond to configurations that produce an overlapping of the objects, and merge those $\mathcal{C}$-vertices that correspond to the same contact configuration (i.e. a contact configuration involving more than three basic contacts).

- **$\mathcal{C}$-edge pruning**: Divide each $\mathcal{C}$-edge into segments, depending on the number of $\mathcal{C}$-vertices where it is involved, and validate each segment. Eliminate those $\mathcal{C}$-edges without valid segments.

- **$\mathcal{C}$-face pruning**: Divide each $\mathcal{C}$-face into patches, depending on the $\mathcal{C}$-edges where it is involved, and validate each patch. Eliminate those $\mathcal{C}$-faces without valid patches.

The detailed algorithms can be found in [8]. As an example Figure 1 shows the $\mathcal{C}$-space of the assembly task of Figure 2.

2.5 $\mathcal{C}$-space partition

Let $e_1(\phi)$, $e_2(\phi)$ and $e_3(\phi)$ be the configurations of three $\mathcal{C}$-edges of the $\mathcal{C}$-space for a given orientation ϕ. A $\mathcal{C}$-prism is defined as the set of configurations $c \in \mathcal{C}_{free}$ that satisfy:

$$c = \alpha e_1(\phi) + \beta e_2(\phi) + \gamma e_3(\phi) \qquad (11)$$

with

$$\begin{aligned}\alpha, \beta, \gamma &\in [0,1] \\ \alpha + \beta + \gamma &= 1 \\ \phi &\in [\phi_{bottom}, \phi_{top}]\end{aligned} \qquad (12)$$

$[\phi_{bottom}, \phi_{top}]$ being the range of orientations were the three $\mathcal{C}$-edges simultaneously exist and c does not belong to any other $\mathcal{C}$-edge.

The $\mathcal{C}$-prisms are a partition of $\mathcal{C}_{free}$, since they are disjoint regions whose union is $\mathcal{C}_{free}$. Figure 2 shows a section of the $\mathcal{C}$-space for a given orientation, where each triangle represents the section of a $\mathcal{C}$-prism. This exact cell partition is based on [1].

2.6 Planning

Let a $\mathcal{C}$-node be the middle configuration of the border between two $\mathcal{C}$-prisms, and let a $\mathcal{C}$-arc be an arc of curve within a $\mathcal{C}$-prism connecting two $\mathcal{C}$-nodes defined as follows. The configuration c with orientation ϕ of a $\mathcal{C}$-arc between two $\mathcal{C}$-nodes n_i and n_g satisfies:

$$\vec{ce_1} = \alpha \, \vec{e_2 e_1} + \beta \, \vec{e_3 e_1} \qquad (13)$$

with

$$\begin{aligned}\alpha &= \alpha_i + (\alpha_g - \alpha_i)\frac{\phi - \phi_i}{\phi_g - \phi_i} \\ \beta &= \beta_i + (\beta_g - \beta_i)\frac{\phi - \phi_i}{\phi_g - \phi_i}\end{aligned} \qquad (14)$$

where ϕ_g and ϕ_i are the orientations of n_g and n_i, respectively, and α_i, α_g, β_i and β_g are determined by (13) for the values ϕ_g and ϕ_i.

A graph is created whose nodes represent $\mathcal{C}$-nodes and whose arcs represent $\mathcal{C}$-arcs. Then, given a goal node, an initial node and a cost function, the Dijkstra algorithm is applied in order to compute the path of minimum cost. The cost of the arcs is set equal to the length of the $\mathcal{C}$-arcs, but some policies can be defined to modify the cost associated to some given arcs in order to guide the search.

The partition and graph representation of $\mathcal{C}_{contact}$ is obtained in a similar way, and then planning in $\mathcal{C}_{contact}$ is done as in $\mathcal{C}_{free}$.

3 Motion Synthesis

3.1 Contact uncertainty analysis

Once a nominal solution path has been found, its feasibility has to be evaluated considering all the uncertainties affecting the task. Contact uncertainty analysis is not in the scope of this paper, but the main ideas are briefly sketched in this Section. The thorough contact uncertainty analysis can be found in [8].

Modelling and sensing uncertainties have been taken into account, which include: a) manufacturing tolerances, b) imprecision in the positioning of the static objects, c) imprecision in the positioning of the manipulated object in the robot gripper, and d) imprecision in the position and orientation of the robot

Uncertainty is handled by associating a Contact Configuration Domain U^S to the current measured robot configuration c_o for each contact situation S, and maintaining the nominal $\mathcal{C}$-space. Then, a contact situation S is compatible with c_o if the intersection of U^S and the set of the corresponding nominal contact configurations is not empty.

Force measurements can help to solve some ambiguous contact situations [2]. A Generalized Force Domain is defined for each contact situation considering the uncertainties that affect the possible directions of the reaction force. Then, a contact situation is compatible with a measured force if the uncertainty region due to imprecision in the force measurements intersects the corresponding Generalized Force Domains. These tests are done using the dual representation of forces [3].

3.2 Path evaluation

In order to evaluate the $\mathcal{C}$-arcs of the path, a discrete number of configuration of each $\mathcal{C}$-arc is classified as follows. A configuration:

- is *uniquely identifiable* if it is assigned to only one contact situation by the contact identification algorithm.

- is *compliant* at a contact situation for a given commanded velocity if for any possible (contact) realization of this configuration, the resulting motion direction moves the manipulated object towards the nominal path [8].

Then, each configuration is classified as:

- *Compliant*: if a contact can take place due to uncertainties and the manipulated object is compliant at it.

- *Guarded*: if it is not compliant but uniquely identifiable.

- *Ambiguous*: if it is not compliant and nor uniquely identifiable.

And each arc is classified into:

- *Compliant*: if all its configurations are compliant.

- *Guarded*: if none of its configurations are ambiguous, and at least one is guarded.

- *Ambiguous*: if there is at least one ambiguous configuration.

A compliant $\mathcal{C}$-arc can be traversed even if the contact situation changes. A guarded $\mathcal{C}$-arc terminates when a new contact situation occurs; then, an error patch plan must be issued in order to recover from the contact situation. An ambiguous $\mathcal{C}$-arc should not be traversed.

3.3 Path synthesis

Given two configurations, either in $\mathcal{C}_{free}$ or $\mathcal{C}_{contact}$, the following algorithm is used to find the optimum non-ambiguous path if it exists, or the path with less ambiguous $\mathcal{C}$-arcs.

```
Find-path(C-space ){
    find nominal path
    DO{
        evaluate all the arcs of the nominal path
        IF no arc is ambiguous then RETURN
        ELSE{
            eliminate the ambiguous arcs
            find nominal path
        }
    }until no nominal path exists
    RETURN the path with less ambiguous arcs
}
```

The solution path is synthesized using the following algorithm, which patches the path in $\mathcal{C}_{contact}$ whenever a solution in $\mathcal{C}_{free}$ is not feasible.

```
Path-Synthesis( ){
    C-space construction
    Find-path ($\mathcal{C}_{free}$)
    IF the path is no ambiguous RETURN
    ELSE{
        FOR all the sets of ambiguous arcs{
            Find-path ($\mathcal{C}_{contact}$)
        }
    }
}
```

As an example, figure 3 shows a final solution path which includes $\mathcal{C}$-arcs in both $\mathcal{C}_{free}$ and $\mathcal{C}_{contact}$.

4 Command motion synthesis

The generalized damping control mode is assumed. The velocity commands sent to the robot are computed from two velocity components, $\vec{v}_t$ and $\vec{v}_f$ (Figure 4).

The $\vec{v_t}$ component tries to follow the nominal $\mathcal{C}$-arcs of the solution path (either in $\mathcal{C}_{free}$ or $\mathcal{C}_{contact}$). It is computed as the tangent direction of the $\mathcal{C}$-arc at c_o. The expression of the current $\mathcal{C}$-arc of the solution path being traversed is iteratively updated by substituting n_i by the current measured configuration c_o in equations (13) and (14).

The compliant component, $\vec{v_f}$, has as aim to maintain a constant bounded force during motion in $\mathcal{C}_{contact}$. Given a desired reaction force $\vec{F}_d$, a force control loop with a PID controller is used to generate $\vec{v_f}$ as follows. The input of the PID controller is the force error between the desired reaction force and the actual measured reaction force. The output, multiplied by the predefined accommodation matrix, is the compliant component $\vec{v_f}$.

The desired reaction force $\vec{F}_d$ is computed as follows. Given a contact situation involving one basic contact, if the commanded velocity points towards the $\mathcal{C}$-face, a reaction force arises within the generalized friction cone [4]. If the mobile object moves along an instantaneous direction of motion over the tangent plane, the reaction force lies on one of the edges of the generalized friction cone:

$$\vec{e}^{-} = (n_x - \mu n_y, n_y + \mu n_x,$$
$$[r_y(n_x - \mu n_y) - r_x(n_y + \mu n_x)]/\rho)$$
$$\vec{e}^{+} = (n_x + \mu n_y, n_y - \mu n_x,$$
$$[r_y(n_x + \mu n_y) - r_x(n_y - \mu n_x)]/\rho) \quad (15)$$

where (n_x, n_y) and (r_x, r_y) are the normal to the contact edge and the vector from the contact vertex to the manipulated object reference frame, respectively, defined in $\{W\}$, μ is the friction coefficient, and ρ the radius of gyration of the manipulated object. The proposed $\vec{F}_d$ is in the direction of one of the edges of the generalized friction cone, with a given desired module.

For contact situations involving two basic contacts, $\vec{F}_d$ is in the direction of a linear combination of the corresponding edges of the generalized friction cones:

$$\vec{F}_d = \frac{\beta_i \vec{e}_i + \beta_j \vec{e}_j}{|\beta_i \vec{e}_i + \beta_j \vec{e}_j|} F_d \quad \text{with } \beta_i, \beta_j > 0 \quad (16)$$

5 Task Execution

The assembly task used as example was executed by a Staübli RX-90 robot equiped with a JR3 force sensor. A force control loop with a PID controller has been designed to achieve a good performance in the maintenance of a constant bounded force. The PID parameters were $K_P = 1266.0$, $K_D = 920.6$ and $K_I = 20.6$, and the sampling time was 32 ms. The reaction force module of a type-B contact motion involving translation and rotation is shown in Figure 5. Figure 6 shows snapshots of the real task execution that follows the solution path of Figure 3.

6 Conclusions

The paper has presented the development and implementation of a new fine-motion planner based on a two-phase approach for the performance of planar assembly tasks with robots, as a previous step towards its extension to assembly tasks in the space. The potential application of this research is the automation of complex assembly tasks (i.e. which require several motions to be performed), where the clearance between the objects to be assembled is small with respect to the manufacturing tolerances and uncertainties affecting the task.

References

[1] Avnaim F., Boissonnat J. D., and B. Faverjon. "A practical exact motion planning algorithm for polygonal objects amidst polygonal obstacles". *Proc. of the 1988 IEEE ICRA*, pp. 1656-1660.

[2] Basañez L., Suárez R. and J. Rosell. "Contact situations from Observed Reaction forces in Assembly with Uncertainty", *Proc. of the 13th IFAC World Congress, 1996*, Vol A, pp. 331-336.

[3] Brost R. C. and M. Mason, "Graphical Analysis of Planar Rigid-Body Dynamics with Multiple Frictional Contacts, *Fifth International Symposium of Robotics Research*, 1989.

[4] Erdmann M., "On a representation of Friction in Configuration Space", *The International Journal of Robotics Research*, 13 (3), pp. 240-271, 1994.

[5] T. Lozano-Perez, M.T. Mason and R.H. Taylor. "Automatic Synthesis of Fine-Motion Strategies" *The International Journal of Robotics Research*, Vol.3, No.1, pp.3-24, 1984.

[6] Rosell J., Basañez L. and R. Suárez "Determining Compliant Motions for Planar Assembly Tasks in the Presence of Friction" *Proc. of the 1997 Int. Conf. on Intelligent Robots and Systems*, pp. 946-951.

[7] Rosell J., Basañez L. and R. Suárez, "Embedding Rotations in Translational Configuration Space" *Proc. of the 1997 IEEE ICRA*, pp. 2825-2830.

[8] Rosell J. "Fine-motion planning for robotic assembly under modelling and sensing uncertainties" Ph.D. Thesis, Polythecnical University of Catalonia, 1998.

[9] Suárez R., Basañez L. and Rosell J. Using Configuration and Force Sensing in Assembly Task Planning and Execution *Proc. of the 1995 IEEE Int. Symposium on Assembly Task Planning*, pp.273-279.

[10] J. Xiao and R. Volz, "On Replanning for Assembly Tasks Using Robots in the Presence of Uncertainties", *Proc. of the 1989 IEEE ICRA*, pp. 638-645.

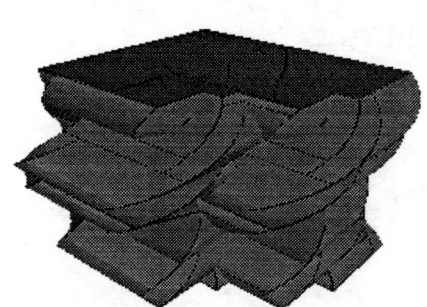

Figure 1: $\mathcal{C}$-space of the assembly task.

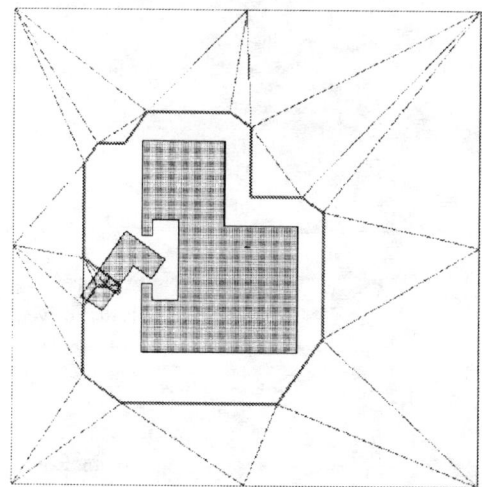

Figure 2: Slice of $\mathcal{C}$-space showing the sections of the $\mathcal{C}$-prisms that partition $\mathcal{C}_{free}$.

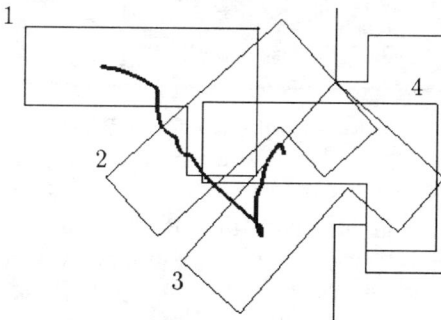

Figure 3: Solutionpath in $\mathcal{C}_{free}$ (from 1 to 2) and $\mathcal{C}_{contact}$ (from 2 to 4).

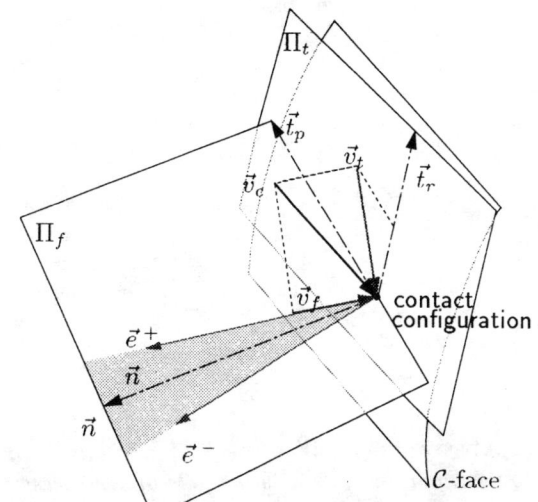

Figure 4: Commanded velocity decomposition.

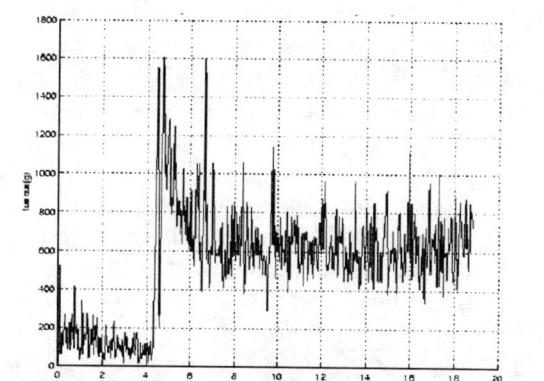

Figure 5: Reaction force module during a type-B contact motion involving translation and rotation.

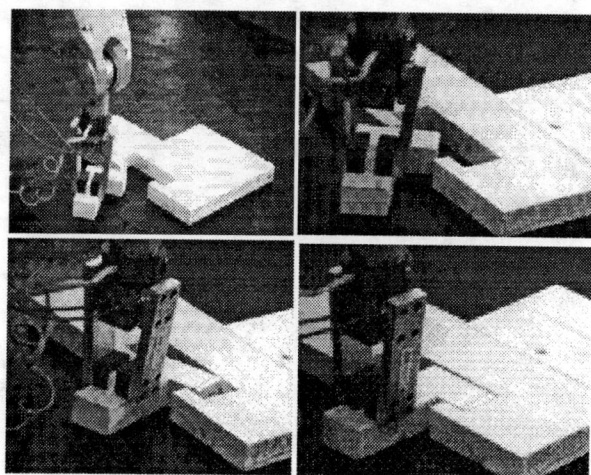

Figure 6: Snapshot of the real task execution.

Learning Friction Estimation for Sensorless Force/Position Control in Industrial Manipulators

V. Zahn, R. Maaß, M. Dapper, R. Eckmiller

University of Bonn, Dept. of Comp. Science VI
Römerstr. 164, D-53117 Bonn, F.R. Germany
Tel.: +49-228-73-4168, FAX: +49-228-73-4425
e-mail: zahn@nero.uni-bonn.de

Abstract - *We present a novel type of friction estimation applied to the field of sensorless force/position control. Friction estimation is used twice in our force/position control: as part of a computed torque controller and in an estimation of external forces exerted by the tool center point (TCP). As a part of a position based Neural Force Control (NFC-P) the estimation friction and external force allows a force/position control without using a force sensor. NFC-P consists of a hybrid force/position controller that accurately generates contact forces to objects with arbitrary flexibility and uncertain distance or shape. NFC-P performs force control by modifying the desired joint angle changes in force direction (position based force control) before they are fed into a computed torque controller. The inverse dynamics of the manipulator is modeled in a computed torque controller. Kinematical mappings guarantee singularity robustness in the entire workspace.*

Results from real time experiments are presented with a 6 DOF industrial manipulator as test bed. The control functions are implemented on a PC based operating system (RCON). All mappings are solved in a cycle time of 0.7msec (Pentium PC 166)[1].

1 Introduction

Many robotic applications involve interaction between manipulator and environment. Stable contact to moving objects and surface tracking with defined contact force typify demanding tasks[5, 10, 11]. In several approaches of force/position control, the contact between the end effector and the environment is assumed to be soft, realized by a flexible tool [1, 4, 6]. Furthermore most approaches involve a very expensive force torque

[1] This work was supported in part by Federal Ministry of Education, Science, Research, and Technology (BMBF) under grant "DEMON"

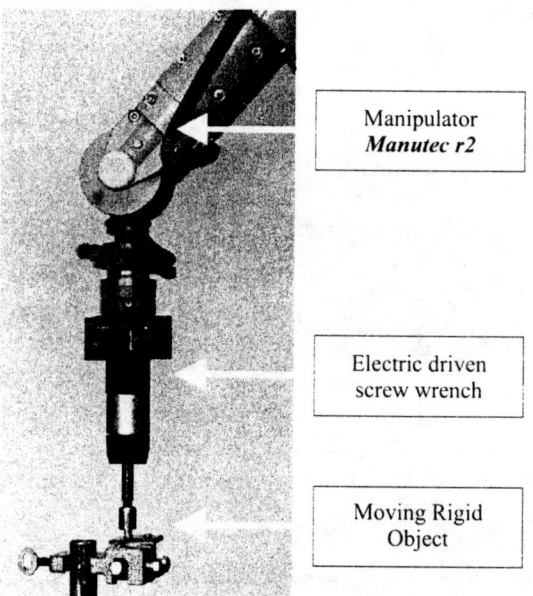

Fig. 1. 6 DOF manipulator (Siemens Manutec r2) with screwdriver tool. Control functions are implemented on a Pentium PC 166 MHz with the RCON-Operating system

sensor. A sensorless approach for a 2 DOF SCARA type robot uses a force estimator build with a structure of disturbance observers[7].

A lot of applications, like unfastening screws in dismounting processes, polishing of unknown surfaces or deburring, only need an approximate contact force within a tolerable range. In such task, an force sensor can be replaced by our developed force estimation.

Our approach is new in two ways: 1) a neural network based force estimation for external forces from known joint torques is developed. 2) an artless tool with no built-in flexibility and without force sensor is used (fig. 1).

2 Results

2.1 Control Concept NFC-P

The position based Neural Force Control (NFC-P) [9] in fig. 2 is a hybrid force/position control system, that can be applied to a high range of surface tracking and force contact tasks for manipulators. This control principle includes neural modules for trajectory optimization, force estimation, and force control as well as a neural computed torque controller.

In a typical application, a planning instance estimates the expected shape and relative position of the target object. The surface tracking task is described by a via point sequence and a desired contact force trajectory. The via point sequence is trained into the neural trajectory optimization that generates a smooth and kinematically valid joint trajectory [8].

NFC-P performs an orthogonal separation of the desired joint position vectors θ_d and velocity vectors $\dot{\theta}_d$:

$$\delta\boldsymbol{\theta}_{d,\parallel} = \mathbf{J}^*\mathbf{S}\mathbf{J}\delta\boldsymbol{\theta_d} \qquad (1)$$
$$\delta\boldsymbol{\theta}_{d,\perp} = \mathbf{J}^*(\mathbf{I}-\mathbf{S})\mathbf{J}\delta\boldsymbol{\theta_d}, \qquad (2)$$

with the Jacobian matrix J, the singularity robust inverse Jacobian matrix J^*, and the separation matrix S.

The desired force F_d, the actual force $F_{contact,a}$, the actual joint positions θ_a, and the desired parallel joint changes $\theta_{d,\parallel}$ are fed into the force controller. This control function computes corrected desired parallel joint changes $\delta\theta_{d,\parallel,corr}$.

The computed torque controller does not see any difference between the tracking component of the desired joint values $\theta_{d,\perp}$ and the force motivated component $\theta_{d,\parallel}$ and performs the inverse dynamics of the manipulator as accurate as possible:

$$\boldsymbol{\tau} = \mathbf{H}(\boldsymbol{\theta})\ddot{\boldsymbol{\theta}} + \mathbf{C}(\boldsymbol{\theta},\dot{\boldsymbol{\theta}})\dot{\boldsymbol{\theta}} + \mathbf{g}(\boldsymbol{\theta}) + \boldsymbol{\tau_F}, \qquad (3)$$

with the inertia matrix H, the Coriolis- and centripetal matrix C and the gravitation vector g. τ_F is the motor torque component that compensates the joint frictions.

In contact to a surface with the tool center point (TCP) the manipulator exerts the contact force

$$\mathbf{F}_{contact} = \mathbf{S}(\mathbf{J}^{-1})^\top(\boldsymbol{\tau_{motor}} - \boldsymbol{\tau_F} - \boldsymbol{\tau_g}), \qquad (4)$$

with the motor torque τ_{motor}, joint friction τ_F, and gravitation component τ_g. This contact force can be controlled by increasing or decreasing the desired parallel joint changes $\theta_{d,\parallel}$.

2.2 Singularity Robust Kinematics During Force Control

Singularities reduce the manipulator's degrees of freedom, thus the workspace is bounded. In or near singularities joint velocities can increase uncontrollable and cause disturbances in force values and in worst case the tool or the work peace can be damaged. Thus in sensor guided manipulator movements singularity robustness (SR) is very important for it is not possible to calculate the movements in advance. NFC-P guarantees singularity robust movements by online trajectory modifications. A more detailed description of the used kinematics can be found in [9].

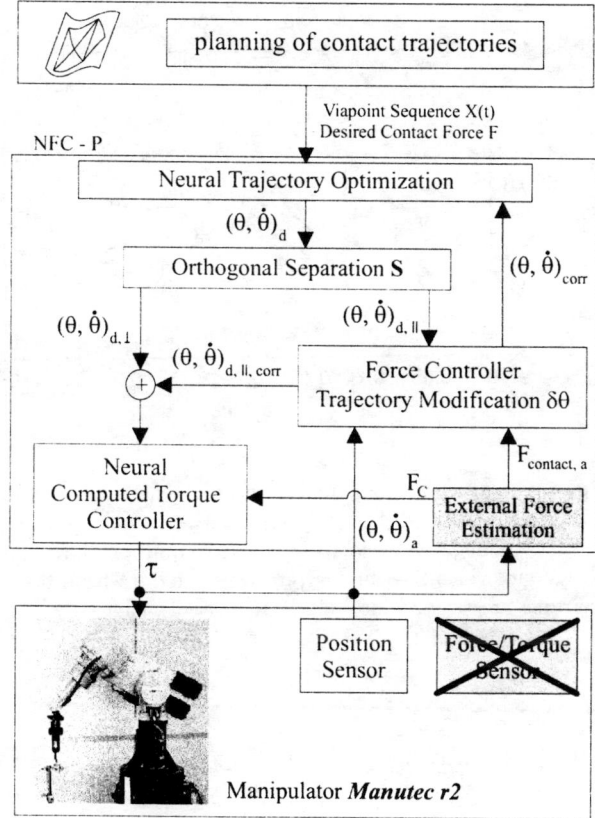

Fig. 2. Control concept NFC-P (Position based Neural Force Control) including neural trajectory optimization, force estimation, force controller, and neural computed torque controller. NFC-P includes singularity robust kinematics in the control loop. The force sensor is replaced by an external force estimation.

2.3 Friction Estimation

Central topic in sensorless force/position control is an accurate friction estimation to calculate the external forces from the joint torques. The model of a n-DOF manipulator with a friction model

$$\tau_F = \mathbf{F}_v + \mathbf{F}_C \cdot sign\,\dot{\boldsymbol{\theta}} \qquad (5)$$

consisting of Coulomb friction $\mathbf{F}_C$ and velocity dependent friction $\mathbf{F}_v$ is described by eq. 3.

In this work a neural computed torque controller is used, that performs a nonlinear decoupling and feedback linearization of the manipulator's dynamics using the plants inverse dynamics model (eq. 3).

The computed torque controller models and calculates the effects of mass matrix $\mathbf{H}$, Coriolis coupling $\mathbf{C}$, and gravitation-caused torques $\mathbf{g}$ in real time from the manipulators state $(\boldsymbol{\theta}, \dot{\boldsymbol{\theta}})$ [2].

position and the sign of joint velocity; the most important effects in the tested manipulator were caused by Coulomb friction.

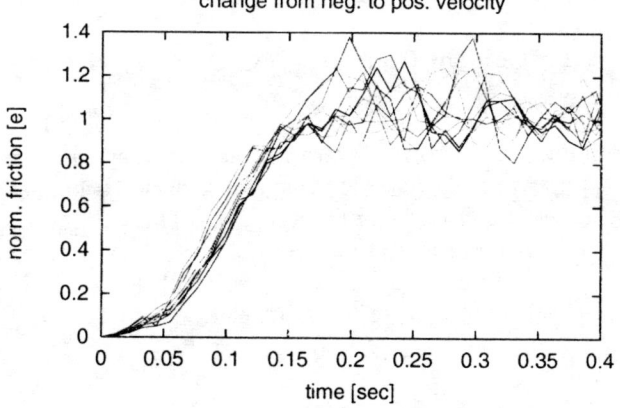

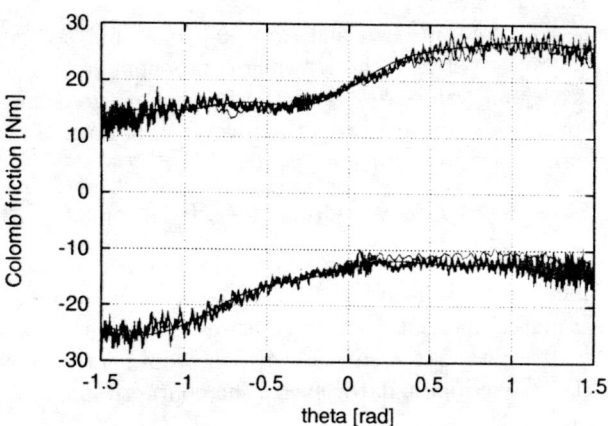

Fig. 3. Coulomb friction of a 6 DOF manipulator (Siemens Manutec r2) at two different velocities. The smooth line in between is the output of the trained RBF networks.

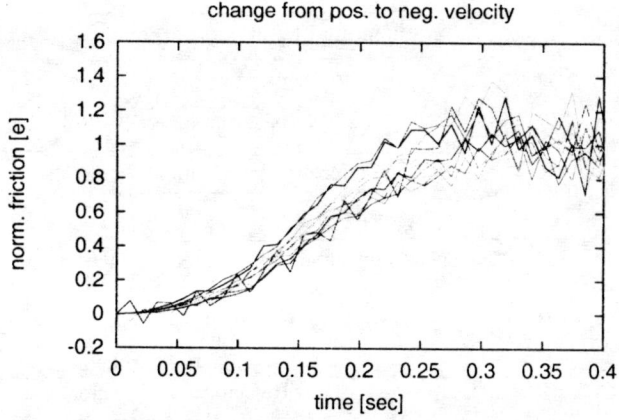

Fig. 4. Change over behavior of axis 2: the normalized signal from several tests show, that the friction change from positive to negative velocities is much quicker than the change vice versa. For fitting purposes the data from both diagrams are trained into one RBF network each.

This approach was extended to a Lyapunov stable controller in another work [3, 12] by using the model of Coulomb and velocity dependent friction (eq. 5) to derive an online learning rule for friction compensating RBF networks.

In this work we decided to go another, more experimental way. We measured the joint friction in free movements of dedicated joints at different, constant velocities. The tests showed, that the friction can be assumed joint specific; no coupling between different axes was measured. The friction terms were mostly constant at different speeds, but depended on the joint

Consequently we decided to model the Coulomb friction per axis and sign of velocity; we trained two small RBF networks (RBF^+, RBF^-), which allow a smooth fitting and fast calculation. Fig. 3 shows a measurement of friction for axis 2, the axis with the largest Coulomb friction and gravitation, at low and high velocity together with the output from the trained networks. The signals of slow and fast movements show the same behavior and are position dependent within a range width of more than 10 Nm.

The most difficult problem in modeling Coulomb

friction is the course of friction at a sign change of velocity. The difference in joint friction between negative and positive velocities in the tested manipulator are up to 50 Nm and the change over behavior is very joint specific, depending on the type of motors and gear boxes. We measured the course of friction in such change overs at different positions. Fig. 4 shows normalized measurements, where 0 means friction from previous movement direction and 1 means new friction. Again it showed to be a direction dependent behavior: for example in axis 2 the change from negative to positive velocity was nearly the same in all positions, but took only $0.17 sec$, where the change vice versa took $0.29 sec$. We trained these axis specific time course into RBF networks to model the change over.

To eliminate destabilizing effects with noisy velocity signals at joint velocities near $0 \frac{rad}{s}$, we use the desired velocities for discrimination, which RBF network to use for friction estimation. The whole friction estimation scheme is shown in fig. 5.

Assuming the robot to move slowly and with small acceleration, the external forces can be calculated using (eq. 4) from the robot equation (eq. 3). The inverse kinematics is represented by the singularity robust inverse Jacobian $\mathbf{J^{-1}}$. $\mathbf{S}$ defines the separation matrix, which separates directions of position control from directions of force control.

2.5 Contact Detection

Another difficult point in sensorless force/position control is to detect a contact. A TCP acceleration has to be distinguished from an external force. Both acceleration and contact cause higher joint torques than the friction and gravitation compensating torques.

Forces in the TCP mean acceleration of the TCP; thus the Cartesian acceleration in force direction is used to detect a contact. If desired and actual Cartesian acceleration do not correspond, a contact is detected (fig. 6).

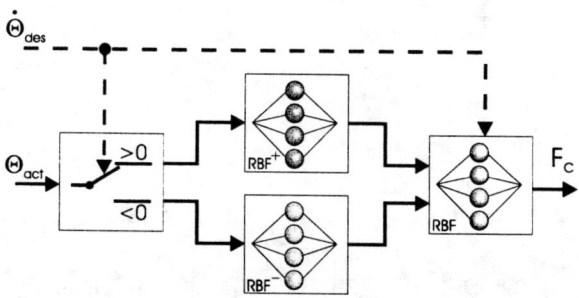

Fig. 5. Coulomb friction estimator(CFE): two trained RBF networks, one for each movement direction are used to estimate the Coulomb friction. Depending on the sign of the actual velocity, the according RBF network is chosen. The change over time course of friction is modeled in an additional RBF network.

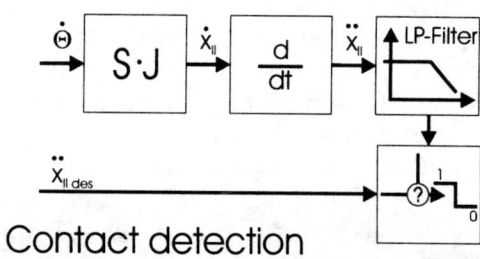

Fig. 6. Contact detection: the Cartesian acceleration in force direction is calculated by transposing the joint velocity into the Cartesian space, derivating the Cartesian velocity and filtering. If desired and actual Cartesian acceleration do not correspond, a contact is detected (output=1).

2.4 External Force Estimator

The most important effects in force/position control are caused by gravitation and Coulomb friction. For velocities and accelerations are small in force/position tasks, especially in directions relevant to force control, effects from mass matrix (see eq. 3) and Coriolis coupling are neglectable. Thus it is sufficient to take only Coulomb friction and gravitation into account to calculate external forces from joint torques.

Fig. 7 shows the complete force estimator, which is implemented in NFC-P.

3 Real time Experiments

As an example task we choose dismounting a screw: the manipulator had to drive along a parallel line of the z axis with no contact and finding an rigid object by contacting it and keeping a contact force of $10N$ to it, even if it moves (see fig. 8). Similar tasks are needed in mounting a screw, polishing or deburring.

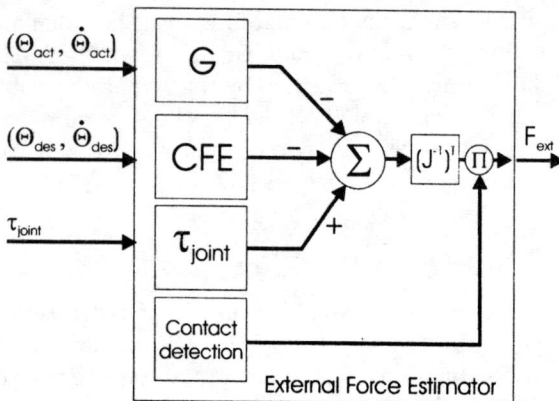

Fig. 7. Force estimator: the force estimator is combined from a calculation of gravitational forces (G), the Coulomb friction estimator (CFE) and the contact detection unit.

Fig. 9 shows results from axis 2, which is most involved in this task, in contacting a stiff surface (steel-steel): the axis starts its movement at $t = 86.6s$ and accelerates to constant velocity. At $t = 88.5$ the contact is detected from an unexpected force peak, force control is activated and after a short overshoot the contact force is established. The external contact force is established with about $9N$, thus 10% to low. This results from not perfectly compensated friction in computed torque controller and friction estimation. In the experiments the estimated external force was used to control the manipulator; an additionally attached force sensor was only used to measure the real external forces and to compare estimation and reality.

We performed many contact experiment with different kinds of surfaces, from soft(foam) to stiff(steel). The contact detection works better with stiff surfaces. In a very soft surface the manipulator moves nearly as in free space and thus only slightly larger torques are applied and these are not distinguishable from friction torques.

If once contact is established, NFC-P enables the manipulator to follow movements of the contacted object by online trajectory generation.

4 Conclusions

We presented a novel approach for sensorless hybrid force/position control. The proposed concept was stable in establishing and keeping a desired contact force.

By learning friction terms a force estimator was developed to estimate external forces from joint torques,

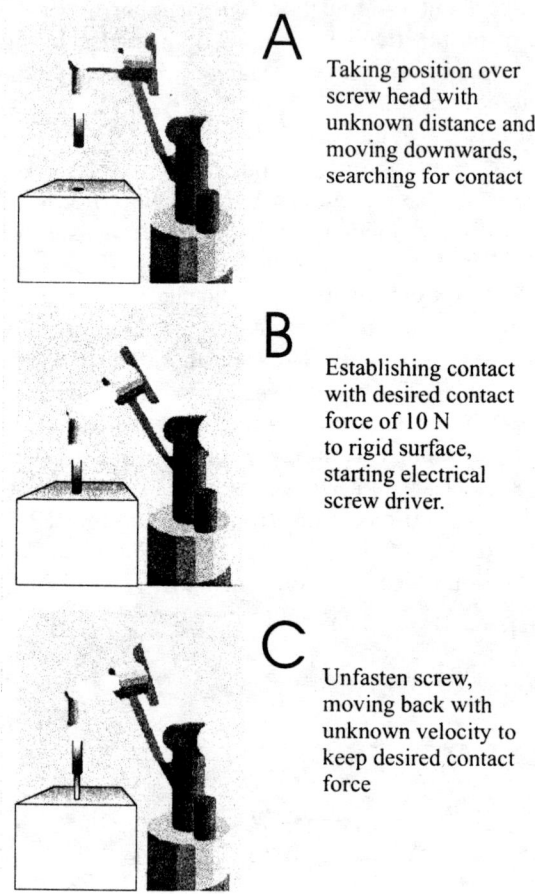

Fig. 8. Dismounting a screw: the tool center point (TCP) is positioned over the screw head(A), the TCP moves towards the screw head and contacts the head (B), after starting the electrical screw driver the screw head moves with unknown velocity and the TCP eludes (C) while keeping a constant contact force.

positions, and velocities of the manipulator joints. For force/position control a developed concept of position based Force Control (NFC-P) was used. By using this concept it is now possible to realize both: sensor based and sensorless force/position control.

The special relevance of the developed force estimator lies in force/position applications, where the precision of an expensive force sensor is not needed or where force sensors cannot be operated. The PC based implementation together with the absence of a force sensor offers solutions for a wide range of interacting tasks of multi joint manipulators and environment, without an extended use of expensive hardware devices.

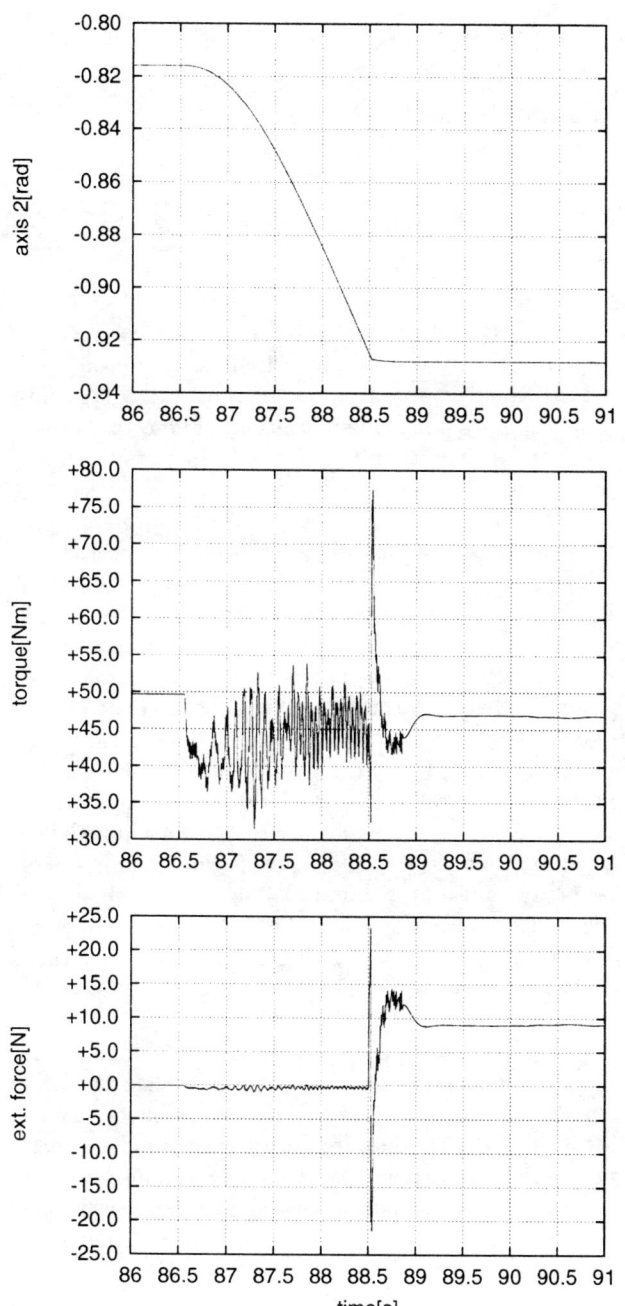

Fig. 9. Experiment: the manipulator drives along a parallel line to the z axis and meets an obstacle. Shown are angle and torque of axis two, which is mainly involved. At $t = 88.5s$ the object is contacted and after short swing the contact is established and the defined contact force is hold. The force measurement in this diagram is done with a force sensor in the tool.

References

1. G. Bonitz and T. Hsia. Robust internal force based impedance control for coordinating manipulators. In *IEEE Proc. International Conference on Robotics and Automation, Mineapolis, Minesota*, pages 622–528, 1996.
2. M. Dapper, R. Maaß, V. Zahn, and R. Eckmiller. Neural force control (NFC) applied to industrial manipulators in interaction with moving rigid objects. In *IEEE Proc. Int. Conf. Robotics and Automation, Leuven, Belgium*, pages 2048–2053, 1998.
3. M. Dapper, V. Zahn, R. Maaß, and R. Eckmiller. How to compensate stick slip friction in neural velocity force control (NVFC) for industrial manipulators. In *IEEE Proc. Int. Conf. Robotics and Automation, ICRA99, Detroit(Mi), USA*, 1999.
4. R.V. Dubey, T. F. Chan, and S. E. Everett. Variable damping impedance control of a bilateral telerobotic system. *IEEE Control Systems Magazine*, 17(1):37–45, 1997.
5. G. Ferretti, G. Magnani, and P. Rocco. On the stability of integral force control in case of contact with stiff surfaces. *Journal of Dynamic Systems, Measurement, and Control*, 117(4):547–553, 1995.
6. B. Heinrichs, N. Sepehri, and A. B. Thornton-Trump. Position-based impedance control of an industrial hydraulic manipulator. *IEEE Control Systems Magazine*, 17(1):46–52, 1997.
7. W.K Chung K.S. Eom, I.H. Suh and S.-R. Oh. Disturbance observer based force control of robot manipulator without force sensor.
8. R. Maaß, M. Dapper, and R. Eckmiller. Neural trajectory optimization (nto) for manipulator tracking of unknown surfaces. In *Springer Proc. Int. Conf. Artificial Neural Networks, ICANN98, Skoevde, Sweden*, pages 893–898, 1998.
9. R. Maaß, V. Zahn, M. Dapper, and R. Eckmiller. Hard contact surface tracking for industrial manipulators with sr position based force control. In *IEEE Proc. Int. Conf. Robotics and Automation, ICRA99, Detroit(Mi), USA*, 1999.
10. J. Steck, K. Rokhsaz, and S. P. Shue. Linear and neural network feedback for flight control decoupling. *IEEE Control Systems Magazine*, 16(4):22–30, 1996.
11. D. Surdilovic. Contact stability issues in position based impedance control: Theory and experiments. In *IEEE Proc. International Conference on Robotics and Automation, Mineapolis, Minesota*, pages 1675–1680, 1996.
12. V. Zahn, R. Maaß, M. Dapper, and R. Eckmiller. How to compensate friction in dynamic hard contact tasks by means of neural networks. In *Int. Conf. on Computational Intelligence for Modelling, Control and Automation, CIMCA99, Vienna, Austria*, Feb. 1999.

Learning of Robot Tasks via Impedance Matching

Suguru Arimoto†, P.T.A. Nguyen†, and Tomohide Naniwa‡

† Department of Robotics, Ritsumeikan University
Kusatsu, Shiga, 525-8577 Japan

‡ Graduate School of Science and Engineering, Yamaguchi University
Tokiwadai 2557, Ube, Yamaguchi, 755 Japan

Abstract

This paper is aimed at presenting a physical interpretation of practice-based learning (so-called "iterative learning control") for robotic tasks from the viewpoint of "bettering impedance matching". At first, the concepts of impedance and impedance matching that are inherent to linear electric circuits are generalized for a class of nonlinear dynamics including robotic tasks by means of passivity. It is then shown in the simplest case when the tool endpoint is free to move that a simple iterative scheme of learning enables robots to make a progressive advance in a sense of zero-impedance matching at every trial of operation. In case of impedance control when a soft and deformable finger-tip presses a rigid object or environment, it is shown that, for a given desired periodic force, physical interaction between the soft fingertip and the rigid object, the robot learns steadily the desired task by monotonously increasing the grade of impedance matching pertaining to dynamics of the robot task with controller dynamics.

1 Introduction

It is already well known as in the literature [1]~[5] that mechanical robots can learn desired motions iteratively via repeated exercises if the next control input is designed carefully by modifying the previous control input by adding an appropriate error modification term. One of the authors [2] showed that a D-type iterative learning control scheme when a numerical differentiation of the angular velocity errors in each joint is used in modification of the control input makes the trajectory of robot motion converge asymptotically to the desired one. In order to avoid rapid accumulation of noises caused by numerical differentiation of measured velocity signals, one of the authors proposed a class of P-type ILC (see [3]) where errors of measured angular velocity versus the desired velocity profile are directly used in modification of the next control input without taking differentiation. These two learning control schemes differ not only in physical units of used signals but also in the proof of convergence of trajectory tracking. The proof of convergence in the case of D-typer learning is based on a contraction mapping in a Banach space originally used in Picard's proof for the existence of a unique solution to a differential equation. On the other hand, that of P-type learning is based on the passivity of residual nonlinear dynamics between the desired motion and the actual one. This paper attempts to present a physical interpretation that a learning control scheme with using an SP-D (Saturated Position and its Derivative) signal enables robots to make progress toward impedance matching. To do this, it is shown that for a class of robot dynamics an SP and D (angular velocity) output satisfies passivity when the robot endpoint is free to move. An iterative learning control scheme is then considered to be a process of acquiring a desired control input realizing the given desired motion by making progress toward zero-impedance matching. In the case of impedance control when a soft finger is pressing a rigid object, an iterative learning control scheme for a desired periodic force of physical interaction is proposed. It is shown by computer simulation that the proposed learning control scheme realizes the ideal force tracking after a certain number of repetitions of periodic operation.

2 Learnability on the Basis of Passivity

Given a desired output $y_d(t)$ over a time interval $[0,T]$ for a nonlinear dynamical system with input u and output y, the problem of iterative learning control (ILC) is to find a recursive form of learning control law $u_{k+1} = F(u_k, \Delta y_k)$ in trial number k that eventually realizes the convergence $\Delta y_k(t) \to 0$ as $k \to \infty$

in some sense, where Δy_k stands for the output error, i.e., $\Delta y = y - y_d$ (see Fig. 1). This section is concerned

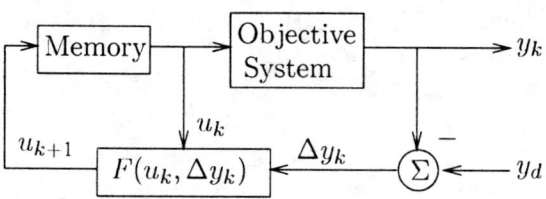

Figure 1: Schematic diagram of Iterative Learning Control

with a problem of what kind of characterizations of dynamical systems is crucial in ensuring learnability (the existence of a function norm $\|\cdot\|$ such that $\|\Delta y_k\| \to 0$ as $k \to \infty$). Among potential candidates of recursive forms, let us pick up one of the simples laws (Fig. 2):

$$u_{k+1} = u_k(t) - \Phi \Delta y_k(t) \qquad (1)$$

where Φ is a constant positive definite matrix. It is

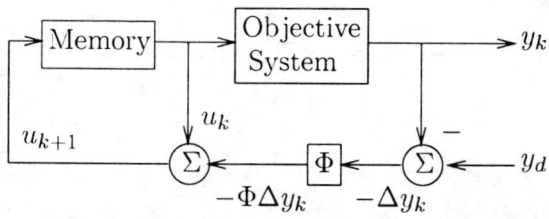

Figure 2: One of the simplest schemes of ILC

assumed implicitly that an ideal input u_d that realizes the prescribed output y_d exists but it can not be calculated in ordinary cases due to the uncertainty of description of system dynamics. Subtraction of u_d from both sides of (1) yields

$$\Delta u_{k+1} = \Delta u_k - \Phi \Delta y_k \qquad (2)$$

where $\Delta u_k = u_k - u_d$. Inner products of both sides of (1) with themselves via the inverse of Φ are expressed as

$$\Delta u_{k+1}^T \Phi^{-1} \Delta u_{k+1}$$
$$= \Delta u_k^T \Phi^{-1} \Delta u_k - 2 \Delta u_k^T \Delta y_k + \Delta y_k^T \Phi \Delta y_k. \qquad (3)$$

Integration of this equation over time interval $[0, T]$ leads to

$$\|\Delta u_{k+1}\|_{\Phi^{-1}}^2 = \|\Delta u_k\|_{\Phi^{-1}}^2 + \|\Delta y_k\|_{\Phi}^2$$
$$- 2 \int_0^T \Delta y_k^T(\tau) \Delta u_k(\tau) d\tau \qquad (4)$$

where $\|\cdot\|$ is defined as

$$\|\Delta u_k\|_{\Phi^{-1}}^2 = \left\{ \int_0^T \Delta u_k^T(\tau) \Phi^{-1} \Delta u_k(\tau) d\tau \right\}^{1/2}. \qquad (5)$$

If the inner product of output error Δy_k with input error Δu_k is always positive with a margin that makes compensation for $\|\Delta y_k\|_{\Phi}^2$ in (4), that is, if the inequality

$$\int_0^T \Delta y_k^T(\tau) \Delta u_k(\tau) d\tau$$
$$\geq \frac{1+\beta}{2} \int_0^T \Delta y_k^T(\tau) \Phi \Delta y_k(\tau) d\tau$$
$$= \frac{1+\beta}{2} \|\Delta y_k\|_{\Phi}^2 \qquad (6)$$

follows with a positive constant β, then it follows from (4) that

$$\|\Delta u_{k+1}\|_{\Phi^{-1}}^2 \leq \|\Delta u_k\|_{\Phi^{-1}}^2 - \beta \|\Delta y_k\|_{\Phi}^2. \qquad (7)$$

The inequality implies that the sequence $\{\|\Delta u_k\|_{\Phi^{-1}}\}$ is monotonously decreasing with increasing k as long as $\|\Delta y_k\|_{\Phi}$ does not vanish. Since $\{\|\Delta u_k\|_{\Phi^{-1}}\}$ is bounded from below, the monotonous decrease of $\{\|\Delta u_k\|_{\Phi^{-1}}\}$ implies the convergence, which proves that $\|\Delta y_k\|_{\Phi} \to 0$ as $k \to \infty$. Since Φ is a fixed positive definite matrix, this means that $\Delta y_k \to 0$ as $k \to \infty$ in the sense of $L^2[0, T]$ norm.

The passivity with a margin as described by (6) can be regarded as "dissipativity" for dynamical systems. In terms of "passivity" and "dissipativity" it is possible to discuss the learnability of such nonlinear systems in a rigorous manner.

3 Learnability for Robot Dynamics

Dynamics of a robotic arm with all rotational joints are expressed in the following form:

$$\left\{ H(q)\frac{d}{dt} + \frac{1}{2}\dot{H}(q) \right\} \dot{q} + S(q, \dot{q})\dot{q} + r(\dot{q}) + g(q) = u \qquad (8)$$

where $q = (q_1, \cdots, q_n)^T$ denotes the joint angle vector, $H(q)$ the inertia matrix, $S(q, \dot{q})$ a skew-symmetric matrix, $r(\dot{q})$ the damping term including frictional forces, $g(q)$ the gravity term, and u the control input vector whose components are torques generated by joint actuators. It should be noted that $H(q)$ is symmetric and positive definite. Moreover, since each entry of $H(q)$ is constant or a sinusoidal or cosine function of components of q, $S(q,\dot{q})$ is linear and homogeneous in $\dot{q}$ with property $S(q,0) = 0$ and each non-diagonal entry of $S(q,\dot{q})$ is a sinusoidal or cosine function. Further, it is natural to assume that $\dot{q}^T r(\dot{q}) \geq 0$ and there exists a potential function $U(q)$ such that $\partial U/\partial q = g(q)^T$. Since $U(q)$ is a linear combination of sinusoidal and cosine functions of q, it is possible to take the constant term of $U(q)$ so that $\min_q U(q) = 0$.

As to the robot dynamics of eq.(8), the pair of input u and output $\dot{q}$ satisfies passivity, i.e.,

$$\int_0^t \dot{q}^T(\tau) u(\tau) d\tau$$
$$\geq E(t) - E(0) + \int_0^t \dot{q}^T(\tau) r(\dot{q}(\tau)) d\tau$$
$$\geq -E(0) = -\gamma_0^2 \quad (9)$$

where E denotes the total energy expressed as $E = \frac{1}{2}\dot{q}^T H(q)\dot{q} + U(q)$. This passivity relation plays a crucial role in building a bridge between the energy conservation law in physics and the operational input-output characterization in system theory. In light of this property, it is possible to make mechanical robots learn a desired motion through repeated practices.

In fact, consider the problem of trajectory tracking for a given joint-trajectory $q_d(t)$ defined for $t \in [0,T]$. Let us conveniently introduce a linear combination of velocity error $\Delta \dot{q}$ and saturated position error $\alpha s(\Delta q)$, i.e.,

$$\Delta y = \Delta \dot{q} + \alpha s(\Delta q) \quad (10)$$

where $\Delta q = q - q_d$, $\alpha > 0$ is a constant, $s = (s_1, \cdots, s_n)^T$, and each $s_i(\Delta q_i)$ is a saturated function with a profile exhibited in Fig. 3. Then, it is already shown in the literature [7] that the pair $\{\Delta u, \Delta y\}$ satisfies passivity concerning the error dynamics between

$$\left\{ H(q)\frac{d}{dt} + \frac{1}{2}\dot{H}(q) \right\} \dot{q} + \{S(q,\dot{q}) + B\}\dot{q}$$
$$+ D\Delta y + g(q) = u \quad (11)$$

and

$$\left\{ H(q_d)\frac{d}{dt} + \frac{1}{2}\dot{H}(q_d) \right\} \dot{q}_d + \{S(q_d,\dot{q}_d) + B\}\dot{q}_d$$
$$+ g(q_d) = u_d. \quad (12)$$

More detailedly, it is possible to show (see [7]) that there exists a storage function $V(\Delta q, \Delta \dot{q})$ positive definite in Δq and $\Delta \dot{q}$ and a scalar-valued dissipation function $\xi(y) > 0$ with $\xi(0) = 0$ such that

$$\int_0^t \Delta y^T \Delta u d\tau \geq V(\Delta q(t), \Delta \dot{q}(t))$$
$$- V(\Delta q(0), \Delta \dot{q}(0)) + \int_0^t \xi(\Delta y(\tau)) d\tau. \quad (13)$$

Further, it is possible to adjust the velocity gain B by damping injection so that $\xi(\Delta y)$ is in the order of a quadratic function of Δy, i.e.,

$$\xi(y) \geq \beta \|\Delta y\|^2 \quad (14)$$

with some $\beta > 0$. The most important result concerning robot's learnability is that robots can acquire the desired motion through repeating practices if the control input is updated iteratively by the law of eq.(1). It has been shown (see [7]) that if $\Phi < 2\beta I$ then $\Delta y_k \to 0$ as $k \to \infty$ in the sense of $L^2(0,T)$. Since $\Delta \dot{q}_k + \alpha s(\Delta q) \to 0$ as $k \to \infty$, $\Delta q_k(t) \to 0$ as $k \to \infty$.

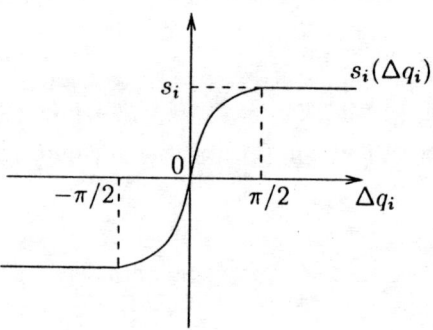

Figure 3: $s_i(\Delta q_i)$ is a saturated nonlinear function.

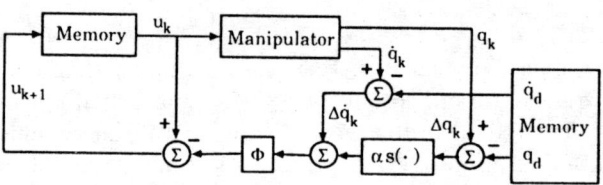

Figure 4: Iterative learning based on SP-D type update law.

The most noteworthy merit of using the output Δy in the learning update law instead of $\Delta \dot{q}$ is that the initial setting at the beginning of every trial is unnecessary if the desired output q_d satisfies $q_d(0) = q_d(T)$ and $\dot{q}_d(0) = \dot{q}_d(T)$. Fig. 6 shows the experimental data for comparison of the use of $\Delta \dot{q}$ in Fig. 6(a) with the use of Δy in Fig. 6(b), where a three d.o.f DD arm was used. The details of the experiment was reported in a dissertation at Yamaguchi University (see Naniwa [8]).

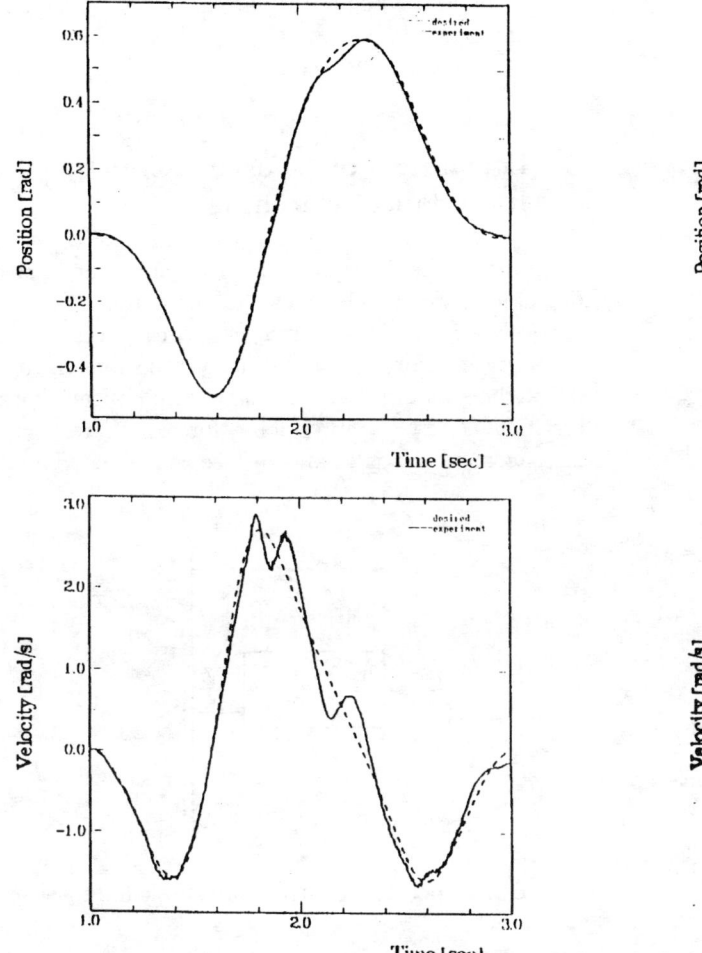

Figure 5(a): Joint angle position and velocity trajectories of link 2 after 30 trials.

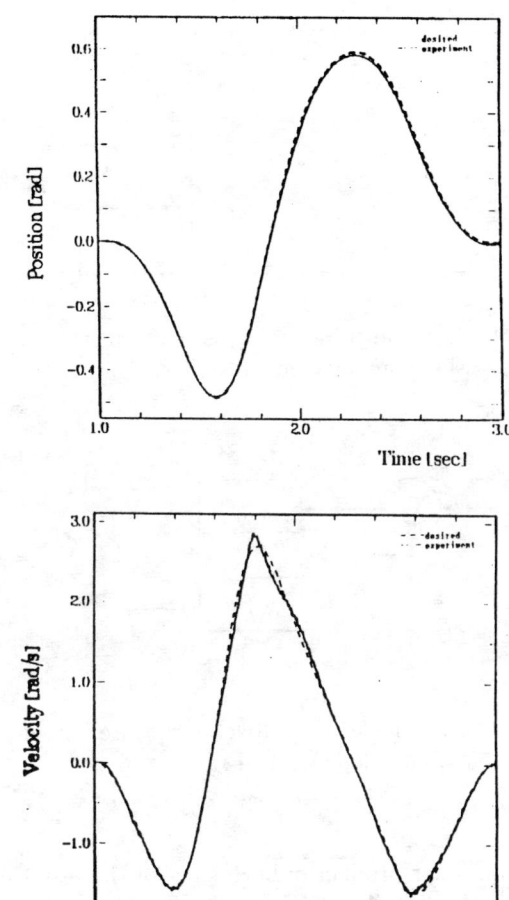

Figure 5(b): Joint angle position and velocity trajectories of link 2 after 30 trials using Δy.

4 Learning toward Zero-Impedance Matching

Iterative learning can be considered to be a process of acquiring a desired control $u_d(t)$ realizing the given desired motion $q_d(t)$ for $t \in [0,T]$ by making progress toward impedance matching through repeated practices. To see this, it is necessary to simplify the argument and gain a physical insight into the problem by treating the simplest case that the objective system is linear and time-invariant, of single input and single output, and satisfies dissipativity. First note that the circuit in Fig. 6 can be rewritten in a form of negative feedback control system as shown in Fig. 7.

Suppose that, given a desired periodic output $y_d(t)$ (corresponding to E in Fig. 7) with period T, i.e.,
$y_d(t) = y_d(t + T)$, the problem is to find the desired u_d (corresponding to V in Fig. 7) that realizes $Zu_d = y_d$. It should be noted that the impedance function of the objective system is strictly positive real and the internal impedance Z_0 must be very close to zero-impedance. Then, as discussed in section 3, the iterative learning control system can be depicted as in Fig. 8, where the learning update law is described by eq.(1). Since the strict positive realness of Z implies the existence of a positive constant $\gamma > 0$ such that

$$\int_0^t \Delta u_k \Delta y_k d\tau \geq V_k(t) - V_k(0) + \int_0^t \gamma \|\Delta y_k\|^2 d\tau \tag{15}$$

where V expresses a storage function that is nonnegative (see [7]), it is possible to obtain the following inequality (subtract u_d from both sides of eq.(1) and

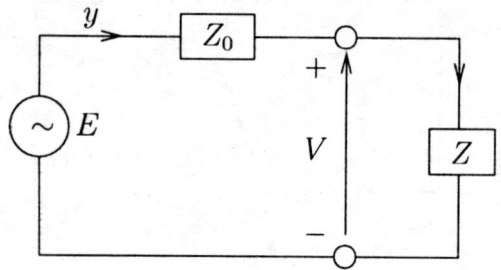

Figure 6: Impedance matching is realized when $Z = Z_0^*$. Z_0^* is the complex conjugate of Z_0.

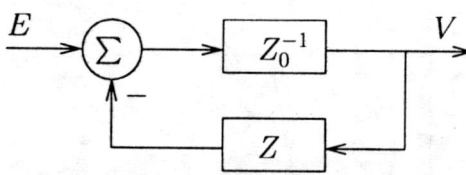

Figure 7: This negative feedback structure is equivalent to the circuit depicted in Fig. 6.

taking an inner product of both sides of the resultant equation through Φ^{-1}):

$$\Phi^{-1}\|\Delta u_{k+1}\|^2 + V_{k+1}$$
$$\leq \Phi^{-1}\|\Delta u_k\|^2 + V_k + (\Phi - 2\gamma)\|\Delta y_k\|^2 \quad (16)$$

where $y_k(t) = y(t + kT)$, $u_k(t) = u(t + kT)$, $V_k(t) = V(t + kT)$, and $\|\Delta u\|$ denotes the norm of Δu in $L^2[0,T]$. Hence, if Φ is chosen so that $0 < \Phi < 2\gamma$, then eq.(16) means that $y_k(t) \to y_d(t)$ in $L^2[0,T]$ as $k \to \infty$. This concludes that the forward path in the

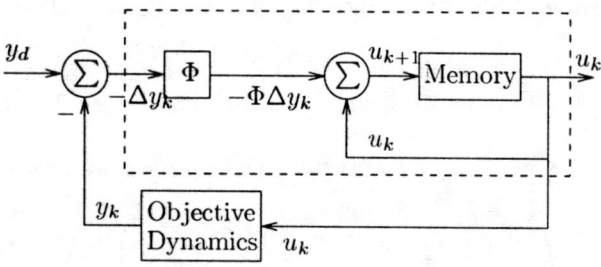

Figure 8: Iterative learning as impedance matching with internal zero-impedance

system of Fig. 8 tends to realize the zero-impedance (the infinitely large admittance), because $\Delta y_k \to 0$ and $u_k \to u_d$ as $k \to \infty$.

5 Learning of Force Control via Impedance Matching

Consider the simplest problem of impedance control depicted in Fig. 9, where a single d.o.f. tool whose end is covered with soft material must press a rigid object at the desired force f_d. In ordinary situations the mass M of the tool is uncertain and the nonlinear characteristics $f(\Delta x)$ of reproducing force with respect to displacement Δx is unknown (see Fig. 10). The dy-

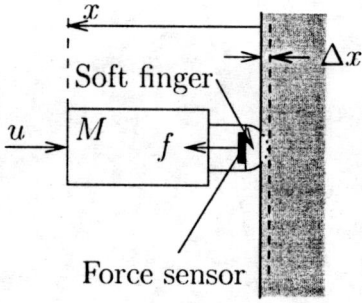

Figure 9: Impedance control for a single d.o.f. system.

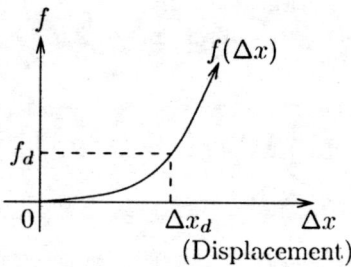

Figure 10: Nonlinear characteristics of reproducing force.

namics of the system can be described as

$$M\Delta\ddot{x} + c\Delta\dot{x} = -f + u \quad (17)$$

where u denotes the control input. A reasonable way to design u is that

$$u = f_d + \hat{M}\dot{r} + \hat{c}r + v \quad (18)$$

where $\hat{M}$ and $\hat{c}$ stand for estimates for M and c, r an appropriate signal defined later, and v is an extra input. Substituting eq.(18) into eq.(17) yields

$$M(\Delta\ddot{x} - \dot{r}) + c(\Delta\dot{x} - r) \\ + \Delta M\dot{r} + \Delta cr = -\Delta f + v \quad (19)$$

where $\Delta M = M - \hat{M}$ and $\Delta c = c - \hat{c}$. By denoting $\Delta y = \Delta\dot{x} - r$ and setting

$$\hat{M}(t) = \hat{M}(0) - \int_0^t \gamma_M^{-1} \dot{r}(\tau) \Delta y(\tau) d\tau, \quad (20)$$

$$\hat{c}(t) = \hat{c}(0) - \int_0^t \gamma_c^{-1} r(\tau) \Delta y(\tau) d\tau, \quad (21)$$

it is possible to see that an inner product between y and eq.(19) yields

$$\frac{d}{dt}\frac{1}{2}\left\{M\Delta y^2 + \gamma_M \Delta M^2 + \gamma_c \Delta c^2\right\} \\ + c\Delta y^2 = -\Delta y(\Delta f + v). \quad (22)$$

This form suggests us the best design of signal Δy through the design of signal r. In other words, if the pair $\Delta y, \Delta f$ satisfies passivity or dissipativity, then the input v and the output Δy of the overall system satisfies passivity or dissipativity. This observation leads to the definition

$$r = -\alpha\Delta x - \beta\Delta F, \quad \Delta F = \int_0^t \Delta f d\tau \quad (23)$$

where α and β are appropriate positive constants. Then,

$$y = \Delta\dot{x} + \alpha\Delta x + \beta\Delta F \\ = \delta\dot{x} + \alpha\delta x + \beta\Delta\bar{F} \quad (24)$$

where $\delta x = \Delta x - \Delta x_d$, $f(\Delta x_d) = f_d$, and $\Delta\bar{F} = \Delta F + \alpha\Delta x_d/\beta$. If v is considered to be an extra damping injection Dy plus an original disturbance n then eq.(24) can be written in the form

$$M\Delta\ddot{y} + (c+D)\Delta y + (\Delta M\dot{r} + \Delta cr) = -\Delta f + n. \quad (25)$$

The pair of eqs.(24) and (25) can be expressed in the same circuit as in Fig. 11 except the current source w_k and ΔF_k (in this case $w_k = 0$ and ΔF_k should be replaced with $\Delta\bar{F}_k$). Note that the circuit of Fig. 11 is just a kind of nonlinear version of an electric circuit depicted in Fig. 6.

Next consider the case that the desired reproducing force is a period function in time, i.e., $f_d(t) = f_d(t+T)$ with $T > 0$. The problem is to find a desired input $n_k(t)(= n(t+kT))$ that drives the system of Fig. 11

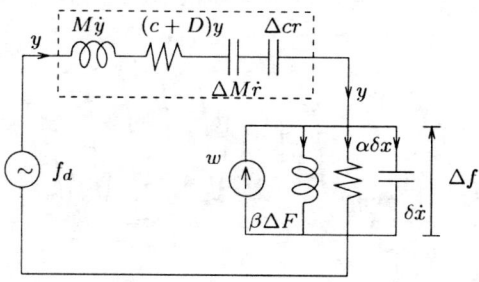

Figure 11: A circuit-theoretic expression of impedance control

to make Δf_k tend to vanish as $k \to \infty$. We design a learning update law by

$$w_{k+1} = w_k - \Phi\Delta f_k \quad (26)$$

and the signal $r_k(t) = r(t+kT)$ and the control input u_k are defined as

$$r_k = -\alpha\Delta x_k - \beta\Delta F_k + w_k, \quad (27)$$
$$u_k = f_d + \hat{M}_k\dot{r}_k + \hat{c}r_k - D\Delta y_k \quad (28)$$

where Δy_k is defined as

$$\Delta y_k = \Delta\dot{x} - r_k = \delta\dot{x}_k + \alpha\delta x_k + \beta\Delta F_k - \Delta w_k \quad (29)$$

where $\Delta w_k = w_k - w_d$ and $w_d = \Delta\dot{x}_d + \alpha\Delta x_d$. Then, it follows that

$$M\Delta\dot{y}_k + (c+D)\Delta y_k + \Delta M\dot{r}_k + \Delta cr_k = -\Delta f_k. \quad (30)$$

It should be noted that eqs.(29) and (30) can be expressed as a circuit depicted in Fig. 11. In the present stage, it seems hard to show passivity of the pair $\{\Delta f_k, \delta\dot{q}_k\}$, but is is possible to show passivity of $\{\Delta f_k, \delta\dot{x}_k + \alpha\Delta x_k\}$ if α is sufficiently large (the details is omitted in this paper).

The computer simulation shows effectiveness of the proposed method (see Fig. 12), which was conducted under the following data:
$M = 0.3$ [kg], $c = 0.3$ [m/s], $f_d = 1.0 + 0.3\sin(\pi t)$, $f(\Delta x) = 3500(\Delta x)^{3/2}$, $\alpha = 80$, $\beta = 0.3$, $D = 24$, $\Phi = 0.05$, $\gamma_M = 0.1$, $\gamma_c = 0.1$, $\hat{M}(0) = 0.0$, $\hat{c}(0) = 0.0$.

6 Conclusion

It has been shown that iterative learning control schemes for a class of nonlinear dynamics pertaining to robotic motions can be interpreted as a type of progressive learning toward impedance matching.

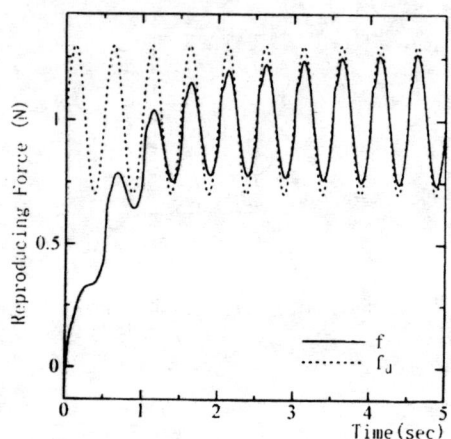

Figure 12(a): Response of Reproducing Force in case of unknown $f(\Delta x)$

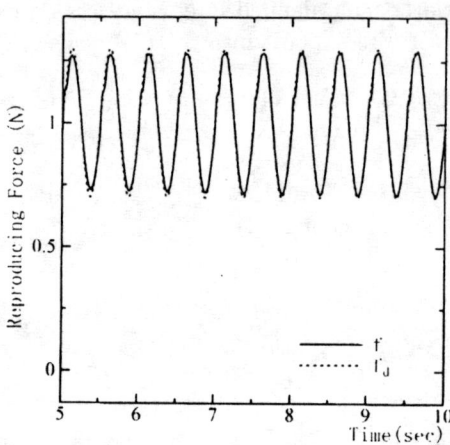

Figure 12(b): Response of Reproducing Force in case of unknown $f(\Delta x)$

References

[1] Arimoto, S., Kawamura, S., and Miyazaki, F. (1984a). Bettering operation of robots by learning, J. of Robotic Systems, vol. 2, pp. 123–140.

[2] Arimoto, S., Kawamura, S., and Miyazaki, F. (1984b). Can mechanical robots learn by themselves? In Robotics Research: 2nd Int. Symp. (ed. H. Hanafusa and H. Inoue), MIT Press, Cambridge, MA, pp. 127–134.

[3] Arimoto, S. (1990). Learning control theory for robotic motion, Int. J. of Adaptive Control and Signal Processing, vol. 4, pp. 543–564.

[4] Moore, K., Dahleh, M., and Bhattacharyya, S. (1992). Iterative learning control, J. of Robotic Systems, vol. 9, pp. 563–594.

[5] Horowitz, R. (1993). Learning control of robot manipulators, ASME J. of Dynamic Systems, Measurement, and Control, vol. 115, pp. 402–411.

[6] Arimoto, S. (1995). Fundamental problems of robot control: Part I Innovation in the realm of robot servo-loops, Robotica, vol. 13, pp. 19–27.

[7] Arimoto, S. (1996). Control Theory of Nonlinear Mechanical Systems: A Passivity-Based and Circuit-theoretic Approach, Oxford University Press, Oxford, UK.

[8] Naniwa, T. (1996). Trajectory control of a DD robot by learning: A dissertation conducted by H. Muraoka at Yamaguchi University (in Japanese).

[9] Arimoto, S. et al. (1998). Extension of impedance matching to nonlinear dynamics of robotic tasks, to be published in Systems and Control Letters.

Releasing Manipulation with Learning Control

Chi Zhu Yasumichi Aiyama and Tamio Arai

Department of Precision Machinery Engineering
Graduate School of Engineering, The University of Tokyo
7-3-1 Hongo, Bunkyo-Ku, Tokyo, 113-8656, Japan
E-mail: zhu@prince.pe.u-tokyo.ac.jp

Abstract

In this paper, the properties of releasing manipulation are given out. To improve the precision of object posture and decrease trial numbers, two iterative learning control schemes, learning control based on convergent condition (LCBCC), and learning control based on optimal principle(LCBOP) are designed in an experiment-oriented way. These two methods are all based on a linearized model. The experimental results show that these methods are effective. After discussing the characteristics of these control methods, we postulate that in the case of where the system does not have enough knowledge, LCBCC is the only choice and to learn system knowledge, after enough experience has been acquired, LCBOP is better than LCBCC, in the view of convergence rate and precision.

Key Words: dynamic manipulation, slide, impact, friction, learning

1. Introduction

For the task of transferring an object from one place to another by a robot, the conventional method is "pick and place". Recently, some dynamic approaches were proposed. Huang et al [1,2] proposed *impulsive manipulation* in which, by striking a circular object to give it an initial velocity, the object slides on a surface and stops at the required destination due to friction. Giving an object an initial velocity and then letting the object slide on a surface to stop at the preassigned destination is called *"releasing manipulation"* [3,4,5]. Obviously, releasing manipulation is different from the conventional ones. It is a discrete process and consists of two phases: acceleration and free motion of the object on the support surface. This manipulation is very simple, fast, and the reachable range of the object can be out of the workspace of manipulator. The last one is a very important feature, different from other manipulations. Since there are uncertainties, and during the sliding phase the motion of object is uncontrollable and completely determined by its initial velocity and environmental conditions such as the smoothness of the surface, its disadvantage is that the posture (stop position and orientation) precision of object can not be very high. Our experiment shows [3,4] that even with the same hitting velocity/position, the object is stochastically scattered over some range and its stop position and orientation fall on a normal distribution.

To improve the posture precision, in impulsive manipulation [1,2] with a circular object, "two-tap" and "multi-tap" planing methods are proposed. Such approaches are effective since the displacement of a circular object is only about several centimeters and the reachable range of the object is within the workspace of the manipulator. However, in releasing manipulation, the manipulator hand manipulates the object only one time, since the object can be out of the workspace of the manipulator, the above methods are not suitable. We adopt learning control to improve the posture precision by iterative trials.

Learning control is an iterative approach to the problem of improving behavior for processes that are repetitive in nature. There has been a lot of literature describing this approach in robotics. For releasing manipulation, different from other learning schemes, there are two essential features. One is that the control result can only be available after the object stops in each trial. The acquired system information is limited, it is impossible to execute an on-line control. The other is that the stop posture of the object is stochastic. This means that to reduce the error of the stop posture, it is necessary to execute control continually even if a "correct system model" has been obtained.

Our previous work [5] proposed an iterative learning control scheme for a circular object in which, based on virtual contact points assumption, a simplified system model is established. By modifying the "target" of the object and using Newton-Raphson method, the hitting velocity/position of manipulator is calculated and trials are processed iteratively. Though satisfactory experimental results are obtained with this approach, there are three problems. The first is that this approach is only suitable for circular objects, since for other objects it is impossible to determine the position and number of virtual contact points. The second is that it is difficult to ensure the convergence of the algorithm in theory. The third is that the computation is very large.

To overcome the above problems, in this paper we propose two learning control schemes for object of arbitrary shape. First, we give out a linearized model to approximate the nonlinear dynamics of releasing manipulation; then based on this model and the convergent condition, we present a learning control law that we call *learning control based on convergent condition (LCBCC)*. In order to utilize the learned experience to accelerate the convergent rate and to further improve precision, a *learning control based on optimal principle(LCBOP)* is adopted with system identification.

In section 2, releasing manipulation is defined and its basic characteristics are given out from our previous work. In section 3, we propose two learning control approaches for positioning task and in section 4, the experimental system is described and experimental results are given out. The discussion is in section 5, and in section 6, we give out the conclusion.

2. Concept of Releasing Manipulation
2.1. Definition and Research Problems

As mentioned above, releasing manipulation is such that a manipulator gives the object initial translational and angular velocities, releases it and makes it slide on a surface by its inertia, slowing down by friction force until coming to rest at the destination. This definition is rather general. Our work makes a number of assumptions. The manipulation is implemented with striking (in this sense, releasing manipulation is very similar to impulsive manipulation). The surface is a plane. Objects are planar of arbitrary shape and have a uniform support distribution, all relevant object properties are known, only Coulomb friction acts on the object after impact. The impact process follows the classical model of two-dimensional impact with friction.

The difficulty in releasing manipulation is the inverse problem, that is, with the desired stop position and orientation of the object, how to determine the required initial velocity of the object and hitting velocity/position of the robot. This is because 1) the motion equations of a sliding object with friction are nonlinear and friction force is non-conservative; 2) the impact process is very complex, and related to factors of restitution coefficient and friction coefficient between the object and the robot hand. It is difficult to determine the hitting velocity and position of robot hand. For such nonlinear, inverse dynamics, there are generally two approaches, one is a neural network and the other is a look-up-table. We adopt the look-up-table method to solve this problem since the look-up-table approach is simpler to implement and more robust than the neural network approach [10]. The other problem is how to transfer the object to the desired position and orientation more precisely. The purpose of releasing manipulation is to transfer object to desired position and orientation as precisely as possible. Since there are some uncertainties, which can not be modeled or modified by parameter adjustment, the hitting velocity/position found from the table may not guarantee that the object reaches the desired posture. Fortunately releasing manipulation is a discrete process; the initial condition, i.e., initial position and orientation of the object, can be easily held the same and iterative learning control is suitable to be used to improve the posture precision of object. This problem is the emphasis of this paper.

2.2 Some Characteristics of Releasing Manipulation

Comparing with the results for a circular object [1,2], our previous work shows that:
(1) The translation motion and rotation motion stop at the same time, regardless of the initial velocities.
(2) The displacement and rotated angles are greater than in pure translation and pure rotation. The time to rest will increase with increasing values of the initial velocities.
(3) In theory, the trajectory of any other object is a curve while the trajectory of a circular object is rectilinear. Though the displacement component, which is in the direction vertical to initial velocity, varies with initial velocity, compared with the one in the direction of initial velocity it is so small that it can be neglected.
(4) Compared with experimental values, the initial velocities of the object are larger than the evaluated with the impact model. We modified this model and obtained satisfactory results [4].

Property (3) is very important. Based on this, we can use the sliding model of a specific object, for instance, a rectangle or circle, to approximate the motion of others. Simulation and experiment show that this approach does not produce big errors. This gives us the convenience to build a reasonably general approach to solve the inverse problem for an object with an arbitrary shape.

3. Learning Control for Positioning Task
3.1 System Model

As we know, when solving linear equations, we have nothing to do other than to derive the inverse matrix of the coefficient matrix of the linear equations, if this inverse matrix exists. However, to solve nonlinear equation(s) or an optimal problem, it generally needs to be given initial values and to be processed iteratively. Learning control is just like this. If a system is linear, known and invertible, its best input is just the output of its inverse system and iterative process is no necessary. Contrarily, for a nonlinear system, unknown or partially unknown system, iterative approach is necessary to generate an optimal system input so that the system output is as close as possible to the desired output, this is just iterative learning control.

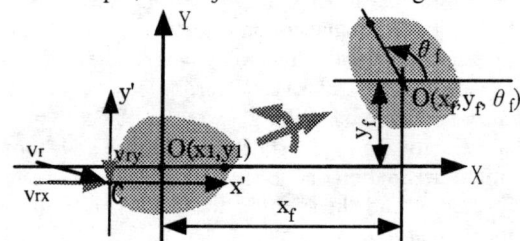

Fig. 1 Transferring Task of Releasing Manipulation

Notations: as shown in Fig.1,
OXY: world coordinate system, the COM (center of mass) of the object is always initially set at its origin at the start of each trial.
cx'y': the tangent-normal system at hitting point c (direction x' is opposite to the normal direction).
x_f, y_f, θ_f: translation displacement component along X, Y direction and rotated angle of object, respectively.
$x_f(k), y_f(k), \theta_f(k)$: respectively the value of x_f, y_f, θ_f at k^{th} trial.
$\mathbf{X_f}$: stop position and orientation vector, $\mathbf{X_f}=(x_f, y_f, \theta_f)^T$.
$\mathbf{X}(k)$: the value of $\mathbf{X_f}$ at k^{th} trial. $\mathbf{X}(k)=(x_f(k), y_f(k), \theta_f(k))^T$.
$\mathbf{X_d}$: desired stop position and orientation vector of the object. $\mathbf{X_d}=(x_d, y_d, \theta_d)^T$.
x_1, y_1: coordinates of the COM of object in cx'y'. Note: y_1 is the eccentricity from point c to COM of object, which is regarded as the hitting position of the robot hand in this paper.
V_r: hitting velocity of robot hand in OXY.
v_{rx}, v_{ry}: components of V_r along X, Y, respectively.

V_r: *hitting velocity vector, $V_r = (v_{rx}, v_{ry}, y_1)^T$.

v_{x0}, v_{y0}, ω_0: initial translation velocity component along X, Y and initial angular velocity, respectively.

V_0: initial velocity of object, $V_0 = (v_{x0}, v_{y0}, \omega_0)^T$.

μ_1, e: friction coefficient and restitution coefficient between object and robot hand, respectively.

μ: friction coefficient between object and support surface.

*: For simplicity, in Fig.1, we select a point c such that the coordinate system cx'y' is parallel to OXY, therefore the components of velocities (including hitting velocity of robot and initial velocity of object) in direction x' and y' are the same as ones in X and Y, respectively. For other case, what has to be done is only to transform the hitting velocity from OXY to cx'y' or from cx'y' to OXY. So such approach does not lose generality.

During the impact phase, the object obtains initial velocities, V_0, which is a function of V_r, e, and μ_1, that is
$$V_0 = g(V_r, e, \mu_1) \quad (1)$$
On the other hand, during the sliding phase, the stop position and orientation X_f of the object is a function of its initial velocity V_0 and friction coefficient μ, that is
$$X_f = f(V_0, \mu) \quad (2)$$
Substituting (1) into (2) and expanding X_f about $(V_{rp}, e_p, \mu_{1p}, \mu_p)$, we can get

$$X_f(Z) = X_f(Z_p) + \left(\frac{\partial X_f}{\partial V_0} \cdot \frac{\partial V_0}{\partial V_r}\right)_{Z_p} \cdot \Delta V_r + \left(\frac{\partial X_f}{\partial V_0} \cdot \frac{\partial V_0}{\partial e}\right)_{Z_p} \quad (3)$$
$$\cdot \Delta e + \left(\frac{\partial X_f}{\partial V_0} \cdot \frac{\partial V_0}{\partial \mu_1}\right)_{Z_p} \cdot \Delta \mu_1 + \left(\frac{\partial X_f}{\partial V_0} \cdot \frac{\partial V_0}{\partial \mu}\right)_{Z_p} \cdot \Delta \mu + h_1$$

where, $Z = (V_r, e, \mu_1, \mu)^T$, $Z_p = (V_{rp}, e_p, \mu_{1p}, \mu_p)^T$, $\Delta V_r = V_r - V_{rp}$, $\Delta e = e - e_p$, $\Delta \mu_1 = \mu_1 - \mu_{1p}$, $\Delta \mu = \mu - \mu_p$, h_1 represents all the higher-order terms and unmodelable components, $X_f(Z_p) = f(g(V_{rp}, e_p, \mu_{1p}), \mu_p)$. By simple experiment, μ and e can be easily measured; though μ_1 can not be determined directly, simulations show that by proper selection (≥ 0.15, for the objects used in our experiment), it almost doesn't affect results; experiments also show that this selection is acceptable. Therefore the affects of e, μ_1, μ can be ignored, and (3) can be revised as a linear model
$$X_f(V_r) = X_f(V_{rp}) + K \cdot \Delta V_r + h \quad (4)$$
where,
$$K = \left(\frac{\partial X_f}{\partial V_0} \cdot \frac{\partial V_0}{\partial V_r}\right)_{V_{rp}} \quad (5)$$

h represents nonlinear terms and unmodelable components. By neglecting h, as shown in Fig.2, we rewrite (4) as
$$X_f(V_r) = K \cdot (V_r - V_a) \quad (6)$$
where, $V_a = V_{rp} - K^{-1} \cdot X_f(V_{rp})$. In our experimental environment as described in next section, within such a range, X_d from (450mm, 0mm, ±90deg) to (800mm, ±200mm, ±450deg), by simulation, the range of found V_{rp} is from (750mm/s, 0mm/s, ±4.35mm) to (1380mm/s, ±200mm/s, ±22mm), while the range of V_a is from (80mm/s, 0mm/s, ±0.46mm) to (185mm/s, ±16mm/s, ±1.75mm), i.e., $V_a/V_{rp} \doteq 1/10$, thus V_a can be ignored. Therefore, (6) can be further simplified as

$$X_f = K \cdot V_r + \xi \quad (7)$$
where, ξ is regarded as model error, K is called *system matrix* in this paper, though in fact, it is a Jacobean matrix. Note that this is a 3-input 3-output system.

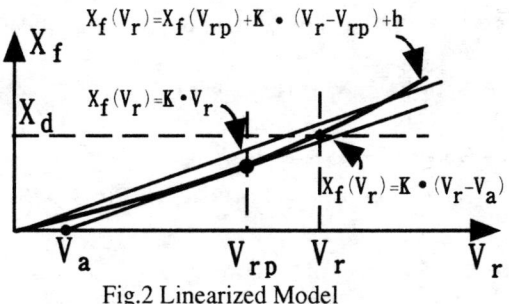

Fig.2 Linearized Model

3.2 Control Strategies

For a single or specific task, based on model (7), according to a convergent condition, we propose a simple learning control method that is called *learning control based on convergent condition(LCBCC)*. To utilize the learned experience and improve the convergence rate, based on a criterion function, which is that the square sum of error of X_f, be minimum, we suggest *learning control based on optimal principle (LCBOP)* by identifying the system matrix K. Note that, in LCBOP, using the error of the previous trial, its input is iteratively modified based upon the system model (7), therefore the control effect is determined by the precision of model. Since it is impossible to get the precise model of releasing manipulation until the experimental results of high precision are obtained, so we first run LCBCC to learn the information of system (in other words, learn the experience), then execute LCBOP to expect to obtain the higher convergent rate and higher precision.

3.3 Learning Control Based on Convergent Condition(LCBCC)

3.3.1 Determination of System Matrix K and Initial Hitting Velocity and Position

Theoretically, system matrix K can be derived from (5), but this is not easy because function (1) and (2) are complex, especially for function (1). Instead, by numeric calculation, we store functions (2), (1) into *Table 1* and *Table 2*, respectively. With known e, μ_1, μ, for a given X_d, by looking up Table 1, we can get V_0, and with V_0 we can find the required V_r from Table 2. In the same way, other two different sets (X_d^1, V_r^1), (X_d^2, V_r^2) can be obtained (but X_d^1 and X_d^2 can not be too close to X_d to avoid K being singular), thus by solving the following linear equation
$$(X_d^1 : X_d : X_d^2) = K \cdot (V_r^1 : V_r : V_r^2) \quad (8)$$
the system matrix K is obtained. The V_r found above from X_d is taken as the initial value of the next iterative learning control. By the way, V_r found is just V_{rp} in Fig.2.

3.3.2 Control Law for LCBCC

For the nth trial, model (7) is rewritten as
$$X_f(n) = K \cdot V_r(n) + \hat{\xi}(n) \quad (9)$$

where n is the number of trials.

The learning law is
$$\mathbf{V_r}(n+1) = \mathbf{V_r}(n) + \mathbf{G} \cdot \mathbf{E}(n) \quad (10)$$
where $\mathbf{G}$ is the gain matrix and $\mathbf{E}(n)$ is the error vector,
$$\mathbf{E}(n) = (x_f(n) - x_d, y_f(n) - y_d, \theta_f(n) - \theta_d)^T \quad (11)$$

3.3.3 Determination of Gain Matrix G

By ignoring $\xi(n)$, we have
$$\mathbf{E}(n+1) = \mathbf{X_f}(n+1) - \mathbf{X_d} = \mathbf{K} \cdot \mathbf{V_r}(n+1) - \mathbf{X_d} \quad (12)$$
From (10), we get
$$\mathbf{E}(n+1) = \mathbf{K} \cdot \mathbf{V_r}(n) - \mathbf{X_d} + \mathbf{K} \cdot \mathbf{G} \cdot \mathbf{E}(n)$$
$$= \mathbf{X_f}(n) - \mathbf{X_d} + \mathbf{K} \cdot \mathbf{G} \cdot \mathbf{E}(n) = \mathbf{E}(n) + \mathbf{K} \cdot \mathbf{G} \cdot \mathbf{E}(n) \quad (13)$$
$$= (\mathbf{I} + \mathbf{K} \cdot \mathbf{G}) \cdot \mathbf{E}(n)$$
Taking norm to two sides of (14), we have
$$\|\mathbf{E}(n+1)\| = \|(\mathbf{I} + \mathbf{K} \cdot \mathbf{G}) \cdot \mathbf{E}(n)\| \le \|(\mathbf{I} + \mathbf{K} \cdot \mathbf{G})\| \cdot \|\mathbf{E}(n)\| \quad (14)$$
With known $\mathbf{K}$, the gain matrix $\mathbf{G}$ is properly chosen to ensure
$$0 < \rho \le \|(\mathbf{I} + \mathbf{K} \cdot \mathbf{G})\| < 1 \quad (15)$$
i.e. $\quad \|\mathbf{E}(n+1)\| < \rho \cdot \|\mathbf{E}(n)\| \quad (16)$
where ρ is called convergence coefficient. (16) means,
$$\|\mathbf{E}(n)\| \to 0 \text{ as } n \to \infty$$
and $\quad \mathbf{X}(n) \to \mathbf{X_d}$ as $n \to \infty$

Of course, the choice of matrix $\mathbf{G}$ is not unique. For simplicity, $\mathbf{G}$ is selected as a diagonal matrix,
$$\mathbf{G} = \begin{pmatrix} g_x & 0 & 0 \\ 0 & g_y & 0 \\ 0 & 0 & g_\theta \end{pmatrix} \quad (17)$$

3.4 Learning Control Based on Optimal Principle(LCBOP)
3.4.1 Estimation Methods

With the above control, the object can be positioned to different positions and orientations corresponding to different $\mathbf{X_d}$ with satisfactory results. We consider whether it is possible to utilize the learned experience to accelerate the convergence rate and further improve the precision. Apparently, this requires the estimation of the system matrix $\mathbf{K}$.

By rewriting model (9) into estimation form, the recursive least-square (RLS) estimator [13] is used to estimate the matrix $\mathbf{K}$. The essential feature of estimation in releasing manipulation is that the estimator can only be updated one time with each trial.

We first used the standard RLS to estimate the system $\mathbf{K}$, and executed LCBOP, we found that the error of the rotated angle couldn't be decreased to less than 10%. For this reason, we respectively use the estimator of RLS with covariance modification and RLS with selective data weighting, for rotated angle component to estimate $\mathbf{K}$. Then the error can be decreased to less than 8% even with initial values found by look-up-table.

As the results of LCBCC, the learned different $\mathbf{V_r}$ and its corresponding $\mathbf{X_d}$ (equal or more than 3 sets) are used to initialize the estimator to accelerate the convergent rate. In this occasion, the sets $(\mathbf{V_r}, \mathbf{X_d})$ used must be different each other to meet PE (persistently exciting) condition. Then, the system is being identified with every each trial to renew the estimated system matrix $\hat{\mathbf{K}}$.

3.4.2 Control Law for LCBOP

Our purpose is to decrease the error of $\mathbf{X_f}$ as small as possible. According to system (9), let the following function be min., i.e., $f(\mathbf{V_r}(n)) \to 0$,
$$f(\mathbf{V_r}(n)) = \frac{1}{2}\left(\hat{\mathbf{K}}(n) \cdot \mathbf{V_r}(n) - \mathbf{X_d}\right)^T \cdot \left(\hat{\mathbf{K}}(n) \cdot \mathbf{V_r}(n) - \mathbf{X_d}\right) \quad (18)$$
The control law is
$$\mathbf{V_r}(n) = \mathbf{V_r}(n-1) + \Delta \mathbf{V_r}(n-1) \quad (19)$$
The Taylor's series expansion about $\mathbf{V_r}(n-1)$ is
$$f(\mathbf{V_r}(n)) = f(\mathbf{V_r}(n-1)) + \nabla f(\mathbf{V_r}(n-1))^T \cdot \Delta \mathbf{V_r}(n-1)$$
$$+ \frac{1}{2} \Delta \mathbf{V_r}(n-1)^T \cdot \nabla^2 f(\mathbf{V_r}(n-1)) \cdot \Delta \mathbf{V_r}(n-1) + 0(\Delta^2) \quad (20)$$
the necessary condition of $f(\mathbf{V_r}(n)) \to 0$ is
$$\nabla f(\mathbf{V_r}(n)) = 0 \quad (21)$$
this means
$$\Delta \mathbf{V_r}(n-1) = -\left(\nabla^2 f(\mathbf{V_r}(n-1))\right)^{-1} \cdot \nabla f(\mathbf{V_r}(n-1)) \quad (22)$$
On the other hand, the sufficient condition is
$$\nabla^2 f(\mathbf{V_r}(n-1)) > 0 \text{ (positive definite)} \quad (23)$$
However, $\nabla^2 f(\mathbf{V_r}(n-1))$ can not be guaranteed to be positive definite. Instead, the DFP method [14] is used to search a positive definite matrix iteratively to approximate $\left(\nabla^2 f(\mathbf{V_r}(n-1))\right)^{-1}$.

4. Experiments
4.1 Experimental System

The experiment system is set up as illustrated in Fig.3. The industrial manipulator Js-2, which is made by Kawasaki Heavy Industries, controlled by a computer connected via RS232C, hits the object. Camera 1 is used to measure the initial velocity of the object right after hitting, Camera 2 is used to measure the position and orientation of the object. The table surface is covered with paper, the hand of the manipulator is an aluminum disk, its diameter is 80mm and thickness is 10mm. With our proposed calibration approach [4] for vision system, the position and orientation can be easily obtained.

The objects used in experiments are a wooden rectangle and an L-shape, their initial position and orientation are shown in Fig.4. The mass of the rectangle is 51.5g, its μ is 0.35, e is 0.49; for the L-shape, its mass is 55g, μ is 0.34 and e is 0.50.

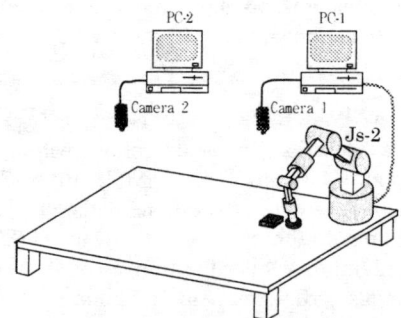

Fig. 3 Experimental System

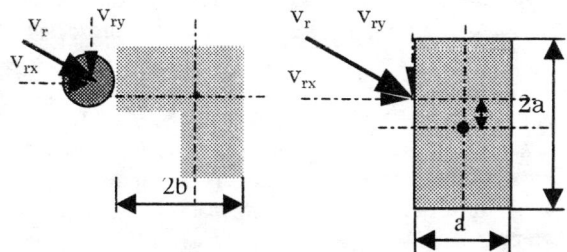

Fig.4 L-Shape (b=40.8mm) and Rectangle (a =50mm)

4.2 Experimental Results
4.2.1 Results from LCBCC

The given $\mathbf{X}_d$ is (500mm, 50mm, 180deg)^T. Here, we only give out the results of experiment for the L-shape as shown in Fig. 5, since the results for the rectangle are almost the same. These results show that the relative errors of all components of $\mathbf{X}_f$ are decreased to within 5% only after 3,4 trials.

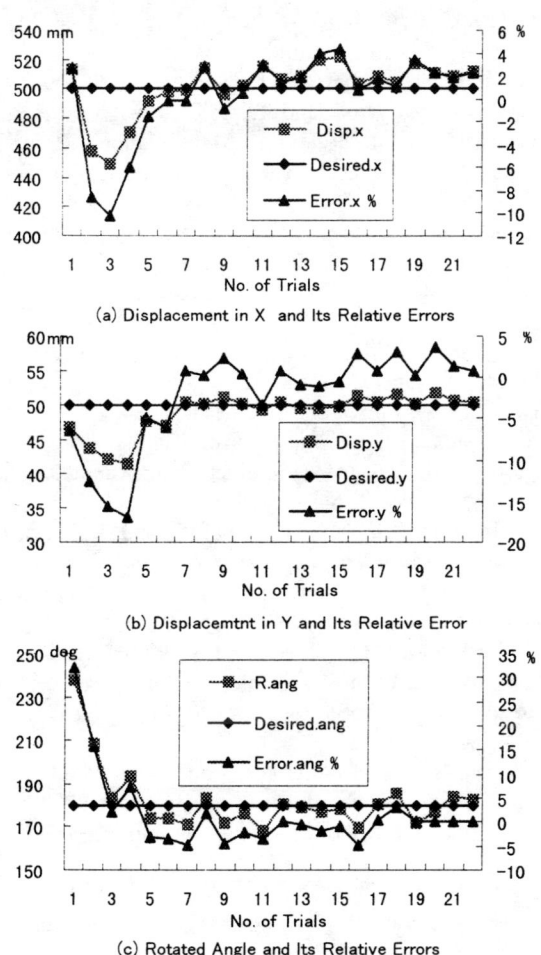

Fig.5 LCBCC for L-Shape

4.2.2 Results from LCBOP

Here we want to arrange three identical rectangles into a row as shown in Fig.6. First, with LCBCC, having learned experience about three points 1,2,3, where their $\mathbf{X}_d$ are (500mm, 50mm, 180deg)^T, (650mm,150mm,-180deg)^T, (550mm,-150mm, 360deg)^T, respectively, the system matrix $\mathbf{K}$ can be obtained with a high degree of precision.

Then, control law (19) is executed and $\hat{\mathbf{K}}$ is further estimated with each trial. After the object is positioned in P_1 and P_2, we transfer the 3rd rectangle to P_3, where $\mathbf{X}_d$ is (600mm, -100mm, -90deg)^T. Here, only the result of the RLS estimator with selective data weighting is given out in Fig.7. The result shows that the convergence rate of this method is faster than LCBCC as we expect; after about 1,2 trials, the errors can be decreased to less than 5%. For other points, this property is also held. Note that in this case the errors of the first trial are smaller than in other cases. This is because the estimated matrix $\hat{\mathbf{K}}$ is well refined by previous tasks. By the way, in this case, because P_1, P_2, P_3 are close each other, the matrix $\hat{\mathbf{K}}$ is in fact, approximately singular, but experiments show that LCBOP is still effective to decrease errors.

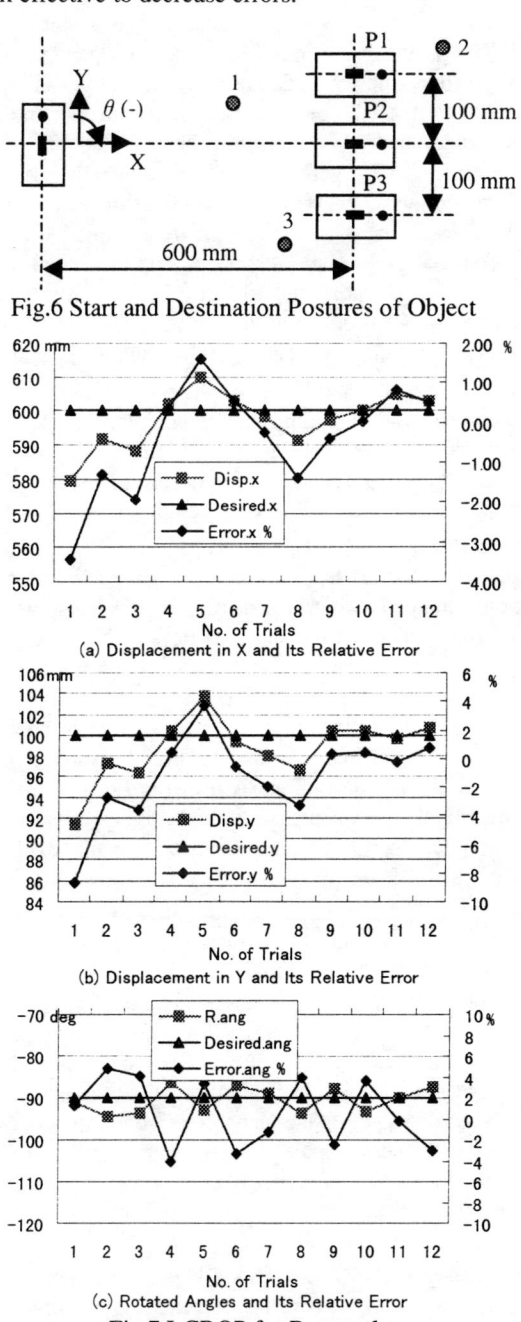

Fig.6 Start and Destination Postures of Object

Fig.7 LCBOP for Rectangle

5. Discussions

With the results from the above and other experiments, we give out the following discussions:

LCBCC is applicable for any X_d, especially under the condition of no "a priori" experience. Its convergence rate is determined by the gain matrix G. For a proper G, after 3,4 trials, it can decrease errors within 5% from initial errors of 20-30% or even more. However, for a new X_d, even if its initial value is given by well refined $\hat{K}$, its convergence rate can not be obviously improved. Its control precision seems unable to be further improved. Moreover, computation is large since it needs to search the required initial values in two large tables.

LCBOP relies strongly on the precision of the system model and the initial value of V_r. If it is initialized by learned experience, its convergence rate is faster than in LCBCC and the precision of the translation components can be further improved. Concretely, only after about 1 or 2 trials, the errors can be reduced to less than 5%, and after some more trials, the position errors can be further decreased to 2-3%, even 1%. If the system K is initialized by learned experience, but the initial V_r is given by a look-up-table, experiments show that errors can not be reduced to 8% until after 14 trials. If both K and the initial V_r are given by look-up-table, up to 30 trials, the errors still do not decrease. We don't think this is acceptable in the view of practical application. The other advantage is that it is still effective when estimated $\hat{K}$ is approximate singular.

Of course, when the approximate singular matrix is met, another method is switching from LCBOP to LCBCC. In this case, the convergence rate and control precision will return to the ones of LCBCC. In addition, it is worth to note that when the estimated $\hat{K}$ is very approximately singular, we do not think LCBOP is still valid. But this case only happens when a series X_d is very close and the size of the object is very small. If not, we can safely say that LCBOP is applicable.

Note that even with learned experience, LCBOP can not further decrease the rotation error (this error is about 3-5%). The main reasons are that the initial setting error of the object at the start of each trial(Fig.1), and that the rotation angle is very sensitive to the hitting position y_1.

Also note that our system model doesn't explicitly include the factors of μ_1, e, and μ, this gives our convenience that we need not to consider these factors when we design learning controller. The affections of these factors is hidden in the initial value of LCBCC.

Generalizing the above discussion, we say that for a new work or experimental condition, such as for a new object, a new table surface, and etc., because of the lack of knowledge, first, LCBCC is used to position the object and learn experience simultaneously. After obtaining enough experience (at least three different points), LCBOP is recommended to be put into use for other new desired postures of the object.

6. Conclusion

In this paper, the properties of releasing manipulation are given out. Based on simulation results, a linear model is established to approximate the real dynamic system. Two control schemes, LCBCC, and LCBOP are designed in an experiment-oriented way and experiments with these schemes are performed with satisfactory results. The characteristics of these control schemes are discussed. Finally, we conclude that under the condition of lack of enough system knowledge, LCBCC is the unique choice to be used to acquire system information. After enough experience has been accquired, LCBOP is a better candidate than LCBCC, in the view of the convergence rate and control precision.

References

[1] W. H. Huang, E. P. Krotkov, M. T. Mason, "Impulsive Manipulator", Proc. IEEE Int. Conf. on Robotics and Automation, 120-125, 1995.

[2] W. H. Huang, M. T. Mason, "Experiments in Impulsive Manipulation", Proc. IEEE Int. Conf. on Robotics and Automation, 1077-1082, 1998.

[3] C. Zhu, Y. Aiyama, T. Chawanya and T. Arai, "Releasing Manipulation", Proc. IEEE/RSJ Int. Conf. On Intelligent Robots and System, 911-916, 1996.

[4] C. Zhu, Y. Aiyama, and T. Arai, "The Experimental Characteristics of Releasing Manipulation", Proc. 3rd Asian Conf. On Robotics and its Application, 911-916, 1997.

[5] Y. Aiyama, T. Chawanya., C. Zhu, T. Arai, "Releasing Manipulation for Object Transferring Task", Proc. IEEE/ASME Int. Conf. On Advanced Intelligence Mechatronics'97 (AIM' 97), CD-ROM, 1997.

[6] S. Arimoto, S. Kawamura, and F. Miyazaki, "Bettering operation of robots by learning", J. of Robotics Systems, Vol. 1, 123-140, 1984.

[7] S. Arimoto, "Learning Control Theory For Robot Motion", Int. J. of Adaptive Control and Signal Processing, Vol. 4, 543-564, 1990.

[8] M. Togai and O. Yamano, "Analysis and design of an optimal learning control scheme for industrial robots: A discrete system approach", Proc. of 24th Conf. on Decision and Control, 1399-1404, 1985.

[9] K. L. Moore, M. Dahleh, and S. P. Bhattacharyya, "Iterative Learning Control: A Survey and New Results", J. Of Robotics Systems, Vol. 9, No.5, 563-594, 1992.

[10] E. Burdet, B. Sprenger, and A. Codourey, "Experiments in Nonlinear Adaptive Control", Proc. Int. Conf. on Robotics and Automation, 537-542, 1998.

[11] Y. Wang, M. T. Mason, "Two-Dimensional Rigid-Body Collisions With Friction", Journal of Applied Mechanics, Sept. 1992, Vol. 59/635-642.

[12] W. Goldsmith, "Impact: The Theory and Physical Behavior of Colliding Solids",1960.

[13] G. C. Godwin, K. S. Sin, "Adaptive, Filtering, Prediction and Control", Prentice-Hall, Englewood, New Jersey, 1984.

[14] W. H. Press et al, "Numerical Recipes in C: the art of Scientific Computing", 2nd edition, Cambridge University Press, 1992.

Learning Method for Hierarchical Behavior Controller

Yasuhisa Hasegawa

Dept. of Micro System Engineering,
Nagoya University
Furo-cho, Chikusa-ku, Nagoya 464-8603, Japan
yasuhisa@mein.nagoya-u.ac.jp

Toshio Fukuda

Center for Cooperative Research in Advanced
Science & Technology, Nagoya University
Furo-cho, Chikusa-ku, Nagoya 464-8601, Japan
fukuda@mein.nagoya-.ac.jp

Abstract

Complex behavior is hardly obtained using any unsupervised leaning methods, because of enormous searching space. In order to reduce the searching space, a hierarchical behavior structure is effective. In this paper, we proposed the hierarchical behavior controller, which consists of three types of modules, behavior coordinator, behavior controller and feedback controller. We also propose a new learning algorithm for the behavior coordinator and the behavior controller that consists of some sub-coordinators and some sub-controllers, respectively. This algorithm selects a deficient one by evaluating each sub-coordinator or sub-controller using multiple regression analysis based on previously obtained evaluation values. This can reduce the searching area and the learning times by avoiding the necessity of trying to tune good sub-coordinators or sub-controllers. The hierarchical behavior controller is applied to the problem of controlling a seven-link brachiation robot, which moves dynamically from branch to branch like gibbon swinging its body.

1. Introduction

Intelligence can be observed to grow and evolve, both through growth in computational power, and through the accumulation of knowledge of how to sense, decide and act in a complex and changing world. There are four system elements of intelligence: sensory processing, world modeling, behavior generation and value judgment. Input to, and output from, intelligent systems are via sensors and actuators[1].

In this paper, we focus on a behavior generating architecture and its acquisition method. The behavior is exhibited by a range of continuous actions that are performed by a robot with multiple degrees of freedom. To make such a robot perform an objective complex task, the controller is required to have the capability for managing multiple inputs and outputs through its behavior. It would be very hard work to design that controller even if some powerful learning algorithms are adopted, e.g., evolutionary algorithm, reinforcement learning algorithm, back-propagation method and so on, because of vast searching space and the nonlinear property among all inputs and outputs. Therefore, we proposed the hierarchical behavior controller that consists of three layer: multiple feedback controllers on the bottom layer, behavior controllers for generating simple behavior on the second layer and behavior coordinators on the top layer for managing the behavior controllers. The object behavior is generally decomposed into some simple behaviors so that the searching space could be divided into some smaller ones. Some learning algorithms above could obtain the behavior controllers for devided simple behaviors. If the complex behavior could be decomposed into some simple behaviors, it is generated by their coordination. The behavior coordinator is adopted to perform the complex behavior, coordinating some simple behavior controllers.

Existing methods [2][3] to coordinate multiple behaviors assume that the behavior state spaces do not interfere with each other or they are completely independent of each other. While, Asada, et al.[4] proposed a method for behavior coordination in a case that the subtask state spaces interfere with each other and that hidden states are involved. It is not however applicable to a more complex behavior generation based on some simple behaviors.

We propose a novel architecture for generating more comples behavior by coordinating the simple behaviors. Furthermore, the learning algorithm for acquisition of the behavior coordinator and behavior controller is proposed in this paper. This hierarchical behavior controller with the learning algorithm is able to acquire more complex behavior, combining some previously obtained behaviors.

The proposed method is applied to obtaining two kinds of behavior coordinators and one behavior controllers in hierarchical behavior controller, for 7-link brachiation robot in order to show its effectiveness.

2. Architecture of Hierarchical Behavior Controller

A robot is often required a complex dynamical behavior as a desired task, but a designing process of its controller is hard and almost impossible work. However, in most case, the complex behavior is decomposed into some fundamental behaviors. Therefore the behavior controllers for each fundamental behavior are designed, and then the complex behavior is performed by cooperation with them. That is the behavior-based controller is easily designed architecture.

The hierarchical behavior controller shown in fig. 1 consists of three layers: feedback controllers, behavior controllers and behavior coordinators. Each feedback controller in the bottom layer makes an actuator follow the desired trajectory indicated by the behavior controllers, using a linear or nonlinear feedback control method. When the desired trajectories from more than one behavior controller are input to a feedback controller, the mean value of them becomes the desired trajectory. The output value yd to actuator k is

$$yd_k = F(x_k, \frac{1}{m}\sum_{i \in m} yb_k^i) \quad (1)$$

x: current actuator state
yb^i: the desired state from behavior controller i
m: all controllers that indicate desired state to the feedback controller k

The behavior controller in the middle layer generates a fundamental behavior that is a sequence of actions and outputs the desired trajectories to feedback controllers to perform the fundamental behavior. The behavior controller consists of multiple sub-controllers that have multiple input variables from sensors or sensor fusion section and one output variable to its feedback controller, as shown in fig. 2. The number of sub-controllers is the same as that of actuators that are needed to generate its fundamental behavior. The output value from behavior controller i to the feedback controller k is

$$yb_k^i = yf_k^i \sum_{j \in n} yc_i^j \quad (2)$$

yf_k^i: output value from sub-controller in behavior controller i to the feedback controller k
yc^j: activation value from behavior coordinator j
n: all behavior coordinators that indicate an activation value to the behavior controller i

The behavior coordinator combines some fundamental behaviors so that more complex behavior could be generated. The behavior coordinator outputs the activation values(gains) to behavior controllers or behavior coordinators on the lower level. The gain means the amplitude of the behavior coordinator or behavior controller output. The behavior coordinator on the top level can perform the desired task by coordinating behavior coordinator or controller on the lower level. The output value yc^j from behavior coordinator j is

$$yc_i^j = yc_j^l yf_i^j \quad (3)$$

yc_j^l: activation value to the behavior coordinator j from

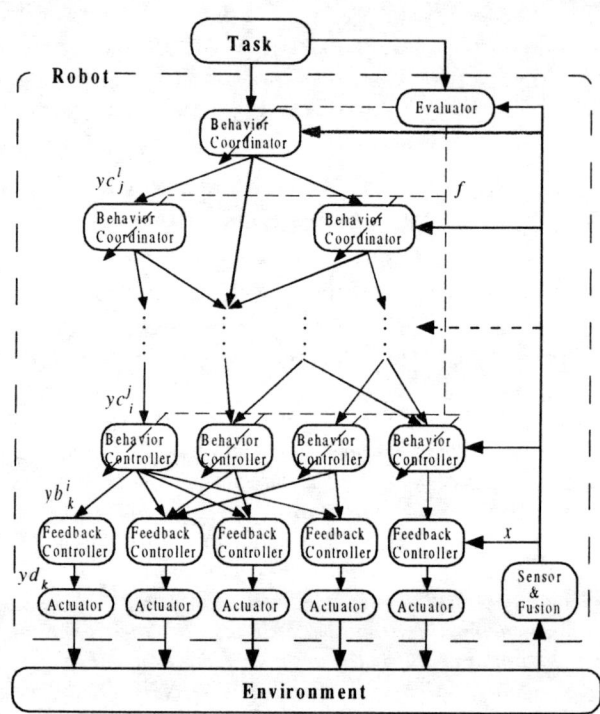

Fig. 1 Architecture of Hierarchical Behavior-based Controller

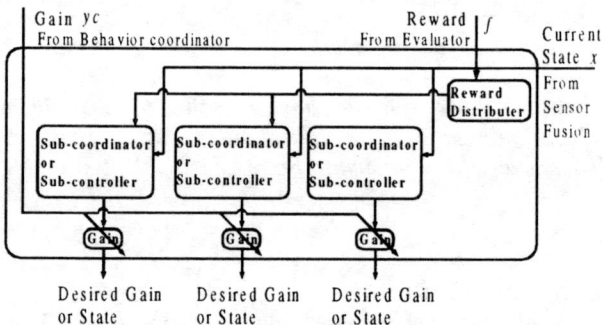

Fig. 2 Behavior coordinator or behavior controller

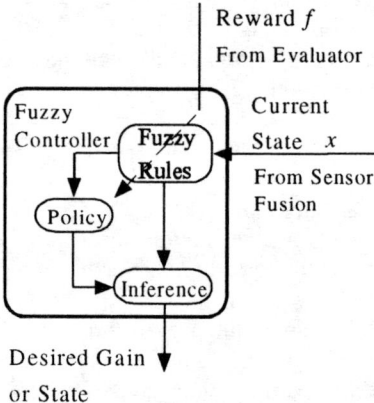

Fig. 3 Sub-coordinator or Sub-controller

the upper behavior coordinator l.

yf_i^j: output value from sub-coordinator in behavior coordinator j.

The environment parameters which are not measurable by the sensors, for example the center of gravity, the figure of object and so on, is calculated in the sensor fusion section and then output to the behavior coordinators, the behavior controllers and the feedback controllers.

3. Selection of Sub-Controller or Sub-coordinator

Most robots have multiple degrees of freedom, which are required to perform various kinds of objective behaviors or tasks. Therefore the controller is required to perceive multiple input variables from robot's sensors and read out multiple output variables to its actuators. In order to handle them easily, we installed the same number of sub-controllers in the behavior controller as that of actuators that are necessary to generate its simple behavior. Each of sub-controllers outputs the desired trajectory to an actuator according to multiple input values. The behavior is performed by the cooperation of all sub-controllers in the behavior controller. Therefore we can independently deal with sub-controller and can fabricate the sub-controllers one by one. When fixing the behavior controller with multiple sub-controllers, we must choose which sub-controller requires tuning in order to improve the whole behavior. However the behavior controller with the multiple sub-controllers is given only the numerical evaluation of the behavior after the robot has finished the behavior. Therefore the selection of the sub-controller that is inappropriate and that should be tuned is very difficult yet very important. We proposed novel algorithm using multiple regression analysis to determine the deficient sub-controller. This algorithm evaluates each sub-controller using multiple regression analysis based on the previously obtained evaluation values, and choose a sub-controller to be tuned in the next trial with the probability of choice in proportion to the evaluation values of the sub-controllers. This can reduce the learning iterations by avoiding attempts to tune good sub-controllers. The method is explained below.

At first, some tuning iterations are done to each sub-controller in order to evaluate sub-controllers. The tuning process is firstly to add some changes to the sub-controller, secondarily to make a trial, then to evaluate its motion and finally to update the sub-controller. After all sub-controllers are evaluated, a candidate for tuning is selected according to the evaluation-based probability. The evaluation probability is updated every tuning. We explain in detail.

After tuning the sub-controller, m, we calculate the improvement of the performance of the robot owing to its

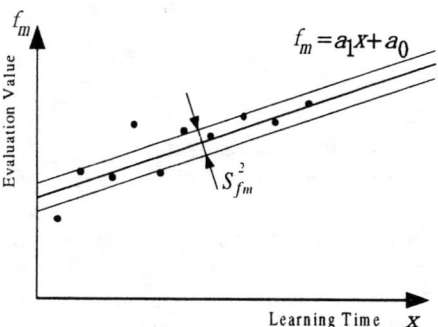

Fig. 4 Relation between learning time of sub-controller and obtained evaluation value

Sub-controller changes. The evaluation value of the sub-controller is,

$$f_{m(x)} = f_{m(x-1)} + f_{(x)} - f' \quad (4)$$

where $f_{m(x-1)}$ is the performance evaluated in the previous trial of sub-controller m, $f_{(x)}$ is the performance score calculated by the given evaluation function, f' is the best evaluation value and x is the learning time about the sub-controller.

We assume that the evaluation value, f_m, of the sub-controller is roughly proportional to its learning time, x (Fig. 4)

$$f_m = a_1 x + a_0 \quad (5)$$

The coefficient values, a_1, and a_0 are calculated by multiple regression analysis as follows.

$$a_1 = \frac{S_{xf_m}}{S_x^2} \quad (6)$$

$$a_0 = \bar{f}_m - a_1 \bar{x} \quad (7)$$

S_{xfm} is the covariance between x and f_m. The bar denotes its mean value.

$$S_{xf_m} = \frac{1}{N} \sum_{i=1}^{N} (x_i - \bar{x})(f_i - \bar{f}) \quad (8)$$

and S_x^2 is the variance of x.

$$S_x^2 = \frac{1}{N-1} \sum_{i=1}^{N} (x_i - \bar{x})^2 \quad (9)$$

a_1 corresponds to the prospect of improvement of the system performance when its sub-controller is tuned in the succeeding trial. The variance of f_m, S^2_{fm}, indicates the reliability of the system improvement. The evaluation function of sub-controllers therefore consists of two terms, a_1 and S^2_{fm} shown below.

$$S_{f_m}^2 = \frac{1}{N-1} \sum_{i=1}^{N} (f_{m(i)} - \bar{f}_{m(x)})^2 \quad (10)$$

$$f_c = \alpha a_1 - \beta S_{f_m}^2 \quad (11)$$

A sub-controller for tuning in the succeeding trial is selected according to the probability, g_c in the following equation, which is based on the evaluation values, f_c.

$$g_c = \frac{f_c}{\sum_{c=1}^{C} f_c} \quad (12)$$

The behavior coordinator also has multiple sub-coordinators that indicate activating values to each behavior coordinator or controller. The output values of the behavior coordinator or controller are changed in proportion to the activating value. The same number of sub-coordinators in the behavior coordinator is embedded as that of simple requisite behaviors to generate complicate behavior. In the same way, when tuning the sub-coordinator, we use the same selection algorithm.

4. Self Scaling Reinforcement Learning

After a robot executes a range of objective tasks or motions, the results are evaluated by a given evaluation function. The evaluated value is assumed to be a value of f. We calculate the past performance a, eq. (13), which is the weighted mean of the past evaluation values. Furthermore we compute an inner reinforcement value using eq. (14) below.

$$a_{(s)} = \frac{\sum_{i=0}^{s} \gamma^{s-i} f_{(i)}}{\sum_{i=0}^{s} \gamma^{s-i}}, \quad (13)$$

$$r' = f_{(s)} - a_{(s-1)} \quad (14)$$

where s is the trial number and gamma is the positive weight<1.

The range of reinforcement value r' is influenced by the evaluation function which is arranged by designer. Therefore it is transformed into r with its range in [-1, 1].

$$r(s) = \frac{1 - \exp\left(-\frac{r'(s)}{\beta}\right)}{1 + \exp\left(-\frac{r'(s)}{\beta}\right)}, \quad (15)$$

where beta is a coefficient value > 0.

We assign an internal reinforcement to each fuzzy rule in fuzzy controller. The internal reinforcement, r_i, assigned to fuzzy rule, i, is updated as follows.

$$r_i \Leftarrow r_i(1 - \max_{t_d} \mu_i(t_d)) + \lambda^{(T^* - T_i)} r \max_{t_d} \mu_i(t_d) \quad (16)$$

where lambda is a positive number < 1, T_i is the time when its fitness value of the membership function is a maximum, and T^* is the time when the reinforcement is received.

The center values and the dispersion values of the consequent parts in the used fuzzy rules during robot motion are updated according to the internal reinforcement calculated by the above equation (16). The equations are

$$w_{ci} \Leftarrow w_{ci} + r_i(w_i - w_{ci}). \quad \text{if } r \geq 0 \quad (17)$$

$$\sigma_i \Leftarrow \sigma_i(1 - r_i) \quad \text{if } r \geq 0 \quad (18)$$

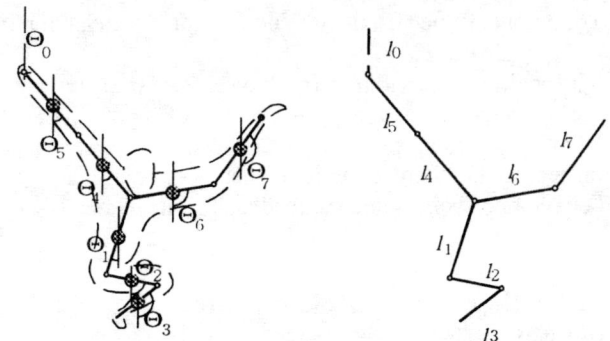

Fig.5 Model of seven-link brachiation robot

$$\sigma_i \Leftarrow \sigma_i + \alpha \quad \text{if } r < 0 \quad (19)$$

where alpha is an increase value > 0. For the detail of the fuzzy controller and of learning algorithm, refer to [5][6][7].

5. Seven-link Brachiation Robot

Figure 5 shows the brachiation mobile robot(BMR) of a seven-link model used for computer simulation in this study. This robot is a mobile robot, which dynamically moves from branch to branch like a gibbon, namely a long-armed ape, swinging its body like a pendulum [8][9]. It has two arms, a body and a leg. The robot has seven degrees of freedom and six control inputs to actuators: elbows and shoulders of both arms, hip and knee. It has a redundant degree of freedom to move from branch to branch. On the tips of the two arms, the grips are set to catch horizontal parallel bars. The grip is automatically closed when the grip is in the seizable region. In this study, the motion of the robot is assumed to be within a vertical plane[10].

6. Simulations and Results

The hierarchical behavior controller is applied to the motion control of the seven-link brachiation robot. The ultimate task is to move from branch to branch. The robot hangs down with one arm as an initial state. While increasing the amplitude of its body, the robot catches the front branch by another arm and repeats to catch the next branch, controlling the body swing. To perform this task, we design the hierarchical structure shown in fig. 6, which has two fundamental behavior controllers on the second layer and three behavior coordinators on the third layer. All fuzzy rules in behavior controllers and coordinators are initially set to output zero. At first, the amplitude controller is obtained using the proposed method and the approach behavior controller to the target branch is PD controller designed by kinematics. In the next step, each of the first and second locomotion coordinators is obtained by the proposed method. In the final step, the continuous locomotion coordinator is obtained by coordinating the first and the second locomotion coordinators.

6.1 Amplitude controller

The amplitude controller is to increase amplitude of the center of gravity of the robot. We assume that all six actuators are necessary for the desired behavior and prepare the six fuzzy controllers in the amplitude controller. Each fuzzy controller has two input variables: an angle of the center of gravity and an angular velocity of the center of gravity, and one output variable: target angle for each joint. The initial value of fuzzy map is zero. We use the proposed algorithm to select fuzzy controller effectively. The selected fuzzy controller is tuned by self-scaling reinforcement learning and its evaluation function, eq.(20) is set based on the energy efficiency for motion energy par energy consumption.

$$f = (\frac{1}{2}\dot{\theta}_g^2 + \frac{1}{2}\omega^2\theta_g)/E \quad (20)$$

θ_g: angle of the center of gravity

$\dot{\theta}_g$: angular velocity of the center of gravity

E: consumption energy

Figure 7 shows the locus of the center of gravity when the amplitude controller obtained in the 320 iterations controls the seven-link brachiation robot. The seven-link brachiation robot increased its amplitude, then saturated around to 70 degree. The behavior coordinator can control the robot amplitude in the way of changing of the activation value that indicates the size of the amplitude controller outputs.

Each of the learning iterations is shown in fig. 8. The number of iterations for the shoulder of holding arm was a maximum and that for the shoulder of free arm was a minimum. The proposed algorithm reduced learning iterations by 60 percent of the learning iterations in case that all fuzzy controllers are tuned one by one.

6.2 Approach controller to a target bar

This behavior is to control the shoulder and elbow of the free arm so that the grip of the free arm could catch the target bar. The target angles for the shoulder and elbow of free arm are calculated using kinematics. We designed this behavior controller mathematically because the target angles are easily obtained without redundant mechanism.

6.3 First and second locomotion behaviors

The first locomotion behavior is that the robot increases the amplitude from the initial state where the robot hangs down from the bar with one arm, and then catches the succeeding bar by the other arm. The second locomotion behavior is that the robot catches the succeeding bar after the initial state where the robot holds

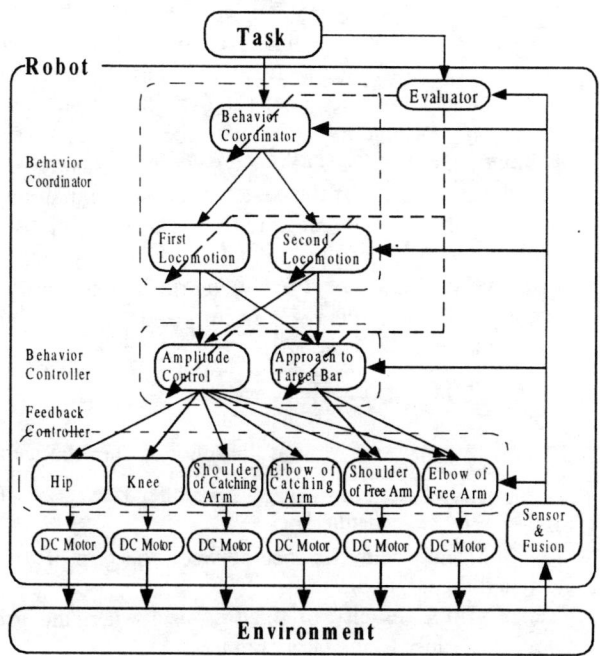

Fig. 6 Hierarchical control architecture for seven-link robot

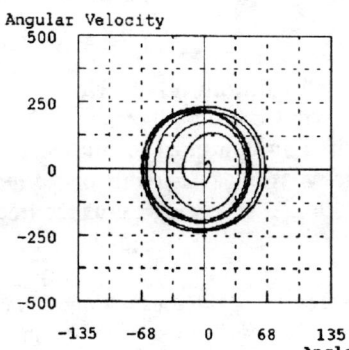

Fig. 7 Locus of center of gravity

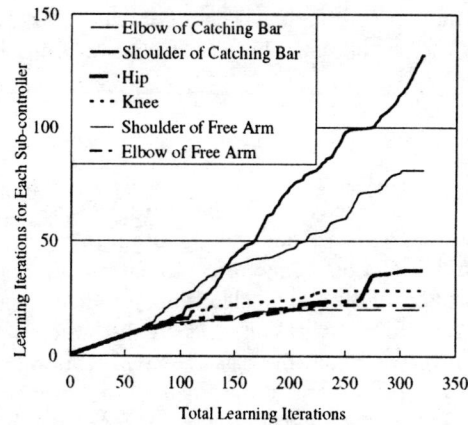

Fig. 8 Learning iterations of each fuzzy controller

two bars in front and behind with each arm. We assume that these behaviors could be accomplished by the two fundamental behaviors above: the amplitude behavior and the approach behavior. We set two fuzzy controllers in behavior coordinators to indicate the gains to these two behavior controllers. The fuzzy controller has two input variables: an angle of the center of gravity and an angular velocity of the center of gravity and one output. We use the proposed algorithm to select the fuzzy controller effectively. The selected fuzzy controller is tuned by self-scaling reinforcement learning and its evaluation functions , eqs. (21) and (22), are

$$f = 20 + \frac{1}{E} + \frac{1}{1 + 20\min d} + \frac{1}{1 + v_t}$$

if the robot catches the bar (21)

$$f = \frac{1}{E} + \frac{1}{1 + 20\min d} + \frac{1}{1 + v_t} \quad \text{otherwise} \quad (22)$$

$\min d$: minimum distance between the grip of the free arm and the target branch

v_t: grip's velocity of the free arm when the grip is close to the target branch

The control results of the obtained first locomotion behavior coordinator are shown in fig. 9, and those of the obtained second locomotion behavior coordinator are shown in fig. 10.

6.4 Continuous locomotion behavior

The continuous locomotion behavior is that the robot increases the amplitude from the initial state where the robot hangs down from a bar with one arm, and then continues to catch the succeeding bar by the free arm. This motion is performed by switching the behavior coordinators from the first locomotion to the second locomotion. That is the continuous locomotion coordinator keeps outputting one to the first locomotion coordinator and zero to the second locomotion coordinator during the first locomotion. After that, it keeps outputting zero to the first locomotion coordinator and one to the second locomotion coordinator. The locomotion is shown in fig. 11. The robot can continue to move from branch to branch.

7. Conclusions

We proposed the architecture of the hierarchical behavior controller and the algorithm to get the behavior controller and the behavior coordinator with multiple sub-controllers. Using this algorithm, we can construct the hierarchical behavior controller by the bottom-up process, in which we obtain the basic behavior controllers at first, then get the organizer for them. We have applied our algorithm to the control problem of seven-link brachiation robot and obtained the proper behavior controllers and

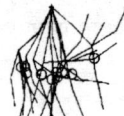
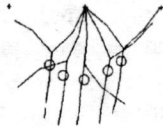

Fig. 9 Stick Diagram of the first locomotion

Fig. 10 Stick Diagram of the second locomotion

(The circle shows center of gravity of the robot at intervals of 180[ms]. The marks "+" mean bar positions)

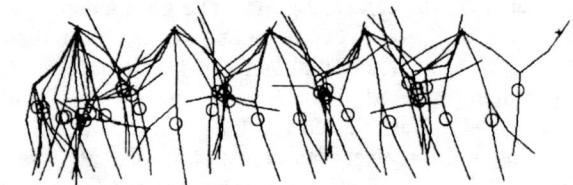

Fig. 11 Stick Diagram of the continuous locomotion

coordinators that can perform the locomotion of the brachiation robot from branch to branch.

Reference

[1] J. S. Albus, Outline for a Theory of Intelligence, IEEE Trans. on Systems, man, and Cybernetics, Vol. 21, No. 3 1991.

[2] J. H. Connel and S. Mahadevan, Robot Learning, Kluwer Academic Publishers, 1993

[3] S. D. Whitehead, J. Karlsson and J. Tenenberg, Learning Multiple Goal Behavior Via Task Decomposition and Dynamic Policy Merging, chapter 3 in Robot Learning, Kluwer Academic Publishers, 1993

[4] E. Uchibe, M Asada and K. Hosoda, Behavior Coordination for a Mobile Robot Using Modular Reinforcement Learning, Proc. IROS, pp.1329-1336, 1996

[5] K. Shimojima, T. Fukuda and Y. Hasegawa, RBF-Fuzzy System with GA Based Unsupervised Learning Methods, Proc. 4th IEEE int. Conf. on Fuzzy Systems/2nd Int. Fuzzy Engineering Symp. Fuzz-IEEE/IFES'95, Vol. 1, pp.253-258, 1995

[6] T. Fukuda, Y. Hasegawa, K. Shimojima and F. Saito, Reinforcement Learning Method for Generating Fuzzy Controller, Int. Conf. on Evolutionary Computation Vol.1, pp.273-278 1995.

[7] V. Gullapalli, A stochastic reinforcement learning algorithm for learning real-valued functions, Neural Net., Vol. 3, pp.671-692, 1990.

[8] T. Fukuda, H. Hosokai and Y. Kondo, Brachiation Type of Mobile Robot, Proc. IEEE Int. Conf. Advanced Robotics, pp. 915-920, 1991.

[9] T. Fukuda, F. Saito and F. Arai, A Study on the Brachiation Type of Mobile Robot(Heuristic Creation of Driving Input and Control Using CMAC), Proc. IEEE/RSJ Int. Workshop on Intelligent Robots and Systems, pp. 478-483, (1991).

[10] T. Fukuda and Y. Hasegawa, Learning Method for Multi-Controller of Robot Behavior, JSME Int. Journal, Series C, Vol. 41, No. 2, 1998.

Feedback Control of a Planar Manipulator with an Unactuated Elastically Mounted End Effector

Mahmut Reyhanoglu

Department of Physical Sciences
Embry-Riddle Aeronautical University
Daytona Beach, FL 32114
reyhanom@db.erau.edu

Sangbum Cho[1] N. Harris McClamroch[2]

Department of Aerospace Engineering
University of Michigan
Ann Arbor, MI 48109
[1]sbcho@engin.umich.edu [2]nhm@engin.umich.edu

Abstract

We study feedback control laws that can stabilize an equilibrium position of an underactuated planar PPR robot manipulator. The manipulator operates in a horizontal plane, and it consists of two actuated prismatic links, an actuated revolute link, and an unactuated end effector that is elastically mounted on the third link. The fact that the end effector is unactuated and is elastically mounted on the manipulator, and thus can be controlled only through coupling with the manipulator dynamics, provides a major obstacle to control design for this problem. The control objective is to design a feedback controller that can achieve arbitrary positioning and arbitrary attitude of the end effector, while suppressing any motion of the end effector relative to the third link. A nonlinear control model is constructed for this underactuated manipulator and a time-invariant discontinuous feedback law is constructed that achieves the objectives with exponential convergence rates. The effectiveness of the proposed feedback law is illustrated through simulations.

1 Introduction

Many examples of underactuated mechanical systems have provided interesting and challenging models of nonlinear control systems. Nonlinear control theory has provided considerable insight into means for achieving various control objectives for underactuated systems. A large literature now exists on the use of nonlinear control theory to treat the control of underactuated mechanical systems; a number of different underactuated mechanical examples have been studied, e.g. see ([9], [10], [11], [12], [14], [15]). In particular, interesting and challenging examples of underactuated robot manipulators have been studied. A PPR planar manipulator with unactuated revolute joint was studied in [7]. A 2R planar robot with a single actuator was studied in [5]. An underactuated 3 DOF manipulator was studied in [1], [2].

The manipulator control problem studied in this paper is motivated by manipulators with unactuated internal dynamics, e.g. due to flexible link dynamics or due to flexible load dynamics. Simplifying assumptions are made so that the problem is tractable, while still reflecting the important coupling between the unactuated internal dynamics and the dynamics of the manipulator. Our control objective is to control the complete robot configuration; in particular, we desire to suppress any relative motion of the unactuated internal dynamics. The particular problem that we study in this paper is a PPR manipulator with an unactuated end effector that is elastically mounted on the revolute link. This problem is substantially different from the previously studied underactuated manipulator examples. In this case, it is an elastic degree of freedom for the relative end effector dynamics that is unactuated, rather than the PPR robot itself.

The problem considered here can be viewed as an extension of our previous work [14], where we constructed a feedback law that achieved position control of a similar underactuated mechanical system; however, the controller developed here has significantly improved properties when compared with the controller in [14]. In addition, the development here is presented explicitly in a robotics context.

For the nonlinear control system that we formulate, we construct a discontinuous feedback law. The methodology followed in the construction of the discontinuous feedback law is based on first transforming the system into a discontinuous one in which the design of a feedback law is easily carried out. Then, transforming back into the original coordinates yields a discontinuous feedback law which controls the original system to the desired equilibrium with exponential convergence rates. The discontinuous coordinate transformation employed here constitutes an example of a σ-process [3], which has proved useful in the stabilization of a special class of nonholonomic systems ([4]) and underactuated mechanical systems ([12], [13]).

2 Model Formulation

In this section, we formulate a mathematical model for a planar robot manipulator with three actuated joints and a single unactuated degree of freedom that describes the relative motion of an end effector that is elastically mounted on the manipulator. This is indicated in Figure 1. Let (x,y) denote the inertial position of the the revolute joint and let θ denote the attitude of the third link, which is also the attitude of the end effector. External control forces, described in an inertial frame by X, Y act on the two prismatic joints, and an external control moment T acts on the revolute joint. The end effector is modelled as a single degree of freedom mass that is constrained to move axially with respect to the third link; the relative position of the end effector is denoted by s. There is an elastic restoring force on the end effector with respect to the third link. The constant c denotes the length of the third link.

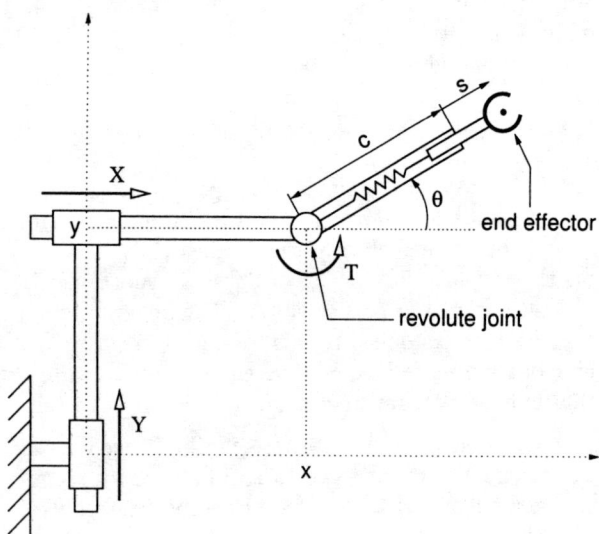

Figure 1: Model of a Robot Manipulator with an Unactuated Elastically Mounted End Effector.

By defining control transformations from (X, Y, T) to new control inputs (u_1, u_2, u_3) the equations of motion can be written in a nonlinear control system form

$$\ddot{\theta} = u_1 , \tag{1}$$

$$\ddot{y} = u_2 , \tag{2}$$

$$\ddot{x} = u_3 , \tag{3}$$

$$\ddot{s} = -u_2 \sin\theta - u_3 \cos\theta + (c+s)\dot{\theta}^2 - k_s s , \tag{4}$$

where $k_s > 0$ is the elastic stiffness of the spring connecting the end effector with the third link.

Let $q = (\theta, y, x, s)$. Then equations (1)-(4) define a nonlinear control system with control (u_1, u_2, u_3) and state $(q, \dot{q}) \in \mathbf{M} = \mathbf{S}^1 \times \mathbf{R}^7$. The set of equilibrium solutions corresponding to $(u_1, u_2, u_3) = (0, 0, 0)$ is given by

$$\mathbf{M}_e = \{(q, \dot{q}) \in \mathbf{M} \mid \dot{q} = 0\} .$$

It is possible to achieve our control objective if we impose the design constraint

$$-\ddot{y}\sin\theta - \ddot{x}\cos\theta + (c+s)\dot{\theta}^2 - k_s s + \lambda^2 s + 2\lambda\dot{s} = 0 , \tag{5}$$

which implies satisfaction of

$$\ddot{s} = -\lambda^2 s - 2\lambda\dot{s} .$$

Thus the constant $\lambda > 0$ denotes the desired decay rate of the unactuated relative end effector dynamics. This simple argument is a key to our development. Clearly, this constraint can be equivalently written as

$$u_3 = -u_2 \tan\theta + [(c+s)\dot{\theta}^2 - k_s s + \lambda^2 s + 2\lambda\dot{s}]\sec\theta , \tag{6}$$

so long as the singularities at $\theta = \pm\frac{\pi}{2}$ are avoided.

The equations of motion take the form

$$\ddot{\theta} = u_1 , \tag{7}$$

$$\ddot{y} = u_2 , \tag{8}$$

$$\ddot{x} = -u_2 \tan\theta + [(c+s)\dot{\theta}^2 - k_s s + \lambda^2 s + 2\lambda\dot{s}]\sec\theta , \tag{9}$$

$$\ddot{s} = -\lambda^2 s - 2\lambda\dot{s} . \tag{10}$$

Clearly, these equations can be used to design feedback controls $u_1(q, \dot{q})$, $u_2(q, \dot{q})$. Once these feedback controls are designed, the feedback control $u_3(q, \dot{q})$ can be determined using relation (6), thereby obtaining a feedback law $u(q, \dot{q}) = (u_1(q, \dot{q}), u_2(q, \dot{q}), u_3(q, \dot{q}))$ for the system (1)-(4).

3 Discontinuous Feedback Control Law

In this section, we consider the problem of designing a feedback control law of the form $u = u(q, \dot{q})$ based on the nonlinear control system (7)-(10). We restrict our consideration to designing a time-invariant discontinuous feedback law.

Note that the problem of controlling the system to a given equilibrium $(q^e, 0)$ can be reduced to the problem of controlling the system to the origin via an appropriate state transformation. Hence, without loss of generality, we assume that $q^e = 0$.

We first study the problem of controlling the following reduced order system, which is obtained by considering equations (7)-(9) and letting $(\dot{\theta}, \dot{y})$ be the input variables (v_1, v_2):

$$\dot{\theta} = v_1 , \tag{11}$$

$$\dot{y} = v_2 , \tag{12}$$

$$\dot{x} = w - v_2 \tan\theta , \tag{13}$$

$$\dot{w} = v_1 v_2 \sec^2\theta + [(c+s)\dot{\theta}^2 - k_s s + \lambda^2 s + 2\lambda\dot{s}]\sec\theta , \tag{14}$$

where

$$w = \dot{x} + \dot{y}\tan\theta . \tag{15}$$

Note that the s and $\dot{s}$ terms in equation (14) are exponentially decaying functions of time according to (10).

Control of the Reduced System

The idea that will be employed is based on first transforming the reduced system (11)-(14) into a discontinuous one by applying a discontinuous coordinate transformation, e.g. by applying a σ-process (see e.g. [3]). From the analytical point of view, the σ-process, also referred to as the process of resolution of singularities, consists of a rational coordinate transformation.

Consider the reduced system (11)-(14). Restricting consideration to $\theta \neq 0$, apply the σ-process definitions

$$\eta_1 = y, \ \eta_2 = \frac{x}{\theta}, \ \eta_3 = \frac{w}{\theta}$$

to obtain

$$\dot{\theta} = v_1 , \tag{16}$$

$$\dot{\eta}_1 = v_2 , \tag{17}$$

$$\dot{\eta}_2 = \eta_3 - \frac{v_1}{\theta}\eta_2 - \frac{\tan\theta}{\theta}v_2 , \tag{18}$$

$$\dot{\eta}_3 = -\frac{v_1}{\theta}\eta_3 + [(c+s)v_1^2 - k_s s + \lambda^2 s + 2\lambda\dot{s}]\frac{\sec\theta}{\theta}$$

$$+ \frac{v_1 v_2}{\theta}\sec^2\theta . \tag{19}$$

The feedback control law

$$(v_1, v_2) = (-k\theta, -l\eta) , \tag{20}$$

where $\eta = (\eta_1, \eta_2, \eta_3)^T$, and $0 < k < \lambda$ and $l = (l_1 \ l_2 \ l_3)$ are feedback gains, yields the reduced closed-loop system

$$\dot{\theta} = -k\theta , \tag{21}$$

$$\dot{\eta}_1 = -l_1\eta_1 - l_2\eta_2 - l_3\eta_3 , \tag{22}$$

$$\dot{\eta}_2 = k\eta_2 + \eta_3 + \frac{\tan\theta}{\theta}(l_1\eta_1 + l_2\eta_2 + l_3\eta_3) , \tag{23}$$

$$\dot{\eta}_3 = k\eta_3 + [k^2\theta^2(c+s) - k_s s + \lambda^2 s + 2\lambda\dot{s}]\frac{\sec\theta}{\theta}$$

$$+ k\sec^2\theta(l_1\eta_1 + l_2\eta_2 + l_3\eta_3) \tag{24}$$

The η-dynamics can be rewritten as

$$\dot{\eta} = (A_1 + A_2(t))\eta + h(t) , \tag{25}$$

where

$$A_1 = \begin{pmatrix} -l_1 & -l_2 & -l_3 \\ l_1 & k+l_2 & 1+l_3 \\ kl_1 & kl_2 & k+kl_3 \end{pmatrix} , \tag{26}$$

$$A_2(t) = \begin{pmatrix} 0 & 0 & 0 \\ l_1 r_1(t) & l_2 r_1(t) & l_3 r_1(t) \\ kl_1 r_2(t) & kl_2 r_2(t) & kl_3 r_2(t) \end{pmatrix} , \tag{27}$$

$$h(t) = \begin{pmatrix} 0 \\ 0 \\ \{k^2\theta^2(c+s) - k_s s + \lambda^2 s + 2\lambda\dot{s}\}\frac{\sec\theta}{\theta} \end{pmatrix} ,$$

$$r_1(t) = \frac{\tan\theta}{\theta} - 1 , \ r_2(t) = \sec^2\theta - 1 .$$

The spectrum of the matrix A_1 can be assigned arbitrarily through the gain matrix l. Clearly, the solutions of the θ-dynamics converge exponentially to $\theta = 0$. Moreover, since $h(t)$ and the matrix $A_2(t)$ given by (27) go to zero as $t \to \infty$ and

$$\int_0^\infty \|A_2(t)\| dt < \infty , \ \|h(t)\| \leq c_1 e^{-c_2 t}, \ \forall t \geq 0 ,$$

for some positive constants c_1 and c_2, the η-dynamics can also be guaranteed to have exponential convergence to the origin $\eta = 0$ by selecting $l = (l_1 \ l_2 \ l_3)$ such that the matrix A_1 given by (26) is a Hurwitz matrix.

Note that in the (θ, y, x, w) coordinates the control (20) takes the form

$$(v_1, v_2) = (-k\theta, -l_1 y - l_2 \frac{x}{\theta} - l_3 \frac{w}{\theta}) , \tag{28}$$

and the reduced closed-loop system is given by

$$\dot{\theta} = -k\theta , \tag{29}$$

$$\dot{y} = -l_1 y - l_2 \frac{x}{\theta} - l_3 \frac{w}{\theta} , \tag{30}$$

$$\dot{x} = w + \tan\theta(l_1 y + l_2 \frac{x}{\theta} + l_3 \frac{w}{\theta}) \tag{31}$$

$$\dot{w} = k\theta \sec^2\theta(l_1 y + l_2 \frac{x}{\theta} + l_3 \frac{w}{\theta})$$

$$+ [(c+s)\dot{\theta}^2 - k_s s + \lambda^2 s + 2\lambda\dot{s}]\sec\theta . \tag{32}$$

We now present the following result:

Proposition 1: *Consider the reduced closed-loop system (29)-(32) with $0 < k < \lambda$ and $l = (l_1 \ l_2 \ l_3)$ selected such that the matrix A_1 given by equation (26) is a Hurwitz matrix. Let $(\theta_0, y_0, x_0, w_0)$ denote an initial condition with $\theta_0 \neq 0$ and $-\frac{\pi}{2} < \theta_0 < \frac{\pi}{2}$. Then the following hold.*
(i) The trajectory $(\theta(t), y(t), x(t), w(t))$ is bounded for all $t \geq 0$ and converges exponentially to zero.
(ii) The control $(v_1(t), v_2(t))$ is bounded for all $t \geq 0$ and converges exponentially to zero.

Remark 1: The above result demonstrates that for initial conditions satisfying $\theta_0 \neq 0$ and $-\frac{\pi}{2} < \theta_0 < \frac{\pi}{2}$, the feedback control law (28) is well-defined for all $t \geq 0$. Moreover, it drives the system (29)-(32) to the origin, while avoiding the set

$$N = \{(\theta, y, x, w) \mid \theta = 0, \ (\theta, y, x, w) \neq 0\} .$$

Note that one can use a finite time feedback control law [6] to move the system away from N.

Control of the Complete System

We now return to the problem of controlling the system (1)-(4) with (u_1, u_2, u_3) as inputs. Note that $(u_1, u_2) = (\dot{v}_1, \dot{v}_2)$ and u_3 is determined by the relation (6). It should be remarked that the integrator backstepping approach developed for smooth systems [8] cannot be directly applied here due to the discontinuous nature of the system.

Again restrict consideration to $\theta \neq 0$, $-\frac{\pi}{2} < \theta < \frac{\pi}{2}$ and consider the following controller

$$u_1 = -K(\dot{\theta} + k\theta) + s_1(q, \dot{q}) ,\qquad(33)$$

$$u_2 = -L(\dot{y} + l_1 y + l_2 \frac{x}{\theta} + l_3 \frac{w}{\theta}) + s_2(q, \dot{q}) ,\qquad(34)$$

$$u_3 = -u_2 \tan\theta + [(c+s)\dot{\theta}^2 - k_s s + \lambda^2 s + 2\lambda \dot{s}] \sec\theta ,\qquad(35)$$

where $w = \dot{x} + \dot{y}\tan\theta$ is defined in (15) and K and L are control gains. The functions $(s_1(q,\dot{q}), s_2(q,\dot{q}))$ correspond to the time derivative of the feedback control (28) for the reduced system along the trajectories of the complete system:

$$s_1(q, \dot{q}) = -k\dot{\theta} ,$$

$$s_2(q, \dot{q}) = -l_1 \dot{y} - l_2 \frac{\dot{x}\theta - x\dot{\theta}}{\theta^2} - l_3 \frac{\dot{w}\theta - w\dot{\theta}}{\theta^2} .$$

Now assume that the control parameters are selected such that $\lambda, k, l_1, l_2, l_3$ satisfy the conditions of Proposition 1 and $K > k$, $L > 0$. The main idea behind the proposed control law is to implement the control law (28) through the integrators by choosing the gains K and L appropriately, while avoiding the set

$$N' = \{(q, \dot{q}) \in \mathbf{M} \mid \theta = 0 ,\ (q, \dot{q}) \neq 0\} .\qquad(36)$$

Define

$$\xi_1 = \dot{\theta} + k\theta,\ \xi_2 = \dot{y} + l_1 y + l_2 \frac{x}{\theta} + l_3 \frac{w}{\theta}.$$

It can be shown that in the (θ, ξ, η) coordinates the closed-loop system can be written as

$$\dot{\theta} = -k\theta + \xi_1,\qquad(37)$$

$$\dot{\xi}_1 = -K\xi_1\qquad(38)$$

$$\dot{\xi}_2 = -L\xi_2\qquad(39)$$

$$\dot{\eta} = (A_1 + \tilde{A}_2(t))\eta + \tilde{h}(t) ,\qquad(40)$$

where A_1 is the matrix given by (26) and

$$\tilde{A}_2(t) = \begin{pmatrix} 0 & 0 & 0 \\ l_1 r_1(t) & l_2 r_1(t) - \frac{\xi_1}{\theta} & l_3 r_1(t) \\ -l_1 r_3(t) & -l_2 r_3(t) & -l_3 r_3(t) - \frac{\xi_1}{\theta} \end{pmatrix},$$

$$\tilde{h}(t) = \begin{pmatrix} \xi_2 \\ (r_1(t) - 1)\xi_2 \\ [(\xi_1 - k\theta)^2(c+s) + (\lambda^2 - k_s)s + 2\lambda\dot{s}]\frac{\sec\theta}{\theta} \\ +(k + r_3(t))\xi_2 \end{pmatrix},$$

$$r_3(t) = \frac{\xi_1}{\theta} + (\frac{\xi_1}{\theta} - k)(\sec^2\theta - 1).$$

The solutions of the (θ, ξ)-dynamics converge exponentially to $(\theta, \xi) = (0, 0)$. Moreover, it can be easily shown that if $\theta_0 \xi_{10} \geq 0$ (or, equivalently, $\theta_0(\dot{\theta}_0 + k\theta_0) \geq 0$), then $\tilde{A}_2(t)$ and $\tilde{h}(t)$ go to zero as $t \to \infty$; and

$$\int_0^\infty \|\tilde{A}_2(t)\|dt < \infty ,\ \|\tilde{h}(t)\| \leq c_3 e^{-c_4 t},\ \forall t \geq 0 ,$$

for some positive constants c_3 and c_4. It follows that for any initial condition $(\theta_0, \xi_0, \eta_0)$ satisfying $\theta_0 \neq 0$, $-\frac{\pi}{2} < \theta_0 < \frac{\pi}{2}$ and $\theta_0 \xi_{10} \geq 0$, both the trajectory $(\theta(t), \xi(t), \eta(t))$ and the control $(u_1(t), u_2(t), u_3(t))$ are bounded for all $t \geq 0$ and converge exponentially to zero.

We now present the following result.

Proposition 2: *Consider the system (1)-(4) with the feedback controls (33)-(35), where the control parameters are selected such that $\lambda, k, l_1\ l_2$ and l_3 satisfy the conditions of Proposition 1 and $K > k$, $L > 0$. Let $(q_0, \dot{q}_0)$ denote an initial condition satisfying $\theta_0 \neq 0$, $-\frac{\pi}{2} < \theta_0 < \frac{\pi}{2}$ and $\theta_0(\dot{\theta}_0 + k\theta_0) \geq 0$. Then the following hold.*

(i) The trajectory $(q(t), \dot{q}(t))$ is bounded for all $t \geq 0$ and converges exponentially to zero.

(ii) The control $u(t)$ is bounded for all $t \geq 0$ and converges exponentially to zero.

Remark 2: Note that the above choice of the feedback control guarantees that

$$\theta(t) = e^{-kt}\theta_0 + \frac{e^{-kt} - e^{-Kt}}{K - k}(\dot{\theta}_0 + k\theta_0) .$$

It can be easily seen that if $\theta_0 \neq 0$ and $\theta_0(\dot{\theta}_0 + k\theta_0) \geq 0$, then $\theta(t) \neq 0$, $\forall t \in [0, \infty)$. Thus, for all initial conditions satisfying $\theta_0 \neq 0$, $-\frac{\pi}{2} < \theta_0 < \frac{\pi}{2}$ and $\theta_0(\dot{\theta}_0 + k\theta_0) \geq 0$, the feedback control law (33)-(35) is well-defined for all $t \geq 0$. Moreover, it drives the system (1)-(4) to the origin, while avoiding the set (36). Clearly, one can use a finite time feedback control law to move the system to a state satisfying the conditions of Proposition 2. For example,

$$u_1 = -|\theta - \epsilon|^a sign(\theta - \epsilon) - |\dot{\theta}|^b sign(\dot{\theta}) ,\qquad(41)$$

$$u_2 = u_3 = 0 ,\qquad(42)$$

where $b \in (0, 1)$, $a > b/(2-b)$ and $\epsilon \neq 0$ are constants, can be used to transfer the system to a state satisfying the conditions of Proposition 2 in finite time.

4 Simulations

We illustrate the results of the paper with simulations of a typical manipulation task. The control objective is that the manipulator move to a specified equilibrium position, while not exciting the unactuated end effector dynamics. The end effector stiffness is $k_s = 0.5$. The length of the third link is assumed to be $c = 1$.

The desired attracting equilibrium state is given by $(q, \dot{q}) = (0, 0)$. Simulations are provided for two different cases. In the first case, the initial state is given by $(\theta, y, x, s, \dot{\theta}, \dot{y}, \dot{x}, \dot{s}) = (\pi/4, 1, 1, 0, 0, 0, 0, 0)$ so that there is no initial relative motion of the end effector. In the second case, the initial state is given by $(\theta, y, x, s, \dot{\theta}, \dot{y}, \dot{x}, \dot{s}) = (\pi/4, 1, 1, 0.2, 0, 0, 0, 0)$ corresponding to nontrivial initial motion of the end effector.

A discontinuous feedback control law has been constructed, using the construction indicated in the paper; the control parameters are

$$k = 1, \lambda = 2, l = (3.0, -15.0, 11.5), K = 5, L = 5.$$

This choice of gains locates the eigenvalues of the matrix A_1 at (-1.0,-1.5,-2.0).

The results of the simulations are shown in the following figures. Figure 2 shows the time responses of the variables θ, y, x and s for the first case. The time responses for the controls u_1, u_2 and u_3 are shown in Figure 3. Exponential convergence of the closed-loop state and control trajectories can be observed. Moreover, the end effector can be seen not to be excited, i.e. $s(t) = 0$, $\forall t \geq 0$. Figures 4 and 5 show the time responses for the state and control variables for the second case. We can see that the initially perturbed relative end effector motion converges to the origin while the other states converge to the desired values.

5 Conclusions

In this paper, a time-invariant discontinuous feedback control law has been derived to control the planar motion of a PPR robot manipulator with an unactuated elastically mounted gripper to a desired configuration with exponential convergence rates. This feedback control law achieves the control objective without exciting the elastic end effector dynamics.

The proposed feedback control is a time-invariant discontinuous control law, the discontinuity arising from the artificial singularity in the end effector attitude that has been introduced. It is desirable to make a few comments about the strengths and weaknesses of the resulting closed loop systems. We note that if $k_s = 0$, the given manipulator control model is not linearly controllable but the presented control approach still provides good closed loop properties. Otherwise, the manipulator control model is linearly controllable so that more traditional control approaches, including use of linear feedback, may be used. The advantages of the discontinuous controller presented in this paper are: (1) it provides a large guaranteed domain of attraction consistent with large manipulator motions, (2) it provides a good framework to make tradeoffs between the speed of response and the control levels, and (3) it provides excellent guaranteed suppression of the elastic end effector dynamics. On the other hand, this control design approach has not been widely studied except for several specific nonholonomic control problems. It is known that the equilibrium of the closed loop system is not necessarily stable in the sense of Liapunov [4]. This deficiency is not serious in terms of practical implementation of the results. In summary, the time-invariant discontinuous feedback law that is constructed achieves the desired manipulator control objectives.

It is possible to develop similar feedback laws for robot manipulator with more than a single unactuated degree of freedom. Another class of generalizations, involving a formulation in three dimensions, is possible.

Acknowledgements

M. Reyhanoglu wishes to acknowledge the support provided by Embry-Riddle Aeronautical University. N.H. McClamroch acknowledges the partial support provided by NSF Grant ECS-9625173.

References

[1] H. Arai, "Controllability of a 3-DOF Manipulator with a Passive Joint under a Nonholonomic Constraint," *Proceedings of IEEE International Conference on Robotics and Automation*, 1996, pp.3707-3713.

[2] H. Arai, K. Tanie, and N. Shiroma, "Nonholonomic Control of a Three-DOF Planar Underactuated Manipulator," *IEEE Transactions on Robotics and Automation*, 14(5), 1998, pp.681-695.

[3] V.I. Arnold, *Geometrical Methods in the Theory of Ordinary Differential Equations*, Springer-Verlag, New York, 1983.

[4] A. Astolfi, "Discontinuous Control of Nonholonomic Systems," *Systems and Control Letters*, 29, 1996, pp.91-95.

[5] A. De Luca, R. Mattone, and G. Oriolo, "Stabilization of Underactuated Robots: Theory and Experiments for a Planar 2R Manipulator," *Proceedings of IEEE International Conference on Robotics and Automation*, 1997 Albuquerque, NM, pp.3274-3280.

[6] V.T. Haimo, "Finite Time Controllers," *SIAM J. Control and Optimization*, 24(4), 1986, pp.760-770.

[7] J. Imura, K. Kobayashi, and T. Yoshikawa, "Nonholonomic Control of 3 Link Planar Manipulator with a Free Joint," *Proceedings of IEEE Conference on Decision and Control*, 1996, Kobe, Japan, pp.1435-1436.

[8] M. Krstic, I. Kanellakopoulos and P. Kokotovic, *Nonlinear and Adaptive Control*, John Wiley and Sons, Inc., New York, 1995.

[9] N. H. McClamroch, C. Rui, I. Kolmanovsky, S. Cho, and M. Reyhanoglu, "Planar Maneuvers of a Rigid Body with an Unactuated Internal Degree of Freedom," *Proceedings of American Control Conference*, 1998, Philadelphia, PA, pp.229-233.

[10] M. Reyhanoglu, A.J. van der Schaft, N.H. McClamroch and I. Kolmanovsky, "Nonlinear Control of a Class of Underactuated Systems," *Proceedings of IEEE Conference on Decision and Control*, 1996, Kobe, Japan, pp.1682-1687.

[11] M. Reyhanoglu, A.J. van der Schaft, N.H. McClamroch and I. Kolmanovsky, "Dynamics and Control of a Class of Underactuated Mechanical Systems," *IEEE Transactions on Automatic Control*, **44**(9), September 1999.

[12] M. Reyhanoglu, 'Exponential Stabilization of an Underactuated Autonomous Surface Vessel," *IFAC Journal Automatica*, **33**(12), 1997, pp.2249-2254.

[13] M. Reyhanoglu, "Control of a Super-Articulated Robot Manipulator with Joint Elasticity," *Proceedings of International Conference on Control, Automation, Robotics and Vision*, 1996, Westin Stamford, Singapore, pp.172-176.

[14] M. Reyhanoglu, S. Cho, N.H. McClamroch, and I. Kolmanovsky, "Discontinuous Feedback Control of a Planar Rigid Body with an Unactuated Degree of Freedom," *Proceedings of IEEE Conference on Decision and Control*, 1998, Tampa, FL, pp.433-438.

[15] C. Rui, I. Kolmanovsky, P. McNally, and N.H. McClamroch, "Attitude Control of Underactuated Multibody Spacecraft," *13th IFAC World Congress*, 1996, San Francisco, CA, Volume F, pp.425-430.

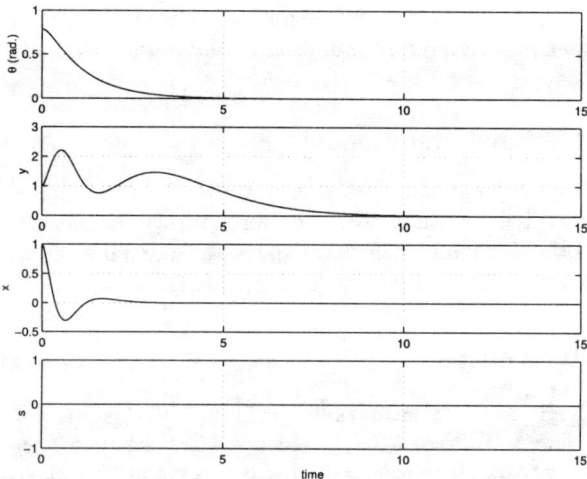

Figure 2: State Variables θ, y, x, and s (Case 1).

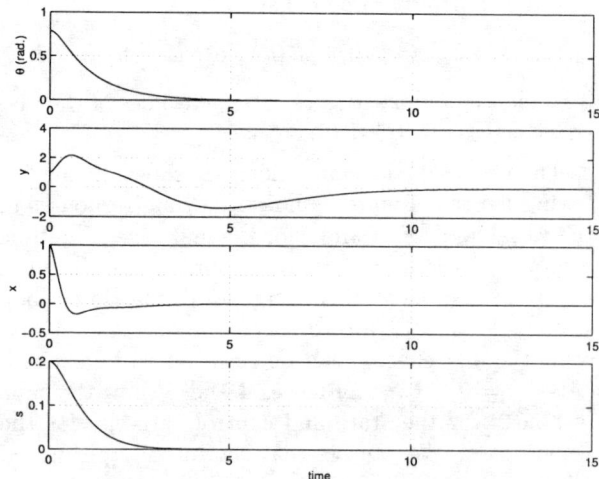

Figure 4: State Variables θ, y, x, and s (Case 2).

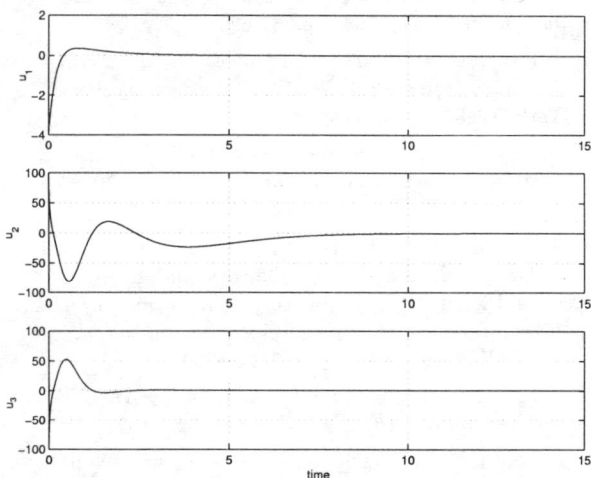

Figure 3: Controls u_1, u_2, and u_3 (Case 1).

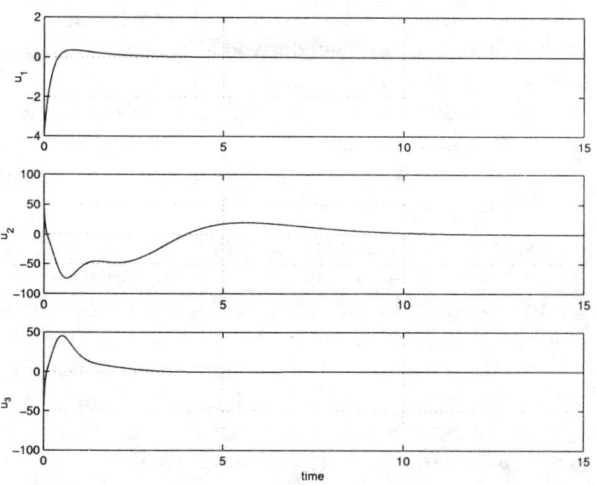

Figure 5: Controls u_1, u_2, and u_3 (Case 2).

Posture Control of Casting Manipulation

Hitoshi Arisumi and Kiyoshi Komoriya

Mechanical Engineering Laboratory, AIST, MITI
Namiki 1-2, Tsukuba Science City, Ibaraki, 305-8564 Japan
Tel +81-298-58-7282   Fax +81-298-58-7201   e-mail arisumi@mel.go.jp

Abstract

We have been developing a casting manipulator which includes a flexible string in the link mechanism to enlarge the work space of the manipulator. During casting manipulation, the gripper is thrown to the target by releasing the string at a suitable time after generating enough motion by swinging. In this paper we deal with the posture control of the gripper after the throwing motion. First, we point out some problems of planning for throwing the gripper while considering its posture when catching. Then we propose a method of applying an impulsive force to the gripper in the air through the string by restricting the motion of the string. Investigating the characteristics of the string through experiments, we confirmed the effectiveness of the impulsive force. This method was evaluated by comparing the results of numerical simulations with those of our experiments using the experimental casting manipulator hardware.

1 Introduction

One of the basic requirements of manipulators is a large work space. For conventional manipulators consisting of rigid links, it is not easy to enlarge the work space because of their mechanical limitations, such as link length. Extending the link length or adding extra degrees-of-freedom[1, 2] produces other problems, such as an increased weight or inefficient operation. Mobile manipulators are a possible means of enlarging the work space[3], however the extent of their reach may be limited, for example a deep depression or a steep slope.

To solve these problems, we noted the human skills of fly fishing or throwing a lasso as a simple way of catching a distant target. For applying these skills to robotic manipulation, we proposed a new type of manipulator which we call a casting manipulator[4]. The casting manipulator consists of rigid links, a flexible string, and a gripper as shown in Figure 1. By throwing the gripper to a target by releasing the flexible string, a distant target can be reached without changing the location of the manipulator. With its small size, simple mechanism and small power, the manipulator realizes a large work space by making more effective use of its dynamics, compared with the conventional manipulators. Since the manipulator uses substantially less energy to move and does not suffer from tumbling by moving, it might be possible to use it to remove rubbles in a disaster-stricken district, or to investigate the nature of the soil on Mars. Furthermore, it can move its main body to an inaccessible place by catching some fixed object and winding a string.

We have investigated the method of generating stable swing motion and the planning the throwing of the gripper to a target as these are the basic motions of casting manipulation. In this paper we discuss the mid-air posture control of the gripper. The paper is organized as follows: Section 2 introduces the process of casting manipulation and presents the periodic swing motion of a casting manipulator model with two links. In Section 3, we propose a method of controlling the posture of the gripper while the gripper is flying in the air. We also describe the results of numerical simulations. In Section 4, we describe the results of our experiment and discuss the effectiveness of the proposed method. Concluding remarks are given in the final section.

2 Casting manipulation

As shown in Figure 1, casting manipulation consists of the following five phases.

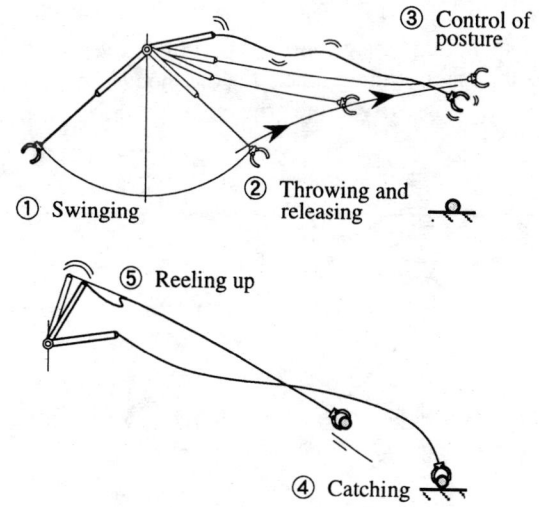

Figure 1: Phases of casting manipulation

(1) Swing phase: To swing the rigid link until the desired motion of the gripper to reach the target is generated.
(2) Throwing and releasing phase: To throw the gripper to the target by releasing the string at a suitable time.
(3) Posture control phase: To control the posture of the gripper while it is flying to the target in order to catch it.
(4) Catching phase: To let the gripper grasp the target with the impulsive force.
(5) Reeling up phase: To reel up the string with the swing motion of the rigid link, in order to retrieve the target.

We proposed a method of generating a desired swing motion[5], that is phase (1), and a planning method for releasing the gripper[6], phase (2). In this paper, we deal with the posture control of the gripper by conducting experiments on the tension of a flexible string.

Similar research on controlling an object with a string has been done. For example, the motion control of a crane system[7], the kendama (Japanese ball-in-cup game) robot system[8] were investigated. As for throwing motion planning, the kendama robot system is similar to our casting manipulator system. But in that system the length of the string is fixed and the posture of the ball while it is flying in the air was not considered.

2.1 Model of two-link casting manipulator

The analytical model of the casting manipulator is shown in Figure 2. The model is a two-link planar manipulator with one actuator only at joint 1 but no actuator at joint 2. Link 1 is a rigid link, and link 2 consists of a string and a gripper. We consider the negative direction of the Y axis in Figure 2 as the direction of gravitational force. Point A at the end of the gripper is the connection between the string and the gripper. The distance between point A and the center of gravity of the gripper is d, L_i is the length of link i. Since we consider that the gripper is a part of link 2, $L_2 = \overline{J_2 A} + d$. M is the mass of the gripper: I is the moment of inertia about the center of gravity of the gripper: θ_i is the angle of joint i:

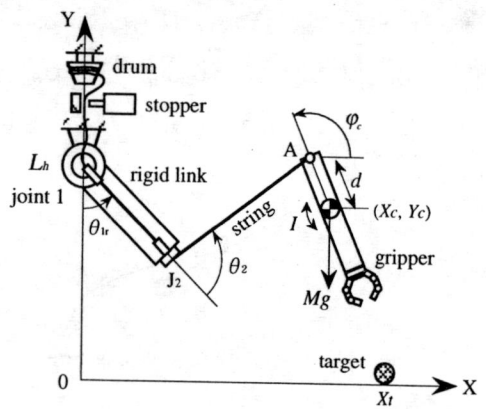

Figure 2: Casting manipulator with two links

and g is the acceleration of gravity. L_h is the Y co-ordinate of joint 1 ($> L_1 + L_2$). To analyze the motion of the gripper, we used the specifications of this system shown in Table 1.

Table 1: Specification of casting manipulator

L_1	0.337 m	L_2	0.61 m
M	0.13 kg	I	0.0006 kgm²
d	0.113 m	L_h	1.17 m

2.2 Periodic swing motion

As a human skill of fly fishing, a stable swing motion should be generated before throwing the gripper. Because the manipulator can accumulate the energy needed to throw the gripper further by swinging, and a stable periodic swing makes it simple to decide the throwing timing.

However in this system the input for the swing motion is only the torque of joint 1. It is thus impossible to control the angles of joint 1 and joint 2 independently. Therefore, we proposed a method of generating a periodic pendulum swing motion by changing and fixing the angle of joint 2 as mentioned in reference[4]. For the excitation or suppression of the swing, we used the periodic reference of joint 2. As the swing comes close to the desired pendulum swing, the reference of joint 2 becomes zero.

When θ_2 approaches zero, the motion of the swing is shown by the following equation which is given with non-holonomic restriction.

$$\ddot{\theta}_1 = -a \sin \theta_1, \quad (1)$$

where
$$a = \frac{MgL_2}{M(L_2^2 + L_1 L_2) + I}.$$

Differential equation (1) can be rewritten as a canonical equation. The phase portrait of this equation is shown in Figure 3. In the figure there are two kinds of motion at joint 1, namely the periodic swing, or mode (a), and the rotation in the constant direction, or mode (b). The motion on the elliptic orbit in the gray area represents mode (a), and the motion on the orbit outside the gray area represents mode (b). Here, we select mode (a) before throwing the gripper. Multiplying both sides of equation (1) by $\dot{\theta}_1$ and integrating it, we obtain the following equation:

$$\frac{1}{2}(\dot{\theta}_1)^2 = a \cos \theta_1 + C,$$

where C is a constant value. Rewriting θ_1 by θ_{1r}, the above equation becomes the following equation:

$$\dot{\theta}_{1r} = \pm \sqrt{2a(\cos \theta_{1r} - \cos \theta_{1\max})}, \quad (2)$$

where θ_{1max} is the maximum swing angle of joint 1.

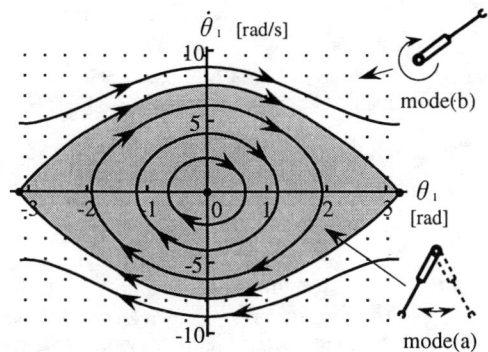

Figure 3: Phase portrait of periodic swing motion

3 Posture control

3.1 Throwing motion planning

Here we consider how to decide the throwing timing. Letting t_c be the time from the throwing of the gripper, the relationship between the configuration of the gripper at the throwing point (X_r, Y_r, φ_r) and the configuration of the gripper (X_c, Y_c, φ_c) at time t_c is generally given by,

$$\dot{X}_r t_c + X_r = X_c \tag{3}$$

$$\dot{Y}_r t_c - \frac{g}{2} t_c^2 + Y_r = Y_c \tag{4}$$

$$\dot{\varphi}_r t_c + \varphi_r = \varphi_c. \tag{5}$$

The position and the velocity of the gripper at the throwing point in the pendulum swing are described by angle θ_{1r} and angular velocity $\dot{\theta}_{1r}$ at the throwing. Rewriting equations (3), (4) and (5) by $(\theta_{1r}, \dot{\theta}_{1r})$ we obtain

$$L_{12} \dot{\theta}_{1r} C_1 t_c + L_{12} S_1 = X_c \tag{6}$$

$$L_{12} \dot{\theta}_{1r} S_1 t_c - \frac{g}{2} t_c^2 + L_h - L_{12} C_1 = Y_c \tag{7}$$

$$\dot{\theta}_{1r} t_c + \theta_{1r} - \frac{\pi}{2} = \varphi_c, \tag{8}$$

where $C_1 = \cos\theta_{1r}$, $S_1 = \sin\theta_{1r}$, $L_{12} = L_1 + L_2$.

Assuming (X_t, Y_t) is the position of the target, and φ_t is the desired posture of the gripper for catching the target shown in Figure 4, we obtain the following equations as a condition for catching a target at time t_c.

$$X_c = X_t + d \cos\varphi_c \tag{9}$$

$$Y_c = Y_t + d \sin\varphi_c \tag{10}$$

$$\varphi_c = \varphi_t, \tag{11}$$

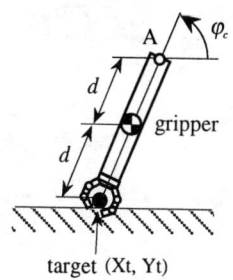

Figure 4: Catching posture of gripper

Eliminating t_c from equations (6) and (7) by using equation (8) and substituting (9), (10), (11), we obtain

$$L_{12}\{(\varphi_t - \theta_{1r} - \frac{\pi}{2})C_1 + S_1\} = X_t + d\cos\varphi_t \tag{12}$$

$$2\dot{\theta}_{1r}^2 [L_{12}\{(\varphi_t - \theta_{1r} - \frac{\pi}{2})S_1 - C_1\} + L_h - Y_t - d\sin\varphi_t]$$
$$= g(\varphi_t - \theta_{1r} - \frac{\pi}{2})^2. \tag{13}$$

The possible range of catching posture φ_t is given by the structure of the gripper and the shape of the target. Here we give φ_t as follows:

$$(60 + 360*n) \deg < \varphi_t < (120 + 360*n) \deg, \tag{14}$$

where n, which is an integer, represents the number of rotations of the gripper until it reaches the target.

According to equation (12) and the inequality (14), in the case that $n = 0$ and 1, the relationship between θ_{1r} and X_t is shown in Figure 5. If $(X_t, Y_t) = (1.3\text{m}, 0)$, throwing angle θ_{1r} is restricted to the region between upper two curves. Calculating the intersections of the dotted line ($X_t = 1.3$) and upper two curves, the inequality, $84.15 \deg < \theta_{1r} < 86.6 \deg$, is obtained as a necessary condition. In the case of $84.15 \deg < \theta_{1r} < 86.6 \deg$, and $420 \deg < \varphi_t < 480 \deg$, the relation

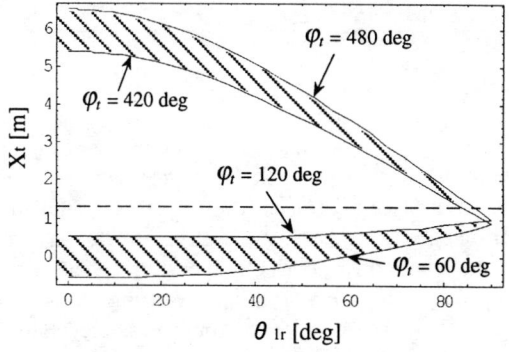

Figure 5: Necessary condition of θ_{1r} for Xt = 1.3 m

between θ_{1r}, φ_t and $\dot{\theta}_{1r}$ is shown in Figure 6 which is calculated numerically using equation (13). As Figure 6 shows, $\dot{\theta}_{1r}$ is restricted by the inequality, 4.1 deg $<\dot{\theta}_{1r}<$ 4.7 deg at most.

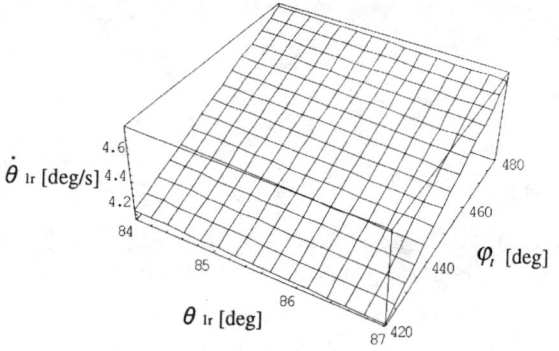

Figure 6: Restricted region of $\dot{\theta}_{1r}$

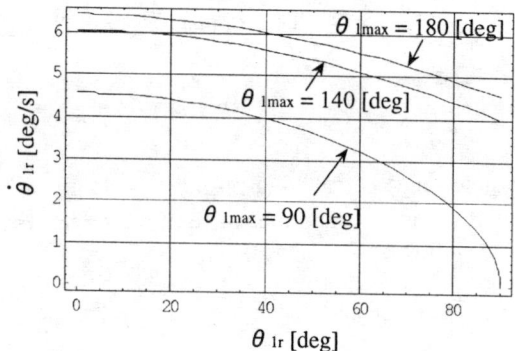

Figure 7: Parts of elliptic orbit

Throwing point ($\theta_{1r}, \dot{\theta}_{1r}$), which is a point on the elliptic orbit in Figure 3, is calculated using equation (2). Figure 7 shows part of the elliptic orbits. In the case of 4.1 deg $<\dot{\theta}_{1r}<$ 4.7 deg, θ_{1max} is required to be around 140 deg. But to prevent the string from slacking, θ_{1max} should be under 90 deg. In the case when swing angle θ_{1max} is under 90 deg, we found it possible to reach the target, but not possible to catch the target. Consequently only throwing the gripper is not enough to enlarge its practical work space. The control of the gripper after throwing is essential.

3.2 Impulsive force

The first question we had to answer is how to control the posture of the gripper to catch the target. As one solution, we considered applying force on the gripper through the string by pulling the string during the flight of the gripper. This control motion corresponds to the third phase of casting manipulation in Figure 1. Pulling the string by swinging the rigid link can be considered one possible solution for applying force on the gripper. But the flight time of the gripper is so short that a rapid motion is required to control the rigid link. However it is difficult to reduce the vibration of the rigid link or error from the reference trajectory. Therefore a vibration problem or error problem may have bad effects on the control of the gripper. And also it is difficult to control the rigid link in real time while sensing the tension of the string, or the position and posture of the gripper. Thus we proposed a simpler way of applying force on the gripper. Here we assume that the rigid link is fixed at the throwing point by braking the string which is drawn out of the drum during the flight of the gripper, at a suitable time instead. To brake the string, we used the solenoid shown in Figure 8. Just after braking the string, the gripper movement is restrained in the direction of the string movement. At that time, the gripper rebounds as if there were a wall perpendicular to the string as shown in Figure 8. Using the impulsive force transmitted by the string, we tried to control the posture of the gripper.

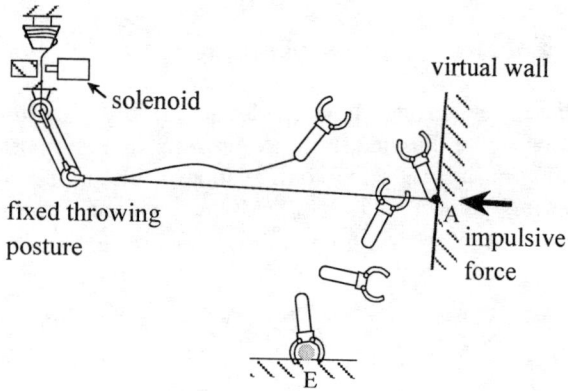

Figure 8: Applying impulsive force to gripper

3.3 Rebound characteristics of string

Before we discuss posture control, it will be useful to clarify the physical characteristics of the string and to examine the possibility of realizing a virtual wall as shown in Figure 8.

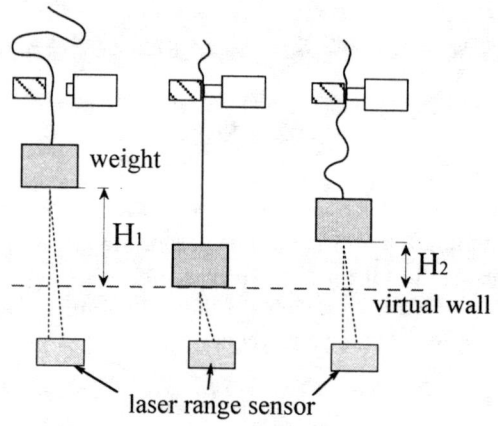

Figure 9: Free-fall experiment

Figure 9 shows our free fall experiment. The horizontal broken line represents a virtual wall. First a weight falls from height H_1 above the broken line. Just as the weight reaches the broken line, the motion of the string is braked by the solenoid. After the weight rebounds, the highest position of the weight is sensed by the laser range sensor. H_2 represents the highest height above the broken line after rebound. Figure 10 shows the result in the case where $H_1 = 0.5$m. The horizontal axis represents the mass of the weight. From this result, the rebound is almost independent of the mass of weight. Here we assume that V_1 represents the velocity just before rebound, and V_2 represents the velocity just after rebound. V_1 and V_2 are given by following equations.

$$V_1 = \sqrt{2gH_1}$$
$$V_2 = \sqrt{2gH_2} \quad (15)$$

Then, rebound coefficient e is given by

$$e = \frac{V_2}{V_1} = \sqrt{\frac{H_2}{H_1}} \quad (16)$$

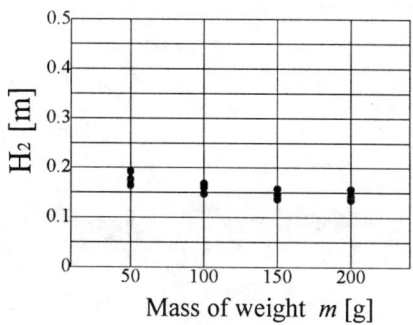

Figure 10: Height of rising weight

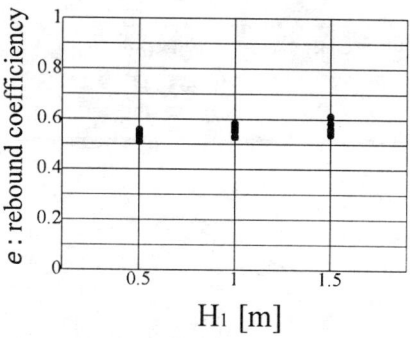

Figure 11: Rebound coefficient

Using equation (16), we examined the rebound coefficient with changing H_1. Figure 11 shows that the rebound coefficient of the string is nearly constant. Generally speaking, the string expands to a greater or less extent when it is stretched. The expansion of the string is due to its elasticity which depends on the material, the cross-sectional shape, the thickness, and the length of the string. Then when H_1 becomes lager, the rebound coefficient will not be constant. But in the region of $0.5m < H_1 < 1.5$ m, we can consider the rebound coefficient to be about 0.56. Also the experiment shows the repeatability of the virtual wall position.

3.4 Gripper motion after collision

We assume the virtual wall, which is vertical to the string, is generated when the string is braked. Using a rebound coefficient, we thus examined the gripper motion after the collision between the gripper and the virtual wall. In Figure 12 (a), point A becomes a collision point, which is the origin of the original coordinate system, Σ_A O_A-$X_A Y_A$. We located an orthogonal coordinate system, Σ_A^* O_A-$X_A^* Y_A^*$, on the virtual wall as the surface of the wall contains Y_A^* axis as shown in Figure 12 (a). When we assume that the angle between X_A axis and the string is β, the coordinate system Σ_A^* is given by rotating the original coordinate system Σ_A with $\pi - \beta$ in the clockwise direction. Here we assume that (Vx, Vy, ω) represents the velocity of the gripper at the center of gravity in Σ_A, and (Vx^*, Vy^*, ω^*) represents the velocity of the gripper at the center of gravity in Σ_A^*. Then the relationship of the velocities which are represented by the two coordinate systems are given by the following equations.

$$\begin{bmatrix} Vx^* \\ Vy^* \end{bmatrix} = Rot(\pi - \beta) \begin{bmatrix} Vx \\ Vy \end{bmatrix} \quad (17)$$
$$\omega^* = \omega$$

Where $Rot(\theta) = \begin{bmatrix} \cos\theta & -\sin\theta \\ \sin\theta & \cos\theta \end{bmatrix}$.

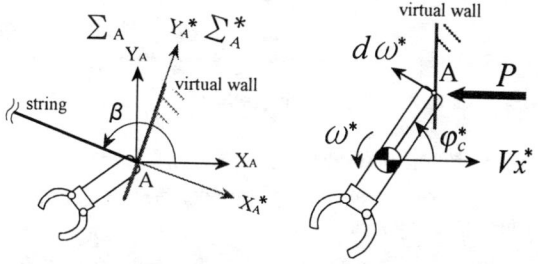

(a) Transformation of coordinates (b) Change of velocity

Figure 12: Collision of gripper

Since a collision between the gripper and the virtual wall occurs in the direction of X_A^* axis, Vx^* and ω^* are changed by the collision. But in the Y^* direction, the gripper does not receive frictional force, and this is different from an actual wall, so Vy^* is not changed by the collision. Therefore, we let $(Vx^{*'}, Vy^{*'}, \omega^{*'})$ be the velocity of the gripper in Σ_A^* just after a collision, and $Vy^{*'} = Vy^*$. Next we mention $(Vx^{*'}, \omega^{*'})$ using Figure 12 (b). Assuming P represents an impulse which is applied to the gripper, equations with respect to momentum and angular momentum are given as follows:

$$P = M(Vx^{*'} - Vx^*)$$
$$Pd\sin\varphi^* = -I(\omega^{*'} - \omega^*) \quad (18)$$

Translational velocity at collision point A is obtained by

$$V_{Ax}^* = Vx^* - d\omega^*\sin\varphi^*$$
$$V_{Ax}^{*'} = Vx^{*'} - d\omega^{*'}\sin\varphi^{*'} \quad (19)$$

And still using rebound coefficient e, we obtain the following equations.

$$V_{Ax}^{*'} = -eV_{Ax}^* \quad (20)$$

From equations (18),(19),(20), and $\varphi^{*'} = \varphi^*$, we obtain

$$\omega^{*'} = \frac{I\omega^* + Md\sin\varphi^*(eV_{Ax}^* + Vx^*)}{Md^2\sin^2\varphi^* + I}$$
$$Vx^{*'} = (\omega^{*'} + e\omega^*)d\sin\varphi^* - eVx^* \quad (21)$$

Finally we obtain the velocity of the gripper after braking the string as the following equations by performing coordinate transformation.

$$\begin{bmatrix} Vx' \\ Vy' \end{bmatrix} = Rot(\beta - \pi)\begin{bmatrix} Vx^{*'} \\ Vy^{*'} \end{bmatrix} \quad (22)$$
$$\omega' = \omega^{*'}$$

3.5 Planning for catching a target

Letting t_s be the time from the throwing the gripper to the braking of the string motion, we consider how to decide these three parameters $(\theta_{1r}, \dot{\theta}_{1r}, t_s)$ in order to catch a target.

Assuming (X_e, Y_e, φ_e) and $(V_{ex}, V_{ey}, \omega_e)$ represent the configuration and the velocity of the gripper at the landing point E as shown in Figure 8, they are determined by three parameters $(\theta_{1r}, \dot{\theta}_{1r}, t_s)$. Each of them, $(X_e, Y_e, \varphi_e, V_{ex}, V_{ey}, \omega_e)$, is expressed by a function of $(\theta_{1r}, \dot{\theta}_{1r}, t_s)$.

Let us consider the case, for example, in which the target position $X_t = 1.3$ m, the conditions for catching the target are given as follows:

$$(X_e, Y_e) = (1.3 \text{ m}, 0 \text{ m}) \quad (23)$$

$$60 \text{ deg} < \varphi_e < 120 \text{ deg} \quad (24)$$

In this case the set of solutions which satisfies equation (23) and the inequality (24) is calculated numerically as shown in Figure 13.

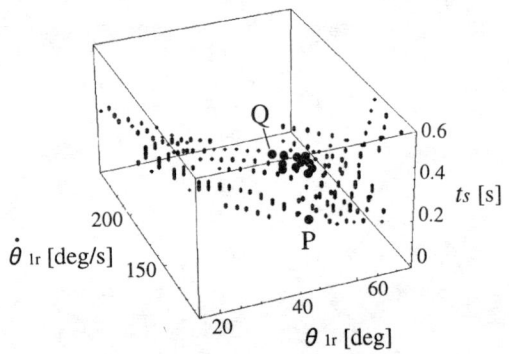

Figure 13: Set of solutions

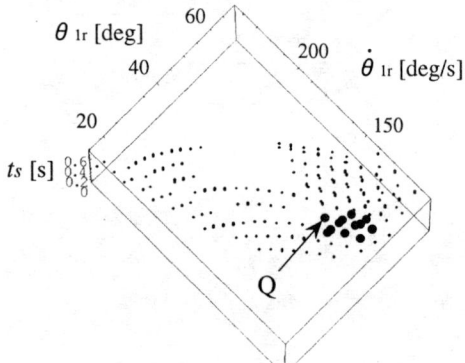

Figure 14: Top view

In the case where the fingers of the gripper are closed passively by collision with a target, or to reduce an impact generated by collision of the gripper and a target, we need to satisfy the following inequality based on the hardware specification.

$$2 \text{ m/s} < \sqrt{V_{ex}^2 + V_{ey}^2} < 2.5 \text{ m/s}$$
$$|\omega_e| < 30 \text{ deg/s} \quad (25)$$

Adding the restriction such as the inequality (25) to the calculation of $(\theta_{1r}, \dot{\theta}_{1r}, t_s)$ for catching the target, the set of solutions is given by the large black points in Figure 13. Observing from the top of Figure 13, we obtain Figure 14. As shown in this figure, each point locates on each elliptic orbit which corresponds to the parts of the orbits in Figure 3. When we consider the time delay when throwing the gripper, we chose point Q as the throwing point because the rotational direction of each elliptic orbit is clockwise.

3.6 Simulation results

Based on the results in Section 3.5, we examined the planning of throwing the gripper and braking the string with numerical simulations.

Figure 15 shows the trajectory of the gripper, when the gripper is thrown at point P in Figure 13 in the case where the inequality (25) is not considered. In this case, the time history of the velocity is shown in Figure 16. The velocity is reduced by impulsive force. However the terminal velocity at the time of catching the target does not satisfy equation (25). As Figure 15 shows, the gripper rotates more than one time. So the string may possibly get tangled.

Figure 17 shows the trajectory, when the gripper is thrown at point Q in Figure 13. The time history of the velocity is shown in Figure 18. By braking the string, we can reduce the velocity and angular velocity within the permitted limit. Therefore, we can reduce the impact at the time of catching the target.

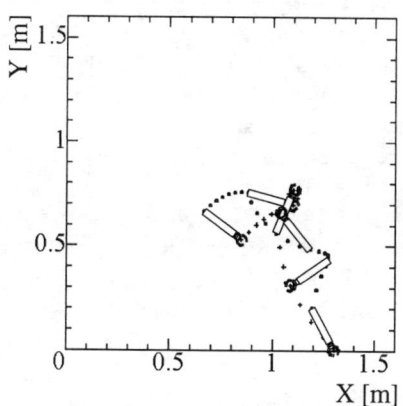

Figure 15: Trajectory of gripper for point P

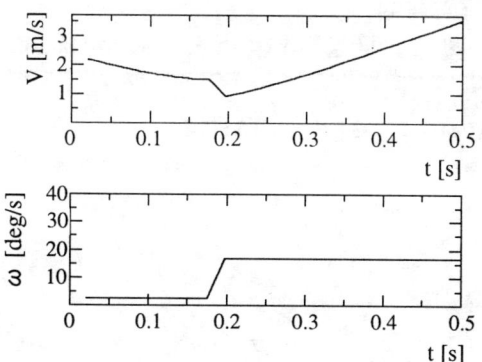

Figure 16: Time history of position and posture (at point P)

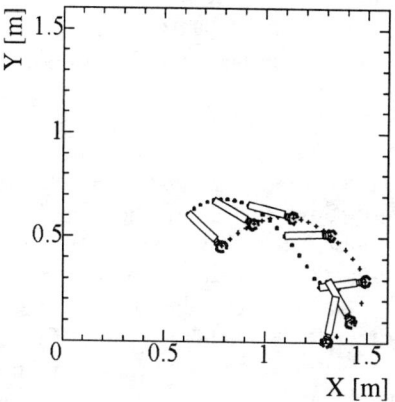

Figure 17: Trajectory of gripper for point Q

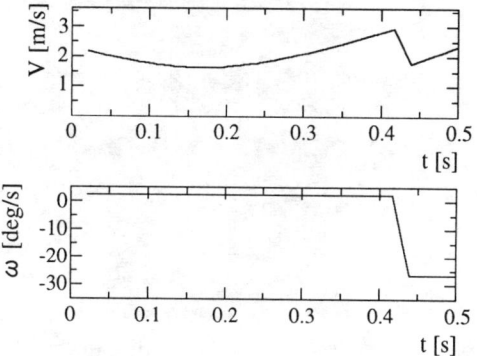

Figure 18: Time history of position and posture (at point Q)

4 Experiments

4.1 Hardware

The casting manipulator with two links, shown in Figure 19, was constructed. The first link made of aluminum plate is driven by the DD motor attached at its base. On the other end of the first link, a potentiometer is attached to estimate joint angle of the second link, a string which is a fishing line. To throw the gripper by releasing the string, a stopper is installed at the base plate above the DD motor and is actuated by the solenoid to brake the string and to release it. The gripper is set at the tip of the string. An optical encoder with a resolution of 81,000 pulse/rev is installed on the DD motor to measure joint angle of the first link. The specifications of the casting manipulator is given in Table 1.

4.2 Experimental results

To investigate the possibility of our proposed method, we conducted an experiment of throwing a rigid pipe, which we regard as the gripper. Instead of catching a target, we gave it the task of throwing the pipe into a box with a small hole.

The diameter of the hole is 1.5 times as large as that of the pipe. We used the Tracking Vision system to monitor the configuration of the gripper. Figure 20 shows a sequence of photographs of the gripper which is flying in the air. As you can see in the photo, the posture of the gripper was changed by braking the string, and the gripper went straight into the box. Comparing with the simulation results shown in Figure 17, Figure 20 shows that the rotational speed of the gripper after braking was a little smaller. Because the string was braked a little earlier due to the time delay of throwing the gripper. But we found that the experimental results almost agreed with our simulation results. Consequently, the effectiveness of our proposed method of posture control was proved through the experiment.

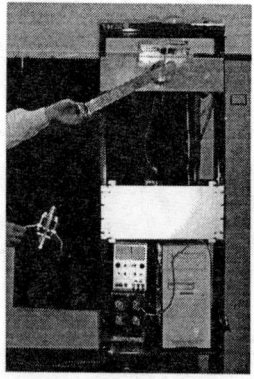

Figure 19: Overview of casting manipulator system

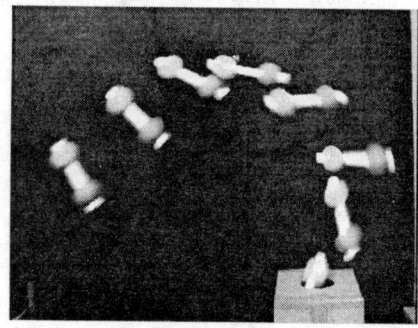

Figure 20: Sequence photographs of gripper

5 Conclusions

In this paper, we discussed the posture control phase of casting manipulation. As a simple way of posture control, we proposed a method of applying impulsive force to the gripper through a string. First, we discussed the possibility of this method by examining the characteristics of the string. According to our simulation results, the gripper could catch a target within the desired approach velocity by controlling the time of braking the string. We also demonstrated the effectiveness of our proposed method through experiments, and the results coincided well with our numerical simulation.

Our proposed method also might help to reduce the impact when the gripper collides with the target.

References

[1] H. Mochiyama, E. Shimemura and H. Kobayashi, "Shape Correspondence between a Spatial Curve and a Manipulator with Hyper Degrees of Freedom," in *Proceedings of the IEEE/RSJ International Conference on Intelligent Robots and Systems (IROS'98)*, pp.161-166, 1998.

[2] N. Takanashi et al, "Simulated and Experimental Results of Dual Resolution sensor Based Planning for Hyper-Redundant Manipulators," in *Proceedings of the IEEE/RSJ International Conference on Intelligent Robots and Systems (IROS'93)*, pp.636-643, 1993.

[3] O. Khatib, K. Yokoi, K. Chang, D. Ruspini, R. Holmberg and A. Casal, "Vehicle/Arm Coordination and Multiple Mobile Manipulator Decentralized Cooperation," in *Proceedings of the IEEE/RSJ International Conference on Intelligent Robots and Systems (IROS'96)*, pp.546-553, 1996.

[4] H. Arisumi, T. Kotoku and K. Komoriya, "A study of Casting Manipulation (Swing Motion Control and Planning of Throwing Motion)," in *Proceedings of the IEEE/RSJ International Conference on Intelligent Robots and Systems (IROS97)*, pp.168-174, 1997.

[5] H. Arisumi, T. Kotoku and K. Komoriya, "Swing Motion control of Casting Manipulation (Experiment of Swing Motion Control)," in *Proceedings of the IEEE International Conference on Robotics and Automation (ICRA98)*, pp. 3522-3527, 1998.

[6] H. Arisumi, T. Kotoku and K. Komoriya, "A study on Casting Manipulation (Experiment of Swing Control and Throwing)," in *Proceedings of the IEEE/RSJ International Conference on Intelligent Robots and Systems (IROS98)*, pp.494-501, 1998.

[7] Y. Hashimoto, T. Tsuchiya, I. Sugioka and T. Matsuda, "Transversal Load-Swing suppression Control of Travelling Crane," *The trans. of RSJ*, vol. 11, No.7, pp. 1073-1082, 1993. (in Japanese)

[8] T. Sakaguchi and F. Miyazaki, "Dynamic Manipulation of Ball-in-Cup Game," in *Proceedings of the IEEE International Conference on Robotics and Automation (ICRA94)*, pp. 2941-2948, 1994.

Achieving Fine Absolute Positioning Accuracy in Large Powerful Manipulators

Marco A. Meggiolaro, Peter C. L. Jaffe and Steven Dubowsky
Department of Mechanical Engineering
Massachusetts Institute of Technology
Cambridge, MA 02139

Abstract

Important robotic tasks could be most effectively done by powerful and accurate manipulators. However, high accuracy is generally unattainable in manipulators capable of producing high task forces due to such factors as high joint, actuator, and transmission friction and link elastic and geometric distortions. A method called Base Sensor Control (BSC) has been developed to compensate for nonlinear joint characteristics, such as high joint friction, to improve system repeatability. A method to identify and compensate for system geometric and elastic distortion positioning errors in large manipulators has also been recently proposed to improve absolute accuracy in systems with good repeatability using a wrist force/torque sensor. This technique is called Geometric and Elastic Error Compensation (GEC). Here, it is shown experimentally that the two techniques can be effectively combined to enable strong manipulators to achieve high absolute positioning accuracy while performing tasks requiring high forces.

1 Introduction

Large robotic manipulators are needed in nuclear maintenance, field, undersea and medical applications to perform high accuracy tasks requiring the manipulation of heavy payloads. Hydraulic robot's high load carrying capacity is attractive for such applications, but high joint friction and actuator nonlinearities make them difficult to control. The nozzle dam positioning task for maintenance of a nuclear power plant steam generator is an example of a task that requires a strong manipulator with very fine absolute positioning accuracy [14]. Absolute accuracy, rather than simple repeatability, is required for autonomous operation or for teleoperation with advanced virtual aides, such as virtual viewing.

A number of approaches exist for improving fine motion manipulator performance through friction compensation. Some of these require modeling of the difficult to characterize joint frictional behavior [1, 10]. Some require the use of specially designed manipulators that contain complex internal joint-torque sensors [11].

A simple, yet effective control method has been developed that is modeless and does not require internal joint sensors [5, 9]. The method, called Base Sensor Control (BSC), estimates manipulator joint torques from a self-contained external six-axis force/torque sensor placed under the manipulator's base. The joint torque estimates allow for accurate joint torque control that has been shown to greatly improve repeatability of both hydraulic and electric manipulators.

Even with improved repeatability, high absolute positioning accuracy is still difficult to achieve with a strong manipulator. Two principal error sources create this problem. The first is kinematic errors due to the non-ideal geometry of the links and joints of manipulators. These errors are often called geometric errors. Task constraints often make it impossible to use direct end-point sensing to compensate for these errors. Therefore, there is a need for model-based error identification. Research has been done in this area, commonly referred to as robot calibration [4, 12].

The second error source that often limits the absolute accuracy of a large manipulator is the elastic errors due to the distortion of a manipulator's mechanical components under large task loads. Methods have been developed to deal with this problem [13]. These methods depend upon detailed and difficult to obtain analytical models of the manipulator.

Recent work has resulted in methods that can correct for errors in the end-effector position and orientation caused by geometric and elastic errors in large manipulators [2, 7]. The similar methods, called Geometric and Elastic Error Compensation (GEC), yield measurement based error compensation algorithms that predict the manipulator's end-point position and orientation as a function of the configuration of the system and the task forces. Given the task loads from a conventional wrist force/torque sensor and the joint angles of the manipulator, the algorithm compensates for the combined elastic and geometric errors. They do not require detailed modeling of the manipulator's structural properties. Instead they use a relatively small set of offline end-point experimental measurements to build a "generalized error" representation of the system [6]. These methods can substantially reduce the absolute errors in manipulators with good inherent repeatability.

In this research, an approach is developed that substantially improves the absolute accuracy in strong powerful manipulators lacking good repeatability and having significant geometric and elastic errors. The method uses base force/torque sensor information to apply BSC in concert with GEC, which uses wrist sensor information to achieve greatly improved absolute accuracy in a strong manipulator exerting high task loads. Its effectiveness is shown experimentally on a large powerful hydraulic industrial manipulator. While strong, robust and reliable, this manipulator does not inherently have fine repeatability and absolute accuracy. The algorithm does not require joint velocity or acceleration measurements, a model of the actuators or friction, or the knowledge of manipulator mass parameters or link stiffnesses, yet it is able to substantially improve its absolute positioning accuracy.

2 Analytical Background

2.1 Base Sensor Control (BSC)

Here the basis for BSC is briefly reviewed. The complete development is presented in [9]. A simplified version of the algorithm sufficient and effective for fine-motion control is formulated in [5].
As shown in Figure 1, the wrench, $\mathbf{W_b}$, exerted by the manipulator on its base sensor can be expressed as the sum of three components:

$$\mathbf{W_b} = \mathbf{W_g} + \mathbf{W_d} + \mathbf{W_e} \quad (1)$$

where $\mathbf{W_g}$ is the robot gravity component, $\mathbf{W_d}$ is caused by manipulator motion, and $\mathbf{W_e}$ is the wrench exerted by the payload on the end-effector. Note that joint friction does not appear in the measured base sensor wrench. In the fine-motion case, it is assumed that the gravity wrench is essentially constant, and this wrench can be approximated by the initial value measured by the base sensor. Hence, the complexity of computing the gravitational wrench, such as identification of link weights and a static manipulator model, is eliminated. Under this assumption, the Newton Euler equations of the first i links are:

$$\begin{cases} \mathbf{W}_{0 \to 1} = -\mathbf{W_b} \\ \mathbf{W}_{1 \to 2} = \mathbf{W}_{0 \to 1} - \mathbf{W}_{d_1} \\ \vdots \\ \mathbf{W}_{i \to i+1} = \mathbf{W}_{i-1 \to i} - \mathbf{W}_{d_i} \\ \vdots \\ -\mathbf{W_e} = \mathbf{W}_{n-1 \to n} - \mathbf{W}_{d_n} \end{cases} \quad (2)$$

where $\mathbf{W}_{i \to i+1}$ is the wrench exerted by link i on link i+1, and $\mathbf{w}_{d_i}$ is the dynamic wrench for link i.
For fine tasks it is assumed that the manipulator moves very slowly so that $\mathbf{W_d}$ can be neglected. Therefore, for slow, fine motions, only the measured wrench at the base

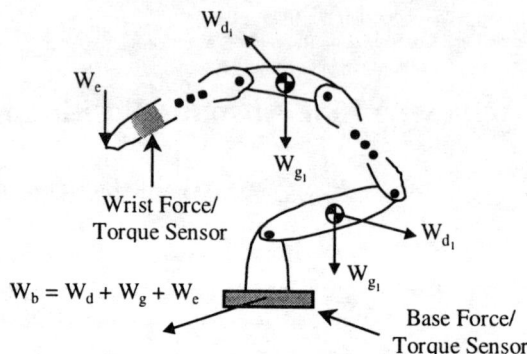

Figure 1 - External and Dynamic Wrenches

is used to estimate the torque in joint i+1. The estimated torque in joint i+1 is obtained by projecting the moment vector at the origin O_i of the ith reference frame along the joint axis z_i:

$$\tau_{i+1} = -\mathbf{z}_i^T \cdot \mathbf{W_b}^{O_i} \quad (3)$$

The value of τ_{i+1} depends only on the robot's kinematic parameters, joint angles and base sensor measurements.
With estimates of the joint torque, it is possible to perform high performance torque control that can greatly reduce the effects of joint friction and nonlinearities. This results in greatly improved repeatability. This method will not compensate for sources of random repeatability errors, such as limited encoder resolution. In addition, a manipulator with good repeatability may not have fine absolute position accuracy.

2.2 Geometric and Elastic Error Compensation (GEC)

The main sources of absolute accuracy errors in a manipulator with good repeatability are mechanical system errors (resulting from machining and assembly tolerances), elastic deformations of the manipulator links, and joint errors (bearing run-out). These can be grouped into geometric and elastic errors. Although these physical errors are relatively small, their influence on the end-effector position of a large manipulator can be significant. A brief review of the error compensation method used here is presented below.
The end-effector position and orientation error, $\Delta \mathbf{X}$, is defined as the 6x1 vector that represents the difference between the real position and orientation of the end-effector and the ideal or desired one:

$$\Delta \mathbf{X} = \mathbf{X}^r - \mathbf{X}^i \quad (4)$$

where $\mathbf{X}^r$ and $\mathbf{X}^i$ are 6x1 vectors composed of the three positions and three orientations of the end-effector reference frame in the inertial reference system for the real and ideal cases respectively.
The error compensation method assumes that physical errors slightly displace manipulator joint frames from

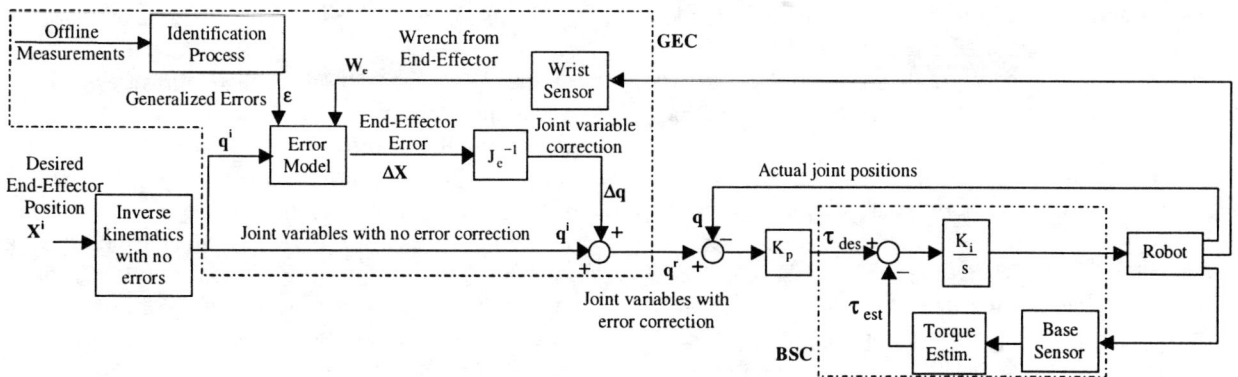

Figure 2 - Base Sensor Control and Error Compensation Scheme

their expected, ideal locations [7]. The real, or actual, position and orientation of each frame with respect to its ideal location is represented by three consecutive rotations and three translational coordinates. These 6 parameters are called here "generalized error" parameters. For an nth degree of freedom manipulator, there are 6n generalized errors represented by a vector ε. When the generalized errors are included in the model,

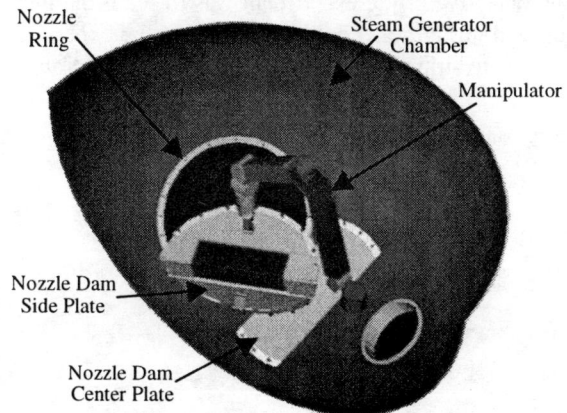

Figure 3 - Simulated Robotic Nozzle Dam Task

the six coordinates of the real end-effector position vector $\mathbf{X}^r$ can be written in a general form:

$$\mathbf{X}^r = \mathbf{f}^r(\mathbf{q}, \varepsilon, \mathbf{s}) \quad (5)$$

where $\mathbf{f}^r$ is a vector non-linear function of the configuration parameters $\mathbf{q}$, the generalized errors ε, and the structural parameters $\mathbf{s}$. In general, the generalized errors depend on the manipulator configuration $\mathbf{q}$ and the end-effector wrench $\mathbf{W}_e$, or $\varepsilon(\mathbf{q}, \mathbf{W}_e)$. To predict the behavior of the manipulator in a given configuration, the task wrench is necessary to calculate the generalized errors from previous offline measurements. For simplicity, the ith element of vector ε is approximated by a polynomial series expansion of the form:

$$\varepsilon_i = \sum_j \varepsilon_i^{(j)} \cdot (q_1^{a_1^{(j)}} \cdot q_2^{a_2^{(j)}} \cdot \ldots \cdot q_n^{a_n^{(j)}} \cdot w_m^{b^{(j)}}) \quad (6)$$

where $q_1, q_2, \ldots, q_n$ are the manipulator joint parameters, w_m is an element of the task wrench, and $\varepsilon_i^{(j)}$ are the polynomial coefficients. It has been found that good accuracy can be obtained using only a few terms in the above expansion. The coefficients of these $\varepsilon_i^{(j)}$ terms are constants and become the unknowns of the problem.

Since the generalized errors are small, $\Delta \mathbf{X}$ can be calculated by the following linear equation in ε:

$$\Delta \mathbf{X}(\mathbf{q}, \mathbf{W}_e) = \mathbf{J}_e \cdot \varepsilon \quad (7)$$

where $\mathbf{J}_e$ is the 6x6n Jacobian matrix of the function $\mathbf{f}^r$ with respect to the elements of the generalized error vector ε. The matrix $\mathbf{J}_e$ depends on the system configuration, geometry, and task wrench.

Once the generalized errors, ε, are identified, the end-effector position and orientation error can be calculated using Equation (7). Assuming all six components of $\Delta \mathbf{X}$ can be measured, for an nth degree of freedom manipulator, its 6n generalized errors ε can be calculated by fully measuring vector $\Delta \mathbf{X}$ at n different configurations. To increase the accuracy of the calculated generalized errors, additional measurements are made and a least mean square procedure is used. All repeatable errors are identified regardless of their source. Figure 2 summarizes how an error model of the type of Equation (7) can be used in an error compensation algorithm, and how the corrected joint angles can be commanded in a Base Sensor Control scheme.

3 The Task and Experimental System

3.1 The Task

The precision control algorithms presented in this paper are being developed for a task in the nuclear power industry. In order for workers to inspect and repair a nuclear power plant's steam generator, two very large pipes (1 meter in diameter) must be sealed with a device called a nozzle dam. The center section of the nozzle dam weighs approximately 60 kg and it must be inserted into a ring with clearances of a few millimeters. In this operation, workers receive high doses of radiation. Hence, performing this task with a robotic manipulator would be very desirable. A simulated robotic nozzle dam placement can be see in Figure 3, where the manipulator is moving the nozzle dam side plate into its position in the nozzle ring. The center plate will then be inserted within the side plate.

Attempts to place the dam with a manipulator have taken too long because of the combination of poor operator visibility and lack of manipulator accuracy. It costs tens of thousands of dollars per hour to keep a nuclear power plant offline. Improving manipulator accuracy is a key to shortening this time. The typical repeatability of manipulators capable of handling the required load is in the range of 10 to 20 mm. The absolute accuracy can be several times these amounts. The automation of this task would require absolute accuracy of a few mm. In this work, the combined BSC/GEC method was experimentally evaluated for this application.

3.2 Experimental System

Figure 4 shows the experimental test-bed constructed for this study. The manipulator chosen for this system is a Schilling Titan II, a six DOF hydraulic robot capable of handling payloads in excess of 100 kg. Its position accuracy is approximately 40 mm (RMS), many times the specification of a few mm. A good part of its lack of accuracy is due to its underlying lack of repeatability. This can be traced to high seal friction in its joints. It has been found that this friction is very difficult to characterize [3, 8]. Hence, model based friction methods are difficult to apply successfully. This system is a good candidate for BSC to improve its repeatability. For this experimental system, the achievable repeatability is limited by the particular control electronics used for the experimental system. The joint resolver signals, standard on the Schilling, are converted to quadrature encoder waveforms using a special purpose Delta Tau Data/PMAC controller design. The joint angle resolution of this configuration is limited to ±0.087 degree, which leads to as much as ±5 mm errors in the end-effector positioning.

A 6-axis force/torque base sensor is mounted under the manipulator to provide wrench measurements for the BSC algorithm. A 15 kg replica of the nozzle dam center-plate was built along with an adjustable plate receptacle that permits the clearances to be varied from interference to several cm. An algorithm to successfully place the rectangular center plate within the receptacle would be easily extendable to perform the other high precision tasks necessary to complete the entire nozzle dam installation, either through teleoperation or as an autonomous subtask.

A pair of Pentax optical theodolites were used to accurately locate the end-effector in 3D space to generate the correction matrix, evaluate weight dependent deflections, and verify the algorithm performance. The resolution of the theodolites was 30 arc seconds, leading to measurement errors of 0.29 mm.

A fixed reference frame, F_0, is used to express the coordinates of all points. The origin of this reference lies at the intersection of the top of the base sensor and the joint 1 axis. Its z-axis is vertical and its x-axis is defined by a specific horizontal reference direction.

A PC based graphical user interface provides the operator with workspace visualization as well as manipulator control functionality. For all experiments, the sampling rate was ten milliseconds, which was sufficiently fast for the experiments.

4 Results

The objective of the experiment was to see if the method outlined in Figure 2 could be applied to the experimental system to improve its repeatability and its absolute

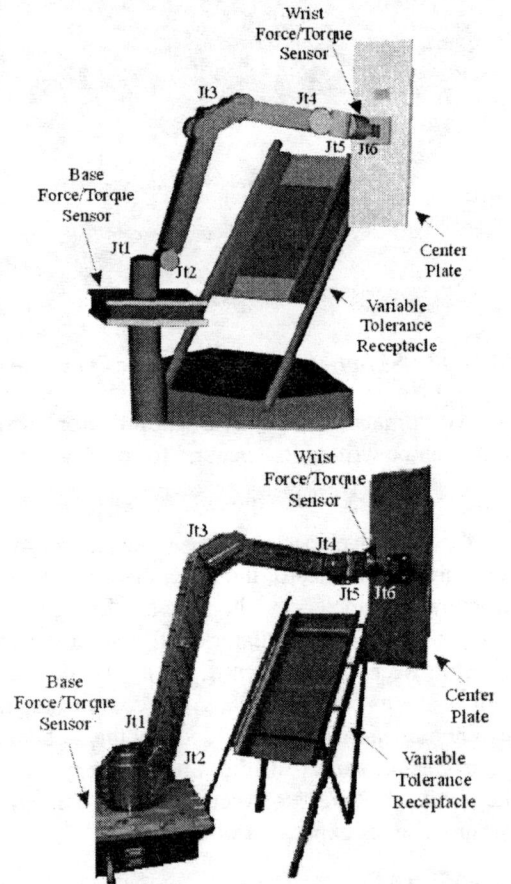

Figure 4 - Simulated and Real Experimental System

accuracy. The object was to have the residual error approach the limit set by the position sensing resolution of the system. In this work, 400 measurements were used to evaluate the basic accuracy of the Schilling. Different payloads were used, with weights up to 45 kg. Most of the measurements focused on two specific payloads: one with no weight and another with a 18 kg weight (the replica nozzle dam plate).

End-effector measurements of the manipulator under PI control determined the baseline uncompensated system repeatability and accuracy. The relative positioning root mean square error was used as a measure of the system repeatability. Recall that the 12-bit discretization of the resolver signal leads to random errors up to 5.0 mm, and imposes a lower limit of 2.0 mm (RMS) on the system repeatability, which sets the accuracy limit of any error compensation algorithm.

The results show that the BSC algorithm was able to reduce the repeatability errors by a factor of 4.73 over PI control. Data was taken by moving the manipulator an arbitrary distance from the test point and then commanding it back to its original position. Figure 5 shows the distributions of the repeatability error with and without BSC. The maximum errors without BSC were 21.0 mm, and the repeatability was 14.3 mm (RMS). BSC reduced the maximum errors to 5.5 mm with a repeatability of only 3.0 mm (RMS).

Although the BSC algorithm greatly reduced the repeatability errors, there are still 35 mm (RMS) errors in absolute accuracy. Since BSC reduced the system repeatability to 3.0 mm, a model based error correction method can be applied to reduce the accuracy errors.

In order to implement GEC, the geometric and elastic deformation correction matrix was calculated using approximately 350 measurements of the end-effector in different configurations and with different payloads. The remaining points were used to verify the efficiency of the GEC method.

From the system kinematic model with no errors, the

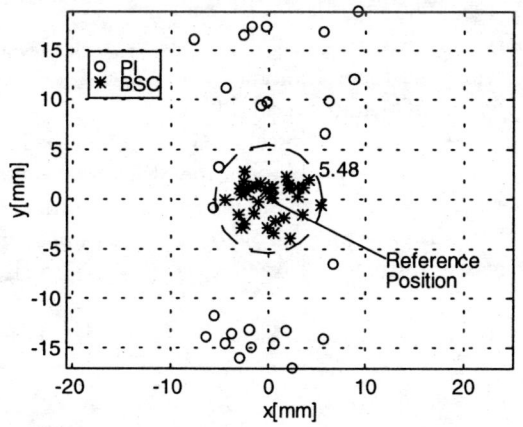

Figure 5 - Repeatability with and without BSC

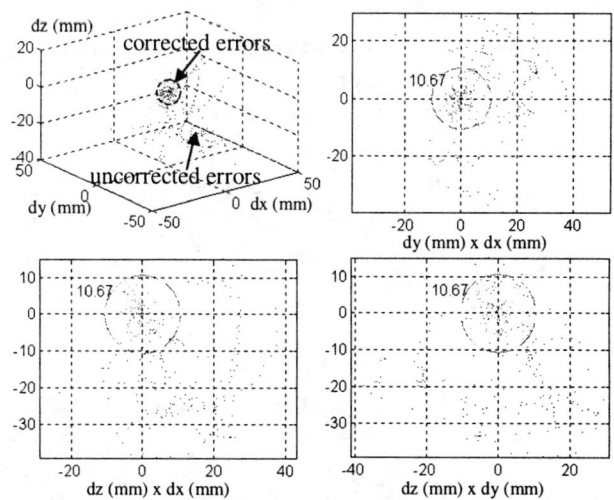

Figure 6 - Measured and Residual Errors After Compensation

ideal coordinates of the end-effector were calculated and subtracted from the experimentally measured values to yield the vector $\Delta \mathbf{X}(\mathbf{q}, \mathbf{W}_e)$ in Equation (7). By treating generalized errors as constant in their respective frames, the system absolute accuracy was improved to 13.4 mm (RMS). Since the GEC method allows for the use of polynomials to describe each generalized error, second order polynomials achieved an absolute accuracy of 7.3 mm (RMS), an additional 100% improvement.

Figure 6 shows the convergence of original positioning errors as large as 55.1 mm (34.3 mm RMS) to corrected errors of less than 10.7 mm (7.3 mm RMS) with respect to the base frame F_0. This demonstrates an overall factor of nearly 4.7 improvement in absolute accuracy by using the GEC algorithm.

An experiment was conducted to demonstrate the application of the joint BSC and GEC algorithm. The Schilling was commanded to a series of 11 points in the same plane under pure BSC control and then with the addition of two forms of the GEC method. The uncorrected data showed absolute accuracy errors of 29.5 mm (RMS), which are of the same order as the 34.3 mm (RMS) error found from the theodolite measurements. The implementation of GEC with constant generalized errors in their frames resulted in errors being reduced to 11.4 mm (RMS). By expanding the GEC algorithm to include second order polynomials, absolute positioning errors were reduced even further to a RMS value of 8.2 mm.

Figure 7 shows the dramatic improvement in absolute position tracking by using a polynomial GEC algorithm over the uncorrected method. Each ideal point is enclosed by a 5 mm radius circle, since the absolute position accuracy is limited by the resolution of the position sensors. The GEC algorithm also corrected for errors perpendicular to the plane of the points, and these values were measured and included in the error calculations. It can be seen that residual errors are approaching the levels of the resolver electronics. With this improvement in performance, it should make feasible such tasks as the nozzle dam insertion.

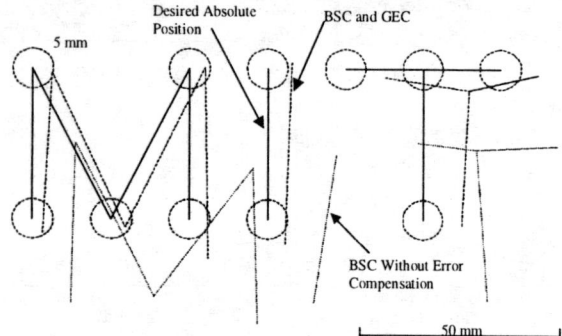

Figure 7 - Uncorrected and Corrected Accuracy

5 Conclusions

In this paper, the simplified, model-free form of Base Sensor Control (BSC) is applied to a hydraulic manipulator. The BSC uses a base force/torque sensor to accurately control joint torques, thereby compensating for joint friction. This in turn, substantially improves the manipulator's poor position repeatability. The BSC controller is then combined with a method, called GEC, that compensates for geometric and elastic errors that degrade the absolute positioning accuracy in large manipulators with inherently good repeatability. The results showed that applying the combined error compensation algorithm improved the absolute accuracy of the manipulator by a factor of 4.7 over pure BSC.

Acknowledgments

The assistance and encouragement of Dr. Byung-Hak Cho of the Korean Electric Power Research Institute (KEPRI) and Mr. Jacque Pot of the Electricité de France (EDF) in this research is most appreciated, as the financial support of KEPRI and EDF.

References

[1] Canudas de Wit, C., Olsson, H, Astrom, K.J. and Lischinsky, P., "A New Model for Control of Systems with Friction." *IEEE Transactions on Automatic Control* **40**, 3, pp. 419-425, 1996.

[2] Drouet, P., Dubowsky, S. and Mavroidis, C., "Compensation of Geometric and Elastic Deformation Errors in Large Manipulators Based on Experimental Measurements: Application to a High Accuracy Medical Manipulator." *Proceedings of the Sixth International Symposium on Advances in Robot Kinematics,* June, Strobl, Austria, 1998.

[3] Habibi, S.R., Richards, R.J., and Goldenberg, A.A., "Hydraulic Actuator Analysis for Industrial Robot Multivariable Control." *Proceedings of the American Control Conference,* **1**, 1003-1007, 1994.

[4] Hollerbach, J., "A Survey of Kinematic Calibration." *Robotics Review*, Khatib O. et al editors, Cambridge, MA, MIT Press, 1988.

[5] Iagnemma, K., Morel, G. and Dubowsky, S., "A Model-Free Fine Position Control System Using the Base-Sensor: With Application to a Hydraulic Manipulator." *Symposium on Robot Control, SYROCO '97,* **2**, pp. 359-365, 1997.

[6] Mavroidis, C., Dubowsky, S., Drouet, P., Hintersteiner, J. and Flanz, J., "A Systematic Error Analysis of Robotic Manipulators: Application to a High Performance Medical Robot." *Proceedings of the 1997 IEEE International Conference of Robotics and Automation,* Albuquerque, New Mexico, 1997.

[7] Meggiolaro, M., Mavroidis, C. and Dubowsky, S., "Identification and Compensation of Geometric and Elastic Errors in Large Manipulators: Application to a High Accuracy Medical Robot." *Proceedings of the 1998 ASME Design Engineering Technical Conference,* Atlanta, 1998.

[8] Merritt, H., *Hydraulic Control Systems.* John Wiley and Sons, New York, USA, 1967.

[9] Morel, G. and Dubowsky, S., "The Precise Control of Manipulators with Joint Friction: A Base Force/Torque Sensor Method." *Proceedings of the IEEE International Conference on Robotics and Automation,* **1**, pp. 360-365, 1996.

[10] Popovic, M.R., Shimoga, K.B. and Goldenberg, A.A., "Model-Based Compensation of Friction in Direct Drive Robotic Arms." *Journal of Studies in Information and Control,* **3**(1), pp. 75-88, 1994.

[11] Pfeffer, L.E., Khatib, O. and Hake, J., "Joint Torque Sensory Feedback of a PUMA Manipulator." *IEEE Transactions on Robotics and Automation,* **5**(4), pp. 418-425, 1989.

[12] Roth, Z.S., Mooring, B.W. and Ravani, B., "An Overview of Robot Calibration." *IEEE Southcon Conference*, Orlando, Florida, 1986.

[13] Vaillancourt, C. and Gosselin, G., "Compensating for the Structural Flexibility of the SSRMS with the SPDM." *Proceedings of the International Advanced Robotics Program, Second Workshop on Robotics in Space,* Canadian Space Agency, Montreal, Canada, 1994.

[14] Zezza, L.J., "Steam Generator Nozzle Dam System." *Transactions of the American Nuclear Society,* **50**, pp. 412-413, 1985.

Performance of Linear Decentralized $\mathcal{H}_\infty$ Optimal Control for Industrial Robotic Manipulators

Jonghoon Park
Graduate student

Wankyun Chung
Associate Professor

Youngil Youm
Professor

Robotics Lab., Department of Mechanical Engineering
Pohang University of Science & Technology (POSTECH)
Pohang, 790-784, Republic of Korea
Email: {coolcat,wkchung}@postech.ac.kr

Abstract

Industrial manipulators are under various limitations against high quality motion control, for example friction and dynamic uncertainties, and simple control structure. A robust linear PID motion controller, called the reference error feedback (REF), is proposed, which solves the nonlinear L_2-gain attenuation control problem. Making use of the fact that the single parameter L_2-gain γ controls the performance and robustness, we propose a simple and stable method of tuning the controller called the "square law". The analytical results are verified through experiments of six degrees of freedom industrial manipulator.

1 Introduction

As many researchers have previously observed, there are two sources of uncertainty in designing an industrial manipulator motion control system. The first is the dynamic interaction between links, and the second is the friction. To deal with these uncertainties, many robust control algorithms were developed, e.g. [1, 2]. However, since industrial manipulator control system still depends on the conventional PID servo in many parts, they experience some difficulties in applying those advanced control theories. In these regards, there were many approaches in designing a simple independent joint control system producing a set of desired performance. With these control, the performance heavily depends on the set of tuning parameters. If the gain tuning is too sophisticated, then it is also difficult to be accepted in industrial community.

Hence, our objective in this article is to develop a robust linear PID motion controller, easily tunable, guaranteeing a desired performance under the dynamic uncertainty, the friction uncertainty, and et al, for a wide range of motion speed. The authors have recently proposed a set of nonlinear $\mathcal{H}_\infty$ motion control for the Euler-Lagrange system including the manipulator system. We refine and simplify the control to a suitable form for industrial motion control, and analyze its property focusing at the performance and robustness. The theoretical results are verified by a six degrees of freedom industrial manipulator.

2 Nonlinear $\mathcal{H}_\infty$ Motion Control: Reference Error Feedback(REF)

For a robot manipulator with $q \in \Re^n$ as the generalized coordinate and $\tau \in \Re^n$ as the input whose equations of motion are given by

$$\tau = M(q)\ddot{q} + C(q,\dot{q})\dot{q} + g(q) + d(q,\dot{q},t), \quad (1)$$

where $M(q)$ is the inertia matrix, $C(q,\dot{q})$ denotes the centripetal and Coriolis matrix, $g(q)$ the gravitational force, and $d(q,\dot{q},t)$ represents all unidentified dynamic forces including friction forces, we are to design a nonlinear $\mathcal{H}_\infty$ controller which tracks $q^{\text{DES}}(t)$ without knowledge of the exact dynamics. For the error $e = q^{\text{DES}}(t) - q$, the control input is given by

$$\tau = \widehat{M}\left(\ddot{q}^{\text{DES}} + k_v\dot{e} + k_p e\right) - u \quad (2)$$

for a constant positive definite matrix $\widehat{M}$. Then the dynamics (1) can be arranged as

$$\begin{aligned} u + w = &M(q)(\ddot{e} + k_v\dot{e} + k_p e) \\ &+ C(q,\dot{q})\left(\dot{e} + k_v e + k_p \int e\right) \end{aligned} \quad (3)$$

where the disturbance $w \in \Re^n$ is put as

$$w = \widetilde{M}(q)(\ddot{q}^{\text{DES}} + k_v\dot{e} + k_p e) \\ + \widetilde{C}(q,\dot{q})\left(\dot{q}^{\text{DES}} + k_v e + k_p \int e\right) + \widetilde{g}(q) + d(t) \quad (4)$$

and $\widetilde{M}(q) = M(q) - \widehat{M}$, $\widetilde{C}(q,\dot{q}) = C(q,\dot{q})$, and $\widetilde{g}(q) = g(q)$ currently.

Let us set the state $x = \left(\int e^T, e^T, \dot{e}^T\right)^T \in \Re^{3n}$. The objective is to design a state feedback controller u such that

$$\int_0^T z^T(x,u,t)z(x,u,t)dt \leq \gamma^2 \int_0^T w^T(x,t)w(x,t)dt \quad (5)$$

for all $T \geq 0$, with $z(x,u,t) \in \Re^z$ being the cost variable defined by

$$z^T(x,u,t)z(x,u,t) = x^T Q(x,t)x + u^T R(x,t)u \quad (6)$$

where $Q(x,t) > 0$ and $R(x,t) > 0$ will be defined later. Note that (5) means the induced L_2-gain (nonlinear analogue of $\mathcal{H}_\infty$ gain for a linear system) from w to z is equal to or less than γ.

The nonlinear $\mathcal{H}_\infty$ control u can be designed by solving the so-called Hamilton-Jacobi-Isaccs (HJI) nonlinear partial differential equation, simply called the HJI equation, as shown in [3, 4, 5]. The following theorem states that the L_2-gain attenuation (5) by an arbitrarily [1] specified $\gamma > 0$ is possible by the simple linear state feedback.

Theorem 2.1. *Let the reference acceleration generation gain k_v and k_p satisfy*

$$k_v^2 > 2k_p > 1. \quad (7)$$

Then for a given $\gamma > 0$ the reference error feedback

$$u = -K\left(\dot{e} + k_v e + k_p \int e\right) \quad (8)$$

satisfies the L_2-gain attenuation requirement for

$$Q = \begin{bmatrix} k_p^2 K_\gamma & 0 & 0 \\ 0 & (k_v^2 - 2k_p)K_\gamma & 0 \\ 0 & 0 & K_\gamma \end{bmatrix}, \quad R = K^{-1} \quad (9)$$

if

$$K_\gamma = K - \frac{1}{\gamma^2}I > 0. \quad (10)$$

PROOF. See Appendic A. □

To appreciate the importance of the theorem, note that the $\mathcal{H}_\infty$ control (8), called the reference error feedback (REF), does not depend on any dynamic parameters. It only depends on the required level of L_2-gain γ.

---
[1] The capability of specifying an arbitrary γ does not imply that one can set $\gamma \approx 0$ due to physical constraint of a control system, e.g. boundedness of control input, and some system noise. For most control systems, the gains can not be arbitrarily big.

Interestingly, the control (2) with the REF (8) which is a solution to the nonlinear $\mathcal{H}_\infty$ control problem, has a form of conventional PID control structure. If k_v and k_p is fixed so that the gain inequality (7) holds, the overall tuning of the controller can be done only by γ.

REMARK 2.1. *As shown in Park et al. [5], the proposed $\mathcal{H}_\infty$ control (8) was derived exploiting only the Euler-Lagrange property. Hence, the equivalent task space dynamics [6] has also similar form of $\mathcal{H}_\infty$ control, i.e. reference error feedback, since it is also Euler-Lagrange system.*

3 Performance Analysis of REF

The L_2-gain attenuation can be further interpreted as follows: The L_2-gain attenuation stated in Thm. 2.1 can be described by

$$\lambda_{min}(Q_{cl})\int_0^T x^T x\, dt \leq \int_0^T x^T Q_{cl} x\, dt \leq \gamma^2 \int_0^T w^T w\, dt, \quad (11)$$

where $Q_{cl} > 0$ is

$$Q_{cl} = \begin{bmatrix} k_p^2(K_\gamma + K) & k_p k_v K & k_p K \\ k_p k_v K & (k_v^2 - 2k_p)K_\gamma + k_v^2 K & k_v K \\ k_p K & k_v K & (K_\gamma + K) \end{bmatrix}$$

which means that the L_2-gain of the error is attenuated by $\gamma\sqrt{\dfrac{1}{\lambda_{min}(Q_{cl})}}$.

Let us analyze how much smaller the error x will be if the L_2-gain is halved. Although exact analysis is impossible, a rough analysis can be easily done due to L_2-gain attenuation property (11). This rough analysis will help to tune the controller performance, as will be verified through experiments.

Noting that if γ is fixed, the above Q is constant denoted by Q, i.e. $Q = \sqrt{\dfrac{1}{\lambda_{min}(Q_{cl})}}$. Let us denote the solution trajectory x_1 and x_2 corresponding to $\frac{1}{2}\gamma$ and γ, respectively. The above inequality about the L_2-norm of error can be written

$$\int \|x_1\|^2 dt \leq \frac{1}{4}Q\gamma^2 \int \|w_1\|^2 dt$$

and

$$\int \|x_2\|^2 dt \leq Q\gamma^2 \int \|w_2\|^2 dt$$

where w_1 and w_2 are disturbances in each case. If $w_1 \approx w_2$, then we get

$$\int \|x_1\|^2 dt \leq \frac{1}{4}\int \|x_2\|^2 dt.$$

Hence, it may be approximated that $\|x_1\| \leq \frac{1}{2}\|x_2\|$. However, as the error norm becomes small, the norm of the disturbances w also tends to be small. Note that

$$w = \widetilde{M}(q)\ddot{q}^{\mathrm{DES}} + \widetilde{C}(q,\dot{q})\dot{q}^{\mathrm{DES}} + \tilde{g}(q) + d(t)$$
$$+ \widetilde{M}(k_v\dot{e} + k_p e) + \widetilde{C}\left(k_v e + k_p\int e\right).$$
$$= a(x,t) + B(x,t)x,$$

and that the magnitude of $a(x,t)$ and $B(x,t)$ does not directly depend on error if $\|x_1 - x_2\| \leq \epsilon$ for small ϵ, since M, C, G continuously depend on q and $\dot{q}$. Hence, we can see that $w^T w$ is dominated by $x^T x$. Therefore, the disturbance norm $\int w^T w dt$ has a tendency to decrease by quarter if x is reduced to its half. Hence, if γ is halved, we have

$$\int \|w_1\|^2 dt \leq \frac{1}{4}\int \|w_2\|^2 dt$$

and, therefore we can roughly say that

$$\|x_1\| \leq \frac{1}{4}\|x_2\|$$

Therefore, a rough rule of thumb is that if γ is halved, the error is reduced to its quarter, and it is called the *square law*. Although it seems too rough to approve, the experimental results verify this square law.

4 Disturbance estimating property of REF

So far, we have shown that the REF with RMC guarantees robustness of the system. Then, the question arises where the robustness comes from. We present somewhat actual interpretation of the robustness of the REF with RMC. That is, the control has the property of estimating the disturbances arising from the control design.

The idea of the disturbance observer proposed by Ohnishi et al. [7] can be reformulated as follows. The disturbance observer begins at parameterizing the system dynamics (1) by linear nominal model plus the disturbance due to model mismatch. That is, the dynamics (1) is arranged as

$$\tau = \widehat{M}\ddot{q} + \tau_{dist} \quad (12)$$

where the disturbance torque is defined by

$$\tau_{dist} = (M - \widehat{M})\ddot{q} + C\dot{q} + G + d. \quad (13)$$

Next, the control input to the system is composed of the reference control to track a desired trajectory $q^{\mathrm{DES}}(t)$ for the nominal model and the estimated disturbance torque

$$\tau = \tau_{ref} + \widehat{\tau}_{dist}, \quad \tau_{ref} = \widehat{M}(\ddot{q}^{\mathrm{DES}} + k_v\dot{e} + k_p e). \quad (14)$$

Specifically, if $\widehat{\tau}_{dist}$ is equal to τ_{dist}, this amounts to reshaping the dynamics (1) to

$$\tau_{ref} = \widehat{M}\ddot{q}. \quad (15)$$

However, it is impossible to obtain the exact value of τ_{dist}.

A promising alternative is to use the estimated one by (12) using a low-pass filter with cutoff frequency ω_i for each joint as

$$\widehat{\tau}_{dist} = \mathrm{diag}\left\{\frac{1}{s/\omega_i + 1}\right\}\left(\tau_{ref} + \widehat{\tau}_{dist} - \widehat{M}\ddot{q}\right) \quad (16)$$

with the abuse of notation $'s'$, the Laplace transform variable. Then by resolving the additional feedback regarding $\widehat{\tau}_{dist}$, there follows

$$\widehat{\tau}_{dist} = \mathrm{diag}\{\omega_i\}\frac{1}{s}\left(\tau_{ref} - \widehat{M}\ddot{q}\right)$$
$$= \mathrm{diag}\{\omega_i\}\int \left(\tau_{ref} - \widehat{M}\ddot{q}\right) dt. \quad (17)$$

The use of integral in the formulation can be eliminated, since

$$\widehat{\tau}_{dist} = \mathrm{diag}\{\omega_i\}\int \widehat{M}(\ddot{e} + k_v\dot{e} + k_p e)$$
$$= \mathrm{diag}\{\omega_i\}\widehat{M}\left(\dot{e} + k_v e + k_p\int e\right). \quad (18)$$

Therefore, noting that the form of the disturbance observer (18) is the same as that of the reference error feedback $\mathcal{H}_\infty$ control (8), one can conclude that the $\mathcal{H}_\infty$ control in the form of reference error feedback has the property of estimating the disturbance due to model mismatch in case of linear nominal model, or linear reference motion compensation. It can be said that the effective closed-loop dynamics is reshaped as the linear nominal model. Unfortunately, this equivalence ceases to hold when the nominal model is nonlinear. Based on this equivalence, we can interpret that the $\mathcal{H}_\infty$ gain γ and the cutoff frequency of the disturbance observer ω are deeply related. Especially, when $\omega_i = \omega$, and $\widehat{M} = mI$, we can conclude that

$$\gamma < \sqrt{\frac{1}{\omega m}}, \quad (19)$$

since

$$K = \omega m I > \frac{1}{\gamma^2}I.$$

Fig. 1: Fara AS1 industrial manipulator

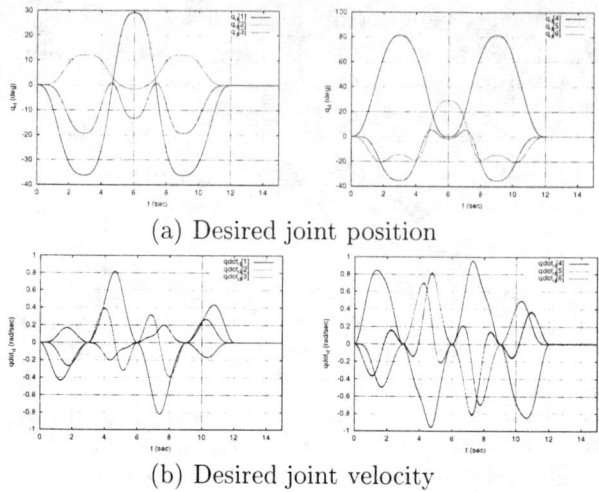

(a) Desired joint position

(b) Desired joint velocity

Fig. 2: Joint trajectory ($t_f = 12.0$) (Left: proximal links, Right: distal links)

5 Experimental Verification

The employed industrial manipulator is called "FARAMAN AS1", shown in Fig. 1, manufactured by Samsung Electronics Co. It has six degrees of freedom whose second and third joints are driven by well known five-bar linkage system. The gear-ratio of each joint is 1/160 for the first, second, and third joints, 1/100 for the fourth and sixth, and 1/108 for the fifth. [2]

5.1 Performance Tuning of REF by L_2-gain γ

The proposed control of the following extremely simple form of nominal model with $\widehat{M} = I$, $\widehat{C} = 0$, and $\widehat{g} = 0$ is applied to motion control of the manipulator. Rewriting the controller (2) with (8), it has the form of

$$\tau_i = \ddot{q}_i^{\text{DES}} + k_v \dot{e}_i + k_p e_i + \frac{1}{\gamma_i^2}\left(e_i + k_v e_i + k_p \int e_i\right)$$
$$= \ddot{q}_i^{\text{DES}} + \left(k_v + \frac{1}{\gamma_i^2}\right)\dot{e}_i + \left(k_p + \frac{k_v}{\gamma_i^2}\right)e_i + \frac{k_p}{\gamma_i^2}\int e_i. \quad (20)$$

where $e_i = q_i^{\text{DES}} - q_i$ for $i = 1, 2, \cdots, 6$. Note that it has a simple structure of PID control with reference acceleration feedforward. The control frequency is set to $1000(Hz)$, and only the angular positions are measured by incremental encoders capable of 2048 pulses per revolution. The angular velocities are obtained through the simple backward difference method. The desired joint trajectory is generated, so that the end-effector traces four straight line segment with simultaneous orientation change. The joint position trajectory is shown in Fig. 2 (a), and the corresponding joint velocity is shown in Fig. 2 (b).

For the controller tuning, we set $k_v = 20$, and $k_p = 100$ such that $k_v^2 > 2k_p > 1$. Controller tuning is indeed easy to do, since the one which one should care is the required L_2-gain $\gamma > 0$. Starting from $\gamma_i = 0.2$ for all i, the control results are shown in Fig. 3 (a). By halving γ_i for the first three joints with γ_j ($j = 4, 5, 6$) fixed to 0.2, we have the results shown in Fig. 3 (b). Further halving γ reduces the error. Note two things: The first is that to tuning of some joint control system does not affect significantly the performance of the others, as shown in the error profile of the distal links, shown in the right column of Fig. 3. They are almost invariant, although a little decrease in magnitude is observed as the proximal joint control system gets better performance. The second thing is that the error profile reduces in quarter way when the L_2-gain is reduced by half, as the rough analysis predicted. This kind of halving process can be stopped if the error signal is too noisy as shown in Fig. 3 (e), where the second and third joints become noisy. One can understand this happens because, as the L_2-gain gets smaller, the bandwidth of the closed-loop system becomes higher. As a consequence, a kind of noise which lives in high frequency region are also magnified, which results in noisy error signal as shown in the figure.

After the first three joint control systems have been tuned, the next three joint control systems can be tuned. Since the remaining three servo systems are not so good as the first three, the achievable L_2-gains without much noise amplification are limited. Finally, we choose the set of L_2-gain attenuatable for the current system as $\gamma = (0.0125, 0.017, 0.017, 0.1, 0.14, 0.14)$.

---

[2] Note that the fifth joint is driven by timing belt system, so that its bandwidth property is lower than the others.

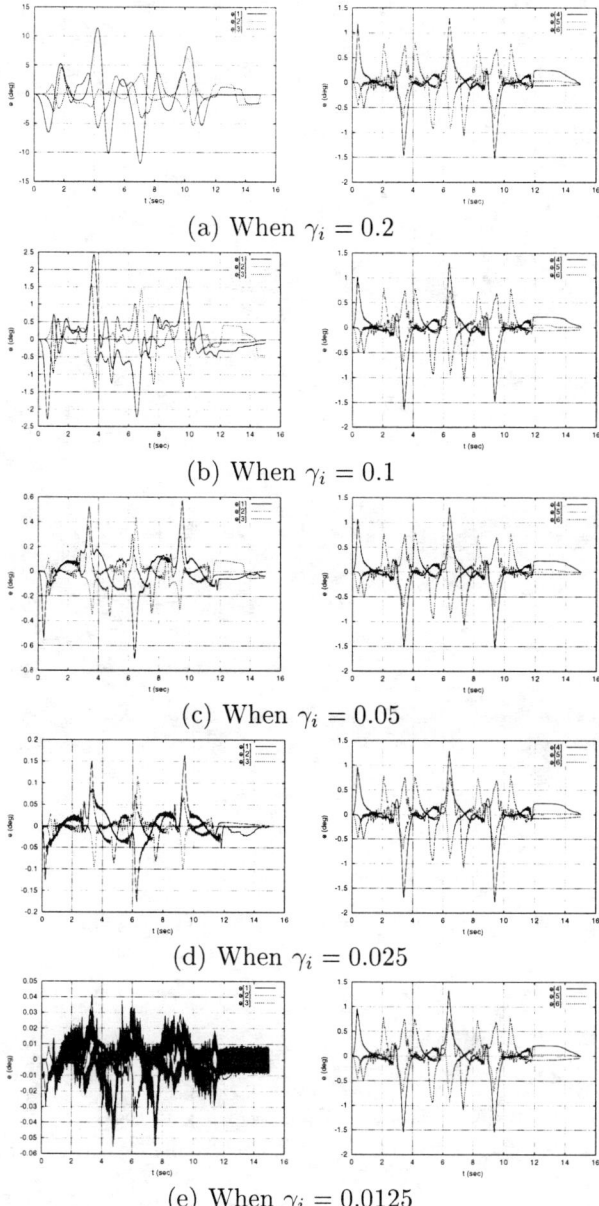

(a) When $\gamma_i = 0.2$

(b) When $\gamma_i = 0.1$

(c) When $\gamma_i = 0.05$

(d) When $\gamma_i = 0.025$

(e) When $\gamma_i = 0.0125$

Fig. 3: Error trajectory of the proposed REF control by L_2-gain γ_i for $i = 1, 2, 3$ and $\gamma_j = 0.2$ for $j = 4, 5, 6$ (Left: proximal link, Right: Distal link)

It should be remarked that the gain tuning based on the REF with RMC, that is (20), is very easy and stable. If the stability of a system is not guaranteed while tuning a set of gains, the controller is hard to apply with ease.

5.2 Performance of REF for a wide range of motion speed

Since conventional PID controller is based on linearization approach [8], a valid region of operation should be small, and does not change too fast. Also, the disturbance rejection capability is much dependent

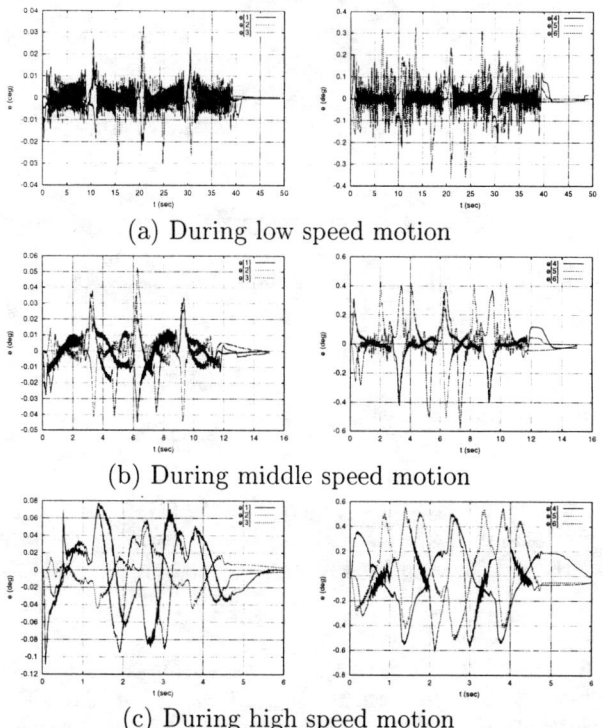

(a) During low speed motion

(b) During middle speed motion

(c) During high speed motion

Fig. 4: Error trajectory of the proposed REF control for a wide range of motion speed (Left: proximal link, Right: distal link)

on the operation region. Therefore there arises a need to change gains region by region. However, this kind of gain planning does not work well if the operating speed becomes faster, and the property of dominating disturbance varies much, e.g. from friction to dynamic uncertainties.

To the contrary, the proposed control has the ability to attenuate the L_2-gain of the error on the global region in a uniform way [3], it should not malfunction as the conventional PID controller does. The following experiment tries to verify this property. In this experiment, all other settings are equal but the operating speed. The speed is classified to three cases, low, middle, and high speed, where the profile of joint position trajectory is same but the time scale. Note that the employed manipulator has really big friction, as you can see this in Fig. 3 (a). At the trailing region of trajectory planning. i.e. from $t = 12.0$, the desired motion is to stay, but the arm still swings back and forth in a stepwise fashion. This implies that the arm does not move as the control input reaches some threshold value which can overcome static friction. Once the move begins, the arm moves relatively fast. The low speed motion is much slower than the case of Fig. 2 because the trajectory completion time denoted as t_f

---

[3]The globality and uniformity of the L_2-gain attenuation property was proved in [5].

is 40.0(*sec*). On the contrary, the high speed motion has $t_f = 4.8(sec)$. Indeed the dynamic uncertainty dominates during the high speed motion, whereas the friction dominates during the low speed motion.

The control results are summarized in Fig. 4. From the results, it can be said that the proposed control performs well for a wide range of motion speed, although during high-speed motion a minor performance degradation is observed. This experimental results imply that one can tune the controller very easily once in a single region for a single motion speed.

6 Conclusion

The robust motion controller, named the reference error feedback (REF), guaranteeing a desired level of performance and robustness in the form of a simple PID form was proposed, and verified through experiments by the six degrees of freedom industrial robot manipulator. Note that the single parameter γ actually controls the error gain attenuation and the magnitude of error norm as well by the square law. The rule of thumb, called the square law, estimating the error norm attenuation was applied to controller tuning, and verified through experiments. The global and uniform disturbance rejection and command tracking of the control were also verified.

A Proof of Thm. 2.1

Using the state $\boldsymbol{x} = \left(\int \boldsymbol{e}^T, \boldsymbol{e}^T, \dot{\boldsymbol{e}}^T\right)^T = (\boldsymbol{x}_1^T, \boldsymbol{x}_2^T, \boldsymbol{x}_3^T)^T$ the closed-loop dynamics (3) is expressed as

$$\dot{\boldsymbol{x}}(t) = \boldsymbol{A}(\boldsymbol{x},t)\boldsymbol{x}(t) + \boldsymbol{B}(\boldsymbol{x},t)\boldsymbol{u}(t) + \boldsymbol{D}(\boldsymbol{x},t)\boldsymbol{w}(t) \quad (21)$$

where

$$\boldsymbol{A} = \begin{bmatrix} 0 & \boldsymbol{I} & 0 \\ 0 & 0 & \boldsymbol{I} \\ -k_p\boldsymbol{M}^{-1}\boldsymbol{C} & -k_v\boldsymbol{M}^{-1}\boldsymbol{C} - k_p\boldsymbol{I} & -\boldsymbol{M}^{-1}\boldsymbol{C} - k_v\boldsymbol{I} \end{bmatrix}$$

$$\boldsymbol{B} = \boldsymbol{D} = \begin{bmatrix} 0 \\ 0 \\ \boldsymbol{M}^{-1} \end{bmatrix}.$$

Then it is to be shown that the following nonlinear quadratic function

$$V(\boldsymbol{x},t) = \frac{1}{2}\boldsymbol{x}^T \boldsymbol{P}(\boldsymbol{x},t)\boldsymbol{x} \quad (22)$$

$$\boldsymbol{P}(\boldsymbol{x},t) = \begin{bmatrix} k_p^2\boldsymbol{M} + k_pk_v\boldsymbol{K}_\gamma & k_pk_v\boldsymbol{M} + k_p\boldsymbol{K}_\gamma & k_p\boldsymbol{M} \\ k_pk_v\boldsymbol{M} + k_p\boldsymbol{K}_\gamma & k_v^2\boldsymbol{M} + k_v\boldsymbol{K}_\gamma & k_v\boldsymbol{M} \\ k_p\boldsymbol{M} & k_v\boldsymbol{M} & \boldsymbol{M} \end{bmatrix} \quad (23)$$

solves the associated HJI equation for

$$\boldsymbol{Q} = \begin{bmatrix} k_p^2\boldsymbol{K}_\gamma & 0 & 0 \\ 0 & (k_v^2 - 2k_p)\boldsymbol{K}_\gamma & 0 \\ 0 & 0 & \boldsymbol{K}_\gamma \end{bmatrix}, \quad \boldsymbol{R} = \boldsymbol{K}^{-1}. \quad (24)$$

First, we simplify the HJI equation, which is indeed a partial differential equation, to a matrix ordinary differential equation, i.e. matrix differential Ricatti's equation. Noting that

$$V_t = \frac{1}{2}\boldsymbol{x}^T \frac{\partial \boldsymbol{P}}{\partial t}\boldsymbol{x}, \quad V_x = \frac{1}{2}\boldsymbol{x}^T \frac{\partial \boldsymbol{P}}{\partial \boldsymbol{x}^T}\boldsymbol{x} + \boldsymbol{x}^T \boldsymbol{P}$$

and $\boldsymbol{P}(\boldsymbol{x},t)$ is not a function of $\boldsymbol{x}_3 = \dot{\boldsymbol{e}}$, we have

$$V_x = \frac{1}{2}\boldsymbol{x}^T \begin{bmatrix} \frac{\partial \boldsymbol{P}}{\partial \boldsymbol{x}_1^T}\boldsymbol{x} & \frac{\partial \boldsymbol{P}}{\partial \boldsymbol{x}_2^T}\boldsymbol{x} & 0 \end{bmatrix} + \boldsymbol{x}^T \boldsymbol{P}.$$

Then

$$\boldsymbol{V}_x\boldsymbol{f} = \boldsymbol{V}_x\boldsymbol{A}\boldsymbol{x} = \frac{1}{2}\boldsymbol{x}^T\left\{\boldsymbol{P}\boldsymbol{A} + \boldsymbol{A}^T\boldsymbol{P} + \sum_{k=1}^{2}\frac{\partial \boldsymbol{P}}{\partial \boldsymbol{x}_k^T}\dot{\boldsymbol{x}}_k\right\}\boldsymbol{x}$$

and

$$V_t + \boldsymbol{V}_x\boldsymbol{f} = \frac{1}{2}\boldsymbol{x}^T\left\{\dot{\boldsymbol{P}} + \boldsymbol{P}\boldsymbol{A} + \boldsymbol{A}^T\boldsymbol{P}\right\}\boldsymbol{x}.$$

Also

$$\boldsymbol{V}_x\left\{g(k^T k)^{-1}g^T - \frac{1}{\gamma^2}\boldsymbol{p}\boldsymbol{p}^T\right\}\boldsymbol{V}_x^T = \boldsymbol{x}^T\boldsymbol{P}\boldsymbol{B}\left(\boldsymbol{K} - \frac{1}{\gamma^2}\boldsymbol{I}\right)\boldsymbol{B}^T\boldsymbol{P}\boldsymbol{x}.$$

Finally, we obtain the following matrix differential Ricatti's equation

$$\dot{\boldsymbol{P}} + \boldsymbol{A}^T\boldsymbol{P} + \boldsymbol{P}\boldsymbol{A} - \boldsymbol{P}\boldsymbol{B}\boldsymbol{K}_\gamma\boldsymbol{B}^T\boldsymbol{P} + \boldsymbol{Q} = 0. \quad (25)$$

Using

$$\dot{\boldsymbol{P}} = \begin{bmatrix} k_p^2\dot{\boldsymbol{M}} & k_pk_v\dot{\boldsymbol{M}} & k_p\dot{\boldsymbol{M}} \\ k_pk_v\dot{\boldsymbol{M}} & k_v^2\dot{\boldsymbol{M}} & k_v\dot{\boldsymbol{M}} \\ k_p\dot{\boldsymbol{M}} & k_v\dot{\boldsymbol{M}} & \dot{\boldsymbol{M}} \end{bmatrix},$$

$$\boldsymbol{P}\boldsymbol{A} = \begin{bmatrix} -k_p^2\boldsymbol{C} & k_pk_v\boldsymbol{K}_\gamma - k_pk_v\boldsymbol{C} & k_p\boldsymbol{K}_\gamma - k_p\boldsymbol{C} \\ -k_pk_v\boldsymbol{C} & k_p\boldsymbol{K}_\gamma - k_v^2\boldsymbol{C} & k_v\boldsymbol{K}_\gamma - k_v\boldsymbol{C} \\ -k_p\boldsymbol{C} & -k_v\boldsymbol{C} & -\boldsymbol{C} \end{bmatrix}$$

$$\boldsymbol{P}\boldsymbol{B}\boldsymbol{R}_\gamma^{-1}\boldsymbol{B}^T\boldsymbol{P} = \begin{bmatrix} k_p^2\boldsymbol{K}_\gamma & k_pk_v\boldsymbol{K}_\gamma & k_p\boldsymbol{K}_\gamma \\ k_pk_v\boldsymbol{K}_\gamma & k_v^2\boldsymbol{K}_\gamma & k_v\boldsymbol{K}_\gamma \\ k_p\boldsymbol{K}_\gamma & k_v\boldsymbol{K}_\gamma & \boldsymbol{K}_\gamma \end{bmatrix}$$

shows that the matrix $\boldsymbol{P}(\boldsymbol{x},t)$ is the solution of (25) thanks to skew-symmetry of $\dot{\boldsymbol{M}} - 2\boldsymbol{C}$.

Then, the $\mathcal{H}_\infty$ control $\boldsymbol{u}$ is defined as

$$\boldsymbol{u} = -\boldsymbol{K}\boldsymbol{B}^T\boldsymbol{P}\boldsymbol{x} = -\boldsymbol{K}\begin{bmatrix} k_p\boldsymbol{I} & k_v\boldsymbol{I} & \boldsymbol{I} \end{bmatrix}\boldsymbol{x} \quad (26)$$

$$= -\boldsymbol{K}\left(\dot{\boldsymbol{e}} + k_v\boldsymbol{e} + k_p\int \boldsymbol{e}\right). \quad (27)$$

The fully general solutions and its proof can be found in [5]. □

References

[1] F. L. Lewis, C. T. Abdallah, and D. M. Dawson, *Control of Robot Manipulators*. Macmillan Pub. Co., 1993.

[2] Z. Qu and D. M. Dawson, *Robust Tracking Control of Robot Manipulators*. IEEE Press, 1996.

[3] A. Isidori, "Feedback control of nonlinear systems," *Int. Jr. Robust and Nonlinear Control*, vol. 2, pp. 291–311, 1992.

[4] A. J. V. D. Schaft, "L_2-gain analysis of nonlinear systems and nonlinear state feedback $\mathcal{H}_\infty$ control," *IEEE Transactions on Automatic Control*, vol. 37, no. 6, pp. 770–784, 1992.

[5] J. Park, W. Chung, and Y. Youm, "Analytic nonlinear $\mathcal{H}_\infty$ optimal control for robotic manipulators," in *Proc. 1998 IEEE Int. Conf. on Robotics and Automation*, pp. 2709–2715, 1998.

[6] O. Khatib, "A unified approach for motion and force control of robot manipulators: The operational space formulation," *IEEE Transactions on Robotics and Automation*, vol. RA-3, no. 1, pp. 43–53, 1987.

[7] M. Nakao, K. Ohnishi, and K. Miyachi, "A robust decentralized joint control based on interference estimation," in *Proc. 1987 IEEE Int. Conf. on Robotics and Automation*, pp. 326–331, 1998.

[8] J. Y. S. Luh, "Conventional controller design for industrial robots-a tutorial," *IEEE Transactions on Systems, Man, and Cybernetics*, vol. SMC-13, no. 3, pp. 298–316, 1983.

Development of a Lightweight Torque Limiting M-Drive Actuator for Hyper-Redundant Manipulator Float Arm

Shigeo Hirose, Richard Chu
Tokyo Institute of Technology
Department of Mechano-Aerospace Engineering
2-12-2 Ookayama, Meguro-ku, Tokyo 152, Japan
hirose@mes.titech.ac.jp, chu@mes.titech.ac.jp

Abstract

In order to realize a practical hyper-redundant manipulator, a new lightweight wire driven transmission mechanism named M-Drive has been developed. This mechanism not only takes advantage of the lightweight properties of a wire and pulley system but also has a unique torque keeping function. The mechanism keeps the joint solid without compliance and not only allows the joint to deform when in contact with an external torque over a threshold torque but is also able to maintain that threshold torque while deforming. Furthermore, M-Drive also has a feature to automatically eliminate wire elongation, the fundamental difficulty of a wire and pulley system. Combining two M-Drive mechanisms in a differential configuration, a two DOF coupled drive actuator that maximizes power has also been developed for a proposed gravity compensated hyper-redundant manipulator named Float Arm. Experimental results for the M-Drive mechanism and the differential M-Drive actuator are also included.

1. Introduction

In the field of robotics, many researchers have endeavored to realize a practical hyper-redundant manipulator because of its ability to enter complicated environments while avoiding obstacles. However, two main difficulties stand in the way of that realization. First, a hyper-redundant manipulator must be able to move and function while supporting its own mass. Thus, methods of reducing or compensating the weight of the manipulator are highly desirable. One method of attempting to overcome this problem, we proposed a hyper-redundant manipulator named Float Arm [1] that has a unique function to be able to compensate for the weight of each link regardless of manipulator posture. A model was constructed using a bevel gear differential actuator to maximize actuator efficiency, but the design still remains too heavy.

Second, the principal problem of existing hyper-redundant manipulators [2,3,4,5] is the inability to withstand unexpected external forces due to excessive moments at the base joints as shown in Fig.1.

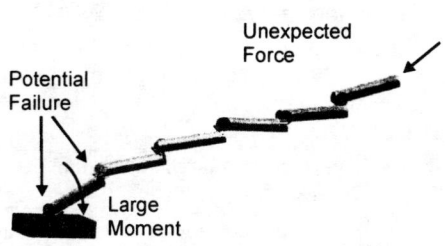

Fig.1 Potential Failure of Base Joints from Excessive Moments Due to Unexpected Forces

One method of overcoming this problem is to use torque limiters, but conventional torque limiters are bulky and impossible to use with a practical hyper-redundant manipulator. Another conventional approach is to introduce compliance in the joints, but this greatly deteriorates the servo control ability of the joints. We thus require a solid joint mechanism that incorporates a torque keeping function to allow the mechanism to deform in accordance with an external load over a set threshold but is also able to maintain that threshold torque while deforming. Therefore, with these considerations in mind, a new transmission mechanism and actuation unit are proposed.

2. Hyper-Redundant Manipulator Float Arm

For hyper-redundant manipulators, unless gravity compensation can be incorporated into every joint, the load will be too large for the actuation system. In addition, the compensation torque required at the joints will heavily depend on the posture of the hyper-redundant manipulator, especially at the base joints. Therefore, a method of gravity compensation independent of posture is needed but difficult to realize. For these reasons, the robotics community has resisted the possibility of using springs for gravity compensation of hyper-redundant manipulators.

2.1 Gravity Compensation

After studying this problem, a hyper-redundant manipulator named "Float Arm" has been developed that has complete gravity compensation for every link and with any posture. This is achieved by the following combination of two mechanisms:

1. A parallel four-bar linkage mechanism that decreases the necessary gravity compensation torque by absorbing the torque component that is due to the posture of the hyper-redundant manipulator
2. A gravity compensation spring mechanism for each link.

In a conventional linkage as shown in Fig.2a, the support torque at the base of the link differs as the load position differs. However, by using a parallel four-bar linkage, the necessary support torque at the base of the link is independent of where a load is placed on the plate attached at the end of the link. This characteristic is because there is a uniform linear displacement Δx on all the plate for a small rotation $\Delta \theta$ as shown in Fig.2b and the parallel four-bar linkage is able to absorb the torque component due to load position.

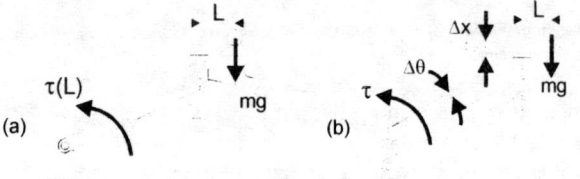

Fig.2 Characteristics of a Parallel Four-Bar Linkage

Float Arm stacks these parallel four-bar linkages to form a hyper-redundant manipulator. With this configuration, the torque τ_i necessary at the pitching joint θ_i is

$$\tau_i = \left(\sum_{j=i}^{n} m_i g \right) \cdot L_i \cos \theta_i \qquad (1)$$

Where $m_i \sim m_n$ are the masses of links $i \sim n$ and L_i is the length of link i. From Equation (1), τ_i is only dependant on mass of link i and the links above it as well as the posture of link i but not on the posture of links $i+1 \sim n$. By utilizing this characteristic, gravity compensation for each link can be realized through the use of springs as shown in Fig.3. Optimized design of the spring gravity compensation has already been studied and determined [1].

Fig.3 Spring Gravity Compensation Mechanism

Furthermore, for Float Arm, ϕ_i the yawing joint axis is always perpendicular maintained by the parallel four-bar linkage and the link may rotate about this perpendicular axis. Combining all these features, Float Arm has been designed with gravity compensation and two degrees of freedom for each link using the parallel four-bar linkage and spring compensation mechanisms. A skeleton model of Float Arm without an actuation system has been developed as shown in Fig.4.

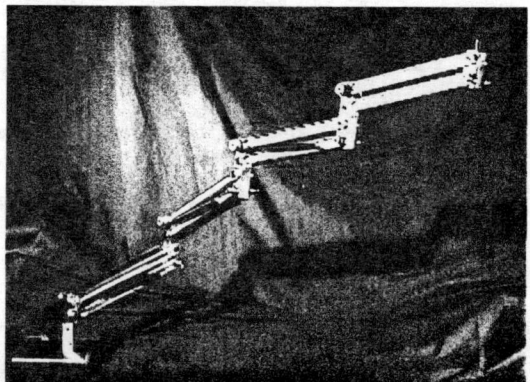

Fig. 4 Skeleton Float Arm

2.2 Driving System
In order to maximize power, a 2 DOF "coupled drive" [6] floating differential actuation system was developed for Float Arm by using a set of bevel gear differentials and two DC motors as shown in Fig.5a. The motors are mounted to the frame of Float Arm while the motor shafts are attached to the bevel gears. With this configuration, three couple drive modes can be obtained as shown in Fig.5b:

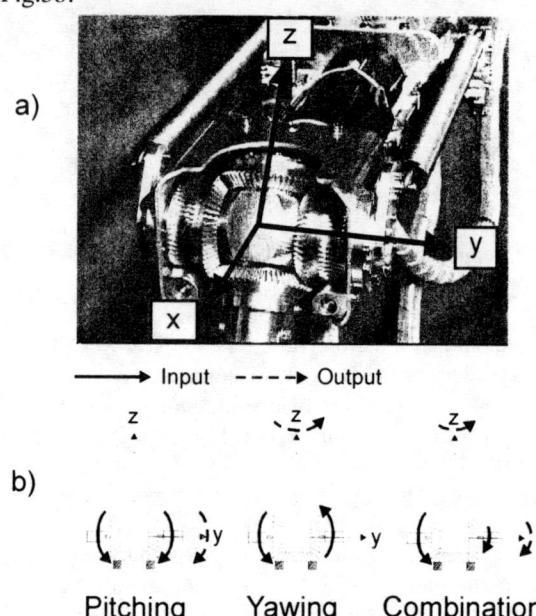

Fig. 5 Coupled Drive Floating Differential Mechanism

1. Pitching: When the motors rotate the bevel gears in the same direction at the same rotational speed, the friction between the bevel gears constrains their rotation. The resulting output is through the frame and the corresponding motion is a rotation about the y-axis.
2. Yawing: When the motors rotate the bevel gears in opposite directions at the same rotational speeds the bevel gears are free to rotate. The resulting output is through the gears and the resulting motion is a rotation about the z-axis.
3. Combination: The previous two modes are for equivalent angular velocities and any difference in angular velocity will produce a combination of yawing and pitching motion.

Although this system is able to maximize the efficiency of the actuators by coupling the outputs, the two considerations for a practical hyper-redundant manipulator, a lightweight design and the ability to withstand unexpected forces are not realized. By using bevel gears, the design is still heavy and as the manipulator travels in 3D space the resultant torsion about the x-axis deforms the frame. Furthermore, a practical torque limiting safety mechanism cannot be equipped with the differential gear configuration. Overcoming these two difficulties is the motivation for the design of the new mechanism.

3. M-Drive

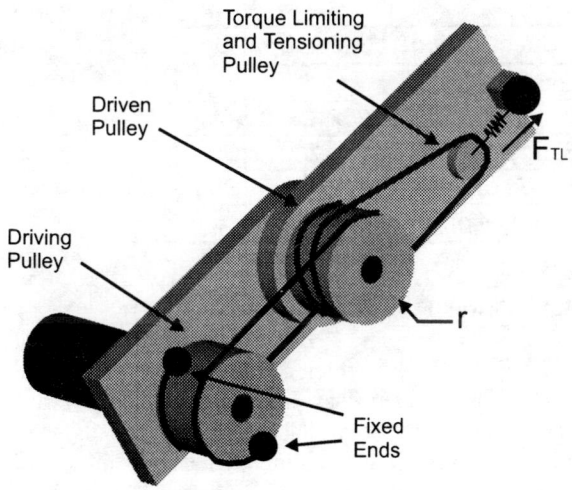

Fig.6 M-Drive mechanism

A new flexible mechanical element power transmission system named M-Drive, shown is Fig.6, is proposed that has the following three features:

1. Is solid and does not have any compliance in normal operation mode, ensuring good servo-control
2. Is compliant when the threshold torque is exceeded such that the system is protected from overloading
3. Automatically adjusts the wire tension by absorbing the elongation

The M-Drive mechanism uses the friction property, like that of a band brake, to control the torque necessary for slip between a flexible mechanical element and a pulley. For the wire and pulley system shown in Fig.7, the relationship between wire forces, friction coefficient μ, and angle of wrap ϕ is expressed by Eytelwein's equation, where F_1 and F_{end} are the forces at the ends of the wire.

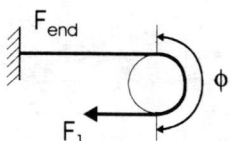

Fig.7 Simple Wire and Pulley System

$$\frac{F_1}{F_{end}} = e^{\mu\phi} \qquad (2)$$

The friction is the difference between F_1 and F_{end} and can be controlled by varying either the angle of wrap ϕ or F_1.

M-Drive uses a wire with both ends fixed at the driving pulley and is wrapped an odd number of times around the driven pulley. The system uses only one force to realize both torque limiting and wire tensioning. That torque limiting and tensioning force is applied in the middle of the wire at the center turn, allowing an equivalent torque to be limited for both clockwise and counterclockwise rotation of the driven pulley. Using only one mid-tensioning force is key and hence, the name, Mid-tensioning Drive or M-Drive for short. A small pulley is also used where the torque limiting and tensioning force is applied in order to allow infinite slip rotation.

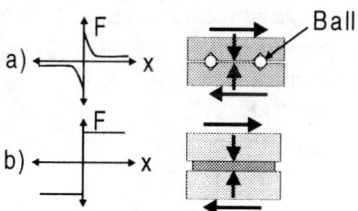

Fig.8 Torque Limiter Slip Profiles

Two main types of conventional torque limiting or guarding mechanisms are used in industry and their respective slip profiles are shown in Fig.8. The first profile is for most torque limiters and shows that after reaching the peak threshold torque, slip occurs. The slip torque is extremely low and for practical robotics that type of profile creates numerous control problems. The second type of profile is for torque keeper mechanisms that are able to maintain the slip torque at the same level as the threshold torque. This is the desired slip profile type, but unfortunately conventional torque limiters with this slip profile are exceedingly heavy and are not feasible for practical applications. However, the wire and pulley torque limiter proposed for the M-Drive mechanism does not only attain the desired slip torque profile but also can be extremely lightweight.

Two points of comparison should be noted between M-Drive and the conventional wire transmission approach [7] as shown in Fig.9. First, the tensioning force may act from numerous locations and the conventional approach would place springs at the ends of the wire. By doing so the tensioning forces needed are extremely high and compliance is introduced into the system, which results in poor servo control. However, M-Drive overcomes this by applying only one force in the middle of the wire at a small pulley. Second, the conventional approach attaches the middle of the wire to the driving pulley. In such a system, if an unexpected force is encountered that is greater than the strength of the wire then the wire will break and the joint will fail. M-Drive's wire configuration, however, allows slip to occur when a defined threshold torque is reach and the slip rotation is infinite.

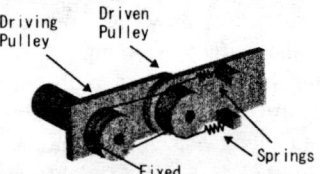

Fig.9 Conventional Wire Transmission System

A model of the M-Drive was made to investigate the torque limiting mechanism. The relationship between the torque limiting force F_{TL} and the limit torque T_L, where μ is the friction coefficient, ϕ is the angle of wrap, and r is the pulley radius using Equation (2) is

$$T_L = F_{TL}\left(e^{\mu\phi} - 1\right)r \qquad (3)$$

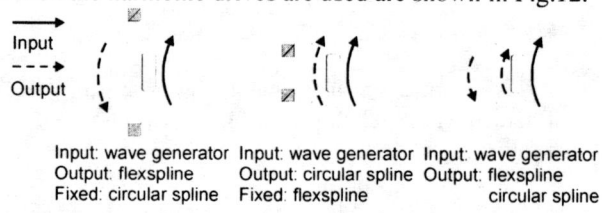

Fig.10 Torque Limit Experimental Results

Experimental results from varying the angle of wrap ϕ and the limiting torque force F_{TL} are as expected and shown in Fig.10.

4. Differential M-Drive Actuator

Utilizing the advantages of the new M-Drive mechanism, a two DOF prototype actuation unit for Float Arm, consisting of two M-Drive mechanisms in a differential configuration, two modified harmonic drives, two 6W DC motors, and two tensioning springs has been developed as shown in Fig.11. The mechanical specifications of the Differential M-Drive Actuator (DMA) are listed in Table.1.

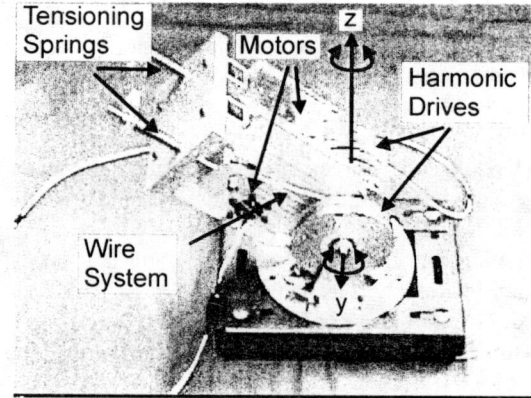

Fig.11 Differential M-Drive Actuator

Specification	Value	Unit
Total Weight	2.2	[kg]
Dimensions	25x12x10	[cm]
Range of Pitch	±70	Deg.
Range of Yaw	±180	Deg.
Gear Ratio	1200:1	

Table 1: Actuator Specifications

4.1 DMA Power Transmission System

The DMA uses Harmonic Drives as a floating differential and gearing mechanism which can be defined as a type of differential mechanism that has one input, the wave generator, and two controllable geared outputs, the flexspline and the circular spline. The three modes in which the harmonic drives are used are shown in Fig.12.

Input: wave generator   Input: wave generator   Input: wave generator
Output: flexspline   Output: circular spline   Output: flexspline
Fixed: circular spline   Fixed: flexspline   circular spline

Fig.12 Harmonic Drive Modes

The wiring system shown in Fig.13 is the mechanical equivalent of a bevel gear differential except that instead of gears, two harmonic drives and a wire and pulley system are used. The base pulley is fixed while the rest of the system rotates about it. Four guide pulleys are also used to avoid wire entanglement as well as two tensioning pulleys for tensioning and torque limiting. Furthermore, by combining two harmonic drive floating differential mechanisms and two motors to drive them, coupled drive can be achieved. This coupling effect is created by using a wire that wraps around the two flexspline output pulleys of the harmonic drive floating differential mechanisms and the base pulley. The actual driving mechanism is shown in Fig.14a and the three couple drive modes corresponding to the three harmonic drive modes are as follows and shown in Fig.14b:

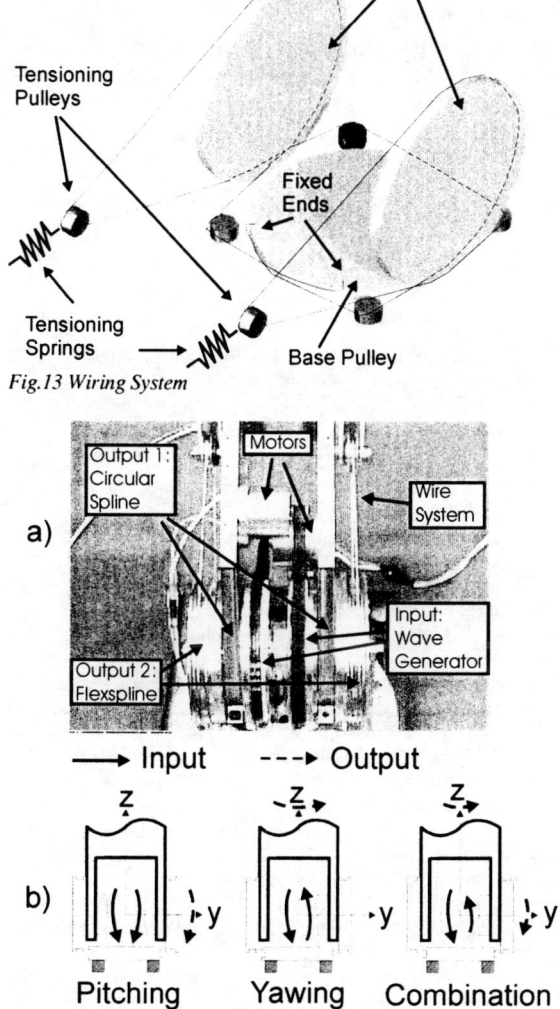

Fig.13 Wiring System

Fig.14 Wire Coupled Drive Floating Differential Mechanism

1. Pitching: When the wave generator inputs rotate in the same direction at the same rotational speed, the friction between the wire and flexspline output pulleys constrains the rotation of these pulleys. The resulting output is through the circular splines and the corresponding motion is a rotation about the y-axis.
2. Yawing: When the wave generator inputs rotate in opposite directions at the same rotational speeds the circular spline outputs are constrained. The resulting output is through the flexspline pulleys and the resulting motion is a rotation about the z-axis.
3. Combination: The previous two modes are for equivalent angular velocities and any difference in angular velocity will produce a combination of yawing and pitching motion.

4.2 DMA Torque Limiting Safety Mechanism

The DMA utilizes a pair of pulleys that are each supported by a sliding shaft and a spring respectively, in order to pretension the wire as well as control the friction between the wire and the flexspline output pulleys. However, with two M-Drive mechanisms and one wire in a differential configuration, the torque limit becomes much more complicated to control. For each pitching and yawing the torque limit is different because effective angle of wrap ϕ is different. Even though for both cases the wire and the pulley configurations are the same, depending on how the actuator makes contact with external forces, the distribution of forces in the wire are different resulting in different effective angles of wrap. Furthermore, the pitching angle also influences the effective angle of wrap but will not be considered at this time and will be investigated in future work.

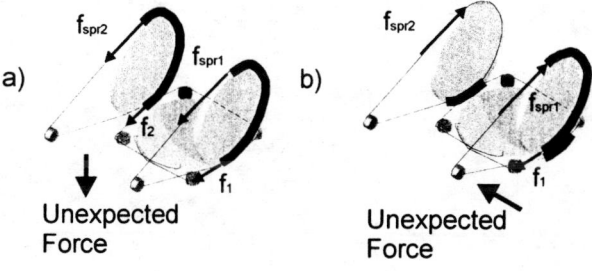

Fig. 15 Effective Angle of Wrap

For both the pitching and the yawing cases, the torque limit is equal to the amount of friction between the wire and pulley. For pitching, the effective angle of wrap is shown in Fig.15a and the corresponding torque is

$$T = (f_1 - f_{spr1} + f_2 - f_{spr2}) \cdot r \qquad (4)$$

substituting in Ertelwein's equation (2) results in

$$T = \frac{f_{spr1} + f_{spr2}}{2}(e^{\phi\mu} - 1) \cdot r \qquad (5)$$

For yawing, the effective angle of wrap is shown in Fig.15b and the corresponding torque is

$$T = (f_1 - f_{spr2}) \cdot r \quad (6)$$

and substituting in Ertelwein's equation results in

$$T = (f_{spr1} e^{\phi\mu} - f_{spr2}) \cdot r \quad (7)$$

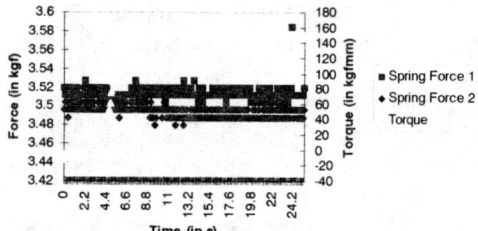

a) Pitch Slip Torque Profile

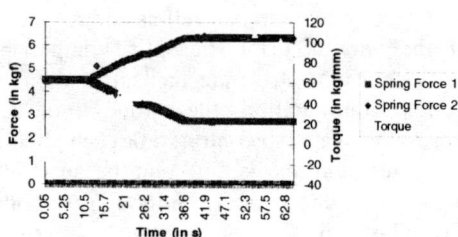

b) Yaw Slip Torque Profile

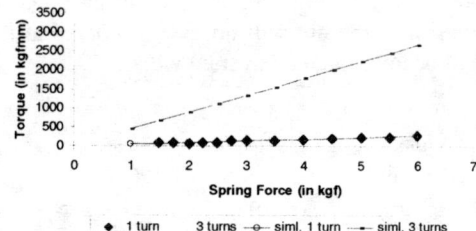

c) Pitch Torque Limit

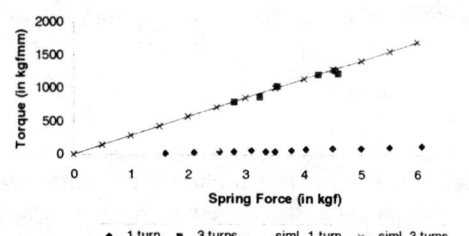

d) Yaw Torque Limit

Fig.16 DMA Experimental Results

Experiments were performed to verify the torque limiting safety mechanism of the DMA unit. A force sensor was used to measure the torque limit and two linear potentiometers were used to measure the two tensioning spring lengths from which the tensioning forces were determined. Several wire configurations and spring forces were used and the general slip characteristics in the pitch and yaw directions are shown in Fig.16a and 16b. The difference in slip characteristic is a result of distribution of wire tension as mentioned above. The torque limits are as expected, increasing linearly versus spring force and exponentially versus contact angle as shown in Fig.16c and 16d. Note that the pitching torque limit is double that of the yawing torque because of the effective angle of wrap. In the pitching case, the wire is tight on both pulleys whereas in the yawing case, the wire is only tight on one pulley.

5. Conclusions

A new M-Drive mechanism that allows slip of the joints when torque exceeds a threshold, using the principles of wire and pulley friction, has been developed. Then by employing two M-Drive mechanisms, a two DOF wire driven actuator that maximizes actuator efficiency by using a coupled floating differential drive is also developed. Experimental results show the feasibility of the mechanisms by demonstrating that friction properties can be controlled and be used for torque limiting. This actuator can be made lightweight and compact, two considerations extremely attractive in the field of robotics. Future work will be to optimize the actuator and to incorporate it with the skeleton hyper-redundant manipulator Float Arm.

Acknowledgement

This research is supported by The Grant-in Aid for COE Research Project of Super Mechano-Systems by The Ministry of Education, Science, Sport and Culture.

6. References

1. S. Hirose: "Design of Hyper-Redundant Arm", The 7th Int. Symposium of Robotics Research, pp.548-557 (1996)
2. K. Asano el al.: "Multijoint Inspection Robot", Proc. IEEE. Trans. On Industrial Electronics, pp.277-281 (1984)
3. P.K.J. Smith: "The "WARRIOR" Welding Manipulator", Proc. Int. Advanced Robotics Programme, 1st Workshop on Manipulators, Sensors and step towards Mobility, pp.107-113 (1987)
4. S. Ma, S. Hirose, H. Yoshinada: "CT Arm-I: Coupled Tendon Driven Manipulator Model I – Design and Experiments", Proc. IEEE Int.Conf. on Robotics and Automation, pp.2094-2100 (1992)
5. E. Paljug, T. Ohm, S. Hayati: "The JPL serpentine Robot: a 12 DOF System for Inspection", Proc. IEEE Int. Conf. Of Robotics and Automation, pp.3143-3148 (1995)
6. S. Hirose, M. Sato: "Coupled Drive of the Multi DOF Robot", Proc. IEEE Int. Conf. Of Robotics and Automation, pp.1610-1616 (1989)
7. K. Arikawa, S. Hirose: "Development of a Quadruped Walking Robot TITAN VIII", Proc.IROS 96, pp.208-214 (1996)
8. Joseph E. Shigley, Charles R. Mischke: Standard Handbook of Machine Design, McGraw-Hill,pp31.1-31.14

The Shape Jacobian of a Manipulator with Hyper Degrees of Freedom

Hiromi Mochiyama

School of Information Science
Japan Advanced Institute of Sci. and Tech.
Tatsunokuchi, Ishikawa 923-1292, Japan

Hisato Kobayashi

Dept. of Electrical and Electronic Eng.
Hosei University
Kajinocho, Koganei, Tokyo 184, Japan

Abstract

The Shape Jacobian which is the counterpart of the manipulator Jacobian plays a key role to control the shape of a manipulator with extraordinarily many degrees of freedom. In this paper, we show some significant properties of the Shape Jacobian; the structure, boundedness, determinant and singularity, in geometric flavor.

1 Introduction

A kinematic degrees of freedom of a manipulator is one of the important factors which indicates task capability of the manipulator. For this reason, a manipulator with extraordinarily many degrees of freedom has been considered from all angles [1, 2, 3, 9]. In this paper, such a manipulator is called a *Hyper Degrees of Freedom manipulator* (HDOF manipulator, for short).

The manipulator Jacobian in one of the most important tool in conventional robot control theory. It is defined as the Jacobian matrix of forward kinematics which describes the relation between joint angles and tip position and orientation. When a manipulator has a lot of kinematic degrees of freedom, it is reasonable to control the shape of a manipulator, that is, we need to change our viewpoint from tip control to shape control. In shape control, it is the Shape Jacobian that corresponds to the manipulator Jacobian [3]. In this paper, we show some important properties of the Shape Jacobian.

In Section 2, we give descriptions of a kinematics of an HDOF manipulator and a parameterized spatial curve. In Section 3, we review the definition of the Shape Jacobian proposed by the authors in order to control the shape of a manipulator. In Section 4, we show significant properties of the Shape Jacobian in geometric flavor. In Section 5, we summarize the results in this paper.

2 Preliminaries

Suppose that an HDOF manipulator satisfies the following assumption (Figure 1):

Assumption 1 (HDOF Manipulators)
An HDOF manipulator has

1. a serial rigid chain structure, and
2. 2-degree-of-freedom(2DOF) revolute joints.

□

The above assumption is satisfied by manipulators developed in [6] and [7], [8].

Under Assumption 1, by the coordinate setting method shown in [4], the kinematics of the manipulator can be expressed as

$$\boldsymbol{\Phi}_i = \boldsymbol{\Phi}_{i-1} \boldsymbol{R}_{w,i}, \quad (1)$$
$$\boldsymbol{R}_{w,i} = \boldsymbol{R}(\boldsymbol{a}_{s,i}, \theta_{s,i}) \boldsymbol{R}(\boldsymbol{a}_{m,i}, \theta_{m,i}), \quad (2)$$
$$\boldsymbol{p}_i = \boldsymbol{p}_{i-1} + l_i \boldsymbol{\Phi}_i \boldsymbol{e}_x, \quad , \quad i = 1, \cdots, n \quad (3)$$

where $\boldsymbol{\Phi}_i \in SO(3)$ is the frame attached to the i-th link, $\boldsymbol{R}_{w,i} \in SO(3)$ is the rotation matrix of the i-th 2DOF revolute joint, $\boldsymbol{R}(\boldsymbol{a}, \theta) \in SO(3)$ the rotation matrix about a unit-length axis $\boldsymbol{a} \in \Re^3$ through an angle $\theta \in [-\pi\ \pi)$, $\boldsymbol{a}_{s,i}, \boldsymbol{a}_{m,i} \in \Re^3$ are the unit-length and constant rotational axes of the joint, $\theta_{s,i}, \theta_{m,i} \in [-\pi\ \pi)$ are the rotaional angles of the joint, $\boldsymbol{p}_i \in \Re^3$ is the position of the link, l_i is the constant link length and $\boldsymbol{e}_x := [1\ 0\ 0]^T$. Define $\boldsymbol{\theta} := [\theta_{s,1}\ \theta_{m,1}\ \cdots\ \theta_{s,n}\ \theta_{m,n}]^T \in \Re^{2n}$. See [4] for more detail.

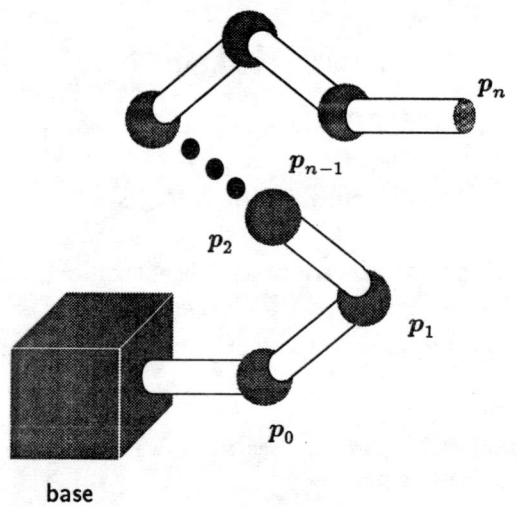

Figure 1: Manipulator with hyper degrees of freedom

In order to prescribe a desired shape for an HDOF manipulator, we use a parametrized spatial curve $c : \Re \to \Re^3$ with the following assumption.

Assumption 2 (Spatial Curves)

A spatial curve $c : \Re \to \Re^3$ has the following properties:

1. Mapping c is of class C^2 in $\Re$.

2. Any tangent vectors on a curve are normalized, i.e., $\forall \sigma$, $\left\| \frac{dc}{d\sigma}(\sigma) \right\|_2 = 1$ where $\| \cdot \|_2$ denotes the euclidean vector norm.

3. The curvature of a curve is bounded.

□

Frenet-Serret formula expresses the relationship between a parametrized curve and geometric values, the curvature and torsion, as follows:

$$\frac{d\Phi}{d\sigma}(\sigma) = \Phi(\sigma)[\omega(\sigma)\times], \quad (4)$$

$$\omega(\sigma) := \begin{bmatrix} \tau(\sigma) \\ 0 \\ \kappa(\sigma) \end{bmatrix}, \quad (5)$$

$$\frac{dc}{d\sigma}(\sigma) = \Phi(\sigma)e_x, \quad (6)$$

where $\Phi(\sigma) \in SO(3)$ is the Frenet frame at σ, $\kappa : \Re \to \Re_+$ and $\tau : \Re \to \Re$ is the curvature and torsion of the curve respectively. For a vector $\boldsymbol{a} = [a_x\ a_y\ a_z]^T \in \Re^3$, $[\boldsymbol{a}\times]$ is defined as

$$[\boldsymbol{a}\times] := \begin{bmatrix} 0 & -a_z & a_y \\ a_z & 0 & -a_x \\ -a_y & a_x & 0 \end{bmatrix}. \quad (7)$$

Note that

$$[\boldsymbol{a}\times]\boldsymbol{b} = \boldsymbol{a} \times \boldsymbol{b}, \quad (8)$$

where the symbol '$\times$' denotes an outer product in euclidean space.

We make the following assumption on the relation between an HDOF manipulator and a spatial curve:

Assumption 3

The given curve $c : \Re \to \Re^3$ passes through the position of the base link, p_0, i.e., there exists a real number σ_0 such that $c(\sigma_0) = p_0$. Without loss of generality, we set $\sigma_0 = 0$. □

3 Definition

Define $p(\theta) \in \Re^{3n}$ by arranging all the link positions in a row as

$$p(\theta) := \begin{bmatrix} p_1(\theta) \\ \vdots \\ p_n(\theta) \end{bmatrix}. \quad (9)$$

Vector $p(\theta)$ includes complete information on all the corners of the chain of line segments which expresses the kinematic feature of an HDOF manipulator. Define also $p_d(\sigma) \in \Re^{3n}$ for a given curve $c : \Re \to \Re^3$ as

$$p_d(\sigma) := \begin{bmatrix} c(\sigma_1) \\ \vdots \\ c(\sigma_n) \end{bmatrix}, \quad (10)$$

where $\sigma_i \in \Re$ ($i = 1, \cdots, n$) is a curve parameter corresponding to the position of the i-th link, and $\sigma := [\sigma_1 \cdots \sigma_n]^T \in \Re^n$. Further define $e(\theta, \sigma) \in \Re^{3n}$ as the difference between them, i.e.,

$$e(\theta, \sigma) := p(\theta) - p_d(\sigma). \quad (11)$$

We can express that the manipulator forms the shape of a given curve by $e(\theta, \sigma) = \mathbf{0}$. Vector e is interpreted as the shape error between the manipulator and curve [†].

[†]Strictly speaking, $e(\theta, \sigma) = 0$ is a necessary condition in order that an HDOF manipulator is 'fitted' to a curve. See [3] or [5] for more detail.

Suppose that θ and σ are time functions and differentiable with respect to time t in $\Re_+$. Then, the derivative of e becomes

$$\dot{e}(q,\dot{q}) = J(q)\dot{q}, \quad (12)$$

where $q := [\theta^T \sigma^T]^T \in \Re^{3n}$ is called the extended joint angles and $J(q) \in \Re^{3n \times 3n}$ is the *Shape Jacobian* which is defined as

$$J(q) := \left[\frac{\partial p}{\partial \theta}(\theta) \quad -\frac{\partial p_d}{\partial \sigma}(\sigma) \right]. \quad (13)$$

The Shape Jacobian contains information on a given curve. Due to this, the Shape Jacobian becomes square. That is a key that we can obtain important properties of the Shape Jacobian we will see next.

4 Properties

First, we see an essential structure of the Shape Jacobian. Next, boundedness of the Shape Jacobian is considered. Then, by use of the structure, the determinant and singularity are analyzed.

4.1 Lower Triangular Structure

Let $I_{i,j} \in \Re^{3n \times 3n}$ be the elementary matrix exchanging the i-th line for the j-th. This matrix has properties such that

$$I_{i,j}^{-1} = I_{i,j}^T, \quad (14)$$
$$\det I_{i,j} = -1. \quad (15)$$

Using this elementary matrix, define $P \in \Re^{3n \times 3n}$ as

$$\begin{aligned} P &:= (I_{2n,2n+1} I_{2n-1,2n} \cdots I_{3,4}) \\ &\quad \cdot (I_{2n+1,2n+2} I_{2n,2n+1} \cdots I_{6,7}) \\ &\quad \cdots (I_{2n+i-1,2n+i} I_{2n+i-2,2n+i-1} \cdots I_{3i,3i+1}) \\ &\quad \cdots (I_{3n-2,3n-1} I_{3n-3,3n-2}) \\ &= \prod_{i=1}^{n-1} \prod_{j=1}^{2(n-i)} I_{2n+i-j, 2n+i-j+1}. \quad (16) \end{aligned}$$

The number of the elementary matrices appeared in the definition P is

$$\sum_{i=1}^{n-1} 2(n-i) = 2\left\{\sum_{i=1}^{n-1}(n-i)\right\}, \quad (17)$$

which is always even. Thus, P has properties:

$$P^{-1} = P^T, \quad (18)$$
$$\det P = 1, \quad (19)$$

i.e., $P \in SO(3n)$. Furthermore, using this P, define $\bar{J}(q) \in \Re^{3n \times 3n}$ as

$$\bar{J}(q) := J(q)P. \quad (20)$$

Then, $\bar{J}(q)$ can be represented as the following lower-triangular 3×3-block matrix:

$$\bar{J} = \begin{bmatrix} J_{11} & & & \\ J_{21} & J_{22} & & \mathbf{O} \\ \vdots & \vdots & \ddots & \\ J_{n1} & J_{n2} & \cdots & J_{nn} \end{bmatrix}, \quad (21)$$

where $J_{ij} \in \Re^{3 \times 3}$ is defined as

$$J_{ij} := \begin{cases} \left[\dfrac{\partial p_i}{\partial \theta_{s,j}}(\theta) \quad \dfrac{\partial p_i}{\partial \theta_{m,j}}(\theta) \quad -\dfrac{dc}{d\sigma}(\sigma_i) \right], \\ \qquad\qquad\qquad\qquad\qquad\qquad i = j, \\ \left[\dfrac{\partial p_i}{\partial \theta_{s,j}}(\theta) \quad \dfrac{\partial p_i}{\partial \theta_{m,j}}(\theta) \quad \mathbf{o} \right], \\ \qquad\qquad\qquad\qquad\qquad\qquad i > j. \end{cases} \quad (22)$$

Property 1 (structure)
The Shape Jacobian can be expressed as a 3×3-block lower-triangular matrix by an even-numbered sorting of its columns. □

It is important to note that partial derivatives of link positions in joint angles are represented by the outer products of vectors as

$$\frac{\partial p_i}{\partial \theta_{s,j}} = (\Phi_{j-1} a_{s,j}) \times (p_i - p_{j-1}), \quad (23)$$

$$\frac{\partial p_i}{\partial \theta_{m,j}} = (\Phi_j a_{m,j}) \times (p_i - p_{j-1}). \quad (24)$$

4.2 Boundedness

Let $\|\cdot\|_2$ denote the euclidean norm for vectors, or the spectral norm for matrices (i.e., the matrix norm induced by the euclidean vector norm). The spectral norm of the Shape Jacobian is evaluated as

$$\begin{aligned} \|J(q)\|_2 &\leq \sqrt{3n}\, \|J(q)\|_1 \\ &= \sqrt{3n} \max\left\{ \max_{x,j} \sum_{i=j}^{n} \left\|\frac{\partial p_i}{\partial \theta_{x,j}}(\theta)\right\|_1, \right. \\ &\qquad\qquad \left. \max_j \left\|-\frac{dc}{d\sigma}(\sigma_j)\right\|_1 \right\}, \quad (25) \end{aligned}$$

where $x \in \{s, m\}$ and $\|\cdot\|_1$ means the 1-norm for vectors or its induced matrix norm. From (23) and

(24), we obtain

$$\left\| \frac{\partial p_i}{\partial \theta_{x,j}}(\theta) \right\|_1 \leq \sqrt{3} \left\| \frac{\partial p_i}{\partial \theta_{x,j}}(\theta) \right\|_2$$
$$\leq \sqrt{3} \| p_i - p_{j-1} \|_2$$
$$\leq \sqrt{3} \sum_{k=j}^{i} l_k. \quad (26)$$

Assumption 2-2 allows us to have the following inequality:

$$\left\| -\frac{dc}{d\sigma}(\sigma_i) \right\|_1 \leq \sqrt{3} \left\| \frac{dc}{d\sigma}(\sigma_i) \right\|_2$$
$$\leq \sqrt{3}. \quad (27)$$

Therefore,

$$\| J(q) \|_2 = \sqrt{3n} \max \left\{ \max_j \sum_{i=j}^{n} \left(\sqrt{3} \sum_{k=j}^{i} l_k \right), \sqrt{3} \right\}$$
$$= 3\sqrt{n} \max \left\{ \max_j \sum_{i=j}^{n} \sum_{k=j}^{i} l_k, 1 \right\}$$
$$\leq 3\sqrt{n} \max \left\{ \sum_{i=1}^{n} \sum_{k=1}^{i} l_k, 1 \right\}. \quad (28)$$

Next we see the bound of the time-derivative of the Shape Jacobian which is defined as

$$\dot{J}(q, \dot{q})$$
$$:= \frac{d}{dt} J(q)$$
$$= \left[\frac{\partial J}{\partial \theta_{s,1}}(q)\dot{q} \quad \frac{\partial J}{\partial \theta_{m,1}}(q)\dot{q} \quad \cdots \quad \frac{\partial J}{\partial \theta_{s,n}}(q)\dot{q} \right.$$
$$\left. \frac{\partial J}{\partial \theta_{m,n}}(q)\dot{q} \quad \frac{\partial J}{\partial \sigma_1}(q)\dot{q} \quad \cdots \quad \frac{\partial J}{\partial \sigma_n}(q)\dot{q} \right]. \quad (29)$$

The spectral norm can be evaluated as

$$\left\| \dot{J}(q, \dot{q}) \right\|_2 \leq \sqrt{3n} \left\| \dot{J}(q, \dot{q}) \right\|_1$$
$$= \sqrt{3n} \max \left\{ \max_{k,x} \left\| \frac{\partial J}{\partial \theta_{x,k}}(q)\dot{q} \right\|_1, \max_k \left\| \frac{\partial J}{\partial \sigma_k}(q)\dot{q} \right\|_1 \right\}$$
$$\leq 3n \max \left\{ \max_{k,x} \left\| \frac{\partial J}{\partial \theta_{x,k}}(q) \right\|_1, \max_k \left\| \frac{\partial J}{\partial \sigma_k}(q) \right\|_1 \right\} \| \dot{q} \|_2. \quad (30)$$

By differentiating (23) and (24), the second-order partial derivatives of p_i with respect to joint angles can be evaluated as

$$\left\| \frac{\partial^2 p_j}{\partial \theta_{y,k} \partial \theta_{x,i}}(\theta) \right\|_1 \leq 2\sqrt{3} \sum_{p=j}^{i} l_p, \quad (31)$$

Thus, we obtain

$$\max_{k,x} \left\| \frac{\partial J}{\partial \theta_{x,k}}(q) \right\|_1$$
$$= \max_{j,k,x,y} \sum_{i=\max\{j,k\}}^{n} \left\| \frac{\partial^2 p_j}{\partial \theta_{y,k} \partial \theta_{x,i}}(\theta) \right\|_1$$
$$\leq 2\sqrt{3} \max_{j,k} \sum_{i=\max\{j,k\}}^{n} \sum_{p=j}^{i} l_p$$
$$\leq 2\sqrt{3} \sum_{i=1}^{n} \sum_{p=1}^{i} l_p. \quad (32)$$

By Frenet-Serret formula and Assumption 2-3, we obtain

$$\max_k \left\| \frac{\partial J}{\partial \sigma_k}(q) \right\|_1 = \max_k \left\| -\frac{d^2 c}{d\sigma^2}(\sigma_k) \right\|_1$$
$$\leq \max_k \sqrt{3} \| \kappa(\sigma_k) \Phi(\sigma_k) e_y \|_2$$
$$\leq \sqrt{3} \sup_\sigma \kappa(\sigma). \quad (33)$$

Therefore,

$$\left\| \dot{J}(q, \dot{q}) \right\|_2$$
$$\leq 3\sqrt{3}n \max \left\{ 2 \sum_{i=1}^{n} \sum_{p=1}^{i} l_p, \sup_\sigma \kappa(\sigma) \right\} \| \dot{q} \|_2. \quad (34)$$

Property 2 (boundedness)
The spectral norm of the Shape Jacobian is bounded from a positive real constant related to the link length and the number of degrees of freedom, i.e.,

$$\forall q \in \Re^{3n} \quad \| J(q) \|_2 \leq J_M, \quad (35)$$

where

$$J_M := 3\sqrt{n} \max \left\{ \sum_{i=1}^{n} \sum_{k=1}^{i} l_k, 1 \right\}. \quad (36)$$

Moreover, the spectral norm of the time-derivative of the Shape Jacobian, $\dot{J}(q, \dot{q}) := \frac{d}{dt} \{ J(\theta) \}$, is bounded from the product of the spectral norm of the extended joint angular velocities $\dot{q}$ and a positive real constant related to the link length, the number of degrees of freedom and the supremum of curvature of a given curve, i.e.,

$$\forall q, \dot{q} \in \Re^{3n} \quad \|\dot{J}(q, \dot{q})\| \leq J_H \|\dot{q}\|, \quad (37)$$

where

$$J_H := 3\sqrt{3} n \max \left\{ 2 \sum_{i=1}^{n} \sum_{p=1}^{i} l_p, \sup_{\sigma} \kappa(\sigma) \right\}. \quad (38)$$

□

4.3 Determinant

The lower-triangular structure of the Shape Jacobian allows us to calculate its determinant as follows:

$$\det \boldsymbol{J}(q)$$
$$= \det \bar{\boldsymbol{J}}(q)$$
$$= \det \begin{bmatrix} \boldsymbol{J}_{11} & & & \\ \boldsymbol{J}_{21} & \boldsymbol{J}_{22} & & \\ \vdots & \vdots & \ddots & \\ \boldsymbol{J}_{n1} & \boldsymbol{J}_{n2} & \cdots & \boldsymbol{J}_{nn} \end{bmatrix}$$
$$= \prod_{i=1}^{n} \det \boldsymbol{J}_{ii}$$
$$= \prod_{i=1}^{n} \det \left[\frac{\partial \boldsymbol{p}_i}{\partial \theta_{s,i}} \quad \frac{\partial \boldsymbol{p}_i}{\partial \theta_{m,i}} \quad -\frac{d\boldsymbol{c}}{d\sigma}(\sigma_i) \right]$$
$$= \prod_{i=1}^{n} \left\{ \left(\frac{\partial \boldsymbol{p}_i}{\partial \theta_{m,i}} \times \frac{\partial \boldsymbol{p}_i}{\partial \theta_{s,i}} \right)^T \frac{d\boldsymbol{c}}{d\sigma}(\sigma_i) \right\}.$$

Taking $j = i$ in (23) and (24), the outer product of the partial derivatives of $\boldsymbol{p}_i$ with respect to $\theta_{s,i}$ and $\theta_{m,i}$ is calculated as

$$\frac{\partial \boldsymbol{p}_i}{\partial \theta_{m,i}} \times \frac{\partial \boldsymbol{p}_i}{\partial \theta_{s,i}}$$
$$= \{\boldsymbol{\Phi}_i \boldsymbol{a}_{m,i} \times (\boldsymbol{p}_i - \boldsymbol{p}_{i-1})\} \times$$
$$\quad \{\boldsymbol{\Phi}_{i-1} \boldsymbol{a}_{s,i} \times (\boldsymbol{p}_i - \boldsymbol{p}_{i-1})\}$$
$$= \det [\ \boldsymbol{\Phi}_i \boldsymbol{a}_{m,i} \quad (\boldsymbol{p}_i - \boldsymbol{p}_{i-1}) \quad (\boldsymbol{p}_i - \boldsymbol{p}_{i-1})\ ]$$
$$\quad \boldsymbol{\Phi}_{i-1} \boldsymbol{a}_{s,i}$$
$$\quad - \det [\ \boldsymbol{\Phi}_i \boldsymbol{a}_{m,i} \quad (\boldsymbol{p}_i - \boldsymbol{p}_{i-1}) \quad \boldsymbol{\Phi}_{i-1} \boldsymbol{a}_{s,i}\ ]$$
$$\quad (\boldsymbol{p}_i - \boldsymbol{p}_{i-1})$$
$$= \det [\ (\boldsymbol{p}_i - \boldsymbol{p}_{i-1}) \quad \boldsymbol{\Phi}_i \boldsymbol{a}_{m,i} \quad \boldsymbol{\Phi}_{i-1} \boldsymbol{a}_{s,i}\ ]$$
$$\quad (\boldsymbol{p}_i - \boldsymbol{p}_{i-1})$$
$$= \det [\ l_i \boldsymbol{e}_x \quad \boldsymbol{a}_{m,i} \quad \boldsymbol{R}^T(\boldsymbol{a}_{m,i}, \theta_{m,i}) \boldsymbol{a}_{s,i}\ ]$$
$$\quad (\boldsymbol{p}_i - \boldsymbol{p}_{i-1}). \quad (39)$$

Therefore, we obtain

$$\det \boldsymbol{J}(q)$$
$$= \prod_{i=1}^{n} \left\{ \det [\ l_i \boldsymbol{e}_x \quad \boldsymbol{a}_{m,i} \quad \boldsymbol{R}^T(\boldsymbol{a}_{m,i}, \theta_{m,i}) \boldsymbol{a}_{s,i}\ ] \right.$$
$$\left. (\boldsymbol{p}_i - \boldsymbol{p}_{i-1})^T \frac{d\boldsymbol{c}}{d\sigma}(\sigma_i) \right\}, \quad (40)$$

Property 3 (determinant)
The determinant of the Shape Jacobian can be expressed as

$$\det \boldsymbol{J}(q) = \prod_{i=1}^{n} \{v_i(\theta_{m,i}) w_i(\boldsymbol{\theta}, \sigma_i)\}, \quad (41)$$

where

$$v_i(\theta_{m,i}) := \det [\ l_i \boldsymbol{e}_x \quad \boldsymbol{a}_{m,i} \quad \boldsymbol{R}^T(\boldsymbol{a}_{m,i}, \theta_{m,i}) \boldsymbol{a}_{s,i}\ ], \quad (42)$$

$$w_i(\boldsymbol{\theta}, \sigma_i) := (\boldsymbol{p}_i - \boldsymbol{p}_{i-1})^T \frac{d\boldsymbol{c}}{d\sigma}(\sigma_i). \quad (43)$$

□

Note that v_i and w_i have geometric meanings. The value $|v_i|$ can be interpreted as the volume of the parallelopiped with edges $l_i \boldsymbol{e}_x$, $\boldsymbol{a}_{m,i}$ and $\boldsymbol{R}^T(\boldsymbol{a}_{m,i}, \theta_{m,i}) \boldsymbol{a}_{s,i}$. The value w_i is the inner product of difference vector, $(\boldsymbol{p}_i - \boldsymbol{p}_{i-1})$, and tangent of a given curve at σ_i, $\frac{d\boldsymbol{c}}{d\sigma}(\sigma_i)$.

4.4 Singularity

Since the determinant of the Shape Jacobian has a product form, we immediately obtain useful expression of the singularity by solving equation $\det \boldsymbol{J}(\boldsymbol{\theta}) = 0$.

Property 4 (Singularity)
The Shape Jacobian is singular if and only if there exists a positive integer, $i \in \{1, \cdots, n\}$, such that at least one of the following two conditions holds:

1. Consider the three directions in euclidean space; the directions of the length of the i-th link and the two rotational axes of the i-th joint. At least two among the directions align, i.e.,

$$\det [\ \boldsymbol{e}_x \quad \boldsymbol{a}_{m,i} \quad \boldsymbol{R}^T(\boldsymbol{a}_{m,i}, \theta_{m,i}) \boldsymbol{a}_{s,i}\ ] = 0. \quad (44)$$

2. The directions of the i-th link length and the tangent at the point corresponding to the i-th link position are orthogonal, i.e.,

$$e_x^T \left(\Phi_i^T \frac{dc}{d\sigma}(\sigma_i) \right) = 0. \quad (45)$$

□

There are three remarks on this property. First, the singularity condition of the Shape Jacobian can be described by completely separated n conditions for each link, joint and curve parameter. This is very helpful for the calculation. Second, each completely separated condition is further devided into two distinguished parts. One part, described by (44), is only related to the mechanical structure and joint angles, while the other, (45), depends on the tangent of the curve. Finally, the derived conditions (44) and (45) have geometric meanings in euclidean space. That is, they can be explained by the geometric relation of vectors, inner and outer products in euclidean space. The first is alignment condition among three vectors, e_x, $a_{m,i}$ and $R^T(a_{m,i}, \theta_{m,i})a_{s,i}$. Note that these vectors correspond to the directions of the link and the two rotational axes of the joint respectively (see Figure 2). The second is orthogonality condition between two vectors, e_x and $\Phi_i^T \frac{dc}{d\sigma}(\sigma_i)$, which are coresspond to the link length direction and the tangent on a curve at σ_i.

5 Conclusion

In this paper, we have shown the following properties of the Shape Jacobian:

1. The Shape Jacobian has a structure of lower-triangular of 3×3-block matricies by appropriate sorting of the columns.

2. The spectral norm of the Shape Jacobian is bounded from a positive constant. Moreover, the norm of the time-derivative is also bounded from the product of the norm of the extended joint anglular velocities and a positive constant.

3. The determinant of the Shape Jacobian consists of the product of the signed volume of a parallelopiped and an inner product of vectors.

4. The singularity of the Shape Jacobian can be interpreted as an alignment and an orthogonality of vectors.

These properties of the Shape Jacobian are very useful for the dynamics-based shape control of HDOF manipulators, which can be seen in [3].

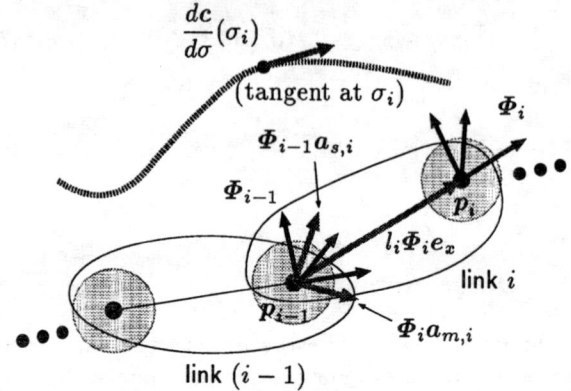

Figure 2: Geometric meanings of singularity

References

[1] G.S. Chirikjian: *Theory and Applications of Hyper-Redundant Robotic Mechanisms*, Ph.D thesis, Dept.of Applied Mechanics, California Inst. of Tech., 1992.

[2] S. Hirose: *Biologically Inspired Robots: Snake-like Locomotors and Manipulators*, Oxford Science Publications, 1993.

[3] H. Mochiyama: *Shape Control of Manipulators with Hyper Degrees of Freedom*, Doctoral Dissertation, School of Information Science, Japan Advanced Institute of Science and Technology, 1998.

[4] H. Mochiyama et al. : "Direct Kinematics of Manipulators with Hyper Degrees of Freedom and Frenet-Serret Formula," *Proc. 1998 IEEE Int. Conf. Robotics and Automation*, Vol.2, 1653/1658, 1998.

[5] H. Mochiyama et al. : "Shape Correspondence between a Spatial Curve and a Manipulator with Hyper Degrees of Freedom," to appear in *1998 IEEE/RSJ Int. Conf. on Intelligent Robots and Systems*, 1998.

[6] M. Nilsson, "Snake Robot Free Climbing," *Proc. 1997 IEEE Int. Conf. Robotics and Automation*, Vol.3, 3415/3420, 1997.

[7] E. Paljug et al.: "The JPL Serpentine Robot: a 12 DOF System for Inspection," *Proc. 1995 IEEE Int. Conf. Robo. and Automation*, Vol.3, 3143/3148, 1995.

[8] N. Takanashi: "Complete Modular Link with the Active Universal Joint Mechanism for 3D Hyper Redundant Robot," *Proc. the 34th SICE Annual Conf.*, Vol.2, 843/844, 1995. (in Japanese)

[9] K.E. Zanganeh et al.: "A Discrete Model for the Configuration Control of Hyper-Redundant Manipulators," *Proc. 1997 IEEE Int. Conf. Robotics and Automation*, Vol.1, 167/172, 1997.

A Geometric Approach to Anguilliform Locomotion: Modelling of an Underwater Eel Robot*

Kenneth A. McIsaac   James P. Ostrowski

General Robotics Automation, Sensing and Perception (GRASP) Laboratory
University of Pennsylvania, 3401 Walnut Street, Philadelphia, PA 19104-6228
E-mail: {kamcisaa, jpo}@grip.cis.upenn.edu

Abstract

In this paper, we approach the modeling, simulation, and control of snake and eel-like robots from a geometric perspective. We propose basic viscous and fluid drag models to capture the effect of external forces acting on the bodies of land-based snakes and aquatic eels, respectively. Using these models, we show that the dynamics can be decomposed into motions in the position and orientation of the overall system and changes in its internal shape. This allows us to write reduced dynamics for the serpentine and swimming motions that isolate the effect of shape changes on the momentum of the system. We explore in simulation several types of locomotive gaits, and report qualitative experimental results that provide validation for these simulations.

1 Introduction

Mobile robots continue to challenge researchers with new applications in a variety of environments. Of recent interest has been the application of robotic technology to underwater exploration, monitoring, and surveillance. In this paper, we explore the modeling, simulation, and design of controllers for snake-like robotic systems capable of both crawling overland and underwater swimming.

Previous research in the area of underwater robotic systems has emphasized larger vehicles used for undersea exploration, such as Woods Hole's JASON [17]. More recently, *biomimetic* (biologically based) approaches to underwater locomotion have been pursued. The biomimetic approach to locomotion systems has several potential advantages, including increased efficiency and agility. Recent research has explored various size ranges of robots, including the RoboTuna [16], and smaller fish-like projects [6, 9]. Less work has been done in the area of *anguilliform* (eel-like) locomotion, though recently Ekeberg [11] has simulated the motion of such systems when controlled by biologically based neural networks. We use simplified models that are similar to Ekeberg's with the goal of understanding analytically the impact of various gait patterns on anguilliform motion. We also emphasize the study of gaits for turning and stopping as well as forward motion.

Land-based *snake-like* robots have also been studied in a variety of contexts, primarily for exploring different environments, including fallen or severely damaged buildings (e.g., after an earthquake), narrow and winding nuclear waste storage facilities, or even the internal organs of human beings [5, 7]. One of the original studies of snake robots, taken from a biological perspective, was conducted by Hirose [7], who performed a series of experiments on live snakes. Based on his observations, he successfully built a snake-like robot capable of propelling itself forward using only internal torques (that is, without directly driving the wheels). Another important body of work on snake robots has been performed by Chirikjian and Burdick, who coined the term **hyper-redundant** robots to describe such systems [5]. They focused on algorithms for controlling snake robots, but also worked on locomotion schemes such as inchworming, serpentine motions, and sidewinding [4, 5, 15].

One of the interesting characteristics of snake-like locomotion is the ability to transition from motion over land to motion underwater. There have been several robotic platforms that have been announced in the past year or so with a similar goal, including the crab-like Ariel system developed by ISRobotics [8] and a lobster robot developed by Ayers [1]. These systems, however, are generally restricted to motion only along the ocean floor.

Finally, we note that this work is motivated by previous work in studying the fundamental geometric principles of locomotion. Our past work has drawn together several areas of research to show how a wide variety of systems, from paramecia to inchworm robots to blimps to snakeboards, can be modeled within a single, unifying framework [10, 12, 14]. We discuss briefly the formulation of the problem and show how the snake and eel dynamics can be modeled within this framework by treating the friction and drag forces as invariant external forces. We also point out potential methods for using recent work on controlling dynamic systems that may be applicable to determining the appropriate choices of gait input patterns for anguilliform robots [3, 13].

---

*This work was partially supported by NSF grant IRI-9711834 and DARPA MURI DAAH04-96-1-0007.

2 Physical Model

For the purposes of simulation, we developed a simple physical model of a snake (we use the term "snake" interchangeably with "eel") to be used as a platform to test various locomotive gaits. The model is outlined in this section.

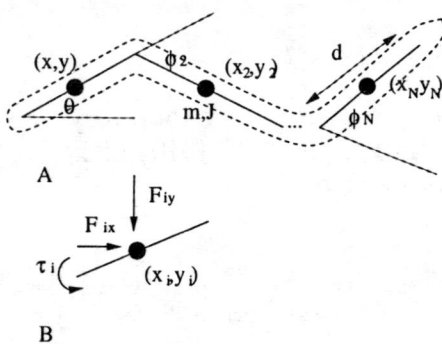

Figure 1: A. Model of snake as a planar, serial chain of links. B. Forces and torques on link i

2.1 Theoretical Development

The model used for simulation of the snake dynamics directly parallels the mechanical model used by Ekeberg [11]. The snake (Figure 1A) consists of a planar, serial chain of N identical links of length d, mass m and inertia J. Independent actuators exist at each joint. The following discussion will refer to the quantities indicated in Figure 1, referenced to an inertial frame.

Taking advantage of the $2N-2$ holonomic constraints enforced by the requirement that the joints do not separate, we define the set q of $N+2$ independent variables called the **configuration variables**. The configuration variables can then be divided into two subspaces. The position and orientation of the first link are called the **group variables**, defined to be $g = (x, y, \theta)$. The equations of motion determine the time evolution of the group variables. The vector of joint angles $r = (\phi_2, ..., \phi_N)$ (defined in Figure 1) are called the **shape variables**. The shape variables are the control inputs to the system.

For convenience, we define $s = (x_2, y_2, ..., x_N, y_N)$ to represent the set of dependent variables. These variables can be readily eliminated using holonomic constraints of the form $s = h(q)$.

We also define two vectors of forces. $F^1 = (F_{1x}, F_{1y}, 0, \tau_2, ..., \tau_N)$ are the forces that act in the dimensions defined by the configuration variables q. F_{ix} and F_{iy} are the forces acting at the centre of link i in the x- and y-dimensions, and τ_i is the torque around joint i. $F^2 = (F_{2x}, F_{2y}, ..., F_{Nx}, F_{Ny})$ contains the forces that act in the dimensions defined by the dependent variables s. Section 2.2 defines the relationship between these forces and the configuration variables.

The Lagrangian (L^s) of the system using the redundant set of coordinates is straightforward:

$$L^s = \frac{1}{2}\dot{q}^T M^1 \dot{q} + \frac{1}{2}\dot{s}^T M^2 \dot{s} \quad (1)$$

where M^1 and M^2 are symmetric, positive definite mass-inertia matrices. If we define the matrix $\beta = \frac{\partial h}{\partial q}$, we can rewrite the Lagrangian using only the configuration variables q as:

$$L = \frac{1}{2}\dot{q}^T \left(M^1 + \beta^T M^2 \beta\right) \dot{q} \quad (2)$$

The expression (2) for the Lagrangian leads to equations of motion of the form:

$$(M^1 + \beta^T M^2 \beta)\ddot{q} = F^1 + F^2 \beta^T - \beta^T M^2 \dot{\beta}\dot{q} \quad (3)$$

where the vector F^2 transforms through β^T into the configuration space.

To simulate the time evolution of the snake's path, we assume the actuators at each joint allow full control of the shape variables ϕ_i, $i \geq 2$. Thus the equations of motion (Eq. 3) reduce to a set of 3 coupled non-linear second-order differential equations in g.

We can further reduce the equations of motion by noticing that this system exhibits **Lie group symmetries** (invariance to changes in position and orientation), also called "pick-and-place" symmetries, since both the kinetic energy of the system and the frictional forces acting on the system are invariant with respect to the group variables g. To exploit this invariance, we first define the vector $\xi \in se(2) \in \mathfrak{g}$ as the body velocity (Lie algebra element) corresponding to the group vector $g \in SE(2)$ where $\xi = \Gamma^{-1}\dot{g}$ represents the tangent map of g acting on velocities (this is often expressed by $g^{-1}\dot{g}$) [14]. We then define a reduced Lagrangian $\ell(\xi, r, \dot{r}) = L(g^{-1}g, \Gamma^{-1}\dot{g}, r, \dot{r})$ and a reduced force vector τ where:

$$\tau(\xi, r, \dot{r}) = \begin{aligned}(F^1(g^{-1}g, r, \Gamma^{-1}\dot{g}, \dot{r}) \\ +\beta^T F^2(g^{-1}g, r, \Gamma^{-1}\dot{g}, \dot{r})) \cdot \Gamma^{-1}\end{aligned} \quad (4)$$

We can now express the equations of motion using the reduced, or **Euler-Poincaré equations** [12]:

$$\frac{d}{dt}\left(\frac{\partial \ell}{\partial \xi}\right) - ad_\xi^* \frac{\partial \ell}{\partial \xi} = \tau(\xi, r, \dot{r}), \quad (5)$$

where ad_ξ^* denotes the dual of the **adjoint action** of the Lie algebra, $\mathfrak{g}$, on itself (this is a constant, linear map—see [2, 12] for more details). Next, we define momenta $p_\alpha = \frac{\partial \ell}{\partial \xi^\alpha}$ and rewrite Eq. 5 in the form:

$$\dot{p} = ad_\xi^* p + \tau(\xi, r, \dot{r}) \quad (6)$$

When τ is quadratic in ξ and $\dot{r}$ (as in our swimming model below), the results of [12] apply directly, so that Eq. 6 can be restated in a quadratic form, called the **momentum equation** [2] (see Eq. 13 in Section 5), which relates the changes in momentum to functions of r only. We will return to this geometric formulation in the discussion that follows.

2.2 Frictional Forces

Locomotion is generated through the frictional force terms, defined by F_{1x}, F_{1y} and F^2 in Eq. 3. The choice of friction models allows us to model different types of undulatory locomotion. Using a fluid drag model, we can simulate the effect of an eel swimming. Using a viscous friction model, we can approximate the motion of a snake travelling over a smooth surface.

To simulate the forces in the water, we adopt a simple fluid mechanical model used by Ekeberg [11]. We assume that the Reynolds number is high enough that inertial forces dominate over viscous effects—a reasonable approximation for smooth bodies in an inviscid fluid. We also assume that the fluid is stationary, so the force of the fluid on a given link is due only to the motion of that link. The pressure differential created by an object moving in a fluid causes a drag force opposing the motion. Under the assumptions above, the drag force developed takes the form $F = \mu_w v^2$. Here, v is the forward speed of the link and μ_w is a drag coefficient for the water, determined by the formula $\mu_w = \rho A C / 2$, where A is the effective area of the object, ρ is the density of water, and C is a shape coefficient.

Following Ekeberg's development, we assume that pressure differentials in the directions parallel to the moving body are de-coupled from pressure differentials perpendicular to the body, to yield:

$$F_i^\perp = -\mu_w (v_i^\perp)^2 \qquad (7)$$

where $v_i^\perp$ is the projection of the vector $(\dot{x}_i, \dot{y}_i)$ along a direction perpendicular to the link. We exclude drag forces parallel to the link because they were determined in simulation to have negligible effects.

To simulate a snake gliding over dry land we assume a proportional viscous drag force of the form:

$$F_i^\perp = -\mu_v v_i^\perp \qquad (8)$$

where μ_v is a coefficient of viscous drag. Once again we have ignored frictional forces parallel to the body—in this case due to Hirose's assertion [7] that the scales of snakes provide much lower coefficients of friction in the parallel direction (and also because, again, simulation indicated that their effect is minimal).

For both of these friction models, the frictional forces are invariant to changes in the group variables, and lead to reduced dynamics of the form of Eq. 6.

2.3 Locomotive gaits

As a starting point we choose a gait corresponding to the gait seen in the biological literature on central pattern generators (CPG's), where motor neurons between segments of a snake introduce a constant phase delay in a travelling wave propagating down the body of the animal. We express the joint angles as:

$$\phi_i = A \sin(\omega t + (i-2)\phi_s) \qquad (9)$$

where ω is the wave frequency, A is the maximum amplitude of the oscillation, and ϕ_s is a constant phase difference between links. Defined in this way, a negative ϕ_s means a wave travelling towards the end of the body (yielding forward propulsion) and a positive ϕ_s has the opposite effect. The phase relationship between two links in the forward gait is shown for $t > 4$ in Figure 2. This choice of phasing leads to motions that closely resemble the serpenoid curve proposed by Hirose to model motions of real snakes [7, 12].

We also tested two turning gaits. In order to turn beginning with zero forward momentum a constant offset is applied to each ϕ_i down the length of the snake:

$$\phi_i = A \sin(\omega t + (i-2)\phi_s) + \phi_{\text{off}} \qquad (10)$$

The bias in the thrust generated during each period of the travelling wave causes the snake to rotate in place. Larger values of ϕ_{off} increase angular velocity.

The second turning gait, in which the snake coils around itself, more closely approximates the actual behaviour of swimming organisms which have some initial momentum before entering a turn. The resulting path is essentially circular. The gait waveform used is displayed in Figure 2 along with one period of the standard forward gait. In this case a "hold" is propagated down the length of the snake's body in addition to the standard sinusoidal gait. To create this gait: i) the angular velocity of each link is held at its maximum for a time T_1 ($3 < t < 3.2$ of the upper curve in Figure 2); ii) the joint angles vary sinusoidally until angular velocity reaches zero ($3.2 < t < 3.3$); iii) the joint angles are held at their maximum for a time T_2. After the hold, the reverse of operations ii) and i) are performed. The value of T_1 controls the degree of coiling, with larger values of T_1 causing the snake to coil into a smaller shape. The value of T_2 is the length of time the coil is maintained and determines the angle through which the snake turns.

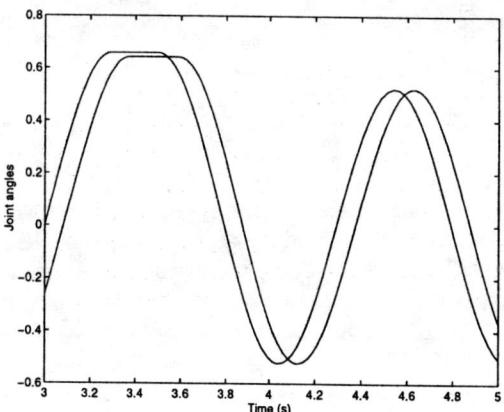

Figure 2: Plot of ϕ_2, ϕ_3 vs. time for the coiling gait

3 Results of Simulation

The physical parameters used in simulation are based on the dimensions of the REEL (Robotic EEL) robot (see Section 4), used to qualitatively test the gaits described in this paper. The value of μ_w was determined as outlined in Section 2.2, for elliptical snake cross-section

($C = 1$). We chose a coefficient of viscous friction (μ_v) to yield dynamic behaviour similar to that of the liquid drag case. The stopping time of an object of mass m and speed v_0 with viscous force $-\mu_v v$ is $t_s = m/\mu_v$. The equivalent time for an object resisted by liquid drag $-\mu_w v^2$ is $2m/\mu_w v_0$. Therefore, for $v_0 \approx 1 m/s$, $\mu_v = \mu_w$ gives similar dynamic behaviour. The numerical values for the simulation parameters are:

$$
\begin{aligned}
m &= 250 \text{ g} & \mu_w &= 4.5 \text{ Ns}^2/\text{m}^2 \\
d &= 15 \text{ cm} & \mu_d &= 4.5 \text{ Ns/m} \\
J &= 2.1 \text{ g} \cdot \text{m}
\end{aligned}
$$

In simulation, we used a 10 segment snake. Except where stated otherwise, a 1Hz gait frequency was used, and simulations were performed for 5 seconds.

3.1 Starting and Stopping

The basic path traced by the snake (initially at rest) using the forward gait is shown in Figure 3 for the liquid drag model. Based on our prediction about the two friction models we expect (for small velocities) to see the same basic behaviour in the case of viscous friction, with a slightly larger stopping time (leading to a wider trace with more "spread" around the centre of mass) This prediction was verified in simulation.

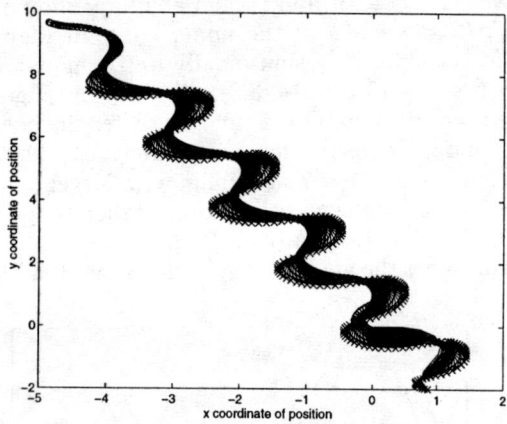

Figure 3: A forward path using the liquid drag model

Of particular interest is the relationship between the design parameters (A and ϕ_s) and the speed of the snake. Our experience with locomotion systems suggests that this relationship should obey an **area rule**, with changes in momentum proportional to an area enclosed by the shape variables. Thus, we expect to see a quadratic dependence between the velocity of the snake and the amplitude of the joint angles (see Section 5 for details). Figure 4 demonstrates that acceleration does vary quadratically with gait amplitude for small joint angles. For a given gait, the acceleration is roughly constant over several cycles, although the velocity saturates as $t \to \infty$. For large amplitudes, the snake accelerates quickly to an average limiting velocity (v_{lim}), and oscillates around that velocity in a straight line (Figure 5). We have not yet determined the dependence of v_{lim} on the design parameters, but we suspect that there is an optimal gait amplitude, where the frictional forces caused by a wide profile balance the thrust provided over a large period of oscillation.

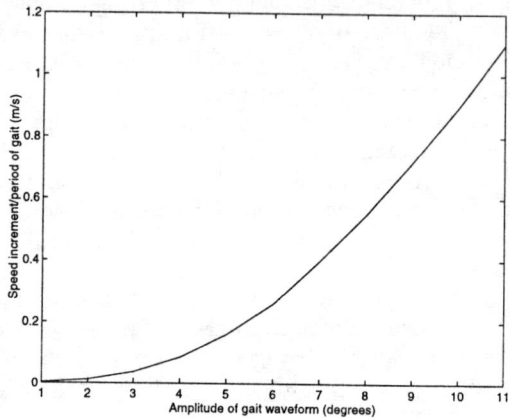

Figure 4: Rate of acceleration for small gait amplitudes.

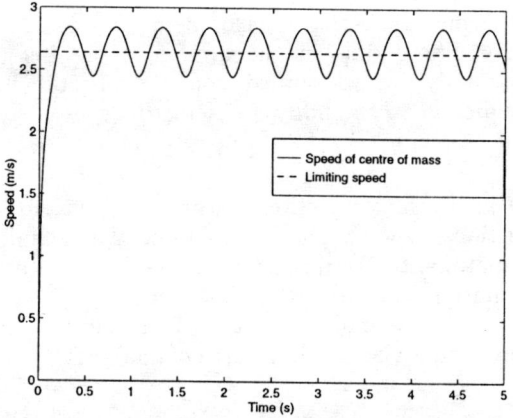

Figure 5: Time profile of the speed of the centre of mass

The reverse gait can be used to move the snake in the "backwards" direction, or to stop the snake given an initial velocity. Figure 6 shows the magnitude of the speed of the centre of mass given a large initial forward velocity in the x-direction. The velocity rapidly decays toward zero, then reverses in the characteristic sinusoidal pattern.

3.2 Turning Gaits

The turning in place gait causes a change in heading with zero initial angular velocity (Figure 7). The rate of change of the snake's heading depends on the magnitude of the offset added to the gait waveform. In simulation this rate can be made quite large, but the constraint that the links cannot intersect imposes an upper bound.

We used the coiling gate to follow an approximately square path (Figure 8). The snake uses the forward gate for several cycles, then turns to emerge on a new heading at right angles to the first. The box is not closed because the turns are not precisely 90°.

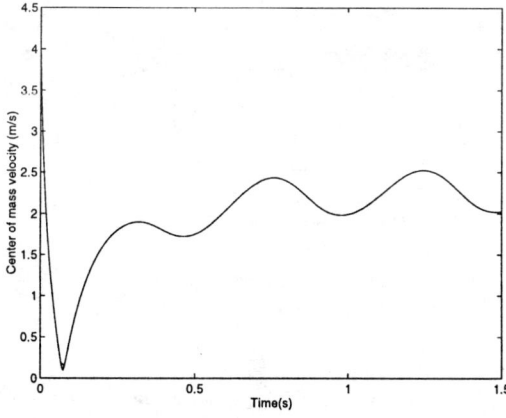

Figure 6: Plot illustrating stopping using the reverse gait

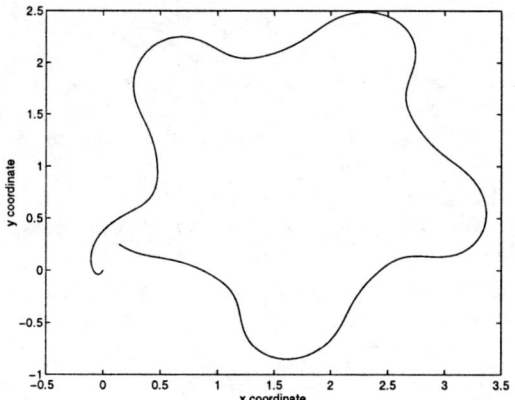

Figure 7: Turning in place for 5 seconds

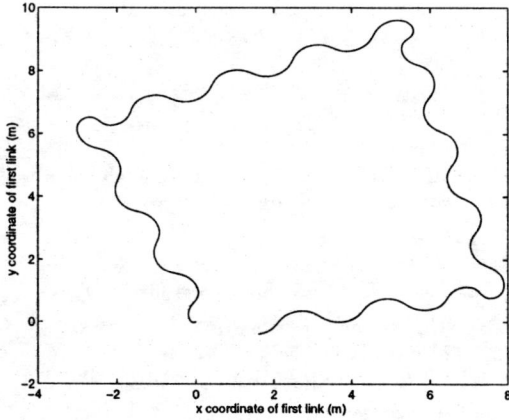

Figure 8: The coiling gait used to travel in a square path

4 The REEL Robot

To test the gaits described in this paper, and to provide a platform for future development of control algorithms for undulatory locomotion, we designed and built the REEL robot (Figure 9). The REEL robot consists of four rigid links (aluminum plates) with servo-motors as the joint actuators. The set of joint angles is transmitted to the robot by radio control. The robot is untethered and contained in a waterproof casing.

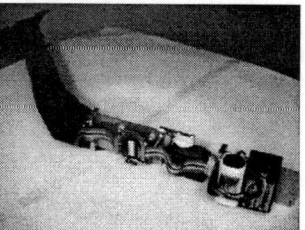

Figure 9: The REEL eel and its skin (black tubing)

Using the REEL robot we have verified the forward, backward, and turning in place gaits. We have also demonstrated stopping using the reverse gait. We have not (as yet) made a successful test of the coiling gait due to technical factors. At present the robot has no position sensing capability or telemetry, so we have only been able to qualitatively observe the robot's behaviour.

Testing the robot on land has been more difficult because the robot (lacking the scales of a land snake) is resisted by equal frictional forces in both the parallel and perpendicular direction. We have, however, been able to move the snake in an inchworm gait, shown in the left half of Figure 9. The inchworm gait is the same basic gait as the forward/backward swimming gait with the robot turned on its side, except that in this mode of locomotion, a *positive* phase shift ϕ_s leads to forward motion instead of the backward motion seen in the water.

5 Discussion

The gaits proposed in this paper are biologically and experimentally motivated. We also use a perturbation analysis (similar to those done in [3, 13]) to understand the factors contributing to the locomotion. After developing a perturbation expansion based on Eq. 13, we examine the terms that most affect the *momentum*. Starting with Eq. 6, we express the reduced Lagrangian, ℓ as:

$$\ell = \begin{pmatrix} \xi^T & \dot{r}^T \end{pmatrix} \begin{pmatrix} I & IA \\ A^T I^T & m \end{pmatrix} \begin{pmatrix} \xi \\ \dot{r} \end{pmatrix}, \quad (11)$$

where I, A, and m are matrix functions of the shape vector r, *only*. Based on this definition for ℓ, we can use the definition of the momenta $p_\alpha = \frac{\partial \ell}{\partial \xi^\alpha}$ to get an expression for ξ in terms of the momenta and shape variables:

$$\xi = I(r)^{-1} p - A(r) \dot{r}. \quad (12)$$

We then substitute this expression for ξ into Eq. 6 to get the momentum equation (see Section 2.1); a non-linear first order differential equation for $\dot{p}_\alpha$ in terms of p_α, r and $\dot{r}$ only:

$$\dot{p}_\alpha = \sigma_\alpha^{\beta\delta}(r) p_\beta p_\delta + \alpha_{\alpha i}^\beta(r) p_\beta \dot{r}^i + \gamma_{\alpha ij}(r) \dot{r}^i \dot{r}^j. \quad (13)$$

In this equation, we have expanded $ad_\xi^* p$ into quadratic expressions of p and $\dot{r}$ and then collected all terms. This equation lends itself to a perturbation analysis, by assuming an expansion in r, $\dot{r}$, and p which models small amplitude periodic shape inputs. In other words, let

$$\begin{aligned} r^i(t) &= r_0^i + \epsilon r_1^i, \quad \dot{r}^i = \epsilon \dot{r}_1^i \\ p_\alpha &= p_{\alpha 0} + \epsilon p_{\alpha 1} + \epsilon^2 p_{\alpha 2} + ..., \end{aligned} \quad (14)$$

where r_0 is constant and all others are functions of t. We then substitute these expressions for r and p into Eq. 13 and equate powers of ϵ to determine the response of the momentum to small amplitude shape inputs. For the body initially at rest we find immediately that $p_{\alpha 0} \equiv 0$.

In the case of the fluid drag model, the forces are proportional to the square of the link velocities (quadratic in p and $\dot{r}$). Because of this, we also find that $p_{\alpha 1} \equiv 0$, and that the equation governing $p_{\alpha 2}(t)$ is given by:

$$\dot{p}_{\alpha 2}(t) = \gamma_{\alpha ij}(r_0) \dot{r}^i(t) \dot{r}^j(t). \quad (15)$$

Eq. 15 suggests that the initial changes in momentum due to periodic inputs will be quadratic in the amplitude of the inputs, as shown in Figure 4. For the viscous friction case, we find that a term linear in p and $\dot{r}$ appears in the momentum equation and $p_{\alpha 1}$ becomes:

$$\dot{p}_{\alpha 1}(t) = \delta_\alpha^\beta(r_0) p_{\beta 1}(t) + \zeta_{\alpha i}(r_0) \dot{r}^i(t).$$

6 Future Work

There are a number of topics of interest that remain to be studied. We will develop a path-planning algorithm using the forward propulsion and coiling gaits to steer the snake from one point to another. We will investigate the dynamics from an optimal control perspective, to verify that the serpenoid curve is the lowest energy gait for an organism (or robot) of this design propelled by undulatory motion. We are also developing a continuum model of the snake, similar to the method of Chirikjian and Burdick [5], to compare the dynamics of the discrete system with models for the infinite dimensional case. Finally, we chose an approximate viscous friction model over dry friction in this work because discontinuous forces do not lend themselves to our geometric approach or to a perturbation analysis. Future work will examine dry friction as a special case.

On the experimental front, we plan to give the robot further functionality. We will add the capability to move in the third dimension, either using an elevator fin, or by adding a degree of freedom to the joints. We also intend to provide the snake with the ability to extract position and orientation with respect to an inertial frame using visual feedback. This will aid us in designing a more complete controller, capable of using position feedback to steer the robot from point to point.

References

[1] J. Ayers and J. Crisman. The lobster as a model for an omnidirectional robotic ambulation control architecture. In R. Beer, R. Ritzman, and T. McKenna, editors, *Biological Neural Networks in Invertebrate Neuroethology and Robots*, pages 287–316. Academic Press, 1992.

[2] A. M. Bloch, P. S. Krishnaprasad, J. E. Marsden, and R. M. Murray. Nonholonomic mechanical systems with symmetry. *Archive for Rational Mechanics and Analysis*, 136:21–99, December 1996.

[3] F. Bullo, N. E. Leonard, and A. D. Lewis. Controllability and motion algorithms for underactuated Lagrangian systems on lie groups. Submitted to IEEE *Transactions on Automatic Control*, February 1998.

[4] J. W. Burdick, J. Radford, and G. S. Chirikjian. A 'sidewinding' locomotion gait for hyper-redundant robots. *Advanced Robotics*, 9(3):195–216, 1995.

[5] G. S. Chirikjian and J. W. Burdick. The kinematics of hyper-redundant locomotion. *IEEE Trans. on Robotics and Automation*, 11(6):781–793, 1995.

[6] K. Harper, M. Berkemeier, and S. Grace. Decreasing energy costs of swimming robots through passive elastic elements. In *Proc. IEEE Int. Conf. Robotics and Automation*, pages 1839–1844, Albuquerque, NM, April 1997.

[7] S. Hirose. *Biologically Inspired Robots: Snake-like Locomotors and Manipulators*. Oxford University Press, Oxford, 1993. Translated by Peter Cave and Charles Goulden.

[8] IS Robotics Research Division. Autonomous legged underwater vehicle. Information available electronically at http://www.isrobotics.com/research/ariel.html, 1999.

[9] S. D. Kelly, R. J. Mason, C. T. Anhalt, R. M. Murray, and J. W. Burdick. Modelling and experimental investigation of carangiform locomotion for control. In *Proc. of the American Control Conference (ACC)*, 1998. (submitted).

[10] S. D. Kelly and R. M. Murray. Geometric phases and locomotion. *J. Robotic Systems*, 12(6):417–431, June 1995.

[11] Örjan Ekeberg. A combined neuronal and mechanical model of fish swimming. *Biological Cybernetics*, 69:363–374, 1993.

[12] J. P. Ostrowski. *The Mechanics and Control of Undulatory Robotic Locomotion*. Ph.D. thesis, California Institute of Technology, Pasadena, CA, 1995. Available electronically at http://www.cis.upenn.edu/~jpo/papers.html.

[13] J. P. Ostrowski. Steering for a class of dynamic nonholonomic systems. Submitted to IEEE *Transactions on Automatic Control*, February 1998.

[14] J. P. Ostrowski and J. W. Burdick. The geometric mechanics of undulatory robotic locomotion. *International Journal of Robotics Research*, 17(7):683–702, July 1998.

[15] J. P. Ostrowski and J. W. Burdick. Gait kinematics for a serpentine robot. In *Proc. IEEE Int. Conf. on Robotics and Automation*, pages 1294–9, April 1996.

[16] M. S. Triantafyllou and G. S. Triantafyllou. An efficient swimming machine. *Scientific American*, pages 64–70, March 1995.

[17] L. L. Whitcomb and D. R. Yoerger. A new distributed real-time control system for the JASON underwater robot. In *Proc. IEEE Intl. Workshop on Intelligent Robots and Systems (IROS)*, Yokohama, July 1993.

Continuum Robots - A State of the Art

G. Robinson, J.B.C. Davies

Department of Mechanical & Chemical Engineering,
Heriot-Watt University, Edinburgh, Scotland, UK

Abstract

Like the human limbs which inspired them most robots are discrete mechanisms with rigid links connected by single degree of freedom joints. In contrast 'continuum' and 'serpentine' robot mechanisms move by bending through a series of continuous arcs producing motion which resembles that of biological tentacles or snakes. This paper provides a single reference to the expanding technology of continuum robot mechanisms. It defines the fundamental difference between discrete, serpentine and continuum robot devices; presents the 'state of the art' of continuum robots and outlines their areas of application; and introduces some control issues. Finally some conclusions regarding the continued development of these devices are made.

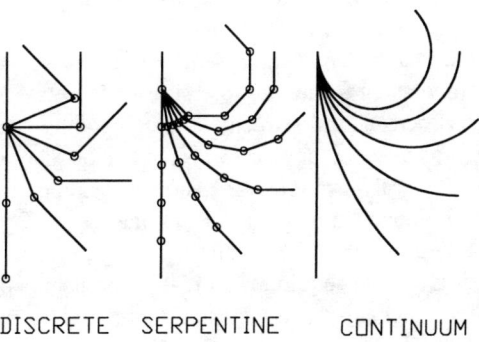

Fig. 1 Robot Motion

1 Introduction

Most robots are discrete mechanisms constructed from a series of rigid links. These are connected by discrete single degree of freedom joints, and controlled movement is only generated at these joints. The intermediate links may be considered infinitely stiff and essentially exist to ensure a known relationship between the centres of consecutive joints. The need for stiff links creates heavy robot mechanisms, large sections of which are passive supporting structures. Although this approach might be essential for many industrial applications where speed of operation and accuracy are paramount, there are many situations where different attributes may provide improved performance.

Serpentine robots also utilise discrete joints but combine very short rigid links with a large density of joints. This creates highly mobile mechanisms which appear to produce smooth curves, similar to a snake.

Continuum robots do not contain rigid links and identifiable rotational joints. Instead the structures bend continuously along their length via elastic deformation and produce motion through the generation of smooth curves, similar to the tentacles or tongues of the animal kingdom [1].

It should be clear that there are fundamental differences in form between conventional discrete, serpentine, and continuum robots (fig. 1). This paper only considers continuum robots.

2 Continuum Robots

Several forms of continuum robot have been developed. These can be broadly classified as 'intrinsic', 'extrinsic', or 'hybrid', according to the method and location of mechanical actuation. In an intrinsic device the actuators are located on and form part of the animated mechanism. Extrinsic devices use remote actuation and motion is transferred into the mechanism via a mechanical linkage. Hybrid devices use a combination of both intrinsic and extrinsic actuation.

Each of these groups can be sub-divided into 'planar' or 'spatial' devices according to the type of motion produced. Planar devices only move in a single plane of bending, whilst spatial devices can bend in any direction perpendicular to their longitudinal axis.

2.1 Intrinsic Planar

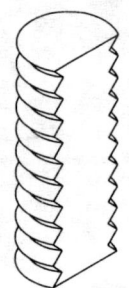

Fig. 2 Intrinsic Planar Actuator

The simplest form of continuum actuators are fluid operated planar devices (fig. 2), which require a single pressure input to produce bending in a single plane. The physical construction of the actuator walls determines the motion produced. Typically one side of the elastic actuator has a different axial stiffness to the remaining walls. Increasing internal pressure generates strain variations across the structure and bending occurs. Elasticity causes the device to straighten on decreasing pressure. Several variations exist and a typical application is for compliant fingers, increasing the versatility and error tolerance of grasping end-effectors, whilst minimising control requirements [2-5]. The use of planar continuum mechanisms to transport simple objects within a plane has also been investigated [6-7].

2.2 Intrinsic Spatial

The basic construction for intrinsic, fluid operated, spatial continuum mechanisms combines both the actuator and the supporting structure into a single unit (fig. 3) [8]. This produces a simple, compact, lightweight mechanism which contains no moving parts and yet can generate motion with up to three degrees of freedom, allowing both the direction and magnitude of tip movement to be controlled. Operation relies on the elastic deformation of parallel actuator chambers placed at equal intervals about a central longitudinal axis. Internal pressures are controlled to generate extension forces and the structure deforms according to constraints provided by the end forms (fig. 4). It is essential for operation that internal pressures generate axial extension rather than radial expansion and several approaches have been developed to ensure anisotropic elasticity. Devices with many degrees of freedom can be obtained by placing several independent actuators in series.

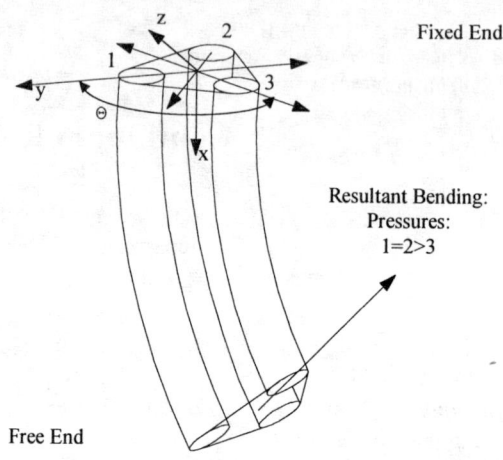

Fig. 4 Actuator Motion

These actuators exhibit passive elastic compliance to external forces in the directions normal to their longitudinal axis. External forces cause compliant motion of the actuator until the force balance within the structure is restored. The resulting increase in strain energy causes the actuator to return to its undisturbed configuration on removal of the external force.

Suzumori et al. described the construction and operation of small low pressure silicon rubber actuators (flexible micro-actuator or FMA) ranging from 1 - 20 mm in diameter [9]. Three parallel chambers were moulded directly into a single cylindrical unit. Integral nylon reinforcing fibres around the circumference of the cylinder resisted radial expansion whilst allowing the actuator to stretch longitudinally. Single stage devices were used as the fingers of a four fingered gripper, whilst a 120 mm long 7 degree of freedom manipulator (including a 1 degree of freedom gripper) was produced using multiple stages. Subsequently FMAs have been used in the construction of mini 6 legged walking robots only 15 mm long [10]. Also stereolithographic techniques have enabled arrays of parallel FMA actuators to be made in a single process. The objective is to manoeuvre objects across the surface of a large FMA array by the distributed actuation of the parallel actuators, in a similar manner to the villi of animal intestines [11].

Davies described a much larger 744 mm long pneumatic device using three parallel thin walled polymeric bellows connected at each end by rigid plates

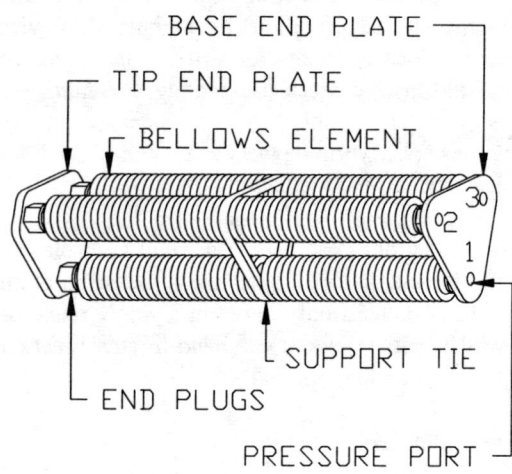

Fig. 3 Spatial Bellows Actuator

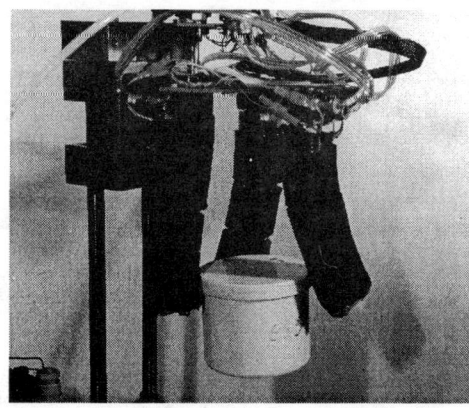

Fig. 5 Polymeric Gripper

[12-13]. The convoluted bellows walls ensured that longitudinal extension due to internal pressure was much larger than radial expansion. This approach also produced a pneumatic gripper which exploited its intrinsic compliance to perform stable open loop manipulations on a range of objects [14] (fig. 5). Compliant finger motion both reduces the risk of damage to grasped objects or fixtures due to excessive contact force, and increases grasp stability by absorbing external disturbances and positional errors which might otherwise lead to grasp failure [15].

Subsequently the Amadeus project is developing continuum end-effector technology and control techniques to improve the sampling and manipulation capabilities of underwater systems [16-17]. Fig. 6 shows the first hydraulic laboratory prototype which used phosphor bronze bellows fingers [18-20]. Successful laboratory operation of this device ensured progression to

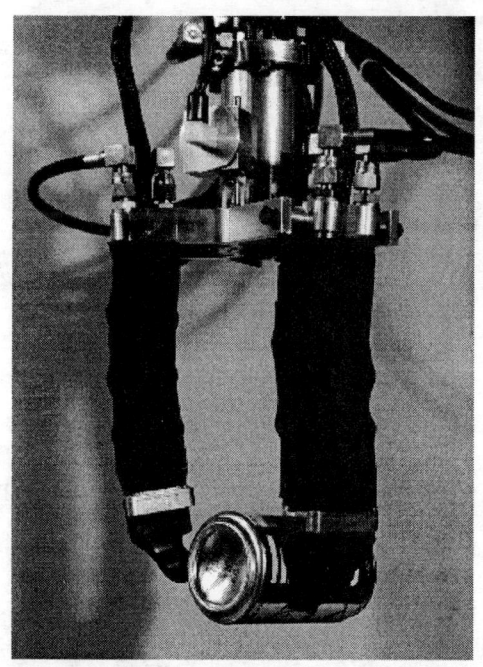

Fig. 6 Amadeus Phase 1 Prototype

the second phase of the project. This is developing a robust system providing enhanced functionality by embodying the benefits of continuum technology, and which is suitable for trials in the marine environment. Robinson et al. discuss the stringent design and operational constraints, and the technical challenges associated with the development of a dextrous robot end-effector for operation in harsh unstructured environments [21].

As the research effort into intrinsic continuum devices increases new methods of actuation are being proposed.

Kallio et al. described the control of a micro-manipulator using a similar parallel bellows construction [22]. Three small nickel bellows support and control the orientation of a platform, which in turn moves a micro-manipulator arm. An innovative system of piezohydraulic actuation operates the bellows producing a system with sub-micrometer resolution in a workspace several hundred micrometers across. The continuum approach was ideal for this application because the bellows are able to both extend and bend, enabling a compact structure to be produced by removing the need for spherical joints.

Ivanescu and Stoin suggested using electrorheological fluids to control fluid filled devices [23]. These fluids change viscosity in the presence of DC electric fields, and hence the provision of suitable electrodes would enable continuum devices to be stiffened once a desired configuration had been reached.

Kornbluh et al. described the use of Electrostrictive Polymer Artificial Muscle (EPAM) actuators in robotic applications and suggested the construction of a multi-sectioned redundant continuum arm [24]. EPAM actuators contract when a voltage is applied across a polymer dielectric and may provide a lightweight compact actuator. Since force is produced by contraction additional supporting elements are required to prevent collapse of the overall continuum structure. Obviously fluid powered artificial muscles such as the McKibben muscle [25] could also be used to construct similar devices.

Shape memory alloys lend themselves to the production of small actuators and applications where low efficiency and slow response are unimportant, and several actively controlled continuum endoscopes have been described for medical intervention [26-27].

2.3 Extrinsic

Fig. 7 Extrinsic Actuator

One technique to construct lightweight continuum devices with many degrees of freedom is to utilise remote actuation, and transfer motion into the structure via groups of antagonistic tendons (fig. 7). Independent tendon pairs or triads control the orientation of rigid plates placed at intervals along the structure. Springs connect successive plates and ensure that the tendons remain in tension. Spatial continuum motion occurs since the springs bend according to sequential plate orientations. Various constructions have been described with different numbers of sections, tendon arrangements, and degrees of freedom [26, 28]. These devices are hollow and suggested applications include the transportation of liquids or bulk materials.

2.4 Hybrid

Essentially hybrid structures follow a similar approach to extrinsic tendon devices, except that the passive springs are replaced by actively controlled bellows (fig. 8). Hemami proposed a potential construction for a redundant hybrid manipulator with many two degree of freedom sections [29]. More recently the commercially available KSI tentacle manipulator has been developed [30]. A central pneumatic bellows is operated in conjunction with two sets of extrinsic tendon triads. One triad inserts halfway along the bellows structure and controls the shape of the proximal half. The other inserts at the free end and controls the form of the distal portion of the device. The internal bellows pressure opposes the operation of the tendons, ensuring they always remain in tension. By varying the tendon lengths and the bellows pressure both the length and stiffness of the structure can be controlled.

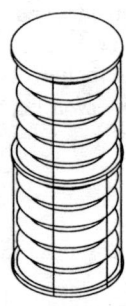

Fig. 8 Hybrid Actuator

3 Applications

Current developments focus on three main areas of application; grasping and manipulation; locomotion; and positioning.

3.1 Grasping and Manipulation

The fluid operated continuum actuators exhibit a useful range properties which ensure their application as the fingers of robot end-effectors. Inherent structural compliance increases grasp stability and minimises contact damage, whilst the absence of moving parts eliminates problems of stiction and reduces the risk of environmental contamination. The devices may also be easily cleaned for applications involving hazardous materials, food or medical production.

Both planar and spatial designs are available depending upon the level of dexterity required. Planar fingers only require a single control input to produce motion and hence the range of available grasps is limited. Spatial fingers require three input pressures per finger but are capable of omni-directional movement. This significantly increases the dexterity of devices using these fingers allowing asymmetric grasps on irregularly shaped objects, and object manipulation (translation and rotation) to be performed solely through finger actuation.

3.2 Locomotion

Spatial fluid actuators have also been investigated as the legs of walking robots. However the payload of such devices is limited by the buckling capacity of the legs and until a compact lightweight actuation system is developed the range and functionality of these robots will be limited by the need to provide services via an umbilical link [8, 10].

The converse application using a static array of parallel actuators for object transportation requires the development of manufacturing techniques, miniature valve technology and distributed control methods [11].

3.3 Positioning

All three forms of continuum robot have been investigated for positioning applications (fig. 9). The simple structures enable lightweight mechanisms with long reaches to be considered for operations within constrained environments with restricted access, such as storage tanks or nuclear cells.

However any mechanism utilising an elastic structure exhibits compliance and will deflect when subject to a disturbing force. This could be due to a collision with an

Fig. 9 Two Stage Continuum Robot

external constraint, the payload, or the self-weight of the device. Thus long continuum mechanisms are suited to positioning tasks requiring relatively low positional accuracy with small payloads, such as inspection, cleaning, or bulk transport.

This structural compliance can also be exploited when working within constrained, unstructured or delicate environments, since devices will deflect rather than generate large contact forces during collisions with its environment.

4 Control Issues

Continuum devices present a novel control problem in that the entire structure undergoes elastic deformation and there are no joints to control or measure. Methods of direct measurement of the device end-point are required for reliable position control. Indirect position models which relate internal bellows pressures to deflected position via physical parameters have been proposed for single stage fluid operated devices [31-32]. Similarly extrinsic tendon operated devices might utilise position models based on tendon tension and length to infer position. The micro-manipulator system used proximity sensors to measure the orientation of the supporting plate and hence obtain the micro-manipulator position [22].

However low accuracy tasks utilising tele-operation may exploit device compliance to absorb positional and contact errors and provide an alternative solution to rigid systems requiring accurate position control.

Conclusions

Continuum robots posses fundamental characteristics which distinguish them from either discrete or serpentine robots.

A number of different construction methodologies have been developed based on intrinsic, extrinsic or hybrid actuation, providing either planar or spatial functionality.

The actuators display a range of useful properties including compliant motion and simple construction.

Used appropriately continuum devices may enhance the performance of other robot systems or enable new and challenging applications to be investigated. Currently described applications include subsea manipulation; micro-manipulation; inspection; jet-washing; walking robots; and endoscopic intervention.

The development of continuum robot technology continues to present many technical challenges and research opportunities.

Acknowledgements

Amadeus phase 2 is supported by the European Community under the Marine Science and Technology (MAST III) research directive grant number: mas3-ct95-0024. The authors would like to acknowledge collaboration with their colleagues in the Ocean System Laboratory, Department of Computing and Electrical Engineering, Heriot-Watt University, Edinburgh; the Departmento Informatica Sistemistica Telematica, University of Genova, Genova, Italy; and the Instituto Per L'Automazione Navale, National Research Centre, Genova, Italy.

References

[1] W. Kier, K. Smith, "Tongues, Tentacles and Trunks; the Biomechanics of Movement in Muscular Hydrostats", Zoological Journal of the Linnean Society, vol. 83, pp. 307 - 324, 1985.

[2] R.L. Orndorff, "Gripping Device", US Patent No. 3601442, 1971.

[3] P. Andorf, D. Franz, A. Lieb, G. Upper, W. Guttropf, "Robot Finger", US Patent No. 3981528, 1976.

[4] U. Tsach, R. Melamed, T.J. Garrison, "Development of a Versatile Multiple Prehension,

Compliance Controlled Robotic End-Effector", ASME Advances in Design Automation, 13th Conference, Boston, MA, Sept. pp 25 - 29, 1987.

[5] K. Khodabandehloo, "Robot Handling and Packaging of Poultry Products", Robotica, vol. 8, pp. 285-297, 1990.

[6] J.F. Wilson, U. Mahajan, "The Mechanics and Positioning of Highly Flexible Manipulator Limbs", Transactions of the ASME Journal Of Mechanisms, Transmissions, and Automation in Design. vol 111, pp 232 - 237, 1989.

[7] D.C. Nemir, "Preliminary Results of the Design of a Robotic Tentacle End-Effector", Proc. 1989 American Control Conference, Pittsburgh, vol. 3, pp. 2374 - 2376. 1989

[8] G. Robinson, J.B.C. Davies, "The Parallel Bellows Actuator", Robotica '98, Brasov, Romania, Nov. 5 - 7, pp. 195 - 200, 1998.

[9] K. Suzumori, S. Iikura, H. Tanaka, "Development of Flexible Micro-Actuator and its Application to Robot Mechanisms", IEEE Conference on Robotics and Automation., Sacremento CA, April, 1991.

[10] K. Suzumori, F. Kondo, H. Tanaka, "Micro-Walking Robot Driven by Flexible Microactuator", Journal of Robotics and Mechatronics, vol. 5, no 6, pp. 537 - 541, 1993.

[11] K. Suzumori, K. Akihiro, F. Kondo, R. Haneda, "Integrated Flexible Micro-Actuator Systems", Robotica, vol 14, pp 493 - 498, 1996.

[12] J.B.C. Davies "An Alternative Robotic Proboscis", Proc NATO Advanced Research Workshop on Traditional and Non-Traditional Robots, Maratea, Italy, Aug 28 - Sep 2, pp 49-55, 1989.

[13] J.B.C. Davies, "Elephants Trunks an Unforgettable Alternative to Rigid Mechanics", Industrial Robot, vol 28, pp29-30, 1991.

[14] J.B.C. Davies, "Dextrous Manipulator for Complex Objects", 5th World Conference on Robotics Research, Sept 27-29, Cambridge, Massachusetts, pp 17-15 - 17-28, 1994.

[15] K.B. Shigoma "Robot Grasp Synthesis Algorithms: A Survey", The International Journal of Robotics Research, vol. 15, no 3, pp 230 - 266, 1996.

[16] D.M. Lane, J. Sneddon, D.J. O'Brien, J.B.C. Davies, G.C. Robinson, "Aspects of the Design and Development of a Subsea Dextrous Grasping System", IEEE OCEANS '94, Brest, France, Sept 13-16, vol II, pp 174-181, 1994.

[17] G. Robinson, J.B.C. Davies, "The Amadeus Project: An Overview", Industrial Robot, vol 24, no 4, pp 290-296, 1997

[18] D.M.Lane, M. Pickett, "Supervisory Planning for Blind Grasping for the Amadeus Gripper", International Journal of Systems Science, vol. 29, no. 5, pp. 513 - 527, 1998.

[19] D.O'Brien, D.M. Lane, "Force and explicit slip sensing for the Amadeus underwater gripper", The International Journal of Systems Science, vol. 29, no. 5, pp. 471 - 483, 1998.

[20] G. Robinson, J.B.C. Davies, E. Seaton, "Mechanical Design, Operation and Direction Prediction of the Amadeus Gripper System", The International Journal of Systems Science, vol. 29, no. 5, pp. 455 - 470, 1998.

[21] G. Robinson, J.B.C. Davies, J.P.P. Jones Development of the Amadeus Dextrous Robot End-Effectors, IEEE Oceans '98, Nice, France, Sept. 29 - Oct 1. vol. 2, pp. 703 - 707, 1998.

[22] P. Kallio, M. Lind, Q. Zhou, H.N. Koivo, "A 3DOF Piezohydraulic Parallel Manipulator", IEEE International Conference Robotics and Automation, Belgium, pp 1823 - 1828, 1998.

[23] R. Kornbluh, R. Pelrine, J. Eckerie, J. Joseph, "Electrostrictive Polymer Artificial Muscle Actuators", IEEE International Conference Robotics and Automation, Belgium, pp. 2147 - 2154, 1998.

[24] B. Tondu, P. Lopez, "The McKibben Muscle and its use in Actuating Robot-Arms Showing Similarities with Human Arm Behaviour", Industrial Robot, vol. 24, no. 6. pp. 432 - 439, 1997.

[25] S. Hirose, "Biologically Inspired Robots", Oxford University Press, 1993.

[26] G. Lim, K. Minami, K. Yamamoto, M. Sugihara, M. Uchiyama, M. Esashi, "Multi-link active catheter snake-like motion", Robotica, vol 14, pp 499 - 506, 1996.

[27] R. Cieslak, A. Moreki, "Technical Aspects of Design and Control of Elastic's Manipulator of the Elephants Trunk Type", 5th World Conference on Robotics Research, Sept 27-29, Cambridge, Massachusetts, pp 13-51 - 13-64, 1994.

[28] A. Hemami, "Design of Light Weight Flexible Robot Arm", Robots 8 Conference Proceedings, Detroit, June 4-7, pp 16-23 16-40, 1984.

[29] G. Immega, K. Antonelli, "The KSI Tentacle Manipulator", IEEE International Conference Robotics and Automation, Japan, pp 3149 - 3154, 1995.

[30] M. Ivanescu, V. Stoian, "A Variable Structure Controller for a Tentacle Manipulator", IEEE International Conference Robotics and Automation, Japan, pp 3155 - 3160, 1995.

[31] J.B.C. Davies, "A Flexible Motion Generator", PhD Thesis, Heriot-Watt University, Edinburgh, 1996.

[32] V. Arrichiello, G. Bartolini, M. Coccoli, "Modelling and Control of New Elements of the Amadeus Gripper Designed for Fast Manipulative Tasks", IEEE Oceans '98, Nice, France, vol. 2, pp. 708 - 712, 1998.

Proceedings of the 1999 IEEE
International Conference on Robotics & Automation
Detroit, Michigan • May 1999

NEW SENSORS FOR NEW APPLICATIONS: FORCE SENSORS FOR HUMAN/ROBOT INTERACTION

William Andrew Lorenz
Michael A. Peshkin
J. Edward Colgate

Department of Mechanical Engineering
Northwestern University
Evanston. IL 60208-3111

Abstract

Conventional force sensors are overdesigned for use in measuring human force inputs, such as is needed in research and application of human/robot interaction. A new type of force sensor is introduced that is suited to human-robot interaction. This sensor is based on optoelectronic measurements rather than strain gauges. Criteria for material selection and dimensioning are given, and results for linearity, noise, and drift are reported.

1. Introduction

Human-robot interaction is an increasingly important field in robotics. Often this interaction is mediated by a force sensor which is the primary control input for the human. Existing force sensor designs are intended for use at the end effector of a robot to monitor assembly or machining forces. In these application very high stiffness may be needed. Commercial multi-axis force sensors typically measure all six axes (forces and torques).

These requirements are unnecessarily stringent for force sensors to be used in human-robot interaction, and cause force sensors to be so expensive as to restrict the development of human/robot applications.

Since the compliance of the human hand and arm are so great, there is no need for the force sensor itself to be orders of magnitude stiffer. Using optoelectronic devices in place of strain gauges allows larger compliances. The force sensor presented here has far lower stiffness than commercial sensors, allowing approximately 0.60 millimeters of deflection at its full scale applied force of 170N.

Other force sensor designs using optoelectronic devices exist. ([8]) The larger compliance allowed by such sensors leads to larger tolerances in the dimensions of the design, easing construction and assembly of the device.

The force sensor described here is a two axis device designed to measure x-y forces in the plane and to ignore the other four forces and torques. This further simplifies the design, as well as offering benefits for certain applications.

The force sensor can be utilized in human-robot coordination, ([3],[4]) teleoperation ([5]), and collaborative robots ([1],[2]).

2. Benefits of new design

Human-robot interaction requires different properties of a force sensor than typical robot applications such as machining and assembly. These differences have substantial impact on how a force sensor can be designed.

2.1 Larger compliance

Humans are relatively insensitive to small displacements on the order of a millimeter. These same displacements would cause trouble if they occurred at the end effector of a robot, which can be thought of as needing (or benefiting from) the accuracy of a machine tool such as a mill. So what advantages are obtained by loosening the restrictions on how much displacement is allowed?

Presently, strain gauges are used in most commercially available force sensors. Strain gauges measure the very slight bend of a flexure element caused by applied forces. However, they are difficult to install and calibrate and are easy to break. While the strain gauge elements themselves are inexpensive, in practice their difficulty of application results in sensors costing thousands of dollars.

In the design presented here, infrared LED/photodiode pairs are used as the key sensor element in place of strain gauges. Strain gauges can measure deflections on the order of microns, and so are applied to very stiff flexure elements. Photosensors can be used to measure deflections on the order of millimeters, and thus are applied to much more compliant flexures. Photosensors suffer from some of the same problems as strain gauges, including nonlinearity and temperature sensitivity. However, these sensors have distinct advantages. Photosensors are cheap, easy to mount, are a non-contact sensor, and are relatively hard to break.

In this design, due to properties of the photosensors and the flexure, the axes are intrinsically decoupled. Separate sensor elements are used to measure the x and y force components. Since the axes are decoupled, less electronics is needed, and calibration issues are much simpler.

2.2 Fewer degrees of freedom

In an instance where a human controls a robot with fewer than six degrees of freedom, the force sensor need not have six degrees of freedom. Fewer degrees of freedom means a simpler mechanical design, less electronics and wires, fewer sensor elements, and less calibration.

One benefit of the design, which follows partially from having fewer degrees of freedom, is the resistance of the force sensor to large undesired force components.

For example, consider a person moving a heavy object suspended from an overhead rail system. Where does one put a force sensor to read the forces applied by the human? If one uses a handle, then the person must grab the device by the handle, even if it might easier for the person to grasp the object directly. If one puts the force sensor between the object and the rail, then manipulating the object directly is allowed, but the force sensor must withstand the weight of the object. The force sensor presented here would be ideal for this problem, as it can withstand large forces out of the plane to be measured.

3. Design

3.1 Overview

The force sensor is shown in figure 1. The outer piece is the housing, while the handle is connected to the inner piece. Connecting the outer and inner pieces is a flexure element, best seen in the upper part of figure 2. The flexure itself is shown in the lower part of figure 2. As forces are applied to the handle, the flexure allows a displacement to occur between the two pieces. Due to the 30:1 (typical) aspect ratio of the flexure, it does not bend significantly in response to forces in the z-direction or torques about the x and y axes.

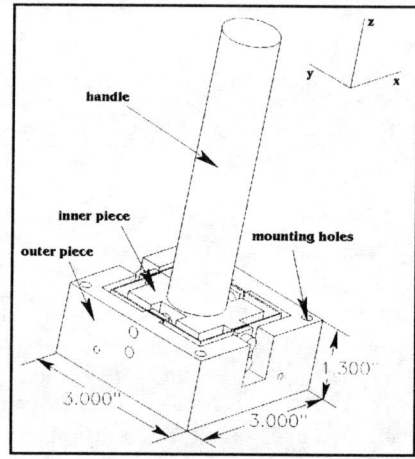

Figure 1

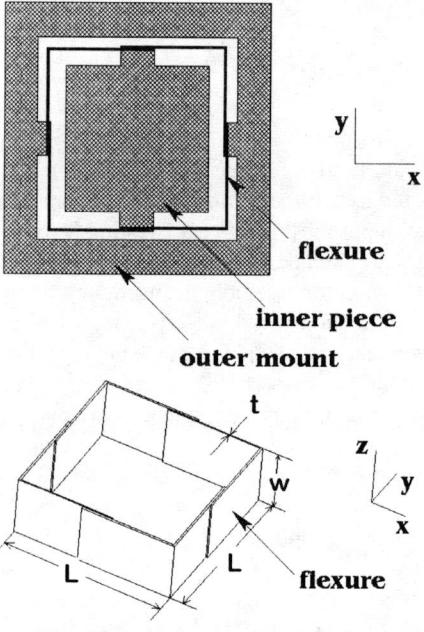

Figure 2

The flexure is designed to withstand one millimeter of motion between the inner and outer pieces, at which point the two pieces physically make contact preventing further motion. This protects the flexure from being broken.

The displacement of the inner piece is measured using what is equivalent to infrared "reflective object" sensors. These sensors are mounted on a printed circuit board which is attached to the inner piece. Light from an LED reflects off the inner wall of the outer piece and is detected by a photodiode, as shown in figure 3.

For each axis, there are two photosensors measuring the distance to the two walls of the outer object. The photosensors labeled A and B in figure 3 for example, are for measuring the x direction. These correspond to the photodiodes labeled A and B in figure 4.

Each LED actually shines on two photodiodes. Each photodiode produces a current proportional to the amount of light it receives. One photodiode, the reference, does not move with respect to LED, and is used to regulate the output of the LED. This significantly reduces the drift and temperature sensitivity of the circuit. This is the photodiode shown in the top part of figure 4.

The other photodiode receives light from the LED off the outer wall, as in figure 3. There are two of these photodiodes per axis. They are shown in the circuit in the bottom half of figure 4. These photodiodes measure the distance to the wall using the constant LED output. The difference in the amount of light received by these diodes causes a current, resulting in the voltage output. The output voltage is approximately linearly related to the displacement, for relatively small displacements. For the small deflections discussed here, the displacement is proportional to the force applied.

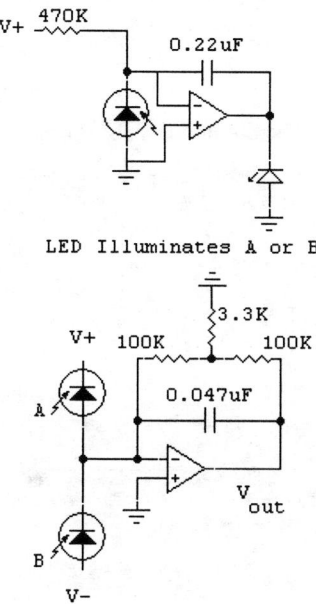

Figure 4

3.2 Compliance matrix

The shape of the square flexure can be used to determine its compliance matrix. The flexure can be considered to be constructed of four identical L's (See figure 2). The compliance matrix of these L's is determined, assuming small deflections and simple stress distributions within the cross-section of the beam, as in [7]. The compliance matrix for one L is given below.

$$\underline{A} = \begin{pmatrix} 2 & -\frac{3}{4} & 0 & 0 & 0 & -\frac{9}{4} \\ -\frac{3}{4} & \frac{1}{2} & 0 & 0 & 0 & \frac{3}{4} \\ 0 & 0 & p+\frac{3}{8}k & \frac{3}{4}p+\frac{3}{8}k & -\frac{3}{4}p & 0 \\ 0 & 0 & \frac{3}{4}p+\frac{3}{8}k & \frac{3}{2}p+\frac{3}{8}k & 0 & 0 \\ 0 & 0 & -\frac{3}{4}p & 0 & \frac{3}{2}p+\frac{3}{8}k & 0 \\ -\frac{9}{4} & \frac{3}{4} & 0 & 0 & 0 & 3 \end{pmatrix} \quad (1)$$

$$\Delta \vec{x} \equiv \begin{pmatrix} \Delta x \\ \Delta y \\ \Delta z \\ \Delta \theta_x \\ \Delta \theta_y \\ \Delta \theta_z \end{pmatrix} = \frac{L^3}{Et^3 w} \underline{A} \begin{pmatrix} f_x \\ f_y \\ f_z \\ \tau_x \\ \tau_y \\ \tau_z \end{pmatrix} \equiv \frac{L^3}{Et^3 w} \underline{A} \vec{f} \quad (2)$$

where $\Delta \vec{x}$ is the displacement, L is the length of one side of the square sheet, E, t, and w are the modulus of elasticity, thickness, and width of the material,

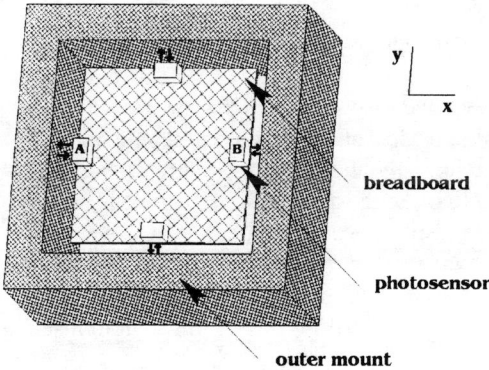

Figure 3

$p=t^2/h^2$, $k=\dfrac{E}{G}$ is the ratio of the modulus of elasticity and the modulus of rigidity of the material, and $\vec{f}$ is the applied force. These compliance matrices are translated and rotated so that they are positioned as in figure 2. Then the compliance matrix, C, is found by combining these four matrices by the following equation

$$\underline{C} = \left(\sum_i \underline{A}_i^{-1}\right)^{-1} \quad (3)$$

For the flexures described as above, we find that

$$\underline{C} = \begin{pmatrix} \frac{1}{20} & 0 & 0 & 0 & 0 & 0 \\ 0 & \frac{1}{20} & 0 & 0 & 0 & 0 \\ 0 & 0 & p\frac{k+p}{4(k+4p)} & 0 & 0 & 0 \\ 0 & 0 & 0 & p\frac{3(k+p)(k+4p)}{12k^2+80kp+40p^2} & 0 & 0 \\ 0 & 0 & 0 & 0 & p\frac{3(k+p)(k+4p)}{12k^2+80kp+40p^2} & 0 \\ 0 & 0 & 0 & 0 & 0 & \frac{3}{112} \end{pmatrix} \quad (4)$$

$$\Delta\vec{x} = \frac{L^3}{Et^3w}\underline{C}\vec{f}. \quad (5)$$

A number of insights can be gained from this compliance matrix. First, we notice that it is diagonal, which tells us that forces and torques create only their corresponding motions. This is clear in view of the symmetry of the part. It also tells us how the choice of our aspect ratio parameter p affects the design. If p is small, then the flexure moves significantly only in response to the forces f_x, f_y, and τ_z. This is how the present design is made. However, if p is large, then we have a flexure that responds to f_z, τ_x, and τ_y. Also we see how the dimensions of the flexure, L, t and w, matter.

3.3 Material selection

The flexure must be able to deflect a desired distance, x_d when the full scale force F is applied in the x (or y) direction. In addition, the flexure must not break at this deflection. The deflection as a function of force can be obtained from equation 5. It is

$$x_d \equiv \Delta x = \frac{1}{20}\frac{FL^3}{Et^3w} \quad (6)$$

where $F=f_x$ from equation 1. From simple mechanics of materials [7] type analysis, we find that the maximum moment M_{max}, is related to the applied force by

$$M_{max} = \frac{3}{40}FL \quad (7).$$

The moment of inertia, I, of the flexure when bent about the z-axis is

$$I = \frac{1}{12}wt^3 \quad (8)$$

We want the maximum stress to be a factor of safety less than the yield stress. The equation for the maximum stress is then

$$\sigma_{max} = \frac{M_{max}c}{I} = \frac{9}{20}\frac{FL}{wt^2} < \frac{\sigma_y}{F.S.} \quad (9)$$

where c =t/2 is the maximum distance from the normal axis of the flexure, σ_y is the yield stress, and F.S. is the factor of safety desired. From equations (6) and (9), we can find equations restricting the length L and thickness t of the material.

$$L < \frac{9}{\sqrt[3]{20}}\frac{(F.S.)x_d^{2/3}F^{1/3}}{w^{1/3}}\frac{E^{2/3}}{\sigma_y} \quad (10)$$

$$t = \sqrt[3]{\frac{F}{20x_dEw}}L \quad (11)$$

We want to be able to make a sensor as small as is possible. This implies minimizing L. Thus we should choose a material that will minimize $E^{2/3}/\sigma_y$, or will maximize

$$\frac{\sigma_y^{3/2}}{E}. \quad (12)$$

Certain materials score well by this criteria ([6]). One is high tensile strength steel, such as the "spring steel" we have used. Other high scores are nylon and certain rubbers. The rubbers are probably not reasonable, as the thickness, t, would have to be so large as to make the design unreasonable, but nylon or other plastics are possible alternatives. Spring steel has good fatigue properties (when a F.S. of 2 or greater is used), and this may be a problem with plastics. However, spring steel is difficult to machine and bend, as its hardness is similar to that of machine tools and it is rather brittle outside its elastic range.

For spring steel, F.S.=2, y_d=1 mm, w=1.9 cm, F=66.3 lb., we find L_{min}=4 cm=1.6 in. This is the value used in the prototype.

3.4 Sensor considerations

As mentioned above, infrared photosensors are used to detect displacement. The arrangement of the sensors in the design are shown in figure 3. One pair of sensors detects motions in the x direction, while the other pair of sensors detects motions in the y direction. The photosensors are reflective, and shine on a uniform reflective flat surface on the outer piece. The sensors detect when the wall comes closer or farther, and are immune to motions parallel with the wall.

This causes the measurements of the two pairs of sensors to be independent, and to measure only their own axis. In addition to this, it causes the sensors to be relatively insensitive to small torques about the z axis. Thus only large z torques that cause range problems for the device need to be worried about. (Due to the compliance matrix, other unwanted force components, f_z, τ_x, and τ_y, do not have this concern.)

It was found for the photosensors used, Panasonic photo diode PN334 and infrared LED LN175, that a decent trade-off between sensitivity and linearity of the sensor response could be obtained if the maximum displacement allowed by the flexure was ±1mm.

4. Three-dof force sensor design

Another design for a similar force sensor is shown in figure 5. In this design, there are three pairs of beams in series. An inner piece is attached at the 2 locations labeled B (One is hidden.). An outer piece is attached at the 4 locations labeled A. The heavily shaded beams are thick and do not bend significantly. The beams labeled α, β, and γ allow flexing in the x, y and z directions respectively. Each pair of beams allows only motion in one direction. All together this design allows motion only in x, y and z, and allows no twisting of the device. As before, photosensors could be used to measure displacements. There would then be 3 pairs of sensors instead of two.

This design is more complicated. In addition, it lacks some of the compactness of the first design, which can be relatively short in the z direction. However, this design is insensitive to all torques and can withstand large torques without breaking or affecting readings.

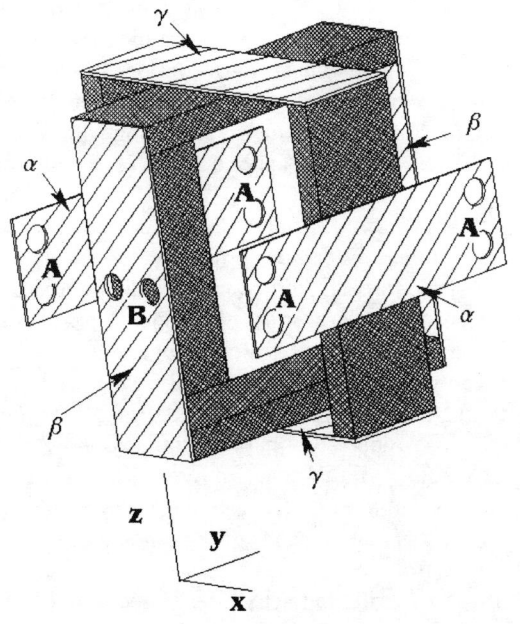

Figure 5

5. Experimental Results

The force sensor described above has been built and calibrated. It has been compared to an existing force sensor, the ATI Gamma model 6 axis force sensor. Results comparing these two sensors are summarized in the table below.

	ATI	Two-axis optical
Max force (Newtons)	65	170
Sensitivity, Volts/N	0.0734	0.0018
Long term drift, % of full scale in 24 hours	0.10%	1.6%
Short term drift, % of full scale in 5 minutes	0.025%	0.23%

The maximum deflection of the two-axis sensor was 0.60 mm (in either direction). This required a force of 170 N. This is very close to the force expected from the theory, which is 177N.

All drift terms above are in terms of peak to peak measurement. Short-term noise was significantly less than the drift terms given in the above table, and was approximately half a millivolt peak to peak. This is after both digital and analog filtering using a time constant of approximately 0.1 seconds. This time constant is reasonable for human interaction, but may be too large for other applications. In this case, the noise contribution would have to be reevaluated.

The five minute drift is the amount the measured force varies in five minutes for a constant applied force, and is much less than the long term drift. For applications involving humans interacting with the sensor, forces tend to be applied for short periods of time.

Forces (and torques) applied to the device in the z, τ_x, and τ_y directions caused no measurable change in the measured force. A torque in the τ_z direction did cause a measured force in the x-y plane. A torque of 0.5 Nm caused a recorded force change of 9.8 N. This seems tolerable for a hand-held device, where the characteristic length is on the order of 2 to 3 cm, but could be troublesome in other applications.

The response of the device to forces was linear for relatively small forces. (See figure 6.) The applied force here was determined from the amount of weight suspended from the device. Larger forces on the order of the maximum force of the device were not as linear. A force of 167 N, almost full scale, applied to the device caused a measured force of 156 N, a 6.7% error.

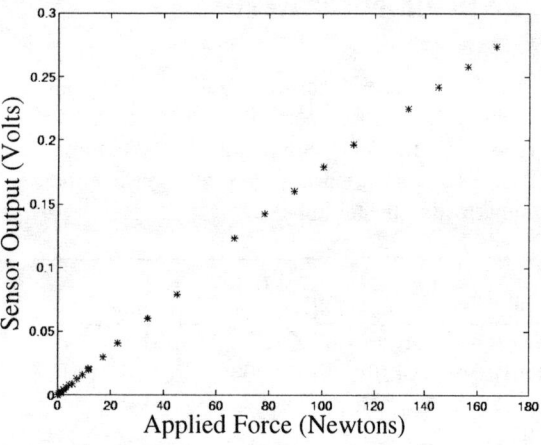

Figure 6

The ATI force sensor clearly outperforms the two-axis sensor, however in many applications its high performance may not be needed. We intend to further increase the sensitivity and reduce the noise and drift levels of the two-axis sensor.

6 Summary

A new force sensor design was described in this paper. It is much cheaper and easier to construct than existing commercial force sensors.

The reduced degrees of freedom (two or three in contrast to six) allows a different class of flexures to be used, two of which are mentioned above. These flexures have diagonal compliance matrices.

The greater allowable compliance in human/robot applications makes the force sensor easier to construct. In addition it allows the use of new types of sensors elements.

The force sensor described above allows measurement of forces in a plane. It is insensitive to other force components. These other force components do not significantly bend the flexure, thus they do not have to be measured and then cancelled. The designs can be modified so that these force components can be very large, and still not break or significantly bend the flexure.

The combination of this class of flexures and photosensors allows the axes to be orthogonal, simplifying sensor placement, electrical design, and calibration.

The sensor is still in the process of being improved. Check our web site at http://lims.mech.nwu.edu/~lorenz/sensor/ for more recent results.

REFERENCES

[1] J. E. Colgate, W. Wannasuphoprasit, and M. A. Peshkin, "Cobots: Robots for Collaboration with Human Operators," Proc. 1996 ASME Int. ME Cong. and Exhib., Atlanta, GA, DSC-Vol. 58, pp. 433-39.

[2] W. Wannasuphoprasit, R. B. Gillespie, J. E. Colgate, M. A. Peshkin, "Cobot Control," Proc. 1997 IEEE Int. Conf. on R&A, Albuquerque, NM, pp. 3571-3577.

[3] O. M. Al-Jarrah and Y. F. Zheng, "Arm-Manipulator Coordination for Load Sharing Using Compliant Control," Proc 1996 Int. Conf on R&A, Columbus, OH, pp. 1000-1005.

[4] K. I. Kim and Y. F. Zheng, "Human-robot Coordination with Rotational Motion," Proc. 1998 IEEE Int. Conf. on R&A, Leuven, Belgium, pp. 3480-3485.

[5] A.K. Bejczy and Z.F. Szakaly, "A harmonic motion generator(HMG) for telerobotic applications," Proc. 1991 IEEE Int. Conf. on R&A, Sacramento, CA, 1991, pp. 2032-2039.

[6] M.F. Ashby, "Material Property Charts" ASM Handbook Vol. 20, pg. 266-280, ASM International, 1997.

[7] F. P. Beer and E. R. Johnston, Jr., Mechanics of Materials, McGraw-Hill, 1976

[8] John A. Hilton, "Force and Torque Converter" United States Patent 4,811,608, 1989

Construction of a Human/Robot Coexistence System Based on A Model of Human Will - Intention and Desire

YAMADA Yoji, UMETANI Yoji
Toyota Technological Institute,
2-12-1, Hisakata, Tempaku, Nagoya 468-8511, Japan
yamada@toyota-ti.ac.jp, umetani@toyota-ti.ac.jp

DAITOH Haruyoshi
Toray Engineering Co. Ltd.,
1-1-45, Ohe, Ohtsu 520-2141, Japan

SAKAI Takayuki
Toyota Motor Co. Ltd.,
1, Motomachi, Toyota, Aichi 471-0854, Japan

Abstract

With the aging problem of manufacturing industries as a social background, we propose a human-will-based approach to constructing a coexistence system in which a robot is assumed to assist the power of a subject while reflecting his intention and desire. Before introducing the system, we discuss a model of human will associated by the work on psychological mindreading, and identifies the advantages of the system in which the robot guides the endtip the subject manipulates along the path intended by him. Then, we describe two major techniques with experimental results: 1) Estimation of the human-intended path pattern from the previously assigned in the early stage of the human motion by proposing MHMM (a modified algorithm of Hidden Markov Model). 2) Merge of the human's force feeling pattern into the desired by proposing FIE (Field Impedance Equalizer) which allows the human to express his desirable field-dependent force/speed feeling and the robot to display the impedance (force divided by speed) along the previously estimated path. Finally, we conclude the study with some comments on future works.

1 Introduction

1.1 Background

The problem of the aging society takes a severe effect to manufacturing industries where the age of workers become higher. However, supplementary labor force cannot be easily expected because the overseas productivity has been enlarged and people in younger generation show less interest in manufacturing industries. And, it is desired to introduce easily manipulable tools such as human power assisting robots to processes of transferring/assembling works, as well as conveying tools [4] because of the above worker's aging problem.

Taking this social background into consideration, we are engaged in the development of coexistence systems comprising humans and power assisting robots for realizing their cooperative motions. Basically, ensuring human safety is prerequisite to such human/robot coexistence systems. We have conducted some research on human-safety-oriented robot design based on human pain tolerance limit [16]. And, it is also important, following the human safety issue, to discuss how to improve operational performance of such power assisting robots. In fact, extremely high performance is required such as to match the desire of each worker in production lines. To meet this requirement, we need to take the mechanism of human mind into consideration. In the next subsection, we discuss a model of human will which involves intention and desire and functions before any action is conducted.

1.2 Model of human will

Let us consider a simplified example of typical work handling operations in production lines. For the study, we pay our attention to the fact that, in many cases, such operations have several fixed task patterns. **Figure 1** illustrates a simplified operation in which a worker grabs a handle mounted at the endtip of a power assisting robot and pulls the handle closer to himself in one of the three preassigned directions.

Throughout the experiment, we frequently en-

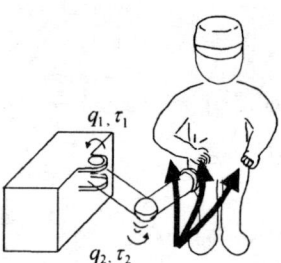

Figure 1: A task simulating Human/Robot coexistence systems in production lines

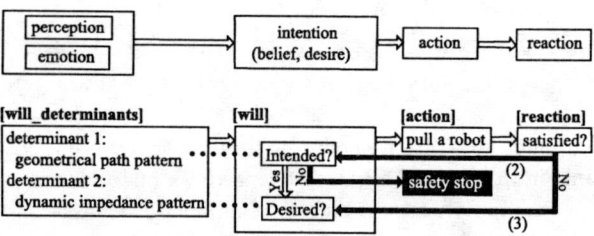

Figure 2: Model of human will in association with mindreading

counter the phrases of the subjects such as " intended (not intended)" path pattern or "desired (not desired)" impedance pattern. The phrases associate us with the psychological study on mindreading where a model for a theory of mind is proposed [1, 15] as represented by the above flow in **Figure 2** where $\Longrightarrow$'s indicate the processes in the subject's mind. Closer examination into the model applied to our study provides us with the system advantages for supporting the operation of the subject as follows: Preceding an action of the subject, let us locate human *will* [1] which should be determined in consideration of all causal factors (called will determinants in the paper). In our study, there are two principal determinants that at least we have to take in the model as those for the subject to pay his attention to; determinant1-geometrical path pattern, and determinant2-dynamic impedance pattern. He expresses his will so that he will be satisfied with the resultant motion pattern composed of the

---

[1]While 'human intention' is used generally in psychology to represent that plans satisfy goals by integrating desires and beliefs [3], we utilize the term 'human will' because as will be explained in section 3, the human has to decide, in cooperation with the robot, the desired impedance pattern which he has not experienced.

two determinants through the experiment of pulling a robot closer. Again, $\Longrightarrow$'s in the below of Figure 2 indicate the processes of human mind. We understand from the simple model that the robot needs to adapt itself properly to satisfy the subject's will. Such an interactive process should be continued until he attains both intended path pattern and desired impedance pattern as shown by $\Longleftarrow$'s in the same figure, with the two numbers (2) and (3) corresponding to the section numbers in the paper for further technical issues.

Moreover, when we learn that geometric trajectory formation is followed by motor command adjustment in general visuomotor sequence learning [8], we can expect that automatic trajectory guidance by the robot along the path intended by the subject directly leads him to enter into the impedance adjustment stage as shown by the $\Downarrow$ in Figure 2. Finally, even if the resultant motion is different from the estimated intended-path or there occurs any unexpected marked deviation from the path, the system can command the robot to make a safety stop as shown in the figure by the white "safety stop" in the black box, because those situations should be regarded as hazardous motion conditions.

The remainder of the paper describes the following two major topics:

1) Estimation of the path pattern intended by the subject from the previously assigned in the early stage of the subject's motion by proposing MHMM (a modified algorithm of Hidden Markov Model).

2) Merge of the human's force feeling pattern into the desired by the proposal of FIE (Field Impedance Equalizer) which allows the subject to express his desirable field-dependent force/speed feeling and the robot to display the impedance (force divided by speed) while leading along the previously estimated path.

Throughout the study, we utilize the experimental setup of a human/robot coexistence system as shown in **Figure 3**. An LED is attached at the subject's hand for detecting its 2D horizontal position by a PSD camera mounted over his head. The robot with 2 DOF's exhibits horizontally planar motions and can be either position- or torque- controlled with two wire-driven DD motors. The joint angles as well as the joint torques are monitored. A 6-axis force sensor is also mounted at the endtip of the robot.

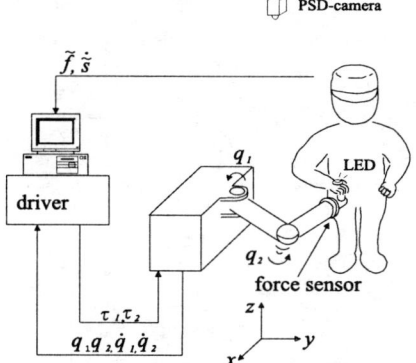

Figure 3: Experimental human/robot coexistence system with signal flows

2 Estimation of human-intended path patterns using Modified Hidden Markov Model

The term intention was used for the first time by Inagaki et al, in the development of a human intention inference system based on a subject's behavior [7]. The importance of understanding human intention in relation to multiple communication channels including unconscious communication was enphasized by [12]. Possibility of identifying the input variables as the candidate factors for human intention was stressed in [13] where the interpretative structural modelling was applied as the identification method.

However, the above works have neither examined the human intention in the model of human mind, nor presented specific methods of estimating the intention. In this section, we present an algorithm of estimating the path pattern intended by a subject from the pre-assigned. Here, the problem of estimating an intended path pattern is defined as a pattern matching problem from incomplete data sequence of a human motion in its early stage. In the next section, we will see that estimating the intended path from an incomplete sequence of the subject's motion even *after* it is started holds technical utility.

2.1 Modified Hidden Markov Model

Since it is necessary to handle data on human behavior which includes the problem of motion rate variation, we pay our technical attention to a Hidden Markov Model (abbreviated as HMM). HMM basically solves three problems [11], but we need to solve two more problems for our study in addition to those three:

- Problem 4: In the original model, each pattern accepts only one trained sequence of data.
- Problem 5: Pattern matching function is conducted after all data of the observed sequence are presented.

In HMM, pattern $\#h$ is modeled by three probability measures; the state transition probability distribution $\mathbf{A} = \{a_{ij}\}$, the observation symbol probability distribution $\mathbf{B} = \{b_{jk}\}$, and the initial state distribution $\boldsymbol{\pi}_h = \{\pi_{h_i}\}$ as in (1).

$$\lambda_h = \{\mathbf{A}_h, \mathbf{B}_h, \boldsymbol{\pi}_h\} \quad h = 1, \cdots, H \quad (1)$$

HMM specifies the model λ_h with the trained data prepared for each pattern and observed symbol sequence $\mathbf{O}_T = \{o_1, \cdots, o_T\}$ (T: terminal point) generates the probability of the sequence, given the model λ_h.

Solution to the Problem 4 accepts multiple observation sequences to reestimate the model parameters and is formulated as (2).

$$\overline{a}_{ij} = \frac{\sum_{k=1}^{K} \sum_{t=1}^{T_k-1} \alpha_t^{(k)}(i) a_{ij} b_j\left(o_{t+1}^{(k)}\right) \beta_{t+1}^{(k)}(j)}{\sum_{k=1}^{K} \sum_{t=1}^{T_k-1} \alpha_t^{(k)}(i) \beta_{t+1}^{(k)}(i)}$$

$$\overline{b}_j(l) = \frac{\sum_{k=1}^{K} \sum_{\substack{t=1 \\ s.t.o_t = v_l}}^{T_k-1} \alpha_t^{(k)}(j) \beta_{t+1}^{(k)}(j)}{\sum_{k=1}^{K} \sum_{t=1}^{T_k-1} \alpha_t^{(k)}(j) \beta_{t+1}^{(k)}(j)} \quad (2)$$

where

$$\alpha_t^{(k)}(i) = P\left(o_1^{(k)} o_2^{(k)} \cdots o_t^{(k)}, s_t = S_i | \lambda\right)$$
$$\beta_t^{(k)}(i) = P\left(o_{t+1}^{(k)} o_{t+2}^{(k)} \cdots o_T^{(k)} | s_t = S_i, \lambda\right) \quad (3)$$

Properly taking initial parameter value $\lambda = \{\mathbf{A}, \mathbf{B}, \boldsymbol{\pi}\}$ which is substituted into the right hand side of (2), we compute λ, and the computation is repeated until it converges, and the converged parameter gives the reestimated model in a way similar to the original HMM algorithm.

On the other hand, the solution to the Problem 5 is given by (4), because we need the probability of obtaining the output $O_t = o_1 \cdots o_t$ and staying at $s_t = S_i$ (only constrained at t) under the $\#h$-th model λ_h.

$$P(\mathbf{O}_t | \lambda_h) = \sum_{j=1}^{N} \alpha_{t_h}(j) \quad (4)$$

which is proved and can be computed iteratively by:

$$\sum_{j=1}^{N} \alpha_{t_h}(j) = \sum_{j=1}^{N} \left[\left\{ \sum_{i=1}^{N} \alpha_{t-1_h}(i) a_{ij} \right\} b_j(o_t) \right] \quad (5)$$

After the above modified algorithm which we refer to as modified HMM (abbreviated as MHMM) is formulated, the probability of being matched as pattern $\#h$ computed at each sampling period by (6).

$$p_h(i) \equiv \frac{P(\mathbf{O}_t|\lambda_h)}{\sum_{h=1}^{H} P(\mathbf{O}_t|\lambda_h)} \quad (6)$$

2.2 Experimental results

We use the system described in section 1.2. The 2D hand position data are sampled at $100ms$ interval for $3s$. As a preliminary setup, robot motion area is divided into the measure of 7×10 squares. By labeling each square, the position data obtained in discrete time domain are coded by symbols beforehand.

Firstly, we show in **Figure 4** the result of the intended path estimation under the condition that only one trained data sequence for each path is given *i.e.* the original HMM algorithm is used. We set three pre-assigned path patterns which are schematically drawn as trained paths $\#1$ through $\#3$ at the bottom of the figure. The trajectory destined to pattern $\#2$ is considerably complicating because the trajectory shares the route with the other patterns $\#1$ and $\#3$ on its way to the destination. The estimated result shows that the calculated probability to all patterns ends in 0% at $t = 10$ when the trajectory heads for pattern $\#2$ or $\#3$. This means that there is no candidate as the intended path and the estimation was not successful because the model cannot accept variously trained paths. Secondly, we demonstrated the same example by using MHMM in (2), the result of which is shown in **Figure 5**. The estimated result gives a small possibility of going to patterns $\#2$ or $\#3$ even at $t = 8$ when the trajectory is on pattern $\#1$. At $t = 12$ when the trajectory is on pattern $\#1$ and moves toward pattern $\#3$, the probability of the intended path pattern $\#2$ is computed highest, while there is still a fairly high possibility of going to pattern $\#3$. Finally at $t = 21$, the probability of being pattern $\#2$ is computed as 100%. The above sequence of estimation verifies that the proposed MHMM gives a correct estimation of the intended path by absorbing multiple data sequence in the training stage.

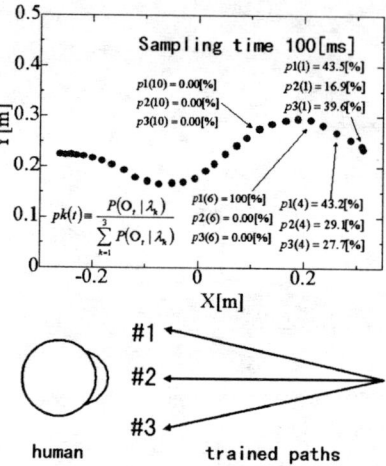

Figure 4: Estimated result of the intention candidate without the proposed method installed

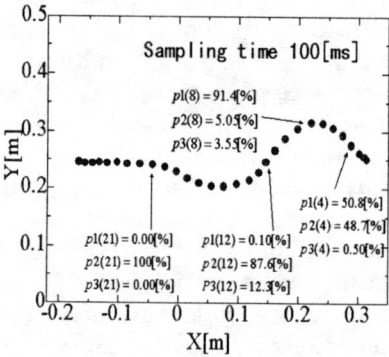

Figure 5: Estimated result of the intention candidate with the modified HMM algorithm installed

The objective of the study is to show the effectiveness of intention estimation by reducing the burden of determining the impedance pattern after the intended path is generated from an incomplete data sequence of human motion. **Figure 6** shows that the intended path estimation is successfully made in the early stage of the human action in flame A, and the robot is controlled to follow the intended path for the rest part of the trial. In the MHMM algorithm, the function of directing from current position toward the destinations in constructing the HMM models is added so that even in the cases where any of the destinations move, the algorithm achieves successful intention estimation for wider applications .

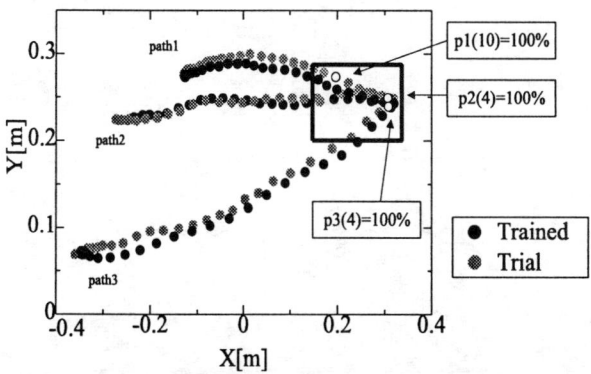

Figure 6: Intended path estimation in the early stage even in the case where target is moving

In our algorithm, the torques supplied to the joint actuators are cut off whenever any excessive operational force exerted to take the actual path off the preassigned is detected by the force sensor at the endtip of the robot. It actually supplies a new concept of ensuring human safety which could not be gained by conventional safety measures based only on physical risk analyses [2], because, in some of the practical cases, even dangerous tools have to be dealt with by the robots in coexistence with workers.

3 Equipment of the system with Field Impedance Equalizer

Now that the system determines the path pattern intended by a subject, all that is left is to determine the impedance pattern along the path. There have been a number of studies conducted for controlling a robot typically in a linear or planar motion which is commanded by an operator who grabs the handle at its endtip. The study was originally proposed by Morasso [10] and the jerk minimum criterion was proposed [5], followed by the torque variation minimum criterion [14]. Besides the scientific interest in human arm movements, we can find some proposals on control strategies [9, 6]. However, the proposal in [9] does not necessarily take the human desire into account, while the applicability of variable damping control in [6] is considered to be limited.

3.1 Proposal of Field Impedance Equalizer

As was discussed in section 1.2, the robot must possess learning capability in order to reflect the desire of a subject. To accept his verbal commands based on his desire, the impedance pattern at the robot endtip should be easily modified. In the last section, we developed an algorithm of estimating the path pattern intended by the subject. Once the intended path is estimated, the robot tries to go along the path, which allows us to assign the desired impedance pattern in a scalar way as $\tilde{f}$ and $\dot{\tilde{s}}$, based on the discussion in subsection 1.2:

$$\tilde{\mathbf{f}} = \tilde{f}\mathbf{e}^T, \quad \dot{\tilde{\mathbf{s}}} = \dot{\tilde{s}}\mathbf{e}^T \qquad (7)$$

where $\mathbf{e} = \mathbf{e}(\mathbf{q}(t))$ is the unit tangent vector defined along the preassigned path, and $\mathbf{q}(t)$ is the joint angular vector. Furthermore, the desired impedance (damping coefficient D in the study) is assigned as a function of the field $\mathbf{s}$

$$\tilde{f}(\mathbf{s}) = \dot{\tilde{s}}(\mathbf{s})D(\mathbf{s}) \qquad (8)$$

where $\mathbf{s} = \mathbf{s}(\mathbf{q}(t))$. Dynamic equation of the robot motion is given by

$$\mathbf{I}(\mathbf{q})\ddot{\mathbf{q}} + \mathbf{h}(\mathbf{q}, \dot{\mathbf{q}}) = \boldsymbol{\tau} + \mathbf{J}^T(\mathbf{q})\mathbf{f} \qquad (9)$$

where

$\mathbf{I}(\mathbf{q})$: inertia tensor,
$\mathbf{h}(\mathbf{q}, \dot{\mathbf{q}})$: Coriolis, centrifugal and viscous friction force vector,
$\mathbf{J}(\mathbf{q})$: Jacobian matrix,
$\boldsymbol{\tau}$: joint torque vector,
$\mathbf{f}$: external force vector (exerted by the subject)

And when a desired impedance pattern $D(\mathbf{s})$ in (8) is obtained, we can calculate external force $\mathbf{f}$ by using the actual velocity deta $\dot{s}(\mathbf{s})$ and the desired output torque can be computed by (9). Formula (8) is the Field Impederce Equalizer (abbreviated as FIE) which makes the desired input to the joint torques $\boldsymbol{\tau}$.

3.2 Experimental results

Firstly, we evaluated the performance of the damping control by the robot system which is shown in **Figure 7**. The evaluation result shows that there was a good proportional relationship between the preassigned force and that actually measured by the force sensor. Similarly, proportionality was obtained in the other experiment where the speed was kept constant.

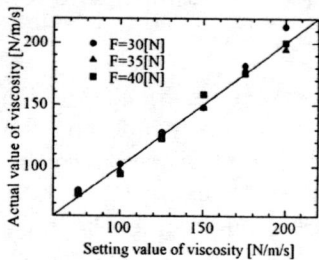

Figure 7: Comparison of velocity at constant force: preassigned and actually measured

Secondly, we verified the usefulness of the proposed FIE. Each of the pre-assigned path track is divided into 10 sections and named as field1 through field10. Each subject is asked to grab the handle of the robot endtip and to pull it toward himself. After each trial is over, he is asked to tell how he wants to change the force and speed patterns at every field along the path toward a more desired impedance pattern by showing e.g. "a little heavier at field1", "much more faster at field5", and so on. The robot operator changes the impedance as the subject desires and repeat the trial until he is satisfied with the impedance pattern along the path track. **Figure 8** serially contrasts the five velocity-and-force patterns in three stages; in the beginning ($n = 1 \sim 5$), in the middle ($n = 8 \sim 12$), and in the last ($n = 14 \sim 18$) stages of the experiment. When the results in the three stages are compared, it is obvious that the five velocity-and-force patterns in the last stage have the least variations. The convergence in velocity and force patterns was seen in most of the experiments and it means that the desired impedance pattern merged through the cooperative adjustment processes between the subject and the robot. It was also evident that the proposed FIE was effective in the sense that subjects felt easy to point out the undesirable velocity or force depending on the field.

4 Conclusion

With the aging problem of manufacturing industries as a social background, we developed a coexistence system comprising a human and a robot which is assumed to assist his power while reflecting his intention and desire. The following three items summarize the study.

1) We examined a model of human will involving intention and desire associated by the work on psychological mindreading, and identified the advan-

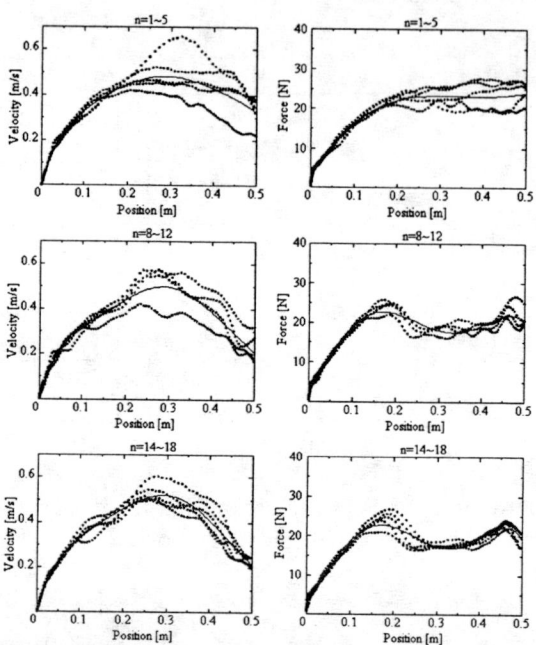

Figure 8: Comparison of force and velocity patterns for the first, middle and the last 5-time cooperative motions

tages of the system in which the robot guided the endtip human subject manipulated along the human-intended path, in relation to motor command adjustment and safety.

2) We developed a function to estimate the human-intended path pattern among the preassigned by using the modified Hidden Markov Model (MHMM). Experimental results demonstrated that MHMM enables the intended-path estimation in the early stage of human motion. MHMM was proved to have wider applicability to the cases where the target points for positioning are variously moved in every motion sequence.

3) We proposed a Field Impedance Equalizer (FIE) which assists the subject to assign along the path an impedance pattern as he desires. Using the FIE made it easier for subjects to assign the desired impedance pattern and the experimental results showed that the desired impedance pattern merged through human/robot cooperative adjustment.

We need to study further on each of the above topics as follows:

1) Not simply the position sequence but also velocity/power sequences may give earlier and more accurate human-intended path estimation. The system should be capable of heuristically recognizing various trajectories as new patterns with human intention or not.

2) Throughout the experiments, subjects were not asked to terminate his action at some definite points. There should be other specific criteria in positioning operations which will be applicable to phase-dependent impedance control. Quick convergence to a desired impedance pattern in each of the intended path should be statisticaly verified.

Acknowledgment

We thank Dr. Morizono Tetsuya, Mr. Konosu Hitoshi, and Mr. Yamamoto Takahisa for their useful comments. This study was partly supported by the Ministry of Education, Science, and Culture under Grant-in-Aid for Scientific Research No.08235224 and No.09221225.

References

[1] S. Baron-Cohen, Chapter 4: Developing Mind Reading: The Four Steps, *Mindblindness*, The MIT Press, 1995

[2] British Standard Code of Practice for Safety of Machinery, B# 5304, pp.9-23, 1988

[3] R. J. Bogdan, Chapter 6: The Big Little Step, *Interpreting Minds*, The MIT Press, 1997

[4] Eri, Y., et al., "The development of working conditions taking the lead an epoch # 2, Proc. of the Japanese Society of Quality, pp.57-60, Aug., 1997 (in Japanese)

[5] T. Flash, et al., "The Coordination for Arm Movements: An Experimentally Confirmed Mathematical Model", The Journal of Neuroscience, Vol.5, No.7, pp.1688-1703, 1985

[6] Ikeura, R., et al., "Cooperative Motion Control of a Robot and a Human", Proc. of the 3rd IEEE Int. Workshop on Robot and Human Communication, pp.112-117, Oct., 1994

[7] Inagaki, y., et al., "A study of a method for intention inference from human's behavior", Proc. of IEEE Int. Workshop on Robot and Human Communication, pp.142-145, 1993.

[8] Hikosaka O., Miyashita K., Miyachi S., Sakai K., Lu X., "Differential roles of the frontal cortex, basal ganglia, and cerebellum in visuomotor sequence learning", Neurobiology of Learning and Memory, No. 70, pp.137-149, 1998

[9] Kosuge K., et al., "Control of a Robot Handling an Object in Cooperation with a Human", Proc. of the 6th IEEE Int. Workshop on Robot and Human Communication, Tohoku, pp.142-147, Sep., 1997

[10] P. Morasso, "Spatial Control of Arm Movements", *Experimental Brain Research*, Springer-Verlag, Vol.42, pp.223-227, 1981

[11] L. R. Rabiner, "A Tutorial on Hidden Markov Models and Selected Applications in Speech Recognition", Proc. of IEEE, Vol.77, No.2, pp.257-286, 1989

[12] Sato, T., et al., "Active Understanding of Human Intention by a Robot through Monitoring of Human Behavior", Proc. of IEEE Int. Symp. on Intelligent Robots and Systems, pp.405-414,1995

[13] Takahashi H., et al., "A Study on an Identification Model for Inferring the Driver's Intentions - When Decelerating on a Downhill Grade -", Trans. on the Society of Instrument and Control Engineers, Vol.32, No.6, pp.904-911, 1996 (in Japanese)

[14] Uno, Y., et al., "Formation and Control of Optimal Trajectory in Human Multijoint Arm Movement: Minimum Torque Change Model", *Biological Cybernetics*, Springer-Verlag, Vol.61, pp.89-101, 1989

[15] Wellman, *The Child Theory of Mind*, The MIT Press, 1990

[16] Yamada, Y., et al., "Fail-Safe Human-Robot Contact in the Safeguarding Space", IEEE/ASME Trans. on Mechatronics, Vol.2, No.4, pp. 230-236 , Dec.,1997

Emergence of Emotional Behavior through Physical Interaction between Human and Robot

Takanori Shibata [*1]   Toshihiro Tashima [*2]   Kazuo Tanie [*1]

*1 Bio-Robotics Division, Robotics Department
Mechanical Engineering Laboratory
Ministry of International Trade and Industry
{shibata, tanie}@mel.go.jp

*2 Fuzzy Technology and Business Promotion Division
OMRON Corporatio

Abstract: Recent advances in robotics have been applied to automation in industrial manufacturing, with the primary purpose of optimizing practical systems in terms of such objective measures as accuracy, speed, and cost. This paper introduces research on artificial emotional creatures that seeks to explore a different direction that is not so rigidly dependent on such objective measures. The goal of this research is to explore a new area in robotics, with an emphasis on human-robot interaction. There is a large body of evidence that shows the importance of the interaction between humans and animals such as pets. We have been building pet robots, as artificial emotional creatures, with the subjective appearance of behaviors that are dependent on internal states as well as external stimuli from both the physical environment and human beings. The pet robots have multi-modal sensory system, actuators, and bodies with artificial skin for physical interaction with human beings.

1. Introduction

A human understands people or objects through interaction. The more and longer they interact, the deeper the human understands the other. Long interaction can result in attachment and desire for further interactions. It may also result in boredom. Interaction stimulates humans, and generates motivations for behaviors. There can be cases in which behaviors are not rational.

Objects with which humans interact include natural objects, animals and artifacts. Studies on interaction between human beings and animals show positive effects on psychology, development of children, and so on [1]. Artifacts that affect people in mentally can be called "aesthetic objects". Such effects are subjective and could not be measured simply in terms of objective measures such as accuracy, energy and time.

Machines are also artifacts. Different from the aesthetic objects, machines have been designed and developed as tools for human beings while being evaluated in terms of objective measures [2]. Machines are passive basically because human beings give them goals. Machines will not be active as long as they are tools for human beings.

However, if a machine were able to generate its motivation and behave voluntarily, it would have much influence to an interacting human. At the same time, the machine should not be a simple tool for humans nor be evaluated only in terms of objective measures. Subjective evaluation is important. For a human, multi-modal stimulation should be influential. People interacting with the machine or observing the interaction may consider the machine as an artificial creature. Behaviors of the machine may be interpreted as emotional.

There are many studies on human-machine interaction. Here, we don't discuss studies on human factors in controlling machines used as tools. In other studies, machines recognize human gestures or emotions by sensory information, and then act or provide some information to the human. However, modeling gestures or emotions is very difficult because these depend on the situation, context and cultural background of each person.

Concerning action by a machine toward a human, an artificial creature in cyber space can give only visual and auditory information to a human. A machine with a physical body is more influential on human mind than a virtual creature.

Considerable research on autonomous robots has been carried out. Their purposes are various such as navigation, exploration and delivery in structured or unstructured environments while the robots adapt to the environments. Also, some robots have been developed to show some emotional expressions by face or gestures. However, even though such robots have physical bodies, most of them are not intended to interact physically with a human.

We have been building pet robots as examples of artificial emotional creatures [1, 3, 4]. The pet robots have physical bodies and behave actively while generating motivations by themselves. They interact with human beings physically. When we engage physically with a pet robot, it stimulates our affection. Then we have positive emotions such as happiness and love or negative emotions such as anger, sadness and fear. Through physical interaction, we develop attachment to the pet robot while evaluating it as intelligent or stupid by our subjective measures.

The chapter 2 discusses subjectivity and objectivity. The chapter 3 discusses emergence of emotional behaviors through physical interaction. The chapter 4 introduces newly developed pet robots [5]. Finally, the chapter 5 concludes this paper.

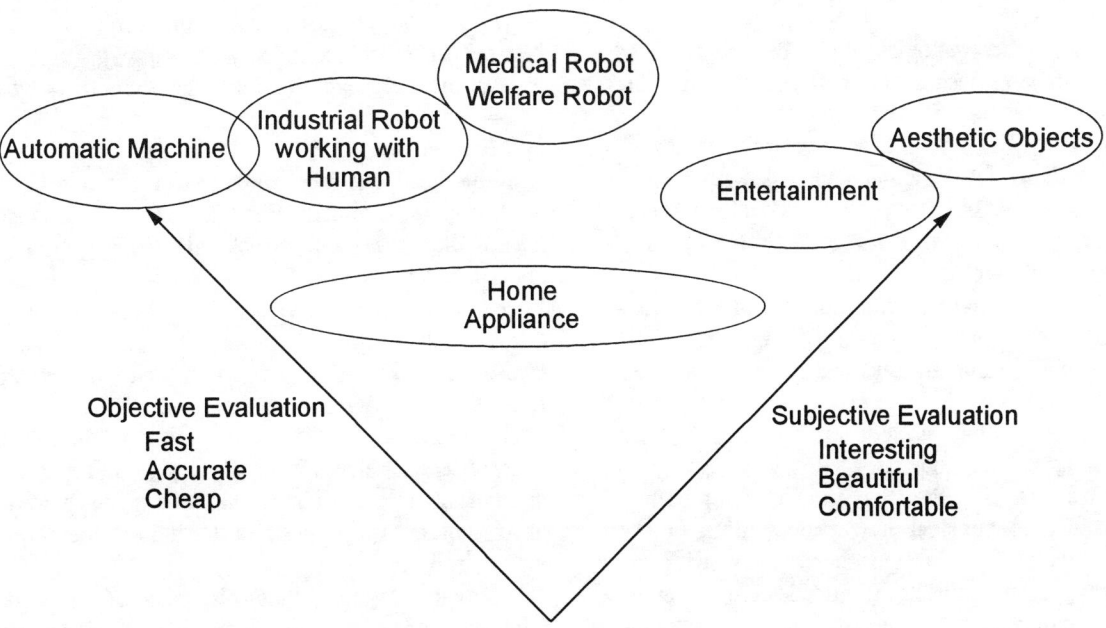

Fig. 1 Objective and Subjective Evaluations based on Application Fields

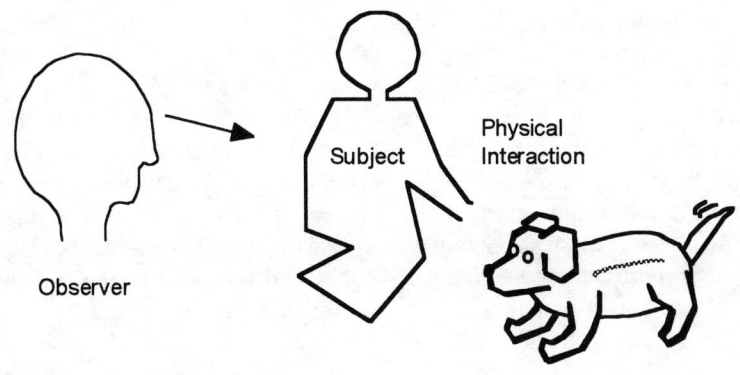

Fig. 2 Physical Interaction between Subject and Pet Robot in a View of Observer

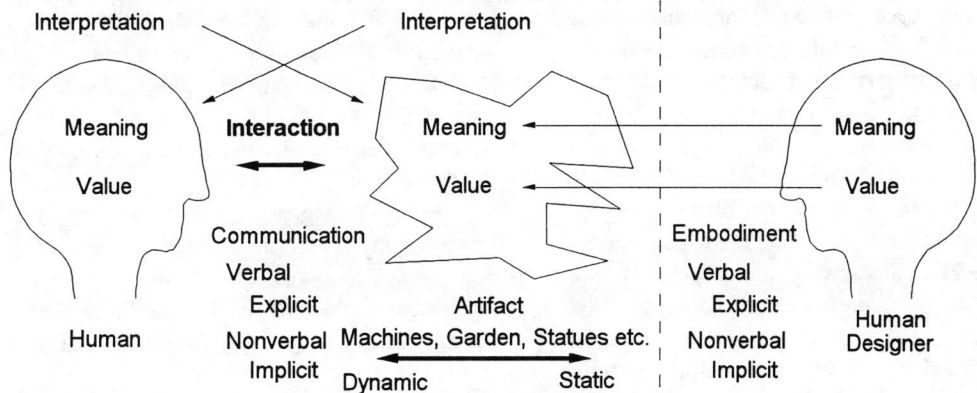

Fig. 3 Subjective Interpretation and Evaluation through Interaction

2. Objectivity and Subjectivity

Science and technologies have been developed through objectivism. Because of this, people can share and use their scientific and technological knowledge in common. When we design machines, we need to use such objective knowledge. A machine which has high value evaluated in terms of objective measures such as speed, accuracy, and cost is useful as a tool for human beings, especially for automation.

A machine that interacts with a human is not always evaluated by such objective measures (Fig. 1). People evaluate a machine subjectively. Even if some machines were useless in terms of objective evaluation, some people put high subjective value on them. Such machines could be considered as aesthetic objects.

When we design robots that interact with human beings, we have to consider how people think of the robots subjectively (Fig. 2). This paper deals with pet robots to investigate subjectivity for designing robots friendly to human beings.

3. Emergence of Emotional Behavior through Physical Interaction

There is an enormous number of studies on emotions. Also, we can make models of emotions by observing many people. However, we can not say which model is correct or even the best. We have many words to express our own emotions, but we don't have the same definition of internal states of our bodies. Emotions are evoked in some situation, and depend on context and cultural background. Therefore, it is difficult to establish a general model of emotions. For example, if a subject and another person are interacting, the person's interpretation of emotions of the subject is not always the same as that of the subject himself. Even if they had long relationship, they would interpret the emotions in different ways.

We are taking a position that emotions emerge through interaction with the environment as Picard classified research on emotions [1, 3, 4, 6]. There is some research on emergent emotions. Toda emphasized the importance of studying whole systems including perception, action, memory, and learning [7]. He proposed a scenario with a fungus eater to illustrate how emotions would emerge in a system with limited resources operating in a complex and unpredictable environment. Toda's robot has the goal of collecting as much uranium ore as possible, while regulating its energy supply for survival. The robot has rudimentary perceptual, planning, and decision-making abilities. Toda proposed urges that are motivational subroutines linking cognition to action, and argued the robot would be emotional with the urges. The urges are triggered in relevant situations and subsequently influence cognitive processes, attention, and bodily arousal. An observer of the robot would interpret that the robot's behaviors are emotional. Pfeifer implemented urges in a mobile robot and showed emergence of emotional behaviors through interaction between the robot and its environment [8].

Braitenberg explained emergent emotions by means of his simple mobile robot, which had two light sensors and two motors [9]. When the robot sees a light source straight ahead, the robot moves toward it, and bangs into it, hitting it frontally. When the source is not straight ahead, then the robot turns and moves so that it still approaches the source and hits it. An observer could interpret the robot's behavior as aggressive as if the robot felt a negative emotion toward the light source. When sensors and motors are wired so as to linger near the source and not damage the robot, the behavior could be interpreted as a more favorable emotion. The robot could have different behaviors in the same environment when its internal system is changed. These mean that emotions have emergent properties and depend on interaction with the environment.

Brooks argued that situatedness, embodiment, intelligence and emergence are key ideas of behavior-based robots [10]. The key idea of situatedness is that the world is its own best model. The key idea of embodiment is that the world grounds the regress of meaning-giving. The key idea of intelligence is that intelligence is determined by the dynamics of interaction with the world. The key idea of emergence is that intelligence is in the eyes of the observer.

When a human and a pet robot interact with each other, they stimulate and affect each other. We call this 'coupling' [4]. When the human evaluates the robot, the human is an observer and a subject at the same time. Following Brooks' ideas, the intelligence of the robot could be determined by the dynamics of interaction with the subject and environment. Also, the subject would interpret or measure intelligence of the robot with his own eyes. The subject's interpretation depends on his knowledge and experiences related to the robot and its designer (Fig. 3). Therefore, the robot's intelligence depends on the subject's intelligence.

At this point, we don't have an explicit definition of intelligence. However, as Minsky suggested, we doubt whether machines can be intelligent without emotions (which doesn't mean emotion models but emotional appearance of behavior in observer's view) [11] Therefore, we consider that emergent emotions are key for intelligence.

As we design pet robots as artificial emotional creatures, an interacting human does not give them goals nor tasks. Pet robots are allowed to generate their own goals and motivations for survival in the world. Therefore, contrary to Asimov's "The Three Laws of Robotics" [12],

1) *They would protect themselves.*
2) *They would not obey human beings.*
3) *They would injure human beings.*

These allowances are the key to let people interpret that pet robots are like living creatures.

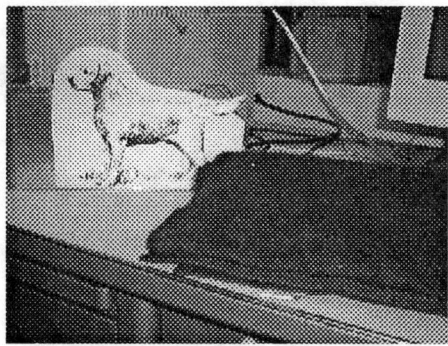

Fig. 4 System for Psychological Experiment

Fig. 5 Prototype of Pet Robot

4. Pet Robot as Artificial Emotional Creature

4.1 Previous Research

We investigated subjective interpretation of robot's behaviors in psychological experiments, in which a picture of a dog was equipped with a 1 DOF tail and subjects were asked to interpret emotions of the dog by wagging tail [4] (Fig. 4). Then, a simple tactile sensor was added to the system and the tail wagged depending on stroking the tactile sensor by subjects. In the first experiment, subjects interpreted meaning of wagging by visual and auditory information. In the second one, subjects had tactile information as well as vision and audition. Interpretations of emotions were various because of knowledge of dogs; for example, some had experience of owning dogs. However, the second experiment was much more impressive for most subjects because of physical interaction with tangibility.

With this result, we developed a pet robot that had visual, auditory, and tactile sensors, a tail with 1 DOF, and mobility by three wheels [4] (Fig. 5). We emphasized "tangibility" for physical interaction between a human and a robot, different from other research [13]. For this purpose, we developed a new tactile sensor that consisted of a pressure sensor and a balloon covered with synthetic skin. The robot could sense pushing, stroking, and patting. It behaved depending on its internal state that consisted of current input from sensors and regressive input from itself. A human interacting with the robot by touching or stroking got visual, auditory and tactile information. The human felt softness and nice texture like real creatures. Depending on the information, the human changed his behavior. This loop was considered as coupling between the human and robot. Even though the robot didn't have explicit emotion model, people interacting with the robot interpreted the robot's behaviors were emotional.

From viewpoints of physical interaction and subjectivity, there were deficits that the robot was too heavy to hug or hold, and didn't look like real animals in its appearance. Therefore, we developed two pet robots: one is a seal robot (Fig. 6), and the other is a cat robot (Fig. 7) that has more complex structure and more functions.

4.2 Small and Animal Like Pet Robots
4.2.1 Seal Robot

A seal robot has a simple structure in order to investigate emergent emotions through physical interaction. The robot has two legs with two servomotors. Each leg has a clutch bearing at a contacting point with floor. At front and back of its body, the robot has two supports. The front support is a caster. The back support has a clutch bearing at the contacting point with floor. There are four contacting points in total and three have clutch bearings in the same direction. When the robot moves two legs back and forth at the same time, it moves forward like crawling. When it moves two legs back and forth alternately, it doesn't move forward but it shakes its body. When one leg was fixed at back and the other leg moves back and forth, the robot turn to a direction of the fixed leg's side. Concerning on sensory system, the robot has two whiskers which sense contact with its environment, and two pressure sensors with balloons which sense pushing, patting and stroking on its body. The robot has 6811 CPU inside to control itself. Its weight is about 1.0 [kg]. We assumed those, as innate characteristics, the robot likes stable pressure on its body and dislikes being touched its whiskers.

The robot has internal state depending on sensory information and recurrent information. A neural oscillator with two neurons generates motion of each leg, and the internal state was input to a neuron. Phase of the two oscillators was controlled by sensory information.

The robot had some rules to generate its motivation, to change its attention, and to control movement of legs, but it didn't have explicit model of emotions. When people interacted with the robot, they interpreted the robot's behavior differently with some words of emotion to express what the robot was doing. As the movement of the robot depended on context, internal state made robot's behavior more complex in

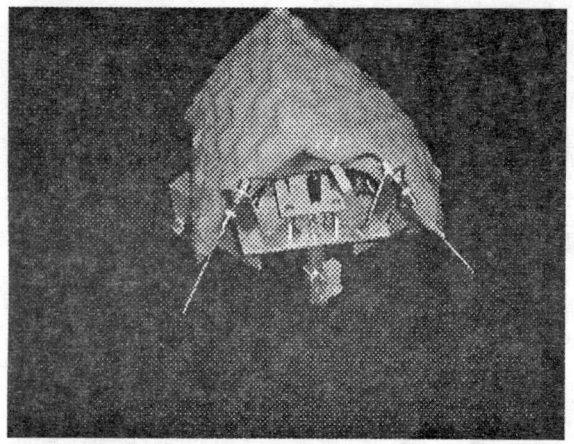

Fig. 6 (a) Front View without Costume of Seal Robot

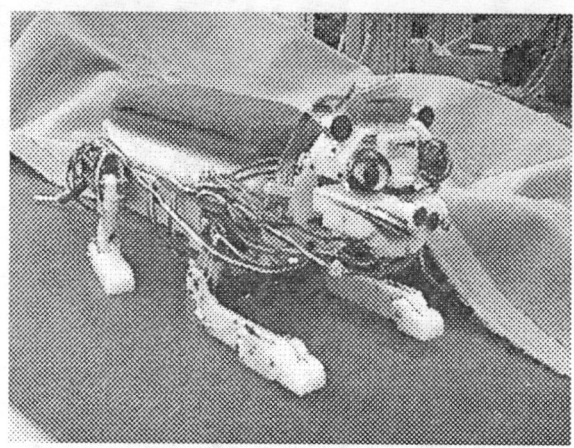

Fig. 7 (a) Cat Robot without Costume

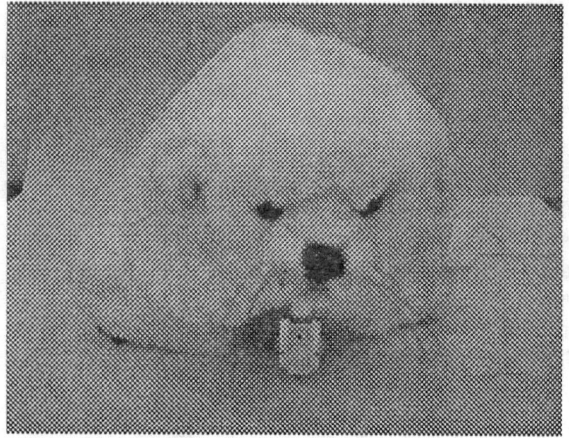

Fig. 6 (b) Front View of Seal Robot with Costume

Fig. 7 (b) Active State

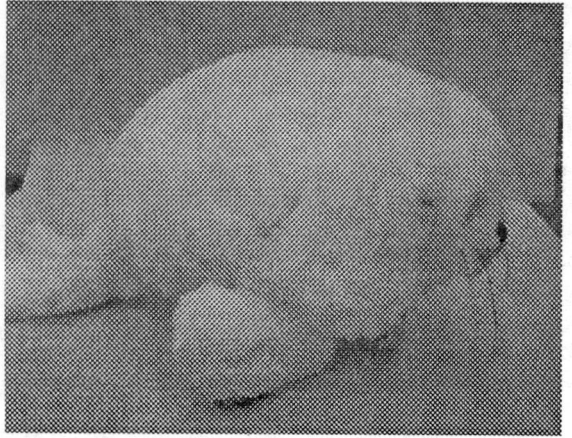

Fig. 6 (c) Side View
Fig. 6 Seal Robot

Fig. 7 (c) Sleeping
Fig. 7 Cat Robot

interpretation than the number of given functions. This was the effect of emergent emotions. Complexity of interpretation depended on subjects because it was very subjective view.

4.2.2 Cat Robot

Cat robot has more complex structure than the seal robot. It was built by OMRON Corp. [5] The robot has tactile, auditory, and postural sensors to perceive human action and its environment. For tactile sensors, it has four piezo-electric sensors on its back and one on its head, and one micro switch at its chin and two at its cheek. The robot can recognize stroking, hitting and touching. For audition, the robot has three microphones for sound localization and one for speech recognition. For posture, the robot has acceleration sensors for three axes. Information is processed one CPU with two DSPs for sensors and one IC for speech recognition. Its weight is 1.5 [kg] including battery.

For movement, the cat robot has one actuator of eyelid, two for neck, two for each front leg, and one for a tail. In total, it has eight actuators. There are one passive joint at front leg's ankle and three passive joints at each rear legs.

As the robot has complex structure, an emotion model is implemented. Therefore, the robot has some patterns and states of emotion for emotional expression, and some model to recognize its environment and human's action. For example, when the cat robot is called by name, it recognizes the word, detect direction of sound source, turn its face to the direction, blink its eyes for eye contact with a human. Also, when robot's internal state is happy, it cries while moving actively. However, in order to keep emergent property in emotional behavior through physical interaction, motion patterns, such as trajectory of a leg, are generated depending on internal states of the robot. Therefore, it is difficult for a subject to predict robot's action.

4.3 Discussions

Each robot has a costume of seal or cat. Without a costume, almost nobody would associate a seal with Fig. 6 (a). A costume gives quite various associations to subjects even without dynamic interaction. For example, in the case of cat robot, people who have experience owing a cat, expect the robot to react in the same way as a real cat. Therefore, when they interpret the robot's behavior, they always compare it with that of real cat.

At this point, we are evaluating the system by statistical data of subjects' impression. As the subjective data is varied, it is difficult to deduce out general principles for designing pet robots with high subjective value. We have to investigate a method to analyze subjective evaluation. A way to evaluate how much people liked a pet robot would be introducing the robot in a commercial market, and counting the number of sales.

5. Conclusions

We developed pet robots as artificial emotional creatures that generated their motivation and behavior through physical interaction with human beings by means of multi-modal information, especially tactile information. These robots have been developed to investigate emotional behaviors emerging through physical interaction with human beings.

We discussed subjectivity in interpretation and evaluation of robot's behavior in physical interaction. Evaluation methods will be investigated with more psychological experiments in the future.

Acknowledgment

This research had been done partially at the AI Lab., MIT. The authors appreciate to Prof. R. Brooks. We also appreciate to Prof. R. Pfeifer, the AI Lab., Univ. of Zurich, for discussion on emergence of emotion.

References

[1] T. Shibata, et al., Emotional Robot for Intelligent System - Artificial Emotional Creature Project, Proc. of 5th IEEE Int'l Workshop on ROMAN, pp. 466-471 (1996)

[2] H. Petroski, Invention by Design, Harvard University Press (1996)

[3] T. Shibata and R. Irie, Artificial Emotional Creature for Human-Robot Interaction - A New Direction for Intelligent System, Proc. of the IEEE/ASME Int'l Conf. on AIM'97 (Jun. 1997) paper number 47 and 6 pages in CD-ROM Proc.

[4] T. Shibata, et al., Artificial Emotional Creature for Human-Machine Interaction, Proc. of the IEEE Int'l Conf. on SMC, pp. 2269-2274 (1997)

[5] T. Tashima, S. Saito, M. Osumi, T. Kudo and T. Shibata, Interactive Pet Robot with Emotion Model, Proc. of the 16th Annual Conf. of the RSJ, Vol. 1, pp. 11, 12 (1998)

[6] R. Picard, Affective Computing, The MIT Press (1997)

[7] M. Toda, Design of a Fungus-Eater, Behavioral Science, 7, pp. 164-183 (1962)

[8] R. Pfeifer, 'Fungus Eater Approach' to emotion: A View from Artificial Intelligence, Cognitive Studies, The Japanese Society for Cognitive Science, 1 (2), pp. 42-57 (1994)

[9] V. Britenberg, Vehicles: Experiments in Synthetic Psychology, The MIT Press (1984)

[10] R. Brooks, Intelligence without Representation, Mind Design II (John Haugeland ed.) The MIT Press, pp. 395-420 (1997)

[11] M. Minsky, The Society of Mind, Simon & Schuster (1985)

[12] I. Asimov, I, Robot, Doubleday (1950)

[13] M. Fujita and K. Kageyama, An Open Architecture for Robot Entertainment, Proc. of Agent'97 (1997)

A Human-Robot Interface Using an Extended Digital Desk

Maho Terashima

Seiko Epson Corporation
Hirooka Office:80, Harashinden
Hirooka, Shiojiri-shi,
Nagano-ken, 399-0785 JAPAN

Shigeyuki Sakane

Faculty of Science and Engineering
Chuo University
1-13-27 Kasuga, Bunkyo-ku,
Tokyo, 112-8551 JAPAN

Abstract

Much attention has recently been paid to Augmented Reality (AR) systems, which can enhance a human's daily life by blending multi-modal information with the real world. Most existing AR systems provide such information for humans by using monitor displays or see-through HMDs (head mounted displays) as interface devices. In applying AR to human-robot interfaces in teaching tasks, however, alternative systems should be developed to overcome the limitations of the existing interfaces and to allow for more flexible interaction.

The paper presents an attempt to make such interface using an extended digital desk for guiding and teaching robot tasks. A prototype system consists of a projector subsystem for information display and a realtime tracking vision subsystem for recognizing human's action. Two levels of interaction have been developed: (1) VOP, virtual operational panel for primitive robot operations and (2) IIP, interactive image panel for intuitive teaching using projected live image of the task environment. Experiments of teaching robot tasks demonstrate usefulness of the proposed system.

1 Introduction

Much attention has recently been paid to Augmented Reality (AR) systems. In contrast to Virtual Reality which completely absorbs the user inside a virtual and synthetic work space, AR allows the user to see the real world with virtual objects superimposed[1],[2]. AR-based environment appears to the user that the virtual and real objects are blended in the same work space. Many application fields have been explored including medical visualization[3], manufacturing[4], annotation[5], and robotics [6],[7].

Most existing AR systems provide information for the user by using monitor displays or see-through HMD (head mounted display) for the interface devices. In applying AR to human-robot interface, however, alternative systems should be developed to overcome the constraints and limitations of the existing interfaces and to allow for more flexible interaction between the robot and the operator.

The paper presents an attempt to create such interface using an extended digital desk for guiding and teaching robot task. A prototype system consists of a projection subsystem for information display and a realtime vision subsystem for recognizing the user's action. Two levels of interaction have been developed: (1) VOP, virtual operational panel for primitive robot operations and (2) IIP, interactive image panel for intuitive teaching using projected live image of the task environment. Experiments of teaching robot tasks demonstrate usefulness of the prototype system.

2 Augmented reality and human-robot interface

2.1 AR-based systems for robotics

Typical AR-based human-robot interfaces provide human operator with useful information such as geometric models of the task environment blended with the live images. Matsui and Tsukamoto [6] developed a multimedia display which provides realtime overlay of 3D geometric models on the live images of the task environment. The system provides stereoscopic views of the scene and the model. Milgram et al.[7] developed ARGOS featured with a display employing 3D graphical overlays on an image of the real scene. Both of the systems, however, require a special hardware for supporting the 3D views.

Sato, Inoue, and Mizoguchi developed a system to assist human's assembly task[8]. Their system uses a projector to provide necessary information for the user in the course of assembly. Although it is aimed for assistance in fabrication rather than for robot teaching, a digital desk approach including recognition of the operator's action plays an important role.

2.2 An extended digital desk approach

We adopt an approach of "digital desk"[9] to establish bilateral communication between a human operator and a robot [10]. The original digital desk provided a virtual desktop GUI by projecting images on a desk and detecting the user's action. Taking the flexibility of the framework into account, we extend the digital desk to form a human-robot interface, especially for guiding and teaching robot tasks.

We have developed two levels of interaction: VOP (virtual operational panel) and IIP (interactive image panel), corresponding to lower and and higher levels of teaching robot tasks. We use an LCD projector for projection of VOP on a desk or on a paper on the desk. We have a choice of hardware systems to provide such panels for the operator. Compared with see-through HMDs, use of a projector has the following advantages:

- The operator needs no special equipment to wear and to adjust himself to the devices. Even when the operator changes to another person, there is no need to setup the system for the replacement.

- More than one operator can share the same information panels without additional equipment.

- Precise tracking of the operator's head position is not required.

Compared with hand-held LCD devices with touch screen, the approach of using a projector has the following advantages:

- The operator does not need to carry such devices.

- Projector has a larger projection area than that of the LCD devices.

- It is easily extended to project much information to objects and the task space.

There are some issues to be developed in using projectors. For instance, the information space is still limited to that of the projection area. However, if intelligent lights with functions of computer-controlled

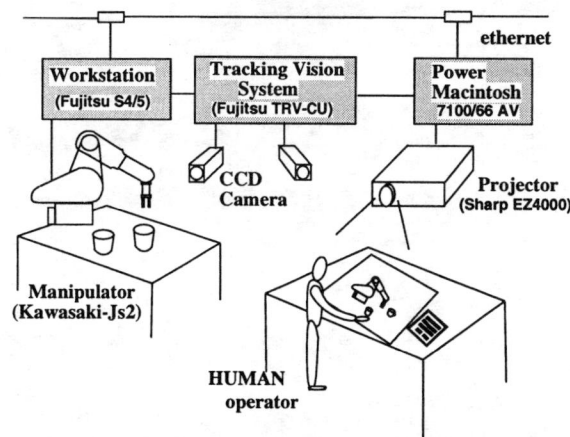

Figure 1: Hardware of the prototype system

projection were available in future, we could enlarge the projection area. Another issue is reliability in detection of finger's touch action. An LCD with touch screen is superior to projectors in terms of the reliability. For our purpose, we consider that extendability in projection to objects and task environment is important.

3 A prototype system

3.1 Hardware and software

Figure 1 shows hardware of the prototype system. An LCD (liquid crystal device) projector (EZ4000, Sharp) is used to provide information for the human operator by projecting VOP and IIP on a desk. A color tracking vision system (TRV-CU, Fujitsu) [11] works for fast tracking subimages based on a block matching algorithm. It is used for recognition of the user's touch action to the panels. Live images of the task environment are captured through a CCD camera with pan, tilt, and zoom control (EVI-G30, Sony).

The video output of the tracking vision system is to video input of a computer (PowerMacintosh 7100/66AV), which is also used as an X-window server for projection of VOP. A CCD camera (EVI-G20, Sony) is used for capturing images of the task environment. A 6-DOF manipulator (Js2, Kawasaki Heavy Industry) is used for manipulating tasks.

To implement software for the prototype system, we used EusLisp [12], an object-oriented lisp language developed for robotic applications at the Electrotechnical Laboratory. EusLisp allows the user to call X-window library functions.

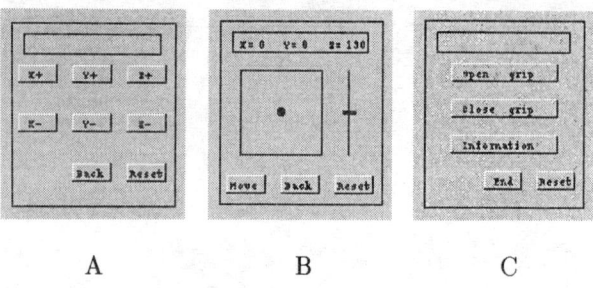

Figure 2: Examples of the virtual operational panels

3.2 VOP (virtual operational panel)

It would be quite useful if task dependent operational panels are available at the real environment and are adjusted their positions and orientations for the operator's requirements. The prototype system provides a flexible interface. It eases the restrictions on the position by projecting VOP on a desk used for robot tasks.

As an example, when a robot needs more precise instruction of the coordinates, the system projects a VOP which has buttons as shown in Figure 2. It allows to give numerical inputs through a virtual teaching pendant. In cases where an instruction of two-dimensional input is required, a VOP with a joystick function could be used. VOP permits to create items of button, slider, joystick, and information window. Examples of these items are as follows:

- The VOP of Fig.2(A) is a combination of button item and information window. User can specify numerical values of x,y,z coordinates. The information window can be used to show current coordinates.

- The VOP of Fig.2(B) shows a combination of slider, joystick, and information window items. This panel is used to set x and y coordinates using joystick and z coordinates using the slider item, respectively. The specified values are shown in the information window.

- The VOP of Fig.2(C) shows an example of information windows. More complex instructions may be given through the window.

VOP returns results of the recognition of the touch action by moving bars on the panel or by changing shade of sides of the buttons as if a real button were pushed down.

Nonverbal communication like a hand gesture helps to infer intentions of a person. In recent years, much

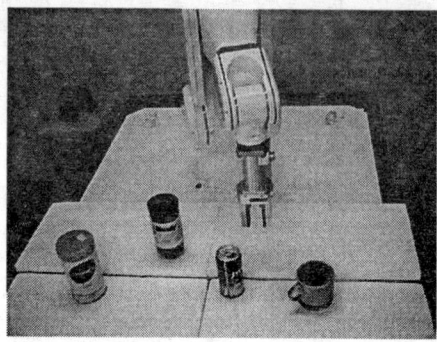

Figure 3: Projected live image of the task environment

effort has been devoted to visual recognition of human's gesture. Since we deal with operator's touch action to virtual panels, the action is constrained to a two-dimensional surface of a desk. So the recognition is more reliable than that of general three-dimensional cases.

3.3 IIP (interactive image panel)

If a robot allows the human operator to use pointing gesture to give instructions to move objects from here to there in the task space, such interface would be closer to the style of human communication. It is a higher level teaching method from a viewpoint of robot task. Moreover, we could superimpose additional information such as marks or geometric information directly on the object or somewhere close to that object as used in AR-based annotation systems. Our prototype system utilizes projection of live images of the robot task environment to a desk as shown in Figure 3. The operator gives his instructions on the projected image by his hand and fingers. In case another view of the robot task is required, teleoperation of a camera is available to the operator. When the operator moves his hand up and down or left-and-right on the desk, the CCD camera with pan-and-tilt function moves its view direction accordingly.

Optical flows are extracted to estimate motion of the operator's hand. That is, when some region monitored in the input image detects the hand's motion, the camera starts to move. When the operator needs to see some part of the scene, he only touches his finger to the object in IIP. Then the system estimates position of the pointed part and performs pan-and-tilt control to direct the camera to the object and then zooms in that part.

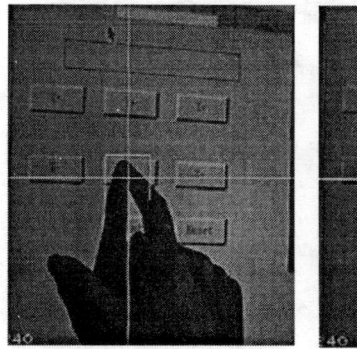
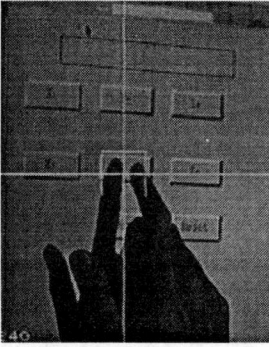

Figure 4: Real image of fingers and the shadows

4 Experiments

To validate effectiveness of the prototype system, we conducted experiments: (1) recognition of touch action for VOP, (2) evaluation of the position error in projection of VOP, and (3) teaching tasks using IIP.

4.1 Recognition of touch action for VOP

The button item is a base for all other items with recognition of the operator's touch action. We made experiments with the button item. To recognize the operator's action for VOP, we use the tracking vision system based on template matching. We utilize difference of the two images before and after the action. Figure 4 (left) shows image when the operator is touching the panel. Figure 4 (right) is an image before the touch action. A template image is stored when the operator is touching. The tracking vision system accumulates, in realtime, the matching error of the template image in the search area.

To reliably distinguish the template image, placement of cameras to monitor the task space is adjusted to give a sidelong view. In matching the template image with subimages in the search area, there are two methods: (i) to search the template at the button area and (ii) to search the template close to the operator's fingertip while tracking. We adopt the first method and set search areas at multiple button items in VOP. The positions of the search areas are determined relative to the corner of the button items.

We made experiments to measure how much the matching error values change during the operator's touch action. Figure 5 shows some of the results. The horizontal axis is time and the number along the axis corresponds to the number of image frames. The vertical axis is the matching error which we call "distor-

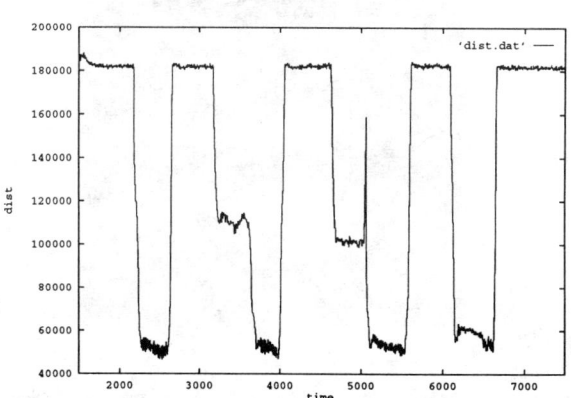

Figure 5: Distortion values of the touch action

Table 1: Distortion values

	touch	shadow only	finger only
average	52811.3	110023.7	100577.6
std.	8293.6	7235.4	8352.8
num. of data	27159	120	120
avg-2std		95552.9	83872.0

avg-2std = $(average - 2 \times std.)$

tion value". As shown in the figure, the value of error about 180,000 corresponds to a steady state when the operator takes no touch action on the panel. On the other hand, the value about 60,000 corresponds to a state of operator's touch action where a subimage was found to have very small error with the template image. Some parts of the graph have error value about 110,000. They correspond to matching with only a finger or shadow of the finger. We made experiments 80 seconds (data of 5 seconds was repeated) with each item in Fig.4. The statistics are summarized in Table 1. Concerning with the average values in the table entry, we take normal average for the touch action while we take average of the minimum values for detecting a finger or shadow of the finger since these minimum values are critical for discrimination of the matching.

To set a threshold with some margin for reliable discrimination of the matching status, we take into account two values: $(average + 2 \times std)$ and $(average - 2 \times std)$, where "std" denotes standard deviation. The former value is 69,398.6 and the latter is 83,872.0. Therefore we use 70,000 for the threshold and obtained good results of recognition.

Figure 6: Tracking a paper for projection of VOP

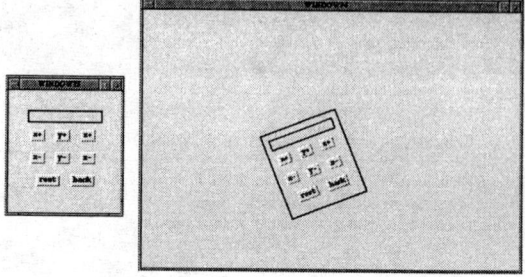

Figure 7: Window for the rotated image

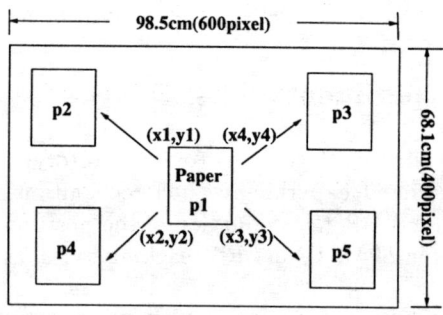

Figure 8: Projection of the virtual panel

4.2 Evaluation of position error in projection of VOP

To generate projected images we used weak perspective transformation [13] and affine transformation. To cope with translation and rotation of a paper on the desk, the system tracks three of the four corners to adaptively project VOP on the paper(Fig.6). The processing steps to generate VOP are the following:

1. Initial position of a paper is used to obtain a matrix (*prj-matrix*) to describe geometric transformation between the projector's image plane and the coordinates on the paper. We use a weak perspective model for the transformation.

2. Similarly geometric relation (*cam-matrix*) between the camera's image plane and the paper is obtained.

3. After the paper was moved, new position of the corners are observed, the system transforms the coordinates, using *cam-matrix*, to those with respect to the initial coordinates of the paper.

4. An affine transform matrix (*afn-matrix*) is calculated with respect to paper's coordinates using the the corners obtained by the step 3.

5. The system transform the original panel image to the new position and orientation using *afn-matrix*.

6. The new panel image obtained by the step 5 is transformed to projector's window coordinates using *prj-matrix*.

To evaluate position error of the projection, we measure the error at various positions of the projection.

The projector is 275[cm] far from the desk and the camera is located with a slant of 30 degrees to the orthogonal direction of the desk. Figure 8 shows the projected positions and measured 500 times for each. Table 2 shows the results at each position. Since the projection area is 98.5 x 68.1 [cm] and projected window has a size of 600 x 400 [pixels], one pixel of the window corresponds to 0.164cm on the desk. The maximum error in the table is 5.61 [pixels] which corresponds to 0.92 [cm]. The experimental results show that the position error of the projection is permissible for our purpose.

4.3 Teaching tasks using IIP

The system allows the operator to use IIP (interactive image panel) which projects live images of the task environment for teaching robot tasks as shown in Figure 9. To identify target objects which the operator is touching his finger in the image, template matching by the tracking vision subsystem is used. Such templates and their names are stored as a database for the task. Recognition of the operator's touch action is required on the projected live images. In contrast to VOP, it is more difficult since variation of the target

Table 2: Error of position

		p1	p2	p3	p4	p5
upper-left	x1	1	0.65	-1	-1.02	-2
	y1	1.97	-0.87	2.88	1.55	2.71
upper-right	x2	0	-1.94	-1.21	-4.0	4.04
	y2	-0.52	-2.24	1.12	3.23	4.59
lower-left	x3	-2.42	-5.03	-3.29	-3.53	1.52
	y3	3.89	0.66	2.85	3.65	4.94
lower-right	x4	-0.42	-1.68	-3.53	-1.55	-2.03
	y4	3.37	2.10	5.61	3.06	5.06

p1:center, p2:upper-left, p3:upper-right, p4:lower-left, p5:lower-right (with respect to the projected coordinates)

Table 3: Distortion values of the items

	without touch		with touch		
	avg	std	avg	std	a+2s
Desk	65812.0	0	26224.9	610.78	27446.5
Coffee	65803.6	16.13	30974.3	623.20	32220.7
Cup	64638.1	305.74	33204.0	8390.68	34883.3
Pot	65380.9	118.71	27582.5	1115.97	29814.5
Robot	65722.6	70.15	28352.2	716.09	29784.4

avg=average, std=standard deviation,
a+2s=average + 2× std

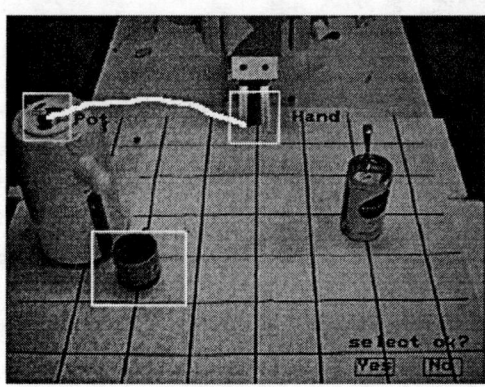

Figure 9: Sample view of the teaching robot task

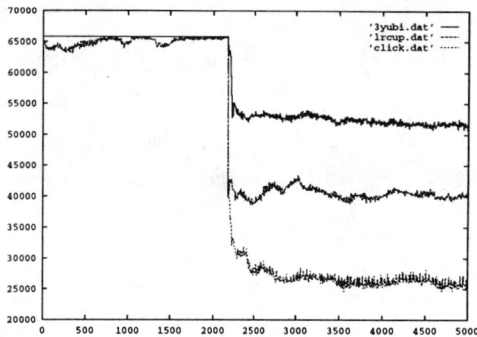

Figure 10: Distortion values

object image affects reliablity of the template matching. To suppress such variation we use high iris control for the camera which monitors operator's action.

We made experiments to measure how much the variation of the background can be suppressed. Table 3 shows the difference of the distortion values of target images with and without the touch action. The average values are obtained over measurements of 2000 times for each entry. Figure 10 shows a part of the experimental results. The horizontal axis is time and the vertical axis is distortion value. The right part of the figure shows three levels. The graph with the highest level corresponds to a case when the finger is 3 [cm] far from the desk. The graph with the middle level corresponds to a case when the finger is touching a cup at the bottom-right of the image. The graph with the lowest level corresponds to a case when the finger is touching the desk. The graphs change to low values when the time is over 2000. And they correspond to the touch action.

We made measurements of 2000 times and obtained average values: 52,225.5 and 44,855.2 when the finger is 3 [cm] and 1 [cm] above the desk, respectively. So a threshold value 43,000 is used and we obtained good recognition results. Fundamental flow of teaching tasks is as follows:

(1) In the projected live images, the operator touches a target object or indicates rectangular region as shown in Fig.9.

(2) The system identifies the target object using the tracking vision subsystem and displays the results superimposed on the projected image (ex. the character Hand or Pot in Fig.9).

(3) The operator verifies the results of the identification in terms of the projected information (as shown in the bottom-right of Fig.9).

(4) Indicate the target object just like to move to a desired place by finger's touch to the object and moving the finger to the place. The system simultaneously superimposes the indicated trajectory on the projected image as shown in Fig.9.

Figure 11: Overview of the experiment

(5) When the operator departs his finger from the image, the position is regarded as the destination of the movement and the system move the object to that place.

The robot task includes the following steps: (1) move a coffee cup under the pot, (2) push the top of the pot, (3) move the cup to the center of the desk, (4) grasp a spoon, (5) put some sugar into the cup, (6) put down the spoon, and (7) serve the cup. Through the experiments, we verified effectiveness of the teaching method using the projected live images. Figure 11 shows an overview of the experiments. In the experiments, the projector is set with horizontal orientation instead of setting at the ceiling and the images are projected on a white board. Since LCD projectors with light weight and compact size are recently developed, we will use the proposed setup in the near future.

5 Conclusions

Human-robot interface is a key for application of future robot systems to various fields. We proposed an extended digital desk as a flexible and gentle communication tool between human and robots. The system supports operator's teaching tasks to robots. Two levels of interaction have been developed: (1) VOP, virtual operational panel for primitive robot operations such as a teaching pendant and (2) IIP, interactive image panel which projects live images of the task environment and allows intuitive instructions for teaching tasks. Experimental results show usefulness of the basic interactive functions of the prototype system.

Future subjects include to extend the prototype system with a function of pointing objects directly in the real environment and to integrate various levels of human-robot interaction.

References

[1] Computer-Augmented Environments: Back to the Real World, *Communication of ACM*, pp.24-97, 1993.

[2] Azuma,R.T.: "A Survey of Augmented Reality," *PRESENCE*, Vol.6, No.4, pp.355-358, 1997.

[3] Bajura,M., Fuchs,H., and Ohbuchi,R.:"Merging virtual objects with the real world: Seeing ultrasound imagery within the patient," *SIGGRAPH*, pp.203-210, 1992.

[4] Feiner,S., MacIntyre,B., and Seligmann,D.: "Knowledge- based Augmented Reality," *Comm. of the ACM* 36(7), pp.52-62, 1993.

[5] J.Rekimoto: "Augmented interaction: Toward a new human-computer interaction style based on situation awareness," *Interactive Systems and Software 2 (WISS'94 Proceedings)*, pp.9-17, 1994.

[6] Matsui,T. and Tsukamoto,M.: "An integrated robot teleoperation method using multimedia display," *In Proc. of 5th Int. Sym. on Robotics Research (ISRR)*, pp.156-163, 1989.

[7] Milgram,P. et al.: "Application of Augmented Reality for human-robot communication," *Proc. of Int. Conf. on Intelligent Robots and Systems*, pp.1467-1472, 1993.

[8] Sato,T., Inoue,E., and Mizoguchi,H.: "Support in situ for operation of fabrication," *JSME Annual Conf. on Robotics and Mechatronics (ROBOMEC '95)*, pp.535-538, 1995 (in Japanese).

[9] Wellner,P.: "Interacting with paper on the DigitalDesk," *Comm. of the ACM*, 36(7), pp.86-96, 1993.

[10] Terashima,M. and Sakane,S.: "Augmented reality and human-robot interface, -An interface using a digital desk approach," *14th Annual Conf. of Robotics Society of Japan*, pp.691-692, 1996 (in Japanese).

[11] Morita,T., Sawasaki,N., Uchiyama,T., Sato,M.: "Color tracking vision system," *14th Annual Conf. of Robotics Society of Japan*, pp.279-280, 1996 (in Japanese).

[12] Matsui,T. and Inaba,M.: "EusLisp: An object-based implementation of LISP," *Int. Journal of Information Processing*, vol.13, no.3, pp.327-338, 1990.

[13] R. Cipolla:"Uncalibrated stereo vision with pointing for a man-machine interface," *In Proc. IAPR Workshop on Machine Vision Applications*, Kawasaki, pp.163-166, 1994.

Proceedings of the 1999 IEEE
International Conference on Robotics & Automation
Detroit, Michigan • May 1999

AUTOMATIC ORIENTING OF 3D SHAPES BY USING A NEW DATA STRUCTURE FOR OBJECT MODELING.

A. ADÁN*, C. CERRADA**, V. FELIU***
* Escuela Técnica Superior de Ingeniería Informática
Universidad de Castilla La Mancha. SPAIN (aadan@inf-cr.uclm.es)
** Escuela Técnica Superior de Ingenieros Industriales
UNED. SPAIN (ccerrada@ieec.uned.es)
*** Escuela Técnica Superior de Ingenieros Industriales
Universidad de Castilla La Mancha. SPAIN (vfeliu@ind-cr.uclm.es)

Abstract

This work presents a solution to the problem of determining the orientation of arbitrarily located 3D parts. The proposed algorithm is based on a new data structure specifically designed to store spherical representation models of 3D objects. This data structure, called Modeling Wave Set (MWS), allows keeping redundant information about local and global features of an object under a unique framework. It involves a different approach to deal with the modeling process providing new methods for solving generic problems like object recognition or pose determination.

The algorithm described in this work is based on the fact that a rotation of a solid can be seen as a change in the wave used to model it, because of the invariance to spatial transformation of the MWS structure. This property derives to an algorithm simpler and with lower computational cost than any other method based on spherical representations.

The algorithm has been tested over a wide set of polyhedral and free-form shapes. Success ratio and 3D orienting errors obtained with the proposed algorithm can be considered as acceptable, as it is shown in the experimental results section.

1. Introduction.

3D object pose determination is required in many robotics and computer vision applications and consequently it has been a problem studied from a great diversity of viewpoints. Obtained solutions are very much related with the working constraints, such as the kind of sensor used or the shapes of objects considered, conducting to different approaches to cope with the problem.

Model based approach is the common way of dealing with the problem. Nevertheless, there are methods that deal with the problem of rotation matrix estimation without using any previously defined object model [1], [2]. Most of them are based on features extraction and matching algorithms [3], [4], but they usually assume that the correspondence between three-dimensional points is known in advance.

Spherical representation model based systems have been developed from the eighties. Orientation problem has been solved for convex objects using the EGI by means of a combination of volumes like in [5], deriving to methods with little robustness and many shape constraints. The use of CEGI improves the solutions because position can also be estimated by means of least squares methods [6], [7].

The SAI model [8], [9] defines an invariant representation valid for concave and convex objects. In these works a distance function extended to all the possible combinations of angles (α,β,γ) for the rotation matrix is minimized to estimate the orientation. Because of the very high computational cost involved in this process a coarse-to-fine approach is suggested to reduce the search space. Potential minima are found using a coarse discretization of the (α,β,γ) space and then a searching is performed around these potential minima using a higher resolution [15].

In this work we present an algorithm for determining the 3D orientation developed from the LSR model, our spherical representation model. The LSR model collects local features of a 3D object over a new discrete data structure. Like other spherical models it can be used in environments that provide range information from the closed surface of an individual object in the three dimensional space. It is suitable for many polyhedral and free-form shapes, except those with abrupt surfaces or non-zero genus, as it is usual in this kind of models.

The construction process of the LSR model is significantly simpler and faster than in its predecessors because usual iterative steps are removed. The use of the data structure called *Modeling Wave (MW)* allows a new approach to the mesh-to-solid fitting process and a more efficient mesh regularization stage. The orientation determination method is built over the LSR model and it benefits of the MW also. The key point of the algorithm is to use the MW as a fast searching tool to estimate the rotation matrix.

This paper is organized as follows. Section 2 establishes the general statement of the problem we are dealing with. Then the data structure called *Modeling Wave Set* is defined. How it is used to build the LSR model is explained in section 4. Section 5 describes the orienting estimation algorithm, while section 6 shows the experimental setup and the main obtained results. Advantages and disadvantages of our method are discusses in the last section.

2. General statement of the problem.

Let us assume an initial situation in which H represents a set of available points from the surface of a given closed object, denoted as Θ. Let H' be another set of points belonging to the same object in a different spatial pose. We deal with the problem of finding a rotation matrix R such that the following spatial transformation for Θ is verified:

$$H' = R \cdot H$$

In a real case, the points of H' usually never verify to be the same points of H rotated, and even $ORD(H) \neq ORD(H')$. Then the solution will try to find the best estimate for R. On the other hand, it is assumed that the object has been identified in advance and it belongs to our database.

In what follows, the **original** object (i.e., that one stored in our database, or any of its spatial transformations) and its model will be symbolized by ($\clubsuit$). The same way, the **unknown** object (i.e., that one whose relative orientation with respect to world coordinate system {W} is unknown) and its model will be marked as ($\heartsuit$). Figure 1 shows a representation of the previous conditions.

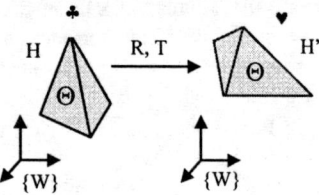

Figure. 1. General statement of the problem of 3D object orientation.

3. Data structures definition.

Our representation model is defined over a mesh built of nodes coming from the tessellation of the unit sphere. Let T_i be this initial that includes an implicit coordinate system. Basically, T_i is implemented as a data structure valid to represent the 3D object in a simplified and normalized manner and, on the other hand, used for mapping some features of the same object.

A way to obtain T_i is by projecting the vertices of a regular polyhedron over the unit sphere. Semi-regular patterns can be achieved by dividing each spherical patch. In our case we are considered a tessellated sphere formed by 320 nodes and 162 polygonal patches, where three connectivity is imposed (see Figure 2).

Let $\{N_{i1}, N_{i2}, ... N_{in}\}$ be the T_i set of nodes. Two relationships based on vicinity and belonging properties can be devised over this set of nodes. As every node N_{ij} has exactly three neighbors in the mesh, a standard vicinity relationship $V(N_{ij}) = \{N_{ij1}, N_{ij2}, N_{ij3}\}$ can be defined. Reciprocally, N_{ij} is one of the neighbors of each of these three nodes. On the other hand, it also can be stated that each node N_{ij} belongs exactly to three polygons. These two relationships can be arranged in tables.

These two arrangements can be considered as typical in the spherical representation systems and they maintain a local characterization. We were interested in building new structures to maintain wider relationships among subsets of T_i nodes. In that sense we have introduced a new elemental structure. This structure organizes the nodes of T_i in disjoint subsets following a belonging relationship. Each subset contains a group of nodes spatially disposed over the sphere as a closed quasi-circle. The number of nodes of each subset grows up from one to a maximum value and then decreases again up to one (or two) for the last subset. Because of this organization resembles the behavior of a wave we call it *Modeling Wave (MW)* (see [12]). Consequently we call *Wave Front (WF)* to each of the disjoint subsets. Observe that to build a MW a node of the tesselated sphere must be chosen as the origin of the structure. This node, called *Initial Focus*, constitutes the first WF of the MW and will be used to identify it.

Formally, a *Modeling Wave* of *Initial Focus* N can be defined as a partition $\{F^1, F^2,...F^q\}$ of T_i verifying:

1. $F^1 = \{N\}$
2. $F^{j+1} = V(F^j) - \bigcup_{m=1}^{j-1} F^m \quad \forall \; j:1..q-1$

where V is the typical neighborhood relationship of 3 connectivity imposed for T_i, F^j is the j-th WF generated in the process being *contiguous* to F^{j+1}. Then two new relationships are established for the nodes: a neighborhood relation called *contiguity* defined among WF, and a belonging relation of each node to one and only one WF. Both aspects allow defining a novel topology over the initial structure that manipulates more resources and can be useful in more applications.

Three major properties can be enumerated to clarify conceptually the powerful of this structure. Except by F^1 and F^q, it is verified that: I) Each node has no neighbor in its own WF. II) All the WF are disjoint. III) Each node has all its neighbors in its two contiguous WF. Figure 2 illustrates the 7 first WF of a given MW.

A more complex structure can be introduced from the elemental one. It can be observed that any node of T_i can be Initial Focus and, therefore, can generate a MW. Then $ORD(T_i)$ different MW can be generated. While all the MW should be identical in an ideal case slight variations appear in practical because of the quasi-regular nature of the discrete mesh. So, there is different number of WF as well as different number of nodes per WF depending on the chosen Initial Focus. We call *Modeling Wave Set (MWS)* to all the possible MW that can be generated over a given tessellated sphere. Figure 3 shows a schematic representation of this structure.

initial mesh T_i is deformed to fit the normalized surface of the object. The deformation phase can be subdivided in two other conceptual steps. In the first step, the original range data is firstly normalized by applying a scaling factor to make the object be bounded by the unit sphere. Then the T_i nodes are approximated to the normalized object surface. In the second step the deformed mesh is regularized. This regularization is based on the propagation of movements experimented by the nodes following the direction implicitly supplied by the Wave Fronts of the Modeling Wave [13]. We call $T_r(S)$ to the resulting mesh at the end of this phase, being S the normalized surface of the considered object.

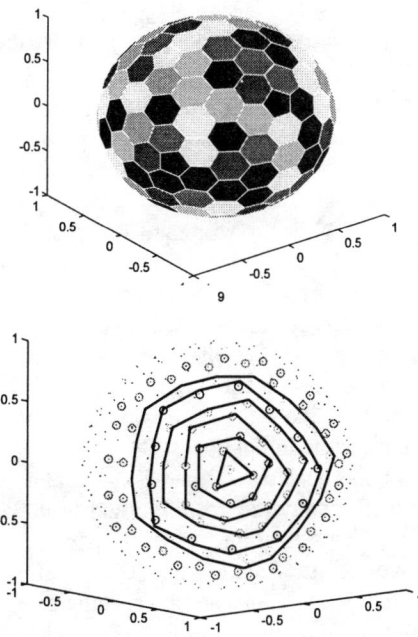

Figure. 2. Initial tessellation T_i (up) and construction of a Modeling Wave (down).

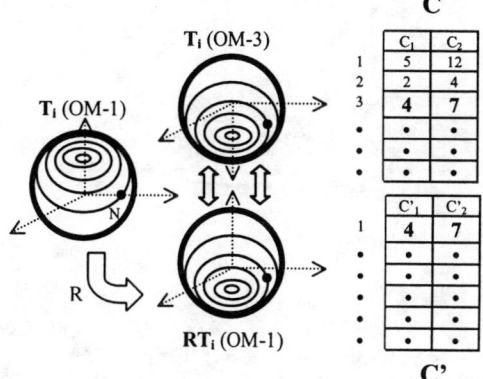

Figure. 4 Equivalence between T_i rotation and change of MW showing the reference systems associated to the two considered MW.

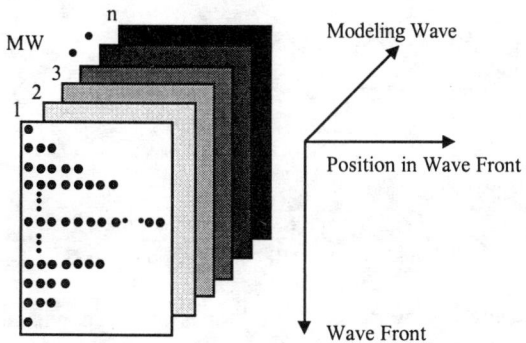

Figure. 3. Illustration of the data structure Modeling Wave Set

Given a MW each node N of T_i can be located by the WF it belongs to and by an ordinal number inside this WF. It means that two coordinates (c1, c2) determine the node N once the MW is chosen. Because n possible MW could be chosen there are n possible coordinate pairs for node N, Therefore, node N can be completely determined in the MWS by a matrix C of dimension $nx2$. It can be shown that C values remain invariant for any rotation of T_i. That is so because each MW defines an implicit coordinate system and a rotation of the main reference system can be explained as a change of the MW. Figure 4 shows graphically this effect.

4. LSR Representation Model

The process of building our 3D representation model of an object is carried out in two phases. In the first phase an

The second phase of model building consists on computing discrete values of some three-dimensional object features and storing them conveniently in the WMS structure. In the present case, the discrete curvature angle (*Simplex Angle* in [9]) is evaluated at each node N_{rj} of $T_r(S)$ and associated to the corresponding N_{ij} node of T_i. We denote as $L_S(N_{ij})$ this local feature and we call LSR (*Local Spherical Representation*) to the overall spherical distribution, denoting by LSR(S) the specific model for the considered object.

Due to the multidimensional nature of our MWS structure when we map the discrete curvature onto it we obtain n different curvature subspaces for the same object, i.e. one for each MW. Each subspace supplies a curvature map of the object organized by the corresponding WF.

Let us remark that MWS structure provides information containing an intrinsic spatial meaning. Our representation model allows to present information about the curvature of the object surface from as many viewpoints as MW exist. In fact, each MW defines an observer viewpoint. We call *LSR-MW-j* to the j-th curvature subspace generated by the j-th MW.

5. Algorithm to determine orientation

The Figure 5 shows schematically the modelization process of a given object Θ in two different locations. To determine the rotation matrix $R(\alpha,\beta,\gamma)$ we first use a fast and robust algorithm that provides initial values $(\alpha_1,\beta_1,\gamma_1)$ near to (α,β,γ) such that the remaining work consists on refining these rough values. We call R_1 to the rotation matrix defined by $(\alpha_1,\beta_1,\gamma_1)$. With this strategy the costly searches extended to 2π radians intervals are significantly reduced.

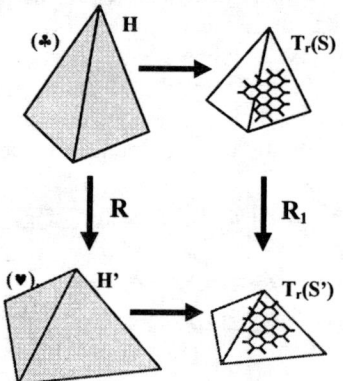

Figure. 5. Modelization process of an object in two different locations.

R_1 determination is carried out by means of the representation models obtained from H and H'. The algorithm uses the MWS structure to establish in a robust way a point to point correspondence between $T_r(S)$ and $T_r(S')$.

Two corresponding nodes, one of $T_r(S)$ and other of $T_r(S')$ must be found in a first stage. Let us have the MW-1 from the LSR(S) and LSR(S') models. It is obvious that the nodes N_{r1} and N'_{r1} are not equivalent (dual) concerning their location at the object surface due to the rotation existing between the two object configurations. The algorithm must find a node N'_{rx} of $T_r(S')$ that should be dual of N_{r1}, the Initial Focus of $T_r(S)$, in the sense that the associated curvature subspaces LSR(S)-MW-1 and LSR(S')-MW-x must be analogous. Therefore it can be said that N_{r1} and N'_{rx} are dual nodes and MW-1 and MW-x are dual Modeling Waves for the locations ♣ and ♥.

In order to obtain N'_{ix}, the Initial Focus of the MW-x, a new parameter called *Curvature Distance* between two subspaces is introduced. With this parameter all the curvatures of two MW are compared node by node following the order established by the Wave Fronts, providing a single accumulative value. It is evident that if the object S and the rotated object S' are traversed following the same path the values of curvature obtained in both WF will be analogous. Formally, we call *Curvature Distance* between two meshes $T_r(S)$ and $T_r(S')$ with respective Modeling Waves MW-1 and MW-j to the function:

$$^jD_C = \sum_{k=1}^{q}\left[\sum_{p=1}^{ORD(^jF^k)}\left[L_s(^1N_{ip}^k) - L_{s'}(^jN_{ip}^k)\right]^2\right]$$

where $^jN_{ip}^k$ is the p-th node of the k-th Wave Front of the MW-j for T_i and $L_S(^jN_{ip}^k)$ is the discrete curvature associated to that node. We assume that the values of discrete curvature are ordered for all the WF and all the possible MW $^jF^k$, j=1...n, k=1..q.

Finally, we say that N'_{rx}, with $N'_{rx} \in T_r(S')$, is the *Dual Node* of N_{r1}, with $N_{r1} \in T_r(S)$, or equivalently that MW-1 and MW-x are *Dual Modeling Waves* in S and S' respectively if $N'_{ix} = {}^hF^1$ being $^hD_C = \min\{^jD_C\}$, j=1,...n.

Notice that this correspondence between two nodes that are Initial Focuses implies the correspondence of all the Wave Fronts for the object in the original and unknown locations, because do not exist two different Modeling Waves with the same Initial Focus.

Once N'_{rx} has been determined as the dual node of N_{r1} a rotation of $T_r(S)$ should be accomplished to make N'_{rx} and N_{r1} coincident. This rotation matrix R_1 is computed in the second stage of the algorithm. The rotation can be decomposed in two rotations around specific axes, as shown in the Figure 6. Formally it can be expressed as $R_1(\varphi,\theta) = R_v(\varphi) \cdot R_u(\theta)$. Conventional methods can be used to compute φ and θ angles and to obtain from them the corresponding rotation angles $(\alpha_1,\beta_1,\gamma_1)$ with respect to the X-Y-Z coordinate axes (see [14]).

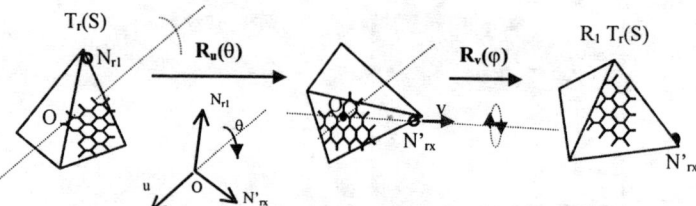

Figure. 6. Representation of the R_1 determination process.

The third stage of the algorithm is involved with the refinement process for the rotation angles around the X-Y-Z axes. The refinement performs a sweep of very small angle intervals and considers the normalized data points at S and S'. The amplitude δ of the intervals will be empirically determined. A least square minimization method is used to obtain the definitive rotation matrix $R(\alpha,\beta,\gamma)$.

6. Experimental Results

The method proposed in this article has been verified using synthetic as well as real range data coming from a *Gray Range Finder*. By using synthetic data a more exact measurement of the errors in the R calculation has been possible. Errors when using real data are measured with less precision because true values only can be roughly determined. Anyway, especial attention has been paid to test the effectiveness of the method for determining the orientation with the help of our models.

The algorithm has been checked over 40 real objects, 20

polyhedral shaped and 20 free-form shaped (see Figure 7) at several locations. The experimentation has been addressed to obtaining the rotation matrix (R_1) between models as the key step to solve the overall problem. We have worked with low resolution meshes, i.e. 320 nodes per mesh, what means an increment in the error of R_1 but a reduction of the computational cost.

Figure. 7. Sets of polyhedral and free-form objects used to test the algorithm.

Figure 8 shows intermediate plots of the algorithm applied to a polyhedral object. Range images corresponding to the object at two different orientations are plotted in the first part. Then the respective deformed meshes $T_r(S)$ and $T_r(S')$ are shown. Node N_{r1} is drawn over the first mesh while three candidates to be N'_{rx} are represented over the second one. Finally the last subplot shows the original object rotated accordingly to the calculated R_1.

Tables 1 and 2 gather the main results of the experimentation for both polyhedral and free-form objects. In both of them, the first column contains an identification code for each object. The second column presents the absolute angular differences in degrees for the three angles (α,β,γ) between the values computed from R_1 and the true values corresponding to R. The third column labeled as "Place" indicates the order of the true (optimum) node N_{rx} in the list of dual node candidates. Notice that the dual node identification procedure explained in the previous section provides an ordered list of candidates. Those cases of wrong Dual Node selection are filtered before performing the R_1 computation. Next candidates of the list are tested consecutively until the true node appears.

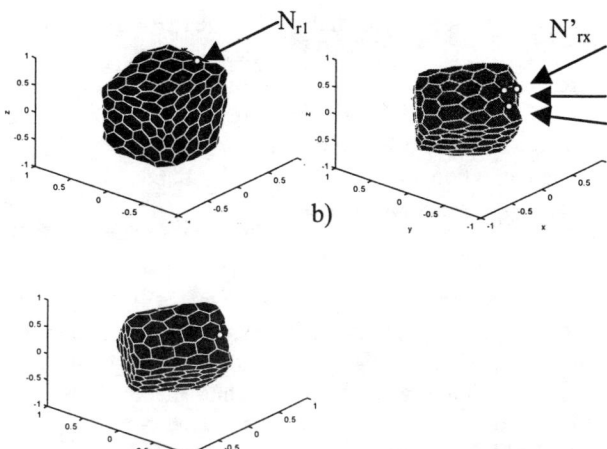

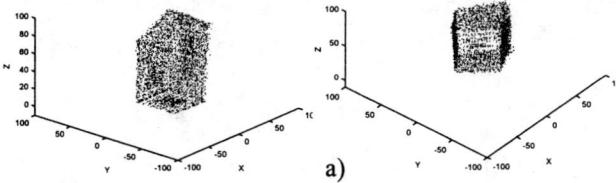

Figure. 8. a) Range image for an object at original (H) and unknown (H') locations. b) Associated models $T_r(S)$ and $T_r(S')$. c) Object positioning with the computed R_1.

Object	Error (°) R_1	Place	Object	Error (°) R_1	Place
Z0	Symmetries(OK)	1°	z20	11 1 13	1°
Z1	Symmetries(OK)	1°	z21	5 8 8	1°
Z2	2 3 1	1°	z22	9 3 2	1°
Z3	4 3 4	3°	z23	Symmetries(OK)	1°
Z4	1 2 1	1°	z24	3 3 2	2°
Z5	8 1 11	3°	z25	0 3 5	2°
Z6	0 3 1	1°	z26	1 2 7	1°
Z7	5 0 9	1°	z27	5 12 1	2°
Z8	3 4 4	2°	z28	No convergence	>3
Z9	3 2 2	3°	z29	10 1 5	1
Z10	9 5 9	2°	z30	0 0 1	2°
Z11	2 4 1	1°	z31	4 9 2	1°
Z12	3 0 0	3°	z32	3 2 3	1°
Z13	1 8 7	3°	z33	1 5 5	2°
Z14	0 2 2	2°	z34	Symmetries(OK)	1°
Z15	9 8 0	3°	z35	Symmetries(OK)	2°
Z16	2 8 4	1°	z36	4 4 9	2°
Z17	7 3 1	3°	z37	4 2 3	1°
Z18	7 8 2	2°	z38	3 2 0	1°
Z19	No convergence	>3	z39	11 9 4	1°

Table 1. Polyhedral objects   **Table 2.** Curved objects

From the examination of the results gathered in Tables 1 and 2 the following comments can be done:
- The method has successfully worked in the 95% of the cases, what proves its robustness for a significant multiplicity of shapes
- The average place of the true Dual Node in the lists of candidates has been 1.7, what proves the robustness of the Dual Node identification procedure.
- In general, errors in R determination reach greater values for free-form objects. The average absolute error in (α,β,γ) estimation over the true values has been 3.7°

for polyhedral shapes, and 4.3° for the free-form ones.
- Certain 3D shapes exhibit some kind of symmetries that can provoke the existence of more than one valid *Dual Node* N'_{rx}. Nevertheless, for all of the considered cases it has been possible to achieve acceptable approximation matrixes R_1. In Tables 1 and 2 these cases are marked as *Symmetries(OK)*.

It is important to remark that the application of our complete algorithm reduces by a factor of 18 times the amplitude of the searching intervals, i.e. the refinement is applied to +/-10° intervals. This fact implies a negligible cost compared with the conventional searching methods based on least squares [10], [11]. Most of this reduction is due to the use of the MWS structure. Notice that the inclusion of MWS in the process does not signify additional computational cost because this information is contained in our model database.

7. Conclusions

A solution to the problem of 3D shape orienting has been presented. The algorithm developed to solve this problem is based on a new spherical representation model that can be applied to 3D closed shapes. The model incorporates a new general data structure called *Modeling Wave Set* (MWS) built from a tessellated sphere. This structure allows representing an object from as many viewpoints as nodes the initial tessellation has.

The concept MW supplies an implicit spatial direction of searching. Each node can be found from *n* different MW defined in the model and, therefore, the model is defined n a multidimensional space by means of a *nx2* matrix of components invariant to rotations. This circumstance makes conceive a rotation of the object as a change of the MW used to represent it.

The method provides a reduction of the computation time with respect to its precedent approaches. In fact, after the node correspondence stage is solved, only one angle searching is required instead of three. On the other hand, our algorithm exhibits a great robustness because the searching is performed over a structure that completely defines the solid, instead of looking for local features with local criteria. This fact allows making posterior refinement tasks over small searching intervals of 5 to 10 degrees.

Another important characteristic of our method with respect to others is that it starts from a situation in which no *a priori* knowledge of the object location is available and no interaction with the user is required. In that sense, it can be labeled as an automatic method.

A verification test to validate the algorithm with polyhedral and free-form three-dimensional shapes has been performed. After analyzing the experimental results it can be concluded that the success rate as well as the error margins obtained in 3D orienting determination with our algorithm can be considered as satisfactory.

Acknowledgments

This research has been supported by the Spanish CICYT under the project labeled as TAP95-0129.

References

[1] T.Masuda N.Yokoya ."A Robust Method for Registration and Segmentation of Multiple Range Image". *Proceedings IEEE Workshop on CAD-based Vision*". Feb. 1994.

[2] K. Kemmotsu, T.Kanade. "Uncertainty in Object Pose Determination with Three Light- Stripe Range Measurements". *IEEE Trans. on Robotics and Automation.* Vol 11. No 5. pp 741-747. October 1995.

[3] P. W. Smith, N. Nandhakumar, C.H. Chien. "Object Motion and Structure Recovery for Robotic Vision Using Scanning Laser Range Sensor". *Proc. IEEE/RSJ Int.Conf.Robots &Sistems.*pp 117-122. 1995.

[4] J.Azarbajani, T.Galyean, B.Horowitz, A.Pentland. "Recursive Estimation for CAD Model Recovery". *Proc.IEEE Workshop on CAD-based Vision".* Feb 1994.

[5] J. Little. "Determining Object Attitude From Extended Gaussian Image". *Proc. Intern. Conf. on Art. Intelligence.* pp 960-963. August 1985.

[6] S. Bing Kang, K. Ikeuchi. "Determining 3-D Object Pose Using the Complex Extended Gaussian Image". *CVPR,* 1991

[7] S. B. Kang, K.Ikeuchi. "The Complex EGI: A New Representation for 3-D Pose Determination". *IEEE Transactions On PAMI*, Vol 15, No 7 , July 1993

[8] K. Higuchi, H. Delingette, M. Hebert, K. Ikeuchi "Merging Multiple Views Using a Spherical Representation". P*roceedings IEEE Workshop on CAD-based Vision".* Feb 1994.

[9] H. Delingette, M. Hebert, K. Ikeuchi. "A Spherical Representation for the Recognition of Curved Objets" *International Conference Computer Vision,* Berlin 1993

[10] K. Ikeuchi, N. Hebert "Spherical Representation: From EGI to SAI" *Technical Report. CMU.* October 1995.

[11] M. Hebert, K. Ikeuchi, H. Delingette." A Spherical Representation for Recognition of Free-Form Surfaces". *IEEE Transactions on PAMI",* vol. 17, no 7. July 1995.

[12] A. Adán. "Modelado 3D de Formas Libres Orientado a Reconocimiento y Posicionamiento de Objetos Tridimensionales". PhD. Thesis. Nov. 1997

[13] A.Adán, C.Cerrada, V.Feliu. "A Fast Mesh Deformation Method to Build Spherical Representation Model of 3D Objects". *ACCV.* Hong Kong. Enero 1998.

[14] J.J. Craig "Introduction to Robotics". *Addison Wesley,* 1986.

[15] H. Shum, M. Hebert, K. Ikeuchi, R. Reddy. "An Integral Approach to Free-Form Object Modeling". *PAMI* Vol 19 No 12, December 1997.

Toward Development of a Generalized Contact Algorithm for Polyhedral Objects

Christopher A. Tenaglia and David E. Orin

Department of Electrical Engineering
The Ohio State University
Columbus, OH 43210

Robert A. LaFarge and Chris Lewis

Intelligent Systems and Robotics Center
Sandia National Laboratories
Albuquerque, NM 87185

Abstract

This paper presents a contact model for polyhedral objects. For a given geometrical description, body state, and viscoelastic properties, it is possible to compute the contact wrench between polyhedral bodies of arbitrary shape and complexity. The C-Space Toolkit (CSTk) [1] geometry engine is used to determine interpenetration distances of the multiple points in contact between the bodies. Kinematic equations describing motion of the multiple points of contact are developed. A compliant contact force model is developed which models both impact and sustained contact, as well as elastic effects such as wedging and jamming. As objects rotate relative to one another, rolling effects are included as well. Forces along the surface at the contact model microslip as well as sliding conditions. Results are obtained for the example of a torus jammed on a cone.

1 Introduction

Dynamic simulation of robotic systems often requires modeling contact between two or more objects. A contact model is required to simulate common cases of contact, such as manipulators grasping parts or mobile robots operating on rough terrain. For the purposes of general simulation (where the nature of object interaction is not known a priori), the contact model should *efficiently* model sustained contact and impacts among multiple bodies while still providing realistic results. A desirable contact model should also work with arbitrary polyhedral models like those created with existing CAD packages.

One of the major tasks in contact simulation is to compute the contact forces, and basically two approaches exist for this purpose. One is the hard contact approach, like that presented in [2] and [3]. With the hard contact approach, bodies are assumed to be rigid, and sustained, persistent contact is treated as a set of algebraic constraints. Often, with a complex system of multiple interacting polyhedral objects whose contact conditions are changing, these constraints inhibit the ability to derive a single set of governing equations for a dynamic simulation. With the assumption of Coulomb friction, the rigid body model can result in nonexistent, or multiple solutions for the governing equations. It has also been shown that even when one consistent set of governing equations exist, the forces that maintain the constraints may be unstable [4].

The compliant contact approach, like that presented in [5], [6], and [7], models compliance and a localized deformation between bodies at the points of contact. This approach allows for non-rigid, elastic effects to be modeled, such as wedging and jamming, which may be important in a number of applications such as power grasp [8]. In addition, the compliant contact approach breaks kinematic loops and simplifies the dynamics computation because the kinematic chains are "open". Because of the importance of compliant effects in many applications, the compliant contact approach will be taken in this work.

Simulations using the compliant contact approach must quickly detect collisions and also determine spatial characteristics of collisions, such as object interpenetration distances. The C-Space Toolkit (CSTk), developed by Sandia National Laboratories, provides a geometry engine which efficiently determines distances between objects [1], and the CSTk will be used as the basis for the generalized contact algorithm for polyhedral objects which is developed here. Computational efficiency is obtained in the CSTk by maintaining a hierarchical ordering of the three-dimensional wrappers which contain the constituent polygons of a polyhedron. In addition, robust collision detection is provided by generating swept hulls of moving polygons. The CSTk provides functionality for a set of arbitrary polyhedral models, making it an attractive choice for detecting collisions when developing a generalized contact algorithm.

This paper develops a contact model for multiple interacting polyhedral bodies. A method for computing the kinematics of points in contact is first presented. A compliant contact force model is then developed which uses a nonlinear damping term to model impacts. Microslip and sliding conditions are modeled during sustained contact. Results are obtained for the case of a torus jamming on a cone.

2 Contact Modeling for Polyhedral Objects

2.1 Geometric Models

Polyhedral models are often used in graphical simulations of physical systems and in particular robotic systems. There are a number of reasons for using polyhedral models, including: fast simulation, simplified model construction, and reduced computation for collision detection. In this work, polyhedral objects of any geometry may be specified, including concave or convex shapes. A rich set of interesting objects

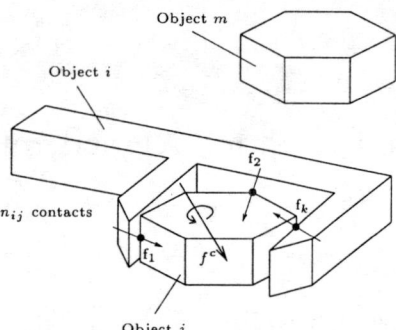

Figure 1: A scene consisting of several bodies which experience intermittent contacts. Of the m polyhedral objects, a pair of objects i and j contact each other at n_{ij} locations. Only a force acts at each location. Each force contributes to the total contact wrench f^c.

may therefore be simulated, such as manipulation and grasping systems, 3-D parts and feeders, and legged vehicles moving on uneven terrain.

An example of a scene with several polyhedral objects appears in Fig. 1. The contact algorithm developed in this paper will compute the force and moment components of the resultant contact wrench, f^c from one object onto another. The wrench is computed for all contacts between m objects (note that $m = 3$ in Fig. 1), any of which may sustain intermittent contact. The resultant contact wrench for any pair of objects i and j is determined as the combined effect of the forces transmitted at the n_{ij} contact points between the objects.

For the purposes of this paper, objects will be modeled as a polyhedral rigid core with a surrounding viscoelastic shell of uniform thickness. From all pairs within the set of m objects, the CSTk is used to compute the interpenetration distances of the polygonal faces of one polyhedral object into the elastic shell of another polyhedral object. In particular, the closest points are obtained on pairs of faces which are in contact. To simplify this computation, the combined shell depth, d_s, is used:

$$d_s = d_{ei} + d_{ej} , \qquad (1)$$

where d_{ei} and d_{ej} are the thicknesses of the elastic shells of objects i and j, respectively.

Figure 2 shows an example of two polyhedral objects coming into contact. Note that the elastic shells of the two objects have been combined to give a shell depth d_s per Eq. (1), and that d_s is shown completely on object j for simplicity. The CSTk returns the coordinates of the closest points on the rigid cores of objects i and j. Specifically, these points are denoted as c and l in Fig. 2. In order to compute the forces at the contact, the relative position and velocity must first be determined, and this problem is briefly addressed in the next section.

2.2 Contact Kinematics

First, some notation must be established. The world coordinate system is fixed in space and is denoted as $[x_0 \; y_0 \; z_0]$ and also as $\{0\}$. For some object i, a coordinate system $[x_i \; y_i \; z_i]$ or $\{i\}$ is fixed to and moves with it. Similarly, $\{j\}$ is defined for object j.

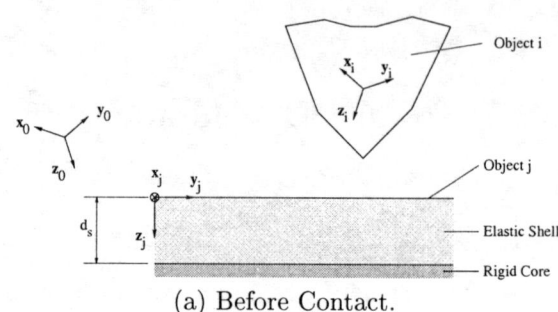

(a) Before Contact.

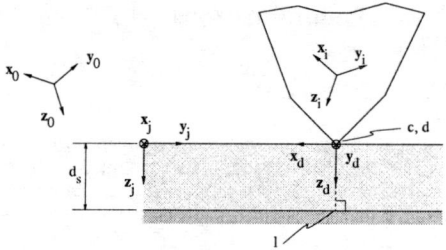

(b) Making Contact.

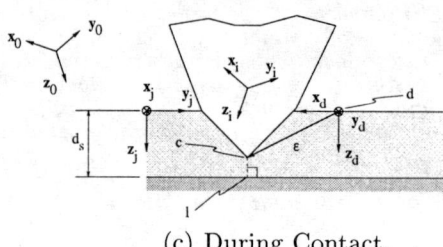

(c) During Contact.

Figure 2: Example of two objects coming into contact.

The position of the origin of $\{j\}$ with respect to $\{0\}$, with components expressed in $\{i\}$, is denoted as $^i(^0p_j)$. For convenience, $^i(^ip_j)$ is denoted as ^ip_j. The rotation matrix (3x3) that transforms vectors in $\{i\}$ to $\{j\}$ is denoted as ^jR_i.

Figure 2 shows objects i and j in successive stages of contact. If the magnitude of the distance between the closest points (c and l) on the rigid cores of the objects is greater than the elastic shell depth d_s, then there is no contact between the elastic shells. Specifically, $|^cp_l| > d_s$, for the case of no contact.

Initial contact of the objects occurs when $|^cp_l| = d_s$. Figure 2(b) illustrates this situation. In order to track the penetration of object i into the elastic shell, a point d is set on the surface of the elastic shell and remains fixed to object j. At initial contact, points c and d are coincident, and the contact state variable ^jp_d is initialized:

$$^jp_d = {}^jp_i + {}^j(^ip_c) . \qquad (2)$$

In order to track the penetration of the object into and along the surface, the frame $\{d\}$ is defined with its origin at d and its z_d coordinate vector normal to the surface. See Fig. 2. It is assumed that the normal to the contact surface lies in the direction of the shortest distance between the rigid cores, so that ^0z_d may be computed as follows:

$$^0z_d = \frac{^0(^c p_l)}{|^c p_l|} \ . \qquad (3)$$

The orientation of ^0x_d and ^0y_d on the contact surface is arbitrarily chosen to form a right-hand set of unit vectors. The rotation matrix for the contact coordinate system is simply $^0\mathbf{R}_d = [^0x_d \ ^0y_d \ ^0z_d]$.

When $|^c p_l| \leq d_s$, there is penetration. Let the displacement of the penetration be denoted as $\boldsymbol{\epsilon}$, which may be computed as follows:

$$^0\boldsymbol{\epsilon} = {}^0\boldsymbol{p}_c - {}^0\boldsymbol{p}_d \ . \qquad (4)$$

Note that the magnitude of $\boldsymbol{\epsilon}$ typically is small when the elastic shell is relatively stiff. The penetration velocity $\dot{\boldsymbol{\epsilon}}$ is computed as the relative velocity of c with respect to d. For the purpose of computing the velocities $^0\boldsymbol{v}_c$ and $^0\boldsymbol{v}_d$, the points are considered fixed to objects i and j, respectively.

3 Contact Force Model

Computing the stresses and strains at the contacts between two bodies is a complex function of their elastic properties, geometry, surface characteristics, and contact state (rolling, sliding, wedging, etc.). However, there are many cases in which the total effects are captured with simple models which account for the most dominant contact characteristics [9]. In this paper, the following two assumptions will be made so as to simplify the computational analysis:

1. The force at each point of contact may be computed as a function of the interpenetration distance and the velocity into, and along the surface of, the elastic shells of the two objects.

2. Only a force is transmitted at each contact point (moments occurring at individual points of contact are negligible), and each point's force may be computed independently of the other contact points.

The first assumption is consistent with Hertzian theory for two spheres whose contact area dimensions are small with respect to their radii. The first assumption is also consistent with Gesley's model, which predicts system behavior when the deformation is small, such as in a clock mechanism [10], [11]. The second assumption allows us to compute the force at each of the contact points independently. Once the contact states are known, it is possible (using superposition) to determine the contact wrench between two objects as a function of their relative position and velocity. In general, a moment is generated from one object onto another when the forces at the spatially distributed contact points are combined.

These assumptions are valid in many cases in which the dimensions of the objects themselves are large compared to the dimensions of the contact area so that the stresses in this region are not critically dependent upon the shape or contact of the objects distant from the contact area [9].

For the case of contacting spheres, the contact area and the normal contact force are essentially independent of tangential shear [9]. This is particularly true when the coefficient of friction between the two surfaces is appreciably less than unity, which is often the case. This approach is taken for the normal force so that it is separated out such that

$$^d\mathbf{f}^n = h^n(\epsilon_z, \dot{\epsilon}_z) \ , \qquad (5)$$

and

$$^d\mathbf{f}^t = h^t(\epsilon_x, \epsilon_y, \dot{\epsilon}_x, \dot{\epsilon}_y, {}^d\mathbf{f}^n) \ , \qquad (6)$$

and the total contact force is an augmentation of tangential and normal components:

$$^d\mathbf{f}^c = \begin{bmatrix} ^d\mathbf{f}^t \\ ^d\mathbf{f}^n \end{bmatrix} \ . \qquad (7)$$

In the next section, the model to compute the normal force is given, while the frictional force in both the microslip and sliding regions is given after that. Section 3 ends by providing an equation for computing the total contact wrench on an object.

3.1 Normal Contact Force

A method to compute contact force with a nonlinear damping term is presented in [6] and [12]:

$$^d\mathbf{f}^n = h^n(\epsilon_z, \dot{\epsilon}_z) = -K_n{}^d\epsilon_z - \lambda_n{}^d\epsilon_z{}^d\dot{\epsilon}_z \ , \qquad (8)$$

where K_n is the stiffness and λ_n is the damping coefficient in the normal direction. This method is computationally simple, provides a good model for impact, and accurately reflects the variation of the coefficient of restitution due to impact velocity. Later results by Marhefka and Orin [6] show that nonlinear damping is consistent with the case of spheres with sustained contact. The contact area grows (and therefore the damping) with the interpenetration.

3.2 Microslip Region

To model the friction force along the surface between two contacting objects, the tangential component of the contact force is separated into two terms: a position-dependent term and a velocity-dependent term. For small displacements at the contact or for the case of velocity reversals, the friction force is often mainly a function of the displacement [9],[13]. This characteristic accounts for Dahl's effect in ball bearings as well [14].

The position-dependent term is also appropriate for modeling contacting spheres in the microslip region before the frictional traction force reaches the limit where full sliding occurs [9]. The contact develops an annulus of slip along the periphery of the (elliptical) contact area as the tangential force increases. Within this annulus, the ratio of the tangential to normal stress values is equal to the limiting coefficient of friction (microslip) μ, while it is lower in the center region of the contact. As the tangential force increases, the annulus grows inward toward the center of the contact. The limiting elastic tangential force (before sliding) occurs when all the points within the contact region are (micro)slipping, and the tangential force is limited by

$$\mathrm{F}_s = \mu_s \mathrm{f}_p^n \ , \qquad (9)$$

where μ_s is the coefficient of static friction, and $f_p^n = K_n{}^d\epsilon_z$.

The relationship between the tangential displacement δ and the tangential force $\mathbf{f}^t$ in the microslip region is [9]:

$$\delta = \mu_s\, \mathrm{f}^n\, c \left\{ 1 - \left[1 - \frac{|\mathbf{f}^t|}{\mu_s \mathrm{f}^n} \right]^{2/3} \right\}, \quad (10)$$

where c is a constant which may be computed from the elastic constants for the spheres. A good linear approximation to this expression is [9]:

$$\delta = |\mathbf{f}^t|/K_t. \quad (11)$$

Again, K_t may be computed from the elastic properties of the spheres. One interesting characteristic of the tangential stiffness, K_t, is that the ratio of the normal stiffness, K_n, to the tangential stiffness varies from 1.17 to 1.5 for typical materials [9]. For the contact model presented here, the position-dependent term is simply computed as:

$$^d\mathbf{f}_p^t = -K_t\, \epsilon_{xy}\, {}^d\mathbf{u}_1, \quad (12)$$

where $\epsilon_{xy} = \sqrt{\epsilon_x^2 + \epsilon_y^2}$ is the magnitude and ${}^d\mathbf{u}_1$ is the direction of the tangential displacement.

3.3 Sliding Contact

Once the contact moves out of the microslip region, kinetic friction becomes important. The friction force is predominantly a function of the relative velocity at the contact. A comprehensive seven-parameter function model is used in [15], with good results, to develop a compensation for the control of an X-Y table. Besides the linear displacement-dependent term, they include Coulomb, viscous, and Stribeck friction terms. With Stribeck friction, the friction force decreases with increasing velocity within the low-velocity sliding region. The decrease is attributed to partial fluid lubrication in solid-to-solid contact [13].

In the present work, investigation is continuing in using terms to account for the Stribeck effect in stick-slip motion. For simplicity, at present only a viscous term is included for the velocity-dependent term:

$$^d\mathbf{f}_v^t = -B_t({}^d\mathrm{f}^n)\, \dot{\epsilon}_{xy}\, {}^d\mathbf{u}_2, \quad (13)$$

where B_t is the coefficient of viscous friction, and $\dot{\epsilon}_{xy}$ is the magnitude and ${}^d\mathbf{u}_2$ is the direction of the tangential velocity. Note that B_t is computed as a function of the normal force ${}^d\mathrm{f}^n$, and for simplicity, at present a linear function is used. The velocity-dependent force always acts in the opposite direction of the tangential velocity, and this term is also present during microslip. While the elastic term dominates the resistance in the microslip region, the damping can account for the energy loss that has been measured in a vibrating contact between two objects [9].

Figure 3 shows a close view of one point of contact between two bodies with the compliant contact model. As was the case in Fig. 2, d_s is shown on object j while interpenetration occurs. The displacement ϵ is

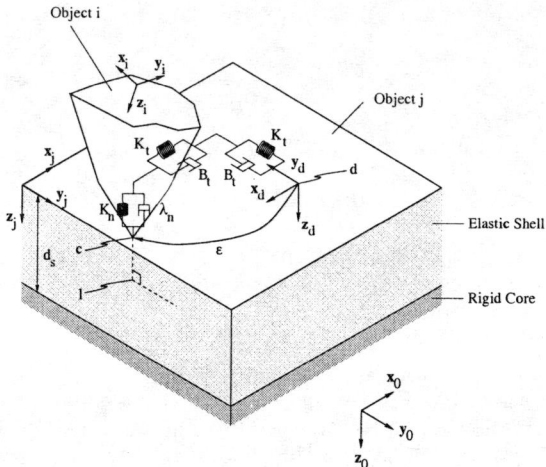

Figure 3: A close view of one point of contact between two bodies, showing the compliant contact model.

resolved into components along and into the surface of object j. The tangential displacement, ϵ_{xy}, is zero at first contact. Inspection of Eqs. (8), (9), and (12) then shows that the contact initializes in the microslip region.

When the tangential displacement becomes large, the contact moves from the microslip region to that of a full sliding contact. To adjust for sliding, the position of point d (see Fig. 3) is changed such that the center of the tangential "spring" is reset in order that the position-dependent force does not exceed the static friction value $\mathbf{F}_s$.

An algorithm that maintains the state of position d, to adjust for sliding, is given in Table 1. This algorithm has several important characteristics. First, the elastic displacement ϵ_{xy} is limited so that the tangential force does not exceed the limiting static friction value. Second, the location of d is continually adjusted so that it follows c during a sliding movement. Also, a query of the contact state on each iteration allows for consistent transitions between the microslip and sliding regions. Finally, since the position-dependent and velocity-dependent terms are considered separately, the tangential force is properly oriented when velocity reverses direction.

3.4 Resultant Contact Wrench Computation

Once the contributing forces from all the individual contacts with all other bodies are known, the total force and moment imparted on an object due to contact must be determined. The total contact wrench on object i, $\boldsymbol{f}_i^c$, is computed as the superposition of the contact force and moment due to contact with all the other objects:

$$^0\boldsymbol{f}_i^c = \sum_{j=1, j\neq i}^{m} {}^0\boldsymbol{f}_{ij}^c. \quad (14)$$

Note that $\boldsymbol{f}_{ij}^c$ denotes the contact wrench generated from contact between objects i and j.

Recall that n_{ij} contacts exist between objects i and

$$\begin{aligned}
&F_s = \mu_s K_n{}^d\epsilon_z \\
&\mathbf{f}_p^t = -K_t \epsilon_{xy} \mathbf{u}_1 \\
&\mathbf{f}_v^t = -B_t \dot{\epsilon}_{xy} \mathbf{u}_2 \\
&\text{IF } (|\mathbf{f}_p^t| > F_s) \\
&\qquad \epsilon_{xy} = \frac{F_s}{|\mathbf{f}_p^t|} \epsilon_{xy} \\
&\qquad \mathbf{f}_p^t = -F_s \mathbf{u}_1 \\
&\qquad \mathbf{p}_d = \mathbf{p}_c - \epsilon \\
&\mathbf{f}^t = \mathbf{f}_p^t + \mathbf{f}_v^t
\end{aligned}$$

Table 1: Algorithm to maintain a contact's state through microslip and sliding. Note that $\mathbf{u}_1$ is the direction of the tangential displacement, and $\mathbf{u}_2$ is the direction of tangential velocity.

j. The contact wrench $\mathbf{f}_{ij}^c$ is computed as a superposition of all of the n_{ij} contact forces and moments between objects i and j. Since $|\epsilon|$ typically is small, the force at a single contact is considered to act at point c. The contact wrench $\mathbf{f}_{ij}^c$ is then computed as:

$$^0\mathbf{f}_{ij}^c = \sum_{k=1}^{n_{ij}} \begin{bmatrix} {}^0({}^i\mathbf{p}_c) \times ({}^0\mathbf{R}_d{}^d\mathbf{f}_{ijk}) \\ {}^0\mathbf{R}_d{}^d\mathbf{f}_{ijk} \end{bmatrix}, \quad (15)$$

where $^d\mathbf{f}_{ijk}^c$ is the contact force generated from some contact k. Note also that c and d are used here to denote c_k and d_k for brevity. Recall also that $^d\mathbf{f}_{ijk}^c$ is an augmentation of normal and tangential components (see Eq. (7)). Combining Eqs. (14) and (15), the total contact wrench on object i may be computed as:

$$^0\mathbf{f}_i^c = \sum_{j=1,j\neq i}^{m} \sum_{k=1}^{n_{ij}} \begin{bmatrix} {}^0({}^i\mathbf{p}_c) \times ({}^0\mathbf{R}_d{}^d\mathbf{f}_{ijk}) \\ {}^0\mathbf{R}_d{}^d\mathbf{f}_{ijk} \end{bmatrix}. \quad (16)$$

4 Results

A standard dynamic simulation program (see Fig. 4) was developed using C++ on a Windows NT workstation with a 300 MHz Pentium II CPU. The simulation uses an interface to the CSTk API to determine contact penetration and normal direction. The simulation uses geometrical models of polyhedra that are compatible with standard CAD packages like ADAMS.

Two polyhedral models, a torus and a cone, are used to test the contact simulation program. In this example, the cone is inverted and is fixed while the torus is shoved upward onto the cone. This action is mechanically similar to the Morris taper, which is used in machinery to hold a collet in place.

In order to wedge the torus on the cone, the elastic force at the contact location must support the weight of the torus. The applied force f_A therefore obeys a constraint that is a function of the geometry of the system:

$$f_A > mg \left(\frac{\tan\gamma + \alpha}{\tan\gamma - \alpha} + 1 \right), \quad (17)$$

where m is the torus mass, and α and γ are the cone angle and friction cone angle, respectively. For this example, $\alpha = 20°$, and a friction coefficient of 0.6 is chosen ($\gamma \approx 30°$). The torus has a mass $m = 0.25$kg,

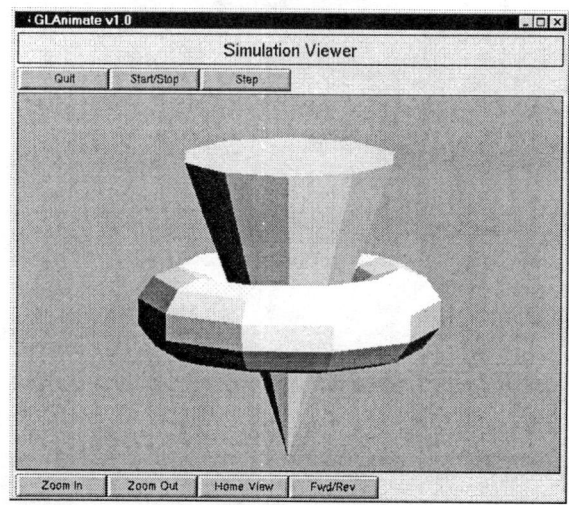

Figure 4: A graphical simulation of a torus jammed on a cone.

an outer radius of 0.07m and an inner radius of 0.03m. In this case, Eq. (17) requires $f_A > 7.4mg$.

A force of $8mg$ is applied at the time of the initial contact, and is removed once the torus stops moving upward. Results for this experiment appear in Fig. 5. The torus accelerates upward immediately after contact due to the applied force. When the torus stops moving upward, the vertical contact force reaches a value of $-8mg$. When the torus is released, the vertical contact force settles to a value of mg and completely supports the weight of the torus. The multiple points of contact are equally spaced around the cone, approximately 0.03m away from the cone axis. Simulation output predicts an outward force of 13.9N, which results in 74N/m of outward horizontal loading around the ring of contact.

Several experiments were conducted to verify that the model obeys the prediction of Eq. (17). Specifically, failure to wedge the torus was observed for three cases: an inadequate applied force of $7mg$, a blunt cone angle $\alpha = 25°$, and an insufficient $\mu = 0.4$.

5 Summary and Conclusions

The contact model developed in this paper uses the C-Space Toolkit (CSTk) to give interpenetration distances of one object into the elastic shell of another. CSTk uses a hierarchical representation of geometric features to efficiently compute the distances. The spatial features of a distributed set of multiple contact points are then determined.

Kinematic equations for the spatial motion of points in contact have been derived. These equations can be used along with any contact force computation model in which the contact forces are assumed to be a function of interpenetration distance and velocity. A contact state is also defined so that the frictional force is a function not only of the relative position and velocity, but the contact history as well.

The normal contact force computation employs a nonlinear damping term to simulate impacts as well as sustained contact. For more highly faceted polyhedra, the contact at the deepest penetration supplies the greatest normal force. This effect is consistent with

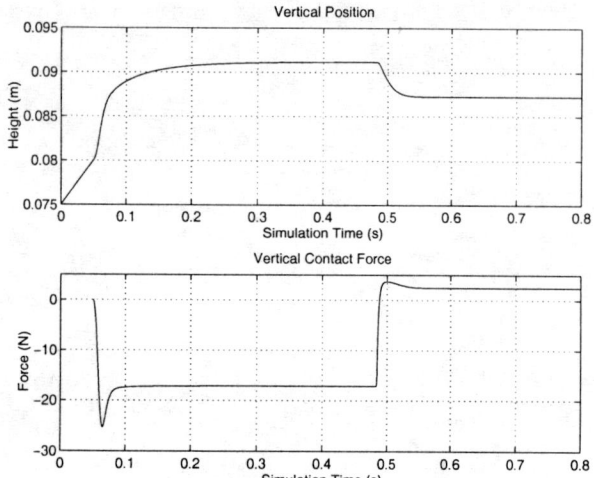

Figure 5: The vertical position and contact force applied to the torus shown in Fig. 4.

Gesley's model [10], [11]. A viscoelastic model along the surface models friction in the microslip and sliding regions. In general, a position-dependent term dominates the microslip region, while a velocity-dependent term dominates the sliding region.

Results show the conforming behavior of a torus wedging onto a cone. The effect of brief impact, as well as sustained contact, is evident in results obtained within the same simulation. CAD packages for creating dynamic simulations, such as ADAMS, allow for direct and simple incorporation of a contact model like the one presented in this paper. Hopefully, dynamic simulation packages that use polyhedra to model a rich set of objects will eventually incorporate contact models such as the one presented here to better simulate objects experiencing intermittent contacts.

6 Acknowledgments

Thanks to Mark Hanes and Duane Marhefka of The Ohio State University, who collaborated in the evaluation of the contact algorithm. Thanks also to Patrick Xavier of Sandia National Labs, whose work greatly contributed to the implementation of the contact force computation model presented here. Support for this work was provided in part by Sandia National Laboratories, under contract #AV-8825 to The Ohio State University. This work was supported in part by the United States Department of Energy under Contract DE-ACO4-94AL85000 to Sandia. Sandia is a multiprogram laboratory operated by Sandia Corporation, a Lockheed Martin Company, for the United States Department of Energy.

7 References

[1] P. G. Xavier, "Fast swept-volume distance for robust collision detection," in *Proc. 1997 ICRA*, pp. 1162–9, Albuquerque, NM, April 1997.

[2] J. Trinkle and J. Pang, "Dynamic multi-rigid-body systems with concurrent distributed contacts," in *Proc. 1997 ICRA*, pp. 2276–81, Albuquerque, NM, April 1997.

[3] B. Mirtich and J. Canny, "Impulse-based simulation of rigid bodies," in *Proc. Monterey Symp. on Real-Time Interactive Graphics*, April 1995.

[4] P. Dupont and S. P. Yamajako, "Stability of frictional contact in constrained rigid-body dynamics," *IEEE Transactions on Robotics and Automation*, vol. 13, pp. 230–236, April 1997.

[5] S. Goyal, E. N. Pinson, and F. W. Sinden, "Simulation of dynamics of interacting rigid bodies including friction I: General problem and contact model," *Engineering with Computers*, vol. 10, pp. 162–174, 1994.

[6] D. W. Marhefka and D. E. Orin, "A compliant contact model with nonlinear damping for simulation of robotic systems." EE Dept. Technical Report, The Ohio State University, July 1998.

[7] Y. T. Wang, V. Kumar, and J. Abel, "Dynamics of rigid bodies undergoing multiple frictional contacts," in *Proc. 1992 ICRA*, pp. 2764–69, Nice France, May 1992.

[8] K. Mirza, M. D. Hanes, and D. E. Orin, "Dynamic simulation of enveloping power grasps," in *Proc. 1993 ICRA*, pp. 430–5, Atlanta, May 1993.

[9] K. L. Johnson, *Contact Mechanics*. Cambridge University Press, 1987.

[10] A. Gesley, *Knowledge-Based Simulation: Methods and Applications*. Springer-Verlag, 1989.

[11] E. Rimon and J. W. Burdick, "Mobility of bodies in contact–part II: How forces are generated by curvature effects," *IEEE Trans. on Robotics and Automation*, vol. 14, pp. 709–17, Oct. 1998.

[12] Hunt and Crossley, "Coefficient of restitution interpreted as damping in vibroimpact," *ASME J. of Applied Mechanics*, pp. 440–5, June 1975.

[13] B. Armstrong-Hélouvry, *Control of Machines with Friction*. Kluwer Academic Press, 1991.

[14] P. R. Dahl, "Measurement of solid friction parameters of ball bearings," in *Proc. 6th Annual Symp. on Incremental Motion, Control Systems and Devices*, pp. 49–60, Univ. of Illinois, 1977.

[15] J. Kim, H. Chae, J. Jeon, and S. Lee, "Identification and control of systems with friction using accelerated evolutionary programming," *IEEE Control Systems*, pp. 38–47, August 1996.

Estimation of Mass and Center of Mass of Graspless and Shape-Unknown Object

Yong Yu   Kenro Fukuda   Showzow Tsujio

Dept. of Mechanical Engineering, Faculty of Engineering

Kagoshima University

Kagoshima 890-0065, Japan

Abstract

In many cases of manipulating an object stably and accurately by robot, it is required to know the mass and center of mass of the object. For the case when the weight or shape of an object is over the grasp capacity of a robot hand, this paper proposes a technique that can estimate the mass and center of mass of the graspless and shape-unknown object. A plane called Gravity Equi-Effect Plane is first defined, which contains the center of mass and a contact line where the object is in line-contact with an environment. If three or over three orientation-different gravity equi-effect planes of an object are obtained, the center of mass of the object can be estimated by the intersect point of the planes. In order to estimate the gravity equi-effect plane, Tip Operation by robot finger, which tips the object repeatedly, is proposed. Then an algorithm to estimate the gravity equi-effect plane and an algorithm to estimate the mass and center of mass of the object are addressed by using the fingertip position and force information measured from tip operations.

1 Introduction

For manipulating an object stably and accurately by robot, it is required to know the mass and center of mass of the object. Otherwise, it will be difficulty to make an accurate estimation for the object motion, so that the operation may be unstable and hard to control in some cases. About the estimation of the mass and center of mass of an unknown object, some researches have been carried on by firmly grasping and operating an object in robot hand [1] [2], and by holding up and handling an object with a space robot hand [3]. However, these researches did not consider how to estimate the mass and center of mass of an unknown object which is beyond robot hand's grasp.

In many work fields such as factories, mines, construction works, there are many heavy and/or big objects which can not be handled by using robot grasp operations. In order to move or operate a heavy and/or big object well, people usually take advantage of the contact between the object and its environment. In the same way, this operation can be applied to robot manipulation, where the operation is called as *Graspless Manipulation* since the object is hard to be firmly grasped and wholly held up. The robot pushing operation [4], robot tumbling operation [10], robot pivoting operation [5], and robot sliding operations [6], for example, are the graspless manipulations. For performing these graspless operations, it is also necessary to know the mass and the center of mass of the unknown objects.

Providing that the weight and/or shape of an object are over the grasping capacity of a robot hand, this paper proposes a technique that can estimate the mass and center of mass of the graspless and shape-unknown object. A plane called *Gravity Equi-Effect Plane* is first defined, which contains the center of mass and a contact line where the object is in line-contact with an environment. If three or more than three orientation-different gravity equi-effect planes of an object are obtained, the center of mass of the object can be estimated by the intersect point of the planes. For estimating the gravity equi-effect plane, *Tip Operation* by robot finger, which tips the object repeatedly, is proposed. Using the fingertip position and force information from tip operations, an algorithm to estimate the gravity equi-effect plane is addressed. Then an algorithm to estimate the mass and center of mass of the object is given by several independent gravity equi-effect planes. Lastly, experimental verification on the proposed approach is performed.

2 Gravity Equi-Effect Plane and Tip Operation

2.1 Problem

The discussion of this paper is on the bases that:
- The object is a shape-unknown rigid polyhedron. And the mass and center of mass of the object are also unknown. The environment is a plane whose position and normal are unknown.
- A robot hand will touch the object with point contact (touching point hereafter).
- The object is graspless because its weight and/or shape are over the grasping capacity of robot hand.
- When the object is tipped and leaned by a robot hand, the object will be in line (or 2 points) contact with the environment (contact line hereafter).
- There is enough friction between the object and environment so that their contact position can be kept not to move during an object operation.

On these bases, the purpose of this paper is to estimate the mass and center of mass of the graspless and shape-unknown object by dexterous object manipulation on an environment. Note that if there is not enough friction, it is conceivable, for example, dexterously putting one (or more) finger at the contact part between the object and the environment to avoid the object slipping on the environment (for example see [11] or Fig.1).

2.2 Gravity Equi-Effect Plane

Under the condition that an external force acting on an object is momentally balancing the gravity of the object around an axis, there exists a plane containing the axis and the center of mass (see Fig.1). If the plane and the object have the same mass and the same position of the center of mass, the external force acting on the object will have an equivalent effect on the plane. Hence, an object can be simplified as an equivalent plane when the moment

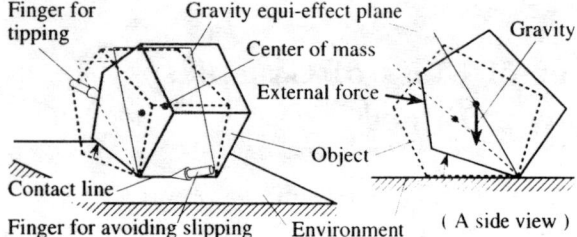

Fig.1 Gravity equi-effect plane

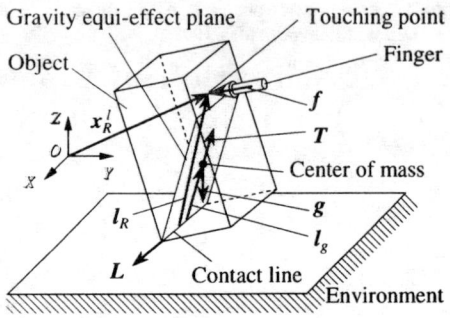

Fig.3 A leaned posture

effect around an axis is considered. The equivalent plane containing an axis and the center of mass is referred to as *Gravity Equi-Effect Plane* in this paper.

The different axes on an object correspond to different gravity equi-effect planes. Three or more than three orientation-different gravity equi-effect planes intersect at one point, i.e., the center of mass of the object. Hence, the center of mass of object can be estimated by estimating three or more than three orientation-different gravity equi-effect planes of an object.

2.3 Parameters of Gravity Equi-Effect Plane

Since the object is beyond robot hand's grasp, the contact with the environment will be used also for operating the object. The contact condition between the object and an environment plane can be classified into point contact, line contact and plane contact (see Fig.2). For the graspless object, a point contact is unstable and will be easy to change to a line contact or a plane contact. This paper, therefore, will use line contact and plane contact to operate the object. The contact line of a line contact will be treated as an axis of a gravity equi-effect plane.

When an object is leaned at rest and is in line-contact with an environment (Fig.3), the equilibrium condition of the moment about the contact line can be expressed as

$$l_R \times f + T \times g = o, \quad (1)$$
$$T \triangleq m l_g, \quad (2)$$

where l_R is a distance vector from the contact line to the fingertip which keeps in touch with the object, f is the contact force at the fingertip measured by finger force sensor, g is the gravity. T is defined as such a vector that crosses the contact line at right angles and points toward the center of mass, m is the mass of the object, l_g is a distance vector from the contact line to the center of mass. And L in Fig.3 is an unit vector representing the direction of contact line. l_R, m, l_g, T and L are unknown since the shape of the object, mass and center of mass are unknown. According to eq.(2), vectors T and l_g have the same direction but their magnitudes are different. In this paper, T, l_g and l_R are called respectively as *Toward-C.M. Vector* from a contact line, *C.M.-Position Vector* from a contact line, and *Fingertip-Position Vector* from a contact line.

As mentioned above, a gravity equi-effect plane contains a toward-C.M. vector T and a contact line, which meet each other at right angles. If we know the toward-C.M. vector T and the contact line, we can determine the gravity equi-effect plane. According to eq.(1), the unknown fingertip-position vector l_R must be estimated first for getting the vector T. If there is enough friction between the object and environment, the contact line will not displace while the object is suitably leaned and turned around the contact line by a robot finger. Moving the object with this condition, a set of fingertip displacement and force information when the finger is tipping the object can be measured. In the next section, we will show that the vectors l_R, T and contact line can be estimated from the measured fingertip displacements and forces.

2.4 Tip Operation by Robot

For estimating a gravity equi-effect plane, therefore, several independent fingertip displacements and fingertip contact forces should be obtained firstly.

To obtain the necessary information, we consider using robot finger to tip the object repeatedly at different fingertip positions under the condition that the contact line between the object and environment is kept not displacing. This object operation is referred to as *Tip Operation* in this paper (see Fig.4).

The procedure of tip operations can be set as:

Step 1 Set the object posture before tipped as an initial object posture where the object is in plane-contact with the environment. From the initial posture, use robot finger to tip and lean the object around a contact line which is kept not displacing, under the condition that the fingertip is displaced along the inverse direction of the sensed fingertip force if possible.

Step 2 Measure the force and position of fingertip when the object is kept in a leaned and still posture (leaned posture hereafter).

Step 3 Make the object return slowly to the initial posture with keeping the current contact line not displacing. Then measure the fingertip position on the object which was touched by the fingertip at Step 2.

Step 4 With changing the touching point position of the fingertip on the object, repeat Step 1 to Step 3 n ($n \geq 2$) times around the same contact line.

On performing these steps, the force values of finger sensor will also be checked to avoid tumbling down the object. For example, if the value is less than the half

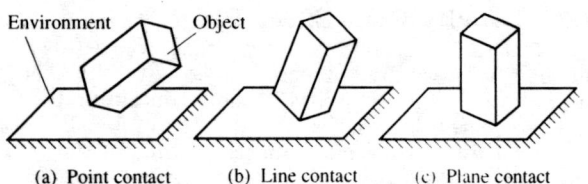

Fig.2 Three kind of contact conditions

of the value when a finger just starts to tip an object, a judgement can be given to stop tipping the object.

3 Estimation of Gravity Equi-Effect Plane

3.1 Fingertip-Position Vector from Contact Line

As mentioned in subsection 2.3, at first, let us consider estimating the fingertip-position vector which is from the contact line to a touching point of fingertip. As shown in Fig.4, let $O-XYZ$ denote the base frame on the environment. Around a contact line j, let $L_{Rij} \in R^3$, $l_{Rij} \in R^3$ denote two fingertip-position vectors respectively for an initial object posture and a leaned object posture of ith tip operation, $\Delta l_{Rij} \in R^3$ denote ith fingertip displacement from the initial object posture to ith leaned object posture. Let $\Delta l_{ij} \in R^3$ denote the distance vector of fingertip positions from ith leaned object posture to $(i+1)$th leaned object posture (see Fig.5). Note that the angle between Δl_{Rij} and $\Delta l_{R(i+1)j}$, the magnitude difference between the two vectors, the distance of touching points in an initial posture corresponding to the two vectors should be made as large as possible for making estimated errors as small as possible.

Since vectors L_{Rij}, l_{Rij}, Δl_{Rij} meet contact line j at right angles (see Fig.4 and 5), let us project the vectors on a plane, which crosses contact line j at right angles, to analyze the position relations among the vectors. For ith and $(i+1)$th tip operations, we have

$$l_{Rij} - L_{Rij} = \Delta l_{Rij}, \quad (3)$$

$$l_{R(i+1)j} - l_{Rij} = \Delta l_{ij} - \frac{[\Delta l_{Rij} \times \Delta l_{R(i+1)j}]^T}{\|\Delta l_{Rij} \times \Delta l_{R(i+1)j}\|} \Delta l_{ij} \frac{[\Delta l_{Rij} \times \Delta l_{R(i+1)j}]}{\|\Delta l_{Rij} \times \Delta l_{R(i+1)j}\|}, \quad (4)$$

where $\|*\|$ denotes the norm of vector $*$. Also, since fingertip-position vectors l_{Rij}, L_{Rij} cross contact line j at right angles and have same magnitude, we can get

$$[\Delta l_{Rij} \times \Delta l_{R(i+1)j}]^T l_{Rij} = 0, \quad (5)$$

$$[\Delta l_{Rij} \times \Delta l_{R(i+1)j}]^T L_{Rij} = 0, \quad (6)$$

$$l_{Rij}^T l_{Rij} - L_{Rij}^T L_{Rij} = 0. \quad (7)$$

By using eq.(3), eq.(7) can be rewritten as

$$-2\Delta l_{Rij}^T L_{Rij} = \Delta l_{Rij}^T \Delta l_{Rij}. \quad (8)$$

To conveniently use the above relation equations for the fingertip-position vector estimation, we define the following matrices when the tip operation is repeated $n(\geq 2)$ times around contact line j:

$$C_{1j} \triangleq [\; I_{3n} \quad -I_{3n}\;] \in R^{3n \times 6n}, \quad (9)$$

$$C_{2j} \triangleq [\; A_{2j} \quad O_{3(n-1) \times 3n}\;] \in R^{3(n-1) \times 6n}, \quad (10)$$

$$A_{2j} \triangleq \mathrm{diag}[\;[-I_3 \;\; I_3],\cdots,[-I_3 \;\; I_3]\;] \in R^{3(n-1) \times 3n}, \quad (11)$$

$$C_{3j} \triangleq \mathrm{diag}[\; A_{3\{1j\}}^T, A_{3\{2j\}}^T, \cdots, A_{3\{(n-1)j\}}^T, A_{3\{(n-1)j\}}^T, $$
$$A_{3\{1j\}}^T, A_{3\{2j\}}^T, \cdots, A_{3\{(n-1)j\}}^T, A_{3\{(n-1)j\}}^T\;]$$
$$\in R^{2n \times 6n}, \quad (12)$$

$$A_{3\{ij\}} \triangleq [\Delta l_{Rij} \times \Delta l_{R(i+1)j}] \in R^3, \quad (13)$$

$$C_{4j} \triangleq [\; O_{n \times 3n} \quad -2A_{4j}\;] \in R^{n \times 6n}, \quad (14)$$

$$A_{4j} \triangleq \mathrm{diag}[\Delta l_{R1j}^T, \cdots, \Delta l_{Rnj}^T] \in R^{n \times 3n}, \quad (15)$$

$$c_{1j} \triangleq [\Delta l_{R1j}^T \;\cdots\; \Delta l_{Rnj}^T]^T \in R^{3n}, \quad (16)$$

$$c_{2j} \triangleq [\; a_{2\{1j\}} \;\cdots\; a_{2\{(n-1)j\}}\;]^T \in R^{3(n-1)}, \quad (17)$$

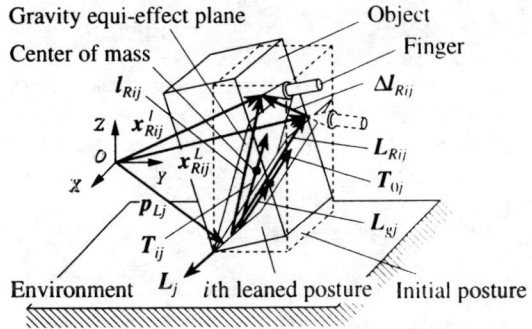

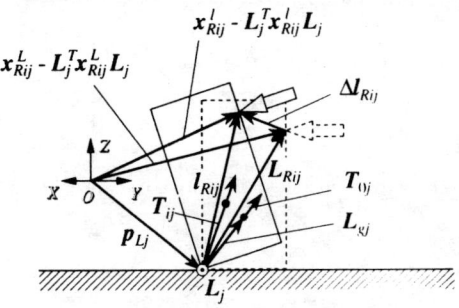

(A projection on a plane crossing L_j at right angles)

Fig.4 ith tip operation

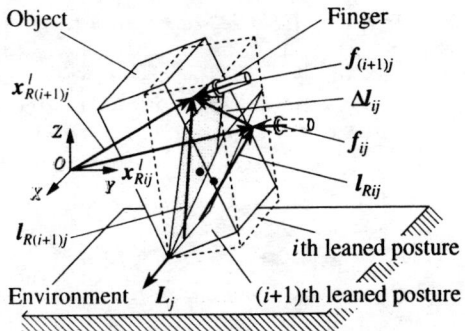

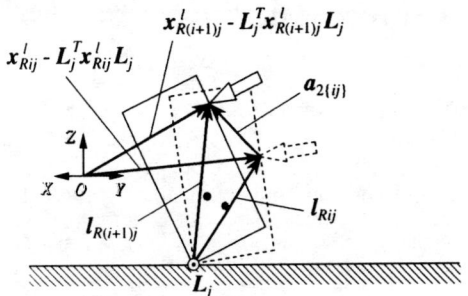

(A projection on a plane crossing L_j at right angles)

Fig.5 ith and $(i+1)$th leaned postures

$$a_{2\{ij\}} \triangleq \Delta l_{ij} - \frac{A_{3\{ij\}}^T}{\|A_{3\{ij\}}\|} \Delta l_{ij} \frac{A_{3\{ij\}}}{\|A_{3\{ij\}}\|} \in \mathbf{R}^3. \tag{18}$$

$$c_{3j} \triangleq O_{2n \times 1} \in \mathbf{R}^{2n}, \tag{19}$$

$$c_{4j} \triangleq \begin{bmatrix} \Delta l_{R1j}^T \Delta l_{R1j} & \cdots & \Delta l_{Rnj}^T \Delta l_{Rnj} \end{bmatrix}^T \in \mathbf{R}^n. \tag{20}$$

By eqs.(3)(4)(5)(6)(8) and the above defined matrices, we can get

$$C_j l_j = c_j \in \mathbf{R}^{3(3n-1)}, \tag{21}$$

$$C_j \triangleq \begin{bmatrix} C_{1j}^T & C_{2j}^T & C_{3j}^T & C_{4j}^T \end{bmatrix}^T \in \mathbf{R}^{3(3n-1) \times 6n}, \tag{22}$$

$$l_j \triangleq \begin{bmatrix} l_{R1j}^T & \cdots & l_{Rnj}^T & L_{R1j}^T & \cdots & L_{Rnj}^T \end{bmatrix}^T \in \mathbf{R}^{6n}, \tag{23}$$

$$c_j \triangleq \begin{bmatrix} c_{1j}^T & c_{2j}^T & c_{3j}^T & c_{4j}^T \end{bmatrix}^T \in \mathbf{R}^{3(3n-1)}. \tag{24}$$

where I_* denotes a $*$-dimensional identity matrix and $O_{*_1 \times *_2}$ denotes a $(*_1 \times *_2)$ zero matrix. When $n \geq 2$ and $\text{rank} C_j = 6n$, the unique solution (if no measured error) or the optimal approximate solution (if measured errors exist) to l_j can be obtained from eq.(21) since $3(3n-1) > 6n$. Therefore, considering that some measured errors may exist in Δl_{Rij} and Δl_{ij}, the vectors l_{Rij}, L_{Rij} ($i = 1, 2, \cdots, n$, $n \geq 2$) for contact line j can be estimated by

$$l_j = C_j^+ c_j, \tag{25}$$

where $C_j^+ \in \mathbf{R}^{6n \times 3(3n-1)}$ is the pseudoinverse of C_j.

3.2 Estimation of Contact Line

This subsection considers estimating the direction of a contact line j and giving the equation of the line, based on the estimated vectors l_{Rij}, L_{Rij} ($i = 1, 2, \cdots, n$, $n \geq 2$) and the known fingertip displacements.

Since Δl_{Rij} and contact line j cross at right angles,

$$[\Delta l_{Rij} \times \Delta l_{R(i+1)j}]^T L_{ij} = k_{Sij} \|\Delta l_{Rij} \times \Delta l_{R(i+1)j}\|, \tag{26}$$

$$k_{Sij} \triangleq \text{sgn}\left\{ [\Delta l_{Rij} \times g]^T [\Delta l_{Rij} \times \Delta l_{R(i+1)j}] \right\} \tag{27}$$

can be obtained, where $L_{ij} \in \mathbf{R}^3$ represents the unit vector of contact line j to ith tip operation, whose direction is given by $[\Delta l_{Rij} \times g]^T L_{ij} > 0$. Because L_{ij} is parallel to $\Delta l_{Rij} \times \Delta l_{R(i+1)j}$, L_{ij} has the minimum norm among the vectors satisfying eq.(26). After repeated the tip operation $n (\geq 2)$ times about contact line j, there are

$$D_{Lj} \hat{L}_j = d_{Lj} \in \mathbf{R}^{(n-1)}, \tag{28}$$

$$D_{Lj} \triangleq \text{diag}\left[[\Delta l_{R1j} \times \Delta l_{R2j}]^T, \cdots, [\Delta l_{R(n-1)j} \times \Delta l_{Rnj}]^T \right]$$
$$\in \mathbf{R}^{(n-1) \times 3(n-1)}, \tag{29}$$

$$\hat{L}_j \triangleq \begin{bmatrix} L_{1j}^T & \cdots & L_{(n-1)j}^T \end{bmatrix} \in \mathbf{R}^{3(n-1)}, \tag{30}$$

$$d_{Lj} \triangleq \begin{bmatrix} K_{S1j} \|\Delta l_{R1j} \times \Delta l_{R2j}\| & \cdots \\ K_{S(n-1)j} \|\Delta l_{R(n-1)j} \times \Delta l_{Rnj}\| \end{bmatrix}^T \in \mathbf{R}^{(n-1)}. \tag{31}$$

The solution with minimum norm of eq.(28) can be obtained by using the pseudoinverse of D_{Lj}. Thereby, considering that some measured errors can exist in Δl_{Rij}, the direction L_j of contact line j can be estimated by

$$L_j = \frac{E_{Lj} \hat{L}_j}{\|E_{Lj} \hat{L}_j\|} \in \mathbf{R}^3, \qquad \|L_j\| = 1, \tag{32}$$

$$\hat{L}_j = D_{Lj}^+ d_{Lj} \in \mathbf{R}^{3(n-1)}, \tag{33}$$

$$E_{Lj} \triangleq \begin{bmatrix} I_3 & \cdots & I_3 \end{bmatrix} \in \mathbf{R}^{3 \times 3(n-1)}. \tag{34}$$

where $D_{Lj}^+ \in \mathbf{R}^{3(n-1) \times (n-1)}$ is the pseudoinverse of D_{Lj}.

Using the obtained L_j and l_{Rij}, L_{Rij} ($i = 1, 2, \cdots, n$, $n \geq 2$), we describe the equation of contact line j as

$$x = p_{Lj} + k_{Lj} L_j \in \mathbf{R}^3, \qquad \|L_j\| = 1, \tag{35}$$

$$p_{Lj} \triangleq \frac{1}{2n} \left\{ \sum_{i=1}^{n} \left(x_{Rij}^l - L_j^T x_{Rij}^l L_j + l_{Rij} \right) \right.$$
$$\left. + \sum_{i=1}^{n} \left(x_{Rij}^L - L_j^T x_{Rij}^L L_j + L_{Rij} \right) \right\}, \tag{36}$$

where $p_{Lj} \in \mathbf{R}^3$ is taken as a representative point on contact line j, which is estimated by the arithmetical mean since some measured errors can exist. k_{Lj} denotes the distance along the line form the representative point to a point x. And x_{Rij}^l, x_{Rij}^L denote the touching point positions from the origin of $O - XYZ$ with ith initial or ith leaned object posture respectively (see Fig.4).

3.3 Estimation of Toward-C.M. Vector from Contact Line

Now let us consider estimating the toward-C.M. vector T_j according to the estimated l_j, and the fingertip forces f_{ij} ($i = 1, 2, \cdots, n$, $n \geq 2$) which are measured when the object is on leaned postures.

By repeating the tip operation $n (\geq 2)$ times, the following matrices can be derived from eq.(1),

$$G_{1j} T_j = b_{1j} \in \mathbf{R}^n, \tag{37}$$

$$G_{1j} \triangleq \begin{bmatrix} O_{3n \times 3} & B_{1j} \end{bmatrix} \in \mathbf{R}^{3n \times 3n}, \tag{38}$$

$$B_{1j} \triangleq \text{diag}\begin{bmatrix} -[g \times], \cdots, -[g \times] \end{bmatrix} \in \mathbf{R}^{3n \times 3(n-1)}, \tag{39}$$

$$T_j \triangleq \begin{bmatrix} T_{0j}^T & T_{1j}^T & \cdots & T_{nj}^T \end{bmatrix}^T \in \mathbf{R}^{3(n+1)}, \tag{40}$$

$$b_{1j} \triangleq \begin{bmatrix} [l_{R1j} \times f_{1j}]^T & \cdots & [l_{Rnj} \times f_{nj}]^T \end{bmatrix}^T \in \mathbf{R}^{3n}, \tag{41}$$

where $[g \times]$ represents a skew symmetry matrix defined by $[g \times] l_{Rij} \triangleq g \times l_{Rij}$, T_{0j} and T_{ij} correspond to the initial or ith leaned object posture respectively.

When the object is leaned and turned about contact line j with ith tip operation, the angle between toward-C.M. vectors T_{ij} and T_{0j} is the same as the angle between vectors l_{Rij} and L_{Rij}, and the magnitudes of T_{ij} and T_{0j} are the same. Accordingly, there are

$$T_{0j} - R_{Rij} T_{ij} = O_{3 \times 1} \in \mathbf{R}^3, \tag{42}$$

$$R_{Rij} \triangleq r_{ij} R_{ij} r_{ij}^T \in \mathbf{R}^{3 \times 3}, \tag{43}$$

$$r_{ij} \triangleq \begin{bmatrix} \dfrac{L_{Rij} \times L_j}{\|L_{Rij} \times L_j\|} & \dfrac{L_{Rij}}{\|L_{Rij}\|} & L_j \end{bmatrix} \in \mathbf{R}^{3 \times 3}, \tag{44}$$

$$R_{ij} \triangleq \begin{bmatrix} \dfrac{L_{Rij}^T l_{Rij}}{\|L_{Rij}\| \|l_{Rij}\|} & -\dfrac{\|L_{Rij} \times l_{Rij}\|}{\|L_{Rij}\| \|l_{Rij}\|} & 0 \\ \dfrac{\|L_{Rij} \times l_{Rij}\|}{\|L_{Rij}\| \|l_{Rij}\|} & \dfrac{L_{Rij}^T l_{Rij}}{\|L_{Rij}\| \|l_{Rij}\|} & 0 \\ 0 & 0 & 1 \end{bmatrix} \in \mathbf{R}^{3 \times 3} \tag{45}$$

Repeating the tip operation $n (\geq 2)$ times, we have

$$G_{2j} T_j = O_{3n \times 1} \in \mathbf{R}^{3n}, \tag{46}$$

$$G_{2j} \triangleq \begin{bmatrix} E_{2j} & -B_{2j} \end{bmatrix} \in \mathbf{R}^{3n \times 3(n+1)}, \tag{47}$$

$$B_{2j} \triangleq \text{diag}\begin{bmatrix} R_{R1j}^T, \cdots, R_{Rnj}^T \end{bmatrix} \in \mathbf{R}^{3n \times 3n}, \tag{48}$$

$$E_{2j} \triangleq \begin{bmatrix} I_3 & \cdots & I_3 \end{bmatrix}^T \in \mathbf{R}^{3n \times 3}. \tag{49}$$

By eqs.(37) and (46), we can get

$$G_j T_j = b_j \in R^{6n}, \quad (50)$$

$$G_j \triangleq \begin{bmatrix} G_{1j}^T & G_{2j}^T \end{bmatrix}^T \in R^{6n \times 3(n+1)}, \quad (51)$$

$$b_j \triangleq \begin{bmatrix} b_{1j}^T & O_{3n \times 1}^T \end{bmatrix} \in R^{6n}. \quad (52)$$

Same as the solution to l_j, when $n \geq 2$ and $\text{rank} G_j = 3(n+1)$, the optimal approximate solution of T_j can be obtained from eq.(50) since $6n > 3(n+1)$. Accordingly, considering that some measured errors may exist, the toward-C.M. vector T_j can be estimated by

$$T_j = G_j^+ b_j, \quad (53)$$

where $G_j^+ \in R^{3(n+1) \times 6n}$ is the pseudoinverse of G_j.

3.4 Estimation of Gravity Equi-Effect Plane

From the contact line $x = p_{L_j} + k_{L_j} L_j$ and the toward-C.M. vector T_{0j}, we can estimate the gravity equi-effect plane corresponding to the contact line j as

$$[L_j \times T_{0j}]^T (x - p_{L_j}) = 0, \quad (54)$$

where $[L_j \times T_{0j}]$ is a normal vector of the gravity equi-effect plane, x represents a point on the gravity equi-effect plane. And the representative point p_{L_j} of contact line j is also used as a representative point for the gravity equi-effect plane.

As the above discussion, the different gravity equi-effect planes can be estimated respectively by performing the tip operation to the different contact lines of the same initial object posture.

4 Estimation of Mass and Center of Mass
4.1 Estimation of Center of Mass

Since a gravity equi-effect plane contains the center of mass of the object, the intersection point of the three or more gravity equi-effect planes, which have different orientations and do not intersect on one straight line in a same object posture, is the center of mass. Let us consider estimating the center of mass by using the plural gravity equi-effect planes in the initial object posture where the object is in plane-contact with the environment. Note that the angle between two plane's normals should be larger enough than 0° and smaller enough than 180° (for example, within 20°~160°), for making estimated errors as small as possible.

Since a gravity equi-effect plane contains the center of mass as mentioned previously, we have

$$[L_j \times T_{0j}]^T (P_g - p_{L_j}) = 0 \quad (55)$$

from eq.(54), where $P_g \in R^3$ denotes the position of the center of mass. For the obtained gravity equi-effect planes $(j = 1, 2, \cdots, k, k \geq 3)$, there are

$$H P_g = h \in R^k, \quad k \geq 3, \quad (56)$$

$$H \triangleq \begin{bmatrix} [L_1 \times T_{01}] & \cdots & [L_k \times T_{0k}] \end{bmatrix}^T \in R^{k \times 3}, \quad (57)$$

$$h \triangleq \begin{bmatrix} p_{L1}^T [L_1 \times T_{01}] & \cdots & p_{Lk}^T [L_k \times T_{0k}] \end{bmatrix}^T \in R^k. \quad (58)$$

When $k \geq 3$, $\text{rank} H = 3$, the optimal approximate solution of P_g can be obtained from eq.(56). If an object is in plane-contact with an environment in its initial posture, the contact plane between the two polyhedrons will be a polygon. Hence for an initial object posture, at least 3 polygon edges can be used as 3 independent contact lines, so that $k \geq 3, \text{rank} H = 3$ can be guaranteed. Accordingly, we can estimate the center of mass of the object by

$$P_g = H^+ h, \quad (59)$$

where $H^+ \in R^{3 \times k}$ is the pseudoinverse of H.

4.2 Estimation of Mass of Object

In an initial posture, let L_{gj} denote a C.M.-position vector which is from a contact line j to the position P_g of the center of mass (see Fig.4). From the obtained P_g, L_{gj} can be described by

$$L_{gj} = (P_g - p_{Lj}) - L_j^T (P_g - p_{Lj}) L_j \in R^3. \quad (60)$$

For the obtained gravity equi-effect planes, by substituting L_{gj} and T_{0j} $(j = 1, 2, \cdots, k, k \geq 3)$ in the initial posture into eq.(2), we get

$$\|T_{0j}\| = m \|L_{gj}\|, \quad j = 1, 2, \cdots, k, \ k \geq 3. \quad (61)$$

Considering that some measured errors may exist, from eq.(61) the object mass m can be estimated by

$$m = \frac{1}{k} \sum_{j=1}^{k} \frac{\|T_{0j}\|}{\|L_{gj}\|}, \quad k \geq 3. \quad (62)$$

5 Verification by Experiment

This section demonstrates the validity of the proposed method by three experiments. The unknown object for the first and second experiments is a wooden box shown in Fig.6(a) and (b). For the second experiment, a weight is added on the bottom of the box in order to change the mass and center of mass of the box. The unknown object of the third experiment is a shape-complex wooden table shown in Fig.6(c), whose center of mass is difficult to know in general. The experimental environment is an unknown plane and there is enough friction between the objects and the environment plane.

In the experiments, the contact lines between the objects and environment plane are kept not to move because of the enough friction, while the touching positions

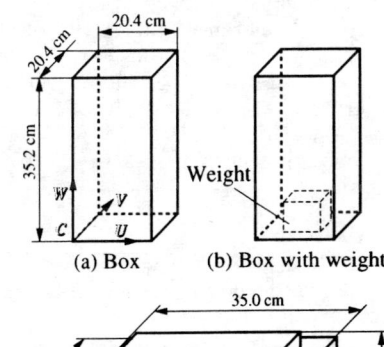

(a) Box   (b) Box with weight

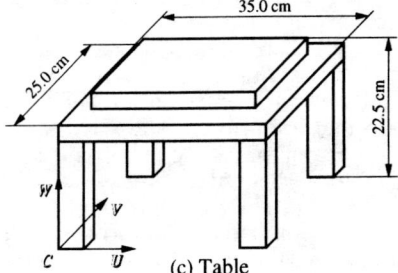

(c) Table

Fig.6 Experimental objects

Table 1 Result of experiment 1

	P_{gu} [cm]	P_{gv} [cm]	P_{gw} [cm]	m [kg]
Reference value	10.20	10.40	17.95	1.872
Estimated value	10.28	10.18	17.42	1.888
Error (%)	0.8	2.1	2.9	0.85

Table 2 Result of experiment 2

	P_{gu} [cm]	P_{gv} [cm]	P_{gw} [cm]	m [kg]
Reference value	10.20	10.38	16.08	2.115
Estimated value	10.16	10.16	15.89	2.131
Error (%)	0.4	2.1	1.2	0.76

Table 3 Result of experiment 3

	P_{gu} [cm]	P_{gv} [cm]	P_{gw} [cm]	m [kg]
Reference value	-	-	-	1.771
Estimated value	17.81	12.50	18.52	1.780
Error (%)	-	-	-	0.51

between the fingertip and objects are allowed to move. After an object is leaned and the touching position and force of the fingertip are measured, the touching position will be marked on the object. Then, after the object is slowly returned to its initial posture, the marked point will be positionally measured again by fingertip to get the touching position at the initial posture, so that fingertip displacements Δl_{Rij} can be obtained.

The estimated results and the reference values about the masses and centers of mass of the objects are shown in Table 1 to Table 3, where the centers of gravity are given with regard to the object frame $C-UVW$ for convenience. The reference values are obtained by human skill for the centers of mass and a dial scale for masses. But the center of mass of the table is unable to be measured by human skill because of its complex shape. Comparing the estimated and reference values, the relative errors of the position of the centers of mass are less than 3%, and the relative errors of masses of the objects are less than 1%. Considering that the accuracy of force sensor is 2% and the position error of robot arm is 1%, the estimated results have enough accuracy and the validity of proposed approach is shown.

6 Conclusion

This paper has proposed a technique that can estimate the mass and center of mass of a graspless and shape-unknown object. A plane called *Gravity Equi-Effect Plane* was defined in this paper, which contains the center of mass and a contact line where the object is in line-contact with an environment. Three or more than three orientation-different gravity equi-effect planes will intersect at one point, i.e., the center of mass of the object. Therefore, the center of mass of object can be estimated by estimating three or more than three orientation-different gravity equi-effect planes of an object. For estimating the gravity equi-effect plane, *Tip Operation* by robot finger, which tips the object repeatedly, was proposed. Using the fingertip position and force information measured from tip operations, an algorithm to estimate the gravity equi-effect plane was described. Then an algorithm to estimate the mass and center of mass of the object was given by several independent gravity equi-effect planes. Lastly, experimental verification on the proposed approach was performed and its results were outlined.

The proposed method can also be used for the object which is not a polyhedron but can be in contact on its environment with 3 (or more than 3) different line-contacts.

References

[1] C. G. Atkeson, C. H. An and J. M. Hollerbach, "Rigid Body Load Identification for Manipulators," *Proc. of IEEE Int. Conf. on Decision and Control*, Vol.2, pp.996-1002, 1985.

[2] T. Arai and Y. Nakamura, "Estimation of Mass and Center of Gravity for Unknown Object Using 6 Axis F/T Sensor Data," *Proc. of IEEE Int. Conf. on Advanced Robotics*, pp.547-552, 1995.

[3] Y. Murotsu, K. Senda, M. Ozaki and S. Tsujio, "Parameter Identification of Unknown Object Handled by Free-Flying Space Robot," *AIAA Jour. of Guidance, Control, and Dynamics*, Vol.17, No.3, pp.488-494, 1994.

[4] K. M. Lynch, "The Mechanics of Fine Manipulation by Pushing," *Proc. of IEEE Int. Conf. on Robotics and Automation*, Vol.3, pp.2269-2276, 1992.

[5] Y. Aiyama, M. Inaba and H. Inoue, "Proviting: A New Method of Graspless Manipulation of Object by Robot Fingers," *Proc. of IEEE Int. Conf. on Intelligent Robots and Systems*, Vol.1, pp.136-143, 1993.

[6] Y. Yokokohji, Y. Yu, N. Nakasu and T. Yoshikawa, "Quasi-Dynamic Manipulation of Constrained Object by Robot Fingers in Assembly Tasks," *Proc. of IEEE Int. Conf. on Intelligent Robots and Systems*, Vol.1, pp.144-151, 1993.

[7] Y. Yu, Y. Yokokohji and T. Yoshikawa, "Two Kinds of Degree of Freedom in Constraint State and Their Application to Assembly Planning," *Proc. of IEEE Int. Conf. on Robotics and Automation*, Vol.3, pp.1993-1999, 1996.

[8] T. Yoshikawa, Y. Yu and M. Koike, "Estimation of Contact Position between Object and Environment Based on Probing of Robot Manipulator," *Proc. of IEEE Int. Conf. on Intelligent Robots and Systems*, Vol.2, pp.769-776, 1996.

[9] Y. Yu and T. Yoshikawa, "Evaluation of Contact Stability between Objects," *Proc. of IEEE Int. Conf. on Robotics and Automation*, Vol.1, pp.695-702, 1997.

[10] N. Sawasaki, M. Inaba and H. Inoue, "Tumbling Objects Using a Multi-Fingered Robot," *Proc. of IEEE Int. Symp. on Industrial Robots*, pp.609-616, 1989.

[11] M. Kaneko, M. Kessler, A. Weigl and H. Tolle, "Capturing Pyramidal-like Objects," *Proc. of IEEE Int. Conf. on Robotics and Automation*, Vol.4, pp.3619-3624, 1998.

[12] Y. Yu, K. Takeuchi and T. Yoshikawa, "Optimization of Robot Hand Power Grasps," *Proc. of IEEE Int. Conf. on Robotics and Automation*, Vol.4, pp.3341-3347, 1998.

4-Axis Electromagnetic Microgripper

Arianna Menciassi*, Blake Hannaford**, Maria Chiara Carrozza*, Paolo Dario*,

*Scuola Superiore Sant'Anna - MiTech Lab, Pisa (Italy)
**University of Washington, Department of Electrical Engineering, Seattle (USA)

Abstract

This paper describes a novel 4-axis microgripping system consisting of two fingers, each driven by a 2-axis, moving coil actuator taken from a CD-lens assembly. These electromagnetic actuators are small, very linear, virtually frictionless and low cost. We measured electrical actuator parameters and characterized actuator performance in terms of displacement vs. current, force vs. current and resonant frequency. Experimental results indicate that the proposed microgripping system can be an attractive solution to the problem of micromanipulating small objects for precision manufacturing and biotechnology with high accuracy in a relatively large workspace.

1. Introduction

There has been a strong increase in interest in micromanipulation technology motivated by increased demands from precision manufacturing and biotechnology, as well as emerging microfabrication technologies [1, 2, 3, 4, 5]. Earlier work has produced planar microgrippers with one degree of freedom for opening and closing [6, 7, 8]. The resulting dexterity has been limited. There is a need to get micromanipulators "out of the plane".

Micromanipulators have been available as mechanical devices for some time from manufacturers such as Leica. These devices are tailored to the biology market and typically control the 3-axis position of a pipette or needle. One axis of motion is typically aligned with the needle direction since micromanipulators are often used for experiments in which a probe is inserted into biological tissue. Among the limitations of existing micromanipulator technology are, low backdrivability / poor force control, limited dexterity, lack of computer control and sensor guidance, and poor human interface. Several previous approaches to solve this problem will be briefly reviewed. Bhatti et. al. [9] used a pair of 2.5 cm needle-like fingers actuated by a piezo-electric bi-morph element to give a single opening-closing degree of freedom. The jaw faces were polished flats. The maximum opening size of the grasper was 0.5 mm. The fingers were mounted on a three degree of freedom mini direct drive robot, and demonstrated the ability to pick and place micro-objects (grains of sand).

Kim et al. [10] built a microfabricated polysilicon gripper actuated by an electrostatic comb actuator. Carrozza et al. [11] built a planar microgripper using the LIGA technique composed of flexure elements. The gripper was driven by an external piezoelectric stack.

Additional research in this area, involving other micromanipulation approaches such as optical trapping, and micro-planar stepper motors has been summarized in workshops at the International Conference on Robotics and Automation for the past several years.

Consumer electronics technology has the benefit of huge market revenues to support extensive engineering efforts. Newer consumer media such as compact disk players, 8-mm video, and digital video formats, have produced exceptional electromechanical technology at very low cost. An educational project at the University of Washington [12] has made use of these devices to teach hands-on engineering skills and concepts of productization and manufacturability. One of the laboratories in that course studies the electromagnetic actuation and control problems of CD-player read head optical servo units in detail. As a result of this educational project, a lot is known about the precision microactuators used. These actuators are small, very linear, electromagnetic actuators which rapidly displace the laser pickup lens in two directions to follow data-track motions. The motion range of these devices is approximately +/- 1mm in each of two directions.

In this paper, we describe a novel 4-axis microgripping system consisting of two fingers, each driven by a CD-lens actuator assembly. Lens actuator assemblies were supplied from "8x" CD-rom drives (LG Electronics Corp.).

2. Design

The CD-optical tracking actuator assembly is about 10x15x5 mm^3 and is positioned along the disk radius by a coarse positioning motor and drive screw providing 20-30 mm of linear motion. The head actuator consists of a set of small coils suspended in a magnetic field created by two permanent magnets. The lens and coils are suspended on springs in the center of the magnetic gap which allow motion "up and down" (lens optical axis) and "in and out" (along the radial direction of the disk). Sometimes the springs are the electrical conductors of current to the coils (our case), and sometimes they are made of a plastic material.

A schematic diagram and a photo of the actuator are given in Figure 1.

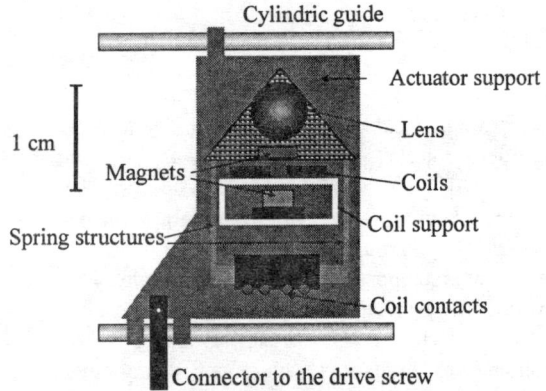

Figure 1. Scheme (above) and photograph (below) of the actuator.

We exploit two head actuators to independently drive two fingers; each finger can move in the vertical and in the horizontal direction in order to get 4 degrees of freedom. We integrated two head actuators in the assembly that tracks CD platter motion: one of the actuators already belonged to the assembly; the second one was inserted in the cylindrical guides of the first one. We modified the second head actuator in order to reduce the distance between the two head actuators.

Afterwards, we glued the two actuators and we connected them by a metal strip, so that it was possible to carry the second actuator exploiting the coarse positioning motor of the first actuator. The final arrangement is visible in Figure 2.

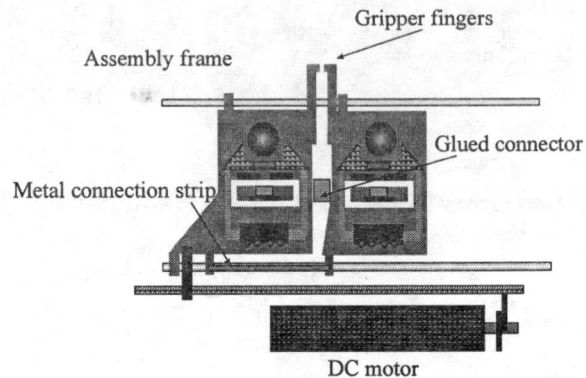

Figure 2. Final arrangement of two actuators.

Because it was impossible to remove the lenses from the head actuators without damaging the spring structures, a finger was designed to fit directly on the lens. The design is visible in Figure 3.

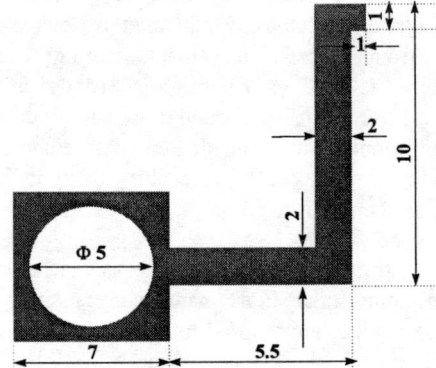

Figure 3. Finger design (dimensions in millimeters).

The hole has the same diameter as the lens. To assure the positioning of the finger, we used a bit of cyanoacrylate glue. Two identical fingers made out of PEEK (Polietereterchetone, $\rho = 0.074$ gr/cm^3) were built; each finger has a mass of 0.03 g. Figure 4 shows the motion range of the system. Each fingertip can move in a vertical plane 1x2 mm. The horizontal spacing between lens centers (currently about 20 mm) determines the trade-off between gripper opening and horizontal displacement.

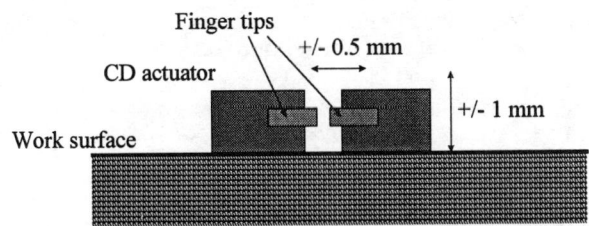

Figure 4. Side view of the system and motion range.

3. Results

Before testing, we measured electrical actuator parameters (resistance, inductance, elastic constant, etc.).

The resistances of the CD head actuator coils are the following: up/down coil resistance = 7 Ω, left/right coil resistance = 7.1 Ω. The inductance, L, of the actuator was measured by using a simple circuit including the coil, a variable resistance and a waveform generator. The result is L = 1.6 µH.

On the mechanical side, the differential equation for the system is

$$K_m I + F_{ext} - (M d^2x/dt^2 + B dx/dt + K \cdot x) = 0,$$

where M is the mass, B is the damping of the support, K_m is the torque constant of the actuator, K is the spring constant of the support and F_{ext} is the external force (if present).

When velocity, acceleration, dI/dt and I are all equal to zero, the coil behavior is the same of a normal spring behavior, so the following relation is valid

$$F_{ext} = K \cdot x,$$

where x is the displacement associated to the external force. To apply an external force, some tiny weights were placed on the lens and the correspondent displacement was measured by a laser triangulator (Mel GmbH). This device is able to measure displacements between -2 mm and +2 mm with an accuracy of 1 µm, by comparing the output laser beam with the light reflected by the object surface.

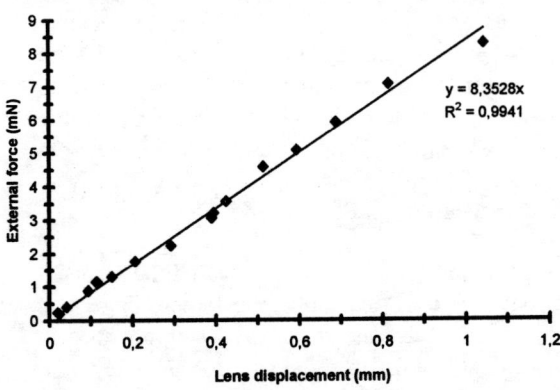

Figure 5. External force versus lens displacement.

The tiny weights were made out of soldered tin and were weighted by a precision balance. Figure 5 shows the obtained results.

The slope of the graph indicates the elastic constant of the up/down actuator spring: K = 8.35 mN/mm.

By setting position, velocity and acceleration equal to zero, we get:

$$F_{ext} = - K_m \cdot I$$

where I is the current and K_m is the actuator motor constant. Using this relation, it is possible to measure K_m by varying F_{ext} and by adjusting I so that the lens displacement is equal to zero. Specifically, we added some tiny weights and adjusted the current until the lens came back to "zero". In Figure 6 the weight vs. current diagram is shown.

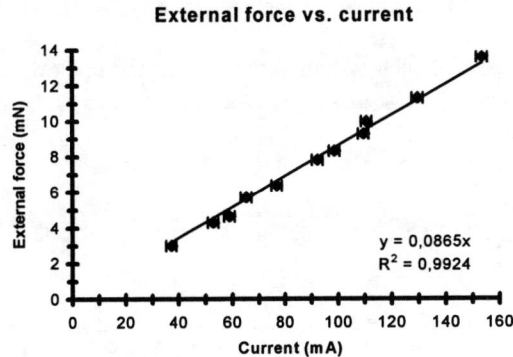

Figure 6. External force versus current.

The slope of the diagram indicates the motor constant of the up/down actuator: Km = 0.087 mN/mA.

The tests performed on the system concern displacement and force vs. current measurements and natural frequency determination. The test bench used to perform actuator displacement vs. current measurements is composed by a voltage supply, a current gage and a laser triangulator. The laser beam was focused on the side and on the top of the head actuator in order to measure respectively the left/right and the up/down displacement. For each current value, a displacement measurement in microns was obtained. Figure 7 shows the results for the left/right displacement.

From the diagrams, a slight difference is observed in the position gain between the right and the left displacement; the displacement is more effective towards the left side (2.2 µm/mA) than towards the right side (1.5 µm/mA). The mean position gain results 1.87 µm/mA.

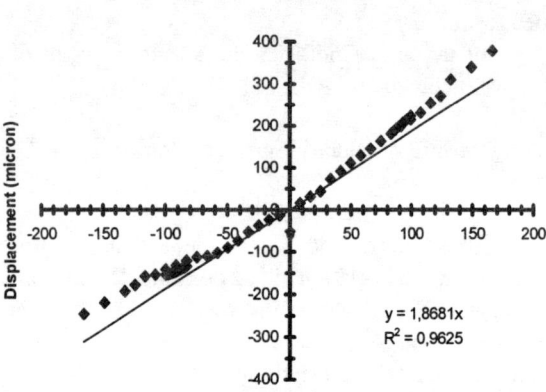

Figure 7. Left/Right displacement vs. current.

The same test bench was used to measure the up/down displacement. The results are shown in Figure 8.

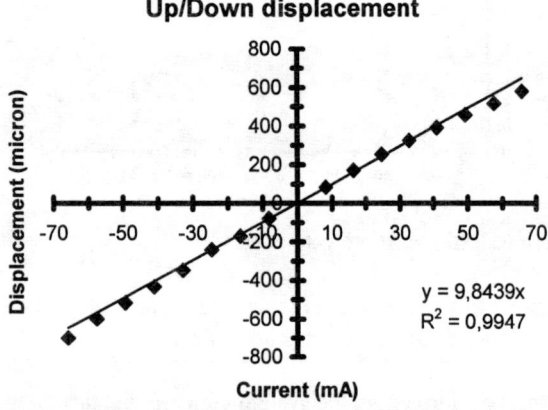

Figure 8. Up/down displacement vs. current.

Also in this case, a slight difference of the position gain between the up and the down displacement is present. However, the position gain is much higher for the up/down motion than for the left/right motion; the average gain is 9.8 µm/mA.

The test bench used to perform force vs. current measurements is composed by the voltage supply and a load cell (PTC Electronics; working range: ±0.3 N; accuracy: 10^{-3} N). The load cell was used in compression and the forces during the left, the right and the up displacement were measured.
The results are shown in Figure 9.

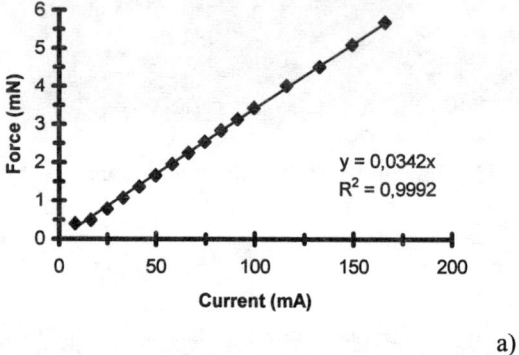

a)

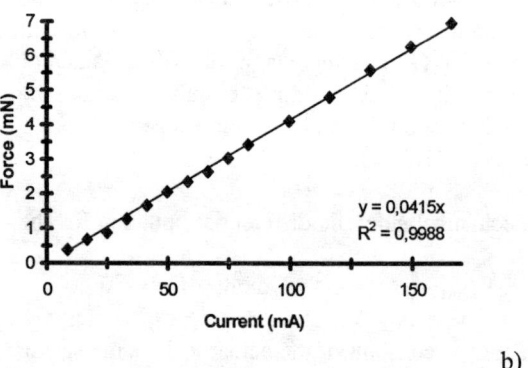

b)

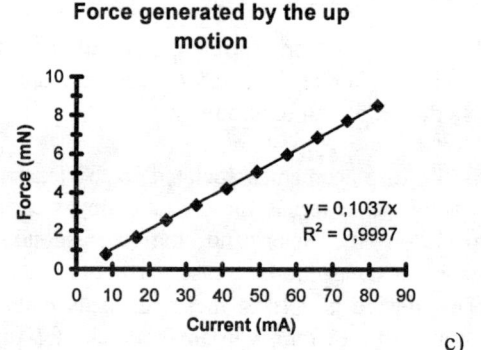

c)

Figure 9. Force vs. current.

As for the displacement, the force generated during the up/down motion is higher than that generated during the left/right motion. It was impossible to measure the force generated during the down displacement, but it can be considered approximately equal to the force generated during the up motion. The "force gain" results 0.1 mN/mA for the up motion, 0.042 mN/A for the right motion and 0.034 mN/A for the left motion. This is an independent measurement from that of Figure 6.

To gauge dynamic response, we measured the resonant frequencies of two CD actuators: one without finger and one with finger. Resonant frequencies were obtained in two ways. First, we dropped the actuator assembly on a hard surface, causing the lens to vibrate, and recorded the voltage signal on the oscilloscope. Since the motion of the lens is highly under-clamped, the frequency of the excited voltage is close to the natural frequency of the actuator. Then, we excited the actuators with a voltage of 200 mV$_{pp}$ at variable frequency and observed the motion in a microscope: when we reached the natural frequency, the oscillation amplitude is amplified.

For the CD actuator without finger, we obtained an average frequency of 26 Hz for the up/down motion, and of 55 Hz for the left/right motion. For the CD actuator with finger, we obtained an average frequency of 26 Hz for the up/down motion, and of 49 Hz for the left/right motion.

Theoretically the resonant frequencies of the actuator without the finger should be larger than the ones of the actuator with the finger. In fact the resonant frequency is:

$$f = (2\pi)^{-1} \cdot (K/M)^{1/2}$$

where K is the elastic constant of the system and M is the mass. On the other side, our measurements do not give a clear difference related to the different mass of the two systems.

The finger mass is 0.03 g and the actuator mass is about 1g; if f is the frequency of the actuator without finger and f' is the frequency of the actuator with finger, we obtain:

$$\Delta F = f - f' = f \cdot 0.015 ;$$

if f = 27.7 Hz, then

$$\Delta F = 0.42 \text{ Hz}.$$

This difference is not resolvable with our type of measurement.

4. Conclusions

We have built and tested a new mini gripper based on a 2-axis moving coil actuators directly driving each finger. The measured performance is summarized below:

	"up/down"	"left/right"
F$_{Max}$	± 12 mN	± 6-7 mN
Disp$_{Max}$	± 1 mm	± 0.5 mm
ω$_n$	26 Hz	49 Hz

The "left/right" displacement does not include the 25 mm range of the coarse positioning lead screw. The force/current relationship, as well as the force/displacement relation are highly linear. Stiction, backlash, and deadband were un-measurable with our equipment. The actuator assemblies were obtained at low cost in very compact form by dissection of CD-Rom drivers. Further reduction of the actuator volume may be possible by additional dissection. Control of actuation force is straightforward due to the absence of friction and the linear relationship between force and current.

Future work on this device includes feedback control of position and mechanical impedance through position sensing. We will evaluate both video image processing and Hall effect sensing for this purpose. We also intend to further reduce the actuator volume (without affecting motion and force output) and to mount the device on a dexterous platform for manipulation experiments.

Acknowledgment

This work has been performed during the stay of B. Hannaford at the Scuola Superiore Sant'Anna of Pisa as visiting professor. B. Hannaford should like to thank the Scuola Superiore Sant'Anna and the LG Corp. Production Engineering Research Lab. A. Menciassi, M.C. Carrozza and P. Dario thank the Scuola Superiore Sant'Anna for the financial support of basic research on micromanipulation.

5. References

[1] I.W. Hunter, S. Lafontaine, P.M.F. Nielsen, P.J. Hunter, J.M. Hollerbach, "Manipulation and Dynamic Mechanical Testing of Microscopic Objects Using a Tele-Micro-Robot System", Proc. of the International Conference on Robotics and Automation, Scottsdale, Arizona, May 14-19, 1989, pp. 1553-1558.

[2] F. Arai, K. Morishima, T. Kasugai, T. Fukuda, "Bio-Micro-Manipulation (New Directions for Operation Improvement)", Proc. of the International Conference on Intelligent Robotic Systems IROS'97, Grenoble, France, September 7-11, 1997, pp. 1300-1305.

[3] R. Hollis, "Whither Microrobots ?", Proceedings of the International Symposium on Micro Machine and Human Science, Nagoya, Japan, October 2-4, 1996, pp. 9-12.

[4] K. Koyano and T. Sato, "Micro Object Handling System with Concentrated Visual Fields and New Handling Skills", Proc. of the 1996 IEEE International Conference on Robotics and Automation, Minneapolis, Minnesota, April 22-28, 1996, pp. 2541-2548.

[5] R.S. Fearing, "Survey of Sticking Effects for Micro Parts Handling", Proc. of the International Conference on Intelligent Robotics and Systems, Vol.2, pp. 212-217 (1995).

[6] Y. Ando, "Micro Grippers", Journal of Robotics Mechatronics, Vol. 2, pp. 214-217 (1992).

[7] F. Arai, T. Fukuda, H. Iwata, K. Itoigawa, "Integrated Micro Endeffector for Dexterous Micromanipulation", Proc. of the International Symposium on Micro Machine and Human Science, Nagoya, Japan, October 2-4, 1996, pp. 149-156.

[8] P. Chu, K. Pister, "Analysis of Closed-Loop Control of Parallel-Plate Electrostatic Microgrippers", Proc. of the International Conference on Robotics and Automation, San Diego, USA, May 8-13, 1994, pp. 820-825.

[9] P. Bhatti, P.H. Marbot, B. Hannaford, "Microscopic Pick and Place with the Mini-Direct Drive Arm", SPIE Telemanipulation Symposium, Boston, November 1992.

[10] C. Kim, A.P. Pisano and R. Muller, "Silicon-Processed Overhanging Microgripper", Journal of Microelectromechanical Systems, Vol.1, N°.1, pp. 31-36 (1992).

[11] M.C. Carrozza, P. Dario, A. Menciassi, A. Fenu, "Manipulating Biological and Mechanical Micro-Objects with a LIGA-Microfabricated End Effector," Proceedings of the International Conference on Robotics and Automation ICRA'98, Leuven, Belgium, May 16-20, 1998, pp. 1811-1816.

[12] B. Hannaford, K. Kuhn, "A Hands-On Course in Consumer Electronics Design", International Journal of Mechatronics, Vol. 5, pp. 759-762, October, 1995.

Goal Directed Reactive Robot Navigation with Relocation Using Laser and Vision *

J. R. Asensio   J. M. M. Montiel   L. Montano

Dpto. de Informática e Ingeniería de Sistemas
Centro Politécnico Superior, Universidad de Zaragoza
María de Luna 3, E-50015 Zaragoza, SPAIN
e-mail : {jrasensi,josemari,montano}@posta.unizar.es

Abstract

This paper presents a method to perform a goal directed reactive navigation in unknown indoor environments. Two sensors cooperate to accomplish this task: trinocular vision and 3D laser rangefinder. Trinocular vision selects the initial goal location for the navigation task. Laser is used to accomplish a reactive navigation to avoid the obstacles. Laser is also used to periodically relocate the goal with respect to the robot, so the dead-reckoning drift is compensated. An Extended Kalman Filter is used to solve the data association problem and to perform the goal relocation while the robot navigates. Experimental results involving a real mobile robot are presented, validating the proposed method.

Keywords: *Mobile Robot Navigation, Robot Relocation, Vision and Laser Cooperation, Extended Kalman Filter.*

1 Introduction

Potential field based techniques are often used for mobile robot navigation purposes. In [2] some limitations of potential based methods are described, in particular the difficulty to go through narrow spaces between obstacles (e.g., door frames). Recent works [4] improve and adapt the classical reactive navigation potential field techniques. Moreover, several recent works solve the problem of accurately locating a mobile robot in indoor environments, by continuously fusing the information of several kinds of sensors to build geometric maps [3]. In [9] a technique based on evidence grids is proposed for continuous robot location, correcting the odometry information, but using an a priori map of the environment. There have been

---
*This work was partially supported by spanish CICYT project TAP97-0992-C02-01

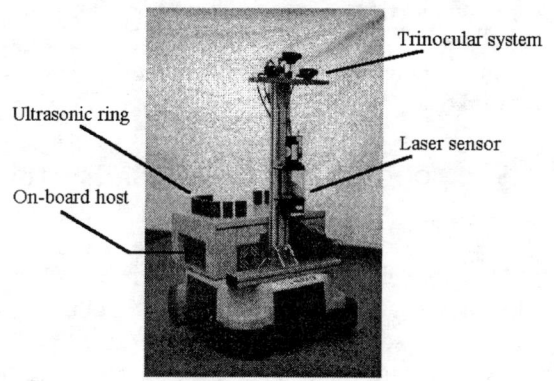

Figure 1: Location of laser and trinocular vision systems on our Labmate robot.

efforts to perform safe autonomous navigation using only trinocular stereo vision systems [7], but using specialized processors (DSP, transputers). In [10] a neural networks based robot location from landmarks –such as doors– is described, but a time consuming learning step is highlighted as its inconvenient.

In this work a cooperation between complementary sensors to achieve a real time navigation is presented. No a priori information about the environment is used, and no dense maps are built while the robot moves (only relevant landmark locations are included in the map). A stereo vision system obtains a 3D geometric reconstruction of the scene, useful to locate several significant features, which allow the robot to plan a navigation task. The features considered in this paper are the doors present in the scene. Additionally, a laser rangefinder sensor provides simple and precise information that is processed in real time. We use the laser information for: a) performing a reactive navigation using a potential field technique [5]; and b) periodically reestimating the goal location using data association, while the robot moves. The problem of passing through a narrow passage has been solved by estimating its center to accurately place an intermediate goal. The goal is reestimated with respect to a reference associated to the robot, which is enough to perform the navigation. Moreover, these periodic reestimations allow to reduce the accumulated location errors due to the odometry drift. Hereinafter, the process to reestimate the goal with respect to a reference associated to the robot will be simply referred as *the relocation process*.

In section 2 we present the robot platform and its sensorial system. The technique to detect and locate a high-level feature from the vision system is briefly presented in section 3. The proposed technique to obtain the relocation of the feature while avoiding obstacles is explained in section 4. In section 5 experimental results are presented, which show how the cooperation between the two sensors improves the navigation task. Some conclusions and future works are related in section 6.

2 The robot and its sensorial system

The Labmate robot is a differentially driven mobile robot (developed by Helpmate Robotics, Inc). It has two active and four passive wheels. Its maximum speed is 1 m/sec. A control software has been developed to have a controller from an on-board host computer.

The stereo trinocular system is composed of 3 CCD monochrome cameras connected to each of the RGB inputs of a color frame grabber, so the three images are simultaneously taken. The 3D *lidar* laser rangefinder (Helpmate), radially scans the environment around the robot. The maximum range is 6.5 meters, and the accuracy of the distance measurement is 2.5 cm irrespective of the distance to the target. The location of the trinocular vision system and the 3D laser sensor on the robot structure can be viewed in Fig. 1.

3 High-level feature location

Due to the limited range of the laser sensor, the only way to locate some features placed far from the robot is using the vision system. Using the trinocular system on board the robot, a straight segment based 3D reconstruction of the scene is achieved. The reconstruction technique uses a probabilistic model to represent the location of the segments. Each segment is represented by a reference system attached to it, and by a covariance matrix representing its location uncertainty. To find a high level feature, such a door as presented in this paper, the system looks for a door pattern in the 3D reconstruction. The uncertainty information associated to each segment is used to find the pattern, also defined up to an uncertainty level. In the matching process to find the door, several unary and binary geometric constraints are verified, using tests based on the Mahalanobis distance to accept or reject hypotheses, following the formalism proposed in [8]. The whole process is summarized in the following steps:

1. Segments in the three gray-level images are detected.

2. Matching and 3D reconstruction are computed using the trinocular algorithm proposed in [6].

3. Small or non-vertical 3D segments are removed.

4. Pairs of 3D vertical segments whose distance belongs to an interval compatible with the width of a typical door are selected; thus, each selected pair is a possible door.

5. Possible doors which have a horizontal 3D segment over it are definitely considered as a door in the scene.

6. To test if the door is open or closed, the existence of a horizontal 3D segment at the door bottom is considered. Since this step is no so reliable as previous ones, the laser sensor is used to verify the door status when the robot is close to it.

This algorithm has been tested in indoor environments including corridors and rooms. As a result, the system locates 50% of the visible doors. The rest of the doors have not been detected due to bad illumination conditions, or have been imprecisely located and are considered as *bad* doors. Precision in the door location is about 10 cm in depth and 4 cm in its perpendicular direction. The precision decreases with the distance from the robot to the feature.

Finally, the system obtains one open door in the room, whose location x_0 and location uncertainty covariance matrix P_0 are used as initial values in the following navigation and relocation processes, explained below. It must be noted that other high-level features, such as corridors, intersection between corridors, etc., could be located from the 3D reconstruction, and then used for path planning purposes.

4 Continuous feature relocation

The task the robot must accomplish can be defined as follows: from any initial location in a room, it must pass through an open door, while avoiding obstacles

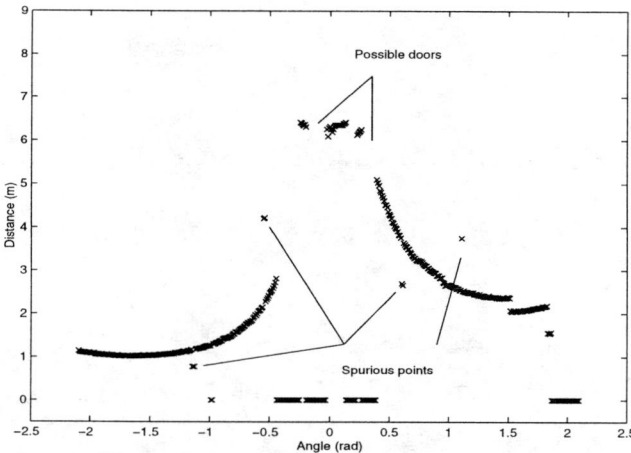

Figure 2: Laser scan information ρ vs. θ

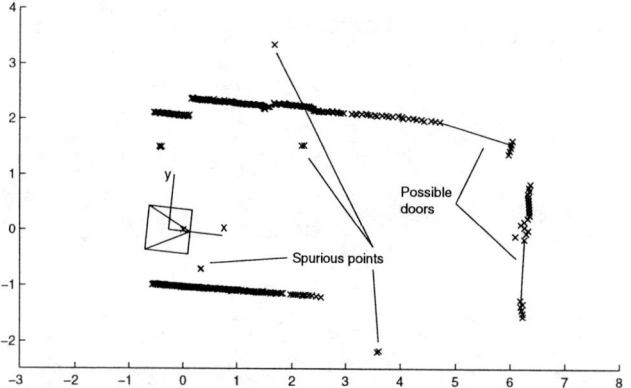

Figure 3: Scan obtained from an experimental location.

and relocating the door. There is no *a priori* information of the environment. The first step is to use the trinocular vision system to detect and obtain a first estimation for the door location, as described in the previous section.

Once the door location has been estimated, a goal beyond the door is computed, and a reactive navigation based on the laser information is performed, driving the robot towards the goal. The reactive navigation is based on artificial potential fields techniques: an attractive force is exerted by the goal, while repulsive ones are exerted by obstacles (and since there is no *a priori* information, walls are also considered as obstacles). The information provided by the laser is used both to perform the reactive navigation and to periodically relocate the feature with respect to the robot. To speed up the information processing required for navigation, only some points around the robot are used to compute the repulsive forces: the environment is scanned every 0.1 sec., this scan is sectored, and one representative point for each sector is selected. Since the scans are taken while the robot moves, the location of the selected points are updated using: a) the robot internal state information (location, and both linear and angular velocities) estimated from the odometry; and b) the time at which the scan has been taken (t_{scan}). This updating process is made in two steps: 1) selected points are referred to the laser location at t_{scan}; and 2), an integration of the selected points of the last 10 scans is performed, transforming the relative location of the sensed points to the current sensor location. See [5] for details about this technique. A wall-following technique is used to exit from potential field minima.

For relocation purposes, a set of possible features –doors, in this case– are detected from the laser information, and the nearest to the previous location estimation is selected by means of a probabilistic data association method. Thus, there are two main steps in the process to relocate the feature. First, a set of *holes* are detected in a planar scan (section 4.1). Then, a data association method is applied to perform the matching process (section 4.2).

4.1 The hole detector

The first step is to obtain a set of holes from the available environmental information. The aim in this phase is to make the simplest laser information processing that could serve to obtain a set of holes, from which a new door location could be obtained in the second phase. The low computation load of this phase allows to implement this method in real time on the mobile robot. This process is carried out as follows:

Scan acquisition. The laser sensor scans the environment at different elevation angles. In spite of this, the laser controller gives the data projected on a horizontal plane. One out of ten scans has an elevation angle between 0 and 7 degrees, which makes it useful to obtain the set of holes; this kind of scans will be called *quasi-planar* scans. The laser provides a quasi-planar scan every second, and all its points are corrected for the feature relocation task, in a similar way to the correction made for navigation purposes –in this case only selected points are corrected (see above).

Scan filtering. Fig. 2 shows a typical quasi-planar scan. Each scan consists of a set of points in polar coordinates (ρ, θ) with respect to the laser sensor reference; when the laser beam does not return, it sets $\rho = 0$; these points can be considered to be more distant than the maximum laser range. First a distance filter is applied: this avoids to take into account the environment beyond the door, because this can confuse the *hole detector*. The distance used as reference

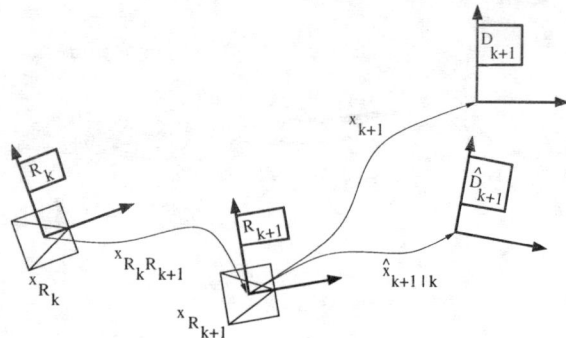

Figure 4: References used in the relocation process. R $\Rightarrow$ robot, D $\Rightarrow$ door

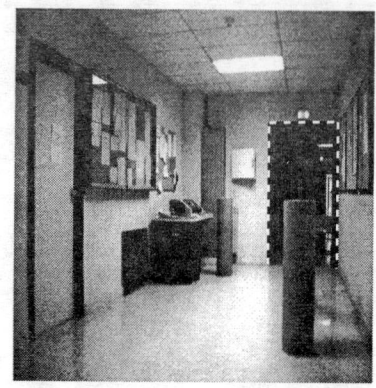

Figure 5: One image from the trinocular system.

is the distance between the robot and the door edges, which is estimated from the previous door location relative to the robot, and from the odometry system (though the odometry system has a drift, the travelled distance between two consecutive relocations is short, and the precision obtained is enough to accomplish the task). Since one of the scan component available is the distance ρ, this algorithm is extremely fast.

Due to reflections and other problems, isolated points can appear in the scan (see Fig. 3). These isolated points are considered spurious and are removed using a median filter applied to the distance component ρ of the scan, in a similar way as the one used in computer vision to filter the *salt-and-pepper* noise.

Hole detection. A hole is defined as a set of consecutive measurements with $\rho = 0$. Obviously, not all the holes are valid candidates to be a door, and distance between the extreme points that delimit each hole is computed to reject all the holes that are not within a width interval. It must be noted that with this simple method there is not needed to apply any kind of segmentation process to the sensorial information.

4.2 The relocation process

Once the set of possible doors has been detected, there is a data association problem that must be solved: only one hole must match the previous estimation for the door –it is even possible that none of the holes match the door. The data association problem is solved using an Extended Kalman Filter (EKF), with its three typical phases: prediction, matching and updating [1]. The variable estimated by the EKF is the location of the door relative to the robot, $\mathbf{x} = \mathbf{x}_{RD} = (x, y, \phi)^T$. In Fig. 4 the relative locations involved in the algorithm are shown. The state equation is:

$$\mathbf{x}_{k+1} = \ominus \mathbf{x}_{R_k R_{k+1}} \oplus \mathbf{x}_k \quad (1)$$

being $\oplus$ and $\ominus$ the location vector composition and its inverse, and where $\mathbf{x}_{R_k R_{k+1}}$ is taken from the odometry (see Fig. 4). And the measurement equation is:

$$\mathbf{x}_{R_k D_k} = \mathbf{x}_k \quad (2)$$

where the measurement $\mathbf{x}_{R_k D_k} = \mathbf{x}_{R_k L_k} \oplus \mathbf{x}_{L_k D_k}$, $\mathbf{x}_{R_k L_k}$ is the location of the laser sensor relative to the robot reference –that is constant and known–, and $\mathbf{x}_{L_k D_k}$ is the measured location for the door, relative to the laser reference.

EKF initialization. The vision system provides both an initial estimated location for the door $\mathbf{x}_0$, and an estimation for the geometric uncertainty with which the door has been detected –that is represented through a covariance matrix, $\mathbf{P}_0$ (see section 3).

Prediction. The prediction stage follows the next equations:

$$\hat{\mathbf{x}}_{k+1|k} = \ominus \mathbf{x}_{R_k R_{k+1}} \oplus \hat{\mathbf{x}}_{k|k} \quad (3)$$

$$\mathbf{P}_{k+1|k} = \mathbf{P}_{k|k} + \mathbf{F}_k \mathbf{Q}_k \mathbf{F}_k^T \quad (4)$$

where $\mathbf{P}_k$ is the covariance matrix that represents the uncertainty of $\mathbf{x}$, $\mathbf{Q}_k$ is the covariance matrix that represents the uncertainty of $\mathbf{x}_{R_k R_{k+1}}$, and $\mathbf{F}_k$ is the matrix that translates this uncertainty to the robot reference at $k+1$,

$$\mathbf{F}_k = \mathbf{J}^{-1}\left(\hat{\mathbf{x}}_{k+1|k}\right) \mathbf{J}^{-1}\left(\mathbf{x}_{R_k R_{k+1}}\right) \quad (5)$$

where $\mathbf{J}$ is the jacobian matrix to deal with the change in the base reference:

$$\mathbf{J}\{\mathbf{x}\} = \begin{pmatrix} \cos\phi & -\sin\phi & y \\ \sin\phi & \cos\phi & -x \\ 0 & 0 & 1 \end{pmatrix} \quad (6)$$

Matching and updating. A nearest neighbor matching process is performed in order to select the *hole* –if one exists– that corresponds to the door which location must be estimated. The matching process uses a probabilistic test: the hole with smallest Mahalanobis distance is matched with the predicted door location (only if this distance is less than a threshold).

From the measurement equation 2, the innovation is computed as follows:

$$\boldsymbol{\nu} = \mathbf{x}_{R_{k+1}D_{k+1}} - \hat{\mathbf{x}}_{k+1|k} \qquad (7)$$

where the operator $-$ has been used in place of the operator $\ominus$ since the equation has been linearized. From the innovation, the Mahalanobis distance is computed as follows:

$$\mathcal{D}^2 = \boldsymbol{\nu}^T \mathbf{S}_{k+1}^{-1} \boldsymbol{\nu} \qquad (8)$$

where $\mathbf{S}_{k+1}$ is the covariance matrix associated to the measurement prediction, computed as:

$$\mathbf{S}_{k+1} = \mathbf{P}_{k+1|k} + \mathbf{G}_{k+1}\mathbf{R}_{k+1}\mathbf{G}_{k+1}^T \qquad (9)$$

In this equation, $\mathbf{R}_{k+1}$ is the covariance matrix associated to the measurement –that is obtained from the sensor measurement characteristics–, and

$$\mathbf{G}_{k+1} = \begin{pmatrix} \cos\phi & -\sin\phi & 0 \\ \sin\phi & \cos\phi & 0 \\ 0 & 0 & 1 \end{pmatrix} \qquad (10)$$

is used to translate the measurement uncertainty from the associated to the door location (D in the Fig. 4) to the robot location.

Since –under linear-Gaussian assumptions– the Mahalanobis distance follows a chi-square probability density with 3 d.o.f., the statistical test for accepting the data association is that $\mathcal{D}^2 \leq a$, where a can be found in the chi-square tables ($a = \mathcal{D}^2_{n=3,\alpha}$).

Only if a hole matches the door location estimation $\hat{\mathbf{x}}_{k+1|k}$, the rest of computations needed for the door relocation are done. This avoids to do these computations for *bad* holes, reducing the overall computation load. The new door location estimation $\hat{\mathbf{x}}_{k+1|k+1}$ after an observation $\mathbf{x}_{R_{k+1}D_{k+1}}$ has been matched is:

$$\hat{\mathbf{x}}_{k+1|k+1} = \hat{\mathbf{x}}_{k+1|k} + \mathbf{W}\boldsymbol{\nu} \qquad (11)$$

and its associated covariance matrix is:

$$\begin{aligned} \mathbf{P}_{k+1|k+1} &= (\mathbf{I}-\mathbf{W})\mathbf{P}_{k+1|k}(\mathbf{I}-\mathbf{W})^T + \\ &\quad \mathbf{W}\mathbf{G}_{k+1}\mathbf{R}_{k+1}\mathbf{G}_{k+1}^T\mathbf{W}^T \end{aligned} \qquad (12)$$

where $\mathbf{W} = \mathbf{P}_{k+1|k}\mathbf{S}_{k+1}^{-1}$.

If there is no valid measurement to relocate the door, the best possible estimation is

$$\hat{\mathbf{x}}_{k+1|k+1} = \hat{\mathbf{x}}_{k+1|k} \qquad (13)$$
$$\mathbf{P}_{k+1|k+1} = \mathbf{P}_{k+1|k} \qquad (14)$$

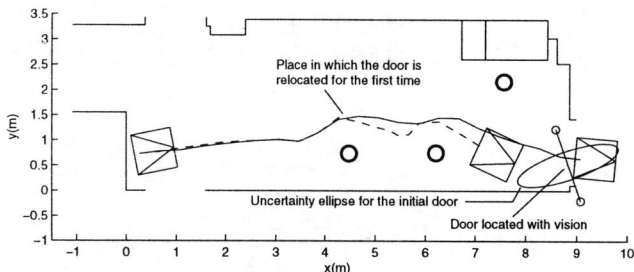

Figure 6: Experimental trajectories: — with feature relocation, and − − without it.

This implies a bigger uncertainty, because the robot has moved.

5 Experimental results

To validate the proposed method, the following experiment has been done. The robot is placed in a room (9×3.5 meters), close to one of its corners, and it must exit through the door located at the other end (see Fig. 6). At the initial robot location, the trinocular vision system takes three images (one of them is shown in Fig. 5), and from these images a segment based 3D reconstruction of the environment is computed. Thus, the door location shown in Fig. 6 is obtained, with the error that can be seen in the same figure, in which its uncertainty ellipse for the x and y components is represented.

Then, the robot performs a reactive navigation in order to cross the initial door provided by the vision system. The reactive nature of the navigation allows to avoid the obstacles that appear in the trajectory – like the cylinders that can be shown in the image and in Fig. 6. When the robot is close enough to *see* the door with the laser sensor (Fig. 7), the implemented algorithm is able to relocate it. For this purpose, one planar scan is obtained each second, and used to relocate the door as the robot moves at an average speed of 0.25 m/s. The location door is relocated in 16 of the 41 planar scans made in this experiment run.

Being all the relocation steps similar, let us analyze the first for which a new location door is obtained from laser readings. This is represented in Fig. 7, where the accumulated odometry drift causes the differences between the environment and the plotted laser readings. This is the most critical step on the trajectory, due to the big uncertainty associated to the door prediction, since the robot has moved ≈ 4.5 meters from the only door estimation available until this moment: the one obtained from the vision system. At this step the covariance matrix $\mathbf{P}_{k+1|k}$ associated to the prediction has grown due to the odometry drift. In spite of this,

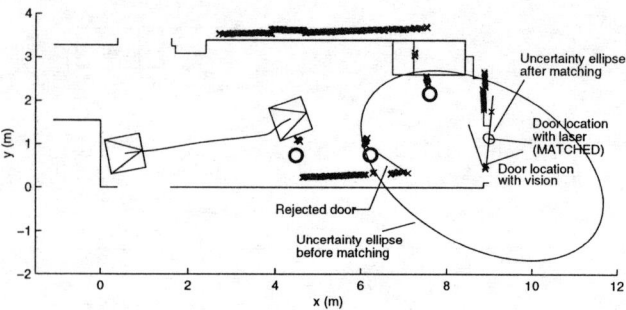

Figure 7: First door relocation.

the *bad* door is rejected, and the *good* door is matched with the previous door estimation. The Mahalanobis distance for the rejected door represented in Fig. 7 is high because of the door orientation. In this sense, a well tuned covariance model is essential to reject the measured holes that are slightly misoriented. These kind of holes, if matched, can cause the robot to cross the door from a non perpendicular direction. The robot can even bump against the door frame or can fall in a potential field minimum.

When the match between the previous door estimation and the correct hole has been done, the door location uncertainty is highly reduced –it is represented through the little ellipse in Fig. 7. Then, a goal beyond the relocated door is computed and passed to the reactive navigation task executor.

In Fig. 6 the door estimation obtained by the vision system is shown, with two experimental trajectories: one without periodic door relocation, and the other with the relocation process active. In the first the robot cannot reach the next room because it has become confused by the growing odometry drift (it has fallen into a potential field minimum) and stops.

6 Conclusions

Navigation of a mobile robot without a priori information in partially cluttered indoor environments is presented in this work. The cooperation between a fast and precise sensor (laser, with a limited scan range) with a sensor that provides geometric and semantic information (trinocular vision system, with a higher visibility range) is carried out. The vision system provides the goal location from a high level feature –a door– that is recognized and located from the 3D segment based reconstruction of the scene. The laser information is used both to perform a safe reactive navigation to the goal, and to periodically relocate the feature. The data association problem is solved using a probabilistic method (Extended Kalman Filter) that takes into account the geometric uncertainties. In this way, relocation notoriously improves the navigation task allowing the robot to pass through narrow ways, where a navigation based only on dead-reckoning fails due to the monotonically growing drift.

The proposed technique will be generalized to other typical indoor features used as landmarks, such as corridors or intersections in corridors. Thus, located features will be used as absolute references to the robot location when robot moves in the next room. In this way, a special high-level feature locations map of the environment will be built and used to reference the robot location in each step of the navigation process.

References

[1] T. Bar-Shalom and T. E. Fortmann. *Tracking and Data Association*. Academic Press Inc., 1988.

[2] J. Borenstein and Y. Koren. Potential field methods and their inherent limitations for mobile robot navigation. In *IEEE Int. Conf. on Robotics and Automation*, pages 1398–1404, 1991.

[3] J. A. Castellanos, J. M. Martinez Montiel, J. Neira, and J. D. Tardós. Simultaneous map building and localization for mobile robots: A multisensor approach. In *IEEE Int. Conf. on Robotics and Automation*, pages 1244–1249, 1998.

[4] H. Haddad, M. Khatib, S. Lacroix, and R. Chatila. Reactive navigation in outdoor environments using potential fields. In *IEEE Int. Conf. on Robotics and Automation*, pages 1232–1237, 1998.

[5] L. Montano and J. Asensio. Real-time robot navigation in unstructured environments using a 3d laser rangefinder. In *IEEE-RSJ Int. Conf. on Intelligent Robots and Systems*, pages 526–532, 1997.

[6] J. M. M. Montiel and L. Montano. Probabilistic structure from camera location using straight segments. *Image and Vision Computing*. To appear.

[7] D. Murray and C. Jennings. Stereo vision based mapping and navigation for mobile robots. In *IEEE Int. Conf. on Robotics and Automation*, pages 1694–1699, 1997.

[8] J. Neira, L. Montano, and J.D. Tardós. Constraint-based object recognition in multisensor systems. In *IEEE Int. Conf. on Robotics and Automation*, 1993.

[9] A. C. Schultz and W. Adams. Continuous localization using evidence grids. In *IEEE Int. Conf. on Robotics and Automation*, pages 2833–2839, 1998.

[10] S. Thrun. Finding landmarks for mobile robot navigation. In *IEEE Int. Conf. on Robotics and Automation*, pages 958–963, 1998.

Vision-based Self-Localization of a Mobile Robot Using a Virtual Environment

M. Schmitt, M. Rous, A. Matsikis, K.-F. Kraiss
Department of Technical Computer Science
Aachen University of Technology (RWTH)
Ahornstr. 55, 52074 Aachen, Germany
{schmitt, rous, matsikis, kraiss}@techinfo.rwth-aachen.de

Abstract

This paper presents a method for position estimation of mobile robots based on the comparison of real camera snapshots taken by an on-board camera and images taken by a virtual camera in a virtual environment. We propose a technique for texturing planar walls of a 3D model of the operating environment and make use of this model for improving operator situation awareness as well as robot self-localization. By applying texture created from camera snapshots a more realistic impression of the environment can be obtained, so that the virtual environment could even be used for inspection tasks. Furthermore, the texture provides additional structure which is especially useful in hallways that contain only few hints for robot navigation.

1 Introduction

The vision-based self-localization technique proposed here is embedded in our control centre VERONA and tested with a mobile robot called TAURO, that are both sketched first. We then discuss how the environment model is created and especially how the texture is composed from the robot's on-board camera snapshots. Subsequently, we present the principle of the position estimation and some experiments we performed.

1.1 Control Centre VERONA

Mobile robots mostly do not operate fully autonomously, except when they are designed and programmed for one special task. Universally usable service robots have to be advised, programmed and controlled. The results of a robot mission are to be reviewed and evaluated - especially when they are used for inspection and monitoring tasks. Therefore, operators are needed and they should be supported by an integrated man-machine interface, that is easy to use and allows an intuitive handling of one or several robots. Virtual Environments improve the operator's survey, enable a free choice of viewpoints and give independancy of real view conditions and transmission quality of an on-board camera. These advantages have already been proved in similar projects, as e.g. [1].

VERONA (Virtual Environment for Robot Navigation) is a control centre that is based on a 3D environment model. It facilitates mission and data management by providing the operator with a visual interface for all interaction between operator and mobile system(s). All sensor data are stored at the corresponding recording position in the model to allow a visual database search regarding times and places, to improve operator's survey and for and an easy review or comparison of robot missions. Furthermore, a visual goal designation is possible [2].

1.2 Mobile Robot TAURO

The control centre is being tested together with our semi-autonomous mobile robot TAURO shown in figure 1. TAURO is a mobile service robot developed for inspection, stocktaking and documentation tasks in indoor environments. It is designed to assist human staff through the automation of uniform and repeating monitoring trips. Possible application environments include museums, office buildings and industry.

The system is based on an electric wheelchair with a differential drive and unsteered wheels. It is equipped with 14 ultrasound sensors, two wheel encoders and two on-board PCs running QNX realtime operating system. On top a CCD camera is mounted with a pan/tilt-unit. The robot is connected to the control centre via two radio links: an analog one for continuous video transmission and a digital one for all control

Figure 1: Mobile robot TAURO

information. It is capable of performing autonomous exploration trips, route planning and collision avoidance. The software architecture is based on the Blackboard concept for communication of several individual modules [3].

2 Model construction

Existing models from CAD data are often not up to date or do not reflect the real visible sizes of a building. Therefore it seems to be more appropriate to construct the model from the robot's sensor data. We are investigating two different approaches: a hybrid technique based on ultrasound sensors together with odometry for ground plan generation with subsequent texturing from camera snapshots and a second approach based on photogrammetry that is not discussed here. The aim of our research is to create the environment model without manual input and to rely on the robot's sensor data only.

2.1 Exploration, ground plan and 3D modelling

The first approach can be divided into three steps: ground plan construction from distance measurements, 3D modelling by 'pulling up' walls from the ground plan and finally texturing the walls. The concept is to first explore an indoor environment by following the walls and logging all sensor data. When no initial model is available, localization of the robot depends on odometry only in this phase. As this is a source of errors, algorithms for an iterative model refinement have to be considered [3]. During these exploration trips, camera snapshots are taken in short periods. The data is transmitted to the control centre, where line segments are extracted and a wireframe model is built. Since this part still shows some errors, a manually created environment model is used to ensure correctness for the following steps.

2.2 Texture composition

In order to create a texture from a sequence of images recorded during the exploration trips, the translation between each of the neighboring snapshots has to be computed. Knowing this translation of overlapping images we can merge them together exactly. We apply a blending to get rid of sharp transitions at the borders of the single parts and finally place the image as texture onto the wall.

2.2.1 Cepstrum Analysis

With *Cepstrum Analysis* it is possible to determine the translation between two images very precisely [8]. Some conditions have to be fulfilled, though. First, the images should represent a similar scenery with an overlapping area of more than 50% - otherwise the algorithm could fail in some cases. Second, the snapshots may not show distortions or different scales, but should be taken from viewpoints perpendicular to the walls in similar distances. This precondition has to be met when exploring a room by wall following.

The *Power-Cepstrum C* of a function is defined as the *Power-Spectrum* of the logarithm of the Power-Spectrum of this function. The Power-Spectrum S itself is defined as the squared absolute value of the *Fourier* transformed function. The general Power-Cepstrum can be transformed easily into the two-dimensional case for digital image proccessing (1).

$$\mathcal{C}\{f(x,y)\} = \left| \mathcal{F}\left\{ \log \left| \mathcal{F}\{f(x,y)\} \right|^2 \right\} \right|^2 \quad (1)$$

To determine the translation between two images $r(x,y)$ and $t(x,y)$, we calculate the *Cepstrum C* of the image $g(x,y) = r(x,y) + t(x,y)$ and one of the original images r or t as reference. Subtraction of the reference $\mathcal{C}\{r(x,y)\}$ from $\mathcal{C}\{g(x,y)\}$ reduces the noise and we obtain an image with several *Dirac* peaks (2). The displacement (x_0, y_0) of the peaks closest to the middle depicts the image translation we are looking for (figure 2).

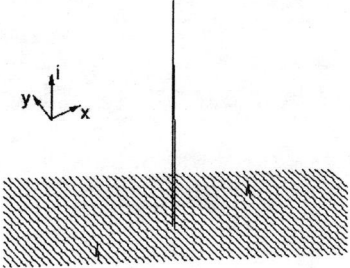

Figure 2: Peaks in the Cepstrum image used for detecting the image translation

$$\begin{aligned}\mathcal{C}\{g(x,y)\} &= \mathcal{C}\{r(x,y)\} + A\delta(u,v) + \quad (2)\\ &\quad B\delta(u \pm x_0, v \pm y_0) + \\ &\quad C\delta(u \pm 2x_0, v \pm 2y_0) + \ldots\end{aligned}$$

A, B, C are the weights of the dirac peaks δ. The peak with weight A is called the main maximum and it is located in the middle of the *Cepstrum* image at (u, v). The translations x_0 and y_0 are the searched values.

To speed up these calculations, the camera images (767 x 575 x 24 Bit) are temporarily translated to graylevels.

2.2.2 Image blending

With the calculated translations x_0 and y_0 we can extract the required image parts and stick them together exactly. We use a modified blending algorithm to smooth the transitions between the single parts, as each snapshot might differ a slightly in brightness.

As lens distortions and brightness differences are growing from the middle of a snapshot to the border, we use strips extracted from the middle of each image. This is sketched in figure 3 for the case of a horizontal image sequence. m_1, m_2 and m_3 depict the middle of each image; x_0 is the translation between image 1 and image 2, x_1 between image 2 and image 3. We start at m_1 and extract a strip from m_1 to $m_1 + x_0$ from image 1 and add the strip from $m_2 - x_0$ to m_2 out of image 2 to the sequence (as image 1 is the first image of the sequence, we also take the left half of it). In contrast to usual blending, we adjust the intensity of the image contribution to the new strip over the image size, as shown in figure 4.

Figure 5 shows a texture generated with the discribed technique. It is composed from a sequence of 35 camera snapshots, covering an area of 15 meters.

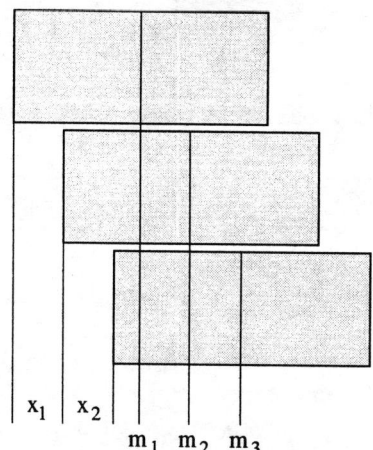

Figure 3: Extracting strips from an image sequence

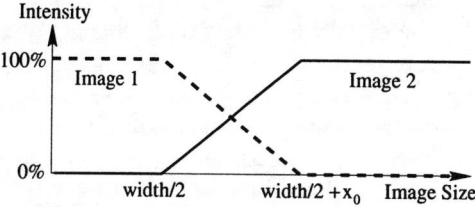

Figure 4: Horizontal image blending with adaptive intensity

The error when merging the single snapshots together is one pixel at most and due to the subsequent blending, the image borders can hardly be detected.

The starting position where the texture is placed in the model is still being determined by the robot position (known from odometry) when taking the first image of the sequence, but is planned to be derived from matching doorways between texture and ground plan in order to automate the process.

The technique presented is usable for planar walls like in this example of a hallway. These hallways can often be found in office buildings. In rooms with a more complex structure (e.g., with furniture), it is less useful. Therefore, we are investigating another 3D reconstruction method based on camera images, inspired by [4]. These optical techniques also provide independancy of odometry errors on the initial exploration trips.

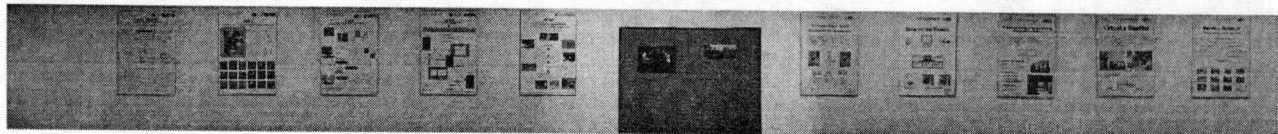

Figure 5: Example of a piecewise composed texture from 35 single snapshots

3 Vision-based Self-Localization

3.1 Using several camera snapshots and the periphery angle

Texturing the virtual environment is not only useful for the operator in front of the control monitor by providing him with a more realistic impression of the operating environment, but also for position estimation, since it yields additional structure that would otherwise be missing on planar walls. Our approach is to compare the real camera view with the virtual camera view from an estimated position in the virtual environment. The robot position has to be roughly known from odometry to ensure that both cameras look in a similar direction. Since the texture is obtained from real camera snapshots, the real and the virtual camera view look very similar and can be compared. We do not need to identify objects in the snapshots but can use the whole image as a landmark. We obtain a landmark position by computing the intersection of the center of the view with the wall in the model.

Figure 6 sketches the concept: by comparing real and virtual camera image and computing the relative translation, we get a position in the model, i.e., the position of the real image and the heading the real camera is looking at. With the pan/tilt-unit we can measure the angle between two of these headings. This yields the periphery angle of a circle intersecting the two positions in the model derived from the image centers and the current viewpoint. The circle can be derived from the periphery angle and the two positions. Since the periphery angle is identical for all positions on this circle, these are possible locations for our robot. With a third image and another periphery angle we can construct a second circle intersecting the first one. From this we get the robot position as one of the two possible intersections of these circles. Since one of these two intersections lies in the observed wall, the point to choose is determined. Main advantage of the proposed method is, that it does not need artificial landmarks. In contrast to a similar approach (using vertical lines in a single image) presented by [9], we make use of the image positions without identifying

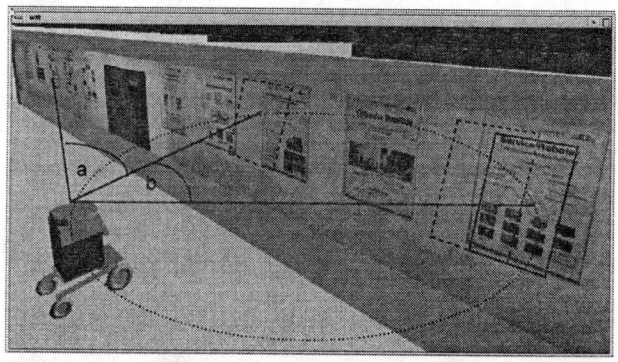

Figure 6: Position estimation using the periphery angle

specific objects respectively knowing the landmarks.

3.2 Comparing real and virtual camera view

What we need is a match of the real and the virtual camera snapshots. We cannot use the Cepstrum technique in this case, since the images might not be similar enough. Thus, we compare the vertical line patterns of the image pairs. First, the second derivative edge filter is applied for binarization and edge detection. The vertical lines are then searched with the *Hough* transformation [5, 6] and are matched between each image pair in the Hough space. Figure 7 gives an example of the detected vertical lines in camera snapshots used to determine the translation between real and virtual view.

We examined different approaches for image matching: cross correlation (3) and the minimisation of absolute differences (4), introduced by [7]. We then developed a more adapted method described below (5), that we call *Positive-Gaussian-Differencesum (PGD)*.

Cross correlation of two graylevel images G and H with dimensions M, N is defined in equation (3). The correlation value at the point (m', n') is stored and after all correlation values are calculated the maximum can be determined, which represents the best correlated translation of the two images or it's line patterns,

Figure 7: Detected vertical lines in real and virtual camera view used for image matching

resp. This technique works for small translations up to the half image width.

$$\mathcal{C}(m',n') = \frac{1}{MN} \sum_{m=0}^{M-1} \sum_{n=0}^{N-1} [\, G(m',n') \\ \star H(m'+m, n'+n)\,] \quad (3)$$

The second method we tested is the minimisation of the absolute difference of the translated accumulators. In [7] the absolute difference $d(m',n')$ of two graylevel images $G(m',n')$ and $H(m',n')$ is defined as:

$$d(m',n') = \sum_{m=0}^{M-1} \sum_{n=0}^{N-1} (\,|\, G(m',n') \\ -H(m'+m, n'+n)\,|\,) \quad (4)$$

The result is similar to the correlation technique. But both technique have a common disadvantage: if the distances between the extracted lines are not similar in both images (caused, e.g., by different scales, perspectives or rendering reasons), the best match is at the point where the 'longest' line is located, i.e., where the most points are lying on the line. If another line is two pixel beside the corresponding line in the other image, it has no effect on the result when using the correlation or absolute difference. But the correct translation is supposed to be the mean value of the translations of the two lines.

Therefore we developed a modified algorithm (5), which considers the neighborhood of a pixel with the weight $\alpha(i)$. We use $\alpha(\pm 3) = 0.1$, $\alpha(\pm 2) = 0.4$, $\alpha(\pm 1) = 0.8$ and $\alpha(0) = 1$. We presume that the result after subtraction is positiv or zero. The minimum of all PGD represents the best matched translation. We achieved the best results with this technique as compared to the two others mentioned. The correct match could be found even in cases where correlation and absolute difference sum were not able to detect a correct match (e.g., due to small distortions between the images to be compared).

$$PGD(m',n') = \sum_{m=0}^{M-1} \sum_{n=0}^{N-1} (\, G(m',n') \\ - \sum_{i=-\varepsilon}^{i=+\varepsilon} \alpha(i) H(m'+m+i, n'+n+i)\,) \quad (5)$$

Two different strategies have been investigated to find the best match of the line patterns: an iterative technique that computes the translation several times to approach to the best fit and a direct match with the most prominent vertical line that is searched near the image center. With the second alternative we obtained more accurate positions and much better performance.

3.3 Experiments

In several experiments we proved the feasibility of the proposed techniques. We built a 3D model of our department floor and textured a wall of the hallway as described above. To ensure correctness for the subsequent position estimation, the texture has been placed manually at the corresponding recording position. To eliminate some possible sources of errors (concerning the robot and pan/tilt-unit), the snapshots were taken with the camera mounted on a tripod and the tripod position was measured with reference to the surrounding walls.

Figure 8 shows the results of our tests. We invoked the position estimation from a viewpoint that was gradually moved away from the true position to investigate the robustness of the method and the calculated results. This initial displacement has been set separately and in combination for x, y and the orientation to positions in a range of $\pm$ 10 centimeters and a deviation of up to $\pm$ 5 degrees. As depicted in the diagram, most calculated positions lie in a small region near the origin. In some cases, the deviations add in such a way that real and virtual images do not overlap enough any more.

The calculated positions can be divided into two groups: 88 of the 92 results lie in a circular region of 2 cm radius (with an orientation error of $\pm$ 1 degrees), thus reducing an assumed initial uncertainty (from odometry) of $\pm$ 10 centimeters to 1/30 and 4 results are scattered in the original area. The mean distance of all calculated positions from the average position is about 0.87 cm. Tests without combination of x, y, ϕ deviations (not shown in the diagram) showed to be even more robust. Displacement between average und true position (0,0) might be due to a systematic error in angle measurement or model calibration. The time needed for these computations is about 2 minutes on a Workstation with a MIPS R10000, 150 MHz processor.

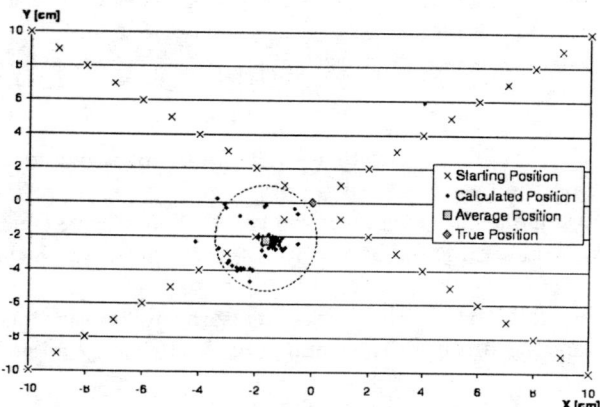

Figure 8: Results of position estimation for combinations of x, y, ϕ initial deviations

We found that the accuracy depends mostly on the cutting angle of camera view and walls. If the angle is too small, little deviations cause big variances in the position of a landmark. Other influences are the resolution of the pan/tilt-unit and the resolution of camera images and textures.

Instead of using a precise pan/tilt-unit, the developed technique of matching vertical line patterns can be applied to an image sequence recorded during panning to compute the angle between the images used for position estimation, if the scenery respectively all images contain enough structure. In the current state, such calculations show an accuracy of ± 0.5 degrees.

4 Conclusion

Virtual environments can help to improve operator's situation awareness as well as robot localization. Placing textures composed from pictures taken by the on-board camera adds structure, that otherwise would be missing on planar walls, as they can often be found in office buildings. They can be used to compare real and virtual camera views. Thus, we do not need to identify objects but can use the whole images as landmarks.

The presented techniques have been tested in a hallway, where environment features detectable by ultrasound sensors, e.g., are rare. With the developed self-localization, position uncertainty due to cumulating odometry errors can be reduced significantly, if invoked periodically before real camera view and model view diverge too much.

In simple environments, sticking together the camera snapshots taken during exploration trips yields quite realistic views. For more complex environments, camera-based 3D reconstruction techniques, that will be investigated in future work, promise better results.

References

[1] Piguet, L., Fong, T., Hine, B., Hontalas, P. and Nygren, E.: VEVI: A Virtual Reality Tool For Robotic Planetary Explorations. In: Proceedings of 'Virtual Reality World 95', Stuttgart, Germany, 1995.

[2] Schmitt, M. and Kraiss, K.-F.: A Mobile Robot Control Centre for Mission and Data Management. In: Proceedings of the 24th Annual Conference of the IEEE Industrial Electronics Society (IECON '98), Aachen, Germany, 1998, Vol. 4/4, pp. 2174–2179.

[3] Pauly, M.: Ferninspektion mit mobiler Sensorik. PhD Thesis, Dept. of Technical Computer Science, Aachen University of Technology (RWTH), Logos, Berlin, 1998.

[4] Pollefeys, M., Koch, R., Vergauwen, M. and Gool, L. V.: Flexible 3D Acquisition with a Monocular Camera. In: Proceedings of the IEEE International Conference on Robotics & Automation, Leuven, Belgium, 1998, Vol. 4/4, pp. 2771–2776.

[5] Hough, P. V. C.: Method and Means for Recognizing Complex Patterns. U.S. Patent 3.069.654, Dec. 1962.

[6] Duda, R. O. and Hart, P. E.: Use of the Hough Transformation to Detect Lines and Curves in Pictures. Communications of the ACM, 1972, Vol.15, No.1, pp. 11–15.

[7] Rembold, D., Zimmermann, U., Längle, T. and Wörn, H.: Detection and Handling of Moving Objects. In: Proceedings of the 24th Annual Conference of the IEEE Industrial Electronics Society (IECON '98), Aachen, Germany, 1998, Vol. 3/4, pp. 1332–1337.

[8] Yeshurun, Y. and Schwartz, E. L.: Cepstral Filtering on a Columnar Image Architecture: A Fast Algorithm for Binocular Stereo Segmentation. In: IEEE Transactions PAMI, Vol.II, No. 7, July 1989, pp. 759–767.

[9] Muñoz, A.J. and Gonzalez, J.: Two-Dimensional Landmark-based Position Estimation from a Single Image. In: Proceedings of the IEEE International Conference on Robotics & Automation, Leuven, Belgium, 1998, Vol. 4/4, pp. 3709–3714.

Continuous Mobile Robot Localization: Vision vs. Laser

J. A. Pérez, J. A. Castellanos, J. M. M. Montiel, J. Neira, J. D. Tardós

Universidad de Zaragoza
Departamento de Informática e Ingeniería de Sistemas
c/María de Luna, 3, E-50015 Zaragoza, SPAIN
email: {jperez,jacaste,josemari,jneira,tardos}@posta.unizar.es

Abstract

In this work we describe a comparative study of the performance of map-based robot localisation processes based on diverse sensing devices such as monocular and trinocular vision systems, and laser rangefinders. We study both the precision (error with respect to the true values) and robustness (sensor measurements correctly paired with map features) of each localisation process. The experiment design that we have followed allows us to compare these processes under exactly the same conditions. We conclude that comparable precision levels can be attained with each of the three sensors. With respect to robustness, monocular and trinocular vision pose more complex matching problems than laser, requiring more elaborate solutions to make the process robust.

Keywords

Robot Location, Monocular Vision, Trinocular Vision, Laser Rangefinder

1 Introduction

The problem of precisely and robustly locating a mobile robot following a trajectory in a known environment has been treated extensively in the literature [1, 2, 3, 4]. In general, the basic idea is always the same: at each location of the robot trajectory the localisation system perceives the environment with some sensor, compares the sensorial information with the expected values (predicted from an *a priori* map), and corrects the available robot location, generally given by an odometric system, to make the perceived data match better with the expected data. The precision and robustness of this process depends on several factors, being the most significant: the sensor used, the available map, and the matching process. Some researchers have reported studies of the precision of localisation processes using some sensors. However, these works have not established what the *true* error is, because the true location of the robot is not known (only error bounds are obtained).

We have studied this problem to try to answer the following questions: under exactly the same conditions, which sensor performs better in carrying out the localisation process? Can we attain with monocular vision comparable precision levels than with laser? Or is stereo vision (in this case, trinocular vision) necessary to attain sufficient precision levels? In order to adequately answer these questions, the following must be taken into account:

1. In order to be able to measure and compare the precision of the results in each case, we need to know the *true* robot location at each step of the trajectory, and have a precise description of the environment.

2. We must be able to test the location processes using the same *a priori* information and for the same trajectory. We may even want to try out different matching processes for the same sensor data, to compare robustness.

We have designed an experiment that allows us to make this comparative study guaranteeing that the former requirements are met. Due to their popularity, we have concentrated in comparing laser rangefinders and vision systems, monocular and trinocular. Vision is a more versatile sensor, so we are basically interested in knowing whether it can achieve comparable precision and robustness levels. We have included both monocular and trinocular vision to see whether only one camera, where there is no depth information, may attain satisfactory precision and robustness levels. We use trinocular instead of stereo vision because the redundancy given by the third camera makes the correspondence problem simpler.

This paper is organised as follows: in section 2 we describe the experiment design. Section 3 contains a description of the location process carried out for each sensor. The obtained results are discussed in section 4. Finally, in section 5 we draw the main conclusions of this work. In order to have sufficient space for presenting our results and discussing them, we have avoided including any mathematical detail related to the representation and fusion of sensorial information in the localisation processes. All mathematical details of these processes have been described extensively and can be found in [4, 5].

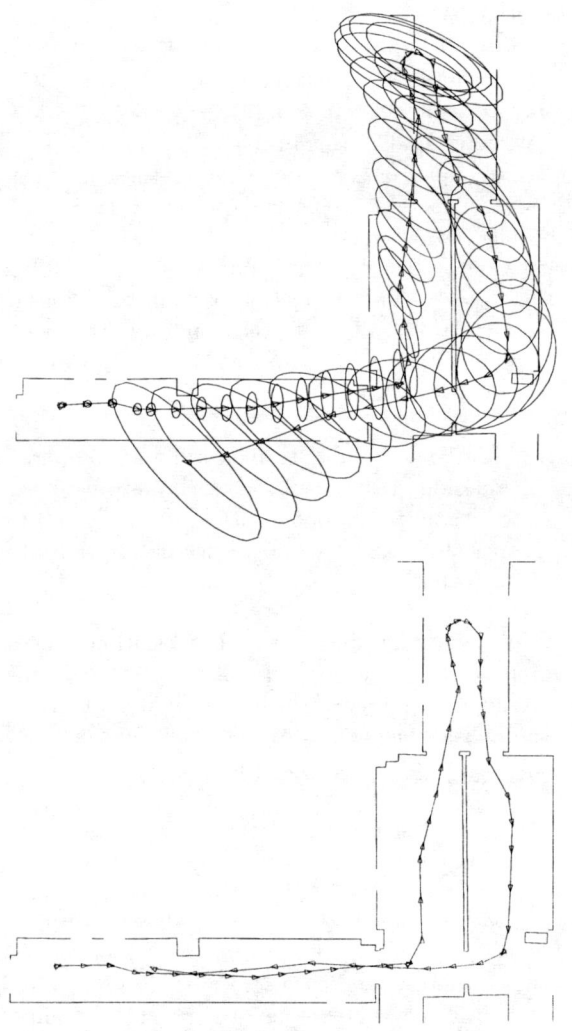

Figure 1: Robot location according to odometry (top) and the theodolites (bottom).

2 The Experiment

The experiment was carried out using the following equipment: a Labmate mobile robot, a laser range system and a trinocular vision system, both mounted on the mobile robot, and a pair of theodolites, used as a precise location measurement equipment, independent from the sensors and the mobile robot.

The *environment model* is a set of vertical edges, corresponding to wall corners and door frames, whose location was measured with the theodolites. The location of vertical walls was calculated using this information. The resulting 2D environment map is shown in fig. 1.

The robot was programmed to follow a trajectory in our laboratory. At each step of the path, the robot location according both to odometry, and measured with the theodolites, was obtained. Being the most precise measurement independent from sensor observations, it is considered the *true* robot location. In fig. 1, both the odometric (with uncertainty ellipsoids) and the theodolite robot location are shown, highlighting the cumulative nature of odometric errors that eventually make the measurement useless.

At each step of the robot trajectory, the environment was sensed using both trinocular vision (fig. 2), and the laser rangefinder. Fig. 3 shows the environment information obtained at step 22 of the robot trajectory. Large vertical edges (more than 100 pixels) are extracted from the three images: the ones corresponding to the central image are considered monocular infinite projection rays of corners and door frames, and the rays of the three images are matched to obtain the location of trinocular vertical edges also corresponding to corners and door frames [7]. The laser scan is segmented to obtain segments corresponding to walls [5]. In the next section we describe the three experiments that we carried out with this information. A detailed description of this experiment, including complete robot and sensor specifications, as well as all data sets, can be found in [6].

3 Robot Localisation

The process of refining the estimated location of the mobile robot given a sensor measurement E of a certain environment feature M, which is applicable to the three types of sensor information that we consider (monocular vision rays, trinocular vision points and laser segments), is carried out in the following way:

1. The imprecision in both the estimated robot lo-

Figure 2: Trinocular Image Set at step 22 of the robot trajectory.

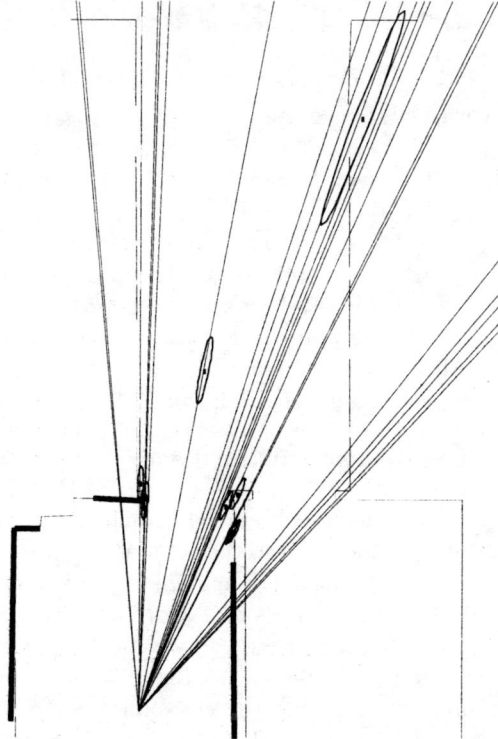

Figure 3: Features observed at step 22 of the robot trajectory : laser segments (thick segments) corresponding to walls, monocular rays (infinite projection rays) corresponding to corners and door frames, and trinocular points (thick points) corresponding also to corners and door frames.

cation and the sensor measurement is represented using a probabilistic model, the SPmodel [8].

2. At each step in the robot trajectory, the sensor acquires a set of measurements of the environment features. A matching process must decide either to pair each of these measurement with some map feature, or to label the measurement as spurious. This matching process is different for each type of sensor, and it is explained in each case below.

3. The possible correspondence between a certain sensor observation and a map feature is verified using a statistical test (the chi-square test) based on the squared Mahalanobis distance between the observation and the feature.

4. Having found that sensor observation E is compatible with map feature M, we establish a nonlinear measurement equation that states that the location of E and M must coincide, and fuse this information using a specialised version of the Kalman Filter, the SPfilter [4], to refine the estimated robot location.

3.1 Localisation using Monocular Vision

As it can be seen in fig. 3, the matching process between projection rays given by monocular vision and vertical edges corresponding to corners and door frames is very complex, due to several facts:

- There are many environment visual features which are not represented in the map.

- Some corners represented in the map cannot be detected by the vision system due to illumination conditions.

- The projection ray is infinite, so there is no information about the distance to the observed feature.

- The proximity of different points of the map and the uncertainty in the initial odometry estimation of the robot position causes that multiple map corners match as candidates for pairing with an observation.

To minimise the risk of the robot getting lost, secure pairings (projection rays that have a unique map corner as pairing candidate) are integrated first. Next, if there are non-secure pairings, a limited number of them is integrated. The trajectory resulting from this process is shown in 4, top.

3.2 Localisation using Trinocular Vision

The trinocular vision system gives us the location of vertical edges in 3D space (which correspond to points in 2D space). These points are obtained by a correspondence algorithm [7]. At each point of the trajectory, the predicted location of potentially visible corners is calculated, and again the chi-square compatibility test is applied to each of the trinocular points with each of the predicted corners. If more than one corner matches a trinocular point, the corner with the smallest square Mahalanobis distance is selected. Because the correspondence algorithm has eliminated most of the spurious information, this process is much simpler than the one of monocular vision, although the depth precision of trinocular vision is rather low. The resulting trajectory is shown in 4, middle.

3.3 Localisation using Laser

Laser segments are obtained by application of a segmentation technique [5] to the set of data gathered by the laser rangefinder. The matching process is quite simple. Again, at each point of the trajectory, the predicted location of potentially visible walls is calculated, and a suitable pairing for each laser segment is obtained by applying the chi-square compatibility test with each of the predicted walls. The segment is paired to the wall that satisfies the test, and whose extension includes the segment location. If there is more than one, the wall with the smallest Mahalanobis distance is selected. The trajectory resulting from applying this process is shown in 4, bottom.

4 Discussion

In this section we compare the results obtained in the localisation process using the three sensor systems, analysing the precision, robustness and practicality of each approach.

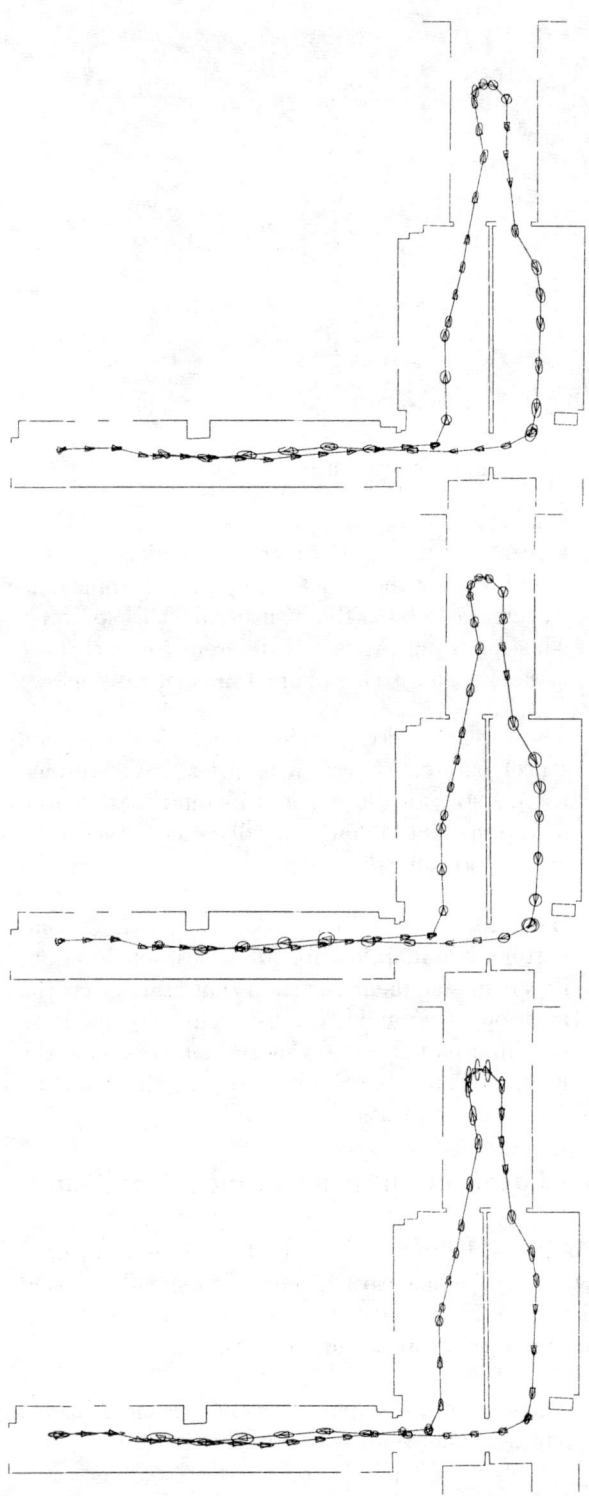

Figure 4: Estimated trajectory using Monocular Vision (top), Trinocular Vision (middle) and Laser (bottom). Uncertainty × 3.

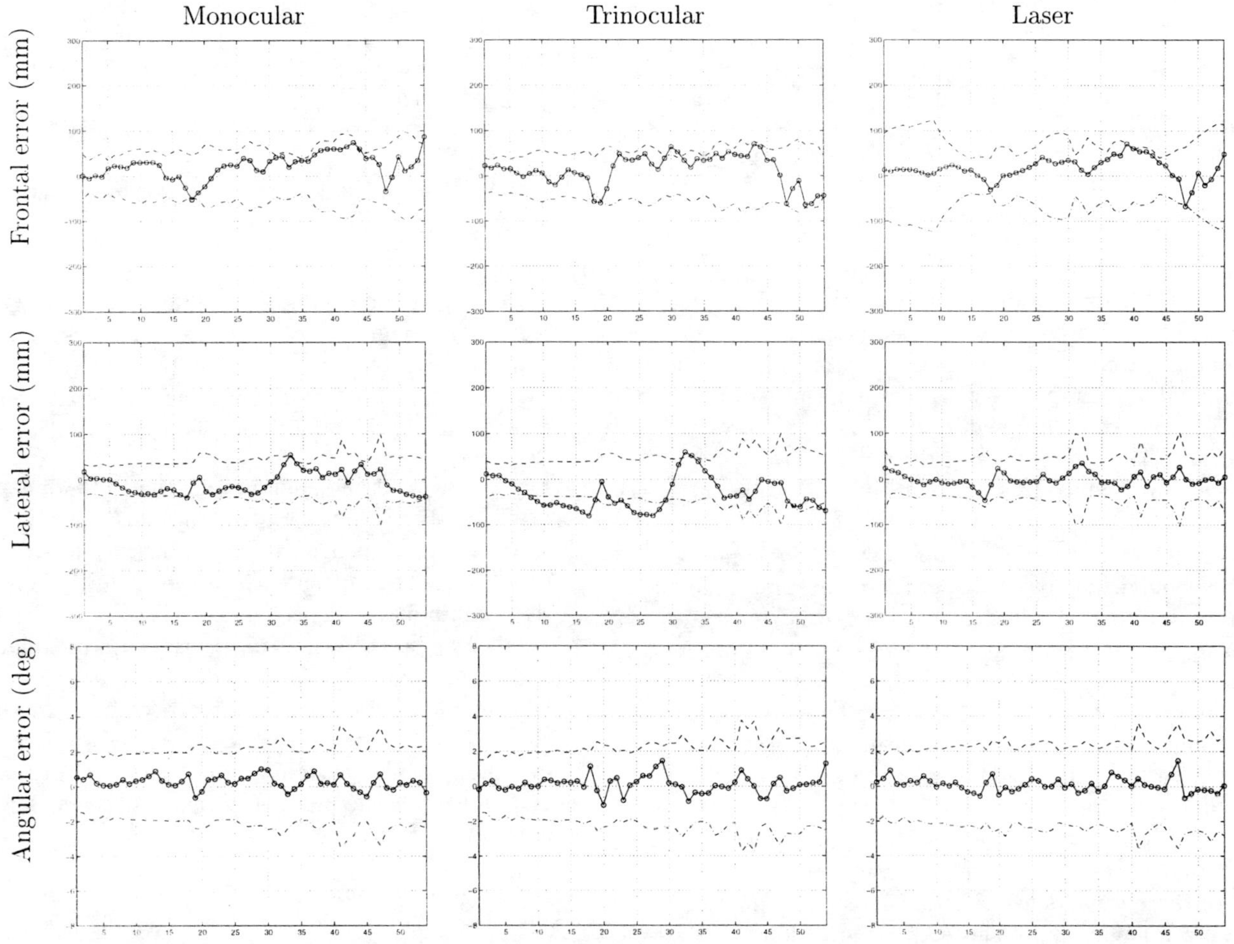

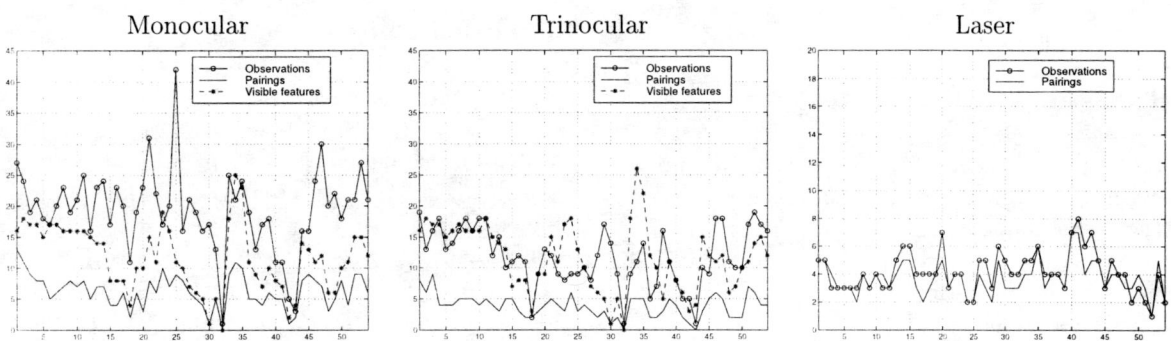

Figure 5: Real errors and 95% uncertainty interval during the localization process for each sensor.

Figure 6: Number of observations and pairings with Monocular Vision (left), Trinocular Vision (middle) and Laser (right).

4.1 Precision

Figure 5 shows the errors obtained in the robot location estimation relative to the *true* location, measured with the theodolites, and the computed location uncertainty, drawn as the 95% confidence interval. A summary of the precision obtained with each sensor is given in Table 1. All three sensors allow to estimate the robot location with a worst-case position error of

	Frontal (mm)		Lateral (mm)		Angular (deg)	
	avg	max	avg	max	avg	max
Mono.	30	87	21	55	0.4	1.0
Trino.	32	71	41	81	0.4	1.5
Laser	23	71	11	46	0.3	1.5

Table 1: Average and worst case localization errors

	Acq.	Proc.	Match. and Loc.	Tot.
Mono.	0.04	0.15	0.03	0.22
Trino.	0.04	0.55	0.02	0.61
Laser	0.10	0.06	0.01	0.17

Table 2: Processing times for each sensor measured in seconds on a SPARCstation 20

less than 10cm, and an orientation error of less than 2deg. This precision can be adequate for successful autonomous navigation in most environments. The computed uncertainty gives a realistic measurement of the true errors obtained in the localisation process, allowing the system to stop the robot and launch a recovery procedure in the event that the robot gets lost.

The lateral precision obtained with trinocular vision is somewhat lower than with monocular vision, mainly due to the fact that a given feature must appear in the three images in order to be located by the trinocular vision systems, and thus in some cases less features may be fused. Precision can be increased using a longer baseline (it was 40cm in our experiment), at the cost of a small loss in field of view, and a bulkier installation.

It is important to highlight that the resulting error plots in the three cases are very similar. That is, the three sensors are coherent in their estimations of the robot location. This shows the validity of the approach based on extracting and fusing geometrical information from the sensors, and supports the plausibility of a multisensor fusion approach based on geometric information.

4.2 Robustness

The robustness of the localisation process is mainly related with the reliability of the matching between observations and map features: incorrect pairings can make the estimation converge to an incorrect solution with covariances under-estimating the real errors obtained. This forces us to adopt a conservative approach: it is better to reject a correct matching than risking to accept an incorrect one. In this work, we have used the chi-square test based on the square Mahalanobis distance as the basic matching tool, with the following results:

- In the case of monocular vision, the lack of depth information produces many potential matches for a given observation. This leads easily spurious pairings. In order to limit this possibility, observations with one candidate map feature are fused first, and the number of observations with several candidate map features that are fused next is limited. This makes the number of accepted pairings somewhat low in some cases (see fig. 6, left).

- In the case of trinocular vision, the redundancy provided by the third camera allows to successfully reject nearly all incorrect pairings between vertical edges in each image. Additionally, the availability of depth information reduces the number of potential pairings between trinocular points and map corners and door frames, giving as result a much more robust procedure. There are however many environment features observed by the trinocular system but not included in the map, and thus the number of pairings is also low in some cases (see fig. 6, center).

- The information provided by the laser rangefinder allows to locate with precision the walls around the robot, most of which can be matched in without errors (fig. 6, right).

4.3 Practicality

Taking into account processing time, laser is superior (Table 2). However, given the simple image processing techniques that we have used (only long vertical edges are extracted), both monocular and trinocular vision can be used to localise in real time a robot moving at 1m/s, with precision less than 10cm.

In the case of laser, all range information is obtained in one horizontal scan, which can constitute a problem if there happens to be many obstacles at this particular working height. If the scan were performed with the robot in motion, either an additional error would be introduced, making the results less precise, or additional processing would be required to compensate for this distortion. Apart from that, the information provided is simple, robust, and easy to process. By contrast, visual information is much richer and versatile, at the cost of more processing time, and more difficulties in finding correct pairings.

Another difficulty for the practical use of vision is the criticality of the calibration procedure, mainly in the case of trinocular. In our experiment the system was calibrated using a pattern close to the robot, resulting in a poor precision at middle and long distances. On the other hand, the higher cost, size and weight of laser rangefinders make them less suitable for some applications like autonomous wheeled chair navigation.

5 Conclusions

Map-based robot localisation based on vision, both monocular and trinocular, achieve levels of precision equivalent to those attained by the use of laser sensors. Laser is computationally simpler, and it gives precise and robust results. It is however more limited, in the sense that no additional information may be obtained from a scan. Since the main drawback of using vision has been robustness, the development of robust matching techniques for vision is one of the subjects of future work. Another important factor affecting the precision and robustness of vision is the fact that the *a priori* map contains information that includes many features detectable with laser (walls), but few visual features (only corners and door frames). It is costly and difficult to obtain a precise and complete *a priori* map. This suggests another line of future work: using the sensor to locate the robot and build a map simultaneously.

References

[1] J.J. Leonard and H.F. Durrant-Whyte, *Directed Sonar Sensing for Mobile Robot Navigation*, Kluwer Academic Publishers, London, 1992.

[2] X. Lebègue and J.K. Aggarwal, "Significant line segments for an indoor mobile robot," *IEEE Trans. on Robotics and Automation*, vol. 9, no. 6, pp. 801–815, 1993.

[3] J. Horn and G. Schmidt, "Continuous localization of a mobile robot based on 3d-laser-range-data, predicted sensor images and dead-reckoning," *Journal of Robotics and Autonomous Systems, Special Issue "Research on Autonomous Mobile Robots"*, vol. 14, pp. 99–118, 1995.

[4] J. Neira, J. Horn, J.D. Tardós, and G. Schmidt, "Multisensor mobile robot localization," in *Proc. 1996 IEEE Int. Conf. on Robotics and Automation*, Minneapolis, USA, 1996, pp. 673–679.

[5] J.A. Castellanos and J.D. Tardós, "Laser-based segmentation and localization for a mobile robot," in *Robotics and Manufacturing vol. 6 - Procs. of the 6th Int. Symposium (ISRAM)*, F. Pin, M. Jamshidi, and P. Dauchez, Eds., pp. 101–108. ASME Press, New York, NY, 1996.

[6] J.A. Castellanos, J. M. Martínez, J. Neira, and J.D. Tardós, "Testbed for mobile robot navigation: The CPS Experiment," Technical Report RR-97-11, CPS, Zaragoza, Spain, 1997.

[7] J.M Martínez, Z. Zhang, and L. Montano, "Segment-based structure from an imprecisely located moving camera," in *IEEE Int. Symposium on Computer Vision*, Florida, November 1995, pp. 182–187.

[8] J.D. Tardós, "Representing partial and uncertain sensorial information using the theory of symmetries," in *Proc. 1992 IEEE Int. Conf. on Robotics and Automation*, Nice, France, 1992, pp. 1799–1804.

Assistance System for Crane Operation with Haptic Display
– Operational Assistance to Suppress Round Payload Swing –

Mitsunori Yoneda, Fumihito Arai, Toshio Fukuda

Nagoya University
Department of Micro System Engineering
Furo-cho 1, Chikusa-ku, Nagoya 464-8603, Japan
E-mail: yoneda@robo.mein.nagoya-u.ac.jp

Keisuke Miyata

Komatsu Ltd.
1200, Manda, Hiratsuka,
Kanagawa 254-0913, Japan

Toru Naito

Kinjo Gakuin University
Omori, Moriyama-ku,
Nagoya 463-0021, Japan

Abstract

It is a difficult operation to suppress round payload swing of a rough terrain crane. We propose a control rule to assist the operation. The stability by the control rule is confirmed with Lyapunov stability criterion. We propose an operational assistance method with haptic display based on the rule. The system can adapt the strength of assistance to operator's state using Recursive Fuzzy Inference. Some experiments on a crane simulator are performed and show the effectiveness of the proposed system.

1 Introduction

The rough terrain crane (Fig.1) has excellent operativity and mobility, yet the operation of the rough terrain crane is very difficult because of complexity of control system, swing of payload, or difficulty of control on underactuared system. Thus skilled operators are needed for safe and efficient operation. Many researchers have studied on crane system to reduce the operating load. Such system can be divided into two types: automatic (or semi-automatic) control system [1][2] and operational assistance system. The aim of most conventional studies is to optimize various control rules for the automatic control system. Although these researches represent good results, those studies of automatic crane system have some problems for practical application such as; difficulty of a countermeasure for various and changeable operation demands in actual construction sites, safety of workers around the payload, responsibility of a countermeasure of the system against a sudden trouble or accident. Thus it is very difficult to make practicable the complete automatic control crane system without operator's assistance.

To reduce the operating load, we have studied another method, an operational assistance system. The system does not perform automatic control directly.

The system just represents various types of information which is useful for an operator to perform more efficient and safer operation[3].

The studies on such operational assistance system have recently been performed actively by the development of VR technology. It is very worthwhile that the system can change information which is hard or impossible for the operator to understand directly into information which is easy to understand for the operator. Such transformation of information is easily realized by applying VR technology to the system. For example, a study of a tele-machining system by Mitsuishi et al.[4] represents a machining state by cutting sound and vibrating grip of control lever to improve operativity of the system.

In this paper, we propose a control rule for swing suppression of a payload. The rule can apply to three-dimensional swing, such as rotary swing. The rule is also used to infer operator's skill level to suppress payload's swing. The system adapts the strength of operational assistance to the skill level using Recursive Fuzzy Inference. We perform some experiments on a VR crane simulator based on the proposal system, and show the effectiveness of the system.

2 Proposal Crane System

The concept of proposal crane system is schematized as Fig.2. This system has two major functions to improve operativity, Multi-modal Display and Time Constant Regulator. The latter regulates time constant according to operation state. The improvement of operation results are confirmed by simple operation experiments of swing suppressing[5]. This paper deals with the former one, Multi-modal Display. It represents various types of information which enable an operator to make safe and efficient control decision

Fig. 1 Rough Terrain Crane

(a) Outside Appearance   (b) Driver's View

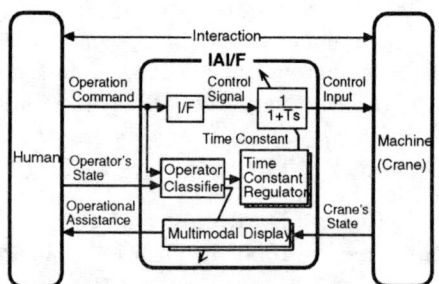

Fig. 2 System configuration

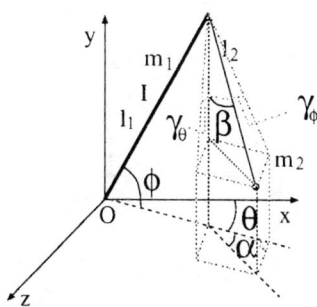

Fig. 3 3-D Crane Model

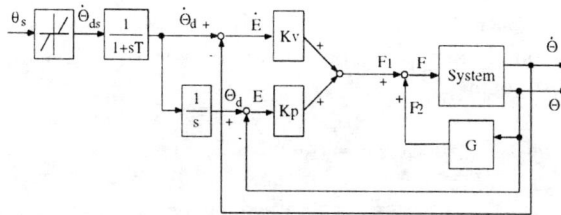

Fig. 4 Block Diagram

easily. This paper shows operational assistance system for swing suppressing operation of a rough terrain crane system with Multi-modal Display.

3 A Control Rule for Swing Suppression

In this section, we propose a control rule to suppress payload's swing in both host and rotation directions, and shows the effectiveness of the control rule We also propose a proper operation pattern to suppress payload's swing for operational assistance.

3.1 Crane Model and Block Diagram

We use a simple crane model shown as Fig.3. Modeling assumption and derivation of the equation of motion are described in the previous paper[5]. In this model, payload's swing can be expressed not only α, β but also $\gamma_\phi, \gamma_\theta$. We call the latter two as swing in the hoist direction and swing in the rotation direction. The control system is schematized as Fig.4. The system controls the crane with velocity control, namely changes operating angle of each control lever (θ_s) into target angular velocity of each degree of freedom.

3.2 Control Rule for Suppression

A control rule to suppress payload's swing is expressed as the following simple equations:

$$\dot\phi_d = -k_\phi \gamma_\phi \quad (1)$$

$$\dot\theta_d = -k_\theta \gamma_\theta \quad (2)$$

Equation 1 and 2 represent target angular velocities for swing suppression. Where k_ϕ and k_θ are positive real numbers, $\dot\phi_d$ and $\dot\theta_d$ represent target angular velocities by operator's lever control. Because equation 1 and 2 are included in the following equation, we deal with the following equation as a basic control rule for swing suppression.

$$\dot\Theta_d = -K\Theta \quad (3)$$

Where $\Theta = (\phi, \theta, \alpha, \beta, l_2)^T$ represents a vector of joint angles, K is a 5×5 square matrix of which all elements are positive numbers or zeros.

We apply Lyapunov stability criterion and confirm that the crane system is stabilized with the equation. Equation of motion of crane model is expressed by the following equation:

$$M(\Theta)\ddot\Theta + V(\Theta,\dot\Theta) + G(\Theta) = F \quad (4)$$

Where M represents a 5×5 matrix of inertia, Θ is a vector of joint angles, V is a vector of Coriolis force, G is a vector of gravity, and F is a vector of external force to control joint angles.

As the block diagram (Fig.4) shows, external force for this system is given as:

$$F = K_p E + K_v \dot E + G(\Theta) \quad (5)$$

Where E represents errors of joint angles and is calculated as $E = \Theta_d - \Theta$. K_p and K_v are 5×5 matrices. Since we consider velocity control only in this paper, K_p is set to a zero matrix and K_v takes the form:

$$K_v = \begin{pmatrix} k_{v11} & 0 & 0 & 0 & 0 \\ 0 & k_{v22} & 0 & 0 & 0 \\ 0 & 0 & 0 & 0 & 0 \\ 0 & 0 & 0 & 0 & 0 \\ 0 & 0 & 0 & 0 & k_{v55} \end{pmatrix} \quad (6)$$

Where $k_{v11}, k_{v22}, k_{v55}$ are positive real constants. Substituting equation 5 into equation 4, equation of motion is expressed as:

$$M(\Theta)\ddot\Theta = K_v \dot E - V(\Theta,\dot\Theta) \quad (7)$$

We apply the following Lyapunov function:

$$v = \frac{1}{2}\dot\Theta^T M(\Theta)\dot\Theta + \frac{1}{2}\Theta^T K_v K \Theta \quad (8)$$

Where K is a matrix which was previously defined as Equation 3. Differential of equation 8 is following.

$$\dot v = \dot\Theta^T M(\Theta)\ddot\Theta + \frac{1}{2}\dot\Theta^T \dot M(\Theta)\dot\Theta + \Theta^T K_v K \dot\Theta$$

Substituting equation 6 to the above equation is following:

$$\dot v = \dot\Theta^T K_v \dot E - \dot\Theta^T V(\Theta,\dot\Theta) + \frac{1}{2}\dot\Theta^T \dot M(\Theta)\dot\Theta + \Theta^T K_v K \dot\Theta$$

Substituting an identity equation derived from Lagrange equation of motion is following:

$$\frac{1}{2}\dot\Theta^T \dot M(\Theta)\dot\Theta = \dot\Theta^T V(\Theta,\dot\Theta), \quad (9)$$

$$\dot v = \dot\Theta^T K_v \dot E + \Theta^T K_v K \dot\Theta$$
$$= -\dot\Theta^T K_v \dot\Theta + \dot\Theta^T K_v \dot\Theta_d + \Theta^T K_v K \dot\Theta$$

Substituting equation 3 is following:

$$\dot v = -\dot\Theta^T K_v \dot\Theta - \dot\Theta^T K_v K \Theta + \Theta^T K_v K \dot\Theta$$

Hence last two terms are eliminated as follows:

$$\dot v = -\dot\Theta^T K_v \dot\Theta \quad (10)$$

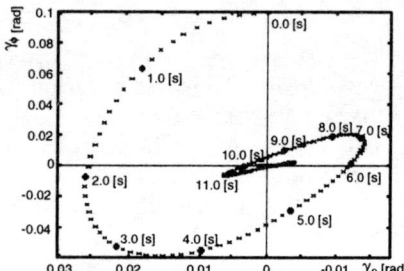

Fig. 5 Automatic Suppression

Equation 10 is clearly a negative number or zero. It is confirmed that the crane system is stabilized by equation 3. Because $\dot{v}$, a function of $\dot{\Theta}$, is zero only where $\dot{\Theta}$ is zero, and decrease in the other case. Thus this system has asymptotic stability.

We have a simulation experiment to confirm the effectiveness of the proposal control rule for swing suppression. Automatic suppressing control is performed to suppress initially given rotary swing of a payload. Where initial states are given as:

$\dot{\alpha} = 0.2[rad/s], \beta = 0.1[rad],$
$k_{v11} = k_{v22} = 10^7, k_{v55} = 10^5$

The result of the experiment is schematized as Fig.5. This figure represents a locus of the payload from top view. Smooth and rapid suppression of swing is observed by automatic suppressing control.

3.3 Operation Pattern for Suppression

Proper control rule to suppress payload's swing is given by equation 1 and 2. In this section, we describe operation pattern of a control lever in the hoist direction only. Operation pattern of a control lever in the rotation direction is given by the same process. Proper operation angle of a control lever to suppress payload's swing is given as following:

$$\xi_{\phi d} = \begin{cases} -\dfrac{k_\phi \gamma_\phi (T_{\phi max} - T_{\phi cut})}{\dot{\phi}_{max}} + T_{\phi cut} & (\gamma_\phi < 0) \\ -\dfrac{k_\phi \gamma_\phi (T_{\phi max} - T_{\phi cut})}{\dot{\phi}_{max}} - T_{\phi cut} & (\gamma_\phi > 0) \end{cases} \quad (11)$$

Where ξ_{ϕ_d} is a proper operating angle of a control lever, which is calculated by proposal control rule, k_ϕ is a positive constant, $T_{\phi cut}$ represents dead zone of the control lever, $T_{\phi max}$ is maximum operating angle of the control lever, $\dot{\phi}_{max}$ is maximum angular velocity of the boom hoist motion.

Although payload's swing will be suppressed by operating the control lever according to equation 11, the operation pattern may cause a bad influence when an operating angle is small. Then the operator must move the control lever largely because each control lever has dead zone. Thus it is hard to follow smooth change of operating angle of control lever, such as a sine curve. Because the operator cannot follow the indicated operation pattern, we observed some bad influences in previous experiments. For example, an operator enlarged payload's swing by late operation

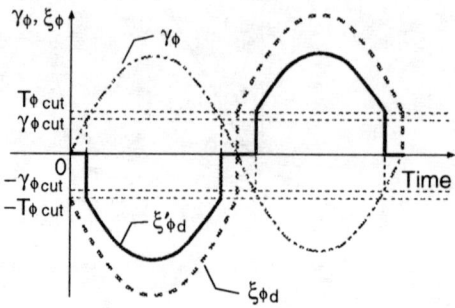

Fig. 6 Operation Pattern of Control Lever

from proper operation pattern. Another operator complained that the operation pattern is too quick to follow.

We explain the reason why the proper operation pattern caused bad influences by Fig.6. Where γ_ϕ is a payload swing angle in the hoist direction, $\xi_{\phi d}$ is the proper operating angle given as the proper operation pattern. An operator must operate the control lever largely when swing angle approaches to zero because of dead zone, namely in case of $\xi_{\phi d} < |T_{\phi cut}|$.

We improve the operation pattern ($\xi'_{\phi d}$) as:

$$\gamma'_\phi = \begin{cases} \gamma_\phi + \gamma_{\phi cut} & (\gamma_\phi \leq -\gamma_{\phi cut}) \\ \gamma_\phi - \gamma_{\phi cut} & (\gamma_{\phi cut} \leq \gamma_\phi) \end{cases} \quad (12)$$

$$\xi_{\phi d} = \begin{cases} -\dfrac{k_\phi \gamma'_\phi (T_{\phi max} - T_{\phi cut})}{\dot{\phi}_{max}} + T_{\phi cut} & (\gamma_\phi \leq -\gamma_{\phi cut}) \\ 0 & (-\gamma_{\phi cut} < \gamma_\phi < \gamma_{\phi cut}) \\ -\dfrac{k_\phi \gamma'_\phi (T_{\phi max} - T_{\phi cut})}{\dot{\phi}_{max}} - T_{\phi cut} & (\gamma_{\phi cut} \leq \gamma_\phi) \end{cases} \quad (13)$$

Where $\gamma_{\phi cut}$ is an important parameter. If $\gamma_{\phi cut}$ is large, the payload will be hard to swing because target angular velocity becomes zero in large range. However payload's swing by operator's wrong operation will be decreased, it takes long time to suppress payload's swing. If $\gamma_{\phi cut}$ is constant, proper operation pattern may become zero when payload's swing is small.

Thus we set a parameter $\gamma_{\phi cut}$ from expected maximum swing angle ($\gamma_{\phi max}$) for conservation of energy. Maximum swing angle $\gamma_{\phi max}$ is calculated by the following equation.

$$\gamma_{\phi max} = \cos^{-1}(\dfrac{1}{2g}\cos\gamma_\phi(1 - l_2\cos\gamma_\phi\dot{\gamma}_\phi^2)) \quad (14)$$

$$\gamma_{\theta max} = \cos^{-1}(\dfrac{1}{2g}\cos\gamma_\theta(1 - l_2\cos\gamma_\theta\dot{\gamma}_\theta^2)) \quad (15)$$

4 Inference of Operator's Skill Level

The operational skill level is inferred with Recursive Fuzzy Inference[6] from two states : deviation between proper operation pattern and operator's operation pattern, and operation time needed for swing suppression.

(a) Rule Map

Deviation from Proper Control Pattern	Operation Time: Short	Operation Time: Middle	Operation Time: Long
Small	0	0.1	0.2
Middle	0.1	0.3	0.6
Large	0.3	0.7	1

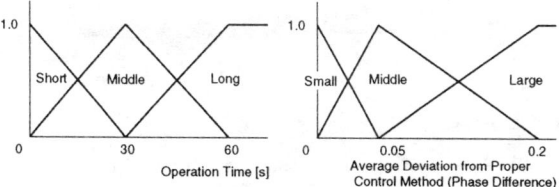

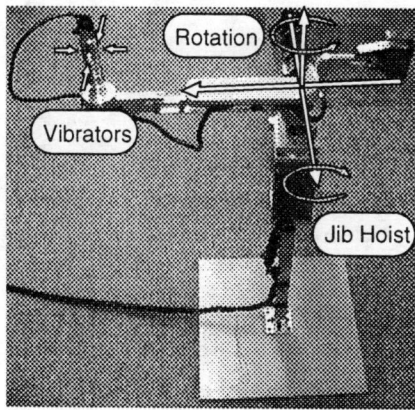

(b) Membership Functions

Fig. 7 Fuzzy Rule for Evaluation of Assistance Level

Proper Control Pattern	Control Input: +	Control Input: 0	Control Input: -
+	0	0.2	1
0	0.5	0	0.5
-	1	0.2	0

Fig. 8 Evaluation Value of Phase Difference

4.1 Recursive Fuzzy Inference

Recursive fuzzy inference (RFI) uses previous inferred data with some weights to infer a new evaluation value. The system with RFI can avoid sudden change of the system responses caused by a single observation. We confirmed that the determination of system parameters with RFI is more effective than that of general fuzzy inference in designing adaptive interface[7]. RFI is expressed as following:

$$f^{(1)} = \frac{\sum_{i,j}(\mu_a^{(1)}\mu_b^{(1)})K_R(a_i^{(1)}, b_j^{(1)})}{\sum_{i,j}(\mu_a^{(1)}\mu_b^{(1)})} \quad (n = 1) \tag{16}$$

$$f^{(n)} = \frac{\sum_{k=0}^{n}\sum_{i,j}\rho^k(\mu_a^{(n-k)}\mu_b^{(n-k)})K_R(a_i^{(n-k)}, b_j^{(n-k)})}{\sum_{k=0}^{n}\sum_{i,j}(\mu_a^{(n-k)}\mu_b^{(n-k)})} \quad (n \geq 2) \tag{17}$$

Where n is a number of recursive data for inference, $\mu_a(\mu_b)$ is the fitness value of the membership function for input a(b), K_R is the inference result based on μ_a and μ_b, and ρ is a fading factor. ρ is 0.5 in this paper.

4.2 Inference of Operational Skill

We set fuzzy rules for this system as Fig.7. Evaluation value of deviation from proper operation pattern, which is shown as Y axis of the rule map in Fig.7(a), is represented as:

$$E_{val} = \frac{E'_{val}(N-1) + Epd}{N} \tag{18}$$

Fig. 9 Control Lever

Where E_{pd} represents an evaluated value of a deviation between proper operation pattern for swing suppression and operator's operation pattern schematized as Fig.8, $Eval$ is an evaluated value by last inference, and $Eval$ represents the average value of previous E_{pd} values. We set $N = 200$ in this system.

Since the evaluated value of operational skill level approaches 0 for skilled operators and 1 for beginner operators, we use this value as "evaluated value of operational assistance level".

5 Operational Assistance

The proposed system is not an automatic control system but an operational assistance system. That is quite different from most conventional studies on crane system. In this section, we focus on the haptic display to assist the suppressing operation of payload swing, and state how the display assists in detail.

5.1 Control Lever

In the conventional crane system, an operator uses many 1 DOF control levers or pedals to control hoist angular velocity, rotation angular velocity, boom length, and rope length. There are some problems on such control lever. One is that it is hard to understand how to operate the levers by intuition. Another is that the operation requires great technical skill for safe and efficient operation when the operator uses many control levers at once. Although some crane systems apply 2 DOF control lever, these operational problems are not solved. Moreover there is another problem by interference of these two motions.

Thus we made a new control lever schematized as Fig.9. The design of the control lever is based on actual motion of crane, and enables operators to understand the operation method easily. An operator changes hoist angular velocity by hoisting his/her arm around his/her elbow, and rotate angular velocity by rotating his/her arm around his/her elbow. Since each motion of the arm is independent, interference of each motion of the arm is small.

We call a grip with built-in actuators and a lever as the grip part, and the other as the base part. An operator puts the arm on the control lever, grasps grip, and operates control levers for boom's hoist and rotation motion by motion around the elbow.

5.2 Haptic Operational Assistance

The system has a function named as Multi-modal Display that presents various operational assistance. We explain a haptic display only in this paper. Haptic devices are included in the grip of the control lever. The haptic device is a small DC motor with eccentric weight, which are often used for a pager or a portable phone.

- Haptic display: proper operation pattern

The information of operational assistance should be able to change its strength for adaptive operational assistance to adapt various operators and situations. Thus haptic display or force display are good for operational assistance.

We explain how to calculate the strength of assistance of the vibrator which is set at the farthest position from the elbow only. We set the strength of operational assistance p based on both proper operation pattern of control lever ξ_ϕ and system's operational assistance level (operator's operational skill level) f on proposal system,

$$p = \begin{cases} \dfrac{\xi_\phi}{T_{max}} f & (\xi_\phi \geq \xi_{\phi cut}) \\ 0 & (\xi_\phi < \xi_{\phi cut}) \end{cases} \quad (19)$$

The strength of assistance of the other vibrator is easily calculated by similar process.

The system adjusts the strength of haptic display by changing interval time of intermittently vibrations shown as Fig.10. The interval time ($Interval$) is represented by the following equation.

$$Interval = (1 - p) \cdot Interval_{max} \quad (20)$$

Where T_{max} represents a maximum displacement angle of control lever, $Interval_{max}$ represents maximum interval time of vibration.

The similar adaptive operational assistance can be realized by force display with DC motors in the base part. We have already confirmed effectiveness of such system[5]. However, this force display forces to move the control lever for proper operation pattern directly. The system makes some problems such as operator's discomfort and interfere with operator's control. Thus we divide the control lever into two parts, the grip part for indirect operational assistance and the base part for operation and direct operational assistance. We apply a haptic display to our operational assistance system because the haptic display has little influence to user's operation.

6 Operational Experiment
6.1 Experimental System

We made a rough terrain crane simulator on a graphic workstation based on the proposal system (Fig.2). VR crane system is schematized as Fig.11.

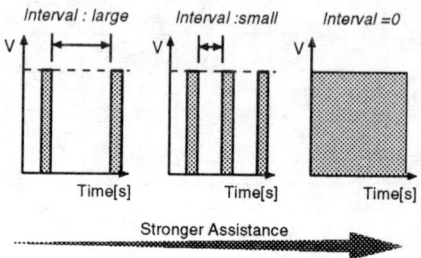

Fig. 10 Strength of Operational Assistance

6.2 Experimental Method

Operational experiments are performed on VR simulation system of rough terrain crane to confirm the effectiveness of proposal system. Aim of the experiments are the suppression of payload's rotary swing by operation of the 2 DOF control lever. The payload's swing is initially given as $\beta = 1.0[rad], \dot{\alpha} = 0.3[rad/s]$.

We observe these two states: operation time to suppress payload's swing throughly and evaluated value of operator's skill level (equation 14). The number of subjects is six. The operational assistance mode is changed as the following order:

(a) Haptic display : proper control method

(b) No haptic display

Measurement of experimental data is repeated five times in each operational assistance mode after adequate operational tests. First and third subjects perform the experiments in reverse order.

6.3 Experimental Results

Experimental results are shown as Table 1 and 2. These tables represent average and variance of operation time and inferred operational skill level in five operational experiments. Improvement of operational performance is observed from these results by haptic display. Comparing both average and variance of operational performance and skill level which is inferred by recursive fuzzy inference, we can convince the validness of the inferred skill level.

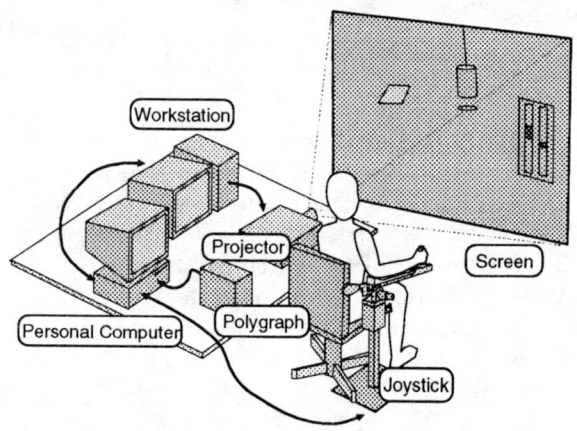

Fig. 11 Experimental System

7 Conclusion

We proposed a control rule to suppress the payload's swing of both hoist and rotation directions, inference method of user's operational skill level based on the control rule, and operational assistance with haptic devices. We confirm the effectiveness of the proposal system from experimental results on a VR simulator system of a rough terrain crane.

8 Acknowledgements

This research is partially supported by the Scientific Grants, Japanese Ministry of Education (No.90001628).

Table 1 Average Operation Time

Assistance Method	Subject 1 Ave.[s] (Var.)	Subject 2 Ave.[s] (Var.)	Subject 3 Ave.[s] (Var.)
Without Assistance	37.31 (20.30)	48.18 (25.06)	61.576 (47.65)
With Assistance	23.45 (18.05)	38.35 (35.90)	49.30 (13.427)
	Subject 4 Ave.[s] (Var.)	Subject 5 Ave.[s] (Var.)	Subject 6 Ave.[s] (Var.)
	60.10 (46.12)	36.85 (33.32)	106.6 (65.09)
	58.12 (39.28)	34.47 (19.79)	48.40 (24.47)

Table 2 Average Assistance Level

Assistance Method	Subject 1 Ave.[s] (Var.)	Subject 2 Ave.[s] (Var.)	Subject 3 Ave.[s] (Var.)
Without Assistance	0.619 (0.197)	0.750 (0.203)	0.741 (0.190)
With Assistance	0.409 (0.219)	0.502 (0.260)	0.696 (0.136)
	Subject 4 Ave.[s] (Var.)	Subject 5 Ave.[s] (Var.)	Subject 6 Ave.[s] (Var.)
	0.715 (0.314)	0.657 (0.298)	0.831 (0.255)
	0.668 (0.152)	0.620 (0.115)	0.569 (0.099)

References

[1] G.A. Forster Alsop.C.F. and F.R.Holmes. Ore unloader automation — a feasibility study. *Proc. of the IFAC-Tokyo Symposium on Systems Engineering and Control Systems Design*, Vol. 2, No. 22, pp. 295–305, 1965.

[2] Y.Sakawa and Y.Shindo. Optimal control of container crane. *Automatica*, Vol. 18, No. 3, pp. 257–266, 1988.

[3] Fumihito Arai, Mitsunori Yoneda, Toshio Fukuda, Keisuke Miyata, and Toru Naito. Assistance system for crane operation using multimodal display. *Proc. IEEE Int. Conference on Robot and Automation*, pp. 40–45, 1997.

[4] Mamoru Mitsuishi, Toshio Hori, Tomoharu Hikita, Masao Teratani, and Takaaki Nagao. Using predictive information display for operational environment transmission in a tele-handling/machining system. *Proc. of IEEE Int. Workshop on Robot and Human Communication*, pp. 45–52, 1995.

[5] Fumihito Arai, Mitsunori Yoneda, Toshio Fukuda, Keisuke Miyata, and Toru Naito. Operational assistance of the crane system by the interactive adaptation interface. *Proc. IEEE Int. Workshop on Robot and Human Communication*, pp. 333–338, 1995.

[6] Kouhei Nomoto and Michimasa Kondo. Autotuning controller with recursive fuzzy inference (in japanese). *Journal of the Society of Instrument and Control Engineers*, Vol. 25, No. 10, pp. 1126–1133, 1989.

[7] Toshio Fukuda, Fumihito Arai, Takahiro Yamamoto, Toru Naito, and Teruyuki Matsui. Interactive adaptation on interface monitoring and assisting operator by recursive fuzzy criterion. *Proc. IEEE Int. Workshop on Robot and Human Communication*, pp. 448–453, 1993.

Haptic Exploration of Fine Surface Features

Allison M. Okamura and Mark R. Cutkosky
Dextrous Manipulation Laboratory
Stanford University
Stanford, CA 94305
{allisono, cutkosky}@cdr.stanford.edu

Abstract

In this paper we consider the detection of small surface features, such as ridges and bumps, on the surface of an object during dextrous manipulation. First we review the representation of object surface geometry and present definitions of surface features based on local curvature. These definitions depend on the geometries of both the robot fingertips and the object being explored. We also show that the trajectory traced by a round fingertip rolling or sliding over the object surface has some intrinsic properties that facilitate feature detection. Next, several algorithms based on the feature definitions are presented and compared. Finally, we present simulated and experimental results for feature detection using a hemispherical fingertip equipped with an optical tactile sensor.

1 Introduction

Haptic exploration is an important mechanism by which humans learn about the properties of unknown objects. Through the sense of touch, we are able to learn about characteristics such as object shape, surface texture, stiffness, and temperature. Unlike vision or audition, human tactile sensing involves direct interaction with objects being explored, often through a series of "exploratory procedures" [6]. Dextrous robotic hands are being developed to emulate exploratory procedures for the applications of remote planetary exploration, undersea salvage and repair, and other hazardous environment operations.

This paper focuses on a subset of haptic exploration: the sensing of small surface features such as cracks and bumps. Brown, et al. [2], define features as "application and viewer dependent interpretations of geometry." As we will show, the identification of a feature is not only dependent on the geometric properties of the object being explored, but also on the properties of the robotic finger performing the exploration.

Some of these features cannot be sensed accurately through static touch; motion is required. To excite the fast-acting, vibration sensitive mechanoreceptors embedded in the fingertips, humans roll and/or slide the fingertips over the surface [6]. Emulating this behavior presents a challenge in that manipulation must accomplish two tasks: (1) move the object to a desired location with a stable grasp and (2) move the fingers and sensors over the features.

In this work we consider first how features affect the path of a round fingertip rolling and/or sliding over an object surface. Although tactile sensors may be needed for manipulation control, we show that features can be detected using only the path of the fingertip center. We then present several algorithms for identifying surface features, with and without contact information from tactile sensors.

1.1 Previous Work

A number of investigators have addressed the problem of using robotic fingers in exploratory procedures. Examples include [1, 15, 16]. This work has mainly focused on the sensing of global object shapes and on fitting shapes to object models. The integration of tactile sensing and dextrous manipulation with rolling or sliding has also been addressed in recent work [9, 10, 15].

Much work on the identification of surface features has been done by researchers in the vision community. In early work, the definitions of features in applications such as topography were often ambiguous because they were based on natural language. More recently, researchers have developed definitions based on various mathematical models, including local maxima of pixels in a discrete 2D image, height or intensity graphs, and differential geometry (local extrema of principal curvatures or curvature properties of level sets of smooth functions) [4]. In another approach, Kunii, et al. [7] extended the idea of the Medial Axis Transform to develop skeletons of object shape and used caustic singularities to determine the locations of ridges. Applications of ridge detection include medical imaging [13] and the analysis of topographic data [8].

2 Defining Surface Features

This paper will take a differential geometry based approach to surface feature definition. In this section, we will briefly review a mathematical description of an object surface and define the surfaces associated with the path of a round finger rolling or sliding over the object surface. Next, we define several features in the context of robotic haptic exploration. Finally, we provide an example of a three-dimensional surface feature.

2.1 Mathematical Background

Montana [14] defines a *Gauss map* for a manifold S as a continuous map $g : S \to S^2 \subset \mathcal{R}^3$ such that for every $s \in S$, $g(s)$ is perpendicular to S at s. The pair (f, U), where f is the invertible map $f : U \to S \subset \mathcal{R}^3$, is a coordinate system for S. The normalized *Gauss frame* at a point $\mathbf{u} = (u, v) \in U$ is defined as the coordinate frame with origin at $f(\mathbf{u})$ and coordinate axes

$$\mathbf{x}(\mathbf{u}) = \frac{f_u(\mathbf{u})}{\|f_u(\mathbf{u})\|}$$
$$\mathbf{y}(\mathbf{u}) = \frac{f_v(\mathbf{u})}{\|f_v(\mathbf{u})\|}$$
$$\mathbf{z}(\mathbf{u}) = g(f(\mathbf{u})). \quad (1)$$

The curvature form K is the 2×2 matrix

$$K = \begin{bmatrix} \mathbf{x}(\mathbf{u})^T \\ \mathbf{y}(\mathbf{u})^T \end{bmatrix} \begin{bmatrix} \frac{\mathbf{z}_u(\mathbf{u})}{\|f_u(\mathbf{u})\|} & \frac{\mathbf{z}_v(\mathbf{u})}{\|f_v(\mathbf{u})\|} \end{bmatrix}. \quad (2)$$

The principal curvatures $k_1(S)$ and $k_2(S)$ of the surface are the elements of the diagonalized curvature matrix K_d. The principal curvatures correspond to the $\mathbf{x}(\mathbf{u})$ and $\mathbf{y}(\mathbf{u})$ directions of the Gauss Frame.

Now consider a robotic finger with a spherical fingertip (surface S_f) with radius r_f following the contours of the surface. Using the formulation above, the curvature form for the fingertip is

$$K_f = \begin{bmatrix} \frac{1}{r_f} & 0 \\ 0 & \frac{1}{r_f} \end{bmatrix}. \quad (3)$$

As the finger traces over the surface of an object, the center point of the finger creates a *parallel surface*, S_p. This concept is also related to the offset surface as defined in CNC toolpath planning and other applications [17]. Figure 1 shows a 2D slice of a surface with several features and the parallel surface created by a spherical fingertip tracing over it.

The parallel surface is defined as the envelope of spheres whose centers are on the surface. A simple way to construct it is to draw a normal line (determined by the z-axis of the Gauss frame, $\mathbf{z}(\mathbf{u})$) with length r_f at each point on the surface. The locus of the end points is the parallel surface. $\mathbf{s}_p(\mathbf{u}) \in S_p$ maps U to $\mathcal{R}^3$:

$$\mathbf{s}_p(\mathbf{u}) = f(\mathbf{u}) + r_f \mathbf{z}(\mathbf{u}). \quad (4)$$

where $\mathbf{z}(\mathbf{u})$ is the z-direction of the Gauss frame, r_f is the radius of the fingertip, and $\mathbf{u}$ parameterizes the surface.

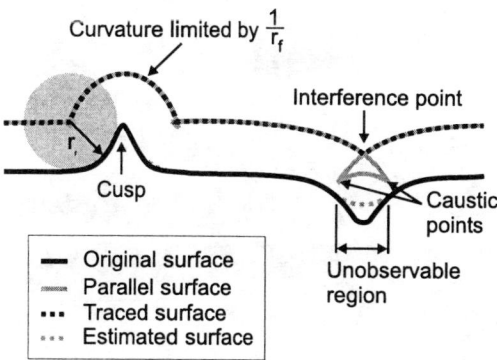

Figure 1: 2D slice of a spherical fingertip with original, parallel, traced, and estimated surfaces

In some cases, however, the actual surface created by moving a finger over the surface is not described by this equation. When the curvature of the surface is greater than that of the finger, the parallel surface has singularities known as *caustics* [8], even if the original surface is smooth. The presence of caustics indicates that the finger cannot exactly follow the contours of the object.

As shown in Figure 1, the caustic points bracket the area of the parallel surface where one of the original surface curvatures, $k_i(S)$, is greater than that of the finger, $\frac{1}{r_f}$. Due to interference, however, the path of an actual finger tracing over the surface is limited before the region bracketed by the caustics. The *interference point* is defined as the point where the finger can no longer follow the parallel surface when entering a region of $k_i(S) < -\frac{1}{r_f}$. Any curve on the parallel surface that travels in and out of a continuous area with $k_i(S) < -\frac{1}{r_f}$ must travel through two interference points. We define the *traced surface*, S_t, as the parallel surface without the portions that the fingertip cannot access due to curvature and interference limitations. A traced surface will always have a discontinuity at the interference point. Figure 1 shows interference points (discontinuities in the traced surface) and caustic points (discontinuities in the parallel surface), as well as the regions of the traced surface that round or fillet the object surface.

2.2 A New Feature Definition

Using the surface and parallel surface descriptions developed above, a new feature definition can be created for the purposes of feature identification and location. The concept of a feature in the context of haptic exploration with robotic fingers is not only dependent on the surface of the object, but also on the size and shape of the finger. In this paper, we assume a spherical fingertip.

We begin by defining a *curvature feature*, then use this as a building block to define *macro features*. Sup-

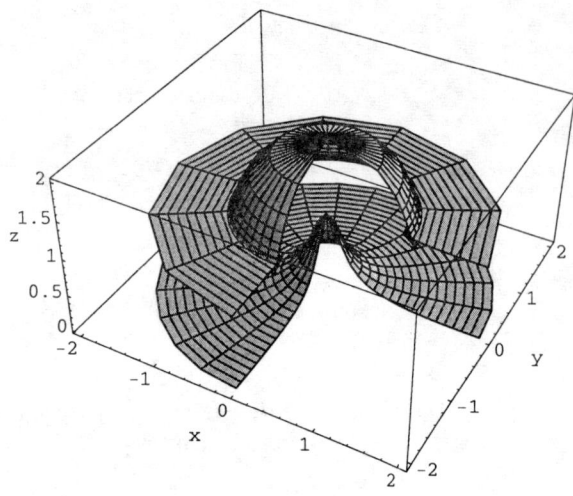

Figure 2: A bump feature (bottom), with parallel surface (top)

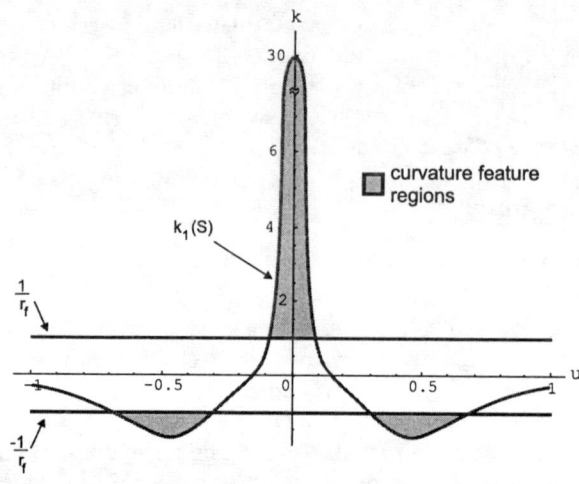

Figure 3: Curvature features are identified where $|k_i(S)| > \frac{1}{r_f}$

pose that an object can be locally fit with a surface S with principal curvatures $k_1(S)$ and $k_2(S)$. The basic criterion for defining a feature is maximum principal curvature:

Definition 1 *A curvature feature, as detected by a spherical robotic fingertip with radius r_f tracing over a surface S with curvature K, is a region where one of the principal curvatures satisfies $k_i > \frac{1}{r_f}$ or $k_i < -\frac{1}{r_f}$. These are* positive curvature *(convex) and* negative curvature *(concave) features, respectively.*

This means that the magnitude of the radius of curvature of the fingertip is larger than that of the feature. The simple definition above can be extended to define macro features, which consist of patterns of curvature features. For example:

Definition 2 *A bump feature is an area with one of the principal curvatures k_i following the pattern {negative curvature feature}{positive curvature feature}{negative curvature feature} as the finger travels over the surface.*

Definition 3 *A ridge feature is an area with principal curvature k_1 following the pattern {negative curvature feature}{positive curvature feature}{negative curvature feature} and $k_2 < \frac{1}{r_f}$. The ends of the ridge feature are defined when the second principal curvature k_2 follows the pattern {positive curvature feature}{negative curvature feature}.*

These definitions can be extended to additional compound features, however that is beyond the scope of this paper.

2.3 Example: A Bump Feature

Consider the three-dimensional parabolic surface of revolution "bump" shown in Figure 2. The bump surface is defined by the set

$$U = \{(u,v) | -1 < u < 1, 0 < v < \pi\} \quad (5)$$

and the map

$$f : U \to \mathcal{R}^3,$$
$$(u,v) \mapsto (u \cos v, u \sin v, \frac{1}{1+au^2}) \quad (6)$$

for some $a \in \mathcal{R}$.

It can be shown that (f, U) is an orthogonal coordinate system and thus the normalized Gauss frame exists for all $\mathbf{u} = (u, v) \in U$. For the bump example, the Gauss frame is

$$c = \sqrt{1 + \frac{4a^2 u^2}{(1+au^2)^4}} \quad (7)$$

$$d = 1 + au^2 \quad (8)$$

$$\mathbf{x}(\mathbf{u}) = \begin{bmatrix} \frac{1}{c} \cos v & \frac{1}{c} \sin v & \frac{2au}{cd^2} \end{bmatrix}^T$$
$$\mathbf{y}(\mathbf{u}) = \begin{bmatrix} -\sin v & \cos v & 0 \end{bmatrix}^T$$
$$\mathbf{z}(\mathbf{u}) = \begin{bmatrix} \frac{2au}{cd^2} \cos v & \frac{2au}{cd^2} \sin v & \frac{1}{c} \end{bmatrix}^T. \quad (9)$$

The curvature matrix is

$$K = \begin{bmatrix} \frac{2ad(1-3au^2)}{c(d^4+4a^2u^2)} & 0 \\ 0 & \frac{2a}{cd^2} \end{bmatrix}. \quad (10)$$

Using the simplification variables c and d defined in Equations (7) and (8), the parallel surface is defined

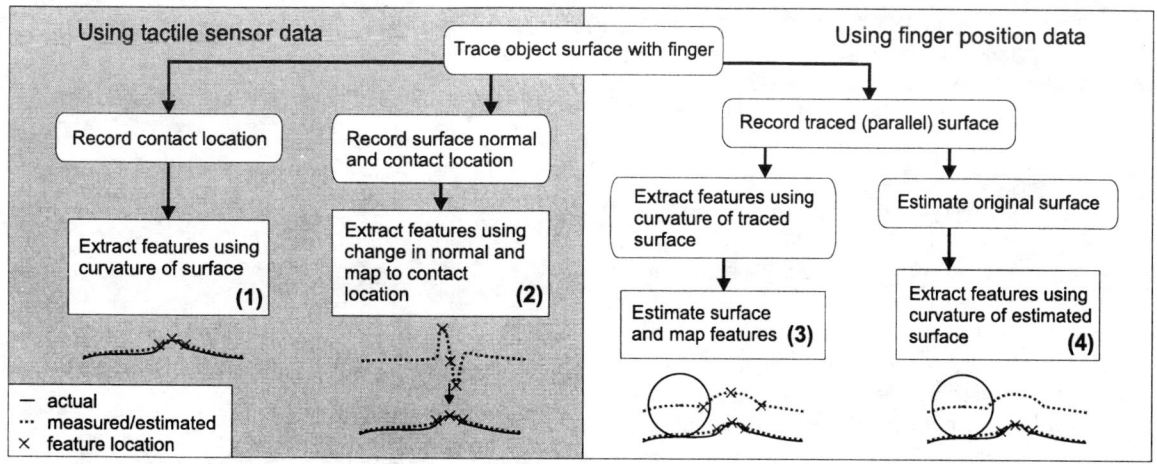

Figure 4: Algorithms for feature detection

by points

$$\mathbf{s}_p(\mathbf{u}) = \begin{bmatrix} (u + \frac{2ar_f u}{d^2 c}) \cos v \\ (u + \frac{2ar_f u \sin v}{d^2 c}) \sin v \\ \frac{1}{d} + \frac{r_f}{c} \end{bmatrix}. \quad (11)$$

To determine the locations of the curvature features, we can plot the principal curvatures of the object surface and the robotic fingertip with respect to the variable u. Figure 3 shows the principal curvature $k_1(S)$ (the K_{11} element of the matrix in Equation 10) and $\pm \frac{1}{r_f}$, where $r_f = 1$, plotted against u for $a = 15$. There are three curvature features: two negative, and one positive. The pattern of curvature features can be identified as a bump as presented in Definition 2.

3 Feature Detection

In this section we present algorithms for detecting features using the surface feature definitions.

Figure 4 shows how position and tactile sensor data from the fingertip can be used to determine the features and global shape of the surface. Different chains on the diagram result in different identification algorithms. The algorithms can be divided into two groups: those that use tactile sensor data to determine the object surface and features, and those that use finger position data.

3.1 Using Tactile Sensor Data

Using the tactile sensor data and a model of the robot finger, one can compute the trajectory of the finger/object contact point in space and build a model of the surface, which we will call the *estimated surface*. This method works for any fingertip shape, as long as the contact point can accurately be sensed.[1]

[1]In fact, most tactile sensors are somewhat noisy and of low resolution and bandwidth compared to joint angle and force sensors, so the contact location data are not highly accurate.

The first algorithm (1) in Figure 4 calculates the curvature at each point on the estimated surface using any numerical difference method. Depending on the noise in the data, the points on the estimated surface may first need to be smoothed or fit to an analytical model before differentiation. Points on the smoothed or modeled surface with a curvature greater that that of the fingertip may then be identified as features.

The second algorithm (2) in Figure 4 requires that the normal of the surface be recorded over time. With a spherical fingertip, this is easily determined from the contact point on the finger. Again, due to noise in the sensor or position of the robot finger, the normal may need to be smoothed. The normal is then differentiated with respect to arclength, providing a spike at the location of the feature. This location can then be mapped to the estimated surface. This method uses the same information as the first algorithm, but improves feature detection because it uses a directly sensed quantity (the surface normal) rather than than the derivative of a noisy surface estimation.

3.2 Using Fingertip Center Position Data

Feature detection can also be accomplished without tactile sensor data, using the traced surface. If the fingertip is spherical (in 3D) or circular (in 2D), the traced surface created by the center of the fingertip can be used to estimate the original surface without tactile sensor data.

Estimation of the original surface can be done in several ways. Just as the theoretical parallel surface can be calculated from the original surface as in Equation (4), the estimated surface points $\mathbf{s}_e(\mathbf{u}) \in \mathcal{R}^3$ can be determined from the traced surface using

$$\mathbf{s}_e(\mathbf{u}) = \mathbf{s}_t(\mathbf{u}) - r_f \mathbf{z}_t(\mathbf{u}), \quad (12)$$

where $\mathbf{s}_t(\mathbf{u})$ are points on the traced surface, $\mathbf{z}_t(\mathbf{u})$ is the z-direction or the Gauss frame of the traced surface, r_f is the radius of the fingertip, and **u** parameterizes the surface. This method, however, is very susceptible to noise in $\mathbf{z}_t(\mathbf{u})$. Before calculating $\mathbf{z}_t(\mathbf{u})$, an improved estimation of the surface can be obtained by fitting a curve to or smoothing the traced surface.

The third algorithm (3) in Figure 4 extracts the features based only on the curvature of the traced surface. This has an advantage over the previous methods because the traced surface is likely to be less noisy than the estimated surface which used tactile sensor data. In addition, the traced surface will show interference points, which are a good indicators of features because they are discontinuities in the traced surface. This serves to enhance negative curvature features.

The fourth algorithm (4) in Figure 4 is similar to the previous algorithm, although the order of surface estimation and feature detection is switched. Now we use the traced surface to estimate the original surface first, and then perform feature detection using the curvature of the estimated surface.

3.3 Smoothing Data Using Curvature Limitations

For algorithms that use only the position of the fingertip, two nonlinear smoothing algorithms based on curvature limitations can be used to smooth noisy data in the traced and estimated surfaces. The algorithms apply the observations in Section 2.1 that the traced surface must round sharp cusps or corners on the object surface and fillet sharp indentations.

At each point on the traced and estimated surfaces, the principal curvatures are calculated using a numerical difference method. On the traced surface, any points for which $k_i(S_t) > \frac{1}{r_f}$ are invalid (i.e., they cannot correspond to motion along an object surface) and should be deleted. After removing these points, the traced surface can be smoothed by a standard smoothing method. After the traced surface is smoothed, the estimated object surface is calculated using the *envelope* of r_f circles centered on the traced surface. In the estimated surface, the principal curvature $k_i(S_t)$ must be at least $-\frac{1}{r_f}$. Although the actual object surface could have regions with larger negative curvature, it is impossible for the fingertip to detect such regions. Again, such points should be deleted and the remaining surface can be smoothed using any standard method.

4 Simulation and Experiments

4.1 Simulated Data

A realistic simulation of a spherical fingertip traveling over a step shows the curvature limitations in surface estimation. For the purposes of simulation, a realistic traced surface must be calculated. First, the

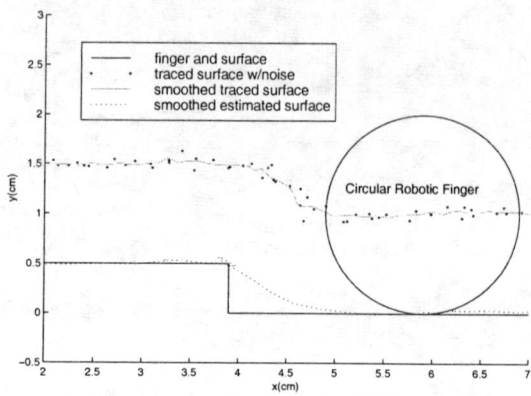

Figure 5: Simulated data of a finger tracing over a step

parallel surface is calculated from the original surface, then the interference points are identified and the unreachable points are removed to form the traced surface. Next, because points on the parallel surface are not spaced equally, the surface is re-calculated using equal arclengths between the points. Finally, Gaussian noise with a variance of 0.01 is added for realism.

Now we invoke the nonlinear smoothing algorithms to limit the curvature to realistic values. The first curvature limitation algorithm is used to remove unreachable points from the traced surface, then it is smoothed. Then, the original surface is estimated using the envelope of circles with centers on the smoothed traced surface. Next, the second curvature limitation algorithm is used to remove unreachable points from the estimated surface. As a final step, the surface can be fit to create an analytical model. Figure 5 shows a 2D slice of four surfaces in this example: the object surface, the traced surface with noise, the smoothed traced surface, and the smoothed estimated object surface.

4.2 Experiments

4.2.1 Apparatus

Manipulation experiments were performed using two degree-of-freedom robotic fingers and tactile sensors developed by Maekawa, et al. [11, 12]. The optical waveguide tactile sensors on the fingers form a hemispherical fingertip and provide analog signals that can be used to calculate the intensity and centroid of the contact point(s). The 20 mm diameter sensors can be sampled at 5 kHz, and have a field of detection of ±75 degrees from the sensor pole. The error is aproximately ±1 degree, although this can change when the contact area increases during sliding. A calibration was performed to characterize and remove the nonlinearity of the sensor.

As is typical of many robotic fingers, the workspace was limited and thus a combination of rolling and slid-

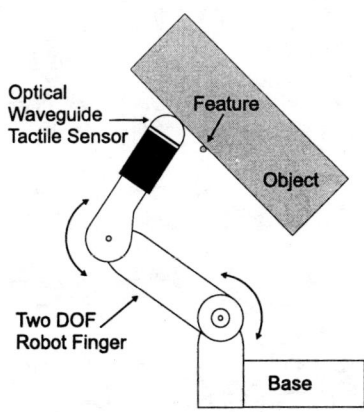

Figure 6: Experimental Apparatus

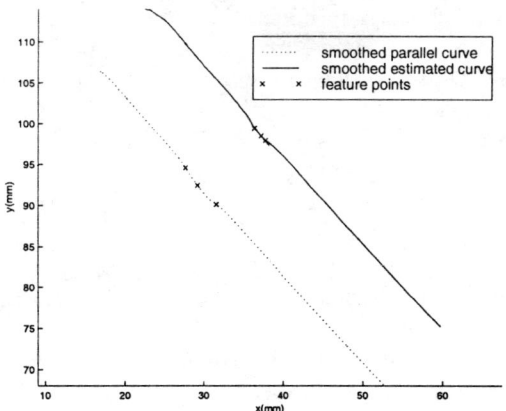

Figure 7: Feature detection with a robotic finger rolling/sliding on a 45° surface with a 0.65 mm bump feature

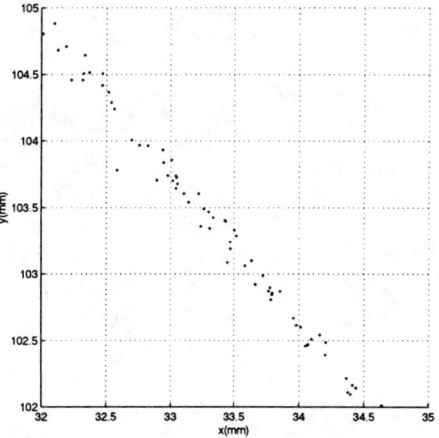

Figure 8: Close up of estimated object surface data

ing was necessary to move the fingers over the surface of the object. A hybrid force/velocity control was used to obtain smooth sliding over flat surfaces with bump features (0.5-1.5 mm wires placed on the surface).

During motion, joint angle potentiometers were used to determine the position of the center of the hemispherical fingertip. The location of the contact point on the fingertip was then used to determine the direction of the contact normal, which was filtered to reduce noise. The center of the fingertip can be sensed to within ± 0.2 mm and the contact location can be estimated within ± 0.5 mm. A tension differential-type torque sensor is used to measure torque in the joints and calculate the cartesian force at the fingertip.

Because the fingertip is spherical, the contact location on the finger gives the tangent and normal of the rigid surface. The velocity of the fingertip tangent to the surface and the force normal to the surface were controlled using a simple proportional law. The finger moved with an average speed of 0.03 m/sec and a normal force of 1 ± 0.01 N.

4.2.2 Results

Each of the algorithms outlined in Figure 4 was tested using the data from these experiments. Algorithm (1) often resulted in false negative identifications because the estimated surface was obtained directly from the contact point. Concave curvature features on the object were overlooked because of curvature limitations; the data are these points are automatically smoothed to the radius of curvature of the fingertip. Using algorithm (2), spikes in the contact normal indicated the presence of negative curvature (concave) features. This method does not provide good detection of positive (convex) features, because the normal direction does not change as quickly at those points.

Using algorithm (3), features were extracted using the curvature of the traced surface, then mapped to an estimated surface. This algorithm performed the best for our application of small bump features on a flat surface. Figure 7 shows the traced and estimated surfaces and features detected for a bump on a flat surface angled at 45 degrees, with a 0.65 mm diameter wire stretched across the surface. Figure 8 shows a close up view of the unsmoothed estimated object surface data in a region with no features. The orientation of the object is the same as that in Figure 6. Algorithm (4) resulted in false negatives for the same reason as algorithm (1): using the estimated surface to determine locations with high curvature is not feasible because of the curvature limitations.

5 Conclusions and Future Work

In this work, we have defined features for the context of robotic haptic exploration with curved fingertips. Features may be identified using the curvatures of traced or estimated surfaces, without tactile sensor data. In particular, the traced surface described by a rounded fingertip accentuates concave features on the object surface. This is an interesting result because, while we may need contact location data for stable manipulation control, we do not need them to recreate the object shape or to identify small surface features with round fingertips. We also presented definitions for curvature and macro surface features in this context and methods for smoothing data based on the inherent curvature limitations of the physical system.

One area for future work is the development of 3D haptic exploratory procedures that take advantage of the feature definitions. For example, if a fingertip encounters a ridge, some specific strategies may be used to determine the size and extent (length) of the feature. Another possible direction for this work is fitting the features onto a global object model. If the general shape of the object is fit to some simple surface, it should be possible to add the details of fine surface features using a simple data structure. Finally, we are studying the display the object global shape and fine surface features using force- and vibration-feedback.

Acknowledgments

This work was supported by NSF dissertation enhancement award 9724763 and STRICOM grant M67004-97-C-0026 through Immersion Corporation. Special thanks are given to Dr. Kazuo Tanie for hosting Ms. Okamura's stay at the Mechanical Engineering Lab in Tsukuba, Japan, and Dr. Hitoshi Maekawa for providing the experimental apparatus.

References

[1] P.K. Allen and P. Michelman, "Acquisition and Interpretation of 3-D Sensor Data from Touch," *IEEE Trans. on Robotics and Automation*, Vol. 6 No. 4, pp. 397-404, 1990.

[2] K.N. Brown, C.A. McMahon, and J.H. Sims Williams, "Features, aka The Semantics of a Formal Language of Manufacturing," *Research in Engineering Design*, Vol. 7, No. 3, pp. 151-172, 1995.

[3] R.D. Howe and M.R. Cutkosky, "Touch Sensing for Robotic Manipulation and Recognition," *The Robotics Review 2*, O. Khatib, J. Craig and T. Lozano-Perez, eds., M.I.T. Press, Cambridge, MA, pp. 55-112, 1992.

[4] D. Eberly, et al., "Ridges for Image Analysis," *Journal of Mathematical Imaging and Vision*, Vol. 4, pp. 353-373, 1994.

[5] R.E. Ellis, "Extraction of Tactile Features by Passive and Active Sensing," *Intelligent Robots and Systems*, D.P. Casasent, Ed., 1984.

[6] R.L. Klatzky and S. Lederman, "Intelligent Exploration by the Human Hand," Chapter 4, *Dextrous Robot Manipulation*, S.T. Venkataraman and T. Iberall, eds., Springer-Verlag, 1990.

[7] T.L. Kunii, et al., "Hierarchic Shape Description via Singularity and Multiscaling," *Proc. of the 18th Annual Intl. Comp. Software and Applications Conf. (COMPSAC 94)*, Taipei, Taiwan, pp. 242-251, 1994.

[8] I.S. Kweon and T. Kanade, "Extracting Topographic Terrain Features from Elevation Maps," *CVGIP: Image Understanding*, Vol. 59, No. 2, pp. 171-182, 1994.

[9] Z.X. Li, Z. Qin, S. Jiang and L. Han, "Coordinated Motion Generation and Real-time Grasping Force Control for Multifingered Manipulation," *Proc. IEEE Intl. Conf. on Robotics and Automation*, pp. 3631-3638, 1998.

[10] H. Maekawa, K. Tanie, and K. Komoriya, "Tactile Sensor Based Manipulation of an Unknown Object by a Multifingered Hand with Rolling Contact," *Proc. 1995 IEEE Intl. Conf. on Robotics and Automation*, pp. 743-750, 1995.

[11] H. Maekawa, et al., "Development of a Three-Fingered Robot Hand with Stiffness Control Capability," *Mechatronics*, Vol. 2, No. 5, pp. 483-494, 1992.

[12] H. Maekawa, K. Tanie, and K. Komoriya, "A Finger-Shaped Tactile Sensor Using an Optical Waveguide," *Proc. 1993 IEEE Intl. Conf. on Systems, Man and Cybernetics*, pp. 403-408, 1993.

[13] J.B.A. Maintz, P.A. van den Elsen and M.A. Viergever, "Evaluation of Ridge Seeking Operators for Multimodality Medical Image Matching," *IEEE Trans. of Pattern Analysis and Machine Intelligence*, Vol. 18, No. 4, pp. 353-365, 1996.

[14] D.J. Montana, "The Kinematics of Contact and Grasp," *Intl. Journal of Robotics Research*, Vol. 7, No. 3, pp. 17-32.

[15] A.M. Okamura, M.L. Turner and M.R. Cutkosky, "Haptic Exploration of Objects with Rolling and Sliding," *Proc. 1997 IEEE Intl. Conf. on Robotics and Automation*, Vol. 3, pp. 2485-2490, 1997.

[16] K. Pribadi, J.S. Bay, and H. Hemami, "Exploration and Dynamic Shape Estimation by a Robotic Probe," *IEEE Trans. on Systems, Man, and Cybernetics*, Vol. 19, No. 4, pp. 840-846, 1989.

[17] J.R. Rossignac and A.A.G. Requicha, "Offsetting Operations in Solid Modeling," *Computer Aided Geometric Design*, Vol. 3, No. 2, pp. 129-148, 1986.

Passive Implementation for a Class of Static Nonlinear Environments in Haptic Display

Brian E. Miller[1] * J. Edward Colgate[1][†] Randy A. Freeman[2][‡]
Department of Mechanical Engineering[1]
Electrical and Computer Engineering[2]
Northwestern University
2145 Sheridan Rd
Evanston, IL 60208 USA

Abstract

This paper derives conditions for the absence of oscillations for a parameterized class of nonlinear environments. This class includes discrete-time environments that can exhibit non-passive behavior. The motivation for considering non-passive discrete-time environments is based on the fact that an interesting class of passive continuous-time environments have non-passive discrete-time counterparts.

A design methodology is introduced that provides relationships between the haptic device, virtual coupling and maximum negative stiffness exhibited by the environment.

1 Introduction

Envision for a moment that you are a virtual environment design engineer for haptic systems. Your job is to develop virtual perceptions for human users, with the most important design criteria being human safety. You ask yourself, what can I expect to render that satisfies this criteria? Obviously, unstable simulations are unacceptable, however another concern are simulations that exhibit oscillations. Controlled oscillations present a safety hazard and destroy any illusion the percept is attempting to convey. The current work derives conditions that are necessary to guarantee the absence of oscillations. An important contribution of this work is that it outlines a design methodology that allows environments that exhibit non-passive behavior. The interest in non-passive discrete-time environments is illustrated by the following example.

*bemiller@nwu.edu
†colgate@nwu.edu
‡freeman@ece.nwu.edu

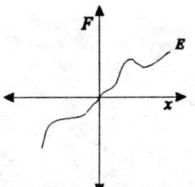

Figure 1: Force/displacement history.

Consider the force/displacement history of nonlinear stiffness (Figure 1).

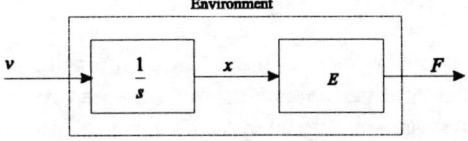

Figure 2: Continuous-time block diagram representing nonlinear stiffness. The environment block consists of an integrator and static nonlinear operator E.

In continuous-time, the block diagram representing this environment is shown in Figure (2). The continuous-time representation results in a passive environment. Next, consider a discrete-time version:

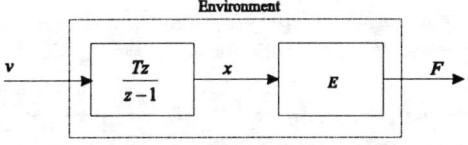

Figure 3: Discrete-time block diagram representing nonlinear stiffness. The environment block consists of an integrator and static nonlinear operator E.

Unless E is restricted to be monotonically increasing,

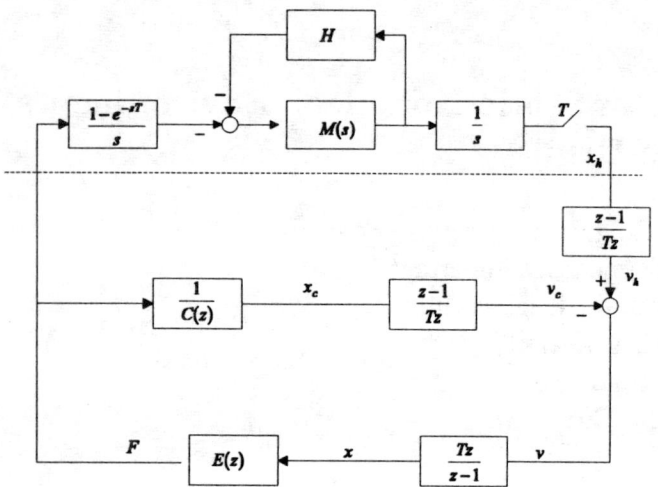

Figure 4: Block diagram for a one DOF haptic display.

the environment block is non-passive.

It is important to note that the loss of passivity is not a physical feature of the environment, rather a byproduct when obtaining the discrete-time approximation. Therefore, it seems plausible that careful design of the system should allow these discrete-time approximations to be displayed passively.

2 Previous Work

Previous work has made the assumption that the operator and virtual environment are passive, focusing on other aspects of the system to ensure stable haptic interaction. Work by Adams et al. has presented a comprehensive approach to virtual coupling design for all possible network causality. The approach taken in the current work is closely related to the results presented in [3]. These results are briefly summarized.

Consider a one degree-of-freedom (DOF) haptic display pictured in Figure (4) with operator H, haptic device M, virtual coupling C and environment E.

To achieve the goal of deriving conditions under which the haptic display appears passive, as seen by the operator, a multiplier was added to the block diagram. Using this interpretation, along with the assumption that the operator and environment are passive, a condition was placed on the amount of physical damping necessary in the device.

The current work also makes the assumption that the operator must behave in a passive manner, however we demonstrate that it is not necessary that the discrete-time environment be passive.

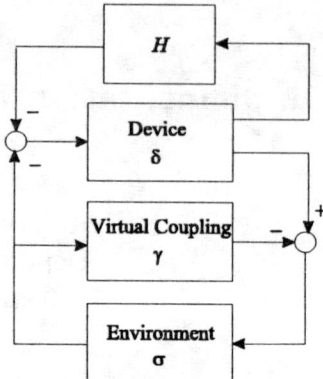

Figure 5: Graphical representation of the analysis.

Several steps in the analysis require the nyquist plot of a transfer function to be located in a particular region. A transfer function is positive real (PR) or passive if the nyquist plot lies in the closed right half plane. A transfer function is μ output strictly passive (μ-OSP) if the nyquist plot is located in a disk of finite radius.

3 Problem Statement

The remainder of this paper will derive conditions for stable haptic interaction in the sense that it guarantees the absence of oscillations.

A graphical interpretation of this argument is presented in Figure (5). Each component in the haptic structure has an associated parameter that will be used in the subsequent text to derive conditions that must be satisfied to achieve stability. The main result of this paper is that if the condition

$$\delta > \frac{\sigma T}{2} \quad (1)$$

is satisfied, and the virtual coupling is designed such that

$$\gamma = \frac{\delta \alpha}{\delta - \alpha} \quad (2)$$

holds, then a class of nonlinear non-passive environments can be displayed stably, in the sense described above, for all passive H. Equation (1) can be interpreted and used in a variety of ways.

- For a particular device parameterized by δ and fixed sample rate, a class of nonlinear environments can be identified.

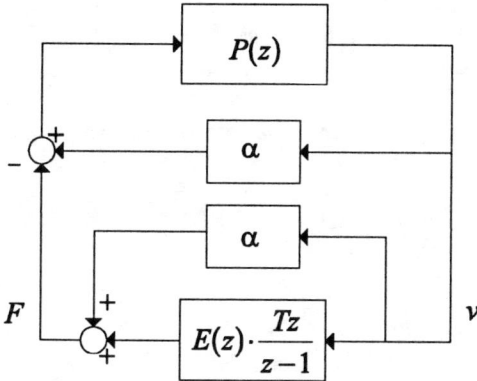

Figure 6: Coupled stability block diagram.

- For a particular device parameterized by δ and desired class of environments, a suitable sample time can be calculated.

- To guide hardware design, for a class of environments and fixed sample time, the amount of necessary physical damping can be determined.

Methods to determine values for parameters δ, γ and σ will be addressed in Section 4.

We begin by making some modifications to the coupled stability block diagram (Figure 6).

The parameter α is added to measure the amount of *excess* passivity in the top sub-system, referred to as the operator/device sub-system. It will be used to provide a connection between the physical characteristics of the haptic device (δ) and conditions imposed on the environment (σ). This results in the coupled stability block diagram shown in Figure (6). In this form the following relationship is defined

$$R(z) = \frac{P(z)}{1 - \alpha P(z)} \qquad (3)$$

where $P(z)$ represents the $\mathcal{ZOH}$ equivalent of the continuous-time portion of Figure (4). Define

$$N(s) = \frac{M(s)}{1 + M(s)H(s)} \qquad (4)$$

$$G(z) = \frac{z-1}{Tz}\mathcal{ZOH}\left\{\frac{1}{s}\cdot N(s)\right\}$$

$$= \frac{(z-1)^2}{Tz^2}\mathcal{Z}\left\{\frac{1}{s^2}\cdot N(s)\right\} \qquad (5)$$

$$= \frac{z-1}{Tz}\mathcal{Z}\left\{\frac{1-e^{-sT}}{s^2}\cdot N(s)\right\}$$

$$P(z) = G(z) + \frac{z-1}{Tz}\cdot\frac{1}{C(z)} \qquad (6)$$

4 Storage Function

This section will develop a storage function for use with the discrete-time version of LaSalle's invariance principle [7], to guarantee the absence of oscillations. This will be accomplished by considering the operator/device and nonlinear sub-systems separately. The positive real lemma guarantees the existence of a storage function in the operator/device sub-system, while a candidate storage function is proposed for the nonlinear sub-system. The desired storage function is obtained by adding these two functions.

4.1 Operator/Device Sub-system

The operator/device sub-system includes the haptic device, sample and hold operation and the virtual coupling network. The signal being sampled is position to accurately reflect the fact that encoders, collocated with the motors, are the only sensors present in the system. Along with obtaining a valid storage function, the main contributions of this section are as follows:

- General device model is accommodated that will allow us to characterize the haptic device with a single parameter, δ, which will be a function of device damping.

- A single transfer function (Q), independent of H, subject to a passivity condition will allow us to design the virtual coupling such that $R \in \text{PR}$.

- The parameter α provides us with a measure of the excessive passivity possessed by the operator/device sub-system.

The goal of this section is to show that $R(z) \in \text{PR}$, then from the positive real lemma a valid storage function, $V_L(\chi)$, exists. We begin with the ascertain that if $M(s)$ is δ-OSP, for some $\delta > 0$, and $H \in \text{PR}$, then

$$G(e^{j\omega T}) \in \frac{1}{\delta}e^{-j\omega T}\cdot D \qquad (7)$$

where $D = \{x \in \mathbb{C} : |2x-1| \leq 1\}$, a unit disk centered at $\frac{1}{2}$, is defined for convenience. Adding α in positive feedback, along with the δ-OSP condition on $M(s)$, requires the Nyquist plot of $zG(z)$ to lie within a disk intersecting the origin and a finite positive point $\frac{1}{\delta}$.

One difficulty in analyzing this sub-system is related to the fact that every transfer function that includes H really represents a class of transfer functions

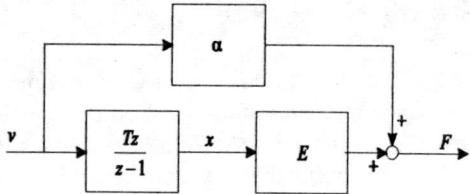

Figure 7: Environment block with parameter α. The mapping from velocity to force must be passive.

defined to include each passive H. To eliminate this difficulty we define the following expression,

$$Q(z) = \frac{z-1}{Tz}\left[\frac{1}{C(z)} - \frac{T}{2\delta}\right] \quad (8)$$

with (7) allowing it to be expressed independent of H, and require it to be γ-OSP. The utility of $Q(z)$ is that we are able to design the virtual coupling, and plot the Nyquist curve of $Q(z)$ to find γ. Then, if $M(s)$ is δ-OSP with $\delta > \alpha$ we can compute

$$\alpha = \frac{\gamma\delta}{\gamma + \delta} \quad (9)$$

Satisfying this relationship states that $R \in$ PR for every $H \in$ PR. It follows from the positive real lemma that a storage function, $V_L(\chi)$, exists that is quadratic in the states.

The next section takes α and derives a relationship with the environment parameter σ to identify a set of nonlinear environments that can be displayed passively.

4.2 Nonlinear Sub-system

This section is concerned with the block diagram in Figure (7). We demonstrate that it is possible to obtain a storage function for a class of nonlinear environments that exhibit non-passive behavior. Furthermore, a condition is derived that allows us to determine how non-passive the environment can be while still achieving stability, as outlined in Section 3. A candidate storage function will be proposed for the nonlinear environment, V_E, that will be used along with V_L to define a storage function for the coupled system. Recall that α is a measure of the *excess* passivity in the operator/device sub-block and has units of damping, therefore it helps to stabilize the environment (Figure (7)). It is important to note that α is a tool, including it in the environment analysis provides us with insight on how the overall system will behave once the two sub-systems are connected. Although α is not directly implemented, the amount of damping it represents is always available from the operator/device sub-system.

We begin the analysis by deriving the state equations associated with Figure (7).

$$x_k = x_{k-1} + Tv_k \quad (10)$$
$$F_k = E(x_k) + \alpha v_k \quad (11)$$

Defining a new state $q_{k+1} = x_k$ the state equations become:

$$q_{k+1} = q_k + Tv_k \quad (12)$$
$$F_k = E(q_{k+1}) + \alpha v_k \quad (13)$$

We propose the following storage function:

$$V_E(q) = \frac{1}{T}\int_0^q E(\xi)d\xi \quad (14)$$

It will be evident in the subsequent text, to achieve the desired goals the nonlinearity must meet certain conditions.

- $\sigma \geq \sup\{\frac{E(b)-E(a)}{a-b} : a \neq b\}$, where $\sigma \geq 0$ is the magnitude of the largest negative slope of the nonlinearity, with a and b points on the curve described by E. Note, we are only concerned with negative slope, therefore if E is non-decreasing we take $\sigma = 0$.

- $V_E(q) \geq E_{min}$, where E_{min} is the minimum energy exhibited by the nonlinearity.

- The nonlinearity described by E causes $V_E(q)$ to become radially unbounded.

We now show that the storage function in (14) is valid while providing some physical insight on the conditions imposed on the nonlinearity.

The environment with static nonlinearity $E(z)$, is passive if the environment slope

$$\sigma \leq \frac{2\alpha}{T} \quad (15)$$

For discrete-time systems the passivity condition yields

$$\Delta V_E = V_E(q_{k+1}) - V_E(q_k) \quad (16)$$
$$= \frac{1}{T}\int_{q_k}^{q_{k+1}} E(\xi)d\xi \leq F_k v_k \quad (17)$$

Physical reasoning suggests that negative stiffness is a non-passive behavior. Therefore, we seek a condition that bounds ΔV_E which will eventually lead to the constraint on the environment slope.

Consider two points **a** and **b** located on the curve defined by **E**. We assume a bound on these points:

$$\sigma \leq \frac{E(b) - E(a)}{a - b} \quad (18)$$

We define the following equation that represents a line with slope $-\sigma$ passing through the point q_{k+1}.

$$L(r) = E(q_{k+1}) - \sigma(r - q_{k+1}) \quad (19)$$

The above expression will be used to bound the storage function in Equation (14).

$$\frac{1}{T}\int_{q_k}^{q_{k+1}} E(r)dr \leq \frac{1}{T}\int_{q_k}^{q_{k+1}} L(r)dr \quad (20)$$

To continue the analysis we evaluate the right hand side of (20).

$$\frac{1}{T}\int_{q_k}^{q_{k+1}} L(r)dr = F_k v_k - (\alpha - \sigma\frac{T}{2})v_k^2 \quad (21)$$

For passivity to hold the above expression must produce a result that is $\leq F_k v_k$. To achieve this, the second term on the right hand side must be positive. Equation (15) follows when solving for σ.

For the radially unbounded condition to be satisfied, the nonlinearity must cause (14) to approach infinity as $q \to \infty$.

4.3 Coupled System

In previous sections we have derived conditions under which the equation

$$V(\chi, q) = V_L(\chi) + V_E(q) \quad (22)$$

is a valid storage function. The discrete-time version of LaSalle's invariance principle [7] can be directly applied to demonstrate that in steady state the velocity input to the environment is exactly zero. Assuming that E is defined such that (14) is both bounded from below by E_{min} and radially unbounded, this storage function guarantees the compactness of the set

$$\Omega = \{V(\chi, q) \leq V(0)\} \quad (23)$$

Since V is a non-increasing function the solutions of the coupled system are bounded and $V \to c$, c being some arbitrary constant. The fact that (23) is compact proves the existence of the positive limit set L^+ that is compact and positively invariant. Furthermore, the

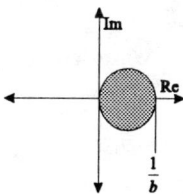

Figure 8: Nyquist region for one DOF haptic display.

solutions of the coupled system tend toward L^+ as $k \to \infty$ and $V(L^+) \to c$.

Now we wish to show that $L^+ \subset \{v(k) = 0\}$. Since $V(L^+) \to c$ and L^+ is a positively invariant set, $\Delta V(L^+) = 0$. From the fact that $\Delta V \leq -\epsilon v^2(k)$ we can state:

$$\Delta V(L^+) \subset \{\Delta V = 0\} \subset \{v(k) = 0\} \quad (24)$$

It follows that the solutions tend toward $L^+ \subset \{v(k) = 0\}$.

To prove that $F(k)$ approaches a constant, we use the fact that $v = 0$ in steady state. The projection of this set for the following relationship

$$x = \frac{Tz}{z-1}v \quad (25)$$

results in a constant x. It follows that $E(x)$ is a constant in steady state.

Recall that this is achieved without requiring that the environment be passive, thus providing the designer with a larger class of environments that can be rendered while addressing human safety.

5 Results

Application of this design methodology will be illustrated using a one DOF haptic display (Figure (4)) assumed to behave like a rigid body (m) with viscous friction (b). A spring/damper virtual coupling is considered along with fixed sample time T.

$$C(z) = \frac{(KT+B)z - B}{Tz} \quad (26)$$

$$M(s) = \frac{1}{ms+b} \quad (27)$$

The task is, for a given level of physical damping, to design a virtual coupling to obtain the largest class of discrete-time nonlinear environments, based on our parameterization, that can be displayed passively. Assuming a passive H we first draw the Nyquist region of the device/human feedback loop (Figure 8).

From the Nyquist plot we observe that the disk has a radius of $\frac{1}{2b}$ and is centered at $\frac{1}{2b}$, therefore we conclude that $\delta = b$. We begin with (1) which immediately tells us the maximum negative stiffness allowed in the environment. Next, we use (15) to determine how much excess passivity is available.

To design the virtual coupling we use (8) and the fact that it must be γ-OSP. Recall,

$$Q = \frac{z-1}{Tz}\left[\frac{1}{C(z)} - \frac{T}{2b}\right] \quad (28)$$

$$= \frac{z-1}{Tz}\left[\frac{Tz}{(KT+B)z - B} - \frac{T}{2b}\right] \quad (29)$$

For choices of K and B the following inequality must hold to ensure that Q is γ-OSP.

$$\frac{(KT)^2}{2b} < KT + 2B \leq 2b \quad (30)$$

Lets take a step back and think about the role we want the virtual coupling to take. Ideally, we would like to eliminate the virtual coupling because it limits the stiffness of the environment. Unfortunately, when it is removed no stability result can be achieved. To minimize the effect it has on the environment we want to minimize the magnitude of the virtual coupling transfer function $\frac{1}{C(z)}$ over the entire range of frequencies. From the bounds on the virtual coupling parameters (30), this is achieved by setting the virtual damping $B = 0$, and choosing $K = \frac{2b}{T}$. From this $Q \equiv 0$ making it γ-OSP for any $\gamma > 0$. If virtual damping is desired in the coupling the analysis would use (9) to solve for γ. Then the values of K and B need to be determined such that Q is γ-OSP.

6 Summary and Conclusions

This work has developed a design methodology that allows passive implementation for a class of non-passive discrete nonlinear environments. The motivation for interest in non-passive discrete-time environments was laid out in the introductory section.

Although this document targeted nonlinear environments the majority of the analysis has wide-spread application. Using the operator/device analysis to produce values for δ and α has given us stability results for linear environments with computational delay. We have also been successful in adding nonlinear damping to the environment block, which adds another condition to the nonlinear sub-system analysis restricting the environment damping to a sector.

Finally, the analysis has made it unnecessary to approximate the device behavior using a model (even though this was done in Section 5). We have plans to determine the parameter δ experimentally, which is all that is needed to proceed with the development.

Acknowledgments

Research supported in part by NSF Grant ECS-9703294.

References

[1] R. Adams and B. Hannaford, "A Two-Port Framework for the Design of Unconditionally Stable Haptic Interfaces," *Proc. IEEE/RSJ International Conference on Intelligent Robots and Systems*, 1998.

[2] R. Adams and B. Hannaford, "Stability Performance of Haptic Displays: Theory and Experiments," *Proc. International Mechanical engineering Congress and Exhibition*, ASME, Anaheim, CA, Vol.SAX-64,p.227-234, 1998.

[3] J.M. Brown and J.E. Colgate, "Passive Implementation of Multibody Simulations for Haptic Display," *Proc. International Mechanical Engineering Congress and Exhibition*, ASME, Dallas, TX, Vol.DSC-61, p.85-92, 1997.

[4] J.E. Colgate and G.G Schenkel, "Passivity of a class of sampled-Data systems: Application to Haptic Interfaces," *Journal of Robotic Systems*, Vol. 14(1), p.37-47, 1997.

[5] J.E. Colgate, M.C. Stanley and J.M. Brown, "Issues in the Haptic Display of Tool Use ," *Proc. IEEE/RSJ International Conference on Intelligent Robots and Systems*,1995.

[6] G.C. Goodwin and K.S. Sin, *Adaptive Filtering Prediction and Control* , Englewood Cliffs, 1984.

[7] J.P. Lasalle, *Stability and Control of discrete Processes*, Springer-Verlag, New York, 1986.

[8] D. Lindorff, *Theory of Sampled-Data Control systems*, New York, 1965.

A Three-Dimensional Touch/ Force Display System for Haptic Interface

Tsuneo YOSHIKAWA and Akihiro NAGURA
Faculty of Engineering, Kyoto University
Kyoto 606-8501, Japan
yoshi@mech.kyoto-u.ac.jp

Abstract

It is widely recognized that haptic display is one of the key elements for making the virtual reality technology more powerful and various force and/or tactile display devices have been developed. However, many of these devices are always in touch with fingers or hands of operators and it is impossible for an operator to feel difference between contact and non-contact state with a virtual object directly from sensory channel of touch in his/her finger or hand. This paper presents a new three-dimensional haptic display device we have developed for providing both touch and force feeling to the operator's fingers. This is an extension of our previous device for two-dimensional use. Control algorithm for the new device and some experimental results are also presented.

1 Introduction

Researches in the field of virtual reality technology at the initial stage were mainly focused on displaying visual and auditory information to an operator by, for example, a head mounted display. However, it soon became obvious that, in cases where the operator wants to touch, grasp and manipulate objects in the virtual space, haptic display of force and/or tactile information is also very important for providing him a more realistic feeling.

Various force display devices have been developed to provide this haptic feeling to the operator [1]. Most of them are, however, held or worn by the operator and the operator's finger or hand is always in touch with the display device. In such systems it is impossible for an operator to feel difference between touch and non-touch states with the virtual object directly from tactile sensory channel in his/her finger or hand. One way to cope with this problem is to develop a display system that does not have any contact with the operator when the operator is not in touch with the virtual object and that can display force when the operator is in touch with it.

Along this line, we developed a display device that consists of a two-joint link mechanism and uses a set of optical glass-fiber on-off sensors for measuring the position of the operator's finger [2]. However, this device was just for use in two-dimensional space. In this paper we present a new three-dimensional touch and force display device which is an extension of our previous two-dimensional device. Control algorithm of the new device and some experimental results are also presented.

Note that several devices with similar principle have already been developed [3] [4]. However, since the device we have developed has simple mechanism using optical sensors, it is not necessary for an operator to wear anything and we hope that it is easy to extend to multiple-finger case. By using this kind of haptic display devices, it will become possible to display a clear feel of force and touch in various tasks involving contact and separation between the fingers and virtual objects such as picking up and manipulating a virtual object with two or more fingers.

Note also that a few encounter-type haptic display devices [5] [6] have been developed. These devices are different from the one developed here in the sense that they do not track the operator's finger or hand closely.

2 Outline of Display System

Fig 1 shows the basic structure of the proposed device. The device consists of two arms and a cap attached to the tips of the arms. The shape of the cap is a combination of a half sphere and a cylinder. The operator can insert his/her finger into the cap to interact with the virtual world. During the finger is in free

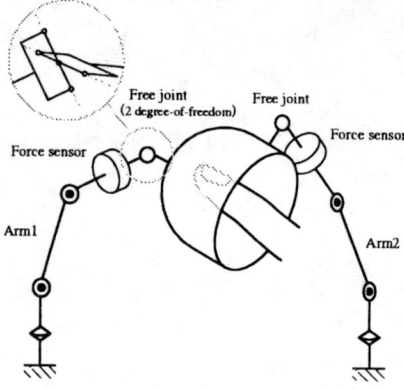

Figure 1: Basic Structure of Device

space the cap is controlled to track the fingertip without touching the finger and when the fingertip gets into touch with a virtual object, the cap is controlled so that the inner surface of the cap touches the finger from the direction of the virtual object and the contact force is displayed to the operator.

The two arms that support the cap have three joints driven by DC servo motors. The first and second arms are, respectively, connected to the cap by a two-degree-of-freedom passive joint and a three-degree-of-freedom passive joint. This way the cap is prevented from rotating around the axis connecting the two passive joints. Hence the position and orientation of the cap can be freely changed by moving the tips of the two arms appropriately.

Rotary encoders are attached to the joints of the arms. Force sensors are placed just behind the tips of the arms to measure the force applied to the finger. A position/orientation sensor system is attached to the cap for measuring the relative position and orientation of the finger with respect to the cap.

The position/orientation sensor system of the cap consists of optical glass-fiber on-off sensors. It is of course desirable to be able to obtain an accurate finger position and orientation measurement, but usually a large and heavy sensor system is necessary to do this. Hence we constructed a new sensor system consisting of multiple light-weight, on-off type optical sensors shown in Fig.2. This sensor system enables us to roughly estimate the three-dimensional position and orientation of the finger in the cap.

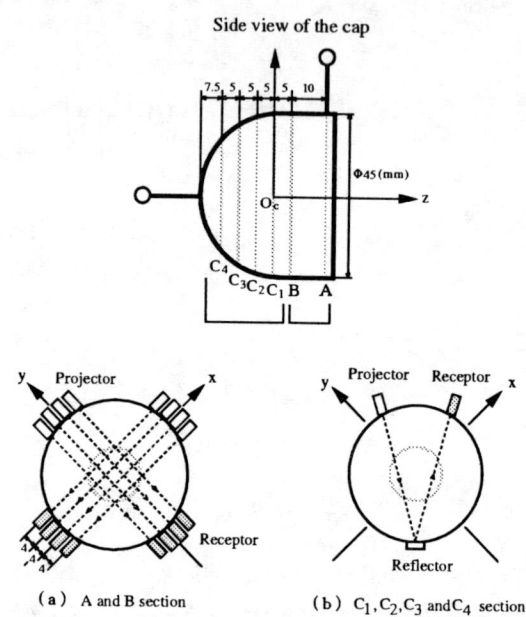

Figure 2: Sensor System of Cap

3 Estimation of Fingertip Position and Orientation

The optical sensors are put on the cap as shown in Fig.2 so that the fingertip position and orientation can be determined from their outputs. Since the sensors are placed only discretely in space, the information on the fingertip position and orientation obtained from this sensor system is discrete in time. Therefore, we first determine an instantaneous estimate of the fingertip position and orientation with respect to the cap and transform this estimate to that with respect to the reference coordinate frame. Then we obtain a smoothed estimate by using a simple filter so that the estimate is continuous in time. This procedure is described in detail in the following.

3.1 Instantaneous Estimate

From the measurements of the two arrays of sensors placed on A and B sections of the cap shown in Fig2(a), we can obtain the positions of the central axis of the finger on both A and B sections and estimate the inclination of the finger. Besides, from the measurements of the sensors placed on C_1, C_2, C_3 and C_4 sections shown in Fig2(b), we can see whether the finger is crossing each section and estimate the fingertip position.

The orthogonal coordinate frame Σ_c attached to

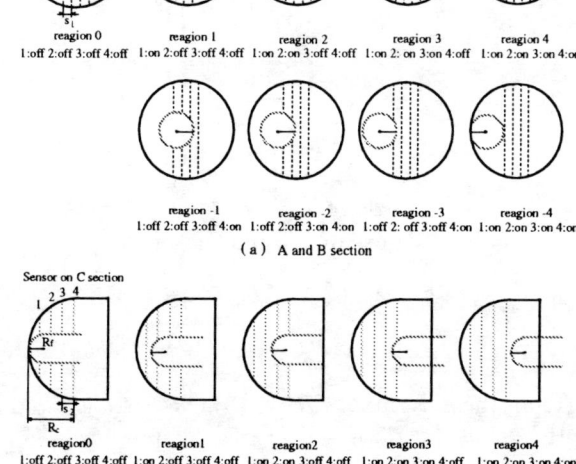

Figure 3: Relation Between Finger and Cap

the cap is placed such that the z axis directs to the cap center axis and the x axis and the y axis to the sensor lines on A and B sections. We approximate finger sections on A and B sections to circles and let the instantaneous estimates of their center positions with respect to Σ_c be $^c r_A$ and $^c r_B$. We approximate the tip of the finger to a hemisphere and define its center as the fingertip position. Let the instantaneous estimate of the fingertip position with respect to Σ_c be $^c r_f$.

The relative position between the finger and the parallel on-off sensors in the x- and y-directions of A and B sections can be classified into nine regions shown in Fig.3(a). Here R_f is the radius of the finger, R_c is the radius of the cap, and s_1 is the distance between sensors, which is smaller than $2/3 R_f$. Based on this information of region, we can determine the center position $^c r_i = [^c r_{ix}, {}^c r_{iy}]^T$ of the finger in section $i (i = A, B)$ as follows:

$$^c r_{ij} = 0, \pm(R_f - s_1), \pm(R_f), \pm(R_f + s_1),$$
$$\text{or } \pm(R_c - R_f) \quad (1)$$

if the sensor state is in region $0, \pm 1, \pm 2, \pm 3,$ or ± 4, respectively.

Since the finger cannot go out of the cap, if $\sqrt{{}^c r_{ix}^2 + {}^c r_{iy}^2} > R_c - R_f$ holds, this $^c r_{ij}$ is replaced by

$$\frac{R_c - R_f}{\sqrt{{}^c r_{ix}^2 + {}^c r_{iy}^2}} {}^c r_{ij} \quad (j = x, y) \quad (2)$$

The z-component $^c r_{fz}$ of $^c r_f$ is defined as follows based on the state shown in Fig3(b). Let s_2 be the distance between C sections, then

$$^c r_{fz} = R_f - R_c, R_f - \frac{5}{2}s_2,$$
$$R_f - \frac{3}{2}s_2, R_f - \frac{1}{2}s_2, R_f + \frac{1}{2}s_2 \quad (3)$$

in region 0, 1, 2, 3, 4, respectively.

Let the instantaneous estimate of the unit vector representing the inclination of the finger be $^c d_f$. We define $^c d_f$ as follows.

$$^c d_f = \frac{{}^c r_A - {}^c r_B}{\| {}^c r_A - {}^c r_B \|} \quad (4)$$

We assume $^c r_f$ is on the line that links $^c r_A$ to $^c r_B$ and it is given by

$$^c r_C = {}^c r_A + \frac{{}^c r_{fz} - {}^c r_{Az}}{{}^c d_{fz}} {}^c d_f \quad (5)$$

From the condition that the finger cannot go out of the cap, if $^c r_{fz} < 0$ and $\| {}^c r_f \| > R_c - R_f$ hold, $^c r_f$ is replaced by

$$\frac{\sqrt{(R_c - R_f)^2 - {}^c r_{fz}^2}}{\sqrt{{}^c r_{fx}^2 + {}^c r_{fy}^2}} {}^c r_{fj} \quad (j = x, y) \quad (6)$$

If $^c r_{fz} \geq 0$ and $\sqrt{{}^c r_{fx}^2 + {}^c r_{fy}^2} > R_c - R_f$ hold, $^c r_f$ is replaced by

$$\frac{R_c - R_f}{\sqrt{{}^c r_{fx}^2 + {}^c r_{fy}^2}} {}^c r_{fj} \quad (j = x, y) \quad (7)$$

In these cases, $^c d_f$ is replaced by

$$^c d_f = \frac{{}^c r_A - {}^c r_f}{\| {}^c r_A - {}^c r_f \|} \quad (8)$$

3.2 Instantaneous Estimate with Respect to Reference Frame

The Procedure of transforming the instantaneous estimates $^c r_f$, $^c d_f$ relative to the cap to those relative to the reference coordinate frame is described in this subsection. Fig.4 shows the reference frame Σ_U, the fingertip frame Σ_h attached to arm 1, and the cap frame Σ_c.

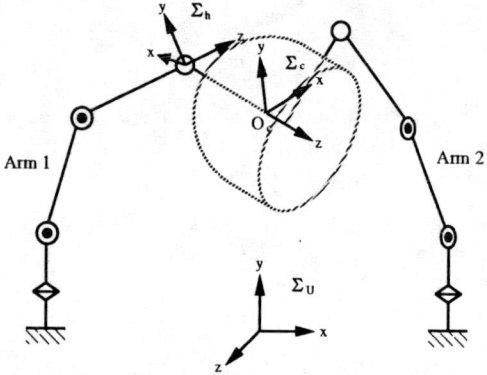

Figure 4: Coordinate Frames

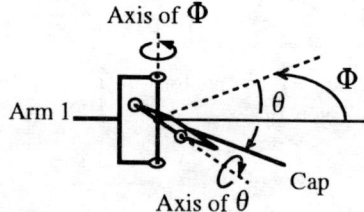

Figure 5: Angle Variables of Free Joint

Let the rotation matrices between Σ_U and Σ_h, Σ_h and Σ_c, and Σ_U and Σ_c be $^U\boldsymbol{R}_h$, $^h\boldsymbol{R}_c$, and $^U\boldsymbol{R}_c$, respectively. Then $^U\boldsymbol{R}_h$ is given as a function of the joint vector $\boldsymbol{q}_1$ of arm 1. By introducing the two variables ϕ, θ shown in Fig.5, rotation matirx $^h\boldsymbol{R}_c$ is given by

$$^h\boldsymbol{R}_c = \frac{1}{\sqrt{2}} \begin{bmatrix} C_\phi + S_\phi S_\theta & -C_\phi + S_\phi S_\theta & \sqrt{2} S_\phi C_\theta \\ C_\theta & C_\theta & -\sqrt{2} S_\theta \\ -S_\phi + C_\phi S_\theta & S_\phi + C_\phi S_\theta & \sqrt{2} C_\phi C_\theta \end{bmatrix} \quad (9)$$

Variables ϕ and θ is determined uniquely for any given tip positions $\boldsymbol{r}_{h1}, \boldsymbol{r}_{h2}$ of arms 1 and 2. Let the vector connecting the two armtips be $\boldsymbol{h}(= \boldsymbol{r}_{h2} - \boldsymbol{r}_{h1})$, and that represented in Σ_c be $^c\boldsymbol{h}$, then $^c\boldsymbol{h}$ is a constant vector given by

$$^c\boldsymbol{h} = \left[\frac{b}{\sqrt{2}}, \frac{b}{\sqrt{2}}, a\right]^T \quad (10)$$

where a and b are constants shown in Fig.6. Since the relation

$$(^U\boldsymbol{R}_h)^T \boldsymbol{h} = {^h\boldsymbol{R}_c}{^c\boldsymbol{h}} \quad (11)$$

holds, by letting $[h_1, h_2, h_3]^T$ denote the components of $(^U\boldsymbol{R}_h)^T \boldsymbol{h}$, the values of ϕ and θ are given by

$$\phi = atan2(h_1, h_3) \quad (12)$$

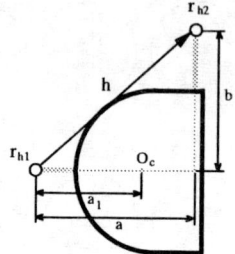

Figure 6: Parameters of Cap

$$\theta = atan2(b\sqrt{(h_1^2 + h_3^2)} - ah_2, a\sqrt{(h_1^2 + h_3^2)} + bh_2) \quad (13)$$

From these ϕ and θ, $^h\boldsymbol{R}_c$ in (9) can be calculated and the value of $^U\boldsymbol{R}_c$ can also be calculated using

$$^U\boldsymbol{R}_c = {^U\boldsymbol{R}_h}{^h\boldsymbol{R}_c} \quad (14)$$

The position of origin of Σ_c with respect to Σ_U be $^U\boldsymbol{p}_c$ is given by

$$^U\boldsymbol{p}_c = \boldsymbol{r}_{h1} - {^U\boldsymbol{R}_c}{^c\boldsymbol{r}_{h1}} \quad (15)$$

where $^c\boldsymbol{r}_{h1} = [0, 0, -a_1]^T$ is the arm-tip position of arm 1 relative to Σ_c (a_1 is a constant distance shown in Fig.6).

With these preparations, the instantaneous estimates $^c\boldsymbol{r}_f$ and $^c\boldsymbol{d}_f$ with respect to Σ_U (a reference frame fixed to the base of the display device) is given by

$$\boldsymbol{r}_f = {^U\boldsymbol{R}_c}{^c\boldsymbol{r}_f} + {^U\boldsymbol{p}_c} \quad (16)$$

$$\boldsymbol{d}_f = {^U\boldsymbol{R}_c}{^c\boldsymbol{d}_f} \quad (17)$$

where $^U\boldsymbol{R}_c$ is the rotation matrix from Σ_U to Σ_c and $^U\boldsymbol{p}_c$ is the position vector of the origin of the cap frame with respect to Σ_U.

3.3 Smoothed Estimate

Since the instantaneous estimate is discontinuous in time, we define the smoothed estimate $\hat{\boldsymbol{r}}_f$ of the fingertip position and the smoothed estimate $\hat{\boldsymbol{d}}_f$ of the inclination of the finger by the following equation.

$$\hat{\boldsymbol{r}}_f(t) = (1 - e^{\frac{t_0 - t}{T_1}})(\boldsymbol{r}_f(t_0) - \hat{\boldsymbol{r}}_f(t_0)) + \hat{\boldsymbol{r}}_f(t_0) \quad (18)$$

$$\hat{\boldsymbol{d}}_f(t) = \frac{\boldsymbol{k}}{\|\boldsymbol{k}\|} \quad (19)$$

$$\boldsymbol{k} = (1 - e^{\frac{t_0 - t}{T_2}})(\boldsymbol{d}_f(t_0) - \hat{\boldsymbol{d}}_f(t_0)) + \hat{\boldsymbol{d}}_f(t_0) \quad (20)$$

where t is the time variable, t_0 is the last time instant of change of the instantaneous estimate, and T_1, T_2 are constants.

4 Algorithm for Non-Contact Tracking and Force Display

An algorithm for realizing non-contact tracking and force display is as follows. The motion of the virtual object caused by the measured finger force is calculated. The desired cap position is determined from the relative position between the finger and the virtual object. For realizing the desired cap position, two arms are controlled.

4.1 Dynamics of Virtual Object

Let the force applied to the cap by the finger be $\boldsymbol{f}_h$. In case of non-contact state, $\boldsymbol{f}_h$ equals 0. In case of contact state, $\boldsymbol{f}_h$ is given by

$$\boldsymbol{f}_h = \boldsymbol{f}_1 + \boldsymbol{f}_2 - M_c \boldsymbol{g} \tag{21}$$

where $\boldsymbol{f}_1$ and $\boldsymbol{f}_2$ are the force sensor measurements, M_c is the mass of the cap, $\boldsymbol{g}$ is the gravitational acceleration vector.

On the assumption that there is no friction between the finger and the virtual object and that the force applied to the virtual object by the finger is only the perpendicular component $\boldsymbol{f}_{hv}$ of $\boldsymbol{f}_h$ to the surface of the virtual object at the contact point, the dynamics of the virtual object is given by

$$M_o \ddot{\boldsymbol{r}}_G = M_o \boldsymbol{g}_v + \boldsymbol{f}_{hv} \tag{22}$$

$$\boldsymbol{I}\dot{\boldsymbol{w}} + \boldsymbol{w} \times (\boldsymbol{I}\boldsymbol{w}) = \boldsymbol{r}_{Gh} \times \boldsymbol{f}_{hv} \tag{23}$$

where M_o and $\boldsymbol{I}$ are the mass and the inertia matrix of the virtual object, and $\boldsymbol{g}_v$ is the gravitational acceleration vector in the virtual space. $\boldsymbol{r}_G$ is the position of the center of the mass of the virtual object, $\boldsymbol{r}_{Gh}$ is the vector from $\boldsymbol{r}_G$ to the contact point, and $\boldsymbol{w}$ is the angular velocity vector.

Note, however, that it is possible to display a force $\boldsymbol{f}_h$ with tangential frictional component also, if it is within the limit of real frictional constraint between the finger and the cap. To implement this, we need to adopt a friction model and to modify the desired cap position and orientation in a similar way to [1]. This could be a next step of this research.

4.2 Desired Cap Position

We define the origin of the cap frame as the cap position and a unit vector directing to the cap center axis as the inclination of the cap. Let the desired cap

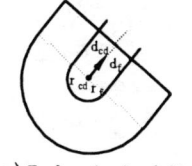

(a) Far from the virtual object

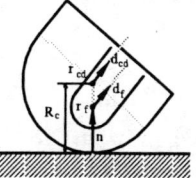 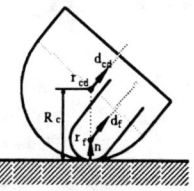

(b) Near the virtual object   (c) Contact with the virtual object

Figure 7: Desired Cap Position

position be $\boldsymbol{r}_{cd}$ and the desired inclination of the cap $\boldsymbol{d}_{cd}$.

When no virtual object is near the finger, the finger position is set to the cap position(Fig.4(a)), i.e. $\boldsymbol{r}_{cd} = \hat{\boldsymbol{r}}_f$

When a virtual object comes closer to the finger than the cap radius, the cap is made to move at the closest surface of the virtual object to the finger(Fig4(b)). So we determine $\boldsymbol{r}_{cd}$ by the following equation. Let the vector from the surface of the virtual object to the finger be $\mathbf{n}$.

$$\boldsymbol{r}_{cd} = \hat{\boldsymbol{r}}_f + \frac{R_c - \|\mathbf{n}\|}{\|\mathbf{n}\|} \boldsymbol{n} \tag{24}$$

Note that $\boldsymbol{n}$ and R_c are functions of the object's configuration determined by (22) and (—refeq31).

While the finger is in contact with the virtual object, $\boldsymbol{r}_{cd}$ is defined similar to (17) so that the cap is kept at the surface of the virtual object(Fig4(c)).

The desired inclination of the cap is determined as follows so that it always equals to the inclination of the finger.

$$\boldsymbol{d}_{cd} = \hat{\boldsymbol{d}}_f \tag{25}$$

4.3 Control Algorithm

A control algorithm for making the arm positions $\boldsymbol{r}_{hi}(i = 1, 2)$ track the desired arm positions $\boldsymbol{r}_{hdi}$ that correspond to the desired cap position and inclination will be given.

When an external force $\boldsymbol{F}_i$ is applied to the arm, the dynamics of the arm is given by

$$\boldsymbol{\tau}_i = \boldsymbol{M}_i(\boldsymbol{q}_i)\boldsymbol{J}_i^{-1}(\boldsymbol{q}_i)[\ddot{\boldsymbol{r}}_{hi} - \dot{\boldsymbol{J}}_i(\boldsymbol{q}_i)\dot{\boldsymbol{q}}_i]$$
$$+ \hat{\boldsymbol{h}}_i(\boldsymbol{q}_i, \dot{\boldsymbol{q}}_i) - \boldsymbol{J}_i^T(\boldsymbol{q}_i)\boldsymbol{F}_i \tag{26}$$

where τ_i is the joint driving force vector of the arm, q_i is the joint vector, $M_i(q_i)$ is the inertia matrix of the arm, $\hat{h}_i(q_i, \dot{q}_i)$ is the term representing the centrifugal and Colioris forces, viscose friction, and gravity forces, and $J_i(q)$ is the Jacobian matrix of r_{hi} with respect to q_i.

Adopting the two-stage control scheme, we propose to use the control algorithm consisting of a linearizing compensator

$$\tau_i = M_i(q_i)J_i^{-1}(q_i)[u_{ri} - \dot{J}_i(q_i)\dot{q}_i] + \hat{h}_i(q_i, \dot{q}_i) - J_i^T(q_i)F_i \quad (27)$$

and a servoing compensator

$$u_{ri} = -K_{rv}\dot{r}_{hi} + K_{rp}(r_{hdi} - r_{hi}) \quad (28)$$

where K_{rv} and K_{rp} are velocity and position feedback gains, respectively.

5 Experiment

5.1 Outline of Experimental System

The picture of the device is shown in Fig.5.

The arms are made of Mini-Robot (Daikin Industries, Ltd.). The force sensor can measure up to 4.9 N(=500gf). The encoder is of 500P/R type with a reduction gear of ratio 1:8. The maximum torque of the DC motor is 35N·mm(=3600gf·mm) and a reduction gear of 1:8 is attached to the motor. The optical sensors of the cap are KEYENCE fiber sensors (ϕ3mm, approximately 1g, response time of the amplifier is 0.5msec).

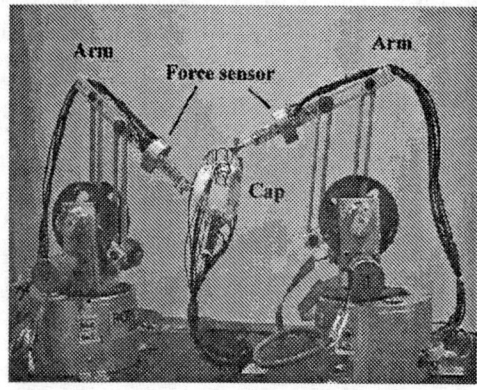

Figure 8: Touch/Force Display Device

5.2 Task Description and Parameter Values

The following four experiments have been performed.

Experiment 1: The operator moves his/her finger in a region far from the virtual object. Tracking ability of the device is examined.

Experiment 2: The operator touches a horizontal floor. It is examined if the cap gets into contact with the fingertip just when the fingertip touches the horizontal floor.

Experiment 3: The operator throws up a virtual object. It is examined if the dynamics of the virtual object is displayed.

Experiment 4: The operator touches a cube from three orthogonal directions. It is examined if the device is effective in displaying objects in three dimensional space.

Parameter values for the experiment are as follows. The cap radius is R_c=22.5[mm], the finger radius is R_f=7.5[mm]. The values of $\{T_1, T_2\}$ in (18)–(20) are $\{0.01, 0.02\}$ for Experiment 1, $\{0.02, 0.03\}$ for Experiment 2, and $\{0.03, 0.06\}$ for Experiments 3 and 4, respectively. The position and velocity feedback gains in servo compensators are $K_{rp} = 1500$ [1/s^2] and $K_{rv} = 35$ [1/s]. The gravitational acceleration in the virtual space is $9800 \times 0.03 = 294$[mm/s^2]. The sampling time is approximately 2.5ms for Experiments 1 and 2, and 4ms for Experiments 3 and 4. Note that the reference coordinate frame is taken in such a way that the x axis directs to the right and the y axis is upward and the z axis is forward when the operator is facing the device.

5.3 Results

It has been confirmed by using human operators' fingers that non-contact tracking and force/touch display are achieved by the developed display system. However, since it is difficult to measure the real position and orientation of the operator's fingertip, we used the tail of the stylus of force display device PHANToM (SensABLE Technologies, type 101AG) instead of human fingertip for obtaining data of Experiments 1 and 2 (Fig.9–12), that is, the operator inserted the tail of stylus into the cap and the position and orientation of the stylus is determined by using the encoders of PHANToM.

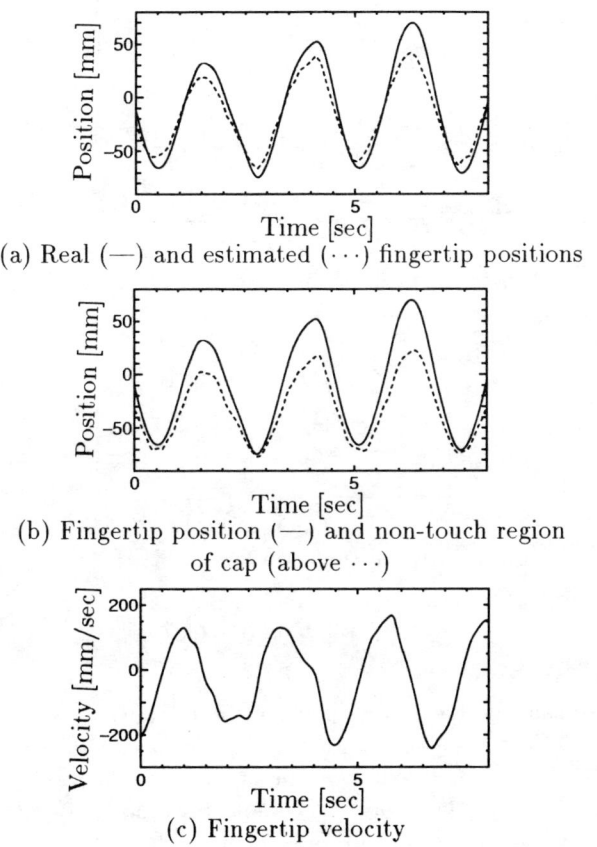

Figure 9: Experiment 1-a

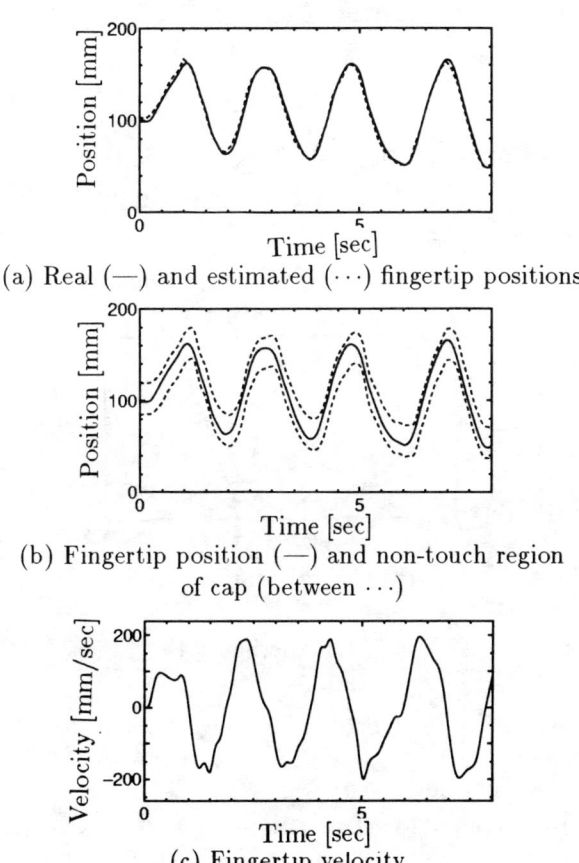

Figure 10: Experiment 1-b

Experiment 1: Fig.9 and Fig.10 are time trajectories of several key variables for the case where the fingertip is waved in the direction of cap center axis and in the vertical direction, respectively, with the orientation of the fingertip kept parallel to the $-x$ direction. In (a) of the figures, the solid line is the real trajectory of fingertip and the dotted line is its smoothed estimate. Although in Fig.10 (a) they coincide rather well, they do not in Fig.9 (a). This difference will be due to the difference of resolution in the two directions. In (b) the solid line is the real trajectory of fingertip and the dotted lines show the boundary of the region in which the fingertip does not touch the cap (above the dotted line in 9 (b) and betweeen the two dotted lines in Fig.10 (b)). From the figure it can be seen that the finger is actually not in touch with the cap. From (c) which shows the fingertip velocity, it can be seen that even when the fingertip velocity is 200mm/s, a non-contact tracking is realized.

Fig.11 is for the case where only the orientation of fingertip is changed in the $x-y$ plane. Solid line in (a) is the angle between the fingertip axis and the x axis, and the dotted line is its smoothed estimate. (b) shows the orientation of the cap. From (a) it is seen that the difference between the real and estimated angles sometimes becomes very large. This will be due to a low resolution of the angle measurement caused by a short distance between the sections A and B in Fig.2. Since the angle of cap in (b) tracks the estimated angle in (a), the cap angle also deviates from the real fingertip angle. However, no contact between the fingertip and the cap has occured in the experiment.

Experiment 2: A virtual horizontal floor is placed at $y=50$ [mm] of the refernce frame and the operator moved his finger into this floor. Fig.12 shows trajectories of some variables in y direction. The solid line in (a) is the real fingertip position and the dotted line is the cap position. When the fingertip is far away from the floor, the fingertip and cap positions agree very well. As the fingertip comes within the distance of R_c from the floor, the fingertip and cap positions begin to separate. When the fingertip touches the floor, the distance between the fingertip and the cap is almost equal to $R_c - R_f$ and the fongertip force appears as

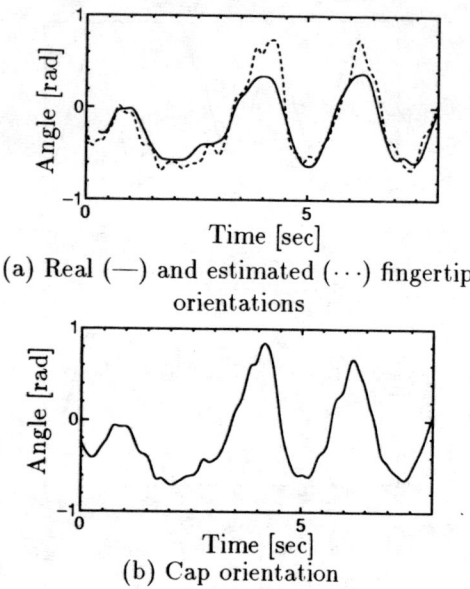

(a) Real (—) and estimated (···) fingertip orientations

(b) Cap orientation

Figure 11: Experiment 1-c

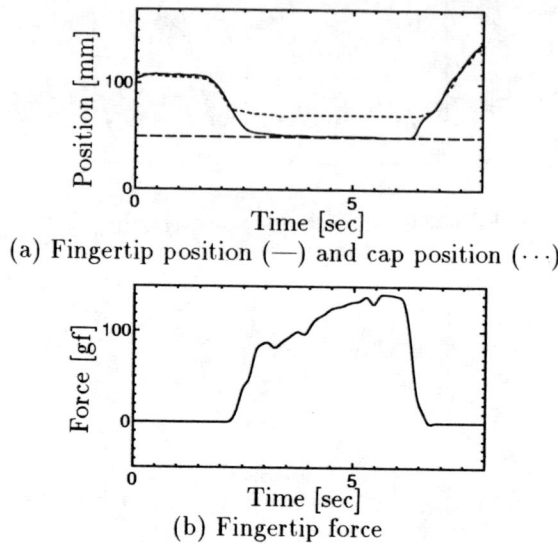

(a) Fingertip position (—) and cap position (···)

(b) Fingertip force

Figure 12: Experiment 2

shown in (b), implying that the fingertip and the cap are in contact to each other. The operator could feel the flat floor fairly well.

Experiment 3: It is suppose that there is a virtual cube with sides 50[mm] long on a virtual horizontal floor at $y=50$[mm]. The motion is restricted to the translation in the vertical direction (y axis). It is also assume that there is no interaction between the fingertip and the floor. Fig.13 shows the motion and force in the vertical direction. The solid line in (a) is the real fingertip position, the dotted line is the cap position, and the broken line is the floor location. The cap position reaches the surface of the cube at time A and begins to deviate from the fingertip position. When the fingertip makes a contact with the cube at time B, the cap also makes a contact with the fingertip. After receiving an impact force shown in (c) the cube flies up in the air as shown in (b). Sometime later, the cube falls down and bounces on the floor as can be seen from (b).

Experiment 4: It is supposed that there is a cube fixed to the floor as shown in Fig.14 (a). The operator tries to touch the cube from side, top, and front, to feel the surface of the cube in three dimensional space. The sphere in the figure represents the fingertip and its shadow is projected on the floor to show the fingertip position better in three dimensional space. The straight line extended from the sphere to the cube shows the direction and magnitude of the force applied on the cube by the fingertip. From the figure we can see that fingertip is kept on the surface of the cube and the contact force is displayed to the operator during the contact. The operator could feel the direction of the cube surface not only from the sense of force but also from the sense of touch.

6 Conclusion

A new three-dimensional haptic display device has been developed for providing both touch and force feeling to the operator's fingers. This device can follow the motion of the finger of the operator without contact when the finger is in free space and gets into contact with the finger and displays contact force on the operator's finger when the finger touches a virtual object. This device was built using light-weight, optical on-off sensors for detecting the finger position and orientation. Some experimental results have been presented to show the performance of the device.

References

[1] T. Yoshikawa and H. Ueda: "Haptic Virtual Reality: Display of Operating Feel of Dynamic Virtual Objects", Proc. of the 7th International Symposium of Robotics Research, pp.191–198, 1995

[2] T. Yoshikawa and A. Nagura, "A Touch and Force Display System for Haptic Interface," Proc.

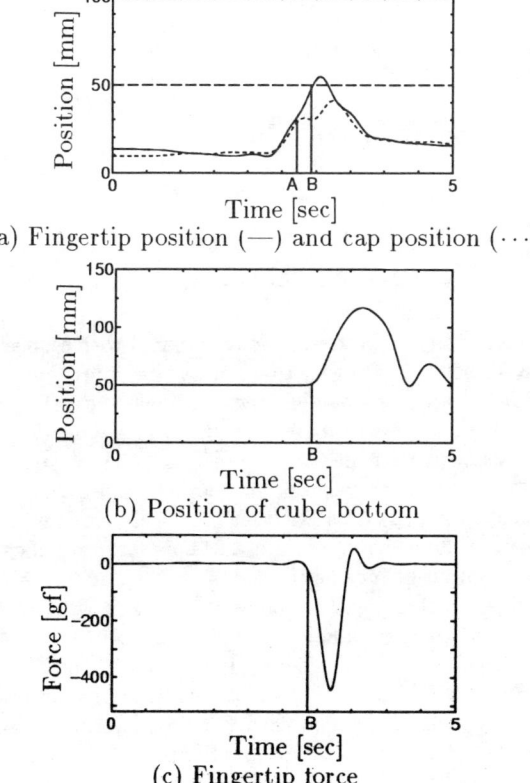

(a) Fingertip position (—) and cap position (···)

(b) Position of cube bottom

(c) Fingertip force

Figure 13: Experiment 3

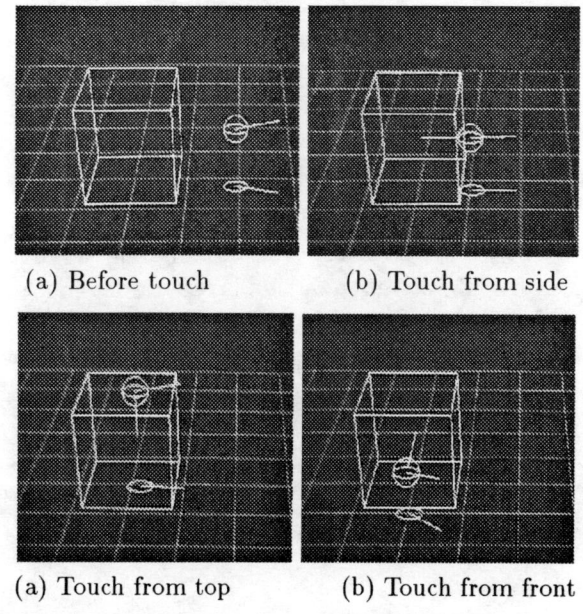

Figure 14: Experiment 4

of 1997 Int. Conf. on Robotics and Automation, pp.3018–3024, 1997.

[3] K. Hirota, J. Saito, and M. Hirose: "Force Display Depending on Mock Surface Method", Proceedings of the Annual Conference of the Robotics Society of Japan, pp.355–356, 1993.

[4] M. Inami, N. Kawakami, S. Hasegawa, and T. Hatakeda: "Design and Application of a Non-Contact Tracking Force Display "ARMS–III" ", Proc. of the 11th Symposium on Human Interface, pp.505–510, 1995.

[5] K. Hirota and M. Hirose: "A Curved Surface Presentation Device for Surface Display", Proc. Tenth Sympo. on Human Interface, 1994; 193–196 (in Japanes).

[6] H. Hoshino, R. Hirano, T. Maeda, and S. Tachi: "Study on Virtual Haptic Space (II) — VR System Displaying the Shape of an Object —", Proc. 33th Annual Conference of the Society of Instrument and Control Engineers in Japan, 1994, 691–692 (in Japanese).

Proceedings of the 1999 IEEE
International Conference on Robotics & Automation
Detroit, Michigan • May 1999

Numerically Efficient Trajectory Tracking Control of Polynomic Nonlinear Systems[+]

Raj Madhavan

Department of Mechanical and Mechatronic Engineering, The University of Sydney &
The Co-operative Research Centre for Mining Technology and Equipment (CMTE), Australia.
E-mail: raj@mech.eng.usyd.edu.au, Fax: (+61 2) 9351 3760.
WWW: http://mecharea.mech.eng.usyd.edu.au & http://www.cmte.org.au

Abstract— This paper proposes a numerically efficient tracking controller for polynomic nonlinear systems by a combined Volterra series and an extended Kalman filtering approach. The developed state estimation approach is applied for the state estimation of the states of a pendulum model and the tracking control of the pendulum is accomplished by utilizing the internal model principle and the feedback linearization method. It is demonstrated that this approach is more numerically efficient as compared to the standard extended Kalman filtering approach and thus also leads to a numerically efficient control scheme..

I. Introduction

The Kalman filter is an optimal estimator that can be characterized as an algorithm for computing the conditional mean and covariance of the probability distribution of the state of a linear dynamic finite-dimensional stochastic system with uncorrelated Gaussian process and measurement noise [1], [2]. The extended Kalman filtering extends the principles for the design of a linear Kalman filter to a nonlinear problem. The main computational burden in the standard hybrid EKF (hereafter referred to as the EKF1 approach) algorithm is in the state estimate and error covariance propagations. The error covariance propagation involves computing the Jacobian, using current estimates at each time instant. Also the state estimate propagation (for example via Euler integration) has to be done n times (where n being the number of propagations in each sampling interval). As the propagations are done every ΔT_s seconds, the number of propagations are much more than what would have been in the case of propagating every T_s seconds. Also the numerical complexity and thus the computational time is increased by a significant factor. To overcome this disadvantage, the state estimate extrapolation and the error covariance extrapolation are desired to be done every T_s seconds. If a closed-form input-output relation of the model at discrete time instants is obtained, then this is possible instead of propagating every ΔT_s seconds. Two extended Kalman filters, EKF2 and EKF3, are developed for this purpose by utilizing the measurements from three sensors - an encoder, a rate-gyro and a tilt-sensor, that allow five possible combinations, for the estimation of the states of a pendulum model. The two types of extended Kalman filters are then compared with a linear Kalman filter (LKF) in order to quantify the improvement in estimation accuracy at higher swing-angles. Finally, it will be shown how the estimates that are obtained using the various estimation schemes discussed, can be employed in an effective way to achieve desired robotic tracking control. In particular, it will be shown that the discrete-estimator based controller will not require any hardware modifications nor significant increase in computational power.

The paper is structured as follows: In section II, a brief description of the pendulum model citing reasons for nonlinear state estimation is included. In section III, the Zero-Order-Hold (ZOH) equivalence of polynomic nonlinear systems and the extended generalized Wiener model structure are introduced and it is shown how the pendulum can be modeled to fall within this structure. State estimation results via the proposed approaches are detailed here. Tracking control of the pendulum is then effected by using the internal model principle and the feedback linearization technique where the estimated states obtained by the proposed methods are employed (section IV) and section V provides the conclusions.

II. System Description

A. Pendulum Model

Consider the pendulum shown in Figure 1 which rotates about a pivot at one end where ℓ indicates the length of the massless rod and m the mass of the bob. Let θ denote the angle subtended by the rod and the vertical axis through the pivot point. The pendulum is considered in the down position (corresponding to the equilibrium position). Using Newton's second law of motion, the equation of motion of the pendulum can be written as

$$J\ddot{\theta}(t) + K_1\dot{\theta}(t) + K_0 \sin\theta(t) = G_0\tau(t) + \omega_0(t) \quad (1)$$

where $K_0 = mg\ell$. When nonlinearities are not severe, local linearization may be employed to arrive at linear models which are approximations of the nonlinear equations in the neighborhood of an operating point. Unfortunately, this pendulum model (like most problems in robotics) is not well suited to this approach since the effect of nonlinearities becomes severe at larger swing-angles, and consequently the $\sin\theta$ term must be included in the model description.

[+]This research was done when the author was a M.Eng. (Research) student with the Dept. of Engg., The Aust. Natl. Univ., Canberra, Australia. The author received financial assistance in the form of a Dept. of Engg. tuition fee waiver and also an ANU ME (R) scholarship and wishes to thank Prof. D. Williamson for his suggestions.

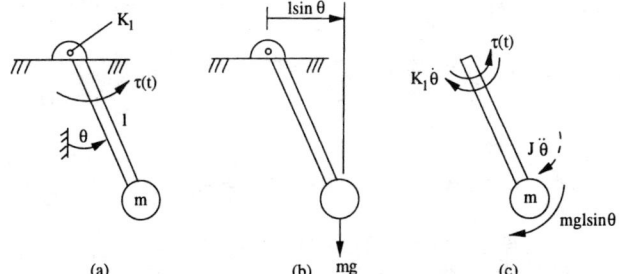

Fig. 1. (a) Pendulum model (b) Partial diagram to determine the torque produced by the weight (c) Free-body diagram showing all torques about the pivot point

B. Input-output Representation via Volterra Series

Many physical systems exhibit a nonlinear behavior that requires a more accurate analysis than that afforded by the linearization of the system model. The Volterra series provides a powerful way of portraying the input-output relationship of a nonlinear model. Functional representations of single-valued nonlinear functions were first introduced by Volterra and they provide a generalized non-parametric method of expressing the response of a nonlinear system. A Volterra model can describe nonlinear behavior such as asymmetric output changes in response to symmetric changes in the input. A controller based on this model can yield improved performance over a linear model-based controller [3]. Using the power series expansion of $\sin\theta$, (1) becomes,

$$\ddot{\theta}(t) + \left(\frac{K_1}{J}\right)\dot{\theta}(t) + \left(\frac{K_0}{J}\right)\left(\theta(t) - \frac{\theta^3(t)}{6} + \frac{\theta^5(t)}{120}\right)$$
$$= \left(\frac{G_0}{J}\right)u(t) + \left(\frac{1}{J}\right)\omega_0(t)$$

where $u(t) \equiv \tau(t)$ is used for notational simplicity. Considering (1) and writing a Volterra series expansion [4],

$$\theta(t) = \theta_1(t) + \theta_2(t) + \cdots = \sum_{i=1}^{\infty} \theta_i(t) \quad (2)$$

where $\theta_1(t)$ is defined as the output of the linear system, $\theta_2(t)$ as the output of the second-order system, and so forth. Equating terms of the same order, we find that all higher order even terms are equal to zero. Thus the closed form expressions for the linear and the non-linear (cubic) part with the additive measurement noise up to the third order are:

$$\ddot{\theta}_1(t) = -\left(\frac{K_0}{J}\right)\theta_1(t) - \left(\frac{K_1}{J}\right)\dot{\theta}_1(t) + \left(\frac{G_0}{J}\right)u(t)$$
$$+ \left(\frac{1}{J}\right)\omega_0(t)$$
$$\ddot{\theta}_3(t) = -\left(\frac{K_0}{J}\right)\theta_3(t) - \left(\frac{K_1}{J}\right)\dot{\theta}_3(t) + \left(\frac{K_0}{3!J}\right)\theta_1^3(t)$$
$$\theta(t) \approx \theta_1(t) + \theta_3(t)$$
$$\dot{\theta}(t) \approx \dot{\theta}_1(t) + \dot{\theta}_3(t)$$

III. Extended Kalman Filtering of Polynomic Nonlinear Systems

A. ZOH Equivalence

Consider a linear time-invariant system with the forcing term $u(t)$

$$\dot{\mathbf{x}}(t) = \mathbf{A}\mathbf{x}(t) + \mathbf{b}u(t); \quad \dot{u}(t) = 0;$$

followed by a polynomial nonlinearity of the type as shown in Figure 2, where $\mathbf{x}(t) \in \mathcal{R}^n$, $\mathbf{y}(t) \in \mathcal{R}^m$, $u(t) \in \mathcal{R}^\ell$ and

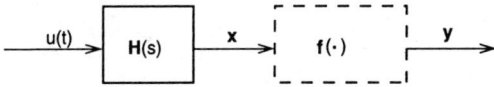

Fig. 2. Non-homogeneous polynomic nonlinear system

$H(s) = \mathbf{C}(s\mathbf{I} - \mathbf{A})^{-1}\mathbf{b}$ and $\mathbf{C}$ is the output matrix. The continuous input $u(t)$ is given in terms of the discrete input $u(kT_s)$ by

$$u(t) = u(kT_s); \quad kT_s \leq t < (k+1)T_s; \quad (3)$$

where $u(t)$ is switched at discrete time instants kT_s. Assuming $\mathbf{A}^{-1}$ exists, the discrete input-output relationship is given by [5]

$$\begin{aligned}
\mathbf{y}(k) &= \mathbf{C}^T\mathbf{x}(k) + \mathbf{C}^{[p]T}\tilde{\mathbf{x}}^{[p]}(k) \text{ where} \quad (4)\\
\mathbf{x}(k+1) &= e^{\mathbf{A}T_s}\mathbf{x}(k) + u(k)\mathbf{A}^{-1}\left[e^{\mathbf{A}T_s} - \mathbf{I}_n\right]\mathbf{b}\\
\tilde{\mathbf{x}}^{[p]}(k+1) &= e^{\tilde{\mathbf{A}}_{[p]}T_s}\tilde{\mathbf{x}}^{[p]}(k)
\end{aligned}$$

where $\mathbf{I}_n$ is the identity matrix of order n and T_s is the sampling interval.

B. Extended Generalized Wiener Model

The analysis and estimation of a large class of nonlinear systems that arise in control and communication networks can be represented as the inter-connection of linear subsystems and polynomial type nonlinearities and is dependent on a mathematical description between the system input and the output. Such systems fall under what is called the *extended generalized Wiener model* structure [6]. These systems constitute a class of nonlinear systems admitting finite Volterra series representation. In estimating the states of such nonlinear systems, the tensor techniques along with the Volterra series and the discrete extended Kalman filtering serve as a powerful tool. These systems can be modeled from the input-output point of view as a combination of a linear system $H(s)$ followed by a polynomial-type nonlinearity and another linear system $G(s)$. A case where the nonlinearity is a cubic law is shown in Figure 3 which has the following state-space representation:

$$\begin{aligned}
\dot{\mathbf{x}_1}(t) &= \mathbf{A_1}\mathbf{x_1}(t) + \mathbf{B_1}\mathbf{u}(t); \quad \sigma(t) = \mathbf{C_1}^T\mathbf{x_1}(t); \quad (5)\\
\gamma(t) &= \mathbf{d_3}\sigma^{[3]}(t)\\
\dot{\mathbf{x}_2}(t) &= \mathbf{A_2}\mathbf{x_2}(t) + \mathbf{B_2}\gamma(t) \quad (6)\\
\mathbf{y}(t) &= \mathbf{C_1}^T\mathbf{x_1}(t) + \mathbf{C_2}^T\mathbf{x_2}(t)
\end{aligned}$$

where $\mathbf{x_1}(t) \in \mathcal{R}^n$, $\mathbf{x_2}(t) \in \mathcal{R}^m$, $\mathbf{u}(t) \in \mathcal{R}^\ell$, $\gamma(t) \in \mathcal{R}^r$, $\sigma(t) \in \mathcal{R}^q$, $\mathbf{y}(t) \in \mathcal{R}^k$, $H(s) = \mathbf{C_1}(s\mathbf{I} - \mathbf{A_1})^{-1}\mathbf{B_1}$ and $G(s) = \mathbf{C_2}(s\mathbf{I} - \mathbf{A_2})^{-1}\mathbf{B_2}$ and the matrices have appropriate dimension and without loss of generality constitute minimal realizations of $H(s)$ and $G(s)$. The term $\sigma^{[3]}$ represents a vector homogeneous form. There exists a non-

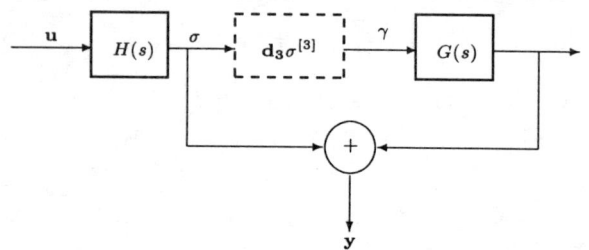

Fig. 3. Extended generalized Wiener model representation

linear difference equation from $\mathbf{y} \to \mathbf{u}$ in terms of $\mathbf{u}(k)$ where $\mathbf{u}(k)$ is a PAM signal as in (3). This equation will then simplify the state and error covariance extrapolations in the discrete extended Kalman filtering (EKF2 and EKF3) approaches. For some $m \times N$ matrix $\mathbf{M}(T_s)$ and assuming $\mathbf{A_1^{-1}}$ exists [5],

$$\mathbf{x_2}(k+1) = e^{\mathbf{A_2}T_s}\mathbf{x_2}(k) + \mathbf{M}(T_s)\hat{\mathbf{x}}_1^{[3]}(k)$$
$$\mathbf{x_1}(k+1) = e^{\mathbf{A_1}T_s}\mathbf{x_1}(k) + \mathbf{u}(k)\mathbf{A_1}^{-1}\left[e^{\mathbf{A_1}T_s} - \mathbf{I}_n\right]\mathbf{B_1}$$
$$\mathbf{y}(k) = \mathbf{C_1}^T\mathbf{x_1}(k) + \mathbf{C_2}^T\mathbf{x_2}(k)$$

define the behavior of (5) and (6) at the time instants $t = kT_s$.

Application of EKF2 and EKF3 to the Pendulum Model - Define

$$\mathbf{x_1} = \begin{pmatrix} \theta_1 \\ \dot{\theta}_1 \end{pmatrix} = \begin{pmatrix} x_{11} \\ x_{12} \end{pmatrix}; \quad \mathbf{x_2} = \begin{pmatrix} \theta_3 \\ \dot{\theta}_3 \end{pmatrix} = \begin{pmatrix} x_{21} \\ x_{22} \end{pmatrix};$$

Then in (5), (6)

$$\mathbf{A_1} = \mathbf{A_2} = \begin{pmatrix} 0 & 1 \\ -\frac{K_0}{J} & -\frac{K_1}{J} \end{pmatrix}; \quad (7)$$

$$\mathbf{B_1} = \begin{pmatrix} 0 & \frac{G_0}{J} \end{pmatrix}^T; \quad \mathbf{B_2} = \begin{pmatrix} 0 & \frac{K_0}{6J} \end{pmatrix}^T;$$

$$\mathbf{h}^{[3]T} = \begin{pmatrix} 1 & 0 & 0 & 0 & 0 & 0 & 0 & 0 & 0 & 0 \end{pmatrix};$$

$$\mathbf{C_1}^T = \mathbf{C_2}^T = \begin{pmatrix} 1 & 1 \end{pmatrix};$$

and thus $\mathbf{B_2}\mathbf{h}^{[3]T}$ is of order 2×10. The $\mathbf{M}(T_s)$ matrix is dependent on the sampling time, T_s (the argument T_s is used to indicate the explicit dependence of $\mathbf{M}(T_s)$ on T_s). Components of $\mathbf{M}(T_s)$ which are small relative to others can be neglected in order to reduce the number of nonlinear terms. Considering only the six elements $\{m_{1,1}, m_{1,2}, m_{1,3}, m_{2,1}, m_{2,2}, m_{2,3}\}$ of the $\mathbf{M}(T_s)$ matrix instead of the twenty elements and setting the value of the other elements to zero, the $\mathbf{M_a}$ matrix is obtained where the additional subscript a indicates the approximated matrix such that

$$\mathbf{M_a}(T_s)\hat{\mathbf{x}}_1^{[3]} \approx \begin{pmatrix} m_{1,1} & m_{2,1} \\ m_{1,2} & m_{2,2} \\ m_{1,3} & m_{2,3} \end{pmatrix}^T \begin{pmatrix} x_{11}^3(k) \\ \sqrt{3}x_{11}^2(k)x_{12}(k) \\ \sqrt{3}x_{11}^2(k)u(k) \end{pmatrix}$$

When the $\mathbf{M_a}$ matrix is employed for the estimation, this type of discrete extended Kalman filtering will be termed as the EKF3 approach.

C. Results

The setup that is used for the experiments exists in the automated systems laboratory. The experimental data is collected using the real-time *Vx-Works* operating system. Compiled programs are downloaded to a Motorola 68040 microprocessor which is located on a VME board. Three sensors are utilized for providing the measurements of the pendulum from which the states of the pendulum can be estimated. In addition to the three cases pertaining to the measurements provided by the individual sensors, two additional cases with the combination of encoder and rate-gyro and tilt-sensor and rate-gyro are also considered. The states that are to be estimated are θ and $\dot{\theta}$ (a third state is introduced when the tilt-sensor is included) [7]. The estimation error variances for all the sensor combinations are shown in Table I.

Performance of the Extended Kalman Filters - In estimating the time required by matrix computations, it is traditional to estimate the number of multiplications. It gives a rough idea of the relative complexities of alternative algorithms. The complexities will be functions of the problem size, which can be represented by the dimensions of the matrices involved. EKF2 and EKF3 perform as well as the EKF1 but the main difference is the computational speed as shown in Table II.

TABLE II
COMPARISON OF THE DIFFERENT FILTERS

Filter	No. of Muls.
EKF1	906
EKF2	192
EKF3	152

IV. Tracking Control of the Pendulum

A. The Internal Model Principle

The problem of controlling a fixed plant in order to have its output track a reference signal is one of the most important problems in control theory [8]. The work of [9] has shown that, in the case of error feedback, any regulator which solves the problem in question, incorporates a model of the dynamical system generating the reference signal which must be tracked. This property is commonly known as the *internal model principle*. Consider a continuous-time plant governed by

$$\dot{\mathbf{x}}_\mathbf{p}(t) = \mathbf{A_p}\mathbf{x_p}(t) + \mathbf{b_p}u_p(t) + \omega_p(t) \quad (8)$$
$$y(t) = \mathbf{C_p}\mathbf{x_p}(t)$$

where $\mathbf{x_p}(t) \in \mathcal{R}^{n_p}$ is the state vector, $u_p(t) \in \mathcal{R}^{m_p}$ denotes the input, $y(t) \in \mathcal{R}^{\ell_p}$ is the output which is required

TABLE I
ESTIMATION ERROR VARIANCES IN THE EKF1, EKF2 AND EKF3 CASES

Sensor	EKF1	EKF2	EKF3
Enc.	0.0219, 0.3287	0.0229, 0.3296	0.0218, 0.3267
RG	0.0220, 0.3115	0.0232, 0.3200	0.0219, 0.3113
Enc.+RG	0.0217, 0.3114	0.0220, 0.3120	0.0217, 0.3082
TS	0.0222, 0.3268, 0.0000	0.0232, 0.3298, 0.0000	0.0220, 0.3269, 0.0000
RG+TS	0.0220, 0.3114, 0.0000	0.0231, 0.3200, 0.0001	0.0220, 0.3113, 0.0000

to track a reference trajectory r and $\omega_p(t)$ represents the process noise. The internal model principle is used for achieving tracking which possibly involves the design of a pre-compensator. If the plant (8) does not contain all the modes of the reference signal, a pre-compensator is added which is represented by

$$\dot{x}_a(t) = A_a x_a(t) + b_a u(t); \quad u_p(t) = C_a x_a(t);$$

where $x_a(t) \in \mathcal{R}^{n_a}$ is the state vector, $u_p(t) \in \mathcal{R}^{m_a}$. The combined plant and pre-compensator can be represented as

$$\begin{pmatrix} \dot{x}_p(t) \\ \dot{x}_a(t) \end{pmatrix} = \begin{pmatrix} A_p & b_p C_a \\ 0 & A_a \end{pmatrix} \begin{pmatrix} x_p(t) \\ x_a(t) \end{pmatrix}$$
$$+ \begin{pmatrix} 0 \\ b_a \end{pmatrix} u(t) + \begin{pmatrix} \omega_p(t) \\ 0 \end{pmatrix} \quad (9)$$
$$y(t) = \begin{pmatrix} C_p & 0 \end{pmatrix} \begin{pmatrix} x_p(t) \\ x_a(t) \end{pmatrix}$$

This can be written more succinctly as

$$\dot{x}(t) = A x(t) + b u(t) + \omega(t); \quad y(t) = C x(t);$$

where the definitions of A, b, C and $\omega(t)$ follow from (9).

Control Objective - The desired objective is to make the output of the pendulum model follow a pre-specified reference trajectory in the presence of sensor (measurement) additive white noises in the system. A ramp reference trajectory is used as it generates a profile through which a robot link moves.

Reference Model - The reference signal r can be modeled as the response of

$$\dot{x}_r(t) = A x_r(t); \quad x_r(0) \in \mathcal{R}^n; \quad r(t) = C x_r(t); \quad (10)$$

The solution of (10) is

$$x_r(t) = e^{At} x_r(0); \quad r(t) = C e^{At} x_r(0); \quad (11)$$

subject to which the initial conditions are calculated.

B. Selection of Control Gains

For a closed-loop continuous-time linear system with system matrices A_c and b_c where A_c is $n \times n$ and a gain matrix K, the resulting closed-loop characteristic equation is given by

$$\det[sI - (A_c - b_c K)] = 0 \quad (12)$$

where det refers to the determinant. The control law design consists of picking the gains K so that the roots of (12) are in desirable locations. Let the desired locations be $s = s_1, s_2, s_3, \cdots$ and then the corresponding desired control characteristic equation is

$$\alpha_c(s) = (s - s_1)(s - s_2) \cdots (s - s_n) = 0 \quad (13)$$

Hence the required elements of K are obtained by matching coefficients in (12) and (13).

C. Feedback Linearization

Feedback linearization is a method of designing control for a class of feedback linearizable nonlinear systems. The basic idea is to first transform the nonlinear system (either completely or partially) into a linear system, and then use the well-known and powerful linear design techniques to complete the control design for the class of nonlinear systems. The main advantage of feedback linearization over point-wise linearization is that once such a control is found, linearization is achieved independently of the operating point.

C.1 Controller-Estimator Design via Feedback Linearization

Consider the pendulum model equation

$$J\ddot{\theta}(t) + K_1 \dot{\theta}(t) + K_0 \sin\theta(t) = u(t) + \omega(t)$$

where $u(t)$ refers to the control torque applied to the pendulum. Let

$$u(t) \equiv K_0 \sin\hat{\theta}(t) + V(t) \quad (14)$$

where $\hat{\theta}(t)$ refers to the estimate obtained by either the LKF or the EKF2 or the EKF3 approach and $V(t)$ to the control law to be proposed. Substituting (14) into the pendulum model equation,

$$J\ddot{\theta}(t) + K_1 \dot{\theta}(t) + K_0 \sin\theta(t) = K_0 \sin\hat{\theta}(t) + V(t) + \omega(t)$$
$$\text{where } V(t) = -k_1\left(\hat{\theta}(t) - \theta_r(t)\right) - k_2\left(\dot{\hat{\theta}}(t) - \dot{\theta}_r(t)\right).$$

$\theta_r(t)$ and $\dot{\theta}_r(t)$ are generated by the reference model based on the internal model principle which generates the necessary reference signals and k_1 and k_2 are the controller gains to be determined. Assuming that $\hat{\theta}(t) = \theta(t)$, the pendulum equation becomes

$$J\ddot{\theta}(t) + K_1 \dot{\theta}(t) = V(t) + \omega(t)$$

The state-space representation after feedback linearization is

$$\dot{x}(t) = A_f\, x(t) + b_f\, V(t) + d_f\, \omega(t) \tag{15}$$

$$A_f = \begin{pmatrix} 0 & 1 & 0 \\ 0 & \frac{-K_1}{J} & \frac{1}{J} \\ 0 & 0 & 0 \end{pmatrix};\; b_f = \begin{pmatrix} 0 \\ 0 \\ 1 \end{pmatrix};\; d_f = \begin{pmatrix} 0 \\ \frac{1}{J} \\ 0 \end{pmatrix};$$

A model for the constant ramp function is $y_r(t) = \alpha t$. The initial conditions of the ramp reference signal, $x_r(0) \in \mathcal{R}^3$ subject to (11) are

$$x_r(0) = \begin{pmatrix} 0 \\ -\alpha \\ -\alpha K_1 \end{pmatrix}$$

Control Gains - The characteristic control equation after feedback linearization is

$$det\,[sI - (A_f - b_f K_f)] = 0 \tag{16}$$

where the subscript f indicates that these matrices pertain to that after feedback linearization defined in (15) and K_f indicates the controller gain matrix and is such that $K_f = (k_1\; k_2\; k_3)$. (16) yields

$$s^3 + \left(\frac{K_1}{J}\right)s^2 + \left(\frac{k_2}{J}\right)s + \left(\frac{k_1}{J}\right) = 0;\; k_3 = 0;$$

from which the gains were calculated using the pole-placement technique.

D. Results

The structure of the discrete-estimator based controller structure is shown in Figure 4. $\omega(t)$ and $v(t_k)$ refer to the process noise and the sampled measurement noise respectively. The encoder is the sensor that is used for the measurement of the output (θ) of the pendulum system which is to be tracked. LKF, EKF2 and EKF3 are employed for providing the estimates for the tracking control. The corresponding error variances are tabulated which provides a basis for analyzing the errors associated with both the tracking and the estimation. The estimation error is also a means by which the effectiveness of the proposed feedback linearization can be studied. The estimation error is the error that has occurred due to the usage of $\hat{\theta}(t)$ instead of $\theta(t)$.

Reference Signals - Two reference ramp signals, $r_1(t)$ and $r_2(t)$ are employed for the tracking control. Both of these reference trajectories are periodic. i.e., $r_i(t+T_i) = r_i(t), i = 1, 2$ where T_i is the period of the corresponding trajectory.

$$r_1(t) = \begin{cases} \alpha t; & 0 \leq t \leq 20 \text{ sampling seconds} \\ -\alpha(t - 2T_1); & 20 < t < 60 \text{ sampling seconds} \end{cases}$$

$$r_2(t) = \begin{cases} \alpha t; & 0 \leq t \leq 40 \text{ sampling seconds} \\ -\alpha(t - 2T_2); & 40 < t < 120 \text{ sampling seconds} \end{cases}$$

The results are presented for both the reference signals. Note that $r_1(t)$ varies with time, twice as fast as $r_2(t)$. The error variances are calculated in the steady-state since the transient error variances are of little significance. For the reference signal $r_1(t)$, the error variances are calculated for the range [25 − 55] sampling seconds and for the reference signal $r_2(t)$, for [50 − 110] sampling seconds. The error plots for the tracking control with LKF for the reference signals, $r_1(t)$ and $r_2(t)$, are shown in Figures 5 & 6 and that with the nonlinear estimates are shown in Figures 7 through 10. The estimation and tracking error variances are tabulated in Table III. It is evident that the nonlinear estimates are to be preferred and the employment of the proposed approaches thus result in a numerically efficient control scheme.

TABLE III
ERROR VARIANCES FOR THE TRACKING CONTROL

Filter	Ref. Signal	EEV	TEV
LKF	r_1	0.0021	0.0027
	r_2	0.0016	0.0026
EKF2	r_1	0.000095997	0.00022774
	r_2	0.000075406	0.00015585
EKF3	r_1	0.00010773	0.00041290
	r_2	0.000099043	0.00023775

V. Conclusions

Two numerically efficient extended Kalman filtering approaches were proposed in this paper. These approaches were employed for the state estimation of a simple pendulum model and the estimated states were in turn employed for achieving the tracking control of the pendulum in a numerically efficient manner which thus demonstrates that these approaches perform as good as the standard hybrid approach but are computationally efficient and does not demand any additional hardware configurations.

References

[1] R.F. Stengel, *Stochastic Optimal Control: Theory and Application*, John Wiley & Sons Inc., 1986.
[2] A. Gelb, *Applied Optimal Estimation*, MIT Press, 1974.
[3] B.K. Maner et al., "Nonlinear Model Predictive Control of a Simulated Multivariable Polymerization Reactor using Second-order Volterra Models," *Automatica*, pp. 1285-1301, Vol 32 No. 9 1996.
[4] H.L. VanTrees, "Functional Techniques for the Analysis of the Nonlinear Behavior of Phase-Locked Loops," *IEEE Transactions on Communication*, pp. 542-556, Vol. Com-27, No.3 March 1979.
[5] R. Madhavan and D. Williamson, "State Estimation of Polynomic Nonlinear Systems," in *Proc. of the IEAust CONTROL97 Conf.*, 1997, pp. 598-603.
[6] J. Sandor and D. Williamson, "Identification and Analysis of Nonlinear Systems by Tensor Techniques," *International Journal of Control*, pp. 853-878, Vol. 27 No. 6 June 1978.
[7] R. Madhavan, "Control of a Pendulum using Extended Kalman Filtering," *M.E. (Research) Thesis, The Australian National University*, 1997.
[8] A. Isidori and C.I. Byrnes, "Output Regulation of Nonlinear Systems," *IEEE Transactions on Automatic Control*, pp. 131-140, Vol 35 No. 2 1990.
[9] B.A. Francis and W.M. Wonham, "The Internal Model Principle of Control Theory," *Automatica*, pp. 457-465, Vol.12 1976.

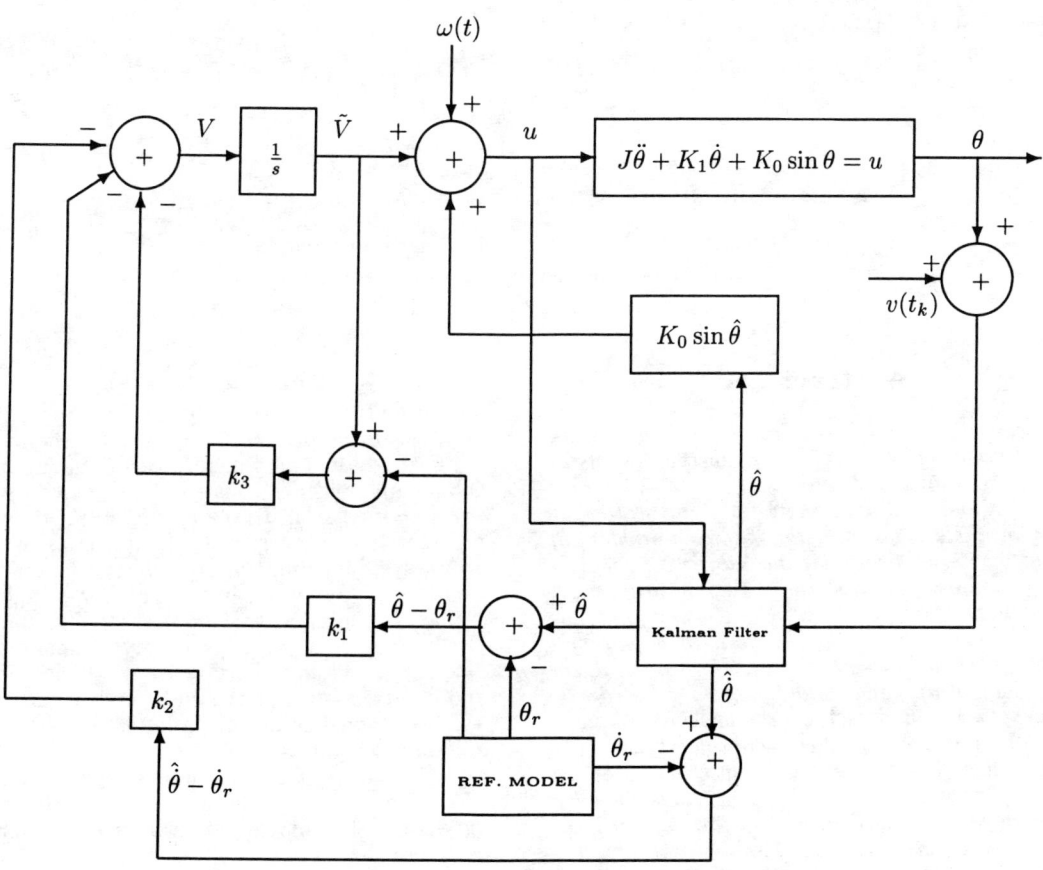

Fig. 4. Controller-estimator based tracking control structure

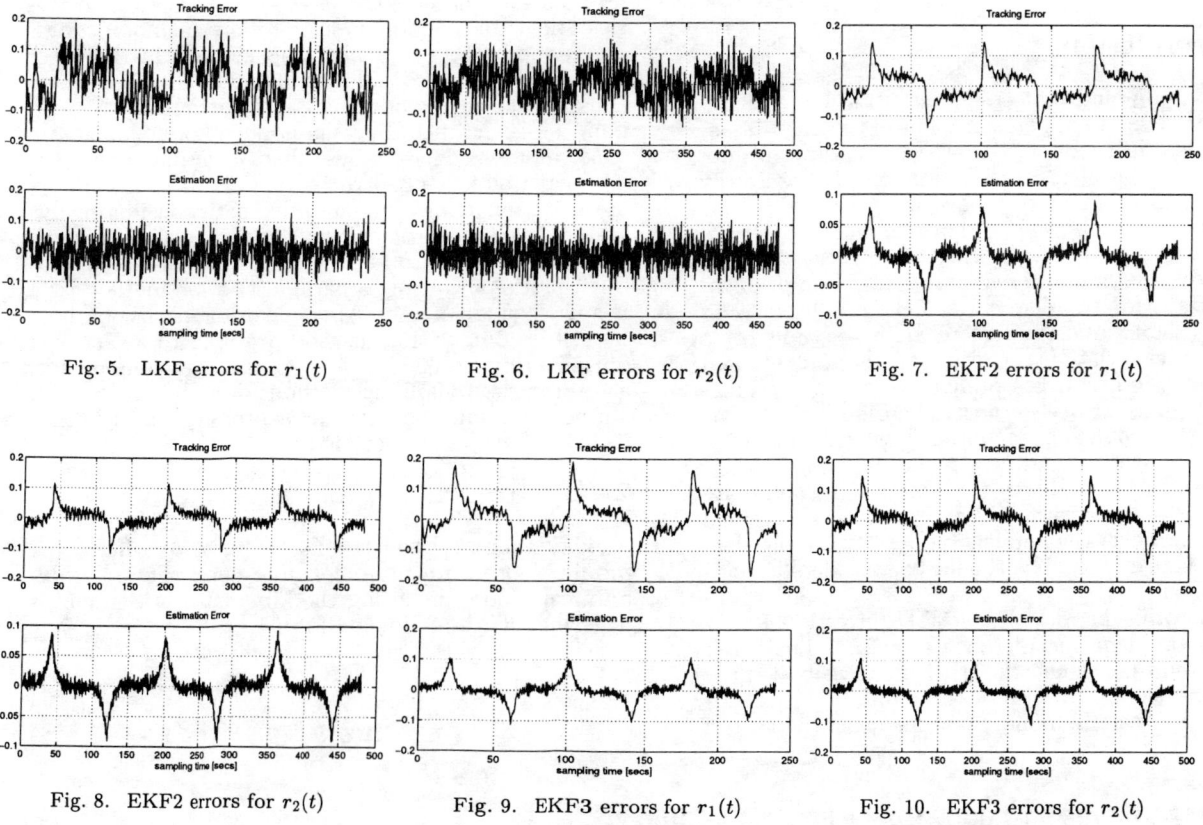

Fig. 5. LKF errors for $r_1(t)$

Fig. 6. LKF errors for $r_2(t)$

Fig. 7. EKF2 errors for $r_1(t)$

Fig. 8. EKF2 errors for $r_2(t)$

Fig. 9. EKF3 errors for $r_1(t)$

Fig. 10. EKF3 errors for $r_2(t)$

Quasi-Time-Optimal Motion Planning of Mobile Platforms in the Presence of Obstacles

Motoji Yamamoto, Makoto Iwamura, *and* Akira Mohri

Department of Intelligent Machinery and Systems
Faculty of Engineering, Kyushu University
Fukuoka 812-8581, JAPAN

Abstract

This paper addresses a problem of optimal motion planning of mobile platforms amidst obstacles considering mobile platform's dynamics. Due to nonholonomic constraints, actuator constraints, and state constraints by obstacle avoidance, the planning problem of mobile platform with two independently driven wheels is a complicated one. In this study, a dynamical model for the mobile platform is presented including nonholonomic kinematic constraints. An idea of a path parameter is introduced to simplify the planning problem considering dynamics and nonholonomic constraints. Using the path parameter, the optimal motion planning problem is divided into two sub-problems, i.e.,(1) time-optimization of trajectory along specified path, (2) search for optimal path. Then two methods to solve each problem are proposed, using an idea of path parameter and parametrization by B-spline function. Finally, quasi-time-optimal solution for the original problem are planned by combining the two methods. Numerical examples show effectiveness of the motion planner.

1 Introduction

Motion planning that design a path to connect initial position and final position is one of the most essential problem in robotics. So far, various types of motion planning problems such as time-optimal or energy consumption minimization have been studied. This paper treats the optimal motion planning problem of mobile platforms in the presence of obstacles.

Particularly time-optimal motion planning problem is important for the demand of high productivity in factories, therefore many researchers have studied this problem. Reister et al.[1] and Renaud et al.[2] studied the problem to find time-optimal motions of two independently driven wheels type mobile robots. They formulated the problem as an optimal control problem of a non-linear system, and proposed a planning method using the structure of bang-bang solution by Maximum Principle[3]. This method is attractive in theoretical meaning, however the application of practical problem where obstacle avoidance should be considered seems to be difficult.

On the other hand, Jiang et al.[4] investigated the planning problem considering obstacles. They first generate multiple geometrical path using visibility graph[5], then plan velocity profiles of the multiple paths. Final path is selected among the paths using the cost of traveling time. However, their solution is not optimal because of various assumptions and approximations.

Most of the proposed methods for the motion planning are based on kinematics of the mobile robots. They only consider constraints of velocities and acceleration of wheels. To plan more accurate optimal motion for realistic application, dynamics of the mobile robot system should be considered. As for the studies considering dynamical model, Shiller et al.[6] proposed a planning method based on a search technique. They use a point mass model as the dynamical model of mobile robot. The model may be insufficient for realistic mobile robot such as two independently driven wheels type and car-like mobile robots. Moreover, nonholonomic constrains, which is essential kinematic characteristics for most mobile robots, should be considered. However, consideration of dynamics including the nonholonomic kinematic constraints in motion planning problem is no so easy.

The current paper deals with time-optimal motion planning problem for two independently driven wheels type mobile platforms (robots) which are commonly used as automated guided vehicles (AGV) in factories and experimental mobile robots in laboratories. The paper proposes a planning method in the presence of obstacles considering dynamic and kinematic constraints. By introducing Lagrange multiplier, a dynamical model including the nonholonomic kinematic constraints is concisely described. The planning algorithm is divided into two parts. The first is velocity optimization along a pre-specified path considering dynamics and kinematic constraints using an idea of path parameter. The second is a global optimization of mobile robot's path using a parametrization method by a B-spline control points. Multiple initial paths obtained by cell decomposition method are used to avoid local optimization problem. The two parts are combined as a total motion planning method for mobile platforms. Numerical simulations are presented to confirm validity of the proposed method.

2 Dynamical model

The kinematic constraints for most of wheel type mobile robots are nonholonomic constraints. A dynamical model including the kinematic constraints is described as the following form[7];

$$M(q)\ddot{q} + c(q,\dot{q}) + g(q) = B(q)u + A^T(q)\lambda \quad (1)$$

$$A(q)\dot{q} = 0 \quad (2)$$

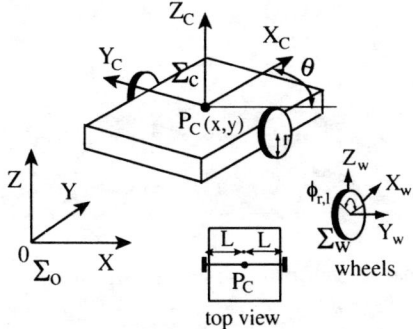

Fig.1 Geometry of mobile platform.

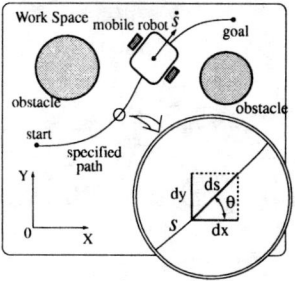

Fig.2 The mobile robot motion along a specified geometric path.

where, $q \in R^n$ is generalized coordinate vector, $M(q) \in R^{n \times n}$ is inertia moment matrix, $c(q, \dot{q}) \in R^n$ is centrifugal and Coriolis force vector, $g(q) \in R^n$ is gravitational force vector, $B(q) \in R^{n \times r}$ is input transformation matrix, $u \in R^n$ is control vector, $\lambda \in R^m$ is Lagrange multiplier vector, and $A(q) \in R^{m \times n}$ is constraints matrix with full rank.

In the followings, a dynamical model is derived for two independently driven wheel type mobile platform shown in Figure 1.

The platform is assumed to move on a flat floor and to have the center of gravity in the middle of the wheel axis (the point P_C in the figure). By assuming that the radius of the wheel r is a small value compared with the length of axle $2L$, we consider the inertia moment of the platform only around Z_c axis of the platform coordinate system Σ_c. In the followings, $\sin\theta \equiv S_\theta$, $\cos\theta \equiv C_\theta$, and x, y are coordinates of the reference point P_c in the world coordinate Σ_o, θ is posture of robot, ϕ_r, ϕ_l are angles of right and left wheels, m_1, m_2 are mass of platform and wheel, I_{z1} is inertia moment of the platform around Z_c-axis, and I_{z2} and I_y are inertia moments of the wheel around Z_w-axis and Y_w-axis in Σ_w coordinate system respectively.

2.1 Kinematics of mobile platform

Assuming no slip along longitudinal direction of wheel (pure rolling) yields the following kinematic relations for each wheel.

$$\dot{x}C_\theta + \dot{y}S_\theta + L\dot\theta = r\dot\phi_r \quad (3)$$

$$\dot{x}C_\theta + \dot{y}S_\theta - L\dot\theta = r\dot\phi_l \quad (4)$$

Assuming no slip along lateral direction of wheel (no side slip) yields

$$-\dot{x}S_\theta + \dot{y}C_\theta = 0 \quad (5)$$

By defining a generalized coordinates $q = [x, y, \theta, \phi_r, \phi_l]^T$, equations (3)~(5) are summarized in the form of (2), where A is

$$A(q) = \begin{bmatrix} C_\theta & S_\theta & L & -r & 0 \\ C_\theta & S_\theta & -L & 0 & -r \\ -S_\theta & C_\theta & 0 & 0 & 0 \end{bmatrix} \quad (6)$$

2.2 Dynamics of mobile platform

By introducing Lagrange multiplier $\lambda_1, \lambda_2, \lambda_3$ the dynamics of mobile platform satisfying the kinematic constraints (3)~(5) can be expressed by

$$m\ddot{x} = C_\theta \lambda_1 + C_\theta \lambda_2 - S_\theta \lambda_3 \quad (7)$$
$$m\ddot{y} = S_\theta \lambda_1 + S_\theta \lambda_2 + C_\theta \lambda_3 \quad (8)$$
$$I_z \ddot\theta = L\lambda_1 - L\lambda_2 \quad (9)$$
$$I_y \ddot\phi_r = u_1 - r\lambda_1 \quad (10)$$
$$I_y \ddot\phi_l = u_2 - r\lambda_2 \quad (11)$$

where $m = m_1 + 2m_2, I_z = I_{z1} + 2I_{z2} + 2m_2L^2$ and u_1, u_2 are both input torques for left and right wheels. Equations (7)~(11) are summarized in the form of (1) as;

$$M(q) = \begin{bmatrix} m & 0 & 0 & 0 & 0 \\ 0 & m & 0 & 0 & 0 \\ 0 & 0 & I_z & 0 & 0 \\ 0 & 0 & 0 & I_y & 0 \\ 0 & 0 & 0 & 0 & I_y \end{bmatrix}, \quad \begin{array}{l} c(q, \dot{q}) = 0, \\ \\ g(q) = 0, \end{array}$$

$$B(q) = \begin{bmatrix} 0 & 0 \\ 0 & 0 \\ 0 & 0 \\ 1 & 0 \\ 0 & 1 \end{bmatrix}, \quad u = \begin{bmatrix} u_1 \\ u_2 \end{bmatrix}, \quad \lambda = \begin{bmatrix} \lambda_1 \\ \lambda_2 \\ \lambda_3 \end{bmatrix} \quad (12)$$

3 Optimization along a specified path

This chapter describes a method to plan optimal trajectories for a specified mobile platform's path as shown in Figure 2, satisfying dynamical constraints. Using an idea of path parameter which is path length along the specified path, the optimization problem is simplified.

3.1 Dynamics described with path parameter

Firstly, eliminating $\lambda_1, \lambda_2, \lambda_3$ in (7)~(11) yields

$$\frac{mr}{2}C_\theta \ddot{x} + \frac{mr}{2}S_\theta \ddot{y} + \frac{I_z r}{2L}\ddot\theta + I_y \ddot\phi_r = u_1 \quad (13)$$

$$\frac{mr}{2}C_\theta\ddot{x} + \frac{mr}{2}S_\theta\ddot{y} - \frac{I_z r}{2L}\ddot{\theta} + I_y\ddot{\phi}_l = u_2 \quad (14)$$

Using (3) and (4), $\ddot{\phi}_r, \ddot{\phi}_l$ are eliminated as

$$m_d r C_\theta \ddot{x} + m_d r S_\theta \ddot{y} + I_d \ddot{\theta} - \frac{I_y}{r}S_\theta \dot{x}\dot{\theta} + \frac{I_y}{r}C_\theta \dot{y}\dot{\theta} = u_1 \quad (15)$$

$$m_d r C_\theta \ddot{x} + m_d r S_\theta \ddot{y} - I_d \ddot{\theta} - \frac{I_y}{r}S_\theta \dot{x}\dot{\theta} + \frac{I_y}{r}C_\theta \dot{y}\dot{\theta} = u_2 \quad (16)$$

where

$$m_d = \frac{I_y}{r^2} + \frac{m}{2}, \quad I_d = \frac{I_y L}{r} + \frac{I_z r}{2L}$$

The path of mobile platform is assumed to be described by a parameter s as;

$$x = x(s), \quad y = y(s) \quad (17)$$

where the parameter s, called "path parameter", is defined as length along a specified robot's path (see Fig.2). Then time derivatives of x, y, θ can be described using the parameter s as

$$\begin{cases} \dot{x} = x'\dot{s} = C_\theta \dot{s} \\ \dot{y} = y'\dot{s} = S_\theta \dot{s} \\ \dot{\theta} = \theta'\dot{s} = \kappa \dot{s} \\ \ddot{x} = x''\dot{s}^2 + x'\ddot{s} = -S_\theta \theta'\dot{s}^2 + C_\theta \ddot{s} \\ \ddot{y} = y''\dot{s}^2 + y'\ddot{s} = C_\theta \theta'\dot{s}^2 + S_\theta \ddot{s} \\ \ddot{\theta} = \theta''\dot{s}^2 + \theta'\ddot{s} = \eta \dot{s}^2 + \kappa \ddot{s} \end{cases} \quad (18)$$

where $(\cdot)'$ expresses derivatives on s and $\kappa \equiv \theta', \eta \equiv \theta''$. Substituting (18) into (15) and (16) yields

$$(m_d r + I_d \kappa)\ddot{s} + I_d \eta \dot{s}^2 = u_1 \quad (19)$$

$$(m_d r - I_d \kappa)\ddot{s} - I_d \eta \dot{s}^2 = u_2 \quad (20)$$

By summing both sides of (19) and (20), we get a following dynamic equation along the specified path.

$$m_d \ddot{s} = \frac{u_1 + u_2}{2r} \quad (21)$$

where $\dot{s}, \ddot{s}$ are velocity and acceleration along the path.

3.2 Constraints of driving torques

Due to a limit of actuator, the driving torques of both wheels are constrained by

$$u_{\min} \leq u_1, u_2 \leq u_{\max} \quad (22)$$

Using (19) and (20), the above inequality is rewritten as

$$u_{\min} - I_d \eta \dot{s}^2 \leq (m_d r + I_d \kappa)\ddot{s} \leq u_{\max} - I_d \eta \dot{s}^2 \quad (23)$$

$$u_{\min} + I_d \eta \dot{s}^2 \leq (m_d r - I_d \kappa)\ddot{s} \leq u_{\max} + I_d \eta \dot{s}^2 \quad (24)$$

The inequalities (23),(24) can be dscribed by the form of constraints for $\dot{s}$ as follows;

$$\dot{s}_{\min} \leq \dot{s} \leq \dot{s}_{\max} \quad (25)$$

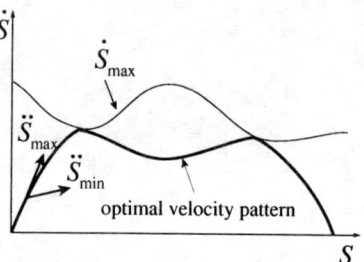

Fig.3 Optimal velocity pattern computed in $s - \dot{s}$ space.

The condition of the existence for $\ddot{s}$ in (23),(24) can be written in the form of the inequality on $\ddot{s}$,

$$\ddot{s}_{\min} \leq \ddot{s} \leq \ddot{s}_{\max} \quad (26)$$

3.3 Optimization of velocity pattern

In this chapter we have investigated how the dynamic equation and constraints are described using the path parameter s for the case of specifying spatial path by $(x(s), y(s))$. The constraints of (23) and (24) contain $\kappa = \theta'$ and $\eta = \theta''$. By noting that $\theta(s) \equiv \tan^{-1}\frac{dx(s)}{dy(s)}$, we find κ and η are described by the specified path which is a functions of the parameter s. As a result, the dynamic equation including nonholonomic (kinematic) constraints and dynamical constraints are described by the specified path $(x(s), y(s))$ and the path parameter s.

To minimize traveling time, velocity $\dot{s}$ along the specified path should be as fast as possible within constraints. The optimum velocity pattern is obtained using a technique by projecting the constraints (25) on $\dot{s}$ onto the phase plane $s - \dot{s}$ as shown in Fig.3. Shin et al.[8],[9] and Ozaki et al.[10] proposed such algorithms for manipulator planning problem. Basically the algorithm can be applied also for the mobile robot motion planning problem. In the following, we call the method MTTPSP (Minimum Time Trajectory Planning with Specified Path) algorithm. Detaisl of the algorithm is omitted. Using the MTTPSP algorithm, we can get exact optimum velocity pattern effectively. Once we get optimum velocity pattern $\hat{\dot{s}}$, then minimum traveling time for the specified path is calculated by

$$t_e = \int_0^{t_e} dt = \int_0^{s_e} \frac{dt}{ds} ds = \int_0^{s_e} \frac{1}{\dot{s}} ds \quad (27)$$

where t_e is minimum traveling time and s_e is total path length. (For details of a procedure to get optimum velocity pattern $\hat{\dot{s}}$, see [11]).

4 Optimization of spatial path

This chapter describes an optimization method by searching B-spline control points using traveling time as cost. The cost is calculated by MTTPSP algorithm. Generally it is difficult to get global optimum solution by local modify of paths. Thus, we first generate multiple initial paths based on global structure of free space. We use cell decomposition method to get the multiple initial paths. These initial paths are optimized by searching the B-spline control points. Consequently this method is one of the parameter optimization method.

4.1 Path by B-spline function

We represent a mobile robot spatial path $(x(s), y(s))$ using control points of B-spline function.

First, we divide the path length s into m sections $[s_0(=0), \cdots, s_k, s_{k+1}, \cdots, s_m(=s_e)]$. Then the mobile platform's path $\boldsymbol{P}(s) \equiv (x(s), y(s))^T$ by B-spline with order three at the section $s_k \leq s \leq s_{k+1}$ is described by control points $[\hat{x}_k, \hat{y}_k](k=0,\cdots,m)$ as

$$\boldsymbol{P}(s(l)) = \sum_{j=0}^{3} \boldsymbol{N}_j(l) \hat{\boldsymbol{Q}}_{k+j-1} \quad (28)$$

$$s(l) = s_k + l\Delta s_k, \ \ 0 \leq l \leq 1 \quad (29)$$

where $\boldsymbol{N}_j(l)$ is base function of B-spline and $\Delta s_k \equiv s_{k+1} - s_k$. We adopt B-spline function with order three considering the path's continuity on curvature.

Once a path $(x(s), y(s))$ is described as a set of B-spline control points, we can calculate minimum traveling time for the path by MTTPSP algorithm. The B-spline control points are modified based on the cost (traveling time). As a result, the original minimum time trajectory planning problem is reduced to a parameter optimization problem by the set of control points. The parameter optimization problem can be solved using various searching techniques[12].

4.2 Search of initial path

The parameter optimization problem formulated in the previous section has $2m$ parameters. The problem is easier to treat than the original problem. However, it is still not so easy to search global optimal solution due to the dimension and local minimum problem. We take a combined approach of global search for multiple initial solutions and their local optimization to tackle the difficulties.

The method consists of two steps. At first, several paths which does not collide with obstacles are used as initial solutions $[\hat{\boldsymbol{x}}_{int}, \hat{\boldsymbol{y}}_{int}]$, generated by cell decomposition method [5]. These initial paths are obtained based on global structure of free space. Then these paths are optimized locally using the B-spline control points starting with $[\hat{\boldsymbol{x}}_{int}, \hat{\boldsymbol{y}}_{int}]$. This section presents an outline of the method to generate multiple global initial feasible (collision-free) paths (see Fig.4).

Step 1. Approximate the robot by a cylinder which encloses it entirely, and extend the obstacles by its radius. As a result, the robot condenses to a point. Then, approximate collision regions of the workspace with polygons. (a)

Step 2. Connect nearest vertices of the polygons except one of it's own. This operation decomposes the free space of workspace into a set of convex polygons (cells). (b)

 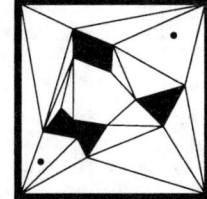

(a) Approximate collision regions with polygons.
(b) Cell decomposition by convex polygones (cells).

(c) Search all sequence of the cells and generate connectivity graph.
(d) Regenerate the polygonal lines by B-spline curve.

Fig.4 Searching of global initial feasible paths by exact cell decomposition method.

Table 1 Parameters of mobile platform.

L=0.75 m	m_1=50.0 kg	I_{z1}=26.04 kg·m^2
b=2.00 m	m_2=1.00 kg	I_{z2}=0.0025 kg·m^2
r=0.10 m		I_y=0.0050 kg·m^2

Step 3. Search all sequences of the cells from a start cell including the initial points to a goal cell including the final points. (c)

Step 4. For each sequence of cells obtained in Step 3, connect midpoints of the boundary of the adjacent cells of the sequence, then we get polygonal lines. Place the control points of B-spline on the polygonal line. Generate path using the B-spline curve. (d)

5 Numerical examples

This chapter shows numerical simulation results by applying the proposed planner. Kinematical parameter and dynamical parameters in this example are shown in Table 1. The constraints on driving torques of two wheels are given as

$$-1.0 \leq u_1, u_2 \leq 1.0 \quad (\text{N·m}) \quad (30)$$

Figure 5 shows an example of planning time-optimal trajectory of mobile robot form initial point (x, y, θ) = (3m, 3m, 0rad) to target point $(21m, 21m, \frac{\pi}{6}\text{rad})$ in 24m × 24m flat floor containing three obstacles. Figure (a) and (c) in Fig.5 are two initial paths obtained by cell decomposition method and B-spline smoothing. Figure (b) and (d) are optimized paths by parameter optimization method using B-spline function presented in section 4.1. In this example, we use a hill-climbing method[13] as a parameter optimization technique. In the figures, the obstacles are polygons daubed with black. The traveling time for each path is presented by T. We select the path of (d) as final quasi-optimum solution for this example. Driving torques of wheels are shown in Fig.6 for the initial path (c) and final path (d).

Another more complicated example is shown in Fig.7 which is including 5 obstacles. Initial point is (x, y, θ) = (3m, 3m, 0rad) and final point is $(33m, 33m, \frac{\pi}{4} rad)$ in a 36m × 36m flat floor. Four candidates of initial path are selected by cell decomposition method for this example. The four paths in Figure 7 are optimized ones by parameter optimization starting with each initial path. Final quasi-optimum path is (b) for this example. The total calculation time to get the final paths (Fig.7(b)) was about 10 minutes including the calculation of initial paths (Intel 266MHz Pentium, gcc compiler).

6 Conclusions

In this paper, a quasi-time-optimal trajectory planning method for mobile platform have been presented. The main points of the paper are summarized as follows.

1. A method using Lagrange multiplier to treat nonholonomic kinematic constraints in a dynamical equation is presented. Using the method, a dynamical model including kinematic constraints for two independently driven wheel type mobile platform is derived.

2. Using an idea of path parameter, the dynamical constraints are simplified for the case of specified path. Time-optimum trajectory along the specified path is easy to plan by optimization of scalar function in $s - \dot{s}$ plane.

3. To get multiple initial paths based on structure of free space, cell decomposition method is used for local minimum problem in parameter optimization step.

4. The original planning problem results in a parameter optimization method by B-spline representation of spatial path.

This paper focuses on time-optimal problem. Basically, MTTPSP algorithm can treat another kind of cost function than traveling time. The optimization method presented in this paper is based on the cost calculated by MTTPSP algorithm. Thus, general optimization problem such as minimum energy consumption problem will be also applicable using the same method proposed in this paper.

References

[1] David B.Reister, and Francois G.Pin : Time-Optimal Trajectories for Mobile Robots With Two Independently Driven Wheels, The International Journal of Robotics Research, Vol.13, No.1, 38-54, 1994.

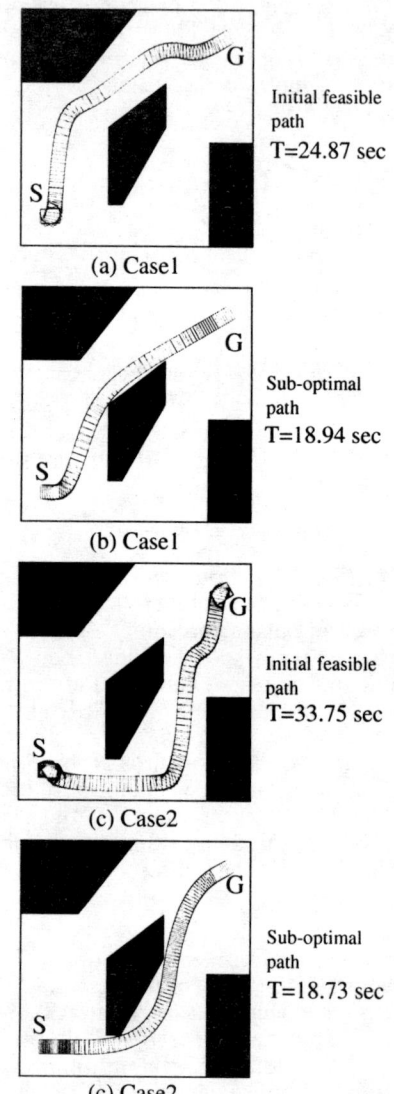

Fig.5 Simulation result (example 1).

[2] Marc Renaud, and Jean-Yves Fourquet : Minimum time motion of a mobile robot with two independent, acceleration-driven wheels, Proc. of 1997 IEEE ICRA, 2608-2613, 1997.

[3] A.E.Bryson and Y.-C.Ho : Applied Optimal Control, Talor and Francis, 1975.

[4] K.Jiang, L.D.Seneviratne and S.W.E.Earles : Time-optimal smooth-path motion planning for a mobile robot with kinematic constraints, Robotica, Vol.15, 547-553, 1997.

[5] J.C.Latombe : Robot Motion Planning, Kluwer Academic Publishers, 153-355, 1991.

[6] Z.Shiller and Y.Gwo : Dynamic Motion Planning of Autonomous Vehicles, IEEE Trans. on Robotics and Automation, Vol.7, No.2, 241-249, 1991.

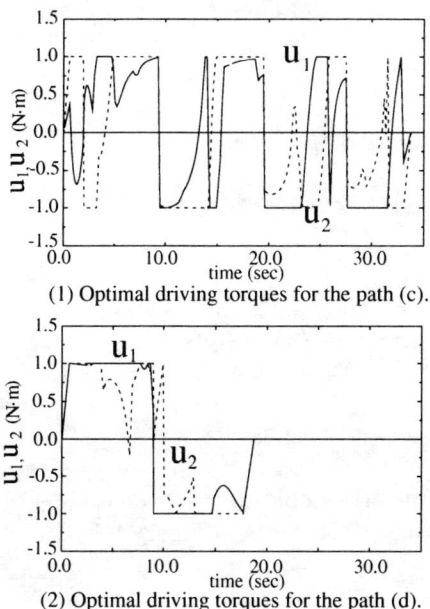

(1) Optimal driving torques for the path (c).

(2) Optimal driving torques for the path (d).

Fig.6 Torques and velocities of wheel (example 2).

[7] Herbert Goldstein : Classical Mechanics, Addison-Wesley Publishing Company, 1980.

[8] Kang. G. Shin, Neil D. Mckay : Minimum-Time Control of Robotic Manipulators with Geometric Path Constraints, IEEE Transaction of Automatic Control, Vol.AC-30, No.6, 531-541, 1985.

[9] Kang. G. Shin, Neil D. Mckay : A Dynamic Programming Approach to Trajectory Planning for Robotic Manipulators, IEEE Transaction of Automatic Control, Vol.AC-31, No.6, 491-500, 1986.

[10] H.Ozaki, M.Yamamoto, A.Mohri : Optimal and Near-Optimal Manipulator Joint Trajectories with a Pre-planned Path, Proc. of the 26th IEEE Conf. on Dec. and Contr., 1029-1034, 1987.

[11] M.Yamamoto, M.Iwamura, and A.Mohri : Time-Optimal Motion Planning of Skid-Steer Mobile Robots in the Presence of Obstacles, Proc. of the 1998 IEEE/RSJ Intl. Conf. on Intelligent Robots and Systems, 32-37, 1998.

[12] M.S.Bazaraa, H.D.Sherali, C.M.Shetty : Nonlinear Programming Theory and Algorithms, John Wiley and Sons, Inc., 1979.

[13] A.Mohri, M.Yamamoto, G.Hirano : A Trajectory Planning Algorithm with Path Search for Cooperative Multiple Manipulators, 1995 IEEE Int. Conf. on Systems, Man and Cybernetics, 898-903, 1995.

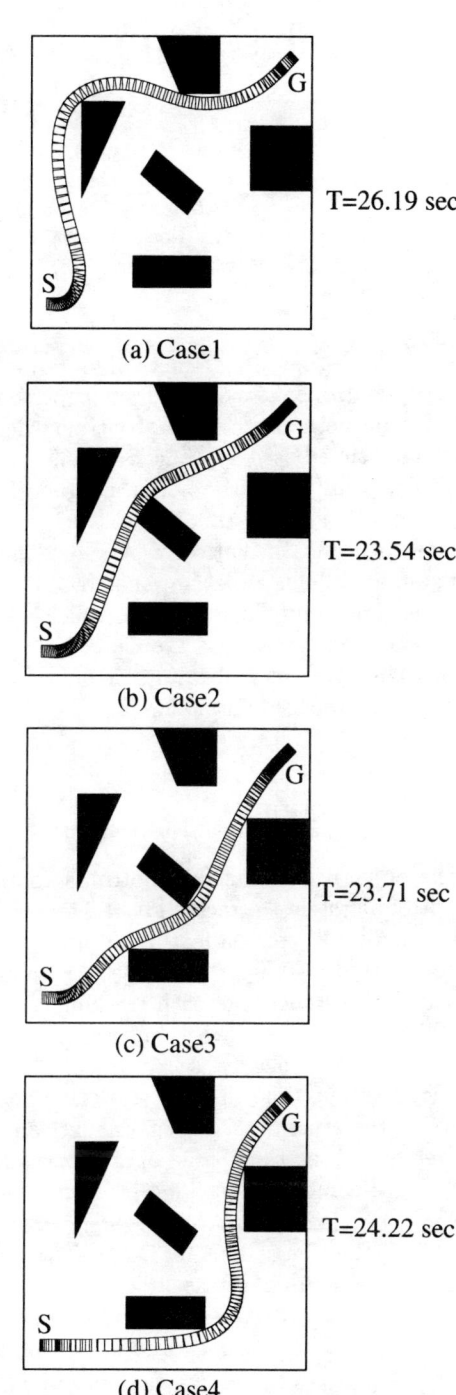

Fig.7 Simulation result (example 2).

Analysis and Design of Non-Time Based Motion Controller for Mobile Robots *

Wei Kang
Mathematics Department
Naval Postgraduate School
Monterey, CA 93943, U. S. A.

Ning Xi and Jindong Tan
Department of Electrical and Computer Eng.
Michigan State University
East Lansing, MI 48824, U.S.A.

Abstract A new design method for non-time based tracking controller of mobile robots is presented. The new design method converts a controller designed by traditional time-based approaches to a non-time based controller using any given action reference. The stability condition of non-time based tracking controller is developed and theoretically proved. The analysis and design methods are exemplified by a mobile robot tracking control problem. The controller has been implemented and tested in a Nomadic XR4000 mobile robot. The experimental results demonstrate the advantages of proposed method.

1 Introduction

The objective of tracking control is to ensure the output of a system to track a given reference input or desired path. A common feature of many path tracking feedback design is a state trajectory following approach. For instance, the path tracking controller for a mobile robot is required to ensure the convergence of the its states to desired states which is a prescribed function of time. The time plays a role of action reference in the system. While mathematically elegant, the well developed time-based approaches may not be the best for some path-tracking problems.

Non-time based controller has attracted researchers' attention of different fields. In [1], the event-based controller design was first introduced. Since then, it has been successfully applied to robot motion control [2], multi-robot coordination [3], force and impact control [4], robotic teleoperation [5] and manufacturing automation [6].

The basic idea of non-time reference is to introduce the concept of an action reference parameter which is directly relevant to the sensory measurement and the task. The event-based planning and control scheme and the traditional time-based planning and control scheme are compared in Figure 1.

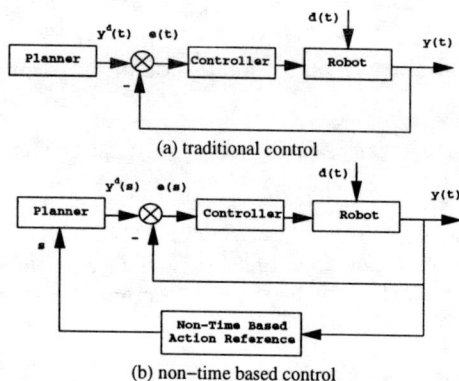

Figure 1: The comparison

In the event-based planning and control scheme, the function of the Action Reference block in Figure 1, (b) is to compute the action reference parameter s on-line, based on sensory measurements. The planner generates the desired value to the system, according to the on-line computed action reference parameter s. The action reference parameter is calculated near or at the same rate as the feedback control. In other words, the action plan is adjusted at a very high rate which enables the planner to handle unexpected or uncertain discrete and continuous events. For example, the event-based planning and control scheme has been successfully applied to deal with unexpected obstacle in a robot motion [2]without extra sensors which are necessary with traditional methods. Furthermore, once the obstacle is removed, the traditional methods require replanning to complete the task. But the event-based planning and control in this case does not require any extra force-torque or contact sensing, or replanning. The experimental results in [2] clearly demonstrate the feature.

---
*Research Partially supported under NSF Grant IIS-9796300 and IIS-9796287.

Since the system no longer uses the time as a reference, the traditional time-based controller cannot be directly applied. In this paper, we develop a new method which is different from all the non-time based approaches mentioned above. We focus on the problem of finding controllers for the tracking of a desired path described as a function of non-time based action reference. Instead of developing new design algorithms based on the non-time based action reference, we use the existing design methods for time-based controller. Then, the time variable in the feedback is substituted by a special transformation called state-to-reference(STR) projection. The resulting feedback becomes non-time based controller which also can drive the system asymptotically approaching the desired path. Our approach has the following four advantages. First, the model of the system is a general nonlinear dynamical control system. The method is applicable to a wide range of tracking control problems. Second, a transformation is provided to transform a time dependent controller into a non-time based controller. It bridges the traditional state trajectory tracking with the non-time based design. The results enable us to design non-time based feedback using the existing, well known methods of feedback design in time domain such as LQR and H_∞ control, and feedback linearization. The third advantage is the flexibility in choosing motion references. In our method, the motion reference can be any parameter. The fourth advantage is that the transformation from state space to the action reference is not necessarily the orthogonal projection. For many curves, orthogonal projection requires on-line numerical solution of minimization. Our method allows the designer to use any mapping satisfying a projection condition. A good choice for the projection mapping can dramatically reduce the on-line computational time.

2 Motion Reference Projection

Dynamic systems are modeled by differential equations in which the free variable is the time variable t. A desired trajectory is often modeled as a function of time. We denote it by $x_d(t)$, where x represents the state of the system. Controllers can be designed so that the trajectory of the plan system $x(t)$ asymptotically approaches the desired trajectory $x_d(t)$. This is a typical tracking control problem. There are a lot of control design methods in the literature of tracking control. Since the design is based on the model driven by t, the controller is time dependent. However, in the control of mobile robots, we often prefer a feedback controller which is independent of time. Furthermore, the desired path is not necessarily defined as a function of time. It is defined by parametric equations. The parameter is called *motion reference* or *action reference*, which is denoted by s. For instance, a desired path can be defined by a parametric equation $x_d(s)$, where s is the arc length, or s is the projection of x_d to a coordinate axis. In this section, we prove the feasibility of transforming a time dependent feedback controller into a time independent controller for a desired path driven by motion reference.

A system is defined by the equation

$$\dot{x} = f(x, u), \quad x \in {I\!\!R}^n, u \in {I\!\!R}^m, y = h(x), \quad y \in {I\!\!R}^k \tag{2.1}$$

where u is the control input, and x is the state of the system. Suppose that $y = h(x)$ represents the output of the system, the desired path is given by a parametric equation $y = y_d(s)$, where s is the motion reference.

The first step in the controller design is to define a corresponding path in time domain. If s is not time, then we assign an strictly increasing function

$$s = v(t).$$

How to pick the function $v(t)$ depends on the desired speed. For example, if the problem require that the system is operated so that s is increasing at a constant speed v_0, then $v(t) = v_0 t$.

The feedback satisfies

$$\lim_{t \to \infty} (h(x(t)) - y_d(v(t))) = 0.$$

Specifically, there exists an initial condition of the system x_0 such that $h(x(t)) = y_d(t)$. Denote this path by $x_d(s)$ or $x_d(v(t))$.

The third step is to find a suitable transformation, $s = \gamma(x)$, from the state space to the reference s. The transformation satisfies

$$\gamma(x_d(s)) = s \tag{2.2}$$

For example, given any state x_0, let $x_d(s_0)$ be the orthogonal projection from x_0 to $x_d(s)$. If we define $\gamma(x_0) = s_0$, then it satisfies (2.2). The transformation satisfying (2.2) is called a state-to-reference projection (STR projection).

The last step is to construct the feedback. Let

$$u(x) = \tilde{u}(x, v^{-1}(\gamma(x))) \tag{2.3}$$

where $\tilde{u}(x,t)$ is the feedback found in the second step, $\gamma(x)$ is a state-to-reference projection. The closed-loop system is

$$\dot{x} = f(x, u(x)). \tag{2.4}$$

In the following, we prove that the non-time based control law (2.4) drives the vehicle asymptotically approaching the desired path. Given any state x and any time t, the distance from x to the point $x_d(t)$ in the desired path is denoted by $d(x,t)$. So, $d(x,t) = \|x - x_d(t)\|$ is a function from $\mathbb{R}^n \times \mathbb{R}$ to $\mathbb{R}$. The open set $U(x_d, r)$ consists of $(x,t) \in \mathbb{R}^n \times \mathbb{R}$ such that $d(x,t)$ is less than r, i.e.

$$U(x_d, r) = \{(x,t) \in \mathbb{R}^n \times \mathbb{R} \mid d(x,t) < r\}.$$

The set $W(\delta)$ is the δ neighborhood of the curve x_d in the state space. In this paper, we assume that the vector field $F(x,u)$, the feedback $u(x,t)$, and their derivatives are bounded. Assume that the derivatives of $v(t)$ and $v^{-1}(s)$ are bounded. It can be seen that these assumptions are true for mobile robots.

Let $\tilde{\phi}(t, t_0, x_0)$ be the trajectory of the closed-loop system

$$\dot{x} = f(x, \tilde{u}(x,t)) \qquad (2.5)$$

with the initial time t_0 and initial condition $x(t_0) = x_0$. We assume that there exists a neighborhood $U(x_d, r)$ so that $(x_0, t_0) \in U(x_d, r)$ implies that $\tilde{\phi}(t, t_0, x_0)$ approaches x_d exponentially. Furthermore, the trajectory satisfies

$$me^{-\alpha_1(t-t_0)}d(x_0, t_0) \leq d(\tilde{\phi}(t, t_0, x_0), t)$$
$$d(\tilde{\phi}(t, t_0, x_0), t) \leq Me^{-\alpha_2(t-t_0)}d(x_0, t_0) \qquad (2.6)$$

for some $m > 0$, $M > 0$, $\alpha_1 > 0$ and $\alpha_2 > 0$. The following function is used as our Lyapunov function,

$$V(x_0, t_0) = \int_{t_0}^{\infty} d(\tilde{\phi}(t, t_0, x_0), t) dt. \qquad (2.7)$$

We assume that all derivatives of $V(x,t)$ of order less than or equal to three are bounded in a neighborhood $U(x_d, r)$. From (2.6), it is easy to show that

$$md(x,t) \leq V(x,t) \leq Md(x,t). \qquad (2.8)$$

Theorem 2.1 *There exists a neighborhood $W(\delta)$ of the desired trajectory $x_d(s)$ such that the trajectory $x(t)$ of (2.4) satisfies*

$$\lim_{t \to \infty} d(x(t), v^{-1}(\gamma(x(t)))) = 0. \qquad (2.9)$$

provided $x(0)$ is in $W(\delta)$. This implies that the trajectory of the closed-loop system with a non-time based feedback asymptotically approaches the desired path.

Proof. The trajectory of (2.5) with initial time t_0 and initial state $x(t_0) = x_0$ is denoted by $\tilde{\phi}(t, t_0, x_0)$. Similarly, a trajectory of (2.4) is denoted

by $\phi(t, t_0, x_0)$. By the definition of $V(x,t)$, it is easy to check that the derivative of $V(x,t)$ along $\tilde{\phi}(t, t_0, x_0)$, a trajectory of (2.5), satisfies

$$\frac{d}{dt}V(x(t), t) = -d(x(t), t).$$

Therefore,

$$\frac{\partial V(x,t)}{\partial x}f(x, \tilde{u}(x,t)) + \frac{\partial V(x,t)}{\partial t} = -d(x,t). \qquad (2.10)$$

The desired trajectory $x_d(v(t))$ is a solution to the equation (2.5). So, $\dot{x}_d(v(t)) = f(x_d, \tilde{u}(x_d, t))$. Since $s = \gamma(x_d(s))$, we have $t = v^{-1}(\gamma(x_d(v(t))))$. The derivative of this equation with respect to t yields

$$\frac{\partial v^{-1}}{\partial s}\frac{\partial \gamma}{\partial x}f(x_d, \tilde{u}(x_d, t)) = 1. \qquad (2.11)$$

Now, let's consider the closed-loop system (2.4) with the feedback (2.3). The derivative of $V(x, v^{-1}(\gamma(x)))$ in the direction of (2.4) is

$$\dot{V}(x, v^{-1}(\gamma(x)))$$
$$= \frac{\partial V}{\partial x}f(x, u(x)) + \frac{\partial V}{\partial t}\frac{\partial v^{-1}}{\partial s}\frac{\partial \gamma}{\partial x}f(x, u(x)).$$

From (2.3) and (2.10), we have

$$\dot{V}(x, v^{-1}(\gamma(x)))$$
$$= -d(x, v^{-1}(\gamma(x)))$$
$$+ (\frac{\partial v^{-1}}{\partial s}\frac{\partial \gamma}{\partial x}f(x, u(x)) - 1)\frac{\partial V}{\partial t}(x, v^{-1}(\gamma(x))). \qquad (2.12)$$

By the definition of $V(x,t)$, it is easy to check that $V(x,t) > 0$ for all x in a neighborhood of $x_d(t)$ and $V(x_d(v(t)), t) = 0$. Therefore, $V(x,t)$ has a minimum value at $x_d(t)$. It can be proved that

$$V(x,t) = (x - x_d(t))^T Q(x,t)(x - x_d(t))$$

for some positive definite matrix $Q(x,t)$. Therefore, $\frac{\partial V}{\partial t}(x,t) = P(x,t)(x - x_d(t))$ for some row vector $P(x,t)$. In a neighborhood $U(x_d, r)$, $P(x,t)$ is bounded. So,

$$\left|\frac{\partial V(x,t)}{\partial t}\right| \leq M_1 d(x,t) \qquad (2.13)$$

in a neighborhood $U(x_d, r)$ for some $M_1 > 0$. Meanwhile, (2.11) implies

$$\frac{\partial v^{-1}}{\partial s}\frac{\partial \gamma}{\partial x}f(x, u(x)) - 1 = 0$$

if x is a point on $x_d(s)$. Therefore,

$$\left|\frac{\partial v^{-1}}{\partial s}\frac{\partial \gamma}{\partial x}f(x,u(x))-1\right|<\frac{1}{2M_1} \quad (2.14)$$

in a neighborhood $W(\delta)$. From (2.12), (2.13) and (2.14), we have

$$\dot{V}(x,v^{-1}(\gamma(x)))<-\frac{d(x,v^{-1}(\gamma(x)))}{2}.$$

From (2.8),

$$\dot{V}(x,v^{-1}(\gamma(x)))<-\frac{1}{M}V(x,,v^{-1}(\gamma(x)))$$

in a neighborhood of x_d. By Gronwall's inequality, $V(x,v^{-1}(\gamma(x)))$ approaches zero exponentially along a trajectory $\phi(t,t_0,x_0)$ if x_0 is in a neighborhood of x_d. From (2.8),

$$md(x,t)\leq V(x,t).$$

Therefore, $d(x,v^{-1}(\gamma(x)))$ approaches zero exponentially along a trajectory $\phi(t,t_0,x_0)$ if x_0 is in a neighborhood, $W(\delta)$, of $x_d(s)$. ◁

3 Path tracking control for mobile robots

In this section, tracking control of a mobile robot using state feedback is discussed. Considering the mobile robot in Figure 2, the rear wheels are aligned with the vehicle while the front wheels are allowed to spin about the vertical axes. The constraints on the system arise by allowing the wheels to roll and spin, but not slip. Let (x,y,ϕ,θ) denote the configuration of the robot, parameterized by the location of the rear wheels. The dynamics model of the mobile robot can be represented as ([8])

$$\begin{aligned}\dot{x}&=u_1\cos\theta\\ \dot{y}&=u_1\sin\theta\\ \dot{\theta}&=\frac{u_1}{l}tan\phi\\ \dot{\phi}&=u_2,\end{aligned} \quad (3.1)$$

where u_1 corresponds to the forward velocity of the rear wheels of the car and u_2 corresponds to the angular velocity of the steering wheel, the angle of the car body with respect to the horizontal is θ, the steering angle with respect to the car body is ϕ, (x,y) is the location of the rear wheels, l is the length between the front and the rear wheels. In the following, we

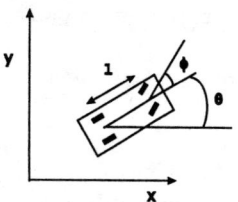

Figure 2: The configuration of a mobile robot.

will derive a time dependent feedback for an arbitrary path given by $y=f(x)$.

Suppose that the desired path is the following function of time

$$x=x_d(t),\quad y=y_d(t).$$

Let's first find a feedback for u_1. Since

$$\frac{d}{dt}(x-x_d)=u_1\cos\theta-\dot{x}_d,$$

we define

$$u_1=\frac{1}{\cos\theta}(\dot{x}_d-(x-x_d)). \quad (3.2)$$

Under (3.2), the first equation of (3.1) becomes

$$\frac{d}{dt}(x-x_d)=-(x-x_d).$$

Its solution asymptotically approaches zero as $t\to\infty$.

We defined the following change of coordinates to achieve linearization of the last three equations in (3.1),

$$\begin{aligned}z_1&=y,\quad z_2=\sin\theta,\quad z_3=\frac{1}{l}\cos\theta\tan\phi\\ \alpha_1&=-\frac{u_1}{l^2}\sin\theta(\tan\phi)^2,\quad \beta_1=\frac{\cos\theta}{l\cos^2\phi}\\ v&=\alpha_1+\beta_1u_2\end{aligned} \quad (3.3)$$

The last three equations in (3.1) are transformed into

$$\dot{z}_1=u_1z_2,\quad \dot{z}_2=u_1z_3,\quad \dot{z}_3=v \quad (3.4)$$

Define

$$e_1=z_1-y_d,\quad e_2=z_2-\frac{\dot{y}_d}{u_1},\quad e_3=z_3-\frac{1}{u_1}\frac{d}{dt}(\frac{\dot{y}_d}{u_1}) \quad (3.5)$$

Then,

$$\dot{e}_1=u_1e_2,\quad \dot{e}_2=u_1e_3,\quad \dot{e}_3=v-\frac{d}{dt}(\frac{1}{u_1}\frac{d}{dt}(\frac{\dot{y}_d}{u_1})) \quad (3.6)$$

The right side of the last equation in (3.6) satisfies

$$\frac{d}{dt}(\frac{1}{u_1}\frac{d}{dt}(\frac{\dot{y}_d}{u_1})) = \frac{\dddot{y}_d}{u_1^2} - \frac{3\ddot{y}_d\dot{u}_1}{u_1^3} + \frac{3\dot{y}_d\dot{u}_1^2}{u_1^4} - \frac{\dot{y}_d\ddot{u}_1}{u_1^3} \quad (3.7)$$

Notice that $\ddot{u}_1$ contains the term $\dot{\phi}$, which depends on u_2. Therefore, the expression (3.7) has the form

$$\alpha_2(x,y,\theta,\phi,t) + \beta_2(x,y,\theta,\phi,t)u_2 \quad (3.8)$$

To stabilize (3.6), we define

$$v = \alpha_2(x,y,\theta,\phi,t) + \beta_2(x,y,\theta,\phi,t)u_2 \\ + u_1(a_1e_1 + a_2e_2 + a_3e_3) \quad (3.9)$$

Then, (3.6) is equivalent to

$$\dot{e}_1 = u_1 e_2, \quad \dot{e}_2 = u_1 e_3, \quad \dot{e}_3 = u_1(a_1e_1 + a_2e_2 + a_3e_3) \quad (3.10)$$

Select the values of $a_i, i = 1, 2, 3$, so that all the eigenvalues of

$$A = \begin{bmatrix} 0 & 1 & 0 \\ 0 & 0 & 1 \\ a_1 & a_2 & a_3 \end{bmatrix}$$

are on the left half plane. Assume that

$$x_d(t) \to \infty \text{ and } -\frac{\pi}{2} < \theta(t) < \frac{\pi}{2}. \quad (3.11)$$

From (3.2), we can prove that $\int_0^t u_1(\tau)d\tau \to \infty$ as $t \to \infty$. The solution of (3.10) is

$$e(t) = e_0 \exp(A \int_0^t u_1(\tau)d\tau).$$

Since A is stable, we know $\lim_{t \to \infty} e(t) = 0$. Therefore, $y - y_d(t) \to 0$. To find a formula for u_2, we solve for u_2 from (3.9) and the last equation in (3.3). The feedback is

$$u_2 = \frac{\alpha_2 - \alpha_1 + u_1(a_1e_1 + a_2e_2 + a_3e_3)}{\beta_1 - \beta_2} \quad (3.12)$$

Theorem 3.1 *Under the feedback (3.2,3.12), $(x(t), y(t))$ of (3.1) asymptotically approaches a desired path $(x_d(t), y_d(t))$, provided that (3.11) is satisfied.*

Remark. A circle does not satisfy (3.11). Feedbacks to track circles can be found in [7]. ◁

The feedback (3.2,3.12) depends on time. However, it can be converted into a time-invariant feedback using state-to-reference projection as proved in § 2. In the following, we use the method developed in § 2 and the feedback (3.2,3.12) to design time-invariant feedbacks for the tracking of the curve $y = \sin x$. The key step is to find a motion reference for the task. The design method proposed in this paper is not based on a specific motion reference. It works for arbitrary choice of the reference. Different from circles and lines, arc length and orthogonal projection are not the best choice in this case for real time computation.

If the desired path is

$$x_d(v(t)) = t, \quad y_d(v(t)) = \sin t \quad (3.13)$$

the non-time based control u_1 and u_2 can be computed as following

$$u_1 = \frac{\dot{x}_d - (x - x_d)}{\cos \theta}, \\ u_2 = \frac{\alpha_2 - \alpha_1}{\beta_1 - \beta_2} + \frac{u_1}{\beta_1 - \beta_2}(a_1e_1 + a_2e_2 + a_3e_3) \quad (3.14)$$

The coefficients a_1, a_2 and a_3 are obtained from the linear control design.

4 Experimental results

The new non-time based controller has been implemented on a Nomadic XR4000 mobile robot. The mobile robot has four wheels. Each wheel can be individually steered. However, in order to implement the above control scheme, only two front wheels are steered in the experiment. The other two wheels maintain straight forward configuration. The distance between front and rear wheels is $0.3053\ m$. The velocity along a given path in the following experiments is always $0.1\ m/sec$. The given desired path is $y = 0.5 \sin(2x)$. In the controller the closed loop poles are assigned to the locations, $-3, -5 - i$ and $-5 + i$.

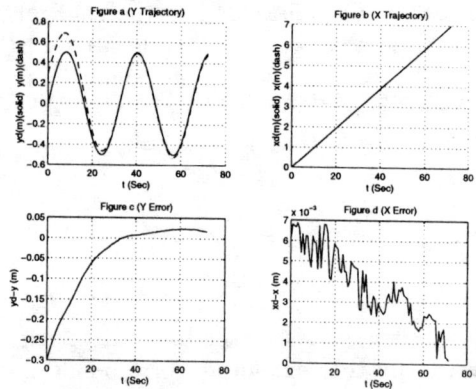

Figure 3: Tracking a sine curve with initial error.

If the initial position of the robot is not on the desired path as shown in Figure 3, the controller was able to command the robot approach the desired path. There existed $0.3m$ initial error in the experiment. But the tracking error converged to zero once the robot started moving. This experiment demonstrates the ability of the controller to rapidly overcome initial tracking error.

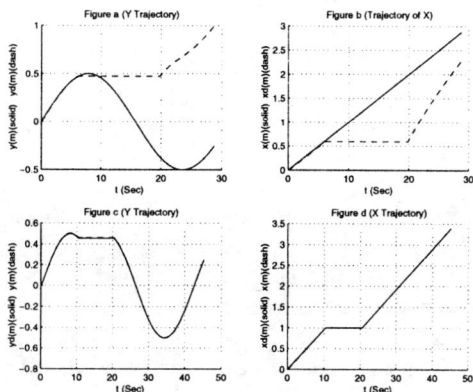

Figure 4: Time based control.

The interesting results of experiments are presented in Figure 4. While tracking a pre-planned desired path, the robot was stopped by an obstacle unexpectly at the $10th$ sec. The Figure 4a,4b presents a results of a time-based controller. Its desired input was a function of the time. Therefore the desired position would keep changing even the robot was blocked. As a result, the tracking error kept increasing until the system became unstable. However, for the non-time based controller, the results are completely different. As shown in Figure 4c and 4d, once the robot is blocked the desired position remained constant. So did the tracking errors. Furthermore, after the obstacle was removed at the $20th$ sec, the robot automatically resumed the original motion without re-planning. This experiment demonstrates the capability of the non-time based controller for coping with unexpected events. It should be noticed that the non-time based controller is not simply replacing the time by a non-time based motion reference. The new motion reference is determined by sensory measurements. Therefore it contains the information related to the states of the system.

5 Conclusion

A new design method for non-time based tracking controller has been developed. It can convert a controller designed by traditional time-based approach to a new non-time based controller. The method has been applied to design a tracking controller of a unmanned vehicle. The experimental results have clearly demonstrated the advantage of non-time based controller. More importantly, the results have provided a efficient and systematic approach to design non-time based tracking controller for a general nonlinear dynamic system which is an important for developing a theoretical foundation for sensor-referenced intelligent control.

References

[1] Ning Xi, "Event-Based Planning and Control for Robotic Systems", Doctoral Dissertation, Washington University, December, 1993.

[2] T.J. Tarn, A.K. Bejczy and Ning Xi, "Intelligent Planning and Control for Robot Arms," Proceedings of the IFAC 1993 World Congress, Sydney, Australia, July 18-23, 1993.

[3] Ning Xi, T.J. Tarn and A.K. Bejczy, "Event-Based Planning and Control for Multi-Robot Coordination," Proceedings of the 1993 IEEE /ICRA, Atlanta, Georgia, pp. 251-258, May 2-6, 1993.

[4] Y. Wu, T.J. Tarn, Ning Xi, "Force and Transition Control with Environmental Uncertainties", The Proceedings of the 1995 IEEE/ICRA , Nagoya, Japan, May 1995.

[5] T.J. Tarn, Ning Xi, C. Guo, A.K. Bejczy, "Fusion of Human and Machine Intelligence for Telerobotic Systems", The Proceedings of the 1995 IEEE /ICRA, Nagoya, Japan, May, 1995.

[6] Ning Xi, T.J. Tarn, "Integrated Task Scheduling and Action Planning/Control for Robotic Systems Based on a Max-Plus Algebra Model", the Proceedings of 1997 the Proceedings of IEEE/RSJ International Conference on Intelligent Robotics and Systems, Sept. 1997, Grenoble, France.

[7] R. M. DeSantis, *Path-tracking for a tractor-trailer-like robot*, The International Journal of Robotics Research, Vol. 13, No. 6, 533-544, 1994.

[8] R. M. Murry and S. S. Sastry, *Nonholonomic motion planning: steering using sinusoids*, IEEE Trans. Automat. Contr., Vol. 38, pp. 700-716, 1993.

Sensor Fusion Based on Fuzzy Kalman Filtering for Autonomous Robot Vehicle

J.Z. Sasiadek and Q. Wang

Dept. of Mechanical & Aerospace Engineering
Carleton University
1125 Colonel By Drive, Ottawa, Ontario, K1S 5B6, Canada

Abstract

This paper presents the method of sensor fusion based on the Adaptive Fuzzy Kalman Filtering. This method has been applied to fuse position signals from the Global Positioning System (GPS) and Inertial Navigation System (INS) for the autonomous mobile vehicles. The presented method has been validated in 3-D environment and is of particular importance for guidance, navigation, and control of flying vehicles. The Extended Kalman Filter (EKF) and the noise characteristic has been modified using the Fuzzy Logic Adaptive System and compared with the performance of regular EKF.

It has been demonstrated that the Fuzzy Adaptive Kalman Filter gives better results (more accurate) than the EKF.

Introduction

When navigating and guiding an autonomous vehicle, the position and velocity of the vehicle must be determined. The Global Positioning System (GPS) is a satellite-based navigation system that provides a user with the proper equipment access to useful and accurate positioning information anywhere on the globe [1]. However, several errors are associated with the GPS measurement. It has superior long-term error performance, but poor short-term accuracy. For many vehicle navigation systems, GPS is insufficient as a stand-alone position system. The integration of GPS and Inertial Navigation System (INS) is ideal for vehicle navigation systems. In generally, the short-term accuracy of INS is good; the long-term accuracy is poor. The disadvantages of GPS/INS are ideally cancelled. If the signal of GPS is interrupted, the INS enables the navigation system to coast along until GPS signal is reestablished [1]. The requirements for accuracy, availability and robustness are therefore achieved.

Kalman filtering is a form of optimal estimation characterized by recursive evaluation, and an internal model of the dynamics of the system being estimated. The dynamic weighting of incoming evidence with ongoing expectation produces estimates of the state of the observed system [2]. An extended Kalman filter (EKF) can be used to fuse measurements from GPS and INS. In this EKF, the INS data are used as a reference trajectory, and GPS data are applied to update and estimate the error states of this trajectory. The Kalman filter requires that all the plant dynamics and noise processes are exactly known and the noise processes are zero mean white noise. If the theoretical behavior of a filter and its actual behavior do not agree, divergence problems will occur. There are two kinds of divergence: Apparent divergence and True divergence [3][4]. In the apparent divergence, the actual estimate error covariance remains bounded, but it approaches a larger bound than does predicted error covariance. In true divergence, the actual estimation covariance eventually becomes infinite. The divergence due to modeling errors is critical in Kalman filter application. If, the Kalman filter is fed information that the process behaved one way, whereas, in fact, it behaves another way, the filter will try to continually fit a wrong process. When the measurement situation does not provide enough information to estimate all the state variables of the system, in other words, the computed estimation error matrix becomes unrealistically small, and the filter disregards the measurement, then the problem is particularly severe. Thus, in order to solve the divergence due to modeling errors, we can estimate unmodeled states, but it add complexity to the filter and one can never be sure that all of the suspected unstable states are indeed model states[3]. Another possibility is to add process noise. It makes sure that the Kalman filter is driven by white noise, and prevents the filter from disregarding new measurement. In this paper, a fuzzy logic adaptive system (FLAS) is used to prevent the Kalman filter from divergence. The fuzzy logic adaptive controller (FLAC) will continually adjust the noise strengths in the filter's internal model, and tune the filter as well as possible. The FLAC performance is evaluated by simulation of the fuzzy adaptive extended Kalman filtering scheme of Fig.1.

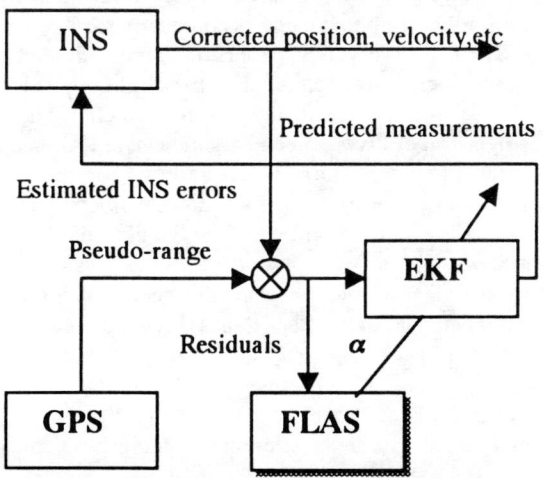

Fig.1. Fuzzy adaptive extended Kalman filter

Weighted EKF

Because the processes of both GPS and INS are nonlinear, a linearization is necessary. An extended Kalman filter is used to fuse the measurements from the GPS and INS. To prevent divergence by keeping the filter from discounting measurements for large k, the exponential data weighting [4] is used.

The models and implementation equations for the weighted extended Kalman filter are:

Nonlinear dynamic model

$$\mathbf{x}_{k+1} = f(\mathbf{x}_k, k) + \mathbf{w}_k \qquad (1)$$

$\mathbf{w}_k \sim N(0, \mathbf{Q})$;

Nonlinear measurement model

$$\mathbf{z}_k = h(\mathbf{x}_k, k) + \mathbf{v}_k \qquad (2)$$

$\mathbf{v}_k \sim N(0, \mathbf{R})$;

Let us set the model covariance matrices equal to

$$\mathbf{R}_k = \mathbf{R}\alpha^{-2(k+1)} \qquad (3)$$

$$\mathbf{Q}_k = \mathbf{Q}\alpha^{-2(k+1)} \qquad (4)$$

where, $\alpha \geq 1$, and constant matrices $\mathbf{Q}$ and $\mathbf{R}$. For $\alpha > 1$, as time k increases, the $\mathbf{R}$ and $\mathbf{Q}$ decrease, so that the most recent measurement is given higher weighting. If $\alpha = 1$, it is a regular EKF.

By defining the weighted covariance

$$\mathbf{P}_k^{\alpha-} = \mathbf{P}_k^{-} \alpha^{2k} \qquad (5)$$

The Kalman gain can be computed:

$$\mathbf{K}_k = \mathbf{P}_k^{-} \mathbf{H}_k^T (\mathbf{H}_k \mathbf{P}_k^{-} \mathbf{H}_k^T + \mathbf{R}\alpha^{-2(k+1)})^{-1}$$

$$= \mathbf{P}_k^{\alpha-} \mathbf{H}_k^T \left(\mathbf{H}_k \mathbf{P}_k^{\alpha-} \mathbf{H}_k^T + \frac{\mathbf{R}}{\alpha^2} \right)^{-1} \qquad (6)$$

The predicted state estimate is:

$$\hat{\mathbf{x}}_{k+1}^{-} = f(\hat{\mathbf{x}}_k, k) \qquad (7)$$

The predicted measurement is:

$$\hat{\mathbf{z}}_k = h(\hat{\mathbf{x}}_k^{-}, k) \qquad (8)$$

The linear approximation equations can be presented in form:

$$\Phi_k \approx \left. \frac{\partial f(x, k)}{\partial x} \right|_{x = \hat{x}_k^{-}} \qquad (9)$$

The predicted estimate on the measurement can be computed:

$$\hat{\mathbf{x}}_k = \hat{\mathbf{x}}_k^{-} + \mathbf{K}_k (\mathbf{z}_k - \hat{\mathbf{z}}_k) \qquad (10)$$

$$\mathbf{H}_k \approx \left. \frac{\partial h(x, k)}{\partial x} \right|_{x = x_k^{-}} \qquad (11)$$

Computing the *a priori* covariance matrix:

$$\mathbf{P}_{k+1}^{-} = \Phi_k \mathbf{P}_k \Phi_k^T + \mathbf{Q}\alpha^{-2(k+1)} \qquad (12)$$

Re-writing (12) gives:

$$P_{k+1}^{\alpha-} = \alpha^2 \Phi_k P_k^{\alpha} \Phi^T + Q \qquad (13)$$

Computing the *a posteriori* covariance matrix gives:

$$P_k^{\alpha} = (I - K_k H_k) P_k^{\alpha-} \qquad (14)$$

The initial condition is:

$$P_0^{\alpha-} = P_0$$

In equation (10), the term $z_k - \hat{z}_k$ is called residuals or innovations. It reflects the degree to which the model fits the data.

INS and GPS

The inertial navigation system (INS) consists of a sensor package, which includes accelerometers and gyros to measure accelerations and angular rates. By using these signals as input, the attitude angle and three-dimensional vectors of velocity and position are computed [5]. The errors in the measurements of force made by the accelerometers and the errors in the measurement of angular change in orientation with respect to inertial space made by gyroscopes are two fundamental error sources, which affect the error behavior of an inertial system. The inertial system error response, related to position, velocity, and orientation is divergent with time due to noise input [6]. There are biases associated with the accelerometers and gyros. In order to correct the errors of INS, the GPS measurements are used to estimate the inertial system errors, subtract them from the INS outputs, and then obtain the corrected INS outputs. There is number of errors in GPS, such as ephemeris errors, propagation errors, selective availability, multi-path, and receiver noise, etc. By using differential GPS (DGPS), most of the errors can be corrected, but the multi-path and receiver noise cannot be eliminated.

Fuzzy Logic Adaptive System

It is assumed that both, the process noise w_k, and the measurement noise v_k are zero-mean white sequences with known covariance Q and R in the Kalman filter. If the Kalman filter is based on a complete and perfectly tuned model, the residuals should be a zero-mean white noise process. If the residuals are not white noise, there is something wrong with the design and the filter is not performing optimally [4]. The Kalman filters will diverge or coverage to a large bound. In practice, it is difficult to know the exact value for Q and R. In order to reduce computation, we have to ignore some errors, but sometimes those unmodeled errors will become significant. These are the instrument bias errors of INS. Generally the w_k does not always have zero mean. In those cases, the residuals can be used to adapt the filter. In fact, the residuals are the differences between actual measurements and best measurement predictions based on the filter's internal model. A well-tuned filter is that where the 95% of the autocorrelation function of innovation series should fall within the $\pm 2\sigma$ boundary [7]. If the filter diverges, the residuals will not be zero mean and become larger.

There are very few papers on application of fuzzy logic to adapt the Kalman filter. In [8], fuzzy logic is used to the on-line detection and correction of divergence in a single state Kalman filter. There were three inputs and two outputs to fuzzy logic controller (FLC), and 24 rules were used. The purpose of our fuzzy logic adaptive system (FLAS) is to detect the bias of measurements and prevent divergence of the extended Kalman filter. It has been applied in three axes — East (x), North (y), and Altitude (z). The covariance of the residuals and the mean of residuals are used as the inputs to FLAS for all three fuzzy inference engines. The exponential weighting α for three axes are the outputs. As an input to FLAS, the covariance of the residuals and mean values of residuals are used to decide the degree of divergence. The value of covariances relates to R. The equation for covariance of the residual is:

$$P_z = H_k P_k^- H_k^T + R \qquad (15)$$

When the Kalman filter is performing optimally, the mean values of residuals are near zero. Generally, when the covariance is becoming large, and mean value is moving away from zero, the Kalman filter is becoming unstable. In this case, a large α will be applied. A large α means that process noises are added. It can ensure that in the model all states are sufficiently excited by the process noise. When the covariance is extremely large, there are some problems with the GPS measurements, so the filter cannot depend on these measurements anymore, and a small α will be used. The perfect measurements are given more weighting. By selecting appropriate α, the FLAS will adapt the Kalman filter optimally and try to keep the innovation sequence acting as zero-mean white noise.

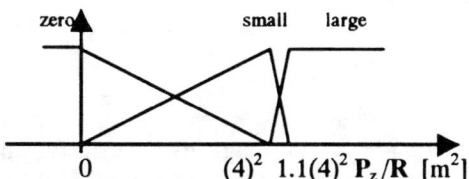

Fig.2. Covariance Membership Functions

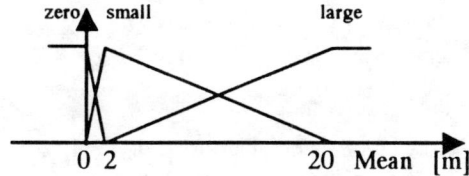

Fig.3. Mean Value Membership Functions

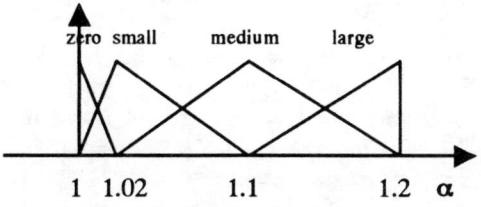

Fig.4. α Membership Functions

The FLAS uses 9 rules, such as:

*If the covariance of residuals is large **and** the mean values are zero **Then** α is large.*

*If the covariance of residuals is zero **and** the mean values are large **Then** α is zero.*

The fuzzy adaptive Kalman filtering has been used for guidance and navigation of mobile robots, especially for 3-D environment. The navigation of flying robots requires fast, on-line control algorithms. The "regular" Extended Kalman Filter requires high number of states for accurate navigation and positioning. The FLAC requires smaller number of states for the same accuracy and therefore it would need less computational effort. Alternatively, the same number of states (as in "regular" filter) would allow for more accurate navigation.

Table. 1. Rule Table for FLAS

α		Mean Value		
		Z	S	L
P	Z	S	Z	Z
	S	Z	L	M
	L	L	M	Z

S --- Small;   M --- Medium;
L --- Large;   Z --- Zero;

Simulation

MATLAB codes developed by authors has been used to simulate and test the proposed method.

The state variables used in simulation are:

$$\mathbf{x}_k = [x_k, \dot{x}_k, y_k, \dot{y}_k, z_k, \dot{z}_k, c\Delta t, c\dot{\Delta}t] \quad (16)$$

The states are position, and velocity errors of the INS East, North, Altitude, GPS range bias and range drift. The covariance of GPS measurement $\mathbf{R}$ is 5 [m^2]. It is assumed that the measurements of INS have some biases. In the first simulation (Fig. 5), the mean values of INS are 0.0014 meter, 0.00035 meter, and 0.0007 meter for the East (x), North (y), and Altitude (z) respectively. A white noise with a standard deviation of 3 meter is added to GPS measurements. The sample period is 1 second. The first row in Fig. 5 is the innovations of fuzzy adaptive EKF and EKF in East (x). The innovation of EKF had a large drift, and was stable at a high mean value. The fuzzy adaptive EKF clearly improved the performance of EKF, and the mean value was much smaller than that of EKF. Other figures present the corrected position (first column) and velocity (second column) errors. The corrected error is the current INS error minus estimated INS error. The dashed lines are the corrected errors of EKF, and the solid lines are the corrected errors of fuzzy adaptive EKF. The fuzzy adaptive EKF significantly reduced the corrected position and velocity errors. In the second simulation (Fig. 6), the same measurements as in the first simulation for INS were used. A white noise with a standard deviation of 2 meter from 0 s to 1000 s and 1500 s to 2000s was applied to GPS measurements. From 1000 s to 1500 s, the standard deviation of 6 meter with mean value of 6 meter was added to GPS measurements. Although, the GPS measurement noises features were changed, the fuzzy adaptive EKF still worked well. Those simulations also showed that the corrected errors of EKF were proportional to the mean values of INS measurements. In other word, the more errors are not modeled, the worse the EKF performs.

Conclusions

In this paper, a fuzzy adaptive extended Kalman filter has been developed to detect and prevent the EKF from divergence. By monitoring the innovations sequence, the FLAS can evaluate the performance of an EKF. If the filter does not perform well, it would apply an appropriate weighting factor α to improve the accuracy of an EKF.

The simulation results show that the FLAS significantly reduces the corrected position and velocity errors when the EKF results diverge. In FLAS, there are 9 rules and therefore, little computational time is needed. It can be used to navigate and guide autonomous vehicles or robots [9] and achieved a relatively accurate performance. Also, the FLAS can use lower order state-model without compromising accuracy significantly. Another words, for any given accuracy, the fuzzy adaptive Kalman filter may be able to use a lower order state model. The FLAS makes the necessary trade-off between accuracy and computational burden due to the increased dimension of the error state vector and associated matrices.

References

[1] Brown, R. G. and Hwang, P.Y.C. " Introduction to Random Signals and Applied Kalman Filtering ", John Wiley & Sons, Inc., 1992.

[2] Abidi, Mongi A. and Gonzalez, Rafael C. " Data Fusion in Robotics and Machine Intelligence ", Academic Press, Inc., 1992.

[3] Gelb, A., " Applied Optimal Estimation ", The MIT Press, 1974.

[4] Lewis, F. L., " Optimal Estimation with an Introduction to Stochastic Control Theory ", John Wiley & Sons, Inc., 1986.

[5] Jochen, P., Meyer-Hilberg, W., and Thomas, Jacob. "High Accuracy Navigation and Landing System using GPS/IMU System Integration", 1994 IEEE Position Location and Navigation Symposium, pp.298 – 305.

[6] Kayton, M. and Fried, W. R., " Avionics Navigation Systems ", John Wiley & Sons, Inc. 1997.

[7] Cooper, S. and Durrant-Whyte H., " A Kalman Filter Model for GPS Navigation of Land Vehicles ", 1994 IEEE Int. Conf. on Intelligent Robot and Systems, pp. 157 – 163.

[8] Abdelnour, G., Chand, S., Chiu, S., and Kido T., " On-line Detection & Correction of Kalman Filter Divergence by Fuzzy Logic ",1993 American Control Conf., pp 1835 – 1839.

[9] Sasiadek, J. and Wang, Q., "3-D Guidance and Navigation of Mobile and Flying Robot using Fuzzy Logic", 1998 AIAA Guidance, Navigation and Control Conference, Boston, MA, August 1998, Paper No. AIAA-98-4126.

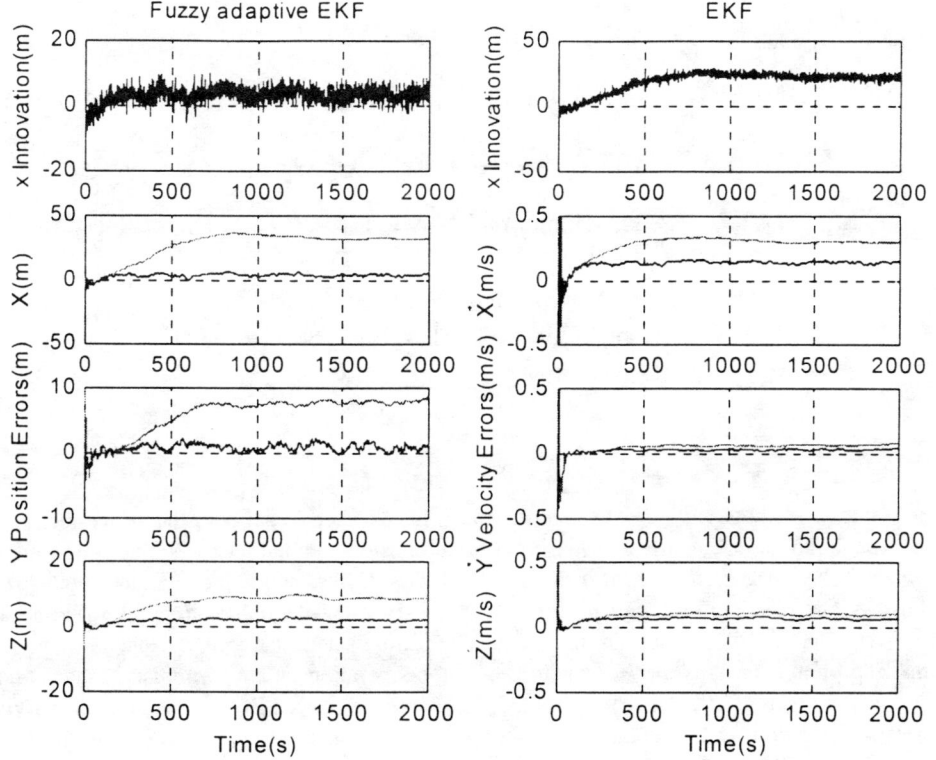

Fig.5. Simulation 1

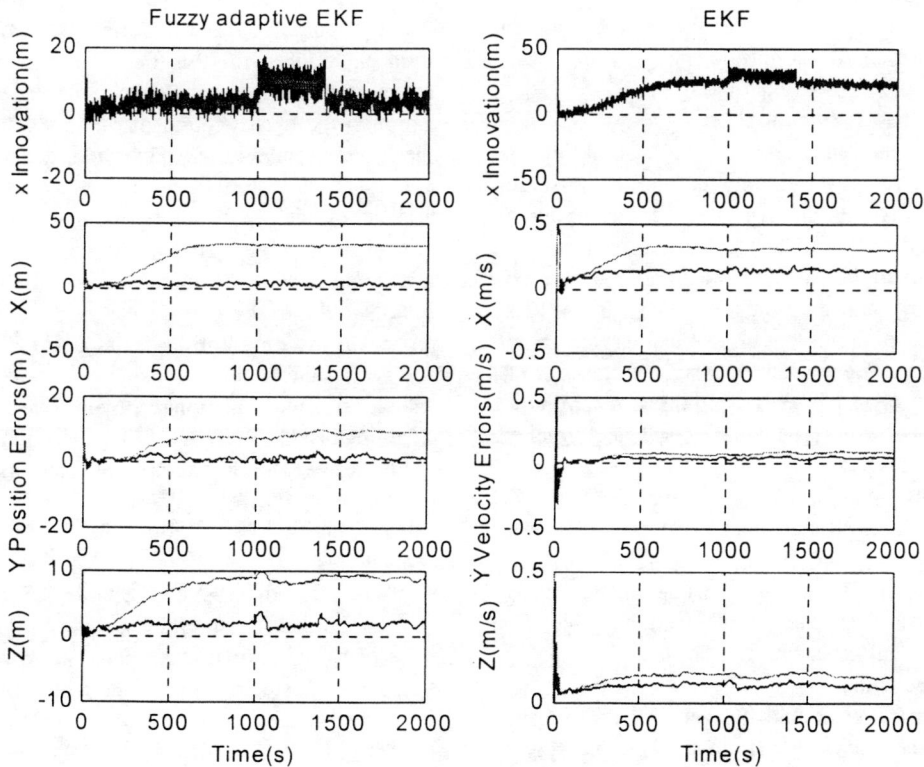

Fig.6. Simulation 2

Environmental Support Method for Mobile Robots Using Visual Marks with Memory Storage

Jun OTA   Masakazu YAMAMOTO   Kazuo IKEDA
Yasumichi AIYAMA   Tamio ARAI

Dept. of Precision Machinery Engineering, The University of Tokyo
7-3-1 Hongo, Bunkyo-ku, Tokyo 113-8656, JAPAN
E-mail: ota@prince.pe.u-tokyo.ac.jp

Abstract

This paper presents a methodology of environmental support for autonomous mobile robots using visual marks with memory storage. The mark makes up for insufficient function of sensing and recognition; self-positioning, positioning of objects, and deciding methods to operate objects. The mark proposed in this paper consists of landmark part and memory part. The landmark part is to be estimated the relative position and orientation between robots and the mark, and the memory part is to have information about what it is, what tasks there are, and how to conduct the tasks. Task execution using the marks with real robots is described to show the effectiveness of the proposed methodology.

Key Words: service robots, environmental support, artificial landmark, mark recognition

1. Introduction

Robots have been used mainly in the industrial fields for many years, where they did only pre-defined assembly sequences. They have been very effective for conducting simple, fixed and repetitive tasks. Nowadays, robots are being extended to applications in which they must co-exist with human-beings and they must conduct various tasks, such as in office or home environments, service sectors and medical fields. The concept of service robots includes assistive robots for people with disabilities, rehabilitation robots, healthcare robots, personal robots and fetch & carry hospital robots.

Now I discuss characteristics of service robots in comparison with industrial robots. Two factors can be considered.
1. Many kinds of tasks should be implemented to the robot because there are many functional objects in the working environment, and because they may be replaced frequently.
2. The geometrical position and orientation (pose) of the objects in the working environments can be slightly changed from the map information of the robot. In other words, lightly structured environments should be assumed.

We need to get over the above-mentioned difficulties. A considerable number of studies have been done to improve the robot's intelligence against them. However, there still exist a lot of troubles to apply the robot to the real worlds with its current ability. Hence we believe that we need cooperation among robots, human users, and working environments. Both robots and working environments should be carefully designed. About the latter case, many approaches have already been adopted in industrial field [8], and it should be extended to office and home environments where human beings and robots coexist.

In order to examine how working environments can support robot's behavior, we discuss what kind of task robots should do. DeVAR project [1], which is one of the pioneer studies in the field of rehabilitation robots, enumerated the tasks that the robot system could realize. Through the classification of them, we define the role of the service robots in this paper as follows: an intermediate agent among functional objects including electric appliances such as refrigerators, furniture such as tables, human beings and so on. Many of them are single-functional except for human beings. Tasks of the service robot would be classified into two: (a) transmission of physically existing entities from one functional object to another (we call this "transmission task"), and (b) state change of one functional object (we call this "state-changing task").

Figure 1 shows an example of relationship between the functional objects (I call this "FO") and the physically existing entities (I call this "PE"). The followings are example tasks of the robot.
- Put the book (PE) on the table (target FO) that is on the bookshelf (initial FO).
- Push the button on the microwave oven (target FO).
- Clean the dishes (PE) at the table (initial FO) and put them into the cupboard (target FO) via the washer & drier (target FO then initial FO).

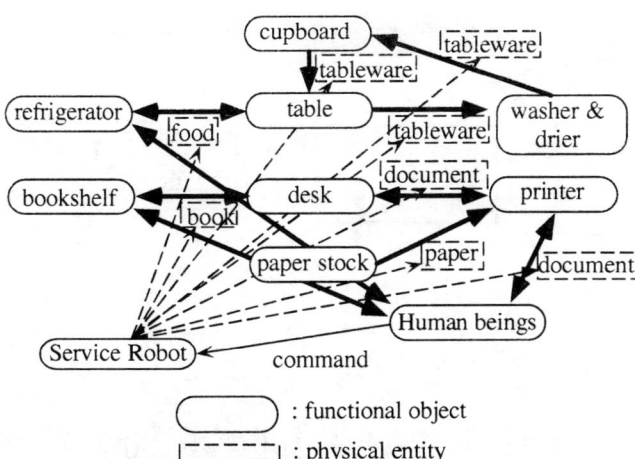

Figure 1: Service robot as an intermediate agent

The first example is included in the transmission task. The second example is included in the state-changing task. The last example is a compounded one of the two tasks. The contents of the tasks are relatively simple. The tasks can be expressed with three factors at most: an initial functional object, a target functional object, and a physical entity.

Process of the transmission task consists of five steps: (a) getting a command from human beings, (b) approaching the initial object, (c) grasping the physical entity, and (d) transmitting it to the target object, and (e) positioning it. As for the state-changing task, it consists of three steps: (a) getting a command from human beings, (b) reaching the target object, and (c) changing the state of the target object. The robots should follow these steps.

Support of robots should be made for the above-mentioned steps. The first one is to support self-positioning ability in the working environments, which is necessary for navigation. The second one is to support positioning ability of robots from FOs. The last one is to support understanding ability of task contents.

Several former studies were made to solve the problems. Sato et al. [9] proposed a new concept of the human symbiosis robots called a robotic room that consists of multiple surrounding sensors and actuators to support human beings. Coen [10] proposed a distributed software agent system that controls a certain room called intelligent room. Takeda et al. [11] aimed at creating cooperative space between humans and artificial agents by using appropriate ontology. These studies aimed at supporting human beings in certain environments, but not for supporting robots. Hada et al. [12] tried to facilitate object recognition by means of code information based on mark-base vision concept. Tashiro et al. [13] proposed how to dispose artificial landmarks in the working environments based on error analysis. Takano et al. [14] proposed RECS (Robot-Environment Compromise System) concept and developed indoor mobile robots with new wheel mechanism to run over steps and with navigation system using indoor lighthouses. These studies did not deal with the understanding ability of task contents. Fujii et al. [15] extended capacity of data carriers and proposed an intelligent data carrier embedded in the working environment. The contents of data are changed dynamically depending on task process. However, the first and the second (robots' positioning) problems were not dealt with in the paper.

We propose environment design technique by designing and disposing visual marks with memory storage as shown in Figure 2. The system will be confronted with a kind of "frame problems" if all the task information is gathered to the robot when working environments are more and more complicated. Task information should be located at the place the tasks are happened. We set the purpose of the study as follows: (1) to propose visual marks with memory storage consisting of landmark part and memory part. The landmark part is to be estimated the relative pose between a camera on the robot (mobile manipulator) and the mark, and the memory part is to have information about what it is, what tasks there are, and how to conduct the tasks. (2) To propose an environmental support methodology using the marks and robot motion strategy in the designed environment.

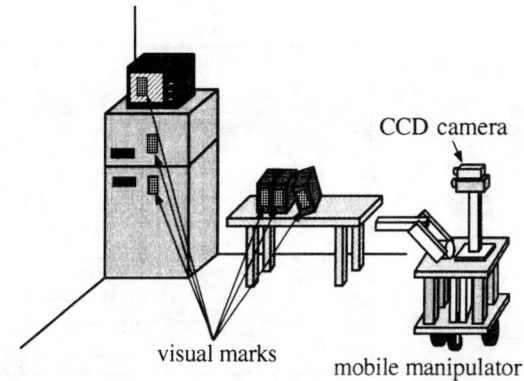

Figure 2: Working environments with marks

Table 1: Utilization of the mark with memory storage

Function	Contents of the memory
Mark for self-positioning	Absolute position/orientation (pose) of the mark
Mark for Manipulation	Contents of task process

2. Environmental Support through Visual Marks with Memory Storage

2.1 Specification of the mark

We propose the design methodology of the mark

discussed at Chapter 1. The landmark part supports positioning ability of robots. The memory part has information of the object to which the mark is attached. There are two kinds of usage of the mark as shown in Table 1. The mark should be compatible to both functions.

1.2 Design of landmark part

Many kinds of landmarks have been proposed (for example, [16][17]). In this paper, we adopt square-shape marks because

- Two-dimensional marks are better than three-dimensional ones from the viewpoint of easiness of attachment to objects.
- Point features are easy to calculate relative pose.
- Four points are the minimal number to get one solution for pose estimation.
- Closed shape marks are better to distinguish from the natural scene.

We adopted the method of [18] to calculate relative pose of the square-shape mark from a camera because this approach is relatively robust from sensing errors, and calculation cost is very low.

Table 2: An example data at the memory storage

	Object ID	1234
	Object name	sliding door
	Object shape	H955 x W800 x D255[mm]
	Center of gravity	(90,890)[mm]
	Number of operation	2
1	Operation name	Open
	Number of primitives	5
	Sequence	Seq.1	Move with (-60,-65,30)
		Seq.2	Move with (0,0,-55) Stop if the contact information is acquired
		Seq.3	Move with (150,0,0)
		Seq.4	Move with (0,0,50)
		Seq.5	End of task
2	Operation name	Close
	Number of primitives	…

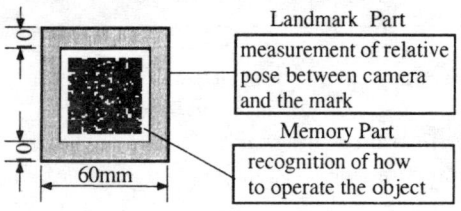

Landmark Part: measurement of relative pose between camera and the mark

Memory Part: recognition of how to operate the object

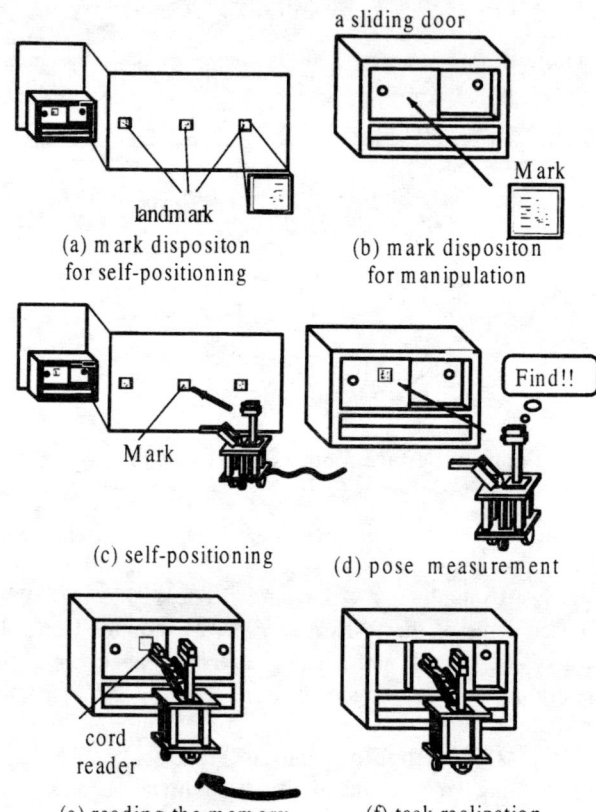

Figure 3: The proposed mark

(a) mark disposition for self-positioning
(b) mark disposition for manipulation
(c) self-positioning
(d) pose measurement
(e) reading the memory
(f) task realization

Figure 4: An example of task realization

1.3 Design of memory part

What kind of information is necessary for the contents of the task process when we attach the mark against the functional object? Here, we consider

(a) ID (what is the object?)
(b) Pose of the mark with respect to the object coordinate
(c) Kind of possible operations against the object and the task process of each operation

We need the second information in order to know the relative pose of the object from the robot by measuring relative pose of the mark from the robot. As for the third information, the following contents are written: (1) the place of the task, (2) finite number of sequence of the robot's motion primitives. The primitive is expressed as a target velocity and angular velocity of the robot's tip and the condition for transition from a certain primitive to the next one (arrival at the target position or change of contact information). Table 2 is an example data of the memory storage at a sliding door.

2.4 Proposal of the mark and task realization process

The designed mark is shown in Figure 3. The memory part consists of QR code developed by Denso Co., which is a kind of two-dimensional bar codes. QR code is selected from the viewpoint of the cost and easiness of writing information as shown in Table 3.

Table 3: Candidates for the memory storage

name	readable range	Price	quality	Memory capacity
Magnetic card	0 (contact)	○	×	150 (Byte)
Flush card	0 (contact)	△	○	2.8~3.4M
IC card (contact)	0 (contact)	×	○	8000
IC card (non-contact)	~ 1 m	×	○	8000
RF-ID	~ 1 m	△	○	200
2D barcode	Variable	○	○	3000

Environmental support methodology and task realization process with the use of the proposed mark is explained in Figure 4. First, as environmental support, the marks for self-positioning are adequately disposed in the working environment (Figure 4(a)) [13]. The marks for manipulation are attached to all the functional objects (Figure 4(b)). Here, the robot has the map information of location of all the marks and all the functional primitives. I assume that this map information does not have large geometrical errors from a real world. Next human operator indicates the target functional object (in this example, a sliding door) and task contents (in this example, open) to the robot. When the robot gets the command, It sets the target position of the task based on the map information, and navigates to the target with refining its position by watching the self-positioning marks (Figure 4(c)). Then it finds the mark for manipulation on the target functional object, and calculates relative pose of the object (Figure 4(d)). Next it approaches to the mark, and reads the memory information by using the bar code reader (Figure 4(e)). It realizes task operation based on the acquired information (Figure 4(f)).

Merits of this system are as follows:
- All the task specific information is written in the marks. The robot does not need deep inference for task realization, which means there is a possibility for avoiding "frame problem" in some degree.
- We can replace the functional objects if we attach the mark with task information to a newly introduced object.
- This system can be applied to lightly structured environments because environmental changes can be measured by watching poses of the marks.

3. Experimental system and evaluation of the landmark
3.1 Experimental system

Architecture of the proposed system is shown in Figure 5. The pose measurement part consists of a CCD camera, a lightning system and image processing system. A mobile robot system with two driving wheels is utilized as a mobile part. An end effector with a touch sensor and a proximity sensor are attached to the tip of a three-degree-of-freedom manipulator. A code reader is utilized to read bar code data.

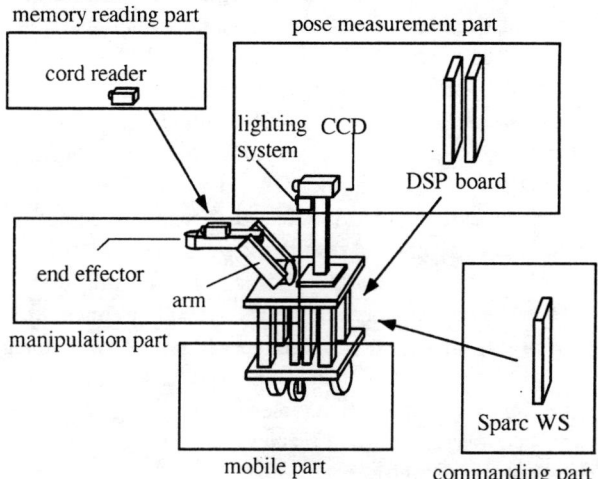

Figure 5: System architecture

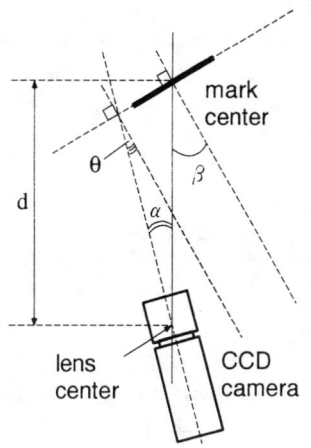

Figure 6: Relative pose of CCD camera and the mark

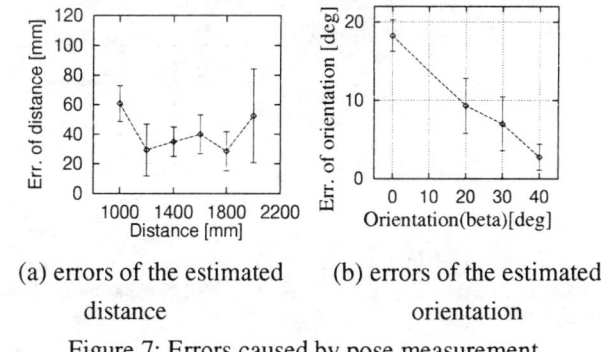

(a) errors of the estimated distance

(b) errors of the estimated orientation

Figure 7: Errors caused by pose measurement

3.2 Measuring experiments of the landmark

First, we measure the accuracy of the proposed

landmark through experiments. Parameters α and β are expressed as shown in Figure 6. Here, orientation θ can be calculated as $\beta - \alpha$.

During the experiment, pose (position and orientation) of the mark is measured by setting d = 1000, 1200, 1400, 1600, 1800, 2000 [mm], α = 0.5 [deg], β = 0, 20, 40, 60 [deg] respectively. Ten-time measurements are made for each set of parameters. As part of experimental results, Figure 7(a) shows the relationship between d and distance errors when α = 0 [deg], β = 20 [deg]. Figure 7(b) shows the relationship between β and orientation errors when α = 5 [deg], d = 1200 [mm]. Vertical lines along measured points mean range of values based on standard deviation of data. The experimental results show

- We can measure the distance with affordable errors when the distance is between 1200 [mm] and 1600 [mm]. This is because measurement with nearer distance causes camera images to be spread.
- As for orientation measurement, error values becomes large when β is less than 30 [deg]

3.3 Sensing strategy

Experimental results at Section 3.2 show that there is possibility for getting insufficient values for pose measurement of the mark. We take the following sensing strategies in order to solve the problem.

1. When the robot run along two marks without rotation like a corridor environment, robot gets the values of d and α from two marks respectively while moving, and can measure the pose by integrating them without utilizing β which has an large error.
2. When we cannot get enough accuracy in operating some objects, n-times touching operations (exploring behavior) of the objects by robot manipulators make it possible to know the pose of the object being touched. Placement of touching operation can be made with [19] by doing error analysis and solving a non-linear optimization problem.
3. Off-line path of the robot are made to keep the distance from 1200 [mm] to 1600 [mm] from the marks for self-positioning.

4. Navigating and door opening experiments

Experiments are made to verify the effectiveness of the proposed system.

In this experiment, in order to extract the mark information from the natural scene with high reliability, (a) reflective sheet is used for the material, and (b) a differential image of the same environment is utilized, with turning a 50W-halogen lamp on and off.

The task is conducted to open a sliding door located several meters away from the initial position of the robot.

Figure 8 shows the experimental environment and planned trajectory of the robot. Size of the sliding door is shown in Figure 9. Please be careful that the trajectory of the robot is planned in advance in this experiment. The necessary accuracy for this door-opening task is about ±5mm in the vertical direction. Figure 10 shows an example of the process of the experiments.

Motion of the robot is described as follows:
(1) Command is given to the robot at point A.
(2) The robot moves to point B and observes the nearby mark for self-positioning.
(3) The robot observes another mark at point C. Two measured values at point B and C are integrated for estimating absolute pose of the robot.
(4) The robot refines its own position based on the self-positioning process at (3) and goes to point D.
(5) The robot goes to point E, where it can observe the mark attached to the sliding door.
(6) It observes the mark at point E (Figure 10(a)), and approaches the mark for manipulation, reads the memory part of it (Figure 10(b)), and conducts the task sequence written at the memory part. Consequently the task of opening sliding door is realized successfully (Figure 10(c)(d)).

Task success ratio is 5 times for 6 trials, which is a high reliability. Because total running length of the robot is about 10 [m], navigation process using the marks is shown to be effective. Total task realization time is 8 minutes and 20 seconds in average. Most of the time is taken by overhead of communication in the computer processes. The time can be decreased with the improvement of the computer architecture. Other processes such as self-positioning and object positioning process with the landmark part, task recognition process with the memory part, and navigation process can be made with real time. This result shows the effectiveness of the proposed system.

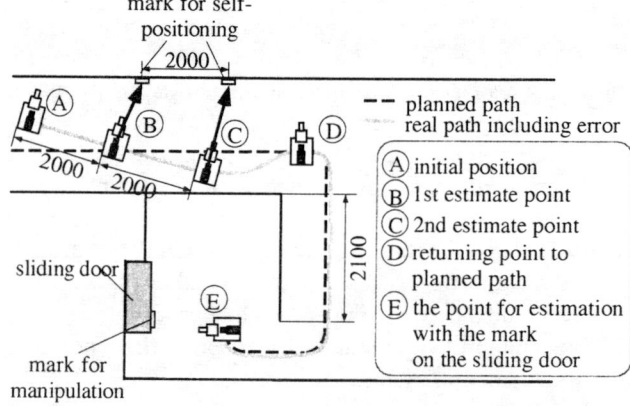

Figure 8: Experimental environments for navigation

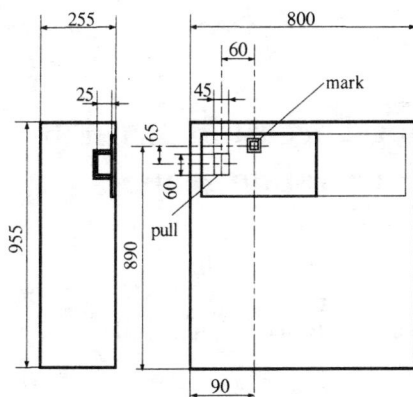

Figure 9: The sliding door used in the experiment

4. Conclusion

This paper proposes the environmental support methodology for service robots.
- We design the visual mark with memory storage. The mark consists of square-shaped landmark part and memory part with 2D bar code.
- Task description methodology for service robots is discussed and embedded information to the memory part is determined.
- Basic experiments of the landmark part were made, and sensing strategies are proposed to assure restricted pose errors.
- Navigation and door opening experiments show the robustness of the proposed methodology.

References

[1] J. Hammel, K. Hall, D. Lees, L. Leifer, M. Van der Loos, I. Perkash, R. Crigler, "Clinical evaluation of a desktop robotic assistant," J. Rehabilitation Research and Development, 26, 3, 1/16(1989).

[2] K.G.Engelhardt, R.A.Edwards, "Human-robot Integration for Service Robotics," Human-Robot Integration, Eds. M.Rahimi, W.Karwowski, Taylor & Francis, 315/346(1992).

[3] K.Kawamura, M.Iskarous, "Trends in Service Robots for the Disabled and Elderly," Prod. Int. Conf. Intel. Robots and Syst., 3, 1647/1654(1994).

[4] P.Dario, E.Guglielmelli, B.Allotta, C.Carrozza, "Robotics for Medical Applications," IEEE Robotics and Automation Magazine, September, 1996, 44/56(1996).

[5] R.M.Mahoney, "Robotic Products for Rehabilitation: Status and Strategy," Proc. Int. Conf. Rehabilitation Robotics, 12/17(1997).

[6] G.Engelberger, "HelpMate: a Service Robot with Experience," Industrial Robot, 25, 2, 101/104(1998).

[7] The ProVAR Homepage: http://provar.stanford.edu

[8] T. Tsumura, "AGV in Japan," Proc. 1994 Int. Conf. Intel. Robots and Syst., 3, 1477/1484(1994).

[9] T. Sato, Y. Nishida, H. Mizoguchi, "Robotic Room: Symbiosis with Human through Behavior Media," Robotics and Autonomous Systems, 18, 185/194(1996).

[10] M.H. Coen, "Building Brains for Rooms: Designing Distributed Software Agents," Proc. 9th Conf. Innovative Applications of Artificial Intelligence, 971/977(1997).

[11] H. Takeda, N. Kobayashi, Y. Matsubara, T. Nishida, "Towards Ubiquitous Human-Robot Interaction," Working Notes for IJCAI-97 Workshop on Intelligent Multimodal Systems, 1/8(1997).

[12] Y. Hada, K. Takase, "Task-Level Feedback Control of a Robot Based on the Integration of Real-time Recognition and Motion Planning", Proc. 28th International Symposium on Robotics, 1997.

[13] K. Tashiro, J. Ota, Y.C. Lin, T. Arai, "Design of the Optimal Arrangement of Artificial Landmarks," Proc.1995 IEEE Int. Conf. Robotics and Automation, 1, 407/413(1995).

[14] M. Takano, T. Yoshimi, K. Sasaki, H. Seki, "Development of Indoor Mobile Robot System Based on RECS Concept," Proc. 4th Int. Conf. on Contr., Automation, Robotics and Vision, 868/872(1996).

[15] T. Fujii, H. Asama, T. Fujita, Y. Asakawa, H. Kaetsu, A. Matsumoto, I. Endo, "Knowledge Sharing among Multiple Autonomous Mobile Robots through Indirect Communication using Intelligent Data Carriers," Proc. 1996 IEEE/RSJ Int. Conf. Intell. Robots and Syst., 1466/1471(1996).

[16] J.H. Kim, H.S. Cho, "Experimental Investigation for the estimation of a mobile robot's position by linear scanning of landmark, Robotics and Autonomous Systems, 13, 39/51(1994).

[17] P. Bison, G. Trainito, S. Venturini, "A Barber Pole Beacon for Mobile Robot Cooperation," Distributed Autonomous Systems 3, T. Lueth, R. Dillmann, P. Dario, H.Wörn Eds., 193/202(1998).

[18] A. Takahashi, I. Ishii, H. Makino, M. Nakashizuka, "A Measuring Method of Marker Position/Orientation for VR Interface by Monocular Image Processing," Trans. Institute of the Electronics, Information and Communication Engineers, J79-A, 3, 804/812(1996).

[19] J. Ota, M. Van der Loos, L. Leifer, "Adaptive Motion Generation with Exploring Behavior for Service Robots," Proc. 1998 IEEE Int. Conf. Robotics and Automation, 1, 77/82(1998).

(a) watching the sliding door   (b) reading memory part

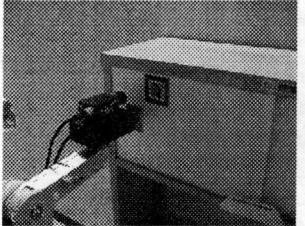 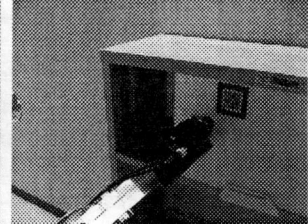

(c) inserting the tip into a pull   (d) opening the door

Figure 10: Process of the experiments

A Real Time Detection Algorithm for Direction Error in Omnidirectional Image Sensors for Mobile Robots

Dong Sung Kim, Young Shin Kim, Wook Hyun Kwon

Engineering Research Center for Advanced Control Instrumentation
School of Electrical Engineering
Seoul National University
Seoul, 151-742, Korea

Abstract

A novel algorithm for the real time detection of the direction error in heading of the mobile robot is proposed. The proposed algorithm is designed for an omnidirecional image sensor. For the omnidirectional image sensor, the hemispherical mirror, one CCD camera, and a vision board are used. This configuration has the low cost and the efficiency in collecting omnidirectional data with one image compared with other types of image sensors. The proposed algorithm is verified through computer simulations and a practical mobile robot.

1 Introduction

For the AMR to be able to move around in a narrow environment without collision, high resolution data and detection of closer object are necessary. It may be more necessary in FMS plant. In this case, slight error in heading of the AMR will lead to a collision with the wall and other machines.

In general, ultrasonic sensors are used in navigation systems for mobile robots, because it gives cost efficiency and an easy way to find the distance to obstacles. But the ultrasonic sensors have two major problems. One is that ultrasonic sensors have low resolution. The other is that ultrasonic sensors often fail to detect an object which lies close to the mobile robot. To overcome these problems, many configurations using multiple ultrasonic sensors have been suggested[1,2,3,4].

There are two kinds of errors in the movement of the AMR. One is a distance error can be detected and compensated by mutiple ultrasonic sensors in most situation. The other is a direction error. It is hard to detect direction error if the mobile robot is equipped only with ultrasonic sensors. This is more likely to cause serious problems. These kinds of errors will eventually result in a serious navigation error to the target point.

For these reasons, various image sensors have been developed and implemented for mobile robots[5,6,7]. But these systems have a high complexity in computation and additional mechanical parts. To overcome these problems, omnidirectional image sensors have been studied[8,9,10]. The image-based homming system was proposed by Hong[8]. A sensing system with a hyperboloidal mirror was proposed by Yagi[9,10]. But these researches were focused mainly on the structure of omnidirectional sensor without practical application to mobile robots. Previous works lack from experimenting with real data to show how omnidirectional image sensors operated well. For practical applications, the detection of the direction error are demanded.

Our aim is to develop a omniderectional image sensor capable of performing accurate and reliable detection of direction error for the mobile robot. The proposed omnidirectional image sensor can detect natural landmarks in real time, because it has low complexity in computations. The proposed configuration has the low cost and the efficiency of collecting omnidirectional data with one image compared with other types of image sensors. Using the proposed image sensor, the AMR detects direction error easily, and estimates the present location efficiently.

In this paper, an algorithm for real time detection of the error in heading of the mobile robot is suggested. The developed algorithm is applied and tested to the omnidirectional mobile robot developed. Section 2 gives a brief description of an image sensor. In Section 3, a filtering algorithm for preprocessing data and the detection algorithm for the direction error is suggested. Section 4 presents he performance of detection algorithm through computer simulations for a practical mobile robot. Finally, the conclusion is drawn in Section 5.

2 Image Sensor

Figure 1: Image Sensor with Experimental Mobile Platform

The developed mobile robot is shown in Figure 1. The image sensor is mounted on the center of the top plate, as shown in Figure 1. The image sensor is composed of a camera, a image processing board and a sensor housing. The design guideline of the image sensor is as follows. First, the image sensor must be able to collect data from all directions around the mobile robot. This property is necessary for the developed omnidirectional mobile robot. Secondly, it must be appropriate for the detection of the heading angle error in real-time processing. And thirdly, cost must be as low as possible. In this configuration, data from all directions can be extracted with one image effectively. So, potential errors in the image sensor are minimized because the camera is fixed on the mobile robot. And the building cost is virtually minimized by it's simple architecture. Based on the idea explained above, the image sensor has been built. Figure 2 shows a prototype of the image sensor.

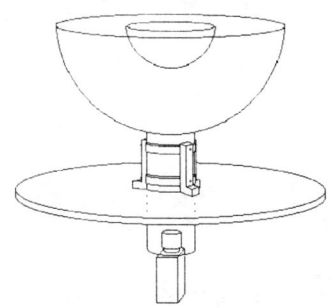

Figure 2: Configuration of Image Sensor

3 Detection Algorithm for Direction Errors

3.1 Filtering Process

The raw image taken from the image sensor is in Figure 3. For the preprocessing data in the each check point, moving average filter(LPF filtering)was applied(Figure 4). The moving average filter has been applied in order to reduce sensor noises. In first, high frequency noises are removed by using moving average filter. In this algorithm, the value of $M = N = 2$ is used for the 5-point moving average filter.

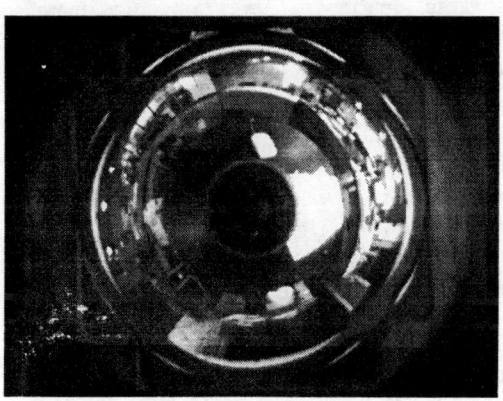

Figure 3: Raw image from Image Sensor

$$y[n] = \frac{1}{N+M+1} \sum_{k=-N}^{M} x[n-k] \quad (1)$$

And then, neighboring data is connected for vector sets. Local maximum values and minimum values are calculated and pointed in the next step. During the calculation, variations within the threshold value

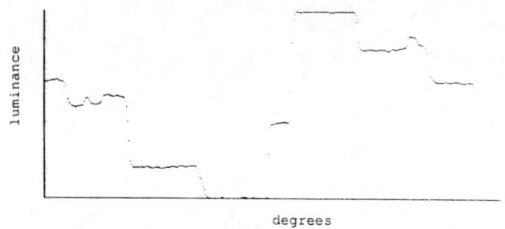

Figure 4: Low Pass Filtered Data

are ignored. Figure 5 represents the procedure of the detection algorithm.

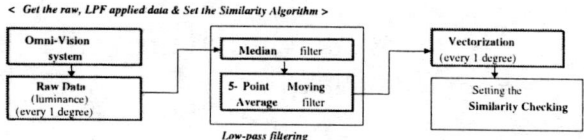

Figure 5: Detection Algorithm for Direction Error

3.2 Detection Algorithm using Similarity Check

The indoor environment has some features different from an outdoor one. That is, almost all structures are vertical in indoor environments. Thus, the feature of a specific location can be determined from series of luminance data(gray level) that are sampled from a horizontal plane. The horizontal plane, on any position in vertical axis between the floor and the ceiling, will give virtually same series of luminance data. Series of luminance data are sampled on one circle which corresponds to a virtual horizontal plane is used for the detection algorithm. Such sampled data can be plotted on a direction angle versus luminance in the coordinate plot, and an example is in Figure 6. In Figure 6, the distance from the center to a point means the luminance value of the present direction. In Figure 6, we can easily observe that the highest luminance value is in directions approximately 5° and 75°. The scheme for the image sensor as a heading error detector is as follows. The PC connected with the image sensor stores series of the luminance data at the every predetermined checkpoint. When the AMR reaches that check point again, the processor gets series of the luminance data from the image sensor system, and then compares it with the one stored before. If there is some shift between these two sets of data, the PC notifies the motor controller about the direction and amount of the heading error.

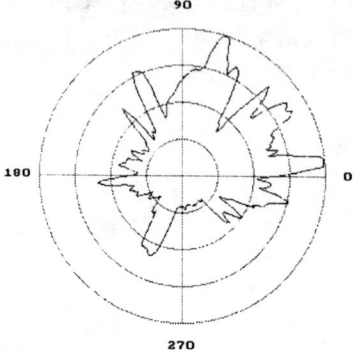

Figure 6: Direction angle versus luminance plot

The core of this scheme is 'How to compare two sets of data that have the same pattern, but there is some angular shift'. First, we consider what it would be like if a person did this process by the human's perception scheme. A human brain will focus on the global shape of the pattern. That is, small fluctuations in data will be ignored, and the luminance data at a direction will be perceived as a part of the 'recognizable pattern'. This means that the relations among neighboring luminances are more important than the individual luminance value in a specific direction. Having this concepts in mind, we try on composing an array that contains the signed difference between the neighboring luminance values. For formal representation, it will be described as follows

$$L = (l_0 \ l_1 \ l_2 \ \ldots \ l_{359}) \quad (2)$$

$$D = (l_0 - l_1 \ l_1 - l_2 \ \ldots \ l_{358} - l_{359} \ l_{359} - l_0). \quad (3)$$

where L is an array of the luminance values, D is an array of the signed difference values between neighboring luminance data. It is easier to compare two sets of data, with arrays denoted as D, of each data sets. The comparison scheme is as follows

$$(angular\ shift) = n \quad (4)$$

where n is given as

$$r_n = \min(r_0, r_1, \ldots, r_i, \ldots, r_{359}) \quad (5)$$

$$r_i = \sum_{m=0}^{359} |(D_{stored} - D_{current.i})_m|. \quad (6)$$

$(D_{stored} - D_{current.i})_m$ means the m-th element of $D_{stored} - D_{current.i}$, $D_{current.i}$ means a shifted version of $D_{current}$ by the amount of ith with wrap-around.

But it still needs many calculations(360 × 3 subtractions and 359 additions), which will result in low comparison speed in computation time. Thus, there must be some modifications as follows

$$(angular shift) = n' \quad (7)$$

where n' is given as

$$s'_n = min(s_0 \; s_1 \ldots s_i \ldots s_{359}) \quad (8)$$

$$s_i = S_{stored} S_{current.i}^T \quad (9)$$

$$S = [sgn(l_0 - l_1) \; sgn(l_1 - l_2) \ldots sgn(l_{359} - l_0)]. \quad (10)$$

It is natural that modified equation seems a little more complicated and time-consuming, because there are multiplications and signum functions in the formula. It needs 360 × 2 subtractions, 360 × 2 signum functions, 360 multiplications and 359 additions. But the time spent for signum functions and the multiplications can be neglected, because the signum functions can be substituted with a combination of simple and fast bitwise operations. And multiplications can also be substituted with a simple look up table.

4 Experiment

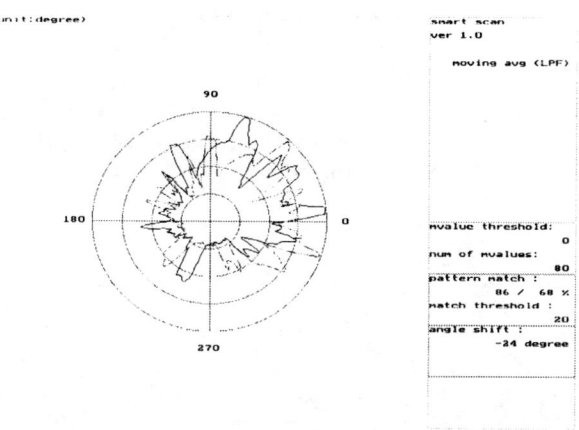

Figure 8: $-24°$ rotation

The minus sign means the data sampled is rotated counterclockwise from the reference data. The calculated value of the angle shift is almost stationary while the sampling was being repeated, with some rare fluctuations which were in the range of $\pm 1°$ maximum. The second experiment shows the case of 60° rota-

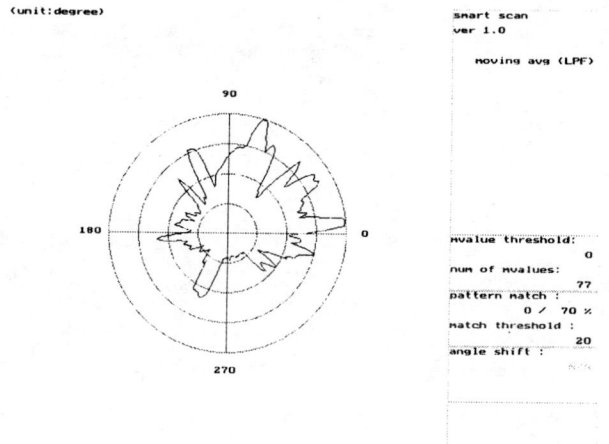

Figure 7: Reference(stored) data

Following experiments are chosen to test the algorithm presented previously. For some typical situation, a rotation up to 180° and a translation of the AMR is tested. Figure 8 shows the result of $-24°$ rotation.

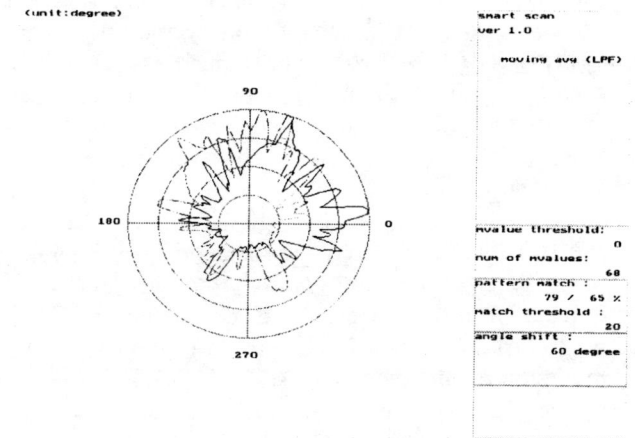

Figure 9: $+60°$ rotation

tion. The calculated rotation angle (i.e., the angle shift value) is as stationary as before. The third experiment shows results for the cases of 120°. The fluctuation of calculated angle shift value was still in the range of $\pm 1°$ maximum, though the value was a less stationary. We observe that the details of the luminance curve become more and more different from the reference curve as the angle shift value increases. The reference data super-imposed on +60° on from the error in manufacturing the hardware part (i.e., mirror, holder, etc.)shown in Figure 2. It is easy to see that the angle shift value is very precise despite of the variation of the luminance data.

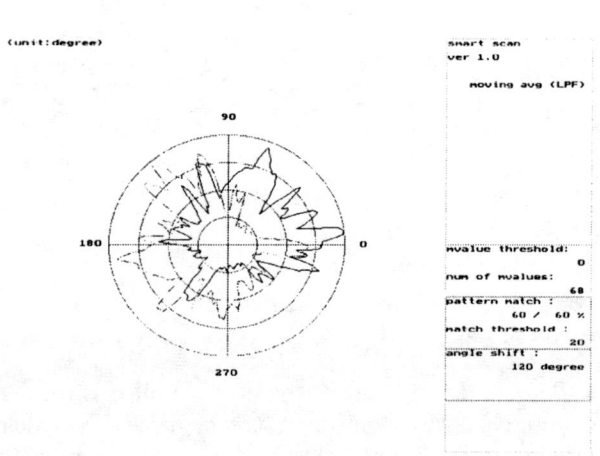

Figure 10: +120° rotation

In Figure 10, we observe that the variation of the data is more eminent, but the algorithm gives a right answer, as long as the global characteristic of the curve is preserved. It is used a value named as the pattern match in Figure 8-Figure 12. This value is represented as current value/minimum requirement value. When this value becomes to smaller than the minimum requirement, the angle shift becomes non-acceptable case(N/A). It means that the global characteristics of the two luminance curves are significantly different, so the mobile robot is not on the exact position of the checkpoint in the navigation. If so, calculating the error in heading angle of the AMR is meaningless. The minimum requirement value is set to decrease linearly, as the absolute value of the angle shift increases. In formal representation, these values are given as follows

$$(current\ value) = (m/360) \times 100 \quad (11)$$

$$(min_{req}) = -(15/180) \times (angleshift) + 70. \quad (12)$$

where m is the number of data which satisfies $|(reference\ data) - (current\ data)| < \theta_{th}$. ($min_{req}$), θ_{th} represents minimum requirement value and threshold value. Eq. 12 is determined empirically, to form a tight linear boundary of the minimum requirement value. That is, rotations up to ±180° at the reference position may not yield the result that angle shift is N/A. Following experiments(Figure 11, 12) are about cases of the translation errors.

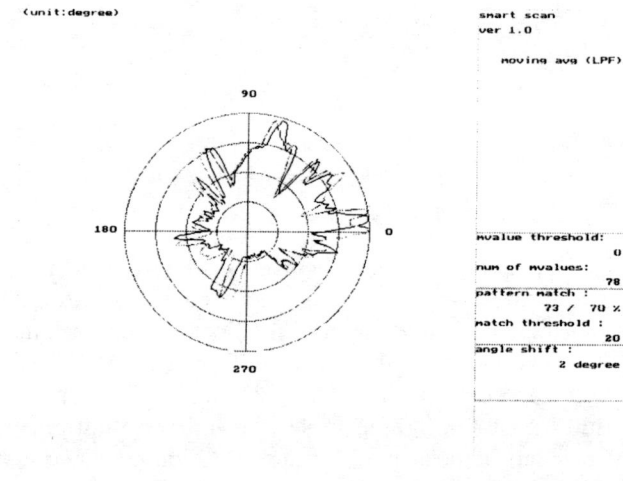

Figure 11: 20cm translation

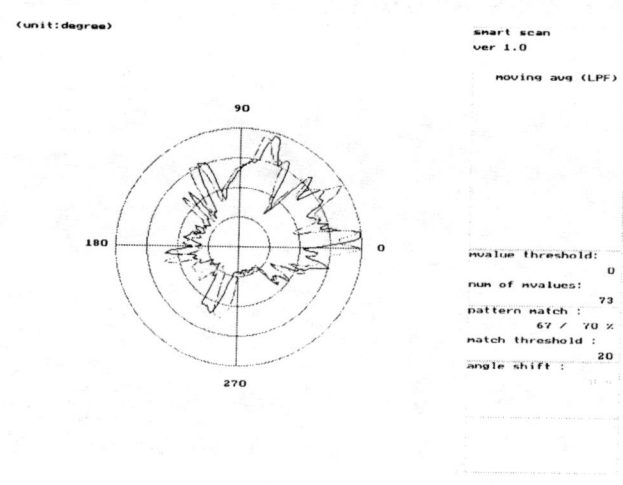

Figure 12: 40cm translation

In Figure 11, we observe that the translation by 20cm is interpreted as 2°rotation. It can be found that this kind of misinterpretation will persist until the distance of translation reaches approximately 40cm, in Figure 12. Figure 12 shows the case of 40cm translation, and the angle shift value is now displayed as N/A. Some more experiments showed that the 40cm is the threshold value of the misinterpretation. That

tion by the amount of $40cm$ or farther, the angle shift value becomes N/A. We expect this threshold value will decrease because these experiments are performed in a hall which was a lot bigger and wider than the narrow aisle where the proposed image sensor system had been designed to operate. The image sensor can be used to detect the objects which lie closer than the detection limit of the ultrasonic sensors. This is the reason that we used a hemispherical mirror(Figure 13). We need a slight modification of the algorithm, which means taking a surplus sample on a circle of radius $r2$ along with that of on the circle of radius $r1$(where $r2 < r1$). The surplus data contains the information about objects near the AMR. It can be used to detect an obstacle, which is too close for the ultrasonic sensors to detect.

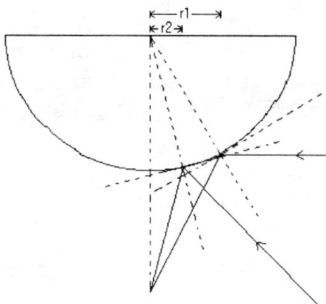

Figure 13: Model of hemispherical mirror

5 Conclusion

In this paper, an algorithm for the real time detection of the direction error in heading of the AMR is proposed. The proposed algorithm is designed for the omnidirecional image sensor for mobile robots. For the omnidirectional image sensor, the hemispherical mirror, one CCD camera, and a vision board are used. The current position and the heading direction of the mobile robot are obtained from a single image projected on a hemispherical mirror.

The proposed algorithm is proved to have efficiency in the computation time. This configuration has the low cost and the efficiency in collecting omnidirectional data with one image, because it uses commercially available projection bulb instead of mirrors. The proposed algorithm also eliminates the need to rotate the camera for tracking obstacles and landmarks.

The proposed image sensor and algorithm can be applied to the case of unmanned moving vehicles and automobiles.

References

[1] Claude Fennema, Allen Hanson, Edward Riseman, J. Ross Beveridge, and Rakesh Kumar, "Model-Directed Mobile Robot Navigation," *IEEE Trans. System, Man, and Cybernetics, vol. 20, no. 6, 1990, pp.1352-1369.*

[2] K. Song, C. Chang "Ultrasonic Sensor Data Fusion for Environmental Recognition" *IEEE International Conference on Intelligent Robots and Systems: Yokohama, Japan, 1993.*

[3] O. Patrouix, B. Jouvencel."Range Information Extraction Using U-BAT: an Ultrasonic Based Aerial Telemeter" *IEEE International Conference on Robotics and Automation, 1993.*

[4] Prabir K.Pal and Asim Kar. "Sonar-Based Mobile Robot Navigation t hrough Supervised Learnning on a Neural Net" *Autonomous Robot, Vol.3, 1996, pp.355-374*

[5] David J. Kriegman, Ernst Triendl, and Thomas O. Binford, "Stereo Vision and Navigation in Buildings for Mobile Robots," *IEEE Trans. Robotics and Automation, vol. 5, no. 6, 1989, pp. 792-803.*

[6] T. D'Orazio, M. Ianigro, E. Stella, F. P. Lovergine, and A. Distante, "Mobile Robot Navigation by Multi-Sensory Integration," *IEEE International Conference on Robotics and Automation, 1993, pp. 373-379.*

[7] Vitor Santos, Joao G. Goncalves, and FRancisco Vaz, "Perception Maps for the Local Navigation of a Mobile Robot: a Neural Network Approach," *IEEE International Conference on Robotics and Automation, 1994, pp. 2193-2198.*

[8] Jiawei Hong, Xiaonan Tan, Brian Pinette, Richard Weiss, and Edward M. Riseman "Image-Based homing" *IEEE Magazine of Control Systems, Feb. 1992, pp.53-59.*

[9] Yasushi Yagi, Shinjiro Kawato, and Saburo Tsuji, "Real-Time Omni-Directional Image Sensor(COPIS) for Vision-Guided Navigation" *IEEE Trans. Robotics and Automation, vol. 10, no. 1, February, 1994, pp. 11-22*

[10] Kazumasa Yamazawa, Yasushi Yagi, and Masahiko Yachida, "Omnidirectional Imaging with Hyperboloidal Projection" *IEEE International Conference on Intelligent Robots and Systems, 1993, pp. 1029-1034*

Modeling and Classification of Rough Surfaces Using CTFM Sonar Imaging

Z. Politis and P. J. Probert

zpol,pjp@robots.ox.ac.uk
Robotics Research Group
Department of Engineering Science
University of Oxford
Parks Road, Oxford OX1 3PJ, UK

Abstract– *The typical use of ultrasonic sensors has been limited to estimation of the location of targets in a robot workspace. CTFM sonars have also been used successfully in classifying primitive targets. In this paper the classification is extended to include textures typical of these found in pathways the robot may need to follow or identify. The pathway classes examined are considered to be plane surfaces of various roughness corresponding to hard smooth floor, carpet, and asphalt. Each class is modeled using an extension of the Kirchhoff approximation method describing the scattering of the acoustic wave on rough surfaces. The CTFM sonar image corresponding to each class is derived and compared with the experimental one. Then a feature is extracted that exploits the differences between the three surface models. A neural network is trained for recognition with excellent results.*

Keywords– *CTFM sonar, acoustics, Kirchhoff approximation method, surface recognition, robot navigation.*

1 Introduction

The most frequent application of ultrasonic sensors in robotics is in areas like map building, obstacle avoidance and primitive target identification. The CTFM sonar system [1] has been used in these areas successfully [2]. However, the clear interpretation of the environmental information that provides, enables the classification of more complex targets [3]. One example could be the classification of rough surfaces, where the characteristics of the surface could be extracted from the information included in the sonar signal [4]. These characteristics, like degree of roughness, or returned energy, could be associated with the features allowing the classification of the surface.

In this paper we developed models for three different surface types that correspond to a smooth (hard smooth floor), a rough (asphalt) and a moderately rough (carpet) surface. Then a set of features is extracted to discriminate them by training a classifier. The advantage of the CTFM sonar is that it needs only a single measurement to assign a surface to a class, while other methods [5] based on pulsed sonar systems require a scan reading of the surface instead.

In section 2 we describe the CTFM sonar system and the parameters on which the output signal depends. In section 3 a general model for rough surfaces is presented using an approximation of the Helmholtz-Kirchhoff integral which determines the scattered field. Section 4 evaluates the theoretical results by comparing them with equivalent measurements. An application to pathway recognition follows in section 5.

2 Sensing with CTFM Sonar

The CTFM sonar system is based on a continuous transmission of a modulated frequency operating over a frequency range of f_l to f_h, using the saw-tooth frequency pattern of Fig. 1. When an echo, s_R, is re-

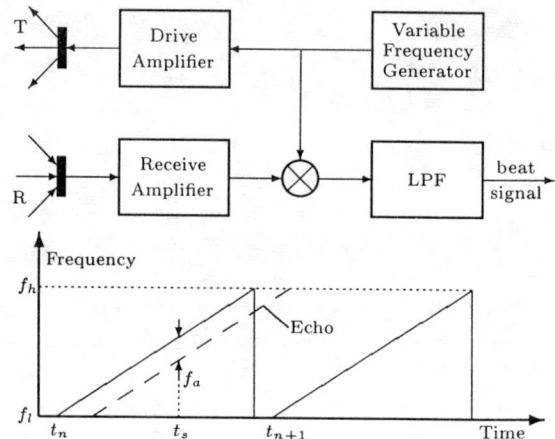

Figure 1: The used CTFM sonar and the frequency pattern of the transmitter. (f_l=45kHz, f_h=90kHz, and T=184ms).

ceived it is mixed with the transmitted signal at the time being. Because the received signal follows the same frequency pattern as the transmitted one, but shifted in time due to the propagation delay, the product of these two signals will have a constant frequency equal to the difference of their frequencies at each moment[1]. The output signal, s_α, (called *beat signal*) consists of the difference frequency, f_α, which corresponds to the range of the reflected object.

Under the assumption that the sawtooth pattern of the modulated frequency does not affect the bandwidth of the transmitted signal s_T, we get [1]

$$s_T(t) = A\mathrm{Re}\left\{e^{2\pi j\left(f_l t_s + \frac{m}{2} t_s^2\right)}\right\} \quad (1)$$

where $t_s = t - t_n$ is the time during a sweep cycle, i.e. $0 \le t_s \le T_s$, $T_s \simeq t_{n+1} - t_n$ is the sweep period, m is a constant ($m = \frac{f_h - f_l}{T_s}$), and A is the amplitude. In the case of a single reflection at distance r from an angle θ, the returned signal has the form

$$s_R(t) = \frac{1}{\mathcal{J}}\alpha(\lambda, r)\varrho(\lambda, r, \theta)\mathcal{D}(\lambda, \theta)s_T(t - \tau) \quad (2)$$

where

τ is the time delay between s_T and s_R, equal to $\frac{r}{c}$, c is the speed of sound.
$\alpha(\lambda, r)$ is the attenuation of the signal in the air.
$\varrho(\lambda, r, \theta)$ describes the reflection of the signal on the object.
$\mathcal{D}(\lambda, \theta)$ is the directivity of the transducer.
$\mathcal{J}$ is the sensitivity of the transducer.
All these terms depend on the wavelength, which means they are varying with time.

After mixing the received and transmitted signals, the beat signal will be

$$s_\alpha(t) = \frac{A^2}{\mathcal{J}}\alpha(\lambda, r)\varrho(\lambda, r, \theta)\mathcal{D}(\lambda, \theta)$$
$$\cdot \mathrm{Re}\left\{e^{2\pi j\left(f_l \tau - \frac{m}{2}\tau^2 + \frac{m}{c} r t_s\right)}\right\} \quad (3)$$

which is a sinusoidal signal of varying amplitude, but of constant frequency f_a given by

$$f_a = \frac{1}{2\pi}\frac{d(\phi_R(t) - \phi_T(t))}{dt} = \frac{2mr}{c} \quad (4)$$

where $\phi_T(t)$ and $\phi_R(t)$ are the transmitted and received phases of the signal respectively.

The signal s_α may include multiple reflections from various ranges. In that case the previous equation will take the more general following form

$$s_\alpha(t) = \frac{A^2}{\mathcal{J}}\sum_i \alpha_i(\lambda, r)\varrho_i(\lambda, r, \theta)\mathcal{D}_i(\lambda, \theta)$$
$$\cdot \mathrm{Re}\left\{e^{2\pi j\left(\phi_i + \frac{m}{c} r_i t_s\right)}\right\} \quad (5)$$

[1] The sum of their frequencies has been eliminated by the LPF after the mixer.

where ϕ_i is the phase shift corresponding to each reflection. The amplitude spectrum of the beat signal is called CTFM *sonar image* and corresponds to a range map in the direction of the sonar sensor.

The sonar image from a rough surface is consisted of a sum of tones from each reflective point, randomly distributed over a wide band, forming a spectrum uniquely related to its geometry and reflective characteristics. In the next section rough surfaces are modeled by evaluating the backscattering coefficient for each surface type.

3 Surface scattering model

The scattering of the incident wave on to a rough surface is described by the Kirchoff approximation method (KAM) [6], which is an approximate solution to the Helmholtz-Kirchhoff integral, a mathematical formulation of Huygen's principle.

We define first the Cartesian coordinate system of Fig. 2. An emitted wave from T with angles β_1 and γ_T is scattered on S to the directions given by the angles β_2 and γ_R to reach the receiver R. The surface

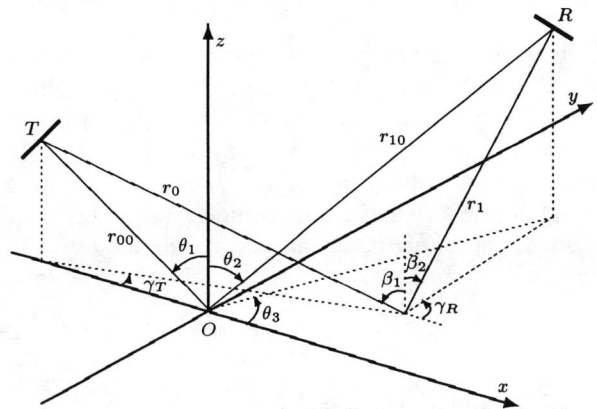

Figure 2: The geometry of scattering of a wave emitted from T and received from R.

height is characterized as a random process, described by the function $\zeta(x, y)$, and it is assumed that the mean value of ζ is on the x-y plane. We assumed that ζ follows a Gaussian distribution of zero mean and standard deviation σ. The origin is chosen at the intersection of the line-of-sight of T and R, while the positions of the transmitter and the receiver are at the points $(x_T, 0, z_T)$ and (x_R, y_R, z_R) respectively. θ_1 is the angle between the line-of-sight of T and the z axis and θ_2 and θ_3 are the angles between the line-of-sight of R with respect to the z and x axis respectively.

If E_{inc} is the pressure field of the incident wave and E_s is the scattered pressure field on the surface, then the pressure field E_R at the receiver including all the reflected and scattered components on the whole surface S, is given by

$$E_R = \frac{1}{4\pi} \int \int \left(E_{inc} \frac{\partial E_s}{\partial n} - \frac{\partial E_{inc}}{\partial n} E_s \right) dS \quad (6)$$

where the partial derivatives are evaluated in the direction of surface normal, and the integration is performed over the surface S. The direction of the surface normal is given by

$$\mathbf{n} = \frac{1}{\sqrt{\zeta_x^2 + \zeta_y^2 + 1}} \begin{pmatrix} -\zeta_x \\ -\zeta_y \\ 1 \end{pmatrix} \quad (7)$$

where ζ_x and ζ_y are the partial derivatives of $\zeta(x,y)$ with respect to x and y respectively. dS can be expressed in terms of dx and dy by

$$dS = \sqrt{\zeta_x^2 + \zeta_y^2 + 1}\, dx dy \quad (8)$$

The integral is calculated on the following approximate boundary condition since it is very difficult to do so on S.

$$E_s = (1+\mathcal{R})E_{inc}, \quad \frac{\partial E_s}{\partial n} = (1-\mathcal{R})\frac{\partial E_{inc}}{\partial n} \quad (9)$$

where $\mathcal{R}$ is the reflection coefficient between the two media for a smooth plane boundary. These conditions are exact only for a smooth surface, and their accuracy decreases as the irregularities on the surface increase.

If in the expressions of E_{inc} and E_s we consider only the wave equation, by eliminating the attenuation of the signal in the air and the directivity of the transducer them we get

$$E_{inc} = e^{i\mathbf{k}_1 \cdot \mathbf{r}_0}, \quad E_s = e^{i\mathbf{k}_2 \cdot \mathbf{r}_1} \quad (10)$$

where $\mathbf{k}_1$ and $\mathbf{k}_2$ are the propagation vectors of the incident and scattered waves and $|\mathbf{k}_1| = |\mathbf{k}_2| = k$, k is the wave number.

$$\mathbf{k}_1 = k \begin{pmatrix} \sin\theta_1 \\ 0 \\ -\cos\theta_1 \end{pmatrix}, \quad \mathbf{k}_2 = k \begin{pmatrix} \sin\theta_2 \cos\theta_3 \\ \sin\theta_2 \sin\theta_3 \\ \cos\theta_2 \end{pmatrix}$$

Under these assumptions the pressure field E_R is dependent only on the scattered waves on the surface.

The distance that the wave travels from the transmitter to receiver can be approximated as follows, when the area of illumination is sufficiently small.

$$r_0 + r_1 \approx r_{00} + r_{10} + [s_x x + s_y y + s_z \zeta(x,y)] \quad (11)$$

where

$$\begin{aligned} s_x &= \sin\theta_1 - \sin\theta_2 \cos\theta_3 \\ s_y &= -\sin\theta_2 \sin\theta_3 \\ s_z &= -\cos\theta_1 - \cos\theta_2 \end{aligned} \quad (12)$$

After some algebraic manipulations [7], equation 6 takes the form

$$E_R = \frac{-ike^{ik(r_{00}+r_{10})}}{2\pi} \mathcal{R} G \int dx \int dy\, e^{ik[s_x x + s_y y + s_z \zeta(x,y)]} \quad (13)$$

where

$$G = \frac{1 + \cos\theta_1 \cos\theta_2 - \sin\theta_1 \sin\theta_2 \cos\theta_3}{\cos\theta_1 \cos\theta_2} \quad (14)$$

In this expression the factor $e^{i\omega t}$ has been supressed and the only dependence on time comes from the wave number k.

If we view the channel from the transmitter to the receiver as a stochastic linear system, then if the input signal $s_t(t)$ is aperiodic, the received signal $s_R(t)$ will have the form

$$s_R(t) = \frac{1}{2\pi} \int_{-\infty}^{+\infty} S_T(\omega) E_R(\omega) e^{i\omega t} d\omega \quad (15)$$

where $S_R(\omega)$ is the Fourier transform of $s_R(t)$. If the surface is random, the expression for $E_R(\omega)$ will also be random, so a form of averaging is needed

$$<s_R(t)s_T^*(t)> = \frac{1}{4\pi^2} \int_{-\infty}^{+\infty} d\omega_1 \int_{-\infty}^{+\infty} d\omega_2$$
$$\cdot S_T(\omega_1) S_T^*(\omega_2) <E_R(\omega_1) E_R^*(\omega_2)> e^{i(\omega_1 - \omega_2)t} \quad (16)$$

where the term $<E_R(\omega_1) E_R^*(\omega_2)>$ corresponds to the frequency correlation of the channel. For simplicity we can assume that $\omega_1 = \omega_2$. Also, the Fourier transform of the transmitted signal can be approximated by the following function

$$S_T(\omega) = \begin{cases} 1 & w \in [k_l, k_h], \\ 0 & \text{else.} \end{cases} \quad (17)$$

Then equation 16 corresponds to the average scattered energy on the surface S received at R and is analog to the scattering coefficient associated with all the reflections from the circle, given by

$$\mathcal{P}(r) \propto \frac{\mathcal{R}^2 G^2}{4\pi^2} \int_{k_l}^{k_h} dk \int_{-\infty}^{+\infty} dx_1 \int_{-\infty}^{+\infty} dx_2 \int_{-\infty}^{+\infty} dy_1 \int_{-\infty}^{+\infty} dy_2$$
$$\cdot k^2 e^{i2k(r_{00}+r_{10})} e^{iks_x(x_1-x_2)} e^{iks_y(y_1-y_2)} <e^{iks_z(\zeta_1-\zeta_2)}> \quad (18)$$

The term $<e^{iks_z(\zeta_1-\zeta_2)}>$ is the two dimensional characteristic function of the surface given by

$$<e^{iks_z(\psi_1-\psi_2)}> = \Phi(s_z k, -s_z k) = e^{-\sigma^2 k^2 (1-C)} \quad (19)$$

where $C = e^{-\tau_1^2/T_1^2 - \tau_2^2/T_2^2} \approx -\frac{\tau_1^2+\tau_2^2}{T^2}$, $\tau_1 = x_1 - x_2$ and $\tau_2 = y_1 - y_2$. T_1 and T_2 are the correlation lengths along the two axis and we use $T = T_1 = T_2$ because we have assumed a uniform surface.

Since the transmitter and receiver are very close to each other, we can assume that $\theta_1 = -\theta_2 = \theta$, $\theta_3 = 0$, and $r_{00} = r_{10}$. Then we also have that $x_T = x_R = r_{00} \sin\theta$, $z_T = z_R = R$, and $y_R = 0$. The backscattering coefficient must be associated with the distance r, so the integration on the surface has to be done under the constrain that the distance $r_0 + r_1 = $ const. This means that the integral has to be calculated on circular zones centered at the T/R projected point on the surface. If the width of these zones is set to be very small, the integration on a surface is decayed on a circle. Note that for constant distance r, the wave has the same angle of incident. So the reflectivity of the surface for a certain distance corresponds to the reflectivity of the surface only on a circle of radius $r_a = \sqrt{r^2 - R^2}$ and under an incident angle ξ given by $\xi = \mathrm{acos}\frac{R}{r}$.

Changing to circular coordinates (r_a, ϕ_a) using the following variable substitutions helps to define properly the integral limits since the integration has to be done on a small circular surface zone.

$$\begin{aligned} x_1 &= r_{a1} \cos\phi_{a1} - r_{00} \sin\theta \\ x_2 &= r_{a2} \cos\phi_{a2} - r_{00} \sin\theta \\ y_1 &= r_{a1} \sin\phi_{a1} \\ y_2 &= r_{a2} \sin\phi_{a2} \end{aligned} \quad (20)$$

The expression for $\mathcal{P}$ of equation 18 takes the form

$$\mathcal{P}(r) = \frac{K\mathcal{R}^2}{8\pi(k_h - k_l)\cos^2\theta} \int_{k_l}^{k_h} dk \int_{r_a}^{r_a+\Delta r_a} dr_{a1} \int_{r_a}^{r_a+\Delta r_a} dr_{a2} \int_0^{2\pi} d\phi_{a1}$$

$$\cdot \int_0^{2\pi} d\phi_{a2} k^2 r_{a1} r_{a2} e^{i2k \sin\theta(r_{a1}\cos\phi_{a1} - r_{00}\sin\theta)}$$

$$\cdot e^{i2k\sin\theta(r_{a2}\cos\phi_{a2} - r_{00}\sin\theta)}$$

$$\cdot e^{\frac{-4\cos^2\theta\sigma^2 k^2(r_{a1}^2 + r_{a2}^2 - r_{a1}r_{a2}\cos(\phi_{a1}-\phi_{a2}))}{T^2}} \quad (21)$$

where K is a scaling constant to make $\mathcal{R}$ equal to the expression on the right. The calculated value of the backscattering coefficient is associated only with the specified transmitted signal, and the assumed surface which is characterized by the triplet $(\mathcal{R}, \sigma, T)$.

In Fig. 3 we plotted the values of the backscattering coefficient for a smooth and a rough surface with $R = 0.7$ and $\theta = 45°$. In the case of a smooth surface all the reflected waves come from the area close to

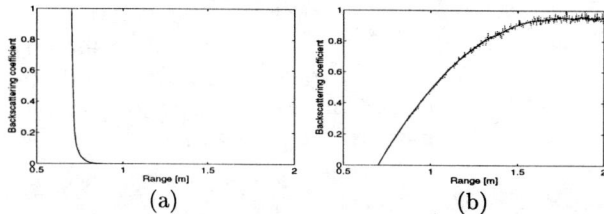

(a)   (b)

Figure 3: The plot of the values of $\mathcal{P}$ with respect to the distance r for (a) smooth ($\mathcal{R} = 1$, $\sigma/T = 4 \cdot 10^{-3}$) and (b) rough ($\mathcal{R} = 1$, $\sigma/T = 8.3 \cdot 10^{-4}$) surface. The integral was calculated using the Monte Carlo integration method (dotted line), and the results were smoothed by averaging (solid line).

the normal of the T/R to the surface plane, so the backscattering coefficient becomes very big for small angles of incidence. In contrary, when the surface is rough the waves are scattered at all directions and the backscattering coefficient is not so much dependent on the angle of incidence.

4 CTFM sonar image of surfaces

As we show in section 2, the CTFM sonar image can be determined by equation 5. The parameters involved in this equation are α, ϱ, and $\mathcal{D}$. The attenuation of the signal in the air and the directivity of the transducer can be expressed by the following equations [2]

$$\alpha(r) = \frac{e^{-a_{\lambda_m} r}}{r}, \quad \mathcal{D}(\eta) = \left(\frac{2\pi a J_1(ka \sin\eta)}{ka \sin\eta}\right)^2$$

where a_{λ_m} is the mean value of the absorption coefficient (≈ 1.3), a is the radius of the transmitter and receiver (0.01m), and finally η is the angle between the transmitted/received ray and the line-of-sight. It is calculated by intersecting a cone centered at the transducer's line-of-sight towards the surface with the plane ($z = 0$) and is given by

$$\eta = \mathrm{atan} \frac{\sqrt{(r_a \cos\phi_a - r_{00}\sin\theta)^2 \cos^2\theta + r_a^2 \sin^2\phi_a}}{((r_a \cos\phi_a - r_{00}\sin\theta)\sin\theta + r)^2} \quad (22)$$

r_a is the radius of the circle on the surface, ϕ_a is the angle of the scattered wave on the circle, and θ is the angle between the line-of-sight of the transducer and the vertical to the surface.

In both expressions for α and $\mathcal{D}$ the wavelength is assumed to be constant and equal to its mean transmitted value. This approximation is very close to the real solution.

When a rough surface is examined, the amplitude of the image at a specific distance will be the sum of all the scattered rays on a circle on the surface plane with center the projection point of the T/R on the surface received by R. For all these scatterings, the terms α and $\mathcal{D}$ are constant in terms of time (wave number) and r, so we have

$$S_a(r) = \frac{A^2}{2\mathcal{J}}\alpha(r)\int_0^{2\pi}\mathcal{D}(r,\eta)\varrho(r,\theta)d\phi_a \quad (23)$$

where the angles η and θ are related with the angle ϕ_a of the integral from equation 22. Because the angle θ is constant on the integration line and since we know that the backscattering coefficient for a rough surface calculated in section 3 corresponds to the scattering of all the rays emitted from the T and ended to R we can write

$$\mathcal{P}_{rough}(r) = \int_0^{2\pi}\varrho(r,\theta)d\phi_a = 2\pi\varrho(r,\theta) \quad (24)$$

Then equation 22 takes the form

$$S_a(r) = \frac{A^2}{4\pi\mathcal{J}}\alpha(r)\mathcal{P}_{rough}(r)\int_0^{2\pi}\mathcal{D}(r,\eta)d\phi_a \quad (25)$$

For a smooth surface $\mathcal{P}_{smooth}(r)$ is given by a similar expression to that of equation 24 while $S_a(r)$ will have the same form of equation 25.

A different surface pattern is the moderately rough surface, where the characteristics of both the smooth and rough surfaces are evident. In this case the received energy from the scattering on the circle of the surface can be decomposed in the sum of scattered waves caused by a smooth surface and those by a rough. Then equation 25 can be rewritten as

$$S_a(r) = \frac{A^2\alpha(r)}{4\pi\mathcal{J}}\left(\mathcal{P}_{smooth}(r) + \mathcal{P}_{rough}(r)\right)\int_0^{2\pi}\mathcal{D}(r,\eta)d\phi_a \quad (26)$$

As a result the CTFM sonar image of a smooth, a moderately rough and a rough surface has the shape illustrated at Fig. 4, where the theoretical results are compared with experimental measurements.

5 Rough surface classification

Based on the modeling of the CTFM sonar images of smooth, moderately rough, and rough surfaces, we will try to point out the physical differences between these three classes.

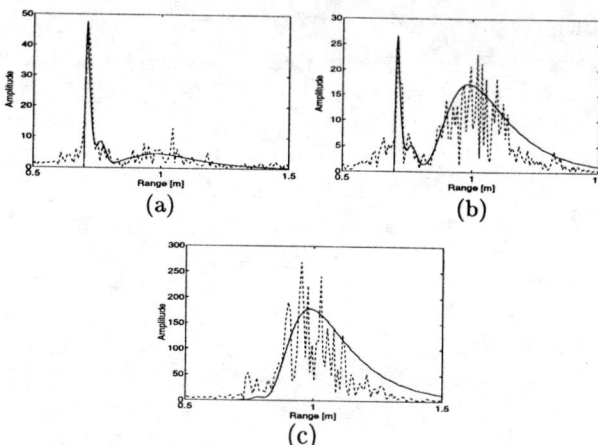

Figure 4: Comparison of the theoretical (solid lines) and experimental (dashed lines) sonar images of (a) a smooth (hard smooth floor), (b) a moderately rough (carpet), and (c) a rough (asphalt) surface. The values for $\mathcal{R}$ and σ/T are the same as in Fig. 3.

The smooth surface (hard smooth floor) behaves like a mirror, so we get back only specular reflections. This means that wherever the angle between the line-of-sight and the normal to the surface is, the only reflections we will get back will come from the normal point of the T/R to the surface. So the image will be composed of only one peak at the normal distance whose amplitude will depend on the declination of the line-of-sight with the surface normal.

In contrary to the smooth surface, every point of the rough surface (asphalt) behaves as a diffuse reflector. This practically means that R can receive a scattered ray from any point on the surface at any distance. The final image will be dependent mostly on the directivity of the transducer.

A moderately rough surface (carpet) is the combination of the previous two surfaces. So the image will be composed of a specular and diffuse reflections from various distances.

By observing the images of Fig. 4 we can discriminate the three classes very easily by using the following feature

$$\mathcal{F} = \frac{E_{smooth}}{E_{rough}} \quad (27)$$

where E_{smooth} and E_{rough} are the energies of the smooth and rough component of the image respectively. The values of $\mathcal{F}$ can be assigned to each class according to the following scheme

$$if\ \mathcal{F}\begin{cases}\gg 1 & \longrightarrow \omega_s \\ \approx 1 & \longrightarrow \omega_m \\ \ll 1 & \longrightarrow \omega_r\end{cases} \quad (28)$$

	ω_s	ω_m	ω_r
Mean($\mathcal{F}$)	11.815	0.779	0.033
Std($\mathcal{F}$)	3.502	0.436	0.014

Table 1: The parameters of each class.

	ω_s	ω_m	ω_r
Max($\mathcal{F}$)	24.158	1.743	0.067
Min($\mathcal{F}$)	7.559	0.232	0.017

Table 2: The feature's range for the testing data.

where ω_s is the class corresponding to a smooth surface, ω_m to a moderately rough and ω_r to a rough.

Experiment verification

All the measurements were taken using a sensor with $R = 0.7$m and $\theta = 45°$. Since for every image the values of R and θ are the same, we expect the peak corresponding to the specular reflection to be around 0.7m, while the components corresponding to the diffuse reflections to be around 1m. In order to estimate the energies related to E_{smooth} and E_{rough} we use the following expression

$$E_{smooth} = C \int_{0.7}^{0.8} S_a^2(r)dr, \quad E_{rough} = \int_{0.8}^{1.4} S_a^2(r)dr.$$

where $S_a(r)$ is the image amplitude at a distance r, and C is a scaling constant.

To classify the three classes we used a neural network with two layers of 4 hidden units and 2 output units. The neural was trained by a set of 60 measurements from each class using the back-propagation learning rule. The mean value and standard deviation of each class are presented in Table 1. Since the classes are very well separated we expect a very high recognition rate. We tested the system using a set of 60 images, 20 from each class. The recognition rate achieved was 100%. The maximum and minimum values of the feature for each class of the testing data are presented at Table 2.

6 Conclusions

A method for classifying different texture patterns of rough surfaces was presented based on geometrical scattering modeling. It has been proven that surfaces such as hard smooth floor, carpet and asphalt can be very easily identified by using a single feature from the CTFM sonar image which incorporates the basic physical differences between them. It does not require a scan of the surface, complicated device structures, or complex preprocessing algorithms. The method can be extended to include more surface classes, which can be modeled by developing equation 13 according to the statistical characteristics of the surface heights and classified by extracting any proper additional features.

Although the results were preliminary, they exploit the potentials of this method for applications in robotics, like recoginizinging and/or tracking pathways of particular textures and estimating the transient points between them. Also its advantages in terms of simplicity and speed makes it very useful in real time applications.

Acknowledgements

One author, Z. Politis, is supported by the State Scholarships Foundation (SSF) of Greece.

References

[1] L. Kay and M. A. Do. An artificially generated multiple object auditory space for use where vision is impaired. *Acustica*, 36(1):1–8, 1976/77.

[2] Z. Politis and P. Probert. Perception of an indoor robot workspace by using ctfm sonar imaging. *Proceeding of the IEEE International Conference on Robotics and Automation*, May 1998.

[3] Z. Politis. *Perception of a Robot Workspace by Extracting Navigational Information Using CTFM Sonar Imaging*. Transfer report, University of Oxford, Trinity Term 1997.

[4] G. Kao. *FM Sonar Modelling for Navigation*. Transfer report, University of Oxford, July 1996.

[5] O. Bozma and R. Kuc. A physical model-based analysis of heterogenous environments using sonar – ENDURA method. *IEEE Transactions on Pattern Analysis and Machine Intelligence*, 16(5):497–506, May 1994.

[6] C. Eckart. The scattering of sound from the sea surface. *The Journal of the Acoustical Society of America*, 25:566–570, 1953.

[7] O. Bozma and R. Kuc. Characterizing pulses reflected from rough surfaces using ultrasound. *Journal of the Acoustical Society of America*, 89(6):2519–2531, June 1991.

Laser Based Pose Tracking

Patric Jensfelt

Signal, Sensors & Systems
Royal Institute of Technology
SE-100 44 Stockholm

Henrik I. Christensen

Autonomous Systems
Royal Institute of Technology
SE-100 44 Stockholm

Abstract

The trend in localization is towards using more and more detailed models of the world. Our aim has been to answer the question "How simple a model can be used to provide and maintain pose information in an in-door setting?". In this paper a Kalman filter based method for continuous position updating using a laser scanner is presented. By updating the position at a high frequency the matching problem becomes tractable and outliers can effectively be filtered out by means of validation gates. Presented experimental results show that the method performs very well in an in-door environment.

1 Introduction

A notorious problem in mobile robotics is localization and maintenance of a position estimate. Traditionally position feedback is provided by odometry, but due to slippage etc., a drift that is proportional to the maneuvers is introduced. To circumvent this problem it is necessary to complement the odometry based estimation with input from other sensory modalities. Most mobile systems use ultra-sonic sonars for localization. An excellent overview of such systems can be found in [1]. The spatial resolution of sonars is limited, which requires significant post-processing of data to provide accurate position updating. An alternative to sonars is laser based position estimation. Traditionally laser scanners have been quite expensive and they have thus primarily been used in hostile or space applications, where cost is of secondary importance. Recently a new generation of laser scanners has been introduced into the market, which allow for use of such scanners even for regular applications. Some of the advantages of laser scanners are that they have the potential to deliver data at a high rate (over 25Hz) and that their spatial resolution typically is as good as 1mm to 1cm at an angular resolution of 0.5°.

With laser based scanners it is possible to provide very detailed models of the environment and utilize these for localization and pose tracking. One problem with the detailed models is that the computational complexity often limits the achievable updating frequency. In general the trend has been towards use of more and more detailed models. In this paper the opposite problem is addressed. The question raised here is "How simple a model can be used to provide and maintain pose information in an in-door setting?".

In the discussion of models two different approaches have been explored in the literature: i) grid-based [2, 3] and ii) feature-based [4, 5]. The grid-based approach is attractive as the semantic interpretation is very limited and it is straight forward to perform on-line updating. Unfortunately the model complexity can be very large, especially for large scale in-door environments, like our laboratory (10 by 70 meters). In the feature based approach laser scans are segmented into a set of features, e.g., line segments, corners, inflection points, etc. The features from the sensor is then matched to the model. The matching is potentially NP-hard, which can result in prohibitive processing requirements, when using detailed world models. In this paper we will use a feature based approach and investigate the use of "simple" world models to allow for fast updating.

Initially a model of the uncertainty for the sensor is developed, as a basis for uncertainty weighted fusion of readings. The problems of absolute localization and pose tracking are then introduced. From this discussion a least square estimator for pose tracking is developed using a Kalman filter approach. A key problem addressed is the filtering of data to provide a robust basis for matching. To simplify matching over time and suppress outliers, validation gates are computed and used for initial filtering of data. The developed technique is tested in an in-door environment where a significant amount of clutter is present.

The experiments involve evaluation in multiple rooms, to demonstrate operation in varying settings. The experiments also involve pose tracking while passing between rooms. Finally the results are summarized and avenues for future research are outlined.

2 Characteristics of the Sensor

In this work a proximity laser scanner, the PLS 200 from SICK Electro-Optics, is used. The SICK sensor can scan the environment at a rate of 25Hz. Due to hardware limitations (max serial speed 38.4 kBaud) we get a maximum sampling rate of 3Hz. The sensor is placed approximately 93 cm above ground on a Nomad200 platform as shown in Figure 1.

Figure 1: The Sick sensor is placed on top of a Nomad200 platform equipped with an array of sensors.

The SICK scanner uses a Time of Flight (TOF) ranging principle that is driven by a 6 GHz clock, which provides a ranging resolution of 50 mm. The laser scanning is performed using a rotating mirror that rotates at 25Hz. In practice the sensor provides a polar range map of the form (α, r), where α is the angle from 0° - 180° discretized into 0.5° bins. Analyzing the output from the sensor it is apparent that the uncertainty of the data is uniformly distributed over the ranges $[-25, 25]$ mm and $[-0.25°, 0.25°]$, respective. Converting the polar data into a Cartesian frame of reference we obtain:

$$\vec{c} = \begin{bmatrix} a \\ b \end{bmatrix} = r \begin{bmatrix} \cos\alpha \\ \sin\alpha \end{bmatrix}. \quad (1)$$

Let Δr and $\Delta \alpha$ define the size of the area over which the parameters r and α are distributed. The density functions can be written as $f_r(r) = \frac{1}{2\Delta r}$ for $r \in [\bar{r} - \Delta r, \bar{r} + \Delta r]$ and $f_\alpha(\alpha) = \frac{1}{2\Delta \alpha}$ for $\alpha \in [\bar{\alpha} - \Delta \alpha, \bar{\alpha} + \Delta \alpha]$. The variance in the a and b directions, respectively, can be derived from straight forward calculations assuming that the uncertainty in r and α is independent. The result being:

$$\sigma_{aa} = \tfrac{1}{2}(\bar{r}^2 + \tfrac{1}{3}\Delta r^2)\left(1 + \cos(2\bar{\alpha})\cos(\Delta\alpha)\tfrac{\sin(\Delta\alpha)}{\Delta\alpha}\right)$$
$$-\bar{r}^2 \cos^2(\bar{\alpha}) \left(\tfrac{\sin(\Delta\alpha)}{\Delta\alpha}\right)^2 \quad (2)$$

$$\sigma_{bb} = \tfrac{1}{2}(\bar{r}^2 + \tfrac{1}{3}\Delta r^2)\left(1 - \cos(2\bar{\alpha})\cos(\Delta\alpha)\tfrac{\sin(\Delta\alpha)}{\Delta\alpha}\right)$$
$$-\bar{r}^2 \sin^2(\bar{\alpha}) \left(\tfrac{\sin(\Delta\alpha)}{\Delta\alpha}\right)^2 \quad (3)$$

$$\sigma_{ab} = \tfrac{1}{4}(\bar{r}^2 + \tfrac{1}{3}\Delta r^2)\cos(2\bar{\alpha})\cos(\bar{\alpha})\tfrac{\sin(\Delta\alpha)}{\Delta\alpha}$$
$$-\bar{r}^2 \sin(\bar{\alpha})\cos(\bar{\alpha}) \left(\tfrac{\sin(\Delta\alpha)}{\Delta\alpha}\right)^2 \quad (4)$$

In our experiments we settled for using a, from a mathematical point of view, much simpler model of the uncertainty. We assumed the uncertainty to have an independent Gaussian distribution in a and b direction with a standard deviation of 50 mm. This a very crude model, but one that has proven to be adequate in our experiments.

3 Localization

Localization is an essential component in most systems. It has been studied extensively in the literature, using a variety of different sensors. As already mentioned in the introduction the most common sensor is simple odometry using for example quadrature encoders. The odometry is, however, subject to drift etc.

3.1 Previous Work

The ultrasonic sensor has been, is and will probably remain the most widely used complement to odometry for mobile robot localization. Many techniques has been developed, both grid-based techniques (e.g. Moravec and Elfes [3] and Borenstein & Koren [6]) and feature based techniques (e.g. Crowley [4] and Leonard & Durrant-Whyte [5]).

3.2 Absolute Localization vs Pose Tracking

Localization is here interpreted as re-initialization of the position estimate and a process that is carried out at a set of discrete instances, for example upon entering a room. Given the large uncertainty that is associated with localization, i.e. the position estimate might be off by meters, the process is often slow. This is, however, often of secondary importance as the process only is carried out at discrete instances. Given availability of frequent (and reliable) sensory information it is possible to change the localization process into a continuous tracking task in which landmarks in the environment are tracked over time and used for maintenance of the position and orientation (together termed the pose) estimate. Localization is then used for boot-strapping of the tracking and given limited motion between pose updates, the matching process is suddenly a tractable problem.

In an in-door environment the physical boundary (the walls) of a room can often be approximated as a rectangle. An interesting research issues is then, "is it possible to perform localization and pose tracking using a simple rectangular model of a room?".

4 Position Tracking

4.1 Idea

The idea behind the pose tracking method presented here is simple. By using the information extracted from detecting walls, a good estimate of the pose can be generated. Let $\mathbf{x}_k = \begin{pmatrix} x & y & \theta \end{pmatrix}_k^T$ represent the pose of the robot at time k. In the state space formulation of the system that follows, this is the state vector. By modeling the odometry as a deterministic input (with some noise) we get:

$$\mathbf{x}_{k+1} = \mathbf{x}_k + \mathbf{g}_k + \mathbf{w}_k \qquad (5)$$

where $\mathbf{g}_k$ is the input from the odometry and $\mathbf{w}_k$ represents the process noise. $\mathbf{w}_k$ is a measure of the odometric quality of the system. The characteristics of the odometry has been studied through a series of tests (see Appendix). The odometric noise is modeled as dependent on two factors, the distance traveled and the amount of steering. The uncertainty in distance traveled is 1%, and each meter traveled adds 0.6° in standard deviation to the orientation estimate. Steering the wheels increases the standard deviation in orientation by 0.1% of the amount of steering. Borenstein [7] points out one danger with having a statistical model for the odometry "they cannot anticipate the unpredictable and potentially "catastrophic" effect of larger bumps or objects encountered on the floor". This is a point well worth considering and it will have to be taken into consideration in the future.

The measurement equation can be described by

$$\mathbf{z}_{i,k} = \mathbf{h}_i(\mathbf{x}_k) + \mathbf{v}_{i,k}, \ i = 1, \ldots, N \qquad (6)$$

where $\mathbf{h}_i$ represents a possibly non-linear measurement function and $\mathbf{v}_{k,i}$ is the corresponding measurement noise. In our case the measurement update is performed in multiple steps, one for each wall, where each wall will give information about the position in either x or y direction plus the orientation. The transformation from the pose of the robot to the measurements from the wall is a non-linear function, but we have chosen to model all nonlinearities as part of the noise. This is possible as we have chosen the walls parallel to the coordinate axes and therefore a distance to a wall corresponds to a position in x or y direction, respectively (see Figure 2).

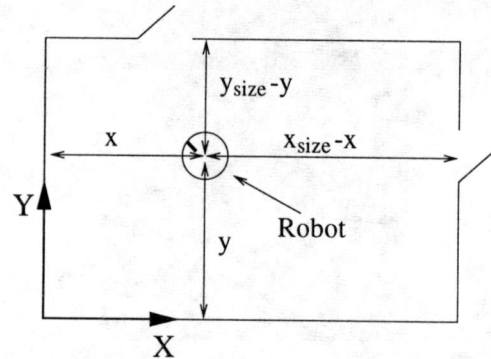

Figure 2: Each wall in a rectangular room will give information about either the x or y position in the room.

$$\mathbf{z}_{i,k} = \begin{pmatrix} x \text{ or } y \\ \theta \end{pmatrix}_{k,i} + \begin{pmatrix} \sigma_x \text{ or } \sigma_y \\ \sigma_\theta \end{pmatrix}_{k,i}, \ i = 1, \ldots, N \qquad (7)$$

Let $\hat{\mathbf{x}}_{k|k}$ be the estimate of the pose at time k given past measurements up until time k. The prediction stage of the tracking is then:

$$\hat{\mathbf{x}}_{k+1|k} = \hat{\mathbf{x}}_{k|k} + \mathbf{g}_k. \qquad (8)$$

The measurements are used in sequence to update the pose.

$$\begin{aligned}\hat{\mathbf{x}}_{k+1|k,i} &= \hat{\mathbf{x}}_{k+1|k,i-1} \\ &+ \mathbf{K}_{i,k}\left(\mathbf{z}_{i,k+1} - C\hat{\mathbf{x}}_{k+1|k,i-1}\right)\end{aligned} \qquad (9)$$

where $\hat{x}_{k+1|k,0} = \hat{x}_{k+1|k}$, $\hat{x}_{k+1|k,N} = \hat{x}_{k+1|k+1}$, $\mathbf{K}_{i,k}$ is the Kalman gain and C is either $\begin{pmatrix} 1 & 0 & 0 \\ 0 & 0 & 1 \end{pmatrix}$ or $\begin{pmatrix} 0 & 1 & 0 \\ 0 & 0 & 1 \end{pmatrix}$ depending on if x or y is measured.

4.2 Filtering of Data

The filtering of data is performed as follows (see dashed rectangle in Figure 3).

1. Data is run through four validation gates, one for each wall. This could be extended to be n walls, as well as other features.

2. A local Range Weighted Hough Transform[1] [8] is performed on each set of validated data.

3. A second stage validation is carried out. The validation uses position and orientation estimates that have been updated using the RWHT.

4. Finally Least Squares fits are made using the points that are close to each wall, corresponding to peaks in the RWHT.

Based on the filtering a set of zero to four wall hypotheses are available for parameter updating (using a rectangular room model). Each wall will have an uncertainty attached, which depends on the data that was used to hypothesize the location of the wall. We have used the model from [9] for the least square algorithm, which also provides estimates of the uncertainty for the line parameters. The representation of the walls is (ρ, φ) where $\rho \in [0, \infty)$ and $\varphi \in [0, 360°)$. ρ is a measure of the distance to one of the walls, i.e. it will give either the x or y position in the room. φ is a measure of the orientation of the robot.

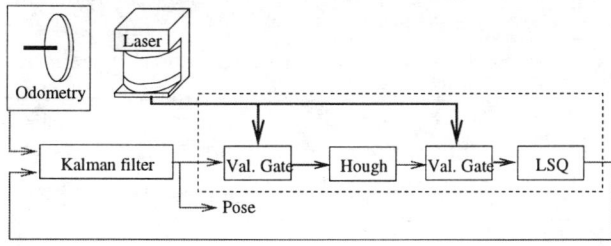

Figure 3: Signal flow in the localization module.

---

[1]Like a standard Hough transform except that the vote of a point far away carries more weight than a point close to the sensor. The idea is that points close to the sensor tend to be closer together and therefore they will influence more unless a range weight is introduced.

4.2.1 The Validation Gates

Given an initial estimate of the pose of the robot, it is possible to predict where the wall(s) will be in the data. The uncertainty in the pose as well as the quality of the data will determine the size of the validation gates. The locations of the validation gates are based on the prediction of the pose of the robot and the map of the current room. The size of each validation gate is defined by two parameters (see Figure 4), the size of the initial opening, δ and the opening angle, γ.

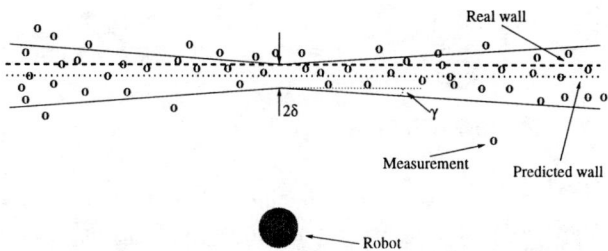

Figure 4: Validation gate where the error in the predication of the wall and the size of the validation gate were scaled for illustration purposes.

4.3 Managing the Size of the State Uncertainty

In a Kalman filter the so called innovation process, or the difference between what is predicted and what is actually measured, ensures convergence towards the true states. The amount of effect the innovation will have on the state estimate depends on the uncertainty both in the state estimate and the measurement. The lower the noise level is on the state estimate the less effect the innovation will have. This will cause a problem in the case of a mobile robot tracking its position since we want the measurements to compensate for wheel slippage etc. One way of handling this is simply to set a lowest allowed level for the uncertainty in the state estimate. This has proven most effective in our experiments.

In passing from one room to another the risk for unmodeled errors to occur is larger than otherwise as the robot might slip on the threshold and/or bump into things. In our experiment the uncertainties are increased with 100 mm in both x and y direction and the angular uncertainty is increased by 2° (the values refer to an increase in the standard deviation).

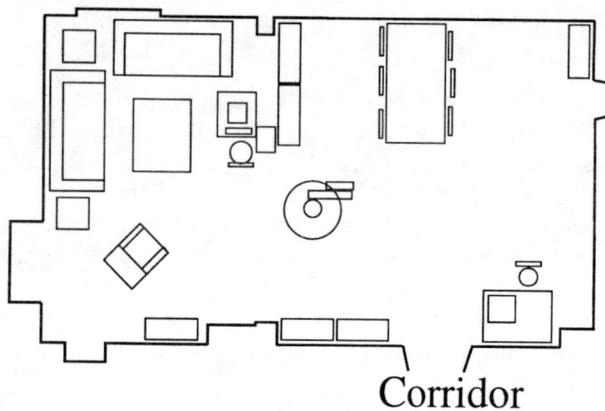

Figure 5: Sketch of the living-room seen from above.

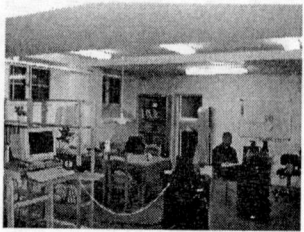

Figure 6: View of the south-east and north-west corners of the living room.

5 Experimental Results

In this section we will show that the presented method can handle rooms that are far from being sparsely furnished.

5.1 General Description of Experiments

The experiments have been performed on the Nomad200 platform shown in Figure 1. The localization module is a part of a larger system that already has capabilities to do obstacle avoidance, pass through doors and go between points, etc. No active sensing strategy has been used. The platform is equipped with a lift mechanism that makes it possible to pick up simple objects. The obstacle avoidance behavior in the system makes sure that the lift as well as the the platform itself do not hit anything. This means that even though no active sensing strategy has been implemented the direction of the sensor will change as a result of the obstacle avoidance.

5.2 Different rooms

At a first glance it might appear as though the method presented will fail as soon as there are no clear wall to be seen by the sensor. To prove that this is not the case we performed experiments in many different rooms in our laboratory. Here we will present the results from three different rooms.

Living-room Our mobile robot lab is setup as a living-room with IKEA furniture. The room is approximately 8.6 by 5 m. A top view of the room can be found in Figure 5 and two pictures of the same room from different view points are shown in Figure 6. As can be seen from the figures above the living room is quite cluttered. The only wall that provides a clear view is the wall on the right in Figure 5. Many experiments were performed in the living-room and only once did the robot make a significant error in the estimation of the pose. This occurred after having moved for minutes without having had any input from neither the top or the lower wall (refers to Figure 5). As the robot moved the uncertainty in that direction grew. As the uncertainty increased the size of the validation gates increased and more data were left unfiltered. Eventually the validation gates were open enough to let data that came from the bookshelves and the pillar on the lower wall form a hypothesis strong enough to be interpreted as the wall. When this occurred the estimate in the position was offset by approximately 300 mm, corresponding to the depth of the bookshelves. To tackle this situation the uncertainty in all directions were increased manually by 400 mm (standard deviation) and the robot was given a clear view of three of the walls (lower, left and top walls) which resulted in the robot returning to the correct pose again.

Office This is a typical office in our laboratory. The room is so small that the allowable movements of the robot are very limited. The rectangular model of this office room is approximately 5 by 3 meters. The office is divided by a cubicle divider into two parts. This room posed no problem either, partly because the limited movements in the room ensured this, even though two orthogonal wall were not seen all the time, the uncertainty never grew very much.

Corridor The corridor at the ground floor of our laboratory is approximately 55 m long and the width is about 2.3 m (see Figure 8). The corridor has been modeled as a single room in our experiments. It is obvious that the problem in the corridor is going to be maintenance of a good estimate of the position in the length direction of the corridor. Since the corri-

Figure 7: Office   Figure 8: Corridor

dor is modeled as a rectangle, the only two possible sources of information about the position along the corridor comes from the two short walls. When moving far away ($> \approx 15$ m) from the short walls they can no longer be used, as too few points are accumulated. This implies that almost half the length of the corridor has to be driven using only odometric information for position updating along the corridor. The biggest problem with the odometry is usually that rotational errors result in big translational error, especially over long distances. This is true for our platform too. In the corridor this could potentially lead to very big errors, but the fact that the orientation of the platform is estimated with high accuracy compensates for the rotation of the platform. The biggest error left is the error that comes from mapping the number of wheel rotations to distance traveled.

Many experiments were performed and in most cases ($> 75\%$) the robot were able to drive from one end of the corridor to the other while maintaining an accurate estimate of its pose. When coming close enough ($< \approx 15$ m) to the other end, the short wall would be tracked and the uncertainty would go down.

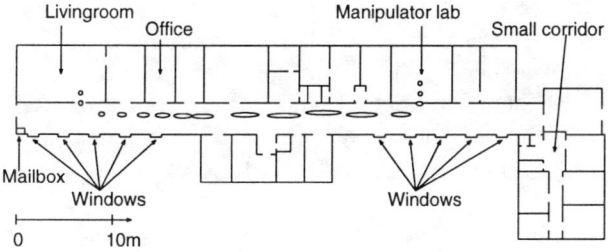

Figure 9: Result of moving from the living-room to the manipulator lab. The size of the uncertainty ellipses are made bigger to be seen easier.

5.3 Navigation Between Rooms

The experimental results in the previous section demonstrated that the pose tracking method worked very well in different types of rooms. ¿From the experiment in the corridor it is evident that the estimate of the position along the corridor might cause problems when moving in the mid area of the corridor. One way to solve this problem would be to extend the model and track the position of doors as well. As we are investigating the limit of using a "simple" model alternatives are more interesting. In Figures 9 and 10 the result of driving from the living room, through the corridor and into the manipulator laboratory is shown. It can be seen how the uncertainty grows large in the mid area where no input in the direction along the corridor is given. As the robot approaches the manipulator laboratory the short wall at that end of the corridor becomes strong enough and the uncertainty decreases. Two approaches for keeping the uncertainty can easily be identified i) use active sensing to maximize the use of the short walls and ii) go into a room on its way to update the position in the direction of the corridor.

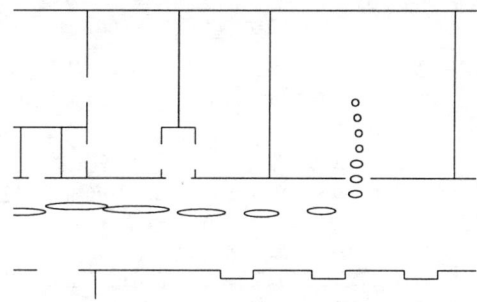

Figure 10: A close up of the situation when passing from the corridor to the manipulator lab.

6 Conclusions

It this paper we have shown that it is possible to continuously track the pose of a robot, modeling each room as a "simple" rectangle. The answer to the question posed at the beginning of the paper must thus be that the model can be this simple in a typical in-door office setting. It has been shown by experiments that the method can handle a variety of room types and that moving between rooms poses no problem either. On the contrary, when moving from one room to another, the position of the robot could potentially be updated very accurately as the door has a well defined position in the world. Having structures that form a line that are close to and parallel with a real wall can result in errors in the data association as was seen in the experiments in the living-room.

An active sensing strategy is still missing. Such a strategy might significantly improve the performance as in many situations the robot could have gotten much better information by looking in another direction. The living-room is a perfect example of this.

Future research will also address the questions: i) Allowable clutter before the tracking fails? ii) How non-rectangular must a room be before the method fails? I.e. does it fail if applied in an old building with non-orthogonal walls?

7 Acknowledgement

This research has been sponsored by the Swedish Foundation for Strategic Research through the Centre for Autonomous Systems. The funding is gratefully acknowledged.

References

[1] J. Borenstein, H. Everett, and L. Feng, *Navigating Mobile Robots: System and Techniques*. A K Peters, Ltd., 1996.

[2] S. Thrun, A. Bücken, W. Burgard, D. Fox, T. Fröhlingshaus, D. Henning, T. Hofmann, M. Krell, and T. Schmidt, *Map Learning and High-Speed Navigation in RHINO*, ch. 1, pp. 21–54. 445 Burgess Drive, Menlo Park, CA 94025: AAAI Press/ The MIT Press, 1998.

[3] H. Moravec and A. Elfes, "High resolution maps form wide angle sonar," in *Proc. of International Conference on Robotics and Automation*, pp. 116–121, IEEE, 1985.

[4] J. L. Crowley, "World modeling and position estimation for a mobile robot using ultrasonic ranging," in *Proc. of International Conference on Robotics and Automation*, 1989.

[5] J. Leonard and H. Durrant-Whyte, "Mobile robot localization by tracking geometric beacons," *IEEE Transactions on Robotics and Automation*, vol. 7, no. 3, pp. 376–382, 1991.

[6] J. Borenstein and Y. Koren, "Real-time obstacle avoidance for fast mobile robots," *IEEE Transactions on Systems, Man, and Cybernetics*, vol. 19, pp. 1179–1187, Sept. 1989.

[7] J. Borenstein and L. Feng, "Gyrodometry: A new method for combining data from gyros and odometry in mobile robots," in *Proc. of International Conference on Robotics and Automation*, (Minneapolis, Minnesota), pp. 423–428, IEEE, Apr. 1996.

[8] J. Forsberg, P. Åhman, and Å. Wernersson, "The hough transform inside the feedback loop of a mobile robot," in *Proc. of International Conference on Robotics and Automation*, vol. 1, pp. 791–798, IEEE, 1993.

[9] R. Deriche, R. Vaillant, and O. Faugeras, *From Noisy Edges Points to 3D Reconstruction of a Scene : A Robust Approach and Its Uncertainty Analysis*, vol. 2, pp. 71–79. World Scientific, 1992. Series in Machine Perception and Artificial Intelligence.

Appendix

A series of tests were performed in order to characterize the odometric noise. One of the test was the UMBmark [1]. This test was designed for a differential drive platform, but it still gives clues to the performance of the odometry on our synchrodrive Nomad200 platform. In theory the platform base should have the same direction at all time, but moving the platform changes the direction of the base. We decided to use the distance traveled and the amount of steering to parameterize the odometry model.

When driving the robot in our long corridor it was found that the scaling was wrong. In order to get a zero mean estimate of the distance traveled, the value given by the odometry has to be scaled by a factor 1.006. This is part of the odometric model. The standard deviation was about 1% of the distance traveled.

Another test that was performed was to look at how the steering effects the orientation of the platform. By placing the robot in front of a wall with a laser pointer at the wall, the change in angle can be measured when the wheels move. The tests indicated a bias of 0.1% of the amount of steering with a standard deviation of about the same magnitude.

Translating the robot without steering will still effect the orientation of the platform. For each meter traveled the platform will rotate about 0.6%. The problem here is that it is not perfectly predictable, i.e. a change in orientation is not always be observed. This was modeled as a bias of 0.6% with a standard deviation of the same magnitude.

Further tests have to be conducted to get a more truthful model of the odometry. It is good to have an accurate model as long as nothing unforeseen happens. When something unforeseen occurs trusting in the odometric model might prove fatal. It is therefore of great importance to be able to detect situations where the odometry fails.

Limbless Locomotion: Learning to Crawl

Kevin Dowling[1]

The Robotics Institute, Carnegie Mellon University

Abstract

This research develops a general framework for teaching a complex electromechanical robot to become mobile where sequences of body motions alone provide progression. The framework incorporates a learning technique, physical modeling, metrics for evaluation, and the transfer of results to a snake-like mobile robot. The mechanism and control of a 20 degree of freedom snake robot is described and multiple gaits are demonstrated including novel non-snake-like gaits. This research furthers the design and control of limbless robots.

Introduction

Biological snakes are pervasive across the planet; their diverse locomotion modes and physiology make them supremely adapted for the wide variety of terrains, environments and climates that they inhabit. It would be wonderful to capture these broad features of movement and capability in man-made equivalents. While wheeled and walking machines have undergone decades, even centuries, of development, they are still limited in the types of terrain they can traverse. A snake-like device that could slide, glide and slither could open up many applications in exploration, hazardous environments, inspection and medical interventions. This research addresses robots that crawl and slither without the use of wheels or legs; *where body motions alone enable progression.*

Potential advantages of serpentine locomotion include high stability (you can't easily fall over), good terrainability, high traction due to surface contact (can exceed product of mass and friction coefficient), redundancy (tolerance of joint failure), and a sealed exterior. Disadvantages include a large number of degrees of freedom to control, limited payload capability, and thermal control issues due to the high surface area to volume ratio.

As shown in Figure 1, while several works have bridged physical simulation and learning and others have bridged learning for robots, little has been done in bridging all three areas. As systems grow in complexity and as classical modeling techniques become more intractable, then connecting these areas will be critical. This work takes a step towards bridging the areas of mechanism, learning and simulation.

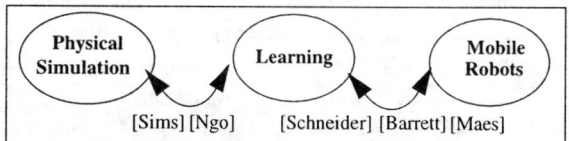

Figure 1: Bridging physical simulation, learning and robot mechanism for complex mobile robots.

---

1. Currently at PRI Automation, Billerica, MA email: kdowling@pria.com

Framework

A snake robot mechanism is relatively complex; the design is a repeated structure of many identical links, all of which need to be coordinated and each of which have to be controlled. The fundamental issue is determining the sequences of motion of the elements that move the mechanism in a particular direction. While it may be possible to construct sequences by hand that move the system, this is fairly tedious, inefficient, and likely to miss a variety of interesting gaits. Schemes tried in some previous works appear inadequate for determining and expressing a variety of gaits for this mechanism; you need to know the gait before attempting it. Ideally, the robot would move forward after learning how to move. The problem then becomes: how to teach the robot to move.

Figure 1 shows the development of a structure or framework to allow both representation and development of gaits. It is a modular framework with the following elements: learning, simulation, and evaluation. Other elements include techniques or protocol for communicating and passing information from one element to another and selection of parameters and the form of the parameters that are passed back and forth within the framework.

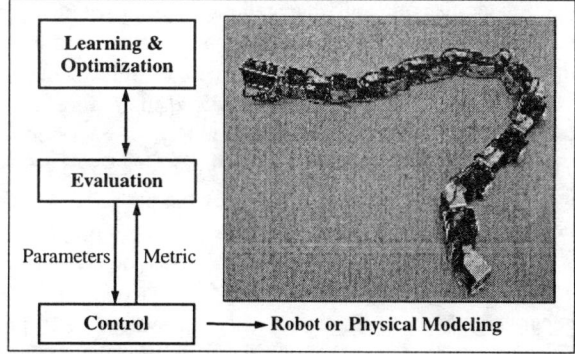

Figure 2: Learning framework

Physical modeling is incorporated into a larger scheme where control and evaluation take place. This scheme provides a means of testing the results of various control methods. Expanding on this idea further, a machine learning technique, using an appropriately chosen metric, interface to the modeling and control techniques for its own use.

The impetus for modeling at all is that testing and evaluating motions on a real system will be both time consuming and poses problems for mechanism reliability and lifetime. By developing gaits in simulation, mechanism wear and tear and much time can be saved.

In particular, control parameters are selected and are operated on by a learning function. These parameters, in turn, become the inputs to the physical simulation. As the simulation learns to behave in a desired fashion, using an appropriate metric, then these results can be ported to a physical device, the robot, for testing and further refinement. This testing can still be done in the context of this original framework, by substituting the robot for the modeled device.

The framework of optimization and learning, evaluation, control, and physical modeling provides a loop of test and iteration that culminates in the performance in physical testing. The results in each set of tests provide input into the next. In the following sections we will examine each component of the framework.

Learning and Optimization

Learning provides the generation and selection process for new sets of parameters to determine the 'best' set of these values. These are used in the course of determining and tuning efficient gaits for the mechanism.

While the technique selection is important, knowing what to optimize, and knowing how to evaluate a solution are even more critical. *What to Optimize* is a metric, specific resistance. *How to Evaluate* is determining this from simulation or measurements from the robot. With the metric and the learning method combined with physical modeling a complete framework can be constructed.

Learning provides improved performance with frequent testing and evaluation. After examining and testing a number of learning methods, I selected a stochastic technique, genetic algorithms or GAs, and, later, probabilistic-based incremental learning or PBIL.

Baluja introduced Probabilistic-Based Incremental Learning (PBIL) as a means to provide the same functionality as other stochastic methods such as GAs but in a more efficient manner. Rather than implicitly maintaining statistics within a population, as GA's do, PBIL explicitly maintains the statistics for the genome. PBIL's compactness, ease-of-implementation, effectiveness and simplicity made it a good choice for evaluating metrics in physical simulation. [Baluja]

This encoding of solutions as statistics can, in many cases, be far more efficient than traditional GA methods. The executions are more efficient, use fewer cycles, and converge more quickly. It is also very compact, the evaluation code in my implementation is not written as a separate function but simply integrated into the evaluation loop. In this system, PBIL provides similar, if not better, performance than GA's with lower overhead.

There still remains the problem of selecting a way of representing the information presented to the learning technique. The general problem is to represent a wide variety of progressing waveforms that provide a systematic displacement of the robot. The waveform parameters are adjusted during the learning process to provide efficient motions of the snake joints and thus provide locomotion. Representation issues include compactness, calculation overhead, complexity of implementation and of interpretation, as well as the correspondence between learning representation and parameters values used in control. Representations tried included trigometric waveforms, fourier coefficients and tabular arrays:

- A trigonometric waveform approach to representation using parameters of magnitude, frequency and phase and offset is a powerful one and this was used for a number of earlier gait applications on a smaller 2D caterpillar robot. Magnitude and frequency are obvious and phase represents the shift of the waveform along the body. (Joint values, not body shape) An additional parameter, offset, can be used to provide a nominal starting configuration. For example, the body may form a helix which is deformed to provide locomotion. Sinusoidal patterns are also easy to represent and parameterize. The problem is they are too restrictive in representing arbitrary time-varying waveforms.

- The Fourier series can represent any periodic function as a sum of exponential terms, usually sine curves. A series can be constructed where angular values of the angle between links are a function of time and link. Since all gaits, by definition, are periodic sequences, the Fourier series can represent them. The mapping of time and joints correspond to the coefficients of the Fourier series which are magnitude and frequency values. A table of values, in turn, can be represented in a binary string which facilitates operation within the learning framework. The values in the string can be the coefficients of the values in the array.

- Finally, rather than trying to explicitly represent all manner of waveforms or trying to identify parameters to refine for different modes of locomotion, a conceptually simpler route is to directly represent the joint angles over time. By explicitly representing the angular position of the links within a one-dimensional tape, the joints can be essentially 'masked' off and the snake shifted through the tape, adjusting the positions to reflect the tape values as shown in Figure 3. By adjusting the set of tape values in physical simulation and looking for effective modes, a variety of locomotion modes can be represented, including those that are not exhibited in natural snakes. This representation then forces the time-histories of the individual joints to be identical but shifted in time.

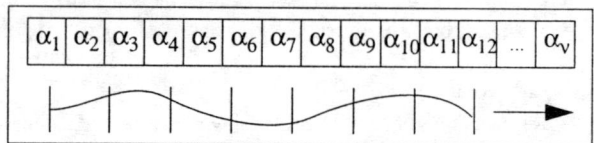

Figure 3: A snake 'tape' defining joint angles at each time step for a periodic waveform.

A more general extension of the 1-D array approach is to represent the joint angles in a column of a 2-dimensional array and, with each timestep, march across the columns where the column entries represent time steps for a particular waveform. This works well not only for periodic forms, but also for locomotion forms such as concertina, a accordion-like locomotion mode, where the time history of different joints is not the same. The array becomes the representation of each joint angle or, even more compactly, the difference between the angles in adjacent time slices. See Figure 4.

The values in the table can also be constrained to represent the likely sequences. This constrains movement between slices and joints. This has a two-fold benefit: first it prevents abrupt jumps and it culls unlikely gait

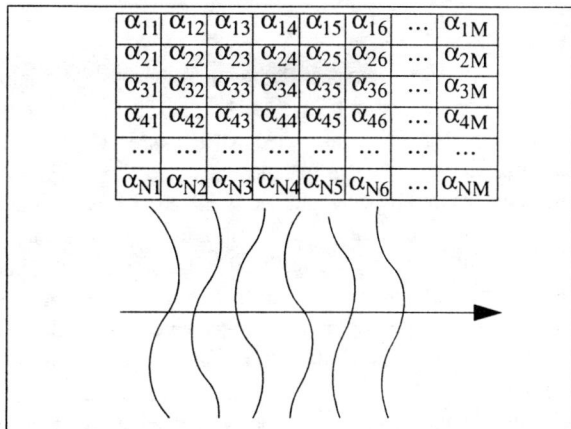

Figure 4: A 2D array representing joint motions over time.

patterns and body contortions, and finally, it reduces the gait space significantly.

The representation structure chosen for the framework is a tabular array with entries that directly represent desired positions of body segments. This representation is the most general and does not depend on combinations of mathematical functions to attempt all possible configurations; it represents those configurations directly. The danger in the tabular approach is that it opens the search space further, but the generality is worth the risk.

A random walk doesn't work for generating table values because the distance of a random walk is related to the square root of the number of time steps. Array values are generated by a fourier technique where magnitude and phase are defined to create a waveform with the right properties. The magnitude function is defined as 1/f, the power spectrum for noise. This creates a rolling cutoff to prevent high spatial frequencies where the joint angles change abruptly. The phase is then seeded with random values and the coefficients are calculated and then the series is created.

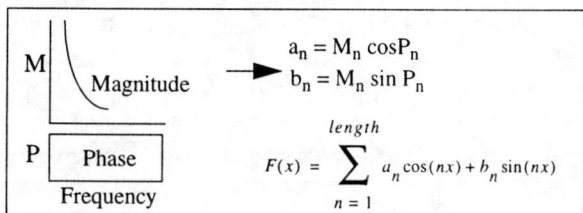

Figure 5: A fourier series is used to generate joint positions for tabular method.

Another issue is the size of the array. For example, if a gait sequence takes 2 seconds before repeating and the robot has 20 DOF, then using a minimum 10Hz update rate gives 400 parameters. Each of the parameters might use a full 8 bits or, if differences are used, 2-3 bits. This gives a a total of a few thousand total bits, which creates a very large set of parameters to adjust. However, while the space of configurations described in this model is large, the representation is simple, easily understood and easily mapped to the robot.

Evaluation and Metrics

The general question is: *how to evaluate the performance of mobile robots?* Specific questions for evaluating performance include: how fast does it move? how much energy does it use? how far can it go? how well does it carry out a task?

Teaching a serpentine robot to crawl requires evaluation of locomotion strategies or the use of metrics. Metrics are measures of performance and the selection of good metrics is critical to learning methods because they provide a means to compare the results, techniques and methods of locomotion. In addition to providing assessment and comparison, they should be easily calculated and measured. Metrics are used to evaluate and drive the learning. Thus, the simpler and more straightforward the metric, the easier it is to track and guide learning.

One example measure is velocity, the maximum distance the device moves in a given period of time. The metric can take time, energy, distance and other measures into account in determining the efficacy of that set of parameters. In a sense, this is a measure of the efficiency of the particular sequence of body motions that effect forward movement.

After generation and evaluation of many such metrics, I selected specific resistance as the measure of performance for learning locomotion. It provides a non-dimensional notion of energy, time and weight of the robot. Power/(Weight x Velocity). Weight is obviously a constant, and it utilizes two measurements that can be determined from both the robot and from the physical model. It is easily and quickly calculated and provides a clean and understandable metric for evaluation during the learning process [Gabrielli].

Metrics reveal only how well a vehicle did on a particular performance measure. It does not reveal why, although it can provide clues, and, finally, it does not directly reveal how to make the performance better. It can be used as a tool to ascribe trends through the use of small changes in the control techniques and hence, develop a better understanding of what makes a better gait. The metric is used as part of the learning process and placed into the overall framework for teach the snake robot to locomote.

Whatever the particular metric value, it is not a good idea to draw too many conclusions or provide close comparisons to other robots. It is too easy to contrive a metric that favors a particular robot. It is also too easy to draw conclusions about vehicles that don't take environment and task into account.

Control

From the optimization, a set of parameter values is chosen and these are used to run a program or controller to implement and execute a run. Although Control is shown as a distinct box within the framework, it is really two separate components; one to interact with Physical Modeling and the other to interact with the physical robot. The Control module handles interaction with Parameters generated by learning and maps those parameters to either the control and sequencing of the physical model or the robot.

Physical Modeling.

The key element for the framework is the ability to simulate the geometry of the physical robot and its interactions with external surfaces and contacts. Physical simulation allows the modeling of complex shapes and their interaction. Physical quantities that are modeled include torques, forces, velocities, acceleration, both kinetic and potential energies. Modeled quantities in physical simulation also include gravity, dynamics, masses, and material properties such as friction and density.

Physical modeling is a useful tool for the configuration and control of complex mechanisms and allows observation of these robots in a simulated environment. The rationale for physical modeling and simulation is to represent a physical device such that inputs and resultant outputs are reflected accurately and, in turn, provide an understanding of mechanism behavior. The user configures geometries, defines physical relationships between geometries, sets forces and any time-dependent relationships. The physical modeler then creates the simulated world and allows the user to observe the subsequent behaviors and interactions.

Coriolis, a toolkit developed at CMU to support interactive simulation, was used for modeling. The toolkit is implemented as a set of C++ classes that are instantiated to represent relationships and forces in the simulated world. Class types within Coriolis include Bodies, used to represent physical objects and their geometries, Constraints, used to describe relationships between different bodies and, finally, Influences, used to describe the forces that act upon Bodies. Coriolis does not provide graphics, I/O formats or an interface and these are written by the user [Baraff].

The framework is also designed to be used on the robot mech-

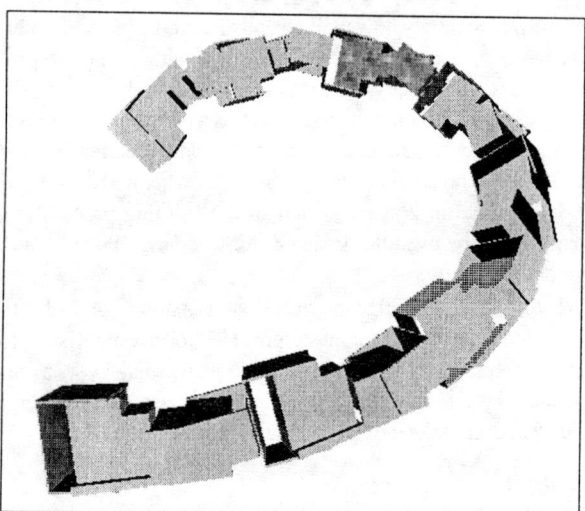

Figure 6: An accurate 3D model of the physical snake.

anism. The physical modeler is replaced by the robot mechanism, as shown in Figure 7, and the same criteria and optimization techniques can be used on the real mechanism after the physical simulation produces useful results.

However, even with a great deal of effort to correctly model the physical mechanism, the simulation does not model the robot perfectly; the vagaries of the real world prohibit accurate and high fidelity predictions of behavior. However, the initial physical model is used to provide general gait forms that can be refined, through the same framework, on the real system.

Implementation

A first generation robot, an 8DOF caterpillar, provided a useful tool for testing and refining the framework. The snake robot, show below, is constructed of 10 links, each with 2DOF. Each joint uses two high performance servos for actuation. Although each joint was capable of subtending excursions of almost 180 degrees, this range of motion wasn't necessary for the modes of locomotion developed in this work. In fact, there are diminishing returns even for threading though tight areas because motion limitations are due to aspect ratio and size of links rather than joint excursion. Joint excursions past 20 degrees show little advantage; short joints show greater efficacy in moving through tight areas. This geometry issue is further developed in [Dowling].

Figure 7: The 20 DOF snake robot

Specifications for the 20 DOF serpentine robot:
- Mass: Mechanism is 1.32 kg. Total mass, including wiring and controllers is 1.48kg.
- Length: 102 cm (10 links, each 1.02cm long)
- Diameter: 6.5 cm
- Power: 24.2W max total mechanical output. 75W max total electrical input.
  Quiescent: 1.15W (9.5mA @ 6VDC)

Additional work in the area of snake 'skins' was also done and a wide variety of materials were tried and evaluated. The resulting skins made from stretchable fabrics provide protection, a low-pass filtering of the mechanics akin to a spline, and a smooth ground contact interface. This is accomplished while being faithful to the shape and movement of the underlying mechanism.

Electronics and control utilized a serial bus and offboard computing and power for communications and energy.

Experimental setup

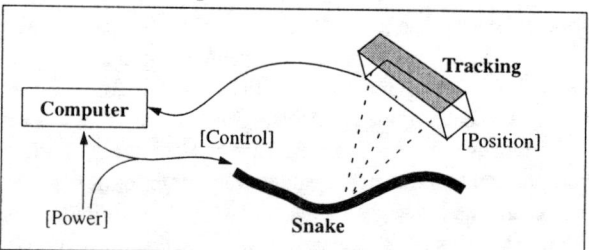

Figure 8: Experimental runs used tracking system and power monitoring to determine metric values.

Measuring the metric value on the real system was accomplished through power monitoring (current and voltage) and a tracking device. The tracking device provides data rates of up to 1000Hz and tracking accuracies to 0.1mm. Ideally, the position data would be for the center of mass of the robot, but this would require a tracking LED at each link and wires to connect them to the strobers. A single tracking LED was used near the middle of the robot to provide travel distance over the experimental time. Although a single LED does not provide the true center of mass, it gives a reasonable approximation to movement of the robot. The maximum distance, or error, that the center of mass can be from the center link is one-quarter of the length of the snake robot; this is a pathological configuration. Most configurations, especially symmetric ones, result in small differences between center of mass and center link position.

Results

Experiment began with physical modeling and then were transferred to the robot. One of the impetuses for physical modeling was to evaluate configurations and gaits much faster than would be possible on a real robot. However, the physical simulation speed is a function of computing power and the number of ground contacts (of which a snake has many). Speed was slower than real-time. Useful simulations could take several hours for a particular set of parameters and parallelism, running on multiple computers, helped.

A variety of interesting patterns did eventually emerge and some of these gaits are shown here: Sidewinding, ventral wave (akin to a rectinlinear motion), lateral rolling, the wheel and others.

Sidewinding is a relatively efficient mode of locomotion with little sliding ground contact but with an odd means of moving laterally. Sidewinding is really two waves, one ventral and one lateral that are out of phase. Together they produce a motion that moves a body section to the side and the rest moves along and settles the body down to form successive parallel tracks as it moves. Figure 9 shows the sidewinding result from physical modeling.

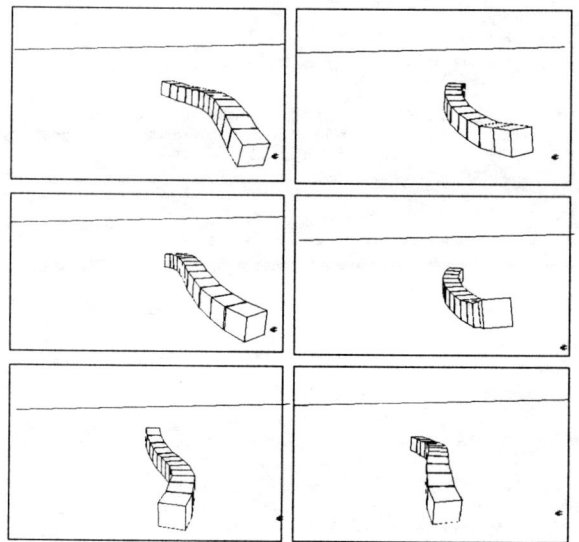

Figure 9: Sidewinding locomotion

Additional gaits demonstrated, shown in Figure 10, included rectinlinear motion, lateral undulation, and a varient of the lateral rolling I term the smoke ring. which could be used in climbing structures such as pipes by wrapping about the object and rolling upwards. The fourth gait is a form of traveling wave where a small amplitude wave is propogated around the ring; akin to the traveling wave ultrasonic motor. The final one is flapping motion, similar to the arm motions in a butterfly swimming stroke.

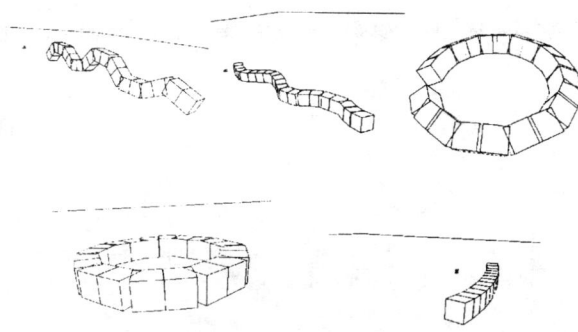

Figure 10: Additional generated gaits included trectinlinear, lateral undulation, 'smoke ring', rotary traveling wave and flapping locomotion.

An intriguing gait is formed by a U-shaped body and providing oscillating motions about each joint. I term this lateral rolling and this sequence in shown in Figure 11.

It appears, at first glance, that there is a continuous rotation of the joints but this, of course, is not possible and unnecessary. The gait is similar to sidewinding in that it uses two waves out of phase: a lateral sine wave and a ventral cosine wave. However, in this case the phase of the waves is zero; all joints on similar axes move together.

The development of the gaits did not proceed in the manner originally envisioned; In many cases these gaits can be described by elementary formulations. However, it is also clear that natural snakes do not use these formulations. A number of these gaits use a simple family of forms to describe joint and subsequent body motions.

```
Amp * Sine(Time/Period + Phase(link)) + Offset
```

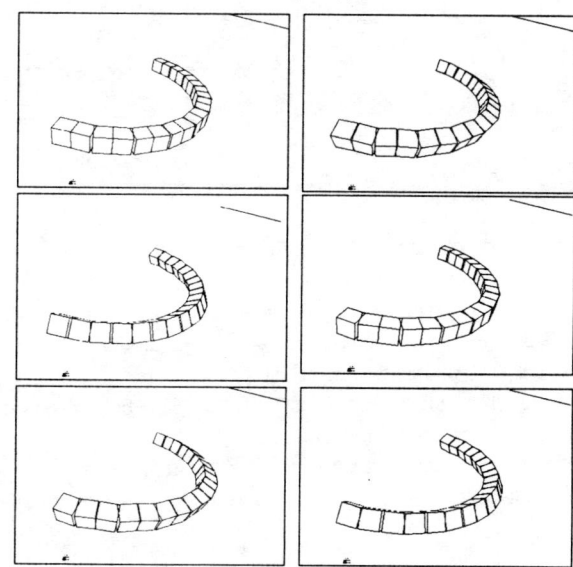

Figure 11: Lateral rolling

The phase is usually the product of a phase value and the joint number. This produces a traveling or propagating wave down the body of the robot. Zero phase, obviously, produces the same motion at all joints simultaneously, whereas a phase of Pi phase produces alternating and opposite motions in adjacent links. Offset can also be particular to a specific joint. For example, the body could describe a nominal 3D shape, such as a helix and use that as a base from which to propagate other waveforms.

Note that this formulation describes the motion of the *links* and not the waveform of the body. In fact, it is the *phase* that most nearly describes the body form. This formulation can also be scaled to be proportional to the link number of the inverse. Additional features of these formulations include an amplitude factor that is proportional to the link number. This allows heading changes in the gait so that the system can be 'steered' in a desired direction.

Run-times in physical simulation for longer snakes went up quadratically and, in some cases, even worse. The runtime depends mostly on the amount of contacts which, for a snake, are numerous. In addition, there is a lot of coupling between links that are closed through the ground or surface. These closed kinematic chains, plus the large number of contacts, contribute to substantially longer run-times.

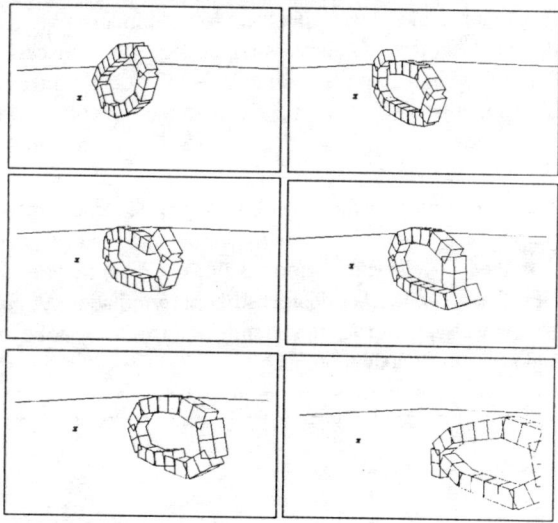

Figure 11: Reinventing the wheel

These, and other results, were ported to the real robots and shown to work by using the framework as well. Specifically, the wheel and rectilinear motion were shown on the caterpillar robot and lateral undulation, rectilinear and lateral rolling were shown on the 20 dof snake robot. Transitions to the real system show close form of the derived gait, but substantial ground slip-stiction interaction as well. The electrical system of the first generation snake robot exhibited noise problems and joint interactions that were filtered and reduced but never completely eliminated.

Conclusion

The old aphorism, "you have to crawl before you can learn to walk", has not applied to mobile robotics. There have been many walking machines over the decades but few crawling machines. In fact, crawling appears to be a harder problem. In this research, a snake robot and framework demonstrates learning to crawl and crawl in several different ways.

Robots can learn to locomote even when they have no wheels or legs. A general framework is used teach a complex electromechanical robot to become mobile that includes a learning method, metrics for evaluation, physical simulation and the transfer of results to two robots.

A number of interesting gaits were developed and, surprisingly, many of these were non-snake-like. The gaits in simulation were able to provide a good measure of performance relative to specific resistance. It was also found that the ideal path from simulation coupled to learning and then learning coupled with the robot did not provide the ideal path for gait development; the ideal serial process became an iterative one.

The confluence of new generation design tools, physical modelers, recent advances in learning and a novel mechanism promise to bring advances in robot design. Physical modeling, a relatively new tool primarily developed for use in the graphics community, can provide the designer feedback and provide a tool in a larger context; evolutionary design. The framework shown here, comprises modeling, simulation, evaluation and control of the robot.

Is this research extendable to other mechanisms and forms? The framework and loop of learning, testing and evaluation is certainly applicable to a wide variety of domains for physical control. For locomotion, all patterns of motion, gaits, can be described in terms of cyclical or periodic forms and this architecture lends itself well to learning those modes of locomotion.

References

Baraff, David, "Coriolis Documentation," Carnegie Mellon University and Physical Effects, Inc., 1997.

Barrett, D., "Optimization of Swimming Locomotion by Genetic Algorithm," Recent Trends in Robot Locomotion Workshop, ICRA, 1997, Albuquerque, NM. April 21. 1997.

Baluja, S., Caruana, R., "Removing the Genetics from the Standard Genetic Algorithm," Proceedings of the Twelfth International Conference on Machine Learning, Lake Tahoe, CA. pp. 38-46, July, 1995.

Chirikjian, G.S., *Theory and Applications of Hyper-Redundant Manipulators.* Ph.D. Thesis, California Institute of Technology, Pasadena, CA 1992.

Dowling, K., "Limbless Locomotion: Learning to Crawl," Ph.D. Thesis, Carnegie Mellon University,, Pittsburgh, PA 1997.

Gabrielli, G., von Kármán, T.H., "What Price Speed?," Mechanical Engineering. 72: 775-781, 1950.

Hirose, S., *Biologically Inspired Robots: Snake-like Locomotors and Manipulators,* Oxford University Press, 1993, ISBN 0 19 856261 6.

Sims, K., "Evolving Virtual Creatures," Computer Graphics, Annual Conference Series, SIGGRAPH '94 Proceedings, July 1994, pp. 15-22.

Maes, P., Brooks, R., "Learning to Coordinate Behaviors," Machine Learning, 1990, pages 796-802.

NEC Corporation, "Orochi 12DOF Snake Like Robot," Press Release, NEC Corp. Melville, NY. Jan, 1996, 6 pp. Orochi was also featured in advertisements and several short mentions in popular magazines.

Ngo, J.T., Marks, J., "Spacetime Constraints Revealed," SIGGRAPH 93, Anaheim, CA August 1993. pp. 343-350.

Gabrielli, G., von Kármán, T.H., "What Price Speed?," Mechanical Engineering. 72: 775-781, 1950.

Schneider, J., Gans, R.F., "Efficient Search for Robot Skill Learning: Simulation and Reality," International Conference on Intelligent Robots and Systems, 1994

Yim, M., "New Locomotion Gaits," IEEE Robotics and Automation Conference, 1995, pp. 2508-2514.

Analysis of Snake Movement Forms for Realization of Snake-like Robots

Shugen MA

Department of Systems Engineering
Ibaraki University, Faculty of Engineering
4-12-1 Nakanarusawa-Cho, Hitachi-Shi, 316-8511 JAPAN
Tel: +81-294-38-5209   Email: shugen@dse.ibaraki.ac.jp

Abstract

This research arms to discover mechanism and principle for the emergence of the snakes' movement in order to realize a snake-like robot. In this study, we elucidated the standard creeping movement form of a snake, which is the typical locomotive motion shown by snakes. The called Serpentine curve *in the constant steady-state velocity was derived for the uniform creeping locomotion of the snake, through analyzing physiologically its muscle characteristics. Muscular force was then discussed for this uniform locomotive curve. We, in this study, also compared the locomotive efficiencies for various creeping movement curves of snake locomotion, by analyzing the ratio of the tangential force to the normal force and the power required for snake locomotion. The results showed that the proposed Serpentine curve is more valid as the snake creeping locomotion shape than the formerly suggested curves.*

1 Introduction

Snakes perform many kinds of movement adapted to their environment. Instead of being handicaps, the lack of limps, the considerable elongation of the body, and the particular mode of locomotion have allowed snakes to expand into diverse environments. One of snakes' assets is their ability to immediately respond to a new environment by changing reptantion modes. This research arms to utilize the snake (*its forms and motion*) as a model to develop a snake-like robot, which behaves the snakes' function. For this purpose, we first discover mechanism and principle for the emergence of the snakes' movement.

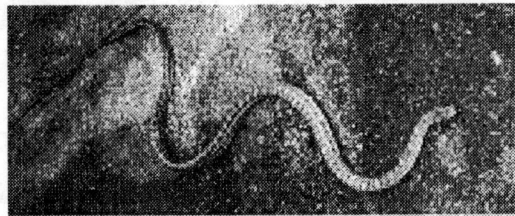

Figure 1: Creeping locomotion of a snake

Shakes are so diversified that they include many examples of specialization, in which one reptantion mode is favored. When these reptantions are broadly classified, there exist four gliding modes[1]: 1) Serpentine movement, 2) Rectilinear movement, 3) Concertina movement, and 4) Sidewinding movement. However, the Serpentine movement (*Creeping locomotion*) is the movement to be seen typically in almost all kinds of snake, and is a gliding mode whose characteristic is that each part of the body makes similar tracks. Figure 1 shows the form that the snake is during creeping locomotion and figure 2 shows the gliding curve of the snake creeping locomotion, respectively.

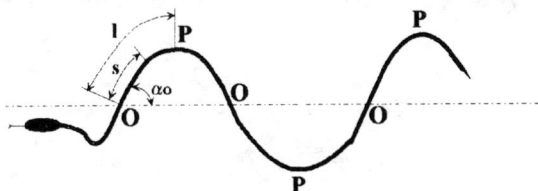

Figure 2: Gliding curve of the snake creeping locomotion

In this study, we elucidate the standard creeping movement form (*S-shaped curve*) of a snake. The form in the constant steady-state velocity is derived, through analyzing physiologically its muscle characteristics. Muscular force is also discussed for this uniform locomotive curve, and the locomotive efficiencies are compared for various forms of the snake creeping locomotion as well, by analyzing the ratio of the tangential force to the normal force and the power required for locomotion. The results proposed are demonstrated by computer simulation to agree with previously reported experimental results, and showed that the proposed Serpentine curve is more valid as the snake creeping locomotion shape than the previously-suggested Clothoid spiral, the Serpenoid curve, and others.

2 Physiological Consideration on Gliding Forms

The motion of snake comes from the contraction of the muscles. Before to discuss the Gliding curves of the snake creeping locomotion, we first discuss the muscle characteristics.

2.1 Physiological characteristics of muscle

Each muscle is made up of muscle fibers. They are what create the striated appearance in skeletal muscle. The contraction of muscle fibers is based on the interaction between two different types of filaments, known as think and thin. Muscular force (*muscle contractive force*) depends upon the contractive velocity; it becomes smaller while the contractive velocity become larger. The relation among the contractive force f_m (*Maximum value* f_{m0}), the contractive velocity v_m (*Maximum value* v_{m0}), and power output P (*Maximum value* P_0) is shown in figure 3 (a) [2] [3] [4] [5].

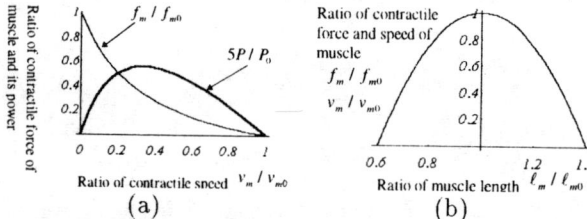

(a)   (b)

Figure 3: Characteristics of muscle expressed in 2 dimension; wherein, (a) is shown the relation among contractive force, contractive velocity and power output, and (b) is relation among contractive force, contractive velocity and muscle length

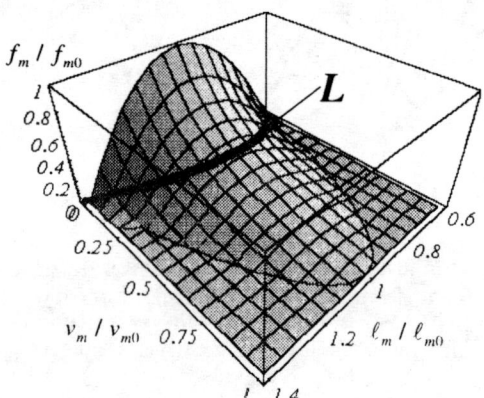

Figure 4: Characteristics of muscle expressed in 3 dimension

Based on Hill equation, we have the following muscle-contraction equation,

$$(3\frac{f_m}{f_{m0}} + 1)(3\frac{v_m}{v_{m0}} + 1) = 4. \quad (1)$$

While the contractive force f_m and the contractive velocity v_m are a third of their maximum value, the power output of the muscle is largest and loss of the power or energy becomes minimum [2].

On the other hand, the muscle contractive force is also affected by contractive length of the muscle. The contractive force and velocity of muscle are largest in a muscle contractive length ℓ_{m0}, and they become smaller while the muscle length is ever prolonged or shortened from the mentioned length. The relation between the muscle contractive force f_m and the muscle length ℓ_m, and that of the muscle contractive velocity v_m and the length ℓ_m are both shown in figure 3 (b) [2] [3] [4] [5]. The exact equation to describe these relation is hard to find out. In this study, we assumed them to be

$$f_m \simeq f_{m0}(1 - K(\frac{\ell_m}{\ell_{m0}} - 1)^2),$$
$$v_m \simeq v_{m0}(1 - K(\frac{\ell_m}{\ell_{m0}} - 1)^2), \quad (2)$$

where, K is the coefficient to express the characteristics of muscle, and determined from the measurement and the figure 3 (b). Therefore, the relation among the muscle contractive force, the muscle contractive velocity, and the muscle contractive length can be described as

$$(3\frac{f_m}{f_{m0}} + 1 - K(\frac{\ell_m}{\ell_{m0}} - 1)^2)$$
$$\times (3\frac{v_m}{v_{m0}} + 1 - K(\frac{\ell_m}{\ell_{m0}} - 1)^2)$$
$$= 4(1 - K(\frac{\ell_m}{\ell_{m0}} - 1)^2)^2 \quad (3)$$

and its distribution map is shown in figure 4.

As stated-above, the loss of the muscle power (*or energy*) for a muscle length ℓ_m is minimum, in the case that the contractive force and the contractive velocity are a third of their maximum value in the muscle length ℓ_m. That is,

$$f_m \simeq \frac{1}{3}f_{m0}(1 - K(\frac{\ell_m}{\ell_{m0}} - 1)^2),$$
$$v_m \simeq \frac{1}{3}v_{m0}(1 - K(\frac{\ell_m}{\ell_{m0}} - 1)^2). \quad (4)$$

This relation among the muscle contractive force, the muscle contractive velocity, and the muscle contractive length is the curved line **L** shown in figure 4.

2.2 Musculo-skeletal structure of Snake

Snakes have at least 130 vertebrae between head and cloaca, and thus number can exceed 300 [6]. For the snakes that are lack of limps, the long body is necessary for them to perform possible locomotion. Moreover, only a few movements are possible between the two vertebtaes, and these have limited amplitude. The long body is thus only one function for snakes to execute possible locomotion. Figure 5 (a) shows Musculo-skeletal structure of the snake. The lateral muscles of a snake lie on either side of a frame formed by the vertebrae and costae, and stretch from one rib to the next, or from robs to vertebrae. The musculoskeletal mechanism by which bending movements are produced is by no means a simple one. However, to permit the kinematic and static analysis, we consider a schematic model shown in figure 5 (b). If the range of movement of the single-joint $\delta\theta$ is small (*actually* about $\pm 4^0$ at most), the displacement between two rib (or costa) can be given by $\delta\ell_m = a\delta\theta$.

Therefore, the curvature of the snake shape κ and its time-differential $\dot{\kappa}$ can be derived from continuous curve theory, and expressed as

$$\kappa = \frac{d\theta}{ds} \simeq \frac{\delta\theta}{\delta s} = \frac{\delta\ell_m}{a\delta s} = \frac{\ell_m - \ell_{m0}}{a\delta s},$$
$$\dot{\kappa} = \frac{d\dot{\theta}}{ds} \simeq \frac{\delta\dot{\theta}}{\delta s} = \frac{v_m}{a\delta s}. \quad (5)$$

If we describe the muscle contractive force by the curvature κ and its time-differential $\dot{\kappa}$, we have, through Eq.(3),

$$f_m = f_{m0} \frac{1 - \frac{Ka^2\delta s^2}{\ell_{m0}^2}\kappa^2 - \frac{a\delta s}{v_{m0}}\dot{\kappa}}{3\frac{a\delta s}{v_{m0}}\dot{\kappa} + 1 - \frac{Ka^2\delta s^2}{\ell_{m0}^2}\kappa^2}(1 - \frac{Ka^2\delta s^2}{\ell_{m0}^2}\kappa^2). \quad (6)$$

Further on, the muscle contractive force and velocity in case of the minimum loss of muscle power can be also described by the curvature κ and its time-differential $\dot{\kappa}$. They are given by

$$f_m \simeq \frac{1}{3}f_{m0}(1 - \frac{Ka^2\delta s^2}{\ell_{m0}^2}\kappa^2),$$
$$v_m \simeq \frac{1}{3}v_{m0}(1 - \frac{Ka^2\delta s^2}{\ell_{m0}^2}\kappa^2). \quad (7)$$

As a result, we know that, the muscle contractive force is determined by the curvature κ and its time-differential $\dot{\kappa}$, but the muscle contractive force and velocity in case of the minimum loss of muscle power are determined only by the curvature κ of the snake-body shape.

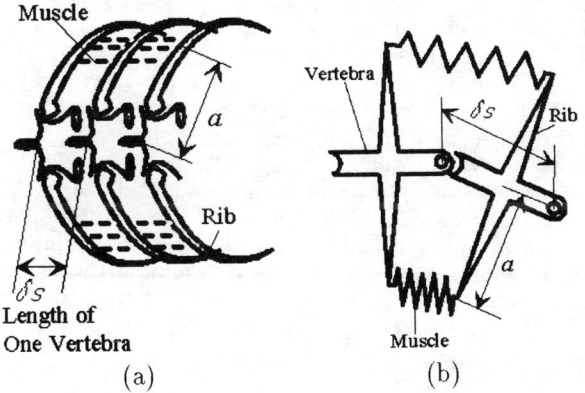

Figure 5: Musculo-skeletal structure of the snake and its schematic model; wherein, (a) is the Musculo-skeletal structure, and (b) is the Schematic model of musculo-skeletal structure

2.3 Discussion on the previously proposed curves

There have in the past been various discussions of the curves of gliding forms during creeping locomotion. Considering creeping movements on land and

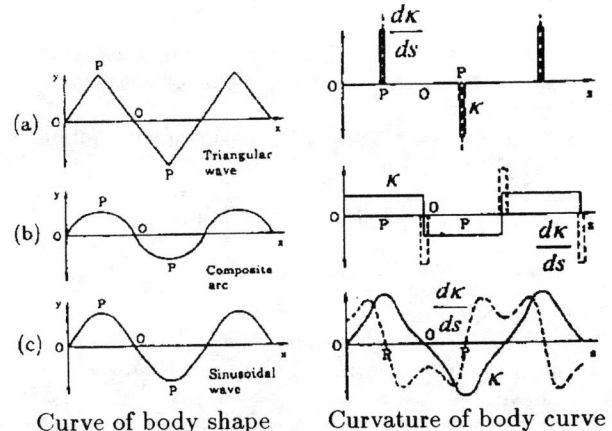

Figure 6: Relation of gliding body-shape curve to curvature along the body axis

through water, N. Rashevsky thought that they approximated to a triangular wave [7], while J. Gray and G. Taylor saw them as resembling sine curves [1] [8]. H. Hertel suggested that gliding curves might be approximated by a curve called a 'Meander', composed of sine functions in the $x-$ and $y-$ axes [9]. However, these proposed curves are chiefly presented as one possible premise for mechanical analysis of creeping movements, and while they do to some extent bear a resemblance to the gliding form of an actual snake, their physiological basis has not been considered. We next show how they are not valid on the basis of the physiological characteristics of muscle and the musculo-skeletal structure of snakes, that stated in subsection 2.1 and 2.2. As shown in figure 6 (a), (b), if we use a triangular wave and/or an arc as the gliding curve of creeping locomotion of snake, the curvature and its time-differential are not continuous, also show a pulse. From Eq.(6) we know that the muscle contractive force becomes uncontinuous. Thus, a triangular wave and/or an arc are physiologically not valid as the gliding curve for creeping locomotion of snake. Moreover, if we use a sine curve as the gliding curve of snake creeping locomotion, it is known from figure 6 (c) that the curvature and its time-differential show complex change, the muscle contractive force becomes very complex and is not valid for snake creeping locomotion. As a result, it is necessary to find out a curve where a curvature and the time-differential of curvature are continuous, and the muscle contraction is natural. Y. Umetani and S. Hirose suggested to use the Clothoid spiral and the Serpenoid called curve as the gliding curve of snake creeping locomotion [10] [11]. These two curves were proposed based on the physiologically qualitative analysis, and are thought valid somewhat as snake creeping locomotion curve. Their curvatures are given by

$$\kappa(s) = -2\alpha_0 \frac{s}{\ell^2}, \quad (8)$$

$$\kappa(s) = -\frac{\pi\alpha_0}{2\ell}\sin(\frac{\pi s}{2\ell}), \qquad (9)$$

where α_0 is the initial bending angle, named as winding angle, and ℓ is the curve length of **OP** along the body axis (figure 2).

In this study, we base on the physiologically quantitative analysis of the muscle contractive force and contractive velocity to elucidate the standard creeping movement form (*S-shaped curve*) of a snake.

3 Proposal of a Creeping Locomotion Curve

From the relation of the muscle contractive velocity and the curvature of body-shape curve (7), and time-differential of the curvature of the body-shape curve (5), we have

$$v_m = a\delta s\dot{\kappa} \simeq \frac{1}{3}v_{m0}(1 - \frac{Ka^2\delta s^2}{\ell_{m0}^2}\kappa^2) \qquad (10)$$

and thus we can obtain

$$\frac{3a\delta s}{v_{m0}}\dot{\kappa} \simeq 1 - \frac{Ka^2\delta s^2}{\ell_{m0}^2}\kappa^2. \qquad (11)$$

In case of the constant steady-state velocity, $\dot{s} =$ *Constant*. The curvature of the body-shape curve is thus derived through solving the differential equation (11), and given by

$$\kappa = A\frac{e^{Bs}-1}{e^{Bs}+1}, \qquad (12)$$

where the coefficients of A and B are

$$A = \frac{\ell_{m0}}{a\delta s\sqrt{K}}, \quad B = \frac{2v_{m0}\sqrt{K}}{3\ell_{m0}\dot{s}}.$$

The body-shape curve whose curvature is given by Eq. (12) is derived for the case of the constant steady-state velocity locomotion by physiologically quantitative analysis of the muscle contraction. It is named as "Serpentine Curve".

The bending angle α (*angle of the tangential line with $x-$axis*) at point of the curve, takes the maximum value α_0 ($\alpha(0) = \alpha_0$) at the point **O** and the minimum value 0 ($\alpha(\ell) = 0$) at the point **P** in the curve segment of OP (figure 2). The bending angle α is integration of the curvature of the snake-shape curve, and given by

$$\alpha(s) = \frac{\alpha_0}{\ell + 2/B\log(2/(1+e^{B\ell}))}\{\frac{2}{B}\log(\frac{1+e^{Bs}}{1+e^{B\ell}})+\ell-s\}, \qquad (13)$$

where ℓ is the curve length of OP along the body axis, same as that stated in subsection 2.3. One example where $B = 2$ is shown in figure 7 (a). In addition, the coordinates of $x-$axis and $y-$axis at point of the curve can be given by

$$x(s) = \int_0^s \cos(\alpha(u))du, \quad y(s) = \int_0^s \sin(\alpha(u))du. \qquad (14)$$

One example where $B = 2$ is shown in figure 7 (b). For comparison, the bending angle and the curve of shape for the Clothoid spiral and the Serpenoid curve are also drawn in figure 7, respectively. As seen, we know that the proposed Serpentine curve is very near the the Clothoid spiral and the Serpenoid curve in shape.

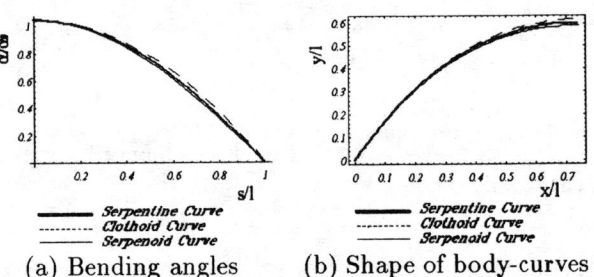

(a) Bending angles (b) Shape of body-curves

Figure 7: Comparison among the Clothoid spiral, the Serpenoid curve, and the proposed Serpentine curve ($\alpha_0 = 60^0$)

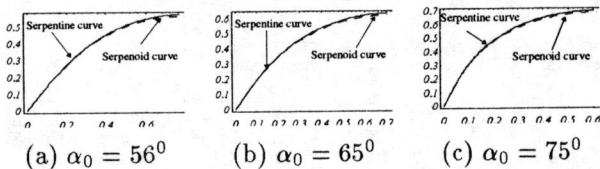

(a) $\alpha_0 = 56^0$ (b) $\alpha_0 = 65^0$ (c) $\alpha_0 = 75^0$

Figure 8: Comparison of shape between the Serpenoid curve and the proposed Serpentine curve

Comparison with an experiment result

Y. Umetani and S. Hirose were allowed a specimen of *Elaphe quadrivirgata* to glide across artificial turf with different friction properties, and used a series of consecutive photographs of its gliding form taken from vertically above [10] [12]. They proved that the Serpenoid curve best resembles the actual form of the snake in shape while creeping and gliding steadily. When one compares this curve with other curves previously-proposed, the Clothoid spiral follows it in closeness of similarity. However, these two curves were not based on the physiologically quantitative analysis. Even they best resemble the actual form of the snake, they are only in shape. In this study, we have derived the Serpentine curve on the basis of the physiologically quantitative analysis of muscle characteristics. If the proposed Serpentine curve resembles very much the Serpenoid curve in shape, we thus can say that the Serpentine curve is more valid as the snake creeping locomotion form. In figure 7, we showed one example in which the winding angle $\alpha_0 = 60^0$. Other examples of $\alpha_0 = 56^0$, $\alpha_0 = 65^0$, $\alpha_0 = 75^0$ are shown in figure 8. From the results, we know that, not only in the case of the winding angle $\alpha_0 = 60^0$ but also in the cases of $\alpha_0 = 56^0$, $\alpha_0 = 65^0$, $\alpha_0 = 75^0$, and others, the Serpentine curve resembles best the Serpenoid curve in shape. Therefore, it is considerable to say that the Serpentine curve would be more valid as the snake creeping locomotion form.

4 Distribution of Muscle Contractive Force

The following conditions have been assumed with regard to the distribution of muscle contractive force in the segment OP [12].

1) Since the muscles are at their natural length is at the midpoint of its range of motion at point **O**, from a physiological point of view it may be assumed that the greatest muscular force within the segment OP is generated at the point **O**, and that muscular force diminishes monotonically towards the point **P**;

2) Point **P** is a transition point, and the muscles change about it from contraction to extension (*or the reverse*). Muscular force at the point **P** is assumed to be zero;

3) Since the changes in muscular force during regular creeping locomotion are of a cyclical nature, even at the transition points **O** and **P** (*especially* **O**, *at which the greatest muscular force is exerted*), the change in muscular force is assumed to be smooth.

S. Hirose suggested the function

$$f_m(s) = F_{m0}\left(1 - (\frac{s}{\ell})^\sigma\right) \quad (15)$$

to express the muscular force that satisfied the above-stated conditions 1) and 2) [12]. Wherein, F_{m0} is the muscular force at point **O**, and is the maximum one in the segment OP. For satisfying the condition 3), the coefficient σ was assumed to be $\sigma > 1$ [12]. The function (15) is estimated from the physiologically qualitative analysis. Even though it agrees with the real muscular force somewhat, but it is not derived from the physiologically quantitative analysis. Here we base on the physiological characteristics of muscle and the musculo-skeletal structure of snakes to discuss the muscular force by quantitative analysis.

Eq.(7), what shows the muscle contractive force in case of the minimum loss of muscle power can be rewritten as

$$f_m(s) = \frac{1}{3}f_{m0}\left(1 - C\kappa(s)^2\right), \quad (16)$$

where f_{m0} is the maximum muscular force that muscle can produce ($\frac{1}{3}f_{m0} = F_{m0}$ *in the body segment* OP), C is the scalar factor to guarantee no negative force be generated and be zero at the point **P**. It is shown that, the muscle contractive force in case of the minimum loss of muscle power is proportional to the square of the curvature of body-shape curve. In the case of the proposed Serpentine Curve, the muscular force is

$$f_m(s) = F_{m0}\left(1 - (\frac{e^{Bs}-1}{e^{B\ell}-1})^2(\frac{e^{B\ell}+1}{e^{Bs}+1})^2\right). \quad (17)$$

Moreover, from the curvature of the Clothoid spiral and the Serpenoid curve, we have

$$f_m(s) = F_{m0}\left(1 - (\frac{s}{\ell})^2\right) \quad (18)$$

for the Clothoid spiral, and

$$f_m(s) = F_{m0}\left(1 - \sin^2(\frac{\pi s}{2\ell})\right) \quad (19)$$

for the Serpenoid curve.

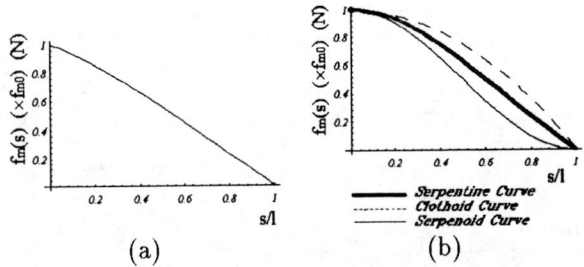

Figure 9: Distribution of muscular force ($\alpha_0 = 60^0$); wherein, (a) is the assumed one in qualitative analysis, and (b) is the derived one in quantitative analysis

Figure 9 (a) shows the profile of the muscular force distribution by Eq. (15) in case of qualitative analysis where $\sigma = 1.2$, and figure 9 (b) shows one by Eqs.(17) (18) (19) in case of quantitative analysis where $B = 2$, respectively. The distribution of muscular force shown in figure 9 (a) does not depend on the form of body-shape and is a same line, because only Eq.(15) is used. In addition, if we compare two distribution of muscular force, one obtained by quantitative analysis shown in figure 9 (b) and another obtained by qualitative analysis shown in figure 9 (a), both satisfied the condition 1) and 2), but the one by quantitative analysis shown in figure 9 (b) is more fitting to the condition 3). As a result, the distribution of muscular force obtained by quantitative analysis of muscle characteristics, makes the snake's muscle to have more natural expansion and contraction, and more fitting to the one of real snakes.

5 Comparison of Locomotive Efficiencies

As shown in figure 5, a schematic model is used to kinematic and static analysis, where δs is the length of one vertebra and a is the distance from joint to muscle. In case of static analysis, a moment of rotation $T(s) = (af_m(s))$ is generated by the force of contraction $f_m(s)$, when one side of the muscles contract. From distribution of the moment $T(s)$ in the segment OP, the tangential force along the body (*or propulsive force*), F_t^{OP}, the normal force perpendicular to the body, F_n^{OP}, and the power density function while the snake is moving at a fixed speed v along the body axis s were derived and given by [12]

$$Tangential\ force: \quad F_t^{OP} = \int_0^\ell \frac{dT(s)}{ds}\kappa(s)ds,$$

$$\text{Normal force}: \quad F_n^{OP} = \int_0^\ell \left| \frac{d^2 T(s)}{ds^2} \right| ds,$$

$$\text{Power density}: \quad P(s) = vT(s) \frac{d\kappa(s)}{ds}.$$

The kinematic conditions required to produce creeping locomotion are: first, that friction with the gliding surface be overcome and that sufficient propulsive force be generated in the direction tangential to the body; and second, in order that the pattern of motion peculiar to creeping, in which the whole trunk follows the same path, can be adopted, slippage in the normal direction must be prevented. Along with these two conditions, it would readily be seen that it is desirable that propulsive force should be maximized and normal force be minimized. In other words, the greater the ratio of propulsive force to normal force, the greater the efficiency of the gliding locomotion. Therefore, we utilize the F_t/F_n as one of the performance criteria for snake creeping locomotion, and called **Locomotive Efficiency**. Another one used is the total power required while the snake is in locomotion, it is described as $P^{OP}(= \int_0^\ell P(s)ds)$.

Table 1: Locomotive efficiency F_t/F_n ($\alpha_0 = 60^0$)

Serpentine	Clothoid	Serpenoid
Case of qualitative analysis		
0.945	0.952	0.942
Case of quantitative analysis		
0.790	0.698	0.349

Table 2: Total power required for creeping locomotion ($\alpha_0 = 60^0$)

Serpentine	Clothoid	Serpenoid
Case of qualitative analysis		
1.130	1.142	1.134
Case of quantitative analysis		
1.226	1.396	1.097

Table 1 shows the locomotive efficiency F_t/F_n, table 2 shows the total required power for snake locomotion, respectively, in the case that the winding angle $\alpha_0 = 60^0$. As seen, the Clothoid spiral shows the highest locomotive efficiency in case of using the force distribution shown in figure 9 (a). However, the force distribution shown in figure 9 (a) was obtained not by the quantitative analysis of muscle characteristics, it is not much valid to use. In case of using the force distribution shown in figure 9 (b), the proposed Serpentine curve gives the highest locomotive efficiency. While we see the total power required for snake locomotion, one by the proposed Serpentine curve is not smallest, but the difference by all three curves is quit small. It is thus not important factor to compare the efficiency of snake locomotion for each curve.

Table 3 shows the locomotive efficiency F_t/F_n, and the total required power for snake locomotion in the case of the other winding angles. As seen, we have same results as the case of $\alpha_0 = 60^0$.

Table 3: Locomotive efficiency (F_t/F_n) and total power for the snake creeping locomotion in the different winding angles (α_0), in case by quantitative analysis

Winding angle α_0	50^0	55^0	65^0	70^0	80^0
Locomotive efficiency F_t/F_n					
Serpentine	0.658	0.724	0.856	0.922	1.053
Clothoid	0.582	0.640	0.756	0.814	0.931
Serpenoid	0.291	0.320	0.378	0.407	0.465
Total power					
Serpentine	1.021	1.124	1.328	1.430	1.634
Clothoid	1.164	1.280	1.513	1.629	1.862
Serpenoid	0.914	1.005	1.188	1.279	1.462

Comparison with an experiment result

S. Hirose and Y. Umetani used an EMG (Electromyogram) and its electrodes to measure the amount of normal muscular force [11] [12]. The statistics of garter snake used in experiment shows in table 4, and one of the experimental result in figure 10.

Table 4: Statistics of garter snake used in experiment [12]

Weight	220 g
Body length	128 cm
(Head to anus)	(100 cm)
Width of belly	1.8 cm
Length of single vertebra	5 mm
Position of measuring devices	50 cm
(Position of earth terminal)	(78 cm)
Body length in OP	12.5 cm
Initial winding angle	60^0

From the experimental result shown in figure 10, we have the average value of the total normal force in the segment **OP**, it is $F_n^{OP} = 16[g]$. Moreover, from table 4 we know that the length of the segment **OP** is 12.5[cm]. We, thus, can obtain the sum total of normal force F_n of whole trunk (*this is defined as the 100[cm] from head to anus*). It is

$$F_n = \frac{100}{12.5} F_n^{OP} = 128 \ [g] \quad (20)$$

On the other hand, the propulsive force required to guarantee regular creeping is equivalent to the product of the coefficient of friction between the snake's trunk and the gliding surface, and the weight of the snake itself. As shown in table 4, the friction coefficient in tangential direction between the belly of the snake and the artificial turf is about 0.46, and the weight of the snake is 220[g]. Accordingly, the tangential force that should be exerted is

$$F_t = 0.46 \times 220 = 101 \ [g] \quad (21)$$

The locomotive efficiency F_t/F_n is thus equal to 0.789. The locomotive efficiency by the proposed Serpentine curve in case of using the force distribution shown in figure 9 (b), is 0.79, and is almost same as the experiment result. This is actually an index to show the validity of the Serpentine curve, even though the normal force and friction coefficients are roughly estimated.

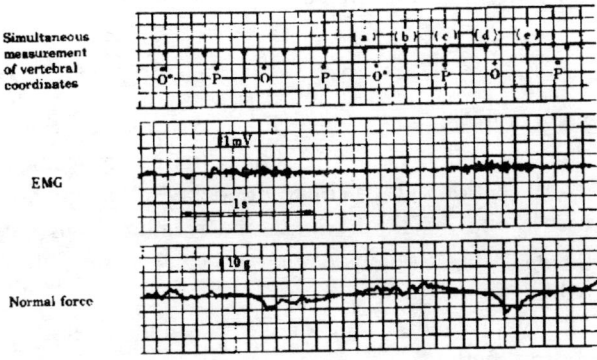

Figure 10: Example of simultaneous experimental data [12]

6 Conclusion and Discussion

In this study, we elucidated the standard creeping movement form (*S-shaped curve*) of a snake. The form in the constant steady-state velocity was derived, through analyzing physiologically its muscle characteristics, and found out that the called **Serpentine curve** as the creeping locomotion form of snake is valid. Muscular force was also discussed, and the locomotive efficiencies were compared for various forms of the snake creeping locomotion as well, by analyzing the ratio of the tangential force to the normal force and the power required for locomotion. It is found out that the proposed Serpentine curve resembles the actual form of the snake in shape and shows the highest locomotive efficiency, thus is more valid as the snake creeping locomotion shape than the Clothoid spiral, the Serpenoid curve, and others.

The discussion up to now is limited on a quarter cycle of the snake-shape [13] [15]. For elucidating exactly the form of snake creeping movement, it is necessary to execute the discussion cross whole snake body. We are, just, continuing this discussion and started to develop a mechanical model of snake-like robot to emerge the snakes' movement [14] [16].

This research is partially supported by the Ministry of Education, Science, Sports, and Culture of Japan, Grands-in-Aid for No. 10751081. At last, I would like to thank Professor S. Hirose of Tokyo Institute of Technology for his valuable advise.

References

[1] J. Gray, *Animal Locomotion*, pp.166–193, Norton, 1968

[2] Y. Yoshifuku, *The Science for Improvement of Sports — Biomechanics for becoming Strong & Smart*, pp.38–48, Koudansha Ltd, 1990 (in Japanese)

[3] I. Numano, *Modern Biology 9 — Physiology and Biochemistry*, pp.353–384, Nakayama Shoten Ltd., 1967 (in Japanese)

[4] R. M. Alexander, *Exploring Biomechanics — Animals in Motion*, W. H. Freeman and Company, 1992

[5] M. J. Adrian and J. H. Cooper, *Biomechanics of Human Movement*, Wm. C. Brown Communications, Inc., 1995

[6] R. Bauchot, *Snakes — A Nature History*, pp.60–75, Sterling Publishing Company., Inc. New York, 1994

[7] N. Rashevsky, *Mathematical Biophysics*, Vol.2, pp.256–261, Diver Publication Inc., 1960

[8] G. Taylor, *Analysis of the Swinging of Long and Narrow Animals*, in Proc. Royal Society of London, Vol.214, pp.158–183, 1952

[9] H. Hertel, *Structure Form Movement*, pp.182–184, Reinhold Publishing Corp., 1966

[10] Y. Umetani, *Mechanism and Control of Serpentine Movement — Kinematics of Locomotion and Model Experiment*, in Proc. 1st Biomechanism Symp., pp.253–260, 1970

[11] Y. Umetani and S. Hirose, *Biomechanical Study on Serpentine Locomotion — Gliding Shape of Snake in Stationary Straightforward Movement*, Trans. of the Society of Instrument and Control Engineers, Vol.10, No.4, pp.513–518, 1974

[12] S. Hirose, *Biologically Inspired Robot —- Snake-like locomotors and manipulators*, Oxford University Press, 1993

[13] S. Ma, *Analysis of Snake Movement Forms — Report 1: Standard creep movement forms of snake*, Trans. of the Japan Society of Mechanical Engineers, Vol.62, No.593(C), pp.230—236, 1996 (in Japanese)

[14] S. Ma and S. Hirose, *Analysis of Snake Movement and Development of Snake-like Robots — Creep Movement of a Snake-like Robot*, in Proc. 8th JSME Annual Conference on Robotics and Mechatronics, Vol.A, pp.251—254, 1996 (in Japanese)

[15] S. Ma and A. Mashiko, *Analysis of Snake Movement Forms — Report 2: Comparison of the locomotive efficiencies for various movement forms of snake creep locomotion*, Trans. of the Japan Society of Mechanical Engineers, Vol.64, NO.617(C), pp.273—278, 1998 (in Japanese)

[16] A. Naito and S. Ma, *Analysis of Creeping Locomotion of Snake-like Robots*, in Proc. 3rd Asian Conf. on Robotics and its Application (3rd ACRA), pp.393—398, 1997

GMD-SNAKE2: A Snake-Like Robot Driven by Wheels and a Method for Motion Control

Bernhard Klaassen,   Karl L. Paap
GMD-AiS, D-53754 St. Augustin, Germany

Abstract

After getting some experience with our first GMD-Snake, we can now present the next generation: GMD-SNAKE2, a robot for inspection tasks in areas difficult to access by humans. Here we tried to imitate the natural scale-driven propulsion of a snake by wheels around the body.

After a description of our robot we concentrate on a method for motion control. It allows a very flexible and consistent method of path planning and calculation where smooth curves are desired. It is applicable not only to snake-like movement but to any wheeled robot which is able to control the curvature of its path.

1. Introduction

The elegance and efficiency of a real snake's movements can hardly be reproduced by a robot but it's still a task to learn from nature.
Watching a snake's path (on normal ground) we may observe that the whole body is following the head in a trace like a railway train. The main propulsion here comes from a source not to be observed at first glance: hundreds of tiny scales are moving ahead and back on the bottom side of the snake.

To imitate such a motion we have to build a mechanism following two main principles for the parts (or sections) of a mechanical snake:
- each section should be actively moveable in direction of its longitudinal axis
- each joint should be bent according to the movement of its predecessor with a certain delay

The first principle can be realized by adding wheels around our snake's body driven by a DC motor in each section. The second principle is an application for distributed computing: Each section has its own processor and a communication unit to talk to its neighbour sections. Hence, if it is possible to estimate the speed of the snake's actual forward movement, every section can calculate the delay after which its own joint position must be identical to the predecessors former position and has to send these data to its successor.
This gives a first idea of our distributed real-time control method.

2. Snake Robot Construction

Our first GMD-Snake (for details see [Paa96], [Wor96] or http://www-set.gmd.de/RS/snake) was built in a very elastic way to allow flexible bending of the parts. A disadvantage was an uncontrolled torsion effect which occured when the snake lifted some of its parts (e.g. to climb on a step).
From this experience we decided to construct the next generation in a more rigid way using universal (or cardanic) joints.

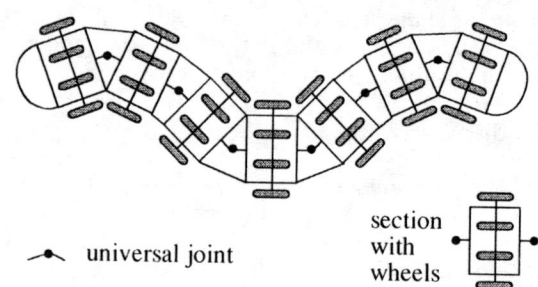

Fig.1: Schematic view of GMD-SNAKE2 with wheels

Each section is basically built like an aluminum cylinder with some holes in its surface for sensors (e.g. we implemented 6 infrared distance sensors per section). Within each cylinder there are three 5W-DC motors to control the position of one universal joint by way of ropes. Between these motors all electronic parts have been integrated, including one C167 processor per section, each with a CAN interface for bus communication. We achieved good results with this bus concept and it is used as well in the joint project MAKRO (with FZI Karlsruhe) [MAK98].

Often snake-like robots are designed for studying control problems with high number of degrees-of-freedom. (E.g. [Chi92] or [Des95]). Our SNAKE2 is intended for practical tasks (inspection of areas hardly to reach by men, e.g. narrow sewerage pipes or destroyed buildings) [Lew94].

We have studied the numerous types of movement performed by real snakes [Bau94] and found one essential technique which is usually not implemented in robot snakes: the forward forces generated by hundreds of active scales under the snake's body. These forces allow a normal land snake to slide within its curved track.

To realize a technically equivalent movement we implemented a ring of wheels around each section of our snake. Each ring of wheels is driven (via special joints) by one additional DC motor per section such that every section can control its own forward force.

By the active joints we can obviously adjust the curvature of the snake's body. This is only limited by the maximal angle (here 45°) and by the discretization of the path induced by the straight rigid parts between the joints.

The diameter of the snake (including wheels) is 18cm while the length of each section is 13.5cm. Hence, with 10 sections plus head we end up with a total length of about 1.5m and 15kg weight.
So, in contrast to several existing snake robots, here the length of one section is clearly smaller than its diameter which is important for a flexible behaviour.

For test runs we can drive our snake with external power through a wire, but for the final version we use an extra section at the tail, filled with batteries and without any motors. So this last joint and ring of wheels will not be driven actively.

The following picture shows four sections of our snake. The joints are covered with rubber bellows to protect them while running on dirty ground.

For more snake robots see the collection [Wor98]. There exist several snake-like robots in research institutes around the world. Two big problems for those of them which are intended as inspection systems (not as fixed manipulators) are weight and autonomy, which are related by the weight of batteries. So all developers of autonomous robots are waiting for lighter batteries.

3. Path Calculation

There exist several 'classical' approaches for the calculation of path curves for vehicles and/or wheeled robots. Of course, the choice depends on the task and on the abilities of the robot. Here we concentrate on two rather general cases:

A) The robot should follow a given path curve or a sequence of points with a certain accuracy.

B) The robot has a given position on a path (with certain direction and curvature) and should calculate the next piece of its path with given end position (and direction).

An example for B) is the situation that the robot's vision system has detected a door nearby while the robot is moving and the navigation system must find a path through this door fitting to the robot's already performed path. Here it is desired that the changes of direction should be smooth in some sense. Also the possible angles of the steering mechanism are limited and it should be known (before the new piece of path is started) if the robot is able to perform it.

The term 'wheeled robot' is not meant in the sense of robots who can turn on the spot and have a more or less piece-wise linear path. In this paper we treat problems with (at least twice) continuously differentiable paths, which occur with the most car-like driven robots when a smooth path curve is desired.

In most cases it is possible to divide case A) into a sequence of cases B) by choosing some points on the planned curve thus defining a sequence of starting end ending points. So a reliable and fast method for case B) solves also A).

In cases when the planned curve in A) is in some sense irregular for the robot's movements (e.g. too sharp bendings) we propose to modify the path into a sequence of 'approved' path pieces.

4. Mathematical Background

Before the task of path planning for robots came into the scope of scientific interest, there existed a long tradition of path planning for streets and motorways. The standard mathematical approach to this task is to use a curve called 'clothoid' (see e.g. [Hir93] chapt. 3) which has some nice properties:
1.1. plain curve parametrized in t
1.2. arclength s is proportional to t
1.3. curvature k is proportional to s and t

curvature here means reciprocal of the radius of curvature $r = (x'^2 + y'^2)^{3/2} / (x'y'' - y'x'')$

This is fulfilled by the well-known formula

$$x = a\int_0^t \cos\frac{\tau^2}{2} d\tau \qquad y = a\int_0^t \sin\frac{\tau^2}{2} d\tau \qquad (1)$$

which defines a clothoid curve starting at zero with start direction to the x-axis.

From this curve normally the plans for motorways are derived since it is desired to have a linear movement of the steering wheel (if constant speed is assumed along the curve).

In mathematical and robotics literature (e.g. [Mee92]) this curve is often used to solve problem B (see introduction). But since there is only one parameter a and an optional scaling of t, it is not possible to apply (1) directly to solve B.
Therefore often rather complicated combinations with other suitable curve pieces are introduced (e.g. circle parts), which makes the resulting algorithms not quite simple.

To overcome this we propose the following enhancement: For robotics it is not really essential to keep property 1.3. One could replace it by 'curvature changes according to a simple continuous function in s within given bounds'. Indeed for a robot it is only important to have a simple function to control its curvature, and e.g. a polynomial function is as good as a linear one. (In [Hir93] sine terms are used but we prefer polynomials for simpler implementation.)

This motivates the definition of a 'second-order clothoid':

$$x = a\int_0^t \cos\left(\frac{u\tau^3}{8} + \frac{v\tau^2}{2} + w\tau\right) d\tau \qquad (2)$$

$$y = a\int_0^t \sin\left(\frac{u\tau^3}{8} + \frac{v\tau^2}{2} + w\tau\right) d\tau$$

with real parameters a, u, v, w and t.

The curvature of this path is (omitting some calculus here) $k(t) = (ut^2 + vt + w)/a$

By this formula it is very easy to know in advance the maximum of curvature for each path segment.

The arclength is $s(t) = at$.
The angle of direction α is
artan(y'/x') =
$$\alpha(t) = \left(\frac{ut^3}{8} + \frac{vt^2}{2} + wt\right)$$

With these nice properties we can define our task B more precisely:
Given a start point (x0,y0), and an end point (x1,y1) together with curvature $k0$ at the start and two directions $\alpha0$ and $\alpha1$ for start and end point respectively. Let us set (x0,y0) to (0,0) and $\alpha0 = 0$, x1 = 1 for convenience (and adjust the other inputs accordingly).

The parameters a, u, v, w and t cannot be assumed to be independent, so we have to fix one of them and set $t = 2$. (A value of 1 would seem more natural but 2 yielded a better numerical behaviour of the solution procedure described below.)

The task is now to find the values a, u, v, w to fulfil the following equations:

$$x(2) = x1 = 1 \qquad y(2) = y1$$
$$k(0) = k0 \qquad \alpha(2) = \alpha1 \qquad (3)$$

which means in words: Find the second-order clothoid piece that fits with angle and curvature to the existing path at (x0,y0) and meets (x1,y1) under a given angle.

From the formula for $k(t)$ we find $w = a\,k0$.
From $\alpha(t)$ we can conclude $u = (\alpha1 - 2v - 2w)$.
So from (3) there remain the first two equations (with x and y from (2)), which can be solved numerically to determine a and v.

The terms in these two equations revealed no big numerical problems while Newton's method was applied.

5. Implementation and examples

On the first glance it seems to be a drawback of our approach that the resulting variables cannot be found explicitly by a formula but have to be calculated by an iterative method. This can be easily circumvented by using table models (e.g. [Kla94]) for $a(y1,k0,\alpha1)$ and $v(y1,k0,\alpha1)$. This is possible since all inputs are limited and the sensitivity of the outputs on variation of input is moderate.

With such a table model we can make sure that the execution time for path calculation is predictable and equal for each path segment. This is important for path predictions during the robot's run-time. Before each model evaluation the local coordinate system must be transformed such that the x-axis coincides with the actual driving direction at the start point of the new path segment. After the evaluation only a rescaling with the original factor (x1-x0) is to be done and t must be scaled to adjust it to the desired velocity.

Now we give some examples for members of our above-defined (see (2)) family of curves:
One can easily observe that it includes circle arcs and straight lines. (We get the first if u and $v = 0$ and $w \neq 0$, the latter if $u, v, w = 0$.)

With $v=1$ and $u,w = 0$ we have the classical clothoid defined in (1) with curvature changing linearly with arclength. All other cases with $u \neq 0$ are changing curvature in a parabolic way, which gives the higher freedom for fitting to boundary values. Usually a path piece of such a kind is not looking much different from the classical clothoid:

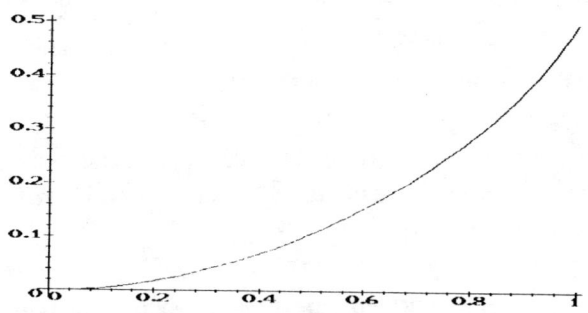

Fig 2. Second-order clothoid path with $\alpha1=1$

But it can also have a turning point as in the following example where $\alpha 1$ is set to 0:

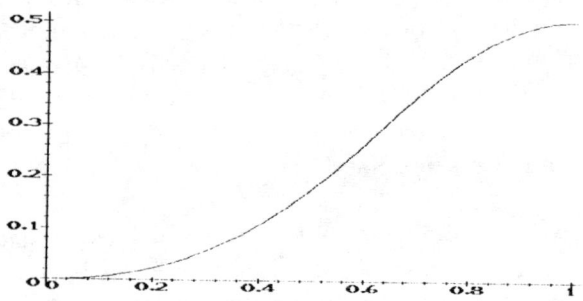

Fig 3. Second-order clothoid path with $\alpha 1=0$

Of course, an approximation of such curves with pieces of first-order clothoids is possible, but our approach has by construction a higher flexibility and includes the cases 'circle arc' and 'straight line' within its parameter range (and not only asymptotically). So as one consequence the length of each path piece can be kept longer, which makes it simpler to look ahead.

A thus generated sequence of second-order clothoids has the following properties:
- continuous in direction and curvature
- controllable by curvature (parabola in t)
- linear relation of arclength and time
- maxima of curvature easily to be found

6. Application to Snake Robot

Now we can apply the method of second-order clothoids to the control of our snake robot. The 2D path planning algorithm performs the following steps:
a) By sensors (infrared distance measurement in each section and camera in the head) build a rough map of the surrounding.
b) Define a sequence of points to be traced (if necessary with a defined direction at each point), e.g. around an obstacle.
c) Fit pieces of second-order clothoids through these points as described above.
d) Check if maximal curvature lies within given bounds. If not, goto b) to simplify path.
e) If path is ok, start moving the snake by controlling speed and curvature. Goto step a) to update the map and the planned path.

It is obvious that such a technique of controlled curvature is ideal for a snake-like robot of this kind. As long as it is possible to get a good calculation of the path length by speed estimation, it is easy to adjust the angles of the following joints according to their position on the path. Here the simple relation of path length and curvature in our method is essential (see section 4).

But also for car-like robots the above path algorithms is applicable since curvature and speed are the 'natural' control parameters and it is useful to know in advance if a proposed path is a possible one (concerning maximal curvature).

7. Conclusion and Future Work

A new snake-like robot and an algorithm for curvature controlled path calculation has been presented. Mathematically it is based on an enhancement of the well-known clothoid curve. It generates smooth curves (continuous in direction and curvature). The method was developed for application to a our snake robot driven by wheels to emulate the snake-typical movement by active scales.

It is planned to apply this robot to areas difficult to reach by men. So this technique of movement will be enhanced to 3D. A simple 3D technique will be a control by 'joystick' and a camera picture. But the robot also has the capability to work in a completely autonomous mode by using a battery tail section.

Each section of our GMD-SNAKE2 is equipped with a microprocesser unit to control its movement in coordination with its neighbour sections. The motors are strong enough to lift the two front sections plus camera head (e.g. to climb over an obstacle).

It should be noted that our method of path calculation is useful not only for such a special kind of robot but also for any wheel-driven vehicle which is able to control speed and direction of its movement.

References

[BAU94] R. Bauchot, (Editor), 'Schlangen', Naturbuch Verlag, Augsburg 1994, ISBN 3-89440-075-7

[CHI92] Chirikjian, G.S. and J.W.Burdick. 'Design implementation, and experiments with a thirty-degree of-freedom `hyper-redundant' robot.' ISRAM 1992, November 1992

[DES95] R.S. Desai, C. J. Rosenberg, J.L.Jones, 'Kaa: An Autonomous Serpentine Robot Utilizes Behaviour Control', IROS '95, vol. 3, pp. 250-255, 1995

[HIR93] Shigeo Hirose, 'Biological inspired Robots', Oxford University Press, 1993

[Kla94] B. Klaassen, K.L. Paap, 'Component table models for analog system simulation', In: Micro System Technologies '94. Berlin: vdi-verlag, 1994.

[LEE94] T. Lee, T. Ohm, S. Hayati, 'A Highly Redundant Robot System for Inspection', Conf. on Intelligent Robotics in the Field, Factory, Service, and Space (CIRFFSS '94), Houston, Texas 1994, pp. 142-149

[LEW94] M. A. Lewis, D. Zehnpfennig, 'R7: A Snake-Like Robot for 3-D Inspection', IROS 1994, Munich

[MAK98] Project MAKRO see http://www.gmd.de/FIT/KI/CogRob/Projects/Makro/makro.html

[Mee92] D. S. Meek and D. J. Walton, "Clothoid spline transition spirals", Mathematics of Computation 59(1992), p.117-133.

[Paa96] K.L. Paap, M. Dehlwisch, B. Klaassen, "GMD-Snake: A Semi-Autonomous Snake-like Robot", In: Distributed Autonomous Robotic Systems 2, Springer-Verlag, Tokyo, 1996

[WOR96] Rainer Worst, Ralf Linnemann, "Construction and Operation of a Snake-like Robot", In: Proceedings IEEE International Joint Symposia on Intelligence and Systems, IEEE Computer Society Press, Los Alamitos, CA, 1996

[Wor98] see http://borneo.gmd.de/~worst/snake-collection.html

Why to use an Articulated Vehicle in Underground Mining Operations?

Claudio Altafini *
Optimization and Systems Theory
Royal Institute of Technology, Stockholm, Sweden
altafini@math.kth.se

Abstract

Engineers in the underground mining industry have known for a long time that an articulated vehicle is preferable to a normal truck for the navigation in the narrow environments of an underground mine because of its higher maneuverability. The two advantages of the articulated configuration can be explained in terms of degrees of freedom in the selection of the directions needed to span the whole tangent space and of the reduced gap between the trajectories followed by the wheels of the truck when steering. Both characteristics can be used in a constructive way for path planning and navigation purposes.

1 Introduction

The action of transporting the material from the stope to the dumping point of an underground mine is performed by a truck called LHD (Load-Haul-Dump). The LHD is an articulated vehicle composed of two bodies connected by a kingpin hitch. Each body has a single axle and the wheels are all non-steerable. The steering action is performed on the joint, changing the angle between the front and rear part by means of hydraulic actuators. Both the shape and the steering mechanism are intended to improve the maneuverability of the vehicle. The effect of the actuated articulation is twofold: the truck can steer on place i.e. the orientation of the vehicle changes varying the steering angle alone and the width spanned by the vehicle when turning is reduced with respect to, for example, a car-like vehicle. Both properties descend directly from the geometry of the articulation and can be explained considering the kinematic model of the vehicle.

The articulated truck is an underactuated drift-free nonlinear system with two inputs, which was proven to be controllable in [6]. The *steering on place* can be justified looking at the vector fields associated with the two inputs. In particular, when comparing with a car-like vehicle, it turns out that the articulated configuration has as consequence a richer range of possible maneuvers that can be explained in terms of independence of the higher order Lie brackets of the vector fields i.e. with the presence of an extra degree of freedom in the selection of the directions needed to complete the tangent space of the configuration space. This degree of freedom can be used to optimize a cost function in a path planning problem. Here we provide a simple example where the cost to minimize is the number of elementary maneuvers required to reach a given point in the configuration space.

Also the second characteristic can be explained considering the geometry of the vehicle. In general, in a multiaxis wheeled vehicle, a nonnull steering angle implies that the wheels (or better the midpoints of the axles) follow different trajectories. This fact becomes a relevant problem when the free space in which the vehicle is allowed to move is limited, like in a narrow road or in an underground tunnel. It is intuitively easy to understand that a vehicle without articulation, say a car-like vehicle, would be more cumbersome i.e. would span a larger area than the LHD when turning. For the articulated vehicle case, the *off-tracking* between the trajectories can be easily calculated in some situations (see also [4]). The main problem for the autonomous navigation of the mining truck is to be able to follow the tunnel keeping a safety margin from both walls. Navigation of the truck requires proper interaction with the environment: in this case the environment (the tunnel) can be modeled as a path to follow and the proper criterion (keeping the middle of the tunnel) can be reformulated as reducing the off-tracking of the *whole vehicle* from the path. This idea,

---
*This work was supported by the Swedish Foundation for Strategic Research through the Center for Autonomous Systems at KTH

formulated in [3], finds here its most suitable application because of the limited width of the workspace of the mining truck. All the several approaches proposed in the literature to solve the path following problem for wheeled vehicles are essentially based on the selection of a single point of the vehicle and on the definition of a tracking criterion for this guidepoint. The task of the controller, then, is to have the corresponding tracking error converging to zero. Here the proposed solution consists in redefining the tracking error of the path following problem not based only on one single distance but on the *sum* of the signed distances of the midpoints of both axles of the vehicle from their orthogonal projections on the path. Stability can be proven locally for paths of constant curvature.

Finally, these being the two advantages of having an articulated configuration, it is also easy to see a drawback: as consequence of the central joint there is no direction of motion in which the open-loop system has a stable equilibrium point: in a sense that will be clarified below the system always behaves as a car moving backwards. Therefore such a configuration is useful only for applications in which the speed range is quite low, like mining, earth moving, forest industry or similar.

2 Steering on place

A typical configuration for a mining truck is the one shown in Fig. 1, where (x_i, y_i), $i = 0, 1$ are the

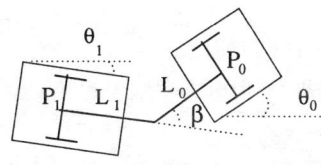

Figure 1: Two-unit articulated vehicle.

cartesian coordinates and θ_i the orientation angles of the midpoints P_0 and P_1 of the axles of the vehicle. At kinematic level, the inputs of the system can be taken to be the speed vector v of the point P_1 and the steering speed $u = \dot{\beta}$ where $\beta \triangleq \theta_0 - \theta_1$.

A set of variables that describes the configuration space of the truck is given by $\mathbf{q} \triangleq [x_1, y_1, \theta_1, \beta]$ with the equations:

$$\dot{\mathbf{q}} = \begin{bmatrix} \cos\theta_1 \\ \sin\theta_1 \\ \frac{\sin\beta}{L_0 + L_1\cos\beta} \\ 0 \end{bmatrix} v + \begin{bmatrix} 0 \\ 0 \\ -\frac{L_0}{L_0 + L_1\cos\beta} \\ 1 \end{bmatrix} u$$

$$= g_1(\mathbf{q})v + g_2(\mathbf{q})u \quad (1)$$

The system (1) is well-defined in the domain

$$D = \left\{ \mathbb{R} \times \mathbb{R} \times \mathbb{S}^1 \times \right] - \arccos\left(-\frac{L_0}{L_1}\right), \arccos\left(-\frac{L_0}{L_1}\right) \left[\right\}$$

If $L_0 > L_1$ the system presents no singularity.

The system (1) was proven to be controllable in [6] using tools from differential geometry, like the rank of the Control Lie Algebra generated by the vector fields associated with the inputs. For the definition of nonlinear controllability, as well as Lie bracket, filtration, distribution etc. refer to a standard textbook on nonlinear control systems like [5].

When $L_0 = 0$, we obtain a car-like vehicle with the well-known system:

$$\dot{\mathbf{q}} = \begin{bmatrix} \cos\theta_1 \\ \sin\theta_1 \\ \frac{\tan\beta}{L_1} \\ 0 \end{bmatrix} v + \begin{bmatrix} 0 \\ 0 \\ 0 \\ 1 \end{bmatrix} u_c$$

$$= g_{1_c}(\mathbf{q})v + g_{2_c}(\mathbf{q})u_c \quad (2)$$

where u_c stands for the steering input of the car. The domain of definition is:

$$D_c = \left\{ \mathbb{R} \times \mathbb{R} \times \mathbb{S}^1 \times \right] -\frac{\pi}{2}, \frac{\pi}{2} \left[\right\}$$

The difference between the two models (1) and (2) is that in the articulated truck the steering input is entering also into the equation for the orientation angle θ_1. This allows to change θ_1 by means of the steering actuator alone, whereas, in the car, θ_1 can be varied only through a sequence of the two inputs. It is equivalent to say that θ_1 is locally accessible by u without need of Lie bracketing the two inputs. The steerability on place can be checked using linearization of the equations around the origin: for the system (1) the two vector fields are $g_1(0) = [1\ 0\ 0\ 0]^T$, $g_2(0) = \begin{bmatrix} 0 & 0 & -\frac{L_0}{L_0 + L_1} & 1 \end{bmatrix}^T$, whereas for the car-like (2) $g_{1_c}(0) = [1\ 0\ 0\ 0]^T$ and $g_{2_c}(0) = [0\ 0\ 0\ 1]^T$.

We can use the original nonlinear system to analyze more in depth this difference. As in all nonholonomic systems, g_1 and g_2 (or for the car g_{1_c} and g_{2_c}) are noncommuting vectors and the distribution they generate $span\{g_1, g_2\}$ is not involutive: the combination of the two vector fields gives a new direction (here the "wriggle"). For the articulated vehicle, the wriggle has the following expression:

$$[g_1, g_2] = \begin{bmatrix} \frac{L_0 \sin \theta_1}{L_0 + L_1 \cos \beta} \\ \frac{-L_0 \cos \theta_1}{L_0 + L_1 \cos \beta} \\ \frac{L_0 \cos \beta + L_1}{(L_0 + L_1 \cos \beta)^2} \\ 0 \end{bmatrix}$$

If $\cos \beta = -\frac{L_0}{L_1}$, then $[g_1, g_2]$ is aligned with g_1 and further bracketing does not generate any independent vector. Therefore, defining $\mathcal{L} \triangleq \min \left\{ \frac{L_0}{L_1}, \frac{L_1}{L_0} \right\}$, we limit our analysis to the subset of D given by:

$$D_a = \{ \mathbb{R} \times \mathbb{R} \times \mathbb{S}^1 \times\,]-\arccos(\mathcal{L}), \arccos(\mathcal{L})[\}$$

which, $\forall \mathcal{L}$, contains an open neighborhood of the origin. In practice, this is not a critical limitation because in a real mining truck the two semichassis have lengths L_0 and L_1 that are similar, when not equal. Both systems (1) and (2) are completely nonholonomic with a filtration that grows regularly in their respective domains D_a and D_c and both distributions that contain vector fields and the Lie brackets of vector fields become involutive (i.e. a Lie Algebra) already at the second level of Lie bracketing (the degree of nonholonomicity of the system is 2 in both cases) implying therefore controllability. The difference between the two cases lies in the arbitrariness of the selection of the fourth vector that complete the distribution to the whole tangent space. If we consider the possible combinations of the vector fields for the two generator case, stopped at the degree 2, we have the five vectors:

$$g_1, g_2, [g_1, g_2], [g_1, [g_1, g_2]], [g_2, [g_1, g_2]]$$

and among those we have to choose the basis of the Control Lie Algebra.

For the car-like system, we obtain that there is an unique way to complete the basis. In fact, $[g_2, [g_1, g_2]]$ results always aligned with the wriggle $[g_1, g_2]$ $\forall \mathbf{q} \in D_c$. In the articulated vehicle instead, both $[g_1, [g_1, g_2]]$ and $[g_2, [g_1, g_2]]$ can be used to complete the basis which means that both inputs can be used to move along the new direction.

The richer behavior of the steering actuator on the articulation joint can be interpreted substituting the articulation with a virtual steering wheel (see [1]). In fact, at kinematic level, the articulated truck is equivalent to a tricycle model or to a couple of car-like models having the steering wheel in common. The steering action on the articulation is equivalent to a combination of steering plus translation on the corresponding car-like models.

We want to exploit now this peculiarity of the articulated truck comparing the two systems in a given maneuver. For an autonomous system it is reasonable to generate motion using the simplest possible sequence of inputs functions, for example piecewise constant inputs, and to give them in a decoupled way, i.e. our input functions have to be:

- piecewise constant
- $u(t) \cdot v(t) = 0 \ \forall\, t \in [0, T]$

We take as maneuver a rotation of 90° around the midpoint of the rear axle (x_1, y_1). This movement can correspond to a docking maneuver, a typical maneuver that has to be performed to place the truck in the correct load/unload position. In the more realistic case of a limitation in the steering angle, say $|\beta| \leq \pi/3$, this has to be accomplished with a sequence of input commands since both vehicles have also to satisfy to the nonholonomic constraints. If we call ψ the phase angle of the rear (actuated) wheels, then $v = \dot{\psi}\rho$ with ρ the radius of the wheels. In order to have a clear picture of how is the evolution of the integral curves of the system, we consider ψ as an extra state and split the state vector into *base* variables (β, ψ) which are directly controlled by the inputs and *shape* variables (x, y, θ) which evolve in $SE(2)$, the special planar Euclidean group. In both cases, simple geometric considerations allow to compute a path between the two shape points $(0, 0, 0)$ and $(0, 0, \frac{\pi}{2})$. Controllability on $SE(2)$ as a subspace of D_a and D_c assures that such a path exists. The corresponding path on the base space, with the requirement that $\beta = 0$ at the end point (again the existence is guaranteed by the nonlinear controllability), and therefore the corresponding piecewise constant time-varying input functions can be calculated using the inverse kinematic of the system. In Fig. 2 the paths that realize the desired motion with a *minimal* sequence of input commands are shown in the two cases. The minimal sequence is not unique in the sense that other combinations of base movements can lead to the same endpoint. On the real system, the controllability of ψ is of no practical relevance, therefore we are not interested in achieving closed paths on the base space but only in the final value of β. Comparing (a) and (b) of Fig. 2, we notice that the minimal sequence for the articulated vehicle is shorter than the other one: this is again a consequence of the direct accessibility of θ from the steering input. Loosely speaking, we can say that for the car-like case the impossibility of generating the whole tangent space only by repeated Lie bracketing g_2 with g_1 is reflected here in the need to apply g_1 three times (instead of the two of the articulated configuration) in order to reach the

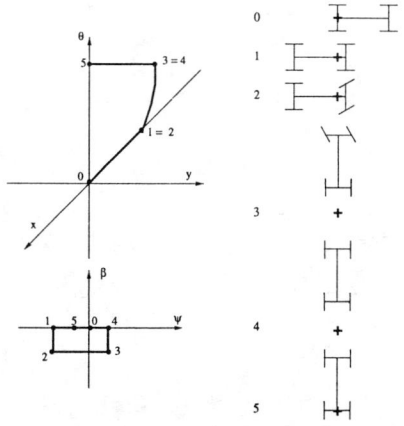

(a) Car-like vehicle

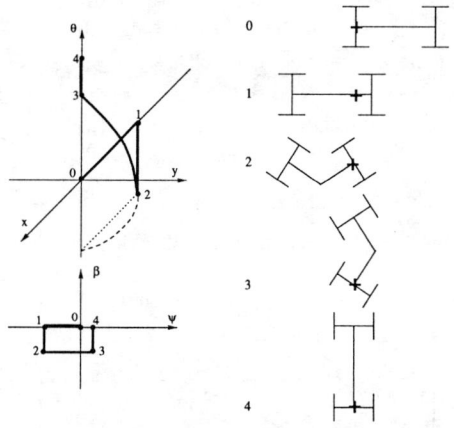

(b) Articulated vehicle

Figure 2: Change of orientation.

desired endpoint, i.e. the arbitrariness in the choice of the fourth vector needed to span the whole tangent space reflects in the larger range of combinations of input commands and implies here the possibility of finding simpler combinations that generate a path between two given points.

3 Off-tracking

The typical work environment for a mining truck is an underground tunnel of limited width. Navigating in such an environment implies an high risk of crashing against the walls of the gallery. It is common characteristic of multiaxis vehicles that during a bend the midpoints of the axles tend to follow different trajectories. The difference between these trajectories can be taken as a measure of how much cumbersome a vehicle is. Comparing again the mining truck with the car-like vehicle, it is intuitively clear that the articulation helps reducing the gap between the trajectories of the midpoints of the two axles. Detailed calculations for the articulated vehicle are reported in [4]. In synthesis, in the two cases, the distance between the two trajectories can be easily computed for a motion with constant steering angle $\beta \neq 0$. In fact, for $v \neq 0$, the two midpoints follow concentric circles whose radius r_0 and r_1 (respectively for the front axle and the rear axle) can be calculated using the geometry of the vehicle. For the car-like vehicle, the off-tracking is (see Fig. 3):

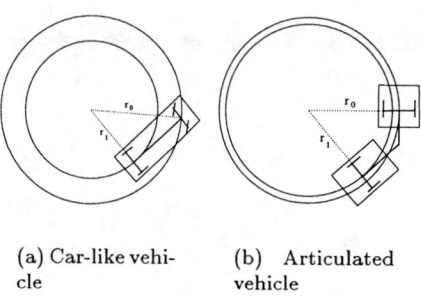

(a) Car-like vehicle

(b) Articulated vehicle

Figure 3: Off-tracking margins.

$$r_0 - r_1 = L \frac{1 - \cos \beta}{\sin \beta}$$

whereas, for the articulated vehicle:

$$r_0 - r_1 = L_1 \frac{\cos \left(\frac{L_0}{L_0+L_1} \beta \right) - \cos \left(\frac{L_1}{L_0+L_1} \beta \right)}{\sin \left(\frac{L_1}{L_0+L_1} \beta \right)}$$

When the two units are symmetric, then $L_0 = L_1$ and the off-tracking is zero i.e. the two trajectories overlay.

3.1 A path tracking criterion

The rest of the paper is dedicated to the formulation of an algorithm for the navigation of the articulated vehicle, aiming at reducing the off-tracking of the whole vehicle from a given path. The idea is adapted from [3].

The underground tunnel in which the truck is navigating is usually represented in terms of a curvature function associated with the length of a trajectory representing for example the middle of the tunnel. Translating this into the cartesian coordinates of an inertial frame is not possible analytically because of the absence, except for trivial cases, of a closed form in the line integral expressing the length of the path covered. Therefore, a particularly convenient local representation is given by a frame moving on the path to follow (see [7, 8]). Under the assumption that the path is at least C^1 and that the curvature has an upper bound (see [9] for the details), a Frenet frame can be used to locally describe the motion of the point with respect to the reference path γ of known curvature κ_γ. The continuity of the curvature function is not required and so also simple paths, composed of straight lines and arcs of circle, can be considered. The set in which the local coordinates are well defined is essentially a "tube" around the path.

In our case, we consider two Frenet frames moving on the curve to follow, corresponding to the projections on γ of the two points P_0 and P_1 of our vehicle. In our frames, we assume to have chosen a base

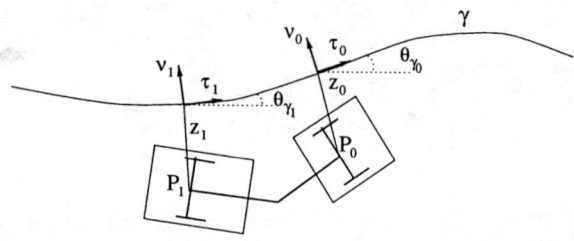

Figure 4: Frenet frames associated with P_0 and P_1.

with the conventions of Fig. 4. Each of the curvilinear frames is represented by two coordinates $(s_{\gamma_i}, \theta_{\gamma_i})$ where s_{γ_i} is the curvilinear abscissa and θ_{γ_i} is the orientation of the frame with respect to the inertial frame. In the Frenet frame, the point P_i is represented by the signed distance z_i between the point itself and its projection and by the relative orientation angle $\tilde{\theta}_i \triangleq \theta_i - \theta_{\gamma_i}$. The equations describing the dynamics of the point P_i in the local frame can be found for example in [7].

For this system, the main issue of the autonomous navigation is by far to *keep the middle* of the tunnel, i.e. to keep control of the lateral dynamics. The main advantage of the Frenet frame is that it provides a natural way to describe the lateral displacement of a point from the path, since z_i represents the signed distance of P_i from its orthogonal projection on γ. This property has been used by several authors as tracking criterion for the path following problem. We will also use it, but redefining the tracking error as the sum of the two signed distances corresponding to the two points P_0 and P_1. In other words, we replace the tracking criterion normally used $z_1 \to 0$ (or an equivalent one) with

$$z_0 + z_1 \to 0 \qquad (3)$$

Using the Frenet frame, the control of the longitudinal dynamics, i.e. how fast the path is covered, becomes an independent problem and relatively secondary with respect to the lateral control. Therefore, it will be neglected in this study. This is equivalent to drop one of the degrees of freedom of the control design that is to say, in our case, to consider the input speed v as a given (open-loop) function. Furthermore, also the dynamic equations for the length covered along the path can be neglected: in fact s_{γ_0} and s_{γ_1} enter into the model only when the curvature κ_γ is varying. Since the stability analysis of our path tracking algorithm will be carried out along an arc of circle, the model is completely independent from s_{γ_0} and s_{γ_1}. In order to keep track of both the distances z_0 and z_1 simultaneously, we must increase the dimension of the state. The new state vector is $\mathbf{p} = \begin{bmatrix} z_0 & z_1 & \tilde{\theta}_0 & \tilde{\theta}_1 & \beta_1 \end{bmatrix}^T$ with the dynamic equations (see [2] for details):

$$\dot{\mathbf{p}} = \begin{bmatrix} \frac{L_1 + L_0 \cos\beta}{L_0 + L_1 \cos\beta} \sin\tilde{\theta}_0 \\ \sin\tilde{\theta}_1 \\ \frac{\sin\beta}{L_0 + L_1 \cos\beta} - \frac{(L_1 + L_0 \cos\beta)\kappa_\gamma(s_{\gamma_0})\cos\tilde{\theta}_0}{(L_0 + L_1 \cos\beta)(1 - \kappa_\gamma(s_{\gamma_0})z_0)} \\ \frac{\sin\beta}{L_0 + L_1 \cos\beta} - \frac{\kappa_\gamma(s_{\gamma_1})\cos\tilde{\theta}_1}{1 - \kappa_\gamma(s_{\gamma_1})z_1} \\ 0 \end{bmatrix} v$$

$$+ \begin{bmatrix} -\frac{L_0 L_1 \sin\beta \sin\tilde{\theta}_0}{L_0 + L_1 \cos\beta} \\ 0 \\ \frac{L_1 \cos\beta}{L_0 + L_1 \cos\beta} + \frac{L_0 L_1 \sin\beta \kappa_\gamma(s_{\gamma_0})\cos\tilde{\theta}_0}{(L_0 + L_1 \cos\beta)(1 - \kappa_\gamma(s_{\gamma_0})z_0)} \\ -\frac{L_0 \cos\beta}{L_0 + L_1 \cos\beta} \\ 1 \end{bmatrix} u$$

or, in more compact form:

$$\dot{\mathbf{p}} = \mathcal{A}(\mathbf{p}) + \mathcal{B}(\mathbf{p})u \qquad (4)$$

Clearly, considering both reference systems in P_0 and in P_1 gives a redundant description of the system. In order to complete this overparameterized state representation, one has to introduce three constraints expressing the fact that the vehicle is a rigid body. These three constraints are given by line integrals that depend on the geometry of the truck and on the curvature of the path between the two projections of P_0 and P_1 on the path. They are obviously holonomic i.e. they reduce the configuration space of the system down to the original number of variables. In what follows we will simply drop them and continue working with the overparameterized model.

Stability analysis For a car-like vehicle, the path following with positive speed implies that the open loop equilibrium point is "naturally" stable whereas backward motion implies that the same equilibrium is open loop unstable. For a mining truck, the path following problem has always an unstable equilibrium point due to the steering action performed on the articulation joint.

We use Lyapunov linearization method to show that the system can be locally asymptotically stabilized to a path of constant curvature. The fact that linearization does not provide global results is not a limitation in our case since the mining truck has to navigate into a tunnel of reduced width and also the local frames are isomorphically defined only in a region around the path.

For a path of constant curvature κ_γ, the equilibrium point $\mathbf{p}_e$ can be calculated from the geometry of the problem. The Jacobian matrix calculated at $\mathbf{p}_e$ $A = \frac{\partial \mathcal{A}}{\partial \mathbf{p}}\big|_{\mathbf{p}=\mathbf{p}_e}$ is a Hurwitz matrix i.e. the closed loop state matrix $A - vKB$ with $B = \mathcal{B}(\mathbf{p}_e)$ can be made stable by choosing an appropriate gain K. Therefore, the system (4) can be locally asymptotically stabilized to a circular path by means of a linear state feedback.

Simulations In Fig. 5, the path to track is an arc of circle. It can be seen that convergence is achieved from a generic admissible initial condition. A similar behavior is obtained with a negative speed.

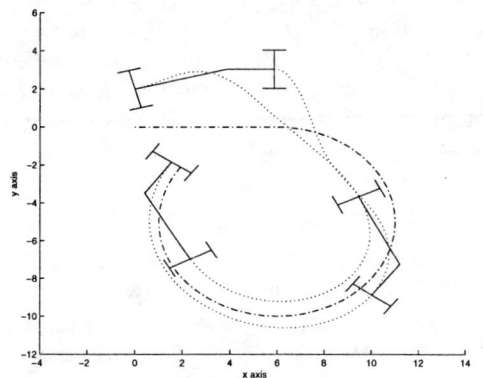

Figure 5: Following an arc of circle of curvature $\kappa_\gamma = -0.2$ (dash-dotted line) from a wrong initial posture.

4 Conclusion

In this paper we have given some mathematical insight into the "higher maneuverability" concept that makes an articulated vehicle more suitable than a car-like vehicle in an environment characterized by a limited free space like it can be an underground mine. How to explain and exploit this difference is the subject of this paper.

References

[1] C. Altafini. The general n-trailer problem: conversion into chained form. In *Proc. of the Conference on Decision and Control*, Tampa, FL, December 1998.

[2] C. Altafini. A path tracking criterion for an LHD articulated vehicle. *International Journal of Robotics Research*, Summer 1999.

[3] C. Altafini and P. Gutman. Path following with reduced off-tracking for the n-trailer system. In *Proc. of the Conference on Decision and Control*, Tampa, FL, December 1998.

[4] L. Bushnell, B. Mirtich, A. Sahai, and M. Secor. Off-tracking bounds for a car pulling trailers with kingpin hitching. In *Proc. of the Conference on Decision and Control*, pages 2944–2949, Lake Buena Vista, FL, 1994.

[5] A. Isidori. *Nonlinear Control System*. Springer-Verlag, 3rd edition, 1995.

[6] J.-P. Laumond. Controllability of a multibody mobile robot. *IEEE Trans. on Robotics and Automation*, 9:755–763, 1993.

[7] A. Micaelli and C. Samson. Trajectory tracking for unicycle-type and two-steering-wheels mobile robots. Technical Report 2097, INRIA, November 1993.

[8] M. Sampei, T. Tamura, T. Itoh, and M. Nakamichi. Path tracking control of trailer-like mobile robot. In *Proc. IEEE/RSJ Inter. Workshop on Intelligent Robots and Systems*, pages 193–198, Osaka, Japan, November 1991.

[9] C. Samson. Control of chained systems: application to path-following and time-varying point stabilization of mobile robots. *IEEE Trans. on Automatic Control*, 40:64–77, 1995.

Simulating Dextrous Manipulation of a Multi-fingered Robot Hand Based on a Unified Dynamic Model*

Joseph C. Chan and Yun-hui Liu
Department of Mechanical and Automation Engineering
The Chinese University of Hong Kong, Shatin, N.T., Hong Kong, China
(ckchan,yhliu)@mae.cuhk.edu.hk

Abstract

A dynamic simulator can facilitate developments and applications of a multi-fingered robot hand. Existing dynamic simulators cannot effectively simulate dextrous manipulation of a multi-fingered robot hand due to the lack of capability to cope with frequent changes in contact constraints and grasping configurations as well as impulsive collision occurring during manipulation. In this paper, we propose a unified framework to model free motions, collisions, and different contact motions including sticking, rolling, and sliding. Furthermore, a new transition model is also proposed to handle transitions between these contact motions. Based on our unified dynamic model, a 3D dynamic simulator has been developed to simulate dextrous manipulation tasks involving combination of different contacts. The simulation results presented will confirm the validity of the dynamic model as well as efficiency of the developed simulator.

1 Introduction

High dexterity and superior ability of fine manipulation offer multi-fingered robot hand extensive potential applications such as those in robot-aided surgery. A dynamic simulator that is able to simulate dynamic motion of multi-fingered hands can facilitate the transfer of research results into applications. For example, it can be served as a testing platform to evaluate different control algorithms. Dextrous manipulation using multi-fingered robot hand is accompanied by frequent changes in contact constraints and grasping configurations. When manipulating an object, different fingers are usually in different kinds of contact with the object. Some of them may be in sticking contacts and others may slide or roll on the surface of the object.

Different contact motions are determined by the frictional properties of the system. Furthermore, fingers may be also required to re-position to other faces of the object. If these situations are modeled case by case, we will suffer from the problem of combinatorial explosion of models. On the other hand, impulsive collision frequently occurring during manipulation causes discontinuity in velocity of the system, which cannot be handled by ordinary contact models.

Although tremendous efforts have been made to robot dynamic simulation, only a few simulators can be found for multi-fingered robot hand. In the study of human-machine coordination, Hashimoto et al. [2] and Kunii and Hashimoto [3] developed a dynamic force simulator for a multi-fingered robot hand in a virtual environment. Reznik and Laugier [7] used a deformable fingertip model to determine the interaction force between fingers and the grasped object. These two simulators are designed to simulate dynamic response in grasping but do not address the relative contact motions. Mirza et al. [6] used the compliant contact model to model friction, rolling, sliding as well as the changing grasp configurations in his 2D dynamic simulator. However, dynamic and geometric properties of 3D manipulation are complicated than those of 2D manipulation. Furthermore, stiff equations may be introduced when parameters are tuned to mimic the rigid body model.

In this paper, we propose a unified framework to model the dynamics of a multi-fingered hand performing dextrous manipulation tasks that involve a combination of free motion, sticking, rolling, and sliding between fingers and the grasped object so that we can easily cope with the frequent changes in contacts and grasping configurations. The key idea lies in the introduction of a *contact mode selection matrix*. With the contact mode selection

---

*This work is supported in part by the Hong Kong Research Grant Council under the grant CUHK 4151/97E.

matrix, one can effectively select dynamic models for different contact motions between the grasped object and the fingers. To determine collision response between fingers and the grasped object, we extended Mirtich's general collision model [5]. Our extended model is able to handle single collision, multiple collisions, and mixed contacts and collisions occurred in dextrous manipulation. A new model is also proposed to model transitions between free motion and different contact motions. Furthermore, we have developed a 3D dynamic simulator for a five-fingered robot hand based on the unified dynamic model. To the best of our knowledge, this dynamic simulator is the first one that can sophisticatedly simulate dextrous manipulation of a multi-fingered robot hand.

2 Kinematic and Dynamic Modeling

Assume that only the fingertips of the fingers can be in contact with the object. The fingers can be either hard fingers or soft fingers. The fingers and the object are assumed to be rigid bodies. Furthermore, we assume that each finger is in one of the three elementary contacts: sticking, sliding, and rolling with the grasped object. In rolling, we suppose that no sliding occurs. It should be noted that different finger may be in different contact motion when manipulating the object. A finger may change the contact motion, i.e. transition between the elementary motions, during the manipulation.

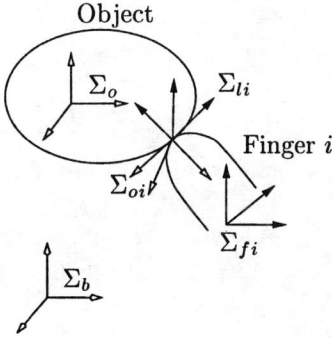

Figure 1: Kinematic and dynamic modeling.

Consider Figure 1. Let Σ_b be the inertial reference frame. Σ_o is the object frame. Σ_{fi} is a frame attached to the ith fingertip. Σ_{oi} and Σ_{li} are the ith contact frames on the surfaces of the object and the fingertip, respectively. The contact kinematics between the ith finger and the object can be writ-

ten as follows [4]

$$B_i^T G_i^T \mathcal{V}_o = B_i^T J_{fi} \dot{\theta}_i + B_i^T \mathcal{V}_{ri} \quad (1)$$

The term $G_i^T \mathcal{V}_o$ represents the velocity of Σ_{oi} with respect to Σ_b. Jacobian J_{fi} transforms the joint velocities $\dot{\theta}_i$ to the velocity of the contact point. Therefore, the term $J_{fi}\dot{\theta}_i$ represents the velocity of Σ_{li} with respect to Σ_b. The relative motion between the finger and the object is defined by $\mathcal{V}_{ri}$, the relative velocity between Σ_{oi} and Σ_{li}. B_i is called the *finger model matrix* which imposes additional constraints to the kinematic equation for different finger models. For example, a frictional hard finger can only exert forces upon the object at the contact point. Thus, the torques exerted by the finger are constrained and the corresponding B_i is in the form of $\begin{bmatrix} I & 0 \end{bmatrix}^T$ where I and 0 are the 3×3 identity and zero matrices. Similarly, for the frictionless hard finger and the soft finger, the finger model matrices are in the form of $B_i = \begin{bmatrix} 0 & 0 & 1 & 0 & 0 & 0 \end{bmatrix}^T$ and

$$B_i = \begin{bmatrix} 1 & 0 & 0 & 0 & 0 & 0 \\ 0 & 1 & 0 & 0 & 0 & 0 \\ 0 & 0 & 1 & 0 & 0 & 0 \\ 0 & 0 & 0 & 0 & 0 & 1 \end{bmatrix}^T \quad (2)$$

For two smooth surfaces in contact, the constraints of sticking, rolling or sliding contact between them can be defined in terms of the relative velocity components, $\mathcal{V}_{ri} = \begin{bmatrix} v_x & v_y & v_z & \omega_x & \omega_y & \omega_z \end{bmatrix}^T$. For sticking contact, all components are zero except ω_z. For rolling contact, all translational components and ω_z are zero. For sliding contact, v_z and all rotational components are zero. In order to incorporate these different contacts into the dynamic model, the *contact mode selection matrix* C_{Mi} is introduced so that the relative velocity $\mathcal{V}_{ri}$ for a particular kind of contact could be rewritten as $C_{Mi}\mathcal{V}_{ri}$ where C_{Mi} is a 6×6 zero matrix except

$$\begin{cases} \text{sticking contact: } C_{Mi}(6,6) = 1 \\ \text{rolling contact: } C_{Mi}(4,4) = C_{Mi}(5,5) = 1 \\ \text{sliding contact: } C_{Mi}(1,1) = C_{Mi}(2,2) = 1 \end{cases} \quad (3)$$

where $C_{Mi}(k,k)$ represents the kth row, kth column element of C_{Mi}. Thus, the generalized contact kinematics can be expressed as

$$B_i^T G_i^T \mathcal{V}_o = B_i^T J_{fi} \dot{\theta}_i + B_i^T C_{Mi} \mathcal{V}_{ri} \qquad (4)$$

Combining all generalized contact kinematics for the robot hand and having some algebraic manipulations, we can obtain the velocity constraint of the entire system

$$A\dot{q} = 0 \qquad (5)$$

where

$$A = \begin{bmatrix} -\tilde{J}_{f1} & \cdots & 0 & \tilde{G}_1^T \\ \vdots & \ddots & \vdots & \vdots \\ 0 & \cdots & -\tilde{J}_{fm} & \tilde{G}_m^T \end{bmatrix} \qquad (6)$$

$\dot{q} = \begin{bmatrix} \dot{\theta} & \mathcal{V}_o \end{bmatrix}^T$ is the generalized velocity of the system. m is the total number of fingers. $\tilde{J}_{fi}$ and $\tilde{G}_i^T$ represent $B_i^T(I - C_{Mi})J_{fi}$ and $B_i^T(I - C_{Mi})G_i^T$ where I is a 6×6 identity matrix. For the ith contact, since both $\tilde{J}_{fi}$ and $\tilde{G}_i^T$ involve $I - C_{Mi}$, they become zero matrices if C_{Mi} is set to be a 6×6 identity matrix. In this case, the rows composed by $\tilde{J}_{fi}$ and $\tilde{G}_i^T$ in A become all zero. As a result, the contact constraint between the ith finger and the object is released and free motion is represented. In other words, different grasping configurations can be achieved in our dynamic model by setting corresponding C_{Mi} to be an identity matrix if there is no contact between that finger and the object.

Finally, the dynamics of the multi-fingered robot hand manipulating an object can be expressed as

$$M_s \ddot{q} + N_s = F_s + A^T \lambda \qquad (7)$$

where M_s is the inertial matrix, N_s and F_s represent the nonlinear terms and the external joint torques, respectively. Matrix A is defined in (6) and λ is the Lagrangian multiplier which actually represents the internal force between the robot hand and the object. Different contact motions such as sticking contact, rolling or sliding can be represented by selecting appropriate contact selection matrix shown in equation (3) and the free motion can be achieved by setting the contact contact selection matrix to be an identity matrix.

3 Modeling of Collisions between Fingers and Object

Collisions between two rigid bodies cannot be handled by the equations derived in Section 2. Collision is completed in infinitesimal time so the impulse introduced only changes the velocity of the system but not the configuration. Traditionally, the pre-collision velocity and the post-collision velocity of the colliding system are related by some algebraic equations. However, it is proved that a differential approach should be used if friction is involved in collision [5]. In this section, we will review the basic equations for our extended differential Mirtich's collision model. Detailed descriptions for its application to the dextrous manipulation simulation can be referred to [1].

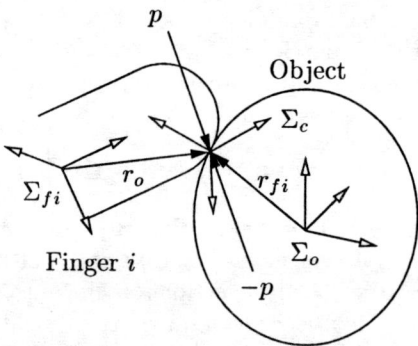

Figure 2: Collision modeling.

The coordinate frames shown in Figure 2 are assigned similar to those in Figure 1. During the collision period, it is assumed that there is no relative motion between two colliding bodies. Thus, frames Σ_{oi} and Σ_{li} are replaced by a contact frame Σ_c located at the contact point. r_o and r_{fi} are the position vectors, and p is the corresponding impulse. These vectors are all with respect to the contact frame. Dynamics of the ith finger is

$$M_{fi} \ddot{\theta}_i + N_{fi} = \tau_i + J_i^T f_c \qquad (8)$$

where M_{fi}, N_{fi} and τ_i are the inertial matrix, nonlinear term and joint input. θ_i is the joint angle and f_c is the collision force. The transpose of Jacobian J_i transforms the force from the contact frame to the joint space. Since N_{fi} and τ_i are negligible because they are much smaller than the collision force f_c, integrating equation (8) through time results an expression in differential joint velocities $\delta \dot{\theta}_i$

$$\delta \dot{\theta}_i(\gamma) = M_{fi}^{-1} J_i^T \, p(\gamma) \qquad (9)$$

where p is the impulse. Since the collision period is very short, it is not appropriate to use t as the integration parameter. t is replaced by γ in equation (9), where γ is the generalized collision parameter which monotonically increases during the course of collision.

We can relate the contact point velocity u_{fi} to the joint velocities by using Jacobian J_i as $u_{fi} = J_i \dot{\theta}_i$. Its differential form with respect to γ is

$$\delta u_{fi}(\gamma) = J_i \delta \dot{\theta}_i(\gamma) \qquad (10)$$

Combining equations (9) and (10), we obtain the relationship between impulse and change of contact point velocity

$$\delta u_{fi}(\gamma) = K_{fi} p(\gamma) \qquad (11)$$

where $K_{fi} = J_i M_{fi}^{-1} J_i^T$.

Meanwhile, the object dynamics can be represented by the Newton-Euler equations. Because the inertial forces can also be negligible during the collision, the differential forms of the Newton-Euler equations are

$$\begin{cases} -p(\gamma) = m_o \delta v_o(\gamma) \\ -r_o \times p(\gamma) = I_o \delta \omega_o(\gamma) \end{cases} \qquad (12)$$

m_o and I_o are the mass and inertial tensor of the object. v_o and ω_o are the translational and rotational object velocities.

The differential contact point velocity of the object is

$$\delta u_o(\gamma) = \delta v_o(\gamma) + \delta \omega_o(\gamma) \times r_o \qquad (13)$$

Combining equations (12) and (13), we obtain

$$\delta u_o(\gamma) = -K_o p(\gamma) \qquad (14)$$

where $K_o = \frac{1}{m_o} I_{3\times 3} - \tilde{r}_o I_o^{-1} \tilde{r}_o$. $\tilde{r}_o$ is a 3×3 skew-symmetric matrix defined as

$$\tilde{r}_o = \begin{bmatrix} 0 & -r_{oz} & r_{oy} \\ r_{oz} & 0 & -r_{ox} \\ -r_{oy} & r_{ox} & 0 \end{bmatrix}$$

r_{ox}, r_{oy}, and r_{oz} are the elements of r_o.

Using equations (11) and (14), the differential relative contact velocity between finger i and the object is

$$\delta u(\gamma) = \delta u_{fi}(\gamma) - \delta u_o(\gamma) = K p(\gamma) \qquad (15)$$

where $K = K_{fi} + K_o$. K is called the collision matrix which is kept constant during collision. The differential equation governing the collision is

$$\frac{d}{d\gamma} u(\gamma) = K \frac{d}{d\gamma} p(\gamma) \qquad (16)$$

Integrating this first order differential equation, we can track through the change of relative contact velocity u during collision. The value of u at the end of collision can be used to determine the corresponding impulse using the Poisson's restitution hypothesis and the Coulomb friction law. Finally, the post-collision joint velocities of the colliding finger and the object post-collision velocity can then be calculated.

Because collision does not involve any change in configurations of the colliding bodies, multiple collisions can be partitioned and solved one by one. If the object is grasped or manipulated by other fingers when a finger is colliding with the object, the grasping forces can be ignored because they are much smaller than the impulsive force. Accordingly, the collision response in this case is also determined by equation (16). After that, the change of the object velocity must be propagated to the contacting fingers and thus instantaneous changes occur in their joint velocities.

4 Dynamic Simulation

The architecture of the dynamic simulation system is shown in Algorithm 1. Interference detection subroutine is implemented to detect whether there are contacts between fingertips and the object. The geometric models of the fingertips and the object are employed for this purpose. When a contact occurs, contact mode is firstly determined and the corresponding contact constraint is then imposed to the kinematics by choosing corresponding contact mode selection matrix. The dynamic response is computed by equation (7). When a collision detected, the change of velocity is computed by the collision model derived in Section 3.

Since we do not know the type of a contact when it is detected, a contact transition model [8] is proposed to handle the transition of different situations such as free motion or rolling contact happened in manipulation. Determination of the contact mode in simulation depends on the information in last and current integrations. In Figure 3, circles represent states of the system and rectangular boxes represent the conditions of transition. For example, when a contact is detected after an impulsive collision, it is classified to be sliding contact if the tangential approaching velocity is larger than zero.

A dynamic simulator based on the full kinematics and dynamics of a five-fingered robot hand has been developed. It was constructed by using C language in the Silicon Graphics *Indigo*² workstation. The corresponding equations were firstly implemented in the MATLAB and then converted to C codes. The graphical output was developed by the IRIS graphical library. The object is restricted to rectangular objects for the ease of implementating the interference detection algorithm. Snapshots of rolling contact and sliding contact are shown in Figure 4 and 6. The corresponding contact point trajectories are shown in Figure 5 and 7. Figure 8 demonstrates the mixed contacts and collisions. The velocity propagation can be identified in Figure 9. Sampling time of the simulation is $1ms$ and the fixed step Runge-Kutta integration method was applied. The friction coefficient used in simulation is 0.6. More simulation results can be found in [1].

5 Conclusions

In this paper, we proposed a new dynamic model for dextrous manipulation using multi-fingered robot hands. Our model is capable of modeling free motion and different contacts in a unified framework. This is achieved by imposing the contact mode selection matrix into the kinematics. By combining this formulation with the new contact transition model described in Section 4, we can efficiently simulate motions of robot hands with different grasping configurations and different contact constraints. An improved collision model is also developed to determine the instantaneous change of velocity caused by collision impulsive force. This model is able to deal with single collision, multiple collisions, and mixed contacts and collisions. Based on our proposed models, a 3D five-fingered robot hand dynamic simulator has been developed and its performance is verified by the simulation results. In future, experiments will be done in manipulating objects using real robot hand. Results will be used to compare the simulated one so that the performance of the dynamic simulation system can be evaluated analytically.

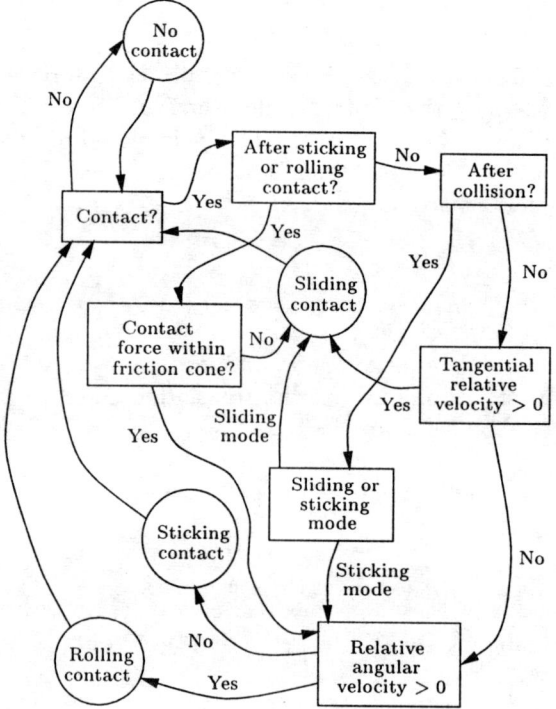

Figure 3: Contact transition model.

References

[1] Chan, J. C. (1998). Dynamic modeling and simulation of a multi-fingered robot hand. Master's thesis, The Chinese University of Hong Kong.

[2] Hashimoto, H., Buss, M., Kunii, Y., and Harashima, F. (1994). Intelligent cooperative manipulation system using dynamic force simulator. In *Proc. of IEEE Int. Conf. on Rob. and Auto.*, pages 2598–2603.

[3] Kunii, Y. and Hashimoto, H. (1996). Dynamic force simulator for multifinger force display. *IEEE Transaction on Industrial Electronics*, 43(1):74–80.

[4] Li, Z., Hsu, P., and Sastry, S. S. (1989). Grasping and coordinated manipulation by a multifingered robot hand. *International Journal of Robotic Research*, 8(4):33–49.

Algorithm 1 Dynamic simulation architecture.

Input: initial system state
1: **loop**
2:    interference detection
3:    determine current contact mode using contact transition model in Figure 3
4:    compute velocity constraint using eq. (5)
5:    compute system dynamics using eq. (7)
6:    numerically integrate system dynamics to obtain the new system state
7:    **if** a collision is detected **then**
8:       **for** each collision **do**
9:          determine post-collision velocity using eq. (16)
10:       **end for**
11:       propagate change of object velocity to other contacting fingers
12:    **end if**
13: **end loop**

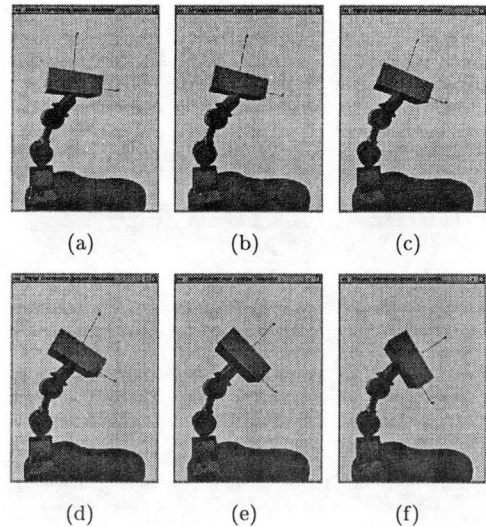

Figure 4: Rolling contact.

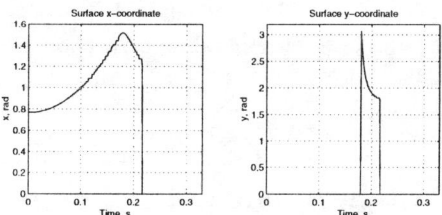

Figure 5: Contact trajectories on the surface of fingertip.

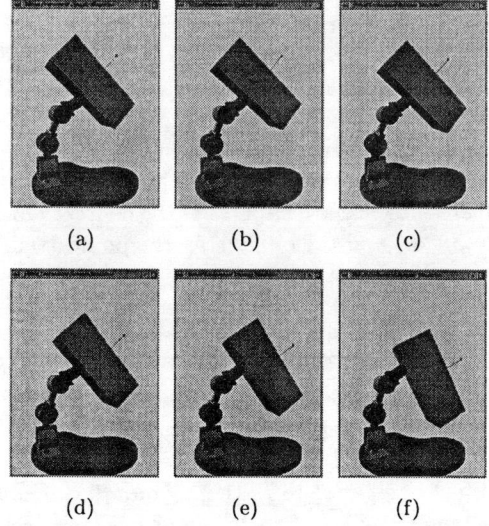

Figure 6: Sliding contact.

[5] Mirtich, D. V. (1996). *Impulse-based dynamic simulation of rigid body systems.* PhD thesis, University of California at Berkeley.

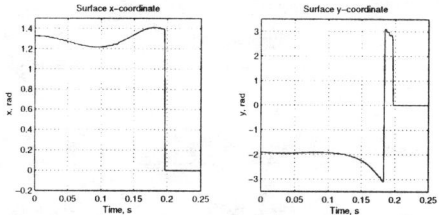

Figure 7: Contact trajectories on the surface of the object.

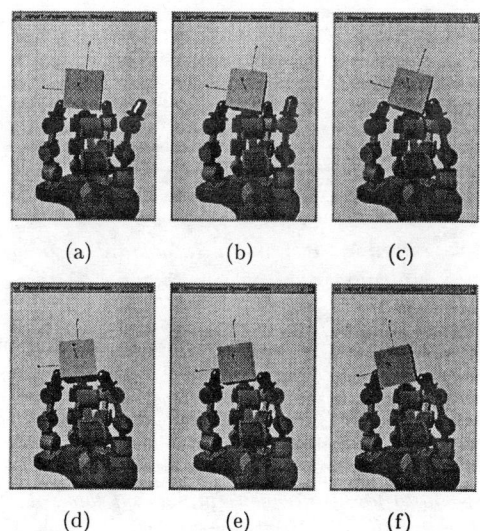

Figure 8: Mixed contacts and collisions.

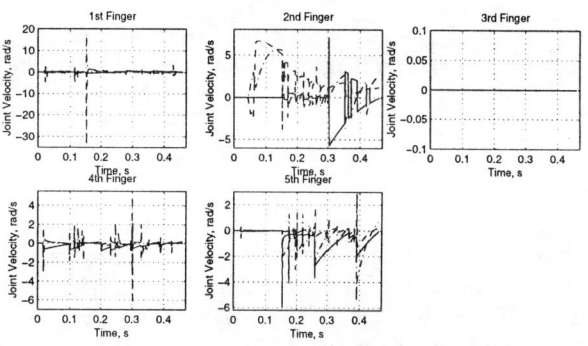

Figure 9: Joint velocities of fingers.

[6] Mirza, K., Hanes, M. D., and Orin, D. E. (1993). Dynamic simulation of enveloping power grasps. In *Proc. of IEEE Int. Conf. on Rob. and Auto.*, pages 430–435.

[7] Reznik, D. and Laugier, C. (1996). Dynamic simulation and virtual control of a deformable fingertip. In *Proc. of IEEE Int. Conf. on Rob. and Auto.*, pages 1669–1674.

[8] Wendlandt, J. M. (1997). *Control and simulation of multibody systems.* PhD thesis, University of California at Berkeley.

Coordinated Motion Generation for Multifingered Manipulation Using Tactile Feedback

S.L. Jiang, K.K Choi and Z.X. Li*
Dept. of Electrical and Electronic Engineering
Hong Kong University of Science and Technology
Clearwater Bay, Kowloon, Hong Kong
(e-mail: eejiangl@ee.ust.hk, fax (852)2358-1485)

Abstract

Dextrous manipulation is an important issue in the study of multifingered robotic hands. Determining the contact velocities in dextrous manipulation with rolling contact is the key problem. In this paper, we propose a coordinated manipulation scheme in which the contact velocities are determined by tactile sensor feedback. We address coordinated motion generation which maintains or improves the grasp quality in dextrous manipulation. We implement several experiments using the HKUST three-fingered robotic hand. The experimental results illustrate the effectiveness of the proposed scheme.

1 Introduction

Over the years, many researches have investigated dextrous manipulation with rolling contact constraints ([6, 7]). Kerr formulated the kinematics of manipulation with rolling contact, namely the relationships between the motion of the fingers, manipulated object and the contact location on both surfaces of the fingertips and the object ([1]). Cole derived the kinematics of rolling contact for two surfaces of arbitrary shape rolling on each other ([14]). Maekawa investigated a new motion control system using tactile feedback for the manipulation of an object by a multifingered hand where the fingertip and the object make rolling contact ([9]). Montana formulated the kinematics of contact between the fingertips and object, which relate the contact velocities to the change rates of the fingertips and the object by using their geometric parameters ([3, 4, 5]). Based on Montana's contact equations, there are two methods to determine the contact ve-

---

*This research is supported in part by RGC Grant No. HKUST 685/95E, HKUST 555/94E and HKUST 193/93E.

locities. One is to specify the change rate of the local coordinate of the manipulated object. The other is to calculate the change rates of the local coordinates of the fingertips by virtue of feedback of tactile sensors. In this paper, we use the second method to address coordinated motion generation for multifingered manipulation.

2 Mathematical Preliminaries

Consider a k-fingered robotic hand grasping an object as shown in Figure 1. We denote by P the palm frame, O the object frame, and F_i the fingertip frame of finger $i, i = 1, \cdots k$. Without loss of generality, we assume that each finger contacts the object at a point around the fingertip (i.e., the surface of the last link) with a contact model described by either a point contact with friction (PCWF) or a soft-finger contact (SFC) ([8]). We parameterize the fingertips and the object with orthogonal coordinate charts $\alpha_{f_i} \in \mathbb{R}^2$ and $\alpha_{o_i} \in \mathbb{R}^2$, which are the local coordinates of the contact point between finger i and the object. At the point of contact, we define local coordinate frames L_{f_i} and L_{o_i}, where L_{f_i} is fixed relative to F_i and L_{o_i} is fixed relative to O. The z-axis of a local frame coincides with the outward unit normal of the respective surfaces. The angle of contact between finger i and the object is denoted by ψ_i. The five parameters $\eta_i := (\alpha_{f_i}, \alpha_{o_i}, \psi_i) \in \mathbb{R}^5$ that describe the contact state between finger i and the object is referred to as the *coordinates of contact*.

We denote by $\theta_i \in \mathbb{R}^{n_i}$ the joint position vector of finger i, where n_i is the degree of freedom of finger i. By viewing each contact as a virtual joint with five degrees of freedom, we let $q_i = (\theta_i, \eta_i) \in \mathbb{R}^{n_i+5}$ be the extended joint coordinates of finger i.

We denote by $g_{ab} = (R_{ab}, p_{ab}) \in SE(3)$ the Euclidean transformation from a coordinate frame B to

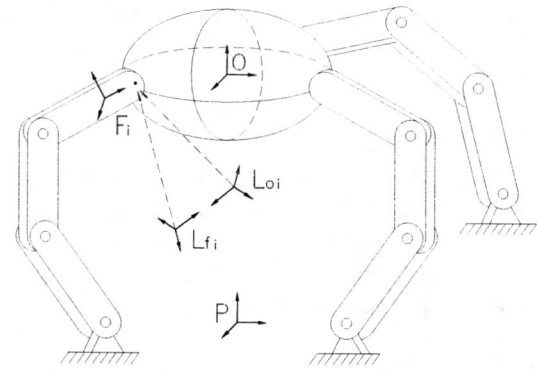

Figure 1: A multifingered hand manipulation system

another coordinate frame A, where $p_{ab} \in \mathbb{R}^3$ and $R_{ab} \in SO(3)$ are the position and orientation of B relative to A respectively, and $SE(3)$ is known as the special Euclidean group. The velocity of B relative to A is denoted by $V_{ab} = (v_{ab}, \omega_{ab}) \in \mathbb{R}^6$. $Ad_{g_{ab}}$ is the adjoint transformation associated with g_{ab}. We refer to [8] for further notation and the kinematic relations embedded in the multifingered manipulation system.

Given the extended joint coordinates $q_i = (\theta_i, \eta_i)$, the position and orientation of the grasped object is given by:

$$g_{po} = g_{pf_i}(\theta_i) g_{f_i o}(\eta_i), \qquad i = 1, \cdots k \quad (1)$$

where $g_{pf_i}(\theta_i)$ is the forward kinematic map of finger i, and $g_{f_i o}(\eta_i)$ consists of three transformations by the contact frames L_{f_i} and L_{o_i} of the fingertip and the object, respectively,

$$g_{f_i o}(\eta_i) = g_{f_i l_{f_i}}(\alpha_{f_i}) \cdot g_{l_{f_i} l_{o_i}}(\psi_i) \cdot g_{l_{o_i} o}(\alpha_{o_i}) \quad (2)$$

The velocity relation of (1) is found to be

$$\begin{aligned} V_{po} &= \begin{bmatrix} Ad_{g_{f_i o}^{-1}} J_{pf_i}(\theta_i), & J_{f_i o}(\eta_i) \end{bmatrix} \begin{bmatrix} \dot{\theta}_i \\ \dot{\eta}_i \end{bmatrix} \\ &:= J_{po}^i(q_i) \dot{q}_i \end{aligned} \quad (3)$$

where $J_{pf_i}(\theta_i) \in \mathbb{R}^{6 \times n_i}$ and $J_{f_i o}(\eta_i) \in \mathbb{R}^{6 \times 5}$ are the Jacobian of finger i and the ith contact, respectively. Equating the velocity relation (3) yields the velocity constraint

$$J_{po}^1(q_1) \dot{q}_1 = \cdots = J_{po}^k(q_k) \dot{q}_k. \quad (4)$$

Note that the velocity relation in (3) can also be expressed in terms of the contact velocities and the result is given by

$$V_{po} = Ad_{g_{f_i o}^{-1}} V_{pf_i} - Ad_{g_{l_{f_i} o}^{-1}} V_{l_{o_i} l_{f_i}}, \quad (5)$$

where $V_{l_{o_i} l_{f_i}} = (v_x^i, v_y^i, v_z^i, \omega_x^i, \omega_y^i, \omega_z^i)^T$ is the contact velocity. Montana's kinematic equations of contact relate the contact velocity to the rate of change of the coordinates,

$$\begin{aligned} \dot{\alpha}_{f_i} &= M_{f_i}^{-1} K_i^{-1} \left(\begin{bmatrix} -\omega_y^i \\ \omega_x^i \end{bmatrix} - \tilde{K}_{o_i} \begin{bmatrix} v_x^i \\ v_y^i \end{bmatrix} \right) \\ \dot{\alpha}_{o_i} &= M_{o_i}^{-1} R_{\psi_i} K_i^{-1} \left(\begin{bmatrix} -\omega_y^i \\ \omega_x^i \end{bmatrix} + K_{f_i} \begin{bmatrix} v_x^i \\ v_y^i \end{bmatrix} \right) \\ \dot{\psi}_i &= \omega_z^i + T_{f_i} M_{f_i} \dot{\alpha}_{f_i} + T_{o_i} M_{o_i} \dot{\alpha}_{o_i} \end{aligned} \quad (6)$$

where $(M_{f_i}, K_{f_i}, T_{f_i})$ and $(M_{o_i}, K_{o_i}, T_{o_i})$ are, respectively, the geometric parameters of the fingertip and the object expressed in terms of local coordinate charts,

$$R_{\psi_i} = \begin{bmatrix} \cos \psi_i & -\sin \psi_i \\ -\sin \psi_i & -\cos \psi_i \end{bmatrix},$$

$\tilde{K}_{o_i} = R_{\psi_i} K_{o_i} R_{\psi_i}$, and $K_i = (K_{f_i} + \tilde{K}_{o_i})$ is the relative curvature between the fingertip and the object viewed from the local frame of the fingertip.

3 The Coordinated Manipulation Scheme Using Tactile Sensor Feedback

Rearranging (5) gives an equation for $\tilde{V}_{pf_i}$, the local expression of V_{pf_i}, in terms of V_{po} and the contact velocity

$$\begin{aligned} \tilde{V}_{pf_i} &= Ad_{g_{o l_{o_i}}^{-1}} V_{po} + V_{l_{f_i} l_{o_i}} \\ V_{pf_i} &= Ad_{g_{f_i l_{o_i}}} \tilde{V}_{pf_i} \end{aligned} \quad (7)$$

A straight forward approach, known as *individual joint control law* ([13]), for computing the fingertip velocity is to first multiply Eq. (7) by B_i^T, the transpose of the wrench basis of the assumed contact model, and then utilize the velocity contact constraint of

$$B_i^T V_{l_{f_i} l_{o_i}} = 0$$

to obtain the equation

$$B_i^T \tilde{V}_{pf_i} = B_i^T Ad_{g_{o l_{o_i}}^{-1}} V_{po} \quad (8)$$

Finally, solving (8) for $\tilde{V}_{pf_i}$ gives a solution of the fingertip velocity V_{pf_i}.

Unfortunately, while the *individual joint control law* is simple to implement, it suffers from a serious drawback as suggested by the following example.

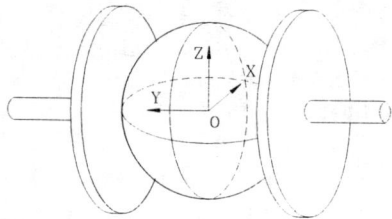

Figure 2: A two-fingered hand manipulating a ball

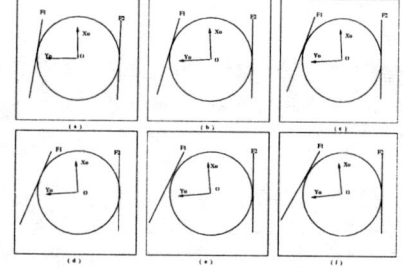

Figure 3: A simulated sequence of intermediate configurations showing undesirable consequence of uncoordinated manipulation.

Example 3.1. Consider a two-fingered robotic hand manipulating a ball as shown in Figure 2. The fingertips are both disk-shaped and the initial grasp configuration is antipodal which is considered optimal. Let the desired trajectory of the object be a rotation about the z-axis through a point in space, i.e., a screw motion generated by a twist of the form $\xi = (0, \rho, 0, 0, 0, 1)^T$, where $\rho \neq 0$. Computing the fingertip velocity using (8) and a soft-finger contact model, we find that rolling motion indeed occur at the contact points and the fingers move closer to each other at one end. Eventually, the force-closure condition would fail to hold and the object drops. A sequence of intermediate configurations illustrating this undesirable phenomena is shown in Figure 3. □

The preceding example shows that as manipulation proceeds the grasp quality would degrade under the individual joint control law ([10, 12]). To overcome this problem, we address coordinated motion generation using tactile feedback as following.

For multifingered manipulation, each contact point can be visualized as a virtual joint which can not be controlled directly. But fortunately, the change rate of the local coordinate of the fingertips can be detected by tactile sensors. Utilizing tactile feedback, the motion of joints can be specified so that each fingertip force always stays inside the friction cone and the fingertip maintains firm contact with the object without slipping nor detaching.

Consider again Eq. (7), rewritten in the form

$$\tilde{V}_{pf_i} = Ad_{g_{olo_i}^{-1}} V_{po} - Ad_{g_{lo_il_{f_i}}} V_{l_{o_i}l_{f_i}}.$$

Note that the kinematic equations of contact given in (6) are invertible. Under pure rolling (or soft-finger) contact, tactile sensor readings for $\dot{\alpha}_{f_i}$ can be used to compute rolling velocities and the rate of change of the remaining contact coordinates,

$$\begin{bmatrix} -\omega_y^i \\ \omega_x^i \end{bmatrix} = (K_{f_i} + \tilde{K}_{o_i}) M_{f_i} \dot{\alpha}_{f_i} \qquad (9)$$
$$\dot{\alpha}_{o_i} = M_{o_i}^{-1} R_{\psi_i} M_{f_i} \dot{\alpha}_{f_i}$$
$$\dot{\psi}_i = (T_{f_i} + T_{o_i} R_{\psi_i}) M_{f_i} \dot{\alpha}_{f_i}.$$

The contact velocity $V_{l_{o_i}l_{f_i}} = (v_x^i, v_y^i, v_z^i, \omega_x^i, \omega_y^i, \omega_z^i)^T$ can be expressed in terms of the rolling velocity which in turn, through the kinematic equations of contact, can be expressed in terms of $\dot{\alpha}_{f_i}$, i.e. we have

$$\boxed{\tilde{V}_{pf_i} = Ad_{g_{olo_i}^{-1}} V_{po} - T_i(\eta_i) \dot{\alpha}_{f_i}} \qquad (10)$$

where

$$T_i(\eta_i) = Ad_{g_{lo_il_{f_i}}} B_i^c R_0 (K_{f_i} + \tilde{K}_{o_i}) M_{f_i}$$

and

$$B_i^c = \begin{bmatrix} 0 & 0 & 0 & 1 & 0 & 0 \\ 0 & 0 & 0 & 0 & 1 & 0 \end{bmatrix}^T, \quad R_0 = \begin{bmatrix} 0 & 1 \\ -1 & 0 \end{bmatrix}.$$

Assume the grasped object is convex and to be a known geometric model. Let $\alpha_o = (\alpha_{o_1}, \cdots, \alpha_{o_k}) \in \mathbb{R}^{2k}$, where $\alpha_{o_i} \in \mathbb{R}^2, i = 1, \cdots, k$. We can define a cost function:

$$f : \mathbb{R}^{2k} \longrightarrow \mathbb{R} \mid \alpha_o \longmapsto f(\alpha_o) \qquad (11)$$

on the contact coordinates of the object for measuring the quality of a grasp. Conditioning of the grasp map $G(\alpha_o)$ is improved if $f(\alpha_o)$ is maximized (or minimized). For example, (1). considering a two-fingered hand manipulation a unit ball, we can define the grasp quality function f as the length of the chord jointing the two contact points on the sphere. Note that f attains maximum when the grasp is antipodal and minimum when the two contact points coincide. (2). for a there-fingered hand manipulating a unit ball, we can define the function f as the square of area formed by the three contact points. Note that the function f is maximized when the three points are located 120° apart along an equator. See([10, 12]) for

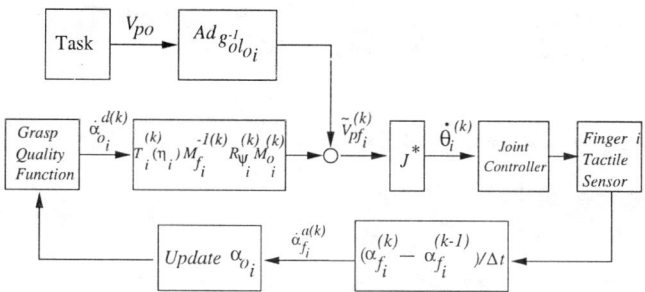

Figure 4: The block diagram of coordinated motion generation using tactile sensor feedback

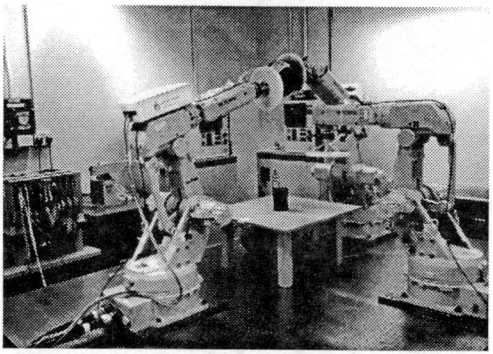

Figure 5: The HKUST three-fingered robotic hand manipulating a ball

details. Other geometric properties of the object surface such as curvature can be incorporated by defining more sophisticated cost functions. See ([2, 4, 12]) for more details.

A simple and sensible solution for $\dot{\alpha}_{o_i}$ is given by

$$\dot{\alpha}_{o_i} = \lambda_i \nabla_i f(\alpha_o) \quad (12)$$

where $\nabla_i f(\alpha_o)$ is the gradient of $f(\alpha_o)$ with respect to α_{o_i} and $\lambda_i \in (0,1)$ is a step-size. If minimizing the objective function is desired then negative of the gradient can be used in (12).

We model each finger as a position controlled device. The fingertip velocity is treated as pseudo input of the system. We have the following algorithm.

Coordinated motion generation using tactile feedback algorithm:

1. Generate the desired object velocity for a given task

2. Calculate the desired $\dot{\alpha}_{o_i}^{(k)}$ by (12) from the constructed grasp quality function

3. Calculate the contact velocity by Montana's contact equations

4. Calculate the fingertip velocity and feed it to the joint controller

5. Detect the contact location at fingertip $\alpha_{f_i}^{(k)}$ from the readings of tactile sensor and calculate the change rate of the local coordinate of the fingertip

$$\dot{\alpha}_{f_i}^{(k)} = (\alpha_{f_i}^{(k)} - \alpha_{f_i}^{(k-1)})/\Delta t$$

where Δt is sampling period

6. Update the contact coordinates and the local geometric parameters of the fingertips and the object

7. Iterate from 2.

The coordinated motion generation block diagram can be described as Figure 4, where $J^* = J_{pf_i}^\# Ad_{g_{f_i l_{o_i}}}$ and $J_{pf_i}^\#$ is the Moore-Penrose inverse of J_{pf_i}, the Jacobian for finger i.

4 Experiments

In this section, we present the details on the experimental platform and coordinated manipulation experiments.

4.1 Experiment Platform

The HKUST dextrous robotic hand developed at the Hong Kong University of Science and Technology for study of multifingered manipulation is shown in Figure 5. Each robot finger consists of a Motoman K3S robot. The fingers are arranged $120°$ apart along a circle of 1.5 meters. High performance AC servo motors with harmonic gears of gear ratios ranging from 30:1 to 100:1 are used to drive the robot. Each finger has 6 degrees of freedom (DOF) but some of its joints can be locked to give a 3-DOF or 4-DOF finger. Disk-shaped capacitive tactile sensors with radius $104mm$ are installed at the fingertips (refer [11, 12] for details).

4.2 Experimental Results

Consider the preceding example again. The two fingers of the HKUST manipulate a ball as shown in Figure 5. The desired trajectory of the ball is a rotation about the z-axis through a point $(10, 0, 0)$, i.e., a screw motion generated by a twist of the form $\xi = (0, 10, 0, 0, 0, 1)^T$.

Figure 6 shows the trajectories of the local coordinates of the contact point at the fingertips for two fin-

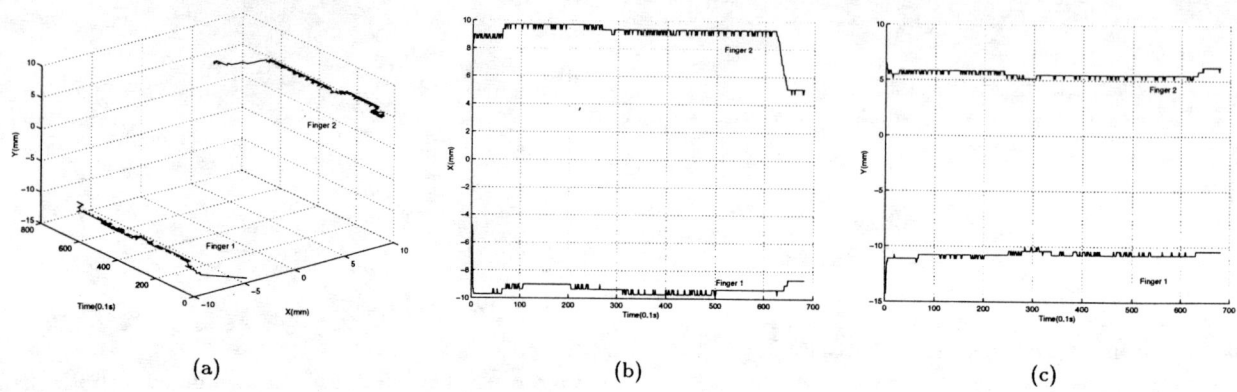

Figure 6: The trajectories of the α_{f_1} and α_{f_2} for two fingers manipulating a ball under the *individual control law* with the optimal (antipodal) initial grasp configuration.

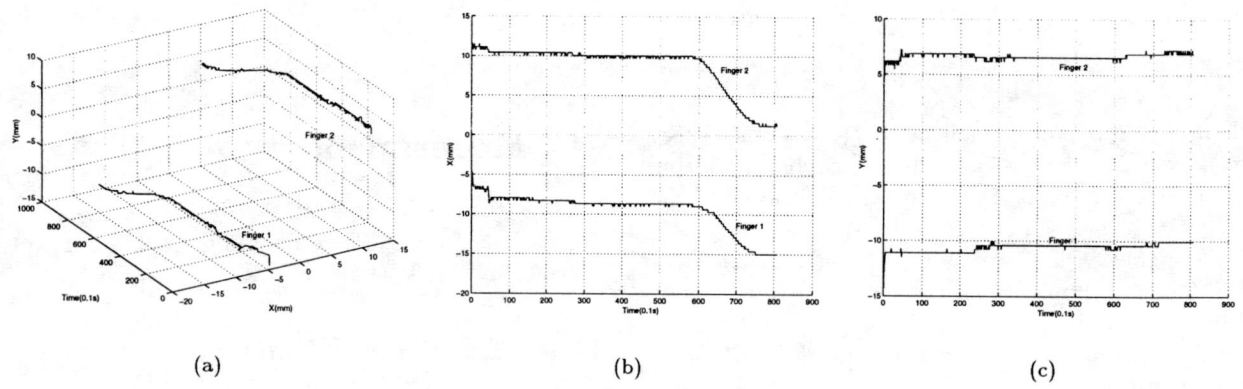

Figure 7: The trajectories of the α_{f_1} and α_{f_2} for two fingers manipulating a ball under the *tactile feedback control law* with the optimal (antipodal) initial grasp configuration.

gers manipulating the ball under *the individual joint control law*. The initial grasp configuration is optimal (antipodal). From $t = 0 \sim 620$ units, the hand grasped and lifted the ball. From $t = 620 \sim 700$ units, the hand implemented the given task. The results in Figure 6 (b) show that fingertip 2 has bigger motion in its x-direction but fingertip 1 has very small motion. The stable grasp is eventually lost. The results in Figure 6 (c) show that fingertips have very small motion in y-direction.

Figure 7 shows the trajectories of the local coordinates of the contact point at the fingertips for two fingers manipulating the ball under *the tactile feedback control law*. The initial grasp configuration is optimal (antipodal). The function of the grasp quality f is defined as the length of the chord joining the two contact points on the sphere. The desired change rate of the local coordinate of the object $\dot{\alpha}_{o_i}$ is determined by Eq.(12). From $t = 0 \sim 580$ units, the hand grasped and lifted the ball. From $t = 580 \sim 800$ units, the hand implemented the given task. The results in Figure 7 (b) and (c) show that the fingertips have almost same motion in their x- and y- directions. The results show that the fingertips maintain the optimal grasp during manipulation.

Figure 8 shows the trajectories of the local coordinates of the contact point at the fingertips for two fingers manipulating the ball under *the tactile feedback control law*. The initial contact coordinates are: $\eta_1 = (0, 0, 8°, 90°, 0)$ and $\eta_2 = (0, 0, 6°, -90°, 0)$. From $t = 0 \sim 700$ units, the hand grasped and lifted the ball. From $t = 700 \sim 950$ units, the hand implemented the given task. The results show that the fingertips first roll on the sphere in y-direction (Figure 8 (c)) so as to achieve the optimal grasp configuration and the fingertips then roll on the sphere in x-direction (Figure

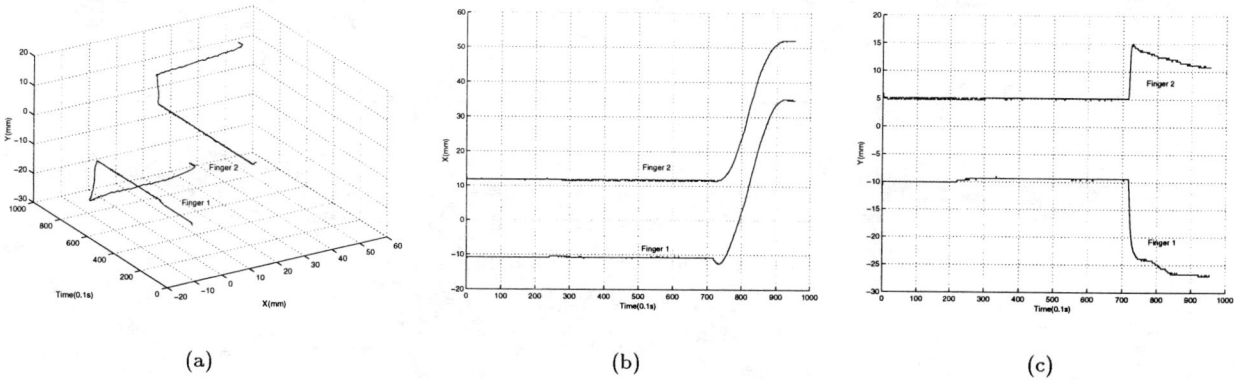

Figure 8: The trajectories of the α_{f_1} and α_{f_2} for two fingers manipulating a ball under the *tactile feedback control law* without the optimal (antipodal) initial grasp configuration.

8 (b)) so as to maintain this optimal grasp during manipulation.

5 Conclusion

In this paper, we address coordinated motion generation using tactile feedback and present the details of the coordinated manipulation scheme. Given a desired object velocity, not only the object velocity can be realized but also the grasp quality(configuration) can be optimized by utilizing tactile feedback. Experimental results show the effectiveness of coordinated motion generation using tactile feedback. Future works include construction of grasp quality function for an object with known geometric model and determination of optimal grasp for an arbitrary object by utilizing tactile feedback.

References

[1] J. Kerr and B. Roth. Analysis of multifingered hands. *International Journal of Robotics Ressearch*, 4(4):3–17, 1986.

[2] W. Howard and V. Kumar. On the stability of grasped objects. *IEEE Transaction on Robotics and Automation*, pages 904–917, December 1996.

[3] D. Montana. The kinematics of contact and grasp. *International Journal of Robotics Ressearch*, 7(3), 1988.

[4] D. Montana. The condition of contact grasp stability. In *Proceedings of IEEE International Conference on Robotics and Automation*, pages 412–417, 1991.

[5] D. Montana. The kinematics of multi-fingered manipulation. *IEEE Transactions on Robotics and Automation*, 11(4):491–503, 1995.

[6] J. K. Salisbury. Kinematics and force control of articulated robot hands. In M. Mason and J. K. Salisbury, editors, *Robot Hands and the Mechanics of Manipulation*. MIT Press, 1985.

[7] K.B. Shimoga. Robot grasp synthesis algorithms: A survey. *Int J. on Robot. Res.*, vol.15(no. 3):pp230–266, 1996.

[8] R. Murray, Z.X. Li, and S. Sastry. *A Mathematical Introduction to Robotic Manipulation*. CRC Press, 1994.

[9] H. Maekawa, K. Tanie, K. Komoriya. Tactile sensor based manipulation of an unknown object by a multifingered hand with rolliing contact. In *Proceedings of IEEE International Conference on Robotics and Automation*, pages 743–750, 1995.

[10] Z.X. Li, Z. Qin, S. Jiang, and L. Han. Coordinated motion generation and real-time grasping force control for multifingered manipulation. In *Proceedings of IEEE International Conference on Robotics and Automation*, pages 3631–3638, 1998.

[11] K.K. Choi, S.L. Jiang and Z.X. Li. Multifingered robotic hands: contact experiments using tactile sensors. In *Proceedings of IEEE International Conference on Robotics and Automation*, pages 2268–2273, 1998.

[12] Z. Qin, Z.X. Li and S. Jiang. $CoSAM^2$: A unified control system architecture for multifingered manipulation (submitted). *IEEE Transactions on Robotics and Automation*, 1997.

[13] R. Murray. Control experiments in planar manipulation and grasping. Technical report, University of California at Berkeley, ERL Memo. No. UCB/ERL M89/3, 1989.

[14] A. Cole, J. Hauser and S. Sastry. Kinematics and Control of multifingered hands with rolling contact. In *Proceedings of IEEE International Conference on Robotics and Automation*, pages 228–233, 1988.

A MULTI-FINGERED END EFFECTOR FOR UNSTRUCTURED ENVIRONMENTS

R M Crowder, V N Dubey, P H Chappell, and D R Whatley
Department of Electrical Engineering
University of Southampton
Southampton, SO17 1BJ, UK

ABSTRACT

In a wide range of robotic tasks, the unstructured nature of the environment may preclude the use of a conventional end effector. This paper considers one such approach to the design a re-configurable three-fingered end effector. The finger mechanism used in this design is based on the previous work at the University of Southampton[1]. The end effector design offers considerable flexibility, particularly in the range of objects that can be handled, both in shape and mass, over the previous designs.

This paper discusses the mechanical structure of the end effector and its control. In order to achieve satisfactory control a novel sensor capable of detecting both slip and applied force has been developed. The proposed controller includes a set of high level commands to orient the gripper, together with a fuzzy logic approach to control the fingertip force, once contact has been made with the object.

1. Introduction

The development of robotic systems to operate in hazardous, unstructured environments is of considerable importance, with applications typified by operations in the nuclear, subsea, and chemical industries. In these environments, a robot can either be fitted with a general purpose gripper, or a special purpose system. Both approaches are restrictive as they may not be capable of undertaking a task, or not provide a cost effective solution. In order to undertake general purpose handling operations, a fully flexible end effector will be required. The objective of the work discussed in the paper is the design of a robotic end effector to operate in an unstructured environment in real time.

A number of anthropomorphic, non-prosthetic end effectors have been developed and reported. Among the most significant designs are the Stanford/JPL[2] and Utah/MIT hands[3]. The significant feature of these and other similar hands is the actuation of the individual finger joints, by flexible tendons from an external actuator. This has a number of significant advantages, in particular low hand mass and high speed of response. However, the location of the actuators external to the fingers necessitates tendons between the finger and the actuator. This approach is not only complex, but prevents the use of such hands in environments where there is limited access space or high robustness is required. In addition a number of non-anthropomorphic hands have been reported, a notable design being the three-fingered hand developed at the University of Pennsylvania[4] and later marketed as the BarrettHand. This compact design, uses four actuators on a worm drive with cable and breakaway clutch to provide finger motions. The design of the end effector discussed in this paper is significantly different from the other end effectors that are based on the used of tendons or other flexible linkages. The philosophy used in this design to locate the finger actuators within the base, and transfer motion to the finger mechanism using direct mechanical linkages.

2. Manipulator Specification

The end effector design considered in this paper is based on three-fingers operated through solid drive-links. This is considered to be the minimum required to achieve a satisfactory stable grip over a wide range of objects to be grasped.

The design specifications for the end effector can be summarised as:

- The end effector has three finger units arranged at 120^0 intervals, on circular mounting structure.

- The end effector is required to have a maximum clearance diameter of 120 mm, with the fingers fully closed.

- The hand has the capability of rotating the plane of finger action relative to the base of the end effector. As discussed later in the paper, the rotation of the fingers gives a considerable number of gripping configurations.

- The end effector should be capable of picking a range of objects directly from a flat surface. In practice there will be a limiting height, below which the end effector will not be able to locate on to an object. This distance will be determined by the design of the fingertips.

- The maximum load capability is 5 kg, with object cross sections ranging between 6 mm and 200 mm.

3. Finger Structure

The end effector was designed to be attached directly to a robot's or a manipulator's tool interface and consists of three equi-spaced finger modules. The key feature of the finger mechanisms, is their capability to adopt a posture dependant on the object being grasped at the time. The advantage of this approach is that only a limited number of actuators are required, compared with a fully articulated and controlled finger.

Each module is identical and is mounted within the body of the end effector. The finger module has two parts; the finger mechanism and the finger. The finger mechanism design splits the single input motion into three different components to drive the fingers in different modes. The mechanism has a central differential unit driving two leadscrews supported on a frame. A motor connected to the differential unit can drive the two leadscrews as well as the frame of the mechanism, providing three different components of a motion. These motions are determined by the suitable use of electromagnetic brakes, Figure 1.

As shown in the figure, brakes 1 and 2 control two leadscrews, while brake 3 is used to lock the finger's orientation relative to the end effector's frame. In the current implementation of the design a d.c. brushed motor, fitted with an integral encoder is used.

The articulated finger is mounted on top of the mechanism, which has three sections (Figure 2) which are jointed at points C and F. Sections 1 and 3 are connected by the control link 4 at pivot points E and G. The control links have the following functions;

- Control link 1 connects the leadscrew nut and the bell crank at points M and K.

- Control link 2 connects the section 1 at points B with second leadscrew nut at point L.

- Control link 3 connects the section 2 with the bell crank at points D and I.

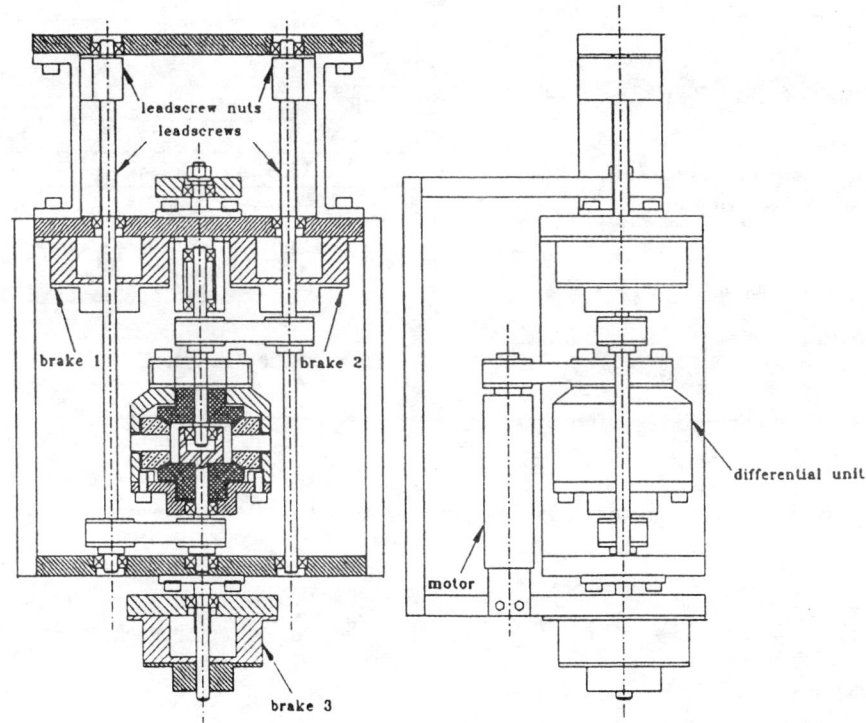

Figure 1 The finger actuation module

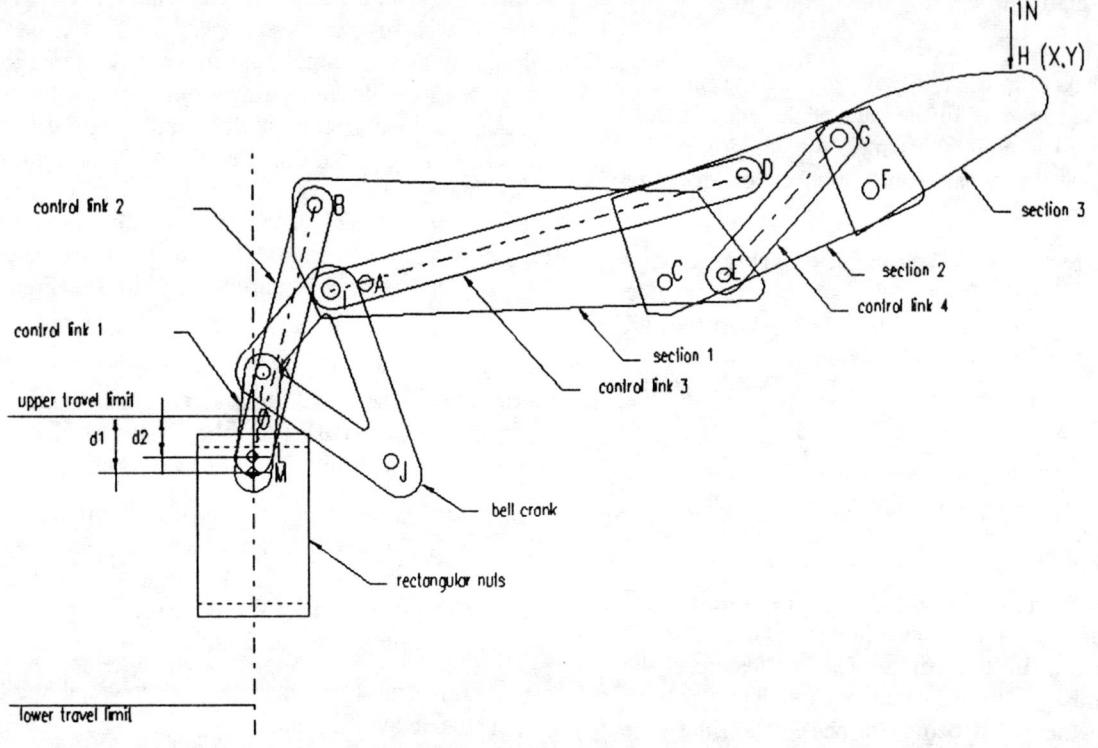

Figure 2: Design of the finger

Joints A and J are fixed to the frame. The input to the finger mechanism is provided by the leadscrew nuts as vertical displacements of pivot points L and M, d1 and d2 respectively. In the figure the two nuts are shown overlapping.

The finger mechanism provides three degrees of freedom to a finger:

- *Curling*, where displacement of control link 1, curls the two upper finger segments about joint C.
- *Bending*, where displacement of control link 2, bends all the three finger segments about joint A.
- *Rotation*, of the whole module about its central axis relative to the end effector body.

The structure of the finger offers independent curl motion while the bend motion is only partially independent as it does result in a slight curling effect in the finger. Each component of motion is driven by a single motor through the differential gearing mechanism. By controlling the three brakes, the three components of motions can be controlled individually or in combination. The truth table for the brakes is given in table 1.

The arrangement of one motor and three brakes per finger offers easier control options than using three different motors for the three components of the motion.

Brake 1	Brake 2	Brake 3	Finger motion
R	R	R	curl, bend and rotate
R	-	-	curl
-	R	-	bend
-	-	R	rotate
R	R	-	curl and bend
-	R	R	bend and rotate
R	-	R	curl and rotate
-	-	-	none

Table 1. Finger motion using brake control, R signifies the brake is released

4. Kinematic Analysis

The forward kinematic relationship was developed to relate the leadscrew displacements with the fingertip position. The continuity of the plot, Figure 3, indicates that all points within the region are attainable. The non-intersecting fingertip loci indicate that unique leadscrew displacements are required to move the fingertip to a specified location. The graph shows that for a range of leadscrew displacements, the inverse kinematics has a unique solution. This has been verified by deriving the inverse kinematic relations

6. Control

The control of the end effector needs to be considered within three operating regimes:

- Configuration of the end effector.
- Closing the end effector
- Controlling the applied force as the manipulator moves the end effector, in order to apply minimum force, while retaining a satisfactory grip.

The arrangement of three brakes and a motor allows a range of gripping control strategies to be provided, as the three components of motions can be driven individually or in combination. Three brakes can be operated in various modes to provide different finger motions, Table 1. However, if it is required to drive each finger to a precise point, it is appropriate to drive each component separately. This will allow maximum control over the position and the orientation of each finger to be maintained. If only a minimum amount of information is available, which is normally the case when operating in an unknown environment, it is preferable to operate both the curl and bend components together. In this mode the fingers will be able to adapt to the geometry of the object. This is achieved by the equalising effect of the differential unit, which distributes the drive torque and hence finger motion according to the resistance offered by the individual motions.

In order to grip a specific object, the end effector must first be positioned and oriented suitably in relation to the object such that the gripping can be possible. If the object is not totally enclosed by the envelope of the three fingers and not sufficiently restrained, the effect of the first contact might simply be to push the object. The control-approach of the end effector can be undertaken by visual control, or pre-programming depending on the application. In the initial control strategy developed, it has been assumed that the end effector is located at the optimum position over the object to be grasped.

The finger mechanism offers various options from a tip to tip pinch, to a full wrap around grasp. In addition the fingers can be operated either in a concentric mode or in the two fingers and an opposing thumb mode. Due to the flexibility of the end effector, the system can also be configured to operate just as a simple two opposing finger gripper, with a third finger disabled. The end effector's design presented in this paper offers a considerable number of configurations, Figure 6. This figure shows the type of the objects that can be grasped. In the figure a 'T' shape has been used to represent the tip-operation of the finger and '|' to represent the full length of the finger wrapping around the object. It can be seen that each configuration in tip grasp has a similar configuration in the wrap around grasp and is operating in either in concentric or in opposing thumb modes. It should be noted that there is a large number of configurations possible between these symmetrical arrangements. This is due to the independent motion of finger's positional axis.

The objects to be handled by the end effector have been defined in various object-class primitives, for example, BLOCK, CYLINDER, SPHERE, etc. These primitives invoke a set of subroutines defined within the controller. The input to the end-effector's controller will be the shape of the object and the mode of grasping (tip or wrap), the output will be the finger's orientation.

In the first step of grasping, the fingers will be oriented according to the given grasp configuration or the object class (for example BLOCK, CYLINDER, etc.). Active fingers (which participate in grasp) are then bent till they register a contact with object, detected by the fingertip

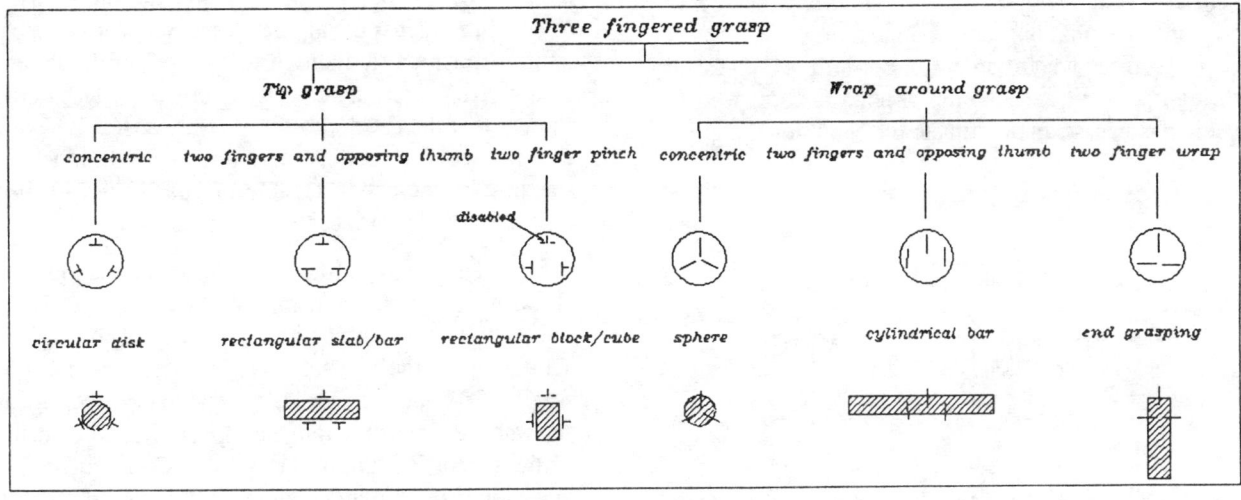

Figure 6 : Available gripping strategies

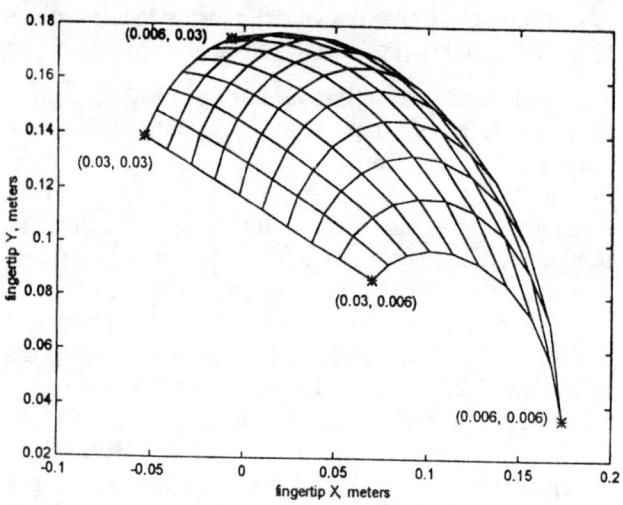

Figure 3: Loci of the finger

5. Sensor Provision

In the object handling operation by a robotic end effector, it is required to ensure a stable grasp without causing excessive gripping force to the object. The grasp stability can be assessed by modelling the various material properties of the finger-object surface in contact and the applied force. This leads to complicated solutions[5], that are difficult to apply in real time. In practice these are not valid for a different object-finger combination and for the changes in the fingertip force and moments during the handling operation. The other approach to estimate the grasp stability is to detect the contact parameters at the finger-object interface using tactile sensors, which could offer the information on changes in the contact parameters directly. In order to perform a successful handling operation, detection of applied force as well as object slip is necessary to estimate the condition the grip.

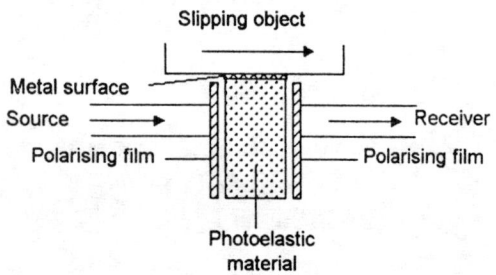

Figure 4: Outline design of the photoelastic slip and force sensor

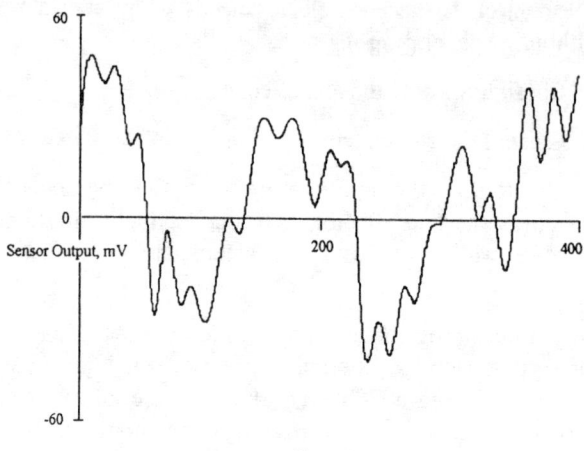

Figure 5: Slip signal from the sensor

A photoelasticity based sensor has been developed for use with the end effector, which is capable of detecting both applied force and the object slip.

The sensor's operation is based on the principle that a photoelastic material undergoing a change in applied stress, produces a corresponding change in the material's angle of polarisation[6,7]. By passing polarised light through the photoelastic medium an effective change in the light intensity received can be observed. The outline design of the sensor is shown in Figure 4.

To enhanced performance a metallic strip has been provides on the contact point[8]. The purpose of the strip is to set up self excited vibration between the object and the sensor. It is this vibration the indicates the magnitude of the slip. In operation, the sensor is used to initially detect

the contact between the sensors and the object. As the object slips against the applied force, the sensor detects a cyclic slip signal in addition to the applied force, as the sensor stick-slips against the object. Measurements showed that the developed sensor was sensitive down to a slip rate of 0.1 mm s^{-1} and provides continuous slip signal. The sensors is capable of operating with highly finished metallic surfaces

A typical result from the slip sensor for an aluminium block moving with a slip rate of 0.6 mm s^{-1} is shown in Figure 5. The magnitude and frequency of the signal are a function of the contact materials and the rate of the slippage.. Though the sensor is capable of detecting small slip rates, the dynamic range of the applied force detected by the sensor is found to be narrow. Consequently, only touch and a minimum applied force can presently be measured.

tactile sensor (photoelastic sensor in the present case). Once this is done, the gripping force on the fingers is gradually increased and tested, this procedure will be repeated until the object slip can be arrested. The detection of the object slip is done by the photoelastic sensor. The actual control action for the motor gripping force is inferred from a fuzzy control algorithm[8], which receives the input of applied force and the object slip from the sensor. The implementation of the fuzzy control has been experimentally found to apply optimal fingertip force for objects of different mass, just above the minimum required for a stable grasp in each case. It also responds dynamically to the external disturbances to retain the grasp.

Fuzzy logic is a suitable control approach to the non-precise and non-well defined information[9,10]. The operation of the end effector in an unstructured environment requires optimal force control as it handles object of different shapes and mass. The use of fuzzy logic forms a suitable control strategy for this specific application and for the real time operation of the end effector.

Currently the controller, sensor and mechanism for a single finger have been implemented, and the full end effector fully modelled. In addition to the fuzzy logic, the controller has a number of routines implemented that optimise the object's position relative to the end effector. These are designed to minimise the induced moments, leading to a more stable grip.

On-going work at Southampton is directed towards the optimisation of the end-effector's trajectory control and enhancing the operational characteristics of the complete system.

7. Discussion

The paper has discussed the design of a three-fingered end effector, photoelastic sensor for slip and applied force detection and the overall control of the system using fuzzy logic. These systems have been designed to operate in an unstructured environment.

We consider that the use of solid linkages removes a number of the significant problems experienced with tendon based designs of robotic end effectors. As currently implemented the use of brakes and motor to control finger position will restrict, to some degree, the operational speed of an individual finger. We consider this limitation to be outweighed by the increased robustness of the design.

Experimental work with the sensor has shown that a single point contact system combined with the finger design is able to achieve the satisfactory, and controllable application of force.

Based on our current studies it is our view that the end effector technologies presented in this paper are capable of offering a practical solution for handling tasks in the unstructured remote environment when compared to a wide range of other gripper designs.

8. Acknowledgement

This work has been supported by the Faculty of Engineering and Applied Science, University of Southampton, and the Overseas Research Student award from the UK Committee of vice-chancellors and Principals.

9. References

1. R. M. Crowder, "An anthropomorphic robotic end effector" *Robotics and Autonomous Systems*, vol. 7 p. 253-268, (1991).

2. M. T. Mason, and J. K. Salisbury, *Robot hands and mechanics of manipulation* The MIT press., (1985).

3. S. C. Jacobsen, J. E. Wood, D. F. Knutti, and K. B. Biggers, "The Utah/MIT dextrous hand: work in progress" in D.T. Pham and W.B. Heginbotham edited *Robot Grippers*, p. 341-389, (1986).

4. N. Ulrich, R. Paul and R. Bajcsy "A medium-complexity compliant end effector" Proc. IEEE Intl. Conf. on Robotics and Automation, p. 434-436, (1988).

5. M. R. Cutkosky, *Robotic Grasping and fine Manipulation,* Kluwer Academic Publishers, Boston, (1986).

6. F. Kvasnik, B. E. Jones, and M. S. Beck "Photoelastic slip sensors with optical fiber links for use in robotic grippers" Institute of Physics Conf. on sensors, Southampton, UK. (1985)

7. P. Dario, D. Femi and F. Vivaldi "Fiber-optic catheter-tip sensor based on the photoelastic effect" *Sensors and Actuators,* vol. 12, p. 35-47, (1987).

8. V. N. Dubey, "Sensing and control within a robotic end effector", PhD thesis, University of Southampton, UK, (1997).

9. A. Nedungadi and D. J. Wenzel, "A Novel approach to Robot Control using Fuzzy Logic" *Proc. of IEEE International Conference on Systems Man and Cybernetics,* vol. 3, p. 1925-1930, (1991).

10. T. Dorsam, S. Fatikow, and I. Streit, "Fuzzy-based grasped-force-adaptation for multi-fingered robot hands" *Proc. of the 3rd IEEE conference on fuzzy System,* p. 1468-1471, (1994).

An Off-Line Iterative and On-Line Analytical Force Distribution Approach for Soft Multi-Fingered Hands

Bing-Ran Zuo[*], Günther Seliger[#] and Wen-Han Qian[*]

[*] Research Institute of Robotics, Shanghai Jiao Tong University, Shanghai 200030, P. R. China
[#] Institute of Machine Tools and Factory Management, Pascalstr. 8-9, D-10587 Berlin, Germany

Abstract

This paper formulates the force distribution for soft multi-fingered hands as choosing a suitable internal force such that the total contact force satisfies the frictional constraints. As a first step, the direction of the optimal internal force is defined based on the grasp locations. Implementation of the algorithm is divided into two phases: (1) Determine the direction of the internal force during grasp planning; (2) Determine the magnitude of the internal force during task manipulation. The total computation cost consists of off-line numerical iterations and on-line analytical computations. Only the latter is related to the real-time control. An example shows that the proposed algorithm can efficiently deal with the soft frictional constraints and well overcome the temporal discontinuity.

1 Introduction

Force distribution is one of the most important topics for multi-fingered hands. The goal is to determine a suitable contact force that balances the external load and at the same time satisfies the frictional constraints under a user specified objective function. In general, the contact force exerted by a multi-fingered hand can be decomposed into two parts [1]: (1) the manipulation force, which is determined by the specified manipulation task, and (2) the internal force, which has no effect on the equilibrium but can be used to modify the contact force to achieve firm fingertip contact. Since the manipulation force has already balanced the external load, the key issue of force distribution is reduced to choosing a suitable internal force such that the total contact force satisfies the frictional constraints.

Many attempts have been made to solve this problem. Early works include various linear programming (LP) [2]-[4] and quadratic programming (QP) [6]-[7] methods. The most serious difficulty inherent in LP methods is that they may introduce undesirable temporal discontinuities, as pointed out in [5]. In addition, all the LP and QP methods are based on the linearized frictional constraints, which are only approximations of the Coulomb model. The nonlinear programming (NLP) [8] and manifold optimization technique [9]-[10] can be used to overcome this inconvenience. The former seems difficult to be real-time implemented with the current computer resources. The latter formulates the force distribution as an optimization problem on the smooth manifold of linearly constrained positive definite matrices, where there are known convergent solutions via gradient flows. It is noted that performance of such optimization needs a valid initial condition at each setpoint, which may involve a nonlinear programming problem and in return increase the computational cost. To make real-time implementation possible, a dynamic grasping force control algorithm using tactile feedback was put forward [11]. By dynamically adjusting the grasping force according to the tactile feedback, all the fingertips can make firm contact with the object. The key issue of this algorithm is to determine the compensation grasping force, which was not detailed in [11]. Some analytical and/or suboptimal methods [12][13] are available for certain particular grasps. Similar issues also exist in walking machines and multiple manipulators.

The above approaches are mainly concentrated on frictional point contact. Some [8]-[10] can be extended to soft finger contact, where the fingertip exerts a spin moment about the contact normal in addition to a pure force. Since these methods are essentially numerical, large iterations are still required at each setpoint to find the optimal solution.

The aim of this paper is to present a fast force distribution algorithm for soft multi-fingered hands. As a first step, the direction of the optimal internal force is defined based on the grasp locations. Implementation of the algorithm is divided into two phases: (1) Determine the direction of the internal force during grasping planning; (2) Determine the magnitude of the internal force during task manipulation. The total computation cost consists of off-line numerical iterations and on-line analytical computations. Only the latter is related to the real-time control. Therefore, the proposed algorithm is hopefully applied to real-time occasions.

2 Problem Formulation

Our approach is under the following assumptions: (1) Each fingertip makes a soft finger contact with the object. (2) The grasp locations relative to the object keep constant during manipulation. (3) All the vectors are specified with respect to the object coordinate frame.

Fig. 1 shows a soft multi-fingered grasp. At each contact, the fingertip exerts a spin moment about the contact normal in addition to a pure force, as shown in Fig. 2.

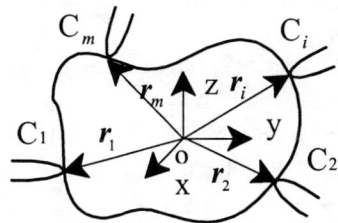

Fig. 1. A soft multi-fingered grasp on an arbitrary object.

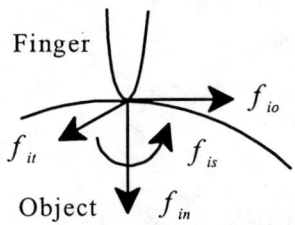

Fig. 2. Soft finger contact

The equilibrium equation for the grasped object is expressed as
$$p = Gf, \quad (1)$$
where p is the resultant wrench, $G = [G_1 \ G_2 \ \cdots \ G_m] \in R^{6 \times 4m}$ is the so-called grasp matrix, $G_i = \begin{bmatrix} n_i & o_i & t_i & 0 \\ r_i \times n_i & r_i \times o_i & r_i \times t_i & n_i \end{bmatrix}$ is the wrench matrix at C_i, $n_i = [n_{ix} \ n_{iy} \ n_{iz}]^T$ is the inward unit normal, o_i and t_i are the two unit tangent vectors, $n_i = o_i \times t_i$, $r_i = [r_{ix} \ r_{iy} \ r_{iz}]^T$ is the position vector, $f = [f_1^T \ f_2^T \ \cdots \ f_m^T]^T \in R^{4m}$ is the total contact force, $f_i = [f_{in} \ f_{io} \ f_{it} \ f_{is}]^T$ is the contact force at C_i, f_{in}, f_{io}, f_{it} and f_{is} are the normal, two tangent force, and spin moment components, respectively.

For any $p \in \text{Im}(G)$, the general solution to eq. (1) has the form
$$f = G^+ p + Wy, \quad (2)$$
where G^+ is the pseudo-inverse of G, $W \in R^{4m \times (4m - \text{rank}(G))}$ is a matrix, whose columns constitute a basis of the null space of G, and $y \in R^{4m - \text{rank}(G)}$ is an arbitrary vector.

The manipulation and internal forces are given by
$$g = G^+ p \quad (3)$$
and
$$h = Wy \quad (4)$$
respectively. To achieve firm contact, the total contact force must satisfy the frictional constraints, where there are two approximate models [9] [14] as follows.

(i) Linear Model
$$\frac{\sqrt{f_{io}^2 + f_{it}^2}}{\mu_i} + \frac{|f_{is}|}{\mu_{si}} \le f_{in}, \ f_{in} > 0 \quad (5)$$

(ii) Elliptical Model
$$\frac{f_{io}^2 + f_{it}^2}{\mu_i} + \frac{f_{is}^2}{\mu_{si}'} \le f_{in}^2, \ f_{in} > 0 \quad (5')$$

where μ_i is the coefficient of friction, μ_{si} and μ_{si}' are the coefficients of spin moment for the linear and elliptical models, respectively.

In the following discussion, we will only concentrate on the linear friction model (actually nonlinear). But the corresponding results can be extended to the elliptical model similarly.

3 Feasible Internal Force

Definition 1: A contact force is called a *feasible internal force* if and only if: (i) it satisfies the equilibrium equation, i.e. $Gf = 0$. (ii) it *strictly* satisfies all the frictional constraints, i.e. $\frac{\sqrt{f_{io}^2 + f_{it}^2}}{\mu_i} + \frac{|f_{is}|}{\mu_{si}} < f_{in}$, $f_{in} > 0, (i = 1, 2, \cdots, m)$.

As shown in [15], the feasible internal force has great effect on the closure properties of grasping. Only when the set of feasible internal force is not null, can the grasp closure be guaranteed. In this sense, the force distribution problem is meaningful. The algorithm to determine the existence of feasible internal forces is derived as follows.

Let
$$h = [h_1^T \ h_2^T \ \cdots \ h_m^T]^T,$$
$$W = [W_1^T \ W_2^T \ \cdots \ W_m^T]^T.$$

Then the internal force at the i-th contact can be written as
$$h_i = W_i y. \quad (6)$$

Let $W_i = \begin{bmatrix} W_{in} & W_{io} & W_{it} & W_{is} \end{bmatrix}^T \in R^{4\times(4m-6)}$,
where $W_{in}, W_{io}, W_{it}, W_{is} \in R^{4m-6}$. The internal force components are given by
$$h_{in} = W_{in}^T y, \; h_{io} = W_{io}^T y,$$
$$h_{it} = W_{it}^T y, \; h_{is} = W_{is}^T y. \quad (7)$$
The feasible internal force exists if and only if
$$\frac{\sqrt{h_{io}^2 + h_{it}^2}}{\mu_i} + \frac{|h_{is}|}{\mu_{si}} < h_{in}. \quad (8)$$
Substituting (7) into (8), we obtain
$$y^T A_i y < 0, \; y^T B_i y < 0, \; a_i^T y < 0, \; b_i^T y < 0, \quad (9)$$
where
$$A_i = \frac{W_{io}W_{io}^T + W_{it}W_{it}^T}{\mu_i^2} - \left(W_{in} + \frac{W_{is}}{\mu_{si}}\right)\left(W_{in} + \frac{W_{is}}{\mu_{si}}\right)^T$$
$$B_i = \frac{W_{io}W_{io}^T + W_{it}W_{it}^T}{\mu_i^2} - \left(W_{in} - \frac{W_{is}}{\mu_{si}}\right)\left(W_{in} - \frac{W_{is}}{\mu_{si}}\right)^T$$
$$a_i = -W_{in} + \frac{W_{is}}{\mu_{si}}, \; b_i = -W_{in} - \frac{W_{is}}{\mu_{si}}.$$

Consequently, the problem has been transformed into determining whether there exists a vector $y \in R^{4m-6}$ complying with (9).

Now let us minimize the maximum value among $y^T A_i y, y^T B_i y, a_i^T y$ and $b_i^T y$ (for all $i = 1, 2, \cdots, m$). At the same time the norm of vector y must be limited. Then we obtain the following nonlinear programming problem (NLP).
$$\begin{cases} \min \eta \\ y^T A_i y \leq \eta \\ y^T B_i y \leq \eta \\ a_i^T y \leq \eta \\ b_i^T y \leq \eta \\ -1 \leq y_j \leq 1 \end{cases} (i=1,\cdots,m, j=1,\cdots,4m-6), \quad (10)$$

where y_j is the component of vector y. The last constraint is complemented to make sure the nonlinear programming problem has finite optimal solution. Considering the *cone* property of internal force set, we only concern the relative magnitudes of the components of vector y.

Suppose the optimal solution to problem (10) is (y^*, ε^*), then the feasible internal force exists if and only if $\varepsilon^* < 0$.

When the feasible internal force does not exist, the grasp robustance is not guaranteed [15]. In this case, a grasp planner is required to adjust the contact positions.

4 Optimal Internal Force

It is noted that the optimal solution to problem (10) implies a measure of the maximum safety margin on frictional constraints, and the product of matrix W and vector y^*, Wy^*, indicates the optimal ratio of the internal force components. Therefore, the *optimal internal force* is defined as
$$h^* = \sigma W y^*. \quad (11)$$
Define a *direction vector* as
$$d = Wy^*. \quad (12)$$
Then
$$h^* = \sigma d, \quad (13)$$
where $\sigma \geq 0$ is a *compensation factor* implying the magnitude of the internal force.

The total contact force is given by
$$f = g + \sigma d. \quad (14)$$
Now we are in a position to determine the compensation factor σ. The following method is based on the *convexity* of soft frictional constraints [15].

Let
$$d = c + \delta, \quad (15)$$
where
$$d = \begin{bmatrix} d_1^T & d_2^T & \cdots & d_m^T \end{bmatrix}^T \in R^{4m},$$
$$d_i = \begin{bmatrix} d_{in} & d_{io} & d_{it} & d_{is} \end{bmatrix}^T,$$
$$c = \begin{bmatrix} c_1^T & c_2^T & \cdots & c_m^T \end{bmatrix}^T \in R^{4m},$$
$$c_i = \begin{bmatrix} d_{in} - \delta_i & d_{io} & d_{it} & d_{is} \end{bmatrix}^T,$$
$$\delta = \begin{bmatrix} \delta_1^T & \delta_2^T & \cdots & \delta_m^T \end{bmatrix}^T \in R^{4m},$$
$$\delta_i = \begin{bmatrix} \delta_i & 0 & 0 & 0 \end{bmatrix}^T,$$
$$\delta_i = d_{in} - \left(\frac{\sqrt{d_{io}^2 + d_{it}^2}}{\mu_i} + \frac{|d_{is}|}{\mu_{si}}\right). \quad (16)$$

Since d strictly satisfies the frictional constraints, it follows
$$\delta_i > 0. \quad (17)$$
From (14)-(15), we get
$$f = (g + \sigma\delta) + \sigma c. \quad (18)$$
Note that the second term σc has already satisfied the frictional constraints. The total contact force f will

satisfy the frictional constraints if the first term $(g+\sigma\delta)$ does.

Substituting each component of $(g+\sigma\delta)$ into (5), we obtain

$$\frac{\sqrt{g_{io}^2+g_{it}^2}}{\mu_i}+\frac{|g_{is}|}{\mu_{si}}\leq g_{in}+\sigma\delta_i. \quad (19)$$

Let

$$\sigma_i=\frac{1}{\delta_i}\left(\frac{\sqrt{g_{io}^2+g_{it}^2}}{\mu_i}+\frac{|g_{is}|}{\mu_{si}}-g_{in}\right). \quad (20)$$

Then the compensation factor is given by

$$\sigma=\max\{\sigma_1\ \sigma_2\ \cdots\ \sigma_m\ 0\}. \quad (21)$$

It is noted that $\max(\cdot)$ is a continuous function, which guarantees the smoothness of the exerted forces during task manipulation. This implies that the proposed algorithm itself can overcome the temporal discontinuity.

5 Force Distribution Algorithm

Fig. 3 shows the diagram of the force distribution algorithm for soft multi-fingered hands. The shaded blocks indicate on-line implementation, while others are off-line implemented. Computation of the direction of the internal force will involve a nonlinear programming problem, see (10) and (12). Expressions of the manipulation force calculator, contact force calculator, and compensation factor calculator are stated in (3), (18), (19)-(21) and (16), respectively.

As we see, determination of the contact force is divided into two phases: (1) Determine the direction of the internal force during grasp planning; (2) Determine the manipulation force and compensation factor during task manipulation. Although the former may involve large numerical iterations, it is off-line implemented. The latter only relates to analytical computations. Therefore, the algorithm can be regarded as off-line numerical and on-line analytical, which makes force distribution fast enough for real-time applications.

6 Simulation Results

Fig. 4 (a) depicts a rectangle object manipulated by a soft four-fingered hand. The coefficient of friction $\mu=0.4$ and spin moment $\mu_s=0.2\text{m}^{-1}$. The dimensions of the object are $40\times40\times40\text{mm}^3$, and the mass $M=0.2$kg. The position vectors of contact points are: (unit: mm)

$$r_1=\begin{bmatrix}0 & 20 & 0\end{bmatrix}^T, r_2=\begin{bmatrix}-20 & 0 & 0\end{bmatrix}^T,$$
$$r_3=\begin{bmatrix}20 & 0 & 0\end{bmatrix}^T, r_4=\begin{bmatrix}0 & 0 & 10\end{bmatrix}^T.$$

The desired trajectory consists of two segments: a line and a semi-circle. The radius of the trajectory $R=500$mm, and the motion path is along $P_1\to O\to P_2\to P_3\to P_1$, see Fig. 4 (b). At each segment, the velocity of object will first accelerate, then keep constant, and finally decelerate to zero, see Fig. 4 (c). Expressions of the velocity are given by

$$v=\begin{cases}0.75(1-\cos 5t), & 0\leq t\leq 0.2\\ 1.5, & 0.2\leq t\leq T-0.2\\ 0.75(1-\cos 5(t-T+0.4)), & T-0.2\leq t\leq T\end{cases}$$

where $T_l=0.87$s for the line segment, and $T_c=1.25$s for the semi-circle segment.

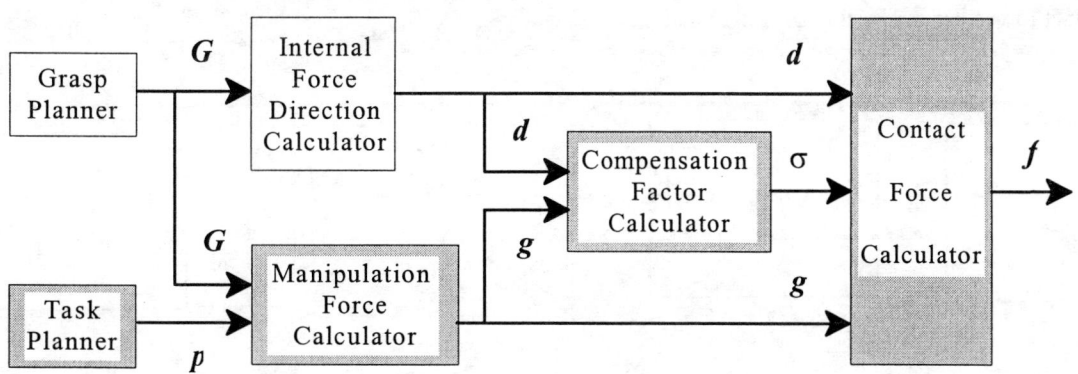

Fig. 3. The diagram of the off-line numerical and on-line analytical force distribution algorithm.

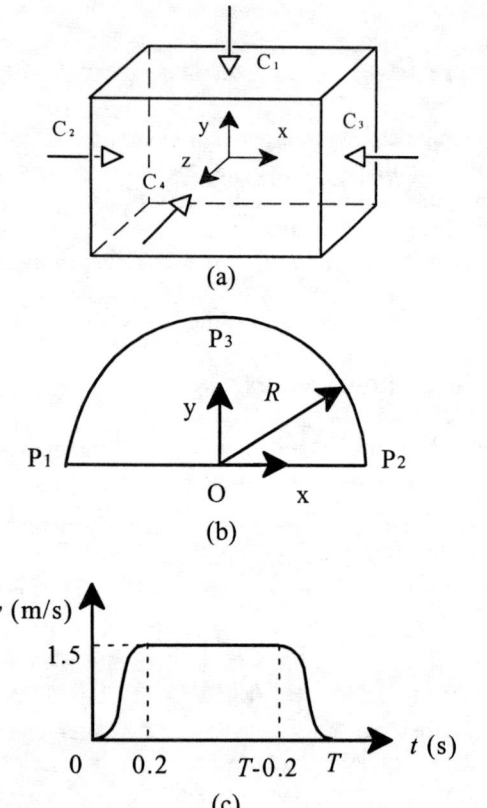

Fig. 4. A soft four-fingered manipulation example. (a). Grasp locations. (b). Desired trajectory. (c). Manipulation velocity.

Using the optimization toolbox of Matlab, we obtain the optimal solution to problem (10) as follows.

$$y^* = [0.2110\ -0.2388\ 1.0000\ 0.2001\ 0.2708$$
$$0.2390\ 0.5055\ 0.1814\ -0.2367\ 0.7234]^T$$
$$\varepsilon^* = -0.1923.$$

Since $\varepsilon^* < 0$, the given grasp is feasible.

For simulation results, Fig. 5 shows the trajectories of the resultant force and compensation factor, and Fig. 6 shows the contact forces exerted by the soft finger hand.

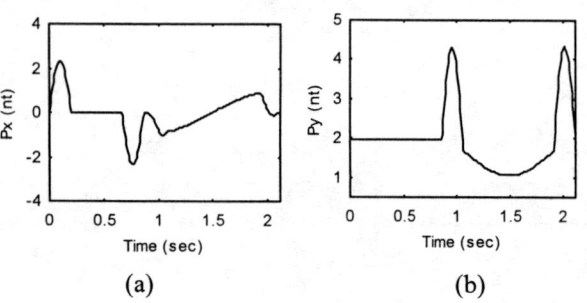

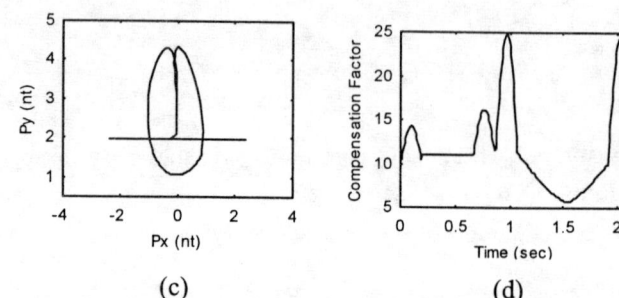

Fig. 5. Trajectories of the resultant force and compensation factor. (a) x-component of the resultant force. (b) y-component of the resultant force. (c) Trajectory of the resultant force. (d) Trajectory of the compensation factor.

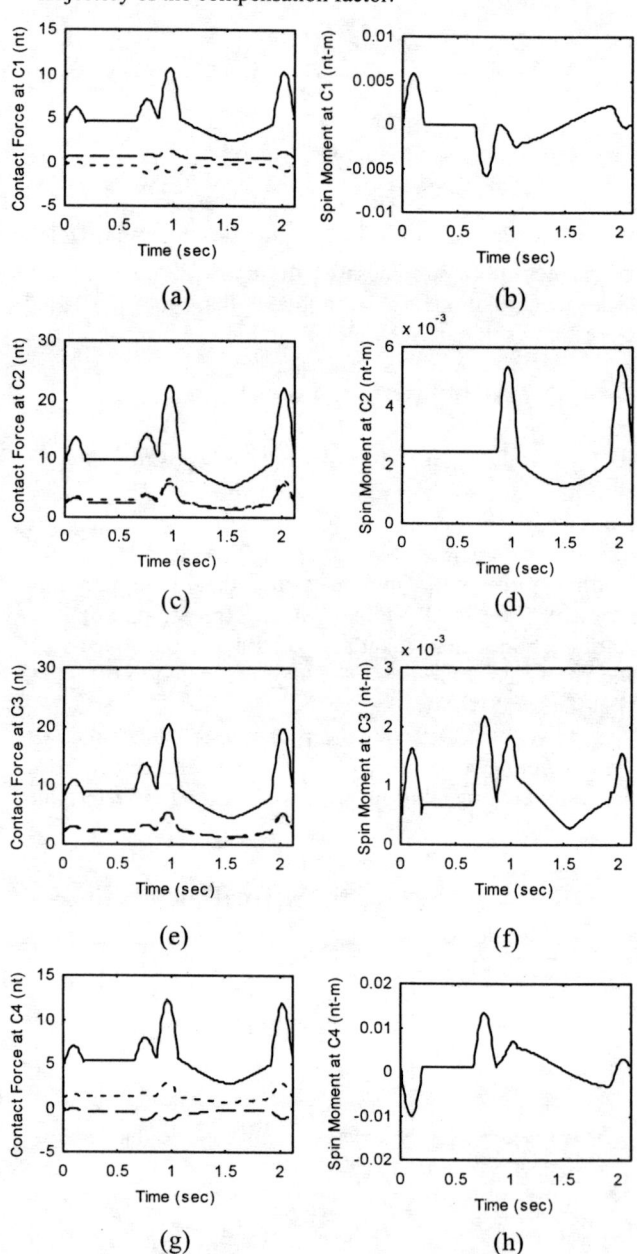

Fig. 6. Contact forces and spin moments exerted by the soft finger hand. —f_{in} ---f_{io} --f_{it} —f_{is}

It is verified that the contact forces satisfy the equilibrium equation as well as the frictional constraints. Thus the grasp robustance is guaranteed. The temporal discontinuity is overcome due to the continuous function adopted to determine the compensation factor, see Fig. 6. The computation cost at each setpoint is 120 add., 148 mult., 8 sqrt., 8 abs., and 4 comp. Hence, the algorithm is fast enough for real-time manipulation.

7 Conclusions

This paper presents an off-line numerical and on-line analytical force distribution algorithm for soft multi-fingered hands. The main contributions are: 1) The feasible internal force for soft multi-fingered grasp is introduced, along with an algorithm given to determine its existence. 2) The optimal internal force is defined as a feasible internal force that satisfies the maximum safety margin criterion on the frictional constraints. Determination of the direction of the internal force involves a nonlinear programming problem, which can be implemented during grasp planning. The manipulation force and the magnitude of the internal force are analytically computed during task manipulation. The total computation cost consists of off-line numerical iterations and on-line analytical computations, which makes force distribution fast enough for real-time manipulation. 3) It is shown that so long as the direction of the internal force strictly satisfies the frictional constraints, there always exists a compensation factor such that the total contact force satisfies the frictional constraints. It is this distinction that makes the proposed algorithm possible.

Acknowledgment

This work was supported by the National Natural Science Foundation of China and the Collaborative Research Center 281 "Disassembly Factories" funded by the German Science Foundation.

References

[1] T. Yoshikawa and K. Nagai, "Manipulating and grasping forces in manipulation by multifingered robot hands," *IEEE Trans. Robotics Automat.*, vol. 7, no. 1, pp. 67-77, Feb. 1991.

[2] J. Kerr and B. Roth, "Analysis of multifingered hands," *Int. J. Robotics Res.*, vol. 4, no. 4, pp. 3-17, 1985.

[3] J. Jameson and L. Leifer, "Automatic grasping: An optimization approach," *IEEE Trans. Sys., Man, Cybernet.*, vol. 17, no. 5, pp. 806-814, 1987.

[4] F. T. Cheng and D. E. Orin, "Efficient algorithm for optimal force distribution-the compact-dual LP method," *IEEE Trans. Robotics Automat.*, vol. 6, no. 2, pp. 178-187, Apr. 1990.

[5] C. A. Klein and S. Kittivatcharapong, "Optimal force distribution for the legs of walking machine with friction cone constraints," *IEEE Trans. Robotics Automat.*, vol. 6, no. 1, pp. 73-85, Feb. 1990.

[6] M. Nahon and J. Angeles, "Real-time force optimization in parallel kinematic chains under inequality constraints," *IEEE Trans. Robotics Automat.*, vol. 8, no. 4, pp. 439-450, Aug. 1992.

[7] P. Sinha and J. Abel, "A contact stress model for multifingered grasps of rough objects," *IEEE Trans. Robotics Automat.*, vol. 8, no. 1, pp. 7-22, Aug. 1992.

[8] Y. Nakamura, K. Nagai and T. Yoshikawa, "Dynamics and stability in coordination of multiple robotic mechanisms," *Int. J. Robotics Res.*, vol. 8, no. 2, pp. 44-61, 1989.

[9] M. Buss, H. Hashimoto and J. B. Moore, "Dextrous hand grasping force optimization," *IEEE Trans. Robotics Automat.*, vol. 12, no. 3, pp. 406-418, June 1996.

[10] M. Buss, L. Faybusovich, and J. B. Moore, "Dikin-type algorithms for dextrous grasping force optimization," *Int. J. Robotics Res.*, vol. 17, no. 8, pp. 831-839, 1998.

[11] H. Maekawa, K. Tanie and K. Komoriya, "Dynamic grasping force control using tactile feedback for grasp of multifingered hand," in *Proc. IEEE Int. Conf. Robotics Automat.* (Minneapolis, Minnesota), April, 1996, pp. 2462-2469.

[12] Z. Ji and B. Roth, "Direct computation of grasping force for three-finger tip-prehension grasps," *ASME J. Mechanisms, Transmissions Automat. Design*, vol. 110, pp. 405-413, Dec. 1988

[13] V. R. Kumar and K. J. Waldron, "Suboptimal algorithms for force distribution in multifingered grippers," *IEEE Trans. Robotics Automat.*, vol. 5, no. 4, pp. 491-498, Aug. 1989.

[14] R. D. Howe, I. Kao and M. R. Cutkosky, "The sliding of robot fingers under combined torsion and shear loading," *Proc. IEEE Int. Conf. Robotics Automat.* pp. 103-105, Philadephia, April, 1988.

[15] B. R. Zuo and W. H. Qian, "A force-closure test for soft multi-fingered grasps," *Science in China*, Ser. E, vol. 41, no. 1, pp. 62-69, 1998.

Assessment of Feedback Variables for Through the Arc Seam Tracking in Robotic Gas Metal Arc Welding

Jianming Tao and Peter Levick
Fanuc Robotics North America
Rochester Hills, Michigan

INTRODUCTION

Through the arc seam tracking for robotic arc welding is used by many robot manufacturers and automated arc welding system builders to compensate for weld joint location variance relative to the taught path of the robot manipulator. The method typically requires the robot manipulator to weave the weld torch assembly across the weld joint and to subsequently monitor the variation in welding arc current as a function of torch position. New weld power supply designs incorporating various proprietary control mechanisms require changes in the feedback variable selection traditionally used for through the arc sensing. Understanding and quantifying the control techniques used by the subject weld power source is required in order to provide robust seam tracking.

Experiments were conducted to assess the impact of weld power source control technique on feedback values of welding current and voltage relevant to seam tracking. Several pulsed output power sources with different control philosophies were evaluated and the resulting data is presented

The results of these experiments are presented along with conclusions regarding the optimum choice of feedback variable for through arc seam tracking.

BACKGROUND

Through the arc seam tracking in Gas Metal Arc Welding (GMAW) has been practiced for approximately 25 years. The basic technique is an outgrowth of Automatic Voltage Control (AVC) for arc length regulation developed for non-consumable electrode processes such as Gas Tungsten Arc (GTA) or Plasma Arc Welding (PAW). The method relies on monitoring of the unregulated variable in the welding power circuit in synchronization with phased transverse oscillation of the welding electrode. In GTAW and PAW the unregulated variable is voltage as the power source output is current controlled. The traditional GMAW power source for DC arc welding uses constant potential control and thus the unregulated variable is current. As such, through the arc seam tracking systems have relied on monitoring the weld current as a function of weld torch position in the weave cycle.

Pulsed output power supplies for GMAW extend the range of usable deposition rates and also provide benefits in the form of reduced average heat input, lower spatter emission, and improved process stability. Pulsed power sources are generally current controlled although some variations of constant current and constant potential exist, and thus the unregulated variable should be voltage. However, the arc length regulation method used with pulsed power sources varies with different equipment manufacturers and this introduces some variability in the signal selection that provides the optimum sensitivity for seam tracking. The measurements reported here are based on correlation of the actual values of weld current and voltage with respect to the Contact Tip to Work Distance (CTWD) for two different power supply designs. The two power sources employed for this testing are the Lincoln Electric Power Wave 450[1] and the Miller Electric AutoInvision[2]. Both units are inverter designs with on-board microprocessor control and incorporate synergic control algorithms that correlate output with filler wire speed.

EXPERIMENTAL PROCEDURE

The experimental method used in this study is based on work presented by Wu and Richardson[3] as well as Shepard[4] and Cook[5] where the GMAW torch is oscillated in a direction normal to the workpiece at varying amplitude and frequency while monitoring the weld current and voltage. The tests were performed with electrode feed rates ranging from 150-450 inches per minute (6.35 cm/sec-19.05 cm/sec) in 50 IPM increments. Torch height with respect to the workpiece is monitored by a laser rangefinder mounted on the torch body which provides an analog output voltage proportional to distance. The torch is carried by a six axis articulated robot arm, and welding is conducted in the flat position on mild steel plate. The welding conditions are as follows:

Welding Filler Metal: Mild Steel, AWS classification ER70S-3, Lincoln L-50
Electrode diameter: 0.045" (1.2mm)
Shield Gas: 90% Argon / 10% CO_2
Weld Travel Speed: 20 inches per minute (50.8 cm/min)
Nominal Contact Tip to Work Distance (CTWD): 5/8" (15.9 mm)
Torch normal to the workpiece

Testing was carried out with the torch CTWD varied +/- 2 mm in a sinusoidal pattern about the nominal torch height. Data was taken with dwell in the increasing stick out direction of 0.5 seconds and in the decreasing stick out direction of 0.3 seconds. A repeat series of the tests were conducted with no dwell at the extremes of oscillation. A sample of the data table follows:

Wire feed (IPM)	Trim	Travel speed	CTWD amp.	CTWD freq.	Dwl Up	Dwl Down
150	88	20	2 mm	1 Hz	0.5	0.3
200	88	20	2 mm	1 Hz	0.5	0.3
250	88	20	2 mm	1 Hz	0.5	0.3
300	92	20	2 mm	1 Hz	0.5	0.3
350	92	20	2 mm	1 Hz	0.5	0.3
400	92	20	2 mm	1 Hz	0.5	0.3
450	92	20	2 mm	1 Hz	0.5	0.3

Shown below is a representative sample of the data record from the file labeled "newpw_04". The weld current and voltage are sampled at 20 kHz. after passing through a signal conditioning module with a 10kHz. anti-aliasing filter. At this time scale, there is very little that may be determined about the control feedback that is most directly proportional to variations in CTWD. However, the overall envelope of the pulsed output may be observed with respect to the torch position, and is included here for reference.

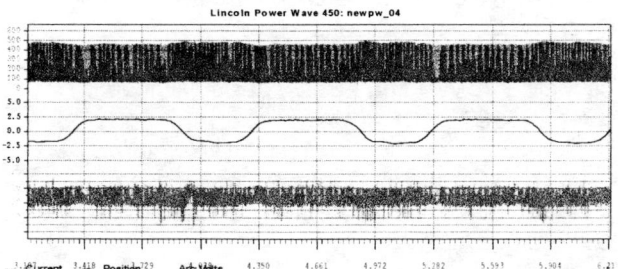

RESULTS AND DISCUSSION

The primary objective of this evaluation is to identify the feedback variable that is most appropriate to achieve maximum sensitivity for seam tracking. As discussed by Shepard[4], the process sensitivity correlates to a value of amperes/mm change in CTWD or volts/mm change in CTWD. Each of the weld power sources evaluated here use different control techniques for maintaining constant arc length despite changes in torch to work distance. The Lincoln Power Wave 450 employs a technique which varies several of the pulsed current variables (Peak Current, Base Current Time, etc.) to regulate arc length relative to a preset "Trim" reference. As a result, the average current is modulated as a secondary effect based on monitoring the Peak Voltage. The Miller AutoInvision varies only the Base Current Time, so the effective duty cycle is changed to maintain arc length constant with respect to a "Trim" reference.

Shown here is a representative plot of the Lincoln Power Wave 450 weld power source output current and voltage with respect to position while welding at 350 IPM filler wire speed.

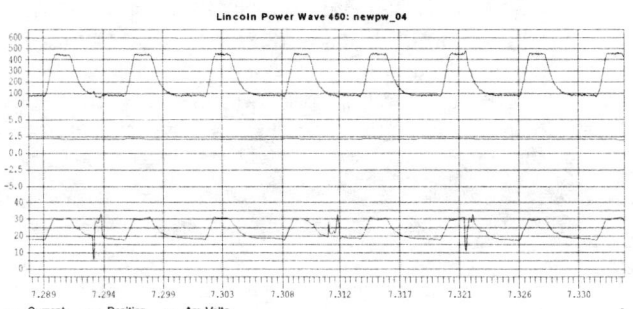

A corresponding plot from the Miller AutoInvision power source is shown below.

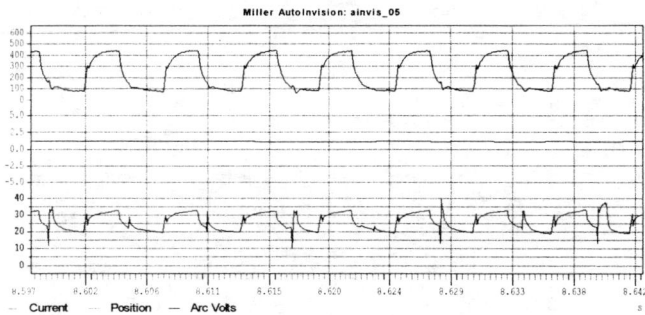

The different output current and voltage pulse shape is evident from comparison of these two plots, although a conclusion regarding process sensitivity cannot be seen here. Plotting this data provides a clue to the process sensitivity, which is the adaptive arc length control sensitivity to changes in CTWD. The following plot indicates the relative sensitivity of pulse frequency to changes in CTWD for both of the subject power supplies.

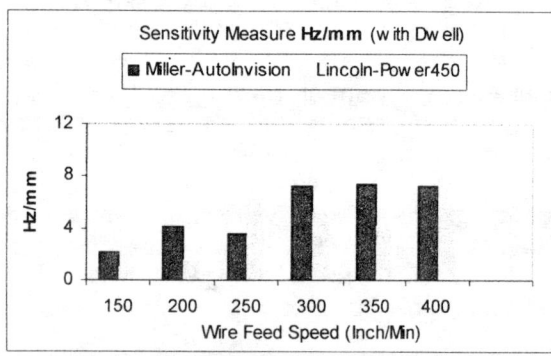

Several researchers have identified the relationship between frequency of CTWD oscillation and process sensitivity. The figure above indicates the frequency in Hertz/mm change in CTWD for oscillation of 2 mm amplitude and 1 Hz. frequency. The figure below illustrates the increased sensitivity that results when oscillation frequency is increased to 2 Hz by eliminating dwell from the extremes of oscillation.

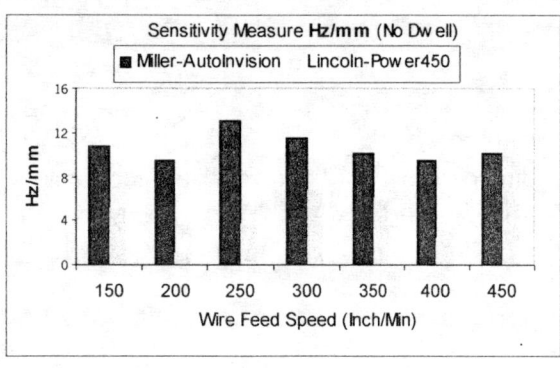

It is evident from evaluation of the changes to the current and voltage while oscillating the CTWD that the adaptive arc length control algorithm employed by the Lincoln Power Wave 450 modulates the Peak Current and Pulse Frequency as the primary variables. This means that the overall current waveform shape is modulated in order to maintain arc length constant while keeping metal transfer consistent.

The data indicates that the most significant variable for changes in CTWD with the Miller AutoInvision is the base current time, as indicated in the overall pulse frequency. This change in duty cycle is directly reflected in the average weld current, so the traditional feedback variable used for constant potential non-pulsed DC power sources remains a good analog to CTWD with the Miller AutoInvision due to the adaptive arc length control algorithm.

SUMMARY

The evaluation determined that the process variable most sensitive to CTWD variation is dependent on the power supply adaptive arc length control algorithm. In both of the subject power sources arc length is regulated by modulating the weld current to maintain a reference voltage. As such average current is an effective feedback variable for seam tracking purposes.

Each of these power sources exhibits different sensitivity for changes in CTWD amplitude, oscillation frequency, and operating level which indicates that the gain of the seam tracking function should be altered based on these characteristics.

---

[1] Lincoln Electric Company, PowerWave 450 product literature.
[2] Miller Electric Company, AutoInvision product literature.
[3] Wu, G.-D., Richardson, R.W.; "The Dynamic Response of Self-Regulation of the Welding Arc"; CWR Tech Report #529501-84-23, 1984; The Center for Welding Research, Ohio State University, Columbus, Ohio.
[4] Shepard, M.E.; "Modeling of Self-Regulation in Gas Metal Arc Welding", PhD Thesis, Vanderbilt University, 1991.
[5] Cook, G.E.; "Modeling of Electric Welding Arcs for Adaptive Feedback Control"; Conf. Record, IEEE/IAS 1983 Annual Meeting; pp.1234-1240; Mexico City, Oct. 1983.

"THROUGH-ARC" PROCESS MONITORING TECHNIQUES FOR CONTROL OF AUTOMATED GAS METAL ARC WELDING

Darren Barborak, Chris Conrardy, Bruce Madigan, and Troy Paskell
Weldware, Inc.[1]
1165 Chambers Road
Columbus, OH 43221 USA

Abstract - Improved sensors are needed to enable automatic GMAW to accommodate a greater range of disturbing inputs. A low-cost, non-intrusive sensing technique, known as Through-Arc sensing, involves collecting and analyzing welding current and voltage signals. Through-Arc sensing can be used to detect various GMAW quality indicators. It is shown that Through-Arc sensing can detect arc-start quality, steady-state arc stability, and the mode of metal transfer. Furthermore, the technique shows promise for detecting the onset of GMAW process disturbances including variations in tip-to-work distance, insufficient shielding gas coverage, electrode feeding problems, joint fit-up problems, and contact tip wear. Additional work is needed to refine and implement these techniques into a production-worthy system.

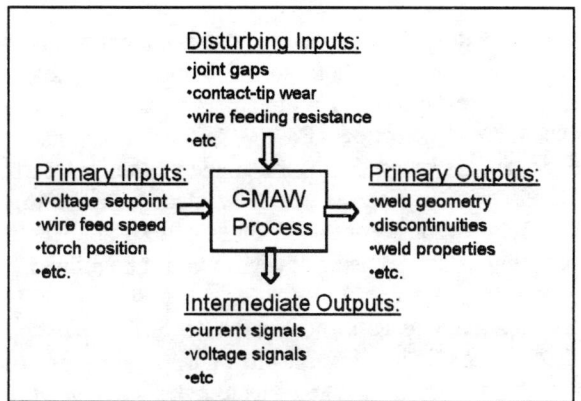

Figure 1 GMAW Process Diagram

1 INTRODUCTION

The basic objective of applying control in automated Gas Metal Arc Welding (GMAW) applications is to achieve the desired weld characteristics. These characteristics are represented as the primary process outputs of Figure 1 and include:
- Maintaining desired weld geometry (weld size, fusion zone shape, and surface profile)
- Achieving required weld mechanical properties
- Limiting weld discontinuities and defects to within acceptable limits

Compared with other manufacturing processes, welding has a reputation for being more difficult to control and less likely to achieve consistent quality. This is the result of the multiplicity of interrelated control parameters (i.e., primary inputs) the complexity of control relationships and the difficulty of in-process sensing.

The traditional method of ensuring adequate weld quality for automated GMAW applications involves establishing satisfactory welding parameters (primary inputs) by procedure trials and testing; and then adhering to these procedures in production. This approach assumes that if all the process inputs remain fixed within acceptable limits then a satisfactory output in terms of weld quality is consistently obtained.

Unfortunately, even if the primary inputs are carefully controlled, disturbing inputs can adversely impact the weld characteristics. Common disturbing inputs include the following:
- Part fit-up variations
- Misalignment of the welding electrode from the joint
- Insufficient shielding gas coverage due to accumulation of spatter in nozzle
- Disruption of wire feeding due to wire liner clogging
- Variation in gun-to-workpiece distance due to variation in fixturing and distortion
- Excessive wear of contact tips.

For many applications such disturbances often result in

---

[1] WeldWare, Inc. is currently working with Accudata, Inc. (Jackson, MI) to develop a next generation automatic GMAW process monitoring system.

sub-standard weld quality, necessitating costly repair. Typical problems caused by GMAW disturbing inputs include incomplete weld fusion, poor weld shape, burn-through, excessive spatter, porosity, and electrode burn-back.

The degree to which disturbances can be tolerated depends to a large degree on the application. In general, automated GMAW welding applications are less tolerant to such disturbances than are manual welding applications. Manual operations are more tolerant because manual welders are trained to identify process disturbances and compensate for them.

A means of automatic feedback control is sought to allow automated GMAW applications to accommodate a greater range of disturbing inputs. To implement such a control, a sensor capable of detecting GMAW disturbing inputs or process outputs is needed. Unfortunately, the presently available sensor technology is generally inadequate for this purpose. Before robust control schemes can be applied in manufacturing, process sensing of GMAW must be refined.

In this paper, the results of research to further develop Through-Arc techniques for sensing and monitoring of GMAW are discussed. These techniques involve the measuring of two GMAW process intermediate outputs (welding current and voltage signals) and using this information to infer GMAW disturbing inputs and primary process outputs.

In the sections, which follow Through-Arc sensing techniques are first described. Then, the application of these techniques to the detection of various GMAW process characteristics is described.

2 THROUGH-ARC SENSING

Through-Arc Sensing is a non-contact sensing method that relies on detecting changes in arc voltage and/or current during welding. The raw arc voltage and welding current signals are digitized using computerized data acquisition instrumentation. Various signal-processing techniques are then applied to extract arc stability information from which weld quality may be inferred.

This technique is attractive for GMAW process monitoring for several reasons:
- Through-Arc sensing is non-intrusive since voltage and current signals can be captured remotely from the point of welding
- Through-Arc sensing makes use of inexpensive and robust sensors
- Through-Arc sensing is amenable to automation.

2.1 DATA ACQUISITION RATES

Data acquisition rates vary depending on the objectives of the analysis and methods of signal processing utilized. Much of the previous work has utilized sampling rates ranging from 1,000 to 10,000 Hz. The sampling rate has a direct effect on the number of calculations that can be performed in real-time, as well as the amount of data that can be stored for later analysis. The lower the sampling rate, the lower the performance requirements (and hence the cost) of the computer-based instrumentation. However, at low sampling rates higher frequency components in the signal are lost, which may reduce the effectiveness of the quality determination for certain applications.

Sampling rates of 1000Hz or less are adequate for determining several coarse measures of quality, such as averages and standard deviations of current and voltage. However, higher sampling rates are needed for more detailed analysis. Observation of spectrographs for spray, short-circuiting and globular transfer modes indicate that most of the critical process information lies below approximately 400 Hz for all modes. While most events occur at frequencies below 400Hz, a sampling rate of at least 10 times or 4kHz is required to get good resolution of the events. For example, GMAW short-circuiting typically occurs at frequencies of less than 400 Hz, however the sampling rate must be an order of magnitude higher to get an accurate measure of waveform attributes, such as the short-circuiting duration.

Because high frequency noise is present in most Through-Arc signals, a low-pass or anti-aliasing filter with a cut-off frequency of 400 Hz should be used.

2.2 DATA ANALYSIS TECHNIQUES

A variety of analysis techniques have been investigated to identify the most robust means of detecting GMAW quality problems. A robust analysis technique will maximize detection of sub-standard welds ("true-positive" condition) while minimizing the rejection of acceptable welds ("true-negative" condition) and the misdetection of substandard welds ("false-positive" condition).

Common time-domain analysis parameters include mean, minimum, maximum, and standard deviation of voltage or current, as well as duration's of various events. A common frequency-domain analysis technique is the frequency spectrum of the voltage or current signal. It is also common to apply threshold techniques or traditional statistical process control (SPC) techniques. In most cases it is necessary to tune the various analysis parameters for each particular application. A change in gas mixture, electrode type, or weld quality requirements can easily cause an algorithm to indicate "false-positive" or "true-negative" conditions.

3 GMAW QUALITY DETERMINATION

Depending on the requirements of a particular application, different GMAW quality characteristics may be important. In this section the approaches used for determining various aspects of weld quality are reviewed.

3.1 MEASURING ARC START QUALITY

For applications requiring the production of high-quality short-duration GMAW welds, the quality of the arc start becomes critical. Proper fusion is usually not achieved until the welding arc has stabilized. For short duration welds this arc stabilization period represents a significant portion of the total welding time, so a poor start could result in inadequate weld fusion[1]. This is particularly the case for automated applications where the weld duration is fixed regardless of the quality of the arc start.

The objective of this portion of the work was to develop a reliable means of determining the quality of a GMAW arc start using current and voltage signals. This would provide a useful quality control tool by allowing the quality of the arc start to be determined automatically. Integration of the arc start quality sensing technique into automated welding equipment could also allow a closed-loop control to wait for the arc to stabilize prior to continuing the weld.

Melton[2] briefly describes a method for quantifying the quality of an arc start using the time integral of the square of the current during the initial short-circuiting period, referred to as the "Action Integral".

$$Action\ Integral = \int_{\tau=0}^{t} i^2(\tau)d\tau$$

This work was extended by Farson, et.al.[3] Using high-speed video images, the Action Integral was correlated to the arc starting mode. It was found that high values of Action Integrals were correlated to arc starts where the wire extension disintegrated at mid-extension. The Action Integral was very consistently found to be at a high value whenever the wire first disintegrated at any point other than the base metal contact point. Lower values of the Action Integrals (although more variable) were found correlate to arc initiations occurring at the base metal contact point.

This work was extended to short duration GMA welds using solid steel (ER70S) electrodes under a variety of conditions[4]. Welding parameters and the electrode end condition (e.g., cut sharp, cut blunt, melted) were systematically varied to achieve a variety of arc starting characteristics. High-speed (2,000-40,000 Hz) video with laser backlighting was employed to capture images of the arc starts to evaluate the quality of the arc starts. Welding current and voltage signals were also captured at high-speed (5,000 Hz) using computer-based data acquisition equipment.

The experimental investigations established the use of the Action Integral for quantifying the quality of arc starts. It was found that high action integral values were associated with wire extension expulsion and resultant arc extinction. Two sets of designed experiments were then used to identify the relationship between weld process parameters and Action Integral for a particular application. From the results of the first screening experiment, three parameters were selected for detailed evaluation in the second experiment. The variables studied in detail were run-in wire feed speed, starting voltage, and contact-tip-to-work distance (CTWD). Relatively low run-in speeds, low CTWDs and high starting voltages were found to result in low Action Integral arc starts, characterized as "good" starts with little or no expulsion. Thus, the Action Integral was found to be a reliable indicator of GMAW arc start quality under a variety of conditions.

3.2 STEADY STATE ARC STABILITY ANALYSIS AND IDENTIFYING DISTURBING INPUTS

The purpose of steady state arc stability analysis is to detect the on-set of disturbances during the steady-state portion of the weld, (i.e., after the arc has been established). While a number of techniques have been reported for determining GMAW arc stability[5,6,7,8,9,10], much of the prior work did not adequately test robustness in recognizing common disturbing inputs.

Additional work was performed to identify which process disturbing inputs can be reliably identified using various Through-Arc sensing algorithms, and to determine the techniques which work best for extracting important waveform attributes[11,12].

Experimentation was conducted on lap joints in 2.5-mm (0.1-in.) thick mild steel sheet. A constant voltage-type power source was used. Wire feed speed and voltage were varied to achieve short-circuiting, globular, spray, and shorting-spray (i.e., "mixed mode") transfer. Travel speed was varied inversely with wire feed speed to maintain a constant deposited weld volume throughout the testing. The power source inductance setting was varied to produce different degrees of short-circuiting transfer stability. Welding procedures were selected which produced nominally acceptable welds for three combinations of wire feed speed and travel speed. The following disturbances were then introduced:
- CTWD was varied by +/- 6 mm (0.25-in.) from the nominal
- Joint gap was varied to three levels: 0, 0.75 mm (0.030-in.), and 1.5 mm (0.060-in.)

- Shielding gas flow rate was varied from 40 cfh down to 0 cfh
- Wire electrode drive roll tension was reduced until electrode feeding problems were encountered
- Electrode wire was offset from the weld joint by +/-2 mm (0.08-in) in 1 mm (0.040-in.) increments

The following techniques were used to analyze the voltage and current waveforms:
- Simple statistics
- Histograms (Probability Density Distributions)
- Waveform factors
- Frequency analysis.

Methods categorized as "simple statistics" included calculating the RMS (root-mean-square) and standard deviation of current (I), voltage (V), power (V*I), and conductance (I/V).

Figure 2 shows a typical voltage histogram. Methods that utilize histograms provide a summary of data in which the individual values are sorted by the range of values into which they fall (also known as bins), and in which the number of individual values fall within each bin is counted. The data is plotted by showing the number of values in each bin.

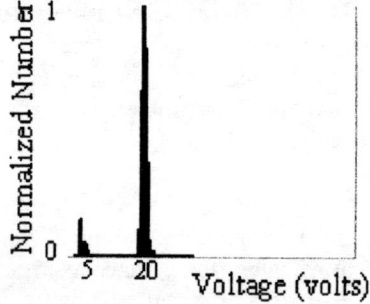

Figure 2 Typical Voltage Histogram for Short-Circuiting GMAW.

Waveform Factors that were calculated include:
- Mean arcing and short-circuiting times
- Standard deviations of arcing and short-circuiting times
- Arcing and short-circuiting frequency
- Mean voltage and current during arcing and during short circuiting
- Standard deviation of voltage and current during arcing and during short circuiting.

Short-circuiting conditions are determined when the voltage drops below a threshold (Figure 3) while the current exceeds a threshold. Likewise, the arcing conditions are determined when the voltage maintains above a threshold (Figure 3) while the current remains below a threshold. The threshold was calculated as a percentage deviation from the mean.

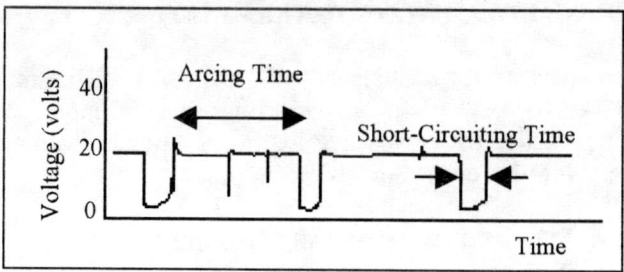

Figure 3 Illustration of Arcing and Short-Circuiting Times

Welding parameters were selected to produce stable short-circuiting and spray modes of metal transfer. GMAW process disturbing inputs were then systematically introduced, and current and voltage waveforms were analyzed.

It was found that no single waveform attribute could be used to detect the onset of all disturbances. Rather, specific attributes of the current, voltage, power, and conductance waveforms show merit for detecting specific types of disturbances for a given transfer mode as discussed below.

<u>Contact Tip-to-Work Distance Variations</u>

The CTWD was varied over a range of 12 mm for both short-circuiting and spray modes of metal transfer. A strong correlation was found between CTWD and a number of waveform attributes including:
- The mean and standard deviation of current
- The mean short-circuiting time.

<u>Insufficient Shielding Gas Coverage</u>

The shielding gas flow rate was systematically reduced from 40 to 0 cfh. It was found that arc instabilities caused by reduced flow rate could be detected before the occurrence of visible surface porosity. Waveform attributes showing promise include:
- The standard deviation of the current
- The standard deviation of the voltage and current during arcing
- The location and amplitude of the 2nd peak in the conductance histogram.

<u>Electrode Feeding Problems</u>

The wire feeder drive roll tension was reduced until wire feeding became erratic. It was found that a number of waveform attributes reflect wire feeding disruption. Unfortunately, the degree of wire feeding disruption was not controlled with sufficient sensitivity to allow selection of the most telling waveform factors. Additional testing is needed to further quantify this disturbance.

Joint Fit-Up Problems

Joint gaps of 0.75 and 1.5 mm (0.030 and 0.060 in.) were introduced. No clear trends were identified when welding in the spray mode of metal transfer. However, trends in several waveform attributes were identified for short-circuiting mode, including:
- The standard deviation of voltage
- The mean voltage during arcing
- The amplitude of the second peak in the power histogram.

Electrode Alignment Errors

The electrode was offset from the joint until weld shape became unacceptable. While several waveform attributes showed trends with varying electrode offset, none correlated strongly for all welding conditions. Additional work is planned to further evaluate waveform attributes for detecting electrode offset. However, it is unlikely Through-Arc sensing alone will provide robust detection of offset without weaving of the electrode.

Contact-Tip Wear

As the electrode passes through the copper contact tip, welding current is transferred from the tip to the electrode. Contact tip wear occurs due to the mechanical abrasion as the electrode passes through it, affecting the electrical contact. As the electrical contact deteriorates, more arcing is present between the electrode and tip causing exponential growth in wear, and eventually tip failure. Contact tip life is sometimes unpredictable from one batch of tips or electrode wire to another. Tip failure during welding can necessitate costly weld repair and downtime for automated equipment.

While several researchers have investigated Through-Arc techniques for monitoring contact tip wear, the most promising technique has been developed by Siewert et.al.[13,14]. This technique looks at the power spectral density of the welding voltage or current, depending on the type of power supply used. If the integral of the power spectral density of the normalized process signal exceeds a predetermined threshold, the life of the contact tip is expected to end shortly and preventative maintenance can then be planned to change the tip.

3.3 IDENTIFYING METAL TRANSFER MODE

There are several modes in which the molten metal can be transferred from the end of the electrode to the weld pool. Commonly used modes include short-circuiting mode, globular mode, and spray mode. "Shorting-spray" mixed mode is also often used for high speed welding of sheet metal.

The mode of metal transfer directly affects the shape, depth of penetration, depth of fusion, and amount of spatter of the resulting weld. Depending on the application, one mode is typically considered optimal. Several researchers have investigated Through-Arc sensing of transfer modes and have devised several techniques to discriminate between modes[15,16,17,18,19].

To verify these results, experiments were conducted to determine which waveform attributes best identify and analyze the various transfer modes during GMAW of thin sheet. Wire feed speed, voltage, and inductance were varied to produce the various transfer modes. Results indicate that the shape of the instantaneous voltage-current oscillogram (V-I plot), voltage and current histograms, and duration's of arcing and short-circuiting are good indicators of transfer mode.

Figure 4 compares V-I plots for globular and short-circuiting modes. The distribution of the data on the V-I plots clearly differentiates the various modes.

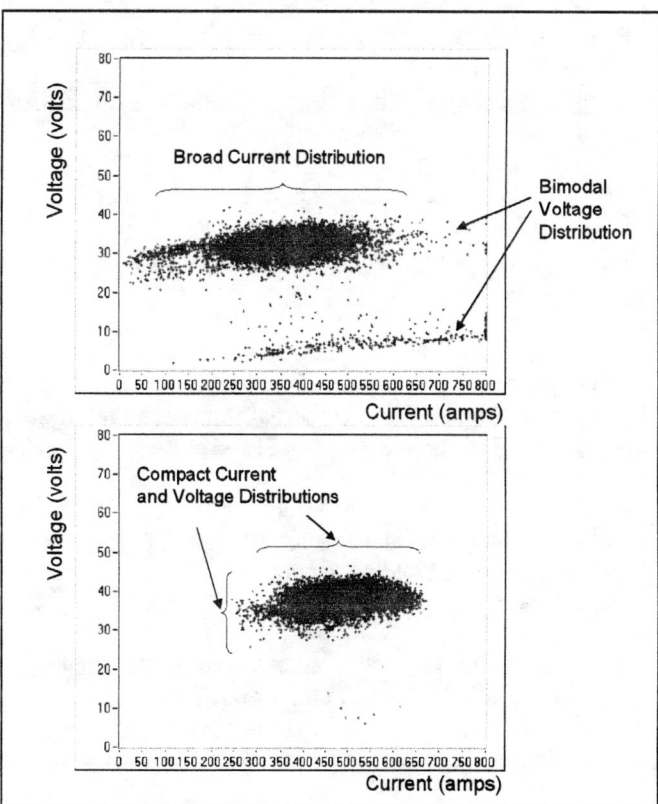

Figure 2 V-I Plots for Short-Circuiting (top) and Globular (bottom) Transfer Modes

The Voltage Histogram differentiates spray from short-circuiting transfer by the presence of one concentrated peak Vs two distinct peaks. The Current Histogram differentiates between transfer modes through distribution and magnitude of the peak.

Transfer mode can also be identified through average arcing time. The average arcing time for globular transfer was nearly double that for short-circuiting transfer. Likewise, the frequency of short circuits is lower for globular transfer than for short-circuiting transfer.

4 CONCLUSIONS

It is shown that Through-Arc sensing can detect arc-start quality, steady-state arc stability, and the mode of metal transfer. Furthermore, the technique shows promise for detecting the onset of GMAW process disturbances including variations in contact tip-to-work distance, insufficient shielding gas coverage, electrode feeding problems, joint fit-up problems, and contact tip wear. Additional work is on going to refine and implement these techniques into a production-worthy system.

5 REFERENCES

[1] Farson, D., Barborak., D., and Conrardy, C. " GMAW Arc Starting; Experimental and Theoretical Study", Proc. 5th Int. Conf. On Trends in Welding Research, ASM International. June, 1998.

[2] Melton, G.B., "Factors Affecting MIG Arc Initiation in Mechanized Systems" The Welding Institute Research Bulletin, The Welding Institute, Cambridge England, July, 1987, p. 229-230.

[3] Farson, D., Conrardy, C., Talkington, J., Baker, K., Kerschbaumer, T., and Edwards, P., "Arc Initiation in Gas Metal Arc Welding", Welding Journal, American Welding Society, vol.77, no.8. June, 1998. p.315s-321s.

[4] Farson, D., Barborak., D., and Conrardy, C. " GMAW Arc Starting; Experimental and Theoretical Study", Proc. 5th Int. Conf. On Trends in Welding Research, ASM International. June, 1998.

[5] Dorn, L., and Roppl, P. "Process Analysis of Gas-Shielded Metal-Arc Welding - Stability of the Arc, Spatter, and Instantaneous Value Curves for the Electrical Variables with Varying Process Parameters". Schweissen Schneiden, vol. 36, no. 5, pp. 219-224, May, 1984.

[6] Shinoda, T., Nishikawa, H., and Shimizu, T., "The Development of Data Processing Algorithms and Assessment of Arc Stability as Affected by the Titanium Content of GMAW Wires During Metal Transfer", Paper 57, Proc. 6th Int. Conf. on Computer Technology in Welding", The Welding Institute. June 1996.

[7] Mita, T., Sakabe, A., and Yokoo, T. "The Estimation of Arc Stability on CO_2 Gas Shielded Arc Welding". Paper 24, Proc. 1st Int. Conf. in Advanced Welding Systems. ISBN 0-85300191-X. The Welding Institute. 1987.

[8] Siewert, T., et. al. "Through-the-Arc Sensing for Monitoring Arc Welding". Proc. 3rd Int. Conf. on Trends in Welding Research. ASM International. 1993.

[9] Rehfeldt, D., et. al.. "Computer-Aided Quality Control by Process Analyzing, Monitoring, and Documentation", Proc. Int. Conf. Joining/Welding 2000, International Institute of Welding, Pergamon Press ISBN#0-08-041280-7, 1992.

[10] Norrish, H., "Computer-Based Data Acquisition for Welding Research and Production" Proc. 39th Welding Conf.: Welding Towards 2000, Welding Technology Institute of Australia, 1991.

[11] Conrardy, C., and Barborak. D. " Monitoring of Weld Quality During Thin-Sheet GMA Welding", Proc. Sheet Metal Welding Conference VII. American Welding Society. 1996.

[12] Conrardy, C., Barborak. D., and Paskell, T. "Through Arc Monitoring of Gas Metal Arc Spot Weld Quality", Presented at the American Welding Society 50th Int. Welding Convention, Detroit. 1998.

[13] Siewert, T., et. al. "Contact Tube Wear Detection in Gas Metal Arc [GMA] Welding". Welding Journal, 74(4)115s-121s. American Welding Society. April, 1995.

[14] Siewert, T., Madigan, B., and Quinn, T. "Prevention of Contact Tube Melting in Arc Welding". United States Patent #5514851. 1996.

[15] Johnson, J., Carlson, N., Smartt, H., and Clark, D. "Process Control of GMAW: Sensing of Metal Transfer Mode", Welding Journal, 70(4)91s-99s. American Welding Society. April, 1991.

[16] Adam, G., and Siewert, T. "Sensing of GMAW Droplet Transfer Modes using ER100S-1 Electrode", Welding Journal, 69(3)103s-108s. March, 1990.

[17] Lui, S. et. al. "Metal Transfer Mode in Gas Metal Arc Welding", Proc. 2nd Int. Conf. on Recent Trends in Welding Science and Technology TWR'89. ASM International. 1989.

[18] Wang, W. et. al. "Flux Cored Arc Welding: Arc Signals Processing and Metal Transfer Characterization". Welding Journal, 74(11)369s-377s. American Welding Society. November, 1995.

[19] Ogunbiyi, B., and Norrish, J. "GMAW Metal Transfer and Arc Stability Assessment using Monitoring Indices". Proc. 6th Int. Conf. on Computer Technology in Welding. The Welding Institute. June, 1996.

Dynamic Modeling of GMAW Process

Zafer Bingul and George E. Cook

Department of Electrical and Computer Engineering
Vanderbilt University
Nashville, Tennessee, 37235

Abstract

A model has been developed correlating the anode temperature profile with the dynamic melting rate in gas metal arc welding. Components of this model are identified, as the electrode melting rate, temperature-dependent resistivity of electrode and arc voltage. The differential equations describing the dynamic behavior of the electrode extension were derived from mass continuity and energy relations. The temperature of the electrode extension was determined by convective heat transfer and joule heating. One-dimensional solutions of temperature and heat content were used to obtain the dynamic melting rate equation. The purpose of the present paper is to provide quantitative analyses, concentrating on the thermal behavior and the electrical characteristics of the arc welding system, to aid in a fundamental understanding of the process, and to develop a dynamic model that may be used in adaptive control. The model is tested by comparing simulations to experimental results.

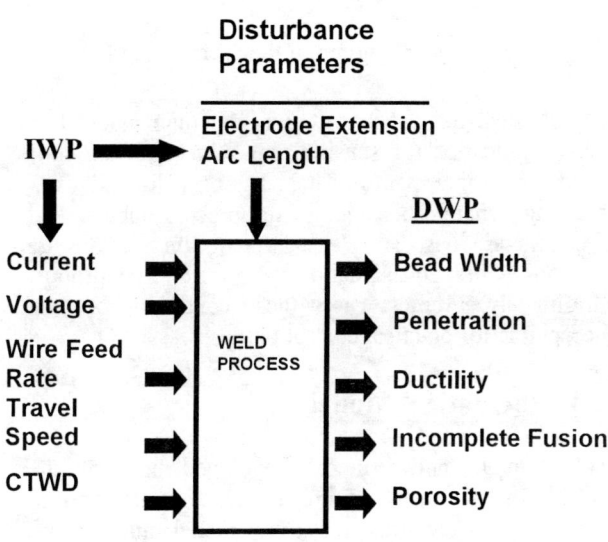

Figure 1. Input and output variables of welding process.

1 Introduction

Improved automation of welding processes has become increasingly important in the push for higher weld quality, safety and reduced manufacturing cost. Robotic welding offers the reduced manufacturing cost sought, but its widespread use demands as means of sensing and correcting for inaccuracies in the part, the fixturing, and the robot.

Changes in the welding environment can result in defective welds (inadequate penetration and undercut). The quality of a weld is a function of many factors: selecting satisfactory welding parameters, maintaining the same parameters in production, monitoring internal and external changes in the process and correcting them. Weld quality is usually assessed and controlled manually. To reduce the reliance on manual inspection, developing a real time quality monitoring and control strategy for automated welding systems is needed.

Cook [1] separates the variables of the welding process into two categories: direct weld parameters (DWP) and indirect weld parameters (IWP) as shown Figure 1. The disturbance parameters can be varied on line during welding to affect change in the DWP so they must be controlled by the other indirect welding parameters. Since a nonlinear relationship exits between the highly coupled direct and indirect weld parameters, the coefficients of the differential equations that relate the direct and indirect weld parameters vary as a function of operating conditions. Hence, the feedback control systems with the added capability of self-adjustment are necessary to maintain stable control over the operating range of the process.

There have been many attempts to model the gas metal arc welding (GMAW) process. These models that have been developed by using different methods can be divided into three groups: (1) the model is based on finite element and finite difference methods, (2) the model is derived empirically or statistically from experimental data, and (3) the model is based on physical reasoning with analytical solution. Models in the first group typically require excessively time-consuming numerical computations that make them unusable directly in a real-time control system. The empirical models have only a limited range of applicability and do not lend themselves to real-time "tuning" in a multivariable control system application. In addition, they contain little information

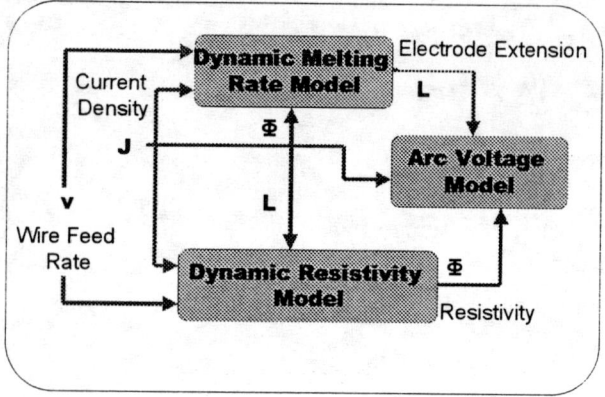

Figure 1. Schematic diagram of the GMAW model.

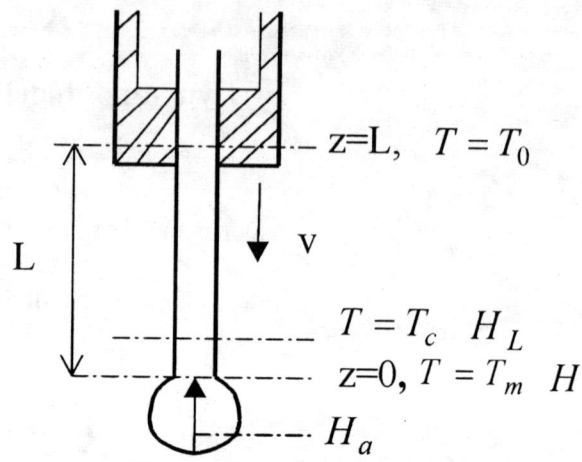

Figure 1. Temperature and energy along electrode extension.

about physical details. An accurate dynamic model that provides guidance for sensing and corrective action is needed for on-line control. There has been limited work describing dynamic behavior of the process but most of the efforts are based, at least in part, on steady-state conditions or quasi-stationary state conditions. Unfortunately, in most cases these conditions become unacceptable for practical control purposes.

2 Mathematical Model

The mathematical model developed here of the GMAW process consists of three components, the melting rate model, the electrode resistivity model, and the arc voltage model. A schematic diagram of the model is shown in Figure 2.

2.1 Melting Rate Model

Electrode melting rate is one of the most important parameters in the GMAW process because arc stability depends on how closely the electrode feed rate and the electrode melting rate are balanced to maintain a constant arc length. It is determined by the balance between energy and mass entering and leaving the electrode. The arc can be considered to be a moving energy (heat) source in the z direction. Therefore, energy conservation and mass conservation theories can be applied by specifying the boundary conditions and initial conditions. The difficulties arising in the modeling of the electrode are mainly due to the complex nature of boundary conditions. However, treating the electrode and the arc column as a separate block helps to eliminate these boundary conditions [2]. This is illustrated in Figure 3. The assumptions made for the model are that the heat dissipation losses due to convection and radiation, and the influence of metal droplets are negligible. These assumptions don't affect the model accuracy's because many authors [3-6] have shown that the convective heat produced by the electric current is the major part of the energy transferred to the anode from the welding arc.

Based on the above assumptions, the conservation equations are expressed in terms of the z-axis as follows:

Mass continuity:

$$\frac{\partial \rho}{\partial t} + v \frac{\partial \rho}{\partial z} = 0 \quad (1)$$

Conservation of Energy:

$$\frac{\partial}{\partial z}\left(\lambda(T)\frac{\partial T}{\partial z}\right) + J^2 \Phi(T) = C_p(T)\rho(T)\left(\frac{\partial T}{\partial t} + v\frac{\partial T}{\partial z}\right) \quad (2)$$

The energy equation contains a Joule heating term. The boundary conditions are:

$$\begin{array}{ll} T = T_0 & z = L \\ T = T_m & z = 0 \\ \frac{\partial T}{\partial r} = 0 & \end{array} \quad (3)$$

Combining the mass continuity equation and the energy equation, the *damped nonlinear heat equation* can be obtained as follows.

$$\frac{\partial}{\partial z}\left(\lambda(T)\frac{\partial T}{\partial z}\right) + J^2 \Phi(T) = 0 \quad (4)$$

Heat transfer and temperature distributions during welding are complex and a solution to the equations is dependent on the thermal conductivity, specific heat and density of the mass as a function of temperature. To find analytical solutions to the equations, it is therefore helpful to eliminate the relationship between temperature and the above variables by means of defining a new

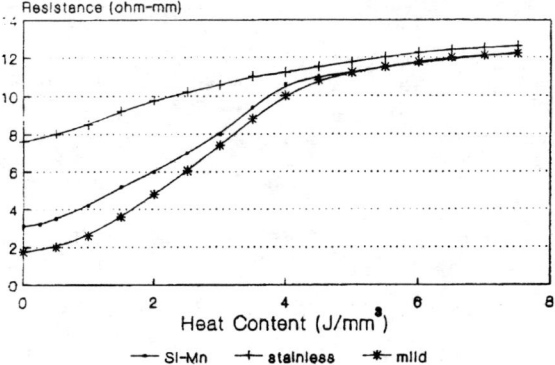

Figure 2. Resistivity vs. heat content for typical wires.

variable which serves as a bridge between them. The heating effect, called the heat content is defined as:

$$H = \frac{\lambda(T)}{v}\frac{\partial T}{\partial z} \qquad (5)$$

Replacing the parenthesis expression with $H v$ in Equation 4 yields an equivalent first order nonlinear differential equation given by:

$$\frac{\partial H}{\partial z} = -\frac{J^2 \Phi(T)}{v} \qquad (6)$$

The resistivity versus heat content is shown for three representative electrode wire materials in Figure 4. The curves may be approximated with two straight lines about the knee of the curve. The resistivity can be written in terms of heat content as,

$$\Phi(H) = \begin{cases} a_1 H + b_1 & H \leq H_c \\ a_2 H + b_2 & H > H_c \end{cases} \qquad (7)$$

where H_c (4 J/mm^3) is heat content at the Curie temperature.

From Equation 6, the heat content as a function of time can be written as:

$$\frac{\partial H}{\partial t} = -J^2 \Phi(T) \qquad (8)$$

The desired relationship between heat content, time and current density can be obtained by rearranging and integrating Equation 8 over time.

$$\int_0^{H_L} \frac{\partial H}{\Phi(H)} = \int_{t-\frac{z}{v}}^{t} J^2 \partial t \qquad (9)$$

where H_L is the Joule heating energy per unit volume of wire. The left side of Equation 9 is called the *action integral* [16]. It is dominant in the dynamic behavior of the process. Equation 9 shows that the heat content based on temperature and temperature history is related to the current density and current density history. Substitution of Equation 7 into Equation 9 gives:

$$\int_0^{H_L} \frac{\partial H}{a_2 H + b_2} = W \qquad H > H_c \qquad (10)$$

This equation can be solved for the Joule heating energy:

$$H_L = \begin{cases} \frac{b_1}{a_1}\left(e^{a_1 W} - 1\right) & H \leq H_c \\ \frac{b_2}{a_2}\left(k e^{a_2 W} - 1\right) & H > H_c \end{cases} \qquad (11)$$

The total heat content along the electrode extension is composed of the following elements:

$$H(z,t) = H_a(0,t) + H_L(z,t) \qquad (12)$$

where H_a is called the anode heat content and is derived using Equation 5. This equation is still valid at the electrode tip, but the melting rate instead of wire feed rate must be used in the equation. Therefore, Equation 5 can be written as:

$$\frac{\partial T}{\partial z} = \frac{v_m H}{\lambda(T)} \qquad (13)$$

By using the chain rule, the temperature as a function time and z can be expressed in the following form:

$$\frac{\partial T(z,t)}{\partial z} = \frac{\partial T}{\partial H}\frac{\partial H}{\partial z} + \frac{\partial T}{\partial t}\frac{\partial t}{\partial z} \qquad (14)$$

Melting wire is assumed to have constant melting temperature at the anode tip so that $\partial T/\partial t$ is considered to be zero. After substituting Equations 6 and 13 into Equation 14, the heat content at the anode tip becomes:

$$H = -\frac{\Phi(T)\lambda(T)J^2}{v_m^2}\frac{\partial T}{\partial H} \qquad (15)$$

In order to formulate the anode heat content in terms of current density and melting rate, the above equation associated with the boundary conditions is rearranged and integrated over H and T.

$$-\int_{H_L}^{H} H \partial H = \frac{J^2}{v_m^2}\int_{T_c}^{T_m} \lambda(T)\Phi(T) \partial T \qquad (16)$$

The solution to Equation 16 under these conditions is given by:

$$-H_a = \frac{J}{v_m}\sqrt{\int_{T_c}^{T_m}\frac{\lambda(T)\Phi(T)}{0.5+\delta(T)}\partial T} = \frac{J\phi(T)}{v_m} \quad (17)$$

where $0 \leq \delta(T) \leq 1$ is the relative proportion of solid in the two-phase zone and $\phi(T)$ is called the *work function*. The physical meaning of the negative sign is that the droplet should have to supply thermal energy to decrease the temperature difference. After substituting for H_a and H_L from Equations 11 and 17, the final form of the heat content is:

$$H = \begin{cases} \dfrac{J\phi(T)}{v_m}+\dfrac{b_1}{a_1}(e^{a_1 W}-1) & H \leq H_c \\ \dfrac{J\phi(T)}{v_m}+\dfrac{b_2}{a_2}(ke^{a_2 W}-1) & H > H_c \end{cases} \quad (18)$$

The results of wire melting experiments show that the heat content can be described by the exponential.

$$H = H_0(1-e^{-ct}) \quad (19)$$

Differentiating this equation gives the dynamic nature of the heat content as represented in the following equation.

$$\frac{\partial H}{\partial t} = c(H_0 - H) \quad (20)$$

The solution of Equation 20 gives the dynamic melting rate.

$$v_m = \begin{cases} \dfrac{c\phi(T)J + \phi(T)\dfrac{\partial J}{\partial t} + J\dfrac{\partial \phi(T)}{\partial t}}{-b_1 e^{a_1 W}(\dfrac{c}{a_1}+\dfrac{\partial W}{\partial t})+c(\dfrac{b_1}{a_1}+H_0)} & H \leq H_c \\ \dfrac{c\phi(T)J + \phi(T)\dfrac{\partial J}{\partial t} + J\dfrac{\partial \phi(T)}{\partial t}}{-b_2 k e^{a_2 W}(\dfrac{c}{a_2}+\dfrac{\partial W}{\partial t})+c(\dfrac{b_2}{a_2}+H_0)} & H > H_c \end{cases} \quad (21)$$

Shepard's burn-off rate model is also derived from this equation under appropriate simplifying assumptions. He assumed that the work function is constant and the heat content for the wire is above H_c (4 J/mm^3). It is concluded from his assumptions that the power is dissipated in a constant potential drop in the arc. By using his assumptions, the equation reduces to:

$$v_m = \frac{k_1 J}{1 - k_2 W} \quad H > H_c \quad (21)$$

Under steady state conditions, this equation reduces to Lesnewich's equation [7].

$$v_m = k_1 J + k_2 L J^2 \quad (22)$$

2.2 Electrode Voltage Model

The current changes in the GMAW process cause significant temperature changes along the electrode extension. These changes lead to a temperature-dependent resistance which produce the electrode voltage drop. The basic equation for the resistance is given by:

$$R_L = \frac{\Phi(T)}{A}L \quad (23)$$

The voltage drop is the product of the welding current and resistance of the electrode extension.

$$V_L = I R_L \quad (24)$$

2.2 Arc Voltage Model

There have been a number of models for the arc voltage model in the gas tungsten arc welding (GTAW) and the GMAW processes. These models show that as the welding current is increased, the relationship between the voltage and current is approximately linear. The arc voltage drop can be modeled as

$$V_{arc} = V_0 + R_a I + E_a l \quad (25)$$

where R_a defines the current dependency of the arc voltage and E_a defines the length dependency of the arc. The total voltage can be obtained by adding the electrode voltage drop and arc voltage drop.

3 Experimental Procedures

Welding experiments were structured to examine the dynamic nature of the electrode extension and arc length from both electrical signal measurements and high-speed video system (2000 frames/s). A 1.2mm (0.045") diameter ER705-2 steel wire and Argon+5%CO^2

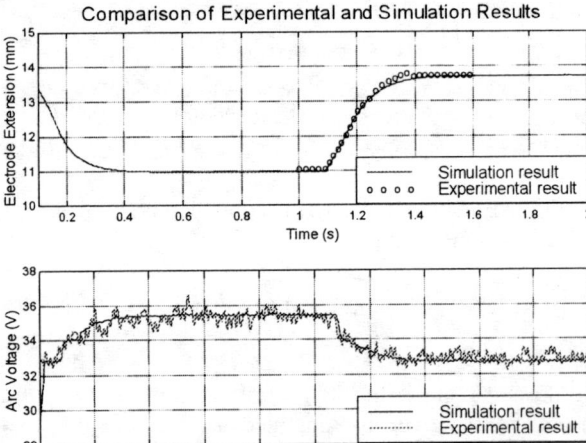

Figure 5. Experimental and simulation results for step increase in the current.

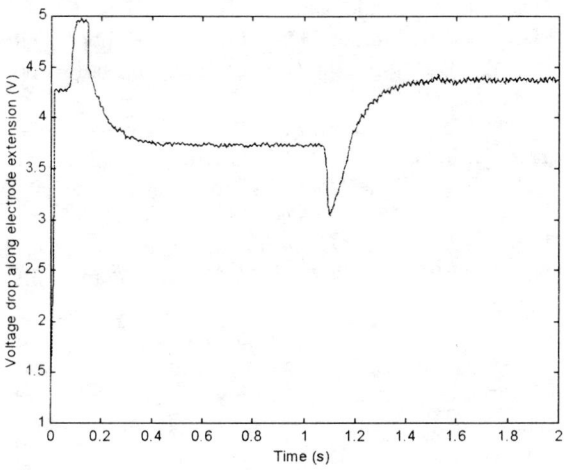

Figure 6. Electrode voltage drop changes

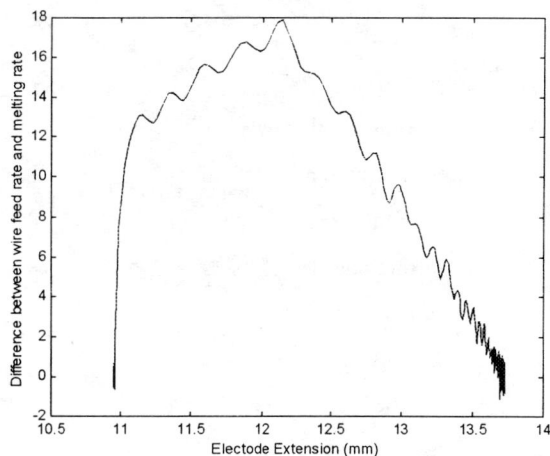

Figure 7. Phase plane analysis

shielding gas were used. The contact-tip-to-work distance (CTWD) was 20mm (3/4") and the travel speed was set at 6 mm/s. The current and voltage signals were collected at 5000Hz. Using a constant current power source with electrode positive, both square wave and sinusoidal perturbations, of variable amplitude (20 A, 30 A and 40 A peak-to peak), were superimposed on the current to allow measurement of changes in the electrode extension, arc length, and total voltage. The square wave perturbations provided a direct measure of the time response of the electrode melting.

4 Results and Discussion

The model was tested by introducing 20A step changes to the arc current, as the wire feed rate was constant at 140 mm/s. Comparison of the experimental and simulation results is shown in Figure 5. The electrode extension settles from an initial value of 13.5mm to 11mm in 0.4s. Similarly, the arc voltage changes

The effect of electrode voltage drop changes on the dynamic behavior of the process can be seen in Figure 6. As shown, the voltage suddenly decreases to about 3V and increases to about 4.4 V, during the current transition. This shows that the electrode voltage drop is a significant factor in the dynamic response of the process.

In order to examine motion trajectories corresponding to various initial conditions and information concerning stability, phase plane analysis was made. As seen in Figure 7, the initial point (13.5mm) for the electrode extension is moved to another point (11mm) in the phase plane. Oscillatory behavior of the curve results from the dynamic resistance changes in the electrode extension.

An interesting aspect of the high-speed video system used in capturing the experimental electrode extension dynamics is its ability to capture the power source and droplet dynamics as well. Power spectrum analysis of a single pixel corresponding to the high-speed video is illustrated in Figure 8. As seen, the frequency components are at 120Hz, 240Hz and 320Hz. These results match results of the same analysis for the arc current. The high-speed video system has been shown to be capable of capturing the dynamics of the power supply and the process itself, and it can also be used to detect droplet transfer frequency.

5 Conclusions

A dynamic model of the GMAW process is developed and tested by comparing simulations to experimental results. The model is applicable to both constant potential and constant current modes (including pulse current) of the GMAW process. The dynamic model developed in this work is essential to develop truly adaptive arc length regulation with pulsed current GMAW.

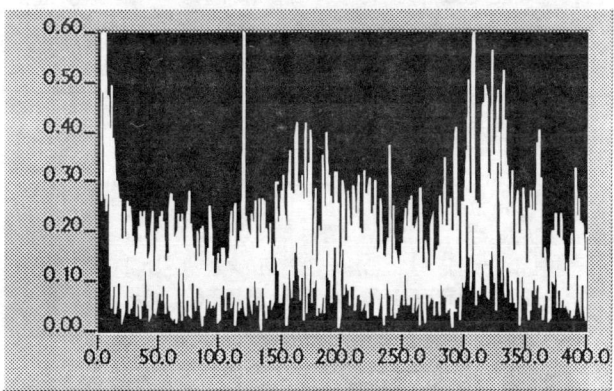

Figure 8. Power spectral density of the pixel of high-speed films

Nomenclature

ρ	Density (g/mm^3)
λ	Thermal conductivity ($W/(mm\,K)$)
ϕ	Work function (V)
Φ	Resistivity of wire ($\Omega\,mm$)
J	Current density (A/mm^2)
v	Wire feed rate (mm/s)
v_m	Melting rate (mm/s)
C_p	Specific heat capacity ($J/(g\,K)$)
H	Heat content (J/mm^3)
H_0	Melting heat content (J/mm^3)
H_a	Anode heating content (J/mm^3)
H_L	Joule heating content (J/mm^3)
T	Temperature (K)
T_m	Melting temperature (K)
T_0	Initial temperature (K)
W	Action integral ($J/(\Omega\,mm^4)$)
V_L	Voltage drop along electrode extension (V)
V_{arc}	Arc voltage drop (V)
k_{12}	Coefficients of Lesnewich's equation
a_{12}, b_{12}	Coefficients of the resistivity
I	Current (A)
c	Time constant
l	Arc length (mm)
L	Electrode extension (mm)

References

[1] Cook, G. E., Andersen K. and Barrett R.J., "Keynote Address Feedback and Adaptive Control in Welding," *International Trends in Welding Science and Technology*, ASM International, pp 891-903, 1992.

[2] P. Zhu and J.J. Lowke, " Theoretical study of the melting of the cathode tip of a free burning arc in argon for various conical angles", *J. Phys. D: Appl Phys*, (26) ,1073-6, 1993.

[3] Halmoy, E., *Pulsating Welding Arc*, SINTEF Report STF16-A79021, Trondheim, Netherlands.

[4] Halmoy, E., "Wire melting rate, droplet temperature, and effective anode melting potential," *Proc. Int. Conf. On Arc Physics and Weld Pool Behavior*, The Welding Institute, Cambridge, UK, May 1979.

[5] Wilson, J. L. and Claussen, G. E., "The Effect of I^2R Heating on Electrode Melting Rate," *Welding Journal*, Vol. 35(1), pp 1s-8s, 1956.

[6] Waszink, J. H. and van den Heuvel, G. J. P. M., "Heat Generation and Heat Flow in the Filler Metal in GMA Welding," *Welding Journal*, Vol. 61(8), pp 269s-282s, 1982.

[7] Lesnewich, A., "Control of Melting Rate and Metal Transfer in Gas-Shielded Metal-Arc Welding; Part 1 – Control of Electrode Melting Rate," *Welding Journal*, vol. 37 (8), pp 343s-353s, 1958.

[8] Bingul, Z., *Stability Considerations for the Gas Metal Arc Welding Process*, M.S. Thesis, Vanderbilt University, 1996.

[9] Bingul, Z., Cook, G. E., Barnett, R. J., Strauss, A. M., and Wells, B. S., "An Investigation of Constant Potential GMAW Instability Behavior," 5th International Conference on Trends in Welding Research, June 1-5, 1998.

[10] Cook, G. E., "Decoupling of Weld Variables for Improved Automatic Control," 5th International Conference on Trends in Welding Research, June 1-5, 1998.

[11] Quintino, L. and Allum, C. J., "Pulsed GMAW Interactions Between Process Parameters, Part 1," *Welding and Metal Fabrication*, March 1984.

[12] Amin, M., "Pulse current parameters for arc stability and controlled metal transfer in arc welding," *Metal Construction*, May 1983.

[13] Amin, M., "Microcomputer control of synergic pulsed MIG welding," *Metal Construction*, April 1986.

[14] Smati, Z., "Automatic pulsed MIG welding," *Metal Construction*, January 1986.

[15] Richardson, M. I., Bucknall, P. W., and Stares, I., "The Influence of Power Source Dynamics on Wire Melting Rate in Pulsed GMAW Welding," *Welding Journal*, Vol. 73(2), pp 32s-37s, 1994.

[16] Shepard, M. E., *Modeling of Self-Regulation in Gas-Metal Arc Welding*, Ph.D. Dissertation, 1991.

[17] Quinn, T. P. and Madigan, R. B., "Dynamic Model of Electrode Extension for Gas Metal Arc Welding," *International Trends in Welding Science and Technology*, ASM International, pp 1003-1008, 1993.

[18] Mao, W. and Ushio, M., "Measurement and theoretical investigation of arc sensor sensitivity in dynamic state during gas metal arc welding," *Science and Technology of Welding and Joining*, Vol. 2(5), pp 191-198, 1997.

Development of Impulsive Object Sorting Device with Air Floating

Shinichi Hirai, Masaaki Niwa, and Sadao Kawamura

Dept. of Robotics,
Ritsumeikan Univ., Kusatsu, Shiga 525-8577, Japan
E-mail: hirai@se.ritsumei.ac.jp

Abstract

A new object sorting device using impulsive manipulation and air floating will be developed. First, two methods for impulsive object sorting are evaluated using computer simulation for the design of a sorting device. Second, an object sorting system with air floating will be developed. Finally, the developed system is evaluated experimentally.

1 Introduction

Object handling has been one of fundamental operations in manufacturing. Electrical parts and mechanical parts are aligned and are provided to assembly stations. Products are sorted and are guided to an appropriate storage in mixed production. Moreover, recycling of used products requires disassembly of the products into parts and sorting of the disassembled parts. Object handling is an important operation in food industry and in agriculture as well. Various daily foods are handled in food production. Many fruits and vegetables should be sorted according to their size and weight to be packed into boxes.

Recently, fast and dexterous object handling is needed more. For example, disassembled parts in product recycling should be sorted according to their properties such as material and condition. It is necessary to detect the properties and to sort the parts according to the properties. Food industry needs fast handling of plenty of food materials so that the foods can be supplied into market quickly. From these observation, we have found that fast and dexterous handling of objects is a challenging issue in many fields. Especially, fast and dexterous sorting of various light-weighted objects is required in recycling and in food production.

Object sorting includes 1) detection of object properties, 2) determination of an object category corresponding to the sorted object, and 3) guidance of an object to a corresponding storage. In many sorting devices, the above three functions are tied into one and cannot be separated clearly. These hard-wired devices are simple and reliable but their capacity of object sorting is limited. We have to distinguish the above functions to develop faster and more dexterous sorting devices. For dexterous handling, sensors such as machine vision and weight sensor as well as control strategy connecting sensory information and the motion of a sorting device should be developed. For fast handling, we have to improve mechanical devices for object guidance.

Object guidance involves collision between an object and a sorting device. Collision among rigid objects has been analyzed theoretically and experimentally in mechanics. Models of impulsive forces during collision have been proposed [1, 2]. In robotics literature, dynamic simulation with collision has been studied and efficient algorithms for the simulation have been proposed [3, 4]. In relation to object guidance, a concept of impulsive manipulation has been proposed and its operation strategy has been derived [5]. Collision between an object and a device has been discussed experimentally [6] and a device for orienting parts using collision has been developed [7]. Use of computer simulation has been applied to variational design of parts feeders [8].

Many researches have focused on modeling of friction and collision between objects and on control strategies of handling devices. On the other hand, little attention is paid to the structure of handling devices that enable the object guidance to be fast and reliable. The motion of a handled object depends on the collision and the friction between an object and a handling device. Since precise motion control under the friction among objects is difficult due to its uncertainty, it is effective to reduce the friction between an object and a handling device so that the object motion can be controlled in a simple manner. Thus, we will design a handling device with little friction instead of introducing many sensors or complex control strategies.

In this paper, we will propose a new object sorting system with impulsive manipulation and air floating. Using air floating, we can reduce uncertainty caused by friction between an object and a guiding table. This yields simplicity in the motion control of a sorted object. Moreover, impulsive manipulation in air floating will realize fast object sorting. First, an overview of a proposed device will be briefly explained. Two methods for object sorting are then introduced. We will investigate their performance in object sorting using computer simulation for the design of a sorting device. Second, an object sorting system with air floating will be developed. Finally, the developed system is evaluated experimentally. Experimental results are compared with simulation results to examine the effectiveness of the simulation.

2 Design of Object Sorting Device using Computer Simulation

We will propose an object sorting system illustrated in Figure 1. This object sorting system consists of 1) an air floating table, 2) an object sorter, 3) a vision system, and 4) a system controller. Objects with various shapes, colors, and materials are conveyed to the sorting system from a shooter. Conveyed objects are floated on a table by air pressure so that their motion can be controlled without uncertainty caused by friction. A vision system can measure the shape of a conveyed object and its color as well as the motion of the object. The motion of an object sorter is then determined from the measurements so that the object can be guided to an appropriate area in a storage. A system controller regulates the sorter motion.

We will propose two sorting mechanisms; 1) *wall hitting method* and 2) *pin hitting method*. Wall hitting method utilizes a rotating plate perpendicular to the table top plane, as illustrated in Figure 2. The collision angle between an object and a rotating plate is determined appropriately according to the object. Consequently, objects can be sorted by controlling the rotational angle of a plate. Pin hitting method uses a pin moving in the table top plane, as shown in Figure 3. The position of collision between an object and a movable pin is adjusted according to the object. Thus, objects can be sorted by controlling the position of a pin. Let us examine possibility of these two methods and investigate the control strategy of a better method using computer simulation. It is necessary to simulate the motion of a planar rigid body with impact and friction between the rigid body and a sorter. Here we will use a physical simulation software, Working Model 2D. This software is capable of simulating the motion of rigid bodies with impact and friction among them.

Let us consider planar objects with three shapes to be sorted, as illustrated in Figure 4. The shape of an object shown in Figure 4-(a) is a square with a side 50[mm] long. The shape of an object shown in Figure 4-(b) is an equilateral triangle with a side 50[mm] long. The shape of an object shown in Figure 4-(c) is a circle with a diameter 50[mm] long. All objects are 5[mm] in thickness. The initial orientation of a square object is

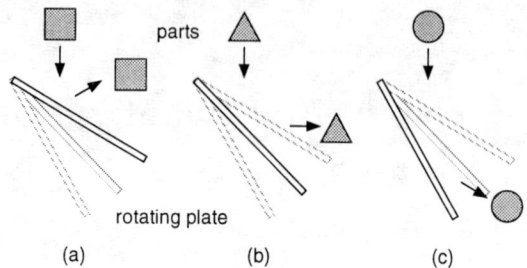

Figure 2: Wall hitting method

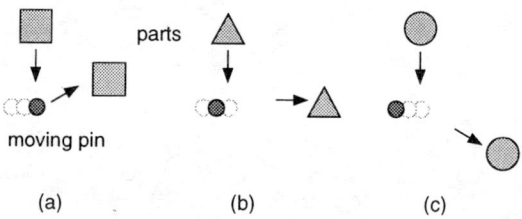

Figure 3: Pin hitting method

assumed to be 0° through 90° at intervals of 15°. The initial orientation of a triangular object is supposed to be 0° through 120° at intervals of 15°. The initial velocity velocity is assumed to be 1.0[m/sec].

Figure 5 shows a simulation result in wall hitting method. The angle between the direction of the initial object velocity and the normal of a rotating plate is equal to 30° in Figure 5-(a). The angle coincides to 45° in Figure 5-(b) and is equal to 60° in Figure 5-(c). Coefficient of restitution between an object and a plate is assumed to be 1.0 and coefficient of friction between them is assumed to be 0.0. As shown in the three figures, the trajectory of an object after collision with the rotating plate mainly depends on the angle between the initial object velocity and the normal of a plate. Dependency of the trajectory on the shape of an object and its orientation is smaller. Moreover, a set of trajectories after the collision in Figure 5-(a), that in Figure 5-(b), and that in Figure 5-(c) can be separated linearly one another. This implies that objects with these three shapes can be sorted by controlling the rotational angle of a plate according to the object shape.

Figure 6 shows a simulation result in pin hitting method. The distance between the center of an object and a pin is equal to 125[mm] in Figure 6-(a). The distance coincides to 188[mm] in Figure 6-(b) and is equal to 250[mm] in Figure 6-(c). As shown in the

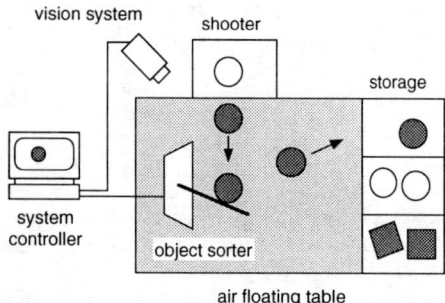

Figure 1: Overview of impulsive object sorting system with air floating

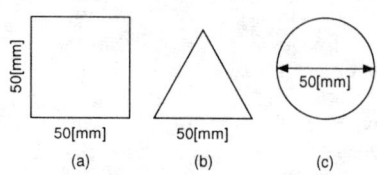

Figure 4: Three shapes of objects in simulation

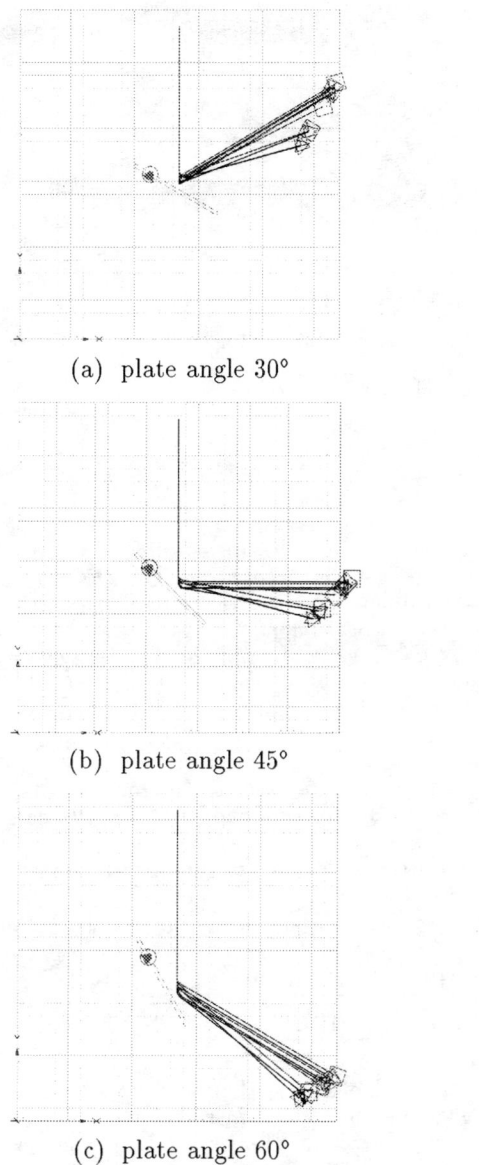

(a) plate angle 30°

(b) plate angle 45°

(c) plate angle 60°

Figure 5: Simulated trajectories in wall hitting method

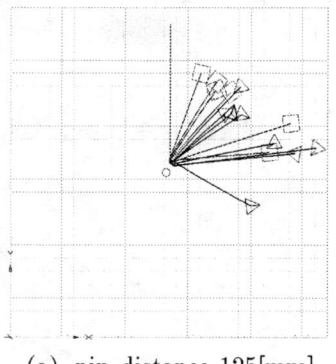

(a) pin distance 125[mm]

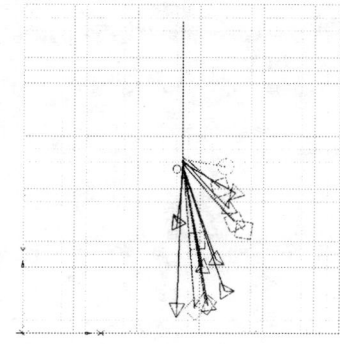

(b) pin distance 188[mm]

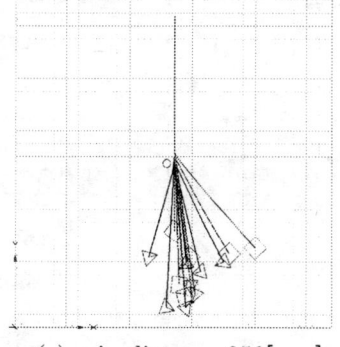

(c) pin distance 250[mm]

Figure 6: Simulated trajectories in pin hitting method

three figures, the trajectory of an object after collision with the pin strongly depends on the position of a pin as well as the shape of the object and its orientation. A set of trajectories after the collision in Figure 6-(a), that in Figure 6-(b), and that in Figure 6-(c) overlap one another. This implies that objects with these three shapes cannot be sorted by controlling the position of a pin alone. Pin hitting method requires to control the pin position according to the shape of an object as well as its orientation just before the collision. Namely, it is necessary to measure the orientation of an object just before the collision with a pin. It is difficult to measure the orientation of a moving object precisely and an additional measuring equipment may be needed.

Comparing the two methods in computer simulation, we have found that the ability of object sorting is better in the wall hitting method than in the pin hitting method. Consequently, we will select the wall hitting method as the mechanism in an object sorter.

3 Development of Object Sorting Device with Air Floating

We have developed an object sorting device, which consists of an air floating table, an air distribution system, a rotating plate, a vision system, and a system controller.

Figure 7 shows the air floating table and the air distribution system. The air floating table has 1000 ×

Figure 7: Air floating table and air distribution system

Figure 9: Rotating plate for object sorting

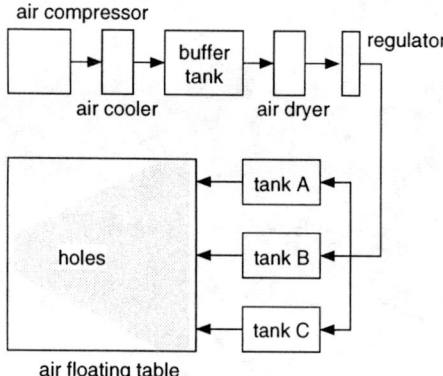

Figure 8: Diagram of air distribution system

1000[mm] area. 1543 holes of diameter 1[mm] are bored through the table at intervals of 20[mm]. Diagram of the air distribution system is illustrated in Figure 8. An air compressor, Hitachi 1.5U-7V6, produces compressed air, which is stored in a buffer tank, SMC AT11C-06, of storage 200[ℓ]. Compressed air in the buffer tank is then distributed to three tanks, Iwata SAT-33H-100, of storage 33[ℓ]. Air in the three tanks is supplied to the air floating table via three different entrances on the table. Air through three different entrances is mixed in the table and provides air flow through the holes on the table top face. This air distribution system helps to keep air pressure at individual holes uniform.

Figure 9 shows the rotating plate for object sorting through wall hitting method. The rotating plate is made of aluminum and is 240[mm] in length, 12[mm] in height, and 12[mm] in width. The plate is rotated by a DC servo motor, SANYO R506T-002, which is driven by a motor driver, SANYO PDT-A03, and a gear of reduction ratio 60. Input voltage to the motor driver is given by a PC, Gateway2000 GP6-333, through a D/A conversion board, CONTEC AS12-16. Motor angle is measured by a rotary encoder, OM-RON E6B2, and its measurement is sent to the PC through a counter board, CONTEC CNT24-4. We have implemented a PID control law to the control of the plate rotation.

A CCD camera captures successive images of an object to be sorted. Images from the CCD camera are sent to a vision board, Fujitsu Tracking Vision, which is installed on the PC. This vision board has a capacity of finding some patterns in an image captured by a CCD camera. Patterns are memorized in the board in advance. Using this capacity, the PC can determine a storage number corresponding to an object and can compute the velocity of the object. According to this information, the angle of a rotating plate is determined so that the object can be guided to an appropriate storage. The plate is rotated to this determined angle before the object collides to the plate.

4 Experimental Evaluation of Object Sorting Device

We will evaluate the capacity of the object sorting device developed in the previous section. Planar objects shown in Figure 4 are dropped from a slant shooter to the air floating table, where objects are floated by air pressure. All objects are made of acrylic. A square object, a triangular object, and a circular object are 6.8[g], 3.0[g], and 5.5[g] in weight, respectively. Moving objects collide with a rotating plate to be guided to the storage. Successive images of a moving object are captured by a CCD camera and are recorded on a VTR tape. The trajectory of the object can be plotted by analyzing the recorded images.

Let us measure the trajectories of the three objects corresponding to three plate angles, 30°, 45°, and 60°. A set of trajectories of a square object is shown in Figure 10. The angle of a plate is set to 30°, 45°, and 60° in Figure 10-(a), in Figure 10-(b), and in Figure 10-(c), respectively. Nine trajectories have been measured at each plate angle. A set of trajectories of a triangular object is shown in Figure 11 and that of a circular object is shown in Figure 12. The plate angle coincides to 30°, 45°, and 60° in Figure 11-(a), in Figure 11-(b), and in Figure 11-(c), respectively, and is equal to 30°,

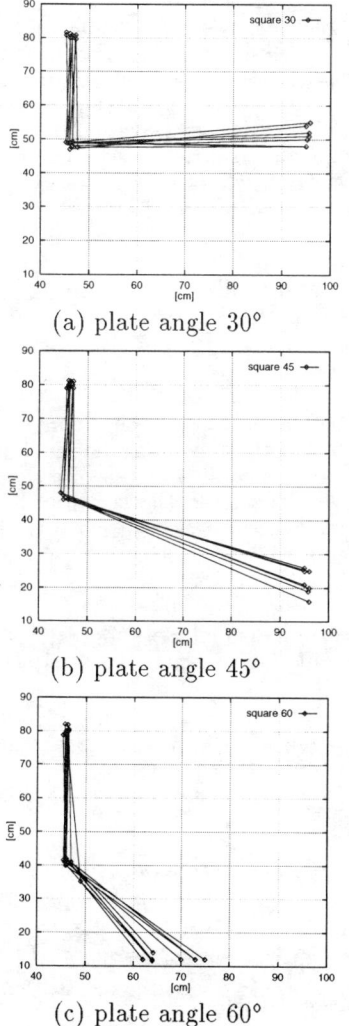

Figure 10: Measured trajectories of square object

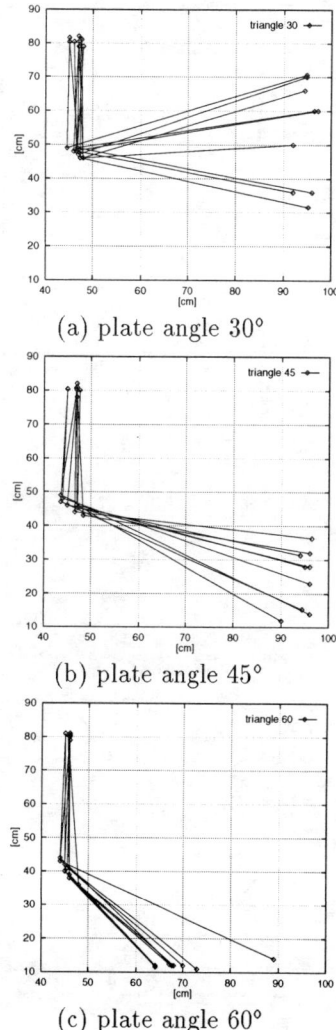

Figure 11: Measured trajectories of triangular object

45°, and 60° in Figure 12-(a), in Figure 12-(b), and in Figure 12-(c), respectively.

Comparing the above figures, we have found that a set of trajectories of a triangular object scatters more than that of a square object or that of a circular object at each plate angle. A set of trajectories of a circular object scatters least. Note that the moment acting on an object during the impact with a rotating plate strongly depends on the orientation of the object just before the impact. The trajectory of an object after the impact is affected by the moment during the impact. The moment acting on a triangular object varies most according to its orientation while the moment acting on a circular object remains almost constant despite of its orientation. This variation of the moment during the impact causes different scattering of trajectories.

Friction between an object and a table, which may disturb the object trajectory, is small due to the air floating. As shown in Figure 12, a set of trajectories of a circular object scatters little after the impact with a rotating plate. This proves that the friction between an object and a table is reduced through the air floating. Consequently, we have found that air floating is effective for precise motion control of objects to be sorted.

The three objects can be separated almost linearly. For example, a set of trajectories in Figure 10-(a), that in Figure 11-(b), and that in Figure 12-(c) are separated linearly one another except a few trajectories. This shows that the three objects can be sorted by controlling the rotational angle of a plate to 30°, 45°, or 60° corresponding to the object shape. For more precise control of object motion, it is necessary to reduce uncertainty in shooter or to measure signals including velocity of an object just before the collision and its angular velocity. Introducing a vision system will enable to measure these signals.

Comparing experimental results and simulation results, we find that the direction of trajectories of three

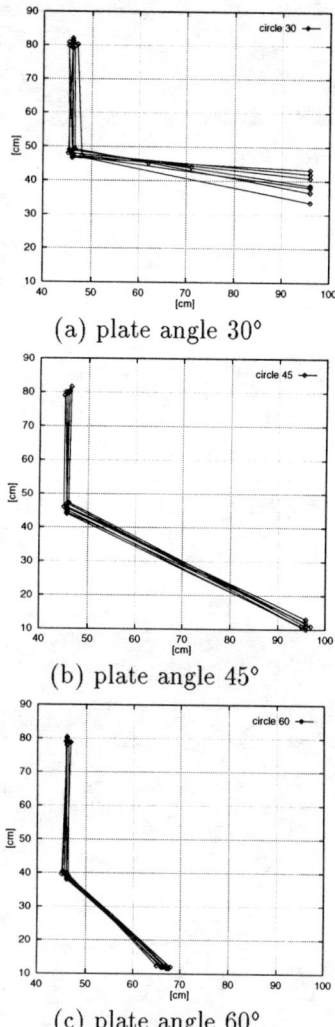

(a) plate angle 30°

(b) plate angle 45°

(c) plate angle 60°

Figure 12: Measured trajectories of circular object

objects after the collision in the experiment differs from that in the computer simulation. This discrepancy is caused by the difference of coefficient of restitution between an object and a plate as well as that of coefficient of friction between them. Despite of this discrepancy, the relative direction of trajectories of three objects in the experiment is similar to that in the computer simulation. Note that the separatability of objects is depend on the relative direction of objects rather than their absolute direction. Thus, we can conclude that the computer simulation is useful for the design of an object sorting device.

5 Concluding Remarks

An object sorting device using impulsive manipulation and air floating has been proposed for a fast and reliable sorting of objects. First, wall hitting method and pin hitting method have been proposed for object sorting. The two methods are evaluated using computer simulation and we have found that the wall hitting method shows better performance in object sorting. Second, an object sorting system with air floating was developed. Based on the computer simulation, we have implemented the wall hitting method. Finally, the developed system was evaluated experimentally. From the experimental result, we have found that the developed sorting system is capable of sorting a set of objects with three different shapes. Moreover, we have found that the computer simulation in design phase of sorting devices is useful despite of discrepancy between the experimental result and the simulation result.

Discussion in this paper has the following limitations. First, we have ignored the incident angle between an object and a rotating plate. Since we have explored a sorting strategy robust to the incident angle, the developed system does not control the incident angle. Control of the incident angle is, however, effective to the precise control of the object motion. Second, we have focused on the sorting of objects with three shapes. It is necessary to investigate a device to sort objects with more than three shapes. Third, air flow is exhausted in a few minites since the air is always distributed to all holes. Thus, we have found that future problems include 1) detection of the velocity and the angular velocity of an object just before the collision with a rotating plate for more reliable sorting, 2) investigation on the use of multiple rotating plates so that objects with more than three shapes can be sorted, and 3) analysis and redesign of the air distribution system for high efficient air flow.

References

[1] Brach, R. M., *Mechanical Impact Dynamics: Rigid Body Collisions*, John Wiley, 1991

[2] Hunt, K. H. and Crossley, F. R. E., *Coefficient of Restitution Interpreted as Damping in Vibroimpact*, ASME J. of Applied Mechanics, Vol.42, pp.440–445, 1975

[3] Joukhadar, A., Deguet, A. and Laugier, C., *A Collision Model for Rigid and Deformable Bodies*, Proc. IEEE Int. Conf. on Robotics and Automation, pp.982–988, 1998

[4] Ullrich, C. and Pai, D. K., *Contact Response Maps for Real Time Dynamic Simulation*, Proc. IEEE Int. Conf. on Robotics and Automation, pp.1950–1957, 1998

[5] Huang, W. H., Krottkov, E. P., and Mason, M. T., *Impulsive Manipulation*, Proc. IEEE Int. Conf. on Robotics and Automation, pp.120–125, 1995

[6] Huang, W. H., and Mason, M. T., *Experiments in Impulsive Manipulation*, Proc. IEEE Int. Conf. on Robotics and Automation, pp.1077–1082, 1998

[7] Salvarinov, A. and Payandeh, S., *Flexible Part Feeder: Manipulating Parts On Conveyer Belt by Active Fence*, Proc. IEEE Int. Conf. on Robotics and Automation, pp.863–868, 1998

[8] Berkowitz, D. R. and Canny, J., *Designing Parts Feeders Using Dynamic Simulation*, Proc. IEEE Int. Conf. on Robotics and Automation, pp.1127–1132, 1996

Integrated Computer Tools for Top-Down Assembly Design and Analysis

R. Mantripragada, J. D. Adams, S. H. Rhee, Dr. D. E. Whitney
Department of Mechanical Engineering
Massachusetts Institute of Technology, Cambridge, MA 02139.

Abstract

This paper describes a prototype software system that implements a top-down approach to the concept stage of assembly design and analysis. The software implements a number of concepts and techniques introduced by the authors in previous publications. Its research contributions comprise a definition of a top-down approach to assembly design, development of several new assembly analysis and synthesis algorithms, and integration with existing algorithms. New assembly analysis and synthesis tools include: a graphical layout tool for defining an assembly level dimensional control plan, mathematical models of assembly features, assembly constraint analysis, variation propagation analysis in the presence or absence of adjustments, and an optimal control algorithm for assembly feature synthesis. Existing tools include assembly sequence constraint generator and assembly sequence generator-editor. A common assembly database is created that is used by the different analysis tools. Some of the functionalities are still evolving and hence are implemented approximately to give a flavor of the potential benefits of the approach.

1. Introduction

Assembly Oriented Design (AOD) uses a number of assembly analysis tools in order to help an assembly designer plan out and analyze candidate assembly schemes prior to having detailed knowledge of the geometry of the parts. In this way, many assembly schemes can be inexpensively evaluated for their ability to deliver the important characteristics of the final product. This ability is important for a product design group to improve quality while shortening the product development time. The tools used in AOD are best applied to new product design, but can also be used to analyze an existing assembly and suggest areas for improvement.

1.1 Need

Most modern CAD systems are "part-centric" in that they are adept at helping designers create detailed geometric models of single parts on the computer screen. The datuming scheme used to create the part and features on it is normally determined by finding the scheme that allows the part to be most easily created geometrically. Often these geometric datums do not correspond to the assembly datums of the assembly into which the part fits. By datuming scheme we mean the determination of all six degrees of freedom of each part in a conclusive way, based on how the parts mate to each other. Assembly design involves coordinating assembly-level and part-level datums in a top-down way. Thus there is a need for a computer-based tool which allows designers to implement a top-down design process. Such a process would start at an assembly or final product level, define datuming schemes consistent with the important aspects of that product or assembly, and coordinate those datums down the assembly tree to the part level. In order to do this, the computer tools must support the design process with analysis that can help the designer select among the various schemes and quantitatively rank them according to defined performance criteria. The toolset and approach described here differ from those in [1], which implemented a parts-first bottom-up approach.

1.2 Desirable properties of an Integrated System

In the initial stages of design, the exact design geometry is often only approximately defined. Thus an analysis system, intended to be used during this stage of design, must be able to give valid results which are independent of detailed geometry. Assembly designers are interested in evaluating how well their chosen assembly scheme will deliver important characteristics of the product such as dimensional accuracy and performance. Assemblability is also an important issue, along with the assembly time, cost of assembly, tolerances required on the parts, assembly system requirements, and others specific to each assembly. An ideal system would thus support quantitative evaluation of all such criteria. Aside from the analysis capabilities, the system should be easy to use and integrate easily with other existing analysis programs such as CAD systems.

2. AOD Approach

In the course of this research we have defined a top-down approach to modeling and analyzing assemblies and their assembly processes. This approach is presented using a flowchart in Figure 1. The method presents several techniques to represent and analyze assemblies at a conceptual stage of design in an attempt to satisfy some of the needs outlined above. The process starts by carefully identifying the assembly requirements from the top level customer requirements down to the fabrication of individual parts using a method called Key Characteristics (KCs) [2]. KCs are intended to capture a few important characteristics of the product, differentiating them from the large number of un-prioritized tolerances that normally appear on engineering drawings. These KCs are expressed first as customer requirements and then translated to supporting engineering specifications for assemblies and parts. The KCs are then used to identify important datums on parts and subassemblies, and to define relationships between them. A detailed description of the different types of KCs can be found in [3].

Even though only sketches of parts might be available early in the design process, AOD can be used to define the interfaces between parts that will permit the KCs to be delivered. AOD accomplishes this by permitting the designer to define assembly features on parts. Assembly features are the local geometry regions on parts that assemble to like regions on other parts. These features are linked by a graph that defines a hierarchy among the parts, defining how the parts are positioned in space and which parts have responsibility for locating which other parts.

We use a technique called the Datum Flow Chain to assign this hierarchy to the datums and to define which parts locate which other parts in the assembly. A DFC is a graphical representation of the designer's strategy to locate the parts with respect to each other, which amounts to specifying the underlying structure of dimensional, constraint, and datum references on parts constituting the assembly. The theory and algorithms underlying the DFC can be found in [4, 5].

The DFC is also used to plan assembly sequences. First the complete feasible set of assembly sequences for the assembly is identified based on geometric constraints. Then the DFC is used to prune the set to a smaller set of assembly sequences based on the order of establishment of these dimensional references and the need to have only dimensionally coherent and fully constrained subassemblies. We call this smaller set a family of assembly sequences.

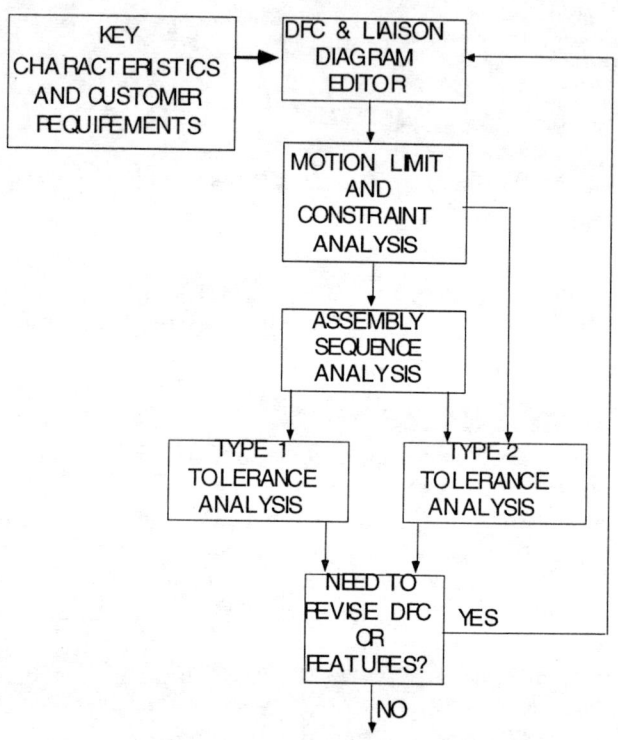

Figure 1: A Flowchart of the Assembly Oriented Design System

Candidate assembly feature sets are designed that implement the structure of dimensional and datum references imposed by a DFC. These feature sets are evaluated for complete, over-, or under-constraint.

Candidate DFCs are then evaluated and compared using criteria such as KC deliverability and tolerance analysis.

We have defined two types of assemblies, each of which require a different kind of analysis. Type 1 assemblies are typical mechanical products like engines and gearboxes. The parts for these assemblies are given their assembly features during fabrication. Assembly of Type-1 assemblies is accomplished simply by "putting together" the parts by joining the assembly features. Type-2 assemblies are typical sheet metal items like car bodies and aircraft fuselages. The parts for these assemblies do not have all of their assembly features when they arrive at the assembly line. Typically they are first assembled to fixtures which provide dimensional location and constraint. Once mated to the fixtures, these parts are fastened to each other. During this process, they are given the remainder of their assembly features in the form of drilled and riveted holes, spot welds, and so on.

Different design and analysis procedures are employed for type-1 and type-2 assemblies. For the former, the selection of mating features is determined by considering function and is not a part of the assembly planning process. For the latter, feature type and location selection is a crucial part of the assembly process design. The design of these mating features in both types of assemblies places requirements on the assembly and fabrication processes. The DFC is then used to construct tolerance chains to perform a three dimensional tolerance analysis and choose between assembly sequences and feature sets within the most promising family identified. The end result of the exercise is a location strategy reflected in the choice of dimensional datums and their hierarchy (DFC), an assembly feature set, and an assembly procedure that satisfies the DFC hierarchy.

In this view, design of the assembly process is driven directly by customer requirements and is implemented by selecting datum flow chains while only a skeleton of the assembly's logic and sketches of the parts exist. Detailed part geometry plays almost no role (except in the neighborhood of the assembly features), even though assembly sequence and tolerance analysis are performed. The following sections describe the different elements of the assembly oriented design approach in detail. This flowchart represents our current best understanding of the AOD approach. It is still at an evolving stage and as new modeling techniques are developed, pieces of this flowchart too will evolve to incorporate these techniques. Section 3 describes the several modules in the flowchart and their characteristics. A typical design session using this AOD approach is presented in Section 4.

3. Capabilities of Current Prototype System

3.1 Summary Capabilities

Using the programs in this toolkit, a designer is able to accomplish the following tasks:
1. Design a dimensional control plan consistent with the important characteristics of the product.
2. Lay out an indexing and datuming scheme that is consistent through the assembly, sub-assembly, part, and feature levels.
3. Choose assembly features to realize the constraints between parts and evaluate the performance of the chosen feature set.
4. Generate, view, and edit all possible assembly sequences according to geometrical considerations.
5. Generate, view, and edit all desirable assembly sequences that support the logic of the DFC.
6. Perform variation propagation analysis on any or all assembly sequences.
7. Rank the assembly sequences according to the ability of each to achieve the objectives of the KCs.

3.2 Synthesize an optimal set of assembly features using the DFC Editor

This is an interactive conceptual CAD tool that provides the front end of the AOD system and integrates the different modules. It has a graphical user interface that lets the designer interactively construct the DFC and liaison diagram for the assembly in the same session. Properties of the DFC and underlying theory can be found in [3]. The constructed DFC can be checked for violation of any DFC properties by performing consistency checks. These checks include detection of loops, presence of only one root node, over and under constraint, etc. The DFC editor will be linked to a program called MLA (Motion Limit Analysis), described in the next section. The DFC editor also creates an assembly database where information about the DFC, liaison diagram and feature information is stored. The DFC editor is linked to the assembly sequence analysis (ASA) module. The different modules extract the necessary information directly from the assembly database. This enables generation of both the complete set and a family of assembly sequences of the assembly and DFC under design. The graphical user interface of the DFC editor is shown in Figure 2.

3.3 Motion limit analysis (MLA)

Motion Limit Analysis (MLA) provides mathematical models of assembly features from which the ability of a feature to position one part relative to another in space can be calculated. The underlying theory can be found in [6],[12], and [13]. A user of this theory is able to obtain three major types of information about an assembly:

1. Knowledge of the directions and quantitative amounts of possible part motions of a part that is being added to an assembly at a given assembly station via connection of a defined set of assembly features.

2. Knowledge of whether or not the defined feature set over-, under-, or fully-constrains the location and orientation of the part.
3. Knowledge that the defined feature set can establish the desired location of the part within the assembly it is being added to.

The MLA software receives some input from the DFC editor and some input from the user. The DFC editor provides information about the parts and interconnections between parts in an assembly. The MLA software is then used to choose features that realize these interconnections and perform calculations about the properties of the chosen set of features. The user performs these steps:

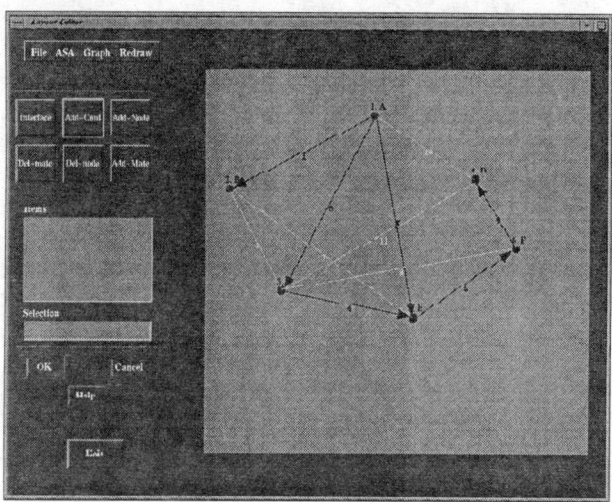

Figure 2: The interactive DFC editor. The designer uses this window to define the DFC by placing nodes and connecting them with arrows to define the logic of the datum scheme.

1. Define the location and orientation in global coordinates of all parts in the assembly under study.
2. Choose assembly features to physically realize these connections between parts that are inferred from the DFC diagram of the assembly.
3. Define the location and orientation of these assembly features on each part.
4. Specify geometric parameters defining the feature, and/or specify numerical limits on the motions that each feature will allow acting individually.

The software first analyzes all mates in the assembly to check if the mates do indeed constrain all six degrees of freedom of each part as intended. The mates are also checked to see if they over-constrain the parts. Over-constraint means that a given degree of freedom of a part is being controlled by more that one assembly feature. Next the contacts are analyzed. If the assembly feature accomplishing the contact, acting independently, would allow relative motion between the two parts it connects, the limits on this rigid body motion are calculated. These limits are provided as inputs to the Type-2 tolerance analysis routine as physical limits on positional adjustments that can be made to the parts at the assembly station where the contact corresponding to each vector of limits is made. The front end interface of this program is shown in Figure 3.

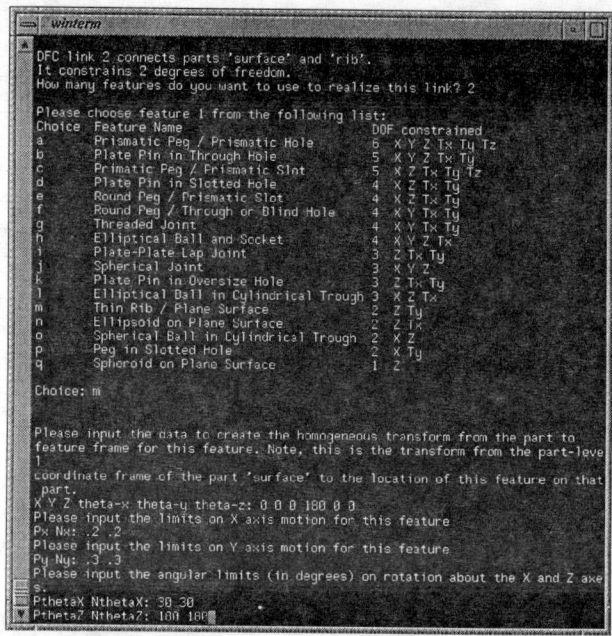

Figure 3: Graphical user interface of the MLA software. The designer uses this window to define the kind of feature at each arc in the DFC and to choose which degrees of freedom are constrained by each feture.

3.4 Assembly Sequence Analysis software (ASA)

This software refers to a suite of tools used to generate and evaluate assembly sequences. These tools can be used both as stand alone tools and integrated with the DFC editor. They are described as follows:

3.4.1 Assembly Sequence Generation

Assembly sequence generation involves first generating the assembly precedence constraints and then using the precedence constraints to generate the assembly sequences. Two types of precedence constraints are considered: geometric and DFC related.

3.4.1.1 Geometric Precedence Constraints (SPAS) [7, 8, 9]

This software generates assembly precedence relations for the given assembly from the geometric constraints imposed by the shape and size of the different parts in the assembly. The analysis used to generate these precedence relations is based on a graphical representation of the joints between parts called the "liaison diagram".

Based on this information, geometric assembly precedence constraints of the form

$$(i \ \& \ j) >= (k \ \& \ l) \ ;$$

are generated. The operator ">=" means "must precede or concur with". The above constraint is read as: liaison i and j must be completed before or concurrently with, completion of (both) liaisons k and l (but not necessarily before or concurrently with either liaison k or l). A typical session of this program is illustrated in Figure 4:

3.4.1.2 DFC Precedence Constraints (DFCPR)

A program called **DFCPR** takes the liaison diagram and the DFC under evaluation as inputs and generates additional constraints that eliminate subassemblies that are incompletely constrained. The precedence constraints generated by DFCPR are of the same form as the geometric constraints described above.

These assembly precedence constraints are then used by another software program to generate the set of assembly sequences that is represented using an assembly sequence graph [9]. If only the geometric precedence constraints are fed to the program, an assembly sequence graph representing the complete set of assembly sequences is created. On the other hand, if both geometric and DFC precedence constraints are fed to the program, then an assembly graph representing only a family of assembly sequences for the DFC is generated.

3.4.2 Assembly Sequence Evaluation (EDIT)

A software program called **EDIT** represents the generated assembly sequence graph in a graphics window where the user can interactively query, inspect, evaluate and delete assembly states and transitions based on several criteria. This software gives the designer the ability to visualize the available assembly sequences . Editing based on conditions such as: deletion of moves where a particular set of liaisons are made, specification that a particular move must immediately precede another, subassemblies hard to assemble due to accessibility problems, etc. can be applied. These editing techniques quickly reduce the number of sequences to a handful that can be subject to more detailed analysis. More details about the capabilities and functions of the EDIT program can be found in [8]. A sample session is illustrated as follows in Figure 5.

3.5 Control System Analyzer

The algorithms described in [10] have been implemented as computer programs and can be used to evaluate assembly sequences and mating feature combinations.

Two programs called TYPE1.tol and TYPE2.tol have been developed to study variation propagation in type-1

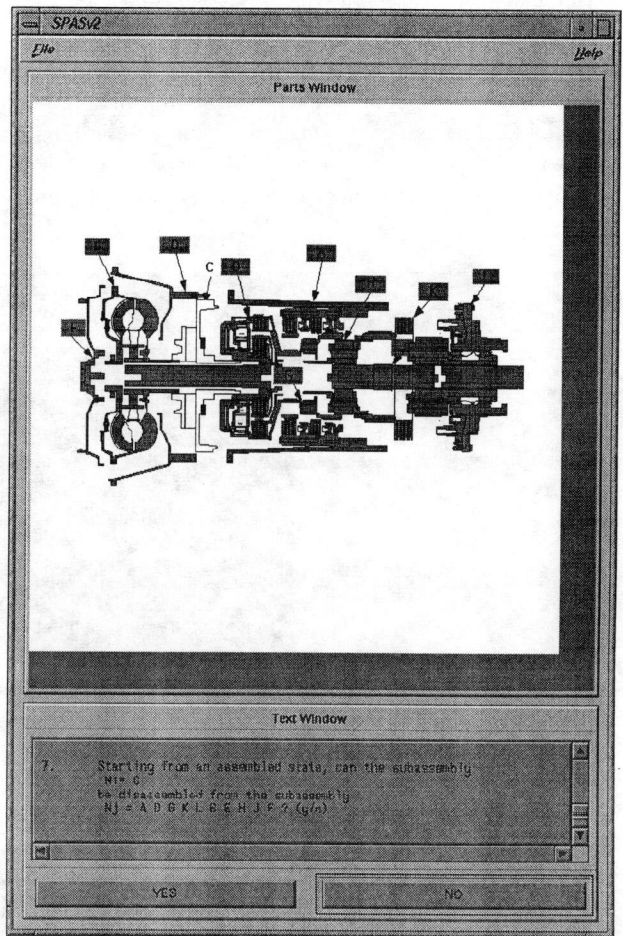

Figure 4: Graphical user interface of the SPAS software for generating geometric assembly precedence constraints. The designer uses this interface to answer precedence questions, using a diagram of the product as a guide.

and type-2 assemblies and are described in sections 3.5.1 and 3.5.2. Interactive software implementation of the optimal controller design is described in 3.5.3. Currently these tools are developed as stand alone tools and require manual input to use them. In the future we foresee these tools being fully integrated within the AOD framework so that they can be used automatically by the DFC editor to analyze DFCs and assembly sequences.

3.5.1 Type-1 tolerance analysis module (TYPE1.tol)

TYPE1.tol is used to study variation propagation in type-1 assemblies. The input to the software is a tolerance chain for the KC that we wish to analyze. The inputs are currently provided interactively by the user and include nominal and variant transforms between coordinate frames on mating features on parts. The input and output information for this software are identical to that for TOLA which is described in [11]. We envision that the input information would eventually be extracted automatically from an assembly database for the design. The assembly database would have this information when the DFC and associated mating features describing the liaisons are designed. Currently the input window that takes this information interactively as shown in Figure 6.

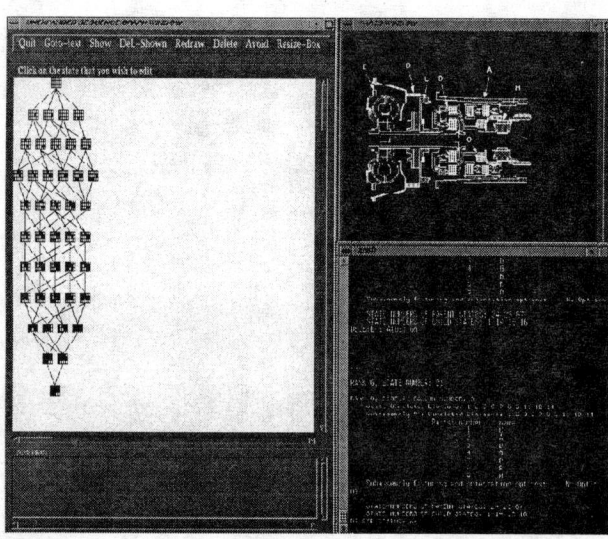

Figure 5: An interactive session with the EDIT software for evaluating assembly sequences. The designer uses this interface to edit the assembly sequence network, which appears in the left window.

From the input information provided, the software computes the $A(k)$ and $F(k)$ matrices and computes the variation associated with the position and orientation of Nth frame with respect to the base frame. Covariance matrices for the position (V_p) and orientation vector (V_θ) are constructed and their Eigen values are computed to determine the resulting variation distribution. The Eigen values of the V_p covariance matrix are variances in the principal directions of the probability density ellipsoid. Standard deviations in each of the three principal directions are the square roots of the variances. The full lengths of these ellipsoids are equal to six times the standard deviation in the principal directions.

3.5.2 Type-2 tolerance analysis module (TYPE2.tol)

TYPE2.tol is a computer program that performs variation propagation analysis in the presence of a set of adjustments. It codes the algorithms described for type-2 assemblies in [10]. The inputs to this program are identical to those for TYPE1.tol except that the 6×1 adjustment vectors are also provided for each station by MLA. In places where adjustments are not possible a zero vector is input. The output from this program is identical to that of TYPE1.tol and is a description of the position and orientation error associated with the Nth frame.

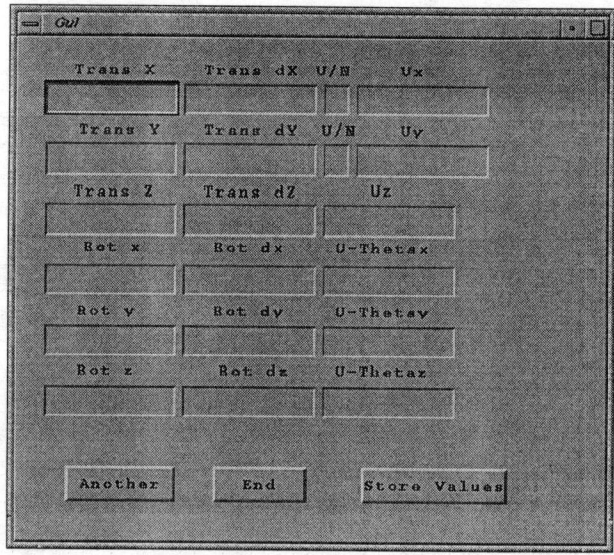

Figure 6: Graphical user interface of the input window that interactively takes information for every coordinate frame and provides it to TYPE1.tol

3.5.3 Optimal Controller design module

The computer program to generate an optimal set of contact features for type-2 assemblies is developed as a part of TYPE2.tol. The inputs to this module are the same as that for TYPE2.tol except for the fact that no adjustment vectors are provided as inputs In addition the user interactively provides the weighting matrices to define the optimization function. The significance of these matrices is described in [10]. Based on the inputs and weighting matrices, the program computes an optimal set of adjustment vectors which are to be interpreted by the user to design contact features. The program also determines the resulting variation distribution associated with the Nth frame given this set of optimal adjustments. An interactive session for the program shown in Figure 7.

4. Design Session

In this section we present a description of how we envision a designer would use AOD system. A step through the flow chart presented in Figure 1 is presented below. Many data flows that would be seamless in a commercial system are manual at present.

1. The first step in the design process is the construction of the assembly's liaison diagram and DFC. We assume that identification and classification of KCs is provided to us as input. Based on the KCs for the assembly, the designer interactively constructs a DFC and liaison diagram for the assembly using the DFC editor. The DFC editor creates an assembly database and stores information about the DFC and the liaison diagram for the assembly.

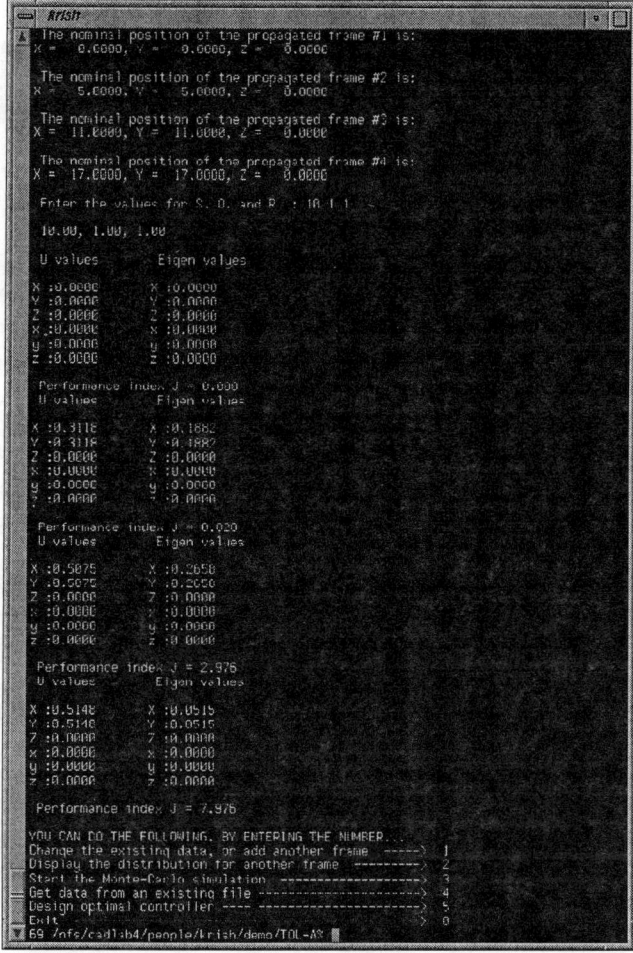

Figure 7: Example interactive session with the control theory analyzer. The designer uses this to provide information about mates and contacts and relative costs of different kinds of errors.

2. The next step is to design the mating features for the mates and contacts in the assembly. This is done using the MLA software. MLA reads information about the parts and fixtures, connectivity between parts, and DOFs constrained by mates directly from the assembly database. Using MLA the designer defines the features that constitute the mates and contacts for the assembly. Different kinds of analyses such as determination of over and under constraint conditions and motion limit analysis for contact features are performed using the MLA program. Motion limit analysis determines the absorption zone for every contact feature set in the assembly and stores them in the assembly database. For type-2 assemblies, these are the adjustment vectors used by the TYPE2.tol to perform variation propagation analysis in the presence of adjustments.

3. The next step involves planning assembly sequences for the assembly.

4. The next step is to evaluate the family of assembly sequences. As described in [3], the family can be evaluated using several criteria. The one of prime importance is variation propagation. For type-1 assemblies, TYPE1.tol is used to study variation propagation to address assemblability type problems. In type-1 assemblies all assembly sequences in a family of assembly sequences yield identical results in variation propagation analysis. Hence only one assembly sequence from the family need be analyzed to evaluate a complete family of sequences.

5. For type-2 assemblies, TYPE2.tol is used to perform variation propagation in the presence of adjustments. Here different sequences within a family need to be evaluated with respect to each other for any given set of contact features.

6. For type-1 assemblies, based on the results obtained from TYPE1.tol, the designer may choose to redesign the elements of the DFC or tolerances on mating features. In the case of type-2 assemblies, the designer can choose to redesign the DFC, redistribute tolerances on mating features or select an alternate set of contact features. The designer may also choose to run the optimal controller design module to design an optimal set of contact features that will repeatedly deliver the KCs.

7. The resulting family of assembly sequences and feature sets can then be evaluated for other assembly planning criteria such as desirable subassembly states, desirable assembly transitions, existence of multiple subassemblies, sequence of establishment of liaisons, etc using the EDIT software.

8. The above process can be iterative and is performed until the designer is satisfied with the assembly design. At the end of the whole process, the designer will be left with a DFC for the assembly, a set of assembly features, a desirable assembly sequence and a measure of the probability of the design and assembly process delivering the KCs repeatedly.

9. For assemblies involving multiple assembly stations, a DFC is constructed for each assembly station. In such cases, the above seven steps are performed for each assembly station. The entire assembly process is thus modeled using a series of clusters of assembly operations.

5. Conclusion and Future Work

5.1 Distributed design environment

With product development activity becoming increasingly global these days, there is a general shift towards network based design tools to permit dispersed teams to work together. We foresee that CAD tools in the future will be a lot more web based and network centric. Some of the applications described above have been developed to run under a distributed type of environment. The ASA software has been developed to use a WWW browser such as Netscape or Internet Explorer as a front end and can be used by remote users to analyze assemblies. This work was done as a part of the ACORN (Advanced Collaborative Open Resource Network) project.

In future, we shall see the applications having a lot more web based content to make the Assembly Oriented Design environment a true distributed design environment so that dispersed teams can be involved in different aspects of the assembly design.

6. Acknowledgements

The authors wish to thank the following:
- The USAF Wright Laboratory Manufacturing Technology Directorate, administering ARPA Contracts F33615-94-C-4428 and F33615-94-C-4429
- National Science Foundation Grant DMI-9610163
- Boeing Commercial Airplane Group
- The Ford Motor Company
- Vought Aircraft Division of Northrop-Grumman
- The Charles Stark Draper Laboratory, Inc., for use of the assembly sequence analysis software.

7. References

[1] T. L. De Fazio et al, "A Prototype of Feature-based Design for Assembly," ASME J. Mech. Des., v 115 no 4, Dec, 1993, pp 723-34

[2] D. J. Lee and A. C. Thornton, "Key Characteristics for Agile Product Development and Manufacturing," presented at Agility Forum 4th Annual Conference Proceedings, 1995.

[3] T. Cunningham, R. Mantripragada, D. Lee, A. Thornton, and D. Whitney, "Definition, Analysis, and Planning of a Flexible Assembly Process," presented at Japan/USA Symposium on Flexible Automation, Boston, USA., 1996.

[4] R. Mantripragada and D. E. Whitney, "Datum Flow Chain: A Systematic Approach to Assembly Design and Modeling," Research in Engineering Design, Oct, 1998.

[5] R. Mantripragada, "Assembly Oriented Design: Concepts, Algorithms and Computational Tools," in Ph.D. Thesis, Mechanical Engineering. Cambridge: MIT, 1998.

[6] J. D. Adams, in M.S. Thesis, Mechanical Engineering. Cambridge: MIT, 1998.

[7] T. L. DeFazio and D. E. Whitney, "Simplified generation of all mechanical assembly sequences," IEEE Journal of Robotics and Automation, vol. RA-3, pp. 640-658, 1987.

[8] D. F. Baldwin, "Algorithmic methods and software tools for the generation of mechanical assembly sequences," in S. M. Thesis, Mechanical Engineering. Cambridge: MIT, 1989.

[9] D. F. Baldwin, T. E. Abell, M.-C. M. Lui, T. D. Fazio, and D. Whitney, "An Integrated Computer Aid for Generating and Evaluating Assembly Sequences for Mechanical Products," IEEE Transactions on Robotics and Automation, vol. 7, pp. 78-94, 1991.

[10] R. Mantripragada and D. E. Whitney, "Modeling and Controlling Variation Propagation in Mechanical Assemblies using State Transition Models," Accepted for IEEE Trans. on Robotics and Automation, 1997.

[11] O. L. Gilbert, "Representation of Geometric Variations using Matrix Transforms for Statistical Tolerance Analysis in Assemblies," in S. M. Thesis, Mechanical Engineering. Cambridge, MA: MIT, 1992.

[12] R. Konkar, "Incremental Kinematic Analysis and Symbolic Synthesis of Mechanisms," Ph.D. Dissertation, Stanford University, Stanford, CA 94305, USA, 1993.

[13] R. Konkar and M. Cutkosky, "Incremental Kinematic Analysis of Mechanisms," Journal of Mechanical Design, Vol. 117, December 1995, pp. 589-596.

The Development of a Rapid Prototyping Machine System for Manufacturing Automation

Ren C. Luo and Wei Zen Lee

Intelligent Automation Laboratory

Department of Electrical Engineering, National Chung Cheng University

160 Shang-Shing, Ming-Hsiung, Chia-Yi, Taiwan 621, R.O.C

Email: luo@ia.ee.ccu.edu.tw , wlee@ia.ee.ccu.edu.tw

http://www.ia.ee.ccu.edu.tw

Abstract

A rapid prototyping machine system which combines three axes PC-based controller with thermal extrusion method is presented. The main features of this system are to build up a physical object model through the extrusion of thin thermal plastic material and built-up layer by layer. User can use this system to turn CAD (Computer Aided Design) model into physical object model. Hardware design includes three axes working table, a thermal extrusion head, and a PC-based control system.

The rapid prototyping system software includes CAD-model Slicing Process and 2D-layer Cross Sectional Fill Generation Algorithms. The process begins with the design of a conceptual geometric CAD model of an object. The cross sectional STL format data is interfaced to the rapid prototyping machine system. The machine has three axes working table and move relative to each other along XYZ axes in a predetermined pattern to create three-dimensional model by extruding ABS (Acrylonitrile-Butadiene-Styrenes) material. The system can extend to use other plastic materials from the thermal extrusion head and build up model layer by layer.

1. Introduction

Since the advent of computer-assisted design (CAD) tools, manufacturers have taken significant steps toward utilizing these tools for new product design and development. Traditionally, engineers have created three-dimensional models and prototypes using conventional fabrication methods, such as CNC based machining. This has limited the flexibility and the time to market for the products. In recent years, design Rapid Prototyping (RP) has been developed to improve the new product design and development to accelerate time for new product to the market significantly. The manufacturers can use rapid prototyping system to develop a model or prototype within minutes or hours from a CAD design. Often, engineers can model much more complex geometries with an RP system than they can produce using conventional methods.[1]

ABS materials provides the impact resistance, toughness, heat stability, chemical resistance and rigidity to perform functional tests on sample parts. So far, only STRATASYS Company's RP system - Fused Deposition Modeling (FDM) offered ABS as a modeling material. Material filament of ABS is pulled into the FDM head by the drive wheels. It enters the liquefier where it is heated until the material reaches a near-liquid state[2][3][4][5]. The filament material requires additional process and costs. We have developed a thermal extrusion based Rapid Prototyping system which uses ABS powder as object building material.

The system we have developed includes three independent axes platform, a thermal extrusion head, a temperature controller, and a PC-based control system. Figure 1 presents the flow diagram for the process and Figure 2 illustrates the configuration of the system. The thermal extrusion mechanism is mounted on top of workbench and is driven in X-Y direction horizontally to build the object model layer by layer. The description of the design for the RP machine is in Section 2 including the hardware design of the extrusion head. The structure of the software algorithms for slicing 3D CAD model and data transfer to drive the RP machine is depicted in Section 3. Section 4 presents the experimental results and future applications for the RP machine.

2. Overall System Architecture

A 3D object model is designed on a CAD system. The model represents the physical part to be built. And it must be represented as closed surfaces. The solid or surface model to be built is then converted into layer slicing format called STL file format. Basically, the STL file approximates the surfaces of the 3D object model by triangular polygons. The layer slicing algorithm and driving software serve as a bridge between STL files and the rapid prototyping machine.

During the model building process, physical model be stabilized during layer fabrication to prevent the distortion of unsupported areas. Therefore, the fabrication processes require that a separate support structure be designed, added to the model geometry file, and then fabricated as a part of the physical model concurrently.

To start the model building process the three dimensional representation of the object is converted to a series of thin cross sectional layers by intersecting a series of planes with the STL file representation in a process called "Slicing". Each two dimensional cross section is a set of closed areas that must be filled with a solid material. A physical model is formed by sequentially building and stacking the thin two dimensional layers to form a complete three dimensional object.

The cross sections data are the processed with the 2D-layer Fill Generation Algorithm to generate working path of the thermal extrusion head. Attributes can be assigned to each slice curve, such as path width and fill patterns. A 3 axes motion control program of rapid prototyping system will read working path file and control the motion of rapid prototyping machine via the PC-based control system.

In order to mechanically build-up each successive layer, three AC servo motors are provided to selectively move the base platform and the thermal extrusion head relative to each other in a predetermined pattern along "X" and "Y" axes while the model building material is being thermally extruded. Relative vertical movement along "Z" axis is carried out after the completion of each layer to achieve desired layer shape and thickness.

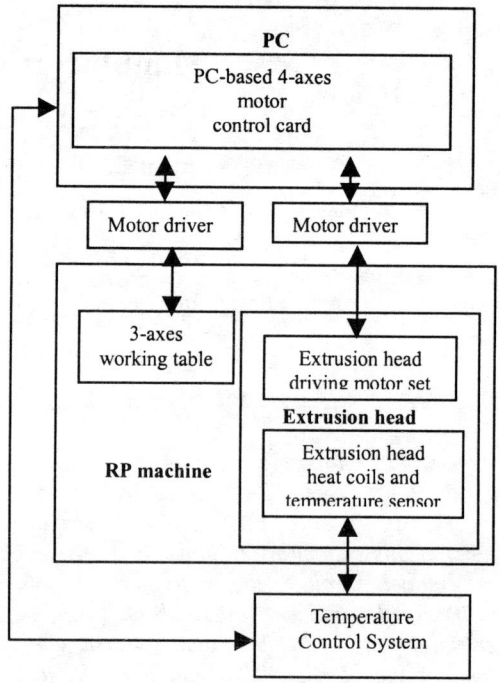

Figure 1. Thermal Extrusion Based Rapid Prototyping System Architecture.

3. Design of the Rapid Prototyping System

3.1 Rapid Prototyping Machine Structure

The structure of the Rapid Prototyping machine structure is illustrated in Figure 2. The hardware configuration of this system includes three independent axes working table, a thermal extrusion head, a temperature control system, and a PC-based control system. The high precision ball screw equipped X, Y, and Z-axes with 0.001 inch accuracy are driven through AC servo motor.

We have used a motion control card to control the 3 axes motion of the working table. It consists of an intelligent motherboard populated with four intelligent plug-in function modules. There are on-board linear interpolation, circular contouring, continuous path motion functions and dual-ported memory for high-speed communications in this card.

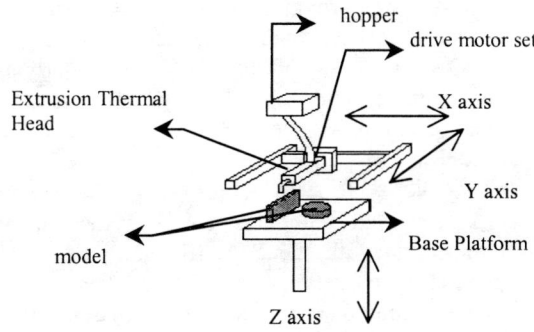

Figure 2. Configuration of the thermal extrusion based Rapid Prototyping Machine

3.2 Thermal Extrusion Head

The structure of the thermal extrusion head is illustrated in Figure 3. The structure includes a extrusion screw, a barrel, a hopper, a temperature sensor, a digital temperature controller, and a drive motor set.

After the plastics move from the barrel to form a thin film of melted plastic on the barrel surface, the relative motion of the barrel and screw drags the melt, which is picked up by the leading edge of the advancing flight of the screw. This edge flushes the polymer down in front of it, forming a circulating pool. The extrusion head is first conducted from the barrel through the film of plastic attached to it. Heat then enters the plastic by shearing action, the shear energy being deriven from the screw turning. The width of the melted polymer increases as the width of the solid bed decreases. Melting is completed at the point where the width of the solid bed is zero.

An extrusion screw as shown in Figure 4 which consists of three sections: feed, melting, and metering. The feed section, which is generally fed by gravity from the hopper into this section. Heating is accomplished by conduction and mechanical friction. The melting section is the area where the softened material is transformed into a continuous melt and compressed to avoid bubbles in the melt. In the metering section, the material is smeared and sheared to give the melt a uniform composition and a uniform temperature for delivery to the tip.

The following formula is a simplified approach to the output of the metering section.

$$R = 2.3D^2 hgN$$

Where R = output rate in pounds per hour
   D = extrusion screw diameter in inches
   h = depth of the metering section in inches
   g = specific gravity of the resin
   N = extrusion screw rpm

By the extrusion screw turning speed, we can predict the output rate of the screw and determine the velocity of thermal extrusion head motion.

A temperature controller which is responsive to the temperature sensor on the extrusion mold head is to closely control the temperature of the barrel and the material. The heater coils wrapped around the extrusion mold head is responsive to the temperature controller which provides the material temperature being controlled in a fluid state slightly above its solidification temperature.

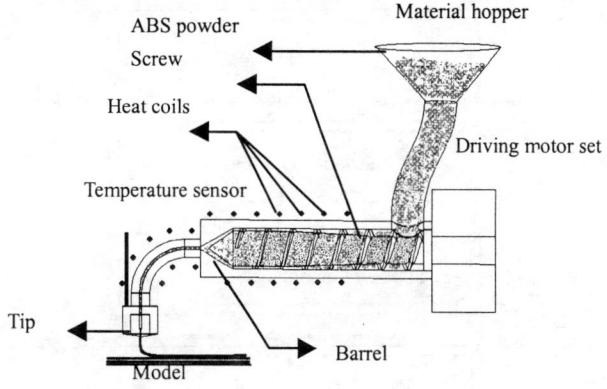

Figure 3. Extrusion Head of Rapid Prototyping System

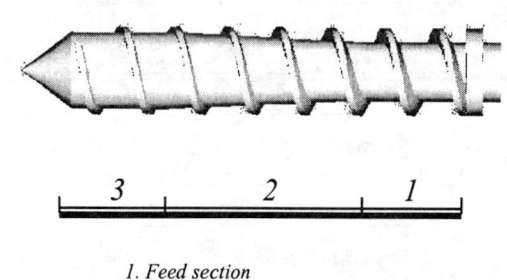

1. Feed section
2. Melting (transition) section
3. Metering section

Figure 4. The Design of the Extrusion Screw

3.3 Software Design

Based on the hardware configuration, the rapid prototyping software includes Model Support Generation, Slicing Process, Cross Section Fill Generation, and Machine Codes Translation. The translation procedure between 3D CAD model and model manufacturing is shown in Figure 5.

The model is built by first creating a 3D CAD model using CAD software like PRO/ENGINEER, AUTOCAD, ARIES, IDEAS etc. The CAD model is converted to the StereoLithography (STL) format acceptably by the model slicing process. This file consists of a patch work of facets, representing the outside skin of the object. Facets are created using a process known as tessellation or triangulation which generates triangles to approximate the CAD model.

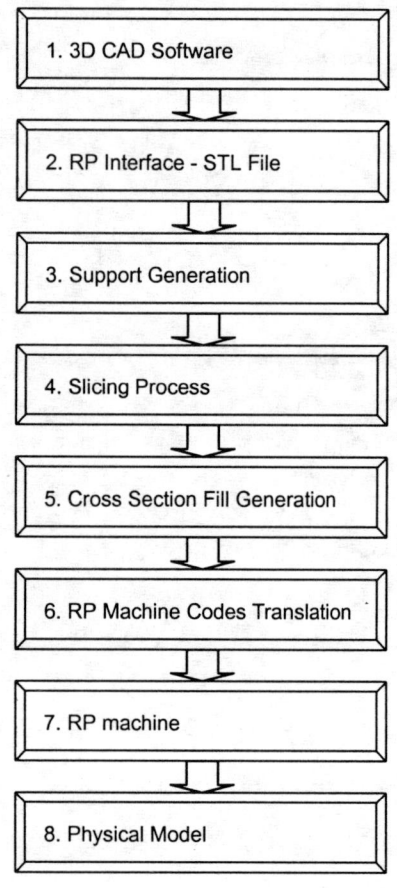

Figure 5. The Data flow between 3D CAD models and models manufacturing.

Un-supported features are considered to be model structural features which may fail or collapse because of the lack of underlying supporting structure. Model Support Generation can generate the support needed to build the model.

A Slicing Process Algorithm can generate cross-sectional boundaries based on the a plane intersection with the STL file object. From a geometric modeling perspective, the slicing process should result in the formation of a boundary representing the wall of the object at that layer.

The Slicing Algorithm works as follows. The layers are defined as the geometric intersections of planes horizontal to the platform of rapid prototyping machine with object. As the plane moves up from the base of the object, successive layers are defined and the thermal extrusion head traces out a contour defining the cross section on each layer. Basic analytical geometry is used to determine the intersections, which are then stored in a dynamic array. Each facet that intersects the plane will form a directed line segment on the planar slice. These intersection lines will define the contour. The intersections are not found in any particular order, so the vectors must now be sorted in head-to-rail fashion in order to derive a closed contour on that slice.

The following is an outline of the plane-object intersection approach:

procedure plane_obj_intersection
for *plane at each incremental z level from bottom to top* **do**
  begin
  segment array = 0;
  find intersections of plane with object_facets;
  update segment-array;
  sort segments in head-to-tail fashion
    while *there are possible starting vectors* **do**
    for *an arbitrary starting vector* **do**
      begin
      contour_array(i) = 0;
      find all line_segments forming a closed contour;
      update contour_array(i);
      delete all used segments from segment_array;
      i++
    end;
end;
end;

Cross Section Fill Generation Algorithm makes up working path of thermal extrusion head, which can fill the cross-sectional boundaries. We can see an example illustrated in Figure 6.

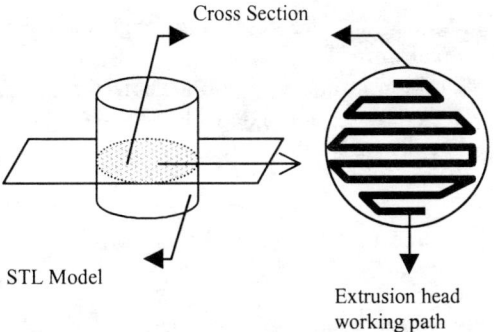

Figure 6. The Slicing and Fill Generation Process.

The Slicing Process Algorithm and the Cross Section Fill Generation Algorithm are described as follows. Figure 7 illustrates the parameters for the material extrusion process. Layer thickness L is controlled by the distance between the tip of the orifice and the extrusion surface. Width of extrusion material W is controlled by the translation speed of the tip and material flow rate through the tip.

The motion control consists of Applications Programming Interface (API) which provides a high level function call interface to PC-based motion control card. The API can be programmed in C or C++. We have used a program with API to read the file, which includes working path and extrusion of thermal extrusion head, and control the rapid prototyping machine by this motion control card.

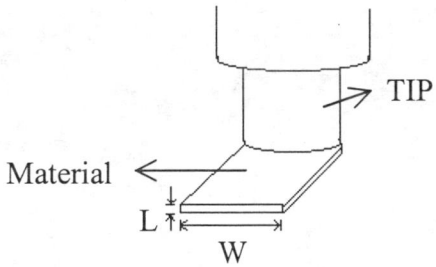

Figure 7. Illustration of Material Extrusion.

4. Applications:

We can build physical RP model with a variety of thermal plastic materials by the RP machine, such as PVC, PE, PP, PS etc. First we must examine the property of the materials, which include softening temperature, specific gravity, and melt flow index etc. Then we can set the parameters of the RP machine based on the material properties. These parameters include thermal extrusion head temperature, thermal extrusion screw speed, and the motion velocity of 3-axes working platform. After these physical models are built up, we can examine some typical properties of the object model, which include tensile strength, tensile elongation, flexural modulus, and Rockwell strength etc. Based on these test data we can find optimal material properties to build an optimal physical model as required.

5. Experimental Results and Concluding Remarks:

Rapid Prototyping System has extended to a variety of applications. We have developed an architecture in Rapid Prototyping system using thermal extrusion method.

The Rapid Prototyping machine has been built and tested. Some sample models are shown in Figure 8 and in Figure 9. The performance of this RP machine is satisfactory however, continuing improvement in accuracy and build speed is still in progress.

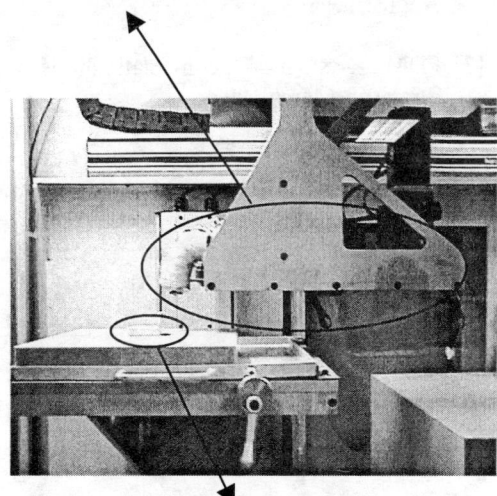

Figure 8. The Thermal Extrusion Head and the Physical Model built up by the RP Machine.

Figure 9. The Physical Model have been built by the RP Machine.

Acknowledgement

The authors thanks for the support of this research by National Science Council, Taiwan, R.O.C, under the reference number, NSC 86-2622-E-194-008R.

Reference:

[1] Chua Chee Kai, "Three-dimensional rapid prototyping technologies and key development areas", IEE Computing & Control Engineering Journal, pp. 200~206, August 1994.

[2] FDM System Documentation Manual, Stratasys, March 1997.

[3] U.S. Patent 5402351, Model generation system having closed-loop extrusion nozzle positioning, International Business Machines Corporation, Armonk, N.Y, 1995.

[4] U.S. Patent 5121329, Apparatus and method for creating three-dimensional objects, Stratasys, Inc., Minneapolis, MN, 1992.

[5] U.S. Patent 5503785, Process of support removal for fused deposition, Stratasys, Inc., Eden Prairie, MN, 1996.

[6] Injection molding handbook second edition, Donald V. Rosato, New York, 1995.

[7] Dharmaraja S. Rajan, Ren C. Luo, "CAD model slicing and surface smoothing for building rapid prototyping parts", IECON'96 vol.3, pp.1496~1501,Taipei, Taiwan, 1996.

[8] YaMei Ma, "Representation and Slicing of Freefrom 3-D Geometric Models for Desktop Prototyping and Manufacturing", A dissertation submitted to the Graduate Faculty of North California State University, 1995.

[9] Jerome L. Johnson, *Principles of Computer Automated Fabrication,* Palatino Press Inc, 1994.

Qualification of Standard Industrial Robots for Micro-Assembly

Michael Höhn *
Institute for Machine Tools
and Industrial Management
Professor G. Reinhart
Technische Universität München

Christian Robl [†]
Laboratory for Process Control
and Real-time systems
Professor G. Färber
Technische Universität München

Abstract

The precision and micro assembly requires special tools, in order to fulfill the high demands in accuracy and control of the assembly process. A precise gripper tool, the micro positioning system (MPS) has been developed, that makes micro assembly with industrial robots possible. The whole assembly process can be surveyed and controlled with a vision system. A compact micro manipulator based on piezo actuators makes visual servoing with a high accuracy possible. An integrated 3D-accelerometer system senses vibrations of the handling device and disturbances of the environment, that can be dynamically compensated by the piezo actuators. The MPS can be adapted to different assembly applications due to changeable grippers. The combination of the big workspace of an industrial standard robot and the high precision of the MPS results in cost advantages comparing to precision handling devices. A sample application, the assembly of radio-controlled wrist-watches, shows the performance potential of the MPS.

1 Introduction

Micro-technology enables new products with a performance, that has not been reached yet (see Figure 1). In electronics there is a trend towards high integrating direct chip bonding. Micro optical components are becoming more important in growing markets like communication networks. But also the miniaturization of consumer products with mechanical units has become more and more an argument for saling. In order to achieve a positive market trend, an efficient and cost saving production of hybrid microsystems in addition to the development of suitable design methods is necessary. Especially the assembly of miniaturized components causes production technology problems

*e-mail: hn@iwb.mw.tu-muenchen.de
[†]e-mail: robl@lpr.ei.tum.de

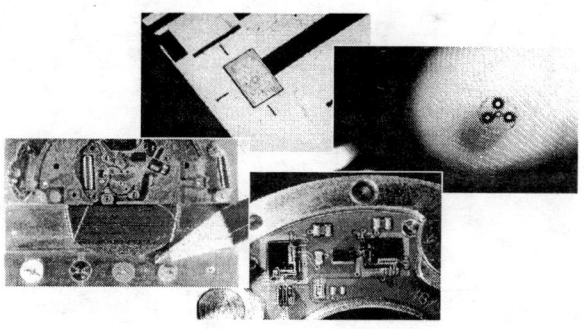

Figure 1: micro mechanical, micro optical and micro electronical products

for many manufacturers. In this application field there are specialized tools necessary to fulfill the strong requirements in accuracy and control of the assembly process.

In section 2 we discuss different positioning strategies and present in section 3 a suitable tool concept for micro assembly. In section 4 we show the necessary control system and show experimental results of the realized MPS in section 5. A sample application is briefly described in the last section.

2 Vision based positioning strategies

The required assembly accuracy in precision and micro assembly is today about $20\mu m$ and will be about $5\mu m$ in the near future. Positioning systems for precision assembly must be able to sense and compensate the positioning error referring to the target position, that results from tolerances of the components and the handling device. This problem can nowadays only be solved by combining several measurement systems with an expensive precision robotic system.

The integration of image processing into the assembly process allows to sense the positional offset between the component to be joined and the target position. Conventional positioning systems locate first the position of the object being handled with a fixed camera and then the target position with a second camera, that is attached on the manipulator [6]. The component's position and the target position can be determined within the view of the referring camera. The results are transformed into the tool coordinates in order to compute the resulting position displacement, that will then be corrected with a movement of the handling device (see Figure 2). The positional off-

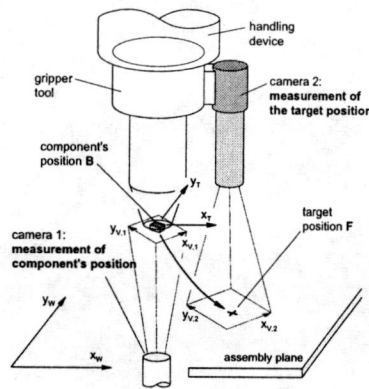

Figure 2: conventional positioning strategy

set between the component being handled (B) and the target position (F) can be written in tool coordinates as

$$\Delta c_T = B_T - F_T \quad (1)$$

The transform of the vision coordinates into the tool coordinates can be written as

$$B_T = (^V M_T)_{cam1}(B_V) \quad (2)$$

$$F_T = (^V M_T)_{cam2}(F_V) \quad (3)$$

with $(^V M_T)_{cam1}$ and $(^V M_T)_{cam2}$ as transformation matrices for the transforms of the component's coordinates measured with camera 1 and the target's coordinates measured with camera 2, respectively.

The positional offset in tool coordinates, in which the corrective movement is processed, can with equ.1, 2, 3 be written as

$$\Delta c_T = (^V M_T)_{cam1}(B_V) - (^V M_T)_{cam2}(F_V) \quad (4)$$

The assembly accuracy in this case is limited by the errors of the position measurement and the position correction: The errors of the position measurement consists of the inaccuracies of the sensors and the calibration errors of the coordinate transform with the matrices $(^V M_T)_{cam1}$ and $(^V M_T)_{cam2}$. The error of the position correction depends on the absolute positioning accuracy of the corrective movement. The spatial offset between joining axis and optical axis results in a separation of measurement position and joining position. Therefore the real position after the corrective movement cannot be determined by the vision system. The position correction is done once by an open-loop movement and depends on the accuracy of the handling device. According to that this positioning strategy requires an expensive handling device with a high position accuracy within a big workspace.

A better assembly accuracy is reached with a centric integration of the camera in the gripper tool (see Figure 3). If the joining axis is within the field of view

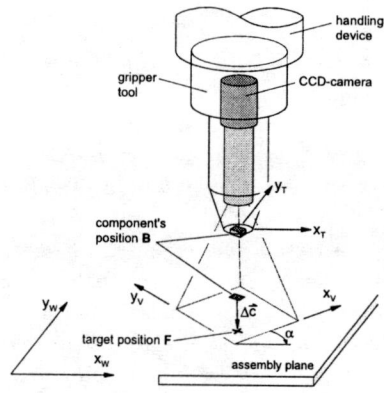

Figure 3: measuring positional offset with a centric integrated camera

of the camera, then it is possible to see both, the component being handled and the target position in one image. The positional offset can then be determined with

$$\Delta c_T = B_T - F_T \quad (5)$$

Both, the position of the component being handled and the target position are measured in vision coordinates as

$$B_T = (^V M_T)(B_V) \quad (6)$$

$$F_T = (^V M_T)(F_V) \quad (7)$$

with $(^V M_T)$ is the transformation matrix vision coordinates into tool coordinates. Due to the corrective movement the positional offset between B and F can be compensated. This can be written as

$$\Delta c_T = B_T - F_T = (^V M_T)(B_V - F_V) \stackrel{!}{=} 0$$

$$\Rightarrow (B_V - F_V) \stackrel{!}{=} 0 \quad (8)$$

Considering equ. 8, the assembly accuracy is only limited by the inaccuracies of the sensor system. Calibration errors based on the tranformation matrix $(^V M_T)$

do not influence the assembly accuracy. This configuration enables a closed control loop for the positioning process due to the possibility to measure the corrective movement with the vision system. Therefore the accuracy is only limited by the motion resolution and not by the absolute positioning accuracy.

3 Tool concept

The positioning of the above mentioned gripper tool with an integrated CCD-camera requires only a small range to compensate the measured positional offset. In opposite to conventional systems (Fig. 2) far movements to cover the offset between measurement and joining axis are not necessary. The corrective movement can therefore be done with a high resolution actuating system that is integrated in the gripper tool. A suitable micro manipulator therefore is a piezo-actuating system.

Such a gripper tool with an integrated fine motion device can be attached to a standard industrial robot, because it requires only a less accurate handling device. First the gripper tool is positioned by the industrial robot with an accuracy that is within the range of the fine motion device. For this step repeatability of conventional robots of about 50 microns is sufficient. In the second step, the corrective movement is done at a high precision with the fine motion device of the gripper tool. This positioning strategy combines the big workspace of standard robots with the high accuracy of the fine positioning gripper tool. This is an cost saving advantage compared to high precision robotic systems.

However, the use of standard robots causes structural vibrations on the end-effector, so that the accuracy of the assembly process is affected. But these vibrations can be measured with an accelerometer system and, due to its good dynamics, the fine motion device can be used to compensate them. Figure 4 shows the structure of the realized precision gripper tool (MPS) for industrial robots with integrated CCD-sensor, accelerometer and piezo-system. The field of view of the CCD-sensor can be varied from 12x9mm to 2.9x2.2mm, using different application dependent objectives. The resulting optical resolution in the range between 18 and 4 microns can be improved with subpixel-image processing. The 3D-accelerometer system [2] is integrated in the movable platform. The combination of micro-mechanical sensor elements with a full-custom ASIC makes miniaturization at a high degree, low power consumption and a high resolution ($2\mu g/\sqrt{Hz}$) possible. The accelerometer system consists of a sensor head with a digital interface (see Fig-

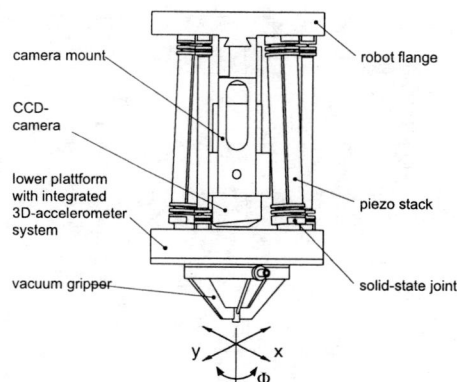

Figure 4: tool concept for precision assembly

ure 5 and a decimation filter, that is located on the process interface board (section 4.2). The corrective

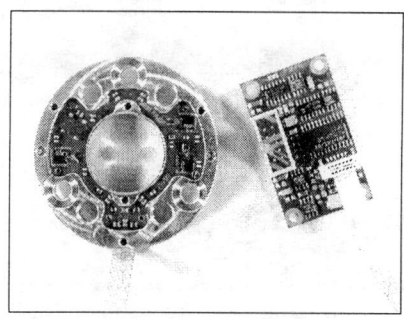

Figure 5: integrated accelerometer system

movement is done by an integral piezo actuator, consisting of four piezo stacks, in two dof [7]. Thus reaching a natural frequency of about 135Hz and a range of $\pm 250\mu m$. In Figure 6b a prototype of the presented precision gripper tool can be seen. In order to see the accelerometer system, the lower cover with the vacuum gripper has been removed.

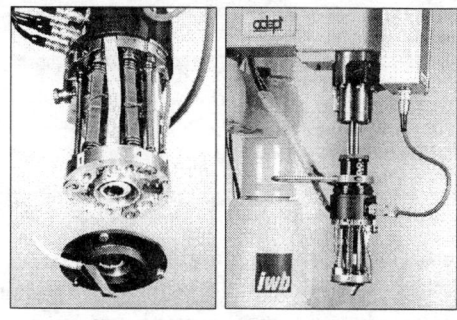

Figure 6: a; gripper tool for precision assembly b; scara-robot with MPS

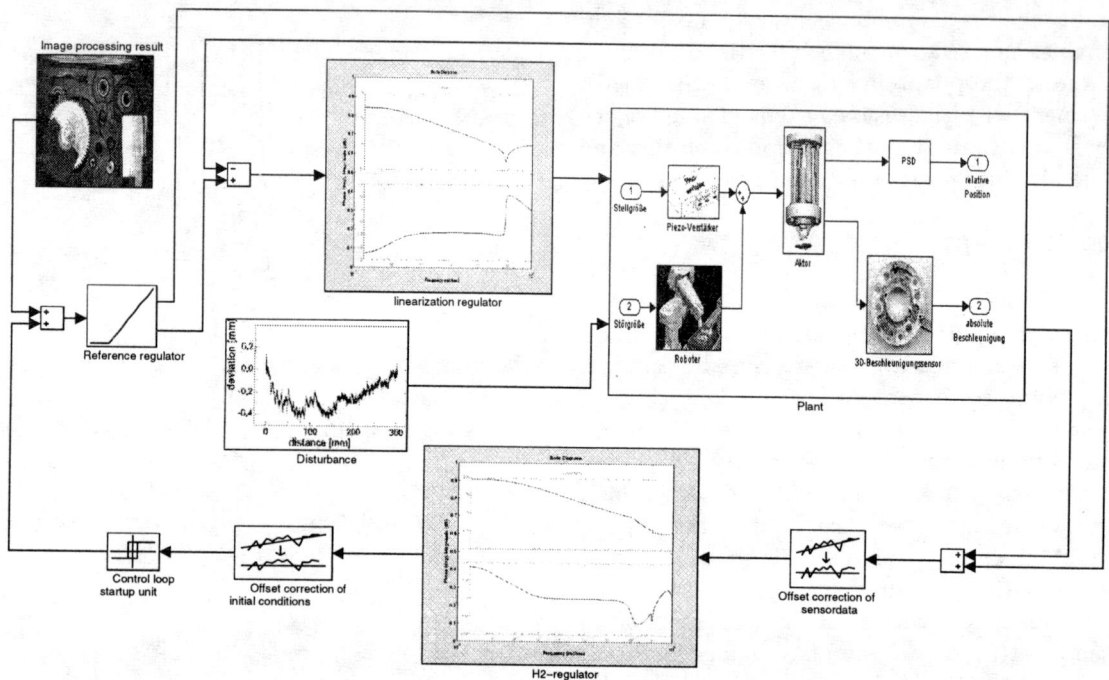

Figure 7: control system

4 Control system

The control system consists of two closed-loop controllers and one open-loop controller (see Figure 7). The goal of the control system is on the one hand to suppress the robot's vibrations and on the other hand to compensate asynchronous changing static displacements. The two degrees of freedom are decoupled, if the system is well calibrated. Therefore for simplicity only one degree of freedom is regarded in this section. The control system is running on a heterogeneous distributed multiprocessor architecture.

4.1 Control design

The task of the inner control loop is to linearize and damp the piezo-driven actuator. Since all piezo-actuators are non-linear and show hysteresis effects, the creation of an accurate model is only possible with a huge computational effort, that has to be done online [5]. In our approach the linearization regulator transforms the piezo-actuator with position feedback (PSD-sensor) into a linear and time invariant PT_2-model, whose damping factor can be adjusted in a wide range. That is why the plant can be described with the actuator's PT_2-system, the PT_1-time delay caused by the piezo amplifier and the PT_2-coupling of the robot's vibrations as disturbance. The dynamic behavior of both sensors can be neglected in the band limited frequency range of the robot's vibrations [9]. With an enhanced model of the plant (weighted states as additional outputs) an H_2-regulator with acceleration feedback for the second control loop can be designed. Since the acceleration is not a state of a mechanical PT_2-system, an lqr-regulator with a state estimator is difficult to develop [4]. H_2-regulators are robust controllers and are used with acceleration feedback for the control of buildings for aseismic protection [3]. The piezo-actuator can be seen as a one-story building with an active mass damper, the robot's vibrations as an earthquake. With this analogy the H_2-development process presented in [3] can be used with a few adaptations also in our case for suppressing the unknown robot vibrations. The open-loop reference regulator is necessary to compensate static displacements, that are determined with the MPSs own image processing [1]. This regulator is based on a parallel model and manipulates the H_2-regulator input in a way, that vibrations caused by the step answer of the actuator are not visible to the H_2-regulator. Two poles of the H_2-regulator represent a double integrator, that is necessary to create a position output of the regulator with an acceleration input. That is the reason why the offset of the sensor data has to be removed. This is done with the value of the discrete

Fourier-transform at 0Hz with a spectral density of 1Hz. This reflects the mean value of the windowed sensor-data. This method obtains an offset suppression of about 130db. Due to the unknown initial conditions of the double integration, the output of the H_2-regulator is drifting away. Therefore the aforementioned offset suppression algorithm has to be applied in a three stage configuration. During the startup time of the offset suppression, the H_2-regulator is driven in open-loop. As a consequence the output is over controlled and the control loop start unit is necessary for a smooth closing of the control loop. The whole control flow is managed with a state machine, that generates all signals necessary for startup, handles the communication with the image processing and controls the calibration of both sensor-systems.

4.2 System architecture

An overview of the system architecture can be seen in Figure 8. The main part is the control PC running VxWorks as RTOS. The whole control system including the state machine is processed on this unit. The image processing is done on a separate PC running Linux, that is connected to the control PC via RS232. The PCI-process interface with an onboard DSP reads and decimates the data coming from the sigma-delta converters of the 3D-accelerometer system [2]. Additionally the piezo-amplifier is driven by the analog outputs of this unit. The data of the internal position measurement system is acquired with a PCI analog I/O-card. The trigger events of the robot, for example starting of a new cycle, are linked to the interrupt controller of the process interface. These events are served within a PCI-interrupt service routine on the control PC. This is the only link that is necessary to the robot control system.

5 Experimental Setup and results

Our experimental setup was necessary, because the measurement of the robot's vibrations during an assembly task with the required accuracy is very difficult. We attached the MPS on a table, whose movements can be measured with an external position measurement system. The results of the image processing unit, the reference input, are simulated with a step function within the state machine. The different coordinate systems have only been roughly calibrated. The left of Figure 9 shows a movement of the upper actuator platform without any control system through the points (0/0; 0/-0.05; -0.05/-0.05). On the right side the same movement with activated linearization regulator can be seen. The adjusted damping factor is 0.7 (aperiodic PT_2-system). The actuator is successfully damped and the non-linearities as well as the hysterisis effects are compensated. The coordinate system of the internal position measurement system determines the coordinate axes of the actuator.

In order to test the whole control system, the table with the actuator was stimulated with vibrations of 60 microns at its natural frequency of about 5Hz. In Figure 10a the measured movement of the table and the corrective movement of the upper actuator platform referring to the table are shown. It can be seen, that the disturbances can be successfully damped. The maximum error of the correction movement is less than 30 microns. Figure 10b shows the result, when a new position of the image processing, a corrective movement of 100 microns for both degrees of freedom, is simulated after 5 and 7 seconds, respectively. It can be seen, that the new position is reached after a short overshooting. This effect can be reduced, when the used models are more refined. The step in the x-axis influences the y-axis and vice versa, because the accelerometer-system has not been calibrated and corrected yet, as suggested in [1].

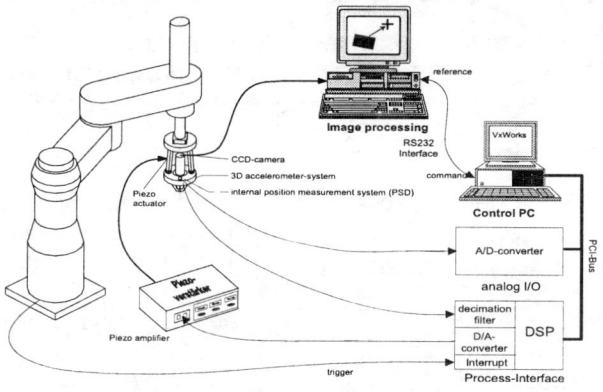

Figure 8: system architecture

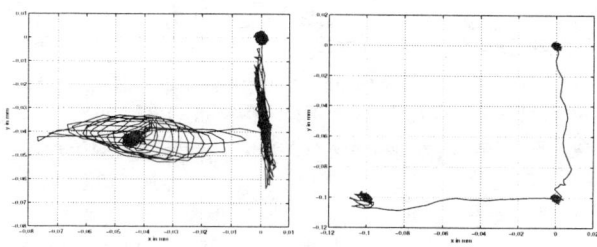

Figure 9: results of the linearization regulator

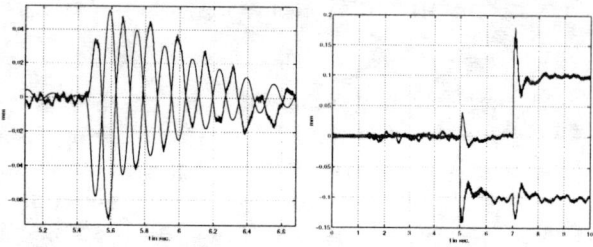

Figure 10: a; disturbance and corrective movement (noisy curve) b; step answer of the control system

6 Assembly of watch movement

We realized a pilot plant for the assembly of miniature gear trains of radio-controlled wrist-watches based on the presented gripper tool. In this application miniaturized gearwheels must be accurately placed in the bearing points of the watch baseplate [8]. The diameter of the bearing points is 300 microns and the required positioning accuracy is about 20 microns. A scara-robot adept 550 is used as handling device (Figure 6a). The gearwheels are handled with a vacuum gripper, that only obscures parts of the component. Thus tracking of characteristic features like the center of the gearwheel in the field of view of the camera is possible (Figure 11).

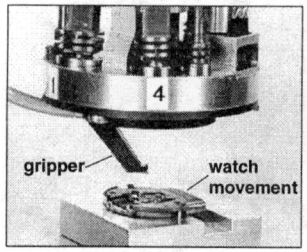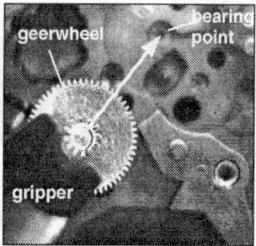

Figure 11: assembly of gearwheels

7 Conclusion

The results show, that the presented gripper tool is suitable for micro assembly with standard industrial robots. It combines the advantages of the robot (workspace) and fine motion device (accuracy) to a cost saving high precision handling device. Positioning errors as well as vibrations during joining can be sensed. The vision system in combination with the control system is able to calibrate the system online for each assembly cycle. Thus compensating environmental influences like temperature drift is possible. The robust control system determines the necessary corrective movements, without complicated sensor data fusion and synchronization. Future work will be focused on refining the plant's models in order to further improve the assembly accuracy.

Acknowledgments

The work presented in this paper was supported by the *Bayerische Forschungs-Stiftung* as a part of the Bavarian Research Union on microsystem technologies FORMIKROSYS.

References

[1] S. Blum, C. Robl, "Sensor data processing and control of a micro positioning system" Workshop on European Scientific and Industrial Collaboration on promoting Advanced Technologies in Manufacturing WESIC '98, Girona, Spain, 1998.

[2] S. Wüstling, "Hochintegriertes triaxiales Beschleunigungssensorsystem", Phd-thesis, Forschungszentrum Karlsruhe, Germany, 1997.

[3] B. Spencer, J. Suhardjo and M. Sain, "Frequency Domain Optimal Control Strategies for Aseismic Protection", Journal of Engineering Mechanics, Vol. 120, No. 1, pp 135-159, 1994.

[4] S. Baba et al., "Active Optimal Control of Structure Using Optimal Observer", Journal of Engineering Mechanics, Vol. 115, No. 9, pp 2564-2581, 1989.

[5] M. Goldfarb, N. Celanovic, "Behavioral Implications of Piezoelectric Stack Actuators for Control of Micromanipulation", IEEE International Conference on Robotics and Automation ICRA, Minneapolis, USA, 1996.

[6] T. Ando et. al, "Chip mounting apparatus", U.S.-patent 5084959, Matsushita Co., Japan, 1992.

[7] U. Reiländer, "A New Design Of A Three Axis Micromanipulator", Acuator '98, 1998.

[8] G. Reinhart, M. Höhn, "Growth into Minitrization - Flexible Micro-Assembly Automation", CIRP Annals Vol. 46/1, 1997

[9] C. Robl et al., "Micro Positioning System with 3dof for a Dynamic Compensation of Standard Robots", IEEE International Conference on Intelligent Robots and Systems IROS, Grenoble, France, 1997.

Understanding of Mechanical Assembly Instruction Manual by Integrating Vision and Language Processing and Simulation

Norihiro Abe †, Kazuaki Tanaka †, Hirokazu Taki ‡

† Kyushu Institute of Technology, Kawazu 680-4, Iizuka, Fukuoka, 820-8502, Japan

‡ Wakayama University, sakae-dani 930, wakayama, 640-8510, Japan

Abstract

The paper reports a system which builds assemblies by simulating in virtual space assembly operations specified in an assembly manual. A manual consists of both illustration and explanation.

Illustration consists of not only figures but also phrases and symbols which relate the phrases to the figures. This research uses the three dimensional models of mechanical parts generated with a modeling tool. A noun phrase in illustration specifies the name of a mechanical part and determines the model used. A verb phrase imposes an initial condition to be held before the execution of an assembly operation.

Auxiliary lines in an illustration are virtual lines, which are used to mate the parts at both ends of them. Both the assembly way and the terminal condition of the operation are elucidated in explanation.

The state of a subassembly obtained is hard to infer because an operation often occurs a side effect, which is affected with the geometries of the parts involved in the operation. Configuration obtained with simulation is propagated to inference mechanism and this allows the system recognize the state of the machine assembled.

1 Introduction

We have explored the methods for automatically assembling/disassembling mechanical objects. The methods are basically divided into two classes.

The first class is to find all possible assembling procedures by disassembling a given assembly [1],[8]. Algorithms proposed so far add only one part to a subassembly each step. As a result, it is difficult to divide one object into several groups which we refer to as subassemblies.

The second class has no problem to use the concept of subassembly but it takes too much time to find all assembly procedures with computer algorithm, that is, it is computationaly impossible. Consequently, the assembling procedures have to be given by experts or designers. The first one is considered more appropriate than the second one under the automatic assembling process because robots only need attach a part to a subassembly on a moving belt conveyer. However, when considering the case where an assembly must be disassembled for finding and repairing parts broken down, we find the second one preferable to the first one.

In fact, an instruction manual is an example of the second method. Realizing a system which understands an instruction manual and constructs the assembly is interesting to both robotics and artificial Intelligence. In this research, we intend to come up with a system which takes the similar manner that we take when reading an instruction manual.

Instruction manuals consist of illustration and explanation for people to read/see on demand. An expert can assemble a target machine even without reading explanation. But the purpose of this research is to make a model of novices who have no expertise of mechanics. Nevertheless, there are dissimilar points between human and computers. We can know no more than what we have acquired by watching how novices read/see an instruction manual.

Since no standard way of writing an instruction manual is yet admitted, we use the manual written by Health INC. [14]. Two typical examples are shown in Figures 1 and 2. The illustration contains not only figures but also description like noun and verb phrases. An auxiliary line is depicted between a part and an object to which the part is attached in order to make it easy to recognize the relation to be set between them. Consequently, illustration prescribes the initial configuration of each part or subassembly which should be held before an assembly operation is performed.

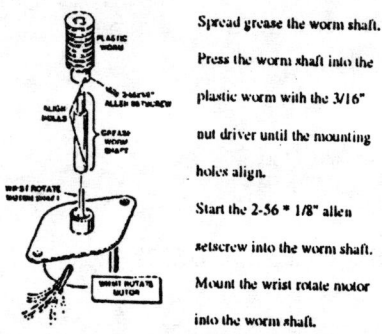

Figure 1: The first part of the manual.

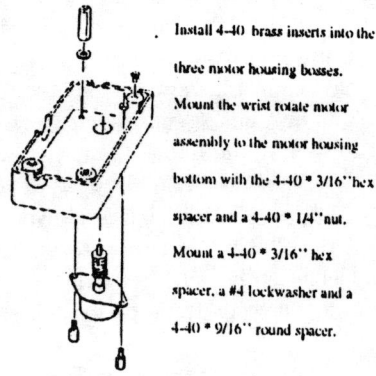

Figure 2: The next part of the manual.

On the other hand, explanation specifies condition which each operation must satisfy according the assembly procedure. Each sentence prescribes which objects participate in the assembly operation and which terminal condition must hold. These facts allow the system to perform an operation at first by reading a sentence and then by determining the position and orientation of each object after recognizing it and the corresponding auxiliary line from the illustration.

It is needless to say that the system must interpret sentences in explanation and recognize the each object in illustration [5], [12]. The papers reported so far by us concerns the bottom-up method for understanding an assembly instruction manual. This research is based on them but the behavior of an assembled object is difficult to know from them because the parts depicted in the illustration are too incorrect for the system to simulate the operation specified in explanation. The system reported in this paper makes it possible to have true understanding of the instruction manual. Researches using vision and language cooperation system are proposed in [12], and this research is naturally included in them but treats not only static information but also dynamic behavior and side effects caused with assembly operations or alteration of behavior. The dynamic behavior of machinery is studied in [9], [10], but they assumed that machinery has been constructed and the representation is given in advance which is suitable for inferring behavior. On the other hand, our system constructs representation of machinery by interpreting both illustration and explanation and finds behavior of the assembly exploiting not only assembly relation built so far but also the state description discovered in the simulation of mechanism done using virtual space.

2 Presupposition of the system

At the beginning of an instruction manual there is a nominal list of mechanical parts used in an assembly, their names (hereafter, we call it p-name) are registered into the system dictionary. The set of three dimensional models of the parts used are given to the system. The models are defined using a modeling system, both of a surface model and a wire frame model are available to the system. These models are given their p-names but every component of each model is not always given a p-name in advance. Components corresponding to mounting holes, bosses and so on are not given their name in advance, though they are referred from explanation or illustration. When the system encounters such names, it must resolve what they are by comparing illustration to their parent models. The names of tools and adhesive agents and so on are given to the system but all operations concerning them are supposed to always succeed. Under these preparation, interpretation of both explanation and illustration is performed by simulating the corresponding operation using three dimensional models of parts. The resultant state of relation between parts or their surfaces is recorded in the representation referred from an inference engine.

3 Separating character strings from image

Each page is treated as image and is separated roughly into two regions, explanation and illustration. Illustration is then examined to extract character strings and figure symbols like an arrow, using the method in [3], [5]. Fig. 3 shows the result obtained from Fig. 1. The characters are recognized using OCR and sentences are analyzed with a syntactic parser using pattern matching method. In this process, all of characters are not successfully recognized, but we want to proceed to the detailed explanation of the system.

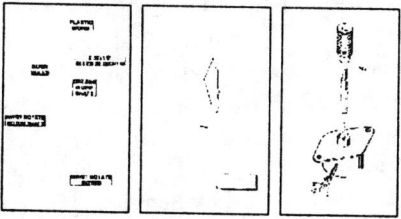

Figure 3: (1)Phrases (2)Symboles (3)Parts.

4 Interpretation of phrases and sentences

Concerning the detail of the syntactic parser used, please see [2], [7]. Concerning interpretation of illustration, at first, phrases extracted from it are examined because the noun phrases are much the same as p-names and the verb phrases specify the initial condition to be held between subparts/components of two parts. Here, the word 'subpart' means the substructure of a parent part whose name properly includes the name of the subpart. For an example, a wrist rotate motor shaft is a subpart of a wrist rotate motor.

On the other hand, the world 'component' means the substructure of a parent part whose name does not include the name of the component. Examples are mounting holes, motor housing bosses. Although the initial state of objects to be assembled is not prescribed in explanation, it is indispensable to determine their attitudes for beginning the assembly operation.

The p-name plays an important role because it allows the system to determine the model of the part depicted in the illustration. Names except for p-names are divided into the following two classes: The first one is a set of abbreviated forms of p-names or noun phrases corresponding to their subparts. Their examples are motor, wrist rotate motor shaft and so on. The second one is a set of names corresponding to the component of a part/subassembly.

When finding a noun phrase corresponding to a subpart, the parser instructs the image analyzer to match the three dimensional model of the parent of the subpart against a figure after finding the figure relevant to the phrase from the illustration. In this case, figure symbols are used to show which figure region is related to the phrase. If the figure region has been matched to some three dimensional model, the phrase is determined to be a name of the substructure of the model. For an example, a verb phrase "align holes" in Fig. 1 means that there is a hole at the end of each arrow, and that the two holes must be aligned. Consequently, before mating the plastic worm and the worm shaft, their holes must be arranged.

In this manner, a rough initial arrangement is found by analyzing phrases given in illustration. But the precise arrangement can be unknown until the configuration of parts in illustration have been determined by matching them to their three dimensional models.

5 Determining attitude and location of parts

When a p-name appears in a sentence, its three dimensional model must be found from illustration. In the case, one of auxiliary lines attached to the model is superimposed onto one of those detected from the figure.

For an example, consider the case that auxiliary line 1 of a plastic worm in Fig. 4 (a) is selected. The figure in Fig. 4 (b) is generated by transforming the coordinate system of the original figure so as to have the auxiliary line to agree with y axis and then constructing a circumscribed rectangle which is symmetric with respect to the auxiliary line. Next one of auxiliary lines attached to the model is selected as y axis and a projected figure of the model is generated by viewing it from the same line of sight as that used when the illustration has been drawn. Then a circumscribed rectangle is generated for a figure obtained by rotating the figure around the axis. An appropriate rotation will allow the figure shown in Fig. 4 (c) to be identical to the one in Fig. 4 (b). This means that an attitude of the plastic worm has been found.

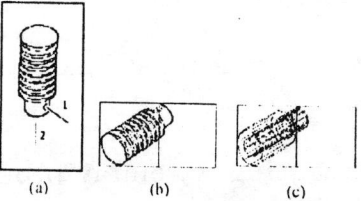

Figure 4: (a)Plastic worm, (2)A circumscribed rectangle, (c)That for the model (a) in Fig. 5.

Here we need a method to decide if two figures are

one and the same one because their scales are not the same. The scale factor is defined to be the ratio of the longitude to the width of the circumscribed rectangle. If two figures are equal then their scale factors must be also equal, and their shapes enclosed by contours must be nearly equal after the original rectangle is magnified/reduced so that it is equal to that constructed from the model. The value of magnification/reduction must be constant value k for all parts depicted in an illustration.

There may be several models passing this inspection if the external appearance is equal to the part except for the number of holes or textures. The wrong model will be surely rejected by examining assemblability along the auxiliary line between itself and the partner to be mated. If the selected model should be wrong, the system would have to backtrack. When the first object appearing in the first sentence has been correctly processed, the subsequent analysis can exploit the resultant value k and attitude of the object recognized so far.

In our example, a worm shaft is to be mated to the plastic worm through the auxiliary line2 shown in Fig. 4 (a). This means that a model of the worm shaft should be arranged so that the one of the auxiliary lines is collinear to that of the plastic worm. An attitude of the worm shaft is to be restricted using the similar way as mentioned above. But it is symmetric with respect to the auxiliary line, no rotation is needed.

At this point, it seems that an initial arrangement of the operation is completed, but the constraint from the verb phrase in the illustration must be examined. It requires two holes to be aligned, then the corresponding holes (it is hard to find them from the illustration) are found from their models and one of them is rotated around the auxiliary line until the holes align.

Here the initial arrangement has been completed. The sentence requires two operations: press the worm shaft into the plastic worm, and align the mounting holes. A virtual assembly operation required to perform the operation specified in a sentence needs only translation except for particular cases including engagement of worms or gears and so on.

6 Implementing Assembly process

The first operation is satisfied by translating the worm shaft toward the plastic worm along the auxiliary line. The sentential form [[Main][until][Sub]] instructs to attain the state of SUB by performing an operation of Main. Consequently, the system has only to translate the worm shaft by the distance between the holes. Note here that "with the 3/16" nut driver" is neglected.

Next, a setscrew must be started to the worm shaft. Another auxiliary line concerning the worm shaft is the line1 in Fig. 4 (a). As there is just one auxiliary line attached to a setscrew, the allocation of it is easy. The setscrew is located at the entrance of the hole of the plastic worm then the setscrew is inserted into the mounting hole of the worm shaft. As the result of this operation these three parts are connected and the effect is represented by merging their scene graphs into a new one of INVENTOR. This means that they are transferred together by the rule of INVENTOR if one of them is transferred.

A new object 'worm assembly' appears in the next sentence, which corresponds clearly to the assembled object. The assembly is mounted by inserting the motor shaft into the hole at the bottom of the worm shaft. Five auxiliary lines are given to the model of the motor as shown in Fig. 5, they are examined which of them corresponds to the auxiliary line in the illustration. In the similar way stated above, a circumscribed rectangle of a projected figure of the motor is generated by selecting one of them as an axis. It is then rotated around the axis to examine if the revised rectangle generated from the illustration is nearly equal to that. This revision is based on the scale factor computed so far.

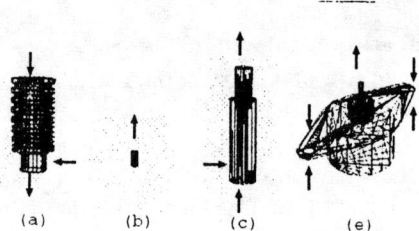

Figure 5: The models used in Fig. 1.

The subordinate clause constraints the quantity of insertion and it is determined by detecting collision between the assembly and the motor. In this case, the condition specified in the subordinate clause is easily confirmed by examining the scene graph and the subassembly obtained so far: surely, the assembly was moved with the setscrew. Note here that the motor is not yet merged into the scene graph of the assembly. By tightening the setscrew, they are merged into a new scene graph in Fig. 6 (a). Here mb, sf, and msa mean motor body motor shaft and motor subassembly.

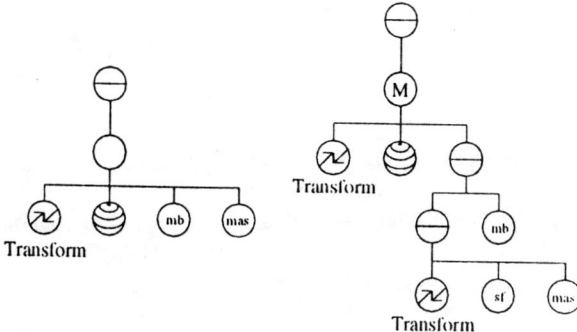

Figure 6: A wrong scene graph (a) and a correct scene graph of the assembly constructed from Fig. 1.

7 Behavior of subassembly

Though the explanation tells nothing about the function of this subassembly, let consider what kind of function has been attained. The four parts discovered from their p-names constitute this subassembly but strictly speaking, not the motor itself but its shaft participates in the subassembly whose members are mutually locked, and there is one freedom between the motor and its shaft. Consequently the plastic worm is to rotate together with the motor shaft.

At the same time, the motor shaft is transferred if the motor is transferred. For representing these relations, the definition of a motor must be defined as shown in Fig. 6 (b). This implies that some of subparts also need their definitions in the same way as their parents are defined. Under this establishment, the result we expect is realized by the system.

Another point noted here is that the operation specified in the last sentence is wrong because the motor dose not work if the worm assembly is moved until it collides with the motor. We have common sense knowledge that a motor shaft should rotate and been taught to examine if a correct function is not violated when we encounter an object possessing the useful function. To find erroneous instruction, the equivalent scheme to that we have is necessary for the system. It is, however, difficult to prepare knowledge for correcting erroneous instruction.

To simulate physical phenomena in a virtual space, there are many problems to be solved including the one shown above. In this paper, we will assume that the assembly obtained here works correctly by keeping the motor from colliding against the worm assembly. As the first subassembly is completed, it is added to the set of models together with the wire frame model of it. The model inherits auxiliary lines from constituents constructing the assembly except for them used.

8 Treatment of missing operations

There are eleven auxiliary lines in the next illustration. Models corresponding to them are shown in Fig. 7. Note here the model (d) in Fig. 7. This corresponds to the subassembly just built but there is no auxiliary line around the worm assembly. It is not impossible for the system to add it to the model because the worm assembly inherits the function of the motor shaft, but there remain several questions which constituents of the model should be given auxiliary lines to which direction.

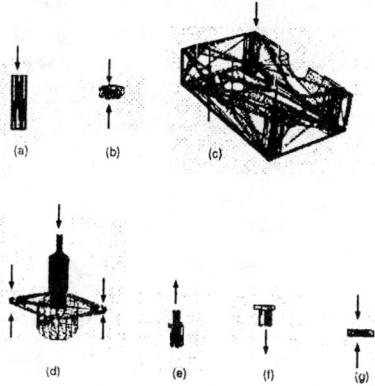

Figure 7: The models used in Fig. 2.

The first operation has relation to the boss and motor housing bottom. It is not easy to determine the angle of incidence from the model of a boss, as the size is small, which corresponds to (f) in Fig. 7. On the other hand, as the bottom has a hole inside it, the incident angle is computable. The auxiliary line between the brass inserts and bottom in the illustration is to be selected to generate a circumscribed rectangle for comparing the figure of the bottom to the model, but the hole is also available to determine the attitude of the bottom. Fig. 8 shows the rectangle generated for the model(c) in Fig. 7 with respect to the line L. In the same way as the previous one, the model of the bottom is rotated along the axis corresponding to L.

The insert operation succeeds but the instruction requires of three motor housing bosses to be inserted with brass inserts. By matching the figure and model of both the motor housing bottom and the brass insert, the location of a boss is discovered. Three bosses and inserts should be depicted in the illustration, but simplification like this is well applied to make it easy

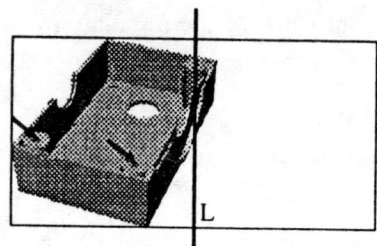

Figure 8: Discovery of remaining bosses.

to recognize illustration. Note here that models of primitive parts must have been given all of auxiliary lines at their substructures which may be connected with something.

As a heuristic knowledge for treating this type of simplification, we know that substructures with the same function which are arranged systematically on a plane or cylinder of a parent part will be repeatedly applied with the same operation. This rule allows the system find remaining two bosses from the model. Consequently, the operations are successfully realized.

9 Conclusion

We have reported the system which integrates vision, language, and simulation in virtual environment for inferring both assembly relation between mechanical parts and behavior of an assembly constructed. There are problems to be solved for realizing the final assembly specified in the instruction manual used in this research.

References

[1] S. Yamada, N. Abe, and S. Tsuji, "Construction of Consulting System from Structural Description of a Mechanical Object," *Proc. of IEEE on Robotics & Automation*, pp.1412-1418, (4 1987).

[2] N. Abe and S. Tsuji, "Robot task specification in Natural Language," *Proc. of Conf. & Ex. on AI*, pp.586-595, (10 1987).

[3] S. He, N. Abe, and T. Kitahashi, "Understanding Assembly Illustrations in an Assembly Manual without Any Model of Mechanical Parts," *Proc. of Int. Conf. on Computer Vision*, pp.573-576, (12 1990).

[4] N. Abe, K. Ohno, S. He, and T. Kitahashi, "Task Specification Using Technical Illustration," *Proc. of Robotics and Automation*, Vol.2, pp.58-64, (5 1993).

[5] S. He, N. Abe, and T. Kitahashi, "Assembly Plan Generation by Integrating Pictorial and Textual Information in an Assembly Illustration," *AAAI-94 Workshop on Integration of Natural Language and Vision Processing*, Seattle Washington USA, pp.66-73, (1994).

[6] N. Abe, J. Y. Zheng, K. Tanaka and H. Taki, "A Training System using Virtual Machines for Teaching Assembling/Disassembling Operations to Novices," *International Conference on Systems, Man, and Cybernetics*(1996).

[7] N. Abe, K. Tanaka, J.Y.Zheng, S. He and H. Taki, "Integration of Language and Image Processing for Assembly Instruction Manual Understanding," *Proceedings of International Symposium on Artificial Intelligence*, 135-140(1996).

[8] L. S. Homem de Mello and A. C. Sanderson, "AND/OR graph representation of assembly plan," *IEEE Trans. Robotics Automat.*, vol.7,no.2, pp.188-199(1991)

[9] L.Joskowies, E.P.Sacks, "Computational Kinematics," *Artif. Intell.* 51,1991,pp381-416.

[10] A.Gelsey, "Automated reasoning about machines," *Artif. Intell.* 74,1995,pp1-53.

[11] J. Wernecke, "The Inventor Mentor," *Addison Weseley Publishing Company*,1994.

[12] P. McKevitt, "Integration of Natural Language and Vision Processing," *AAAI-94 workshop program*.

[13] K.D.Forbus and J.D.Kleer, "Building Problem Solvers," *MIT press*, 1993.

[14] "HERO ROBOT arm accessary manual," *HEALTH COMPANY*, MICHIGAN, 1982.

An Integrated Interface Tool for the Architecture for Agile Assembly

Jay Gowdy and Zack J. Butler

The Robotics Institute
Carnegie Mellon University

Abstract

Developing automated assembly systems normally happens in two distinct stages: first an "off-line" stage in which the system is designed and programmed in simulated and then an "on-line" stage in which the simulation results are used to minimize the deployment and integration time of the physical machines. The distinction is so great that usually completely different software environments are used in the design phase than are used in the deployment and operation phase. We are developing The Architecture for Agile Assembly (AAA): a comprehensive, integrated framework that is designed to blur these stages together and ease the transitions between them. We have used the protocols of AAA to create an integrated interface tool which can be used throughout the life-cycle of a developing AAA factory, from its design to its operation. We have tested the integrated interface tool both in simulation and with our prototype hardware, which is designed for high precision four-degree-of-freedom assembly.

1 Introduction

Off-line robot programming promises to reduce the time necessary to produce automated assembly systems, since much of the design and programming work can be done in simulation, where experimentation is usually cheap and mistakes are generally not catastrophic[5, 9, 13]. Unfortunately, a problem inherent in off-line programming is evident in its name: for an "off-line" program to be truly useful, it must become an "on-line" program running on a physical assembly system. If the physical factory does not match the simulated factory to a sufficient degree in either geometric or functional characteristics, then the transition from the world of pixels and bits to the world of actuators and metal can be arduous.

This difficult transition has long been recognized as a problem with off-line programming.[3] Approaches to easing the transition include

- Calibration: sensing the factory configuration and adapting the programs to match[8].
- High fidelity modeling: using standardized models of components and component assembly con-

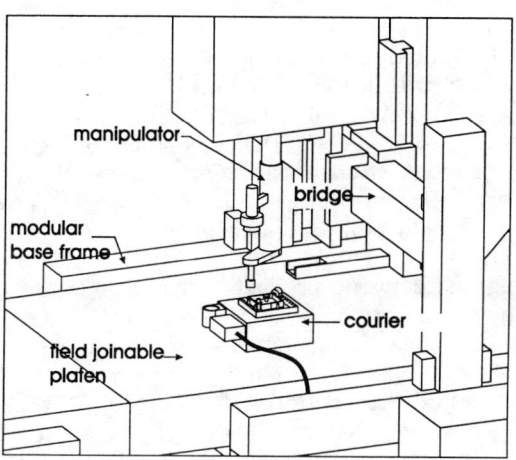

Figure 1: A minifactory segment

straints combined with physical modeling to reduce the number of surprises that can occur in the transition[4].

- Sensor feedback: using sensors such as machine vision or force feedback to overcome discrepancies between simulation and reality[1].

These techniques can ease the transition between simulation and reality, but they need to be embedded in a system which can move back from reality to simulation. If they are used in a "one-pass" approach, in which the transition from simulation to reality is done just once, then the full potential of the simulation will be lost. For example, if contact with the physical world necessitates large changes in programming and factory configuration, all of those changes must be made in the relatively unforgiving environment of the physical machines. Similarly, if the product or processes change, since the original simulated factory is not the same as the implemented physical factory it is difficult to meaningfully address those changes in simulation. To truly achieve *agility*—the ability to rapidly deploy factories to deliver a product to market quickly and to rapidly reconfigure factories to adapt to changing technologies and market needs—simulation and physical implementation need to be more closely coupled and done iteratively.

The overall goal of the Architecture for Agile Assembly (AAA)[12] is such agility: AAA is designed to allow real factories to be built incrementally and reconfigured often as manufacturing processes are perfected and needs change. Rather than having a strict separation in methods and mechanisms between design and execution, we have developed a single *interface tool* which bridges both. This integrated interface tool provides a 3D graphical environment that guides an evolving AAA factory through its various stages, allowing the user to view and interact with the factory as it is assembled, simulated, programmed, implemented, and operated.

AAA systems are composed of a set of modular robust robotic *agents*. Each agent operates in a deliberately limited domain, but possesses a high degree of capability within that domain. For example, our instantiation of AAA, *minifactory* (Fig. 1) is focused on four-degree-of-freedom (DOF) assembly of high-value, high-precision electro-mechanical systems. In a minifactory there are courier agents that are "experts" in moving products in the plane of the factory floor (the platen), and manipulator agents that are "experts" in lifting and rotating products. The agents are physically, computationally, and algorithmically modular, and only when acting cooperatively can they perform the 4-DOF operations required to produce a product. In order to perform this kind of cooperation, each agent knows its own characteristics, abilities, and state, and can use the protocols of AAA to advertise that information to its partners.

It is this self-representation that is the key to the integrated interface tool. Not only do agents share their self-knowledge with each other, they also share it with external entities such as the graphical user interface. This paper describes how the AAA approach and protocols enable such a unified interface tool to shepherd a minifactory throughout its life cycle.

2 Loading Factory Components

A minifactory is made up of *components*, which include active agents such as manipulators and couriers, and passive support elements such as bridges, modular base frames, and field joinable platens (Fig. 1). Every component in a minifactory has a *description*, which is a database containing all of the pertinent information about that component, including how to view it, how to model its geometry and behavior, and how to constrain its assembly to other components.

In order for the simulated minifactory to quickly become a working minifactory it is vital that the component descriptions loaded into the interface tool match the real attributes of the physical components. An agent which cannot fit where its description says it can fit or cannot do what its description says it can do will drastically increase the time spent implementing and integrating the minifactory.

Fortunately, a fundamental aspect of AAA agents

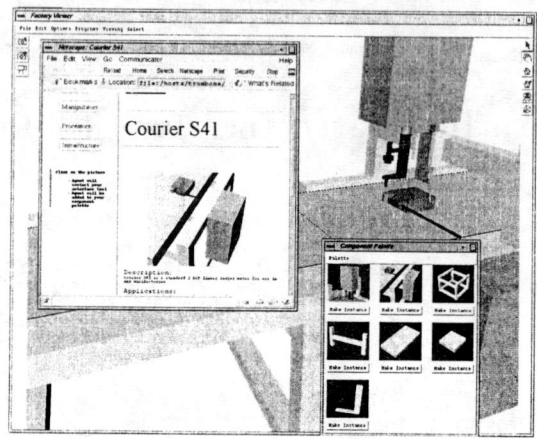

Figure 2: Interface tool interacting with a web browser

is that they all maintain and publish descriptions of themselves. If a factory designer wants to minimize the likelihood of inaccurate information, that designer should directly access the agent itself rather than any catalog or other mediator. There is no absolute guarantee that the agent's self-description is accurate, but it is more likely to contain accurate, up-to-date information specific to that agent than is a generic description of an agent of the appropriate class, or even a passive catalog of agent descriptions.

To demonstrate the direct-access approach, we have implemented a simple example of loading an agent's self-description into the interface tool (Fig. 2). Factory designers can use their World Wide Web (WWW) browser to find a catalog of our prototype agents. Each catalog entry is associated with one particular agent in our laboratory. Clicking on a button next to the catalog entry starts up a helper application which sends an agent "URL" to the interface tool. The interface uses the agent URL to contact the actual agent, and if it is active puts a reference to that agent in a *component palette*. The component palette is a list of iconified representations of the components along with buttons which allow the user to reserve a component and insert a representation of that component into the design environment. Designers can also use the component palette to insert "clones" of components into the design environment, *i.e.* the remote components will not be reserved for use, but simulated representations of them will be available for design experiments.

We have also implemented simple component servers which distribute and reserve descriptions of passive components, such as our prototype base frames and platens. These components do not have any active computational element as the agents do, so there is nothing uniquely associated with a passive component which the interface tool can contact. Thus, the component servers act as active catalogs which the user interface can access. This approach should suffice, since passive components should not change often, and thus

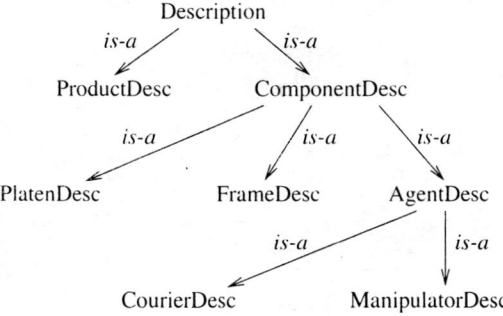

Figure 3: Class diagram for factory descriptions

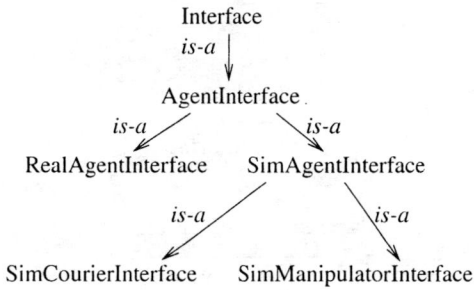

Figure 4: Class diagram for agent interfaces

can have stable, well cataloged geometric and functional attributes.

3 Constructing Factories

The interface tool enables factory designers to quickly take the component descriptions they have found and assemble them together into a variety of different factory configurations, snapping them together and pulling them apart as they design the final factory.

Each component description contains a specification of how it snaps together with other components. Currently we use a short cut which takes advantage of the object oriented capabilities of our implementation language, C++, to "hard-code" this specification. As the class diagram in Fig. 3 shows, all factory infrastructure component descriptions are descendents of the *ComponentDesc* class. All such subclasses must implement an **assembleTo** method, which takes another *ComponentDesc* as an argument and determines first if the target component is a valid candidate for component assembly, and if it is, performs the appropriate manipulations on the description to attach the original component to the target.

For example, if the user desires to assemble a given platen to a given base frame, then the user simply selects them both in the interface tool's 3D rendering and directs that they be assembled together. The **assembleTo** method of the platen description will be invoked with the base frame description as an argument. This **assembleTo** method will verify that the base frame is a valid type for assembling itself to, and will extract from the base frame's description where platens should be mounted. Finally, the platen description's **assembleTo** method will move the rendered platen representation so that it is mounted appropriately on the chosen simulated base frame.

4 Simulation and Programming

The interface tool lets a factory designer write and debug simulated agent programs while the factory is being constructed. The factory designer views a 3D rendering of the running simulated factory as a whole, and interacts with individual agents through virtual control panels, using them to observe their state variables and debug the individual agent programs.

Every agent description must be a subclass of *AgentDesc* which contains a field named **interface**. The **interface** field is itself a database which encapsulates the actual implementation of the agent—whether it is a simulated agent running in the interface tool or it is a physically instantiated agent which the interface tool is interacting with remotely. The entries in the **interface** database will be items such as state variables which can be monitored or parameters which can be changed to affect the simulated or physical agent operation.

The value of the **interface** field must be a subclass of *Interface* (Fig. 4). Any subclass of *Interface* must implement an **update** method. The interface tool maintains a list of agent descriptions that it is monitoring or simulating, and calls the **update** method of each agent description's interface as often as possible. The particular implementation of the **update** method appropriately moves the rendered parts of the agent description, such as the graphical representation of a manipulator's end effector.

Currently, the user specifies an agent's run-time behavior through the use of a program written in Python (a byte-coded, object-oriented programming language). These programs are similar to the applets used in WWW programming in that they instantiate objects with required methods rather than simply being a sequence of commands. The program objects specifically have a *bind* method which is used to indicate all of the global factory components the agent will use, and a *run* method which is the actual script that determines the agent run-time behavior[6].

The architecture for simulating agent behavior is shown in Fig. 5. When the user runs a script, the interface tool creates or contacts a scripting manager process which launches the thread that will actually execute the script. This script performs the necessary inter-agent cooperation and sets up and monitors *controllers*—the specification of the time-varying behavior of the agent. The controllers are actually simulated in the interface tool within the **update** method of the simulated agent interfaces, even though which

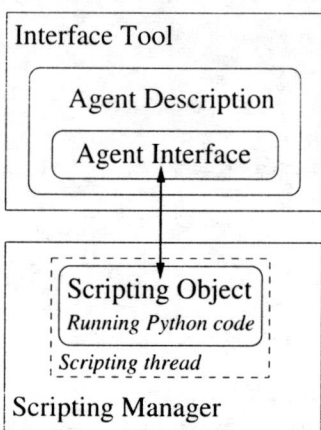

Figure 5: Architecture for simulated agents

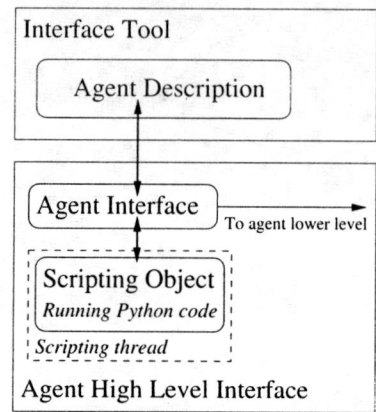

Figure 6: Architecture for physically instantiated agents

controllers are being run is specified by the external scripting thread. All such simulated agent interfaces are subclasses of *SimAgentInterface*, and have similar implementations of their `update` method: first the rendered position and orientation of the end effectors are used to generate current state variable values, then the simulated controller specified and parameterized by the scripting thread is run, and finally, the new set of estimated state variables is transformed into modified 3D renderings of the end effectors.

The simulations currently only model the motions of the agent end-effectors. These simulations are fundamentally kinematic, using a motion model with bounded velocity and acceleration ($\ddot{x} = u$, $\|\dot{x}\|_\infty \leq V_{max}$, $\|u\|_\infty \leq U_{max}$). The different controllers are sequenced using the same efficient hybrid control strategies for robust motion execution that are used in the physical agents[11].

5 Physical Instantiation

Simulation never matches reality, a fact which often limits the utility of off-line programming, since much of the effort is expended in the details of making a system work on physical machines well after the simulations are finished. AAA agents first address this gap by being able to calibrate themselves and explore their physical environments. As long as the simulated world and the real world are topologically compatible and the real world allows the same physical motions as the simulated agents, programs should be able to run in the newly calibrated environment unchanged.

The same interface tool that was used to program a simulated minifactory initiates the transformation into a real factory and monitors this calibration and exploration process. After exploration, the virtual factory rendered by the interface tool is in essentially perfect correspondence with the real factory. We expect users to then switch back to simulation to test the programs using the registered virtual environment before actually trying them out on the real equipment. Any fatal discrepancies, such as swapping manipulator positions or moving components too close together to allow safe passages for couriers, should be detectable and corrected by the user at this point.

We do not expect that this geometric registration will make the transition between simulation and reality seamless: simulations will always miss some aspect of reality that may necessitate reprogramming and possibly reconfiguration. We believe that lowering the cost of moving back and forth between simulation and reality and using the same development environment for both will naturally encourage an iterative approach to developing an automation system, in which major changes imposed by any unmodeled surprises or changes in requirements and capabilities can be addressed first in simulation, and then transferred to reality.

Switching from a simulation of an agent to using a real agent is a simple matter of changing the agent description's interface to one on a physical agent and transferring the program to that new interface. In simulation, the interface tool "owns" the agent description's interface, *i.e.* it completely resides within the interface tool process. The interface to a physical agent will be owned by the high level agent process running on the physical agent's computer, as shown in Fig. 6. Thus, for a physically instantiated agent, the agent description's `interface` field will contain only a reference to a remote database existing on the agent hardware itself. The same script will run on the physical agent as on the simulated agent, except that the scripting thread will be run by the agent rather than by any independent scripting manager and the controllers will be run using the actual control algorithms of the agent rather than being simulated by the interface tool.

Before programs run on the agents, the agents attempt to calibrate themselves and their environment. Part of every agent's calibration process is introspective, *i.e.* establishing and confirming their own physical and behavioral characteristics such as joint-limits or acceleration abilities. Additionally couriers, being the mobile factory agents, explore the exact geometric relationships between themselves and the factory components around them. Each courier is equipped with an optical sensor that can locate precisely placed LEDs to sub-micron resolution[7]. Multiple LEDs mounted on relevant components such as manipulator end-effectors or bridges are used to precisely determine their actual positions and orientations relative to the exploring courier's frame of reference.

We are currently experimenting with two scenarios to accomplish this exploration. In the first scenario, each courier is given the estimated locations of the relevant landmarks (*i.e.* components with locator LEDs mounted on them) by the interface tool. It then negotiates with other couriers for the right to confirm the positions of these landmarks, and builds up an accurate local map from the inaccurate one. The second scenario assumes no initial local map, but rather has each courier run a geometrically complete algorithm that covers its workspace. It will then have discovered the exact locations of all of the relevant components around it[2].

In either case, the interface tool audits the results of each courier's exploration and uses those local maps to generate a new accurate global map. In order to maintain consistency within the global map, the interface tool simply creates a tree of the geometric relationships between the factory components. When a courier precisely identifies the geometric transform between two components, the interface tool immediately adds this information to the tree if possible. If the new information describes a significant inconsistency with previous information, the user is notified. Whereas this solution is not capable of resolving redundant calibration information in an intelligent way, it is sufficient for the current implementation. More capable reconciliation algorithms based on geometric graph matching are under investigation.

6 Monitoring

Once a factory is instantiated, the interface tool should no longer be strictly necessary, since the programs run on each agent without need for any central control or resources. However, the interface tool can still provide a central place for monitoring factory operations in order to keep records, detect problems, and discover means of optimizing the factory.

The primary mechanism for monitoring the factory operations is watching the agents' state and monitoring the products that the agents are manipulating. The interface tool provides a 3D rendering of the whole minifactory as it performs assembly tasks, as well as providing virtual control panels to monitor less visible elements of an individual agent's state.

The 3D rendering of the physical agents' states is achieved through their interface's **update** method,

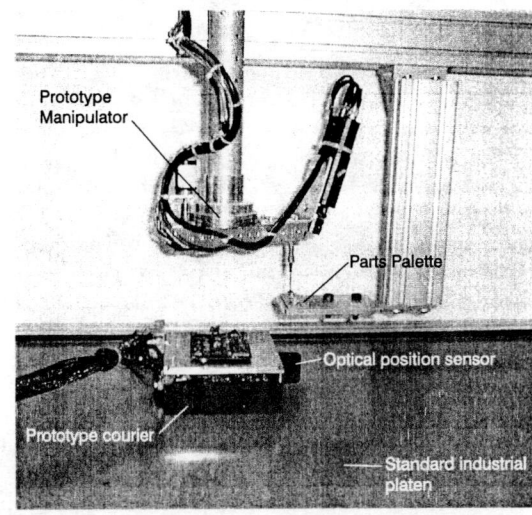

Figure 7: Prototype minifactory hardware

just as the 3D rendering of the virtual agents' states was. All physical agent interface field values are subclasses of the *RealAgentInterface* class, and their **update** method simply examines the current state values and converts those values into updates of the position and orientation of the rendered agent's end effector.

The interface tool monitors products flowing in the system by registering with each of its agents as a "product auditor," *i.e.* the interface tool will be notified when the agents pick up products, when they transfer products, and when they drop products. Thus the interface tool can track the progress of products through the factory, both visually, by presenting a 3D representation of the products as they flow through the factory, and statistically, by maintaining records on the products as they flow through the factory.

The interface tool may prove to be a communications bottleneck for monitoring factory operations, *i.e.* as a factory gets large, sending product information and state values to the interface tool may prove overwhelming. We have attempted to be efficient about the amount of information that is transmitted from the agents to the interface tool, but no matter what is implemented, it is inevitable that a single, central monitoring point will become overwhelmed for some large factory size. Fortunately, since the interface tool is not the central brain of the factory, overwhelming it does not cripple the factory. Furthermore, since the interface tool does not need to serve as the central coordinating resource, there is no need to have only one. For a very large factory, it may in fact make sense to have two or more interface tools monitoring sections of the factory to eliminate any potential bottlenecking problems.

7 Conclusions

We have tested the integrated interface tool with our prototype minifactory test-bed, which uses a fully instrumented courier and an overhead manipulator running on a standard platen and frame (Fig. 7). The integrated interface tool can contact these prototype agents and load their descriptions into its environment. The courier can explore its environment, locate landmarks which contain LEDs with a repeatability of ± 2 μm (which is comparable to the limit of repeatability of our current open-loop courier motion control) and transmit the updated positions of those landmarks to the interface tool, which adjusts the 3D rendering of the minifactory. Once closed-loop control algorithms[10] have been integrated onto the AAA couriers, the repeatability of the localization is expected to move closer to the sub-micron resolution of our optical and magnetic sensors. We have successfully developed programs in simulation within the interface tool and transferred them to the physical agents and run them, after exploration, with no change.

The most pressing deficiency in our current interface tool is the hard-coding of the assembly constraints and behavioral specifications as C++ methods, which means that no fundamentally new agents can be introduced to the system at run time. Another major problem with our current interface tool is that specifications of components and products must go through a tedious and somewhat error prone conversion process from their original CAD model to our current custom factory description format. We are investigating addressing both of these problems through use of the STEP standard (Standard Exchange of Product Model Data)[14]. Adopting STEP will give a means of importing models using an industry standard mechanism while also providing a means of adding the run-time flexibility we need, as extensible elements of STEP can be used to specify behaviors and assembly constraints of novel agents at run-time.

There are many other areas of research and development involving the interface tool, such as using geometric and physical modeling in the simulations or implementing better programming interfaces. We are attempting to build the foundation from which we can experiment in these various areas.

AAA is not only a comprehensive vision, it is an integrated vision which is involved in the development of an automation line throughout its life cycle, from conception to implementation. In fact, the goal of AAA is to not only be involved at the various stages of development, but to blur the distinction between the stages to allow incremental design and modification of automated factories. Such an integrated vision demands an integrated interface tool such as the one we have implemented. Fortunately, at the same time as AAA demands an integrated interface tool, it provides the mechanisms that make such a tool if not trivial, at least straight-forward to implement.

Acknowledgements

This work was supported in part by NSF grant DMI-9523156. Zack Butler was supported in part by an NSF Graduate Research Fellowship. The authors would like to thank Ralph Hollis, Alfred Rizzi, Arthur Quaid, and Patrick Muir for their invaluable work on this project and support for this paper.

References

[1] B. Brunner, K. Arbter, and G. Hirzinger. Task directed programming of sensor based robots. In *Proceedings of IEEE Int'l. Conf. on Intelligent Robots and Systems*, pages 1080–1087, 1994.

[2] Z. J. Butler. CC_R: A complete algorithm for contact-sensor based coverage of rectilinear environments. Technical Report CMU-RI-TR-98-27, Robotics Institute, Carnegie Mellon Univ., 1998.

[3] S. F. Chan, R. H. Weston, and K. Case. Robot simulation and off-line programming. *Computer-Aided Engineering Journal*, pages 157–162, August 1988.

[4] J. J. Craig. Simulation-based robot cell design in AdeptRapid. In *IEEE Int'l. Conf. on Robotics and Automation*, pages 3214–3219, Albuquerque, NM, April 1997.

[5] S. Derby. GRASP from computer aided robot design to off-line programming. *Robotics Age*, 5(2):11–13, 1984.

[6] J. Gowdy and A. A. Rizzi. Programming in the architecture for agile assembly. In *IEEE Int'l. Conf. on Robotics and Automation*, 1999.

[7] J. W. Ma. Precision optical coordination sensor for cooperative 2-DOF robots. Master's thesis, Carnegie Mellon, 1998.

[8] R. S. McMaster and F. M. Ribeiro. Cell calibration and robot tracking. In *IEE Colloquium on Next Steps for Industrial Robots*, 1994.

[9] J. S. Mogal. IGRIP - a graphics simulation program for workcell layout and off-line programming. In *Robots 10 Conf. Proc.*, pages 65–77, 1988.

[10] A. E. Quaid and R. L. Hollis. 3-DOF closed-loop control for planar linear motors. In *Proc. IEEE Int'l Conf. on Robotics and Automation*, pages 2488–2493, Leuven, Belgium, May 1998.

[11] A. A. Rizzi. Hybrid control as a method for robot motion programming. In *IEEE Int'l. Conf. on Robotics and Automation*, pages 832–837, Leuven, Belgium, May 1998.

[12] A. A. Rizzi, J. Gowdy, and R. L. Hollis. Agile Assembly Architecture: An agent-based approach to modular precision assembly systems. In *IEEE Int'l. Conf. on Robotics and Automation*, pages 20–25, Albuquerque, NM, April 1997.

[13] SILMA Division, Adept Technology Inc., San Jose, CA. *The Adept-Rapid Users Manual*, 1996.

[14] STEP: Standard for the exchange of product model data. ISO 10303, 1994.

Programming in the Architecture for Agile Assembly

Jay Gowdy and Alfred A. Rizzi

The Robotics Institute
Carnegie Mellon University

Abstract

The goal of the Architecture for Agile Assembly (AAA) is to enable rapid deployment and reconfiguration of automated assembly systems through the use of cooperating, modular, robust, robotic agents. AAA agent programs must be completely distributed and specify cooperative precision behavior in a structured, well known environment. Thus, the structure of agent programs is carefully designed to allow packaging of all the information necessary for coordinated execution when downloaded to a physical agent. To make the specification and execution of the potentially complex and fragile cooperative behaviors robust, our programs define ordered sets of control strategies and allow a low-level real-time hybrid control system to sequence the strategies rather than burdening the agent program with the management of this critical detail. This novel approach to programming automation systems has been tested both in simulation and on prototype hardware.

1 Introduction

The overall goal of the Architecture for Agile Assembly (AAA) is agility, — to enable both the rapid deployment of factories to deliver a product to market quickly and the rapid reconfiguration of factories to adapt to changing technologies and market needs. As described in [7], AAA achieves such agility by depending on modular robust robotic *agents*. Each agent operates in a deliberately limited domain, but possesses a high degree of capability within that domain. For example, our instantiation of AAA, minifactory, is focused on four degree-of-freedom (DOF) assembly of high-value, high-precision electro-mechanical systems (Fig. 1). In a minifactory there are agents (called couriers) that are "experts" in moving products in the plane of the factory floor, and other agents (called manipulators) that are "experts" in lifting and rotating products. The agents are physically, computationally, and algorithmically modular, and thus only when acting cooperatively in groups can they perform the 4 DOF operations required to produce a product.

In AAA, specifying factory behavior presents some

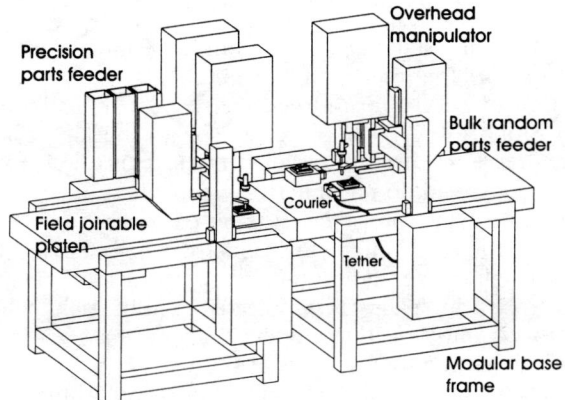

Figure 1: A minifactory segment

unique challenges, since there is no central factory "brain", and thus there is no single program for an entire AAA factory. Instead each agent has its own program which must reliably execute without access to any central or global database. AAA does provide an integrated interface tool, described in [3], which allows centralized design, simulation, and monitoring of the factory, but this centralized tool need not be present for a factory to operate.

In practice an agent's execution in AAA is divided into two layers. A higher-level discrete layer is responsible for the semantics of factory operation and the associated discrete events. This layer must deal with such issues as resource negotiation, factory scheduling, and product flow decisions. In general these tasks require minimal communications bandwidth between agents and are not concerned with true real-time operation of the agent. A lower-level continuous layer is responsible for sequencing and executing the specific control laws used to effect the physical environment of the agent. This continuous layer not only executes individual parameterized control strategies, but also manages the transitions between a carefully selected set of parameterized controllers. The continuous layer may require high communications bandwidth, since often the states that it must monitor will be on other agents, and true real-time operation is critical. This notion of automatically managing the transition between controllers was introduced in [2] and applied, theoretically, to the domain of minifactory in [6].

The programs written by the user and downloaded to the agent form the upper half of this program hierarchy. The lower half is "hard-coded" in the form of a palette of real-time control strategies and a manager which executes them and administers their sequencing. The discrete layer programs parameterize and deploy the control strategies used by the continuous layer, and then views the continuous state of the agent through the discrete "lens" of monitoring transitions between controllers.

1.1 Programming Model

Most industrial robot programming languages are based on standard computer languages, with the addition of special primitives, constructs, and libraries to support the physical control of a robot[5]. These languages are usually targeted at the control of a single robot, and do not inherently provide support for a program which must be distributed across many different robots. While this model can be effective for "trade-show" or "laboratory" demonstrations of a single robot, it leads to significant complications when a robot must be integrated and coordinate with its neighbors in an actual factory.

More abstract programming models are available[1, 4]; typically these are either "task" or "process" based and are often utilized for programming of work cells — which may contain multiple robots. While such approaches can eliminate some of the problems associated with developing coordination strategies for arbitrary machines, they require a central system "controller," and are thus vulnerable to single point failures and bottlenecks.

Fundamentally, these approaches to robot programming make a distinction between the continuous domain of control theory and the discrete domain of event management. We choose to place this distinction at a slightly higher level and make it a more formal abstraction barrier than most. Traditionally, the continuous, state based view is relegated to running controllers, with all decisions about which controllers to run and when to run them made by systems using a discrete, event based view. Instead we make use of continuous mechanisms to guide the transitions between controllers as well as to run the controllers themselves, freeing the agent programs to deal with the more relevant and abstract problem off deciding what to do and how to do it.

2 Distributed Programming

As there is no central controlling program for a minifactory, the operation of the system results from the cooperation of programs running on each agent. The agents interact with each other and with the factory infrastructure to perform the desired assembly task in an efficient and reliable manner. The distributed, but cooperative nature of the program content has considerable implications upon the program form.

Currently, our agent programs are completely text based and written in Python, an object oriented language which can be interpreted or byte-compiled[8]. An agent program is not simply a script, but rather defines an instance of a class which has a number of specific methods. The program can define a new class to be instantiated, or subclass from a preexisting one, but the class must provide a standard set of methods to be valid. This concept is very similar to the Java applets that are used in world-wide web programming.

A key to writing and distributing an agent program is that even though each agent's program must execute without access to any central database, each individual agent program will necessarily reference other parts of the factory. For example, a courier must be able to know it will be interacting with a particular manipulator much as a manipulator needs to know it will get parts of a specific type from a specific parts feeder.

```
# Agent class definition
class Program(ManipProgram):
    # Binding method
    def bind(self):
        # bind a bulk feeder
        self.feeder = self.bindDescription("ShaftFeeder")

        # bind product information
        self.product = self.bindPrototype("ShaftB")

    # Execution method
    def run(self):
        while 1:
            # convenience function for getting a
            # product from a feeder
            self.getPartFromFeeder(self.product, self.feeder)

            # Wait for a courier to rendezvous
            # with the manipulator for feeding
            partner = self.acceptRendezvous("Feeding")

            # and transfer the product to the courier
            self.transferGraspedProduct(partner)

# instantiate the applet
program = Program()
```

Figure 2: A simple manipulator program

In order to reference these factory components within the text of an agent program, users refer to these factory elements by names. Currently, the names must be unique in the factory, *i.e.* if a manipulator references a parts feeder named *ShaftFeeder* then there must be only one parts feeder with that name in the factory. A running physical agent can not resolve the name *ShaftFeeder* with a central resource, so the agent program is split into two segments, a "bind" step and a "run" step, which means that any valid program instance must have two methods, bind and run. The bind method declares what "global" entities the agent will utilize during its execution. The run method is the script which actually runs during execution, implementing the "high-level" discrete logic of the agent which initiates and coordinates the agent behavior. Figure 2 shows the definition of these methods for a

```python
# "Applet" class definition
class Program(CourierProgram):
    # Binding method
    def bind(self):
        # superclass has some binding to do
        CourierProgram.bind(self)

        # Bind to a particular manipulator
        self.source = self.bindAgent("FeederManip")

        # Bind to a particular factory area
        self.corridor = self.bindArea("CorridorA")

    # Execution method
    def run(self):
        # initialize the movement
        self.startIn(self.corridor)

        # block until manipulator is ready
        self.initiateRendezvous(self.source, "Feeding")

        # move into the workspace
        self.moveTo(self.sourceArea)

        # coordinate with manipulator to
        # get product from it
        self.acceptProduct()

        # The coordinated maneuver is done
        self.finishRendezvous("Loading")

        # move out of the workspace
        self.moveTo(self.corridor, blocking=1)

# instantiate the applet
program = Program()
```

Figure 3: A simple courier program

sample manipulator program.

When executing a simulated agent program, the bind step is simply a matter of looking up the relevant items in the simulation database and proceeding to execute the run method. Before an agent program can be downloaded from the simulation environment to the physical agent it must be first be "bound" with all of the global factory information the agent will need to run the program. To bind a program, the interface tool executes that program's bind method, and uses the results to construct a small database of all the information necessary for the agent to locate, both geometrically and logically, all of the factory elements needed to run the program without reference to any global resources. For example, in the sample courier program (Fig 3) the bind method calls bindAgent("FeederManip"), which declares that the agent program wants to know about the manipulator named *FeederManip* and assigns the result of that binding to a local member variable for use in the run method. As a result of the invocation, the interface tool will add the relative position of *FeederManip* in the courier's frame of reference as well as the network address of *FeederManip* to the local database which is sent to the courier with the program text. After being downloaded with this database the physical agent can execute the run method, which in-turn interacts with the lower-level continuous layer to execute the desired physical behavior.

3 High level protocols

Just as there is no central database that agents can rely on during factory operations, there is no central coordinator to organize and direct the agents. Each agent must be programmed to coordinate with its peers to effect the appropriate product flow and assembly operations. In order to achieve this coordination, agents must know and understand common communications protocols. We identify several different types of protocols within our factory, such as built-in protocols that every agent must provide in order to make possible safe factory operations, and extensible protocols that are specific to a particular instantiation of AAA or particular solution approach.

3.1 Built-in protocols

A built-in protocol is one that will be necessary for any agent in any AAA system to produce and understand. For example, there can not be any central arbiter parceling out resources in any AAA system, so every agent has built-in the ability to negotiate with other agents over resource reservation. Agents must have this ability to negotiate for resources in order to ensure safe factory operations.

The primary shared resource that our agents negotiate over currently is space on the platen. We assume that a courier will only go where it says it will go, and that there are no "outside" influences which fail to reserve the resources they consume. These assumptions — which are reasonable in the highly structured, very stable, and well known minifactory environment — allow us to dispense with the inter-agent perception systems that would be necessary to implement completely "reactive" motion, in which agents would be required to observe other agents' positions and intentions (either with sensors or by querying) prior to taking action. The low cost and predictable behavior obtained through the use of a reservation system far outweighs the risk of our assumptions being violated and the minor efficiency losses which will inevitably be incurred. We foresee using a similar distributed reservation system to arbitrate the consumption of more abstract resources such as vibration, noise, thermal, or optical emissions.

Another example of a built-in information protocol is seen during a rendezvous, *i.e.* when a courier and a manipulator cooperate to perform some process on the products: When agents cooperate to perform the manufacturing process, they also must exchange information about that process. In AAA, there is no central database of product information, so product information must flow with the products themselves. Our products have two levels of information, prototype information — information that is true about all products of a certain type, such as nominal geometry, and instance information — information that applies only to a particular instance of a product, such as serial numbers or dimension variations. AAA provides built in protocols for passing product instance information between agents and for acquiring product prototype information either from peers at run-time or from a database at program binding time.

3.2 Extensible protocols

One more protocol that all agents share is the protocol for defining and extending semantic protocols. A particular semantic protocol may not be in use by all agents, but agents can negotiate to confirm whether they share the same semantic protocols before proceeding with operations.

For example, in our current approach to programming agents in minifactory, we view the agent programs as having two types of interactions, the rendezvous between a courier and a manipulator in which the manufacturing process is performed and information about the process is exchanged, and the gross courier motion, in which couriers move from rendezvous to rendezvous without colliding. Keeping the couriers from colliding into each other results from using the built-in geometry resource negotiating protocols, but deciding in what order couriers may rendezvous with manipulators, *i.e.* distributed factory scheduling, is the domain of an extensible protocol. An example of this protocol can be seen in the sample courier program (Fig. 3), which initiates a rendezvous to be accepted as specified in the sample manipulator program (Fig. 2).

This protocol is particular to minifactory, and particular to our current approaches to programming minifactory agents. It may not be useful in other AAA instantiations, and almost certainly will be significantly changed or augmented over time in our minifactory instantiation. Thus, a courier and a manipulator need to negotiate to ensure that they share a common rendezvous protocol before they can work together.

4 Low Level Programming

As outlined in Section 1 an agent program in a minifactory has two distinct but related run-time responsibilities: *i)* it must carry out semantic negotiations with its peers to perform the goal of the factory; *ii)* it must properly parameterize and sequence the application of low-level control strategies to successfully manipulate the physical world. The programming model we are utilizing attempts to simplify the relationship between these two responsibilities and minimize their impact upon one another. The basics of how to accomplish high-level programming tasks was the topic of Section 3. Here we turn our attention to the low-level tasks and the programming of physical motion.

Specifically, in an effort to reduce the complexity associated with writing programs for agents we have adopted the notion of allowing the low level control strategies to become responsible for their own switching and sequencing. Thus the problem of deciding exactly when and how to switch between low level control strategies is removed from the agent program and is thus isolated from the high-level semantic negotiations that are the primary domain of the agent program.

4.1 Underlying Model

The fundamental model for the execution of control strategies was presented in [6]. Briefly, rather than ask the program to generate trajectories through the free configuration space of the agent, the program will be responsible for decomposing the free configuration space into overlapping regions and parameterizing controllers associated with each region. A *hybrid control* system is then responsible for switching or sequencing between the control policies associated with this decomposition to achieve a desired overall goal.

This scheme describes the behavior of any one agent in terms of a collection of feedback strategies based on the state of both the individual agent and its immediate peers. The result is a hybrid on-line control policy (one that switches between various continuous policies) which makes use of the entire collection of available policies to systematically make progress toward a goal based on an agent's estimate of both its own and its peers' state. To provide the desired level of system flexibility the selection of goals and the associated prioritized decomposition of the free space is left to the agent program.

More formally, given a set of controllers, $U = \{\Phi_1, ..., \Phi_N\}$, each with an associated goal, $G(\Phi_i)$, and domain, $D(\Phi_i)$ — where it is presumed that under the action of Φ_i any state that starts in $D(\Phi_i)$ will be taken to $G(\Phi_i)$ without leaving the set $D(\Phi_i)$. We then say that controller Φ_1 *prepares* controller Φ_2, denoted $\Phi_1 \succeq \Phi_2$, if its goal lies within the domain of the second $G(\Phi_1) \subset D(\Phi_2)$ [2, 6]. For an appropriately parameterized set of controllers, U, this relation induces a generally cyclic directed graph. Assuming that the overall goal, G, coincides with the goal of at least one controller, $G(\Phi_i) = G$, then by starting with Φ_i and recursively tracing the relation backwards through the corresponding graph, one arrives at $U_G \subset U$ — the set of all controllers from whose domains the overall goal might be eventually reached by switching between control policies. The domain of a properly conceived composite controller, should then be $\bigcup_{\Phi \in U_G} D(\Phi)$, and thus we have an "automatic" method by which to guide the system from any state in this union of domains to the goal.

Consider, for example, the trivial planar configuration space depicted in Fig. 4. Note that the free space has been decomposed into four separate regions with the overall goal located in the upper right corner of the configuration space (G_4). Here, Φ_1 is responsible for for taking all states in the lower convex region to

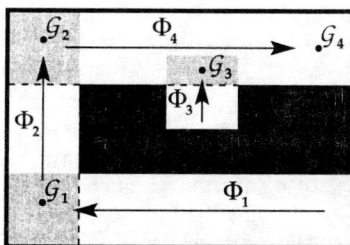

Figure 4: Example decomposition of a trivial planar configuration space.

G_1, and thus prepares Φ_2. Similarly the placement of G_2 and G_3 allow both Φ_2 and Φ_3 to prepare Φ_4 which regulates the state to G_4, the overall goal. It is the responsibility of the underlying hybrid control system, discussed in detail in [2], to switch between the individual Φ_i to achieve this overall goal. While this trivial example is illustrative it is important to note that we have only considered the configuration of the system in this example, while in general the actual domains, $D(\Phi_i)$, for the constituent controllers are defined over the state space of the system — the positions and velocities of the agent as well as those of its peers with which it is closely coordinating.

4.2 Programmatic Interface

Given the underlying model for executing physical action described in Section 4.1, it becomes the responsibility of the agent program (specifically the script defined by its run method) to create, parameterize, and manage the currently active set of controllers, U, along with the associated sets of goals, $G(\Phi_i)$, and domains, $D(\Phi_i)$. Thus the script is only responsible for choosing the current "overall" goal along with appropriate intermediate sub-goals, and providing parameterizations of control strategies to accomplish those goals. The complex and error prone problem of making real-time changes to the underlying control system is left to the hybrid control structure outlined above.

The interface between the script and this *controller manager* is quite straightforward. The class from which a particular agent program instance is derived provides standard tools for creating and parameterizing controllers and their associated domains. Having constructed a controller the script can then insert into an ordered list of active controllers from which the controller manager will select the appropriate instance to execute in real-time. The details of high-bandwidth monitoring and coordination of an agent and its peers state is performed by these lower levels, and utilizes a dedicated local network. This local network is used to pass relevant information between agents only about those variables that effect their execution, resulting in efficient utilization of the available communication bandwidth in a manner that is transparent to the agent program.

Communication of progress and completion of tasks back to the script is accomplished by use of either call-back functions or direct polling of the actual state of the agent. In general the expectation is that scripts will submit a moderately sized list of control actions along with a set of fail-safe and fall-back strategies capable of responding to the most dire circumstances, then *sleep* (wait for a call-back) until either progress has been made or a failure has been detected. When appropriate progress has been made the script will, while motion is still executing, append additional control actions to the "top" of the active controller list indicating new goals and delete those control actions which are no longer useful. If a failure has been detected the program will proceed in a similar fashion, only the actions added to the list will most likely attempt to recover from the problem.

By parameterizing (setting the goal, defining the domain of applicability, specifying gains, etc.) the specific controllers and ordering of their placement on the list of active controllers a script is able to specify complex and efficient physical motion that is fundamentally robust. This provides a rich and expressive method for programs to specify physical motion while at the same time minimizing the risks associated with writing those programs.

```
# submit actions to move from self.current to area
def moveTo(area):
    # get the goal at boundary of area
    # and self.current in self.current
    x,y = self.getBoundaryGoal(area)

    # create and submit action
    controller = self.goTo(x,y)
    domain = self.inArea(self.current)
    self.submit(controller, domain)

    # reserve area, blocking if necessary
    self.reserve(area)

    # get goal at boundary of area and
    # self.current in area
    x,y,overlap = self.getOverlapGoal(area)

    # create and submit action to cross into
    # the new area
    self.submit(self.goTo(x,y), self.inRegion(overlap))

    # create and submit action to drive to the
    # goal in area
    # note that a callback class is invoked when
    # this action starts which unreserves self.current
    self.submit(self.goTo(x,y), self.inArea(area),
                start=Unreserve(self.current))

    # keep track of current area
    self.current = area
```

Figure 5: Code fragment for moveTo.

In practice the details of this interface are hidden from the programmer by a set of standard "convenience functions." For example the moveTo(...) call in Fig. 3, would actually expand to the code fragment shown in Fig. 5. It is here that the specific resource reservation protocol mentioned in Section 3 is implemented and where a "standard" set of controllers are parameterized and placed on the list of active controllers. Note the registration of a call-back method to indicate exit from the initial area and to free the reservation held on it.

5 Conclusion and Future Work

The requirements of AAA have led us to a new model for programming assembly systems. AAA agent programs must be completely distributed and specify cooperative precise behavior in a structured, well known environment. Thus, the structure of agent programs is carefully designed to allow packaging of all the information needed to execute when downloaded to a physical agent. The programs must use standard high-level protocols to initiate the required cooperative behavior. To make the specification and execution of the potentially complex and fragile cooperative behaviors robust, our programs define ordered sets of control strategies and allow a low-level real-time hybrid control system to sequence the strategies rather than burdening the agent program with the management of this critical detail.

We have tested this approach in simulation by constructing virtual factories with several couriers and manipulators cooperating to perform part of the assembly of small (2 millimeter) transducers. In addition, we have written agent programs which, both in simulation and hardware, exercise our prototype courier. We are currently integrating our prototype manipulator in order to implement multi-agent pick-and-place tasks. We will continue to validate the programming approach with real tasks as we develop additional hardware that supports those tasks.

There is much yet to do to address some of the practical implications of our programming model. For example, in order to produce a working factory, users must generate many correct cooperating agent programs. Fortunately, an individual agent's scope is fairly limited, and it has powerful tools for working within its scope so our hope is that each agent program will be relatively simple and short. Unfortunately, no matter how short or simple the programs, the fact remains that some factory programmer has to generate an individual program for each agent in the system. In addition to the potential tedium of generating dozens of programs, the user is essentially writing large, very distributed programs, with all of the known pitfalls of that domain, such as deadlocks or livelocks.

We could address this problem through the use of graphical programming techniques to ease the production of the individual agent programs, but we feel that any advantage gained would be purely cosmetic. Fundamentally, what is required is a method of presenting the factory programmer with a different way of looking at the programming problem. For example, users may want a factory-centric view of the problem, in which they can specify the factory behavior as a whole by inputting a work-flow model, *i.e.* what processes have to occur and in what order. Ultimately, users may want to take a product-centric view, in which they enter product models annotated with some process information. The AAA programming environment would have to provide semi-automatic, user-guided methods of transforming such centralized user views into factory layouts and distributed agent programs.

Regardless of what view the user has of factory programming, agent-centric, factory-centric, or product-centric, ultimately an actual AAA factory must execute that user program as a set of completely distributed programs on a set of agents interacting with each other and with the product components to perform the assembly task. This paper has documented the programming model and protocols we have designed as a basic building block for future systems which can bring the vision of rapid deployment, reconfiguring, and reprogramming of automated assembly systems closer to reality.

Acknowledgements

This work was supported in part by NSF grant DMI-9523156. The authors would like to thank Ralph Hollis, Arthur Quaid, Zack Butler, and Patrick Muir for their invaluable work on the project and support for this paper.

References

[1] G. Berry and G. Gonthier. The ESTEREL synchronous programming language: design, semantics, implementation. *Science of Computer Programming*, 19(2):87–152, November 1992.

[2] R. R. Burridge, A. A. Rizzi, and D. E. Koditschek. Sequential composition of dynamically dexterous robot behaviors. *International Journal of Robotics Research*, 1998. (to appear).

[3] J. Gowdy and Z. J. Butler. An integrated interface tool for the architecture for agile assembly. In *IEEE Int'l. Conf. on Robotics and Automation*, 1999.

[4] R. W. Harrigan. Automating the operation of robots in hazardous environments. In *Proceedings of the IEEE/RSJ Int'l Conf. on Intelligent Robots and Systems*, pages 1211–1219, Yokohama, Japan, July 1993.

[5] T. Lozano-Perez. Robot programming. *Proceedings of IEEE*, 71(7):821–841, 1983.

[6] A. A. Rizzi. Hybrid control as a method for robot motion programming. In *IEEE Int'l. Conf. on Robotics and Automation*, pages 832–837, Leuven Belgium, May 1998.

[7] A. A. Rizzi, J. Gowdy, and R. L. Hollis. Agile Assembly Architecture: An agent-based approach to modular precision assembly systems. In *IEEE Int'l. Conf. on Robotics and Automation*, pages 20–25, Albuquerque, NM, April 1997.

[8] G. van Rossum. *Python Tutorial*. Corporation for National Research Initiatives, Reston, VA, August 1998.

An Approach to Automated Programming of Industrial Robots

Donald R. Myers
Intelligent Automation, Inc.
2 Research Place
Rockville, MD 20855
email: dmyers@i-a-i.com

Abstract

This project develops a "teach-by showing" approach that permits unskilled users to program robots by physically demonstrating examples of a task to a trainable controller. Each example may be shown in one of three ways: by direct demonstration, by placing the robot in a zero-force mode and physically leading it through the task, or by operating the robot in a master/slave mode. During each demonstration, both the sensor inputs to the robot controller and the control outputs from the controller are presented to the training system. Following the demonstrations, the training system then produces a prototype procedural program suitable for downloading into a commercial robot controller. Unlike the conventional program development process, the operator need not postulate about the value of sensor feedback variables required at a particular level of the controller. In fact, the operator need not be familiar with the syntax and semantics of the programming language at all. Examples of successful, but perhaps quite different task strategies can be demonstrated by the human teacher until the robot system is capable of successful autonomous operation.

Background

For certain applications it is not possible for humans to develop fully functional computer programs to control complex systems. In some applications, domain knowledge is incomplete or insufficient. In other applications human time or resources may be insufficient. This work is part of an on-going effort to develop systems that can automatically program military and industrial robots.

Conventional programming methods require the user to learn the syntax and semantics of a proprietary robot programming language and to debug the program by iteratively editing, downloading, and executing the program. If the task to be performed by the robot requires sensor feedback (e.g., force or touch), the programmer must speculate on the proper sensor values and guardedly wait by the "emergency-stop" button in case of error.

Hannaford and Lee (1991) were among the first to investigate teach-by-showing methods. They used hidden Markov models (HMM) to segment force readings from one-degree-of-freedom gripper teleoperated "peg-in-hole" assembly manipulations. The operator manually indicated the beginning and completion of each subtask and its name.

Ikeuchi, et al. (1993) develped an approach that uses video cameras to monitor a human teacher performing a task. The vision system recognizes object relations and transition between relations. It then maps relation transitions to the assembly tasks required to cause such transitions. A program is then generated that instructs a robot to reproduce the series of movements originally performed by the human.

Takahashi, et al.(1993) require the teacher to lead the end-effector of the robot through successful completion of a task. The method then classifies the teacher's movements into assembly states, represents each state as a set of primitive movements, and uses this set to drive the robot. The assembly states are derived, however, from a geometric model of the parts to be manipulated and the fixture. An off-line CAD model from which an assembly graph can be constructed is required.

Skubic and Volz (1997) use a robot under teleoperated control to demonstrate a contact-type task. End-effector forces and moments were sampled throughout the duration of the task. A fuzzy classifier was constructed to separate force/moment data into classes. The membership functions for the fuzzy classifier were constructed from the mean and standard deviation from half the demonstrated examples representative of each class.

Partially supported by US Ballistic Missile Defense Organization, Contract Nos. DASG60-97-M-0105 and DASG60-99-C-0008

Approach

The approach used here is to treat the task level of a robot controller as an unknown system and characterize the intended behavior of that level using a set of fuzzy inference rules. With a human operator demonstrating the task, the inputs to and outputs from the level can be sampled. By treating the problem as a classical system identification problem, a model of the task level can be found. In our case, we choose to characterize the level using fuzzy inference rules, because the inference rules can be easily translated into a conventional robot-specific procedural program. The resultant program can then be used to control a robot performing a similar task.

As a simple example of the methods to be used, consider a simple one degree-of-freedom (DOF) robot whose function is to maintain contact with an object moving at a variable velocity along the robot coordinate axis. The robot has a force sensor to monitor contact with the object. Consider force to be the sensed variable and velocity as the control variable. As a user demonstrates the task, force and velocity are measured and graphed as shown in Fig. 1.

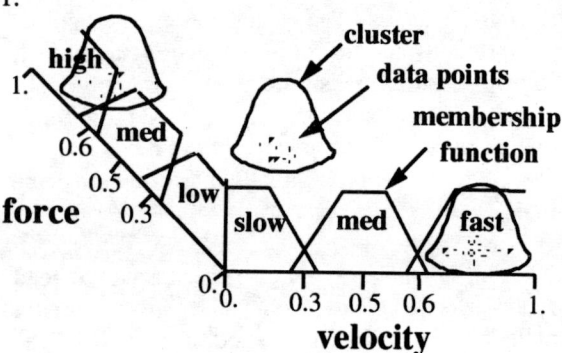

Figure 1. Sensor Cluster Example

Force and velocity are represented as two fuzzy variables whose universe of discourses are normalized. Note that the data appear to cluster in three distinct regions of force-velocity space. The height of the cluster curves (the third dimension in the graph) represents the degree to which any one data point belongs to a cluster.

Membership functions are developed by projecting each cluster onto each of the variable axes and constructing a piece-wise linear trapezoid that encloses the data points for that cluster. In this example, three membership functions are defined for each variable. Fig. 2 shows the projection of data from the top, left cluster of Fig. 1 onto the *force* axis. A trapezoidal curve, containing the data points, represents the membership function for *high force*.

The height of the trapezoids represents the degree to which a data point belongs to the cluster. Linguistic descriptors can then be assigned to each membership function. For the variable *velocity*, the membership functions are *slow, medium,* and *fast*. For the variable *force*, the membership functions are *low, medium,* and *high*. The values of the degree of membership are within the interval [0,1].

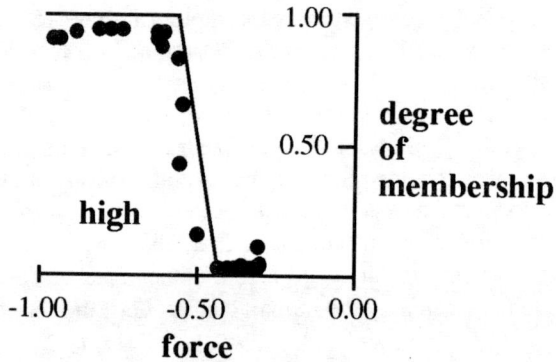

Figure 2. Construction of *High Force* Membership Function

Three inference rules are evident from the cluster projections:

   IF force is high
      velocity is slow

   IF force is medium
      velocity is medium

   IF force is low
      velocity is fast

These inference rules can then be translated into the robot's proprietary programming language and used to control the robot to perform the task demonstrated by the operator.

Experimental Work

The teach-by-showing (TBS) approach was demonstrated and tested by docking a pallet into a fixture. The task was sufficiently unconstrained that sensors were required to accomplish the task.

The robot used was a PUMA 560 equipped with a six-axis force sensor and pneumatically actuated parallel-jaw gripper. The pallet to be acquired was a 30cm x 30cm x 10cm wooden box on which a handle, designed for use with the parallel-jaw gripper, was mounted. A receptacle for the pallet was designed slightly larger than the outer dimensions of the pallet.

For this initial effort, the demonstration task was limited to three translational degrees-of-freedom. The pallet and receptacle and coordinate frames assigned to each are shown in Fig. 3. Forces/moments F_x, M_y, and M_z were considered as the sensed variables and velocities V_x, V_y, and V_z were considered as the control variables. A second parallel-jaw gripper was constructed to serve as a teaching gripper. The teaching gripper was constructed as a scissors mechanism so that internal gripping forces were reflected to the teacher's hand in a 1:1 ratio. Both grippers incorporated six-axis force/torque and position/orientation sensors. The teaching gripper is shown in Fig. 4.

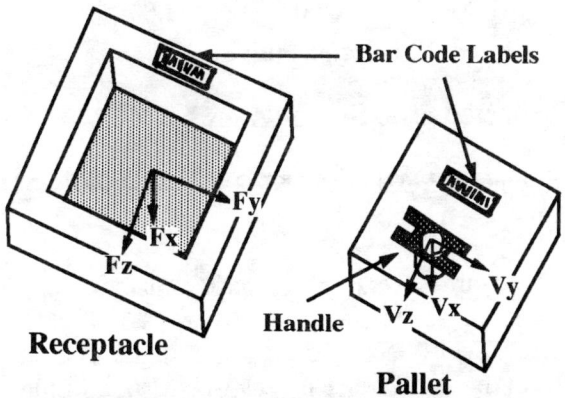

Figure 3. Pallet and Receptacle

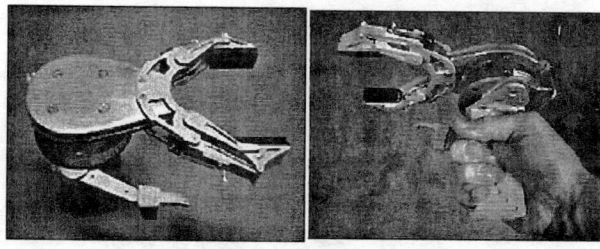

Figure 4. Teaching Gripper

The robot was controlled through a Pentium-based PC. The PC controlled the manipulator, the end-effector, and the data collection from the sensors. A simple program, coded in the robot control language VAL II, allowed the PUMA to accept the velocity commands from the PC, add the velocities to the present position of the robot, and drive the robot to the new desired position.

For the training runs, three teachers manually demonstrated the task to be performed by the robot. Each teacher, using the teaching gripper, grasped the pallet by the handle and placed it into the receptacle. Since the tolerance between the pallet and the receptacle prevented insertion without contacting the sides of the receptacle well, a force-guided insertion strategy was required. The teacher was instructed to make his/her insertion motions slow and deliberate. Data on the position of the pallet and the reaction forces were collected as the human teacher performed the task.

Algorithms to reduce the data to a procedural program were based on qualitative modeling of dynamic systems using fuzzy set theory. Essentially, the velocities and forces are represented in a six-dimensional hyperspace. The data are then examined to find clusters that reveal the underlying relationship between input commands and output response. The algorithms used to cluster the data and determine the degree to which each data point belongs to each cluster are based largely on the fuzzy c-means method (FCM) introduced by Bezdek (1981). FCM is an iterative procedure for finding the center of each cluster and the grade of membership for each data point in each cluster, given the anticipated number of clusters. Since the number of clusters is not known a priori in this work, an additional criteria (as in Sugeno, 1993) was introduced to determine the number of clusters.

As an example of how inference rules are constructed, consider the projection of one of the data clusters (#7) onto the M_y, M_z, V_x, and V_z axes (Fig. 5). The abscissa represents the values of the sensor variables, normalized from -1 to +1. Velocity and force were normalized by dividing each data point by the maximum of the absolute value of the largest force and velocity values within the data set. Positive and negative force and velocity directions are indicated by the direction of arrows of the force coordinate system shown in Fig. 3. The ordinate represents the degree to which the data belongs to cluster 0, from 0 to +1. The dots represent the individual data points. The solid lines forming triangles represent the membership functions that were constructed from the cluster projections.

Triangular-shaped membership functions were constructed to decrease the programming complexity and increase the run-time efficiency during defuzzification of the resultant output

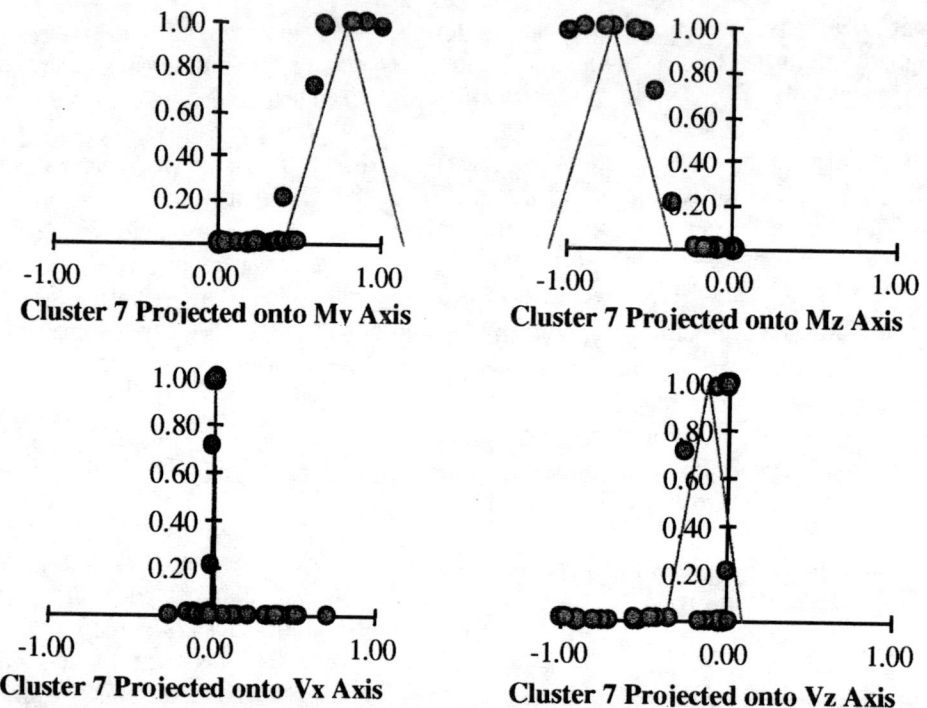

Figure 5. Data from Cluster 7 Projected onto Sensor Axes

velocities. From Fig. 5, it is evident from the values of data on the force axes that the cluster represents the portion of the task where the pallet is resting on the lower right quadrant of the receptacle. Although the descriptive labels have not yet been assigned to the membership functions, it is evident that components of the inference rule corresponding to the cluster will be something like:

$$IF(... \text{ AND } M_y > 0 \text{ AND } M_z < 0)$$
$$V_x \approx 0 \text{ AND } ... \text{ AND } V_z < 0$$

Finally, the linguistic descriptors are assigned to the membership functions. For this initial phase of the work, the linguistic descriptors were selected and assigned manually. Descriptors such as ZERO, SLOW, POSITIVE, and NEGATIVE were chosen to conform to the way in which an operator might describe the task. In future work, assignment of descriptors will be automated.

Using the derived membership functions, the inference rules shown in Fig. 6 were automatically constructed. For readers unfamiliar with fuzzy nomenclature, it should be emphasized that ZERO represents a fuzzy membership function. ZERO does not represent the crisp number zero. Loosely translated, ZERO means a value close to zero.

Similarly, POSITIVE, NEGATIVE, and SLOW are fuzzy membership functions.

There are several observations that are apparent. First, all inputs are explicitly tested and all outputs are explicitly set in each instruction. This follows from the fact that each vector consisting of the three forces and three velocities is represented in a six-dimensional space. TBS simply represents the mapping of vector clusters between the three inputs (F_x, M_y and M_z) and the three outputs (V_x, V_y, and V_z) in the form of fuzzy inference rules. Given this, one might expect that all possible combinations of input and output variables be represented. In the instructions set produced above, there is an omission of negative moments about the y-axis. The lack of NEGATIVE M_y rules can be explained by a physical constraint imposed receptacle. The bar code label on the receptacle was raised about 3 mm above the surface. As a result, all trainers avoided placing the pallet on that portion of receptacle. Consequently, no data sets with negative y moments were collected.

Results

An application program was coded using the conventional, iterative, trial-and-error approach and compared with the TBS-derived program. A simple program was developed to control a PUMA 560

```
BEGIN PROGRAM
    // MATE PALLET INTO RECEPTACLE
    DO FOREVER
        IF( $F_x$ IS ZERO AND $M_y$ IS ZERO AND $M_z$ IS ZERO )
            $V_x$ IS SLOW AND $V_y$ IS ZERO AND $V_z$ IS ZERO
        IF( $F_x > K_1$ AND $M_y$ IS ZERO AND $M_z$ IS ZERO )
            STOP
        IF( $F_x > K_1$ AND $M_y$ IS ZERO AND $M_z$ IS POSITIVE)
            $V_x$ IS ZERO AND $V_y$ IS ZERO AND $V_z$ IS SLOW
        IF( $F_x > K_1$ AND $M_y$ IS POSITIVE AND $M_z$ IS ZERO )
            $V_x$ IS ZERO AND $V_y$ IS SLOW AND $V_z$ IS ZERO
        IF( $F_x > K_1$ AND $M_y$ IS POSITIVE AND $M_z$ IS POSITIVE )
            $V_x$ IS ZERO AND $V_y$ IS SLOW AND $V_z$ IS SLOW
        IF( $F_x > K_1$ AND $M_y$ IS ZERO AND $M_z$ IS NEGATIVE)
            $V_x$ IS ZERO AND $V_y$ IS ZERO AND $V_z$ IS NEGATIVE SLOW
        IF( $F_x > K_1$ AND $M_y$ IS POSITIVE AND $M_z$ IS NEGATIVE )
            $V_x$ IS ZERO AND $V_y$ IS SLOW AND $V_z$ IS NEGATIVE SLOW
    END DO
END PROGRAM
```

Figure 6. Derived Inference Rules

robot using either the conventional program or the TBS inference rules. The pseudo-code for the program consists of:

```
BEGIN PROGRAM DEMO
    DO FOREVER
        Read force sensor
        Normalize force data
        Execute either conventional OR
            TBS inference rules
            to calculate velocity outputs
        Defuzzify velocities
        Denormalize velocities
        Send velocities to robot controller
    END DO
END PROGRAM
```

A simple program, coded in the robot control language, VAL II, was running on the PUMA to accept the velocity commands from the PC, add the velocities to the present position of the robot, and drive the robot to the new desired position. The manipulator controller itself consists of an LSI-11 residing on a Q-bus in which the trajectory and kinematic calculations are performed. The resultant inverse kinematic solutions are the joint set points, which are in turn passed across a 40 bit parallel interface to a J-bus on which six 6503-type processors handle the servo calculations for the joints.

Demonstrations with both the conventional program and the TBS inference rules were effective in placing the pallet into the receptacle. Both demonstrations performed effectively regardless of the starting position of the robot. In this initial phase of the work, no attempt was made to quantify and compare the performance of the robot running under the conventional or TBS programs.

Summary and Future Work

It should be emphasized that the intent of this work is not to develop another control law for mapping sensed forces into robot movements. The purpose of this work is to develop a means of automatically producing code for any level of a generic robot controller – regardless of whether the inputs and outputs are numeric or symbolic. The assembly application presented here was intended to illustrate our method of deriving the program logic.

A pleasing by-product of the fuzzy clustering approach is that the relative importance of all sensor inputs is explicitly quantified. The data from a sensor on which no membership functions have been defined from a cluster projection is not needed for that task. Of course, all sensors must be present and functional during the teach process. The TBS will not indicate the necessity of a sensory modality that was not originally used during teaching. It is anticipated that for typical use of the TBS, a complete complement of available sensors be used during the teach process. Following clustering, the TBS technique will then indicate the relative importance of each sensor.

One of the major problems with the current system is its inability to handle causality. As an example, during the teach process, as contact forces increase, the teacher compensates with slight movements away from the surface. Such a rule does not appear in the TBS program. The reason such a rule could not be produced by TBS is that large contact force causes compensatory movement, but an instantaneous snapshot of the sensor system will not show a large force in one direction along one axis and velocity in the opposite direction at the same instant in time. To handle such causal relationships, we propose to implement the program as a state machine. The input variables will be augmented to include the previous state of the system, and the output variables augmented to include the next state. A state table implementation will serve to include causality into the procedure.

Commercial Applications

The Advanced Manufacturing Technology Development Division within Ford Motor Company is tasked with identifying opportunities and developing improved methods to manufacture Ford products. A long-term target of the group, and of the automobile industry in general, is assembly of power train products, including such components as automatic transmissions. Currently all such assembly is manual. Many groups have tried to automate this process, but to date no one has been able to achieve this automatic assembly with sufficient flexibility or reliability for a practical system. IAI is working with Ford to apply TBS to the programming of robots to perform such tasks.

Two components are illustrated in Fig. 7 below. Although not clearly shown in the Figure, the inside surface of the housing on the left contains several concentric ring gears, each of which must be properly aligned before insertion can be accomplished. In a single transmission there may be over 1000 such parts, and there are may be several transmissions assembled on one line. Without innovative new ways to program these robots, the programming task would be enormous. In addition, a robot with advanced force or compliance control in six DOFs is required to perform the manipulations.

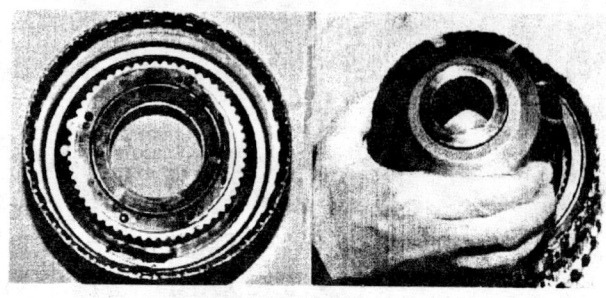

Figure 7. Two Gear Components from Automatic Transmission

References

J.C. Bezdek, *Pattern Recognition with Fuzzy Objective Function Algorithms.* New York: Plenum Press, 1981.

B. Hannaford and P. Lee, "Hidden Markov Model Analysis of Force/Torque Information in Telemanipulation," *Intl. Journ. Of Robotics Research*, pp. 528-538, Oct. 1991.

K. Ikeuchi, M. Kawade, and T. Suehiro, "Assembly Task Recognition with Planar, Curved and Mechanical Contacts," *IEEE Intl. Conf. on Robotics and Automation*, Atlanta, GA,, vol. 2, pp. 688-694, May 1993.

M. Skubic and R.A. Volz, "Learning Force Sensory Patterns and Skills from Human Demonstration," *IEEE Intl. Conf. on Robotics and Automation*, Albuquerque, NM, pp. 284-290, April 1997.

M. Sugeno and T. Yasukawa, "A Fuzzy-Logic-Based Approach to Qualitative Modeling," *IEEE Trans. on Fuzzy Systems,* vol. 1, no. 1, pp. 7-31, Feb. 1993.

T. Takahashi, H. Ogata, and S-Y Muto, "A Method for Analyzing Human Assembly Operations for Use in Automatically Generating Robot Commands," *IEEE Intl. Conf. on Robotics and Automation*, Atlanta, GA, vol. 2, pp. 695-700, May 1993.

An Object-Oriented Realtime Framework for Distributed Control Systems

A. Traub[*]
R. D. Schraft

Fraunhofer Institute
Manufacturing Engineering and Automation
Nobelstraße 12, 70569 Stuttgart (Germany)

This paper presents the 'Realtime Framework', a package of software modules for building distributed real-time control systems in robotics and automation. The Realtime Framework covers the areas of client-server communication, control of program flow and modality through messaging and state machines, and low-level input/output. In addition, it contains a real-time utilities package and wrappers for operating system calls, which shield any operating system dependence from the application built on top of the Framework. Software design patterns were used to document solutions for recurring problems in control systems involving several related classes. The Realtime Framework is mostly aimed at increasing software modularity, portability, and re-usability, thereby reducing software development costs. This paper gives an overview of the major features of the Realtime Framework and summarizes practical experiences from applying it to a mobile robot control system.

1. Introduction

Modern control systems in robotics and automation usually include a large number of different sensors and actuators, and require a considerable amount of data processing capacity. "Intelligent" sensors with micro-controllers help to reduce data transmission rates by pre-processing sensory information. Likewise, "intelligent" actuators can be controlled by high-level commands. Distributed control systems integrate "intelligent" sensors and actuators together with data processing units like computers and micro-controllers, achieving a high level of hardware modularity.

However, the procedural software design techniques which are still widely used in industrial control systems are often unable to cope with the complexities arising from heterogeneous distributed computing. Software, which is not designed in a thoroughly modular way, is hard to develop with large teams of programmers and difficult to adapt to new hardware platforms or new functional requirements. In addition, the robustness of procedural software suffers from the difficulty of testing individual software components independently of others. Besides procedural software design techniques, the second major problem in current control system development is that the communication protocols in distributed systems still vary widely depending on the manufacturer, the operating system and the particular network or bus system used. „Plug and Play" is therefore next to impossible for manufacturers of control system components. All the above factors lead to excessive costs for software development and maintenance.

In order to reduce these costs while maintaining a high level of software quality, the design goals for control software therefore have to include:

- easy and efficient communication between different hardware platforms on bus systems and networks commonly used in automation industry
- re-usability of software components
- portability between operating systems and hardware platforms
- easy familiarization of new team members with the development process and existing software
- robustness towards internal software errors and unforeseen external conditions
- high run-time performance

This paper presents the *Realtime Framework*, a framework of object-oriented software components which specifically addresses the above requirements. The *Realtime Framework* allows an easy and efficient implementation of the communication processes typical for distributed control systems. The hardware components can be connected over a TCP/IP network or a CAN [1] fieldbus, which is widely used in industrial automation systems. The modular structure of the framework increases software re-usability and completely shields any operating system dependence from the application specific software which uses the framework. In addition to

---

[*] Corresponding author; e-mail: *traub@ipa.fhg.de*

software components, it also includes software design patterns [2] which reduce the initial period of new members in the development team. Throughout the framework development, the issues of software robustness and real-time performance have been considered.

In the section 2 of this paper, related approaches to this topic are reviewed. Section 3 describes the overall architecture of an application built with the *Realtime Framework*, and section 4 describes the *Realtime Framework* components in detail. After discussing some implementation details in section 5, we describe a sample application of the Framework for mobile robot control in Section 6. Conclusions are summarized in section 7.

2. Related Work

Several frameworks for control systems have been developed, among them NEXUS [3] and OSACA [4]. They are similar to the *Realtime Framework* presented in this paper in that they use client/server architectures for communication. NEXUS is a decentralized communication system with hierarchical error recovery for robustness. It was developed for mobile robotics but is generic enough for use in other control systems. OSACA is a communication framework for distributed control applications under Windows NT and VxWorks. Both NEXUS and OSACA are limited to TCP/IP protocols and cannot be directly used for communication over industrial fieldbus systems widely used in automation, like CAN, [1], Interbus [10] and Profibus [11].

Commercially available products are ControlShell™ [5] and Rhapsody™ [6]. They feature graphical editors for data flow and state diagrams, from which source code can be generated. Automatically generated source code however tends to be hard to read and understand during debugging. Therefore, this strategy was not adopted for the *Realtime Framework*. ControlShell™ is particular in that it does not use a client/server architecture for communication, but a dissemination system called NDDS. Like NEXUS and OSACA, ControlShell™ [5] and Rhapsody™ cannot be used for communication over industrial fieldbus systems.

ACE [7] is a general communication framework for client/server architectures, mainly focused on TCP/IP networks. It is freely available in source code for several different Windows and UNIX/POSIX platforms and well documented. Among the frameworks presented in this section, it is the only one which explicitly uses software design patterns. In contrast to the *Realtime Framework* however, it does not include convenient classes for the kind of high-level command and data exchange typical in distributed control systems.

Several organizations have developed communication standards for specific fieldbus systems, among them CANOpen [1], DeviceNET [8,9], Interbus [10] and Profibus [11]. A promising new standard is OLE[*] for Process Control (OPC) [12]. Commercial OPC client and server software is available for a variety of network and fieldbus systems. The main drawback of OPC is that it was specifically designed for Microsoft Windows operating systems. The Windows DCOM feature, on which OPC is based, is currently not even included in Microsoft's only real-time operating system, Windows CE [13].

Although not specifically designed for manufacturing environments, the Common Object Request Broker Architecture (CORBA) is gaining popularity in automation industry. A Real-time version of CORBA based on ACE [7] is freely available.

As long as these protocol standardization issues are not resolved, manufacturers of distributed control systems only have the choice of developing software for a specific communication standard or of introducing an extra layer in their software, which can be adapted to different protocols. The latter approach was adopted for the *Realtime Framework*. As shown in the following sections however, the functionality of the *Realtime Framework* goes far beyond that of protocols like OPC or CORBA in that it also provides mechanisms for very efficient client-server communication between processes and threads running on the same machine, design patterns for integrating re-useable components into control applications, and other things.

3. Architecture of a *Realtime Framework* Application

The *Realtime Framework* is designed as a layer of components, on top of which real-time applications and other libraries for more specific areas can be built (Fig. 1). At the time of writing, a library for mobile robotics based on the *Realtime Framework* was available, including components like self-localization and path planning. With this architecture, the amount of application specific code can be kept small by using the framework and other libraries.

Another advantage of this architecture is that all operating system dependent code is concentrated in a thin layer of the *Realtime Framework*, which greatly facilitates porting the application to a new operating system. The runtime overhead of this extra layer can be kept small by using C++ inline functions for operating system calls.

---

[*] Object Linking and Embedding

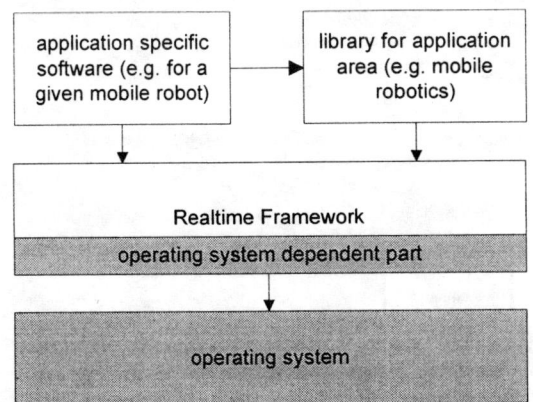

Fig. 1 Architecture of a typical *Realtime Framework* application. Arrows symbolize a 'uses'-relationship. Operating system dependent parts are shaded.

4. Components of the *Realtime Framework*

As shown in Fig. 2, the *Realtime Framework* is organized into several packages of classes which cover the areas common in most distributed control systems. The arrows indicate a "uses"-relationship, which is equivalent to "depends on". Note that circular dependencies have been avoided, so that for example the I/O package and the Utilities packages can be tested and used without the Client/Server and the Program Structure/Program Flow packages.

The I/O package mainly provides an interface to operating system dependent I/O operations for TCP/IP, serial communication and CAN fieldbus. Low-level functions like diagnostic output are concentrated in the Utilities package.

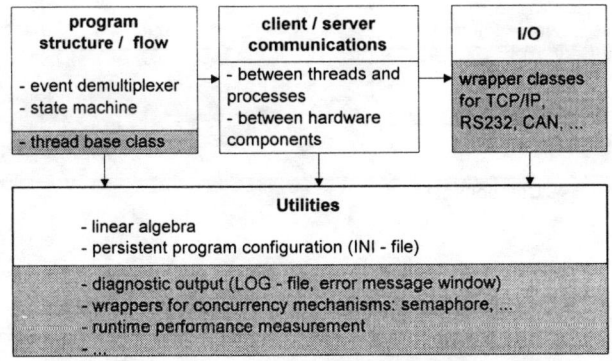

Fig. 2: Structure of the *Realtime Framework*. The Framework classes can be grouped into four packages. Arrows symbolize a 'uses'-relationship. Operating system dependent parts are shaded.

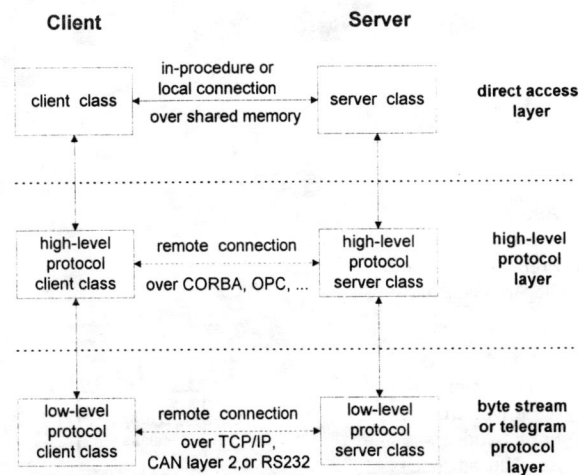

Fig. 3: Layers of client-server communication. If the client-server connection is between two threads of the same process (in-process) or between two processes running on the same machine (local), the communication can be handled in the topmost layer using shared memory. Otherwise, the data is passed on to the middle communication layer, which handles remote connections over high-level protocols like OPC or CANOpen, or to the lower layer, which implements communication over byte stream or telegram protocols.

As mentioned earlier, a major area of concern in current distributed control system development is communication between different hardware and software units. The client/server package implements the major types of client-server communication like data or command exchange and event notifications. In addition, a remote procedure call mechanism was developed. The communication can be either synchronous (blocking) or asynchronous (non-blocking). A priority can be assigned to each data package, which determines the order in which it will be handled by the recipient. In this way, emergency notifications can be handled very fast by assigning a high priority. Due to the layered structure of the client/server package (Fig. 3), the communication can take place between two threads of the same process (in-process), between two processes running on the same machine (local) or between different machines (remote) using several high-level or low-level protocols. In all cases, the same simple and flexible client/server class interface can be used to send or retrieve data. For example, the type of data transmitted between a client and a server can be conveniently specified by a C++ template parameter. Both simple data types like integer or floating point numbers and complex types (data structures) are allowed. Also, arrays of simple and complex data types can be declared. The client/server package thus allows separation of the communication content from the specific kind of communication protocol used, which is a major step towards software modularity and re-usability.

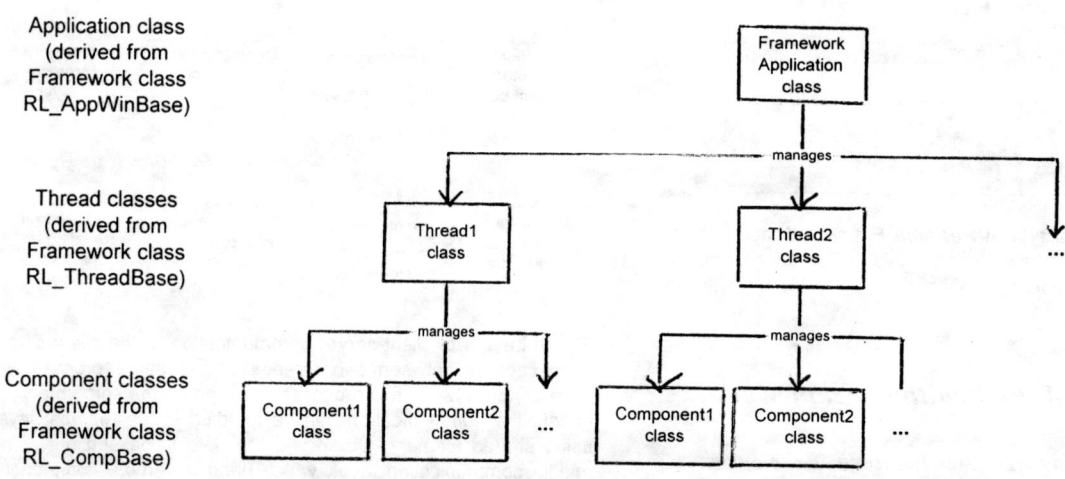

Fig. 4: *Management Tree* design pattern. Application specific classes (rectangles) are derived from corresponding Framework classes, which implement the functionality needed for management.

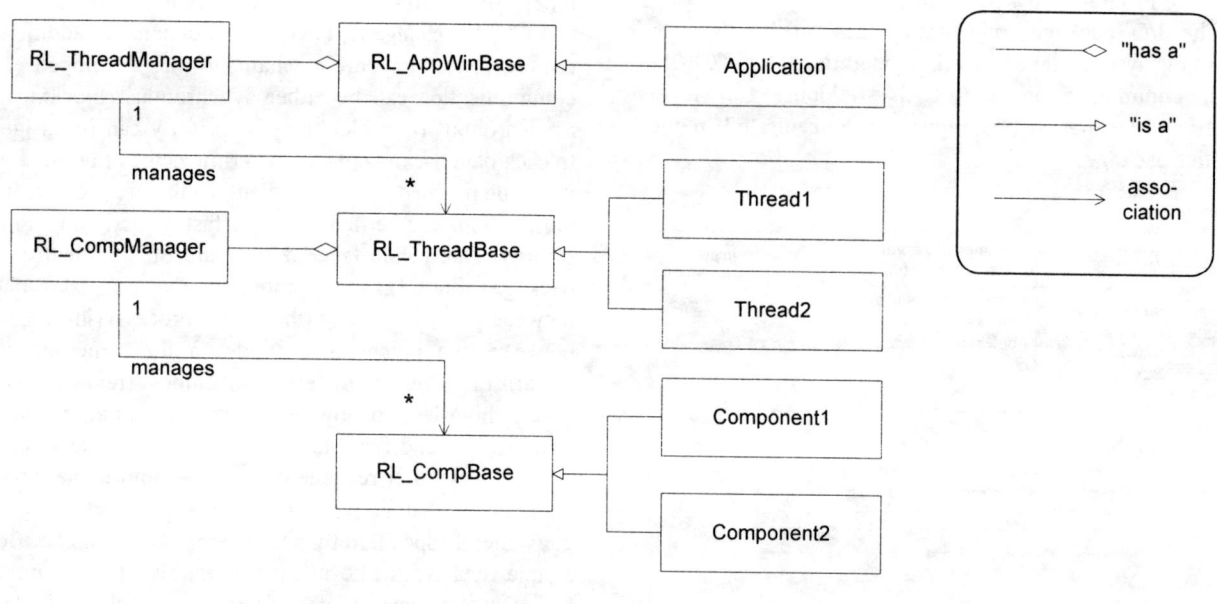

Fig. 5: UML diagram of the classes involved in the *Management Tree* design pattern (simplified). The left and center column of rectangles are *Realtime Framework* classes. Application specific classes are on the right. The application base class RL_AppWinBase contains an object of type RL_ThreadManager, which encapsulates the functionality needed to manage an arbitrary number of thread classes (derived from the thread base class RL_ThreadBase). Similarly, each thread class contains a RL_CompManager object to manage components.

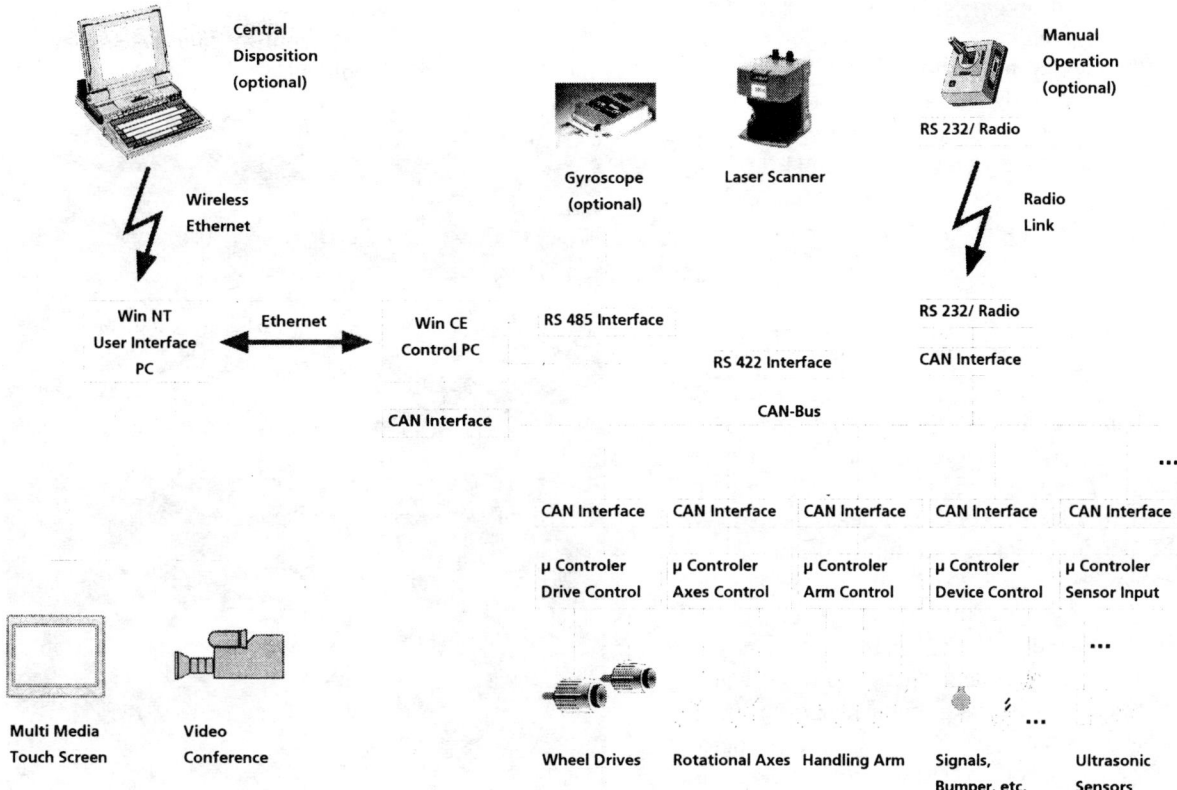

Fig. 6: Distributed control system of Care-O-bot™. Care-O-bot™ is a mobile home care system for elderly and disabled people with a two-wheel driving kinematics and a several axes for object manipulation. Its distributed control system consists of three Pentium PCs and several micro-controllers connected over TCP/IP and a CAN fieldbus.

Since in-process and local connections were implemented efficiently using shared memory, there is very little runtime overhead in using the standardized client and server class interfaces even within a process. In chapter 6, an example is given of how they can be used to structure the control and data flow in an application.

Another recurrent problem in control software development which is often underestimated is fluctuation in the development team. Modern control systems are so complex that it takes a considerable amount of work to understand modules and applications written by someone else. This leads to the commonly known phenomenon that software components are rather re-invented than re-used. Besides a good documentation, software design patterns [2] can help to alleviate this problem. Design patterns describe the organization and interaction of several related classes in an informal way. For example, they are useful for documenting recurrent micro-architectures of classes that have proven useful to build control software.

An important design pattern developed in conjunction with the *Realtime Framework* is the *Management Tree* (Fig. 4), which is used as a skeleton for most applications built with the *Realtime Framework*. In this pattern, application specific classes, which are derived from corresponding *Realtime Framework* classes, are arranged in a tree-like hierarchical structure (all *Realtime Framework* classes start with RL_ to avoid namespace pollution). Note that the arrows in Fig. 4 do not symbolize an inheritance relationship, but rather a "manages" association[*], where "manages" refers primarily to initialization, start/execution, reset and de-initialization. The management associations are created during program start-up with function calls, which register the components with their threads and the threads with the application class. Fig. 5 shows how the design pattern is supported by the *Realtime Framework*. If, for example, one component detects an error that it cannot handle, it notifies the RL_AppBase-derived class, which then resets all other threads and components, allowing for a robust response to unforeseen conditions. The usage of the *Management Tree* design pattern greatly facilitates understanding the critical phases of program start-up,

---

[*] Association is used here in the sense defined by UML [15]

reset and shutdown in *Realtime Framework* applications developed by other people.

Besides the classes used in the *Management Tree* design pattern, the program structure/program flow package includes components like a configurable state machine class and the *Reactor* design pattern [14] with an event demultiplexer for flexible, priority-based handling of simultaneous events within a program thread.

5. Implementation Details

The *Realtime Framework* has been designed with the Unified Modeling Language (UML) [15] and was implemented in C++ on Windows NT and Windows CE. During its development, portability to POSIX-compliant operating systems like VxWorks has been ensured. For example, the Windows WaitForMultipleObjects() system call has been avoided, since it is not supported by POSIX.

Realtime Framework applications are created by statically or dynamically linking *Realtime Framework* libraries with the rest of the application and by calling an initialization function during program start-up. Most applications based on the *Realtime Framework* also derive several application specific classes from Framework classes according to the *Management Tree* and other design patterns (see section 4).

The real-time capabilities of operating systems like Windows CE are preserved in the *Realtime Framework* by guaranteeing deterministic response times for Framework function calls. In addition, the runtime performance and memory efficiency of the Framework were increased by avoiding excessive use of classes with automatic type conversion and memory allocation functions, like string classes.

6. Application of the *Realtime Framework* on a Mobile Robot

The *Realtime Framework* was used as basis for the control software of Care-O-bot™ (Fig. 7, [16]), a mobile home care system for elderly and disabled people which is being developed at the Fraunhofer Institute IPA. Care-O-bot™ was designed with a two-wheel driving kinematics and a multimedia touch screen as a mobile communications terminal and walking aid. A future model with a manipulator arm will also be able to perform simple household tasks like bringing medicine to a bed. The control system of Care-O-bot™ (Fig. 6) includes three PCs and several „intelligent" sensors and actuators connected over a TCP/IP network and a CAN fieldbus. The *Realtime Framework* was used as the base layer of the control software architecture. It manages communication between the different hardware units and provides other real-time control functionality for the applications running on the PCs.

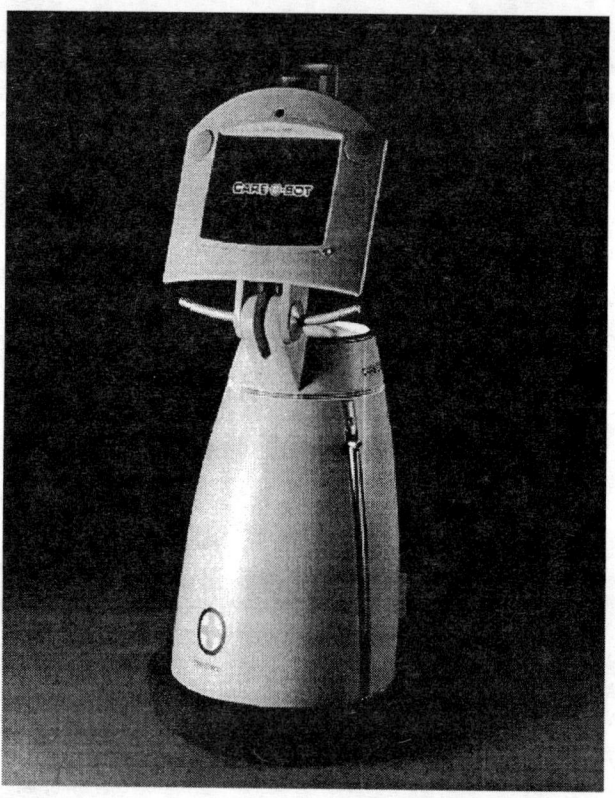

Fig. 7: Care-O-bot™. Below the moveable multimedia touch screen (top) are two handles for use as a walking aid.

The user interface PC accepts input from the user via touch screen or speech recognition and provides audiovisual feedback. It is connected to the control PC over a client-server connection, which is used to write user commands and to read robot status data (Fig. 8). Similarly, the control application is divided into several threads (tasks) which communicate over in-process client-server interfaces. This causes very little runtime overhead, since in-process and local connections are implemented very efficiently using shared memory.

Each thread contains several components according to the *Management Tree* design pattern (Fig. 4). The Motion Control Thread for example contains re-useable components for position interpolation, position control, inverse kinematics, and so on.

Besides the Control Process, a simple Watchdog Process is also running on the control PC, which supervises the correct operation of the safety-relevant part of the control application (motion control including obstacle detection) through a local client-server connection. Since these two processes are running in different address spaces under protected mode, the Watchdog Process cannot be corrupted even when there are serious programming flaws in the control application.

During the software development with up to six people working in parallel, the client-server interfaces of the *Realtime Framework* have proven very useful in integrating and re-using software components written by different people. Also, the *Management Tree* and *Reactor* design patterns were used to implement and document micro-architectures recurring in control software. They greatly facilitated the integration of developers familiar with other *Realtime Framework* applications into the team.

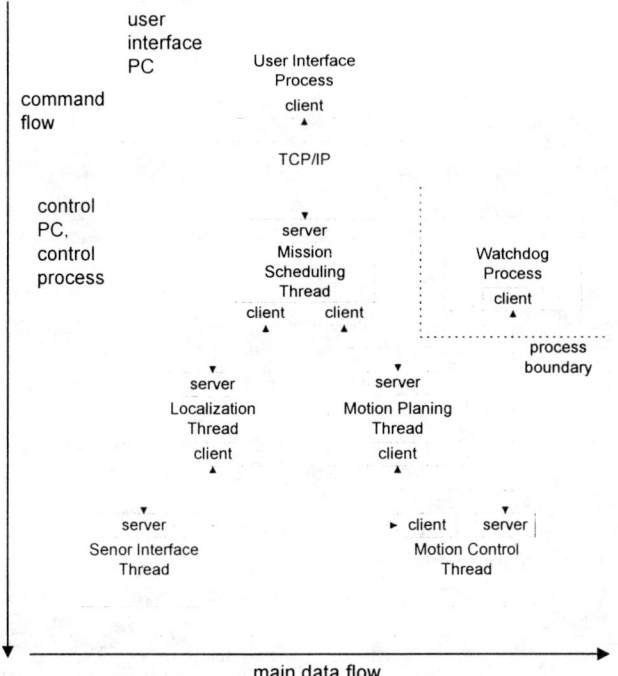

Fig. 8: Using client-server interfaces to structure the control and data flow between the processes and threads running on the user interface PC and the control PC of Care-O-bot™. The command flow is implemented through chains of client-server connections (top to bottom). The data flow is mostly directed from left to right. It can go from client to server or vice versa using either client write or client read operations.

7. Conclusions

The *Realtime Framework* presented in this paper addresses the need for new software architectures and design strategies in distributed industrial control systems. It provides a simple interface for client-server communication not only between different hardware units but also between processes and threads running on the same machine. In the latter case, the data exchange is extremely efficient, since it is implemented through shared memory. Applications based on the *Realtime Framework* are shielded from any operating system dependence, which enhances portability and re-use of software components. Design patterns were developed in order to express and document common structures on a micro-architectural level between *Realtime Framework* applications.

Practical experiences with the *Realtime Framework* were collected during the development of a mobile robot with several data processing units connected over a TCP/IP network and a CAN fieldbus. It was found that object-oriented frameworks and design patterns can significantly reduce software development and maintenance costs while maintaining a high software quality.

Acknowledgments

This work is supported by the Bundesministerium für Forschung und Technologie, Germany, within the project IQ2000.

References

[1] CAN in Automation e.V. Home Page, *http://can-cia.de*
[2] E. Gamma, R. Helm, R. Johnson, J. Vlissides, „Design Patterns: Elements of Reusable Object-Oriented Software", Addison-Wesley, Reading, 1995
[3] J. A. Fernandez, J. Gonzalez, „NEXUS: A Flexible, Efficient and Robust Framework for Integrating Software Components of a Robotic System", Proceedings of the 1998 IEEE ICRA, Leuven, Belgium, 1998
[4] OSACA Association, „The OSACA Project at a Glance", http://www.osaca.org/
[5] Real-Time Innovations Inc (RTI) and Stanford University, „Control Shell: Object-Oriented Framework for Real-Time System Software", http://128.102.240.17/products/cs.html, 1996
[6] I-Logix Inc., Rhapsody product overview, http://www.ilogix.com/fs_prod.htm
[7] Schmidt, D. C, „ACE: an Object-Oriented Framework for Developing distributed Applications", in Proceedings of the 6th USENIX C++ Technical Conference, Cambridge, Massachusetts, USENIX Association, April 1994; see also: http://www.cs.wustl.edu/~schmidt/ACE.html
[8] The Open DeviceNET Vendors Association, http://208.147.96.36/mrop/Organizations
[9] „interesting links and information to DeviceNET", http://www.infoside.de/infida/wissen_devicenet.htm
[10] INTERBUS Online, http://www.ibsclub.com/index.htm
[11] Profibus Home Page, http://www.profibus.com
[12] OPC Foundation Home Page, http://www.opcfoundation.org/
[13] Microsoft Windows CE product information, http://www.microsoft.com/windowsce/default.asp
[14] Schmidt, D. C, „The Object-Oriented Design and Implementation of the Reactor: A C++ Wrapper for UNIX I/O Multiplexing (Part 2 of 2)", C++ Report, vol. 5, Sept. 1993
[15] G. Booch, J. Rumbaugh, I. Jacobsen, „Unified Modeling Language User Guide", Addison Wesley, Longman, 1997
[16] R. D. Schraft, C. Schaeffer, T. May, „The Concept of a System for Assisting Elderly or Disabled Persons in Home Environments", Proceedings of the 24th IEEE IECON, Vol. 4 Aachen (Germany), 1998

Neural Adaptive Control of Two-Link Manipulator with Sliding Mode Compensation

Wen Yu[†], Alexander S. Poznyak[†] and Edgar N. Sanchez[‡]

[†]CINVESTAV-IPN, Seccion de Control Automatico,
Av.IPN 2508, A.P.14-740,
Mexico D.F., 07000, Mexico,
FAX: +525-747-7089, e-mail: yuw@ctrl.cinvestav.mx.

[‡] CINVESTAV, Unidad Guadalajara,
Aparatado Postal 31-438, Gaudalajara, Jalisco, C.P. 45091, Mexico

Abstract— In this paper we developed a new neuro controller for robot manipulators. A simple dynamic neural network is used to estimate the unknown robot manipulators, then the direct linearization controller is derived via this neuro identifier. Because the approximation capability is limited, another robust sliding mode compensator is addressed. Our main contributions are: first we give a bound for the identification error of the parallel neuro identifier; second we establish a bound for the tracking error of the hybrid controller.

I. Introduction

Recently many researchers manage to use modern elegant theories for robot control. Adaptive Control is a popular and powerful approach to control systems with unknown parameters [17]. Sliding Mode Control [18] consists a hypersurface switching surface which leads to the asymptotic trajectory convergence to this sliding surface. In spite of that this control is robust with respect to external disturbances, its implementation is never perfect because of "chattering effects" (state oscillation around sliding surface). Robust Feedback Control [3] is usually designed to guarantee the stability and some quality of control in the presence of parametric or unparametric uncertainties. Robust Adaptive Control can be realized by the following two ways: by adding minimax control or saturation-type control to the existing adaptive control [15] or by changing the adaptation law so there is a negative defined term (leakage-like adaptation) [10]. Adaptive-Robust Control (see [1] and [16]) estimates on-line the size of the uncertainties and uses these estimates in the traditional robust procedures [1]. Unfortunately, the corresponding theoretical study is still not completed. Optimal Control is applied to design a robust control for manipulators with some uncertainties [8].

Neural Networks (NN) control is a very effective tool to control the robot manipulator when we have no complete model information or, even, consider a controlled plant as "a black box" [7]. Neurocontrol is model-free, it is based on the NN model. To get a neuro model there exists two kinds of structure: *serial-parallel model* and *parallel model* [11]. Serial-parallel model can ensure all the signals bounded if the plant is BIBO stable for both multilayer perceptrons (MLP) [11] and dynamic neural networks (DNN) [14], [5]. Most of published papers used the *series-parallel model* because of its stability result. On the other hand, parallel model is very useful when deals with noisy systems, because it avoids problems of bias caused by noise on the real system output [19]. If the identification model is to be used off-line, obviously the parallel model is more suitable. However *parallel model* lacks theoretical verification, it is difficult to enjoy its advantages. In [5] a high order parallel NN can ensure that the identification error converges to zero, but they need the regressor vector is persistently exciting, for closed-loop control it is not reasonable. Neurocontrol may be classified as indirect (identification-based) and direct control [4]. The direct neurocontrol of MLP [9] suffers from some problems, such as lack of knowledge of the plant Jacobian, local minima and requiring specified training data pairs (off-line training). Many efforts are made to overcome these disadvantages. In [12] a on-line estimation of plant Jacobian is presented. A modified continuous-time version of backpropagation algorithm is given in [6] which does not need off-line learning and plant Jacobia. These "static" networks are based on the theory of function approximation which are sensitive to the training data. DNN can successfully overcome this disadvantage as well as demonstrate workable behavior in the presence of unmodeled dynamics, because their structure incorporate feedback. The direct and indirect [14] neural adaptive controls use linear two-layer DNN. Because this DNN has a poor approximation capability, if the neurocontrol is derived from this neuro identifier, the control results are not satisfied. In order to improve the identification results, a high-order DNN is proposed in [5], but the neurocontrol is not easy to realize because they use high-order multiply of control input.

In this paper, a simple DNN similar as [14] is used to identify robot manipulator, instead of serial-parallel model we use parallel identification structure. This neuro identifier cannot give a good approximate accuracy. So our controller for the robot manipulator has two parts: a direct linearization neuro controller and a sliding mode compensator. The main contribution of this paper is connected with the extension of our previous results [13] in the following directions: we consider parallel model of dynamic neural network to identify the unknown robot manipulator, the new learning laws assure stability of the identification error; then a bound for the tracking error of robot manipulator is established.

II. ROBOT MANIPULATOR DYNAMICS

The dynamics of an $n-$link robot manipulator may be expressed in the Lagrange form [6]

$$M(q)\ddot{q} + V(q,\dot{q})\dot{q} + G(q) + F_d(\dot{q}) = \tau \quad (1)$$

where q consists of the joint variables, τ is the generalized forces, $M(q)$ is the intertia matrix, $V(q,\dot{q})$ is centripetal-Coriolis matrix, $G(q)$ is gravity vector, $F_d(\dot{q})$ is the friction vector. $M(q)$ represents inertia matrix. A scheme of the two-link robot manipulator is shown in Fig.1. For

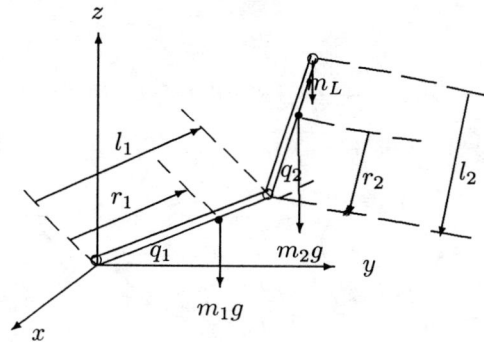

Fig. 1. A scheme of two-links manipulator

the case of two links, the elements can be represented as
$M(q) = \begin{bmatrix} M_{11} & M_{12} \\ M_{21} & M_{22} \end{bmatrix},$

$M_{11} = m_1\left(r_1^2/4 + l_1^2/3\right) + m_2\left(l_1^2 + r_2^2/4 + l_2^2/3 + l_1l_2\cos q_2\right)$
$\quad + m_2\left(l_1^2 + 3/4l_2^2 + l_1l_2\cos q_2\right)$
$M_{12} = m_2\left(r_2^2/4 + l_2^2/3 + \frac{1}{2}l_1l_2\cos q_2\right) + m_Ll_2^2 = M_{21}$
$M_{22} = m_2\left(r_2^2/4 + l_2^2/3\right) + m_Ll_2^2$
$V(q,\dot{q}) = \begin{bmatrix} -V_m\dot{q}_2\sin q_2 & -V_m\sin q_2\left(\dot{q}_1 + \dot{q}_2\right) \\ V_m\dot{q}_1\sin q_2 & 0 \end{bmatrix}$
$V_m = \left(\frac{1}{2}m_2l_1l_2 + m_Ll_1l_2\right),$
$G(q), = \begin{bmatrix} \left(\frac{1}{2}m_1 + m_2\right)gl_1\cos q_1 + \frac{1}{2}m_2gl_2\cos(q_1+q_2) \\ \frac{1}{2}m_2gl_2\cos(q_1+q_2) \end{bmatrix}$
$F_d(\dot{q}) = \begin{bmatrix} \nu_1\dot{q}_1 + \kappa_1 sign(\dot{q}_1) \\ \nu_2\dot{q}_2 + \kappa_2 sign(\dot{q}_2) \end{bmatrix}$

The robot dynamics have following standard properties
- $M(q)$ is a positive symmetric matrix bounded by $m_1 I < M(q) < m_2 I$.
- The norm of matrix $V(q,\dot{q})$ is bounded an known function $v_b(q)$.

So (1) can be rewritten as

$$\dot{x} = f(x_t,t) + g(x_t,t)u_t \quad (2)$$

where $x_t = \begin{bmatrix} q_1, q_2, \dot{q}_1, \dot{q}_2 \end{bmatrix}^T = [x_1, x_2, x_3, x_4]^T$, $f(x_t,t) = \begin{bmatrix} \dot{x}_3, \dot{x}_4, -M^{-1}\left(V\dot{q} + G + F_d\right) \end{bmatrix}^T$, $g(x_t,t)u = [0, 0, M^{-1}(q)\tau]^T$.

The two-link manipulator system (2) has 4-dimension state and 2-dimension input. In order to design a suitable neuro controller, we define following auxiliary variable

$$\bar{q} = \dot{q} + \Lambda q \quad (3)$$

where $\bar{q} \in R^2$, Λ is positive defined diagonal matrix. The system can be rewritten as

$$\dot{\bar{q}} = F(\bar{q},t) + G(\bar{q},t)u$$

where F and Gu are 2-dimension vector functions. Or we can even rewrite it as

$$\dot{\bar{q}} = F'(q,t) + G'(q,u,t) \quad (4)$$

When we design a neurocontrol we do not need to transfer the original system (1) into the standard form (2) or (4). We give the form of (4) in order to illuminate the neuro identifier. Now we only assume the state outputs q is available.

III. ADAPTIVE CONTROL VIA DYNAMIC NEURAL NETWORK

In this section we derive a neuro controller for the robot dynamics in Section II. This controller has two new contribution: first we use parallel dynamic NN to identify the manipulator, a robust learning algorithm is proposed. Second, a hybrid neuro controller is presented. We construct the following parallel dynamic neural network:

$$\dot{\hat{x}}_t = A\hat{x}_t + W_t\sigma(\hat{x}_t) + u_t \quad (5)$$

where $\hat{x}_t \in \Re^n$ is the state of the neural network, $u_t \in \Re^n$, $W_t \in \Re^{n \times n}$ is the weight of neural network. $A \in \Re^{n \times n}$ is a stable matrix. The vector functions $\sigma(\hat{x}_t)$ is assumed to be n-dimensional with the elements increasing monotonically, such as sigmoid functions

$$\sigma_i(x_i) = \frac{a_i}{1 + e^{-b_i x_i}} - c_i. \quad (6)$$

Remark 1: This neural network does not contain any hidden layers. We use this simplified neural network because we want to make the neuro controller more reliable. This model will cause more unmodeled dynamic, but we may compensation the modelling error.

For the two-link manipulator $n = 2$. The robot manipulator to be control is given in (4), it can be presented as a DNN plus a modelling error, i..e. there exists weight W^* such that the system (4) is complete described by

$$\dot{\bar{q}} = A\bar{q} + W^*\sigma(\bar{q}) + u_t + \Delta f(\bar{q}, u_t) \quad (7)$$

where W^* is bounded as

$$W^*\Lambda_\sigma^{-1}W^{*T} \leq \overline{W}. \quad (8)$$

Here Λ_σ and $\overline{W}$ are priory known matrices.

Let define the identification error as,

$$\Delta_t := \widehat{x}_t - q. \quad (9)$$

Because $\sigma(\cdot)$ is chosen as sigmoid functions, the following general Lipschitz condition is fulfilled.

A1: *The function $\sigma(\cdot)$ satisfies*

$$\widetilde{\sigma}_t^T \Lambda_\sigma \widetilde{\sigma}_t \leq \Delta_t^T D_\sigma \Delta_t,$$

where $\widetilde{\sigma}_t := \sigma(\widehat{x}_t) - \sigma(\overline{q})$, Λ_σ and D_σ are known positive constants.

From (4) and (5), if a bounded control input u_t may stabilize the nonlinear system, the unmodeled dynamics $\Delta f(\overline{q}, u_t)$ is bounded, we can assume that

A2: *There exists positive defined matrix Λ_f such that*

$$\Delta f^T \Lambda_f \Delta f := \|\Delta f\|_{\Lambda_f} \leq \overline{\eta}$$

where $\overline{\eta}$ is the upper bound of the modeling error.

It is well known that, if the matrix A is stable, the pair $\left(A, R^{\frac{1}{2}}\right)$ is controllable and the pair $\left(Q, R^{\frac{1}{2}}\right)$ is observable and the *local frequency condition* fulfills,

$$A^T R^{-1} A - Q \geq \frac{1}{4} \left[A^T R^{-1} - R^{-1} A \right] R \left[A^T R^{-1} - R^{-1} A \right]^T. \quad (10)$$

then the following matrix Riccati equation:

$$A^T P + PA + PRP + Q = 0 \quad (11)$$

has a positive solution. So we can introduce following assumption:

A3: *For a given matrix A. there exists a strictly positive defined matrix Q_0 such that the matrix Riccati equation (11) with the matrices R and Q given by*

$$R := \overline{W} + \Lambda_f^{-1}, \quad Q := Q_0 + D_\sigma,$$

has a positive solution.

This conditions is easily fulfilled if we select A as diagonal matrix. The next theorem presents stable learning procedure of the parallel DNN.

Theorem 1: Let us consider the unknown robot manipulator (4) and parallel neural network (5) whose weights are adjusted as

$$\dot{W}_t = \dot{\widetilde{W}}_t = -s_t K P \sigma(\widehat{x}_t) \Delta_t^T \quad (12)$$

where $s_t = \begin{cases} 1 & \text{if } \|\Delta_t\| > \sqrt{\overline{\eta} \lambda_{\min}(Q_0)} \\ 0 & \text{if } \|\Delta_t\| \leq \sqrt{\overline{\eta} \lambda_{\min}(Q_0)} \end{cases}$, $\widetilde{W}_t := W_t - W^*$, K is a positive define matrix, P is the solution of the matrix Riccati equation given by (11). Assuming also that the assumptions **A1-A3** hold, we conclude that the weight and identification error are bounded, i.e.,

$$\Delta_t, W_t \in L_\infty \quad (13)$$

for any $T \in (0, \infty)$ the identification error Δ_t satisfies the following tracking performance

$$\begin{array}{l} \frac{1}{T} \int_0^T \|\Delta_t\|_{Q_0} dt \leq \overline{\eta} + C_0/T \\ C_0 = \Delta_0^T P \Delta_0 + tr\left[\widetilde{W}_0^T K^{-1} \widetilde{W}_0\right] \end{array} \quad (14)$$

Proof: From (9) and (5) we have

$$\dot{\Delta}_t = A \Delta_t + \widetilde{W}_t \sigma(\widehat{x}_t) + W_t \widetilde{\sigma}_t - \Delta f(\overline{q}, u_t). \quad (15)$$

Define Lyapunov function candidate as

$$V_t := \Delta_t^T P \Delta_t + \frac{1}{2} tr \left[\widetilde{W}_t^T K^{-1} \widetilde{W}_t\right]. \quad (16)$$

So, calculating its derivative, we obtain

$$\frac{d}{dt} V_t = 2 \Delta_t^T P \dot{\Delta}_t + tr\left[\dot{\widetilde{W}}_t^T K^{-1} \widetilde{W}_t\right]. \quad (17)$$

As

$$\Delta_t^T P \dot{\Delta}_t = \Delta_t^T P A \Delta_t + \Delta_t^T P (W_t \widetilde{\sigma}_t - \Delta f(\overline{q}, u_t)) + \Delta_t^T P \widetilde{W}_t \sigma(\widehat{x}_t).$$

Using matrix inequality [13]

$$X^T Y + (X^T Y)^T \leq X^T \Lambda^{-1} X + Y^T \Lambda Y \quad (18)$$

which is valid for any $X, Y \in \Re^{n \times k}$ and for any positive defined matrix $0 < \Lambda = \Lambda^T \in \Re^{n \times n}$.

The assumption **A1** leads to

$$\begin{array}{l} 2\Delta_t^T P W^* \widetilde{\sigma}_t \leq \Delta_t^T P W^* \Lambda_\sigma^{-1} W_t P \Delta_t \\ + \widetilde{\sigma}_t^T \Lambda_\sigma \widetilde{\sigma}_t \leq \Delta_t^T \left(P \overline{W} P + D_\sigma\right) \Delta_t, \end{array} \quad (19)$$

From **A2**

$$-2 \Delta_t^T P \Delta f \leq \Delta_t^T P \Lambda_f^{-1} P \Delta_t + \Delta f^T \Lambda_f \Delta f \leq \Delta_t^T P \Lambda_f^{-1} P \Delta_t + \overline{\eta},$$

we obtain:

$$\dot{V}_t \leq \Delta_t^T \left(PA + A^T P + P\left(\overline{W} + \Lambda^{-1}\right) P + D_\sigma + +Q_0\right) \Delta_t$$
$$\Delta_t^T P \widetilde{W}_t \sigma(\widehat{x}_t) + tr\left[\dot{\widetilde{W}}_t^T K^{-1} \widetilde{W}_t\right] - \Delta_t^T Q_0 \Delta_t + \overline{\eta}.$$

Use **A3**

$$\dot{V}_t \leq \Delta_t^T P \widetilde{W}_t \sigma(\widehat{x}_t) + tr\left[\dot{\widetilde{W}}_t^T K^{-1} \widetilde{W}_t\right] - \Delta_t^T Q_0 \Delta_t + \overline{\eta}. \quad (20)$$

If

$$\|\Delta_t\| > \sqrt{\frac{\overline{\eta}}{\lambda_{\min}(Q_0)}},$$

according to (12),(20) becomes

$$\dot{V}_t \leq -\Delta_t^T Q_0 \Delta_t + \overline{\eta} \leq -\lambda_{\min}(Q_0) \|\Delta_t\|^2 + \overline{\eta} \leq 0. \quad (21)$$

If

$$\|\Delta_t\| \leq \overline{\eta} \sqrt{\frac{1}{\lambda_{\min}(Q_0)}}, \quad (22)$$

W_t is constant matrix and from (22) $\|\Delta_t\|$ is also bounded, so V is bounded. (13) is realized. Integrating (21) from 0 up to T yields

$$V(T) - V(0) \leq -\int_0^T \Delta_t^T Q_0 \Delta_t dt + \overline{\eta} T,$$

from which (14) is achieved. ∎

Remark 2: The updating law (12) is similar with the *series-parallel* structure [14], the differences are *series-parallel* structure uses $\sigma(\overline{q})$ and *parallel* structure uses $\sigma(\widehat{x}_t)$; In *series-parallel* structure P is the solution of following Lyapunov equation

$$PA + A^T P = -Q.$$

in p*arallel* structure P is the solution of a matrix Riccati equation (11).

Remark 3: The updating rates $\overline{K} := KP$ can be achieved by a special selection of the gain matrices K, the solution of matrix Riccati equation (11) does not influence the updating law.

Remark 4: If we have no any unmodeled dynamic ($\Delta f = 0$), we obtain $\overline{\eta} = 0$ and, hence, from (14) the globally asymptotic stability is guaranteed, i.e.,

$$\limsup_{T \to \infty} \frac{1}{T} \int_0^T \|\Delta\|_{Q_0} dt = 0,$$

from which we directly obtain $\|\Delta_t\| \underset{t \to \infty}{\to} 0$.

From (7), we know that the nonlinear system (4) may be modeled as

$$\dot{q} = Aq + W_t \sigma(\widehat{x}_t) + u_t + \Delta f(\overline{q}, u_t) + W^* \sigma(q) - W_t \sigma(\widehat{x}_t) \quad (23)$$

Using the assumptions **A1** and **A2**, from theorem 1 we have $\Delta f(q, u_t) + W^* \sigma(q) - W_t \sigma(\widehat{x}_t)$ is bounded. (23) can be rewritten as

$$\dot{q} = Aq + W_t \sigma(\widehat{x}_t) + u_t + d_t \quad (24)$$

where

$$d_t = \Delta f(q, u_t) + W^* \sigma(q) - W_t \sigma(\widehat{x}_t) \quad (25)$$

is bounded

Based on the neural network identifier (5), we will force the robot manipulator system (4) to track a optimal trajectory $x^* \in \Re^2$ which is assumed to be smooth enough. This trajectory is regarded as a solution of a nonlinear reference model:

$$\dot{x}^* = \varphi(x^*, t) \quad (26)$$

with a fixed initial condition. If the trajectory has points of discontinuity in some fixed moments, we can use any approximating trajectory which is smooth. In the case of regulation problem $\varphi(x^*, t) = 0$, $x^*(0) = c$, c is constant.

Theorem 2: Let the desired trajectory be given by (26), the weights of the parallel DNN (5) is tuned by (12). For the two-link manipulator, if the control input is provided by

$$u_t = u_{1,t} + u_{2,t} \quad (27)$$
$$u_{1,t} = \varphi - Ax^* - W_t \sigma(\widehat{x}_t). \quad (28)$$
$$u_{2,t} = \begin{cases} -kP_c^{-1} \text{sign}(\Delta_t^*) & |\Delta_t^*| \geq \delta \\ -kP_c^{-1} \Delta_t^*/\delta & |\Delta_t^*| < \delta \end{cases}, \quad k, \delta > 0 \quad (29)$$

where P is a solution of the Lyapunov equation $A^T P_c + P_c A = -Q_c$, the tracking error

$$\Delta_t^* = q - x^*$$

is bounded.

Proof: From (23) and (26) we have

$$\dot{\Delta}^* = A\Delta^* + W_t \sigma(\widehat{x}_t) + u_t + d_t - \varphi + Ax^*. \quad (30)$$

As $\varphi(x^*, t)$, x^* and $W_t \sigma(\widehat{x}_t)$ are available, we can select u_t as (27) where $u_{1,t}$ is direct linearization control (28). So

$$\dot{\Delta}^* = A\Delta^* + u_{2,t} + d_t. \quad (31)$$

The Sliding Mode technique can be applied to compensate d_t. Let us define Lyapunov-like function as

$$V_t = \Delta_t^{*T} P_c \Delta_t^*, \quad P_c = P_c^T > 0 \quad (32)$$

where P is a solution of the Lyapunov equation $A^T P_c + P_c A = -Q_c$. Using (31) whose time derivative is

$$\dot{V}_t = \Delta_t^{*T} (A^T P_c + P_c A) \Delta_t^* + 2\Delta_t^{*T} P_c U_{2,t} + 2\Delta_t^{*T} P_c d_t. \quad (33)$$

According to sliding mode technique, we may select $u_{2,t}$ as

$$U_{2,t} = -kP_c^{-1} \text{sign}(\Delta_t^*), \quad k > 0 \quad (34)$$

where k is positive constant,

$$sgn(\Delta_t^*) = [sgn(\Delta_{1,t}^*), \cdots sgn(\Delta_{n,t}^*)]^T \in \Re^n$$

Substitute (34) into (33)

$$\begin{aligned}\dot{V}_t &= -\Delta_t^{*T} Q_c \Delta_t^* - 2k\|\Delta_t^*\| + 2\Delta_t^T P_c d_t \\ &\leq -\lambda_{\max}(Q_c) \|\Delta_t^*\|^2 - 2k\|\Delta_t^*\| + 2\lambda_{\max}(P_c)\|\Delta_t^*\|\|d_t\| \\ &= -\lambda_{\max}(Q_c)\|\Delta_t^*\|^2 - 2\|\Delta_t^*\|(k - \lambda_{\max}(P_c)\|d_t\|)\end{aligned}$$

If we select k is big enough such that

$$k > \lambda_{\max}(P_c)\overline{d}$$

where $\overline{d}$ is upper bound of $\|d_t\|$ ($\overline{d} = \sup_t \|d_t\|$), then $\dot{V}_t < 0$. So

$$\lim_{t \to \infty} \Delta_t^* = 0.$$

Because the sliding mode control $u_{2,t}$ is inserted in the closed-loop system, chattering occur in the control input which may excite unmodeled high-frequency dynamics. To eliminate chattering, the following boundary layer compensator can be used

$$u_{2,i} = \begin{cases} -kP_i^{-1}\text{sign}\left(\Delta_{i,t}^*\right) & \left|\Delta_{i,t}^*\right| \geq \delta \\ -kP_i^{-1}\Delta_{i,t}^*/\delta & \left|\Delta_{i,t}^*\right| < \delta \end{cases} \quad (35)$$

where $u_{2,t} = [u_{2,1},\ldots u_{2,n}]$, δ is small enough positive constant. The above boundary layer controller offers a continuous approximation to the discontinuous sliding mode control law inside the boundary layer and guarantees the output tracking error within any neighborhood of the origin [2]. ∎

Remark 5: Theorem 1 assures W_t is bounded. So $u_{1,t}$ in (28) is bounded. From (35) we know $u_{2,t}$ is bounded. So the hybrid control input is bounded. Although we use the sliding mode control [18], we can avoid the "chattering effects" because our hybrid control has two part, the main part $u_{1,t}$ is derived from neuro identifier, and $u_{2,t}$ uses boundary layer compensator to eliminate the chattering.

Remark 6: Compare (5) and (26)

$$\dot{\hat{x}} - \dot{x}^* = A\left(\hat{x} - x^*\right) + W_t\sigma(\hat{x}_t) + u_t - \varphi + Ax^*.$$

If we select $u_t = u_{1,t}$

$$\dot{\hat{x}} - \dot{x}^* = A\left(\hat{x} - x^*\right)$$

Because A is stable, $\lim_{t\to\infty}(\hat{x} - x^*) = 0$. $u_{1,t}$ is used to make neuro identifier (5) follow the reference model (26). Because the neuro identifier (5) cannot follow the robot dynamic (4) exactly, so $u_{2,t}$ is used to compensate the modeling error in order to make the robot (4) follow the reference model (26).

IV. SIMULATION RESULTS

We take the robot parameters as in [6], we also include friction in (1), i.e. $l_1 = l_2 = 2r_1 = 2r_2 = 1$m, $m_1 = 0.8$kg, $m_2 = 2.3$kg, $v1 = v2 = 0.4$, $k1 = k2 = 0.8$, $g = 9.81$. The initial conditions are $\dot{q}(0) = [\dot{q}_1(0), \dot{q}_2(0)] = [0,0]$, $q(0) = [q_1(0), q_2(0)] = [3.14, 0.6]$. We assume the parameters in (4) are unknown, only the position and the velocity of q are available. The neural network used for identification is as (5), where $\hat{x}_t = [\hat{q}_1, \hat{q}_2]^T$, $A = \text{diag}[-2,-2]$,

$$\sigma(x) = \frac{2}{(1+e^{-2x})} - \frac{1}{2}$$

The initial conditions for the neural network are $\hat{x}_0 = [0,0]^T$, $W(0) = \begin{bmatrix} 1 & 1 \\ 1 & 1 \end{bmatrix}$.

First we check the identification ability of parallel DNN, because the open-loop system is unstable, first we use a PD control as

$$\tau = -10(q - q^*) - 5\left(\dot{q} - \dot{q}^*\right) \quad (36)$$

where the reference inputs are $q_1^*(t) = \sin t$, $q_2^*(t) = \cos t$, $q_1^*(0) = q_2^*(0) = 0$. So in (26) $\varphi(q_t^*, t) = [-\cos t, \sin t]^T$. Let

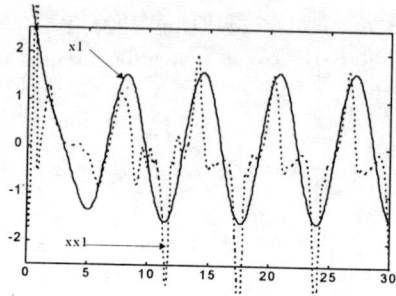

Fig. 2. Identification for q_1

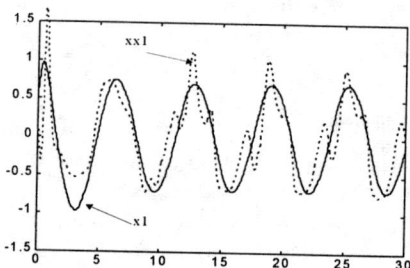

Fig. 3. Identification for q_2

$u_t = \tau$, use (5) to identify the two-link robot. The updating laws are same as (12). we select $KP = \text{diag}[10,10]$, $\overline{\eta} = \lambda_{\min}(Q_0) = 1$. The results are shown in Fig.2 and Fig.3. There exists identification errors because we use second order neural network to model the dynamic of two links robot, there are unmodelled dynamics.

Then we apply the hybrid neurocontrol to the two-link manipulator. For compensation a standard PD control is give as in (36). The result appears in Fig.4 and Fig.5 and is unsatisfactory. If we use parallel DNN, we select $\Lambda = \text{diag}[1,1]$, so the control input is

$$\tau = u_{1,t} + u_{2,t}$$
$$u_{1,t} = \varphi(q^*) - \begin{bmatrix} -2 & 0 \\ 0 & -2 \end{bmatrix} x^* - W_t\sigma(\hat{x}_t)$$
$$= \begin{pmatrix} -3\cos t + \sin t \\ \cos t + 3\sin t \end{pmatrix} - W_t\sigma(\hat{x}_t),$$
$$u_{2,t} \begin{cases} -kP_i^{-1}\text{sign}\left(\Delta_{i,t}^*\right) & \left|\Delta_{i,t}^*\right| \geq \delta \\ -kP_i^{-1}\Delta_{i,t}^*/\delta & \left|\Delta_{i,t}^*\right| < \delta \end{cases}, \quad i = 1,2$$

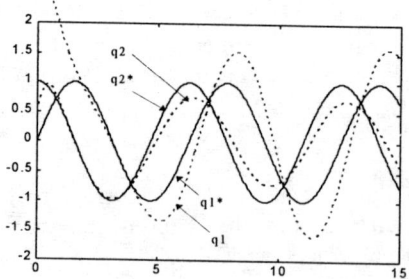

Fig. 4. PD control

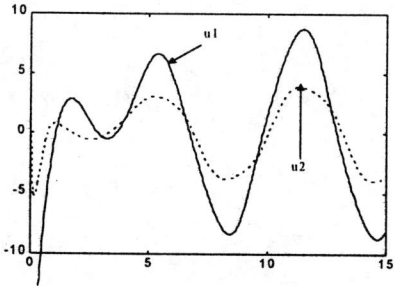

Fig. 5. Control input of PD

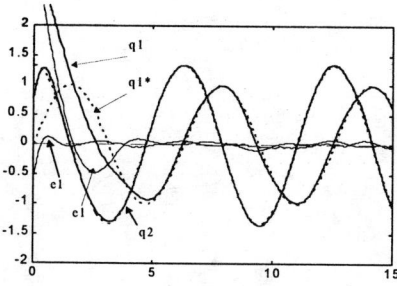

Fig. 6. Response of neurocontrol

where $\delta = 0.1$, $\Delta_t^* = q - q^*$. The neurocontrol presented in this paper are shown in Fig.6 and Fig.7. No initial training or learning phase was need. It is very clear that the addition of DNN makes a significant improvement in the tracking performance.

V. CONCLUSION

A dynamic neural network was developed for the two-link robot manipulator. First we use the parallel DNN to identify the dynamic of robot, then a direct linearization controller is applied base based on this neuro identifier. Because of the modelling error, a sliding mode compensator is presented. In this paper we proof that both of identification error and tracking error are bounded.

REFERENCES

[1] Y.H.Chen, "Adaptive robust model-following control and application to robot manipulators", *Transactions of ASME, Journal of Dynamic Systems, Measurement, and Control*, vol. 109, pp. 209-15, 1987.

[2] M.J.Corless and G.Leitmann, Countinuous Stste Feedback

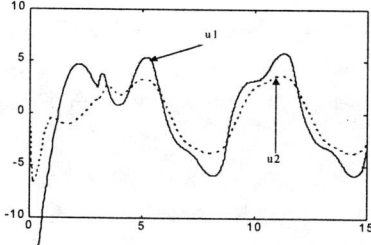

Fig. 7. Control input of neurocontrol

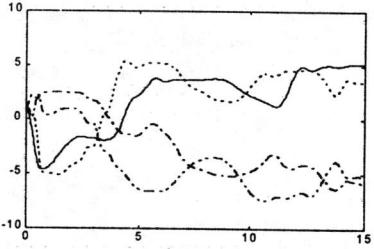

Fig. 8. Representative weightd estimate

Guaranteeing Uniform Ultimate Boundedness fo Uncertain Dynamic Systems, *IEEE Trans. Automat. Contr.* Vol.26, 1139-1144, 1981

[3] D.Dawson, Z.Qu and M.Bridge,"Hybrid adaptive control for the tracking of rigid-link flexible-joint robots", *in Modelling and Control of Compliant and rigid Motion Systems*, 1991 ASME Winter Annual Meeting, Atlanta GA, pp. 95-98, 1991.

[4] K.J.Hunt, D.Sbarbaro, R.Zbikowski and P.J.Gawthrop, "Neural Networks for Control Systems- A Survey", *Automatica*, vol. 28, pp. 1083-1112, 1992.

[5] E.B.Kosmatopoulos, M.A.Christodoulou and P.A.Ioannpu, "Dynamical Neural Networks that Ensure Exponential Identification Errror Convergence", Neural Networks, Vol.10, No.2, 299-314, 1997

[6] F.L.Lewis, A.Yesildirek and K.Liu, Multilayer Neural-Net Robot Controller with Guaranteed Tracking Performance, *IEEE Trans. on Neural Networks*, Vol.7, No.2, 388-399, 1996

[7] F.L.Lewis, Neural Network Control of Robot Manipulators, *IEEE Expert*, Vol.11, No.2, 64-75, 19963

[8] F.Lin, and R.D.Brandt, An Optimal Control Approach to Robust Control of Robot Manipulators, *IEEE Trans. on Robot. Automat.*, Vol.14, No.1, 69-77, 1996

[9] W.T.Miller, R.S.Sutton, and P.J.Werbos, *Neural Networks for Control*, Cambridge: MIT Press, 1991

[10] K. S. Narendra and A. M. Annaswamy, "Stable Adaptive Systems", Prentice Hall, Englewood Cliffs, NJ, 1989.

[11] K.S.Narendra and K.Parthasarathy, "Identification and Control of Dynamical Systems Using Neural Networks", *IEEE Trans. on Neural Networks*, Vol.1, 4-27, 1989

[12] G.W.Ng, and P.A.Cook, On-line Adaptive Control of Nonlinear Plants Using Neural Networks with Application to Liquid Level Control, *Int. J. of Adaptive Control and Signal Processing*, Vol.12, 13-28, 1998

[13] Wen Yu, Alexander S.Poznyak, Indirect Adaptive Control via Parallel Dynamic Neural Networks, *IEE Proceedings - Control Theory and Applications*, accepted for publication.

[14] G.A.Rovithakis and M.A.Christodoulou, "Direct Adaptive Regulation of Unknown Nonlinear Dynamical System via Dynamical Neural Networks", *IEEE Trans. on Syst., Man and Cybern.*, Vol. 25, 1578-1594, 1995.

[15] Zhihua Qu and Darren M. Dawson, "Robust Tracking Control of Robot Manipulators", IEEE Press, New York, 1996.

[16] S.N.Singh,"Adaptive model-following control of nonlinear robotic systems", *IEEE Transactions on Automatic Control*, vol. 30, pp. 1099-1100, 1985.

[17] J.E.Slotine, and W.Lin, Adaptive Manipulator Control: A Case Study, *IEEE Trans. on Automat. Contr.*, Vol. 33, 995-1003, 1988.

[18] V.I.Utkin, "Variable structure systems with sliding modes: A survey", *IEEE Transactions on Automatic Control*, vol. 22, pp. 212-22, 1977.

[19] B.Widrow and S.D.Steans, *Adaptive Signal Processing*, Prentice-Hall, Englewood Cliffs, NJ, 1985

[20] J.C.Willems, "Least Squares Optimal Control and Algebraic Riccati Equations", *IIEEE Trans. Automat. Contr.*, vol.16, 621-634, 1971,

How to Compensate Stick-Slip Friction in Neural Velocity Force Control (NVFC) for Industrial Manipulators

M. Dapper, V. Zahn, R. Maaß, R. Eckmiller

University of Bonn, Dept. of Comp. Science VI
Römerstr. 164, D-53117 Bonn, F.R. Germany
Tel.: +49-228-73-4168, FAX: +49-228-73-4425
e-mail: dapper, maass,zahn,eckmiller@nero.uni-bonn.de

Abstract - *We present a new design of hybrid force/position control NVFC (Neural Velocity Force Control) capable of friction-compensation, which is a mandatory requirement in industrial applications for precise tracking of hard surfaces with desired contact force. The control architecture consists of an outer-loop velocity controller and an inner-loop adaptive hybrid force/position controller, which compensates the friction of the manipulators joints, significantly improving the force tracking performance during slow movements. The cascaded velocity controller ensures the precise approach during neural force/position control with a reduced bounce into the unknown surface, which is a demanding requirement for tasks like deburring or chamfering. The resulting performance of the force control system is illustrated in simulations, and is consistent with first velocity/force control experiments on a 6DOF industrial manipulator controlled with a PC-based realtime controlsystem.*

1 Introduction

Recently, robot manipulators are expected to perform more sophisticated tasks under constrained conditions [10, 13]. To realize tasks like controlling the end-effector orientation and movement in contact with unknown hard surfaces, the control system has got to be flexible enough, to cope with various problems, like joint friction, uncertain model parameters and disturbances between the different controllers respectively [1, 2, 3, 9]. A novel adaptive force control approach satisfying these requirements is presented in this work. We developed an adaptive force control for industrial task as deburring, chamfering or robotic based thermoplastic fibre placement [6, 5, 11], where normal desired forces have to be exerted on unknown shaped surfaces and friction effects due to slow movements or high contact forces are urgent problems: Coulomb friction is uncertain and can vary significantly with load, wear

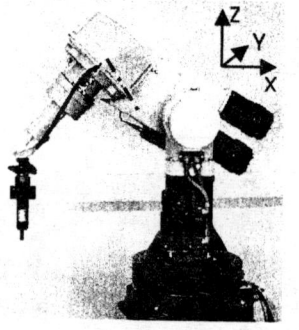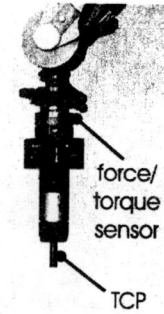

Fig. 1. 6DOF industrial manipulator Siemens Manutec R2 using 2ms control loop equipped with a 6 axis force/torque sensor (50 microseconds sampling time)

and temperature [4]. On hard metal working surfaces, subjected to robotized polishing and finishing operations, adaptive force control like NVFC is an essential requirement, to handle these tasks with sufficient accuracy[1].

2 Results

2.1 Control concept

NVFC is a neural velocity/force controller [1, 3] capable of contacting and tracking hard surfaces and establishes easily stable contact without bouncing while maintaining contact to unknown hard surfaces without any dangerous acceleration of the TCP (tool center point) even if the contact surface is moved away. The friction influence of the manipulator system, like Coulomb friction, which affects very disturbing on the

---
[1] This work was supported in part by Federal Ministry for Education, Science, Research, and Technology (BMBF) under grant "DEMON"

force and therefore needs to be compensated adequately, is adapted via neural networks [7, 8] in the NAC (Neural Adaptive Control) module, which leads to a significant improvement of the force tracking performance especially during slow movements. The manipulator (Siemens Manutec R2) is equipped with a 6-axis force torque sensor ATI FT Mini 100/5, which has been speeded up by a specially developed, parallel, high speed signal processing device, which allows an acceleration factor of 30 in the sampling time, compared to the original system. Thus a sampling rate of 50 microseconds for the complete set for 3 forces and 3 torques is achieved, which allows aquisition and processing of high bandwidth force peaks appearing especially in contacts with hard surfaces like metal or stone.

NAC: For the training of the unknown and changing frictional effects into the RBF nets, a control approach similar to the algorithm of Slotine and Li [12] for parameter adaptation is chosen (eq.1)

$$\begin{aligned}\tau &= \mathbf{M}\dot{\mathbf{v}} + \mathbf{C}\mathbf{v} + \mathbf{g} + \mathbf{F}_v(\dot{\boldsymbol{\theta}}) + \mathbf{F}_C(\dot{\boldsymbol{\theta}}) \cdot sgn(\dot{\boldsymbol{\theta}}) - K_D\mathbf{r} \\ &= \mathbf{Y}\hat{\mathbf{p}} + \hat{\mathbf{F}}_\mathbf{v}(\dot{\boldsymbol{\theta}}) + \hat{\mathbf{F}}_\mathbf{C}(\dot{\boldsymbol{\theta}}) \cdot \mathbf{sgn}(\dot{\boldsymbol{\theta}}) - \mathbf{K_D}\mathbf{r}\end{aligned} \quad (1)$$

$$\mathbf{v} = \dot{\boldsymbol{\theta}}_d - \Lambda(\boldsymbol{\theta} - \boldsymbol{\theta_d}) \quad (2)$$

$$\mathbf{r} = (\dot{\boldsymbol{\theta}} - \dot{\boldsymbol{\theta}}_d) + \Lambda(\boldsymbol{\theta} - \boldsymbol{\theta_d}) \quad (3)$$

with the definition of the used RBF-networks for the viscose friction (eq.4),

$$\begin{aligned}\mathbf{F}_{v,j}(\dot{\boldsymbol{\theta}}_j) &= RBF_j(\dot{\boldsymbol{\theta}}_j)|_{j=1,\ldots,6} \\ &= \sum_{l=1}^{N_{Neurons}} \omega_{jl} \cdot e^{\frac{(\dot{\theta}-\dot{\theta}_c)^2}{\sigma_v^2}} \\ &= \sum_{l=1}^{N_{Neurons}} \omega_{jl} \cdot \Psi_{v,jl} \\ &= \mathbf{w}_j^T \boldsymbol{\Psi}_{v,j}\end{aligned} \quad (4)$$

and the Coulomb friction in the general case (eq.5)

$$\begin{aligned}\mathbf{F}_{c,j}(\dot{\boldsymbol{\theta}}_j) &= RBF_j(\dot{\boldsymbol{\theta}}_j)|_{j=1,\ldots,6} \\ &= \sum_{l=1}^{N_{Neurons}} v_{jl} \cdot e^{\frac{(\dot{\theta}-\dot{\theta}_c)^2}{\sigma_c^2}} \\ &= \sum_{l=1}^{N_{Neurons}} v_{jl} \cdot \psi_{c,jl} \\ &= \mathbf{v}_j^T \cdot \boldsymbol{\Psi}_{c,j}\end{aligned} \quad (5)$$

which can be reduced in the special case with one infinite wide neuron ($\sigma_c \to \infty$) to equation 6.

$$\begin{aligned}\hat{\mathbf{F}}_{c,j} &= RBF_j|_{j=1,\ldots,6} \\ &= v_j^T\end{aligned} \quad (6)$$

Thus neural networks train the hard to model frictional part of the robots dynamic system (eq.7)

$$\tau = \mathbf{M}(\boldsymbol{\theta}) \cdot \ddot{\boldsymbol{\theta}} + \mathbf{C}(\boldsymbol{\theta},\dot{\boldsymbol{\theta}}) \cdot \dot{\boldsymbol{\theta}} + \mathbf{g} + \mathbf{F_v}(\dot{\boldsymbol{\theta}}) + \mathbf{F}_c(\dot{\boldsymbol{\theta}}) \cdot sgn((\dot{\theta})) \quad (7)$$

during a special position control law (eq. 1), where Λ in equation 2 and 3 is a positive definite diagonal matrix, which has got to be chosen, to achieve the desired closed loop performance of the nominal system. Note, that the controller in equation 1 does not perform a linearization of the closed loop system and is different from the control law in equation 17, but stable behaviour can be proved via a Lyapunov function $\mathbf{V}(t)$, whose time derivative $\dot{\mathbf{V}}(t)$ has got to be negativ definite. First substitute equation 7 in equation 1, which delivers the way to find a Lyapunov candidate. The stability condition delivers then the necessary stable adaptation law for the system. First choose a Lyapunov candidate (eq.8) of the form

$$V(t) = \frac{1}{2}\left[\mathbf{r}^T\mathbf{Mr} + \tilde{\mathbf{p}}^T \Gamma^{-1}\tilde{\mathbf{p}} + \sum_{j=1}^{6} \tilde{\mathbf{w}}_\mathbf{j}^T \gamma_\mathbf{j}^{-1} \tilde{\mathbf{w}}_\mathbf{j} + \sum_{j=1}^{6} \eta_j^{-1} \tilde{v}_j^2 \right] \quad (8)$$

with

$$\tilde{\mathbf{p}} = \hat{\mathbf{p}} - \mathbf{p} \quad (9)$$
$$\tilde{\mathbf{w}}_j = \hat{\mathbf{w}}_j - \mathbf{w}_j \quad (10)$$

where $\gamma_\mathbf{j}^{-1}$, Γ^{-1} and η^{-1} are positive definite, diagonal matrices, representing the learning rate for the system parameters and neuron weights. Differentiation of $\mathbf{V(t)}$ with respect to time delivers

$$\begin{aligned}\dot{V}(t) &= \frac{1}{2}\left[\frac{d}{dt}(\mathbf{r}^T\mathbf{Mr}) + \frac{d}{dt}(\tilde{\mathbf{p}}^T \Gamma^{-1}\tilde{\mathbf{p}}) \right. \\ &\quad \left. + \frac{d}{dt}(\sum_{j=1}^{6}\tilde{\mathbf{w}}_j^T \gamma_j^{-1}\tilde{\mathbf{w}}_j) + \frac{d}{dt}(\sum_{j=1}^{6}\eta_j^{-1}\tilde{v}_j^2)\right] \\ &= -\mathbf{r}^T K_D \mathbf{r} + \mathbf{r}^T \mathbf{Y}\tilde{\mathbf{p}} + \sum_{j=1}^{6} r_j \tilde{\mathbf{w}}_\mathbf{j}^T \boldsymbol{\Psi}_\mathbf{j} \\ &\quad + \sum_{j=1}^{6} r_j \tilde{v}_j sgn(\dot{\theta}_j) + \dot{\tilde{\mathbf{p}}}^T \Gamma^{-1}\tilde{\mathbf{p}}\end{aligned}$$

$$+ \sum_{j=1}^{6} \dot{\tilde{\mathbf{w}}}_j^T \gamma_j^{-1} \tilde{\mathbf{w}}_j + \sum_{j=1}^{6} \dot{\tilde{v}}_j \eta_j^{-1} \tilde{v}_j \quad (11)$$

The main part of equation 11 has got to be zero, so that only the negative definite kernel

$$\dot{V}(t) = -\mathbf{r}^T K_D \mathbf{r} \leq 0 \quad (12)$$

remains[12]. This condition delivers the adaptation laws (eq.13,14 and 15) for the parameters respectively.

$$\dot{\tilde{\mathbf{p}}} = -\Gamma \mathbf{Y}^T \mathbf{r} \quad (13)$$

$$\dot{\tilde{\mathbf{w}}}_j = -\gamma_j \, \Psi_\mathbf{j} \, \mathbf{r_j} \quad (14)$$

$$\dot{\tilde{v}}_j = -\eta_j \, sgn(\dot{\theta}_j) \, r_j \quad (15)$$

The training of the RBF networks is performed during 'sufficient rich' movements. Especially slow movements are very suitable for a fast adaptation of the coulomb-friction parameters. After the training phase, the parameters and weights of the system are frozen and then used in the inverse dynamics control law of the NVFC control system. Then the manipulator controller ist switched from the NAC controller to the NVFC controller to perform hybrid position/force control.

NVFC: The NVFC system consists of the modules force mapping (eq. 16)

$$\boldsymbol{\tau}_f = \mathbf{J}^\mathbf{T}(\theta) \cdot \mathbf{S} \cdot \mathbf{F_d} \quad (16)$$

with the separation matrix **S** representing a diagonal matrix with the entries 1 or 0 on the main diagonal and the rest equal to zero. The inverse dynamics controller (eq. 17) inclusive neural friction compensation

$$\boldsymbol{\tau}_p = \hat{\mathbf{M}}(\boldsymbol{\theta})\ddot{\boldsymbol{\theta}} + \hat{\mathbf{C}}(\boldsymbol{\theta},\dot{\boldsymbol{\theta}}) \cdot \dot{\boldsymbol{\theta}} + \hat{\mathbf{G}}(\boldsymbol{\theta})$$
$$+ \hat{\mathbf{F}}_v(\dot{\boldsymbol{\theta}}) + \hat{\mathbf{F}}_c(\dot{\boldsymbol{\theta}}) \cdot sgn(\dot{\boldsymbol{\theta}}) \quad (17)$$

achieves reduced disturbances in force even during slow joint velocities, improving the force tracking performance significantly. Moreover an orthogonal position separation (eq. 18)

$$\boldsymbol{\Delta\theta}_s = \mathbf{J}^{-1}(\boldsymbol{\theta}) \cdot (\mathbf{I} - \mathbf{S}) \cdot \mathbf{J}(\boldsymbol{\theta}) \cdot \boldsymbol{\Delta\theta} \quad (18)$$

inclusive PD controller and nonlinear decoupling via the inertia matrix $\mathbf{M}(\boldsymbol{\theta})$ is used inclusive a modification (eq. 19, 20 and 21)

$$\boldsymbol{\theta}_{dm,i} = \boldsymbol{\theta}_{d,switch} + \boldsymbol{\theta}_{T,i} + \boldsymbol{\theta}_{F,i} \quad (19)$$

$$\boldsymbol{\theta}_{T,i} = \boldsymbol{\theta}_{T,i-1} + \mathbf{J}_\mathbf{i}^{-1} \cdot [\mathbf{I} - \mathbf{S}] \cdot \mathbf{J}_{\mathbf{d,i}} \cdot [\boldsymbol{\theta}_{\mathbf{d,i}} - \boldsymbol{\theta}_{\mathbf{d,i-1}}] \quad (20)$$

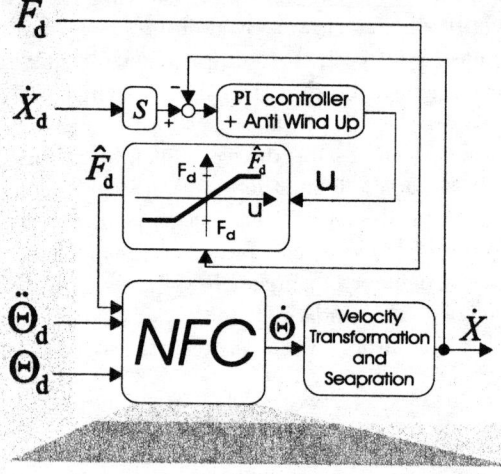

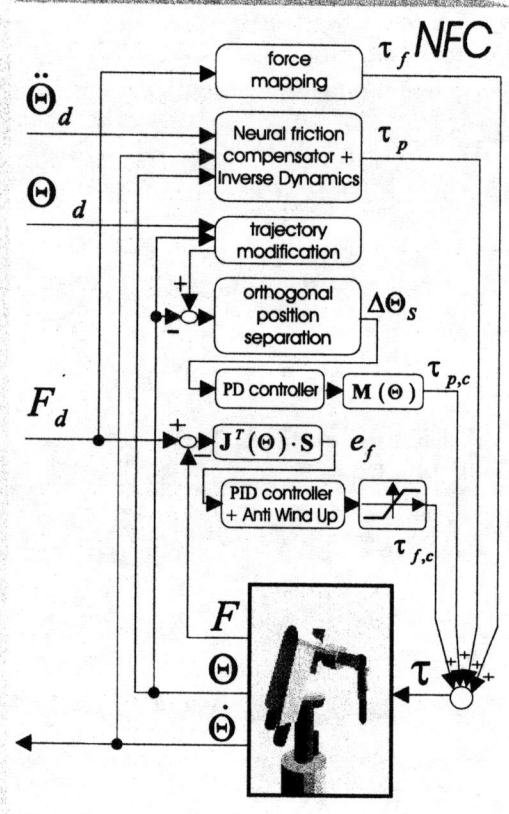

Fig. 2. Neural velocity/force control system running in a 2ms control loop on a PC-based realtime Operationsystem

$$\boldsymbol{\theta}_{F,i} = \boldsymbol{\theta}_{F,i-1} + \mathbf{J}_\mathbf{i}^{-1} \cdot [\mathbf{S}] \cdot \mathbf{J}_\mathbf{i} \cdot [\boldsymbol{\theta}_\mathbf{i} - \boldsymbol{\theta}_\mathbf{i-1}] \quad (21)$$

of the desired trajectory, to keep the angle error $\Delta\theta$ sufficiently small which satisfies the requirements of the differential inverse kinematics of eq. 18. Furthermore a force mapping (eq. 22)

$$\mathbf{e}_f = \mathbf{J}^T(\theta) \cdot \mathbf{S} \cdot \Delta\mathbf{F} \quad (22)$$

of the force control error $\Delta\mathbf{F}$ delivers the control input for the PID force controller, which is equipped with an Anti-Wind-Up module inclusive saturation, to prevent instability of the force control while having no contact to the unknown surface yet, when a constant force error $\Delta\mathbf{F}$ is fed into the PID controller. The four resulting torques

$$\boldsymbol{\tau} = \boldsymbol{\tau}_f + \boldsymbol{\tau}_p + \boldsymbol{\tau}_{p,c} + \boldsymbol{\tau}_{f,c} \quad (23)$$

present the complete control input fed into the manipulators servo controllers. The NFC module [7, 8] is embedded into the NVFC controlsystem consisting of a velocity controller of PI type inclusive Anti-Wind-Up to bound the output of the PI controller to the desired Force $\mathbf{F}_d$, when being in contact to the unknown hard environment. During the approaching phase, when having no force contact or when losing contact by moving away the surface from the TCP, the output of the PI controller is reduced nearly to zero, when having reached the desired TCP velocity. This reduction of the desired force is highly recommended to bound the TCP velocity to an adequate low value, which guarantees a sufficient small impact force peak with extremly reduced bouncing.

2.2 Simulation

The figures 3.(A,B,C) show the adaptation phase of the system for the Coulomb friction of axis 6 during the first 30 seconds. It becomes obvious, that the friction parameter of axis 6 reaches its final value after a very short time, which coincides with the decrease of the joint angle control errror (fig.3.B). The adaptation of the Coulomb friction is performed during slow movements (fig.3.C) to provide a sufficient rich trajectory whereas the viscose friction can only be adapted during a movement with sufficient different joint angle velocities. Figures 4.(A,B,C,D) show the contacting and tracking of an unknown hard surface ($c = 10^6 N/m$) with desired force and desired position trajectory. The NVFC gets a desired trajectory from the user in the xy-plane and a desired force in z-direction and produces the trajectory in Fig.4.A., which makes the manipulator follow the unknown hard surface with constant z-force. Fig.4.B shows the force in z-direction over the position in xy-direction.

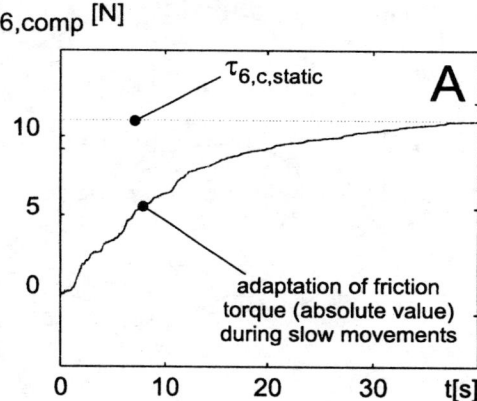

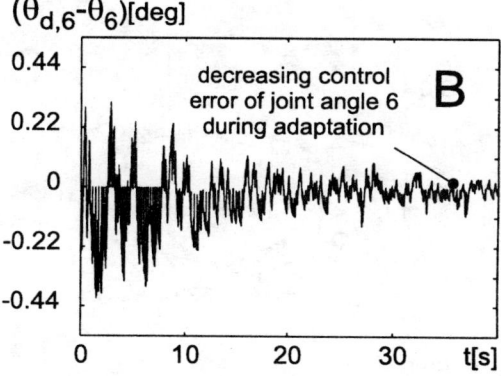

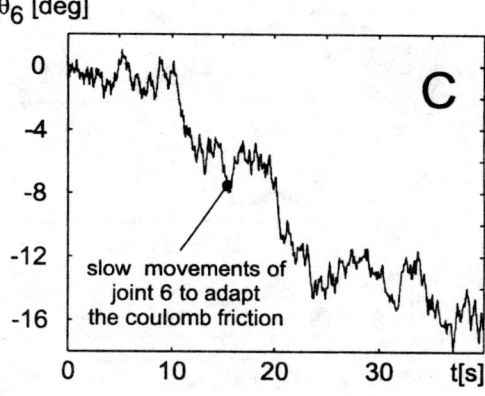

Fig. 3. Simulation results **A**: friction compensation torque during slow adaptation movements, **B**: decreasing joint angle control error during the adaptation phase, **C**: joint angle trajectory for axis 6 during adaptation movements

Fig.4.(C,D) show the velocity in z-direction and the force over time respectively. The force peaks in fig.4.(B,D) result from the position controller, which has a small disturbing influence on the force, since the

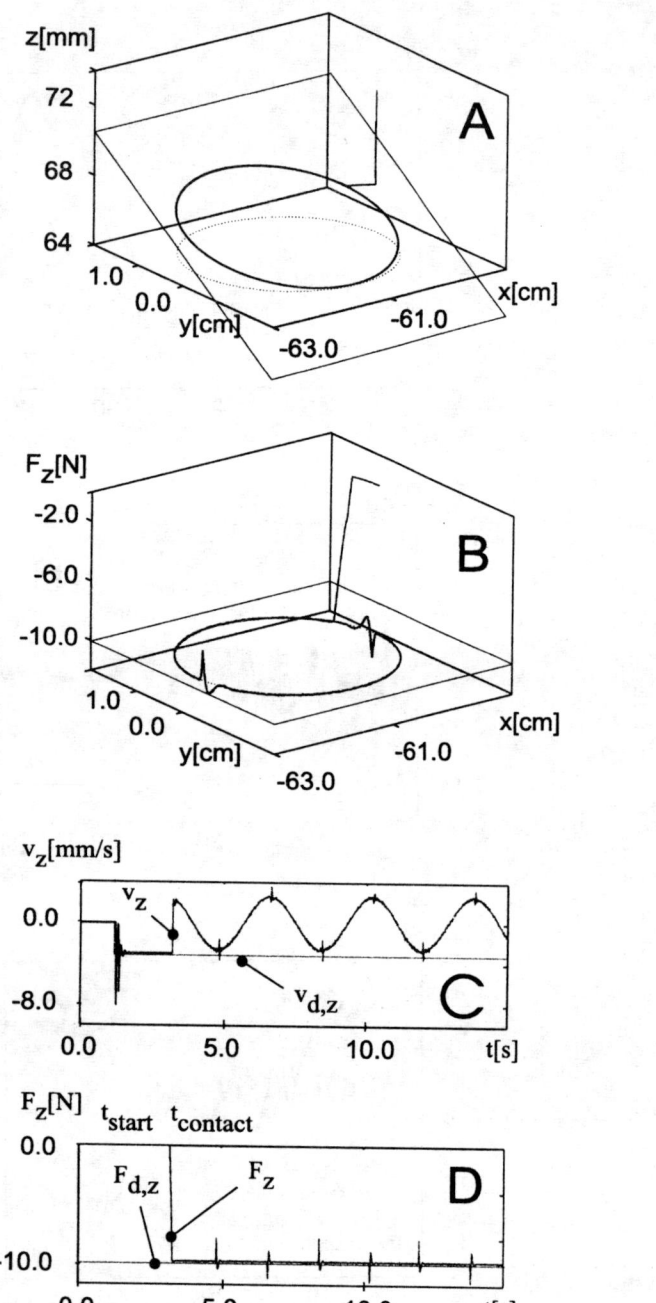

Fig. 4. Simulation Results A: cartesian trajectory generated while tracking an unknown hard surface ($c = 10^6 N/m$), B: contact force in z-direction over xy-position trajectory, C: z-velocity over time, D: contact force in z-direction over time

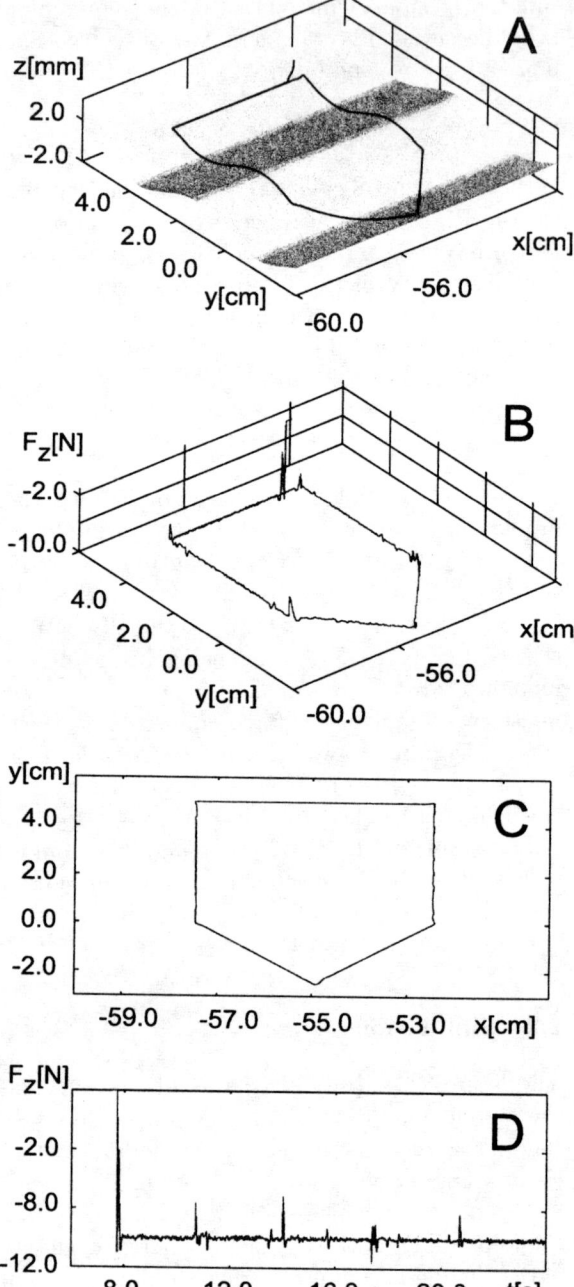

Fig. 5. Simulation Results A: cartesian trajectory, generated while tracking an unknown wave shaped hard surface ($c = 10^6 N/m$) with constant contact force, B: contact force in z-direction over xy-position, C: actual position in xy plane, D: force in z-direction over time

position controller as well as the force controller act on the same joint torques. Small uncompensated friction effects produce position errors, which then cause the PD position controller to produce high torque peaks,

interfering the actual force. So it is necessary, to choose the D term of the PD position controller not to high, to reduce undesired force peaks and interference with the force controlsystem. In Fig.5.(A,B,C,D) this correlation also becomes obvious, where the desired position trajectory in the xy-plane maintains 5 sharp edges ('house', fig.5.A), where the position controller gets a higher error and therefore disturbs the force controller via high torque peaks fig5.B. The unknown contact surface has got a wave shaped structure and nevertheless a good tracking performance in the position is achieved(fig.5.C), as well as a good force tracking performance (fig.5.B). Fig.5.D shows the force in z-direction over time inclusive the contact phase.

3 Discussion

The embedded approach of force control works, in contrast to existing concepts [4] without controller-switching in the contact moment, to reduce bouncing and impact force peak. Our velocity/force control is turned on, when still being quite far away from the contact surface, controlling the desired force, fed into the hybrid position/force controller adequately [1]. The overshooting and oscillation caused by the unknown and/or unmodelled dynamics of a robot manipulator and an environment can be decreased efficiently by using the proposed NVFC concept. Coulomb friction torques (3 to 5 percent of the nominal motor torques) are quite large compared to the torques needed to develop the contact forces, so the influence of friction on force behaviour is quite strong [1, 4]. The NAC adapts the unknown friction parameters of the system and reduces the tracking error providing a better performance of the closed loop system. It is necessary to test differernt learning rates η and γ and to generate sufficient rich adaptation trajectories to achieve a fast learning of the unknown system parameters.

4 Conclusions

The high accuracy in force/position control using NN's offers a wide range of intentional interaction tasks of multi joint manipulators. An adaptive force controlsystem has been developed to ensure that the contact force is regulated to a constant desired value while the end-effector orientation tracks a time varying desired trajectory. In simulations we have demonstrated, that NVFC allows the solution of various hard contact control problems including polishing, deburring, or disassembly, which require a continuous highspeed evaluation of the contact forces.

References

1. M. Dapper, R. Maaß, and R. Eckmiller. Neural velocity force control for industrial manipulators contacting rigid surfaces. In *Springer Proc. International Conference Artificial Neural Networks, ICANN98, Skoevde, Sweden*, pages 887–892, September 1998.
2. M. Dapper, R. Maaß, V. Zahn, and R. Eckmiller. Neural force control (nfc) applied to industrial manipulators in interaction with moving rigid objects. In *IEEE Proc. International Conference on Robotics and Automation, Leuven, Belgium*, pages 2048–2053, May 1998.
3. M. Dapper, R. Maaß, V. Zahn, and R. Eckmiller. Neural force control (nfc) for complex manipulator tasks. In *Springer Proc. International Conference Artificial Neural Networks, ICANN97, Lausanne, Switzerland*, pages 787–792, October 1997.
4. G. Ferretti, G. Magnani, and R. Zavala. Impact modeling and control for industrial manipulators. *IEEE Control Systems Magazine*, 18(4):65–71, August 1998.
5. D. Gorinevsky, Formalsky A., and Schneider A. *Force Control of Robotic Systems*. CRC Press LLC, Boca Raton, Florida, 1997.
6. P. Kraus, V. Kumart, and P. Dupont. Analysis of frictional contact models for dynamic simulation. In *IEEE Proc. International Conference on Robotics and Automation, Leuven, Belgium*, pages 976–981, May 1998.
7. R. Maaß, M. Dapper, and R. Eckmiller. Neural trajectory optimization (nto) for manipulator tracking of unknown surfaces. In *Springer Proc. International Conference Artificial Neural Networks, ICANN98, Skoevde, Sweden*, pages 893–898, September 1998.
8. R. Maaß, V. Zahn, and R. Eckmiller. Neural force/position control in cartesian space for a 6 dof industrial robot: Concept and first results. In *IEEE Proc. International Conference Neural Networks, ICNN97, Houston*, pages 1744–1748, June 1997.
9. C. Natale, B. Siciliano, and L. Villani. Control of moment and orientation for a robot manipulator in contact with a compliant environment. In *IEEE Proc. International Conference on Robotics and Automation, Leuven, Belgium*, pages 1755–1760, May 1998.
10. Y. Oh, W. Chung, Y. Youm, and I. Suh. Motion/force decomposition of redundant manipulator and its application to hybrid impedance control. In *IEEE Proc. International Conference on Robotics and Automation, Leuven, Belgium*, pages 1441–1446, May 1998.
11. F. Pfeiffer and C. Glocker. *Multibody Dynamics with Unilateral Contacts*. Wiley and Sons, New York, 1996.
12. M. W. Spong and M. Vidyasagar. *Robot Dynamics And Control*. Wiley and Sons, New York, 1989.
13. W. Zhu, Z. Bien, and J. De Schutter. Adaptive motion/force control of multiple manipulators with joint flexibility based on virtual decomposition. *IEEE Transactions on Automatic Control*, 43(1):46–60, January 1998.

Transfer of Human Control Strategy Based on Similarity Measure

Jingyan Song[1], Yangsheng Xu[2,3], Michael C. Nechyba[4], Yeung Yam[2]

[1]Department of Systems Engineering & Engineering Management
[2]Department of Mechanical and Automation Engineering
The Chinese University of Hong Kong, Shatin, NT, Hong Kong
[3]The Robotics Institute, Carnegie Mellon University, Pittsburgh, PA 15213 USA
[4]Department of Electrical and Computer Engineering
University of Florida, Gainesville, FL 32611 USA

Abstract

In this paper, we address the problem of transferring human control strategies (HCS) from an expert model to an apprentice model. The proposed algorithm allows us to develop useful apprentice models that nevertheless incorporate some of the robust aspects of the expert HCS models. We first describe our experimental platform – a real-time graphic driving simulator – for collecting and modeling human control strategies. Then, we discuss an adaptive neural network learning architecture for abstracting HCS models. Next, we define a hidden Markov model (HMM) based similarity measure which allows us to compare different human control strategies. This similarity measure is combined subsequently with simultaneously perturbed stochastic approximation to develop our proposed transfer learning algorithm. In this algorithm, an expert HCS model influences both the structure and the parametric representation of the eventual apprentice HCS model. Finally we describe some experimental results of the proposed algorithm.

1 Introduction

Transferring human control strategy (HCS) has potential application in a number of research areas ranging from virtual reality and robotics to intelligent highway systems. In previous work, transfer of human control strategies from human experts to human apprentices has been studied [1]. In this paper, we investigate the transfer of human control strategies, not between human expert and human apprentice, but rather between an expert HCS model and an apprentice HCS model.

In developing an algorithm to accomplish this transfer of control strategies, we rely on a number of related results in human control strategy research. First, we need to be able to successfully abstract a human control strategy to a reliable computational model. Since human control strategies are dynamic, nonlinear stochastic processes, however, developing good analytic HCS models tends to be difficult. Therefore, recent work in modeling HCS has focused on learning empirical models from real-time input-output human control data, through, for example, fuzzy logic [2, 3, 4, 5], and neural network techniques [6].

Second, we need to be able to evaluate our resulting models both with respect to how well they approximate the human control data, and how well they meet certain performance criteria. Since the HCS models are empirical, this is especially important, as few theoretical guarantees exist about their stability or performance. In [7], a stochastic similarity measure, based on hidden Markov model (HMM) analysis, was developed for validating the fidelity of HCS models to the source training data, while in [8, 9], several performance criteria were developed for evaluating the skill inherent in learned HCS models.

This previous work forms the basis for our transfer learning algorithm proposed herein. The algorithm requires that we first collect control data from experts and apprentices. From this data, we can then abstract two types of HCS models, one for the expert and one for the apprentice, using previously developed techniques. Our main goal in this paper is to improve the apprentice model's performance while still retaining important characteristics of the apprentice model. To do this, we look towards the stochastic similarity measure proposed in [7], which is capable of comparing long, multi-dimensional, stochastic trajectories, and has been applied previously towards validating models of human control strategy.

In our transfer learning algorithm, we propose to raise the similarity between an expert HCS model and an apprentice HCS model. Alternatively, we can think of the expert model guiding the actions of the apprentice model. The overall algorithm consists of two steps. In the first step, we let the expert model influence the eventual structure of the HCS model. Once an appropriate model structure has been chosen, we then tune the parameters of the apprentice model through simultaneously perturbed stochastic approximation, an optimization algorithm that requires no analytic formulation, only two empirical measurements of a user-defined objective function.

In this paper, we first describe our experimental platform for collecting and modeling human control data – a real-time graphic driving simulator. We then describe an adaptive neural network learning architecture for abstracting HCS models given human control data. Next, we define the notion of similarity between control trajectories, and develop our transfer learning algorithm for expert and apprentice HCS models. Finally, we describe experimental results of the learning algorithm in the driving domain.

2 HCS modeling

For this work, we collect expert and apprentice driving data through a real-time graphic simulator, whose interface is shown in Figure 1. In the simulator, the

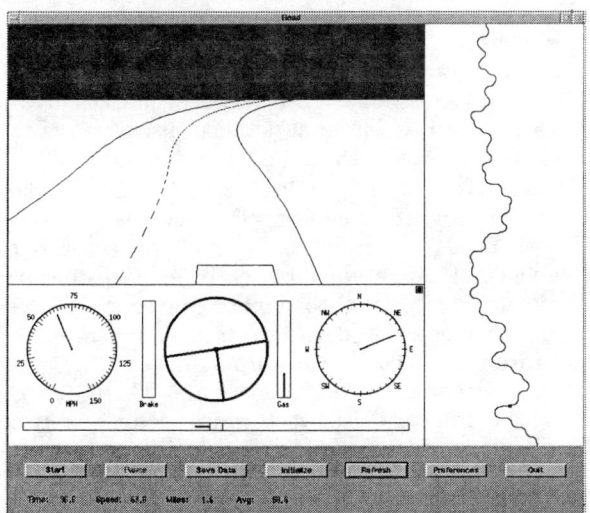

Figure 1: The driving simulator gives the user a perspective preview of the road ahead.

driving operator has independent control of the vehicle's steering as well as the brake and gas pedals. We ask different individuals to navigate across a number of different randomly generated roads, which consist of a sequence of (1) straight-line segments, (2) left turns, and (3) right turns. The map in Figure 1, for example, illustrates one randomly generated 20km road for which human (expert or apprentice) driving data was recorded. Each straight-line segment as well as the radius of curvature for each turn range in length between 100m and 200m. Nominally, the road is divided into two lanes, each of which has width $w = 5$m. The human operator's view of the road ahead is limited to 100m, and the entire simulator is run at 50Hz.

Using the collected data, we choose the flexible cascade neural network architecture with node-decoupled extended Kalman filtering (NDEKF) [10] for modeling the human driving data. We prefer this learning architecture over others for a number of reasons. First, no *a priori* model structure is assumed; the neural network automatically adds hidden units one at a time to an initially minimal network as the training requires. Second, hidden unit activation functions are not constrained to be a particular type. Rather, for each new hidden unit, the incremental learning algorithm can select that functional form which maximally reduces the residual error over the training data. Typical alternatives to the standard sigmoidal function are sine, cosine, and the Gaussian function. Finally, it has been shown that node-decoupled extended Kalman filtering, a quadratically convergent alternative to slower gradient descent training algorithms (such as backpropagation) fits well within the cascade learning framework and converges to good local minima with less computation [10].

The flexible functional form which cascade learning allows is ideal for abstracting human (expert or apprentice) control strategies, since we know very little about the underlying structure of each individual's internal controller. By making as few *a priori* assumptions as possible in modeling the human driving data, we improve the likelihood that the learning algorithm will converge to a good model of the human control data.

In order for the learning algorithm to properly model each individual's human control strategy, the model must be presented with those state and environmental variables upon which the human operator relies. Thus, the inputs to the cascade neural network should include: (1) current and previous state information $\{\nu_\xi, \nu_\eta, \dot{\theta}\}$, (2) previous output (command) information $\{\delta, P_f\}$, and (3) a description of the road visible from the current car position. More precisely, the network inputs are,

$$\{\nu_\xi(k-n_s), \cdots, \nu_\xi(k-1), \nu_\xi(k),$$
$$\nu_\eta(k-n_s), \cdots, \nu_\eta(k-1), \nu_\eta(k), \quad (1)$$
$$\dot{\theta}(k-n_s), \cdots, \dot{\theta}(k-1), \dot{\theta}(k)\}$$
$$\{\delta(k-n_c), \cdots, \delta(k-1), \delta(k),$$
$$P_f(k-n_c), \cdots, P_f(k-1), P_f(k)\} \quad (2)$$
$$\{x(1), x(2), \cdots, x(n_r), y(1), y(2), \cdots, y(n_r)\} \quad (3)$$

where $\nu_\xi(k)$ is the lateral velocity of the car at time k, $\nu_\eta(k)$ is the longitudinal velocity of the car at time k, $\dot{\theta}(k)$ is the angular velocity of the car at time k, $\delta(k)$ is the steering command at time k, $P_f(k)$ is the acceleration command at time k, n_s is the length of the state histories and n_c is the length of the previous command histories presented to the network as input. For the road description, we partition the visible view of the road ahead into n_r equivalently spaced, body-relative (x, y) coordinates of the road median, and provide that sequence of coordinates as input to the network. Thus, the total number of inputs to the network n_i is given by,

$$n_i = 3n_s + 2n_c + 2n_r \qquad (4)$$

The two outputs of the cascade network are $\{\delta(k+1), P_f(k+1)\}$. For the system as a whole, the cascade neural network can be viewed as a feedback controller, whose two outputs control the driving of the vehicle.

3 HCS similarity

In this section, we describe a similarity measure [7], which has previously been applied towards validating HCS models, and which we will use in a subsequent section for our HCS model comparisons. This similarity measure is based on hidden Markov models (HMMs), which are trainable statistical models with two appealing features: (1) no *a priori* assumptions are made about the statistical distribution of the data to be analyzed, and (2) a degree of sequential structure can be encoded by the hidden Markov models. As such, they have previously been applied in a number of different stochastic signal processing applications.

A discrete-output HMM is completely defined by the following triplet,

$$\lambda = \{A, B, \pi\} \qquad (5)$$

where A is the probabilistic $n_s \times n_s$ state transition matrix, B is the $L \times n_s$ output probability matrix with L discrete output symbols $l \in \{1, 2, \cdots, L\}$, and π is the n-length initial state probability distribution vector for the HMM.

Below, we define a stochastic similarity measure, based on discrete-output HMMs. Assume that we wish to compare observation sequences from two stochastic processes Γ_1 and Γ_2. Let $\bar{O}_i = \{O_i^{(k)}\}, k \in \{1, 2, \cdots, n_i\}, i \in \{1, 2\}$, denote the set of n_i observation sequences of discrete symbols generated by process Γ_i. Each observation sequence is of length $T_i^{(k)}$, so that the total number of symbols in set $\bar{O}_i$ is given by,

$$\bar{T}_i = \sum_{k=1}^{n_i} T_i^{(k)}, i \in \{1, 2\}. \qquad (6)$$

Also let $\lambda_j = \{A_j, B_j, \pi_j\}, j \in \{1, 2\}$, denote a discrete HMM locally optimized with the Baum-Welch algorithm to maximize,

$$P(\lambda_j|\bar{O}_j) = \prod_{k=1}^{n_i} P(\lambda_j|O_j^{(k)}), j \in \{1, 2\}, \qquad (7)$$

and let,

$$P(\bar{O}_i|\lambda_j) = \prod_{k=1}^{n_i} P(O_i^{(k)|\lambda_j}) \qquad (8)$$

$$P_{ij} = P(\bar{O}_i|\lambda_j)^{1/\bar{T}_i}, i, j \in \{1, 2\} \qquad (9)$$

denote the probability of the observation sequences $\bar{O}_i$ given the model λ_j, normalized with respect to the sequence lengths $\bar{T}_i$.

Using definition (9), we now define the following similarity measure between $\bar{O}_1$ and $\bar{O}_2$:

$$\sigma(\bar{O}_1, \bar{O}_2) = \sqrt{\frac{P_{21}P_{12}}{P_{11}P_{22}}} \qquad (10)$$

In other words, we first use each set of observation sequences to train a corresponding HMM; this allows us to evaluate P_{11} and P_{22}. We then cross-evaluate each observation sequence on the other HMM to arrive at P_{12} and P_{21}. Finally we take the ratio of these probabilities and take the square root.

For any two sets of observation sequences, the similarity measure σ will range between 0 and 1. For very similar observation sequences, σ will be close to one, while for very dissimilar observation sequences, σ will be close to zero. Additional detailed discussion of the similarity measure's properties can be found in [7].

Finally, we note that in order to apply this similarity measure towards comparing human control strategies, we need to convert real-valued trajectories to a sequence of discrete symbols. To achieve this, we follow a two-step signal-to-symbol conversion process First, we window the data into frames and filter the data through spectral transforms such as the well-known short-time Fourier transform. Then, the resulting spectral coefficient vectors are vector-quantized into discrete symbols. For further details on this processing, please consult [7].

4 HCS transfer

In this section, we develop a learning algorithm for transferring skill from an expert HCS model to an apprentice HCS model. Rather than simply discard the apprentice HCS model, we attempt to preserve important aspects of the apprentice model, while at the same time improving the apprentice model's performance. In the proposed algorithm, the expert model serves as the guide or teacher to the apprentice model, and influences both the eventual structure and parametric representation of the apprentice model. As we will demonstrate shortly, the similarity measure defined in the previous section will play a crucial role in this transfer learning algorithm.

4.1 Structure learning

Recall that in cascade neural network learning, the structure of the neural network is adjusted during training, as hidden units are added one at a time until satisfactory error convergence is reached. Suppose that we train a HCS model from human control data provided by an expert using cascade learning. The final trained expert model will then consist of a given structure, which, in cascade learning, is completely defined by the number of hidden units in the final model.

Now, suppose that we have collected training data from an apprentice – that is, from an individual less skilled than the expert. What final structure should his learned model assume? One answer is that we let the model converge to the "best" structure as was the case for the expert model. Since, we already have an expert model at hand, however, we can let the expert model inform the choice of structure for the apprentice model.

Figure 2 illustrates our approach for structure learning in apprentice HCS models, as guided by an expert HCS model. We first train the expert HCS model in the usual fashion. Then, we train the apprentice HCS model, but impose an additional constraint during learning – namely, that the final structure (i.e. the number of hidden units) of the apprentice model and the expert model be the same.

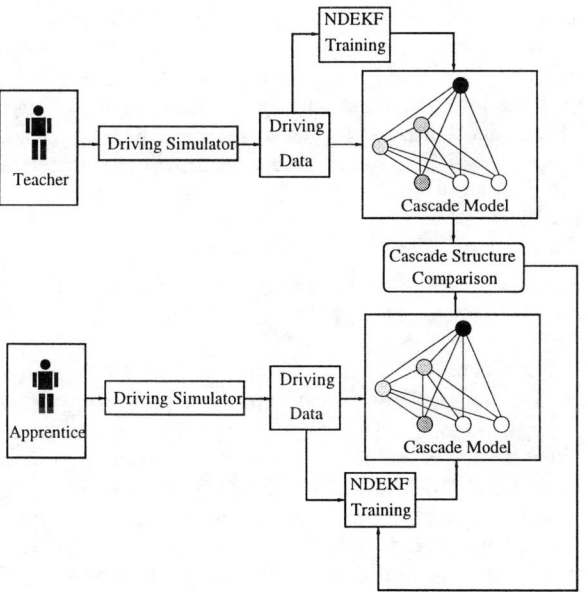

Figure 2: Structure learning in apprentice HCS model.

4.2 Parameter learning

Even though we impose the same structure on the expert and apprentice HCS models, they will clearly converge to different parametric representations. The expert model will be similar to the expert's control strategy, while the apprentice model will be similar to the apprentice's control strategy. We would now like to tune the apprentice model so as to retain part of the control strategy encoded within, while at the same time improving the performance in the apprentice model. Once again we will use the expert model as a guide in this learning, by examining the similarity (as defined previously) between the expert and apprentice models.

Let,

$$\omega = [w_1 \ w_2 \ \cdots \ w_n] \quad (11)$$

denote a vector consisting of all the weights in the apprentice HCS model $\Gamma(\omega)$. Also let $\sigma(O_e, O_p(\omega))$ denote the HMM similarity between the apprentice HCS model and expert HCS model. We would now like to determine a weight vector ω^* which raises the expert/apprentice similarity $\sigma(O_e, O_p(\omega^*))$, while at the same time retaining part of the apprentice control strategy.

Determining a suitable ω^* is difficult in principle because (1) we have no explicit gradient information

$$G(\omega) = \frac{\partial}{\partial \omega} \sigma(O_e, O_p(\omega)) \quad (12)$$

(2) each experimental measurement of $\sigma(O_e, O_p(\omega))$ requires a significant amount of computation. We lack explicit gradient information, since we can only evaluate the similarity measure empirically. Hence, gradient-based optimization techniques, such as steepest-descent and Newton-Raphson are not suitable. Furthermore, because each similarity evaluation is potentially computationally expensive, genetic optimization, which can require many iterations to converge, also does not offer a good alternative. Therefore we turn to *simultaneously perturbed stochastic approximation (SPSA)* to adjust ω.

Stochastic approximation (SA) is a well known iterative algorithm for finding roots of equations in the presence of noisy measurements. Simultaneously perturbed stochastic approximation (SPSA) [11] is a particular multivariate SA technique which requires as few as two measurements per iteration and shows fast convergence in practice. Hence, it is well suited for our application. Denote ω_k as our estimate of ω^* at the kth iteration of the SA algorithm, and let ω_k be defined by the following recursive relationship:

$$\omega_{k+1} = \omega_k - \alpha_k \bar{G}_k \quad (13)$$

where $\bar{G}_k$ is the simultaneously perturbed gradient approximation at the kth iteration,

$$\bar{G}_k = \frac{1}{q} \sum_{i=1}^{q} G_k^i \approx \frac{\partial}{\partial \omega} \bar{\sigma}(O_e, O_p(\omega)) \quad (14)$$

$$G_k^i = \frac{\bar{\sigma}_k^{(+)} - \bar{\sigma}_k^{(-)}}{2c_k} \begin{bmatrix} 1/\triangle_{kw_1} \\ 1/\triangle_{kw_2} \\ \cdots \\ 1/\triangle_{kw_n} \end{bmatrix} \quad (15)$$

Equation (14) averages q stochastic two-point measurements G_k^i for a better overall gradient approximation, where,

$$\bar{\sigma}_k^{(+)} = \sigma(O_e, O_p(\omega_k + c_k \triangle_k)) \quad (16)$$
$$\bar{\sigma}_k^{(-)} = \sigma(O_e, O_p(\omega_k - c_k \triangle_k)) \quad (17)$$
$$\triangle_k = [\triangle_{kw_1} \triangle_{kw_2} \cdots \triangle_{kw_n}]^T \quad (18)$$

where $\triangle_k$ is a vector of mutually independent, mean-zero random variables (e.g. symmetric Bernoulli distributed), the sequence $\{\triangle_k\}$ is independent and identically distributed, and the $\{\alpha_k\}$, $\{c_k\}$ are positive scalar sequences satisfying the following properties:

$$\alpha_k \to 0, \quad c_k \to 0 \text{ as } k \to \infty, \quad (19)$$
$$\sum_{k=0}^{\infty} \alpha_k = \infty, \quad \sum_{k=0}^{\infty} \left(\frac{\alpha_k}{c_k}\right)^2 < \infty \quad (20)$$

For our problem, we define the objective function to be $\bar{\sigma} = 1 - \sigma(O_e, O_p(\omega))$. The weight vector ω_0 is of course the weight representation in the initially stable apprentice model. Finally, we note that while larger values of q in equation (14) will give more accurate approximations of the gradient, in practice, two measurements ($q = 1$) per iteration is often sufficient. Figure 3 illustrates the overall parameter tuning algorithm.

5 Experiment

Here, we test the transfer learning algorithm on control data collected from three individuals, Tom, Dick and Harry. In previous work [8], we have observed that Harry's driving model performs better with respect to certain important performance measures. Therefore, we view Harry as the expert, and Dick and Tom as the apprentices. Furthermore, in order to simplify the problem somewhat, we keep the applied force constant at $P_f = 300N$. In other words, we ask each driver to control the steering δ only.

For this experiment, we first train the expert model on Harry's control data. The final trained model consists of two hidden units with $n_s = n_c = 3$, and $n_r = 15$; because we are keeping P_f constant, the total number of inputs for the cascade network model therefore is $n_i = 42$. Keeping in mind that we want the final structure of the apprentice models to be the same as the expert model's structure, we also train Dick's and Tom's models to two hidden units each, with the same number of inputs ($n_i = 42$).

We now would like to improve the performance of the apprentice models, while still retaining some

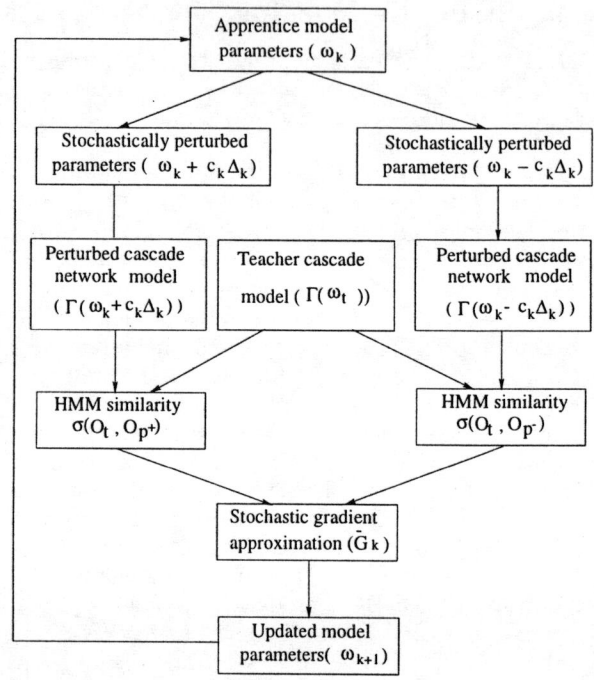

Figure 3: Stochastic transfer learning algorithm for HCS models.

aspects of the apprentice control strategies. In other words, we would like to improve the similarity $\sigma(O_e, O_p)$ defined in equation (10) between Harry's expert model and each of the apprentice models using the SPSA algorithm as discussed in the previous section.

In the SPSA algorithm, we empirically determine the following values for the scaling sequences $\{\alpha_k\}$, $\{c_k\}$:

$$\alpha_k = 0.00001/k^2, \quad k > 0 \quad (21)$$
$$c_k = 0.001/k^{1.25}, \quad k > 0 \quad (22)$$

Furthermore, we set the number of measurements per gradient estimation in equation (14) to $q = 1$. Finally, we denote $\sigma(O_e, O_p^k)$ as the similarity $\sigma(O_e, O_p)$ after iteration k of the learning algorithm; hence, $\sigma(O_e, O_p^0)$ denotes the similarity prior to any weight adjustment in the apprentice models.

Figure 4 plots the similarity measure between Tom's and Dick's apprentice models and Harry's expert HCS model, respectively, as a function of iteration k in the transfer learning algorithm. We observe that for Dick's model, the similarity to Harry's model improves from $\sigma(O_e, O_p^0) = 41.5\%$ to $\sigma(O_e, O_p^{15}) = 78.5\%$. Although for Tom's model the change is less dramatic, his model's similarity nevertheless rises from $\sigma(O_e, O_p^0) = 55.0\%$ to $\sigma(O_e, O_p^{15}) = 72.2\%$. Thus, the transfer learning algorithm improves the similarity of Dick's model by approximately 37% and Tom's model by about 17.2% over their respective initial models.

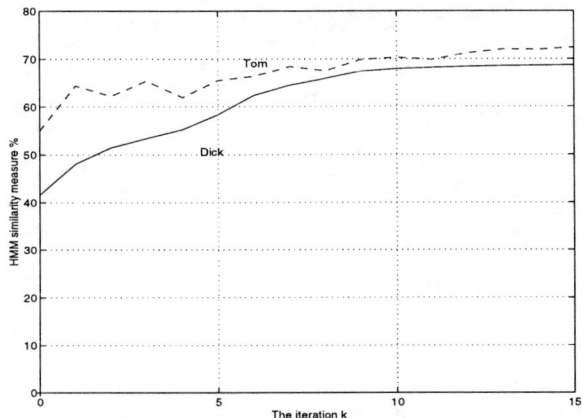

Figure 4: Similarity between Harry's and Dick's (solid) and Harry's and Tom's (dashed) models during transfer learning.

Since the similarity in control strategies improves, we would expect that apprentice performance improves as well. In order to test this, we examine model performance as measured by the obstacle avoidance performance criterion J defined previously in [8]. Let J_e denote the performance criterion value for Harry's expert model. Also, let J_p^0 denote apprentice performance before learning, and let J_p^{15} denote apprentice performance after transfer learning. We note that the performance criterion is defined so that smaller values indicate better obstacle avoidance performance.

Table 1 lists these performance values for Dick and Tom.

	J_p^0	J_p^{15}
$Harry$	0.51	0.51
Tom	1.23	0.87
$Dick$	1.37	1.13

Table 1: Obstacle avoidance performance measure

From Table 1, we note that Harry's performance does not change, of course, since we keep his model fixed. The models for Dick and Tom do, however, improve with respect to the obstacle avoidance performance measure. The adjusted apprentice models can be viewed as hybrid models, which combine the apprentice control strategy with some of the improved techniques of the expert model.

6 Conclusion

In this paper, we have proposed an iterative learning algorithm, based on simultaneously perturbed stochastic approximation (SPSA), for improving the similarity between apprentice and expert models of human control strategy. The transfer algorithm consists of two steps. In the first step, we ensure that the apprentice model reaches the same overall structure as the expert model. In the second step, we tune the parameters (i.e. weights) in the apprentice model to improve its similarity with the expert model. The resulting model will preserve some aspects of the original apprentice model while at the same time improving performance. The proposed algorithm requires no analytic formulation, only two experimental similarity measurements per iteration. In initial experiments, we have demonstrated that apprentice performance improves with the help of the expert model.

Acknowledgments

This work is supported in part by Hong Kong Research Council Grant No. CUHK4138/97E.

References

[1] M. C. Nechyba and Y. Xu, "Human Skill Transfer: Neural Networks as Learners and Teachers," *Proc. IEEE/RSJ Int. Conf. on Intelligent Robots and Systems*, vol. 3, pp. 314-9, 1995.

[2] M. Sugeno and T. Yasukawa, "A Fuzzy-Logic-Based Approach to Qualitative Modeling," *IEEE Trans. on Fuzzy Systems*, vol. 1, no. 1, pp. 7-31, 1993.

[3] C.C. Lee, "Fuzzy Logic in Control Systems: Fuzzy Logic Controller–Part I," *IEEE Trans. on Systems, Man, and Cybernetics*, vol. 20, no. 2, pp. 404-18, 1990.

[4] C.C. Lee, "Fuzzy Logic in Control Systems: Fuzzy Logic Controller–Part II," *IEEE Trans. on Systems, Man, and Cybernetics*, vol. 20, no. 2, pp. 419-35, 1990.

[5] U. Kramer, "On the Application of Fuzzy Sets to the Analysis of the System Driver-Vehicle-Environment," *Automatica*, vol. 21, no. 1, pp. 101-7, 1985.

[6] M. C. Nechyba and Y. Xu, "Human Control Strategy: Abstraction, Verification and Replication," *IEEE Control Systems Magazine*, vol. 17, no. 5, pp. 48-61, 1997.

[7] M. C. Nechyba and Y. Xu, "Stochastic Similarity for Validating Human Control Strategy Models," *IEEE Trans. on Robotics and Automation*, vol. 14, no. 3, pp. 437-51, 1998.

[8] J. Song, Y. Xu, M. C. Nechyba and Y. Yam, "Two Measures for Evaluating Human Control Strategy," *Proc. IEEE Conf. on Robotics and Automation*, vol. 3. pp. 2250-5, 1998.

[9] J. Song, Y. Xu, Y. Yam and M. C. Nechyba, "Optimization of Human Control Strategies with Simultaneously Perturbed Stochastic Approximation", *Proc. IEEE/RSJ Int. Conf. on Intelligent Robots and System, October, 1998.*

[10] M. C. Nechyba and Y. Xu, "Cascade Neural Networks with Node-Decoupled Extended Kalman Filtering," *Proc. IEEE Int. Symp. on Computational Intelligence in Robotics and Automation*, vol. 1, pp. 214-9, 1997.

[11] J. C. Spall, "Multivariate Stochastic Approximation Using a Simultaneous Perturbation Gradient Approximation," *IEEE Trans. on Automatic Control*, vol. 37, no. 3, pp. 332-41, 1992.

Modeling Human Strategy in Controlling Light Source

Jiong Zhang and Yangsheng Xu

Department of Mechanical and Automation Engineering
The Chinese University of Hong Kong, Shatin, NT, Hong Kong

Abstract

In this paper, we present a method of modeling human strategy in controlling light source in dynamic environment. We take a simple example of how to control the light source to avoid a shadow and maintain appropriate illumination condition on the target area of attention to illustrate the procedure and method. The work is valuable to various applications of automatic light control from surgical room, space applications, to inspections.

1 Introduction

Dynamical light source placement is of significance in applications in many areas such as surgical operations, space environment and inspections where illumination condition must be adaptively altered. Traditionally, humans can operate light projection, like surgeon/nurses projects light holder all the time. This observation motivates our research on modeling human light source placement strategy and transfer the strategy to robots, so that the robotic light holder can project an ideal light condition based on modeled strategy. Because human control strategy (HCS) is a dynamic, nonlinear stochastic process, developing valid analytic models of HCS tends to be difficult. We employ cascade neural network to obtain the empirical HCS models.

In the past years, skill acquisition and modeling HCS have been done in many research works. Dillmann and Kaiser [1] studied skill acquisition from human demonstration and its robotic applications in compliant motion, machining, and navigation. Asada and Liu [2, 3] presented the work of transferring human skills to a robot in machining tasks. They used neural network trained from human performance data, to perform on-line adaptation of the stiffness coefficients of a conventional controller in machining. Sukhan Lee and Shimoji [4] proposed a neural network architecture called Hierarchically Self-organizing Learning Network for learning skill mappings. Pomerleau [5] transferred human road following skills to an automated driving system using a single hidden layer neural network which mapped coarsely sampled video images to steering outputs. In [6, 7], Nechyba and Xu used a cascade neural network architecture, trained by node decoupled extended Kalman filtering, to learn skill mappings from human performance data. They also demonstrate human to human skill transfer using the neural network as an intermediary teacher.

In parallel, vision-based robot programming has also received attentions in recently. Ikeuchi et al. [8] and Kuniyoshi et al. [9] implemented this paradigm in task specification. Both systems are based on a visual recognition of human action sequences. This enables the generation of a symbolic description of the task that can be converted into an executable robot program. Ude and Dillmann [10] proposed a method for the specification of robot motions based on stereo vision to tracking hand holding object.

In this paper, we first introduce the problem of dynamical placement of light source. We then discuss the measurement of pose of hand-hold light source, and demonstrate the method of pose estimation. We propose to use cascade neural network to model human control strategy and address on performance measurement in implementation.

2 Simplified experiment

We first consider how to control the pose of the light source to avoid the shadow caused by the occlusion of light. Fig. 1 depicts the simplified concept of the task. Human eyes view the area of interest from the front side and the area of attention is illuminated by a hand-hold light source. The mobile platform can move forward and backward in both directions. When the platform moves, a simple shadow will cast near the area of attention (AOA). When the shadow is too close to AOA, the human operator has to move the light source to avoid the shadow in order to maintain AOA to be appropriately illuminated. We would like to use this simple experiment to demonstrate the way to acquire the human control strategy of shadow avoidance and transfer such a control strategy to a robot. Here we assume that the illumination of AOA is mainly contributed by the hand-hold light source. Collision avoidance is not an issue as the light source is well above the mobile platform.

In the experiment setup, an operator views a picture pattern illuminated by a hand-hold light source. Rigidly attached with the light source is a board with some marks on. A camera (camera 1, not shown in the figure) views the marks for the measurement of the pose of the board. A mobile platform is pulled or pushed

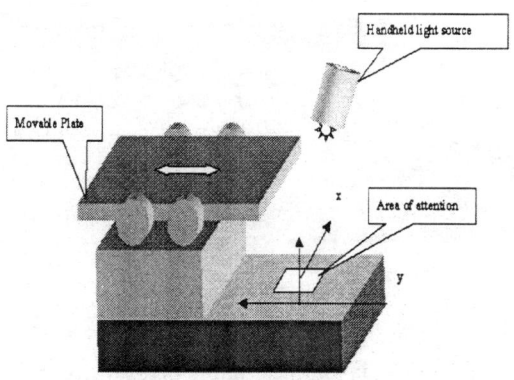

Figure 1: Simplified experiment setup

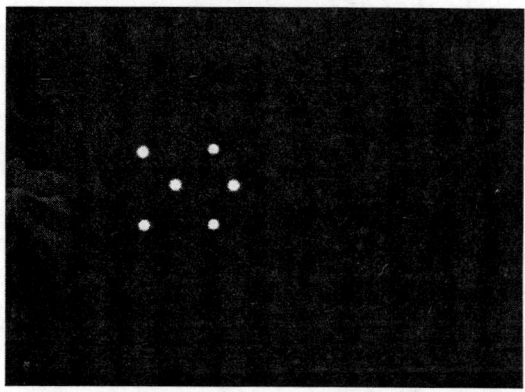

Figure 2: The image of marks with the handhold light off

randomly to produce a shadow. Another camera (camera 2) on other side of the mobile platform is used to obtain the image of AOA and its surrounding area for commutating the location of shadow edge. When the human demonstrator performs the task, images from the two cameras are grabbed into a computer to record the data.

3 Measurement pose of handhold light source

When the demonstrator moves the light source to avoid the shadow for sufficient illumination, the pose of the light source is important. Since the image from the camera 1 inevitably includes the intensive light area generated by the light source, a reliable measurement free of such an effect is needed. We used 6 small torch bulb lights fixed on a board as depicted in experiment. Each bulb is then enclosed in a small tube and the out end of each tube is covered with a piece of white paper. Thus when the camera views the board with the bulb lights on, the bright part of each piece of papers in the image is taken as the tracking mark, as depicted in Fig. 2. Since the brightness of such marks are determined by the bulb lights which are powered by a electric regulator, the marks are free of environmental lighting condition and can be reliably detected. In fact, the marks can be easily detected regardless of the environmental lights on. We set every tube at the same height with respect to the board surface in order that the marks are coplanar. Thus there is no inter occlusion between the marks even when the board has a small angle with respect to the camera optical axis. The components and the board are light in weight, so it is easy for demonstrators to move freely.

The measurement procedure is composed of two steps: marks searching and marks tracking. At searching step, the white blobs (correspond to the marks) are sought in the whole image. If the number of blobs equals to the number of marks, it continues to next tracking step. At tracking step, the program looks for the each blob only in a small area around the position of the mark in the previous image. Thus the marks searching processing only occurs in the part of the mark board area in the image. Since the handhold light is at the back side of the mark board, the intensive light is blocked by the board. This procedure makes it sure that the brightness of the marks is stable. When the marks are being reliably tracked, we turn the handhold light on and then the demonstrating procedure begins. Fig. 3 shows the marks are detected and tracked when the handhold light is on. The program also computes the shadow edge at the same time.

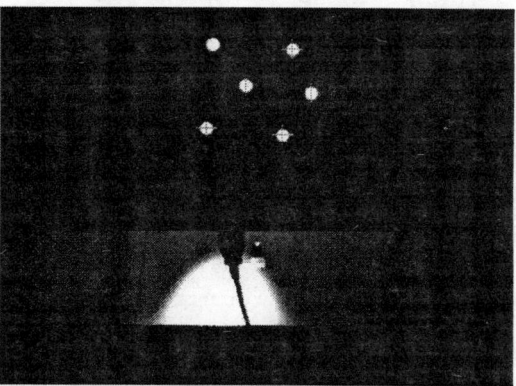

Figure 3: The image of tracked marks with the handhold light on

We take the gravity center of each blob as the each mark position,

$$x_b = \frac{1}{n}\sum x_i, \quad y_b = \frac{1}{n}\sum y_i \quad (1)$$

where x_b and y_b are the center coordinates of a blob, x_i and y_i are the the pixel coordinates in the blob, and n is the number of pixels in the blob. Such obtained positions are of accuracy of subpixel, which is sufficiently high for our requirement.

4 Pose estimation

Once measurements of marks have been obtained, we compute the pose of the light source. As indicated in Fig. 4, when a point is mapped onto a camera image plane, the 3D or 2D mapping can be represented as:

$$\begin{bmatrix} x_c \\ y_c \\ 1 \end{bmatrix} = \begin{bmatrix} 1 & 0 & 0 & 0 \\ 0 & 1 & 0 & 0 \\ 0 & 0 & 1 & 0 \end{bmatrix} \begin{bmatrix} X_c \\ Y_c \\ Z_c \\ 1 \end{bmatrix} \quad (2)$$

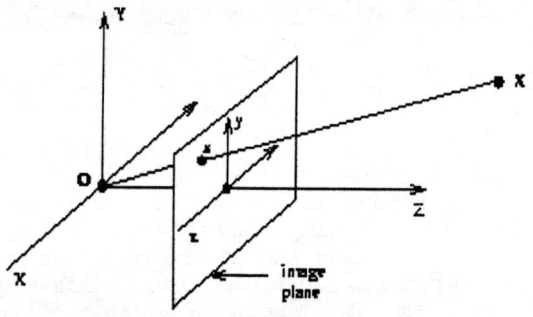

Figure 4: Persecutive projection.

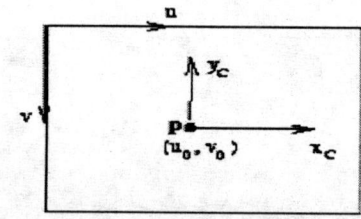

Figure 5: Intrinsic camera parameters.

From Fig. 5, it is understand that $k_u x_c = u - u_0$, $k_v y_c = v - v_0$. We can obtain:

$$x_i = \begin{bmatrix} u \\ v \\ 1 \end{bmatrix} == C \begin{bmatrix} 1 & 0 & 0 & 0 \\ 0 & 1 & 0 & 0 \\ 0 & 0 & 1 & 0 \end{bmatrix} \begin{bmatrix} X_c \\ Y_c \\ Z_c \\ 1 \end{bmatrix} \quad (3)$$

C is a 3×3 upper triangular matrix called the camera calibration matrix:

$$C = \begin{bmatrix} fk_u & 0 & u_0 \\ 0 & -fk_v & v_0 \\ 0 & 0 & 1 \end{bmatrix} = \begin{bmatrix} \alpha_u & 0 & u_0 \\ 0 & \alpha_v & v_0 \\ 0 & 0 & 1 \end{bmatrix} \quad (4)$$

where, $\alpha_u = fk_u$, $\alpha_v = -fk_v$.

Camera calibration matrix C provides the transformation between an image point and a ray in Euclidean 3-space. There are four parameters: the effective focal length of the pinhole camera; the principal point (u_0, v_0), which is the point where the optic axis intersects the image plane; the aspect ratio $sx = k_v/k_u$. The camera is termed calibrated by camera calibration matrix C.

In Tsai's camera model [11], an additional parameter–1st order radial lens distortion coefficient k_1, is introduced.

$$x_c = x_d + k_1 r^2 \quad (5)$$
$$y_c = y_d + k_1 r^2 \quad (6)$$

where (x_d, y_d) is the distorted position on the image plane and $r = \sqrt{x_d^2 + y_d^2}$.

Euclidean transformation between the camera and world coordinates is:

$$X_c = RX_W + P \quad (7)$$

or

$$\begin{bmatrix} X_c \\ Y_c \\ Z_c \\ 1 \end{bmatrix} = \begin{bmatrix} R & P \\ O^T & 1 \end{bmatrix} \begin{bmatrix} X_W \\ Y_W \\ Z_W \\ 1 \end{bmatrix} \quad (8)$$

where R and P are the relative rotation and translation of world coordinates with respect to camera coordinates. They are also extrinsic camera parameters. Therefore there are 11 parameters to be calibrated. In calibration, a linear solution can be obtained from the small set of corresponding points, and this linear solution is then used as the starting point for a nonlinear minimization of the difference between the measured and projected points.

In this paper, we use Tsai's [11] camera model and Tsai's calibration method implemented by Wilson. Basic coplanar calibration requires at least five data points, therefore we use six points in the above measurement. We first use a standard calibration board to obtain the intrinsic parameters in order to exploit the full optimization feature of the algorithm. Then we compute the pose (extrinsic camera parameter) at every time step with the known camera intrinsic parameter.

The obtained above is a series of relative pose of mark board with respect to the camera coordinates. However, since we are interested in the motion of the light source with respect to the area of attention, we need to transform the pose relative to a coordinates fixed on the area of attention. Such a base coordinate frame is shown in Fig. 1. We acquire the relative pose of camera 1 with respect to base coordinates through calibration and measurement. The transformation of light source coordinates with respect to the mark board is measured directly. Therefore we have:

$${}^B_L T = {}^B_C T {}^C_M T {}^M_L T \quad (9)$$

where $T = \begin{bmatrix} R & P \\ O^T & 1 \end{bmatrix}$ represents a 4×4 homogeneous transformation between two coordinates, and

^B_LT, ^B_CT, ^C_MT and ^M_LT represents the 4 × 4 homogeneous transformations between light coordinates and base coordinates, camera coordinates and base coordinates, mark board coordinates and camera coordinates, light source coordinates and mark board coordinates, respectively. The coordinate setup of hand-hold light source is shown in Fig. 6, where the z-axis coincides with the axis of light source.

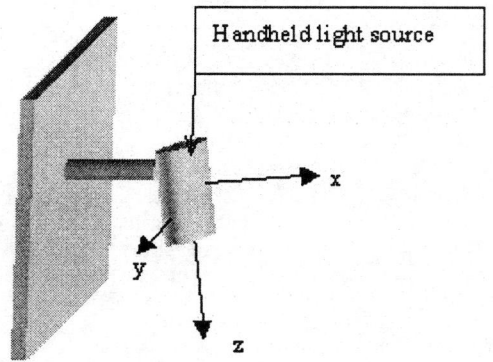

Figure 6: The coordinate setup of light source.

Using Eqn. 8, we obtained a sequence pose measurement of light source represented in the base coordinates. To avoid redundancy, pose parameter is represented as a 6D vector

$$Q = [\alpha\ \beta\ \gamma\ x\ y\ z]^T \quad (10)$$

where α, β, γ are rotation angles around x-axis, y-axis and z-axis respectively. For more convenience, we set the base coordinate origin at the center of AOA. Thus intersection of the z-axis of light source coordinates and the x-y plane of base coordinates can be considered as the center of light project on the AOA plane.

5 Distance between shadow edge and AOA

The human demonstrator moves the light source to react the location changing of shadow due to motion of the mobile platform, the brightness and distribution in the image from camera 2. We first detect the shadow and then compute the shadow's location. As shown in Fig. 7, we take the shadow edge in the image as approximately a straight line considering the small camera distortion.

We first compare the average brightness between a small area (100pels× 100pels) in upper right and one in middle left. If the difference is larger than a threshold there probably is a shadow edge in the image, otherwise the edge is out of the image. The shadow edge as in Fig. 7 has a slow slope, so we use binarization based on the above bright difference to locate the edge points. Since the edge line is a straight line, Hough transform is a good candidate, but it is also computationally intensive

Figure 7: The shadow edge computation

when processing all points in image. We propose a simplified Hough transform to make the detection fast and reliable. We chose one row from every n (n=5 or 6) rows to binarize and then take only one edge point from each binarized row to get a candidate point set. When the point set is transformed to Hough plane the angle is limited in a small range, e.g. 20 grade in our program. If the maximum point number in the transformed plane is larger than a respecified number, we take it as an edge and record the corresponding location. Fig. 7 shows the detected line in the image. We compute the locations of two end points of the straight-line segment in image coordinates, namely (x_1, y_1) and (x_2, y_2).

Similarly to the procedure of calibration in camera 1, we can obtain the intrinsic parameters and 6 external parameters which are also the transformation of base coordinates with respect to the camera coordinates $^{C2}_BT$. From Eqns. 3 and 8, we have:

$$x_{i2} = \begin{bmatrix} u_2 \\ v_2 \\ 1 \end{bmatrix} = C_2 \begin{bmatrix} 1 & 0 & 0 & 0 \\ 0 & 1 & 0 & 0 \\ 0 & 0 & 1 & 0 \end{bmatrix} {}^{c2}_BT \begin{bmatrix} X_b \\ Y_b \\ Z_b \\ 1 \end{bmatrix} \quad (11)$$

Here the radical distortion in camera 2 was not considered.

Plug (x_1, y_1) and (x_2, y_2) into (u_2, v_2) in Eqn. 11, and let $Z_b = 0$, we obtain (X_1, Y_1) and (X_2, Y_2) which are two points on the edge line on the x-y plane in base coordinates. These two points represent the shadow edge line described as

$$y - Y_1 = \frac{Y_2 - Y_1}{X_2 - X_1}(x - X_1) \quad (12)$$

Let $x = 0$, we have

$$y_d = Y_1 - \frac{Y_2 - Y_1}{X_2 - X_1} X_1 \quad (13)$$

When the base coordinates locate at the center of AOA with the x-axis parallel to the occlusion edge of mobile platform as shown in Fig. 1, we take the intersection of the y-axis with shadow edge line to represent the measuring point and its y coordinate is the distance from the shadow edge to the center of AOA. The distance from the right edge to the shadow edge is computed by simply subtracting a constant value from it, i.e.

$$d = y_d - d_{con} \quad (14)$$

where d_{con} is the distance from the center of AOA to its right edge.

6 Modeling human strategy

We use the flexible cascade neural network architecture with node-decoupled extended Kalman filtering (NDEKF) [6, 7] for modeling the human skill in controlling light source motion data. We selected this learning architecture for a number of reasons. First, no a priori model structure is assumed; the neural network automatically adds hidden units to an initially minimal network as the training requires. Second, hidden unit activation functions are not constrained to be a particular type. Rather, for each new hidden unit, the incremental learning algorithm can select that functional form which maximally reduces the residual error over the training data. Typical alternatives to the standard sigmoidal function are sine, cosine, and the Gaussian function. Finally, it has been shown that node-decoupled extended Kalman filtering, a quadratically convergent alternative to slower gradient descent training algorithms (such as backpropagation or quickprop) fits well within the cascade learning framework and converges to good local minima with less computation [12]. The flexible functional form which cascade learning allows is ideal for abstracting human control strategies, since we know very little about the underlying structure of each individual's internal controller. By making as few a priori assumptions as possible in modeling the human skill data, we improve the likelihood that the learning algorithm will converge to a good model of the data. In order for the learning algorithm to properly model each individual's human control strategy, the model must be presented with those state and environmental variables upon which the human operator relies. Thus, the inputs to the cascade neural network should include (1) current and previous state information $\{Q\}$, (2) previous output (command) information $\{S\}$, and (3) the distance of the right edge to the shadow edge $\{d\}$. We define the outputs as

$$S(n) = [\delta_\alpha(n) \ \delta_\beta(n) \ \delta_\gamma(n) \ \delta_x(n) \ \delta_y(n) \ \delta_z(n)] \quad (15)$$

where $\delta_\alpha(n) = \alpha(n) - \alpha(n-1)$, and so forth.

The network inputs are

$$\{Q(k - n_s), \cdots, Q(k-1) m Q(k)\}$$
$$\{S(k - n_c), \cdots, S(k-1) m S(k)\}$$
$$\{d(k - n_r), \cdots, d(k-1), d(k)\}$$

where n_s is the length of the state histories and n_c is the length of the previous command histories presented to the network as input. Taking account of the distance d, the total number of inputs n_i is,

$$n_i = 6n_s + 6n_c + n_r \quad (16)$$

The outputs of the cascade network are the 6 items of $S(k+1)$. In practice, in consideration of output limit of current neural network training software, we take the 4 significant elements, i.e., $[\delta_\alpha(n+1) \ \delta_x(n+1) \ \delta_y(n+1) \ \delta_z(n+1)]$ of $\{S\}$ as the outputs.

When integrating the model in a robot system, the cascade neural network can be viewed as a feedback controller whose outputs control the motion of the light source attached on the manipulator.

In placing the light source, our goal is to maintain appropriate illumination on AOA to allow operator to view a comfortably. As the type of light source plays a important role in operator's movement, we simply assume a model for the light source used in our experiment in Fig. 8. The light source is attached in

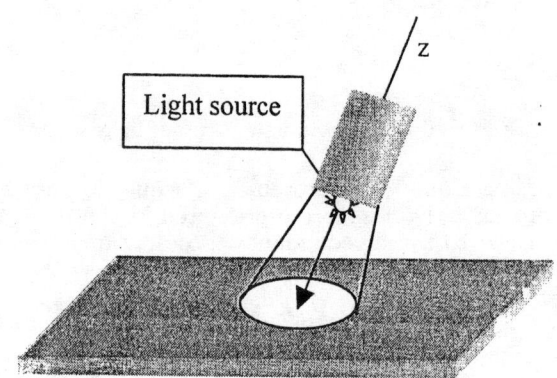

Figure 8: Simple model of light source.

a parabolic reflector, it shapes a circular light cone. The z-axis is coincidence with center axis of the cone. When the light cone shines on plane, an elliptical area is illuminated. The conical surface can be represented by

$$x^2 + y^2 = kz \quad (17)$$

where k is a constant dependent of the angle of the cone.

From Eqn. 8 and 10, we have

$$\begin{bmatrix} X_B \\ Y_B \\ Z_B \\ 1 \end{bmatrix} = {}^B_L T \begin{bmatrix} X_L \\ Y_L \\ Z_L \\ 1 \end{bmatrix} \quad (18)$$

Let $X_L = 0$, $Y_L = 0$ and $Z_B = 0$, we can get X_B, Y_B and Z_L. Here X_B and Y_B approximately represent the center of the light shining on the x-y plane in the base coordinates. Z_L represents the distance from the light source origin to the plane.

Let $D = \sqrt{X_B^2 + Y_B^2}$, then D represents the distance between the light center to AOA on the plane. Furthermore from Eqns. 17 and 18, we can also obtain the equation of the intersection of light cone and the x-y plane and then compute the minimal distance from the elliptical to the center. For simplicity, we only take D into account.

Different demonstrators may give distinct data sets, and data sets from same demonstrator may also vary, especially in a task where there are many control choices to accomplish it, such as a task reported here. We need to know whether the demonstration or the skill model is acceptable or unacceptable. So quality evaluation of demonstrated performance data set is essential.

From the above description of the task, our goal is to make D as small as possible in every movement step. Therefor D can be used as a measure of the demonstrated performance quality.

7 Conclusion

In this paper we present a method to transfer human control strategy in dynamical placement of light source to avoid shadows to a robotic light holder. To obtain the best results, there are several light parameters required to be controlled, such as pose, intensity and distribution of light source. As the first step, we consider how to control the light source to avoid a shadow of simple pattern to maintain proper illumination on the target area of attention. We develop a robust hand pose measurement method based on visual approach. It is composed of a mark board rigidly fixed with the light source and a video camera to measure the marks. Thus human demonstrator can move the light source freely regardless of robot motion limit. We use a simplified hough transform to fast obtain the edge of the shadow. We model the human control strategy through the flexible cascade neural network learning architecture, and propose a task specific performance measure to evaluate the quality of the learned model. Based on the learned model, the robotic light holder is being developed to carry out the same tasks as human light holders may perform. The work is significant in modeling and transferring of human skills in placing light source in dynamical environment and is valuable to variant tasks in automatic control of illumination condition in surgery rooms, space operation, or inspections.

References

[1] M. Kaiser and R. Dillmann. "Hierarchical learning of efficient skill application for autonomous robots." *International Symposium on Intelligent Robotics Systems*, Pisa, Italy, 1995.

[2] H. Asada and S. Liu. "Transfer of human skills to neural net robot controllers" *Proceedings. 1991 IEEE International Conference on Robotics and Automation*, volume 3, pages 2442–8. IEEE Comput. Soc. Press, April 1991.

[3] S. Liu and H. Asada. "Teaching and learning of debarring robots using neural networks." *Proceedings of 1993 IEEE/RSJ International Conference on Intelligent Robots and Systems,* pages 339–45.

[4] Sukhan Lee and Shunichi Shimoji. "Machine acquisition of skills by neural networks." *International Joint Conference on Neural Networks,* volume 2, pages 781–8. IEEE, July 1991.

[5] Dean A. Pomerleau. "Neural network perception for mobile robot guidance." PhD thesis, School of Computer Science, Carnegie Mellon University, 1992.

[6] M. Nechyba and Y. Xu. "Cascade neural networks with node decoupled extended kalman filtering." *Proc. IEEE Int. Symp. on Computational Intelligence in Robotics and Automation,* Monterey, CA, July 1997.

[7] M.C. Nechyba and Yangsheng Xu. "Human skill transfer: neural networks as learners and teachers." *Proceedings of the 1995 IEEE/RSJ International Conference on Intelligent Robots and Systems,* pages 314–19.

[8] K. Ikeuchi, M. Kawade, and T. Suehiro. "Assembly task recognition with planar, curved and mechanical contacts." *Proc. IEEE Int. Conf. Robotics and Automation,* Vol. II, pages 688–694, Atlanta, Georgia, May 1993.

[9] Y. Kuniyoshi, M. Inaba, and H. Inoue. "Seeing, understanding and doing human task." *Proc. IEEE Int. Conf. Robotics and Automation,* pages 2–9, Nice, France, May 1992.

[10] Ales Ude and R. Dillmann, "Robot motion specification: a vision-based approach." *Surveys on Mathematics for Industry,* vol. 5, pp. 109-131, 1995

[11] Roger Y. Tsai, "A versatile camera calibration technique for high-accuracy 3D machine vision metrology using off-the-shelf TV cameras and lenses," *IEEE Journal of Robotics and Automation,* Vol. RA-3, No. 4, August 1987, pages 323-344.

[12] M. C. Nechyba and Y. Xu, "Stochastic similarity for validating human human control strategy," *Proc. IEEE Conf. On Robotics and Automation,* Vol.3 pp. 278-283 1998.

Using Redundancy to Reduce Accelerations near Kinematic Singularities

K. O'Neil and Y.C. Chen

Abstract—Solutions of the inverse kinematic problem for redundant manipulators are studied for workspace paths that pass near, but not through, kinematic singularities. Widely-used algorithms such as pseudoinverse velocity control may compute solution paths that have low joint velocities but very high joint accelerations, which may not be physically desirable or even feasible. In this paper we show the existence of solutions with low jointspace acceleration, which use the redundancy of the mechanism to track the workspace trajectory by means of a more natural joint motion. We derive a redundancy resolution algorithm at the velocity level that is computationally undemanding and tracks the desirable low-acceleration solution path. Simulations illustrating the feasibility of the proposed algorithm are presented.

I. INTRODUCTION

The inverse kinematic problem is a fundamental problem in robotic control: given a desired trajectory $x(t)$ in the position/orientation workspace of a manipulator, find a solution trajectory $\theta(t)$ in the jointspace which is mapped onto $x(t)$ by the forward kinematic function F. As is well known, this problem can be solved effectively for paths which do not contain singular values of F by means of *rate control*, which is simply integration of the differential equation $J(\theta)\dot{\theta} = \dot{x}$ from a valid initial condition; J is the derivative of F. (The use of local coordinates on the Euclidean group allows x and F to be expressed as vectors, so that J is a Jacobian matrix in the usual sense.) The joint velocity is not determined completely by the differential equation if the mechanism is redundant; $\dot{\theta}$ may be given an arbitrary component in the nullspace of J.

If however $x(t)$ passes through a singular posture of the mechanism, where J does not have full rank, this technique breaks down and even the existence of a solution $\theta(t)$ is in question. If there is a feasible solution trajectory, recent work has shown that it is possible to modify the basic control algorithm in a way that takes the nature of the singularity into consideration and allows a smooth solution path to be computed [2,5,7,10-12,14,16,18].

Trajectories which pass near but not through the singular set of F can still pose practical difficulties.

K. O'Neil is with the Department of Mathematical and Computer Sciences, The University of Tulsa, 600 S. College Ave., Tulsa, OK 74104. Y.C. Chen is with Innovative Robotic Solutions, Inc., Santa Clara, CA 95054.

Manipulation in the vicinity of the singular set can result in large jointspace velocities, and even if the solution velocities are reasonable the joint *acceleration* may be unrealistically large.

For example, consider a three-joint planar arm with unit link lengths, tracking the trajectory $(x,y) = (2 - d + cos(t), 0)$ by means of pseudoinverse velocity control, $\dot{\theta} = J^+\dot{x}$. This redundancy resolution scheme minimizes the norm of the joint velocity vector. As can be seen in Figure 1a, which illustrates the case $d = 0.01$, the norm of the acceleration vector has a large increase at the time the end effector reverses direction near the workspace boundary. Direct analysis shows that the peak acceleration norm varies as $d^{-1/2}$, which corresponds to the maximum of $||J^+||$. The situation is qualitatively the same for the singularity-robust or damped version of pseudoinverse velocity control [3,9,19]. Since this scheme does not exactly solve the differential equation $J(\theta)\dot{\theta} = \dot{x}$, a feedback term must be added to obtain a practical controller. Figure 1b shows the result of a simulation with damping factor 0.1 and position error coefficient 1. The acceleration norm peak is reduced by the damping, but a significant error (roughly ten percent of the commanded motion) is introduced.

The preceding discussion illustrates that the presence of singularities in the vicinity of a workspace trajectory can cause some velocity-level redundancy resolution schemes to generate joint solution paths with very large accelerations. For high-speed or high-inertia applications where the manipulator dynamics is important, these joint solutions may not be feasible. Even if the accelerations can be attained, it may be advantageous to find a joint solution that is smoother. The difficulty cannot be avoided merely by shifting to acceleration-level controllers: it is known, for example, that the minimum norm acceleration controller for the arm in Figure 1 is unstable [1,4]. Furthermore, velocity-level control is generally easier to implement than acceleration-level control. So the question presents itself: is there a general velocity-level redundancy resolution scheme which produces joint trajectories with low velocities *and* low accelerations in the vicinity of singularities?

In this paper we present an analysis of the kinematic

function of a redundant mechanism in the neighborhood of a singularity, and study the solutions to the inverse kinematic problem in the context of a "typical" workspace path that passes near, but not through, the singularity. If the singularity is an extremum, such as is found at the workspace boundary, a typical trajectory is a parabolic approach that turns around short of the singularity; if the singularity is an internal one, a typical trajectory is a straight path that passes near but not through the singular point. The analysis proves that solutions to the inverse kinematic problem exist which have neither large velocity nor large acceleration. The phenomenon of large acceleration norm is avoided by using the component of $\dot{\theta}$ in the nullspace of J, rather than joint accelerations, to generate the necessary workspace acceleration components in those directions that correspond to small singular values of J.

We also present a velocity-level redundancy resolution algorithm, the *natural motion algorithm*, that computes these solutions efficiently. The algorithm is exact, in the sense that joint velocities $\dot{\theta}$ are generated that exactly solve the relation $\dot{x} = J\dot{\theta}$. The algorithm can be used for those portions of the commanded motion which lie near the singular set of the manipulator. For the other portions of the motion, the user is free to choose an appropriate redundancy resolution scheme.

II. Existence of Nearly-Linear Motion

This section begins with a mathematical analysis of the geometry of the possible solutions to the path tracking problem for paths that pass near a singular point. This allows us to understand the reason for the large accelerations seen in the examples and prove the existence of solutions which avoid these accelerations.

A *kinematic singularity* is a point θ_s at which the m by n Jacobian matrix $J(\theta_s)$, $m < n$, does not have full rank. It will be convenient to use the singular value decomposition (SVD) of $J(\theta_s)$. This consists of orthogonal matrices U and V, with columns $u_1, \ldots, u_m$; $v_1, \ldots, v_n$; and nonnegative numbers $\sigma_1, \ldots, \sigma_m$ that satisfy the relation $J = \sum_1^m u_i \sigma_i v_i^T$. At a kinematic singularity, one or more of the singular values σ_i equal zero. In this paper for reasons of space we consider only singularities of corank 1, where only one singular value σ_m vanishes. It is clear that the projected kinematic function $u_m^T F(\theta)$ has a critical point at θ_s, since $\nabla(u_m^T F) = u_m^T J = 0$. Thus, we have a quadratic approximation for $u_m^T F$ in a neighborhood of θ_s:

$$u_m^T F(\theta) \approx u_m^T x_s + (1/2)(\theta - \theta_s)^T H_m (\theta - \theta_s) \quad (1)$$

where $x_s = F(\theta_s)$. The difference between the left and right hand sides of (1) is of order $\|\theta - \theta_s\|^3$.

The character of the singular point is determined by the eigenvalues of the hessian matrix H_m of $u_m^T F(\theta)$ at θ_s. If this matrix is definite then the projected kinematic function has an extremum, as would be the case for example at the boundary of the manipulator workspace. A hessian with eigenvalues of both signs would occur only at an internal singularity of the mechanism. An example of this type of singularity is found when the three joint planar arm discussed in the Introduction is folded back on itself: $\theta_2 = \theta_3 = \pi$. If the hessian matrix has some zero eigenvalues and nonzero eigenvalues of only one sign, then the character of the singular point cannot be determined by the hessian matrix alone. It should be noted that the relative signs of the hessian are independent of the local coordinate system used on the Euclidean group. Much recent work has appeared on the classification of manipulator singularities [8,15,17].

Now consider a path $x(t)$ in the workspace that passes near the singular value x_s at time $t = t_s$. We wish to study solutions to the equation $F(\theta) = x(t)$ for t near t_s and θ near θ_s. An equivalent system of equations is

$$u_i^T x = u_i^T F(V\phi), \quad i = 1, \ldots, m \quad (2)$$

where $\phi = (\phi_1, \ldots, \phi_n)$ expresses the joint angles θ in terms of the basis of right singular vectors of $J(\theta_s)$ by the relation $\theta = V\phi$. Consider first the $m - 1$ equations $1 \leq i < m$. The derivative of the right-hand side with respect to ϕ_j is $\sigma_i \delta_{ij}$, so the implicit function theorem implies that these equations implicitly determine $\phi_1, \ldots, \phi_{m-1}$ in terms of $\bar{\phi} := (\phi_m, \ldots, \phi_n)$ and x. Specifically, to first order $(\phi - \phi_s) = J^+(x - x_s) + \bar{V}(\bar{\phi} - \bar{\phi}_s)$, where $\bar{V}$ consists of the last $(n - m + 1)$ columns of V. Now for the last equation in (2), $i = m$, the leading term in $u_m^T F$ is quadratic, and we may use the above expression for $(\phi - \phi_s)$ to obtain:

$$u_m^T(x - x_s) = (1/2)\Theta^T H_m \Theta + O(\|\phi - \phi_s\|^3), \quad (3)$$

$$\Theta = J^+(x - x_s) + \bar{V}(\bar{\phi} - \bar{\phi}_s)$$

which is valid in a neighborhood of ϕ_s.

Now we may investigate solutions to (3) for generic trajectories that pass close to x_s. Let t_s be the time of closest approach to the singularity, and assume that $u_i^T x(t)$, $1 \leq i < m$ is linear in $(t - t_s)$; it follows that $J^+(x - x_s)$ is as well. For the component $u_m^T x(t)$ of the trajectory, we assume that it is linear in $(t - t_s)$, and (in the case that the singularity is an extremum) has a quadratic component that keeps the trajectory inside the workspace. We seek a linear form for $\bar{\phi} - \bar{\phi}_s$ that will satisfy equation (3) to leading order; the implicit function theorem may then be applied again to prove the existence of an exact solution to (3) and thus (2) that is nearly linear, that is, has low acceleration.

If $J^+(x - x_s)$ and $\bar{\phi} - \bar{\phi}_s$ are assumed to be linear expressions in $t - t_s$, then equation (3) may be viewed as a quadratic polynomial in $t - t_s$ and we must find coefficients for $\bar{\phi} - \bar{\phi}_s$ that make the three coefficients of this polynomial vanish. The constant term can be made zero by using a solution to $x = F(\theta)$ at time $t = t_s$. Elimination of the linear (t^1) term amounts to solving an equation that is linear in $d\bar{\phi}/dt$. Finally, the coefficient of t^2 can be set equal to zero by solving the equation

$$u_m^T \ddot{x} = (J^+ \dot{x} + \bar{V}\dot{\bar{\phi}})^T H_m (J^+ \dot{x} + \bar{V}\dot{\bar{\phi}})$$

If H_m is definite, this has a solution so long as $u_m^T \ddot{x} - (J^+ \dot{x})^T H_m (J^+ \dot{x})$ has the correct sign, i.e. the trajectory has the proper curvature at t_s to keep within the workspace. If H_m is indefinite (the internal singularity case), the equation has a solution as long as $\bar{V}^T H_m \bar{V}$ is indefinite. Indefiniteness of $\bar{V}^T H_m \bar{V}$ is a sufficient condition for the existence of self-motion at the internal singularity [12]; this condition is often satisfied in practice. We conclude that there are nearly-linear solutions of (3) in a neighborhood of the singular point.

At this stage it is clear that in general one must use some jointspace velocity $\bar{V}\dot{\bar{\phi}}$ in the nullspace of the Jacobian matrix J near the singularity in order to achieve the desired low-acceleration solution. The pseudoinverse velocity rate control method is a redundancy resolution that always sets this nullspace velocity to zero, and one consequence can be large velocities and/or accelerations near singularities. Consider the generic paths discussed above. If the trajectory is a straight path near an internal singularity, with zero workspace acceleration, then the relations

$$\dot{x} = J(\theta)\dot{\theta}, \quad \ddot{x} = J(\theta)\ddot{\theta} + \dot{J}(\theta)\dot{\theta}$$

imply the relations

$$v_m^T \dot{\theta} = (u_m^T \dot{x})/\sigma_m, \quad v_m^T \ddot{\theta} = -(u_m^T)\dot{x} J^{+T} H_m J^+ (u_m^T \dot{x})/\sigma_m.$$

The small singular value in the denominator makes large values of $\dot{\theta}$ and $\ddot{\theta}$ likely. If instead the trajectory is an out-and-back type trajectory as illustrated in the Introduction, with $\dot{x}(t_s) = 0$, then we have $v_m^T \ddot{\theta} = (u_m^T \ddot{x})/\sigma_m$, showing that large accelerations are possible even when the joint velocity is zero.

III. NATURAL MOTION ALGORITHM

In the previous section the existence of low-acceleration solution trajectories was established. In this section we address the practical issue of computing such solutions.

Straight-line solutions to the path-tracking problem, up to order $\|\theta - \theta_s\|^3$, may be extracted directly from equation (3) by simple algebraic manipulation. However, writing down the equation to begin with requires detailed information about the singular point θ_s, such as the values of all first and second derivatives of the kinematic function. Since the singular point does not lie on the desired trajectory, these data are in some sense extraneous to the problem at hand and may entail unnecessary computation.

To reduce this burden, one may compute instead a solution trajectory which only approximates the ideal straight-line trajectory, but still has low acceleration. This can be done by simplifying equation (3), giving an approximation to the desired nullspace component of $\dot{\theta}$. It is important to note that approximation of this nullspace component does not degrade the accuracy in the workspace of the solution trajectory, since $\dot{x} = J\dot{\theta}$ is still satisfied exactly. A natural simplification of equation (3) is to evaluate the Hessian of the kinematic function using the current joint values rather than values corresponding to the singularity. An approximate velocity vector for the straight-line solution may then be computed by differentiating this equation twice with respect to t, which with the assumption of zero joint acceleration becomes

$$u_m^T \ddot{x} = \dot{\Theta}^T H(u_m^T F)\dot{\Theta}, \quad (4)$$

where this time $\dot{\Theta} = J^+ \dot{x} + \nu$, ν an arbitrary vector in the nullspace of the Jacobian matrix J (at the *current* joint value, not at the singular point.) Thus equation (4) is quadratic in the $n - m$ variables parametrizing the nullspace of J, and will have a solution set of dimension $n - m - 1$ in general. If the degree of redundancy $n - m$ is 1, then $\nu = \lambda v_n$, and (4) reduces to a scalar equation

$$0 = \lambda^2 (v_n^T H_m v_n) + 2\lambda(v_n^T H_m J^+ \dot{x}) + \\ + ((J^+ \dot{x})^T H_m (J^+ \dot{x}) - u_m^T \ddot{x}) \quad (5)$$

There are two discrete solutions. These solutions always exist because of the existence of zero-acceleration solutions to (3). The Hessian matrix in (4) and (5) is easily computed using formulas in [13] for $\dot{J}$ for manipulators with revolute or prismatic joints.

Once the set of nullspace vectors n which cause $v_m^T \ddot{\theta}$ to vanish is determined, the controller must select one vector. If the redundancy is being used to optimize some secondary condition, it should be chosen with this in mind. If minimization of joint acceleration is the only goal, ν should be chosen to minimize the difference between the new joint velocity and the old. In the common case of one degree of redundancy, one of the two possible choices for ν will be consistent with the solution trajectory computed up to the current instant, and one will not be. (The solution trajectory will be tangent to only one of the two straight lines lying in the hyperboloid implicit in equation (3).)

This algorithm has a simple physical interpretation, given the relations

$$u_m^T \dot{J}\dot{\theta} = \dot{\theta}^T H(u_m^T F)\dot{\theta}$$

and

$$\ddot{x}(t) = J\ddot{\theta} + \dot{J}\dot{\theta}. \quad (6)$$

The workspace acceleration is the sum of two contributions. One derives from the joint accelerations, and the other is the acceleration produced by steady joint velocities due to the geometry of the mechanism. (For example, consider a revolute-joint mechanism with only one joint activated: the end effector moves along a circular arc, with non-zero acceleration in the radial direction. This acceleration is due to the geometry of the mechanism only.) The vanishing of jointspace acceleration implies $0 = v_m^T \ddot{\theta} = u_m^T(\ddot{x} - \dot{J}\dot{\theta})$. Hence in the direction of u_m, the workspace acceleration required by the desired trajectory is being produced entirely by this geometric contribution $\dot{J}\dot{\theta}$, and no further joint acceleration is needed. The manipulator is executing what could be called a *natural motion* to follow the trajectory $x(t)$.

To summarize, a practical velocity-level algorithm for a mechanism with a single degree of redundancy that results in a jointspace trajectory with low acceleration that track a workspace trajectory passing close to a singularity is to use joint velocity

$$\dot{\theta} = J^+ \dot{x} + \lambda v_n$$

where λ satisfies equation (5). For mechanisms with higher degrees of redundancy, one solves equation (4) subject to additional conditions or constraints to resolve the remaining $(n-m-1)$ degrees of redundancy.

This natural motion algorithm is appropriate for smooth control only near a singularity of the kinematic function. Therefore, implementation of the algorithm will require it to be switched on as the singularity is approached, possibly by establishing a threshhold value for the smallest singular value σ_m. Abrupt enforcement of the condition in (4) may result in large accelerations, however, which would defeat the purpose of the algorithm. Instead one could replace the relation $u_m^T(\ddot{x} - \dot{J}\dot{\theta}) = 0$ by a linear or exponential decay of the left hand side for the initial period of the algorithm.

IV. SIMULATIONS

In this section we present simulations of two planar redundant manipulators moving near kinematic singularities that illustrate the advantage of the natural motion (NM) algorithm.

The first manipulator is the three link arm discussed in the introduction. This arm has one degree of redundancy, so there are two roots to the scalar equation (5) determining the nullspace component of joint velocity. Figure 2 shows joint positions and the norm of joint acceleration, for the NM controller applied to the same trajectory and initial condition as Figure 1. The acceleration is very near zero at the point of closest approach to the workspace boundary ($t = 0$), and the graphs of the joint angles are nearly straight lines. Thus the controller has successfully computed the solution trajectories discussed in Section 2. The exact nature of the algorithm is evidenced by the fact that the workspace error of the trajectory, computed using a time discretization of Δt, scales as $(\Delta t)^2$. Note also that the acceleration grows as the arm moves away from the boundary: the approximations in the NM algorithm are less accurate there. A practical algorithm for manipulator control would switch to NM control as the boundary was approached, and switch back to the main control algorithm as the arm moves away from the boundary. This simple approach would completely eliminate the acceleration spike seen in Figure 1.

The reader may also note a qualitative difference between pseudoinverse and NM control. The plots of joint positions are symmetric with respect to t for pseudoinverse control, meaning that the arm passes through the same postures as it moves away from the workspace boundary as it did during its approach. In effect, the arm motion is brought to a halt and then reversed to follow the trajectory. The joint plots produced by NM control, however, lack this symmetry, so that the arm postures are different for the two phases of motion. This lack of symmetry allows the joint velocities to remain essentially constant, so that the acceleration of the solution path is small. Thus the NM algorithm brings the same advantage to pick-and-place operations *near* singularities that has been identified for such motions that terminate at a singularity [6].

The second simulation is of a zero-acceleration workspace trajectory passing near the internal folded-arm singularity of the same planar manipulator. Figure 3a shows the joint positions and acceleration norm generated by pseudoinverse velocity control. The initial joint angles are $(\theta_1, \theta_2, \theta_3) = (3.4379, -3.7735, 2.6717)$ and the workspace velocity is the constant $(-1.0, 0)$. The negative of θ_2 is displayed in the position plot for convenience. The acceleration norm generated near the singularity is so large that for clarity its (common) logarithm is displayed; peak acceleration is about 2000 rad/sec^2. Although it is not evident from the position plot, the joint velocities $\dot{\theta}_1$ and $\dot{\theta}_3$ undergo reversals near the singularity. A time-expanded plot of these two joint angles is shown in Figure 3b. The NM algorithm, when started at the same initial joint configuration with the same trajectory, produced the much smoother solution path seen in Figure 4. (Again, $-\theta_2$ is displayed rather than θ_2.) Acceleration is reduced by three orders of magnitude. This reduction, accomplished by judicious use of the redundancy of the manipulator, obscures the fact that the NM trajectory passes even closer to a singular posture for the arm than does the trajectory in Figure 3: the minimum values of the smaller singular value σ_2 are 0.063 and 0.075, respectively.

Our third simulation is of a planar manipulator with one prismatic joint and two revolute joints, and kinematic function

$$F(p, \theta_1, \theta_2) = (c_1 + c_{12}, s_1 + s_{12} + p),$$

where c_1 denotes $\cos(\theta_1)$, etc. (In effect, a prismatic joint is added at the base of a planar two joint arm to create this manipulator.) The workspace trajectory is $(1.99 - t^2, 0)$, which approaches nearest the workspace boundary at time $t = 0$. The behavior of the pseudinverse control (not shown) here is qualitatively similar to Figure 1a; peak acceleration norm is about 20. A plot of joint positions for the NM controller is shown in Figure 5. Again, the NM algorithm successfully computes a solution trajectory with nearly constant joint velocities. The arm is able to move smoothly as it tracks the trajectory in the workspace. The singularity approached in this example is a degenerate extremum, yet the NM algorithm works well.

V. Conclusions

Redundant mechanisms can achieve smooth motion with low acceleration for motions near singularities. These motions are significantly smoother than those sometimes achieved by redundant mechanisms when controlled by pseudoinverse rate control. The reduction in acceleration is obtained by letting the necessary workspace acceleration in the near-singular direction be generated in a natural way by the geometry of the mechanism. A simple algorithm allows these path-tracking solutions to be calculated, as validated in several simulations. This algorithm may prove to be advantageous in high-speed or high-inertia applications.

VI. References

1. T.-H. Chen, F.-T. Cheng, Y.-Y. Sun, and M.-H. Hung, "Torque optimization schemes for kinematically redundant manipulators," *J. Robot. Sys.* v. 11, pp. 257-269, 1994.
2. Y.C. Chen, J. Seng, and K. O'Neil, "Lowest-order rate control of mechanisms near singularities," in *Proceedings, IEEE Int. Conf. Robot. Autom.*, 1997.
3. S. Chiaverini, B. Siciliano, O. Egeland, "Review of the damped least squares inverse kinematics with experiments on an industrial robot manipulator," *IEEE Trans. on Control Systems Technology*, v. 2, no. 2, pp. 123-134, 1994.
4. J.M. Hollerbach and K.C. Suh, "Redundancy resolution of manipulators through torque optimization," *IEEE J. Robot. Automat.* RA-3, pp. 308-316, 1987.
5. J. Kieffer, "Differential analysis of bifurcations and isolated singularities for robots and mechanisms," *IEEE Trans. Robot. Automat.*, v. 20, no. 1, pp. 1-10, 1994.
6. J.C. Kieffer and A.J. Cahill, "Fast pick and place at robot singularities," in *Proceedings, IEEE Int. Conf. Robot. Automat.*, v. 3, pp. 2236-2241, 1995.
7. J. Lloyd, 1996, "Using Puiseux series to control non-redundant robots at singularities," *Proceedings, IEEE Int. Conf. Robot. Automat.*, Minneapolis, pp. 1877-1882, 1996.
8. R. Muszyński and K. Tchoń, "Normal forms of non-redundant singular robot kinematics: three DOF worked examples," *J. Robotic Sys.*, v. 13, no. 12, pp. 765-791, 1996.
9. Y. Nakamura and H. Hanafusa, "Inverse kinematic solutions with singularity robustness for robot manipulator control," *J. Dyn. Syst., Meas., Contr.*, vol. 109, pp. 163-171, 1986.
10. S. Narasimhan and V. Kumar, "A second order analysis of manipulator kinematics in singular configurations," *23rd Biennial Mechanisms Conference*, pp. 477-484, Minneapolis, 1994.
11. D.N. Nenchev, Y. Tsumaki, M. Uchiyama, "Adjoint Jacobian closed-loop kinematic control of robots," in *Proceedings, IEEE Int. Conf. Robot. Automat.*, Minneapolis, pp. 1235-1240, 1996.
12. K.A. O'Neil, Y.C. Chen, and J. Seng, "Removing singularities of resolved motion rate control of mechanisms, including self-motion," *IEEE Trans. Robot. Automat.*, v. 13, no.5, pp. 741-751, 1997.
13. E.D. Pohl and H. Lipkin, "A new method of robotic rate control," *Proc. IEEE Int. Conf. Robot. Automat.*, pp. 1708-1713, 1991.
14. S.K. Singh, "Motion planning and control of non-redundant manipulators at singularities," *Proc. IEEE Int. Conf. Robot. Automat.*, pp. 487-492, 1993.
15. K. Tchoń, "A normal form of singular kinematics of robot manipulators with smallest degeneracy," *IEEE Trans. Robot. Automat.*, v. 11, no.3, pp. 401-404, 1995.
16. K. Tchoń and I. Dulęba, "On inverting singular kinematics and geodesic trajectory generation for robot manipulators," *J. Intelligent and Robotic Systems*, v. 8, pp. 325-359, 1993.
17. K. Tchoń and R. Muszyński, "Singularities of nonredundant robot kinematics," *Int. J. Robotics Res.*, v. 16, no. 1, pp. 60-76, 1997.
18. K. Tchoń and R. Muszyński, "Singular inverse kinematic problem for robotic manipulators: a normal form approach," *IEEE Trans. Robot. Automat.*, v. 14, no.1, pp. 93-104, 1995.
19. C.W. Wampler II and L.J. Liefer, "Applications of damped least-squares methods to resolved-rate and resolved-acceleration control of manipulators," *J. Dynamic Sys., Meas., Contr.*, v. 110, pp. 31-38, 1987.

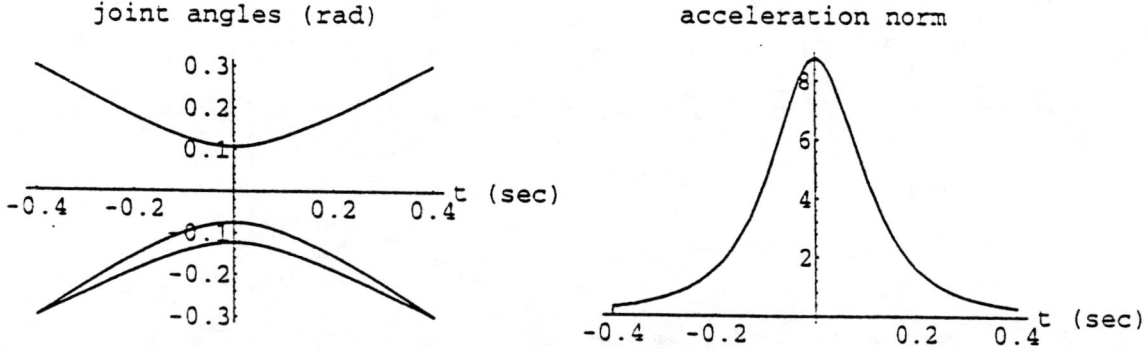

Figure 1. Joint positions and acceleration norm for 3 joint arm tracking trajectory $(1.99 + \cos(t), 0)$ with controller $\dot{\theta} = J^+ \dot{x}$. Initial joint angles$(\theta_1, \theta_2, \theta_3) = (0.3, -0.3, -0.3)$.

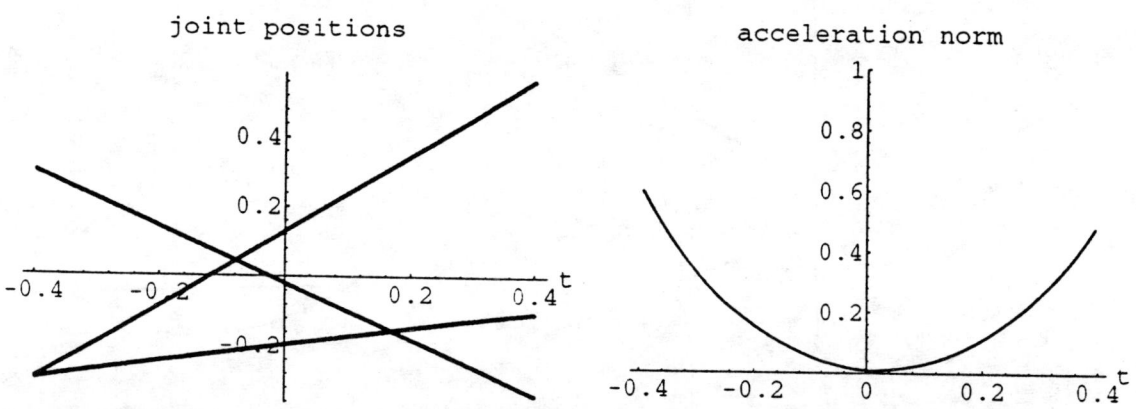

Figure 2. Joint positions and acceleration norm for 3 joint arm tracking trajectory $(1.99 + \cos(t), 0)$ with NM controller. Initial joint angles$(\theta_1, \theta_2, \theta_3) = (0.3, -0.3, -0.3)$.

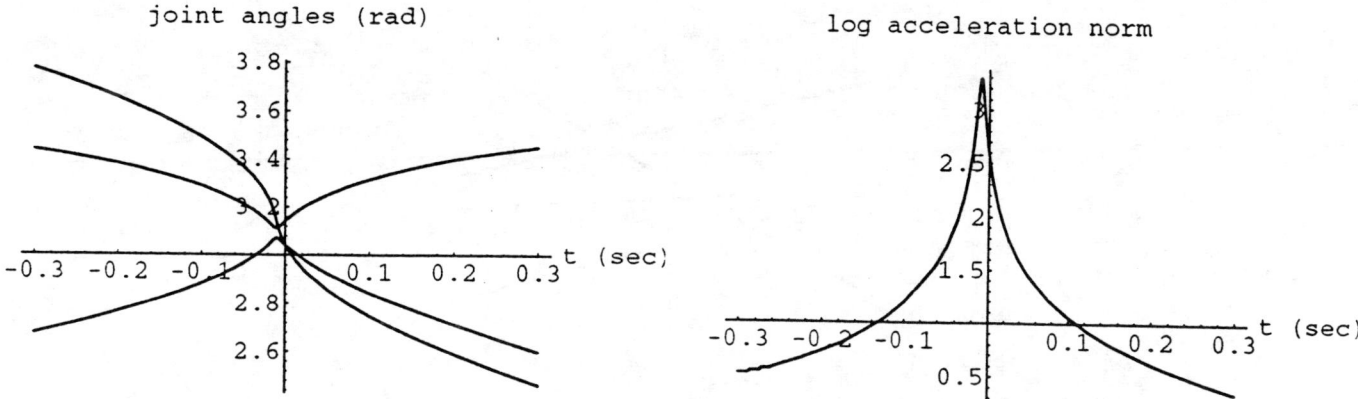

Figure 3. (a) Joint positions and acceleration norm for 3 joint arm tracking linear trajectory with pseudoinverse velocity controller. Negative of θ_2 is shown.

Figure 3. (b) Expanded view of θ_1 and θ_3.

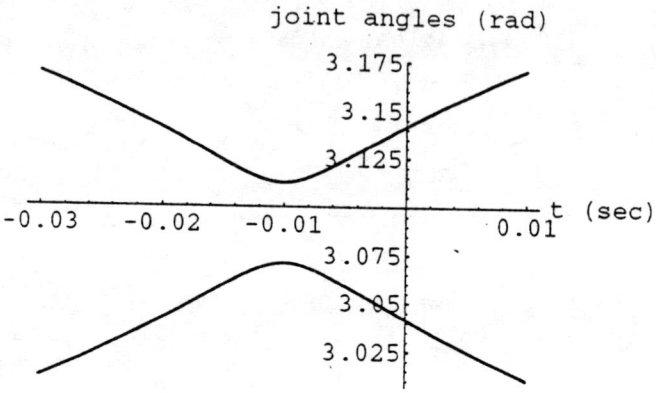

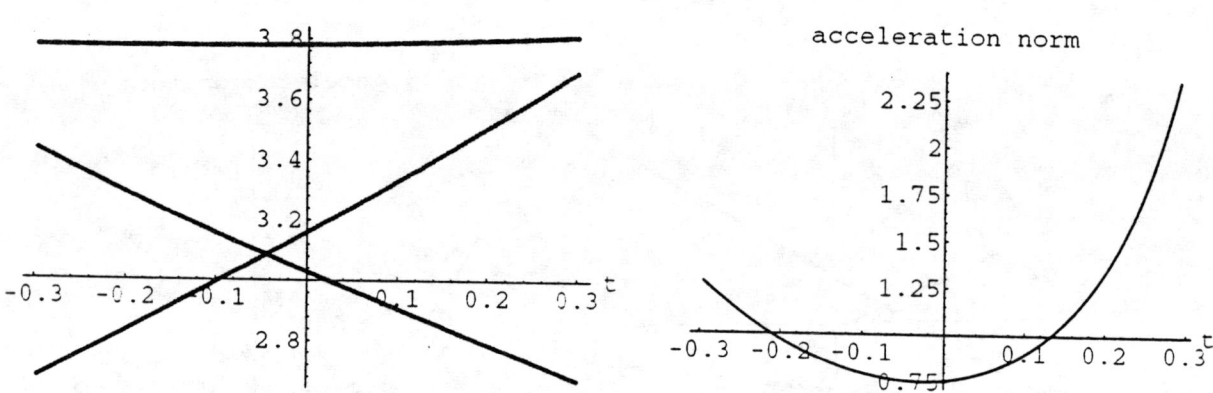

Figure 4. Joint positions and acceleration norm for 3 joint arm tracking linear trajectory with NM controller. Negative of θ_2 is shown.

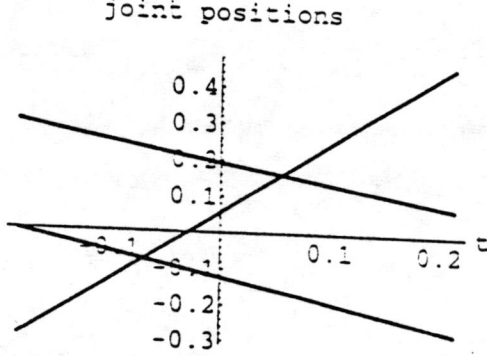

Figure 5. Joint positions for 1P2R arm tracking trajectory $(1.99 - t^2, 0)$ with NM controller. Initial joint values $(p, \theta_1, \theta_2) = (-0.296, 0, 0.3)$

An Improved Trajectory Planner for Redundant Manipulators in Constrained Workspace

Tzu-Chen Liang and Jing-Sin Liu

Institute of Information Science
Academia Sinica
Nankang, Taipei, Taiwan 11529
R.O.C.
email: liu@iis.sinica.edu.tw

Abstract

This paper studies the trajectory planning of redundant robots performing tasks within enclosed workspace. Configuration control of kinematically redundant manipulators using pseudo-inverse with null-space projection method is a well-known scheme. One advantage of this method is that gradient of an objective function can be included in the homogeneous term to optimize the objective function without affecting the motion of end-effector. Different objective functions could achieve redundancy resolution such as obstacle or joint limits avoidance. Along this line of redundancy resolution, a switching objective function is proposed. We modify Liegeois' joint angle availability objective function in [1] so that the midpoints of each joint are switched at a series of prespecified key path points for the end-effector to achieve. These key path points are planned beforehand according to the geometry of constrained workspace. The trajectory planning problem can then be viewed as a series of proper postures determination problems at key path points. The proper postures are determined using a combination of potential field method and elastic model method which take into account joint operating ranges and motion tendency of end-effector. Variable weighting technique to achieve the proper postures effectively is also presented. Simulations of a planar 8-link robot in constrained workspace illustrate the effectiveness of this approach in trajectory planning problems.

Keywords: redundant robot, trajectory planning, null-space projection

1. Introduction

A non-redundant robot manipulator can locate its end-effector to arbitrary configuration within its workspace. A kinematically redundant robot manipulator has more degrees of freedoms (dof) than required for accomplishing a task. For the case of a redundant robot manipulator, infinitely many solutions of joint postures are possible, given any configuration of the end-effector. These extra dof are used to improve the ability of manipulators [23]. The abilities include obstacle avoidance [3], [4], [10], [11], [15], [20], [21], [22], joint limits and singularity avoidance [3], [13], [17], peak torque reduction [16], torque optimization [12], [19], joint failure/fault tolerance [14], [18], etc.. One main method of redundancy resolution for redundant robots is the pseudo-inverse method. However, problems of joint motion drift or repeatability [2], [17] are also raised due to redundancy. A measure of how much drift will occur is given in [6].

The inverse kinematics of redundant manipulators can be solved in the velocity level. For an n-link robot manipulator (depicted in Fig. 1), the relationship between the end-effector variables $X = [x_1, x_2, ..., x_m]^T \in R^m$ and the joint space variables $\theta = [\theta_1, \theta_2, ..., \theta_n]^T \in R^n$ is defined as the forward kinematic equation :

$$X = f(\theta) \quad (1)$$

where $n > m$ and f denotes forward kinematic function.

To solve the inverse kinematics problem, the Jacobian method is a fundamental one. By differentiating the forward kinematic equation (1) with respect to time, the transformation between end-effector velocity and joint velocity is derived as

$$\dot{X} = J\dot{\theta} \quad (2)$$

where J is the $m \times n$ Jacobian matrix. Main advantage of using Jacobian matrix is: If the time interval is short, the finite displacement relation is nearly linear [1], [2], [3], [4]:

$$\delta X \approx J \delta \theta \quad (3)$$

In general, for a redundant robot, an infinite number of joint velocities result in a desired end-effector velocity. The joint velocities correspond to a given end-effector velocity is computed by the null-space projection technique:

$$\dot{\theta} = J^+ \dot{X} + (J^+ J - I_n) z \quad (4)$$

where J^+ is the pseudo-inverse matrix of J, I_n is the $n \times n$ identity matrix, z is an arbitrary vector. $J^+ \dot{X}$ is the minimun-norm solution [2] of equation (2) and $(J^+ J - I_n)$ is the null-space projection matrix [3]. The homogeneous term $(J^+ J - I_n) z$, which is orthogonal to $J^+ \dot{X}$, is referred to as the self-motion of the manipulator since it does not cause end-effector motion.

One redundancy resolution scheme is done by imposing z as the gradient of an objective function P and projecting it to the null space of Jacobian [1]

$$\dot{\theta} = J^+ \dot{X} + \alpha (J^+ J - I_n) \nabla P(\theta) \quad (\alpha > 0) \quad (5)$$

where α is a weighting parameter. By applying various P functions, a secondary criterion can be achieved while the primary goal is tracking the specified end-effector trajectory [3], [4]. In particular, Liegeois [1] has investigated the use of a quadratic objective function, called *joint angle availability*,

$$P(\theta) = \sum_{i=1}^{n} \left(\frac{\theta_i - \theta_{ci}}{\Delta \theta_i} \right)^2 \quad (6)$$

$\Delta \theta_i$: operating range of joint i

to optimize the sum of squares the deviation of the joint displacements from their midpoint θ_{ci}. Yoshikawa [23], Nakamura et al [4] implemented an obstacle avoidance scheme in which the z vector is a specified joint velocity toward a collision-free joint vector. In the case of complex environment, however, no universal collision-free z vector exists. There are other objective functions can be imposed to ensure the approach of a given posture, as the one suggested by A.A Maciejewski [10] and experience in [11].

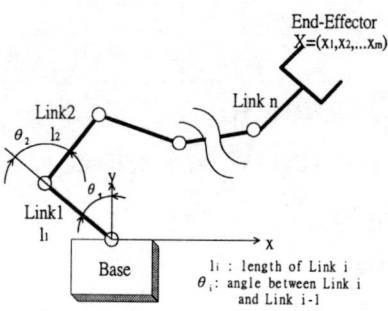

Fig.1. An n – degrees of freedom robot manipulator

One interesting feature drawn from the simulations of [1] and others is that the use of (6) in (5) can cause the joint motions of redundant manipulator to have a tendency to close to θ_{ci}. Based on this observation, a simple but effective obstacle avoidance scheme is proposed. When a manipulator moves in complex workspace, a series of joint postures (called *proper postures*), which are planned according to the current state of end-effector, the future motion and the geometry of constrained workspace, are used. If θ_{ci} in (6) is switched to be proper posture and the α weighting is accordingly tuned at right moment during a task, then motion in constrained workspace will very probably be achieved effectively using the velocity control scheme (5)-(6). The trajectory planning problem then becomes that of proper postures selections at the key path points. The paper is organized as follows. In section 2, an improved obstacle avoidance scheme using switching objective function and variable weightings is proposed. Determination of proper postures in constrained workspace using the elastic model method is presented in section 3. Simulations of a planar 8-link robot are performed in section 4 to demonstrate the effectiveness of the planning method. Finally, conclusions are drawn in section 5.

2. The Switching Objective Function Scheme

2.1 Redundancy resolution using the switching objective function scheme

It is easily proved that $\dot{P} \leq 0$ when $\dot{X}=0$ if $\alpha>0$ in (5). When $\dot{X} \neq 0$, simulations in [1] and others still find that there is a tendency for $\dot{P}$ to be non-positive [4]. Inspired from this observation, the use of velocity control scheme (5)-(6) will drive the joint motion of manipulator to approach the posture θ_{ci} gradually. By suitable plan of the postures and switching θ_{ci} as these postures at the right moment, the redundant manipulators can successfully move in the constrained workspace.

We now state the specific trajectory planning problem which is studied in the paper.

Problem statement (Problem of Trajectory Planning in Constrained Workspace):
Consider the workspace in the form of enclosures with openings. A series of path placement of end-effector and the geometry of constrained workspace are known. It is desired to plan an obstacle avoidance trajectory for joints of the redundant manipulator whose end-effector is required to track a prespecified path.

Typically, a trajectory planning problem may be divided into a geometric path planning phase and a velocity planning phase. Our proposed scheme is the switching objective function scheme, which will be described in details in next subsection. Now refer to Fig. 2. The geometric path planning strategy is:

STEP 1: (Global planning) Plan a series of "Key Path Points" along the path of end-effector off-line. key path points include path points at which the links collide with the obstacles presumably or where the direction of motion substantially changes.

STEP 2: Resolve postures of manipulator at these key path points, and check if these postures are feasible. These postures are called "Proper Postures".

STEP 3: Switch θ_{ci} in (6) successively as the end-effector passes the key path point. Then between the time interval of switching, the joints motion of the robot manipulator will approach next proper posture using the velocity control scheme (5), (6).

STEP4: Repeat STEP 3, until the whole path is completely traversed.

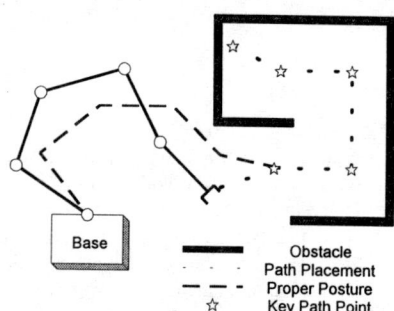

Fig.2 Robot in constrained workspace: Key Path Points and Proper Postures

Applying the switching midpoint (proper postures) strategy, obstacle avoidance at the key path points can be achieved owing to the property of $\dot{P} \leq 0$ tendency along the end-effector path. This completes the collision-free geometric path planning. Although there may not have enough travel time to achieve proper postures between two key path points, it does not affect the accomplishing of collision-free motion due to that the proper postures are mutually independent. Some technique to achieve these proper postures easier will be discussed in next section. Between two successive key path points, the end-effector positions are interpolated and the corresponding joint configurations are calculated. The result is a smooth joint progression from one proper posture to another proper posture. Then STEP 2 of checking if the collision or joint limits exceed will happen can be done by simulations..

2.2 Variable weightings technique to achieve proper postures easier

In STEP 2, "Proper Posture" at a key path point is a "reference arm posture" [4], which serves as a guide for collision-free joint motion while the end-effector is tracking a prescribed trajectory.

There are ways to achieve the subgoal easier, for instance, increase the weighting α in (5). The larger α is, the faster the objective function $P(\theta)$ approaches zero. However, α cannot be increased without upper limit[3], since it also affects the velocity and acceleration of joints motion which are physically bounded. Furthermore, undesirable high velocity caused by large α induces large kinematics computation error. Since each link may have different ranges of motion due to hardware limits, and has different orders of importance for a particular application, it is desirable to penalize the motion of some joints more than the others to embed different ability of approach with joints. This suggests the use of variable weighting α_i for each term in (6), instead of a single fixed α. $\Delta\theta_i$ in (6) now is viewed as a normalizing parameter.

Between two subsequent key path points, a joint is to move from one proper posture θ_{ci} sequentially to its next one, while the primary goal is tracking of a desired path by the end-effector.

Therefore $\Delta\theta_i$ indicates in some sense the difficulty of posture change between two consecutive proper postures. For large displacement $\Delta\theta_i$, α_i should be large. This can be done by normalizing $\Delta\theta_i$ between two key path points, and choose α_i correspondingly, but a threshold minimum value of α_i is used in implementation.

To conclude, suppose the key path points along the end-effector path have been planned as a series of points in Cartesian space:

$$X_0, X_1, X_2, \cdots\cdots, X_f \quad (7)$$

This is the primary goal that redundant robot should achieve. Let the time instants when the end-effector passes the key path points be, respectively,

$$t_0 = 0, t_1, \cdots\cdots, t_f \quad (8)$$

Suppose the proper postures resolved at the key path points are:

$$\theta_c^0, \theta_c^1, \cdots\cdots, \theta_c^f \quad (9)$$

with

$$\theta_c^j = \left[\theta_{c1}^j, \theta_{c2}^j, \cdots\cdots, \theta_{cn}^j\right]^T, \quad j = 0, 1, \ldots, f \quad (10)$$

Define the time-varying function $\theta_c(t)$ as a piecewise constant function that switches its value at the scheduled time instants (7):

$$\theta_c(t) = \begin{cases} \theta_c^0 & if \quad t = 0 \\ \theta_c^1 & if \quad 0 < t \leq t_1 \\ \theta_c^2 & if \quad t_1 < t \leq t_2 \\ \cdots\cdots \\ \theta_c^f & if \quad t_{f-1} < t \leq t_f \end{cases} \quad (11)$$

With the above definitions, the pseudo-inverse with null-space projection method is modified as:

$$\dot{\theta} = J^+\dot{X} + (J^+J - I_n)\alpha(\theta_c(t), t)\frac{\partial P(\theta, \theta_c(t))}{\partial \theta} \quad (12)$$

where the objective function is defined as the switching function:

$$P(\theta, \theta_c(t)) = \sum_{i=1}^{n}\left(\frac{\theta_i - \theta_{ci}(t)}{\Delta\theta_i}\right)^2 \quad (13)$$

and the weighting $\alpha(\theta_c, t)$ is defined as

$$\alpha(\theta_c, t) = \begin{cases} \dfrac{w \cdot diag\{\Delta\theta_{c1}^0, \Delta\theta_{c2}^0, \cdots\cdots, \Delta\theta_{cn}^0\}}{\Delta\theta_{SUM}^0} & if \quad 0 \leq t \leq t_1 \\ \dfrac{w \cdot diag\{\Delta\theta_{c1}^1, \Delta\theta_{c2}^1, \cdots\cdots, \Delta\theta_{cn}^1\}}{\Delta\theta_{SUM}^1} & if \quad t_1 < t \leq t_2 \\ \cdots\cdots \\ \dfrac{w \cdot diag\{\Delta\theta_{c1}^{(f-1)}, \Delta\theta_{c2}^{(f-1)}, \cdots\cdots, \Delta\theta_{cn}^{(f-1)}\}}{\Delta\theta_{SUM}^{(f-1)}} & if \quad t_{f-1} < t \leq t_f \end{cases}$$

$$(14)$$

$$\Delta\theta_{ci}^j = \left|\theta_{ci}^{j+1} - \theta_{ci}^j\right|$$
$$\Delta\theta_{SUM}^j = \Delta\theta_{c1}^j + \Delta\theta_{c2}^j + \cdots\cdots + \Delta\theta_{cn}^j$$

where w is a constant.

In summary, the velocity control scheme given by (12), (13), (14) are the improved obstacle avoidance scheme we propose, which is based on switching objective function with variable weightings.

3. Elastic model method to resolve postures in free space

For a given key path points, the collision-free proper postures of the redundant robot are selected from the self-motion that avoids the obstacles and can be resolved by the well-known potential field method, if the redundancy is not too high. In the potential field method, each link is moved to position of lower potential difference and followed a movement of each distal link. The postures of these links tend to pass through the local minimum of potential field, thus the postures of the links within the constrained workspace are generally collision-free. In the case of highly redundant robot, only the postures for the links that are completely within the constrained workspace can be resolved by the potential field method. There remain links of the robot that are outside the constrained workspace (i.e. at the free space) have not been resolved yet. This situation is shown in Fig.3.

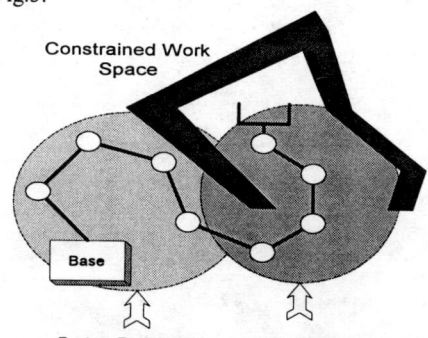

Fig. 3 Postures are not completely resolved by potential field method in highly redundantly case.

One way to resolve the postures for the links at the free space is the elastic model method [5], [7], [8]. Briefly speaking, the elastic model method is to represent the robot by attaching an elastic spring to each joint. Then the postures of manipulator can be determined by solving a set of algebraic equations derived from the requirement of static equilibrium, given the spring constants and the location and orientation of end-effector.

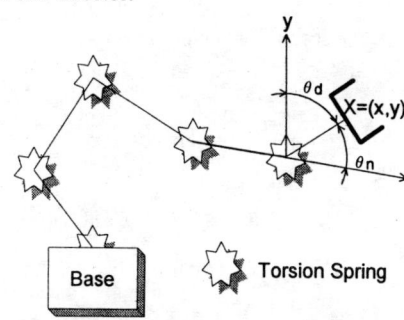

Fig. 4 Illustration of the elastic model method.

3.1 Elastic model method

Consider the planar articulated robot depicted in Fig. 4. The elastic model method represents each revolute joint of the robot as an elastic spring (torsion spring). Let the joint configuration be θ_i with unloaded reference spring configuration θ_{i0}. Let k_i be the spring constant associated with joint i and V_i be the elastic potential. Then

$$V_i = -\frac{1}{2}k_i(\theta_i - \theta_{i0})^2, i = 1, 2, \ldots, n \quad (15)$$

and the joint driving torque T_i is

$$T_i = -k_i(\theta_i - \theta_{i0}), i = 1, 2, \ldots, n \quad (16)$$

Let the position (x_n, y_n) and orientation θ_d (where θ_d denotes the angle between n-th link and y-axis) of the proximal end of the robot be given. Then the following relations hold:

$$\begin{cases} l_1\sin\theta_1 + l_2\sin(\theta_1+\theta_2) + \cdots + l_n\sin(\theta_1+\theta_2+\cdots+\theta_n) = x \\ l_1\cos\theta_1 + l_2\cos(\theta_1+\theta_2) + \cdots + l_n\cos(\theta_1+\theta_2+\cdots+\theta_n) = y \\ \theta_1 + \theta_2 + \cdots + \theta_n = \theta_d \end{cases} \quad (17)$$

where, as defined in Fig. 1, l_i is the length of i th link and θ_i is the joint angle of i th joint. Define the Lagrange L for the elastic system defined by the elastic potential energy (15) subject to the constraints (17) as

$$\begin{aligned} L = & \tfrac{1}{2}k_1(\theta_1-\theta_{10})^2 + \tfrac{1}{2}k_2(\theta_2-\theta_{20})^2 + \cdots + \tfrac{1}{2}k_n(\theta_n-\theta_{n0})^2 \\ & + \lambda_1\bigl(l_1\sin\theta_1 + l_2\sin(\theta_1+\theta_2) + \cdots + l_n\sin(\theta_1+\theta_2+\cdots+\theta_n) - x_n\bigr) \\ & + \lambda_2\bigl(l_1\cos\theta_1 + l_2\cos(\theta_1+\theta_2) + \cdots + l_n\cos(\theta_1+\theta_2+\cdots+\theta_n) - y_n\bigr) \\ & + \lambda_3(\theta_1+\theta_2+\cdots+\theta_n - \theta_d) \end{aligned} \quad (18)$$

where $\lambda_1, \lambda_2, \lambda_3$ are Lagrange multipliers associated with the constraints imposed.
Differentiating (18) with respect to θ_i, and setting the equations equal to zero, we have

$$\begin{cases} \dfrac{\partial L}{\partial \theta_1} = k_1(\theta_1-\theta_{10}) + \lambda_1[l_1\cos\theta_1 + l_2\cos(\theta_1+\theta_2) + \cdots + l_n\cos(\theta_1+\theta_2+\cdots+\theta_n)] \\ \qquad -\lambda_2[l_1\sin\theta_1 + l_2\sin(\theta_1+\theta_2) + \cdots + l_n\sin(\theta_1+\theta_2+\cdots+\theta_n)] + \lambda_3 = 0 \\ \dfrac{\partial L}{\partial \theta_2} = k_2(\theta_2-\theta_{20}) + \lambda_1[l_2\cos(\theta_1+\theta_2) + \cdots + l_n\cos(\theta_1+\theta_2+\cdots+\theta_n)] \\ \qquad -\lambda_2[l_2\sin(\theta_1+\theta_2) + \cdots + l_n\sin(\theta_1+\theta_2+\cdots+\theta_n)] + \lambda_3 = 0 \\ \cdots \\ \dfrac{\partial L}{\partial \theta_n} = k_n(\theta_n-\theta_{n0}) + \lambda_1 l_n[\cos(\theta_1+\theta_2+\cdots+\theta_n)] \\ \qquad -\lambda_2[l_n\sin(\theta_1+\theta_2+\cdots+\theta_n)] + \lambda_3 = 0 \end{cases} \quad (19)$$

Combining (17) and (19), there are totally ($n+3$) nonlinear algebraic equations. To simplify the above equations, define

$$\beta_1 = \theta_1,\ \beta_2 = \theta_1+\theta_2,\ \ldots,\ \beta_n = \theta_1+\theta_2+\cdots+\theta_n$$

and subtract the consecutive equations to obtain:

$$\begin{cases} \dfrac{\partial L}{\partial \theta_1} - \dfrac{\partial L}{\partial \theta_2} = 0: \\ k_1\beta_1 - k_2(\beta_2-\beta_1) + \lambda_1[l_1\cos(\beta_1)] - \lambda_2[l_1\sin(\beta_1)] = c_1 \\ \dfrac{\partial L}{\partial \theta_2} - \dfrac{\partial L}{\partial \theta_3} = 0: \\ k_2(\beta_2-\beta_1) - k_3(\beta_3-\beta_2) + \lambda_1[l_2\cos(\beta_2)] - \lambda_2[l_2\sin(\beta_2)] = c_2 \\ \cdots \\ \dfrac{\partial L}{\partial \theta_{n-1}} - \dfrac{\partial L}{\partial \theta_n} = 0: \\ k_{n-1}(\beta_{n-1}-\beta_{n-2}) - k_n(\beta_n-\beta_{n-1}) + \lambda_1[l_{n-1}\cos(\beta_{n-1})] - \lambda_2[l_{n-1}\sin(\beta_{n-1})] = c_{n-1} \\ \dfrac{\partial L}{\partial \theta_n} = 0: \\ k_n(\beta_n-\beta_{n-1}) + \lambda_1(l_n\cos(\beta_n)) - \lambda_2(l_n\sin(\beta_n)) + \lambda_3 = c_n \end{cases} \quad (20)$$

where $c_1,\ldots,c_n$ are constants generated from the elastic force at the unload position of springs. The equations (20) and (17) constitute a total of ($n+3$) nonlinear algebraic equations in the ($n+3$) unknowns ($\theta_1,\ldots\theta_n,\lambda_1,\lambda_2,\lambda_3$).

The solution of (17),(20) yields a posture that satisfies the end-effector requirement and is a posture of static equilibrium.

3.2 More about elastic model method

Since the potential field method can resolve parts of the postures, the location and orientation of proximal link whose posture has been determined by potential field method provides the constraints (17) for the elastic model method to resolve the remaining postures of manipulator. Combining the postures resolved by the potential field method and those by the elastic model method, one get the whole postures at the key path points. In application of elastic model method to the problem of resolving the postures for links outside the constrained workspace, it is noted that:
(i) If some joint has a configuration which is most suitable for motion, we can set the unload configuration (natural position of spring) to be this configuration. Then the elastic model method will come up with a solution near this configuration for that joint.
(ii) The spring constants k_i are selected according to the operating range of each joint. Larger k_i can be assigned
to joints of wide operating range, then the joint displacement solved by the elastic model method is larger.
(iii) The nonlinear algebraic equations (17), (20) can be solved by numerical iteration method, starting from a good initial guess of joint configuration for convergence. Since θ_d and (x_n,y_n) are given, (x_{n-1},y_{n-1}) is known. Using θ_d and (x_{n-1},y_{n-1}), instead of (x_n,y_n), as constraints are more convenient for numerical solution.

3.3 Potential field method to resolve postures in constrained workspace

Kinds of potential field method are applied in trajectory planning problem in constrained workspace. Now we do not use this method to plan the trajectory of manipulator, but produce statical proper postures at key path points. Assume that the manipulator has abundant dof and working range to handle this task, and there exists only one way, which is through key path points in turn, for end-effector to achieve the goal. Under the presented hypothesis, we can only select a potential field so that it has a maximum value at the obstacle and the potential descends properly along the key path points from goal to the entrance of constrained workspace. Put joints at the entrance of constrained workspace in turn and fix the end-effector position, then we can resolve the collision-free proper posture using minimum number of dof. When the workspace is not so complex compared with the ability of manipulator, this is a simple but effective method.

4. Simulations

To show the effectiveness of the method, two simulations are performed for an 8-dof planar redundant manipulator to move in constrained workspace. The first task is the end-effector moves along the outline of a vertical wall which has two recessions. The second task requires the robot to move into the corner of an enclosed workspace. In both simulations, ten key path points are chosen in advance. The path placement of the end-effector is set as the piecewise straight lines whose connecting points are set to be the key path points. The velocity profile along each straight-line segment is planned as a trapezoid (maximum velocity is 0.05; acceleration/deceleration is 0.01 in simulation). To avoid the occurrence of high joint velocities at the key path points, the velocity of end-effector at all key path points is zero. Note that although we set such a limit, it doesn't means that the end-effector velocity at key path points is necessary zero. If we adjust the path to be a smooth curve and select key path points on it, the velocity plan passing these points can keep nonzero and need not to use the trapezoid velocity profile.

Additionally, a feedback correction term in added (i.e. the Newton's Method [9]) in simulation to obtain a closed-loop inverse kinematics algorithm to reduce the error:

$$\dot{\theta}_k = J^+[\dot{X}_k - (X - X_k)] + (J^+J - I_n)\alpha\frac{\partial P}{\partial \theta} \quad (21)$$

where

$\dot{\theta}_k$: Joint velocity at k-th time interval.

$\dot{X}_k$: Planned end-effector velocity k-th time interval

X_k : Planned end-effector position k-th time interval

X : End-effector position

α, P are defined as (13),(14) with $w = 0.04$.

In addition, unload position of each spring is set as $\theta_{i0} = 0$ except the first link is set as the y axis.

Example 1

The first task is to manoeuvre the robot within the constrained workspace by requiring the end-effector to follow a contour from a given start position along the walls to a destination position and back to the start position. The contour and key path points are shown in Fig. 5. The length of the robot links is $l_1=1$, $l_2=9$, $l_3=0.8$, $l_4=0.7$, $l_5=0.6$, $l_6=0.5$, $l_7=0.5$, $l_8=0.5$. The proper postures resolved are shown in Fig.6 and the simulation result is shown in Fig. 7. A time sequence plot of the whole motion is provided in Fig. 8. It is seen that the objective function is monotonically decaying to near zero at each switching time, as Fig. 9 shows.

Key path point	X	Y
X_0	0.1	2.5
X_1	2	2.4
X_2	3	2.4
X_3	3	1.6
X_4	2	1.6
X_5	2	0.9
X_6	3	0.9
X_7	3	0.1
X_8	2	0.1
X_9	0.1	2.5

Fig. 5 The constrained workspace and the key path points for example 1.

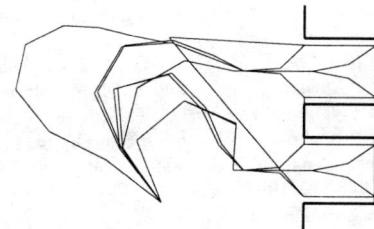

Fig. 6 Proper postures resolved by the potential field method and the elastic model method.

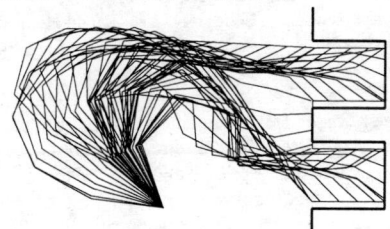

Fig. 7 Simulation result.

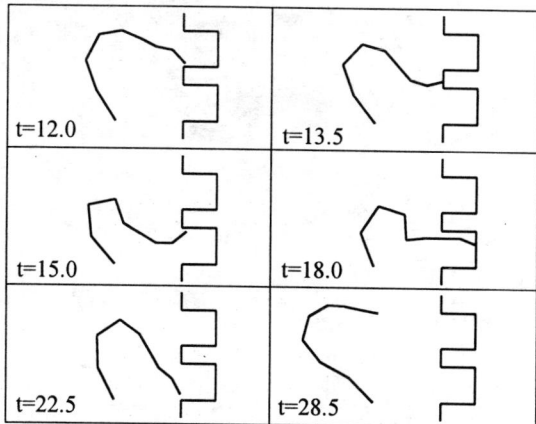

Fig.8 Time sequence plot of simulated joint motion

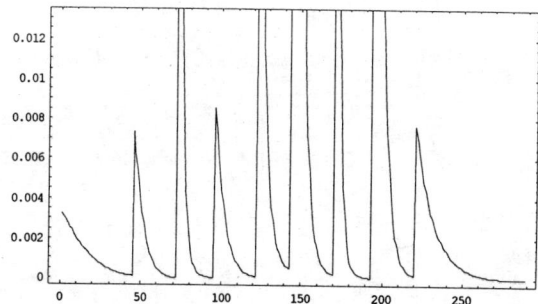

Fig. 9 Objective function-time plot

Example 2

In second task the robot move from a start position into a goal position (the corner) of constrained workspace shown in Fig. 10, where the key path points are also shown in the table. The length of each link is l_1=2.1, l_2=1.9, l_3=1.4, l_4=1, l_5=0.8, l_6=0.7, l_7=0.6, l_8=0.6. The resolved proper posture is shown in Fig. 11 and a collection of simulation results is shown in Fig.12. A sampled time sequence plots of the joint motion are provided in Fig. 13. The objective function is decaying to near zero at each switching time, as Fig. 14 shows.

Key path point	X	Y
X_0	4	3
X_1	3.5	1.5
X_2	3.5	0
X_3	2.75	-0.5
X_4	2.3	-1
X_5	2	-2.5
X_6	3	-2.6
X_7	3.5	-2.65
X_8	4	-2.7
X_9	4.25	-1.7

Fig. 10 The constrained workspace and the Key path points for example 2.

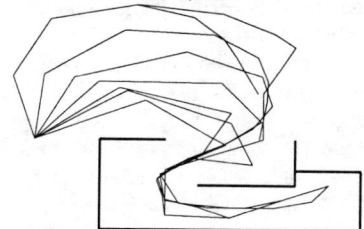

Fig. 11 Proper postures resolved by potential field method and elastic model method.

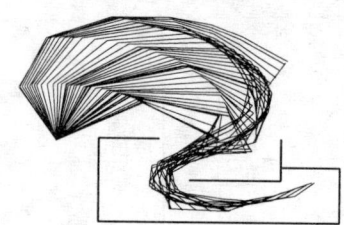

Fig. 12 Simulation result.

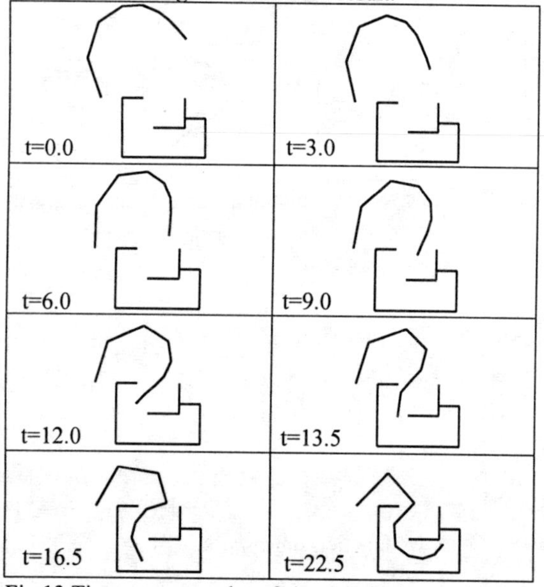

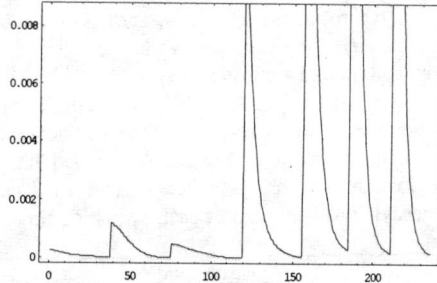

Fig. 13 Time sequence plot of simulated joint motion.

Fig. 14 Objective function-time plot

Simulation results of the above two examples show that the planar 8-dof redundant manipulator does move along the outline of the wall in example 1 or enter the constrained workspace in example 2 without colliding between the links and the constrained workspace.

5. Conclusion

We have presented an improved null-space projection method for trajectory planning problem of redundant robots. The improvement lies in the higher flexibility introduced by the use of switching the midpoints of each joint in joint angle availability objective function with variable weightings and the guide of joint motion by proper postures at the key path points. Regarding to resolve proper postures, we have presented a method of combining the potential field method and the elastic model method to get the whole postures. From the simulation results, the improved planning method is effective to manoeuvre the redundant robot to achieve simultaneously the primary goal of desired end-effector trajectory tracking and the subgoal of achieving collision avoidance if key path points are arranged properly..

References

[1] A. Liegeois, "Automatic Supervisory Control of the Configuration and Behavior of Multibody Mechanisms," *IEEE Trans. Syst., Man, Cyberetics.*, vol. SMC-7, no. 12, Dec., pp.868-871, 1977.

[2] C.A. Klein and K. Kee, "The Nature of Drift in Pseudoinverse Control of Kinematically Redundant Manipulators," *IEEE Trans. on Robotics and Automation*, vol. 5, no. 2, April, pp.231-234, 1989.

[3] C.A. Klein and C. Huang, "Review of Pseudoinverse Control for Use With Kinematically Redundant Manipulators," *IEEE Trans. Sys.,Man, Cybern*etics, vol. SMC-13, no. 3, March/April, pp.245-250, 1983.

[4] Y. Nakamura, H. Hanafusa and T. Yoshikawa, "Task-Priority Based Redundancy Control of Robot Manipulators," *Int. J. Robotics Res.* vol. 6, no. 2, pp.3-15, 1987.

[5] F.A. Mussa-Ivaldi and N. Hogan, "Integrable Solutions of Kinematic Redundancy via Impedance Control," *Int. J. Robotics Res.* vol. 10, no. 5, pp. 481-491, 1991.

[6] S. Luo and S. Ahmad, "Predicting the Drift Motion for Kinematically Redundant Robots," *IEEE Trans. Syst., Man, Cybernetics.*, vol. 22, no. 4, pp.917-728, 1992.

[7] Y.S. Chung, M. Griffis, and J. Duffy, "Repeatable Joint Displacement Generation for Redundant Robotics Systems," *Trans. ASME Journal of Mechanical Design* vol. 116, no. 6, pp.11-16, 1994.

[8] J.A. Kuo and D.J. Sanger, "Task Planning for Serial Redundant Manipulators," *Robotica*, vol. 15, pp. 75-83, 1997.

[9] R. Featherstone, "Accurate Trajectory Transformations for Redundant and Nonredundant Robots," *IEEE Int. Conf. Robotics and Automation*, pp.1867-1872, 1994.

[10] A.A Maciejewski and C.A Klein, "Obstacle Avoidance for Kinematically Redundant Manipulators in Dynamically Varying Environment," *Int. J. Robotics Res.*, vol. 4, no. 3, pp.109-117, pp.109-117, 1985.

[11] J. Wunderlich and C. Boncelet, "Local Optimization of Redundant Manipulator Kinematics within Constrained Workspaces," *IEEE Int. Conf. On Robotics and Automation*, Minneapolis, Minnesota, pp. 127-132, April 1996.

[12] J.M. Hollerbach and KI C. Suh, "Redundancy Resolution of Manipulators through Torque Optimization," *IEEE J. of Robotics and Automation*, vol. RA-3, no. 4, Aug. pp. 308-316, 1984.

[13] S. Chiaverini, "Singularity-Robust Task-Priority Redundancy Resolution for Real-Time Kinematic Control of Robot Manipulators," *IEEE Trans. on Robotics and Automation*, vol. 13, no. 3, June , pp. 398-410, 1997.

[14] R.G. Roberts and A.A. Maciejewski, "A Local Measure of Fault Tolerance for Kinematically Redundant Manipulators," *IEEE Trans. on Robotics and Automation*, vol. 12, no. 4, Aug. , pp.543-552, 1996.

[15] N. Rahmanian-Shahri and I. Torch, "A New On-line Method to Avoid Collisions with Links of Redundant Articulated Robots," *Robotica*, vol. 14, pp. 611-619, 1996.

[16] Degao Li, A.A. Goldenberg and Jean W. Zu, "Peak Torque Reduction with Redundant Manipulators," *IEEE Int. Conf. on Robotics and Automation*, Minneapolis, Minnesota, April, pp. 1775-1780, 1996.

[17] R.G. Roberts and A.A. Maciejewski, "Singularities, Stable Surface, and the Repeatable Behavior of Kinematically Redundant Manipulators," *Int. J. of Robotics Research*, vol. 13, no. 1, Feb. pp. 70-81, 1994.

[18] J.D. English and A.A. Maciejewski, "Fault Tolerance for Kinematically Redundant Manipulators Anticipating Free-Swinging Joint Failures," *IEEE Int. Conf. on Robotics and Automation*, Minneapolis, Minnesota, April, pp.460-467, 1996.

[19] P. Chiacchio, Y. Bouffard-Vercelli and F. Pierrot, "Evaluation of Force Capabilities for Redundant Manipulators," *IEEE Int. Conf. on Robotics and Automation*, Minneapolis, Minnesota, April, pp.3520-3525, 1996.

[20] A. McLean and S. Cameron, "The Virtual Springs Method: Path Planning and Collision Avoidance for Redundant Manipulators," *Int. J. of Robotics Research*, vol. 15, no. 4, Aug. pp. 300-319,1996.

[21] E.S. Conkur and R. Buckingham, "Manoeuvering Highly Redundant Manipulators," *Robotica*, vol. 15, pp. 435-447,1997.

[22] J.Z. Li and M.B. Trabia, "Adaptive Path Planning and Obstacle Avoidance for a Robot with a Large Degree of Redundancy," *J. Robotic Systems*, vol.13, no.3, pp.163-176, 1996.

[23] T.Yoshikawa, "Analysis and Control of Robot Manipulators with Redundancy," *The First Int. Symposium on Robotics Research*, M.Brady and R.Paul Eds., MIT Press, pp. 735-748,1984.

On the quantification of robot redundancy

Jadran Lenarčič

Department of Automatics, Biocybernetics and Robotics
The Jožef Stefan Institute
1111 Ljubljana, Slovenia (jadran.lenarcic@ijs.si)

Abstract

In this paper, we evaluate the measure of kinematic flexibility as an index that quantifies the manipulator's redundancy. Initially we propose a general algorithm for its calculation, then present two illustrative numerical examples of redundant mechanisms and discuss the obtained results. We show that the distribution of the kinematic flexibility throughout the workspace can be a discontinuous function. There exist different sub-domains of joint angles corresponding to the same primary task. In some poses, the manipulator is able to pass from one sub-domain to another continuously, while in other cases the manipulator cannot change the current sub-domain without violating the primary task constraints. This fact may have important implications in robot control.

1 Introduction

The kinematic redundancy notably increases the mobility of a robot manipulator since it enables to execute the designated task in an infinite number of ways [4]. A widely recognised index to quantify redundancy is the degree of redundancy that is defined as the number of superfluous degrees of freedom of the manipulator with respect to the minimum necessary number required by the primary task. However, from the viewpoint of the manipulator's mobility and flexibility, the degree of redundancy has a very limited importance. It does not give much insight in the mechanism's motion ability in a given point or throughout its workspace.

In robotic practice, especially in designing and programming of robot manipulators, we must be able to distinguish between different kinematic structures of redundant mechanisms or to determine regions of the workspace where a manipulator can be more flexible in executing the primary task [3,7,12]. By having more redundancy, the manipulator possesses more resources to solve different secondary tasks, such as the collision avoidance or the minimisation of torques.

Performances of redundant manipulators can be measured by various indices, such as the manipulability [8,14], isotropy [15], workspace [13], or dexterity [5]. However, these are general-type indices that cannot directly be utilised to validate the amount of robot redundancy. In our previous works we have adopted the measure of kinematic flexibility [9,10] to quantify the robot redundancy. It was originally defined for non-redundant manipulators as the number of inverse kinematics solutions. These outline in how many ways a manipulator can solve a specific task, so that the more inverse kinematics solutions (number of configurations) the manipulator possesses, the more flexible it is. Since the number of inverse kinematic solutions of a redundant mechanism is infinite, the kinematic flexibility for redundant robots was associated with the hyper-volume of the whole set of values of joint angles corresponding to a given primary task. For instance, the kinematic flexibility of a 3R planar redundant manipulator with one degree of redundancy is the length of the curve produced during the self-motion as a series of triplets of joint angles in the joint space that correspond to a predefined position of the end effector. This curve was also studied and plotted in [1]. The length of the curve clearly quantifies the redundancy of the manipulator and directly demonstrates the entire domain of joint angles that is available for executing various (in principle not yet known) secondary tasks.

The present article is dedicated to the computational problems of the kinematic flexibility of redundant manipulators. We show that, theoretically, one can develop very simple numerical algorithms. However, we demonstrate that each example must be treated as a unique problem, in particular when the degree of redundancy exceeds one. It is shown that the distribution of the kinematic flexibility throughout the workspace may not be a continuous function. The discontinuity arises from the fact that there exist different sub-domains of joint angles (groups of self-motions) corresponding to the same primary task. In some cases, the manipulator is able to pass from one sub-domain to another continuously, while in other cases the manipulator cannot alternate from one sub-domain to another without violating the primary task constraints. The property can be explained by studying the sub-chains of the mechanism on the basis of a precise and analytic information on the kinematics of the mechanism's self-motion. To confirm and illustrate these conclusions we present two quite different numerical examples, the planar 3R and the spatial 4R anthropomorphic mechanism.

2 Computation of kinematic flexibility

The background of calculating the kinematic flexibility of a redundant manipulator is based on the determination of the self-motion. It can be done by the use of a differential approach [2]. Let $\mathbf{p}$ be the vector of primary task coordinates given as a function of joint angles $\mathbf{q}$

$$d\mathbf{p} = \mathbf{J}_P d\mathbf{q} . \quad (1)$$

$\mathbf{J}_P$ is the primary $m \times n$ Jacobian matrix, $n > m$. A standard reverse relationship is

$$d\mathbf{q} = \mathbf{J}_{PA}^+ d\mathbf{p} , \quad (2)$$

where

$$\mathbf{J}_{PA}^+ = \mathbf{A}^{-1}\mathbf{J}_P^T(\mathbf{J}_P\mathbf{A}^{-1}\mathbf{J}_P^T)^{-1} \quad (3)$$

is the weighted pseudoinverse of $\mathbf{J}_P$ whose dimension is $n \times m$ [6], and $\mathbf{A}$ is a $n \times n$ weight matrix whose role isn't important in our case.

Let the joint increment corresponding to the secondary tasks be collected in vector $d\mathbf{q}_S$ and let the secondary tasks be entirely subordinated to the primary task, so that the joint increments that do not interfere with the primary task can only be searched in the null space of the primary Jacobian

$$d\mathbf{q}_N = \mathbf{N}_{PA} d\mathbf{q}_S . \quad (4)$$

Here, matrix $\mathbf{N}_{PA}$ is of dimension $n \times n$ and is an orthogonal complement of the primary Jacobian matrix, so that $\mathbf{J}_P \mathbf{N}_{PA} = \mathbf{0}$. According to [11], matrix

$$\mathbf{N}_{PA} = \mathbf{I} - \mathbf{J}_{PA}^+ \mathbf{J}_{PA} \quad (5)$$

projects a vector in the null space of $\mathbf{J}_P$. The degree of redundancy $D < l$ can mathematically be determined as $D = \text{rank}\{\mathbf{N}_{PA}\} = n - \text{rank}\{\mathbf{J}_P\}$. Thus, the complete joint increments corresponding to the primary task and the secondary tasks are

$$d\mathbf{q} = d\mathbf{q}_P + d\mathbf{q}_N . \quad (6)$$

It is clear that $d\mathbf{q}_N$ does not produce any change in the primary task coordinates $\mathbf{p}$ since $\mathbf{J}_P d\mathbf{q}_N = \mathbf{0}$. It is, therefore, a vector of the self-motion increments. Hence, the central point of computing the self-motion and, consequently, the kinematic flexibility is to delimit the range of $d\mathbf{q}_N$ in the vector space of $d\mathbf{q}_S$.

The kinematic flexibility F of articulated redundant manipulators (no translations) is defined as the D-dimensional volume of the space of joint vectors that are pertinent to the self-motion of the mechanism (normalised by π^D and the joint angles are within the module of 2π).

$$F = \frac{1}{\pi^D} \iiint d\delta_1 d\delta_2 \cdots d\delta_D , \quad (7)$$

The kinematic flexibility thus quantifies the range of joint motions that are not in contrast with the execution of the primary task. When $D = 1$, it is defined as the total length of the self-motion locus $\psi(\mathbf{q})$ in the space of n-dimensional vectors of joint angles $\mathbf{q}$ that correspond to the selected m-dimensional vector of primary task coordinates $\mathbf{p}$
while the joint angles vary within a module of 2π. Similarly, when $D = 2$, $\psi(\mathbf{q})$ is a surface in the n-dimensional space of $\mathbf{q}$ and the kinematic flexibility is its area. When $D = 3$, $\psi(\mathbf{q})$ is a 3-dimensional space in the n-dimensional space of $\mathbf{q}$ and the kinematic flexibility is its volume.

The form and the size of ψ-curves vary if we change the values of the primary task constraints (the position of the tip, for instance) and if we modify the parameters of the mechanism (the relative lengths of the links, for instance). An optimum design problem would be to find the ratio between the link lengths that assures the highest amount of redundancy throughout the workspace. In [9] we obtained the ψ-curves based on analytical inverse kinematic formulas sweeping through all possible joint values. It must be pointed out, however, that this kind of computation can only be used with very simple manipulators and with very simple primary task constraints, such as the end effector position.

The range of joint increments that produce a self-motion displacement is configured by the amount of vectors $d\mathbf{q}_N$, which are the projections of vectors $d\mathbf{q}_S$ in the null space of Jacobian $\mathbf{J}_P$ through the projection matrix $\mathbf{N}_{PA}$. In general, the vector space $d\mathbf{q}_N$ is spawn by eigenvectors $\mathbf{u}_1, \mathbf{u}_2, ..., \mathbf{u}_D$ of matrix $\mathbf{N}_{PA}\mathbf{N}_{PA}^T$ that correspond to the D non-zero eigenvalues (they are all real since $\mathbf{N}_{PA}\mathbf{N}_{PA}^T$ is symmetric). We can write

$$d\mathbf{q}_N = \mathbf{u}_1 \gamma_1 + \mathbf{u}_2 \gamma_2 + \cdots + \mathbf{u}_D \gamma_D , \quad (8)$$

$\gamma_1, \gamma_2, ..., \gamma_D$ is an arbitrary combination of scalars. The self–motion ability of the manipulator is associated with the whole amount of combinations of joint vectors

$$\mathbf{q}^{(k)} = \mathbf{q}^{(k-1)} + d\mathbf{q}_N^{(k-1)} , \quad (9)$$

$k = 1,2,...,K$, where $\mathbf{q}^{(k-1)}$ is known and $d\mathbf{q}_N$ is computed in accordance to equation (8). There are D possible perpendicular directions for vectors $d\mathbf{q}_N$ in each instant k.

The algorithm to compute the self-motion begins with a know vector $\mathbf{q}^{(0)}$ and consists of D nested loops to search for new vectors $\mathbf{q}$. We set the initial scalars $\gamma_1, \gamma_2, ..., \gamma_D$ to zero and then in each loop gradually increment theirs values by adding a small $d\gamma$, so that

$$\gamma_i^{(k)} = \gamma_i^{(k-1)} + d\gamma , \quad (10)$$

$i = 1,2,...,D$. The corresponding new joint angles are

$$\mathbf{q}^{(k)} = \mathbf{q}^{(k-1)} + \mathbf{u}_i^{(k-1)} d\gamma , \quad (11)$$

Note that this algorithm is not operational in kinematic singularities since they increase the degree of redundancy but these can easily be avoided. More critical point of the algorithm, when $D > 1$, is to compute eigenvectors pertinent to each $\mathbf{q}^{(k)}$ in a way that $\mathbf{q}^{(1)}, \mathbf{q}^{(2)}, ...$ will produce smooth families of paths that enable to sweep through the whole space of the manipulator's self-motion joint angles. Suppose that the set of eigenvectors $\mathbf{u}_1^{(k-1)}, \mathbf{u}_2^{(k-1)}, ..., \mathbf{u}_D^{(k-1)}$ corresponding to $\mathbf{q}^{(k-1)}$ is known and that we have obtained the new values $\mathbf{q}^{(k)}$ based on equation (11). By the use of a standard numerical procedure, we can then compute the new eigenvectors corresponding to $\mathbf{q}^{(k)}$ that are denoted as $\mathbf{e}_1^{(k)}, \mathbf{e}_2^{(k)}, ..., \mathbf{e}_D^{(k)}$. In general, however, $\mathbf{u}_i^{(k-1)}$ and $\mathbf{e}_i^{(k)}$ are not directly related and may point in entirely different and unpredictable directions. What we know for sure is that any desired vector $\mathbf{u}_i^{(k)}$ can always be found in terms of the following linear combination

$$\mathbf{u}_i^{(k)} = \alpha_1 \mathbf{e}_1^{(k)} + \alpha_2 \mathbf{e}_2^{(k)} + ... + \alpha_D \mathbf{e}_D^{(k)} , \quad (12)$$

where $\alpha_1, \alpha_2, ..., \alpha_D$ are scalars constrained by

$$\alpha_1^2 + \alpha_2^2 + ... + \alpha_D^2 = 1 . \quad (13)$$

To follow a smooth path $\mathbf{q}^{(1)}, \mathbf{q}^{(2)}, ...$, we have to derive $\mathbf{u}_i^{(k)}$ as the maximum projection of $\mathbf{u}_i^{(k-1)}$ in the space of the new eigenvectors $\mathbf{e}_1^{(k)}, \mathbf{e}_2^{(k)}, ..., \mathbf{e}_D^{(k)}$. The objective, therefore, is to

$$\text{Maximise } (\mathbf{u}_i^{(k-1)})^T \mathbf{u}_i^{(k)} , \quad (14)$$

with respect to (13). The associated Lagrangean can be written as

$$L = (\mathbf{u}_i^{(k-1)})^T (\alpha_1 \mathbf{e}_1^{(k)} + \alpha_2 \mathbf{e}_2^{(k)} + ... + \alpha_D \mathbf{e}_D^{(k)}) - \\ - \lambda(\alpha_1^2 + \alpha_2^2 + ... + \alpha_D^2 - 1) , \quad (15)$$

where λ is the Lagrangean multiplicator. The partial derivatives with respect to every α_r ($r = 1,2,...,D$) must vanish, giving

$$\frac{\partial L}{\partial \alpha_r} = (\mathbf{u}_i^{(k-1)})^T \mathbf{e}_r^{(k)} - 2\lambda \alpha_r = 0 . \quad (16)$$

The derivative with respect to λ must also vanish, so that equation (13) will hold. But instead of introducing equation (13) directly in the set of equations (16), we impose $\lambda = \frac{1}{2}$ and get

$$\hat{\alpha}_r = (\mathbf{u}_i^{(k-1)})^T \mathbf{e}_r^{(k)} \\ \Rightarrow \hat{\mathbf{u}}_i^{(k)} = \hat{\alpha}_1 \mathbf{e}_1^{(k)} + \hat{\alpha}_2 \mathbf{e}_2^{(k)} + \cdots + \hat{\alpha}_D \mathbf{e}_D^{(k)} \quad (17)$$

To satisfy equation (13) we compute

$$\mathbf{u}_i^{(k)} = \frac{\hat{\mathbf{u}}_i^{(k)}}{\sqrt{(\hat{\mathbf{u}}_i^{(k)})^T \hat{\mathbf{u}}_i^{(k)}}} , \ i = 1,2,...,D. \quad (18)$$

If $\mathbf{u}_1^{(k-1)}, \mathbf{u}_2^{(k-1)}, ..., \mathbf{u}_D^{(k-1)}$ and $\mathbf{e}_1^{(k)}, \mathbf{e}_2^{(k)}, ..., \mathbf{e}_D^{(k)}$ are groups of mutually orthogonal vectors it is easy to prove, by taking into account equations (17) and (18), that the resulting vectors $\mathbf{u}_1^{(k)}, \mathbf{u}_2^{(k)}, ..., \mathbf{u}_D^{(k)}$ are also mutually orthogonal and therefore can form a base of the vector space of joint increments $d\mathbf{q}_N^{(k)}$. The use of vectors $\mathbf{u}_1^{(k)}, \mathbf{u}_2^{(k)}, \ldots, \mathbf{u}_D^{(k)}$ instead of vectors $\mathbf{e}_1^{(k)}, \mathbf{e}_2^{(k)}, \ldots, \mathbf{e}_D^{(k)}$ assures that the space of joint vectors of the manipulator's self-motion will be swept with a D-dimensional matrix of approximately equidistant points.

It is useful to observe that the self-motion locus ψ is a periodic function of joint angles. If we keep the values of joint angles within a module of 2π (when a value of one joint angle is greater, we subtract 2π), ψ will (after certain number of iterations) return to its initial value. Hence, the algorithm's iterations are stopped when

$$(\mathbf{q}^{(k)} - \mathbf{q}^{(0)})^T (\mathbf{q}^{(k)} - \mathbf{q}^{(0)}) < \varepsilon. \qquad (19)$$

Here, ε is a positive constant whose value cannot usually (because of the cumulated error) be chosen smaller than about $3d\gamma$. The smaller is constant $d\gamma$ the smaller will be the cumulated error but more iterations will be needed.

The calculation of the kinematic flexibility is now facilitated by the fact that the obtained vectors in the joint space of the self-motion $\mathbf{q}^{(k)}$, $k = 1, 2, \ldots, K$, represent a number of hyper-cubes with approximately equivalent edges. The kinematic flexibility of the robot manipulator is thus given by the following formula

$$F \approx \frac{d\gamma^D}{\pi^D} K. \qquad (20)$$

3 Planar 3R mechanism

Consider the well know and the most simple planar mechanism possessing three parallel rotations $\mathbf{q} = (q_1, q_2, q_3)^T$ and three links of lengths d_i. Its primary task is to position the tip in the plane, $n = 3$, $m = 2$, and $D = 1$. We assume here that the joint angles are unlimited. We study here the distribution of kinematic flexibility along axis $x, y = 0$ which depends on the ratio between the link lengths.

We observed that there are three critical points that separate the distribution of the kinematic flexibility in four different regions. Assuming that $d_1 < d_2 < d_3$ we refer to the first region as A and is associated with interval $x \in (0, d_1 + d_2 - d_3)$, the second region B is the interval $x \in (d_1 + d_2 - d_3, d_1 - d_2 + d_3)$, region C is the interval $x \in (d_1 - d_2 + d_3, -d_1 + d_2 + d_3)$, and the last D is the interval $x \in (-d_1 + d_2 + d_3, d_1 + d_2 + d_3)$. It must be

pointed out that the self-motion functions ψ achieve different forms in different regions. The visual distinction is associated with the configuration of the last two links, "elbow up" when $q_3 < 0$ and "elbow down" when $q_3 > 0$. The manipulator can cross from one group of configurations to another group of configurations only when the tip is inside regions B or D. In these two regions, the self-motion locus ψ is a single curve and is completed because the manipulator passes through both groups of configurations. In regions A and C, the manipulator cannot change the configuration without violating the primary task constraints and the self-motion locus ψ is composed of two disconnected curves. This is shown in Fig. 1.

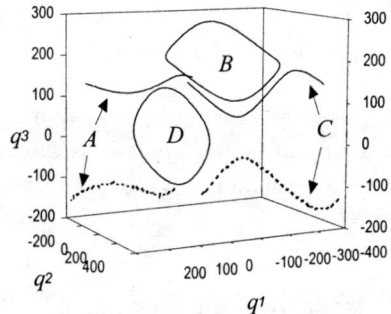

Fig. 1: *Self-motion curves of 3R mechanism*

Hence, the kinematic flexibility in regions A and C can be conceived in two ways, either as the length of both curves or the length of only one curve. Obviously, in the first case the kinematic flexibility is two times higher as in the second case (Fig. 2).

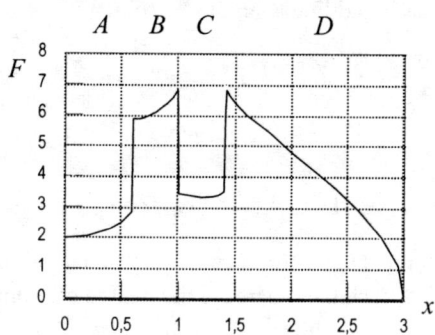

Fig. 2: *Distribution of kinematic flexibility of 3R mechanism*

A special example is the manipulator with equal link lengths, $d_1 = d_2 = d_3$. This manipulator has only one switching point at $x = d_1$. We can then divide the distribution of kinematic flexibility in only two regions, where $x \in (0, d_1)$ and where $x \in (d_1, d_1 + d_2 + d_3)$. In the first region we have pairs of disconnected ψ curves, each corresponding to one group of configurations, while in the second region we have only single curves incorporating both configurations.

4 Anthropomorphic 4R mechanism

The mechanism of the 4R anthropomorphic manipulator is shown in Figure 3. The first three joints, q_1 (humeral abduction), q_2 (humeral flexion), q_3 (humeral rotation) are concentrated in the shoulder complex, the fourth joint q_4 is the elbow flexion. The length of the upper arm is d_1, and the length of the forearm is d_2. The self-motion, when the primary task is to position the tip in Cartesian space ($n = 4$, $m = 3$, and $D = 1$), consists of the rotation of the centre of the elbow joint about axis **r**. We found in [10] that there are two kinematically different manners of the self-motion - depending on whether the circle produced by the elbow intersects axis **x** of the first joint (named manner II) or not (named manner I).

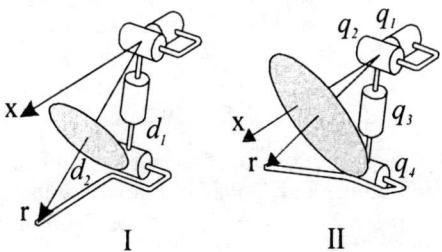

Fig. 3: *The two manners of the self motion*

The self-motion when the tip of the mechanism is positioned along a general skew axis **r** switches between the first and the second manner (let the distance of the tip from the centre of the shoulder complex be r). Typically, when the mechanism is completely stretched and $r \approx r_{max}$ ($r_{max} \approx d_1 + d_2$) the self motion is of the first manner (except when **r = x**). By decreasing r, the self-motion can change (not necessarily) from the first to the second manner. Further decreasing of the value of r (when $r \approx r_{min}$, $r_{min} \approx |d_1 - d_2|$) causes that the first manner of the self-motion is restored. The self-motion curves corresponding to different values of the radial distance r of the tip from the shoulder joint are presented in Figure 4. Here, the parameters of the mechanism are $d_1 = 0.7$, $d_2 = 1.3$ and $r = (0.6, 2.0)$.

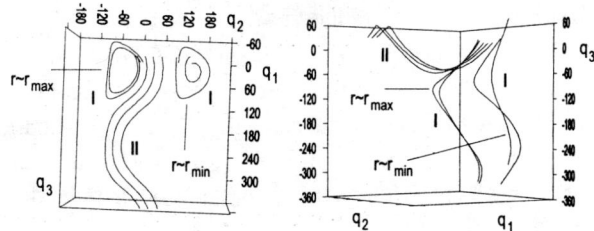

Fig. 4: *Self-motion curves for different positions of the tip*

Figure 5 presents the kinematic flexibility F as a function of the position of the tip along the given skew axis and for different combinations of the link lengths. The manipulator is more kinematically flexible along the well visible edge that delimits the two manners of the self-motion. In absolute terms, however, the highest kinematic flexibility can be achieved in those regions of the workspace where the mechanism produces only the first manner of the self-motion [10]. It is possible to see that such a region of the workspace is plane $x = 0$. In a practical application, we would therefore prefer to position the end effector of the manipulator in this plane and thus provide more ability to the manipulator to solve various secondary tasks.

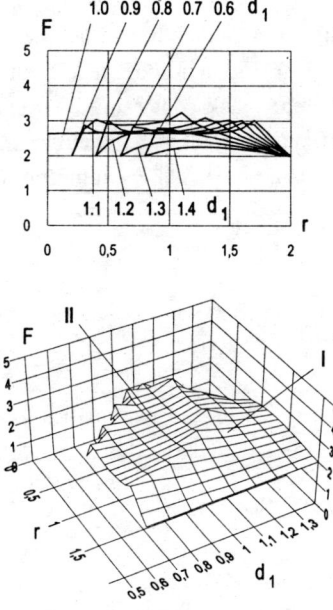

Fig. 5: *Kinematic flexibility for different positions of the tip*

In addition, it is important to notice that there are eight groups of configurations, all corresponding to the same primary task of positioning the tip [9]. During the self-motion, except in singular configurations, the analysed mechanism cannot pass from one group of configurations to another. This is so without regard to the position of the tip in the workspace. Because of this fact, the kinematic flexibility of the mechanism is a continuous function and rather restricted in comparison to some other redundant manipulators.

7 Summary

In the present paper, we introduced the kinematic flexibility to measure the manipulator's redundancy and proposed a general algorithm for its calculation. We presented two numerical examples of redundant mechanisms possessing articulated joints and one degree of redundancy. We demonstrated based on these examples that the distribution of the kinematic flexibility can be a discontinuous function. We showed that there exist different sub-domains of joint angles that correspond to the same primary task constraints. An example is the 3R planar manipulator. For certain values of these constraints, the manipulator can change the sub-domain of joint angles continuously and the kinematic flexibility is higher, but in other cases the manipulator cannot switch from one sub-domain to another without violating the primary task constraints. An example of a mechanism that isn't capable of changing configurations only by its self-motion is the anthropomorphic 4R manipulator.

References

[1] J.W. Burdick, "On the kinematics of redunfant manipulators: Characterization of the self-motion manifolds", in *Proc. IEEE Int. Conf. on Robotics and Automat.*, Scottsdale, Arizona, USA, 1989, pp. 264-270.

[2] J.A. Euler, R.V. Dubey, S.M. Babcock, "Self motion determination based on actuator velocity bounds for redundant manipulators," *J. Robotics Systems*, vol. 6, no. 4, 1989, pp. 417-425.

[3] J. Hollerbach, "Optimum kinematic design for a seven degree of freedom manipulator," in *2nd Int. Symp. on Robotics Res.* (MIT Press, 1985, pp. 215-222.

[4] O. Khatib, "The impact of redundancy on the dynamic performances of robots," *Laboratory Robotics and Automation*, vol. 8, no. 1, 1996, pp. 37-48.

[5] C.A. Klein, B.E. Blaho, "Dexterity measures for the design and control of kinematically redundant manipulators," *Int. J. Robotics Res.*, vol. 6, np. 2, 1987, pp. 72-83.

[6] C.A. Klein, C.H. Huang, "Review of pseudoinverse control for use with kinematically redundant manipulators," *IEEE Trans. on Syst., Man, and Cybern.*, vol. 13, no. 3, 1983, pp. 245-250.

[7] K.P. Lam, "Criteria optimization using self-motion of a 3-R redundant robot," *J. Robotic Systems*, vol. 6, no. 3, 1989, pp. 254-267.

[8] S. Lee, "Dual redundant arm configuration optimization with task-oriented dual arm anipulability," *IEEE Trans. Robotics and Automat.*, vol. 5, no. 1, 1989, pp. 78-97.

[9] J. Lenarčič, "The kinematic flexibility of articulated arms containing one degree of redundancy", in *Proc. 8th Int. Conf. on Advanced Robotics*, Monterey, CA, USA, 1997, pp. 467-472.

[0] J. Lenarčič, "The self motion of an anthropomorphic manipulator", in *Advances in Robot Kinematics: Analysis and Control* (J. Lenarčič and M. Husty, Eds.), Kluwer Academic Publishers, Dordrecht/Boston/London, 1998, pp. 571-578.

[11] A. Liegois, "Automatic supervisory control for the configuration and behaviour of multibody mechanisms," *IEEE Trans. Sys. Man, Cybern.* vol. 7, no. 12, 1977, pp. 868-871.

[12] K.J. Waldron, J. Reidy, "A study of a kinematically redundant manipulator structure," in Proc. *IEEE Conf. On Robotics and Automation*. USA, 1986, pp. 1-8.

[13] D.C. Yang, T.W. Lee, "Heuristic combinatorial optimisation in the design of manipulator workspace," *IEEE Trans. on Syst. Man and Cybern.*, vol. 14, no. 4, 1984, pp. 571-580.

[14] T. Yoshikawa, "Manipulability of robotic systems," *Int. J. Robotics Res.*, vol 4, no. 2, 1985, pp: 3-9.

[15] K.E. Zanganeh, J. Angeles, "On the isotropic design of general six-degree-of-freedom parallel manipulators," *Computational Kinematics*, Kluwer Academic Publishers, Dordrecht, 1995, pp. 213-220.

Learning of Inverse Kinematics Behavior of Redundant Robot

Goran S. Đorđević, Milan Rašić, Dragan Kostić,

University of Niš, Faculty of Electronic Engineering,
Beogradska 14, Niš, Yugoslavia
[gorandj / mrasic / dkostic] @elfak.ni.ac.yu

and

Dragoljub Šurdilović

Fraunhofer IPK, Department Robot & Systems Control,
Pascalstr. 8-9, 10587 Berlin, Germany
dragoljub.surdilovic@ipk.fhg.de

Abstract

Acceleration in redundancy resolution has been achieved by model-based learning. Parameterized joint motions generated by pseudoinverse of parameterized motion primitives in operational space are taken as learning examples. The Method of Successive Approximations as a Function Approximator generalizes given examples resulting in a parameterized model, termed redundancy resolution skill. The SA method results in an analytical solution of joint motions. This particular property enables extension of the universality of skills to tasks never experienced by learning. Moreover, a proper addressing of skill's output results in paths completely different from learned ones. In practice, this means that a segment as a motion primitive suffices in acquiring skill in the learning domain. The example emphasis this property of SA-based redundancy resolution. Highly compressed skill, acquired on segments only, is applied in RR along two test-trajectories – an 'H'-character and a rosette. The gained acceleration of RR recommends this procedure for on-line redundant robot control and easy robot programming.

1 Introduction

The kinematic redundancy in robotics has two sources: task definition in space of lower dimensionality than joint space, and robot construction with more than six degrees-of-freedom (DoF), intentionally designed to manage unstructured and changing environments. Either way, this redundancy can improve the manuevering capabilities of the robot, by increased kinematic dexterity and dynamic performance. In practice, however, these benefits of the redundancy are usually not used if one has to program the robot on daily basis. The main reason is that common redundancy resolution (RR) procedures are cumbersome. Most robot programmers are unfamiliar with them. In our opinion, they do not have to be. Particularly in changing environments the numerical complexity of RR techniques, even with expert knowledge, becomes a limiting factor. All this essentially undermines the benefits redundancy has brought. Generally, we are facing a trade-off between the quality of RR and cost of redundancy application. In this paper we present a tool and a way of its application to the RR problem that unburdens the user, a robot programmer, from dealing with redundancy, enabling him to draw benefits from the redundancy. The idea behind this research is a supervised model learning of RR from numerically generated examples. Such a model of the RR is denoted Redundancy Resolution Skill (RRS).

We assume that the robot will perform one class of tasks in its usage. This consistency of robot tasks brings up some motion primitives (MPs) to focus on, as common ingredients of all necessary paths. In digression, robot programming languages are mostly related to segment and arc movement commands. Therefore, we will confine to segments and arcs as MPs, generating RRS along segments and arcs. Transformations of the primitives sufficient for task description are defined by skill parameters (SPs). Usually, SPs are chosen among affine transformations. A suitable IK algorithm applied on the parameterized primitives yields parameterized joint motions, i.e. learning examples. The RRS should also be parameterized. However, even a reasonable number of MPs combined by a small number of SPs, produces an enormous set of learning examples.

One of the results presented in this paper is a reduction of the model size gained by applying a suitable skill representation. Instead of Neural Networks [3, 5, 7], Fuzzy [8] and other Function Approximators (FA) applied to the RR problem, we use the Successive Approximations (SA) method [2] for skill representation. It is a parametric, multivariate, non-linear FA. Besides enormous reduction of model size, resulting in accelerated RR, it also features several advantageous properties [2]. It enables easy upgrading to new skill parameters, it provides extrapolation of examples along with interpolation and it produces analytic solution. However, it only produces parameterized primitives used in skill learning . So, the rather sparse knowledge contained in RRS enables a coarse representation of robot's IK within the learning domain. The ultimate goal of our research is to extract RR knowledge from already acquired skills. Random addressing of the skill spans its output over a much wider domain. Besides it milding the dimensionality problem of learning space, an effect similar to an introduction of a completely new skill parameter is achievable. Further, instead of a sub-path, we can address a single point of a generalized MP. An array of such points on consecutive MPs leads to a completely new trajectory. Indexing of a parameterized skill's output essentially determines the nature of a new trajectory. For example, skill of segments as MPs can produce an arbitrary curve. This means that one MP with one skill parameter is sufficient for generating an arbitrary path within the learning domain. In such a way, the interpretation of skill-based RR has morphed RRS to IK behavior within the learning domain.

A brief review of the SA procedure applied on RR is given in Section 2. A modification of the SA method that spans universality of skills and enables morphing of skills to IK behavior is presented in detail in Section 3. Finally, Section 4 contains an Example with a 5DoF planar redundant robot in a RR task. Skill is acquired on segments in xOy plane, parameterized by affine transformations. The RRS is successfully applied on circular paths, an 'H'-character and a rosette, in the learning domain.

2 SA – based Redundancy Resolution

The SA-based RR skill is essentially a method that generalizes learning examples. Since neither model size, nor computational burden of the skill application are not affected by the complexity of example generation procedure, various RR algorithms can be adopted in order to achieve the best performance. Such algorithms include various global or local optimization techniques. However, the standard procedure of RR is application of pseudoinverse or some of its modifications [1]. Being algorithmically simple and highly accurate, pseudoinverse is applied in generation of learning examples.

Denote the MP Π, along with its temporal interpretation, termed elementary trajectory (ET), $\Pi_{ET}(t)$. The set of skill parameters denoted S_{SP}, consists of affine and temporal transformations. Precise definitions of SP could be found, for example, in [6]. Each SP applied to $\Pi_{ET}(t)$ results in a new trajectory $\Pi_{ET}^s(t)$. Arbitrary combinations of SPs ($\mathbf{s} = s_1 \circ s_2 \circ \ldots$) lead to a multidimensional field of examples $\Pi_{ET}^\mathbf{s}(t)$. Finally, learning examples, i.e. corresponding joint motions $\mathbf{q}_{ET}(t)$ based on pseudo-inverse solution, are also parameterized on the set S_{SP}. Assuming that examples are selected and generated, we further elaborate the SA procedure.

The Method of SA [2] is a nonlinear, global, parametric, multivariate function approximator. Essentially, it is a procedure of successive least-squares (LS) fitting of approximant's coefficients on the set of S_{SP}. Polynomial LS approximation on a natural basis is used.

SA method starts with normalization in time on the interval $\tau \in [-1,1]$. This yields normalized joint trajectories:

$$\{\mathbf{q}_{ET}(j,t)\} \mapsto \{\theta_{ET}(j,\tau)\}, \quad j=1,\ldots n, \quad (1)$$

where n is the number of joints. We drop out index j, considering that the following procedure is independent and identical for each joint. Normalized trajectories $\theta_{ET}(\tau)$ are approximated by a n_0-th order polynomial

$$Q_0(\tau) = \sum_{k_0=0}^{n_0} \alpha_{k_0} \cdot \tau^{k_0}, \quad (2)$$

where coefficients α_{k_0}, $k_0 = 0,\ldots n_0$ minimize the norm

$$\left\| \theta_{ET}(\tau) - \sum_{k_0=0}^{n_0} \alpha_{k_0} \cdot \tau^{k_0} \right\|, \quad (3)$$

in the sense of least squares. Arranged coefficients α_{k_0} give the matrix $A = [\alpha_0 \; \alpha_1 \ldots \alpha_{n_0}]^T = [\alpha_{k_0}]_{(n_0+1) \times 1}$.

The first SP denoted by s_1 takes ng_{s_1} values on the interval $s_1 \in [s_{1_m}, s_{1_M}]$. SP s_1 is defined as

$$s_1(i_{s_1}) = s_{1_m} + i_{s_1} \Delta s_1,$$
$$\Delta s_1 = \frac{s_{1_M} - s_{1_m}}{ng_{s_1} - 1}, \; i_{s_1} \in \{0,\ldots,ng_{s_1} - 1\}. \quad (4)$$

In other words, ng_{s_1} learning examples, $^{s_1}\theta_{ET}$, are produced by the first SP. Approximation of the examples in time yields matrices of coefficients $[^{i_{s_1}}\alpha_{k_0}]_{(n_0+1) \times 1}$, further arranged in a two-dimensional array $^{s_1}A = [^{i_{s_1}}\alpha_{k_0}]_{(n_0+1) \times ng_{s_1}}$. Naturally, every coefficient $\alpha_{k_0}, k_0 = 0,\ldots,ng_{s_1}$ depends on the value of s_1, thus it can be fitted by a polynomial of m_{s_1}-th order. SP s_1 is taken as an independent variable, and polynomial coefficients $\beta_{k_0 k_1}$ minimize the norm

$$\left\| ^{i_{s_1}}\alpha_{k_0} - \sum_{k_1=0}^{m_{s_1}} \beta_{k_0 k_1} \cdot s_1^{k_1} \right\|, \; k_0 = 0,\ldots,n_0. \quad (5)$$

Solution coefficients $\beta_{k_0 k_1}$ are represented by the matrix

$$^{s_1}B = [\beta_{k_0 k_1}]_{(n_0+1) \times (m_{s_1}+1)}, \quad (6)$$

where k_0 denotes approximation in time and k approximation over the first SP.

The next step is derivation of a new SP s_2, defined on the interval $s_2(i_{s_2}) \in [s_{2_m}, s_{2M}]$ as a discrete function

$$s_2(i_{s_2}) = s_{2_m} + i_{s_2} \Delta s_2,$$
$$\Delta s_2 = \frac{s_{2_M} - s_{2_m}}{ng_{s_2} - 1}, \; i_{s_2} \in \{0,\ldots,ng_{s_2} - 1\}. \quad (7)$$

Combination of elementary skills, i.e. their grafting in a new skill $s_2 = s_2 \circ s$, can be achieved in the following way. For each value of SP s_2 the procedure (1)–(6) is repeated, yielding an array of coefficients matrices:

$$^{s_{21}}B = [^{i_{s_2}}\beta_{k_0 k_1}]_{(n_0+1) \times (m_{s_1}+1) \times ng_{s_2}}. \quad (8)$$

It represents a two-dimensional RRS for the given ET. Eq. (8) is continuous in time and in the first SP but discrete in the second SP. Application of the procedure of m_{s_2}-th order polynomial fitting with coefficients $\gamma_{k_0 k_1 k_2}$, and regarding SP s_2 as an independent variable, minimizes the norm

$$\left\| ^{i_{s_2}}\beta_{k_0, k_1} - \sum_{k_2=0}^{m_{s_2}} \gamma_{k_0 k_1 k_2} \cdot s_2^{k_2} \right\|, \; \begin{array}{l} k_0 = 0,\ldots,n_0, \\ k_1 = 0,\ldots,m_{s_1}. \end{array} \quad (9)$$

It yields a matrix of coefficients

$$^{s_{21}}\Gamma = [\gamma_{k_0 k_1 k_2}]_{(n_0+1) \times (m_{s_1}+1) \times (m_{s_2}+1)}. \quad (10)$$

The matrix of coefficients $^{s_{21}}\Gamma$ is a skill that represents RR along $\Pi_{ET}(t)$ modified by two SPs – s_{21}.

Similarly, approximation procedure generalized on a N-dimensional case, gives a $N+1$-dimensional matrix of coefficients

$$^{s_{N\ldots 21}}\Psi = [\psi_{k_0 k_1 \ldots k_N}]_{(n_0+1)\times(m_{s_1}+1)\times\cdots\times(m_{s_N}+1)}. \quad (11)$$

The matrix of coefficients $^{s_{N\ldots 21}}\Psi$ represents RR skill of one robot joint along the elementary trajectory $\Pi_{ET}(t)$, modified by a combination of skill parameters $s_1 \circ s_2 \circ \ldots \circ s_N$. This matrix stands for a skill $s_{N\ldots 21}$. RR skill (11) represented by the SA method features several prestigious properties.

First, applied generalization yields an enormous rate of model compression along with a high acceleration in calculation. The size of the model is directly proportional to the complexity and/or the size of the MP. Second, besides interpolation only, also an extrapolation of the examples [2] is achieved, due to the use of polynomials in the SA-based RR skill. Third, SA-based RR generally requires only summation and multiplication, as long as polynomials are approximants. Oppositely, pseudoinverse demands, by far, more operations. Fourth, robustness on initial position errors, discussed in [2], gives more flexibility in the selection of parameter values. This means that a trajectory learned in one part of the working space could be reproduced without significant distortion in other, adjacent parts of the space. Fifth, and the most important for the further discussion, SA-based RR is an analytic function of joint variables in time. The succession of MPs results in concatenated polynomials, representing joint motions. Being so, joint velocities and accelerations are easily achievable. Finally, SA-based RR is aware of limits in the robot's working area, even if the skill is not trained on them. From a user point of view, the power of this method is reflected by the fact that the complexity of trajectory generation does not depend on the complexity of RR applied in the learning procedure. Furthermore, a control system augmented with such a skill enables robot programming without actual knowledge of exactly how redundancy should be resolved. On the other hand, the speed of RR computation recommends this procedure for on-line control of redundant robots.

A major drawback of the SA method is its batch nature. Hence, it is more suitable in off-line skill acquisition. In real-time, it should be additionally treated by incremental learning algorithms. A modification of the SA procedure, presented in Section 3, will reduce the side effects of SA-based IK skills. It will expand the universality of IK skill to the IK behavior.

3 Interpretation of SA-based skills

Robot's tip path is defined by $\mathbf{X} = \Pi(\sigma)$, where $\mathbf{X} \in R^m$, $m \leq 6$, and $\Pi(\sigma)$ is a parameterized spatial curve. Variations of the parameter σ on the interval $[\sigma_0, \sigma_f]$ result in a reposition of the robot's tip from the starting to the end point along the path. Then, variations of σ in time $\sigma = \sigma(t)$ determines a profiled motion and defines a trajectory along the path $\Pi(\sigma)$. The choice of the execution interval $[t_0, t_f]$, determines the mapping

$$[t_0, t_f] \xrightarrow{\sigma(t)} [\sigma_0, \sigma_f] \quad (12)$$

termed motion profile. It is clear that the choice of time interval $[t_0, t_f]$ affects the velocity of motion (smaller interval, faster movement), while the shape of $\sigma(t)$ function determines the motion profile.

The function $\sigma = \sigma(t)$ is a continual, single-valued, and monotone function. If it is increasing, the tip motion starts at σ_0 and ends at σ_f; if decreasing the tip will go in the opposite direction. The latter case can be formalized in a similar manner to the former one, so it will not be discussed. This property will be used in the Example.

It is a common case that the motion profile is given by a velocity profile, i.e. by the function $\dot{\sigma}(t)$. Some well-known profiles are given below.

$$\dot{\sigma}_0(t) = \frac{\sigma_f - \sigma_0}{t_f - t_0}, \quad (13)$$

$$\dot{\sigma}_1(t) = \begin{cases} \dfrac{(\sigma_f - \sigma_0)(t - t_0)}{\Delta t_a (t_f - \Delta t_a - t_0)}, & t_0 \leq t \leq t_0 + \Delta t_a, \\ \dfrac{(\sigma_f - \sigma_0)}{(t_f - \Delta t_a - t_0)}, & t_0 + \Delta t_a \leq t \leq t_f - \Delta t_a, \\ \dfrac{(\sigma_f - \sigma_0)(t_f - t)}{\Delta t_a (t_f - \Delta t_a - t_0)}, & t_f - \Delta t_a \leq t \leq t_f, \end{cases} \quad (14)$$

$$\dot{\sigma}_3(t) = \frac{\sigma_f - \sigma_0}{t_f - t_0}\left[1 - \cos\left(\frac{2\pi(t - t_0)}{t_f - t_0}\right)\right], \quad (15)$$

with $\sigma_i(t_0) = \sigma_0$, $\sigma_i(t_f) = \sigma_f$, $i = 0, \ldots, 3$, satisfied. Pseudo-constant velocity profile is represented by Eq. (13), trapezoidal profile by Eq. (14), and cosine profile is given by Eq. (15). The term Δt_a in (14) represents the acceleration/deceleration period. Triangular velocity profile is a special case of the trapezoidal profile. Each motion profile can be transformed into a pseudo-constant profile by inverse substitution $t = \sigma^{-1}(\sigma)$. This procedure is termed regularization.

Let us discuss two paths $\Pi(\sigma)$ and $^1\Pi(\upsilon)$, where $^1\Pi(\upsilon)$ is a sub-path of the path $\Pi(\sigma)$, as shown in Fig. 1.

Then, a set of parameters

$$^1\sigma \in [^1\sigma_0, ^1\sigma_f] \subset [\sigma_0, \sigma_f] \quad (16)$$

exists, and the relation

$$^1\Pi(\upsilon) \equiv \Pi(^1\sigma), \\ \forall \upsilon \in [\upsilon_0, \upsilon_f], \forall ^1\sigma \in [^1\sigma_0, ^1\sigma_f] \subset [\sigma_0, \sigma_f] \quad (17)$$

holds true. Values of the parameters $^1\sigma_0$ and $^1\sigma_f$ are derived from conditions

$$^1\Pi(\upsilon_0) = \Pi(^1\sigma_0), \quad ^1\Pi(\upsilon_f) = \Pi(^1\sigma_f). \quad (18)$$

The existence of parameters $^1\sigma_0$ and $^1\sigma_f$ enables realization of a sub-path of the path used in example generation. This procedure can be further generalized to the case of sub-path degeneration to a single-point.

Let us assume that the parameters of the trajectory during the motion are given, i.e. the $\Pi(t)$ function is known. If the velocity profile and time interval $[t_0, t_f]$ are given, we can determine the representation of $\Pi(t)$ function uniquely, by taking the inverse velocity profile $t = \sigma^{-1}(\sigma)$ function. Thus, the parametric definition of the path could be associated with an arbitrary interpretation of the velocity profile by taking $\sigma = \sigma(\tau)$. The shape of the function σ represents the motion profile, its monotonicity determines the direction of motion, and the interval $[\tau_0, \tau_f]$ determines the velocity of the motion.

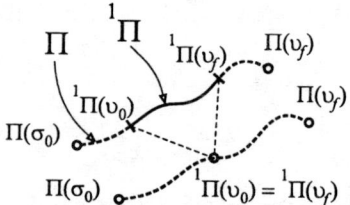

Fig. 1. *Example of path and sub-path composition*

In such a way, the benefits of the interpretation of SA-based RRS is twofold. First, partial addressing of a skill, and second, arbitrary shaping of the velocity profile along the path are gained. This procedure can generate trajectories different from the path used in the skill acquisition procedure. The problem of minimizing the set of SPs is partially milded, which directly results in a reduction of the learning space dimensionality. This facilitates learning, reduces the model size, and accelerates IK resolution. However, the best we can achieve is a reproduction of trajectories that match the nature of those applied in the skill acquisition.

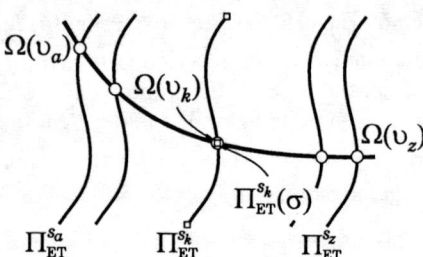

Fig. 2. *Intersection of arbitrary path Ω and parameterized ETs in hypothetical case of translation as skill parameter*

To remedy this problem we must change a part of the skill interpretation procedure discussed above. The idea is explained in Fig. 1, where a sub-path reduces in a point: $^1\Pi(\upsilon_0) = {}^1\Pi(\upsilon_f)$. Consider an arbitrary path $\Omega(\upsilon)$, different from the ETs and within the learning domain. Points along $\Omega(\upsilon)$ shown as circles in Fig. 2, are determined by the motion profile. Each value of υ, say $\upsilon_a, \upsilon_b,...$, corresponds to a certain value of the SP s, say $s_a, s_b,...$ Each value of SP determines related Π_{ET}^s.

Consider the intersection point between $\Omega(\upsilon)$ and Π_{ET}^s, for $\upsilon = \upsilon_k$, i.e. $s = s_k$, shown as a box in Fig. 2. The value of parameter σ at ET $\Pi_{ET}^{s_k}$ along with SP $s = s$ determines RR for the path point $\Omega(\upsilon_k)$. In such a way, knowledge of RR contained in the skill could be extended to the arbitrary path that belongs to the learning domain. This interpretation of SA-based skill promotes the significance of the learning domain, overcoming the nature of learning examples. In other words, the question "where to learn" dominates over "what to learn". Note that redundancy is resolved at every point of a new path generated from the RRS by summation and multiplication operations, only.

4 Example

Redundant planar 5 DoFs robot, with the first three revolute joints and the last two prismatic joints, given in Fig. 3, is discussed. The lengths of robot's links are: 0.5; 0.5; 0.2; 0.1; 0.1 m. In example, two characteristic tasks will be elaborated.

At first, the robot is trained on two orthogonal segments (0.5 m) as MPs in operational space. Skill parameters are translations along axes, resulting in two RR skills: s_{tx} and s_{ty}. Every SP has 11 values, generating 11 equidistant (5 cm) segments of each MP. Trajectories are generated with a triangular velocity profile, with a duration time of 2.5 s. Learning examples, i.e. joint motions, are generated by pseudoinverse along the MPs, with joint limit criteria imposed on the last two segments.

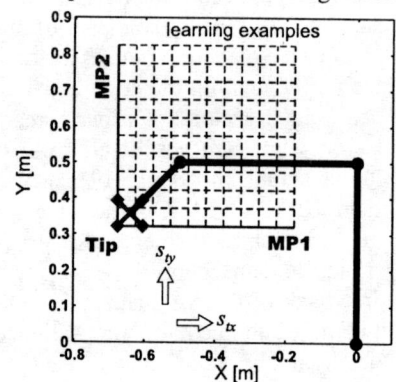

Fig. 3. *Planar 5DoF robot. Examples are forming a grid. Velocity profile is triangular. Duration of examples is 2.5s.*

The method of SA is applied on each joint, for both SPs. Preprocessing regularizes every example to a pseudo-constant velocity profile, and normalizes it. The first step of SA procedure, i.e. approximation in time, was taken with a polynomial of 20-th order. The second step, approximation in the first SP, was taken with a polynomial of 10-th order. The two approximation procedures result in a skill model contained in $(20+1)(10+1)$ i.e. 231 coefficients, per skill for each joint. In total, this is 18.05 Kb. This number of coefficients is the RR skill of the planar robot performing two distinct trajectories, each translated along related axes in a given domain. As discussed in Section 3, conventional SA-based skill could reproduce only the MP that match the nature of those used in learning. However, the interpretation of the skill, proposed in Section 3, results in an effect similar to an introduction of a new skill parameter – scaling. This results in a full range of paths from very short to 0.5 m long. Also, velocity profiles could be varied on the basis of functions σ, Eqs. (13–15). This reduction of the dimensionality of the learning space facilitates learning efforts. In other words, we have learned how to resolve redundancy along arbitrary orthogonal lines, maintaining joint limit criteria.

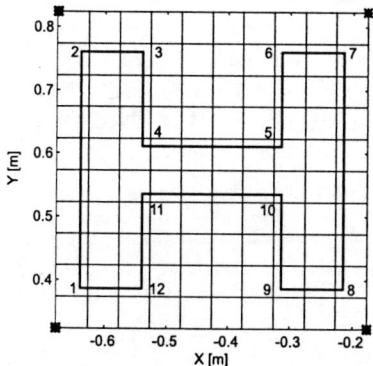

Fig. 4. *Test trajectory in the shape of* H *character, sequence of points (1-2-...-12-1) and learning domain, marked by four stars.*

A complex test trajectory in the shape of 'H'-character is made of twelve segments, Fig. 4. Total execution time is 12 s. Velocity profiles are trapezoidal, with 20% of acceleration/ deceleration. These segments are chosen in such a way that they could be derived from s_{tx} or s_{ty}. None of the segments belong to the example set, and each segment is a sub-path of parameterized MPs.

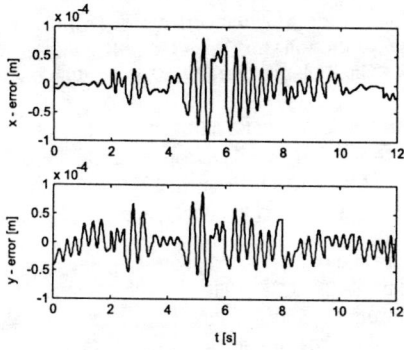

Fig. 5. *Cartesian errors along Ox and Oy during execution of test trajectory (1-2-...-12-1); in both cases it is below 0.1 mm.*

Time histories of the Cartesian errors, shown in Fig. 5, are less than 0.1 mm. Such errors correspond to the accuracy of most industrial robots. So, we may conclude that the SA-based redundancy resolution provides acceptable results. The number of flops along the test trajectory for SA-based RR is about 10% of the flops used with pseudoinverse augmented with joint limit criteria (used in training).

From the former discussion, it is clear that skill-based RR is justified. Maintaining the quality of IK solution that pseudo-inverse offers, SA applied on skill acquisition provides advantages such as small computational effort, and easy managing of redundancy. This promotes such FA as efficient, fast and cheap RR tool, particularly suitable for real-time tasks.

In the second task, the robot is trained on one segment Π in operational space,

$$x = -0.308, \quad y \in [0.4605, 0.6605] \quad (19)$$

given in Fig. 3 is considered as MP. Skill parameter s_r, is a point-rotation on the interval $s_r \in [0, 2\pi]$ defined by

$$s_r(i_{s_r}) = i_{s_r} \Delta s_r, \quad (20)$$

where, $\Delta s_r = 2\pi/72$, $i_{s_r} = \{0,...,71\}$. The point of rotation is the first point of the path $\wp : (-0.308, 0.4605)$. Transformation of the path based on (20) results in a set of 72 segments Π^{s_r}. Triangular velocity profile with a duration of 1 s, results in an elemental trajectory, $\Pi^{s_r}_{ET}(t)$, along which IK is to be learned. IK is resolved by pseudoinverse augmented with joint limit criterion imposed to the last two prismatic joints. This procedure, for each trajectory $\Pi^{s_r}_{ET}(t)$, results in joint motions ${}^j\mathbf{q}^{s_r}_{ET}(t)$, $j = 1,...5$. The first step of the preprocessing procedure is regularization in time that maps a triangular velocity profile to a pseudo-constant velocity profile, given by Eq. 15. Next, normalization in time takes place. At the end of preprocessing, obtained parameterized and normalized joint motions are used as learning examples. RR skill of the robot consists of a set of five skill models, independently generated. By using the SA method, a skill model is generated by a 9-th order polynomial for approximation in time and an 8-th order polynomial for approximation regarding the SP s_r, for each joint. In total, for the whole robot, a set of 450 coefficients represents the skill. This takes only 3.5Kb of memory, it resolves redundancy of the robot fast and accurately along a segment of 0.2 m in length, arbitrarily rotated in the learning domain. However, available outputs are only segments, sharing the same starting point. Additional interpretation could result in shortened segments in radial directions. This has the same effect as introducing SPs for rotation and scaling. Still, skill output is only a coarse set of available trajectories. Furthermore, curvilinear trajectories could not be generated without a new learning procedure. The interpretation procedure (Section 3) further extends the SA-based skill model. It enables generation of paths of arbitrary nature, regardless of those used in learning. The only limitation is that newly generated paths should belong to the learning domain. In such a way, the importance of the learning domain dominates over the nature of the learning trajectory.

To evaluate the procedure of interpretation of SA-based skill of redundancy resolution we choose a test-trajectory of completely different nature compared to the MP. The test-trajectory is a rosette (Fig. 6) with equation

$$\rho = 0.2 \sin 3\phi, \quad \phi \in [0, 2\pi], \quad (21)$$

translated to the center of learning domain ($\wp$), and rotated for $\delta = +5\pi/6$. The rosette (21) is within the learning domain. Motion of the robot's tip along the rosette is 5 s with a triangular velocity profile.

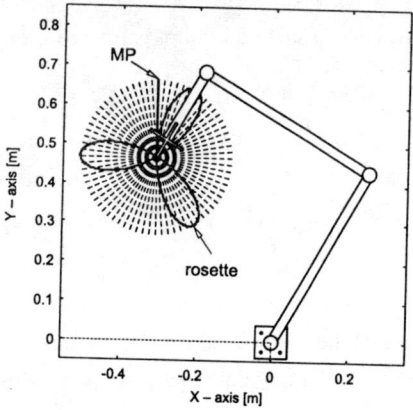

Fig. 6. *Planar 5DoF robot (3R+2T). Examples are segments forming radial lattice. Test-trajectory is rosette.*

Redundancy resolution based on pseudoinverse, augmented with joint limits criterion imposed on the last two joints, for the test trajectory defined by 500 points, demands a total of approximately 1.6 Mflops. On the other hand, RR based on the SA-method with interpretation given in Section 3, with results of joint motions depicted in Fig. 4, demands approximately 1 Mflops. Compared to the pseudoinverse, it is approximately a 40 % reduction of flops used. In addition, the joint limits criterion is maintained. Also, it is obvious that the robot takes the same poses at the end of the rosette as at the beginning of the motion, with maintained cyclicity of joint motions.

Operational space errors, shown in Fig. 5, are less than 1 mm along both axes, which is quite acceptable compared to the whole motion range of 0.4 m. Additional fine-tuning of the SA procedure could result in a partial reduction of the external errors. However, this can only be estimated. This interpretation procedure does not result in an analytical solution of joint motions and computation of joint velocities should be performed separately.

5 Conclusion

In this paper we have discussed the RR problem on motion primitives (MPs) in operational space as ingredients of future robot trajectory planning. We introduced some parameters that spans MPs over the whole learning domain. Along parameterized MPs redundancy is resolved yielding learning examples. Application of the Successive Approximations method generates a model, i.e. skill of RR along parameterized examples. In such a way we built an efficient system capable of fast and competent RR. At this point we can only generate a limited number of parameterized MPs. This is overcome by appropriate interpretation of the skill output. Fragmental addressing of the skill produces joint motions resulting in a new trajectory of similar nature as learned ones within the learning domain. Also, velocity profiles can be generated arbitrarily, overcoming learning example's profiles. Further interpretation of skill addresses only a single point of the MP. Concatenating those points, a completely different path can be generated with resolved redundancy. Two suitable examples shows the effectivity of the method proposed. Redundancy of a 5DoF planar robot is resolved accurately and with significant acceleration. Further, the RR along an essentially different trajectory, a rosette, is obtained from the skill based on segments as examples. The proposed method enables fast and cheap RR suitable for on-line control. The acceleration in RR computation time can be further achieved by parallel processing of the skills, each of which is responsible for a single joint. Also, the RR skill reduces redundant robot programming to a path planning problem only, since such skilled controller will come along with previously resolved redundancy.

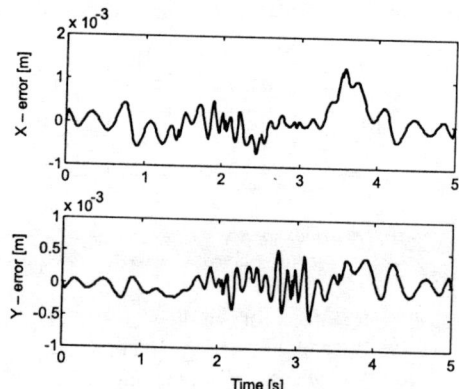

Fig. 7. *Operational space errors along rosette.*

6 References

[1] S. Chiaverini, "Singularity Robust Task-Priority Redundancy Resolution for Real-Time Kinematic Control of Robot Manipulators", *IEEE Trans. Robot. and Automat.*, vol. 13, no. 3, pp. 398–410, 1997.

[2] G. Dordević, V. Potkonjak, M. Rašić and D. Kostić, "Skill-Based Inverse Kinematics of Redundant Robots", 3rd *ECPD ICARIAAS*, Bremen, Sept. 1997, pp. 301–306.

[3] M. Eldracher, D. Hernàndez, and M. Kinder, "Neural Modeling of the Kinematics of Robot Trajectories", Statusseminar des BMFT, Neuro-Informatik, Maurach, Germany, pages 227-236, 1992.

[4] S. Kawamura, and N. Fukao, "Interpolations for Input Torque Patterns obtained through Learning Control", *Proc. ICARCV'94*, Singapore, Nov. 9-11, 1994., pp. 2194–2198.

[5] S. Kieffer, V. Morellas, and M. Donath, "Neural Network Learning of the Inverse Kinematic Relationships for a Robot Arm", *Proc. ICRA.*, 1991, pp. 2418–2425.

[6] K. Takayama, and H. Kano, "A New Approach to Synthesizing Free Motions of Robotic Manipulators Based on a Concept of Unit Motions", *IEEE Trans. Syst. Man Cybernet.*, vol. 25, no. 3, pp. 453–463, March, 1995.

[7] S.G. Tzafestas, and S.N. Raptis, "Neural Networks for Inverse Robot Kinematics", In 3rd *Int. Conf. on ARIAAS*, Bremen, Sept. 15-17.1997, pp. 398–403.

[8] K.-Y. Young, and S.-J. Shiah, "An Approach to Enlarge Learning Space Coverage for Robot Learning Control", *IEEE Trans. Fuzzy Systems*, vol. 5, no. 4, pp. 511–522, Nov. 1997.

INTERACTIONS AND MOTIONS IN HUMAN-ROBOT COORDINATION*

J.Y.S. Luh Shuyi Hu
Clemson University
Clemson, SC 29634
U.S.A.

ABSTRACT

In the human-robot cooperative tasks, the robot is required to memorize different trajectories for different assignments, and retrieve a proper one from them in real-time for the robot to follow when an assignment is repeated. A three-point representation of the space curves is translation/rotation/scaling invariant. Schemes of matching positions by vectors, and orientations by matrices are efficient in that computing time is short enough for real-time implementations. Physical experiments are performed to illustrate and verify the analytical results.

INTRODUCTION

Consider the task of carrying a rigid object jointly by a human and a robot. Since the human has all the essential sensors, he/she mainly makes decisions. The robot, which has a sensing wrist, carries the physical load. To start the task, the human leads the robot through the rigid object to travel along a suitable trajectory and thereby achieves the desired goal. During the process, the trajectory is recorded and stored in memory. When the task is repeated, the stored trajectory is played back along which the robot travels. The human need not participate in the action. This process is well known. It can be achieved by storing and then playing back the joint trajectories.

This paper, however, deals with tasks, which requires the robot having trajectory learning and matching capabilities. The tasks are described as follows.

The human-robot team is required to perform a finite number of different categories of tasks. Each category involves a different trajectory. For every new task, the human is required to lead the robot. During the process, the trajectories are recorded and stored in memory as "skillful trajectories". The representation of the stored trajectories is translation/rotation/scaling invariant. This feature is needed for handling the situation in which the relative translation/rotation/scaling between the robot's end-effector and the goal location is altered.

For completely or partially repeated tasks, the robot searches for a matching skillful trajectory in memory. If the robot finds one, it travels along this trajectory without human assistance. Because of the invariance property of the stored trajectories (the matching and skillful ones), the on-going trajectories could be different by a proper translation, rotation and scaling. Once the matching process is completed, the robot follows the matched trajectory without human assistance. However, when the human notices that the path of the object is not, or will not be, acceptable, he/she interrupts the motion and applies compliance control [1] to lead the robot until another acceptable, skillful trajectory is identified and followed by the robot. The human then withdraws his/her assistance. However, interruption by the human is repeated intermittently whenever it is needed until the desired goal is accomplished.

It is seen that two major subjects are involved in the operation: compliance control for human-robot interactions, and trajectory learning and matching for motions. They are presented in the following sections.

COMPLIANCE CONTROL

Compliance control is concerned with the interactions between the human and robot when they make a physical contact through the rigid object. By means of the sensing wrist, the robot measures the forces/moments that are exerted by the human, and thereby understands and follows human's intended motion. By a proper coordinate transformation, the measured forces/moments may be expressed with reference to the end-effector's coordinates. Let

$$F_e(kT) = \begin{bmatrix} f_{ex} & f_{ey} & f_{ez} & m_{ex} & m_{ey} & m_{ez} \end{bmatrix}^T$$

be the total forces/moments vector exerted on the end-effector at time kT, and

---

*This work was supported by NSF Grant No. IRI 9405276 through Ohio State University Research Foundation Project No. 306130

Let
$$F_o(kT) = [f_{ox} \quad f_{oy} \quad f_{oz} \quad m_{ox} \quad m_{oy} \quad m_{oz}]^T$$
be the forces/moments exerted on the end-effector due to gravity on the tool and object at time kT. Then
$$D_p(kT) = [f_{ex} - f_{ox} \quad f_{ey} - f_{oy} \quad f_{ez} - f_{oz}]^T,$$
and
$$D_\varphi(kT) = \begin{bmatrix} 0 & -m_{ez} + m_{oz} & m_{ey} - m_{oy} \\ m_{ez} - m_{oz} & 0 & -m_{ex} + m_{ox} \\ -m_{ey} + m_{oy} & m_{ex} - m_{ox} & 0 \end{bmatrix}$$
represent the forces and moments, respectively, exerted by the human at kT. Consequently, the compliance motion strategy may be expressed by the recursive 3×1 position vector of the end-effector's coordinate frame $P_E^0(kT)$ at time kT and 3×3 orientation matrix $R_E^0(kT)$:
$$P_E^0((k+1)T) = P_E^0(kT)$$
$$+ R_E^0(kT)\{K_f D_p(kT)T + K_{df}[D_p(kT) - D_p((k-1)T)]\}$$
$$R_E^0((k+1)T) = R_E^0(kT)$$
$$+ R_E^0(kT)\{K_m D_\varphi(kT)T + K_{dm}[D_\varphi(kT) - D_\varphi((k-1)T)]\}$$
where K_f and K_m are corresponding 3×3 stiffness matrices while K_{df} and K_{dm} are corresponding 3×3 damping matrices.

TRAJECTORY LEARNING AND MATCHING

End-effector's trajectories are kinematically equivalent to the joint trajectories. However the joint trajectory representation is not orientation invariant, and hence not suitable for the intended purpose.

Since the end-effector's coordinate frame consists of both position and orientation, trajectories of each of them are discussed separately as follows.

Positional Trajectory

Conventionally, space curves may be described by their tangent vectors, principal normal vectors and/or the positive valued curvatures at points on the curves [2]. But tangent and normal vectors are also not rotation invariant. Although the curvature does not fall into this category, the disadvantage is its proportionate to the derivative of tangent vector with respect to the arc length. Thus it is very sensitive to the disturbance and hence not a suitable representation in the trajectory matching process. Consequently, it is desirable to search for a new representation of the space trajectories for learning and matching.

It is noticed that the relative locations among points on a space curve do not change when the curve is translated and/or rotated [3,4]. When the curve is scaled with a factor λ, their relative locations is scaled with the same factor λ. Thus the information about the relative locations among points on the trajectory is chosen to be its representation. based on this observation, the process of trajectory learning and matching proceeds as follows. When a new trajectory starts to appear, the process of comparing it against those existing skillful trajectories that are in memory starts immediately. Suppose that the entire time duration of the new trajectory that has appeared is $[0, t_c]$, while that of the stored trajectory being compared is $[0, t_{sc}]$, $t_{sc} \geq t_c$. Select three different past time instants t_o, t_m and t_l with $0 \leq t_o < t_m < t_l < t_c$. The three corresponding points on the new trajectory are
$$P_o = P(t_o) = [x_o, y_o, z_o]^T = [f(t_o), g(t_o), h(t_o)]^T,$$
$$P_m = P(t_m) = [x_m, y_m, z_m]^T = [f(t_m), g(t_m), h(t_m)]^T \text{ and}$$
$$P_l = P(t_l) = [x_l, y_l, z_l]^T = [f(t_l), g(t_l), h(t_l)]^T.$$
They are all known. Let t_p be a future time instant, $t_p \in (t_c, t_{sc}]$. Corresponding to t_p, the point
$$P = [x_p, y_p, z_p]^T = [f(t_p), g(t_p), h(t_p)]^T$$
has not appeared on the new trajectory yet.

The points on the skillful trajectories at time instant t_p, t_o, t_m and t_l, are denoted by P_s, P_{so}, P_{sm}, and P_{sl} respectively. All these four points are known.

Assume that for $t \in [t_o, t_l]$, the portion of new trajectory has matched that of the skillful trajectory. It is desirable to have a method of predicting the point P on the new trajectory based on the information on three known points P_o, P_m and P_l on the new trajectory and four known points P_{so}, P_{sm}, P_{sl} and P_s on the skillful trajectory. If the predicted point P matches the physical point P when it appears on the new trajectory at time t_p, then the new trajectory matches the skillful one in memory for $0 < t \leq t_p$.

Counstruct a new coordinate system $o' - x'y'z'$ for the new trajectory with the conditions:

(1) origin o' is at P_o,

(2) $\overrightarrow{P_o P_l}$ is in the positive x' direction,

(3) P_m is on the (x', y')-plane, and

(4) projection of P_m onto y'-axis is positive.

Figure 1 depicts the $o' - x'y'z'$ coordinate system. Then, with reference to this coordinate system, $P_o = \begin{bmatrix} 0 & 0 & 0 \end{bmatrix}^T$, $P_m = \begin{bmatrix} x'_m & y'_m & 0 \end{bmatrix}^T$, $P_l = \begin{bmatrix} x'_l & 0 & 0 \end{bmatrix}^T$ and $P = \begin{bmatrix} x'_p, y'_p, z'_p \end{bmatrix}^T$. The $o' - x'y'z'$ system relates to the original $o - xyz$ system through a translation $\begin{bmatrix} x_o, y_o, z_o \end{bmatrix}^T$ and then a rotation

$$R = \begin{bmatrix} r_{11} & r_{12} & r_{13} \\ r_{21} & r_{22} & r_{23} \\ r_{31} & r_{32} & r_{33} \end{bmatrix}$$

so that

$$\begin{bmatrix} x_l & x_m & x_o \\ y_l & y_m & y_o \\ z_l & z_m & z_o \end{bmatrix} = \begin{bmatrix} 0 & 0 & x_o \\ 0 & 0 & y_o \\ 0 & 0 & z_o \end{bmatrix} + R \begin{bmatrix} x'_l & x'_m & 0 \\ 0 & y'_m & 0 \\ 0 & 0 & 0 \end{bmatrix}$$

or,

$$\begin{bmatrix} x_l & x_m & 0 \\ y_l & y_m & 0 \\ z_l & z_m & 0 \end{bmatrix} = \begin{bmatrix} r_{11}x'_l & r_{11}x'_m + r_{12}y'_m & 0 \\ r_{21}x'_l & r_{21}x'_m + r_{22}y'_m & 0 \\ r_{31}x'_l & r_{31}x'_m + r_{32}y'_m & 0 \end{bmatrix}.$$

The left side contains known quantities, but the right side are unknown values. Since R is orthogonal and each column vector is a unit vector, then, when o' coincides with o,

$$x'_l = \sqrt{x_l^2 + y_l^2 + z_l^2}$$

$$x'_m = \frac{x_l x_m + y_l y_m + z_l z_m}{x'_l}$$

$$y'_m = \sqrt{x_m^2 + y_m^2 + z_m^2 - x'^2_m}$$

So that

$$r_{11} = \frac{x_l}{x'_l}, \quad r_{21} = \frac{y_l}{x'_l}, \quad r_{31} = \frac{z_l}{x'_l},$$

$$r_{12} = \frac{x_m - r_{11}x'_m}{y'_m}, \quad r_{22} = \frac{y_m - r_{21}x'_m}{y'_m},$$

$$r_{32} = \frac{z_m - r_{31}x'_m}{y'_m}, \quad r_{13} = r_{21}r_{32} - r_{31}r_{22},$$

$$r_{23} = r_{31}r_{12} - r_{11}r_{32} \quad r_{33} = r_{11}r_{22} - r_{21}r_{12}.$$

Thus R is determined. It relates any point $\begin{bmatrix} x, y, z \end{bmatrix}^T$ in $o - xyz$ system to the same point $\begin{bmatrix} x', y', z' \end{bmatrix}^T$ in $o' - x'y'z'$ system by

$$\begin{bmatrix} x & y & z \end{bmatrix}^T = \begin{bmatrix} x_o & y_o & z_o \end{bmatrix}^T + R \begin{bmatrix} x' & y' & z' \end{bmatrix}^T.$$

Since the new trajectory matches the skillful one for $t \in [t_o, t_l]$, then

$$\frac{x'_l}{x'_{sl}} = \frac{y'_l}{y'_{sl}} = \frac{z'_l}{z'_{sl}} = \frac{x'_m}{x'_{sm}} = \frac{y'_m}{y'_{sm}} = \frac{z'_m}{z'_{sm}}.$$

If the new trajectory is a scale-down (or –up) version of the skillful trajectory with a scale factor λ, then the above ratios equal λ, and hence λ is determined. When $\lambda=1$, the two trajectories have the same size.

If, for $t_p \in (t_c, t_{sc}]$, the new trajectory segment would match the skillful trajectory segment, then

$$\frac{x'_p}{x'_{sp}} = \frac{y'_p}{y'_{sp}} = \frac{z'_p}{z'_{sp}} = \lambda$$

would be true. By the relation

$$\begin{bmatrix} x_p & y_p & z_p \end{bmatrix}^T = \begin{bmatrix} x_o & y_o & z_o \end{bmatrix}^T + R \begin{bmatrix} x'_p & y'_p & z'_p \end{bmatrix}^T$$

$$= \begin{bmatrix} x_o & y_o & z_o \end{bmatrix}^T + \lambda R \begin{bmatrix} x'_{sp} & y'_{sp} & z'_{sp} \end{bmatrix}^T,$$

the computed $\begin{bmatrix} x_p & y_p & z_p \end{bmatrix}^T$ would be the future point on the new trajectory at time t_p. If the physical point $P = \begin{bmatrix} x_p & y_p & z_p \end{bmatrix}^T$ when it appears on the new trajectory at time t_p is indeed equal to the computed $\begin{bmatrix} x_p & y_p & z_p \end{bmatrix}^T$ shown above, then the two trajectories do match for $0 < t \leq t_p$.

In the above discussion, the relation between the end-effector's coordinate frame and the coordinates of the goal is fixed. If, however, the relation is altered, the matching process between the new and the stored skillful trajectories requires the consideration of appropriate translation and rotation. With the coordinates of the goal as a reference, the new trajectory should be transformed consecutively by a rotation $\hat{R}$ first, then a scaling λ, and a translation p_o. Let P be the position vector in the end-effector's coordinate frame, and $\tilde{P}$ be the same after the above mentioned transformation, then

$$\tilde{P} = \lambda \hat{R} P + p_o.$$

In this transformation, λ has been determined previously. An empirical method of determining $\hat{R}$ and p_o can be performed as follows. If the coordinates of the goal remains unchanged but the home location (position and orientation) of the end-effector is altered, then lead the end-effector physically from its original home location to its new home location. Consequently $\hat{R}$ and p_o can be computed from the changes of all the joint variables (angles and/or displacements) through kinematics relations. On the other hand, if the end-effector's home location remains the same but the goal's coordinates are altered, then lead the end-effector physically from the goal's original coordinates to its new coordiantes to measure the changes of all the joint variables for the purpose of computing $\hat{R}$ and p_o.

Alternatively, $\hat{R}$ may be determined using an orientation matrix in the end effector's coordinate frame when it reaches the goal at a fixed orientation. Note that the oirientation matrix is orthogonal and its inverse equals its transpose. Let R_E be that matrix before the relative location between the end-effector's coordinate frame and the coordinates of the goal is altered. Let $\widetilde{R}_E$ be the corresponding orientation matrix after the relative location is altered. Then $\widetilde{R}_E = \hat{R} R_E$ so that

$$\hat{R} = \widetilde{R}_E \left(R_E\right)^{-1} = \widetilde{R}_E \left(R_E\right)^T.$$

Once $\hat{R}$ is known, then take any test position vector P_T from a fixed point on the end-effector to a fixed point on the goal when they both are at their original locations, and the corresponding position vector $\widetilde{P}_T$ after the relative locations of the end-effector and goal are altered, so that one may compute

$$p_o = \widetilde{P}_T - \lambda \hat{R} P_T.$$

Finally the computed $\begin{bmatrix} x_p & y_p & z_p \end{bmatrix}^T$ for identification and matching is modified as

$$\begin{bmatrix} \widetilde{x}_p & \widetilde{y}_p & \widetilde{z}_p \end{bmatrix}^T = \lambda \hat{R} \begin{bmatrix} x_p & y_p & z_p \end{bmatrix}^T + p_o.$$

Orientational Trajectory

Note that the orientation matrix is not affected by scaling or translation. Therefore only the rotation should be considered. Let the orientation matrix of the end-effector in its coordinate frame at time t be $R_E^0(t)$. If a rotation $\hat{R}$ is operated on the orientational trajectory, the rotated orientation matrix of the end-effector at time t becomes $\widetilde{R}_E^0(t) = \hat{R} R_E^0(t)$. Then for any two time instants t_1 and t_2, $\widetilde{R}_E^0(t_1)$ and $\widetilde{R}_E^0(t_2)$ represent the rotated orientation matrices with reference to the base coordinates at t_1 and t_2, respectively. Again these matrices have orthogonality property. Consequently the orientation at t_2, with reference to the coordinates corresponding to t_1 is

$$\left[\widetilde{R}_E^0(t_1)\right]^T \widetilde{R}_E^0(t_2) = \left[\hat{R} R_E^0(t_1)\right]^T \hat{R} R_E^0(t_2) = \left[R_E^0(t_1)\right]^T R_E^0(t_2).$$

Thus

$$\widetilde{R}_E^0(t_2) = \widetilde{R}_E^0(t_1) \left[R_E^0(t_1)\right]^T R_E^0(t_2).$$

Let $R_E^0(\sigma)$ and $R_E^0(s)$ be the orientation matrices at time σ and s, respectively, of the skillful trajectory in memory for $t_o \leq \sigma \leq t_l$ and $t_o \leq s \leq t_l$. During that time interval, suppose that the rotated orientation of the new trajectory (created by the moving robot and object) $\widetilde{R}_E^0(s)$ matches the skillful one. Then

$$\widetilde{R}_E^0(s) = \widetilde{R}_E^0(\sigma) \left[R_E^0(\sigma)\right]^T R_E^0(s).$$

For $t \in (t_c, t_{sc}]$, $R_E^0(t)$ is in memory and known, but $\widetilde{R}_E^0(t)$ does not appear yet. If the orientation $\widetilde{R}_E^0(t)$ of the new trajectory matches that of the stored skillful one, then

$$\widetilde{R}_E^0(t) = \widetilde{R}_E^0(\sigma) \left[R_E^0(\sigma)\right]^T R_E^0(t)$$

must be satisfied. This condition serves as the matching criterion for the orientational trajectory.

This matching condition requires the comparison of matrices on both sides of the equation which is time consuming. The matrices, which essentially express the consecutive rotations about the three orthogonal axes, is also the representation of a rotation of an angle θ about an axis **k** in the cartesian space [5]. By converting the rotation matrix to these two paramenters (one scalar and one vector), the comparasion requires much less time.

EXPERIMENTS

Experiments were performed with a Zebra-Zero robot manipulator. It is chosen for its size and strength so that the risk of injuring the human co-worker is low should an unexpected accident occur.

The robot has a forces/moments sensing wrist. Besides the controller for the robot, the system is controlled by a computer with

CPU:	Intel 486 DX2-66
System:	DOS
Programming:	TC++ 3.1

The description of the general experiments is stated in the INTRODUCTION. The goal is to move a 12"×22" wooden board, which has two hocks at its top edge, from some location to a fixed frame, which has two rings held by short chains. The human-robot team is required to hang the board under the frame by threading the hocks into the rings. Figure 2 depicts the set-up of the experiments.

The CPU time for computing one representing point on the trajectory is 0.05 ms for position and 0.19 ms for orientation. The time required for searching and matching the stored skillful trajectory is not exactly linearly proportional to the number of trajectories under comparison. They are tabulated as:

Number of Skillful Trajectories	Time Required
1	0.015 ms
2	0.027 ms
3	0.038 ms
5	0.063 ms
10	0.121 ms
50	0.500 ms

The communication between the human and the robot is done in two different ways. The human commands are delivered by tapping the end-effector. The robot's responses for acknowledging the human's messages are vocal. The commands include "open gripper", "close gripper", "begin", "continue", "compliance", "set start position", "stop", "ready", etc.

A three-minute segment of video tape showing the experiments was produced and included in the 1999 IEEE International Conference on Robotics and Automation.

As one expected, different human co-workers gave different performances. It was observed that human with more experience on the experiments yielded more smooth trajectories, and required less time to complete the experiments. No noticeable time delay was observed in robot's response/action to human's command.

CONCLUSION

The three-point representation of space curves is translation/rotation/scaling invariant. The learning and matching process by comparing the position vector and orientation matrix of the newly appeared point on the new trajectory is efficient in that the computing time is short, which makes the real-time operation possible.

Fine motion at the neighborhood of the goal location is still under investigation for matching-accuracy improvement. Human command by tapping the end-effector is workable. Experiments on voice command as a replacement are in progress. Difficulties caused by vocal accents of different co-workers are expected.

REFERENCES

[1] O.M.Al-Jarrah and Y.F.Zheng, "Arm-Manipulator Coordination for Load Sharing Using Compliant Control", Proc. 1996 International Conference on Robotics and Automation, April 22-28, 1996, Minneapolis, Minnesota, pp. 1000-1005.

[2] D.J.Struck, "Lectures on Classical Differential Geometry", 2nd edition, Addison-Wesley, 1950, Chapter 1.

[3] J.Y.S.Luh and Shuyi Hu, "Real-Time Trajectory/Profile Learning for Robots in Human-Robot Interactions", Proceedings of the 1997 IEEE International Conference on Robotics and Automation, 1997, 901-906.

[4] J.Y.S.Luh and Shuyi Hu, "Real-Time Orientaion-Invariant Trajectory Learning in Human-Robot Interactions", Proc. 1998 IEEE/RSJ International Conference on Intelligent Robots and Systems (IROS), pp. 918-923.

[5] Richard Paul, Robot Manipulators, 1981, MIT Press, pp. 25-34.

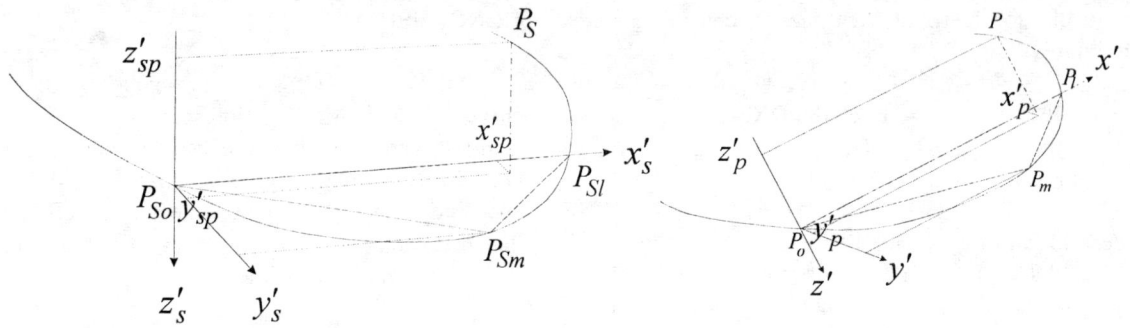

Skillful Trajectory   New Trajectory

Figure 1

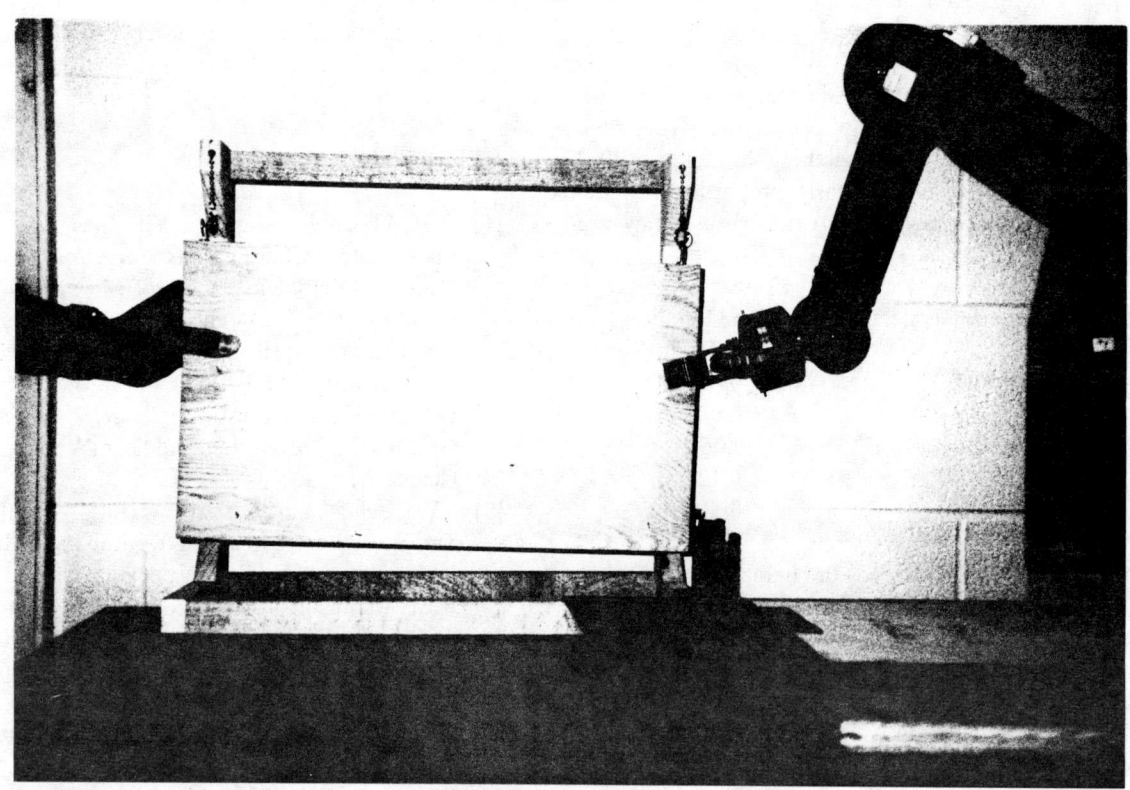

Figure 2

Emotional Communication Between Humans and the Autonomous Robot Which has the Emotion Model

Tetsuya OGATA *and* Shigeki SUGANO

Department of Mechanical Engineering, Waseda University
3-4-1, Okubo, Shinjuku-ku, Tokyo 169-8555, Japan
Phone: +81-3-5286-3264, Fax: +81-3-5272-0948
E-mail: {ogata,sugano}@paradise.mech.waseda.ac.jp
URL:http://www.sugano.mech.waseda.ac.jp/wamoeba/index.html

ABSTRACT

This study discusses the communication between autonomous robots and humans through the development of a robot which has an emotion model. The model refers to the internal secretion system of humans and it has four kinds of the hormone parameters to use to adjust various internal conditions such as motor output, cooling fan output and sensor gain. We surveyed 126 visitors at '97 International Robot Exhibition held in Tokyo, Japan (Oct. 1997) in order to evaluate psychological impressions of the robot. As a result, the human friendliness of the robot was confirmed and some factors of the human-robot emotional communication were discovered.

1.INTRODUCTION

In recent years, interactive simulation games have become very popular. Human beings enjoy communication with machinery which is not controlled by operators but behaves autonomously in ways which the observers cannot anticipate. These machines will be successful in the future.

Robot hardware and virtual agents have been developed to date using the above considerations. Bates developed a virtual agent that mimics the behavior of living things using animation [1]. SONY proposed "Robot Entertainment," and developed a pet-robot has with 4 legs [2]. Moreover, there is some research which has applied robot behavior to human-machine interface. F.Hara proposed "Active Human Interface (AHI)," and made a robot which makes the facial expressions to indicate the conditions at the machinery [3].

Most of these robots or agents have original feeling models. The feeling model not only improves the autonomy of the robot but also affects the smoothness of human-robot communication. In general, feeling models are structured based on the taxonomy of feeling in psychology, and is constructed using a static rule such as finite-state-automatons etc. However, there is no common theory on the classification of feeling, because all theories depend on individual human experiences.

R.Pfeifer noted that the simple behavior of robots can be interpreted as "feelings" by observers [4] referring to the "Fungus Eater" proposed by M.Toda [5]. He regards feelings as emergent phenomena.

We focused on the "internal secretion system" as a biological mechanism which generates the body conditions recognized as feelings. Research shows that feelings accompany various bodily changes caused by the autonomic nervous system and the internal secretion systems, e.g. tension in muscles, shrinkage of pupils, and rise in temperature, etc. We call these objective phenomena an "emotion" in order to distinguish the feelings experienced subjectively. In other words, "emotion" is the subordinate concept of feelings.

This study proposes a new model of feelings based on the basic concept, "emotion," in order to realize the human-robot emotional communication for use in human-machine interface fields. To date, the model of the internal secretion system using fuzzy set theory has been proposed and implemented in an autonomous robot, WAMOEBA-1R (Waseda Amoeba, Waseda Artificial Mind On Emotion BAse) [6]. This paper describes the three functions of an independent autonomous robot WAMOEBA-2, which improved version of WAMOEBA-1R. The first function is hardware for the human-interface, the second is a method for emotional expressions generated by the proposed model of the internal secretion system, and the third is a behavior generation algorithm based on human psychology in communication. Moreover, the factors in emotional communication of robots are described by referring to the results of a questionnaire at the '97 international robot exhibition held at Tokyo Big-Site, Japan, Oct., 1997.

2. AUTONOMOUS ROBOT: WAMOEBA-2

WAMOEBA-2 has been developed to investigate a robot emotional communication with humans, and it focused on the emergence of intelligence for communication. It is a completely independent robot which has the batteries and the control computers (Pentium 233 MHz and 200 MHz) into the body, respectively. The dimensions are 983 (L) x 862 (W) x 1093 (H) [mm], and the weight is around 60 [kg]. The characteristics of the robot hardware are 1) communication functions, 2) model of the internal secretion system, and 3) behavior generation algorithm.

2.1 COMMUNICATION FUNCTIONS

From the perspective of being "human friendly," the arrangement of the sensors and the motors refers to the morphologies of the creatures represented by human beings. WAMOEBA-2 has two arms for emotion expressions using gestures. In addition, it has two LCDs on the head and the chest to indicate internal conditions. It is important for WAMOEBA-2 to detect visual, sound, and tactile information, because human beings can generate this information directly. WAMOEBA-2, therefore, has various sensors, e.g. four ultrasonic range sensors, two color CCD cameras, two microphones (right and left), two touch panels (the head and the chest), and eight tactile sensors on the vehicle. The motor chair is adapted to the vehicle part, so that WAMOEBA-2 can acquire a wide activity area, so it does not have to stay indoors. WAMOEBA-2 can driven for about 30 [min.] using the battery in the motor chair.

2.2 MODEL OF INTERNAL SECRETION SYSTEM

The original characteristic of WAMOEBA-2 is internal mechanism architecture for modeling the internal secretion system of humans. The internal secretion system controls the entire state of the living organism using hormones. It is thought that, for robot hardware, these organisms correspond to the control mechanisms of electric power consumption and circuit temperature, etc. WAMOEBA-2 receives the battery voltage and the driving current. Moreover, using temperature sensor IC, it can

Fig. 1 WAMOEBA-2 (Photgraph)

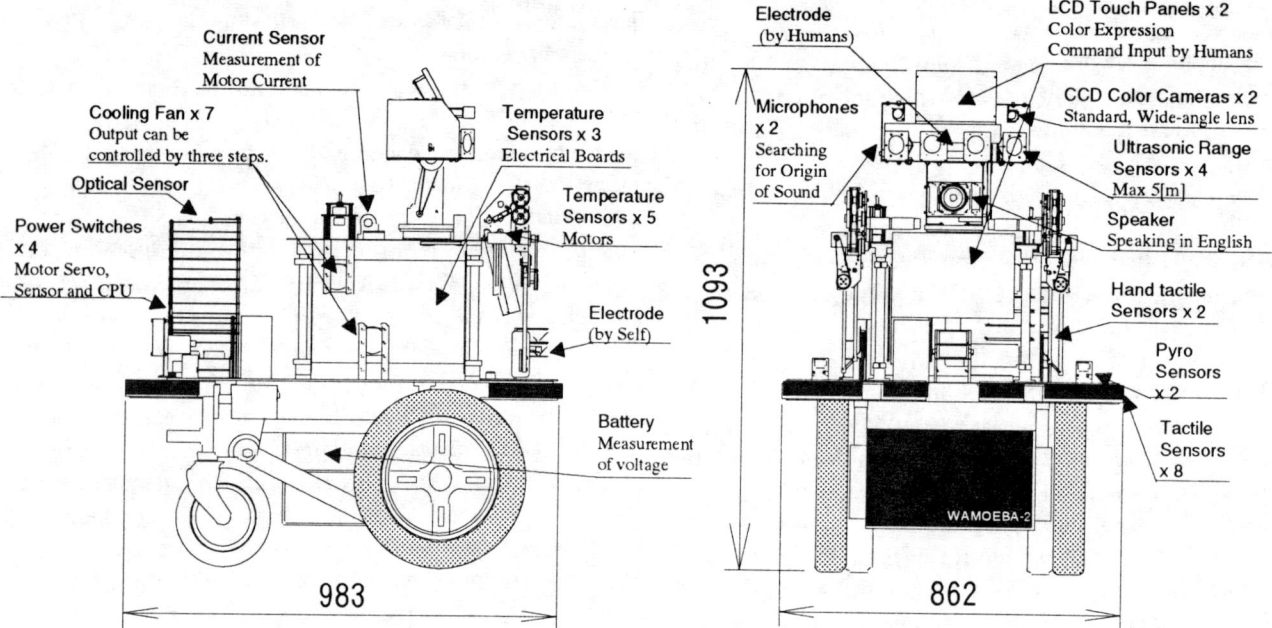

Fig. 2 Configuration of WAMOEBA-2 functions

acquire eight positions of temperature, which are the motors (the head, the neck, the shoulder, the elbow, and the motor chair) and the circuits (the image processing board and A/D boards etc.) It can control the output of the cooling fans, and the switch the power supply of each motor on or off by itself.

WAMOEBA-2 controls these internal mechanism using the hormone parameters calculated by the original algorithm "Self-Preservation Evaluation Function." This function is one kind of fuzzy set membership function which converts sensor input into the evaluation values of durability (breakdown rate) of robot hardware between 0-1. Each function consists of two sigmoid function in order to simulate the properties of human senses, and it has one minimum value which stands for the best state for the self-preservation of the robot. When this value is close to zero, the state of self-preservation is positive, and if this value gets close to one, the state is negative. WAMOEBA-2 has seven kinds of self-preservation evaluation functions that correspond to eleven internal and external sensors. The shapes of these functions are chosen depending on the basic hardware specs. For example, the evaluation function of the voltage of battery is shown in Fig.3. In this case, the shape of the function is decided depending on the lowest voltage of the circuit drive, the standard voltage of the battery. WAMOEBA-2 calculates the output of the hormone parameters using total value P of all self-preservation evaluation functions in every program cycle as follows;

$$\frac{dH_i}{dt} = \alpha_i \cdot (P - P^{th}) + \beta \cdot \sigma_i\left(\frac{dP}{dt}\right) + \gamma \cdot (H_i - H^{th}) \quad (1)$$

where, α_i, β and γ are coefficients that correspond to the potential, the change quantity, and the stabilization. α_i represents if the hormone output is continuous. P^{th} and H^{th} represent the standard values about P and H. $\sigma_i(x)$ is the sigmoid function which suppresses dP/dt within the range of 0-1. There are four kinds of the hormone parameters [$H1$ to $H4$] corresponding to four conditions: if the evaluation value P is positive or negative (mood), and

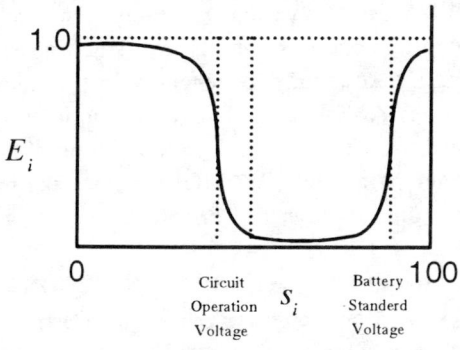

Fig.3 Evaluation Function of Voltage Battery

Table 1 Affects of the Hormone Parameters of WAMOEBA-2

		H1	H2	H3	H4
Actuator Output		Up	Down	Down	Up
Cooling Fan Output		Down	Up	Up	Down
CCD Camera Viewing Angle		Decrease	Increase	Increase	Decrease
Ultrasonic Sensors Sensing Area		Narrow	Wide	Wide	Narrow
Sound	Volume	Up	Down	Down	Up
Sound	Speed	Up	Down	Down	Up
Sound	Loudness	Down	Down	Up	Up
LCD Color		Red	Blue		Yellow

Table 2 Outline of the Affects of a Hormone model

Radical Unpleasantness	cause	Bumper switches, Ultra-sonic range sensors (radical approach)
Radical Unpleasantness	expression condition	Decrease of the viewing angle, Increase of the motor output, Red color expression on the head LCD and Low voice
Unpleasantness	cause	Temperature of the motors and the electrical circuits, Ultra-sonic range sensors
Unpleasantness	expression condition	Increase of the viewing angle, Decrease of the motor output, Blue color expression on the head LCD
Pleasantness	cause	Charge
Pleasantness	expression condition	Decrease of the viewing angle, Decrease of the motor output, Yellow color expression on the head LCD and High voice

if P changes dynamically or not (arousal). These hormone parameters affect many hardware conditions such as sensor gains, the motor outputs, the temperatures of the circuits and energy consumption in parallel. The affects of each hormone are decided referring to the physiology [7] shown in Table1. Table2 shows examples of the correspondences between the morphologies of the emotional expressions caused by the hormone parameters, however, these are not fixed but are changed by the mixture condition of the four hormone parameters.

2.3 MOTOR AGENT

Next, the methodology by which WAMOEBA-2 generate its behavior for emotional communication should be discussed. We considered some algorithms of robot behavior as follows.

A conventional model-based robot behaves based on the environmental model given a priori. It requires accurate sensor input, an optimal environment, and a large amount of calculation. R.Brooks proposed a "behavior based approach" [8] which has some behavior modules that correspond to the tasks. Each behavior module does not require higher level behavior planning.

However, there is a limitation on the variety of behavior, because there are only the combinations of each behavior module which are fixed a priori. In communication, humans can easily forecast robot behavior through the experiments, and would then become tired. It is an extremely difficult problem to design a behavior module for communication with humans.

We thought that the behavior should be described not at the level of the "task" but at "motor activity" in order to generate the diversity of the behavior. R.Brooks has developed a humanoid robot, "Cog," which moves its arms based on oscillators in the motors [9].

As the first step, we proposed the behavior reflection system of WAMOEBA-2 named "motor agent." In the motor agent algorithm, each motor acquires all sensor information and other motor drive conditions through the network in the robot hardware. Based on this information, each motor decides its actions autonomously. Motion command M_i of motor i is calculated as follows:

$$a_i = \sum_p w_{ip}^s S_p + \sum_{j \neq i} w_{ij}^m M_j \quad (2)$$

Here, the input value of sensor p is defined as S_p (ex. the sound volume and the visual moving area etc.), the output of motor j is M_j, and the activity of motor i is a_i. The commands for motor i are generated using the absolute value and the positive and negative values of a_i. In this architecture, the morphology of the behaviors depends on weight value w, in which descriptions are not explicit. The initial value of w depends on the physical arrangement of the motors and the sensor; i.e., w is a large value when the distance between the sensors and the motors is small. In this stage, w is adjusted by a designer who observes the behaviors of WAMOEBA-2. Table 3 shows some connections, which have a large w value between the sensors and the motors.

Table 3 The Outline of "Motor Agent" in WAMOEBA-2

Motor Part	Sensor Input
Head (2 DOF)	Vision (Moving Area, Color Area), Sound, Ultra-sonic Range sensors, Bumper switches
Shoulder (x 2)	Degree of Head Motor, Hand touch sensors
Elbow (x 2)	Degree of Shoulder Motor, Hand touch sensors
Vehicle (2 DOF)	Degree of Neck Motor, Vision, Sound, Ultra-sonic Range sensors, Bumper switches, Hand touch sensors

Table 4 Speaking Words of WAMOEBA-2

WORDS	Stimulus
"Hello, my name is WAMOEBA-2."	Start of the program
"Auch!"	Bumper Touch Sensors
"Delicious"	Charging
"I'm sleepy."	Increase of H2 Hormone Parameter
"Good Morning!"	Decrease of H2 Hormone Parameter
"Good bye!"	End of the program

Based on only implicit expressions, the "motor agent," WAMOEBA-2 generates the behavior using the whole body, e.g. imitation of the movement area, the sound origin, and avoidance behavior, etc. Although the w value is fixed a priori, the hormone parameters affect all motor activity: a_i. That is, hormone parameters H_1 and H_4 (which cause the exciting conditions in WAMOEBA-2) increase the a_i, and, on the other hand, H_2 and H_3 (which cause the calm conditions) decrease a_i. As the results, the forms of the behavior change as follows: for example, when WAMOEBA-2 is quiet due to the low battery etc., it imitates moving object by the head only. In the exciting conditions however, it follows the object using the arms and/or the vehicle. The behaviors based on motor agent are expected to surprise observers.

3. COMMUNICATION USING WAMOEBA-2

3.1 EMOTIONAL EXPRESSION OF WAMOEBA-2

This chapter outlines the communication between WAMOEBA-2 and humans in actual experiments. Humans can communicate with WAMOEBA-2 by hand waving, clapping, calling, touching the tactile sensors etc. WAMOEBA-2 makes various reactions such as approaching, escaping, making sounds, eye tracking, and arm stretching. These actions are generated by the motor agent. In addition, WAMOEBA-2 changes the motion speed, volume/speed/loudness of sounds, and color output on LCD by hormone parameters. WAMOEBA-2 speaks some English according to the behavior shown in Table 4.

3.2 CHARACTERISTICS OF COMMUNICATION

Most conventional emotion models have a limited ability to communicate with humans. Usually, a human being observes and judges the expressions of the emotion model, and the recognition rate is the evaluation of the model. In communication between WAMOEBA-2 and humans, there is no scenario like this. The psychological impressions at humans change dynamically according to the behavior of the robot and/or the humans. The characteristics of WAMOEBA-2 communication are as follows.

1) Adaptability in real world

Since WAMOEBA-2 is an independent and behavior based robot, it is not necessary to standardize its environment. Moreover, there is no limitation for humans in the standing position and/or motion, etc.

2) Diversity of the ways to communicate

Human beings can communicate without special interface tools. Moreover, neither "words" nor "gestures" etc. for communication are specified, and preliminary knowledge is not needed.

3) Development of communication

Communication is developed according to the behavior of humans and WAMOEBA-2 in real-time. There is no "story" and/or "scenario" set beforehand by a designer.

It is believed that the "freedom degree" mentioned above (where humans are not restrained in communication with robots) is an important element in order to realize robot-human emotional communication.

4. EXPERIMENTS

In order to evaluate WAMOEBA-2, demonstrations and questionnaires experiments were done at the '97 international robot exhibition held in "Tokyo Big Site," Oct. 1997, where about 150,000 people attended. The objective of this experiment is to investigate the general psychological impressions of human beings that communicated with the autonomous robot.

To make the questionnaires, we referred to research [10] about an evaluation of human subjectivity from robot behavior. The questionnaires consisted of free comments and a seven-step evaluation of five adjective pairs. Answers were obtained from 126 visitors, which involve many men (88%), engineers (17%) and people in there twenties (50%). The reason for the about percentages is the exhibition focused on industrial robots.

Experimental procedure was as follows: first, we explained the research background and the functions of WAMOEBA-2 to the testee operating the WAMOEBA-2. Next, we had the testee actually make interactions, and did the questionnaire. Fig.4 shows the results of the adjective pairs in the questionnaire, "Goodwill or Hostile", "Hate or Like", and "Untame or Tame". These adjective pairs are related to the emotional impressions of testees. WAMOEBA-2 was given the impressions of "Goodwill", "Like", and "Tame" generally.

About the free comments, since there were no orders to

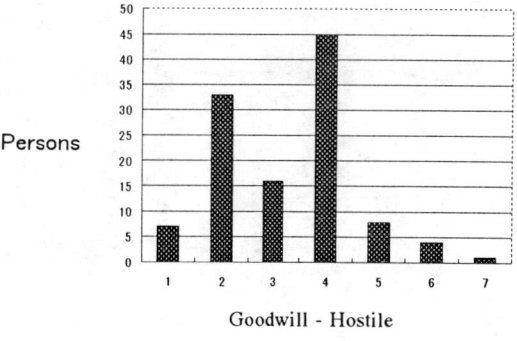

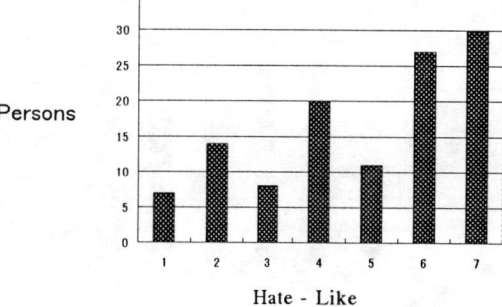

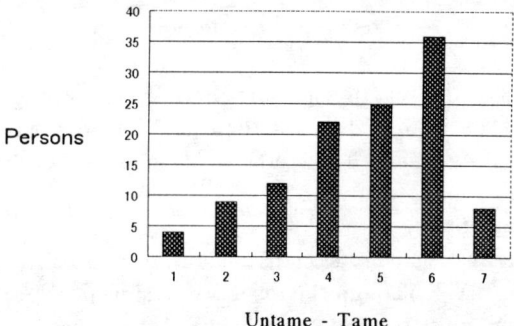

Fig.4 Results from Questionnaire 1

testees, the contents of the comments involve various topics. We categorized these topics as follows. The ratio for the entire answer is described in parentheses.

a) External design (20.2 %)
b) Mechanical / Electrical hardware structure (9.3 %)
c) Software involving Algorithms (39.5 %)
d) Objective and/or methodology of the research, etc. (21.7%)

As the results of the arrangement of the questionnaires, the factors that affects human psychological impressions are as follows:

1) Robot Design and Form

There was a resemblance between the robot and living things caused by the arrangement of the sensors and motors influences the psychological impressions of the testee. Moreover, the testee drew various opinions about the design and the covering of WAMOEBA-2.

2) Behavior

There were many testees who were interested in the

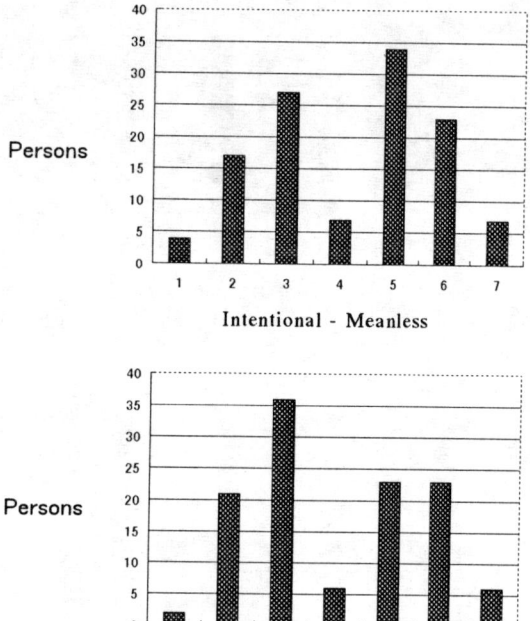

Fig.5 Results from Questionnaire 2

changes of the reaction of the robot. There were more testees who try to understand the English words, which WAMOEBA-2 said. The importance of language was clarified again.

3) Responsibility

Many testees were sensitive to the responsibility of WAMOEBA-2. Though the robot was designed to search for the moving area and sound origin that are thought to represent human beings, there was a lot of sensory noise in the exhibition hall. It is important for robots, who communicate with human beings to detect the stimulus caused by human intentions.

In addition, Fig.5 shows the results of the adjective pairs, "Intentional - Meanless" and "Animal like - Machine like." The answers here showed a tendency to separate into two groups largely. This tendency was not related to whether the testee is male or female, the age of the testee, and the job. These questionnaire results are thought to show the important and inevitable problem when human beings interprets these concepts and/or words in robotics.

5. CONCLUSION AND FURTHER RESEARCH

This study examined the emotional communication between an autonomous robot and human beings. The autonomous robot WAMOEBA-2 has been developed by implementing the emotion model by referring to the internal secretion system, and the motor agent which generates various behavior based on implicit descriptions in the network. The questionnaire examined the human psychological impressions from robot and done at the '97 International robot exhibition, and the human-friendliness of the robot was confirmed. Further, from the questionnaire, some factors and problem in the robot's emotional communication were clarified, represented by "humans freedom of degree,"

The communication between an autonomous robot and human being discussed in this paper is a new concept which is different from conventional robot-human communication which only focused on the cost and the efficiency. However, the autonomy and/or the intelligent of robots will advance, and they will be used in homes and hospitals, etc. There, the communication discussed here will be indispensable, and has many problems, e.g. methods to maintain human-friendliness and the human empathy to robots etc.

In the future, the emotion model of WAMOEBA-2 should be optimized from the perspective of the kinds of hormone parameters and their effects. Further, there are many comment concerning with the "will" of WAMOEBA-2 in the questionnaire. It is thought that WAMOEBA-2 should equip more intelligent functions for human-robot emotional communication.

REFERENCES

[1] J.Bates: "The Role of Emotion in Belivable Agents," Communications of the ACM, pp.122-125, 1994.
[2] M.Fujita and K.Kageyama: "Robot entertainment," in Proc. of the 6th Sony Research Forum, pp.234-239, 1996.
[3] F.Hara and H.Kobayashi: "Computer graphics forexpressing robot artificial emotions," in Proc. of IEEE Int. Workshop on Robot and Human Communication, pp.155-160, 1992.
[4] R.Pfeifer: "Emotions in Robot Design," in Proc. of IEEE International Workshop on Robot and Human Communication (ROMAN'93), pp.408-413, 1993.
[5] M.Toda, "The Urge Theory of Emotion and Cognition., School of Computer and Cognitive Sciences," Chukyo Univ., Technical Report, 1994.
[6] S.Sugano, T.Ogata, "Emergence of Mind in Robots for Human Interface - Research Methodology and Robot Model," IEEE Int. Conf. on Robotics and Automation, pp.1191-1198, 1996.
[7] R.Nieuwenhuys, J.Voogd, and Chr.Huijzen: "The Human Central Nervous System-A Synosis and Atlas," Springer-Verlag, 1988.
[8] Brooks,R.A. : "A robust layered control system for a mobile robot, IEEE Journal of Robotics and Automation, RA-2, April, and pp.14-23, 1986.
[9] R.A.Brooks, C.Breazeal, R.Irie, C.C.Kemp, M.Marjanovic, B.Scassellati and M.M.Williamson, Alternative Essenses of Intelligence, American Association for Artificial Intelligence (AAAI), 1998.
[10] T.Nakata, T.Sato, T.Mori, and H.Mizoguchi: "Generation Familiarity by Robot Behavior toward a Human Being," Journal of Japan Robotics Society, Vol.15, No.7, pp.1068-1074, 1997. (in Japanese)

Development of Human Symbiotic Robot: WENDY

Toshio MORITA Hiroyasu IWATA Shigeki SUGANO

Dept. of Mechanical Engineering, Waseda University
3-4-1 Ookubo, Shinjuku, Tokyo 169-8555, Japan
E-Mail: morita@paradise.mech.waseda.ac.jp
URL: http://www.sugano.mech.waseda.ac.jp

Abstract

An objective of this study is to find out design requirements for developing human symbiotic robots, which share working space with human, and have the ability of carrying out physical, informational, and psychological interaction. This paper mainly describes design strategies of the human symbiotic robots, through the development of a test model of the robots, WENDY (Waseda ENgineering Designed sYmbiont).

In order to develop WENDY, mobility and dexterity of a humanoid robot Hadaly-2, which was developed in 1997, are improved on. The performances of WENDY are evaluated by experiments of object transport and egg breaking, which requires high revel integration of whole body system.

1. Introduction

The development of the next generation robots capable of supporting human labor through the use of several interaction channel are expected to be a measure against labor shortages in aging societies. This study calls these robots "human symbiotic robots," in distinction from conventional industrial robots, because they must share behavioral space, working space, and thinking space with human.

Design requirements of the human symbiotic robots are similar to that of the humanoid robots [1], since the final target is almost the same. Recently, several studies on humanoid robots have been conducted, however, these results are not applicable to the human symbiotic robots directly, because these researches mainly focus on bipedal walking problems. In the case of human symbiotic robots, the capability of appropriately adapting motion to humans while ensuring safety, must be given top priority.

According to the circumstances, the Humanoid Project at the Advanced Research Institute for Science and Engineering (RISE), Waseda University has developed Hadaly-1 and Hadaly-2 as prototypes of the human symbiotic robots. Among these robots, Hadaly-1 [2] was developed as an informational assistant robot in 1995. Hadaly-1 can realize informational interaction with human by combining with audio-visual information, voice dialogue, and gesture motion using manipulator.

Hadaly-2 was developed in 1997, from motivations for improving physical interaction ability of Hadaly-1 [3]. In designing Hadaly-2, an anthropomorphic shape is adopted, since the human symbiotic robots must work in human's living space which is designed to suit with daily life. Hadaly-2 can carry blocks in reference to the operator's demands, by using its informational and physical interaction abilities. From the development of Hadaly series, design requirements and system integration strategy were clarified.

Succeeding to these results, this study targets to construct a new human symbiotic robot named WENDY (Waseda ENgineering Designed sYmbiont) by improving each subsystem of Hadaly-2. First, this paper describes a design strategy of head, waist, vehicle, and finger tip in reference to remarks which are given by development of Hadaly-2. Then, the developed limb mechanisms are integrated into human symbiotic robot: WENDY. Finally, dexterity and mobility are evaluated by experimental tasks, such as picking up object on the floor and breaking eggs.

2. Design Strategy for Human Symbiotic Robots
2.1 Assurance of Safety and Operability

In this study, assurance of safety and operability is regarded as essential requirements for designing human symbiotic robot. The details are listed as follows.

(1) Safety

Safety measures of human symbiotic robot can categorized into two phases, such as impact safety and after collision safety. In order to ensure impact safety, this study has proposed a double safety measure [4] which combining viscoelastic structure and reflex movement. It is the reason that Hadaly-2 employs the shock absorbing cover and the passive impedance controlled manipulators.

In addition to the measurement of impact safety, adaptive motions after collision by coordinating whole limb systems (such as a arm, a body and a vehicle) are important for carrying out cooperative tasks with human. For realizing safety after collisions, capabilities of object recognition by using visual and tactile information, and coordinated motion of whole body are required.

(2) Operability

Operability of human symbiotic robots can be categorized into mobility and dexterity. As regards mobility, motion space in both horizontal and vertical plane is required, because human's living space has unevenness. However, most of conventional mobile manipulators are designed in consideration of mobility in horizontal plane.

Moreover, dexterity of manipulators is also required to operate daily necessities. Assistance of cooking, for example, requires ability of handling of cookware and foods. In order to realize operability and mobility, abilities of object recognition and coordination of whole body motion are also effective.

Considering the requirements of safety and operability, Hadaly-2 is improved with focus on the uniformity of the whole body system. The following chapters describes a new design strategy of human symbiotic robots.

2.2 Mechanisms of Head and Neck

A head subsystem of human symbiotic robots should act as not only visual information sensor, but also interface for collaborative tasks with human. For example, it was confirmed that operators infer the working condition of Hadaly-2 from the motions of its eyes. So, this study improves resolution of visual information processing, by developing a new head mechanism which employs two color CCD cameras for detecting objects and humans. An assembly drawing of the seven D.O.F. head mechanism is shown in figure 1.

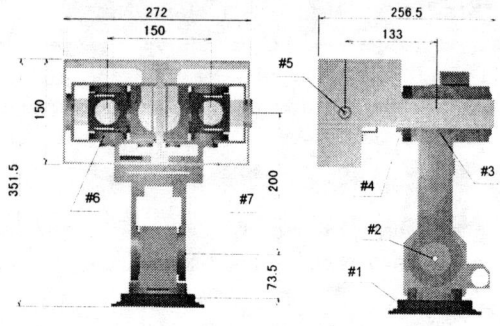

Figure 1 Head Mechanisms

The head mechanism consists of a three D.O.F. eye mechanism and a four D.O.F. neck mechanism. In regard to the eye part, locations of the CCD cameras are important, because the distance between two eyes affects both image processing and human's impression. Two cameras are separated as possible, under the conditions that the vision system can catch objects which are located at 10 [cm] from the face. And, in order to observe objects which are located on the floor, the neck mechanism employs two degrees of freedom in the direction of flexion and extension.

2.3 Mechanisms of Waist

Waist mechanisms are utilized for expanding the working spaces in vertical plane. Moreover, evacuating motions using redundancy of whole body are effective for suppressing the impact forces, which generate in the case of collision between humans and robots. In order to realize the motions stable, a waist mechanism which employs three degrees of freedom is developed.

The waist mechanisms consist of one rotation axis (joint #1) and two pitching axes (joint #2 and #3). By rotating joint #2 to the opposite direction of joint #3, the waist can bend without stumbling over. Consequently, the end point of the WAM-10 manipulator [2] which is attached to the body mechanism can reach to the point on the floor, at a distance of 260[mm] from the rotation axis of the body. An overview of the developed waist mechanism is shown in figure 2.

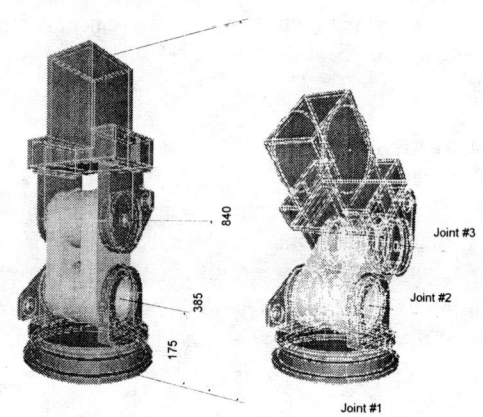

Figure 2 Waist Mechanisms

2.4 Mechanisms of Vehicle

Mobility in horizontal plane is necessary for human symbiotic robots to extend working space. In that case, however, stability of the mobile mechanisms is the most essential, because human-robot collaboration requires insurance of operator's safety. From this remarks, this study adopted wheeled vehicle mechanisms among many forms of the mobile mechanisms.

Figure 3 shows the assembly drawings of the developed two wheeled vehicle mechanisms. In designing the mechanisms, the eccentric casters with suspensions are located at each vertex of rectangle, since vehicle must climb over bumps without stumbled over.

The developed vehicle has enough power for dragging the cables which connects the internal actuators to the external controllers, by utilizing combination of high power motor (300[W], AC) and harmonic drive gear (reduction ratio 1/33). Specification of the vehicle is as follows; maximum velocity, 5[km/h]; maximum payload, 120[kg]; climbable height, 1[cm]; weight, 53[kg]; maneuvering error, 1[%].

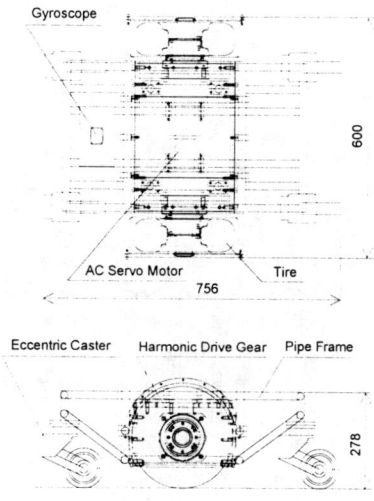

Figure 3   Vehicle Mechanisms

2.5 Mechanisms of Arm and Hand

Manipulators of the human symbiotic robots are utilized as interface between humans and robots, while carrying cooperative tasks. So, safety and adaptive manipulation is essential for constructing human symbiotic robots. In consideration of these problems, the authors have developed the WAM-10 (Waseda Automatic Manipulator #10), and design requirements for carrying out several tasks which involves physical interactions between robots and humans were clarified [5]. In this study, the hand mechanisms of the WAM-10 are improved by focusing on dexterity of pinching motion.

Human changes the poses of finger tips with adapting to the shapes of objects. This motion can be regarded as control of friction by adjusting pressure, which is shown in figure 4. Coefficient of adhesive friction μ is changeable by adjusting areas of contact, in reference to following equation [6].

$$\mu = \frac{As}{W} = \frac{A(s_0 + \alpha\ p)}{W} = \frac{s_0}{p} + \beta \qquad (1)$$

$$\theta = \tan^{-1} \mu \qquad (2)$$

where, A: contact Area, W: contact force, s, s_0: shearing stress, p: pressure, α, β: constants, and θ: acute angle of friction cone.

From these remarks, this study focuses on the design parameters of finger tips, such as shape, structure, material, and size. For continuously adjustment ability of contact conditions from point to area, curved shapes of the finger tip are designed by using rubber and bone. A nail is also attached to the finger tip, in order to realize extremely high pressure on the contact point. Figure 5 shows a structure of finger tips.

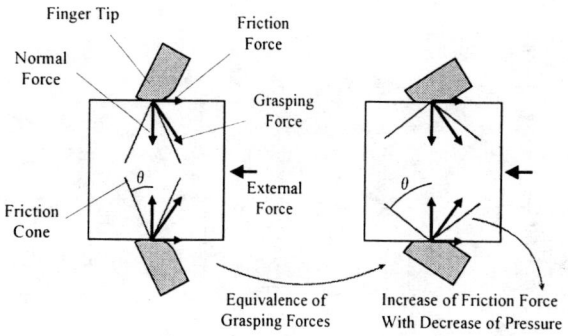

Figure 4   Adjustment of Friction by Pressure Control

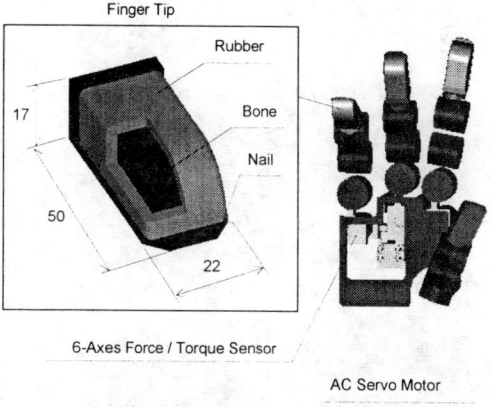

Figure 5   Structure of Finger Tip

3. Construction of WENDY
3.1 Mechanical Hardware and Functions

By combining the head, the waist, the vehicle and the arm-hand system, a limb mechanism of WENDY (Waseda ENgineering Designed sYmbiont) is constructed. An assembly drawing and an overview are shown in Figure 6, and 7. Specifications are arranged in table 1.

WENDY improves Hadaly-2 in the human interaction ability, since whole body motion control and visual-auditory information processing are implemented into design concept. In the system, image processing part is utilized for measuring the relative position from the base of WENDY to the target objects, by using stereo vision. In the process, shapes of the objects is recognized by visual images (512×480[pixel], 8×3[bit]RGB) through GPB-J board, which is on the market. In addition, possibilities of speech recognition is also realized by using AUDIO-98VOICE board.

The limb mechanisms of WENDY has the possibility of handling small sized object which is located in human's living space. And it can turn around on the position while maintaining the end point positions of manipulators.

Table 1 Specifications of WENDY

DOF	Finger	13×2
	Arm	7×2
	Neck	4
	Eye	3
	Waist	3
	Vehicle	2
	Total	52
Size [mm]	Height	1471.5
	Width	930.0
	Depth	756.0
Weight [kg]	Hand	2.0×2
	Arm	25.0×2
	Neck + Eye	8.0
	Waist	46.0
	Vehicle	53.0
	Total	161.0
Actuators	Neck, Eye, Lower Arm	DC Servo Motor×17
	Hand, Upper Arm, Waist, Vehicle	AC Servo Motor×35
	Compliance Control of Arm	DC Servo Motor×14
	Damping Control of Arm	Electromagnetic Brake×14
	Total	80
Sensors	Optical Encoder (Whole Body)	66
	Limit Switch (Arm, Waist)	17
	6-Axes Force Sensor (Hand)	8
	CCD Camera (Eye)	2
	Gyro Sensor (Vehicle)	1
	Microphone	1
	Total	95
Computers	Host	PC×1
	Arm-Hand Motion Control	PC×2
	Waist and Vehicle Control	PC×1
	Vision Processing	PC + GPB-J×2
	Voice Dialogue Processing	PC+AUDIO-98VOICE×1
	Total	7

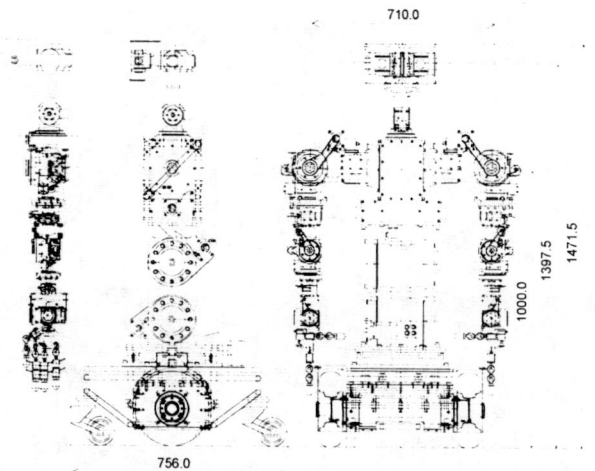

Figure 6 Assembly Drawing of WENDY

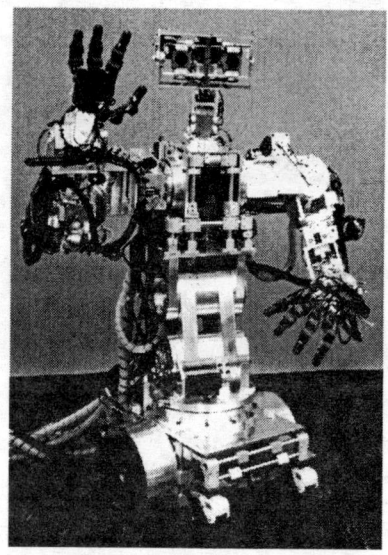

Figure 7 Overview of WENDY

3.2 Control Systems and Algorithms

For development of control systems of WENDY, control architectures of Hadaly-2 is succeeded. The control systems of WENDY is shown in figure 8.

In the system, the host computer generates behavior codes by using visual, and voice dialogue information, through CAAD (Creating Actuator Angle Data) operating system.

The CAAD operating system was developed for arm-hand coordination control of Hadaly-2. The system is similar to finite state machine or petri-net. Therefore, environmental conditions, such as scenarios of dialogues and arrangements of objects, must be previously measured and installed into the behavior models.

Each motion is created by using graphical user interface GL-Model, and implemented into the CAAD. The GL-Model is shown in figure 9. Then, the CAAD produces the limb motions in reference to the environmental situations, by combining the several motion patterns which involve gestures. Each body control system is received the behavior codes from the host computer through DP-RAM.

In this system, mobility are produced by combining straight line and rotation of the vehicle, and maneuvering errors are compensated by combination of the information from the vision and the gyro sensor.

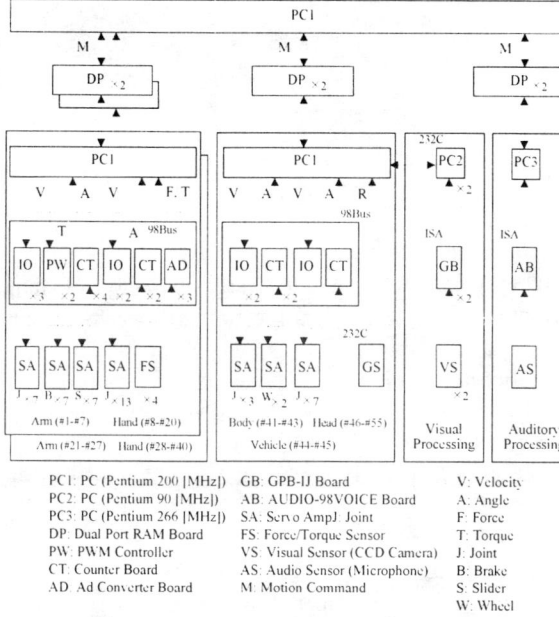

Figure 8   Control System of WENDY

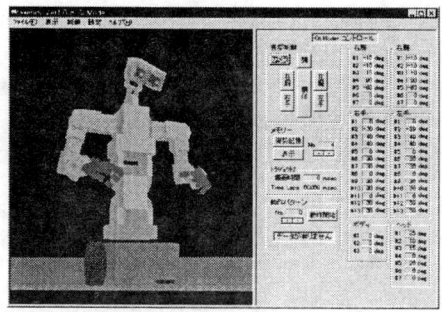

Figure 9   GUI for Motion Teaching (GL Model)

4. Evaluation Experiments
4.1 Image Processing and Head Motion

For evaluation of the head system, experiments on visual image processing and arm-head coordination are conducted, respectively. Figure 10 shows the experimental setup of WENDY.

In regard to the visual image processing, recognition of objects shape, and measurement of distance are adopted for evaluation items. Contents of the experiments are listed as follows;

(1) Mission: Search a target object from any number of objects, which are located on the chair. Then, measure distance between the object and the base of WENDY.

(2) Environment: Objects are located at 1.0[m] (x-axis) from the base. Background color is blue. Sizes, shapes and its color are previously defined. (object #1; sphere; white, object #2; cube; red, object #3; pyramid; yellow)

The experimental conditions and results are shown in figure 10 and 11. From the results, it is confirmed that the ratio of success of object recognition is approximately 90 [%], the measuring errors are 5 [cm] (z-axis) and 1 [cm] (x, y-axis), and the time for processing is 250 [ms].

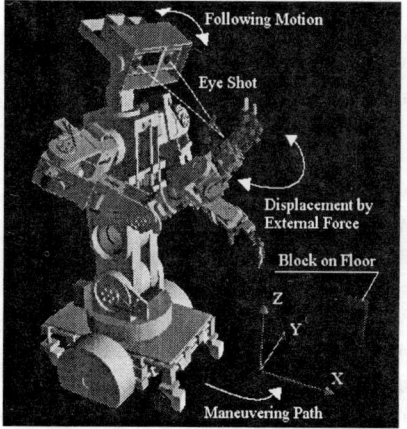

Figure 10   Coordination on Experiments

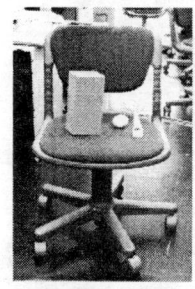 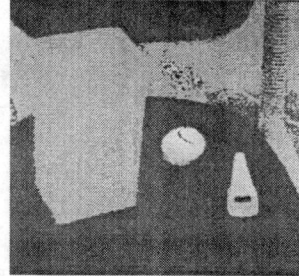

Figure 11   Experimental Results of Image Processing

Next, servo responses of the head mechanisms are evaluated by the head-arm coordination experiments. In the experiments, mechanical compliance coefficient of the manipulator is set to maximum, and operator shakes the manipulator. The applied external force is approximately 0.5 [Hz] sine curve. The eye shot of WENDY is controlled to follow the end point of the manipulator, by calculating DK of the arm and IK of the head. The distance between hand and eye is set to 1.0 [m], and the angle of joint #4 is 0.0 [deg]. The experimental result on x axis, is shown in figure 12. It is confirmed that fine following motions of head mechanisms are achieved by head-arm coordination.

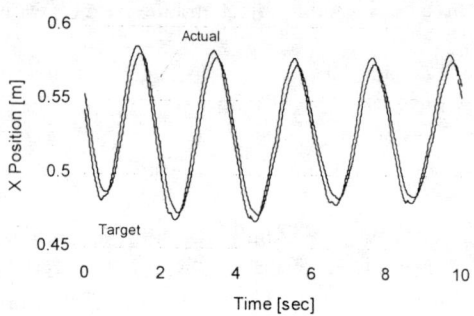

Figure 12 Following Motion of Head System

4.2 Mobility and Dexterity

To evaluate mobility and dexterity of WENDY, object picking up and egg breaking are selected for experimental tasks. In the object picking up task, WENDY approaches to the target object by using the vehicle, and picks the object up. Maneuvering path is generated from the object position, with combination of linear lines and curves. Size of the object is 8 [mm] in width, 10 [mm] in depth, 20 [mm] in height, and 100 [g] in weight. The block is located at 700 [mm] in front, and 300 [mm] to the right.

Figure 13 is the maneuvering path of the vehicle in the experiments. The ratio of success of picking up object is almost 100 [%]. This result shows that WENDY can realize accurate mobile manipulation by using whole body motion.

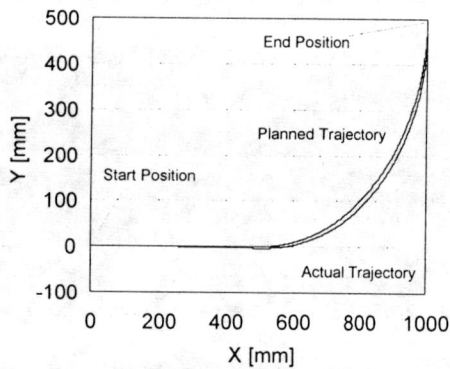

Figure 13 Results of Vehicle Trajectory Control

In regard to the egg breaking tasks, information from force sensor is not utilized, because the experiment mainly focuses on the evaluation of the mechanical characteristics of the finger tips. In the experiment, motion patterns are previously produced by the GL-Model. Figure 14 shows the experimental set up of egg breaking tasks. The ratio of success becomes 20 [%] by using the new finger tip, while the ratio of success is 0[%] by using the conventional hemispheric finger tips. This result shows that the pressure control method using finger tip is effective for improving the dexterity of handling.

Figure 14 Experiments of Egg Breaking

5. Conclusion

This paper proposes a design strategy of human symbiotic robots which has the possibilities for ensuring safety and operability. By using the strategy, a human symbiotic robot, WENDY is developed.

Effectiveness of the design strategy is confirmed from the results of several evaluation experiments, such as object picking up and egg breaking, which require high completeness of whole body system.

Acknowledgments

This study was conducted as a part of the Humanoid Project at the Advanced Research Institute for Science and Engineering (RISE,) Waseda University. This project has been supported by NEDO (New Energy and Industrial Technology Development Organization), and Grant in Aid for Scientific Research (B). The authors would like to thank for the cooperation.

References

[1] K. Hirai, M. Hirose, Y. Haikawa, and T. Takenaka, "The development of Honda Humanoid Robot," Proc. of ICRA'98, pp. 1321-1326, 1998.

[2] S. Hashimoto et al., "Humanoid Robot -Development of an Information Assistant Robot Hadaly-," Proc. of 6th Roman, pp.106-111, 1997.

[3] T. Morita, and S. Sugano, "Design and Control of Mobile Manipulation System for Human Symbiotic Humanoid: Hadaly-2," Proc. of ICRA'98, pp. 1315-1320, 1998.

[4] T. Morita, and S. Sugano, "Safety Materials and Control of Human-Cooperative Robots," Journal of Robotics and Mechatronics, Vol.9, No.1, pp.33-40, 1997.

[5] T. Morita, and S. Sugano, "Development and Evaluation of Seven-D.O.F. MIA ARM," Proc. of ICRA'97, pp.462-467, 1997.

[6] F. P. Bowden, and D. Tabor, "The Friction and Lubrication of Solids," Pt. II, Oxford University Press, 1964.

Reflexive Behavior of Personal Robots Using Primitive Motions

Li Xu
Department of Electrical Engineering
Zhejiang University
Hangzhou, 310027, P.R. China

Yuan F. Zheng
Department of Electrical Engineering
The Ohio State University
Columbus, OH 43210, USA

Abstract— Personal robotics is a new and attractive use of robotic technologies. In this paper we study one of its important topics – real-time motion planning of personal robots. We propose to use primitive motions and their combination to make this possible. A reflexive motion control scheme is proposed to activate appropriate primitive motions for a desired complex motion. Experiments results are presented to show the effectness of the proposed scheme.

I. Introduction

PERSONAL robotics is a new and attractive area in robotics research. The idea of personal robots [12] was proposed with the increasing interests in healthcare robotics and welfare and service applications [4]-[6]. Although there still lacks a widely accepted definition, it is generally agreed that a personal robot should have such features as safety, friendliness, capacity of interaction, flexibility, adaptability, and modularity in autonomy. The development of personal robots provides quite a few challenging topics to the robotics community, one of which is real-time motion planning in unstructured environments. Some of the related fields are teleoperation [1], robotic extender [2], human-skill learning [3], and human-robot coordination [7]-[10].

The objective of this work is to develop a real-time motion planning scheme for personal robots. Unlike industrial manipulators, personal robots usually work in an unconstrained environment, where off-line motion planning is not applicable. We propose to use primitive motions and their combination to make possible the real-time motion planning. The complex motion of a personal robot is decomposed into a series of primitive motions, each of which has a short trajectory and is activated by a human command or a stimulus to the robot sensors. By using the primitive motion, the human needs only to issue discrete and sparse commands to obtain a useful trajectory rather than steering the robot all the time.

The primitive motion was addressed by several researchers. Mujtaba [14] compared several different trajectories including a Bang-Bang, a quintic, a cosine, a

---
This work was conducted while L. Xu was at The Ohio State University. The material is based on work supported by the National Science Foundation under grant IRI-9405276.

sine on ramp, and a critical damping motion. Scheuer and Fraichard [17] and Pinchard [19] studied straight segments connected with tangential circular arcs of minimum radius. Continuous-curvature paths were also studied using quintic polynomials and clothoid [18], cubic spiral [15], intrinsic spline [16], clothoid and anticlothoid [16], polar spline [19], and B-spline [20]. However, all the research efforts were focused on the forms of primitive motions and their concatenation [13], and no literature was found which dealt with the combination of primitive motions.

We propose to utilize the force and moment signals exerted by the human to serve as stimuli. The voice command is also a good means to direct the robot. However, this will increase the complexity because it involves the speech recognition problem. Furthermore, the force and moment signals have strength and can be used to determine the magnitude of the stimulus which will produce a corresponding reflexive motion.

The idea of reflexive motion control is adopted here to direct the personal robot. In biological systems, muscular activities are regulated via reflexes. A reflex response is a distinguished involuntary action in response to a specific stimulus. Study of the neuromuscular system has attracted researchers to mimic some of its aspects in robotics [22][23]. Reflexive control has been used for collision avoidance [24] as well as arm-manipulator coordination [10].

In this work we consider the personal robot as a single point, which could represent a robotic arm installed on a mobile platform or the platform itself. This is to focus the current study on the primitive motion and will not affect the validity of the result. When the dimension of the robot becomes an issue such as in collision avoidance, its physical shape can easily be taken into consideration.

In this paper we will first study the pattern of the primitive motion in Section 2. The approach of the reflexive motion control is introduced in Section 3, and the experimental results are presented in Section 4. A conclusion is given in the final section.

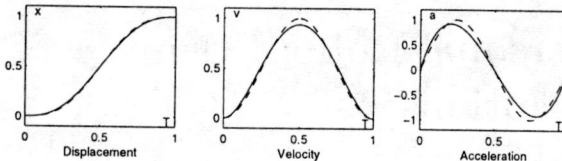

Fig. 1. Patterns of the primitive motion where the solid line represents the path of the point-to-point motion of the human hand as introduced in [11], and the dashed line is the sinusoidal representation.

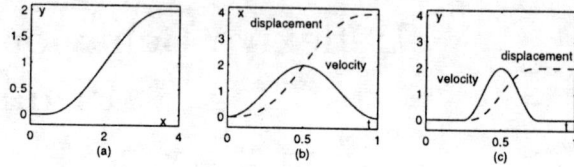

Fig. 2. Combining primitive motions to form a curved path: (a) the path of the robot hand in the two-dimensional space, (b) the primitive motion in the x direction, and (c) the primitive motion in the y direction.

II. PRIMITIVE MOTIONS

The primitive motion plays a very important role in the operation of personal robots. In determining the pattern of the primitive motion, it is necessary to consider the following two aspects. Firstly, the primitive motions should have a unified pattern; otherwise, the computer will have too large a memory space and be busy searching for the desired pattern in real-time. For the same reason, the pattern should be simple yet powerful so that multiple primitive motions can be combined to form complex paths. Secondly, the human compatibility should be taken into consideration. That is, the person should feel comfortable when working with the robot. For this reason, the pattern should be similar to that of the human motion. Ideally, it closely resembles the latter.

A fundamental motion pattern of the human hand is experimentally identified in [11]. It is shown that the point-to-point motion of a human hand is approximately straight with bell-shaped tangential velocity profiles (Fig. 1). As expressed in the following equation, this pattern is obtained by optimizing a smoothness index of the hand motion, which is equated as to minimize the mean-square jerk. The jerk is mathematically defined as the rate of change of acceleration [11].

$$x(t) = x(0) + [x(0) - x(T)](15\tau^4 - 6\tau^5 - 10\tau^3) \quad (1)$$

where $x(0)$ and $x(T)$ are initial and final hand position coordinates, respectively, T the time period for a point-to-point motion, and $\tau = t/T$. Note that $x(t)$ is a general description of the trajectory and can be in any of the six directions of the Cartesian space. It is shown in [11] that the measured hand path of human subjects is very close to the predicted path using the above equation. Because this model is experimentally verified by the human subject, it should be suitable for the personal robot with respect to human compatibility. The minimal jerk of the motion is a plus because rapid changes of acceleration are not desirable for the robotic joints. For the same reason, the rotation of the robot hand may also use this pattern. By using (1), one only needs to specify $x(T)$ and T to activate a primitive

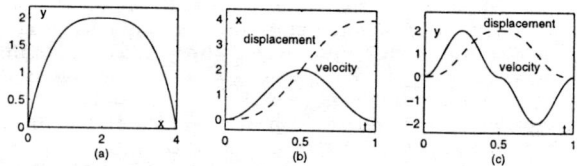

Fig. 3. Combining primitive motions to form a more complex path: (a) the path of the robot hand in the two-dimensional space, (b) the primitive motion in the x direction, and (c) the primitive motion in the y direction.

motion. Note that $x(T)$ is related to the magnitude of the primitive motion, and T affects the speed.

Another pattern considered here is the sinusoidal function. The velocity profile of this pattern can be described by the following equation:

$$v(t) = \frac{1}{T}[x(T) - x(0)](1 - cos2\pi\tau). \quad (2)$$

This velocity profile can be translated into the following position profile corresponding to (1):

$$x(t) = x(0) + [x(T) - x(0)](\tau - \frac{1}{2\pi}sin2\pi\tau). \quad (3)$$

It can be seen from Fig. 1 that the sinusoidal function produces similar position, velocity and acceleration patterns as Eqn. (1). However, the sinusoidal function offers a few advantages: *a)* it is mathematically concise to express, and *b)* it is convenient to fuse multiple sinusoidal functions for forming complex motions. The second advantage is important for the personal robot. Consider that the human wants the robot to execute a curved motion. He or she may do this by issuing a *move right* command first and then *move up*. The *move up* motion starts before *move right* is completed. Assuming that *right* and *up* be along with the x and y axes, respectively, the resulted path of the robot hand is shown in Fig. 2. By alternating $x(T)$ and T of individual primitive motions and time delays between consecutive ones, even more complex trajectories can be formed, such as the one shown in Fig. 3, where there are two primitive motions in the y direction, positive and negative, respectively, and one in the

x direction. By using this concept of combination, the primitive motion becomes a powerful building block of complex motions.

III. Reflexive Motion Control

Reflex [21] is a simple, unlearned, yet specific behavioral response to a specific stimulus. Reflexes are exhibited by virtually all animals from protozoa to primitives. Generally, the strength of a reflexive response depends upon the magnitude of the stimulus. We propose to apply this idea of reflexive control to the real-time motion planning of personal robots. That is, each of the primitive motions carried out by the personal robot is trigged by a stimulus to the robot, and the stimulus may be exerted by the human.

To implement the reflexive motion control, we need to formulate the instantaneous sensation that triggers the primitive motion and then to describe the primitive motion which is activated by the stimulus.

The sensation signal is the force and moment exerted to the manipulator. When the human wants the personal robot to move *forward*, *backward*, *left*, or *right*, he would normally push or pull the robot for a short period of time; therefore, the stimulus received by the robot is impulsive rather than constant and may be expressed as

$$F_\delta(t) = \int_t^{t+\eta} F_m(\tau) d\tau \quad (4)$$

where $F_\delta(t)$ is the stimulus received by the the robot at time t, F_m is the force and moment exerted to the robot, and η is the period of time for the robot to sample force and moment signals for each stimulus.

The relation between the reflexive motion or the deviation of the robot position, X_f, and the sensation signal may be described by

$$X_f = H(F_\delta - F_{th}) \quad (5)$$

where H is the coefficient matrix, and F_{th} a thresholding value. The above equation reflects the fact that the strength of the reflex response is proportional to the magnitude of the stimulus.

When the sinusoidal velocity pattern is employed, we may rewrite Eqn. (2) as,

$$v(t) = v_m(1 - cos2\pi\tau) \quad (6)$$

where $v_m = [x(T) - x(0)]/T$ is the magnitude of the velocity profile, and the maximum velocity is $2v_m$. Let

$$X_f = X(T) - X(0). \quad (7)$$

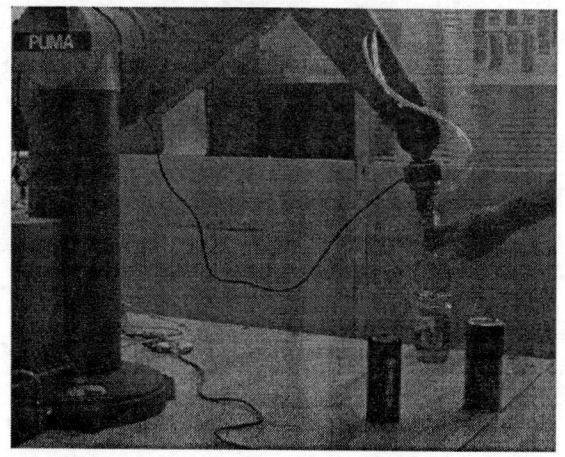

Fig. 4. The robot arm is guided by the human to avoid obstacles.

It is obvious that the deviation caused by this pattern is $v_m T$. Therefore, we have two parameters to determine for each primitive motion. One is the velocity magnitude v_m, and the other is the time duration T.

It is important that the human should feel comfortable when working with the personal robot. One of the factors which affects the compatibility is the speed with which the robot moves. If the robot moves too fast, the human may feel dangerous because he may be hit by the robot if he issues an incorrect command. For this reason, we set a limit, v_{lim}, to the velocity magnitude of individual primitive motions.

Based on the above consideration, the two parameters may be determined as follows,

$$v_m = a\sqrt{|F_\delta - F_{th}|} \quad (8)$$

$$T = b\sqrt{|F_\delta - F_{th}|} \quad (9)$$

where a and b are the weighting coefficients. Larger force and moment signals will result in larger $|v_m|$ and T. However, if $|v_m| > |v_{lim}|$, we set $|v_m|$ equal to $|v_{lim}|$.

If the person pushes or pulls the robot for a period of time longer than η, a second stimulus may result. If the time is longer than 2η, a third stimulus may be formed, but this will not become valid if the first one is not completed and if the three primitive motions are along the same direction. This is to prevent the robot from moving too fast.

By using this proposed approach, the robot can understand the intention of the human and work with the person friendly.

IV. Experimental Results

We have conducted experiments to demonstrate and verify the proposed schemes. The system consists of a supervisor computer, a Unimation PUMA 560 robot and a multi-axis force/torque sensor system [25]. The

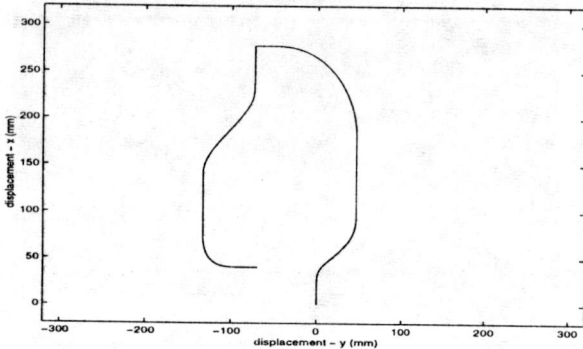

Fig. 5. Four combined motions formed by primitive motions.

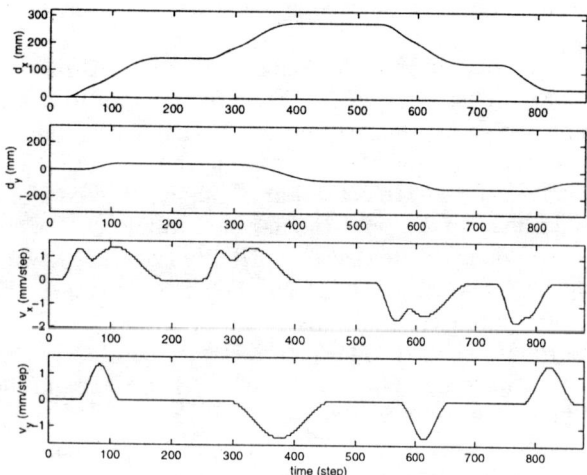

Fig. 6. Decomposed displacement and velocity profiles contributing to the four combined motions as shown in Fig. 5. The upper two figures represent the displacement in x and y axes, respectively, and the lower two are the corresponding velocity profiles.

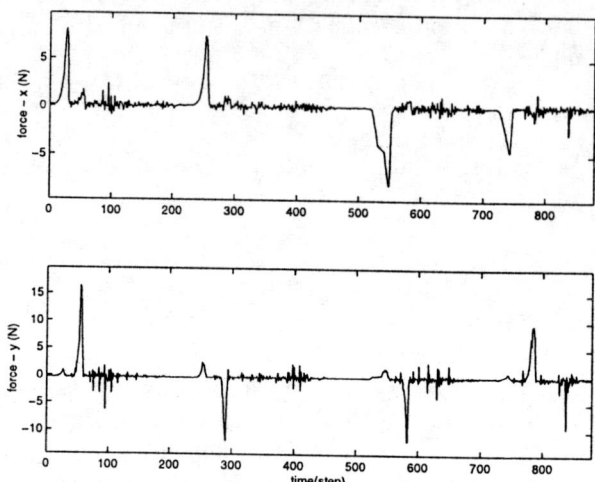

Fig. 7. Force signals corresponding to the combined motions as shown in Fig. 5.

joint controller of the robot arm receives a command every 28 ms, which is the same sampling period for the force/torque sensor system. It is also referred to as one $step$. The force/torque sensor is mounted between the gripper and the mounting flange of the robot wrist. The communications between the robot, the computer and the sensor controller are via serial ports. The robot is expected to understand the human intention when it is pulled or pushed and to move accordingly. Fig. 4 illustrates the experiment system.

The experiments were carried out in a 3-dimensional space, that is, the robot could move in x, y and z directions. However, only the results in the x-y plane are presented here.

When the robot moves, the sensor signals are quite noisy due to the vibration. We have devised the following measures to distinguish the human command from the noise: a) A human command is composed by a package of force signals, the length of the package is proportional to the sampling time η. In each of the six axes, a package of signals of human intention should all be greater than a certain threshold and be monotonic or bell-shaped; otherwise, they are considered noise. If the signals in all axes are noise, this package is discarded; and b) For a valid package of signals, only one primitive motion is activated. By using the above measures, the robot can understand the human intention quite well in most cases. However, it sometimes moves in undesired directions with small displacement due to occasionally large magnitude noises.

The first experiment was to demonstrate how different paths could be generated by combining primitive motions along different axes with different starting time and different time durations. As shown in Fig. 5, the robot started from the origin and generated four combined motions with the forms of an S-$shape$, an arc, an S-$shape$, and an arc, and occurred between $step$ intervals $[0, 200]$, $[200, 500]$, $[500, 700]$, and $[700, 880]$, respectively. For the first combined motion, the human first pushed the robot for a relatively longer time along the $+x$ axis, and then the $+y$, thus activating two primitive motions along the $+x$ and $+y$ axes, respectively. Consequently, the primitive motion along the $+y$ started later but ended earlier than the one along the $+x$ axis. As a result, an S-$shape$ path was formed. For the second combined motion, the primitive motion along the $-y$ started later and ended later than the one along the $+x$ axis, thus generating an arc. The other two combined motions were formed accordingly. The detailed displacement and velocity profiles as well as force sensor signals can be found in Figs. 6 and 7, respectively.

The second experiment was conducted to show how

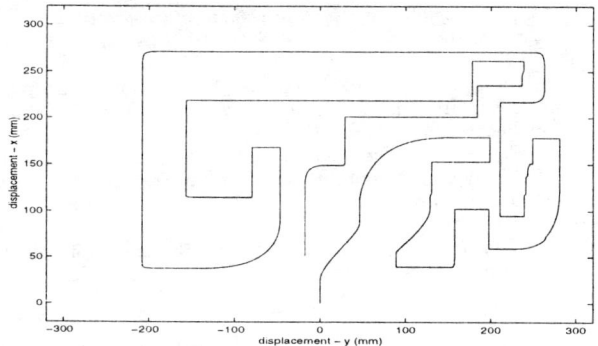

Fig. 8. A complex trajectory was formed when the person continuously pulled or pushed the robot.

the robot moves in the entire space when the robot was pulled or pushed continuously. As demonstrated in Fig. 8, the robot again started its motion at the origin, and ended at a spot very close to the starting point. The entire trajectory was composed by a series of primitive motions and their combination. Some primitive motions were activated before the previous ones were completed, thus forming smooth transitions, while some started after the previous one ended, thus forming a sharp turn when these two were along different axes.

The third experiment was to show how the robot could avoid obstacles and reach target points. In this case, there are three tea pots serving as the obstacles and a small metal piece serving as the target point, and the gripper was hung with a screw through a rope (Fig. 4). As displayed in Fig. 9, the robot first moved from the origin to the target point, lowered down to let the screw hit the metal piece, rose up again, and then moved around each tea pot. Finally it returned back to the target point and hit the metal piece again. The tea pots were approximately 75 mm in diameters, and the radius of the metal piece and the screw were 11 mm and 10 mm, respectively. The detailed displacement along the x and y axes and the corresponding velocity profiles were illustrated in Fig. 10. As can be seen in Fig. 11, the force signals were very noisy.

V. Conclusions

Reflexive behavior of personal robots using primitive motions was studied in this paper. We proposed to combine primitive motions to form complex trajectories. We also proposed to use the reflexive control to activate appropriate primitive motions. The primitive motion scheme is an important contribution to the development of personal robots. By using this scheme, the robot needs no constant steering while interacting with human beings. On the other hand, complex motion trajectories of the robot can still be generated by

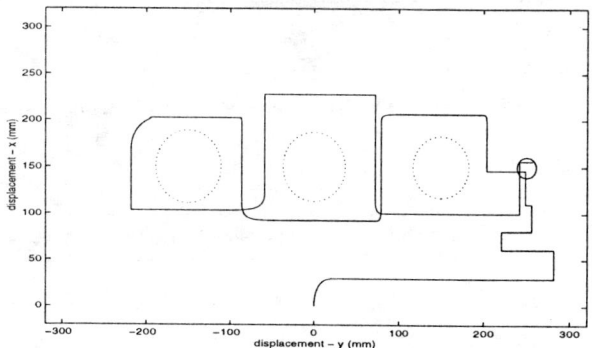

Fig. 9. Avoiding obstacles and reaching the target point. The circles in dotted lines represent the obstacles, and the one in solid line is the target.

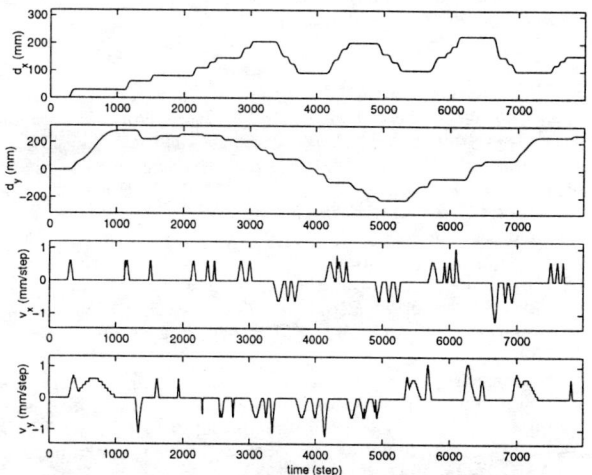

Fig. 10. Decomposed displacement and velocity profiles contributing to the trajectory as shown in Fig. 9.

the personal robots. The experimental results show that the proposed approach provides an effective and flexible way for the motion generation of of personal robots.

The proposed scheme is a realistic approach based on the current state-of-the-art in robotics. Recall that machine intelligence has been studied for a long time, but robust autonomous robots still remain a dream for researchers in the robotics community. On the other hand, the need for personal robots becomes more and more urgent by many sectors of the society. We hope the primitive motion scheme proposed in this paper represents a constructive effort in meeting that need.

References

[1] B. Bon and H. Seraji, "On-line collision avoidance for the ranger telerobotic flight experiment," *Proc. 1996 IEEE Int. Conf. on Robotics and Automation*, Minneapolis, MN, April 22-28, 1996, pp. 2041-2048.

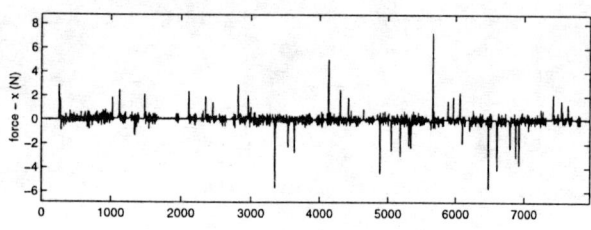
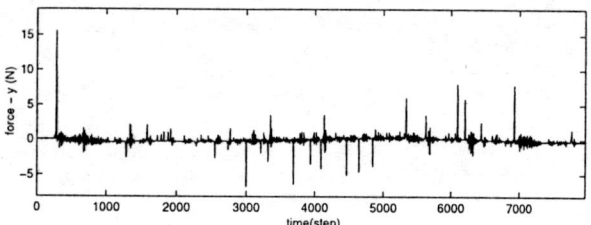

Fig. 11. Force signals corresponding to the trajectory as shown in Fig. 9.

[2] H. Kazerooni and M. Her, "The dynamics and control of a haptic interface device," *IEEE Trans. on Robotics and Automation*, Vol. 10, No. 4, August, 1994, pp. 453-464.

[3] C. Lee and Y. Xu, "Online, interactive learning of gestures of human/robot interface," *Proc. 1996 IEEE Int. Conf. on Robotics and Automation*, Minneapolis, MN, April 22-28, 1996, pp. 2982-2987.

[4] A. Agah and K. Tanie, "Human interaction with a service robot: mobile-manipulator handling over an object to a human," *Proc. 1997 IEEE Int. Conf. on Robotics and Automation*, Albuquerque, NM, April 20-25, 1997, pp. 575-580.

[5] S. Mascaro, J. Spano, and H. Asada, "A reconfigurable holonomic omnidirectional mobile bed with unified seating (RHOMBUS) for bedridden patients," *Proc. 1997 IEEE Int. Conf. on Robotics and Automation*, Albuquerque, NM, April 20-25, 1997, pp. 1277-1282.

[6] P. Fiorini, K. Ali and H. Seraji, "Health care robotics: a progress report," *Proc. 1997 IEEE Int. Conf. on Robotics and Automation*, Albuquerque, NM, April 20-25, 1997, pp. 1271-1276.

[7] K. Kim and Y. Zheng, "Human-robot coordination with rotational motion," *Proc. 1998 IEEE Int. Conf. on Robotics and Automation*, Leuven, Belgium, May 16-21, 1998.

[8] O. Al-Jarrah and Y. Zheng, "Intelligent compliant motion control," *Proc. 1996 IEEE Int. Conf. on Robotics and Automation*, Minneapolis, MN, April 22-28, 1996, pp. 2610-2615.

[9] O. Al-Jarrah and Y. Zheng, "Arm-manipulator coordination for load sharing using compliant control," *Proc. 1996 IEEE Int. Conf. on Robotics and Automation*, Minneapolis, MN, April 22-28, 1996, pp. 1000-1005.

[10] O. Al-Jarrah and Y. Zheng, "Arm-manipulator coordination for load sharing using reflexive motion control," *Proc. 1997 IEEE Int. Conf. on Robotics and Automation*, Albuquerque, NM, April 20-25, 1997, pp. 2326-2331.

[11] T. Flash and N. Hogan, "The coordination of arm movements: an experimentally confirmed mathematical model," *J. of Neuroscience*, Vol. 5, No. 7, July, 1985, pp. 1688-1703.

[12] P. Dario, et al, "Huamn-robot interface for welfare and services", *1997 IEEE International Conference on Robotics and Automation*, Workshop on Human-Centered Robotics, Albuquerque, NM, April 20-25, 1997.

[13] M. Brady, J.M. Hollerbach, T.L. Johnson, T.Lozano-Perez, and M.T. Mason (eds), *Robot motion : planning and control*, The MIT Press, 1982.

[14] M.S. Mujtaba, "Discussion of trajectory calculation methods", in *Exploratory study of computer integrated assembly systems*, Binford, T.O. et al, Stanford University, Artificial Intelligence Laboratory, AIM 285.4, 1987.

[15] Y. Kanayama and B.I. Hartman, "Smooth local path planning for autonomous vehicles", *Proc. 1989 IEEE Int. Conf. on Robotics and Automation*, Vol. 3, pp. 1265-1270.

[16] S. Fleury, P. Souerse, J.-P. Laumond, and R. Chatile, "Primitives for smoothing mobile robot trajectories", *Proc. 1993 IEEE Int. Conf. on Robotics and Automation*, Vol. 1, pp. 832-839;

[17] A. Scheuer and Th. Fraichard, "Planning continuous-curvature paths for car-like robots", *Proc. IROS'96*, Vol. 3, pp. 1304-1311;

[18] M. Zefran and V. Kumar, "Planning of smooth motions on SE(3). *Proc. 1996 IEEE Int. Conf. on Robotics and Automation*, Vol. 1, pp. 122-126;

[19] O. Pinchard, "Generalized polar polynomials for vehicle path generation with dynamic constraints", *Proc. 1996 IEEE Int. Conf. on Robotics and Automation*, Vol. 1, pp. 915-920;

[20] H. Ozaki and C.-J. Lin, "Optimal B-spline joint trajectory generation for collision free movements of a manipulator under dynamic constraints", *Proc. 1996 IEEE Int. Conf. on Robotics and Automation*, Vol. 4, pp. 3592-3597.

[21] McGraw-Hill, *Encyclopedia of Science & Technology*, No. 15, RAB-RYE, pp. 285-287, 8th edition.

[22] R. Tomovic, "Transfer of motor skills to machines", *Robotics and Computer-Integrated Manufacturing*, Vol. 5, No. 2/3, pp. 261-267, 1989.

[23] C.H. Wu, K.Y. Young, K.S. Hwang, and S. Lehman, "Voluntary movements for robotic control", *IEEE Control Systems Magazine*, Vol. 12, February, pp. 8-14, 1992.

[24] T.S. Wikman, M.S. Branicky, and W.S. Newman, "Reflexive collision avoidance: a generalized approach", *Proc. 1993 IEEE Int. Conf. on Robotics and Automation*, Altanta, GA, May 2-6, 1993, pp. 31-37.

[25] ATI Industrial Automation, "Installation and operation manual for intelligent multi-axis force/torque sensor system", Garner, NC, 1994.

# High Speed Grasping Using Visual and Force Feedback

Akio Namiki, Yoshihiro Nakabo, Idaku Ishii, and Masatoshi Ishikawa

Department of Mathematical Engineering and Information Physics
University of Tokyo
7-3-1, Hongo, Bunkyo-ku, Tokyo 113-8656, Japan

E-mail : namik@k2.t.u-tokyo.ac.jp
URL : http://www.k2.t.u-tokyo.ac.jp/~sfoc/

Abstract

In most conventional manipulation systems, changes in the environment cannot be observed in real time because the vision sensor is too slow. As a result the system is powerless under dynamic changes or sudden accidents. To solve this problem we have developed a grasping system using high-speed visual and force feedback. This is a multi-fingered hand-arm with a hierarchical parallel processing system and a high-speed vision system called SPE-256. The most important feature of the system is the ability to process sensory feedback at high speed, that is, in about 1ms. By using an algorithm with parallel sensory feedback in this system, grasping with high responsiveness and adaptiveness to dynamic changes in the environment is realized.

Key Words: *grasping, high-speed visual and force feedback, sensor fusion, hierarchical parallel architecture, multi-fingered hand-arm, active vision*

1 Introduction

A grasping motion is one of the most important processes for control of multi-fingered hands. To complete the grasping process multiple types of sensor information are needed such as visual information and force/haptic information, and grasping motion should be controlled by fusing this sensory information. Then the sensory information should be acquired at a rate as the cycle time of grasping control so that the grasping motion has responsiveness to dynamic changes in the environment.

In conventional research, a delay of visual feedback is a serious problem for realizing the above requirements. In most cases a CCD camera is used to acquire visual information, which requires over 33ms to acquire an image. As a result the acquired image is always behind the real world action and prediction is necessary to compensate for the delay. As research based on such an approach, Allen et al. realized grasping of a moving model train on stereo triangulation of optic-flow field [1], and Hong et al. realized grasping of a flying object using stereo vision [2]. In these researches it was assumed that motion of an object was known beforehand.

For this reason there was a problem in that the system could not complete a task in an unknown environment.

On the other hand, vision chip systems have recently been developed for high-speed visual processing [3, 4]. Because they have a parallel processing architecture in which each photodetector is directly connected to a corresponding processing element, visual processing is realized at a high rate [5, 6]. The use of such high-speed vision in manipulation can solve the problem of the delay of visual feedback and realize high responsiveness and flexibility in an unknown environment.

Against this background our objective is to develop a grasping system in which high-speed visual feedback by a vision chip system is fused with force/haptic feedback and to realize responsive and flexible grasping in the real world environment.

2 Architecture for High Speed Grasping

In this section a system architecture suitable to fuse sensor information is discussed by analyzing grasping process in the real world. There are two important features to realize grasping in the real world, as follows:

(A) Flexibility under multiple conditions

A grasping system should have flexibility to complete various tasks under various conditions. The grasping process should be suitably changed according to the grasping condition, for example an object's position, an object's shape and an object's motion.

To implement this, a hierarchical parallel processing architecture with several types of sensor is valid. Multiple types of sensory-motor fusion processing coexist in one system based on it. As a result, flexibility in multiple grasping environments is realized.

(B) Responsiveness to dynamic changes

In the real world, the grasping environment changes dynamically and is possible that the object moves at high speed and sudden accidents happen during grasping.

To overcome this, grasping control based on high-speed sensory feedback is effective. High-speed sensory feedback means to return feedback of external sensor information at a rate higher than the rate of control. Because the system can recognize an external environment in real time, responsiveness to dynamic changes in the grasping environment is realized.

We adopt an architecture in which both flexibility and responsiveness are realized. This is a hierarchical parallel architecture in which each element consists of high-speed sensory feedback within 1ms as shown in Figure 1. Because each feedback process is completed within 1ms, adjustment to various conditions is realized at high speed.

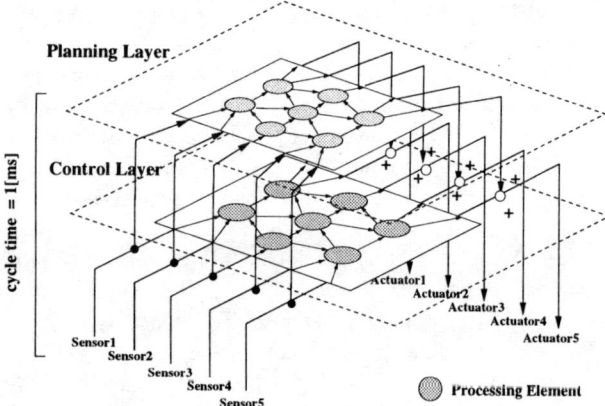

Figure 1. Hierarchical Parallel Architecture Based on High-speed Sensory Feedback

In general the cycle time of 1ms is necessary to prevent mechanical resonance in robotic control. In our architecture we decided that the cycle time of each sensory feedback should be 1ms to ensure stable grasping control.

As a related research Albus proposed a hierarchical parallel architecture based on the model of humans [7], and Brooks proposed a behavior-based hierarchical architecture consisting of layered sensory feedback modules [8]. We adopt a similar hierarchical parallel architecture, but responsiveness based on high-speed sensory feedback is not considered in these architectures.

3 1ms Sensory-Motor Fusion System

Using the idea of a hierarchical parallel architecture based on high-speed sensory feedback described in Section 2, we have developed a system called the "1ms Sensory-Motor Fusion System" to realize high-speed sensory feedback and fusion of sensory information. This system exhibits high performance processing of all sensory feedback, including visual feedback, with a cycle time of 1ms. Because the processing result is directly used to control the manipulator, each task is realized with high responsiveness.

Figure 2 shows the system components and Figure 3 shows a photograph of the system. This system consists of three parts: (1) a DSP subsystem for high-speed sensor fusion processing; (2) an active vision subsystem for high-speed visual processing; and (3) a multi-fingered hand-arm subsystem for high-speed manipulation.

3.1 DSP Subsystem

The DSP subsystem is the main part for fusion processing of sensory feedback within 1ms. It has a hierarchical parallel architecture consisting of 7 DSPs connected to each other, and many I/O ports are installed for inputing various types of information in parallel.

In this system we use a floating-point DSP TMS320C40 which has high performance (275 MOPS) and 6 I/O ports (20 Mbytes/sec). By connecting several C40 processors, a low bottle-neck hierarchical parallel architecture is realized. In our system the following I/O ports are prepared; ADC (12 bit, 64 CH), DAC (12 bit, 24 CH), and Digital I/O (8 bit, 8 ports). These I/O ports are distributed on several DSPs to minimize the I/O bottleneck so that sensor signals are input in parallel.

A parallel programming development environment has been prepared in which multi-process and multi-thread programming is easily realized. This function is useful to program parallel sensory feedback.

3.2 Active Vision with SPE

The active vision subsystem consists of a vision chip system called SPE-256 and a 2-axis actuator moved by DC servo motors. SPE-256 consists of a 16×16 array of processing elements (PE) and PIN photo-diodes (PD). The output of each PD is connected with a corresponding PE. Each PE is a 4-neighbor connected SIMD based processor which has a 24 bit register and a bit-serial arithmetic logic unit capable of AND, OR, and XOR operations etc. Because the visual processing is perfectly executed in parallel, high-speed visual feedback is realized within 1ms [5].

The SPE-256 is a scale-up model of an integrated vision chip and the next generation is currently being developed in which all elements of the vision chip architecture are integrated in one chip [4].

The actuator part of the active vision subsystem has two degrees of freedom; pan and tilt. This is used to move the sensor platform and this is controlled by a DSP assigned for active vision control.

3.3 7-axis Manipulator with Dextrous Multi-fingered Hand

The hand-arm subsystem is a 7-axis manipulator with a dextrous multi-fingered hand.

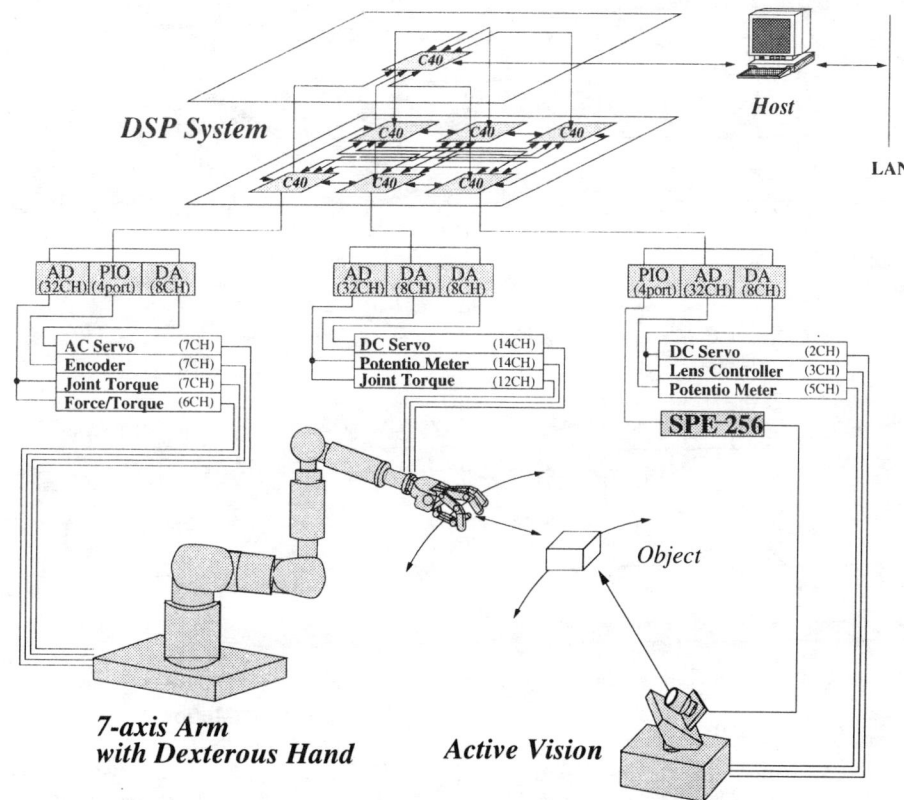

Figure 2. Architecture of 1ms Sensory-Motor Fusion System

The multi-fingered hand has 4 fingers and 14 joints. Its structure is similar to a human hand, in which a thumb finger is installed opposite to the other three fingers. Each joint is controlled by DC servo motors in a remote place using a control cable consisting of an outer casing and an inner wire. Each joint of the hand has a potentiometer for position control and a strain gage for force control.

The arm has 7 joints controlled by AC servo motors. An encoder is installed in each joint and a 6-axis force/torque sensor is installed at the wrist.

Two DSPs are assigned to control the hand and the arm.

4 Algorithm of High Speed Grasping

Using the idea of a hierarchical parallel architecture based on high-speed sensory feedback described in Section 2, we have developed an algorithm to realize responsiveness to dynamic changes of a grasping object.

Figure 4 shows a block diagram of the algorithm. In this algorithm 4 sensory feedbacks and one sensory information processing are executed in parallel; tracking control of the active vision, tracking control of the arm, reaching control of the arm, grasping control of the hand, and visual object recognition. These processes are distributed to 3 DSPs which are respectively assigned to the active vision, the arm, and the hand.

Though no prediction methods are used in this algorithm, our system has a high performance and sufficient responsiveness based on high-speed sensory feedback whose cycle time is less than 1.5ms.

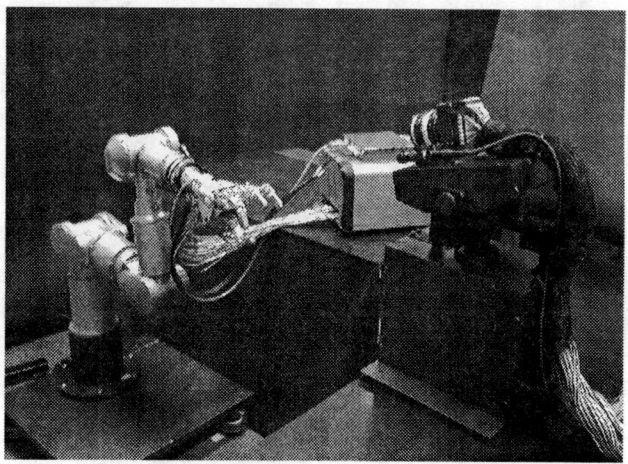

Figure 3. 1ms Sensory-Motor Fusion System

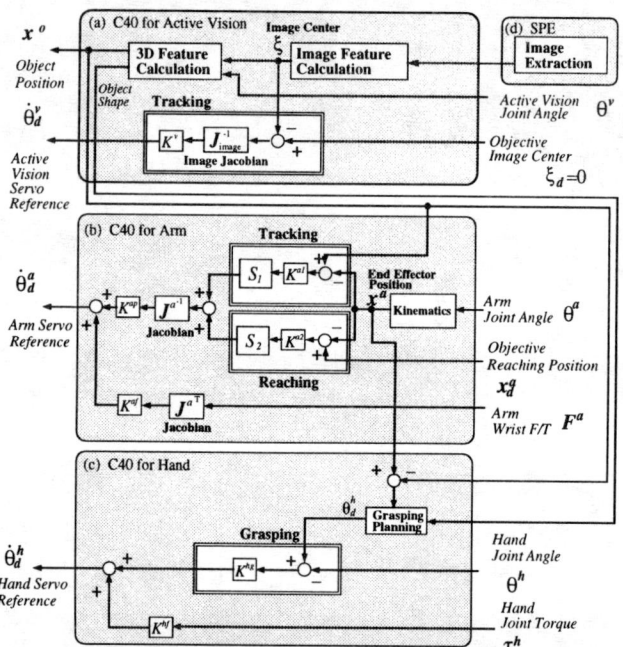

Figure 4. Algorithm of High Speed Grasping

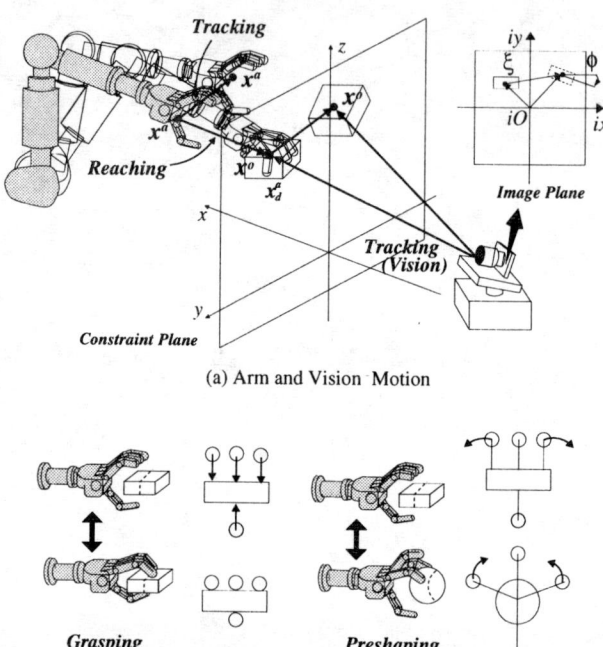

(a) Arm and Vision Motion

(b) Hand Motion

Figure 5. Motion of High Speed Grasping

4.1 Processing for Active Vision

In the DSP for active vision, two processes are executed in parallel: (1) object recognition by visual information, and (2) target tracking control.

4.1.1 Object Recognition by Visual Information

First, in SPE-256 a manipulated object is extracted in the image plane using an algorithm called Self Windowing which is realized by utilizing features of high-speed vision [5]. Next some image features are calculated from the extracted image in the DSP: the center of the image $\xi \in \mathbf{R}^2$, and the angle of rotation of the image $\phi \in \mathbf{R}$.

Lastly, some 3-D geometrical parameters are calculated in the DSP: the 3-D position and the orientation (role, pitch, and yow angles) of the object: $x^o \in \mathbf{R}^6$, and the object size. Then the object shape is detected based on these parameters. In the present configuration we assume that the motion of an object is limited to a constraint plane given beforehand. Using parameters of both the plane and image features, 3-D parameters are calculated.

4.1.2 Tracking Control of Active Vision

In this system the objective of active vision control is to acquire reliable object information by keeping an object in sight. To realize this we adopt feedback control using the image feature ξ written as,

$$\dot{\theta}_d^v = K^v J_{\text{image}}^{-1}(\xi_d - \xi) - K^{vv}\dot{\theta}^v, \qquad (1)$$

where $\dot{\theta}_d^v \in \mathbf{R}^2$ is the control input to the active vision servo, $\xi_d \in \mathbf{R}^2$ is the objective position on the image plane, $\theta^v \in \mathbf{R}^2$ is the joint angle vector of the active vision, and $K^v \in \mathbf{R}^{2\times 2}$ and $K^{vv} \in \mathbf{R}^{2\times 2}$ are diagonal gain matrices. The matrix $J_{\text{image}} \equiv \frac{\partial \xi}{\partial \theta^v} \in \mathbf{R}^{2\times 2}$ is called the "image Jacobian".

4.2 Processing for Arm

On the DSP for arm control, two sensory feedback controls are executed: (1) tracking control to object motion, and (2) reaching control to the grasping position.

The objective of tracking control is cancelation of the object motion by maintaining the relative position between the hand and the object. In the present configuration it is realized as a servo control so that the relative position error is kept to zero on the constraint plane, as shown in Figure 5(a).

The objective of reaching control is to reach the hand to the object by control of the relative position. In the present configuration this is controlled in the direction orthogonal to the constraint plane, as shown in Figure 5(a). In this control the objective trajectory of the hand is given beforehand.

By integrating tracking motion and reaching motion, the arm control scheme can be written as,

$$\begin{aligned}\dot{\theta}_d^a = &\ K^{ap}J^{a-1}SK^{a1}(x^o - x^a) \\ &+ K^{ap}J^{a-1}(I-S)K^{a2}(x_d^a - x^a) \\ &- K^{av}\dot{\theta}^a + K^{af}J^{aT}F^a\end{aligned} \qquad (2)$$

where $\dot{\boldsymbol{\theta}}_d^a \in \boldsymbol{R}^6$ is the control input to the arm servo, $\boldsymbol{\theta}^a \in \boldsymbol{R}^6$ is the joint angle vector of the arm, and $J^a \in \boldsymbol{R}^{6\times 6}$ is the jacobian matrix of the arm. The vector $\boldsymbol{x}^o \in \boldsymbol{R}^6$ is the position and the orientation of the object observed by vision, $\boldsymbol{x}^a \in \boldsymbol{R}^6$ is the position and the orientation of the hand obtained by haptic sensors, and $\boldsymbol{x}_d^a \in \boldsymbol{R}^6$ is the objective trajectory for reaching. The matrix K^{ap}, K^{a1}, K^{a2}, K^{av}, and $K^{af} \in \boldsymbol{R}^{6\times 6}$ are diagonal gain matrices. The matrix $I \in \boldsymbol{R}^{6\times 6}$ is the unit matrix and $S \equiv \mathrm{diag}\{s_i\}(i=x,y,z,\mathrm{role},\mathrm{pitch},\mathrm{yow})$ is a task partition matrix defined as,

$$s_i \equiv \begin{cases} 1 & \text{if } i = y, z, \text{role} \\ 0 & \text{otherwise} \end{cases} \quad (3)$$

In Eqn.(2) tracking motion and reaching motion respectively correspond to the first term and the second term. Because reaching motion is orthogonal to the tracking motion, there is no interaction. Then the fourth term is force feedback of the wrist force/torque sensor for compliance control.

4.3 Processing for Hand

In this system the objective of grasping control is to fix the object with the hand for manipulation. Using the compliance control method the hand is controlled as follows:

$$\dot{\boldsymbol{\theta}}_d^h = K^{hg}(\boldsymbol{\theta}_d^h - \boldsymbol{\theta}^h) - K^{hv}\dot{\boldsymbol{\theta}}^h + K^{hf}\boldsymbol{F}^h \quad (4)$$

where $\dot{\boldsymbol{\theta}}_d^h \in \boldsymbol{R}^{14}$ is the control input to the hand servo, and $\boldsymbol{\theta}^h \in \boldsymbol{R}^{14}$ is the joint angle vector of the hand. Matrices K^{hg}, K^{hv}, and $K^{hf} \in \boldsymbol{R}^{14\times 14}$ are diagonal gain matrices, and $\boldsymbol{F}^h \in \boldsymbol{R}^{14}$ is the joint torque vector. The vector $\boldsymbol{\theta}_d^h \in \boldsymbol{R}^{14}$ is the objective trajectory for grasping and is planned according to reaching motion $\boldsymbol{x}_d^a$.

Furthermore, according to the object shape, preshaping motion is executed to set the appropriate hand shape for grasping. In the present configuration the grasping shape is changed by distinguishing a circle and a rectangle in the 2D image-plane, as shown in Figure 5(b).

5 Experimental Results

We have performed experiments of grasping an object on the 1ms Sensory-Motor Fusion System.

The experimental result is shown in Figure 6 as a continuous sequence of pictures. All sensory feedback is executed in parallel according to the object motion at high speed: tracking motion of the active vision, tracking and reaching motion of the arm, and grasping motion of the hand. In Figure 7 a close-up view of the same motion is shown. In this figure tracking is executed from 0.0ms to 0.5ms and both reaching and grasping motion start at 0.5ms and all motion is completed at 0.8ms. Then in Figure 8 a close-up view of the grasping motion of a spherical object is shown. It is shown that the shape of the hand is changed to a suitable shape for grasping of a sphere.

In Figure 9 the trajectory of the hand is shown when grasping and releasing are alternately executed. In this figure, the Y axis position of the hand and the object show the tracking motion, and the X axis position of the hand and objective trajectory for reaching motion show the reaching motion. This figure shows that both responsive tracking by visual feedback during the releasing phase and stable grasping by visual and force feedback during the grasping phase are realized.

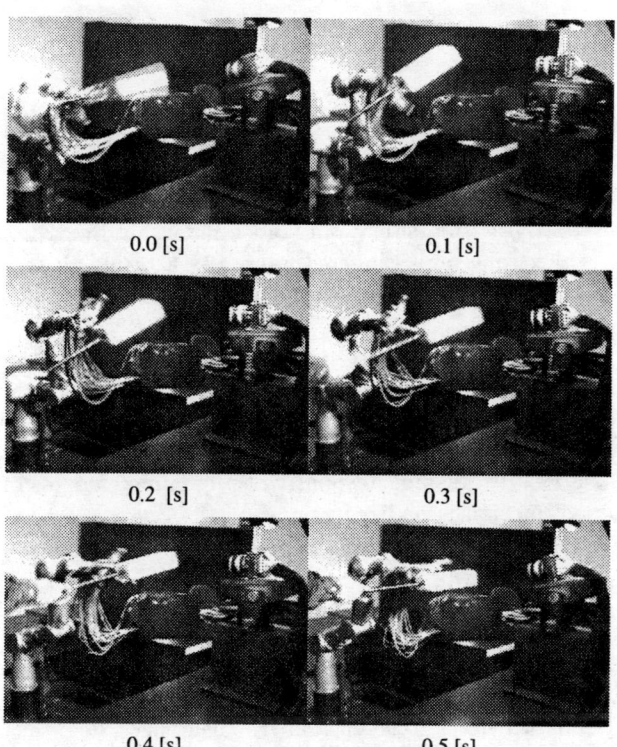

Figure 6. Experimental Result: Grasping of a Hexahedron

In these experiments, because the object is moved by a human hand, its trajectory is irregular and difficult to predict. Using the speed of the sensory feedback this problem is solved.

6 Conclusion

In this paper we describe a grasping system using high-speed sensory feedback with visual and force feedback. The system consists of two parts.

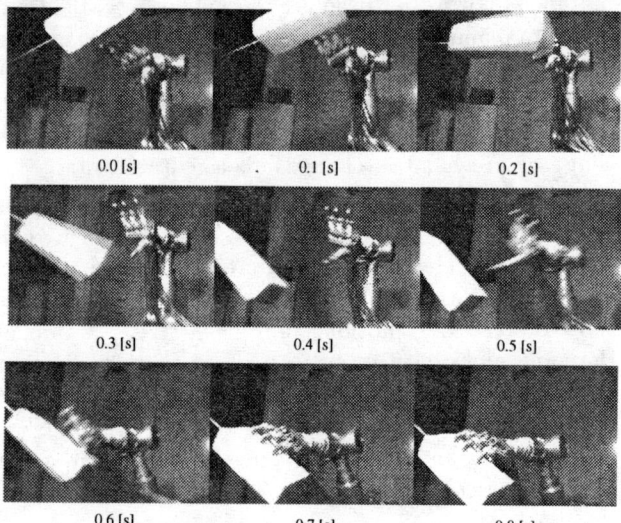

Figure 7. Experimental Result: Grasping of a Hexahedron

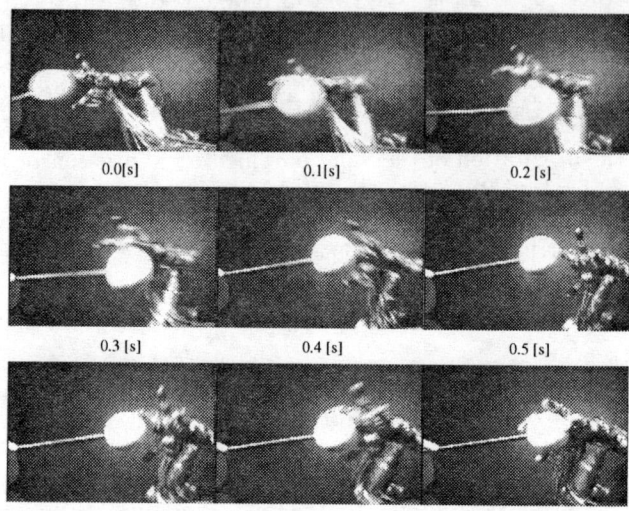

Figure 8. Experimental Result: Grasping of a Sphere

1. 1ms Sensory-Motor Fusion System has been developed to process and fuse sensory information at high speed. This consists of a hierarchical parallel processing subsystem with DSPs, a high-speed active vision subsystem and a manipulator with a dextrous multi-fingered hand. As a result all sensory feedback can be realized in about 1ms.

2. An algorithm for high-speed grasping is proposed. In this algorithm a grasping task is decomposed into some subtasks and each subtask is executed by high-speed sensory feedback in parallel.

As a result, grasping responsive to dynamic changes of object motion is realized.

Now we are developing various types of application on our system to realize responsive and flexible manipulation in the real world environment.

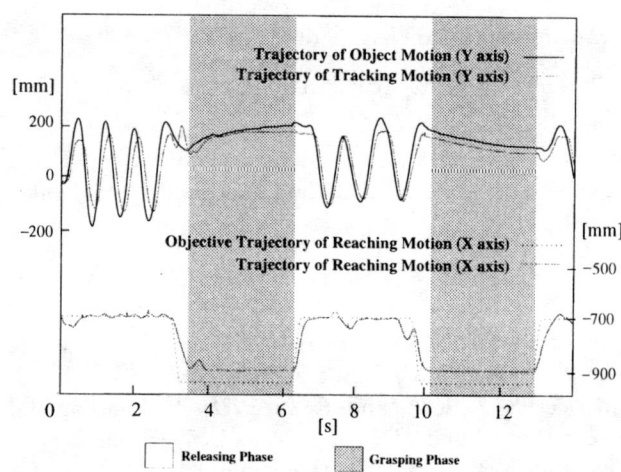

Figure 9. Feedback Response

References

[1] P.K. Allen, B. Yoshimi, and A. Timucenko. Real-time visual servoing. *Proc. IEEE Int. Conf. Robotics and Automation*, pages 2376–2384, 1991.

[2] W. Hong and J.E. Slotine. Experiments in hand-eye coordination using active vision. *Proc. 4th Int. Symp. on Experimental Robotics (ISER'95)*, 1995.

[3] M. Ishikawa, A. Morita, and N. Takayanagi. High speed vision system using massively parallel processing. *Proc. IEEE Int. Conf. on Intelligent Robotics and Systems*, pages 373–377, 1992.

[4] T. Komuro, I. Ishii, and M. Ishikawa. Vision chip architecture using general-purpose processing elements for 1ms vision system. *Proc. 4th IEEE Int. Workshop on Computer Architecture for Machine Perception (CAMP'97)*, pages 276–279, 1997.

[5] I. Ishii, Y. Nakabo, and M. Ishikawa. Target tracking algorithm for 1ms visual feedback system using massively parallel processing vision. *Proc. IEEE Int. Conf. on Robotics and Automation*, pages 2309–2314, 1996.

[6] Y. Nakabo and M. Ishikawa. Visual impedance using 1ms visual feedback system. *Proc. IEEE Int. Conf. on Robotics and Automation*, pages 2333–2338, 1998.

[7] J.S. Albus. Outline for a theory of intelligence. *IEEE Trans. on Systems, Man, and Cybernetics*, 21(3):473–509, 1991.

[8] R.A. Brooks. A robust layered control system for a mobile robot. *IEEE Journal of Robotics and Automation*, RA-2(1):14–23, 1986.

Vision-Aided Object Manipulation by a Multifingered Hand with Soft Fingertips

Yasuyoshi YOKOKOHJI, Moriyuki SAKAMOTO*, and Tsuneo YOSHIKAWA
Department of Mechanical Engineering, Graduate School of Engineering
Kyoto University, Kyoto 606-8501, Japan
(*currently with KAWASAKI HEAVY INDUSTRIES, LTD.)
yokokoji@mech.kyoto-u.ac.jp   moriyuki@tech.khi.co.jp   yoshi@mech.kyoto-u.ac.jp

Abstract

Soft fingertips are easy to achieve high friction contacts but their frictional moment around the contact normals makes the manipulation control difficult. In this paper, we propose a control scheme to manipulate an object by a multifingered hand with soft fingertips. The proposed scheme cancels the frictional moment according to the direction of the relative rotation between the fingertip and the object.

Whatever the fingertip contact type is, accurate pose measurement is very important for the object manipulation. In this paper, we also propose a systematic method to estimate the object pose with a complementary use of camera images and finger joint information.

The proposed control scheme and the pose estimation method were combined and implemented to an experimental system. Results of the experiment show the effectiveness of the proposed pose estimation method and the overall control system for manipulating an object by a soft-fingered hand.

1 Introduction

Dexterous manipulation by a multifingered hand is one of the essential technologies that improve robot task flexibility. Typical three contact types of multifingered hands are (i) fixed point contact with friction, (ii) hard rolling contact with friction, and (iii) soft-finger (or pure rolling) contact[6]. Hard rolling contact assumes point contact and thus generates no twist moment around the contact normal. Point contact assumption of (ii) is adequate only when both the fingertip and the object are rigid bodies. Since friction coefficient at each contact point should be large enough to prevent undesirable slip, however, we usually make fingertips by rubber or other soft materials. This means that we need to consider the soft-finger contacts and the frictional twist/spin moment around the contact normal.

To always keep the soft-finger contact, each finger must have enough degrees of freedom (more than four DOF). Sarkar et al.[7] showed a control scheme that can control the object pose and contact point locations simultaneously using two arms, each of which has 6 DOF and a flat plane at the endpoint. Robot hands, however, should have round fingertips rather than flat planes and DOF of each finger should be as small as possible. If we could cancel the frictional twist/spin moment at the contact points, and allow the fingertips to rotate relatively around the contact normal, we can use conventional three DOF fingers, while keeping a high tangential friction with soft fingertips. In this paper, we propose a control scheme for object manipulation by three-fingered hand with soft fingertips, where each finger has at least three joints.

Whatever type of fingertip contact, measuring the pose of the grasped object is also an important problem. When the contact constraint is nonholonomic like our case, it is difficult to estimate object pose only from the joint sensor information. In addition, any friction modeling contains some uncertainties, and it is difficult to predict the actual slipping phenomena accurately even with a precise friction modeling[3][4]. Thus, it is reasonable to avoid the precise friction model and use external sensors, such as video cameras, to measure the object pose directly.

Resolution of vision sensors, however, is lower than joint sensors. In addition, video images are usually updated 30 times a second, which is too slow for the servo control frequency. In this paper, we also propose a systematic method to estimate the pose of the manipulated object with a complementary use of camera images and finger joint information. The proposed estimation method updates the object pose and velocity at servo frequency.

To verify the proposed control scheme for soft fingertips and the proposed pose estimation method, we built a camera-hand system and conducted some experiments. The experimental results show the effectiveness of the proposed control scheme and the pose estimation method.

2 Target System

The target system of this paper is shown in **Fig.1**. The hand has three fingers, each of which has three joints and one soft fingertip. The given task is to manipulate an object by the hand along with a desired trajectory. Each fingertip is equipped with a force sensor to measure the force vector applied to the grasped object. The object is rigid and its contact surface is smooth. As shown in **Fig.1**, some landmarks are attached to the object. A single camera tracks these

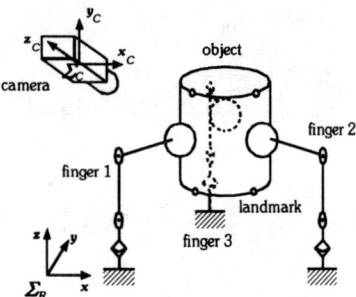

Figure 1: Target system

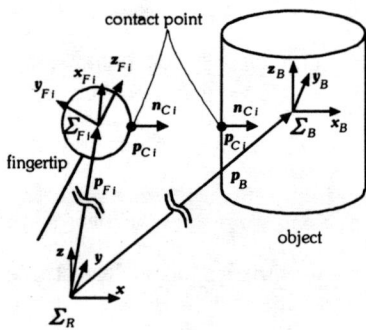

Figure 2: Fingertip and object

landmarks to estimate the object pose.

Regarding the soft fingertips, the following assumptions are made: (i) As shown **Fig.2**, the shape of fingertips is sphere; (ii) Contact area is small and will be regarded as point contact in the following analysis, but the frictional twist/spin moment is considered; (iii) Energy dissipation due to the fingertip deformation[1] is negligible; (iv) Tangential frictional force and frictional twist/spin moment around the contact normal are independent and they can be approximated by the Coulomb model.

3 Control Scheme for Soft Fingertips
3.1 Basic strategy

The proposed scheme in this paper is based on the formulations by Cole et al.[2] and Yoshikawa[8]. But we consider the frictional twist/spin moment due to the soft-finger type contact. The basic strategy to compensate frictional twist/spin moment is as follows:

1. If the fingertip and the object are relatively rotating around the contact normal, the control scheme will cancel the dynamic frictional spin moment.

2. If no relative rotation but the servo controller commands relative rotational acceleration, the control scheme will cancel the static frictional twist moment to realize the commanded acceleration.

3. If no relative rotation and the commanded acceleration is zero, the control scheme will not apply any moment for the friction compensation.

3.2 Constraints of rolling contact

As shown in **Fig.1** and **Fig.2**, a base coordinate frame, Σ_R, is attached to the hand base and an object coordinate frame, Σ_B, is placed at the gravity center of the object. A fingertip coordinate frame, Σ_{Fi}, is located at the center of fingertip sphere of the i-th finger ($i = 1, 2, 3$). Position and orientation of Σ_B with respect to Σ_R are represented by a vector $\boldsymbol{p}_B \in \Re^3$ and a rotation matrix $\boldsymbol{R}_B \in \Re^{3\times 3}$, respectively. In the same way, position and orientation of Σ_{Fi} are given by $\boldsymbol{p}_{Fi} \in \Re^3$ and $\boldsymbol{R}_{Fi} \in \Re^{3\times 3}$. The i-th finger's contact point is denoted by $\boldsymbol{p}_{Ci} \in \Re^3$.

If no tangential slip occurs at the contact point, we get the following constraint between the object velocity and the fingertip velocity[2][8]:

$$\boldsymbol{D}_{Bi}\begin{pmatrix}\dot{\boldsymbol{p}}_B \\ \boldsymbol{\omega}_B\end{pmatrix} = \boldsymbol{D}_{Fi}\begin{pmatrix}\dot{\boldsymbol{p}}_{Fi} \\ \boldsymbol{\omega}_{Fi}\end{pmatrix} \triangleq \boldsymbol{v}_{Ci}, \qquad (1)$$

where $\boldsymbol{\omega}_B \in \Re^3$ and $\boldsymbol{\omega}_{Fi} \in \Re^3$ denote angular velocities of the object and the i-th fingertip, respectively. Matrices $\boldsymbol{D}_{Bi} \in \Re^{3\times 6}$ and $\boldsymbol{D}_{Fi} \in \Re^{3\times 6}$ are defined as follows:

$$\boldsymbol{D}_{Bi} = [\boldsymbol{I}_3 \ -[(\boldsymbol{p}_{Ci} - \boldsymbol{p}_B)\times]], \qquad (2)$$
$$\boldsymbol{D}_{Fi} = [\boldsymbol{I}_3 \ -[(\boldsymbol{p}_{Ci} - \boldsymbol{p}_{Fi})\times]], \qquad (3)$$

where $[\cdot\times]$ denotes a skew symmetric matrix equivalent to the cross product operation. $\boldsymbol{v}_{Ci} \in \Re^3$ in eq.(1) denotes a vector of the contact point velocity, where the components of the contact point movement due to the rolling are excluded.

A fingertip velocity and the corresponding joint velocities are related by

$$\begin{pmatrix}\dot{\boldsymbol{p}}_{Fi} \\ \boldsymbol{\omega}_{Fi}\end{pmatrix} = \boldsymbol{J}_{Fi}\dot{\boldsymbol{\theta}}_i, \qquad (4)$$

where $\boldsymbol{\theta}_i \in \Re^3$ denotes a joint vector of the i-th finger and $\boldsymbol{J}_{Fi} \in \Re^{6\times 3}$ is the Jacobian matrix of the i-th finger. From eq.(4), we get a kinematic relationship between the contact point velocity and the joint velocity as follows:

$$\boldsymbol{v}_{Ci} = \boldsymbol{D}_{Fi}\boldsymbol{J}_{Fi}\dot{\boldsymbol{\theta}}_i \triangleq \boldsymbol{J}_{CFi}\dot{\boldsymbol{\theta}}_i. \qquad (5)$$

Instead of a 3×3 orientation matrix, object orientation can be represented by three variables, such as roll, pitch and yaw angles. Hence, the object pose can be expressed by a six dimensional vector, $\boldsymbol{r}_B = \text{col}^*[\boldsymbol{p}_B, \boldsymbol{\phi}_B]$. Then $\dot{\boldsymbol{r}}_B$ can be related to the original velocity expression, $\text{col}[\dot{\boldsymbol{p}}_B, \boldsymbol{\omega}_B]$, by [2][8]

$$\begin{pmatrix}\dot{\boldsymbol{p}}_B \\ \boldsymbol{\omega}_B\end{pmatrix} = \boldsymbol{T}\dot{\boldsymbol{r}}_B, \qquad (6)$$

---

*"col" means a column vector or matrix formed by the following elements.

where $T(r_B) \in \Re^{6\times 6}$ is a matrix that transforms $\dot{\phi}_B$ to the corresponding ω_B. The object velocity and the contact point velocity are related by

$$D_B \begin{pmatrix} \dot{p}_B \\ \omega_B \end{pmatrix} = v_C, \qquad (7)$$

where $D_B = \text{col}[D_{B1}, D_{B2}, D_{B3}] \in \Re^{9\times 6}$ and $v_C = \text{col}[v_{C1}, v_{C2}, v_{C3}] \in \Re^9$.

From eqs.(5), (6) and (7), the constraint between the joint velocity, $\dot{\theta} = \text{col}[\dot{\theta}_1, \dot{\theta}_2, \dot{\theta}_3]$, and the object velocity, $\dot{r}_B$, is given by

$$D_B T(r_B) \dot{r}_B = J_{CF} \dot{\theta}, \qquad (8)$$

where $J_{CF} = \text{diag}^\dagger[J_{CF1}, J_{CF2}, J_{CF3}] \in \Re^{9\times 9}$. Eq.(8) will be used in section 4.3 to estimate the object velocity from the joint velocities.

3.3 Control scheme for frictional twist/spin moment compensation

In this section, we derive a control scheme that can cancel the frictional twist/spin moment at the contact point of soft fingertips. Using T in eq.(6), dynamics of the object is given by

$$M_B T \ddot{r}_B + h_{Br}(r_B, \dot{r}_B) = t_B, \qquad (9)$$

where $M_B \in \Re^{6\times 6}$ denotes an inertia tensor of the object, $h_B \in \Re^6$ includes centrifugal, Coriolis and gravitational force terms as well as an inertial term associated with $\dot{T}$. $t_B \in \Re^6$ is a total force/moment vector applied to the object by the hand.

Dynamics of each finger can be united into one equation:

$$M_F(\theta)\ddot{\theta} + h_F(\theta,\dot{\theta}) = \tau - (J_{CF}^T + J_m^T)f_C, \qquad (10)$$

where $\tau = \text{col}[\tau_1, \tau_2, \tau_3] \in \Re^9$ ($\tau_i \in \Re^3$ denotes a joint torque vector of the i-th finger), $f_C = \text{col}[f_{C1}, f_{C2}, f_{C3}] \in \Re^9$ ($f_{Ci} \in \Re^3$ denotes a finger force vector applied to the object by the i-th finger). $M_F = \text{diag}[M_{F1}, M_{F2}, M_{F3}] \in \Re^{9\times 9}$, where $M_{Fi} \in \Re^{3\times 3}$ is an inertia matrix of the i-th finger, and $J_m = \text{diag}[J_{m1}, J_{m2}, J_{m3}] \in \Re^{9\times 9}$, where $J_{mi} \in \Re^{3\times 3}$ relates the fictional twist/spin moment at the i-th contact to the joint force of the i-th finger. J_{mi} is defined as a transposed form by

$$J_{mi}^T = J_{Fi}^T \begin{pmatrix} \Phi_3 \\ N_i \end{pmatrix}, \qquad (11)$$

where Φ_3 denotes 3×3 zero matrix. $N_i \in \Re^{3\times 3}$ specifies the mode of rotational frictions at the i-th contact point, which is defined by

$$N_i = \alpha_i \mu_i n_{Ci} n_{Ci}^T, \qquad (12)$$

where $n_{Ci} \in \Re^3$ denotes an outward unit normal vector of the fingertip surface at the i-th contact point,

$\dagger$ "diag" means a block diagonal matrix.

Table 1: Direction of spin and frictional moment

current spinning direction β_i		spin at the next instant γ_i	α_i	μ_i	frictional twist/spin moment m_{ci}
spinning	+1	—	+1	$\bar{\mu}_i$	dynamic frictional moment $+\bar{\mu}_i n_{ci} n_{ci}^T f_{ci}$
	−1	—	−1	$\bar{\mu}_i$	dynamic frictional moment $-\bar{\mu}_i n_{ci} n_{ci}^T f_{ci}$
not spinning	0	will spin	+1	$\tilde{\mu}_i'$	maximum static frictional moment $+\tilde{\mu}_i' n_{ci} n_{ci}^T f_{ci}$
			−1	$\tilde{\mu}_i'$	maximum static frictional moment $-\tilde{\mu}_i' n_{ci} n_{ci}^T f_{ci}$
		will not spin	0	0	0

and μ_i is the moment friction coefficient. μ_i may be either the static friction coefficient, $\tilde{\mu}_i'$, or the dynamic friction coefficient, $\bar{\mu}_i$, according to the condition of the relative rotation around the contact normal. $\alpha_i(=1, -1,$ or $0)$ shows the direction of the frictional spin moment compensation, which is determined from the current spinning direction, $\beta_i(=1, -1,$ or $0)$, or the commanded spin direction, $\gamma_i(=1, -1,$ or $0)$. The current spinning direction at the i-th contact is obtained by

$$\beta_i = \text{sign}(n_{Ci}^T(\omega_{Fi} - \omega_B)). \qquad (13)$$

ω_{Fi} is obtained by eq.(4), while ω_B can be estimated from eqs.(6) and (8) (or by eq.(45) given later). γ_i can be determined in the same manner. Practically speaking, signs of β_i and γ_i should be determined by appropriate threshold values, ε_β and ε_γ, respectively. **Table 1** summarizes how α_i and μ_i are determined.

The total force/moment applied to the object and the fingertip forces are related by the following equation:

$$f_C = A^+ t_B + (I - A^+ A) E \tilde{f}, \qquad (14)$$

where $A \in \Re^{6\times 9}$ denotes an extended grip matrix defined by the following equation:

$$A = D_B^T + \begin{pmatrix} \Phi_3 & \Phi_3 & \Phi_3 \\ N_1 & N_2 & N_3 \end{pmatrix}, \qquad (15)$$

where D_B^T is the original grip matrix and the second term is related to the contribution by the frictional twist/spin moments. In eq.(14), the second term in the right hand side represents internal forces, and $E \in \Re^{9\times 3}$ is defined by

$$E = \begin{pmatrix} \mathbf{0} & -e_{31} & e_{12} \\ e_{23} & \mathbf{0} & -e_{12} \\ -e_{23} & e_{31} & \mathbf{0} \end{pmatrix}, \qquad (16)$$

where $e_{ij} \in \Re^3$ is a unit vector directing from the contact point of the i-th finger to the j-th one. $\tilde{f} \in \Re^3$ specifies the magnitude of the internal forces.

From eqs.(8), (9), (10), and (14), we can derive the dynamics of the total hand/object system as follows:

$$\boldsymbol{\tau} = \boldsymbol{M}_W \boldsymbol{T} \ddot{\boldsymbol{r}}_B + \boldsymbol{B}\tilde{\boldsymbol{f}} + \boldsymbol{Q}, \qquad (17)$$

where

$$\boldsymbol{M}_W = \boldsymbol{M}_F \boldsymbol{J}_{CF}^{-1} \boldsymbol{D}_B + (\boldsymbol{J}_{CF}^T + \boldsymbol{J}_m^T)\boldsymbol{A}^+ \boldsymbol{M}_B, \quad (18)$$

$$\boldsymbol{B} = (\boldsymbol{J}_{CF}^T + \boldsymbol{J}_m^T)(\boldsymbol{I} - \boldsymbol{A}^+\boldsymbol{A})\boldsymbol{E}, \qquad (19)$$

$$\boldsymbol{Q} = \boldsymbol{M}_F \boldsymbol{J}_{CF}^{-1}\{(\dot{\boldsymbol{D}}_B\boldsymbol{T} + \boldsymbol{D}_B\dot{\boldsymbol{T}})\dot{\boldsymbol{r}}_B - \dot{\boldsymbol{J}}_{CF}\dot{\boldsymbol{\theta}}\}$$
$$+ \boldsymbol{h}_F + (\boldsymbol{J}_{CF}^T + \boldsymbol{J}_m^T)\boldsymbol{A}^+ \boldsymbol{h}_{Br}. \qquad (20)$$

Let us consider the following control scheme, which linearize the hand/object system of eq.(17):

$$\boldsymbol{\tau}_d = \boldsymbol{M}_W \boldsymbol{T} \boldsymbol{u}_B + \boldsymbol{B}\boldsymbol{u}_I + \boldsymbol{Q}, \qquad (21)$$

where $\boldsymbol{u}_B \in \Re^6$ and $\boldsymbol{u}_I \in \Re^3$ are new inputs for the object pose and the internal force, respectively.

Letting $\boldsymbol{r}_{Bd}$ and $\tilde{\boldsymbol{f}}_d$ be the desired trajectories of the object pose and the internal force, respectively, $\boldsymbol{u}_B$ and $\boldsymbol{u}_I$ are given by the following servo control:

$$\boldsymbol{u}_B = \ddot{\boldsymbol{r}}_{Bd} + \boldsymbol{K}_V(\dot{\boldsymbol{r}}_{Bd} - \dot{\boldsymbol{r}}_B) + \boldsymbol{K}_P(\boldsymbol{r}_{Bd} - \boldsymbol{r}_B), \quad (22)$$

$$\boldsymbol{u}_I = \tilde{\boldsymbol{f}}_d + \boldsymbol{K}_I \int_0^t (\tilde{\boldsymbol{f}}_d - \tilde{\boldsymbol{f}})dt', \qquad (23)$$

where $\boldsymbol{K}_V, \boldsymbol{K}_P \in \Re^{6 \times 6}$ and $\boldsymbol{K}_I \in \Re^{3 \times 3}$ are gain matrices. The internal force components are calculated from the measured fingertip forces $\boldsymbol{f}_C$ by

$$\tilde{\boldsymbol{f}} = ((\boldsymbol{I} - \boldsymbol{A}^+\boldsymbol{A})\boldsymbol{E})^+ (\boldsymbol{I} - \boldsymbol{A}^+\boldsymbol{A})\boldsymbol{f}_C. \qquad (24)$$

When all fingertips are relatively rotating around the contact normals (i.e., $\beta_i \neq 0$ for $i = 1, 2, 3$), and if

$$\det(\boldsymbol{D}_B \boldsymbol{M}_B^{-1} \boldsymbol{A} + \boldsymbol{J}_{CF}\boldsymbol{M}_F^{-1}(\boldsymbol{J}_{CF}^T + \boldsymbol{J}_m^T)) \neq 0, \quad (25)$$

then the position error, $\boldsymbol{e}_P = \boldsymbol{r}_{Bd} - \boldsymbol{r}_B$, and the force error, $\boldsymbol{e}_f = \tilde{\boldsymbol{f}}_d - \tilde{\boldsymbol{f}}$, satisfy the following equations:

$$\ddot{\boldsymbol{e}}_P + \boldsymbol{K}_V \dot{\boldsymbol{e}}_P + \boldsymbol{K}_P \boldsymbol{e}_P = \boldsymbol{0}, \qquad (26)$$
$$\dot{\boldsymbol{e}}_f + \boldsymbol{K}_I \boldsymbol{e}_f = \boldsymbol{0}. \qquad (27)$$

Thus, with appropriate feedback gain matrices, $\boldsymbol{K}_V$, $\boldsymbol{K}_P$ and $\boldsymbol{K}_I$, the actual object pose and the actual internal force will converge to the desired trajectories asymptotically. Note that when some of the fingers fall into the state of static frictional moment ($\beta_i = 0$), eq.(26) is no longer satisfied, because the object looses some degrees of freedom. The proposed control scheme, however, cancels the static frictional twist moment immediately and prevents that finger from stacking by the static friction. Therefore, falling into the static friction state can be regarded as an impulsive disturbance for eq.(26).

In eq.(22), we implicitly assumed that the current object pose, $\boldsymbol{r}_B$, and its velocity, $\dot{\boldsymbol{r}}_B$, can be obtained at the servo frequency. In the next section, we will discuss how to estimate the object pose and its velocity for the servo control.

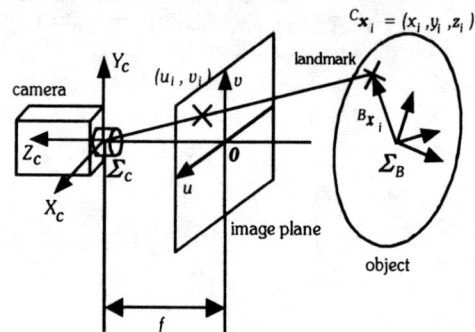

Figure 3: Camera model

One more point that should be mentioned here is about the conditions for the frictional moment cancellation when $\beta_i = 0$ for some i. We have no space to explain the detail, but the conditions, which are obtained by extending the condition of eq.(25), largely depend on the finger joint configurations. For example, when $\beta_i = 0$ for all i, the finger configuration in **Fig.1**, which is not singular in a usual sense, becomes singular in a sense that static frictional moments are indeterminate. The configuration in **Fig.6**, on the other hand, is not singular in this sense.

4 Object Pose Estimation by Sensor Fusion

4.1 Object tracking by single camera with least-squares minimization

Suppose n landmarks are attached to the object. A single camera is fixed on the ground base. As shown in **Fig.3**, a camera coordinate frame, Σ_C, is placed in such a way that its origin is at the lens center and the z-axis is along with the lens axis. The camera is approximated by a pinhole camera model and its focal length is f. Then, the i-th landmark at $^C\boldsymbol{x}_i = (x_i\ y_i\ z_i)^T$ $(i = 1, 2, \cdots, n)$ expressed by Σ_C is projected to a point on the image plane by the following equations:

$$u_i = -\frac{fx_i}{s_u z_i}, \qquad v_i = -\frac{fy_i}{s_v z_i}, \qquad (28)$$

where $\boldsymbol{u}_i = (u_i\ v_i)^T$ denotes a position vector of the projected landmark in pixel values and s_u and s_v are pixel ratios in u and v directions, respectively.

Position and orientation of Σ_B with respect to Σ_C is given by a six dimensional vector $^C\boldsymbol{r}_B = \mathrm{col}[^C\boldsymbol{p}_B, ^C\boldsymbol{\phi}_B]$, where $^C\boldsymbol{p}_B = (x_B\ y_B\ z_B)^T$ denotes the position vector of the origin of Σ_B and $^C\boldsymbol{\phi}_B = (\phi_B\ \theta_B\ \psi_B)^T$ denotes orientation angles of Σ_B, such as roll, pitch and yaw angles. The object pose can also be represented by a position vector $^C\boldsymbol{p}_B \in \Re^3$ and a rotation matrix $^C\boldsymbol{R}_B \in \Re^{3 \times 3}$.

Letting the position vector of the i-th landmark

from Σ_B be $^B x_i$ (= const), $^C x_i$ is given by

$$^C x_i = {}^C R_B {}^B x_i + {}^C p_B. \quad (29)$$

If the object pose has an infinitesimal displacement $\Delta^C r_B$, the landmarks on the image plane move Δu accordingly. This relation can be described by using the well-known image Jacobian, $^C J_u \in \Re^{2n \times 6}$, as follows:

$$\Delta u = {}^C J_u \, \Delta^C r_B, \quad (30)$$

where

$$\Delta u = (\Delta u_1 \quad \Delta v_1 \quad \cdots \quad \Delta u_n \quad \Delta v_n)^T, \quad (31)$$

$$^C J_u = \begin{pmatrix} \frac{\partial u_1}{\partial x_B} & \frac{\partial u_1}{\partial y_B} & \frac{\partial u_1}{\partial z_B} & \frac{\partial u_1}{\partial \phi_B} & \frac{\partial u_1}{\partial \theta_B} & \frac{\partial u_1}{\partial \psi_B} \\ \frac{\partial v_1}{\partial x_B} & \frac{\partial v_1}{\partial y_B} & \frac{\partial v_1}{\partial z_B} & \frac{\partial v_1}{\partial \phi_B} & \frac{\partial v_1}{\partial \theta_B} & \frac{\partial v_1}{\partial \psi_B} \\ \vdots & \vdots & \vdots & \vdots & \vdots & \vdots \\ \frac{\partial u_n}{\partial x_B} & \frac{\partial u_n}{\partial y_B} & \frac{\partial u_n}{\partial z_B} & \frac{\partial u_n}{\partial \phi_B} & \frac{\partial u_n}{\partial \theta_B} & \frac{\partial u_n}{\partial \psi_B} \\ \frac{\partial v_n}{\partial x_B} & \frac{\partial v_n}{\partial y_B} & \frac{\partial v_n}{\partial z_B} & \frac{\partial v_n}{\partial \phi_B} & \frac{\partial v_n}{\partial \theta_B} & \frac{\partial v_n}{\partial \psi_B} \end{pmatrix}, \quad (32)$$

$$\Delta^C r_B = (\Delta x_B \quad \Delta y_B \quad \Delta z_B \quad \Delta \phi_B \quad \Delta \theta_B \quad \Delta \psi_B)^T. \quad (33)$$

When $n \geq 3$ and these landmarks are not in a singular configuration, $\mathrm{rank}(^C J_u) = 6$ and we get

$$\Delta^C r_B = {}^C J_u^+ \Delta u, \quad (34)$$

where $^C J_u^+$ is the pseudo inverse of $^C J_u$.

Usually camera images are updated by video rate (30Hz). If the object motion is too fast, the estimated pose, which is updated by adding $\Delta^C r_B$ once at each video cycle, might not be close enough to the true value. In such a case, we can iterate the update by eq.(34) at each camera sampling until the estimation error becomes small enough, just like the Newton-Raphson method.

The estimated object pose by eq.(34) is given in camera coordinates. One can transform it to the base frame representation using known camera frame position and orientation.

4.2 Fusing finger joint information and visual information

Object pose estimation from camera images tends to be inaccurate. Especially estimating by a single camera may result in a large error in the depth direction[5]. Thus, we will fuse the finger joint information and the visual information in a complementary manner.

Before going to the formulation of fusing two disparate sensor measurements, let us consider an iterative method of pose estimation from the joint information alone. Suppose each fingertip moved a little and caused a small change of the object pose, Δr_B. If the estimated object pose is not updated, i.e., if the object pose is virtually moved $-\Delta r_B$ from the current true pose while keeping each fingertip at the current position, the contact point on the object surface, p_{Ci}^B,

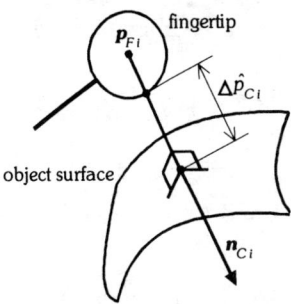

Figure 4: Distance from the fingertip to the estimated object surface

may apart from the current contact point on the fingertip surface, p_{Ci}^F. Hence, we can update the object pose in such a way that the estimated contact points on the object surface and the corresponding contact points on the fingertip surface coincide.

From eqs.(6) and (7), a small displacement of the i-th contact point on the object surface, $\Delta p_{Ci} = p_{Ci}^B - p_{Ci}^F$, corresponding to a small change of the object pose, $-\Delta r_B$, is given by

$$\Delta p_{Ci} = D_{Bi} T(-\Delta r_B) = -D_{Bi} T \Delta r_B. \quad (35)$$

Δp_{Ci} is a three dimensional vector but what we can know from the finger joint information is just the minimum distance between the object surface and the fingertip. Thus, we extract the surface normal component from eq.(35) as follows:

$$\Delta \hat{p}_{Ci} = -n_{Ci}^T D_{Bi} T \Delta r_B, \quad (36)$$

where $\Delta \hat{p}_{Ci}$ is a scalar value representing the minimum distance between the fingertip surface and the objet surface as shown in **Fig.4**. If the object surface is smooth and non-concave, it is possible to calculate this minimum distance by projecting a perpendicular line from the fingertip center p_{Fi} to the object surface based on the current pose estimation. Unifying eq.(36) for all three contact points, we get

$$\Delta \hat{p}_C = -\hat{D}_B T \Delta r_B, \quad (37)$$

where $\Delta \hat{p}_C = (\Delta \hat{p}_{C1}, \Delta \hat{p}_{C2}, \Delta \hat{p}_{C3})^T$ and $\hat{D}_B = \mathrm{col}[n_{C1}^T D_{B1}, n_{C2}^T D_{B2}, n_{C3}^T D_{B3}] \in \Re^{3 \times 6}$.

From eq.(37), we cannot fully recover Δr_B, because $\mathrm{rank}(\hat{D}_B T) \leq 3$. For example, suppose that the grasped object is a cylinder. If all three fingers are contacting with the side surface of the cylinder, we cannot know the rotational error around the cylinder axis and translational error along with the same axis from the finger joint information alone.

Now, we are ready to formulate the pose estimation by sensor fusion. Remember that we could obtain the infinitesimal displacement of the object pose, $\Delta^C r_B$, from Δu by eq.(30). Transforming $\Delta^C r_B$ to the base frame expression Δr_B, we get

$$\Delta u = {}^C J_u {}^C T^{-1} U_C^T T \Delta r_B \triangleq J_u \Delta r_B, \quad (38)$$

where $^C T \in \Re^{6\times 6}$ is a transformation matrix (defined by eq.(6)) in camera coordinates and $U_C \in \Re^{6\times 6}$ is defined by

$$U_C = \begin{pmatrix} R_C & \Phi_3 \\ \Phi_3 & R_C \end{pmatrix}, \quad (39)$$

where $R_C \in \Re^{3\times 3}$ denotes an orientation matrix of Σ_C.

From eqs.(37) and (38), Δu and $\Delta \hat{p}_C$ corresponding to Δr_B are given by

$$\begin{pmatrix} \Delta u \\ \Delta \hat{p}_C \end{pmatrix} = \begin{pmatrix} J_u \\ -\hat{D}_B T \end{pmatrix} \Delta r_B. \quad (40)$$

Using a weighting matrix $W \in \Re^{(2n+3)\times(2n+3)}$, we can get Δr_B as follows:

$$\Delta r_B = \left(W \begin{pmatrix} J_u \\ -\hat{D}_B T \end{pmatrix} \right)^+ W \begin{pmatrix} \Delta u \\ \Delta \hat{p}_C \end{pmatrix}. \quad (41)$$

Δu and $\Delta \hat{p}_C$ may contain some measurement errors and these errors may affect the estimation result by eq.(41). Thus, it is reasonable to set each diagonal element of W to the inverse of the standard deviation of the corresponding measurement error and normalize the effect of errors. If the variance of the landmark measurement on the image plane is given by σ_u^2, its standard deviation is σ_u. The variance of $\Delta \hat{p}_C$, on the other hand, should include the effects of the encoder measurement errors and uncertainties of the fingertip shape due to the deformation and the fabrication error. Representing the variances of the encoders of all joints by a matrix $\Gamma_\theta = \text{diag}[\sigma_{\theta_1}^2, \sigma_{\theta_2}^2, \sigma_{\theta_3}^2]$, the variance of the fingertip center along with the contact normal n_{Ci} is given by

$$\sigma_{FCi}^2 = n_{Ci}^T (J_{Fi} \Gamma_\theta J_{Fi}^T) n_{Ci}. \quad (42)$$

Letting the variance of the fingertip shape error be σ_s^2, the standard deviation of the i-th fingertip position along with the direction of n_{Ci} is given by

$$\sigma_{pCi} = \sqrt{\sigma_{FCi}^2 + \sigma_s^2}. \quad (43)$$

Thus, we can set the weighting matrix as follows:

$$W = \text{diag}(\underbrace{1/\sigma_u, \cdots, 1/\sigma_u}_{2n}, 1/\sigma_{pC1}, 1/\sigma_{pC2}, 1/\sigma_{pC3}).$$

$$(44)$$

4.3 Velocity estimation

For the servo control of eq.(22), we need the information of the object velocity. Since the camera image is updated every 33msec, however, it is not adequate to estimate the object velocity by differentiating the sequence of the object pose obtained by eq.(41). Fortunately, we can estimate the object velocity directly from the joint velocities.

We have derived the velocity constraint between $\dot{\theta}$ and $\dot{r}_B$ by eq.(8). Solving it for $\dot{r}_B$, we get

$$\dot{r}_B = T^{-1} D_B^+ J_{CF} \dot{\theta}. \quad (45)$$

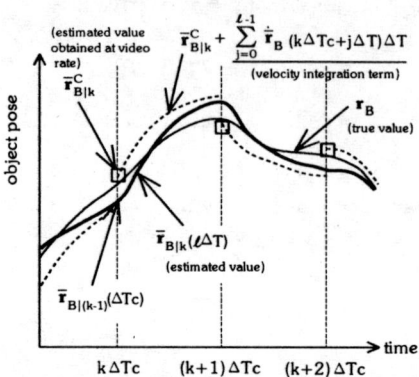

Figure 5: Estimation of object position/orientation

Eq.(45) can be calculated at servo rate and the estimated object velocity $\dot{r}_B$ can be used in eq.(22).

4.4 Estimation at inter-camera-sampling

Although we could estimate the object pose accurately by sensor fusion in section 4.2, it is updated only thirty times a second, which is too slow for servo control. Although the object velocity is available at the servo rate by eq.(45), continuing to integrate the object velocity may accumulate errors. To prevent this error accumulation, we restart the integration at each camera sample, resetting by a new pose updated by eq.(41).

Now let ΔT be a sampling time of the servo loop, ΔT_C be the camera sampling (i.e., 33.3msec). To make the problem simple, we suppose that $\Delta T_C = l_C \Delta T$, an integer multiple of ΔT. Then let k ($= 0, 1, 2, 3, \cdots$) be a camera sampling number, and l ($= 0, 1, 2, 3, \cdots, l_C$) be a servo control cycle number. Then, the estimated object pose $\bar{r}_B(t)$ at $t = k\Delta T_C + l\Delta T$, which is denoted by $\bar{r}_{B|k}(l\Delta T)$, is given by

$$\bar{r}_{B|k}(l\Delta T) = (1-\eta^l)\bar{r}_{B|k}^C + \eta^l \bar{r}_{B|k-1}(\Delta T_C)$$
$$+ \sum_{j=0}^{l-1} \dot{\bar{r}}_B(k\Delta T_C + j\Delta T)\Delta T, \quad (46)$$

where $\bar{r}_{B|k}^C$ denotes the estimated pose by eq.(41) at the most recent camera sample, i.e., $t = k\Delta T_C$, and $\dot{\bar{r}}_B(k\Delta T_C + j\Delta T)$ denotes the object velocity obtained by eq.(45) at $t = k\Delta T_C + j\Delta T$ ($j = 0 \cdots l-1$). η ($0 < \eta < 1$) is a smoothing factor, by which the weight is shifted from the second term to the first one in the right hand side of eq.(46) as the servo cycle number l increases. **Fig.5** illustrates a graph that conceptually explains the estimation method by eq.(46).

5 Experiment
5.1 Experimental system

Fig.6 shows the system configuration of our experimental setup. This setup has two computer systems,

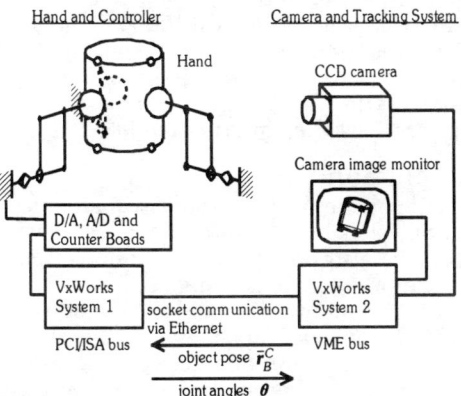

Figure 6: Configuration of the experimental system

one for the hand control and the other for the object pose estimation. Two systems communicate each other via Ethernet.

The hand controller has a Pentium MMX CPU (233MHz) and it is operated by VxWorks. Each finger is a DAIKIN Miniature Robot, three DOF parallel link robot. The length of the second link is 128[mm] and the third link is 169[mm]. A three-axis force sensor is attached to the third link of each finger. An interface board, RIF-01A by FUJITSU, is used for I/O (D/A, A/D, and encoder pulse counter). The object pose estimation system has a CPU board (SPARC 2CE by Force Computers), which is also operated by VxWorks. Tracking Vision, an image-based motion tracker by FUJISTU, is used for the landmark tracking. Tracking Vision has the capability to track more than 100 regions in video rate by template matching. Image size of Tracking Vision is 512×512[pixel]. A CCD camera (TI-24A by NEC) is used for capturing the image.

Each fingertip is made from a soft sponge ball with 20[mm] radius, where the inside of the ball was followed out and a wooden ball with 15.5[mm] radius was inserted as a core. Since this sponge is easily deformed, we set the radius of the fingertip to 18[mm].

The object is a plastic cylindrical cup, 37[mm] radius, 100[mm] height and 105[gw] weight. Four small spheres (3[mm] radius), which are used for landmarks, are attached to the object as shown in **Fig.6**. Since the hand has no joint velocity sensor, joint velocities are obtained by numerically differentiating the sequence of the encoder count values.

5.2 Parameter setting

Object orientations are expressed by roll(ϕ)-pitch(θ)-yaw(ψ) angles. The desired trajectory for the experiment is, as shown in **Fig.7**, picking the object up 40[mm] from the initial position((i) in **Fig.7**), moving and returning linearly $20\sqrt{2}$[mm] along with (1, 1) direction in the x-y plane((ii) and (iii)), rotating and returning $\pi/8$[rad] around the $y(\theta)$-axis ((iv) and (v)), around the $z(\phi)$-axis ((vi) and (vii)), and around the $x(\psi)$-axis ((viii) and (ix)), respectively, and going back to the initial position ((x)). In **Fig.7**, one can see two

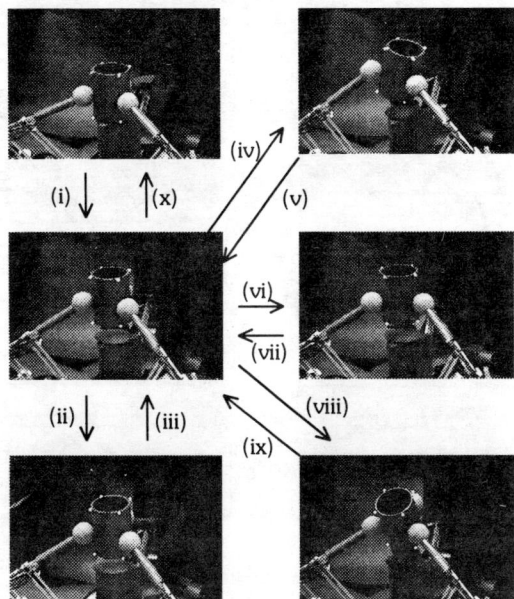

Figure 7: Trajectory in the experiment

more landmarks on the top edge of the cylinder but these points were not used in the experiment.

The estimated dynamic rotational friction coefficient, $\bar{\mu}_i$, was 4.6[mm] and the maximum static rotational friction coefficient, $\bar{\mu}'_i$, was 5.2[mm]. The threshold value to judge the relative rotation, ε_β, was set to 0.5[rad/s] when currently not rotating, and 0.05[rad/s] when currently rotating. This hysteresis setting can avoid the vibrations due to the unnecessary switching of the control input. The threshold value for judging the rotational acceleration, ε_γ, was set to 66[rad/s^2].

Servo gains in eqs.(22) and (23) were set to $\boldsymbol{K}_V = \text{diag}[15, 15, 15, 15, 15, 15][1/\text{s}]$, $\boldsymbol{K}_P = \text{diag}[1800, 1800, 1800, 1800, 1800, 1800][1/\text{s}^2]$ and $\boldsymbol{K}_I = \text{diag}[1, 1, 1][1/\text{s}]$, respectively. The desired internal force was set to $\tilde{\boldsymbol{f}}_d = (1.47[\text{N}]\ 1.47[\text{N}]\ 1.47[\text{N}])^T$.

The standard deviations of the measurements, or diagonal elements of $\boldsymbol{W}^{-1}$, were set to $\boldsymbol{W}^{-1} = \text{diag}(1[\text{pixel}], \cdots, 1[\text{pixel}], 1[\text{mm}], 1[\text{mm}], 1[\text{mm}])$. We assumed that the deformation of the fingertip is dominant over the other factors when estimating the error of the contact points, and its variance was set to a constant value independent to the finger joint angle. η in eq.(46) was set to 0.95.

The proposed control scheme was implemented in the experimental system, together with the proposed pose estimation method. Sampling time, ΔT, was 3.0[msec].

5.3 Experimental result

Fig.8 shows the experimental result. The left side shows the object pose by a control scheme without rotational friction compensation, and the right side shows the result by the proposed scheme with rotational friction compensation. In **Fig.8**, dashed lines

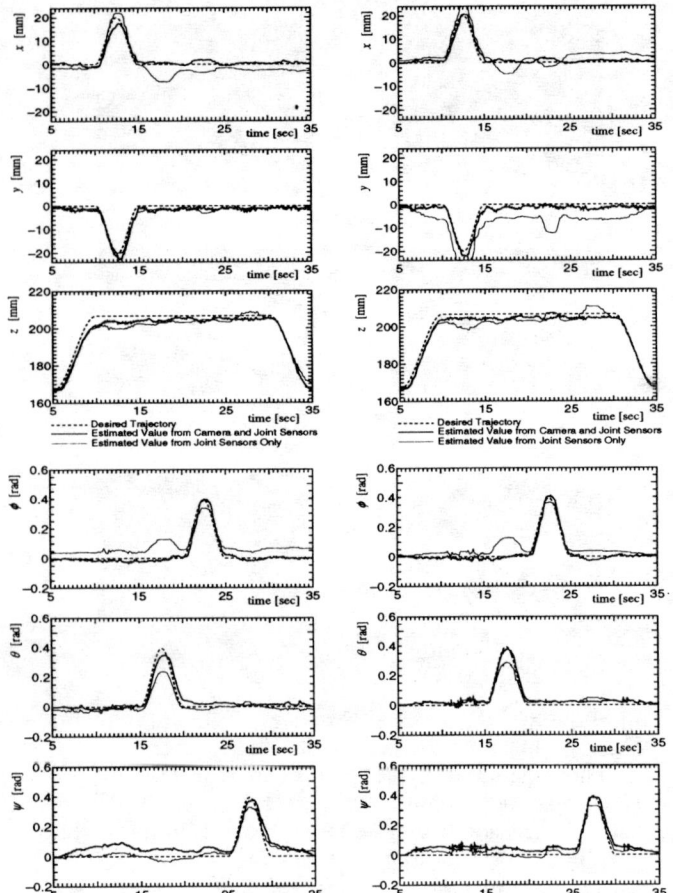

Figure 8: Experimental result (LEFT: object pose without rotational friction compensation, RIGHT: object pose with rotational friction compensation)

show the desired trajectory and thin solid lines are the estimated object pose from the joint sensor information alone (i.e., continued integration of the object velocity obtained by eq.(45)). Bold solid lines show the estimated object pose by the proposed method, fusing visual information and joint information.

Comparing the two results, one can see that the estimated object pose (bold solid line) is closer to the desired trajectory (dashed line) when the frictional moment was compensated. Especially when changing angles θ and ψ, where the frictional moment affects very much, the difference between the two results is clear. As can be seen from **Fig.8**, however, object motion with the proposed scheme tends to be vibratory. This may be due to the inaccurate friction coefficients or the inappropriate threshold values. Anyway, these results show the effectiveness of the proposed control scheme.

In the experiment, we used the proposed object pose estimation method for the servo control. The estimated pose with the joint sensor information alone (thin solid line), however, is far apart from the estimated pose by sensor fusion (bold solid line). This deviation might be due to the accumulation of errors by the velocity integration and unexpected slips. Therefore, if the estimated pose with the joint sensor information alone is used for the servo control, the actual object pose may deviate from the desired trajectory. Thus, this experimental result also shows the effectiveness of the proposed estimation method.

6 Conclusion

In this paper, we proposed a control scheme for object manipulation by a multifingered hand with soft fingertips. We also proposed a systematic method to estimate the pose of the manipulated object by using the vision sensor information and the finger joint information in a complementary manner. We also considered the pose estimation at inter-camera-sampling so that the estimation can be used for the manipulation control.

Our friction model is rather simple and may be inaccurate than those in [3][4]. According to the elliptical limit surface model in [3], where the tangential friction and the frictional spin moment are coupled, we can estimate that the tangential slip caused by a relative rotation at 1[rad/s] would be in the order of [mm/s], assuming the typical friction coefficients of rubber. Thus, it would be reasonable to let the friction model simple and compensate the friction coupling and other modeling uncertainties by vision feedback.

Of course, the proposed pose estimation method can be used not only for the hands with soft fingertips but for other multifingered hands.

References

[1] P.Akella and M.Cutkosky: "Manipulating with Soft Fingers: Modelling Contacts and Dynamics", Proc. IEEE International Conference on Robotics and Automation, pp.764-769, 1989.

[2] A.B.A.Cole, J.E.Hauser, and S.S.Satry: "Kinematics and Control of Multifingered Hands with Rolling Contact", IEEE Trans. on Automatic Control, vol.34, no.4, pp.398-404, 1989.

[3] R.D.Howe and M.R.Cutkosky: "Practical Force-Motion Models for Sliding Manipulation", Int. J. Robotics Res., vol.15, no.6, pp.557–572, 1996.

[4] I.Kao and M.R.Cutkosky: "Quasistatic Manipulation with Compliance and Sliding", Int. J. Robotics Res., vol.11, no.1, pp.20–40, 1992.

[5] B.J.Nelson and P.K.Khosla: "Force and Vision Resolvability for Assimilating Disparate Sensory Feedback", IEEE Trans. on Robotics and Automation, vol.12, no.5, pp.714-731, 1996.

[6] J.K.Salisbury and J.J.Craig: "Articulated Hands: Force Control and Kinematics Issues", Int. J. Robotics Res., vol.1, no.1, pp.4-17, 1982.

[7] N.Sarkar, X.Yun and V.Kumar: "Dynamic Control of 3-D Rolling Contacts in Two-Arm Manipulation ", IEEE Trans. on Robotics and Automation, vol.13, no.3, pp.364-376, 1997.

[8] T.Yoshikawa: "Foundations of Grasping and Manipulation -3. Control", Journal of Robotics Society of Japan, vol.14, no.4, pp.505-511, 1996. (*In Japanese*)

Human Visual Servoing for Reaching and Grasping: The Role of 3-D Geometric Features

Y. Hu[1,2], R. Eagleson[2], and M. A. Goodale[1]

[1]*Vision and Motor Control Laboratory, Department of Psychology*
[2]*Department of Electrical and Computer Engineering*
The University of Western Ontario, London, Ontario, Canada

Abstract

We investigated the kinematics of human visually guided prehension in a situation similar to that typically used in visual servo control of autonomous robots. We found that kinematic parameters of the human grasp, such as transport velocity, path, and grip aperture, were determined by the 3-dimensional geometric structure of the target object, not the 2-dimensional projected image of the object. The results of this study have important implications for the design of reliable and efficient control systems for robots with human-like capabilities.

1. Introduction

Many robots, like human beings, are well-equipped with sensing devices, which provide visual, tactile, and other information about objects in the world. One problem for robots (and for the engineers that design them) is the correct interpretation of the gathered information, so that appropriate actions can be planned and executed. In fact, tasks like object recognition, grasping, and manipulation, which are simple and natural for humans, are a great challenge for current robotic engineering. These human-like capacities are particularly critical for robots that interact with their surroundings in novel and dynamic environments. One way to design such robots is to emulate the control systems that have proved to be so beautifully effective in humans.

The development of robots with human-like capabilities has become a recent trend in robotics and artificial intelligence. Encouraging results have been achieved by some researchers. One example is the integration of image-based visual servo control into autonomous robots [5]. This 'active-vision' approach directly mimics the human visuomotor system, which also uses vision to guide accurate and rapid object-oriented actions. The aim of visual servoing is to control the robot's behavior based on visual features (static and/or dynamic) existing in the workspace. These visual features are acquired through retina-like sensing devices, such as CCD cameras, and are analyzed using modern image-processing technologies. In general, the visual features are geometric primitives and are used to compute the control parameters for grasping or tracking objects in the workspace [1, 2].

Although robots under visual servo control can perform certain tasks, such as grasping objects, quite well, there are inherent problems associated with the particular hardware used to implement the motor actions and the software used to process the "sensory" information. One problem is the heavy computational load involved in action planning. By definition, visual servo control includes an image-processing system in the control loop. Particular image features are assumed to reflect particular spatial relationships between a designated target and the robotic end effectors. To use such a relationship, a Jacobian matrix must be computed to extract the image features, no matter whether the visual sensors are attached to the end effectors or fixed in the workspace [5]. Some difficulties arise due to the need to compute complex Jacobian matrices on-line. To deal with these difficulties, Sanz et.al. [12] employed a two-phase approach in a well-defined workspace. In the first phase, a Jacobian matrix was pre-computed for each point on a grid of possible goal positions in the defined workspace. In the second phase, the visual servo control simply established which goal point was being used and then used the pre-computations to generate the required action. Although this approach will work in a well-defined workspace, it is clearly unsuitable for visual servo control in novel and dynamic environments.

Given that the control of robots in novel environments is not an easy problem to solve, the question arises as to how the human visuomotor system deals with the same problem. As we know, vision enables humans to reach out and grasp objects with great accuracy. Indeed, if one regards the human eye as a visual sensor, the human hand as an end effector, and the human visuomotor system as a visual servo control system, then humans are literally biological robots with capabilities far beyond that of any state-of-the-art robot. Humans can generate successful visually guided reaching and grasping movements within a couple hundred milliseconds of observing the target object [6]. One can ask therefore: how is it that the human visuomotor system can transform visual information into action so efficiently? Answering this question could provide useful insights for the development of efficient control systems for robots and perhaps imbue them with human-like capabilities.

It was this goal that motivated us to conduct an experiment to examine the way in which the human visuomotor system (the human equivalent of the robot's visual servo control system) transforms geometric information about a target object into an accurate visually guided grasp[1]. In short, we examined the relation between the geometric features of target objects and the kinematics of the transport and grip components of a grasping movement. To do this, we varied the dimensions of height and width of the target object and its position in the workspace. During the execution of the movement, subjects could see their hand as well as the target object. This scenario simulates the case of visual servo control, in which the image is analyzed throughout the execution of the movement. We found that kinematic parameters of the human grasp, such as transport velocity, path, and grip aperture, were determined by the 3-D geometric structure of the target object, not the 2-D projected image of the object.

The present paper is organised as follows: In Section 2, we describe the experimental procedure and apparatus, the kinematic parameters we measured, and the statistics we used. In Section 3, we summarize the experimental results. In Section 4, we discuss the implications of these results for the design of robotic control systems. Finally, in Section 5, we present our conclusions and plans for future research.

2. Experimental Methods

All subjects were strongly right handed, as determined by a modified version of the Edinburgh handedness inventory [10]. Subjects had normal or corrected-to-normal vision, with a stereoacuity of at least 40" of arc as determined by the Randot Stereotest (Stereo Optical, Chicago).

Three infrared light-emitting diodes (IREDs) were affixed to the dorsal surface of the right hand. One IRED was placed on the left edge of the nail of the index finger, another on the right edge of the thumbnail, and a third on the left side of the wrist opposite the styloid process. The position of each these IREDs was recorded by a conventional WATSMART system (Northern Digital Inc., Waterloo, Ontario, Canada) at sampling rate of 100 Hz. The accuracy of this system and data analysis techniques have been documented elsewhere [6]. Before initiating a trial, each subject was instructed to bring the tips of her/his index finger and thumb together and to depress a starting button, which was fixed at the subject's midline and at 83.0 cm above the floor. On a given trial, the subject was requested to pick up an object placed in front of her/him on a black working surface, using the index finger and thumb. They were instructed to pick up the object across its width (see Fig. 1). The object was always placed at the subject's midline with its length perpendicular to this midline. A chin rest was employed to restrict the subject's head movements. A pair of liquid-crystal shutter spectacles (Translucent Technologies Inc., Toronto, Ontario, Canada) was used to control stimulus presentation. Data collection began as soon as the shutters opened and stopped when the object was picked up.

All subjects were tested in visual closed loop in which the target object and the moving hand were always visible. As Fig. 1 shows, target objects were divided into three groups, according to the ratios of their dimensions. The six objects in Group A, which

---

[1] The experiment was carried out at the University of Western Ontario in compliance with the Social Sciences and Humanities Research Council (Canada) Guidelines (1981).

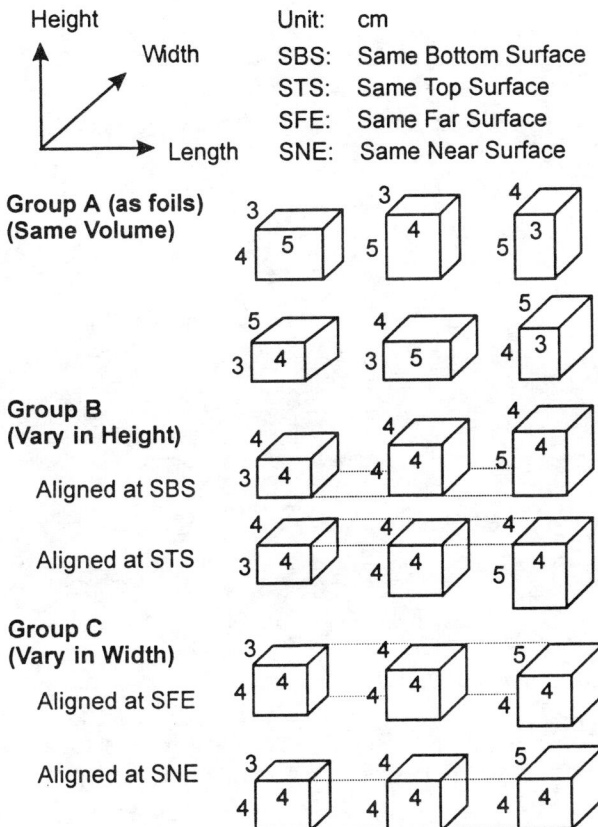

Figure 1 Target objects used in the experiment.

all had the same volume but varied in all three dimensions, served as foils. Group B varied only in height. Group C varied only in width. Length was not manipulated as an experimental variable since it was found not to affect grip aperture in an earlier study [4]. Notice that the middle-sized object in Groups B and C measured 4 × 4 × 4 cm in both cases. This object was placed so that its top surface was 1.0 cm above the plane of the start button and far edge 26.0 cm from the start button. The other two objects in Group B were placed with their bottom surface aligned with the bottom surface of the middle-sized object (Same Bottom Surface) or with their top surface aligned with that of the middle-sized object (Same Top Surface). The other two objects in Group C, were placed so that their far edge was aligned with the far edge of the middle-sized object (Same Far Edge) or with their near edge aligned with that of the middle-sized objec (Same Near Edge). A robotic arm (CRS-A255, Burlington, Ontario, Canada) was used to place the target object in the workspace according to the alignments described above. The order in which different objects and different alignments were presented was completely randomized. Each object was presented four times in each alignment for a total of 72 trials.

The following kinematic parameters were computed from the stored three-dimensional position for each IRED: movement duration, peak velocity, maximum grip aperture, maximum elevation, percent time to peak velocity, percent time to maximum grip aperture, and percent time to maximum elevation. Movement duration was calculated by subtracting the onset time from the time at which the target object was picked up. [Onset time was measured from the moment the resultant velocity of the wrist IRED exceeded a value of 5.0 cm/s over ten consecutive sampling frames.] Peak velocity was peak resultant velocity of the wrist IRED. Maximum grip aperture was the maximum resultant distance that was achieved between the index finger and thumb as the subject reached out to grasp the object. [To reduce intersubject variance, maximum grip aperture was normalized against a standard aperture obtained for each subject by measuring the IRED positions as he or she gripped a standard object with their index finger and thumb.] Maximum elevation was defined as the maximum height that each IRED reached from its initial position over the working surface perpendicularly. Percent time to reach maximum velocity, grip aperture, and elevation was calculated as a percentage of movement duration.

For each kinematic parameter, the means were calculated from the four replications of grasping movements made to each target object (except foils). Only trials for which complete data were available for all parameters were included in the analysis. Less than 1.0% of the trials were eliminated using this criterion and each mean was based on a minimum of three trials. Simple analyses of variance were carried out on the mean values for each kinematic parameter. Post-hoc contrasts were carried out using the Newman-Keuls testing procedure. All tests of significance were based upon an alpha level of 0.05.

3. Experimental results

In general, the wrist velocity and grip aperture profiles for the grasping movements shown by our

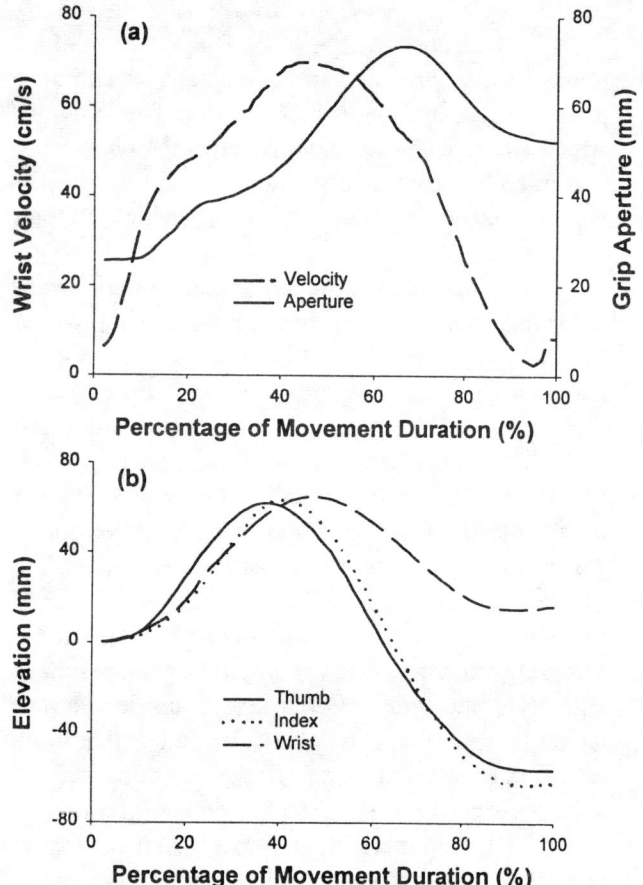

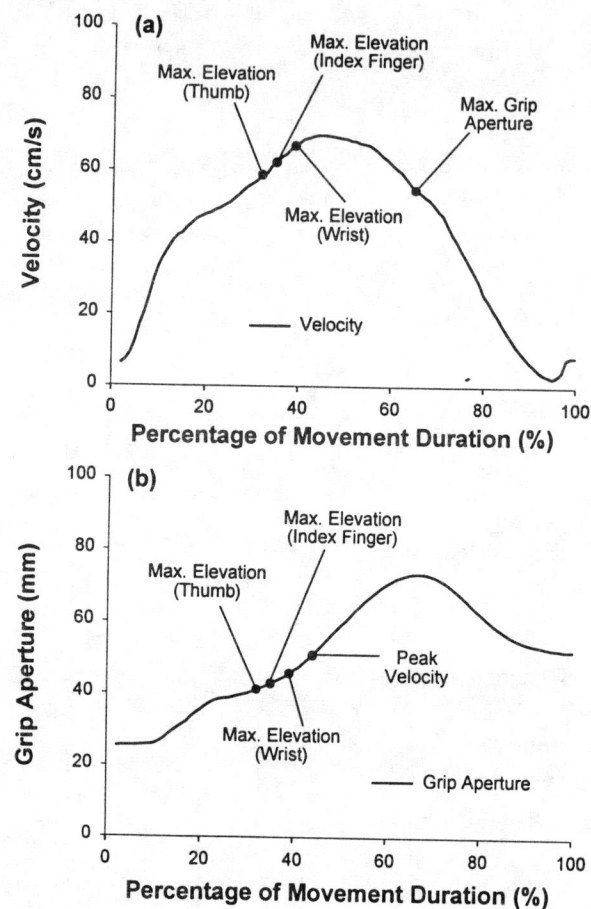

Figure 2 Representative profiles of wrist velocity, grip aperture, and elevation of the thumb, index finger, and wrist.

Figure 3 Description of percent time to each maximum in comparing with peak velocity and grip aperture.

subjects were similar to those typically found in the literature [6]. As illustrated in Fig. 2, wrist velocity, grip aperture, and elevation of the thumb, index finger, and wrist, all showed asymmetrical bell-shaped curves, in which the maximum values were achieved at different times after movement onset. As Fig. 3a indicates, maximum elevation of the thumb, index finger, and wrist was reached just slightly earlier than the peak velocity of the wrist, whereas maximum grip aperture was not achieved until much later, during the phase in which the hand was decelerating on its way to the target object. In fact, when plotted on the grip aperture profile, it is clear that the maximum elevation of the thumb, index finger, and wrist was achieved approximately halfway through the opening phase of the grip (Fig. 3b). The implications of this timing sequence will be discussed later.

Several analyses of variance were carried out to examine the effects of the geometrical properties on the various kinematic parameters that we measured.

Elevation: Separate analyses were carried out to examine the effect of object height and object width on the elevation of the thumb, index finger, and wrist. As Fig.2b illustrates, there was no significant difference in the maximum elevation achieved by the thumb, index finger, and wrist. There were, however, significant differences in the timing of maximum elevation – with the thumb reaching maximum elevation first, followed by the index finger, and then the wrist [$F(2, 22) = 38.27$, $p < 0.001$, $F(2, 22) = 40.22$, $p < 0.001$; for varying object height and width, respectively]. Object height, but not object width, had an effect on the maximum elevation for the thumb, index finger, and wrist [$F(2, 22) = 10.45$, $p < 0.001$; $F(2,22) = 0.10$, $p > 0.05$; for varying object

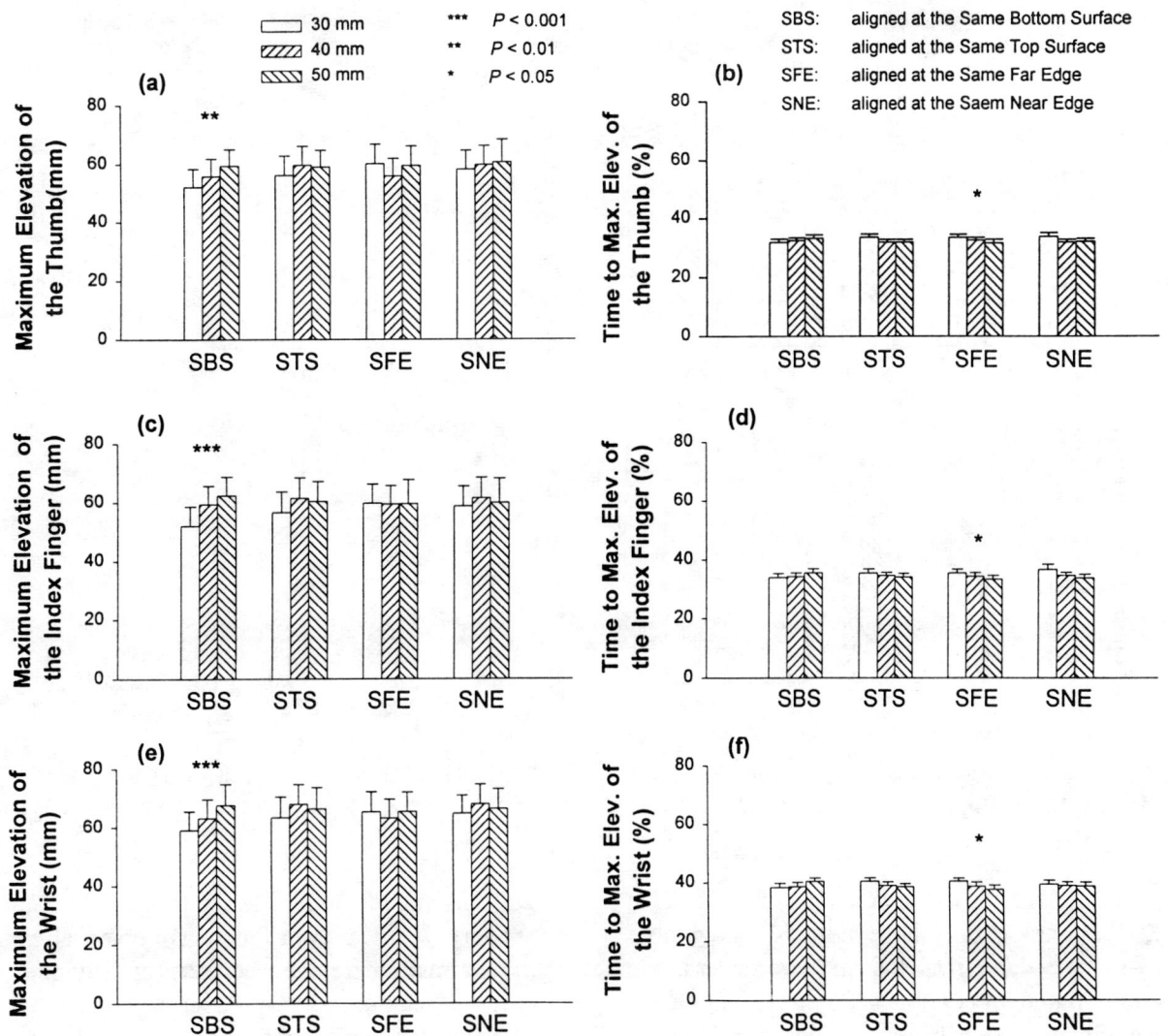

Figure 4 Mean values of maximum elevation and percent time to the maximum as a function of changes of size and alignment: (a) and (b) the thumb; (c) and (d) the index finger; (e) and (f) the wrist. (Error bars represent the standard error).

height and width, respectively]. To further examine how the object's size at each alignment affects the maximum elevation of the thumb, index finger, and wrist, separate analyses of variance were conducted. As Fig. 4 shows, maximum elevation of the thumb, index finger, and wrist increased as a function of object height when the objects in the SBS alignment were compared [$F(2, 22) = 7.41, p < 0.01$; $F(2, 22) = 22.03, p < 0.001$; $F(2, 22) = 11.99, p < 0.001$; for the thumb, index finger, and wrist, respectively]; percent time to these maxima was a decreasing function of object width only when the objects in the SFE alignment were compared [$F(2, 22) = 4.88, p < 0.05$;

$F(2, 22) = 4.69, p < 0.05$; $F(2, 22) = 4.59, p < 0.05$; for the thumb, index finger, and wrist, respectively]. When objects were compared across the other alignment conditions, there was no effect of object's height or width on maximum elevation or percent time to maximum elevation.

Velocity: Separate analyses were carried out to examine the effect of object height and object width on the peak velocity of the reaching component of the grasp. It was found that overall subjects achieved a lower peak velocity for objects in the SBS alignment than for the same objects in the STS alignment [$F(1, 11) = 5.44, p < 0.05$]. As well,

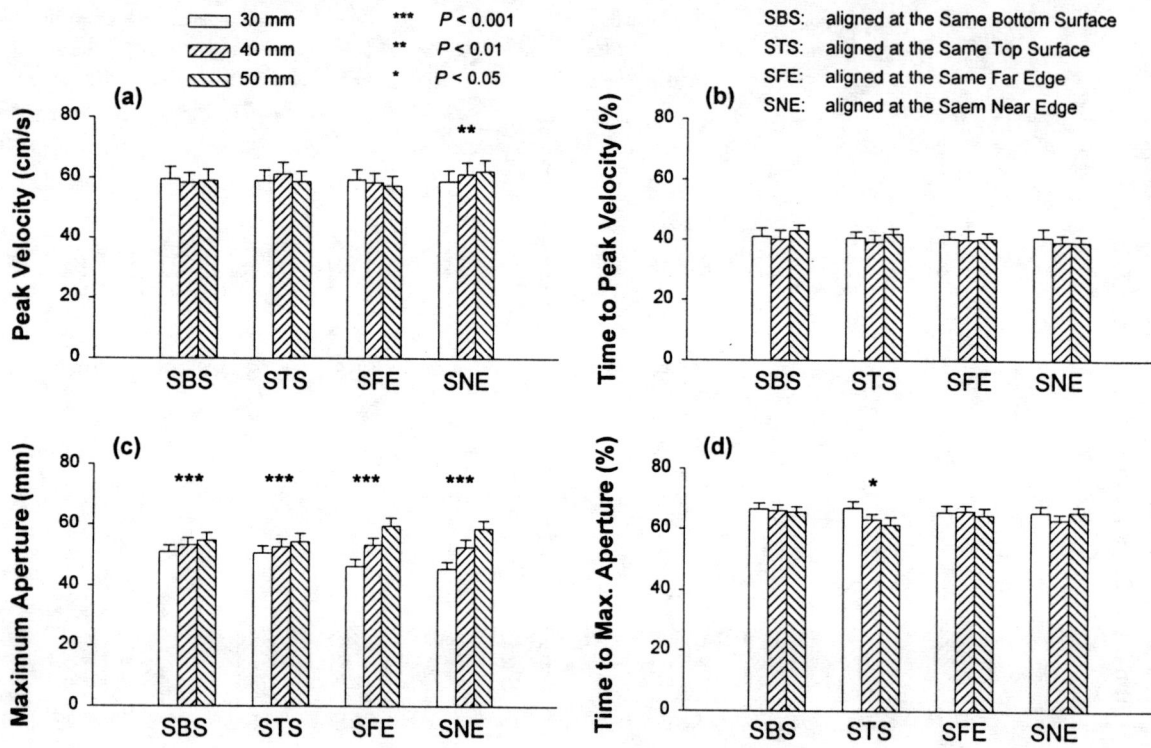

Figure 5 Mean values of kinematic parameters as a function of size and alignment: (a) peak velocity; (b) percent time to peak velocity; (c) maximum grip aperture; (d) percent time to maximum aperture. (Error bars represent the standard error).

subjects achieved a lower peak velocity for objects in the SNE alignment than those in the SFE alignment [$F(1, 11) = 10.09$, $p < 0.01$]. Nevertheless, the different alignment conditions had no effect on the percent time to peak velocity. To pursue the effect of object height and width on peak velocity, percent time to peak velocity and the interaction of these variables with alignment, separate analyses of variance were conducted. As illustrated in Fig. 5, peak velocity increased significantly as a function of object width when target objects were in the SNE alignment [$F(2, 22) = 5.97$, $p < 0.01$], but no significant differences were observed when objects were placed in the three other alignments (although there was a trend for peak velocity to decrease as a function of object width when target objects were in the SFE alignment). Again there was no effect of object height and width on percent time to peak velocity, no matter how objects were aligned.

Grip Aperture: Separate analyses were carried out to investigate the effect of object height and object width on the maximum grip aperture in grasping movements. Maximum grip aperture did not vary significantly as function of object alignment. In contrast, there were effects of alignment on timing: maximum grip aperture was reached significantly earlier when objects were in the STS alignment as compared to when they were in the SBS alignment [$F(1, 11) = 5.62$, $p < 0.05$]. Not surprisingly, as illustrated in Fig. 5, maximum grip aperture increased as a function of object width, no matter how the objects were aligned [$F(2, 22) = 95.18$, $p < 0.001$; $F(2, 22) = 118.26$, $p < 0.001$, for SFE and SNE respectively]. But maximum grip aperture also increased as a function of object height, and again it did not matter how the objects were aligned [$F(2, 22) = 13.45$, $p < 0.001$; $F(2, 22) = 14.11$, $p < 0.001$; for SBS and STS respectively]. It is interesting to note that percent time to maximum grip aperture was affected by object height only when target objects were in the STS alignment [$F(2, 22) = 4.45$, $p < 0.05$]. Object width had no effect on percent time to maximum grip aperture.

4. Discussion

The results of the present experiment show that the 3-D geometric properties of a target object affect the kinematic parameters of human grasping movements. In agreement with earlier studies [6], increasing the size of the 'relevant' dimension (in our case the object's width) led to corresponding increase in maximum grip aperture. What was surprising was the fact that the object's height also had an effect on maximum grip aperture – such that subjects produced wider grasps for higher objects. The increase in grip aperture presumably reflected the fact that the visuomotor system was ensuring that fingers would clear the top surface of the object as the hand closed in for the grasp. This adjustment for object height was also observed in the overall elevation of the hand which also increased as function of object height – but only when the objects in the SBS condition were compared. Of course, when objects of different height are aligned along their bottom surface different clearances are required. When objects of different heights are aligned at their top surface, different clearances do not need to be programmed. Not surprisingly, therefore, we did not observe any effect of object height on the elevation of the hand when the objects were in the STS alignment. In short, the elevation of the hand during the grasp was affected by where the top surface of the object was with respect the hand's initial position.

These aforementioned effects on grip aperture and hand elevation could have only arisen if the human visuomotor system computes a grasping trajectory that takes into account of the 3-D geometric structure of the target objects -- a maneuver that is necessary in achieving a successful and stable grasp. This conclusion is also supported by a number of other studies. Sakata [11], for example, showed that neurons in the parietal cortical area, which is thought to play a critical role in the control of grasping movements, are sensitive to the 3-D structure of objects. In fact, many of these cells show binocularity – and respond to 3-D objects only if both eyes are open. In a recent behavioral study, Mamassian [8] showed that the maximum grip aperture is correlated with the actual size of the target objects rather than the size of its projected retinal image (which would be the case for many visual servo systems that use sensory devices such as CCD cameras). Studies with neurological patients, whose visuomotor systems have been spared despite massive damage to their visual perception system, are now much more dependent on binocular information for computing their grasp than even individuals with normal vision [3, 9]. All these studies, together with the results of the present study, suggest that visual servo control in robots could be significantly improved by using visual sensors that can compute the 3-D geometrical structure of objects – just as the human visuomotor system does.

A number of other interesting findings emerged from this study. The peak velocity of the movement was scaled to the location of the far edge of the target object -- in agreement with an earlier study in our laboratory [13]. A new observation was that the timing of maximum grip aperture varied as a function of the position of the object's lower surface (i.e. the surface on which the target was assumed to rest). The lower the surface with respect to the initial position of the hand, the sooner maximum grip aperture was achieved. A parallel finding was observed in the timing of the elevation of the thumb, index finger, and wrist; namely, time to maximum elevation varied as a function of the distance between the initial position of the hand and the near edge of the target object. The closer the object, the sooner the hand reached maximum elevation. These observations suggest that the location of certain surfaces of the target with respect to the initial position of the hand enters into the computation of grasp kinematics. Although it is not clear how much of a role proprioception plays in this computation, it seems likely that some kind of proprioceptive information is used to compute the initial position of the hand. Such 'proprioception-based' computations might be necessary in robotic systems when visual sensors are separated from the end effectors.

The fact that the thumb always reached maximum elevation before the index finger suggests that the thumb may be used as the end point in the trajectory planning of the grasp. Indeed, this finding is similar to an earlier observation by Wing et al. [14] who suggested that the thumb and index finger may be functioning in a 'master-slave' relationship in the programming of grasp. In fact, the hand begins to open as soon as it leaves the start position. In other words, the 'transport' of the hand and the formation of the 'grasp' occur simultaneously. A number of studies, and the present study was no exception, have shown that the transport component and the grasp component are temporally coupled with the grasp achieving maximum aperture during the deceleration phase of the transport component [6, 7]. These strategies are quite different from those normally used in the visual servo control of a robotic gripper, in which the gripper is first transported to the location over a target object and then the two fingers of the gripper open simultaneously to grasp the target object.

5. Conclusions and future work

In real-world reaching and grasping tasks, humans are able to make use of multiple sensory cues and invoke highly evolved visuomotor strategies. Indeed, the results of the present experiment show that the human visuomotor system takes into account the 3-D geometric features rather than the 2-D projected image of target objects to program and control the required movements. These computations, while more complex than those typically carried out in visual servo control systems, permit humans to operate in an almost limitless range of environments. The development of comparable robotic applications could profit from the importation of the principles used by this highly adaptive biological system.

Additional research is required to determine exactly how the 3-D geometric features to which the human visuomotor system is sensitive are transformed into motor output. Such research could provide the algorithms to the design of reliable and efficient controllers for autonomous robots with human-like abilities.

References

[1] A. Bendiksen and G. Hager, "A vision-based grasping system for unfamiliar planar objects," *IEEE Int. Conf. on Robotics & Automations*, Vol. 3, pp. 2844-2849, Los Angles, CA, USA, May, 1994.

[2] A. Crétual and F. Chaumette, "Image-based visual servoing by integration of dynamic measurements," *IEEE Int. Conf. on Robotics & Automation*, Vol. 3, pp. 1994-2001, Leuven, Belgium, May, 1998.

[3] H.C. Dijkerman, A.D. Milner, and D.P. Carey, "The perception and prehension of objects oriented in the depth plane. 1. Effects of visual form agnosia," *Experimental Brain Research*, Vol. 112, pp. 442-451, 1996.

[4] Y. Hu, R. Eagleson, and M.A. Goodale, "The effect of delay on the kinematics of grasping," *Experimental Brain Research*, in press, 1999.

[5] S. Hutchinson, G. Hager, and P.I. Corke, "A tutorial on visual servo control," *IEEE Trans. on Robotics & Automation*, Vol. 12, No. 5, pp. 651-670, October, 1996.

[6] L.S. Jakobson and M. A. Goodale, "Factors affecting higher-order movement planning: a kinematic analysis of human prehension," *Experimental Brain Research*, Vol. 86, pp.199-208, 1991

[7] M. Jeannerod, "The timing of natural prehension movements," *J. of Motor Behaviour*, Vol. 16, No. 3, pp. 235-254, 1984.

[8] P. Mamassian, "Prehension of objects oriented in three-dimensional space," *Experimental Brain Research*, Vol. 114, pp.235-245, 1997.

[9] J.J. Marotta, M. Behrmann, and M.A. Goodale, "The removal of binocular cues disrupts the calibration of grasping in patients with visual form agnosia," *Experimental Brain Research*, Vol. 116, pp.113-121, 1997.

[10] R.C. Oldfield, "The assessment and analysis of handedness: the Edinburgh inventory," *Neuropsychologia*, Vol. 9, pp.97-112, 1971.

[11] H. Sakata, "Coding of 3-D features of objects used in manipulation by parietal neurons," in *Vision and Movement Mechanisms in the Cerebral Cortex*, R. Caminiti, et al. Eds., Strasbourg: HFSP, 1996, pp 55-63.

[12] P.J. Sanz, A.P. del Pobil, J.M. Iñesta, and G. Recatalá, "Vision-guided grasping of unknown objects for service robots," *IEEE Int. Conf. on Robotics & Automation*, Vol. 4, pp. 3018-3025, Leuven, Belgium, May, 1998.

[13] P. Servos, L.S. Jakobson, and M.A. Goodale, "Near, far, or in between? – Target edges and the transport component of prehension," *J. of Motor Behaviour*, Vol. 30, pp. 90-93, 1998.

[14] A.M. Wing, A. Turton, and C. Fraser, "Grasp size and accuracy of approach in reaching," *J. of Motor Behaviour*, Vol. 18, pp. 245-260, 1986.

Sensor Based Control for the Execution of Regrasping Primitives on a Multifingered Robot Hand

Mohammad Asim Farooqi   Takashi Tanaka   Yukio Ikezawa   Toru Omata
Kazuyuki Nagata*

Tokyo Institute of Technology
4259 Nagatsuda, Midori-ku, Yokohama, Kanagawa 226-8502, JAPAN
*Electrotechnical Laboratory
1-1-4 Umezono, Tsukuba, Ibaraki 305, JAPAN

Abstract

In this paper we demonstrate the robust execution of two regrasping primitives, *rotation* and *pivoting* on a four-fingered hand, based on the real-time sensors data based control. We indicate frequently occurring grasp errors and faults during execution of the two regrasping primitives. The sensors data is obtained from two sources. One is force-torque sensors mounted on the finger tips of the four fingers. The other is a laser range sensor mounted on the palm of the hand. The grasp information acquired through these two sources is useful for both sensing the errors and controlling the fingers. This paper proposes strategies to recover or prevent faulty grasps during manipulation, which may lead to unwanted object movements or grasp failures. The object to be manipulated is a polygonal prism, model of which is known a priori.

1 Introduction

In order to enable a multifingered hand to manipulate an object effectively, elaborate control algorithms are required to be chalked out. In the area of contacts' data feedback control for object manipulation, there are two major areas of research; first the contact sensing, which provide contact data typically in the form of force and torque measurement with respect to the sensor coordinate frame. Various studies relating these sensors and their mathematical descriptions for data evaluation have been conducted [1]-[10]. Second area is the kinematical study and the planning of the regrasping strategies. In this area also, various regrasping primitives have been analyzed [11]-[21].

However, we do find the simultaneous exploitation of these two areas for the practical manipulation of objects yet not very distinct. Such as, most of the work on utilization of contact data is done for object recognition [5], [6], [7], shape interpretation [7], [8], or model acquisition [9], [10], that is to say, towards object perception in general. Similarly, most of the studies regarding regrasp primitives are limited to their kinematical and geometrical analysis and planning only.

Here in this paper, we focus on the execution of regrasping primitives. Regrasping is essentially subjected to errors due to inaccurate control of fingers, such as due to unexpected slips during repositioning of fingers, unstable or insecure grasp and so on.

We have adopted two primitives *rotation* [11] and *pivoting* [12] for changing the orientation of the grasped object. The advantages of developing a manipulation system based on primitives are that (1) they are relatively simple but their sequential execution can achieve three dimensional reorientations [13], (2) errors during execution are analyzed independently for each primitive and control programs are developed independently.

This paper indicates frequently occurring grasp errors and faults during execution of the two primitives and proposes strategies to recover or prevent the faulty grasps, which may lead to unwanted object movements or grasp failures.

For conducting this work we obtain the data about a grasp from two sources. One source of data is force-torque sensors mounted in the last links of the fingers near their tips. The force and torque measurements sensed at the sensor coordinate frame are sufficient to evaluate other global parameters, i.e., the normal, the force and the torque vectors, and position of the contact point with respect to sensor reference frame, given the shape of the finger [3], [7]. The other source of data is a laser-range sensor mounted on the center of the palm of the hand. It measures the linear distance between the object and the palm.

The grasp information acquired through these two sources is useful to estimate the position and orientation of a grasped object. However, some grasps do not give a unique solution to this estimation. This paper discusses how to prevent such grasps as much as possible.

In section 2, we define the *rotation* and *pivoting* primitives and basic conditions for their execution. Section 3 first describes the force-torque sensor and the laser range sensor, then describes estimation of the position and orientation of a grasped object based on data from these two sensors. Section 4 presents grasp errors encountered during execution of *rotation* and *pivoting*, as well the strategies to recover or prevent them. Experimental results are presented.

Assumptions: This work has been conducted on a three jointed four-fingered hand. The hand is installed in inverted position i.e., the fingers point downwards as shown in the picture of Fig. 1.

Figure 1: The robotic hand.

We have certain general assumptions for this work as follows.

1. The hand is provided with at least four fingers.
2. The object is small enough to be manipulated by the hand.
3. The object is a polygonal prism.
4. The geometrical model of the object is known.
5. The finger contact is point contact with friction.

A three-fingered hand could perform regrasping but a four-fingered hand is much better for regrasping from both the sensing and manipulating points of view as we will discuss in this paper. The third assumption is not very restrictive because most of existing objects are a polygonal prism or can be regarded as a polygonal prism. The fingertips in our hand have some softness

and can exert a moment. Since it is not reliable, therefore we develop a manipulation system not relying on it.

2 Primitives

When the object is rotated about the z-axis of the hand reference frame as shown in Fig.2(a), by repositioning the fingers and regrasping the object, we term it as *rotation* primitive [11]. In *rotation* all the finger contacts lie in the same plane parallel to xy-plane surrounding the object. We assume that the *rotation* primitive is executable only about the z-axis of the hand frame, which is reasonable for our hand. Note that this assumption does not mean that the hand can reorient an object only about the z axis of the hand frame. It can also do so about the other axes, such as exhibited in *pivoting* or the non-regrasping strategy described next.

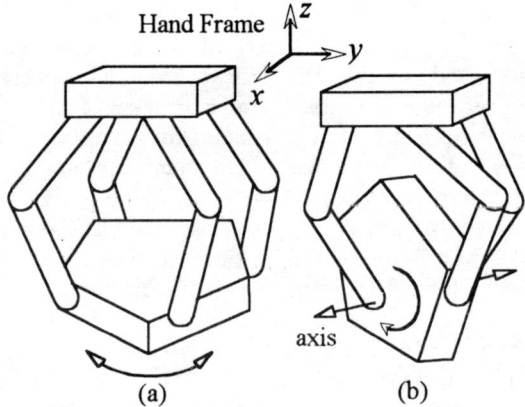

Figure 2: (a) Rotation. (b) Pivoting.

In *pivoting* primitive [12], two fingers form a pivot axis grasping an object at its two parallel and opposite faces as shown in Fig.2(b). The third finger pushes the object so that it pivots about the pivot axis. The two faces through which the pivot axis passes should be parallel or nearly parallel. *Pivoting* is executable only when the pivot axis is lying in a plane parallel to xy-plane of the hand frame as shown.

To reorient the object at a large angle, *rotation* and *pivoting* are effective. On the other hand, the hand can rotate and translate the object without regrasping it as far as none of the fingers reach their workspace boundaries. We implement this non-regrasping strategy to control the position and orientation of an object. If the error between the desired and the current orientation, as well as between the desired and the current position, exceeds some given threshold, the non-regrasping strategy is invoked to generate trajectories for the fingers to eliminate the error.

3 Components of sensors data based system

3.1 An intrinsic force-torque sensor

An intrinsic force-torque sensor mounted on the fingertips (last link) measures the force and moment at that finger in the sensor frame. Using these measurements, the force and the moment at the contact, and the position of the contact can be estimated. Mathematical models and their solutions for soft and hard fingers were shown by Bicchi [3]. Such a model for a semi-sphere shaped finger is illustrated in Fig.3, where B is the sensor frame, and p is the location of the center point of the contact with respect to B.

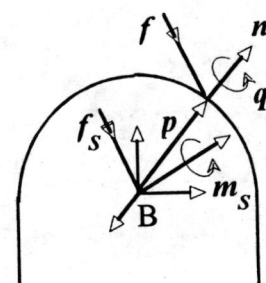

Figure 3: Force and moment at the sensor frame

The relationship between force f_s and moment m_s - the known quantities at B, and the force f and moment q - the unknown quantities at the contact centroid for a soft finger model are given by,

$$f_s = f \quad (1)$$

$$m_s = q + p \times f \quad (2)$$

For the semi-sphere fingertip

$$q = sn \quad (3)$$

$$p = rn \quad (4)$$

where n is the unit normal at the contact point, s is the magnitude of the moment, r is the radius of the semi-sphere. Further details regarding solutions of these equations can be found in [3] and [7].

3.2 Estimation of the position and orientation of a grasped object

Let n_i and p_i; $i = 1, \ldots, N$, represent the normal and position vectors, respectively, at the contact point of finger i on the grasped object with respect to the hand frame. These can be calculated using data from the force-torque sensors and joint angle sensors.

Let F_i be the face of the object finger i makes contact with. The model of F_i is given by

$$m_i^T x = d_i \quad (5)$$

where m_i represents the normal vector of F_i with respect to the model frame and d_i represents the distance from the origin of the model frame. In Fig. 4 the model frame is coincident with the hand frame.

Then the orientation R and the position P of the grasped object with respect to the hand frame are given by rotating and translating the model frame. They are optimally estimated as described by the following two equations respectively [5],

$$\min_R \sum_{i=1}^{N} \|n_i - Rm_i\|^2 \quad (6)$$

$$\min_P \sum_{i=1}^{N} \|m_i^T R^T(p_i - P) - d_i\|^2 \quad (7)$$

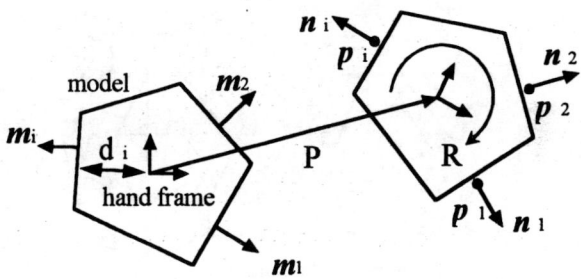

Figure 4: Position P and orientation R of the grasped object w.r.t. the hand frame

We assume that at an initial grasp F_i is known, that is, which face of the object finger i is in contact with is known. The search algorithm in [5] could remove this assumption.

3.3 Laser range sensor

The contact normals and contact positions of the fingers do not necessarily determine the position and orientation of a grasped object uniquely. Fig. 5 (1) through (4) illustrate such grasps. In (1), all four fingers grasp the side faces of the object parallel to z-axis, which is just the grasp form in *rotation*. The translation of the object along the z-axis does not change the parameters of the contact normals and contact positions. Thus this data alone cannot determine uniquely the relative position of the object along the z-axis with respect to the hand.

To solve this, we have mounted a micro-laser range sensor on the palm of the hand. It measures the linear distance of the object from the palm.

The laser sensor, however, still cannot determine the position and orientation of the object in Fig. 5 (2), (3), and (4). In (2), only two fingers are in contact with the object. The two-fingered grasp requires soft fingertips and therefore it is not reliable. We decide not to allow the hand to grasp an object with only two fingers.

During *pivoting* two fingers grasp an object forming a pivot axis and the third finger pushes the object. The third finger needs to be repositioned when it reaches its workspace boundary. If it is lifted, only the two fingers forming the pivot axis grasp the object. To avoid this, we use the fourth finger. The fourth finger makes contact with the object before the third finger is lifted. This makes execution of *pivoting* much more reliable.

In (3) more than one finger is in contact with the same face of the object and in (4) one of the fingers is in contact with the same face as irradiated by the laser. We allow these grasps currently for the best use of the hand. Otherwise the hand cannot manipulate an object well because of the limited workspaces of the fingers.

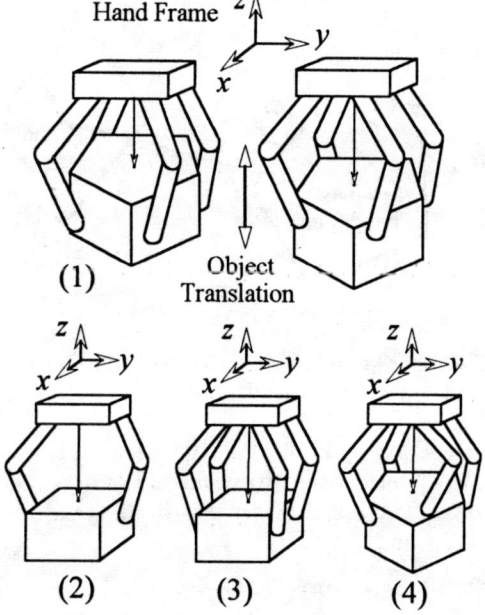

Figure 5: Indeterminate position and orientation of a grasped object

3.4 Irradiated face shifts during the *pivoting*

During *pivoting*, the face irradiated by the laser may shift as the object is being pivoted. The linear distance measurement by the laser sensor is only available if the irradiated face is known. Without the distance information from the laser sensor the position of the object in the direction of l in Fig. 6 is indeterminate. An exceptional case is that the fourth finger touches the object on a face not parallel to the face the pushing finger touches.

Since the model of the object is known, the face shift can be detected as illustrated in Fig. 6. Suppose that initially face 1 is irradiated. As the object is pivoted counterclockwise, either face 1 or 2 is irradiated. If the laser irradiates face 2, then the measured distance must be in between the range of those of points A and B, and if the laser irradiates face 1, it must be greater than the distance of point B. Thus by calculating the distances of points A and B and comparing them with the measured distance, we can determine the irradiated face and in turn evaluate the indeterminate position.

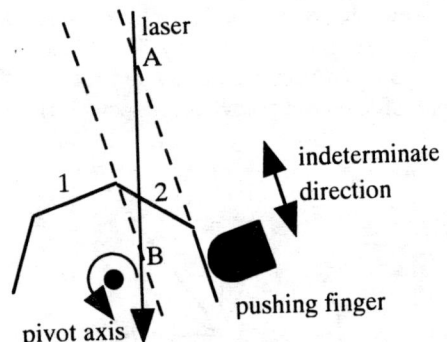

Figure 6: Detection of irradiated face shifting

4 Typical errors and their recovery strategies

In this section we indicate frequently occurring errors and faults encountered during execution of *pivoting* and *rotation*. We simultaneously describe the strategies to eliminate such grasp faults, and other precautionary measures to prevent such errors from growing deep towards grasp failures.

4.1 pivoting

(*a*). During execution of *pivoting*, the pivot axis may slip under the force of the pushing finger as shown in Fig. 7 (1). If the pivot axis is situated far away from the center of gravity of the object, then an uncontrolled *pivoting* may occur under the action of the gravity. Therefore to make the grasp stable, it is desirable that the pivot axis should pass as near as possible to the cog. To ensure it, we calculate the location of the center of gravity and monitor the distance between the cog and the pivot axis. If this distance increases from a certain allowed range, then the pivot axis is brought in the allowed region by lifting or sliding the two fingers forming the pivot axis with the third and fourth fingers. This can be achieved easily if the object has another pair of parallel (or nearly parallel) faces for the third and fourth fingers to grasp it as shown in Fig. 7 (2).

(b). The slip of the pivot axis may also cause it to draw near the edge of the object, thus results in a grasp failure. This error is prevented by keeping the pivot axis at safe distance from each edge of the contact face as shown in Fig. 7 (1). For its recovery, we calculate the position of each edge. The distance between the edge and the pivot axis is constantly checked, and if it decreases to a low value, the pivot axis is re-adjusted in a safe direction with the third and fourth fingers similarly as described in (a).

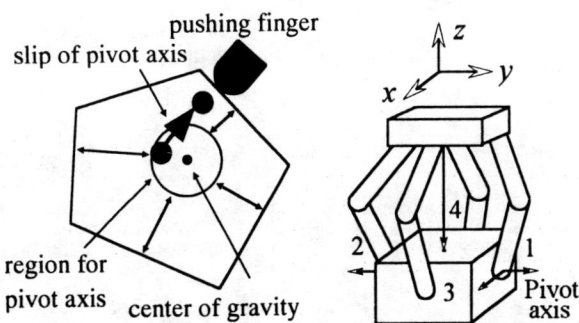

Figure 7: (1) Slip of the pivot axis (2) re-adjustment of the pivot axis

Experiment: Fig. 8 shows an object with which experiments were conducted. Its dimensions are also shown. We select the axis P on the object as a pivot axis. Initially the laser irradiates face 0 and the pushing finger pushes face 1. The pushing finger searches the object surface for a good contact point for *pivoting* by actually making contact with the object.

Fig. 9 shows experimental results: (1) angle of rotation, (2) d_{af}, the distance between the pivot axis and the face the pushing finger pushes, and (3) d_{ac}, the distance between the pivot axis and the cog.

At 20s the irradiated face shift occurs from face 0 to face 1, the one which the pushing finger pushes. The position in the direction of I as in Fig. 8 becomes indeterminate and thus d_{ac} is undefined. But d_{af} is still defined in the duration from 20s to 40s and its value decreases as we expected ((2) C). The angle of rotation increases ((1) A). At 40s the pushing finger is lifted from face 1 and repositioned to face 2 at 80s. The position and orientation of the object is determined uniquely after this. The graph in (2) D shows the distance between the pivot axis and face 2. d_{ac} is greater than $1.5cm$ ((3) F). The recovery routine is invoked and reduces d_{ac} to almost 0 ((3) G). Then the fingers can successfully increase the angle of rotation ((1) B).

From 20s to 40s, in spite of the indeterminate position in the direction of I, the fingers could successfully execute *pivoting* without dropping the object. This is because the pivot axis was situated far enough from both face 0 and 2 at 20s.

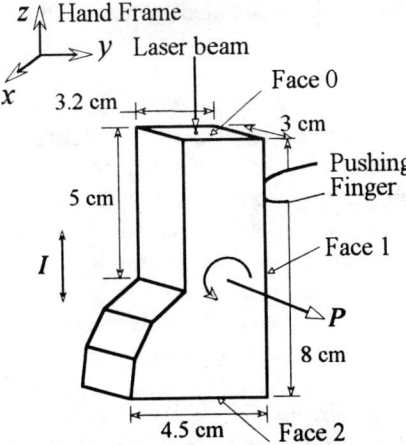

Figure 8: Object for *pivoting*

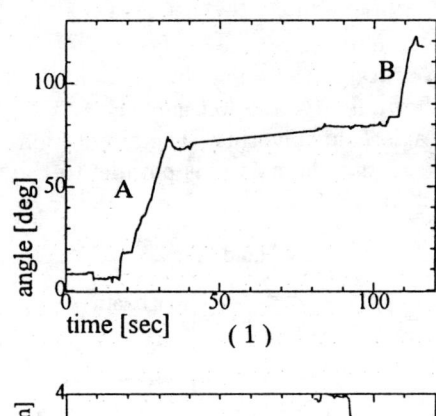

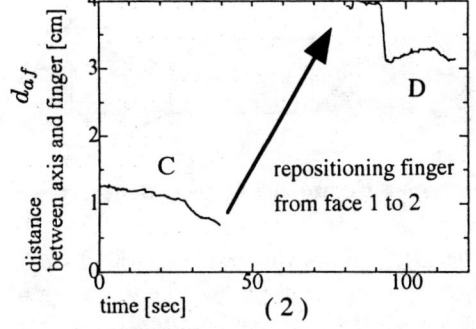

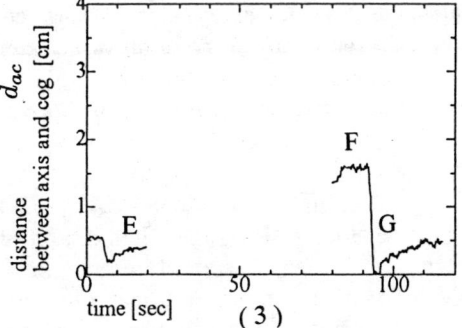

Figure 9: Experimental results

4.2 Rotation

Rotation is performed by simultaneously moving all the fingers along with the held object, and then repositioning them alternatively one by one. Also keeping in view the shape of the hand, the weight of the object always acts downwards in parallel to the direction of the fingers. So the frictional forces are large during *rotation*. These two factors may cause the object to slip downwards. As we discussed in section 3, the force-torque sensors alone are unable to detect this fault. Therefore we program to constantly check the grasp to maintain a constant distance between palm and the object through the laser range sensor.

Experiment: The motion of the fingers is not planned here as we discussed previously in [11]. Instead we only check the security of the grasp before lifting the fingers. One of the fingers is selected as a repositioning finger if the rest of the fingers can maintain equilibrium.

We rotated the same object about a rotation axis R as shown in Fig. 10. Actually the axis R is the same as the axis P for pivoting. The orientation of the object is such that the axis R is parallel to the z-axis of the hand frame.

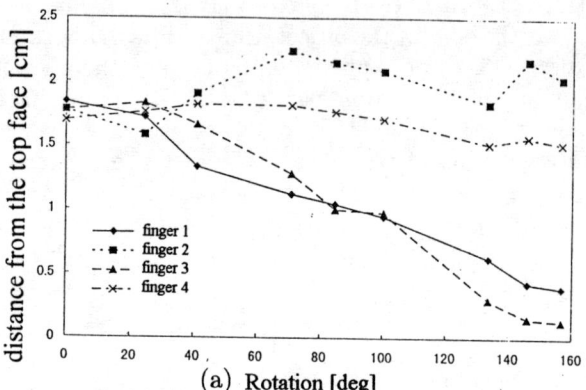

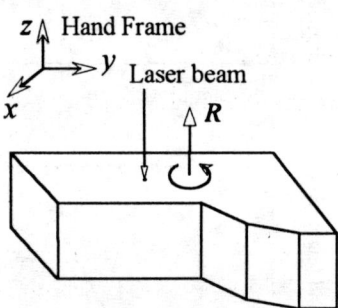

Figure 10: Object for *rotation*

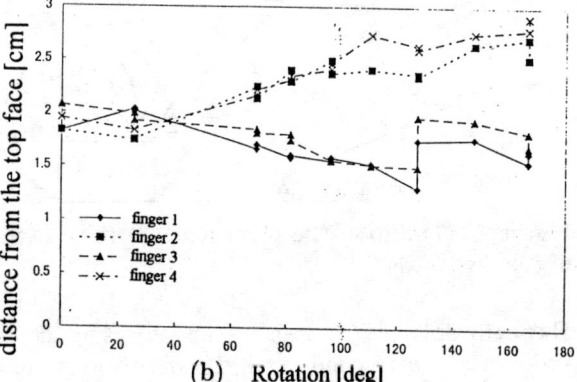

Figure 11: (a). Fingers position measured from top margin of the object *vs.* angle of *rotation*, (b). Recovery from insecure grasp.

Fig.11(a) shows the graph for the distance between fingers' contact position and the top margin of the object, verses the angle of *rotation*. It is seen that the positions of all the four fingers at the start of the *rotation* is near to 1.7 cm. But this distance for finger 1 and 2 decreases to a very small value after 120°. At such occasion the grasp may be lost. Fig.11(b) shows the correction made to this fault at about 125°.

4.3 Other errors

A repositioning finger may mistakenly make contact with a face of the object other than a target face. Such an error is defined as a discrete error in [22]. This could happen during *pivoting* and *rotation* especially for a thin prism. In *pivoting* even a slight misalignment in the contact positions of pivot axis fingers may create large unnoticed deflection in the orientation of the object as shown in Fig. 12 (1). In such cases the pushing finger may be positioned on the same face as one of the fingers forming the pivot axis. In *rotation*, a repositioning finger may mistakenly make contact with one of the base faces of the prism as shown in Fig. 12 (2).

This error can be detected by examining the contact normal of a repositioned finger. It is essential to detect this error as quickly as possible because the grasp is significantly disturbed if a finger makes contact with an undesired face of the object. To conduct experiments for various cases is our future work.

5 Conclusion

In this paper we have indicated grasp errors and faults during execution of the two regrasping primitives, *rotation* and *pivoting*, and proposed strategies to recover or prevent them. We obtain grasp information from two sources, force-torque intrinsic contact sensors and a laser-range sensor. Data obtained from these two sources are used to sense the errors and to control the fingers. Experimental results show that the

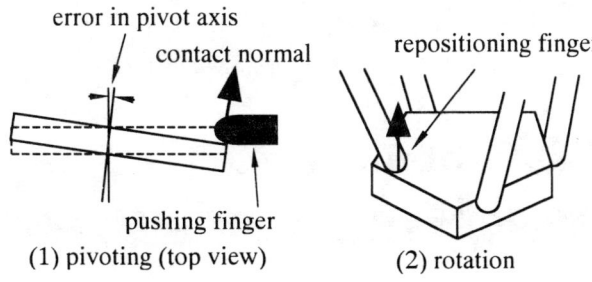

Figure 12: Grasp fault due to misplaced finger contact

proposed strategies successfully execute *rotation* and *pivoting* recovering or preventing faulty grasps during execution. This work paves the way for advance application of manipulation primitives based on real-time hand control.

References

[1] Salisbury, J. K., "Interpretation of Contact Geometries from Force Measurements," Proc. of 1st Int. Symp. on Robotic Research, pp. 134-150, MIT Press, 1984.

[2] Mason, M.T., Salisbury, J. K., "Robot Hands and the Mechanics of Manipulation," MIT Press, Canbridge, MA, 1985.

[3] Bicchi, A., "Intrinsic Contact Sensing for Soft Fingers," IEEE R.A. Conf., pp. 968-973, 1990.

[4] Brock, D., Chiu, S., "Environment Perception of an Articulated Robot Hand Using Contact Sensors," ASME Publication PED-Vol.15, Robotics and Manufacturing Automation, pp. 89-96, 1985.

[5] Grimson, W. E., Lozano-Peres, T., "Model-Based Recognition and Localization from Tactile Data," IEEE R.A. Conf., pp. 248-255, 1984.

[6] Gatson, P. C., Lozano-Peres, T., "Tactile Recognition and Localization using Object Models: The case of Polyhedra on a Plane," IEEE Trans. Pattern Analysis and Machine Intelligence, Vol PAMI-6, No. 3, pp. 257-266, 1984.

[7] Hahn. H., "Recognition of 3D objects by a 3-Fingered Robot Hand Equipped with Tactile and Force Sensors," IEEE Int. Conf. on Multisensor Fusion and Integration for Intelligent Systems, pp. 297-304, 1994.

[8] Fearing R., "Tactile Sensing for Shape Interpretation," Dexterous Robot Hands, Spring-Verlag, pp. 209-238. 1989.

[9] Allen, P. K., Michelman, P., "Acquisition and Interpretation of 3-D sensor Data from Touch," IEEE Trans. Robotics and Automation, Vol. 6, No. 4, pp. 397-404, 1990.

[10] Nagata, K., Keino, T., Omata, T., "Acquisition of an Object Model by Manipulation with a Multifingered Hand," Proc. of IROS Conf., pp. 1045-1051, 1996.

[11] Omata, T., Nagata, K., "Planning Reorientation of an object with a Multifingered Hand," IEEE R.A. Conf., pp. 3104-3110, 1994.

[12] Omata, T., Farooqi, M. A., "Regrasps by a Multifingered Hand based on Primitives," IEEE R.A. Conf., pp. 2772-2780, 1996.

[13] Omata, T., Farooqi, M. A., "Reorientation Planning for a Multifingered Hand based on Orientation States Network Using Regrasp Primitives," Proc. of IROS Conf., pp. 285-290,1997

[14] Fearing, R. S. "Implementing a force strategy for Object Re-orientation," IEEE R.A. Conf., 96-102, 1986.

[15] Cole, A. A., Hsu, P., Sastry, S., "Dynamic Regrasping by Coordinated Control of Sliding for a Multifingered Hand," IEEE R.A. Conf., 781-786, 1989.

[16] Rus, D. "Coordinated Manipulation of Polygonal Objects," IEEE R.A. Conf., 106-112, 1993.

[17] Gupta, K., "Motion Planning for Reorientation using Finger Tracking: Landmarks in SO(3) x w," IEEE R.A. Conf., 446-451, 1995.

[18] Trinkle, J. C., "A Quasi-static analysis of Dexterous Manipulation with Sliding and Rolling Contacts," IEEE R.A. Conf., 788-793, 1989.

[19] Paetsch, W., Wichert, W. V., "Solving Insertion task with a Multifingered Gripper by Fumbling," IEEE R.A. Conf., 173-179, 1993.

[20] Michelman, P., Allen, P., "Forming Complex Dexterous Manipulations from Task Primitives," IEEE R.A., 3383-3388, 1993.

[21] Speeter, T. H., "Primitive based Control of the Utah/MIT Dexterous Hand," IEEE R.A. Conf., 866-875, 1991.

[22] Schegl, T, Buss, M, "Hybrid Closed-Loop Control of Robotic Hand Regrasping," IEEE R.A. Conf., 3026-3031, 1998

VISP: A Software Environment for Eye-in-Hand Visual Servoing

Éric Marchand
IRISA - INRIA Rennes
Campus de Beaulieu, F-35042 Rennes Cedex
Email: Eric.Marchand@irisa.fr

Abstract

In this paper, we describe a modular software that allows fast development of eye-in-hand image-based visual servoing applications (ViSP states for "Visual Servoing Platform"). Visual servoing consists in specifying a task as the regulation in the image of a set of visual features. Various issues have thus to considered in the design of such application: among these issues we find the control of camera motions and the tracking of visual features. Our environment features a wide class of control skills as well as a library of real-time tracking processes. Some applications that used this modular architecture on a six dof cartesian robot are finally presented.

1 Overview

Visual servoing is a very important research area in vision-based robotics (see [4] for an extensive review). Despite all the research in this field, it seems that there are no software environment that allows fast prototyping of visual servoing tasks. The main reason is certainly that it usually requires specific hardware (the robot and, most of the time, dedicated image processing boards). The consequence is that the resulting applications are not portable and can be merely adapted to other environment. Today's software design allows to propose elementary components that can be combined to build portable high-level applications. Furthermore, the increasing speed of micro-processors allows the development of real-time image processing algorithms on a simple workstation.

Chaumette, Rives and Espiau [3] proposed to constitute a *"library of canonical vision-based tasks"* for eye-in-hand visual servoing that contents the most classical linkages that are used in practice. Toyama and Hager [9] describe what such a system should be in the case of stereo visual servoing. The proposed system is specified using the same philosophy as the XVision system [7]. This system (called SERVOMATIC) would have been independent from the robot and the tracking algorithms. Following these ways, ViSP, the software environment we present in this paper features all these capabilities: independence with respect to the hardware, simplicity, extendibility, portability. Moreover, ViSP features a large library of elementary positioning tasks wrt. various basic control features (points, lines, circles, spheres, cylinders, etc.) that can be combined together, and an image processing library that allows the tracking of visual cues (dot, segment, ellipse, spline, etc.). This modular platform has been developed in C++ on Unix Workstations.

The remainder of this paper presents the background dealing with the eye-in-hand visual servoing and the tracking algorithms. Software design is then presented in the light of a simple example. Experimental results carried out on a six dof cartesian robots are finally presented.

2 Image Based Visual Servoing

The *image-based visual servoing* consists in specifying a task as the regulation in the image of a set of visual features [6][8]. Embedding visual servoing in the task function approach allows us to take advantage of general results helpful for the analysis and the synthesis of efficient closed loop control schemes. These control issues are now well known, we just give a rapid overview of the visual servoing process.

2.1 Control issues

Let us denote $\mathbf{P}$ the current value of the set of selected visual features[1] used in the visual servoing task and measured from the image at each iteration of the control law. To ensure the convergence of $\mathbf{P}$ to its desired value $\mathbf{P_d}$, we need to know the interaction matrix (also called image Jacobian) $\mathbf{L}_\mathbf{P}^T$ defined by the classical equation [6]:

$$\dot{\mathbf{P}} = \mathbf{L}_\mathbf{P}^T(\mathbf{P}, \mathbf{p})\mathbf{T_c} \qquad (1)$$

where $\dot{\mathbf{P}}$ is the time variation of $\mathbf{P}$ due to the camera motion $\mathbf{T_c}$. The parameters $\mathbf{p}$ involved in $\mathbf{L}_\mathbf{P}^T(\mathbf{P}, \mathbf{p})$ represent the depth information between the considered objects and the camera frame.

A vision-based task $\mathbf{e}$ is defined by:

$$\mathbf{e} = \mathbf{W}^+\mathbf{C}(\mathbf{P} - \mathbf{P_d}) + (\mathbf{I} - \mathbf{W}^+\mathbf{W})\mathbf{g}_s^T \qquad (2)$$

---

[1]In the reminder of this paper, we will call these features *"control features"* by opposition to the features tracked in the image sequence (*the "visual cues"*) from which the control features are extracted.

where **C**, called combination matrix, has to be chosen such that $\mathbf{CL_P^T(P,p)}$ is full rank about the desired trajectory $q_r(t)$. It can be defined as $\mathbf{C} = \mathbf{WL_P^{T+}(P,p)}$ ($\mathbf{L^+}$ denotes the pseudo inverse of $\mathbf{L}$). In that case, we set $\mathbf{W}$ as a full rank matrix such that $\text{Ker } \mathbf{W} = \text{Ker } \mathbf{L_P^T}$. If the vision-based task does not constrain all the n robot degrees of freedom, a secondary task $\mathbf{g}_s$ can also be performed. $\mathbf{g}_s$ is the gradient of a cost function h_s to be minimized ($\mathbf{g}_s = \frac{\partial h_s}{\partial \bar{r}}$). This cost function is minimized under the constraint that $\mathbf{P} = \mathbf{P_d}$. The two projection operators $\mathbf{W^+}$ and $\mathbf{I} - \mathbf{W^+W}$ guarantee that the camera motion due to the secondary task is compatible with the regulation of $\mathbf{P}$ to $\mathbf{P_d}$.

In order to make **e** exponentially decrease and then behave like a first order decoupled system, we get:

$$\mathbf{T_c} = -\lambda \mathbf{e} - \mathbf{W^+}\widehat{\frac{\partial \mathbf{e_1}}{\partial t}} - (\mathbf{I} - \mathbf{W^+W})\frac{\partial \mathbf{g}_s^T}{\partial t} \quad (3)$$

where:

- $\mathbf{T_c}$ is the camera velocity;
- λ is the proportional coefficient involved in the exponential convergence of **e**;
- $\widehat{\frac{\partial \mathbf{e_1}}{\partial t}}$ (with $\mathbf{e_1} = \mathbf{C(P - P_d)}$) represents an estimation of a possible autonomous target motion.

2.2 A library of visual servoing skills

One of the difficulty in visual servoing is to derived the interaction matrix $\mathbf{L}^T$ corresponding to the selected control features. A systematic method has been proposed to derived analytically the interaction matrix of a set of control features defined upon geometrical primitives [6][3]. Any kind of visual information can be considered within the same visual servoing task (coordinates of points, line orientation, surface (or more generally inertial moments), distance, etc.)

Knowing these interaction matrices, the construction of elementary visual servoing tasks is straightforward. As explained in [3] a large library of elementary skills can be proposed. The current version of VISP allows to define X-to-X feature-based tasks with X = {point, line, sphere, cylinder, circle, etc.}. Using these elementary positioning skills more complex tasks can be considered by stacking the elementary matrices. For example if we want to build a positioning task wrt. to a segment, defined by two points $\mathbf{P_1}$ and $\mathbf{P_2}$, the resulting interaction matrix will be defined by:

$$\mathbf{L_P^T} = \begin{bmatrix} \mathbf{L_{P_1}^T} \\ \mathbf{L_{P_2}^T} \end{bmatrix}$$

where $\mathbf{L_{P_i}^T}$ is defined, if $\mathbf{P_i} = (X, Y)$ and z is its depth, by:

$$\mathbf{L_{P_i}^T} = \begin{pmatrix} -1/z & 0 & X/z & XY & -(1+X^2) & Y \\ 0 & -1/z & Y/z & 1+Y^2 & -XY & -X \end{pmatrix}$$

This way, more feature-based tasks can be simply added to the library.

An other important feature is the capability to introduce a secondary task. This secondary task can be a solution to the impossibility to plan the camera trajectory by introducing some constraints in its motion. We proposed a large library of secondary tasks from trajectory tracking to occlusions avoidance and joint limits/singularities avoidance.

Tracking capabilities have also been integrated. When the target is moving, an estimation of the autonomous target motion is required in order to avoid tracking errors.

3 Tracking Visual Cues

Parallel to the development of the platform with respect to control part of visual servoing, we have to develop algorithms dealing with tracking issues. The visual servoing formalism allows to use more complex features or combination of features. VISP allows to consider, besides the usual "dots", visual cues such as lines, ellipses, or more complex curves. Information available in the description of these shapes are then used to describe the control features.

We have developed fast real-time tracking processes. Few systems features real-time capabilities on a simple workstation. The XVision system [7] is a good example of such systems, however, it does not features all the tracking capabilities we wanted[2]. In our case, we decided to use the ME (moving edges) algorithm [2] adapted to the tracking of parametric curves. It is a local approach that allows to match moving contours. Previous works have been done to use this algorithm to track line segments [1] on a dedicated IP board.

3.1 General Algorithm

One of the advantages of the method is that it does not require any edge extraction, furthermore it can be implemented with convolution efficiency and can therefore ensure a real-time computation [2]. As we want an algorithm that is fast, reliable, robust to partial occlusions and to false matches, we decided to track only a list L^t of pixels along the considered edge and then to determine, by a robust least square approach, the equation of the support primitive that fits these data.

We will not described the initialization process in details, let us just say that we are able to initialize the pixels list L^t. Then, for each pixel we estimate the direction of the tangent to the edge $\hat{\theta}$. The process consists in searching for the correspondent P_i^{t+1} in image I^{t+1} of each pixel $P_i^t \in L^t$. We determine a 1D search area $Q_i^j, j \in [-J, J]$ in the direction of the normal to the contour δ. For each pixel P_i^t of the list L^t, and for each position Q_i^j lying in the direction δ we compute the matching criterion corresponding to a log-likelihood ratio ζ^j. This is nothing but the absolute sum of the convolution values computed at P_i and Q_i^j using a pre-determined mask M_θ function of the orientation of the contour. New position P_i^{t+1} is given by:

$$Q_i^{j*} = \arg\max_{j \in [-J,J]} \zeta^j \quad \text{with} \quad \zeta^j = \mid I(P_i) * M_\theta + I(Q_i^j) * M_\theta \mid$$

---

[2] Even if, in many way, it is far more complete that the system we propose.

providing that the value ζ^{j^*} is greater than a threshold λ. Then pixel P_i^{t+1} given by $Q_i^{j^*}$ is stored in L^{t+1}. A new list of pixels is obtained.

Finally, given the list L^{t+1} the new parameters of the feature are computed using a least squares technique.

3.2 Tracking Visual cues

Line segments. The simplest case we consider is the line segment [1]. The representation considered for the line are the polar coordinates (ρ, θ):

$$x \cos \theta + y \sin \theta - \rho = 0$$

This case is very simple as the direction θ is directly given by the parameters of the features. The choice of the convolution mask is then straightforward. A points insertion processes either in the middle of the segment, to deal with partial occlusions or miss-tracking, and at the extremities of the segment to deal with sliding movements has been introduced in the tracking method.

Ellipses. Dealing with the ellipse many representation can be defined, we chose to use the coefficients K_i that are obtained from the polynomial equation of an ellipse:

$$K_0 x^2 + K_1 y^2 + 2K_2 xy + 2K_3 x + 2K_4 y + K_5 = 0$$

The ellipse correspond to the case $K_2^2 < K_0 K_1$. The parameters K_i can be estimated from the pixels of the list L_t using a least square method. From the parameters K_i we can derived other representations as the parameterization $(X_c, Y_x, \mu_{11}, \mu_{02}, \mu_{20})$ based on the normalized inertial moments.

Splines. A spline is defined by a parametric equation:

$$Q(t) = \sum_{j=-d}^{n-1} \alpha_j B_j(t), \quad t \in [0, 1]$$

where the α_j are the control knots of the spline, d is the degree of the spline ($d = 3$ for a cubic spline) and B_j spline basis function. Since the number p of tracked point is usually greater than the number of desired control knots n, a least square method is used.

Discussion. The proposed tracking approach algorithms based on the ME algorithm allows a real-time tracking of geometric features in an image sequence. It is robust with respect to partial occlusions and shadows. However, as a local algorithm, its robustness cannot be ensure in complex scenes with highly textured environment.

4 Software environment

As already stated, while developing this software, our goal was to allow a portable (independent from the hardware), fast and reliable prototyping of visual servoing applications. The first part of this section presents the internal architecture of the system and how it has been implemented. Describing the full implementation of the software is out of reach in this paper, therefore, we will focus on the notion of extendibility and portability. The second part describes how to use the available libraries from a end-user point of view. Let us note that all the functionalities described in this section have been implemented and are *fully operational*.

4.1 An overview of the ViSP architecture

To fulfill the extendibility and portability requirements we divided the platform into three different parts (the tracking parts, a library of control features, and the controller itself) and we widely use C++ capabilities (templates, derivation, inheritance, virtual classes, etc.).

As can be seen on the network of Figure 1, each library of the ViSP environment is indeed extensible. In the controller library, the CServoAfma class is derived from a CServo virtual class and is specific to our six dof Afma robot. It redefines some pure virtual methods defined in CServo such as robot motion orders (*i.e.*, the CServoAfma::UpdateCameraVelocity specific to a given robot) and inherits all the methods and attributes of CServo (*i.e.*, generic control issues). Adding a new robot is then straightforward. In the same way some specific frame grabber classes (here CSunvideo or CEdixia, that are our frame grabbers) can be derived from a generic CFrameGrabber class. This two examples also show that the platform is independent from the hardware and then portable.

The extendibility can be mainly seen in the control features library. Each specific control feature is derived from a virtual class CBaseFeatures. This class mainly defines a few variables (*e.g.*, a vector that describes the parameters **P** and **p**) and a set of virtual members functions that are features dependent (*e.g.*, the way to compute the image Jacobian **L** or the virtual function that allows the tracking of the feature in the images). It is important to note that all the relations between the controller library and the features library is done through this class. The virtual functions define in CBaseFeatures can be directly used by the controlled even if they are not yet defined. The consequence is that the controller library never know the nature of the manipulated features. Another consequence is therefore that it is absolutely not necessary to modify the controller library when adding a new feature. On the other hand, when adding a new feature in the control features library, the programmer must define the number of visual information, the way to compute the image Jacobian, etc. This is done at a lower level (*e.g.*, CPoint, CLine, ...). Some member functions of these derived classes remain virtual. Indeed a control feature can be defined by many visual cues, member functions like Track(...) cannot be defined at this level. Therefore, a few classes derive from the feature classes in order to supply the users with many tracking capabilities. For example a point can be define as a simple dot (CPointDots in Figure 1) or as a intersection of two lines (CPointTwoLines). In any case, using the inheritance mechanism the controller knows which function it has to use. In conclusion, adding new capabilities within the control features library can therefore be done

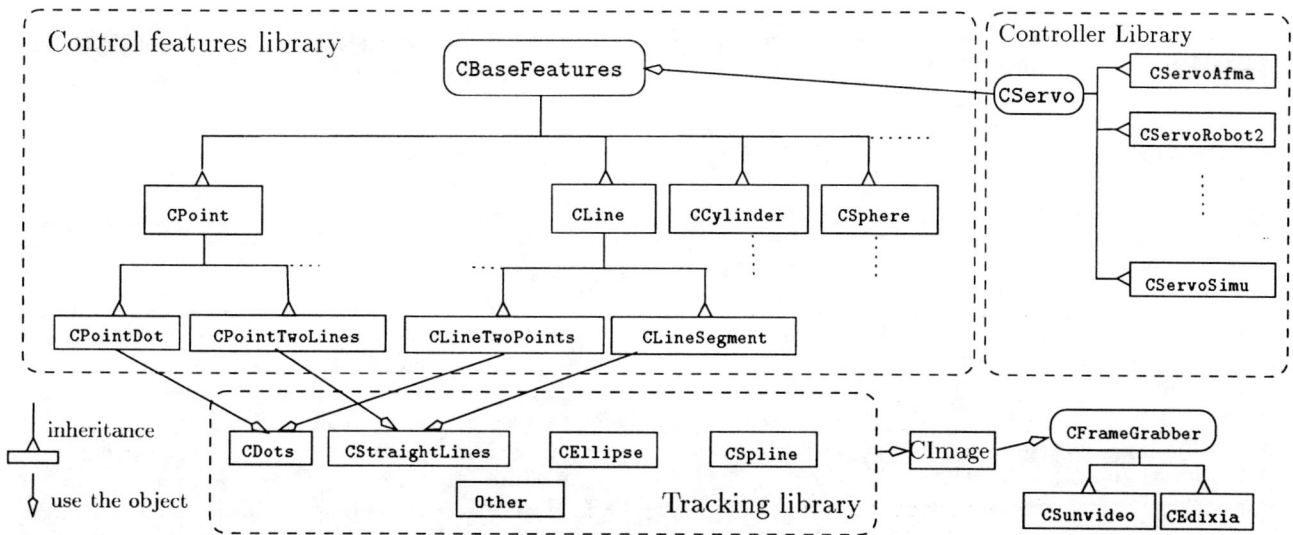

Figure 1: VISP structure of classes

at two levels: adding a new control feature (*e.g.*, a circle in Figure 1) or adding a new way to link this feature to the visual cues (for example, defines a point as the center of an ellipse).

4.2 VISP from a end-user point of view

Our other claim was that VISP is simple to use. We will therefore describe the software environment, from the end-user point of view, in the light of two simple examples implemented using VISP. The first example is a positioning task wrt. a line.

```
0  main() {
1    CImage I ;
2    I.InitAcqImage(Sunvideo()) ;
3    I.AcqImage() ;
4
5    CLineSegment line ;          // or CLineTwoDots line ;
7    line.InitTracking(I) ;
8
9    CLine Ld(0,PI/2) ;  // Init here the desired visual
10                       // features Ld (centered/horizontal)
11
12   CServoAfma task ;
13   task.AddLink(line,Ld) ;
14   task.InitInteractionMatrix() ;
15
16   while(...) {
17     CColVector Tc(6) ;
18
19     I.AcqImage() ;
20     task.GetCurrent(I) ;
21     Tc = -0.2 * task.TaskFunction() ;  // Tc = -l L^+(s-sd)
22     task.UpdateCameraVelocity(CAMERA,Tc) ;
23   }
24 }
```

Line 2 defines the image acquisition protocol (here through the sunvideo frame grabber). The `CImage::AcqImage()` function puts the bitmap in the computer memory where the tracking process will be done. A control feature, a line, is defined (5-6) as a segment (`CLineSegment`). This means that the visual cue is a line segment. Line 9 defines the desired position of the visual feature in the image. The controlled `CServoAfma` derived from the `CServo` virtual class is defined line 12. Line 13 defines a link between the current position (`line`) of the visual feature in the image and the desired position (`Ld`) while line 14 creates the corresponding interaction matrix (here a 2×6 matrix). Note that more complex tasks can be defined by "stacking" other links using the `CServo::AddLink` method. It is straightforward to write the loop itself. It features the image acquisition (19) and the current visual features extraction or tracking (20). Finally the vision based task e is computed using the `CServo::TaskFunction()` method and the new robot velocity orders $\mathbf{T_c} = -\lambda e$ (22).

The task may be more complex. Let us consider, a curves following task (see the experimental results in the next section). This problem can be divided in two subtasks. The primary task consists in servoing on the tangent to the curve (*i.e.*, maintain this tangent horizontal and centered in the image). The positioning skill used in this experiment is therefore a line-to-line link. Indeed, image features used here are $\mathbf{P} = (\rho,\theta)$ where ρ and θ are the parameters of a line that is nothing but the tangent to the curve (2 dof are then controlled). The secondary task is a trajectory tracking at a given velocity in the X direction and is defined by a cost function $h_s = X - X_0 - V_x t$. Image processing consists here in tracking a spline in the image sequence and to compute the equation of the tangent to the curve from which we can control the camera motion. From a control point of view, this task is similar to the previous one. However, in our software environment there are no direct relations between the tangent to a curve (that is, here, the visual cue) and a line (that is the control feature). As explained in the previous example, VISP usually allows to avoid an explicit access to the trackers, but the number of relations visual cues/control features is virtually infinite. Therefore a direct access to the trackers is sometimes necessary. The spline tracker, here, is defined in line 1 whereas the visual feature (a straight line) is associated to this tracker in line 4. Then in the control loop, the spline is tracked in each frame (the CSpline::Track(CImage &I) method in line 10) and the new control feature is computed (line 11) and introduced in the controller (the CServo::NewCurrent(...) method in line 12). This visual servoing task also features the use of a secondary task. Vector **g** is the gradient of h_s, $\mathbf{g} = V_x$ and is combined with the primary task using the projection operation $\mathbf{I} - \mathbf{W}^+\mathbf{W}$ (line 14). These simple examples allow us to show the importance of the three libraries: the trackers library (define by variable in line 1), the control features library (line 3), and the controller library (line 6) and how they interact with each other.

```
main() {
   CImage I ;
1  CSpline S(CUBICSPLINE) ;
2  S.InitTracking(I) ;

3  CLine L ;
4  L = S.Tangent(0,0) ;
5  CLine Ld(0,PI/2) ;

6  CServo task ;
7  task.AddLink(L,Ld) ;
8  task.InitInteractionMatrix() ;

   while(...) {
      CColVector Tc(6), g(6)
9     I.AcqImage() ;
10    S.Track(I) ;         // Track the spline (visual cue)
11    L = S.Tangent(0,0) ; // compute the tangent
12    task.NewCurrent(L) ; // define the new control feature
13    g[0] = Vx ;          // secondary task
14    Tc = -0.2* (task.TaskFunction() + task.I_WpW*g);
15    task.UpdateCameraLocation(CAMERA,Tc) ;
   }
}
```

5 Some Applications

Table 1 sums up the different elementary positioning tasks that have been implemented using VISP. Most of these results are now classic and will not be described here. We just propose results for less classic experiments, *i.e.* positioning wrt. to a sphere and curves following. Other applications implemented with VISP are structure from controlled motion (point, sphere), joint limits and singularities avoidance, occlusions avoidance, ...

The application presented in this section has been implemented at IRISA on a six dof cartesian robot. Image processing (described in Section 3) and control law (Section 2) are performed on a Sun Ultra Sparc 1 (170 Mhz). The frame grabber was a Sunvideo board. Tests performed on the Sun show that a line can be tracked in 3.6 ms (with 40 pixels in the list L^t and a maximum displacement $J = \pm 10$) whereas an ellipse is tracked in 6.5 ms. A spline with 40 pixels in L^t and 15 control knots is tracked in less than 10 ms.

Curves following task. The principle of this experiment has been described in the previous paragraph. Let us just add that the goal was to follow a long pipe (2 meters long) that features three 90° corners. Figure 3 show three images acquired during this task in the first corner (note the tracked spline and the tangent in the image). Figure 2a depicts the translational velocities. The bottom curves is the due to the secondary task. In order to avoid high rotational velocity in high curvature area, the velocity V_x has been defined as $V_x = V_{max} \exp(-\alpha(\theta - \theta^*)^2)$. This explain the noise in this plot and the three "⋂" that can be observed (due to the three corners). Figures 2b and c depict the errors $\theta - \theta_d$ and $\rho - \rho_d$ observed for the selected visual features. The important error in θ while crossing the high curvature area are due to the fact that the curvature is not predicted. This problem is very similar to the tracking errors that can be observed in a target tracking task. Figure 2d depicts the XY camera trajectory that reflects the shape of the pipe.

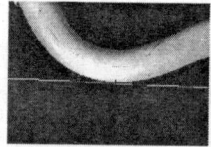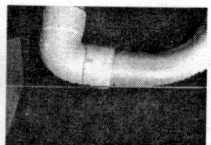

Figure 3: Pipe following task: 3 images acquired during the task and tracked curve (red) and tangent to the curve (green)

Let us note that a similar problem has already been addressed in [5]. But even if the tracking was done using snakes, the visual features used in the control were points.

Positioning wrt. a sphere. In this second experiment, we want to servo on a sphere (a ping-pong ball). The task is to observe the sphere as a centered circle in the image with a given radius and then to move the camera at a constant distance from the sphere center. The projection of a sphere in the image plane is an ellipse. Image processing consists in tracking this ellipse (here on a non-uniform environment). We chose as visual features the inertial moments of this ellipse (*i.e.*, $X_c, Y_x, \mu_{11}, \mu_{02}, \mu_{20}$). Since the real radius of the ping-pong ball is known (19 mm), the secondary motion (along the $\vec{X}$ axis) does not introduce any perturbations in the primary task Each iteration is performed in 40 ms.

6 Conclusion and future work

VISP is a *fully functional* modular architecture that allows fast development of visual servoing applications. It is mainly composed of three C-callable libraries: two dedicated to control issues (one of control processes and one of canonical vision-based tasks that contains the most classical linkage [3]) and one dedicated to real-time tracking (based on the Moving edges algorithm). This

objects	Control features (see [6] for details)		Visual Cues			
			dots	lines	ellipse	B-Spline
point	point	$\mathbf{P} = (X, Y)$	×	×		
line	line	$\mathbf{P} = (\rho, \theta)$		×		× (tangent)
circle	ellipse	$\mathbf{P} = (X_c, Y_c, \mu_{11}, \mu_{20}, \mu_{02})$			×	
sphere	ellipse	$\mathbf{P} = (X_c, Y_c, \mu_{11}, \mu_{20}, \mu_{02})$			×	
cylinder	lines	$\mathbf{P} = (\rho_1, \theta_1, \rho_2, \theta_2)$		×		
square	points	$\mathbf{P}_i = (X_i, Y_i), i = 1..4$	×	×		
	lines	$\mathbf{P}_i = (\rho_i, \theta_i), i = 1..4$		×		
quadrilateral	moments/orientations $\mathbf{P} = (X_c, Y_c, \mu_{00}, \alpha_1, \alpha_2, \alpha_3)$		×			

Table 1: Elementary positioning tasks using VISP: tracked cues and control features.

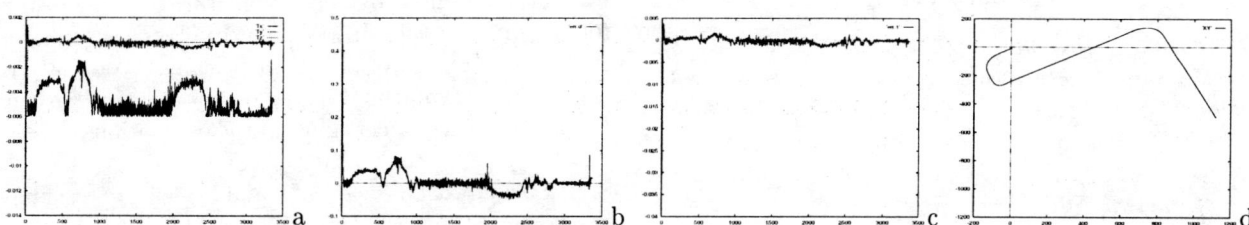

Figure 2: Pipe following task: (a) Translational camera velocity (b) Error $\theta - \theta_d$ and (c) $\rho - \rho_d$ (d) XY camera trajectory

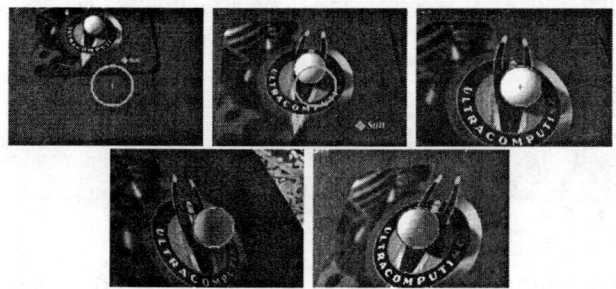

Figure 4: Tracking ellipse and servo on a sphere: $t = 62$, $t = 126$, $t = 416$, $t = 760$, $t = 1602$

software has already been used for the development of a large number of applications. Let us finally note that VISP also features a virtual six dof robot that allows to simulate visual servoing experiments. Dealing with vision-based control, many new features can be added to this software. Only eye-in-hand 2D visual servoing with respect to geometrical features have been proposed. Therefore other visual servoing technics can be introduced such as stereo visual servoing, or visual servoing wrt. dynamic informations or motion, etc. Furthermore, now that a set of basic tools are available, it is necessary to be able to deal with high-level mission that combines many tasks.

References

[1] S. Boukir, P. Bouthemy, F. Chaumette, and D. Juvin. A local method for contour matching and its parallel implementation. *Machine Vision and Application*, 10(5/6):321–330, April 1998.

[2] P. Bouthemy. A maximum likelihood framework for determining moving edges. *IEEE Trans. on PAMI*, 11(5):499–511, May 1989.

[3] F. Chaumette, P. Rives, and B. Espiau. Classification and realization of the different vision-based tasks. *in* K. Hashimoto, editor, *Visual Servoing*, pp. 199–228, World Scientific, Singapor, 1993.

[4] P.I. Corke. Visual control of robot manipulator. K. Hashimoto, editor, *Visual Servoing*, pp. 1–32, World Scientific, Singapor, 1994.

[5] E. Coste-Manière, P. Couvignou, and P. Khosla. Visual servoing in the task-function framework: a contour following task *Journal of Intelligent and Robotics Systems*, 12:1–21, July 1995.

[6] B. Espiau, F. Chaumette, and P. Rives. A new approach to visual servoing in robotics. *IEEE Trans. on Robotics and Automation*, 8(3):313–326, June 1992.

[7] G. Hager and K. Toyama. The XVision system: A general-purpose substrate for portable real-time vision applications. *Computer Vision and Image Understanding*, 69(1):23–37, January 1998.

[8] K. Hashimoto, editor. *Visual Servoing: Real Time Control of Robot Manipulators Based on Visual Sensory Feedback*. World Scientific Series in Robotics and Automated Systems, Vol 7, World Scientific Press, Singapor, 1993.

[9] K. Toyama, G. Hager, and J. Wang. Servomatic: A modular system for robust positioning using stereo visual servoing. In *Proc. of the International Conference on Robotics and Automation*, pp. 2636–2643, Minneapolis, April 1996.

Measurement error estimation for feature tracking

Kevin Nickels
knickels@trinity.edu
Dept. of Engineering Science
Trinity University
San Antonio TX 78212

Seth Hutchinson
seth@uiuc.edu
Dept. of Electrical and Computer Engineering
and The Beckman Institute
University of Illinois at Urbana-Champaign
Urbana IL 61801

Abstract

Performance estimation for feature tracking is a critical issue, if feature tracking results are to be used intelligently. In this paper, we derive quantitative measures for the spatial accuracy of a particular feature tracker. This method uses the results from the sum-of-squared-differences correlation measure commonly used for feature tracking to estimate the accuracy (in the image plane) of the feature tracking result. In this way, feature tracking results can be analyzed and exploited to a greater extent without placing undue confidence in inaccurate results or throwing out accurate results. We argue that this interpretation of results is more flexible and useful than simply using a confidence measure on tracking results to accept or reject features. For example, an extended Kalman filtering framework can assimilate these tracking results directly to monitor the uncertainty in the estimation process for the state of an articulated object.

1 Introduction

Estimating the effectiveness of feature tracking information is a very important topic in image processing today. In correspondence-based object tracking the results from several feature trackers, each tracking salient points or edges of an object, are combined to track a (possibly articulated) object [1]. Point tracking for the purpose of computing image flow requires a metric for the confidence in a motion estimate, so that estimates from regions of high confidence can be used to improve estimates in regions of low confidence [2]. Feature tracking confidence measures have also been used in a visual servo control to increase the robustness of the servo control [3].

We characterize the *accuracy* of a feature tracking result by the accuracy of the location computed by the feature tracking. This characterization leads to the evaluation of the *confidence* of a feature tracking result for a more general purpose than that of accepting or rejecting the feature for use in tracking. We analyze the uncertainty in the tracking process, so that we can keep track of the uncertainty in the estimation process being driven by the feature tracking.

We begin with a review of standard feature tracking methods. Following this, we describe our goals with respect to characterizing the spatial discrimination of features. Then we present a Gaussian approximation and describe how sufficient statistics can be used to characterize this approximation. Finally, we present some results from our implemented tracking system.

2 Correlation and feature templates

In correlation-based feature tracking, a *feature template* is used to detect a feature in an image. A feature template contains some representation of the feature and is compared against portions of an image to locate that feature in the image. This comparison utilizes a similarity metric to rate the similarity of the template and the image patch. The image region found to be the most similar to the template is usually taken to be the location of the feature.

The following sections discuss three areas crucial to correlation based tracking: the content of this template, the definition and use of a specific similarity metric for tracking, and the definition of confidence measures on the tracking results.

2.1 Template content

The content of the template is an important choice in feature tracking. If the template faithfully reproduces the actual appearance of the feature in the image, tracking will work well. However, if a template is oversimplified or does not match the appearance of a feature in the image due to unmodeled effects, feature tracking will perform poorly.

A template can be generated from a canonical view of the feature, and template matching done in a search window centered about the predicted position of the image. Brunelli and Poggio give a good review of this technique in the context of facial feature tracking [4]. The main problem with this straightforward approach is that the simple template is a 2D entity, and the image patch may undergo transformations that the template cannot model, such as rotation.

A more complex algorithm that also works in certain situations is to use an image patch from the previous image, taken from the area around the last computed position of the feature in that image, for the template. Hager [5] uses this approach for visual servoing. Hager and Belhumeur [6] have also used previous tracking information to warp this image patch before use as a feature template, which increases the flexibility of this approach.

If object and scene modeling are part of the tracking framework, it is possible to create templates from this information. Lopez et al. [7] have a 3D registered texture of a face as part of their object model. Computer graphics techniques are used to render the relevant portion of the scene complete with sophisticated texture mapping to estimate the appearance of a feature in the image. This image patch is then used as a template in the feature tracking portion of the system. Our work uses 3D models for complex articulated objects, also in a graphics-based framework [8], to generate feature templates.

2.2 The SSD similarity metric

In correlation-based tracking, a similarity metric is used to compare the feature template described above to areas of the image to locate the feature in the image.

The standard sum-of-squared-differences (SSD) metric for grayscale images is defined as:

$$SSD(u,v) = \sum_{m,n \in N} [T(m,n) - I(u+m,v+n)]^2, \quad (1)$$

where T is the template image and I is the input image. The location (u,v) represents some location in the input image whose content is being compared to the content of the template. Papanikolopoulos [3] uses the SSD measure to generate tracking results that are then used for robotic visual servoing experiments. Anandan [2] and Singh and Allen [9] use this SSD metric for the computation of image flow.

Often, this measure is not computed for the entire input image, but only for some *search window* in the input image. Primarily for computational reasons,

this restriction also serves as a focus of attention for the feature tracking algorithm. Singh and Allen [9] define a fixed size square search window surrounding the previous location of the feature. Kosaka and Kak [10] consider at length the shape and location of the search window. They model the scene and compute a spatial probability density function for the location of each feature, then search the image area corresponding to 85% of the probability mass. We use a constant-velocity model for an articulated object to predict 3D positions for relevant points on the object. Imaging models are then used to project these locations to points on the image plane. A fixed size rectangular search window centered at these locations is established in the input image. See [8] for more details.

3 Confidence, uncertainty estimation, and spatial uncertainty

It has been noted [2] that popular similarity measures often lead to some unreliable matches, particularly in image regions with little textural information. For this reason, it is often helpful to compute a *confidence* on the match found, as well as a location. This confidence measure typically gives information regarding the reliability of the match score. This scalar score often is used to estimate the reliability of the feature, i.e. for use in later tracking operations or to propagate image flow information from one portion of an image to another. Below, we will describe a matrix-valued covariance matrix that contains information both about the overall confidence in a feature measurement, and information about how accurate the measurement is in all directions.

Anandan [2] used the SSD matching scores of a template with a 5×5 image region to develop a match confidence measure based on the variation of the SSD values over the set of candidate matches. Anandan argued that if the variation of the SSD measure along a particular line in the search area surrounding the best match is small, then the component of the displacement along the direction of that line cannot be uniquely determined. Conversely, if there is significant variation along a given line in the search area, the displacement along this line is more likely correct.

Singh and Allen define a *response distribution* based on the SSD metric (1) as

$$\mathcal{RD}_c(u,v) = exp(-kSSD(u,v)), \quad (2)$$

where k is used as a normalization factor. The normalization factor k was chosen in [9] so that the maximum response was 0.95. Singh and Allen then argue that each point in the search area is a candidate for the "true match." However, a point with a small response is less likely to be the true match than a point with a high response. Thus, the response distribution could be interpreted as a probability distribution on the true match location – the response at a point depicting the likelihood of the corresponding match being the true match. This interpretation of the response distribution allows the use of estimation-theoretic techniques.

Under the assumption of additive zero mean independent errors, a covariance matrix is associated with each location estimate:

$$\mathbf{P}_m = \begin{bmatrix} \widehat{\sigma_u^2} & \widehat{\rho\sigma_u\sigma_v} \\ \widehat{\rho\sigma_u\sigma_v} & \widehat{\sigma_v^2} \end{bmatrix} \quad (3)$$

$$\widehat{\sigma_u^2} = \sum_{u,v \in N} \mathcal{RD}(u,v)(u-u_m)^2/TP \quad (4)$$

$$\widehat{\sigma_v^2} = \sum_{u,v \in N} \mathcal{RD}(u,v)(v-v_m)^2/TP \quad (5)$$

$$\widehat{\rho\sigma_u\sigma_v} = \sum_{u,v \in N} \mathcal{RD}(u,v)(u-u_m)(v-v_m)/TP \quad (6)$$

$$TP = \sum_{u,v \in N} \mathcal{RD}(u,v) \quad (7)$$

where u_m and v_m are the estimated locations, in the u and v directions, of the feature. The reciprocals of the eigenvalues of the covariance matrix are used as confidence measures associated with the estimate, along the directions given by the corresponding eigenvectors. To our knowledge, Singh and Allen are the first researchers treat the location of the best match as a random vector, and the (normalized) SSD surface is used to compute the spatial certainty of the estimate of this vector [9]. These confidence measures are used in the propagation of high confidence measurements for local image flow to regions with lower confidence measurements, such as caused by large homogeneous regions.

Our work develops a different normalization procedure for $\mathcal{RD}$ that is useful for the evaluation of isolated feature measurements from template images. As described in Section 4.2, we compute one covariance matrix and one location for each feature, and use this information in a model-based object tracking framework. We do not reject any tracking information, but weight each measurement on the basis of this covariance matrix, using as much information as possible from the feature tracking.

As the SSD measure is used to compare the template to areas of the image near the area generating the minimum SSD score, some measure of the *spatial discrimination* power of the template can be generated [2]. Spatial discrimination is defined as the ability to detect feature motion along a given direction in the image. This concept is quite similar to the confidence measures discussed in Section 3 that estimate the reliability of the location estimate. However, we interpret the confidences as spatial uncertainties in the returned location.

While conclusions about the efficacy of a given template for feature localization can be drawn from the fully computed SSDS, it is expensive both computationally and from a computer memory standpoint to maintain the complete surface for this purpose. In the next section, we derive a approximation for $\mathcal{RD}$ that is more useful.

4 A practical approximation for $\mathcal{RD}$

In order to maintain and use relevant information about the shape of the response distribution, we introduce a mathematical approximation to the distribution given in (2). By suppressing the off-peak response of the feature tracking result, this response distribution function converts the SSDS into an approximately Gaussian distribution that contains the feature tracking information we wish to maintain. Since many object tracking systems (including all Kalman filter-based systems) assume measurements are random vectors with Gaussian probability density functions, we explicitly model and approximate this density.

4.1 Uncertain feature measurements

The measurement vector $\mathbf{z}_k$ is interpreted as an uncertain location in the (u,v) plane, and modeled as a 2D Gaussian random vector. It is illustrative to analyze the behavior of the density function for this vector with respect to the spatial certainty of the feature tracking result as $\mathbf{R}_k$, the covariance matrix for the vector, changes. For example, if $\mathbf{R}_k = \sigma^2 I$, where σ^2 is the variance of the vector, the location is equally certain in each direction. The ellipses of equal probability on the density surface are circles. If $\sigma_u \neq \sigma_v$, where σ_u^2 and σ_v^2 are the variances in the u and v directions, the location is more certain in one direction (given by the minor axis of the ellipses of equal probability) than in the other direction (given by the major axis). As the length of the major axis approaches infinity, complete uncertainty on the location along this dimension is asserted. It is well known that the mean and covariance are sufficient statistics for a Gaussian random variable. Therefore, if this Gaussian density surface is sufficient to model the tracking behavior,

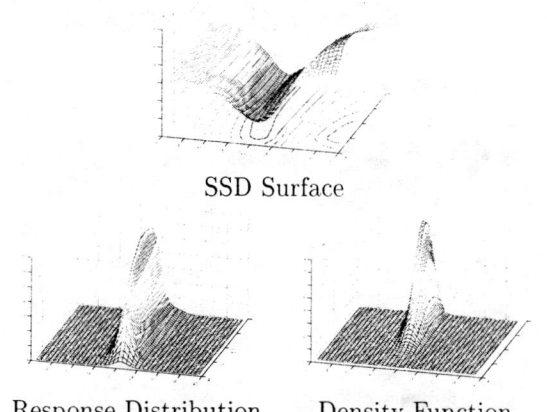

Figure 1: Approximation of response distribution by density function.

it is no surprise that the mean and covariance suffice to maintain this information. In the next section we explain how we estimate these quantities.

4.2 Parameter estimation from the SSDS

This section describes a process for analyzing the SSDS to arrive at estimates for the mean and variance of a Gaussian random vector. The density function of this vector acts as an approximation to the response distribution $\mathcal{RD}$ (see (2)) for the purpose of tracking features.

Our computation of the normalization factor k in (2) differs from that of Singh and Allen [9]. We chose k such that

$$\sum_{u,v \in N} \mathcal{RD}(u,v) \approx 1. \qquad (8)$$

This has the effect of suppressing the off-peak response of the feature detector, when compared with Singh and Allen's normalization. Since we are using correlation between synthetic templates and images, the off-peak response in our situation is more significant than for Singh and Allen. As shown in Figure 1, our normalization makes the response distribution approximate a Gaussian density function with the desired characteristics with respect to feature tracking.

The mode, or most probable value, of a random vector is located at the peak of the density function. We take the location of the minimum of the SSDS as our value for the mode of the vector,

$$\mathbf{z}_k = argmin_{u,v} SSD(u,v). \qquad (9)$$

The variance of u (σ_u^2), the variance of v (σ_v^2), and the covariance between u and v ($\rho_{uv}\sigma_u\sigma_v$) can be estimated directly from the response distribution using Equations (2) and (3)-(7), yielding the desired covariance matrix,

$$\mathbf{R}_k = \begin{bmatrix} \widehat{\sigma_u^2} & \widehat{\rho_{uv}\sigma_u\sigma_v} \\ \widehat{\rho_{uv}\sigma_u\sigma_v} & \widehat{\sigma_v^2} \end{bmatrix}, \qquad (10)$$

which, as described above, contains complete information about the *orientation* and *shape* of the error ellipsoids. Figure 1 illustrates this process for a vertical edge feature.

Of course, as we are only maintaining the mean and variance of the random vector, and not the complete SSDS, this is only an approximation to the complete information about local image structure given by the SSD. However, it does give an indication of both the absolute quality of the match and, in cases where edge features exist, the direction of the edge.

5 Results

5.1 Gripper feature

The feature illustrated in this section is the end-effector of a robot. Figure 2 shows the search region, tracking result, and measurement uncertainty estimates for two different cases. Note that since the SSD measure involves the image area surrounding a pixel, a border around the search region must be retained for each search region.

The results of feature tracking in normal circumstances are shown in Figure 2(a). The cross indicates the location of the minimum point of the SSD surface. The complete SSDS and the Gaussian approximation to this surface are shown in Figure 2(b)-(c), both indicating equal accuracy of the tracking result in all directions. Note the effect in (b) and (c) of our normalization procedure. Even though there is significant off-peak response, the relative certainty in the peak response with respect to the lower responses indicated a single proper match. This fact is evident from the final result shown in (c).

In (d)-(f), we present an illustration of the usefulness of the on-line estimation of template efficacy. This feature has the same template as in the previous case. However, a person has stepped between the camera and the feature, occluding the feature. The feature template thus does not match any portion of the search window well, as shown in (e). This mismatch causes larger values for the variances for this measurement, and (f) indicates high uncertainty of the feature location in all directions.

A 2D measurement, represented by the cross in Figure 2(a) and (d), and a 2 × 2 covariance measurement are the output of the feature tracking, and are used directly in the EKF framework described in [8].

extent: the result is neither endowed with inappropriate confidence due to the good accuracy in the direction orthogonal to the edge nor unduly devalued due to the poor accuracy in the direction along the edge.

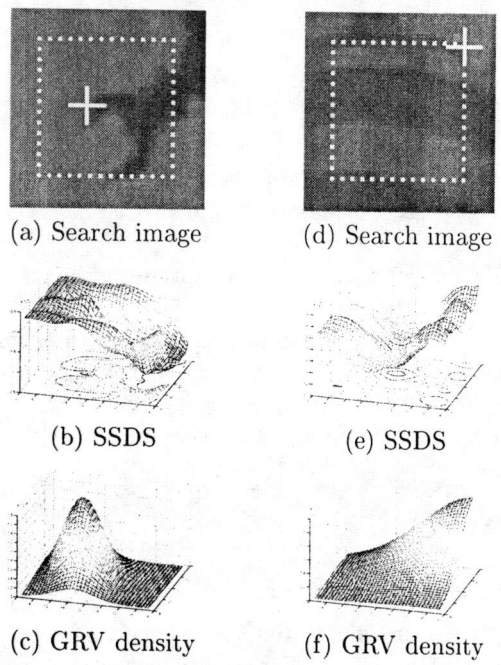

Figure 2: Tracking results for gripper feature (a)-(c) normal (d)-(f) externally occluded

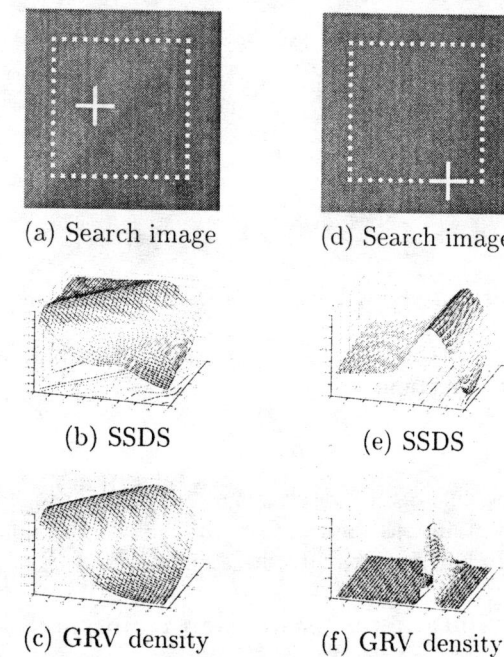

Figure 3: Tracking results for edge feature.

5.2 Edge feature

This case illustrates the usefulness of the measurement uncertainty estimation for tracking features with poor spatial discrimination in one direction. An edge feature can be tracked well only in the direction orthogonal to the edge. This feature arises from a point on the edge of the robotic arm. Thus, the orientation of the edge in the feature depends on the configuration of the robot. As the configuration of the robot changes, the direction of the edge projected onto the image plane will change. In Figure 3 (a), the edge is in a diagonal orientation. The full SSDS shown in (b) has a ridge along this direction, indicating good match scores along the ridge. After normalization, the density function shown in (c) exhibits the same ridge, while suppressing the off-peak match scores on both sides of the ridge. Similarly, the edge in (d) is in a vertical orientation, so the ridges in (e) and (f) are in the vertical direction.

By maintaining this information, the system can exploit the feature tracking information to a greater

5.3 Degenerate point feature

In this section, we illustrate another aspect of the usefulness of on-line estimation of template efficacy. Since our object-tracking system works under widely varying configurations of the object, the appearance of features may change significantly during tracking. A single feature acceptance or rejectance decision will not suffice in this case. Figure 4 shows tracking results for a feature that undergoes such a change in appearance, a point of intersection of a black line on the edge of the robotic arm with the rear edge of the arm.

This feature is a point of high texture in both directions when the arm is approximately parallel to the image plane, as shown in (a), and acts like a point feature. This feature location can be found with high accuracy in all directions, shown in (c).

However, this feature acts like an edge feature in other configurations, such as the configuration shown in (d), where the arm is pointing roughly toward the camera. In this configuration, the feature appears as a vertical edge feature. The location of the feature in this case can be found with high accuracy in only the horizontal direction, as seen in (e) and (f).

3234

Again, the maintenance of the covariance matrix instead of a single confidence measure makes this suboptimal tracking result not only tolerable, but useful.

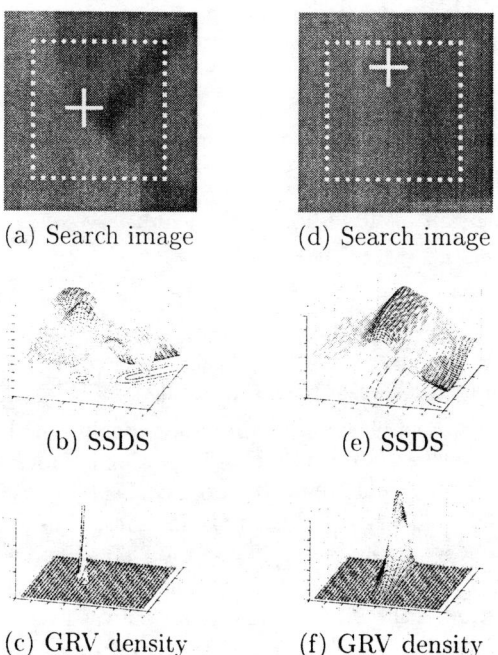

(a) Search image (d) Search image

(b) SSDS (e) SSDS

(c) GRV density (f) GRV density

Figure 4: Tracking results for a point feature (a)-(c) nondegenerate and (d)-(f) degenerate (acting as edge feature)

6 Conclusions

The method presented uses the SSDS, a common intermediate result in correlation-based feature tracking, to compute quantitative estimates for the spatial accuracy of the feature tracking result. This estimate consists of a covariance matrix for a Gaussian random vector. Analysis of this matrix yields information about the directions (if any) in which the template is discriminating the feature from the image background, and provides a quantitative measure of confidence in each direction.

The feature tracking results, combined with this matrix, yields a composite measure that is useful when analyzing the tracking results. An example of the use of this matrix in model-based object tracking can be found in [8]. Analysis of the results can detect templates that do not discriminate effectively in any direction. By associating spatial confidence measures with feature tracking results, those results can be more fully exploited: the fact that some directions may have high confidence does not lead us to accept the entire measurement, and the fact that some directions may have low confidence does not lead us to disregard useful data.

References

[1] D. Lowe, "Robust model-based motion tracking through the integration of search and estimation," *International Journal of Computer Vision*, vol. 8, pp. 113–122, Aug. 1992.

[2] P. Anandan, "A computational framework and an algorithm for the measurement of visual motion," *International Journal of Computer Vision*, vol. 2, pp. 283–310, Jan. 1989.

[3] N. P. Papanikolopoulos, "Selection of features and evaluation of visual measurements during robotic visual servoing tasks," *Journal of Intelligent and Robotic Systems*, vol. 13, pp. 279–304, July 1995.

[4] R. Brunelli and T. Poggio, "Face recognition: Features versus templates," *IEEE Transactions on Pattern Analysis and Machine Intelligence*, vol. 15, pp. 1042–1052, Oct. 1993.

[5] G. Hager, "Real-time feature tracking and projective invariance as a basis for hand-eye coordination," in *Proceedings IEEE Computer Society Conference on Computer Vision Pattern Recognition*, pp. 533–539, 1994.

[6] G. Hager and P. Belhumeur, "Real-time tracking of image regions with changes in geometry and illumination," in *Proceedings IEEE Computer Society Conference on Computer Vision Pattern Recognition*, pp. 403–410, 1996.

[7] R. Lopez, A. Colmenarez, and T. S. Huang, "Vision-based head and facial feature tracking," in *Advanced Displays and Interactive Displays Federated Laboratory Consortium, Annual Symposium*, Advanced Displays and Interactive Displays Federated Laboratory Consortium, Jan. 1997.

[8] K. M. Nickels, *Model Based Articulated Object Tracking*. PhD thesis, University of Illinois, Urbana, IL, 61821, Oct. 1998.

[9] A. Singh and P. Allen, "Image flow computation: An estimation-theoretic framework and a unified perspective," *Computer Vision Graphics and Image Processing: Image Understanding*, vol. 56, pp. 152–177, Sept. 1992.

[10] A. Kosaka and A. C. Kak, "Fast vision-guided robot navigation using model-based reasoning and prediction of uncertainties," *Computer Vision and Image Understanding*, vol. 56, pp. 271–329, Nov. 1992.

Visual servoing of a 6 DOF manipulator for unknown 3D profile following

Jacques A. Gangloff, Michel de Mathelin* and Gabriel Abba

Strasbourg I University – LSIIT UPRES-A CNRS 7005
ENSPS – Parc d'Innovation, Bd. Sébastien Brant, 67400 Illkirch, France
fax: +33 (0)3 88 65 54 89   e-mail: demath@hp1gra.u-strasbg.fr

Abstract

This paper presents the visual servoing of a 6 DOF manipulator for profile following. The profile has an unknown curvature, but its cross-section is known. The visual servoing keeps the transformation constant between a cross-section of the profile and the camera with respect to 6 degrees of freedom. The position of the profile with respect to only five of these degrees of freedom can be measured with the camera, since the image gives no position information along the profile. The kinematic model of the robot is used to reconstruct the displacement along the profile and allow us to control the profile following velocity. Experiments show good accuracy for positioning at a sampling rate of 20ms. Two control strategies are tested: PI control and Generalized Predictive Control (GPC). The visual servoing exhibits better accuracy with the GPC in simulations and in real experiments.

Introduction

In industrial applications, there exist many robotic tasks which consist in following a path along a profile, *e.g.*, to make a soldered joint along an edge, to distribute the glue for a sealing strip on the door of a car [9], to remove the burrs along the edges of a casted piece of metal or polymer. The use of an external vision sensor, like a CCD camera, is very helpful to control the end-effector of the robot, so that it stays in the same position with respect to a cross-section of the profile. Indeed, with visual servoing, there is no need to know the exact curvature of the profile, nor its position with respect to the robot frame of coordinates. This paper describes such a visual servoing application where we assume that only the cross-section of the profile is known, in order to keep the position constant with respect to the profile.

There exist some previous works dealing with profile following. For example, a simulation is shown in [5] by Espiau *et al.*, where the goal consists in following a road modelled by three parallel lines. The camera must stay on the center of the right lane at a defined height with respect to the road and must look in the direction of the road. This simulation was intended to validate an image-based approach to profile following where the features are straight lines instead of points.

An edge following experiment is described by Lange *et al.* in [9]. It is a two dimensional line following experiment where the line can be curved but stays always in a plane parallel to the image plane of the camera.

In our approach, the profile must contain at least 3 parallel distinguishable curves or edges. It is the minimum needed to estimate the pose between the camera frame and the current cross-section of a 3D profile (see section 1.3). We use a position-based control strategy to control the visual loop (see [8] or [4], for a tutorial on visual servoing techniques). We propose a method for the reconstruction of the pose based on the determination of the tangent vectors to the profile at the current cross-section. A problem that we have to solve is the determination of the current cross-section, that is, the cross-section of the profile where the pose is estimated. This cross-section moves along the profile at the velocity required by the profile following task. Unfortunately, the vision system cannot give velocity and position measurements in the direction of the profile. Therefore, we use additional measurements based on the kinematic model of the robot to control the speed of the displacement along the profile. Two controllers were tested for the visual loop: a PI controller and a Generalized Predictive Controller (GPC). The tuning technique used for the GPC is the same than the one described in [6]. The GPC exhibits better perfor-

---

*Author to whom all correspondence should be addressed

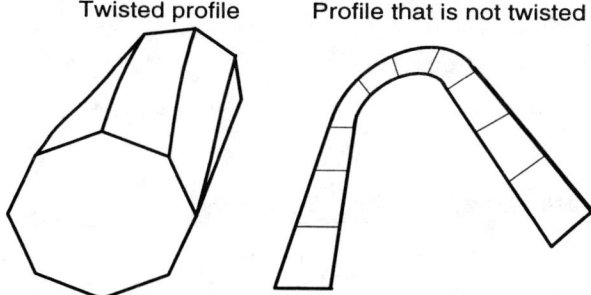

Figure 1: Difference between a twisted profile and a profile which is not twisted.

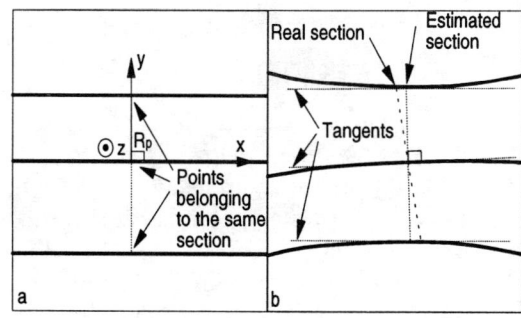

Figure 2: Finding the projection of a cross-section.

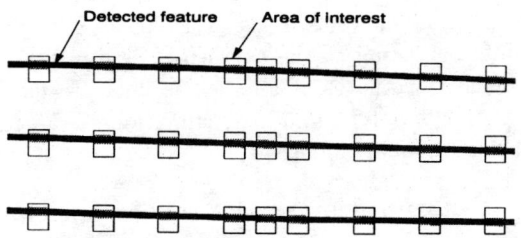

Figure 3: The image processing.

mance than the PI, thanks to the fact that it takes into account the dynamics of the manipulator. The paper is divided into 3 sections. Section 1 describes how to derive the pose between the profile and the camera. In section 2, the whole visual servoing loop model is presented. Finally, simulation and experimental results are presented in section 3.

1 Calculation of the pose vector

1.1 Background material

By definition, a 3D profile is generated by the displacement of a constant cross-section along a 3D reference curve with the cross-section always kept perpendicular to the curve. In other words, the intersection of the profile and a plane perpendicular to the reference curve is identical to the cross-section at every point of this reference curve.

A profile that is *not twisted* can be divided in portions where the transformation between two cross-sections is a translation (for a linear portion) or a rotation (for a curved portion). In this paper, we assume that the profiles are not twisted (see figure 1). Now, in practice, industrial profile cross-sections often contain edges. Given the definition of a profile that is not twisted, the tangent vectors to these edges at points belonging to the same cross-section are parallel together.

At least 3 parallel lines are needed (this point is discussed in section 1.3) to reconstruct the transformation between a frame attached to the camera and a frame attached to the parallel lines in 3D space if only the perspective projection of these lines in the camera frame is known. Let us consider the following general property of perspective projection:

Property: *The perspective projection of a tangent to a differentiable curve in 3D space at a point P of this curve is identical to the tangent to the projection of this curve at a point equal to the projection of P.*

Based on this property, if we can find in the image the projections of points that belong to the same cross-section, then, the tangents at these points to the projection of the profile's edges, are the projection of the tangents to the profile's edges in 3D space which are parallel, since they belong to the same cross-section. Furthermore, if we have at least 3 visible edges in the image, we are able to reconstruct the pose between the cross-section and the camera (see section 1.3). Therefore, the problem consists in finding the points in the image which are projections of points in 3D space belonging to a particular cross-section. In order to solve this problem, an assumption must be made. If we assume a small curvature of the edge projections in the image, then, small error in the determination of these points leads to a negligible error in the detected tangents as shown in figure 2b.

1.2 Modelling of the profile projection

The reconstruction process of the pose between the camera and the profile is based on finding the tangents at points of the profile's edges' projection. Therefore, these tangents must be determined with high accuracy. In order to obtain this type of accuracy, the

3237

curves in the image are modelled with high order polynomials.

Figure 3 shows how the image processing works. After an initial global detection of all the features in the image, a polynomial model of the curves is determined. Then, for each measurement, Areas Of Interest (AOI) are acquired along each curve according to the polynomial model. These AOI are selected at the position occupied by the curve in the last image.

Then, edge detection is performed in these AOI and the center of mass of all the points where the gradient is greater than a threshold value is computed. In theory, this center of mass belongs to the middle of the corresponding curve. So, 9 AOI are processed, giving 9 points belonging to the curve (see figure 3). Then, sub-pixel coordinates of all these points are used to find the next polynomial approximation of the curve with a least-squares method.

1.3 Computation of the target position

In this section, we assume that the camera has a focal length of $f = 1$. This yields no loss of generality since all the measurements in the image can be converted to a virtual image plane where $f = 1$. Let $\mathcal{T}_{pi} = \mathcal{P}(\mathcal{T}_i)$, $i = 1...n$, be the perspective projections in the image plane of the tangents $\mathcal{T}_i$ to the profile at points P_i belonging to the same cross-section:

$$\begin{aligned}
\mathcal{T}_i &= \begin{pmatrix} x_i \\ y_i \\ z_i \end{pmatrix}_{R_c} = k_i \vec{V}_i + P_i \\
&= k_i \begin{pmatrix} v_{xi} \\ v_{yi} \\ v_{zi} \end{pmatrix}_{R_c} + \begin{pmatrix} r_{xi} \\ r_{yi} \\ r_{zi} \end{pmatrix}_{R_c} \\
\mathcal{T}_{pi} &= \begin{pmatrix} x_{pi} \\ y_{pi} \end{pmatrix}_{R_c} \\
&= k_{pi} \begin{pmatrix} u_{xi} \\ u_{yi} \end{pmatrix}_{R_c} + \begin{pmatrix} o_{xi} \\ o_{yi} \end{pmatrix}_{R_c} \quad (1)
\end{aligned}$$

where k_i and k_{pi} are scalars parameterizing the tangents and $\vec{V}_i$ are direction cosines (i.e. $\|\vec{V}_i\| = 1$). We assume that these tangents are expressed in a frame of reference, R_c, attached to the camera. The frame R_c is defined such that the image plane is perpendicular to the z axis, the x axis is vertical in the image, and y is horizontal (see figure 4). The relationship between the parameters of a 3D line and the parameters of its perspective projection are given by (see [7]):
$\forall i = 1...n$,

$$u_{yi} r_{xi} - u_{xi} r_{yi} + (o_{yi} u_{xi} - o_{xi} u_{yi}) r_{zi} = 0 \quad (2)$$
$$u_{yi} v_{xi} - u_{xi} v_{yi} + (o_{yi} u_{xi} - o_{xi} u_{yi}) v_{zi} = 0 \quad (3)$$

Since the 3D lines of interest are parallel:

$$\vec{V}_i = \begin{pmatrix} v_{xi} \\ v_{yi} \\ v_{zi} \end{pmatrix}_{R_c} = \begin{pmatrix} v_x \\ v_y \\ v_z \end{pmatrix}_{R_c} = \vec{V} \quad \forall i = 1...n \quad (4)$$

Furthermore, since $\|\vec{V}\| = 1$:

$$v_x^2 + v_y^2 + v_z^2 = 1 \quad (5)$$

Since the $\mathcal{T}_{pi}$ are known from the image processing (see section 1.2), the equations (3) have 3 unknowns: v_x, v_y and v_z:

$$u_{yi} v_x - u_{xi} v_y + (o_{yi} u_{xi} - o_{xi} u_{yi}) v_z = 0 \quad i = 1...n \quad (6)$$

Solving these equations for $n \geq 2$ and using (5) gives $\vec{V}$, the direction cosines common to all the 3D lines tangent to the profile.

Now, let us introduce the geometrical model of a cross-section. For this purpose, a frame of reference, R_p, linked to the profile must be defined (see figure 2a and figure 4). This frame of reference, R_p, is chosen so that all the points where the tangents intersect the cross-section are in the (yz) plane. Let P_i, $i = 1...n$, be these points:

$$P_i = \begin{pmatrix} 0 \\ P_{yi} \\ P_{zi} \end{pmatrix}_{R_p} \quad (7)$$

Furthermore, we choose the origin of R_p to be one of these points (the central one in the image). Let $P \in \{P_i\}$ be this central point. Then, $P = [0\ 0\ 0]^T_{R_p}$.

We chose also the axis x of R_p to be tangent to the profile at P. Let M_{cp} be the homogeneous transformation between the frame attached to the camera, R_c, and the frame attached to the profile, R_p, defined in R_c:

$$M_{cp} = \begin{pmatrix} r_{11} & r_{12} & r_{13} & T_x \\ r_{21} & r_{22} & r_{23} & T_y \\ r_{31} & r_{32} & r_{33} & T_z \\ 0 & 0 & 0 & 1 \end{pmatrix}_{R_c} \quad (8)$$

The first column of M_{cp} is known since the direction cosines of the 3D lines are known. Indeed, the first column of the rotation sub-matrix in M_{cp} contains the direction cosines of the x axis of R_p, expressed in R_c. Therefore, $r_{11} = v_x$, $r_{21} = v_y$ and $r_{31} = v_z$. Then, P_i can be expressed in R_c using M_{cp}:

$$\begin{aligned}
P_i &= \begin{pmatrix} r_{xi} \\ r_{yi} \\ r_{zi} \end{pmatrix}_{R_c} = \begin{pmatrix} r_{12} R_{yi} + r_{13} R_{zi} + T_x \\ r_{22} R_{yi} + r_{23} R_{zi} + T_y \\ r_{32} R_{yi} + r_{33} R_{zi} + T_z \end{pmatrix}_{R_c} \\
i &= 1...n \quad (9)
\end{aligned}$$

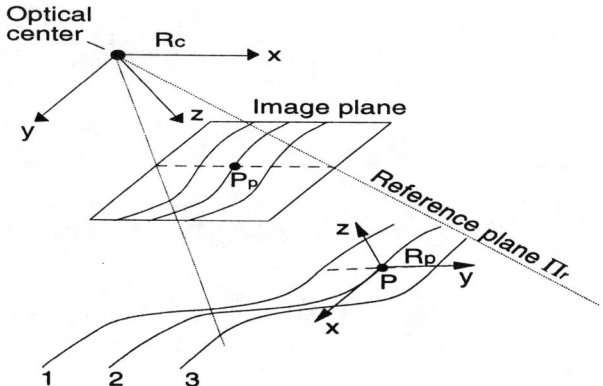

Figure 4: Definition of the reference plane.

Hence, using (2) with (9) yields:

$$u_{yi}(r_{12}P_{yi} + r_{13}P_{zi} + T_x)$$
$$-u_{xi}(r_{22}P_{yi} + r_{23}P_{zi} + T_y)$$
$$+(o_{yi}u_{xi} - o_{xi}u_{yi})(r_{32}P_{yi} + r_{33}P_{zi} + T_z) = 0$$
$$i = 1...n \quad (10)$$

These equations contain nine unknowns: r_{12}, r_{13}, r_{22}, r_{23}, r_{32}, r_{33}, T_x, T_y, T_z. At least three 3D parallel lines are required to compute M_{cp}. Indeed, if $n \geq 3$, then at most six other equations are necessary. The properties of the rotation matrix in M_{cp} supply five equations:

$$r_{12}^2 + r_{22}^2 + r_{13}^2 = 1 \quad (11)$$
$$r_{13}^2 + r_{23}^2 + r_{33}^2 = 1 \quad (12)$$
$$v_x r_{12} + v_y r_{22} + v_z r_{32} = 0 \quad (13)$$
$$v_x r_{13} + v_y r_{23} + v_z r_{33} = 0 \quad (14)$$
$$r_{12}r_{13} + r_{22}r_{23} + r_{32}r_{33} = 0 \quad (15)$$

The sixth equation is obtained by defining a reference plane Π_r. This reference plane Π_r (see figure 4) is defined by the optical center of the camera, which is the origin of the reference frame R_c, and a vertical line in the image plane going through P_p, the projection of P. Therefore, this plane fixes the position P of the current cross-section along the profile. Its equation is as follows:

$$\Pi_r = \begin{pmatrix} \pi_x \\ \pi_y \\ \pi_z \end{pmatrix}_{R_c} \quad \text{with} \quad \pi_y - a\pi_z = 0 \quad (16)$$

where a is a scalar chosen so that the intersection between Π_r and the image plane is within the image and so that the displacement along the profile has a predefined speed, V_f (see section 2.1). Then, since the

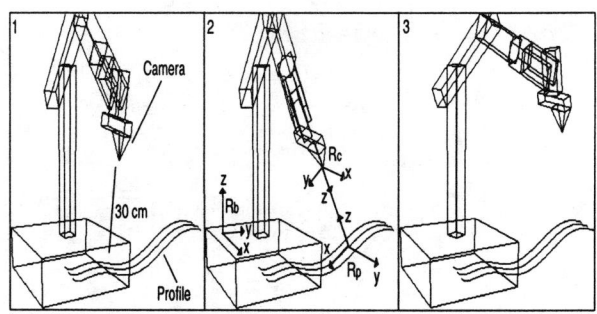

Figure 5: Simulation of the profile following.

origin of R_p belongs to Π_r:

$$T_y - aT_z = 0 \quad (17)$$

Particular case:

An exact solution to this set of equations can be found in the particular case of 3 parallel and coplanar lines. Indeed, in this case, the corresponding profile is flat (see figure 5). Let us define P_1, P_2, P_3 in the frame attached to the profile, then:

$$P_1 = \begin{pmatrix} 0 \\ d_1 \\ 0 \end{pmatrix}_{R_p} P_2 = \begin{pmatrix} 0 \\ 0 \\ 0 \end{pmatrix}_{R_p} P_3 = \begin{pmatrix} 0 \\ -d_3 \\ 0 \end{pmatrix}_{R_p} \quad (18)$$

From (10):

$$u_{y1}(d_1 r_{12} + T_x) - u_{x1}(d_1 r_{22} + T_y)$$
$$+(o_{y1}u_{x1} - o_{x1}u_{y1})(d_1 r_{32} + T_z) = 0 \quad (19)$$
$$u_{y2}T_x - u_{x2}T_y + (o_{y2}u_{x2} - o_{x2}u_{y2})T_z = 0 \quad (20)$$
$$u_{y3}(-d_3 r_{12} + T_x) - u_{x3}(-d_3 r_{22} + T_y)$$
$$+(o_{y3}u_{x3} - o_{x3}u_{y3})(-d_3 r_{32} + T_z) = 0 \quad (21)$$

Using (20) and (17), we obtain:

$$T_x = \frac{u_{x2}a - o_{y2}u_{x2} + o_{x2}u_{y2}}{u_{y2}} T_z \quad (22)$$

It holds from (13), that:

$$r_{12} = -\frac{v_y r_{22} + v_z r_{32}}{v_x} \quad (23)$$

Then, using (17), (22) and (23) in (19) and (21) yields:

$$u_{y1}\left(-\frac{d_1(v_y r_{22} + v_z r_{32})}{v_x}\right.$$
$$\left.+\frac{T_z(u_{x2}a - o_{y2}u_{x2} + o_{x2}u_{y2})}{u_{y2}}\right)$$
$$-u_{x1}(d_1 r_{22} + aT_z)$$
$$+(o_{y1}u_{x1} - o_{x1}u_{y1})(d_1 r_{32} + T_z) = 0 \quad (24)$$

and

$$u_{y3}\left(\frac{d_3(v_y r_{22} + v_z r_{32})}{v_x}\right.$$
$$+\frac{T_z(u_{x2}a - o_{y2}u_{x2} + o_{x2}u_{y2})}{u_{y2}}\right)$$
$$-u_{x3}(-d_3 r_{22} + aT_z)$$
$$+(o_{y3}u_{x3} - o_{x3}u_{y3})(-d_3 r_{32} + T_z) = 0 \quad (25)$$

Only 3 unknowns, r_{22}, r_{32} and T_z, remain in these two equations. Thus, solving (24) and (25) for r_{22} and r_{32} yields two relationships:

$$r_{22} = K_1 T_z \quad (26)$$
$$r_{32} = K_2 T_z \quad (27)$$

With K_1 and K_2 properly defined. Finally, using (26) and (27) in the last equation, (11), gives the value of T_z^2. Since the profile is always in front of the camera, $T_z > 0$. So we can deduce the value of T_z and all the other components of the homogeneous matrix M_{cp}.

2 The profile following task

2.1 Control of the position along the profile

The image gives no information about the position along the profile. In order to follow the profile at a constant velocity V_f, a measurement of the position along the profile must be done. For this purpose, we use the kinematic model of the robot. Let M_{bp}^- the homogeneous transformation between R_b, the frame attached to the base of the robot (see figure 5) and R_p^-, the frame attached to the previous current cross-section, i.e. the cross-section that was used to make the measurement of M_{cp} at the last sampling time, that we will note M_{cp}^- to distinguish it from the current M_{cp}. Then, M_{bp}^- is given by:

$$M_{bp}^- = M_{bc}^- M_{cp}^- \quad (28)$$

where M_{bc}^- is the homogeneous transformation between the frame attached to the base and the frame attached to the camera, obtained by computing the kinematic model of the robot at its last position using axis position measurements. If the profile following velocity is V_f, then, the theoretical position, M_{bp}^0, of the frame R_p attached to current cross-section, is given by $M_{bp}^0 = M_{bp}^- \mathcal{T}_x$ where $\mathcal{T}_x$ is an homogeneous matrix making a translation of a distance $T_s V_f$ (where

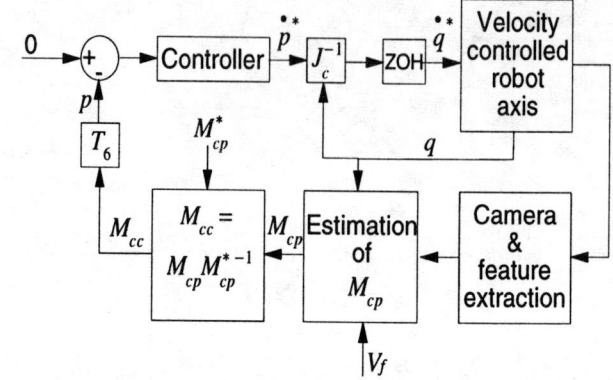

Figure 6: Diagram of the visual servoing loop.

T_s is the sampling time interval) in the x axis direction of the frame R_p^- (the direction of the profile following task, see figure 4). We will note R_p^0, the frame whose position is defined by M_{bp}^0. Now, M_{bp}^0 is a theoretical position assuming that the profile is made of straight lines in the x axis direction, that it does not move, and that the kinematic model of the robot is perfectly known. In reality, these conditions are not fulfilled. So, this position must be updated by the current visual measurement. For this purpose, firstly, the position of R_p^0, is expressed in the camera frame, R_c: $M_{cp}^0 = M_{bc}^{-1} M_{bp}^0$ where M_{bc} is obtained by computing the kinematic model of the robot at its current position. Given the definition of R_p, the homogeneous transformation M_{cp}^0 defines a theoretical plane $(yz)^0$, perpendicular to the direction of the profile following task, where the point P of the current cross-section should be. Secondly, we must find in the current image, the cross-section corresponding to the profile cross-section in 3D space whose point P belongs to the plane $(yz)^0$ so that the profile following speed is approximately V_f. This is done iteratively. The parameter in the iterative algorithm is the position of the reference plane Π_r, given by the scalar a (see equation (17)). The condition that must be fulfilled is $\overrightarrow{P^0 P} . \vec{x^0} = 0$, where P^0 is the origin of R_p^0, P is the origin of R_p, updated by vision, and $\vec{x^0}$ a vector in the direction of the x axis in the frame R_p^0. Once the iterations have converged, P, R_p, P_p and M_{cp} are known thanks to the procedure given in section 1.3. Then, M_{cp} is the measurement used in the visual servoing loop. Finally, at the next sampling time, M_{cp} becomes M_{cp}^-, that is used to compute M_{bp}^- in (28).

2.2 The visual servoing loop

The diagram of the visual servoing loop is described in figure 6. The visual information from the camera gives an estimation of the transformation between the camera frame and the current cross-section, M_{cp}. This homogeneous matrix is multiplied by the inverse of the reference transformation, M_{cp}^*. This reference is initialized at the start of the visual servoing task and corresponds to the pose between the camera and a cross-section whose projection is in the middle of the image frame (so, initially, $a = 0$ in equation (17)). Consequently, even in the case of a bad calibration of the camera, this initial pose will be maintained by the visual servoing. Then, M_{cc} is the transformation between the position of the current camera frame and the desired one: $M_{cc} = M_{cp} M_{cp}^{*-1}$. Let T_6 be the transformation that converts an homogeneous transformation matrix into 6 Cartesian coordinates: 3 translations and 3 rotations (in our case pitch, roll, yaw). Therefore, the pose p of the desired camera frame expressed in the current one is expressed as:

$$p = [T_x\ T_y\ T_z\ \Theta_p\ \Theta_r\ \Theta_y]^T_{R_c} = T_6(M_{cc}) \qquad (29)$$

Assuming that the visual loop is fast enough to maintain M_{cc} small, Θ_p, Θ_r and Θ_y will stay small. So, the computation of p from M_{cc} avoids the problem of the switching between $+\pi$ and $-\pi$. The pose p must be set to zero by the visual servoing control loop. We use a velocity-controlled 6 DOF manipulator for a higher bandwidth instead of a position-controlled manipulator. Therefore, the controller must yield velocity references $\dot{p}^*$ for each DOF. These references are converted into velocity references for the robot joint angles with the Jacobian J_c (see [6], for more details about the computation of J_c). Finally, these references are applied directly to the input of the joint-level analog velocity control loops of the manipulator through a digital to analog converter.

3 Simulations and experiments

3.1 Simulation results

A graphical simulation (see figure 5) was developed in order to validate all the algorithms before their implementation on the real system. The robot dynamics are simulated with a high order dynamic model. Figure 7 shows the error measurement of the pitch angle, Θ_p, for the profile following task along the profile shown in figure 5 at the constant velocity of 10

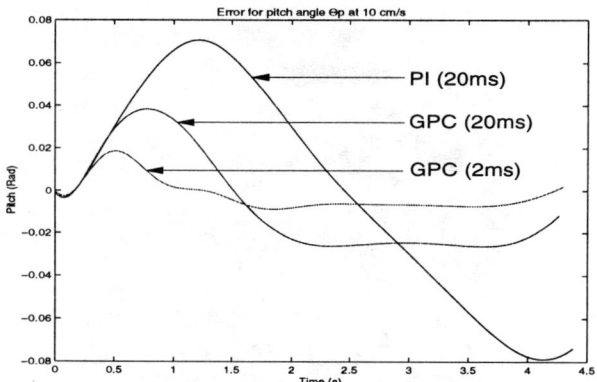

Figure 7: Simulations of the profile following experiment.

cm/s (Θ_p is the angle around the x direction of the frame R_c in figure 5.2). Two controllers were tested: a Proportional-Integral controller (PI) and a Generalized Predictive controller (GPC) (see, [1], [3] and [10]). The tuning of the GPC is explained in [6]. Furthermore, two different sampling rates are simulated with the GPC: $20ms$ which corresponds to the sampling rate of the real experimental setup and $2ms$, corresponding to future systems with a fast CCD camera. These simulations show clearly the superiority of the GPC compared to the PI. This is due to the fact that the GPC use a linearized dynamical model of the velocity controlled manipulator (see [6] for more explanations). Moreover, the simulations show that a decrease in the sampling rate improves the profile following precision by significantly reducing the squared error integral.

3.2 Experimental results

The profile is composed of 3 black lines printed on a white paper tape. Its curvature can be easily altered by an object placed below it. The visual sensor is a regular CCIR CCD black and white camera. By using alternately the odd and the even field of a frame, the sampling rate of the visual measurement can be set at 50 Hz. The whole visual loop is synchronized with the vision, so the sampling rate of the whole loop is 50Hz. The image processing and the control of the loop is performed by a Pentium II^{TM} 300 computer with an Imaging technology^{TM} PCI frame grabber board. The manipulator is a 6 DOF SCEMI 6P01 robot with 6 rotational joints. For all the experiments, we use the same profile which is shown in figure 8. This profile is divided in two parts: a straight one and a curved one. The curvature is created by a rectangular object

Figure 8: Experimental setup.

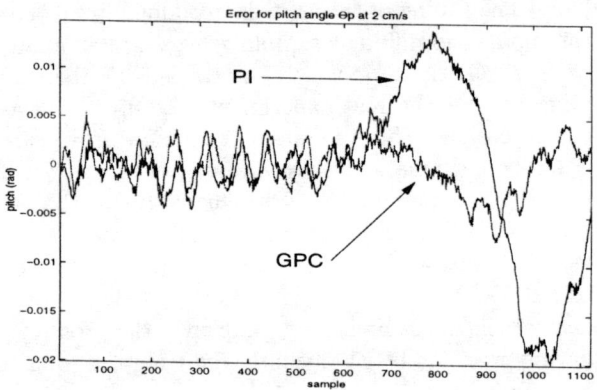

Figure 9: Comparison between PI control and GPC.

of about 5 cm high. Figure 9 shows the comparison of the visual loop error for the pitch angle Θ_p with a PI controller and a GPC at a profile following velocity of 2 cm/s. It clearly shows that the GPC rejects better the perturbations due to the curvature variations than the PI controller. Indeed, the tuning of the GPC takes into account a model of the robot dynamics whereas the PI controller is optimized by manual tuning on the real system. Figure 10 shows the error in the z direction of the visual servo-loop with a GPC at a profile following velocity of 6 cm/s. In the straight part of the profile, the profile following precision is less than 0.5 mm. Whereas in the curved part of the profile, the error is less than 1.5 mm. As it is shown in the simulation, this error could be reduced by increasing the sampling rate.

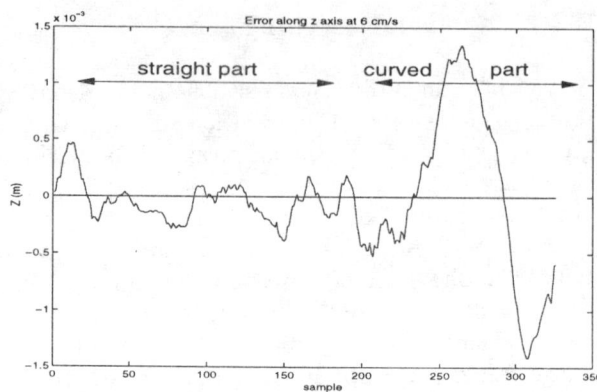

Figure 10: Error along axis z at a speed of 6 cm/s.

References

[1] R. Bitmead, M. Gevers, and V. Wertz. *Adaptive optimal control, the thinking's man's GPC*. Prentice Hall, London, 1990.

[2] M. P. Do Carmo. *Differential geometry of curves and surfaces*. Prentice-Hall, Englewood Cliffs, New Jersey, 1976.

[3] D.W. Clarke, C. Mohtadi, and P.S. Tuffs. Generalized predictive control - part 1. The basic algorithm. *Automatica*, 23:137–160, 1987.

[4] Peter I. Corke. *Visual control of robots*. Research Studies Press Ltd., Taunton, Somerset, U.K., 1996.

[5] B. Espiau, F. Chaumette, and P. Rives. A new approach to visual servoing in robotics. *IEEE Transactions on Robotics and Automation*, 8(3):313–326, 1992.

[6] J. A. Gangloff, M. de Mathelin, and G. Abba. 6 DOF high speed dynamic visual servoing using GPC controllers. In *Proceedings of the IEEE International Conference on Robotics and Automation*, pages 2008–2013, 1998.

[7] R. M. Haralick and L. G. Shapiro. *Computer and robot vision*, volume 2. Addison-Wesley, Reading, Massachusetts, 1993.

[8] S. Hutchinson, G.D. Hager, and P.I. Corke. A tutorial on visual servo control. *IEEE Transactions on Robotics and Automation*, 12(5):651–670, 1996.

[9] F. Lange, P. Wunsch, and G. Hirzinger. Predictive vision based control of high speed industrial robot paths. In *Proceedings of the IEEE International Conference on Robotics and Automation*, pages 2646–2651, 1998.

[10] K. Yamuna Rani and H. Unbehauen. Study of predictive controller tuning methods. *Automatica*, 33(12):2243–2248, 1997.

Accurate Image Overlay on Head-Mounted Displays Using Vision and Accelerometers

Yasuyoshi YOKOKOHJI, Yoshihiko SUGAWARA, and Tsuneo YOSHIKAWA
Department of Mechanical Engineering, Graduate School of Engineering
Kyoto University, Kyoto 606-8501, Japan
{yokokoji|yosihiko|yoshi}@mech.kyoto-u.ac.jp

Abstract

In this paper, we propose a method for accurate image overlay on head-mounted displays (HMDs) using vision and accelerometers. The proposed method is suitable for video see-through HMDs in augmented reality applications but not limited to them. Acceleration information is used for predicting the head motion to compensate the end-to-end system delay and to make the vision-based tracking robust. Experimental results showed that the proposed method can reduce the alignment errors within 6 pixels on average and 11 pixels at the maximum, even if the user moves his/her head quickly (with 10 $[m/s^2]$ and 49 $[rad/s^2]$ at the maximum).

1 Introduction

Head-mounted displays (HMDs) are widely used in VR applications. To give immersive images to the user, the user's head motion should be tracked accurately without delay. In augmented reality (AR) applications, where synthetic images are overlaid on real scenes, head tracking should be more accurate than for VR, because users may easily detect an even small misalignment between the synthetic image and the real image[2].

In AR applications, see-through HMDs are used for image overlay. Optical see-through HMDs have optical combiners through which the user can see real scenes, whereas video see-through HMDs display real scenes through video cameras. Hence, video see-through HMDs are compatible with the vision-based tracking approach. If the target scene includes some landmarked points or lines, it is possible to recover the camera pose (position and orientation in 3D space) with respect to the target object by tracking these points or lines[8]. The vision-based approach can achieve more accurate static registration than using magnetic sensors[2]. However, video images are usually taken at 30 [Hz] (video rate) and the end-to-end system delay (total latency from measuring data until displaying the final image) is not negligible. Hence, if the user moves his/her head quickly, the overlaid image is misaligned from the real image and even swings around (dynamic registration error)[2]. A quick head motion may also cause the vision-based tracker to lose the landmarks.

In this paper, we propose a method for accurate image overlay on HMDs using vision and accelerometers. The proposed method, which is based on the extended Kalman filter (EKF), is suitable for video see-through HMDs in augmented reality applications but not limited to them. Acceleration information is used for predicting the head motion to compensate the end-to-end system delay and to make the vision-based tracking robust. Experimental results show the effectiveness of the proposed method.

2 Related Works

Kalman filters or extended Kalman filters are widely used for motion tracking and motion prediction[6][5][1][4][3]. Some of them used different kinds of sensors simultaneously for tracking the head motion. Azuma and Bishop[1] used linear accelerometers and gyros to reduce the dynamic registration errors of their optical tracker. They used time derivatives of the measured angular velocity for motion prediction.

Merhav and Velger[6] proposed a complementary filter for head tracking. They supposed that the orientation information is obtained at 30 [Hz], while the acceleration is measured at a much higher sampling rate. They assumed that the orientation information is delayed one sampling cycle, like vision. Although the orientation information is delayed, it compensates the drift caused by integrating the acceleration.

Emura and Tachi[3] proposed a head tracking method by multi-sensor integration. They used a magnetic tracker and gyros, each of which samples the data in a different frequency. They tried to predict the head motion assuming a constant angular velocity.

Foxlin[4] developed an inertial head-tracker which consists of three gyros, inclinometers and a compass. They implemented a complementary Kalman filter for the angular motion estimation, where the drift, which is caused by integrating the gyro readings, is compensated by the inclinometers and the compass.

Accurate head tracking or image overlay is a difficult problem and far from solved. None of the previous attempts has shown the performance at a satisfactory level.

3 Estimation of Head Motion

3.1 Target system

Fig.1 illustrates the target system including a user who wears a video see-through HMD, on which accelerometers are mounted. **Fig.2** shows the relation between a landmark and its projection on the image plane, assuming a pinhole camera model. A camera coordinate frame is placed in such a way that its origin is at the lens center and the z-axis is parallel to

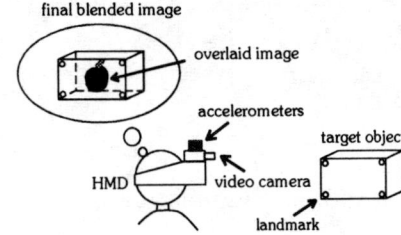

Figure 1: Target system

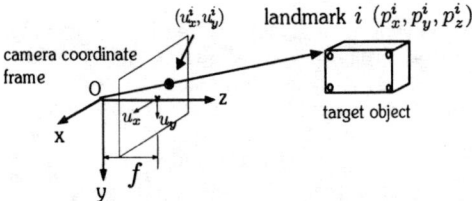

Figure 2: Pinhole camera model

the lens axis. Suppose n landmarks are attached on the target object. The i-th landmark ($i = 1, \ldots, n$) located at (p_x^i, p_y^i, p_z^i) in camera coordinates is projected to a point (u_x^i, u_y^i) on the image plane, and its coordinates are given by

$$u_x^i = \frac{f p_x^i}{s_x p_z^i}, \quad u_y^i = \frac{f p_y^i}{s_y p_z^i}, \quad (1)$$

where f denotes the focal length of the camera, and s_x and s_y are pixel ratios of the image plane.

Since the landmark coordinates (u_x^i, u_y^i) ($i = 1, \ldots, n$) are taken from each frame of the video stream, we get the following discrete-time state equation that describes the motion of the head/HMD system.

$$\begin{aligned} x_{k+1} &= f_k(x_k) + D_k(x_k) u_k, & (2)\\ y_k &= h_k(x_k), & (3) \end{aligned}$$

where x_k denotes a twelve dimensional state vector composed of the camera position and orientation (roll, pitch and yaw angles), and their velocities. u_k means a system input, or accelerations of the camera/HMD system in this case. A measurement vector $y_k = (u_x^1, u_y^1, \ldots, u_x^n, u_y^n)^T$ contains n pairs of landmark coordinates in the image plane. The output equation (3) includes the landmark projection given by eq.(1). The state vector is updated by eq.(2) at video rate (30 [Hz]).

3.2 EKF with acceleration measurement

We use the extended Kalman filter (EKF) to estimate the head motion. The measured acceleration is used as a system input of the estimator. Acceleration is also used to predict the camera motion in the future for the image overlay.

Let u_k and w_{uk} denote a true acceleration of the system and a measurement noise, respectively. Then, the measured acceleration, $\tilde{u}_k$, is given by

$$\tilde{u}_k = u_k + w_{uk}. \quad (4)$$

Considering the process noise w_{pk} in eq.(2) and the measurement noise v_k in eq.(3), we get

$$\begin{aligned} x_{k+1} &= f_k(x_k) + D_k(x_k) \tilde{u}_k \\ &\quad - D_k(x_k) w_{uk} + G_k(x_k) w_{pk} \\ &= f_k(x_k) + D_k(x_k) \tilde{u}_k + G_k(x_k) w_k, \quad (5) \\ \tilde{y}_k &= h_k(x_k) + v_k, \quad (6) \end{aligned}$$

where $\tilde{y}_k$ denotes an actual measurement vector including measurement noise. In eq.(5), we introduced a new process noise, $w_k \triangleq w_{pk} - G_k(x_k)^{-1} D_k(x_k) w_{uk}$. Since the above system is nonlinear, we linearize it and obtain the following extended Kalman filter:

$$\begin{aligned} \hat{x}_{k|k} &= \hat{x}_{k|k-1} + K_k[\tilde{y}_k - h_k(\hat{x}_{k|k-1})], & (7)\\ \hat{x}_{k+1|k} &= f_k(\hat{x}_{k|k}) + \hat{D}_k \tilde{u}_k, & (8)\\ K_k &= \hat{\Sigma}_{k|k-1} H_k^T (H_k \hat{\Sigma}_{k|k-1} H_k^T + \Sigma_{vk})^{-1}, & (9)\\ \hat{\Sigma}_{k|k} &= \hat{\Sigma}_{k|k-1} - K_k H_k \hat{\Sigma}_{k|k-1}, & (10)\\ \hat{\Sigma}_{k+1|k} &= F_k \hat{\Sigma}_{k|k} F_k^T + \hat{G}_k \Sigma_{wk} \hat{G}_k^T, & (11) \end{aligned}$$

where
- K_k : the Kalman gain at time k
- $\hat{x}_{k|k}$: the optimal estimate of x_k
- $\hat{x}_{k+1|k}$: the estimate of x_{k+1} propagated from $\hat{x}_{k|k}$
- $\hat{\Sigma}_{k|k}$: the covariance matrix of $\hat{x}_{k|k}$
- $\hat{\Sigma}_{k+1|k}$: the covariance matrix of $\hat{x}_{k+1|k}$
- Σ_{wk} : the covariance matrix of w_k
- Σ_{vk} : the covariance matrix of v_k

and
$F_k = \partial f_k / \partial x_k |_{x_k = \hat{x}_{k|k}}$, $H_k = \partial h_k / \partial x_k |_{x_k = \hat{x}_{k|k-1}}$, $\hat{D}_k = D_k(\hat{x}_{k|k})$ and $\hat{G}_k = G_k(\hat{x}_{k|k})$.

4 Implementation Tips

To make the image overlay accurate, one must take care of some points regarding video images and timing. In this section, we give some tips to realize accurate image overlay.

4.1 Landmark tracking

Landmark tracking is one of the essential parts of the vision-based tracking. In this paper, we suppose a CCD camera with the conventional interlace scanning and an image-based motion tracker based on the template matching. The template matching is basically to find the location where a pre-registered template matches best in a given image. **Fig.3** shows an example of templates and search area. In this example, basic template size is 8×8 pixels. By skipping one pixel or more, the apparent size can be 16×16 pixels or wider. Search area is usually square, and its size is typically three times of the template size as shown in **Fig.3**. One can specify within which area the tracker

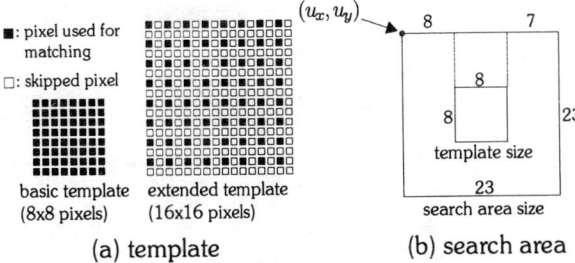

Figure 3: Template and search area

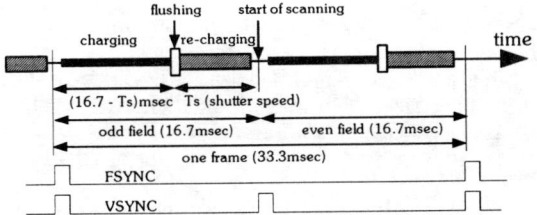

Figure 4: Timing chart of CCD camera

should search by specifying, for example, the location of top-left corner of the search area in display coordinates (i.e., by integer numbers).

The template image is usually taken beforehand from a static scene. Thus, the template has no motion blur. Images taken from a moving camera with a slow shutter speed, however, may have motion blurs, which make the matching difficult. So we first have the following tip.

Tip 1: Shutter speed of the video camera should be as fast as possible. □

A fast shutter speed is also helpful for determining the timing of image overlay.

Most vision-based trackers take images frame by frame and search the landmarks in each frame. A single frame by interlace scanning consists of even and odd fields. **Fig.4** shows a timing chart of CCD cameras. Note that the shutter timings for even field and for odd field are different. Thus, even with a fast shutter speed, images taken from a moving camera may be different from the static image. **Fig.5** conceptually shows the image shift due to the interlace scanning, which also makes the template matching difficult. If we use an extended template by skipping one pixel or an odd number of pixels, the tracker does the matching only on either odd or even field image. In addition, it is possible to let the tracker search all landmarks only from either even or odd fields by giving either even or odd integers when specifying the vertical coordinate of the search area location. Tracking from a single field image gives consistent landmark locations because we can guarantee that all landmark images have been taken at the same timing. To reduce the end-to-end delay, we should use the even field, which is taken later than the odd field in a given frame.

Tip 2: One should use only either even or odd field images for template matching. Even field images are better. □

In theory, registration accuracy of vision-based tracking can be up to the level of a pixel in the display coordinates. Although the accuracy in the depth direction is worse than in other directions, the depth error does not contribute much to the final alignment error in the image. The EKF estimates the pose based on the error $\tilde{y}_k - h_k(\hat{x}_{k|k-1})$ in eq.(7). If we put landmarks apart from the target object, even a small estimation error is magnified and it may result in a large alignment error. Therefore, landmarks should be located near to the target object. At least, landmarks should not be far from the target object; otherwise, they might be out of the camera view. Thus, when the user swings his/her head, the tracker must change the landmarks so that enough number of landmarks is always in the camera sight.

Tip 3: Landmarks should be located near the target scene where the supplemental image is currently overlaid. □

In order to achieve an accurate image overlay, one should know the shutter timing and synchronize the EKF program with it. The easiest way to do so is to use FSYNC or VSYNC signals, which are generated from the camera, as a trigger of hardware interrupts in the EKF program. Usually real-time operating systems, like VxWorks, can handle the hardware interrupt. Synchronizing with the camera FSYNC or VSYNC is also important for knowing the exact timing when overlaying images should be rendered.

Tip 4: The estimation program should be synchronized with FSYNC or VSYNC of the camera. □

4.2 Motion prediction

One of the most difficult problems of the augmented reality systems is the dynamic registration error due to the end-to-end system delay[2]. Since it is impossible to make the system delay completely zero, we have to compensate it by predicting the head motion. One may want to delay the display timing of real scenes by buffering them so that the latency is canceled. Nevertheless, it is not a fundamental solution because the user may feel a motion sickness due to the inconsistency between his/her head motion and the displayed images.

The EKF formulated in section 3.2 propagates an optimal estimate until the next discrete time step (33.3 [msec] ahead) by eq.(8). Besides, we have to predict the pose of the head/HMD system, usually further than 33.3 [msec], for rendering overlaid images. Motion prediction could be a simple extrapolation assuming a constant velocity. However, this method may not be so accurate because the acceleration is neglected. Unlike the camera images, acceleration can be measured immediately and more frequently than 30 [Hz]. Integrating accelerations may accumulate errors and

(a) landmark image in a static scene (b) landmark image when the camera horizontally moved very fast

Figure 5: Image shift due to interlace scanning

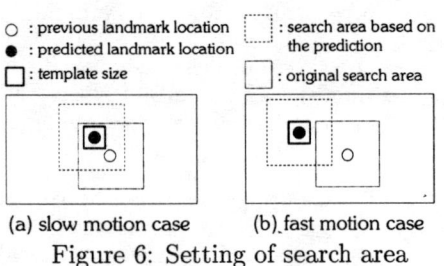

Figure 6: Setting of search area

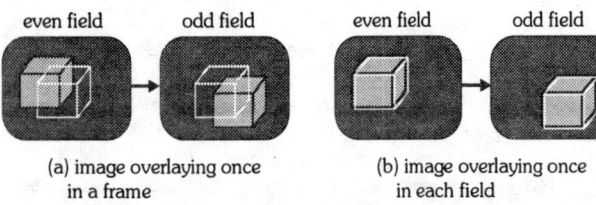

Figure 7: Timing of image overlay

may result in a drift, but it can be compensated by the landmark measurements.

Gyros are often used for head tracking. Sometimes measurements of gyros are numerically differentiated when predicting the motion in the future. Since numerical differentiation is sensitive to noise, it is better to measure the necessary information directly.

Tip 5: Measure acceleration directly rather than numerically differentiating lower ordered data. □

For accurate prediction, one should know exactly when accelerations were measured. Hardware timer interrupts can control the exact timing of measurement at a constant period.

Tip 6: Acceleration should be measured more frequently than video rate and the measurement timing should be controlled by a hardware timer. □

Whichever tracked by hardware or software, the search area is limited by the hardware specification or the computational cost. Often, we set the search area around the location of the landmark found in the previous frame as shown in **Fig.6**. With this setting, however, the tracker tends to lose the landmark when the head motion is fast. To solve this problem, we can set the search area around the predicted location as shown in **Fig.6**. With this new search area setting, the tracker would seldom lose the landmarks as long as they are in the camera view.

Tip 7: Search area for template matching should be set around the predicted landmark location. □

4.3 Rendering overlaid images

What kind of images is overlaid may depend on the target applications. In AR applications, wire framed images are often used. Whether one can render the overlaid image once at each frame or twice on odd and even fields also depends on what kind of rendering hardware we use. Images of real scene captured by the video camera, on the other hand, are updated field by field in the interlaced manner.

When one can render the overlaid image only once at each frame, the prediction time for the overlaid image should be in the middle of the timings of even and

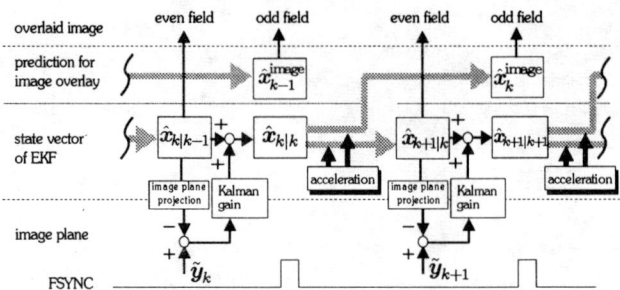

Figure 8: Estimation/prediction flow

odd fields. This means that, when the camera moves fast, the overlaid image is not aligned with the real images in both fields as shown in **Fig.7**(a), and from our experience the overlaid images look flickering. When one can render the overlaid image in both odd and even fields, it is possible to make the overlaid image aligned with the real image in both fields, as shown in **Fig.7**(b), so that the overlaid image has no flickering.

Tip 8: If possible, overlaid images should be predicted and rendered separately for even and odd fields. □

Fig.8 illustrates the timing chart of our prototype system. Note that the horizontal distance does not exactly correspond to the actual time. When propagating the optimal estimation to the next discrete time (i.e., $\hat{x}_{k|k} \rightarrow \hat{x}_{k+1|k}$), accelerations, which are measured more frequently than video rate, are integrated. Since $\hat{x}_{k|k}$ is estimated by using $\tilde{y}_k$ taken from the previous even field, the state vector $\hat{x}_{k+1|k}$, which is propagated 33.3 [msec] ahead, can be used for rendering the overlaid image in the next even field. For the next odd field, we have to predict a pose denoted by $\hat{x}_k^{\text{image}}$, which is 16.7 [msec] ahead from $\hat{x}_{k+1|k}$ or 50 [msec] ahead from $\hat{x}_{k|k}$.

When the prediction time span becomes longer, the overlaid images tend to jitter. This is because we did not consider frame-to-frame or field-to-field continuity. For example, when we render an image based on $\hat{x}_{k+1|k}$, we do not care where we rendered the previous image based on $\hat{x}_{k-1}^{\text{image}}$. We have no space to explain the detail, but one can consider this frame-to-frame (or field-to-field) continuity based on the location where the previous image was rendered.

Tip 9: One should consider frame-to-frame (or field-to-field) continuity of the overlaid images. □

5 Experiment

5.1 Prototype system

After some preliminary experiments by using a conventional CCD camera and a CRT display and mimicking the head motion by manually moving the camera[9], we actually implemented the proposed method in a video see-through HMD. **Fig.9** shows the configuration of our prototype system.

We used a HMD by SHIMADZU Co. (See-Through Vision STV-E). As can be guess from the product name, it can be used as an optical see-through HMD.

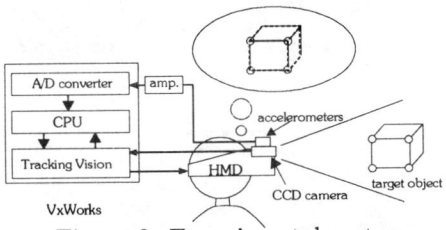

Figure 9: Experimental system

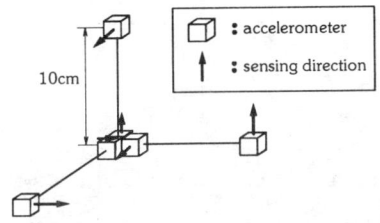

Figure 10: Accelerometer configuration

However, we added a small CCD camera (IK-UM40 by TOSHIBA Co.) and modified the original HMD to a video see-through type. **Fig.10** illustrates the accelerometer configuration proposed by Ohta and Kobayashi[7]. Six accelerometers are used to measure linear/angular accelerations. **Fig.11** illustrates the sensor arrangement on the HMD, and **Fig.12** shows an overview of the HMD with accelerometers.

The target object is a paper box (40 [cm]×30 [cm]×20 [cm]) with four landmarks (one at each corner of the frontal face), and it is located about 2.5 [m] apart from the user. The landmarks are tracked by Tracking Vision (TRV), a motion tracking system by FUJITSU Co. TRV can track more than one hundred regions of 8×8 pixels in video rate (30 [Hz]) by template matching. TRV consists of two VME-bus boards: a video module (VMDL-2) and a tracking module (TMDL-2). TRV can handle monochrome images of 512×512 pixels. The overlaid image is a wire frame of the target box and is drawn in the overlay memory in the video module. The final blended image (also monochrome) is displayed on two small CRTs (for left and right eyes) in the HMD. The displayed images are not stereo.

The TRV is supervised by a CPU board (SPARC 2CE by FORCE Computers) under the VxWorks real-time operating system. The measured acceleration is taken to the CPU board through an A/D converter (PVME-301 by Internix Inc.).

We implemented all tips in section 4, including tip 9. Accelerations are measured exactly every 1 [msec] in a process invoked by an auxiliary timer interrupt. We averaged three accelerations to reduce the measurement noise and got a sequence of the averaged

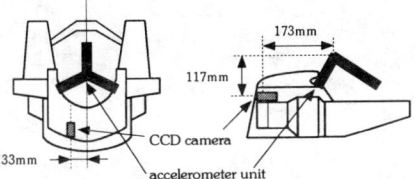

Figure 11: Sensor arrangement on HMD

Figure 12: Video see-through HMD with accelerometers

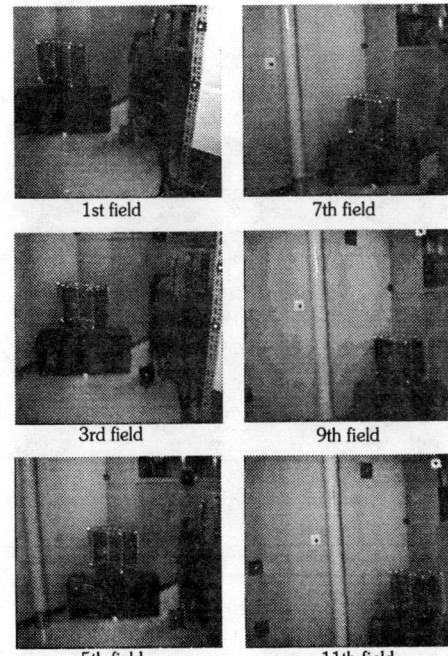

Figure 13: Sequence of the displayed images

accelerations at every 4 [msec]. The timing of the EKF process is synchronized with FSYNC of the video signal by a hardware interrupt. Since 33.3 [msec] is not the multiple of 1 [msec], FSYNC also resets the counter used in the process to measure the accelerations. Shutter speed of the CCD camera was set to 1/500 [sec]. The template size was 16×16 pixels with one pixel skipping. The apparent size of search area for each template was widened to 78×78 pixels by fitting four original search areas (also skipped one pixel) together.

5.2 Experimental result

A user actually wore the HMD and moved his head. **Fig.13** shows the experimental result. There are other landmarks than those on the box, but they were not used in the experiment. In this sequence, the maximum liner/angular accelerations were 10.5 [m/s^2] and 49.3 [rad/sec^2], respectively, and the maximum liner/angular velocities were 0.65 [m/s] and 3.3 [rad/s]. Even with such a quick head motion, the wire frame was overlaid accurately and the tracker did not lose the landmarks. For the sake of comparison, **Fig.14** shows a sequence when no accelerations are used.

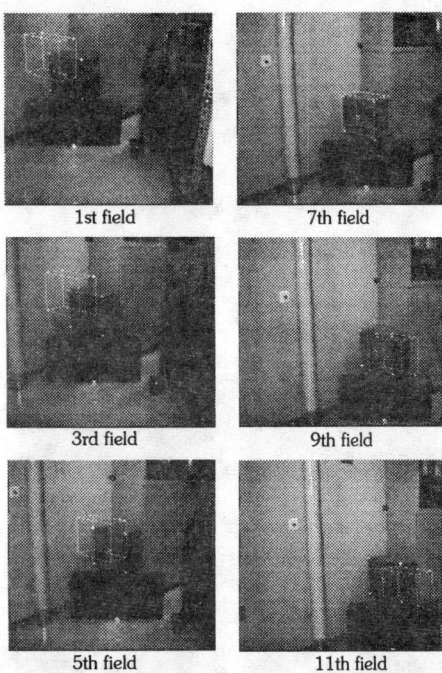

Figure 14: Sequence of the displayed images without acceleration

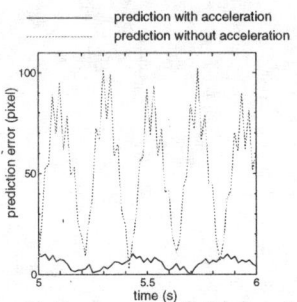

Figure 15: Prediction error of the landmark location

Although we cannot know the true value in real-time, we can evaluate the accuracy of the prediction off-line after the experiment. **Fig.15** shows errors of the predicted landmark locations on the image plane. For evaluating the prediction error, we tracked another landmark located at the center of the frontal face of the target box. This landmark was not used for the pose estimation and prediction. The proposed method could reduce the alignment errors within 5.3 pixels on average and 10.2 pixels at the maximum. The 9th field in **Fig.13** corresponds to the instant of the maximum alignment error. Compared to the result without acceleration (53.3 pixels on average and 102.3 pixels at the maximum), the prediction errors were reduced by 1/10.

Note that **Fig.15** is just an evaluation on the image plane and we do not know how accurate the estimated 3D pose is in base coordinates. To evaluate the accuracy of the estimation in base coordinates, we need to measure the true value of the head pose by a passive linkage or other method. We can imagine that the estimated pose has a larger error in the camera depth direction because the landmark location on the image plane is not sensitive to this direction. It should be noted, however, that as long as the estimated pose is used only for the image overlay, the evaluation on the image plane is enough.

6 Conclusion

In this paper, we proposed a method for accurate image overlay on HMDs using vision and accelerometers. Both linear and angular accelerations are measured and used for predicting the head motion to compensate the end-to-end system delay. Prediction also makes the vision-based tracking robust.

It is impossible to predict the future motion perfectly. In this paper, we tried to predict the head motion as accurate as possible by sampling the acceleration more frequently than video rate.

Experimental results showed that the proposed method can reduce the alignment errors within 6 pixels on average and 11 pixels at the maximum, even when the user moves his/her head quickly (about 10 $[m/s^2]$ and 49 $[rad/s^2]$ at the maximum).

Acknowledgments

The authors would like to express their thanks to Mr. Takayuki Yoshigahara, ex-Visiting Research Scholar at Carnegie Mellon University, for his comments on CCD cameras. They also express their appreciation for the accurate answers and valuable comments by FUJITSU Tracking Vision Technical Support Division. The research in this paper was financially supported by Grant-in-Aid for Scientific Research, Ministry of Education, Science, Sports and Culture, Japan (Research No.09450105).

References

[1] R. Azuma and G. Bishop, "Improving Static and Dynamic Registration in an Optical See-through HMD", *Proceedings of SIGGRAPH '94*, pp.197–204, (1994)

[2] R. Azuma, "A Survey of Augmented Reality", *Presence*, Vol.6, No.4, pp.355-385, (1997)

[3] S. Emura and S. Tachi, "Multisensor Integrated Prediction for Virtual Reality", *PRESENCE*, vol.7, no.4, pp.410–422, (1998)

[4] E. Foxlin, "Inertial Head-Tracker Sensor Fusion by a Complementary Separate-Bias Kalman Filter", *Proceedings of VRAIS'96*, pp.185–194, (1996)

[5] M. Friedmann et al., "Synchronization in Virtual Realities", *PRESENCE*, vol.1, no.1, pp.139–144, (1992)

[6] S. Merhav and M. Velger, "Compensating Sampling Errors in Stabilizing Helmet-Mounted Displays Using Auxiliary Acceleration Measurements", *J. Guid. Control Dyn.(USA)*, vol.14, no.5, pp.1067–1069, (1991)

[7] K. Ohta and K. Kobayashi, "Measurement of Angular Velocity and Angular Acceleration in Sports Using Accelerometer", *Transactions of the Society of Instrument and Control Engineers*, vol.30, no.12, pp.1442–1448, (1994) (*In Japanese*)

[8] M. Uenohara and T. Kanade, "Vision-Based Object Registration for Real-Time Image Overlay", *Comput., Biol., Med.*, vol.25, no.2, pp.249–260, (1995)

[9] Y. Yokokohji, Y. Sugawara and T. Yoshikawa, "Vision-Based Head Tracking for Image Overlay and Reducing Dynamic Registration Error with Acceleration Measurement", *Proc. of 1998 Japan-U.S.A. Symposium on Flexible Automation*, pp.1293–1296, (1998)

Design of a 3R Cobot Using Continuously Variable Transmissions

Carl A. Moore
Michael A. Peshkin
J. Edward Colgate

Dept. of Mechanical Engineering
Northwestern University
Evanston, IL 60208-3111

Abstract

Cobots are capable of producing virtual surfaces of high quality, using mechanical transmission elements as their basic element in place of conventional motors. Most cobots built to date have used steerable wheels as their transmission elements. We describe how continuously variable transmissions (CVTs) can be used in this capacity for a cobot with revolute joints.

The design of an "arm-like" cobot with a three-dimensional workspace is described. This cobot can implement virtual surfaces and other effects in a spherical workspace approximately 1.5 meters in diameter. Novel elements of this cobot include the use of a power disk that couples three CVTs directly.

I. Introduction

Several robotic devices have been proposed for the purpose of creating programmable constraints and virtual surfaces. One such device by Book et. al. [1], called P-TER (Passive Trajectory Enhancing Robot), is a 2-degree of freedom (dof) manipulator designed to guide its end effector along a desired path while being pushed by the user. Clutches are used to vary the coupling between the two major links of the device, while brakes are used to remove energy from the links. Delnondedieu and Troccaz [2] have developed another device, called PADyC (Passive Arm with DYnamic Constraints), intended for guided execution of potentially complex surgical strategies. The prototype system has 2 dof and uses 2 each of a motor, clutch, and free wheel to dynamically constrain each joint.

Neither of these devices is able to provide arbitrarily oriented, smooth, hard virtual surfaces.

II. Scooter Cobot

To illustrate how cobots provide smooth, hard virtual surfaces, we use the example of Scooter (shown in Fig. 1), a cobot with a three-dimensional workspace (x, y, θ) [3]. Three small motors are used to steer the wheels of which two are visible in Fig. 1. The motors cannot cause the wheels to roll; they can only change the wheels' rolling direction. A force sensor on the center post handle measures forces applied by the user.

A cobot's two simplest modes of operation are free mode (in which the user can move the cobot freely in (x.y.θ)-space) and virtual surface mode (in which only motion along a virtual constraint is allowed).

Fig. 1. Scooter three wheel cobot.

Free Mode:

In free mode Scooter operates as if it were supported by casters, like those on an office chair, which permit any desired motion direction. Unlike casters whose shafts are off center, Scooter's wheels are on straight-up shafts and are steered using motors. When the user applies a force to Scooter by pushing on the center handle, the computer monitors the force perpendicular to Scooter's rolling direction and attempts to minimize it by changing Scooter's rolling direction. Scooter's rolling direction is described by a center of rotation (COR). If the COR lies directly in the center of Scooter, the only allowed motion is rotation about the handle. If the COR is infinitely far away, corresponding to all wheel axes being parallel, then Scooter will follow a straight line. Therefore, in free mode, the computer monitors the user's forces, determines the required COR, and turns the wheels to allow that motion.

Virtual Surface Mode:

In virtual surface mode a cobot filters the user's motion. If the user brings Scooter up to a programmed virtual surface, the computer ceases to steer the wheels in a direction that minimizes the perpendicular force. Instead, the wheels are steered such that the allowed motion is tangent to the surface. The computer does continue to monitor the user-applied forces. Forces that would cause Scooter to penetrate the surface are ignored. Forces that would bring Scooter off of the surface and back into the free space are interpreted as before in free mode, and Scooter again behaves as if it were on casters.

When a cobot is in contact with a virtual surface or constraint, it is possible for the user to apply a force into the constraint that is large enough to cause the constraint to collapse. The strength of the virtual constraint is related to the mechanism by which the cobot resists perpendicular forces. With Scooter, coulomb friction forces between the steered wheels and the working planar surface resist forces applied against the constraint. If the applied force becomes greater than the friction force, the virtual surface crumbles and the cobot enters the restricted area.

III. Rotational CVT

Scooter is restricted to a three-dimensional planar workspace because its virtual surface behavior relies on the presence of a flat working surface on which to roll. Revolute arm-like architectures have proven very versatile for robots, and so we now address the problem of creating an arm-like cobot with revolute joints.

The role of the steered wheels in Scooter is to establish a mechanically enforced ratio between the x-velocity and the y-velocity of the steering axis of each wheel. That ratio, v_y/v_x, is given by α, the steering angle of the wheel, which is under computer control. This principle may be considered obvious for a wheel, but it lies at the heart of cobots that have planar workspaces, like Scooter.

To extend cobots to workspaces typical of revolute jointed robots, we require a mechanical element whose function is analogous to that of the wheel in scooter. For revolute joints the mechanically enforced ratio is between two angular velocities, ω_1 and ω_2, rather than two linear velocities as in scooter. Also, the ratio ω_2/ω_1, which is enforced mechanically, must be adjustable under computer control just as the angle of each of Scooter's wheels.

The requirements above call for a continuously variable transmission, or CVT: a device which couples two angular velocities according to any adjustable ratio. Such a CVT is shown in Fig. 2.

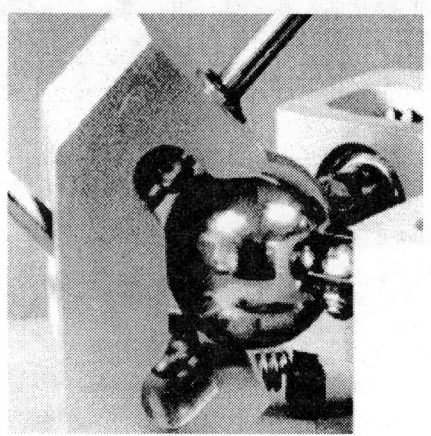

Fig. 2. Rotational CVT

A CVT holds two angular velocities in proportion: $\omega_2/\omega_1 = T$, where T is the continuously variable transmission ratio.

As diagramed in Fig. 3, the CVT consist of a sphere caged by four rollers. The rollers are arranged at the corners of a stretched tetrahedron so that the angle subtended by the contact points is 90° (a regular tetrahedron would have angles of 108°). The two drive rollers with angular velocities ω_1 and ω_2 interface to the joints of the cobot. The other two rollers are steering rollers whose orientation controls the central sphere's rotational axis and thereby the ratio $\omega_2:\omega_1$.

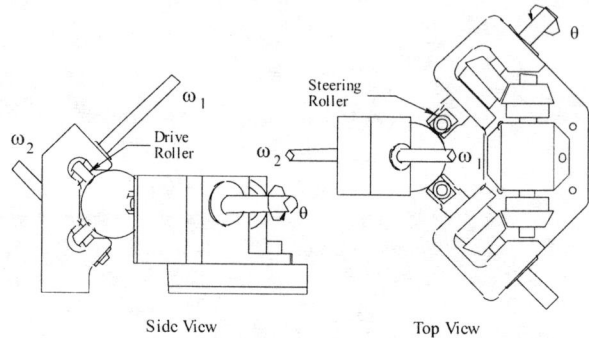

Fig. 3. Rotational CVT

Since both drive rollers are in rolling contact with the central sphere, its rotational axis must lie in the plane containing both drive rollers and pass through the non-translating center of the sphere. If the orientation of the rotational axis is located by an angle γ measured from drive roller 2, the transmission ratio T is

$$T = \tan\gamma = \frac{\omega_2}{\omega_1}. \qquad (1)$$

The angle γ is determined by the steering roller angle θ by

$$\gamma = \tan^{-1}\left(\frac{\sqrt{2}}{2}\tan\theta\right) - 45°. \qquad (2)$$

The transmission ratio $\omega_2:\omega_1$ assumes the full range of values from $-\infty$ to $+\infty$ as the steering rollers are turned from -90° to +90° [4].

IV. Serial Cobot

Fig. 4 shows a hypothetical arm cobot that uses 1 CVT to couple its two joints. We present this diagram to show, in the simplest possible application, how a CVT is used in a revolute jointed cobot. The steering rollers along with the mechanism used to hold the CVT in place on link 1 is not shown.

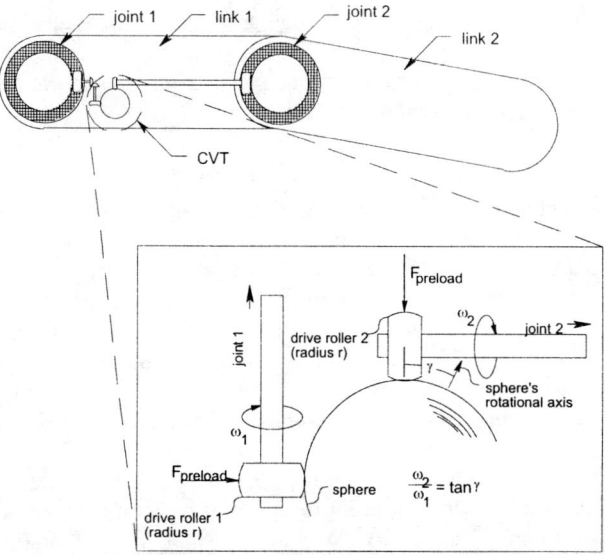

Fig. 4. Arm cobot with 2 joints

The rotation of each joint is coupled to one drive roller of the CVT. The drive rollers are held in rolling contact with the central sphere through a preload force F that is usually applied by springs. Because the ratio of drive roller angular speeds is equal to the tangent of the sphere's spin axis angle γ, the allowed endpoint velocity (v_x, v_y) in task space is also a function of γ:

$$\begin{bmatrix} v_x \\ v_y \end{bmatrix} = \mathbf{J} \begin{bmatrix} \omega_1 \\ \omega_1 \tan \gamma \end{bmatrix}. \quad (3)$$

where $\mathbf{J}$ is the cobot's Jacobian.

The equations governing free caster and constraint control are beyond the scope of this paper, but can be found in [5]. However, a basic understanding of free caster control is helpful to understanding the relationship between a CVT and the user applied endpoint forces.

The user's applied force $\mathbf{F}_{ext}$ is read by a force sensor at the endpoint and used to assess the endpoint acceleration desired by the user. Using this desired acceleration and the endpoint velocity, it is possible to predict a desired C-space path. The principle unit normal for this path is then found and used along with the appropriate Jacobian to determine the necessary CVT steering speeds. The steering speed ω is proportional to the endpoint force perpendicular to the velocity vector $F_\perp$ and inversely proportional to the endpoint speed u times the endpoint mass M or

$$\omega = \frac{F_\perp}{uM}. \quad (4)$$

When the speed of the endpoint is zero, the steering velocity is undefined, and the steering angle is set such that the allowed motion direction is parallel to $\mathbf{F}_{ext}$.

A. Connecting CVTs

There are two methods of connecting CVTs to a cobot's joints. In the first method, each successive pair of joints is coupled to the drive rollers of one CVT. Fig. 5 is a schematic of a 3-joint cobot with 2 CVTs in series.

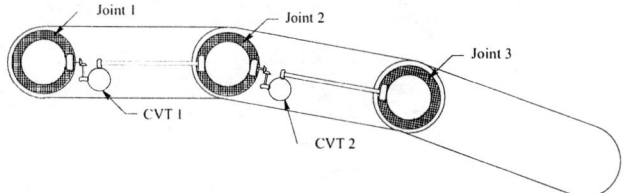

Fig. 5. 3-Link Cobot with 2 CVTs in series

The signal flow between task space velocity $\mathbf{V}$ (v_x, v_y) and the two serially connected CVTs is diagramed in Fig. 6. It is important to remember that the CVTs determine the joint speed ratios, while the joint speeds themselves are a function of the user applied endpoint forces.

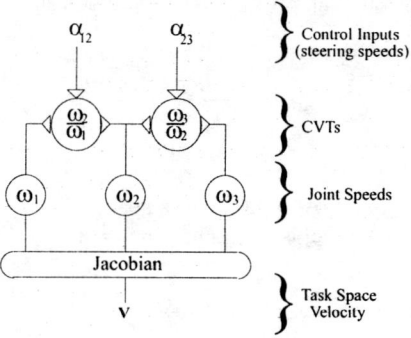

Fig. 6. Signal flow between serial CVTs

The interdependence of joint angular speeds is an important characteristic of the serial model. When an internal joint of the cobot has a near zero angular speed, an extremely high transmission ratio will be required by another CVT in the chain. Take for example the 3-joint, 2-CVT model of Fig. 5. If the desired joint angular velocities are $\omega_1 = \omega_d$, $\omega_2 = \ll \omega_d$, and $\omega_3 = 2\omega_d$ the following CVT transmission ratios are required:

$$T_1 = \frac{\ll \omega_d}{\omega_d} \qquad T_2 = \frac{2\omega_d}{\ll \omega_d} \quad (5)$$

The first transmission ratio is near zero and attainable. The second transmission ratio is near infinite and not possible because friction internal to the CVT bounds the maximum transmission ratio to approximately 20:1. A second issue with the serial chain model is that transmission ratio errors in one CVT are propagated to the others.

CVTs can also be connected in parallel. A three joint example is shown in Fig. 7. The angular velocity of each joint is coupled to a separate drive roller and the remaining three drive rollers are tied to a common shaft.

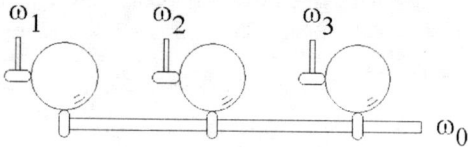

Fig. 7. CVTs connected in parallel

In the parallel configuration, a CVT transmission ratio T_i relates the joint velocity ω_i to the common shaft velocity ω_0

$$T_i = \frac{\omega_i}{\omega_0}, \qquad (6)$$

where the angular velocity of the common shaft is

$$\omega_0 = \frac{\omega_1 + \omega_2 + \omega_3}{T_1 + T_2 + T_3}. \qquad (7)$$

The signal flow between task space velocity **V** and the 3 CVTs in parallel is diagramed in Fig. 8.

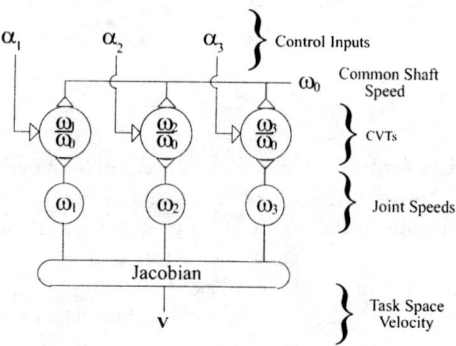

Fig. 8. Signal flow between parallel CVTs

It was shown in the serial case that a near zero angular velocity for one joint required an unattainably high transmission ratio in a neighboring CVT. In contrast, in the parallel configuration CVT, transmission ratios are proportional to the angular speed of their joints. Therefore, a near zero joint speed in the parallel configuration requires merely that its own CVT have a transmission ratio which is also nearly zero. So, the speed ratio between any two joints connected by CVTs in parallel can assume the full range of values (-∞ to +∞) with finite CVT transmission ratios.

Another interesting characteristic of CVTs in parallel is their ability to caster without steering. If the transmission ratios of each CVT are set to infinity, the common shaft has zero angular velocity and each joint can rotate freely without respect to any other joint. The only other time that the speed of the common shaft is zero is when the speed of all joints is zero.

B. Adding Power

Traditionally, cobots are physically passive: their kinetic energy is bounded by the energy supplied by the user's hand. However, when we started to design cobots with larger gear ratios to create harder constraints, the increased friction reflected to the user became an issue. There was also the desire to make cobots more responsive by magnifying the user's force at low speed. The most convenient way to accomplish these goals is to add kinetic energy to the cobot using a motor.

The common shaft of CVTs in parallel provides an elegant attachment point for a power assist motor. When a motor is added to the common shaft, the signal flow changes - the angular velocity of the common shaft is now an input to each CVT (see Fig. 9).

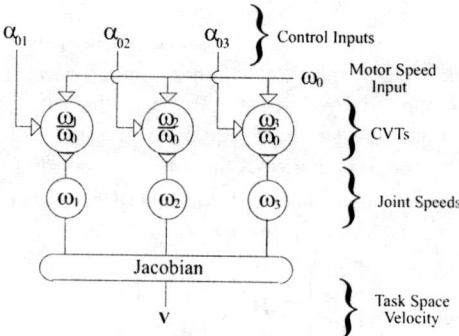

Fig. 9. Signal flow between parallel CVTs with power assist

By equation (5), a joint velocity ω_i is the product of the CVT transmission ratio T_i and the angular velocity of the common shaft ω_0. Therefore, the cobot's task space velocity **V** is

$$\mathbf{V} = \mathbf{J}\mathbf{T}\omega_0. \qquad (8)$$

The same idea holds for an endpoint force. Each joint torque τ_i is a product of the common shaft torque τ_0 and the inverse of the transmission ratio such that the force $\mathbf{F}_{xyz}$ reflected to the endpoint is

$$\mathbf{F}_{xyz} = \mathbf{J}^{T^{-1}} \frac{1}{\mathbf{T}} \tau_0. \qquad (9)$$

The noteworthy result here is that regardless of the dimension of the cobot's taskspace, a single motor coupled to the common shaft of CVTs in parallel produces an endpoint force and speed that is parallel to the allowed motion direction.

V. Arm Cobot with Power Assist

We have designed a cobot with a parallelogram arm configuration as shown in Fig. 10. It has a 3 dimensional (x,y,z) workspace and a 4-link parallelogram configuration which is counter-balanced for gravity. Its 3 CVTs are connected in parallel, and a power assist motor delivers power to each of the three joints through a wheel that is in rolling contact with each of the CVTs. A force sensor is located at the end effector.

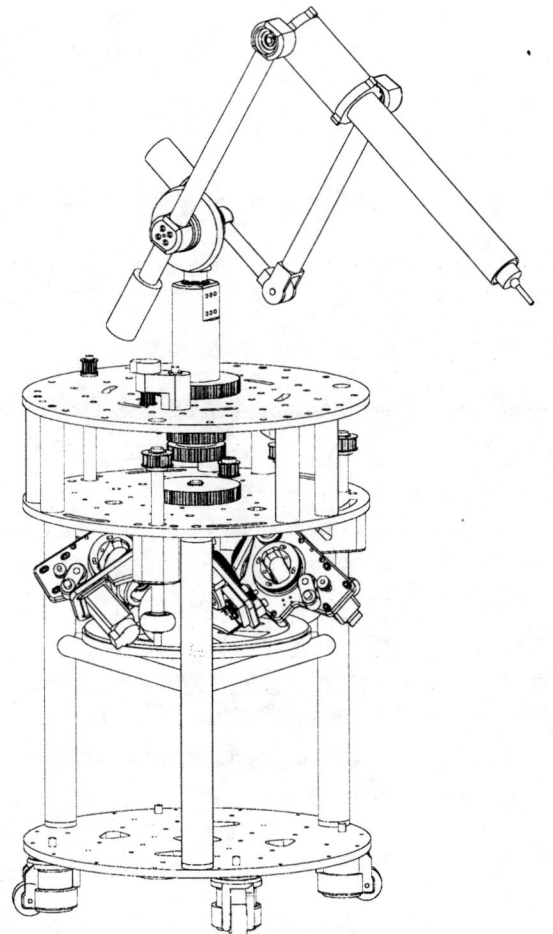

Fig. 10. Parallelogram arm cobot

It was decided early on that the cobot's CVTs would remain stationary (grounded) during motion of the arm. Connecting each CVT to ground lowers the complexity of the design and decouples the mass of the CVT subsystems from the arm's dynamics. The resulting design couples the rotation of the joints to three concentric shafts. Two sets of bevel gears connect the two non-vertical joint axles to the two innermost shafts. The third joint does not require bevel gears because its axis of rotation is already parallel to the CVTs' drive roller shafts.

Graphite was chosen for the links due to its high bending strength (E = 96MPa) and low weight (ρ = 1.66 g/cm^3). The largest diameter link has a 8.26cm OD, and the arm's reach is over 76cm. With the support stand attached, the origin of the three x, y, and z joint rotations is approximately 1.5m above the floor. The workspace of the cobot is diagramed in Fig. 11.

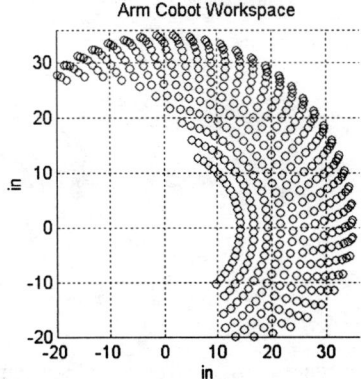

Fig. 11. Arm workspace

The links are counter-balanced against gravity for all arm configurations by two counter weights. The moments provided by the counter weights are approximately 6.6N-m and 3.8N-m.

Considering that there are three CVTs in this design, the size of the CVT dictates the footprint of the entire mechanism. The original CVT design (Fig. 3) with its long steering axes was ruled out for having too much wasted volume. The new design shown in Fig. 12 is much more compact than the original. The steering axles have been replaced with low profile steering hubs that surround the steering rollers. Equal and opposite rotation of the steering hubs is ensured by bevel gears that are synchronized to each hub through timing belts (belts are not shown). There is a 45-watt steering motor with encoder on each CVT assembly.

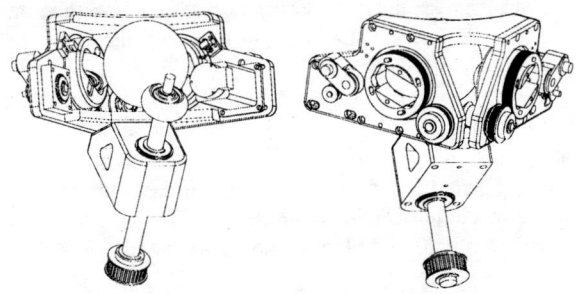

Fig. 12. Front and back view of the compact CVT

As noted earlier, CVTs in parallel have the rotation of one of their drive rollers coupled to a common shaft. The CVT of Fig. 12 has only one drive roller. In order to couple the rotations of each CVT together there is a common wheel (instead of a common shaft) that is in rolling contact with the central sphere of each CVT as shown in Fig. 13. This wheel is called the power wheel and is driven using a motor (the timing belt pulley for the motor and wheel axle is hidden on the bottom side of the support plate).

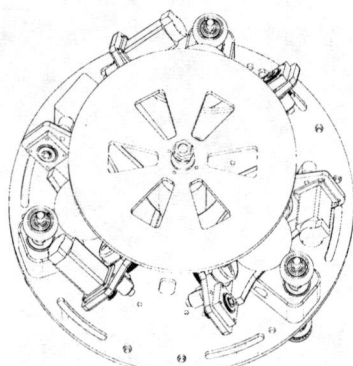

Fig. 13. 3 CVTs and a power wheel

The power wheel is made from an aluminum plate that has a rubber running surface on one side. Sections of material are removed from the wheel to lower its inertia. The power wheel has a diameter of 36.8cm, and its running surface contacts the CVTs 16.5cm from the center post.

There are many benefits to the symmetric arrangement of CVTs. It permits a single spring on the power wheel's axle to apply an equal preload force to all CVTs. The spring currently being used can apply a maximum of 22.7Kg to the rolling elements of each CVT. Also, in this arrangement, the drive roller shafts are parallel to their joint shaft axes allowing power transfer between the two with zero backlash timing belts.

The coulomb friction forces that exist between the rolling elements of the CVTs determine the force of constraint that the arm cobot can display. Taking $\mu_s = 0.8$ as the coefficient of static friction between the CVTs rolling elements, a drive roller radius of 2.84cm, and a normal force of 133.4N, the resulting maximum torque that the drive rollers can resist is 3.0N-m. With a gear ratio of 6:1 between the drive rollers and the joints, the maximum sustainable joint torques are 18.0N-m.

Fig. 14 uses force ellipses to display the static force characteristics for the arm [6]. The major (minor) axis of each ellipse represents the maximum (minimum) endpoint force that can be resisted in the direction of the axis. The largest force that can be supported at a position regardless of direction is recorded in pounds next to each ellipse.

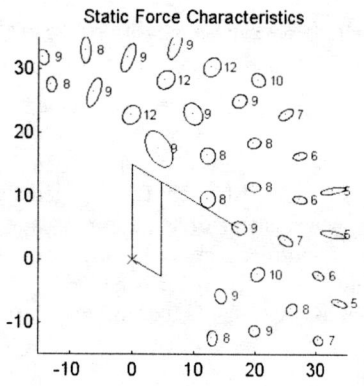

Fig. 14. Static force characteristics

VI. Contribution of Design

The parallelogram arm cobot will be the first 3R cobot. Its ability to move in traditional x,y,z 3-space opens an entire new class of tasks to cobotic solutions. The addition of power assist to the traditional passive cobotic model will result in a cobot that can reduce or magnify the inertia that is reflected to the user making larger cobots or cobots with large transmission ratios possible.

The powered arm will be able to perform tasks autonomously like a traditional robot while remaining backdrivable. A backdrivable arm is attractive to persons that want a powered manipulator that can also be easily positioned by hand, such as those interested in robot-assisted surgery.

VII. References

1 Book, W., Charles, R., H., Davis, Gomes, M., "The Concept and Implementation of a Passive Trajectory Enhancing Robot," *Proceedings of the ASME Dynamics Systems and Control Division*, DSC-Vol 58, 1996.

2 Delnondedieu, Y., Troccaz, J., "PADyC: a Passive Arm with Dynamic Constraints," *Proceedings of the 2nd International Symposium on Medical Robotics and Computer Assisted Surgery*, 1995.

3 Wannasuphoprasit, W., Colgate, J.E., Peshkin, M.A., "The Design and Control of Scooter, a Tricycle Cobot", *Proceeding of the IEEE 1997 International Conference on Robotics & Automation*; March 1997

4 Moore, C.A., "Continuously Variable Transmission for Serial Link Cobot Architectures," Master's thesis, Department of Mechanical Engineering, Northwestern University, March 1997.

5 Wannasuphoprasit, W., Gillespie, R.B., Colgate, J.E., Peshkin, M.A., "Cobot Control", *Proceeding of the IEEE 1997 International Conference on Robotics & Automation*; March 1997.

6 Asada, H., "Direct-drive robots: theory and practice," Cambridge, Mass., MIT Press, 1987.

Development of the Anthropomorphic Head-Eye Robot WE-3RII with an Autonomous Facial Expression Mechanism

*,**Atsuo Takanishi, *Hideaki Takanobu:
*Department of Mechanical Engineering, Waseda University
**Advanced Research Center for Science and Engineering, Waseda University

***Isao Kato, ***Tomohiko Umetsu:
***Graduate School of Science and Engineering, Waseda University

Abstract

In this study, the authors have been developing anthropomorphic head-eye robots, in order to elucidate the human vision system from an engineering point of view, and to develop a new head mechanism and function for a humanoid robot having the ability to communicate naturally with a human. We developed the anthropomorphic head-eye robot "WE-3RII" (Waseda Eye No.3 Refined II), with eyes, eyelids, eyebrows, lips, jaw and a neck. The robot has the ability to express an autonomous facial expression using three independent parameters of a psychological model. With the robot, the mechanism of changing eye movement from smooth pursuit eye movement to saccadic, and the control of scanning eye movement like that of a human's can be achieved. The robot is also able to express more human-like motion in pursuing a target by changing its eyeball control parameters, rather than that of WE-3R which we developed in 1996. Therefore, the authors have confirmed that WE-3RII has the ability to realize a pursuing motion in three dimensional space with its autonomous facial expression adapting to the target position and brightness like a human.

1. Introduction

The human head has five senses, as well as various faculties for the expression of mood or health condition, *etc*. Considering communication, the head is the most important part of the human body. To enable humanoids to actively participate in our society, natural communication with humans will be one of the most important issues. From this point of view, the authors believe it is necessary for humanoids to have the capability of performing human-like motions.

Also, by developing an anthropomorphic head-eye robot, we are attempting to elucidate human perception and the recognition system from an engineering point of view.

We have presented WE-3, at the IROS'97 in France [1]. WE-3 was developed in 1995 to realize the pursuing motion in a three dimensional space using coordinated head-eye motions with V.O.R (Vestiburo-Ocular Reflex) [2][3], and pursuing motions in a depth direction using an angle of convergence. We have also presented WE-3R, at the ICRA'98 in Belgium [4]. WE-3R was developed in 1996 to realize the human-like function of adjusting to brightness by improving WE-3.

Adding eyebrows, lips and a jaw to the basic mechanism of WE-3R, WE-3RII was developed so that a human-like expression could be realized. Further, we realized a facial expression control method founded on three independent parameters of a simple psychological model.

Below are descriptions of hardware, the facial expression control algorithm based on the psychological model, and experiments on WE-3RII.

2. Hardware Configuration

Each part and its velocity of WE-3RII has been designed so that it will give a natural impression to humans. Fig.2.1 shows the robot that we developed in 1997. Fig.2.2 shows the points of action and the direction of motions of the eyebrows, the lips and the jaw.

2.1 Eyebrows and Lips

To actuate the eyebrows and the lips, we adopted the outer tube and inner wire driving mechanism using

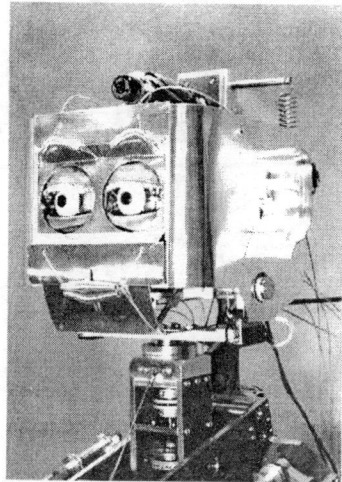

Fig.2.1 Anthropomorphic Head-Eye Robot WE-3RII

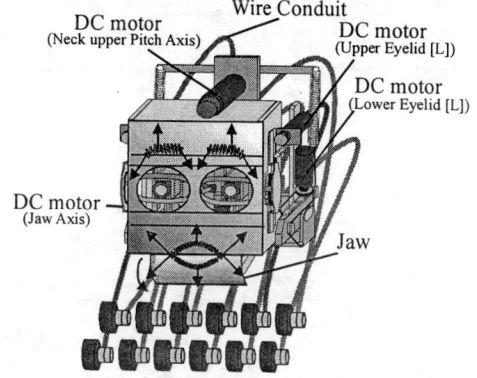

Fig.2.2 The mechanism and the motion direction of the each expression part

stepping motors located in the robot's back to lighten the head part. Considering the eyebrows' and lips' shape, we used a spring which has a swollen part on its center. The point of action and direction of motion were decided based on Ekman's Action Units (AU) [5] and our video analysis. The eyebrows and lips have six D.O.F (Degrees Of Freedom), and each motion speed is similar to a human's. The total D.O.F is 12.

2.2 Jaw

The role of the jaw in the six basic facial expressions is to open and close the mouth. Therefore, we concluded that one D.O.F is enough for the jaw. To actuate the jaw, we used a small and light weight DC servo motor. The jaw is also able to move as a human's like the eyebrows and the lips can.

2.3 Eyeballs, Eyelids and Neck

We used the same mechanism as of WE-3R for the eyeballs, eyelids and neck. The yaw axis of the neck is driven by a harmonic drive DC motor, and the other axes and the eyelid part are driven by a tendon driven mechanism using a DC motor and springs. The maximum angular velocity of each axis is similar to a human's with 600[deg/s] for the eyeballs, 160[deg/s] for the neck and 900[deg/s] for the eyelids. Furthermore, this robot is able to blink within 0.3[s] as fast as a human can.

2.4 The total system

Fig.2.3 shows the total system of WE-3RII. The image processing computer is PC/AT compatible with a single 150[MHz] Intel Pentium CPU. The image processing boards are commercially available boards. The computer controlling the DC motor is PC/AT compatible with a single 133[MHz] Intel Pentium CPU. This PC also controls the jaw. Another computer, which controls twelve stepping motors for the eyebrows and lips, is PC/AT compatible with a single 120[MHz] Intel Pentium CPU. These three PCs are connected to each other by RS-232C cables.

The image processing computer calculates the center position of the object from the input images, obtained from the right and left CCD cameras and the luminance of the object.

The targeting positions are then sent to the DC motor control computer. At the same time, the motion pattern of the facial expression from the state of the object using a psychological model, which is described in chapter 3 in detail, is sent to the stepping motor control computer.

The stepping motor control computer gives the reference input as the basic facial expression pattern to the stepping motors. It gives the reference input for the jaw and the eyelids to the DC motor control computer through the image processing computer.

The DC motor control computer receives the reference input from the stepping motor control computer. Then, the DC motor of the jaw uses the received value as the reference input. Eyelids obtain the reference input prior to the brightness adaptation state from the image processing computer.

The calculation rate of the position using data from the image processing boards is 30[Hz]. The rate of the local position control is 500[Hz]. The interrupt time of the computer for controlling the stepping motor is 20[kHz]. It is necessary for this to be set higher than 1.0[kHz], which is the maximum frequency of the stepping motor drive pulse.

3. Control algorithm

3.1 Smooth pursuit eye movement and saccadic eye movement

Humans use smooth pursuit eye movements and

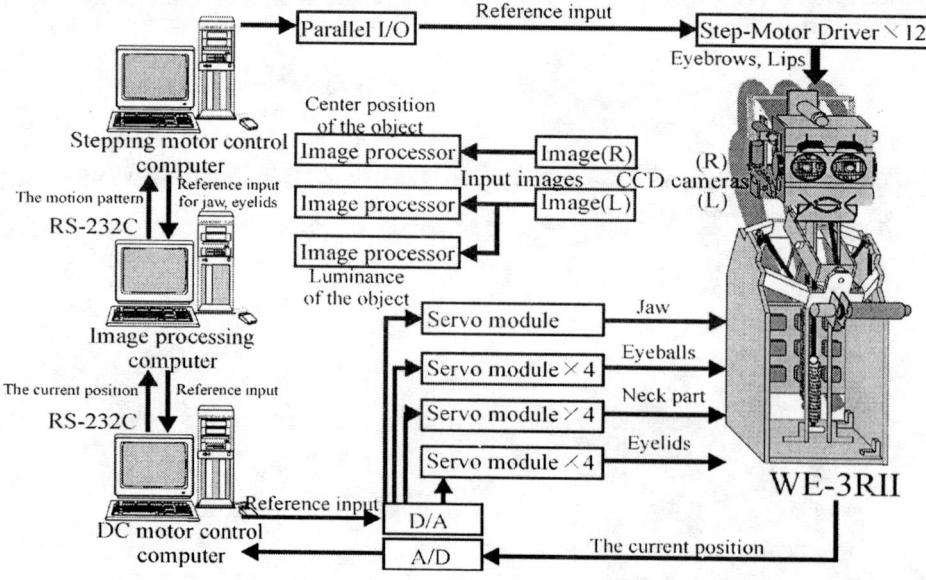

Fig.2.3 The Total System Configuration

saccadic eye movements to pursue a target. Smooth pursuit eye movement happens when pursuing a moving target, and the velocity of the human eye is 45[deg/s] or below. Otherwise it is saccadic eye movement. Therefore, for humans the threshold between smooth pursuing and saccadic eye movement is about 45[deg/s][11].

Like in the flowchart shown in Fig.3.1, we aimed at generating more human-like expressions by changing these eye movements to pursue targets like humans.

The robot measures the eyeball velocity and judges whether or not it is within V1[deg/s]. Here, V1 is the threshold of the changing mode of the robot's eye movement. If it is below V1, the robot uses smooth pursuit eye movement. If it is above V1, the robot switches to saccadic eye movement.

In the case of saccadic eye movement, it gains 200[ms] as the non-responsive period of a human in the Time Loop in Fig.3.1. During this period, if the range of retina error is within RE_A, it returns to smooth pursuit eye movement. This process is considered as the 'Catch Up Saccade', which jumps to attain the target while using saccadic eye movement [11]. After elapsing 200[ms] as the non-responsive period, the eye movement is decided again by the range of retinal error. The range of retinal error means the error that corresponds to 200[ms] in theory. This is caused when the eye movement becomes still at every 200[ms]. The threshold (RE_B in Fig.3.1) must be equivalent to the error of V1. We need to set these values (V1, RE_A, RE_B) to permit smooth pursuit eye movement to shift smoothly saccadic eye movement, and to reverse its motion. These values are obtained as the results of heuristically experiments. In this research, we set the V1 to 60[deg/sec], the RE_A to 4[deg], and the RE_B to 10[deg].

3.2 Scanning eye movement

When a human sees a stationary object, he or she watches the place where the object lays with saccadic eye movement, which is called scanning eye movement. Therefore, we focused on this scanning

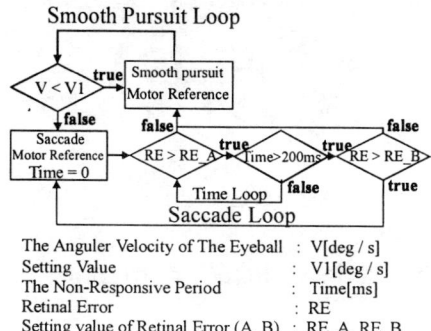

Fig.3.1 Control of Smooth Eye Movement and Saccadic Eye Movement

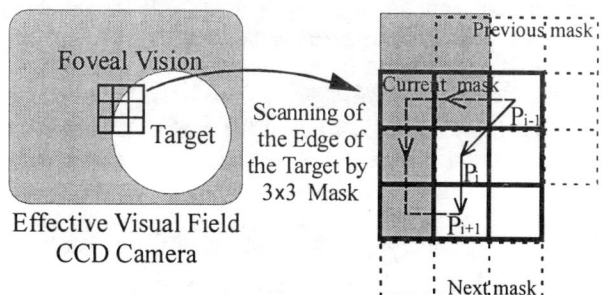

Fig.3.2 Scanning Eye Movement

eye movement. The simplest method of extracting a line from black and white images is by using a mask composed of 3×3 pixels. We introduced the method of scanning eye movement of along the edge of objects. Fig.3.2 shows the method of scanning eye movement. The mask is in the center of the retina. First, the robot judges the area around the center of the mask according to its brightness or darkness counterclockwise from the previous center of the mask. Then, the eyes obtain the next watching point from where the brightness changes from darkness to brightness. By repeating this movement, the eyes scan the edge of a target.

In case of more complex object watching, humans pay close attention when the effective visual field becomes narrow. The area of the mask of 3×3 pixels is assumed to be the effective visual field. The more focused the attention is, the narrower the area of the mask will be, and the shallower it is, the wider it will be.

3.3 Expression by setting the control parameters of the eyes
(1) Setting of a non-responsive area from eyeball velocity input of the retina

The previous robot WE-3R [2] gives velocity references to the eyeballs' controller which are proportional to the retinal error. It suppresses the vibration of the eye movement by setting up a non-responsive area that gives zero reference velocity at some degree of error. If we intentionally enlarge this non-responsive area, the tracking ability of the robot becomes worse, and the robot can express the state of stupefaction. Feveation happens when a human pays attention to an object. Peripheral vision happens when a human pays less attention. For WE-3RII, we realized these behaviors by using the above method.

(2) Setting of the non-responsive area by neck control reference

WE-3RII gives velocity references to the neck that are the current positions of the eyes. It suppresses the vibration from the neck movement by setting up a non-responsive area around the current position. However, if we intentionally enlarge this non-responsive area in that the neck, WE-3RII realizes various human-like head behaviors by making a variable setting value for the non-responsive area.

(3) Variable offset position of the eyeball angle

In WE-3RII, the face direction angle deviates when the current zero position of the eyeball angle is changed. WE-3RII can express the human-like casting of a sideways, upward and downward glance.

3.4 Facial expression control using eyebrows and mouth
(1) Stepping motor control

The stepping motor control computer generates a control pulse through the I/O board using timer interrupts on software. Based on its velocity specifications, the required frequency is 1[kHz], but we set a 20[kHz] interrupt time for making detailed controls.

(2) Setting the intensity of the six basic facial expressions

Based on the AU, we made the six basic facial expressions pattern using the developed eyebrows and mouth. The intensity of each facial expression is variable by a five-grade proportional interpolation of differences in location from the neutral facial expression.

3.5 Facial expression control using three independent parameters from a psychological model

Deciding the internal state of a robot is necessary to control the facial expression using the developed hardware. Thus, we proposed a simple psychological model that has three independent parameters as an early stage for our robot's development as shown in Fig.3.3. In this model, the various conditions of the

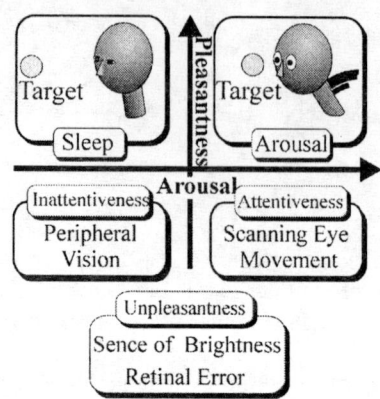

Fig.3.4 Movement of The Eyeballs

target decide the internal state of the robot.

In this psychological model, the conviction axis [4] is added to the two axes of the pleasantness axis and the arousal axis. The six basic facial expressions (happiness, anger, surprise, sadness, fear and disgust) are mapped to this model with a five-grade intensity. Fig.3.4 shows the relationship between this psychological model and the eye movements. Unfortunately, there is no space in this paper for the details of this model. We are planning to report them in another paper in the near future.

4. Experiments
4.1 Smooth pursuit eye movement and saccadic eye movement

We evaluated the effect of smooth pursuit eye movement and saccadic eye movement by the method described in chapter 3. Only eye movement was used in this experiment while the neck was fixed. The result is shown in Fig.4.1, which shows the state change of the smooth pursuit eye movement and saccadic eye movement by the EyeMoveFlag. When the EyeMoveFlag is still, saccadic eye movement

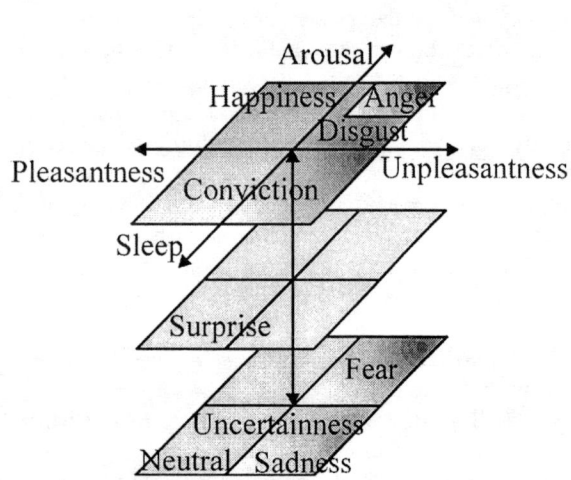

Fig.3.3 Simple Psychological Model

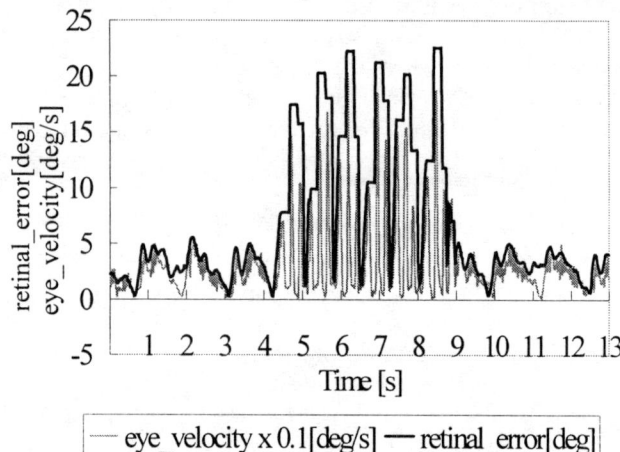

Fig.4.1 Smooth Eye Movement and Saccadic Eye Movement

occurs.

The retinal error indicates the error of the target on the retina. It shows that the eyes are pursuing this error during the smooth pursuit eye movement. There is a non-responsive period of 200[ms] when saccadic eye movement replaces smooth pursuit eye movement.

From these results, we have confirmed the effect of the control algorithm described in chapter 3. We also realize more human-like eye movements as compare to WE-3R. The retina error as the degree of unpleasantness is larger for saccadic eye movement than for smooth pursuit eye movement.

4.2 Evaluation of scanning eye movements

We applied the scanning eye movement algorithm to WE-3RII and examined it. We set the target object about 10[cm] in front of the eyes, and took the data of the 3×3 mask in the scanning. The data of these eyeball angles are shown in Fig.4.2. In these figures, the data were plotted at 200[ms] intervals. The mask range means for pixels are halfway between two cell's center points as in Fig.3.2.

Based on this result, the rougher the mask becomes, the higher the scanning velocity. Also the edge of the bulb is recognized as a circle. We have confirmed the effect of the scanning eye movements' algorithm, as well as that WE-3RII is able to express more human-like eye movements compared to WE-3R.

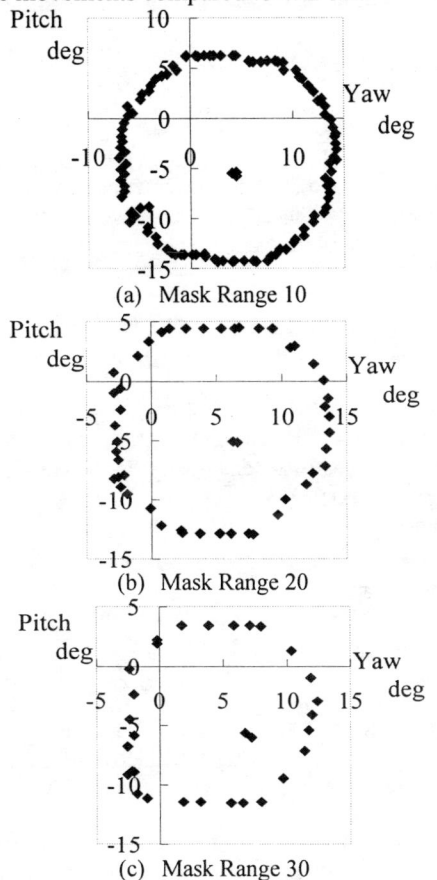

Fig.4.2 Results of Scanning Eye Movement

4.3 Expression by setting the eyeball control parameters

By setting the non-responsive area by the eyeball velocity input of the retina, the robot can express the state of stupefaction. By setting the non-responsive area based on the neck control reference, the robot can convey an expression that the eyeballs will do more used than the neck part. Further, by changing the current zero position of the eyeball angle, the robot can express the human-like casting of a sideways, upward and downward glance as shown in Fig.4.3. Therefore, we have confirmed that WE-3RII can express more human-like movements as compared to WE-3R by adding variable eyeball control parameters.

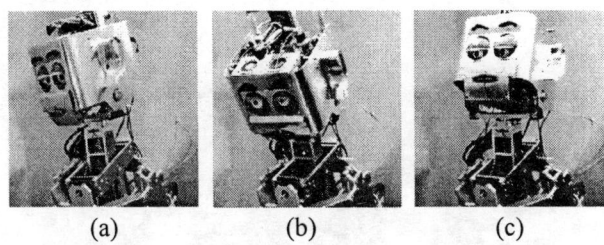

Fig.4.3 Expression by Setting the Control Parameter of the Eyeballs

4.4 Expressing the six basic facial expressions

We did experiments on the six basic facial expressions with WE-3RII. Fig.4.4 shows the six basic facial expressions by WE-3RII. Some of these expressions (happiness, anger, surprise and sadness) were successfully recognized by some students. However, others (fear and disgust) were confused for each other. Based on this result, we need to improve this issue for our future work.

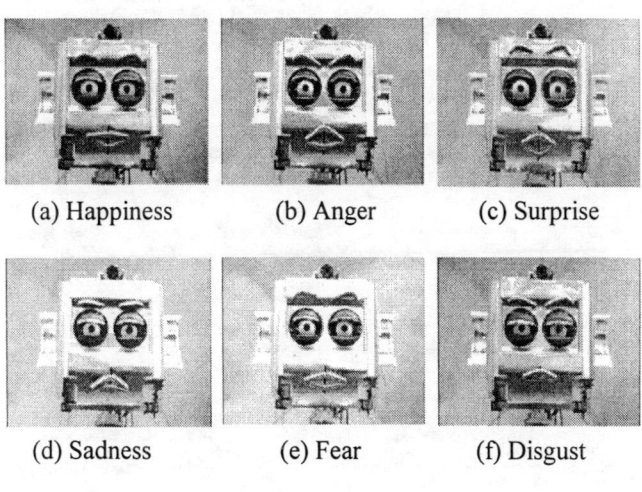

Fig.4.4 Six Basic Facial Expressions

4.5 Target pursuing in a three dimensional space with a facial expression

We evaluated pursuing in a three dimensional space with facial expression using WE-3RII. We have confirmed that WE-3RII realized the movement of pursuing a target in a three dimensional space by adjusting to the target position and the brightness. Fig.4.5 shows the appearance of the experiment. By introducing the changing to the smooth pursuit eye movements, saccadic eye movements and scanning eye movements, we have confirmed that WE-3RII was able to give more human-like expressions than WE-3R. Besides, we have also confirmed the effect of the psychological model.

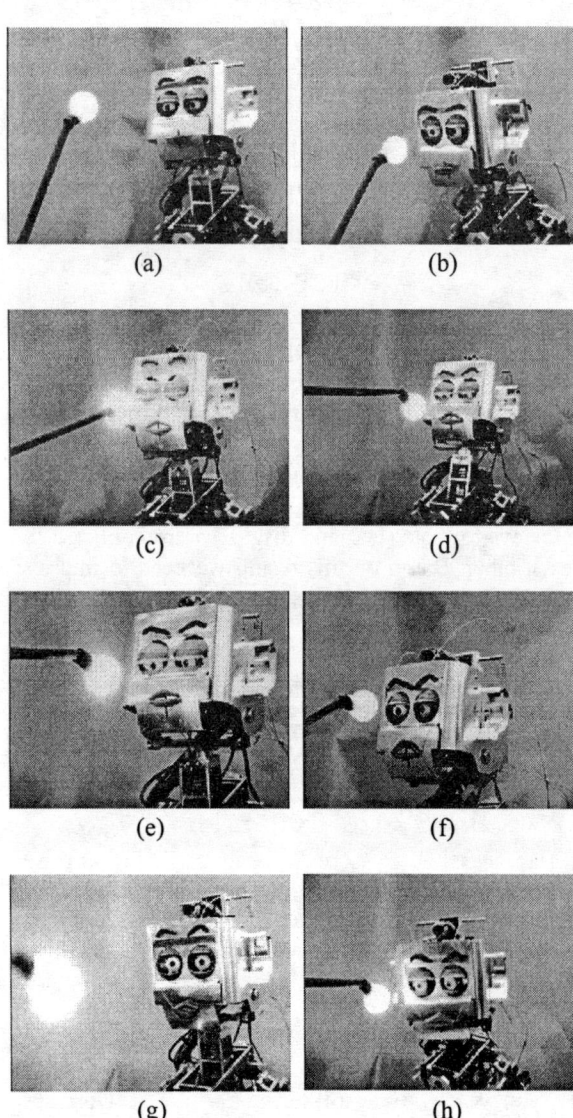

Fig.4.5 Motion of WE-3RII pursuing a light bulb with facial expression

5. Conclusion

We developed the anthropomorphic head-eye robot WE-3RII with an autonomous facial expression mechanism using a simple psychological model by adding a conviction axis on two axes of the pleasantness axis and the arousal axis. The robot realized the control of switching between the smooth pursuit eye movement and the saccade eye movement, and realized the control of the scanning eye movement by the saccadic eye movement. The robot was also able to express more human-like motion in pursuing a target by changing its eyeball control parameters as compared to WE-3R.

From our experiments, we have confirmed that WE-3RII realized the pursuing motion in a three dimensional space with its autonomous facial expression adapting to the target position and the brightness. With these results, it can be said that we have reached a higher step toward the realization of human friendly humanoid robots.

6. Reference

[1] Atsuo Takanishi, Tadao Matsuno, Isao Kato: "Development of an Anthropomorphic Head-Eye Robot with Two Eyes", Proceedings of the IEEE/RSJ International Conference on Intelligent Robots and Systems, pp.799-804, 1997

[2] Kazutaka Mitobe, etc.: "Consideration of Associated Movements of Head and Eyes to Optic and Acoustic Stimulation", The Institute of Electronics, Information and Communication Engineers, Vol.91, pp.81-87, 1992

[3] Laurutis V.P. and Robinson D.A.: "The vestibulo-ocular reflex during human saccadic eye movements", J. Physiol., 373, pp.209-233, 1986

[4] Atsuo Takanishi, Satoshi Hirano, Kensuke Sato: "Development of an Anthropomorphic Head-Eye System for a Humanoid Robot -Realization of Human-like Head-Eye Motion Using Eyelids Adjusting to Brightness-", Proceedings of the IEEE International Conference on Robotics and Automation, 1998

[5] Paul Ekman, Wallace V. Friesen : Facial Action Coding System, Consulting Psychologists Press, Inc., (1978)

[6] Haruhiko Chiba : Face and Mind・Entrance of the facial psychology, pp.110-135, (1993)

[7] Toshiaki Miura : Cognitive Science & Information Processing ex.3, pp.100-141, 1993

[8] Akihiro Yagi : The Psychology Today series 6 : The perception and the recognition, Baifukan, pp.72-83

[9] Hiroshi Kobayashi, Fumio Hara, Tsuyoshi Uchida, Munehisa Ono : Of the Face Robot for Active Human Interface(AHI),RSJ Journal, vol.12 No.1, pp.155-163

[10] Chang Seok Choi, Hiroshi Harashima, Tsuyosi Takebe : 3-Dimensional Facial Model-Based Description and Synthesis of Facial Expressions, Japan.ICICE. A, vol. J73-A No.7, pp.1270-1280. 1990

[11] Atsushi Komatsuzaki, Yoshikazu Shinoda, Toshio Maruo : Neurology of the Oculomotor System, IGAKU_SHOIN Ltd. pp.1-17

Interaction with a Realtime Dynamic Environment Simulation using a Magnetic Levitation Haptic Interface Device

Peter J. Berkelman and Ralph L. Hollis
The Robotics Institute
Carnegie Mellon University
Pittsburgh, PA 15213

David Baraff
Pixar Animation Studios
Richmond CA, 94804

Abstract

A high performance six degree-of-freedom magnetic levitation haptic interface device has been integrated with a physically-based dynamic rigid-body simulation to enable realistic user interaction in real time with a 3-D dynamic virtual environment. The user grasps the levitated handle of the device to manipulate a virtual tool in the simulated environment and feels its force and motion response as it contacts and interacts with other objects in the simulation.

The physical simulation and the magnetic levitation controller execute independently on separate processors. The position and orientation of the virtual tool in the simulation and the levitated handle of the maglev device are exchanged at each update of the simulation. The position and orientation data from each system act as impedance control setpoints for the other, with position error and velocity feedback on each system acting as virtual coupling between the two systems. The setpoints from the simulation are interpolated by the controller at the faster device control rate so that the user feels smooth sliding contacts without chattering due to the slower updates of the simulation. The simple feedback coupling between the two systems enables the overall stiffness and stability of the combined system to be tuned easily and provides realistic haptic user interaction. Sample task simulation environments have been programmed to demonstrate the effectiveness of the haptic interaction system.

1 Introduction

Haptic refers to the tactile and kinesthetic sensing modalities of the hand. The aim of the work described here is to provide high fidelity tool-based haptic interaction with a dynamic simulated rigid-body environment so that a user can grasp a tool handle on the haptic interface device to interact in real time with simulated manual task environments while feeling the detailed reaction forces of the tool in the simulation due to solid contacts, friction, and texture. Magnetic levitation devices are well suited for haptic interaction due to simple control dynamics, full 6-DOF motion with one moving part, high control bandwidths, and the absence of actuator nonlinearities such as backlash and hysteresis. Potential applications of haptic interface technology include prototyping, simulation, training, and teleoperation in areas such as medical, CAD, machine or vehicle operation[1, 2, 3].

A graphic display of the simulated environment is also generated during haptic interaction. A simple coupling method provides a general way to interface a haptic interface device with any realtime simulation to provide the user with realistic, stable interaction with a simulated environment model of a haptic task.

Lorentz force magnetic levitation devices have previously been developed by Hollis and Salcudean [4, 5]. The new device described here has a larger range of motion and was designed specifically for haptic interaction instead of as a fine motion robot wrist. Haptic interaction for solid contacts and friction has been developed using previous magnetic levitation devices by Salcudean and Berkelman [6, 7].

Colgate *et al.* first proposed using a virtual coupling between a simulation and a haptic display to simplify design and ensure stability [8] and Adams and Hannaford generalized the idea and analyzed its stability conditions [9]. The use of interpolated, locally updated intermediate representations to interface haptic devices with simulations at a slower update rate has been developed by Adachi, Mark *et al.*, and Vedula [10, 11, 12], primarily for point force interactions rather than for 6-DOF rigid-body interaction.

2 Maglev Haptic Interface Device

The use of noncontact actuation and sensing with a single lightweight moving part results in control bandwidths and position resolution that would be very difficult to achieve with any conventional 6-DOF force feedback linkage mechanism. The actuation forces

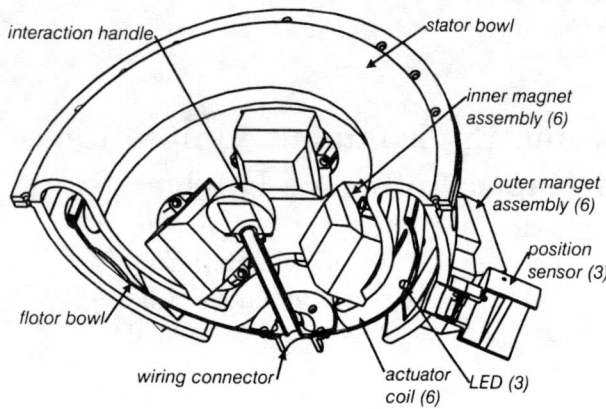

Figure 1: Cutaway View of the Maglev Haptic Device

of the magnetic levitation device are caused by the Lorentz force generated on wire coils carrying electric current in a magnetic field. Six Lorentz force actuators arranged about the levitated bowl enable forces and torques to be generated in all directions. Three sensor assemblies focus light from LEDs on the levitated bowl through lenses onto fixed planar position sensing photodiodes to provide noncontact position sensing. A cutaway schematic of the magnetic levitation device showing the fixed stator with magnet and sensor assemblies and the levitated flotor with coils, LEDs, and the interaction handle is given in Fig. 1. Additional details of the device design are described in a design paper[13].

Our maglev device and its amplifiers, controller, and power supplies are contained in a single desktop-height cabinet shown in Fig. 2. The device cabinet can be easily positioned next to a desk for user operation with a graphical display, as in Fig. 3. The interaction handle of the device is at desktop level and the top rim of the magnet stator bowl mounted into the top of the cabinet so that the user can manipulate the handle with the fingertips while resting his or her wrist on the rounded rim of the permanent magnet housing.

The range of motion of the levitated handle is 25 mm in translation and 15-20° in rotation in all directions to provide a comfortable fingertip motion range. For small motion amplitudes, the ±3 dB attenuation position control bandwidth of the magnetic levitation device is 100 Hz or greater for all axes. The closed-loop position control and sensing resolution of the device is 5-10 μm throughout its motion range. The maximum sample rate of the present 68060 control processor is 1.4 kHz. The peak forces and torques attainable are limited to 50 N and 6 Nm by the current amplifiers in use. The translational stiffness range is 0.005 N/mm to 25.0 N/mm. More results are given in a dynamic performance paper [14].

Figure 2: Haptic Interface Device Enclosure

Figure 3: Using the Haptic Interface Device with a Graphics Display

3 Physical Simulation

The CORIOLIS^{tm} 3-D physically-based dynamic simulation package developed by Baraff, an extension of the 2-D simulation described in [15], is used for the interactive simulation. This package calculates forces and motions of rigid bodies in space due to Newtonian mechanics, motion constraints, collisions, and friction quickly and efficiently.

The Newtonian free rigid-body dynamic state equations given below are integrated for each free body in the simulation:

$$\frac{d}{dt}\begin{pmatrix} \boldsymbol{x}(t) \\ \boldsymbol{R}(t) \\ \boldsymbol{P}(t) \\ \boldsymbol{L}(t) \end{pmatrix} = \begin{pmatrix} \boldsymbol{v}(t) \\ \boldsymbol{\omega}(t)\boldsymbol{R}(t) \\ \boldsymbol{F}(t) \\ \boldsymbol{\tau}(t) \end{pmatrix},$$

$$v(t) = \frac{P(t)}{M}, \quad \omega(t) = I(t)L(t),$$
$$I(t) = R(t)I_{body}R(t)^T.$$

with position x, rotation matrix R, momentum P, angular momentum L, velocity v, angular velocity ω, force F, and torque τ. Body mass M is constant and inertia I is a coordinate transform of constant inertia matrix I_{body}.

When contact is detected, the necessary surface forces to prevent interpenetration are computed and introduced into the simulation. For collisions, the impulsive forces and accelerations are instantaneous so integration must be restarted with new initial conditions since differential equation integration methods assume continuous dynamics. Simulated objects are perfectly rigid and non-interpenetrating and bounding boxes are used for fast collision detection.

The simulation was tuned to run as fast as possible for realistic realtime interaction. The midpoint or second-order Runge-Kutta integration method was selected for speed and simplicity. A 100 Hz update rate was achieved for simulations with up to 10 polyhedral objects of 10 vertices or less on an SGI Indigo 2 workstation. If the multiple rigid body contact states are sufficiently complex, such as when a chain of several objects collides nearly simultaneously, the simulation update rate may occasionally slow down for one or two frames.

The 3-D rendered graphics display is updated asynchronously in the background after the simulation dynamics are updated by an interval timer signal handler. The resulting frame rate of the graphics display is typically 15-30 Hz, depending on the complexity of the simulation and any other processes executing on the workstation.

4 Coupled Simulation and Control

The simulation and the device controller can each operate independently and communicate using TCP/IP sockets over a standard Ethernet network. A schematic representation of the intercommunication between the device, controller, simulation, and graphical display systems is shown in Fig. 4.

Interactive simulations have been implemented on the local control processor but they are limited to static environments with fewer than 10 to 15 total vertices in the models of the tool and its environment. The present control processor is not sufficiently fast to perform dynamic simulation and collision detection on multiple moving objects while calculating the sensor kinematics and feedback control at a rate sufficient for stable stiff contacts.

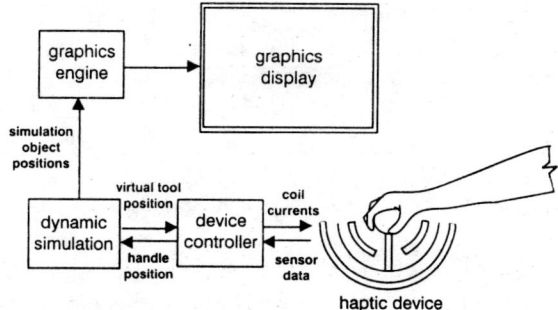

Figure 4: Haptic and Visual Interface System

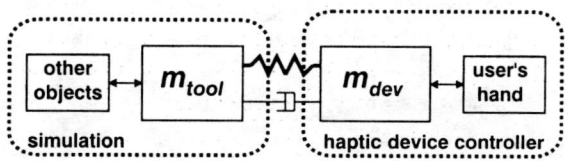

Figure 5: Virtual Coupling of Simulation and Device

The virtual coupling between the realtime simulation and the maglev haptic device controller as proposed by Colgate [8] is shown in Fig. 5. In our implementation, the position and orientation vector of each system is periodically sent to the other system to act as its control setpoint. The stability and responsiveness of the simulation and device can be set by the spring and damper gains of the coupling as seen from either side. The generated forces on the maglev haptic device f_{dev} and the virtual tool in the simulation f_{tool} are given by:

$$f_{dev} = f_g + K_p(x_{tool} - x_{dev}) + K_v r(x_{dev} - x_{devprev}),$$
$$f_{tool} = f_{other} + K_{spring}(x_{dev} - x_{tool}) + K_{damp}v_{tool}.$$

where f_g is gravity feedforward to reduce the weight of the levitated bowl, K are the coupling gains, r is the control rate of the device, and f_{other} are the forces from the other objects in the simulation. When the virtual tool is not in contact with any other objects in the simulation, force feedback to the device is switched off. Realistic and stable performance for most task simulations has been obtained with the following coupling gains, where K_{spring} and K_{damp} are the gains from the simulation side and K_p and K_v are the maglev device control gains:

Gain	Position	Rotation
K_{spring}	100 N/mm	800 Nm/rad
K_{damp}	10 N/mm/s	20 Nm/rad/sec
K_p	4.0 N/mm	25 Nm/rad
K_v	0.1 N/mm/s	0.5 Nm/rad/sec

The time required to send a set of position setpoints from the simulation and receive a reply of setpoints from the control processor is generally approximately 1 ms. The simulated environment takes 10 ms to respond to forces or motions exerted by the user, however, due to its 100 Hz update rate. The response delay due to this latency is perceived by the user as stickiness or sluggishness in haptic interaction.

Interpolation between the setpoints supplied by the simulation eliminates the jittering or chattering feel experienced by the user during sliding contacts between the simulated tool and other objects. The desired position setpoints x_{goal} on the haptic device controller are interpolated from the last setpoint supplied by the simulation x_{simnew} and the differences between the last setpoint and the previous one $x_{simprev}$:

$$x_{goal} = x_{goal} + \frac{x_{simnew} - x_{simprev}}{T_{avg}}$$

where T_{avg} is the average of the last three simulation time intervals since there are variations due to overruns in the simulation calculations, network traffic, or other delays. This interpolation scheme is a first-order hold since the most recent sample and the slope between samples are used to calculate the interpolated setpoints.

This same method of virtual coupling with interpolated setpoints can be used to add haptic interaction to any realtime simulation that calculates motions due to dynamic forces. Independent operation of the simulation and the haptic device controller significantly simplifies development, testing, and debugging of the integrated system.

Sample tasks have been programmed into the simulation to demonstrate rigid-body haptic tool manipulation. Each of these tasks requires 6-DOF haptic manipulation and feedback and could not be performed with 3 DOF only. The first general task world contains several free polyhedral objects and fixed walls, as pictured in Fig. 6. The user tool is a square scoop with a handle. The scoop can be used to feel, strike, push, pick up, or throw and catch other solid objects in the simulation while the user feels its dynamic response. The second task, shown in Fig. 7, demonstrates the classic peg-in-hole manipulation problem. The world contains only a fixed square hole and the user tool is a peg of a slightly smaller cross-section than the hole. The third task, in Fig. 8, is a variation of the peg-in-hole where the task setup includes a fixed keyhole, a movable bolt, and a key as the haptic user tool. The user can insert the key into the hole and rotate it to slide the bolt sideways.

A one-to-one mapping of the levitated handle position to the virtual tool position in the simulated environment provides sensitive interaction for fine fingertip operations such as insertion, but the user cannot

Figure 6: Block Manipulation Task

Figure 7: Peg-in-Hole   Figure 8: Key and Lock

move the tool over larger distances in the simulated environment. Therefore, additional control modes have been added to the haptic user interface to enable the user to operate the haptic tool over arbitrarily large distances and rotations in the virtual environment in a natural, intuitive way. During interaction the user can change the position scaling factors and zero offsets between the levitated handle and the simulated environment, switch to rate control mode, and move the graphics viewpoint for the simulation display.

The mapping from the actual levitated handle position x_{device} and Θ_{device} to the simulation setpoint x_{setp} and Θ_{setp} is given by

$$x_{setp} = x_{offset} + x_{scale} x_{device}, \text{ and}$$

$$\Theta_{setp} = \Theta_{offset} + \theta_{scale} \Theta_{device}.$$

The variable scaling factors x_{scale} and θ_{scale}, and offsets x_{offset} and Θ_{offset} for rotation and translation are set by sliders in a graphical user interface control panel. The offsets determine the position and orientation of the virtual tool when the haptic device handle is in the centered position. These offsets can be set so that the full sensitivity and motion range of the haptic interface device can be made available at any

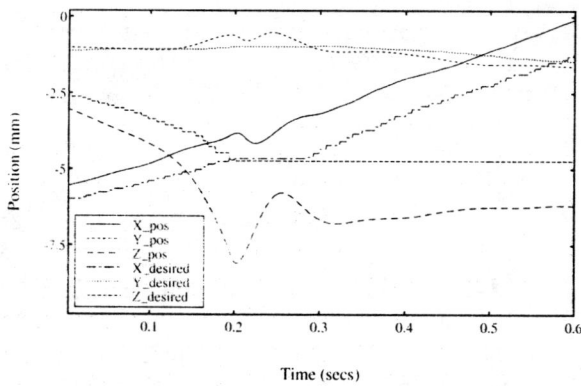

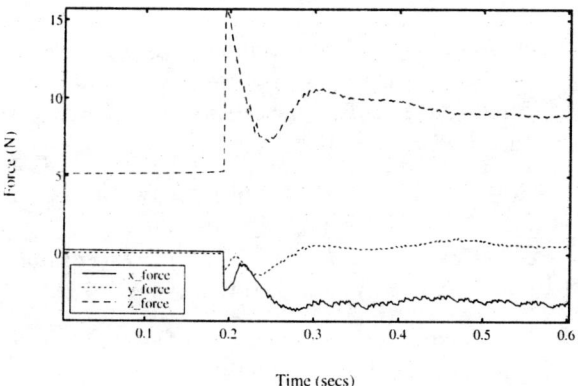

Figure 9: Position and Force Data from Impact and Sliding Motions

point in the simulated environment. The user can easily switch from making large motions across the entire simulation world to fine motions in one spot.

In rate control mode, the position of the haptic device handle determines the velocity instead of the position of the tool in the simulation. Rate control mode can be either manually selected by clicking a button on the user interface control panel window, or automatically invoked as the levitated handle approaches its motion limits.

Viewpoint controls are provided for the user to zoom, rotate, and pan to display areas of interest in the simulated environment. Automatic viewpoint control mode tracks the tool so that the zero offset position of the tool is centered in the display and the major axis of the tool in its zero offset orientation is aligned with the viewing direction.

5 Sample Experimental Results

The data plots in Fig. 9 were obtained from the magnetic levitation controller during typical user interaction with the haptic interface system. Rotation and torque data are not shown. The user brought the virtual scoop of Fig. 6 into contact with the environment floor while moving it in the positive x direction. The data show free tool motion until approximately the 0.2 seconds, then vertical impact and sliding. The desired position setpoints in the controller are obtained from the virtual tool positions in the dynamic simulation.

In unconstrained axes, the desired setpoints lag behind the actual setpoints because there is no force feedback exerted by the system. Due to the slower update rate of the simulation, the desired position setpoint curves have a stair-step appearance. Since objects in the simulation cannot interpenetrate, the z setpoint is pinned to the floor level at -5 mm after impact. There is an inflection in the x position curve after the impact from the user's hand rebounding due to the friction in the simulation.

The force data plot shows the force commands exerted on the flotor by the Lorentz actuators. While the virtual tool is in free space, the user feels no resistance to motions other than the actual inertia and passive eddy current damping in the flotor. No Lorentz forces are generated in the x, y horizontal plane and 5 N are generated upwards in z to partially cancel the weight of the flotor. When the virtual tool in the simulation contacts the floor, the force in y remains close to zero and a negative force in x is generated due to sliding friction modelled in the dynamic simulation. The force in z changes suddenly and rebounds at impact, then settles to support the weight of the flotor and additional disturbance forces from the user's hand.

6 Summary and Conclusions

We have implemented a complete general haptic interface system which demonstrates the effectiveness and practicality of several combined technologies in haptics:

Tool-based haptic interaction using a magnetic levitation device can provide sensitive, high-bandwidth haptic interaction with a dynamic simulation. The virtual spring and damper coupling is a simple yet effective method for coupling a haptic interface device with a simulation to provide realistic, high-performance, 6-DOF haptic interaction. Separation of the simulation and the haptic interface simplifies development and testing and the overall system stiffness and stability can be tuned by adjusting the parameters of the virtual spring and dampers. Setpoint interpolation by the haptic device controller practically eliminates the feel of jitter during sliding contacts caused by the slower update rate of the simulation.

The three tasks set up for the simulation system demonstrate common haptic 6-DOF rigid body tasks that cannot be performed naturally using 3-DOF devices. The added haptic user interface features of rate control mode, variable scaling and offsets, and automatic viewpoint tracking enable the user to navigate in a natural and intuitive manner through an arbitrarily large simulation workspace and also interact with high sensitivity when desired.

7 Planned Work

To further increase the fidelity of haptic interaction, the effects of communication latency and compliance between the simulation and the device must be reduced. To accomplish this, all the virtual tool contact point data will be sent to the device controller from the simulation instead of only its position. The feedback forces generated by the device will be calculated from each contact point rather than simply servoing to the interpolated setpoint.

To add realism and detail to the haptic interaction, surface texture and friction will be modeled on the local control processor and added to the force feedback during sliding contacts. To handle the added computation, the local control processor will be upgraded. Finally, user studies will be carried out to evaluate the effectiveness and ease of use of the haptic interface system.

References

[1] N. Parker, S. Salcudean, and P. Lawrence, "Application of force feedback to heavy duty hydraulic machines," in *IEEE Int'l Conf. on Robotics and Automation*, (Atlanta), pp. 375–381, May 1993.

[2] J. W. Hill, P. S. Green, J. F. Jensen, Y. Gorfu, and A. S. Shah, "Telepresence surgery demonstration system," in *IEEE Int'l Conf. on Robotics and Automation*, (San Diego), pp. 2302–2307, May 1994.

[3] S. Singh, M. Bostrom, D. Popa, and C. Wiley, "Design of an interactive lumbar puncture simulator with tactile feedback," in *IEEE Int'l Conf. on Robotics and Automation*, (San Diego), pp. 1734–1739, May 1994.

[4] R. L. Hollis, S. Salcudean, and A. P. Allan, "A six degree-of-freedom magnetically levitated variable compliance fine motion wrist: design, modeling, and control," *IEEE Transactions on Robotics and Automation*, vol. 7, pp. 320–332, June 1991.

[5] S. Salcudean, N.M. Wong, and R.L. Hollis, "Design and control of a force-reflecting teleoperation system with magnetically levitated master and wrist," *IEEE Transactions on Robotics and Automation*, vol. 11, pp. 844–858, December 1995.

[6] S. Salcudean and T. Vlaar, "On the emulation of stiff walls and static friction with a magnetically levitated input-output device," in *ASME IMECE*, (Chicago), pp. 303–309, November 1994.

[7] P. J. Berkelman, R. L. Hollis, and S. E. Salcudean, "Interacting with virtual environments using a magnetic levitation haptic interface," in *Int'l Conf. on Intelligent Robots and Systems*, (Pittsburgh), August 1995.

[8] J. Colgate, M. Stanley, and J. Brown, "Issues in the haptic display of tool use," in *Int'l Conf. on Intelligent Robots and Systems*, (Pittsburgh), August 1995.

[9] R. Adams and B. Hannaford, "A two-port framework for the design of unconditionally stable haptic interfaces," in *Int'l Conf. on Intelligent Robots and Systems*, (Victoria, B,C.), August 1998.

[10] Y. Adachi, T. Kumano, and K. Ogino, "Intermediate representation for stiff virtual objects," in *Proc. IEEE Virtual Reality Annual Intl. Symposium*, (Research Triangle Park, N.C.), pp. 203–210, March 1995.

[11] W. Mark, S. Randolph, M. Finch, J. V. Werth, and R. Taylor, "Adding force feedback to graphics systems," in *Computer Graphics (Proc. SIGGRAPH)*, pp. 447–452, 1996.

[12] S. Vedula and D. Baraff, "Force feedback in interactive dynamic simulation," in *Proceedings of the First PHANToM User's Group Workshop*, (Dedham, MA), September 1996.

[13] P. J. Berkelman, Z. J. Butler, and R. L. Hollis, "Design of a hemispherical magnetic levitation haptic interface device," in *Proc. of the ASME IMECE Symposium on Haptic Interfaces for Virtual Environment and Teleoperator Systems*, (Atlanta), November 17-22 1996.

[14] P. J. Berkelman and R. L. Hollis, "Dynamic performance of a hemispherical magnetic levitation haptic interface device," in *SPIE Int'l Symposium on Intelligent Systems and Intelligent Manufacturing, SPIE Proc. Vol. 3602*, (Greensburgh, PA), September 1997.

[15] D. Baraff, "Interactive simulation of solid rigid bodies," *IEEE Computer Graphics and Applications*, vol. 15, pp. 63–75, 1995.

Proceedings of the 1999 IEEE
International Conference on Robotics & Automation
Detroit, Michigan • May 1999

GUARANTEED CONVERGENCE RATES FOR FIVE DEGREE OF FREEDOM IN-PARALLEL HAPTIC INTERFACE KINEMATICS

CHRISTOPHER D. LEE
Electrical Engineering
University of Colorado
Boulder, CO 80309-0526
toph@colorado.edu

DALE A. LAWRENCE
Aerospace Engineering
University of Colorado
Boulder, CO 80309-0429
dale.lawrence@colorado.edu

LUCY Y. PAO
Electrical Engineering
University of Colorado
Boulder, CO 80309-0425
pao@colorado.edu

Abstract

Numerical computation of forward kinematics for an in-parallel interface are developed which conform to the needs of real-time haptic rendering. This approach produces a computationally fast algorithm for computing Cartesian position and orientation from actuator position measurements. Explicit bounds on the convergence rates are derived which can be used to design appropriate computational strategies to maximize control loop cycle times, yet preserve acceptable accuracy.

1 INTRODUCTION

Parallel mechanisms have well known advantages for haptic interfaces [1], [2], including low moving inertia, high stiffness, and high bandwidth transmission of forces to the operator. Disadvantages include reduced workspace compared to serial chain mechanisms, and computationally expensive forward kinematics [3], [4].

This paper is motivated by a 5 degree-of-freedom (DOF) parallel interface designed by the authors for scientific visualization using a stylus. As currently configured, the interface provides haptic rendering in 3 Cartesian translations and 2 orientation angles. Orientation changes about the stylus long axis are difficult to accomplish using a pencil grip, and have limited value in data rendering, hence this sixth DOF is neither sensed nor controlled. See Figure 1. The stylus is actuated by 5 prismatic joints, which produce axial forces on rods that connect actuator units mounted on a hexagonal ring to end points of a stylus held in the operator's hand. This produces a workspace that is compatible with the range of motion of an operator's hand and wrist when the elbow is placed on a horizontal rest. Scientific data is rendered haptically (e.g. as a virtual surface) by prescribing particular Cartesian forces and torques on the operator's hand as functions of Cartesian translation and orientation motions. Cartesian position is derived from measurements of the length of each actuator rod via the forward kinematics mapping.

As in other work (e.g. [5]) this paper uses an iterative numerical approach to solve the forward kinematics problem. We discuss how the haptic interface application is particularly well-suited to a numerical solution, and contribute bounds on initial conditions and algorithm parameters that guarantee specific exponential convergence rates. These bounds can be used to design computational schemes which minimize cycle time subject to desired levels of accuracy.

Section 2 derives bounds for the Cartesian location

---

*Support for this work by the National Science Foundation (IIS-9711936) and the Office of Naval Research (N00014-97-1-0354) is gratefully acknowledged.

Figure 1: 5 DOF haptic interface being used in data rendering experiments for scientific visualization.

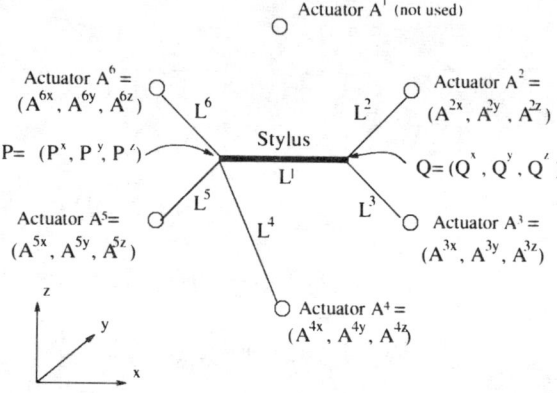

Figure 2: Schematic of the haptic interface showing how actuator rods are connected to points P and Q on the stylus. The actuators lie in the plane of the paper, and the stylus moves in a workspace above this plane.

of the left and right end points of the stylus, which completely describe its position and orientation within the 5 degrees of freedom. The conservativeness of the bounds are examined in Section 3 using data from the haptic interface in operation.

2 STYLUS END POINT KINEMATICS

Cartesian translation and orientation of the stylus in 5 DOF (i.e. excluding roll of the stylus about its long axis) is completely determined from the Cartesian location of the stylus end points P and Q. This section presents a numerical algorithm for calculating the location of P and Q from measured rod lengths connecting the stylus to 5 actuators arranged on a hexagonal base. Figure 2 shows a schematic of the stylus and actuator rod arrangement. Although the inverse kinematics are simple and in closed form, the forward kinematics are needed to convert measured rod lengths to Cartesian position for haptic data rendering. The forward kinematics are not so simple, in general requiring the solution of high order polynomial equations [4]. This is difficult to accomplish in real time when high sampling rates are desired for good haptic rendering (e.g. high virtual stiffnesses). Fortunately, high accuracy is usually not needed in this application, since human perception limits what errors can be detected in data rendering due to inaccurate measurement of Cartesian position. Note however, that perception of edges or jumps in forces is highly acute [6], [7], requiring Cartesian position estimates that are smooth functions of haptic interface motion.

The goal is thus to find a fast online estimation method for calculating the positions of P and Q without actually inverting the inverse kinematics map. Speed and smoothness are of primary importance, provided that bounds on estimation error are within reasonable values.

The inverse kinematic map $h : \mathbb{R}^6 \to \mathbb{R}^6$ from Cartesian stylus end point coordinates X to rod lengths L at any time k is defined by the following five expressions for rod lengths, together with an adjoined stylus length expression.

$$X_k = [P_k^x, P_k^y, P_k^z, Q_k^x, Q_k^y, Q_k^z]^T$$
$$L_k = h(X_k) = [L_k^1, L_k^2, L_k^3, L_k^4, L_k^5, L_k^6]^T$$
$$= [h^1(X_k), h^2(X_k), h^3(X_k), h^4(X_k), h^5(X_k), h^6(X_k)]^T$$
$$L_k^1 = \sqrt{(P_k^x - Q_k^x)^2 + (P_k^y - Q_k^y)^2 + (P_k^z - Q_k^z)^2}$$
$$L_k^2 = \sqrt{(Q_k^x - A^{2x})^2 + (Q_k^y - A^{2y})^2 + (Q_k^z - A^{2z})^2}$$
$$L_k^3 = \sqrt{(Q_k^x - A^{3x})^2 + (Q_k^y - A^{3y})^2 + (Q_k^z - A^{3z})^2}$$
$$L_k^4 = \sqrt{(P_k^x - A^{4x})^2 + (P_k^y - A^{4y})^2 + (P_k^z - A^{4z})^2}$$
$$L_k^5 = \sqrt{(P_k^x - A^{5x})^2 + (P_k^y - A^{5y})^2 + (P_k^z - A^{5z})^2}$$
$$L_k^6 = \sqrt{(P_k^x - A^{6x})^2 + (P_k^y - A^{6y})^2 + (P_k^z - A^{6z})^2}$$

(1)

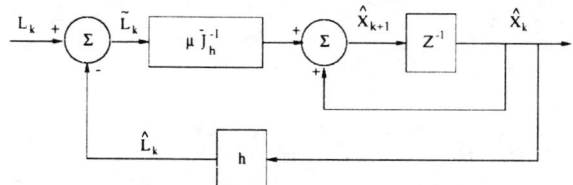

Figure 3: Position Estimation Algorithm

L_k^1 is the length of the stylus, which is fixed. The A^i are Cartesian locations of the 5 actuators, as illustrated in Figure 2.

Since the inverse kinematics $L_k = h(X_k)$ are easy to calculate, our strategy is to iterate on an estimated value $\hat{X}_k$ for the Cartesian position X_k, using a Newton-Raphson method based on the estimation error $h(X_k) - h(\hat{X}_k)$ to improve the estimate at each iteration k. The correct update direction is found via the Jacobian J_f of the forward kinematics map $X_k = f(L_k)$, since $X_k - \hat{X}_k \simeq J_f(h(X_k) - h(\hat{X}_k))$. J_f is found as the inverse of the 6×6 J_h. To maximize speed we wish to use a fixed J_f rather than update it with each iteration. We choose the Jacobian of the inverse kinematics map at a nominal location: $J_f = \bar{J}_h^{-1} = (J_h(X_{nom}))^{-1}$. The resulting algorithm is defined by

$$\hat{X}_{k+1} = \hat{X}_k + \mu \bar{J}_h^{-1}(L_k - \hat{L}_k), \qquad (2)$$

where the step size μ is chosen (below) to obtain fast convergence of $\hat{X}_k$ to X_k. Figure 3 shows this iterative scheme in block diagram form. Observe that this algorithm contains an integration of the rod length prediction error, which amounts to a low pass filtering operation on measured rod length signals. Hence estimated Cartesian positions are inherently smooth functions of rod lengths.

In an implementation, the actual Cartesian position X_k will be constant on intervals of length N, where N is the number of iterations of the estimation algorithm between sensor updates δt of the measured rod lengths L_k. Hence the L_k are also constant on length N intervals. If the initial estimate $\hat{X}_k$ on one such N-interval is in a small enough neighborhood of the actual value X_k, and this X_k is close enough to the nominal location for P and Q (where the Jacobian $\bar{J}_h^{-1}$ is correct), this algorithm converges, i.e. provided that X_k remains constant for all k, $\lim_{k \to \infty} X_k - \hat{X}_k = 0$. In the haptic interface application, a more useful result would be a bound that guarantees a certain reduction factor in $X_k - \hat{X}_k$ over a finite number N of algorithm iterations, where this bound depends explicitly on the size of the initial estimation error and the physical distance from the nominal position. Such a bound would support the selection of computer hardware and the design of software to provide the desired speed/accuracy trade-off.

The Theorem and Lemmas 1 and 2 below develop such a bound. This bound depends on a linearization error Δ and a Jacobian error $\tilde{J}$ defined as follows. Expand $\hat{L}_k = h(\hat{X}_k)$ in a Taylor series around X_k:

$$h(\hat{X}_k) = h(X_k) + J_h(X_k)(\hat{X}_k - X_k) + \Delta(X_k, \hat{X}_k - X_k) \qquad (3)$$

$$\Rightarrow \hat{L}_k - L_k = J_h(X_k)(\hat{X}_k - X_k) + \Delta(X_k, \hat{X}_k - X_k)$$

Now let

$$\tilde{X}_k = X_k - \hat{X}_k$$
$$\tilde{L}_k = L_k - \hat{L}_k$$

Then

$$\tilde{L}_k = J_h(X_k)(\tilde{X}_k) - \Delta(X_k, -\tilde{X}_k) . \qquad (4)$$

Also, define the Jacobian error $\tilde{J}$ by

$$\bar{J}_h^{-1} J_h(X_k) = I + \tilde{J}(X_{nom}, X_k) \qquad (5)$$

where I is the 6×6 identity matrix. With $\|\cdot\|$ the Euclidean norm of a vector, and $\bar{\sigma}(\cdot)$ the maximum singular value of a matrix, we have the following explicit bounds on convergence rate:

Theorem: Suppose that for all estimation errors $\tilde{X}_k$ within a ball of radius d, and for all positions X_k within a ball of radius $\sqrt{2}r$ of the nominal position X_{nom}, there exist bounds $b > 0$ and $\bar{\sigma}(\tilde{J})$ such that

$$\|\Delta(X_k, -\tilde{X}_k)\| \le b\|\tilde{X}_k\| \qquad (6)$$
$$\bar{\sigma}(\tilde{J}) + b\bar{\sigma}(\bar{J}_h^{-1}) \le \alpha < 1 . \qquad (7)$$

where α is the least upper bound of the quantity on the left in (7). Then if the actual position X_k is fixed on the N-interval $k \in [1, N]$, choosing $\mu = 1$ in (2) results in

$$\|\tilde{X}_{k+1}\| \le \alpha \|\tilde{X}_k\|, \ k \in [1, N] \qquad (8)$$

Remark: Cartesian position estimation error $\|\tilde{X}_k\|$ is exponentially decreasing during the N computational cycles of one data sampling interval, resulting in a net reduction factor α^N applied to the initial error on each interval. The low pass filter dynamics of this algorithm therefore have the effective time constant $\tau = -\delta t/(N \ln \alpha)$ sec., where δt is the data sampling period. N and α can then be designed to achieve desired convergence speed, or to manage stability margin erosion in a Cartesian control loop due to phase lag in the kinematics calculation.

Proof: Substituting (4) into (2):

$$\hat{X}_{k+1} = \hat{X}_k + \mu \bar{J}_h^{-1} J_h(X_k)(\tilde{X}_k) - \mu \bar{J}_h^{-1} \Delta(X_k, -\tilde{X}_k)$$

Since the actual position is fixed, $X_{k+1} = X_k$ and

$$\tilde{X}_{k+1} = \tilde{X}_k - \mu \bar{J}_h^{-1} J_h(X_k)\tilde{X}_k + \mu \bar{J}_h^{-1} \Delta(X_k, -\tilde{X}_k) \quad (9)$$

Using (5), we get

$$\tilde{X}_{k+1} = \tilde{X}_k - \mu \tilde{X}_k - \mu \tilde{J} \tilde{X}_k + \mu \bar{J}_h^{-1} \Delta(X_k, -\tilde{X}_k)$$

Thus with (6) and the triangle inequality

$$\|\tilde{X}_{k+1}\| \le |1-\mu|\|\tilde{X}_k\| + \mu\bar{\sigma}(\tilde{J})\|\tilde{X}_k\| + \mu b \bar{\sigma}(\bar{J}_h^{-1})\|\tilde{X}_k\| \quad (10)$$

and with (7) and $\mu \le 1$

$$\|\tilde{X}_{k+1}\| \le [1 - \mu(1-\alpha)] \|\tilde{X}_k\|. \quad (11)$$

Since $\alpha < 1$, it can be seen that the bracketed expression in (11) is minimized for $\mu = 1$, establishing the result (8). ∎

We now take advantage of the special structure of h and its associated J_h for parallel mechanisms to obtain the bounds b and $\bar{\sigma}(\tilde{J})$ for the Jacobian error in terms of the ball radii d and $\sqrt{2}r$.

Let L_{nom} be the vector of rod lengths corresponding to the nominal positio. Let P_{nom} and Q_{nom} be the Cartesian positions of the left and right ends of the stylus in the nominal position.

Lemma 1: Suppose P_k and Q_k each lie within a ball of radius r of P_{nom} and Q_{nom}, respectively. Then $\|X_k - X_{nom}\| \le \sqrt{2}r$ and

$$\bar{\sigma}(\tilde{J}) \le a\bar{\sigma}(\bar{J}_h^{-1}) \quad (12)$$

where

$$a^2 = 8\left(\frac{r}{L^1}\right)^2 + \sum_{i=2,\ldots,6}\left(\frac{r}{L_{nom}^i}\right)^2\left(1 + \frac{L_{nom}^i + r}{L_{nom}^i - r}\right)^2 \quad (13)$$

Remark: Note that the bound on the Jacobian error is directly proportional to r, the Euclidean distance of each end point from its respective nominal position.

Proof: Let $\delta J_h = J_h(X_{nom}) - J_h(X_k)$ then
$$\bar{J}_h^{-1} J_h(X_k) = \bar{J}_h^{-1}(J_h(X_{nom}) - \delta J_h) = I - \bar{J}_h^{-1}\delta J_h$$
$$\Rightarrow \tilde{J} = -\bar{J}_h^{-1}\delta J_h \quad \text{using (5)}$$
$$\Rightarrow \bar{\sigma}(\tilde{J}) \le \bar{\sigma}(\bar{J}_h^{-1})\bar{\sigma}(\delta J_h)$$
$$\Rightarrow \bar{\sigma}^2(\tilde{J}) \le \bar{\sigma}^2(\bar{J}_h^{-1})\bar{\sigma}^2(\delta J_h) \quad (14)$$

Since $\bar{\sigma}(\bar{J}_h^{-1})$ is fixed, we now proceed to bound $\bar{\sigma}(\delta J_h)$ as $P_k - P_{nom}$ and $Q_k - Q_{nom}$ each vary within the ball of radius r. The Jacobian $J_h(X_k)$ of h at X_k is

$$\begin{bmatrix} \frac{P_k^x - Q_k^x}{L_k^1} & \frac{P_k^y - Q_k^y}{L_k^1} & \frac{P_k^z - Q_k^z}{L_k^1} & \frac{Q_k^x - P_k^x}{L_k^1} & \frac{Q_k^y - P_k^y}{L_k^1} & \frac{Q_k^z - P_k^z}{L_k^1} \\ 0 & 0 & 0 & \frac{Q_k^x - A^{2x}}{L_k^2} & \frac{Q_k^y - A^{2y}}{L_k^2} & \frac{Q_k^z - A^{2z}}{L_k^2} \\ 0 & 0 & 0 & \frac{Q_k^x - A^{3x}}{L_k^3} & \frac{Q_k^y - A^{3y}}{L_k^3} & \frac{Q_k^z - A^{3z}}{L_k^3} \\ \frac{P_k^x - A^{4x}}{L_k^4} & \frac{P_k^y - A^{4y}}{L_k^4} & \frac{P_k^z - A^{4z}}{L_k^4} & 0 & 0 & 0 \\ \frac{P_k^x - A^{5x}}{L_k^5} & \frac{P_k^y - A^{5y}}{L_k^5} & \frac{P_k^z - A^{5z}}{L_k^5} & 0 & 0 & 0 \\ \frac{P_k^x - A^{6x}}{L_k^6} & \frac{P_k^y - A^{6y}}{L_k^6} & \frac{P_k^z - A^{6z}}{L_k^6} & 0 & 0 & 0 \end{bmatrix}$$

$$= \begin{bmatrix} A & B \\ C & 0 \end{bmatrix}$$

Thus

$$\delta J_h = \begin{bmatrix} \delta A & \delta B \\ \delta C & 0 \end{bmatrix} = \begin{bmatrix} A_{nom} - A & B_{nom} - B \\ C_{nom} - C & 0 \end{bmatrix}.$$

We now use the result that

$$\bar{\sigma}^2(\delta J_h) \le tr(\delta J_h^T \delta J_h)$$
$$= \sum_{i=1..6} |\delta J_{hi}|^2$$
$$= \sum_{i=1..3} |\delta A_i|^2 + \sum_{i=1..3} |\delta B_i|^2 + \sum_{i=1..3} |\delta C_i|^2 \quad (15)$$

where δJ_{hi} and δA_i are the columns of δJ_h and δA respectively. We now find bounds for $\bar{\sigma}(\delta A), \bar{\sigma}(\delta B), \bar{\sigma}(\delta C)$, starting with $\bar{\sigma}(\delta C)$. Each row of δC, (with $i = 4, 5, 6$) is

$$\begin{bmatrix} \frac{P_{nom}^x - A^{ix}}{L_{nom}^i} - \frac{P_k^x - A^{ix}}{L_k^i} \\ \frac{P_{nom}^y - A^{iy}}{L_{nom}^i} - \frac{P_k^y - A^{iy}}{L_k^i} \\ \frac{P_{nom}^z - A^{iz}}{L_{nom}^i} - \frac{P_k^z - A^{iz}}{L_k^i} \end{bmatrix}^T$$

Letting

$$\frac{1}{L_k^i} = \frac{1}{L_{nom}^i} + \delta_i \quad (16)$$

each row becomes

$$\frac{[P^x_{nom} - P^x_k, P^y_{nom} - P^y_k, P^z_{nom} - P^z_k]}{L^i_{nom}}$$
$$- [P^x_k - A^{ix}, P^y_k - A^{iy}, P^z_k - A^{iz}]\delta_i. \quad (17)$$

Since the real position P_k is within a radius r of P_{nom},

$$L^i_{nom} - r \leq L^i_k \leq L^i_{nom} + r \Rightarrow \frac{1}{L^i_{nom} + r} \leq \frac{1}{L^i_k} \leq \frac{1}{L^i_{nom} - r}$$

We note that
$$\frac{1}{L^i_{nom} - r} = \frac{1}{L^i_{nom}} + \frac{r}{L^i_{nom}(L^i_{nom} - r)}$$
$$\Rightarrow \frac{1}{L^i_k} \leq \frac{1}{L^i_{nom} - r} = \frac{1}{L^i_{nom}} + \frac{r}{L^i_{nom}(L^i_{nom} - r)}$$

and similarly that

$$\Rightarrow \frac{1}{L^i_k} \geq \frac{1}{L^i_{nom} + r} = \frac{1}{L^i_{nom}} - \frac{r}{L^i_{nom}(L^i_{nom} + r)} \quad (18)$$

With (16), we have,

$$\frac{-r}{L^i_{nom}(L^i_{nom} + r)} \leq \delta_i \leq \frac{r}{L^i_{nom}(L^i_{nom} - r)} \quad (19)$$

Using (17), we see that the Euclidean norm of each row of δC is bounded by

$$\frac{r}{L^i_{nom}} + (L^i_{nom} + r)|\delta_i| \leq \frac{r}{L^i_{nom}}\left(1 + \frac{L^i_{nom} + r}{L^i_{nom} - r}\right) \quad (20)$$

Hence,

$$|\delta C_i|^2 \leq \left(\frac{r}{L^i_{nom}}\right)^2 \left(1 + \frac{L^i_{nom} + r}{L^i_{nom} - r}\right)^2 \quad (21)$$

Since Q is within a ball of radius r of Q_{nom}, we obtain a bound for rows 2 and 3 of δB similarly:

$$|\delta B_i| \leq \frac{r}{L^i_{nom}}\left(1 + \frac{L^i_{nom} + r}{L^i_{nom} - r}\right) \quad (22)$$

Since $L^1_k = L^1_{nom}$ (stylus length is fixed), row 1 of δB is:

$$\begin{bmatrix} \frac{Q^x_{nom} - P^x_{nom} - (Q^x_k - P^x_k)}{L^1} \\ \frac{Q^y_{nom} - P^y_{nom} - (Q^y_k - P^y_k)}{L^1} \\ \frac{Q^z_{nom} - P^z_{nom} - (Q^z_k - P^z_k)}{L^1} \end{bmatrix}^T$$

The numerator in each component is the sum of differences between P and P_{nom} and Q and Q_{nom}, so we see that

$$|\delta B_i| \leq \frac{2r}{L^1}$$

Thus,

$$\sum_{i=1..3} |\delta B_i|^2 \leq \left(\frac{2r}{L^1}\right)^2 + \sum_{i=2,3}\left(\frac{r}{L^i_{nom}}\right)^2 \left(1 + \frac{L^i_{nom} + r}{L^i_{nom} - r}\right)^2 \quad (23)$$

Finally, the bound δA is the same as the bound on the first row of δB, so that

$$|\delta A_i|^2 \leq \left(\frac{2r}{L^1}\right)^2 \quad (24)$$

Expression (12) follows from (20), (22) and (23). ∎

We now find a value for the linearization error bound b in (6). Let L^{min} be the minimal rod length over the physical workspace, and let d be the maximal $\tilde{X}_k$ during one data sampling interval δt of N algorithm iterations.

Lemma 2: If $\hat{P}_k$ lies within a ball of radius d_p about P_k, and $\hat{Q}_k$ is within d_q of Q_k such that $\sqrt{d_q^2 + d_p^2} \leq d$, then $\|\hat{X}_k - X_k\| \leq d$ such that $\sqrt{d_q^2 + d_p^2} \leq d$ and

$$\|\Delta(X_k, -\tilde{X}_k)\| \leq b\|\tilde{X}_k\|$$

where

$$b = d\sqrt{\frac{3}{4(L^{min})^2} + \frac{1}{(L^1)^2}} \quad (25)$$

Remark: Note that the linearization error b is proportional to d, the largest initial error between X and the estimate $\hat{X}_k$. Normally, this error is quite small, since if the algorithm has converged to the correct value for X before motion occurs, subsequent initial errors in each data sampling interval δt will be limited by the displacement caused by the maximum voluntary motion speed that an operator can produce multiplied by the sampling interval δt.

Proof: From (3), we have

$$\Delta(X_k, -\tilde{X}_k) = h(\hat{X}_k) - h(X_k) - J_h(X_k)(\hat{X}_k - X_k)$$

So for each row:

$$\Delta^i(X_k, -\tilde{X}_k) = h^i(\hat{X}_k) - h^i(X_k) + \nabla h^i(X_k)(\tilde{X}_k) \quad (26)$$

For rods 2 and 3, we have the following expression for the gradients (rows of J_h):

$$\nabla h^i(X_k) = \left(0, 0, 0, \frac{Q^x_k - A^{ix}}{L^i_k}, \frac{Q^y_k - A^{iy}}{L^i_k}, \frac{Q^z_k - A^{iz}}{L^i_k}\right)$$

and for rods 4, 5, and 6:

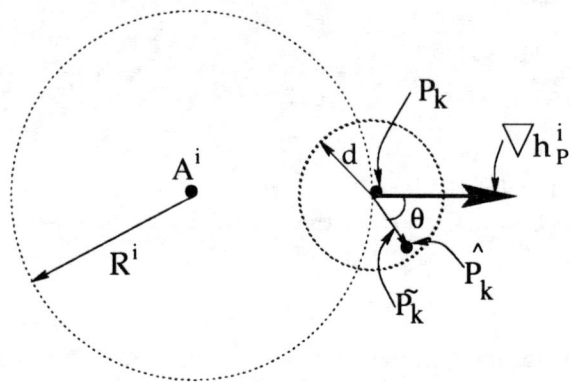

Figure 4: Geometry of the gradient and error calculation for Lemma 2.

$$\nabla h^i(X_k) = \left(\frac{P_k^x - A^{ix}}{L_k^i}, \frac{P_k^y - A^{iy}}{L_k^i}, \frac{P_k^z - A^{iz}}{L_k^i}, 0, 0, 0\right)$$

Designate the P and Q related components of ∇h^i as ∇h_P^i and ∇h_Q^i, respectively, allowing (26) to be written as

$$\Delta^i(X_k, -\tilde{X}_k) = h^i(\hat{X}_k) - h^i(X_k) + \nabla h_P^i(P_k)\tilde{P}_k + \nabla h_Q^i(Q_k)\tilde{Q}_k$$

Define the angle θ as in Figure 4 for P and the corresponding angle ϕ for Q. (Note the gradient direction is always radially outwards as in Figure 4 for rods 2-6.) We will demonstrate that for all X_k on a ball of radius R^i around any actuator A^i, $\Delta^i(X_k, -\tilde{X}_k)$ varies only as a function of θ and ϕ. Write $\Delta^i(X_k, -\tilde{X}_k)$ as

$$\Delta^i(X_k, -\tilde{X}_k)$$
$$= \underbrace{h^i(\hat{X}_k) - h^i(X_k) + \|\nabla h_P^i(P_k)\|\|\tilde{P}_k\|\cos(\theta)}_{} $$
$$+ \underbrace{\|\nabla h_Q^i(Q_k)\|\|\tilde{Q}_k\|\cos(\phi)}_{} \quad (27)$$

with $h^i(X_k) = R^i$, and $\hat{h}^i(X_k) = h^i(X_k) + \tilde{P}_k \cos(\theta)$

The underbraced quantities thus depend only on the magnitude (and not the direction) of P_k and Q_k, relative to the A_i. Since θ and ϕ range over $[0, 2\pi]$, the bound on $\Delta^i(X_k, -\tilde{X}_k)$ will be invariant for any Q_k on a ball of radius R^i around A^i (for $i = 2, 3$), and for any P_k on a ball of radius R^i around A^i (for $i = 4, 5, 6$).

We now address components of $\Delta(X_k, -\tilde{X}_k)$ for rods 2 and 3 in detail. The analysis for rods 4, 5, 6 is analogous, and will be omitted. Rod 1 (the adjoined stylus) is a special case which we will discuss separately.

For rods 2 and 3, we pick the convenient point $X_k = (P_k, Q_k) = (0, 0, 0, R_i + A^{ix}, A^{iy}, A^{iz})$, where the gradient is seen to be $(0, 0, 0, 1, 0, 0)$ since $L_k^i = R_i$. The resulting $\Delta^i(X_k, -\tilde{X}_k)$ in the neighborhood of this point is:

$$\Delta^i(X_k, -\tilde{X}_k) = h^i(\hat{X}_k) - R_i + (0, 0, 0, 1, 0, 0)\tilde{X}_k$$
$$= \sqrt{(\hat{Q}_k^x - A^{ix})^2 + (\hat{Q}_k^y - A^{iy})^2 + (\hat{Q}_k^z - A^{iz})^2} - R_i + \tilde{Q}_k^x$$

Let $\hat{Q}_k$ lie on the ball of radius d_q about this Q_k, i.e. let

$$\hat{Q}_k^x = Q_k^x + \delta x = R_i + A^{ix} + \delta x$$
$$\hat{Q}_k^y = Q_k^y + \delta y = A^{iy} + \delta y$$
$$\hat{Q}_k^z = Q_k^z + \delta z = A^{iz} + \delta z$$

where $\sqrt{\delta x^2 + \delta y^2 + \delta z^2} = d_q$ so that

$$\Delta^i(X_k, -\tilde{X}_k) = \sqrt{R_i^2 + 2R_i\delta x + d_q^2} - R_i - \delta x$$

We seek the maximum of this function for $|\delta x| \leq d_q$:

$$\frac{\partial \Delta^i}{\partial \delta x} = -1 + \frac{R_i}{\sqrt{R_i^2 + 2R_i\delta x + d_q^2}} = 0$$
$$\Rightarrow \delta x = -\frac{d_q^2}{2R_i}$$

We evaluate Δ^i at this point and obtain $\frac{d_q^2}{2R_i}$. The other rods have a similar bound, so that with $R_i \geq L^{min}$, we have the bound

$$i = 2, 3: \quad \|\Delta^i(X_k, -\tilde{X}_k)\|_2 \leq \frac{d_q^2}{2L^{min}}$$
$$i = 4, 5, 6: \quad \|\Delta^i(X_k, -\tilde{X}_k)\|_2 \leq \frac{d_p^2}{2L^{min}} \quad (28)$$

We now find the bound on $\Delta^1(X_k, -\tilde{X}_k)$ for rod 1. As for the other 5 rods, we may pick the direction of P_k and Q_k as we wish, since $\Delta^1(X_k, -\tilde{X}_k)$ only depends on X_k by the distance between P_k and Q_k, which is fixed at L^1 (stylus length). We pick the convenient point $X_k = (-L^1/2, 0, 0, L^1/2, 0, 0)^T$. The gradient here is $(-1, 0, 0, 1, 0, 0)$ and we obtain

$$\Delta^1(X_k, -\tilde{X}_k) = h^1(\hat{X}^k) - L^1 - \tilde{P}_k^x + \tilde{Q}_k^x$$

Once again, let $\hat{P}_k$ travel on a ball of radius d_p and $\hat{Q}_k$ travel on a ball of radius d_q about P_k, and Q_k, respectively, so that

$$\hat{Q}_k^x = \frac{L^1}{2} + \delta Q^x \;;\; \hat{Q}_k^y = \delta Q^y \;;\; \hat{Q}_k^z = \delta Q^z$$
$$\hat{P}_k^x = -\frac{L^1}{2} + \delta P^x \;;\; \hat{P}_k^y = \delta P^y \;;\; \hat{P}_k^z = \delta P^z$$
$$\sqrt{(\delta P^x)^2 + (\delta P^y)^2 + (\delta P^z)^2} = d_p$$
$$\sqrt{(\delta Q^x)^2 + (\delta Q^y)^2 + (\delta Q^z)^2} = d_q$$

Now $h^1(\hat{X}_k)$ becomes:

$$\sqrt{(-L^1 + \delta P^x - \delta Q^x)^2 + (\delta P^y - \delta Q^y)^2 + (\delta P^z - \delta Q^z)^2}$$

Define $\delta x = \delta Q^x - \delta P^x$, $\delta y = \delta Q^y - \delta P^y$, $\delta z = \delta Q^z - \delta P^z$, to obtain

$$\begin{aligned} h^1(\tilde{X}_k) &= \sqrt{(L^1 + \delta x)^2 + \delta y^2 + \delta z^2} \\ &= \sqrt{(L^1)^2 + 2\delta x L^1 + \delta x^2 + \delta y^2 + \delta z^2} \end{aligned} \quad (29)$$

This gives

$$\Delta^1(X_k, -\tilde{X}_k) =$$
$$\sqrt{(L^1)^2 + 2\delta x L^1 + \delta x^2 + \delta y^2 + \delta z^2} - L^1 + \delta P^x - \delta Q^x$$
$$= \sqrt{(L^1)^2 + 2\delta x L^1 + \delta x^2 + \delta y^2 + \delta z^2} - (L^1 + \delta x).$$

Note that the argument of the square root is greater than $(L^1+\delta x)^2$, making the above non-negative. Also, using $\sqrt{\delta x^2 + \delta y^2 + \delta z^2} = \|\tilde{P}_k - \tilde{Q}_k\|$ and the triangle inequality $\|\tilde{P}_k - \tilde{Q}_k\| \leq \|\tilde{P}_k\| + \|\tilde{Q}_k\| \leq d_p + d_q$, yields

$$0 \leq \Delta^1(X_k, -\tilde{X}_k)$$
$$\leq \sqrt{(L^1)^2 + 2\delta x L^1 + (d_p + d_q)^2} - (L^1 + \delta x) = \Delta^1_u$$

We seek the maximum of this quantity over $|\delta x| \leq d_p + d_q$:

$$\frac{\partial \Delta^1_u}{\partial \delta x} = -1 + \frac{L^1}{\sqrt{(L^1)^2 + 2L^1\delta x + (d_p+d_q)^2}} = 0$$
$$\Rightarrow \delta x = -\frac{(d_p + d_q)^2}{2L^1}$$

and we conclude that

$$\Delta^1(X_k, -\tilde{X}_k) \leq \frac{(d_p + d_q)^2}{2L^1}. \quad (30)$$

We are now in a position to establish a bound on $\Delta(X_k, -\tilde{X}_k)$ as a whole, where (30) holds for the first component of $\Delta(X_k, -\tilde{X}_k)$ and (28) holds for the remaining five components:

$$\|\Delta(X_k, -\tilde{X}_k)\|$$
$$\leq \sqrt{3\left(\frac{d_p^2}{2L^{min}}\right)^2 + 2\left(\frac{d_q^2}{2L^{min}}\right)^2 + \left(\frac{(d_p+d_q)^2}{2L^1}\right)^2}$$
$$\leq \sqrt{3\left(\frac{d_p^2+d_q^2}{2L^{min}}\right)^2 + \left(\frac{2d_p^2 + 2d_q^2}{2L^1}\right)^2}$$
$$= \sqrt{(d_p^2+d_q^2)\left(3\frac{d_p^2+d_q^2}{(2L^{min})^2} + \frac{d_p^2+d_q^2}{(L^1)^2}\right)}$$

Noting that $\|\tilde{X}_k\| = \sqrt{d_p^2 + d_q^2} = d$ in this construction, we obtain

$$\|\Delta(X_k, -\tilde{X}_k)\| \leq d\sqrt{\frac{3}{4(L^{min})^2} + \frac{1}{(L^1)^2}} \|\tilde{X}_k\| \quad (31)$$

which establishes (23). ∎

3 COMPARING BOUNDS TO NUMERICAL RESULTS

Our haptic interface has the following physical parameters for rod lengths, where the nominal position was chosen near the center of the workspace.

$$\bar{\sigma}(\bar{J}_h^{-1}) = 5.42,$$
$$L^{min} = 195 \text{mm},$$
$$L^{max} = 534 \text{mm},$$
$$L_{nom} = [202, 363.8, 320.7, 352.2, 323.4, 361.1]^T \text{mm},$$

The constant d is found as follows:

$$d = \sqrt{d_p^2 + d_q^2} = \sqrt{2} V_{max} \delta t \quad (32)$$

where V_{max} is the greatest speed at which a rod may move under voluntary control of an operator, and δt is the data sampling interval. In our case, using $V_{max} = 2$ m/sec and $\delta_t = 0.25$ msec, we obtain $d = 0.71$mm. From Lemma 2 we get

$$b = 0.0047$$

From Lemma 1 and the Theorem, we obtain the relationship between convergence rate parameter α and distance r away from the nominal location as shown in Figure 5. If $\alpha \leq 0.67$ is the desired incremental error contraction factor, then we need $r \leq 6$ mm by the results of the Theorem. That is, this rate of error convergence is guaranteed, as long as the current physical locations of the stylus end points P and Q in the vector X are within $r = 6$ mm ($a = 0.1158$ from Lemma 1) of the nominal location X_{nom} used to obtain the nominal Jacobian $\bar{J}_h^{-1}$ used in (2). Said another way, an upper bound on the convergence rate throughout the workspace can be guaranteed, provided that the nominal Jacobian $\bar{J}_h^{-1}$ used in (2) is changed with stylus motion to keep the corresponding nominal position X_{nom} always within 6 mm of the current physical position. Based on the bound above for motion speed,

$$\frac{r}{\delta_t V_{max}} = 12 \text{ sample periods/update}$$

indicating that we can have 12 sample periods between successive updates of the Jacobian.

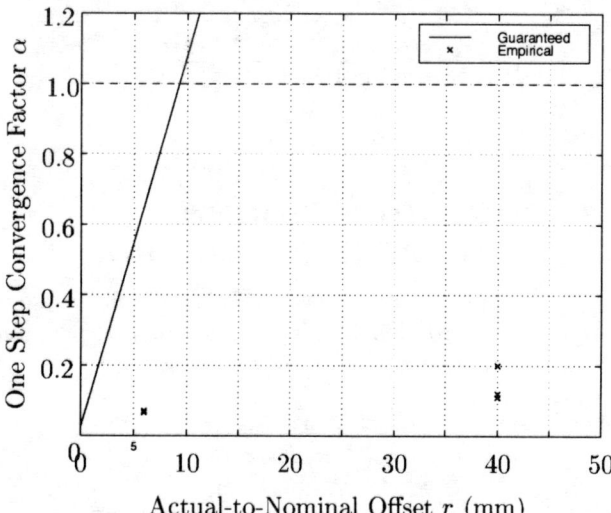

Figure 5: Guaranteed convergence rates α (Line) and empirical convergence rates $\hat{\alpha}$ ($\times$) for the 5 DOF haptic interface.

It has been our experience that the actual rate of convergence of the algorithm is in fact far better than the bounds above. When $r = 6$ mm and $d = 0.71$ mm, moving X_k in a variety of directions around X_{nom}, and moving $\hat{X}_1$ in a variety of directions around X_k, we computed the empirical convergence rates as

$$\hat{\alpha} = \frac{\|\tilde{X}_{k+1}\|}{\|\tilde{X}_k\|}$$

This resulted in values of $\hat{\alpha}$ of around 0.07, which are plotted in Figure 5 as individual points ($\times$). These empirical convergence rates were therefore an order of magnitude faster than the guaranteed rate provided by the theory. We also found that good convergence ($\hat{\alpha} \leq 0.2$) still occurred when r was increased to 40 mm.

4 CONCLUSION

Haptic interfaces have unique requirements that make iterated numerical solutions for kinematic transformations quite suitable, provided that solution speed and accuracy can be quantified. Explicit bounds on algorithm convergence rate were provided in terms of physical parameters of the device and initial estimation errors. These bounds can be used as guidelines for designing appropriate computational schemes. For example, it is clear from the case presented that inversion of the 6×6 Jacobian in the recursion for the forward kinematics is not necessary at each data sample; a single value (perhaps precomputed) can be used over a relatively large region of the workspace. The size of this region is explicitly linked to convergence rate in our results, providing a tool for making engineering trade-offs in algorithm design, sample rate selection, hardware architecture, etc.

Analytic bounds are relatively easy to derive in the case the parallel mechanism studied here, due to the linear form of the Jacobian of the inverse kinematics. More general parallel kinematic configurations may allow similar bounding approaches. This is currently being investigated by the authors.

REFERENCES

[1] M. D. Bryfogle, C. C. Nguyen, S. S. Antrazi, and P. C. Chiou, "Kinematics and Control of a Fully Parallel Force-Reflecting Hand Controller for Manipulator Teleoperation", *J. Robotic Systems*, Vol. 10, No. 5, pp. 745–766, 1993.

[2] V. Hayward, C. Nemri, X. Chen, and B. Duplat, "Kinematic Decoupling in Mechanisms and Application to a Passive Hand Controller Design", *J. Robotic Systems*, Vol. 10, No. 5, pp. 767-790, 1993.

[3] O. Didrit, M. Petitot, and E. Walter, "Guaranteed Solution of Direct Kinematic Problems for General Configurations of Parallel Manipulators", *IEEE Trans. Robotics and Automation*, Vol. 14, No. 2, pp. 259–266, 1998.

[4] M. L. Husty, "An Algorithm for Solving the Direct Kinematics of General Stewart-Gough Platforms", *Mech. Mach. Theory*, Vol. 31, no. 4, pp. 365–379, 1996.

[5] L-C. T. Wang and C. C. Chen, "On the Numerical Kinematic Analysis of General Parallel Robotic Manipulators", *IEEE Trans. Robotics and Automation*, Vol. 9, No. 3, pp. 272–285, 1993.

[6] P. A. Millman and J. E. Colgate, "Effects of Non-Uniform Environment Damping on Haptic Perception and Performance of Aimed Movements", *Int. Mech. Engr. Congress and Exp*, San Francisco, CA, 1995.

[7] D. A. Lawrence, L. Y. Pao, M. A. Salada, and A. M. Dougherty, "Quantitative Experimental Analysis of Transparency and Stability in Haptic Interfaces", *5th Annual Symp. Haptic Int., Int. Mech. Engr. Congress and Exp.*, Atlanta, GA, 1996.

A Muscular-Like Compliance Control for Active Vehicle Suspension

Shih-Lang Chang[*], Chi-haur Wu[+] and D. T. Lee[**]
Department of Electrical and Computer Engineering
Northwestern University
Evanston, IL 60208
[+] chwu@ece.nwu.edu;(847)491-7076

Abstract

Inspired by the natural suspension capabilities in the biological limb system, a design of active suspension system is developed for damping the undesired forces and displacements caused to vehicles by the irregularity on the road surface. The proposed controller for suspension is based on a muscular-like model fitted from the responses of different voluntary and involuntary limb movements. Because the muscular-like model emulates the property of biological damping behavior, the proposed active suspension controller has a very unique feature that will enable the controlled vehicle to adapt to varying loads and sudden impacting forces. This is in contrast to the conventional linear controller that has sensitivity and stability problems when the dynamic system is subjected to varying loads. To realize our controller, a small-scale, quarter-car model was built for our experiments on active suspension.

1. Introduction

The primate limb is an amazing elastic system. By compression and extension of muscle tissues, the limb can store and dissipate potential energy and kinetic energy. And the limb is not just like an assembly of springs and dampers. It has been recognized that a normal leg provides five functions in walking (Shurr'90). The first one is shock absorption that reduces the reaction force generated when the heel hits the ground. The second smoothes the motion of body's center of mass, which is usually located at one thigh length above the knee joint. To reduce the vertical displacement of center of mass and generate a smooth motion, the ankle coordinates with the foot to change the motion of the limb during walking. The third is surface adaptation that helps people walk and stand on a rough or slanting surface. The fourth one improves knee stability. As the foot strikes the ground while walking, there is a corresponding reaction force from the ground against the foot. Whereas no joint moment is required when the reaction-force line passes through the knee joint, such joint moment is opposed by the posterior structure of the knee joint. In this case, the knee is considered to be stable. However, if the force line passes the knee joint posteriorly, the extensor muscle must make an effort to provide an opposing torque to bear the body's weight. In such a case, the knee is considered to be in an unstable situation. By adjusting the angle of the ankle joint, a normal leg keeps the reaction-force line passing or close to the knee joint to minimize the required muscular force. From the analysis of human walking, the ankle is about 20 degrees of plantar flexion at the initial swing phase. To avoid dragging the foot, the last function returns the ankle to the neutral position after the toes leave the ground. In performing these five functions, the limb works not only like an assembly of springs and dampers but also like a self-tuning system.

Paradoxically, the attributes of an ideal suspension reflect the same spectrum of critical functions performed by the primate limb. An ideal suspension should be able to absorb and dissipate the energy generated by road irregularities, smoothen the motion curve of the vehicle, adapt to surface irregularities, even out the anti-stability load force, and regulate suspension movements (Hedrick'81). Supporting these functions is beyond the ability of a conventional suspension system composed of springs and dampers. Like the primate limb whose muscles can store and dissipate energy as well as generate forces to regulate its motion, an ideal suspension system should also include a controllable actuator to improve the comfort of a vehicle ride. Most research effort has been devoted to controlling such an actuator. In short, a good controller used in suspension system must be compliant enough to adapt to road irregularity, yet stiff enough to regulate the vibration of the vehicle. The shock-absorbing property of primate limbs in walking has been widely studied in biomechanics. However, it has not been applied to engineering. Therefore, our research effort has been devoted to designing a controllable active suspension system by emulating the force absorbing capability of the primate limb.

2. Passive and Active Suspension Systems

The components of a conventional suspension system include springs and fluid dampers (shock absorbers). From early 1980s, it was widely recognized that adding an active component to the suspension system would

---

[*] He was a graduate student of Mechanical Engineering at Northwestern University.
[**] Current address: Institute of Information Science, Academia Sinica, Nankang, Taipei, Taiwan. (dtlee@iis.sinica.edu.tw)

dramatically improve the performance of the system. The active component is defined as a power-driven device that can inject, store, transfer or extract energy. Besides active and passive suspensions, semi-active suspension and adaptive suspension are also under extensive studies (Karnopp et al.'74). A semi-active suspension differs from an active suspension in that the body weight is carried by a passive structure and the active component is only conditionally turned on. Many research findings have indicated that the semi-active suspension fulfills to some degree the level of performance achieved by the active suspension at much lower power consumption. The adaptive suspension also uses a passive structure to support the body weight, but the rate of damper is selected from several preset levels upon the detected chassis motion. Usually, the adjustment of stiffness is achieved by controlling the valve in the hydraulic/pneumatic damper.

Darling et. al (Darling'92) outlined the requirements that a suspension system needs to meet as follows: *a)Isolate the passenger from road irregularities (ride); b) Maintain contact between the tire and the road; c) Provide safe handling during maneuvers; d) React changes in load; e) Contain the suspension displacements within the limits of travel; and f) Provide control over the pitch and roll motion of the vehicle body.* In sum, the purpose of the suspension is to minimize the undesired motions during driving. Besides bounces caused by road irregularities, rolls during cornering and pitches during accelerating, braking also reduces driving comfort and safety.

In a vehicle suspension system, the supported mass representing vehicle body is usually called sprung mass and the wheel is called unsprung mass. In a passive suspension system its components can only temporarily store or dissipate energy. Springs need to be soft enough for ride comfort but also hard enough for safety. Therefore, in the past decades, using active suspension concepts to improve ride quality has been under intensive research in both industrial and academic works. The "skyhook" damper might be the simplest and the most well known control method (Pollard&Simons'84; Appleyard&Wellstead'95). In this method, the actuation force is proportional to the absolute velocity of the supported body. A skyhook damper can reduce the low-frequency response without enhancing the high-frequency response. However, it tends to damage the overall response at curvature. An on-off semi-active suspension requiring less power than a full-active suspension was also proposed (Lane'90); however, the force discontinuity tends to create jerks during the switch transition. To investigate the effect of feedback control on suspension performance Hedrick ('81,'84) investigated a control law consisting of the feedback signals of absolute velocity and acceleration of vehicle body. Sharp and Prokop presented another control concept that uses the road preview information to improve the ride quality and decides control input by utilizing LQG (Linear Quadratic Gaussian) and LQR (Linear Quadratic Regulator) optimal control theories (Prokop'95; Sharp'94). Hrovat ('93) used LQ (Linear Quadratic) optimal control to minimize the cost function that combine excessive suspension stroke, sprung-mass acceleration, and sprung-mass jerk in all one-DOF model, two-DOF model, half-car model, and full-car model. To overcome the practical insufficiency of the generally used quarter-car model, Venhovens et al. ('94) focused on the interaction relation of the four suspension systems in a full car model. An optimal control method based on LQG theory was derived in this work. Ulsoy et al. ('94) compared the performance difference between the suspensions designed by using LQ and by LQG control theories and concluded that a good LQ controller should use accompanied passive components that are low in stiffness and damping. Compared to the corresponding LQ controller, the LQG controller pales in both performance and robustness. Salman et al. ('90) proposed two reduced-order control strategies based on the oscillatory factor that there were two distinct oscillatory modes in the dynamics of suspension system corresponding respectively to the natural frequencies of sprung mass and unsprung mass. The main advantage of this approach is its simplified structure. To account for the essentially nonlinear nature of a suspension system, another type of control method based on fuzzy logic, which is inherently nonlinear, was also utilized (Cherry&Jones'95).

As surveyed, there are many approaches trying to improve the performance of suspension systems. But none of the developed approaches has taken advantage of biological damping capabilities as described earlier. In next section, we will describe some biological properties extracted from the muscle-reflex system that will be used to design an excellent active suspension.

3. Muscular-Like Control

The muscle-reflex mechanisms (McMahon'84) form a feedback system called the motor servo, consisting of a muscle, its spindle receptors, and the corresponding reflex pathways back to the muscle (Houk et al.'81; Gielen&Houk'87). Findings from physiological experiments (Gielen et al.'84; Miller'84) further indicated that the motor servo of neuromuscular system has two salient nonlinear dynamic features, a stiffness enhancement at small signals and a fractional velocity-dependent viscosity, for its unique adaptability. The stiffness of the neuromuscular system, which is not constant over the stretching range, has more effect on the initial elastic force response, a property referred to as the *short-range enhancement* (Miller'84). The velocity-dependent viscous force response, however, shows a nonlinear viscosity property that has been postulated to damp limb movements when variable loads are involved.

According to an extensive body of experimental evidence (Miller'84), this nonlinear effect extracted from force-velocity data shows that the force response is proportional to a low fractional power, 0.17, of muscle's velocity. Based on this fractional power in velocity-dependent viscosity, our initial analysis started with a simple second-order system modeled with an estimated nonlinear damping effect of 1/5 power and a linear elastic stiffness (Wu et al.'90). In our study, a very interesting property was observed that this unusual kind of damping property serves as a control function that can enhance the limb motion when high velocity is needed and, conversely, it also helps to stop limb motion promptly when velocity decreases to the low-velocity range. The reason is that this system has low viscosity at the high-velocity range and high viscosity at the low-velocity range. In addition, we also observed that this nonlinear system can maintain stability for a wide range of inertial load for the same damping constant and stiffness. Without adjusting any parameter, the system is stable over a load with range of 1000 fold (Wu et al.'90). By comparison, a linear system would induce the stability problem that is typical in a system with linear damping. This capability of adapting to various or uncertain loads possessed by the nonlinear damping system can greatly reduce the control sensitivity of a dynamic system and promises a design of active damping control for a dynamic vehicle system.

3.1 A Model of Muscular-Like Compliant Controller

Our initial analysis demonstrated that a nonlinear damping system could emulate the adaptability possessed by the muscle-reflex system. After further fitting of experimental data recorded from voluntary and involuntary limb movements, a muscular-like model was developed for controlling various movements of robot manipulators (Wu et al.'90,,'92; Wu &Chang '95; Wu&Chang&Lee'96). For the past few years, our research in modeling the properties of biological muscle-reflex mechanisms has induced a nonlinear model of muscular-like control for a dynamic system. Specifically, this muscular-like control was originally modeled from two muscles producing opposing torques on the hand rotating about the wrist with a single degree of freedom (DOF) (Wu et al.'92). To emulate the operations of both extensor and flexor for voluntary movements (driven by motor commands from Central Nervous System) and involuntary movements (reflex motion by external forces), the developed muscular-like model consists of a pair of agonist and antagonist muscles for flexion and extension.

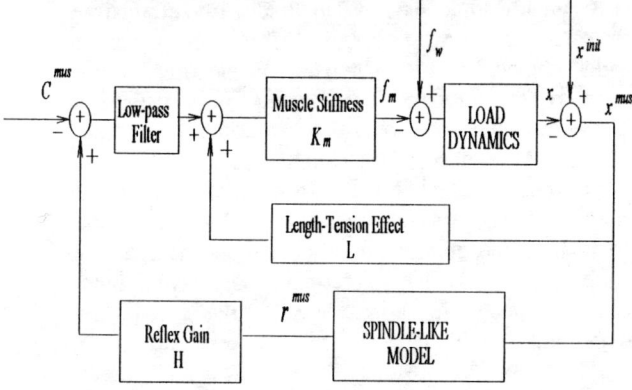

Figure 3.1. Muscular-Like Compliant Control

Each muscle is emulated by a muscular-like model that consists of two major parts: spindle-like mechanism emulating the reflex property of biological muscular system for absorbing impacting forces, and muscle-stiffness mechanism emulating muscle stiffness for tracking various movements. The control blocks of this muscular-like model are depicted in Fig. 3.1. The developed model has two possible inputs, motor command C^{mus} and external disturbing force f_w. The reflex signal of the spindle-like model, r^{mus} (will be explained later), scaled through a reflex gain coefficient, H, is combined with the motor command to produce a reflex-induced command for muscle force. To simulate the smoothing of commands by motor neuron pools such that a bell-shaped velocity profile for limb movements can be induced, a low-pass filter is also included in the model. Based on the experimental data (Gielen & Houk'87), a time constant of 30 ms is extracted for this filter effect. The linear feedback gain coefficient L in Fig. 3.1, represents the effect of muscle length-tension that the muscle length varies with load position (Houk & Rymer'81). This effect shows that any change in muscle length will produce a muscle force f_m through muscle stiffness K_m. In sum, both spindle and muscle stiffness mechanisms will induce muscle forces. As in Fig. 3.1, the control law for generating muscle force f_m is

$$f_m = K_m \cdot \left(L \cdot x^{mus} + LPF(H \cdot r^{mus} - C^{mus}) \right) \quad (3.1)$$

where $LPF(\cdot)$ represents a low-pass filter. The resultant muscle force combined with the disturbing force f_w at the load will then move the load.

Since spindle receptors of the muscle-reflex system will sense both muscle length and the rate of change of muscle length, and produce a firing rate reflecting both measurements, the simulated spindle-like model in Fig. 3.1 must capture the physiological properties of the nonlinear, fractional, damping effect multiplied by the short-range elasticity enhancement. After fitting experimental data in our previous research (Wu et al.'90), a nonlinear spindle-like model with the described properties was constructed as follows:

$$B_p \dot{x}_p^{1/5} \left(|x_p| - x_{p0} \right) = K_r \left(x^{mus} - x_p \right) \quad (3.2)$$

where x_p, $\dot{x}_p$, and x_{p0} are the internal position, velocity and bias position of the simulated spindle; B_p is a scaled damping coefficient; K_r represents the reflex stiffness; and the property of the short-range elasticity enhancement is represented by $\left(|x_p| - x_{p0} \right)$. x^{mus} represents the equilibrium position of the emulated muscle and can be computed as:

$$x^{mus} = x^{init} - x \quad (3.3)$$

where x represents the position variable of the controlled dynamic system and x^{init} is its initial position. With the spindle-like model represented by Eq.(3.2), its internal position, x_p, will be excited by the load position and produce a reflex signal, r^{mus}, as follows:

$$r^{mus} = K_r\left(x^{mus} - x_p\right) \quad (3.4)$$

Therefore, any change of load position will induce a reflex signal r^{mus} by this spindle-like model. And this reflex signal scaled by a reflex gain H will then produce a muscle force to respond to the change. To further understand the structure of this muscular-like control, a mechanical analogy of a single mass system controlled by a simplified muscle-reflex model with no low-pass filter is illustrated in Fig. 3.2. The effect of the motor command can be simulated by exciting an internal spindle position, x_p, to move the load. In this case, the muscular-like model becomes a controllable oscillator to excite the load.

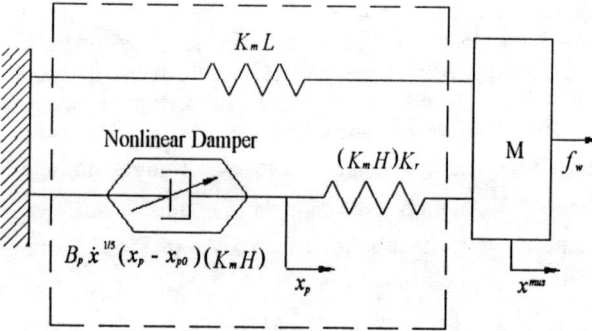

Figure 3.2. A Simplified Mechanical Analogy of Muscular-like Model.

In general, the spindle mechanism is capable of making automatic corrections in muscle toning to respond to external disturbance forces. It has the capability in absorbing impacting forces that cause displacement x^{mus} by generating reflex muscle forces to the motion. It serves as a reflex sensor on muscle force. When the muscle force rises past a threshold level, the spindle receptors will inhibit the motor unit, and turn the muscle activity off to prevent muscle from over-lengthening. Meanwhile, it also regulates the muscle stiffness for various movements.

Using the developed muscular-like compliant controller to control a movement, the motor command C^{mus} will define the equilibrium position. If the movement is imposed by an external force, the transient equilibrium position will be shifted by the reflex signal generated from the spindle-like model to adapt to the constraint and comply with the disturbance force. However, when the disturbance force dissipates, the equilibrium position will return to the desired target defined by the command. In other words, this controller will react to changes of movement sensed from the controlled structure and actively damp out the disturbance force. Based on this capability of adapting to disturbance forces, an active control of suspension can damp out vibrations occurred during the ride in the same manner as muscles.

Our preliminary results show that the stiffness of the controlled system can be adjusted by the muscle stiffness K_m and the length-tension feedback gain L. The strength of reflex signals can be adjusted by the reflex gain H. For a fixed spindle-like model represented by Eqs.(3.2-3.4), the product of the reflex gain and the muscle-stiffness gain provides certain enhancement for muscle force. This effect can be identified from the mechanical analogy diagram shown in Fig. 3.2. With the same amount of force enhancement by $K_m \cdot H$, the higher the reflex gain is, the lower the muscle stiffness. It will cause the system to be more compliant. On the other hand, the higher the muscle stiffness, the lower the reflex gain. It will cause the system to be stiffer. The developed nonlinear controller serves as a controllable damper. It can suppress vibration levels occurred in the structure and absorb various forces or torques occurred between components of mechanical systems and provide compliance needed for mechanical components to reduce fatigue and structural damage upon impact. In other words, this damper controller will control the active suspension system and provide a vehicle with a better ride quality.

The damping control is a key feature in designing a controller for suppressing vibrations or absorbing forces occurred in an active suspension system. The findings from the biological limb system provide a unique muscular-like model for controlling an active suspension system. The structure of this model has an analogy to a biological limb that has flexible muscles. Any change in muscle length will induce muscle forces to move the limb. In other words, this model will provide a unique damping property for the controlled dynamic system to adapt to various loads and suppress various undesired forces and movements. Therefore, it seems natural to design a muscular-like control for an active suspension system.

4. Experimental Results

In our preliminary studies, control of a quarter-car was first analyzed using MATLAB simulations (Chang &Wu'97). Our theoretical analysis and simulation results showed that an active-suspension under the proposed muscular-like damping control has better performance in comparison with some existing methods. Encouraged by our simulated results, we then built a small scale, quarter-car model, depicted in Fig. 4.1, using an electrical motor. The car-body (sprung mass) is connected to the suspension frame that is composed of a motor mount, three round shafts, and a bottom support. The handle simulating the wheel is attached to a robot arm. To simulate the input of road roughness to the suspension model, a PUMA 560 robot is used to hold the quarter-car model as shown Fig. 4.2. While the robot vibrates the quarter-car model as though the quarter-car model were running on a rugged road, the controlled motor will drive the ball screw to adjust the suspension deflection in reduction of the sprung mass vibration. A linear transducer is used to feed back the suspension deflection for our control. By sensing the displacement caused by the disturbance force, the muscular-like controller reflects a compliance force to isolate the car-body (the sprung mass) from the vibration and maintains the ride comfort.

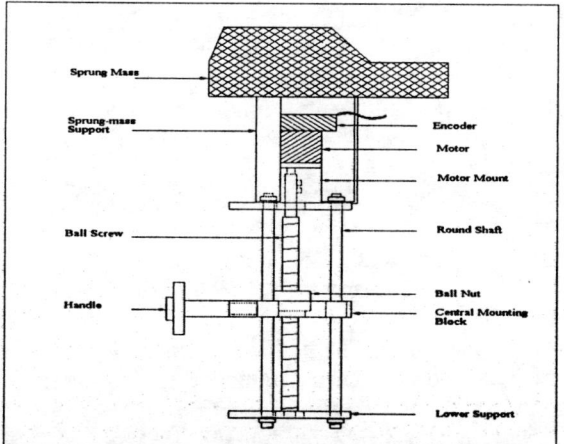

Figure 4.1. Sketch of the designed quarter-car model.

Fig. 4.2 Experimental Set up

Using the muscular-like controller we propose two modes of operation, damper mode and control mode, for actively controlling suspension. The damper mode and the control mode have different goals. In the damper mode, the muscular-like model works as a nonlinear damper to dissipate the disturbance force. The body weight of the vehicle will be mainly supported by passive components. In the control mode, the muscular-like model tries to combine the disturbance force to support the vehicle body weight and the suspension is fully active.

Based on the above two control modes, experiments were performed in both time domain and frequency domain. In the first experiment, two disturbance waves were applied to the model. The first wave has decreasing amplitude but increasing frequency. Following the first wave, the second has decreasing amplitude and decreasing frequency. Figure 4.3(a) shows the response of the control mode active suspension. The dashed lines in the figure show the ground surface simulated by the PUMA robot and the solid lines show the position of the simulated vehicle body. Figures 4.3(b) shows the response of a damper mode active suspension to the same disturbance input. Figures 4.3(a) and (b) show that both the control mode and the damper mode can reduce the effect of the disturbance by different levels.

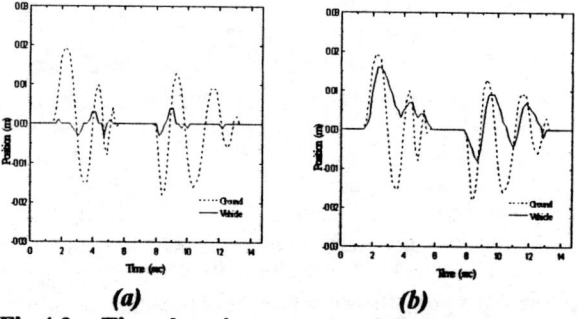

Fig 4.3 Time domain responses of vehicle body position. (a) Control mode. (b) Damper mode.

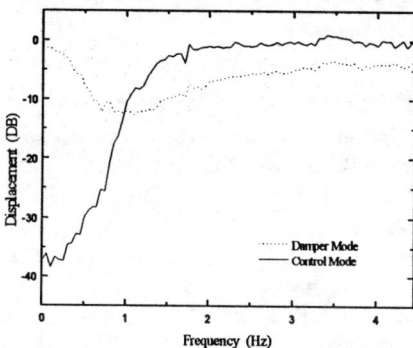

Fig. 4.4 Frequency responses of vehicle body.

Figure 4.4 shows the frequency responses of the quarter-car controlled by both the control mode and the damper mode. The frequency response from the control mode (solid-lines) demonstrates that the magnitude of car-body position response was remarkably reduced at the lower frequency. It means that the control mode can provide ride comfort for low-frequency road roughness. The bandwidth shown is around 1.5 Hz, which corresponds to the selected gains used in our controller. The overall bandwidth is also limited by the structural frequency of the quarter-car model. As for the damper mode (the dotted-lines), it shows that it can reduce the magnitude of disturbance at a higher frequency. In other words, if we combine the control mode and damper mode to actively control a suspension system, we can provide ride comfort for a wide range of road roughness.

5. Discussion and Conclusion

In this paper a muscular-like control is developed for an active suspension system. A physical quarter-car model was also designed and fabricated to verify the design concepts. In our experiments, this quarter-car model was validated in both time domain and frequency domain using two control algorithms, control mode and damper mode. The experimental results showed that both the control mode and the damper mode active suspensions could reduce the effect of road roughness. The control mode has superior performance at low frequency while the damper mode performs better at high frequency. At low frequency, the control mode can efficiently estimate and adapt to the disturbance whereas the damper mode follows the input. At high frequency, the damper mode working as a nonlinear-damper will absorb and reduce the vibration levels but the control mode can't efficiently adjust to the fast input changes.

A conclusion drawn from our experiments is that the control mode is superior for reducing low-frequency vibration, and the damper mode is better for high-frequency vibration. This conclusion suggests that we should combine the control mode and damper mode so that a better active suspension system can be designed. In addition, in reality there are two to four suspension systems in a vehicle. To optimize the performance in comfort and safety, the suspension systems may need to coordinate with each other. Therefore to study the performance of the proposed suspension on a real vehicle, models of half-car and full-car will be studied in the future.

REFERENCES

Appleyard, M. and P. E. Wellstead, "Active Suspension: Some Background," *IEE Proceedings-Control Theory and Applications*, vol. 142, pp. 123-128, 1995.

Chang, S.L. and C.H. Wu, "Design of an Active Suspension System Based on a Biological Model," *Proceedings of 1997 American Control Conference*, Albuquerque, NM, June 1997.

Cherry, A. S., and R. P. Jones, "Fuzzy logic control of an automotive suspension system," *IEE Proceedings-Control Theory and Applications*, vol. 142, pp. 149-160, 1995.

Darling, J., R. E. Dorey, and R.-M. T.J., "A Low Cost Active Anti-Roll Suspension for Passenger Cars," *Journal of Dynamic Systems, Measurement, and Control*, vol. 114, pp. 599-605, 1992.

Gielen, C.C.A.M. and Houk, J.C., ``Nonlinear Viscosity of Human Wrist,'' *J. Neurophys.*, Vol.52, No.3, September 1984.

Gielen, C.C.A.M. and Houk, J.C., ``A Model of the Motor Servo: Incorporating Nonlinear Spindle Receptor and Muscle Mechanical Properties,'' *Biol. Cybern.*, 1987, 57:217:231.

Hedrick, J. K., D. N. Wormley, F. Buzan, G. Celniker, and M. Partridge, *Application of Active and Passive Suspension Techniques to Improve High Speed Ground Vehicle Performance-Phase I*. Washington D.C.: U.S. Dept. of Transportation, OST, 1981.

Hedrick, J. K., D. Cho, R. Barletta, and L. C. Chen, *The Use of Pneumatic Active Suspensions to Improve Lateral Rail Vehicle Ride Quality*. Wahington D.C.: U.S. Dept. of Transportation, OST, 1984.

Houk, J.C. and Rymer, W.Z., ``Neural Control of Muscle Length and Tension,'' *Handbook of Physiology --- The Nervous System II*, Bethesda, MD, American Physiol. Soc., 1981, Sect. 1, Vol. II, Chap. 8, pp. 257-323.

Hrovat, D., "Applications of Optimal Control to Advanced Automotive Suspension Design," *Journal of Dynamic Sys., Measu., and Contr.*, vol. 115, pp. 328-342, 1993.

Karnopp, D., M. J. Crosby, and R. A. Harwood, "Vibration Control Using Semi-Active Force Generators," *Journal of Engineering for Industry*, vol. 96, pp. 619-629, 1974.

Lane, R. and N. M. Charles, "Methods for Eliminateing Jerk and Noise in Semi-Active Suspensions," in *Total Vehicle Ride, Handling and Advanced Suspensions*. Warrendale, PA: Society of Automotive Engineers, 1990.

McMahon, A. Thomas, *Muscles, Reflexes, and Locomotion*, University Press, 1984.

Miller, L.E., ``Reflex Stiffness of the Human Wrist," M.S. Thesis, Dept. of Physiology, Northwestern Univ., 1984.

Pollard, M. G., and N. J. A. Simons, "Passenger Comfort - the Role of Active Suspensions," *Railway Engineer*, vol. 1, pp. 17-31, 1984.

Prokop, P. and R. S. Sharp, "Performance enhancement of limited-bandwith active automotive suspensions by road preview," *IEE Proceedings-Control Theory and Applications*, vol. 142, pp. 140-148, 1995.

Salman, M. A., A. Y. Lee, and N. M. Boustany, "Reduced Order Design of Active Suspension Control," *Journal of Dynamic Systems, Measurement, and Control*, vol. 112, pp. 604-610, 1990.

Sharp, R. S. and C. Pilbeam, "On the Ride Comfort Benefits available from Road Preview with Slow-active Car Suspensions," in *Dynamics of Vehicles on Roads and on Tracks*, Z. Y. Shen, Ed. Lisse: Swets & Zeitlinger, 1994, pp. 437-448.

Shurr, D. G. and T. M. Cook, "Prosthetics and Orthotics," Appleton and Lane, 1990.

Ulsoy, A. G., D. Hrovat, and T. Tseng, "Stability Robustness of LQ and LQG Active Suspensions," *Journal of Dynamic Systems, Measurement, and Control*, vol. 116, pp. 123-131, 1994.

Venhoven, P. J., A. C. M. Knaap, A. R. Savkoor, and A. J. J. Weiden, "Semi-Active Control of Vibration and Attitude of Vehicles," in *Dynamics of Vehicles on Roads and on Tracks*, Z. Y. Shen, Ed. Lisse: Swets & Zeitlinger, 1994, pp. 522-540.

Wu, C.H., J.C. Houk, K.Y. Young, and L.E. Miller, "Nonlinear Damping of Limb Motion," Chap. 13, *Multiple Muscle Systems: Biomechanics and Movement Organization* edited by J. M. Winters and S. L-Y. Woo, Springer-Verlag New York Publishers, 1990. 214-235.

Wu, C. H., K.Y. Young, K.S. Hwang, and S. Lehman, "Analysis of Voluntary Movements for Robotic Control," *IEEE Control Systems Magazine*, 2(1), Feb. 1992, 8-14.

Wu, C. H. and S. L. Chang, "Implementation of a Neuromuscular-like Controlfor Compliance on A PUMA 560 Robot", *Proceedings of the 34th IEEE 1995 Int.l Conf. on Decision and Control*, Dec. 1995, 1597-1602.

Wu, C. H., Chang, S. L., and D. T. Lee, "A Study of Neuromuscular-like Control for Rehabilitation Robots", *Proceedings of the 1996 IEEE International Conference on Robotics and Automation*, Minneapolis, April 1996.

Wu, C.H., K.S. Hwang, and S.L. Chang, "Analysis and Implementation of a Neuromuscular-like Control for Robotic Compliance," *IEEE Trans. on Control Systems Technology*, .5(6), Nov. 1997, pp. 586-597.

A General Contact Model for Dynamically-Decoupled Force/Motion Control

Roy Featherstone [*]
Department of Engineering Science
Oxford University
Parks Road, Oxford OX1 3PJ, England
roy@robots.oxford.ac.uk

Stef Sonck Thiebaut
Aerospace Robotics Laboratory
Department of Aeronautics and Astronautics
Stanford University, CA 94305, USA
ssonck@sun-valley.Stanford.EDU

Oussama Khatib
Robotics Laboratory
Department of Computer Science
Stanford University, CA 94305, USA
khatib@cs.Stanford.EDU

Abstract

This paper presents a general first-order kinematic model of frictionless rigid-body contact for use in hybrid force/motion control. It is formulated in an invariant manner by treating motion and force vectors as members of two separate but dual vector spaces. These more general kinematics allow us to model tasks that cannot be described using the Raibert-Craig model; a single Cartesian frame in which directions are either force- or motion-controlled is not sufficient. The model can be integrated with the object and manipulator dynamics in order to model both the kinematics and dynamics of contact. These equations of motion can be used to design force and motion controllers in the appropriate subspaces. To guarantee decoupling between the controllers, it is possible to apply projection matrices to the controller outputs that depend solely on the kinematic model of contact, not a dynamic one. Experimental results show a manipulation that involves controlling the force in two separate face-vertex contacts while performing motion. These multi-contact compliant motions often occur as part of an assembly and cannot be described using the Raibert-Craig model.

1 Introduction

The modern concept of hybrid control, as described in the work of Raibert and Craig [12], has attracted a

[*]Supported by EPSRC Advanced Research Fellowship number B92/AF/1466.

great deal of interest over the years. In addition to the various improvements, extensions and practical implementations that have been proposed, two theoretical errors in the original formulation have received attention: the non-invariant formulation of the original contact model [3, 9] and the phenomenon of kinematic instability caused by an incorrect filtering of error signals into force and motion components [1, 6]. Another problem with the Raibert-Craig model, which appears to have been neglected, is that it lacks sufficient generality to describe an arbitrary state of contact between two rigid bodies. Other published models vary on this point: some suffer the same problem (e.g. [4]) while others are completely general (e.g. [2, 7, 13, 14]).

This paper is organized as follows. First, we show that the Raibert-Craig model is not general and give two examples of contacts that it cannot handle. Then, after a brief review of dual vector systems, we describe the new model. This model can describe any (nonsingular) state of frictionless contact between two rigid bodies. Force and motion filtering is accomplished by projection matrices that are invariant with respect to choice of units and coordinate systems; and it can be shown that these matrices are kinematically stable, although this subject is not discussed in this paper.

The next section combines the kinematic model with the dynamics of the manipulator and environment. It presents an analysis of the equations of motion of a manipulator in contact using the projection matrices derived earlier. It is possible to guarantee that a force/motion controller exhibits dynamically decoupled behavior (i.e., the actual contact force and acceleration depend only on the outputs of the respective force and motion controller) using projection matrices that depend solely on the kinematic model of contact.

To validate the theory, we present experimental results showing a robot performing a compliant motion task that cannot be described by the Raibert-Craig model. The task involves controlling the contact force between the manipulated object and the environment in two face-vertex contacts, while executing a motion.

2 Generality of Contact Models for Force/Motion Control

A state of contact between a robot's end effector and its environment defines a constraint surface in the robot's operational space. At any given instant, this surface defines two vector spaces: a space of tangent vectors containing all permissible motions (velocities, infinitesimal displacements, or accelerations after compensation for velocity-product effects), and a space of normal vectors containing all permissible contact forces. A mathematical model of contact must include a means of describing these two spaces.

The contact model used by Raibert and Craig was based on the theoretical work of Mason [10], and consists of a Cartesian coordinate frame, called the constraint frame, and a compliance selection matrix. This model is characterized by six geometric parameters, giving the position and orientation of the constraint frame, and six binary parameters to select motion or force control in each direction. Unfortunately, any contact model that uses only six geometric parameters is lacking in generality.

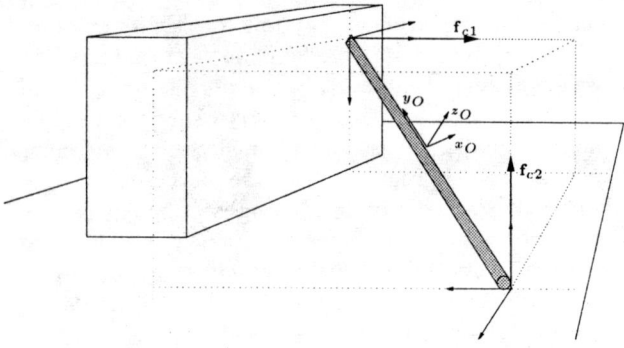

Figure 1: A general two-point contact with skew, non-intersecting contact normals.

Figures 1 and 2 show examples of contact states that cannot be described by the Raibert-Craig model. For example, in figure 1 it is not possible to find a location for the constraint frame such that the space spanned by the two contact normals along forces $\mathbf{f}_{c1}$ and $\mathbf{f}_{c2}$ equals a space spanned by two constraint-frame directions.

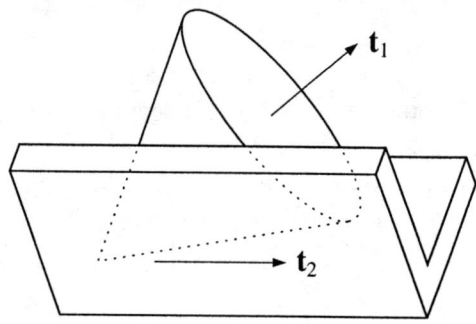

Figure 2: A cone with two motion freedoms: rotation about its axis ($\mathbf{t}_1$) and translation along the groove's axis ($\mathbf{t}_2$). The two axes are neither parallel nor perpendicular.

3 A General First-order Model of Contact

The basic idea of dual vector spaces is that we have two separate vector spaces, one containing force-type vectors, and the other containing motion-type vectors. We shall call them F^6 and M^6. The scalar product defined between them is called the reciprocal product; it is the work done by a force-type vector acting on a motion-type vector.

A Cartesian coordinate frame defines two separate bases: $\{f_1 \ldots f_6\}$ for F^6 and $\{m_1 \ldots m_6\}$ for M^6. The abstract vectors $f \in \mathsf{F}^6$ and $v \in \mathsf{M}^6$ are represented by the column matrices $\mathbf{f}$ and $\mathbf{v}$ in these bases. The basis vectors must satisfy the reciprocity condition

$$f_i \cdot m_j = \begin{cases} 1W & \text{if } i = j \\ 0 & \text{otherwise} \end{cases},$$

where W is the appropriate unit of work. This condition enforces consistency of units and ensures that the scalar product $f \cdot v$ can be expressed as $\mathbf{f}^T \mathbf{v}$. By convention, the two bases comprise three unit forces, three unit couples, three unit linear motions and three unit angular motions.

A general state of contact between two rigid bodies defines a constraint surface in their relative configuration space. At any given instant, this surface defines two vector spaces: an r-dimensional space of contact normal vectors $N \subseteq \mathsf{F}^6$ and a $(6-r)$-dimensional space of tangent vectors $T \subseteq \mathsf{M}^6$, where r is the degree of motion constraint. These spaces can be modeled by means of a $6 \times r$ matrix $\mathbf{N}$ and a $6 \times (6-r)$ matrix $\mathbf{T}$ such that

$$N = \text{range}(\mathbf{N}), \quad T = \text{range}(\mathbf{T}).$$

The columns of $\mathbf{N}$ are any r linearly independent force vectors in N, and the columns of $\mathbf{T}$ are any $6 - r$ linearly independent motion vectors in T. There is no need for any of these vectors to be normalized or orthogonal in any sense.

One of the basic properties of a contact constraint is that a constraint force does no work against an infinitesimal displacement that is consistent with the constraint. In other words, the scalar product of any member of N with any member of T is zero. This property can be expressed as $N \perp T$ or, in terms of matrices,

$$\mathbf{N}^T \mathbf{T} = \mathbf{0}.$$

In many situations, a suitable value for $\mathbf{N}$, $\mathbf{T}$ or both can be obtained by inspection. For the contact shown in Figure 1, a suitable value for $\mathbf{N}$ is $\mathbf{N} = [\mathbf{f}_{c1}\ \mathbf{f}_{c2}]$; and for the contact shown in Figure 2, a suitable value for $\mathbf{T}$ is $\mathbf{T} = [\mathbf{t}_1\ \mathbf{t}_2]$.

4 The Projection Matrices

Let us introduce two more spaces, N' and T', and their matrix representations $\mathbf{N}'$ and $\mathbf{T}'$, satisfying

$$N \oplus N' = \mathsf{F}^6, \quad T \oplus T' = \mathsf{M}^6,$$

where $\oplus$ means direct sum. N' can be any $(6-r)$-dimensional force subspace with no non-zero element in common with N, and T' can be any r-dimensional motion subspace with no non-zero element in common with T. It is important to recognize that N' and T' are not defined by the contact.

Each possible value of N' defines a unique decomposition of a general force vector into components in N and N'. Similarly, each possible value of T' defines a unique decomposition of a general motion vector into components in T and T'. Thus, given $f \in \mathsf{F}^6$ and $v \in \mathsf{M}^6$, we have

$$\mathbf{f} = \mathbf{f}_1 + \mathbf{f}_2 = \mathbf{N}\,\alpha_1 + \mathbf{N}'\,\alpha_2, \quad (1)$$

$$\mathbf{v} = \mathbf{v}_1 + \mathbf{v}_2 = \mathbf{T}\,\beta_1 + \mathbf{T}'\,\beta_2, \quad (2)$$

where α_1, α_2, β_1 and β_2 are all uniquely determined.

We are now in a position to define the projection matrices, $\mathbf{\Omega}_f$ and $\bar{\mathbf{\Omega}}_f$, which split a force vector $\mathbf{f}$ into its components $\mathbf{f}_1 = \mathbf{\Omega}_f \mathbf{f} \in N$ and $\mathbf{f}_2 = \bar{\mathbf{\Omega}}_f \mathbf{f} \in N'$, and the motion projection matrices $\mathbf{\Omega}_m$ and $\bar{\mathbf{\Omega}}_m$ that

act similarly on $\mathbf{v}$. These projections are not uniquely defined by N and T, but also depend on N' and T'. After a little algebraic manipulation, the results are

$$\bar{\mathbf{\Omega}}_f = \mathbf{N}'\,(\mathbf{T}^T\,\mathbf{N}')^{-1}\,\mathbf{T}^T = \mathbf{1} - \mathbf{\Omega}_f, \quad (3)$$

$$\bar{\mathbf{\Omega}}_m = \mathbf{T}'\,(\mathbf{N}^T\,\mathbf{T}')^{-1}\,\mathbf{N}^T = \mathbf{1} - \mathbf{\Omega}_m. \quad (4)$$

Equations 3 and 4 express every possible invariant projection matrix that is consistent with the given contact constraint. Without loss of generality, it is possible to represent N' and T' in the form

$$N' = \mathbf{A}\,T = \text{range}(\mathbf{A}\,\mathbf{T}), \quad T' = \mathbf{B}\,N = \text{range}(\mathbf{B}\,\mathbf{N}),$$

where $\mathbf{A}$ and $\mathbf{B}$ are positive-definite 6×6 matrices representing linear mappings from M^6 to F^6 and F^6 to M^6 respectively. Using this representation, all possible projection matrices for a given contact constraint can be expressed as

$$\mathbf{\Omega}_f(\mathbf{A}) = \mathbf{N}\,(\mathbf{N}^T\,\mathbf{A}^{-1}\,\mathbf{N})^{-1}\,\mathbf{N}^T\,\mathbf{A}^{-1} \quad (5)$$

$$\bar{\mathbf{\Omega}}_f(\mathbf{A}) = \mathbf{A}\,\mathbf{T}\,(\mathbf{T}^T\,\mathbf{A}\,\mathbf{T})^{-1}\,\mathbf{T}^T \quad (6)$$

$$\mathbf{\Omega}_m(\mathbf{B}) = \mathbf{T}\,(\mathbf{T}^T\,\mathbf{B}^{-1}\,\mathbf{T})^{-1}\,\mathbf{T}^T\,\mathbf{B}^{-1} \quad (7)$$

$$\bar{\mathbf{\Omega}}_m(\mathbf{B}) = \mathbf{B}\,\mathbf{N}\,(\mathbf{N}^T\,\mathbf{B}\,\mathbf{N})^{-1}\,\mathbf{N}^T \quad (8)$$

Some useful relationships are independent of the choice of $\mathbf{A}$ or $\mathbf{B}$:

$$\mathbf{\Omega}_f(\mathbf{A}_1)\,\mathbf{\Omega}_f(\mathbf{A}_2) = \mathbf{\Omega}_f(\mathbf{A}_2), \quad (9)$$

$$\bar{\mathbf{\Omega}}_f(\mathbf{A}_1)\,\mathbf{\Omega}_f(\mathbf{A}_2) = \mathbf{0}, \quad (10)$$

$$\mathbf{\Omega}_m(\mathbf{B}_1)\,\mathbf{\Omega}_m(\mathbf{B}_2) = \mathbf{\Omega}_m(\mathbf{B}_2), \quad (11)$$

$$\bar{\mathbf{\Omega}}_m(\mathbf{B}_1)\,\mathbf{\Omega}_m(\mathbf{B}_2) = \mathbf{0}. \quad (12)$$

5 Combining the Contact Model with Dynamics

The basic operational-space equation of motion for the object and the manipulator(s) [8] is

$$\mathbf{\Lambda}_0(\mathbf{x})\,\dot{\vartheta} + \mu_0(\mathbf{x}, \vartheta) + \mathbf{p}_0(\mathbf{x}) + \mathbf{f}_c = \mathbf{f} \quad (13)$$

where $\mathbf{x}$ is a vector of operational-space coordinates, $\vartheta, \dot{\vartheta} \in \mathsf{M}^6$ are end-effector velocity and acceleration vectors, $\mathbf{\Lambda}_0$ is the operational-space inertia matrix, $\mu_0, \mathbf{p}_0 \in \mathsf{F}^6$ are vectors of velocity-product and gravitational terms, $\mathbf{f}_c \in \mathsf{F}^6$ is the contact force applied by the robot to the environment, and $\mathbf{f} \in \mathsf{F}^6$ is the force command to the robot. We model the environment by the equation of motion

$$\mathbf{a}_e = \mathbf{\Phi}_e\,\mathbf{f}_c + \mathbf{b}_e. \quad (14)$$

$\mathbf{a}_e$ is the environment's acceleration, $\mathbf{\Phi}_e$ is its inverse inertia (a positive semidefinite matrix), and $\mathbf{b}_e$ is its bias acceleration (the acceleration it would have in the absence of a contact force) [5]. Eq. 14 models any environment that accepts an arbitrary applied force. A fixed, stationary environment can be modeled with $\mathbf{\Phi}_e = \mathbf{0}$, $\mathbf{b}_e = \mathbf{0}$.

The contact imposes constraints on the relative acceleration between the end effector and environment, and on the contact force:

$$\dot{\vartheta}' = \mathbf{T}\dot{\beta}, \qquad (15)$$

$$\mathbf{f}_c = \mathbf{N}\alpha, \qquad (16)$$

where $\dot{\beta}$ and α are unknown acceleration and force vectors, and $\dot{\vartheta}'$ is the relative acceleration with the velocity-product terms removed:

$$\dot{\vartheta}' = \dot{\vartheta} - \mathbf{a}_e - \dot{\mathbf{T}}\beta.$$

Combining Eqs. 13–15 and rearranging gives

$$\dot{\beta} = (\mathbf{T}^T \mathbf{\Lambda}_{rel}\mathbf{T})^{-1} \mathbf{T}^T \mathbf{\Lambda}_{rel} \mathbf{\Lambda}_0^{-1} \mathbf{f}', \qquad (17)$$

$$\alpha = (\mathbf{N}^T \mathbf{\Lambda}_{rel}^{-1}\mathbf{N})^{-1} \mathbf{N}^T \mathbf{\Lambda}_0^{-1} \mathbf{f}', \qquad (18)$$

where

$$\mathbf{f}' \triangleq \mathbf{f} - \mu_0 - \mathbf{p}_0 - \mathbf{\Lambda}_0 (\dot{\mathbf{T}}\beta + \mathbf{b}_e),$$

and

$$\mathbf{\Lambda}_{rel} \triangleq (\mathbf{\Lambda}_0^{-1} + \mathbf{\Phi}_e)^{-1}.$$

$\mathbf{\Lambda}_{rel}$ is the relative inertia between the manipulator and the environment—the tensor that relates contact force to relative acceleration. Note that if $\mathbf{\Phi}_e = \mathbf{0}$ then $\mathbf{\Lambda}_{rel} = \mathbf{\Lambda}_0$. Substituting Eqs. 17 and 18 back into Eqs. 15 and 16 gives (the factor $\mathbf{\Lambda}_{rel}\mathbf{\Lambda}_0^{-1}$ disappears if $\mathbf{\Phi}_e = \mathbf{0}$)

$$\dot{\vartheta}' = \mathbf{\Lambda}_{rel}^{-1}\bar{\mathbf{\Omega}}_f(\mathbf{\Lambda}_{rel})\left(\mathbf{\Lambda}_{rel}\mathbf{\Lambda}_0^{-1}\mathbf{f}'\right), \qquad (19)$$

$$\mathbf{f}_c = \mathbf{\Omega}_f(\mathbf{\Lambda}_{rel})\left(\mathbf{\Lambda}_{rel}\mathbf{\Lambda}_0^{-1}\mathbf{f}'\right). \qquad (20)$$

The applied force has been split into two components by means of the projections $\mathbf{\Omega}_f(\mathbf{\Lambda}_{rel})$ and $\bar{\mathbf{\Omega}}_f(\mathbf{\Lambda}_{rel})$, with one component responsible for the manipulator's relative acceleration and the other responsible for the contact force. Although there are an infinite number of force decompositions that are compatible with the kinematics of a contact constraint (given by Eqs. 5–8), there is only one that models the dynamic behavior correctly.

Now let us consider the effect of controlling the manipulator via the following dynamic control structure

$$\mathbf{f}' = \mathbf{\Lambda}_0 \dot{\vartheta}'_{comm} + \mathbf{\Lambda}_0 \mathbf{\Lambda}_{rel}^{-1} \mathbf{f}_{ccomm}, \qquad (21)$$

with $\dot{\vartheta}'_{comm}$ the commanded relative acceleration and $\mathbf{f}_{ccomm}$ the commanded contact forces. This results in the equations of motion

$$\dot{\vartheta}' = \mathbf{\Omega}_m(\mathbf{\Lambda}_{rel}^{-1})\dot{\vartheta}'_{comm} + \mathbf{\Lambda}_{rel}^{-1}\bar{\mathbf{\Omega}}_f(\mathbf{\Lambda}_{rel})\mathbf{f}_{ccomm}, \qquad (22)$$

$$\mathbf{f}_c = \mathbf{\Omega}_f(\mathbf{\Lambda}_{rel})\mathbf{f}_{ccomm} + \mathbf{\Lambda}_{rel}\bar{\mathbf{\Omega}}_m(\mathbf{\Lambda}_{rel}^{-1})\dot{\vartheta}'_{comm}. \qquad (23)$$

If the command inputs are filtered by any pair of motion and force projection matrices, so that $\dot{\vartheta}'_{comm} = \mathbf{\Omega}_m(\mathbf{B})\dot{\vartheta}'_u$ and $\mathbf{f}_{ccomm} = \mathbf{\Omega}_f(\mathbf{A})\mathbf{f}_{cu}$, where $\dot{\vartheta}'_u$ and $\mathbf{f}_{cu}$ are the unfiltered command signals, then Eqs. 22 and 23 simplify to

$$\dot{\vartheta}' = \mathbf{\Omega}_m(\mathbf{B})\dot{\vartheta}'_u, \qquad (24)$$

$$\mathbf{f}_c = \mathbf{\Omega}_f(\mathbf{A})\mathbf{f}_{cu}. \qquad (25)$$

(See Eqs. 9–12.) These equations are decoupled in the sense that the relative acceleration depends only on the filtered output of the motion controller, and the contact force depends only on the filtered output of the force controller.

6 Experimental Application of the model

The model can be used in several ways to build force/motion controllers that can deal with contact tasks that go beyond the Raibert-Craig model.

Fig. 3 shows the experimental platform: a dual-arm robotic workcell, consisting of two SCARA-type manipulators and an overhead vision system which tracks the positions of the arms and the objects that they are in contact with. Both arms cooperatively grasp a metal rod (approximately 60 cm long). Hinges in the grippers allow the bar to rotate with respect to the end effectors. Each individual arm has four actuated degrees of freedom; both arms together can move the rod in five degrees of freedom. The two end-points of the rod are in contact with the environment, following the configuration of Fig. 1.

A joint-level controller uses joint torque sensors to compensate for (exaggerated) joint-flexibility and the fact that the actuators are not ideal torque sources [11]. The dynamics of the two arms and the manipulated object are combined in the operational space and are of the form of Eq. 13, where $\mathbf{\Lambda}_0 = \mathbf{\Lambda}_{0l} + \mathbf{\Lambda}_{0r} + \mathbf{\Lambda}_{0o}$ is the total inertia of the two arms and the manipulated object, $\mathbf{f}$ is the total torque from both arms, and $\mathbf{f}_c = \mathbf{f}_{c1} + \mathbf{f}_{c2}$ is the total effect of the two contacts.

Figure 4 shows the global architecture of the force/motion controller. In this implementation, the contact model is represented by a 6×2 matrix $\mathbf{N}$. It

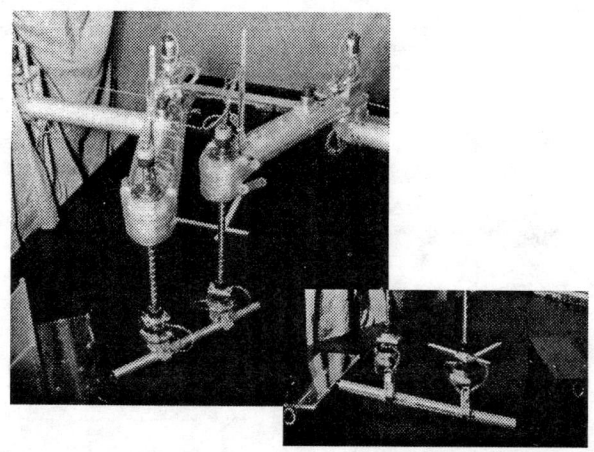

Figure 3: The dual-arm robotic workcell and the two-contact task

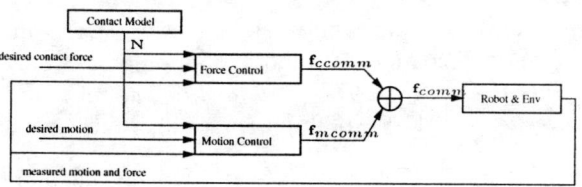

Figure 4: Decoupled Architecture for Force/Motion Control

is continuously updated as the arms and/or objects move.

Fig. 5 shows the basic force control structure. The measured forces are first filtered through the projection matrix $\Omega_f(\Lambda_{0o})$ which depends on the inertia Λ_{0o} of the manipulated object. It is necessary to remove the forces that are caused by accelerations of the manipulated object in the tangential space $\mathbf{T}$, because the 6-dof force-sensors sit between the arm end-points and the grippers holding the manipulated object.

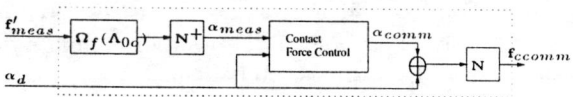

Figure 5: Force Control in the N space

The filtered force is projected onto the two-dimensional space of contact normal magnitudes by an arbitrary generalized inverse of $\mathbf{N}$. The contact force controller compares the measured contact forces, $\boldsymbol{\alpha}_{meas}$, with the vector of desired contact force magnitudes, $\boldsymbol{\alpha}_d$, and computes a command vector that is designed to reduce the difference between the two. Finally, the command vector is combined with the desired force vector (used as a feed-forward term), and the result is transformed back to operational space by multiplying by $\mathbf{N}$. From Eq. 22, it is clear that this controller will not disturb the control of motions.

The experiment of Fig. 6 shows that the reciprocity condition holds when the contact constraints are respected. The top part of the plot is the total instantaneous power $\boldsymbol{\vartheta}^T \mathbf{f}'_{meas}$, which includes the power resulting from object motion. The bottom plot is the power $\boldsymbol{\vartheta}^T \left(\Omega_f(\Lambda_{0o}) \mathbf{f}'_{meas} \right)$ delivered by the forces projected in the contact space. In theory, $\boldsymbol{\vartheta}^T \Omega_f(\mathbf{A})$ should be exactly zero for any admissible $\mathbf{A}$.

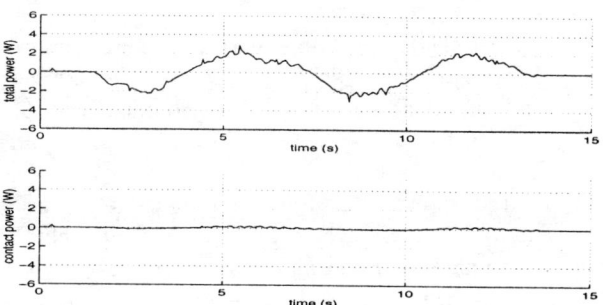

Figure 6: Power with and without projection while contacts are maintained

Figure 7 shows how the reciprocity condition is violated when the constraints are not respected. Every peak corresponds to a collision between the rod and the environment.

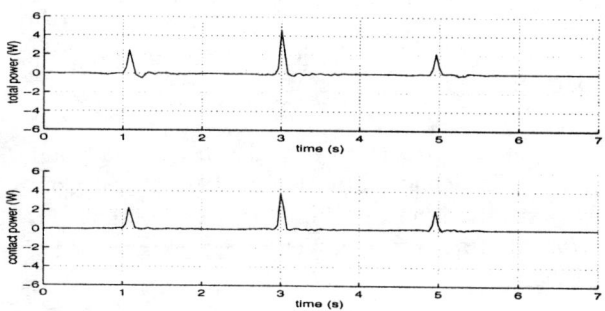

Figure 7: Power with and without projection when contacts are *not* maintained

Figure 8 shows the desired and measured contact forces $\mathbf{f}_{c1}$ and $\mathbf{f}_{c2}$. These forces are controlled in the two-dimensional space N (described by matrix $\mathbf{N}$), which can simply not be described by the Raibert-Craig model.

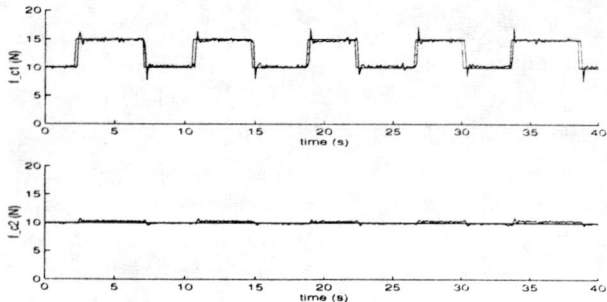

Figure 8: Multi Contact Force Control

7 Conclusion

This paper extends the Raibert-Craig model to a more general framework in which any contact state between the manipulated object and the environment can be described (to first order). This kinematic model can be combined with the dynamics of the arm and the manipulated object. This approach makes it possible to model and control force/motion tasks that include multiple points of contact and cannot be tackled with the Raibert-Craig model. The experimental results show how the reciprocity condition holds for the projected values of the measured contact forces while the state of contact is maintained, and how the model can be used to control forces in multiple points of contact between the robot and the environment.

References

[1] An, C. H., and Hollerbach, J. M., Kinematic Stability Issues in Force Control of Manipulators, IEEE Int. Conf. Robotics and Automation, pp. 897–903, Raleigh, NC, 1987.

[2] Bruyninckx, H., Demey, S., Dutré, S., De Schutter, J., Kinematic Models for Model-Based Compliant Motion in the Presence of Uncertainty, Int. Jnl. Robotics Research, Vol. 14, No. 5, pp. 465–482, 1995.

[3] Duffy, J., The Fallacy of Modern Hybrid Control Theory that is Based on "Orthogonal Complements" of Twist and Wrench Spaces, Jnl. Robotic Systems, Vol. 7, No. 2, pp. 139–144, 1990.

[4] Faessler, H., Manipulators Constrained by Stiff Contact—Dynamics, Control and Experiments, Int. Jnl. Robotics Research, Vol. 9, No. 4, pp. 40–58, 1990.

[5] Featherstone, R., Robot Dynamics Algorithms, Kluwer Academic Publishers, Boston/Dordrecht/Lancaster, 1987.

[6] Fisher, W. D., and Mujtaba, M. S., Hybrid Position/Force Control: A Correct Formulation, Int. Jnl. Robotics Research, Vol. 11, No. 4, pp. 299–311, 1992.

[7] Jankowski, K. P., and ElMaraghy, H. A., Dynamic Decoupling for Hybrid Control of Rigid-/Flexible-Joint Robots Interacting with the Environment, IEEE Trans. Robotics & Automation, Vol. 8, No. 5, pp. 519–534, Oct. 1992.

[8] Khatib, O., Inertial Properties in Robotic Manipulation: An Object-Level Framework, Int. Jnl. Robotics Research, Vol. 14, No. 1, pp. 19–36, 1995.

[9] Lipkin, H., and Duffy, J., Hybrid Twist and Wrench Control for a Robotic Manipulator, ASME Jnl. Mechanisms, Transmissions & Automation in Design, vol. 110, No. 2, pp. 138–144, June 1988.

[10] Mason, M. T., Compliance and Force Control for Computer Controlled Manipulators, IEEE Trans. Systems, Man & Cybernetics, Vol. SMC-11, No. 6, pp. 418–432, June 1981.

[11] Pfeffer, L. E., and Cannon, R. H., Experiments with a Dual-Armed, Cooperative, Felxible-Drivetrain Robot System, IEEE Int. Conf. Robotics and Automation, pp. 601–608, Atlanta, GA, 1993.

[12] Raibert, M. H., and Craig, J. J., Hybrid Position/Force Control of Manipulators, ASME Jnl. Dynamic Systems, Measurement & Control, Vol. 103, No. 2, pp. 126–133, June 1981.

[13] West, H., and Asada, H., A Method for the Design of Hybrid Position/Force Controllers for Manipulators Constrained by Contact with the Environment, IEEE Int. Conf. Robotics and Automation, pp. 251–259, St. Louis, MO, 1985.

[14] Yoshikawa, T., Sugie, T., and Tanaka, M., Dynamic Hybrid Position Force Control of Robot Manipulators — Controller Design and Experiment, IEEE Jnl. Robotics & Automation, Vol. 4, No. 6, pp. 699–705, 1988.